Scientific Foundations of Clinical Biochemistry

BIOCHEMISTRY
IN
CLINICAL PRACTICE

SCIENTIFIC FOUNDATIONS OF CLINICAL BIOCHEMISTRY

Biochemistry in Clinical Practice

Edited by

DAVID L. WILLIAMS

MB, PhD, FRSC, MRCPath

Consultant Chemical Pathologist, Royal Berkshire Hospital, Reading; Visiting Reader in Clinical Biochemistry, University of Surrey, Guildford

and

VINCENT MARKS

DM, FRCP(Ed), FRCP, FRCPath

Professor of Clinical Biochemistry, University of Surrey, Guildford; Consultant Chemical Pathologist, St. Luke's Hospital, Guildford, Surrey

LONDON
WILLIAM HEINEMANN MEDICAL BOOKS LIMITED

First published 1983

© David L. Williams and Vincent Marks 1983

ISBN 0–433–36386–X

Text set in 10/11 pt Linotron 202 Times, printed and bound
in Great Britain at The Bath Press, Avon

PREFACE

The application of analytical chemistry to the furtherance of medical knowledge began in earnest during the late nineteenth century with the work of such pioneers as Thudicum. These workers looked upon themselves as chemical pathologists whose determination it was to elucidate the chemistry of disease and by doing so to increase their understanding. It was only much later that the results of chemical analyses on biological fluids, initially especially on urine, began to be applied to the recognition and differentiation of disease in individual patients and this heralded the beginning of the era of clinical chemistry. Organisationally, these two branches of what can now best be described as clinical biochemistry remained separate; the elucidation and extension of knowledge of the biochemical basis of disease remained firmly in the hands of clinicians, medical investigators and academic biochemists and the application of that knowledge to individual patients developing as a branch of clinical pathology. The artificiality of this division and its detriment to the growth and proper utilisation of chemical and biological knowledge for the general benefit of mankind has not prevented its perpetuation in many parts of the world. Recognition is long overdue that clinical biochemistry, which can be defined as the application of analytical chemistry and a knowledge of biochemistry to the improvement is understanding, diagnosis, treatment and prevention of disease in man, is a single unified scientific discipline. It encompasses a number of subdivisions which have for many years masqueraded as separate academic disciplines such as metabolic medicine, chemical pathology, clinical chemistry and analytical chemistry and biochemistry. All seek to achieve the same objective, namely the improvement of the health of individuals and of the community.

It was with this view of clinical biochemistry in mind that the editors embarked upon the preparation of this textbook on the Scientific Basis of Clinical Biochemistry. For the sake of convenience the work has been divided into two volumes. The first was concerned with the analytical principles and techniques used to obtain the data which are the bedrock of medical knowledge and understanding. The second, here presented, is concerned with the interpretive aspects. No attempt has been made except by way of illustration to describe particular investigative procedure or analytical methods which are usually best obtained from original publications. Nor, in the main, has emphasis been given to listing "reference values" since, despite attempts at international standardisation of unitage, there is still no uniformity of methodology upon which these values so heavily depend.

The subject matter chosen for inclusion in the two volumes is similar to that taught to the M.Sc. students in Clinical Biochemistry at the University of Surrey and to students on other advanced courses in clinical biochemistry.

VM/DLW *July* 1983

CONTENTS

SECTION I
DISORDERS OF FLUID AND ELECTROLYTE BALANCE

		PAGE
1.	DISORDERS OF FLUID AND ELECTROLYTE BALANCE G. Walters	1
2.	DISORDERS INVOLVING CHANGES IN HYDROGEN ION AND BLOOD GAS CONCENTRATIONS Joan F. Zilva	24
3.	POLYURIA AND POLYDYPSIA D. Donaldson	43

SECTION II
DISORDERS OF THE GASTROINTESTINAL SYSTEM

4.	THE FUNCTION OF THE HUMAN GASTROINTESTINAL TRACT AND LABORATORY ASSESSMENT N. W. Read and C. L. Corbett	67
5.	DIFFERENTIAL DIAGNOSIS OF MALABSORPTION AND MALNUTRITION H. G. Sammons	95
6.	MALNUTRITION AND VITAMIN AND MINERAL DEFICIENCIES AND EXCESSES J. W. T. Dickerson	113
7.	THE LIVER Neil McIntyre	139
8.	THE DIFFERENTIAL DIAGNOSIS OF ACUTE ABDOMINAL PAIN Harold Ellis	155
9.	CLINICAL BIOCHEMISTRY OF THE 'ARTIFICIAL LIVER' D. B. A. Silk and Roger Williams	164

SECTION III
DISORDERS OF THE BLOOD CONSTITUENTS

10.	HAEM SYNTHESIS AND THE PORPHYRIAS G. H. Elder	175
11.	HAEMOGLOBINOPATHIES R. W. Carrell and Hermann Lehmann	187
12.	THE ANAEMIAS R. Vaughan Jones	211
13.	ABNORMALITIES OF PLASMA PROTEINS J. T. Whicher	221

		PAGE
14.	THE CLINICAL BIOCHEMISTRY OF BLOOD COAGULATION D. E. G. Austen	251

SECTION IV
DISORDERS OF THE RENAL TRACT

15.	FUNCTIONS OF THE KIDNEY AND TESTS OF THESE FUNCTIONS: ACUTE AND CHRONIC RENAL FAILURE: URAEMIA H. A. Lee	271
16.	EXAMINATION OF URINE R. R. McSwiney, Brenda M. Slavin, Patricia R. N. Kind, R. V. Brooks, I. S. Menzies and J. N. Mount	277
17.	UROLITHIASIS William C. W. Alston	290

SECTION V
DISORDERS OF THE SKELETAL SYSTEM

18.	DISORDERS OF CALCIUM AND SKELETAL METABOLISM J. A. Kanis, A. D. Paterson, R. G. G. Russell	299
19.	DISORDERS OF JOINTS A. B. Myles	325
20.	DISORDERS OF CONNECTIVE TISSUE A. B. Myles	337

SECTION VI
PHARMACOLOGY

21.	THE BIOCHEMISTRY AND TOXICOLOGY OF HEAVY METALS A. Taylor	347
22.	LABORATORY INVESTIGATION OF THE POISONED PATIENT M. J. Stewart	363
23.	REGULATION AND MONITORING OF DRUG THERAPY E. Perucca and A. Richens	379
24.	TRAUMA, SHOCK AND SURGERY A. C. Ames	400

SECTION VII
DISORDERS OF THE NEUROMUSCULAR SYSTEM

25.	THE BIOCHEMICAL INVESTIGATION OF COMA AND STUPOR R. Capildeo, F. R. Smith and F. Clifford Rose	413
26.	PERIPHERAL NEUROPATHY J. P. Patten	426
27.	THE MYOPATHIES B. P. Hughes	430
28.	CHEMISTRY OF CEREBROSPINAL FLUID IN HEALTH AND DISEASE P. T. Lascelles, E. J. Thompson and D. S. Warner	445

CONTENTS

SECTION VIII
DISORDERS OF THE CARDIOVASCULAR SYSTEM

		PAGE
29.	BIOCHEMISTRY OF DEGENERATIVE VASCULAR DISEASE G. S. Boyd and I. F. Craig	457
30.	CHEST PAIN AND CARDIAC FAILURE T. H. Foley	473
31.	HYPERTENSION Vincent Marks	479

SECTION IX
NEUROPSYCHIATRIC DISORDERS

32.	HEREDITARY AND ACQUIRED MENTAL DEFICIENCY J. Stern	489
33.	THE ROLE OF THE LABORATORY IN SOME PSYCHIATRIC ILLNESSES D. M. Shaw	523

SECTION X
DISORDERS OF THE ENDOCRINE SYSTEM

34.	INTRODUCTION TO THE ENDOCRINE SYSTEM D. L. Williams	541
35.	DISORDERS OF THE ANTERIOR PITUITARY GLAND J. W. Wright	548
36.	THE THYROID GLAND D. L. Williams and Richard Goodburn	560
37.	THE ENDOCRINE PANCREAS Vincent Marks	583
38.	GASTROINTESTINAL HORMONES Vincent Marks	606
39.	ADRENAL CORTEX Vincent Marks	616
40.	THE GONADS S. L. Jeffcoate	633

SECTION XI
CLINICAL BIOCHEMISTRY OF PREGNANCY AND THE NEONATAL PERIOD

41.	THE CONFIRMATION AND MONITORING OF PREGNANCY G. R. Wilson and A. I. Klopper	653
42.	PRENATAL DIAGNOSIS OF INHERITED METABOLIC DISEASES, NEURAL TUBE DEFECTS AND LUNG MATURITY J. B. Holton	663
43.	PAEDIATRIC CLINICAL CHEMISTRY: SOME SPECIAL PROBLEMS G. M. Addison	682

INDEX   695

CONTRIBUTORS

G. M. Addison, MA, MB, BChir, PhD
Consultant Chemical Pathologist, Royal Manchester and Booth Hall Children's Hospital, Manchester.

William C. Alston, MB, ChB, BSc, PhD, MRCPath
Consultant Chemical Pathologist, West Surrey and North East Hants District and South West Surrey District, Frimley Park Hospital, Frimley, Surrey.

A. C. Ames, BSc, MB, BS, FRCPath
Consultant Chemical Pathologist, West Glamorgan Health Authority, Neath Hospital.

D. E. G. Austen, BSc, PhD, CChem, FRSC
Principal Scientific Officer, Haemophilia Centre, Churchill Hospital, Oxford.

*G. S. Boyd, PhD, FRSC, FRSE
Professor and Chairman of the Department of Biochemistry, University of Edinburgh.

R. V. Brooks, DSc, PhD, FRSC, CChem, FRCPath
Professor of Chemical Endocrinology, St. Thomas's Hospital Medical School, University of London.

R. Capildeo, MRCP
Consultant Neurologist, Oldchurch Hospital, Romford.

R. W. Carrell, MB, BSc, PhD, MRCPath, FRACP
Professor of Clinical Biochemistry, Christchurch Hospital, Christchurch, New Zealand.

C. L. Corbett, MD, MRCP
Consultant Physician, Bassetlaw Health Authority, Nottinghamshire.

I. F. Craig, PhD
Post-doctoral Fellow, Department of Biochemistry, University of Edinburgh. (At present Medical Student, University of Dundee).

J. W. T. Dickerson, PhD, FIBiol
Professor of Human Nutrition, University of Surrey, Guildford; Consultant Adviser in Clinical Nutrition, South West Thames Regional Health Authority.

D. Donaldson, MB, ChB, MRCP, FRCPath
Consultant Chemical Pathologist, Redhill General Hospital, Redhill, Surrey.

G. H. Elder, BA, MD, FRCPath, MRCP
Professor of Medical Biochemistry, Welsh National School of Medicine, Cardiff.

Harold Ellis, DM, MCh, FRCS
Professor of Surgery, Westminster Hospital Medical School, University of London.

T. H. Foley, MD, FRCP
Consultant Physician, South West Surrey District, St. Luke's Hospital, Guildford.

Richard Goodburn, BSc
Senior Biochemist, Ashford Hospital, Ashford, Middlesex.

J. B. Holton, BSc, PhD, FRCPath
Top Grade Biochemist, Southmead Hospital, Bristol.

B. P. Hughes, BSc, PhD
Former Muscular Dystrophy Group Research Fellow, Neurochemistry Department, The National Hospital, Queen Square, London.

S. L. Jeffcoate, MB, BChir, PhD, MRCPath
Professor of Biochemical Endocrinology, Chelsea Hospital for Women, London.

J. A. Kanis, BSc, MB, ChB, MRCP, MRCPath
Reader in Human Metabolism, University of Sheffield Medical School.

Patricia R. N. Kind, BSc, MSc, MCB, FRCPath
Principal Biochemist, St. Thomas's Hospital, London.

A. I. Klopper, PhD, MD, FRCOG
Professor of Reproductive Endocrinology, University of Aberdeen Medical School.

P. T. Lascelles, MD, FRCPath
Senior Lecturer and Honorary Consultant in Chemical Pathology, The National Hospital, Queen Square and Maida Vale, London.

H. A. Lee, BSc, MB, BS, FRCP, MRCS
Professor of Renal Medicine, University of Southampton; Director and Consultant Physician, Wessex Regional Renal Unit; Co-Director Wessex Regional Transplant Unit, St. Mary's Hospital, Portsmouth.

Hermann Lehmann, CBE, MD, ScD, FRCP, FRSC, FRCPath, FRS
Emeritus Professor of Clinical Biochemistry, University of Cambridge.

* Deceased

LIST OF CONTRIBUTORS

Vincent Marks, DM, FRCP(Ed), FRCP(Lond), FRCPath
Professor of Clinical Biochemistry, University of Surrey, Guildford; Consultant Chemical Pathologist, St. Luke's Hospital, Guildford, Surrey.

N. McIntyre, BSc, MD, FRCP
Professor of Medicine, Royal Free Hospital School of Medicine; Honorary Consultant Physician, Royal Free Hospital, London.

R. R. McSwiney, MB, BS, MRCP, FRCPath
Former Senior Lecturer (Chemical Pathology), St. Thomas' Hospital Medical School, London.

I. S. Menzies, MB, BS, FRCPath, DPath
Senior Lecturer in Chemical Pathology, St. Thomas' Hospital Medical School, London.

J. N. Mount, MSc, MIBiol
Principal Biochemist, St. Thomas' Hospital, London.

A. B. Myles, BA, MB, BChir, FRCP
Consultant Physician, Department of Rheumatology, St. Peter's Hospital, Chertsey, Surrey.

A. D. Paterson, MA, BM, BCh, MRCP
Senior Registrar in General Medicine and Renal Medicine, Hallamshire Hospital, Sheffield.

J. P. Patten, BSc, MB, FRCP
Consultant Neurologist, South West Thames Regional Neurological Unit, Guildford, Surrey.

E. Perucca, MD, PhD
Lecturer, Department of Medical Pharmacology, University of Pavia, Italy.

N. W. Read, MA, MD, MRCP
Senior Lecturer in Physiology, University of Sheffield, Honorary Consultant Gastroenterologist, Royal Hallamshire Hospital, Sheffield.

A. Richens, PhD, FRCP
Professor, Department of Pharmacology and Therapeutics, Welsh National School of Medicine, Cardiff.

F. Clifford Rose, FRCP
Physician in Charge, Department of Neurology, Charing Cross Hospital, London.

R. G. G. Russell, PhD, DM, FRCP, MRCPath
Professor of Human Metabolism and Clinical Biochemistry, University of Sheffield Medical School.

H. G. Sammons, BSc, PhD, DSc, FRSC, FRCPath
Consultant Clinical Biochemist (Retired), East Birmingham Hospital.

D. M. Shaw, PhD, FRCP, FRCPsych
Clinical Tutor, Welsh National School of Medicine, Cardiff, External Scientific Staff, Medical Research Council.

D. B. A. Silk, MD, MRCP
Consultant Physician, Department of Gastro-enterology and Nutrition, Central Middlesex Hospital, London.

Brenda M. Slavin, BSc, MB, BCh, MRCPath, MCB
Senior Lecturer, Chemical Pathology and Metabolic Disorders, St. Thomas' Hospital, London.

F. R. Smith, MB, ChB, MRCP
Registrar, Department of Neurology, Charing Cross Hospital, London

J. Stern, BSc, PhD, FRCPath
Clinical Biochemist, Queen Mary's Hospital for Children, Carshalton, Surrey.

M. J. Stewart, BSc, PhD, MCB, MRCPath, MRSC, CChem
Top Grade Biochemist, Head of the Drug Investigation Unit, Royal Infirmary, Glasgow.

A. Taylor, BSc, MSc, PhD
Senior Biochemist, St. Luke's Hospital, Guildford; Visiting Lecturer, Robens Institute of Industrial and Environmental Health and Safety, University of Surrey, Guildford.

E. J. Thompson, PhD, MD, MRCPath
Senior Lecturer and Honorary Consultant in Special Chemical Pathology and Clinical Neuro-chemistry, The National Hospitals, Queen Square, London.

R. Vaughan Jones, MA, MB, BChir, FRCP, MRCPath
Consultant Haematologist, North West Surrey District Health Authority, St. Peter's Hospital, Chertsey.

G. Walters, MD, FRCP, FRCPath, LRSC
Consultant Chemical Pathologist, Bristol Royal Infirmary.

D. S. Warner, FIMLS
Senior Chief MLSO, The National Hospital, Queen Square, London.

J. T. Whicher, MA, MB, BChir, MSc, MRCPath
Consultant Chemical Pathologist, Bristol Royal Infirmary; Lecturer in Chemical Pathology, University of Bristol.

D. L. Williams, MB, PhD, FRSC, MRCPath
Consultant Chemical Pathologist, Royal Berkshire Hospital, Reading; Visiting Reader in Clinical Biochemistry, University of Surrey, Guildford.

Roger Williams, MD, FRCP
Consultant Physician and Director of the Liver Research Unit, King's College Hospital, London.

*G. R. Wilson, PhD
Scientific Officer, University of Aberdeen Medical School.

J. W. Wright, MSc, MRCP, MRCPath
Consultant in Clinical Biochemistry, St. Luke's Hospital, Guildford, Surrey, Reader in Metabolic Medicine, University of Surrey, Guildford.

Joan F. Zilva, BSc, MD, FRCP, FRCPath
Professor of Chemical Pathology and Honorary Consultant, Westminster Hospital Medical School, London.

* Deceased.

SECTION I

DISORDERS OF FLUID AND ELECTROLYTE BALANCE

		PAGE
1.	DISORDERS OF FLUID AND ELECTROLYTE BALANCE	1
2.	DISORDERS INVOLVING CHANGES IN HYDROGEN-ION AND BLOOD GAS CONCENTRATIONS	24
3.	POLYURIA AND POLYDYPSIA	43

1. DISORDERS OF FLUID AND ELECTROLYTE BALANCE

G. WALTERS

Electrolyte and water composition of the body
The renal handling of water and salt
  Factors influencing the reabsorption of water
    Factors influencing ADH secretion
  Factors influencing the renal tubular reabsorption of sodium
Abnormalities of the sodium and water content of the body
  Methods of assessment
  Sodium depletion
  Hypernatraemia
    Water depletion with relatively little sodium depletion
    Hypernatraemia due to excessive sodium intake
    Effects of hypernatraemia
  Excessive water retention
    The syndrome of inappropriate secretion of ADH
    Hyponatraemia without ectopic ADH
  Retention of water and sodium
    Fluid retention with oedema
  Effects of hyponatraemia
  Other guides
    The anion gap
    Plasma and urine osmolality
    Urinary sodium measurements
Potassium
  Distribution
  The renal handling of potassium
  Hypokalaemia and potassium deficiency
    Causes of hypokalaemia
    Urinary potassium measurements
    Effects of hypokalaemia
    Treatment of hypokalaemia
  Hyperkalaemia
    Reduction in renal potassium excretion
    The effects of hyperkalaemia
    Treatment of hyperkalaemia

ELECTROLYTE AND WATER COMPOSITION OF THE BODY

On average, the total body water constitutes about 60% of the body weight in men and about 50% in women. There are very large differences between individuals, associated particularly with variations in the amount of body fat. About two-thirds of the water is in the cells and one-third outside the cells in the extracellular fluid (ECF), so that in a 70 kg man total body water is about 42 litres, with 14 litres of ECF and 28 litres of intracellular fluid. The ECF comprises several anatomically separate compartments, namely, the blood plasma, the interstitial fluid which lies between the cells outside the vascular compartment, the lymph, and the transcellular fluids, i.e. the small amounts of free fluid in the pleural, pericardial and peritoneal cavities, the cerebrospinal fluid, and the secretions of various glands such as the digestive secretions.

The electrolyte composition of the cells and extracellular fluid are very different and as the transcellular fluids are secreted their composition may differ from that of the rest of the ECF. In the latter, sodium is by far the most abundant cation whereas in the cells it is present only in about 10–20 mmol/l, the most abundant intracellular cations being potassium and magnesium; the relative amounts of the latter two vary in different tissues. Representative values, in meq/l, are given in Table 1.

Only 93% of the plasma volume is water, the other 7% being occupied by the plasma proteins. The concentration of sodium and other ions in the plasma water will therefore be correspondingly higher than their concentration in whole plasma, e.g., the sodium concentration will be 150 mmol/l.

In the interstitial fluid the protein concentration is much lower than in plasma. As a result of the Donnan equilibrium the concentration of sodium in the interstitial fluid water will be less than in the plasma water, amounting to $0.92 \times$ concentration in plasma water.

The large differences between the sodium and potassium concentrations in the cells and the ECF are maintained by active transport of sodium out of the cell. This is mediated by a mechanism dependent on the activity of the sodium- and potassium-dependent ATP-ase in the cell membrane. If the rate of cell metabolism is reduced by cooling or if the cell is damaged the activity of this so-called 'sodium pump' ceases; then sodium accumulates in the cells and potassium escapes.

The relationship between the water content of the cells and the ECF is governed by differences in osmotic pressure across the cell membranes. Cell membranes are very permeable to water, and the intracellular and extracellular osmolalities must be equal; in the event of a change in one of them water will move rapidly across the cell membrane from the lower osmolality to the higher until a new equilibrium is attained. However, the cell membranes are not strictly semi-permeable but are

TABLE 1
ELECTROLYTE COMPOSITION (MEQ/L) OF EXTRA- AND INTRA-CELLULAR FLUID

Extracellular fluid				Intracellular fluid			
Sodium	140	Chloride	100	Sodium	10	Chloride	10
Potassium	4	Bicarbonate	27	Potassium	140	Bicarbonate	10
Calcium	5	Protein	17	Calcium	2	Protein	50
Magnesium	2	Phosphate	3	Magnesium	30	Phosphate	80
		Others	5			Others (mainly sulphate)	30

selectively permeable to solutes. Therefore the osmotic pressure across the cell membrane will be governed by the total concentration of particles to which the cell membrane is impermeable; this is sometimes known as the 'effective' osmotic pressure. Notwithstanding this, in clinical practice it is the total concentration of the osmotically active particles in the plasma water, i.e. osmolality, that is used to assess osmotic changes and not measurements of osmotic pressure.

In the ECF 95% of the osmolality is contributed by the electrolytes and is determined by the sodium concentration. If this changes by a large amount, say 40 mmol/l, there cannot be a compensatory change in any of the other extracellular cations because of their relatively low concentrations. Therefore, there must be a corresponding change in the total concentration of anions in order to maintain cation-anion equivalence; hence, the osmolality will change by twice the change in sodium concentration. In contrast, a large change in the most abundant anion, chloride, can be compensated for by an equal but opposite change in other anions. The sodium concentration therefore does not change and the total number of particles, and hence the osmolality, will also be unchanged. The contribution of urea and glucose to plasma osmolality is normally very small but may be large when pathologically raised.

The relationship between the plasma and the interstitial water is dependent on other factors. Water and electrolytes can all move freely across the capillary wall which separates these two compartments, but the plasma proteins cannot. There will therefore be a slightly higher osmotic pressure or the *oncotic pressure*. The movement fluid due to the plasma proteins; this will attract water into the vascular compartment and is called the colloid osmotic pressure or the oncotic pressure. The movement of fluid across the capillary wall is therefore governed by differences between the oncotic pressure of the proteins and by the hydrostatic forces in the capillaries and in the interstitial fluid. At the arterial end of the capillary the arterial hydrostatic pressure exceeds the sum of the oncotic pressure and the hydrostatic pressure in the tissues, and there is a net movement of water out of the capillary. At the venous end the capillary hydrostatic pressure is lower and the oncotic pressure is higher owing to the loss of fluid at the arterial end. The resultant pressure now favours the movement of fluid back into the capillary. Any fluid which does not return to the vascular compartment in this way will normally be returned via the lymphatics.

In health the plasma volume is maintained relatively constant, though small changes of about 5% do occur with alterations of posture, due to shifts of fluid into and out of the vascular compartment. After haemorrhage the capillary dynamics change so that sufficient interstitial fluid moves into the vascular compartment to restore the blood volume to normal. During this process there is also a shift of protein into the vascular compartment, thus maintaining the plasma oncotic pressure near its normal value. Restoration of the blood volume in this way does not occur if the extracellular fluid is diminished as a result of the loss of water and sodium.

THE RENAL HANDLING OF WATER AND SALT

In health the amounts of water and electrolytes in the body are maintained within relatively narrow limits by the kidney acting under the influence of hormones and other factors. The kidney excretes a water load or conserves water by varying the concentration of the urine within the range of specific gravity 1·001–1·030 corresponding to osmolalities of 40–1400 mmol/kg water. This enables the plasma osmolality to be maintained within the narrow limits of 290 ± 5 mmol/kg.

About 70% of the glomerular filtrate is reabsorbed iso-osmotically in the proximal tubule in the renal cortex. In the interstitial fluid of the medulla the concentration of sodium chloride and urea increases progressively from the cortico-medullary junction to the papillary tip. As the remainder of the glomerular filtrate passes down the descending limb of Henle's loop, water is reabsorbed into the hypertonic interstitium of the medulla. The tubular fluid therefore becomes very concentrated to about 1400 mmol/kg by the time it reaches the bend. The ascending limb of the loop of Henle is impermeable to water but actively extrudes chloride, sodium following passively along the electrochemical gradient thus established. The tubular fluid therefore becomes more dilute as it traverses the ascending limb, and is hypotonic to plasma when it enters the distal convoluted tubule. The thick part of the ascending limb of the loop of Henle is known as the diluting segment. Beyond it the capacity for further reabsorption of solute without water is slight, though some reabsorption of urea and electrolytes can occur in the collecting ducts, so that proper functioning of the diluting segment is necessary if the body is to excrete solute-free water.* The production of a concentrated urine depends on the reabsorption of water in the distal part of the convoluted tubule and the collecting tubule which become permeable to water in the presence of the antidiuretic hormone (ADH). Water is reabsorbed osmotically from these segments into the concentrated medulla, thereby raising the concentration of urea, etc. in the tubular fluid. The more distal part of the collecting tubule in the inner medulla is permeable to urea which diffuses into the medullary interstitium and thus maintains its concentration.

After the reabsorption of sodium in the ascending limb of the loop of Henle, more sodium is reabsorbed in the distal convoluted tubule, under the influence of aldosterone, in exchange for potassium and hydrogen ions. The final adjustment to the amount of sodium excreted in the urine is made in the collecting ducts.[1]

Factors Influencing the Reabsorption of Water[2]

Physiologically a concentrated urine depends on the ADH-mediated final reabsorption of water from the collecting ducts. Excretion of a water load depends on the absence of such concentration and also on the fluid reaching the collecting ducts being hypotonic, i.e. it depends on the reabsorption of sodium chloride in the ascending limb of the loop of Henle. Diuretics which prevent the reabsorption of sodium chloride in the diluting segment will therefore diminish the kidney's ability to form a dilute urine.

The elaboration of maximally concentrated urine is dependent upon ADH rendering the collecting tubules permeable to water, but the reabsorption of water through the permeable epithelium requires a hypertonic medulla. Factors which increase medullary blood flow are said to reduce the concentration of the interstitium by a washout effect, and therefore to diminish the maximal concentration of the urine. Diversion of blood flow to the outer cortex will have a similar effect on urine concentration as the nephrons in this part have short loops of Henle which do not penetrate to the deepest, and therefore the most hypertonic, part of the medulla.

Factors Influencing ADH Secretion

(a) Osmotic. The secretion of ADH from the posterior pituitary gland is normally regulated by changes in plasma osmolality. The probable existence of receptors sensitive to changes in plasma osmolality was demonstrated many years ago in the dog, but further studies were hampered by difficulties in the bioassay of ADH. Since the advent of radioimmunoassay methods for ADH, considerable progress has been made in understanding the mechanism of its secretion. It has been shown that there is a threshold of plasma osmolality, about 285 mmol/kg, below which ADH levels in plasma are too low to measure with present techniques, and above which plasma ADH increases with plasma osmolality (Fig. 1); a change in plasma osmolality of 3 mmol/kg, consequent on a change in the sodium concentration,

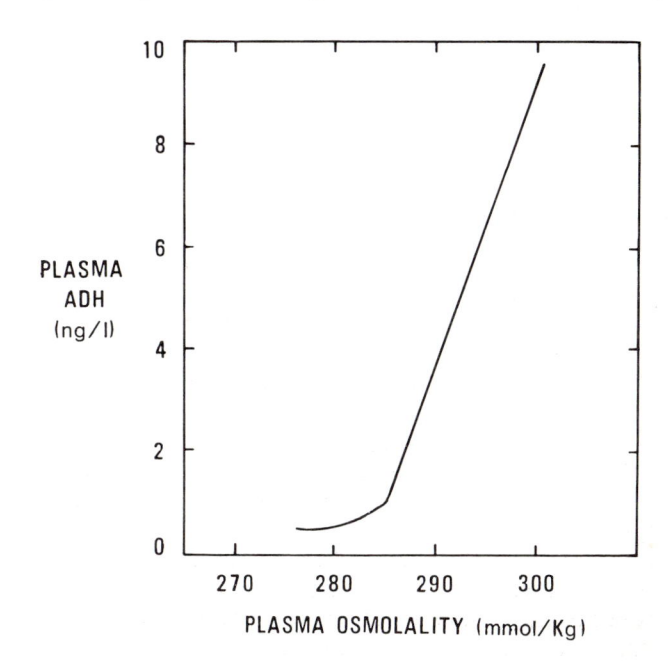

FIG. 1. Diagrammatic representation of the relationship between plasma ADH and plasma osmolality altered by changing the sodium concentration. From the data of Robertson et al.[2]

is sufficient to affect the urine concentration. The osmoreceptors are thus very sensitive, the sensitivity in any individual being indicated by the slope of the line relating osmolality to ADH levels in that individual.

Not all solutes have the same effect on ADH release as sodium. Increasing the plasma urea concentration does not stimulate ADH secretion unless it is done very rapidly. This is because urea readily enters the cells and therefore does not cause an osmotic gradient across the cell membrane so that the cell volume does not change. The uptake of glucose by the somatic cells is insulin-dependent and takes place more slowly than the uptake of urea. High extracellular glucose levels therefore draw water out of the cells, causing hyponatraemia, and are found not to increase plasma ADH levels but, paradoxically, to lower them. This implies that while water is being withdrawn from the somatic cells it actually enters the osmoreceptor cells; the mechanism of this atypical behaviour of the osmoreceptors has not yet been elucidated.

(b) Non-osmotic. ADH release is stimulated physiologically by changing from the supine to the erect posture. In pathological states its release is stimulated or inhibited by a variety of haemodynamic disturbances. These are thought to be mediated through volume (low

---

* The term 'free water excretion' signifies the amount of water that is excreted in excess of that needed to carry the solute load iso-osmotically.
Free water clearance = $V - C_{osm}$
where V = urine flow rate;
C_{osm} = osmolal clearance i.e. $U_{osm} \times V/P_{osm}$

pressure) receptors located possibly in the left atrium, and through the high pressure aortic and carotid baroreceptors; experimentally such responses have been abolished by interruption of the afferent pathways from the baroreceptors and left atrium.

A fall in blood volume of 10–15% in man is a powerful stimulus to ADH release, and will override the inhibition of secretion caused by a fall in plasma sodium concentration. These effects are presumably mediated through the consequent fall in the cardiac filling pressure and hence cardiac output, as impeding the venous return experimentally has similar effects, dependent on the integrity of the baroreceptors.

A rise in left atrial pressure has been shown to inhibit ADH release, and this may explain the diuresis sometimes caused by paroxysmal tachycardia. But a fall in cardiac output will stimulate ADH release through stimulation of the baroreceptors. When such opposing factors co-exist, which of them predominates is variable.

There are other non-osmotic agents which appear to act directly on the brain. The infusion of noradrenaline increases water excretion, an effect that is blocked by α- but not by β-blockade. The effect is independent of haemodynamic changes and is due to suppression of ADH secretion. Conversely, isoprenaline, which has predominantly β activity, stimulates ADH release.

Cortisol influences the effect of ADH; the distal part of the nephron can become maximally impermeable to water only in the presence of an adequate level of cortisol.

Factors Influencing the Renal Tubular Reabsorption of Sodium[1]

The amount of sodium presented to the renal tubules is a function of the glomerular filtration rate (GFR) and the plasma sodium concentration. Each day more than ten times the total extracellular sodium, amounting to 24 000 mmol, is filtered and more than 99% of it is reabsorbed. During everyday activity there are fluctuations in the GFR which do not lead to fluctuations in the amount of sodium excreted because corresponding adjustments are made to the amount of sodium reabsorbed in accordance with large or small variations in the sodium intake. Even in renal disease with a large reduction in the total GFR the individual may be able to regulate his sodium balance satisfactorily. The adjustment of tubular sodium reabsorption in accordance with fluctuations in the amount filtered is referred to as glomerulo-tubular balance.

The reabsorption of sodium in the distal tubule is regulated by aldosterone, secreted by the zona glomerulosa of the adrenal cortex in response to activation of the renin-angiotensin system. The major stimulus to the latter is not hyponatraemia but diminished renal blood flow, such as occurs in states of hypovolaemia in which cardiac output is reduced. Expansion of the blood volume on the other hand inhibits the secretion of renin and hence of aldosterone. The importance of the plasma potassium concentration as a stimulus to aldosterone secretion is discussed later. Normally the distal tubule accounts for only about 8% of sodium reabsorption, most of which occurs in the proximal tubule.

It was once thought that the proximal tubular reabsorption of sodium might be regulated only by alterations in the GFR, but there is strong evidence for the existence of a 'third factor', i.e. in addition to GFR and aldosterone, which regulates sodium reabsorption in the proximal tubule. Thus, patients maintained on fixed doses of cortisol and mineralo-corticoids because of Addison's disease or bilateral adrenalectomy are still able to vary their sodium excretion. Furthermore, the daily administration of a salt-retaining steroid to normal subjects induces a positive sodium balance for only a few days. Then, although the positive balance is maintained, sodium excretion in the urine rises in accordance with the dietary intake despite continuance of the mineralocorticoid.

The nature of this third factor is still a matter of dispute but there is a good deal of evidence to implicate a humoral substance, although its source is unknown. Several workers have claimed to have isolated a peptide natriuretic hormone, but its existence is still not universally accepted. Another hypothesis is that adjustments to proximal tubular reabsorption are mediated through the sodium concentration of the tubular fluid reaching the macula densa, a specialized area of epithelium at the beginning of the distal convoluted tubule which is in contact with the juxtaglomerular body. Whatever the mechanism, the response of the kidneys under different conditions of salt-loading fits with the existence of a natriuretic hormone whose secretion is stimulated and inhibited by expansion and contraction of the ECF volume respectively. Thus, the effect is to promote sodium diuresis by inhibiting sodium reabsorption; the renin-angiotensin-aldosterone system must be switched off to achieve the same effect.

It is to be expected that proximal tubular reabsorption and distal tubular reabsorption of sodium will act in concert and interact with ADH as follows. Expansion of the ECF volume with, say, isotonic saline will expand the plasma volume. To correct this, third factor activity will be stimulated and tubular reabsorption of sodium will decrease. At the same time renin and aldosterone secretion will be suppressed thus inhibiting sodium reabsorption in the distal tubule. The excess sodium is therefore excreted. It will be accompanied by the excess water because the excretion of sodium alone would lower the plasma sodium concentration and hence suppress ADH secretion. Conversely, contraction of the ECF volume will inhibit third factor activation but stimulate aldosterone secretion, both of which will promote sodium reabsorption. The resulting tendency for the ECF sodium concentration to rise will stimulate ADH release, and enough water will be retained with the sodium to restore the volume and composition of the ECF to normal.

ABNORMALITIES OF THE SODIUM AND WATER CONTENT OF THE BODY

Methods of Assessment

There is often a tendency to think about abnormalities of sodium and water only in terms of the plasma concentration of sodium. This is of great importance because any change in osmolality resulting from a change in sodium concentration will cause water to move into or out of the cells and, if severe enough, may be fatal. However, alterations in the *total amount* of water and sodium in the extracellular fluid, which may occur without changes in plasma sodium concentration, are even more important because plasma volume contracts in parallel with the extra-cellular fluid volume. As the integrity of the circulation is dependent on maintenance of the blood volume, reduction in ECF volume will, if large enough, cause hypovolaemic shock. On the other hand, expansion of the extracellular fluid volume, which may also occur without alteration of sodium concentration, will lead to expansion of the plasma volume and, in some circumstances, to heart failure and acute pulmonary oedema. Since the changes in the total body water and the total extracellular sodium cannot be gleaned from measurement of plasma electrolytes alone, other methods must be used.

When abnormalities occur solely during a period of observation the changes can be determined by measuring the intake and the output of water and electrolytes, and calculating the 'balance', a positive balance signifying a net gain of a substance and a negative balance signifying a net loss. This method is of great importance in clinical practice and forms the basis for formulating treatment in patients whose problems arise in hospital. However, when an electrolyte disturbance is already established by the time of admission to the hospital the most direct method of obtaining quantitative data is to measure the size of the cellular and extracellular fluid compartments, and the total body sodium, directly by means of radioisotopes or other suitable compounds.

These methods are based on the dilution principle: If an accurately known amount of a substance is introduced into a closed system in which it will mix uniformly, the volume of the system can be calculated from the dilution of the marker when mixing is complete:
i.e.,

$$\text{volume} = \frac{\text{amount of marker added}}{\text{final concentration of the marker}}$$

For the measurement of a body fluid compartment the final concentration is measured in plasma and, if necessary, allowance is made for the amount of marker lost in the urine during the period of equilibration, so that the formula becomes:-

$$\frac{\text{amount of marker administered} - \text{amount excreted in urine}^*}{\text{concentration in plasma}}$$

* The amount lost in faeces, sweat and expired air are usually very small and are ignored.

This principle has been applied to the measurement of the total body water, the extracellular fluid volume and the plasma and total red cell volumes.

Total body water was originally measured with heavy water, D_2O, but now tritiated water is more commonly used. These substances reach equilibrium with the total body water in about two hours. Other substances which distribute themselves throughout the total body water and have been used for its measurement are urea, antipyrine, and ethanol.

Measurement of the ECF volume is less satisfactory. Early measurements with inulin indicated at least two mixing phases, a rapid first phase followed by a slow phase thought to be due to the slow penetration of the marker into the extracellular fluid in dense connective tissue, bone and cartilage. A number of different substances have been used to determine the ECF volume but they have given values ranging from about 15% of body weight for inulin to about 28% for bromide. These differences are due largely to the variable extent to which markers penetrate the cells and transcellular fluids, and to the difficulties of deciding when equilibration is complete. Since some markers are distributed throughout a volume that does not coincide with either the extracellular volume or the total body water, the term 'space' is used to refer to the volume of distribution of a particular marker. The sulphate space, now determined with ^{35}S-sulphate is widely regarded as the best estimate of the ECF volume, but opinions differ on the ideal equilibration time. Following intravenous injection sodium ^{35}S-sulphate exhibits three mixing phases, and some authors rely on a single sample of plasma collected 20 minutes after the injection. This is probably adequate in many patients without circulatory abnormalities but a preferable procedure is to take serial samples of venous blood from an indwelling cannula and to plot the log of the concentration of radio-sulphate against time; the final exponential part of the curve can then be extrapolated back to zero time to give the theoretical concentration had mixing been instantaneous (Fig. 2); this value may be used to calculate the ECF volume.

The same principles apply to the measurement of the plasma volume. A large number of macromolecular substances which leave the vascular compartment very slowly has been used. The two which have been used most are the blue dye T 1824 (Evans blue) and human albumin labelled with ^{131}I or ^{125}I. Mixing in the vascular compartment is very rapid and a single sample of plasma obtained ten minutes after the injection may be used; alternatively, several samples may be taken over 45 minutes and the plot of the log concentration *versus* time extrapolated back to zero time as for sulphate. The total blood volume can be calculated from the formula:

$$\text{blood volume} = \frac{\text{plasma volume} \times 100}{100 - H}$$

where H = the venous haematocrit, preferably multiplied by 0·91 to allow for the fact that the whole body haematocrit differs from the venous haematocrit.

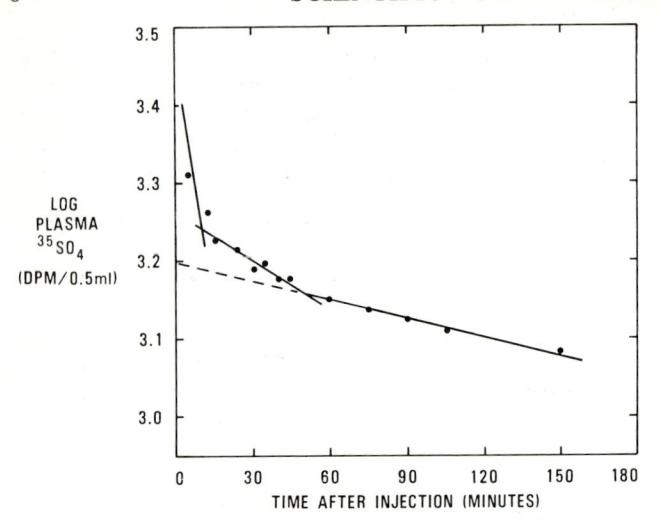

Fig. 2. Disappearance rate of radioactive sulphate from plasma after a single intravenous injection.

Similarly, the sodium and potassium spaces may be determined. Isotopes of sodium need 12 hours to reach equilibrium in normal subjects and longer in some patients with circulatory abnormalities or oedema. The sodium space exceeds the probable extracellular fluid volume by about 15%, indicating that some of the sodium is in the cells. If the sodium space is multiplied by its concentration of sodium, i.e. the plasma sodium concentration,* the total amount of sodium with which the isotope has exchanged will be obtained; this is known as the exchangeable sodium, and on average it amounts to something of the order of 40 mmol/kg body weight in men and slightly less in women. This is considerably less than values for total body sodium found by analysis of cadavers or by whole-body neutron activation. The reason is that about 30% of the body sodium is present in bone, and this does not exchange. This non-exchangeability of bone sodium means that it cannot be used to replace sodium lost from the ECF.

For the determination of the potassium space and the exchangeable potassium a longer equilibration period is required than for sodium, varying, according to different authors, from 24 to 48 hours. Total body potassium can be determined by whole-body counting of the naturally occurring radioisotope ^{40}K, and this is higher than the exchangeable potassium by about 10–15%.

With careful technique and modern counting equipment it is possible to make the above measurements with a precision of 2–6%, total body water and exchangeable potassium measurements being the least precise. It is also possible to adopt a procedure which enables tritiated water and sulphate spaces to be determined simultaneously with the exchangeable sodium and potassium. Such measurements have added greatly to knowledge, but their use in acute fluid and electrolyte disturbances is not practicable because of the long equilibration times required for the electrolyte measurements; although the equilibration times for total body water and ECF volume measurements are acceptable, the counting procedures lengthen the time required, especially if the lowest possible doses of isotope are used.

Another very great limitation to the use of such measurements in clinical practice as opposed to research is the difficulty of interpreting a single measurement in relation to a predicted normal value for the individual. This difficulty derives from the very wide reference ranges, the coefficient of variation being at least 10% when the values are related to body weight, height or surface area.[3] Various formulae have been recommended using higher powers of height and weight combined, but they do not narrow the range sufficiently. Additional difficulties arise in patients with loss of weight.

The above parameters vary most of all, but not solely, with the amount of body fat. When the fat-free body weight was determined by carcass analysis of animals or by whole-body specific gravity measurements in man, it was found that total body water amounted to 71–73% of the fat-free weight. Body fat may be estimated approximately by measuring skinfold thickness with constant-tension calipers, and this enables a better prediction of a patient's normal values to be made. But again, severe loss of weight will disturb this relationship because the non- or poorly-exchangeable components of bone will make a disproportionately large contribution to the lean body mass.

An alternative approach is to analyse the tissues directly. Many reports indicate that the electrolyte content of erythrocytes changes in disease, but they are atypical cells and the changes are not representative of those in the rest of the cell mass. Leucocytes have also been analysed, though less frequently, but there are unresolved difficulties in the estimation of sodium content which is very imprecise; the weight of leucocytes which can be harvested from 20 ml of blood is very small and it is difficult to allow satisfactorily for the suspending medium trapped in the final centrifuged deposit of cells.

Skeletal muscle analyses, though representing the largest part of the cell mass, are also less than ideal. Some of the assumptions on which allowance for the ECF content of the sample was based in the past are now known to be invalid, and the possibility of alterations in cell membrane permeability casts doubt on others. Moreover, variations have been found with age and sex, and differences have been reported in the electrolyte content of different muscles.

These approaches are not applicable to the acutely ill patient. Fortunately in most such cases it is possible to manage the patient satisfactorily without them, by evaluating the problem from a combination of the history, which can yield very useful information on the amount of fluid lost, the physical examination, measurements of the plasma concentrations of electrolytes and urea, and a measurement reflecting total protein concentration such as plasma specific gravity or refractive index. The interpretation of these observations must be

---

* Strictly speaking, the concentration in the plasma water should be used and an allowance made for the Donnan equilibrium, but these corrections approximately cancel out.

based on knowledge of the changes which are known to occur in different pathological states and of the homeostatic responses to them. It must be remembered that changes in extracellular osmolality due to alteration of the sodium and glucose concentrations will result in a shift of water into or out of the cells, but that changes in plasma urea will not unless they occur very rapidly; that the loss or retention of sodium will, if the homeostatic mechanisms are intact, lead to the loss or retention of water, provided it is available; and that the homeostatic responses will be triggered by changes in concentration in some cases and by changes in volume in others.

Sodium Depletion

The normal dietary intake of sodium varies from about 50–150 mmol per day. There is in addition a very large turnover of digestive secretions so that the amount of sodium secreted into the digestive tract daily is five or six times the dietary intake. This is almost all absorbed in the small intestine and an amount equivalent to the dietary sodium is excreted in the urine. A small amount of sodium is absorbed in the colon, probably in exchange for potassium, and a very small amount of sodium is excreted in the faeces and a little in sweat. Sodium depletion is usually caused by the loss of gastrointestinal secretions or by loss through the kidney due to renal or endocrine disease. The loss of other transcellular fluids, for example CSF or pleural fluid, will, if large enough also cause sodium depletion. The normal renal mechanisms for sodium homeostasis are so efficient that a low sodium intake alone does not cause severe sodium depletion except after a prolonged period.

The composition of transcellular fluids is variable so that their loss results in the loss of various proportions of water and sodium (Table 2). The effect on the body fluid compartments and on the plasma sodium concentration is dependent on these proportions and on the nature of any replacement fluid that might be given.

TABLE 2
SOME SOURCES OF ELECTROLYTE LOSS

	Na	K	Cl
Gastric juice (acid)	70	10	125
Bile	130	4	110
Small intestinal fluid	120	10	100
Faeces in diarrhoea	50	30	50
Rectal mucus* (villous papilloma)	100	40	100
Pleural and peritoneal fluids	as for plasma		

* Note:- Although published reports emphasize the potassium loss, the sodium loss is probably of greater importance (see Fig. 4)

These are average values (mmol/l) but are suitable for calculating treatment in a high proportion of cases. When the losses are very large it is better to base treatment on a quantitative analysis of the fluid, since the composition of a given fluid does vary from case to case.

Sodium Depletion without Water Depletion

The sequence of events induced by sodium depletion was studied by McCance[4] by making normal subjects sweat and then replacing only the water which had been lost, so that sodium depletion without water depletion was induced. At first the plasma sodium concentration remained unchanged, but the body weight declined (Fig. 3) because the water ingested was excreted in the urine

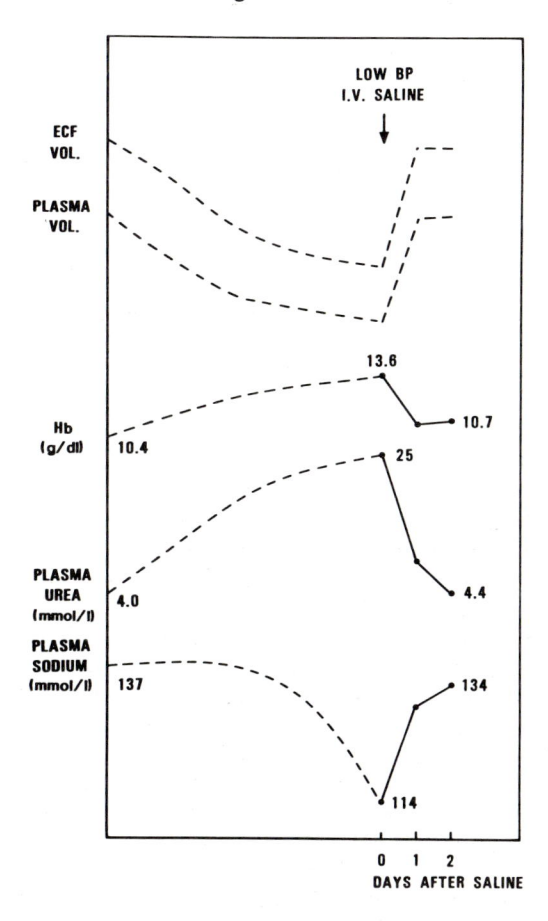

FIG. 3. Changes in extracellular fluid composition caused by the loss of bile and its replacement mainly with water. The numbers are the values found at the times indicated,
○——○ observed changes
————— presumed changes in accordance with McCance's observations.

and a negative water balance ensued. This resulted in contraction of the ECF volume, including a reduction in plasma volume as indicated by increased concentrations of red blood cells and plasma proteins. Plasma urea concentration was also raised by the reduced renal blood flow consequent upon the contracted plasma volume. Giving water in the face of sodium loss resulted at first in excretion of dilute urine, owing to inhibition of ADH secretion, which maintained the isotonicity of the extracellular fluid at the expense of its volume.

However, after several days the weight loss ceased because water retention began and a fall in the plasma sodium concentration ensued. Presumably, release of ADH was stimulated by volume depletion, this stimulus overriding the inhibitory effect of the resultant hyponatraemia. Thus, in the later stages of sodium depletion maintenance of plasma osmolality is sacrificed for the maintenance of plasma volume.

This experiment is sometimes inadvertently repeated in clinical practice when the loss of some secretion containing relatively large amounts of sodium is replaced by the administration of either water alone or 0·18% sodium chloride in 5% glucose, so called fifth-normal saline, which contains only 30 mmol sodium chloride per litre. An example of this is illustrated in Fig. 3.

In a man aged 62, bile, lost through a T-tube after surgery for gallstones, was replaced with fifth-normal saline (see Table 2). After some days a dramatic fall in blood pressure occurred. Plasma urea and electrolyte concentrations at this stage are shown in the figure together with a diagrammatic representation of the events which must have occurred before the onset of hypotension, based on the observations of McCance. Clearly, the plasma volume depletion resulting from the inadequately replaced loss of sodium had stimulated ADH release and also lowered the cardiac output to a level at which the blood pressure could not be sustained. The changes in plasma volume which had occurred are indicated by the changes in the concentrations of blood haemoglobin and plasma urea that occurred with rehydration.

The level of plasma urea found in such cases is extremely variable. It depends not only on the amount by which the glomerular filtration rate is reduced, but also on its duration, on the rate of protein catabolism and on increased tubular urea reabsorption, caused by reduced tubular flow rate. A rapid, large loss of circulating volume will cause a fall in blood pressure, calling for vigorous treatment, before there is time for the plasma urea to rise even though the glomerular filtration rate may fall to zero. Slowly progressive changes on the other hand, associated with maintenance of the blood pressure, may cause the plasma urea to rise by 5–10 mmol/24 hours to reach a level as high as 60 mmol/l entirely due to impairment of the renal blood flow. There will be a small volume of concentrated urine containing a high level of urea but hardly any sodium because the fall in circulating volume resulting from sodium depletion stimulates ADH release and the sodium-retaining mechanisms. Pre-existing renal disease or the supervention of acute tubular necrosis will cause the urea to rise even faster.

The onset of hypotension in such a case is due primarily to plasma volume contraction and not to the low sodium concentration. This is illustrated by the following case (Fig. 4).

A woman aged 62 who presented with a detachment of the retina was found to be hypotensive with a plasma sodium concentration of 99 mmol/l resulting from sodium loss in large volumes of mucus secreted by a villous papilloma of the rectum. Resuscitation was begun immediately with intravenous isotonic saline, before the plasma electrolyte results were known. By the time they became available the blood pressure had been restored to normal by the infusion of 1½ litres of normal saline in about one hour. The 225 mmol of sodium administered would be a small fraction of the total sodium deficit and is therefore most unlikely to have brought about resuscitation by

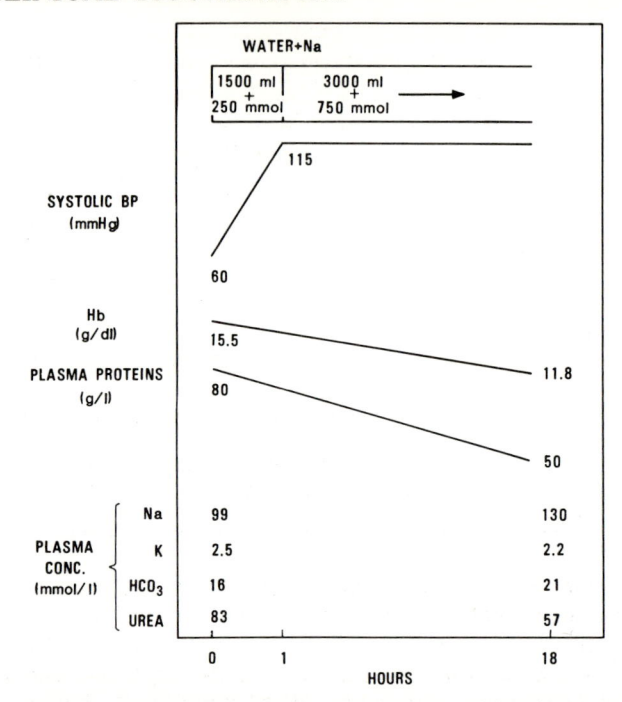

Fig. 4. Circulatory and biochemical changes in a patient with severe sodium depletion due to the loss of mucus from a villous papilloma of the rectum.
Note the large fall in plasma urea in the first 18 hours. A high urine flow rate was maintained by means of a high fluid intake, and the plasma urea returned to normal in five days.

elevating the plasma sodium concentration. The importance of volume expansion, even in patients with hyponatraemia, is further emphasized by the effect of slowing the rate of infusion too much after the initial resuscitation. As hypovolaemic patients with normo-natraemia this induces a recurrence of hypotension as the crystalloid solution equilibrates with the interstitial fluid, but the blood pressure rises again when the rate of infusion is increased to a level that ensures retention of much of the fluid in the vascular compartment.

Sodium Depletion with Water Depletion

The loss of sodium in gastrointestinal secretions or through the kidney, without water replacement, will give a different picture. The loss of any fluid with a sodium concentration approximately the same as that in plasma (Table 3) will obviously deplete the extracellular fluid volume without alteration of the plasma sodium concentration, apart from a tendency for it to rise slightly as a result of insensible water losses. The absence of a change in plasma sodium concentration means that there will be no shift of water between the cells and the ECF, so that the fall in plasma volume will be unabated. Progressive contraction of the plasma volume will lead, as before, to haemoconcentration, to a rising level of plasma urea, and ultimately to hypotension. As before, the sodium retaining mechanisms will be stimulated so that although the plasma sodium concentration is normal, sodium will disappear from the urine and the release of ADH,

TABLE 3

A COMPARISON OF THE CHANGES RESULTING FROM SODIUM DEPLETION (THE CASE IN FIG 3) AND A CASE OF INAPPROPRIATE SECRETION OF ADH. (THE CONCENTRATIONS ARE IN MMOL/L)

	Sodium Depletion	SIADH
Plasma urea	25	3·0
Plasma sodium	115	115
Plasma potassium	3·8	4·0
Plasma chloride	76	75
Plasma bicarbonate	24	27
Urine sodium	2	80
Haemoconcentration	present[1]	absent[2]
Blood pressure	low or tending to fall	normal

1. Haemoconcentration may be masked by preexisting anaemia and hypoproteinaemia.
2. Some degree of haemodilution should be present but the possibility of anaemia makes the interpretation of the haemoglobin concentration difficult.

stimulated by contraction of the plasma volume, will produce a small volume of concentrated urine. *It cannot be stressed too strongly that a normal plasma sodium concentration does not exclude sodium depletion which may be so severe as to make death imminent.* Such a case is illustrated in Fig. 5.

A different pattern develops from the loss of fluid with a much lower sodium concentration than plasma, e.g. acid gastric juice (Table 2). In the absence of appropriate replacement this causes the plasma sodium concentration to *rise*, thus inducing a shift of water out of the cells. Consequently, the rise in sodium concentration is minimized and the contraction of the extracellular fluid volume is less than when sodium and water are lost in equivalent amounts. But if the losses continue without replacement the changes must progress and the extracellular volume will continue to contract as the plasma sodium concentration continues to rise. It is therefore perfectly possible for circulatory failure to occur from sodium and water depletion with elevation of the plasma sodium level. In practice, there is seldom more than a modest elevation of plasma sodium arising from the loss of gastric juice, probably because some replacement is usually administered. The highest levels of plasma sodium in patients with significant sodium depletion are probably seen in infants with diarrhoea.

Some other causes of sodium depletion will be dealt with later when the problem of primary water retention is discussed.

Hypernatraemia

Water Depletion with Relatively Little Sodium Depletion

The loss of water without sodium must cause a rise in the plasma sodium concentration, and this will induce a shift of water out of the cells. The rise in sodium concentration caused by the loss of a litre of water from the ECF is greater than that caused by the loss of a litre of fluid containing 80 mmol of sodium. Therefore the former will result in a larger shift of water out of the cells to re-establish the osmotic equilibrium across the cell membrane, and hence will have the smaller residual deficit of the ECF volume. The more sodium there is in the fluid lost, the larger will be the fall in the ECF volume.

The progressive loss of water therefore leads to a rising plasma sodium concentration with minimal contraction of the ECF volume. The latter will nevertheless decline and, as in the case of sodium loss, there will ultimately be a rise in the plasma urea level though not to the same extent as in sodium depletion. As the ECF volume contracts, sodium retention will be stimulated so that the small volume of concentrated urine due to ADH release will have a low sodium content despite the high plasma levels.

Conditions in which water depletion predominates are as follows:

1. *Inadequate intake*

This occurs either when fresh water is not available or when some physical disability prevents drinking. Examples are:

(a) Carcinoma of the oesophagus.

(b) Acutely ill elderly people living alone may be too weak to get themselves a drink; they sometimes lie undiscovered for several days after a stroke, losing extra water because of hyperventilation.

It is not unusual to see patients in these categories with a plasma sodium level of 150–155 mmol/l on admission to hospital.

(c) Patients with high fever who are unable to

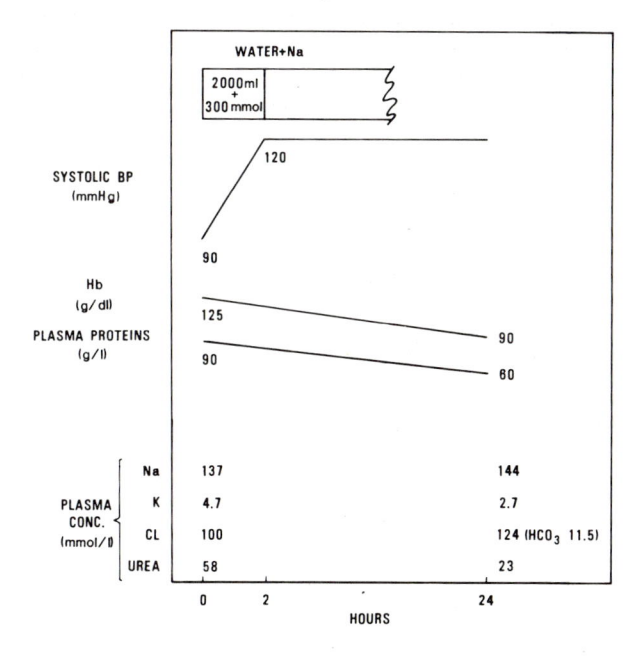

FIG. 5. Acute hypotensive circulatory failure due to sodium depletion secondary to gastric aspiration and diarrhoea. Note the normal plasma sodium concentration. The initial contraction of the ECF volume is indicated by the fall in haemoglobin and plasma protein concentrations with treatment. The final plasma chloride concentration indicates that a mixture of chloride and bicarbonate would have been more appropriate than saline alone; this is not of great importance during rapid resuscitation, when the important point is that the fluid used should be isotonic with respect to sodium. Note how rapidly the plasma urea began to fall, a normal level being attained in 72 hours.

complain of thirst, e.g. babies, or adults in coma. The development of hypernatraemia is a good indication that they are not receiving enough water.

(d) Brain tumour or other lesions which destroy the thirst mechanism.

2. *Excessive water losses despite hypernatraemia*

(a) Diabetes insipidus. Usually thirst ensures an adequate water intake, but when diabetes insipidus develops after a head injury the water intake will almost certainly be inadequate if the urine output is not carefully measured and the fluid balance calculated daily. A rising plasma sodium concentration in the absence of a large sodium intake should direct attention to this diagnosis which, in these circumstances, is usually easy to confirm by measuring the urine concentration.

(b) Resetting of the osmoregulatory centre upwards. This, may account for some cases of hypernatraemia associated with severe injuries to the brain.[5] The plasma sodium level may be maintained anywhere between 150–170 mmol/l, and administering water fails to lower it. This is, of course, what happens in unrecognized diabetes insipidus, but the two conditions should be distinguishable by measurements of the changes in urine concentration and plasma ADH level. In diabetes insipidus the urine concentration will be inappropriately low with any level of plasma sodium. With an osmoregulatory centre reset at, say, a plasma sodium level of 160 mmol/l, the urine will be dilute if the water intake is sufficient to lower the plasma sodium concentration slightly below this. But a slight rise above 160 mmol/l as a result of, say, 8 hours of water deprivation, will induce a 'normal' ADH response and a rise in urine concentration, i.e. the ADH response will be similar to that in Fig. 1 but shifted to the right because the threshold osmolality will be high. (See also Fig. 6.)

(c) Osmotic diuresis. This occurs when the renal tubular fluid contains solute that cannot be reabsorbed and which therefore holds water in the tubule. The intrarenal mechanisms are complex but lead to a large volume of urine, containing about 30 mmol/l of sodium, even if there is ADH secretion. The plasma sodium concentration will therefore rise if the replacement of water is inadequate.

Diabetes mellitus with glycosuria is a common cause of osmotic diuresis. But the resulting tendency to hypernatraemia is opposed by the withdrawal of water from the cells by the high plasma glucose level. Insulin-induced uptake of glucose by the cells will cause water to return to the cells and the plasma sodium level will tend to rise further, especially if isotonic sodium solutions have been administered; this is because much of the sodium is retained while much of the water is lost in a continuing osmotic diuresis, which will persist until control of the diabetic state is re-established.

High nitrogen excretion due to high protein feeding or hypercatabolism is another cause of osmotic diuresis. Because of the limit to urine concentration a very large solute load requires a large volume of urine. The urine output may therefore appear to be 'satisfactory' even though the patient is becoming progressively water depleted (Fig. 7). A rising plasma sodium

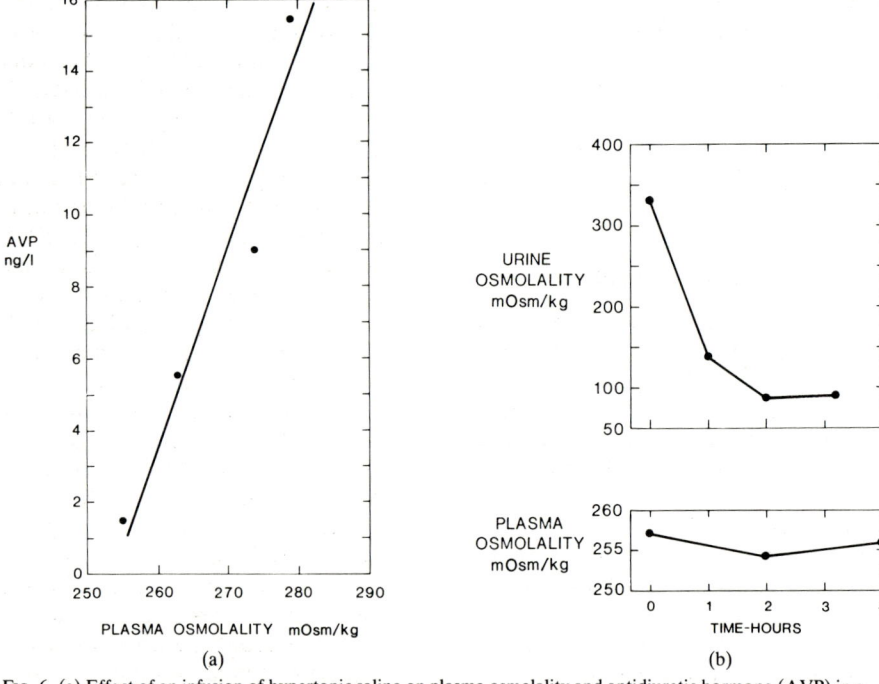

FIG. 6. (a) Effect of an infusion of hypertonic saline on plasma osmolality and antidiuretic hormone (AVP) in a case of infectious polyneuritis with hyponatraemia. (b) Changes in urine and plasma osmolality after an oral waterload of 20 ml/kg body weight given to the same patient when the plasma sodium was 125 mmol/l. Reproduced, with permission from Penny et al.[10]

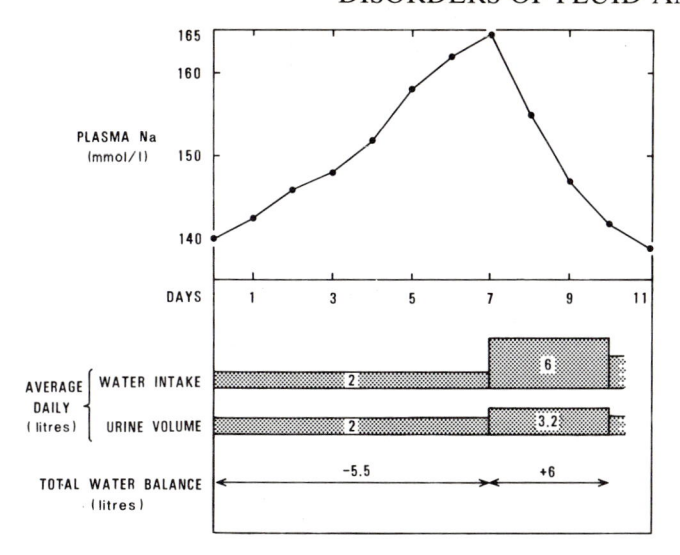

FIG. 7. Plasma sodium concentration in a case of water depletion induced by high protein feeding. A male aged 23 with multiple injuries excreted 45–60 g urea/day when given 80 g protein/day through a nasogastric tube. Note that for the first 7 days when the sodium concentration was rising the urine output was 'good', but that on average it equalled the water intake of 2 litres/day so that the patient was in negative balance by the amount of insensible loss, taken as 800 ml/day. Increasing the fluid intake to 6 litres/day increased the urine output and also led to a positive balance which corrected the abnormality in 3 days.

concentration can be avoided only by ensuring an adequate water intake, making due allowance for insensible losses. It is desirable to allow an intake of about 100 ml of water for every gram of nitrogen (2 g urea) excreted.

There is clearly no sharp dividing line between the changes observed with water loss alone and those due to water loss accompanied by small amounts of sodium, the cumulative effects of which may ultimately cause acute circulatory failure. A striking example of this is provided by infants with gastroenteritis in whom circulatory failure may occur with the plasma sodium concentration as high as 160 to 190 mmol/l. This requires sodium administration as well as relatively more water to restore the extra-cellular fluid composition and its volume to normal.

Hypernatraemia Due to Excessive Sodium Intake

This sometimes arises from the administration of hypertonic sodium solutions intravenously, notably 8·4% sodium bicarbonate to correct acidosis, especially in patients with cardiac arrest. As much as one or two litres have sometimes been given, presumably due to a failure to appreciate that this solution contains 1000 mmol/l of sodium!

Hypernatraemia of the order of 170 mmol/l has been reported due to hypertonic saline enemas, to the administration of high salt-containing feeds to babies, and to the use of a salt solution as an emetic in cases of poisoning. In these the urine will be concentrated as before but with a high sodium content, because the renal tubular reabsorption of sodium will be inhibited as a result of expansion of the ECF volume by the withdrawal of water from the cells. The latter also ensures, in the absence of renal failure, a normal plasma urea level. In practice, of course, the way to distinguish between hypernatraemia due to excessive sodium intake and that due to water depletion is from the history.

Effects of Hypernatraemia and their Therapeutic Implications

Patients with severe hypernatraemia are usually very ill so that clinical manifestations may be due to the primary illness. But where hypernatraemia developed acutely in babies due to the inadvertent administration of excessive amounts of salt, there was a rapid onset of confusion, followed by coma and, in some cases, death: this is attributable to cellular dehydration, especially in the brain. In the fatal cases there were striking changes in the brain, including petechial haemorrhages and subdural haemorrhage, thought to have been due to traction induced by rapid contraction of the brain volume. Although progressive water depletion ultimately proves fatal, the gradual development of a plasma sodium level of 170 mmol/l in adults does not usually cause coma; some form of adaptation thus seems likely. Infants and young children with hypernatraemia due to gastroenteritis frequently exhibit fits and coma; the acute mortality rate is high and residual neurological abnormalities are common, possibly as a sequel to the anatomical lesions mentioned above. In these infants and children fits are also associated with too rapid correction of the hypernatraemia.

The effects of acute and less acute hypernatraemia have been studied experimentally in animals[6] and several differences between the two states have been shown. The rapid induction of hypernatraemia caused coma and death; the osmolality of the brain tissue was increased and its water content was correspondingly reduced. When hypernatraemia persisted for several days the brain osmolality again increased, but the fall in water content was less than expected from the rise in osmolality; it is suggested that in this case the increased osmolality was due to the formation and retention within the brain cells of organic molecules of relatively small molecular weight to which the name 'idiogenic osmoles' has been given, and that this is an adaptive response to prevent too much shrinkage of the brain cells.

The implications of these studies for treatment are that when hypernatraemia occurs acutely, its rapid correction by the administration of water will allow the brain volume to return to normal with benefit. But it is suggested that when hypernatraemia has been present for some days rapid restoration of extracellular fluid tonicity will result in overhydration of the brain cells because of the intracellular idiogenic osmoles which draw water into the cells and which dissipate only slowly. In practice it seems to take adults 24–48 hours to achieve a large enough positive water balance to correct severe hypernatraemia and this does not seem to be associated with adverse effects. In infants with diarrhoea it is generally recommended that, after initial correction of any circulatory failure present with a solution of half-strength plasma or its equivalent, the hypernatraemic state should be corrected at a rate of between 1–2 mosmol/kg plasma water/hr, since these rates have been found not to induce fits. It must be emphasized however that a rate of 2 mosmol/kg plasma water/hr enables a plasma sodium concentration of 165 mmol/l to be corrected to 140 within 24 hours, and even the slower rate allows correction of a sodium level of 155 in a little over 24 hours. Slower rates of correction leave the child exposed longer to the risks of hypernatraemia and their sequelae.

Excessive Water Retention

Although sodium loss gives rise to hyponatraemia in some circumstances, hyponatraemia does not always indicate sodium loss. Conceptually, it merely indicates the presence of proportionately less sodium than water in the ECF. This may be due to the loss of sodium with the loss of relatively less water, or it may be due to the retention of water with the retention of proportionately less sodium. There are in fact many causes of water retention, in some of which retention of sodium also occurs.

It must be remembered that spurious hyponatraemia may be encountered when a blood sample is taken from a vein receiving an infusion of fluid low in sodium. Hyperlipidaemia and occasionally hyperproteinaemia in myelomatosis may also cause an apparently low plasma sodium concentration owing to the volume occupied by the lipoprotein or paraprotein. In such a sample a method which measures the sodium concentration in the plasma water must be used. Alternatively, an approximate value may be derived from the serum osmolality.

In order to excrete a water load efficiently the kidney must be able to elaborate dilute urine. As described above, this requires that ADH secretion be inhibited and also that the tubular fluid be dilute when it enters the collecting tubules, which is achieved by extrusion of sodium chloride in the diluting segment of the tubule. Interference with the normal inhibition of ADH secretion or with the function of the diluting segment may therefore prevent the excretion of solute-free water and hence cause hyponatraemia.

The Syndrome of Inappropriate Secretion of ADH (SIADH)

This term was introduced by Schwarz and Bartter[8] to refer to cases of carcinoma of the bronchus with hyponatraemia. It has since been described in association with other tumours, e.g. carcinoma of the pancreas and various lymphomas, and is thought to be due to the secretion of an ADH-like substance by the tumour. The plasma sodium concentration is often below 120 mmol/l, with levels as low as 100 mmol/l having been encountered. The expanded extracellular volume inhibits the sodium-retaining mechanisms and sodium continues to be excreted in the urine. A typical set of electrolyte concentrations in this condition is shown in Table 3, together with data from a patient with sodium depletion.

The pattern in SIADH differs from that of sodium depletion in that the plasma urea is normal or even low, there is abundant sodium in the urine if the intake is normal, there is no haemoconcentration, the blood pressure is normal and shows no tendency to fall in the upright posture, and thirst is unlikely. The acid-base state is usually normal unless other factors are present. In both cases the urine will be inappropriately concentrated for the plasma sodium concentration though it may not exceed the concentration of plasma. The criteria for diagnosis laid down by Schwarz and Bartter are given in Table 4.

TABLE 4

CRITERIA FOR THE DIAGNOSIS OF THE SYNDROME OF INAPPROPRIATE SECRETION OF ADH[8]

1. Hyponatraemia with corresponding hypo-osmolality of the plasma.
2. Absence of clinical evidence of fluid volume depletion.
3. Concentration of the urine greater than that appropriate for the prevailing tonicity of the plasma.
4. Continued renal excretion of sodium.
5. Normal renal function.
6. Normal adrenal cortical function.

Hyponatraemia Without a Source of Ectopic ADH

The term SIADH is often applied to other cases of hyponatraemia without a tumour but in which the diagnostic criteria of Schwarz and Bartter are present. The number of such conditions is very large (Table 5) and in

TABLE 5

SOME CAUSES OF THE SYNDROME OF INAPPROPRIATE SECRETION OF ADH (EXCLUDING A SOURCE OF ECTOPIC HORMONE PRODUCTION)

Trauma:
    accidental and surgical
Infections:
    pneumonia, peritonitis, septicaemia, urinary infections, tuberculosis
Pulmonary disease:
    chronic bronchitis and emphysema
    artificial ventilation
Endocrine disorders:
    hypopituitarism
    Addison's disease
    hypothyroidism
CNS disorders:
    head injury
    infections
    tumours
    cerebrovascular accidents
    acute psychosis
    acute intermittent porphyria
    acute infective polyneuritis
    acute toxic polyneuritis (vincristine)
Drugs which interfere with the release of ADH or potentiates its action on the renal tubule:
    chlorpropamide
    vincristine, cyclophosphamide
    fluphenazine, carbamazepine, amitriptyline
    clofibrate
    thiazides

some of these ADH levels have been shown to be raised, presumably secreted by the posterior pituitary. In most cases the mechanism of increased ADH secretion is not understood, and sometimes it may be appropriate for the preservation of volume. Some of the conditions in Table 5 are rare but some are common and will be considered in more detail.

(a) After injury or surgery. Hyponatraemia is not uncommon within the first two or three days after surgical operations and may occur after any major injury. It has been attributed to the administration of too much hypotonic fluid and also to a shift of sodium into the cells because of damage to the cell membrane, i.e. a 'sick-cell'. The latter on its own would cause water to enter the cells with sodium leaving the extracellular sodium concentration unchanged. Flear[9] attributed the hyponatraemia in the 'sick cell syndrome' to the escape from the cell of organic solutes, as yet unidentified, through a damaged cell membrane. As a result, water is drawn out of the cells and the extracellular sodium concentration falls. This explanation is based on his observation that the plasma osmolality in such cases was higher than that calculated from the sodium, urea and glucose concentrations.

This sick cell syndrome may occur in any ill patient but it is very uncommon. A far more common cause of hyponatraemia, even in patients with infections, is the administration of too much water or hypotonic saline when the patient is in a water-retaining state due to ADH secretion as part of the metabolic response to injury. Such patients may, surprisingly, excrete more than three litres of urine per day if the volume of fluid administered is very large. But they do not as a rule excrete the whole of the water load so that water retention gradually builds up. The duration of ADH secretion after injury varies with the severity of the injury, and sometimes it may be very prolonged. Hyponatraemia then occurs if too much water is given, but the measured plasma osmolality is not different from the calculated osmolality. As in the SIADH due to tumours, overexpansion of the extracellular fluid volume causes the loss of sodium in the urine.

The mechanism which triggers such prolonged water retention is not understood, but there is evidence that in some cases at least, osmoregulation may be reset at a lower level so that ADH secretion will be switched on and off around a level of plasma sodium below the normal. This is illustrated in Fig. 6 which shows data from a case of infective polyneuritis with hyponatraemia; the plasma sodium settled at a new stable level of 125 mmol/l, but at this level a waterload was excreted normally and the infusion of enough hypertonic saline to increase the plasma sodium concentration into the upper part of the normal range evoked a marked ADH response.

(b) Addison's disease. It might be thought that the inability to excrete a waterload normally in Addison's disease could be adequately explained by the stimulation of ADH release by the contraction of the ECF volume resulting from sodium depletion. Correction of this abnormality by intravenous saline has been reported, but this cannot be the whole explanation since the same phenomenon occurs in hypopituitarism in which aldosterone secretion is unimpaired. A fall in the glomerular filtration rate, frequently indicated by a raised plasma urea concentration, would also impair water excretion by reducing delivery of fluid to the distal part of the nephron. Water handling however is restored to normal in both Addison's disease and hypopituitarism by cortisol. The exact nature of the abnormality is uncertain but it seems that cortisol has an effect on the collecting ducts necessary for maximal impermeability to water.

(c) Hypothyroidism. A small proportion of patients with hypothyroidism develop hyponatraemia with the features of water retention. The mechanism is not understood but is reversed by thyroxine.

(d) Drugs. (i) *Diuretics*.[11] It is uncommon for diuretics to cause hyponatraemia but they may do so in several ways. Those which inhibit sodium reabsorption in the diluting segment will thereby diminish the excretion of free water. Hyponatraemia may then occur due to water retention; the plasma urea remains normal as in the SIADH, but unlike the latter hypokalaemia is often present. Recovery occurs when the diuretic is stopped, often within a few days. More rapid recovery can be achieved if necessary by giving salt with very little water (see below).

There is evidence to suggest that, in some cases at least, hyponatraemia is a consequence of severe potassium depletion, which induces a shift of sodium into the cells and possibly interferes with the regulation of ADH secretion.

A third mechanism by which diuretics may produce hyponatraemia is sometimes seen when vigorous diuretic therapy is continued after the oedema has gone. Sodium depletion and contraction of the ECF volume ensue with malaise, weakness and a rising plasma urea level in the presence of a falling plasma sodium concentration.[12] The cautious administration of salt and water restores the extra-cellular fluid to normal, with a resultant fall in the plasma urea and symptomatic improvement. There is often reluctance to administer salt to such patients because of the fear that it will precipitate a relapse into heart failure. But the therapeutic response can be very dramatic as in the following case:

A 52-year-old woman underwent mitral valve replacement, after which diuretics were continued for six weeks in the same dosage as before. She was admitted to hospital with vomiting, generalized muscle aches and weakness. Plasma sodium concentration was 120 mmol/l, potassium 2·6 mmol/l, chloride 70 mmol/l and bicarbonate 27 mmol/l; plasma urea was 52 mmol/l. There was no elevation of the jugular venous pressure and no peripheral or pulmonary oedema. In order to keep the water administration as low as possible initially, one litre of hypertonic saline containing 600 mmol sodium chloride was given intravenously over 8 hours. This restored the plasma sodium to normal and only 41 mmol of sodium was excreted in the first 12 hours, in contrast to the response observed in hyponatraemia due to water retention (Table 6). The vomiting and muscle pains had ceased by 36 hours after admission and the urea was normal within a few days. There was no recurrence of heart failure.

TABLE 6

EFFECTS OF IV HYPERTONIC SALINE (450 MMOL/L) GIVEN ON DAYS 1 AND 2 FOR WATER INTOXICATION IN A 95-YEAR-OLD WOMAN WITH ALL THE FEATURES OF THE SIADH FOLLOWING A FRACTURED NECK OF FEMUR

Days	1	2	3	4	5	Total
Plasm sodium (mmol/l)	111	121	134	128	123	
Sodium intake (mmol/24 hrs)	225	450	150	0	0	825
Sodium in urine (mmol/24 hrs)	*	300	150	150	120	720+

* Not measured, but the sodium concentration in a random specimen was 80 mmol/l

Note. The data illustrate: (a) the efficacy of the treatment; (b) its safety, given normal cardiovascular and renal function; (c) the rapid recurrence of hyponatraemia when the water intake was not curtailed after the treatment.

It was quite unnecessary to give the litre of normal saline (150 mmol/l) on day 3, and the 5% dextrose in water which followed it 'to keep the drip open' was contraindicated. Unless there is a clear need for continuing i.v. therapy (which must be carefully formulated) it is best to discontinue i.v. fluids as soon as the hypertonic saline has been given.

(ii) *Other drugs.* Some drugs appear to interfere with the action of ADH and induce excessive water retention. These are listed in Table 5.

The intravenous infusion of Syntocin, a synthetic oxytocin with antidiuretic properties, is frequently used to induce labour. When infused in 5% dextrose solution it sometimes induces sufficient water retention to cause severe hyponatraemia and water intoxication, with fits and coma. Proper attention to the urine output will prevent this; should water intoxication occur, stopping the drug will result in spontaneous correction by a rapid diuresis.

Similarly, the treatment of diabetes insipidus with vasopressin also causes hyponatraemia if the water intake is inappropriately high.

Retention of Water and Sodium

This may be entirely iatrogenic due to the administration of so much saline that the patient is unable to excrete it as fast as it is being administered, even with normal kidney function. It is particularly liable to occur in patients with acute oliguric renal failure or with chronic renal failure in which the number of functioning nephrons may be too small to handle the additional load. The plasma sodium concentration will depend on the relative amounts of salt and water that are given. If the water intake is higher than the sodium intake hyponatraemia will develop.

Hyperaldosteronism and excessive cortisol secretion are other conditions in which sodium and water retention occur, sometimes with hypernatraemia.

Fluid Retention with Oedema

Oedema occurs when the net movement of fluid out of the capillaries exceeds the capacity of the lymphatics to remove it. Apart from lymphatic obstruction the factors which promote this are increased capillary hydrostatic pressure, decreased plasma oncotic pressure, and increased capillary permeability.

A small rise in the amount of fluid in the interstitial space evokes a sharp rise in tissue hydrostatic pressure. This will increase the lymphatic drainage and oppose the flow of more fluid out of the capillaries, but above a certain level there is a dramatic fall in the pressure response to a given volume increase, and fluid then accumulates.[13] The mechanisms of fluid retention in some cases of oedema are complex and not fully understood, especially those in which hyponatraemia is common.

(a) Congestive heart failure. With venous congestion there is increased capillary pressure, and oedema forms. In severe cases with massive oedema, hyponatraemia develops and is a bad prognostic sign. The exchangeable sodium is grossly increased, with an increased amount of it in the cells, and there is proportionately more water retention due to ADH secretion. There is usually very little sodium in the urine owing to hyperaldosteronism.

The exact mechanism of these changes is not fully understood. The low cardiac output is thought to be the stimulus for ADH release, perhaps through the baroreceptors, and also for the release of renin and aldosterone through a reduction in renal blood flow. Volume receptors per se should be inhibited as the blood volume and atrial pressures are high, but they are over-ridden by stimulation of the other receptors. A low cardiac output will also reduce the glomerular filtration rate and so further impair water excretion by reducing the delivery of fluid to the distal tubule.

(b) Hepatic cirrhosis with portal hypertension. Patients with advanced cirrhosis and portal hypertension may become oedematous because of a fall in the plasma oncotic pressure due to hypoalbuminaemia, and they also develop ascites. These changes cause the plasma volume to fall thus leading to the release of both ADH and aldosterone and hence water and sodium retention. In advanced cases there appears to be retention of proportionately more water than sodium leading to hyponatraemia, possibly because changes in intrarenal blood flow and impairment of glomerular filtration rate also affect free water excretion adversely. As in heart failure, there is avid sodium retention with very low levels in the urine due to hyperaldosteronism.

(c) The nephrotic syndrome. The severe hypoalbuminaemia which results from the renal loss of albumin lowers the plasma oncotic pressure and so promotes oedema formation. This tends to lower the plasma volume and stimulates sodium- and water-retaining mechanisms. It should be remembered that in this condition hyponatraemia may be spurious owing to the hyperlipoproteinaemia.

(d) Intermittent idiopathic oedema.[14] This is a distressing condition which seems to occur only in women. There is intermittent pitting oedema, and sometimes abdominal swelling, unrelated to the menstrual cycle. The mechanism is unclear and there is almost certainly more than one entity. In some patients a sudden change from supine to an upright posture precipitates a large fall in plasma volume associated with sodium retention, but tthi is not an invariable finding. In others, the intermittent use of diuretics, possibly to lose weight or to remove 'oedema', has been a causal factor; sodium retention and oedema recur rapidly on stopping the diuretics, and are attributed to hyperaldosteronism induced by chronic volume depletion. Permanently discontinuing the diuretics leads to recovery in most of the patients. Subclinical hypothyroidism is also associated with this form of oedema, which is relieved when therapy is instituted.

Effects of Hyponatraemia and Their Therapeutic Implications

It is a common clinical observation that, whereas some patients with a plasma sodium concentration of 115 mmol/l due to water retention will develop the syndrome of water intoxication with anorexia, vomiting, confusion, headaches, muscle cramps, convulsions, coma, and ultimately death, others with a similar plasma sodium concentration may have none of these features. The difference is probably related to the rate of fall of sodium level. A large fall occurring in a few hours will induce water intoxication, whereas development over a number of days probably will not. Some adaptation therefore seems likely in the slowly developing cases.

In animals, the rapid induction of hyponatraemia was found to cause brain swelling due to an increased water content.[15] When the sodium level was allowed to fall over several days there was less brain swelling although the brain tissue osmolality was just as low because of the loss of potassium from the brain cells. Potassium is lost from the cells generally in longstanding hyponatraemia; similar adaptive changes have been observed *in vitro* with cells placed in a hypotonic solution.

These observations are relevant to treatment. If, in hyponatraemia, the brain cells in particular have adapted to the low ECF osmolality by extrusion of potassium, the rapid correction of the extracellular sodium concentration might be expected to cause cellular dehydration. If there are no symptoms or signs of water intoxication, correction may be achieved slowly by restricting the patient's water intake enough to cause a large negative balance. The total negative balance required to increase the plasma sodium concentration to 135 mmol/l is

$$\frac{(135 - \text{Plasma [Na]})}{135} \times \text{total body water}$$

where total body water equals 60% or 50% of body weight (kg) in males females respectively. Increasing the dietary sodium or giving isotonic saline intravenously while the ECF volume is still expanded merely results in increased sodium excretion.

Signs of water intoxication are an indication for more urgent correction of hyponatraemia which can be achieved by the intravenous infusion of *hyper*tonic saline, given over six to eight hours. The amount of sodium, in millimoles, required to raise its plasma level to 135 mmol/l is

$$(135 - \text{observed plasma [Na]}) \times \text{total body water}$$

To keep the water administered to a minimum, the sodium can be given as a molar solution of the chloride or, if there is an acidosis to be corrected, as a mixture of chloride and bicarbonate. This will correct the hyponatraemia and draw water out of the cells, thus transiently expanding the ECF volume still further; the excess sodium is excreted in the urine within 2–3 days and takes with it the excess water that has been withdrawn from the cells (Table 6). Once the plasma sodium concentration has been restored to normal or near normal in the SIADH, the improvement will only be sustained by continued water restriction.

Other Guides

The Anion Gap[16]

The anion gap is defined as the difference between the plasma sodium concentration* and the sum of the chloride and bicarbonate; it normally falls within the range 8–15 mmol/l. Its calculation is sometimes useful diagnostically and it is useful for quality control because an error in estimating one of the components will distort the anion gap.

Since the sum of the cations (meq/l) equals the sum of the anions (meq/l),

$$\text{Na} - (\text{Cl} + \text{HCO}_3) = \text{proteins} + \text{unmeasured anions} - (\text{K} + \text{Ca} + \text{Mg})$$

Hence the anion gap may change because of an alteration in any of the quantities on the right-hand side of the equation.

A rise in the anion gap is the more common abnormality and is usually due to a rise in the unmeasured anions in diabetic ketoacidosis or renal failure, and, less frequently to a large salicylate overdose. A small increase may occur in alkalosis owing to the increased anionic equivalence of the plasma proteins, and this may be combined with a further small increase due to increased protein concentration if the ECF volume is low. A much less common cause is lactic acidosis; penicillin and carbenicillin are anions in plasma and may also cause a high anion gap. Other causes which should be suspected

---

* Sometimes the sum of the sodium and potassium is used.

in obscure cases include certain poisons which are metabolized to acids, such as ethylene glycol and methanol, metabolized to oxalate and formate respectively, and ethanol which may cause lactic acidosis. A fall in the concentration of only one of the cations has too small an effect on the anion gap to be noticed.

A reduced anion gap is caused by a low plasma albumin concentration. Albumin normally contributes about 10 meq of the 15 meq or so of anions contributed by the plasma proteins. A fall in the plasma albumin to 25 g/l will therefore cause a fall of 5 meq/l in the total anions, but this is usually counter-balanced by a corresponding rise in the plasma chloride concentration which lowers the anion gap. A low gap may also be present with a high plasma concentration of potassium, magnesium or calcium, and high levels of lithium in cases of overdose will contribute. Recently it has been recognized that IgG has cationic properties and that high levels of paraproteins may result in an anion gap of only 1 or 2 meq/l, and even a negative one. Exceptionally high levels of polyclonal IgG also may cause the anion gap to be abnormally low.[17]

Plasma and Urine Osmolality Measurements

The measurement of the osmolality of both serum and urine is widely recommended for the study of fluid balance problems, but the significance of such measurements seems often to be misunderstood and their importance grossly exaggerated. More often than not, in clinical practice, they add nothing to the information derived from other routine measurements.

In health, the plasma osmolality depends mainly on the sodium concentration, with small contributions from urea and glucose. In disease the contribution from urea or glucose is sometimes very large, but these three substances together still account for virtually the whole of the osmolality except in the comparatively few instances when substances not normally found in plasma are present, e.g., mannitol or the unidentified metabolites in the 'sick cell syndrome' (see below).

In *hyperosmolal states* what really matters is the *reason* for the high osmolality, because this is what determines treatment. Thus, an osmolality of say 360 mmol/kg might be due to one of the following: (i) a plasma sodium concentration of about 175 mmol/l with a slightly raised urea and a normal glucose; (ii) normal concentrations of sodium and urea, but a glucose level of 70 mmol/l; (iii) normal sodium and glucose concentrations but a urea level of 70 mmol/l. These three conditions are clearly entirely different therapeutic problems, which can only be identified by knowing the concentrations of sodium, urea and glucose. If these are known, the osmolality can be calculated:* measurement of osmolality, in addition, is not helpful. It must be remembered that a high plasma osmolality does not necessarily indicate the need for treatment to dilute the body fluid because some substances, e.g., urea and ethanol, are distributed throughout the total body water and will not cause a change in cell size.

If there is marked elevation of only one of these factors, serial osmolality measurements can be used *instead* of specific measurements to follow the response to treatment, but these serve no purpose if measured *as well as* the specific factor. Moreover, measuring osmolality alone may sometimes be misleading. For example, a patient with diabetic ketoacidosis resuscitated with an infusion of isotonic sodium solutions commonly sustains a rise in plasma sodium concentration as the blood glucose falls in response to insulin. The fall in plasma osmolality will then be less than the fall in the plasma glucose level and could lead, mistakenly, to the administration of more insulin.

Hyperosmolality may be due to large amounts of toxic substances in the plasma, notably ethanol, methanol and ethylene glycol. Smell will direct attention to the first and, since it is usual to measure the electrolytes in coma of obscure origin, the others will be revealed by acidosis with a high anion gap.

With significant *hypo-osmolality* there must be hyponatraemia. However, the latter does not always mean hypo-osmolality because it occurs when a shift of water out of the cells is induced by high levels of some other solute, e.g., mannitol or glucose. This is the mechanism postulated in the 'sick cell syndrome'[9] in which intracellular organic metabolites which are normally confined to the cell are thought to escape into the ECF. The metabolites have not been identified and to diagnose this state it is necessary to show that the measured plasma osmolality exceeds the calculated osmolality. However, this is a very uncommon cause of hyponatraemia, most cases of which are due to the administration of too much hypotonic solution.

The value of measuring *urine osmolality* also seems to be misunderstood. It is a measure of the concentration of the urine just like specific gravity, and in normal urine there is a linear relationship between the two. For most clinical purposes a very accurate assessment of urine concentration is not required and specific gravity and osmolality give equally useful information. In the diagnosis of the syndrome of inappropriate secretion of ADH or diabetes insipidus, urine specific gravity is as useful as osmolality if it is remembered that a specific gravity of 1·010 is approximately equivalent to the osmolality of normal plasma, i.e. about 290 mmol/kg.

Urine osmolality measurements are, however, essential if it is necessary to know the urine concentration when there is reason to suspect the presence in the urine of substances which increase the specific gravity more than the osmolality, e.g., glucose, very large amounts of protein, mannitol, dextran-40, or radiographic contrast media. The quantitative study of free water excretion also necessitates the measurement of plasma and urine osmolality.

---

* There are many methods recommended for calculating the plasma osmolality.[18] The simplest is to double the plasma sodium concentration and then add the millimolar concentrations of urea and glucose if they are abnormally high.

Urinary Sodium Measurements

Reference has been made previously to the changes in urine sodium concentration that occur characteristically in different states. It does not follow, however, that measurements of urine sodium will necessarily contribute to the diagnosis or management of patients. Their usefulness is, in fact, very limited, and they are made far too often because their significance is misunderstood.

The question often asked when such measurements are being interpreted is what is the normal urine sodium concentration? There is no short answer to this because in health the 24-hour excretion of sodium can vary from zero to several hundred millimoles according to the intake; the concentration of sodium depends also on the urine volume. In pathological states the urine sodium levels are always within this range and in that sense are always 'normal', but what matters is whether the level is appropriate for the circumstances. For this to be determined other information is necessary which often, but not always, renders the measurement of little value.

A low urine sodium concentration of less than, say, 10 mmol/l, may, in the absence of extreme urine dilution, be due to sodium depletion. But it may also be due merely to a low sodium intake without clinically significant depletion. Moreover, such levels also occur when the exchangeable sodium is actually increased as in heart failure and cirrhosis of the liver, owing to secondary hyperaldosteronism.

Conversely a urine sodium concentration of more than, say, 40 mmol/l in a patient with hyponatraemia might be taken to indicate the syndrome of inappropriate secretion of ADH rather than sodium depletion. But additional information is necessary before this distinction can be made, because uncontrolled loss of sodium in the urine might be the cause of sodium depletion, as in Addison's disease, renal disease, diuretic therapy or osmotic diuresis.

Clearly, the urine sodium concentration cannot be interpreted on its own, but given other necessary information its measurement can on occasions be useful diagnostically. Thus, in a patient with clinical evidence of sodium depletion in whom the cause is not apparent, the urine content must be measured to determine whether this is the source of sodium loss. Likewise, in a patient acutely ill with diarrhoea and vomiting who, it is thought, might have Addison's disease, the measurement of urine sodium might enable the diagnosis to be refuted or strengthened according to whether the sodium level is very low or not.

In cases of hypernatraemia the cause should be apparent from the history, but in the absence of this measurement of the urinary sodium level will distinguish water depletion, in which it will be low, from sodium overload, in which it will be high. It follows that a low value associated with hypernatraemia should not lightly be taken to indicate that sodium retention is the cause of hypernatraemia.

The measurement of urine sodium excretion is often made ostensibly as a guide to treatment. This is useful in the relatively few patients whose kidneys cannot conserve sodium in the normal manner, as the obligatory threshold can be determined. But in most acutely ill patients the sodium that is in the urine is that which has been given in excess of the requirements, and it is misguided to measure the urinary sodium daily in order to replace it exactly in the following 24 hours.

POTASSIUM

Distribution

On average, a 70 kg man will contain a little less than 4000 mmol of potassium, or 55 mmol/kg body weight, although there are very large differences between individuals. Ninety-eight per cent of this is in the cells, held in proportion to the amount of water, protein and carbohydrate; three-quarters of it is in skeletal muscle and about 7% of it is in the bone. Naturally occurring potassium is mainly ^{39}K with a small proportion of radioactive ^{40}K, so that the total body potassium can be determined by whole-body counting of ^{40}K. The value obtained by this method is some 15% more than the exchangeable potassium measured with ^{42}K or ^{43}K. The ECF contains only 50–60 mmol in all so that large alterations in the extra-cellular concentration may result from the shift of comparatively small amounts of potassium into or out of the cells. The average concentration of potassium in the cells is around 130 mmol/l, but there are large differences between different tissues.

The normal dietary intake of potassium is variable but averages about 100 mmol/day. Both meat and vegetables are rich sources so that a diet adequate in other respects is unlikely to be deficient in potassium. Most of the dietary potassium is absorbed together with the potassium in the digestive secretions, and 90% of it is excreted in the urine. About 8–10 mmol/day are excreted in the faeces, probably as a result of exchange of potassium for sodium in the colon, and there is a little lost in the sweat. The amount in both sweat and faeces is increased by aldosterone.

The Renal Handling of Potassium and its Regulation[19]

The filtered potassium is completely, or almost completely, reabsorbed by the renal tubule, and what is excreted in the urine is added to the tubular fluid in the distal tubule in exchange for sodium. Reabsorption takes place mainly in the proximal tubule and the ascending limb of the loop of Henle, but probably continues by an active process along the whole length of the tubule, including the collecting duct. In the distal tubule, potassium enters the tubular fluid probably by passive diffusion down the electrochemical gradient created by the active reabsorption of sodium under the influence of aldosterone. Both hydrogen and potassium ions are exchanged for sodium ions at this site and the exchange process is dependent on the relative amounts of potassium and hydrogen ion in the tubular cells, as well as on the amount of sodium reaching the distal tubule. A low dietary sodium leads to a low urinary potassium, and a high sodium intake leads to a high potassium excretion, but this is not solely due to the amount of sodium available for reabsorption in the distal tubule. Although the exchange of potassium for sodium is increased by aldosterone, the effects of the hormone on sodium reabsorption and potassium excretion may be dissociated. Thus the administration of a mineralocorticoid causes sodium retention only for several days, after which sodium excretion 'escapes'. This, however, is not accompanied by a fall in potassium excretion.

The renal mechanism for regulating potassium levels is very responsive to changes in potassium status, and an intake as high as 200 mmol/day is easily tolerated by a corresponding increase in the urinary excretion. A low intake leads to renal conservation though it takes a number of days to achieve maximal conservation which, even then, is not as efficient as in the case of sodium, urine potassium levels seldom falling to less than 5 mmol/l.

The major factor influencing the excretion of potassium in the urine is aldosterone, which increases the exchange of potassium for sodium in the distal tubule. This is secreted not only in response to stimulation of the adrenal cortex by angiotensin II, whose activity reflects changes in sodium metabolism, but also in response to a high plasma potassium concentration. Conversely, a low plasma potassium concentration suppresses aldosterone secretion and may do so even with elevated renin levels.

When the potassium intake is high, the plasma level rises transiently during the absorptive phase, and falls in the post-absorptive phase in conjunction with a raised plasma concentration of aldosterone and increased renal excretion. Similar effects are produced by the intravenous infusion of potassium. The extent to which plasma potassium and aldosterone levels rise in response to the infusion of a given amount of potassium varies with the dietary intake of potassium. On a low oral intake of 40 mmol/day, the rise of plasma potassium is higher and that of aldosterone; lower than when the oral intake is 200 mmol/day. If the patient is in sodium balance, variations in the sodium intake do not affect these responses, but if the patient is sodium depleted the response to hyperkalaemia is enhanced several fold.

This interrelationship between sodium and potassium intake is of great clinical importance. The administration of a high sodium intake, say 200 mmol/day, will suppress aldosterone secretion in normal subjects but not in patients with hyperaldosteronism, if the potassium intake is less than the average of 80 mmol/day. But if the potassium intake is also high at 200 mmol/day, the high sodium intake will not suppress aldosterone secretion even in a normal subject, a point of obvious importance when investigating patients for hyperaldosteronism. It is also possible for the administration of very large doses of potassium, given therapeutically, to elevate plasma levels and so stimulate aldosterone secretion and lead to loss of some of the potassium supplement in the urine. This, of course, also acts as a safeguard during potassium therapy provided that renal and adrenal function are normal.

The acid-base state is another important influence on the renal tubular handling of potassium. Alkalosis and acidosis respectively increase and decrease the intracellular potassium concentration and thus increase or decrease potassium secretion by the distal tubule.

Hypokalaemia and Potassium Deficiency

Hypokalaemia sometimes results from the transfer of a small amount of potassium from the ECF into the cells. In practice, however, it usually means potassium depletion, but the plasma level is not a good indication of the total body deficit. Nor does potassium depletion always cause hypokalaemia because when potassium is lost from the cells it may accumulate in the ECF if acidosis or renal failure interferes with normal renal excretion (see below). In such circumstances rehydration of the cells or the administration of insulin and glucose rapidly causes a small shift of potassium back into the cells and the appearance of hypokalaemia.

Causes of Hypokalaemia

1. Starvation. This alone does not usually cause hypokalaemia as potassium is lost from the cells in proportion to the loss of protein and carbohydrate, and is excreted in the urine if there is no acidosis due to starvation. Some potassium depletion may occur if starvation is prolonged, because the kidney is unable to reduce potassium excretion to zero. Vomiting in anorexia nervosa, for example, will cause a metabolic alkalosis and thus lead to increased loss of potassium in the urine and hypokalaemia (see below).

2. Loss through the gastrointestinal tract. All gastrointestinal secretions contain potassium and even though the concentration may not be high (Table 2), the loss of such secretions is a common cause of potassium depletion; an additional factor is stimulation of aldosterone secretion as a result of sodium depletion and hence the loss of potassium in the urine. Urinary potassium loss may be further increased by alkalosis when there is loss of acid gastric juice. Plasma potassium levels frequently fall below 2·5 mmol/l and may be less than 2·0. Losses from the large bowel may be covered by the dietary intake until some adverse factor such as the onset of vomiting, reduces it.

Other causes are:

Acute or chronic diarrhoea including cholera and the watery diarrhoea associated with pancreatic tumours which secrete gastrin or the vasoactive intestinal polypeptide. The loss of mucus from the large bowel is another potent source of potassium loss. A villous papilloma of the rectum may secrete a litre per day or more of mucus rich in potassium and sodium (Table 2). Mucus may also be lost in large amounts in ulcerative colitis and, occasionally, in carcinoma of the rectum. If the sodium intake is inadequate to replace the sodium losses hyperaldosteronism will develop and further increase the loss of potassium.

3. Loss through the kidney. (a) *The metabolic response to injury.* In starvation, potassium is lost in the urine at the rate of 2·7 mmol/g nitrogen. Immediately after injury there is loss of potassium from the cells in excess of this. The amount of potassium lost in this way varies with the severity of the injury. Similar losses may occur in any acute illness, and low plasma levels will persist during convalescence unless adequate supplements are given.

(b) *Alkalosis.* In alkalosis the hydrogen ion that is lost from the cells is replaced partly by potassium ion. The decreased amount of hydrogen ion in the renal tubular cells means that more potassium must be lost from these cells in exchange for sodium reabsorbed from the tubular filtrate; this increases the degree of hypokalaemia.

(c) *Diuretics.*[20] Hypokalaemia has long been recognised as a complication of treatment with diuretics, which are now so widely used to treat hypertension that their use is probably the commonest cause of hypokalaemia. By causing sodium loss, diuretics stimulate aldosterone secretion and, at the same time, increase the amount of sodium at the site of aldosterone action so that potassium secretion is increased. However, the clinical significance of this change has been questioned. It has been shown that when loop diuretics and thiazides cause hypokalaemia in susceptible hypertensives, the change occurs rapidly and is maximal after only one week's treatment, thus rendering frequent laboratory measurements to monitor the change unnecessary. Moreover, studies of exchangeable potassium and measurements of total potassium by whole-body counting have cast doubt on the clinical importance of the hypokalaemia, as some authors have found that the body deficit is so small as not to be a cause for concern. Others, however, have found larger deficits on occasions and there can be little doubt that some patients on long term diuretic therapy do acquire a sizeable potassium deficit; there are reports of some patients becoming so depleted that hyponatraemia developed owing to a shift of sodium into the potassium-depleted cells, being reversed by correcting the potassium deficit.[11]

Notwithstanding the conclusions on the size of the potassium deficit in hypertensives, most physicians do give potassium supplements to hypertensive patients treated with diuretics. But normal plasma levels are restored in only a proportion of them, the others responding to the increased intake by increasing the urinary excretion. The explanation of these changes is not entirely clear. Possibly the induced sodium losses sensitize the renin-aldosterone system to the effects of potassium administration. The potassium-sparing diuretics, spironolactone, triampterene and amiloride, which block sodium/potassium exchange in the distal tubule, seem to be more effective than supplements in maintaining normal plasma potassium levels. But they too do not always prevent hypokalaemia, perhaps because of incorrect dosage, and they pose a risk of hyperkalaemia.

(d) *Other drugs.* Carbenoxolone, a drug used in the treatment of peptic ulcer and oesophageal reflux, potentiates the action of aldosterone on the distal tubule and causes hypokalaemic alkalosis. Liquorice is another substance with similar effects due to the presence of glycyrrhizinic acid. Both of these can cause very severe symptomatic hypokalaemia.

(e) *Corticosteroid excess.* Aldosterone is the major factor regulating potassium excretion. Therefore hypokalaemia occurs in primary hyperaldosteronism and in secondary hyperaldosteronism due to renin-secreting tumours, Bartter's syndrome of hyperplasia of the juxtaglomerular bodies,[21] and renal ischaemia. Excessive secretion of other mineralocorticoids such as corticosterone or deoxycorticosterone also cause hypokalaemia. Because cortisol also has some mineralocorticoid activity hypokalaemia sometimes occurs in Cushing's syndrome and sometimes complicates the therapeutic administration of glucocorticoids.

(f) *Renal disease.* Potassium loss and hypokalaemia occur in several forms of renal disease. The loss may be associated with chronic parenchymal disease, particularly if there is renal ischaemia which gives rise to secondary hyperaldosteronism, which in turn may increase potassium loss through the remaining nephrons and will also increase loss in the faeces. The term 'potassium losing nephritis' is sometimes used very loosely to refer to a variety of renal lesions associated with inappropriately high potassium excretion.

In acute tubular necrosis the problem is more likely to be *hyper*kalaemia, but when recovery ensues and the urine output increases, there may be uncontrolled loss of potassium in the urine. In any osmotic diuresis, the urine potassium concentration may be an 'obligatory' 20 mmol/l.

Excessive loss of potassium also occurs in renal tubular acidosis when it is the only cation available to exchange with sodium in the distal tubule, in view of the inability of hydrogen ion to do so. Hypokalaemia occurs frequently and may cause severe symptoms.

(g) *Diabetes mellitus.* Insulin promotes the cellular uptake of glucose and potassium, and the lack of it leads to the loss of potassium from the cells. Ketoacidosis also promotes the loss of potassium from the cells and interferes with potassium exchange in the distal tubule, so that a high plasma level is maintained. Potassium is nevertheless lost in the urine as a consequence of the osmotic diuresis, and the patient becomes potassium deficient. This becomes manifest as a low plasma potassium level during treatment with insulin coupled with cellular rehydration and correction of the acidosis, all of which promote the re-entry of potassium into the cells.

(h) *Familial periodic paralysis.* In this condition there are intermittent attacks of paralysis of skeletal muscle lasting up to 24 hours, due to a shift of potassium into the cells. With recovery the plasma potassium returns to normal. Total body potassium between attacks has been variously reported as normal or low.

Urinary Potassium Measurements

Not uncommonly hypokalaemia is discovered when there is no immediately obvious explanation. Loss through the kidney must then be considered, together with the possibility of chronic purgation, which the patient may not readily admit. The measurement of urinary potassium excretion may be helpful in these circumstances. The excretion of more than 25 mmol/day when the plasma potassium concentration is less than 3·0 mmol/l is presumptive evidence that the kidney is the route of loss. In cases of purgation, or of potassium deficiency from causes no longer operating, e.g. diuretic treatment or an acute illness in the recovery phase, the excretion may be down to 10 mmol/day or less. In the latter, low levels of potassium in the urine will persist until the patient is repleted but the administration of potassium supplements to patients with Conn's syndrome will cause a dramatic rise in urinary potassium excretion before the plasma level approaches normal (Fig. 8). In alkalosis and in 'potassium-losing nephropathies' the measurement of urine potassium excretion is sometimes useful as a guide to therapy. These conditions apart, it is usually unnecessary to measure the urine potassium excretion daily in order to replace it exactly, as once repletion has been achieved the urine will contain that which is not required.

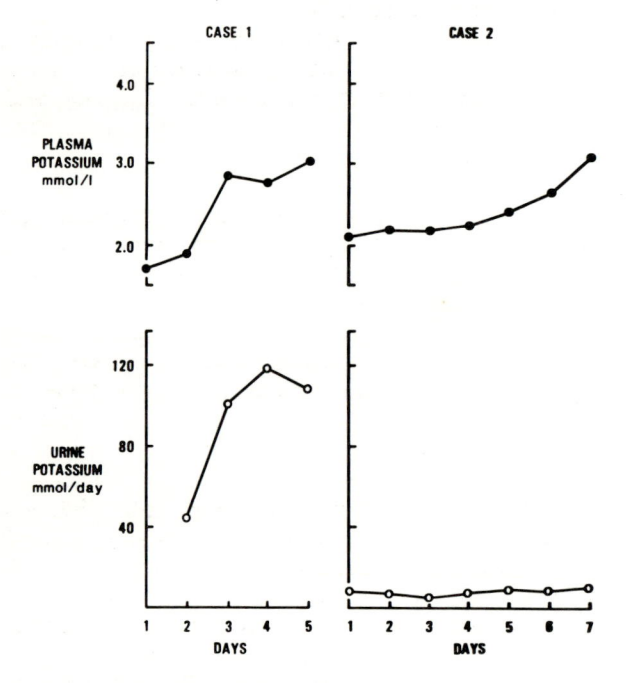

FIG. 8. Effect of potassium supplements on plasma and urine potassium in a case of Conn's syndrome (Case I) and one of potassium depletion due to chronic purgation (Case II). In Case I the supplement was 60 and 120 mmol on days 1 and 2 respectively, and thereafter was 150 mmol/day. In Case II the supplement was 100 mmol/day throughout.

The effects of hypokalaemia. The membrane potential of nerve and muscle cells, and therefore their excitability, is related, among other factors, to the ratio of intracellular to extracellular potassium. Experimentally, potassium depletion with hypokalaemia has been found to cause muscular weakness, depressed tendon reflexes and loss of intestinal motility. Clinically, severe hypokalaemia has effects on skeletal muscle, cardiac muscle and possibly smooth muscle, but in general deficits of less than 300 mmol are unlikely to be manifest by clinical signs, and there is no close correlation between the level of plasma potassium and its effects. Indeed, patients with plasma levels as low as 1·5 mmol/l may have no symptoms whereas higher levels have been accompanied by flaccid quadriplegia and even respiratory paralysis, which is fortunately rare. The reason for such differences is not clear but may be due to the loss of relatively similar proportions of potassium from the intracellular and extracellular compartments, leaving the ratio little altered, or to counteracting changes in other electrolytes, e.g. magnesium or hydrogen ions.

The effect on the heart is more complex because the myocardium and the conduction system are affected differently. There is increased sensitivity to the effect of digitalis and a characteristic sequence of changes in the ECG, although this is not closely related to the plasma potassium concentration. In cases of prolonged depletion there may be morphological changes, including focal necrosis, in the myocardium.

Loss of intestinal motility is observed in experimental potassium depletion, and paralytic ileus is said to be a feature of clinical deficiency. However, there is usually a local cause for ileus and its recovery solely from correcting potassium depletion must be rare.

Chronic hypokalaemia also has adverse effects on the kidney. The ability to concentrate the urine is greatly reduced and the tubular response to ADH is impaired. Thirst and polyuria may occur, the urine volume in some patients exceeding 4 litres per day. Histologically, changes are seen in the proximal convoluted tubule, but not in the collecting tubules. The former become swollen and vacuolated, but the changes have been shown by serial renal biopsies to be reversible. In longstanding depletion in animals a picture resembling chronic interstitial nephritis has been described and the possibility of this happening in man cannot be discounted.

Glucose tolerance is diminished in 50% of patients with Conn's syndrome, and reverts to normal with correction of potassium deficiency. Many publications link hypokalaemia with diminished glucose tolerance and impaired insulin release, but these are not often found in potassium deficiency.

Alkalosis associated with hypokalaemia is described in Chapter 2.

Principles of Treatment of Hypokalaemia

The total dose of potassium required to correct hypokalaemia cannot be calculated from the plasma level. Treatment must therefore be empirical to some extent, but clearly the rate of administration must exceed the rate of loss; the small supplements given to patients

taking diuretics have no place in therapeutic regimes.*
As a general rule, when the homeostatic mechanisms are normal, therapeutic doses of potassium should be at least 100 mmol/day more than the loss. In most cases an approximate value for the latter can be derived from the data in Table 2, but when large volumes are lost of the fluids known to have a high potassium concentration, the fluid should be analysed. Estimation of the urinary potassium excretion is also useful in patients with severe alkalosis.

It is generally acknowledged that whenever possible potassium supplements should be given by mouth, as this is thought to be the best safeguard against overtreatment. However, it is important to establish not simply that the patient can take supplements by mouth but that he *will* take *enough* in this way—if not, there is no option but to give at least some of it intravenously. When the latter route is used there often seems to be great reluctance to give adequate amounts of potassium because of the fear of causing hyperkalaemia. The risk of this is small in depleted patients with normal homeostatic mechanisms because any elevation of plasma potassium above normal stimulates aldosterone secretion, and hence increases the urinary loss. The amount of potassium which needs to be infused to cause sustained hyperkalaemia is very large, and in severely depleted patients, with plasma levels below 2 mmol/l, 200–300 mmol can be infused quite safely in 12 hours, preferably in glucose solutions which will facilitate the uptake of potassium by the cells. In the most severe cases, and especially if there is muscular paralysis, even larger amounts may be given, the infusion of up to 40 mmol/hour being considered safe. During the infusion of such large amounts the ECG should be monitored continuously for large T-waves, which renders unnecessary the measurement of plasma potassium every few hours.

The concentration of potassium in intravenous fluids should not usually exceed 50 mmol/l because if this solution is given too quickly it may cause severe pain along the course of the vein. Higher concentrations may be given, if infused slowly, but should not be administered through a central venous catheter because an inadvertent increase in the drip rate could be fatal.

In severe depletion the plasma potassium level is likely to remain low for days after beginning treatment. Ultimately it will rise into the normal range quickly over one or two days, and to be sure of detecting this when large doses are being given it is necessary to measure the plasma level daily. Once a normal plasma level has been achieved the dose should be reduced, but not to the level of the estimated loss because this may have been underestimated—if the homeostatic mechanisms are normal it is usually better to give a little too much than to give insufficient, but, when the homeostatic mechanisms are impaired, great care is needed during potassium replacement to avoid dangerous hyperkalaemia. (See 'Hyperkalaemia' below.)

Hyperkalaemia

Unlike hypokalaemia, hyperkalaemia is not related to the total amount of potassium in the body and even occurs in the presence of potassium depletion as discussed above. It may be an artefact arising *in vitro* owing to haemolysis or the breakdown of leukaemia cells during sampling. If whole blood is allowed to stand overnight before separating the plasma, levels exceeding 10 mmol/l can occur if the blood is kept at 4°C, which depresses the activity of the sodium pump. In a recently described familial disorder of the red cell membrane the plasma potassium level rose to 6 mmol/l when whole blood was allowed to stand at room temperature for two hours.[22] Such artefacts are usually identified, but lesser changes may merely obscure the presence of hypokalaemia.

Reduction in Renal Potassium Excretion

Normally, any tendency for the plasma potassium concentration to rise *in vivo* stimulates aldosterone secretion and the excess is rapidly excreted. When hyperkalaemia does develop *in vivo*, there is usually a reduced capacity for potassium excretion, which has been exceeded by the potassium intake or by the release of potassium from the cells.

The conditions that impair the renal excretion of potassium are acidosis, renal failure, mineralocorticoid deficiency, the administration of certain diuretics, and a group of conditions in which there appears to be a disorder of tubular transport:

1. Acidosis. In acidosis, potassium is displaced from the cells by hydrogen ions, and the electrochemical gradient between the cells in the distal tubule and the tubular fluid is less favourable for the excretion of potassium. The plasma level may be as high as 7 mmol/l even though the urine volume exceeds 2 litres/day. Not all patients with acidosis show a high plasma potassium level, which is often normal or even low when there is severe potassium depletion.

2. Renal failure. In chronic renal failure hyperkalaemia does not often present a problem until the glomerular filtration rate is very low, unless there is also acidosis. Adaptive changes result in the excretion of more potassium by the remaining nephrons and the excretion into the colon is also increased. Hyperkalaemia is more likely to appear in acute oliguric renal failure. This occurs in conditions in which there is also increased release of potassium from damaged tissues, viz trauma, burns, increased catabolism, and haemorrhage into the tissues or intestine.

The plasma of stored blood contains up to 20 mmol of potassium per litre and although this seems likely to re-enter the red cells after transfusion, a proportion of the transfused cells is rapidly destroyed and the plasma

---

* It cannot be stressed too strongly that the commercially available slow-release preparations of potassium contain only small amounts of potassium, and that as many as 10 tablets per day may provide no more than 65 mmol, which is almost always far too little for a depleted adult.

potassium concentration may rise if there is renal failure. Massive blood transfusions have been reported as a cause of hyperkalaemia but this is rarely a problem if the homeostatic mechanisms are normal.

When oliguric renal failure is associated with increased tissue breakdown the plasma potassium level may rise at an alarming rate, increasing from normal to 8 mmol/l in a few hours and rapidly reaching fatal levels if untreated.

3. Deficiency of aldosterone. As in Addison's disease or as an isolated defect also reduces the efficiency of potassium excretion. The same effect is seen with the use of so called 'potassium-sparing diuretics', amiloride, spironolactone, and triamterene, which oppose the effect of aldosterone on the sodium–potassium exchange mechanism in the distal tubule.

4. Hyperkalaemic periodic paralysis. This is a condition in which a rapid shift of potassium out of the cells causes hyperkalaemia and muscular paralysis episodically. The condition is hereditary but its pathogenesis is unknown, although strenuous exercise appears to be a predisposing factor.

5. Tubular unresponsiveness to aldosterone (pseudohypoaldosteronism).[23] There have been reports of a number of cases of hyperkalaemia with acidosis in the presence of high levels of aldosterone and renin. Many of these cases were diagnosed in infancy; renal sodium wasting was an additional feature, although this disappeared in later childhood. There was no response to exogenous mineralocorticoids and an abnormality of the aldosterone receptors in the renal tubule has been postulated. An accompanying failure of other organs to respond to mineralocorticoids has also been described.

6. Low-renin hypoaldosteronism.[24] This term refers to those patients in whom hyperkalaemia and acidosis are associated with a low level of plasma aldosterone, but also with a low level of plasma renin, which distinguishes the syndrome from a primary mineralocorticoid deficiency. In many of the reported cases glomerular filtration was reduced. The low aldosterone level is thought to be secondary to the low renin levels, because salt restriction or infusion of ACTH or angiotensin have each been shown to increase aldosterone secretion, and some patients have responded to treatment with mineralocorticoids. The pathogenesis of the failure to secrete renin has not been elucidated. In a few reported cases it was shown that extracellular fluid volume and total exchangeable sodium were increased; other patients responded to treatment with thiazide diuretics. It has been postulated that there is a tubular defect which leads not only to a failure to excrete potassium but also to sodium retention. The possibility exists that this is a heterogeneous group.

The Effects of Hyperkalaemia

Hyperkalaemia has effects on cardiac and skeletal muscle which depend on the plasma level. As this increases, cardiac conduction is depressed and arrhythmias occur. There are progressive changes in the ECG, starting with peaked T-waves and culminating in ventricular fibrillation or asystole at levels above 11 mmol/l. In general, the earliest ECG manifestations appear with levels of about 7 mmol/l but this is inconstant and the ECG may not become abnormal until potassium levels exceed 9 mmol/l. High plasma potassium levels also diminish the excitability of skeletal muscle and cause weakness and even paralysis. These potentially fatal effects may be preceded by paraesthesiae but these cannot be relied upon to appear.

Principles of Treatment of Hyperkalaemia

As there may be little warning before the onset of a fatal arrhythmia, an awareness of the circumstances in which hyperkalaemia occurs is vital so that the rate of change can be assessed and corrective measures applied early. A slow rate of rise can be arrested by means of an ion-exchange resin administered orally together with measures to correct acidosis and cellular dehydration.

A dramatic fall in the plasma level can be induced by the intravenous injection of 25 grams of glucose in a 50% solution together with 20 units of insulin. This causes potassium to move into the cells, and can be reinforced by creating an alkalosis with intravenous sodium bicarbonate even in non-acidotic patients. Hyponatraemia should also be corrected as this too will antagonize the effects of a high potassium level.

These measures may lower the serum potassium dramatically from, say, 10 to 7 mmol/l but the fall may not be sustained for more than a few hours. It is of paramount importance therefore to monitor the ECG after such treatment, and if there is further deterioration despite continuing treatment some form of dialysis becomes necessary.

Arrhythmias can be abolished temporarily by the intravenous injection of calcium chloride, as the calcium ions block the effect of the high potassium level on cardiac excitability. This is a short-lived effect which does not affect the level of potassium but will enable procedures with more lasting effects to be instituted.

REFERENCES

1. De Wardener, H. E. (1978), 'The control of sodium excretion.' *Amer. J. Physiol.*, **235**, F163–F173.
2. Robertson, G. L., Shelton, R. L. and Athar, S. (1976), 'The osmoregulation of vasopressin.' *Kidney International*, **10**, 25–37.
3. Skrabal, F., Arnot, R. N. and Joplin, G. F. (1973), 'Equations for the prediction of normal values for exchangeable sodium, exchangeable potassium, extracellular fluid volume, and total body water.' *Brit. Med. J.*, **2**, 37–38.
4. McCance, R. A. (1936), 'Experimental human salt deficiency.' *Lancet*, **i**, 823.
5. Taylor, W. H. (1962), 'Hypernatraemia in cerebral disorders.' *J. clin. Path.*, **15**, 211–220.

6. Arieff, A. I., Guisado, R. and Lazarowitz, V. C. (1977), 'Pathophysiology of hyperosmolar states.' In *Disturbances in Body Fluid Osmolality*, pp. 227–250. (Andreoli, T. E., Grantham, J. J., Rector, F. C. (Eds.) Besthesda, Maryland: American Physiological Society.
7. Kahn, A., Brachet, E. and Blum, D. (1979), 'Controlled fall in natraemia and risk of seizure in hypertonic dehydration.' *Intens. Care Med.*, **5,** 27–31.
8. Bartter, F. C. and Schwartz, W. B. (1967), 'The syndrome of inappropriate secretion of ADH.' *Amer. J. Med.*, **42,** 790.
9. Flear, C. T. G. and Singh, C. M. (1973), 'Hyponatraemia and sick cells.' *Brit. J. Anaesth.*, **45,** 976.
10. Penney, M. D., Murphy, D. and Walters, G. (1979), 'Resetting of osmoreceptor response as cause of hyponatraemia in acute idiopathic polyneuritis.' *Brit. Med. J.*, **2,** 1474–1476.
11. Fichman, M. P., Vorherr, H., Kleeman, C. R. and Telfer, N. (1971), 'Diuretic-induced hyponatraemia.' *Annals of Internal Medicine*, **70,** 853–863.
12. Schroeder, H. A. (1949), 'Renal failure associated with low extracellular sodium chloride. The low salt syndrome.' *JAMA*, **141,** 117–124
13. Guyton, A. C. (1976), *Textbook of Medical Physiology*, 5th Edition, p. 397. Saunders: Philadelphia and London.
14. MacGregor, G. A. and De Wardener, H. E. (1979), 'Idiopathic oedema.' *Lancet*, **ii,** 355.
15. Arieff, A. I., Kleeman, C. R., Keushkerian, A. and Bagdoyan, H. (1972), 'Brain tissue osmolality: Method of determination and variations in hyper- and hypoosmolar states.' *J. Lab. Clin. Med.*, **79,** 334–343.
16. Emmett, M. and Narins, R. G. (1977), 'Clinical use of the anion gap.' *Medicine*, **56,** 38–54.
17. Keshgegian, A. A. (1978), 'Decreased anion gap in diffuse polyclonal hypergammaglobulinaemia.' *N. Engl. J. Med.*, **299,** 99–100.
18. Weisberg, H. F. (1975), 'Osmolality—Calculated, "Delta" and more formulas.' *Clin. Chem.*, **21,** 1182–1185.
19. Kunau, R. T. and Whinney, M. A. (1978), 'Potassium transfer in distal tubule of normal and remnant kidneys.' *Amer. J. Physiol.*, **235,** F186–F191.
20. Morgan, D. B. and Davidson, C. (1980), 'Hypokalaemia and diuretics: an analysis of publications.' *Brit. Med. J.*, **280,** 905–908.
21. Gill, J. R. and Bartter, F. C. (1978), 'Evidence for a prostaglandin-independent defect in chloride reabsorption in the loop of Henle as a proximal cause of Bartter's syndrome.' *Amer. J. Med.*, **65,** 766–772.
22. Stewart, G. W., Corrall, R. J. M., Fyffe, J. A., Stockdill, G. and Strong, J. A. (1979), 'Familial pseudohyperkalaemia: a new syndrome.' *Lancet*, **ii,** 175–177.
23. Oberfield, S. E., Levine, L. S., Carey, R. M., Bejar, R. and New, M. I. (1979), 'Pseudohypoaldosteronism: multiple target organ unresponsiveness to mineralocorticoid hormones.' *J. Clin. Endocrinol. Metab.*, **48,** 228–234.
24. Lee, M. R. and Morgan, D. B. (1980), 'Familial hyperkalaemia responsive to benzothiadiazine diuretic.' *Lancet*, **i,** 879.

RECOMMENDED FURTHER READING

Andreoli, T. E., Grantham, J. J. and Rector, F. C. (Eds.) (1977), Disturbances in body fluid osmolality. Bethesda, Maryland: American Physiological Society.

Brenner, F. M. and Stein, J. H. (Eds.) (1978), *Contemporary Issues in Nephrology, Vol 1: Sodium and Water Homeostasis*. London: Churchill Livingstone.

Brenner, B. M. and Stein, J. H. (Eds.) (1978), *Contemporary Issues in Nephrology, Vol 2: Acid-Base and Potassium Homeostasis*. London, Edinburgh & New York: Churchill Livingstone.

Maxwell, M. H. and Kleeman, C. H. (Eds.) (1980), *Clinical Disorders of Fluid and Electrolyte Metabolism*. 3rd Edn. New York: McGraw-Hill Book Company.

Kramer, H. J. and Kruck, F. (1978), *Natriuretic Hormone*. New York: Springer-Verlag.

Swales, J. D. (1975), *Sodium Metabolism in Disease*. London: Lloyd-Luke.

2. DISORDERS INVOLVING CHANGES IN HYDROGEN ION AND BLOOD GAS CONCENTRATIONS

JOAN F. ZILVA

Introduction

Buffering
   The Henderson–Hasselbalch Equation
   Bicarbonate
   Haemoglobin
   Other proteins
   Phosphate

Control of the components of the extracellular bicarbonate buffer pairs
   Control of CO_2 by the lung
   The erythroyctes and the role of haemoglobin
   Control of bicarbonate by the kidney
   'Re-absorption' of filtered bicarbonate
   Replacement of bicarbonate depleted by buffering
   Secretion of hydrogen ion by the gastric mucosa
   Handling of bicarbonate by the intestinal tract

Assessment of hydrogen ion status
   Plasma total carbon dioxide
   Measured values on arterial blood
   Derived values

Disturbances in hydrogen ion balance
   Acidosis
      Metabolic acidosis
      Respiratory acidosis and hypoxia
   Alkalosis
      Metabolic alkalosis
      Respiratory alkalosis
   Salicylates and hydrogen ion balance

Investigation of hydrogen ion disturbance
   Interpretation of plasma $T\text{CO}_2$
   Anion gap
   Indications for arterial estimations

INTRODUCTION

Complete aerobic metabolism of the skeletons of organic compounds converts the constituent hydrogen, carbon and oxygen, to water and carbon dioxide. The carbon dioxide (CO_2) is central to hydrogen ion homeostasis: the hydrogen is metabolized via reduced coenzyme and does not affect hydrogen ion balance. Other metabolic processes release a net amount of about 50 to 100 mmol of hydrogen ion (H^+) daily into about 15 to 20 litres of extracellular fluid. Homeostatic mechanisms are so efficient that the normal extracellular hydrogen ion concentration is only about 40 nmol/l (pH 7·4), and varies little despite changing loads.

Physiological processes which generate hydrogen ion are summarized in Table 1. They fall into two groups:

1. Complete Catabolism of Organic Constituents other than Carbon, Hydrogen and Oxygen. The two most important processes are conversion of amino nitrogen to urea, and of sulphur in SH groups of amino acids to sulphate. High protein diets, especially those containing a high proportion of sulphur amino acids, lead to the excretion of an acid urine.

2. Incomplete Catabolism of Carbon Skeletons. For example, anaerobic metabolism of carbohydrate yields lactate, and incomplete metabolism of fatty acids and ketogenic amino acids yields ketones; both types of process, either directly or indirectly, produce H^+. In the normal subject exercise produces lactic acid, and periods of fasting increase ketogenesis. These processes may be much increased in pathological states.

Three very simple compounds are essential for hydrogen ion homeostasis.

Water is ubiquitous, and incorporates the hydrogen ions, which have been buffered by bicarbonate, in an unionized (non-toxic) form.

TABLE 1
EXAMPLES OF PHYSIOLOGICAL PROCESSES YIELDING HYDROGEN IONS

Metabolism of non-carbon groups	*Products*
Oxidation of sulphur	Sulphate + $2H^+$ per $-SH$
Metabolism of nitrogen	Urea + H^+ per $-NH_3^+$
Incomplete metabolism	
Anaerobic glucose metabolism	2 lactate$^-$ + $2H^+$ per glucose molecule
Incomplete fatty acid metabolism	Acetoacetate + H^+ ⎫ per acetyl CoA
or	β-hydroxybutyrate + H^+ ⎭ (or per fatty acid molecule)

No attempt has been made to balance these equations.
The production of hydrogen ions is not always a direct consequence of the reaction.
Interested readers should consult the references at the end of the chapter for more detailed information.

Carbon dioxide is constantly produced by metabolism. Water and carbon dioxide combine to form the buffer anion, *bicarbonate* and hydrogen ion.

Two organs are essential to co-ordinate the system.

The *lungs,* controlled by the hypothalamic respiratory centre, adjust the level of CO_2.

The *kidneys* use CO_2 and water to control the level of *bicarbonate*.

A most important reaction in hydrogen ion homeostasis is the very simple, reversible one involving these three components.

$$H^+ + HCO_3^- \overset{a}{\rightleftharpoons} H_2CO_3 \overset{b}{\rightleftharpoons} CO_2 + H_2O$$
(**Equation 1**)

Before I discuss how the mechanism works *in vivo* I shall outline some important aspects of buffer action and explain which metabolites can act as effective buffers.

BUFFERING

A buffer pair is made up of a weak (little dissociated) acid and its conjugate base. If either free H^+ or free OH^- are added to a solution of such a pair they will be partially converted to the unionized form.
Thus

$$B^- + H^+ \rightleftharpoons HB \quad \text{(\textbf{Equation 2})}$$

or

$$HB + OH^- \rightleftharpoons H_2O + B^- \quad \text{(\textbf{Equation 3})}$$

where HB denotes a weak acid and B^- its conjugate base.

The ability to buffer hydrogen ions is more important to the body than the buffering of hydroxyl ions.

Obviously the buffering capacity depends partly on the concentration of the buffer. If the amount of H^+ added exceeds the B^- present the excess cannot be buffered. Moreover, since the reaction is reversible (see **Equation 2**), as $[B^-]$* falls and [HB] increases the reverse reaction increases and buffering becomes progressively less effective. If the initial mixture contains much more HB than B^- it will be a poor buffer for even small amounts of added H^+. The nearer the ratio of [HB] to $[B^-]$ is to unity, the more effectively the buffer system minimizes changes due to addition of either H^+ or OH^-. If there is more B^- than HB it will be a more efficient buffer for H^+ than for OH^-.

The Henderson–Hasselbalch Equation

If we include the equilibrium constant, K_a, in **Equation 2**, we can express the relative concentrations of the reactants at equilibrium thus

$$[B^-][H^+] = K_a[HB] \quad \text{(\textbf{Equation 4})}$$
or
$$[H^+] = K_a \frac{[HB]}{[B^-]} \quad \text{(\textbf{Equation 5})}$$

* Square brackets are used to denote the concentration of the substance within the brackets.

We have shown that buffering is most effective if $[HB] = [B^-]$: the value of $[HB]/[B^-]$ is then 1 and, at equilibrium, $[H^+] = K_a$. A buffer system is therefore most effective at a hydrogen ion concentration near its equilibrium constant.

The Henderson–Hasselbalch equation uses the term pH, rather than $[H^+]$. pH is log $1/[H^+]$ where $[H^+]$ is expressed as mmol/l. If we similarly denote log $1/K_a$ as pK_a, we can rewrite **Equation 5** as

$$pH = pK_a + \log\frac{[B^-]}{[HB]}$$

The antilog of 1 is zero. If $[HB] = [B^-]$, pH = pKa: a buffer pair is therefore most effective in maintaining a pH near its pK_a, although it can be shown to be reasonably effective within about 1·5 pH units of the pK_a.

If we assume the target extracellular pH to be 7·4, effective buffers will have a pK_a between about 5·9 and 8·9—although the nearer the pK_a to 7·4 the better buffers they will be. Since acidosis is more common than alkalosis, a pK_a nearer the lower end of the range (when $[B^-]$ is greater than [HB]) is usually less disadvantageous than one at the upper limit.

We will now examine some of the possible *in vivo* buffer systems with two factors in mind.

1. The concentration of the buffer
2. The pK_a of the buffer

Bicarbonate

Extracellular fluid and glomerular filtrate contain bicarbonate at a concentration of about 25 mmol/l: quantitatively it is an adequate buffer in these fluids. The intracellular concentration is low.

The bicarbonate ion (HCO_3^-) and carbonic acid (H_2CO_3) form the buffer pair, for which the Henderson–Hasselbalch equation can be written.

$$pH = pK_a + \log\frac{[HCO_3^-]}{[H_2CO_3]}$$

We cannot measure the concentration of H_2CO_3, but we can measure the partial pressure of CO_2 (P_{CO_2}). If we inspect the important **Equation 1**—the buffering of H^+ by bicarbonate we can see that H_2CO_3 is in equilibrium with CO_2.

$$H^+ + HCO_3^- \overset{a}{\rightleftharpoons} H_2CO_3 \overset{b}{\rightleftharpoons} CO_2 + H_2O \quad \text{(\textbf{Equation 1})}$$

H_2CO_3 can therefore be replaced by CO_2 in the Henderson–Hasselbalch equation.

$$pH = pK_a' + \log\frac{[HCO_3^-]}{[CO_2]}$$

where pK_a' represents the combined pK_a for reactions a and b in **Equation 1**. This pK_a' has a value of about 6·1.

$[CO_2]$ expressed in mmol/l can be calculated from P_{CO_2} by multiplying by the solubility coefficient for carbon dioxide. This constant has a value of 0·225 if $P\text{co}_2$ is expressed in kPa and 0·03 if it is expressed in mmHg.

Therefore

$$\text{pH} = 6\cdot 1 + \log \frac{[\text{HCO}_3^-]}{P\text{co}_2 \times 0\cdot 225}$$

when $[\text{HCO}_3^-]$ is expressed in mmol/l and $P\text{co}_2$ in kPa.

The pK_a' of 6·1 of the bicarbonate system is near the lower limit of 5·9 for an effective buffer at pH 7·4. If 7·4 is inserted for pH in the Henderson–Hasselbalch equation, $\log [\text{HCO}_3^-]/[\text{CO}_2]$ will be 7·4–6·1 = 1·3; the ratio of $[\text{HCO}_3^-]$ to $[\text{CO}_2]$ at pH 7·4 is therefore 20 (the antilog of 1·3). The concentration of the base is therefore much higher than that of the acid, and buffering for H^+ will be more effective than that for OH^-; this is fortunate, because H^+ rather than OH^- is produced by the body. The effectiveness of the system is further improved by dissociation of H_2CO_3 to gaseous CO_2, the level of which can be kept low by the lungs, and because the interconversion of carbonic acid and CO_2 (**Equation 1**) can be catalysed by the enzyme carbonate dehydratase (carbonic anhydrase). The bicarbonate system is of prime physiological importance, and acts co-operatively with other buffers.

Haemoglobin

The complex buffering effect of proteins depends on the polar groups of the constituent amino acids, the pK_a's of which are affected by neighbouring groups. No accurate figure can be given either for concentration of buffer, or for the pK_a. Haemoglobin is a better buffer than most proteins at pH 7·4 because of the relatively high concentration of imidazole groups (pK_a approximately 7) of the constituent histidine molecules. The erythrocyte haemoglobin concentration is only about 4 mmol/l (about 2 mmol/l in whole blood), but that of buffering groups is much higher.

Haemoglobin is also relatively more effective as a buffer than plasma proteins because it is packaged in erythrocytes containing carbonate dehydratase; as we shall see, this enables the bicarbonate and haemoglobin buffering systems to work co-operatively.

Deoxyhaemoglobin is a better buffer than oxyhaemoglobin. The converse is also true, i.e. that hydrogen ions decrease the affinity of haemoglobin for oxygen.

Other Proteins

Plasma proteins are present at lower molar concentrations than haemoglobin. Their buffering capacity per mole is less, and they do not work co-operatively with the carbonate dehydratase system. They are therefore very unimportant blood buffers. The concentration in interstitial fluid is even lower than in plasma.

Proteins are present in much higher concentrations in cells than in plasma. They are probably important in buffering H^+ before release from cells.

Phosphate

Mono- and di-hydrogen phosphate form a buffer pair with a pK_a of about 6·8.

$$\text{pH} = 6\cdot 8 + \log \frac{[\text{HPO}_4^{--}]}{[\text{H}_2\text{PO}_4^-]}$$

The extracellular concentration of phosphate is only about 2 mmol/l, so that, despite the favourable pK_a, phosphate contributes little to extracellular buffering.

The phosphate concentration rises during passage of glomerular filtrate through the distal renal tubule, and reaches about 20 mmol/l in urine. It is an important urinary buffer, and, like haemoglobin in erythrocytes, works co-operatively with the bicarbonate system.

All other potential buffers are present in very low extracellular and urinary concentrations (for example, urate is about 0·2 mmol/l in plasma and about 2 mmol/l in urine) and have unfavourable pK_a values even at the pH of urine, which is lower and more variable than that of extracellular fluid.

In summary, the three important buffers are:
The *bicarbonate* system in the extracellular fluid and glomerular filtrate, which is central to hydrogen ion homeostasis, and works co-operatively with
haemoglobin in the erythrocytes and
phosphate in renal tubular fluid.

We shall see that *ammonia* in renal tubular fluid, although not an effective buffer, also acts co-operatively with the bicarbonate system.

CONTROL OF THE COMPONENTS OF THE EXTRACELLULAR BICARBONATE BUFFER PAIR

Carbon dioxide, the denominator in the Henderson–Hasselbalch equation, is a 'waste product' of aerobic metabolism. It forms bicarbonate, the numerator in the same equation, and so plays an important part in controlling another toxic product of metabolism, namely hydrogen ion.

Control of Carbon Dioxide by the Respiratory Centre and Lungs

Carbon dioxide diffuses freely from the pulmonary capillaries into the alveolar air. Normal alveoli have a very large capacity to eliminate carbon dioxide, provided that the concentration gradient from blood is maintained by removal from the alveolar sac: the rate of removal depends mainly on the rate and depth of respiration. In metabolic acidosis, in which there is overproduction of CO_2 from bicarbonate, the $[CO_2]$ is maintained at a normal or even low level by hyperventilation, and only rises if alveolar diffusion or lung movements are impaired. Respiratory excursion is controlled by the hypothalamic respiratory centre, which, by responding to the pH of the blood flowing through it, maintains arterial $P\text{co}_2$ at about 5·3 kPa (40 mmHg)—corresponding to about 1·2 mmol/l [CO_2]. This residual CO_2 is of central importance in hydrogen ion homeostasis.

We have seen that CO_2 can combine with water to form carbonic acid, and that carbonic acid can dissociate to bicarbonate and hydrogen ions.

$$CO_2 + H_2O \stackrel{a}{\to} H_2CO_3 \stackrel{b}{\to} H^+ + HCO_3^- \quad \textbf{(Equation 1)}$$

If the HCO_3^- generated by this reaction is to be useful as a buffer for hydrogen ions arising from metabolic processes, the H^+ generated in reaction 1b should be disposed of. The pH of the system depends on the ratio of [HCO_3^-] to [CO_2]: to minimize changes in pH, the reaction should proceed rapidly to the right to produce HCO_3^- as [CO_2] increases at the site of production (tissues), and should reverse rapidly to remove HCO_3^- as [CO_2] falls at the site of removal (lungs).

Two types of cell not only contain high concentrations of *carbonate dehydratase* which speeds the reaction, but can also dispose of H^+, so that HCO_3^- is made available as a buffer.

The Erythrocytes and the Role of Haemoglobin

The erythrocytes are of great physiological importance in minimizing the change in ratio of [CO_2] to [HCO_3^-], and therefore in pH, between arterial and venous blood.

The erythrocyte produces little CO_2 because it cannot metabolize aerobically. As shown in Fig. 1 (step 1), when CO_2 diffuses from tissue cells into the extracellular fluid, the extracellular ratio of [HCO_3^-] to [CO_2] and therefore the pH, falls. The rising venous PCO_2 creates a concentration gradient, between the plasma and erythrocyte. CO_2 therefore diffuses into the red cell (step 2) where its reaction with water is catalysed by carbonate dehydratase (step 3). Haemoglobin (mostly in the deoxy- form after releasing oxygen to tissues) buffers the H^+ produced by the reaction (step 4), and this stimulates further oxygen release. The increase in cellular [CO_2] and the removal of H^+ by haemoglobin keep the net reaction proceeding to the right and the cellular concentration of bicarbonate rises. Diffusion into the extracellular fluid (step 5), brings the ratio of [HCO_3^-] to [CO_2] (and pH) towards normal, the negatively charged HCO_3^- being replaced by chloride diffusing from plasma into the cell along the electrochemical gradient ('chloride shift').

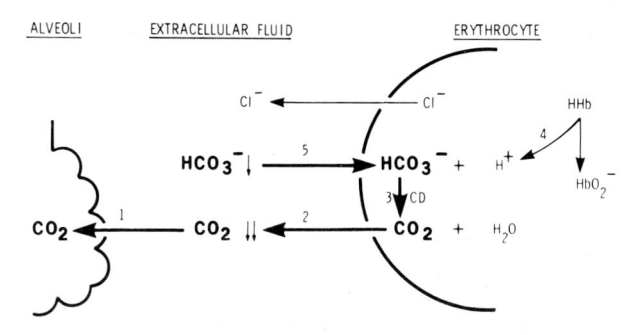

FIG. 2. Role of haemoglobin in controlling extracellular pH during carbon dioxide removal in lungs.

The process is reversed in the lungs. CO_2 diffuses from plasma across the alveolar wall (Fig. 2, step 1). This creates a concentration gradient across the red cell membrane. As CO_2 leaves the cell (step 2) the carbonate dehydratase reaction is reversed (step 3) and the consequent removal of H^+ from haemoglobin (step 4) is accelerated as haemoglobin takes up oxygen to form the weaker buffer, oxyhaemoglobin; the oxygenation of haemoglobin is further stimulated by loss of H^+. Bicarbonate diffuses into the cell along the concentration gradient (step 5) and chloride diffuses out along the electrochemical gradient.

The diffusion of chloride is the rate limiting step in both these processes.

This process is so effective in minimizing the change in ratio between [HCO_3^-] and [CO_2] that the arteriovenous pH difference is only about 0·03 to 0·04 units (a change of hydrogen ion concentration from about 40 to about 43 nmol/l). The haemoglobin buffering capacity is not exceeded because it is alternately in the form of HHb and HbO_2^-; at the same time bicarbonate is alternately formed from, and broken down to, CO_2. If CO_2 cannot be removed at the normal rate by the lungs further bicarbonate generation is impaired because almost all the haemoglobin may be in the form HHb. The erythrocyte can play little part in the correction of an acidosis; the limitations of the concept of the 'standard' bicarbonate depend on this fact (see below).

Control of Bicarbonate by the Kidney

The erythrocyte is important in minimizing the arteriovenous pH difference, but has little effect on net

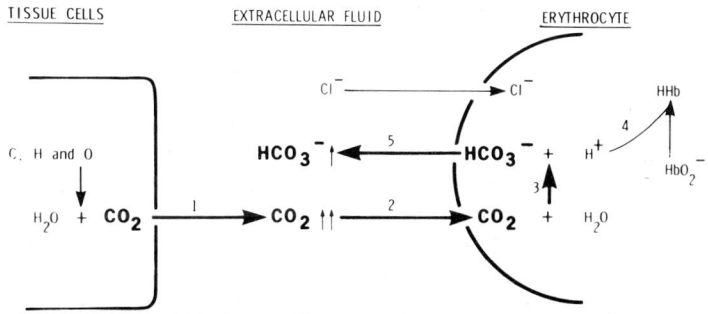

FIG. 1. Role of haemoglobin in controlling extracellular pH during carbon dioxide release from tissues.
HbO_2^- = oxyhaemoglobin
HHb = acid haemoglobin

bicarbonate balance. The kidney is the most important organ in controlling the extracellular bicarbonate level.

Extracellular bicarbonate might be lost from the body by one or more of three mechanisms.

1. Filtered bicarbonate, at plasma concentration (normally about 25 mmol/l), might not all be reabsorbed.
2. Circulating bicarbonate might be depleted when it buffers hydrogen ions produced by metabolism.
3. Bicarbonate secreted into the intestine might not be reabsorbed normally.

The kidney normally prevents the first two, and helps to minimize the effect of abnormal intestinal losses, by the carbonate dehydratase (CD) mechanism.

Bicarbonate is formed within the tubular cell by the carbonate dehydratase mechanism.

$$CO_2 + H_2O \overset{CD}{\rightleftharpoons} H^+ + HCO_3^-$$

There is one important difference between the erythrocyte and renal tubular mechanism. The H^+ formed in the red cell remains in the body, but that formed by the same mechanism in the tubular cell can be secreted into the tubular fluid and lost in the urine. So long as H^+ can be secreted into the urine, HCO_3^- can continue to be formed in the tubular cell; the net reaction therefore continues to the right and does not reverse. The secretion is an active process, and can occur against a moderate hydrogen ion concentration gradient. The luminal cell wall is impermeable to H^+ so that the gradient can be maintained.

The tubular cell also differs from the erythrocyte because it can produce CO_2 by aerobic metabolism. This constant CO_2 production within the cell is another factor helping to keep the net reaction proceeding to the right. CO_2 diffuses out of the cell along a concentration gradient.

The following factors may *increase* the rate of renal production of HCO_3^-.

1. A rise in extracellular or luminal P_{CO_2}. CO_2 can diffuse out of the cell into the tubular luminal fluid as well as into the extracellular fluid. An increase in concentration of CO_2 in either of these fluids reduces the concentration gradient, slows diffusion out of the cell and increases the intracellular $[CO_2]$ which is one of the reactants in the carbonate dehydratase reaction.

2. A fall in the level of extracellular HCO_3^-. Bicarbonate cannot diffuse through the luminal cell membrane, but can diffuse into extracellular fluid along a concentration gradient. The increased gradient caused by a fall in extracellular $[HCO_3^-]$, by reducing the intracellular level of one of the products, stimulates the carbonate dehydratase reaction.

3. An increased rate of secretion of H^+ into the tubular lumen, increases the rate of removal of the other product.

The rate of production of HCO_3^- may be *reduced* by:

1. A fall in extracellular P_{CO_2}
2. A rise in extracellular $[HCO_3^-]$
3. A decreased rate of secretion of H^+

'Reabsorption' of Filtered Bicarbonate

Bicarbonate filtered at the glomerulus (Fig. 3, step 1) is reclaimed by the carbonate dehydratase mechanism. The luminal cell membrane is relatively impermeable to ions and HCO_3^- cannot be reabsorbed directly; an equivalent amount to that filtered can be formed within the tubular cell, and returned to the extracellular fluid through the permeable transluminal membrane. This process occurs in the proximal part of the tubule.

H^+ formed by the carbonate dehydratase mechanism within the tubular cell (Fig. 3, step 2) is pumped into the

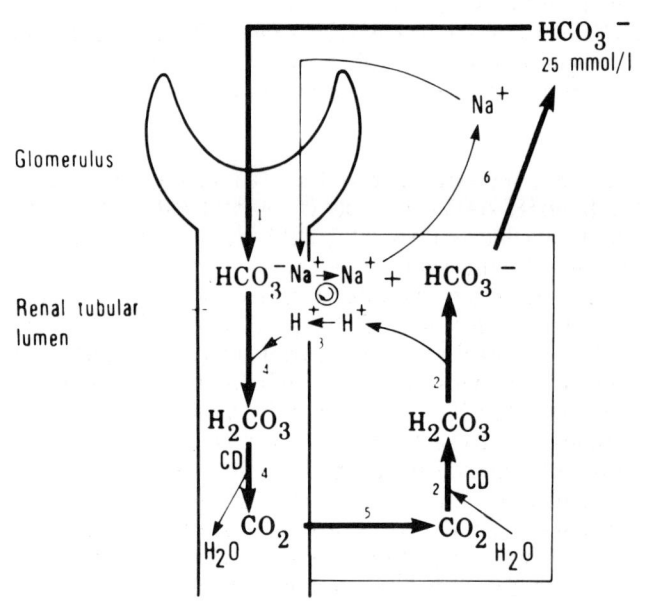

Fig. 3. 'Reabsorption' of filtered bicarbonate by the proximal renal tubular cells. (Reproduced by kind permission from *Clinical Chemistry in Diagnosis and Treatment*. (3rd Edition 1979) by Joan F. Zilva and P. R. Pannall. London: Lloyd-Luke.)

tubular lumen in exchange for filtered sodium (step 3). This active removal of H^+ from the cell continues to stimulate its production. Neither H^+ nor HCO_3^- can diffuse passively from the lumen into the cell across the impermeable membrane, and as luminal H^+ rises above cellular H^+ it combines with HCO_3^- filtered with the sodium. The original cellular reaction is reversed in the lumen (step 4)—a process which is probably catalysed by carbonate dehydratase on the cell surface. The rise in luminal P_{CO_2} is reflected within the cell (step 5), and accelerates further intracellular reaction between CO_2 and water. The hydrogen ion formed is again secreted into the urine. HCO_3^- and Na^+ diffuse through the permeable transluminal membrane as their intracellular concentration rises. Each mmol of HCO_3^- filtered can combine with one mmol of H^+, and each time one mmol of H^+ is exchanged for luminal Na^+ one mmol of HCO_3^-

and Na$^+$ return to the extracellular fluid: thus an amount of HCO$_3^-$ equal to that filtered can be formed in the cell. Urinary pH gradually falls below that of plasma and below about pH 6·5 the urine is bicarbonate free. Bicarbonate reabsorption is usually complete in the proximal tubule.

Replacement of Bicarbonate Depleted by Buffering: the Role of Urinary Phosphate

Bicarbonate 'reabsorption' causes no net change in the amount of either bicarbonate or hydrogen ion in the body: the process is, as shown in Fig. 3, cyclical. Replacement of bicarbonate used in buffering could occur if generation continued after all filtered bicarbonate had been 'reabsorbed'. If secreted H$^+$ were lost in the urine with anion other than bicarbonate the carbonate dehydratase mechanism would continue to return HCO$_3^-$ to the extracellular fluid. The action of the hydrogen ion pump would be limited if the H$^+$ gradient were too high; the H$^+$ should, ideally, be mostly buffered.

Urinary Phosphate and the Carbonate Dehydratase Mechanism. As bicarbonate is 'reabsorbed' less is available to accept secreted H$^+$, until, below pH 6·5 none is present. At the same time the concentration of phosphate is increasing as water is reabsorbed and phosphate secreted in the more distal parts of the tubule. The buffering capacity of the phosphate pair therefore increases (both because of the rise in its concentration, and because of the fall of luminal pH) as that of bicarbonate declines, until the former takes over. As in the 'reabsorption' mechanism the hydrogen ion is derived from water in the tubular cell by the carbonate dehydratase mechanism; HCO$_3^-$ is returned to the extracellular fluid with sodium each time secreted H$^+$ combines with mono- to form di-hydrogen phosphate (Fig. 4). This represents a net gain by the extracellular fluid of one mmol of HCO$_3^-$ per mmol H$^+$ buffered by phosphate.

Urinary Ammonia and the Carbonate Dehydratase Mechanism. As more H$^+$ is secreted more of the monohydrogen phosphate is converted to dihydrogen phosphate, until, below pH about 5·5, almost all is in this form. At times when the tissues are producing a large H$^+$ load there may not always be enough urinary buffering capacity to enable the carbonate dehydratase mechanism fully to replete the bicarbonate losses.

As the urine becomes more acid, it can be shown to contain increasing amounts of ammonium salts. Although ammonia and ammonium ion form a buffer pair of sorts, the pK$_a$ of the system is about 9·2 (an effective buffering range of between pH about 8 and 11): this pair cannot act as a straightforward physiological buffer, especially in the relatively acid urine. It does, however, co-operate with the carbonate dehydratase system to replete plasma HCO$_3^-$.

Ammonium ion is formed in the tubular cell from the two amino groups of glutamine, the two deamination steps being catalysed by glutaminase and glutamate dehydrogenase. The net effect is the production of one mole of 2-oxoglutarate and two moles of ammonium ion from one mole of glutamine.

$$\text{Glutamine} \xrightarrow{\text{glutaminase}} \text{Glutamate}^- + \text{NH}_4^+$$

$$\text{Glutamate} \xrightarrow{\text{glutamate dehydrogenase}} \text{2-oxoglutarate}^{2-} + \text{NH}_4^+$$

The high pKa of the ammonia pair indicates that most of the intracellular ammonium ion will not dissociate: for instance, at pH 7·2 there is 100 times as much NH$_4^+$ as NH$_3$, and at pH 7·4 the factor is 60. However, just as one member of the bicarbonate pair can form gaseous CO$_2$ which, unlike bicarbonate ion, can diffuse through the luminal cell membrane, so ammonia can diffuse into the lumen whilst the ammonium ion cannot. If the ammonia diffusing into the urine were prevented from returning to the cell, more of the ammonium ion formed in the cell would dissociate, and its formation from glutamine would continue.

Figure 5 shows how ammonium production could co-operate with the carbonate dehydratase system so that H$^+$ could continue to be secreted into the urine; equimolar amounts of HCO$_3^-$ would continue to enter the extracellular fluid. Figure 5 should be compared with Fig. 4. The action of glutaminase and glutamate dehydrogenase liberates NH$_4^+$ into the distal tubular cell (Fig. 5, step 1a). This NH$_4^+$ is in equilibrium with a very small amount of NH$_3$ (step 2a). The luminal concentration is even lower than that in the cell, and ammonia diffuses along a gradient into the lumen (step 3), where it combines with H$^+$ generated in the usual way by the carbonate dehydratase mechanism (step 4). The re-formed NH$_4^+$ cannot diffuse back into the tubular cell. This 'trapping' of NH$_3$ and H$^+$ in the lumen removes one of the products of both the carbonate dehydratase and the glutaminase/glutamate dehydrogenase reactions and

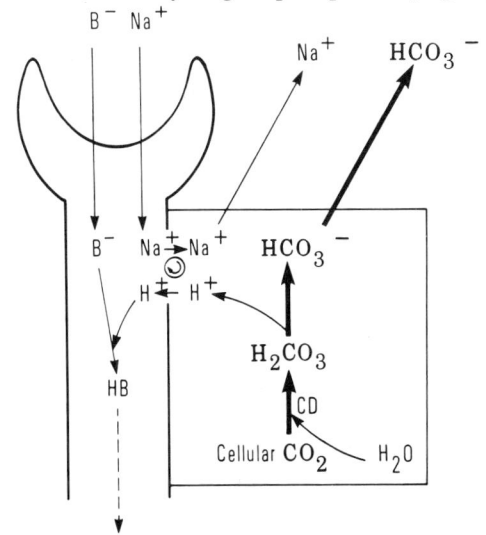

FIG. 4. Role of filtered buffer anions in net generation of bicarbonate by renal tubular cells.
B$^-$ = non-bicarbonate buffer anion (mostly monohydrogen phosphate).
(Reproduced by kind permission from *Clinical Chemistry in Diagnosis and Treatment*. (3rd Edition 1979) by Joan F. Zilva and P. R. Pannall. London: Lloyd-Luke.)

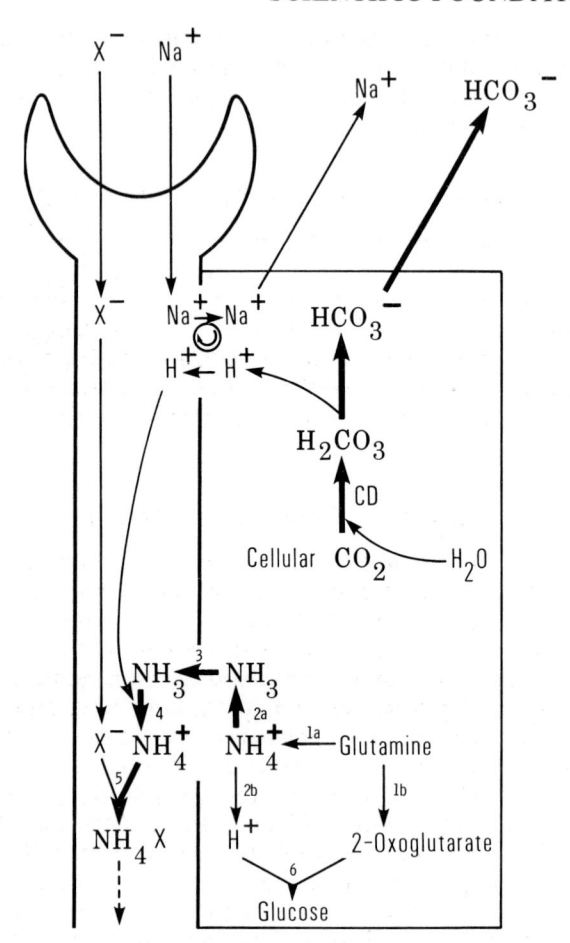

FIG. 5. Role of urinary ammonia in net generation of bicarbonate by renal tubular cells.
X^- = non-buffer anion (mostly chloride).

bicarbonate generation and glutamine breakdown are accelerated. The ammonium is passed in the urine with non-buffer anion (mostly chloride) and the accompanying sodium is exchanged for the secreted H^+, and returned to the extracellular fluid with bicarbonate. By incorporating the H^+ in NH_4^+ the carbonate dehydratase system can continue to generate HCO_3^-, and to replete that used in buffering, even when urinary phosphate buffer has been exhausted (hydrochloric acid is highly ionized, and chloride cannot buffer H^+ directly).

One problem remains. When NH_4^+ dissociates it releases H^+ as well as NH_3 (step 2b). Unless this intracellular H^+ could be disposed of it would have to be buffered by an equivalent amount of bicarbonate, and the excretion of NH_4^+ would have no advantage. The explanation may be that 2-oxoglutarate, the other product of deamination of glutamine (step 1b) is a substrate for gluconeogenesis—a process which can only occur in the renal tubular cell and the liver; formation of glucose by this pathway can be shown to use H^+. Gluconeogenesis is stimulated by acidosis, and H^+ liberated in the tubular cell as shown in Fig. 5 (step 2b) may be incorporated into glucose (step 6). The net effect on the body would then be loss of one H^+ and gain of one HCO_3^- for each NH_3 entering the urine.

Hepatic urea production is inhibited by acidosis. This pathway also uses glutamine and liberates H^+ (Table 1), but at a site at which the H^+ will deplete extracellular bicarbonate. In acidosis more glutamine is used to generate, rather than use, HCO_3^-.

We can now see how ingeniously the renal tubular mechanism repletes the bicarbonate used in buffering hydrogen ion derived from metabolic processes. The H^+ diffuses into the extracellular fluid and is incorporated locally into water in the unionized form; the oxygen in the water comes from the HCO_3^-. The normal respiratory centre responds to the slight rise in $P\text{co}_2$. The falling $[HCO_3^-]$ stimulates the renal carbonate dehydratase mechanism, and the extracellular buffering reaction is reversed in the tubular cell, using CO_2 and water, within the cell. The reformed H^+ will be lost in the urine as dihydrogen phosphate, and as this buffering becomes depleted, increasingly more will appear in ammonium salts. If addition of H^+ to the extracellular fluid stops, the processes will continue until extracellular $[HCO_3^-]$ is normal, when the stimulus ceases, and the former steady state is reached. Urinary phosphate and ammonia are acting in the same way as red cell haemoglobin—as hydrogen ion acceptors—allowing the continued formation of HCO_3^- from CO_2.

Water is the inert carrier of potential hydrogen ions from the site of production to a site at which they can be eliminated from the body (Fig. 6).

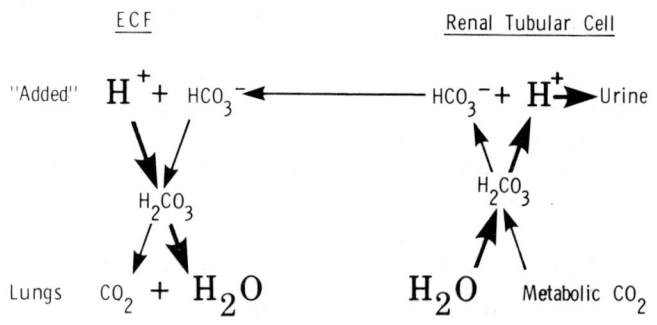

FIG. 6. Hydrogen ion 'shuttle' between site of buffering and kidneys.
(Reproduced by kind permission from *Clinical Chemistry in Diagnosis and Treatment*. (3rd Edition 1979) by Joan F. Zilva and P. R. Pannall. London: Lloyd-Luke.)

Urinary Sodium and Chloride and the Carbonate Dehydratase Mechanism. Hydrogen ion secretion, and therefore bicarbonate generation in the tubular cell, depends on adequate amounts of luminal sodium for exchange. A reduction in the availability of sodium may cause acidosis, and increased availability may contribute to alkalosis.

The amount of chloride in the glomerular filtrate is one of the factors which may affect the amount of sodium reaching distal sites. Most filtered sodium is reabsorbed in the proximal tubule, and it is the small unabsorbed fraction that takes part in hydrogen ion exchange. Proximal reabsorption of cationic sodium depends on the availability of anion to accompany it. The two predominant filtered anions are chloride (about 100 mmol/l) and bicarbonate (about 25 mmol/l).

Hypochloraemia without equivalent hyponatraemia impairs proximal tubular reabsorption, and enough sodium may reach distal sites to aggravate alkalosis. This should be remembered when treating the biochemical disturbance of pyloric stenosis.

Potassium Balance and the Carbonate Dehydratase Mechanism. Potassium within the renal tubular cells competes with hydrogen ion, formed by the carbonate dehydratase mechanism, for exchange with luminal sodium.

Abnormally low potassium levels in these cells favour secretion of relatively more hydrogen ion in exchange for the same amount of sodium. Increased H^+ secretion is one of the factors stimulating the carbonate dehydratase mechanism, and more HCO_3^- is returned to the extracellular fluid causing alkalosis. Any of the other factors stimulating the mechanism lead to acidosis (a high extracellular P_{CO_2} or a low extracellular $[HCO_3^-]$) and make relatively more hydrogen ion available for exchange; less potassium is lost in the urine and hyperkalaemia may result.

Conversely, high intracellular potassium levels might be expected to inhibit HCO_3^- formation; low extracellular P_{CO_2} or high extracellular $[HCO_3^-]$ might be expected to cause potassium depletion.

Secretion of Hydrogen Ion by the Gastric Mucosa

Acid secretion by the gastric mucosa also depends on the carbonate dehydratase mechanism. The empty gastric cavity contains very little sodium or buffer, and the gastric mucosa can tolerate a very low pH. Hydrogen ion is secreted with chloride rather than in exchange for sodium. As usual, HCO_3^- is returned to the extracellular fluid each time H^+ is secreted.

Handling of Bicarbonate in the Intestinal Tract

Some bicarbonate diffuses passively into the duodenal and jejunal lumen, and is probably reabsorbed by an exactly analogous mechanism to that for reabsorption from the renal tubular lumen (Fig. 3).

The cells of the pancreas and biliary tract may carry out the reverse process. The H^+ exchanges for sodium from the extracellular fluid; it is buffered by, and so reduces the level of extracellular bicarbonate. At the same time the bicarbonate formed with the H^+ in the cell enters the duct, and hence the duodenal lumen, with the sodium from the extracellular fluid. The net effect is secretion of sodium bicarbonate from the extracellular fluid into the duodenum.

Bicarbonate is also secreted by ileal and colonic cells, by a mechanism which is probably the reverse of that in the gastric mucosa. Luminal chloride exchanges for bicarbonate formed in the cell, and the hydrogen ion is transported into the ECF, where it reduces the plasma bicarbonate concentration. The net effect is secretion of bicarbonate in exchange for chloride.

The tendency to alkalosis during post-prandial gastric secretion (the 'alkaline tide') is normally offset by bicarbonate secretion as food enters the small intestine. The chloride secreted into the gastric lumen is regained in the ileum and colon.

ASSESSMENT OF HYDROGEN ION STATUS

The components of the bicarbonate buffer system can easily be measured. The system is central to the homeostatic mechanisms, and will be affected if they are disturbed; the availability of haemoglobin as a buffer is of some importance in acute disturbances, and most modern 'blood gas machines' correct automatically for blood haemoglobin concentration, and for its degree of oxygen saturation.

Plasma Total Carbon Dioxide (T_{CO_2})

Determination of plasma T_{CO_2} concentration is the most commonly performed estimation of a component of the buffer pair. It is often the only necessary one if clinical and other findings are taken into account and can be carried out on venous blood at the same time as estimation of plasma sodium, potassium, urea and, if necessary, glucose and chloride. T_{CO_2} concentration includes the sum of bicarbonate, dissolved CO_2 and carbonic acid. At pH 7·4 the ratio of $[HCO_3^-]$ to the other two components is 20 to 1, and if T_{CO_2} were 21 mmol/l, 20 mmol/l of this would be derived from HCO_3^-. Even if the pH were as low as 7·1 the ratio would be 10 to 1, and, at a T_{CO_2} of 22 mmol/l, 20 mmol/l would come from HCO_3^-. T_{CO_2} is therefore a very acceptable estimate of bicarbonate concentration.

Simple precautions should be taken during collection of the specimen to ensure that the result truly reflects the circulating concentration. The commonest cause of artefactually low levels, especially in specimens from small children, is loss of CO_2 from the specimen into the air in the specimen tube: as this happens bicarbonate is converted to CO_2 which in turn is lost. Centrifugation may cause significant losses if a small specimen is put into a larger container with a large dead space between the surface of the blood and the stopper. Tubes should be filled nearly to the stopper and if the specimen is small it should be put into a small tube. If this criterion is not fulfilled the laboratory should request a fresh specimen. Old and haemolysed specimens are also unsuitable.

Measured Values on Arterial Blood

There is a large arteriovenous difference in P_{CO_2}, and measurements other than bicarbonate (as T_{CO_2}) must be made on arterial blood. Arterial puncture is slightly more unpleasant and dangerous to the patient than venous puncture, and estimation of all parameters is often unnecessary. The clinician should only request 'blood gases' when there are clear clinical indications (see p. 42).

Blood should be drawn into, and left in, a heparinized syringe, and should be mixed by rolling the syringe gently between the hands, without allowing air to enter. Results from estimations on capillary specimens are less likely to represent circulating arterial levels especially if there is peripheral vasoconstriction. If they must be used the site of puncture should be warm and pink and heparinized capillary tubes should be completely filled with blood and sealed. All specimens should be sent to the laboratory immediately, preferably on ice to reduce the rate of cellular metabolism. Excess heparin can significantly affect results and the anticoagulant should only wet the surface of the syringe. Plasma sodium estimation should not be requested on specimens taken with sodium heparin.

The measurements are made at 37°C. Corrections may be made for hypoxia and hypothermia.

The pH is measured by passing blood directly through a glass electrode from the syringe or capillary tube. An abnormal result indicates a serious abnormality of hydrogen ion balance, but a normal one does not exclude an abnormality. Compensatory changes may be keeping the pH normal.

The P_{CO_2}, a measure of respiratory disturbance, is measured by passing blood through a CO_2 electrode.

Derived Values

All other values are derived from measured values for pH, P_{CO_2}, and sometimes haemoglobin and its degree of oxygen saturation. The calculations are usually made automatically by modern machines. It should be remembered that the more factors used in the derivation the more suspect the significance of the final result.

Actual bicarbonate concentration is the most useful of these calculated values. It is derived by inserting measured values of P_{CO_2} and pH into the Henderson–Hasselbalch equation for the bicarbonate pair. It is a measure of circulating arterial bicarbonate concentration, and should agree closely with T_{CO_2} measured at the same time.

Standard bicarbonate concentration is an *in vitro* measurement. If circulating P_{CO_2} were abnormal, but were equilibrated *in vitro* with a 'normal' P_{CO_2} of 5·3 kPa (40 mmHg), the pH could again be measured, and the new ('standard') $[HCO_3^-]$ calculated. *In vitro* changes in P_{CO_2} have the same effect on erythrocyte metabolism, and therefore on bicarbonate levels, as those in Figs. 1 and 2. If the circulating P_{CO_2} were high, *in vitro* correction to normal would result in a 'standard' bicarbonate level lower than that of actual bicarbonate; if the circulating P_{CO_2} were low, the 'standard' would be higher than the actual bicarbonate. The difference is therefore a measure of the *in vivo* increment or decrement in actual bicarbonate due to the erythrocyte mechanism. Any *in vivo* change due to the renal mechanism is unaffected. In my experience this derived value rarely gives any more information than the P_{CO_2} level, although it is still widely requested.

I have even greater reservations about the other two derived values, which seem of no more help than direct measurements. *Buffer base* is said to be an indication of the *total* blood buffering capacity. *Base excess or deficit*, is a hypothetical measure of the quantity of base or acid needed per litre of blood to bring all the parameters to normal. These two values will not be discussed further in this chapter, but the reader may consult the reference at the end of the chapter for further information.

In the following discussion I shall describe the changes in pH, P_{CO_2} and actual bicarbonate concentration in each group of disturbances; these changes are summarized in Table 2. Later in the chapter I shall consider which estimations provide clinically useful information.

DISTURBANCES IN HYDROGEN ION BALANCE

The bicarbonate buffer pair is affected in any disturbance of hydrogen ion homeostasis, and its components can easily be measured. It is therefore convenient and valid to discuss disturbances in terms of changes in this pair. pH is determined by the ratio of $[HCO_3^-]$ to P_{CO_2}.

In so-called *metabolic* disturbances a change in the concentration of *bicarbonate* is the primary abnormality.

In so-called *respiratory* disturbances a change in P_{CO_2} is the primary abnormality.

If the level of the other member of the pair changes in the same direction, the ratio of the concentration of bicarbonate to $[CO_2]$, and therefore the pH, is brought towards normal.

In *metabolic* disturbances the *compensatory* change is in P_{CO_2}, in the same direction as the original change in $[HCO_3^-]$.

In *respiratory* disturbances the *compensatory* change is

TABLE 2
FINDINGS IN ARTERIAL BLOOD IN HYDROGEN ION DISTURBANCES

Type	Uncompensated			Partially compensated			Fully compensated		
	pH	P_{CO_2}	$[HCO_3^-]$	pH	P_{CO_2}	$[HCO_3^-]$	pH	P_{CO_2}	$[HCO_3^-]$
Acidosis									
Metabolic	↓↓	N	↓↓	↓	↓	↓↓	N	↓↓	↓↓
Respiratory	↓↓	↑↑	N	↓	↑↑	↑	N	↑↑	↑↑
Alkalosis									
Metabolic	↑↑	N	↑↑	Compensation relatively ineffective					
Respiratory	↑↑	↓↓	N	↑	↓↓	↓	N	↓↓	↓↓

in $[HCO_3^-]$ in the same direction as the original change in Pco_2.

Compensatory changes may correct the pH partially or completely (see Table 2), but buffering capacity remains abnormal unless the original abnormality is corrected.

Acidosis

Acidosis is much commoner than alkalosis. Primary products of metabolism are hydrogen ion and carbon dioxide rather than hydroxyl ion and bicarbonate. Compensatory mechanisms for acidosis are more effective than those for alkalosis.

Either a fall in $[HCO_3^-]$ in metabolic acidosis or a rise in Pco_2 in respiratory acidosis will reduce the ratio of $[HCO_3^-]$ to Pco_2 and therefore cause a fall in pH.

$$\text{pH} = 6 \cdot 1 + \log \frac{[HCO_3^-]}{Pco_2 \times 0 \cdot 225}$$

Certain secondary findings are common in many cases of acidosis, whether respiratory or metabolic in origin.

1. *Hyperkalaemia.* H^+ and K^+ compete for tubular secretion in exchange for Na^+, and acidosis is often accompanied by hyperkalaemia. Exceptions occur when potassium is lost with bicarbonate into the intestine, or when the tubular exchange mechanisms are impaired by carbonate dehydratase inhibitors or by impairment of the tubular function.

2. An acid medium favours ionization of calcium salts. *Very prolonged* acidosis may cause significant bone decalcification, with clinical and radiological findings of *osteomalacia,* and a high plasma alkaline phosphatase activity. In most cases the kidney responds to the acidosis by secreting an acid urine, which prevents renal calcification and stone formation despite the hypercalciuria: in 'classical' renal tubular acidosis (p. 37) the systemic acidosis results from failure to acidify the urine, and precipitation may occur in the alkaline urine. Hypercalcaemia is an exceedingly rare finding in acidosis.

Metabolic Acidosis

The primary abnormality in the bicarbonate buffer pair in metabolic acidosis is a low plasma $[HCO_3^-]$; the other findings in the uncompensated, partially compensated, and fully compensated state are indicated in Table 2. Compensation depends on stimulation of the respiratory centre by the acidosis, followed by deeper than normal respiratory excursions (in severe cases this may be evident clinically as so-called Kussmaul respiration). Compensation is impaired by any disturbance of the hypothalamic-respiratory axis.

A primary fall in $[HCO_3^-]$ may be due to any of the following factors.

1. Increased utilization in buffering at a rate which exceeds the renal ability to replace it.

2. Extrarenal loss of HCO_3^- at a rate which exceeds the renal ability to replace it.

3. Impairment of the tubular mechanisms generating HCO_3^-.

Increased Utilization in Buffering. The production of many, but not all, intermediate anionic metabolites is associated with equivalent H^+ production. These are utilized by further metabolism, but if one of the pathways is interrupted more bicarbonate than normal is used to buffer the extra hydrogen ions. So long as the addition of hydrogen ions is not too rapid, bicarbonate generation and hydrogen ion secretion by the renal tubular cell will prevent a significant fall of plasma $[HCO_3^-]$. Once the rate of secretion of H^+—and therefore the generation of HCO_3^-—has reached a maximum any further addition of H^+ will cause a reduction in plasma bicarbonate levels, the bicarbonate being replaced by an equivalent amount of the anionic metabolite. The pH may be partially or completely corrected by rapid loss of CO_2, but extracellular buffering capacity is reduced.

Some conditions associated with such overproduction of hydrogen ions are listed in Table 3. In the adult *diabetic ketoacidosis* is the commonest; *hypoxic lactic acidosis* is common, but is usually part of a complicated picture which includes many other causes of acidosis.

In infants *inborn errors* should always be considered when an unexpected—and authenticated—low Tco_2, is found. These fall into two groups—those errors of carbohydrate metabolism associated with lactic acidosis, and errors of fatty or amino acid metabolism associated with keto- and other organic acid production.

Ketoacidosis

Acetoacetate is a product of acetyl CoA, the acetyl group of which is oxidized to CO_2 and water via the tricarboxylic acid cycle. If acetyl CoA production exceeds its rate of removal the acetyl groups of two molecules condense to form acetoacetate and coenzyme A (CoASH) may be reused for further metabolism. The net reaction is:

2CH$_3$COSCoA + H$_2$O → CH$_3$COCH$_2$COO$^-$ + H$^+$
acetyl CoA   acetoacetate   + 2CoASH

Initially equimolar amounts of acetoacetate and hydrogen ion are produced. Acetoacetate is converted to β-hydroxybutyrate by reduced coenzyme (NADH + H$^+$) and this reaction does not utilize the H$^+$ generated from acetyl CoA

CH$_3$COCH$_2$COO$^-$ + NADH + H$^+$ →
　　　　　　　　　CH$_3$CHOHCH$_2$COO$^-$ + NAD$^+$
　　　　　　　　　β-hydroxybutyrate

The HCO_3^- utilized in buffering H$^+$ is replaced by a quantitatively equivalent sum of acetoacetate and β-hydroxybutyrate ions. The fall in extracellular $[HCO_3^-]$ stimulates the tubular carbonate dehydratase mechanism and, as HCO_3^- is generated, H$^+$ is secreted in exchange for sodium filtered with β-hydroxybutyrate (pK 4·7) and acetoacetate (pK 3·6); these ketoacids may be detectable

TABLE 3
EXAMPLES OF PATHOLOGICAL CONDITIONS CAUSING HYDROGEN ION OVERPRODUCTION

Conditions	Predominant anion(s)
Lactic acidosis (see Fig. 7)	lactate
Ketoacidosis due to diabetes or starvation	acetoacetate + β-hydroxybutyrate
Long-chain ketoacidosis due to inborn errors of fatty acid metabolism	
β-ketothiolase deficiency	α-methylacetoacetate + α-methyl-β-hydroxybutyrate
propionyl CoA carboxylase deficiency (propionic aciduria)	propionate + associated ketoacids
methylmalonic acidaemias	methylmalonate + associated ketoacids
Inborn errors of branched-chain amino acid metabolism	
maple syrup urine disease (ketoacidosis)	branched-chain amino- and ketoacids.
isovaleric acidaemia	isovalerate.
β-methylcrotonyl CoA carboxylase deficiency (β-methylcrotonic aciduria)	β-hydroxyisovalerate + β-methylcrotonyl glycine
Glutathione synthetase deficiency (5-oxoprolinuria)	? 5-oxoproline
Administration of ammonium chloride	chloride

in the urine. Any renal impairment—whether due to a reduction in glomerular filtration rate, to tubular damage, or to both—will reduce the renal capacity to replace the depleted HCO_3^-.

Acetoacetate overproduction occurs whenever there is a relative increase in the rate of fatty acid compared with carbohydrate catabolism.

During *fasting* the intracellular availability of glucose is reduced in those tissues without significant glycogen stores. Triglyceride catabolism in adipose tissue provides the substrate for acetoacetate production. Ketosis may occur after only short periods of fasting, and is reversed by giving carbohydrate. In prolonged starvation the rate of production may be so great that some acidosis results.

In *diabetes,* ketosis is also due to intracellular glucose deficiency in those cells dependent on insulin (despite high extracellular levels). There may be no detectable acidosis whilst production is only slightly increased but in diabetic ketoacidosis plasma bicarbonate may fall to very low levels; if compensatory loss of CO_2 through the lungs is not rapid enough to bring the ratio back to normal there will be a fall in pH. The acidosis is aggravated if volume depletion lowers the GFR, so reducing the amount of sodium filtered, and therefore its availability for exchange with hydrogen ion. Poor tissue perfusion, also due to fluid depletion, may simultaneously cause lactic acidosis. The importance of adequate rehydration during therapy is therefore obvious.

The cause of the low $T\text{co}_2$ is usually obvious if plasma or blood glucose is estimated, and if urine and perhaps plasma are tested for ketones. Knowledge of the actual level of ketones, or of lactate, is usually of academic interest, since correction of volume depletion coupled with insulin therapy (which aims to increase glucose entry to cells and therefore decrease fatty acid catabolism) should correct both. It is common to request estimation of pH, $P\text{co}_2$, and derived values at frequent intervals. However, unless there is a superimposed respiratory condition, this is unnecessary and the adequacy of therapy can be followed by monitoring $T\text{co}_2$. In my experience overemphasis on unnecessary investigations sometimes distracts attention from such important findings as continued hyperosmolality, despite control of hyperglycaemia, due to the development of hypernatraemia.

Longer chain ketoacid accumulation may be due to some inborn errors of amino- and fatty acid metabolism (Table 3).

Lactic Acidosis

Reduction of pyruvate by NADH produces lactate (compare the production of β-hydroxybutyrate from acetoacetate)

$$\underset{\text{Pyruvate}}{CH_3COCOO^-} + NADH + H^+ \rightarrow \underset{\text{Lactate}}{CH_3CHOHCOO^-} + NAD^+$$

Lactate and hydrogen ion production are almost always equimolar. Lactic acid may therefore accumulate due to overproduction or under-utilization of pyruvate, or to overproduction of NADH.

As shown in Fig. 7, pyruvate is the product of anaerobic glycolysis. If oxygen is available it is usually metabolized to acetyl CoA and hence to CO_2 and water (in the tricarboxylic acid cycle), or to fatty acids. In the liver and kidney pyruvate can also be used for gluconeogenesis, a process needing ATP and GTP mostly generated by the tricarboxylic acid cycle: gluconeogenesis is therefore indirectly dependent on oxygen. If pyruvate is produced anaerobically faster than it can be used by oxygen-requiring pathways it is converted to lactate; lactate is later reconverted to pyruvate and metabolized. Conversion of lactate either to glucose, or to CO_2 and water, uses hydrogen ion. The normal liver uses more lactic acid aerobically than it produces anaerobically.

Lactic acidosis may therefore be due to any of the following factors:

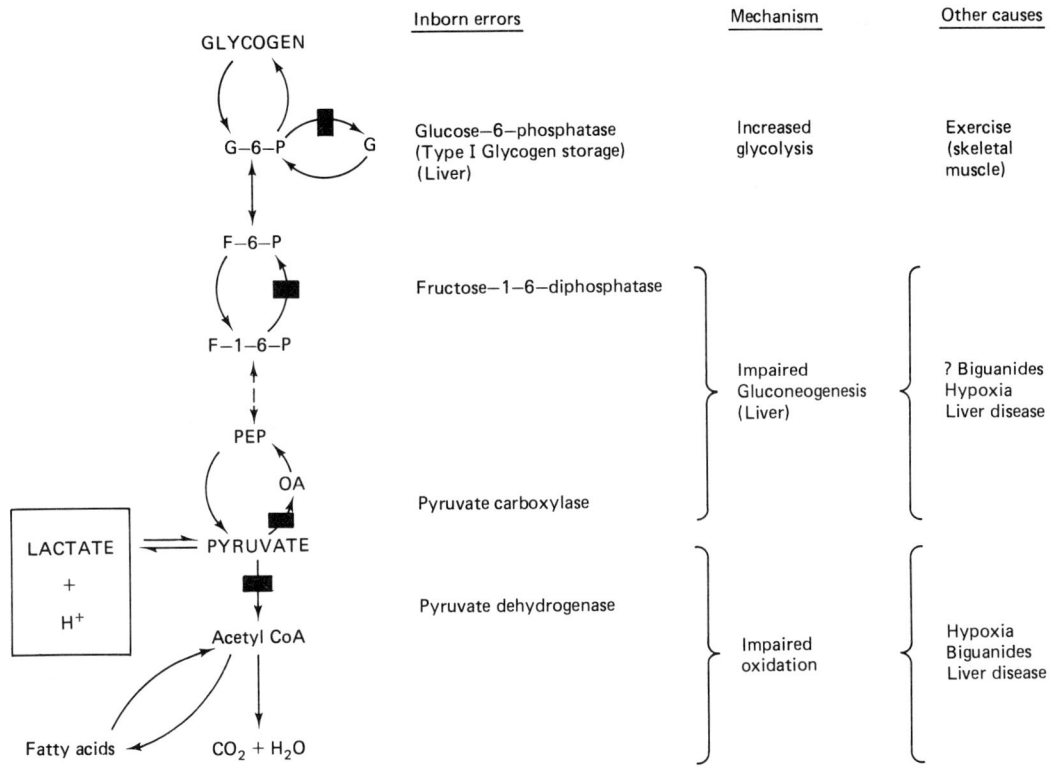

FIG. 7. Schematic representation of some disturbances of pyruvate metabolism which may cause lactic acidosis.
G = glucose
G-6-P = glucose-6-phosphate
F-1-6-P = fructose-1-6-diphosphate
PEP = phosphoenol pyruvate
OA = oxaloacetate

1. Pyruvate Overproduction by Anaerobic Glycolysis

The temporary overproduction following *muscular exercise* is due to rapid glycogenolysis and glycolysis in skeletal muscle.

In the liver, glucose-6-phosphate (G-6-P) derived from glycogenolysis and gluconeogenesis, can, in the presence of *glucose-6-phosphatase*, be converted to glucose and released into the extracellular fluid. In *Type I glycogen storage disease* this enzyme is deficient, and when glycogen breakdown is stimulated (for instance, by adrenaline or glucagon) less of the resultant G-6-P than normal can be converted to glucose; the increment can only be metabolized via pyruvate, and overproduction may precipitate lactic acidosis.

Fructose, sorbitol and xylitol can enter liver cells at a lower extracellular concentration than can glucose: they then enter the glycolytic pathway. Infusion of any of these sugars, or sugar alcohols, especially into a patient with some hepatic hypoxia (and therefore impairment of gluconeogenesis and pyruvate oxidation), can increase hepatic pyruvate production sufficiently to cause lactic acidosis. Glucose is a preferable energy source for parenteral nutrition.

2. Under-utilization of Pyruvate

Pyruvate, and therefore lactate, accumulation may be due to under-utilization despite a normal rate of production. Because the tricarboxylic acid cycle and synthetic pathways are interdependent most factors impairing the tricarboxylic acid cycle also impair hepatic gluconeogenesis and fatty acid synthesis.

Metabolism in the tricarboxylic cycle is impaired if the oxygen supply to the tissues is inadequate. This may occur in states of 'shock' and is especially severe after cardiac arrest. In such conditions the hypoxic liver changes from a predominantly lactic acid consuming organ to a predominantly lactic acid producing one. This type of lactic acidosis is usually superimposed on acidosis due to renal and respiratory impairment, and may sometimes accompany diabetic ketoacidosis.

Some drugs may affect aerobic pathways, and therefore impair pyruvate utilization by all routes. *Biguanides* such as phenformin may act in this way. Lactic acidosis due to these drugs may be severe in patients with impaired renal homeostatic mechanisms: this cause should be remembered in a known diabetic presenting with a low plasma $T\text{CO}_2$ despite a normal, or only slightly elevated plasma glucose or urea level, and without detectable ketosis. *Salicylates* may 'uncouple' oxidative phosphorylation, and in overdosage may so impair pyruvate utilization as to cause lactic acidosis.

Hepatic gluconeogenesis accounts for most of the normal metabolism of lactic acid produced in other tissues (such as skeletal muscle), and *severe liver disease* may rarely cause lactic acidosis.

Pyruvate dehydrogenase is an enzyme catalysing the

conversion of pyruvate to acetyl-CoA—the first step in its entry into the tricarboxylic acid cycle. The inborn error due to deficiency of this enzyme may present with lactic acidosis.

Primary impairment of gluconeogenesis. As shown in Fig. 7, two steps in the glycolytic pathway are not directly reversible. Gluconeogenesis can only proceed in the presence of the enzymes catalysing these two steps. If the level of either is reduced glycolysis, pyruvate oxidation, and fatty acid synthesis can still proceed, but gluconeogenesis is impaired. Inborn errors involving *fructose-1-6-diphosphatase* or *pyruvate carboxylase* are associated with lactic acidosis.

3. Increased Conversion of Pyruvate to Lactate

Elevated $NADH_2$ levels might be expected to increase lactate production. Mild lactic acidosis may be found after ingestion of large amounts of alcohol. Ethanol is metabolized to acetyl CoA via acetaldehyde, each step involving reduction of NAD^+, the net reaction being:

$$C_2H_5OH + CoASH + 2NAD^+ \rightarrow CH_3COSCoA + 2NADH + 2H^+$$

It may be that the usually mild lactic acidosis which may occur after ethanol ingestion is due to utilization of this reduced NAD to convert pyruvate to lactate and that it may be aggravated if lactate utilization is impaired by liver disease. This is an unimportant finding in some alcoholic patients. Methyl alcohol poisoning causes lactic acidosis by a similar mechanism.

Acidosis due to Ethylene Glycol and Paraldehyde Poisoning

Ingestion of these substances is known to cause metabolic acidosis.

Glycollate is an intermediate in ethylene glycol breakdown, and its further metabolism may be rate-limiting. The anion in the acidosis of ethylene glycol poisoning is probably glycollate.

The cause of the acidosis associated with paraldehyde poisoning is less well understood.

Administration of Ammonium Chloride

Ammonium ion is converted to urea in the liver, and this process generates H^+. Large amounts of ammonium chloride may cause acidosis. The anion replacing HCO_3^- is chloride, and ammonium chloride administration is one of the causes of hyperchloraemic acidosis (see Table 6). Ammonium chloride is rarely given therapeutically nowadays, but is used to test the integrity of the renal carbonate dehydratase mechanism (see Renal Tubular Acidosis (p. 37)).

Extrarenal Loss of Bicarbonate into the Intestinal Tract

Pancreatic and small intestinal secretions are very alkaline because of their high bicarbonate content. Normally this loss is balanced by the earlier bicarbonate generation in the gastric mucosa during gastric acid secretion, and by generation in the renal tubular cells. In patients with very severe diarrhoea, or drainage of large amounts of fluid from small intestinal fistulae, the net loss may be much larger than normal. The renal mechanisms, stimulated as usual by a low plasma $[HCO_3^-]$, may nevertheless be prevented from functioning optimally by a low GFR due to volume depletion; the amount of sodium and buffer filtered may then be inadequate for maximal operation of the carbonate dehydratase mechanism. If the sodium lost into the intestine is replaced with fluids of low sodium concentration, the amount of filtered sodium is further reduced.

The cause of the low total CO_2 is usually obvious from the clinical history and examination. If volume depletion is corrected and plasma sodium concentration is kept normal bicarbonate therapy is not indicated in the uncomplicated case: these steps will provide the renal tubule with the wherewithal to replace the lost bicarbonate.

Small intestinal secretions are relatively rich in potassium, and this type of acidosis is one of the exceptions to the rule that acidosis causes hyperkalaemia.

Transplantation of the ureters into the colon or ileum, for instance after total cystectomy, delivers urine into this part of the intestine. The lining cells exchange chloride from the lumen for extracellular HCO_3^-—the reverse process to that in the stomach. This results in *hyperchloraemic acidosis*. The cause is usually obvious, and it should be treated by giving sodium bicarbonate.

Impairment of the Tubular Mechanisms Generating Bicarbonate

In the two groups already discussed the renal homeostatic mechanisms were capable of generating bicarbonate, and the urine was acid. In the group to be discussed the acidosis is due to impairment of bicarbonate generation and hydrogen ion secretion.

The causes are shown in Table 4.

We have already explained that a *low GFR*, by reducing the total amount of sodium and phosphate filtered, can impair the correction of keto- or lactic acidosis, or of the acidosis due to intestinal bicarbonate loss. A significantly low GFR, whether due to renal or pre-renal causes, can so reduce the function of intact tubular cells that, even without complicating factors, Tco_2 falls; such a reduced GFR will be indicated by oliguria, and raised plasma urea and creatinine levels. Acidosis due only to a low GFR is rarely severe. Volume and sodium depletion should be corrected by administration, intravenously if necessary, of the appropriate fluid.

Renal tubular damage with a normal GFR may cause a low Tco_2 with normal plasma urea and creatinine levels. There is often associated potassium and phosphate loss, and the uncommon association of acidosis and hypokalaemia may provide the clue to the cause. Evidence of tubular damage, such as glycosuria and aminoaciduria, should be sought.

TABLE 4
IMPAIRMENT OF RENAL TUBULAR BICARBONATE GENERATION

Cause	Mechanism
Low GFR	Carbonate dehydratase mechanism inhibited because less sodium available for exchange with hydrogen ion and less buffer available to accept hydrogen ion.
Renal tubular damage	Renal tubular function impaired.
Proximal renal tubular acidosis	Deficiency in proximal tubular carbonate dehydratase mechanism. Bicarbonate wastage.
Distal ('classical') renal tubular acidosis	Distal tubular cells abnormally permeable to hydrogen ion. Distal carbonate dehydratase mechanism inhibited by H^+. Inability to acidify urine.
Acetazolamide	Direct inhibition of carbonate dehydratase.

More commonly acidosis is due to *generalized renal failure,* which involves both glomeruli and tubules; plasma urea and creatinine levels will be high and there is variable hyperkalaemia. The acidosis *per se* rarely requires treatment.

Renal tubular acidosis is due to failure of the tubular cells to reabsorb and/or generate HCO_3^-, without equivalent failure of other tubular functions. By contrast with the other causes of renal acidosis, quantitative reabsorption of sodium is relatively normal. However, since less hydrogen ion is available for exchange, a greater proportion of filtered sodium is reabsorbed isosmotically with chloride, or in exchange for potassium. The acidosis of this group of conditions has, therefore two unusual features—*hyperchloraemia* and *hypokalaemia*.

There are two forms of renal tubular acidosis. In both types the capacity to secrete ammonia appropriate to the pH of the urine is retained.

1. A defect in the *proximal tubular* carbonate dehydratase mechanism causes *impairment of bicarbonate reabsorption,* and urinary bicarbonate loss is inappropriate to the plasma level. The distal tubular cells can still secrete H^+ and generate bicarbonate and if the plasma, and therefore filtered, bicarbonate levels fall very low, little bicarbonate reaches the distal tubule despite impaired proximal reabsorption; under these circumstances the distal tubule will function normally, the H^+ secreted mostly combining, as usual, with phosphate and ammonia, rather than with bicarbonate. *The ability to form an acid urine in severe systemic acidosis is retained.*

2. A defect in the *distal tubule* is associated with an *inability to acidify the urine* even in the face of severe systemic acidosis. In this so-called 'classical' type of renal tubular acidosis proximal HCO_3^- reabsorption can be shown to be normal, and the carbonate dehydratase mechanism is probably intact. The abnormality is probably in the luminal membranes of the distal cells. These are normally impermeable to secreted H^+, allowing a hydrogen ion gradient of 100:1 to build up between urine and plasma (at urinary pH 5·4); in 'classical' renal tubular acidosis the membranes may be abnormally permeable. Return of secreted H^+ to the tubular cell would inhibit an otherwise normal carbonate dehydratase mechanism, and acidification of the urine and bicarbonate generation would be impaired. The high filtered load of calcium common to all forms of acidosis meets relatively alkaline urine, and the patient may present with renal calculi or nephrocalcinosis.

The ability to acidify the urine may be *tested* by administering 0·1 mg of ammonium chloride orally per kg body weight. In the normal subject and in patients with generalized renal tubular damage or proximal renal tubular acidosis, the pH of the urine will fall to below 5·2 at between 2 and 8 hours after the dose. Only in cases of 'classical' renal tubular acidosis will the pH remain above this level, reflecting inability to acidify the urine.

Distal renal tubular acidosis may be familial or may be a transient finding in the newborn. However, both types are often secondary to damage to tubular cells: causes include hypercalcaemia, ingestion of various toxins (for instance, amphotericin B or mercury), Bence–Jones proteinuria or hyperglobulinaemic states, and inborn errors such as those listed in Table 5, in which metabolites may be deposited in the renal tubules. Renal

TABLE 5
SOME INBORN ERRORS WHICH MAY CAUSE RENAL TUBULAR ACIDOSIS

Inborn error	Tubular deposition of:
Hereditary fructose intolerance	Fructose-1-phosphate
Galactosaemia	Galactose-1-phosphate
Wilson's disease	Copper
Cystinosis	Cystine
Fabry's disease	Abnormal glycosphingolipids

tubular acidosis has been reported in severe hyperthyroidism. Most of these secondary causes may, later in the course of the disease, be associated with generalized tubular damage (Fanconi syndrome).

Either type responds to early treatment with oral sodium bicarbonate and potassium supplements. Permanent tubular damage from hypokalaemia, and in the primary distal form, nephrocalcinosis, may complicate untreated cases.

Acetazolamide is a carbonate dehydratase inhibitor. Its effect is therefore similar to that of renal tubular acidosis; although the patient on such therapy may develop hyperchloraemic, hypokalaemic metabolic acidosis, the small doses of the drug, used in ophthalmology rarely cause these disturbances. Acetazolamide is not now used as a diuretic because of the danger of acidosis.

TABLE 6
METABOLIC ACIDOSIS ASSOCIATED WITH HYPERCHLORAEMIA AND HYPOKALAEMIA

	Hyper-chloraemia	Hypo-kalaemia
Extrarenal (intestinal) loss of bicarbonate and potassium	−	+
Ureteric transplantation into the intestine	+	±
Impairment of renal tubular function:		
Generalized tubular disease	−	+
Impairment of CD mechanism only		
Renal tubular acidosis	+	+
Acetazolamide therapy		
Administration of ammonium chloride	+	−

The causes of acidosis which may be associated with hyperchloraemia and/or with hypokalaemia are summarized in Table 6.

Inborn Errors Causing Metabolic Acidosis

We have seen that metabolic acidosis is a common finding in inherited metabolic disorders and that they should always be remembered in infants presenting with low T_{CO_2}.

In summary, acidosis associated with inborn errors may be due to:

1. Lactic acidosis due to inborn errors of the gluconeogenic pathway and in pyruvate dehydrogenase and glucose-6-phosphatase deficiency (Fig. 7).
2. Overproduction of other intermediate metabolites with H^+ (Table 3).
3. Renal tubular acidosis due to deposition of metabolites in the renal tubular cells (Table 5).

Respiratory Acidosis and Hypoxia

The primary abnormality in the bicarbonate buffer pair in respiratory acidosis is a high P_{CO_2}; the other findings in the uncompensated, partially compensated, and fully compensated states are indicated in Table 2.

Compensation depends on acceleration of the renal tubular carbonate dehydratase mechanism by the high P_{CO_2} (see p. 28). If renal function is normal the capacity to generate bicarbonate is very high, but full compensation takes time. Thus in chronic respiratory disease the P_{CO_2} may stabilize at a very high level; the kidneys will then generate HCO_3^- until the ratio of $[HCO_3^-]$ to P_{CO_2} and, therefore the pH, is normal and a new steady state is reached. By contrast, a much smaller rise of P_{CO_2} in acute respiratory failure may be too rapid to allow much rise in bicarbonate, and acidosis may be severe.

The Standard Bicarbonate in Respiratory Acidosis

The capacity of the red cell to generate HCO_3^- is limited by the buffering capacity of haemoglobin. The difference between the standard and actual bicarbonate in respiratory acidosis is an index of the increment in circulating bicarbonate due to the red cell mechanism (p. 27). In *acute respiratory failure* the kidney and red cell both generate bicarbonate in response to the high P_{CO_2} and a relatively high proportion of the increment will be of red cell origin (the amounts will depend on the haemoglobin level, and on renal function): the actual bicarbonate will be slightly elevated, but the *standard bicarbonate is usually normal*.

Once the haemoglobin buffering capacity is saturated any further increment is of renal origin: in chronic respiratory disease the standard bicarbonate will be above 'normal', but lower than the actual bicarbonate; the deviation of the standard bicarbonate from 'normal' in the chronic state reflects the proportionally larger contribution by the kidneys. In my experience the standard bicarbonate is no more informative than the P_{CO_2} and pH in the assessment of uncomplicated respiratory failure.

If there is a superimposed metabolic acidosis in a patient with acute respiratory failure, some of the bicarbonate generated from CO_2 will be used in buffering: a low standard bicarbonate reveals this by eliminating part of the respiratory component. In such complicated cases a knowledge of the 'standard' bicarbonate level is occasionally helpful.

Conditions which Impair Alveolar Function

Impaired alveolar function affects oxygen as well as carbon dioxide transport. It is convenient to discuss both effects together.

Gas exchange might be impaired for any of the following reasons:

1. Diffusion between the blood and the alveolar cavity might be reduced by damage to the alveolar walls.
2. The alveolar sac might be partially or completely filled with oedema fluid or exudate.
3. Air entry to the sac might be impaired by:
   blockage of airways;
   reduction of the magnitude of respiratory excursions.

The effect on arterial oxygen and carbon dioxide levels depends on the extent of the disease, and on the integrity of the blood supply to the malfunctioning alveoli.

In conditions such as *acute and chronic bronchitis, broncho- and lobar pneumonia, and pulmonary fibrosis, infiltration or collapse* impaired function of parts of the lung is due to more than one of the above factors. In *pulmonary oedema* the alveoli fill with oedema fluid, and diffusion of oxygen is more impaired than that of carbon dioxide.

Aeration of the whole of both lungs may be reduced by partial blockage of the main airways (as during laryngeal spasm, or after inhalation of a large foreign body). Impaired movements of the rib cage due to *injury, ankylosing spondylitis* and *gross obesity*, may cause underventilation. The normal feed-back response of the respiratory centre may be impaired by *hypothalamic lesions,* or the response may not be transmitted to the chest-wall muscles in, for instance, *poliomyelitis* or *peripheral neuritis.*

In the conditions mentioned so far the blood supply to the affected alveoli remains relatively intact. In *emphysema* reduced blood supply to malfunctioning alveoli may produce different effects on arterial levels of oxygen and carbon dioxide.

Arterial $P\text{CO}_2$ and $P\text{O}_2$ in Respiratory Dysfunction

Any condition causing a rise in $P\text{CO}_2$ will be associated with a low $P\text{O}_2$, but a low $P\text{O}_2$ is not always accompanied by a high $P\text{CO}_2$.

Alveolar dysfunction with a relatively normal blood supply. In conditions in which alveolar blood supply is relatively normal arterial $P\text{O}_2$ is more likely to be abnormal than $P\text{CO}_2$.

The physiological stimulus to the respiratory centre is a fall in pH of the blood perfusing it; hypoxia can also stimulate it. If the function of a significant proportion of alveoli is impaired an increased respiratory tidal volume can neither eliminate retained CO_2, nor maintain a normal $P\text{O}_2$. However, increased aeration of a relatively large proportion of functioning alveoli may be able to compensate for the retention of CO_2, but not for the low O_2, in blood leaving the hypofunctioning sections of the lung. This difference in response is due to differences in diffusibility and mode of transport of the two gases.

Dissolved carbon dioxide can diffuse from the blood into alveolar air at approximately twenty times the rate that dissolved oxygen can diffuse in the reverse direction. CO_2 can also diffuse more rapidly through oedema fluid. In many cases of *pulmonary oedema* the $P\text{O}_2$ is reduced but the $P\text{CO}_2$ is normal; the low $P\text{O}_2$ causes overbreathing by stimulating the respiratory centre, and the elimination of carbon dioxide may be increased enough to reduce the $P\text{CO}_2$ below normal. The $P\text{CO}_2$ may rise in very severe pulmonary oedema.

Very little oxygen is carried in the blood in simple solution. The small amount of dissolved oxygen in pulmonary capillary plasma diffuses into the red cell, and reaches equilibrium with the larger amount in oxyhaemoglobin. At a normal $P\text{O}_2$ (about 13 kPa or 100 mmHg) about only 3% of arterial haemoglobin remains unsaturated, and even at a $P\text{O}_2$ as low as 8 kPa (60 mmHg) only about 10% of haemoglobin is free to take up more oxygen (see Fig. 8): unless the $P\text{O}_2$ of inspired air is above atmospheric, increased respiratory excursion due to stimulation of the respiratory centre by hypoxia cannot significantly increase the oxygen content of arterial blood. By contrast, although a little CO_2 is bound to haemoglobin (as carbaminohaemoglobin), most is in simple solution and it can diffuse faster than oxygen. The capacity of normal alveoli to eliminate this dissolved CO_2 is very large, and overbreathing can reduce the $P\text{CO}_2$ of blood leaving normal alveoli to abnormally low levels. Thus, blood leaving malfunctioning alveoli will have a high $P\text{CO}_2$ and low $P\text{O}_2$ and this will

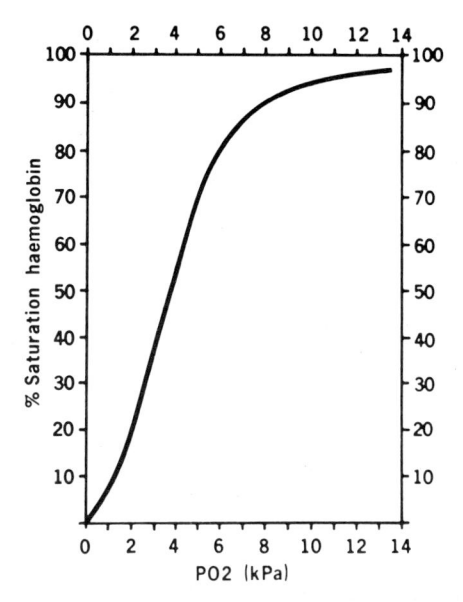

FIG. 8. Oxygen dissociation curve of haemoglobin (pH 7·4; Temp. 37°C). Haemoglobin remains over 90% saturated with oxygen until the $P\text{O}_2$ falls below 8 kPa (60 mmHg). Saturation falls rapidly as $P\text{O}_2$ is further reduced.

mix, as it enters the pulmonary vein, with blood leaving relatively normal alveoli with a low $P\text{CO}_2$ and normal $P\text{O}_2$.

1. *If the proportion of normally functioning alveoli is high* the low $P\text{CO}_2$ in blood leaving them may 'overcompensate' for the high $P\text{CO}_2$ in blood from impaired areas. The normal $P\text{O}_2$ from these areas may prevent a significant fall in $P\text{O}_2$, but it cannot 'overcompensate'.

Therefore *arterial $P\text{CO}_2$ will be low and $P\text{O}_2$ will be normal or low.*

2. *If the proportion of normally functioning alveoli is slightly less,* the arterial $P\text{O}_2$ *will soon fall* below normal levels. The $P\text{CO}_2$ *will remain normal or low.*

3. *If the proportion of malfunctioning alveoli is high* the smaller fraction of blood with a low $P\text{CO}_2$ will not balance that with a high $P\text{CO}_2$. $P\text{O}_2$ *will continue to fall, and $P\text{CO}_2$ will rise.*

If the $P\text{O}_2$ falls below about 8 kPa (about 60 mmHg) haemoglobin saturation falls and the patient becomes cyanosed.

Alveolar dysfunction with impaired blood supply. If the blood supply to damaged alveoli is impaired both arterial $P\text{CO}_2$ and $P\text{O}_2$ are relatively well preserved until late.

The reduction of arterial $P\text{O}_2$ in the relatively localized pulmonary disease described above is due to mixing of blood with a low $P\text{O}_2$ with that of a normal $P\text{O}_2$. If the impaired alveoli were not perfused, arterial blood would only be derived from normal alveoli. The $P\text{O}_2$ would be only slightly reduced, despite quite extensive lung disease. The oxygen dissociation curve of haemoglobin is sigmoid and saturation falls very little until the $P\text{O}_2$ falls below about 8 kPa (60 mmHg) (Fig. 8). Cyanosis does not occur until this type of disease is very extensive. However if the response of the respiratory centre is normal it is stimulated by a slight reduction in $P\text{O}_2$; the $P\text{CO}_2$ falls early. The pink well-oxygenated appearance of these breathless patients (usually with emphysema) has inspired the nickname 'pink puffers'. This contrasts with the less breathless 'blue bloaters' with chronic bronchitis, who are cyanosed and have a high $P\text{co}_2$; (the 'bloating' may be due to the oedema of right heart failure) and the relative absence of dyspnoea to a reduced response of the respiratory centre to chronic elevation of $P\text{co}_2$ and reduction of $P\text{o}_2$.

The findings in these conditions are summarized in Table 7.

Alkalosis

Alkalosis is uncommon and may be induced by the actions of the patient or of his medical attendant.

Either a rise in [HCO_3^-] in metabolic alkalosis, or a fall in $P\text{co}_2$ in respiratory alkalosis will reduce the ratio of [HCO_3^-] to $P\text{co}_2$ and therefore cause a rise in pH

$$\text{pH} = 6.1 + \log \frac{[HCO_3^-]}{P\text{co}_2 \times 0.225}$$

TABLE 7
ARTERIAL BLOOD GASES IN PULMONARY DYSFUNCTION

	Blood leaving				Arterial blood			
	Dysfunctioning alveoli		Normal alveoli		Patchy disease		Extensive disease	
	$P\text{o}_2$	$P\text{co}_2$	$P\text{o}_2$	$P\text{co}_2$	$P\text{o}_2$	$P\text{co}_2$	$P\text{o}_2$	$P\text{co}_2$
Pulmonary oedema	↓	Usually N	–	–	–	–	↓	Usually N
Alveolar blood supply normal	↓↓	↑↑	N	↓↓	↓	N or ↓	↓↓	↑
reduced	–	–	N or ↓	↓	N or ↓	↓	↓	N or ↑

It is important to remember the stimulatory effect of a low $P\text{o}_2$ before giving oxygen to patients with pulmonary hypofunction. The increased respiratory rate due to the low $P\text{o}_2$ may be minimizing the rise in $P\text{co}_2$ and, in chronic respiratory disease, compensation has probably occurred: moreover the respiratory centre may have become less responsive to the chronically high $P\text{co}_2$. Restoration of the $P\text{o}_2$ to normal may so reduce the respiratory drive as to cause dangerous acute CO_2 retention without immediate compensatory changes and lead to a fall in pH. Haemoglobin saturation, and the supply of oxygen to the tissues, falls very little until the $P\text{o}_2$ is below 8 kPa. Oxygen therapy in such cases should aim to maintain $P\text{o}_2$ at just above this level; this ensures continued respiratory drive with adequate tissue oxygenation.

Oxyhaemoglobin is a less effective buffer than haemoglobin. If arterial oxygenation is reduced, haemoglobin buffering capacity, and therefore the ability of the erythrocyte to form bicarbonate from CO_2, is increased. This slight tendency to increase the ratio of extracellular [HCO_3^-] to $P\text{co}_2$ in hypoxic states is relatively unimportant in offsetting the primary effects of CO_2 retention.

Conversely buffering of excess H^+ by haemoglobin impairs its oxygen binding capacity and in hypoxic states a high $P\text{co}_2$ further reduces the ability to form oxyhaemoglobin. Erythrocyte hypoxia increases the 'shunt' pathway by which 1:3-diphosphoglycerate, a metabolite on the glycolytic pathway, is converted to 2:3-diphosphoglycerate (2:3-DPG); 2:3-DPG, like H^+ and carbon dioxide, impairs the oxygen binding by haemoglobin: this is probably an adaptive process that helps to maintain release of oxygen to tissues, in, for instance, anaemic and chronic hypoxic states.

Certain secondary findings are common to many cases of alkalosis, whether respiratory or metabolic in origin. These are the opposite of the effects due to acidosis described on p. 33.

1. Hypokalaemia is a common accompaniment of alkalosis. Alkalosis, by reducing the availability of H^+ for renal tubular exchange for sodium, leads to increased renal loss of potassium. Conversely, potassium depletion causes alkalosis, and it may not always be easy to distinguish cause from effect.

2. Tetany. Alkalosis may reduce the ionized fraction of circulating calcium enough to cause tetany despite a normal total calcium level.

Metabolic Alkalosis

The primary abnormality in the bicarbonate pair in metabolic alkalosis is a high plasma [HCO_3^-]. Although some compensation due to CO_2 retention can be demonstrated, severe alkalosis could only be fully compensated at the expense of serious hypoxia: the tendency of the alkalosis to inhibit the respiratory centre is overridden by the consequent rise in $P\text{co}_2$ and fall in $P\text{o}_2$. Correction therefore depends on inhibition of the renal tubular carbonate dehydratase system by the high extracellular bicarbonate level.

This inhibition impairs both 'reabsorption' and net generation, and the bicarbonate loss in the urine may be adequate to correct the extracellular elevation. A reduced GFR will limit the total amount which can be lost and will hamper correction. It can now be seen that an adequate GFR is an important factor in the correction of either acidosis or alkalosis.

A primary rise in plasma [HCO_3^-] occurs in three situations.

1. Large amounts of sodium bicarbonate given by intravenous infusion, or even by mouth, may exceed the renal capacity to excrete it. This is particularly likely to cause alkalosis in patients with impaired renal function.

2. Potassium depletion. Generation of bicarbonate by the renal tubule in potassium depletion is a common cause of extracellular alkalosis. The fall to normal of the high plasma Tco$_2$ can be used to monitor the adequacy of potassium replacement if there is no other cause of high extracellular bicarbonate concentration. It is important to exclude the possibility of a high Tco$_2$ due to compensated respiratory acidosis before monitoring in this fashion.

3. Pyloric stenosis. Bicarbonate is generated in the parietal cells of the stomach during secretion of gastric acid. The H^+ is secreted accompanied by chloride, and each mmol of HCO_3^- generated by this mechanism is coupled with the loss of a mmol of chloride: the sum of plasma [HCO_3^-] and [Cl^-] remains constant. Normally this process is followed by secretion of HCO_3^- into the duodenum. If vomiting occurs when the pylorus is patent the loss of secreted HCO_3^- counteracts the effect of its generation by the gastric mucosa, and there is little disturbance of hydrogen ion balance. In pyloric stenosis the obstruction between the stomach and duodenum reduces bicarbonate loss. Loss of gastric hydrogen and chloride ions in the vomitus stimulates the carbonate dehydratase mechanism of mucosal cells, and more bicarbonate is generated. Unless renal loss of bicarbonate is equally rapid, alkalosis follows.

Two factors may impair renal correction of the alkalosis in patients with pyloric stenosis. Firstly volume depletion, consequent on vomiting, may reduce the GFR. Secondly, the hypochloraemia due to loss of chloride in the vomitus, results in a low chloride concentration in the glomerular filtrate. Isosmotic sodium reabsorption in the proximal tubule is impaired, and more sodium is available for exchange with H^+ and K^+. The hydrogen ion secretion causes inappropriate bicarbonate 'reabsorption' and generation, and the potassium loss aggravates that due to alkalosis. The consequence is hypochloraemic alkalosis with hypokalaemia. This advanced picture is not found if the volume depletion and chloride loss are corrected early by saline infusion. Once the biochemical picture is established large amounts of iso-osmolar saline with added potassium should be infused: the restoration of normal plasma chloride levels and GFR will enable the kidney to correct the alkalosis.

Respiratory Alkalosis

The primary abnormality in the bicarbonate buffer pair in respiratory alkalosis is a low Pco$_2$. Compensatory renal loss of, and failure to generate, HCO_3^- depends on the inhibition of the renal carbonate dehydratase mechanism by the reduced Pco$_2$.

A primary fall in Pco$_2$ follows abnormally rapid or deep respiration despite relatively normal CO_2 transport across the alveolar wall. It may occur in any of the pulmonary conditions described on p. 39 in which *oxygen exchange is impaired more than that of CO_2*. The slight tendency to alkalosis in such cases is relatively unimportant.

Excessive artificial respiration may occur relatively frequently on intensive care units, and may rapidly be corrected by adjusting the respirator. If this cause of respiratory alkalosis is coupled with metabolic alkalosis following bicarbonate infusion, severe hypokalaemia may follow.

Hysterical overbreathing may cause respiratory alkalosis. The patient may present complaining of tingling in hands due to the reduced level of ionized calcium.

Raised intracranial pressure, and some *brain stem lesions* may overstimulate the respiratory centre.

Salicylates and Hydrogen Ion Balance

An overdose of salicylates initially tends to cause a respiratory alkalosis by direct stimulation of the respiratory centre. The 'uncoupling' effect on oxidative phosphorylation may so impair aerobic pathways that metabolic lactic acidosis develops. Both these disturbances of hydrogen ion homeostasis may result in low Tco$_2$ levels, but the pH may be high (if respiratory alkalosis is predominant), low (if metabolic acidosis is predominant), or normal. The true state of hydrogen ion balance can only be assessed if blood pH is measured. In practice, such knowledge rarely affects therapy.

INVESTIGATION OF HYDROGEN ION DISTURBANCES

Understanding the mechanisms behind disorders is an essential preliminary to critical assessment of how much investigation is necessary. Without such understanding the doctor may overinvestigate, and, even more dangerously, misinterpret the results of overinvestigation.

Interpretation of Plasma Tco$_2$

This estimation, an adequate assessment of plasma bicarbonate concentration, has the advantage that it can be performed on venous blood. Plasma [HCO_3^-] *alone* tells us little about the state of hydrogen ion balance: the pH depends on the *ratio* of [HCO_3^-] to Pco$_2$. Nevertheless, measurement of Tco$_2$, together with the use of the cheapest of all commodities—the brain—usually yields adequate information.

If the Tco$_2$ is low the following procedure should be adopted. As soon as the diagnosis is obvious, further investigations are unnecessary.

1. Exclude artefactual causes due to *in vitro* CO_2 loss from a specimen that has been standing overnight, or which has a large dead space above it.

2. Estimate plasma urea and glucose (if not already available), and test the urine for ketones.

3. Reassess the clinical picture with special reference to the presence of:
  hypotension, dehydration or other causes of poor tissue perfusion;
  diarrhoea or intestinal fistulae;
  transplantation of the ureters into the intestine;
  a drug history, with specific reference to biguanides such as phenformin.

In the great majority of cases the diagnosis will now be obvious. If it is not further estimations may be indicated.

4. Estimate plasma chloride. Hyperchloraemia with a low $T\text{co}_2$, without obvious cause, is most likely to be due to renal tubular acidosis. If the plasma pH is normal the diagnosis of 'classical' renal tubular acidosis may be confirmed by an ammonium chloride loading test.

5. Estimate blood pH and $P\text{co}_2$ to exclude compensated, or partially compensated, respiratory alkalosis.

If the acidosis is confirmed and the cause is still obscure continue as follows.

6. Estimate lactate. Lactate estimation is unhelpful if tissue hypoperfusion is obvious; the acidosis in such cases is multifactorial, and must be treated on its merits. It may help to identify rarer causes of lactic acidosis, especially in infants.

7. Remember inborn errors in infants. If the level of chloride or lactate is insufficient to account for all the fall in bicarbonate, look for other organic acidaemias.

If someone knowledgeable sees the results early, the first 6 steps can be accomplished with no delay. A logical approach is likely to yield an early answer. It is usually unnecessary to ask for everything at the outset.

If the $T\text{co}_2$ is high, look at the plasma potassium result. Reassess the clinical picture, particularly noting whether there is respiratory disease. Take a drug history, with special reference to potassium-losing diuretics, and to bicarbonate ingestion or infusion. If there is severe vomiting, which might indicate pyloric stenosis, estimate plasma chloride.

Anion Gap

The sums of plasma cation and anion concentrations expressed as mEq/l (numerically the same as mmol/l for univalent ions) must be equal if there is electrochemical neutrality. Sodium with potassium make the largest contribution to the cations (about 144 mEq/l) and chloride and bicarbonate to the anions (about 125 mEq/l). The sum of these measured cations usually exceeds the sum of measured anions by about 15 to 20 mEq/l, and this difference has been called the 'anion gap'. It is made up of the difference between the sum of 'unmeasured' cations such as ionized calcium and magnesium, and of 'unmeasured' anions such as phosphate, urate, organic acids, and plasma proteins.

A low plasma bicarbonate is the hallmark of metabolic acidosis. Each mEq of bicarbonate used to buffer a mEq of hydrogen ion is replaced by a mEq of accompanying anion. In the rare hyperchloraemic acidoses (Table 6) this anion is chloride, and the sum of measured anions and measured cations, and therefore of the 'anion gap', is unchanged. In all other causes of metabolic acidosis the anion is neither chloride nor bicarbonate: the chloride level is unchanged as the bicarbonate falls, and the 'anion gap' is increased. It has been suggested that plasma chloride be routinely estimated with sodium, potassium and bicarbonate and the 'anion gap' calculated. In my experience this is unnecessary if the simple procedure recommended above is followed. Other views may be found in references quoted at the end of the chapter.

Indications for Arterial Estimations

Arterial blood estimation of pH and $P\text{co}_2$ is only indicated in a few situations.

1. If there is doubt about the cause of an abnormal $T\text{co}_2$, estimation of blood gases and pH may elucidate the cause.
2. If mixed respiratory and non-respiratory conditions may be present and if therapy may be affected by the result (for instance, after cardiac arrest frequent estimations are needed to monitor the control and estimation is necessary).
3. If the patient is on a respirator.
4. If there has been an acute exacerbation of chronic pulmonary disease, or if acute pulmonary disease is potentially reversible, vigorous therapy or artificial respiration may tide the patient over until the acute condition improves. In patients with no potential for improvement of pulmonary function knowledge of all parameters can only be of academic interest.
5. If arterial blood is taken for estimation of $P\text{o}_2$ it is sensible to estimate other parameters.

FURTHER READING

Those interested in trying to predict the metabolic pathways (and therefore the pathological disturbances) which produce hydrogen ions, and who would like to understand how they do it, might start their search in:

McGilvery, R. W. (1979), *Biochemistry: a functional approach,* 2nd edn. W. B. Saunders Co., Philadelphia. London. Toronto.

The search is interesting, but time-consuming. An example may be found in:

Zilva, J. F. (1978). The origin of acidosis in hyperlactataemia, *Ann. Clin. Biochem.,* **15,** 40–43.

The following references provide background reading on the physiology of gas exchange:

Wagner, P. D. (1977). Diffusion and chemical reaction in pulmonary gas exchange, *Physiol. Rev.*, **57**, 257–312.
Perutz, M. F. (1978). Hemoglobin structure and respiratory transport, *Scientific American*, **239** (Dec) 68–86.

An explanation of the derived values obtained from measurement of pH and $P\text{co}_2$ will be found in:

Astrup, P., Jørgensen, K., Siggaard Andersen, O. and Engel, K. (1960). The acid-base metabolism. A new approach. *Lancet*, **1**, 1035–1039.

Other references:

Park, R. and Arieff, A. I. (1980). Lactic Acidosis. *Adv. Int. Med.* **25**, 33–68.
Seldin, D. W. and Wilson, J. D. (1978). Renal tubular acidosis. In *The Metabolic Basis of Inherited Disease*. Eds. Stanbury, J. B. Wyngaarden, J. B. and Fredrickson, D. S. McGraw-Hill Book Co., 4th edn. N.Y. etc.
Emmett, M. and Narins, R. G. (1977). Clinical use of the anion gap. *Medicine*, **56**, 38–54. This should be read with the editorial and subsequent correspondence in: *Lancet* (1977) **1**, 785, 948, 1054 (see correction for one of these letters in **2**, 50), and 1304 and **2**, 34, 513 and 710.

3. POLYURIA AND POLYDIPSIA

D. DONALDSON

Introduction

Basic principles of water control
   The requirement for water
   Distribution of water in the body
   Intake and output of water
   Dynamic aspects of water
   Water balance in infancy
   Homeostasis of water and sodium

Physiological control of water homeostasis
   Osmolality
   Thirst
   Water absorption
   Renal function
   Antidiuretic hormone
   Consequences of imbalance of water homeostasis

Clinical aspects
   Primary polyuria and secondary polydipsia
   Primary polydipsia and secondary polyuria
   Hypodipsia and adipsia
   Syndrome of inappropriate ADH secretion (SIADH)

Laboratory investigation of polyuria and polydipsia
   Clinical groups
   Scheme of clinical investigation
   General principles of differential diagnosis and interpretation

INTRODUCTION

In clinical medicine, the symptoms of polyuria and polydipsia are commonly encountered together. Nevertheless, from the diagnostic standpoint, it is necessary to attempt to establish which of the two is the primary event, and which is the secondary consequence. It is clear therefore, that one can at the outset divide such disorders into two groups—in one, polydipsia is primary and polyuria follows, whereas in the other, polyuria is the cause of the polydipsia. However, clinically, both usually present simultaneously, and it is not always easy in retrospect to ascertain which came first.

The mechanisms underlying maintenance of normal hydration of the body, necessarily require discussion in this context. Intake and output of water must be considered. There is, moreover, close integration between fluid intake and urinary output. The anatomical structures involved in this integration, and their normal functioning, will all be referred to in this text.

There is, for instance, continuous monitoring of the plasma osmolality by osmoreceptors in the hypothalamic area. In turn, the results of this monitoring are passed on to other areas of the brain, where physiological action is either initiated or inhibited. Elevated osmolality of the plasma, for example, causes stimulation of the thirst centre. In addition there is secretion of antidiuretic hormone (ADH). This has, of course, been formed in the hypothalamus prior to storage in the posterior lobe of the pituitary gland. It is not fluid intake, but actual excretion from the body which is determined by the ADH mechanism. This hormone acts on the distal tubules and collecting ducts in the kidney, thus serving as the final control over loss of water.

The kidney is a vital organ in fluid balance. It is involved not only in controlling excretion of fluid from the body, but also in regulation of water intake. This is by way of the renin-angiotensin mechanism. It has been shown that angiotensin II is a potent stimulus to drinking.

Viewing these observations against the background of the biological system as a whole, it is clear that in intracellular dehydration, the more important aspect is to take in water, rather than merely to conserve what is already there.

The importance of each of the physiological mechanisms available for maintenance of water balance will be very apparent in the discussion on diseases involving the structures concerned. The causes of polyuria and polydipsia will be elaborated upon in turn, from both physiological and clinical aspects. Finally, those investigations which may aid in the elucidation of the clinical diagnosis will be documented and evaluated.

BASIC PRINCIPLES OF WATER CONTROL

The maintenace of water homeostasis depends essentially on a sensitive balance between the release of antidiuretic hormone (ADH) which controls renal output of water, and the presence of an efficient thirst mechanism, which controls input to the body.[1]

In fact, both water intake and output are largely controlled by osmolality of the plasma. This in turn is largely dependent on sodium concentration. Although it is true that the sodium content largely determines the amount of water in the body of the normal individual, this is not true in pathological conditions such as hyperglycaemia, uraemia, or following administration of highly osmotically active substances such as mannitol and amino acids. The mechanisms underlying osmotic diuresis in these disorders will be considered later.

The Requirement for Water

Water serves as the moisture necessary for the easy intake of nutrition to the body, and itself comprises approximately 55%–60% of the body by weight in the adult man, and 50%–55% in the adult woman. There is, nevertheless, a wide range for the water content in the body—from 40% of body weight in the obese person to 70% of body weight in the lean individual. Water thus forms a large part of the body substance, both intracellularly and extracellularly (Table 1).

It also acts as the medium in which metabolic reactions take place, and as the vehicle for excretion of products of metabolism, either in solution (i.e. the urine) or in the form of admixture with insoluble excretory products (i.e. the faeces). In the latter instance, water is needed for plasticity of the intestinal contents; this plasticity is essential for moulding to the continuously varying intestinal shape and pressure. Intestinal transit is thereby permitted, and hence easy expulsion of the faeces from the body follows after a suitable interval of time, during which as much water as possible is reabsorbed. Evaporation of water from the skin surface (i.e. sweat) and from mucous membranes, serves as a cooling mechanism in hot environments. The dynamic flow of water through the body in general, serves to transport substances throughout the tissues, or in certain instances to convey selected compounds to parts of the body where they are required for special metabolic purposes (e.g. gastrointestinal secretions etc.).

TABLE 1
DISTRIBUTION OF BODY WATER IN AVERAGE NORMAL YOUNG ADULT MALE

Source	ml/kg of body weight	% of total body water	
Intracellular	330	55·0	
Extracellular	270	45·0	
Plasma		45	7·5
Interstitial-lymph		120	20·0
Dense connective tissue and cartilage		45	7·5
Inaccessible bone		45	7·5
Transcellular		15	2·5
Total body water	600	100·0	

Reference From: Edelman and Leibman, *Am. J. Med.*, **27**, 256, 1959

Distribution of Water in the Body

A 70 kg man has in his body, 40 l of water. The extracellular component is about 15 l, of which 3 l is plasma and 12 l interstitial fluid. It is increase of this latter portion which leads to clinical presentation of oedema. Intracellular fluid totals about 25 l in a 70 kg man, of which 2 l is present in the erythrocytes. The 3 l of plasma and 2 l of water in the erythrocytes totals 5 l of blood (Table 2).

TABLE 2
APPROXIMATE FLUID CONTENTS OF THE BODY COMPARTMENTS

Source	Approximate % of body weight	Approximate volume in litres
Extracellular fluid	20	15
-Interstitial fluid	15	12
-Blood plasma	5	3
Intracellular fluid	40	25
-Erythrocyte fluid	3	2
-Non-erythrocyte fluid	37	23

In the male subject, body water tends to decrease slightly up to about twenty years of age. In the female, however, there is a steady persistent decrease up to the menopause. This is on account of increased fat deposition—with its low water content.

Different tissues have different amounts of water—the water content, by weight, of muscle is 70%, of adipose tissue 20%, and of bone and cartilage it is as low as 10%.

Intake and Output of Water

The general sequence of movement of water through the body is displayed diagrammatically in Fig. 1. In normal circumstances, the oral route forms the only

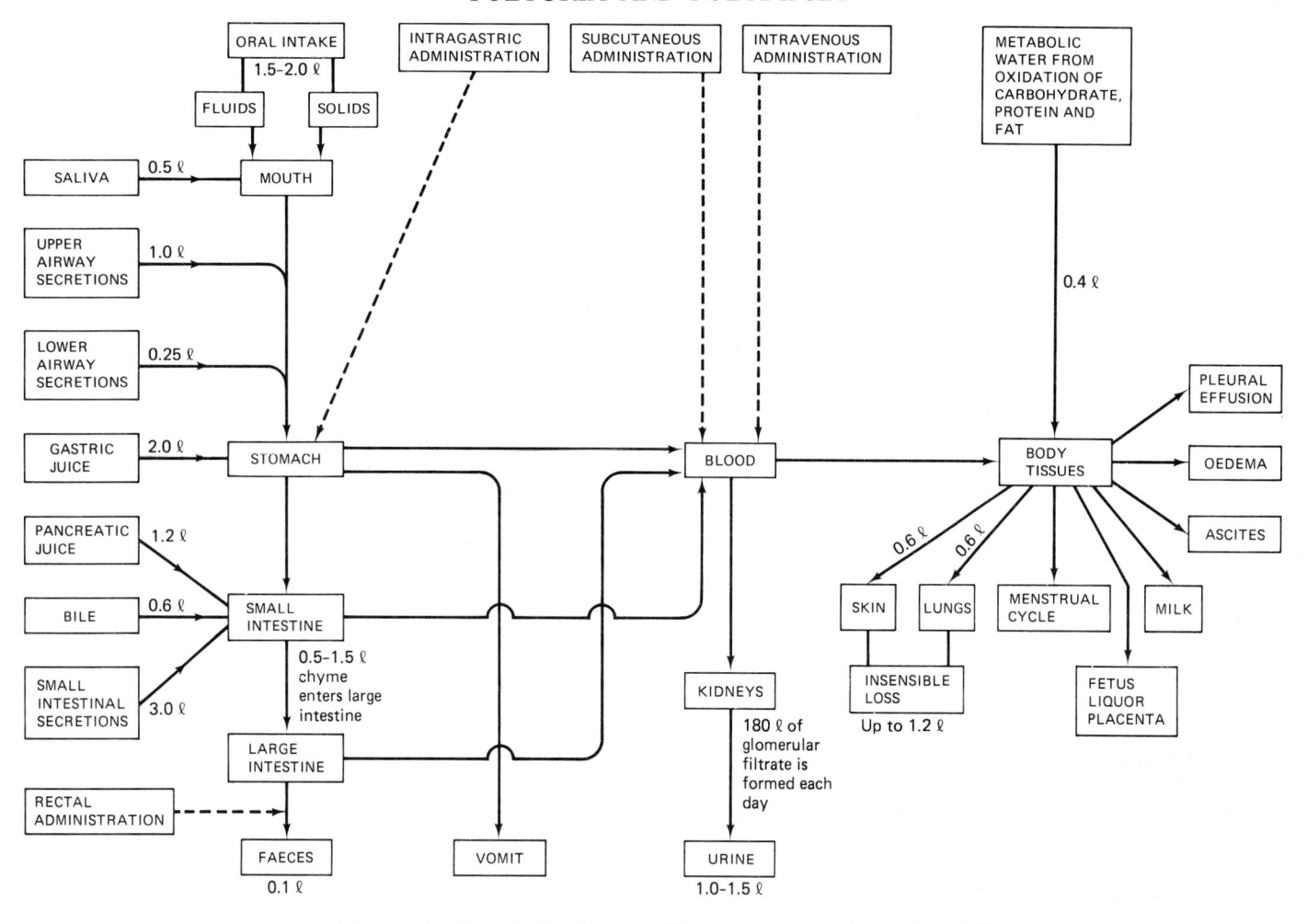

FIG. 1. Summary of the dynamics of water in the adult human. Figures represent approximate volumes in litres per 24 hours.

input other than the contribution from oxidation of ingested food, the latter being derived from carbohydrate, protein and fat. One gram of starch yields 0·60 gm water, 1 gm of protein yields 0·41 gm water, and 1 gm fat yields 1·07 gm water. Output of water can be by several routes—i.e. urine, sweat, respiration, and faeces. In starvation, endogenous breakdown of body tissues further supplements the water supply.

Fluid input

This is determined by the desires of the individual, other than in special circumstances where clinical requirement demands intravenous, subcutaneous, intragastric and other routes to be used. Input is influenced by thirst, by habit and also by the degree of availability of fluid. Social drinking is another aspect sometimes requiring consideration.

Fluid output

Output from the body is determined by several complex mechanisms, thus permitting release of water from the various portals of exit. There is close integration by way of feed-back systems between the factors controlling intake and output, hence maintaining the body water at or close to the optimum concentration and amount.

Once water is within the confines of the gastrointestinal system, there are basically two ways in which it may be lost—by vomiting, or as diarrhoea. Once it has entered the tissues of the body, however, it is then involved in endogenous metabolic pathways and processes; other modes of leaving the body then become available.

In normal daily life, water is lost insensibly through the skin (600 ml per 24 h) and lungs (600 ml per 24 h), and is lost from the body in the urine (1500 ml per 24 h). A small amount is lost in the faeces (100 ml per 24 h). In pathological states water can be sequestered in the tissues as tissue fluid (oedema, ascites, pleural effusion etc.); this may be seen in cardiac failure, nephrotic syndrome, and in hepatic disease, in addition to a large number of other clinical disorders. In females in the reproductive phase of life, a small quantity of water is lost each month in the menstrual flow, and should pregnancy occur, water will of course be incorporated into the products of conception—namely the foetus, and placenta, together with the amniotic fluid.

The post parturition phase is accompanied by further loss of fluid from the mother on account of lactation.

Dynamic Aspects of Water

In the normal state, the body composition of water remains fairly constant. However, in reality, there is an extraordinary degree of movement of water through the tissues. Therefore, although the quantitative composition of the organs and tissues is very constant, the flow through them is vast.

The renal system in particular displays this mobility to a remarkable degree. It is known that approximately 180 l of glomerular filtrate are formed each day. The average urine volume for an adult is of the order of 1000–1500 ml per 24 h. This means that well over 99% of the fluid passing through the tubules is reabsorbed; less than 1% of the glomerular filtrate therefore becomes urine.

The gastrointestinal tract is another system of the body which shows dynamics of water flow in marked degree. In addition to orally ingested fluids, and the water content of 'solid' foods (often around 70% in many foods), both of which enter the upper end of the tract, there are many litres of gastrointestinal secretions which cascade into the system at different levels. The 24 h production of saliva is around 500 ml, gastric juice 2000 ml, pancreatic juice 1200 ml, bile 600 ml, and small intestinal secretion 3000 ml. Almost the whole volume of these secretions, together with ingested fluid and the water content of solid foods, is absorbed in the small intestine. Only 500–1500 ml passes as chyme into the large intestine. All but 100 ml of this is reabsorbed again, in the large intestine. The normal daily fluid loss via the faeces is around 100–150 ml.

With reference to the respiratory system, it is not generally appreciated that upper airway secretion (i.e. above the level of the larynx) is up to 1 l per day, most of which is reabsorbed in the gastrointestinal tract. Moreover, lower airway secretion (i.e. below the level of the larynx) is about 250 ml/day. Expired air also contains water vapour (up to 600 ml/day). The pulmonary loss, together with non-visible loss from the skin, is referred to as 'insensible loss'. In addition, the skin also produces, in certain environmental circumstances, visible sweat. Insensible loss amount to about 1 l per day.

Water Balance in Infancy

Infants are more prone to dehydration than are adults. This is on account of their relatively greater body water content (75%–80% compared with 55%–60% for the adult male and 50%–55% for the adult female). In addition, renal concentrating ability is poorer than in adults. Hence in order to maintain water and electrolyte balance, intake of fluid must be sufficiently high to ensure outflow of a dilute urine. This accounts for the relatively high fluid intake requirement compared with the adult (150 ml per kg of body weight per 24 h for a one-week old infant compared with less than 60 ml per kg per 24 h for an adult).

The premature baby is even more prone to dehydration on account of possible feeding difficulties, the even greater water content compared with that of the full-term child, the even poorer renal concentrating ability, the greater requirement for fluid than the normal full-term infant—i.e. 180 ml per kg body weight per 24 h—the increased liability to infection with consequent vomiting, diarrhoea and high respiration rate, the increased liability to diarrhoea and vomiting following on from digestive problems, and the risk of over-warming the baby.

Urine output in the normal infant of 1–3 days of age is 20 ml/24 h. At 2 weeks the urine volume is around 200 ml/24 h, and at 3 months it achieves 300 ml/24 h.

The % of body water decreases with age; by 1 year it has fallen to 60% of the total body weight.

Homeostasis of Water and Sodium

The mechanisms of homeostasis for water and sodium are interlinked, as shown in Fig. 2.[16] This displays the inter-relationships of ADH and aldosterone. The cells of the osmoreceptor sites in the hypothalamus detect differences in osmolality between the plasma and themselves, and in turn control both ADH secretion and the sensation of thirst. ADH controls water absorption in the distal tubules and collecting ducts of the kidneys thereby affecting urine volume. Renal blood flow is determined by several factors, and circulating volume is one. A reduced flow will lead to activation of the renin–angiotensin–aldosterone pathway. Aldosterone secretion controls absorption of sodium in the distal part of the renal tubular system. Retention of sodium leads to secretion of ADH and water retention follows, thereby restoring intravascular volume.

In essence, the extracellular sodium concentration controls ADH secretion, but the latter controls water; and water controls aldosterone secretion, which in turn controls sodium.[16]

PHYSIOLOGICAL CONTROL OF WATER HOMEOSTASIS

Osmolality

Accurate measurement of the freezing point of a solution (solute dissolved in water) followed by comparison of this with the freezing point of pure water, enables one to make a very accurate assessment of the number of particles in that solution. These particles may be molecules, ions or molecular aggregate. Osmolality refers to the number of dissolved particles in 1 kg solvent. The usual way of determining this in the laboratory is with the aid of an osmometer. The fact that the depression of freezing point so measured is proportional to the osmolality, makes it possible to calibrate the machine to give a direct read-out in mosmol/kg. However, it should not be

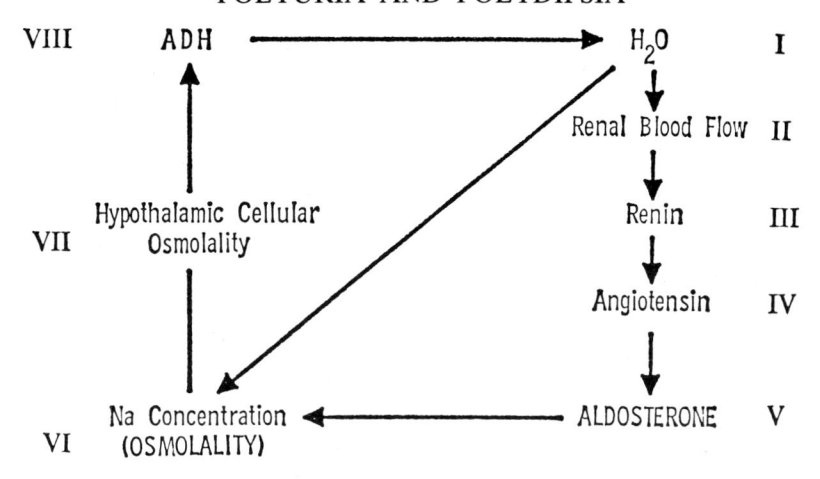

FIG. 2. Simplified cycle of sodium and water homeostasis, assuming there is no influence exerted by other substances such as mannitol or amino acids, or by pathological states such as hyperglycaemia or uraemia.
From: J. F. Zilva and P. R. Pannall.[16]

overlooked that the osmometer is merely a very sophisticated means of measuring the freezing point accurately.

Determination of freezing point will not provide information as to the size of the particles, nor of their shape, nor whether they are charged. It merely informs one as to how many particles there are.

When a mole (molecular weight in grams) of any solute is dissolved in 1 kg of water, the freezing point is depressed by 1.858°C. It should be emphasized, however, that if the solute ionizes during the solution process, the ions so formed will each contribute to the freezing point depression. If, for instance, 1 mole of sodium chloride is dissolved, this will form twice as many particles (ions) when solution is complete as there were particles (molecules) in the original solute. Hence the freezing point depression will be twice that of 1 mole of urea or glucose for example, neither of which ionises on solution.

Plasma is a complex solution containing much water in which are dissolved many different types of particles. Urea and glucose will be present un-ionized, but sodium chloride and potassium chloride, for instance, will be ionized. In addition, plasma contains proteins; these and other large molecules contribute in only a minor way to osmolality. This is because large molecules are vastly outnumbered by the smaller ionic particles and molecules around them, which themselves contribute so much to osmolality.

Osmometry

Cooling pure water. The basis of laboratory determination is briefly summarized in this section. At a constant rate of cooling, pure water shows three phases on the temperature chart. The initial phase shows a straight downward slope (A in Fig. 3). When the temperature reaches 0°C, the slope becomes a plateau on account of crystallization (B in Fig. 3), which in turn indicates that some energy has been released—in the form of heat. This temporarily nullifies the cooling process and so the actual temperature recorded remains stationary for a while. Nevertheless, when the crystallization process is complete, the cooling which continues steadily, then permits the temperature chart to again resume its downward trend. (C in Fig. 3).

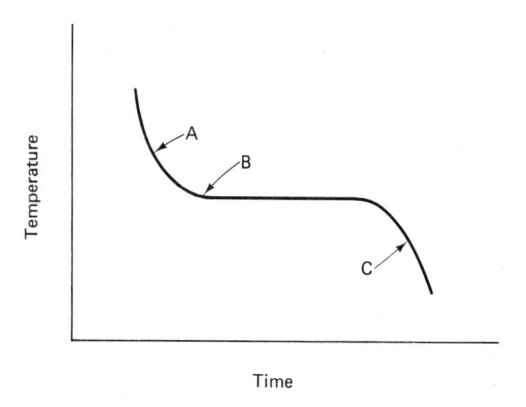

FIG. 3. Freezing curve of a pure solvent. (For explanation see text.)

Cooling a simple solution. In the case of a solution, it is only the solvent which freezes—the solute is already 'solid'. When such a simple solution is exposed to a constant rate of cooling, freezing first occurs far away from the particles surrounded by their shells of solvent. The solvent thus commences to crystallize. This is accompanied by a sharp decrease in the rate of cooling, on account of energy release in the form of heat (D in Fig. 4). The result is that the solute concentration rises as the particles are squeezed into the liquid that remains. As the concentration rises, there is in consequence even greater depression of the freezing point. A hyperbolic curve appears on the cooling chart at this point (E, Fig. 4).

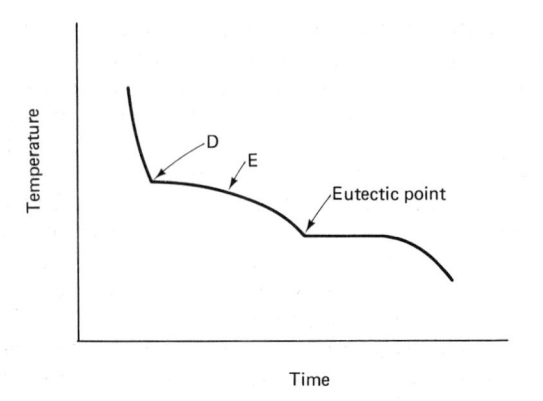

FIG. 4. Freezing curve of a pure solution. (For explanation see text.)

The concentration continues to increase steadily as more and more ice is formed. Eventually the solubility limit is reached (eutectic point), and as more solvent is frozen, the solute continues to precipitate out at a steady rate.

Cooling plasma. Different solutes have different eutectic temperatures. Therefore, the cooling curve of plasma, with its many solutes, is incredibly complex. In laboratory practice the situation is solved, however, by use of the process of 'super-cooling'. This process allows a solution to be cooled a little way below its true freezing point, without it actually freezing—this applies only if the solution is absolutely motionless. As soon as movement occurs, usually evoked by sudden vibration of a stirring rod, freezing commences and heat of crystallization is released. Freezing and thawing rapidly proceed in this blanket of slush, and the practical result is a plateau on the temperature chart. The temperature of this plateau can be used to calculate a close approximation to the osmolality of the solution (i.e. plasma) (Fig. 5).

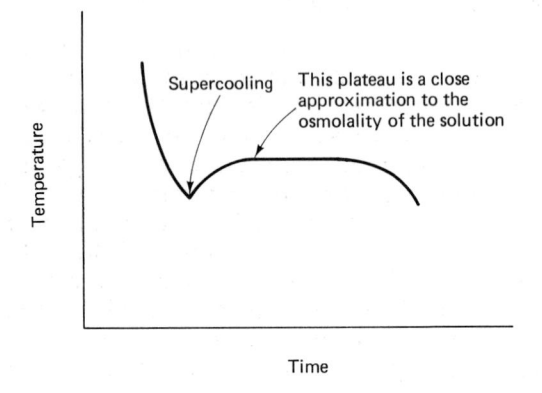

FIG. 5. Freezing curve of plasma containing its many solutes. (For explanation see text.)

Quantitative Aspects of Concentration

In the normal state, about 1000 mosmol of solute must be excreted in the urine each day. This solute is derived from breakdown of ingested nutrients, and from metabolic processes within the body. These end-products are normally excreted as a concentrated solution. Such is the concentrating ability of the renal tubular system, however, that in the normal adult, about 1 l of water is required for this purpose in each 24 h period. This concentration of approximately 1000 mosmol/kg is about 3 times the osmolality of normal serum. In the fasting state, the solute load requiring excretion, may fall to less than half—i.e. to about 400 mosmol. Even this amount demands a urine volume of at least 400 ml for successful excretion. Extrarenal losses of water cannot be reduced to below 500 ml in 24 h.

Lesser degrees of concentrating ability are shown by the immature renal system in infancy. In addition, concentration is impaired in the renal system of elderly people, and in certain disease processes involving the kidney in adult life.

As the osmolality of serum is normally between 280 and 290 mosmol/kg, there is clearly a marked osmotic gradient between it and urine (perhaps 1000 mosmol/kg as discussed above). In fact, physiological extremes of water intake can vary the urine osmolality between 50 and almost 1500 mosmol/kg. This gradient is an expression of concentrating ability, and it diminishes with the onset of renal disease and with ageing. Eventually, in disease of the kidney, no gradient at all may exist, at which point the individual becomes unable to elevate the urine osmolality much beyond the upper limit of the osmolality of normal serum. This level is around 290 mosmol/kg and equates approximately with the oft quoted urinary specific gravity figure of 1·010. The implication behind this is that at least 3 times the normal volume of urine is required in order to perform excretion of the 1000 mosmol of solute by the renal system each day. The increased urine excretion which follows would lead to rapid dehydration of the body, but the onset of thirst prevents this, providing there is free availability of fluid. A mild degree of dehydration has in effect thereby stimulated the hypothalamic osmoreceptors, and polydipsia has averted further dehydration.

In the normal individual, urine achieves greater concentration overnight than during the course of the day. This property is lost early in chronic renal disease. The consequence is the onset of nocturia, simply because the bladder cannot comfortably retain all the urine until the normal waking hour.

There are other pathological states, however, in which the osmotic load presented to the renal tubular system is greater than normal—i.e. in excess of the normal 1000 mosmol of solute per day. Diabetes mellitus is the classical example of this, in which case the urine contains large quantities of glucose, some extra electrolytes, and more water than normal.

Osmolal Clearance and Free Water Clearance

The concept of osmolal clearance and free water clearance is helpful for the understanding not only of the effect of ADH on renal tubular reabsorption of water, but also of the clinical situations which can arise.

In any individual, the urine volume over 24 hours is

equal to the sum of osmolal clearance and free water clearance. Osmolal clearance (C_{osm}) is defined as the volume of water which is required simply to dissolve the solutes present in the urine in a solution isosmotic with plasma. If more water is present in the urine than meets this requirement, then the excess is defined as free water clearance (C_{H_2O}). If, however, less water is present than that able to maintain the urinary solutes isosmotic with plasma, then free water clearance achieves a negative value.

During water diuresis, a high positive figure is obtained for free water clearance. In antidiuresis, however, a negative value indicates that solute-free water has been absorbed, and that the urine is more concentrated than plasma with reference to osmolality. In the situation where an osmotic diuresis is occurring, there will be a rise of osmolal clearance. However, should this occur at the same time as a decrease of free water clearance, through the influence of ADH on the kidney, then total urine flow may be slightly reduced or even continue to rise—thus masking the ADH effect.

Thirst

Several physiological stimuli contribute to production of the sensation of thirst. Increase in osmolality of body fluids is the main one of these. It is equally effective whether there is administration of salt in excess of water, or whether there is loss of water in excess of salt; both situations will produce hyperosmolality. Moreover, in experimental circumstances, rapid onset of thirst occurs when hypertonic solutions of sodium chloride are administered. The high osmotic pressure thus produced in the circulation leads to withdrawal of fluid from the intracellular compartment of the body. Thirst develops, and drinking persists until the osmolality of the extracellular fluid again reverts to normal. The opposite occurs when plasma is diluted—in this case there is inhibition of thirst.

Another factor involved in thirst control is the volume of extracellular fluid. When this is decreased, there is generation of intense thirst. It is the low extracellular fluid volume, rather than lack of water itself that is responsible. Animal experiments confirm this—deprivation of salt over a long period leads to reduction of extracellular fluid osmolality. Thirst is the consequence, and relief is obtained when the normal extracellular fluid volume is restored following administration of adequate sodium chloride.

In the rat, it has been found that more than 30% blood loss must occur before water intake rises above that of control animals. In humans and animals, other studies with lesser degrees of blood loss have produced no consistent effects on thirst. Where drinking does ensue, however, it is found to be accompanied by elevation of plasma angiotensin II. Hence it is via activation of the renin–angiotensin system, rather than by reduced stimulation of the distension receptors in the left atrium and pulmonary veins, that thirst is so stimulated.

Extracellular and intracellular dehydration may lead to a 'dry mouth' which is relieved by drinking. But a 'dry mouth' may also be due to atropine administration. In this instance there is, of course, no change of extracellular or intracellular volume. Nevertheless drinking may occur, although it is not based on the usual sensation known as thirst. It is merely the discomfort of a dry mouth which produces drinking via a non-thirst mechanism.

Other factors influencing thirst should also be referred to. They include exposure of mucosal surfaces of the mouth, pharynx and oesophagus to water, and gastric distension, both of which temporarily reduce further intake. Changes in mood and emotion can also modify physiological stimuli controlling thirst in various ways. Lesions of certain parts of the hypothalamus may induce adipsia or hypodipsia. Moreover, electrical stimulation in the appropriate cerebral areas is known in experimental animals to arouse the sensation of thirst.

Temporary Relief of Thirst by Drinking

In a thirsting individual, there is prompt relief in response to the drinking of water—especially if cold. This relief occurs before water is absorbed into the circulation from the gastrointestinal tract. Thirst may return, however, after a very short while.

Entry of water to the gastrointestinal system, moreover, distends the gut, and the temporary relief of thirst may continue. In fact, distension of the stomach with a balloon can alone relieve thirst for up to thirty minutes.

The basis of the development of such a mechanism must surely be the prevention of over-hydration, in response to thirst. It takes perhaps up to 1 h for water to be absorbed into the circulation and to reach the extracellular fluid. Should excessive drinking occur on account of thirst, there would necessarily be the danger of over-dilution of body fluids.

Experimental Aspects Relating to Thirst

Experimental insertion of a balloon into the inferior vena cava of a dog has thrown more light on the thirst mechanism. If the balloon is inflated 24 h later, when the animal has fully recovered from the anaesthetic, there will be diminution of blood return to the right atrium. Ten minutes later, drinking commences; it has been shown that the greater the fall of arterial pressure, the greater is the amount of water drunk. If the balloon is left inside the inferior vena cava for several days, it can be shown that fluid intake on days when it is inflated is twice as much as on days when it is deflated.

The renin-angiotensin system has been shown to be the physiological mechanism responsible. Moreover, urine flow falls in these circumstances, and the dog, in fact, drinks itself eventually into a state of water excess.

Osmoreceptors

Early experiments yielding proof of the control of water balance by a cerebral sensory mechanism were carried out using dogs. It was suggested at that time that the cerebral receptors were stimulated by hypertonic solutions of sodium salts, sucrose, and fructose, flowing past them, but not by urea or D-glucose. The latter two substances equilibrate with the intracellular fluid, and therefore cause no osmotic difference. However, the first three substances mentioned do not equilibrate quickly. An osmotic difference is thereby created between the interior and exterior of these cells. Dehydration of the intracellular compartment follows the passage of water from the cells towards the newly hypertonic extracellular fluid. Hence the conclusion was that the receptors were stimulated by reduction of their own volume.

The subcellular basis of osmoreception is the presence of vesicles within neural cells of the supraoptic nuclei. They enlarge when the osmolality of extracellular fluid is less than that of intracellular fluid, and they decrease in size when the situation is reversed.

Should the concentration of osmotically active substances in the extracellular fluids rise significantly, these osmoreceptors in the supraoptic nucleus of the hypothalamus will thereby be stimulated. Neural impulses are then sent to the posterior pituitary gland where ADH is released, causing increased water absorption in the distal tubules and collecting ducts of the kidney. Osmotic stimuli are probably amongst the most important factors controlling ADH in physiological circumstances. However, a number of other factors can also influence its secretion.

The Renin-Angiotensin System

Renin is produced by the juxta-glomerular apparatus in the kidney. It circulates and acts on angiotensinogen in the blood to produce angiotensin I. Converting enzyme then causes conversion of angiotensin I into angiotensin II. The latter in turn stimulates production of aldosterone by the adrenal cortex. In addition angiotensin II stimulates drinking and increases the blood pressure.

The first work claiming involvement of the renin-angiotensin system in the production of thirst was reported by Fitzsimons. Further work in the rat showed that intravenous infusion of angiotensin II produced thirst. More adventuresome application in the forms of intracranial injection was found to stimulate short-term drinking even in rats which were well hydrated.

Angiotensin II, however, does not cross the blood-brain barrier, and therefore equilibrates only slowly with the interstitial fluid of brain tissue. The reason that this octapeptide can act so promptly following intravenous administration is that there are certain areas in the brain that lack this barrier. These areas are referred to as the circumventricular organs. In these areas, rapid action is likely on account of the more direct contact with blood constituents.[14]

The circulating angiotensin II is found to act at more than one such blood-brain barrier free locus. These circumventricular organs include the area postrema, the subfornical organs, organum vasculosum and the neurohypophysis. The consequence of exposure of these areas to angiotensin II is elevation of blood pressure, secretion of antidiuretic hormone, ACTH secretion, and drinking. It seems that one of the most sensitive of these areas is the subfornical organ, because the dose of angiotensin II which promotes drinking, is less here than for other areas of the forebrain. Moreover, if this area is damaged selectively, intravenous angiotensin II is without any effect.

In the rat, as little as a few femtomoles of angiotensin II produces a brisk drinking response. This occurs within 1 minute of injection, even if the animal is resting at the moment of the experiment. Drinking may continue up to 10 minutes later, and the effect is so marked that the volume imbibed in that time can reach the normal 24 h intake for the animal.

Saralasin has been shown to block the action of angiotensin II at the receptor site. Injection of this compound leads to reduction of water intake in the hour following administration, compared with the one hour period prior to the dose.

The main site of sensitivity to angiotensin II in the brain is the subfornical organ; the vascular organ of the lamina terminalis is also responsive to it. These two sites are now recognized as the two dipsogenic areas, and are both outside the blood-brain barrier.

Angiotensin II. The molecular structure of angiotensin II is as follows:

$$\text{Asp-Arg-Val-Tyr-Ile-His-Pro-Phe}$$
$$1 \quad 2 \quad 3 \quad 4 \quad 5 \quad 6 \quad 7 \quad 8$$

It has been shown that amino acids 3–8 inclusive comprise the minimum length necessary for full biological activity. The terminal phenylalanine is essential for stimulation of the receptor site. The N-terminal aspartic acid contributes to binding, and also influences duration of action. Arginine also contributes to the binding process. Amino acids 3 (valine), 5 (isoleucine), and 7 (proline), all possess side-chains which help to stabilize the secondary structure of the whole sequence. Amino acid 4 (tyrosine), and 6 (histidine), are functional groups which are essential for binding to the receptor site.

Water Absorption

Absorption of fluid by the intestinal tract is almost complete in normal circumstances. Most of this takes place in the small intestine.

Small Intestinal Absorption

Almost all of the approximately 9 l of fluid passing through the small intestine each day is absorbed there, at varying rates up to about 400 ml per hour. The mechanism of absorption is very similar to that in the proximal tubules of the kidney.

Chyme leaves the stomach and enters the duodenum. It is almost isotonic with plasma at this point. Sodium, monosaccharides and amino acids in particular are actively absorbed, thus creating hypotonicity in the intestinal lumen. Water then rapidly diffuses from this slightly hypotonic solution to the plasma. So rapid is this sequence of events, that the luminal contents remain on the whole essentially isotonic, in spite of the absorption of large amounts of both solute and water.

Nevertheless, the intestinal lumen is permeable to water in both directions. In circumstances where there is high osmolality in the duodenum, water flows rapidly in the opposite direction—i.e. from the blood to dilute this concentrated solution. Such a situation could arise where there is rapid digestion of relatively few large molecules, such as proteins and polysaccharides, to a large number of small molecules such as amino acids and monosaccharides. These small molecules will, of course, create much greater osmotic attraction for water. The result may be a marked fall of blood volume, with cardiovascular problems as the consequence.

Colonic Absorption

The main role of the colon is absorption of water and electrolytes during the onward passage of its contents. Absorption takes place throughout the whole length of the organ, although the major portion occurs on the right side.

There are a number of factors which act on the colon thereby influencing this absorption process; the concentration of sodium in the colonic fluid, antidiuretic hormone, and aldosterone, are three of them.

The transport of water in the intestine is a passive process, and occurs secondary to both osmotic and hydrostatic pressure gradients. These, in turn, are produced by the active transfer of solute.

Both sodium and chloride ions, for instance, are absorbed through the brush border into the columnar cells lining the intestinal tract. This is followed later by transverse passage through the lateral membranes into the hypertonic fluid lying between these cells (i.e. the intercellular space). This hypertonic fluid creates an osmotic gradient, so that water is directly attracted to it from the intestinal lumen. Water therefore forces itself between the cells through the tight junction at the distal ends of the cells, and thereby distends and dilutes the intercellular space. Elevated hydrostatic pressure has thus been created, following which the now isotonic fluid is absorbed through the basement membrane into the capillaries below. The intercellular space later becomes less tense, and the cell walls of adjacent cells return again to close proximity. The ingestion of fluids of varying tonicity can further impose osmotic gradients on this system (Fig. 6).

Adrenal cortical hormones influence electrolyte movement in the colon. Both the proximal and distal colon are able to absorb sodium, but only the distal portion is able to secrete potassium. This is in response to aldosterone.

Antidiuretic hormone also has the effect of decreasing salt and water absorption in the colon in man.

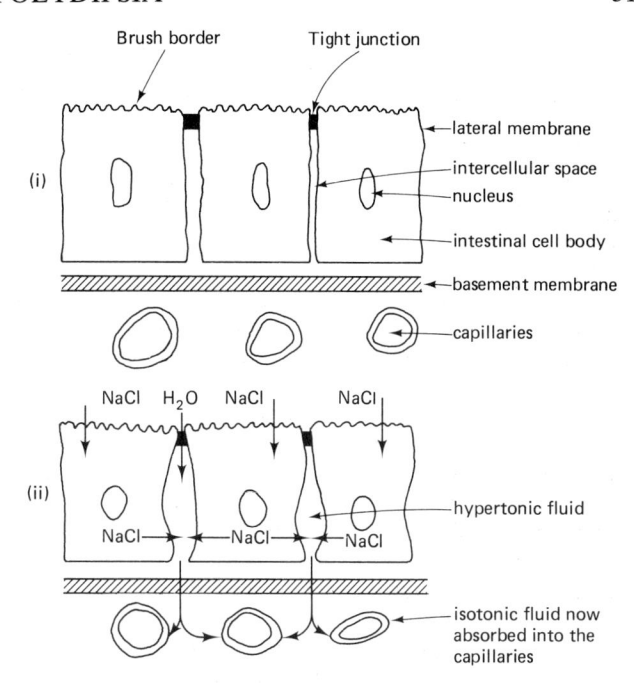

FIG. 6. Colonic absorption of water, showing the normal sequence of events. Sodium chloride is absorbed into the intestinal cells. Following this there is transverse passage through the lateral borders of the cells, thus creating a hypertonic state. Water is now osmotically attracted into this hypertonic fluid, and thus forces itself through the tight junctions between the cells. The intercellular fluid now becomes isotonic, the elevated hydrostatic pressure thus developed aids absorption of water into the capillaries below.
Modified from: Krejs, G. J. and Fordtran, J. A. (1978), Physiology and pathophysiology of ion and water movement in the human intestine, in *Gastrointestinal Disease; pathophysiology, diagnosis and management.*

Renal Function

The renal mechanisms by which urinary concentration is controlled are illustrated in Fig. 7.

The kidney consists basically of two parts—one is the filtration system, and the other the transport passages for the fluid so filtered. The glomerulus is the filtration unit, and there are approximately one million in each kidney. Continuous circulation of blood at the correct pressure through the glomeruli is necessary for their proper function. The filtration process takes place through the capillary network into which the afferent arteriole divides. On account of the fact that the major portion of the plasma proteins does not cross this filter, blood leaving the glomerulus via the efferent arteriole is more concentrated than that in the afferent arteriole.

The composition of the fluid appearing at the commencement of the proximal tubule is almost identical with that of plasma except that it contains little if any protein, and there are no cells. Many substances in the filtrate are required by the body, and are reabsorbed readily in the tubular system. In addition there is excretion directly into the lumen of other substances not needed. As the fluid flows along the tubules, there is change not only of composition, but also of volume. The urine eventually formed, collects in the bladder, which is

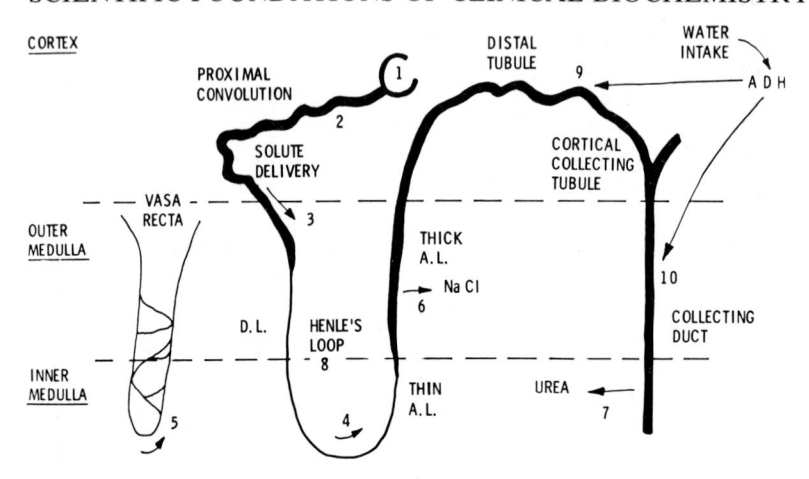

FIG. 7. Diagram of a nephron to illustrate the factors which may influence urinary concentration. GFR (1), total solute concentration in the glomerular filtrate and proximal tubular reabsorption of sodium (2) together determine solute delivery (3) to the medullary counter-current system. The latter's efficiency is reduced by increased flow rate through Henle's loop (4) or the vasa recta (5) solute accumulation in the medulla depends chiefly on sodium reabsorption from the ascending limb (AL) (6) and urea absorption from the collecting duct (7) changes in permeability of the descending limb (DL), distal tubule (9) or collecting duct (10) could affect urinary concentration. (9) and (10) are physiologically regulated by ADH, release of which is affected by water intake.
From: Bissett, G. W. and Jones, N. F. (1975) 'Antidiuretic hormone' in *Recent Advances in Renal Disease* (N. F. Jones, Ed.) p. 350–416. Churchill-Livingstone, Edinburgh.

the temporary storage organ, and from which it is voided periodically.

The membrane lining the proximal tubule and descending limb of the loop of Henle is very permeable to water, which is absorbed here in proportion to the solute which is also reabsorbed. The distal tubule and collecting ducts of the kidney are, however, controlled very differently. They are sometimes permeable to water, and sometimes not—depending on the amount of ADH in the body fluids. ADH causes increased absorption of water at these two sites.

There is control also in the distal part of this system by aldosterone and cortisone. These two hormones act on the collecting ducts to increase sodium absorption, at the same time promoting potassium loss.

Reabsorption of Water in the Proximal Tubule

Reabsorption of water in this part is a passive process along an osmotic gradient. Such a gradient is created in the proximal tubule of the kidney by active reabsorption of solutes, such as sodium and glucose. As referred to earlier, this process is similar to absorption in the small intestine. The walls of the proximal tubule are therefore freely permeable to water. As seen from Table 3, 140–150 l of the total 180 l filtered each day at the glomeruli and are absorbed in this very vascular area; this equates with approx. 80% of the glomerular filtrate. The bloodstream carries the absorbed substances away rapidly, and distributes them as required.

TABLE 3
TO SHOW RENAL BLOOD FLOW AND THE VOLUMES OF FLUID PASSING DIFFERENT POINTS IN THE RENAL TUBULAR SYSTEM OF A NORMAL INDIVIDUAL DURING THE COURSE OF 24 HOURS

	ml per minute	litres in 24 h
Renal blood flow	1200	1800·0
Glomerular filtrate	120	180·0
Distal end of proximal tubule	24	35·0
Distal end of descending limb of the loop of Henle	6	10·0
Distal end of ascending limb of the loop of Henle	6	10·0
Urine	1	1·5

Reabsorption of Water in the Loop of Henle

Of the 180 l of glomerular filtrate formed each day, approximately 35 l passes through to the descending limb of the loop of Henle. It will be seen from Table 3 that 75% of this is absorbed in the descending limb, thus leaving 25% to flow through the ascending limb where there is no water absorption—but where there is absorption of solute.

The Descending Limb

Isosmotic reabsorption of water in the proximal tubule means that fluid of the same osmolality as plasma enters the descending limb of the loop of Henle. This is about 280–290 mosmol/kg. By means of counter current multiplication, high osmolality is achieved at the distal end of the descending limb, on account of passage of sodium

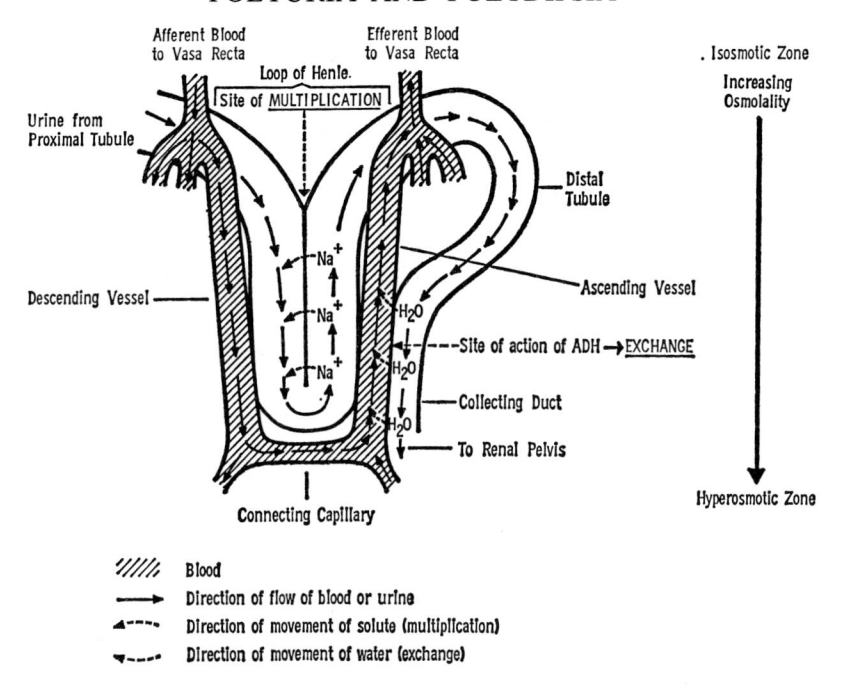

FIG. 8. Diagrammatic representation of the countercurrent mechanism in the kidney.
From: *Clinical Chemistry in Diagnosis and Treatment*, by J. F. Zilva and P. R. Pannall, 2nd edition (1975) Fig. 1 Lloyd-Luke (Medical Books) Ltd., London.

from the ascending to the descending limb during the onward progression of fluid. Furthermore, osmotic equilibrium is achieved between the fluid contents of the distal portion of the descending limb and the surrounding medullary tissue and blood vessels. Fig. 8 displays the typical relationships.

The Ascending Limb

Osmolality of the fluid in the ascending limb is very high at the junction with the descending limb, but decreases rapidly distally. No water is absorbed in the ascending limb, but sodium absorption does occur.

Effects of ADH on the Distal Tubules and Collecting Ducts

In the distal tubules and collecting ducts of the kidney, presence of ADH is necessary for water absorption to take place. Furthermore, the renal tubule at this point must be sensitive to it, for it to work properly. Even when maximal quantities of ADH are present, a large volume passing through the distal tubules and collecting ducts will produce a water diuresis. This may occur, for example, in hyperglycaemia, uraemia or following mannitol administration. Other disorders where there is gross tissue damage, may lead to protein breakdown to amino acids, and consequently to urea elevation. Osmotic diuresis and water depletion may follow. Intravenous feeding may be excessively administered, thereby leading to high production of urea from the amino acids, and hence osmotic diuresis.

Antidiuretic hormone alters the permeability of the distal tubular system to water. Without ADH, permeability is absent, and the consequence is that hypotonic fluid leaves the renal system, and eventually a dilute urine is passed. If, however, ADH is present, there is opportunity for water to be reabsorbed, as the permeability rises. Hence water passes across an osmotic gradient towards the hypertonic interstitial fluid in the medulla, and thence to the bloodstream.

This countercurrent exchange process displayed in Fig. 8, is seen to involve passage of water into the ascending vasa recta along the osmotic gradient previously determined by countercurrent multiplication.

In the normal individual, ADH concentrates the urine, and in return plasma is diluted.

Antidiuretic Hormone

The role of ADH is that of conservation of water. In the intact animal, however, it is clear that in intracellular dehydration, the more important aspect is to take in water, rather than only to conserve what is already there.

The mechanism of increased absorption of water is by increase of pore size in the collecting ducts, sufficiently large for water, but not other solutes, to pass through.

Antidiuretic hormone is produced by the hypothalamus and stored in the posterior pituitary gland. Its release is affected by a number of factors, and furthermore, its peripheral action can actually be impaired in certain circumstances.

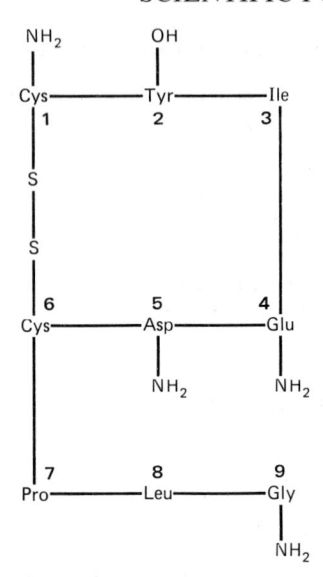

```
      NH2           OH
       |             |
    Cys———————Tyr————————Ile
     1            2         3
     |
     S
     |
     S
     |
     6            5         4
    Cys———————Asp————————Glu
                  |         |
                 NH2       NH2

     |
     7            8         9
    Pro———————Leu————————Gly
                            |
                           NH2
```

FIG. 9. Amino acid sequence of oxytocin.

Chemical Structure

There are similarities in the sequences of amino acid residues of oxytocin, arginine-vasopressin, and lysine-vasopressin (Figs. 9 and 10). Lysine-vasopressin occurs in the pig and some related animals, whereas arginine-vasopressin is the antidiuretic hormone in man. A number of analogues have now been prepared synthetically, and these have been used both for research and therapeutically. Both vasopressin and oxytocin have molecular weights of around 1000 daltons.

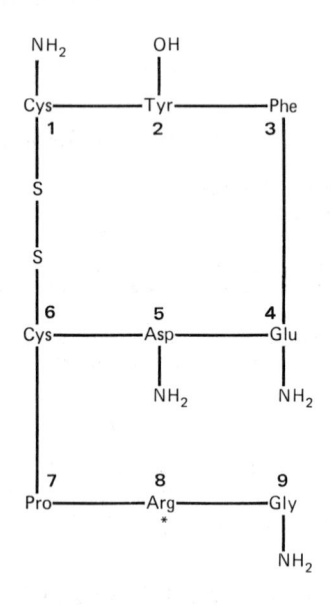

```
      NH2           OH
       |             |
    Cys———————Tyr————————Phe
     1            2         3
     |
     S
     |
     S
     |
     6            5         4
    Cys———————Asp————————Glu
                  |         |
                 NH2       NH2

     |
     7            8         9
    Pro———————Arg————————Gly
                  *         |
                           NH2
```

FIG. 10. Amino acid sequence of arginine vasopressin. In lysine vasopressin, there is merely interchange of lysine with arginine (*).

Neurophysins

The human pituitary gland has been shown to contain at least two neurophysins, H–Np I and H–Np II, both with a molecular weight of just over 10 000 daltons. These proteins bind vasopressin and oxytocin within the neurosecretory granules where they form in the supraoptic and paraventricular nuclei of the hypothalamus. It has been suggested that binding of neurophysins and hormones in the granules, through a weak non-covalent bond, serves as a means of decreasing the osmotic effect of the hormones.

With reference to Fig. 10, it is seen that the sequence of the first 6 amino acid residues in the vasopressins form a ring structure. The ring is closed by means of a disulphide bond (–S–S–) between the two hemicystine (Cys) residues (amino acid residues 1 and 6). It is known that the terminal–NH₂ group of this hemicystine residue is not required for biological activity of the molecule. However, it is the means by which it binds to neurophysin, with the aid also of secondary hydrophobic bonds involving residues two and three.

It seems likely that one neurophysin is associated with each hormone. Also, it is probable that the splitting of a large precursor substance gives rise to neurophysin and hormone.

Neurophysins have also been detected in tumours which have been shown to produce vasopressin, the implication of this being that both are actively produced within the neoplastic cells.

Site of Synthesis

There are neurosecretory cells in the hypothalamus—in the supraoptic nucleus and the magnocellular part of the paraventricular nucleus. These cells secrete vasopressin, oxytocin and the neurophysins. The latter are the carrier proteins for the two hormones. Nerve endings of these neurosecretory cells are found in the median eminence, the infundibular stem, and the infundibular process (the neural lobe of the hypophysis).

Neurosecretion with reference to ADH is not a new concept. Indeed, the brain is now recognized to be one large neurosecretory unit, each neurone within it producing one or more neurotransmitter and/or neuromodulatory substances. These can be of many types and are largely peptide, amino acid, or amine in nature. Secretion of antidiuretic hormone is no longer to be regarded therefore as a strange and unique property of certain neural structures—secretion is the means by which all neurones communicate with each other, and with more distant structures of the body.

Transport and Storage

As the hormones in their neurosecretory granules, and in combination with neurophysins, travel down the axons from the hypothalamus to the pituitary gland, dissociation takes place; hence they become more concentrated.

Mechanisms Controlling ADH Secretion

Stimulation

A multiplicity of factors are involved in release of vasopressin from its storage sites. One of the most sensitive mechanisms, and the one that is of major importance in physiological conditions, is detection of increased plasma osmolality. A very small increase of sodium is detectable by the thirst centre. In fact, a rise of as little as 2 mmol/l (i.e. approximately 1–2%) suffices to produce a marked increase in ADH output.

Other factors are able to influence ADH secretion in different circumstances. Decrease of extracellular volume, hypotension, and decreased circulating volume, are three such factors. The last of these can be a very potent stimulus to ADH secretion—so potent, that it can even override inhibition of ADH secretion caused by low osmolar states. Experiments in man and the primates reveals that a deficit of approximately 10% of the blood volume must occur before elevation of plasma ADH becomes detectable. There is evidence which implicates receptors in the region of the carotid bifurcation and in the left atrium in this process. This was referred to earlier in this review.

Increase of ADH is noted in response to muscle pain, and also during emotional and mental stress. All can be associated with antidiuresis. Physiological studies reveal diurnal variation—i.e. ADH is increased at night, during exercise, and in the erect posture.

The nicotine test for stimulating ADH has been in use for many years, and in smokers, maximal stimulation of ADH occurs immediately after smoking. Many other drugs have now been implicated as interfering with ADH mechanisms, and the list is a growing one. Drugs and pharmacological agents that have been reported to stimulate ADH release include acetylcholine, barbiturates, bradykinin, chlorpropamide, cinchoninic acid, chlorthalidone, clofibrate, ether, ferritin, hydrochlorothiazide, methylclothiazide, morphine, nicotine, and polythiazide.[12] Carbamazepine and thioridazine may also stimulate hypo-osmolar states, but in the former instance there is some controversy regarding the mechanism of action. It is possible that there is potentiation of the action of ADH on the renal tubular system, or perhaps there is increased secretion from the posterior pituitary gland, or there may be a combination of both. However, the hypo-osmolar state reverts to normal on cessation of the drug. Chlorpropamide also potentiates the peripheral action of ADH on the renal tubule (Table 4).

Inhibition

Inhibition of ADH release occurs in hypo-osmolar states, elevation of extracellular fluid volume and also during exposure to ethyl alcohol. Although the effect with alcohol is weak, a marked diuresis usually follows ingestion of large quantities.

Inhibition of the peripheral action of ADH occurs in hypercalcaemia, hypokalaemia, and in states of prolonged overhydration. The peripheral action of ADH on the renal tubule is inhibited also by lithium carbonate and by demeclocycline, both of which have been used therapeutically in the syndrome of inappropriate antidiuretic hormone secretion (SIADH).

TABLE 4
CAUSES OF THE SYNDROME OF INAPPROPRIATE ADH SECRETION (SIADH)

1. *Tumours* (presumed ectopic production of ADH)
   Carcinoma of lung (especially oat cell tumours)
   Carcinoma of pancreas
   Carcinoma of duodenum
   Carcinoma of suprarenal gland
   Carcinoma of prostate
   Carcinoma of ureter
   Thymoma
2. *SIADH from probable neurohypophyseal dysfunction*
   Meningitis (pyogenic and tuberculous)
   Encephalitis
   Brain abscess
   Cerebral tumours
   Subarachnoid haemorrhage
   Subdural haematoma
   Head injury
   Cerebrovascular disease
   Central pontine myelinolysis
   Guillain-Barré syndrome
   Acute intermittent porphyria
   Systemic lupus erythematosus with focal cerebrovascular disease
3. *Thoracic disease*
   Pulmonary tuberculosis
   Pneumonia-bacterial, viral and fungal
   Cardiac surgery
   ? Positive pressure ventilation
4. *Endocrine disorders*
   Addison's disease
   Myxoedema
   Hypopituitarism
5. *Drugs*
   Chlorpropamide
   Thiazide diuretics
   Carbamazepine
   Vincristine
   Cyclophosphamide
   Clofibrate*
   Metformin*
6. *Miscellaneous conditions* (cause unknown)
   Hodgkin's disease
   Acute myeloid leukaemia
7. *SIADH associated with non-osmotic 'physiological' stimuli*
   Hypovolaemia
   Hypotension
   Pain
   Surgery
   Trauma
8. *Oedematous states in which SIADH may play a part*
   Congestive cardiac failure
   Hepatic cirrhosis with ascites
   Nephrotic syndrome

* These drugs have been shown to have actions that could lead to SIADH but the syndrome has yet to be described in patients receiving them.

From Bisset, G. W. and Jones, N. F. (1975), 'Antidiuretic Hormone', in *Recent Advances in Renal Disease* (N. F. Jones, Ed.) Edinburgh, London & New York: Churchill Livingstone.

Molecular Action on Distal Tubules and Collecting Ducts

ADH has been shown to act on certain receptive structures in the renal system, when it is present in the peritubular fluid. It has no effect, however, when introduced directly into the tubular lumen.

The clinical effect of ADH may be detected as soon as 2 to 4 minutes following intravenous injection. It probably binds to receptor cells by means of covalent linkages between its own disuphide bridges (involving the cystine residues of the molecule), and sulphydryl radicals on the cell surface.

In common with many other hormones, ADH acts by way of the cyclic AMP system. Having combined with the receptor site, it then activates membrane bound adenylate cyclase, which promptly converts ATP into adenosine 3', 5'-monophosphate (cyclic AMP, cAMP), and pyrophosphate (PP). In turn, cAMP is the active component which induces membrane permeability for water. Excess cAMP is, however, converted into adenosine 5'-monophosphate (5'-AMP) by means of the enzyme phosphodiesterase (Fig. 11). It is possible to inhibit phosphodiesterase by theophylline and other related xanthines, thereby prolonging the cellular action of cAMP.

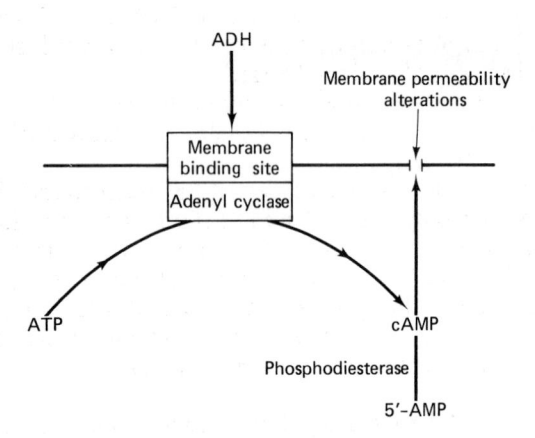

FIG. 11. Molecular action ADH. ADH binds with the membrane binding site, and this leads directly to activation of adenyl cyclase. This in turn converts ATP into cAMP which mediates membrane permeability changes. Phosphodiesterase is required for conversion of cAMP into 5'-AMP.

Consequences of Imbalance of Water Homeostasis

Excessive Water Intake

Where there is a high water load, dilution of the extracellular fluid occurs. In consequence there is a fall of osmolality and this leads to inhibition of ADH release. As the distal tubules and collecting ducts are not now permeable to water, countercurrent multiplication must necessarily occur without countercurrent exchange (Fig. 8). A dilute urine is thereby produced. The increase of blood volume on account of the excessive water intake, causes the renal blood flow to increase, thereby causing some dilution of the normally high osmolality in the medullary region (distal portion of the descending limb of the loop of Henle). This solute is then absorbed into the circulating blood and compensates for the decreased osmolality caused by drinking.

Therefore, at the same time as extra water is lost in the urine, more solute is absorbed into the circulation to compensate for the decreased osmolality of the extracellular fluid caused by the initial high water intake.

Decreased Water Intake

Decreased water intake leads to increased osmolality of the plasma, and decreased volume of the extracellular fluid. Both factors lead to ADH production. Countercurrent exchange is thereby permitted in the distal tubules and collecting ducts of the kidney. The reduced blood volume leads to diminished flow through the vasa recta, and hence permits high medullary tonicity to develop through the process of countercurrent multiplication. The consequence is reduced urine volume and absorption of the water back into the circulation, thereby again tending to increase blood volume.

Osmotic Diuresis

Proximal tubule absorption of water is a passive process. It occurs along an osmotic gradient created by solute absorption; sodium is the main contributor to this. If, however, solute is present in the glomerular filtrate, in such a high amount that it cannot be fully absorbed (e.g. glucose in diabetes mellitus), or not as fully absorbed as usual (e.g. urea in chronic renal failure), then the osmotic effect all the way down the renal tubular system will cause increased water excretion.

Furthermore, if a pharmacological agent such as mannitol is administered, this is passed freely into the glomerular filtrate. Mannitol cannot significantly be reabsorbed actively or passively in the proximal tubule, and hence its osmotic effect causes retention of a greater amount of water than normal in the lumen. The fluid in the tubule is still isosmotic with plasma, however, largely on account of the presence of mannitol, but with a small contribution from sodium. This high volume proceeds along the whole length of the tubular system. It flows through the distal tubules and collecting ducts in greater amounts than can be absorbed there, thus causing a diuresis.

CLINICAL ASPECTS

The preceding account, including discussion on the physiological background, has been provided with the purpose of bringing into perspective each of the clinical disorders to be documented in this section. It has been seen that mechanisms responsible for maintaining water homeostasis involve a sequence of biochemical and physiological pathways utilizing feed-back processes. Abnormalities may occur at any point along the line. Polyuria, polydipsia, adipsia, hypodipsia, and the syndrome of inappropriate antidiuretic hormone secretion (SIADH) all occur. Reference to each of these will be made; the less common disorders are not only interesting, but study of them further aids the more complete understanding of normal homeostasis.

Polydipsia can, of course, occur apart from polyuria. Satisfaction of thirst may not produce polyuria if dehydration is sufficiently gross initially. Moreover, polyuria may exist in the absence of polydipsia. This can be seen, for instance, in a patient over-enthusiastically transfused with intravenous fluid. He may develop polyuria, but his own desire to drink may be suppressed.

However, there are other disorders in which there is both polyuria and polydipsia. There is more than one way in which these conditions may be classified. There is the basic physiological approach, as taken in this text already, in which each of the processes in the circuit is discussed in turn. In some cases, clinical disorders are found to have only one fault in the whole urinary concentrating mechanism, and hence fit this classification neatly. However, many conditions show more than one abnormality, and hence the alternative approach is a more clinical classification. This has merit on account of the fact that it partially overcomes the problem just referred to. The latter classification, therefore, is the one that will be used in this section. Each disorder will be elaborated upon in turn, and opportunity will be taken to draw attention, where required, to the faulty biochemical and/or physiological mechanism(s) responsible.

Primary Polyuria and Secondary Polydipsia

Diabetes Mellitus

Diabetes mellitus is the foremost example of obligatory excretion of a solute load. The glucose content of the proximal tubular fluid exceeds the capacity of the cells to reabsorb it. Osmotic retention of water in the tubular lumen occurs; the high osmotic pressure overrides the capacity of the ADH mechanism more distally to absorb water. Glucose appears in the urine, together with extra sodium, chloride, and water. Dehydration follows the polyuria, and thirst is the consequence. Glycosuria is, of course, a feature not only of diabetes mellitus, but also of nonketotic hyperosmolar coma. Both disorders will lead to osmotic diuresis.

Renal Disorders

The nephron is the basic unit of the kidney, and comprises the glomerulus and renal tubule. Disease processes responsible for chronic renal failure may not only involve different parts of the nephron, but may involve different nephrons in varying degree. Glomerular involvement restricts the filtration process. On account of continued protein catabolism, however, retention of urea occurs. Nevertheless some nephrons may not be damaged at all, and as the plasma urea concentration rises, so does the amount of urea filtered by them into their proximal tubules. Osmotic imbalance is created, and hence retention of water in the tubular system results, with polyuria as the clinical outcome. In addition, some nephrons may display essentially tubular damage, rather than glomerular pathology; these tubules may reabsorb poorly and will also contribute to the osmotic diuresis referred to above.

Late Chronic Renal Failure. In late chronic renal failure, there is reduced ability to conserve sodium. However, if dietary intake of sodium is normal, then no serious consequences ensue. But if there is persistent urinary sodium loss concurrent with restriction of intake, then reduction of extracellular fluid volume could occur. This event would secondarily lead to reduction of the glomerular filtration rate, and uraemia may be exacerbated.

Medullary Cystic Disease. Medullary cystic disease of the kidney, also leads to urinary loss of sodium. However, in this disorder, the sodium wasting is noted to occur even in the presence of normal dietary intake. It is clear then, in this instance, that dietary supplements of sodium chloride should be taken.

Persistent Obstruction of the Genito-urinary Tract. Persistent obstruction of the genito-urinary tract with progressive renal failure, may present with polyuria. In the literature, there are a number of reports in which such individuals having been relieved of their obstruction, obtain alleviation not only of their renal insufficiency but also of their polyuria.[11]

It is likely that there is a defect in the permeability of the collecting ducts to water, thereby simulating nephrogenic diabetes insipidus.

In patients with this disorder, it has been found that relief of the polyuria and renal failure rapidly follows removal of the obstruction. The implication is that no organic damage has occurred, and that merely temporary functional changes are involved. Renal biopsies confirm that glomeruli and tubular epithelial cells possess normal appearances in these patients.

Relief of Genito-urinary Tract Obstruction. Massive polyuria is well recognized following relief of obstruction in the genito-urinary tract.[11] The tubular damage that is present leads to defective sodium reabsorption. It is only when the obstruction has been relieved, however, that the glomerular filtrate, with its contained sodium, is permitted to test the absorption capacity of the proximal tubule. Loss of salt and water occurs, and there is acute reduction of the extracellular fluid volume. Nevertheless, the tubular function usually recovers within a few days, and sodium absorption recommences.[12] The polyuria is therefore of short duration.

Renal Transplantation. Following renal transplantation there is polyuria in some patients.[7] It is recognized that there are several possible causes of this—including: fluid overload; osmotic diuresis related to retention of metabolites on account of uraemia; elevated blood glucose also giving rise to osmotic diuresis; moderate intraoperative tubular damage; and retention of preoperatively administered diuretics (e.g. frusemide).

Husberg and colleagues[7] described a patient treated with 80 mg frusemide per day over a period of several months. During this time his renal function had deteriorated. At this time, the patient's mother agreed to donate a kidney, and the frusemide therapy was stopped 2 days prior to the renal transplant operation. Following surgery, however, massive diuresis occurred, so much so that there was increased risk to the patient's life. On the first postoperative day a diuresis of 40 l occurred. It was only at the 11th day that the urine volume fell below 5 l per 24 h. Preoperative renal failure had presumably led to retention of massive amounts of the diuretic.

Renal Tubular Dysfunction. Tubular dysfunction impairs the normal renal concentration and reabsorption processes. Amongst many functions, the countercurrent mechanisms will be disturbed, together with failure of sodium reabsorption in the proximal tubule. The latter is the essential cause of impaired water reabsorption at that level. Sodium and water are therefore lost in the urine. Nevertheless, depletion of body sodium is not likely to be great if the dietary sodium is adequate.

Predominantly tubular damage is present in the recovery phase of acute oliguric renal failure, which may in turn be due to prolonged untreated shock, or to exposure of the proximal tubule cells to toxic agents such as carbon tetrachloride. More slowly evolving disorders such as hypercalcaemia, hypokalaemia, and heavy metal poisoning, will produce progressive damage to the tubules.

Oral ingestion of a soluble salt of mercury, for instance, leads to severe acute local inflammation at all points of contact. There is rapid onset of oral, pharyngeal, laryngeal and abdominal pain, together with nausea and vomiting within a few minutes. The mercury is absorbed into the body, and is excreted through the kidney where it is concentrated; it quickly damages the renal tubules. The consequence is a rapid onset of diuresis within 3 hours of ingestion. Very soon, however, there is vomiting and dehydration which leads to shock. In association with the progressive tubular damage, anuria and renal failure follow. Furthermore, mercury salts cause severe enteritis with the onset of diarrhoea containing much blood. Uraemia is the usual cause of death.

Nephrogenic Diabetes Insipidus

This is caused by insensitivity of the distal tubules and collecting ducts of the kidney to ADH. It may be hereditary or acquired. Moreover, concentration of urine cannot be achieved following exposure to exogenously administered ADH. This is totally unlike the dramatic effect normally obtained in diabetes insipidus of cranial origin.

Familial nephrogenic diabetes insipidus is a very rare disorder. It is due to insensitivity of the renal tubules to ADH. Elevated levels of ADH have been shown in serum and urine of such patients. The biochemical basis is unknown, but two patients have been described in whom ADH infusion did not lead to the normal rise of urinary cAMP. Moreover, cAMP injections in children with the disorder failed to cause the usual antidiuresis.

Acquired nephrogenic diabetes insipidus has a wide spectrum of causes ranging from naturally occurring diseases to administration of drugs. Renal disorders such as glomerulonephritis and pyelonephritis may cause it, as well as hydronephrosis, polyarteritis nodosa, amyloidosis, and infiltration of the kidney with myeloma tissue. Severe hypokalaemia renders the renal tubules unresponsive to ADH, and pathologically there is noted to be vacuolation of the renal tubular cells. Hypercalcaemia leads to deposition of calcium in the renal tubular cells. Lithium carbonate therapy may also inhibit effective action of ADH on the renal tubules, thus producing a nephrogenic diabetes insipidus. Demeclocycline and methoxyflurane also interfere with ADH action.

Pharmacological agents probably interfere at two biochemical sites, namely with the normal ability of ADH to activate adenyl cyclase-mediated cAMP, and also with the initiation of reactions leading up to increased water permeability of the epithelial cells in the distal tubule and collecting ducts.

Hypercalcaemia. Hypercalcaemia is a particularly important cause clinically of nephrogenic diabetes insipidus. There is often loss of ability of the kidney to concentrate urine, prior to any structural damage being detectable microscopically. Later, calcification of the renal tubules can further add to the problems, by causing inability to absorb water in sufficient amounts. Polyuria is the outcome, and polydipsia follows.

Lithium Salts. Lithium carbonate therapy has been used in the treatment of manic depressive psychosis for a number of years. It is particularly efficacious in alleviating the manic phase. In addition, the drug has also been used in other disorders—carcinoma of the thyroid gland, thyrotoxicosis, Parkinson's disease, and inappropriate antidiuresis.

Over the years there have been reports of severe lithium intoxication leading to renal failure. Nevertheless, if serum levels of the drug are monitored regularly, this complication should not arise. However, experiments in animals, and observations in humans, have both revealed the possibility of histological changes in the kidney following lithium exposure.

Early conclusions from the available evidence, suggested that no serious sife-effects would arise, if the serum lithium concentration was maintained within the therapeutic range. But this seems to be no longer true. It has been reported that a high proportion of patients taking oral lithium carbonate develop impairment of urinary concentration, even when serum lithium levels are within the therapeutic range.

Theoretically, several possible mechanisms underlying this syndrome have been considered. They include nephrogenic diabetes insipidus, cranial diabetes insipidus, and primary polydipsia.

In a series of 48 lithium-treated patients, 17 developed a urinary concentrating defect with serum lithium levels in the therapeutic range. Of these, 10 had nephrogenic diabetes insipidus, 1 had cranial diabetes insipidus, and the remainder could not be classified. There was no evidence of primary polydipsia in any of them.

Animal experiments, involving rats, have revealed that the nephrogenic diabetes insipidus syndrome is associated with a defect in the renal production of cyclic AMP (cAMP) in response to vasopressin.

In summary, therefore, lithium leads to inhibition of action of ADH on the renal tubules.

Sepsis. In patients with severe sepsis, inappropriate polyuria has been described on many occasions.[4] It can lead to severe hypovolaemia and hypotension. Physiological studies have revealed neither abnormalities of glomerular filtration rate, nor of renal blood flow. Exposure of such patients to vasopressin injections has not prevented the polyuria. It seems, therefore, that diminished release of endogenous antidiuretic hormone is not the cause of this disorder. The basis is likely to be that of the effect of a toxin or of a toxic metabolic breakdown product, leading secondarily to impairment of sodium and water conservation. Another possibility suggested by Cortez et al.[4] was that the polyuria could represent self-protection, in the sense that there is requirement for the renal system to excrete an unidentified septic metabolite.

Two patients with Gram-negative pneumonia have been documented in which hypotension was treated with intravenous dopamine hydrochloride.[6] Both patients developed polyuria with excessive loss of sodium in the urine. Dopamine hydrochloride normally has the advantage over adrenaline in that it produces increased cerebral, renal and mesenteric blood flow. However, in these two individuals, it is possible that the dopamine may have enhanced the hypotension by increasing the diuresis and thereby leading to hypovolaemia. The authors suggested that the Gram-negative infection might have played a significant part, by enhancing the effect of the drug on the vasculature of the renal system.

Cranial Diabetes Insipidus

The pathological causes of cranial diabetes insipidus can be further subdivided according to the site of the lesion. This may be in the hypothalamus, in the hypothalamo-hypophyseal tract, or in the posterior lobe of the pituitary gland.

In many cases the cause of the condition is idiopathic, but primary and secondary tumours feature prominently. Craniopharyngiomas, benign and malignant pituitary tumours, and other tumours adjacent to the pituitary gland (e.g. meningiomas) may all be responsible. Metastatic deposits from bronchial and breast tumours sometimes cause diabetes insipidus. Trauma in that region, such as may occur in head injuries, or following surgical intervention, can be the basis of destruction of the ADH-producing and releasing mechanisms. Occasionally diabetes insipidus follows a childhood exanthem, and very rarely the disorder may be congenital or hereditary. Chronic granulomatous lesions that may cause it include eosinophilic granuloma, Hand–Schüller–Christian disease, sarcoidosis, syphilis (gumma), and tuberculosis (tuberculoma). Diabetes insipidus is sometimes seen to complicate basal meningitis.

The symptoms are less severe if the anterior pituitary gland is also involved in the disease process, as cortisol also plays a part in normal water excretion. Conversely, partial diabetes insipidus may be worsened in the presence of exogenous ACTH or adrenal glucocorticoids. Reference to Table 3 indicates that in the normal individual, of every 6 ml of tubular filtrate emerging from the distal end of the ascending limb of the loop of Henle, 5 ml are reabsorbed through the ADH mechanism, and hence only 1 ml becomes urine. In the absence of ADH it is clear therefore that the whole 6 ml entering the distal tubule is likely to become urine. Typically, therefore, the urine volume in diabetes insipidus is up to six times the normal volume—i.e. instead of 1000–1500 ml per 24 h, there may be 6000–9000 ml per 24 h, and it is of low osmolality.

In cases where the dipsogenic areas of the hypothalamus are involved, thirst will not be a feature. The osmostat mechanism is set much higher than normal and hence polyuria occurs alone. This group of disorders is discussed separately under the heading of 'Adipsia and hypodipsia'. In all other instances, however, the usual primary polyuria and secondary polydipsia is characteristic.

Osmotic Diuresis

A number of reports have indicated that blood urea levels may be elevated when babies are fed with overstrength cows' milk. The higher the blood urea, the more likely it is that overstrength feeds are being given. Dehydration is caused by osmotic diuresis.

Another clinical cause is intravenous feeding; excessive administration leads to high production of urea from the contained amino acids. Again osmotic diuresis may follow.

Moreover, excessive protein feeds given orally, or via a tube directly into the stomach, can also lead to osmotic diuresis, and elevated blood urea. It is not always apparent that the diuresis has a nutritional basis in this case, as the medical practitioner may not be directly supervising the feeding process. In any case of unexplained diuresis, enquiry should always be made concerning not only the mode of feeding, but also the quantities being given.

Therapeutically used osmotic diuretics include mannitol, trometamol, and urea. All of them reduce the water content of the body, and some are used to reduce or prevent cerebral oedema.

In addition, excessive radiation of radiosensitive tumours has been observed to produce an osmotic diuresis associated with massive increases in the concentration of circulating uric acid.

Mannitol. Mannitol is a hexahydric alcohol related to mannose. It does not possess significant energy value, but can be given intravenously for the purpose of producing osmotic diuresis in certain patients with oedema and ascites, and also in some cases of oliguria and anuria. In addition it can be used to initiate diuresis in patients who have had self-administered drug overdoses.

It is reported that doses of mannitol in excess of that normally recommended, may cause symptoms of water intoxication. Even normal doses of mannitol may cause the patient to complain of thirst. Moreover, on account of its high osmotic effect, should the mannitol become extravasated, local oedema will develop, followed occasionally by thrombophlebitis.

Trometamol. Trometamol is an organic amine base which causes osmotic diuresis following intravenous infusion. It penetrates into intracellular compartments and has been used as an osmotic diuretic.

Urea. Urea is an osmotic diuretic which is sometimes used to reduce intracranial pressure due to cerebral oedema.

Primary Polydipsia and Secondary Polyuria

Psychogenic Polydipsia

This syndrome is alternatively known as 'compulsive water drinking'. A number of psychiatric disorders can lead to this state of polydipsia, which may itself range from mild to utmost severity. These individuals may have a severe psychosis (e.g. schizophrenia), or they may indeed appear to be relatively normal people, merely presenting with complaints of thirst and polyuria. If the condition is sufficiently severe, water intoxication may occur. Chronic ingestion of large volumes of water can thus impair the renal concentrating mechanism, and sometimes there is difficulty in differentiating this condition from true diabetes insipidus. Those who are severely mentally impaired may well present with polydipsia, involving drinking of any fluid that is freely available, including flower pot water, puddles and even from the toilet etc. Between the ages of 3 and 4 years, psychogenic polydipsia is occasionally seen. Lack of water requirement during the night is a feature, compared with the desperate need for it in cranial diabetes insipidus.

There is ingestion of a larger amount of water than the renal system is able to excrete. The normal kidney cannot excrete more than just over 1 l per h. The consequence is that extracellular volume expands, and dilutional hyponatraemia occurs. The renin–angiotensin–aldosterone system is inhibited by the overloaded circulation, and hence there is renal wasting of sodium. The specific gravity of the urine is very low (SG = 1·001) and the osmolality is similarly low. However, the first urine sample in the morning may be of higher osmolality than samples throughout the day, on account of possible reduction of drinking during the night hours.

Iatrogenic polydipsia

This is not a common cause of polyuria, but it has been observed in an individual who presented with nephrolithiasis.[3] On account of this, his physician had suggested increase of the fluid intake. His somewhat obsessive personality caused him to take this advice to the extreme, and he was found four years later to be taking close to 12 l of fluid orally per day; his urine volume was approximately the same. Advice to reduce to more moderate hydration was adhered to, and the urine volume accordingly diminished.

It is occasionally encountered also, on a more temporary basis in test circumstances, when a patient is asked to drink copiously. The large volume of urine that follows, will, of course, be recorded in the patient's notes. This must not, however, be misinterpreted as representing his normal daily output—but should be recognized as the temporary iatrogenically produced volume that it is.

Hypokalaemia

It is well recognized that there is association between hypokalaemia and polyuria.[2] It is now established that in this state of potassium depletion, there are two possible influences over water homeostasis, one being a central mechanism, and the other renal. The central process involves stimulation of the thirst centre, thus leading to polydipsia, which in turn will account for some of the polyuria observed in hypokalaemia. However, potassium deficiency also leads to an effect on the kidney, thereby diminishing the ability of the renal tubule to achieve maximal urinary concentration. The latter process is not dependent on the former.

Experiments in rats have been largely used to determine the physiological details of the effects of hypokalaemia on water balance. Incontrovertible evidence that potassium depletion could be associated with polyuria and polydipsia via impaired maximal urinary concentration had been well established, but it was only realized later that perhaps the more important effect was stimulation of the thirst centre.

Polyuria and polydipsia may therefore be caused by chronic hypokalaemia from any cause. There is resistance to the normal action of ADH at renal level. In addition, the hypokalaemia leads to diminished renal medullary tonicity, and thirdly there may be stimulation of the thirst mechanism.

Renin-secreting Wilm's Tumour

A patient has been described in whom there was polydipsia of unusual degree, together with polyuria, hyponatraemia and hypertension.[13] The patient was a 16-month-old black male infant, who was subsequently shown to have a Wilm's tumour in the right kidney. Examination of the kidney postoperatively, revealed that the tumour cells contained many secretory granules as seen on electron microscopy. Plasma renin activity was enormously elevated.

Presumably the angiotensin II, stimulated by the renin production, was responsible for activating the dipsogenic centre in the hypothalamus. Postoperatively, serum renin fell markedly by the third day, serum sodium had risen to within the normal range, and urinary aldosterone was normal. Moreover, the urine output had fallen dramatically from 8 l per 24 h, and the specific gravity was reported to have risen from 1·001 to 1·014.

Miscellaneous

Hypodipsia and Adipsia

A number of patients have been described with hypodipsia and hypernatraemia, in the presence of hypothalamic and suprasellar damage.[9] Post mortem studies have revealed a variety of pathological lesions. The mechanism is likely to be that damage to the thirst centre causes disordered release of ADH, through uncoupling of the osmoreception–ADH release system. The consequence is that plasma osmolality may become grossly elevated largely on the basis of hypernatraemia. Sodium retention would be expected, on account of low glomerular filtration rate, and secondary hyperaldosteronism.

Clinically these patients show personality change, drowsiness, impaired concentration and memory, lethargy, apathy, confusion, weakness and low labile blood pressure, but do not show obvious dehydration or polyuria. The urine osmolality is inappropriately low in spite of marked elevation of plasma osmolality. Moreover, on rehydration, plasma electrolyte abnormalities may return to normal, but an abnormal plasma/urine osmolality ratio persists. Typically the serum sodium achieves 150–160 mmol/l or more, and the serum osmolality 310 mosmol/kg or more. The urine osmolality is often less than 200 mosmol/kg. Plasma volume was reduced in several patients in whom it was studied. In one patient extracellular fluid volume was also decreased.

A typical feature is that these patients do not complain of thirst. It is only on direct questioning that it emerges they do not feel thirsty, although some confess to the sensation of dryness of the mouth. It is clear therefore that unless special questions are directed to the possible absence of thirst, the patient will not volunteer it. Thirst may not be totally absent, and it may be that drinking only occurs when the serum sodium is sufficiently high to activate the osmoreceptors—as though the 'osmostat' has been set at a biologically unacceptable high level.

In these patients, rehydration sufficient to bring the serum sodium to within the normal range, may not decrease the osmolality proportionally—it is considered that the high plasma osmolality is due to increased total solute. Moreover, overhydration leaves the urine osmolality inappropriately high—there thus seems to be a failure to detect and respond appropriately to the plasma osmolality. In some patients it was noticed that full rehydration did not produce a urine osmolality as low as is generally seen in classical diabetes insipidus.

A multitude of pathological lesions have been noted as being the basis of the disorder. They include granulomas, (e.g. eosinophilic granuloma), secondary carcinomatous deposits (e.g. from the bronchus), meningioma, craniopharygioma, and germinoma. Lesions in one patient were found in the area of the pituitary stalk, infundibulum of the third ventricle, hypothalamus, mammillary bodies, mammillo–thalamic tracts, and specifically not in the supraoptic and paraventricular nuclei which remained intact. In other patients, the supraoptic and paraventricular nuclei have been infiltrated by tumour. The disorder is therefore a subdivision of diabetes insipidus—but specifically of hypothalamic origin; moreover, it is even restricted to the dipsogenic areas of the hypothalamus.

Syndrome of Inappropriate ADH Secretion (SIADH)

There is a large number of diseases in which the complication of the syndrome of inappropriate ADH secretion may occur (SIADH). The excessive secretion of ADH leads to reduced serum osmolality and hyponatraemia. The urine osmolality is inappropriately higher than the plasma reading. This is due to persistent sodium loss on account of the inhibition of the renin–angiotensin mechanism, which in turn is due to the overloaded circulation. Peripheral oedema is not a feature of the disorder.

Urine osmolality is not low, on account of the unimpaired action of ADH on the distal tubules and collecting ducts of the kidney. In normal circumstances, of course, the feed-back system tends to control ADH production according to requirements. A low plasma osmolality would therefore normally lead to reduction of ADH secretion. Where ADH is autonomously produced, however, or persistently stimulated by some abnormal mechanism, no such feed-back occurs.

The full syndrome is not difficult to diagnose. There may be more difficulty, however, in the case of the incomplete disorder. In these cases the blood shows no changes of sodium or osmolality until a stress test in the form of a water load is applied. If the patient has more circulating ADH than normal, there will be delayed excretion in response. Moreover, the plasma will become hypotonic at this time.

There are now many recognized causes of this syndrome (Table 4). It is well known, for instance, that many malignant tumours are able to synthesize and secrete polypeptide substances, including ADH. Moreover, neurophysins have also been detected in tumour tissue.

In other disorders producing the SIADH syndrome, the ADH originates from its normal site in the hypothalamus. The excessive amounts of it, however, occur simply because of cellular damage in the hypothalamus, or on account of unusual stimulation.

The occurrence of SIADH in respiratory disease has produced more than one theory as to its origin. Reduced filling of the left atrium has been suggested as being the stimulus to ADH production in inflammatory lung disease. However, extraction of the lung in one patient with pulmonary tuberculosis, yielded a large amount of ADH, although no conclusion could be drawn as to whether this was locally produced, or whether it was secreted in the normal way and merely trapped there.

There have also been reports of patients who developed water intoxication following oxytocin infusion, given for the induction of labour or abortion. Indeed fatalities have been recorded. The mechanism is presumably that of the antidiuretic effect of oxytocin, in combination with the large volumes of water in which it is administered.[10]

Drugs which promote the action of ADH could do so in several theoretical ways. They could stimulate ADH release, prolong the half life of the hormone, enhance its action on the kidney, or perhaps the drug itself could have an antidiuretic effect by some other means. Some of the drugs responsible are listed in Table 4, and more are quoted on p. 55.

LABORATORY INVESTIGATION OF POLYURIA AND POLYDIPSIA

When one considers the laboratory investigations of polyuria and polydipsia, it is clear that the clinical disorders discussed in the previous section fall into three groups.

Clinical Groups

Group 1

It is recognized that in certain diseases, polyuria and polydipsia occur as late complications. There is no surprise at the onset of diuresis when a patient enters the recovery phase of acute oliguric renal failure. Diuresis is anticipated, its cause is known, the patient is already in hospital, and laboratory investigations are orientated towards management, rather than establishment of diagnosis. Similarly, the possible polyuric and polydipsic complications of other renal disorders such as known chronic renal failure, medullary cystic disease, persistent obstruction of the genito-urinary tract, relief of genito-urinary obstruction, and renal transplantation are well documented. Nephrogenic diabetes insipidus caused by severe sepsis is referred to in the previous section, and its possibility should be kept in mind. The cause of polyuria will be obvious if a patient is receiving osmotic diuretics such as mannitol or urea. It may not be so apparent if the diuresis is caused by the feeding of overstrength cows' milk in babies, or by excessive protein feeds in adults. This applies, whether it is orally self-administered, or given via a tube directly into the stomach. In these circumstances, enquiry will be necessary regarding the mode of feeding, and also as to how much is being given at each feed, and how often.

Group 2

The clinical presentation of diabetes mellitus, chronic renal failure, hypercalcaemia and hypokalaemia is often with polyuria and polydipsia, in association with other complaints. If the patient does not directly admit to polyuria, routine questioning with reference to the urinary system should establish it. The laboratory investigations required here are basically simple and routine.

Urine glucose, blood glucose, serum urea and creatinine, urine protein and microscopy, serum calcium and serum potassium are usually sufficient initially to confirm, or exclude, these disorders. Should any of them prove abnormal, it will be necessary, of course, to pursue each of them in the conventional way.

It is also necessary to exclude the possibility of iatrogenic polyuria and polydipsia, and self-medication. The patient should be questioned regarding the taking of drugs such as diuretics, hypnotics and opiates, and also asked about alcohol consumption.

Group 3

More discussion is necessary regarding individuals belonging to the third group. These patients may present with polyuria and polydipsia, but none of the obvious causes listed under groups 1 and 2 can be established. Group 3 comprises psychogenic polydipsia, cranial diabetes insipidus, and nephrogenic diabetes insipidus.

Before proceeding to detailed investigations, iatrogenic polydipsia must also be considered and excluded. The patient should be questioned whether he is drinking fluids in abundance on the advice of a physician, perhaps even a long time ago. This may have been suggested initially, for example, following the finding of cystine stones in the renal tract, etc.

In the history taking, some care does need to be observed in establishing polyuria and polydipsia. It is not always volunteered and it is not always divulged on direct questioning. Occasionally it is found that the excessive requirement for fluid has become so much a way of life, that it is now considered to be normal. The nocturnal ritual of placing a large jug of water on the bed-side table, and the daily ritual of taking a bottle of dilute orange squash to work, can easily be missed. Very detailed enquiry about daily habits is essential when there is suspicion of disease involving the hypothalamic-pituitary area.

A chance observation that the 24 h urine volume recorded on a laboratory form is well above the normal range can be an important pointer, particularly if seen on more than one occasion.

The adipsia or hypodipsia of hypothalamic disease involving the dipsogenic centre(s) needs special mention. These patients do not complain of lack of thirst. They just do not drink, as their physiological mechanisms do not recognize it any longer as being a normal requirement, until the serum sodium is grossly elevated. At this high level the patients may drink, but cessation will soon occur so that the serum sodium and osmolality do not fall below the new level determined by the elevated osmostat mechanism.

It is important for the physician to interpret correctly the patient's own history, and not to misinterpret for polyuria, what is really frequency of micturition associated with a urinary tract infection. Furthermore attention focused on the urinary system may cause the patient himself to misrepresent his symptoms. The patient may think he is passing a large volume of urine, and yet all laboratory tests, and detailed history, may be unable to confirm it.

Investigations that may be required in this group of patients will now be elaborated upon.

Scheme of Clinical Investigation

This scheme is suggested to serve as a framework for biochemical investigation of patients with polyuria and polydipsia. It is easily modifiable to suit special circumstances and requirements. Reference to discussion in the previous section on the clinical grouping system is necessary before proceeding.

Patient Preparation and Procedure for Water Deprivation Test and Vasopressin Test

(1) Exclude patients belonging to Group 1. Their future investigations are directed towards management rather than diagnosis of the polyuric state.

(2) Exclude patients belonging to Group 2 by ascertaining urine glucose, blood glucose, serum urea, serum creatinine, urine protein, urine microscopy, serum calcium, and serum potassium. Ask regarding diuretics, hypnotics, opiates, and alcohol.

(3) Patients of Group 3 are now left by exclusion. Observe the patient for 1 day. Put on a fluid intake and output chart. Collect all urine samples passed, and measure volumes and osmolalities of each. If many are presented, then a small random selection need only be analysed. Calculate the total 24-h urine volume.

(4) The water deprivation test now follows. Smoking is forbidden to the patient throughout the whole procedure. Weigh the patient initially. Permit a light breakfast and normal fluid intake on the morning of the test. During the test, exclude all access to fluid for 8 hours, but be prepared to stop test if body weight falls by greater than 3%, or if the patient begins to feel unwell and looks dehydrated. Collect all urine samples passed and measure volumes and osmolality. Collect serum samples 2-hourly, and measure the osmolality of each.

(5) At the end of the 8 hour period of water deprivation, do the vasopressin test using the synthetic vasopressin substance desmopressin, desaminocys[1]–D arg[8]–vasopressin, (DDAVP). $2\mu g$ DDAVP is given intramuscularly. Again collect all urine samples passed over the next 16 hours—i.e. up to the end of the whole 24 h period from the commencement of the water deprivation procedure. Measure urine volume and osmolalities. Collect 3 or 4 serum samples over this period for osmolality determination. During the DDAVP test, food is given as normal. This is the end of the test.

(6) Interpretation of this scheme is given on p. 64 under the section entitled 'General principles of differential diagnosis and interpretation'.

Nicotine Test

Maximal stimulation of ADH occurs immediately after smoking. For many years the smoking of one to three cigarettes rapidly with deep inhalation, or the intravenous nicotine test, was used to test the capacity of the hypothalamo-pituitary area to secrete ADH.

Intravenous nicotine in sufficient dose, leads to an antidiuretic response in the normal person within 30 minutes of administration. There is decrease of urine flow, together with reduction of free water clearance. The urine concentrations of solutes increase. This response is absent in the patient with cranial diabetes insipidus.

The test is time-consuming, not entirely reliable, and moreover, patients may suffer toxic effects such as nausea, vomiting and vertigo. There seems little merit in doing this test now that osmolality is so easily and quickly measured. Nevertheless, knowledge of the principle of it serves as a reminder that patients must not be permitted to smoke before or during investigations of polyuria and polydipsia.

Intravenous hypertonic saline test

This test was used for many years for testing the hypothalamo-pituitary capacity to secrete ADH. Three per cent saline (10 ml per kg) was infused over 45 minutes. The normal antidiuretic response, seen within 60 minutes of infusion, does not occur with cranial diabetes insipidus.

This test is clinically time consuming, and is reported to be not always in agreement with the nicotine test. Now that osmolality tests on the urine and serum are so easily available, this test would seem now to have little value.

Urine Arginine Vasopressin Test (AVP)

Urinary AVP is measured by radioimmunoassay. It is rarely required, however, as osmolality tests nearly always provide the diagnosis, are so easy, and are much less expensive.

Urine levels are dependent on the clinical state of the patient, and certain drugs already referred to will influence the value obtained. The clinical state at the time of collection is important, as secretion depends on plasma osmolality and volume.

Special requirements for collection of urine need to be observed. The local Chemical Pathology laboratory will need to be consulted. In general, however, about 15 ml should suffice, as long as it is collected into a plastic container, and made acid with 1 ml of 1 M HCl per 5 ml urine immediately it is passed. The urine and acid must be mixed well, and then stored at $-20°C$.

General Principles of Differential Diagnosis and Interpretation

Plasma ADH, plasma osmolality and urine osmolality, are theoretically the three most important factors in understanding the differential diagnosis of psychogenic polydipsia, cranial diabetes insipidus, and nephrogenic diabetes insipidus. Plasma ADH is not readily available as a routine test, but the osmolality determinations are.

The Normal Individual

The normal serum osmolality is between 280 and 290 mosmol/kg. It does not matter whether serum or plasma is used for the determination of osmolality, as the figures obtained are virtually identical.

Overnight deprivation of fluid may possibly elevate serum osmolality to 295 mosmol/kg, and urine osmolality to more than 1000 mosmol/kg. The first urine specimen after sleep is likely to be the most concentrated of the whole day.

Intramuscular DDAVP administration following overnight deprivation of water cannot lead to greater osmolality of urine; the endogenous ADH is already being secreted maximally, and exerting its full effect.

Psychogenic Polydipsia

This is characterized by appropriately low plasma and urine osmolalities in the presence of low plasma ADH. In this disorder there is normally fluid intake in excess of the capacity of the renal system to excrete it. The tendency is therefore, for serum sodium and osmolality to be towards, at, or below the normal range for the laboratory concerned. Serum osmolality may even be below 255 mosmol/kg and urine osmolality less than 50 mosmol/kg. If it is possible to collect all samples of urine passed, it is likely that the early morning one will have an osmolality greater than those during the remainder of the day. This is on account of the probably reduced or absent input during sleep. Urine volume and fluid input should both be measured over 24 h, if possible, but in a severe psychiatrically disturbed individual this may not prove possible.

Deprivation of fluid intake may lead to elevation of serum osmolality to the upper normal range, around 290 mosmol/kg. Urine osmolality may rise to the same level or greater, depending on the severity of the condition initially. It is not likely, however, to rise as much as in the normal person, but it may reach 600 mosmol/kg.

Intramuscular injection of DDAVP, following water deprivation causes water absorption in the distal tubules and collecting ducts of the kidney. There is, in consequence, a decrease of serum osmolality to perhaps 255 mosmol/kg, as drinking may not be inhibited. Urine osmolality, however, may not change much from the elevated level achieved following fluid deprivation. It may rise to 750 mosmol/kg. If the condition is severe, it may be that urine osmolality remains low after fluid deprivation. Hence there will be opportunity for elevation of it following the intramuscular DDAVP injection, perhaps to 500 mosmol/kg.

Exposure to DDAVP can be dangerous sometimes in psychogenic polydipsia, as the patient may suffer from overhydration some hours afterwards, if fluid intake is not curtailed.

Cranial Diabetes Insipidus

This is characterized by low or absent plasma ADH in the presence of high plasma osmolality. The disorder is normally accompanied by urine output being in excess of the capacity of the individual to counteract it by drinking. The tendency, therefore, is for serum sodium and osmolality to be near, at, or above the upper limit of the normal range. Urine osmolality of all separate specimens passed throughout the course of the day tends to show a fixed low level. The early morning sample does not have a higher osmolality than the others, and is therefore unlike psychogenic polydipsia. Urine volumes over 24 h periods should be recorded, and a fluid intake/output chart commenced.

Water deprivation causes the serum osmolality to increase beyond 295 mosmol/kg, perhaps even to 305 mosmol/kg. Nevertheless, urine osmolalities remain at the same low level as found initially, maybe around 150 mosmol/kg. Abundant urine continues to be passed. In milder degrees of the illness, the urine can be concentrated to some degree, and the osmolality rises towards or even surpasses that of serum; it may reach 400 mosmol/kg.

Deprivation of fluids in cranial diabetes insipidus may indeed be dangerous if overdone. The patient may become dehydrated rapidly, and may appear sunken eyed, grey and lethargic. Water is relatively heavy, and therefore loss of much urine in the absence of intake, will lead to weight reduction. Great care should be exercised in supervising this test. It should be stopped as soon as sufficient information has been gained, and immediately there is any cause for concern regarding the patient's well being, as vasomotor collapse is possible in severe dehydration. Hourly weighing of the patient will alert the clinician to consider ending the test should the body weight fall by greater than 3%. The test must therefore only be done by day when there is strict supervision by experienced staff.

Accurately timed 1 or 2 hour volumes of urine may also be collected for determination of ADH (AVP, arginine vasopressin) both before and after fluid deprivation, for as long as can safely be maintained. Special requirements for the collection and storage of the urine are quoted under 'Urine arginine–vasopressin test'.

Administration of intramuscular DDAVP leads to a very dramatic fall of urine output, the appearance of the urine changes from colourless to yellow, the osmolality rises, and the patient, not only begins to feel improved, but also looks better. Urine osmolality may reach 800 mosmol/kg in mild cases, and around 450 mosmol/kg when the disorder is more severe.

Nephrogenic Diabetes Insipidus

Patients with this disorder are found to have an appropriately high plasma ADH in the presence of a low urine osmolality.

Findings of the water deprivation test in this disorder are similar to those in cranial diabetes insipidus. Fluid intake and output should be recorded over 24 h periods.

Administration of DDAVP intramuscularly to a patient with nephrogenic diabetes insipidus is without effect on urine output, and the osmolalities do not change from the initial values.

Hypodipsia and Adipsia

Biochemical features include high serum sodium, perhaps between 150 and 160 mmol/l, but may be more. In spite of this, thirst is significantly absent as the 'osmostat' is elevated. Serum osmolality is usually in excess of 310 mosmol/kg. The urine osmolality is inappropriately low, and may be less than 200 mosmol/kg. On rehydration, the serum electrolytes may return to normal, but there may be persistence of an abnormal plasma/urine osmolality ratio.

Syndrome of Inappropriate ADH secretion (SIADH)

Biochemical findings include reduced serum sodium and osmolality. The serum sodium may be down to 120 mmol/l, and the osmolality around 240 mosmol/kg. Urine osmolality is inappropriately higher than plasma osmolality, and this is due to the loss of sodium. This occurs on account of the overloaded circulation producing inhibition of the renin-angiotensin-aldosterone mechanism, and because of the persistent excessive activity of ADH on the renal tubules and collecting ducts.

The incomplete disorder, however, may require a stress test in the form of a water load. In this case delayed excretion of the water occurs on account of the elevated ADH activity. At the same time, the plasma will become hypotonic.

REFERENCES

1. Andersson, B. (1978), 'Regulations of water intake,' *Physiological Reviews*, **58**, 582–603.
2. Berl, T., Linas, S. L., Aisenbrey, G. A. and Anderson, R. J. (1977), 'On the mechanism of polyuria in potassium depletion—the role of polydipsia,' *J. Clin. Invest.*, **60**, 620–625.
3. Berry, E. M., Halon, D. and Fainaru, M. (1977), 'Iatrogenic polydipsia,' *Lancet*, **2**, 937–938.
4. Cortez, A., Zito, J., Lucas, C. E. and Gerrick, S. J. (1977), 'Mechanism of inappropriate polyuria in septic patients,' *Arch. Surg.*, **112**, 471–476.
5. Fichman, M. P. and Brooker, G. (1972), 'Deficient renal cyclic adenosine 3', 5'-monophosphate in nephrogenic diabetes insipidus,' *J. Clin. Endocrinol. and Metab.*, **35**, 35–47.
6. Flis, R. S., Scoblionco, D. P., Basti, C. P. and Popovtzer, M. M. (1977), 'Dopamine—related polyuria in patients with Gram-negative infections,' *Arch. Intern. Med.*, **137**, 1547–1550.
7. Husberg, B., Hellsten, S., Bergentz, S., Hansen, T. and Möller-Jensen, K. (1977), 'Massive diuresis after renal transplantation due to retention of furosemide,' *Transplantation*, **23**, 101–103.
8. Jones, N. F., Barraclough, M. A., Barnes, N. and Cotton, D. G. (1972), 'Nephrogenic diabetes insipidus: effects of 3'-, 5'-cyclic adenosine monophosphate,' *Arch. Dis. in Childhood*, **47**, 794–797.
9. Lascelles, P. T. and Lewis, P. D. (1972), 'Hypodipsia and hypernatraemia associated with hypothalamic and suprasellar lesions,' *Brain*, **95**, 249–264.
10. Morgan, D. B., Kirwan, N. A., Hancock, K. W., Robinson, D. and Ahmad, S. (1977), 'Water intoxication and oxytocin infusion,' *Brit. J. Obst. Gynae.*, **84**, 6–12.
11. Nagar, D., Ferris, F. Z. and Schacht, R. A. (1976), 'Obstructive polyuric renal failure following renal transplantation,' *Amer. J. Med.*, **60**, 702–706.
12. Schwartz, W. B. (1979), 'Disorders of fluid, electrolyte and acid-base balance,' in *Cecil Textbook of Medicine*, Chapter 530, pp. 1950–1969, (Beeson, P. B., McDermott, W. and Wyngaarden, J. B. Eds.), Philadelphia, London, Toronto: W. B. Saunders Co.
13. Sheth, K. J., Tang, T. T., Blaedel, M. E. and Good, T. A. (1978), 'Polydipsia, polyuria, and hypertension associated with renin—secreting Wilm's tumour,' *J. Paediatrics*, **92**, 921–924.
14. Weindl, A. (1973), 'Neuroendocrine aspects of circumventricular organs,' in *Frontiers in Neuroendocrinology* (Ganong, W. F. and Martini, L. Eds.). New York: Oxford University Press.
15. Zilva, J. F. and Pannall, P. R. (1979), 'The kidneys: renal calculi,' in *Clinical Chemistry in Diagnosis and Treatment*, Chapter 1, pp. 1–30, (Zilva, J. F. and Pannall, P. R. Eds). London: Lloyd-Luke (Medical Books) Ltd.
16. Zilva, J. F. and Pannall, P. R. (1979), 'Sodium and water metabolism,' in *Clinical Chemistry in Diagnosis and Treatment*, Chapter 2, pp. 31–63, (Zilva, J. F. and Pannall, P. R. Eds.). London: Lloyd-Luke (Medical Books) Ltd.

FURTHER READING

Beeson, P. B., McDermott, W. and Wyngaarden, J. B. (1979), *Cecil Textbook of Medicine*, Philadelphia, London and Toronto: W. B. Saunders Co.

Dormandy, T. L. (1967), 'Osmometry' *Lancet*, **1**, 267–270.

Fitzsimons, J. T. (1972), *Physiological Reviews*, **52**, 468–560.

Johnson, F. N. and Johnson, S. (1978), *Lithium in Medical Practice*, Liverpool: Medical Technical Press.

Jones, N. F. (1975), *Recent Advances in Renal Disease*, Edinburgh, London and New York: Churchill Livingstone.

Sleisenger, M. H. and Fordtran, J. S., (1978), *Gastrointestinal Disease—pathophysiology, diagnosis and management*, Philadelphia, London and Toronto: W. B. Saunders Co.

Vander, A. J., Sherman, J. H. and Luciano, D. S. (1975), *Human Physiology—the mechanisms of body function*, New Delhi: Tata McGraw-Hill Publishing Company Ltd.

Wintrobe, M. M., Thorn, G. W., Adam, R. D., Bennett, T. L., Braunwald, E., Isselbacher, K. J. and Petersdorf, R. G. (1970), *Harrison, Principles of Internal Medicine*, Tokyo: Kogakusha Company Ltd.

Zilva, J. F. and Pannall, P. R. (1979), *Clinical Chemistry in Diagnosis and Treatment*, London: Lloyd-Luke (Medical Books) Ltd.

SECTION II

DISORDERS OF THE GASTROINTESTINAL SYSTEM

		PAGE
4.	THE FUNCTION OF THE HUMAN GASTRO-INTESTINAL TRACT AND ITS LABORATORY ASSESSMENT	67
5.	DIFFERENTIAL DIAGNOSIS OF MALABSORPTION AND MALNUTRITION	95
6.	MALNUTRITION AND VITAMIN AND MINERAL DEFICIENCIES AND EXCESS	113
7.	THE LIVER	139
8.	THE DIFFERENTIAL DIAGNOSIS OF ACUTE ABDOMINAL PAIN	155
9.	CLINICAL BIOCHEMISTRY OF THE 'ARTIFICIAL LIVER'	164

4. THE FUNCTION OF THE HUMAN GASTRO-INTESTINAL TRACT AND ITS LABORATORY ASSESSMENT

NICHOLAS W. READ AND CHRISTOPHER L. CORBETT

The Stomach
    Acid secretion
    Pepsinogen secretion
    Intrinsic factor
    Gastric secretory tests
    Gastric mucosal barrier
    Gastric emptying
    Vomiting

The Pancreas
    Pancreatic secretion
    Assessment of pancreatic function

Bile
    Bile secretion
    Gallbladder function
    Bile acid metabolism and enterohepatic circulation
    Functions of bile acids
    Disturbances of the enterohepatic circulation
    Assessment of bile acid metabolism

The Small Intestine
    Transport of fluid and electrolytes
    Digestion and absorption of carbohydrate
    Protein digestion and absorption
    Digestion and absorption of fat
    Absorption of calcium, magnesium and phosphate
    Iron absorption
    Water soluble vitamins
    Investigation of malabsorption
    Small intestinal motility

The Colon
    Motility
    Absorption of nutrients
    Salt and water absorption
    Diarrhoea

THE STOMACH

The stomach functions as a large muscular bag modified for the storage and acid digestion of food and the delivery of the acid contents in small aliquots to the duodenum. It is divided into cardia, fundus, body and antrum. The fundus and the body are the portions which secrete acid and are lined with a columnar epithelium punctuated with numerous invaginations called gastric pits into which one or two simple gastric glands open (Fig. 1). The cells lining the surface and the gastric pits secrete alkaline mucus, while the gastric glands contain large triangular oxyntic or parietal cells which secrete gastric acid, chief cells which secrete pepsinogen and some mucus-secreting cells. The glands in the antrum secrete only pepsinogen and mucus into the lumen but contain G cells which release the hormone gastrin into the blood stream.

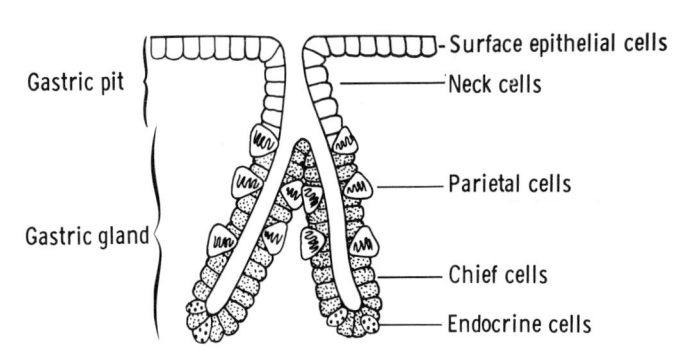

FIG. 1. A diagram of the epithelium of the acid secreting portion (body and fundus) of the stomach. Both surface epithelial and neck cells secrete mucus. The parietal or oxyntic cells secrete acid and the chief cells secrete pepsinogen.

ACID SECRETION

The fundamental reaction in the production of hydrogen ion by the parietal cell is the ionization of water to hydrogen and hydroxyl ions.

$$H_2O = H^+ + OH^-$$

The hydrogen is actively transported across the canalicular membrane by an active process accompanied by the active transport of chloride. The OH^- combines with CO_2 derived from the arterial blood or mucosal metabolism to form HCO_3^- which then enters the blood stream. This accounts for the metabolic alkalosis which occurs shortly after a meal. The hydration of CO_2 to HCO_3^- + H^+ is catalysed by the abundant carbonic anhydrase in the gastric mucosa.

The composition of gastric juice in healthy subjects is dependent upon the rate of secretion. At maximal rates of secretion it contains up to 150 mM HCl, at basal rates it contains a mixture of sodium chloride and sodium bicarbonate. Thus there is an inverse relationship between the concentrations of Na^+ and H^+ in gastric juice.

The source of sodium in the gastric juice is still unresolved, but two hypotheses have been proposed. One, first suggested by Pavlov and Hollander, proposes that gastric juice is an admixture of variable amounts of HCl secreted by the parietal cells and relatively fixed amounts of NaCl and $NaHCO_3$ secreted by the surface and neck cells. At high rates of acid secretion, the parietal component predominates while at lower rates the non-parietal alkaline secretion constitutes the major fraction. The other theory (Teorell) suggests that hydrogen ion exchanges for sodium as it comes into contact with the cells of the gastric pits or surface. At high rates of acid secretion, there will be little opportunity for exchange to occur but at low rates most of the acid will be exchanged for sodium.

Control of Acid Secretion

Acid secretion from the parietal cell can be stimulated by three chemical transmitters; acetylcholine, gastrin and histamine. The recent finding that a histamine H_2-receptor antagonist inhibits acid secretion induced by acetylcholine as well as gastrin, indicates that histamine must play a central role in the control of acid secretion though it is not known whether it has a permissive action or functions as an essential mediator.

For convenience the control of acid secretion during ingestion of a meal is divided into cephalic, gastric and intestinal phases. This does not imply that these phases occur sequentially or are mediated by separate mechanisms. They probably all involve nervous as well as endocrine mechanisms and occur simultaneously to produce the optimum acid response for the meal in question.

The cephalic phase of secretion is stimulated by the sight, smell, taste and mastication of food. This phase is abolished by vagotomy. However, there is evidence to suggest that in patients with duodenal ulceration part of the acid secretory response to sham feeding is mediated by release of gastrin,[95] though the latter is probably released by vagal stimulation.

The presence of food in the stomach induces acid secretion by distension as well as by chemical stimuli. Selective distension of the fundus with a balloon induces acid secretion,[50] which is correlated to the degree of distension. This response can be blocked by atropine and can be reduced but not completely inhibited by vagotomy. This suggests that the acid secretory response to distension is mediated by a short cholinergic reflex within the gastric wall as well as a long reflex involving the vagus nerve. Selective distension of the antrum[8] has been found to induce acid secretion only in patients with duodenal ulcers.

The chemicals in food which stimulate acid secretion are small peptides and amino acids. These release gastrin from the antrum probably by interacting with receptors on the microvillous border of the antral G-cell.

The intestinal phase of gastric secretion can be stimulated by infusion of peptides and amino acids into the duodenum. This induces an acid secretory response which is about 30% of the maximum acid output, induced by pentagastrin,[65] but is not associated with a concomitant rise in serum gastrin level. As this intestinal phase of gastric secretion is enhanced in patients whose portal blood by-passes the liver[116] it may be mediated by an as yet undiscovered hormone which is inactivated by the liver.

Other stimuli which induce acid secretion include caffeine and calcium. The latter probably acts in part by releasing gastrin. This is of little clinical importance because most patients with hypercalcaemia due to hyperparathyroidism do not have hypergastrinaemia or excess acid secretion unless they also have a gastrinoma.[49]

Exposure of the antral mucosa to acid suppresses gastrin release and inhibits acid production. This mechanism helps to ensure that only sufficient acid is produced for acid digestion. Acid secretion is also inhibited by the presence of acid or fats in the duodenum. This response is probably mediated by release of secretin and cholecystokinin-pancreozymin (CCK).

PEPSINOGEN SECRETION

Pepsinogens are secreted in an inactive form and are converted to the active proteolytic pepsins by contact with gastric acid. There are two immunologically distinct groups of pepsinogen.[130] Group I (pepsinogens 1 to 5) are secreted solely by the chief cells and the mucus-secreting cells of the fundus and body whereas those in group II (pepsinogens 6 and 7) are found in the antrum and duodenum as well. The ability of group I pepsins to hydrolyse the synthetic substance, N-acetyl 1-phenyl-alanyl 1-diodotyrosine distinguishes between the two groups. It has been observed that group I pepsinogen levels in blood are strongly correlated with acid secretion.[131] This offers a non-invasive method for studying acid secretion in man.

Pepsinogen secretion is stimulated by the same factors which enhance acid secretion. However, secretin also enhances secretion of pepsinogen but inhibits acid secretion.

INTRINSIC FACTOR

Intrinsic factor is a mucoprotein of molecular weight 44 000, which is secreted by the parietal cell in man. Its secretion is enhanced by the same factors which enhance acid secretion and it is absent, or present in only very small amounts, in stomachs which have lost the capacity to secrete acid. It binds to Vitamin B_{12} and is necessary for its normal absorption in the ileum. In the absence of intrinsic factor, Vitamin B_{12} absorption is inadequate and pernicious anaemia develops.

GASTRIC SECRETORY TESTS

Pentagastrin Test

The pentagastrin test has now superseded the histamine or Betazole test as the standard method for measuring the capacity of the stomach to secrete acid.

After an overnight fast, a wide bore gastric tube (usually equipped with a second small bore tube to act as an air line to facilitate collection) is inserted into the stomach and positioned fluoroscopically so that the tip of the tube lies in the gastric antrum just to the right of the midline while the collection ports lie in the most dependent portion of the greater curvature. After discarding any residual gastric contents, the gastric juice is aspirated for a period of one hour (four 15-minute periods). Then pentagastrin is administered by subcutaneous injection (1 to 6 µg/Kg) or intravenous infusion (1 to 6 µg/Kg/h) and the gastric juice is collected for a further hour. The acid output during each 15-minute period is calculated from the volume and the hydrogen ion concentration (obtained either by titration against NaOH or from pH measurements using the tables of Moore and Scarlata.[107] Basal acid output is calculated by summing the four basal collections (Normal range = 0 to 5 mmol/h). Peak acid output (PAO) is obtained by adding the two highest 15 minute collections after pentagastrin and multiplying by two (normal range = 5 to 40 mmol/h).

Nowadays there are relatively few indications for the pentagastrin test. It is perhaps most useful in the investigation of patients suspected of having a gastrin secreting tumour (Zollinger-Ellison Syndrome). These patients characteristically have abnormally high peak acid outputs, often up to 100 mmol/h and a basal output of at least 60% of the peak. In patients with high blood gastrin concentrations, secretory testing distinguishes between gastric atrophy and a gastrinoma. Finally, the presence of hypersecretion in patients with recurrent ulceration after gastric surgery may suggest the Zollinger-Ellison syndrome or indicate retained gastric antrum.

The pentagastrin test is now of little use in the diagnosis and clinical management of patients with peptic ulceration. Although patients with an ulcer in the pylorus and duodenum tend to secrete acid at higher rates than normal subjects, at least 60% of these patients have a PAO within the normal range. Moreover, it has never been proven that the level of gastric secretion in a patient with duodenal ulceration is any guide to the sort of operation that should be performed. Patients with gastric ulcer secrete acid at lower rates than normal subjects but there is again considerable overlap with the normal range. The distinction between malignant or benign gastric ulcers on the grounds that patients with benign ulcers secrete acid while those with cancer do not is also unreliable as it is now clear that 80% of patients with malignant ulcers secrete acid while few patients with benign gastric ulceration have pentagastrin-fast achlorhydria.

Insulin Test

Induction of hypoglycaemia by insulin stimulates the vagus nerve and induces acid secretion. Thus the insulin test has been used as a test of the completeness of surgical vagotomy for duodenal ulcer.

Insulin is injected at a dose of 0·2 units/kg by rapid intravenous injection of by infusion of 0·1 units/kg/h for two hours. In both cases acid output is measured over two hours.

The basic assumption that hypoglycaemia induces acid secretion solely by stimulating the vagus nerve has never been proven and seems unlikely to be correct. For example, hypoglycaemia stimulates release of adrenaline which in turn releases gastrin.[140] Thus, the finding of some acid secretion on insulin testing may not necessarily indicate an incomplete vagotomy. Moreover, the finding of insulin-induced acid secretion cannot be used to predict which patients are likely to develop an ulcer. Kronborg[77] showed that only 27% of patients who secreted more than 3 mmol/h after insulin stimulation developed an ulcer, while some patients who had no insulin-stimulated acid secretion developed an ulcer. The only value of insulin testing may be in deciding whether it is necessary to reassess the completeness of the vagotomy during operation for recurrent ulcer.

As this test induces all the symptoms of hypoglycaemia from sweating and tachycardia to coma, it is essential for a doctor to be in attendance. Tolbutamide, which stimulates the β-cells of the pancreas to release insulin, has been used as an alternative to insulin infusion, but it is not established that side effects are any less frequent with this drug. The other agent which has been used to stimulate the vagus, 2-deoxyglucose, acts by competitively inhibiting the uptake of glucose by the brain. This is potentially more dangerous than insulin because its side effects are not so easily reversed by injecting glucose. It should therefore not be used for routine clinical testing in man.

Sham Feeding

Sham feeding offers a safer and more physiological method of stimulating the vagus than insulin testing.[125] Moreover, it seems likely that the effect of sham feeding is solely mediated by the vagus nerve (though vagal stimulation may induce some acid secretion indirectly by releasing gastrin).

After an overnight fast, subjects are intubated and a basal collection is carried out. They are then given an appetizing meal and instructed to chew the meal thoroughly without swallowing a morsel and to spit the masticated food into a beaker. Gastric contents are collected continuously during this procedure. Clearly the efficacy of this test depends on the ability of the subject to stop himself from swallowing any of the food.

Calcium Infusion

This test is used to diagnose the Zollinger-Ellison syndrome. Intravenous infusion of 4 to 5 mg/kg/h of calcium for three to four hours gradually stimulates gastric acid secretion to approximately 30 to 50% of the PAO in normal subjects and in patients with duodenal ulceration. The peak response to calcium infusion occurs in the third hour and is associated with a small but significant increase in serum gastrin level and and increase in plasma calcium level of 0·7–0·8 mmol/l

(3 mg/100 ml). In the Zollinger-Ellison syndrome, the secretory response to calcium infusion is 100% of the PAO and the serum gastrin level increases dramatically to over 300 pg/ml.[7]

Secretin Test

Another manoeuvre which is useful in identifying patients with Zollinger-Ellison syndrome is the intravenous injection of secretin (1 to 2 units/kg of GIH secretin, 4 to 8 units/kg of Boots secretin). This produces a decrease in blood gastrin level in normal subjects but a paradoxical increase in patients with Zollinger-Ellison syndrome.[74]

Non-Invasive Techniques for Measuring Gastric Secretion

The capacity of the stomach to secrete acid can be indicated by several non-invasive techniques. The Diagnex Blue test involves the oral administration of an ion-exchange resin (Diagnex Blue) which dissociates in the presence of acid. The dissociated form is absorbed in the small intestine and excreted in the urine which it turns blue. This test does not give a quantitative index of acid secretion, but has been used as a screening test for detection of gastric cancer. The technetium test is more useful. Technetium can be secreted by the gastric epithelium and it has recently been found that following intravenous injection of ^{99m}Technetium, there is a good correlation between PAO and the radioactive counts measured over the surface of the stomach.[144] Serum pepsinogen I levels also show a good correlation with acid output and may be used as a non-invasive test of acid secretion.[131]

GASTRIC MUCOSAL BARRIER

Damage to the stomach wall by gastric acid is prevented by the close apposition of the cells to each other which renders the gastric epithelium virtually impermeable to small ions such as hydrogen and sodium. This gastric mucosal barrier maintains high concentration gradients for hydrogen ions from lumen to plasma and for sodium ions from plasma to lumen and a high transepithelial potential difference (PD). According to the theory of Davenport,[27,28] exposure of the stomach to aspirin, ethanol or bile acid damages this mucosal barrier, leading to an increase in permeability. Acid then enters the submucosa where it causes inflammation and eventually gastric ulceration. This hypothesis is based on the observations that the volume and sodium content of the gastric juice are increased while the hydrogen ion concentration and the PD are reduced in the presence of gastritis or agents which are thought to break the mucosal barrier. While these changes undoubtedly suggest an increase in gastric permeability, they could equally well result from a stimulation of bicarbonate secretion from surface endothelial cells.[145] The alkaline mucus is probably of little importance in protecting the stomach against digestion by its own secretions, although it clearly has some buffering capacity and can inhibit diffusion of acid. If mucus were an efficient barrier, gastric acid would never be able to escape into the lumen in the first place.

Measurement of Gastric Potential Difference

This has been advocated as a convenient clinical method for diagnosing gastritis or damage to the gastric mucosal barrier.[44] The gastric potential difference can be measured by means of an intragastric electrode containing saline or concentrated potassium chloride in agar and a subcutaneous or intravenous reference electrode. The electrodes are connected via calomel half cells to a battery-powered electrometer. The normal resting PD in the body or fundus of the stomach is approximately 40 mV. When the stomach is exposed to agents which disrupt the gastric mucosal barrier, the PD falls to 20 mV or lower. The fall in PD is thought to be due to the decrease in electrical resistance brought about by an increase in permeability to small ions. However, a fall in PD may occur as a result of several other physiological events occuring in the epithelium or the gastric lumen. Of particular importance is the reduction in PD which is observed when the rate of acid secretion is increased. This is largely caused by the formation of a physicochemical junction potential between the electrode solution and gastric acid. It is thus possible to interpret a lowering of gastric PD as due to disruption of the mucosal barrier only if either there is no acid secretion or if appropriate corrections are made for junction potentials.[122]

GASTRIC EMPTYING

The stomach receives and stores food, churns it and mixes it with gastric secretions and delivers it in a controlled manner to the duodenum. Its motor activity is carried out by a proximal reservoir, comprising the fundus and proximal body, and a distal pump incorporating the distal body and antrum.

The fundus and proximal portion of the body of the stomach have no phasic contractile activity, but exhibit the phenomena of receptive relaxation to accommodate the meal, and tonic contraction which gently squeezes food down towards the antral pump. After a meal waves of contraction sweep down through the antrum from a point in the mid body of the stomach towards the pylorus at a maximum rate of three to four a minute. By churning and mixing the food with gastric secretions, the contractions gradually liquefy the gastric contents and deliver them at a steady rate into the duodenum. The pylorus appears to act as a filter allowing small particles of food to pass into the duodenum, but retaining larger particles to be ground further by antral contractions. It is also important in preventing duodeno-gastric reflux.

The rate of gastric emptying is regulated by the volume and the composition of a meal. Larger meals are emptied more slowly than smaller ones. Liquids leave the stomach more quickly than solids that have to be churned into a fluid before being emptied. However, the differential emptying of solids and liquids probably depends upon the particle size of the solids and on the volume of liquid present. A meal containing a small amount of liquid together with food which is quickly reduced to small particles will be mixed rapidly to a slurry in which both liquid and solid components will be emptied at equivalent rates.[56] The acidity, osmolality and amount of fat in duodenal contents and the presence of certain amino-acids delay gastric emptying presumably by interacting with duodenal receptors. On reviewing the results of other studies, Hunt and Stubbs[64] concluded that although there may be a variety of factors which control gastric emptying, the overall rates of gastric emptying for a variety of meals were closely correlated with their nutritive densities. Meals with higher nutritive densities were emptied more slowly. Nevertheless, the delivery of calories to the duodenum was greater with meals with high nutritive densities than with meals with low nutritive densities. These findings are important in terms of the relationship between satiety and obesity.

The inhibition of gastric emptying by certain food substances must be partly mediated by release of hormones. Secretin, CCK and gastric inhibitory peptide (GIP) are all released in the duodenum by the presence of acid or food and can all delay gastric emptying.

Surgical vagotomy increases emptying of liquids by increasing the tone in the fundus of the stomach and decreases emptying of solids by inhibiting antral contractions. The latter does not occur after the proximal vagotomy currently performed for duodenal ulceration. Gastric emptying is often slowed in patients with atrophic gastritis and increased in patients with duodenal ulcer or gallstones. Studies in patients with pancreatic insufficiency have shown that gastric emptying of starch is more rapid than in normal subjects, presumably because there is less stimulation of osmoreceptors if the starch is not digested.[92] The same mechanism presumably explains why gastric emptying of lactose in patients with lactase deficiency is more rapid than in normal subjects. Emptying of triglycerides is more rapid in pancreatic insufficiency than emptying of the equivalent amount of free fatty acids,[73] which suggests that it is the free fatty acid which is largely responsible for slowing of gastric emptying by fat.

Methods for Measuring Gastric Emptying

The methods for measuring gastric emptying, which are most clinically acceptable, involve labelling food with a non-absorbable radioactive marker and estimating the disappearance of the marker from the stomach by means of the gamma camera or a fixed scintillation detector. Because solid and liquid components of certain meals empty at different rates, methods have been devised to label the solid components of the meal independently of the liquid components. In one ingenious method ^{99m}Technetium has been incorporated into chicken liver by injecting the isotope into the wing vein of a chicken.[105] The chicken is killed, the liver removed, cooked, diced into cubes and eaten with the meal. Other methods of labelling solid components include incorporating the marker into the white of an egg or coating ^{99m}Technetium impregnated paper with plexiglas to resemble cornflakes. These methods are only of importance in studying meals that are readily divided into solid and liquid components (i.e. steak and water). Most meals can be quickly turned into a slurry, in which both solids and liquids empty at the same rate.

Other techniques for measuring gastric emptying involve intubation and sampling of gastric contents and can be recommended only for research purposes. The simplest method is to incorporate a marker with the meal and aspirate the gastric contents after a known period of time. The product of the volume of gastric contents and the concentration of the marker indicates the amount of marker remaining in the stomach. To avoid complete aspiration of gastric contents, gastric volume may be measured by injecting a second marker into the stomach and measuring its concentration in a sample taken after completely mixing it with gastric contents.[45] These methods are only really suitable for liquid meals. Malagelada et al,[91] have recently developed a method for solid meals which allows simultaneous measurements of post-prandial gastric secretion, volume of gastric contents and rate of delivery of food into the duodenum. This method, however, is difficult to carry out, as it involves the use of a gastric tube and a triple-lumen duodenal tube as well as two non-absorbable markers.

VOMITING

The vomiting act encompasses an integrated set of physiological events culminating in the forcible expulsion of gastric contents. It consists of three stages: nausea, retching and actual vomiting. Nausea is a psychological experience which is almost impossible to define, but is associated with a loss of gastric tone and contractions, an increase in duodenal tone and often the reflux of duodenal contents into the stomach. Hypersalivation is also usually present. Retching consists of spasmodic contractions of the diaphragm and muscles of the abdominal wall. Observations in the cat[97] have shown that this is associated with repeated thoracic herniations of the abdominal oesophagus and cardia, accompanied by the to and fro passage of gastric contents into a dilated oesophagus. The antral and pyloric portion of the stomach appears contracted during this phase. Finally, vomiting is the passage of gastric contents through the mouth produced by forcible abdominal contractions occurring when the cardia of the stomach is raised and open. Observations in the cat suggest that herniation of the cardia into the chest caused by retching is an important prerequisite for vomiting as it overcomes the antireflux mechanism at the lower oesophagus.

The sequence of events that comprise vomiting is thought to be coordinated by an area in the medulla called the vomiting centre.[15] Stimulation of this area induces vomiting. Ablation renders the animal unresponsive to the factors which normally induce vomiting.

The vomiting centre may be stimulated from a number of sources (Figure 2). Distension of the gut, bile ducts, ureters and Fallopian tubes is neurally mediated as is the

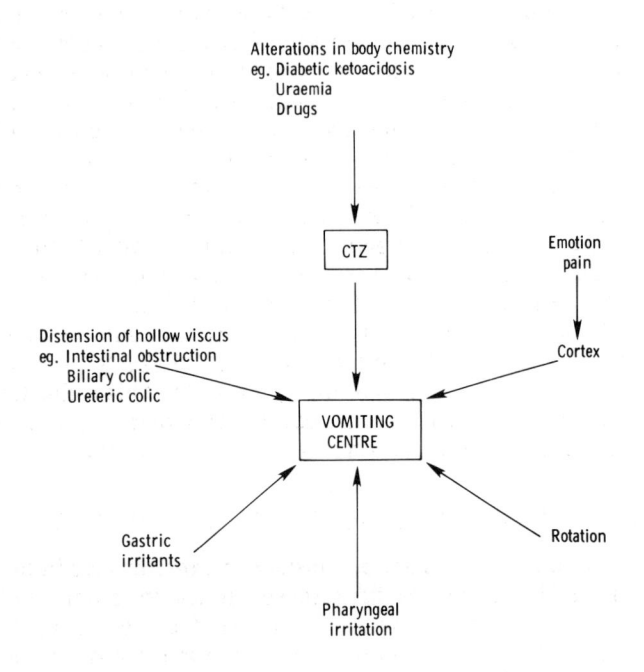

FIG. 2. A diagram of the afferent neural pathways involved in the vomiting reflex.

vomiting induced by irritation of the pharynx and the stomach. The alteration to body chemistry caused by uraemia, diabetic ketoacidosis and drugs causes vomiting by stimulating an area in the floor of the fourth ventricle called the chemoreceptor trigger zone (CTZ) which has nervous connections to the vomiting centre. Ablation of the CTZ renders the animal completely resistant to chemical emetics administered parenterally, but not to the same emetics given intragastrically. Vomiting induced by rotation (motion sickness) is also neurally mediated, by stimulation of the vestibular apparatus. Thus, all the emetic responses are mediated by reflex arcs which pass through the vomiting centre regardless of whether these responses are initiated at peripheral or central sites. It is possible that in some cases of psychogenic vomiting, the sensitivity of the vomiting centre may be enhanced by input from the cerebral cortex.

In many cases, the cause of vomiting can be rapidly diagnosed by clinical examination and simple radiological procedures. Nevertheless, every patient with vomiting for which no cause is apparent should have blood taken for estimation of urea, calcium, corticosteroids, liver function tests and glucose. As a clinical test, measurement of gastric emptying is only useful in patients suspected of having conditions such as diabetic neuropathy causing gastric stasis, as these may have normal barium meals. Finally, a history of repeated vomiting, but no clinical or biochemical evidence of dehydration, alkalosis or weight loss suggests psychogenic origin.

THE PANCREAS

The pancreas is an elongated, flattened gland lying on the posterior abdominal wall. The head lies within the duodenal loop and the tail extends across the abdomen as far as the gastric surface of the spleen. The exocrine portion of the gland comprises a system of acini which form lobules separated by connective tissue. Each acinus is composed of pyramidal cells whose apices point into a central lumen, which is drained via a system of ducts into the main pancreatic duct. This usually joins the common bile duct to open into the second part of the duodenum at the ampulla of Vater.

PANCREATIC SECRETION

Pancreatic juice is an isotonic secretion consisting of an enzyme-rich component secreted by the acinar cells and a fluid and electrolyte component which largely emanates from the ductular cells. Sodium and potassium comprise the major cations of pancreatic juice and are secreted in proportions similar to those found in plasma. Chloride and bicarbonate are the major anions and their concentrations vary reciprocally, the proportion of bicarbonate increasing as the flow rate increases. The most likely explanation for this phenomenon is that a relatively fixed amount of chloride-rich fluid is secreted by the acinus, but varying amounts of bicarbonate-rich fluid are added by the ductular cells according to the degree of neurohumoral stimulation.[86] Recent work suggests that the mechanism underlying bicarbonate secretion is in fact the active transport of hydrogen ion from the ductular lumen into the cell in exchange for sodium.[142] The hydrogen ion combines with bicarbonate inside the cell, increasing the $P{CO_2}$. Carbon dioxide then diffuses into the lumen where it dissociates into hydrogen ion and bicarbonate. The continued transport of hydrogen ion out of the lumen leads to an apparent net secretion of sodium bicarbonate.

The alkaline pancreatic juice is largely responsible for neutralizing the gastric acid and creating the optimum duodenal pH for the activity of pancreatic enzymes. The enzymes secreted by the pancreas include proteases (trypsin, chymotrypsin, elastase and carboxypeptidases), amylase and lipase. They are synthesized in the acinar cells, stored as zymogen granules, and probably leave the cell by exocytosis. The proteases are secreted as inactive precursors. Trypsinogen is converted by duodenal enterokinase to trypsin which then activates the other proteases. Lipase and amylase are secreted in the active form.

Amylase breaks down starch to maltose, maltotriose and α-limit dextrins (short chains of glucose molecules containing C 1–6 branch points). Lipase hydrolyses triglyceride to fatty acid and monoglyceride. Proteases are of two sorts: endopeptidases (trypsin, chymotrypsin and elastase) cleave peptide bonds in the interior of protein or polypeptide molecules, whilst exopeptidases (carboxypeptidase A and B) separate the terminal amino-acids at the carboxyl end of the chain.

Although pancreatic enzymes are predominant in the digestion of carbohydrates, fats and protein, the exocrine function of the pancreas is not essential for survival. Alternative pathways of digestion and absorption exist which allow a limited intake of nutrients under these conditions. Some starch may be digested by salivary amylase, and recent studies suggest that a large proportion of unabsorbed carbohydrate may be salvaged in the colon by bacterial conversion to short chain fatty acids which may then be rapidly absorbed.[13] Fat may be digested in part by pharyngeal and gastric lipases and nonspecific esterases in the intestinal mucosa. Short-chain and medium-chain triglycerides can be absorbed intact without prior digestion. Finally, oligopeptides can be broken down in the intestine by brush border peptidases and some peptides can be absorbed intact.

Control of Pancreatic Secretion

Major roles in the control of pancreatic secretion have traditionally been assigned to the hormones secretin and CCK. Secretin, released in response to duodenal acidification, provokes a high-volume HCO_3^--rich secretion, containing only small amounts of enzymes. CCK, released in response to food, (particularly amino acids, peptides and fatty acids) entering the duodenum, elicits an enzyme-rich secretion. These two hormones potentiate each other's actions.

Other hormones may also stimulate pancreatic secretion. Gastrin is similar in structure to CCK, but is much less potent in eliciting pancreatic enzyme secretion in man. Vasoactive intestinal polypeptide (VIP) is structurally similar to secretin and stimulates output of bicarbonate-rich fluid. It is, however, a weaker agonist than secretin and when the two hormones are given together it may competitively inhibit the actions of secretin. Glucagon and pancreatic polypeptide inhibit both the enzyme and bicarbonate secretion stimulated by CCK and secretin respectively. The physiological role of these inhibitors has not been established.

The precise role of neural mechanisms in the control of pancreatic secretion is unresolved.[137] Vagal stimulation and cholinergic agents induce an enzyme-rich pancreatic secretion and also potentiate the actions of CCK and secretin. Anticholinergic agents and vagotomy reduce the secretory response to these hormones.[99] Adrenergic agents also influence pancreatic secretion. α-adrenergic agents decrease water and electrolyte secretion. β-adrenergic agents are ineffective in man. Dopamine is a pancreatic secretogogue.[132]

There is evidence to suggest the existence of a cephalic as well as a gastric phase of pancreatic secretion in man.[133,155] Both of these phases stimulate a small amount of enzyme secretion and their major role may be to mobilize pancreatic enzymes in the ducts so that they are flushed out by the fluid secretion induced by secretin. The most potent stimulus for pancreatic secretion is the entry of food into the small intestine. The assumption that secretin is largely responsible for the increase in volume of pancreatic juice stimulated by food has been recently questioned by the observations that the pH of duodenal contents after a meal does not fall low enough to stimulate large quantities of pancreatic juice and increases in secretin levels in the blood during a meal are barely detectable.[80] It is possible that the interaction of only small amounts of secretin with the other neural and humoral factors released by the meal is sufficient to cause the normal increase in pancreatic secretion. The release of pancreatic enzymes is stimulated by peptides and amino acids, an action mediated probably both by neurohumoral interactions, as these substances stimulate vagal afferents, and by the release of CCK.

The potency of intestinal contents to stimulate pancreatic secretion depends on the length of small intestine exposed to the products of digestion.[104] This indicates that mucosal receptors mediating pancreatic secretion are arrayed for some distance down the intestine.

ASSESSMENT OF PANCREATIC FUNCTION

Laboratory tests are of considerable importance in the diagnosis of pancreatic disease. Rises in blood and urinary enzyme activities are major features of acute pancreatitis, whilst stimulation tests form the basis of the diagnosis of chronic pancreatic insufficiency.

Acute Pancreatitis

Acute damage to the pancreas leads to the leakage of enzymes into the circulation and a rise in their blood and urinary concentrations. Amylase is the most commonly measured enzyme, and assay by the saccharogenic technique is claimed to be the most accurate method of analysis.[84] Although amylase is usually only measured in the serum or plasma, measurement of urinary amylase concentrations may be particularly useful as hyperamylasuria may persist for several days after blood amylase levels have returned to normal following acute pancreatitis.

Both blood and urinary amylase levels may be increased in conditions other than acute pancreatitis. These include renal impairment, diabetic ketoacidosis, intestinal obstruction and perforation, afferent loop obstruction, perforated duodenal ulcer, cholelithiasis, ruptured ectopic pregnancy, and acute and chronic liver disease. This may cause diagnostic difficulty. Determination of the ratio of urinary amylase clearance to creatinine clearance[85] may help to distinguish the hyperamylasaemia of acute pancreatitis from other causes. This ratio is increased approximately three-fold in acute pancreatitis,[153] but may also be elevated in other conditions.[82]

Blood and urinary amylase levels and the amylase: creatinine clearance ratio may also be abnormal in acute episodes of pain occurring in the course of chronic pancreatitis, but as destruction of the gland proceeds, elevation of the level of blood amylase in acute episodes diminishes. Elevation of blood amylase activity with normal amylase excretion may be seen in the rare condition macroamylasaemia, in which an abnormal amylase isoenzyme, too large to be filtered by the glomeruli, is present in serum. These patients may occasionally have abdominal pain.

Increase of blood lipase activity is also seen in acute pancreatitis, and in most of the other conditions in which raised blood levels of amylase are encountered, with the exception of mumps, hepatitis and macroamylasaemia. Estimation of lipase activity appears to have little advantage over that of amylase and is a more difficult practical procedure.

The recent development of a radioimmunoassay for trypsin may prove a very useful and specific test in acute pancreatitis[36] because unlike lipase and amylase, trypsin is only produced in the pancreas.

Other biochemical abnormalities which may occur in association with acute pancreatitis may be of diagnostic use and include the presence of methaemalbuminaemia, transient jaundice, hyperglycaemia, hypocalcaemia, hypercalcaemia (hypocalcaemia may develop as a consequence of acute pancreatitis, whereas hypercalcaemia is a predisposing cause), hyperlipoproteinaemia and diminished lipoprotein-lipase activity.

Chronic Pancreatitis

Collection and analysis of pancreatic juice secreted in response to standard stimuli is perhaps the best way to assess chronic pancreatic insufficiency. The gland may be stimulated directly by injection of secretin and CCK, or indirectly by a standard meal.

Secretin Test

In the secretin test of pancreatic function,[157] one tube is positioned in the fourth part of the duodenum for continuous aspiration of pancreatic juice, and another is placed in the antrum of the stomach to prevent acid from entering the duodenum and neutralizing pancreatic bicarbonate. Secretin is given either by bolus injection or by continuous infusion. CCK may also be given if information regarding enzyme output is required. Duodenal aspiration is continued for 80 minutes, and complete collection is necessary. Variations in the dosage and potency of stimulating hormones used, and the technique of stimulation (whether secretin is given alone or with CCK, by bolus injection or by continuous infusion) renders a strict definition of normal ranges difficult.[5] Nevertheless, it is perhaps safe to say that in chronic pancreatitis maximum bicarbonate concentration is less than 80 to 90 mmol/l, and volume flow less than 2 ml/kg over 80 minutes. Bicarbonate output may be a better measure of exocrine function than maximum bicarbonate concentration. Volume flow may also be low in carcinoma of the pancreas, especially if it involves the head of the gland, though the bicarbonate concentration is more likely to be normal.

Lundh Test

Indirect stimulation of the pancreas is carried out by feeding a test meal.[67] One that is often used consists of 19 g corn oil, 15 g casilan, 40 g glucose, in 300 ml of water. Tryptic activity is estimated in the duodenal aspirate. This is a simpler test to perform than the secretin test, and is more physiologically relevant as it tests pancreatic response to a meal. The tube is positioned in the fourth part of the duodenum as before, and no gastric tube is necessary. Complete collection of duodenal contents is unnecessary. In chronic pancreatitis mean tryptic activity is reduced from a mean of 17 IU/l to below 6 IU/l in more than two-thirds of cases.[108] Abnormally low results may also occur in coeliac disease, presumably because of impaired release of duodenal hormones. The test is also inappropriate after gastric surgery.

Other Methods

The development of the technique of endoscopic cannulation of the pancreatic duct has allowed the collection of pancreatic juice uncontaminated by food, gastric or biliary secretions. Results of stimulation tests under these conditions have so far, however, proved disappointing, with considerable overlap in volume, bicarbonate and protein output between normal subjects and patients with chronic pancreatitis.[29]

A test of pancreatic function which avoids the discomfort of intestinal intubation involves the oral use of a synthetic peptide (N-benzoyl-L-tyrosyl-p-aminobenzoic acid).[6] This peptide is cleaved by chymotrypsin, releasing p-aminobenzoic acid (PABA) which is absorbed and excreted in the urine. Recovery of PABA in the urine of patients with chronic pancreatitis is about 50% of that in normal subjects. However, false positive results may occur in the presence of renal impairment, intestinal and liver disease[106] and these limit the usefulness of this test.

Faecal enzyme activity is low in chronic pancreatitis, and this has been used as a test of pancreatic function, particularly in children.[35] Chymotrypsin estimation is more reliable than that of trypsin.

Many patients with chronic pancreatitis or pancreatic cancer have abnormal glucose tolerance, and the insulin response to injected glucagon and tolbutamide is diminished.[69]

Choice of tests in chronic pancreatitis

Despite their drawbacks, the most specific and sensitive laboratory tests of pancreatic function are those involving intestinal intubation and stimulation with either secretin-CCK or a test meal. The secretin test has the advantage of offering a more standard and controlled stimulus to the gland, whereas the Lundh test mimics the normal physiological stimulus. In chronic pancreatitis with steatorrhoea, when at least 90% of exocrine function is lost, these two tests give comparable results. In the absence of steatorrhoea the secretin-CCK test is claimed to be more sensitive in the diagnosis of chronic pancreatitis. It is, however, considerably more difficult to perform than the Lundh test, and each laboratory requires to establish a standard method and normal range. For these reasons the Lundh test is performed in many centres.[152]

Faecal enzyme estimation and the PABA test can be regarded only as screening tests at the present time.

BILE

BILE SECRETION

Hepatic bile is an isotonic, slightly alkaline fluid containing bile acids, phospholipids (mainly lecithin), cholesterol, bilirubin, protein, electrolytes and water.[37,41] Approximately 500 to 600 ml is secreted daily. Active secretion of bile acids into the biliary canaliculi provides the osmotic driving force for secretion of water into bile (bile acid-dependent pathway) via gaps between the cells (the paracellular route)[79] (Fig. 3). Electrolytes

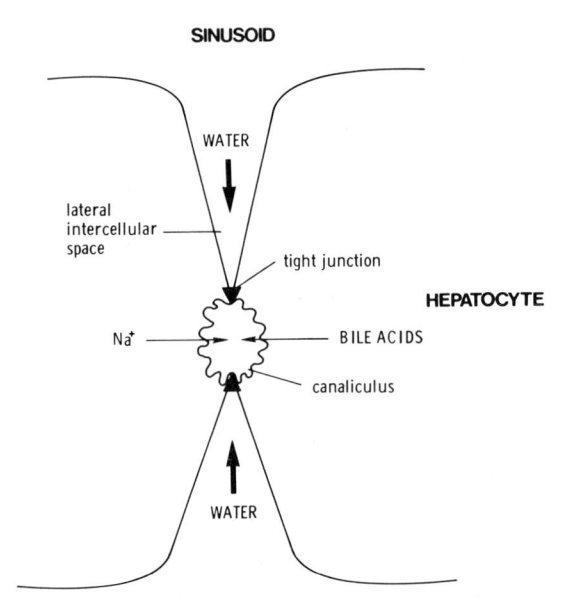

FIG. 3. Canalicular secretion. Secretion of bile acids across the canalicular membrane causes the osmotic flow of water via the lateral intercellular spaces and tight junctions into the canaliculus (bile acid-dependent flow). Bile acid-independent secretion is probably related to active sodium transport, with secondary osmotic water flow by the same route, but the anatomical locations of the active steps in both sodium and bile acid secretion are unresolved.

and some small solutes then accompany water flow by the process of solvent drag. There is also an additional, bile acid independent, canalicular secretion which may be linked to active sodium transport. Substances with larger molecular weights, such as bilirubin, organic acids, oestrogens and drugs, are probably actively transported into bile. The transport of cholesterol and lecithin, the major biliary lipids, appears to be passive. Lecithin secretion is tightly coupled to that of bile acids, cholesterol less so.

Canalicular secretion continues throughout the 24-hours at a low basal level, and is periodically stimulated by meals. The main factor regulating canalicular secretion is thought to be the concentration of bile acids returning to the liver in the portal blood.

After leaving the canaliculi, bile is modified by the secretion of an alkaline fluid, rich in bicarbonate and chloride, from the cells lining the biliary ductules. This secretion is stimulated by secretin and perhaps other hormones released by the presence of food in the duodenum.

GALLBLADDER FUNCTION

During fasting, hepatic bile accumulates in the gallbladder, held back by the resistance of the sphincter of Oddi. Bile is concentrated in the gallbladder by the active reabsorption of sodium chloride which creates an osmotic gradient across the epithelium, encouraging passive water absorption.[42] The entry of food into the duodenum stimulates the release of CCK which relaxes the sphincter of Oddi and contracts the gallbladder, causing the bile in the gallbladder to enter the intestine. At the same time, secretin is released by duodenal acidification and stimulates bile flow by enhancing ductular secretion.

BILE ACID METABOLISM AND THE ENTEROHEPATIC CIRCULATION

Two primary bile acids are synthesized in the hepatocytes from cholesterol. These are cholic acid (3α, 7α, 12α-trihydroxy-5β-cholanoic acid), and chenodeoxycholic acid (3α, 7α-dihydroxy-5β-cholanoic acid)[59] (Fig. 4). After synthesis, conjugation with either glycine or taurine occurs. Conjugation increases the resistance of bile acids to precipitation in the intestinal lumen, and decreases their passive reabsorption in the proximal small bowel.

The total bile acid pool size, measured by isotope dilution, is 5 to 10 mmol. During fasting practically all of the pool is stored in the gallbladder, but it enters the intestine when a meal is eaten. The majority of bile acids are reabsorbed by an active transport process in the terminal ileum and return in the portal blood to the liver. (Fig. 5). Bile acids are efficiently extracted from portal venous blood into the hepatocytes by a sodium-dependent saturable process, and are promptly re-secreted into the canaliculi, thus further stimulating bile flow. Recycling by this enterohepatic circulation continues until the meal is absorbed, and the entire bile acid pool may be recycled two or three times for a single meal. After the meal is absorbed, hepatic bile re-enters the gallbladder, and the reduction in bile acids returning to the liver reduces bile acid secretion and bile flow back to basal levels.

A small proportion of the total bile acid pool escapes ileal reabsorption and is dehydroxylated by colonic bacteria to produce the secondary bile acids; deoxycholic acid (3α, 12α-dihydroxy-5β-cholanoic acid) from cholic

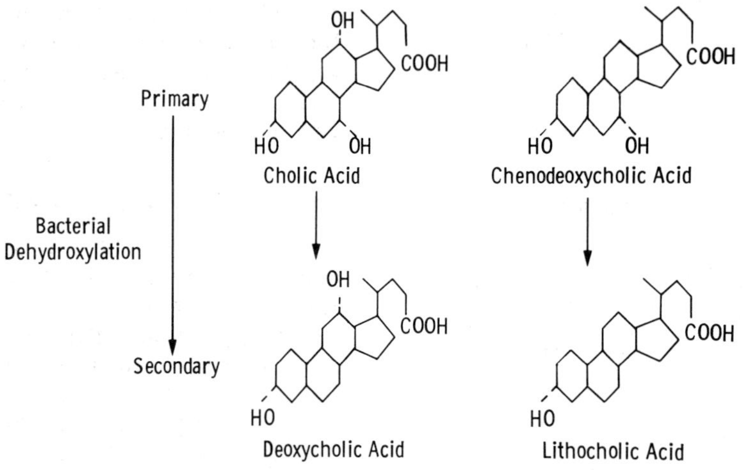

FIG. 4. Major primary and secondary bile acids in man. In normal bile, conjugates of cholic and chenodeoxycholic acids predominate (30–40% each) with lesser amounts of deoxycholic acid (10–30%). Lithocholic acid is present only in small amounts and is mostly sulphated. In some people the tertiary bile acid ursodeoxycholic acid ($3\alpha,7\beta$-dihydroxy-5β-cholanoic acid) is also present in significant amounts. This is formed in the liver from 7-keto-lithocholic acid, a product of bacterial dehydrogenation of chenodeoxycholic acid.

acid, and lithocholic acid (3α-hydroxy-5β-cholanic acid) from chenodeoxycholic (Fig. 4). Some of the deoxycholic acid is lost in the faeces but up to 50% is reabsorbed from the colon, conjugated in the liver, and then circulates like the primary bile acids (Fig. 5). A smaller proportion of lithocholic acid is absorbed in the colon. On reaching the liver this is conjugated and largely sulphated, reducing subsequent ileal reabsorption and promoting its loss in the faeces.

The daily faecal losses of bile acids (15 to 20% of the total pool) are balanced by hepatic synthesis (Fig. 5), which is controlled by the negative feedback of bile acids returning in the portal venous blood on the rate-limiting hepatic enzyme cholesterol 7α-hydroxylase. Bile acids also inhibit the rate-limiting enzyme for cholesterol synthesis, hydroxymethyl-glutaryl Coenzyme A reductase, limiting the availability of cholesterol for bile acid synthesis. Synthesis of bile acids can be increased several fold by bile acid depletion, and is depressed by total fasting, bile acid feeding, and in cholestasis.

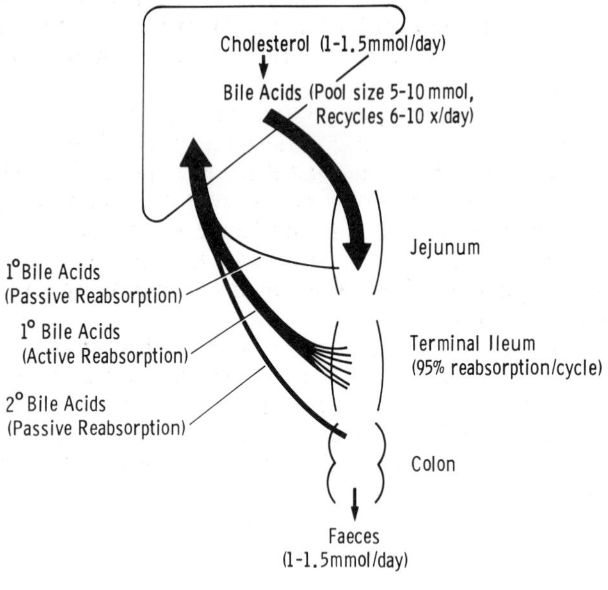

Because of the efficiency of hepatic extraction, very little spillover of bile acids into the systemic circulation occurs, although postprandial rises in blood bile acid levels can be detected.

FUNCTIONS OF BILE ACIDS

Stimulation of Bile Flow

See page 75.

Transport of Lipids in Bile

Bile acids are essential for the transport of insoluble cholesterol in bile. Conjugated bile acid molecules are amphiphilic, and at a lipid-water surface become orientated with their hydrophobic radicals (hydroxyl groups) in the water phase and their hydrophobic radicals in the lipid phase. This detergent property enables bile acids to solubilize cholesterol by dispersing it into small polymolecular aggregates named micelles, with a diameter of approximately 5 nm.[57] Micelles in bile are probably disc- or drum-shaped and consist of a bilayer of lecithin, surrounded by bile acid molecules (Fig. 6). Cholesterol molecules are contained within the hydrophobic interior of the micelle between the fatty acid chains of the lecithin.

FIG. 5. Enterohepatic circulation of bile acids. Bile acid loss in the faeces is balanced by synthesis from cholesterol in the liver. Passive reabsorption of unconjugated bile acids can occur throughout the small intestine; conjugated bile acids are actively reabsorbed in the terminal ileum. Colonic absorption of secondary bile acids is passive.

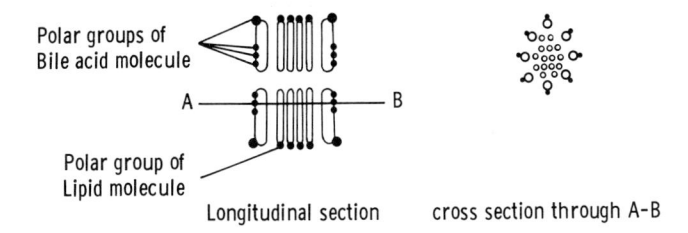

FIG. 6. Structure of a mixed micelle. The polar groups of the bile acid molecules, and those of the relatively water insoluble lipid molecules in the hydrophobic interior of the micelle, face outwards into the water phase. The interior of the mixed micelle in bile contains lecithin and cholesterol, and in the intestinal lumen contains fatty acids, monoglyceride, fat soluble vitamins and cholesterol.

Lipid Absorption

In the duodenum the bile acids form micelles with fatty acids and monoglyceride, released from hydrolysis of triglyceride. The micelle enhances absorption of fatty acids and monoglyceride by facilitating their diffusion through the unstirred water layer immediately adjacent to the microvillous membrane of the enterocytes. The minimum concentration of bile acids necessary for the formation of intraluminal micelles, the critical micellar concentration, is between 2 and 5 mM.

Colonic Fluid Transport

As dihydroxy bile acids (deoxycholic and chenodeoxycholic) inhibit colonic water absorption, and can induce secretion[100] it has been suggested that the entry of small amounts of bile acid into the colon may be one factor regulating colonic water transport.[60]

DISTURBANCES OF THE ENTEROHEPATIC CIRCULATION

Impaired Ileal Transport of Bile Acids

Since bile acids are predominantly reabsorbed in the terminal ileum, resection or disease of this part of the intestine, or excessively rapid transit through this region, can lead to bile acid malabsorption.[58] Bile acid synthesis can increase several-fold, but if faecal losses exceed this synthetic reserve, the concentration of bile acids in the intestine may fall below the critical micellar concentration and lead to malabsorption of fat, cholesterol and fat soluble vitamins. Diarrhoea may occur due to colonic secretion induced by unabsorbed dihydroxy bile acids,[9,100] or by hydroxy fatty acids formed by bacterial degradation of unabsorbed long-chain fatty acids.[4]

Diarrhoea induced by bile acids will occur only if bile acids, particularly dihydroxy bile acids, persist in solution in the colon. As secondary acids produced by bacterial dehydroxylation are relatively insoluble at normal colonic pH, the occurrence of diarrhoea depends on the balance between the entry of bile acids into the colon and the ability of the bacterial flora to dehydroxylate them.

A further consequence of bile acid malabsorption is a reduction in bile acid pool size leading to a decrease in bile acid concentration in hepatic and gallbladder bile, and ultimately cholesterol gallstone formation. This explains the association between gallstones and ileal resection and disease.[53]

Contaminated Small Bowel Syndrome

Contamination of the small intestine by overgrowth of bacteria can produce steatorrhoea.[71] It seems likely that this is related to the bacterial deconjugation of bile acids in the upper small intestine. The reduction in the concentrations of conjugated bile acid may prevent micelle formation and thus impair fat absorption. Unconjugated bile acids may further impair absorption by damaging the small intestinal mucosa.

Cholestasis

Obstruction of bile flow in the liver and bile ducts leads to impaired fat absorption and steatorrhoea. Bile acid synthesis is suppressed. Bile acids accumulate in the systemic circulation and tissues such as the skin, and are thought to cause pruritus. Under these conditions hepatic sulphation of primary bile acids, and their subsequent excretion in the urine, becomes an important route for bile acid elimination from the body.[141]

Pathogenesis of Gallstones

Gallstones are largely composed of aggregates of cholesterol crystals. Cholesterol, which is insoluble in water, is normally maintained in solution in bile by being transported in mixed micelles together with bile acids and lecithin. Decreases in the concentrations of bile acids or lecithin, or an increase in the concentration of cholesterol, lead to the formation of hepatic bile which is supersaturated with cholesterol, and ultimately to cholesterol gallstones (Fig. 7).

Evidence suggests that gallstone patients have an increased rate of hepatic cholesterol synthesis.[22] They also have a smaller than normal bile acid pool,[149] which recycles more frequently than normal.[114] Abnormally rapid small bowel transit is one factor associated with a reduced pool size,[34] though it is as yet not established whether patients with gallstones have abnormally rapid intestinal transit.

Although it seems likely that abnormalities in bile acid pool size and cholesterol synthesis contribute to the formation of lithogenic bile, it is important to note that supersaturated bile can be found in normal subjects, especially during fasting. Thus it seems likely that other factors (such as bacterial infection, diet, obesity and gallbladder function) may be important in initiating the formation of gallstones.

ASSESSMENT OF BILE ACID METABOLISM

Numerous techniques have been developed for investigating the metabolism of bile acids and the integrity of

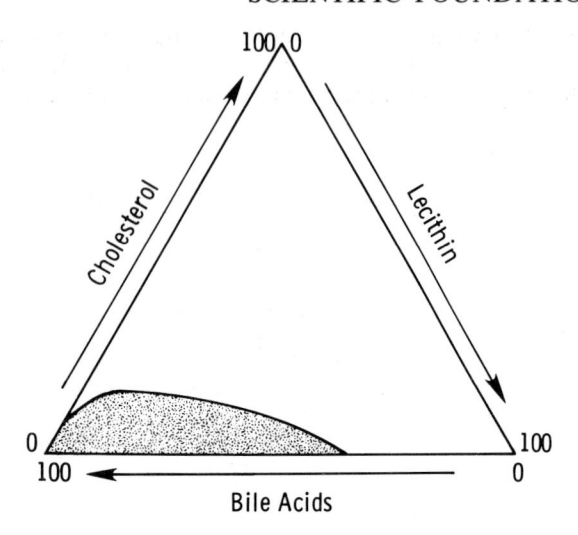

FIG. 7. Triangular co-ordinate method for determining cholesterol solubility in bile. The relative molar proportions of cholesterol, lecithin and bile acids, which together comprise 100%, are plotted on triangular co-ordinates. Bile whose composition lies within the shaded area (micellar zone) is a true single-phase solution. Outside this zone the bile is supersaturated with cholesterol (lithogenic bile) and may exist as liquid, liquid crystal and crystalline phases, predisposing to cholesterol gallstone formation (Admirand and Small,[1]; Holzbach, et al.[63]).

the enterohepatic circulation,[59] but most are used only as research tools. The tests described below perhaps have most application to clinical practice.[54]

Serum Bile Acid Estimation

Serum bile acids can be measured enzymatically, by gas-liquid chromatography or by radioimmunoassay. Elevation of two-hour postprandial serum bile acid concentration is a sensitive index of hepatobiliary disease.[70] The measurement of the clearance of ^{14}C-labelled bile acids from the serum has not proved helpful as a method for detecting mild liver disease.[89] Alteration of the ratio of cholic acid to chenodeoxycholic acid serum has been used to distinguish between biliary obstruction and parenchymal liver disease.[118]

Faecal Bile Acids

Bile acid malabsorption can be detected by measuring bile acid excretion in the faeces. Chemical methods have been developed but are not widely available.[51] Measurement of faecal radioactivity after ingestion of labelled bile acid is a practicable method of diagnosing bile acid malabsorption and may be used in conjugation with the ^{14}C-glycocholate breath test.[134]

^{14}C-Glycocholate Breath Test

This test detects bacterial deconjugation of bile acids occurring in the small intestine or the colon.[43,66] Bacterial action releases ^{14}C-glycine which is then metabolized, giving off $^{14}CO_2$ which is detected in the breath. In the standard method of carrying out this test, 5 μCi of ^{14}C-glycocholate is ingested together with a Lundh Test Meal (see page 74). Breath samples are then taken every hour for four hours. $^{14}CO_2$ is semiquantitatively estimated by collecting exhaled air in Hyamine Hydroxide, which traps a known amount of CO_2. Scintillation fluid is added to the vial and the ^{14}C counted in a scintillation counter. Results are usually expressed as the percentage of the total radioactive dose per mmol CO_2 exhaled. Values below 0·14% dose/mmol CO_2 at any time within four hours of ingesting the meal are normal. Anything above this level indicates excessive bacterial deconjugation of bile acids. This can either occur as a result of bacterial overgrowth in the small bowel, or because of malabsorption of bile acids in the ileum and deconjugation by colonic bacteria. The test is therefore positive in bile acid malabsorption due to resection or disease of the ileum, rapid small intestinal transit, as well as in small intestinal bacterial overgrowth. Concomitant measurement of stool radioactivity may help to identify those patients with bile acid malabsorption.[134]

Detection of Deconjugated Bile Acids in Jejunal Aspirate

See page 86.

THE SMALL INTESTINE

The small intestine is a long tube (200 to 300 cm) modified for the terminal digestion and absorption of food material. Approximately 90% of a meal is absorbed in the small intestine, even though food material remains in contact with the epithelium for only about five hours after ingestion. This prodigious absorptive capacity is made possible by the increase in surface area produced by myriad microscopic villi upon which are situated the absorptive cells or enterocytes, and by the microvilli on the surface of each enterocyte (Fig. 8). Many of the enzymes responsible for the terminal digestion of carbohydrate and protein, as well as the special carrier proteins for the absorption of hexose sugars, amino acids, vitamins, minerals and electrolytes, are situated on the surface of the microvilli, while a core of contractile filaments within the microvilli may aid absorption by pumping absorbed material into the cytosol. The enterocytes are joined together at their luminal poles by dense areas called tight junctions. Although these areas may appear anatomically impermeable, together with the lateral intercellular spaces they comprise the channels for the majority of water and electrolyte movement across the small intestinal epithelium. The depressions between the villi are known as crypts and these are lined by immature enterocytes. The division of the small intestine into jejunum and ileum is made on a functional rather than an anatomical basis. The ileum differs from the jejunum in that it secretes bicarbonate and has special active mechanisms for absorbing vitamin B_{12} and bile acids.

Absorption of solutes across the intestinal wall can take place by several mechanisms.

(i) diffusion can occur via aqueous pores in the

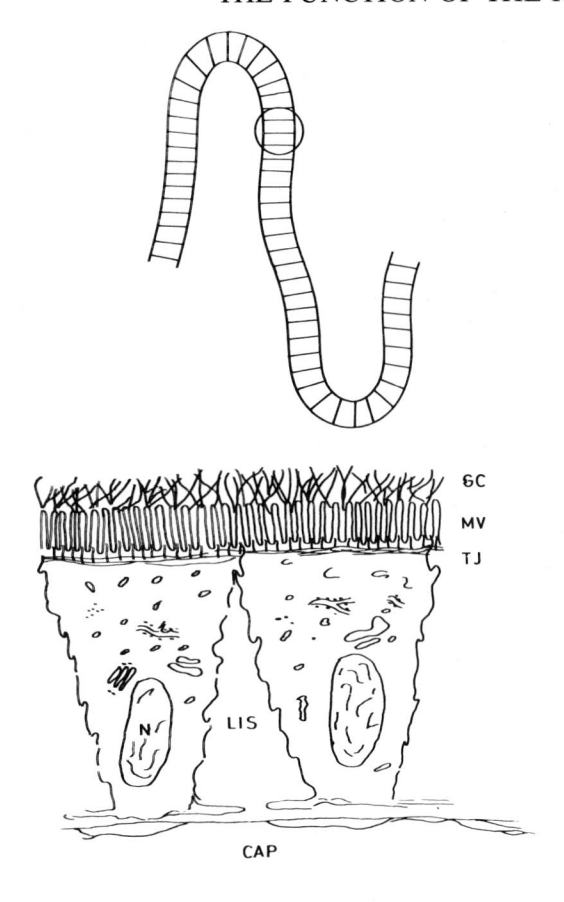

FIG. 8. The small intestinal epithelium. Above: a diagram of a villus and crypt. Below: a diagram of two villous enterocytes showing the glycocalyx (GC), microvilli (MV), tight junctions (TJ) between the cells, lateral intercellular spaces (LIS) and capillary (CAP).

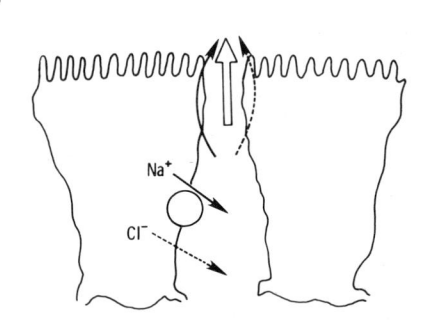

FIG. 9. Possible mechanisms of fluid absorption (above) and secretion (below) across the small intestinal epithelium. Active extrusion of sodium at the basolateral membrane of the enterocyte, followed by passive movement of chloride down the electrical gradient established by sodium extrusion creates a hypertonic zone in the lateral space. When the epithelium is absorbing (above) the tight junction is cation selective and rapid leakage of sodium into the lumen is prevented by the adverse electrical gradient set up as soon as sodium starts to move. Instead, the hypertonic zone encourages the osmotic movement of water into the lateral space via the tight junctions. This expands the lateral space and increases the hydrostatic pressure, which causes water to flow across the path of least resistance towards the capillary. Intestinal secretogogues (below) enhance the junctional permeability to chloride. This removes the electrical brake on sodium movement and allows NaCl to leak passively from the lateral spaces into the lumen, accompanied by water.

membrane if the solute is water soluble, or by partition into the lipid membrane if it is fat soluble.

(ii) low molecular weight solutes can accompany the osmotically induced influx of water by the process of solvent drag.

(iii) the solute may combine with a membrane-bound carrier to form a lipid-soluble complex which can then diffuse through the membrane. This process is known as facilitated diffusion.

(iv) active transport also involves combination with a membrane-bound carrier but in this case the solute can travel against a concentration gradient utilising a source of energy.

TRANSPORT OF FLUID AND ELECTROLYTES

Absorption of Water

Water is absorbed in response to the osmotic forces set up by the active transcellular absorption of sodium, hexose sugars and amino acids. According to the most popular hypothesis,[31,129] the active extrusion of sodium and the transport of solutes from the cell into the lateral intercellular space produces a zone of hypertonicity, which encourages the flow of water from the lumen through the tight junctions between the cells. This then expands the lateral spaces and water passes along the path of least resistance towards the capillaries (Fig. 9).

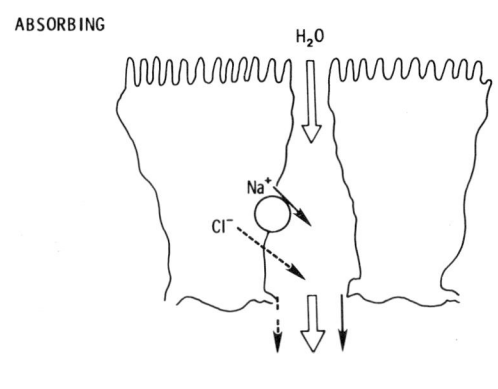

The efficiency of this mechanism depends on the patency of the tight junctions between the cells. In the jejunum where tight junctions are widely patent, rapid absorption of water occurs, whereas in the ileum, where tight junctions are less permeable, water absorption is slower.

In recent years, Lundgren and his colleagues[90] have suggested that the zone of hypertonicity responsible for fluid absorption in the small intestine is situated in the lamina propria at the villous tip. According to this hypothesis, the hypertonic zone is maintained by the operation of a counter-current multiplier, formed by the hairpin vascular loop configuration of the central artery and the subepithelial capillary network (Fig. 10) and operating in a manner analogous to the loop of Henlé in the kidney.

sodium from capillary to artery, while water travels in the reverse direction from artery to capillary (Fig. 10). The effect is multiplied towards the villous tip so that very large osmotic gradients are built up.

Absorption of Electrolytes

As electrolytes carry either a positive or negative charge, their passive movement across the intestinal wall depends on electrical as well as concentration gradients. Furthermore, the active transport of an electrolyte will result in a change in the transepithelial potential difference (electrogenic transport) unless its transport is coupled with the movement of an ion of opposite charge in the same direction or an ion of the same charge in the opposite direction (neutral transport).

Electrolytes can cross the epithelium either by going across the cells (the transcellular route) or going between them (the paracellular route). The relative importance of these two routes depends on the permeability of the tight junctions between the cells. In the jejunum, where the junctions are leaky, most electrolyte absorption accompanies the osmotic influx of water by the process of solvent drag.[40] Transcellular absorption of sodium provides the major osmotic force for water and further electrolyte absorption. This involves carrier-mediated entry into the cell down its electrochemical gradient and active extrusion by a sodium pump situated at the basolateral pole (Figs. 9 and 11). The entry step is coupled to the entry of hexose sugars, amino acids, or peptides (electrogenic) or the entry of chloride (neutral) or exit of hydrogen (neutral).[147] The extrusion mechanism is thought to obtain its energy from ATP by means of a sodium and potassium dependent ATPase.[17]

In the ileum, where the junctions between the cells are tighter, solvent drag is a less efficient mechanism for electrolyte absorption. Most sodium and chloride absorption occurs via a neutral transcellular double-

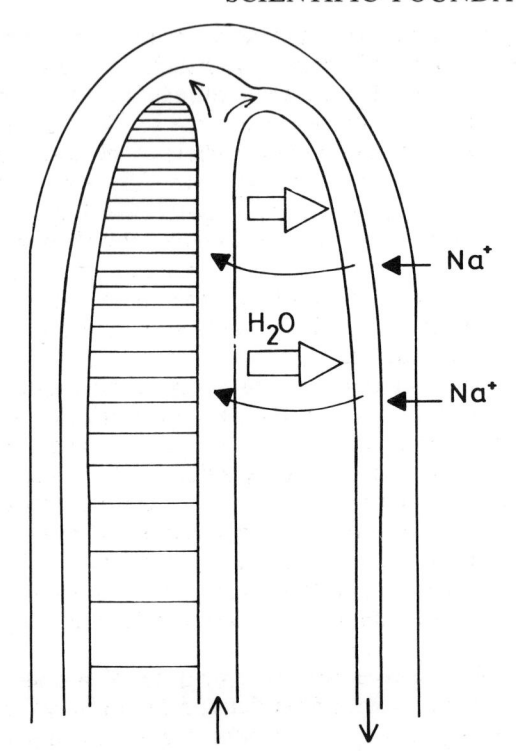

FIG. 10. A diagram of the villous counter current mechanism. The villus is supplied by a central artery which divides at the villous tip into arterioles and capillaries. These run down the outside of the villus adjacent to the epithelium. The active absorption of sodium across the enterocytes increases the sodium concentration in the capillaries. Sodium then diffuses across into the central artery while water passes in the opposite direction driven by osmotic forces. This leads to a high osmolality at the villous tip (indicated by the horizontal lines) which provides the osmotic gradient for water to be absorbed across the epithelium.

Active transepithelial absorption of sodium establishes a sodium gradient between the peripheral capillary and central artery. This results in a cross diffusion of

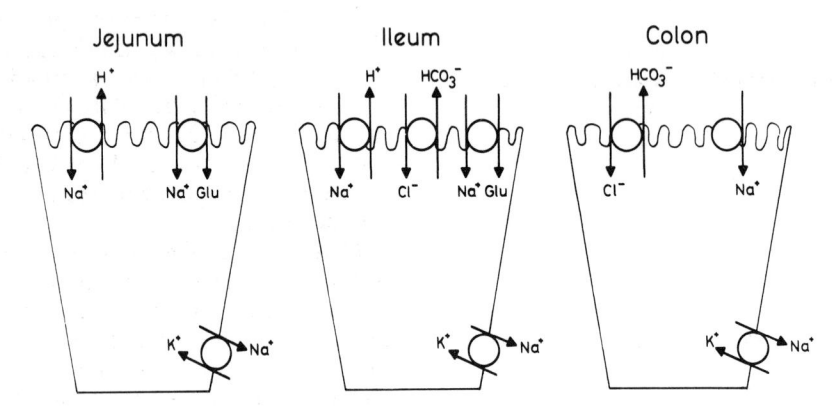

FIG. 11. A diagram of the carrier mediated ion transport processes across the jejunum, ileum and colon in man. All three sites contain a mechanism for the electrogenic entry of sodium into the cell. In the jejunum and ileum this is coupled to the entry of hexose sugar (Glu) or amino acid. The jejunum and ileum also exhibit neutral entry of sodium in exchange for hydrogen ion. Finally, the ileum and colon contain a neutral mechanism for absorption of chloride in exchange for bicarbonate secretion. Coupled sodium and chloride entry, which is so well documented in experiments carried out *in vitro* has not been substantiated *in vivo*. The exit of sodium at the basolateral pole of the enterocyte is mediated at all three sites by an active sodium potassium exchange pump, which is thought to pump 50% more sodium out of the cell than potassium in.

exchange process;[146] sodium is absorbed in exchange for hydrogen ion and chloride is absorbed in exchange for bicarbonate (Fig. 11). The hydrogen and bicarbonate for this exchange are derived from the hydration of CO_2 within the cell. The electrogenic transport of sodium induced by glucose in the ileum does not result in net transport of sodium across the epithelium because the electrical gradient drives the sodium back along the paracellular pathway into the lumen.[39]

Secretion of Water and Electrolytes

Secretion of water and electrolytes can also take place in the small intestine. Indeed, the direction and rate of transport of salt and water across the small intestine is thought to be the net result of those factors which cause absorption and those factors which induce secretion. Secretion probably takes place in the crypts while absorption occurs on the villi. Until recently it was thought that secretion was initiated by the active electrogenic pumping of chloride across the mucosal membrane of the enterocyte. However, there is now evidence to indicate that the major event initiating the secretion is an increase in the permeability to chloride, of the tight junctions (Fig. 9). This allows sodium and chloride ions to leak from high concentrations within the lateral space coupled with the passage of water.[62] Intestinal secretion may be stimulated by a large variety of substances (Table 1). The action of some of these secretogogues is mediated by cyclic AMP.[38] Others may act by increasing calcium entry into the enterocyte.[10]

TABLE 1
AGENTS WHICH INDUCE INTESTINAL SECRETION

Associated with activation of the adenyl cyclase system
  Enterotoxins (cholera toxin, heat labile toxin of *E. Coli*)
  Methylxanthines
  Prostaglandins
  VIP
  Dihydroxy bile acids
  Hydroxy fatty acids

Not proven to be associated with activation of the adenyl cyclase system
  Calcitonin
  Serotonin
  Histamine
  Glucagon
  GIP
  Secretin
  CCK
  Substance P
  Cholinergic agents
  Laxatives
  Bacterial enterotoxins (*Shigella*, *E. coli* (heat stable), *Pseudomonas*, *Klebsiella*, *Staphylococcus*)

DIGESTION AND ABSORPTION OF CARBOHYDRATE

Starch, sucrose and lactose are the major dietary carbohydrates. Dietary starch exists in two forms, amylose, consisting of straight chains of glucose molecules joined by 1,4 linkages and amylopectin, consisting of branched chains of glucose molecules joined at the branch points by 1,6 linkages. Starch is broken down, predominantly by pancreatic amylase to maltose, maltotriose and α-limit dextrins (short chains of glucose molecules containing the 1,6 branch points). These substances are further broken down by oligosaccharidases within the brush border to glucose which is absorbed by the enterocyte. Sucrose and lactose do not undergo pancreatic digestion, but are broken down by brush border disaccharidases to yield the monosaccharides, glucose, galactose and fructose.

The major mechanism for glucose and galactose absorption in the small intestine is carrier-mediated active transport. The active entry of the sugar at the mucosal pole of the enterocyte is thought to be linked to the entry of sodium on a common carrier so that the energy for sugar entry is provided by the inwardly directed electrochemical gradient for sodium.[23,135] The exit of glucose and galactose from the cell is thought to occur by carrier-mediated diffusion not requiring the presence of sodium.[156]

There is evidence for the existence of a transport system for monosaccharides derived from disaccharides which is separate and distinct from the transport mechanism for free monosaccharide. This disaccharidase-related transport system is not dependent on sodium and can continue to transport glucose even when the transport system for free glucose is completely saturated.[121] This mechanism may be responsible for the observation that absorption of glucose is more rapid when presented to the epithelium as maltose rather than the equivalent amount of glucose.[20,98]

Fructose appears to be absorbed by carrier-mediated facilitated diffusion, which is dependent on the presence of sodium.[46,61]

PROTEIN DIGESTION AND ABSORPTION

70 to 100 g of protein are present daily in the average diet of Western man. Endogenous protein, in the form of mucus, biliary and pancreatic secretions, and desquamated cells, may total two or three times the dietary intake. Protein digestion is initiated by gastric pepsin, but this mechanism is relatively unimportant, as patients with pernicious anaemia are able to digest dietary protein adequately.

Most protein digestion occurs in the duodenum and jejunum by pancreatic proteases. The products of digestion are small peptides and amino-acids, the former probably predominating. Some peptides are broken down by oligopeptidases within the brush border to amino acids which are then absorbed by a sodium-coupled active-transport process similar to that described for glucose and galactose. Other peptides are absorbed intact by an active mechanism also coupled to sodium entry and later hydrolysed by peptidases within the cytoplasm.[94] The separate mechanisms for peptide and amino acid absorption confer a nutritional advantage as nitrogen absorption is faster from a protein hydrolysate containing many small peptides than from the equivalent amino acid mixture.[136]

DIGESTION AND ABSORPTION OF FAT

Long-chain Triglyceride

Dietary fat exists mainly in the form of water-insoluble triglyceride. This is emulsified by the motor activity of the stomach and duodenum and the detergent action of bile acids. It is then hydrolysed to fatty acids and 2-monoglycerides by pancreatic lipase, which acts at the oil-water interface of the emulsion particle. Bile acids can inhibit the activity of lipase by displacing it from the oil-water interface. This is prevented by co-lipase[30] which is also secreted by the pancreas. Co-lipase also lowers the pH optimum of lipase (8·5) to the intraluminal pH of the proximal intestine (6·5). Fatty acids and 2-monoglycerides are poorly soluble in water, but are brought into solution by combining with bile acids in micelles. The micelle is important for the absorption of fat because it presents the fat in a water-soluble form, which can readily diffuse to the cell surface through the unstirred layer of water that lies adjacent to the epithelium. In the absence of bile acids, the rate-limiting step for absorption of long-chain fatty acids is diffusion through this unstirred layer, the resistance being greater for those fatty acids which are most rapidly taken up by the enterocyte.[32]

At the brush border the micelle appears to disaggregate, the bile acids pass back into the lumen to form further micelles and the fatty acids and monoglycerides diffuse across the lipid membrane of the enterocyte. Inside the cell fatty acids combine with a specific binding protein.[115] This combination maintains the concentration gradient for the diffusion of more fatty acid into the cell and is also necessary for adequate re-esterification of fatty acids to triglyceride in the endoplasmic reticulum. The triglyceride so formed then combines with lecithin and a specific protein to form chylomicrons and very low density lipoproteins (VLDL).[126] Chylomicrons and VLDL leave the enterocyte by exocytosis and enter the villous lymphatics. In the rare congenital defect abetalipoproteinaemia, there is an inability to synthesize apoprotein B which is necessary for the formation of chylomicrons and VLDL. Triglyceride cannot be absorbed and accumulates in the cell as lipid droplets.

Medium-chain Triglyceride (MCT)

Medium-chain triglycerides contain fatty acids with a chain length of 6 to 12 carbon atoms. Although these contribute little to normal fat intake, they are therapeutically important in patients who cannot digest or absorb the long-chain triglyceride normally present in the diet.[48] MCT is hydrolysed and absorbed more rapidly than long-chain fatty acids even when the concentrations of bile acids and lipase are reduced. At least one-third of MCT can be absorbed intact in the absence of lipase and is then hydrolysed by enzymes within the enterocyte to yield medium-chain fatty acids (MCFA). MCFA are not esterified to triglyceride or packaged into chylomicrons and are largely transported away from the cell in the portal blood.

Thus MCT can be an important source of dietary lipid under conditions of pancreatic insufficiency, biliary obstruction, abetalipoproteinaemia and lymphatic obstruction.

Cholesterol

Cholesterol in the intestinal lumen derives from bile, where it is present in bile acid-lecithin-cholesterol micelles, and from the diet. Cholesterol absorption is incomplete, only 20 to 40% of ingested cholesterol being absorbed in man, mostly in the jejunum.[14] Some hydrolysis of cholesterol occurs in the bowel lumen due to the action of cholesterolesterase, secreted by the pancreas. Incorporation into micelles is important for transport to the cell surface. Thereafter cholesterol diffuses through the cell membrane, though some may combine with a carrier and be actively transported into the cell.

Fat-soluble Vitamins

Absorption of the fat-soluble vitamins A, D, E and K, is bile acid dependent and occurs mainly in the jejunum. Hydrolysis of Vitamin A and possibly other vitamins by pancreatic enzymes occurs before absorption. The mechanism for absorption is probably non-ionic diffusion, but there is evidence that Vitamin K may be absorbed at least in part by a carrier-mediated active mechanism.

ABSORPTION OF CALCIUM, MAGNESIUM, AND PHOSPHATE

Most calcium is absorbed in the jejunum either by passive diffusion along the paracellular route or by an active transcellular mechanism.[109] The latter is the most important route and involves passive carrier-mediated entry at the brush border and active extrusion at the basolateral pole. The entry step is thought to be regulated by a specific calcium-binding protein, the production of which is stimulated by the active form of vitamin D, 1,25-dihydroxycholecalciferol (1,25 OH-D_3). Extrusion of calcium at the basolateral pole is probably mediated by a sodium-calcium exchange pump or a specific calcium ATPase. Active calcium absorption is increased during growth, pregnancy and after a low calcium diet. It is also enhanced by a low intraluminal pH and by certain sugars and amino acids. These solutes may act by increasing solubility, but a recent study suggests that glucose enhances calcium absorption because the glucose-induced water flow concentrates calcium near the cell surface.[112] Fats, oxalates and phytates form insoluble complexes with calcium in the gut lumen and inhibit its absorption.

Magnesium can be absorbed by an active process stimulated by 1,25 OH-D_3 and probably utilises a similar or identical binding protein.

Phosphate is absorbed both by sodium-dependent and by independent routes. The sodium-dependent mechanism may involve interaction with sodium on a common carrier and is stimulated by 1,25 OH-D_3.

IRON ABSORPTION

The availability of iron for absorption is also influenced by intraluminal factors. Gastric acid and pepsin release the inorganic iron in food as soluble ferrous salts. The bicarbonate and perhaps other components of pancreatic juice inhibit iron absorption by forming insoluble precipitates. This, however, may be prevented by bile, which tends to keep iron in solution in the more alkaline pH of the small intestine by forming ligands of low molecular weight. Phytates, oxalates, carbonates and phosphates also bind to iron in the upper small intestine and render it insoluble, and unavailable for absorption. On the other hand, ascorbic acid, some amino acids and some sugars increase its solubility.

Haemoglobin iron is not susceptible to the same intraluminal constraints as other forms of dietary iron. It is broken down by pancreatic enzymes to haem which is absorbed intact releasing the free iron inside the enterocyte.[154]

Iron is absorbed mainly in the duodenum by a specific cellular mechanism, which is very sensitive to body requirements, being increased in iron deficiency states and reduced in iron overload. Uptake into the cell is an active carrier-mediated process and is increased in iron deficiency.[21] The rate of uptake may be determined by the number of mucosal iron receptors at the time of enterocyte formation or maturation in the crypts. Inside the cell the iron may either enter a soluble transport pool or be incorporated into soluble ferritin. Iron not transferred to plasma is retained by the cell, mostly as ferritin, until the cell is sloughed off at the tip of the villus. Transfer across the serosal cell border may also involve receptors, the iron being attached to transferrin for transport in the plasma.

The mechanisms whereby iron stores and the rate of erythropoiesis influence iron absorption are complex and ill-understood.[87] Serosal transfer of iron appears to be the rate-limiting step for iron absorption, although control may also be exerted at other stages. The mediators of these regulatory mechanisms are unknown.

WATER-SOLUBLE VITAMINS

Folate

Folates are present in animal and plant foodstuffs, the daily requirement being approximately 100 μg. Body stores are usually sufficient for only three months' needs. Folates consist of a pteroyl moiety with a variable number of glutamic acid residues. Dietary folates are predominantly polyglutamates, and are probably hydrolysed by brush border enzymes to mono- and diglutamates.[124] The monoglutamate can be as absorbed by a saturatable mechanism in the jejunum and recent evidence suggests that an acid microclimate adjacent to the epithelium is necessary for optimal absorption of the vitamin.[81] The sequence of events within the enterocyte is unclear, but reduction to dihydro- or tetrahydro-folate and methylation occurs. The major biological active folate is 5-methyltetrahydrofolate.

Vitamin B_{12}

Vitamin B_{12} is found only in animal foodstuffs, the daily requirements are approximately 1 to 2 μg and body stores are usually sufficient for several years. Absorption of vitamin B_{12} in man is a complicated procedure requiring the binding to a specific mucoprotein, intrinsic factor (IF) secreted by the gastric parietal cell.

IF-B_{12} complex then travels down the intestine and binds to the enterocytes of the ileum where it is absorbed by a saturatable mechanism. Although the intrinsic factor dependent mechanism accounts for the majority of B_{12} absorption, a small proportion can be absorbed by a process independent of intrinsic factor that resembles facilitated diffusion.

Recent studies have indicated that B_{12}-IF binding occurs predominantly in the upper small intestine.[3] In the stomach B_{12} forms complexes with salivary R-protein in preference to intrinsic factor. Pancreatic proteases are necessary to digest R-protein, leaving the B_{12} free to bind with intrinsic factor. Inability to remove R-protein from B_{12} seems to be the explanation for the reduced B_{12} absorption occuring in pancreatic insufficiency.

Other Water-Soluble Vitamins

Recent evidence suggests that sodium-dependent carrier-mediated transport processes exist for many of the water-soluble vitamins.[127] Absorption is maximal in the jejunum for all those vitamins studied, except ascorbic acid, which is mainly absorbed in the ileum.

INVESTIGATION OF MALABSORPTION

The diagnosis of malabsorption is often suspected in a patient who presents with weight loss, diarrhoea and anaemia which cannot be explained by any other condition. The investigation of such a patient generally proceeds in two stages. Firstly, a range of haematological and biochemical tests is performed to detect nutritional deficiencies (Table 2). If several of these investigations are abnormal, the next step is to carry out a jejunal biopsy. This is the only certain way to diagnose coeliac disease and some of the more obscure conditions causing malabsorption such as Whipple's disease, intestinal lymphagiectasia and abetalipoproteinaemia. If the jejunal biopsy is normal, then other causes of impaired nutrition should be considered, such as an inadequate diet, bacterial overgrowth of the small intestine and an abnormally rapid intestinal transit.

Isolated abnormalities on the nutritional screen may suggest other diagnoses. For example, severe steator-

TABLE 2
LIST OF BLOOD TESTS WHICH MAY BE CARRIED OUT AS A SCREEN TO DETECT MALABSORPTION OR MALNUTRITION

Haemoglobin
Serum and red cell folate
Vitamin B_{12}
Iron, total iron-binding capacity
Albumin
Calcium
Phosphate
Alkaline phosphatase
Prothrombin time
Cholesterol
Leucocyte ascorbic acid
Zinc
Carotene

rhoea may indicate pancreatic insufficiency, low B_{12} may indicate ileal disease or gastric atrophy.

The following section indicates some specific laboratory tests of small intestinal function which may be useful in establishing a definitive diagnosis in patients with suspected malabsorption.

Assessment of Carbohydrate Absorption

Generalized carbohydrate malabsorption occurs in severe mucosal disease, after extensive gut resection, and in association with excessively rapid small intestinal transit. As carbohydrate reaching the colon is partly metabolised by bacteria, faecal measurements are of little value in the assessment of carbohydrate malabsorption.

Malabsorption of specific sugars can occur, most commonly due to deficiency of brush border disaccharidases, particularly lactase. The only specific defect in sugar transport is congenital glucose-galactose malabsorption which occurs in infancy and is thought to result from lack or impairment of the hexose carrier.

Absorption of individual sugars is best studied by intestinal perfusion techniques, but these are not generally available and in any case the wide normal ranges limit their value as clinical tests. The laboratory tests in common use are those related to the detection of disaccharidase deficiency and the D-xylose absorption test. Breath hydrogen tests are also being increasingly applied to the assessment of carbohydrate malabsorption.

Detection of Lactase Deficiency

Lactase Assay. The definitive test for the diagnosis of lactase deficiency is assay of enzyme activity in jejunal mucosa obtained by peroral biopsy.[25] The glucose produced after incubation of the tissue with lactose is measured. Other rarer disaccharidase deficiencies (sucrase-isomaltase and trehalase deficiency) may also be diagnosed by this method.

Lactose Tolerance Test. The fasting subject ingests 50 or 100 g of lactose in water, and blood samples are taken at 15, 30, 60 and 90 minutes after ingestion. A rise in blood glucose of less than 20 mg% (1·1 mmol/l) indicates lactase deficiency. The occurrence of abdominal pain, borborygmi and diarrhoea supports the diagnosis. Similar results may be obtained in patients with deficit in mucosal transport of monosaccharides. Thus abnormal results should be followed by a control test in which the equivalent amounts of glucose and galactose are ingested in order to ensure that the patient has normal absorption of the constituent monosaccharides.

Breath Hydrogen Test. Carbohydrate which is not absorbed in the upper small intestine is fermented by colonic bacteria giving off hydrogen gas. This diffuses rapidly through body tissues and can be estimated in the breath using sensitive detectors. Thus, an increase in breath hydrogen after ingestion of a standard dose (50 g) of lactose is inversely proportional to lactase activity as hydrogen will only be produced if unabsorbed lactose reaches the bacterial pool in the colon.[12,101] False positive results may occur in bacterial overgrowth of the small intestine and in the presence of excessively rapid transit in the small intestine. False negative results can occur if the patient does not possess the appropriate bacterial flora for producing hydrogen gas on exposure to lactose. This may occur after a course of antibiotics.

The breath hydrogen test can also be performed after ingestion of sucrose to detect the rare deficiency of sucrase.[102]

^{14}C-lactose Breath Test. The principle underlying this test differs from the breath hydrogen test in that the test gas, $^{14}CO_2$ is detected if the carbohydrate is absorbed in the upper small intestine and a portion rapidly metabolized. Thus, in theory, excretion of $^{14}CO_2$ in the breath after ingestion of ^{14}C-lactose is directly proportional to lactase activity.

This test is of limited value in practice because $^{14}CO_2$ is also given off when unabsorbed lactose reaches the colon and is metabolized by colonic bacteria. However, reasonable discrimination between normal subjects and patients with lactase deficiency is obtained by measuring $^{14}CO_2$ in the first hour after ingestion.[110]

Choice of Test in Lactase Deficiency. Assay of enzyme activity in jejunal biopsy specimens is the definite diagnostic test but requires intestinal intubation and biopsy. As a screening test, breath hydrogen analysis is the simplest test to perform and is well tolerated by the patient. It is more sensitive than the ^{14}C-lactose breath test and it does not require administration of radioisotopes. However, most units are not equipped to measure breath hydrogen, and in these the lactose tolerance test remains a perfectly adequate and simple method to diagnose lactase deficiency. All of the screening tests may be abnormal in patients with coeliac disease or other conditions causing impairment of monosaccharide transport.

D-xylose Excretion Test

D-xylose excretion is a test of mucosal uptake and transport of sugars by the jejunum. D-xylose is a pentose sugar not normally found in the diet. It is absorbed, probably actively, in the jejunum, and approximately half of the absorbed sugar is excreted unchanged in the urine.

25 g of D-xylose is given in 500 ml of water in the fasting state. An additional 250 ml of water is drunk after 90 minutes. Urine is collected for five hours and complete urine collection is essential. Xylose excretion of more than 5 g is taken as normal. A blood sample taken at 90 minutes can increase the usefulness of the test in the presence of renal impairment, a value greater than 33 mg% (2·2 mol/l) being normal. The test may be performed using a 5 g dose of D-xylose, especially in children, as the larger dose may cause nausea and diarrhoea.

The D-xylose excretion test has been used mainly as a screening test for coeliac disease. However, diminished xylose excretion is seen in other conditions which cause malabsorption, such as bacterial overgrowth of the small bowel, rapid intestinal transit, and administration of neomycin, as well as in renal impairment or dehydration. Unless both blood and urine xylose concentrations are measured, and strict criteria for abnormality applied, the discriminatory power of the test is poor.[138] The ready availability of jejunal biopsy has greatly reduced the need to employ this test.

Glucose Tolerance Test

The oral glucose tolerance test is of little value as an indicator of carbohydrate malabsorption, a 'flat' curve being seen in many healthy young adults. A diabetic curve may point to pancreatic disease.

Assessment of Protein Absorption and Protein Loss from the Gut

No satisfactory test of protein absorption or protein loss is available. Increased faecal nitrogen excretion, like faecal fat, is often present in malabsorptive states, but the origin of the faecal nitrogen is a subject of controversy. Much of it may be endogenous rather than dietary in origin. The low plasma albumin concentration commonly seen in malabsorptive disease may be the product of increased loss from the gut, inadequate intake, defective absorption, or reduced hepatic albumin synthesis.

Enhanced protein loss into the bowel lumen has been described in most inflammatory diseases of the gut, extensive carcinoma or ulceration, and in coeliac disease. It may be the primary manifestation in diseases such as intestinal lymphangiectasia or Ménètrier's disease of the stomach.[150]

Gastrointestinal protein loss is difficult to measure as protein lost into the upper intestine may be digested and reabsorbed lower down. A qualitative assessment can be made by estimating the radioactivity in faecal samples and the decrease in plasma radioactivity after the intravenous injection of ^{51}chromium chloride-labelled albumin, ^{67}Cu-caeruloplasmin, or ^{131}I-polyvinyl-pyrrolidine (PVP). Chromium chloride-labelled albumin is perhaps the best method as ^{67}CU has a short half-life, and ^{131}I may become split from PVP and reabsorbed.[151] Normal subjects clear only about 1% of their plasma pool each day, whereas in patients with protein-losing enteropathy, the clearance may be 30% or more.[150]

Assessment of Fat Absorption

Faecal Fat Excretion

Fat is normally present in faeces, up to 6g per day being accepted as normal. Not all of this is of dietary origin as 1 to 3 g per day is excreted even on a fat-free diet. Although steatorrhoea is clinically associated with pale, bulky offensive stools, which float in the pan, stool appearances may be normal in the presence of mild or moderate fat malabsorption. Moreover, floating of stools depends on their gas content rather than their fat content.[83] Increased faecal fat excretion, in the absence of ingestion of unabsorbable fat, is unequivocal evidence of a defect in digestion or absorption. It is found in the majority of patients with generalized malabsorption, and also in patients with pancreatic insufficiency, biliary obstruction and terminal ileal disease or resection. Gross steatorrhoea (35 g/day or more) is rarely found in malabsorption and strongly suggests pancreatic insufficiency.

Since the intake and output of fat varies appreciably from day to day, collections over several days are recommended[88] while the patient is ingesting a diet containing 100 g fat. Most reliable estimates of fat excretion are obtained using faecal markers. A constant amount of an inert substance is given daily in divided doses. After an equilibration period of several days, stool collections are made. The amount of marker appearing in the stool indicates the number of days or fractions of a day to which the faecal fat in the collection should be referred. Chromic oxide has been most widely used as a marker, but barium sulphate, cuprous thiocyanate, polyethylene glycol or radio-opaque plastic markers may also be suitable.

The most widely used method of chemical estimation of faecal fat is that of Van de Kamer et al.[148] In this method the fatty acid content of the faeces is measured by titration and stool fat expressed as grammes of fatty acid. Various modifications exist to measure medium-chain triglyceride which the original method excludes. Alternatively, total lipid content of faeces can be estimated by extraction with lipid solvent and direct weighing of the extracted lipid. This gives somewhat higher values.

Indirect Tests of Fat Absorption

As faecal fat estimation is an unpleasant procedure, attempts have been made to develop methods to measure fat absorption that do not require chemical analysis of large amounts of stool. Microscopic analysis of stool smears for fat globules, blood vitamin A estimation, vitamin A tolerance tests and fat tolerance tests are abnormal in severe steatorrhoea but less useful in mild fat malabsorption.

Measurement of the faecal excretion of fat labelled with a radio-isotope has been extensively studied, but has proved unsatisfactory when compared with chemical faecal fat determination. More promising are tests based on the measurement of $^{14}CO_2$ in expired air after the ingestion of ^{14}C-labelled triglycerides. ^{14}C-triolein appears to be most suitable and results correlate well with chemical fat measurements.[111]

The association of hyperoxaluria with steatorrhoea[33] has recently been used as the basis of a screening test for fat malabsorption,[120] and awaits further evaluation.

Assessment of Vitamin B_{12} Absorption

Because body stores of this vitamin are substantial, dietary deficiency is uncommon. Malabsorption of vitamin B_{12} occurs in gastric mucosal atrophy, ileal disease, (Crohn's disease and severe coeliac disease), ileal resection, bacterial overgrowth in the small intestine and pancreatic insufficiency. Severe folate deficiency and ingestion of drugs such as neomycin, p-aminobenzoic acid and alcohol, may also interfere with absorption of vitamin B_{12}. Malabsorption may be present for some time before a fall in serum levels of vitamin B_{12} occurs, and absorption tests may therefore be of value in suspected malabsorption even if serum levels are normal.

Schilling Test

This test relies on determination of radioactivity appearing in urine after oral administration of 0·5 to 2·0 μg of radioactive cobalt-labelled vitamin B_{12}.

The oral dose of labelled vitamin B_{12} is given, together with an intramuscular 'flushing' dose of 1 mg unlabelled B_{12}. The latter saturates tissue and plasma binding sites so that the absorbed labelled vitamin is excreted in the urine. Total urine output is collected for 24 hours. Normal subjects excrete more than 10% of the labelled vitamin in the urine over 24 hours.

Nowadays Schilling tests are often carried out using vitamin B_{12} labelled with two different isotopes of cobalt, one free and one bound to intrinsic factor. This allows distinction between B_{12} malabsorption due to gastric atrophy and that caused by ileal disease or resection.

In patients suspected of having reduced B_{12} absorption due to bacterial contamination of the small intestine, repeating the Schilling test after administration of antibiotics may be a useful diagnostic procedure. Similarly, an abnormal Schilling test which returns to normal after administration of pancreatic enzymes may indicate pancreatic insufficiency.

Finally, it should be emphasized that urinary excretion of B_{12} is severely impaired in patients with dehydration or renal disease.

Urinary Methylmalonic Acid Excretion

Vitamin B_{12} is a co-enzyme for the conversion of methylmalonic acid to succinate, and deficiency leads to increased urinary excretion of methylmalonic acid. This test is now rarely used.

Assessment of Folate Absorption

Since body stores of folate are small, dietary deficiency is common. Low plasma levels are commonly seen in severe illness, but red cell levels are a better guide to true tissue depletion. Folate is required for the conversion of formiminoglutamic acid (FIGLU) to glutamic acid and ammonia and increased FIGLU excretion in the urine, particularly after histidine loading, has been used as a test of folate absorption, but has not found wide acceptance. Measurements of urinary excretion and increase in blood levels after the ingestion of unlabelled or tritiated synthetic folic acid (pteroylmonoglutamic acid) have also been of little value in clinical practice. One reason is that this test does not assess the intestinal handling of the polyglutamates which are the normal form of folate in the food.

Tests for the Detection of Bacterial Contamination of the Small Bowel

Direct Examination of Bacterial Flora

Culture of organism from the upper small intestine, which is normally almost sterile, is the most specific method, but the considerable difficulties in sampling, transport and culture techniques required for quantitation of anaerobic bacteria limit its widespread use.[16]

Intraluminal Deconjugation of Bile Acids

Bacterial deconjugation and dehydroxylation of bile acids in jejunal aspirates can be detected by thin-layer chromatography.[113] This is a useful test if the technique is available.

Urinary Indican Excretion

A simple urine test has obvious appeal, and urinary indican excretion is often elevated in bacterial overgrowth syndromes. Indican in the urine derives from bacterial metabolism of tryptophan. Increased indican excretion is not, however, specific for bacterial overgrowth, and is seen in other malabsorptive states.[47] In some studies it has also shown a poor correlation with jejunal bacterial counts.[96]

^{14}C-glycocholate Breath Test (see page 78)

This test is simple to perform and does not require intestinal intubation. Increased excretion of $^{14}CO_2$ in the breath after the ingestion of ^{14}C-glycocholate is seen in bacterial overgrowth because bacteria detach the bile acid from the glycine which is then absorbed and metabolized. However, positive results will also be found in patients with bile acid malabsorption due to ileal disease or resection, or excessively rapid small intestinal transit. Simultaneous measurement of ^{14}C excretion in stool may be of value in identifying patients with bile acid malabsorption.[134] The test may be normal in 30 to 40% of patients with bacterial overgrowth of the small intestine proven by culture of small intestinal juice,[78] probably because the bacterial species implicated do not deconjugate bile acids.

^{14}C-D-xylose Breath Test

Xylose is normally rapidly absorbed in the jejunum and excreted in the urine without being metabolized. Bacteria can, however, metabolize xylose to release CO_2 which can be detected in the breath. Thus excretion of $^{14}CO_2$ following ingestion of ^{14}C-D-xylose indicates bacterial overgrowth in the small intestine and is claimed to provide a more sensitive index than the bile acid breath test.[72]

Schilling Test

Abnormal vitamin B_{12} absorption is a common feature of bacterial overgrowth in the small intestine. Bacteria may detach B_{12} from intrinsic factor and compete with the enterocyte for absorption. Improvement in vitamin B_{12} absorption after antibiotic treatment is strongly indicative of bacterial overgrowth.

Breath Hydrogen Excretion

Hydrogen in expired air is derived solely from bacterial metabolism. An increase in breath hydrogen excretion following ingestion of 50 g glucose, which is rapidly and completely absorbed in the upper small intestine, strongly suggests bacterial overgrowth.[103]

Choice of Tests

There is at present no one test which is perfect for diagnosing bacterial overgrowth of the small intestine. Culture of small intestinal juice is the only method for proving the existence of bacterial overgrowth, but this is uncomfortable and presents considerable technical problems, both in obtaining samples and in the culture of the anaerobic bacteria present in the gut. The other tests assume that the bacterial flora in the small intestine may be able to metabolize the test substrate, but this may not always be the case. In addition, false results can occur in conditions of rapid small-intestinal transit or malabsorption where the test substance may reach the colon and be metabolized by the colonic pool of bacteria. In our opinion the glucose hydrogen test will probably provide the most reliable non-invasive method of detecting bacterial overgrowth because the likelihood of a small dose of glucose reaching the colon is extremely small, but until that is more widely available, the clinical response of an abnormal Schilling test or steatorrhoea to treatment with antibiotics must remain the mainstay of the clinical diagnosis of bacterial overgrowth.

SMALL INTESTINAL MOTILITY

Contractile activity in the small intestine breaks food up, mixes it with digestive juices and propels it gradually towards the colon in preparation for the next meal. The frequency and pattern of contractions are controlled by the myoelectrical slow wave.[18] This is a regular oscillation in the membrane potential of smooth muscle cells. Its frequency is about 12 per minute in the duodenum, but falls to about 8 per minute in the ileum, and for the most part it appears to be propagated aborally down the intestine. When the membrane potential is partially depolarized by various stimuli, the cyclic slow wave activity elicits bursts of spike potentials which initiate contractions of smooth muscle (Fig. 12). As only one

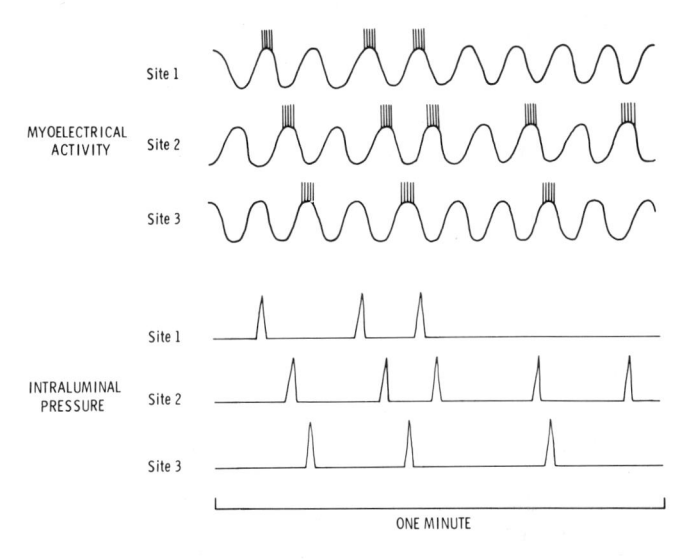

FIG. 12. A diagram of the myoelectrical activity and intraluminal pressure at three sites in the small intestine, separated by 10 cm. The myoelectrical activity is dominated by a regular oscillation of the membrane potential of smooth muscle occurring at a relatively fixed frequency (slow wave) and apparently propagated down the small intestine. Depolarisation of this membrane generates spike potentials which initiate a muscle contraction. As only one spike burst and contraction can occur per one slow wave the frequency and timing of intestinal contractions are set by the underlying slow wave. Note that although some spike bursts and contractions are propagated with the slow wave, this is not always the case.

spike burst or contraction can occur with each slow wave, the frequency of slow waves determines the maximum frequency of contractions at any one site in the intestine. The degree of depolarization of the smooth muscle determines the degree to which each slow wave initiates a contraction.

During fasting the motor activity of the stomach and small bowel is organized into a distinct phasic pattern

INTRALUMINAL PRESSURE

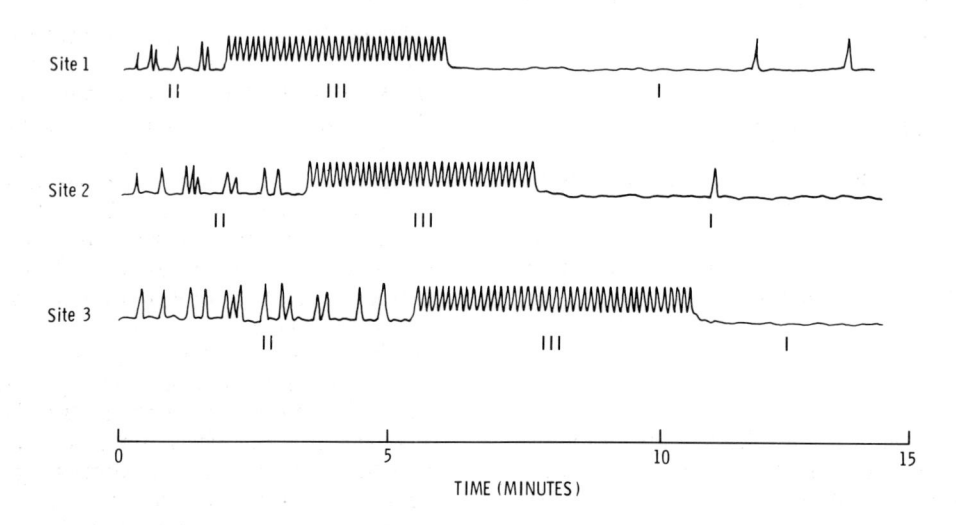

FIG. 13. A diagram of the fasting motor pattern in three sites in the upper small intestine in man, separated by 20 cm. The motor pattern occurs in a recurrent pattern consisting of three distinct phases. These are a period of relative quiescence when there are a few contractions (phase I) followed by a period of intermittent activity where contractions occur frequently but in an irregular manner, (phase II), and finally a period of regular contractile activity in which contractions occur at a regular frequency, determined by the underlying slow wave (phase III). The latter migrates down the small intestine at a slow rate and gives its name to the migrating complex.

called the migrating complex[143] (Fig. 13). This consists of a phase of quiescence during which there are few or no contractions, followed by a phase of intermittent contractile activity, which is followed in turn by a brief burst of regular contractile activity, where contractions occur with every slow wave. Thereafter the cycle begins again with the periods of quiescence. Each complete cycle lasts approximately one and a half hours in man and appears to migrate slowly (less than 10 cm per minute) through the intestine. It is associated with corresponding fluctuations in the rates of gastric, pancreatic and possibly also intestinal secretion, and probably represents the background rhythmic activity in a system controlling small intestinal function. The pattern is disrupted by food leading to a period of prolonged intermittent contractile activity, which lasts for several hours.

Although much work has been carried out in describing the patterns of motor activity, we still do not know how these patterns of contractile activity relate to transit of food through the small intestine. However, as each slow wave is propagated aborally, it seems likely that increased coupling between electrical and mechanical events will not only increase the frequency of contractions, but also the extent to which each contraction is propagated with the slow wave down the intestine. Thus increased contractile activity should lead to increased rates of small intestinal transit. This may be too simple a scheme, because monitoring the passage of food labelled with a radioisotope through the small intestine with a gamma camera has indicated that although food passes relatively quickly through the jejunum, it accumulates in the ileum before emptying into the colon.[68]

The rapid intestinal transit induced by laxative agents and certain enterotoxins can be explained in part by the presence of spike bursts which are propagated rapidly over long distances.[2,93]

Measurement of Small Intestinal Transit Time

For food to be adequately absorbed, it is necessary for it to remain in contact with the small intestinal epithelium for a minimum period of time. Abnormally rapid transit of food through the small intestine may well lead to impaired absorption of a number of dietary constituents, and diarrhoea. Until recently the only method for measuring the small intestinal transit time was to monitor the passage of barium through the small intestine radiologically. Nowadays transit time can be measured more simply by estimating the hydrogen exhaled in the breath following ingestion of a drink of lactulose.[11] As this disaccharide cannot normally be digested and absorbed in the small intestine, it passes quickly to the colon where it is rapidly fermented by bacteria to yield hydrogen. The hydrogen diffuses rapidly through body tissues and is exhaled in the breath where it can be measured by special detectors. The time from ingesting the lactulose to the first peak in hydrogen indicates the transit time of the head of the lactulose bolus through the small intestine. The method has several limitations. Firstly, the transit time of ingested lactulose gives no indication of the time that food takes to pass through the small intestine.[123] Second, the test may give spuriously short transit times in the presence of bacterial overgrowth in the small intestine, and third, the test relies on the presence of an appropriate bacterial flora in the colon. Despite these drawbacks, the lactulose test remains a useful technique for comparing small intestinal transit under different experimental conditions or in different disease states.

THE COLON

The colon can be regarded as a commodious stagnant reservoir containing food residues which have not been absorbed higher up and teeming with vast quantities of bacteria. The epithelial surface is flat, but is punctuated with numerous pits. Mucus secreting goblet cells are liberally interspersed between colonic enterocytes.

MOTILITY

In contrast to the small intestine, the colon is relatively inactive. The myoelectrical slow wave is not usually propagated longitudinally in the colon. Consequently most contractions occur as static rings and are segmenting rather than propagative. Thus, in contrast to small intestinal motility, increased colonic motor activity obstructs flow, slows colonic transit and leads to constipation, while decreased motor activity is associated with diarrhoea.[19] Occasionally the pattern is broken by propagated spike bursts and associated contractions which propel food residues towards the anus. Such propagated events have been observed in response to certain laxative agents.[52]

Colonic motor activity is markedly enhanced by eating.[139] Evidence implicates gastrin and the vagus nerve as leading contenders for the mediation of this gastrocolic reflex.

ABSORPTION OF NUTRIENTS

The stagnant conditions which predominate in the colon are ideal for the bacterial fermentation of food residues. Carbohydrate which is not absorbed in the small intestine is fermented in the colon to volatile fatty acids (VFA) which can then be rapidly absorbed by the colonic epithelium.[128] Production of VFA in the large intestine contributes up to 40% of basal energy requirements in herbivorous species such as the rabbit[117] and between 10% and 20% in omnivorous species such as the pig. Its importance in man has yet to be established, but it has been shown that in patients with jejuno-ileal by-pass up to 40% of carbohydrate unabsorbed in the small intestine can be salvaged by the colon.[13]

The colon may also be able to salvage unabsorbed protein and amino acids by bacterial degradation of ammonia, which is absorbed into the portal blood and converted into amino acids by the liver. However, most of the nitrogen for ammonia production comes from urea, which diffuses into the colon from the blood. Finally, the colonic epithelium is able to absorb some long-chain fatty acids. However, much unabsorbed fat is probably converted to insoluble soaps or hydroxy fatty acids which cause diarrhoea and are poorly absorbed.

SALT AND WATER ABSORPTION

The colon also has an important role in extracting salt and water from the faecal residues. As the tight junctions between the cells are much less permeable than they are in the small intestine, sodium absorption in the colon occurs almost entirely via the transcellular route by an active electrogenic mechanism which is not linked to the cotransport of any other solute. The tightness of epithelium prevents the sodium from leaking back down its electrochemical gradient and allows sodium to be extracted from colonic contents against high electrochemical gradients.

Water is absorbed in response to the high osmolality established in the lateral space by active sodium absorption. The low permeability of the tight juctions means that water is absorbed more slowly than in the small intestine, but absorption occurs against much higher osmotic gradients. The ability to extract water against a transepithelial osmotic gradient is a consequence of the osmolality in the lateral space being higher than in the capillary or the lumen.

Potassium is secreted into the colonic lumen, probably by passive diffusion down the electrical gradient created by active sodium transport. These movements account for the low sodium and high potassium concentrations in normal stool.

Chloride is absorbed actively in the colon via the transcellular route by means of a chloride-bicarbonate exchange pump.

DIARRHOEA

Diarrhoea can be defined as the passage of more than 200 g of liquid motions per day and results from an abnormality in the handling of fluid by the gut. Approximately 9 litres of fluid normally enter the gut within 24 hours (2 litres from the diet and 7 litres of secretions). The majority of this fluid is absorbed in the small intestine (Fig. 14), leaving somewhat less than 200 ml voided in the faeces. This is by no means a rigid scheme and the volumes may vary according to different conditions. For example, it seems likely that, when large volumes of fluid enter from above, the absorptive capacity of the small intestine may be considerably in excess of 8 litres a day, while the absorptive capacity of the colon under similar circumstances can be as high as 7 litres a day.

Diarrhoea occurs if the normal balance between the fluid entering the gut and the fluid leaving the gut is disturbed. This can occur when a) excessive volumes of fluid are ingested, b) abnormally large amounts of fluid are secreted into the gut and c) if there is impairment of fluid absorption. Diarrhoea caused by excessive fluid ingestion is rare but may be found in some hysterical water drinkers and some beer drinkers who consume several litres of fluid per day. Ingestion of large volumes of salt solutions is one method for cleansing the bowel in preparation for colonoscopy or bowel surgery.[26]

Excessive secretion into the gut sufficient to over-

DIET + SECRETION ≃ 9 LITRES

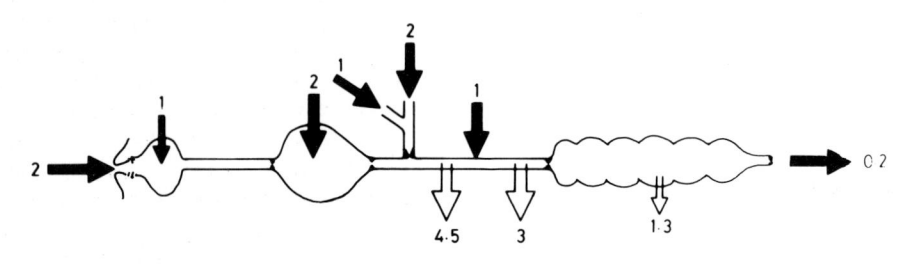

ABSORPTION ≃ 8.8 LITRES

Fig. 14. A diagram of the typical fluid movements into (solid arrows) and out of (open arrows) the gastrointestinal tract of man.

whelm the normal absorptive mechanisms and cause diarrhoea usually comes from the small and large intestines. Diarrhoea due to pathological increases in salivary, pancreatic or biliary secretion has, to our knowledge, never been described and hypersecretion of fluid from the stomach is only one factor in the aetiology of diarrhoea in patients with Zollinger-Ellison syndrome. Profuse small intestinal secretion sufficient to produce copious diarrhoea (secretory diarrhoea) can be caused by infection with enterotoxin-producing bacteria (*Vibrio cholerae, Shigella, Salmonella, Escherichia coli, Staphylococcus*), tumours secreting excessive amounts of polypeptide hormones and prostaglandins, (e.g. medullary carcinoma of the thyroid, pancreatic cholera syndrome) and ingestion of laxative agents (anthracene derivatives, phenolphthalein) (see Table 1). Small intestinal secretion may also be the mechanism whereby idiosyncratic reactions to other drugs (i.e. mefenamic acid, methyldopa, diuretics) and foodstuffs cause diarrhoea. Abnormal secretion from the colon sufficient to cause diarrhoea commonly occurs secondary to malabsorption of bile acids or long-chain fatty acids in the small intestine. These substances are converted by colonic bacteria to potent cathartic agents.

Impaired water absorption can result from:

(1) ingestion of solutes which cannot normally be absorbed or digested and hence retain water in intestinal lumen. These include the osmotic laxatives lactulose and magnesium sulphate and certain indigestible saccharides present in fruit and vegetables;

(2) impaired digestion of normal food constituents (lactose in lactase deficiency, pancreatic insufficiency);

(3) impaired absorption of normal food material. This can occur because of destruction or loss of the absorptive cells or enterocytes (as in coeliac disease or certain viral or bacterial enteritides), impairment of specific transport processes (as in rare congenital glucose-galactose malabsorption and congenital chloridorrhoea) and abnormally rapid transit of food through the small intestine leaving insufficient contact time for normal absorption. Rapid transit may be of importance in some patients with irritable bowel syndrome and is probably also a factor in the diarrhoea caused by enteritides and ingestion of certain laxative agents.[2,93] Diarrhoea occurs because unabsorbed solute traps water in the lumen and the increased bulk of the food stimulates peristalsis leading to rapid delivery of food to the colon. If the colon fails to accommodate the load, diarrhoea will ensue.

Malabsorption of bile acids is of particular importance in the pathogenesis of diarrhoea. This can occur because of ileal resection of Crohn's disease. The unabsorbed bile acids pass into the colon where they stimulate secretion and motility.

Approximately 500 to 1500 ml of unabsorbed food material enters the colon each day;[119] the colon usually retains food residues for approximately three days (at least ten times as long as the food remains in the small intestine). During this time, extraction of water occurs against high transepithelial osmotic gradients, secondary to the active absorption of sodium. This can only take place because spaces between the cells are small and prevent salt and water leaking back across the epithelium. Any factor which increases colonic permeability will impair fluid absorption. An increased colonic permeability is thought to be responsible for the diarrhoea observed in inflammatory diseases of the colon and may also play a role in the diarrhoea caused by unabsorbed bile acids and fatty acids. As extraction of water by the colon occurs slowly, increases in the rate of transit throughout the colon will also impair water absorption and result in the passage of material resembling ileal contents.

Investigation of Diarrhoea

Many cases of diarrhoea can be diagnosed by the standard means of investigations listed in Table 3. However, it is by no means uncommon to find patients for whom these investigations fail to reveal a diagnosis. In this case, the following measurements may elucidate the pathogenesis of the diarrhoea.

Measurement of Stool Weight and Osmotic Gap[75]

This is of particular value in patients with large volume diarrhoea (at least 500 ml/day) and it indicates whether the diarrhoea is caused by abnormal secretion of fluid and electrolytes into the gut lumen (secretory diarrhoea)

TABLE 3
LIST OF CLINICAL TESTS COMMONLY USED TO INVESTIGATE PATIENTS WITH CHRONIC DIARRHOEA

Sigmoidoscopy
Rectal smear and biopsy
Faecal fat estimation
Faecal occult blood
Stool microscopy for ova and parasites
Stool culture
Barium meal and follow through
Barium enema
Jejunal biopsy
Jejunal aspirate (examination for giardia and bacterial culture)
Xylose excretion test
Schilling test
Lactose tolerance test
Gastric acid analysis
Blood eosinophil count
Thyroid function tests
Plasma cortisol
Immunoglobulins
Urinary 5-hydroxyindole acetic acid
Urinary metanephrins or VMA
Bile acid breath test
Nutritional screen (see Table 2)

or whether it is caused by the presence of unabsorbed osmotically active substances in the bowel lumen (osmotic diarrhoea). Ideally faeces are collected for three days while the patient is fasted, all fluid requirements being provided by an intravenous infusion. Each stool collection is weighed and if liquid, a sample is centrifuged, the supernatant analyzed for electrolytes and osmolality and the osmotic gap calculated.

$$\text{Osmotic gap} = \text{Osmolality} - 2(Na + K)$$

Secretory diarrhoeas are characteristically of large volume, usually more than a litre/day. They persist and may even get worse if the patient fasts. Finally, the osmolality of the stool water can almost completely be accounted for by the sum of the electrolytes. Osmotic diarrhoeas may also be of large volume, but are considerably reduced or absent when the patient fasts and the analysis of stool water reveals a large osmotic gap (over 100 mosmol/l).

Not all diarrhoeas can be subdivided into osmotic or secretory categories by this test, particularly if the stool volumes are less than 500 ml/day. Some diarrhoeas, such as those caused by malabsorption of fat or bile acids have both osmotic (small intestinal malabsorption) and secretory (colonic secretion) components. Coeliac disease is also characterized by abnormal small intestinal secretion as well as malabsorption. Nevertheless, identification of secretory or osmotic diarrhoeas provides a useful physiological basis for further investigation.

Investigation of secretory diarrhoea

The causes of large volume secretory diarrhoea are few, but of considerable importance, as they result in severe fluid and electrolyte depletion. They include infection with toxigenic bacteria, polypeptide-secreting tumours and surreptitious ingestion of drugs.

Toxigenic bacteria can be identified by appropriate cultures of small intestinal aspirate or faecal fluid. Development of secretion following injection of a sample of faecal fluid into a loop of rabbit ileum has been used as a method of proving the presence of an enterotoxin.

The easiest way to diagnose surreptitious ingestion of drugs is by a careful search of the patient's room and belongings.[24] However, as this measure cannot always be justified on ethical grounds, physicians often have to rely on biochemical analysis of the stool or urine. Phenolphthalein can be easily identified in urine or stool water by the purple colour which appears after addition of alkali. There are also simple assays available for detection of anthracene derivatives (senna, cascara, dorbanex) and bisocodyl in stool water. Surreptitious ingestion of Epsom salts (which cause osmotic diarrhoea) can be detected by the large amounts of magnesium and sulphate in stool water. However, there are other agents, such as diuretics, which cause severe diarrhoea which cannot be detected except by specific chromatographic procedures.

Intestinal secretion can be caused by several polypeptides but only a few are recognized to be associated with severe diarrhoea when secreted in large amounts by endocrine tumours. The polypeptide-secreting tumours which are commonly associated with diarrhoea include Zollinger-Ellison syndrome (gastrin), pancreatic cholera syndrome (VIP) and medullary carcinoma of the thyroid (calcitonin). The diarrhoea produced by these tumours is not necessarily a direct result of the intestinal secretion produced by high circulating levels of polypeptide. In the Zollinger-Ellison syndrome, it is almost certainly secondary to the acid entering the small intestine, while, in pancreatic cholera syndrome and medullary carcinoma of the thyroid, intestinal secretion is also stimulated by the large amounts of prostaglandins released from the tumour. Diarrhoea has also been reported in patients who have tumours secreting somatostatin, GIP, and glucagon, though these are extremely rare. Although specific immunoassays exist for all of these agents, the reported incidence of high VIP results from some normal subjects and some patients with surreptitious drug ingestion leads us to interpret the results of radioimmunoassay with caution.[76] Certainly, the diagnosis of pancreatic cholera syndrome should never be made in somebody who does not have a large volume secretory diarrhoea.

Perfusion of the small intestine with electrolyte solutions has been used in specialist centres to confirm the existence of intestinal secretion caused by polypeptide-secreting tumours.[76] In patients whose secretory diarrhoea has been caused by the surreptitious ingestion of drugs, perfusion techniques show net absorption of fluid as long as the patients have not ingested the drug immediately before or during the perfusion study.

Investigation of Osmotic Diarrhoea

With the exception of ingestion of osmotic laxatives such as magnesium sulphate or lactulose, most cases of osmotic diarrhoea result from the malabsorption or maldigestion of food and are investigated by the tests of pancreatic and small intestinal function described above.

REFERENCES

1. Admirand, W. H. and Small, D. M. (1968), 'The physicochemical basis of cholesterol gallstone formation in man,' *J. Clin. Invest.*, **47**, 1043–1052.
2. Aitchison, W. D., Stewart, J. J. and Bass, P. (1978), 'A unique distribution of laxative induced spike potentials from the small intestine of the dog,' *Am. J. Dig. Dis.*, **23**, 513–520.
3. Allen, R. H., Seetharam, B., Podell, E., and Alpers, D. H. (1978), 'Effect of proteolytic enzymes on the binding of cobalamin to R protein and intrinsic factor,' *J. Clin. Invest.*, **61**, 47–54.
4. Ammon, H. V. and Phillips, S. F. (1973), 'Inhibition of colonic water and electrolyte absorption by fatty acids in man,' *Gastroenterology*, **65**, 744–749.
5. Arvanitakis, C. and Cooke, A. R. (1978), 'Diagnostic tests of exocrine pancreatic function and disease', *Gastroenterology*, **74**, 932–948.
6. Arvanitakis, C. and Greenberger, N. J. (1976), 'Diagnosis of pancreatic disease by a synthetic peptide. A new test of exocrine pancreatic function,' *Lancet* **i**, 663–666.
7. Basso, N. and Passaro, E. (1970), 'Calcium-stimulated gastric secretion in the Zollinger-Ellison syndrome,' *Arch. Surg.*, **101**, 399–402.
8. Bergegardh, S., Nilsson, G. and Olbe, L. (1976), 'The effect of antral distension on acid secretion and plasma gastrin in duodenal ulcer patients,' *Scand. J. Gastroenterol.*, **11**, 475–479.
9. Binder, H. J., Filburn, C. and Volpe, B. T. (1975), 'Bile salt alteration of colonic electrolyte transport: role of cyclic adenosine monophosphate,' *Gastroenterology*, **68**, 503–508.
10. Bolton, J. E. and Field, M. (1977), 'Ca ionophore-stimulated ion secretion in rabbit ileal mucosa; relation to actions of cyclic 3′,5′-AMP and carbamylcholine,' *J. Membr. Biol.*, **35**, 159–173.
11. Bond, J. H. and Levitt, M. D. (1975), 'Investigation of small bowel transit time in man utilizing pulmonary hydrogen (H_2) measurements,' *J. Lab. Clin. Med.*, **85**, 546–555.
12. Bond, J. H. and Levitt, M. D. (1976), 'Quantitative measurement of lactose absorption,' *Gastroenterology*, **70**, 1058–1062.
13. Bond, J. H., Currier, B. E., Buchwald, H. and Levitt, M. D. (1980), 'Colonic conservation of malabsorbed carbohydrate,' *Gastroenterology*, **78**, 444–447.
14. Borgström, B. (1969), 'Quantification of cholesterol absorption in man by faecal analysis after the feeding of a single isotope-labelled meal,' *J. Lipid Res.*, **10**, 331–337.
15. Borison, H. L. and Wang, S. C. (1953), 'Physiology and pharmacology of vomiting,' *Pharmacol. Rev.*, **5**, 193–230.
16. Borriello, P., Hudson, M. and Hill, M. (1978), 'Investigation of the gastro-intestinal bacterial flora,' *Clin. Gastroenterol.*, **7**, 329–349.
17. Charney, A. N., Kinsey, M. D., Myers, L. Gianella, R. A. and Gots, R. E. (1975), 'Na^+-K^+-activated adenosine triphosphatase and intestinal electrolyte transport. Effect of adrenal steroids,' *J. Clin. Invest.*, **56**, 653–660.
18. Christensen, J. (1971). 'The control of gastrointestinal movements: some old and new views,' *New Eng. J. Med.*, **285**, 85–98.
19. Connell, A. M. (1962), 'The motility of the pelvic colon. II. Paradoxical motility in diarrhoea and constipation,' *Gut*, **3**, 342–348.
20. Cook, G. C. (1970), 'Comparison of the absorption and metabolic products of sucrose and its monosaccharides in man,' *Clin. Sci.*, **38**, 687–697.
21. Cox, T. M. and Peters, T. J. (1980), 'Cellular mechanisms in the regulation of iron absorption by the human intestine: studies in patients with iron deficiency before and after treatment,' *Br. J. Haematol.*, **44**, 75–86.
22. Coyne, M. J., Bonorris, G. G., Goldstein, L. I. and Schoenfield, L. J. (1976), 'Effect of chenodeoxycholic acid and phenobarbital on the rate-limiting enzymes of hepatic cholesterol and bile acid synthesis in patients with gallstones,' *J. Lab. Clin. Med.*, **87**, 281–291.
23. Crane, R. K. (1962), 'Hypothesis of mechanism of intestinal transport of sugars,' *Fed. Proc.*, **21**, 891–895.
24. Cummings, J. H., Sladen, G. E., James, O. F. W., Sarner, M. and Misiewicz, J. J. (1974), 'Laxative-induced diarrhoea: A continuing clinical problem,' *Br. Med. J.*, **1**, 537–541.
25. Dahlqvist, A. (1968), 'Assay of intestinal disaccharidases,' *Anal. Biochem.*, **22**, 99–107.
26. Davis, G. R., Santa-Ana, C., Morawski, S. G. and Fordtran, J. S. (1980), 'Development of a lavage solution associated with minimal water and electrolyte absorption or secretion,' *Gastroenterology*, **78**, 991–995.
27. Davenport, H. W., Warner, H. A. and Code, C. F. (1964), 'Functional significance of gastric mucosal barrier to sodium,' *Gastroenterology*, **47**, 142–152.
28. Davenport, H. W. (1967), 'Salicylate damage to the gastric mucosal barrier,' *N. Engl. J. Med.*, **276**, 1307–1312.
29. Denyer, M. E. and Cotton, P. B. (1979), 'Pure pancreatic juice studies in normal subjects and patients with chronic pancreatitis,' *Gut*, **20**, 89–97.
30. Desnuelle, P. (1976), 'The lipase co-lipase system,' in *Lipid Absorption: Biochemical and Clinical Aspects.* pp. 23–36. (Rommel, K., Goebell, H. and Böhmer, R., Eds.). Lancaster, MTP Press
31. Diamond, J. M. and Bossert, W. H. (1967), 'Standing-gradient osmotic flow. A mechanism for coupling of water and solute transport in epithelia,' *J. Gen. Physiol.*, **59**, 2061–2083.
32. Dietschy, J. M. and Westergaard, H. (1975), 'The effect of unstirred water layers on various transport processes in the intestine,' in *Intestinal Absorption and Malabsorption*, pp. 197–207. (Csaky, T. Z. Ed.). New York, Raven Press.
33. Dobbins, J. W. and Binder, H. J. (1977), 'Importance of the colon in enteric hyperoxaluria,' *N. Engl. J. Med.*, **296**, 298–301.
34. Duane, W. C. and Hanson, K. C. (1978), 'Role of gallbladder emptying and small bowel transit in regulation of bile acid pool size in man,' *J. Lab. Clin. Med.*, **92**, 859–872.
35. Dyck, W. P. (1967), 'Titrimetric measurements of faecal trypsin and chymotrypsin in cystic fibrosis with pancreatic exocrine insufficiency,' *Am. J. Dig. Dis.*, **12**, 310–317.
36. Elias, E., Redshaw, M. and Wood, T. (1977), 'Diagnostic importance of changes in circulating concentrations of immunoreactive trypsin,' *Lancet*, **ii**, 66–68.
37. Erlinger, S. and Dhumeaux, D. (1974), 'Mechanisms and control of secretion of bile water and electrolytes,' *Gastroenterology*, **66**, 281–304.
38. Field, M. (1979), 'Intracellular mediators of secretion in the small intestine,' in *Mechanisms of Intestinal Secretion*, pp. 83–92. (Binder, H. J. Ed.). New York, Alan R. Liss Inc.
39. Fordtran, J. S., Rector, F. C. and Carter, N. W. (1968), 'The mechanisms of sodium absorption in the human small intestine,' *J. Clin. Invest.*, **47**, 884–900.
40. Fordtran, J. S. (1975), 'Stimulation of active and passive sodium absorption by sugars in the human jejunum,' *J. Clin. Invest.*, **55**, 728–737.
41. Forker, E. L. (1977), 'Mechanisms of hepatic bile formation,' *Annu. Rev. Physiol.*, **39**, 323–347.
42. Frizzell, R. A., Dugas, M. C. and Schultz, S. G. (1975), 'Sodium chloride transport by rabbit gallbladder. Direct evidence for a coupled NaCl influx process,' *J. Gen. Physiol.*, **65**, 769–795.
43. Fromm, H. and Hofmann, A. F. (1971), 'Breath test for altered bile-acid metabolism,' *Lancet*, **ii**, 621–625.
44. Geall, M. G., Phillips, S. F. and Summerskill, W. H. J. (1970), 'Profile of gastric potential difference in man,' *Gastroenterology*, **58**, 437–443.
45. George, J. D. (1968), 'New clinical method for measuring the rate of gastric emptying: the double sampling test meal,' *Gut*, **9**, 237–242.
46. Gracey, M., Burke, V. and Oshin, A. (1972), 'Active intestinal transport of D-fructose,' *Biochim. Biophys. Acta*, **266**, 397–406.
47. Greenberger, N. J., Saegh, S. and Ruppert, R. D. (1968), 'Urine indican excretion in malabsorptive disorders,' *Gastroenterology*, **55**, 204–211.
48. Greenberger, N. J. and Skillman, T. G. (1969), 'Medium chain

triglycerides. Physiological considerations and clinical implications,' *N. Engl. J. Med.*, **280**, 1045–1058.
49. Grossman, M. I. (1978), 'Control of gastric secretion,' in *Gastrointestinal Disease*, pp. 640–659. (Sleisenger, M. and Fordtran, J. S., Eds.). Philadelphia, W. B. Saunders Co.
50. Grotzinger, U., Bergegårdh, S. and Olbe, L. (1977), 'The effect of fundic distension on gastric acid secretion in man,' *Gut*, **18**, 105–110.
51. Grundy, S. M., Ahrens, E. H. and Miettinen, T. A. (1965), 'Quantitative isolation and gas-liquid chromatographic analysis of total faecal bile acids,' *J. Lipid Res.*, **6**, 397–410.
52. Hardcastle, J. D. and Wilkins, J. L. (1970), 'The action of sennosides and related compounds on human colon and rectum,' *Gut*, **11**, 1038–1042.
53. Heaton, K. W. and Read, A. E. (1969), 'Gallstones in patients with disorders of the terminal ileum and disturbed bile salt metabolism,' *Br. Med. J.*, **iii**, 494–496.
54. Heaton, K. W. (1979), 'Bile salt tests in clinical practice,' *Br. Med. J.*, **i**, 644–646.
55. Hill, M. J. and Drasar, B. S. (1975), 'The normal colonic bacterial flora,' *Gut*, **16**, 318–323.
56. Hinder, R. A. and Kelly, K. A. (1977), 'Canine gastric emptying of solids and liquids,' *Am. J. Physiol.*, **233**, E335–E340.
57. Hofmann, A. F. and Small, D. M. (1967), 'Detergent properties of bile salts: correlation with physiological function,' *Annu. Rev. Med.*, **18**, 333–376.
58. Hofmann, A. F. (1972), 'Bile acid malabsorption caused by ileal resection,' *Arch. Intern. Med.*, **130**, 597–605.
59. Hofmann, A. F. (1977a), 'The enterohepatic circulation of bile acids in man,' *Clin. Gastroenterol.*, **6**, 3–24.
60. Hofmann, A. F. (1977b), 'Bile acids, diarrhoea and antibiotics: data, speculation and a unifying hypothesis,' *J. Infect. Dis.*, **135**, S126–132.
61. Holdsworth, C. D. and Dawson, A. M. (1965), 'Absorption of fructose in man,' *Proc. Soc. Exp. Biol. Med.*, **118**, 142–145.
62. Holman, G. D. and Naftalin, R. J. (1979), 'Fluid movements across rabbit ileum coupled to paracellular ion movements,' *J. Physiol.*, **290**, 351–366.
63. Holzbach, R. T., Marsh, M., Olszewski, M. and Holan, K. (1973), 'Cholesterol solubility in bile. Evidence that supersaturated bile is frequent in healthy man,' *J. Clin. Invest.*, **52**, 1467–1479.
64. Hunt, J. N. and Stubbs, D. F. (1975), 'The volume and energy content of meals as determinants of gastric emptying,' *J. Physiol.*, **245**, 209–225.
65. Isenberg, J. I., Ippoliti, A. F. and Maxwell, V. (1977), 'Perfusion of the proximal small intestine with peptone stimulates gastric acid secretion in man,' *Gastroenterology*, **73**, 746–752.
66. James, O. F. W., Agnew, J. E. and Bouchier, I. A. D. (1973), 'Assessment of the ^{14}C-glycocholic acid breath test,' *Br. Med. J.*, **iii**, 191–195.
67. James, O. (1973), 'The Lundh test,' *Gut*, **14**, 582–591.
68. Jian, R., Pecking, A., Najean, Y. and Bernier, J. J. (1979), 'Étude de la progression d'un repas dans l'intestin grêle de l'homme par une méthode scintigraphique,' *Gastroenterol. Clin. Biol.*, **3**, 755–762.
69. Joffe, B. I., Bank, S., Jackson, W. P. U., Keller, P., O'Reilly, I. G. and Vinik, A. I. (1968), 'Insulin reserve in patients with chronic pancreatitis,' *Lancet*, **ii**, 890–892.
70. Kaplowitz, N., Kok, E. and Javitt, N. B. (1973), 'Postprandial serum bile acid for the detection of hepatobiliary disease,' *J.A.M.A.*, **225**, 292–293.
71. King, C. E. and Toskes, P. P. (1979), 'Small intestinal bacterial overgrowth,' *Gastroenterology*, **76**, 1035–1055.
72. King, C. E., Toskes, P. P., Guilarte, T. R., Lorenz, E. and Welkos, S. L. (1980), 'Comparison of the one-gram d-(^{14}C) xylose breath test to the (^{14}C) bile acid breath test in patients with small-intestine bacterial overgrowth,' *Dig. Dis. Sci.*, **25**, 53–58.
73. Knox, M. T. and Mallinson, C. N. (1971), 'Gastric emptying of fat in patients with pancreatitis,' *Rendic Gastroenterol.*, **3**, 115–119.
74. Kolts, B. E., Herbst, C. A. and McGuigan, J. E. (1974), 'Calcium and secretin-stimulated gastrin release in the Zollinger-Ellison syndrome,' *Ann. Intern, Med.*, **81**, 758–762.
75. Krejs, G. J. and Fordtran, J. S. (1978), 'Physiology and pathophysiology of ion and water movement in the human intestine,' in *Gastrointestinal Disease*, pp. 297–335. (Sleisenger, M. H. and Fordtran, J. S. Eds.). Philadelphia, W. B. Saunders Co.
76. Krejs, G. J., Walsh, J. H., Morawski, S. G. and Fordtran, J. S. (1977), 'Intractable diarrhoea. Intestinal perfusion studies and plasma VIP concentrations in patients with pancreatic cholera syndrome and surreptitious ingestion of laxatives and diuretics,' *Am. J. Dig. Dis.*, **22**, 280–292.
77. Kronborg, O. (1974), 'Gastric acid secretion and risk of recurrence of duodenal ulcer within six to eight years after truncal vagotomy and drainage,' *Gut*, **15**, 714–719.
78. Lauterburg, B. H., Newcomer, A. D. and Hofmann, A. F. (1978), 'Clinical value of the bile acid breath test: evaluation of the Mayo Clinic experience,' *Mayo Clin. Proc.*, **53**, 227–233.
79. Layden, T., Elais, E. and Boyer, J. L. (1978), 'Bile formation in the rat. The role of the paracellular shunt pathway,' *J. Clin. Invest.*, **62**, 1375–1385.
80. Lee, K. Y., Tai, H. H. and Chey, W. Y. (1976), Plasma secretin and gastrin responses to a meat meal and duodenal acidification in dogs,' *Am. J. Physiol.*, **230**, 784–789.
81. Lei, F. H., Lucas, M. L. and Blair, J. A. (1977), 'The influence of pH, low sodium ion concentration and methotrexate on the jejunal-surface pH: A model for folic acid transfer,' *Biochem. Soc. Trans.*, **5**, 149–152.
82. Levitt, M. D. (1979), 'Clinical use of amylase clearance and isoamylase measurements,' *Mayo Clin. Proc.*, **54**, 428–431.
83. Levitt, M. D. and Duane, W. C. (1972), 'Floating stools—flatus versus fat,' *N. Engl. J. Med.*, **286**, 973–975.
84. Levitt, M. D., Johnson, S. G., Ellis, C. and Engel, R. R. (1977), 'Influence of amylase assay technique on renal clearance of amylase-creatinine ratio,' *Gastroenterology*, **72**, 1260–1263.
85. Levitt, M. D., Rapoport, M. and Cooperbrand, S. R. (1969), 'The renal clearance of amylase in renal insufficiency, acute pancreatitis and macroamylasemia,' *Ann. Intern. Med.*, **71**, 919–925.
86. Lightwood, R. and Reber, H. A. (1977), 'Micropuncture study of pancreatic secretion in the cat,' *Gastroenterology*, **72**, 61–66.
87. Linder, M. C. and Munro, H. N. (1977), 'The mechanism of iron absorption and its regulation,' *Fed. Proc.*, **36**, 2017–2023.
88. Losowsky, M. S., Walker, B. E. and Kelleher, J. (1974), *Malabsorption in Clinical Practice*. Edinburgh and London, Churchill Livingstone.
89. Luey, K. L. and Heaton, K. W. (1979), 'Bile acid clearance in liver disease,' *Gut*, **20**, 1083–1087.
90. Lundgren, O. and Svanik, J. (1977), 'Gastrointestinal circulation,' in *International Review of Physiology*, pp. 1–34. vol. 12. (Crane, R. K. Ed.). Baltimore, University Park Press.
91. Malagelada, J-R., Longstreth, G. F., Summerskill, W. H. J. and Go, V.L.W. (1976), 'Measurement of gastric functions during digestion of ordinary solid meals in man,' *Gastroenterology*, **70**, 203–210.
92. Mallinson, C. N. (1968), 'Effect of pancreatic insufficiency and intestinal lactase deficiency on the gastric emptying of starch and lactose,' *Gut*, **9**, 737.
93. Mathias, J. R., Carlson, G. M., Dimarino, A. J., Bertiger, C., Morton, M. E. and Cohen, S. (1976), 'Intestinal myoelectrical activity in response to live *Vibrio cholerae* and cholera enterotoxin,' *J. Clin. Invest.*, **58**, 91–96.
94. Matthews, D. M. and Adibi, S. A. (1976), 'Peptide absorption,' *Gastroenterology*, **71**, 151–161.
95. Mayer, G., Arnold, R., Feurle, G., Fuchs, K., Ketterer, H., Track, N. S. and Creutzfeldt, W. (1974), 'Influence of feeding and sham feeding upon serum gastrin and gastric acid secretion in control subjects and duodenal ulcer patients,' *Scand. J. Gastroenterol.*, **9**, 703–710.
96. Mayer, P. J. and Beeken, W. L. (1975), 'The role of urinary indican as a predictor of bacterial colonisation in the human jejunum,' *Am. J. Dig. Dis.*, **20**, 1003–1010.
97. McCarthy, L. E., Borison, H. L., Spiegel, P. K. and Friedlander, R. M. (1974), 'Vomiting: radiographic and oscillographic correlates in the decerebrate rat,' *Gastroenterology*, **67**, 1126–1130.
98. MacDonald, I. and Turner, L. J. (1968), Serum fructose levels

after sucrose and its constituent monosaccharides,' *Lancet*, **i**, 841–843.
99. McGregor, I., Parent, J. and Meyer, J. H. (1977), 'Gastric emptying of liquid meals and pancreatic and biliary secretion after subtotal gastrectomy, or truncal vagotomy and pyloroplasty in man,' *Gastroenterology*, **72**, 195–205.
100. Mekhjian, H. S., Phillips, S. F. and Hofmann, A. F. (1971), 'Colonic secretion of water and electrolytes induced by bile acids: perfusion studies in man,' *J. Clin. Invest.*, **50**, 1569–1577.
101. Metz, G., Jenkins, D. J. A., Peters, T. J., Newman, A. and Blendis, L. M. (1975), 'Breath hydrogen as a diagnostic method for hypolactasia,' *Lancet*, **i**, 1155–1157.
102. Metz, G., Jenkins, D. J. A., Newman, A. and Blendis, L. M. (1976), 'Breath hydrogen in hyposucrasia,' *Lancet*, **1**, 119–120.
103. Metz, G., Gassull, M. A., Drasar, B. S., Jenkins, D. J. A. and Blendis, L. M. (1976), 'Breath-hydrogen test for small-intestinal bacterial colonisation,' *Lancet*, **i**, 668–669.
104. Meyer, J. H. and Kelly, G. A. (1976), 'Canine pancreatic responses to intestinally perfused proteins and protein digests,' *Am. J. Physiol.*, **231**, 682–691.
105. Meyer, J. H., MacGregor, I. L., Gueller, R., Martin, P. and Cavalieri, R. (1976), '^{99m}Tc-tagged chicken liver as a marker of solid food in the human stomach,' *Am. J. Dig. Dis.*, **21**, 296–304.
106. Mitchell, C. J., Humphrey, C. S., Bullen, A. W. and Kelleher J. (1979), 'The diagnostic value of the oral pancreatic function test,' *Scand. J. Gastroenterol.*, **14**, 183–187.
107. Moore, E. W. and Scarlata, R. W. (1972), 'Liquid junction potentials: their importance in gastric PD measurements, theoretical estimation and measurement *in vitro*,' *Gastroenterology*, **62**, 786–790.
108. Mottaleb, A., Kapp, F., Noguera, E. C. A., Kellock, T. D., Wiggins, H. S. and Waller, S. L. (1973), 'The Lundh test in the diagnosis of pancreatic disease: a review of five years' experience,' *Gut*, **14**, 835–841.
109. Nellans, H. N. and Kimberg, D. V. (1979), 'Intestinal calcium transport. Absorption, secretion and vitamin D.' in *International Review of Physiology*, pp. 227–262. vol. 19. (Crane, R. K. Ed.). Baltimore, University Park Press.
110. Newcomer, A. D., McGill, D. B., Thomas, P. J. and Hofmann, A. F. (1975), 'Prospective comparison of indirect methods for detecting lactase deficiency,' *N. Engl. J. Med.*, **293**, 1232–1236.
111. Newcomer, A. D., Hofmann, A. F., DiMagno, E. P., Thomas, P. J. and Carlson, G. L. (1979), 'Triolein breath test. A sensitive and specific test for fat malabsorption,' *Gastroenterology*, **76**, 6–13.
112. Norman, D. A., Morawski, S. G. and Fordtran, J. S. (1980), 'Influence of glucose, fructose and water movement on calcium absorption in the jejunum,' *Gastroenterology*, **78**, 22–25.
113. Northfield, T. C., Draser, B. S. and Wright, J. T. (1973), 'Value of small intestinal bile acid analysis in the diagnosis of the stagnant loop syndrome,' *Gut*, **14**, 341–347.
114. Northfield, T. C. and Hofmann, A. F. (1975), 'Biliary lipid output during three meals and an overnight fast. 1. Relationship to bile acid pool size and cholesterol saturation of bile in gallstone and control subjects,' *Gut*, **16**, 1–17.
115. Ockner, R. K. and Manning, J. A. (1976), 'Fatty acid binding protein. Role in esterification of absorbed long chain fatty acid in rat intestine,' *J. Clin. Invest.*, **58**, 632–641.
116. Orloff, M. J., Chandler, J. G., Alderman, S. J., Keiter, J. E. and Rosen, H. (1969), 'Gastric secretion and peptic ulcer following portacaval shunt in man,' *Ann. Surg.*, **170**, 515–524.
117. Parker, D. S. (1976), 'The measurement and production rates of volatile fatty acids in the caecum of the conscious rabbit,' *Br. J. Nutr.*, **36**, 61–70.
118. Pennington, C. R., Ross, P. E. and Bouchier, I. A. D. (1977), 'Serum bile acids in the diagnosis of hepatobiliary disease,' *Gut*, 18, 903–908.
119. Phillips, S. F. and Giller, J. (1973), 'The contribution of the colon to electrolyte and water conservation in man,' *J. Lab. Clin. Med.*, **81**, 733–746.
120. Rampton, D. S., Kasadis, G. P., Rose, G. A., and Sarner, M. (1979), 'Oxalate loading test; a screening test for steatorrhoea,' *Gut*, **20**, 1089–1094.
121. Ramaswamy, K., Malathi, P., Caspary, W. F. and Crane, R. K. (1974), 'Studies on the transport of glucose from disaccharides by hamster small intestine *in vitro*. II. Characteristics of the disaccharidase-related transport system,' *Biochim. Biophys. Acta*, **345**, 39–48.
122. Read, N. W. and Fordtran, J. S. (1979), 'The role of intraluminal junction potentials in the generation of the gastric potential difference in man,' *Gastroenterology*, **76**, 932–938.
123. Read, N. W., Miles, C. A., Fisher, D., Holgate, A. M., Kime, N. D., Mitchell, M. A., Reeve, A. M., Roche, T. B. and Walker, M. (1980), 'The transit of a meal through the stomach, small intestine and colon in normal subjects and its role in the pathogenesis of diarrhoea,' *Gastroenterology*, (In press).
124. Reisenauer, A. M., Krumdieck, C. L. and Halsted, C. H. (1977), 'Folate conjugase: two separate activities in human jejunum,' *Science*, **198**, 196–197.
125. Richardson, C. T., Walsh, J. H., Cooper, K. A., Feldman, M. and Fordtran, J. S. (1977), 'Studies on the role of cephalic-vagal stimulation in the acid secretory response to eating in normal human subjects,' *J. Clin. Invest.*, **60**, 435–441.
126. Riley, J. W. and Glickman, R. M. (1979), 'Fat malabsorption—advances in our understanding,' *Am. J. Med.*, **67**, 980–988.
127. Rose, R. C. (1980), 'Water-soluble vitamin absorption in intestine,' *Annu. Rev. Physiol.*, **42**, 157–171.
128. Ruppin, H., Bar-Mein, S., Soergel, K. H., Wood, C. M. and Schmitt, M. G. (1980), 'Absorption of short chain fatty acids by the colon,' *Gastroenterology*, **78**, 1500–1507.
129. Sackin, H. and Boulpaep, E. L. (1975), 'Models for coupling of salt and water transport,' *J. Gen. Physiol.*, **66**, 671–733.
130. Samloff, I. M. (1971), 'Pepsinogens, pepsins and pepsin inhibitors,' *Gastroenterology*, **60**, 586–604.
131. Samloff, I. M., Secrist, D. M. and Passaro, E. (1975), 'A study of the relationship between serum group I pepsinogen levels and gastric acid secretion,' *Gastroenterology*, **69**, 1196–1200.
132. Sarles, H. R. (1977), 'The exocrine pancreas,' in *International Review of Physiology*, pp. 173–221. vol. 12., (Crane, R. K. Ed.). Baltimore, University Park Press.
133. Sarles, H., Dani, R., Prezelin, G., Souville, C. and Figarella, C. (1968), 'Cephalic phase of pancreatic secretion in man,' *Gut*, **9**, 214–221.
134. Scarpello, J. H. B. and Sladen, G. E. (1977), 'Appraisal of the ^{14}C-glycocholate acid test with special reference to the measurement of faecal ^{14}C excretion,' *Gut*, **18**, 742–748.
135. Schultz, S. G. and Zalusky, R. (1964), 'Ion transport in isolated rabbit ileum,' *J. Gen. Physiol.*, **47**, 1043–1059.
136. Silk, D. B. A., Clark, M. L., Marrs, T. C., Addison, J. M., Burston, D., Matthews, D. M. and Clegg, K. M. (1975), 'Jejunal absorption of an amino acid mixture simulating casein and an enzymatic hydrolysate of casein prepared for oral administration to normal adults,' *Br. J. Nutr.*, **33**, 95–100.
137. Singh, M. and Webster, P. D. (1978), 'Neurohumoral control of pancreatic secretion,' *Gastroenterology*, **74**, 294–309.
138. Sladen, G. E. and Kumar, P. J. (1973), 'Is the xylose test still a worthwhile investigation?' *Br. Med. J.*, **iii**, 223–226.
139. Snape, W. J., Matarezzo, S. A. and Cohen, S. (1978), 'The effect of eating and gastrointestinal hormones on human colonic myoelectrical and motor activity,' *Gastroenterology*, **75**, 373–378.
140. Stadil, F. and Rehfeld, J. F. (1973), 'Release of gastrin by epinephrine in man,' *Gastroenterology*, **65**, 210–25.
141. Stiehl, A. (1977), 'Disturbances of bile acid metabolism in cholestasis,' *Clin. Gastroenterol.*, **6**, 45–67.
142. Swanson, C. H., and Solomon, A. K. (1975), 'Micropuncture analysis of the cellular mechanisms of electrolyte secretion by the *in vitro* rabbit pancreas,' *J. Gen. Physiol.*, **65**, 22–45.
143. Szurszewski, J. H. (1969), 'A migrating electric complex of the canine small intestine,' *Am. J. Physiol.*, **217**, 1757–1763.
144. Taylor, T. V., Holt, S., McLoughlin, G. P. and Heading, R. C. (1979), 'A single scan technique for estimating acid output,' *Gastroenterology*, **77**, 1241–1244.

145. Thjodliefsson, B. and Wormsley, K. G. (1977), 'Backdiffusion—fact or fiction,' *Digestion*, **15**, 53–72.
146. Turnberg, L. A., Bieberdorf, F. A., 'Morawski, S. G. and Fordtran, J. S. (1970), Interrelationships of chloride, bicarbonate, sodium and hydrogen transport in the human ileum,' *J. Clin. Invest.*, **49**, 557–567.
147. Turnberg, L. A., Fordtran, J. S., Carter, N. W. and Rector, F. C. (1970), 'Mechanism of bicarbonate absorption and its relationship to sodium transport in the human jejunum,' *J. Clin. Invest.*, **49**, 548–556.
148. Van der Kamer, J. H., Huinink, H. and Weyers, H. A. (1949), 'Rapid method for the determination of fat in faeces,' *J. Biol. Chem.*, **177**, 347–355.
149. Vlahcevic, Z. R., Bell, C. C., Buhac, I., Farrar, J. T. and Swell, L. (1970), 'Diminished bile acid pool size in patients with gallstones,' *Gastroenterology*, **59**, 165–173.
150. Waldmann, T. A. (1966), 'Protein-losing enteropathy'. *Gastroenterology*, **50**, 422–443.
151. Waldmann, T. A. (1970), 'Protein-losing enteropathy, in *Modern Trends in Gastroenterology*, pp. 125–142. vol. 4., (Card, W. I. and Creamer, B. Eds.). London, Butterworths.
152. Waller, S. L. (1975), 'The Lundh test in the diagnosis of pancreatic disease. A comment from the moderator,' *Gut*, **16**, 657–658.
153. Warshaw, A. L. and Fuller, A. F. (1975), 'Specificity of increased renal clearance of amylase in diagnosis of acute pancreatitis,' *N. Engl. J. Med.*, **292**, 325–328.
154. Weintraub, L. R., Weinstein, M. B., Huser, H-J., and Rafal, S. (1968), 'Absorption of haemoglobin iron: the role of a heme-splitting substance in the intestinal mucosa,' *J. Clin. Invest.*, **47**, 531–539.
155. White, T. T., McAlexander, R. A. and Magee, D. F. (1963), 'The effect of gastric distension on duodenal aspirates in man.' *Gastroenterology*, **44**, 48–51.
156. Wright, E. M., Mircheff, A. K., Hanna, S. D., Harms, V., van Os, C. H., Walling, M. W. and Sachs, G. (1979), 'The dark side of the intestinal epithelium. The isolation and characterisation of basolateral membranes,' in *Mechanisms of Intestinal Secretion*, pp. 117–130. (Binder, H. J. Ed.). New York, Alan R. Liss.
157. Wormsley, K. G. (1972), 'Pancreatic function tests'. *Clin. Gastroenterol*, **1**, 27–51.

5. DIFFERENTIAL DIAGNOSIS OF MALABSORPTION AND MALNUTRITION

H. G. SAMMONS

Introduction
   General information profile

Causes of malabsorption

Category I—Dietary deficiency

Category II—Inadequate digestion
   Loss of pancreatic secretion
   Loss of biliary secretion
   Effects of gastric secretion

Category III—Biochemical/mucosal cell abnormality
   Methods of study
   Protein-losing enteropathy
   Failure of immune response
   Iatrogenic malabsorption

Category IV—Inadequate surface
   Gastrectomy and dumping syndrome
   Small bowel resection

Category V—Contaminated bowel

Category VI—Endocrine disorders
   Thyroid
   Tumours

Conclusion

INTRODUCTION

Patients may be malnourished because they have either a dietary deficiency or malabsorption or a combination of both. The presenting features are in many respects so similar that a painstaking investigation is usually necessary to clarify the diagnosis. The former situation can be found in a wide variety of subjects, for instance, in the elderly person living alone, in a subject with a dietary insufficiency associated with sectarian practices, or in dietary indiscretions, such as found in patients with chronic alcoholism. In addition, anorexia must not be overlooked as a possible contributor to a deficiency state. Consequently a clear, well documented, dietary history is essential, as much of the interpretation of the laboratory investigations will depend on it. For instance, steatorrhoea will not be demonstrated, even when enteropathy is present, if little or no fat is being taken in the diet.

Certain initial general laboratory-based information is fundamental to subsequent investigations. Therefore, in this chapter, a basic information profile will be considered first as it can then form a common ground for the more specific biochemical features associated with particular diseases.

Following this, the chapter covers six categories of malabsorption which are presented on an aetiological basis (Table 1). To avoid confusion with too much detail, only those situations most frequently encountered in general hospital practice have been chosen within each category. More comprehensive texts dealing with the abnormalities of gastrointestinal function may be found in Losowsky et al.[1] McColl and Sladen,[2] Dietschy,[3] Sleisenger and Fordtran[4] and Walker-Smith.[5]

Within each category the most frequently used types of investigation are given (Table 2). Each serves to illustrate the minimal investigations which would be

TABLE 1
MAIN CATEGORIES OF MALABSORPTION (MALNUTRITION*)

Aetiology	Disease state	Methods of investigation
I *Dietary deficiency.	Malnourished and wasting. Failure to thrive.	Observe the effect of replacement.
II Inadequate digestion.	Atrophy of gastric mucosa. Pancreatic insufficiency.	Gastric acid and gastrin levels. Intubation for enzyme and bile salt levels. Fat absorption test. Sweat test.
	Bile salt deficiency.	Faecal analyses.
III Biochemical/Mucosal cell abnormality.	Gluten induced enteropathy (Coeliac disease). Disaccharidase deficiency.	Xylose absorption. Faecal analyses. Sugar absorption test. Intestinal biopsy.
	Specific failure of absorption of calcium, iron, B_{12}, Mg, etc. Regional enteritis. Protein-losing enteropathy. Lymphatic obstruction. Failure of immune response. Ulcerative colitis. Iatrogenic enteropathy.	Specific absorption tests. Faecal $^{51}CrCl_3$ after injection. Fat absorption. Serum immunoglobulins. Faecal analyses.
IV Inadequate surface area.	Gastric operations. Intestinal resection or bypass. Ileostomy.	B_{12} absorption. Analysis of discharges. Faecal analyses.
V Abnormally contaminated bowel: bacteria, viruses or parasites.	Contaminated bowel syndrome. Blind loops. Enteritis.	Urinary indican. Breath hydrogen. Breath $^{14}CO_2$
VI Endocrine abnormality. Tumours.	Thyroid. Apudomas. Carcinoid.	Thyroxine and TBG. Serum gastrin. Urine 5HIAA.

expected to be done. This does not say that the others should not be considered as the investigation proceeds. There is also bound to be some overlap between categories. For instance, reduction in effective surface area (Category IV) is also a feature of coeliac disease (Category III).

Some comment and criticism of the investigational methods is given together with, where applicable, the experience of the author.

The results most frequently obtained with each category are also shown together with some interpretation. To avoid repetition, where a method is used in more than one category, e.g. faecal analysis which may be made in almost all categories, the procedure will be discussed only once and then referred to later where necessary.

General Information Profile

Initially, there are seven areas which need to be assessed when the patient is admitted before proceeding to further investigations.

1. *Water, electrolyte and acid-base status.*
   Plasma levels of sodium, potassium bicarbonate or chloride and hydrogen ion levels are necessary as they can indicate deficiency or imbalance. All can be affected by vomiting or diarrhoea.

2. *Renal status.*
   Plasma urea and creatinine levels.
   Because the body attempts to conserve where it can, some indication of malabsorption is often reflected in 24-hour urine levels. These are themselves dependent on a good glomerula filtration. The plasma urea concentration on its own may be misleading as raised levels may be found after gastrointestinal haemorrhage as well as in prerenal urea syndromes. Renal function also affects the synthesis of such substances as 1,25 dihydroxycholecalciferol and the catabolism of others, such as gastrin.

3. *Liver and nutritional status.*
   Good gastrointestinal function is dependent on or associated with several functions of the liver. These may be affected in cirrhosis, obstructive lesions or hepatitis. Consequently, the activities of either alanine or aspartic aminotransferase and alkaline phosphatase, as well as the level of bilirubin should be known. A measurement of prothrombin time will help in the assessment of vitamin K status. Nutritional status may be reflected in albumin levels which can be half the normal level though low levels will also be found in protein-losing enteropathy. Calcium and phosphate levels may also be abnormal and the interpretation of a cal-

TABLE 2
METHODS OF STUDY

Method	I	II	III	IV	V	VI
Replacement	*					
Gastric acid		*		*		
Gastrin		*				
B$_{12}$ absorption		(*)	*	*		
Xylose absorption			*			
Specific Absorption						
Fat		*	*			
Iron			*			
Disaccharide (lactose)			*			
Faecal analysis		*	*	*	*	*
Intubation, enzymes		*				
Bile salts		*				
Immunoglobulins			*			
Breath analysis				*		
^{51}CrCl$_3$ injection			*			
Thyroxine and TBG						*
Urine analysis electrolytes	*	*				
Indoxyl sulphate					*	
5HIAA						*
CEA						*

* Indicates in which category (see Table 1) the method is most applicable.

cium level is to some extent dependent on the albumin concentration.

A total protein measurement may indicate raised levels of one or more of the globulins; immunoglobulins are often raised in malabsorption, especially IgA. The level of the acute-phase reactive protein orosomucoid is frequently raised in inflammatory bowel disease, so is a useful initial measurement.

4. *General metabolic status.*
Diabetes and thyroid function can be a contributing factor to GI symptoms as for instance in chronic pancreatitis associated diabetes. Fasting blood sugar, serum thyroxine and thyroxine-binding globulin levels are also desirable.

5. *Haematological status.*
Anaemia is a frequently presenting symptom; so the haemoglobin level, a full blood count, plasma and red cell folate concentrations, vitamin B$_{12}$, serum iron, and iron binding capacity should be known. Faecal occult blood is a useful measurement on a random specimen as it may give a pointer to the cause of anaemia, but it must be emphasised that the test is not quantitative and that it takes only about 1 ml of whole blood in a 24-hour stool homogenate to give a positive result.

In many patients much of this preliminary profile can be assessed in the out-patients clinic before, or if necessary, on admitting the patient. The aetiology of malabsorption can have so many facets that it seems unwise to omit any of these fundamental preliminary reference parameters.

CATEGORY I: DIETARY DEFICIENCY

Apart from a general lack of sufficient food, which reaches an extreme in anorexia nervosa, there are examples of specific deficiencies. For instance, protein malnutrition can occur in areas of the world where carbohydrate is the staple diet resulting in kwashiorkor, vitamin deficiencies such as vitamin C arise in the elderly person living alone or B$_{12}$ deficiency in vegans who eat no animal food. These may be clinically demonstrable by loss of weight, oedema or anaemia. Clearly a good dietary history and some knowledge of the environment or condition under which the patient is living are an essential part of the diagnosis. This will be further helped by the basic biochemical and haematological measurements. For instance, folic acid and/or B$_{12}$ deficiency will produce a macrocytic anaemia and vitamin D deficiency a raised activity of alkaline phosphatase of bone origin and a low plasma concentration and urinary output of calcium.

As indicated in Table 2, a useful addition to these basic measurements would be analyses of several consecutive 24-hour urine collections. Analysis of these for electrolytes, calcium and magnesium will give an indication of the body's natural response to conserve as much of these constituents as possible. General malnutrition or protein deficiency will be indicated by a low urine urea or total nitrogen output.

Having established that the deficiencies are dietary in origin, then no further investigation will be necessary prior to oral replacement therapy. If there is a failure to respond, or only a partial response, then the cause of the deficiencies will need to be sought within one or more of the remaining areas of malabsorption listed in Table 1.

CATEGORY II: INADEQUATE DIGESTION

Loss of Pancreatic Secretion

The process of chyme passing from the stomach to the upper small bowel sets off a series of hormonal mechanisms which result in a flow of pancreatic juice, rich in enzymes and electrolytes, especially sodium bicarbonate. This juice, arising from the acinar cells of the pancreas, provides the digestive ferments necessary for the hydrolysis, solubilisation or emulsification of large molecular forms of carbohydrate, protein and lipid. Loss of this secretion may be either a primary defect in the pancreas, such as in chronic pancreatitis or cystic fibrosis or, but much more rarely, destruction of the pancreatic enzymes by acid in the lumen of the intestine as found in patients with a gastrinoma. Comprehensive reviews, devoted entirely to the laboratory study of the exocrine pancreas have been given by Gowenlock[6] and Arvanitakis.[7]

Methods for the Study of Exocrine Pancreatic Function

Four main approaches are available. They involve

1. A direct measure of the components of intestinal fluid, concerned with digestion.
2. By feeding tests and measuring corresponding changes in blood levels.
3. By faecal analyses.
4. By the sweat test in cystic fibrosis of the pancreas.

Assessment of Digestive Efficiency of the Aspirated Intestinal Fluid. Intestinal fluid, comprising secretions from the pancreas, bile and succus entericus, is obtained by intubation of the small intestine. A radio-opaque double lumen tube is passed so that the tip of one lumen lies in the antrum and the other passes into the small intestine to a point distal to the entrance of the common bile duct. The position of the tube is checked fluoroscopically. To begin the test, two baseline 15 minute aspirates are obtained either by syringe or by continuous suction. Following this, pancreatic flow is stimulated. Two methods have been recommended. The first is a direct intravenous stimulation by secretin and/or pancreozymin, which may be administered as a single injection or by infusion. The second is by a standard oral meal.

In the first method, commercial hormone preparations are employed. Secretin (95 Crick-Harper-Raper units) is administered slowly over 4 to 5 minutes in case any untoward reaction occurs, followed 30 minutes later by pancreozymin, 100 units. Two 15 minute samples are collected after each stimulation. The aspirated specimens are analysed for volume, pH, bicarbonate, enzymes and bile salts. The specimens are collected on ice to preserve the carbon dioxide content. Bicarbonate and pH should be measured immediately but for enzyme assay and bile salts, the specimen can be deep frozen until analysis is convenient.

With respect to the second method, the standard meal consists of 18 g corn oil, 40 g glucose and 15 g protein (Casilan). The use of the method was reviewed by James[8] and a modification for use in children described by McCollum et al.[9] Although it is argued that this stimulation is more physiological, the aspirated specimens contain partially digested meal and although enzymes can be estimated, bicarbonate and pH cannot as they are affected by the components of the meal. Furthermore, the interpretation of the results may be misleading since stimulation of exocrine secretion by this means may be deficient when there is disease or dysfunction of the small intestine, for instance in coeliac disease or even duodenal ulcer. One other advantage of the hormone stimulated juice is that it can be seen by its orange/brown appearance after stimulation to contain bilirubin. This is a useful check on the siting of the tube and for subsequent estimation of bile salts.

The enzymes usually measured are amylase, lipase and trypsin, though others may be equally important. At least two enzymes should be measured. As deficiencies of lipase, trypsinogen and enterokinase, which activates trypsinogen, have been described, we prefer to estimate lipase and trypsin. Others have recommended amylase alone but here there is always the danger of contamination with swallowed salivary amylase.

The methods found most helpful in our laboratory are as follows:

For lipase: We use a method which measures the fatty acids released with the aid of bile salts during the hydrolysis of olive oil. This has the advantage that the ability of the mixture to form a fat emulsion can be seen as the digestion proceeds.

For trypsin: The specific substrate p-toluene-sulphonyl-L-arginine methyl ester is used. Both the activity of trypsin, and also the conversion of trypsinogen to trypsin by enterokinase, are enhanced by bile salts.

Total bile salts can be semiquantitatively estimated in the jejunal aspirate by the simple colorimetric Petenkoffer reaction after alcohol extraction. Alternatively the specific enzymatic technique used by McCollum et al.,[9] may be employed.

Results. pH: Normally the pH will, in fresh juice, lie between 6 and 8, the lower pH being largely due to the presence of high P_{CO_2} which can vary between 1 and 30 kPa. Both P_{CO_2} and bicarbonate can be measured on a blood gas apparatus. If the pH of the specimens is below 6, then the measured activities of lipase and trypsin are unreliable as the enzymes are rapidly inactivated at low pH. If the pH is too acid, the brown bilirubin will be converted to green biliverdin. This observation can be of practical value in interpreting low enzyme activities.

If complete collection of pancreatic secretion could be guaranteed, then clearly the total amount of bicarbonate and enzyme activity could be reported. Unfortunately, total recovery is not easily achieved so it is better to report both concentration and total recovered values. In our experience, the normal range is wide.

The bicarbonate concentration can increase after stimulation from a baseline of 10–20 mmol/l to reach values > 100 mmol/l. Activities above one lipase unit/ml or 17 trypsin units/ml can be considered satisfactory, the basal level normally rising after stimulation by at least ten times when expressed as total activity.

Almost complete absence of activity has a good diagnostic reliability for complete pancreatic failure but any value above the values quoted gives no indication of the extent of loss of pancreatic function as there is considerable pancreatic reserve. In cystic fibrosis of the pancreas, the juice will be low in volume and in bicarbonate concentration. It will be thick and viscous and difficult to aspirate. It is often so viscous that it will stay at the bottom end of the test tube when inverted.

The level of bile salts is about 4 mmol/l in the fasting state and increases markedly when the gall bladder is stimulated to contract following the pancreozymin injection as it also has cholecystokinin activity. Alkaline phosphatase activity in the intestinal lumen will also rise about tenfold after stimulation for the same reason but there is some evidence that it may also be released from the intestinal mucosa. In contrast to acute pancreatitis, the measurement of plasma amylase and lipase levels in chronic pancreatic insufficiency is of little value even after hormonal stimulation.

Assessment of Pancreatic Function by Feeding Tests. Absorption tests using starch or olive oil have been used with varying success. We have found the method of Robards[11] to be helpful, using double cream milk and feeding it as ice cream (60 g fat to adult: 30 g to the child). Blood levels are measured by nephelometry and the increase is normally 5 to 10 fold in 2 to 3 hours after the oral dose. A flat curve will be found in severe pancreatic failure. Glucose absorption may be rapid and an abnormal diabetic type glucose absorption test may be found in chronic pancreatitis but the test is non-specific. Xylose is normally absorbed and is a most useful differential test in diagnosis of this condition. A sensitive test utilising exhaled $^{14}CO_2$ after feeding labelled triolein has recently been described[10] but it may be expensive and one is reluctant to use carbon-14 in human subjects and especially in the child.

Faecal Analysis

General Appearance and Volume. Although faecal analysis will not distinguish between malabsorption and maldigestion faecal material in severe pancreatic failure has several characteristic features. The faeces are bulky, pale, semi-formed and offensive, frequently with the odour of butyric acid derived from dietary milk fat. They may float in the pan but this characteristic is due as much to entrapped gases as it is to fat. Though semi-formed, the volume sometimes exceeds 1 litre or 5 to 10 times the normal volume (Table 3).

For quantitative purposes a four day period is recommended. Much has been said about the use of markers. Although theoretically they may reduce the number of faecal analyses, in practice they are not easy to organise. They must be given, for equilibration purposes, several days before the test so lengthening the period of study. They must be given in precise amounts during each day making supervision more necessary. It is much easier to maintain a patient on a normal ward diet and analyse each day's faecal loss. Depending on the volume, the combined stool or each 24-hour specimen is homogenised and aliquots taken for the required analysis.

Faecal Fat. The normal output of fat in the faeces is about 17 mmol/day, but in pancreatic failure the degree of steatorrhoea can be as high as 200 mmol. The fat lost

TABLE 3
FAECAL/DIET COMPOSITION/24 HOURS (MMOL/DAY)

Constituent	Faecal output (normal)	Average intake	Faecal output in malabsorption
Volume (ml)	200	1500	250–2500+
Fat	<17	248	25–200
Nitrogen	<140	900	200–600
Volatile Acids (Acetic)	<20	50–60	34–170
Lactic acid	<3	10–20	3–11
Sodium	7	150	→200
Potassium	15	80	→100
Chloride	3	150	→200
Calcium	6–10	10–25	→20
Phosphate	3·5	39	→10
Magnesium	7	11	→14

in the faeces may equal the dietary intake. In partial pancreatic failure, however, the results are not nearly so convincing and steatorrhoea may not develop until at least 85% of the available lipase has been lost. The extent of the hydrolysis of the faecal fat is not a good indication of pancreatic failure as considerable hydrolysis may take place, presumably by bacteria, even when there is complete pancreatic atrophy. The one exception to this is in the stool of a child with pancreatic cystic fibrosis. Here, under light microscopy, fat globules may sometimes be seen. If the steatorrhoea is severe, calcium soaps may be present and some of the pallor may be due to their presence. However, as with hydroxy fatty acids which are also said to be present, their presence is by no means confined to pancreatic failure.

The estimation of faecal fat as fatty acid depends on their release from protein binding, soaps and triglyceride. This is achieved in the method of Van de Kamer et al.,[12] a method which has now gained universal acceptance. Slight modifications have been introduced in the method we have used in this department. In this modification much of the alcohol is not refluxed, but boiled off during the saponification process. This conveniently avoids the partition solubility problem between alcohol and petroleum ether phases which occurs when the alcohol concentration is too high. If hydroxy acids are suspected, then diethyl ether can replace the petroleum ether but being much more volatile, it is less easy to handle. The amounts of the latter, when present, are relatively low, about 10%, so the quantative contribution is not usually important.

Faecal Nitrogen. Faecal nitrogen excretion is also raised in severe failure but not to the same extent in proportion to the normal value as fat (Table 3). Whereas the normal is up to 140 mmol/day it may be increased to 600 mmol, which represents about 60% of the normal intake.

Nitrogen may be estimated on an aliquot using the normal Kjeldhal sulphuric acid digestion procedure. One or two commercial instruments, such as Kjeltec system (Tecator Ltd), have work-simplified this procedure making it semi-automated and safer.

Increased faecal excretion of nitrogen may also be found in certain systemic diseases, such as diabetes mellitus, amyloid and radiation enterocolitis, the nitrogen arising presumably from increased bacterial flora and mucus.

Although daily faecal fat and nitrogen analyses are not a great deal of value in differential diagnosis of malabsorption, they can be of help in following the effectiveness of replacement therapy with pancreatic extracts in proven cases, though faecal fat is probably the more sensitive index.

Faecal Enzymes. Normally only slight proteolytic activity is found in faeces from adults, but there is marked activity in that from children. Presumably this is because the transit time is so much shorter and for this purpose a fresh random specimen is adequate. Loss of activity in the child's faeces can have useful significance as an indication of pancreatic failure. A simple inexpensive total proteolytic assay is sufficiently sensitive for this purpose. We have found the digestion of azo-albumin as described by Tomarelli[13] adequate for this purpose. If good activity is found, then cystic fibrosis of the pancreas can be virtually ruled out. In the adult, the situation is not nearly so clear cut, though some have found the specific estimation of trypsin and chymotrypsin to be of value, especially if purgation is also used.

However, inactivation by bacteria or contribution by bacterial proteases may confuse the issue. Faecal protease activity should not of course be looked for whilst the patient is on pancreatin replacement therapy.

Further comments on the use of faecal analyses are made in Category III and V.

The Sweat Test in Cystic Fibrosis

Whether or not the pancreas is involved, sweat sodium chloride concentration will almost certainly be abnormal in cystic fibrosis. When the pancreas is affected then such parameters as faecal fat, proteolytic activity and duodenal enzymes will be abnormal but as these may not always occur at first, then other symptoms which lead to the paediatrician to suspect cystic fibrosis need to be carefully investigated.

The abnormality in the sweat gland is such that sodium is not reabsorbed and its concentration in the sweat can approach that of plasma. As the secretion is under parasympathetic control, a stimulant such as acetylcholine or its analogues or pilocarpine injected intradermally can be used to stimulate sweat production.

The very reliable method of Anderson and Freeman[14] uses methylcholine (Mecholyl) as a stimulant. Unfortunately, the procedure takes 3 hours to complete and requires an intradermal injection into the forearm or back and sometimes sedation is needed for this. The iontophoretic method uses pilocarpine as a stimulant but requires meticulous attention to detail and occasionally burning of the skin has been reported. Recently the sweat chloride meter has been introduced which also uses pilocarpine as a stimulant. Price and Spencer[15] have pointed out that the pilocarpine concentration is critical and must be greater than 500 mg/100 ml. This method takes only 10 minutes to complete and avoids the difficulty of the operator personally injecting the stimulant intradermally, not an easy exercise in a young child unless under sedation.

The level of sodium or chloride should be less than 50 mmol/l. Levels between 50 and 70 are suspiciously high. In cystic fibrosis, levels of 70 to 120 may be found. The importance of an accurate result cannot be stressed too much. Smalley et al.,[16] have given good evidence of the unreliability of results in inexperienced hands and Schwarz et al.,[17] found that only 4 patients out of 30 with sweat sodium ranging from 50–75 mmol/l proved to have cystic fibrosis after several years observation. It seems at the present time that the sweat chloride meter is a useful screening method but if high levels are found then a comfirmatory stimulation test for the actual collection of sweat should be employed. At least 70 mg of sweat should be obtained, adequate controls employed and great care and cleanliness taken in the procedure. In any case, the diagnosis of cystic fibrosis is never justified on the basis of the sweat test alone.[17]

With modern treatments for chronic respiratory disease, cystic fibrosis may not become apparent until adolescence or adult life. Sweat sodium and chloride concentrations increase with age and normal adult ranges up to 120 mmol/l may be found but with increasing incidence of higher levels in patients with chronic chest diseases, though not related to any type of chest disease. Malabsorption would appear to be a less frequent finding in the adult form of the disease.

Other Screening Tests for Cystic Fibrosis. Screening for cystic fibrosis in the newborn by demonstration of albumin in the meconium has been recommended, but the test is invalidated if the specimen is contaminated with blood. The technique has been improved by the demonstration that lactase can be detected in meconium only from babies with cystic fibrosis.

Loss of Biliary Secretion

The main role of bile salts would appear to be to assist in the digestion and absorption of triglyceride fats. As bile salts are synthesised from cholesterol in the liver any process which interferes with this production and subsequent secretion into the lumen of the small intestine will interfere with this process. The common causes of this deficiency are therefore common bile-duct obstruction, liver failure, fistulae which drain away the bile, and congenital bile salt deficiency. The degree of steatorrhoea is usually mild, 50 to 70 mmol/day of fatty acid being lost in the faeces. Nitrogen excretion on the other hand, is normal. It has been stated that in obstructive liver disease, the degree of steatorrhoea correlates with the faecal stercobilinogen level but in cirrhosis and hepatitis steatorrhoea can occur in the absence of jaundice.

The pallor of the stools is no indication of bile

deficiency and may be due as much to the reduction of pigment by bacteria and the presence of calcium soaps as to the absence of sterobilin. In contrast to pancreatic insufficiency, fat absorption studies have not been shown to be helpful. The functional integrity of the intestinal mucosa would appear to be unimpaired in bilary deficiency so absorption of xylose and B_{12} is unaffected. The effect of absence of bile on steatorrhoea is well documented in the child with congenital bile-salt deficiency described by Ross et al.[18] Treatment with bile salts reduced the faecal fat from 70 mmol/day to nearly normal levels within a few days.

Effect of Gastric Secretion

In the human subject, digestion in the stomach is normally of minor importance. Some breakdown of tissue protein occurs to enable natural foodstuffs to be adequately mixed before passage into the upper intestine. Nevertheless, abnormalities in gastric secretion can have a pronounced effect on absorption from the small bowel. For instance, hyperacidity can, when stomach emptying occurs, cause the upper part of the small intestine to become too acid and thus destroy the activity of pancreatic enzymes. Conversely, achlorhydria will allow an alkaline reaction to be maintained and thus encourage bacteria to proliferate in an otherwise sterile area of the upper small intestine. Gastric atrophy will reduce the production of the glycoprotein necessary for the absorption of vitamin B_{12} from the normal ileum.

The need to measure gastric function is therefore of considerable importance in the differential diagnosis of malabsorption.

Methods of Study

Acid Production. The questions appropriately to be asked in a situation concerning the malabsorption syndrome are: is there achlorhydria present as in pernicious anaemia? *or* is there hyperchlorhydria leading to inactivation of pancreatic enzymes as in the Zollinger-Ellison syndrome? For this purpose intubation is necessary followed by gastric aspiration under controlled conditions. The procedure adopted by Baron,[19] using a radio-opaque tube, is recommended. The gastric juice is aspirated over a prescribed interval of time before and after stimulation with a parasympathomimetic agent such as pentagastrin. The hydrogen ion concentration is best measured in the juice obtained using a pH meter and back-titrating to pH 7·4. In certain cases, for example in suspected pernicious anaemia, it may be sufficient to demonstrate achlorhydria by foregoing the titration and simply testing the aspirate with Universal pH test paper; if acid (pH 1–2) then pernicious anaemia can be excluded from the diagnosis. The full test is rather prolonged occupying at least two hours and usually requires constant supervision even when using a peristaltic aspiration pump as the pump tube often becomes blocked or collapses under partial vacuum thus created.

The normal rate of hydrogen ion production is less than 10 mmol/hour for the baseline level and less than 35 mmol/hour after stimulation. For hypersecretors the basal level is probably the most informative. Overnight collections are unpleasant for the patient and unless constantly supervised, may be inaccurate. Furthermore, the patient may lose a large volume of water together with its contained electrolytes by this procedure.

Other methods of hydrogen ion detection, such as the tubeless test or dyed-resin technique, are too subject to artifacts to be reliable. The pH-sensitive pill only measures acidity, and gives no idea of the total amount of acid being secreted. It can be used in children. Peptic activity, if required, can be measured by the method of using haemoglobin as substrate,[20] or by a nephlometric technique.[21] Pepsin levels are usually complementary to the acid production.

Gastrin Levels. Plasma gastrin concentration is also increased in pernicious anaemia and markedly so in patients with gastrinoma. Blood (heparinised) for this assay should be taken from a patient in the fasted state as a protein meal will increase blood levels. The plasma should be kept frozen until assayed.

Gastrin may be measured by radioimmunoassay. The normal fasting level is less than 50 pg/ml but, as antisera vary, it is best to establish a normal range for the laboratory carrying out the assay. Levels greater than 400 may be found in patients with pernicious anaemia, gastrinomas, G-cell hyperplasia or hypercalcaemia.

Investigation of B_{12} Deficiency. Apart from genuine dietary deficiency, low levels of vitamin B_{12}, i.e. less than 250 ng/l, are usually associated either with intrinsic factor deficiency or with malabsorption in the ileum, though other rarer causes may sometimes occur. These are listed in Table 4 approximately in their order of occurrence in general hospital practice.

TABLE 4
CAUSES OF B_{12} DEFICIENCY AND ASSOCIATED DISEASE

Intrinsic-factor deficiency	Pernicious anaemia
	Total or partial gastrectomy
	Atrophic gastritis
	Intrinsic-factor antibodies
	Congenital deficiency
Ileal malabsorption	Diseased ileum
	Resected ileum
	Congenital malabsorption of B_{12} — I.F. complex
Other associated causes	Bacterial or parasitic infestation
	Pancreatic dysfunction
	Transcobalamin II deficiency

The subject of vitamin B_{12} absorption and malabsorption was reviewed by Donaldson.[48]

Intrinsic factor (IF) may be measured by radioimmunoassay in gastric juice before and after stimulation with histamine or pentagastrin. It is claimed that the method is as sensitive as the Schilling test (*vide infra*) or whole-body counting for B_{12} absorption. Furthermore, Ardeman and Chanarin[22] found that out of 21 patients with pernicious anaemia, IF was absent in 11 and very low in 10 with no overlap with other megaloblastic anaemias. Our impression is that it is a useful adjunct to the B_{12} absorption (Schilling) test.

Absorption of vitamin B_{12} labelled with the radioactive cobalt isotopes ^{57}Co or ^{58}Co can be assessed by measurements in the blood, in the faeces, in the urine and by whole-body counting. All methods are probably equally effective but there are drawbacks and limitations for each. After an oral dose, absorption into the blood is low, reaching a peak level in 6 to 8 hours. The level of activity reached is only just above background so a counting time of 1 to 2 hours is required. A good counter is necessary, free from background interference. Faecal counting after an oral dose has to be continued for 7 days, to ensure complete excretion. Whole-body counting takes 14 days to complete and requires equipment which is available in only a very few centres. It is more common practice to measure excretion in the urine over a period of 24 hours. In patients with depressed glomerular filtration rate, excretion may be delayed and much of the dose may be excreted in the second 24 hours. It is also necessary to give a non-radioactive flushing dose of 1000 μg of vitamin B_{12} intramuscularly during the test, to ensure that the tissue stores are saturated and excretion into the urine of the oral dose is as complete as possible. This procedure may obscure further similar studies for some weeks. Nevertheless, although the same doses may be used for all the methods and there is much to be said for measurements in the plasma, the urine excretion test seems to be the most popular. A combined preparation, DICOPAC CT 31P, is provided by the Radiochemical Centre, Amersham, UK. The cobalt-labelled vitamin B_{12} is given in two forms; in one ^{57}Co is bound to intrinsic factor and in the other, ^{58}Co is unbound. The test will measure at the same time IF deficiency in the stomach or ileal malabsorption. If IF is deficient, only ^{57}Co vitamin B_{12} will be adequately absorbed. If ileal dysfunction is present, neither will be adequately absorbed.

The reference range for either isotope is that 10–20% of the dose is normally excreted in the urine in 24 hours. If intrinsic factor deficiency (IFD) is suspected, then the ratio of $^{57}Co/^{58}Co$ is helpful. Normally it is 0·8–1·3 but in IFD it could be 1·8–15·0. Using the ratio only has the advantage of being independent of a full 24-hour urine collection. In ileal malabsorption the levels will be only 0–5%. A complete 24 hour urine collection is required for this so any loss will invalidate the test.

It should not be overlooked that in vitamin B_{12} deficiency intestinal function may be impaired in a more general way, e.g. reduced xylose absorption or mild steatorrhoea. All parameters quickly revert to normal on treatment. In contrast folic acid absorption may not be affected.

CATEGORY III: BIOCHEMICAL/MUCOSAL CELL ABNORMALITY

Before considering the effects of mucosal cell dysfunction it is helpful to stress some characteristic features concerning intestinal function.

There is a certain purposefulness about absorption from the small intestine. The absorptive cell and its associated glycocalyx coating seem designed or have evolved to absorb as much nutrient as is available after preparation for absorption has taken place in the lumen of the intestine. Many processes are available to achieve this objective including the use of hydrolases and specific proteins within the brush border or the cell itself. Consequently, many steps are involved in the passage from lumen to blood or lymph, any of which may be rate limiting. Nevertheless, the overall rate in most cases is remarkably rapid, many water soluble substances appearing in the blood within minutes after oral ingestion.

The absorptive process is astonishingly efficient and very little is normally rejected in the faeces (Table 3). Not only is most of the diet normally absorbed but so are digestive secretions especially water and salts in even greater amounts than the dietary intake.

Except for vitamin B_{12} and the constituents of bile which are absorbed in the ileum, most substances are absorbed throughout the small intestine, though perhaps most effectively in the upper half. There is therefore, a fair amount of reserve capacity before malabsorption is pronounced. The large intestine is also responsible for the absorption of residual water and salts not absorbed in the small bowel.

Any or all of these salient characteristics may be affected in malabsorption to a greater or less extent. They will therefore form a background to the functional irregularities associated with the diseases included in this category.

As indicated in Table 1, there are a variety of lesions affecting the mucosal cell but the classical presenting symptoms of diarrhoea, weight loss, steatorrhoea and malnutrition are similar. Passage of the products of digestion from the lumen through the enterocyte and into the blood are in some way affected in all cases. It is the problem of the biochemical laboratory to find out the extent of their abnormality. To see if it is a generalised malabsorption as in coeliac disease, or specific as in lactase deficiency, or may even be introgenic following antibiotic therapy. The problem may be complicated by the coexistence of more than one pathology as in iron malabsorption in renal failure. It is therefore important that the tests should be as sensitive as possible. Furthermore, they should be taken in conjunction with other abnormalities that may be present, such as the nature of the anaemia or clotting factor deficiency or bone lesions such as osteomalacia.

For this category (mucosal cell abnormality), therefore, the tests listed in Table 2 will be described in some detail together with their relative merits.

Methods of Study

Xylose Absorption Test. The measurement of xylose absorption fulfils nearly all the criteria necessary for assessing proximal intestinal function, and has been found to be nearly always abnormal in primary malabsorption. Xylose is water soluble and absorbed as quickly as glucose. This supports the view that it is transported through the cell by an active process, requiring good cellular function. Xylose is not normally present in the blood, is metabolised slowly and is excreted in the urine in proportion to the administered dose if the latter does not exceed 5 g; above this figure xylose may not be completely absorbed and at the 25 g dose level it can be nauseating and produce diarrhoea.

Besides some disagreement about the recommended dose there is also some controversy about whether the test is best done by analysing blood or urine. We have examined this problem and come to the conclusion[23] that although there may be certain cases where urine analyses are helpful, blood levels taken at 1 hour after a 5 g dose differentiate best between normal and abnormal subjects, providing the level is adjusted in the adult to a constant surface area of 1·73 m^2.

Urine analyses to be meaningful require a knowledge of renal function as the excretion depends on the GFR. Furthermore, careful supervision of the patient is essential. Blood measurements require only venous samples to be taken at two prescribed intervals of time. If very low blood levels are obtained (<0·2 mmol/l) then there is always the possibility that this could be due to delay in stomach emptying. In this situation the test should be repeated and measurements made at 0, 1, 2, and 3 hours consecutively. Alternatively consecutive split urine collections for 2 and 3 hours can be analysed.

At one hour after the 5 g oral dose, the normal 95% range in this laboratory, for any age above 14, is 0·65–1·33 mmol xylose per litre of serum. The distribution is skewed with a median at 0·92 mmol. For children up to 8 years (<30 kg body weight) the figure should be greater than 1·35 with a median of 2·8 mmol. The concentration in the blood appears to fall linearly with increasing body weight from 5 kg reaching the upper end of the 'adult' range of 1·3 at about 40 kg.

The percentage dose excreted in the urine in 0–2 hours is 12–32 (95% range) with a median of 19·5. In 0–5 hours it is 21–50 with a median of 35·5. These figures apply to the 20–59 age range; because of possibly falling GFR urine levels are not reliable in patients over the age of 60.[23] It is of interest that the first use of xylose was for measuring renal function after an intravenous injection.

Xylose absorption is almost always abnormal in coeliac disease and tropical sprue with values usually between 0·4 and 0·6 mmol/l. It is most helpful in following the course of treatment with, for instance, gluten-free diet.[23] This test can be used to show the effect of a gluten challenge on coeliac patients or to show the effect of cows' milk protein on the function of the small intestine in children intolerant to this protein. Clearly it is a test that can be used effectively in any suspected protein-induced enteropathy.

There have been criticisms of the xylose absorption test in the past but many of these criticisms can be attributed to a high dose or difficulties in urine collection. In our experience, when a 5 g dose is given and blood levels rather than urine excretion is used, most of those difficulties are overcome. Furthermore, the shorter time of one hour makes the possibility of bacterial destruction of a sufficient degree to affect the blood levels most unlikely except perhaps to a small extent.

Lactose Absorption. Lactose absorption depends on adequate amounts of the lactase in the brush border of the enterocyte. Another lactase within the cell is non-specific and is a galactosidase. It is not concerned in absorption though it is normally measured together with brush border lactase when enzyme activity is determined in a mucosal biopsy. There are three distinct situations affecting the mucosal cell of the small intestine that may affect lactose absorption. Firstly, the brush border lactase is biologically labile and seems to be easily destroyed in chronic villus atrophy such as untreated coeliac disease or in acute conditions such as gastroenteritis. The situation is reversible on treatment but sometimes recovery is rather slow. Some temporary relief of symptoms may be achieved by putting patients on a low-milk diet. Secondly, two other situations arise where lactase activity may be low, but in both of these the cell is not structurally affected. The first is a rare congenital deficiency which may be present at birth and is demonstrable on the initiation of milk feeding. The second has ethnic characteristics: in most ethnic groups the lactase activity decreases during childhood and reaches low levels in the adult; however, in the caucasians of northern and western European extraction lactase activity persists at a high level throughout adulthood. A summary of the incidence of lactase deficiency in ethnic groups as culled from the literature is given in Table 5.

TABLE 5
INCIDENCE OF LACTASE DEFICIENCY OR LACTOSE INTOLERANCE AMONGST DIFFERENT PEOPLES

Group studied	Number studied	% Abnormal lactose absorption test or lactase deficiency
Europeans	797	6–17 Average 10
American whites	363	6–19 Average 13
American negroes	301	70–81 Average 75
African negroes	39	14–95 Average 58
American indians	156	66
Asians	153	54–100 Average 86

So it appears that a low lactase activity may be the normal state in most populations except for those of northern European extraction. It has been calculated that there must be at least 30 million lactase-deficient citizens in America.

A review of the aetiology of low lactase activity is given by Sahi.[24] In addition to these remarkable racial variations the fact must not be overlooked that there are also differences in the quoted normal ranges from various laboratories (Table 6). It is apparent from Table 6

TABLE 6
NORMAL RANGE FOR LACTASE ACTIVITY

u/g wet wt	u/g protein	Investigators
Not given	3–94	Dunne et al. (1977)[27]
Not given	39–258	Haemmerli et al. (1965)[28]
0–11·1	3–83	Newcomer, McGill (1967)[26]
0·2–19·0	Not given	Dahlqvist (1964) (1968)[25]
1·0–25·0	11–280	This laboratory

that the normal range depends on the laboratory concerned and the units used to express the activity. It is, of course, difficult to establish a normal range for this assay as normal healthy subjects are unwilling to submit to a jejunal biopsy. Usually the quoted ranges are based on results on hospital patients who cannot strictly be classified as normal, or on volunteers from prison who are in an unnatural environment. The definition of lactase deficiency is therefore still arbitrary and may depend more on clinical symptoms than actual values. Theoretically, it should be possible to demonstrate that less than a certain lactase activity is rate-limiting to the hydrolysis, and hence the absorption, of the disaccharide, but as far as the author is aware this has not been done.

Despite the fact that low lactase activity is regarded as the norm in such groups as North American Indians, Orientals and black races, milk intolerance is not usually troublesome. They live on their natural diet of yoghurt, cheese and other milk fermented products without dietary problems. It is possible that milk intake levels played a part in the development of lactase levels and that lactose intolerance may be only clinically important among societies in which milk is a vital constituent of the diet. Therefore, although a state of deficiency or malabsorption can be identified by the disaccharidase result or the lactose absorption test these must be interpreted in conjunction with other tests and symptoms and in relation to the race and dietary habits of the patient concerned. Whereas a deficiency of lactase leading to a lactose intolerance may be clinically important in a European, it may have little or no significance in an Asian who has all but excluded milk from the diet.

Lactose Absorption Test. It is important to distinguish between lactose intolerance and lactose malabsorption. The former is a clinical term and refers to the symptoms of pain, borborygmi, flatulence or diarrhoea which may occur after taking lactose, usually in the form of milk.

Lactose malabsorption can be measured by feeding 50 g lactose in 250 ml water to a patient in a fasted state and measuring blood glucose levels at 0, 1, 1½ and 2 hours. In a child the dose will be at a level of 2 g/kg but for an adult 50 g is adequate. This is the amount contained in 1 litre of milk. The blood level normally reaches a maximum in 30–90 minutes. Lactose malabsorption may be defined as the inability of the blood glucose to rise by an amount greater than 1·1 mmol/l after an oral dose of 50 g lactose. The test does not distinguish between primary and secondary lactose deficiency[4] but the latter occurs more frequently in clinical practice as a result of destruction of the integrity of the mucosal cells and the brush border.

Lactase Assessment. Lactase activity can be measured in a jejunal biopsy specimen of 10–20 mg. The activity is stable when kept frozen at $-20°C$. A suitable control for future assays is to homogenise several grams of tissue and store aliquots deep frozen. An acceptable method for the assay is that described by Dahlqvist[25] which is also applicable to the assay of sucrase and maltase using the appropriate buffers and substrates.

The normal ranges for lactase are given in Table 6. The range is wide but this may be associated with geographical areas from which the subjects have been chosen. They can be expressed either as units/g wet weight or units/g protein. There is good correlation between them so to avoid the extra step, the former is recommended.

When taking the biopsy it is important to obtain mucosa from as near as possible to the duodenal-jejunal junction as disaccharidase activity is not uniformly distributed throughout the intestine but highest in the jejunum. It has been recommended that a sucrase: lactase ratio will identify lactase deficiency better, but this is not our experience.

Reports of correlation of lactose absorption with mucosal lactase are controversial. We, as others,[26] find that about 30–35% of patients with an abnormal lactose absorption can have a normal lactase activity.

Lactase activity can be depressed in any mucosal cell defect affecting the jejunum. There is also support for the concept that it may be depressed in more generalised inflammatory disease of the whole bowel. For instance, a decrease in brush border enzymes including lactase has been shown in Crohn's disease[27] and in ulcerative colitis. Although either the absorption of the other commonly occurring disaccharides, sucrose and maltose; or the activity of their respective hydrolases, sucrase and maltase, may also be depressed in mucosal cell damage, lactase seems to be the most sensitive and takes longest to recover.

Although lactase deficiency may be secondary to a disease which has caused damage to the small intestinal mucosa, the enzyme may also be inhibited by certain drugs. For example, there is an association in alcoholism of lactose malabsorption and disaccharidase deficiency. A recent study in this laboratory showed that aspirin and salazopyrin at therapeutic plasma levels could inhibit lactase *in vitro* to an extent of 10 and 16% respectively. There was no effect on sucrase or maltase. These observations emphasise once more the biological lability of lactose.

In conclusion, it would seem that, in adults, assessment of lactose intolerance by an oral lactose absorption test and by measurement of intestinal lactase activity is worthwhile as a guide to treatment or dietary restric-

DIFFERENTIAL DIAGNOSIS OF MALABSORPTION AND MALNUTRITION

TABLE 7
FAECAL APPEARANCE & COMPOSITION
(Normal & Abnormal)

Patient: ER

Date	Appearance	Colour	Vol ml	OB	Fat mmol	VFA mmol	Na mmol	K mmol	N_2 mmol	Ca Soaps
20–29/1/77	Fluid	Brown	570	−ve	6·3	6·5	52	14	120	—
29–30	Insuff									
30–31	Formed	Brown	650	−ve	22·5	44	66	30	180	—
31–1	Formed	Brown	200	−ve	5·0	11	14	12	85	—
1–2	Formed	Brown	150	−ve	0·3	9	4	10	60	—
2–3	Formed	Brown	150	−ve	10·0	10	6	8	80	—
3–4	Semiformed		200	−ve	5	11	16	10	110	—
Mean			274		7·0	13	22	12	91	

Patient: AH

Date	Appearance	Colour	Vol ml	OB	Fat mmol	VFA mmol	Na mmol	K mmol	N_2 mmol	Ca Soaps
16–17/9/76	Formed	Brown	200		48	7	4	14	150	−ve
17–18	Formed	Brown	470	−ve	98	18	8	62	150	
18–19	No stool									
19–20	Formed	Pale	150	+ve	103	11	2	24	115	−ve
20–21	Insuff									
21–22	Formed	Pale	240	+ve	141	11	4	36	135	
Mean			177		65	8	3	23	92	

Patient: TM

Date	Appearance	Colour	Vol ml	OB	Fat mmol	VFA mmol	Na mmol	K mmol	N_2 mmol	Ca Soaps
25–26/4/78	Formed	Lt brown	340	+ve	208	28	12	42	325	+ve
26–27	Formed	Pale	200	−ve	118	15	4	26	160	
27–28	Formed	Pale	350	−ve	175	18	12	46	265	
28–29	Formed	White & cream	430	−ve	126	16	14	36	200	
29–30	Formed	Pale	810	−ve	170	40	28	46	620	
Mean			426		160	23	14	40	314	

OB = Occult Blood, VFA = Volatile Fatty Acids

tions. At the same time, great emphasis should be placed on the clinical symptoms which may develop within 24 hours of an oral dose. This is probably even more applicable to children where a congenital deficiency is possible.

Faecal Analyses. Normal faecal output compared with intake and output in malabsorption is given in Table 3. As in pancreatic deficiency, so also in intestinal disease associated with mucosal cell damage, it is essential for the measurement of other constituents (except for faecal enzymes and sugars in the child) to collect faecal material daily for at least 4 consecutive days to obtain meaningful results. Bowel evacuation varies so much from day to day, possibly even more than in a normal person, that analysis of single random specimens or even a single day collection can be misleading. The need is illustrated in Table 7 which compares the results of consecutive faecal analyses from a patient with no steatorrhoea (ER) with those from one who has untreated adult coeliac disease (AH) and one with severe streatorrhoea due to pancreatic carcinoma (TM).

It can be seen that a significant departure from normality (c.f. Table 3) can occur sporadically even when the average result is normal, possibly as a result of bowel retention. Whether the stool is formed or unformed, the volume is not necessarily related to its consistency. The colour of the stool is more related to the reducing action of bacteria on faecal pigment than absence of stercobilin since there was no evidence of biliary insufficiency in any of these patients. This is a common observation in the malabsorption syndrome.

Although present in only one of these patients, calcium soaps can confer a degree of pallor to the stool suggestive of aluminium paint. These can be easily demonstrated by extraction into ether in which they are dispersed but insoluble. The possibility of a low calcium level, increased alkaline phosphatase activity, or even osteomalacia should be considered when the faecal finding is consistently positive. Steatorrhoea may be present without calcium soaps being demonstrated. In other words, fat and calcium absorption are not necessarily related.

Several other points can be made from Table 7. The degree of steatorrhoea in untreated coeliac disease is not as severe as commonly found in pancreatic insufficiency. This patient (AH) was on a normal ward diet. It must be stressed that between 60 and 80 g of fat per day must be supplied and *known* to be eaten. If less is taken then it could be almost totally absorbed—malabsorption is only a relative term. This may be why some patients when investigated are reported as having no steatorrhoea. Furthermore, it may be difficult to obtain this level of fat in the diet if the patient has a poor appetite or is fastidious—as is often the case in a child with coeliac disease. In such a situation, a proper balance study must

TABLE 8
AN EXAMPLE OF FAECAL COMPOSITION IN ULCERATIVE COLITIS

Date	Appearance	Colour	Vol ml	OB	Fat mmol	VFA mmol	Na$^+$ mmol	Cl' mmol	K$^+$ mmol	N$_2$ mmol
1–2/3	Semi formed and fluid	Brown	760	+ve	7	9	48	54	16	336
2–3	Unformed and fluid	Brown	1340	+ve	12·4	60	124	156	56	320
3–4	Unformed and fluid	Brown	240	+ve	5·0	3	26	30	18	105
4–5	Liquid and formed	Brown	1630	+ve	15·0	87	112	124	68	680
5–6	Liquid and semi formed blood	Brown	1400	+ve	3·75	66	96	108	64	490

OB = Occult Blood, VFA = Volatile Fatty Acids

be done and the faecal fat expressed as a percentage of intake; normally this is about 95%. The type of fat in the diet is important; unsaturated fats are often better absorbed than saturated fats presumably because they may be more easily emulsified. Using gas chromatography an analysis of the different fatty acids found in various patients with steatorrhoea was published by us some years ago.[29] As the fat is absorbed via the lymphatic system a marked steatorrhoea of about 100 mmol/day will be found in diseases affecting the lymphatic system such as Whipples disease and lymphangectasia.

When compared with the normal value steatorrhoea is almost always more pronounced than azotorrhoea as typified by patient TM (Table 7).

Sodium concentrations can be normal in these patients, but small bowel malabsorption may lead to osmotically powerful substances reaching the colon in sufficient quantity to exceed its normal absorptive capacity. These, if acidic, may be accompanied by increased sodium excretion. Slightly increased excretion of potassium is not an uncommon finding and suggests increased colonic secretion.

In lesions affecting the large bowel the faecal presentation is often quite different; a typical faecal excretion pattern in ulcerative colitis is given in Table 8. Normally, about 50 mmol of sodium chloride is absorbed by the large bowel per day, but in ulcerative colitis faecal levels may be 50–150 mmols/day. In addition, potassium may be lost to a similar extent. It will be noted that there is no steatorrhoea in this condition but a fair loss of nitrogen. This latter increase though is not specific, and may also be found in protein-losing enteropathy affecting the small intestine. The volume is always high in lesions affecting the large bowel and there will often be overt blood and leucocytes present, the latter probably accounting for the high lysozyme levels seen in the faeces of these patients.

Faecal Enzymes. As pointed out in category II, faecal enzymes of pancreatic origin are only of value in pancreatic failure in the child. Bacterial proteases and lipases are present but have no diagnostic significance. Lysozyme has already been referred to and when present is a reflection of the number of polynuclear leucocytes present. It is not a factor in the aetiology of ulcerative colitis, but possibly represents an exacerbation of the disease.

Faecal Protein. Most of the faecal protein is provided by bacteria and cellular debris which is measured as part of the total faecal nitrogen. The exception is α_1-antitrypsin which is considered under protein-losing enteropathy.

Faecal Sugars. Sugar malabsorption can sometimes be detected in the diarrhoea stools of a child after feeding. This has been used to detect specific disaccharide malabsorption. A simple test for reducing substances on the supernatant fluid after centrifuging the specimen is often adequate. A random specimen is adequate but it must be fresh to avoid bacterial degradation. Chromatography will help to identify the offending sugar. Diarrhoea has to be very severe to detect sugar in faeces of adults.

Specific Mucosal Defective Absorption

Iron. Although poor absorption of all nutrients can be expected in subjects with upper intestinal mucosal damage, specific absorption defects may also be found. An example of such a situation is iron. Poor absorption of iron can accompany renal failure, rheumatoid arthritis, or other conditions where erythropoesis is depressed, and also in the elderly.[30] Furthermore, we occasionally find a patient who will recover the capacity to absorb xylose normally on iron treatment alone.

The simplest way to test for iron malabsorption is to give ferrous sulphate by mouth and measure the increase in the blood level over the next three hours. In iron-deficiency anaemia, if absorption is satisfactory, normal blood levels should be reached in 1–1½ hours. This is a simple technique we use regularly. Other more sophisticated and expensive methods used are radioactive ^{59}Fe uptake or whole-body counting. An iron-absorption test is useful to be able to assess whether oral iron will be potentially effective.

Calcium. Calcium, like iron, requires the synthesis and presence of a specific protein in the intestinal cell. In the case of calcium, this synthesis seems to be controlled by a metabolite of vitamin D formed from calciferol in the liver and kidney. Calcium absorption is depressed in renal failure and appears to be related to the degree of failure.

Calcium absorption may be assessed using radioactive calcium, ^{47}Ca, by measuring the blood level increase after an oral dose, by a balance technique, or by whole-body counting. A simple non-radioactive absorption technique has been used successfully in this department. Rushton et al.,[31] showed that calcium absorption from a small dose of 50 mg calcium chloride was depressed in patients on haemodialysis. It is possible that any situation where the metabolism of vitamin D is impaired could result in a specific malabsorption of calcium. Over-treatment with vitamin D can result in hyperabsorption of calcium. Increased absorption of calcium has also been implicated in some patients with a tendency to form renal stones.

Calcium absorption by a balance technique measuring faecal and urine output is now rarely used. The study needs to be done for 2 or 3 weeks before the assessment of a change in output consequent upon treatment can be made.

Protein-Losing Enteropathy. Excessive loss of protein has been shown to occur in many gastrointestinal diseases. Alpers[32] gives a list of nearly 70 disorders in which it may be present.

There are two main groups. Firstly, those intestinal diseases in which the process results in ulceration of an area of the bowel and exudation of protein-containing fluid. Secondly, those with obstruction to intestinal lymphatics with loss of chyle-rich fluid containing fat and protein. There will probably be a moderate increase in faecal nitrogen—2–3 times the normal—in all these cases. Steatorrhoea may be absent in patients in the first group, for instance in ulcerative colitis, but quite severe in the second. In Whipple's disease, for instance, the levels may be 6 to 7 times normal. In both these groups there will be hypoproteinaemia, especially a low plasma albumin possibly of the order of 20–30 g/l, but no proteinuria or hepatic disease. There will probably be low immunoglobulin concentrations, especially of IgG. Oedema, though not specific, is a common finding.

The techniques available for the detection of extracellular fluid protein loss are presented in a review by Tavill[33] who also outlines the criteria necessary for an ideal label to detect a leak of large molecules from the plasma.

The technique which most closely fufils these criteria and is most popular is that of Van Tongeren and Majoor.[34] Plasma albumin is labelled with chromium chloride ($^{52}CrCl_3$) by injecting 30–100 μl intravenously in saline. Faeces are collected for 5 days and counted. Normally less than 1% of the dose is excreted but up to 35% may be excreted in cases of intestinal protein loss. Some chromium is excreted in the urine, so contamination of the faeces by urine must be avoided. It is not a test that one would wish to repeat frequently as a large percentage of the chromium is retained and has an isotopic half-life of 28 days.

Most serum proteins on excretion into the gut lumen undergo complete hydrolysis and much of the product is reabsorbed. However, α_1-antitrypsin would appear to be an exception and can be found in gastrointestinal secretions and faeces. Some have recommended its estimation in the faeces as a simple test for exudative enteropathy, but others have shown 25% false negatives when compared with the $^{58}CrCl_3$ technique.

Failure of Immune Response. The measurement of the immunoglobulins in the serum of patients with malabsorption can be helpful, especially IgA, IgM, and IgG. There is, as yet, no immunological pattern which is clearly characteristic of a particular intestinal disease. It would appear that their measurement is most helpful in specific primary immunological disease. In this group IgA is usually low or absent and there is low IgM and IgG.

It is well known that the intestine can provide a local immune response independent of systemic immunity and which protects the mucosal surface. The ability to provide this protection appears to be mediated by means of a secretory protein, synthesised in the intestine and predominantly of the IgA class and which appears to be resistant to intestinal proteolysis.

Malabsorption may also be associated with low levels of immunoglobulins such as may be found in patients with hypogammaglobulinaemia or protein-losing enteropathy.

There is also a group of intestinal diseases such as coeliac disease which may have an immunological aetiology by virtue of hypersensitivity to ingested antigens. Such patients may show raised IgA and low IgM and IgG concentrations.

A specific abnormality of IgA has also been described in which there is severe malabsorption[35] and only the heavy α-chains are produced. On electrophoresis, this protein will be found in the α_2-β region. There will also be a rise in intestinal alkaline phosphatase clearly shown on polyacrylamide electrophoresis. A review of α-chain disease and related lymphomas is given by Doe.[35]

Except in the latter case, immunoglobulin levels are not in general diagnostic as there is considerable overlap between diseases with differing aetiologies. It is beyond the scope of this chapter to deal adequately with the immunological patterns associated with gastrointestinal pathology and the reader is referred to the texts of Asquith[36] and Thomas and Jewell.[37]

Iatrogenic Malabsorption. Drugs may have a primary effect on absorption within the lumen or a secondary effect on the mucosal cells. An example of the first process, cholestyramine, an anion-exchange resin binds organic acids such as the bile acids and long-chain fatty acids. It is used in the chloride form to treat patients with pruritus. Depending on the dose, it will cause a mild steatorrhoea by reducing the bile acid concentration in the lumen and inhibiting micelle formation. Excretion of long-chain fats is increased in the faeces by four to five times but short-chain fats (8–10 carbon chain) are unaffected as they do not depend on micelle formation for absorption. Cholestyramine may also interfere with the absorption of substances such as fat-soluble vitamins, e.g. vitamin K, and hypoprothrombinaemia has been described. Liquid paraffin has also been reported to cause loss of fat-soluble vitamins.

Aluminium hydroxide can affect the absorption of phosphate. As an antacid aluminium hydroxide neutralises hydrochloric acid in the stomach. Subsequently in the small bowel, the chloride ion is replaced by phosphate thus causing its malabsorption. This action has been used therapeutically in the treatment of bone disease in renal failure but may also cause increased absorption of calcium and hypercalcuria leading eventually to osteomalacia. Change in intestinal pH may cause the malabsorption of substances requiring an acid medium such as iron but drug absorption such as the tetracyclins can also be affected. When given together with aluminium hydroxide, the serum level is reduced in 24 h by a factor of ten.

Laxatives can cause malabsorption when used continuously: fat and nitrogen excretion may be increased.

Drugs which affect the intestinal mucosa such as to cause malabsorption may do so in different ways. Antibiotics such as neomycin affect the morphology of the cell or villi if treatment is prolonged. Others such as indomethacin affect intestinal transport for example the xylose absorption test can be depressed despite no mucosal damage. Using the one hour xylose test, methotrexate has been shown to induce functional malabsorption in children with lymphoblastic leukaemia; it has also been suggested that the drug has less effect if doses are spaced more widely and that such drugs may themselves reduce their own absorption or that of other drugs.

A review of drug-induced malabsorption with a comprehensive list of references has been written by Longstreth and Newcomer.[38]

The possibility that malabsorption can be drug induced should always be borne in mind when investigating a patient for malabsorption. This is especially relevant to a patient who has developed diarrhoea subsequent to treatment with an antibiotic for intestinal infection. There is no doubt that investigations such as the 5 g xylose absorption test are most helpful in studying the effectiveness of drug removal in suspected situations.

CATEGORY IV: INADEQUATE SURFACE AREA

Gastrectomy and Dumping Syndromes

The sequelae of permanent loss of part of the gastrointestinal tract by surgery are varied according to the site and extent involved. For instance, following gastrectomy, there are several causes of poor food utilisation. These include rapid emptying of the remaining stomach, increased mobility by the hyperosmolar contents presented to the upper small bowel, inadequate co-ordination and mixing with pancreatic and biliary secretions and bacterial overgrowth, especially when blind loops are left, causing retention of intestinal contents. These actions may cause an early 'dumping' syndrome a few minutes after eating or a late dumping syndrome after one or two hours and caused by reactive hypoglycaemia, due to excessive quantities of insulin liberated by high blood sugar levels. The syndrome may also develop in the gastric-bypass technique for the treatment of obesity. More moderate intake of food especially carbohydrate will usually relieve symptoms. These multiple effects may give rise to malnutrition, weight loss and some steatorrhoea. After gastric surgery, there is always the possibility of B_{12} deficiency developing eventually. Iron-deficiency anaemia is also not uncommon probably caused in part by failure to convert ferric to ferrous ion due to loss of reducing substances such as ascorbic acid and too rapid transit through the absorbing area. Measurements of plasma iron, iron-binding capacity, an iron absorption test and a B_{12} absorption test, with and without intrinsic factor, are worth doing so that future treatment can be confidently recommended. Inadequate absorption of calcium has also been described which may lead to osteomalacia or osteoporosis.

Small Bowel Resection

Resection of the small bowel is sometimes needed for diseased or traumatised parts of the intestine. Inevitably, loss of the distal small bowel will eventually lead to B_{12} deficiency and an impairment of the enterohepatic circulation of bile salts. The malabsorption that ensues results in severe electrolyte disturbance, hypoproteinaemia, and general nutritional inadequacy, though the degree of loss will depend on the extent and nature of the resection. Plasma sodium, potassium, calcium, magnesium, iron and folic acid levels may be low and there may be some streatorrhoea and an acidosis due to lack of bicarbonate. Once the immediate postoperative effects have been controlled and a steady state obtained, it is important to determine the digestive and absorptive capacity of the remaining part of the small bowel. Some compensation and hypertrophy of the remaining bowel is likely to occur so repeated tests to measure the extent of recovery are also desirable. Faecal and urine analyses to measure the extent of loss and deficiency are initially helpful. In addition, the Schilling test to assess the capacity to absorb B_{12}, a xylose test to measure bowel failure, and tests to measure pancreatic function which may be depressed need to be measured.

The course of recovery seems to depend on the extent of the original surgery but patients have been described who have made remarkable recoveries with only a metre or less of intestine available for absorption; indeed some patients have eventually maintained good health with less than 50 cm of gut remaining. In these patients much of their initial malabsorption was associated with a plasma magnesium concentration of not more than 0·25 mmol/l; after correction with oral magnesium hydroxide, the generalised malabsorption improved dramatically.

Patients who have an ileostomy are susceptible to sodium depletion and dehydration The colon normally absorbs about 500 ml water and 50 mmol sodium per

TABLE 9
COMPARATIVE COMPOSITION: ILEOSTOMY FLUID AND FAECAL LOSS IN 24 HOURS

	Vol ml	Na$^+$ mmol	K$^+$ mmol	Cl$'$ mmol	N$_2$ mmol	Calories	Fatty Acid mmol	Cholesterol mmol
Ileostomy content	656	63	3	33	350	105	11	2·6
Faeces	63	3	6	1	150	50	15	Trace

day, so this is the approximate loss per day in ileostomy fluid. A comparative study of two subjects on the same diet, one normal and one with an ileostomy and colectomy is given in Table 9. There was little or no difference in the total or differential fatty acid excretion but a marked difference in cholesterol output. A review of the subject and additional data was given by Kanaghinis et al.[39] It would appear that the large bowel has a marked influence on the excretion of water, electrolytes, and cholesterol.

CATEGORY V: CONTAMINATED BOWEL

In contrast to the lower small bowel, the stomach and upper small intestine are virtually sterile. This is to a large extent probably due to the relatively rapid changes in pH which occur. Consequently, increased contamination will occur in achlorhydria or where mobility is depressed as in scleroderma. Contamination will also occur where there are blind loops, diverticulae—an increasing possibility in the elderly—and chronic infection such as ascending cholangitis. The bacteria can affect dietary components especially carbohydrates within the lumen by breaking them down to short chain fatty acids and thus initiating an osmotic intestinal hurry and diarrhoea. Hydroxy fatty acids can be formed by bacteria especially when steatorrhoea is severe as in the sprue syndrome. It has been postulated that they may play a part in the mechanism of diarrhoea but they are usually hydroxy derivatives of saturated fatty acids[29] and would not necessarily have such a toxic action on the bowel as that of the hydroxy-unsaturated ricinoleic acid, the major component of castor oil.

Bacteria may also affect the mucosa itself by producing an inflammatory process and affecting the absorption of most substances including water. They may also, by virtue of their metabolic action, produce toxic compounds within the gut, for example, the hydrolysis of conjugated bile acids to produce acids such as chenodeoxycholic acid with cathartic properties. On the other hand, they may in certain types of treatment be used therapeutically. For instance, in the treatment of inflammatory bowel disease by salazopyrine. The drug is hydrolysed by bacteria in the lower small bowel or colon to sulphapyridine and 5-amino salicylic acid. The 5-amino salicylic acid is said to be the active fraction.

The important need for the clinical chemistry laboratory in this category of malabsorption, is to try to determine to what extent bacterial overgrowth in the small intestine is contributing to, or indeed causing, the enteropathy.

Some indirect evidence can be obtained from faecal analyses but more direct evidence is obtained from analysis of the breath and urine.

Faecal Analyses. There will probably be a mild steatorrhoea due to the inhibition of micelle formation through the bacterial hydrolysis of conjugated bile acids.

A frequent finding, sometimes without steatorrhoea, is an increase in volatile fatty acids. These are so called because they are steam distillable and consist mostly of acetic proprionic and butyric acids. They can be found in many enteropathies (Table 10) and volatile fatty

TABLE 10
VOLATILE FATTY ACID OUTPUT (24 HR) IN FAECES AND ILEOSTOMY FLUID

	No. of patients	Acetic (mmol)	Proprionic (mmol)	Butyric (mmol)
Ulcerative colitis	4	16·1	3·8	4·8
Gluten-induced enteropathy	5	24·4	9·3	11·7
Other malabsorption	11	28·1	10·9	8·8
Normals	9	14·7	4·9	3·7
Ileostomy fluid	17	15·0	0·6	3·0

acids have been found in the aspirated jejunal contents of patients with stagnant blind loop syndrome. They are formed by the action of bacteria on substances of dietary origin, mostly unabsorbed carbohydrate. The relation of bacterial contamination to the mechanism of diarrhoea has been reviewed by Lowbeer and Read.[40] The mechanism would appear to be largely osmotic with poor absorption of water, sodium and short chain acid. Volatile fatty acids can be measured in an aliquot of faeces using the Kjeldhal distillation equipment to measure the total amount or by gas chromatography to measure the individual acids.

Breath Tests. Calloway[41] first correlated increases in breath hydrogen with certain ingested foods though the presence of combustible gases in the intestine has been known for more than 160 years. There have even been reports of explosions occurring during colonic surgery. The subject of breath analysis tests in gastroenterology has been reviewed by Newman.[42] The gases in the breath usually used to detect inappropriately located excessive bacterial overgrowth are carbon dioxide and hydrogen. All mammalian cells and bacteria produce carbon dioxide as an end-product of metabolism but hydrogen is produced only by bacteria. Consequently, carbon dioxide when produced in the intestine by bacteria needs to be labelled radioactively so that it can be distinguished from the normal endogenous form in the exhaled gases. Hydrogen needs no label and can therefore be analysed directly in the breath by gas chromatography.

In the intestine with a normally distributed bacterial population, the ingested material will be absorbed before reaching the bacteria in the lower bowel but if not then bacterial breakdown to carbon dioxide and hydrogen will take place in the colon and show itself in the breath. This will occur in two to four hours. If, however, the bacterial population is abnormally distributed and present in the more proximal areas of the small bowel, then fermentation will take place and the gases will be exhaled earlier. If serial studies are made, say every 15 minutes, then ideally two peaks will be demonstrated one corresponding to passage through the small bowel, the other through the colon. This is the basis of the breath test. It makes two major assumptions, that the transit times are comparable and the bacteria can catabolise the ingested substances. The substances so far found useful for this test are cholyl-^{14}C-glycine, lactulose[43] and ^{14}C-xylose in low doses.

The breakdown of cholyl-^{14}C-glycine is dependent first of all on the hydrolysis of the conjugate then on the production of carbon dioxide from the amino acid glycine. It has been stressed that to differentiate between bile-acid malabsorption and bacterial overgrowth, ^{14}C analysis should be made in faeces as well as in breath. In comparing lactulose and cholylglycine in 37 patients, Rhodes et al.[43] found 6 false positives with cholylglycine and only 1 with lactulose.

Lactulose (β-1,4 galactosido-fructose) is a synthetic disaccharide and is used for this purpose because owing to an absence of lactulase it is poorly absorbed yet is metabolised by most bacteria occuring in the intestine. There have been reports that not all bacteria will metabolise lactulose, and there is evidence that some absorption of the intact molecule can occur as a result of mucosal damage. The use of xylose as a test of absorption has been criticised because of its possible destruction by intestinal bacteria; as a breath test it is most sensitive and specific when a very low dose (1 g) is used. Xylose is metabolised to carbon dioxide to the extent of 16% when given intravenously.

From a practical point of view, the counting of radioactive specimens is easier to carry out than analysis by gas chromatography, which can be very labour intensive if serial 15' samples are taken as one has to allow about 20 minutes between each analysis to eliminate other gases. The patient must be fasted overnight before the lactulose is given. The normal level of hydrogen in the breath is less than 20 ppm for the first 2 hours of the test. It may rise to 40–50 ppm or higher in a patient with a contaminated bowel[43] though a negative result should not exclude the possibility as it could arise through too rapid transit or failure of the endogenous organisms to produce hydrogen after antibiotic treatment. High fasting levels have been found in patients with intestinal cysts.

Indoxyl Sulphate. Indoxyl sulphate (sometimes incorrectly designated indican which is the glycoside of indoxyl) is normally derived from indole in the liver after production of indole from trytophan by colonic bacteria. The indole is absorbed from the colon in small amounts and excreted in the urine. When colonisation of the small bowel occurs much more is produced and absorbed and the urine excretion rises.

A reliable method of estimation is that of Curson and Walsh[44] in which indoxyl sulphate reacts with dimethyl aminobenzaldehyde to produce a red colour. It is well to preserve the urine during collection with merthiolate unless refrigerated during the collection. A raised urobilinogen content will interfere with the test and urine from patients on Parentrovite or rifampicin will give high blanks. It is also said that excretion products of the drug Salazopyrine interfere with its estimation. The normal range in the literature is unfortunately wide but the consensus of opinion would give as the upper limit about 400 μmol (85 mg)/24 h but between 200–400 (43 mg–85 mg) is suspect. Although the increase in urinary levels is associated with a contaminated bowel it is non-specific; it is probably related to the amount of tryptophan in the dietary protein and the degree of bacterial overgrowth. Miloszewski et al.[45] found no increase in patients with pancreatic insufficiency until pancreatin had been given presumably achieved by the proteolytic release of tryptophan. Examination of the data from this laboratory on patients with enteropathy indicates that there is no obvious correlation with faecal volatile fatty acids or the degree of steatorrhoea. Nevertheless, it is a test which we find helpful as a significant increase (500–1000 μmol) is a genuine indication of bacterial overgrowth.

CATEGORY VI: ENDOCRINE DISORDERS

The intestinal endocrine system is discussed elsewhere in this book but malabsorption is not uncommon in endocrine abnormalities of other organs in the body.

In the course of a general clinical examination, it is useful to be aware that malabsorption may be associated with thyroid, parathyroid, adrenal, pituitary dysfunction or with diabetes. In certain situations such as poor growth and delayed sexual development, the endocrine abnormality may be secondary to gluten-induced enteropathy.

Thyroid

The subject of thyroid hormones and the gut has been reviewed by Middleton.[46]

Alteration in bowel habit may lead to a mild steatorrhoea in both hyper- and hypo-thyroidism. Increase in frequency of bowel action may be a pointer to thyrotoxicosis; diarrhoea, sometimes accompanied with steatorrhoea may be the only complaint in patients with thyrotoxicosis.

The metabolic defects present in thyroid dysfunction do not alter xylose absorption as measured in the blood though renal excretion may be affected in myxoedema.

Tumours

Carcinoid

Diarrhoea is a feature of the rare malignant carcinoid syndrome usually attributed to the release of 5-hydroxy tryptamine (5HT). In this disease metabolism of 5HT may lead to a considerable increase in urinary excretion of 5-hydroxy indole acetic (5HIAA) especially when liver metastases have developed, though a modest increase to twice the normal levels may be found in other malabsorption states.

The normal excretion is 2–10 mg per day but levels of 50–1000 mg may be found in the true carcinoid syndrome.

A suitable method of estimation is that of Udenfriend et al.[47] During the urine collection period the patient should avoid foods containing relatively high levels of 5HT, the precursor of 5HIAA, such as bananas, cheese and tomatoes. Some drugs such as phenothiazines interfere with the test.

Carcinoma

Attention must be drawn to the value of increased blood levels of carcinoembryonic antigen (CEA) in intestinal tumours. CEA has been found to be increased in some patients with gastric carcinoma and especially in those with adenocarcinoma of colorectal or pancreatic origin. Its finding is non-specific and raised levels may be found in many non-malignant states. Measurement of CEA would appear to be of real value when serial estimations are made every 3 months or so postoperatively as an indication of tumour recurrence in patients who have undergone curative resections for colorectal cancer. Normal levels are quoted as being below 12·5 ng/ml rising to 500 ng/ml in carcinoma, but as with most radioimmunoassays it is best for the laboratory carrying out the assays to establish its own normal range.

CONCLUSION

The rational approach to the differential investigation of a patient with malabsorption depends on many factors. Malabsorption cannot be properly investigated by one test alone as the intestine has evolved to perform a myriad of functions, digestive, absorptive, synthetic, immunological, and excretory, any one of which may be defective but the overall presenting symptoms of malnutrition including weight loss, anaemia, and increased faecal excretion can be so very similar. This aetiological diversity may well account for the apparent inconsistencies occasionally found. For instance, a poor xylose absorption may be shown in the absence of steatorrhoea.

The intestine has a tremendous reserve area and absorption tests although extremely valuable as a measure of the ability to absorb a single dose, do not 'titrate' the capacity of the whole intestine.

In the investigation of malabsorption serial studies over weeks or months are of immense value. Not only do they assist in evaluating the effect of treatment but by adding a second temporal dimension to the investigation they can also help to define more clearly the nature of the enteropathy—for instance, in the reintroduction of gluten in suspected gluten-induced enteropathy.

Excluding iatrogenic reasons, there are over 50 possible causes of malabsorption in the human subject.[49] Yet, despite this daunting prospect, a painstaking investigation can be very rewarding as in most patients so much can be done to rectify the abnormality.

REFERENCES

1. Losowsky, M. S., Walker, B. E., Kelleher, J. (1974), 'Malabsorption in clinical practice.' Edinburgh, London: Churchill Livingstone.
2. McColl, I., Sladen, G. E. G. (1975), *Intestinal Absorption in Man*. London, New York, San Francisco: Academic Press.
3. Dietschy, J. M. (Ed.) (1976), *Disorder of the Gastrointestinal Tract, Disorders of the Liver, Nutritional Disorders*. New York, San Francisco, London: Grune and Stratton.
4. Sleisenger, M. H., Fordtran, J. S. (1978), *Gastrointestinal Disease*, 2nd Edn. Philadelphia, London, Toronto: W. B. Saunders Co.
5. Walker-Smith, J. (1979), *Diseases of the Small Intestine in Childhood*, 2nd Edn. London: Pitman Medical.
6. Gowenloch, A. H. (1977), 'Tests of exocrine pancreatic function.' *Annals Clin. Bioch.*, **14**, 61–89.
7. Arvanitakis, C., Cooke, A. R., Greenberger, N. J., (1978), 'Laboratory aids in the diagnosis of pancreatitis.' *Med. Clinics N. Amer.*, **62**, 107–128.
8. James, O. (1973), 'The Lundh test.' *Gut*, **14**, 582–591.
9. McCollum, J. P. K., Muller, D. P. R., Harris, J. T. (1977), 'Test meal for assessing intraluminal phase of absorption in childhood.' *Arch. Dis. Child.*, **52**, 887–889.
10. Newcomer, A. D., Hofmann, A. F., Di Magno, E. P., Thomas, P. J., Carlson, G. L. (1978), 'Triolein breath test: A sensitive and specific test for fat absorption.' *Gastroenterology*, **76**, 6–13.
11. Robards, M. F. (1975), 'Changes in plasma nephelometry after oral fat loading in children with normal and abnormal morphology.' *Arch. Dis. Child.*, **50**, 631–636.
12. Van de Kamer, J. H., fen Bokkel Huinink, Weijers, H. A. (1949), 'Rapid method for the determination of fat in faeces'. *J. Biol. Chem.*, **177**, 347–355.
13. Tomarelli, R. M., Charney, J., Harding, M. L. (1949), 'The use of azo albumin as a substrate in the colorimetric determination of peptic and tryptic activity.' *J. Lab. Clin. Med.*, **34**, 428–433.
14. Anderson, C. M., Freeman, M. (1958), 'A simple method of sweat collection with analysis of electrolytes in patients with fibrocytic disease of the pancreas and their families.' *Med. J. Australia*, 419–422.
15. Price, C. P., Spencer, K. (1977), 'Problems associated with measuring sweat chloride concentration with an ion specific electrode.' *Annal Clin. Bioch.*, **14**, 171–178.
16. Smalley, C. A. Addy, D. P., Anderson, C. M. (1978), 'Does that child have cystic fibrosis?' *Lancet*, **ii**, 415–417.
17. Schwarz, V., Simpson, N. I. M., Ahuja, A. S. (1977), 'Limitation of diagnostic value of the sweat test.' *Arch. Dis. Child.*, **52**, 870–874.
18. Ross, C. A. C., Frazer, A. C., French, J. M., Geerard, J. W., Sammons, H. G., Smellie, J. M. (1955), 'Coeliac disease. The relative importance of wheat gluten.' *Lancet*, **1**, 1087–1091.
19. Baron, J. H. (1978), 'Clinical tests of gastrin secretion.' In *History, Methodology Interpretation*, pp. 5–18, London: MacMillan Press Ltd.
20. Berstad, A. (1970). 'A modified haemoglobin substrate method for the estimation of pepsin in gastric juice.' *Scand. J. Gastroent.*, **5**, 343–348.
21. Gerring, E. L., Allen, E. A. (1969), 'A nephlometric pepsin method.' *Clin. Chim. Acta*. **24**, 437–443.
22. Ardeman, S., Chanrin, I. (1965), 'Assay of gastric intrinsic factor

in the diagnosis of Addisonian pernicious anaemia.' *Brit. J. Haemat.*, **11**, 305–314.
23. Haeney, M. R., Culank, L. S., Montgomery, R. D., Sammons, H. G. (1978), 'Evaluation of xylose absorption as measured in blood and urine: a one-hour blood xylose screening test in malabsorption'. *Gastroenterology.*, **75**, 393–400.
24. Sahi, T. (1978), 'Dietary lactose and the aetiology of human small intestinal hypolactasia.' *Gut*, **19**, 1074–1086.
25. Dahlqvist, A. (1964), 'Assay of disaccharidases.' *Anal. Biochem.*, **7**, 18–25; and (1968), *Anal. Biochem.*, **22**, 99–107.
26. Newcomer, A., McGill, D. B. (1967), 'Disaccharidase activity in the small intestine; prevalence of lactase deficiency in 100 healthy adults.' *Gastroenterology*, **53**, 881–889.
27. Dunne, W. T., Cooke, W. T., Allen, R. N. (1977), 'Enzymatic and morphometric evidence for Crohn's disease as a diffuse lesion of the gastrointestinal tract.' *Gut*, **18**, 290–294.
28. Haemerli, V. P., Kistler, H., Ammann, R., Marthale, R. T., Semenza, G., Auricchio, S., Prader, A. (1965), 'Acquired milk intolerance in the adult caused by lactose malabsorption due to a selective deficiency of intestinal lactase activity. *Am. J. Med.*, **38**, 7–30.
29. Gompertz, S. M., Sammons, H. G. (1963), 'The composition of human faecal fats.' *Clin. Chim. Acta.*, **8**, 591–603.
30. Montgomery, R. D., Haeney, M. R., Ross, I. N., Sammons, H. G., Barford, A. V., Balakrishnan, S., Mayer, P. P., Culank, L. S., Field, J., Gosling, P. (1978), 'The ageing gut, a study of intestinal absorption in relation to nutrition in the elderly.' *Quart. J. Med.*, **47**, 197–211.
31. Rushton, M. L., Sammons, H. G., Robinson, B. H. B. (1971), 'A study of calcium absorption using an automated fluorimetric assay procedure.' *Clin. Chim. Acta*, **35**, 5–16.
32. Alpers, D. H. (1976), 'Differential approach to protein-losing enteropathy.' In *Disorders of the Gastrointestinal Tract, Disorders of the Liver, Nutritional Disorders*, pp. 66–68, Dietschy, J. M. (Ed.). New York, San Francisco, London: Grune and Stratton.
33. Tavill, A. S. (1971), 'Protein losing enteropathy.' *J. Clin. Path.*, **24**, Supp (Roy. Coll. Path) 45–54.
34. Van Tongeren, J. H. M., Majoor, C. L. H. (1966), 'Demonstration of protein losing gastroenteropathy. The disappearance rate of ^{51}Cr from plasma and the binding of ^{51}Cr to difference serum proteins.' *Clin. Chim. Acta.*, **14**, 31–41.
35. Doe, W. F. (1979), 'A-chain disease and related small intestinal lymphomas.' In Immunology of the Gastrointestinal Tract, pp. 306–315, Asquith, P. (Ed.). Edinburgh, London, New York: Churchill Livingstone.
36. Asquith, P. (Ed.) (1979), *Immunology of the Gastrointestinal tract*. Edinburgh, London, New York: Churchill Livingstone.
37. Thomas, H. L., Jewel, D. P. (1979), *Clinical Gastrointestinal Immunology*. London, Oxford: Blackwell Scientific.
38. Longstreth, G. F., Newcomer, A. D. (1975), 'Drug induced malabsorption.' *Mayo Clinical Proc.*, **50**, 284–293.
39. Kanaghinis, T., Lubran, M., Coghill, N. F. (1963), 'The composition of ileostomy fluid.' *Gut*, **4**, 322–338.
40. Lowbbeer, T. S., Read, A. E. (1971), 'Diarrhoea; mechanisms and treatment.' *Gut*. **12**, 1021–1036.
41. Calloway, D. H. (1966), 'Respiratory hydrogen and methane as effected by gas-forming foods.' *Gastroenterology*, **51**, 383–389.
42. Newman, A. (1974), 'Breath-analysis tests in gastroenterology.' *Gut*, **15**, 308–323.
43. Rhodes, J. M., Middleton, P., Jewell, D. P. (1979), 'The lactulose hydrogen breath test as a diagnostic test for small bowel bacterial overgrowth.' *Scand. J. Gastroenterology*, **14**, 333–336.
44. Curson, G., Walsh, J. (1962), 'A method for the determination of urinary indoxyl sulphate (indican). *Clin. Chim. Acta.*, **7**, 657–663.
45. Miloszewski, K., Kellher, J., Walker, B. E., Losowsky, M. S. (1975), 'Increase of urinary indican excretion in pancreatic steatorrhoea following replacement therapy.' *Scand. J. Gastroenterol.*, **10**, 481–485.
46. Middleton, W. R. J. (1971), 'Thyroid hormones and the gut.' *Gut*, **12**, 172–177.
47. Udenfriend, S., Titus, E., Weissbach, H. (1955), 'Identification of 5-hydroxy-3-indoleacetic acid in normal urine and method for its assay.' *J. Biol., Chem.*, **216**, 499–505.
48. Donaldson, R. M. (1975), 'Mechanisms of malabsorption of cobalamin.' In *Cobalamin*, pp. 335–368, Babion, B. B. M. (Ed.). London, New York: John Wiley and Son.
49. Brooks, F. P. (1974), *Gastrointestinal Pathophysiology*, pp. 218–219, New York, Toronto, London: Oxford Univ. Press.

6. MALNUTRITION AND VITAMIN AND MINERAL DEFICIENCIES AND EXCESSES

J. W. T. DICKERSON

Protein-energy malnutrition
  Introduction
  Protein-energy malnutrition in children
    Classification
    Pathogenesis
    Prevalence
    Clinical features
    Prognosis and mortality in severe PEM
    Pathophysiology and biochemical disturbances
    Diagnosis and assessment of severity
    Investigations during treatment
  Protein-energy malnutrition in adults
    Prevalence and pathogenesis
    Clinical importance
    Assessment for PEM in adults
    Treatment
  The malnutrition of malignancy
    Anorexia
    Cachexia

Vitamin deficiency and excess
  Introduction
  Clinical features
    Vitamin A
    Vitamin D
    Vitamin E
    Vitamin K
    Thiamine (vitamin B1)
    Riboflavin (vitamin B2)
    Nicotinic acid (Niacin) and Nicotinamide
      (vitamin B3)
    Pyridoxine (vitamin B6) and related compounds
    Cyanocobalamin (vitamin B12) and related
      substances
    Folate (folic acid, folacin)
    Ascorbic acid (vitamin C)
  Determination of vitamin status

Trace element deficiency and excess
  Introduction
  Clinical features
    Zinc
    Copper
    Manganese
    Chromium
    Cobalt
  Determination of trace element status

Conclusions

PROTEIN-ENERGY MALNUTRITION

Introduction

The human body possesses considerable powers of adaptation to changes in its environment and this is nowhere more marked than in response to a decrease in nutrient availability. It is clear then that we cannot suppose that there is only one 'normal' nutritional state of the organism, and adaptation over a wide range of nutrient availability is compatible with health. Protein-energy malnutrition (PEM) represents a breakdown of adaptive mechanisms which is manifested as clinical illness.[70] PEM is characteristically a disorder of children and as such is the most widespread and serious nutritional problem known to medical science. As we shall see, the causes are complex with a large number of interacting factors but the basic cause is a lack of sufficient food of the right kind. PEM may, however, occur as a secondary consequence of other diseases which interfere with the ingestion of food or the absorption of nutrients, or in which there are losses by abnormal routes (i.e. fistulae). The most severe form of malnutrition in adults is that which occurs in patients with cancer and is known as cancer cachexia.

Protein-energy Malnutrition in Children

Classification

PEM is a term coined originally as 'protein-calorie malnutrition' to cover a spectrum of conditions ranging from marasmus at one extreme to kwashiorkor at the other. The term 'marasmus' is derived from the Greek 'marasmos' meaning wasting and has been recognised for centuries as being, with gastroenteritis, a major cause of infant mortality. The term 'kwashiorkor' was first used by Dr. Cicely Williams in 1933 to describe a condition which she recognised in the Gold Coast. The word is taken from the Ga language of Ghana and literally means 'the disease the first child gets when the second is on the way'.

In marasmus the child is underweight with very little body fat and consequent loosening of the skin. There is also muscle atrophy and, depending on the chronicity of the condition, short stature. Kwashiorkor on the other hand is characterised by skin and hair changes, oedema, 'moon-face', fatty liver, hypo-albuminaemia, and psychomotor changes. The clinical manifestations of this disease differ in different parts of the world as also does the age of development. These differences have tended to lead to some confusion in nomenclature and classification of the different forms of PEM.

A classification of PEM should be suitable for use in (1) the international classification of diseases, (2) prevalence studies and other observations in communities and (3) clinical and research investigations. The simplest classification is that known as the Wellcome classification (Table 1) which is based simply on the deficit in body weight and the presence of oedema. This classification has advantages because of its simplicity, but it cannot be applied when the age of the patient is not

TABLE 1
WELLCOME CLASSIFICATION OF INFANTILE MALNUTRITION[47]

	Percentage of expected weight for age*	Oedema
Marasmus	< 60	absent
Marasmic kwashiorkor	< 60	present
Kwashiorkor	60–80	present
Underweight	60–80	absent

* Taken as the 50th percentile of the Boston standards[69]

known and does not take into account the chronicity of the disease process.

McLaren et al.[52] introduced a simple scoring system for classifying the severe forms only (Table 2). This

TABLE 2
SIMPLE SCORING SYSTEM FOR PROTEIN-ENERGY MALNUTRITION[52]

	Points
Oedema	3
Dermatosis	2
Oedema + dermatosis	6
Hair change	1
Hepatomegaly	1
Serum albumin (g/l)	
< 10·0 >	7
10·0–14·9	6
15·0–19·9	5
20·0–24·9	4
25·0–29·9	3
30·0–34·9	2
35·0–39·9	1
> 40·0	0
Classification	
Marasmic	0–3 points
Marasmic Kwashiorkor	4–8 points
Kwashiorkor	9–15 points

method is precise and provides a means of objectively classifying the type of patients likely to be hospitalized. It does, however, require access to laboratory facilities. Another method, introduced by Kanawati and McLaren[45] requires only the use of a stiff tape measure and is based on the ratio of mid-arm circumference to head circumference which is independent of age at least from 3 months to 48 months and is similar for the two sexes (Table 3). This method is, however, rough, unsuitable for use in individual children and is intended for use only in screening large numbers.

It seems desirable that any classification of PEM should take into account deficits in weight and height for age. One such classification is shown in Table 4. The

TABLE 3
ASSESSMENT OF MARGINAL MALNUTRITION[45]

	Mid-arm circumference /head circumference	Percentage of weight for age
Nutritionally healthy	> 0·310	> 90
Mild-moderate PEM	0·310–0·280	60–90
Severe PEM	< 0·250	< 60

TABLE 4
CLASSIFICATION OF NUTRITIONAL STATUS IN EARLY CHILDHOOD[53]†

Classification	Observed weight as percentage of ideal weight/length/age
Overweight	> 110
Normal range	90–109
Mild PEM	85–89
Moderate PEM	75–84
Severe PEM*	< 75

* Marasmus (no oedema) or kwashiorkor (with oedema)
† Authors give a nomogram for rapid calculation and classification

diagnosis of the kind of severe PEM with which the child presents, kwashiorkor or marasmus, may change quickly for if oedema is lost a kwashiorkor child becomes a marasmic one.

Pathogenesis

Although PEM is a nutritional deficiency disease, it is doubtful if nutrition can ever be separated from the whole variety of social factors which make up the complex environment in which children grow up. Thus, economic stringencies, overcrowding, poor hygiene, and religious and other taboos all contribute to the development of the condition. The exact form of PEM which develops depends on the age of the child, the duration of breast-feeding and also weaning practices. Industrialization has influenced the nature of the most prevalent types of PEM in different communities. From an analysis of these factors in the Lebanon and in Jordan, McLaren[51] put forward a scheme (Fig. 1) to account for the pathogenesis of the extreme forms.

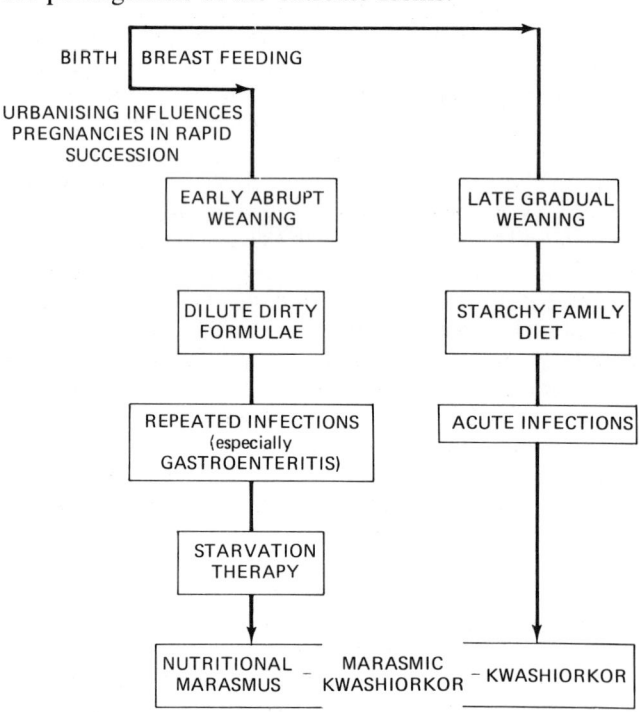

FIG. 1. Paths leading from early weaning to nutritional marasumus and from protracted breast feeding to kwashiorkor.[51]

Classically it was held that marasmus is caused by a shortage of food, predominantly of energy, and that kwashiorkor is caused by a shortage of protein of the right kind but that the energy supply might be adequate. It has become clear that this simplistic view is a misconception which has had important consequences for it has led to ideas that there is a 'protein gap' and this has been described in fact as a 'protein fiasco'. In India children have been described as developing either marasmus or kwashiorkor whilst consuming the same diet. It now seems clear that whether a child who is on the 'knife-edge' of good nutrition develops marasmus or kwashiorkor is determined by a number of non-dietary forces. Diet may be said to have a permissive rather than a determinative role in the human situation. This is clearly not so in animal experiments for a diet may be made to be determinative by eliminating other factors. A kwashiorkor-like condition can be produced in rats or pigs by feeding a diet containing a small amount of protein with excess or adequate energy. Similarly, by reducing the amount of a good diet allowed to the animals, severe stunting or a marasmic type condition can be produced. In the human child infections which reduce intake and impair utilization, together with accompanying social and sensory deprivation, have a profound impact on the child. Furthermore, it may be expected that as the process of urbanization increases in poor susceptible countries these non-dietary factors may be found to play an increasingly important role.

Prevalence

Two terms are used by epidemiologists to describe the distribution of disease in communities. They are 'incidence' and 'prevalence'. The incidence rate is the number of new cases that come into being during a specified period in a specific group of the population. Point prevalence is the proportion of the population that have a disease at a particular instant.

All attempts to estimate the prevalence of PEM are subject to a considerable degree of uncertainty but information has been collected for a number of years by The World Health Organization (WHO) and Bengoa[6] reported data from 77 nutrition surveys in 46 developing countries, totalling nearly 200 000 children mostly under 5 years of age. Based on weight-for-height data extrapolated from these figures it was estimated that about 100 000 000 children throughout the world are suffering from moderate or severe PEM. Of these about 31% would be found in Asia, 26% in Africa and 19% in Latin America. Mortality rates for children during the first year of life and in the period 1–4 years are related to the incidence of PEM and may be used as an indirect guide in assessing nutritional status in a community.

The relative prevalence of marasmus and kwashiorkor differs in different communities. Some of the reasons for these differences have already been suggested (Fig. 1). Of the surveys quoted by Bengoa only 22 distinguished between the two conditions and 17 of these reported more marasmus than kwashiorkor. This is in line with present information which does suggest that marasmus is emerging as the predominant form, although it does in fact tend to be neglected in estimates of PEM prevalence and only kwashiorkor which is a notifiable disease in certain areas is reported. There is no doubt, however, that the predominant need of the world's children is for food, and not specifically for protein.

Clinical features

Kwashiorkor: Oedema is the cardinal feature of kwashiorkor. The severity differs widely from one child to another and may manifest itself by puffiness around the eyes and swelling of the feet and hands or in other cases it may be more generalized. The accumulation of fluid in the extremities may mean that severe dehydration may be present in a child with oedematous legs. Most children with kwashiorkor are underweight for their age. Marked muscle wasting may be masked by overlying well-preserved subcutaneous tissue and by oedema. Deficits in height suggest that the disease manifests itself more acutely than marasmus although bone age and growth in length and head circumference are retarded. Some of the most striking clinical features of kwashiorkor are the skin and hair changes though these are not present in all cases and are reversible on recovery. The skin changes include hypopigmentation and "flaky-paint" dermatosis. In some cases the thin shiny skin is tautly stretched on oedematous limbs or trunk, often the skin breaks down with ulceration at the flexures in the groin and on the buttocks. The hair shows a wide range of abnormalities especially when the disease is of long duration. Hair that is normally dark and curly becomes lighter and straighter. Dispigmented hair may alternate with darker hair according to the nutrient supply at the time when the hair was being formed, this has been described as the 'flag' sign. Hair is dry, thin and silky and is easily plucked from the scalp. In some parts of the world there is marked hepatomegaly which is mainly due to fatty infiltration of the organ. Hepatomegaly, like the hair changes, does not occur in all cases of kwashiorkor and is therefore not diagnostic of the condition. Children with kwashiorkor are characteristically listless, apathetic, miserable and irritable. Anorexia is common and often requires intragastric tube feeding in those children where it persists. The causes of the psychological disturbances are doubtlessly complex and involve the effects of malnutrition on the brain itself and also the effects of sensory deprivation. These, together with the age of the child, play an important role in determining whether the disease process is fully reversible in terms of mental function.

Marasmus: The most striking feature of marasmus is the marked deficiency in weight and to a lesser extent in height. The marasmic infant is grossly emaciated with an apparently large head, staring eyes and shrunken muscles under redundant folds of skin. The skin itself is thin with no subcutaneous fat. The marasmic infant is usually irritable and fretful but may also display the features of apathy and misery which are so characteristic of kwashiorkor. Anorexia is less common in marasmus and the appetite is often good. The performance of these children in mental tests may be improved by sensory stimulation in addition to nutritional rehabilitation. However, the completeness of rehabilitation may well depend upon the age of the child, being less likely in children who have had chronic early malnutrition. Skin and hair changes may occur in marasmus but they are less common and less marked than those occurring in kwashiorkor. Hepatomegaly is found less frequently in marasmus than in kwashiorkor.

Marasmic kwashiorkor: As the name implies children are classified as having marasmic kwashiorkor when they present clinical features of both maramus and kwashiorkor and are regarded as presenting intermediate forms of severe PEM. As shown in Table 1 oedema is present and body weight is less than 60% of expected standard for age. Skin and hair changes, psychological changes and a palpable fatty liver characteristic of kwashiorkor are often found in children with marasmic kwashiorkor.

Prognosis and mortality in severe PEM

Severe PEM is associated with infection especially of the lower respiratory tract and with gastroenteritis. Marked fluid and electrolyte disturbances may occur as a result of the diarrhoea, and mortality in severe PEM is high although the rate differs in different regions. Most workers report a mortality of over 20%. A majority of deaths occurs within the first few days of hospitalization. Factors which influence prognosis include infection, electrolyte and fluid imbalance, the magnitude of weight deficit, marked liver enlargement, hypothermia, hypoglycaemia, severe skin changes, xerophthalmia and evidence of disturbance in liver function. Opinions may differ as to those features which are associated with a grave prognosis but workers in Africa have reported that hypothermia with a rectal temperature of less than 35°C carries a grave prognosis. Profound hypoglycaemia is life-threatening and requires immediate treatment.

Pathophysiology and Metabolic Disturbances

There is a considerable literature dealing with the pathological and biochemical changes in PEM. The following is a brief account of the cardinal features. A more detailed review will be found elsewhere.[2]

Elucidation of the metabolic changes involved in the pathogenesis of PEM has presented considerable difficulties and it is almost certain that the changes are not completely understood. Prospective studies form the only real method by which the changes can be elucidated: Whitehead and Alleyne[72] attempted to do so by studying in depth the first 3 years of life of 60 children living in an area where kwashiorkor was prevalent. The conclusions from this study are summarized in Figs. 2 and 3.

It is suggested that in marasmus there are two fundamental ways in which the body adapts to deficiency of dietary energy. Increased cortisol output induces muscle wasting with gluconeogenesis and the liberation of essential amino acids for the synthesis of proteins essential for homeostasis. At the same time subcutaneous fat is mobilized with fatty acids being used to synthesize various metabolites and participate in the maintenance of homeostasis. Glycerol and fatty acids are also used to provide energy.

In uncomplicated kwashiorkor, tissue catabolism does not occur to the same extent as in marasmus, possibly due to the provision of a nearly sufficient amount of dietary energy. As a result only small amounts of essential amino acids are made available for necessary homeostatic processes. This would account for the fall in plasma levels of essential amino acids found in kwashiorkor (see later) and for the decreased albumin and β-lipoprotein synthesis.

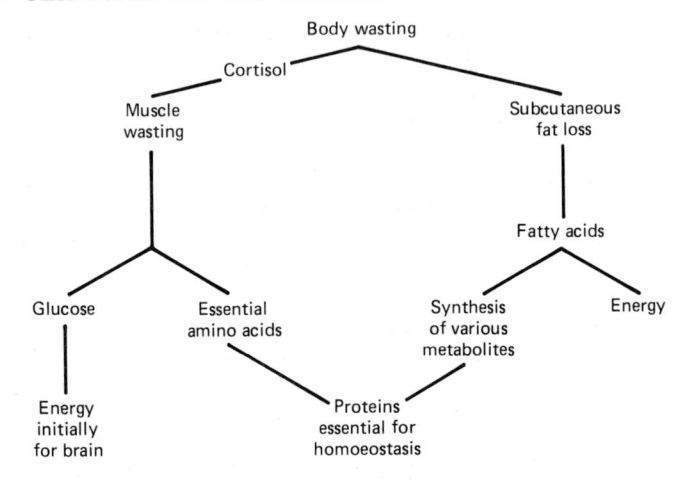

FIG. 2. Adaptation to energy deficiency.[72]
(Reproduced by permission of the Medical Department, The British Council).

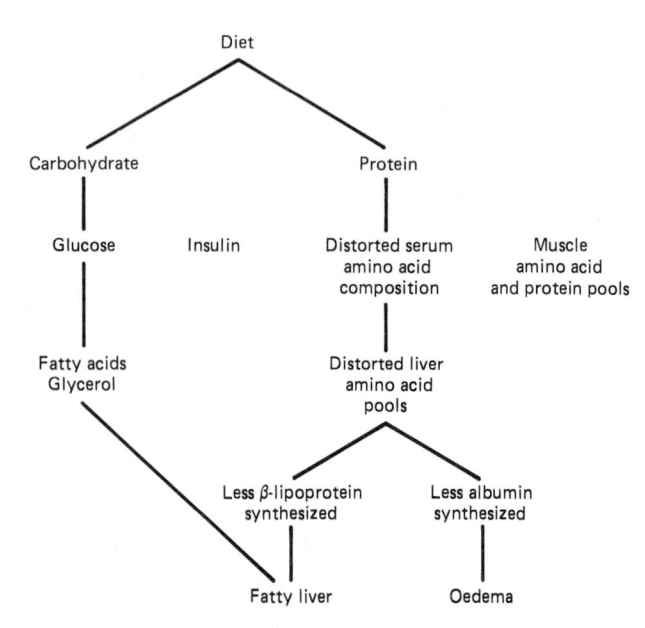

FIG. 3. Disturbances in homeostasis during the development of kwashiorkor.[72]
(Reproduced by permission of the Medical Department, the British Council).

In these schemes it is suggested that the response of the adrenal gland plays a key role and that this is more marked in maramus than in kwashiorkor. However, a number of workers have reported that adrenal function is well preserved in most cases of kwashiorkor. It thus seems clear that uncomplicated kwashiorkor such as that depicted in Fig. 3 rarely occurs in real life and that complications do modify the metabolic changes in various ways.

Body composition. Slowing or cessation of growth in PEM is associated with a gradual increase in total body water (TBW) relative to body weight. This occurs as a result of the mobilization of fat and the wasting of muscle and other tissues. Hansen et al.[38] showed that there is a direct relationship between the deficit in body weight and the proportion of water in the body. The increase in total body water is due to an expansion of the volume of extracellular fluid and this is particularly marked in the presence of oedema.

A decrease in the cell mass accounts in part for the lower level than normal of potassium found in the body in PEM. At first the decrease in potassium is proportional to the loss in body nitrogen, but diarrhoea causes a disproportionate loss of potassium with the result that there is a greater fall in potassium than in nitrogen in the cells. Potassium depletion plays an important role in metabolic disturbances at the cellular level particularly in muscle, kidney and pancreas. Brain potassium is also depleted in patients with diarrhoeal losses. Studies with the whole body counter based on determinations of the naturally occurring isotope ^{40}K have shown that in kwashiorkor the mean concentration of potassium in the body was 31 mM/Kg and in marasmus 39 mM/Kg.[54] On recovery values rose to 45 mM/Kg (normal range 45–55 mM/Hg).

In children dying of severe PEM, chemical analysis of whole bodies[34] showed deficits of sodium, calcium and phosphorus of 7%, 23% and 21% respectively of expected values. There is evidence that the cellular depletion of magnesium, like that of potassium, may be greater than that of nitrogen in muscle of children dying of PEM.

Total body protein is severely reduced in PEM and total body analysis of a limited number of cases showed deficits of 20–45% when compared with expected values. Compared with children of the same height those with PEM had a greater deficit of total protein than of body weight. The principal loss is of non-collagen protein and on a whole-body basis collagen is relatively little affected. However, this is probably not true of all tissues and in particular not of the skin from which collagen is lost to a marked extent as a result of dietary deficiencies of energy or of protein[14].

The muscle mass is greatly diminished with degeneration and loss of fibres and shrinkage of remaining fibres.

In marasmus, body fat may be as little as 5% of the body weight (normal value approximately 19%). Subcutaneous fat is well preserved in kwashiorkor and in this condition the liver may contain up to 50% of its wet weight as fat. Liver fat may in fact contribute 0.4% to 40% of the total body fat but there is no constant relationship between the amount of fat in the liver and that in the subcutaneous tissues.

The macroscopic, yellow greasy appearance of the fatty liver of PEM is very striking. Microscopically the fat appears first in the periportal area and spreads to the central vein area of the liver lobules and distends the liver cells. The excess fat is mainly triglyceride. Liver cirrhosis does not develop as a consequence of fatty infiltration, but is more likely to be due to viral or tropical parasitic infections, alcohol or other hepatotoxins. Aflatoxin may also be a cause in some areas.

The lipid that accumulates in the liver in PEM is triglyceride and this has a similar fatty acid composition to that of adipose tissue. It seems likely that the accumulation is due to a combination of an increased flux of fatty acids from adipose tissue together with decreased hepatic synthesis of β-lipoprotein which normally transports triglycerides from the liver. In untreated kwashiorkor the concentration of β-lipoproteins in the plasma is reduced but rises on treatment. Synthesis of the apoprotein part of the lipid transporting mechanism seems to be particularly sensitive to dietary protein intake. The results of a study of plasma and liver lipids (Table 5) are in agreement with these suggestions.

Pancreatic function is abnormal in PEM and histologically there is atrophy of the acinar cells though not of the islets. Exocrine function is depressed in untreated cases but responds rapidly to treatment. The endocrine function of the pancreas has attracted considerable attention and in particular that of the beta cells and insulin production. Fasting plasma insulin levels are low in kwashiorkor and rise on recovery. Insulin response to oral glucose is low in kwashiorkor and in most marasmic

TABLE 5
RELATION OF SERUM LIPIDS AND LIPOPROTEINS TO FATTY LIVER IN KWASHIORKOR[68]

	Liver fat grading	Serum albumin g/l	Total cholesterol	β-Lipoprotein cholesterol	α-Lipoprotein cholesterol	Triglycerides	Total phospholipids
			mg/100 ml				
A. Severe fatty liver	3 to 4+	16.4	89	52	37	78	136
B. Mild fatty liver	0 to 2+	21.4	104	69	34	103	134
A/B × 100		77	87	76	107	76	101
Controls			163	123	41	104	199

children. Intravenous glucose yields a greater incidence of normal responses. When the response is impaired, the rate of return to normality is variable with many patients taking 2–10 months after leaving hospital to recover normal responses in insulin secretion.

Disturbances in the structure and function of the gastro-intestinal tract play an important role in the development of PEM and in the potential for ultimate recovery.

Macroscopically at autopsy the bowel appears atrophic. In the jejunum the most severe mucosal changes probably occur in kwashiorkor rather than in marasmus but the changes range from almost normal villi to severe villous atrophy with only convolutions or ridges being seen. The crypts between the villi are not affected so severely. These changes are not specific to PEM but occur in a number of small intestinal pathologies, notably sprue and coeliac disease. Moreover, the initial changes occur as part of the normal mucosal pattern in most tropical areas of the world. The change from the normal finger-like villi to spade villi with the occasional convolution or ridge is probably an adaptation to the 'luminal milieu' as distinct from atrophy of existing villi. The appearance of intestinal villi is also affected by bacterial contamination. Histologically, the epithelial cells lose their columnar shape and become cuboidal and there is atrophy of the brush border. Paneth cells may be reduced and inflammatory cells (usually plasma cells) infiltrate the underlying lamina propria.

Electron microscope studies have concentrated on the appearance of lipoid droplets which were more numerous in the cells of the villous tip than in those of the crypt. The appearance was reported as being similar to that found in β-lipoprotein deficiency in which there is a failure of chylomicron formation and accumulation of lipid droplets in the mucosal cells. Other studies on a small number of cases of severe kwashiorkor have reported no lipid in villous cells. In contrast, these workers describe considerable disorganization of intracellular architecture, not only of epithelial cells but also of endothelial cells lining the villous capillaries. They suggested that these changes could result in marked malfunction and malabsorption. They did note, however, that marked improvement occurred in a few days when rehabilitation was attempted, even in those children who subsequently died. Differences in diet prior to obtaining the samples could well account for the differences in these two reports.

Enzyme activity within the mucosal cells is reduced in PEM. The enzymes most particularly affected are the disaccharidases, peptidases, adenosine triphosphatase and alkaline phosphatase, all of which are normally located in the brush border. Of these enzymes lactase has received the most attention and lactose intolerance is common in PEM, particularly in kwashiorkor. Although feeds containing lactose increase stool weight during the initial recovery phase this phenomenon is often short-lived and does not require a lactose-free diet. In severe cases sucrose and glucose intolerance also occur. Lactose intolerance may persist after recovery from PEM but the children usually tolerate milk well.

The large intestine is hypotonic with thinning and increased vascularity of the walls. Histologically the surface epithelial cells are flattened and the mucous glands show cystic dilatation with increased numbers of inflammatory cells, particularly plasma cells, in the sub-mucosa.

The long-term effects of PEM on the alimentary tract obviously have a bearing upon the prognosis for recovery of the child. Some workers[18] have reported that cellular improvement in the villi occurs only slowly. Other workers[7] reported reversion to a normal appearance within 9 months. However, most of these reports have been from areas where tropical sprue is endemic and it could well be that the mucosal changes are not primarily the result of malnutrition but of some other factor which is responsible for both malnutrition and the mucosal changes. This other factor could be bacterial colonization, for Tompkins et al., have shown that persistent colonization within the mucosa by enterobacteria is related to continued mucosal abnormality. Genetic differences may also play a role.

In acute PEM the mucosal atrophy of the small intestine and the exocrine pancreas, the abnormalities at sub-cellular level, the depression of enzyme activities, and the reported decreased concentration of conjugated bile salts in the upper jejunum, all work together to reduce absorption. It would appear, however, that total mucosal function is still adequate for the absorption of the necessary nutrients to allow for rapid and complete recovery of all but the most severely marasmic children. Problems arise mainly from the entry of non-absorbed residue into the lower gut with resulting bacterial fermentation and altered osmotic pressures within the intestinal lumen, and an invasion of the upper gut by bacteria from the colon. In spite of all the apparent problems and difficulties there seems to be no evidence of any persisting malabsorption after recovery from PEM apart from a lactose intolerance which is common in the ethnic groups amongst which PEM is common.

Protein and Amino Acid Metabolism. Changes in body proteins occur in all forms of PEM. Those in certain tissues have already been mentioned because of their role in the pathogenesis of the disorder. Because of the ease with which they can be determined and the accessibility of the material as well as their potential as a diagnostic tool, considerable attention has been devoted to the plasma proteins. The level of total protein tends to be reduced in all forms of PEM but the levels are lowest in kwashiorkor with the greatest reduction occurring in the albumin fraction. In a study of malnourished and recovered children receiving high and low protein diets, James and Hay[43] showed that the rate of synthesis of albumin increased or decreased in response to a high or low protein diet respectively. Furthermore, corresponding changes in the catabolic rate occurred after a lag period of about a week. They further showed that adaptation to a low protein diet involved a decrease in the extravascular albumin mass with an increase in the intravascular albumin content. These observations have been confirmed in a study of experimental malnutrition.[20]

There is experimental evidence that at least two separate factors may be involved in determining whether hypoalbuminaemia appears in PEM. Thus, experiments in rats fed diets containing different concentrations of protein showed that whilst the lower the protein concentration of the diet, the greater was the extent of hypoalbuminaemia which developed, dietary restriction with an increase in plasma glucocorticoid concentration and body wasting could initially delay the development of the hypoalbuminaemia. However, in the final stages of wasting which ensued a low plasma albumin concentration could develop because of failure of the mechanisms which had earlier been able to preserve them at normal levels. It was suggested that these two separate mechanisms might have a parallel in the human situations in developing countries.

Gamma globulin metabolism, unlike that of albumin, is unaffected by PEM and in the presence of infection the rate of synthesis is increased.[17] The concentration of transferrin in the plasma is markedly reduced in severe PEM. In a study in which serum albumin and transferrin levels were measured in children suffering from marasmus, kwashiorkor and marginal undernutrition[60] it was concluded that plasma transferrin levels provide an index of severity in severely malnourished children. The plasma transferrin concentrations showed a significant linear relationship with deficits in length for age and weight for length. Moreover, in severely undernourished children who died the plasma transferrin levels, in contrast to the albumin levels, were lower than in those who survived.

Techniques are now available by which it is possible to measure total body protein turnover. Picou and Taylor-Roberts[58] used these techniques in malnourished children who were fed an adequate diet. They showed that at the onset of treatment the malnourished children had higher rates of protein synthesis, catabolism and turnover than recovered children. A reduction of protein intake was followed by an increase in the catabolic rate with no change in protein synthesis. These studies clearly demonstrated the adaptive mechanisms, which have been confirmed in experimental animals, by which the body conserves nitrogen in response to a dietary deficiency. It was shown that the fall in the excretion of urinary nitrogen which follows the consumption of a diet low in protein is brought about by a marked reduction in the proportion of the amino-acid flux converted to urea. Thus, when a high protein diet was fed, 23% of the amino-acid flux was converted to urea, whereas after consumption of a low protein diet the amount converted was reduced to 2·4%.

An important aspect of this shift is the changes in the body whereby protein synthesis in certain tissues, particularly skeletal muscle, is reduced, whereas that in other tissues, such as the liver, is well maintained. As already indicated it has been suggested that glucocorticoids play an important role in bringing about these tissue adaptations in protein metabolism. Confirmation of these has been obtained experimentally[50] in an experiment in which weanling rats were given a diet containing 3·1% protein by weight and received at the same time 1·25mg of cortisone acetate per day. Under these conditions the plasma albumin concentration rose and there was a rise in the liver weight and in its protein content whereas the protein content of skeletal muscle fell. These changes in muscle and liver proteins were accompanied by changes in the corresponding free amino-acid pools, for the free amino-acids of muscle generally decreased in concentration whilst those in the liver rose.

Plasma amino-acids: In untreated severe PEM the amounts of total amino acids may be reduced to half of their normal value. In kwashiorkor there is a fall of the concentrations of most of the essential amino acids particularly the branched chain amino-acids, threonine and tryptophan. The concentrations of lysine and phenylalanine are less affected and the concentrations of the non-essential amino-acids are fairly well maintained or even increased. The concentration of histidine is raised, probably due to decreased activity of hepatic histidinase. Another consequence of the impaired metabolism of histidine is the excretion of imidazole acrylic acid.

Urea metabolism: In severe PEM the plasma concentration and urinary excretion of urea are reduced.

Carbohydrate metabolism: There is general agreement that the blood glucose concentration is lower in malnourished than in recovered or normal children. The hypoglycaemia is of two types. Thus, it may be asymptomatic, noted in blood sugar determinations, or profound and almost invariably fatal. The clinical history of the children showing the latter type of hypoglycaemia frequently shows that they have been anorexic and have not fed for 12–24 hours.

Glucose tolerance tests show delayed disappearance of blood glucose. Plasma insulin levels are low in PEM and insulin and glucagon response to intravenous glucose administration is impaired. The response returns to normal but only after some 2–10 months of rehabilitation. It has been suggested that the carbohydrate intolerance found in PEM predisposes the children to diabetes. The reasons for the insulin insensitivity in PEM are probably complex. It has been suggested that the raised cortisol and growth hormone levels found in the condition play a role, for both of these hormones increase peripheral resistance to insulin. Chromium deficiency may also contribute to the condition and potassium deficiency has also been implicated because deficiency of this element results in impaired insulin release in response to glucose infusion.

There is evidence of impairment of peripheral glycolysis, which persists after recovery and which occurs at a stage of glycolysis after the phosphofructokinase step.

Gluconeogenesis is in part dependent on amino-acid availability and liver function. Tissues like the brain and

TABLE 6
METABOLIC RATE OF MALNOURISHED, RECOVERING, RECOVERED AND CONTROL CHILDREN EXPRESSED IN TERMS OF DIFFERENT TERMS OF REFERENCE [12]

	$\dfrac{Kj}{Kg^{-1}d^{-1}}$	$\dfrac{Kj}{(Kg^{0.375})^{-1}d^{-1}}$	$\dfrac{Kj}{100\,cm^{-1}d^{-1}}$	$\dfrac{Kj}{(m^2)^{-1}d^{-1}}$	$\dfrac{Kj}{(m.molTBK^{0.75})^{-1}d^{-1}}$
Malnourished	210	315	1,654	3,450	17.9
Recovering	284	453	2,852	5,010	26.1
Recovered	251	417	2,772	4,827	24.2
Controls	236	409	2,914	4,857	22.9

erythrocytes which usually use glucose become dependent on hepatic gluconeogenesis in PEM and for this alanine from skeletal muscle is the principal precursor. With prolonged fasting the brain adapts to the use of ketones as fuel. It is not known to what extent this may occur in children with PEM. Hepatic glycogen levels are generally depressed in PEM though there have been reports of elevated levels in kwashiorkor.

Energy metabolism: The basal metabolic rate depends on the composition of the body, on cellular activity or both. A number of bases of reference have been used in studies on children with PEM, but regardless of the reference used, the metabolic rates of malnourished children have been found to be lower than in recovering or control children (Table 6). Energy intake is usually the limiting factor in controlling the rate of recovery from uncomplicated PEM. With a high energy intake of the order of 490 kj/kg^{-1}d^{-1} (119 kcals/kg/d) and a moderate protein intake rapid rates of weight gain can be achieved. It is reckoned that for each gram of tissue gain 20 kj (5 kcals) is required. A post-prandial increase in metabolic rate is probably related to the energy cost of growth.

Thermoregulation: Rectal temperatures of 35.5°C (96°F) or less occur in patients with PEM and in some parts of the world such low temperatures have been regarded as indicative of a poor prognosis. The degree of hypothermia is related to the severity of malnutrition. Moreover, hypothermia is more common in marasmic children with their marked reduction in skinfold thickness than it is in children with kwashiorkor who have relatively well-preserved subcutaneous tissues. Another factor in the pathogenesis of hypothermia is an inadequate supply of substrates for energy production. This applies particularly to the child with marasmus. There is a marked rise in the temperature when food is provided. Hypothermia may be an indication of asymptomatic hypoglycaemia in the severely malnourished infant.

Oedema. The aetiology of oedema in kwashiorkor is complex. It cannot occur in the absence of sodium and water retention, and hypoproteinemia is a predisposing factor. Albumin is the most important substance contributing to plasma colloidal osmotic pressure, but in studies in Uganda it has been found that the serum plasma colloidal osmotic pressure did not fall significantly until the albumin concentration fell to values of between 25.1 and 27.5 gl^{-1}. A significant correlation between the two did not appear until the plasma albumin concentration was low.[11] Increases in the concentration of globulins could account for the lag in fall of the colloidal osmotic pressure. An increased activity of ADH may also play an important role in the aetiology of oedema in kwashiorkor. It has also been suggested that increased aldosterone production is a causative factor but there is no good evidence to support this suggestion.

Diagnosis and Assessment of Severity

The diagnosis of severe PEM is usually straightforward; marasmus and kwashiorkor are easily diagnosed by the clinical features already mentioned. However, it is important that the diagnosis is correct before treatment is commenced. The identification of the child at risk of developing clinical PEM, and the detection of mild or moderate PEM and its proper assessment can make an important contribution to preventive medicine. True preventive medicine requires the identification of social, economic and other factors which pre-dispose children to PEM. However, for the present purpose we will concentrate on those methods with which the clinical biochemist is most likely to be concerned—anthropometry and laboratory investigations. The latter fall into two groups; those used to detect malnutrition and those necessary for treatment.

Anthropometry is the main method by which the severity of malnutrition and the chronicity of deprivation can be assessed. It can only be used, however, if the age of the child is known, for the interpretation of the measurements depends upon reference to standards for age.

The standardization of techniques for anthropometry is of prime importance. The operator needs to be practised at the techniques so that the various measurements can be made in a strictly reproducible fashion. Moreover, he needs to have equipment that will enable the measurements to be made accurately and quickly. Without sound equipment and training, anthropometric measurements are likely to be of little value.

The ratio of weight to ideal weight for age is a satisfactory expression of deficit of body size. Weight for height is probably a better index of wasting than weight for age and is a measure of 'acute' marasmus.

Percentage weight for height is calculated as follows:

$$\frac{\text{Weight of child} \times 100}{\text{Weight of standard child of the same height}}$$

A deficit in height for age usually indicates that the child has suffered chronic nutritional deprivation as stunting of the skeleton takes longer to become manifest than does a deficiency in body weight.

Ideally, 'normal' reference growth data should be available for each ethnic group in order to control for the genetic contribution to growth patterns. However, the most used standards[69] are the Harvard or Boston standards and in the U.K. those produced by Tanner and his colleagues.[65]

Fat can be conveniently assessed by making use of the fact that in man much of the total body fat, probably about 50%, is subcutaneous and the 'thickness' of this can be measured with a skin-fold caliper. If convenient, the skin-fold thickness at more than one site should be measured in order to allow for individual differences in the distribution of sub-cutaneous fat. The four sites suggested by Durnin and Womersley[31] are the biceps and triceps at the mid-point between the acromion and olecranon processes, the sub-scapular at the point of the shoulder blade and the supra-iliac measured at the tip of the iliac crest. The sum of these skin-fold thicknesses may then be used to calculate the total body fat as a percentage of the body weight using a nomogram.[45]

Mid-upper arm circumference can be used to provide useful information about muscle mass if combined with measurement of the triceps skinfold. The measurement has an added attraction in that it requires only a steel or non-stretchable plastic tape measure. The arm circumference represents a summation of the bone, muscle, fat and skin of the arm. What has become known as the 'mid-upper-arm muscle circumference' can be calculated as follows:–

Muscle circumference = arm circumference – (π × triceps skinfold)

Hair morphology. The hair root is a site of active protein synthesis throughout life. Moreover, samples are easily obtained with little or no discomfort to the subject and their examination requires only the availability of a simple microscope. Details of the method of obtaining hair samples together with interpretation guidelines were presented by Bradfield.[11] These guidelines are shown in Table 7. The hair root is affected differently in kwashiorkor and marasmus; this is not surprising since hair changes are a characteristic feature of the former condition. In kwashiorkor, the number of growing roots (anagens) is significantly less than normal, and usually they show severe atrophy. The mean bulb diameter may be reduced to one-third of normal values. The number of bulbs in the resting (telogen) phase is significantly increased. In marasmus, there is almost complete lack of bulbs in the growing phase and more broken hairs are found in marasmus than in kwashiorkor. Bradfield has concluded that the method is useful for the detection of the presence of malnutrition and not for the assessment of its severity.

Plasma Albumin Concentration. Marked hypoproteinaemia, in particular hypoalbuminaemia, is a characteristic of severe kwashiorkor. There has been considerable debate as to whether this determination is a sufficiently sensitive indicator of marginal malnutrition, but Whitehead et al.[73] have shown that when the level reaches $30 gl^{-1}$ all the metabolic changes associated with kwashiorkor develop. This information provided a basis in which to make the following interpretation guidelines: Concentrations $(g^{-1}) >$ 35 Normal; 30–34 Subnormal; 25–29 Low; <25 Pathological.

Other Plasma Proteins. The iron-carrying globulin transferrin is much more reduced in percentage terms than albumin. However, its use as an index of malnutrition requires further investigation, for the levels are evidently affected by iron, as well as protein status. Furthermore, more needs to be known about differences in the values in kwashiorkor compared with those in marasmus.

Plasma Amino-acid Patterns. It has been suggested that the pattern of amino-acids in the plasma which is abnormal in kwashiorkor could be used as a means of detection of children at risk of this form of PEM[74]. Simple semi-quantitative determinations of non-essential and essential amino-acids enables the calculation of a ratio:

$$\text{Ratio} = \frac{\text{glycine} + \text{serine} + \text{glutamine} + \text{taurine}}{\text{valine} + \text{leucine} + \text{isoleucine} + \text{methionine}}$$

In normal children the ratio was 1·5; in subclinical malnutrition values ranged from 2·0 to 4·0 whilst in frank kwashiorkor they were above 3·5, with a mean value of 5·0. The main attraction of this method, which was developed in Uganda, was its sensitivity to malnutrition at an earlier stage than plasma albumin. However, there are a number of problems with the interpretation, for the ratio very rapidly returns to normal when a good diet is given and it is clearly very sensitive to the protein content of the diet. An interpretation of the values is shown in Table 8.

Urinary Creatinine-height Index. Skeletal muscle is vulnerable to malnutrition and an estimate of muscle mass would provide a useful assessment of nutritional

TABLE 7
SUGGESTED STANDARDS FOR THE NUTRITIONAL INTERPRETATION OF HAIR ROOT SPECIMENS[11]

Measurement	Normal	Moderate Malnutrition	Severe malnutrition
		Values	
Mean bulb diameter (mm × 10^{-2})	>11	6–11	<6
Atrophy (percentage of anagen)	0–25	26–50	>50
Anagens (percentage)	>50	30–50	<30
Telogens (percentage)	<20	20–45	>45

TABLE 8
INTERPRETATION OF FASTING SERUM AMINO ACID RATIOS IN CHILDREN[2]

Ratio	Interpretation
<2·0	Adequate diet
2·1–3·5	Low protein/high carbohydrate diet being eaten at insufficient levels of protein to meet requirements; serum albumin levels probably decreasing leading to sub-clinical kwashiorkor
>3·5	Indicative of severe dietary deprivation and impending kwashiorkor

status. Biochemical assessment is based on the fact that creatinine is a breakdown product of creatine, nearly all of which is contained in muscle. The 24 hour excretion of creatinine is therefore related to the muscle mass, with 1g of creatinine excreted being equivalent to 20 kg of muscle in children. The creatinine-height index is the ratio between the creatinine excreted by the subject in 24 hours divided by the daily creatinine output of the average normal child of the same height. Normal values are given in the original publication.

Values determined by this method showed a good correlation with estimates of musculature obtained from the total body potassium measurements made in a whole body counter, and with estimates of musculature derived from soft tissue radiography.

Other methods: Other methods that have been suggested as suitable for the detection of malnutrition in children include the urinary urea/creatinine ratio, the hydroxyproline-creatinine index, the urinary excretion of 3-methylhistidine and determinations of both plasma and urinary ribonuclease. Further validation of each of these methods is required before they can be used with confidence.

Investigations During Treatment

The metabolic disturbances that occur in patients hospitalized for PEM are complex. However, a few simple determinations are all that are really necessary for their day to day care.

Anaemia is common and blood smears should be examined in order to determine the type that is present. The prothrombin index is also commonly depressed, particularly in septicaemic patients. Septicaemia is often associated with jaundice. The plasma bilirubin level should be determined in order to detect mild jaundice. Plasma electrolyte levels should also be determined as very low plasma potassium levels (approx. 1·5 mmol/l) are commonly seen. The low plasma urea level of kwashiorkor is raised by an adequate protein intake. The plasma albumin level is a sensitive guide to recovery.

Blood glucose determinations are necessary for the detection of asymptomatic hypoglycaemia though these can be done roughly with Dextrostix (Ames Company).

Bacteriological investigations are necessary to exclude the presence of infections and the tuberculin test should always be done on admission.

Discussion of the dietary approach to the rehabilitation of children with PEM is beyond the scope of this chapter and will be found elsewhere.[71]

Protein-energy Malnutrition in Adults

Prevalence and Pathogenesis

Broadly speaking, there are three groups of adults who may develop PEM. These are hospital patients, the elderly and those taking multiple drugs. Clearly, these groups are not mutually exclusive for some hospital patients are elderly, whilst others both old and young, receive a number of drugs.

The possibility that PEM may occur in hospital patients is now well documented as a result of careful studies carried out in the U.S.[9] and the U.K.[40] The medical or surgical condition necessitating hospitalization is a prime factor and patients may be malnourished on admission. However, the nutritional status of some patients may decline during their stay in hospital particularly if this is of more than two weeks duration. Again, this situation could be related to the nature of the disease process or to its treatment. The causes of a reduction in voluntary food intake are complex, but include the psychological effect of strange surroundings and stress on appetite and the effects of large scale cooking and presentation on the acceptability and nutritional value of food. The adequacy of hospital diets cannot be assessed from measurements of what is provided. The actual intakes of patients must be weighed.

Other factors involved in the aetiology of PEM in hospital patients include the use of a menu system which requires that the patient may have to choose a meal two days in advance. An intervening change in well-being may then make the choice unacceptable. Alternatively, a choice may be made on the patient's behalf which proves unacceptable. In some cases, there may be an additional problem with the food provided being totally unsuitable for the patient's condition. In such a situation an inflexible system of catering may not permit nutritious alternatives to be readily available. In the case of the elderly patient in particular, physical handicap, such as arthritis, may prevent the manipulation of cutlery, or mental handicap, confusion or dementia may lead to the neglect of any food presented.

In the elderly there are two main groups of factors that lead to malnutrition and these are summarized in Table 9. Primary and secondary factors often operate together to produce malnutrition in a particular individual. Thus,

TABLE 9
CAUSES OF NUTRITIONAL DEFICIENCIES IN OLD PEOPLE[33]

Primary	Secondary
Ignorance	Impaired appetite
Social isolation	Masticatory inefficiency
Physical disability	Malabsorption
Mental disturbance	Alcoholism
Iatrogenic	Drugs
Poverty	Increased requirements

limited mobility, loneliness, social isolation and depression are all found in some house-bound old people and make them particularly prone to malnutrition when they are receiving insufficient support from others. The recently bereaved are another vulnerable group. This is particularly true of widowers who may not have been used to providing for themselves. Ignorance, as well perhaps as poverty, then become factors which tend to inadequate provision of food.

A nutrition survey of the elderly based on random samples of old people living at home in six areas of the UK showed that malnutrition occurs in about 3% of the elderly population[26]. Iron deficiency and various vitamin deficiencies have received particular attention, but protein energy malnutrition may also be found. In most cases, the malnutrition occurs as a direct or indirect consequence of physical or mental disorders. Nevertheless, in some cases primary causes may be of importance.

As far as protein is concerned measurements of endogenous N metabolism suggest that 0.42 g protein/kg body weight is a safe practical allowance for healthy elderly women. However, for the normal utilization of dietary protein sufficient energy must be available and the protein-energy relationship is more important than the absolute amount of protein. Another factor that must be borne in mind is the quality and amino-acid composition of the dietary proteins; the requirement for the essential amino-acids, methionine and lysine, which cannot be synthesized in the body, may rise sharply after the age of 50 years.

The plasma albumin concentration falls with advancing age and the total albumin pool in the body is about 20% lower in the elderly as compared with the younger subjects. In the DHSS survey of the elderly of 1972[26], it was found that of the old people examined, 13% had plasma albumin levels of less than 35 gl^{-1} and 8% had serum cholinesterase activities of less than 50 units dl^{-1}. These findings bore some relationship to the clinical assessment of the state of health and the low activity of pseudocholinesterase was correlated with skinfold thickness. There was a slight correlation in the subjects between protein and bulk of arm muscle as determined by measurements of arm circumference and skinfold thickness. The plasma albumin level is less sensitive to protein intake than is the protein content of hair bulbs which seems therefore to be a more sensitive indicator of dietary protein deficiencies. However, aging changes in the hair bulb irrespective of dietary intake probably make hair roots an unreliable index of nutritional status in old people.

Drugs interact with nutrition in a variety of ways, some of which are desirable and related to the purpose for which the drug has been given, others undesirable.[27,28] Nutritional deficiencies (proteins, energy, minerals and vitamins) impair the biotransformation of drugs and their consequent removal from the body because of the relationship of nutrients to the activity of the mixed function oxidase drug-metabolizing enzyme system. Obviously, the effects of these interactions are likely to be more severe in severely depleted individuals. However, the most severe nutritional effects of drugs are likely to be those which occur in the elderly for in old people the administration of multiple drugs is common and these may interact with each other as well as with nutrients. In addition, the drug-metabolizing enzyme activity of the liver decreases with age and the ability of the kidney to excrete the end products of drug metabolism is also diminished in elderly people (Fig. 4).

Clinical Importance of Adult PEM

We have already seen that the body can adapt to lower than normal intakes of both energy and protein. However, in the face of disease the capacity for adaptation may be reduced and a series of changes occur which lead to the individual passing into negative nitrogen balance as the result of tissue cannibalization in order to meet the necessary energy requirement for essential bodily functions. The consequences of this negative nitrogen balance are summarised in Table 10. Loss of body weight can have a serious psychological effect on the patient and cause a lack of well-being and enhanced fear of a poor prognosis. Impaired immunocompetence also occurs and this is related to increased susceptibility to infection. In the patient undergoing surgery there is no evidence of impaired wound healing but PEM is associated with increased incidence of wound breakdown. Oedema may occur principally due to a low albumin level and the patient's mental state may be marked by severe apathy.

In the elderly the existence of sub-clinical malnutrition

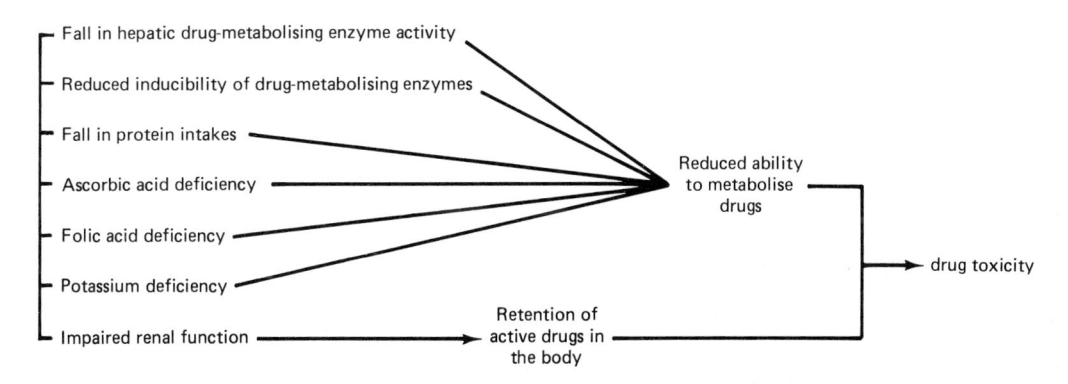

FIG. 4. Nutritional and other factors which may lead to increased drug toxicity in the elderly.[28]

TABLE 10
CONSEQUENCES OF NEGATIVE NITROGEN BALANCE[49]

Loss of body weight
Impaired immunocompetence—humoral and cellular
Increased susceptibility to infection
Increased incidence of wound dehiscence
Hypoproteinaemic oedema
Apathy
Increased mortality

including protein deficiency may have a greater importance than at other ages since homeostatic mechanisms are impaired and stress, associated with a variety of pathological processes, may disturb the precarious physiological balance. It remains, however, to be definitely established whether seeking to restore an adequate state of nutrition in an individual with sub-clinical depletion actually benefits the patient to a marked degree.

Assessment for PEM in Adults

In PEM, one or more of the major tissue compartments of the body is depleted. Whilst a fall in body weight gives some indication of malnutrition, measurements are necessary in order to evaluate the fat stores, skeletal muscle protein, visceral proteins and immune function. A list of useful measurements is given in Table 11. Most of these have been discussed previously and it is necessary only to add brief notes here.

TABLE 11
DETECTION OF PEM IN ADULTS

Anthropometric measurements	Biochemical determinations	Others
Body weight	Serum albumin	Hair roots
Height	Serum transferrin	Immunocompetence:
Triceps skin-fold thickness	Creatinine height index	Total lymphocyte count
Arm circumference		Skin tests

Body Weight. The patient should be weighed daily, preferably after emptying the bladder and before breakfast, on a beam or lever balance. The weight for height index or ratio is calculated by comparing the actual weight of the individual with the ideal value for a person of the same height and sex. For ideal weights, the modified Metropolitan Life Insurance company charts (Table 12) can be employed. This index can then be classified according to the criteria indicated for skinfold thickness and muscle circumference (Table 13). Values over 90% indicate that the individual is not depleted, values between 80 and 90% mildly depleted, those between 70 and 80% moderately depleted and those 60% or less severely depleted.

Fat Stores: The mass of body fat can be assessed by the measurement of the triceps skinfold thickness. Severe depletion of fat stores in a malnourished subject represents a significant nutritional problem for it represents

TABLE 12
IDEAL WEIGHT (KG) FOR HEIGHT (CM) FOR MEN AND WOMEN[10]

Height	Weight		Height	Weight	
	Men	Women		Men	Women
145	51·9	47·5	166	64·0	60·1
146	52·4	48·0	167	64·6	60·7
147	52·9	48·6	168	65·2	61·4
148	53·5	49·2	169	65·9	62·1
149	54·0	49·8	170	66·6	
150	54·5	50·4	171	67·3	
151	55·0	51·0	172	68·0	
152	55·6	51·5	173	68·7	
153	56·1	52·0	174	69·4	
154	56·6	52·5	175	70·1	
155	57·2	53·1	176	70·8	
156	57·9	53·7	177	71·6	
157	58·6	54·3	178	72·4	
158	59·3	54·9	179	73·3	
159	59·9	55·5	180	74·2	
160	60·5	56·2	181	75·0	
161	61·1	56·9	182	75·8	
162	61·7	57·6	183	76·5	
163	62·3	58·3	184	77·3	
164	62·9	58·9	185	78·1	
165	63·5	59·8	186	78·9	

TABLE 13
REFERENCE STANDARDS FOR INTERPRETATION OF TRICEPS SKINFOLD THICKNESS AND ARM MUSCLE CIRCUMFERENCE[10]

Nutritional status (% standard)	Triceps skinfold (mm)		Arm muscle circumference (cm)	
	Men	Women	Men	Women
Standard (100%)	12·5	16·5	25·3	23·2
Not depleted (90%)	11·3	14·9	22·8	20·9
Mildly depleted (80%)	10·0	13·2	20·2	18·6
Moderately depleted (70%)	8·8	11·0	17·7	16·2
Severely depleted (60%)	7·5	9·9	15·2	13·9

a depletion of energy stores and this means that there is less fuel for the individual to draw upon during a state of starvation or semi-starvation. Reference standards for interpretation of triceps skin-fold thickness are shown in Table 13.

Lean Body Mass. The bulk of the protein reserves which are mobilized during starvation are located in the skeletal muscle mass and the size of this can be evaluated using two different parameters; the arm muscle circumference (AMC) and the creatinine/height index (CHI). Reference standards for the interpretation of the arm muscle circumference are shown in Table 13. Because creatinine is a product of muscle metabolism, the 24 hour excretion of creatinine is related to the total muscle mass. A complete urine collection over 24 hours is required to determine the daily excretion. The actual value measured is then compared to the creatinine excretion expected for an individual of the same height, at the ideal weight (Table 14). The creatinine height index is then calculated as a ratio of the actual excretion

Table 14
URINARY CREATININE/24 HOURS/CM HEIGHT FOR MEN AND WOMEN OF IDEAL WEIGHT FOR THEIR HEIGHT (IN SHOES AND INDOOR CLOTHING)[8]

Men				Women			
Height	Ideal weight	Total creatinine	Creatinine mg/cm height/24 hours	Height	Ideal weight	Total creatinine	Creatinine mg/cm height/24 hours
cm	Kg	mg		cm	Kg	mg	
157.5	56	1288	8.17	147.3	46.1	830	5.63
160	57.6	1325	8.28	149.9	47.3	851	5.68
162.6	59.1	1359	8.36	152.4	48.6	875	5.74
165.1	60.3	1386	8.40	154.9	50.0	900	5.81
167.6	62.0	1426	8.51	157.5	51.4	925	5.87
170.2	63.8	1467	8.62	160.0	52.7	949	5.93
172.7	65.8	1513	8.76	162.6	54.3	977	6.01
175.3	67.6	1555	8.86	165.1	55.9	1006	6.09
177.8	69.4	1596	8.98	167.6	58.0	1044	6.23
180.3	71.4	1642	9.11	170.2	59.8	1076	6.32
182.9	73.5	1691	9.24	172.7	61.6	1109	6.42
185.4	75.6	1739	9.38	175.3	63.4	1141	6.51
188	77.6	1785	9.49	177.8	65.2	1174	6.60
190.5	79.6	1831	9.61	180.3	67.0	1206	6.69
193	82.2	1891	9.80	182.9	68.9	1240	6.78

of creatinine divided by the ideal excretion × 100. This index is then interpreted as shown in Table 13.

Visceral Proteins. The level of plasma proteins and particularly albumin falls during a stress situation and low levels are found in protein deficiency. Albumin is therefore usually measured although it is not known whether it is a true index of nutritional status in adults, as for instance after surgery, or whether the low levels simply result from a movement of albumin from the intravascular to the extravascular space. Recent studies indicate that proteins with relatively shorter half-lives, such as transferrin, react to nutritional depletion and repletion more rapidly.

Protein Synthesis. As already indicated, the hair bulb is a site of active protein synthesis throughout life and the hair bulb changes very quickly in response to changes in protein nutrition. Classification of hair bulbs or alternatively the measure of their protein content therefore give a good indication of the state of protein synthesis in the individual.

Immunocompetence: Cellular immune function is severely compromised by malnutrition. Some indication of the level of immunocompetence can be obtained by measurements of the total lymphocyte count. This is calculated from the percentage of lymphocytes multiplied by the white cell count and divided by 100. The value obtained should then be expressed as a percentage of the standard value, i.e. 1200 per cubic millimeter, and the index thus calculated is again used to assess the state of depletion (Table 13). Lymphocyte counts are an unreliable index in cancer patients receiving radio- or chemo-therapy. Cell-mediated immunity is an important part of the host defence mechanisms against infection and its depression is associated with increased morbidity and mortality. The state of this system can be further evaluated by means of skin tests using commonly recalled antigens such as mumps and Candida. The chemical compound dinitrochlorobenzene (DNCB) can also be used. These compounds are individually injected subcutaneously at different sites on the patient's arms. The diameter of the wheal at each site is then measured after 48 hours and a result greater than 5 mm on any one reaction site is considered a normal response.

Ideally, all these measurements should be made systematically on at risk hospital patients on admission and at intervals during their stay. If the stay is two weeks or less they should simply be repeated on discharge. It is important that these records should be kept with the patient's notes and if possible collated by the same individual on to a chart. Haphazard measurements made under varying conditions are of little value.

Treatment

The method of rehabilitation of adults with PEM depends upon the capability of the individual to ingest, absorb and retain nutrients. Sometimes this can be accomplished by encouraging the individual to eat normal foods. In other conditions the nature of the food must be changed so that it is soft and can be swallowed and digested without difficulty. In still others a liquid diet is necessary and this may have to be passed down a naso-gastric tube. When enteral feeding is not possible, resort must be had to total parenteral nutrition. Discussion of each of these methods of feeding will be found elsewhere.[29]

The Malnutrition of Malignancy

PEM can be a serious problem in the patient with cancer. It may interfere with the response to anti-tumour treatment by exacerbating unpleasant side-effects and reducing tolerance to radiotherapy and drugs, or by increasing the likelihood of infection because of its effect on immunocompetence. Furthermore, because of the psychological effect of the accompanying fall in body weight, PEM may decrease the patient's will to live.

Anorexia

Anorexia is one of the most prevalent causes of PEM. The causes of anorexia are often complex, for whilst a decreased food intake may result from mechanical interference with food ingestion, as in patients with tumours of the head and neck, oesophagus or intestinal tract, it may also result from changes in intermediary metabolism, aberrations of taste and smell or from psychological disturbance. If PEM is allowed to continue the adaptive response of the body to malnutrition fails and the patient passes into a cachectic condition which becomes increasingly difficult to redress.

The assessment of anorexia presents a problem, for patients will often deny changes in appetite when they have already reduced their intake. Indirect evidence may be obtained from changes in body weight, and body weight at the time of presentation. However, a fall in body weight may simply indicate that food intake has not kept pace with energy expenditure or that the patient has some degree of malabsorption. A weighed dietary intake may be the only reliable method of assessing the actual amount of food consumed.

Appetite Control. A reduction in appetite, and consequently of voluntary food intake, may occur as a result of effects upon those mechanisms which actually control appetite. Briefly, two kinds of control system have been described—hierarchical and integrated.[25] In hierarchical systems, the central nervous system is usually considered to be at the top of the hierarchy and able to overide other cues. This is in contrast to an integrated system in which each cue has its own impact on the control of food intake and is not overridden by other cues. De Wys favours an integrated system as being the most satisfactory one available at present in helping to explain the anorexia of the cancer patient. However, additional factors which may operate must also be considered. In the rat, lesions placed in the ventromedial nucleus of the hypothalamus cause hyperphagia whereas if they are placed in the lateral nuclei they cause hypophagia. Furthermore, hypothalamic function can be sub-divided according to the type of catecholamine involved, thus a system involving β-adrenergic and dopaminergic receptors suppresses feeding behaviour and it is known that these systems can be manipulated by drugs. Amphetamine, for instance, can cause anorexia by acting on the β-adrenergic and dopaminergic system. A number of hormones also affect appetite. Removal of the ovaries increases eating and insulin may be given to increase appetite. In contrast, glucagon, adrenalin, enterogastrone and cholecystokynin all decrease appetite. Food intake is also influenced by changes in the concentrations of metabolites such as glucose and by the size of the body fat stores.

These effects form the basis of the 'glucostatic' and 'lipostatic' theories respectively of appetite control. The concentrations and patterns of amino-acids in the blood and extracellular fluid also have a complex and incompletely understood effect on appetite.

Psychological Aspects. The extent of fear and pessimism in response to the presence of a suspected or known cancer depends very much on attitudes and beliefs about malignant disease. It is very common for a diagnosis to lead the patient to assume the worst and the resulting worry to lead to weight loss. In such circumstances it is difficult to know whether the weight loss is the direct result of fear of cancer, or of the cancer itself. The origin of feelings of hopelessness and lack of motivation in the patients with cancer is not at present clear. It may be that input is received by the hypothalamus from the higher cerebral cortical centres which in turn help to regulate bodily functioning through neuroendocrine and autonomic nervous system connections. These mechanisms would help to explain the clinical observations that if feelings of hopelessness persist for any length of time there will be loss of appetite, anorexia, malaise and weight loss. There is little information at present available as to the advantage to the patient of encouragement to eat a high energy diet. In considering the effect of psychological state on food intake it seems clear that we are dealing with an interaction in which it is very difficult to decide which is the chicken and which is the egg. It would seem likely that motivation, hope and general mental activity affect cellular metabolism in some way that we do not at present understand. The consequences of these cellular changes may well affect the chance of survival.

Taste and Feeding Behaviour. Patients with cancer often complain of abnormalities of taste; it seems that these are a distant effect of the tumour. Lowering of the recognition threshold for bitter, assessed with urea, has been noted to be associated with raised thresholds for sour and for salt. Longitudinal studies indicate a correlation between changes in taste sensation and changes in disease status. Patients with limited tumour growth have been found to have normal taste thresholds whereas those with advanced disease have a high probability of an abnormality of taste sensation, particularly the recognition of sucrose or urea, or both. It seems possible that these abnormalities of taste sensation are both a cause and a consequence of a reduced energy intake and that they decrease in their severity as the energy intake is increased. However, changes of taste and oral stimulation cannot be considered in isolation for these sensations may trigger physiological reflexes. Stimulation of receptors within the oral cavity is essential for the initiation of swallowing and foods which appeal to the palate result in a copious flow of gastric secretions.

Role of Vitamins and Minerals. Patients with cancer show evidence of multiple nutrient deficiencies and disturbances of tissue and plasma concentrations of a number of minerals. These may interact to reduce taste acuity and thus play a role in reducing food intake. There is some evidence that zinc may play a key role in this respect.

Cachexia

Cancer cachexia is characterized clinically by anorexia, early satiety, weight loss, anaemia and marked asthenia. Whilst the exact pathogenesis of this disorder is unknown it seems clear that both local and systemic adverse effects of the disease play a role and basically only treatment of the disease can completely reverse the syndrome, though some improvement may occur as the result of aggressive nutritional rehabilitation. It has for some time been considered likely that toxic materials produced by the tumour enter the blood stream, are taken up by normal cells, with a consequent adverse effect on their metabolism, and play a role in the genesis of cachexia (Fig. 5).

The changes in metabolism which occur in cancer cachexia differ somewhat in certain respects from those which result from simple starvation (Table 15). The cachectic individual is quite unable to adapt to the reduced intake and increased energy expenditure and as a consequence progressively deteriorates. Fortunately, severe cachexia is only rarely seen in cancer patients today.

Assessment of nutritional status in the cancer patient and the rehabilitation of malnourished patients is carried out as described in the section above of PEM in adults.

VITAMIN DEFICIENCY AND EXCESS

Introduction

Vitamins are organic substances that are essential to health. They cannot be made in the body, at least not in amounts sufficient to meet metabolic requirements, and must thus be supplied in the diet. With one notable exception (vitamin D, or cholecalciferol), vitamins function as co-factors for enzyme systems. Broadly speaking, the vitamins can be divided into two groups according to whether they are fat, or water-soluble. Ill effects from the ingestion of large quantities of water-soluble vitamins are rare and transient. In contrast, ill-effects resulting from the consumption of large doses of some fat soluble vitamins, particularly retinol (vitamin A) and cholecalciferol (vitamin D) can be serious and even fatal. As with protein and energy, the effects of deficiencies and excesses are more serious in the rapidly growing foetus and child than in adults. However, as we shall see, deficiencies of some vitamins present a public health problem in certain ethnic minority groups and amongst old people.

It is important to remember that deficiency diseases are the end result of a series of changes (Fig. 6) and that they take time to develop. The fact that an individual does not show clinical evidence of disease does not mean

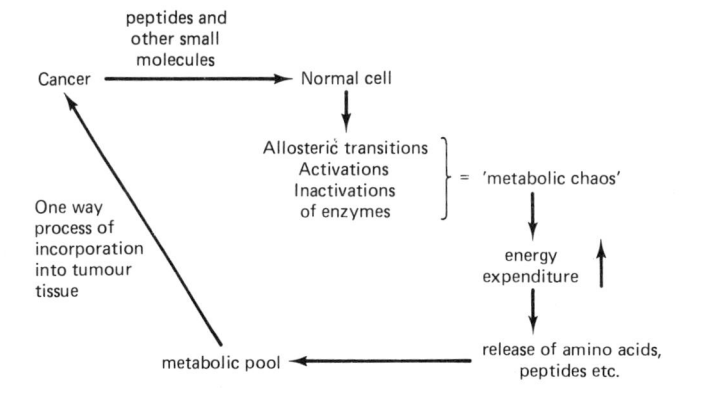

FIG. 5. Pathogenesis of cancer cachexia.[66]

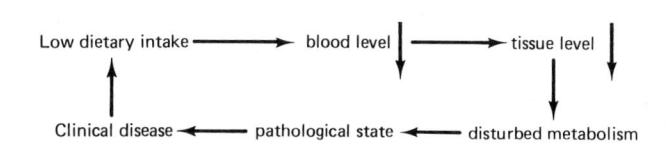

FIG. 6. Sequence of changes leading to the development of a deficiency disease.

TABLE 15
METABOLIC EFFECTS OF STARVATION AND CACHEXIA

	Starvation	Cachexia
Appetite	Good	Poor: early satiety
Energy expenditure: basal metabolic rate	↓	↑
oxidative metabolism	↓	↑
Nitrogen metabolism: blood amino acid N	↓	↑
Carbohydrate metabolism: glycogen	↓	↓ (diabetic type glucose tolerance curve)
Weight: body	↓	↓
liver	↓	No change
Body fat: stores	↓	↓
ketosis	↑	None
Anaemia	None unless severe and chronic	Present; multifactorial

that there is no deficiency of a particular vitamin. Biochemical indicators of circulating vitamin levels may yield low values even in the absence of overt disease. There is considerable debate as to the significance of such values and they are generally said to indicate a state of "sub-clinical deficiency". There are two situations in which such deficiencies may be serious—in the sick, in whom there may be an increased requirement, and in the elderly, in whom, as we have seen already, homeostatic mechanisms may be somewhat brittle.

It is also important, in this connection, to recognise that a level of deficiency which is not sufficient to produce disease may, nevertheless, result in some impairment of function. Many signs of impaired function are somewhat non-specific and only detected by experienced observers.

The situation is made more complex by interactions between certain vitamins whereby excess of one may "protect" or "conserve" another. In addition, blood levels of vitamins which circulate in the plasma bound to specific proteins may be low due to a deficiency of the binding protein. Thus, retinol circulates in the plasma bound to retinol-binding protein (RBP) which forms part of the prealbumin fraction. The synthesis of RBP is dependent on the availability of zinc so that low zinc status can be responsible for a low level of retinol. Both retinol and cholecalciferol are stored in the liver, and in the case of retinol, it is clearly possible for the liver store to be adequate and yet the circulating level, and consequently the amount reaching the tissues, to be low.

Clinical Features

Vitamin A

Vitamin A deficiency causes xerophthalmia. A mild form of this disorder, confined to the conjunctiva, is common in may countries and is not in itself serious. However, it serves as a warning for if the deficiency becomes more severe and spreads to the cornea, there is danger of corneal ulceration and a permanent defect of vision. In severe cases there is softening of the cornea, keratomalacia, which if not treated immediately leads to permanent blindness. Keratomalacia is often associated with PEM in young children. Indeed, the peak age at which blindness is most likely to occur is between two and five years when kwashiorkor is common, but it can occur in association with marasmus in the first year of life.

Vitamin A deficiency is most prevalent in countries where rice is the dietary staple. The diet contains practically no milk or butter and very small amounts of fresh vegetables and fruit. It therefore lacks both retinol and carotenoids which can be converted into retinol in the body. Xerophthalmia and keratomalacia occur in the first year of life amongst artificially fed infants but are rare amongst the breastfed. It is estimated that 250 000 young children become permanently blind every year as the result of keratomalacia.

In experimental animals, vitamin A deficiency causes squamous metaplasia in the cornea, but there are no good reports of similar histological studies in man.

Five clinical manifestations of vitamin A deficiency have been described by WHO and these have been given code numbers to assist reporting (Table 16).

A number of other conditions, e.g. exposure, trauma, bacterial infection, measles and trachoma, cause corneal lesions and keratomalacia and vitamin A deficiency must therefore be distinguished from them.

Populations at risk of vitamin A deficiency should be given a prophylactic oral dose of 66 000 μg vitamin A every 6 months.

Vitamin A can be toxic and both acute and chronic poisoning have been described. Acute poisoning has followed the administration of single large doses of about 100 000 μg to children. There are symptoms of raised intracranial pressure with headache, vomiting and restlessness. Chronic poisoning takes several weeks or years to develop depending on the age of the child and the dosage. Infants can become hypervitaminotic within a few weeks on 20 000 μg/day. A bizarre clinical picture is presented and vitamin A toxicity may not be suspected. The hair becomes coarse and sparse, the skin dry and rough, the lips cracked and the eyebrows denuded. Hepatosplenomegaly develops with headache and gener-

TABLE 16
FORMS OF XEROPHTHALMIA AND ASSOCIATED CONDITIONS[75]

Code	Form and associated condition	Description
X1A	Conjunctival xerosis	Bulbar conjunctiva dry, thickened, wrinkled, and pigmented with a 'smoky' appearance. Similar condition caused by long periods of exposure to glare, dust, and infections. In children under 1 year likely to be due to dietary deficiency
X1B	Conjunctival xerosis with Bitot's spots	
X2	Corneal xerosis	Cornea dry, dull, hazy and lacks lustre. Insensitive to touch
X3A	Corneal xerosis with ulceration	Corneal ulceration may occur from many causes. Erosion occurs which in the absence of infection heals by scarring
X3B	Keratomalacia	Develops from xerosis, little chance of preventing blindness
XN	Night blindness	An early symptom of vitamin A deficiency
Xf	Xerophthalmia fundus	Ophthalmoscopic examination shows white or yellow scattered spots along sides of blood vessels
XS	Corneal scars	White, opaque patches on the cornea. Effect on vision depends on size of scar. Likely to be due to vitamin A deficiency in areas where this is prevalent

alised weakness. The bones also show deformities and osteoporosis.

Carotenaemia occurs in individuals who consume an abnormal quantity of carrots or carrot juice. Death may result from the following of such a fad regimen.

Vitamin D

Rickets and osteomalacia are the result in children and adults respectively of a deficiency of vitamin D. The child with rickets is often restless, fretful and pale with flabby, toneless muscles. Excessive sweating of the head is common, the abdomen is distended as the result of the weak abdominal muscles. Diarrhoea is a common feature and the child is also prone to respiratory infections. Development is delayed and in severe cases growth may be severely impaired.

The bony changes are the most characteristic and easily identifiable sign of rickets. There is widening of the epiphyses where cartilage meets bone. The earliest recognizable lesion is craniotabes, small round unossified areas in the skull bones. Another early sign is enlargement of the epiphyses at the lower end of the radius and at the costrochondral junctions of the ribs—described as the "rachitic rosary". "Bossing" of the frontal and parietal bones may occur with delayed closure of the anterior fontanelle. If allowed to continue the child with rickets develops deformities. These include the so-called "pigeon-chest" in which there is undue prominence of the transverse depression of the chest which makes breathing difficult. Kyphosis of the spine may develop in the second or third year as a result of gravitational and muscular strains, caused by sitting or crawling. There may also be enlargement of the lower ends of the femur, tibia and fibula and when the child begins to walk the weight may cause deformities of the shafts—knock-knees (genuvalgum) or bow-legs (genuvarum). Rickets is most likely to be seen in Britain in children whose parents come from India and Pakistan.[4] Congenital rickets may occur in babies of osteomalacic mothers, usually presenting as neonatal hypocalcemic seizures. Pre-term rickets, formerly called the rickets of prematurity, is thought to be due to the very rapid linear growth of the pre-term infant. This contrasts with the rapid weight gain of the low birth-weight dysmature (light-for-dates) term infant whose main problem is hypoglycaemia. Infantile rickets was formerly common in Britain due to the feeding of unfortified substitutes for breast milk for some time after birth. The earliest sign is craniotabes and costrochondral beading. Toddler rickets affects mainly those aged 9 months to 3 years who have received an adequate intake of vitamin D for the first four to six months but whose intake then falls due to a switch from fortified to doorstep milk. Toddler rickets may be severe enough to cause pelvic contracture and genuvarum in babies already walking. Rickets is all too common amongst Asian immigrant adolescents who are usually Muslims adhering to their religion, dress, language and dietary habits. The diet of many of these immigrants contains a large proportion of chappatis and other bread made from whole flour. Arneil points out that despite pain in limbs, awkward and deteriorating mobility, and marked genuvalgum the condition is often overlooked by teachers, nurses and doctors. Modesty and purdah play an important role in veiling the deformed limbs of pubertal Asian females.

Osteomalacia occurs in women of child-bearing age who have been depleted of calcium by repeated pregnancies. It is common amongst Muslims who practise purdah and is due to the consumption of a poor diet and the absence of exposure to sunlight. The disease is also not uncommon in non-immigrant old women in the U.K. and occurs particularly in those who, because of disability, cannot go out easily. Indeed recent evidence[48] suggests that lack of exposure to sunlight is the main aetiological factor. Osteomalacia often co-exists with osteoporosis. Distinctive features of these two conditions are shown in Table 17.

Osteomalacia occurs in up to 25% of subjects who have undergone gastrectomy with signs of bone disease appearing 10 to 14 years after surgery. A failure to absorb vitamin D or a dietary deficiency of vitamin D and calcium have been suggested as possible reasons. Osteomalacia also commonly occurs as a consequence of chronic diseases of the digestive system such as cystic fibrosis and coeliac disease. A complex bone disease known as renal osteodystrophy occurs in patients with chronic renal failure. This disease is osteomalacia with varying degrees of osteitis fibrosa, osteoporosis, osteosclerosis and, in children growth retardation. The disease is resistant to treatment with all but very high doses of vitamin D. Due to the renal disease there is a failure to synthesize 1,25-dihydroxycholecalciferol [1,25-$(OH)_2D_3$] from the parent vitamin.

Rickets and osteomalacia have been described in patients receiving long-term anticonvulsant drug therapy, particularly with dilantin and phenobarbitone. One possible explanation is that these drugs induce hepatic microsomal enzymes which convert 25- hydroxycholecalciferol to inactive metabolites. Another possible explanation is that the anticonvulsant drugs disturb mineral metabolism perhaps by inhibiting vitamin D function. These explanations are not mutually exclusive and evidence has been obtained in favour of both of them. Practically, it is important to note that the requirements of such patients are raised and that preventive measures against deficiency are adequate to meet this raised requirement. Children on these drugs probably require 25 μg per day, compared with 10 μg per day for children not on these drugs.

Genetically determined forms of vitamin D-resistant rickets occur. Hypophosphataemic vitamin D-resistant rickets is inherited as an X-linked dominant genetic defect although new cases occur sporadically, presumably as new mutations. Vitamin D-dependent rickets is a rare disorder with autosomal recessive heritability.

There is a narrow gap between the nutrient requirement and the toxic dose of vitamin D. As little as five times the recommended intake i.e. 50 μg per day, over prolonged periods can lead to hypercalcaemia and

TABLE 17
THE DIFFERENTIAL DIAGNOSIS OF OSTEOMALACIA AND OSTEOPOROSIS[21]

	Osteomalacia	Osteoporosis
Clinical features		
Skeletal pain	Major complaint, usually persistent	Episodic, usually associated with fracture
Muscle weakness	Usually present, resulting in characteristic 'waddling' gait	Absent
Fractures	Relatively uncommon, healing delayed	The usual presenting feature, heals normally
Skeletal deformity	Common, especially kyphosis	Only occurs with fracture
Biochemical features		
Plasma Ca & P	Often low	Normal
Plasma alkaline phosphatase	Often high	Normal
Plasma vitamin D	Low	Normal
Urinary calcium	Often low	Normal or high
Radiographic features		
Low bone density	Widespread	Irregular, most marked in the spine
Loss of bone detail	Characteristic	Not a feature
	Diagnostic	Absent
Biopsy		
Histological changes	Excess osteoid, normal quantity of bone	Reduced quantity of bone, fully mineralized

nephrocalcinosis in adults. Toxicity can occur in children due to a mother mistaking her instructions in administering a concentrated vitamin D preparation. The earliest toxic symptom in children is usually loss of appetite with nausea and vomiting. Thirst and polyuria develop and constipation may alternate with diarrhoea. Pains in the head frequently occur, and the child becomes thin, irritable and depressed and gradually stuporose. Metastatic calcification has been described in fatal cases.

The syndrome of supravalvular aortic stenosis has been presumed to be due to hypersensitivity to vitamin D. In this condition there is aortic stenosis with left ventricular hypertrophy, myocardial focal necrosis, fibrosis and calcification. Affected children have a close craniofacial similarity to those with severe idiopathic hypercalcaemia and like these may be mentally defective. Patients with this condition seldom received more than 75 μg of dietary vitamin D per day and the question arises as to whether they received too much vitamin D before birth.

There is a wide variation in the 24-hour excretion of calcium between individuals in different parts of the world. In Britain the upper limit of normal is taken as 400 mg for men and 300 mg for women. Hypercalciuria on a normal diet is most commonly related to an abnormally high absorptive capacity for calcium and individuals with the condition have an increased tendency to form urinary calculi. Clearly, the enhanced absorption of calcium in indiopathic hypercalciuria could be due to stimulation of the conversion of vitamin D to its active forms. It might also be due to increased sensitivity of the intestinal mucosa to vitamin D though it seems difficult to explain changes on this basis.

Although hypercalcaemia occurs in less than 20% of patients with sarcoidosis, hypercalciuria is found more often and evidence of hyperabsorption of calcium is an almost constant finding in the active disease. As with idiopathic hypercalciuria, there is a seasonal change in calcium absorption, and this is related to a hypersensitivity to vitamin D which can be corrected with glucocorticoids.

Vitamin E.

Primary vitamin E deficiency has been reported only in premature infants who are born with an inadequate reserve of the vitamin. If given a dietary formula with a high content of polyunsaturated fat or iron medication they may develop a haemolytic anaemia with an associated thrombocytosis, generalized irritability and oedema. Evidence of vitamin E deficiency has been described in patients with malabsorption of fat due to cystic fibrosis, abetalipoproteinemia and other diseases. Red cell survival is reduced but anaemia and clinical manifestations are unusual. Vitamin E toxicity, manifested as multiple haemorrhages, has been described in a patient taking warfarin with high doses of the vitamin. Warfarin and vitamin E antagonize the action of vitamin K in provoking the clotting of blood. Vitamin E is now being taken in large doses (up to 1000 i.u. per day) to prevent clotting problems and also to prevent senile changes in the brain. There is no real evidence that it does either in the human. There is some evidence, however, that high doses may be beneficial in the treatment of intermittent claudication.

Vitamin K

Primary deficiency of vitamin K is rare in adults but has been seen in infants. Some cases of hypoprothrombinaemia in infants are cured by giving vitamin K.

As vitamin K is fat-soluble any disease process which interferes with the absorption of fat from the gut is likely to lead to a deficiency of the vitamin. Severe bleeding after an operation for relief of jaundice due to obstruction of the common bile duct used to be feared by surgeons. It is now customary to give vitamin K by injection before the operation in order to reduce this danger. Bleeding due to vitamin K deficiency may occur in coeliac disease and other conditions in which fats are not collectively absorbed. Vitamin K deficiency with hypoprothrombinaemia and bleeding can also follow the administration of antibiotics which reduce colonic flora in individuals who have been eating poorly.

As already indicated warfarin antagonizes the action of vitamin K and overdosage with warfarin can be overcome by an adequate dose of phytomenadione (Vitamin K_1).

Thiamine (Vitamin B1)

Beriberi due to thiamine deficiency was formerly common amongst the rice-eating peoples of the East. Three forms of the disease occur: (i) wet beriberi characterized by oedema often associated with high-output cardiac failure; (ii) dry beriberi, a polyneuropathy and (iii) the infantile form. Thiamin deficiency is also involved in three conditions which are not uncommon in chronic alcoholics and occur in all parts of the world. These are (i) alcoholic polyneuropathy, clinically indistinguishable from dry beriberi, (ii) a thiamine-responsive cardiomyopathy and (iii) an encephalopathy, the Wernicke-Korsakoff syndrome.

Early symptoms and signs are common to both wet and dry beriberi. The onset is usually insidious with anorexia and ill-defined malaise associated with weakness of the legs. Oedema of the legs and face may become manifest. Tenderness of the muscles occurs on pressure and there are complaints of 'pins and needles' (paraesthesiae) and numbness of the legs. Anaesthesia of the skin particularly over the tibiae is common. The patients are only slightly incapacitated but at any time the disease may develop into either of the severe forms.

Oedema is the most characteristic feature of *wet beriberi* and may develop rapidly to involve legs, face, trunk and serous cavities. Palpitations are marked and there may be breathlessness. Anaemia and dyspepsia are common. Often there is pain in the legs after walking similar to that resulting from ischaemia of muscle. Neck veins become distended and show visible pulsations. High systolic and low diastolic pressures occur in the arteries. If the circulation is well maintained, the skin is warm to the touch owing to associated vasodilation. When the heart begins to fail the skin becomes cold and cyanotic. The mind is usually clear. Sudden increase in oedema may occur with acute circulatory failure extreme dyspnoea and death.

The essential feature of *dry beriberi* is polyneuropathy. Early symptoms and signs are as described above. Muscles become progressively wasted and weak, and walking increasingly difficult. The thin emaciated patient becomes progressively weaker and may succumb to an infection. The *infantile form of beriberi* is always a more or less acute disease due to a deficiency of thiamine in breast milk. The early signs are mild and easily missed by the mother but the disease progresses rapidly and is fatal if not treated at between 2–4 months of age. It is milder

TABLE 18
PRESENTATION OF INFANTILE BERIBERI[13]

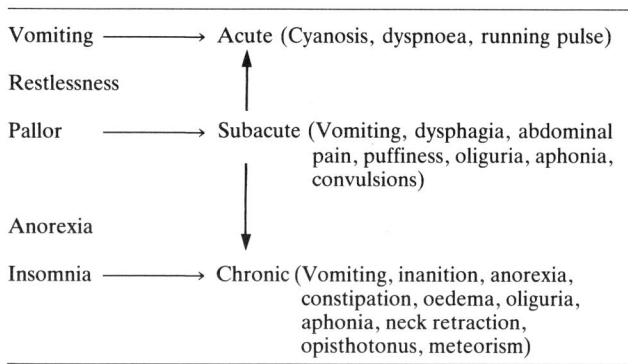

Vomiting ⟶ Acute (Cyanosis, dyspnoea, running pulse)

Restlessness

Pallor ⟶ Subacute (Vomiting, dysphagia, abdominal pain, puffiness, oliguria, aphonia, convulsions)

Anorexia

Insomnia ⟶ Chronic (Vomiting, inanition, anorexia, constipation, oedema, oliguria, aphonia, neck retraction, opisthotonus, meteorism)

and less common after 6 months. The mothers show few signs of the disease themselves.

The clinical features are summarised in Table 18. Three main syndromes have been described:[44] (i) The acute cardiac variety is of sudden onset at between 2 and 4 months of age with physical signs of acute cardiac failure and death may occur within minutes or hours. (ii) Aphoria occurs between 5 and 7 months and coughing and choking suggest a respiratory infection. The child emits a characteristic noiseless cry with laryngeal nerve paralysis, or oedema of the larynx. (iii) The pseudomeningeal variety simulates meningitis in the older infant 8–10 months of age with apathy, drowsiness, head retraction and signs of raised intracranial pressure.

Beriberi is rare after the age of 1 year. In older children it manifests itself in a way similar to that of adults.

Alcoholics who have eaten little or no food for many weeks may develop a disorder of peripheral nerves indistinguishable from dry beriberi. More than one nerve may be involved with the lesions being symmetrical and the nerves of the lower limbs being affected more severely than the upper limbs. Usually there is dysfunction of both sensory and motor fibres. The effects on the sensory fibres may be paresthesiae or sometimes severe nerve pains. Motor nerve involvement manifests itself as foot drop, muscle wasting and impaired knee and ankle jerks.

Wernicke-Korsakoff syndrome is characterized by weakness of eye muscles (ophthalmoplegia), accompanied sometimes with jerky rhythmical movements of the eyes (nystagmus). If the patient can stand he is unsteady (ataxic). The psychosis which is part of the syndrome is characterized by a severe defect in memory and learning. Confabulation is a characteristic feature though not always present. Memory of distant events is good but that for recent events is poor. However, the individual tends to provide a superficially convincing tale about these events rather than admit that he has forgotten them.

Subclinical thiamine deficiency occurs in old people and may indeed manifest itself as a cardiomyopathy. It can also be a cause of confusion in old people, particularly when there is the added stress of an infection or surgery. In the elderly, thiamine deficiency may be caused by interference with the absorption of the vitamin by antacids. The cytotoxic drug, 5-fluorouracil (5-FU) causes sub-clinical deficiency probably due to interference with the phosphorylation of thiamine.[1]

Riboflavin (Vitamin B_2)

Ariboflavinosis can be a cause of angular stomatitis, cheilosis and nasolabial seborrhoea in malnourished subjects. It is not unknown amongst old people.

Nicotinic acid (Niacin) and Nicotinamide (Vitamin B_3)

Pellagra, the disease caused by nicotinic acid deficiency, is endemic amongst poor peasants who subsist chiefly on maize. The typical clinical features are loss of weight, increasing debility, an erythematous dermatitis, gastrointestinal disturbances especially glossitis and diarrhoea, and mental changes. It is said to be characterized by the three D's—dermatitis, diarrhoea and dementia. However in mild cases depression may be the main presenting feature with possibly some skin changes.

The erythema has a characteristic distribution, appearing symmetrically over the parts of the body exposed to sunlight, especially the back of the hands, wrists, forearms, face and neck (Casall's collar). Diarrhoea is common but not always present, the mouth is sore and often shows angular stomatitis and cheilosis. The tongue characteristically has a 'raw beef' appearance—red, swollen and painful.

In mild cases the symptoms in the nervous system consist of weakness, tremor, anxiety, depression and irritability. In severe acute cases, delerium is common and in chronic cases dementia develops. In chronic cases there may be peripheral symptoms with increased sensation in the feet to touch. Spasticity and exaggerated tendon reflexes occur. There may also be foot drop and the impairment of tendon reflexes. Some of these features are those of combined subacute degeneration of the cord and may be due to associated vitamin B_{12} deficiency. Loss of position sense may give rise to ataxia.

Pellagra may be precipitated by the anti-tuberculosis drug isoniazide in patients who have been given poor diets. This occurs because isoniazide acts as an antagonist to pyridoxine which is a necessary coenzyme for the conversion of tryptophan to nicotinamide. In Western Society pellagra is ocasionally seen in alcoholics and in patients with malabsorption syndrome. It can also occur in patients receiving low protein diets such as those used in the treatment of chronic renal failure when prophylactic vitamins are not given. Patients on intermittent dialysis are particularly at risk as the vitamin supplements may be inadvertently forgotten. Pellagrous skin lesions have been described in Hartnup disease in which there is an aminoaciduria including tryptophan, and a defect of tryptophan absorption in both the renal tubules and the intestinal mucosa. The pellagrous skin rash found in this condition responds promptly to nicotinamide.

Pyridoxine (Vitamin B_6) and related compounds

Dietary deficiency of pyridoxine is rare because of the wide distribution of the vitamin in food. Convulsions have occurred in infants in the USA due to feeding a formula which provided little B_6 due to a manufacturing error. Biochemical evidence suggesting pyridoxine deficiency has been found in patients with malabsorption and in PEM. Several drugs interact with pyridoxine and increase their requirement fo the vitamin, notably isoniazid, hydrolazine, penicilline and oestrogens. A proportion of women taking oral contraceptive agents develop nausea, headaches and depression and in some of these women these side effects can be traced to an induced deficiency of pyridoxine which depresses the conversion of tryptophan to 5-hydroxytryptamine (serotonin).

Convulsions cured by comparatively large doses of pyridoxine have been described in otherwise well nourished children. These have been attributed to a pyridoxine dependency which appears to be an inborn error of metabolism. Rarely, a hypochromic sideroblastic anaemia occurs in adults which responds to large doses (20–100 mg/day) of pyridoxine.

Cyanocobalamin (Vitamin B_{12}) and related substances

Megaloblastic anaemia due to B_{12} deficiency has been described in a few very strict vegetarians (i.e. vegans) due to the fact that this vitamin is restricted in its distribution in foods to foods of animal origin. Vitamin B_{12} is stored in the liver and at birth there is sufficiency of the vitamin for about 12 months. The liver of adults usually contains sufficient for about 5 years.

Kwashiorkor with accompanying megaloblastic anaemia due to B_{12} deficiency, has been described in children of around 13 months of age who have been reared on fad diets.

The absorption of B_{12} from the distal part of the ileum requires its conjugation with a glycoprotein secreted by the parietal cells in the stomach (Castle's intrinsic factor). Failure to produce this factor in pernicious anaemia is due to an autoimmune reaction which destroys the secretory cells of the stomach. Secretion is also reduced in atrophic gastritis, after partial gastrectomy, particularly of the Polya type and, of course, after total gastrectomy. In a variety of intestinal diseases, e.g. malabsorption syndrome, especially the blind loop syndrome, an increased bacterial flora may reduce the absorption of B_{12}. The vitamin is also not absorbed in ilial disease. A number of drugs e.g. β-aminosalicylic acid, biguanides, slow-release K, and colchicine, interfere specifically with the absorption of the vitamin.

The onset of symptoms in B_{12} deficiency is insidious with a history of increasing lassitude over perhaps 14–17 months. The longest history is found in patients with neurological disease. The common symptoms are weakness, tiredness, dyspnoea on exertion, paresthesiae and a sore tongue. With severe anaemia, evidence of cardiac failure may appear such as oedema of the feet. Patients with pernicious anaemia are often slightly jaundiced and pyrexial and show premature greying of the hair. The tongue is often smooth, clean and shiny.

B_{12} plays an important role as a coenzyme in the reactions which result in the removal of cyanide from the body by its conversion to thiocyanate. Optic atrophy due to cyanide occurs in heavy smokers and is often effectively treated with hydroxycobalamine. Retrobulbar neuropathy and spinal ataxia occur together in Western Nigeria among people for whom cassava is the staple diet. The syndrome, called tropical ataxic neuropathy, appears to be caused by chronic ingestion of small amounts of cyanide from cassava. It is believed that the disease arises because the diet does not contain sufficient sulphur-containing amino acids and B_{12} to metabolize the cyanide.

Vitamin B_{12} neuropathy, or sub-acute combined degeneration of the cord, occurs as a complication of pernicious anaemia. It affects mainly the dorsal and lateral columns and early symptoms include tingling, coldness, and numbness in the extremities due to peripheral neuropathy. Motor weakness and ataxia appear later and become severe as the cord is progressively involved.

Folate (Folic Acid, Folacin)

Megaloblastic anaemia due to a simple dietary deficiency of folate occurs amongst poor people in the tropics. It can occur in Western societies during pregnancy due to the increased demands for the vitamin. Absorption from the upper small intestine is often impaired in gastrointestinal diseases.

Anticonvulsant and other drugs including alcohol reduce the blood levels of the vitamin but rarely do they cause megaloblastic anaemia. Low blood levels of folate can also occur in patients being given certain intravenous amino acid solutions. This is due to the fact that they contain a high concentration of methionine which requires folate for its metabolism via the trans-sulphuration pathway to cysteine.

Ascorbic Acid (Vitamin C)

The diagnosis of scurvy, due to ascorbic acid deficiency, is often suggested by the gingivitis which is characteristic of the disease. The gums are spongy, swollen and red and bleed at the slightest touch. Cutaneous bleeding also occurs, often initially on the lower thighs just above the knees. The perifollicular haemorrhages may subsequently appear on the buttocks, abdomen, legs and arms and are often followed by petechial haemorrhages not associated with hair follicles. Corkscrew hairs may appear, particularly on the abdomen, due to a heaping up of keratin-like material on the surface and around the mouth of the follicle. Later, large cutaneous bruises (ecchymoses) can occur spontaneously on the skin. The patient will sometimes present with painful joints.

Scurvy has been produced experimentally in volunteers[41] and these presented some additional features including ocular haemorrhages (particularly in the bulbar conjunctiva), Sjögren's syndrome i.e. loss of secretion of salivary and lacrimal glands and swelling of the parotid glands, femoral neuropathy, oedema of the lower limbs and psychological disturbances, hypochondriasis and depression. Early signs are that the patient complains of feeling feeble and listless for some weeks. These signs precede the changes in the gums and skin. Scurvy can occur in infants due to a reduction in the small amount of vitamin C present in milk. The first sign of bleeding in scorbutic infants is usually a sub-periosteal haemorrhage immediately overlying one of the long bones—frequently the femur—producing the characteristic frog-like position. This gives rise to intense pain, and the infant may cry agonizingly, and scream even louder when lifted.

A subclinical deficiency of ascorbic acid is common amongst the elderly and should always be suspected in old people particularly widowers who may not be accustomed to providing for themselves. Only a few foods are rich in the vitamin and it is easily lost during the preparation of meals. Use of the meals-on-wheels service is no guarantee that provision of ascorbic acid is satisfactory.

The requirement for ascorbic acid is increased by stress and low blood levels occur in patients undergoing surgery. Certain drugs, notably prednisone, increase the requirement for the vitamin and may produce scurvy.

Assessment of Vitamin Status

Clinical examination combined with a dietary history and haematological or radiological examination, depending on the nature of the vitamin, may in some cases be sufficient for the diagnosis of a vitamin deficiency. However, the early detection of vitamin deficiencies at the so-called 'sub-clinical' or 'biochemical' level relies on the determination of the levels of the various vitamins in body fluids and tissues. The significance to be attached to low levels of vitamins in the body in the absence of clinical or other manifestations of deficiency depends on the age of the subject, the environment, and stresses that may be experienced. Generally speaking, low levels are of more potential importance in the sick, in whom they may interfere with metabolic processes involved in recovery, and in the elderly.

Biochemical methods that are used for the determination of vitamin status are shown in Table 19. For some vitamins, ascorbic acid and folic acid, for example, plasma or serum levels can be measured. These levels are probably unreliable indicators of vitamin status for they change readily in response to recent intake. For other vitamins, niacin for example, blood levels are difficult to measure and status is more readily assessed from measurements of the urinary excretion of the vitamin or a metabolite, in the case of niacin, N'-methylnicotinamide. Such determinations usually require the collect of a 24-hour urine sample. Alternatively, it may be sufficient to express excretion as a ratio to creatinine. However, the excretion of creatinine is re-

TABLE 19
BIOCHEMICAL METHODS FOR THE DETERMINATION OF VITAMIN STATUS

Vitamin	Methods	Comments	References
A	Plasma retinol and β carotene	Normal levels of vitamin A taken as 50–100 μg/100 ml.	39
	Plasma retinol-binding protein	Low levels alone do not indicate deficiency. Deficiency should be confirmed by assessment of dietary intake and measurement of dark adaptation	36
D	Plasma alkaline phosphatase	Raised levels of alkaline phosphatase occur in rickets and osteomalacia but are non-specific for vitamin D deficiency.	32
	Plasma 25-hydroxycholecalciferol	Normal values for 25-hydroxycholecalciferol vary with the method used	
E	Plasma tocopherol	Plasma level varies with total plasma lipids. Representative lower limits of normal taken as 5 μg/ml or 0.8 mg/g lipid.	39 42 57
	Erythrocyte haemolysis	High peroxide haemolysis not specific to vitamin E deficiency.	
Thiamine	Blood pyruvate and lactate	Raised levels non-specific for thiamin deficiency	
	Red cell (or whole blood) transketolase and stimulation (TPP % value)	Red cell transketolase and particularly raised TPP % value is most reliable index.	30
	Urinary excretion of thiamin	Values > 15 probably indicate deficiency.	5
Riboflavin	Red cell glutathione reductase (GR) and GR stimulation ratio	Stimulation or activation ratios > 1.30 indicative of riboflavin deficiency.	61 5
	Urinary excretion of riboflavin		35
Nicotinic acid	Urinary excretion of N'-methyl nicotinamide (NMN)	Excretion of < 0.6 mg/6 h is subnormal; excretion of 0.2 mg or less/6 h indicative of deficiency. Ratio of NMN to creatinine in 2 h urine collected between 10 AM and 12 AM similar to values obtained on 24 h urine.	22 37
	Urinary excretion of NMN after an oral dose of nicotinamide.	For load test, 50–200 mg nicotinamide given. Normally nourished subjects excrete 20% of dose as NMN in 24 h.	
Pyridoxine	Red cell aspartate transaminase (AST) and red cell AST stimulation	The red cell AST stimulation test is a convenient method of determining pyridoxine status.	55
	Enzymological assay of pyridoxal phosphate with tyrosine decarboxylase. Tryptophan load test.	After a tryptophan load, deficient individuals excrete an increased amount of xanthurenic acid and an increased ratio of hydroxykynurenine:hydroxyanthranilate.	62 5
Folate	Serum folate	Normal range 3–25 μg/l. Concentrations reduced by loss of appetite. Low value not sufficient to diagnose deficiency. Indicates negative folate balance at the time.	16
	Red cell folate	Normal range 150–600 μg/l packed cells. Normal range varies considerably in different laboratories. Low levels always indicative of folate deficiency. Diagnosis confirmed by satisfactory treatment with 200 μg/d.	64
	Histidine load test	15 g L-histidine given. Folate deficiency indicated by excretion of > 18 mg formimino-glutamic acid (FIGLU) in first 8 h.	46
B12	Serum concentration	Normal range quoted as 170–1000 μg/ml but varies with the method used.	
	Methylmalonic acid (MMA) excretion.	Normal range 0.2–15.3 mg MMA/24 h. Level of excretion increased in B12 deficiency particularly if urine is collected after giving valine or isoleucine. Less sensitive and less convenient than measurement of B12.	15
Ascorbic acid	Plasma and leucocyte ('buffy coat') concentrations.	Plasma levels reflect recent intake but repeated values < 0.2 mg/100 ml probably indicate deficiency. Leucocyte concentrations reflect tissue concentrations. Values < 25 μg/10^8 cells indicate deficiency.	23
	Saturation test	Adequately nourished individuals excrete most of the load. Depleted individuals retain a proportionately greater amount.	

lated to muscle mass, as well as to renal function, and changes with age so that different 'normal' or reference values must be used for subjects of different ages.

Vitamins function as coenzymes at various stages in metabolic pathways. Deficiency of a vitamin will lead to a reduction in the activity of a particular enzyme and may cause both a build-up of metabolites prior to that point in the metabolic pathway and increased excretion of one or more metabolites. Thus, for example, folic acid is necessary for the complete conversion of histidine to formic acid and glutamic acid (Fig. 7), and specifically for the last stage of this pathway, that of formiminoglutamic acid (FIGLU) to formic and glutamic acids. It follows therefore that when a histidine load is given to a subject who is folate deficient, a greater than normal amount of the histidine will be excreted as FIGLU. The tryptophan load test for pyridoxine deficiency is based on a similar principle.

Determination of status with respect to a number of vitamins in the B group (e.g. thiamine, riboflavin and pyridoxine) can be made by measuring the activities of enzymes in the red cell for which they act as cofactors. A more reliable index is given by the 'activation coefficient' which is the rise in enzyme activity when the vitamin is added to deficient red cells expressed as a ratio, or percentage, of that already present.

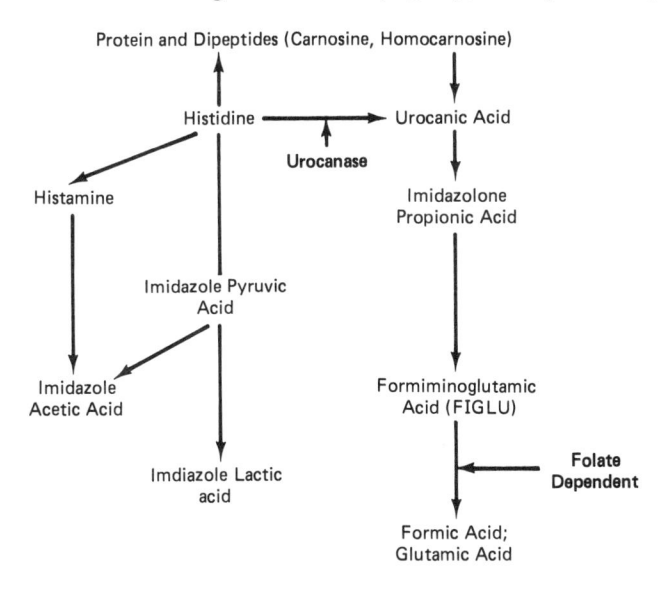

FIG. 7. Metabolic pathways of histidine. The oxidation pathway yields formic and glutamic acids. This pathway is blocked at the urocanase step in kwashiorkor and histidinaemia. Folate deficiency results in elevated excretion of FIGLU.

TRACE ELEMENT DEFICIENCY AND EXCESS

Introduction

Metal ions are essential for the activity of about one-third of all enzymes. These enzymes may be divided into two groups, the metallo-enzymes and the metal-ion activated enzymes. A metallo-enzyme contains the mineral as an integral part of the structure of the molecule which does not dissociate under physiological conditions. In contrast, the metal is only loosely bound to metal-ion activated enzymes; it may be replaced by other metals, and may show some activity when removed altogether. The six most important of these metals are iron, copper, zinc, manganese, molybdenum and cobalt. Other metals present in the human body in still smaller concentrations include nickel, chromium, selenium, tin and vanadium. Other minerals known to be important in man are iodine and fluorine and there have been suggestions that silicon may play an important role in relation to atherosclerosis. This review will be restricted to those trace minerals that are known to be involved in human disease and which are therefore of particular interest to the clinical biochemist—zinc, copper, manganese, chromium and cobalt. Information about iron and other trace elements will be found in the review by Prasad.[59]

Clinical Features

Zinc

A syndrome of zinc deficiency has been described in young men in Iran and Egypt. Affected patients were dwarfed and sexually retarded and showed endocrine abnormalities which resembled hypopituitarism. Multiple deficiencies cannot be ruled out in these patients but they did respond clinically to zinc supplementation. Zinc deficiency may retard wound healing but zinc is not a panacea to aid this process. Zinc deficiency may be a cause of dermal ulcers in patients on long term treatment with corticosteroids and play a role in delaying the healing of leg ulcers. In the latter disorder zinc supplements are effective in promoting healing only in those patients who have low plasma zinc levels. Zinc deficiency may diminish or pervert the senses of taste and smell. Zinc plays an important role in tissue growth and maternal zinc deficiency may play a role in human teratogenicity.

Increased urinary excretion of zinc resulting in zinc depletion occurs in patients with alcoholic cirrhosis, and after surgery. Malabsorption syndromes, pancreatitis, diabetes and sickle-cell anaemia are associated with a depletion of body zinc and low plasma levels are found in women taking oral contraceptives.

Zinc deficiency occurs in patients receiving long-term parental nutrition if they are not being given appropriate mineral supplements. The condition presents as a red blotchy dermatitis of the face, axillae and other parts of the body. The appearance of the skin lesions is similar to that found in the inborn error of zinc metabolism known as acrodermatitis enteropathica. This disease is inherited as an autosomal recessive characteristic and the disease presents at weaning, or earlier if the infant is not breast-fed. Severe dermatitis of the extremities is associated with severe diarrhoea and the children die in infancy unless treated. The cause of the disease is an inability to absorb zinc from the intestine but the mechanism is unknown.

Symptomless elevation of plasma zinc concentration

has been reported as a characteristic of an inherited disorder. Zinc intoxication has occurred following domestic renal haemodialysis using water stored in a galvanized tank.

Copper

Three syndromes of copper deficiency have been described in human infancy, two of them being acquired and the other an inborn error of metabolism. The first is a condition manifesting itself mainly as anaemia and found in infants maintained largely on cow's milk. The condition responds to the administration of both copper and iron. In the second syndrome, anaemia is associated with diarrhoea and skeletal rarefaction and responds well to copper but not iron. The inborn error is known as Menkes' kinky hair syndrome.[56] It is inherited as a sex-linked recessive disorder characterised by growth retardation, progressive mental deterioration, degeneration of aortic elastin, metaphyseal abnormalities in the skeleton with changes similar to scurvy, hypothermia, and a peculiar defect in hair keratin and pigmentation ('Kinky hair'). Most of these abnormalities reflect the known activities of copper-containing enzymes. The disease is due to a failure to absorb copper from the intestine and for this reason oral copper supplements are ineffective whereas parenteral copper improves the condition.

The second inborn error of copper metabolism, that of Wilson's disease or hepatolenticular degeneration, is associated with copper excess. As in Menkes' syndrome, serum copper and caeruloplasmin levels are low, but in contrast copper is absorbed and deposited in steadily increasing amounts in the tissues, particularly in the liver, brain, iris and kidneys. This steady accumulation results in toxicity with cirrhosis of the liver and mental deterioration with ataxia and choreiform movements. Treatment of Wilson's disease relies on the use of chelating agents, such as penicillamine, which increase the urinary excretion of copper (and also of zinc).

Low plasma levels of copper may occur in malabsorption syndromes and in the nephrotic syndrome. High plasma levels occur in patients with certain kinds of tumour, e.g. Hodgkin's disease and in Addison's disease. The clinical significance of these changes is not known.

Manganese

The drug hydralizine (apresoline) used to be used extensively in the treatment of hypertension. In common with a number of other drugs (including procainamide) long continued use of hydralazine can produce a syndrome identical to idiopathic lupus erythmatosus. Patients suffering from both the drug-induced and naturally occurring syndrome have benefited from the administration of manganese salts. A link between rheumatoid arthritis and manganese metabolism has been suggested.

Manganese deficiency in animals gives rise to ataxia and convulsions. Intoxication in humans can give rise to Parkinsonism, extropyramidal syndromes, hallucinations and psychomotor instability.

Chromium

A link has now been established between chromium and glucose metabolism with improvements in glucose tolerance being claimed in elderly subjects as a result of the administration of supplements of 150 μg/day of chromium chloride. The hair chromium of diabetic children in markedly reduced.

Cobalt

Cobalt is an important constituent of vitamin B_{12}. It has not been definitely established whether the metal has a role in human nutrition apart from this. However, there is a report of a single child in a cobalt-deficient area in Scotland with marked geophagia who responded to cobalt by mouth.[63] Pharmacological doses of inorganic cobalt salts induce polycythaemia in many species, including man and doses of 20–30 mg daily have been used with iron in the treatment of iron deficiency anaemia. However, they should not be used in this condition unless cobalt deficiency is suspected, as suppression of thyroid activity, goitre and cardiomyopathy can occur as a result of cobalt toxicity, particularly in infants. Since cobalt does stimulate production of renal erythropoietin it has been used to raise haemoglobin levels in the anaemia of chronic renal failure. An undesirable complication of such treatment is that the blood lipid concentrations are also raised.

Determination of Trace Element Status

Plasma or serum levels are commonly measured for the assessment of trace element nutriture. However, there is increasing interest in the use of hair as a biopsy material. This has obvious advantage in that the taking of the sample is non-invasive and hair contains higher concentrations of many trace elements than those in plasma. Moreover, the results of hair analysis may be more informative of tissue levels and yield a better background of a patient's nutriture because they reflect it over a greater time span. Furthermore, much of the developing interest in the possible role of trace elements in clinical conditions such as psychiatric disorders centres not so much on deficiency or excess of a single mineral but on relationships between different minerals for it is known that relative excesses or deficiencies and ratios between related elements, such as copper and zinc, may be more important clinically than the absolute concentrations of either independently. Such considerations mean that more information is likely to be gained from a trace element profile than from a single plasma determination and a profile is more easily determined on a hair sample.

CONCLUSIONS

In our present state of knowledge the limits of what may be considered 'normal' nutritional status are purely arbitrary. It is only when metabolic processes are disturbed and overt disease develops that we can refer to under- or overnutrition. Be this as it may, nutritional status is an important aspect of patient welfare. Disease may precipitate nutrient deficiencies by increasing requirements, interfering with intake, causing abnormal losses or resulting in malabsorption. Furthermore, nutritional deficiencies may have important interactions with treatment whether by drugs, surgery or radiotherapy. Assessment and maintenance of nutritional status is therefore an important part of the holistic approach to the management of the sick.

ACKNOWLEDGEMENTS

I have drawn freely on published material in writing this chapter. The source of the material is gratefully acknowledged and my thanks go to Authors and Publishers who have readily given their permission.

REFERENCES

1. Aksoy, M., Basu, T. K., Brient, J. and Dickerson, J. W. T. (1980), 'Thiamin status of patients treated with drug combinations containing 5-fluorouracil'. *European J. Cancer*, **16**, 1041–1045.
2. Alleyne, G. A. O., Hay, R. W., Picou, D. I., Stanfield, J. P. and Whitehead, R. G. (1977), *Protein-energy Malnutrition*. London, Arnold.
3. Alleyne, G. A., Trust, P. M., Flores, H. *et al* (1972), 'Glucose tolerance and insulin sensitivity in malnourished children.' *Br. J. Nutr.*, **27**, 585–592.
4. Arneil, G. C. (1979). 'Progress in the prevention and treatment of nutritional rickets,' in *The Importance of Vitamins to Human Health*, pp. 151–62. (Taylor, T. G., Ed.). Lancaster: MTP.
5. Bayoumi, R. and Rosalki, S. B. (1976), 'Evaluation of methods of co-enzyme activation of erythrocyte enzymes for detection of deficiency of vitamins B_1, B_2 and B_6.' *Clin. Chem.*, **22**, 327–335.
6. Bengoa, J. M. (1974), 'The problem of malnutrition.' *WHO Chronicle*, **28**, 3–7.
7. Berkel, I., Kiran, O. and Say, B. (1970), 'Jejunal mucosa in infantile malnutrition,' *Acta Paediat. Scand.*, **59**, 58–64.
8. Bistrian, B. R., Blackburn, G. L., Sherman, M. *et al* (1975), 'Therapeutic index of nutritional depletion in hospitalized patients,' *Surg. Gyn. Obstet.*, **141**, 512–516.
9. Bistrian, B. R., Blackburn, G. L., Vitale, J., Cochran, D. and Naylor, J. (1976), 'Prevalence of malnutrition in general medical patients,' *J.A.M.A.*, **235**, 1567–1570.
10. Blackburn, G. L. and Schlamm, H. T. (1979), 'Nutritional assessment and treatment of hospital malnutrition,' in *Nutrition and Cancer*, pp. 1–30. (Van Eys, J., Seelig, M. S. and Nichols, B. L., Eds.). London, S.P. Med and Scientific Books.
11. Bradfield, R. B. (1972), 'A rapid technique for the field assessment of protein-energy malnutrition,' *Am. J. Clin. Nutr.*, **25**, 720–729.
12. Brooke, O. G. and Cocks, T. (1974), 'Resting metabolic rate in malnourished babies in relation to total body potassium,' *Acta Paediat. Scand.*, **63**, 817–825.
13. Burgess, C. R. (1958), 'Infantile beriberi,' in *Nutritional Diseases*, Proceedings of a conference on beriberi, endemic goiter and hypervitaminosis A. *Fed. Proc.*, **17**, No. 3, Part II, Supplement No. 2, p. 39.
14. Cabak, V., Dickerson, J. W. T. and Widdowson, E. M. (1963), 'Response of young rats to deprivation of protein or of calories,' *Brit. J. Nutr.*, **17**, 601–616.
15. Chanarin, I., England, J. M., Mollin, C. and Perry, J. (1973), 'Methylmalonic acid excretion studies,' *Brit, J. Haemat.*, **25**, 45–53.
16. Chanarin, I., Kyle, R. and Stacey, J. (1972), 'Experience with microbiological assay for folate using a chloramphenicol-resistant L. casei strain,' *J. Clin. Pathol.*, **25**, 1050–1052.
17. Cohen, S. and Hansen, J. D. L. (1962), 'Metabolism of albumin and γ-globulin in kwashiorkor,' *Clin. Sci.*, **23**, 351–359.
18. Cook, G. C. and Lee, F. D. (1966), 'The jejunum after kwashiorkor,' *Lancet*, **ii**, 1263–1267
19. Coward, W. A. (1975), 'Serum colloidal osmotic pressure in the development of kwashiorkor and in recovery: its concentrations and oedema,' *Br. J. Nutr.*, **34**, 459–467.
20. Coward, W. A. and Sawyer, M. B. (1977), 'Whole-body albumin mass and distribution in rats fed on low-protein diets,' *Br. J. Nutr.*, **37**, 127–134.
21. Davidson, S., Passmore, R., Brock, J. F. and Truswell, A. S. (1979), *Human Nurition and Dietetics*. 7th Edition. Edinburgh, Churchill Livingstone.
22. De Lange, D. J. and Joubert, C. P. (1964), 'Assessment of nicotinic acid status of population groups,' *Am. J. Clin. Nutr.*, **15**, 169–174.
23. Denson, K. W. and Bowers, E. F. (1961), 'The determination of ascorbic acid in whole blood cells. A comparison of W.B.C. ascorbic acid and phenolic acid excretion in elderly patients,' *Clin. Sci.*, **21**, 157–162.
24. De Wys, W. D. (1974), 'Abnormalities of taste as a remote effect of a neoplasm,' *Ann. N.Y. Acad. Sci.*, **230**, 427–434.
25. De Wys, W. D. (1979), 'Anorexia as a general effect of cancer,' *Cancer*, **43**, 2013–2019.
26. Department of Health and Social Security (1972), 'A nutritional survey of the elderly.' Report on Public Health and Medical Subjects, No. 3, London. Her Majesty's Stationery Office.
27. Dickerson, J. W. T. (1978), 'The interrelationships of nutrition and drugs,' in *Nutrition in the Clinical Management of Disease*, pp. 308–331. (Dickerson, J. W. T. and Lee, H. A., Eds.). London, Arnold.
28. Dickerson, J. W. T. (1980), 'Nutrition and drugs,' in *Symposium on Nutrition*, pp. 42–62. (Davies, S. H., Ed.). Edinburgh, Royal College of Physicians of Edinburgh.
29. Dickerson, J. W. T. and Lee, H. A. (Eds.) (1978), *Nutrition in the Clinical Management of Disease*. London, Arnold.
30. Dreyfus, P. M. (1962), 'Clinical application of blood transketolase determinations,' *New Eng. J. Med.*, **267**, 596–598.
31. Durnin, J. V. G. A. and Womersley, J. (1974), 'Body fat assessed from total body density and its estimation from skinfold thickness: measurements on 481 men and women aged 16 to 72 yrs,' *Br. J. Nutr.*, **32**, 77–97.
32. Edelstein, S., Charman, M., Lawson, D. E. M. and Kodicek, E. (1974), 'Competitive protein-binding assay for 25-hydroxycholecalciferol,' *Clin. Sci. Mol. Med.*, **46**, 231–240.
33. Exton-Smith, A. N. (1978), 'Nutrition in the elderly,' in *Nutrition in the Clinical Management of Disease*, pp. 72–104. (Dickerson, J. W. T. and Lee, H. A., Eds.). London, Arnold.
34. Garrow, J. S., Fletcher, K. and Halliday, D. (1965), 'Body composition in severe infantile malnutrition,' *J. Clin. Invest.*, **44**, 417–425.
35. Glatzle, D., Korner, W. F., Christeller, S. and Wiss, O. (1970), 'Method for the detection of a biochemical riboflavin deficiency. Stimulation of NADPH2-dependent glutathione reductase from human erythrocytes by FAD In vitro. Investigations on the vitamin B_2 status in healthy people and geriatric patients,' *Int. J. Vit. Nutr. Res.*, **40**, 166–183.
36. Glover, J. (1973), 'Retinol binding proteins,' *Vitam. Horm.*, **31**, 1–42.
37. Goldsmith, G. A. and Miller, O. N. (1967), 'Niacin,' in *The Vitamins*, Vol. VII, pp. 136–167. (Gyorgy, P. and Pearson, W. N., Eds.). London, Academic Press.
38. Hansen, J. D. L., Brinkman, G. L. and Bowie, M. D. (1965), 'Body composition in protein-calorie malnutrition,' *S. Afr. Med. J.*, **39**, 491–495.

39. Hansen, L. G. and Warwick, W. J. (1969), 'A fluorometric micromethod for serum vitamins A and E,' *Techn. Bull. Registr. Med. Techn.*, **39**, 70–73.
40. Hill, G. L., Blackett, R. L. and Pickford, I. et al (1977), 'Malnutrition in surgical patients,' *Lancet*, **i**, 689–692.
41. Hodges, R. E., Hood, J., Canham, J. E. et al (1971), 'Clinical manifestations of ascorbic acid deficiency in man,' *Am. J. Clin. Nutr.*, **24**, 432–443.
42. Horwitt, M. K., Harvey, C. C. Dahm, C. H. Jr. and Searcy, M. T. (1972), 'Relationship between tocopherol and serum lipid levels for determination of nutritional adequacy,' *Ann. N.Y. Acad. Sci.*, **203**, 223–236.
43. James, W. P. T. and Hay, A. M. (1968), 'Albumin metabolism: effect of the nutritional state and the dietary protein intake,' *J. Clin. Invest.*, **47**, 1958–1972.
44. Jelliffe, D. B. (1968), 'Infant nutrition in the subtropics,' World Health Org. Monograph Series No. 29, 2nd Edition, p. 98. Geneva, WHO.
45. Kanawati, A. A. and McLaren, D. S. (1970), 'Assessment of marginal malnutrition,' *Nature*, **228**, 573–574.
46. Kohn, J., Mollin, D. L. and Rosenbach, L. M. (1961), 'Conventional voltage electrophoresis for formiminoglutamic acid determinations in folic acid deficiency,' *J. Clin. Path.*, **14**, 345–350.
47. Lancet (1970), 'Classification of infantile malnutrition,' *Lancet*, **ii**, 302–303.
48. Lawson, D. E. M., Paul, A. A., Black, A. E. et al (1979), 'Relative contributions of diet and sunlight to vitamin D state in the elderly,' *Br. Med. J.*, **2**, 303–305.
49. Lee, H. A. (1978), 'Parenteral nutrition', in *Nutrition in the Clinical Management of Disease*, pp. 349–76. (Dickerson, J. W. T. and Lee, H. A., Eds.). London, Arnold.
50. Lunn, P. G., Whitehead, R. G., Baker, B. A. and Austin, S. (1976), 'The effect of cortisone acetate on the course of development of experimental protein-energy malnutrition in rats,' *Br. J. Nutr.*, **36**, 537–50.
51. McLaren, D. S. (1966), 'A fresh look at protein-calorie malnutrition,' *Lancet*, **ii**, 485–488.
52. McLaren, D. S., Pellet, P. L. and Read. W. W. C. (1967), 'A simple scoring system for classifying the severe forms of protein-calorie malnutrition in early childhood,' *Lancet*, **i**, 533–535.
53. McLaren, D. S. and Read. W. W. C. (1972), 'Classifications of nutritional status in early childhood,' *Lancet*, **ii**, 146–148.
54. Mann, M. D., Bowie, M. D. and Hansen, J. D. L. (1972), 'Potassium in protein-calorie malnutrition,' *S. Afr. Med. J.*, **46**, 2062–2064.
55. Marsh, M. E., Greenberg, L. D. and Rhinehart, J. F. (1955), 'The relationship between B_6 investigation and transaminase activity,' *J. Nutr.*, **56**, 115–127.
56. Menkes, J. H., ALter, M., Steigleder, G. K. et al (1962), 'A sex-linked disorder with retardation of growth, peculiar hair and focal cerebral and cerebellar degeneration,' *Paediatrics*, **29**, 764–779.
57. Nitowsky, H. M. and Tildon, J. T. (1956), 'Some studies of tocopherol deficiency in infants and children. III. Relation of blood catalase activity and other factors to hemolysis of erythrocyte in hydrogen peroxide,' *Am. J. Clin. Nutr.*, **4**, 397–401.
58. Picou, D. and Taylor-Roberts, T. (1969), 'The measurement of total protein synthesis and catabolism and nitrogen turnover in infants in different nutritional states and receiving different amounts of dietary protein,' *Clin. Sci.*, **36**, 283–296.
59. Prasad, A. S. (1978), *Trace Elements and Iron in Human Metabolism*. Chichester, John Wiley and Sons.
60. Reeds, P. J. and Laditan, A. A. O. (1976), 'Serum albumin and transferrin in protein-energy malnutrition,' *Br. J. Nutr.*, **36**, 255–263.
61. Sauberlich, H. E., Judd, J. H. Jr., Nichoalds, G. E. et al (1972), 'Applications of the erythrocyte glutathione reductase assay in evaluating riboflavin nutritional status in a high school student population,' *Am. J. Clin. Nutr.*, **25**, 756–762.
62. Sauberlich, H. E., Canham, J. E., Baker, E. M. et al (1972), 'Biochemical assessment of the nutritional status of vitamin B_6 in the human,' *Am. J. Clin. Nutr.*, **25**, 629–642.
63. Shuttleworth, V. S., Cameron, R. S., Alderman, G. and Davies, H. T. (1961), 'A case of cobalt deficiency in a child presenting as "Earth Eating",' *Practitioner*, **186**, 760–764.
64. Tabor, H. and Wyngarden, L. (1958), 'A method for the determination of formimino-glutamic acid in urine,' *J. Clin. Invest.*, **37**, 824–828.
65. Tanner, J. M., Whitehouse, R. H. and Takaishi, M. (1966), 'Standards from birth to maturity for height, weight, height velocity, and weight velocity: British children, 1965,' *Arch. Dis. Childh.*, **41**, 613–635.
66. Theologides, A. (1977), 'Cancer cachexia,' in *Nutrition and Cancer*, pp. 75–94. (Winick, M. Ed.). Chichester, John Wiley and Sons.
67. Tomkins, A. M., Drasar, B. S. and James, W. P. T. (1975), 'Bacterial colonisation of jejunal mucosa in acute tropical sprue,' *Lancet*, **i**, 59–62.
68. Truswell, A. S., Hansen, J. D. L., Watson, C. E. and Wannenburg, P. (1969), 'Relation of serum lipids and lipoproteins to fatty liver in kwashiorkor,' *Am. J. Clin. Nutr.*, **22**, 568–576.
69. Vaughan, V. C., McKay, R. J. and Behrman, R. E. (Eds.) (1979), *Nelson Textbook of Pediatrics*, 11th Edition, pp. 10–46. London, W. B. Saunders.
70. Waterlow, J. C. (1968), 'Observations on the mechanism of adaptation to low protein intakes,' *Lancet*, **ii**, 1091–1096.
71. Waterlow, J. C., Golden, M. H. N. and Patrick, J. (1978), 'Protein-energy malnutrition: treatment, in Nutrition in the clinical management of disease,' pp. 49–71. (Dickerson, J. W. T. and Lee, A. H., Eds.). London, Arnold.
72. Whitehead, R. G. and Alleyne, G. A. O. (1972), 'Pathophysiological factors of importance in protein-calorie malnutrition,' *Br. Med. Bull.*, **28**, 72–78.
73. Whitehead, R. G., Coward, W. A. and Lunn, P. G. (1973), 'Serum-albumin concentration and the onset of kwashiorkor,' *Lancet*, **i**, 63–66.
74. Whitehead, R. G. and Dean, R. F. A. (1964), 'Serum amino acids in kwashiorkor I and II,' *Am. J. Clin. Nutr.*, **14**, 313–319; 320–330.
75. World Health Organisation (1976), 'Vitamin A deficiency and xerophthalmia.' Tech. Rep. Ser. Wld. Hlth. Org. No. 590, Geneva, WHO.

7. THE LIVER

NEIL McINTYRE

Introduction to liver function tests
  Sensitivity, specificity and predictive accuracy
  Screening and profiling
  Diagnostic tests
  Monitoring the patient's progress
  Research and the evaluation of therapy

Individual tests of liver function
  Bilirubin, bilirubin esters and urobilinogen
  Tests indicating hepatocellular damage
  Tests indicating obstruction to biliary flow
  Plasma proteins
  Bile acids
  Cholesterol and other lipids

The value of liver function tests for screening and profiling

The value of liver function tests in the diagnosis of liver disease
  Classification of jaundice
  Primary and secondary hepatic malignancy
  Tests used for research

INTRODUCTION TO LIVER FUNCTION TESTS

Biochemical investigations are widely used in the management of patients with diseases of the liver and biliary tract, and the term 'liver function tests' has been given, in various laboratory and clinical services, to the small battery of investigations which they use regularly for this purpose. The individual tests used may vary from laboratory to laboratory. Unfortunately, the term is a poor one as 'liver function tests' are not particularly good tests of liver function *per se* and may give abnormal results with disorders of other systems. They do, however, give abnormal results with various kinds of liver disease. Like other special investigations they may be used for several different purposes:

(1) Screening or profiling.
(2) Diagnosis (by confirmation or by exclusion) and Prognostication.
(3) Monitoring of a patient's progress.
(4) Research.
(5) Evaluation of new forms of therapy.

Sensitivity, Specificity and Predictive Accuracy

Before considering the value of liver function tests (and other biochemical investigations) for these various purposes we must first consider what is meant by the terms 'sensitivity', 'specificity', and 'predictive accuracy.'

Let us assume that in a large population N, $P \times N$ people have hepatobiliary disease, P being the proportion with and $(1 - P)$ the proportion without hepatobiliary disease. If a particular test is applied to the whole population then a certain number, a, of those with the disease will give an abnormal or positive result and others with the disease, b, will give a normal or negative result. Of those without the disease, $[(1 - P) N]$, c will give a positive result and d will give a negative result. We call a true positives and c false positives; b represent false negatives and d true negatives.

We can measure the sensitivity and specificity of a test used to detect hepatobiliary disease according to definitions introduced by Yerushalmy.[20] He defined *sensitivity* as the number of true positive results, a, divided by the total number of subjects who have the disease, $P \times N$, (i.e. true positives and false negatives, or $a + b$). A test is therefore considered sensitive to the extent that it detects patients with the disease and excludes false negatives. *Specificity* on the other hand is calculated by dividing the true negatives, d, by the number of subjects without the disease, $(1 - P) N$, (i.e. true negatives plus false positives, or $d + c$). A test is specific to the extent that it identifies those who do not have the disease and particularly in that it minimizes the number of false positives.

Unfortunately these indices are of relatively little use to the practising clinician who needs to know not the sensitivity or specificity of a test but its *predictive accuracy*. In other words he needs to know the probability that a positive test indicates the presence of disease, i.e. $a/(a + c)$ or that a negative test indicates the absence of disease, i.e. $d/(b + d)$.

The predictive accuracy of a test is markedly dependent on the type of population studied and the relative proportion of those who have or do not have the disease. When there are very few subjects with the disease even a small percentage of false positives in the large 'healthy' subgroup could mean that a positive result would have little predictive value; this would be true even if the test was highly sensitive and highly specific. The predictive value of a negative result, however, would be very high under these circumstances. By contrast a positive result of a test with a sensitivity and specificity no better than tossing a coin might have a high predictive accuracy if there was a very high prevalance of the disease.

Unfortunately few of the studies done to evaluate tests used in hepatobiliary disease have taken account of the many problems involved in dermining the sensitivity, specificity and predictive accuracy (indeed it is rarely clear what is meant when the terms sensitivity and specificity are used). Often the studies are done on selected groups of patients in whom the prevalence of various diseases is high. This is particularly unsatisfactory if the tests are then used in another population (e.g. if a test evaluated on inpatients is transferred to a quite different outpatient population). Another problem is that relatively little attention is paid to the purposes to which the test or tests may be put.

Screening and Profiling

In this discussion I shall use the terms, 'screening' and 'profiling' in the way suggested by Whitby in 1974.[19] Screening tests are those done on an apparently healthy population, while profiling tests are those done routinely on patients. Both screening and profiling tests are done in the hope that we might detect unsuspected conditions. Profiling tests are also useful as positives may shed light on the cause of the patient's presentation while negative results allow us to exclude certain relatively common conditions. If, for example, the haemoglobin, red cell indices, and blood film were normal we might do no further haematological investigations; if a test for urine glucose proved negative we would not normally be concerned about the possibility of diabetes mellitus. Profiling tests are also used to provide background information which might subsequently be used to monitor the patient's progress and to assess the response to treatment.

For screening or profiling we require a test with reasonably high sensitivity. If the disease is present then we want to find it and we want to be confident that a negative test excludes the disease. The proportion of false negatives must be low. It does not matter if we get a relatively large number of false positives as these can be sorted out subsequently by the use of more specific diagnostic tests, always providing that these diagnostic tests are relatively harmless, or that the risks involved in performing them are commensurate with the likely benefits.

If screening or profiling is to be worthwhile then the overall benefits must outweigh the overall costs. Such tests are done on a very large number of patients. They must therefore be cheap, and must be acceptable to the population to be studied. They should not be unpleasant or inconvenient.

Diagnostic Tests

At some stage in the work up of a patient we tend to contruct a list of diagnostic hypotheses—of conditions which are probable or possible explanations for the clinical presentation. At this stage we order further tests in the hope that they will confirm a particular diagnosis or rule out alternative diagnoses.

Confirmatory diagnostic tests should have a high specificity. In other words there should be few false positives; this is particularly true if false positives may do psychological or economic harm to the patient and if the disease is serious but not treatable or curable. If the consequences of making the diagnosis are serious (e.g. if an operation would be performed, or if the patient would be treated with a potentially toxic drug) it is clearly important that the test should have a high predictive accuracy.

If the presence of disease is virtually certain when a test is positive (that is if it has a very high predictive accuracy) then it does not matter too much if there are a moderate number of false negatives, particularly if it is possible to make the diagnosis in these patients by other means. The order of diagnostic investigations can be chosen on the basis of cost, inconvenience, etc.

In order to *exclude* a condition with high degree of certainty, however, we need to use a test with a very high sensitivity, i.e. one with a very small number of false negatives. This is particularly true if the disease is serious and if it is treatable; otherwise the risk of a false negative result would not allow us to be confident that we have excluded the disease. Diagnostic tests which are aimed at excluding likely causes of a clinical presentation are therefore analogous to screening and profiling tests in that sensitivity is the most important criterion (i.e. a small number of false negatives is required) while specificity is of less importance. When false negatives and false positives are both equally undesirable we need a test with very high efficiency. (i.e. the ratio true positive + true negative/total tests is very high). Unfortunately not many tests of this kind are available in the field of hepatobiliary disease. The question of the stringency which should be applied in the field of diagnosis has been considered by Galen and Gambino[6] and by Murphy.[11]

In differential diagnosis one tries to match the patient's clinical picture with the 'classical' presentation(s) of the disease(s) in question. In most instances the classical presentation reflects the simultaneous finding of a number of different abnormalities which may be elicited by history taking, physical examination and laboratory investigation. A correct diagnosis may therefore depend on the finding of an appropriate pattern and one can extrapolate the earlier argument about the sensitivity and specificity of individual tests to combinations of findings. Unfortunately this greatly increases the complexity of the evaluation of the tests and the application of discriminant analysis and of computer technology in the evaluation of diagnostic tests is in its infancy.

Monitoring the Patient's Progress

In many clinical situations, particularly in hospital practice, it is necessary to check the patient's progress and to decide whether his condition is improving or getting worse or whether there are any complications of therapy. Monitoring a patient in this way implies that measurements will be done on more than one occasion and in practice much of the work of chemical pathology and haematology laboratories involves repeated tests on individual patients. Surprisingly, in view of the high cost of this work, there is very little written on the use of tests for monitoring, and there seems to be little theoretical basis for deciding on the most efficient and effective strategy for their use.

Certain points are clear however. As most monitoring tests are done to follow progress they should, as far as possible, be either quantitative, i.e. measurements of continuous variables, or semi-quantitative, i.e. giving points on a scale with sufficient intervals to allow adequate documentation of change. When repeated measurements are done it may then be possible to identify clinically significant trends even when all of the results from a single patient remain within the conventional normal range, e.g. a gradual fall in haemoglobin concentration from 17 to 14 g/dl, or an increase in the level of blood urea from 3·0 to 6·5 mmol/l. The methods used for the purposes of monitoring should, of course, be reasonably precise, to ensure that observed changes are, as far as possible, real ones and not due to experimental error. Unfortunately the precision of laboratory results is rarely known to clinicians when they are making their decisions. Tests used for monitoring should be relatively cheap as they will be used on a number of occasions; more expensive tests may be justifiable if the results will be used to make clinical decisions which have particularly expensive consequences.

Research and the Evaluation of Therapy

Workers involved in research on liver disease or those involved in the evaluation of new forms of therapy often use standard liver function tests and other tests designed to evaluate more specific aspects of hepatic function. Standard tests may be used for several reasons. They are useful in classifying patients into different experimental or treatment groups; they ensure that others will understand the basis of the classification as the experimental subjects can be matched with patients from their own experience. This allows studies to be reported on essentially similar groups. In therapeutic studies, standard tests are useful in assessing the results or complications of therapy; they would be expected to improve with successful treatment while deterioration of several standard test results would normally be seen as an undesirable effect of therapy even if certain special indices seemed to have improved.

For research purposes standard tests have some disadvantages. The pathophysiological basis of some of them is not fully understood, e.g. aspartate aminotransferase and alkaline phosphatase elevations, and the results may correlate poorly with other indices of hepatocellular function. The purpose of a therapeutic agent may be to improve one or more relatively specific aspects of hepatocellular function, e.g. the synthesis of an individual protein, or change in membrane lipid composition, and it is obvious that highly specific investigations must be designed to study such effects. But in some cases the investigator may wish to check that 'liver function' has improved; for this purpose standard tests are relatively unhelpful because none of them test liver function *per se*. Abnormalities in the levels of plasma bilirubin, aspartate aminotransferase, alkaline phosphatase and even bile acids may all be influenced by factors other than liver function and this lack of specificity has important implications for research projects designed to answer specific questions. For this reason investigators have turned to tests which reflect individual functions of the liver and which are not affected by disturbances of other organs. Unfortunately there are few good tests for this purpose in the field of hepatology. Details of some of those which have been used are presented in a later section. They are all tests which are relatively complicated to perform and are usually costly both in terms of materials used, laboratory time, and in terms of convenience to the patient. These disadvantages are not necessarily serious if the results of the investigations provide important answers, as such investigations are not usually repeated, except when other workers wish to confirm the results. It is important that studies done using such investigations should be conducted and evaluated as rigorously as possible. It is also important that the methods used should be accurate and precise and should be valid (i.e. they should test the question which the investigator is particularly concerned to answer).

INDIVIDUAL TESTS OF LIVER FUNCTION

Bilirubin, Bilirubin Esters and Urobilinogen[1]

Estimation of the concentration in the plasma or serum of total bilirubin, which includes free bilirubin and bilirubin esters, is a common procedure in clinical laboratories. (The term 'free bilirubin' in this chapter means unconjugated or unesterified bilirubin). Free bilirubin is derived from the catabolism of haem and serum levels rise with haemolytic states and when there is over production of haem for other reasons. It is relatively insoluble in water and high serum levels are not accompanied by hyperbilirubinuria. Free bilirubin is taken up by the liver, conjugated with glucuronic acid, and then excreted by the liver. Normally only a tiny fraction re-enters the blood stream but when liver cells are damaged, or when there is an obstruction to the flow of bile, bilirubin esters enter the blood and make up a variable proportion of the total bilirubin.

The estimation of bilirubin is based on the red colour which develops when Ehrlich's diazo reagent is added to serum. Bilirubin esters give an immediate colour change, the 'direct' reaction, while the presence of free bilirubin causes further change upon the addition of alcohol, the 'indirect' reaction. Unfortunately the estimation of free and esterified bilirubin in a mixture is inaccurate because a small percentage of free bilirubin appears to react directly with the reagent. The 'normal' range for the concentration of total bilirubin in serum or plasma is between 3 and 15 μmol/l and it is claimed that in 99% of normal subjects the figure is below 17 μmol/l (but see later). The upper limit for direct reacting bilirubin is approximately 3 μmol/l and levels above 5 μmol/l, measured in a good laboratory, can be taken as evidence of hepatobiliary disease. Even modest elevations of bilirubin esters are accompanied by bilirubinuria and for practical purposes these elevations may be inferred from the presence of bilirubin in the urine even when the total bilirubin is in the normal range. Bilirubinuria is now usually detected with dipsticks or tablets containing diazo-reagents.

The bilirubin esters which enter the intestine undergo

bacterial hydrolysis and degradation in the ileum and colon with production of 'urobilinogen' (this term refers to a mixture of isomers). Increased amounts of urobilinogen are made when there is over-production of bilirubin, as with haemolysis, and with constipation or with bacterial contamination of the small bowel. Some of this material is excreted in the faeces. A small amount is absorbed and travels in the portal vein to the liver which removes most of it and excretes it in the bile. A small amount escapes hepatic uptake and is excreted in the urine. In the presence of liver damage a greater proportion of urobilinogen escapes hepatic uptake and more is therefore excreted via the kidney; when biliary obstruction is more or less complete, however, urinary levels fall as less bilirubin is available in the intestine for conversion into urobilinogen and less is absorbed. Urobilinogen production is also decreased with diarrhoeal states and following treatment with antibiotics.

The urinary excretion of urobilinogen is markedly dependent on urine pH; tubular reabsorption increases with acidity of the urine which also renders urobilinogen unstable. Estimation of urinary urobilinogen in an acid urine is thus an unreliable index of plasma urobilinogen levels. The peak urinary output of urobilinogen tends to occur between 12 noon and 1600 hours, probably in association with the 'alkaline tide', and urinary estimations of urobilinogen should be done during this time. Urinary urobilinogens give a purple reaction with Ehrlich's aldehyde reagent. A dipstick containing this reagent allows rough and ready quantification, and can be used to monitor urinary urobilinogen on repeated occasions. The test should be done on freshly voided urine.

Tests Indicating Hepatocellular Damage

Aspartate and Alanine Aminotransferases (SGOT; SGPT)[1]

The enzymic activities of serum aspartate and alanine aminotransferases have been measured in a variety of clinical disorders. Aspartate aminotransferase is found in high concentrations in liver and in heart muscle, skeletal muscle, kidney and pancreas. When any of these tissues are damaged they release aspartate aminotransferase into the blood, and serum levels rise. Alanine aminotransferase is present in low concentration in tissues other than liver and high serum levels of this enzyme are relatively specific for hepatic damage.

Aspartate aminotransferase activity is present in the cytoplasm and in the mitochondria of the hepatocyte and a different co-enzyme is present at each site; alanine aminotransferase is confined to the cytoplasm. It has been suggested that reversible liver cell injury results only in a loss of soluble enzyme from the cell sap but that mitochondrial enzymes are also released when there is more extensive damage and necrosis. This led to the use of the aspartate:alanine aminotransferase ratio as a diagnostic tool for the differentiation of hepatitis, cirrhosis, and obstructive jaundice but it has not gained wide support, although some workers consider it useful in the diagnosis of alcoholic liver disease (see later). The ratio of the cytoplasmic and mitochondrial aspartate aminotransferases has also been proposed as a diagnostic test on a similar basis.

Other Tests Indicating the Presence of Hepatocellular Damage or Necrosis

The plasma or serum concentrations of other substances rise in association with hepatocellular damage. Some are enzymes which are presumably released from the damaged hepatocytes. They include lactate dehydrogenase, α-hydroxybutyrate dehydrogenase, glutamate dehydrogenase, isocitrate dehydrogenase, β-glucuronidase, ornithine carbamyl transferase, glutamine deaminase, xanthine oxidase and urocanase. The pattern of response of these serum enzymes does not necessarily mirror that of the aminotransferases; for example, in acute viral hepatitis the lactate dehydrogenase returns to normal much more quickly. Few of these enzymes have been widely used in hepatology and it is difficult to assess their merits relative to the aminotransferase tests which are commonly used. Plasma vitamin B_{12} levels increase in patients with various kinds of liver disease and there is good evidence that it is released from damaged hepatocytes. There is an increase in the amount bound to transcobalamin I, the major B_{12} carrier in normal blood, and a more striking rise in the amount bound to transcobalamin II. Dialysable B_{12} levels are markedly increased and there is an accompanying rise in urinary B_{12} excretion.

Plasma iron levels are high in many patients with hepatocellular necrosis. In acute hepatitis maximum values are not attained until both aminotransferase and bilirubin levels are past their peak; this suggests that simple release of ferritin from damaged liver cells is not an adequate explanation. Furthermore, although ferritin levels rise with hepatic necrosis, even strikingly elevated levels could not account for the magnitude of the rise in plasma iron which occurs in viral hepatitis. Transferrin iron must also be raised but the reason for the elevation is not clear. Diminished utilization of iron for haemoglobin production, haemolysis, impaired reticuloendothelial uptake and impaired hepatic storage of iron may all be contributory factors in patients with liver disease.

Tests Indicating Obstruction to Biliary Flow

Alkaline Phosphatases[15]

Alkaline phosphatases are zinc metallo-enzymes which release inorganic phosphate from a variety of organic orthophosphates. They are found in many tissues as well as in serum where the activity is due to the presence of iso-enzymes derived from bone, liver and intestine (and in late pregnancy from placenta). The normal range (and units) for alkaline phosphatase activity depends on the laboratory performing the test. In the liver, histochemical evidence of alkaline phosphatase can be found in the microvilli of the bile canaliculus and on the sinusoidal surface of hepatocytes. The level of plasma alkaline phosphatase activity rises in many types of liver disease, the highest levels being seen in patients with obstruction to the flow of bile, either intrahepatic or extrahepatic, or with intrahepatic space occupying lesions such as metastatic tumours.

The elevation of activity of plasma alkaline phosphatase seen with liver disease is not due to failure of the liver to excrete circulating enzyme. The phosphatase which accumulates is produced in the liver and it has been shown in animal studies that hepatic synthesis of alkaline phosphatase is increased following biliary ligation. There appear to be two hepatic isoenzymes, one produced by hepatocytes and the other a biliary isoenzyme which presumably regurgitates into plasma from the obstructed biliary tree. The latter enzyme seems to be a better marker of biliary obstruction than total serum alkaline phosphatase activity. Intestinal phosphatase may be the major phosphatase of serum in patients with cirrhosis (normally it does not exceed 20% of the total phosphatase activity of plasma) but the reason for this increase is not clear. Identification of alkaline phosphatase isoenzymes might help to distinguish various kinds of liver disease, and can certainly help to decide whether an elevated alkaline phosphatase activity is due to liver disease or bone disease. In those hospitals where the identification of alkaline phosphatase isoenzymes is not available, it is usual to look for raised levels of other markers of biliary obstruction such as 5′-nucleotidase or γ-glutamyl transferase to confirm a hepatic origin for raised alkaline phosphatase levels.

Alkaline phosphatase is present in many tumours which may secrete the enzyme into plasma. In some tumours specific isoenzymes are produced, e.g. the Regan and Nagao isoenzymes; they may be of value not only for diagnosis but also because they allow monitoring of anti-tumour therapy; successful treatment is associated with a fall or with disappearance of the iso-enzyme from the plasma.

5′-Nucleotidase[15]

5′-nucleotidase is an alkaline phosphatase which attacks nucleotides with a phosphate at the 5′ position of the pentose. It is present in all human tissues but only liver disease appears to cause significant elevation of serum 5′-nucleotidase activity. The normal range of activity in plasma is from 1–15 i.u./l (measured at 37°C).

The highest levels are found with either intrahepatic or extra-hepatic obstruction to bile flow, but increases are also seen with chronic active hepatitis, cirrhosis, hepatitis and with other hepatocellular disorders. In routine practice measurement of 5′-nucleotidase activity has been used to confirm that liver disease is indeed the cause of an elevated alkaline phosphatase but its use for this purpose has now been largely superseded by measurement of gamma glutamyl transferase. It still has a role in infancy and in pregnancy. Both bone alkaline phosphatase and gamma glutamyl transferase levels are high for physiological reasons in infancy; in pregnancy the placental iso-enzyme causes an elevated alkaline phosphatase in the third trimester, but gamma glutamyl transferase levels may fall. Under both circumstances alkaline phosphatase iso-enzyme studies are a better method of establishing the presence of obstructive liver disease.

Leucine Aminopeptidase[15]

Leucine aminopeptidase hydrolyses peptides in which the free amino group is a l-leucine residue. This enzyme is widely distributed in the body and is present in bile and in the bile ducts and canaliculi. Blood levels are highest when there is intrahepatic or extrahepatic obstruction to bile flow. Levels also increase in acute hepatitis, in cirrhosis, in hepatic malignancy, and in the last three months of pregnancy. Like 5′nucleotidase its major use has been the confirmation that increased alkaline phosphatase activity is due to liver disease. This test is rarely used in routine practice.

Gamma-glutamyl Transferase (γ GT)[14]

Gamma-glutamyl transferase catalyses the transfer of gamma-glutamyl groups from γ-glutamyl peptides to other peptides, to aminoacids and to water. Large amounts are found in the kidneys, pancreas, liver and prostate and the enzyme is also found in many other tissues. γGT is present in bile where the activity is approximately 100 times greater than in normal serum. The normal range for serum γGT is higher in men than in women.

Unfortunately an increase in γGT activity is seen in several conditions other than liver disease. It increases in acute pancreatitis and in 50–70% of cases of acute myocardial infarction. The source of the increase in patients with myocardial infarction is not clear; it may be due to secondary effects on the liver, as high levels are also found in congestive cardiac failure. Serum γGT activity increases with the administration of enzyme-inducing drugs such as phenobarbitone, phenytoin or glutethimide. A high ethanol intake also causes raised γGT activity; there is an increase in over 80% of chronic alcoholics but there is a poor correlation between the alcohol intake and the serum γGT activity. Part of the effect of alcohol seems to be due to the liver damage produced by ethanol and part to enzyme induction. With cessation of drinking, γGT levels, when they are elevated, tend to revert to normal over 2–3 weeks but if chronic liver disease is present elevated levels may persist.

Elevated blood γGT activity is found in about 20% of patients with uncomplicated diabetes mellitus but levels are rarely more than twice the upper limit of normal. The source of the enzyme is thought to be the liver but other clinical or biochemical evidence of liver disease is not usually found.

Serum γGT levels rise in almost all kinds of liver

disease and abnormal levels are found in about 90% of patients suffering from hepatobiliary disease. The highest levels of γGT (averaging 10–20 times the upper limit of normal) are found with biliary obstruction or with primary or secondary hepatic malignancies. In acute viral hepatitis serum γGT levels rise to reach a peak in the second or third week of the illness but in some patients the level of γGT is still up at 6 weeks. Levels remain high with the development of chronic active hepatitis or cirrhosis.

Plasma Proteins[17]

The liver is the source of many of the proteins circulating in plasma and it is not surprising that hepatobiliary diseases are often associated with striking changes in the concentration of these compounds. The effect of liver disease on the levels of plasma proteins is complex and depends not only on its effect on protein synthesis, but also on its effects on the volume and distribution of extracellular fluids, on the half-life of the individual proteins and on protein catabolism by various routes. There may also be changes in the metabolism of proteins produced outside the liver. The estimation of 'total plasma protein' is of relatively little value. It is often normal even when there are gross disturbances of the individual components, such as albumin, and is of little value for screening, for diagnosis or for monitoring a patient's progress. When the total level is high or low its significance can only be interpreted by measurement of the major fractions.

Serum protein electrophoresis is often performed in conjunction with measurement of total protein, albumin and globulin. In patients with jaundice the electrophoretic pattern is rarely normal and certain types of electrophoretic strip are characteristic of certain types of liver disease. For example patients with several types of cirrhosis have a low albumin and high gamma globulin concentration, while patients with chronic active hepatitis often have a marked elevation of gamma globulin. Patients with biliary obstruction may show an excess of α_2- and β-globulins due to the accumulation of certain lipoproteins. Pateints with intravascular haemolysis may have a low α_2 band (due to decreased haptoglobin concentration) while α_1-antitrypsin deficiency may be suspected if the strip shows a poor α_1 band.

Despite these characteristic patterns the electrophoretic strip is generally of little value. Various abnormalities are found in many conditions other than liver disease and the changes found in jaundiced patients are of little help in diagnosis.

Plasma Albumin[17]

Plasma albumin, a molecule consisting of a single chain of 575 amino-acids, is the most abundant circulating protein. It is largely responsible for the colloid osmotic pressure of plasma, and has binding sites which have great affinity for a number of naturally occurring compounds, including bilirubin, and for many drugs.

The liver appears to be the only site of synthesis, producing 120–200 mg/kg body weight daily in normal individuals (approximately 12 g in a 70 kg person). The total exchangeable pool of albumin is normally about 3·5–5·0 g/kg body weight, and of this 38–45% is present within the intravascular space. The plasma albumin concentration is normally between 35 and 45 g/l.

About 1 g of albumin is lost each day into the gastrointestinal tract. About 11 g/day appear to be degraded within the body and only a small amount is broken down by the liver. The precise mechanism of degradation is not clear but it does seem to be under fairly precise metabolic control as the fractional rate of albumin degradation can change markedly. When synthesis is reduced plasma albumin levels also fall but the fall is minimized because there is a reduction in the fractional catabolic rate for albumin.

The plasma albumin level has been considered by many as a valuable test of liver function, because it is thought to reflect hepatic protein synthesis. Unfortunately hepatic synthesis is only one of several factors affecting plasma albumin and even when synthesis is markedly reduced the effect on plasma albumin takes many days to become apparent because of the long half-life of albumin (about 20 days).

It is true that albumin synthesis falls in many patients with severe chronic liver disease and they tend to have a low serum albumin level. But in many patients with hypoalbuminaemia and ascites matters are complicated. Although plasma levels are low the plasma volume may be large, and there may be much more albumin than normal in the extravascular space. Although the fractional catabolic rate is low the absolute rate of degradation, and therefore synthesis, may be normal or even high. In cirrhotics with ascites the hepatic secretion of albumin is disturbed; instead of albumin being secreted directly into the bloodstream *via* the sinusoids much of it is diverted into thoracic duct lymph or into the ascites. Finally it must be realized that plasma albumin levels might be expected to fall with prolonged reduction of protein intake or with a variety of infectious or toxic states. Clearly alterations in the concentrations of plasma albumin must be interpreted with caution.

Proteins Involved in Blood Coagulation (including factors I, II, V, VII, IX)

Most clotting factors are synthesized in the liver. Vitamin K plays a key role in the synthesis of factors II, VII, IX and X. The first 10 residues on the amino-terminal end of prothrombin (factor II) are an amino-acid called γ-carboxyglutamic acid. The second carboxyl group is added to these residues after the polypeptide chain of prothrombin has been assembled; vitamin K is a cofactor in this reaction. This chain of γ-carboxyglutamic acid residues appears to be important in the calcium-dependent binding of prothrombin to platelets and to the membranes of disintegrated cells. It seems likely that a similar Vitamin K-dependent process is important in the synthesis of factors VII, IX and X; they also bind to membranes and to calcium, and proximity to each other on the platelet surface would clearly facilitate the coagulation cascade.

The clotting factors have a short half-life (indeed that of factor VII is only a few hours) and so plasma levels fall rapidly when synthesis is impaired. This happens with any severe hepatocellular disturbance. Synthesis of normal II, VII, IX and X is impaired with obstructive jaundice because Vitamin K absorption is reduced due to impaired micelle formation in the intestinal lumen. Individual plasma clotting factors are not usually measured routinely. Instead a number of tests tend to be used which depend on the presence of reasonably normal amounts of various combinations of clotting factors. They are relatively insensitive to even moderate reductions of individual factors. The best established test is the one-stage prothrombin time (PT) which tests the 'extrinsic' coagulation pathways and depends on the presence of factors I, II, V, VII and X. It is normal in patients with deficiencies of VIII, IX, XI and XII.

The partial thrompoplastin time (PTT) tests the 'intrinsic' pathway and is abnormal in patients with deficiencies of factors VIII, IX, XI, XII, as well as in patients with deficiencies of I, II, V, IX.

In evaluating plasma clotting factors in jaundiced patients it is important to recognize that an abnormal PT and PTT may be due mainly to deficiency of Vitamin K. One or two injections of 10 mg of Vitamin K would then correct these tests within 24 to 48 hours. Failure to correct suggests the presence of hepatocellular damage; this has important prognostic implications as the results of tests of clotting (after correction of Vitamin K deficiency) seem to correlate well with the overall severity of disease particularly in patients with acute hepatitis.

The estimation of the PT and PTT are of value in determining whether a liver biopsy can be performed in patients with suspected liver disease but there is no hard and fast rule on the degree of abnormality which should preclude biopsy. Different groups have their own criteria. Some would argue that tests of coagulation are of less importance for this purpose than the platelet count or bleeding time. The latter is certainly an important measurement and percutaneous biopsy should not be performed if the bleeding time is greater than 6 minutes (Duke).

Patients with liver disease may have more complex disorders of coagulation than those resulting from impaired synthesis of clotting factors. Increased fibrinolytic activity has been described, particularly in patients who have recently been injured or undergone surgery, whilst decreased fibrinolytic activity has been described in acute hepatic failure and in obstructive jaundice. The increased fibrinolysis following surgery or trauma may be due to failure of the liver to clear plasminogen activator released into the blood stream; failure of the liver to synthesize antiplasmin may also be important.

In patients with chronic liver disease increased fibrinolysis may be secondary to the presence of disseminated intravascular coagulation (DIC), usually triggered by infection. DIC is suggested by a fall in the platelet count, by marked prolongation of the PT, PTT and thrombin time, by shortening of the fibrinogen half-life and by the finding of elevated levels of fibrin degradation products in plasma. Both DIC and increased fibrinolysis are serious complications and require expert investigation and management which can rarely be provided outside special units.

Bile Acids[9]

Bile acids are derived from the catabolism of cholesterol. The two main primary bile acids of man, cholic acid and chenodeoxycholic acid, are produced in the liver by partial cleavage of the cholesterol side-chain and by α-hydroxylation of the cholesterol nucleus at positions 3, 7 and 12, or positions 3 and 7 respectively. Cholic and chenodeoxycholic acids are conjugated in the liver with either glycine or taurine and these conjugates are excreted in the bile. In the upper small intestine the conjugated bile salts play an important role in the digestion and absorption of fat and fat soluble compounds; they are reabsorbed in the terminal ileum by a very efficient active transport mechanism. A small amount of bile acid escapes into the colon where bacteria cause deconjugation and/or dehydroxylation of the molecule at the 7α-position. As a result of the latter reaction cholic acid is converted to deoxycholic acid, and chenodeoxycholic to lithocholic acid. These newly formed bile acids are absorbed from the colon and travel to the liver via the portal vein. The liver removes bile acids very efficiently from the portal blood and excretes them rapidly into the bile. This completes the enterohepatic circulation of bile acids (i.e. liver—bile—intestine—portal vein—liver). Because of the enterohepatic circulation deoxycholate, produced in the colon, accumulates in the bile acid pool and is one of the major bile acids found in human bile (as a taurine or glycine conjugate). The lithocholic acid absorbed from the colon is largely sulphated by the liver and, as sulphated bile acids are poorly absorbed from the intestine, most lithocholic acid is lost from the body when it returns to the colon in sulphated form. Unlike deoxycholic acid it does not accumulate in the bile salt pool.

Liver disease might be expected to interfere with bile salt metabolism in a number of ways. There might be a change in the amount of cholic and chenodeoxycholic acids synthesized or a reduction in their relative proportions (normally about twice as much cholic acid is produced); unusual bile acids might be produced. There might be a reduction in the amount of bile acid conjugated by the liver or a change in the taurine: glycine ratio. A decrease in the amount of primary bile acid entering the intestine might lead to a reduction in secondary bile acid synthesis with a reduction in plasma levels of deoxycholate. Impaired liver function or diversion of portal blood might lead to a failure of the liver to clear bile acids carried in the portal circulation (or following injection into the systemic circulation) and this would result in increased levels of plasma bile acids, particularly for a period after meals. Finally obstruction to the flow of bile might also lead to a rise in plasma bile acids due to their regurgitation into the blood stream from the obstructed biliary tract. With high plasma bile acid levels one might expect increased excretion of urinary bile acids.

All of these expectations are, in fact, met in patients with liver disease. Most of them are of little value in the clinical assessment of patients. The simplest of the bile acid tests is the measurement of the blood total bile acid concentration, either in the fasting state or after a meal, but even this test is rarely available in routine laboratories. The normal fasting serum bile acid level goes up to about 15 μmol/l but the figure is dependent on the method of measurement used and varies from laboratory to laboratory. The fasting serum bile acid level is elevated in only about two-thirds of patients with a variety of types of liver disease. It is therefore of relatively limited value in screening for patients who might have liver disease.

Blood bile acid levels rise after a meal and usually return to normal after two hours. Levels remain high in almost all patients with significant liver disease and the two hour post-prandial bile acid level is therefore a valuable screening test for liver disease.

Cholesterol, Other Lipids and Lipoproteins[5]

The measurement of total plasma cholesterol has long been used as a 'liver function' test but it is of relatively little practical value. The total cholesterol level tends to rise in patients with either extrahepatic or intrahepatic biliary obstruction, and sometimes rises to very high levels in patients with primary biliary cirrhosis; it falls in patients with severe parenchymal liver disease but it is a crude index and in most patients with liver disease the total cholesterol is within the normal range.

However, there are striking changes in the plasma lipids and lipoproteins in most patients with liver disease but they are only revealed by the use of relatively complex methods. Measurement of free and esterified cholesterol in plasma shows that the rise in total cholesterol in obstructive jaundice is due mainly to a rise in free cholesterol and that this is associated with a corresponding increase (on a one-to-one molar basis) of phospholipid; triglyceride levels also rise. The fall in total cholesterol in parenchymal liver disease is due to a fall in cholesteryl ester; there tends to be a reciprocal rise in triglyceride levels and free cholesterol levels usually rise a little.

In order to understand these lipid changes one must look at the changes which occur in the plasma lipoproteins in liver disease. In normal subjects there are four main classes of lipoproteins; chylomicrons, very low density lipoproteins (VLDL), low density lipoproteins (LDL) and high density lipoproteins (HDL). Chylomicrons are large triglyceride-rich particles which transport fat from the intestine to the blood stream; VLDL are also triglyceride-rich and enter the blood stream from the liver and intestine. LDL particles are rich in cholesteryl ester and are formed in the blood as a result of the catabolism of VLDL. HDL particles are particularly rich in protein; they enter plasma from the liver and gut.

Several enzymes play key roles in the metabolism of lipoproteins. Lipoprotein lipases (LPL) present in tissues are responsible for the hydrolysis of the triglyceride of chylomicrons and VLDL, and the free fatty acids are taken up by the tissues for utilization or storage. There is also a hepatic triglyceride lipase the function of which is not yet clear. Lecithin cholesterol acyl transferase (LCAT) is a plasma enzyme which catalyses the transfer of a fatty acid from the β-position of lecithin (phosphatidyl choline) to the hydroxyl group of free cholesterol with the production of cholesteryl ester and lysolecithin. This enzyme seems to account for nearly all the turnover of plasma cholesteryl ester.

LCAT is produced in the liver and plasma LCAT activity falls in most patients with significant liver disease. LCAT activity is normal or increased in a small number of patients; in them the plasma lipoproteins are of normal structure and composition although patients with biliary obstruction and high LCAT do have high plasma lipid levels, which are probably secondary to the regurgitation of biliary phospholipid. When LCAT levels are low there are striking abnormalities of plasma lipoproteins. In patients with parenchymal liver disease and low LCAT activity fasting levels of VLDL are markedly reduced: LDL levels tend to be normal but although the LDL particles appear normal on electron microscopy they have less cholesteryl ester than normal and a corresponding increase in triglyceride. The HDL fraction has an abnormal composition and small disc-like particles are found which are thought to represent 'nascent' HDL. These discs, which are also seen with familial LCAT deficiency, are converted into normal spherical HDL by the action of LCAT. In obstructive jaundice with low LCAT levels similar changes are seen in the HDL fraction but there is a different pattern of changes in the LDL fraction. Normal sized LDL particles are present, (rich in triglyceride and poor in cholesteryl ester, like those found in parenchymal liver disease) but in addition two other particles are found: one is a large triglyceride rich particle, which may represent an intermediate in the catabolism of VLDL; the other is a large disc called LP-X, which is composed mainly of free cholesterol and lecithin.

It has been claimed that LP-X is a valuable pointer to the presence of biliary obstruction but in fact its clinical usefulness is limited. Not only is it undetectable in patients with high LCAT levels but even when it is present it does not allow us to make the important distinction between extrahepatic and intrahepatic obstruction.

Unfortunately the demonstration of the lipoprotein changes occurring in liver disease is beyond the scope of routine clinical laboratories. Electrophoresis on paper or agarose, which allows the typing of common hyperlipoproteinaemia (Fredrickson types I–V) is of no value in liver disease. When LCAT levels are normal the electrophorectic pattern is normal; when LCAT activity is low the only abnormality seen is a loss or reduction of VLDL (pre-β) and HDL (α) bands with preservation of the β-band. This simplification of the electrophoretic pattern is in striking contrast to the heterogeneity of lipoproteins found by ultracentrifugation and electron microscopy, particularly in patients with obstructive jaundice.

THE VALUE OF LIVER FUNCTION TESTS FOR SCREENING AND PROFILING

Biochemical screening of the 'well' population is an uncommon procedure in the United Kingdom and is usually restricted to routine check-ups done at places like the British United Provident Association (BUPA) medical centre. Their experience, and that of others who have engaged in biochemical screening, is that diabetes mellitus hypercholesterolaemia and liver disease are the three conditions most commonly detected. Unsuspected liver disease, usually on the basis of excessive alcohol intake, is found in approximately 1% of the 'well' population studied, and this would appear to justify the incorporation of 'liver function tests' into the screening battery.

Liver function tests are chosen by some doctors in the profile of tests which they request routinely on patients. I suspect that many more doctors are given the results of these investigations without asking for them because the tests are included in the battery of tests done by automated multi-channel analysers. There is little information on the proportion of hospital inpatients or outpatients found to have unsuspected liver disease. One would anticipate a figure higher than the 1% found with screening of the 'healthy' population but it would clearly depend on the type of clinic attended and the type of population from which the patients were drawn.

For screening purposes the use of a small battery of tests clearly constitutes a more sensitive procedure than the use of any of the individual tests as the number of false negatives must be minimized by this strategy; fewer of the subjects with liver disease would be missed. In addition the use of a battery adds considerable specificity; when more than one test is abnormal liver disease is virtually certain and false positives are uncommon. It is unlikely that there would be serious consequences if patients with liver disease, but with several normal liver function tests were missed. Such patients would tolerate surgical procedures, drugs, etc., far better than those in whom there was clear evidence of hepatic damage.

The individual tests vary considerably in their value as tests for screening or profiling. When an apparently healthy population of adults was studied at the BUPA Centre levels of total plasma bilirubin above the commonly accepted upper limit of 17 μmol/l were found in approximately 5% of the population. Levels were higher in men; about 2% had a bilirubin concentration greater than 25 μmol/l while in only 0·6% of women was this figure exceeded. Because high levels of bilirubin are usually associated with overt jaundice only slight or mild elevations of plasma bilirubin are found in this way. In most cases bilirubin is absent from the urine and the results of other liver function tests are normal. Such hyperbilirubinaemia is almost always benign even though it goes under the name of Gilbert's disease or syndrome. Indeed the overall consequences of finding a high bilirubin level in this way may be deleterious as subjects with high bilirubin levels are sometimes subjected to unnecessary and potentially hazardous medical or surgical intervention.

When 'healthy' populations are screened the proportion found to have elevated levels of aspartate aminotransferase is around 4–6%; although many of these abnormal results are unsuspected it would appear that a definitive new diagnosis of liver disease is made in only a small number of them. It is likely that alcoholic liver disease is the major cause of new cases and mild elevations of asparate aminotransferase may fall with cessation of drinking. The aspartate aminotransferase level appears to rise with excessive alcohol intake which occurs shortly before the test is performed; at the BUPA centre there was a distinct peak in the mean result found on New Year's Day!

Levels of γ-glutamyl transferase (γGT) above the usually accepted upper limits of normal (30 U/l in women, 50 U/l in men) were found in approximately 15% of those screened at the BUPA centre and this led them to adjust their upper limits of normal to 50 U/l for women and 80 mu/l for men.

In a study of 11 000 men, 6% had a γGT activity of greater than 80 and in 4% the level was greater than 100 U/l. Although estimation of the γGT level is useful in screening it lacks specificity for the detection of liver disease. Several other conditions cause elevation of γGT (see earlier). Its main value would appear to be in the detection of people who are at risk for the complications of alcoholism. γGT levels do not appear to increase in non-alcoholic subjects as a result of an evening's drinking. In alcoholics the γGT activity appears to rise even when there is no significant underlying liver disease and to return to normal on cessation of drinking: if it does not return to normal then continued alcohol intake is likely or it may indicate that liver damage has occurred. Unfortunately about 1/3 of heavy drinkers show no elevation of γGT activity. In these subjects a suspicion of heavy ethanol intake is raised if there is a high mean corpuscular volume, or a high plasma concentration of urate or triglyceride.

Alkaline phosphatase levels are found to be elevated in about 2–5% of apparently healthy adults, but in only a small number of these does the finding lead to the establishment of a new diagnosis. Bone disease, especially Paget's disease, makes up a significant proportion of the abnormal findings particularly in the elderly. When an elevated alkaline phosphatase activity is found, further diagnostic tests are indicated and these are discussed in the next section.

THE VALUE OF LIVER FUNCTION TESTS IN THE DIAGNOSIS OF LIVER DISEASE

Some patients are found to have serious liver disease (e.g. well compensated cirrhosis of the liver) even though the results of 'standard' liver function tests are normal. Usually the diagnosis is made only when clinical features, such as splenomegaly, spider naevi and gastrointestinal bleeding, point clearly to the true nature of the patient's problems and lead to further diagnostic effort.

When only one liver function test is abnormal it is necessary to establish that hepatobiliary disease is indeed the cause of the abnormality. When only the total plasma bilirubin concentration is elevated liver disease can be assumed if there is also an increase in the level of plasma bilirubin esters or if there is accompanying bilirubinuria. Unfortunately we cannot assume that liver disease is absent when there is a simple unconjugated hyperbilirubinaemia, as this is found in some patients with liver disease; in some of them of course, it may be due to an associated Gilbert's syndrome, and largely unrelated to the underlying liver disease.

When the only abnormality is a significantly elevated aspartate aminotransferase liver damage is the most likely cause and this can usually be confirmed with another test for hepatocellular damage such as alanine aminotransferase, γGT or serum ferritin etc. A hepatobiliary cause of an elevated alkaline phosphatase level is more or less certain if there is an associated increase in γGT or 5-nucleotidase, or if it can be shown that the high alkaline phosphatase activity is due to an increase in hepatocellular or biliary isoenzyme.

Classification of Jaundice

When two or more standard liver function tests give abnormal results it is highly probable that some form of hepatobiliary disorder is present. The problem then facing the clinician, and that facing him when the confirmatory tests mentioned in the above paragraph are abnormal, is to decide on the nature of the underlying disease. On the basis of the liver function tests, many clinicians divide patients with jaundice into those with *haemolytic jaundice, obstructive jaundice,* or *hepatocellular disease.* This classification has been used for many years. Haemolysis is inferred if the jaundice is acholuric (i.e. if bilirubin is absent from the urine) and if there is a predominantly unconjugated hyperbilirubinaemia. Obstructive jaundice is said to be present if a large proportion of the bilirubin is conjugated, if aspartate transaminase levels are only moderately elevated and if there is a marked increase in alkaline phosphatase activity (with a level above 30 K.A. units/dl or the equivalent in other units). Jaundice is considered to be hepatocellular when the rise in bilirubin concentration is due to both unconjugated bilirubin and bilirubin esters, when the alkaline phosphatase activity is normal or only slightly increased, and when there is a marked increase in aspartate aminotransferase.

Unfortunately the classification of jaundice on this basis is unsatisfactory for diagnostic purposes. In an ideal classification, classes should be collectively exhaustive and mutually exclusive, and it is disconcerting to realize that with the conventional classification the following conditions appear under the heading of both hepatocellular and obstructive jaundice.

TABLE 1
CONDITIONS WHICH MIGHT BE CLASSIFIED AS 'HEPATOCELLULAR' OR 'OBSTRUCTIVE' JAUNDICE ON THE BASIS OF LIVER FUNCTION TESTS

Viral hepatitis
Drug induced hepatitis
Alcoholic liver disease
Chronic active hepatitis
Cirrhosis

A more useful classification for diagnostic purposes is to divide jaundiced patients into those with:

(1) Unconjugated hyperbilirubinaemia
(2) Extra-hepatic biliary obstruction
(3) Parenchymal liver disease

(1) Unconjugated Hyperbilirubinaemia

The preliminary diagnosis of an *unconjugated hyperbilirubinaemia* can be made on the basis of routine liver function tests (including urinary bilirubin) and confirmed by measurement of bilirubin esters. Haemolytic disorders must clearly be excluded as a cause but by far the largest number of patients falling into this sub-group are those with a benign hyperbilirubinaemia. With the conventional classification the benign hyperbilirubinaemias are grouped under hepatocellular jaundice, even though they have few features in common with the other conditions in the group and it seems more logical for diagnostic purposes to group them with haemolytic disorders on the basis of the initial biochemical findings.

(2) Extra-Hepatic Biliary Obstruction

When conjugated hyperbilirubinaemia is present the important distinction to be made is whether or not a patient has an extrahepatic obstruction to the biliary tree. If there is a blockage of the major ducts, surgery will be needed in most cases, but if jaundice is due to parenchymal liver disease surgery must be avoided, not only because it will not relieve the jaundice but because it may exacerbate the patient's condition.

Liver function tests are of little help in making this decision which may also be difficult to make on the basis of the history and physical examination. Sometimes clinical features such as the age of the patient, the presence of an enlarged gallbladder, ascites, a hard irregular liver, etc, may provide valuable clues. In many patients however, the clinical picture is equivocal or even misleading. The classical presentation of obstructive jaundice (with pale stools, dark urine and pruritus) may be seen with the parenchymal liver diseases listed in Table 1, while patients with proven extra-hepatic obstruction may present without pruritus, hepatomegaly or pale stools and occasionally without dark urine. In neither situation are liver function tests of much value. Patients with 'cholestasis' due to parenchymal liver disease may give results typical of those found in patients with extrahepatic obstruction. With extrahepatic obstruction on the other hand serum bilirubin levels may range widely and aspartate aminotransferase levels are frequently greater than 200 iu/l and occasionally exceed 1000 iu/l (normal range 5–15 iu/l). Alkaline phosphatase levels are less than 30 K.A. units/dl in approximately 20% of cases and in a small number of cases, variously estimated between 3 and 10%, alkaline phosphatase levels remain normal despite deep jaundice.

It cannot be stressed too strongly that the precise diagnosis of an extrahepatic block can only be made on the basis of 'anatomical' evidence of biliary obstruction; this can be obtained using special investigations such as ultrasound, computerized axial tomography and cholangiography. When these investigations are not available there must always be serious doubt about the appropriateness of surgical intervention even when the presence of extrahepatic biliary obstruction is probable on the basis of the overall clinical picture.

(3) Parenchymal Liver Disease

Liver function tests alone are clearly of limited value in deciding whether a patient has parenchymal liver disease, for reasons outlined in the previous section. If the clinical picture is equivocal and ultrasound shows no enlargement of the biliary tree then the diagnosis of parenchymal liver disease is usually confirmed by percutaneous liver biopsy, a technique which may also allow more precise identification of the underlying condition. In many patients, however, the clinical picture is helpful, and with liver function tests, and other laboratory investigations, it may be possible to establish the precise diagnosis with a considerable degree of probability. In trying to reach a conclusion it is worth bearing in mind that most patients with parenchymal liver disease are suffering from one of a relatively small number of conditions—namely acute hepatitis, due to a virus or drug, chronic hepatitis, some form of cirrhosis, alcoholic liver disease, primary biliary cirrhosis or hepatic malignancy. Other causes of parenchymal liver disease are relatively uncommon.

Acute Hepatitis[13]

The clinical features of acute hepatitis are similar whether it is caused by a virus or by a drug. With both causes the plasma aspartate aminotransferase level is markedly elevated at the onset of clinical symptoms, falling quickly over the following weeks. In patients with viral hepatitis the aminotransferase level may rise to its peak before the onset of symptoms and striking elevation of the aminotransferase level occurs even in asymptomatic and anicteric cases. In most patients in whom the diagnosis of acute hepatitis is made, the plasma bilirubin level is increased and there tends to be slight elevation of alkaline phosphatase activity. Marked elevation of the level of aspartate aminotransferase may be seen with shock or acute heart failure. These conditions should be readily diagnosed on clinical grounds, but occasionally marked elevations are seen in patients with extrahepatic obstruction or with chronic liver disease, between which conditions there may be considerable diagnostic difficulty. It must be stressed that aminotransferase levels should be measured at the first suspicion of liver disease (usually because of constitutional upset and/or slight jaundice with bilirubinuria). If testing is delayed the aminotransferase level may fall to levels found with many other causes of jaundice, and diagnostic confusion may result which could have been avoided if the early marked risk of aminotransferase had been detected.

To distinguish between drug- or toxin-induced acute hepatitis and acute viral hepatitis it is essential to check the history of drug ingestion with great care. If no drug can be identified as the case, then it is customary to attribute the hepatitis to a virus. It is now possible to make a fairly confident diagnosis of acute hepatitis due to viruses A and B by assessing whether viral components, or antibodies to them, are present in the blood at appropriate times. With type B hepatitis the surface antigen is usually found with the onset of symptoms and tends to persist for several weeks. Antibodies to the virus core appear while the surface antigen is present and later still antibodies to the surface of the virus can be found. Recent infection with hepatitis A virus can be diagnosed with some confidence if IgM antibodies to the virus are found. These tools are becoming more widely available but a number of patients with viral hepatitis will be suffering from Non-A, Non-B hepatitis or from other forms of viral hepatitis such as infectious mononucleosis, cytomegalovirus etc.

Some patients with viral hepatitis run a much more serious course. Fulminant hepatitis is often fatal within a matter of days. Subacute hepatitis is more prolonged with death from liver failure occurring within weeks or months. In both conditions the bilirubin concentration often rises to high levels but in individual patients the bilirubin, aminotransferase and alkaline phosphatase levels are of little help for diagnosis or prognostication, and do not necessarily reflect the severity of the clinical picture. The plasma albumin concentration may fall to very low levels and there is a marked impairment of tests of clotting due to failure of synthesis of the proteins involved in coagulation. Hypoglycaemia occurs in a large proportion of patients with fulminant hepatic failure and may require large amounts of intravenous glucose. There is a general and marked elevation of plasma amino acids, except for the branched chain amino acids (leucine, isoleucine and valine) which may be present at normal or low concentrations. Plasma ammonia usually rises in fulminant hepatic failure but may be normal even in patients with severe encephalopathy. Fluid and electrolyte disturbances tend to be profound with hypokalaemia, hyponatraemia and hypocalcaemia and there may be a marked respiratory alkalosis. The mechanism of these changes is poorly understood and treatment is very difficult. The blood urea level is usually low but rises with the oliguria which often supervenes.

Cirrhosis[16]

The diagnosis of cirrhosis of the liver can often be made confidently on purely clinical grounds. When the classical clinical features are absent biochemical investigations are usually of little diagnostic value, and a firm diagnosis is only possible as a result of liver biopsy. Liver biopsy also allows identification of the cause of cirrhosis when this is due to haemochromatosis, Wilson's disease, alpha-1-antitrypsin deficiency, etc.

In patients with haemochromatosis the plasma iron concentration is usually elevated (to more than 30 μmol/l) while the total iron binding capacity of the plasma is usually reduced. The percentage saturation is usually very high (above 75%) and there is usually more or less complete saturation of transferrin. The total iron binding capacity of plasma is often reduced in other forms of cirrhosis and its percentage saturation may be higher than normal even when the plasma iron level is normal and when there is no excess of iron deposition in the tissues. Some patients with cirrhosis show features of haemochromatosis, including skin pigmentation and diabetes mellitus, and may have variable degrees of iron deposition in the tissues. The diagnosis of haemochromatosis cannot rest, therefore, on the basis of clinical features and an elevation of the iron binding saturation. It is necessary to demonstrate grossly increased iron stores and this can be done, semi-quantitatively, by measuring serum ferritin, the iron content of liver tissue or urinary iron excretion following intramuscular injection of desferrioxamine.

The diagnosis of Wilson's disease is easy when Kayser-Fleischer rings are found in the cornea. When these are absent we must rely on tests of copper metabolism. In 95% of patients with Wilson's disease the serum caeruloplasmin level is below 200 mg/l at the time of diagnosis (caeruloplasmin is an acute phase reactant which may explain the normal levels of caeruloplasmin seen in some patients with Wilson's disease). Unfortunately a low level of caeruloplasmin is far from diagnostic as low levels are found in patients with other forms of cirrhosis in whom hepatic synthesis of the protein is depressed.

Almost all symptomatic patients with Wilson's disease have high levels of 'free' copper in plasma and excrete excessive amounts of copper in the urine. This is also true of some patients with primary biliary cirrhosis or other chronic forms of biliary obstruction but in these subjects the level of caeruloplasmin is usually increased. The concentration of copper in the liver is markedly elevated in patients with Wilson's disease. It is also raised in patients with chronic biliary obstruction but histological examination of the liver usually allows differentiation between the two conditions. When clotting abnormalities render biopsy potentially hazardous then the diagnosis of Wilson's disease can be made by measuring the incorporation of ingested radioactive copper into caeruloplasmin. Patients with Wilson's disease incorporate virtually no label into the newly synthesized protein.

Chronic Active Hepatitis[16]

Chronic active hepatitis is a common clinical problem. The diagnosis can only be made with confidence following biopsy of the liver. The characteristic histological changes may be seen following viral infections of the liver, in association with the administration of certain drugs, with Wilson's disease and alcoholic liver disease, and in 'lupoid' hepatitis, an autoimmune disorder sometimes associated with other autoimmune diseases.

Standard liver function tests are of little value for diagnosis as plasma levels of bilirubin, aspartate transaminase and alkaline phosphatase may range widely. There may be striking elevation of γ-globulin levels, particularly in lupoid hepatitis, but increases in γ-globulins are also seen in a number of other chronic parenchymal diseases. Measurement of IgG, IgA and IgM levels may improve discrimination between different types of liver disease (IgG elevation being common in chronic active hepatitis) but such tests do not obviate the need for liver biopsy.

The clinical picture may be helpful in the diagnosis of autoimmune chronic active hepatitis. Laboratory studies are also useful and the diagnosis is questionable unless there is a high titre (1:80 or more) of antinuclear factor or smooth muscle antibodies. In autoimmune chronic hepatitis, antinuclear antibodies to double-stranded DNA and smooth muscle antibodies to actin are of diagnostic value. The high titre is important as smooth muscle antibody titres up to 1:40 may be seen with acute or chronic viral hepatitides and some normal subjects show an antinuclear factor of low titre (1:10), particularly in the older age group. Recently it has been claimed that 80% of patients with autoimmune chronic active hepatitis have antibodies in serum to a liver membrane antigen. While this antibody is also found in about 20% of patients with primary cirrhosis it is apparently not found in other forms of chronic active hepatitis.

Chronic active hepatitis due to viral infections can usually be diagnosed on immunological grounds. Patients with hepatitis B disease usually have HBsAg in their blood though not always in high titre and in those with no detectable HBsAg it is possible to find antibody to the viral core. Hepatitis A infection is not followed by serious chronic liver disease but chronic active hepatitis does occur with 'non-A, non-B' infections. Precise identification of these cases may soon be possible in routine laboratories. The diagnosis of other forms of chronic active hepatitis is not usually helped by biochemical investigations, except in the case of Wilson's disease.

Primary Biliary Cirrhosis[16]

Primary biliary cirrhosis is a parenchymal liver disease with some clinical and biochemical findings similar to those of extrahepatic biliary obstruction, and other clinical features, such as the presence of associated autoimmune disorders, which suggest a relationship to the autoimmune type of chronic active hepatitis.

The disease is sometimes diagnosed in its early stages, before the onset of jaundice; patients with generalized pruritus may have an isolated elevation of alkaline phosphatase and liver biopsy reveals the characteristic granulomata and bile duct damage. Later in the course there may be deep jaundice, moderate elevation of aminotransferases, marked elevation of alkaline phosphatase activity and other markers of biliary obstruction, and striking hypercholesterolaemia. The diagnosis is sometimes made easily on clinical grounds but the condition may mimic extrahepatic biliary obstruction and it is then essential to exclude disease of the biliary tract.

The finding of a mitochondrial antibody is of great value in the diagnosis of primary biliary cirrhosis. It is found in almost 100% of patients with primary biliary cirrhosis and if it is absent it is essential to exclude other possible causes for the clinical picture. Conversely the presence of mitochondrial antibody throws considerable doubt on the diagnosis of extrahepatic biliary obstruction as the cause of jaundice. Mitochondrial antibodies are also found in about 20% of patients with autoimmune chronic active hepatitis but in specialized laboratories it is possible to distinguish them from the type found in primary biliary cirrhosis.

There is usually an elevation of immunoglobulins in patients with primary biliary cirrhosis. IgM levels tend to be higher than IgG levels, the reverse being true in patients with autoimmune chronic active hepatitis, but this finding is of little diagnostic value as there is considerable variation.

Alcoholic Liver Disease

The diagnosis of alcoholic liver disease may be obvious when there is a history of heavy alcohol intake and other clinical features of alcoholism are present. Liver biopsy provides valuable support for this diagnosis when there is micronodular cirrhosis, clear evidence of alcoholic hepatitis or an alcoholic pattern of fibrosis. Unfortunately in some alcoholics, particularly in the later stages of cirrhosis, histological evidence is equivocal; the biopsy may show chronic active hepatitis or a macronodular cirrhosis.

The standard liver function tests are of little use in identifying alcoholic liver disease. The plasma levels of bilirubin, aspartate aminotransferase, and alkaline phosphatase may all range widely. In the United States great emphasis is placed by some workers on the relative proportions of serum aspartate aminotransferase (SGOT) and alanine aminotransferase (SGPT).[4] They use an SGOT/SGPT ratio greater than 2 as evidence of alcoholic hepatitis and cirrhosis, claiming that it distinguishes alcoholics from patients with extrahepatic obstruction or with viral hepatitis in whom the SGOT/SGPT ratio is usually less than 1·0. Not one of 37 patients with obstructive jaundice, and only 2 out of 52 patients with viral hepatitis, had a ratio greater than 2·0. Unfortunately the ratio was sometimes more than 2 with either postnecrotic cirrhosis or chronic hepatitis. The comparisons appear to have been made between groups of patients in whom the diagnosis was relatively straightforward and we cannot be certain that the ratio would be helpful in patients with a puzzling clinical presentation. It is of interest that the high ratio seems to be associated with a reduction in the alanine aminotransferase content of liver tissue in patients with alcoholic liver disease.[10]

Serum levels of γ-glutamyl transferase are useful for screening for excessive alcohol intake (see earlier) and it might be assumed that the γGT/AST ratio would be useful in the diagnosis of alcoholic liver damage. Unfortunately quite striking elevations of γGT occur in a number of diseases of the liver and the relative proportions of γGT and other enzymes are quite variable.

Primary and Secondary Hepatic Malignancy

Metastases in the liver are commonly found with a number of primary tumours including carcinomas of the stomach, colon, pancreas, breast etc. When the primary tumour has already been identified there is rarely any difficulty in establishing the cause of jaundice or hepatomegaly due to secondary invasion of the liver. When the primary is still occult, however, hepatic involvement may cause diagnostic difficulty.

Liver function tests can be normal even with extensive metastatic disease. When biochemical abnormalities are found they tend to be increases in alkaline phosphatase activity or in the other markers of biliary obstruction—γGT, 5-nucleotidase and leucine aminopeptidase. There are often changes in the plasma bilirubin concentration and the activity of aminotransferases, but in many patients these may be normal. In some patients with intrahepatic metastases there are striking increases in plasma lactate dehydrogenase but this test is relatively non-specific. It is a poor diagnostic test when metastases are suspected and it is not sufficiently sensitive to be worth using as a screening test for liver disease in general.

The two main primary carcinomas of the liver are the hepatocellular carcinoma and the bile duct carcinoma. The latter causes the biochemical changes of extrahepatic biliary obstruction, although the plasma bilirubin concentration may be normal when only one hepatic duct is involved. The hepatocellular carcinoma gives biochemical changes similar to those found with hepatic secondaries, but in addition there may be an increase in alpha-fetoprotein levels in blood which may be of considerable diagnostic value. Modest increases in alpha-fetoprotein concentration may be found in several diseases of the liver including viral hepatitis, HBsAg-negative chronic active hepatitis and alcoholic liver disease but levels greater than 5000 ng/ml are virtually diagnostic of primary hepatocellular carcinoma. This tumour is often found in association with cirrhosis of the liver and its presence may be suspected when there is a marked deterioration of standard liver function tests.

Patients with a primary hepatocellular carcinoma may also demonstrate a number of other biochemical abnormalities, such as hypoglycaemia, hypercalcaemia and hyperlipidaemia, but the pathogenesis of these changes is poorly understood.

Tests Used for Research[3]

Galactose Elimination Capacity[18]

Galactose is a naturally occurring monosaccharide. It is a product of the hydrolysis of ingested lactose, a reaction catalysed in the small intestinal mucosa by the enzyme lactase. Following absorption galactose is removed from blood only by the liver and the kidneys. After entering hepatocytes it is rapidly phosphorylated by galactokinase to galactose-1-phosphate; this prevents the accumulation of free galactose in liver cells and its return to the circulation. The initial phosphorylation of galactose is the rate-limiting step in galactose metabolism (except in galactosaemia in which there is a failure of conversion of galactose-1-phosphate to UDP-galactose). When the plasma galactose level is fairly high the galactokinase reaction is saturated and there is a constant rate of removal of galactose by the liver.

Tygstrup[18] has described a relatively simple test to measure hepatic galactose uptake which involves the intravenous injection of a single dose of 0.5 g galactose per kg body weight over a period of about 5 minutes. Plasma galactose levels are then measured at approximately 5 minute intervals for 60 minutes following the start of the injection. Urinary galactose elimination is also measured. From 20–40 minutes after the onset of injection the disappearance of galactose from the blood is linear with time and if one makes several assumptions and takes urinary excretion into account it is possible to calculate the rate of removal of galactose by the liver. The galactose elimination capacity in normal subjects is approximately 270 ± 40 (S.D.) mg/min per m^2 of body surface area or 6.7 ± 1.0 mg/min per kg body mass.

The galactose elimination capacity correlates well with other indices of hepatocellular function, such as prothrombin time, plasma albumin concentration and antipyrine removal rate but has the advantage that it measures only one aspect of hepatocellular function. It is affected only by the number of individual hepatocytes or by their functional capacity. In the same subject the results of repeated estimations vary by about 10% while the coefficient of variation in the control population is about 16%. There is considerable overlap between normal subjects and patients with several kinds of liver disease and the test is most useful for repeated studies done on individual subjects.

Aminopyrine Removal and Breath Test

The liver plays a key role in drug metabolism and the removal and catabolism of drugs are hepatic functions which can be assessed in several ways. Plasma levels can be measured following a single injection, or the decay in plasma radioactivity estimated after injection of a labelled drug. The production of drug metabolites can be measured in blood or urine. One important reaction in the metabolism of several groups of drugs is N-demethylation. The methyl groups removed are oxidized and their carbon atoms appear in the breath as CO_2.

Labelled aminopyrine (^{14}C-Dimethyl aminoantipyrine) has been used in many studies of hepatic drug metabolism. The rate of disappearance of label from plasma can be measured fairly simply but the main attraction of this compound lies in its use in breath tests. ^{14}C-aminopyrine is given orally to non-ambulant subjects. At various intervals a standard amount of CO_2 (2 mmol) is collected from the breath by getting the subject to blow into scintillation vials containing a trap for CO_2; an indicator changes colour when the desired amount of CO_2 has been taken up. The specific activity of the breath CO_2 is measured and if one assumes constant endogenous production of CO_2 (at 9 mmol kg body wt^{-1} l^{-1}) the amount of labelled CO_2 produced by demethylation of aminopyrine can be calculated.

If the specific activity of CO_2 is plotted against time the disappearance curve over about 12 hours is roughly exponential and a decay constant, K_b, can be estimated. This seems to be the best index which can be obtained from breath analyses[2] and correlates well with the decay constant for disappearance from plasma (K_p), with BSP disappearance and with the galactose elimination capacity. Unfortunately it is rather tedious to sample the breath over many hours. Several groups of workers have modified the breath test to obtain results after a period of 2 hours. Some have taken the mean value for CO_2 specific activity during the first two hours after the ingestion of the drug.[8] Others have simply sampled the breath at 2 hours and compared the specific activity values obtained from normal subjects with those obtained from patients with various kinds of liver disease.[7] Values for patients with hepatocellular dysfunction are generally lower than those found in normal subjects but normal values may be found in patients with biliary obstruction. The aminopyrine 2-hour breath test is not a particularly useful diagnostic test but it does allow several measurements of one aspect of hepatic function to be made in the same patient, the coefficient of variation of duplicate tests being about 6%. There is, of course, some reluctance to use radioactive material with a long half-life for repeated studies in man, but the radiation dosage of the test is small. Theoretically the test could be performed using the stable isotope ^{13}C but this would require mass spectrometry for measurement of the isotope in breath.

Hepatic Removal of Bromosulphthalein, Dibromosulphthalein and Indocyanine Green[3]

Bromosulphthalein (BSP) has been widely used to measure various aspects of hepatic function. It is removed from the blood by hepatic parenchymal cells, bound by intracellular proteins such as ligandin and Z protein, and excreted in bile either unconjugated (approximately 30%) or following conjugation with glutathione. It is easy to measure in blood and bile. In the United States a simple test of BSP excretion is widely used to detect hepatobiliary disease in non-jaundiced subjects. 5 mg of BSP per kg body weight is injected intravenously over approximately 30 seconds. A single blood specimen is taken 45 minutes later and the plasma concentration of BSP measured spectrophotometrically. It is assumed that the dye was distributed in a plasma volume of 50 ml/kg body mass and that the initial BSP concentration was therefore 10·0 mg/dl. The retention of BSP at 45 minutes is easily calculated. In normal subjects the proportion of BSP remaining in plasma at 45 minutes is up to 7% of the original dose. Higher values are found with many types of liver disease, obesity, etc.

The conventional BSP removal test has never been popular in Britain, partly because it provides very limited information and partly because it is a potentially dangerous test. Serious damage is done if the BSP extravasates at the site of injection; anaphylaxis has been reported on numerous occasions, and if the BSP is not properly prepared for injection there may be neurological problems associated with the injection of microcrystals.

More complex studies of BSP removal have been proposed by those who believe that they provide details about several different aspects of the liver's handling of BSP. Following a single intravenous injection of BSP the disappearance curve of the dye from plasma can be represented by two exponentials. If it is assumed that one is dealing with a two compartment model (i.e. plasma and liver), and that loss of BSP occurs only in the urine and in bile, then it is possible to calculate the size of the plasma compartment, hepatic storage, the transfer of dye between the compartments, urinary excretion and the rate of biliary excretion which has a theoretical maximum level, Tm. Unfortunately such studies do not take into account the conjugation of BSP and a number of other problems affect interpretation. These have been fully considered by Carson and Jones[3] who also discuss other methods of interpreting the data. Although there are problems, repeated measurements of BSP removal by this method do offer a useful method of evaluating changes in a fairly clearly defined function of the liver.

Attempts have also been made to estimate the Tm and hepatic storage capacity (S) for BSP from the plasma levels found during constant infusion of the dye at different rates. Unfortunately the method is based on some false assumptions. It can easily be shown that the resulting values for S must be wrong; they appear to depend not only on the liver's capacity to concentrate BSP, but also on blood flow.[12]

Dibromosulphthalein (DBSP) is a compound handled by the liver in the same way as BSP. It has the important advantage that it is not conjugated with glutathione which simplifies the assumptions involved in analysis of plasma disappearance.

Indocyanine green (ICG) is also rapidly removed by the liver and excreted into the bile without conjugation. It is much safer than BSP (though preparations may be contaminated by small amounts of sodium iodide), is simple to measure, and can be used instead of BSP to study hepatic uptake, storage and transport of an exogenous dye. ICG retention at 20 minutes can be measured and plasma disappearance curves can be analysed in the same way as BSP disappearance curves. The major disadvantages of using ICG are its high cost relative to BSP and its instability in plasma which means that plasma levels have to be measured soon after withdrawal of the sample.

Both BSP and ICG have been used to measure hepatic blood flow on the basis of the Fick principle.[16] If plasma levels of either BSP or ICG are held constant during a continuous infusion of the dye then hepatic blood flow can be calculated by dividing the rate of infusion of dye by the difference in concentration between arterial and hepatic venous plasma. The blood flow through the liver is clearly an important functional characteristic and it is unfortunate that its measurement involves hepatic venous catheterization.

Dynamic Studies of Bilirubin Metabolism[3]

Following the injection of radioactive unconjugated bilirubin the label is rapidly removed from the plasma and its disappearance curve can be represented as the product of three exponentials. A number of estimates can be made directly from the analysis of this curve including the mass of unconjugated bilirubin in the initial pool, the initial distribution volume of the injected material, the quantity of unlabelled bilirubin which enters the plasma per unit of time, the hepatic clearance of bilirubin and the plasma bilirubin turnover rate. If it is assumed that the bilirubin enters three compartments then it is also possible to derive various parameters which indicate the rate of transfer between these compartments and the size of the compartments, but the calculations depend on the nature of the assumptions made in constructing the model. The analyses which have been made using this approach have been of great interest in the interpretation of disorders of the metabolism of unconjugated bilirubin such as Gilbert's syndrome and the Crigler-Najjar syndrome but the method, which is both expensive and time consuming, does not allow interpretation of the findings in patients with conjugated hyperbilirubinaemia. It is therefore of relatively little value in the study of patients with most types of liver disease.

Dynamic Studies of Bile Acid Metabolism

Attempts have been made to measure the fractional disappearance rate of bile acids from plasma following the injection of radio-labelled bile acids. These disappear very rapidly (within a matter of minutes) from the plasma and the interpretation of the data is complicated as it seems likely that there is considerable removal of the label before there is adequate mixing. While it may be possible to show differences in removal rate with impairment of hepatic function there are problems in deciding the physiological meaning of the changes.

It is possible to follow the fate of a labelled bile acid through the various compartments of the total bile acid pool. Mathematical interpretation of the results is, of necessity, extremely complicated because of the enterohepatic recirculation, because of the conversion of primary to secondary bile acids, and because of the occurrence of both conjugation and deconjugation with taurine and glycine. At present this method is of relatively little value for repeated study of patients with liver disease.

Acknowledgements

I am grateful to Dr. Howard Thomas and to Dr. Alan Bailey of the BUPA Medical Centre for many helpful discussions during the preparation of this chapter.

REFERENCES

1. Billing, B. H. (1978), 'Twenty-five years of progress in bilirubin metabolism (1952–77)', *Gut*, **19**, 481–491.
2. Bircher, J., Kupfer, A., Gikalov, I., Preisig, R. (1976), 'Aminopyrine demethylation measured by breath analysis in cirrhosis,' *Clin. Pharm. Therap.*, **20**, 484–492.
3. Carson, E. R. and Jones, E. A. (1979), 'Use of kinetic analysis and mathematical modelling in the study of metabolic pathways *in vivo*,' *N. Engl. J. Med.*, **300**, 1016–1027, 1078–1086.
4. Cohen, J. A. and Kaplan, M. A. (1979), 'The SGOT/SGPT ratio—an indicator of alcoholic liver disease,' *Dig. Dis. Sci.*, **24**, 835–838.
5. Day, R. C., Harry, D. S. and McIntyre, N. (1979), 'Plasma lipoproteins and the liver'. Chapter 4 in *Liver and Biliary Disease* (Eds) Wright R., Alberti, K. G. G. M., Karran, S., Millward-Sadler, G. H. London: W. B. Saunders, pp. 63–82.
6. Galen, R. S. and Gambino, S. R. (1975). *Beyond Normality—the Predictive Value and Efficiency of Medical Diagnosis* New York: John Wiley & Sons.
7. Galizzi, J., Long, R. G., Billing, B. H., Sherlock, S. (1978), 'Assessment of the (14C) aminopyrine breath test in liver disease,' *Gut*, **19**, 40–45.
8. Hepner, G. W. and Vesell, E. S. (1975), 'Quantitative assessment of hepatic function by breath analysis after administration of (14C) aminopyrine,' *Ann. Int. Med.*, **83**, 632–638.
9. Heaton, K. W. (1979). 'Bile Salts' Chapter 12 in *Liver and Biliary Disease* (Eds) Wright, R., Alberti, K. G. G. M., Karran, S. and Millward-Sadler, G. H. London: W. B. Saunders, pp. 233–254.
10. Matloff, D. S., Selinger, M. J. and Kaplan, M. M. (1980), 'Hepatic transaminase activity in alcoholic liver disease,' *Gastroenterology*, **78**, 1389–1392.
11. Murphy, E. A. (1976). *The Logic of Medicine*. Baltimore: Johns Hopkins Press.
12. McIntyre, N., Mulligan, R. and Carson, E. R. (1973). BSP Tm and S: 'A critical re-evaluation,' In: *The Liver: Quantitative Aspects of Structure and Function* (Eds) Paumgartner, G., Preisig, R. Karger Basel. pp. 417–427.
13. McIntyre, N. and Heathcote, J. (1974), 'The laboratory in the diagnosis and management of viral hepatitis,' *Clin. Gastroent.*, **3**, 317–336.
14. Rosalki, S. R. (1975), 'Gamma-glutamyl transpeptidase,' *Adv. Clin. Chem.*, **17**, 53–107.
15. Rosalki, S. R. (1976), 'Enzyme tests in diseases of the liver and biliary tract.' Chapter 10 in *Principles and Practice of Diagnostic Enzymology* (Ed) Wilkinson, S. H. London: Edward Arnold.
16. Sherlock, S. (1975). *Diseases of the Liver and Biliary System*. Oxford: Blackwell.
17. Tavill, A. S. (1979), 'Protein metabolism and the liver.' Chapter 5 in *Liver and Biliary Disease* (Eds) Wright, R., Alberti, K. G. G. M., Karran, S. and Millward-Sadler, G. H. London: W. B. Saunders, pp. 83–107.
18. Tygstrup, N. (1964), 'The galactose elimination capacity in control subjects and in patients with cirrhosis of the liver,' *Acta. Med. Scand.*, **175**, 281–289.
19. Whitby, L. G. (1974), 'Screening for disease: Definitions and Criteria,' *Lancet*, **ii**, 819–822.
20. Yerushalmy, J. (1947), 'Statistical problems in assessing methods of medical diagnosis, with special reference to x-ray techniques,' *Pub. Health. Rep.*, **62**, 1432–1449.

FURTHER READING

Feinstein, A. R. (1977). *Clinical Biostatistics*. Saint Louis: C. V. Mosby. (An excellent account of the traps into which amateur and professional statisticians are likely to fall when evaluating clinical and laboratory information).

Jones, E. A. and Berk, P. D. (1977). 'Liver Function.' Chapter 10 in *Chemical Diagnosis of Disease* ed. Brown, S. S., Mitchell, F. L. and Young, D. S. North Holland: Elsevier. (A long and detailed chapter on almost all aspects of the use of the laboratory in the management of liver disease).

Sherlock, S. (1975). *Diseases of the Liver and Biliary System*. Oxford: Blackwell (A classic).

Thomas, H. C. and Jewell, D. P. (1979). *Clinical Gastrointestinal Immunology*. Oxford: Blackwell. (An excellent account of the use of immunological tests in hepatology).

Wright, R., Alberti, K. G. G. M., Karran, S., Millward-Sadler, G. H. (1979). (Eds) *Liver and Biliary Disease—Pathophysiology, Diagnosis, Management*. London: W. B. Saunders. (A recent book with much useful information on laboratory investigations scattered throughout).

8. THE DIFFERENTIAL DIAGNOSIS OF ACUTE ABDOMINAL PAIN

HAROLD ELLIS

Introduction
Steps in diagnosis
Acute appendicitis
Acute intestinal obstruction
Perforated peptic ulcer
Acute pancreatitis
Medical conditions which may simulate the acute abdomen

Introduction

The patient with acute abdominal pain remains one of the last bastions of Clinical Medicine. There are three reasons for this statement. First, in no other common situation is reliance on clinical features, immediate decision and accurate diagnosis of such paramount importance. Second, it is the clinical diagnosis that counts, since ancillary investigations, both from the laboratory and the X-ray Department, are rarely diagnostic, often misleading and can never be used for more than confirmation of the clinical diagnosis. Third, in no other condition commonly encountered is there such a wide differential diagnosis which, indeed, covers the whole of medical practice.

In most other fields, an initially tentative or even incorrect clinical diagnosis is not necessarily harmful; we can wait until it is confirmed or refuted by laboratory and radiological investigations and even await the progression of the natural history of the disease. Provided we take sensible and appropriate measures, a misdiagnosis at our first clinical consultation is readily corrected. For example, a woman is seen by her General Practitioner because she has a lump in the breast. His initial diagnosis is that the lump is malignant; he takes the wise appropriate step and arranges an urgent consultation with his surgical colleague. The surgeon agrees that this tentative diagnosis may indeed be true and makes urgent provision to admit the patient to hospital for excision and histological examination of the lump. When the lump is put under the microscope it proves to be completely benign. The diagnostic process may be reversed; that is to say, an initial diagnosis of a benign breast lump is replaced, after histological examination, with the diagnosis of malignant disease of the breast. Once again, as long as appropriate sensible steps have been taken, no harm has come to the patient. But in acute abdominal conditions, delays of even a few hours in initiating the correct line of treatment may make the difference between a smooth or stormy course; indeed, may even place the patient's life in danger. For example, the General Practitioner who labels a child with acute appendicitis as a case of gastroenteritis and prescribes a kaolin mixture and a sedative overnight may be faced, the next morning, with a desperately ill patient suffering from a generalized peritonitis. The Casualty Officer who sends an abdominal injury home labelled 'slight bruising of the abdominal wall' may be confronted, a few hours later, with a moribund example of haemoperitoneum due to a ruptured spleen. Sir Heneage Olgilvie, of Guy's Hospital, summed this up perfectly years ago when he wrote that 'in the acute abdominal emergencies the difference between the best and the worst surgery is infinitely less than that between early and late surgery, and the greatest sacrifice of all is the sacrifice of time'. Put into less elegant but more practical English, this can be paraphrased 'it is better for you to have your inflamed appendix removed a few hours after the onset of your illness by the House Surgeon than to have your gangrenous perforated appendix removed a couple of days later by the Professor of Surgery'.

In dealing with the patient with acute abdominal pain, the clinician must steel himself to realize that he must rely almost entirely on clinical features rather than laboratory and radiological investigations. These days he has become so used to being able to skimp on history and examination that it comes as a severe shock to realize that, in the middle of the night, in the patient's home or in the Casualty Department, he is going to be forced to rely on his own five senses. It is a very good aphorism that, in the diagnosis of the acute abdomen, the special investigations can only be used to reinforce a clinical diagnosis; seldom if ever can they refute it. We shall be developing this theme as we consider specific examples of acute abdominal pain but a few examples may now be used to underline this statement. Thus, if the clinical diagnosis is one of acute appendicitis it is common practice to ask for a white blood count to be performed. Indeed, in the United States of America, the textbooks list a leucocytosis as one of the cardinal features of this condition. Certainly, one is used to seeing a raised white count in such cases but every clinician will have also encountered many examples of classical acute appendicitis, confirmed at operation, where the white cell count was within normal limits. It is true that many cases of perforated peptic ulcer are associated, radiologically, with the presence of free gas under the diaphragm on a plain X-ray of the abdomen in the erect position. Yet in about a quarter of the cases X-rays fail to reveal free gas, particularly when the perforation has sealed or when fluid rather than gas has principally escaped into the peritoneal cavity through the perforation. It is true that most cases of intestinal obstruction are associated with distended loops of intestine and with the presence of multiple fluid levels on X-ray of the abdomen. However, once again, it is well known that, in some 10% of cases, intestinal obstruction, indeed strangulation of the bowel, may be associated with perfectly normal radiographic appearances. This occurs when a closed loop of intestine containing fluid and free from gas becomes obstructed. Failure to observe our cardinal principle of relying on clinical features and being prepared to discount special investigations may put the patient into the gravest peril.

Acute abdominal pain is not only a vitally important diagnostic problem but it is also often a difficult one, and therefore of particular fascination to the clinician. It is no exaggeration to say that abdominal pain may result from disease of almost any organ in the body. Apart from the abdominal and the retroperitoneal viscera themselves, and these include, of course, the pelvic organs, one must also consider the chest, the central nervous system, and even the ears and the throat. Basal pneumonia, acute cardiac ischaemia and spinal root irritation can all mimic the acute abdomen. We have recently transferred a patient admitted to the Coronary Unit as a case of massive myocardial infarction to the operating theatre to deal with his ruptured aortic aneurysm; many a patient with prodromal herpes zoster has been labelled 'biliary colic' or 'pain of renal origin' before the typical rash appears, and children with otitis media are notorious for complaining of belly-ache. Metabolic disorders, particularly diabetes, but also acute porphyria, may give acute abdominal symptoms. Finally, those extraordinary psychopaths suffering from the Munchausen syndrome provide us with teasing diagnostic difficulties and manage to simulate a wide variety of acute abdominal catastrophes to perfection.

It is a good discipline, therefore, when approaching any patient with abdominal pain to consider its anatomical source; is it intra-abdominal?, is it arising from a retroperitoneal structure (pancreas, kidney, aorta)? or is it of extra-abdominal origin, neurological or functional?

Steps in Diagnosis

The essense of clinical diagnosis is, of course, a careful history and full examination; neither is easy when the patient has acute abdominal pain.[2]

The History

Great skill and a fine economy must be employed in taking a history from a patient with acute abdominal symptoms. There is nothing to gain and much to lose by exhausting and exasperating an ill patient with irrelevant questions. However, such valuable gems of information as a missed menstrual period which suggests ectopic pregnancy, or shoulder-tip pain of diaphragmatic irritation in a perforated peptic ulcer or in haemoperitoneum following abdominal trauma, may only be revealed by direct questioning. We need to know about the speed of onset, the severity, the exact duration of the pain and its accompaniments. Most serious intra-abdominal conditions stop the patient from sleeping and make it impossible for him to continue at his work. Of course there are always exceptions to any rule; I shall never forget as a House Surgeon being called to the Admission Ward to see an obese lady admitted with acute abdominal pain. On arrival, I found that she was in a deep sleep and had a mind to going back to bed myself. However, I roused her, elicited rigidity and exquisite tenderness in the right iliac fossa and assisted at the removal of her gangrenous and perforated appendix. Perforation, torsion and stone impaction all have a sudden acute onset and indeed the patient can describe the exact moment at which his agony commenced. Inflammatory conditions and intestinal obstruction, in contrast, tend to come on more insidiously.

Sometimes it is difficult to be sure whether the patient's symptoms represent some quite minor abdominal upset or warrant more serious consideration. This is a problem that particularly confronts the General Practitioner when called to the patient's home. Here I have always found Sir Zachary Cope's aphorism most helpful; that a previously healthy patient who complains of severe abdominal pain for more than about six hours duration is *usually* suffering from an acute surgical condition. The description of pain given by the patient is rarely diagnostic. Since it is usually a unique experience for the patient, he can only describe it as 'severe' or 'agony' or 'not too bad'. A very florid clinical description—'like a piece of barbed wire heated to a dull red heat and slowly tightened up around my abdomen'—is very likely to be functional in origin. It is the accompaniments of the pain which are far more valuable than any description by the patient of the pain itself. A patient with renal or biliary colic, or with an intestinal obstruction, can usually hardly hold still with the pain and may develop all sorts of strange contortions in an effort to find a position of relief. In contrast, the patient with peritonitis finds the slightest movement agonizing and usually simply wants to be left alone with his knees drawn up, clutching his abdomen and keeping as still as possible. Jaundice or haematuria during the present or a previous attack of pain are important pointers to gallstones or to renal calculi respectively. While on this topic it is worth remembering the important clinical observation by French and Robb,[4] who found that, unlike the usual textbook description, biliary and renal colics are only very rarely intermittent in nature. Usually the pain in these conditions is constant and continuous or else continuous with exacerbations. In contrast, intestinal colic does indeed come on with waves of intermittent pain.

One last point in the history; most patients are more than anxious to tell the doctor about their bowel actions but they will usually only discuss passage of flatus on direct questioning. In spite of the layman's views to the contrary, a few days of constipation are of no great relevance. However, 24 hours without the passage of wind denotes either complete intestinal obstruction or paralytic ileus.

Clinical Examination

The clinical examination requires particular skill. The sick, anxious patient is apprehensive that the doctor is going to produce still more pain and yet it is vital to get the patient to co-operate and to relax. The problem is particularly difficult, of course, with a small child where, in addition, there is the presence of tense anxious parents, hoping for reassurance and yet fearing the worst.

The common mistake in an emergency situation is to rush ahead and palpate the abdomen at once. Time spent in careful observation of the patient as a whole and of the abdomen in particular is time well spent indeed. It is a good plan to start by taking the temperature and pulse rate; useful in themselves, they also give the clinician useful time for general observation. One should be suspicious of the diagnosis of acute appendicitis without pyrexia; it does occur, but it is unusual. In contrast, a pyrexia is rare with a perforated peptic ulcer. Atrial fibrillation in a patient with acute abdominal pain with the features of ileus will suggest a mesenteric embolism. A coated tongue is certainly a common finding in peritonitis and acute appendicitis but it is not an invariable finding; indeed it is a very good rule that there is no such thing as 'always' or 'never' in the acute abdomen. Look at the patient's posture. The restlessness of colic is in sharp contrast to the fear of movement which is so typical of peritoneal irritation. The girl sent into hospital as a case of acute appendicitis who is pale and fainting is more likely to have a ruptured ectopic pregnancy with haemoperitoneum.

Attention is now turned to the abdomen itself. It should be carefully inspected for distension (either local or generalized), for visible peristalsis or for the presence of previous laparotomy scars. A previous abdominal operation, carried out no matter how many years previously, always raises the suspicion of intra-abdominal adhesions or bands. It is a good rule that this should be the first diagnosis in a patient with intestinal obstruction who has a previous scar on the abdominal wall. Although the anatomical boundaries of the abdomen are the xiphoid above to the pubis below, the wise clinician will extend these borders to the nipples above and the knees below and he will ensure that all the territory between is laid bare for his inspection. Perhaps the commonest sign to be missed is a small strangulated inguinal, or particularly, femoral hernia, particularly in an obese female patient. It may seem unbelievable that a patient with intestinal obstruction does not know that he has an irreducible hernia but this is certainly the case. Indeed, I have known of a distinguished doctor who correctly diagnosed himself as having intestinal obstruction but who realized he had a strangulated inguinal hernia only when his surgical colleague pointed this out to him.

An absolute rule, therefore, is always to inspect and palpate the hernial orifices in every patient with acute abdominal pain. At the same time take this opportunity to feel the femoral pulses; their absence is a good clue to the rare dissecting aortic aneurysm.

Palpation is a delicate art which requires constant practice to perfect. The heavy-handed tyro may think that the abdomen is rigid, yet the more experienced clinician will find the muscles to be relaxed under his gentler touch. There is all the difference between voluntary and protective guarding against rough palpation and the involuntary and reflex rigidity of muscles due to underlying peritoneal irritation or inflammation which can be detected with the lightest touch. An abdomen that will relax completely and will allow the posterior abdominal wall to be massaged under the examiner's hand rarely harbours any acute abdominal disorder. However, constant localized tenderness and muscle spasm and the presence of release tenderness are signs which can be trusted and which almost invariably indicate serious mischief beneath. The same applies to tenderness in the pelvic peritoneal pouch on rectal or vaginal examination, which naturally must never be omitted. Digital examination of the pelvis may also reveal such important features as faecal impaction or the presence of a tender pelvic mass. Finally, auscultation may reveal the typical noisy bowel sounds of mechanical intestinal obstruction, the transmitted heart sounds of free fluid in the peritoneal cavity or the ominous silence of peritonitis.

Particular skill is required in the examination of the abdomen in children. Once the patient is hurt or frightened, he cries, the abdominal muscles are voluntarily contracted into iron bands and any opportunity of further examination is lost. Utmost gentleness is essential. The young child is best held in his mother's arms and the examiner's warmed hand then insinuated beneath the blanket and under the cover of soothing words. If all this fails, the young patient must be examined again after suitable sedation.

Having finished clinical examination, a specimen of urine is obtained and examined, especially for blood, pus, bile and sugar.

Differential Diagnosis

It may not, and indeed often cannot, be possible to make an exact pathological diagnosis in a case of acute abdominal pain but at least the clinician should attempt to diagnose the type of disease process which is going on. He must consider whether the lesion is situated within the abdomen (including the retroperitoneum and pelvis) or whether possibly the pain is referred from the chest or is of neurological or functional origin. It is useful to consider the following check-list of the causes of acute intra-abdominal pain:-

1. Inflammation (for example, appendicitis, cholecystitis or sigmoid diverticulitis).

2. Obstruction of a hollow viscus—the intestine (which may or may not be strangulated), the bile duct or the ureter.

3. Perforation of a hollow viscus.

4. Haemoperitoneum (for example, ruptured spleen or ruptured ectopic pregnancy).

5. Torsion of a viscus.

6. A vascular catastrophy (mesenteric occlusion or dissecting aneurysm).

An especial difficulty in late cases is that one pathological process may by now have merged into another. In particular, an inflamed or strangulated viscus may by now have ruptured so that the immediate features are those of acute peritonitis. Here reliance must be placed quite heavily on the history of the events leading up to the patient's present clinical condition. A rigid abdomen with all the features peritonitis within a few hours of onset suggests a primary perforation, of which the commonest cause is a perforated peptic ulcer. However, generalized rigidity in a patient seen at 48 hours who gives a typical history of central abdominal pain which moves to the right iliac fossa and which is now generalized suggests that a gangrenous appendix has by now ruptured.

Having determined the type of abdominal emergency, we then usually have to rely on the laws of probability to guess at the underlying disease. For example, there is nothing at all on history or examination to distinguish between an acutely inflamed appendix and an acute Meckel's diverticulitis, and certainly no laboratory investigation will distinguish between the two. The only difference is that the first is extremely common and the second rather rare. Of course, occasionally good luck or experience will allow the clinician the satisfaction of correctly diagnosing the uncommon, often because some small extra fact in the history or an additional physical sign has been elicited. Thus, pigment spots on the lips and buccal mucosa might suggest that a young patient with intestinal obstruction may have an intussusception of small intestine produced by a Peutz-Jegher polyp.

Special Investigations—an Overview

What help can we obtain from the laboratory and from the radiological department in the diagnosis and the differential diagnosis of acute abdominal pain? Unfortunately, although occasionally invaluable, these special investigations often yield little, if any, more information than that obtained by careful history and examination and, indeed, every now and then they may be frankly misleading. Both the clinician and his colleagues in the ancillary departments should have the following aphorism written in capital letters of gold in the ward and the laboratory:- 'If the special investigations fit in with the diagnosis made after careful clinical study, then they provide useful confirmatory evidence of the diagnosis. If, however, there is total disagreement between a laboratory or radiological test and the carefully elicited clinical features in an acute abdominal emergency, then it is wise to side with the clinical diagnosis rather than that obtained on laboratory findings'. Let us take a very common clinical example. A patient is admitted with all the clinical features of acute appendicitis; he has a typical history, a raised temperature and the signs of local peritoneal inflammation. If a white cell count is now obtained and is found to be raised with a marked polymorph leucocytosis, then this is an interesting finding and reinforces the diagnosis. If, however, the white cell count is normal, the doctor should certainly not budge from his clinical diagnosis of acute appendicitis which can occur, of course, in the presence of an entirely normal white cell count. Another example is a raised serum amylase which is extremely good supporting evidence to the clinical diagnosis of acute pancreatitis, but an elevation of the amylase may occur in perforated peptic ulcer and after morphine; moreover an overwhelming pancreatitis may be associated with a perfectly normal amylase level which is often normal, in addition, in acute pancreatitis following abdominal trauma.

The electrocardiogram and the estimation of the serum transaminases are certainly helpful in making the differential diagnosis between the pain of myocardial ischaemia and an acute upper abdominal crisis such as cholecystitis. However, one must bear in mind that a patient with a recent ischaemic episode may have thrown off an embolism and now be suffering from a mesenteric arterial embolus.

Urine examination, we have already noted, hardly falls into the ambit of a 'special investigation' and is perhaps better regarded merely as an extension of the clinical evaluation of the patient presenting as an acute abdominal emergency. It should certainly never be omitted but, once again, needs cautious interpretation. Pyuria at microscopic level may be seen in patients with acute appendicitis, particularly in a low lying pelvic appendix which is abutting against the bladder or in a retrocolic inflamed appendix lying against the ureter or even on the renal pelvis. The differential diagnosis between appendicitis and right-sided pyelitis under such circumstances requires very careful clinical judgement. The presence of sugar in the urine is, of course, a vital finding and will need to be followed up by blood sugar estimation. Hyperglycaemia may well mimic an acute abdominal episode but equally well one must remember that the diabetic patient is not immune from any of the common abdominal emergencies.

X-ray films of the chest and abdomen, the latter taken with the patient in the erect and lying position, may yield considerable information. The presence of free gas beneath the diaphragm is pathognomonic of perforation of a hollow viscus, except in the postoperative period following laparotomy when the air which has been admitted into the opened abdomen remains visible for some days. However, once again, one must not rely too heavily on this radiological sign; in perhaps 25% of perforations, especially those which have sealed rapidly or where fluid rather than gas has escaped into the peritoneum, the X-rays appear perfectly normal. Intestinal obstruction on X-ray examination of the abdomen usually demonstrates one or more distended loops of bowel with fluid levels seen on the erect films. In theory it would appear to be easy enough to distinguish these findings from the generalized gaseous distension of the intestine in peritonitis or in paralytic ileus (and the latter often accompanies ureteric colic); in practice, however, the two are confused with surprising ease. Furthermore, if the obstructed loop of bowel happens to be distended entirely with fluid, the characteristic air distension is absent and the involved segment of intestine, which may even be entirely gangrenous, is radiologically invisible. Fortunately renal and ureteric stones are usually radio-opaque and can often be identified on plain abdominal

X-rays. Yet about 10% of these are radiolucent and in other cases they may be confused with calcified phleboliths or faecal debris. Unfortunately only about 10% of gallstones are radio-opaque and, again, they may be confused with other opaque intra-abdominal shadows. X-rays are certainly helpful in acute abdominal emergencies but the quality of films which can be obtained in patients who are seriously ill and restless often leave much to be desired, particularly when taken in the middle of the night or at the week-end by junior radiographers. One must really weigh up the value likely to accrue from these investigations against the delay they cause in initiating treatment and the disturbance necessary to an already exhausted patient.

Having now considered the special investigations in general terms, we will now turn to specific entities presenting as acute abdominal pain and consider the laboratory tests in these instances.

Acute Appendicitis

Acute appendicitis is the commonest cause of the acute abdomen in this country. The exact incidence is not known because the disease is not notifiable, but it is estimated that about 125 000 patients are treated yearly in the United Kingdom for appendicitis. Several hundred deaths are recorded yearly from this cause.

The vast majority of patients present with localized pain and tenderness in the right iliac fossa. Nausea and vomiting usually follow the onset of the attack and anorexia is almost invariable. Typically the pain starts centrally and shifts to the right iliac fossa after about six hours; more accurately, the pain moves to the site of the inflamed appendix, so that it may be experienced in the supra-pubic region if the appendix lies in the pelvic position, or in the right loin if the appendix is tucked away behind the ascending colon. Usually the pain is aggravated by movement. When the appendix perforates, the pain may temporarily remit, only to return in a more severe and generalized form with profuse vomiting as general peritonitis develops.

On examination the patient is flushed and in pain; usually the temperature and pulse are elevated. It is uncommon for the tongue to be clean and for there not to be foetor oris. In the region of the inflamed appendix there is usually guarding with localized tenderness on palpation and on sudden release of pressure of the examiner's hand. On rectal examination tenderness is only found when the appendix is in the pelvic position or when there is pus in the pelvic cul-de-sac. In late cases with generalized peritonitis the abdomen is diffusely tender and rigid, the bowel sounds are absent and the patient is obviously very ill. Later still the abdomen is distended and the patient shows the features of advanced peritonitis. If, however, the perforated appendix becomes walled off by surrounding structures into an appendix mass, the examiner can palpate the tender swelling in the right iliac fossa but the rest of the abdomen is soft with no evidence of a generalized peritonitis.

Nothing can be as easy, nor as difficult, as the diagnosis of acute appendicitis. The straightforward case is obvious but difficulties are especially likely to be encountered in young children, the elderly, the pregnant, those with an appendix in an unusual site, those who are poor historians and the obese. Even today many children do not reach hospital until the appendix has perforated and gangrene and perforation occur five times more commonly in the elderly than in the young adult with appendicitis.

Appendicitis in children is particularly worrying. Because in the infant the appendix has a relatively wide lumen, appendicitis is rare below the age of 12 months but after this the incidence increases quite rapidly. Both the mortality and the morbidity of appendicitis are higher in preschool children than in those over the age of five and this undoubtedly reflects delays in diagnosis. The picture is often not the classical one seen in older age groups. There may not be the early central abdominal colic and the shift of pain to the right iliac fossa, the child frequently only complains of generalized abdominal pain. Vomiting is usual but is by no means inevitable. There may be diarrhoea, constipation, or the bowels may remain perfectly normal. Extreme patience is required when it comes to examining the abdomen of the ill child. If there is localized tenderness and muscle guarding in the right iliac fossa in a previously healthy child, then the chances are very strong indeed that the appendix is acutely inflamed.

Differential Diagnosis of Acute Appendicitis

Other intra-abdominal conditions which commonly mimic appendicitis are perforated peptic ulcer, acute cholecystitis, acute intestinal obstruction (especially in the elderly), acute diverticulitis of the colon, acute regional ileitis and, in children, acute Meckel's diverticulitis and mesenteric adenitis. Renal colic and acute pyelonephritis may also be confused with acute appendicitis and in women one includes the common pitfalls of acute salpingitis, ruptured ectopic pregnancy and a ruptured cyst of the corpus luteum.

Other conditions outside the abdomen may mimic acute appendicitis. It is well known that pneumonia may produce acute abdominal pain in children. Jona and Belin[7] found that 12 of 250 children who presented with acute abdominal pain had a basal pneumonia as the only cause. It is interesting that of these 12 children, eight had only mild respiratory symptoms, four had none at all and only two had abnormal physical findings when the chest was examined. However, the abdominal pain was severe, sustained and was associated with abdominal tenderness and even, on occasion, with absent bowel sounds. These authors point out that a chest X-ray is valuable and should include a good lateral film because consolidation may be hidden by the diaphragm in the usual postero-anterior film. However, any experienced surgeon will remember a case of acute appendicitis occurring in a child who has a respiratory tract infection or even pneumonia and I can still remember one such case that I dealt with as a Resident Surgical Officer in Sheffield a quarter of a century ago. Other conditions include diabetes mellitus, infectious hepatitis, gastro-enteritis and sickle cell crisis (which should always be thought of in black children).

An important paper by Valerio[12] describes three children with acute abdominal pain which was the presenting feature of their undiagnosed diabetes. He stresses three important clinical clues which are, first a history of polyuria, polydipsia and anorexia which precede the abdominal pain, second deep sighing rapid respirations and third, severe dehydration. The abdominal pain is usually generalized in contrast to the localized right iliac fossa pain and tenderness of acute appendicitis. It may be difficult to get a urine specimen to test for sugar because of the dehydration and an important step in confirming the suspected diagnosis is to arrange an immediate blood sugar estimation. The abdomen will become pain free and soft within a few hours of appropriate treatment for the diabetes but obviously very close observation is required during this crucial period.

Special Investigations

Acute appendicitis is essentially a clinical diagnosis and there is no laboratory or radiological test which is diagnostic of the condition.

White Blood Count. This is raised above 12 000 in about three-quarters of patients and is only slightly raised or normal in the remainder. Smith[10] in a study of 100 consecutive cases of confirmed acute appendicitis, found that six had entirely normal white cell counts.

Urine Examination. This should of course be routine in every case of acute abdominal pain. Graham[6] quantitatively analysed midstream urine samples in 71 patients operated on with a diagnosis of appendicitis. Of these, 62 had acute appendicitis and the remaining nine had a normal appendix removed; three of these had mesenteric adenitis, the other six had no abnormality detected. In the whole group of patients microscopic pyuria was found in nine, all female, one of whom also had haematuria. One male with acute appendicitis had microscopic haematuria also. The distribution of microscopic pyuria, this author points out, was about that expected in the normal population. Significant haematuria should point to a urinary tract lesion, but, as indicated in this study, the presence of unequivocal clinical features should not deter the surgeon from proceeding to appendicectomy.

Radiography of the Abdomen. There are a number of radiological signs which have been described in plain X-rays of the abdomen in patients with acute appendicitis.[1]

These include:
1. Fluid levels localized to the caecum and to the terminal ileum, indicating localized inflammation in the right lower quadrant of the abdomen.
2. Localized ileus, with gas in the caecum, ascending colon or terminal ileum.
3. Increased soft tissue density in the right lower quadrant.
4. Blurring of the right flank stripe, the radiolucent line produced by fat between peritoneum and transversus abdominis.
5. A faecolith in the right iliac fossa (which may be confused with a ureteric calculus or a calcified mesenteric lymph node and which may also be found in a normal subject).
6. Blurring of the psoas shadow on the right side.
7. A gas-filled appendix.
8. Free intraperitoneal gas (extremely rare).
9. Deformity of caecal gas due to an adjacent inflammatory mass. This is difficult to interpret because there may be disturbance of caecal gas from intraluminal fluid or faeces.

In a review of 200 patients undergoing laparotomy for acute appendicitis, these authors[1] found that 54% of patients with acute non-perforated appendicitis had one or more of these signs positive and the incidence rose to 80% in patients with advanced appendicitis. Fifteen out of 41 patients who did not have acute appendicitis showed one or more of these X-ray appearances. Eight of these had another acute lesion in the right iliac fossa and three had no abnormality discovered at the time of operation.

Acute Intestinal Obstruction[3]

Acute intestinal obstruction is subdivided into simple and strangulated. In the first, there is purely obstruction to the passage of intestinal contents, but in the second there is, in addition, an obstruction to the blood supply of the involved segment of the gut which, left untreated, will inevitably lead to gangrene and perforation of the infarcted bowel. Stragulation may be produced by a band or adhesion, strangulated hernia, volvulus or intussusception.

Intestinal obstruction is characterized by a classical quartet of pain, vomiting, constipation and distension. They may all exist in a particular case or occur in any combination. The pain is a typical colic, cramp-like and intermittent, but may have been disguised by morphia. The vomit, which typically becomes dark brown and faeculent, may be late or even absent in colonic obstruction. Distension of the abdomen is more marked the lower the obstruction and may be imperceptible in a high small-bowel obstruction. If the occlusion is incomplete, flatus may still pass and in any case a loaded colon below an obstruction may still provide one or two bowel actions before constipation becomes total. Examination of the abdomen may reveal distension and the presence of visible peristalsis. Careful inspection should be made for the presence of an abdominal scar, which always suggests an underlying band or adhesion, and a specific search should be made for a strangulated external hernia. Palpation of the abdomen usually reveals tenderness and release tenderness and these, together with muscle guarding, tend to be more marked in the strangulated case. A mass may be detected on palpation which might be a carcinoma of the colon, diverticulitis of the sigmoid or an intussusception. Typically in an intestinal obstruction the rectum is ballooned and occasionally a low-lying obstructive tumour or an impacted mass of faeces may be found on rectal examination. Auscultation of the abdomen often reveals characteristic noisy sounds which accompany each wave of colicky peristalsis.

Special Investigations

There are no specific laboratory tests for intestinal obstruction. Leucocytosis is suggestive of strangulation but is not at all a reliable sign. The laboratory may, however, give valuable help in the important task of fluid replacement in these often dehydrated patients. Elevation of the haemoglobin and of the packed cell volume indicate haemoconcentration and act as guides to fluid replacement. Severe electrolyte depletion owing to loss of gastrointestinal fluid will be reflected by lowered serum sodium, potassium, chloride and bicarbonate concentrations with a raised blood urea.

Plain X-ray films of the abdomen are helpful in cases of intestinal obstruction but it must be stressed that the X-rays may be entirely negative in some 5 to 10% of cases. The physical basis of the radiological signs is that gas and fluid accumulate in the bowel above the obstruction, so that, with the patient in the erect position, the association of gas and fluid give rise to a series of fluid levels. When the patient is in the supine position, the X-rays show the amount and distribution of gas in the gut. The distended loops of small intestine generally lie transversely in a step-ladder fashion across the central abdomen. The distribution of gas may enable the surgeon to differentiate between an obstruction in the small or large bowel.

Two specific forms of intestinal obstruction may give quite typical X-ray appearances on the plain films. An obstruction due to a gall stone lodged in the small intestine may show air in the biliary system (as a result of a cholecyst-duodenal fistula) together with direct visualization of the stone and X-ray evidence of small bowel occlusion. Volvulus of the sigmoid colon usually demonstrates a tremendously distended sigmoid loop which may extend up to the diaphragm and may even fill the right side of the abdomen.

In some cases X-rays may be taken with radio-opaque material. A micropaque barium-sulphate suspension can be given as a barium meal and follow-through study. There is little danger of its impaction above a small bowel obstruction because the considerable fluid accumulation above the block rapidly dilutes the medium. A barium enema X-ray is particularly helpful in cases of colonic obstruction due to carcinoma or diverticular disease.

Perforated Peptic Ulcer

This is yet another example of a common acute abdominal emergency where clinical features are all-important and where laboratory investigations have virtually no part to play in diagnosis. The clinical presentation in the great majority of patients will present no serious difficulty in diagnosis. The sudden onset of the pain is very characteristic. It is agonizing, and involves the whole abdomen. Radiation of the pain to one or other shoulder (referred via the phrenic nerve) is highly characteristic, but is so often masked by the intense abdominal pain that it may not be revealed unless sought by direct questioning. About 70% of patients will give a history of previous symptoms suggestive of peptic ulcer and, indeed, there may have been an actual previous perforation.

The clinical findings are typical; the patient is cold and clammy, in obvious pain and does not wish to move. Respirations are rapid and shallow but, certainly in the early hours after perforation, the pulse is only slightly raised and the blood pressure normal. The rigid abdomen is quite typical—beautifully described by Moynihan as 'every part offers the most inflexible opposition to pressure, the rigidity is obdurate, persistent and unyielding'. Bowel sounds are usually absent and diminution of liver dullness can sometimes be demonstrated by percussion over the lower chest. However, this is not a sign to be relied upon.

Special Investigations

Haematological and other laboratory tests are of no value in the diagnosis of perforated peptic ulcer, although they may be needed to help in the differential diagnosis in some obscure cases.

The X-ray appearances, however, can be helpful although it is important to remember that the X-rays may be entirely negative in about a third of the patients.[5] The characteristic appearance is of free gas between the liver and the right side of the diaphragm. Gas beneath the left leaf of the diaphragm may be confused with gas within the stomach fundus but this may be resolved by inspecting a lateral decubitus film. A generalized 'ground glass' appearance suggests excessive free peritoneal fluid.

The differential diagnosis of the causes of free gas beneath the diaphragm includes perforation of any other part of the gut (for example, perforation of an acutely inflamed diverticulum of the colon), laparotomy within the previous couple of weeks, spontaneous rupture of a gas cyst of the intestine, subphrenic abscess, tubal insufflation within the previous 24 hours and introduction of gas at peritoneal dialysis.

Acute Pancreatitis

Acute pancreatitis is interesting in that its aetiology (or, more probably, aetiologies), remains the subject of debate. It is well recognized that there is an association with the presence of gallstones, acute alcoholism, and also a large number of rarer causes, although in many cases no precipitating factor can be identified.

Among the less common associations may be mentioned hyperparathyroidism,[9] mumps, pregnancy, trauma to the pancreas or surgery on the biliary or upper gastrointestinal tract, and a large number of drugs, particularly steroids (well reviewed by Nakashima and Howard[8]).

Trapnell and Duncan[11] in a 20 year review of 590 cases of acute pancreatitis in the Bristol area, found 54% were associated with biliary tract disease, and 4·4% with chronic alcoholism; however, no less than 34·4% were idiopathic. However, in areas where alcohol consumption is far greater, e.g. Glasgow, New York and Sweden, between a quarter and two-thirds of cases are associated with high alcohol consumption. The reasons for these associations remain conjectural but at least suggest that clinical pancreatitis is the end result of a number of quite different precipitating factors.

The importance of accurate diagnosis in acute pancreatitis is that the majority of these cases can be treated non-operatively. However, in a proportion of cases neither the clinical features nor the laboratory investigations make the diagnosis certain and, under such circumstances, it may be necessary to perform a diagnostic laparotomy.

Clinical Features

Although pancreatitis is occasionally seen in children and more often in young adults, it only becomes common after the age of 50. Onset is usually sudden and severe, situated in the upper abdomen and radiating, in some 65% of cases, into the thoraco-lumbar region of the spine. The great majority of patients also have repeated vomiting.

On examination, the severe case is pale and shocked. The local examination of the abdomen is characterized by rather less marked physical signs than would be expected by the severity of the pain. The pancreas, being retroperitoneal, produces less peritoneal irritation than a ruptured or inflamed intra-abdominal viscus. Usually, therefore, although there is quite marked upper abdominal tenderness, rigidity and guarding are not usually marked features. However, these may be present and may closely mimic a perforated peptic ulcer.

Special Investigations

Acute pancreatitis is unique among the common acute abdominal emergencies in that there is a fairly reliable laboratory confirmatory investigation, the serum amylase.

Serum Amylase. The normal value of the serum amylase activity ranges from 80 to 150 Somogyi units. In the majority of case of acute pancreatitis this level rises to 1000 units or more. However, in a small number of cases the amylase activity may be normal, and this is particularly so when blood is taken two or three days after the onset of the pancreatitis by which time, in the majority of patients, the level has fallen to normal. Serum amylase activity may also be increased in other conditions associated with severe abdominal pain. These include a perforated peptic ulcer, acute cholecystitis, and intestinal obstruction. The amylase level may also be elevated in renal failure and following the administration of morphia and its derivatives which may cause spasm of the sphincter of Oddi. Although raised in these various conditions, the amylase activity very rarely reaches the high level seen in acute pancreatitis.

A number of other biochemical abnormalities may be found in this condition:-

Plasma calcium concentration may fall and the degree of hypocalcaemia is related to the severity of the pancreatitis. Occasionally there may be clinical tetany and in such cases the prognosis is poor.

Blood sugar concentration may be elevated transiently in acute pancreatitis, with associated glycosuria.

The plasma bilirubin concentration may be elevated due to oedema of the pancreatic head, indeed there may be mild clinical jaundice.

Serum methaemalbumin level may be raised in severe pancreatitis and indicates a severe degree of inflammation of the gland.

Plain X-rays of the abdomen are not diagnostic. There may be a C-shaped loop of gas filled duodenum due to localized ileus, absence of the psoas shadows due to retroperitoneal oedema and small fluid levels due to a generalized ileus. The frequent association with biliary disease may be demonstrated by the presence of radio-opaque gallstones. A chest X-ray occasionally shows small left pleural effusion.

Although these special investigations are undoubtedly of value, the surgeon, faced with the patient in whom he suspects the possibility of a ruptured abdominal viscus, will be wise to submit the patient to laparotomy, even in the face of a raised serum amylase. The biochemical tests are, however, of considerable value in reinforcing a clinical diagnosis of acute pancreatitis and will encourage the surgeon, under these circumstances, to manage the patient conservatively.

Medical Conditions which may Simulate the Acute Abdomen

Although many essentially medical diseases may be associated with, or present as, acute abdominal pain, it is equally important to remember that a patient under treatment with some medical condition may develop an acute abdominal emergency. Indeed, it is a well known aphorism that the most dangerous place to perforate is in a medical ward! I have personally operated upon a child with acute appendicitis who was under treatment with pneumonia, dealt with another who was in a plaster bed with severe spinal tuberculosis and seen a perforated peptic ulcer in a patient with post-traumatic paraplegia. Patients may develop a mesenteric embolism while recovering from a myocardial infarct and, much more commonly get faecal impaction or acute retention of urine while on strict bed rest with this condition.

In clinical practice, a careful history combined with meticulous clinical examination will usually enable the practitioner to decide whether the situation represents an acute abdominal emergency in a patient with an associated medical condition, or whether the acute abdominal pain is due to the medical disease itself. For example, a known diabetic patient, who is well stabilized, develops an acute abdominal pain with classical features of intra-abdominal pathology. His stable diabetic condition should not give rise to any confusion. In contrast, a patient with uncontrolled diabetes in precoma with marked ketonuria who is also complaining of abdominal pain should alert the clinician to a diagnosis of prediabetic coma with associated abdominal symptoms. In this latter case, the rapid disappearance of the abdominal pain over the next few hours as the diabetes comes under control will clinch the diagnosis. Perhaps the most important thing to bear in mind is to beware of a 'label' attached to a patient; just because he has had several attacks of pain in the past due to cardiac ischaemia does not mean that his present attack may not be one of acute cholecystitis. It is always best to bring quite a fresh mind to the problem and consider each patient and each painful episode on its own merit.

Chest Diseases

Upper abdominal pain, tenderness and guarding may be associated with acute pulmonary or pleural infection (*see above* under Acute Appendicitis). The local abdominal tenderness secondary to pulmonary disease is not usually increased by abdominal palpation, the respiratory rate is usually increased, associated with shallow breathing and the pain is frequently aggravated by taking a deep breath. Chest radiography is valuable and a lateral film should also be taken since basal shadowing may be hidden behind the diaphragm on the postero-anterior film.

Upper abdominal pain is not unusual in an acute coronary thrombosis but there is seldom associated rigidity or tenderness and there are the other clinical features suggesting cardiac origin of the pain. The electrocardiogram and serum transaminase estimations are certainly valuable confirmatory aids in this condition.

Diabetes

We have already discussed the abdominal pain and vomiting which may accompany prediabetic coma. The increasing thirst, polyuria, drowsiness and dehydration should alert the surgeon. A vital requirement is to test the urine as a routine and to perform a blood sugar estimation if there is any doubt at all. Even if there is an associated abdominal condition, the first necessity is to correct the diabetes. If the abdominal features clear over the next two or three hours as the diabetes comes under control then obviously they were due to the medical condition. If they do not improve, the possibility of some surgical lesion must be further considered.

Sickle Cell Anaemia

This is more or less confined to the black races and is not uncommon in the West Indian community in this country. Abdominal pain with associated tenderness and guarding may occur during an acute haemolytic episode; the nature of the pain is not known. The blood film is characteristic and having determined that the patient has sickle cell anaemia, close observation must be initiated to decide whether or not the pain is due to a haemolytic crisis or whether there is an associated abdominal catastrophe. It is important to remember that patients with any haemolytic anaemia have a raised incidence of gall stones.

Other Medical Conditions

Among other conditions which are listed as being associated with abdominal pain one can mention Herpes zoster affecting the lower thoracic segments, epidemic myalgia (Bornholm disease), the abdominal crises of tabes dorsalis, acute lead poisoning and acute porphyria, but these are all extremely rare. Perhaps under the heading of medical conditions is a convenient place to mention the *Munchausen syndrome*; patients who are either mentally abnormal or who are drug addicts and who can mimic to perfection many diseases including acute abdominal emergencies. Ureteric colic is a favourite and the patient is often shrewd enough to explain to the doctor that he cannot undergo intravenous pyelography because he is iodine sensitive! A bizarre previous history, no fixed address, antecubital injection marks and the long memory of the Casualty Sister are all helpful in making the diagnosis.

REFERENCES

1. Brooks, D. W. and Killen, D. A. (1965), 'Roentgenorgraphic findings in acute appendicitis,' *Surgery*, **57**, 377–380.
2. Ellis, H. (1968), 'Diagnosis of the acute abdomen,' *Brit. Med. J.*, **1**, 491–493.
3. Ellis, H. (1974), *Maingot's Abdominal Operations* 6th edition Vol. 2, 1842–1901, New York: Appleton-Century-Crofts.
4. French, E. B. and Robb, W. A. T. (1963), 'Biliary and renal colic,' *Brit. Med. J.*, **2**, 135.
5. Gough, M. H. and Gear, M. W. L. (1971), *The Plain X-ray in the Diagnosis of the Acute Abdomen*. Oxford: Blackwell Scientific.
6. Graham, J. A. (1965), 'Urinary cell counts in appendicitis,' *Scot. Med. J.*, **10**, 126–128.
7. Jona, J. Z. and Belin, R. P. (1976), 'Basal pneumonia simulating acute appendicitis in children,' *Arch. Surg.*, **III**, 552–554.
8. Nakashima, Y. and Howard, J. M. (1977), 'Drug induced acute pancreatitis,' *Surg. Gynec. Obstet.*, **145**, 105–109.
9. Rosin, R. D. (1976), 'Pancreatitis and hyperparathyroidism,' *Postgrad. Med. J.*, **52**, 95–101.
10. Smith, P. H. (1965), 'The diagnosis of appendicitis,' *ibid.*, **41**, 2–5.
11. Trapnell, J. E. and Duncan, E. H. L. (1975), 'Patterns of incidence in acute pancreatitis,' *Brit. Med. J.*, **2**, 179–183.
12. Valerio, D. (1976), 'Acute diabetic abdomen in children,' *Lancet*, **i**, 66–67.

9. CLINICAL BIOCHEMISTRY OF THE 'ARTIFICIAL LIVER'

D. B. A. SILK AND ROGER WILLIAMS

Introduction

Fulminant hepatic failure (FHF)
   Aetiology
   Clinical features
   Initial investigations

The multisystem abnormalities of FHF
   Changes in acid-base balance
   Renal disturbances
   Metabolic disturbances
   Haematological disturbances
   Haemorrhage
   Infections

Supportive therapy in FHF

Pathogenesis of acute hepatic encephalopathy
   Role of water soluble toxins
   Pathogenic importance of middle molecular weight compounds

Use of charcoal haemoperfusion in acute liver failure

INTRODUCTION

The term 'artificial liver' refers to the liver support devices such as haemoperfusion and haemodialysis systems used in patients with liver failure. To date, use of the artificial liver has been restricted mainly to patients with fulminant hepatic failure (FHF), the clinical syndrome which arises from massive necrosis of liver cells in a patient in whom there has been no previous evidence of liver disease.[1] FHF is characterized by the onset of a severe and progressive encephalopathy, and mortality, which is closely related to the severity of the encephalopathy, is as high as 80–90% in patients who show signs of Grade IV coma.[1-4] Since the liver has remarkable regenerative capacity, patients who recover usually do so completely, and despite the extensive initial damage to the liver, cirrhosis rarely occurs.[5]

FHF is associated with many different biochemical disturbances which arise from the impairment of the hepatic synthetic processes and also through failure of the normal detoxification and excretory functions of the liver. Apart from the biochemical disturbances in the pathogenesis of the encephalopathy, these changes also give rise to multisystem disorders which are an integral part of the FHF syndrome.

The aim of this chapter is to highlight these changes and to discuss their possible relationships to the development of the encephalopathy and to the multisystem disorders. We also describe the role of biochemical testing in the management of these patients and in artificial liver support.

FULMINANT HEPATIC FAILURE

Aetiology

FHF is not common in the UK and current estimates suggest that fewer than 400 patients a year die from the disease. It is often young and previously healthy individuals who suffer from it. During the three-year period 1976–79, 108 patients with FHF, showing signs of Grade IV coma (unresponsive to verbal commands) were admitted to the Liver Failure Unit, King's College Hospital. The commonest causes were acute viral hepatitis and paracetamol self-poisoning (Table 1).

TABLE 1
THE AETIOLOGY OF FHF IN 108 PATIENTS ADMITTED TO THE LIVER UNIT, KING'S COLLEGE HOSPITAL, 1976–79

Cause	Number of patients
Viral hepatitis	52
Paracetamol overdose	50
Halothane	4
Other drug hypersensitivity:	
Rifampicin	1
Aldomet	1

Clinical Features

In many patients clinical diagnosis is not difficult, especially if the patient is deeply jaundiced and has shown a rapid deterioration in consciousness. In others however the initial diagnosis may be missed because clinically apparent jaundice is preceded by signs of hepatic encephalopathy (Table 2). In general, the time between the onset of symptoms (including jaundice) and development of coma in the acute viral hepatitis group is longer than in patients who have taken an overdose of paracetamol. The rapid progression of encephalopathy can be striking in patients with hepatic necrosis caused by paracetamol overdosage and only 12 hours may elapse between Grade I and Grade IV coma. The course

TABLE 2
CLINICAL STAGES IN DEVELOPMENT OF HEPATIC COMA

Grade	Mental state
I	Euphoria, occasionally depression
	Fluctuant, mild confusion
	Slowness of mentation and affect
	Untidy. Slurred speech
	Disorder in sleep rhythm
II	Accentuation of I. Drowsy
	Inappropriate behaviour
III	Sleeps most of time but is rousable
	Incoherent speech. Marked confusion
IV	Unrousable. May or may not respond to noxious stimuli

of the illness may be so short in some patients that jaundice never becomes apparent.

On examination, signs of hepatic encephalopathy are present (Table 2) and those related to acid-base, cardiac, renal, metabolic and haematological disturbances may also be detected. With respect to underlying hepatic damage, the degree of jaundice may be variable, at least at first. Foetor hepaticus may be present. Cutaneous stigmata, including spider naevi, so characteristic of chronic liver disease, are rarely seen in FHF. The liver is usually not palpable and may be demonstrably small to percussion.

Initial Investigations

The standard liver function tests, including prothrombin time, usually provide early confirmation of the diagnosis (Table 3). Serum alanine aminotransferase and

TABLE 3
INITIAL BIOCHEMICAL AND HAEMATOLOGICAL SCREENING IN PATIENTS WITH FHF

Liver function tests:	
Bilirubin	Variable
Alanine aminotransferase (ALT)	>1000 IU/l
Aspartate transaminase (AST)	>1000 IU/l
Alkaline phosphatase	⩾200 IU/l
Albumin	Usually ⩾30–35 g/l
Full blood count	Leucocytosis, anaemia if bleeding has occurred
Platelet count	Low, usual range 30–120 × 10^9/l
Prothrombin time	Prolonged 25–120 secs
Blood glucose	Often low
Serum electrolytes and urea	Hyponatraemia, hypokalaemia, raised blood urea
Serum creatinine	Often goes to >200 μmol/l

aspartate aminotransferase activities are markedly raised, often greater than 1000 IU/1. Alkaline phosphatase levels are often only slightly raised. The half-life of albumin is 13–16 days, so that if FHF has developed rapidly, serum albumin will often be within the normal range. At the onset, levels of less than 30 g/l are rare, and, if present, should alert the clinician to the presence of underlying chronic liver disease rather than FHF. Values will, however, fall during the course of the illness and in our experience, levels of less than 30 g/l are quite common in those patients who recover consciousness and subsequently survive.

One of the most striking features of patients with FHF is their markedly prolonged prothrombin times, and in the initial series of our patients treated by polyacrylonitrile membrane dialysis,[6] the prothrombin time (PT) was 83·4 ± 7·0 (SEM), n = 24, with half the patients having a PT greater than 2 minutes long. In addition to prolongation of PT, these patients may have a profound thrombocytopenia often to levels of less than 20 × 10^9/l.

Blood glucose estimations are important, not from the point of view of diagnosis, but because hypoglycaemia is common in the early stages of FHF. Many patients develop renal failure after a few days of illness, so serum electrolyte and creatinine measurements are an integral part of both the initial and subsequent daily screening.

All patients should be tested for the hepatitis B surface antigen as its presence may not only have significance for the patients in relation to an aetiological factor, but also for the staff who look after them and for those who analyse their blood samples.

Subsequent biochemical testing is best categorized into those measurements primarily concerned with the development and management of the many body systems which become affected in FHF and those which are more specifically related to the encephalopathy and its management by artificial liver support.

THE MULTISYSTEM ABNORMALITIES OF FHF
(Table 4)

Changes in Acid-Base Balance

Arterial blood gas and serum electrolyte measurements are necessary throughout the course of the illness.

Hyperventilation is a constant feature during the early stages of coma. It usually results in a respiratory alkalosis, and the Paco$_2$ falls with increasingly deep coma. In our series, hydrogen ion concentration was low in 44 of 65 observations concerning 26 patients.[7] In 30 of these a respiratory alkalosis was present, in 16 there was a metabolic alkalosis, and in 3 we found a mixed abnormality.

Correction of respiratory alkalosis by either inhalation of CO_2 or infusion of the carbonic anhydrase inhibitor acetazolamide has not been successful and is often associated with further clinical and biochemical deterioration.[8] Sometimes an improvement is seen with intravenous administration of sodium bicarbonate[9] but this benefit may only be temporary and the therapy is limited by the adverse effect of a large sodium load.

Hypoxia is often found in FHF, resulting from decreased respiration due to central depression of the respiratory centre, pulmonary oedema, infection or intrapulmonary shunting.[10] Oxygen therapy should be given and if there is any doubt that the Pao$_2$ cannot be maintained in this way, the patient should be intubated and receive ventilatory support. Since intubation can often prove difficult the position of the endotracheal tube must be checked radiologically. The pathogenesis of the pulmonary oedema which develops in up to 37% of cases is similar to that of the so-called shock-lung syndrome.[11] Severe cases should be treated by intermittent positive-pressure ventilation with positive end-expiratory pressure. Diuretic therapy has not been shown to be effective.

Renal Disturbances

The abnormalities in plasma electrolyte concentration seen in FHF[12] include both hyponatraemia and hypo-

TABLE 4
MULTISYSTEM ABNORMALITIES IN FHF

	Disorder	Appropriate investigations
Acid-base balance	Respiratory and/or metabolic alkalosis Hypoxia	Blood gases, electrolytes
Renal disturbances	Plasma electrolyte abnormalities Pre-renal uraemia Functional renal failure Acute tubular necrosis	Plasma creatinine Plasma electrolytes Plasma osmolality Urinary electrolyte excretion Urinary osmolality
Haematological changes	Haemorrhage	Platelet count Prothrombin time Gastric acid pH during cimetidine therapy
Infection	Septicaemia Chest infection Urinary tract infection	Microbiological Analysis of all biological fluids
Metabolic abnormalities	As in Table III	

kalaemia. These occur in about half of the patients at some time in their illness. The mechanism for the hyponatraemia is probably dilutational, at least in part, as its occurrence has been related to a markedly impaired renal capacity for excretion of free water. Wilkinson et al.[13] have also shown, using peripheral blood leucocytes, that hyponatraemia is associated with increases in intracellular sodium concentrations, which, in turn, has been related to reduced leucocyte-membrane-associated Na,K-ATPase activity. In patients who survive, plasma sodium concentrations rise, and intracellular sodium concentrations fall, observations which are accompanied by return to normal of Na,K-ATPase activity. Toxins in the circulation of these patients are likely to be directly responsible for mediating the electrolyte shifts which are the basis of the plasma electrolyte abnormalities. The renal failure so common in FHF may be classified into pre-renal uraemia, functional renal failure and acute tubular necrosis. In a recently published series, 22 of 40 (55%) patients developed renal failure.[14] An increase in plasma creatinine concentration is detectable before oligoria or anorexia develops. In those patients with functional renal failure, the glomerular filtration rate was greater than 3 ml/min, urinary sodium concentration less than 16 mmol/l, urine:plasma osmolality ratio greater than 1·09 and renal histology normal at post-mortem. In contrast, those with a glomerular filtration rate of less than 3 ml/min had a urinary sodium concentration higher than 16 mmol/l, urine:plasma molality ratio less than 1·09 and evidence of tubular necrosis at post-mortem.

In our experience[15] the development of renal failure in FHF carries a high mortality. When polyacrylonitrile membrane dialysis was used as artificial liver support, however, the incidence of complicating renal failure was similar in survivors and non-survivors,[6] showing that the prognosis of acute renal failure *per se* in FHF can be improved with early institution of renal support.

Metabolic Disturbances

Hypoglycaemia can develop rapidly and may lead to irreversible brain damage unless it is prevented by parenteral administration of 10% glucose. Care must be taken, therefore, to monitor blood glucose concentrations. Several factors are likely to be responsible for the hypoglycaemia associated with FHF, including the high concentrations of circulating endogenous insulin, and failure of glycogenolysis and gluconeogenesis which arise directly from the massive hepatic necrosis.

Haematological Disturbances

Disorders of coagulation are common in FHF. As liver function deteriorates, the synthesis of clotting factors decline. This is manifested as a prolongation of prothrombin time and a depression of circulating levels of fibrinogen and factors II, VII, IX and X. Except in cases where there is also intravascular coagulation, monitoring of the prothrombin time only is necessary. The degree of prolongation of prothrombin time is a good indicator of prognosis and is related to the extent of hepatic necrosis as judged by morphoactive analysis of liver biopsy specimens.[16]

Platelet counts should also be determined for thrombocytopenia which is common in FHF and at least half the patients have a platelet count of less than 80 000. Thrombocytopenia occurs consequent on bone marrow hypoplasia and disseminated intravascular coagulation, the latter, except in certain cases of hepatic failure developing post partum is usually not severe and although heparin administration may be of value in these patients, it is not to be recommended routinely. The use of fresh frozen plasma is followed by some correction of the coagulation disturbances although experience has shown that it is of little value prophylactically in the prevention of bleeding.

Haemorrhage

The usual sites of bleeding are the nasopharynx, oesophagus, stomach, gastrointestinal tract, retroperitoneal space and the bronchial tree. Overt bleeding may not occur, however, despite prothrombin times in excess of 150 seconds (normal value 12 seconds). In an analysis of mortality in an early series of patients, death was directly attributable to this complication in 19% of 132 patients.[16] Upper gastrointestinal endoscopy has now clearly shown that these patients bleeding from superficial erosions in the oesophagus, stomach and duodenum, and in a prospective controlled clinical trial prophylactic cimetidine therapy (300 mg i.v. 6-hourly) was shown to prevent this complication in almost all patients.[17] Careful monitoring of gastric aspirate pH is, however, essential as the pH must be maintained at >5·0, otherwise treatment will not be effective.

Infections

Patients in FHF show an increased susceptibility to bacterial infection. This may be due partly to inhibition of the hexose monophosphate shunt activity in polymorphonuclear leucocytes by circulating toxins, and partly to impaired bacterial opsonization caused by complement deficiency. Most disturbing from the clinical point of view has been the development of fatal infections in some patients after successful liver support at a time when recovery of consciousness has taken place and hepatic function is improving. Infection may occur at all locations, so that frequent microbiological analysis of sputum, blood and urine must be carried out in order to apply the correct antibiotic therapy rapidly when necessary.

SUPPORTIVE THERAPY IN FHF

A full description of the general and supportive therapy is outside the scope of this chapter. The measures currently adopted at King's College Hospital have been reviewed in detail elsewhere.[10] Although improvements in supportive therapy have reduced the incidence of many of the above complications, overall, survival figures have remained remarkably constant. The reason for this is that the standard measures have little effect on either the encephalopathy or cerebral oedema which occur in many patients. At present the cause of the encephalopathy of FHF still remains speculative. However, knowledge in this area is expanding rapidly and it is becoming clear that the pathogenesis of the encephalopathy is wholly, or in part, due to the accumulation of toxic water-soluble and protein compounds in the circulation of these patients. As build-up of toxic concentrations occurs because of the failure of the detoxification processes of the liver, it has seemed for some time that a rational approach to improving the hepatic encephalopathy and thence survival would be to provide a means of lowering the circulating levels of these toxic compounds while hepatic regeneration restores hepatic detoxification and excretory processes. The known adsorptive properties of sorbents such as activated charcoals and polymeric resins of the Amberlite series (XAD 2, 4, 7) as well as the development of novel haemodialysis systems have provided the basis for the development of several support systems which are being applied in the treatment of acute liver failure.

PATHOGENESIS OF ACUTE HEPATIC ENCEPHALOPATHY[18]

Role of Water-Soluble Toxins

(a) Ammonium

Ammonium is normally produced in large amounts in the liver and intestine, but small quantities are produced in other tissue such as the kidney, muscle and brain. Ammonium (which has a pK_a of 9·0) is thought to cross cell membranes predominantly in the un-ionized lipid-soluble form NH_3, although it is known that the ionized form (NH_4^+) is able to enter cells by competition with the potassium ions for the exchange with sodium ions on the sodium pump. For over forty years, accumulation of high circulating levels has been incriminated in the pathogenesis of hepatic encephalopathy. Possibly the strongest evidence to support these claims is the repeated observation that neurological changes leading to a reversible coma can be induced in normal animals by large injections of ammonium. Further evidence is gained from the fact that feeding of urea, protein, ammonium salts or ion-exchange resins in the ammonium phase will raise blood ammonium levels, impair the electroencephalogram and precipitate coma in animals with portocaval shunts and in patients with chronic liver disease.

The principle objections to the importance of high circulating levels of ammonium in relation to the pathogenesis of hepatic encephalopathy stem mainly from the repeated observations that venous and arterial blood levels often correlate poorly with the degree of hepatic encephalopathy.[19] Although these findings argue somewhat persuasively against the importance of ammonium, it is necessary to appreciate that the total intracellular ammonium concentration may be up to 18 times that in the extracellular fluid so that blood vessels are only distantly related to absolute concentrations inside the brain cells. There is also the possibility that the effect of high blood levels of ammonium may be cumulative rather than immediate. Conversely, clinical recovery may be delayed after biochemical improvement so that at both times clinical signs and blood levels may be out of step. Furthermore, Zieve et al.[20] have now shown that the coma-producing potential of ammonium in normal rats increases several-fold when presented with fatty acids and mercaptans.

(b) Amino Acids

Much interest is currently being directed towards the possible role that abnormalities of plasma amino acid

TABLE 5

PLASMA AMINOACID CONCENTRATIONS (μMOL/L) IN PATIENTS WITH FULMINANT HEPATIC FAILURE SHOWING SIGNS OF GRADE IV COMA

Aminoacid	FHF patients median (range) (n = 26)	Controls mean ± SE (n = 11)	Significance of difference*
Aspartate	53 (10–167)	22.4 ± 3.72	NS
Threonine	351 (90–1239)	145 ± 12.7	p < 0.01
Serine	157.5 (72–1152)	139 ± 5.64	NS
Glutamate	123.5 (11–625)	50 ± 7.29	P < 0.01
Glutamine	1439.5 (540–7155)	846 ± 46	p < 0.01
Proline	749 (243–2795)	203 ± 19.2	p < 0.01
Glycine	551.5 (172–1796)	281 ± 16.7	p < 0.01
Alanine	583.5 (215–2165)	491 ± 33	NS
Citrulline	79 (18–421)	43 ± 7.7	p < 0.05
Valine	215 (84–655)	348 ± 23	p < 0.02
Cystine	148 (52–301)	107 ± 17	NS
Methionine	433 (154–929)	35 ± 2	p < 0.01
Isoleucine	50.5 (15–164)	85 ± 4.4	P < 0.02
Leucine	108 (22–255)	172 ± 6	p < 0.02
Tyrosine	270.5 (78–573)	88 ± 3.7	p < 0.01
Phenylalanine	310 (77–716)	83 ± 3.3	p < 0.01
Histidine	176.5 (84–503)	139 ± 17	p < 0.05
Ornithine	177.5 (34–635)	79 ± 7.8	p < 0.01
Lysine	528 (120–1430)	216 ± 15	P < 0.01
Arginine	136 (15–723)	159 ± 38	NS

profiles play in the pathogenesis of the encephalopathy in FHF. As Table 5 shows, the plasma levels of most individual acids were markedly raised when measured in a series of our patients on admission in grade IV coma.[21]

(i) *Methionine*
In the series of patients with FHF referred to in Table 5, methionine was the only amino acid significantly raised in the plasma of *all* patients on admission in grade IV coma. This amino acid is especially toxic to normal animals and, if ingested in large amounts, may produce psychotic signs, not only in patients with schizophrenia but also in normal subjects. In experimental studies high blood concentrations of methionine are associated with raised levels of the methyl donor S-adenosylmethionine in the brain and as Thompson[19] has suggested, onset of coma could be related to excessive methylation of brain amines such as 5-hydroxytryptamine, noradrenaline and dopamine. An alternative way in which raised blood levels of methionine could promote the onset of coma is by increased synthesis of the mercaptans methanethiol and dimethylsulphide, both of which have been isolated in large amounts from the breath of patients with hepatic failure and both of which are capable of inducing coma in normal rats.[20] In our clinical experience, however, the serial plasma levels of methionine correlate poorly with changing coma grade in FHF.[21]

(ii) *Phenylalanine, tyrosine and the branched chain amino acids*
We have shown that in patients with FHF concentrations of the two aromatic amino acids phenylalanine and tyrosine are markedly raised (up to 10-fold, Table 5), whereas levels of the branched chain amino acids leucine, isoleucine and valine are either normal or actually reduced.[21] As these amino acids share a common carrier-mediated transport system across the blood-brain barrier,[22] Fischer et al.[23] have suggested that the net result is an increase in brain uptake of phenylalanine and tyrosine.

It has been suggested that, once inside the brain, the high levels of phenylalanine saturate tyrosine hydroxylase, the enzyme which normally catalyses the conversion of tyrosine to dihydroxyphenylalanine (DOPA), the rate-limiting step in the synthesis of the putative excitatory neurotransmitters dopamine and noradrenaline. Under such circumstances, tyrosine, when present in excess, will undergo preferential decarboxylation to tyramine and β-hydroxylation to octopamine and the remaining phenylalanine will be converted to β-hydroxy phenylethanolamine. Coma ensues either because octopamine and β-hydroxy phenylalanine exhibit inhibitory rather than excitatory post-synaptic actions, or because on a pre-synaptic basis, the two compounds operate as weak excitatory neurotransmitters and when present in excess are released instead of noradrenaline.

We have examined the relationship between serial levels of tyrosine and phenylalanine and changes in conscious levels in the patients with FHF and treated by daily repeated 4-hour periods of haemodialysis.[21] There was a trend, which reached statistical significance, for plasma levels to return towards normal values in the 24 hours before recovery of consciousness in those patients who survived, whereas the levels did not change significantly in those patients who remained in grade IV coma throughout the course of their illness. Moreover, in 3 out of the 6 survivors who were studied in detail, improvement in consciousness was preceded by, and then associated with, a sustained return to normal of plasma phenylalanine and tyrosine levels.

(iii) Tryptophan

Tryptophan, one of the essential amino acids and the precursor for brain synthesis of the putative neurotransmitter 5-hydroxytryptamine (5-HT), exists in the plasma either bound to proteins or in the free state. In FHF total tryptophan levels are normal, but free tryptophan levels are markedly raised[24] and it is this form which is translocated across the blood-brain barrier by the common carrier mediated transport system shared by the aromatic amino acids tyrosine and phenylalanine and the branched chain amino acids leucine, isoleucine and valine. Biopsy specimens of frontal cortex obtained immediately after the death of patients with FHF in grade IV coma contained normal amounts of 5-HT but increased amounts of tryptophan and 5-hydroxy indole acetic acid.[24] Although it is possible to conclude with some certainty that rises in plasma free tryptophan in FHF lead to increased brain 5-HT turnover, it is unclear whether this has a special role in the development of the encephalopathy.

(c) Protein-bound Compounds of Possible Pathogenic Importance

In the circulation of patients with FHF there are raised levels of several low molecular weight compounds, many of which have been shown to be toxic, which are either totally or partially bound to plasma proteins. These compounds include mercaptans, phenolic acids, short-chain fatty acids and bile acids.

(i) Mercaptans

Challenger and Walshe[25] were the first to establish the association of mercaptans with hepatic failure by isolating methylmercaptan from the urine of a patient in deep hepatic coma. Subsequently, by breath analysis, Chen et al.[26] have shown increased levels of methanethiol, ethanethiol, and dimethylsulphide in hepatic failure. Mercaptans have proved difficult to quantitate precisely in the plasma of patients, but increased levels probably play an important role.

(ii) Phenolic acids

Phenolic acids represent a group of compounds whose levels are now known to be increased in the serum of patients with acute liver failure and are known to inhibit enzymes in brain homogenates even at low concentrations. Most of the methods currently available are capable of measuring only the concentrations of the free forms in serum, but since they are also known to be bound to plasma proteins, removal of only the free, biologically active forms by artificial liver support systems would not be expected to have any beneficial effect because the protein-bound moiety would rapidly re-equilibrate with the free plasma pool.

(iii) Fatty acids

Despite previous claims, we have found little evidence to support a major role for short-chain fatty acids in the pathogenesis of acute hepatic encephalopathy. Only acetic, propionic and butyric acids were raised in the blood of our patients with FHF.[27] Moreover, the absolute levels were at least two orders of magnitude less than the concentrations needed to produce coma and inhibit brain Na, K-ATPase activity in animal studies.

(iv) Bile acids

Conjugated and unconjugated bile salts have been shown to have an inhibitory effect on brain respiration *in vitro* and since it is known that in patients with hepatic coma brain oxygen uptake is reduced, it seemed that bile salts could be one of the substances involved in the pathogenesis of the coma of FHF. We have shown that bile acids can be detected in brain tissue and cerebrospinal fluid obtained from patients dying of FHF, whereas none was found in samples obtained from patients dying without evidence of liver disease.[28] The concentrations found, however, were much less than those needed to have a significant inhibitory effect on brain respiration *in vitro*.

Pathogenic Importance of Middle Molecular Weight Compounds

Opolon et al.[29] have suggested that a group of compounds within the molecular weight range 500–5000 may be involved in the pathogenesis of the coma associated with acute liver failure. The evidence upon which their conclusions are based is clinical and biochemical data derived from haemodialysis studies performed on pigs with surgically-induced liver failure. Haemodialysis using a cuprophane membrane was followed by faster deterioration in conscious level and in EEG impairment and shorter survival time than haemodialysis using the permeable polyacrylonitrile membrane. These clinical observations were later mirrored by marked differences in the regional brain neurotransmitter levels in the groups of animals treated by the two systems. As the cuprophane membrane is only freely permeable to compounds of molecular weight up to 500, whereas the polyacrylonitrile membrane permits rapid transfer of substances up to molecular weight 5–7000, it seems that improvement in consciousness was directly associated with removal of compounds in the middle molecular weight range.

Conclusion

No unified concept is ever likely to adequately explain the pathogenesis of the acute hepatic encephalopathy that characterizes FHF. Persuasive experimental evidence favours a contributory role of many low molecular weight toxins existing in the circulation both in the native state and bound to plasma proteins, and in addition, compounds in the middle molecular weight range are also likely to be important. Successful reversal of the coma in acute liver failure will thus require a means of support that is quite capable of removing this wide variety of potentially toxic compounds. The use of one single system can never achieve this and the ultimate aim must be a reproducible combination support system

USE OF CHARCOAL HAEMOPERFUSION IN ACUTE LIVER FAILURE

Background

The broad adsorptive spectrum of activated carbon for water-soluble metabolites has been appreciated for many years. Because of this, as well as its high surface area, capacity and its ready availability, haemoperfusion over activated carbon has been conceived as a rational means of removing endogenous and exogenous toxins from the circulation. Two major problems associated with direct haemoperfusion over activated carbon arose. One was that the charcoal granules gave off charcoal powder, resulting in embolism to various organs. The second was the severe depletion of platelets and leucocytes from circulating blood. Furthermore, the haemoperfusions resulted in sludging and channelling of blood within the columns and reduced clearances. One of the means of overcoming the platelet and leucocyte losses during charcoal haemoperfusion has been physically to separate the charcoal from the formed elements of the blood by coating with a biocompatible polymer (microencapsulation) which is permeable to the desired toxins but not platelets and leucocytes. The coating materials so far tested include albumin-colloid ion, colloid ion alone, poly-hydroxy ethyl methacrylate cellulose (polyhema) and several copolymer preparations.

Adsorptive Properties of the Encapsulated Charcoals

In general low molecular weight water-soluble compounds (molecular weight less than 500) are adsorbed by all the preparations and, with respect to the specific compounds in this molecular weight range held to be important in the development of acute hepatic encephalopathy, the encapsulated charcoal preparations are capable of binding amino acids and free fatty acids. Ammonium, however, which at physiological pH is heavily ionized, appears to be poorly adsorbed to these preparations.

Compounds in the middle molecular weight range (500–5000) are also known to be adsorbed to encapsulated charcoal. Compared with water-soluble compounds, those bound to plasma proteins are relatively poorly adsorbed.

Charcoal Haemoperfusion in FHF

The first report of the use of charcoal haemoperfusion in the treatment of a patient in hepatic coma due to chronic alcoholic liver disease appeared in 1972.[30] The first large series of patients with FHF treated by repeated daily charcoal haemoperfusion was reported from this hospital.[31] Ten of 22 patients with FHF, all of whom showed signs of grade IV coma before treatment, survived to leave hospital. Coconut-based charcoal encapsulated with 4% polyhema was used, and the only side-effect reported was a transient hypotension in some patients at the start of perfusion which was usually reversed by infusion of plasma or blood. Platelet losses across the column were less than 30% in 35 of 42 instances. The effect of charcoal haemoperfusion on biochemical parameters was variable. Thus, no significant extraction of ammonium across the column could be demonstrated. Significant extraction of amino acids by the column was shown. By the time that 37 patients had been treated there were 14 survivors (37·8%). However, many of the subsequent 34 patients treated developed severe unresponsive hypotension during treatment and only three of these survived (8·8%). These initial hypotensive reactions were associated with marked losses in circulating platelets, so detailed studies of platelet function in patients treated by charcoal haemoperfusion were carried out.[32] The results were interesting not only because the patients with FHF had a greater percentage of small platelets in the total count, but also because there appeared to be an additional loss of larger and haemostatically more active platelets during haemoperfusion. Later studies using a specifically designed *in vitro* haemoperfusion test circuit, have confirmed that the platelet losses across the charcoal columns are due to formation of platelet aggregates and that the cause of the associated hypotensive reactions were most likely due to the release of vasoactive amines from the damaged platelets.

In order to test various agents that might protect susceptible platelets from further activation during haemoperfusion we developed an *in vitro* test circuit using paired 10 g charcoal columns perfused with heparinized blood. After two hours of haemoperfusion, rises in Swank screen filtration pressure, which is used to detect the presence of circulating platelet aggregates, are evident. Experiments using sulphinpyrazone infusions to protect the platelets were disappointing, but both prostaglandins PGE_1 and PGI_2 (prostacyclin), when used in the *in vitro* test circuit, were able to prevent rises in Swank screen filtration pressure which occurred in the control column. These drugs are thought to prevent platelet aggregation by stimulating adenyl cyclase. PGE_1 was also effective when bound to the polymer surface and PGI_2 reduced the rise in screen filtration pressure when given as a bolus into the arterial line.

Encouraged by these results, we have recently carried out an *in vivo* assessment of the effectiveness of PGI_2. In this study a new column was used containing 100 g of carbon of a smaller particle size with an improved internal design and fluting which exposes the blood to lower shear forces.

All patients in the study had grade IV encephalopathy due to FHF. Six were given a single 4-hour period of haemoperfusion with the new column and acted as controls. Six others had an identical perfusion with the addition of a PGI_2 infusion. The groups were matched with respect to age, sex, aetiology and preperfusion parameters but not prothrombin time which was significantly more prolonged in the PGI$_2$ group. PGI$_2$ was infused in a glycine buffer at an increasing dose before haemoperfusion and then at 16 ng/kg/min during perfusion. The initial PGI$_2$ infusion was well tolerated with no significant side-effects and subsequently the blood pressure was stable during perfusion in all 6 patients. On the other hand, two patients in the control group developed hypotension. There were no rises in screen filtration pressure in the PGI$_2$-treated group. Platelet damage was measured during perfusion by estimation of arterial platelet counts and by the release from platelet α-granules of the platelet-specific protein β-thromboglobulin. In the PGI$_2$-treated group there was no loss of circulating platelets and only a small rise in β-thromboglobulin of 23 ng/ml, maximal at one hour. In contrast, in the control group, although the fall in arterial platelet count observed was not significant, there was a statistically greater rise in β-thromboglobulin at one hour of 341 ng/ml ($p < 0.05$). These results showed that although platelet damage may be less with the improved column, it was not completely eliminated, as shown by the severe hypotension which developed in one patient. Further platelet protection was needed and this was afforded by the continuous PGI$_2$ infusion.

Development of Sorbents for Removing Protein-bound Toxins

The uncharged resins Amberlite XAD-2 and 4 have a specific adsorptive attraction for lipid-soluble, protein-bound molecules, including barbiturates, glutethamide, echlorvinyl, methaqualone, digitalis and cephaloridine.[33] These resins have been used in haemoperfusion systems in the treatment of intoxication secondary to overdose of protein-bound and poorly dialysable drugs. Subsequent *in vitro* studies have shown that these resins bind bilirubin and bile salts, protein-bound plasma components which may be related to hepatic failure.[34]

Early use of the native resins in an animal model of acute hepatic failure was associated with unacceptable platelet losses and several techniques for circumventing the severe bio-incompatibility problems of resin haemoperfusion have been developed, including the perfusion of platelet-poor plasma directly over the resins and development of coating techniques. We have shown that human serum albumin can be bound tightly to XAD-7 resin, and the first clinical trials have shown that the albumin coating has substantially improved the biocompatibility properties of the resin with respect to platelet damage and losses in patients with FHF. Results also show that the albumin-XAD-7 material has a marked binding capacity for bilirubin and bile salts.

Haemodialysis

Theoretically, low molecular weight, water-soluble toxins important in the pathogenesis of acute hepatic encephalopathy should be removed by standard dialysis methods. In a series of experiments in a pig model of acute liver failure, however, Opolon *et al.*[29] have not only found that conventional cuprophane haemodialysis has no effect on deterioration of consciousness, EEG impairment and survival time, but also that this dialysis system has no effect on regional concentrations of brain neurotransmitters and their precursor amino acids. In contrast, haemodialysis using the much more permeable polyacrylonitrile membrane had a marked beneficial effect on levels of consciousness and EEG recordings, and at the end of treatment regional brain neurotransmitter concentrations were within normal limits.

The differential effects of the two membranes on neurotransmitter metabolism suggested that removal of compounds in the middle molecular weight range (1500–5000 daltons) could be beneficial with respect to improvement of conscious levels in patients with acute liver failure. When polyacrylonitrile haemodialysis was applied to the treatment of patients with FHF, significant improvement in conscious levels was reported, although final survival figures are not appreciably better than those achieved by conservative therapy alone. Our own experience with polyacrylonitrile haemodialysis has been encouraging.[4,6] Of 108 patients with FHF showing signs of grade IV coma before treatment who were treated by repeated daily (4–6 hour) haemodialysis, thirty survived to leave hospital. In contrast to the experience of Opolon *et al.*[35] all but one of the patients who recovered consciousness survived to leave hospital.

In the polyacrylonitrile haemodialysis system, dialysis fluid is circulated in a closed system, and it has been possible to accurately quantitate the total amounts of each individual amino acid removed from the circulation of patients with FHF during the course of each treatment period.[21] With the knowledge of the pre-treatment levels as well as the plasma volume, removal can be expressed in terms of the number of pre-treatment plasma pools of amino acids removed during the course of treatment. As Fig. 1 shows, the equivalent of at least five plasma pools of each amino acid was removed in 4 hours' haemodialysis. This gives an indication of the rapidity with which amino acids are transferred from tissue to plasma pools in patients with FHF, and explains how, despite the substantial removal by sorbents and dialysis membranes, circulating amino acid levels after treatment may not be substantially different from those measured before treatment is started. If one accepts that it is the raised plasma levels of toxic metabolites that forms the basis of the encephalopathy, then tissue to plasma re-equilibration of metabolites during treatment could be of key importance with respect to the efficacy of artificial liver support systems in this condition. At present most patients are being treated by sorbent haemoperfusion of haemodialysis for daily periods ranging from 3–6 hours. There appears to be no rational reason for this choice of treatment and if tissue to plasma movement of other potentially toxic metabolites is as rapid as amino acids, then circulating plasma levels may only be effectively lowered by substantially prolonging each treatment period.

Haemodialysis removes circulating glucose and blood levels need to be checked. Also, removal of blood

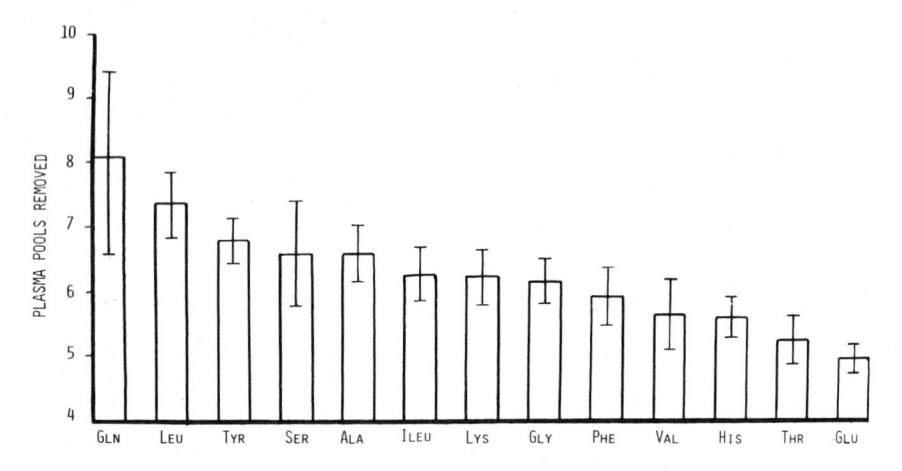

FIG. 1. Number of pretreatment plasma pools removed during 4 hours of haemodialysis with the polyacrylonitrile membrane. Values are mean ± 1SE of data obtained during treatment of 14 patients (7 survivors and 7 non-survivors).

glucose as well as amino acids may result in lowering of the plasma osmolality. It is important to monitor for this, for when large amounts of water solute are removed, for example during a 6-hour period of polyacrylonitrile membrane haemodialysis, small but significant rises in intracranial pressure occur. As many patients have coexisting renal failure, plasma electrolyte, urea, creatinine and urinary electrolyte concentrations should be measured before and after treatment.

Biochemical Selection and Timing of Liver Support in FHF

The availability of a reliable biochemical test which would be of considerable help in predicting prognosis at different stages of the clinical course would be of considerable value in practical management. Considerable efforts have been made in this direction and, for example, recent reports have indicated that serum α-fetoprotein, ^{14}C-cholic acid kinetics, the galactosamine elimination test and Factor VII levels may provide some guide to prognosis. In all these tests, however, overlap between survivors and non-survivors has occurred, which severely limits their usefulness in predicting outcome for the individual patient. We have also assessed the value of twice daily detailed neurological testing and have found absence of the oculo-vestibular reflex to be of absolute prognostic value.[36] Thus none of the patients who lost their reflex during the course of their illness survived. This is not to say that all patients with a positive reflex survive, but that as long as the oculovestibular is present, active liver support therapy should be pursued.

The final comment must be concerned with the question of referral to a specialist unit and the time to do this. Our view is that there is a strong case for early rather than late referral. With very few exceptions, patients who show signs of grade III encephalopathy do continue to deteriorate and it certainly seems more logical to start liver support at this stage when the patients have a positive oculovestibular reflex, rather than leave this until the patient has grade IV encephalopathy with the greater likelihood of cerebral oedema at an irreversible stage. It is at the early stages of encephalopathy that the patient is able to be transferred safely even over long distances.

REFERENCES

1. Trey, C. and Davidson, C. S. (1970), *Progress in Liver Diseases*, Vol. 3 p. 282 (Popper, H. and Schaffner, F. Eds). New York: Grune & Stratton.
2. Ritt, D. J., Whelan, G., Werner, D. J., Eigenbrodt, E. H., Schenker, S. and Combes, B. (1969), *Medicine*, **48**, 151.
3. Saunders, S., Hickman, R., Macdonald, R. and Terbanche, J. (1972), *Progress in Liver Diseases*, Vol. 4, p. 333 (Popper, H. and Schaffner, F., Eds). New York: Grune & Stratton.
4. Silk, D. B. A. and Williams, R. (1978), *International Journal of Artificial Organs*, **1**, 29.
5. Karvountzis, G. G., Redeker, A. G. and Peters, R. L. (1974), *Gastroenterology*, **67**, 870.
6. Silk, D. B. A., Trewby, P. N., Chase, R. A., Mellon, P. J., Hanid, M. A., Davies, M., Langley, P. G., Wheeler, P. G. and Williams, R. (1977), *Lancet*, **ii**, 1.
7. Record, C. O., Iles, R. A., Cohen, R. D. and Williams, R. (1975), *Gut*, **16**, 144.
8. Posner, J. B. and Plum, F. (1960), *J. Clin. Invest.*, **39**, 1246.
9. James, I. M., Nashat, S., Sampson, D., Williams, H. S. and Gavassini, M. (1969), *Lancet*, **ii**, 1106.
10. Ward, M. E., Trewby, P. N., Williams, R. and Strunin, L. (1977), *Anaesthesia*, **32**, 228.
11. Trewby, P. N., Warren, R., Contine, S., Crosbie, W. A., Wilkinson, S. P., Laws, J. W. and Williams, R. (1978), *Gastroenterology*, **74**, 859.
12. Wilkinson, S. P., Blendis, L. M. and Williams, R. (1974), *Brit. Med. J.*, **1**, 186.
13. Alam, A. N., Wilkinson, S. P., Poston, L., Moodie, H. and Williams, R. (1977), *Gastroenterology*, **72**, 914.
14. Wilkinson, S. P., Arroyo, V. A., Moodie, H., Blendis, L. M. and Williams, R. (1976), *Gut*, **17**, 501.
15. Wilkinson, S. P., Weston, M. J., Parsons, V. and Williams, R. (1977), *Clin. Nephrol.*, **8**, 287.
16. Gazzard, B. G., Portmann, B., Murray-Lyon, I. M. and Williams, R. (1975), *Quart. J. Med.*, **44**, 615.
17. Macdougall, B. R. D., Bailey, R. J. and Williams, R. (1977), *Lancet*, **i**, 617.

18. Silk, D. B. A. and Williams, R. (1980), *Sorbents and Their Clinical Application*, p. 415. Academic Press.
19. Thompson, R. P. H. (1976), *Hepatic Support in Acute Liver Failure*, p. 13 (Kuster, G. G. R., Ed.). Springfield, Illinois: Thomas.
20. Zieve, L. (1975), *Artificial Liver Support*, p. 11 (Williams, R. and Murray-Lyon, I. M., Eds). London: Pitman Medical.
21. Chase, R. A., Davies, M., Trewby, P. N., Silk, D. B. A. and Williams, R. (1978), *Gastroenterology*, **75**, 1033.
22. Pardridge, W. M. and Odendorf, W. H. (1977), *J. Neurochem.*, **28**, 5.
23. Fischer, J. F. (1975), *Artificial Liver Support*, p. 31 (Williams, R. and Murray-Lyon, I. M., Eds). London: Pitman Medical.
24. Record, J. O., Buxton, B., Chase, R. A., Curzon, G., Murray-Lyon, I. M. and Williams, R. (1976), *European J. Clin. Invest.*, **6**, 387.
25. Challenger, F. and Walshe, J. E. (1955), *Lancet*, **i**, 1239.
26. Chen, S., Zieve, L. and Mahadevan, V. (1970), *J. Lab. Clin. Med.*, **75**, 628.
27. Lai, J. C., Silk, D. B. A. and Williams, R. (1977), *Clin. Chim. Acta*, **78**, 305.
28. Bron, B., Waldram, R., Silk, D. B. A. and Williams, R. (1977), *Gut*, **18**, 692.
29. Bloch, P., Delmore, M. L., Rapin, J. R., Granger, A., Boschat, M. and Opolon, P. (1978), *Surg. Gynaec. Obstet.*, **146**, 551.
30. Chang, T. M. S. (1972), *Lancet*, **ii**, 1371.
31. Gazzard, B. G., Weston, M. J., Murray-Lyon, I. M., Flax, H., Record, C. O., Portmann, B., Langley, P. G., Dunlop, E. H., Mellon, P. J., Ward, B. M. and Williams, R. (1974), *Lancet*, **i**, 1301.
32. Weston, M. J., Langley, P. G., Rubin, M. H., Hanid, M. A., Mellon, P. J. and Williams, R. (1977), *Gut*, **18**, 897.
33. Rosenbaum, J. L. (1975), *Artificial Liver Support*, p. 118 (Williams, R. and Murray-Lyon, I. M., Eds). London: Pitman Medical.
34. Willson, R. A. (1975), *Artificial Liver Support*, p. 109 (Williams, R. and Murray-Lyon, I. M., Eds). London: Pitman Medical.
35. Denis, J., Opolon, P., Nusinovici, V., Granger, A. and Darnis, F. (1978), *Gut*, **19**, 787.
36. Hanid, M. A., Silk, D. B. A. and Williams, R. (1978), *Brit. Med. J.*, **1**, 1029.

SECTION III

DISORDERS OF THE BLOOD CONSTITUENTS

		PAGE
10.	HAEM SYNTHESIS AND THE PORPHYRIAS	175
11.	HAEMOGLOBINOPATHIES	187
12.	THE ANAEMIAS	211
13.	ABNORMALITIES OF PLASMA PROTEINS	221
14.	THE CLINICAL BIOCHEMISTRY OF BLOOD COAGULATION	251

10. HAEM SYNTHESIS AND THE PORPHYRIAS

G. H. ELDER

Clinical chemistry

Haem biosynthesis
   The pathway of haem biosynthesis
   The regulation of haem biosynthesis

Excretion of haem precursors

Enzyme defects in the porphyrias

Clinical features of the porphyrias
   Porphyrias in which acute attacks occur
   Porphyrias in which skin lesions occur

Laboratory differentiation of the porphyrias
   Patients with symptoms which suggest acute porphyria
   Patients who present with skin lesions

Laboratory tests for the definitive diagnosis and management of the porphyrias
   Acute hepatic porphyrias
   Variegate porphyria and hereditary coproporphyria
   Porphyria cutanea tarda
   Protoporphyria and congenital erythropoietic porphyria

The porphyrias are a group of disorders of haem biosynthesis in which characteristic clinical features are associated with increased formation of porphyrins and porphyrin precursors (Table 1). The primary abnormality in each type of porphyria is a decrease in the activity of a particular enzyme of the biosynthetic pathway. Each type is defined by a specific pattern of accumulation and excretion of haem precursors which reflects an increase in the intracellular concentration of the substrate of the defective enzyme. Recognition of these patterns by the selection and correct interpretation of appropriate biochemical investigations is essential for the accurate diagnosis and the proper management of the porphyrias since the clinical features alone are not sufficiently distinctive.

CLINICAL CHEMISTRY

Porphyrins are cyclic tetrapyrroles which contain a central porphyrin macrocycle and differ according to the number and nature of the β-substituents at positions 1–8 (Fig. 1). In natural porphyrins derived from haem biosynthesis at least six of these positions are substituted. For porphyrins that contain more than one type of β-substituent, there are a number of different position isomers according to the order in which the substituents are arranged around the macrocycle. When only two types of substituent are present, and each pyrrole ring contains both, there are four possible position isomers, types I–IV. Of these, only the symmetrical type I isomer, in which the substituents alternate around the macrocycle, and the asymmetrical type III isomer (Fig. 1) occur in nature.

Porphyrins in solution have characteristic absorption spectra with a strong absorption band, the Soret band, around 400 nm. Absorption in this region is greatest when the porphyrin is in the di-cation form, and measurement of the absorption at the Soret band max-

TABLE 1
CLASSIFICATION AND MAIN CLINICAL FEATURES OF THE PORPHYRIAS

Condition	Main clinical features	
	Acute attacks	Skin lesions
A. Porphyrias with increased red cell porphyrins		
1. Congenital erythropoietic porphyria	0	+
2. Protoporphyria	0	+
B. Porphyrias with normal red cell porphyrins: hepatic porphyrias		
1. Acute intermittent porphyria	+	0
2. Hereditary coproporphyria	+	+
3. Variegate porphyria	+	+
4. Porphyria cutanea tarda	0	+

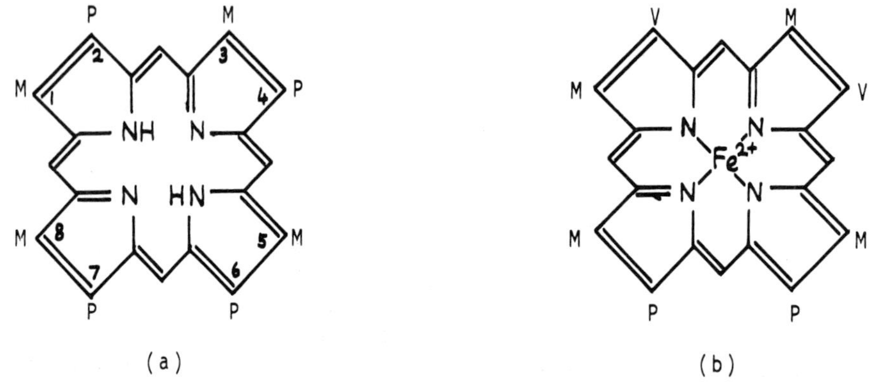

Fig. 1. The structure of (a) coproporphyrin III and (b) haem IX. M, P and V denote methyl, propionic acid and vinyl substituents respectively.

imum in dilute acid is a widely used and sensitive method for determining porphyrin concentration. Porphyrins also have characteristic fluorescence spectra, showing intense red fluorescence when irradiated with long wavelength ultraviolet light. This property provides a convenient method for their detection in solution and on chromatograms and has also been extensively used for quantitative determination, being more sensitive than spectrophotometry.[20,24]

Porphyrins with acidic substituents are soluble in dilute mineral acids and bases. They are least soluble at their iso-electric points around pH 3·5 and at this pH are readily extracted into polar organic solvents, such as diethyl ether, ethyl-acetate or butan-l-ol. Porphyrins can be extracted from ether and ethylacetate by dilute mineral acids. Partition between these solvents and the acid phase depends on the number and type of β-substituents on the porphyrin and the strength of the acid. Thus protoporphyrin, which has two carboxylic-acid side chains, is more ether-soluble than the tetracarboxylic coproporphyrin and requires stronger hydrochloric acid for its complete extraction from ether. Porphyrins with more than five carboxyl groups are still more hydrophilic, uroporphyrin being ether-insoluble. In contrast, haem, which contains no ring nitrogen atoms that can be protonated (Fig. 1) is more soluble in ether and ethyl-acetate than porphyrins and remains in these solvents at acid strengths which extract all natural porphyrins. These differences in the solubility of porphyrins and haem form the basis of the solvent extraction techniques, such as those devised by Rimington,[51] for the fractionation of mixtures of porphyrins from urine, faeces and blood, which are in widespread current use. Because some natural porphyrins have very similar solubility properties, the fractions obtained by these techniques often contain more than one porphyrin and a porphyrin other than that after which the fraction is named may predominate. For example, coproporphyrin fractions may contain mesoporphyrin, deuteroporphyrin or isocoproporphyrin as the main component.[24]

In recent years, solvent extraction techniques have been partly superseded by chromatographic methods which enable individual porphyrins to be isolated. Various techniques using thin-layer chromatography (TLC) or high-performance liquid chromatography (HPLC) have been described,[20,24,35] most of which require prior conversion of porphyrins to their methyl ester derivatives which are more soluble in organic solvents. In general, separation of porphyrins or their methyl ester derivatives according to the number of carboxylic acid or other hydrophilic groups is easily obtained by HPLC or TLC, while separation of position isomers is more difficult but can be achieved by either technique.

HAEM BIOSYNTHESIS

Haem biosynthesis takes place in all cells that contain functioning mitochondria. Adult human subjects synthesize about 0·45 mmol of haem each day. Some 70–80% of this is produced in erythroid cells for the formation of haemoglobin. Most of the remaining 20–30% is synthesized in the liver where a major fraction is used for the assembly of microsomal haemoproteins of the cytochrome P-450 series which act as catalysts for the oxidative metabolism of many drugs, chemicals and endogenous compounds.[42]

The Pathway of Haem Biosynthesis

The pathway of haem synthesis is outlined in Fig. 2. The enzymology of each reaction has been reviewed[42] and will not be described here. Although most of the reactions take place in the cytosol fraction of the cell, the first and final stages take place in the mitochondrion (Fig. 2) and the enzymes catalysing these reactions are not present in mature erythrocytes. The first part of the pathway consists of a series of irreversible condensation reactions, initiated by the formation of 5-aminolaevulinate (ALA) from glycine and succinyl-CoA which is catalysed by ALA-synthetase (Fig. 2). These reactions lead to the formation of the asymmetrical porphyrinogen, uroporphyrinogen III. The polymerization of four molecules of porphobilinogen (PBG) to uroporphyrinogen III is catalysed by two enzymes, uroporphyrinogen-I-synthase and uroporphyrinogen-III-cosynthase. Recent work suggests that the first of these

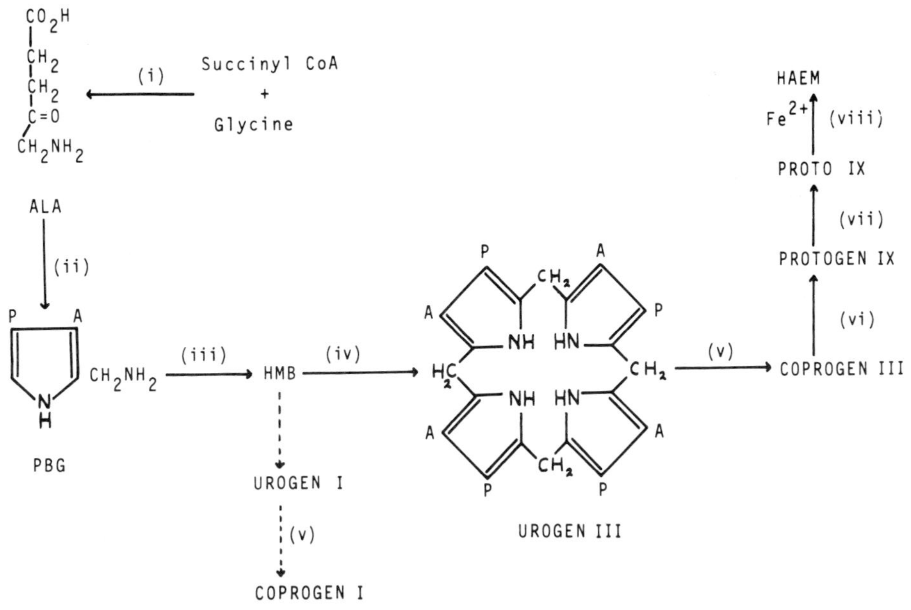

FIG. 2. The pathway of haem biosynthesis. Reactions are catalysed by (i) ALA-synthetase (ii) ALA-dehydratase (iii) uroporphyrinogen-I-synthase (PBG deaminase) (iv) uroporphyrinogen-III-cosynthase (v) uroporphyrinogen decarboxylase (vi) coproporphyrinogen oxidase (vii) protoporphyrinogen oxidase (viii) ferrochelatase. HMB, hydroxymethyl bilane, the linear tetrapyrrole intermediate between PBG and uroporphyrinogen III; A, acetic acid substituent; other abbreviations as in Fig. 1.

enzymes produces an unstable, symmetrical linear tetrapyrrole which serves as a substrate for the second enzyme which catalyses its rearrangement and cyclization to uroporphyrinogen III.[3] In the absence of uroporphyrinogen-III-cosynthase, the linear tetrapyrrole undergoes chemical cyclization to the symmetrical isomer, uroporphyrinogen I.

In the second part of the pathway a series of side-chain decarboxylation reactions produces protoporphyrinogen IX, which then undergoes aromatization to protoporphyrin IX with final insertion of ferrous iron to form haem (Fig. 2).

The porphyrinogens that are the intermediates in the conversion of PBG to protoporphyrin IX are unstable, colourless compounds that readily oxidize to porphyrins which are not metabolized. Thus increased formation of porphyrinogens leads to accumulation in tissues and increased excretion of both porphyrinogens and porphyrins. Since standard techniques for the determination of porphyrins in urine and faeces include procedures for the oxidation of porphyrinogens, these compounds are included in the values that are obtained for porphyrin excretion.

The Regulation of Haem Biosynthesis

The rate of haem synthesis is determined by the activity of the first enzyme of the pathway, ALA-synthetase, which under normal circumstances has a lower activity than the subsequent enzymes. Changes in the rate are brought about by alterations in the activity of this enzyme. The regulation of ALA-synthetase activity has been investigated extensively and much of this work has been reviewed.[30,42]

In liver, ALA-synthetase activity appears to be under negative feedback control by haem. Thus induction of ALA-synthetase by drugs is prevented by administration of haem, while compounds that decrease the concentration of haem in the liver enhance its activity. Immunochemical measurements of enzyme protein in avian hepatocytes have shown that haem inhibits the synthesis of ALA-synthetase in response to inducing chemicals, but there is only indirect evidence for a similar mechanism in mammalian liver. Regulation by end-product repression of synthesis allows effective short-term control because the turnover of ALA-synthetase is rapid, the half-life in rat liver being about 60 minutes. The ALA-synthetase of liver differs from that of other tissues in that it is induced by a wide variety of lipophilic drugs, such as barbiturates and foreign chemicals, many of which are metabolized by the cytochrome P-450-mediated mono-oxygenase system. There is now evidence that these compounds primarily stimulate the synthesis of apocytochromes of the P-450 series, which then combine with pre-formed haem and thus, by depleting the intracellular concentration of 'regulatory' haem, secondarily stimulate ALA-synthetase activity.

ALA-synthetase activity in erythroid tissue is similarly decreased by haem, but here the mechanism appears to involve direct inhibition and fewer factors are known that increase activity. Thus, erythroid ALA-synthetase activity is increased by erythropoietin and hypoxia, which do not affect the hepatic enzyme, but is not altered by the majority of drugs and chemicals; certain

TABLE 2
REFERENCE RANGES FOR HAEM PRECURSORS IN BLOOD, URINE AND FAECES

Sample	Haem precursor	Concentration or excretion rate
Erythrocytes	Total porphyrin (greater than 95% protoporphyrin)	less than 1600 nmol/l erythrocytes[1]
Plasma	Total porphyrin	less than 0·5 nmol/dl
Urine	PBG	less than 8·8 μmol/day
	ALA	" " 46 " "
	Total porphyrin	" " 410 nmol/day (320 nmol/l)
	Coproporphyrin fraction	" " 280 " "
	Uroporphyrin fraction	" " 40 " "
	Coproporphyrin	23–115 nmol/l
	Penta	less than 5 " "
	Hexa	" " 3 " "
	Hepta	" " 4 " "
	Uroporphyrin	" " 24 " "
Faeces	Coproporphyrin fraction	" " 46 nmol/g dry wt
	Protoporphyrin fraction	" " 134 " " "
	Ether-insoluble porphyrin	" " 24 " " "

Figures from Day et al.,[17] Doss,[20] Eales et al.,[21] Elder et al.,[25] Fogstrup & With,[28] Chisholm & Brown.[14] Abbreviations as in Fig. 2.
[1]By a fluorometric micromethod; macroscale solvent extraction procedures give values that are about 50% lower.

5β-H steroids are the only compounds that are active in both systems. The co-ordination of haem and globin synthesis is achieved through initiation of globin chain synthesis by haem.

EXCRETION OF HAEM PRECURSORS

In relation to the quantities used for haem synthesis, only small amounts of haem precursors accumulate in tissues and are excreted in the urine and faeces under normal circumstances (Table 2).

ALA and PBG are excreted exclusively in the urine. The distribution of porphyrins between urine and faeces is determined by their chemical structure and the excretory capacity of the liver. In general, urinary excretion becomes increasingly favoured as the number of carboxyl groups increases. Thus uroporphyrin is excreted mainly in the urine, while protoporphyrin excretion is restricted to the bile. Coproporphyrin I is taken up and excreted by the liver in performance to the series III isomer. Thus faecal coproporphyrin is mainly coproporphyrin I while both isomers are present in about equal amounts in urine. When the excretory function of the liver is impaired, coproporphyrin is diverted from the biliary route and both the total amount of coproporphyrin and the percentage of the series I isomer increases in the urine. The only exception to this finding is in the Dubin-Johnson syndrome where a characteristic decrease in the urinary excretion of coproporphyrin III is found.[61]

Normal faeces contain more protoporphyrin and other dicarboxylic porphyrins than can be accounted for by biliary excretion. It is probable that much of this dicarboxylic porphyrin comes from the action of bacteria on haem derived from the lining of the alimentary tract and from the diet.

ENZYME DEFECTS IN THE PORPHYRIAS

In recent years primary enzyme defects have been identified in all the main types of porphyria (Table 3), although there is some controversy over the location of the defects in congenital erythropoietic[43] and variegate porphyria.[5,10] For each disorder, the observed abnormality of haem precursor excretion can be explained by accumulation of the substrate of the defective enzyme.

With the exception of some types of porphyria cutanea tarda, each form of porphyria has a well-defined pattern of inheritance. Apart from congenital erythropoietic porphyria, all are inherited as autosomal dominant disorders (Table 3). In acute intermittent porphyria, hereditary coproporphyria and variegate porphyria, activities of the defective enzymes are close to 50% of the mean values for subjects without porphyria; this finding is consistent with inheritance of a mutant gene which is not expressed as active enzyme. In protoporphyria, the activity of ferrochelatase is decreased by 75–90%; this has been attributed to an alteration in the microenvironment or change in structure of the residual enzyme. In contrast, porphyria cutanea tarda appears to be a syndrome in which the primary abnormality is a decrease in the activity of hepatic uroporphyrinogen decarboxylase that can arise in several different ways. In some patients, many of whom have a family history of the disorder, the decreased activity is inherited as an autosomal dominant characteristic[6] and can be detected by measuring the enzyme in erthyrocytes.[37] In rare instances, the enzyme defect may be acquired by exposure to chlorinated aromatic hydrocarbons as in the outbreak of porphyria caused by hexachlorobenzene in Turkey between 1956 and 1960 or following accidental industrial exposure to 2,3,7,8-tetrachlorodibenzo-p-dioxin. However, most patients with porphyria cutanea

TABLE 3
ENZYME DEFECTS IN THE PORPHYRIAS

Condition	Enzyme defect	Inheritance	Reference
Congenital erythropoietic porphyria	Uroporphyrinogen-III-cosynthase	Autosomal recessive	Levin[40]
Protoporphyria	Ferrochelatase	Autosomal dominant	Bottomley et al.[9] Bonkowsky et al.[8]
Acute intermittent porphyria	Uroporphyrinogen-I-synthase	Autosomal dominant	Strand et al.[57]
Hereditary coproporphyria	Coproporphyrinogen oxidase	Autosomal dominant	Elder et al.[26]
Variegate porphyria	Protoporphyrinogen oxidase	Autosomal dominant	Brenner and Bloomer[10]
Porphyria cutanea tarda	Uroporphyrinogen decarboxylase	At least two types: autosomal dominant in one type, inheritance not established in others, some may be acquired	Kushner et al.[37]

tarda have neither a family history nor evidence of poisoning with halogenated hydrocarbons. In this group, the enzyme defect is restricted to the liver[27,58] and its pathogenesis is uncertain.

In all forms of porphyria where inheritance of the enzyme defect has been demonstrated, decreased activity has been found in all tissues that have been studied. Thus, in acute intermittent porphyria, uroporphyrinogen-I-synthase activity is decreased in the liver, erythrocytes, cultured skin fibroblasts and amniotic cells, and transformed lymphocytes.[54]

The increase in intracellular substrate concentration that accompanies each enzyme defect appears to be a compensatory adjustment that enables the rate of haem synthesis to be maintained. Under normal circumstances the concentrations of the substrate of the pathway are lower than the Michaelis constants for the corresponding enzymes. Thus any decrease in the rate of haem synthesis that would follow a decrease in the activity of one of the enzymes of the pathway can be compensated by increasing the rate of formation of ALA, and subsequent intermediates, through operation of the negative feedback mechanism for the control of ALA-synthetase activity. Reports of increased ALA-synthetase activity in all forms of porphyria[42] and indirect evidence that the rate of haem synthesis is normal, except during attacks of acute hepatic porphyria,[1] are consistent with this hypothesis.

In contrast to the tissue distribution of the enzyme defects, the compensatory changes are restricted to certain tissues, Thus, in the hepatic porphyrias (Table 1) they are evident only in the liver and there is no detectable increase in the concentration of porphyrins or porphyrin precursors in erythroid cells. In practice, measurement of erythrocyte porphyrin concentration enables the porphyrias to be divided into two groups which show important clinical differences and has been used as the basis of the simple classification shown in Table 1. This classification does not include some very uncommon types of porphyria that have been described nor porphyria caused by production of porphyrins by hepatic tumours, a syndrome that is clinically indistinguishable from porphyria cutanea tarda.[32]

CLINICAL FEATURES OF THE PORPHYRIAS

Patients with porphyria present in one of three ways: with some or all of the features of an acute attack of porphyria, with skin lesions alone, or with acute attacks accompanied by skin lesions. None of these modes of presentation is confined to a single type of porphyria but not all occur in all disorders (Table 1). In variegate porphyria and hereditary coproporphyria acute attacks are not necessarily accompanied by skin lesions or *vice versa*.

In most families with porphyria, individuals with latent porphyria, who have inherited the enzyme defect but have not developed any clinical features, are commoner than those with symptoms.

Porphyrias in Which Acute Attacks Occur

Acute attacks of porphyria occur only in acute intermittent porphyria, variegate porphyria and hereditary coproporphyria (Table 1), and are clinically identical in each of these disorders. In acute intermittent porphyria episodic acute attacks are the only clinical feature, but they are accompanied by skin lesions in about half of the patients with variegate porphyria[46] and in about a third of those with hereditary coproporphyria[11] that present in this way.

The clinical features of acute porphyria have been reviewed.[42,56] The commonest symptom is severe abdominal pain, which may be accompanied by neurological and mental disturbances. About two-thirds of patients develop muscular weakness which may progress to quadriparesis and respiratory paralysis. Acute attacks are commoner in women and are most frequent in the third and fourth decades, being very uncommon before puberty.

Hyponatraemia is common, when vomiting is severe or renal function deteriorates. Occasionally it may be caused by inappropriate secretion of antidiuretic hormone.[56] Hypothalamic neuro-endocrine dysfunction has been postulated to explain this finding and asymptomatic abnormalities of growth hormone and ACTH control.[42,56] Various abnormalities of thyroid function have also been described but patients are usually euthyr-

TABLE 4
SOME FACTORS THAT MAY PRECIPITATE ACUTE ATTACKS OF PORPHYRIA

A. Drugs*

   Barbiturates, chlordiazepoxide, chlorpropamide, diazepam, griseofulvin, glutethimide, hydantoins, methyldopa, oestrogens, oral contraceptives, phenylbutazone, phensuximide and related drugs, pyrazolones, sulphonamides, tolbutamide.

B. Other factors

   Low calorie diet
   Cyclic-menstrual
   Alcohol
   Infection
   Emotional stress

*A more complete list and a list of drugs that are considered safe is given by Moore.[44]

oid, although transient hyperthyroidism has been reported.[13,42]

Most acute attacks are precipitated by drugs, especially barbiturates, endocrine factors or calorie restriction (Table 4). Many of the drugs that provoke acute attacks stimulate the synthesis of, and are metabolized by, cytochrome P-450. The increase in hepatic ALA-synthetase activity and PBG formation that accompanies acute porphyria suggests that, in these disorders, the short-term demand for increased haem synthesis imposed by these drugs cannot be met, because the rate of haem synthesis becomes limited by uroporphyrinogen-I-synthase activity. The activity of this enzyme is substantially lower than the activities of the other enzymes that convert ALA to haem and, even when it is not the site of the primary enzyme defect as it is in acute intermittent porphyria, it may become rate-limiting before other enzymes of the pathway.[26] Thus, during the acute attack, there appears to be hepatic haem deficiency with sustained derepression of ALA-synthetase. Although the relation between these biochemical changes and the neurological disturbance that underlies all the clinical features of the acute attack has not been defined,[4] two manoeuvres that are known to prevent the induction of hepatic ALA-synthetase in laboratory animals appear to be effective in the treatment of acute porphyria in humans. Thus a number of reports have shown that either administration of carbohydrate[12] or haematin[59] decreases PBG formation and may produce rapid remission.

Porphyrias in Which Skin Lesions Occur

Porphyria Cutanea Tarda, Variegate Porphyria, Hereditary Coproporphyria, Congenital Erythropoietic Porphyria

The skin lesions in these disorders occur in sun-exposed areas and are fundamentally similar, although their severity varies. The most prominent abnormality is increased mechanical fragility with displacement of the epidermis and the formation of erosions in response to trivial injury. Subepidermal bullae are frequent and are the usual presenting feature. Pigmentation and hirsutism are common.

The commonest disorder in this group, and the most frequent form of porphyria in Europe and N. America, is porphyria cutanea tarda. The clinical features have been described by Grossman et al.[33] Most cases occur in association with liver cell damage, particularly when caused by alcohol, but synthetic and natural oestrogens, including the contraceptive pill, are also important precipitating factors. Biochemical evidence of liver dysfunction, especially minor increases in serum aspartate transaminase and ferritin levels, is common. Jaundice or severe liver disease is uncommon at presentation. Needle biopsy of the liver usually reveals inflammatory changes often with some fatty infiltration, with cirrhosis being present in less than a third of the patients in most series. The liver contains an increased concentration of uroporphyrin and needle biopsy cores frequently show red fluorescence when viewed in long-wave ultraviolet light. Almost all patients have hepatic siderosis and total body iron stores are increased in about two-thirds of European patients, although rarely to the extent found in haemochromatosis. Depletion of iron stores by repeated venesection leads to clinical and biochemical remission in the majority of patients[50] as does treatment with chloroquine. Acute attacks of porphyria do not occur in porphyria cutanea tarda and there is no reaction to drugs such as barbiturates.

When variegate porphyria presents with skin lesions alone, it cannot be distinguished clinically from porphyria cutanea tarda. Neither the family history nor the usual absence of liver disease provide reliable means of differentiation. The proportion of patients that present in this way appears to vary from country to country[46] but is probably of the order of 60% in the United Kingdom. In contrast, hereditary coproporphyria almost always presents as acute porphyria, and is accompanied by skin involvement in about one-third of cases.[11]

Congenital erythropoietic porphyria is a very rare disorder in which severe skin lesions caused by photosenzitization by porphyrins usually develop in infancy and may be accompanied by haemolytic anaemia. Recently a milder form of this condition, with onset in adult life, has been described that clinically resembles porphyria cutanea tarda.

Protoporphyria

The characteristic clinical features of this form of cutaneous porphyria have been described by De Leo et al.[19] Onset usually occurs during childhood. Photosensitivity with an oedematous reaction in sun-exposed skin is the most prominent feature. Bullae are unusual and increased mechanical fragility is not present. Residual scarring from healed lesions is much less prominent than in the other cutaneous porphyrias. The most important complications are cholelithiasis, with gall-stones that contain protoporphyrin, and liver disease. The latter develops in a small proportion of patients and usually progresses rapidly to death in liver failure with massive

TABLE 5
LABORATORY DIFFERENTIATION OF THE PORPHYRIAS

Condition	Main haem precursor present in excess		
	Erythrocytes	Urine	Faeces
Congenital erythropoietic porphyria	Uroporphyrin I Coproporphyrin I	Uroporphyrin I	Coproporphyrin I
Protoporphyria	Protoporphyrin IX	–	Protoporphyrin may be increased
Acute intermittent porphyria	–	Porphobilinogen	–
Hereditary coproporphyria	–	Coproporphyrin III. Porphobilinogen during acute attack. Both usually normal during remission	Coproporphyrin III
Variegate porphyria	–		Coproporphyrin III, protoporphyrin IX, 'X-porphyrins'.
Porphyria cutanea tarda	–	Uroporphyrin I and III heptacarboxylic and other acetate-substituted porphyrins	Isocoproporphyrin, heptacarboxylic porphyrin and other acetate-substituted porphyrins

accumulation of protoporphyrin in the liver.[5,15] At present this complication cannot be predicted by assay of porphyrins or by other investigations.

LABORATORY DIFFERENTIATION OF THE PORPHYRIAS

All the porphyrias can be accurately differentiated by measurements of haem precursors (Table 5) provided that fresh or adequately preserved[24] samples of urine, faeces and erythrocytes are examined. Failure to investigate the faeces is an important source of error.

Many of the techniques used to measure haem precursors are complex and time-consuming. Most laboratories, therefore, rely heavily on screening tests for the rapid exclusion of porphyria as, for example, in the investigation of abdominal pain. This section describes a scheme for the investigation of suspected porphyria in which simple screening tests are employed as the first stage. A similar approach has been outlined by With.[60] Descriptions of appropriate methods have been given by Doss[20] and Elder.[24]

Patients with Symptoms Which Suggest Acute Porphyria

Acute attacks of porphyria are always associated with increased excretion of PBG. The urine may be dark reddish-brown, due to polymerization of PBG to porphyrins and other compounds, or normal in colour, perhaps later darkening on standing.

In this group, therefore, the essential test is examination of the urine for PBG. This can be done by a screening test in which fresh urine (2·5 ml) and Ehrlich's reagent (0·24 g/dl of p-dimethylaminobenzaldehyde in 7 M-HCl) (2·5 ml) are mixed, allowed to stand for one to two minutes, mixed with saturated sodium acetate (5 ml) and shaken with 2·5 ml of amyl alcohol-benzyl alcohol (3:1, v/v) or butanol. The red colour formed by PBG appears on addition of Ehrlich's reagent and is not extracted into the organic phase. Extraction must be repeated until the organic phase is colourless before the test can be regarded as positive. Urobilinogen and some other compounds also react with Ehrlich's reagent but are extracted into the organic phase (Table 6). If chloroform is used as the extractant in this test, false positive reactions are more frequent.[48] A positive screening test must always be confirmed by a quantitative determination of PBG using a specific method such as that described by Mauzerall & Granick.[41]

Demonstration that the positive screening test is due to increased PBG excretion strongly supports the diagnosis of an acute attack of acute intermittent porphyria, hereditary coproporphyria or variegate porphyria. The only other condition in which PBG excretion may be sufficiently increased is severe lead poisoning.

A negative screening test does not exclude increased PBG excretion. The lower limit of sensitivity of the screening test is about 27 μmol/l which is about three times the upper limit of normal (Table 2). However, if the abdominal pain of a patient is severe enough to

TABLE 6
COMPOUNDS IN URINE WHICH GIVE A RED COLOUR WITH EHRLICH'S REAGENT

A. Compounds that react with p-dimethylaminobenzaldehyde
  (a) Urobilinogen
  (b) Porphobilinogen
  (c) Unidentified metabolites of:
    (i) Methyldopa
    (ii) Levomepromazine
    (iii) Cascara sagrada back extract

B. Compounds that give a red colour with 7M HCl
  (a) Food additives: methyl red
  (b) Phenazopyridine HCl (pyridium)
  (c) Indoles: indoleacetic acid

require admission to hospital and is caused by acute porphyria, it is very unlikely that the screening test will be negative, provided fresh urine is examined. In the later stages of an acute attack, PBG excretion may decrease so that the screening test becomes negative. Thus quantitative measurement is essential when investigations are carried out later in the course of an attack, perhaps when only residual neurological symptoms are present, or when attempting to make a retrospective diagnosis of an acute attack.

In variegate porphyria and hereditary coproporphyria the urinary PBG concentration may fall rapidly as the acute attack subsides and often returns to normal during remission. In acute intermittent porphyria, PBG excretion usually remains increased during remission. The extent of the increase varies markedly from patient to patient and in one patient may be higher during remission than in another during an acute attack. For this reason, measuring the concentration of PBG in the urine is often unhelpful in differentiating acute porphyria from other causes of abdominal pain in a patient who is known to have acute intermittent porphyria, unless serial measurements show that a marked increase occurs in association with symptoms.

Although the urine in acute porphyria usually contains excess porphyrins, measurements of urinary porphyrins are of little diagnostic value in this group of patients. In practice, screening tests for excess PBG and porphyrins are often carried out together. Provided the sample is fresh, a positive screening test for porphyrin and negative test for PBG on urine from a patient with abdominal pain is almost always due to one of the causes of secondary coproporphyrinuria (Table 7). If there is any suspicion that the sample is not fresh, and that any excess PBG present may have polymerized to form porphyrin, fresh urine must be obtained.

TABLE 7
CAUSES OF COPROPORPHYRINURIA OTHER THAN PORPHYRIA

A. Toxic conditions
　(a) Alcoholism
　(b) Lead poisoning

B. Impaired biliary excretion
　(a) Hepatocellular disease
　(b) Cholestasis
　(c) Pregnancy
　(d) Oestrogen therapy, oral contraceptives

C. Others (? impaired biliary excretion)
　(a) Dubin-Johnson syndrome
　(b) Miscellaneous systemic diseases: severe infections, rheumatic fever, reticuloses

The other eseential investigation for this group of patients is examination of the faecal porphyrins. This test enables acute intermittent porphyria, variegate porphyria and hereditary coproporphyria to be differentiated and also detects those patients with variegate porphyria or hereditary coproporphyria who present, or are investigated, after the urinary screening test for PBG has become negative.

Quantitative analysis of faecal porphyrins is time-consuming and a simple screening test can be used to detect samples that do not require further investigation. A small lump of faeces (about 0·5 to 1 cm in diameter) and 0·5 ml glacial acetic acid are ground together with a glass rod and the resulting suspension is mixed thoroughly with 5 ml ether. After centrifugation, the supernatant is decanted and shaken with 0·5 ml 1·4 M HCl, which extracts porphyrins related to haem synthesis but leaves those derived from dietary chlorophyll in the organic phase. A positive test is shown by a faint pink to bright red fluorescence when the acid phase is examined in ultraviolet light (Wood's filter, peak emission 365 nm). Intensity of fluorescence is proportional to the porphyrin content of the sample. Absence of pink or red fluorescence in the acid phase indicates that faecal porphyrin excretion is within normal limits and, in a patient with acute porphyria, confirms a diagnosis of acute intermittent porphyria.

Interpretation of a positive screening test is less straightforward. In a patient with symptoms suggestive of acute porphyria it usually indicates increased faecal porphyrin excretion and thus a diagnosis of variegate porphyria or hereditary coproporphyria. Quantitative determination of faecal porphyrins is then essential both to differentiate these conditions (see below) and to eliminate other causes of a positive screening test. Thus occasionally the test may be positive in subjects with normal porphyrin excretion, particularly those that are constipated. Patients with gastro-intestinal bleeding often have increased faecal porphyrin excretion because haem is converted by gut bacteria to protoporphyrin and other dicarboxylic porphyrins. In such cases, tests for occult blood in faeces are usually positive and measurement of individual porphyrins shows an increase in dicarboxylic porphyrins with a normal concentration of coproporphyrin. Finally, patients with cutaneous porphyria, particularly porphyria cutanea tarda, may present with abdominal or neurological symptoms due to some cause other than porphyria, for example, alcoholism, and may be misdiagnosed unless the appropriate quantitative measurements are made on faeces and urine.

Patients Who Present With Skin Lesions

The full investigation of a patient who presents with skin lesions that might be due to porphyria entails measurement of porphyrins in blood, faeces and urine. In practice, the order in which these investigations are carried out is determined by the clinical presentation. Thus, if protoporphyria is suspected, erythrocyte porphyrin measurement is the first and most important investigation, whereas if skin lesions of the type seen in porphyria cutanea tarda are present, determination of urinary and faecal porphyrins takes precedence. Unless all three types of sample are examined, the rarer types of porphyria such as porphyria cutanea tarda in children, congenital erythropoietic porphyria[34] and hepato-erythropoietic porphyria[16] may be misdiagnosed.

An increased concentration of porphyrin in red blood cells may be detected by screening tests based on solvent extraction or by fluorescence microscopy of erythrocytes.[52] In the solvent extraction procedure, 0·1 ml of blood is added to 2·5 ml of ether: acetic acid (5:1, v/v) and macerated with a glass rod. The supernatant is transferred to a thin-walled glass tube and shaken with 0·5 ml of 3M HCl which extracts porphyrin but leaves haem in the organic phase. When examined in ultraviolet light, the presence of a faint pink to bright red fluorescence in the acid phase indicates excess porphyrin. If the test is positive, it should be repeated with washed red cells to ensure that the excess porphyrin is in these cells and not in the plasma, which should be examined separately. This manoeuvre is essential, since the plasma porphyrin concentration may be sufficiently increased in porphyria cutanea tarda, variegate porphyria or hereditary coproporphyria to give a positive screening test when whole blood is used. Small increases in red cell porphyrin content may not give a positive result with this screening test and occasional cases of protoporphyria may be missed. The second type of screening test, in which a saline-diluted unfixed blood smear is examined for red-fluorescent erythrocytes by fluorescence microscopy[52] is more reliable and is specific for erythrocyte porphyrin.

In recent years, a number of fluorometric micromethods for the measurement of erythrocyte porphyrin have been introduced.[24] These methods are rapid and simple and, in many laboratories, are replacing screening tests and more cumbersome quantitative methods. Some, which employ acid extraction, measure 'free erythrocyte porphyrin' because any zinc-protoporphyrin present in the cell is converted to protoporphyrin under acidic conditions. Others allow both protoporphyrin and its zinc-chelate to be determined because the measurement is carried out under neutral conditions.[38]

The red cell porphyrin concentration is increased in protoporphyria and in congenital erythropoietic porphyria, and may be increased in lead poisoning, iron deficiency and some other disorders (Table 8). In normal subjects and patients with iron deficiency or lead poisoning the red-cell protoporphyrin is mainly present as zinc-protoporphyrin. Protoporphyria differs from these conditions in that the red cell porphyrin is protoporphyrin rather than its zinc chelate, and demonstration of the presence of protoporphyrin by fluorescence emission spectroscopy of a neutral red extract is a useful confirmatory test for this disorder (Fig. 3). Conversely, methods

TABLE 8
CONDITIONS IN WHICH RED CELL PROTOPORPHYRIN IS INCREASED

A. Protoporphyrin
   (a) Protoporphyria

B. Zinc-protoporphyrin
   (a) Iron deficiency
   (b) Lead poisoning
   (c) Sideroblastic anaemias
   (d) Haemolytic anaemias

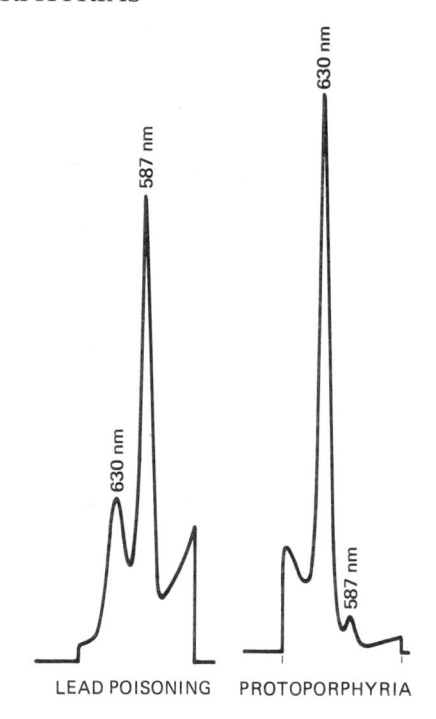

FIG. 3. Fluorescence emission spectra of ethanol extracts of whole blood. The excitation wavelength was 415 nm.

which are designed to measure only zinc-protoporphyrin will not detect protoporphyria unless measurements can be made at the appropriate wavelength.

In other types of porphyria that present with skin lesions, the red cell porphyrin concentration is normal (Tables 1 and 5). But in these conditions, as in congenital erythropoietic porphyria, urinary porphyrin excretion is likely to be increased. Again screening tests for excess porphyrins are widely used.[20,21] A simple method is to shake urine (4 ml), adjusted to pH 3 to 4 with glacial acetic acid, with amyl alcohol (1·0 ml) which extracts all urinary porphyrins. The presence of excess porphyrins is shown by a pink or red fluorescence when the upper phase is examined in ultraviolet light. This test will not detect porphyrin concentrations lower than about 1·0 μmol/l and may be negative in patients whose skin lesions are subsiding and when the urine contains compounds which interfere with porphyrin fluorescence. In a patient with skin lesions, a positive test supports a clinical diagnosis of porphyria and indicates that further investigations are required. Secondary coproporphyrinuria (Table 7) may also produce a positive screening test but can be distinguished from all types of porphyria with increased urinary coproporphyrin excretion by showing that the faecal porphyrin concentration is normal.

The faecal screening test for excess porphyrins should also be carried out on all patients who present with skin lesions. It will be positive in variegate porphyria, hereditary coproporphyria and congenital erythropoietic porphyria and in some patients with porphyria cutanea tarda or protoporphyria. Even when all screening tests are negative it may be necessary to undertake some of the more detailed investigations described below if the clinical evidence for cutaneous porphyria is strong. For example, the total amount of porphyrin excreted in urine or faeces may not be sufficiently increased in porphyria cutanea tarda during remission or when only healed or resolving skin lesions are present to give positive screening tests, yet measurement of individual porphyrins will enable a certain diagnosis to be made.

LABORATORY TESTS FOR THE DEFINITIVE DIAGNOSIS AND MANAGEMENT OF THE PORPHYRIAS

Consideration of the clinical history and the results of the simple tests described in the proceding sections should enable patients who require further investigation to be identified and indicate the type of porphyria that is likely to be present. This section describes tests that are required to establish the diagnosis of a particular type of porphyria, to identify gene carriers and to monitor treatment.

Acute Hepatic Porphyrias

Acute Intermittent Porphyria

This condition can be differentiated from all other forms of porphyria by demonstrating that the increase in urinary PBG excretion is accompanied by a normal or near normal concentration of porphyrin in the faeces. The excretion of ALA is also increased but to a lesser extent than PBG. This is the reverse of the situation in lead poisoning, in which PBG excretion is only increased in severe cases, and then never exceeds that of ALA.

Erythrocyte uroporphyrinogen-I-synthase activity is decreased in the majority of patients with acute intermittent porphyria but, since there is a small amount of overlap between the porphyric and reference ranges,[39] an activity within the reference range does not exclude the diagnosis. Measurement of this enzyme in erythrocytes has now replaced determination of urinary PBG as the method for detecting latent porphyria in the relatives of affected individuals because it has two important advantages. First, it enables latent porphyrics to be identified, and advised to avoid known precipitants of acute attacks, before puberty when PBG excretion is invariably normal and acute attacks very uncommon. Second, it allows the substantial percentage of adult latent porphyrics with normal PBG excretion to be detected. In carrying out family studies, it is useful to measure the enzyme activity in red cells from both the affected and unaffected parent. This helps to interpret enzyme activities that lie in the overlap zone as the level in carriers of the gene tends to be half that of the normal parent.[39] Uroporphyrinogen-I-synthase activity is increased in haematological disorders in which the proportion of reticulocytes and young erythrocytes is increased.[1] For this reason, activities are difficult to interpret in young infants and measurements on the children of affected individuals are best delayed until after the age of four months.[47]

Various methods, in addition to urinary PBG determinations, have been used to monitor the effectiveness of treatment of acute attacks with haematin or carbohydrate infusion. The measurement of serum PBG and ALA levels has been described by Watson et al.[59] and leucocyte ALA-synthetase activity by Brodie et al.[12]

Variegate Porphyria and Hereditary Coproporphyria

In these conditions the diagnosis is established by measuring faecal porphyrin concentrations (Table 5). Solvent extraction procedures which measure porphyrin fractions are usually adequate for this purpose[11,22,46] but measurement of individual porphyrins by t.l.c.[11] or h.p.l.c.[31] may occasionally be required, particularly to distinguish variegate porphyria from porphyria cutanea tarda.[23]

In variegate porphyria the concentration of porphyrin in faeces is increased, often ten-fold or more, with the amount of protoporphyrin being about twice that of coproporphyrin. The ether-insoluble porphyrin fraction from faeces contains large quanties of a hydrophilic porphyrin-peptide conjugate or 'X-porphyrin,'[53] but measurement of this fraction is of little diagnostic value, even in family studies.[46] Recently an unidentified porphyrin which is characteristic of variegate porphyria has been found in plasma.[18,49] Detection of this porphyrin by fluorescence emission spectroscopy of plasma appears to be a useful, rapid technique for distinguishing this condition from other porphyrias, at least when skin lesions are present.[49]

At present, detection of latent variegate porphyria depends on measurement of faecal porphyrin concentrations, as excretion of urinary haem precursors is usually normal. A small proportion of latent cases have normal faecal porphyrin concentrations[46] and it is likely that measurement of the activity of the defective enzyme will eventually prove to be a better method of detection. However, systematic studies of the activities of ferro-chelatase and protoporphyrinogen oxidase in families with variegate porphyria have not yet been reported.

In hereditary coproporphyria, the concentration of coproporphyrin III is increased in the faeces, while the excretion of other porphyrins remains within or close to normal limits. Latent hereditary coproporphyria is best detected by measuring coproporphyrinogen oxidase in lymphocytes or cultured skin fibroblasts as inheritance of the gene is not always accompanied by increased excretion of coproporphyrin III.

Porphyria Cutanea Tarda

The complex pattern of overproduction of acetate-substituted porphyrins in porphyria cutanea tarda allows this condition to be distinguished with certainty from other forms of porphyria (Table 5). Urinary PBG excretion and red cell porphyrin concentrations are normal.

The urine contains excess uroporphyrin and heptacarboxylic porphyrin with smaller quantities of porphyrins with four to six carboxyl groups. When measured by solvent extraction techniques this pattern is reflected by an increased uroporphyrin fraction with a coproporphyrin fraction that may be either increased or normal, but which is always less than the uroporphyrin fraction. Measurement of individual porphyrins shows a characteristic pattern with uroporphyrin (45–80% of the total) and heptacarboxylic porphyrin (15–35% of the total) predominating.[20] The total concentration of porphyrins in faeces is usually increased but may range from normal levels to levels that overlap those seen in variegate porphyria. Solvent partition methods usually show an increased coproporphyrin fraction as the main abnormality with a lesser increase in the ether-insoluble porphyrin fraction. Measurement of individual porphyrins reveals diagnostic increases in the proportion of isocoproporphyrin and heptacarboxylic porphyrins.[22,23] The porphyrin patterns of urine and faeces are sufficiently characteristic to enable subclinical porphyria cutanea tarda to be diagnosed in some patients during prolonged remission and in family studies, even when total porphyrin excretion is close to, or within, normal limits.[6]

The main problem in the differentiation of the cutaneous porphyrias is the separation of variegate porphyria from porphyria cutanea tarda. Provided both urine and faeces are examined, this is usually straight forward but occasionally patients with cutaneous variegate porphyria have urinary porphyrin abnormalities that cannot be distinguished from those of porphyria cutanea tarda by solvent partition methods, perhaps because any PBG present in the urine initially has polymerized to uroporphyrin on standing. In such patients the correct diagnosis is readily made by measurement of individual porphyrins in the faeces[23] or by fluorescence emission spectroscopy of plasma.[49]

Treatment of porphyria cutanea tarda by repeated venesection or chloroquine may be monitored by urinary or plasma porphyrin measurements.[36,45,50]

Protoporphyria and Congenital Erythropoietic Porphyria

An increased concentration of erythrocyte protoporphyrin, as distinct from zinc-protoporphyrin, confirms a diagnosis of protoporphyria. Raised plasma and faecal protoporphyrin levels are less constant findings. Urineary porphyrin excretion is normal, unless severe liver disease develops. At present there is no test that will predict the possible onset of liver disease. Family studies by enzyme or porphyrin measurements[7] identify gene carriers but, due to the variability of clinical expression, cannot identify those who will develop symptoms.

Congenital erythropoietic porphyria is characterized by marked over-production of coproporphyrin I and uroporphyrin I (Table 5). The typical form of this very rare disorder, which presents in infancy, is rarely confused with other types of porphyria.[34]

REFERENCES

1. Anderson, K. E., Alvares, A. P., Sassa, S. and Kappas, A. (1976), 'Studies in porphyria. V. Drug oxidation rates in hereditary hepatic porphyria,' *Clin. Pharm. Ther.*, **19**, 47.
2. Anderson, K. E., Sassa, S., Petersen, C. M. and Kappas, A. (1977), 'Increased erythrocyte uroporphyrinogen-I-synthetase, δ-aminolaevulinate acid dehydratase and protoporphyrin in haemolytic anaemias,' *Amer. J. Med.*, **63**, 359.
3. Battersby, A. R., Fooker, C. J. R., Matcham, G. W. J. and McDonald, E. (1980), 'Biosynthesis of the pigments of life: formation of the macrocycle,' *Nature*. **285**, 17.
4. Becker, D. M. and Kramer, S. (1977), 'The neurological manifestations of porphyria. A review,' *Med.*, **56**, 411.
5. Becker, D. M., Viljoen, J. D., Katz, J. and Kramer, S. (1977), 'Reduced ferrochelatase activity: a defect common to porphyria variegata and protoporphyria,' *Brit. J. Haem.*, **36**, 171.
6. Benedetto, A. V., Kushner, J. P. and Taylor, J. S. (1978), 'Porphyria cutanea tarda in three generations of a single family,' *New Engl. J. Med.*, **298**, 358.
7. Bloomer, J. R., Bonkowsky, H. L., Ebert, P. S. and Mahoney, M. J. (1976), 'Inheritance in protoporphyria: comparison of haem synthetase activity in skin fibroblasts with clinical features,' *Lancet*, **2**, 226.
8. Bonkowsky, H. L., Bloomer, J. R., Ebert, P. S. and Mahoney, M. J. (1975), 'Haem synthetase deficiency in human protoporphyria,' *J. Clin. Invest.*, **56**, 1139.
9. Bottomley, S. S., Tanaka, M. and Everett, M. A. (1975), 'Diminished erythroid ferrochelatase activity in protoporphyria,' *J. Labor. Clin. Med.*, **86**, 126.
10. Brenner, D. A., and Bloomer, J. R. (1980), 'The enzymatic defect in variegate porphyria,' *New Engl. J. Med.*, **302**, 765.
11. Brodie, M. J., Thompson, G. G., Moore, M. R., Beattie, A. D. and Goldberg, A. (1977), 'Hereditary coproporphyria,' *Quart. J. Med. New Series*, **46**, 229.
12. Brodie, M. J., Moore, M. R., Thompson, G. G. and Goldberg, A. (1977), 'The treatment of acute intermittent porphyria with laevulose,' *Clin. Sci. Molec. Med.*, **53**, 365.
13. Brodie, M. J., Graham, D. J. M., Goldberg, A., Beastall, G. H., Ratcliffe, W. A., Ratcliffe, J. G. and Yeo, P. R. B. (1978), 'Thyroid function in acute intermittent porphyria: a neurogenic cause of hyperthyroidism?' *Horm. Metab. Res.*, **10**, 327.
14. Chisholm, J. J. and Brown, D. H. (1975), 'Micro-scale photofluorometric determination of "free erythrocyte porphyrin" (protoporphyrin IX),' *Clin. Chem.*, **21**, 1169.
15. Cripps, D. J. and Goldfarb, S. S. (1978), 'Erythropoietic protoporphyria: hepatic cirrhosis,' *Brit. J. Dermatol.*, **98**, 349.
16. Czarnecki, D. B. (1980), 'Hepatoerythropoietic porphyria,' *Arch. Dermatol.*, **116**, 307.
17. Day, R. S., De Salamanca, R. E. and Eales, L. (1978a), 'Quantitation of red cell porphyrins by fluorescence scanning after thin-layer chromatography,' *Clin. Chem. Acta.*, **89**, 25.
18. Day, R. S., Pimstone, N. and Eales, L. (1978b). 'The diagnostic value of blood plasma porphyrin methyl ester profiles produced by quantitative TLC,' *Internat. J. Biochem.*, **9**, 897.
19. De Leo, V. A., Pho-Fitzpatrick, M., Matthews-Ruth, M. and Harber, L. C. (1976), 'Erythropoietic protoporphyria. 10 years experience,' *Amer. J. Med.*, **60**, 8.
20. Doss, M. (1974), 'Porphyrins and porphyrin precursors,' in *Clinical Biochemistry Principles and Methods*, (Curtius, M. and Roth, M. Berlin: Walter de Gruyter. Eds.). 1323.
21. Eales, L., Levey, M. J. and Sweeney, G. D. (1966), 'The place of screening tests and quantitative investigations in the diagnosis of the porphyrias, with particular reference to variegate and symptomatic porphyria,' *S African Med. J.*, **40**, 63.
22. Eales, L., Grosser, Y. and Sears, W. G. (1975), 'The clinical biochemistry of the human hepatocutaneous porphyrias in the light of recent studies of newly identified intermediates and porphyrin derivatives,' *Ann. New York Acad. Sci.*, **244**, 441.
23. Elder, G. H. (1975), 'Differentiation of porphyria cutanea tarda symptomatica from other types of porphyria by measurement of isocoproporphyrin in faeces,' *J. Clin. Pathol.*, **28**, 601.

24. Elder, G. H. (1980), 'The porphyrias: clinical chemistry, diagnosis and methodology,' *Clin. Haem.*, **9**, 371.
25. Elder, G. H., Magnus, I. A., Handa, F. and Doyle, M. (1974), 'Faecal "X-porphyrin" in the hepatic porphyrias,' *Enzyme*, **17**, 29.
26. Elder, G. H., Evans, J. O., Thomas, N., Cox, R., Brodie, M. J., Moore, M. R., Goldberg, A. and Nicholson, D. C. (1976), 'The primary enzyme defect in hereditary coproporphyria,' *Lancet*, **2**, 1217.
27. Elder, G. H., Lee, G. B. and Tovey, J. A. (1978), 'Decreased hepatic uroporphyrinogen decarboxylase in porphyria cutanea tarda,' *New Engl. J. Med.*, **229**, 274–278.
28. Fogstrup, J. and With, T. K. (1979), 'Urinary total porphyrins by ion exchange analysis: reference values for the normal range and remarks on performed porphyrins in acute porphyria urine,' *Clin. Path.*, **32**, 109.
29. Fuhrhop, J. M. and Smith, K. M. (1975), *Laboratory Methods in Porphyrin and Metalloporphyrin Research*. Amsterdam, Elsevier, p. 243.
30. Granick, S. and Beale, S. I. (1978), 'Hemes, chlorophylls and related compounds: biosynthesis and metabolic regulation,' in *Advances in Enzymology*, (Meister A. Ed.). **46**, 33.
31. Gray, C. H., Lim, C. K. and Nicholson, C. C. (1977), 'The differentiation of the porphyrias by means of high pressure liquid chromatography,' *Clin. Chem. Acta*, **77**, 167.
32. Grossman, M. E. and Bickers, D. R. (1978), 'Porphyria cutanea tarda: a rare cutaneous manifestation of hepatic tumours,' *Cutis*, **21**, 782.
33. Grossman, M. E., Bickers, D. R., Poh-Fitzpatrick, M. B., De Leo V. and Harber, L. C. (1979), 'Porphyria cutanea tarda. Clinical features and laboratory findings in 40 patients,' *Amer. J. Med.*, **67**, 277.
34. Ippen, H. and Fuchs, T. (1980), 'Congenital porphyria,' *Clin. Haem.*, **9**, 323.
35. Jackson, A. H. (1977), 'Modern spectroscopic and chromatographic techniques for the analysis of porphyrin on a microscale,' *Sem. Hematol.*, **14**, 193.
36. Kordač, V. and Semrádorá, M. (1974), 'Treatment of porphyria cutanea tarda with chloroquine,' *Brit. J. Dermatol.*, **90**, 95.
37. Kushner, J. P., Barbuta, A. J. and Lee, G. R. (1976), 'An inherited enzymic defect in porphyria cutanea tarda. Decreased uroporphyrinogen decarboxylase activity,' *J. Clin. Invest.*, **58**, 1089.
38. Lamola, A. A., Joselow, M. and Yamane, T. (1975), 'Zinc protoporphyrin (ZPP): a simple, sensitive, fluorometric screening test for lead poisoning,' *Clin. Chem.*, **21**, 93.
39. Lamon, J. M., Frykholm, B. G. and Tschudy, D. P. (1979), 'Family evaluations in acute intermittent porphyria using red cell uroporphyrinogen-I-synthetase,' *J. Med. Genet.*, **16**, 134.
40. Levin, E. Y. (1975), 'Comparative aspects of porphyria in man and animals,' *Ann. New York Acad. Sci.*, **244**, 481.
41. Mauzerall, D. and Granick, S. (1956), 'The occurence and determination of δ-aminolaevulinic acid and porphobilinogen in urine,' *J. Biol. Chem.*, **219**, 435.
42. Meyer, U. A. and Schmid, R. (1978), 4th edn. 'The Porphyrias,' in *The Metabolic Basis of Inherited Disease*, (Stanbury, J. B., Wyngaarden J. B. and Fredrickson, D. S., Eds.). New York: McGraw-Hill, 1166.
43. Miyagi, K., Petryka, Z. J., Bossenmaier, J., Cardinal, R. and Watson, C. N. (1976), 'The activities of uroporphyrinogen synthase and cosynthase in congenital erythropoietic porphyria (CEP),' *Amer. J. Haem.*, **1**, 3.
44. Moore, M. R. (1980). *International Journal of Biochemistry*, **12**, 1089.
45. Moore, M. R., Thompson, G. G., Allen, B. R., Hunter, J. A. A. and Parker, S. (1973), 'Plasma porphyrin concentrations in porphyria cutanea tarda,' *Clin. Sci. Molec. Med.*, **45**, 711.
46. Mustajoki, P. (1980), 'Variegate porphyria,' *Quart. J. Med. New Series*, **49**, 191.
47. Nordmann, Y., Grandchamp, B., Grelier, M., Phung, N'G. and Verneuil, H. de (1976), 'Detection of intermittent acute porphyria trait in children,' *Lancet*, **ii**, 201.
48. Pierach, C. A., Cardinal, R., Bossenmaier, J. and Watson, C. J. (1977), 'Comparison of the Hoesch and the Watson-Schwartz tests for urinary porphobilinogen,' *Clin. Chem.*, **23**, 1666.
49. Poh-Fitzpatrick, M. B. (1980), 'A plasma porphyrin fluorescence marker for variegate porphyria,' *Arch. Dermatol.*, **116**, 543.
50. Ramsay, C. A., Magnus, I. A., Turnbull, A. and Baker, H. (1974), 'The treatment of porphyria cutanea tarda by venesection,' *Quart. J. Med.*, **43**, 1.
51. Rimington, C. (1971), 'Quantitative determination of porphobilinogen and porphyrins in urine and porphyrins in faeces and erythrocytes,' *Broadsheet No. 70 (Revised Broadsheet 36)*. Association of Clinical Pathologists.
52. Rimington, C. and Cripps, D. J. (1965), 'Biochemical and fluorescence-microscopy screening tests for erythropoietic protoporphyria,' *Lancet*, **i**, 624.
53. Rimington, C., Lockwood, W. H. and Belcher, R. V. (1968), 'The excretion of porphyrin-peptide conjugates in variegate porphyria,' *Clin. Sci.*, **35**, 211.
54. Sassa, S., Zacar, G. L. and Kappas, A. (1978), 'Studies in Porphyria. VIII. Induction of uroporphyrinogen-I-synthase and expression of the gene defect of acute intermittent porphyria in mitogen-stimulated human lymphocytes,' *J. Clin. Invest.*, **61**, 499.
55. Smith, K. M. (1975). *Porphyrins and Metalloporphyrins*. Amsterdam: Elsevier.
56. Stein, J. A. and Tschudy, D. P. (1970), 'Acute intermittent porphyria: a clinical and biochemical study of 46 patients,' *Medicine*, **49**, 1.
57. Strand, L. J., Felsher, B. F., Redeker, A. G. and Marver, H. S. (1970), 'Heme biosynthesis in intermittent acute porphyria: decreased hepatic conversion of porphobilinogen to porphyrins and increased hepatic conversion of porphobilinogen to porphyrins and increased δ-aminolaevulinic acid synthetase activity,' *Proc. Nat. Acad. Sci.*, U.S.A., **67**, 1315.
58. Verneuil, H. De, Atiken, G. and Nordmann, Y. (1978), 'Familial and sporadic porphyria cutanea. Two different diseases,' *Hum. Gen.*, **42**, 145.
59. Watson, C. J., Pierach, C. A., Bossenmaier, I. and Cardinal, R. (1978), 'Use of hematin in the acute attack of the "inducible" hepatic porphyrias,' *Adv. Internal Med.*, **23**, 265.
60. With, T. K. (1978), 'Clinical porphyrin analyses: indications and interpretations,' *Scand. J. Clin. Labor. Invest.*, **38**, 501.
61. Wolkoff, A. W., Cohen, L. E. and Arias, I. M. (1973), 'Inheritance of the Dubin-Johnson Syndrome,' *New Engl. J. Med.*, **288**, 113.

11. HAEMOGLOBINOPATHIES

ROBIN W. CARRELL AND HERMANN LEHMANN

Introduction

Haemoglobin structure and function
   Haem and the haem pocket
   Globin structure: solubility and stability
   Cooperativity and the haemoglobin tetramer

Haemoglobin genetics and expression
   Haemoglobins of embryo, fetus and adult
   α-Gene duplication
   Globin gene structure

Haemoglobin variation and disease
   The abnormal haemoglobins
   Non-genetic variation: HbA_{1c}
   The haemoglobinopathies
   The thalassaemias
   Pathology and haematology

Laboratory diagnosis
   Routine screening

INTRODUCTION

The structure of haemoglobin is typified by the adult human haemoglobin, HbA. This is formed of four globin polypeptide chains each with its own haem group, the globin chains being in the form of two unlike pairs, i.e. $\alpha_2\beta_2$.

The term *haemoglobinopathies* is a collective one for the genetic defects affecting haemoglobin structure or synthesis. These have been intensively studied over the past 20 years and a major conclusion has been the truth of the familiar laboratory law, that everything that can go wrong will go wrong! The defects can be divided into two categories: (1) those that affect the structure of the haemoglobin, usually due to the substitution of one amino acid by another, i.e. the *abnormal haemoglobins*, and (2) those that primarily affect the rate of synthesis of one of the globin chains causing an imbalanced production of the haemoglobin subunits, the *thalassaemias*. There are many different types of thalassaemia, but there are two major classifications: (1) the α-thalassaemias in which there is a defect in synthesis of the α-chain with consequent excess of β-chains and (2) the β-thalassaemias in which there is deficit in β-chain synthesis and hence an excess of α-chains.

The structural variants of haemoglobin, the abnormal haemoglobins, are each individually characterised and may be accompanied by functional alterations to give disease as in sickle cell anaemia, where there is a substitution of the 6th amino acid of the β-chain, glutamic acid, by valine, i.e. HbS $\alpha_2\beta_2^{6Glu\rightarrow Val}$. Many other abnormal haemoglobins, however, are just chance mutations that do not noticeably affect function or the health of the carrier.

Although the distinction between the thalassaemias and the abnormal haemoglobins is quite clear in terms of their definitions, it is not nearly so clear at the level of laboratory-bench diagnosis. The haemoglobinopathies usually present to the clinical biochemist in the form of a blood sample accompanied by some sketchy clinical details ending with . . . '? haemoglobinopathy'. If this question is to be answered competently two things are required of the clinical biochemist. The first is a clear understanding of the basic principles of the structure, function and genetics of haemoglobin; the second is a systematic diagnostic approach which will cover the likely possibilities. These topics and an approach to diagnosis are reviewed in this chapter at a level relevant to the general clinical laboratory; more detailed background and reference information is given in the sources appended under Further Reading.

HAEMOGLOBIN STRUCTURE AND FUNCTION

Barcroft[1] pointed out in 1928 that, physiologically, haemoglobin was required to be

(i) capable of transporting large quantities of oxygen;

(ii) very soluble;

(iii) capable of taking up oxygen at suitable velocity and in sufficient amounts in the blood and of releasing it to the tissues;

(iv) able to buffer a bicarbonate solution.

This list is still relevant, though now it could be re-stated to give a greater emphasis on structure, i.e.

1. Haem and the Haem Pocket: Oxygenation.
2. Globin Structure: Solubility and Instability.
3. The Haemoglobin Tetramer: Cooperativity.
4. Modifiers of Oxygen Transport.

Haem and the Haem Pocket

Haem

Oxygenation occurs at the haem group which is a coordinate of iron and protoporphyrin. The full structure of protoporphyrin is given in the previous chapter, but essential features are its near-planar nature and its highly mobile electron flux. This flux is readily excited by light in the 400 nm range to give the characteristic red, Soret, absorption spectrum. Free protoporphyrin will emit light to give fluorescence but this fluorescence is quenched in haem by the coordinated iron. However, haemoglobin fluorescence may exceptionally occur when iron is prevented from coordinating with the protoporphyrin and is replaced by zinc. This forms the basis of a simple and useful screening test[20] for iron deficiency or lead poisoning, by measurement of whole blood fluorescence.

The electron flux of the protoporphyrin readily allows transfer of charge to the haem iron. This iron has four coordinate bonds to the porphyrin, a fifth bond to a histidine of the globin and a sixth bond to various ligands such as oxygen. The iron is in the ferrous, Fe^{2+}, form in deoxyhaemoglobin but in oxyhaemoglobin there is a partial transfer of an electron from the iron to the oxygen so the iron is in a transition state between Fe^{2+} and Fe^{3+}. The oxygen of oxyhaemoglobin can be readily displaced by small anions such as Cl^- or OH^- (water) to give the ferric methaemoglobin. The changes in charge distribution that result are reflected by characteristic changes in absorption spectra (Fig. 1).

As well as changes in charge distribution there are also changes in the spin state of the iron; these are of significance in triggering the cooperative changes in haemoglobin on oxygenation as well as resulting in spectral modifications. Both ferrous and ferric iron can exist in high- or low-spin states according to the pairing of their outer orbital electrons. The radius of the iron atom increases in the high-spin state with a consequent lengthening of iron-porphyrin bonds. This causes a movement of iron out of the plane of the haem when the low-spin oxyhaemoglobin is converted to high-spin deoxyhaemoglobin (Fig. 2). This movement of the iron will be transmitted to the linked histidine with resultant distortion of the globin and triggering of changes in the other globin subunits of the haemoglobin molecule.

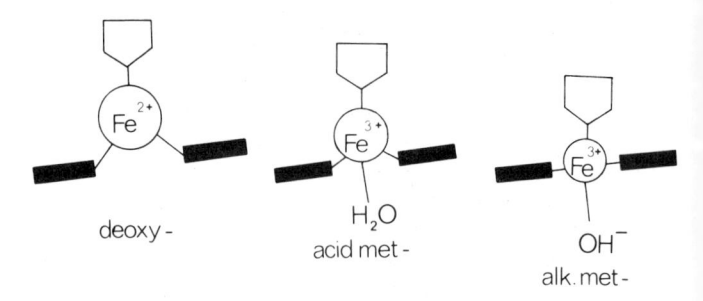

FIG. 2. Deoxyhaemoglobin and acid and alkaline methaemoglobins showing the changes in charge and relationship of iron to the haem plane that produce changes in adsorption spectra. Note how the iron shrinks and moves into the plane of the haem with linked movement of the proximal histidine (cf. Fig. 3.) (Reproduced with permission, from Carrell and Lehmann.[6])

Haem Pocket

The partial transfer to oxygen of an electron from the haem iron makes it very susceptible to displacement by OH^- or Cl^-. This is highly disadvantageous as it results in the physiologically inert methaemoglobin, $Fe^{3+}OH^-$, and the release of the toxic superoxide radical, O_2^-. A consequent priority for reversible oxygenation is that the haem should be buried in a non-polar, hydrophobic environment which prevents access of polar ligands, such as water. This is achieved by placing the haem in a hydrophobic pocket in the globin, in a position structurally comparable to that of a coin pushed deep into a soft bun.

The 20 amino acids lining the pocket form a close steric fit with the haem group, limiting access to the iron and contributing to the overall stability of the globin. The structure of the haem pocket was tightly defined by evolution several hundred million years ago; since then it has remained relatively static whereas the rest of the molecule has changed to meet the requirements of different species. Consequently chance mutations that occur to haem-pocket residues are likely to produce pathology; either by allowing globin bonding to the sixth ligand position, i.e. abnormal methaemoglobins, or due to a distortion of globin structure to give instability of the molecule with consequent red cell haemolysis.

Two of the amino acid residues lining the haem pocket are of particular importance; these are the two haem-

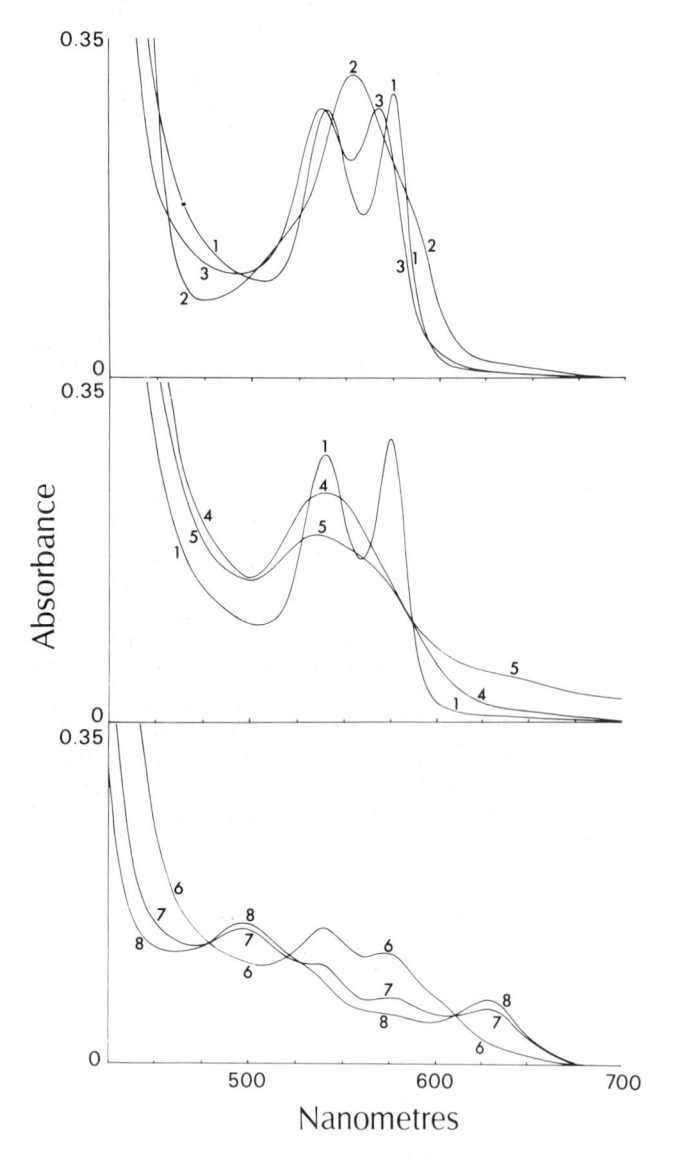

FIG. 1. Absorption spectra of haemoglobin derivatives measured in phosphate buffer, pH 7·4, except as indicated below. (Beckman model 25 spectrophotometer, Miss B. M. McGrath.) 1 = oxyhaemoglobin; 2 = deoxyhaemoglobin, 3 = carboxyhaemoglobin; 4 = cyanmethaemoglobin; 5 = haemichrome; 6 = methaemoglobin in pH 6·0 phosphate buffer; 7 = methaemoglobin; 8 = methaemoglobin in pH 11·0 glycine buffer. (Reproduced with permission, from Carrell and Lehmann.[6]).

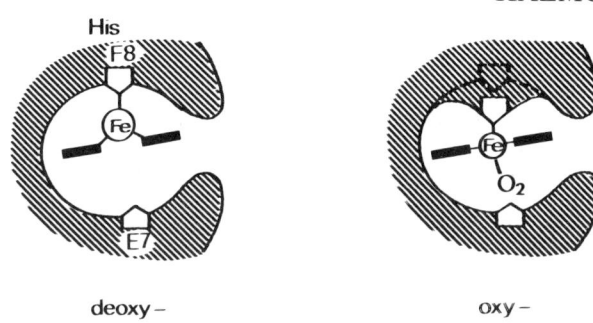

FIG. 3. Schematic representation of the haem pockets of deoxy- and oxyhaemoglobins. The porphyrin ring with its four bonds is shown in planar section. The proximal histidine (F8) forms the fifth coordinate, the sixth coordinate being formed by the ligand, oxygen in this case, which lies between the iron and the distal histidine (E7). Note the linked movement of the globin and the iron as changes in the radius of the iron cause it to move into and out of the haem plane. (Reproduced with permission, from Carrell and Lehmann.[6])

linked histidines illustrated in Fig. 3. The proximal histidine (F8) forms the fifth coordinate of the iron and the distal histidine (E7) occupies a position where it can form a polar bond with the partially-negative oxygen of oxyhaemoglobin. Distortion of the globin, as occurs in partial denaturation, may allow the distal histidine close enough to the iron to bond to it to form the unstable globin haemichromes, initially haemichrome I. If further distortion occurs, other side chains in the globin may displace the histidine to form the irreversible globin haemichrome II, with resultant denaturation and precipitation.

Globin Structure: Solubility and Stability

If one prime task of the globin molecule is to provide a hydrophobic environment for the haem group, the other is the need to give solubility to the molecule as a whole. This need for solubility is a demanding requirement as the red cell is packed with haemoglobin which forms a third of its wet weight. Some idea of the packing can be given by comparison of the red cell and the molecules to a football stadium filled with tennis balls, with only the interstices available for the water of solution. It is not surprising then that one of the likely consequences of globin mutations is a change in solubility, or stability, that results in intracellular precipitation of haemoglobin with consequent haemolysis.

The globin molecule is highly soluble because it is folded in such a way that its ionised and polar amino acid sidechains are externally situated. As shown in Fig. 4, globin is formed of eight helices (A–H) linked by short interhelical segments (AB, CD, etc.). The non-polar aspect of each helix points internally to provide a hydrophobic environment for the haem, the polar aspect of each helix is externally orientated to provide solubility.

Cooperativity and the Haemoglobin Tetramer

Cooperativity

The globin monomer, as in myoglobin or in an isolated haemoglobin subunit, has a hyperbolic oxygen dissocia-

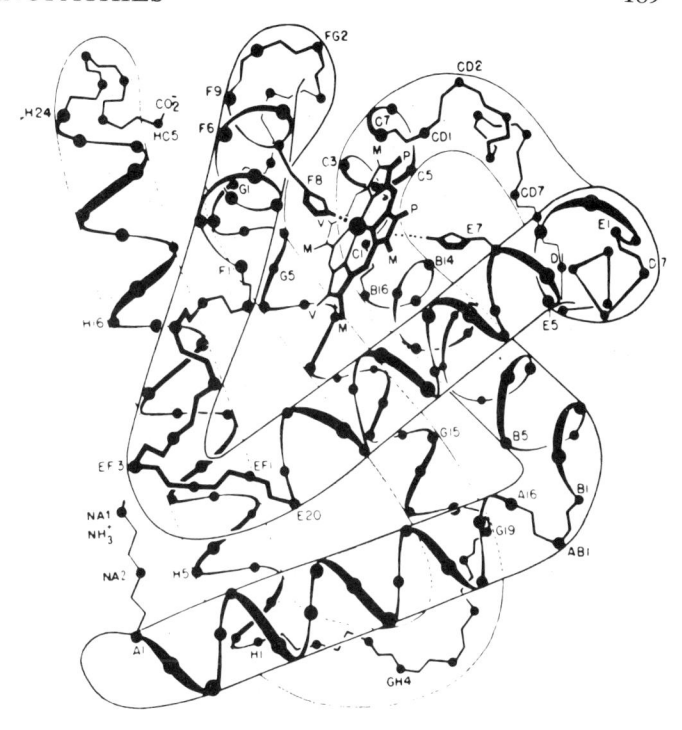

FIG. 4. Structure of myoglobin, which exemplifies those of all the globins: 80% of the molecule is in the helical form, each helix being designated by a letter and number, and non-helical portions by two letters and a number according to their position (thus A1–A16, helical; CD1–CD7, non-helical). (Reproduced by courtesy of Dr. R. E. Dickerson: published in *The Proteins*, Vol. II, edited by Hans Neurath, Academic Press, Fig. 15, p. 634).

tion curve (Fig. 5). Oxygen is bound with high affinity and is released only at very low oxygen tension. The requirement for a physiological oxygen carrier is that it should release a substantial proportion of its oxygen at the tissue tension of 40 mmHg and the rest well before reaching zero oxygen tension. This need for a sigmoid oxygen dissociation curve is met by cooperativity, the term used to describe the way in which the deoxygenation of one globin subunit of haemoglobin decreases the oxygen affinity of the other subunits. This cooperative interaction of subunits involves a change in structural conformation of the haemoglobin molecule: the R to T transformation.

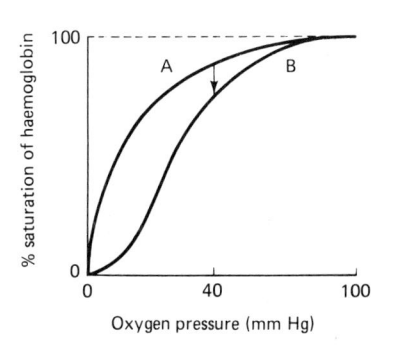

FIG. 5. Comparison of the hyperbolic oxygen dissociation curve of myoglobin (A) with the sigmoid curve of haemoglobin (B). Note the increased efficiency of release of oxygen of the sigmoid curve at the mean venous tension of 40 mm Hg. (Reproduced, with permission, from Lehmann and Casey.[2])

The affinity of fully oxygenated haemoglobin for oxygen is the same as that of its isolated subunits and for this reason oxyhaemoglobin is said to be in the relaxed, R, state. However, on deoxygenation, cooperative effects between the globin subunits constrain the molecule in a low affinity, tense, T state. A simple analogy is to that of a sponge which in its relaxed (R) form will readily take up water but if held in the tense (T) form, will readily lose, and less readily take up, water.

Hill Equation

Cooperativity can be treated quantitatively. The oxygenation of myoglobin or of a single haemoglobin subunit follows first order kinetics:

$$Mb + O_2 \rightleftharpoons MbO_2$$
$$[1-Y] \quad [O_2] \quad [Y]$$

where Y is the fractional saturation and $[O_2]$ the partial pressure of oxygen. If K is the equilibrium constant, then the fractional (percentage) saturation is the hyperbolic function

$$Y = \frac{K[O_2]}{1+K[O_2]}$$

Hill showed that for the haemoglobin tetramer, the oxygen dissociation could be represented by the sigmoid function

$$Y = \frac{K[O_2]^n}{1+K[O_2]^n}$$

where n, the Hill number, indicates the degree of subunit interaction. Normally its value is 2·8, but it will be lower in mutant haemoglobins with decreased cooperativity. In myoglobin and the isolated haemoglobin subunits, n has a value of 1, i.e. no cooperativity.

Haemoglobin Tetramer

The cooperativity of haemoglobin requires a tetramer formed of two pairs of unlike globins, as in HbA, $\alpha_2\beta_2$, adult human haemoglobin. Together the four subunits form an ellipsoid with a molecular weight of 67 000 and dimensions of 64 × 55 × 50A°. Cooperativity occurs due to small changes in subunit shape, i.e. tertiary changes, that result in much larger changes in the overall molecular shape, i.e. quaternary changes. The detailed molecular changes have been elucidated and reviewed by Perutz.[27] He points out that the molecule undergoes 'paradoxical breathing', decreasing in size on oxygenation and increasing on deoxygenation.

Oxyhaemoglobin is in the relaxed, R, state with each of its subunits having the same conformation as they would in the free state. The loss of oxygen from one or more subunits results in a change of conformation of the molecule as a whole to the tense, T, state with consequent further loss of oxygen. The R to T transformation involves a sliding rotation of one $\alpha\beta$ pair about the other; the movement taking place at the $\alpha_1\beta_2$ junction (Fig. 6). This movement, on partial deoxygenation, allows ionic

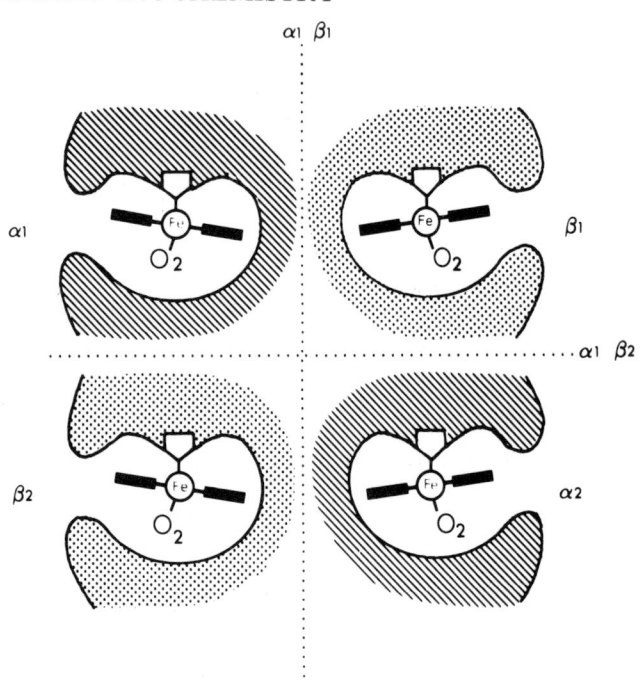

FIG. 6. The haemoglobin tetramer diagrammatically represented in the oxy-, R, state to show the nomenclature of the $\alpha\beta$-interfaces. The change to the deoxy-, T, state involves a sliding movement at the $\alpha_1\beta_2$ junction, the α-chains moving closer together and a gap opening between the β-chains that allows entry of 2,3-diphosphoglycerate. (Reproduced with permission, Carrell and Lehmann.[6])

bonds to form between subunits which in turn induce a change in shape of the individual subunits. This tertiary change, although small, results in a movement of the F-helix away from the haem; in doing so the haem-linked histidine pulls the iron out of the plane of the haem with a change to the high-spin state, with consequent release of any bound oxygen (Fig. 3). This is a reciprocal process; loss of oxygen from one subunit will cause tertiary changes that favour ionic bonding with other subunits and hence the overall R to T transformation. Similarly the change to the quaternary T conformation will induce each subunit to take up the tertiary T conformation, with loss of oxygen from individual globin subunits.

An important conclusion for the clinical biochemist is that mutations which cause gross changes in oxygen affinity are likely to involve only a few residues; those of the $\alpha_1\beta_2$ interface, those forming ionic bonds between subunits and those directly involved in the tertiary, T, change.

Modifiers of Oxygen Transport

It is a mistake to think of oxygen delivery primarily in terms of haemoglobin concentration. It is true that gross anaemia will affect peripheral oxygenation but, in most situations, changes in the oxygen affinity of haemoglobin are of much greater importance. Normally only 25% of the oxygen carried in the blood is released to the tissues, but this can be greatly increased by small alterations in the shape of the oxygen dissociation curve (Fig. 7). This

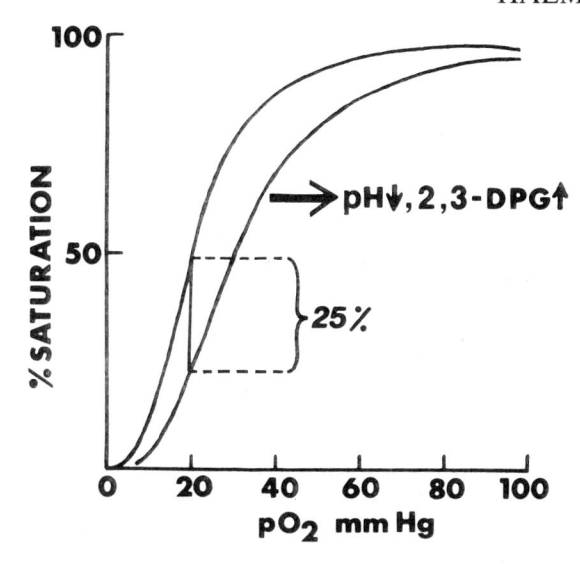

FIG. 7. The oxygen dissociation curve of haemoglobin moves to the right (decreased affinity) with a drop in pH or a rise in 2,3-diphosphoglycerate. As shown, even a small shift, as in a pH change from 7·6–7·2, will result in a large increase (25%) in oxygen delivery at the partial pressure of working tissues.

comes about by the action of small molecules that provide additional ionic bonds to stabilise the T quaternary structure of deoxyhaemoglobin.

An appreciable decrease in oxygen affinity, with consequent release of oxygen, occurs in working tissues due to accumulation of carbon dioxide and lactic acid. The decrease in affinity that occurs with increased acidity, in the physiological range, is known as the alkaline Bohr effect. The release of lactic acid in the periphery gives increased hydrogen ion concentration, this causes protonation of key basic residues in haemoglobin which then form ionic bridges to stabilise the deoxy configuration. This results in both the release of oxygen to the tissues and the buffering of pH, as hydrogen ions are taken up at the periphery and released to the lungs where they are neutralized by combination with bicarbonate ions.

Similarly carbon dioxide can combine with the N-terminal amines of globin subunits to give carbamino groups, which can form stabilizing bridges in deoxy- but not oxy-haemoglobin. The overall result is that haemoglobin acts as a transport agent that selectively releases oxygen to working tissues and removes hydrogen ions and carbon dioxide for excretion in the lungs.

As well as adaptations that influence the release of oxygen to specific tissues there is also an overall control of oxygen dissociation by the red-cell metabolite, 2,3-diphosphoglycerate. This binds stoichiometrically to haemoglobin forming ionic bonds with groups in a pocket between the β-chains in the deoxy-, T, conformation. This stabilizes the deoxy- conformation to give a decrease in oxygen affinity. Consequently, the red cell can partially compensate for the decrease in oxygen transport that would otherwise occur in anaemia or with reduced environmental oxygen as occurs at high altitudes. It also explains the difference between the oxygen affinities of fetal and adult blood that facilitates placental exchange of oxygen.

Both HbF and HbA have the same dissociation curves in simple solution, but in the red cell the affinity of HbA is decreased due to the 2,3-diphosphoglycerate effect. However, the γ-chains of HbF lack one of the 2,3-diphosphoglycerate binding sites and consequently fetal red-cell haemoglobin has a higher affinity for oxygen than that of the adult.

2,3-diphosphoglycerate is a glycolytic pathway side product, formed in the Rapoport–Luebering cycle which by-passes the phosphoglycerate kinase step of the main

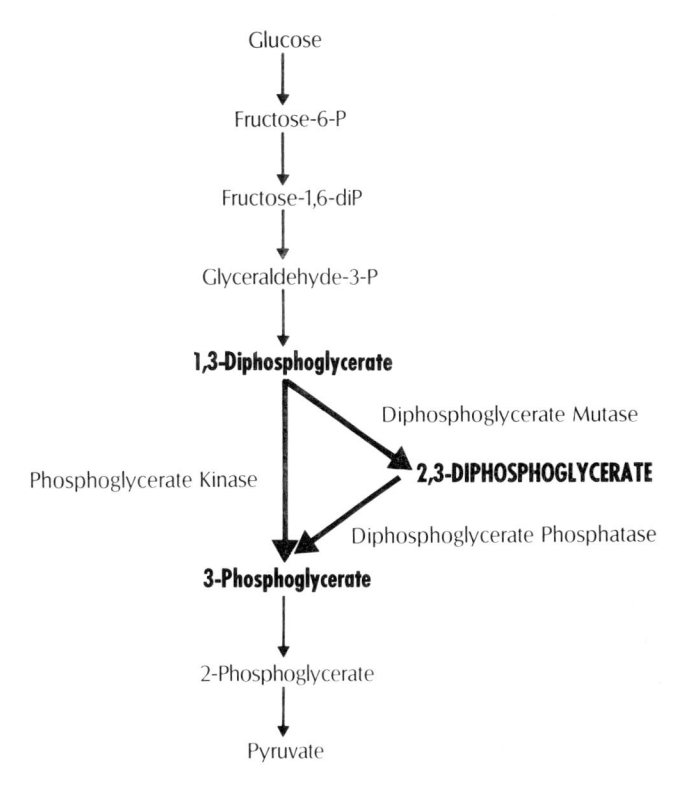

FIG. 8. The Rapoport–Luebering pathway for the production of 2,3-diphosphoglycerate.

pathway (Fig. 8). The synthesis of 2,3-diphosphoglycerate is controlled by a number of factors which compensate for any changes in oxygen delivery.[2] For example, the decrease in oxygen affinity that occurs with increased acidity will be compensated by an equivalent increase in affinity due to inhibition of 2,3-diphosphoglycerate synthesis by the lowered pH. This gives an automatic correction of oxygen dissociation in acid-base disturbances, but, whereas the Bohr effect acts immediately, adjustment of the 2,3-diphosphoglycerate concentration takes place over some eight hours. For this reason the correction of acid-base disturbances is best carried out over a period of hours rather than precipitately. In prolonged acidosis, 2,3-diphosphoglycerate concentration will be low and a sudden correction of the blood pH will cause a drastic increase in the oxygen affinity of the haemoglobin with decreased oxygen release. The only way the body can meet the

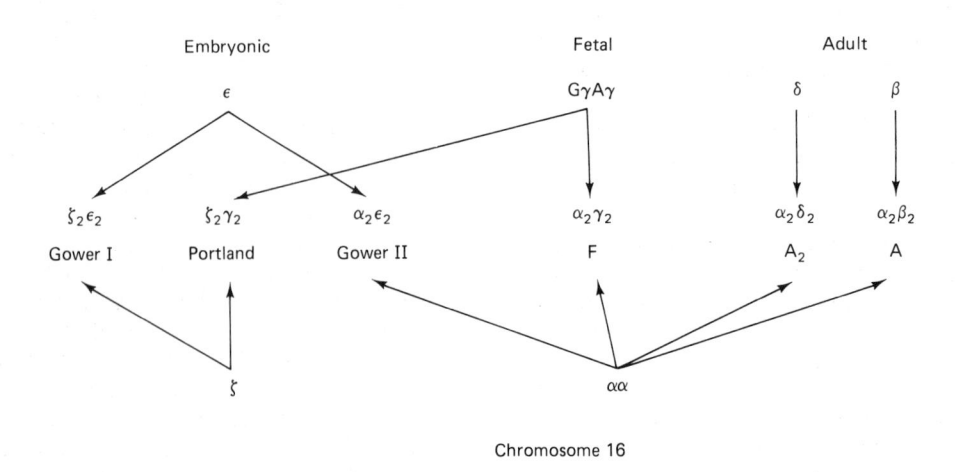

Fig. 9. Human haemoglobins and the genetic control of their production.

HAEMOGLOBIN GENETICS AND EXPRESSION

Haemoglobins of Embryo, Fetus and Adult

Human haemoglobins can be divided into three groups, each designed to meet a different physiological requirement for oxygen: the first are those required for the uptake of oxygen in the free-floating embryo, the second are those required for the uptake of oxygen across the placenta, and the third are those required in the mature human for atmospheric respiration. These haemoglobins are summarised in Fig. 9, and their chronological appearance is shown in Fig. 10. Diagnostically, the important haemoglobins are HbF, $\alpha_2\gamma_2$; HbA, $\alpha_2\beta_2$; and HbA$_2$, $\alpha_2\delta_2$. All of these are present in the mature human: HbA forming 97% of haemoglobin of the adult and HbA$_2$ some 3%. HbF which forms about 60% of the haemoglobin of the newborn comprises less than 0·8% of that of the normal adult. This trace amount of fetal haemoglobin in the adult appears to arise from the occasional switching-on of new lines of erythroid cells which go through an initial phase of production of fetal haemoglobin. Usually the bone marrow is dominated by mature cell-lines but stressed erythropoiesis, as in the leukaemias or severe congenital anaemias, will give a large increase in *de novo* synthesis with significant increases in HbF production. In these situations the HbF is predominantly found in those cells that have arisen from new lines. A blood film stained to show fetal haemoglobin will give a heterogeneous pattern showing a mixed pattern of cells; a few containing a high proportion of fetal haemoglobin but the majority of cells, from mature erythroid lines, having no fetal haemoglobin.

As shown in Fig. 9, there are two fetal haemoglobins which differ, almost imperceptibly, in the structure of their γ-chains; one having an alanine at position 136, the other a glycine. This slight structural difference is diagnostically useful as it enables a better definition of the various genetic abnormalities that lead to persistence of fetal haemoglobin into adult life; these may result in persistent expression of one, or both, of the γ-genes.

α-Gene Duplication

The duplication that has occurred with the γ-globin genes has also occurred with the α-globin genes; in this last instance to give identical duplicates. The presence of paired α-chain genes means that an abnormal α-globin will involve only one quarter of the total haemoglobin, i.e. one of the four α-chain alleles; an abnormal β-globin will involve 50% of the HbA, i.e. one of the two β-chain alleles. The presence of identical duplicates also opens the possibility of unequal crossing-over to give chromosomes with either three or one α-chain genes.[24] The loss of one gene will decrease, by a half, the α-globin produced from that chromosome to give, overall, a small deficit of α-chains; this is a mild α-thalassaemia, previously known as α-thalassaemia-2 trait. This deletion of a single α-globin gene provides a slight compensatory advantage in malarial areas. As a consequence a significant proportion of individuals who originate from Africa, South Asia, or the Mediterranean areas, have

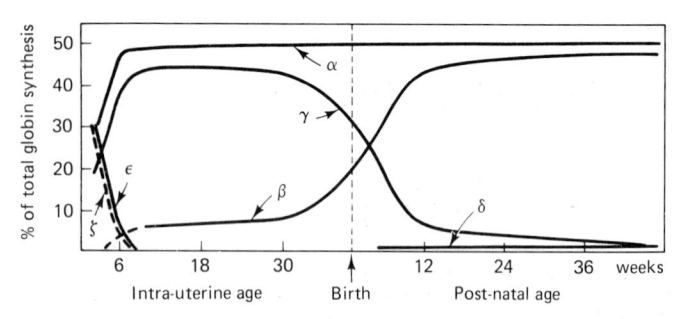

Fig. 10. Changes in globin chain synthesis during human development. (Adapted from Wood.[37])

oxygen demand is by increasing cardiac output; thus cardiac failure may be the end-result of a sudden correction of acidosis in an elderly or debilitated diabetic.

only three, or even two, α-chain alleles compared to the four alleles present in most of the human population.

The frequency of variations in the numbers of α-genes explains the variability observed in the relative proportions of α-chain variants. The same variant may form less than 20% of the total HbA in an individual, with five α-globin alleles, whereas in another individual, with two alleles, it will form 50% of the total haemoglobin. A similar variation is seen in the proportion of β-chain variants for the same reason. A heterozygote for HbE will have some 30% of the abnormal haemoglobin since the abnormal β-chain competes less effectively than the normal β-chain for available α-chains. If the synthesis of α-chain is decreased due to the loss of one α-gene then the proportion of HbE decreases to 24%, with the loss of two α-genes the proportion of the HbE falls to 17%. Careful measurement of variant proportions may, in this way, provide evidence of the α-thalassaemic carrier states which can otherwise be detected only by the more complex techniques of genomic mapping.

Globin Gene Structure

The duplication of the α-chain genes is, in terms of geological time, a relatively recent event but it illustrates the way in which the globin genes evolved; by the process of duplication, mutation and then specialisation.

The globin genes are closely linked, although at some time the α-globin genes on chromosome 16 have segregated from the non-α-globins on chromosome 11. The structure and relation of human globin genes has been described by Proudfoot[29] and others as illustrated in Fig. 11.

The order of the genes in each of the chromosomes represents the order of their expression during, development. The pattern of gene duplication is clearly evident, but as well as the formation of the functional duplicates (α2 & α1, Gγ & Aγ, δ & β), there has also been the formation of non-functional 'pseudo-'genes from duplicated loci which have undergone mutations that prevent transcription (ψβ2, ψβ1, ψα1). The role of these non-functional genes is obscure; their occurrence in each instance between the fetal (or embryonic) and adult genes gives a hint of some role in gene expression. But it is more likely that they are dormant remnants, increasing the potential for genetic diversity and evolution.

The genes that code for globin structure are overwhelmed in size by massive amounts of non-coding, flanking-sequences. Some of the flanking sequences are involved with the initiation and control of transcription, but the function of the bulk of the material is not known. Comparisons between species suggests that only small areas have a direct role in the control of globin expression which is thought to extend over a much larger region of chromosomal DNA than that shown in Fig. 11.

As well as the non-coding flanking sequences there are also non-coding intervening sequences or introns, that interrupt the structural coding of the globin genes. These introns are consistently found in globin genes, homologously placed between codons (amino acids) 30 and 31, and 104 and 105, of the β-globin messenger. The introns are transcribed into nascent m-RNA but are excised to give the messenger ready for ribosomal translation.

An important practical observation is that although there is absolute conservation of the structural portion of the gene there is considerable variation in flanking and intron sequences, even within a single species. It seems certain that the degree of this non-expressed polymorphism within the human will be sufficient to allow much more precise gene linkage studies using DNA from non-specific sources such as amniotic fibroblasts or white cells. The actual structural abnormality, such as HbS,

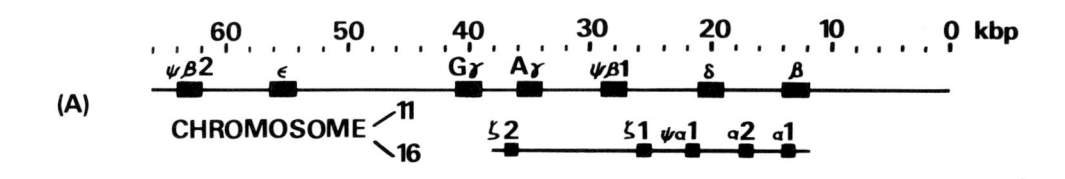

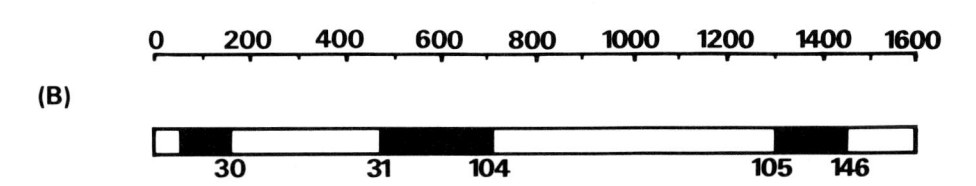

FIG. 11(A). The positioning of the functional and non-functional globin genes; non-α on chromosome 11, and α on chromosome 16. The scale in thousands-of-base-pairs indicates how the coding sequence is overwhelmed in size by the non-structural flanking sequences. The ζI though not a typical pseudo-gene (ψζ) is not functional because a mutation has introduced a stop codon early in its structural sequence. (B) A magnified depiction of the β-globin gene showing how the structural gene is interrupted by large non-coding introns. The structural sequences are in black.

may be difficult to detect from this material but an affected carrier is likely to be readily identifiable by the gross polymorphisms in the linked flanking sequence.

HAEMOGLOBIN VARIATION AND DISEASE

Haemoglobin is the best studied of all proteins in terms of both its structure and of its variant forms in the human. The study of these abnormal haemoglobins has greatly increased our knowledge of the relation of structure to function in proteins; it has also provided a model of protein variation which is being rediscovered again and again in other genetic diseases. The adage of the turn of the century with respect to pathology—'Know syphilis and you will know disease'—could now be re-stated with respect to molecular pathology—'Know the haemoglobinopathies and you will know genetic disease'.

The Abnormal Haemoglobins

Nearly 400 abnormal haemoglobins have now been identified and named after their place of discovery. Initially, the majority of these were found as the result of electrophoretic screening of population samples. Most examples found as a result of this approach turned out to be relatively harmless substitutions of one amino acid for another on the exterior of the molecule where a change in charge alters electrophoretic mobility but is not likely to produce gross molecular pathology. More recently, most new variants have been found as a result of an elective investigation of individuals with evidence of haemoglobin malfunction, usually haemolysis or dysfunction of oxygen transport. These malfunctions mainly arise from lesions that affect the internal amino acids of the globin or involve gross changes such as deletions or insertions of sections of the squence. A third category of variants has also been recently identified; these are the structural variants that result in diminished synthesis to give a thalassaemic-like picture as seen in the chain elongation mutants. Together all these abnormalities illustrate the complexity of variation that can occur in the amino acid sequence of the globins[23] and as shown more recently,[14] in the nucleotide sequence of the globin m-RNA.

Point Mutations

Most abnormal haemoglobins have single amino acid substitutions due to a point mutation, the substitution of a single base in a triplet codon (see genetic code, Fig. 12). For example, the codon GGU which codes for $\beta24$ glycine has been substituted: by CGU (arginine) in Hb Riverdale-Bronx, by GUU (valine) in Hb Savannah and by GAU (aspartic acid) in Hb Moscva. Although nearly all of the identified point substitutions have been compatible with the known nucleotide sequence, exceptions are beginning to be found that suggest some polymorphism may exist at the level of m-RNA. The codon for valine $\beta67$ is GUG which is compatible with two sub-

First Base	Second Base				Third Base
	U	C	A	G	
U	UUU Phe UUC Phe UUA Leu UUG Leu	UCU UCC UCA Ser UCG	UAU Tyr UAC Tyr UAA Stop UAG Stop	UGU Cys UGC Cys UGA Stop UGG Trp	U C A G
C	CUU CUC CUA Leu CUG	CCU CCC CCA Pro CCG	CAU His CAC His CAA Gln CAG Gln	CGU CGC CGA Arg CGG	U C A G
A	AUU AUC Ile AUA AUG Met	ACU ACC ACA Thr ACG	AAU Asn AAC Asn AAA Lys AAG Lys	AGU Ser AGC Ser AGA Arg AGG Arg	U C A G
G	GUU GUC GUA Val GUG	GCU GCC GCA Ala GCG	GAU Asp GAC Asp GAA Glu GAG Glu	GGU GGC GGA Gly GGG	U C A G

FIG. 12. The messenger-RNA triplet code for amino acids. The code for mitochondrial RNA differs slightly. UGA codes for tryptophan, and AUA for methionine.

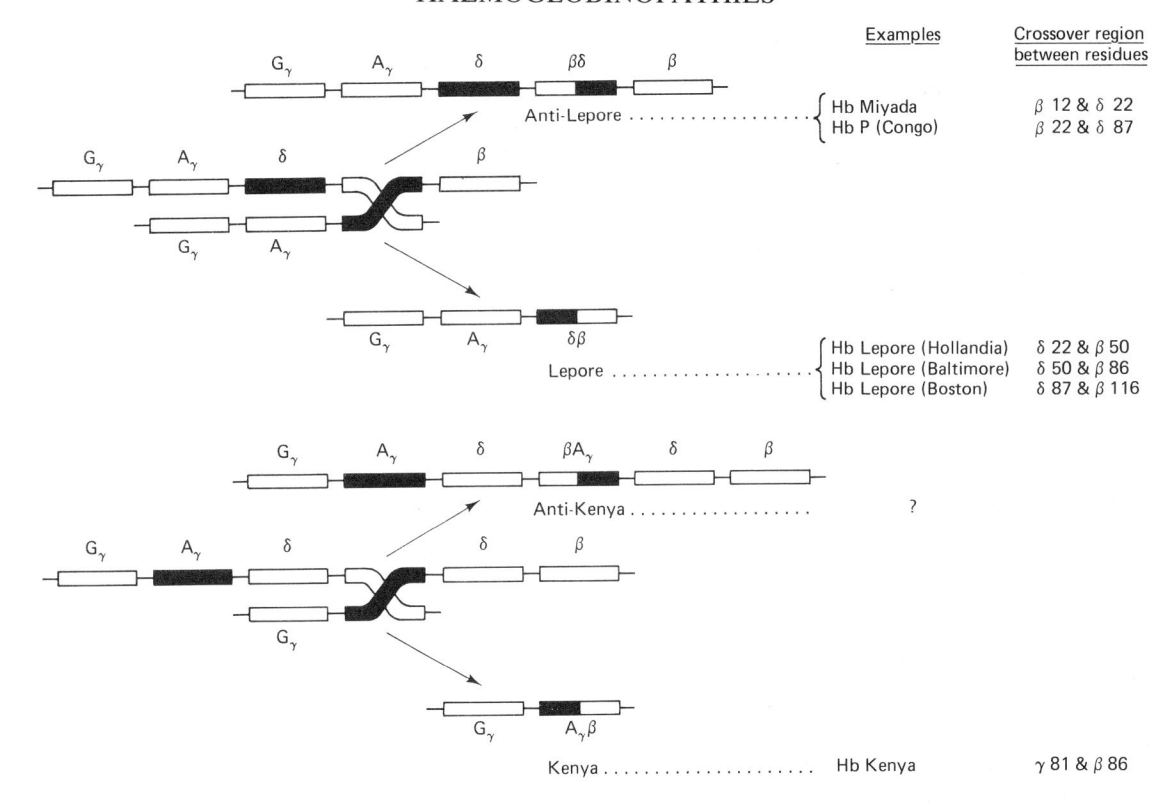

Fig. 13. Mechanism of prediction of the crossover variants Hb Lepore and Hb Kenya. There are 3 different forms of Hb Lepore resulting from crossovers between the δ and β chain loci at different sites. (Reproduced with permission from Weatherall, D. J. and Clegg, J. B. (1979), *Cell*, **16**, 467–479.)

stitutions found at this position, alanine (GCG) in Hb Sydney and glutamic acid (GAG) in Hb Milwaukee. However, a third variant, Hb Bristol, has an aspartic acid at position β67 which requires a codon GAU or GAC, incompatible with a single mutation from that of valine GUG. In other words, some individuals exist with either GUC or GUU, as the alternative codons for valine to GUG, at position β67.

Deletions and Insertions

A number of variants have been identified that have lost or gained small portions of their amino acid sequence. Usually deletions occur between small areas of repetitive sequence as in Hb Gun Hill where there is a deletion between two Leu-His sequences in the β-chain:

Leu-His-*Cys-Asp-Lys*-Leu-His
91   92   93   94   95   96   97

An example of an insertion is the α-chain variant, Hb Grady, in which the tripeptide sequence Glu-Phe-Thr is repeated. It is likely that these deletions and insertions have arisen from unequal crossing-over during meiosis though it is possible that they could result from a defect in DNA repair or replication.

Fusion Variants

These variants have formed by non-homologous crossovers between adjacent genes to give δβ or γβ hybrid or fusion forms. The identification of these fusion variants allowed the deduction of the arrangement of the globin genes on chromosome 11 even before it was confirmed by genomic mapping (Fig. 11).

The most important of the fusion variants are the Hbs Lepore, illustrated in Fig. 13. These arise from a fusion of the δ and the β-chains to give several varieties of δβ products, depending on the level of junction of the δ and β fragments. The δβ product is produced at only one-fifth the rate of that of the β-chain it replaces and this leads to a surplus of α-chains with mild thalassaemic findings. As a consequence, the Hbs Lepore have a similar distribution to the thalassaemias and are a relatively common finding in Mediterranean and other populations from endemic malarial areas. The other fusion globins, the converse anti-Lepore variants (Fig. 13), and the γβ Hb Kenya, are less frequently found.

Termination, Frame-shift and Nonsense Errors

A mutation affecting a single base in a globin gene may produce effects other than a simple amino acid substitution. A mutation in the codon that signals termination of the globin polypeptide may result in elongation products. Most examples have occurred with the α-chain, the reason almost certainly being that the resultant product is synthesised at a much slower rate than the normal α-gene product. This results in a mild thalassaemia which once again has advantages for the heterozygote in malarial areas.

The α-chain termination mutants have an elongation of some 30 residues which represents translation of the normally untranslated 3′ m-RNA up to the next termination codon. The most common example, present in many South-East Asians, is Hb Constant Spring: an α-chain variant in which the stop codon UAA has mutated to CAA the codon for glutamine.

Another type of variation is the frame-shift error in which one or two nucleotides are inserted into, or lost from the structural gene. These frameshift mutations will result in an alteration of the sequence beyond the mutation and are unlikely to produce a viable globin unless they occur near the terminus of the globin gene. An example is the α-chain of Hb Wayne which is unchanged up to position 138 but then has a sequence of eight new amino acids. As shown, this is explicable by the loss of an adenine from the codon for α139 Lys.

```
α-A      137  138  139  140  141  142  143
         Thr  Ser  Lys  Tyr  Arg  Stop
         ACC  UCC  AAA  UAC  CGU  UAA  GCU
α-Wayne  137  138  139  140  141  142
         Thr  Ser  Asn  Thr  Val  Lys
         ACC  UCC  AAU  ACC  GUU  AAG
```

Similarly, Hb Tak and Hb Cranston have elongated β-chains which are explicable by frame-shift mutations.

Yet another type of mutation is the nonsense error in which there is a conversion of the codon for an amino acid to a stop or nonsense codon. An example is Hb McKees Rock where the two C-terminal β-chain residues, β145–146, are absent. This can be explained by the mutation of the β145 tyrosine codon, UAU, to a termination-nonsense codon UAA or UAG. Once again these mutations will only produce a viable globin product if they are near the COOH-terminus. Nonsense or frame-shift mutations earlier in the nucleotide sequence will produce a non-viable product to give, phenotypically, a thalassaemia.

Distribution of Variants in Man

Some of the abnormal haemoglobins, such as Hbs S, C, D and E, affect hundreds of thousands of people while others, such as G-Philadelphia and O-Arab, have an increased frequency within relatively limited groups. Knowledge of the distribution of variants[21] is useful both diagnostically and anthropologically. Diagnostically, the knowledge that HbE is common in South-East Asians will usually allow its provisional differentiation from HbC, a variant with similar electrophoretic mobility but a distribution largely confined to West Africans. Anthropologically, the variants are of interest as they illustrate the migration and mixture of peoples; for example, the not infrequent finding of HbD Punjab in Europeans is likely to be one of the longest lasting reminders of the British Empire.

However, there needs to be caution in the plotting of migrations on the basis of variants as undoubtedly, many examples arise from new mutations. It would be a bold anthropologist who could link the first three observations of HbM Hyde Park; in a Japanese, a black American and a European New Zealander!

The establishment of chance mutations within a population is much more likely when there is a balanced polymorphism, i.e. a situation where the disadvantages of the mutant are balanced by other advantages. An example is HbS or sickle cell haemoglobin (β6 Glu→Val) where the heterozygote has the advantage of protection against malignant malaria. HbS is mostly associated with tropical Africa where it may affect up to 40% of some East African tribes. It is present at a lesser frequency in non-tropical Africa, and is found in North but not South Africa. There are foci of sickling in Italy, the Middle East and in India. Sickle cell anaemia has become an important public health problem in a much wider area due to African settlement in the USA, the West Indies, tropical Latin America, and more recently, in Europe.

Haemoglobin C (β6 Glu→Lys) is a West African genetic marker with its highest incidence in Northern Ghana and Upper Volta. As mentioned previously, HbE (β26 Glu→Lys), which has the same electrophoretic charge as HbC, is frequently found in South-East Asia including Vietnam and also in South-West China. Haemoglobin D (β121 Glu→Gln) is present in 3% of Punjabis and is a not uncommon finding in the increasingly mixed populations of Westernised countries. This relatively common haemoglobin is of diagnostic importance as it moves on electrophoresis with HbS and consequently the HbSD mixed heterozygote can easily be mislabelled as a HbS homozygote.

Two other, less frequent, West African haemoglobins are Hb Korle-Bu (β73 Asp→Asn), found predominantly in Ghana, and HbG Philadelphia, found predominantly in Nigeria. Hb O-Arab (β121 Glu→Lys) first found in Arabs, is in fact at its highest incidence in Macedonia; it occurs occasionally in the Middle-East and in the Nothern Sudan. Hb Hasharon (α47 Asp→His) at first found occasionally in Ashkenazi Jews is now also seen with some regularity in Italy, particularly the Ferrara region.

Non-Genetic Variation: HbA$_{1c}$

Haemoglobin can undergo several types of modification subsequent to its synthesis. One example is the minor HbF fraction that results from acetylation of the amino-terminus of some of the circulating γ-chains. Another example is the proteolytic loss from the α-chains of the terminal arginine that takes place when haemoglobin is released into the plasma, to give the fast electrophoretic component, Hb Koellicker. But by far the most important of the modified forms are the glycosylated haemoglobins, a series of glucose adducts to haemoglobin. The range of modified forms of HbA is illustrated in Fig. 14 which shows the isoelectric-focused electrophoresis of an adult haemolysate after three days refrigerated storage.

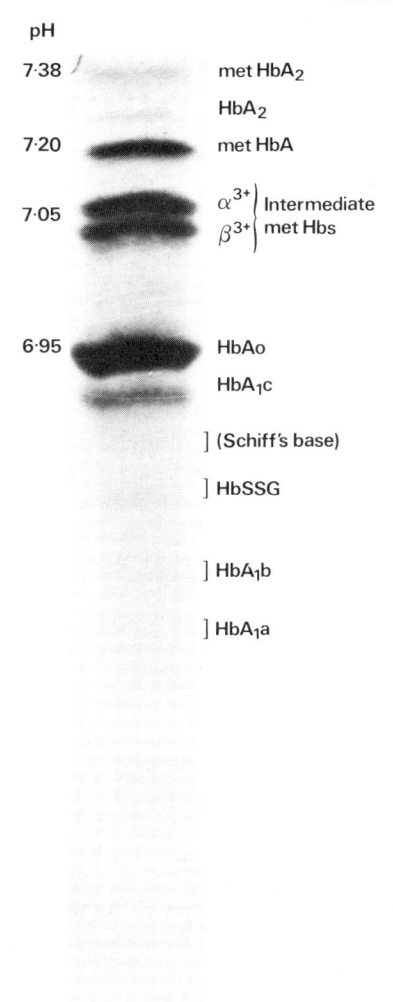

FIG. 14. Isoelectric focusing of haemolysate stored at 4° for 5 days. The pattern is complicated by oxidation to methaemoglobin including the intermediate oxidised hybrids. A number of acidic derivatives are formed; most prominent here is HbA_{1c} but its Schiff base precursor and the haemoglobin glutathione adduct can be major components. The identity and mobility of HbA_{1b} and HbA_{1a} are still a matter of debate. Their position is indicated in the Figure.

HbA_{1c}: The Glycosylated Haemoglobins

Early studies had shown that ion-exchange chromatography of HbA yielded several minor fractions with increased negative charge the most significant of which were HbA_{1a}, HbA_{1b} and HbA_{1c}, the group being collectively labelled HbA_1. Rahbar[30] in 1968 showed that the amount of these fast components was considerably increased in diabetic subjects and later work showed that the increase was primarily due to an increase in HbA_{1c}. This is produced by the formation of a Schiff base between the aldehyde of glucose and the terminal amine acid of the β-chain to give a ketoamine linkage.[4] The level of HbA_{1c} was shown to correlate with the degree of hyperglycaemia to which the haemoglobin had been exposed over its average life of one to three months. Thus the HbA_{1c} gives a potential measure of the efficiency of long term control of diabetes; the normal individual has some 9% of his haemoglobin in glycosylated form whereas a diabetic with bouts of severe hyperglycaemia may have a level of 14% or more. This enables an assessment of overall control of blood glucose levels as opposed to the intermittent levels available from individual blood glucose analyses. The potential value of the measurements is enhanced by the demonstration that a similar glycosylation is occurring in other proteins including those of basement membranes and crystallin, the principal protein of the lens. The inference is that protein glycosylation is at least a contributory factor to the long-term degenerative changes associated with poorly controlled diabetes.

Unfortunately at this stage clinical enthusiasm for the test has moved far ahead of analytical techniques[11] and as a consequence much of the vast literature on the topic is of dubious value except as an object lesson in how not to conduct clinical biochemistry. In the first place the analyses are ill-defined and analytical standards are not available. HbA_{1c} is certainly a single chemical species, but most current results published as HbA_{1c} are really column eluates that include all of HbA_1. Some of the other A_1 components can also reflect hyperglycaemic metabolism since HbA_{1a} includes components in which the linkage is with fructose 1,6-diphosphate (A_{1a1}) and glucose 6-phosphate (A_{1a2}). But the A_1 or 'fast' chromatographic fraction also includes the glutathione adduct of haemoglobin whose concentration varies with the age of the haemolysate and its mode of preparation. There is also a labile component of HbA_1 that reflects recent, short-term hyperglycaemia. This appears to be due to the preliminary formation of the glucose-amine Schiff base without ketoamine linkage. It is therefore reversible and can be eliminated by incubation *in vitro*.

An additional complicating factor is that glycosylated haemoglobins can also be measured colorimetrically. Curiously the results correlate well with the chromatographic technique, even though the colorimetric method is not specific for HbA_{1c}, but also measures other glucose adducts as well as non-specific complexes.

Clearly, what is required is a reproducible technique, measuring defined compounds, with standards that allow interlaboratory reproducibility. In the meantime, there is no doubt as to the potential usefulness of glycosylated protein measurements but the clinical biochemist should be well aware of the limitations of current techniques and endeavour to communicate these limitations to the enthusiastic clinician.

The Haemoglobinopathies

Used in its strictest sense, the term 'haemoglobinopathies' covers those variants that result in clinically apparent disease. The identification of their abnormalities has enabled the correlation of the molecular defect, with the disease process, to open up the new field of molecular pathology.[28]

Sickle Cell Disease

This is due to the enhanced tendency of HbS (β6 Glu→Val) to become insoluble in deoxygenated solutions due to the formation of unidirectional crystals.[3] These crystals are apparently formed in a transitory manner by HbA but the replacement of the hydrophilic glutamic acid, at position β6 in HbA by the hydrophobic valine in HbS, stabilises the crystals. Crystallisation is due to the formation of strands by end-to-end association of haemoglobin molecules, six or more of these strands winding about a central core to form tubular fibrils.

The unidirectional haemoglobin crystals form spikes which deform the shape of the red cells and make them unduly rigid. These rigid, sickle-shaped cells are susceptible to haemolysis and have a shortened half-life. They are also liable to form plugs in the microcirculation causing ischaemia and further sickling, with local infarction. Initially the cells are reversibly sickled and, on oxygenation, can revert to their normal shape. However, after several bouts of sickling, the red cell develops membrane damage and becomes an irreversibly sickled cell, giving rise to an increased viscosity even in fully oxygenated blood.

Sickling is primarily a problem that affects homozygotes and it rarely causes difficulties for the AS heterozygote. The severity of the disease may be greatly modified by interaction with other haemoglobins.[25] Haemoglobins C and O-Arab will co-polymerise with HbS but Hb Korle-Bu (β73 Asp→Asn) is a protective agent against sickling, either by itself or as the double mutant Hb Harlem (β6 Glu→Val, β73 Asp→Asn). Haemoglobin F does not readily take part in filament crystallisation and is therefore a powerful antisickling agent. Even in homozygous HbS disease, a fetal haemoglobin level above 10% usually ensures a mild clinical course and a value above 20% protects against most symptoms.

The important point for the clinical biochemist is that diagnostically the task is not completed by the identification of the presence of HbS along with a positive sickle cell test. Is there a coincidental HbD? Is there also β or α thalassaemia? Is there another haemoglobin such as O-Arab or C Harlem? What is the HbF level? All of these factors are important in the assessment of apparent sickle cell disease and their significance should be recognised and reported to the clinician.

The Unstable Haemoglobins

Abnormalities that disrupt the shape of the globin molecule will decrease its solubility and cause precipitation of the haemoglobin within the red cell to give particulate inclusions—Heinz bodies. Some eighty of these unstable haemoglobins have been identified,[23] almost always as the result of the investigation of the accompanying Heinz body haemolytic anaemia.[7,36]

Most of the unstable haemoglobins have amino acid abnormalities which cause either spatial disruption of the haem pocket or of the α1β1 interface, or produce a major lesion, such as a deletion, that grossly distorts globin conformation. The largest proportion are due to substitutions within the haem pocket (Fig. 15) where even minor steric alterations will allow entry of anions to give methaemoglobin formation. The associated distortion allows bonding with globin sidechains to give haemichrome formation with irreversible denaturation and precipitation. The other lesions of the unstable

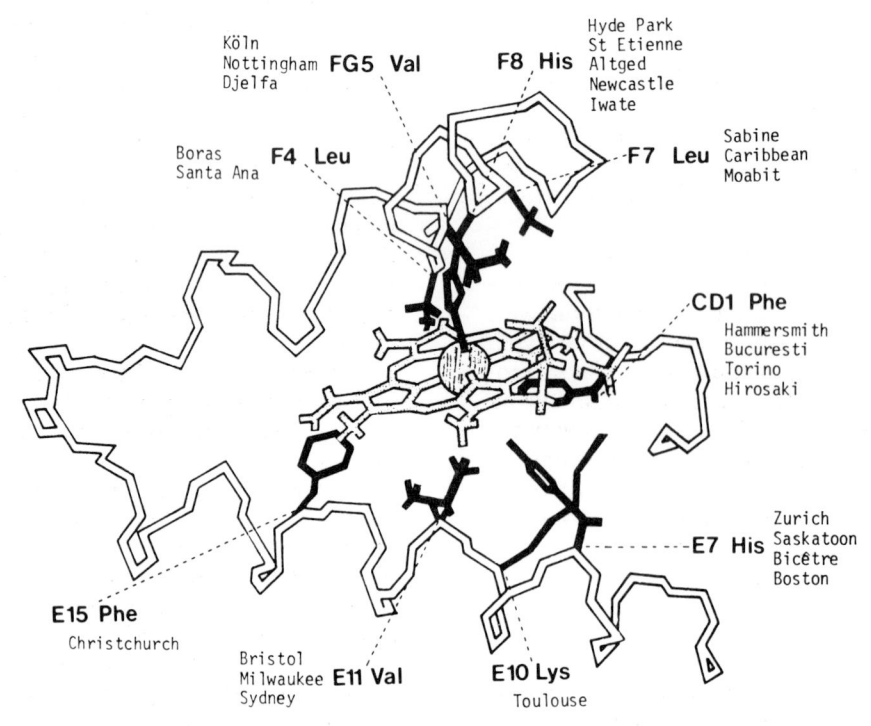

Fig. 15. The haem group (shaded) with some of the amino acid residues in contact with it in the haem pocket. Substitutions of the 9 residues shown account for a third of the unstable haemoglobins, as indicated by the associated names.

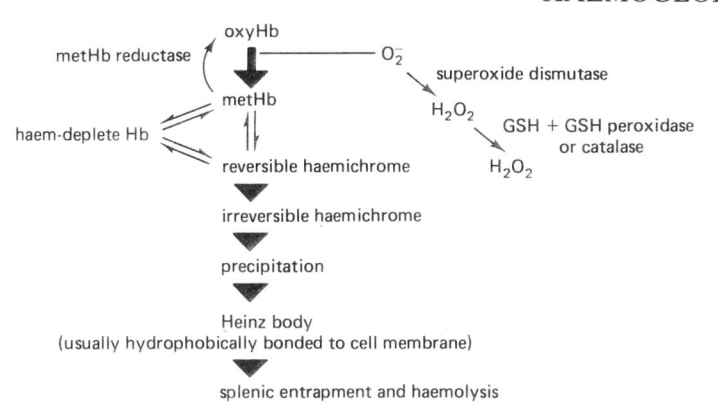

FIG. 16. Sequence of events leading to precipitation and Heinz body formation by the unstable haemoglobins.

haemoglobins will also produce haem pocket disruption with the consequent course of events summarised[7] in Fig. 16.

The haemolytic anaemias occur in the heterozygous carriers of unstable haemoglobins, the most common and severe forms being due to β-chain abnormalities. These give rise to initial symptoms in early infancy following the disappearance of HbF. Globin instability has two main effects, one in the bone marrow, the other in the circulation. The newly-formed, unstable globin is liable to misfold in shape and be eliminated in the red cell precursors of the bone marrow by proteolysis. With major instability this may cause a gross disturbance of erythropoiesis to give a dyserythropoietic bone marrow picture. Usually some half of the unstable globin survives to give circulating levels of 10–30% of the total haemoglobin. This circulating, unstable haemoglobin is readily precipitable either by drugs or by the increased temperature that accompanies even minor childhood infections. The result is precipitation of the haemoglobin as Heinz bodies which are removed by the spleen. This removal process, descriptively called pitting (Fig. 17), results in loss of cell membrane, distortion of the red cell and haemolysis.

The diagnosis of the unstable haemoglobins is almost entirely dependent on the demonstration of their presence by a positive stability test. The details of these tests are discussed later, but they are simple to perform and the important point is that they should be routinely used in the workup of any suspected haemoglobinopathy or unexplained congenital anaemia. Many of the abnormalities do not cause a readily discernible change in mobility on electrophoresis or chromatography and the use of stress tests for stability is the only means of demonstrating the presence of an abnormal component and of isolating it for further investigation.

Oxygen Affinity Variants

As discussed, the R to T transformation in haemoglobin primarily involves residues in the $\alpha 1\beta 2$ interface, the haem pocket and the 2,3-diphosphoglycerate binding site. Mutations at these sites are likely to alter

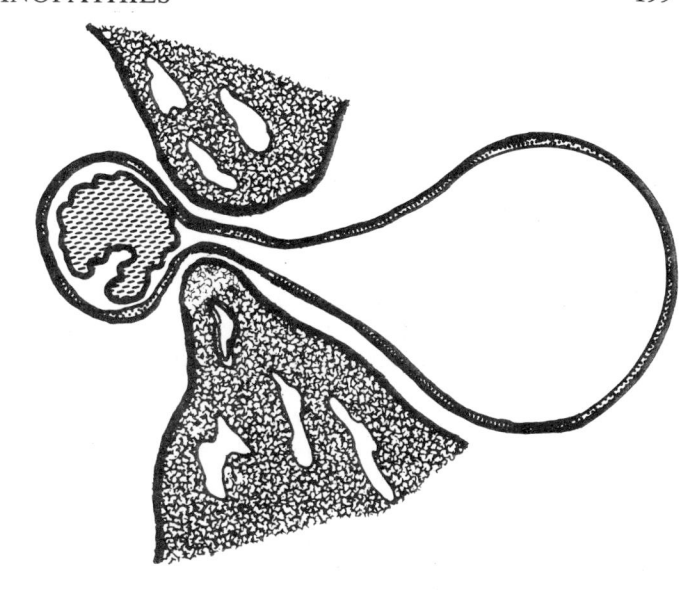

FIG. 17. Inclusion, or Heinz, bodies are removed from the red cell by the process of pitting, a plucking of the rigid body and associated red-cell membrane as the erythrocyte moves through the interstices of the splenic sinusoids.

cooperativity,[19] usually to decrease it to give increased oxygen affinity. Increased affinity will result in a lessened release of oxygen at the periphery and the body will respond to this by increasing red cell production to give an erythraemia which may present as a polycythaemia, but the haematocrit is often within the upper limit of normal. Less frequently, mutations may result in a partial locking of the molecule in the deoxy-, T-form, to give decreased affinity and an apparent cyanosis.

The possibility of an abnormal haemoglobin should always be considered in the investigation of polycythaemia or cyanosis, particularly if there is a family history of the condition or if it has been present from an early age. It is not excluded by the absence of any electrophoretic abnormality since, as with the unstable haemoglobins, the mutations may not involve any change in charge, e.g. the high-affinity variant Hb Heathrow ($\beta 103$ Phe→Leu). Similarly, although it was thought that a normal white cell count with erythraemia indicated the likelihood of a high affinity variant, experience has shown that the white cell count may well be raised.

The lesson is that wherever there is a suspicion of an abnormality of oxygen affinity it should be checked by measurement of the P_{50} of the whole cells along with that of a normal control. A deviation of the P_{50} by more than 3 mm Hg from that of the control is of significance and should be checked. This requires measurement of the dissociation curve of haemolysates, both in the presence of 2,3-diphosphoglycerate and when stripped of phosphates.[10,31] The demonstration of a consistent change in oxygen affinity confirms the presence of an abnormal haemoglobin. A stability test should always be performed as lesions gross enough to cause a change in oxygen affinity will also usually produce a decrease in stability. The corollary is also true; the unstable haemoglobins usually have some alteration in oxygen affinity.

Clinically, the combination of the two effects may be apparent with evidence of haemolysis, due to the instability, and an unexpectedly high haemoglobin concentration, due to increased oxygen affinity.

Over 80 abnormalities, producing clinically evident changes of oxygen affinity, have been identified.[19] The majority have occurred at the $\alpha 1\beta 2$ interface. Examples are the substitutions that occur at position $\beta 99$ which is normally occupied by an aspartic acid residue that forms a hydrogen bond important in stabilizing the deoxy-conformation. This aspartic acid is replaced in four variants associated with polycythaemia; Hb Yakima ($\beta 99$ His), Kempsey ($\beta 99$ Asn), Radcliffe ($\beta 99$ Ala) and Ypsilante ($\beta 99$ Tyr). An increase in affinity also results from the replacement of the penultimate tyrosine ($\beta 145$) which is responsible for the locking of the tertiary structure in the T-form. This occurs with Hb Bethesda ($\beta 145$ Tyr→His) and Rainier ($\beta 145$ Tyr→Cys). Alteration at a 2,3-diphosphoglycerate binding site will also give increased affinity as in Hb Rahere ($\beta 82$ Lys→Thr) or Helsinki ($\beta 82$ Lys→Met).

Methaemoglobinaemia

The iron of the haem group is maintained in its ferrous state by the protective environment of the globin haem pocket. Fluctuations in the conformation of this pocket will occur from time to time and, even with normal HbA, this is sufficient to allow a daily conversion of some 3% of red cell haemoglobin to the ferric methaemoglobin. Several reductive mechanisms exist in the red cell, the most important of which is the NADH-linked methaemoglobin reductase. This enzyme is more correctly labelled as cytochrome-b_5 reductase since it acts indirectly by reducing cytochrome-b_5 which then directly reduces the ferric iron of methaemoglobin.[16]

An increased level of methaemoglobin may result[6] from increased oxidation, as in nitrite-induced methaemoglobinaemia; or from decreased reduction, as in methaemoglobin-reductase deficiency; or from globin abnormalities that result in internal oxidation as in the Hbs M.

It is also useful to remember that the reductive mechanisms are confined to the red cell and consequently haemolysates stored for more than a day or two will contain appreciable levels of methaemoglobin. For this reason blood samples for haemoglobin investigation should always be transported as red cells and haemolysed just prior to testing.

The presence of methaemoglobin A can be demonstrated using a lysate divided into two samples. To one is added a few crystals of sodium dithionite. After mixing and a wait of five minutes this sample is used as a blank for a scanned spectrum of the other sample. The appearance of a peak of 632 nm confirms the presence of methaemoglobin and allows its quantitation.

Acquired Methaemoglobinaemia

Methaemoglobin formation can occur as a result of exposure to a number of drugs and chemicals such as aromatic amines. It is interesting that much of the best work on the drug-induced methaemoglobinaemias is found in rather obscure German papers. The reason is that the most intensive studies of red cell metabolism were made in Germany during the 1939–45 war. This was prompted by the occurrence of methaemoglobinaemia in workers in the explosive industry. This was a serious problem that resulted from the need in Germany to use benzene-based derivatives rather than the toluene-based explosives produced in other countries. Nowadays the biochemist is more likely to see acquired methaemoglobinaemia resulting from nitrate contamination of water supplies or as a side-effect of therapeutic drugs. These are particular problems in the newborn whose red-cell reductive capacity is not fully developed.

The diagnosis of acquired methaemoglobinaemia can usually be made from the history of exposure along with a demonstration *in vitro*, of a normal rate of reduction by the red cells of nitrite-induced methaemoglobin. Red cells are oxidised with nitrite, any excess being removed by further washing. The cells are then incubated in the presence of lactate along with similarly treated cells from two normal controls. In the acquired, as opposed to the congenital methaemoglobinaemia, about 10% of the methaemoglobin will be reduced each hour.

Congenital Methaemoglobinaemia

This arises from a deficiency of the NADH-linked cytochrome b_5 reductase. The result, in the homozygote, is an accumulation of some 10–20% methaemoglobin in the blood. The problem presents in the infant as cyanosis but, unfortunately, methaemoglobinaemia is frequently not thought of until after extensive cardiac investigations, including catheterization, have proved negative.

The enzyme cytochrome b_5 reductase exists in two forms, the full-sized molecule consists of the enzymic portion plus a hydrophobic tail that binds it to the microsomal membrane. This form is found in all cells of the body except the red cell which has no microsomes and consequently its cytochrome b_5 has lost the hydrophobic tail and is soluble in the cytoplasm. A defect which prevents loss of the tail will result in a simple red cell deficiency, i.e. methaemoglobinaemia, whereas a defect which affects the enzymic portion will affect other cells as well as the red cell. Consequently, a proportion of the congenital methaemoglobinaemias are accompanied by general changes, notably, mental retardation.

The presence of a congenital methaemoglobinaemia can be confirmed by the nitrite-reduction test described in the previous section which differentiates it from an acquired methaemoglobinaemia. Enzymic methaemoglobinaemia can be differentiated from an abnormal HbM by the oral administration of methylene blue (1 to 3 mg/kg body weight) or ascorbic acid (0·25–0·5 g/day). This will decrease methaemoglobin levels to near normal in individuals with the reductase deficiency since ascorbate gives direct reduction of methaemoglobin A and methylene blue gives reduction utilizing NADPH thus bypassing the cytochrome b_5 reductase system. There will be no appreciable improvement in patients whose methaemoglobinaemia is due to an abnormal haemoglobin.

The Haemoglobins M

Cyanosis may be due to the presence of an abnormal haemoglobin with a substitution in the haem pocket that allows ligand formation to the ferric iron. This will occur if a negatively charged group can closely approach the haem iron. The replacement of either the proximal or distal histidine by tyrosine allows binding by the hydroxyl of the tyrosine to give a stable ferric haemoglobin as in Hb M Saskatoon (β63 His→Tyr), Hb M Boston (α58 His→Tyr), Hb M Hyde Park (β92 His→Tyr) and Hb M Iwate (α87 His→Tyr). It also occurs with Hb M Milwaukee in which a glutamic acid is substituted for valine β67, in a position that allows direct binding of the carboxylic groups to the iron, again converting it to the ferric form.

Cyanosis will be apparent from birth in the case of an α-chain HbM and from about six months in the case of a β-chain Hb M. The condition is dominant, in that all known cases have been heterozygotes. This may help in its differentiation from a methaemoglobin reductase deficiency which gives rise to cyanosis only in the homozygote. However caution has to be used with this guideline, as an appreciable proportion of observed cases of Hb M, as with unstable haemoglobins and oxygen affinity variants, have arisen as new mutations, and can be absent in both parents.

Differentiation of the Methaemoglobinaemias

The preliminary differentiation of the methaemoglobinaemias[6] can be made, as described above, by use of the *in vivo* administration of methylene blue and the *in vitro* reduction of nitrite-induced methaemoglobin in red cells. An acquired methaemoglobinaemia will respond to methylene blue administration and will give a normal *in vitro* reduction of induced methaemoglobin. A congenital methaemoglobinaemia will respond to methylene blue but will give a reduced rate of reduction on incubation of nitrite-oxidised red cells. A patient with Hb M will not respond to methylene blue but there will be a near normal rate of reduction of oxidised red cells. Electrophoresis at pH 6·5–7·0 will identify the Hbs M which have a different mobility from Hb A at this pH. Additional evidence may be provided by the spectrum of a dilute haemolysate since the Hbs M have characteristically altered spectra though this may be most evident if a preliminary purification of the abnormal component can be carried out.

The Thalassaemias

Pathology and Haematology

The stability of normal haemoglobin is dependent on the presence of an equal number of α and β subunits to give stable, $\alpha_2\beta_2$ tetramers. The underproduction of one subunit will result in an excess of the other with consequent instability and red cell pathology. This imbalance in globin chain production is the basic defect present in the very heterogeneous group of diseases, the thalassaemias.[35] Each type of thalassaemia is named according to the subunit whose synthesis is suppressed and is further classified as to whether it is a heterozygous or homozygous suppression. Thus β-thalassaemia is the suppression of β-chain synthesis; β-thalassaemia minor, or trait, is the heterozygous suppression of one β-allele; β-thalassaemia major is the homozygous suppression of both β-alleles.

There are two factors involved in the pathology of the thalassaemias: the consequences of the presence of an excess of one of the subunits, and the effect on the cell of an overall reduction of haemoglobin.

The presence of excess globin subunits, particularly in the homozygous thalassaemias, causes considerable disruption of marrow red-cell synthesis. This is due to precipitation of the excess chains with consequent intramedullary haemolysis and distortion of red cell morphology. Peripheral haemolysis occurs, partly due to pitting of inclusions by the spleen (Fig. 17) with this in turn leading to splenomegaly. There is a considerable increase in erythropoiesis to give marrow overgrowth with bone malformation. The accompanying increased haemolysis combined with intermittent transfusion may lead to iron overload. Assessment of iron stores provides one means of differentiating the thalassaemia traits from the morphologically similar iron-deficiency anaemias: the thalassaemias will characteristically have a normal or raised iron-binding saturation and plasma ferritin concentration.

The underproduction of total haemoglobin in the heterozygous thalassaemias results in a decrease in mean cell volume (MCV <80 fl) and a decrease in mean cell haemoglobin (MCH <27 pg). There is usually a compensatory increase in red cell numbers (RBC >5 × 10^{12}/l) so that anaemia is not usually a striking feature of thalassaemia minor. In the β-thalassaemias there may be an increase in synthesis of HbF, as γ-chain production attempts to compensate for β-chain suppression. In β-thalassaemia trait the HbF level is usually 0·6–3·0% (normal <0·8%) and in homozygous β-thalassaemia HbF forms 10–90% of the total haemoglobin. The other compensatory increase is in the level of HbA$_2$ ($\alpha_2\delta_2$). This provides the single most useful laboratory test for β-thalassaemia heterozygotes (Fig. 18) who usually have a raised proportion of HbA$_2$, greater than 3·5% (normal <3·3%).

The pathology of the α-thalassaemias is somewhat different from the β-thalassaemias because whereas free α-chains are highly unstable and predominantly precipitate in the red-cell precursors of the bone marrow, β and γ chains are able to form tetramers capable of surviving for some time in the peripheral circulation. These tetramers β_4 (HbH) and γ_4 (Hb Barts) provide diagnostic evidence of the presence of α-thalassaemia. The tetramers are not efficient oxygen carriers but otherwise they

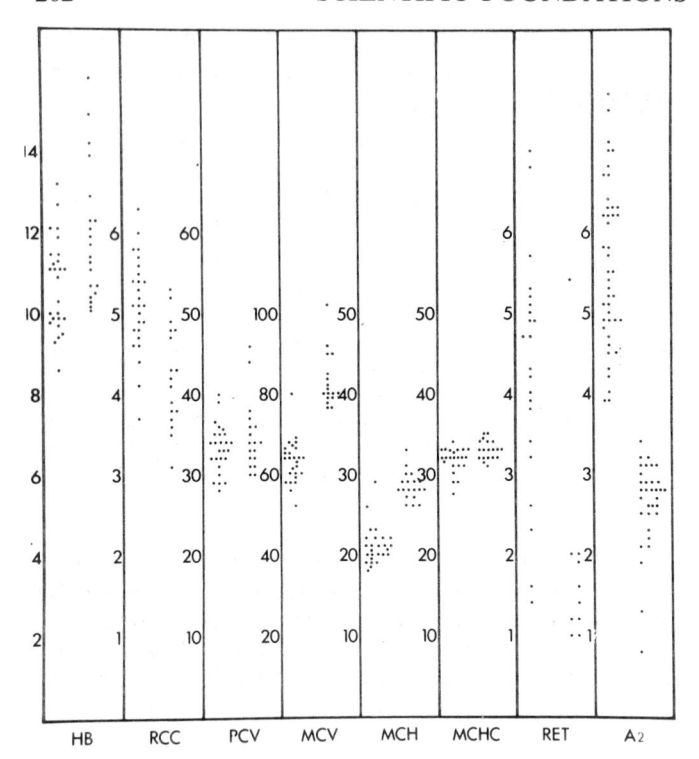

FIG. 18. Red cell indices in β-thalassaemia minor (left) and in non-thalassaemic subjects (right) obtained from one laboratory. The ordinates are as follows: haemoglobin (Hb): g/dl; red cell count (RCC): $\times 10^{12}$/l; packed cell volume (PVC): %; mean corpuscular volume (MCV): fl; mean corpuscular haemoglobin (MCH): pg; mean corpuscular haemoglobin concentration (MCHC): g/dl; reticulocyte count (ret) %; haemoglobin A_2: %. (Reproduced, with permission from Cauchi and Tauro.[8])

peripheral inclusion-body formation, and haemolysis. Of course in the severe α-thalassaemias, i.e. HbH disease, the primary problem is the gross underproduction of haemoglobin with a consequent failure in oxygen transport.

Molecular Defects

The introduction of the techniques of genomic mapping have allowed at least a partial definition of the molecular lesions involved in the thalassaemias.[18] These have proved to be extraordinarily diverse and it is clear that many independent mutations have become established at a population level all resulting in decreased expression of globin genes. These deleterious mutations have been preserved because of their balancing protective advantage in malarial areas. Thus multiple thalassaemic lesions must be added to the already diverse range of other genetic red-cell defects that have arisen in malarial regions and are discussed in a succeeding section.

The lesions producing thalassaemia are of the three types; structural globin variants, structural gene deletions, and mutations of the DNA that controls expression of the structural genes. The best example of a globin structural variant causing thalassaemia is Hb Constant Spring, the α-chain elongation variant commonly found in South-East Asians. Other examples exist of mutations within the structural gene that result in a non-viable and therefore non-expressed globin.

Complete deletion of a gene has been observed in a minority of the β-thalassaemias but is the most common cause of the α-thalassaemias; the loss of one α-gene being a common finding in Africans and Southern Asians. Deletions may extend beyond the structural gene into the area controlling gene expression as shown in Fig. 19. This illustrates why a deletion of the δ and β genes, if limited, will result in a δβ-thalassaemia with a poorly compensated increase in HbF. If, however, the deletion extends towards the γ-genes it apparently removes the area preventing their expression and there is a full expression of HbF to give the harmless Hereditary Persistence of Fetal Haemoglobin (HPFH).

behave like a moderately unstable haemoglobin, halfway in stability between HbE and a grossly unstable variant, e.g. Hb Köln.

The two types of thalassaemia therefore each have a characteristically different pathophysiology. In the β-thalassaemias the main problem is a disruption of bone marrow synthesis due to the early precipitation of the grossly unstable free α-chains. In the α-thalassaemias the pathology is that of a typical unstable haemoglobin with

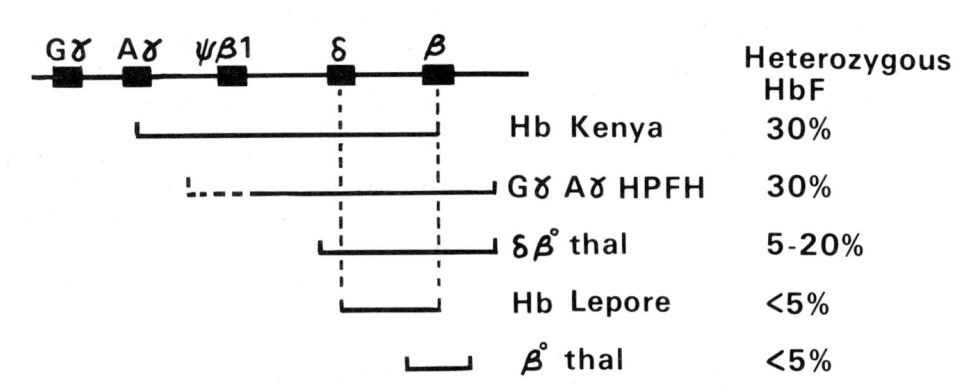

FIG. 19. A diagrammatic representation of the size of deletions responsible for thalassaemic-like lesions of the non-α globin. Note that the further the deletion extends towards the γ-genes the greater is the de-repression of HbF synthesis. (Adapted from Jackson and Williamson.[18])

β-Thalassaemia

There is considerable heterogeneity in the β-thalassaemias. Gene mapping has shown that though in some instances there have been deletions of portions of the β-gene, in most instances the gene is intact. The problem in most cases appears to be due to a failure in expression, either transcription or processing, of the β-globin mRNA. If this failure is complete, and no β-globin is produced, then there is a β^0-thalassaemia. In some cases there is only a partial reduction in mRNA to give the less severe β^+-thalassaemia.

Although the genetic lesions are diverse it does seem that some four or five mutations will account for the majority of the β-thalassaemias. Those identified include a nonsense or stop mutation at codon 39, a frameshift mutation at codon 8, and a mutation at the commencement of an intron that results in non-splicing of the coding areas. An interesting variation of this last mutation has been found that provides an explanation for β^+-thalassaemia. This has been shown to arise from partial suppression of β-mRNA formation due to a false splicing codon within an intron. Splicing can then occur at either the correct position to give a viable messenger or at the incorrect position to give an incorrectly spliced, non-viable messenger. The effect of this β^+ mutation is to give a reduction in β-mRNA whereas the β^0 mutation gives a complete cessation of viable β-mRNA production.

δβ-Thalassaemia

In δβ-thalassaemia there is a complete lack of HbA and HbA$_2$. The lesion (Fig. 19) has been shown to be a deletion of the whole of the β-globin gene and part or all of the δ-globin gene.

Hb Lepore

Here there is a deletion of part of the β and part of the δ-globin genes. The area controlling δ gene expression is unaffected and the fused δβ globin, Hb Lepore, is consequently expressed and synthesised similarly to δ-globin. The result is a mild thalassaemia.

Hereditary Persistence of Fetal Haemoglobin

This term covers a variety of conditions, all associated with increased HbF production beyond infancy. Gene mapping has shown that there is a deletion of both the δ and β genes which extends further into the flanking DNA than does the deletions in δβ thalassaemia. This extension apparently includes a control section for the γ-globin gene causing de-repression of the γ-locus.

α-Thalassaemias

The presence of four α-globin genes results in the α-thalassaemias having a graded severity, from the clinically insignificant suppression of one-gene to the invariably fatal suppression of four genes. These phenotypes

α-Genes		Deleted	Name	Clinical	Endemic	Cord Bart's
|	|	1 gene	α-thal-2 heterozygote	Minimal effect	Africa Asia	1–2%
*	|					
*	|	2 cis	α-thal-1 heterozygote	α-thal minor	S-E Asia	1–5%
*	|					
|	|	2 trans	α-thal-2 homozygote	α-thal minor	Africa Asia	1–5%
*	*					
*	|	2 cis 1 trans	mixed α-thal-1 α-thal-2	Hb H disease	S-E Asia	15%
*	*					
*	*	4	α-thal-1 homozygote	Hydrops foetalis	S-E Asia	90%
*	*					

FIG. 20. α-Thalassaemia classification according to the number of non-functional, *, genes. The crossed genes will usually be due to a deletion but virtually the same phenotypic result would follow if the cross indicated a Constant Spring α-globin gene.

and their rather confusing terminology are illustrated in Fig. 20. Almost all the α-thalassaemias are caused by gene deletions though other lesions occur, particularly the structural variant Hb Constant Spring. To all practical purposes this last can be equated in effects with a gene deletion and can be substituted as such in Fig. 20 in order to determine its clinical effect.

As shown in the figure, each chromosome can have either a single or double α-gene deletion. The single gene deletion is common in those of African descent and by itself causes no clinically evident changes through an increased (1–2%) amount of Hb Barts (γ_4) is present in cord blood. This mild loss of α-globin production is termed α-thalassaemia-2 trait. Its combination with a similar (trans) deletion on the other chromosome will give a clinically evident, but still mild, α-thalassaemia minor.

In Southern Asia there are not only individuals with a single gene deletion but also individuals with two (cis) deletions on the one chromosome, α-thalassaemia-1. The Southern Asian races therefore can have two genotypically distinct but phenotypically similar forms of α-thalassaemia trait (2 gene deletions, cis or trans). It also follows that the combination of lesions may lead in these Asian races, but not in the black races, to a 3-gene deletion, i.e. HbH disease or the fatal 4-gene deletion, Hydrops foetalis.

Diagnostically, the main problem is the recognition of α-thalassaemia minor, or trait, with 2-gene deletion. The red-cell indices are those of a typical thalassaemia minor and the diagnosis is often made by exclusion. In blood films incubated with brilliant cresyl blue a careful search will demonstrate the presence of an occasional red cell containing inclusions, the golf-ball-like HbH cell. The diagnosis can be confirmed by gene mapping or prospectively, but less reliably, by examination of cord blood for Hb Barts.

Malaria and Red Cell Abnormalities

An attempt has been made here to show the heterogeneity of the thalassaemias, but these diseases are only a part of an extraordinarily heterogeneous group of genetic defects that have accumulated as protective devices against malaria.[13] The clinical biochemist should be aware of the range of red cell defects involved, as they must all be considered in the differential diagnosis of individuals with suspected haemoglobinopathies; particularly those with family links to the malaria areas of the world. The range of defects is best illustrated by an outline of the common mechanisms by which they all provide a defence against fatal malarial infection.

Part of the malarial parasite's life-cycle occurs within the red cell and, true to its parasitic nature, it utilises the red cell's metabolism to reduce released oxidants. The parasite produces H_2O_2 which is reduced by NADPH produced by the red cell's hexose monophosphate shunt. A defect of the major enzyme of this shunt, glucose-6-phosphate dehydrogenase (G-6-PD), will result in inadequate reduction with oxidation and elimination of the affected cell.

Even in the presence of fully active red-cell reductive mechanisms the production of oxidants by the malarial parasite is liable to oxidize the red cell haemoglobin to methaemoglobin. Normally this is a reversible process, but mildly unstable haemoglobins such as HbE, HbF, HbH and Hb Bart's will more readily form insoluble haemichromes to give inclusion body formation and, again, preferential elimination of infected cells. The sensitivity of HbF to oxidation explains one reason for the protection afforded by the thalassaemias, but the main reason for the protection is probably the inherently increased susceptibility of the thalassaemic cell to oxidative haemolysis. Thalassaemic cells are exposed to oxidative damage both from superoxide release from isolated haemoglobin subunits and due to the presence of free iron. The additional oxidative load of an intracellular parasite appears *in vitro* to lead to membrane changes that are incompatible with parasite multiplication. *In vivo* there will again be an increased elimination of affected cells.

The range of red cell defects therefore can be seen as mechanisms that give preferential elimination of cells exposed to the additional oxidative stress of the malarial parasite. The defects include G-6-PD deficiency, HbE, the thalassaemias, the Hbs Lepore, and HPFH. Two factors that could also be included under this heading are the external protective mechanisms of the antimalarial drugs and extreme pyrexia—both of these will similarly give additional oxidative stresses favouring elimination of infected cells.

The other outstanding protective mechanisms, HbS, probably, though not certainly, functions through a different mechanism. The maturation of *Plasmodium falciparum* into schizonts takes place in the latter part of the intra-erythrocytic stage of parasite development. At this stage the red cell moves into deeper tissues with a lower oxygen tension. Under these circumstances initial crystallisation of HbS occurs, but not sickling. This initial presickling stage is incompatible with parasite survival, possibly because of membrane electrolyte leakage similar to that occurring in the oxidised thalassaemic cell. It is this protective advantage against *P. falciparum* malaria in the HbS heterozygote that has balanced the gross disadvantage to the homozygote of sickle cell disease.

LABORATORY DIAGNOSISIS
Routine Screening

Approach

The possibility of a haemoglobinopathy usually arises from haematological findings and it is desirable that there should be cooperation between the haematologist and biochemist in its further investigation. It is useful to have a preliminary screen established covering both the haematological and biochemical aspects of the investigations. Often the diagnosis seems obvious, e.g. sickle cell disease, and it is tempting to haemolyse all the sample and move directly to electrophoretic confirmation of the presence of HbS. However, subsequently the possibility of more subtle complications often arise, such as HbS/β-thalassaemia or HbS/HPFH, that require studies on fresh whole cells.

To avoid these problems a full screen should be performed wherever possible on each sample. This would routinely consist of: haematological studies (red cell indices, prolonged incubation for inclusion bodies, iron studies); HbA_2, HbF, sickling test, stability test and electrophoresis at alkaline pH. This provides a good general screen for the haemoglobinopathies though other tests may be added such as staining of red cells for HbF, globin chain electrophoresis, isoelectric focusing, G-6-PD screening test, and spectrophotometric scanning for methaemoglobin or other derivatives. Details of the techniques involved are given in the standard texts in the bibliography.

Specimen Collection

Some 5–10 ml of anticoagulated blood is adequate for almost all investigations. Haemoglobin rapidly deteriorates in haemolysates and it is preferable to transport the sample as whole blood. If a few mg of a 99:1 mixture of glucose and chloramphenicol is added this should keep the sample in a condition to survive a week of airmail transport or two weeks of storage in the refrigerator. Red cell indices and morphology should be performed on a fresh EDTA-sample.

Haematology

Accurate red cell indices are available from automated equipment. Iron studies should be performed, as iron deficiency gives a haematological picture similar to that of the thalassaemias minor and concomitant iron deficiency may also distort the findings in other haemoglobinopathies.

A test for red cell inclusions[10] should be routinely performed to help in the diagnosis of α-thalassaemia and the unstable haemoglobin diseases. The inclusion of Heinz bodies may already be present in the cells and can be seen by staining with methyl violet. However, they are removed by the spleen and in the unsplenectomised patient they are best demonstrated by prolonged incubation of the red cells with a redox dye such as brilliant cresyl blue. The inclusions can vary from the large round Heinz body to the multiple punctate inclusions of the HbH cell.

Tests for Sickle Cell Haemoglobin

The presence of HbS can be demonstrated by either a microscopic sickling test or a solubility screening test. The last is the most suitable for the biochemical laboratory, but either should be confirmed by subsequent electrophoresis. The solubility screening test is based on the comparative insolubility of reduced HbS in concentrated phosphate buffer. It is best performed[12] using haemolysate or washed cells since misleading results may be obtained with whole blood; increased γ-globulins can give rise to false positives, and false negative results may be obtained when there are reduced red cell:plasma ratios, as in severe anaemia.

Haemolysate Preparation

Haemoglobin studies are best carried out using a carefully prepared 10% (10 g/dl) haemolysate. The red cells are washed to remove plasma proteins and lysis is performed by use of a detergent or, more usually, hypotonically by addition of water. Cell debris can be removed by vigorous shaking with carbon tetrachloride.

Haemolysates can be stored indefinitely in liquid nitrogen but longterm refrigerated storage is best[17] in the form of carbon monoxy-haemoglobin with addition of cyanide and EDTA.

Measurement of HbA_2

The careful measurement of HbA_2 as a percentage of the total haemoglobin is a key part of the diagnostic screen.[17] It is the single most useful criterion for the diagnosis of β-thalassaemia minor where it is raised above the normal range (1·8–3·3%), usually being above 3·6%. Normal or low values are obtained with $\delta\beta$-thalassaemia, Hb Lepore, and the α-thalassaemias.

The difference between the upper limit of normal and the level diagnostic of β-thalassaemia minor is slight and there are few laboratory tests which require greater precision and accuracy in their performance. Measurement can be made either by cellulose acetate electrophoresis with elution and comparison of the A and A_2 bands, or by column chromatography. The cellulose acetate technique has the advantage that it will detect HbH and also other abnormal haemoglobins that may give false results with the column technique. On the other hand the column technique is more suitable for handling large numbers of samples.

In either case it is important that the measurements are made regularly, preferably by the same technician, as it is only with constant practice that precision is obtained. Standards ought to be utilised to calibrate the accuracy of the measurements.

Measurement of HbF

The presence of a raised HbF level indicates a life-long anaemia and is a characteristic finding in the β and $\delta\beta$ thalassaemias. It also occurs in the defects of γ-chain switching i.e. HPFH, and the measurement of HbF levels is a necesary prerequisite for the assessment of the severity of sickle cell disease.

Measurement of the total fetal haemoglobin is based on the γ-chain lacking an internal cysteine present in the β-chain. This cysteine in the β globin is ionised at alkaline pH causing instability and precipitation. The γ-chain, lacking this cysteine, is stable at alkaline pH and is said to be alkali-resistant. Haemolysate is added to an alkaline ammonium sulphate solution and, after a fixed period to allow the precipitation of adult haemoglobin, a filtrate is prepared whose concentration gives a measure of HbF. The normal adult HbF level is less than 0·8% of the total haemoglobin, a result above 1·5% indicates haematological disease.

Two procedures for the estimation are in common usage;[12] the Singer technique which provides good precision with higher levels of HbF, i.e. above 5%, and the Betke method, which is the recommended approach as it gives precise results in the 0·2–4% range—the important decision range. Recently a radial immunodifussion technique has been introduced commercially but it has the defect of lack of precision in the range of 2% and below.

The question of whether the HbF is distributed heterogeneously or homogeneously throughout the red cells is of importance in differentiating HPFH, with usually a homogeneous distribution, from $\delta\beta$-thalassaemia with a heterogeneous distribution. The determination of the distribution[12] is based on the work of Kleihauer who showed that an acid buffer adjusted to the pK of HbF will wash out all the non-fetal haemoglobin from an alkali-fixed film, allowing staining of the remaining HbF.

Stability Tests

As with the measurements of HbA_2 and HbF, a stability test should be routinely performed on all samples being assessed for a haemoglobinopathy. Unless this is done unstable haemoglobin variants may be undetected as there is often no associated change in electrophoretic mobility. Furthermore, the finding of mild instability is a valuable aid in the detection and identification of other variants, particularly HbE and HbH.

The most commonly performed test is the isopropanol test[5] in which fresh haemolysate is incubated at 37° with a 17% isopropanol solution for 30 minutes. The change in polarity of the solvent provides a denaturation stress that causes precipitation, with obvious flocculation, of unstable variants. An apparent positive result may be given in the presence of HbF levels above 5% (HbF is mildly unstable as compared to HbA). For this reason positive results should be checked by the less sensitive 50°C heat stress test. Methaemoglobin A is also unstable and it is important that any of the stability tests should be carried out using freshly prepared haemolysate, though the red cells can be stored for up to 10 days in the refrigerator prior to preparation of the haemolysate.

Oxygen Affinity

One of the most important, but yet difficult, groups of haemoglobinopathies to diagnose is the group of variants with altered oxygen affinity. These are mostly high-affinity variants that are suspected because of a polycythaemia, present from youth, and often with a family history. The identification of the underlying abnormality is usually a time-consuming and expensive task. The first priority is to establish if there really is an inherent change in oxygen affinity. The P_{50}, the partial pressure at half saturation, should be determined on whole blood and stripped (phosphate-free) haemolysate. The Hill coefficient should be calculated and the effect of added 2–3 diphosphoglycerate to the stripped haemolysate should be assessed. Commercial oxygen-electrode equipment specially designed for plotting oxygen-dissociation curves is available and is useful for screening samples.

The approach to screening for changes in oxygen affinity has been summarised elsewhere[31] but in practice it is difficult unequivocally to exclude a change in affinity and if there is a real suspicion of an abnormality the sample should be sent to a reference laboratory for full oxygenation studies.

Electrophoresis

Electrophoresis of freshly haemolysed blood at alkaline pH should be part of the routine haemoglobinopathy screen. Electrophoresis can be carried out on either cellulose acetate or starch gel (Fig. 21). Cellulose acetate[32] is more convenient for routine screening but starch gel[12] more readily gives an indication of minor fractions, such as Hb Constant Spring. For a full examination electrophoresis should also be carried out at acid pH, in citrate-agar, and also in dissociating buffers containing 6M urea to visualise the isolated globin chains (Fig. 22). Isoelectric focusing provides an additional approach (Fig. 14) though its great selectivity may be confusing to the worker who is not familiar with the mobility of the various derivatives of normal haemoglobin.

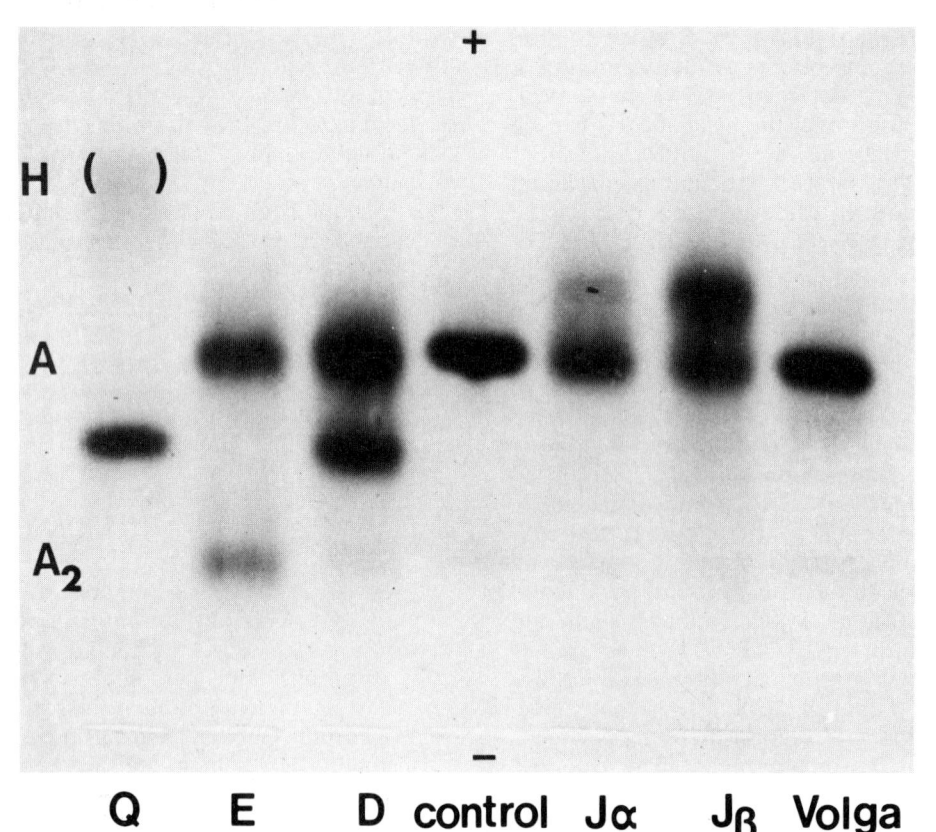

FIG. 21. Starch gel electrophoresis (pH 8·6) of haemoglobin. Note (i) that HbQ an α-variant is linked to α-thalassaemic gene; (ii) HbE in conjunction with α-thalassaemia may mimic a raised HbA$_2$; (iii) HbD, moves in the same position as HbS; (iv) the Jα and Jβ variants show the typical difference in proportions between α and β variants; (v) Hb Volga, an unstable variant shows no change in charge (cf. Fig. 22) but there is a noticeable HbF band.

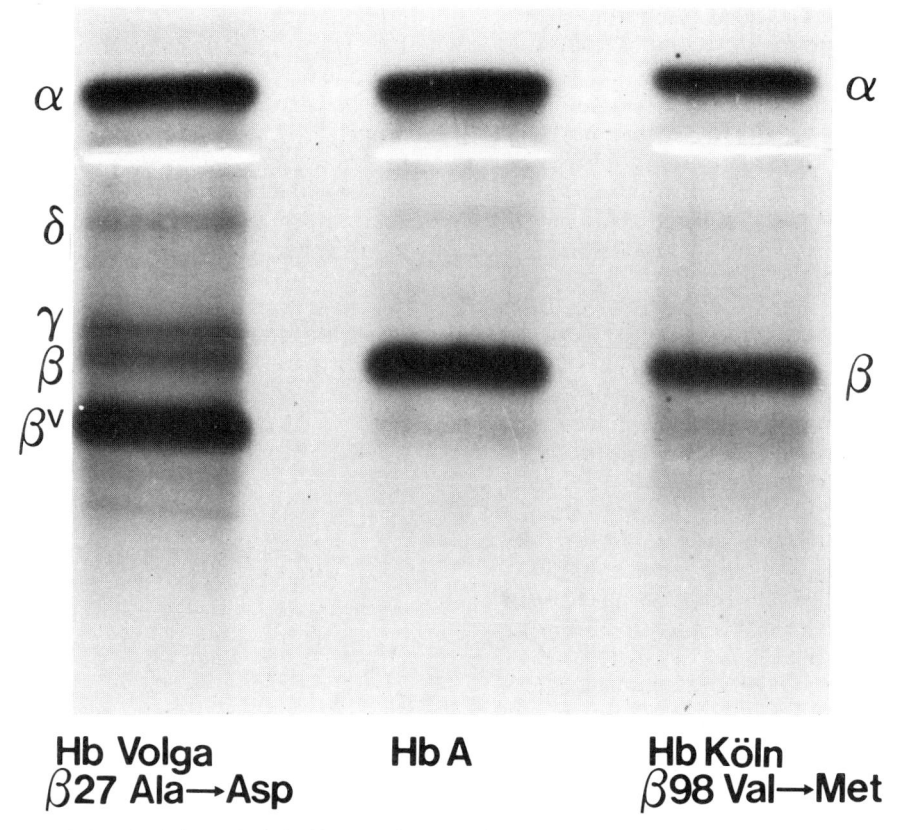

FIG. 22. Starch gel electrophesis in 6 M urea of two unstable haemoglobins compared to a control HbA. The true charge change is revealed in the unfolded globin of Hb Volga (cf. Fig. 21).

Interpretation of Results

The investigation of suspected haemoglobinopathies is seldom routine as there are usually multiple factors to consider. Nevertheless, the scheme[15] outlined in Fig. 23 allows the categorisation of the majority of the diagnostically confusing thalassaemia traits. It brings together the haematological indices[8] and individual tests outlined in the preceding sections. The diagnosis of the thalas-

SCREEN FOR THALASSAEMIA

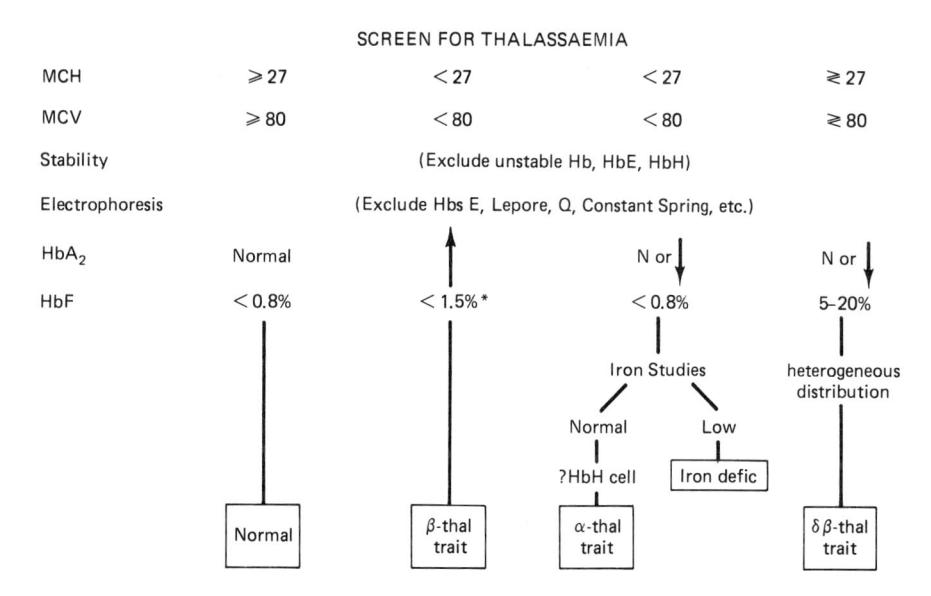

MCH	≥ 27	< 27	< 27	≥ 27
MCV	≥ 80	< 80	< 80	≥ 80
Stability		(Exclude unstable Hb, HbE, HbH)		
Electrophoresis		(Exclude Hbs E, Lepore, Q, Constant Spring, etc.)		
HbA$_2$	Normal	↑	N or ↓	N or ↓
HbF	< 0.8%	< 1.5%*	< 0.8%	5–20%
	Normal	β-thal trait	Iron Studies → Normal (?HbH cell) → α-thal trait; Low → Iron defic	heterogeneous distribution → δβ-thal trait

FIG. 23. A summarised screen to give initial differentiation of the thalassaemia traits. *The level of HbF in β-thalassaemia trait is usually normal but 30% are raised level, usually below 1.5% and seldom more than 3%. HbF levels quoted are measured by the Betke technique,[12] (normal < 0.8%).

saemias major, is usually self-evident on morphological evidence plus altered haemoglobin composition, e.g. HbH in 3-gene α-thalassaemia.

It should be remembered that the haematological findings of a thalassaemia trait can also be due to a heterozygous HbE or Hb Lepore. Both should, however, be readily identifiable by electrophoresis: HbE giving a fraction of some 30% of the total haemoglobin in the position of HbA_2, plus a mildly positive isopropanol stability test: Hb Lepore giving a fraction of some 10% of the total that moves in a position between HbA and HbA_2, close to that of HbS or HbD. An α-thalassaemia may be indicated by the finding of HbQ (Fig. 21) or, more commonly, by Hb Constant Spring which can be seen on electrophoresis as minor bands running between HbA_2 and the origin.

HbS will be readily diagnosable because of the associated sickling but beware of a concomitant HbD; the SD heterozygote will mimic electrophoretically the SS homozygote. HbC which runs in the same position as HbE can usually be distinguished from it on the basis of racial background but an additional differentiation is the stability of HbC to isopropanol stress.

This outline covers the major, clinically-associated haemoglobin abnormalities that can be readily handled by the non-specialist laboratory. The finding of a grossly positive stability test or of a change in oxygen affinity requires the more detailed investigations available in reference laboratories.

Variant Identification

Isolation of Variants

The standard method of preparative isolation of abnormal haemoglobins is by DEAE-Sephadex chromatography as illustrated in Fig. 24. The column is equilibrated and the haemoglobin is applied in alkaline buffer. As the pH is decreased the more positively charged components such as A_2, C and E are eluted, followed by Hbs S and D, the main HbA_0 peak and then its derivative HbA_1 peak along with HbF. It is a useful practice routinely to pool each peak to allow calculation of their relative proportions from the Soret band absorbance (420 nm).

The isolation of unstable haemoglobins is most readily achieved by isopropanol precipitation, using a scaled-up modification of the diagnostic technique.

Chain Separation and Labelling

The development of the technique for preparative globin chain separation by Clegg, Naughton and Weatherall,[9] represented a major advance both in the isolation of abnormal components and in the diagnosis of thalassaemia. Freshly prepared globin is applied to a carboxymethyl-cellulose column and equilibrated with an alkaline buffer in 8M urea with added SH-reducing agent. As the pH is lowered to below 7 there is a precisely reproducible elution of the β-chain followed by the α-chain; the presence of a mutant chain being shown by an extra peak whose position is directly related to the magnitude of charge change involved in the mutation. The technique provides preparative yields but can also be used diagnostically[35] to measure synthesis ratios of α and β globins. Globin is prepared from reticulocyte-rich preparations that have been incubated with ^{14}C-labelled amino acids. Counting of the separated globin peaks allows calculation of the relative $\alpha{:}\beta$ synthesis rates. In the normal, the ratio is usually close to 1·00 but it is proportionately altered in the α and β-thalassaemias.

Peptide Mapping

The identification of the amino acid abnormality in a haemoglobin is made by the process of peptide mapping or fingerprinting.[22] The whole globin or isolated globin chain is broken down into a limited number of peptides using a specific enzyme, usually trypsin. Cleavage occurs at lysine residues to give some 15 tryptic peptides from each globin chain. These peptides can be separated on paper in two dimensions; by electrophoresis at pH 6·4 in the first dimension followed by chromatography with a relatively non-polar solvent in the second dimension. As shown in Fig. 25 the only peptides that will be altered in mobility are those with altered amino acid composition. A change in charge, as with the substitution of a negatively charged glutamic acid by a neutral valine (HbS), will cause the abnormal peptide to move more towards the cathode, whilst the change in polarity will give the peptide increased mobility with respect to the chromatography front. These characteristic alterations in mobility, along with the ability to stain specifically for residues such as histidine, tyrosine and arginine, allows the ready identification of the substitution involved in many variants. Confirmation is provided by amino acid analysis of the peptide eluted from the peptide map.

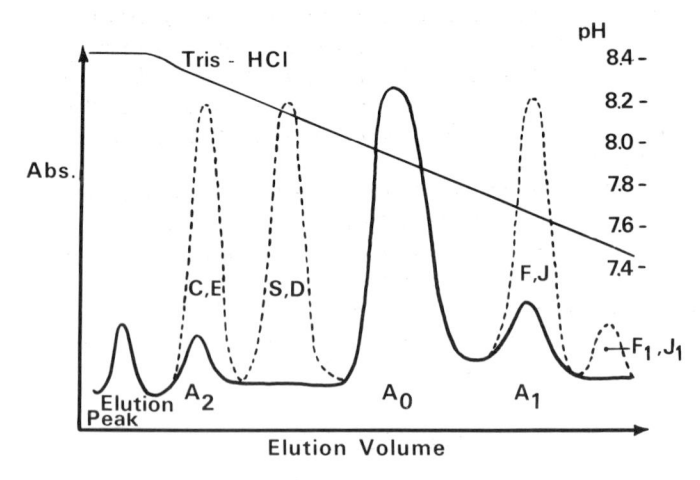

FIG. 24. DEAE-Sephadex column chromatography of a normal haemolysate using a Tris-HCl pH elution gradient. The positions of some common variants are superimposed; note that all will have an 'A_1' artefact as shown in the F_1 or J_1 of the last peak.

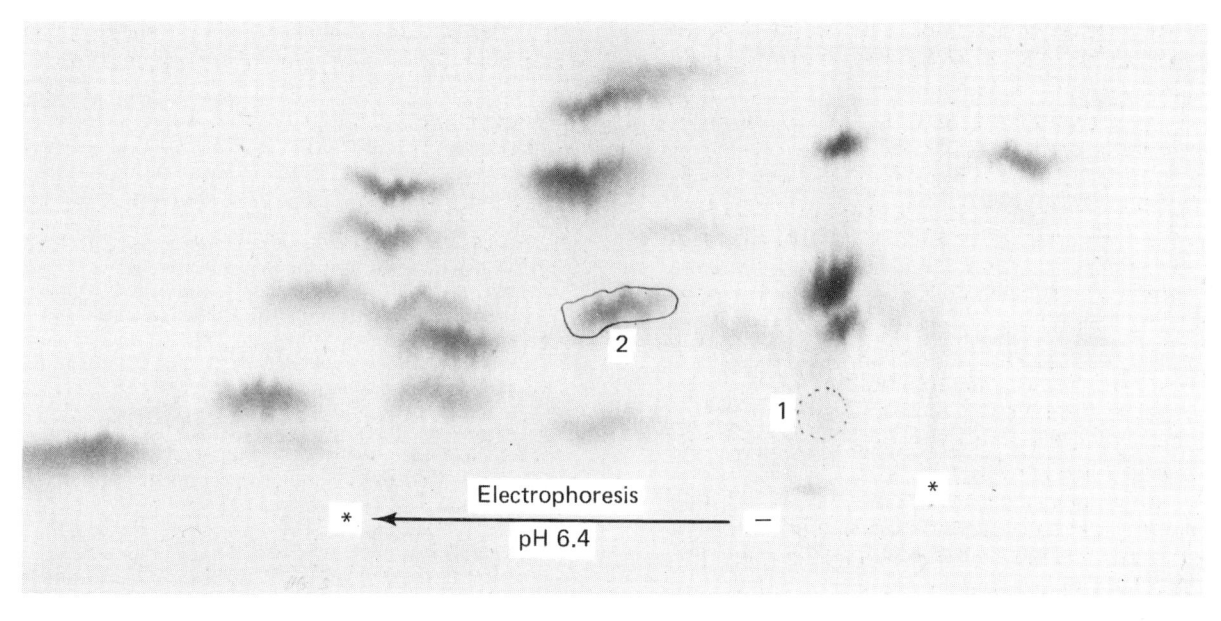

Fig. 25. 'Fingerprint' of haemoglobins. * Point of origin. The dotted line encircles the area (1) where the tryptic peptide $\beta^A 1-8$ is normally found. It is absent and a new peptide $\beta^S 1-8$ is present instead (2).

High Performance Liquid Chromatography

This relatively new technique (HPLC) gives rapid and sensitive separations of peptides and polypeptides on the basis of both their charge and hydrophobicity. Its extreme sensitivity is illustrated by its ability to separate the Gγ from the Aγ fetal haemoglobin[34], the difference between the two being solely in a glycine or an alanine residue in position γ136.

HPLC gives excellent separations of the tryptic peptides of haemoglobin with high sensitivity and good recovery yields.[33] The speed is such that several peptide profiles can be completed in a day. The technique has the potential to replace peptide mapping as the standard means of variant identification but much greater experience is needed before the profiles produced can be interpreted with the confidence of the paper peptide fingerprint.

Gene Mapping

The development of the techniques of molecular biology have allowed a comparable approach[18] to the mapping of genomic DNA to that used for the peptide mapping of proteins. A major difference is that the DNA relating to globin synthesis is present in every cell and is capable of amplification; thus globin gene analysis can be potentially carried out on cells from limited sources and independent of the chronology of expression. For example, amniotic fibroblasts can be utilised to obtain information about the genes for adult, embryonic or fetal globin.

The basis of gene mapping is the availability of specific enzymes, the restriction endonucleases, that give precisely defined cleavage of DNA sequences to yield, from the total DNA, a large, but finite, number of fragments. These DNA fragments can be separated by agarose gel electrophoresis and then transferred by blotting to nitrocellulose sheets. If stained, these sheets would show a blurred packing of fragments, but the unique properties of DNA are utilised to develop only those fragments containing the globin gene under study. A radioactively-labelled DNA copy of the globin gene is hybridised to the DNA on the nitrocellulose sheet and an autoradiograph prepared. This will show only the DNA fragments containing globin genes. The use of several restriction enzymes has allowed the production of characteristic gene maps with further specificity being added by the use of DNA-copies specific for particular globin genes.

In this way the globin genes and their surrounding DNA can be mapped and, with overlaps, the structural relations shown in Fig. 11 can be deduced. The technique can be used to demonstrate the number and extent of gene deletions with particular relevance to α-thalassaemia but also to the $\delta\beta$ deletions (Fig. 19). It has the potential to identify gene inheritance by changes in flanking sequences allowing the genotype of the fetus to be determined, and may, for instance, identify if the fetus has inherited a β-thalassaemic gene from each of its heterozygous parents.

It should be emphasised that these applications are still in their early stages and there is a way to go before they are able to provide reliable and consistent diagnostic evidence. There can be no doubt however, of the potential of the techniques and that in this direction, as in others, the haemoglobinopathies will be the pioneer model for study of genetic disease in general.

REFERENCES

1. Barcroft, J. (1928), *The Respiratory Function of the Blood*, Cambridge University Press.
2. Bellingham, A. J. (1974), 'The red cell in adaptation to anemic hypoxia,' *Clinics in Hematology*, **3**, 577–594.

3. Bunn, H. F., Forget, B. G. and Ranney, H. M. (1977), 'Sickle cell anemia and related disorders,' in *Human Hemoglobins*, pp. 228–281, W. B. Saunders.
4. Bunn, H. F., Shapiro, R., McManus, M. *et al.* (1979), 'Structural heterogeneity of human hemoglobin A due to non-enzymatic glycosylation,' *J. Biol. Chem.*, **254**, 3892–3898.
5. Carrell, R. W. and Kay, R. (1972). 'A simple method for the detection of unstable hemoglobins', *Brit. J. Haematol.*, **23**, 615–619.
6. Carrell, R. W. and Lehmann, H. (1979), 'Abnormal Haemoglobins,' in *Chemical Diagnosis of Disease*, pp. 879–966, (Brown, S. S., Mitchell, F. L. and Young, D. S., Eds) Amsterdam: Elsevier North Holland Publ. Co.
7. Carrell, R. W. and Winterbourn, C. C. (1981), 'The unstable hemoglobins,' in *Human Hemoglobin and Hemoglobinopathies: A Current Review to 1981*, (Schneider, R. G., Charache, S. and Schroeder, W., Eds.). *Texas Rep. Biol. Med.*, **40**, 431–445.
8. Cauchi, M. N. and Tauro, G. (1979), 'The quantitation of haemoglobin A_2 and F.' Report of the techniques review group to the Thalassaemia Society of Victoria. *Pathology*, **11**, 175–179.
9. Clegg, J. B., Naughton, M. A. and Weatherall, D. I. (1966), 'Abnormal human haemoglobin. Separation and characterization of the α and β chains by chromatography,' *J. Molec. Biol.* **19**, 91–108.
10. Dacie, J. V. and Lewis, S. M. (1975), in *Practical Hematology*, 5th Ed., pp. 227–233, Edinburgh: Churchill Livingstone.
11. Editorial (1980), 'Haemoglobin A_1 and diabetes: a reappraisal,' *Brit. Med. J.*, **281**, 1304–1305.
12. Efremov, G. D. and Huisman, T. H. J. (1974), 'The laboratory diagnosis of the hemoglobinopathies,' *Clinics in Hematology*, **3**, 527–570.
13. Etkin, N. L. and Eaton, J. W. (1975), 'Malaria-induced erythrocyte oxidant sensitivity,' in *Erythrocyte Structure and Function*, pp. 219–230. (Brewer, G. J., Ed.). New York: Alan R. Liss.
14. Forget, B. G. (1977), 'Nucleotide sequence of human β globin messenger RNA,' *Hemoglobin*, **1**, 879–881.
15. Galanello, R., Melis, M. A. Ruggeri, R., Addis, M., Cao, A. *et al.* (1979). 'β^0 thalassemia trait in Sardinia,' *Hemoglobin*, **3**, 33–46.
16. Hultquist, D. E., Douglas, R. H. and Dean, R. T. (1975), 'The methemoglobin reduction system of erythrocytes,' in *Erythrocyte Structure and Function*, pp. 297–300. (Brewer, G. J., Ed.). New York: Alan R. Liss.
17. Huntsman, R. G., Carrell, R. W. and White, J. M. (1978), 'Recommendation for selected methods for quantitative estimation of HbA_2 and for HbA_2 reference preparation,' *Brit. J. Haematol.*, **38**, 573–578.
18. Jackson, I. J. and Williamson, R. (1980), 'Annotation: Mapping of the human globin genes,' *Brit. J. Haematol.*, **46**, 341–349.
19. Jones, R. T. and Shih, T.-B. (1980), 'Haemoglobin variants with altered oxygen affinity,' *Hemoglobin*, **4**, 243–261.
20. Lamola, A. A. and Yamane, T. (1974) 'Zinc protoporphyrin in the erythrocytes of patients with lead intoxication and iron deficiency anemia,' *Science*, **186**, 936–938.
21. Lehmann, H. and Casey, R. (1982), 'Human Haemoglobin', in *Comprehensive Biochemistry*, **19B, Part II**, pp. 347–417, (Neuberger, A. and van Deenan, L. L. M., Eds.) Amsterdam: Elsevier.
22. Lehmann, H. and Huntsman, R. G. (1974), *Man's Haemoglobins*, 2nd Ed., pp. 431–451. Amsterdam: North-Holland Publ. Co.
23. Lehmann, H. and Kynoch, P. A. M. (1976) *Human Haemoglobin Variants and their Characteristics*, Amsterdam: North-Holland Publ. Co.
24. Lehmann, H. and Lang, A. (1974), 'Various aspects of α-thalassaemia,' *Ann. N.Y. Acad. Sciences*, **232**, 152–158.
25. Milner, P. F. (1974), 'The sickling disorders,' *Clinics in Hematology*, **3**, 289–331.
26. Pasvol, G., Weatherall, D. J. and Wilson, R. J. M. (1978), 'Cellular mechanism for the protective effect of hemoglobin S against *P. falciparum* malaria,' *Nature*, **274**, 701–703.
27. Perutz, M. F. (1978), 'Hemoglobin structure and respiratory function,' *Scientific American*, **239**, 68–86.
28. Perutz, M. F. and Lehmann, H. (1968), 'Molecular pathology of human haemoglobin,' *Nature*, **219**, 902–909.
29. Proudfoot, N. J., Shander, M. H. M., Manley, J. L. *et al.* (1980), 'Structure and *in vitro* transcription of human globin genes,' *Science*, **209**, 1329–1342.
30. Rahbar, S. (1968). 'An abnormal hemoglobin in the red cells of diabetics,' *Clin. Chim. Acta*, **22**, 296–298.
31. Rosa, J., Beuzard, Y., Thillet, G. M. C. *et al.* (1975), 'Testing for hemoglobins with abnormal oxygen affinity curves,' in *Abnormal Haemoglobins and Thalassaemia. Diagnostic Aspects*, pp. 79–116. (Schmidt, R. M. Ed.). New York: Academic Press.
32. Schneider, R. G. (1980), 'Electrophoretic methods in hemoglobin identification,' *Hemoglobin*, **4**, 521–526.
33. Schroeder, W. A., Shelton, J. B. and Shelton, J. R. (1980), 'Separation of hemoglobin peptides by HPLC,' *Hemoglobin*, **4**, 551–560.
34. Shimizu, K., Wilson, J. B. and Huisman, T. H. J. (1980), 'The distribution of the percentages of $G\gamma$ and $A\gamma$ chains in human hemoglobin by HPLC,' *Hemoglobin*, **4**, 487–496.
35. Weatherall, D. J. and Clegg, J. B. (1981), *The Thalassaemia Syndromes*, 3rd Ed. Oxford: Blackwell.
36. White, J. M. and Dacie, J. V. (1971), 'The unstable hemoglobins—molecular and clinical features', *Progress in Hematology*. Vol. VII, pp. 69–109.
37. Wood, W. G. (1976), 'Haemoglobin synthesis during human fetal development', *Brit. Med. Bull.*, **32**, 282–287.

FURTHER READING

Bunn, H. F., Forget, B. G. and Ranney, H. M. (1977), *Human Hemoglobins*, W. B. Saunders: Philadelphia.

Huisman, T. H. J. and Jonxis, J. H. P. (1980), *The Hemoglobinopathies. Techniques of Identification*. Marcel Dekker: New York.

Lehmann, H. and Huntsman, R. G. (1974), *Man's Haemoglobins*, North-Holland Publ. Co.: Amsterdam.

Schneider, R. G., Charache, S. and Schroeder, W. (1981), 'Human hemoglobin and hemoglobinopathies: a current review to 1981, *Texas Rep. Biol. Med.* 40 (In Press).

Weatherall, D. J. and Clegg, S. B. (1981), *The Thalassaemia Syndromes*, 3rd Ed. Blackwell: Oxford.

12. THE ANAEMIAS

R. VAUGHAN JONES

Erythropoeisis
  Iron metabolism
  Vitamin B_{12}: Folic Acid

The Anaemias
  Introduction
  Anaemias associated with deficiencies
  Acute haemorrhage
  Anaemia due to haemolysis
  The anaemia of chronic disorders
  Anaemia of marrow disorders

Diagnosis of the type of anaemia

A brief description of the physiology of erythropoiesis is a necessary prelude to an understanding of the causes and investigation of anaemia.

ERYTHROPOIESIS

In fetal life erythropoiesis, the formation of red cells, takes place progressively in the yolk-sac, liver and bone marrow but in the child or adult it is confined to the bone marrow.

From stem cells, proerythroblasts are produced by mitosis and in turn, by further mitoses, basophilic, polychromatophilic and orthochromatic normoblasts. The later cells possess increasing amount of haemoglobin in the cytoplasm and the nucleus progressively shrinks as its chromatin clumps together until it becomes pyknotic and finally disappears. Red cell maturation takes approximately seven days, the early part largely occupied by cell divisions, producing probably sixteen cells from one stem cell, and the later part occupied mainly in maturation and haemoglobinization. After the nucleus has been extruded the cell still contains a cytoplasmic reticulum of residual RNA and is known as a reticulocyte. Such a cell takes two or three days to lose its reticulum, one or more of which is normally spent in the marrow. Some 10–15% of the erythroblasts fail to mature and subsequently die—this is known as ineffective erythropoiesis.

When nuclear division lags behind haemoglobinization larger red cells (macrocytes) are produced: when haemoglobinization lags there is failure to trigger a presumed feed-back mechanism which, once a critical haemoglobin concentration is reached, normally shuts off further nucleic acid synthesis and hence cell division. In iron deficiency, failure to achieve, or delay in reaching the critical haemoglobin concentration results in extra mitotic divisions and the formation of small cells or microcytes.

The control of erythropoiesis is regulated at least partly by a circulating substance, erythropoietin, produced in response to anoxia. Probably a glycoprotein, this substance is made mainly in the kidney but continues to be produced in nephrectomized subjects. The kidney appears to elaborate a substance, erythrogenin, which acts on a plasma substrate produced in the liver to release the active erythropoietin.

Erythropoietin is capable of many effects on the bone marrow including stimulating the proliferation and maturation of erythroid stem cells. Its major route of action is stimulation of messenger-RNA production resulting in increased amounts of δ-aminolaevulinic acid synthetase, the rate limiting enzyme in haem synthesis, as well as increased iron incorporation and stromal synthesis in red cells.

There is evidence that the erythrocytosis-promoting effect of androgens is at least partly a direct effect upon the marrow.

Certain raw materials are necessary for the formation of mature red cells and they include amino-acids, iron, folates, vitamin B_{12}, pyridoxine, riboflavin and probably vitamins C and E. Amino-acid shortage is not a constraint except in gross malnutrition or protein starvation—kwashiorkor, a condition with multiple nutritional deficiencies producing anaemia is relatively common in Africa but virtually unknown in Europe.

Iron Metabolism

The total amount of iron in the human body normally amounts to about 4 grams, varying with haemoglobin level and weight. About $\frac{2}{3}$ of this is in the form of haemoglobin and up to 30% is storage iron. The tiny remaining fraction is made up of iron in enzymes and myoglobin. The daily breakdown of red cells results in the release and re-utilization of some 20 mgm of iron each day, corresponding to about 40 ml of blood. Iron is very abundant on the surface of the earth and poisonous in excess, leading to haemosiderosis and haemochromatosis. Very little excretion of iron takes place from the body—estimates of daily loss by a man or a non-menstruating woman are between 0·5 and 1·0 mg. Mean menstrual blood loss in the normal woman has been estimated at 30–40 ml with a range of 3–87 ml, a mean loss of 15–20 mgm of iron per period; in women with iron deficiency it is higher, mean figures being 70–100 ml. Many women lose much greater amounts than these.

The amount of iron in the body is controlled by its absorption which in turn is believed to be regulated by the epithelial cells of the small intestine, probably by the amount of apoferritin which these cells contain when they are formed. In iron deficient subjects little apoferritin is present and iron proceeds from the gut lumen into the intestinal villus to be bound to transferrin and make its way to the marrow. In normal subjects the intestinal columnar cells contain quantities of apoferritin, which are larger in iron-overloaded subjects, and which combine with some or all of the iron from the intestinal lumen to form ferritin; this is retained in the cells and later desquamated with them so that iron absorption is reduced or prevented.

Normal iron loss is of the order of 0·5–1·0 mg per day in men and up to 1·5 mg per day in menstruating women and it is balanced by iron absorption. It is usually agreed that western diets contain about 10–15 mg of iron per day and absorption varies between 5–10% of this.

Excessive amounts of iron are required for growth, especially in infancy and in childhood, in pregnancy, and when there is excessive blood loss, most commonly in menstruating women but also from any other cause. From most sites blood loss is apparent to the patient but from the gastro-intestinal tract it is frequently unnoticed, indeed there may be nothing abnormal visible to be noticed.

Vitamin B_{12}: Folic Acid

Vitamin B_{12} and folic acid will be considered together as deficiency of either may cause a megaloblastic anaemia. In this condition the red cells are large and often oval, with granulocyte nuclei being hypersegmented; marrow examination reveals characteristic features best summarised as a relative delay in nuclear maturation compared with that of the cytoplasm seen in both erythroblasts and granulocyte precursors. Tritiated thymidine studies show an arrest of cells at the stage of DNA synthesis and accumulation of cells in the premitotic resting phase as well as of cells lacking sufficient DNA to divide.

Folates are involved in one reaction in the synthesis of pyrimidines, the methylation of deoxyuridylate monophosphate to thymidylate monophosphate catalysed by thymidylate synthetase. Only the fully reduced form, tetrahydrofolate can be utilized and the enzyme dihydrofolate reductase reduces folic acid and dihydrofolate to the tetrahydro-form. This stage limits the rate of synthesis of DNA in mammalian tissue and probably explains why folate deficiency produces megaloblastic haemopoiesis.

Vitamin B_{12} is involved as methylcobalamin in the regeneration of tetrahydrofolate from methyl tetrahydrofolate; this appears to be its role in haemopoiesis, as apparently the only way in which this methyl group can be removed is by the conversion of homocysteine to methionine in the presence of vitamin B_{12}. Although synthesized only by microorganisms it is found in animal products especially liver and an adult requires to absorb about 2 μgm of the average daily dietary supply of 5–10 μgm. The mechanism of absorption of physiological amounts involves coupling with a glycoprotein, intrinsic factor, which is produced by the gastric parietal cells. The complex is then absorbed in the ileum, not more than about 1·5 μgm from any one dose, after which the ileum remains refractory for several hours. Normal daily gastric intrinsic factor output is greatly in excess of what is needed to complex as much vitamin B_{12} as the ileum can absorb. Body stores of vitamin B_{12} are normally sufficient for several years.

Folates are required in daily amounts of about 100 μgm. Occurring in most foodstuffs, especially in liver, spinach, and other greens, a normal diet provides about 600 μgm of folate daily. This may be destroyed by cooking, especially in large volumes of water at high temperatures. Absorption takes place rapidly through the duodenum and jejunum and the absorptive capacity is ample.

THE ANAEMIAS

Introduction

It is not easy to give a wholly logical and connected account of how to diagnose a given case of anaemia and it is necessary to approach the subject tentatively, first with a study of the causes and diseases, next of the screening tests and finally of the more detailed investigation which should make it possible to clinch a diagnosis.

Red cells normally survive for about 120 days in the peripheral blood before they are destroyed. Anaemia results sometimes from failure of production of red cells in sufficient quantity, sometimes because red cell life is reduced, or quite often for both reasons. The cell life may be shortened because of external loss of blood or because some external cause or some abnormal innate property of the red cells leads to their inability to survive for the normal length of time. Failure of production may arise because of shortage of essential raw materials, notably iron, vitamin B_{12} or folates, because of disease of the marrow or of the process of erythropoiesis or because of inability to utilize available raw materials, as in the anaemia of chronic disorders. Intermingling of shortened life and production failure is very common, e.g. in iron deficiency, megaloblastic anaemia and leukaemia and it is only among the haemolytic anaemias that one is able to find types of anaemia apparently wholly due to the reduction of cell life-span.

Anaemias Associated with Deficiencies

Iron Deficiency

First place must go to iron deficiency as this is much the commonest and most important. Deficiency is due to a negative balance between iron absorption and iron losses. This may result from defective intake, defective absorption or increased losses from chronic haemorrhage.

It is not particularly easy to choose a diet lacking iron and some skill may be required. In the UK flour has its iron content regulated to 1·65 mg for 100 g, rather more in the USA. Meat, fish, eggs and vegetables all contain significant amounts of iron differing in availability and in the percentage which may be absorbed. Dietary iron deficiency is more likely to develop where the diet is lacking in meat and offal as in general these foods contain more iron and absorption is more efficient than from vegetable iron.

Iron is usually poorly absorbed in coeliac disease in children and adults and after partial gastrectomy or jejunal resections and iron deficiency may result. In achlorhydria, although iron absorption may be impaired, the clinical significance is doubtful. Excess mucosal desquamation with iron loss may accelerate iron deficiency in coeliac disease.

Chronic haemorrhage is the most important cause of iron deficiency. When blood loss exceeds about 15 ml per day, it is barely possible for iron absorption to keep pace and iron deficiency ensues when body iron stores have become exhausted. World-wide, hookworm infestation is perhaps the most common cause of chronic haemorrhage, hotly pursued by excessive menstrual losses. Gastrointestinal haemorrhage is very common and is most commonly due to peptic ulceration and haemorrhoids—also important are oesophageal varices, hiatus hernia, aspirin ingestion, diverticulosis, gastrointestinal neoplasms, ulcerative colitis and hereditary haemorrhagic telangiectasia.

When iron balance becomes negative, a greater percentage of dietary iron is absorbed as iron stores become depleted and with further iron loss latent iron deficiency develops—an absence of significant iron stores, as judged by marrow aspirates stained for iron, and a reduction in the amount of saturation of serum transferrin, i.e. of the ratio serum iron divided by serum iron binding capacity. This ratio is normally above about 16% and a fall below this figure is taken as indicative of iron deficiency. At this stage no anaemia is present and there is debate as to whether clinical symptoms are associated with it—certainly a sore tongue may be present; lassitude may be associated also, but is a very common symptom and difficult to assess.

Further loss of blood leads to overt iron deficiency with anaemia—when developed there is a microcytic hypochromic picture in the blood film, and target cells may be present. Red cell measurements show a lower than usual mean cell volume (MCV) and mean cell haemoglobin (MCH). Marrow puncture reveals normoblasts with rather pyknotic nuclei and cytoplasm which is scanty, tattered and often vacuolated—stainable parenchymal iron is absent or very scanty. Serum iron concentration is low and iron binding capacity is high with saturation usually below 10%. Erythrocyte protoporphyrin is raised. When iron deficiency is severe certain clinical features often accompany it, koilonychia (spoon shaped nails), glossitis, angular stomatitis, and web formation in the pharynx or oesophagus. Pica may be present and the items selected contain no iron but often have in common a certain crunchiness, e.g. celery, cucumber, blackboard chalk, toilet paper and, especially, ice.

The foregoing account may suggest that the diagnosis of iron deficiency is exceedingly simple, which is often so, but sometimes the diagnosis is very difficult. The ultimate test may well be a therapeutic trial with an iron preparation.

Vitamin B_{12} Deficiency

The commonest cause of vitamin B_{12} deficiency is pernicious anaemia, a disease involving atrophy of the gastric mucous membrane, under production of intrinsic factor and hence an inability to absorb vitamin B_{12} in sufficient amounts. When the disease is established gastric achlorhydria is invariable. Gastric biopsy usually shows gastric atrophy with a loss of glandular elements, replacement by mucous cells or with intestinal metaplasia and an inflammatory cell infiltrate. Occasionally some gastric glands persist and the appearances are then known as atrophic gastritis. It seems probable that the disease is a genetically determined auto-immune process. The serum and gastric juice of patients with pernicious anaemia often contains antibodies to the parietal cells of normal gastric mucosa and to intrinsic factor, the latter either preventing attachment of the B_{12}-intrinsic factor complex to the ileal mucous membrane or blocking the combination between B_{12} and intrinsic factor. Intrinsic factor antibodies are rarely found except in pernicious anaemia but parietal cell antibodies are much less specific. Free intrinsic factor antibody in gastric juice would seem to represent the last straw to a patient with diminished gastric production as what little intrinsic factor remains may rapidly be neutralized.

A patient with severe pernicious anaemia is commonly grossly anaemic but relatively little distressed by it due to the slow development and compensatory mechanisms. The complexion is lemon yellow in colour due to slight jaundice. The tongue is devoid of papillae, often sore and has been likened to raw beef steak, albeit sometimes a rather pale beef steak, or veal steak. There may be complaints of paraesthesiae or weakness and examination may reveal the characteristic subacute combined degeneration of the spinal cord with impairment of deep pain and position sense and pyramidal tract signs. Optic atrophy or psychiatric abnormalities may also be seen. The spleen is usually palpable. Blood examination reveals an anaemia, indeed often a pancytopenia, with a film showing macrocytes and macro-ovalocytes and considerable variation in shape and size of the red cells. Granulocytes show nuclear hypersegmentation. Nucleated red cells may be present. The mean cell volume is raised, often to very high levels. Marrow puncture reveals a cellular specimen with megaloblastic appearances in which large red cell precursors of all stages have disproportionately immature nuclei, as indeed do myelocytes and metamyelocytes.

Estimation of serum vitamin B_{12} by microbiological assay methods using *Euglena gracilis* or *Lactobacillus leichmanii* reveals markedly low values with normal or raised values for serum folate and normal or low values for red cell folate. Isotopic methods of assay of serum vitamin B_{12} which do not utilize intrinsic factor as a binder may fail to detect up to 10% of low values because of non-specific protein binding. When deficiency is severe, raised urinary excretion of methylmalonic acid, especially after a loading dose of L-valine, may be detected. A significant reticulocyte response after daily injections of 5 μg of vitamin B_{12} is proof of deficiency.

By appropriate techniques gastric involvement may be proved—this may be by demonstrating intrinsic factor antibodies in serum or gastric atrophy on biopsy or finding that histamine or pentagastrin fail to lead to acid production, or more specifically intrinsic factor production, in gastric juices in significant amounts. Failure to absorb radioactive vitamin B_{12} and correction by simultaneous administration of intrinsic factor may be shown by tests involving measurement of whole body or hepatic uptake, faecal excretion, urinary excretion (Schilling test) or plasma radioactivity. Tests should demonstrate both vitamin B_{12} malabsorption and gastric involvement. The numbers and complexity of tests carried out tend to reflect the wealth of the institution and the hardiness of the patient and to vary inversely with the age of the initiator.

Gastric operations, partial and total gastrectomy, and resections of significant lengths of the jejunum may lead to vitamin B_{12} malabsorption. Disease of the ileum may also do so, especially Crohn's disease, as may colonization of the upper intestine by colonic bacteria as in internal gut fistulae, jejunal diverticolosis or blind loops due to surgery or Crohn's disease.

The fish tapeworm, *Diphyllobothrium latum*, appears to destroy or inhibit intrinsic factor. This tapeworm is very common in Finland and other Scandinavian countries but anaemia is rare and this is thought to relate to the level in the intestine at which the worm is attached—when attached high in the intestine the worm appears to absorb vitamin B_{12} at the expense of the host.

The anaemia in these patients with B_{12} deficiency other than pernicious anaemia is also megaloblastic but clearly they have their own distinctive features, not least scarring of the abdomen drawing attention to previous surgery. Barium studies both by meal and enema and occasionally laparotomy may be required to unravel the enteric group—notable clues are given by vitamin B_{12} levels which are low, absence of proof of gastric involvement and failure of intrinsic factor to correct B_{12} malabsorption.

Stores of vitamin B_{12} are considerable in relation to daily requirements and are usually sufficient for several years. Dietary deficiency is unusual except in very strict vegetarians—vegans, although subnormal levels of vitamin B_{12} are common in those Hindus who avoid all meat, fish and dairy products.

Folate Deficiency

As stores of folate are usually sufficient for only weeks of normal usage and daily requirements relatively high compared with dietary supply, dietary insufficiency may lead rapidly to deficiency. This is most often seen in the old, the poor, and in psychiatric and institutionalized patients. Definite malabsorption occurs in coeliac disease, tropical sprue, often in association with dermatitis herpetiformis and may be seen in many conditions such as after partial and total gastrectomy (although anorexia may be a factor also) extensive jejunal resection, alcoholism (anorexia, direct suppression of erythropoiesis and increased urinary excretion may also contribute) and in congestive heart failure.

Diphenylhydantoin and primidone with or without phenobarbitone produce low levels of plasma and red cell folate and megaloblastic anaemia in epileptics. Various explanations have been suggested, including interference with absorption. Administration of drugs which block dihydrofolate reductase such as methotrexate cause megaloblastic anaemia by blocking thymidylate synthesis; the antimalarial daraprim has a similar but weaker inhibitory action.

Folate deficiency may be diagnosed by measurement of serum and red cell folate levels, by the FIGLU test and by a therapeutic trial with 100–200 μg of folic acid daily. The serum folate level, usually between 3–20 μg/l is lowered in deficiency although values between 3 and 6 μg/l are regarded as borderline. The serum level is rapidly influenced by a reduction in dietary intake of folic acid and therefore is very sensitive, perhaps too sensitive. The red cell folate level is much higher and much less sensitive to transient dietary changes, so it is probably a better test in discriminating between genuine deficiency and transient reductions in dietary intake. Unlike the serum folate which tends to rise, in vitamin B_{12} deficiency the red cell folate usually falls substantially so B_{12} deficiency may need to be excluded.

In folate deficiency an oral loading dose of histidine leads to an increased amount of formiminoglutamic acid (FIGLU) in the urine. The test is moderately sensitive but unfortunately is often also positive in vitamin B_{12} deficiency, in liver disease and in many malignancies in the absence of folate deficiency.

Acute Haemorrhage

Whereas chronic haemorrhage leads to iron deficiency anaemia, acute haemorrhage may not produce any anaemia for a number of hours, depending upon its magnitude. After haemorrhage, reflex vasoconstriction takes place, reducing the size of the vascular bed which is to be filled at the expense of perfusion of the skin and splanchnic circulation. If this fails to maintain blood flow through the remaining vital organs the result is often death, although transfusion of blood or other fluids may maintain the necessary circulation. In the event of blood loss which is large but not too large for constriction to deal with, haemodilution takes place over the next 12–24 hours and the missing blood volume is restored by dilution of the plasma so that by the next day the haemoglobin level has fallen.

Anaemia due to Haemolysis

In these diseases the main cause of anaemia is a shortening of the red cell survival. Often there is an attempt at compensation by the marrow, which can increase its output up to 6 or 8 times that of normal in young adults, although less at later ages, and sometimes compensation is successful, resulting in a normal haemoglobin level—a compensated haemolytic state.

Haemolysis is most clearly recognised by its effects on the breakdown products of red cells. There is an increased output of unconjugated bilirubin in the serum, usually not to very high levels in the absence of liver disease or obstruction, and an increased amount of stereobilinogen in the faeces. Characteristically bile is absent from the urine. These findings, the common ones, are those seen where haemolysis is predominantly extravascular. Where it is intravascular, free haemoglobin appears in the plasma and is bound by the plasma haptoglobins; when the plasma concentration of haemoglobin is about 1·25 g/l the haptoglobins are saturated and excess haemoglobin is excreted via the glomeruli, resulting in haemoglobinuria. When haemoglobinuria is chronic some of the pigment is taken up by the kidney and free iron is excreted in the urine as haemosiderin granules which may be stained by a Prussian-blue method. A few hours after haemoglobin is released into plasma it can be demonstrated as methaemalbumin, a brown pigment identified by Schumm's test in which it forms a haemochromogen band at 558 nm.

Red cell regeneration, measured by reticulocytosis, increased numbers of polychromatic red cells, cells with basophil stippling and even nucleated red cells in the blood film, is evidence of compensation rather than of haemolysis but not to be despised for all that. When a reticulocytosis is sustained over some days or weeks and haemorrhage can be excluded, haemolysis is virtually proved. Regeneration after a single brief episode of haemolysis, without obvious pigment changes may not be easy to distinguish from regeneration after a single acute bleed into the intestine when this is not clinically detectable.

In cases of chronic haemolysis the marrow may be expanded enough to be apparent on X-rays as widening of the medulla and thinning of the cortex of long bones such as the femur and the metacarpals and phalanges, and widening of the skull vault.

Many types of haemolytic anaemia are characterized by distinctive abnormalities of the red cells as seen on a blood film, notably spherocytosis, elliptocytosis, target cell formation, fragmentation and contraction and the presence of Heinz bodies when stained by methyl violet.

In chronic haemolytic states where the haemoglobin is stable at levels high enough not to threaten the patient, studies of red cell survival may be carried out. These involve labelling red cells with ^{51}Cr (or diisopropyl-fluorophosphate $DF^{32}P$) and although radioactivity disappears due to elution as well as cell destruction, normally the time for half the ^{51}Cr to disappear is 25–32 days. In haemolytic states this time may be greatly shortened.

Congenital Haemolytic Anaemias

The haemolytic anaemias may be divided broadly into those in which the defect is in the red cell, usually inherited, and those, usually acquired, in which it is in the environment of the red cells. The red cell defects may be further subdivided into disorders of the red cell membranes, and disorders of enzyme production. In this congenital group a history will often be obtained of neonatal jaundice, jaundice on one or more occasions in childhood or adult life, symptoms of an anaemia on occasions, and splenomegaly. On the whole, membrane defects run in families, i.e. they are inherited as autosomal dominants and enzyme defects do not, i.e. they show sex-linked or autosomal recessive inheritance.

The commonest membrane defect is *hereditary spherocytosis*. The precise abnormality is unknown but it appears that the spherocytic shape of the red cells in this disorder may be due to an abnormality in the structure or function of spectrin, a filamentous protein present in the internal surface of the lipid bilayer. The red cells appear spherocytic on blood films and, when exposed to varying concentrations of saline below isotonic, they begin to lyse at concentrations below that at which lysis normally occurs. In borderline cases the difference may be very slight and incubation for 24 hours before testing may help to exaggerate the discrepancy. The autohaemolysis test, in which red cells are incubated at 37°C with and without added glucose for 48 hours before the amount of haemolysis is measured, shows that the cell metabolises rapidly and requires more than normal amounts of glucose for its integrity. Haemolysis takes place in the spleen where the cells become trapped due to their rigid shape—splenectomy corrects the clinical abnormalities although the red cells do not become normal.

Hereditary elliptocytosis often produces no anaemia or haemolysis but in a few cases a relatively severe chronic haemolytic anaemia with splenomegaly is present. The red cells are oval or elliptocytic with many pencil-like forms. Splenectomy cures the clinical manifestations but the elliptocytosis remains.

The enzyme defects of the red cells are numerous but mostly very rare. For reasons of space only the two commonest conditions will be considered at all, glucose-6-phosphate dehydrogenase deficiency (G-6-PD), a very common and important condition and pyruvate kinase deficiency, an interesting but much rarer condition.

Pyruvate kinase deficiency is least rare in populations of North European origin. There is considerable clinical variation in severity so that the condition may be unnoticed or characterized by neonatal and recurrent or continuous jaundice later with the possibility of pigment stones forming in the gall bladder (as in so many of the congenital haemolytic anaemias) which may cause obstructive jaundice. Anaemia may be severe enough to require repeated transfusions. The red cells are not distinctively abnormal; the reticulocyte count may be markedly raised, e.g. to 50% or more. The autohaemolysis test is markedly abnormal and glucose fails to correct it. Direct enzyme assay is necessary to prove the diagnosis.

The red cell is normally subject to oxidant threat, especially if oxidant drugs are given, and if adequate reducing power is not present some of the haemoglobin is oxidized to methaemoglobin. This may be reduced to haemoglobin but if oxidation proceeds further the haemoglobin becomes denatured and cannot then be reduced. The denatured protein forms a Heinz body inside a red cell and usually either the Heinz body or the whole red cell are removed by the spleen. Reduced glutathione in the red cell acts as a reducing agent and protects against such denaturation. The pentose phosphate pathway in the red cell produces NADPH which, coupled to glutathione reductase, reduces glutathione and also, coupled to methaemoglobin reductase is one mechanism which keeps methaemoglobin reduced to haemoglobin. Central to the pentose phosphate pathway is the enzyme glucose-6-phosphate dehydrogenase. Disorders due to *glucose-6-phosphate dehydrogenase deficiency* have been described in almost all races, especially in the countries of the Mediterranean littoral, the Middle East, West Africa and South China. More than 50 abnormal variants of the enzyme are now recognized with varying ranges of defective activity. Inheritance is sex-linked, i.e. the genes are carried in the X-chromosome, and it has been found that in female heterozygotes the individual red cells either have normal function or that of the particular enzyme mutant. On average, carrier females have activities midway between that of hemizygous males and normal individuals. This has practical significance as some tests cannot easily distinguish between female heterozygotes and normals although they can detect hemizygotes and homozygotes.

Clinically, patients with G-6-PD deficiency may present with neonatal jaundice, drug-induced haemolysis after being given one of a number of oxidant drugs (aminoquinolines, analgesics, sulphonomides and sulphones, nitrofurans and miscellaneous drugs), chronic haemolytic anaemia, and favism. The latter is an acute haemolytic anaemia provoked by exposure to the pollen of the bean Vicia faba, or by eating the uncooked bean. An early victim is believed to have been Pythagoras. It occurs in the Mediterranean and Middle Eastern types of the deficiency; acute intravascular haemolysis may follow within a few minutes of inhaling the bean pollen, although there may be a latent period of 6–24 hours after eating the beans. Attacks last from 2–6 days and, although often severe and even fatal in children under six years of age, they are usually self-limiting.

Drug-induced haemolytic anaemia occurring in all races with G-6-PD deficiency varies in severity but acute onset with mild or gross haemoglobinuria is usual. The oldest cells (with least G-6-PD) are destroyed and the younger cells have enough enzyme to prevent their oxidative denaturation and haemolysis, although larger drug doses will destroy younger cells. The red cells of affected Negroes contain more enzyme than those of affected Caucasians and so the latter have more severe haemolysis.

Diagnosis is based on suspicion and testing for the defective enzyme, testing best not done until some weeks after a haemolytic episode lest the higher enzyme levels of young red cells confuse the result. The brilliant cresyl blue dye test is widely used as a screening test, incubating haemolysate, G-6-P, NADP and brilliant cresyl blue. NADPH is normally formed and decolourizes the blue dye, the time varying with the activity of G-6-PD. The test may fail to detect female heterozygotes.

Unstable haemoglobins such as Hb Zurich and Hb Köln may produce haemolysis only after oxidant drug ingestion (Zurich) or chronically, but perhaps worse, after drug administration (Köln). A large number have been described. They and thalassaemia and abnormal haemoglobins are covered in Chapter 11 and will not be discussed further here.

Acquired Haemolytic Anaemias

Perhaps the commonest group of acquired haemolytic anaemias is the auto-immune group, that in which an antibody forms to the red cells and leads, by a little understood mechanism, to their destruction.

Cold Antibody Type

Some of these antibodies attach to red cells in the cold and haemolysis is provoked by exposure to cold, for example *paroxysmal cold haemoglobinuria*; often within a few minutes a rigor develops with pains in back and legs and abdomen and there may be urticaria and cyanosis. This condition is often associated with a virus infection and does not recur when the antibody disappears after a few months. In vitro testing demonstrates the Donath-Landsteiner antibody which attaches to red cells at low temperatures, fixing complement, and lyses them when warmed.

In other cases an antibody can be detected in high titre which agglutinates cells at low temperatures but trouble develops only when exposed parts of the body reach these low temperatures. When the antibody is effective at more nearly physiological temperatures less cold exposure will provoke haemolysis. In this *chronic cold haemagglutinin disease*, especially seen in elderly ladies, fingers and nose and ears may become discoloured in cold weather as a result of circulatory interference by agglutinates of red cells and often also there is a chronic intravascular type of haemolytic anaemia especially in cold weather with haemoglobinaemia and variable haemoglobinuria. The antibody may occur in association with a lymphoproliferative disorder but more often its sources remain obscure and although cytotoxic treatment to reduce the antibody titre is logical and may be used, central heating and appropriate clothing may be more effective.

More commonly cold agglutinins develop in association with either infectious mononucleosis or Mycoplasma pneumoniae infection, usually in the second week of the disease. Rarely these antibodies may cause a very severe type of haemolytic anaemia with haemoglobinuria, threatening life by the very rapid fall in haemoglobin and occasionally causing oliguric renal failure. Transfusion may be necessary and the cells frequently appear to be incompatible. Some comfort may be sought in *in vivo* compatibility testing with a few of the cells to be transfused which have been labelled with ^{51}Cr. In the chronic cold haemagglutinin disease and after M. pneumoniae infection, antibodies have an anti-I specificity, and after infectious mononucleosis they show anti-i specificity. In the infections the cold antibody titre frequently wanes rapidly so the disease tends to spontaneous cure with the passage of time.

Warm Antibody Type

Whereas cold antibodies are usually easy to demonstrate by the agglutination which they produce, warm antibodies are most easily detected, attached to the surface of red cells, by the direct antiglobulin test, using antibodies to human complement or the fractions of globulin, the presence of which on the surface of the red cell is not normal. Whereas the presence of globulins attached to red cells is usually only seen in association with haemolytic states, complement may often be present in patients who show no evidence of haemolysis.

Such antibodies, causing warm antibody type haemolytic anaemia, occur in a variety of disorders, notably in association with autoimmune diseases such as systemic lupus erythematosus and less commonly others especially rheumatoid arthritis, ulcerative colitis and chronic active hepatitis. Drugs, especially α methyldopa and very rarely several others may cause warm type haemolytic anaemia. However, in nearly half of the cases no cause can be found at presentation, although one may emerge months or years later.

The onset may be gradual with pallor or may be more sudden with dyspnoea, jaundice, and even haemoglobinuria. The spleen is often enlarged and signs appropriate to an associated disease may be present. The blood film commonly shows marked spherocytosis and reticulocytosis and pigment changes and a positive antiglobulin test are found. Treatment is usually by steroids, may well be urgent and the disease is often life-threatening. Splenectomy is sometimes necessary and control of haemolysis cannot always be achieved.

Mechanical and Microangiopathic Haemolytic Anaemias

These diseases have in common mechanical damage to circulating red cells which may result from physical violence, heart valve abnormalities and a variety of conditions which probably have in common red cell destruction by strands of fibrin in the lumina of small blood vessels or red cell adherence to damaged endothelial cells. The blood film shows the appearance of distorted red cells, some tiny fragments and microspherocytes and cells lacking parts of their circumference as if a fragment had been torn off.

March haemoglobinuria is a type of exertional haemolytic anaemia in which haemoglobinuria follows a route march or a cross-country run. Sorbo-rubber heels frequently prevent this and it is believed that direct trauma to red cells when the heel hits the road is the cause, possibly in part a function of the technique of running or walking. Similar haemoglobinuria has been described in karate students.

A proportion of patients develop a mechanical haemoly tic anaemia, apparent on the blood film after heart valve replacement. There is usually haemoglobinaemia and haemosiderinuria and occasionally the disorder is severe enough to require transfusion. Some minor degree of haemolysis may be an almost inescapable sequel of a valve replacement but clinical effects are most often seen after aortic valve replacements and, when severe, and when the patient's condition permits, may lead to the need for a further replacement valve. Exercise or severe anaemia may make haemolysis worse, presumably by increasing the cardiac output. Only rarely does natural disease of a heart valve lead to clinical anaemia although reduced cell survival is not uncommon.

Microangiopathic haemolytic anaemias represent a group of diseases showing features resembling mechanical haemolytic anaemias but without obvious mechanical causes for haemolysis. Such conditions, often associated with thrombocytopenia, low levels of fibrinogen and some evidence of increased amounts of fibrinogen breakdown products in the plasma may be seen in a variety of clinical presentations. In childhood, especially in the first year of life, often in association with a febrile illness, there may develop suddenly the *haemolytic uraemic syndrome* with obvious evidence of a haemolytic anaemia and rapidly worsening renal failure. A similar disease in adults is *thrombotic thrombocytopenic purpura*, with haemolysis, thrombocytopenia, possibly uraemia and a variety of central nervous symptoms, especially fits. In both diseases involvement of the micro-vasculature is common, frequently with surface fibrin deposition. *Eclampsia* in pregnancy toxaemia is thought to have a somewhat similar pathogenesis with platelet emboli precipitating fits.

Carcinomatosis may lead to a similar type of haemolytic anaemia and this is known to be associated with metastases from adenocarcinomata, usually mucin secreting. Renal involvement is absent and local coagulation round mucin forming deposits is thought to produce fibrin strands explaining the commonly depressed levels of platelets and factors VIII and fibrinogen. Sometimes evidence of intravascular coagulation can only be obtained, here and in some of the other conditions, by measurement of the disappearance rate of radio-active iodine-labelled fibrinogen.

Paroxysmal Nocturnal Haemoglobinuria

This condition is often diagnosed late and offers many traps in its diagnosis. Once it has developed, although haemoglobinaemia is present, there may be no haemoglobinuria, but, when there is, it may be thought to be haematuria. There is a haemolytic anaemia with a largely normal blood film although leucopenia and thrombocytopenia are common. Haemosiderinuria is present even in the absence of haemoglobinuria and sometimes iron deficiency is present (which in association with a haemolytic anaemia should suggest this diagnosis). The acid-serum test (Ham's test) is diagnostic but correct performance needs attention to detail.

Some patients have a predominantly aplastic course in

place of the more usual attempts at compensation for haemolysis but thrombotic phenomena in both veins and arteries are very common and are important reasons for the rather short average life span in these patients.

Infectious Causes of Haemolytic Anaemia

The malarial parasite destroys red cells and malaria, especially when due to *Plasmodium falciparum*, is an important cause of haemolysis, exacerbated when a malarial patient has glucose-6-phosphate dehydrogenase deficiency and is treated by primaquine.

More important, although rare, in non malarial areas, are the anaemias occasionally associated with *streptococcus pyogenes* infections, usually mild and compensated, and the overwhelming haemolysis associated with the frequently fatal *clostridium perfringens* infections, e.g. in septic abortions. Here haemolysis may be so severe that one cannot tell plasma from red cells in sedimented blood, a phenomenon which the writer happens not to have seen in any other condition.

The Anaemia of Chronic Disorders

This type of anaemia is mild, haemoglobin level not usually below 9 or 10 gm per 100 ml, and non-progressive. It tends to reach a certain level within a few weeks of the onset of the associated condition and to remain unchanged until its cure. It may be seen in association with rheumatoid arthritis and related diseases such as polymalgia rheumatica, with chronic infections such as large chronic abscesses and osteomyelitis, with lymphomata and carcinomata (even in the absence of metastases), and with fractures, severe tissue injury and many other disorders. Severity often relates to the clinical severity of the rheumatoid arthritis or other causative conditions.

The anaemia is usually normocytic and normochromic, or slightly hypochromic, and is characterized by a low serum iron concentration, low serum iron-binding capacity, low percentage saturation of the iron-binding capacity, but not as low as in iron deficiency, little iron in marrow normoblasts (i.e. few sideroblasts) and plentiful iron in the marrow reticulo-endothelial cells.

The cause of the anaemia is in part a defective release of iron from reticulo-endothelial cells to transferrin, but in addition there is a failure to release erythropoietin in amounts appropriate to the degree of anaemia, and a moderately reduced red cell survival due to external causes.

Treatment is that of the disease although where iron deficiency co-exists, as it may, e.g. in rheumatoid arthritis, a trial of oral iron therapy will establish how much of the anaemia is due to iron deficiency and correctable.

Anaemia Due to Marrow Disorders

Under this heading will be discussed very briefly those blood and other disorders in which marrow involvement occurs and in which this appears to be the major reason for anaemia. It is a gross oversimplification to assume that this is because the normal marrow is crowded out in the physical sense, but competition for essential metabolites may place constraints as great as those of dispossession.

First to be considered are *the leukaemias*. Anaemia is usual in all the acute types of leukaemia, myeloid, lymphatic, monocytic, and erythroleukaemia. Additional features may arise because of pancytopenia, sepsis, purpura or spontaneous bleeding from mucous membranes especially of nose and mouth, or because of leukaemic infiltration of bone marrow, skin, central nervous system, testes and indeed most parts of the body. Diagnosis is not usually very difficult and abnormal cells are often to be found in blood films and in large numbers in marrow aspirates.

In chronic leukaemias, anaemia may be a comparatively late manifestation—in chronic lymphatic leukaemia there may be obvious signs of many lymphocytes in the peripheral blood for a decade in some cases before anaemia develops if it ever does. Prolymphocytic leukaemia and hairy cell leukaemia are morphologically distinct and behave somewhat differently clinically. In all these diseases when developed splenomegaly is usual and in chronic lymphatic leukaemia, enlargement of lymph nodes in the neck axillae and groins is common.

In chronic myeloid leukaemia anaemic symptoms are the usual cause of presentation and the patient usually has splenomegaly, often of considerable size. The blood picture shows anaemia, often with an increased number of blood platelets and large numbers of granulocytes and their precursors, myelocytes, metamyelocytes and even small numbers of myeloblasts and promyelocytes. Characteristically there is an absence of alkaline phosphatase activity in the cytoplasm of neutrophils in this condition, contrasting strongly with the increased amounts in association with a neutrophil leucocytes due to a pyrogenic infection. Chromosome studies of the marrow reveal, in most cases, the presence of a small chromosome, the Philadelphia chromosome, in which there is deletion of part of the long arm of one of the four small acrocentric chromosomes.

Other rarer types of leukaemia also may cause anaemia—neither they nor the treatment of the above conditions by cytotoxic drugs will be discussed.

Myelomatosis is a disease in which the bone marrow is infiltrated with plasma cells and presents with anaemia, bone pain which may be very severe, fractures, or symptoms related to the high output of paraproteins (monoclonal immunoglobins) or light chains (Bence-Jones protein) which may be produced. The latter may produce renal failure by tubular blockage as a result of precipitation from the urine, and the former may result in symptoms due to hyperviscosity of the blood—bleeding from mucous membranes, especially of the gastrointestinal tract, purpura, lethargy, mental confusion and even coma, and visual disturbances often due to retinal haemorrhages.

A typical blood picture is a moderate anaemia, sometimes with immature red and white cell precursors (leuco-erythroblastic anaemia), obvious rouleaux formation on the film and commonly a background bluish haze which may enable the diagnosis to be considered when the stained film is picked up. Marrow puncture usually reveals an excess of plasma cells, protein estimation usually an excess of a monoclonal globulin (a paraprotein), the amount being more, often much more, than 20 g per litre of plasma. Urine examination, often only after concentration by dialysis, reveals an excess of light chains. Immunological confirmation is necessary.

Metastatic deposits in the bone marrow may occur, usually from carcinomata of the breast, bronchus, prostate, kidney or thyroid. The classical blood picture is a leuco-erythroblastic anaemia but in many cases anaemia is the only abnormality. Other cases, as discussed above, show features of a micro-angiopathic haemolytic anaemia with fragmented red cells.

Myelofibrosis is usually a chronic disease with anaemia, variable white cell and platelet pictures, and very commonly splenomegaly. The red cells commonly include the characteristic tear drop cells and when such appearances are combined with a leuco-erythroblastic anaemia the diagnosis is in little doubt. Usually marrow puncture yields only a dry tap and marrow trephine is necessary to obtain a core of marrow which when processed and sectioned shows the characteristic increase in reticulin, fibrous tissue and megakaryocytes.

Sometimes the marrow appears to become *aplastic*. Here anaemia is usually accompanied by pancytopenia with very few reticulocytes in view of the degree of anaemia and no primitive or abnormal cells. The red cells are usually normocytic and normochromic but may be macrocytic. There may be a history of exposure to a relevant drug or chemical, or to radiation, but physical examination usually shows signs associated with pancytopenia, such as purpura, pallor and rarely agranulocytic ulcers in the mouth. Marrow puncture may yield very hypoplastic material with very little except a few lymphocytes or blast cells but occasionally it is cellular. Often a trephine in such cases shows that there are a few islands of surviving marrow in a sea of aplastic marrow.

In *acquired sideroblastic anaemia*, not uncommon in elderly people, a certain delay in making the diagnosis is rather common and after some months or years of anaemic symptoms, physical examination being unhelpful, the patient is found to have mainly normocytic normochromic red cells although nearly always there is a second population of hypochromic red cells of varying size. Bone marrow examination reveals erythroblastic hyperplasia, and iron staining shows many sideroblasts with a perinuclear ring of iron granules, the so-called ring sideroblasts. Sometimes associated with antituberculous drug therapy or chloramphenicol and some diseases such as intestinal malabsorption and rheumatoid arthritis, the condition is usually primary.

Rarely in obscure anaemias the cause is found, by examination of marrow aspirates, to be dyserythropoiesis with megaloblastoid erythroblasts and intranuclear chromatin bridges in erythroblasts, or erythroblastic multinuclearity. The group is complex and rare and will not be discussed further.

DIAGNOSIS OF THE TYPE OF ANAEMIA

The method employed in diagnosis will vary depending on who is making it. The problem is different for a general practitioner confronted by a pallid teenage girl with heavy menstrual periods and for a haematologist to whose department a sample of blood from such a patient is sent.

Some clinical notion is usually formed at least that anaemia is present because of tiredness, breathlessness, a falling-off in overall performance, and pallor. Some effort will have been made to decide whether this is physiological due to growth in childhood or accompanying pregnancy, whether it is due to haemorrhage or to iron deficiency because of chronic haemorrhage. Sometimes a sudden onset with jaundice or haemoglobinuria or after the prescription of potentially haemolytic drugs will suggest that the anaemia is haemolytic. The lemon yellow colour described in pernicious anaemia is not a very valuable feature to judge from its apparent presence in many cases of iron and folic acid deficiency anaemias.

Splenomegaly may be present in megaloblastic and haemolytic anaemias, myelofibrosis, acute and chronic leukaemias, liver disease and myelomatosis so that while it may be a valuable guide to the presence of disease, the guidance is non-specific.

Enlarged lymph nodes may be found in lymphatic leukaemia and rarely other blood disorders. Thrombocytopenia may lead to purpuric haemorrhages especially into the skin and mucous membranes and leukaemic infiltration to skin nodules, bone tenderness, hepatomegaly and other rare signs.

Blood examination is much facilitated by modern cell counters which enumerate the haemoglobin level (in g/dl), white cell count, red cell count, packed red cell volume and the derived red cell parameters, mean cell haemoglobin (MCH) in picograms, mean cell volume (MCV) in femtolitres, mean cell haemoglobin concentration (MCHC) in per cent.

Iron deficiency anaemia is characterized by cells which are small and have less haemoglobin in them than normal and both MCV and MCH are subnormal, frequently markedly so. A blood film shows small cells and cells with only a peripheral ring of cytoplasm and undue central pallor, hypochromic cells. Rather similar blood findings are seen in thalassaemia and heterozygotes may cause confusion. Some of these latter are not anaemic and a normal haemoglobin level with low MCV and MCH strongly suggests thalassaemia. Anaemic heterozygotes present more problems but a raised level of haemoglobin A_2 and sometimes HbF is helpful in diagnosing β-thalassaemia heterozygotes, as is the normal transferrin saturation.

Confirmation of the diagnosis of iron deficiency may not be necessary in some circumstances but in others it may, and here help may be sought from measurement of serum iron concentration, the iron-binding capacity and the percentage saturation of transferrin, which is reduced below 16%. Serum ferritin levels are consistently low. Bone marrow aspirates, stained for iron, show a virtual absence of haemosiderin, but this method is non-quantitative and varies somewhat in usefulness with the cellularity of the marrow sample obtained. Correlation between absent haemosiderin and the presence of iron deficiency is, however, good.

Megaloblastic anaemia results in macrocytosis and the mean cell volume is increased. The blood film shows a proportion of macrocytes, especially oval macrocytes and considerable variation in size and shape of the red cells with some tiny cells present. Bone marrow aspirates show megaloblastic features. Serum vitamin B_{12} levels and serum and/or red cell folate levels are measured (note that concurrent presence of antibiotics in the serum may interfere with microbiological assays). The Schilling test, of urine ^{57}Co B_{12} excretion after oral ingestion and a parenteral dose of non-radioactive cyanocobalamin, is the easiest test of vitamin B_{12} absorption. Gastric secretion of intrinsic factor may be measured by aspiration along a gastric tube before and after pentagastrin stimulation—measurement of this and of serum intrinsic factor antibodies depend upon an *in vitro* assay with radioactive vitamin B_{12}. Gastric biopsies may be taken via a flexible fibreoptic gastroscope and show atrophic gastritis or gastric atrophy. Investigation of intestinal function in association with megaloblastic anaemia is in part radiological, by barium meal, follow-through and enema and by jejunal biopsy and other tests of absorption where these are deemed relevant.

The diagnosis of *haemolytic anaemias* depends in large part upon evidence of pigment changes and blood regeneration. The plasma bilirubin will need to be measured and may be raised and it will be mainly of indirect or non-conjugated type. Other liver function tests are usually normal. Lactate dehydrogenase activity is markedly raised in haemolytic (and megaloblastic) anaemias. Plasma may contain free haemoglobin or methaemalbumin and urine may contain haemoglobin or haemosiderin. Plasma haptoglobin levels are often lowered; there is however a fairly wide normal range and as many acute inflammatory and other conditions lead to raised levels, low levels are not invariable and may in any case be due to hepatic disease.

The reticulocyte count is raised when there is compensatory erythropoiesis. However in anaemic patients reticulocytes are released earlier from the marrow and hence persist longer in the peripheral blood and there are less normal red cells to dilute them so in anaemic patients the reticulocytes give a misleading picture of erythroid compensation. If a normal non-anaemic individual produces about 1% of reticulocytes (and of course such low levels have large statistical counting errors) a figure of 7·5% in a patient with a packed cell volume of 15% really means the same daily red cell production, not $7\frac{1}{2}$ times greater production. The blood film often shows nucleated red cells, macrocytosis due to the presence of bluish (polychromatic) large young red cells and often an associated neutrophil leucocytosis. Marrow puncture usually reveals actively proliferating bone marrow with erythroblastic hyperplasia.

Some guide to the type of haemolytic anaemia often comes from examination of the blood film which may show spherocytosis, elliptocytosis, target cells (in association with liver disease), fragmented red cells, cells with basophil stippling and autoagglutinates.

A direct antiglobulin test, which often needs to be done with reagents against IgA, IgG, IgM and complement, and a search for cold acting agglutinins and haemolysins would be a screen for autoimmune haemolytic anaemia.

A red-cell osmotic fragility test serves to demonstrate the presence of spherocytes and the same tests in relatives serves to establish the hereditary nature of the disease although some cases do not show this feature. This red cell autohaemolysis test is most valuable as a screen for pyruvate kinase deficiency and the 48-hour lysis is not reduced by glucose addition in this condition. Screening tests for glucose-6-phosphate dehydrogenase deficiency have been described above.

Evidence of red cell fragmentation in the blood film will lead to the possibility of mechanical causes and, if these have been excluded, a search for evidence of generalized or localized intravascular coagulation. Features of a consumption coagulopathy may be found with low levels of platelets Factors V and VIII and raised levels of fibrinogen degradation products. There may be associated clinical evidence of hypertension, renal failure, eclampsia or carcinomatosis.

Although haemoglobinaemia or haemoglobinuria should be a clue, it is always wise to perform Ham's acid serum test for haemolysis in unexplained cases of haemolytic anaemia. In those cases linked to acute infections blood culture or serological tests will be crucial.

The anaemia of chronic infection may produce either normal or hypochromic microcytic red cells and counter printout and blood film will show either of these. Transferrin saturation may be low normal or normal, marrow iron stain shows plenty of storage iron, and serum ferritin is normal or raised.

The diagnosis of leukaemia, myelofibrosis, myeloma, sideroblastic anaemia and dyserythropoietic anaemia is based on the presence of specific (and gross) abnormalities in blood or bone marrow or both and these need not be enumerated in detail.

It is important to stress that a great deal may be discovered about the cause of a patient's anaemia by study of the above mentioned red cell parameters, counting of reticulocytes, examination of a blood film and if necessary of a bone marrow aspirate. Special tests are required only in a minority of cases.

FURTHER READING

Hardisty, R. M. and Weatherall, D. J. (1974), *Blood and its Disorders*. Oxford: Blackwell.
Harris, J. W. and Kellermeyer, R. W. (1970), *The Red Cell*. Cambridge, Massachusetts: Harvard University Press.
Hoffbrand, A. V. and Lewis, S. M. (1972), *Haematology*. London: Heinemann Medical Books Ltd.

13. ABNORMALITIES OF PLASMA PROTEINS

J. T. WHICHER

Introduction
  History
  Clinical applications

Metabolism of plasma proteins
  Synthesis
  Distribution
  Catabolism

Functions of plasma proteins
  Immune defence
  Inflammation
  Transport proteins
  Signal proteins
  Blood clotting
  Tissue derived proteins and oncofetal proteins

The proteins of immune defence
  Immunoglobulins

Proteins associated with inflammation
  Acute-phase reactant proteins
  Protease inhibitors
  Haptoglobin
  Orosomucoid
  Other proteins of inflammation

Transport proteins
  Albumin
  Transferrin
  Caeruloplasmin
  Other transport proteins

Oncofetal and tissue proteins
  β_2-microglobulin
  α-fetoprotein
  Placental and pregnancy-associated proteins

Plasma precursors of amyloid

Interpretation of electrophoresis
  Serum and plasma
  Cerebrospinal fluid
  Urine

Specific protein measurements in disease
  Inflammation
  Immune-complex disease
  Infectious disease
  Liver disease
  Gastrointestinal disease
  Renal disease
  Neoplasia
  Skin disease

INTRODUCTION

Plasma contains a large number of different proteins of very varying structural characteristics. The function of only a relatively small proportion of them is known but it must be presumed that they are present either with a specific role in the plasma or that they represent cellular proteins shed into the circulation as a result of degradative processes. Many plasma proteins show characteristic changes in concentration or structure either due to genetic defects which may give rise to disease or as a result of secondary changes reflecting pathological or physiological processes.

All the transcellular fluids such as CSF, urine, synovial fluid and saliva contain a considerable number of proteins, many of them derived from plasma but some secreted specifically into the fluid.

History

Plasma has always provided biochemists with an easily available material for the study of proteins, some of which are present in extremely high concentration. Detailed study of proteins did not really start until a series of workers between 1853 and 1869 studied the insolubility of some serum components upon acidification and the addition of water. It was in 1862 that Schmidt instituted the term globulin for the substances from serum that formed a precipitate of tiny spheres on the addition of water. The idea of salt fractionation then appeared and gave rise to a series of separation techniques still widely used today and upon which much of the subsequent work has depended. Thus in 1894 Gurber used ammonium sulphate to purify one of the first proteins ever to be crystallized—horse albumin.

Despite the advent of such techniques little real advance was made as methods were not available to study the structure of proteins separated in this way. However, it was appreciated that the albumin/globulin ratio showed characteristic changes in disease.

It was not until the advent of analytical ultracentrifugation and moving boundary electrophoresis in the 1930's that the real characterization of serum proteins started. The electrophoretic pattern of human serum comprising albumin and the alpha, beta and gamma globulins was described by Tiselius, and in 1939 Tiselius and Kabat showed that antibody was a gamma-globulin. In 1950 paper electrophoresis was described and by the late 1950's had spread into many laboratories giving rise to innumerable clinical studies. In 1953 Grabar and Williams described the use of antibody raised against human serum to characterize the proteins present after electrophoresis. Thus the essential facets of protein biochemistry had now been discovered; techniques for fractionation, separation and identification. The measurement of protein concentrations using specific antibody raised against a single purified protein appeared some ten years later with the work of Mancini and of Laurell.

Clinical Applications of Protein Studies

Changes in proteins in plasma or other body fluids may be genetically determined and result directly in disease due to an abnormality or deficiency in a specific protein. More commonly changes are secondary to a disease process affecting either protein synthesis, distribution or catabolism.

Genetic variants though not common individually form an important group for the clinical laboratory to investigate as they may be the diagnostic key to a disease process. Secondary changes vary enormously in clinical usefulness; they may be important to the diagnosis such as the finding of a paraprotein in myeloma, or simply be a part of the jigsaw of diagnostic information which allows us to assess the probability of a particular disease. More recently we are beginning to appreciate the value of sequential protein measurements in the monitoring of disease activity and response to treatment.

The characteristics of plasma proteins which are used in clinical practice are changes in their concentration, function or structure. It is thus important that both qualitative and quantitative techniques are available in the laboratory. Simple electrophoresis on cellulose acetate or agarose is a most useful initial investigation providing a considerable amount of information upon which the choice of specific protein measurements can be made It also allows the identification of a number of genetic variants which result in changes in electrophoretic mobility. The identification of proteins separated by electrophoresis by the use of immunofixation or immunoelectrophoresis further increases the discrimination of qualitative examination.

For proteins present in plasma at concentrations above 1 mg/l quantitation may be performed by gel diffusion, electroimmunodiffusion or nephelometry. Below this concentration radioimmunoassay and enzyme immunoassay are the most widely used techniques. Functional assays are essential for proteins which are present in normal concentrations but whose function may be defective.

METABOLISM OF PLASMA PROTEINS

Synthesis

Many plasma proteins, including albumin, are synthesized in the liver although there is increasing evidence for the production of some plasma proteins in peripheral tissues. Thus we know that immunoglobulins are only synthesized by B-cells while cell-surface proteins such as β_2-microglobulin are made by many different cells.

The general mechanism of protein synthesis is common to all cells. Intracellular proteins are usually translated on cytoplasmic polyribosomes whereas proteins destined for the cell membrane or for secretion are translated on the membrane-bound ribosomes of the rough endoplasmic reticulum. After synthesis, most proteins probably contain two additional peptides at the N-terminus. A hydrophobic pre-peptide acts as a 'leader' directing the molecule through the membrane of the endoplasmic reticulum. This is cleaved almost immediately leaving the protein with a pro-peptide which is not removed until maturation occurs in the golgi vesicles or in storage granules immediately prior to secretion. Glycoproteins have carbohydrate residues added by sugar transferases in the lumen of the endoplasmic reticulum and these may control folding of the polypeptide chain and act as recognition sites for subsequent metabolism. For example, it is probable that the carbohydrate moiety of the glycoprotein hormones may be the key for interaction with target cell membrane receptors. Control of protein synthesis may be exerted on the rate of transcription of M-RNA from DNA or at the level of translation. Both mechanisms have been described for a number of proteins (Fig. 1).

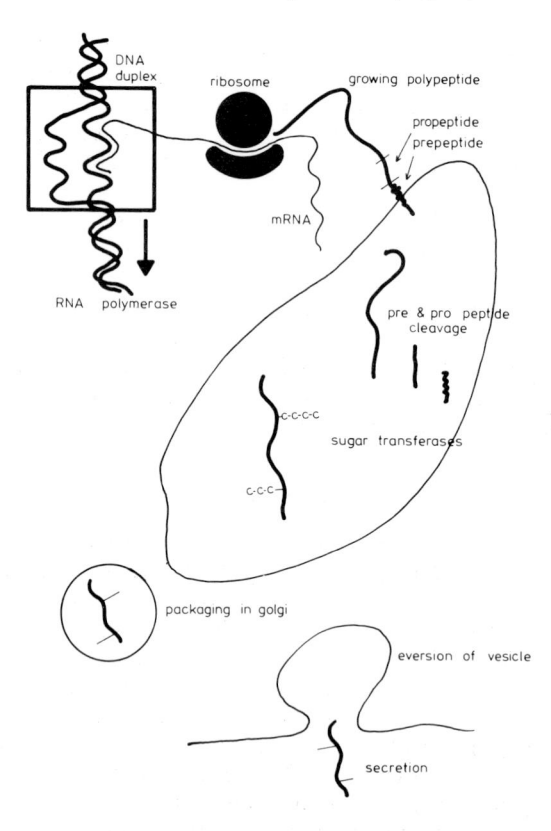

FIG. 1. Protein synthesis. The gene for a polypeptide may exist as a number of separate segments on the chromosome. M-RNA transcription skips intervening segments resulting in a single strand of message for an individual peptide chain. Most proteins probably possess two additional N-terminal peptides, the pre-peptide, responsible for insertion into, and passage across, the membrane of the endoplasmic reticulum and the pro-peptide for maturation and transport to the golgi apparatus. The pre-peptide is probably cleaved shortly after entry into the lumen of the endoplasmic reticulum while the pro-peptide may be retained until just prior to secretion from the cell. Carbohydrate side chains may be added by sugar transferases within the endoplasmic reticulum.

Factors affecting protein synthesis have been largely studied in the liver. The most important regulatory factor in hepatic plasma protein synthesis is amino acid availability. Toxic factors such as ethanol may inhibit

TABLE 1
THE EFFECT OF STEROIDS ON PLASMA PROTEIN CONCENTRATIONS

Protein	Corticoid	Androgen	Oestrogen
Pre-albumin	−	+	−
Albumin	−	N	−
α_1-lipoprotein	N	−	++
Orosomucoid	+	+	−
α_1-antitrypsin	N	+	++
Haptoglobin	+	++	−
Caeruloplasmin	N	N	+++
Transferrin	N	+	+
β-lipoprotein	++	+	N
IgG	−	N	−

These changes are due to a combination of changes in volume of distribution and changes in the rate of synthesis

+ denotes increase; −, decrease; N, no change.

synthesis and hormones have an important but complex effect. In clinical practice the effects of androgens and oestrogens are the most important (Table 1). The synthesis of many proteins by the liver is increased in protein-losing states when the plasma albumin level is decreased.

Specific synthetic regulatory mechanisms are also being increasingly recognized. It is notable that the synthesis of acute phase proteins, most of which have specific functions in the inflammatory reaction, is increased in the liver during inflammation of many tissues. It is possible that proteolytic fragments derived from lysosomal enzyme activity may increase the synthesis of proteins from which they are derived.

Distribution

Plasma proteins pass continuously from the vascular to the extra-vascular space. The passage across capillary walls is both through interendothelial junctions and by pinocytic transport across endothelial cells. In both cases proteins must pass across the basement membrane. It is probable that some of the pinocytic vesicles fuse with lysosomes resulting in enzymatic degradation and it is in this way that many plasma proteins are catabolized by the capillary endothelium. The protein content of the transcellular fluids and extracellular fluid is thus dictated by a combination of these two non-specific mechanisms as well as by specific transport mechanisms which may apply to some of the proteins crossing the endothelial cells in pinocytic vesicles. In all such fluids low molecular weight proteins predominate owing to the molecular sieving effect of the capillary basement membrane. It is important to appreciate that the protein content of such transudates will vary from tissue to tissue according to the nature of the capillary interendothelial junctions and also that the fluid, once formed, may be rapidly changed by subsequent metabolic processes. For example, the protein content of urine is very different from that of the glomerular filtrate owing to the reabsorption of proteins in the proximal tubule. Ventricular CSF is very different from lumbar CSF due to the progressive equilibration of the fluid with plasma during its passage down the spinal column. The effect of changes in posture in altering the concentration of proteins in plasma is also of practical importance.

Catabolism

Plasma protein breakdown probably occurs to a greater or lesser extent in most cells of the body, and degraded plasma proteins may provide an important source of amino acids for cellular protein synthesis. In particular, degradation occurs in the capillary endothelial cell during the process of pinocytic transfer from lumen to basement membrane. Capillary endothelial cells have few lysosomes and while their catabolic activity is limited it is worth noting that the muscles of an adult man contain some 60 000 miles of capillaries with an area of about 6000 square metres. Thus even if only a small proportion of the proteins pinocytosed by each endothelial cell are catabolized the catabolic potential is still enormous.

Proteins that have passed through the endothelium may then be catabolized by the tissue cells they come into contact with, again by pinocytosis and lysosomal degradation.

The role of the liver in protein catabolism has been extensively studied and a number of interesting ideas have emerged. The hepatic sinusoids lack a continuous basement membrane, the endothelial cells have marked intercellular fenestrations and the pericapillary cellular covering is far from complete. This has led to the proposal that the space of Disse and the hepatic sinusoid form a single mixing pool of proteins; plasma proteins thus enter and leave the hepatocyte without hindrance. This of course explains how high molecular weight proteins and particles synthesized in liver cells, such as α_2-macroglobulin (MW 725 000) and very low density lipoprotein, enter the plasma. Most plasma proteins are glycoproteins and the evidence suggests that the carbohydrate side chains may have a key role in controlling degradation. De-sialation of many glycoproteins results in increased catabolism and it is possible that an intact carbohydrate moiety protects circulating proteins from hepatic catabolism. Removal of carbohydrate residues as a result of the action of circulating or membrane-bound enzymes may act as the catabolic initiating process for binding of the protein to hepatocyte membrane receptors with subsequent pinocytosis and intracellular degradation by lysosomal enzymes. It is possible that other molecular changes such as polymerization or complex formation may also act as catabolic initiators.

The kidney has long been known to be the site of breakdown of low molecular weight proteins such as immunoglobulin light chains. Between 2 and 4 g of plasma proteins are filtered by the glomerular capillaries each day but only about 100 mg appear in the urine. Proteins pass through the glomerular filter in inverse proportion to their molecular size and are largely reabsorbed in the proximal tubule by pinocytosis and degraded by lysosomal enzymes. Pinocytosis by tubular cells is probably competitive and there may be selective mechanisms for some types of proteins.

We are thus left with the concept that plasma proteins may be taken up by various cells either by bulk pinocytosis or by receptormediated selective uptake and subsequently catabolized by lysosomal enzymes. This process probably occurs in most tissues but the capillary endothelial cell, the hepatocyte and the proximal tubule are of considerable importance.

FUNCTIONS OF PLASMA PROTEINS

Despite the enormous expansion in knowledge about plasma proteins in the last ten years, many proteins still have no known biological function. A functional classification of plasma proteins is however useful in understanding the changes which occur in disease as proteins of similar function often form interacting systems (e.g., immunoglobulins and complement). It is probable that most of the plasma proteins present at relatively high concentrations have a functional role to play in circulating blood but it is clear that many of those present in trace amounts represent cell surface or intracellular proteins which have been shed into the bloodstream.

Proteins of Immune Defence

Immunoglobulins provide the adaptive immune response of higher animals for the elimination of antigen. *Complement* provides both effector mechanisms for antibody and an additional non-adaptive defence mechanism. *C-reactive protein* may represent a primitive non-adaptive mechanism activating complement.

Proteins Associated with Inflammation

A number of plasma proteins increase in concentration during inflammation in many tissues. It is tempting to speculate that the so-called 'acute phase response' represents the switching on of synthetic mechanisms for proteins which are involved in the inflammatory process either as mediators, participants or inhibitors.

Transport Proteins

Plasma proteins are concerned in the transport of a wide range of substances from the site of production to their site either of action or of catabolism. Another important aspect of their role is the maintenance of a pool of biologically inactive substance in equilibrium with the pool of free active substance. It is being increasingly recognized that the carrier proteins may have a complex role in the metabolism of the molecules with which they are associated such as interacting with enzymes or cellular receptors.

Signal Proteins

It is clear that many plasma proteins are in fact messengers of one sort or another. Most of the classical protein hormones are present at low concentrations but some such as human placental lactogen and the placental protein Sp 1 are present at high enough concentrations to have been classed as 'plasma proteins'.

Proteins of Blood Clotting

The interacting proteins of the blood clotting system form a quantitatively important part of the plasma proteins. It is important to realize that the inhibitors and active proteolytic enzymes of this system interact closely with the complement and kallikrein-kinin systems. They will not be further considered in this review.

Tissue Derived Proteins and Oncofetal Proteins

Some of the proteins present in low concentrations in plasma are cell membrane proteins shed into the blood during cell membrane turnover or as a result of cell death. A number of other trace proteins, the oncofetal proteins, are produced by tumours as a result of depression of genes coding for fetal proteins or proteins not normally produced by the tissue of origin of the tumour.

THE PROTEINS OF IMMUNE DEFENCE

Immunoglobulins

The immune system can be conveniently divided into two functionally cooperative but developmentally independent pathways of lymphoid differentiation. T-lymphocytes derived from the thymus represent a functionally heterogenous group of cells concerned with immune regulation and antigen elimination while B-lymphocytes synthesise and secrete antibody.

The immunoglobulin molecules which mediate antibody activity are so far unique among proteins both in their remarkable structural heterogeneity and in that their synthesis is an adaptive response triggered by the antigenic configuration with which they will interact. Functionally they have the capacity to complex with specific antigens and interact with both cells and other circulatory plasma proteins to give rise to a series of biological effector functions which result in the elimination of antigen (Table 2).

Immunoglobulin Structure

All immunoglobulins consist of one or more basic units comprised of two identical heavy (H) chains and two identical light (L) chains joined together by a variable number of disulphide bonds. Each polypeptide chain falls into a series of globular regions called domains which have considerable amino acid sequence homology (Fig. 2). The N-terminal domains of both H- and L-chains contain the variable amino acid sequence (V region) which determines antigenic specificity, thus allowing each molecule to have two antigen combining sites. The remaining domains of the H chains (constant or C region) have certain structural and thus antigenic

TABLE 2
BIOLOGICAL PROPERTIES OF HUMAN IMMUNOGLOBULINS

Class	Subclass	Complement activation	Binding to macrophage Fc receptors	Transport across placenta	Transport into secretions	Fixation to mast cells and basophils
IgG	1	+	+	+	−	−
	2	±				
	3	+				
	4	−				
IgA	1 and 2	−	−	−	+	−
IgM		+	−	−	−	−
IgD		−	−	−	−	−
IgE		−	−	−	−	+

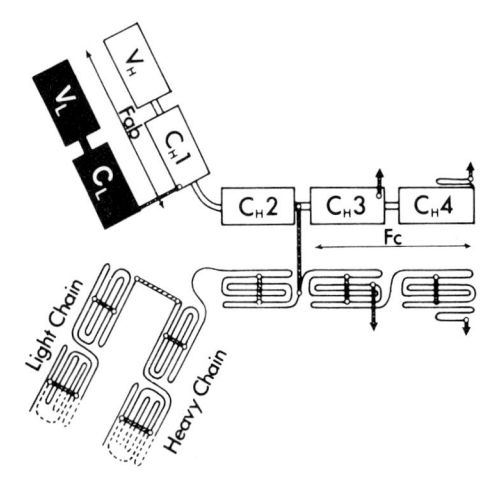

FIG. 2. Diagram of one IgM subunit. The discontinuous lines in the V-domains in the lower half of the diagram mark the hypervariable regions. The thick cross-hatched bars represent disulphide bonds, those with arrows interconnect the IgM subunits in the J chain. (Reproduced with permission from R. S. J. Pumphrey, 'Structure and function of the immunoglobulins' in *Immunochemistry in Clinical Laboratory Medicine*. Milford Ward, A. and Whicher, J. T. Eds. MTP Press, Lancaster, 1978).

differences which allow their classification into 5 classes and a number of subclasses (Table 3). They also contain the effector sites which allow the molecules to interact with cells and complement.

Immunoglobulin synthesis

Each immunoglobulin-producing cell (B-lymphocyte) has some 10^5 immunoglobulin molecules embedded in the phospholipid bi-layer of the cell membrane by a terminal hydrophobic peptide attached to the Fc region. There is strong evidence that the surface immunoglobulin acts as the receptor with which antigen interacts to trigger B-lymphocyte proliferation and maturation to the immunoglobulin-producing plasma cell (Fig. 3). H- and L-chains are synthesized on different polyribosomes present on the rough endoplasmic reticulum and molecular assembly occurs within the lumen of the endoplasmic reticulum. Plasma cells synthesize a slight excess of L-chains most of which are normally catabolized within the lysosomes though a small amount of free L-chains are secreted by the cells. Being of low molecular weight these are filtered at the glomerulus and predominantly reabsorbed in the proximal tubule. It is however normal to find some 10–20 mg/l of polyclonal free L-chains in urine.

Functions of the Immunoglobulin Classes

A classification of the immunoglobulins is summarized in Table 3.

IgG (MW 150 000)

IgG is a monomer of the basic immunglobulin unit and is present in high concentrations in plasma and in low concentrations in all extracellular fluids. It accounts for 75% of the plasma immunoglobulins and may be thought of as the antibody which protects the tissue spaces and, in particular, aggregates or coats small soluble proteins such as bacterial toxins. IgG is pumped across the placenta by an active transport process and provides protection for the fetus and for the newborn infant in the first few months of life before its own antibody synthesis is adequate.

IgA (MW 160 000)

Most plasma IgA molecules are monomeric though a small proportion form polymers and complex with other plasma proteins, especially albumin. The bulk of IgA synthesis occurs in plasma cells located beneath the mucosa of the gastrointestinal tract, the respiratory tract, the skin and in exocrine glands. IgA enters the epithelial cells of the mucosa which synthesize and attach

TABLE 3
ANTIGENIC HETEROGENEITY OF IMMUNOGLOBULINS

Immunoglobulin class	Subclass	Allotype
IgA	α_1	—
	α_2	Am$^+$, Am$^-$
IgG	γ^1	Gm 1, 2, 3, 4, 17
	γ^2	Gm 23
	γ^3	Gm 5, 10, 13, 21, 6, 11, 14
	γ^4	—
IgM	—	Mm
IgD	—	—
IgE	—	—

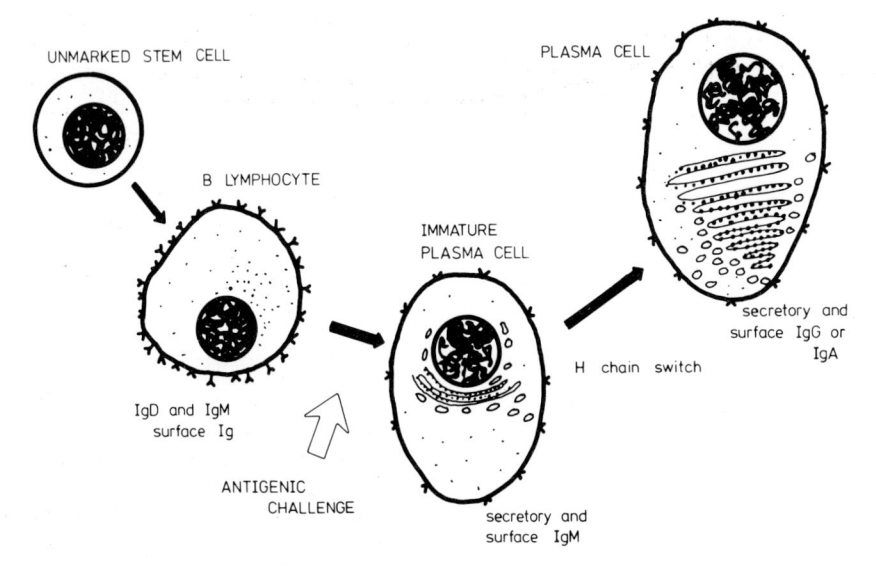

FIG. 3. Maturation of B-cell following antigenic challenge. The mature unstimulated B lymphocyte synthesizes surface IgM and IgD with variable region sequence specificity for a specific antigen for which they act as the receptor. Differentiation to form the immature plasma cells results in switching off IgD H chain synthesis with increasing production of secretory IgM with the same variable region specificity as the surface IgM and IgD. Later in maturation H chain constant region synthesis switches from IgM to either IgG or IgA with transcription of the same V-region gene. In this way the type of immunoglobulin produced by a maturing clone of B-cells changes from low affinity IgM to high affinity IgG or IgA. It is important to appreciate that the gene segments for the H chain constant region and V region sequences may be repeated on the chromosome. Similarly the V-region gene is probably made up of a number of segments which are combined at random to form the heterogeneity of V-region sequences necessary for all the antibody specificities which occur.

a secretory piece which binds to two IgA molecules forming a dimer that is secreted onto the luminal surface of the mucosa. This secretory IgA protects mucosal surfaces against bacterial invasion and is resistant to proteolysis in the gut. It is present in colostrum where it protects the newborn infant from gastrointestinal infection.

IgM (950 000)

Plasma IgM consists of a pentamer of basic immunoglobulin units and because of its very high molecular weight is largely confined to the vascular space. IgM is present with IgD as one of the surface receptor immunoglobulins of the mature unstimulated B-lymphocyte. IgM is the first immunoglobulin to be secreted after stimulation of the B-lymphocyte to differentiate to form the plasma cell. Later in differentiation, H chain C region transcription switches to either IgG or IgA but with the same variable region gene and hence the same antibody specificity. This is why IgM is the first immunoglobulin to be synthesized following antigenic challenge. IgM synthesis is however maintained in response to certain antigens, in particular organisms present in the vascular space. The blood group isohaemagglutinins belong to this immunoglobulin class.

IgE (MW 190 000)

IgE is a monomeric immunoglobulin synthesized by plasma cells throughout the lymphoid system but in particular under the mucosae of the gut and respiratory tract. Plasma IgE is rapidly bound to the cell membranes of mast cells and basophils and circulating levels are low. The combination of antigen with this cytophilic antibody causes degranulation of these cells with release of kinins, histamine and prostaglandins which cause vascular permeability and smooth muscle contraction. Such phenomena are responsible for asthma and hay fever. The beneficial effect of this interaction remains to be convincingly elucidated.

IgD (MW 175 000)

IgD together with IgM constitutes the membrane receptor of the B-lymphocyte. Occasionally cells go on to synthesize IgD but the function of the plasma IgD is unknown.

Immunoglobulin Normal Ranges

Normal ranges for immunoglobulins are not distributed in a gaussian fashion and the shape of the distribution differs for each immunoglobulin class. The levels of IgG and IgM and the rates at which they mature to adult levels from childhood are dependent on antigen challenge. In the tropics mean IgG and IgM levels are often 200% of levels in England. If Africans are domiciled in England IgM levels usually fall but IgG levels often remain elevated reflecting a genetic influence. IgA levels are unaffected by environmental factors.

The newborn child has a high level of IgG of maternal

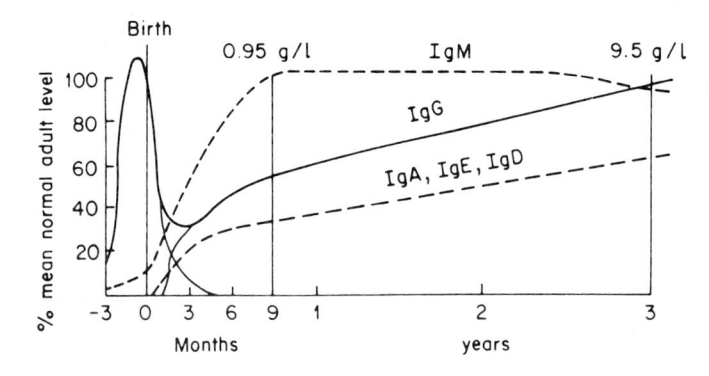

FIG. 4. The normal maturation of serum immunoglobulin levels. IgM reaches adult levels by 9 months, IgG by 3 years and IgA, IgD and IgE by puberty, Salivary IgA matures by about 6 weeks. (Reproduced with permission from Hobbs, J. R., *Developmental Immunology*, Adinolfi, A. Ed. Spastics International Press, London, 1969, p. 114 as modified by Keyser, J., *Human Plasma Proteins*, John Wiley and Sons, Chichester and New York, 1979, p. 2).

origin which declines to a very low level by about six months *post-partum*. The child's own immunoglobulin synthesis begins with IgM and is followed by IgG at about three months after birth (Fig. 4). IgM reaches adult levels at about nine months, IgG at 3 years and IgA, IgE and IgD at about 14 years. IgA in saliva reaches adult levels at about 6 weeks after birth.

Immunoglobulin Deficiency

Genetic Deficiencies. Primary genetic immunoglobulin deficiencies are rarer than secondary deficiencies. They are most often sought in children with recurrent infections or a combination of infections with allergy. They are, however, occasionally fortuitously recognized in the laboratory during electrophoresis performed for other purposes. A number of classifications have been proposed but the WHO classification[1] is to be recommended. In view of their complexity it is not proposed to discuss them in detail (Table 4).

Decreased synthesis of most or all immunoglobulin classes has been described in a number of the familial syndromes, the most well known of which are infantile X-linked agammaglobulinaemia (Bruton form) and the severe combined immunodeficiencies (SCID). SCID is a combined B- and T-lymphocyte deficiency and patients usually die of viral infections. Agammaglobulinaemia is usually associated with pyogenic bacterial infections.

Selective immunoglobulin deficiency may affect all classes but primary IgA deficiency is by far the commonest with an incidence of about 1/500 of the population. It is usually symptomless but may be associated with gastrointestinal, respiratory or renal infections. Affected people have a higher incidence of immune complex disease.

The early diagnosis of immune deficiency is important if adequate treatment is to be instituted. Simple electrophoresis is inadequate as selective immunoglobulin deficiencies may be difficult to recognize; quantitative immunoglobulin measurements are essential. In newborn infants the low levels of serum immunoglobulins make the diagnosis difficult and if panhypogammaglobulinaemia or IgA deficiency are suspected salivary IgA measurements are very useful after 6 weeks of age when this immunoglobulin matures to adult levels.

Secondary Deficiencies. Secondary deficiencies are 10–100 times more common than primary deficiencies and may occur in about 4% of hospital patients.

Transient immunoglobulin defiency in the newborn is important in premature babies. As can be seen from Fig. 4 the majority of maternal IgG crosses the placenta in the last trimester of pregnancy and the baby's own IgG does not reach acceptable levels until 6 months after birth. Those babies born before 22 weeks will have severe hypogammaglobulinaemia and those born before 34 weeks may develop it within two months. Some 4% of full-term infants show delayed maturation of IgG synthesis and may have a temporary hypogamma-

TABLE 4
THE PRIMARY IMMUNOGLOBULIN DEFICIENCIES

Deficiency	Immunglobulins affected	T cells affected	Comments

1. Severe combined immunodeficiency
   Swiss type
   Sex linked form
	all classes	+	die of severe fungal and viral infections in infancy

2. Combined immunodeficiency
   with thymoma
   with achondroplasia
   with thrombocytopenia and eczema
	all classes
	
	variable
		+
		
		deficient response
			severe viral and fungal infections
			
			?defect in afferent limb of immunity

3. Hypogammaglobulinaemia
   X linked, Bruton type
   common and variable type
	all classes	normal	pyogenic infections

4. Selective immunoglobulin deficiency (see Ref. 6)
   selective IgA and IgM deficiency (Type I)
   selective IgG and IgA deficiency (Type II)
   selective IgG deficiency (Type III)
   selective IgA deficiency (type IV)
   selective IgM deficiency (Type V)
   IgG subclass deficiency
	
	IgA and IgM
	IgG and IgA
	IgG
	IgA
	IgM
	IgG subclasses
		
		normal
		normal
		normal
		normal
		normal
		normal
			
			Giardiasis common
			respiratory tract infections
			pyogenic infections
			common—1/500 of the population
			septicaemia occurs after splenectomy
			pyogenic infections

globulinaemia lasting several months. The maturation of salivary IgA to normal levels by 6 weeks after birth and a rising serum IgM is a good indication that a baby will eventually acquire normal immunoglobulin synthesis. Careful follow-up and treatment of infection is essential during the first year of life.

IgG has the longest half-life of the plasma immunoglobulins (22 days) and thus any factors increasing immunoglobulin catabolism will affect IgG most. Protein loss results in an increased endogenous catabolism of most plasma proteins and while proteins synthesized in the liver also show a marked increase in synthesis immunoglobulins do not. The result is a progressive decrease in IgG which may be reduced to levels of 1 g/l even despite small external losses, as for example in a selective glomerular proteinuria. Conditions giving rise to catabolic hypogammaglobulinaemia are shown in Table 5. This type of hypogammaglobulinaemia rarely gives rise to serious infection because the adaptive immune response is still intact with normal production of IgM and IgG in response to antigenic challenge.

Suppressed synthesis of immunoglobulins affects IgM most, then IgA and IgG least. The commonest cause is lymphoid neoplasia which may take months or years to manifest its effect. 'Toxic' factors may give rise to a similar picture (Table 5). There is a very real risk of infection in this type of immune deficiency as B-cell proliferation is suppressed and the immune response to antigenic challenge is progressively lost.

Polyclonal Hypergammaglobulinaemia

In most situations when an immune response is occurring, whether they are infections, immune complex diseases, or auto-immune conditions, a polyclonal increase in immunoglobulins is seen. In certain conditions the immunoglobulin classes respond in a selective manner and may provide some clinically useful information. In general, however, immunoglobulin measurements provide little more information than is available on examination of the electrophoretic strip. For a detailed discussion see Hobbs.[6] An example of the patterns which are seen is shown in Table 6. In general terms auto-immune and immune-complex diseases give rise to a selective increase in IgG; skin, gut, respiratory and renal disease, particularly if associated with infection, stimulate the mucosal IgA system with increased serum IgA; organisms invading the bloodstream give rise to an increase in IgM.

TABLE 5
SECONDARY IMMUNOGLOBULIN DEFICIENCIES

1. *Disorders of immunoglobulin synthesis* (IgM ↓↓, IgA ↓, IgG N or ↓)
    (a) Lymphoid neoplasia—especially
        myelomatosis
        chronic lymphatic leukaemia
        lymphosarcoma
    (b) Toxic factors—
        uraemia
        corticosteroid therapy
        cytotoxic therapy
        coeliac disease
        diabetes mellitus
    (c) Delayed maturation of immunoglobulin production in the neonate (all immunoglobulins affected)
2. *Excessive immunoglobulin catabolism* (IgG ↓↓, IgA ↓, IgM N or ↓)
    (a) Exogenous loss
        nephrotic syndrome
        protein-losing enteropathy
        burns
    (b) Endogenous hypercatabolism
        most causes of exogenous protein loss
        malnutrition
        dystrophia myotonica (IgG only affected)
3. *Decreased IgG transfer to the fetus*
    prematurity

TABLE 6
SOME DISORDERS ASSOCIATED WITH POLYCLONAL HYPERGAMMAGLOBULINAEMIA

1. *Predominant increase in IgM*
    congenital toxoplasma, rubella, syphilis
    primary biliary cirrhosis   marked ↑ in IgM
    Q fever endocarditis
    brucellosis
    malaria, trypanosomiasis, filariasis
    most primary viral infections
2. *Predominant increase in IgG*
    systemic lupus erythematosus   IgM often also ↑
    hypergammaglobulinaemic purpura
    chronic aggressive hepatitis   IgM often also ↑
    kala-azar
    leprosy   IgM often also ↑
3. *Predominant increase in IgA*
    liver cirrhosis   maybe associated with ↑ IgG
    Crohn's disease
    ulcerative colitis
    fibrocystic disease
    bronchiectasis
    pulmonary tuberculosis
    sarcoidosis
    dermatomyositis
    rheumatoid arthritis
    pyelonephritis

Immunoglobulin measurements are useful in diagnosing infection *in utero* and in the first six months of life. The rise in normal IgM after birth is due to the first antigenic challenges met by the infant. Intrauterine infection may give rise to IgM production before birth with high levels of cord blood IgM. This is particularly useful in aiding the diagnosis of congenital toxoplasmosis, rubella, cytomegalovirus, herpes simplex and syphilis, the so-called 'TORCH' agents.

One of the areas in which the differential pattern of immunoglobulin response may be useful is in liver disease. A dominant and often very high level of IgM is typical of primary biliary cirrhosis. A mild IgM response is also seen early on in acute infective hepatitis where it represents the response to primary antigenic challenge. It is also seen in infectious mononucleosis with or without liver involvement as this is usually a primary infection. IgG is raised in chronic aggressive hepatitis while IgA is typically high in the macronodular Laennec type of cirrhosis, regardless of the cause.

IgE measurement has a role in the management of asthma and allergy, particularly in children where high IgE levels are associated with extrinsic asthma and a good response to certain therapeutic regimes. The measurement of antigen specific IgE by radioallergo-absorption techniques (RAST) gives similar information to skin tests in the identification of substances to which a patient is allergic.

Oligoclonal Hypergammaglobulinaemia

Not infrequently a raised gammaglobulin may be of restricted heterogeneity due to the response of relatively few B cell clones to antigenic challenge. Such oligoclonal gammopathies show faint bands or zones in the gamma region on electrophoresis (see Fig. 17). These are often multiple and not clear-cut and are usually distinguishable from true paraproteins (the product of a single clone of B-cells) by the presence of both types of light chain on immunoelectrophoresis or immunofixation. Such restricted heterogeneity of immune response is a characteristic of chronic immune stimulation where there is progressive switching-off of B-cell clones leaving only those with the highest affinity receptors for the antigen still producing immunoglobulin. It is seen in chronic active hepatitis, chronic infections, immune complex disease, and in viral infections in young children, particularly in the presence of partial immune deficiency where fewer B-cell clones are available to respond.[2]

Monoclonal Hypergammaglobulinaemia

Immune stimulation usually results in a more or less heterogenous population of B-lymphocytes undergoing multiplication and differentiation to form plasma cells producing immunoglobulin directed against a wide range of antigenic configurations. This is because almost all macromolecular antigens have multiple antigenic sites. Under special conditions a restricted number of clones or a single clone may respond provoking an oligoclonal gammopathy, or occasionally monoclonal immunoglobulin production. Malignant proliferation of a single clone of immunoglobulin-producing B-cells may also result in the production of a homogenous population of immunoglobulin molecules with reference to class, subclass, light chain type and charge. Such 'paraproteins' form discrete bands on electrophoresis.

Aetiology and Incidence. The incidence of paraproteins depends both on age and on the population under consideration. In a healthy population under the age of 60 years the incidence is about 0·2% and most of these are apparently benign and presumably reflect restricted clonal response to antigenic challenge. In the age group over 90 years the incidence becomes something of the order of 17–19% with a higher incidence of malignant B-cell tumours. Hospital populations show a majority of cases of malignant origin.[3]

The causes of malignant paraproteinaemia are listed in Table 7.

The malignant proliferation of B-lymphocytes may be thought of as representing populations of cells which have undergone neoplastic transformation at various stages during the maturation process (Fig. 5). The B-cell lymphomas and chronic lymphatic leukaemias are typified by small lymphocytes with little capacity for immunoglobulin synthesis except for small quantities of IgM. Some 15% of such tumours produce detectable amounts of immunoglobulin. Wäldenstrom's macroglobulinaemia is a lymphoma comprising intermediate B-lymphocytes with the capacity to synthesise considerable quantities of IgM, producing high serum levels which often give rise to the hyperviscosity syndrome. Skeletal manifestations are rare and like the lymphomas and chronic lymphatic leukaemia there is anaemia and lymphadenopathy. Myeloma is a malignant proliferation of more or less mature plasma cells which most often secrete IgG and less commonly IgA and rarely IgD, IgM or IgE. IgA and IgG_3 may polymerize to cause hypercosity but the disease usually presents with bone involvement, pain or fracture, infection due to immune paresis and renal failure.

Benign paraproteins usually occur in the absence of any obvious predisposing factor but may be associated

TABLE 7
MALIGNANT B-CELL TUMOURS RESULTING IN PARAPROTEINS

Disease	Paraprotein type	Incidence of paraprotein	Incidence of Bence-Jones
Myelomatosis	IgG	50%	60%
	IgA	25%	70%
	IgM	0·5%	100%
	IgD	1·5%	100%
	IgE	0·1%	most
	Bence Jones protein only	20%	all
Soft tissue plasmacytoma	similar distribution to myelomatosis		
Waldenstrom's macroglobulinaemia	IgM	all	80%
Non-Hodgkin's lymphoma	IgM occasionally IgG or IgA	20%	20%
Chronic lymphatic leukaemia	IgM occasionally IgG or IgA	15%	15%

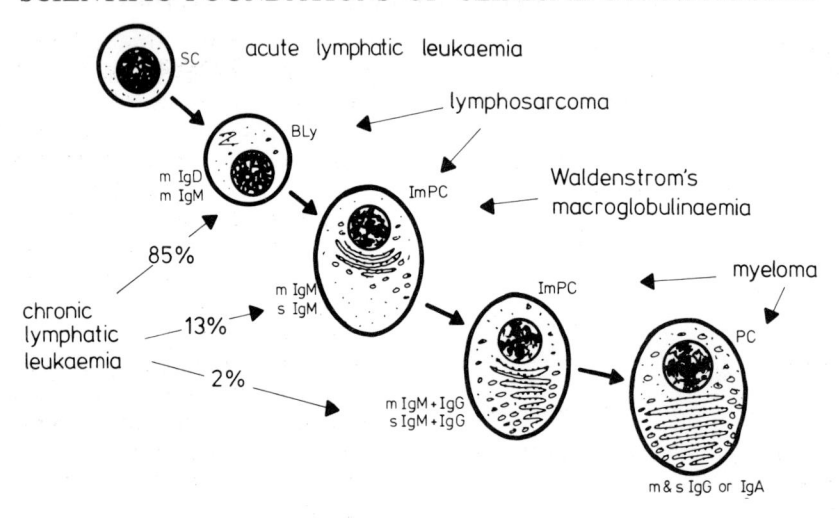

FIG. 5. B-cell tumours classified according to their immunoglobulin production and cell type in relation to the normal maturation pathway of B-lymphocytes; SC-lymphocytes stem cell; BLy-mature B lymphocyte; ImPC-immature plasma cell; PC-plasma cell; M-membrane bound immunoglobulin; S-secretory immunoglobulin.

with the conditions which give rise to the oligoclonal gammopathies.

Laboratory Investigations. The investigation of paraproteinaemia is one of the most important protein studies carried out in laboratories.[4] The aim of the investigation is to establish the monoclonal nature and type of the paraprotein, provide biochemical evidence of the likelihood of a malignant origin, and perform precise estimation of paraprotein for monitoring therapy.

1. Establishment of monoclonality. The existence of a paraprotein must first be established by simple electrophoresis. If it is not detectable by these means then there is little likelihood of detection by immunoelectrophoresis unless it is hidden beneath another fraction on the electrophoretic separation. Immunofixation however increases the sensitivity of electrophoretic detection and may reveal paraproteins not discernible by simple electrophoresis; this is particularly true of Bence-Jones protein in serum or urine.[5] Immunofixation or immunoelectrophoresis against monospecific antisera to H- and L-chains will enable identification and confirmation of monoclonality by the presence of a single class of H-chain and one type of L-chain. Immunofixation (Fig. 6) is by far the most reliable technique and will rapidly replace immunoelectrophoresis for these purposes.

The H-chain type of paraprotein is clinically very relevant as its nature may allow a search for the

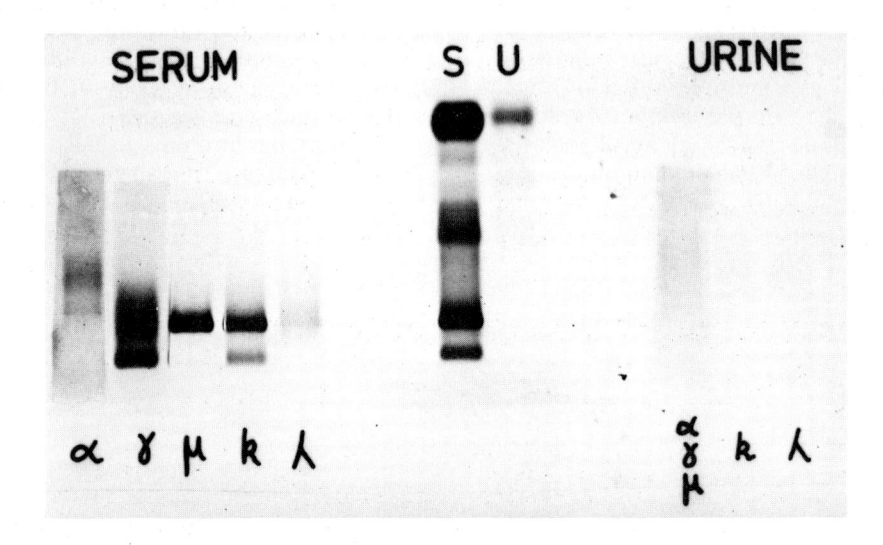

FIG. 6. Immunofixation of a biclonal paraproteinaemia. S and U represent simple agarose electrophoresis of serum and urine: two paraproteins are present in the serum. Immunofixation of the serum with antisera to immunoglobulin H chains (α, γ and μ) and k and λ light chains reveal an IgM k, and IgG k paraprotein. This probably represents a 'switch cell clone' transcribing both IgM and IgG H chains with the same V-region and L chain segments. There is no Bence-Jones protein in the urine.

appropriate type of B-cell tumour in a previously undiagnosed case. For example, an IgM paraprotein is most unlikely to be due to myeloma so that skeletal X-rays are unimportant but examining for enlarged lymph nodes is essential. In established myeloma the complications of the disease are very different with different immunoglobulin classes (Table 8).

TABLE 8
COMPLICATIONS ASSOCIATED WITH DIFFERENT IMMUNOGLOBULIN CLASSES IN MYELOMA

1. *IgG Myeloma*
   (a) Mean ½ life of paraprotein 13 days, therefore paraprotein reaches high levels
   (b) If paraprotein level exceeds 80 g/l polymerization and hyperviscosity is likely, especially if IgG_3
   (c) Immune paresis common
   (d) Amyloid less common than other types
   (e) Cryoglobulinaemia in about 2% of cases
2. *IgA Myeloma*
   (a) Mean ½ life of paraprotein 5 days; lower levels reflect the same tumour mass as higher levels of IgG
   (b) Hypercalcaemia is common
   (c) Amyloid occurs in 5–10% of cases
   (d) Renal failure common even in the absence of Bence-Jones protein
3. *Bence-Jones protein Myeloma*
   (a) Probably faster growing tumour
   (b) Amyloid most common
   (c) Renal failure and hypercalcaemia commoner than other types

2. The presence of immunoglobulin fragments is one of the most important criteria of malignancy (Table 9).[6] Their production by a B-cell clone suggests de-differentiation and failure of normal immunoglobulin synthesis. Such fragments include free immunoglobulin L-chains (Bence-Jones protein), IgM monomer, half-molecules and molecules with deletions.

Bence-Jones protein (BJP), monoclonal free light chain, is by far the most important of the fragments and its presence in urine is a strong but not invariable indication of a malignant B-cell tumour. However, it is important to remember that all plasma cells produce a slight excess of light chains which are released from the cells and pass through the renal glomerulus to be largely reabsorbed in the proximal tubule. Increased amounts of polyclonal light chain may be detectable in the urine in conditions associated with polyclonal hypergammaglobulinaemia. It is thus essential to establish the monoclonality of the light chains by immunofixation of adequately concentrated (usually 300×) urine.[5] It is essential to look for BJP not only as an indication of malignancy but also in the investigation of a suspected B-cell tumour in the absence of a serum paraprotein, as some 20% of myelomas have only BJP production from the tumour. Such tumours may be more de-differentiated and aggressive.

Bence-Jones protein is usually found in the plasma only in the presence of glomerular failure. It may damage the renal tubule causing obstruction and 'myeloma kidney' and it may be responsible for amyloid formation.

3. Paraprotein measurement is important for a number of reasons. The absolute level provides some indication of the probability of malignancy, while a rising level is a strong indicator of a proliferating malignant clone. The paraprotein is of course a tumour marker *par excellence* and the level is extremely useful for monitoring therapy of all types of B-cell tumour. Immunochemical quantitation is unsatisfactory as it provides very inaccurate results which vary enormously between batches and between manufacturers of antisera. The most satisfactory method is to quantitate on the basis of densitometric scanning of an electrophoretic strip.

4. Measurement of the immunoglobulins other than the paraprotein is also useful in assessing the probability of malignancy (Table 9) and in anticipating the complication of immune deficiency in known B-cell tumours.

The Heavy-chain Diseases. Some monoclonal B-cell tumours may produce only immunoglobulin H-chains. The most important of these, α-heavy chain disease, is associated with gut lymphomas in people predominantly of Mediterranean and middle eastern origin. The diagnosis depends upon the demonstration of free H-chains by immunoselection techniques.[7]

Cryoglobulins. These may be simple monoclonal immunoglobulins which precipitate in the cold. Many paraproteins have this characteristic when cooled to 4°C but

TABLE 9
CRITERIA OF VALUE IN ASSESSING PROGNOSIS OF PARAPROTEINS

Criterion	Benign	Malignant
1. Immunoglobulin fragments		
Bence-Jones protein	trace amounts detectable occasionally	often present
Monomeric IgM	rarely present	often present
Heavy chain fragments	may occur due to degradation	present in certain conditions
2. Suppression of normal immunoglobulins	rare—unless paraprotein associated with immune deficiency	common and progressive—especially IgM
3. Paraprotein level	IgG usually not > 20 g/l IgA usually not > 10 g/l	high concentrations common
4. Progressive increase in paraprotein concentration	transient increase may occur	almost invariable

are likely to be clinically significant only if they precipitate above 22°C (Figs. 7, 8). Their presence is hence an important reason for failing to detect a paraprotein because it has precipitated in transit to the laboratory and has been spun down with the cells when the sample has been separated!

Cryoglobulins may also comprise immunoglobulin-anti-immunoglobulin complexes. These are most often monoclonal IgM with anti-IgG activity complexed to IgG (monoclonal rheumatoid factor) but may be polyclonal (polyclonal rheumatoid factor). In all types of cryoglobulin the aggregated immunoglobulin may activate complement and give rise to vasculitis.

If the diagnosis is suspected the blood must be collected and separated at 37°C. If a precipitate then appears on cooling it may be washed in saline at 4°C and subjected to immunofixation to establish whether it is monoclonal or polyclonal. Monoclonal rheumatoid factors may be dissociated from their IgG antigens by treatment with mercaptoethanol or dithiothreitol prior to electrophoresis. The immunoglobulin present in the cryoprecipitate may be quantitated.

The Complement System

Physiology. The complement system comprises a group of proteins which, following activation, interact with each other in a sequential fashion to produce biological effector molecules which facilitate the elimination of antigens by lysis and phagocytosis. The sequence of events may be summarized as follows (Fig. 9).

1. Tissue is invaded by antigen which is bound by specific antibody with activation of C1 by the Fc region of the immunoglobulin molecule. Complement may also be activated directly via the alternative pathway by bacterial lipopolysaccharide. Complement effector molecules are produced (the anaphylatoxins C3a and C5a) which cause histamine release from mast cells and result in smooth muscle contraction and vascular permeability. Oedema and stasis result with passage of further antibody and complement into the infected extravascular space.

2. The spread of infection is limited by thrombosis of surrounding blood vessels due to complement-induced platelet aggregation and intravascular coagulation.

3. Complement-derived chemotaxins (C3a, C5a and C567) result in migration of phagocytes into the area.

4. Phagocytes and killer cells adhere to the antigen by receptors for immunoglobulin Fc and complement (C3b, C3d and C4b) with phagocytosis and proteolytic destruction of the foreign material.

5. Cellular antigens may be lysed by complement (C5–9 complex).

The proteins of the system share a number of important properties. Proteolytic cleavage of some components by the ones preceding them in the pathway results in sequential activation producing low molecular weight activation or conversion products which possess biological activities. Such cleavage may also result in the formation of binding sites for other components, allowing assembly of component complexes, and binding sites for cell membranes, resulting in transfer from the fluid phase to the cell surface. Control is exerted by the short half life of active sites and by the presence of inhibitors at various points in the pathway.

A simple view of the classical and alternative pathways is presented in Fig. 10. For more detailed description see references. It is probable that the alternative pathway is continuously cycling at a low level and that input

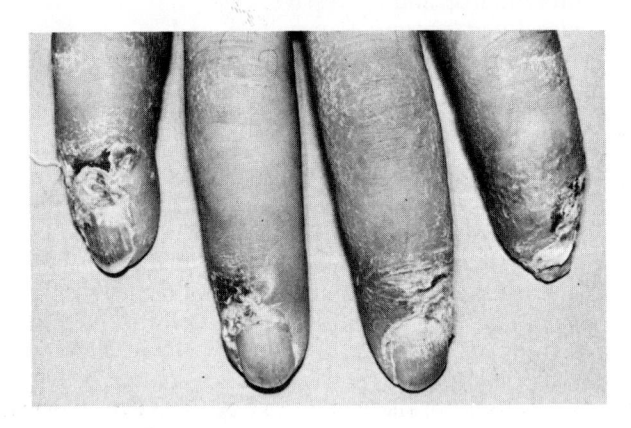

FIG. 7. Digital gangrene in a patient with cryoglobulinaemia. (Kindly supplied by Dr P. Copeman.)

FIG. 8. Serum from a patient with cryoglobulinaemia as seen after standing at 4°C. (Kindly supplied by Professor J. R. Hobbs.)

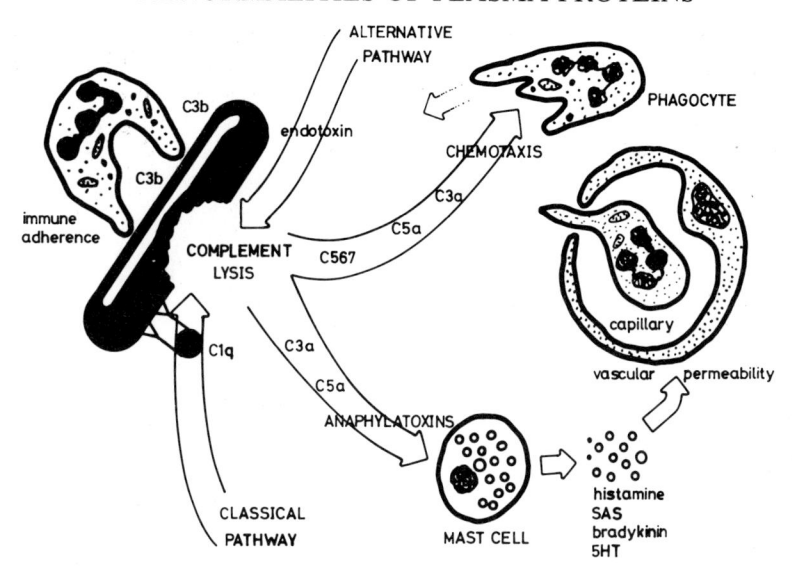

FIG. 9. The role of complement in the defence against micro-organisms. Only the most important functions are shown.

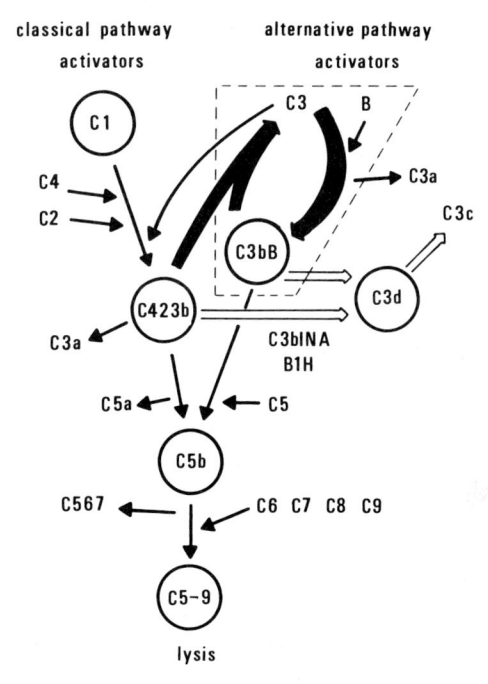

FIG. 10. Complement may be activated by either the classical or alternative pathways. Classical pathway activation usually results from antibody-antigen complexes activating C1. This initiates a proteolytic cascade generating C5b onto which C6–C9 assemble to form the lytic complex. The alternative pathway is a cycle which is activated directly by the protection of C3bB from the inhibitors C3INA and β1H (signified by a broken line) allowing the positive feedback loop to accelerate. The cycle is maintained by the low level spontaneous generation of C3bB from C3 and B in the presence of Mg^{++} and Factor D. C423b will also activate the cycle thus amplifying classical pathway activation. Both pathways produce the lytic complex, the opsonins C3b and C3d, the chematoxins C3a, C5a and C567, and the anaphylatoxins C3a and C5a.

either results from the generation of C423b by classical pathway activation or by protection of spontaneously generated C3bB from inactivation by C3b inactivator (C3b 1NA) and β_1H; alternative pathway activation. In both cases the feedback loop accelerates and C5 activation occurs. This pathway thus acts as a positive feedback loop.

The Value of Complement Measurements in Disease. The complement pathway is thus an important part of the antibody-mediated immune defence mechanism. It is also able to interact directly, via the alternative pathway, with certain antigens resulting in effector mechanisms causing phagocytosis and antigen elimination. This may in fact be the more important aspect of complement function resulting in the immediate non-adaptive elimination of antigens.

Genetic deficiencies or defects in complement proteins may result in decreased resistance to infection and possibly to persistence in the host of viruses which may be responsible for chronic immune complex disease. Deficiencies of inhibitors in the complement pathway may result in spontaneous activation of complement with the production of effector molecules causing disease. In these conditions laboratory investigations are aimed at measuring the concentration or function of the affected proteins. Complement may also be the mediator of tissue damage in diseases associated with the deposition of antibody-antigen complexes within the vascular system. Complement is activated by the Fc zone of antibody molecules within complexes which have adhered to vascular endothelium. Effector molecules are produced which cause vascular permeability and chemotaxis of phagocytes which damage endothelial cells by releasing lysosomal enzymes. The involvement of complement in the disease process results in consumption of some components whose concentration in the plasma falls with the production of detectable breakdown products. Laboratory investigation may use both of these characteristics to indicate the presence of active immune-complex deposition in the vascular system and to monitor therapy for such immune-complex disease.

Inherited deficiencies of complement components. Genetic deficiencies of most of the complement components have now been described and fall into a number of clinically defined groups.

(a) Deficiencies of the early part of the classical pathway, C1, C4 and C2, are associated with immune-complex diseases, especially systemic lupus erythematosis (SLE)-like syndromes. It would seem that the early classical pathway while being unimportant in the defence against bacteria may be important in the case of viruses which may be implicated in such immune-complex diseases as SLE. C2 deficiency is by far the commonest and immunochemical assays for this protein are difficult owing to its low concentration in plasma.

(b) Deficiencies of the alternative pathway have been identified by functional assays and are associated with severe bacterial infections.

(c) Deficiencies of the C5–C9 sequence have shown a striking association with Neisserial infections suggesting that lysis is important in the elimination of these organisms. Some of these defects are functional in nature with normal immunochemically measured component concentrations.

(e) Inherited deficiency of the inhibitor of activated C1, C1-esterase inhibitor (C1 INA), or hereditary angioneurotic oedema, is the commonest clinically important inherited deficiency within the complement system. Patients suffer from recurrent attacks of peripheral, bronchial and gastrointestinal oedema (Fig. 11). Spontaneous activation of C1 occurs with consumption of C4 and C2 but not of C3 or later components. It is probable that C1 activation is caused by proteolytic digestion of C1 by plasmin for which C1 INA is also an important inhibitor. A peptide that induces vascular permeability, released from C2, causes the oedema. Eighty per cent of patients have low levels (10–20% of normal) of C1 INA and are probably heterozygotes for a dominant gene. Twenty per cent have normal immunochemically measured levels of C1 INA but with a non-functional protein. A functional inhibitor assay is thus essential in investigating this disease. The diagnosis is particularly important as there is a high mortality untreated and treatment with the anabolic steroid danazol is very effective.

Complement as an indicator system in immune-complex disease. Immune complexes in plasma may be measured directly by a number of methods varying enormously in their sensitivity and specificity. Evidence of complement activation may be used to infer the active deposition of immune complexes.

During the process of complement activation by immune complexes several complement components are consumed, lowering their plasma levels, and activation products are produced which may be detected in plasma. Either or both of these facets of complement activation may be used to infer immune-complex deposition but a number of factors must be taken into consideration.

(1) Many complement components are acute-phase reactive proteins, their plasma levels rising in inflammatory conditions and perhaps masking the decrease caused by consumption.

(2) Proteolytic enzymes from damaged tissue may give rise to complement breakdown products similar to those of immune activation.

(3) Immune complexes may be produced in showers of short duration, the clinical features of vascular damage appearing after complement has returned to normal.

(4) The low-molecular-weight complement-activation products may result in overestimation of some components by radial immunodiffusion and Laurell rocket immunoelectrophoresis.

(5) Activation of complement occurs easily *in vitro* and in the presence of Ca^{2+} and Mg^{2+}. This can be largely avoided by simple precautions including the use of EDTA in samples.

With these provisos, complement measurement is useful in investigating symptoms which could be due to immune-complex deposition. It is also of value in the

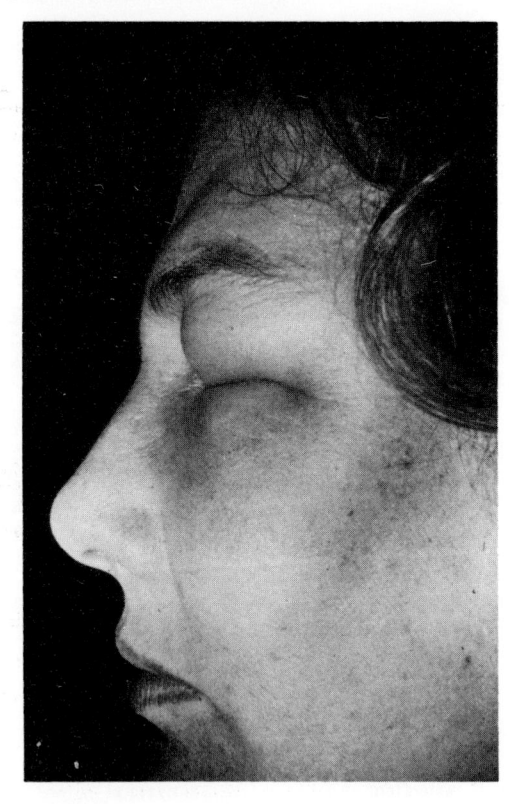

FIG. 11. The typical features of facial oedema due to C1 esterase inhibitor deficiency. (Kindly supplied by Dr R. P. Warin).

TABLE 10
DISEASES IN WHICH COMPLEMENT ASSAYS ARE OF VALUE IN DIAGNOSIS OR TREATMENT IN EUROPE*

	Disease	Investigations
Inborn errors of complement metabolism	C1 esterase inhibitor deficiency	Functional and immunochemical assay of C1 inhibitor
	Monocomponent deficiencies	Total haemolytic complement functional and immunochemical assays of components
Systemic immune complex disease	Systemic lupus erythematosus	Decreased C4 level with C3 conversion products present are the most reliable tests
	Rheumatoid vasculitis	Decreased C3 with C3 conversion
	Polymyalgia rheumatica	C3 conversion products
	Mixed cryoglobulinaemia	C3 conversion products
	Subacute bacterial endocarditis	Decreased total C3 and C4 with conversion products
	'Shunt' nephritis	Decreased total C3 and C4 with conversion products
Glomerulonephritis	Post-streptococcal	C3 decreased with conversion products. Returns to normal in two months
	Mesangio capillary	Persistent low C3 with conversion products and normal C4. Factor B conversion products may be found
Shock syndromes	Gram-negative bacteraemia	Decreased total C3. C3 conversion products present. Normal C4 level Decreased Factor B with conversion products

* Complement assays may be very useful in a number of tropical diseases with immune complex deposition or alternative pathway activation. (Reproduced with permission from J. T. Whicher, 'The Complement System' in *Immunochemistry in Clinical Laboratory Medicine*. A. Milford Ward and J. T. Whicher, Eds. MTP Press, Lancaster. 1978.)

diagnosis and management of a few diseases where the changes are clear-cut (Table 10).

Bacterial shock. Complement activation occurs during bacteraemia especially with gram-negative organisms which activate the alternative pathway. Complement effector molecules may be implicated in the shock syndrome and the detection of breakdown fragments has been used in the rapid diagnosis of this condition.[8]

Laboratory Measurement of Complement Components. Immunochemical assays for many components are now widely available and useful, but some techniques such as radial immunodiffusion suffer severe inaccuracy in the presence of breakdown products. Overestimation, especially of C3, may thus occur in the situation where low levels are being sought as an indication of immune-complex disease.

Function assays for the enzymatic activity of individual complement components are essential when investigating genetic defects in the complement pathway as some of them are of a functional nature with normal component concentrations. All components may be assayed by using the end-product of complement activation, lysis, as the indicator system.

The activation or breakdown products of C3 and factor B are separable from the native molecules on the basis of their different sizes and charges and may thus be detected by immunoelectrophoresis, crossed immunoelectrophoresis or immunofixation.

C-Reactive Protein

C-reactive protein is a complex molecule of molecular weight about 140 000 composed of six identical subunits. It is normally present in plasma with a concentration of about 6 mg/l but rises to high levels in inflammation. It is in this capacity as an acute phase reactive protein (see below) that it is widely measured in clinical practice. Little is known of its function but it is known to bind strongly to the C-polysaccharide of the pneumococcus and other glycopeptides containing N-acetylgalactosamine 6-phosphate as a terminal sugar residue. After complexing with the polysaccharide it is able to activate complement via C1 with the production of effector molecules which could encourage phagocytosis. This has led to the suggestion that its function is that of a non-adaptive defence protein.

Lysozyme

Lysozyme is a low molecular weight protein (MW 15 000) found in lysosomes, plasma and most extracellular fluids, especially secretions. It is normally catabolized by reabsorbtion in the proximal tubules of the kidney and appears in the urine in tubular damage. It is bacteriolytic and probably functions as a bactericidal agent in secretions. Production may be greatly increased in monocytic leukaemia when the concentration may exceed the tubular reabsorbtive capacity and it will appear in the urine as a band in the slow gamma region on electrophoresis. Serum levels like those of other lysosomal enzymes may reflect phagocytic activity in areas of inflammation.

PROTEINS ASSOCIATED WITH INFLAMMATION

The Acute-phase Reactant Proteins

The acute-phase reactants are a group of proteins whose concentration increases in plasma as a result of tissue damage and inflammation (Table 11). They are mainly glycoproteins, synthesized in the liver, and the evidence suggests that they have roles to play as mediators, participants or inhibitors in the inflammatory process (Fig. 12). It is probable that substances released from damaged tissue such as lysozomal enzymes or prostaglandins pass in the circulation to the liver and switch on synthetic mechanisms for these proteins. A

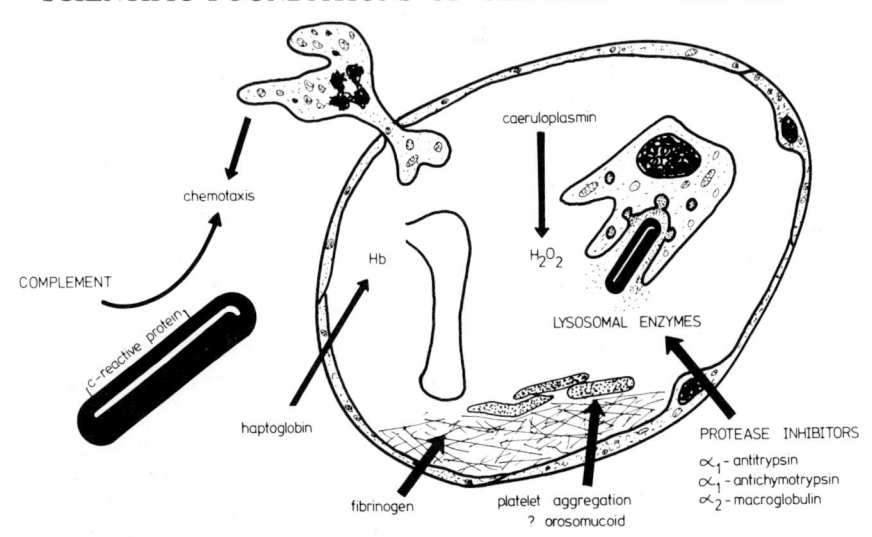

FIG. 12. The roles played by some of the acute proteins in inflammation. Complement activated directly by the alternative pathway, lysosomal proteins, interaction with specific antibody, or C-reactive protein results in the production of chemotaxins and opsonins which aid phagocytosis and cause an accumulation of mononuclear phagocytes and polymorphonuclear neutrophils in the inflamed area. Complement anaphylatoxins interact with mast cells to produce permeability-inducing factors which facilitate the passage of proteins and cells across the capillary wall. Lysosomal enzymes released from tumour cells and phagocytes damage capillary endothelium resulting in platelet adhesion and coagulation. This is kept in check by the circulating protease inhibitors. Caeruloplasmin may destroy peroxide free radicals produced during phagocytosis. Haptoglobin binds free haemoglobin produced by rupture of erythrocytes in fibrin obstructed capillaries.

TABLE 11
ACUTE PHASE PROTEINS

Protein	Possible role in inflammation	Response time
C-reactive protein	opsonin	6–10 hours
C3		
C4	antigen elimination	2–4 days
Factor B		
C1 INA		
α_1-antichymotrypsin	protease inhibition	10 hours
α_1-antitrypsin		24 hours
Orosomucoid	?	24 hours
Haptoglobin	? protease inhibition	24 hours
Caeruloplasmin	? free radical elimination	2–4 days
Fibrinogen	coagulation	24 hours
Serum amyloid A protein	?	6–10 hours

macrophage-derived protein, interleukin I, may also be an important messenger.

Clinically their measurement is of considerable value in the detection, prognosis and therapeutic monitoring of patients with tissue damage. Elevation will occur in almost any disease which involves tissue damage, in particular infections, trauma, burns, immune-complex and auto-immune diseases, and malignancy. Most commonly the concentration of all the acute-phase proteins changes roughly in parallel but in some conditions selective increases in certain proteins are seen. Acute-phase protein changes are often more sensitive indicators of disease activity than ESR, viscosity or leucocyte count. For example, C-reactive protein levels correlate better with joint damage than other parameters in rheumatoid arthritis.[9] Caeruloplasmin reflects disease activity in Hodgkin's disease where other indices may be normal. C1 INA levels are a sensitive indicator of response to therapy in some lymphomas (Fig. 13).

The Protease Inhibitors

Plasma contains a number of protease inhibitors (Table 12).[10] They have varying degrees of specificity and their function is to inhibit circulating proteases such as those involved in the complement and clotting systems or lysozomal proteases which may be released during phagocytosis or cell damage and death. Most but not all of these proteins are acute-phase reactants. It is also clear that a number of interactions between cells such as macrophages depend upon tissue fluid or membrane-bound proteases. This has led to the suggestion that the acute-phase reactive protease inhibitors may have an important role in immune regulation.

α_1-antitrypsin (MW 54 000)

It seems clear that while antithrombin III and C1 INA are mainly concerned with specific inhibition of circulating proteases involved in the dynamic systems of the coagulation, complement and kinin pathways, α_1-antitrypsin, α_1-antichymotrypsin and α_2-macroglobulin inhibit lysozomal proteases. They perform an important role in neutralizing these enzymes which are released in particular during phagocytosis of particulate material by polymorphonuclear leucocytes. In this way tissue damage in areas of inflammation is limited. The consequences of

TABLE 12
PROTEASE INHIBITORS OF HUMAN PLASMA

	Mean concentration (g/l)	Molar concentration (mol/l)	Enzymes inhibited in vivo
α_1-antitrypsin	2·0	36	elastase, collagenase etc
α_1-antichymotrypsin	0·4	7	?
α_1-antithrombin (Antithrombin III)	0·3	4·5	thrombin
Inter α-trypsin inhibitor	0·4	3	? mucosal proteases, acrosine
α_2-macroglobulin	2·0	2·7	many proteases
C1-esterase inhibitor (α-neurominoglycoprotein)	0·25	2·5	activated C1, activated Hageman factor

Adapted from A. M. Ward, 'α_1-antitrypsin' in *Immunochemistry in Clinical Laboratory Medicine*, (A. Milford Ward and J. T. Whicher, Eds. MTP Press, Lancaster, 1978).

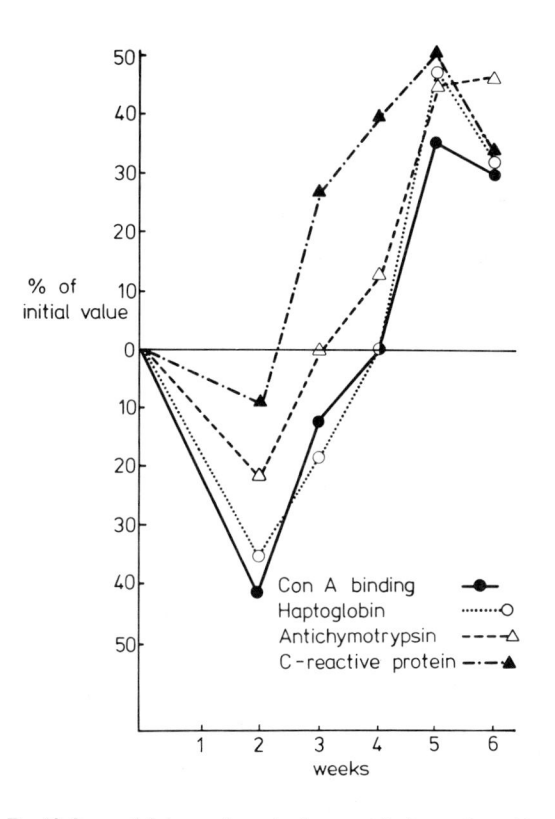

FIG. 13. Sequential changes in acute phase proteins in a patient with rheumatoid arthritis. Con A binding is a method of measuring total acute phase proteins. (Reproduced from Warren, C., Whicher, J. T. and Kohn, J. (1980) *J. Immunol. Methods*, **32**, 141–150.)

failure of this inhibition are seen in α_1-antitrypsin deficiency (see below).

α_1-antitrypsin is a protein with relatively low molecular weight which is able to pass into all body fluids. It is most active against serine proteases such as trypsin and elastase but is known to bind and inactivate a number of lysozomal proteases such as collagenase (metal protease) cathepsin B1 (thiol protease) and cathepsin D (carboxyl protease).

A small peptide is cleaved from the molecule upon binding to the protease. It seems probable that α_1-antitrypsin scavenges proteases in the tissue fluids and returns to the plasma where the protease is handed on to α_2-macroglobulin which traps proteases by enfolding them. The α_2-macroglobulin-protease complex has a very much shorter half-life than the α_1-antitrypsin-protease complex, being rapidly catabolized in the reticulo-endothelial system.

Genetic Variation. α_1-antitrypsin shows marked genetic polymorphism, some 30 distinct allotypes having now been described in this protease inhibitor (Pi) system. The genotypes are designated by the prefix Pi and letters such as B, C, D, E, E_2, F, G, I, L, M, M_2, M_3, M^{Malton}, N, P, S, V, W, W_z, C, Y, Y_2, Z and null. However it is important to appreciate that this is simply an electrophoretic classification and several structural variants may give rise to the same 'electrophoretic allotype'. Thus other evidence such as disease association and population distribution is required to indicate genetic homogeneity. The various phenotypes may be identified by separating the different α_1-antitrypsins by electrophoretic techniques of which isoelectric focusing is the most widely used. The protein shows considerable microheterogeneity on isoelectric focusing which probably reflects post-transcriptional differences in the carbohydrate moiety of the molecule. Differences between allotypes may reflect the carbohydrate microheterogeneity superimposed upon charge differences caused by inherited amino-acid substitutions in the polypeptide backbone.

The enormous clinical interest in α_1-antitrypsin polymorphism has centred around genetic variants associated with disease. Certain allotypes, notably those that are clearly established as polymorphisms, result in decreased secretion of α_1-antitrypsin from the liver. The Pi W gene has 80% of the production of PiM, PiS 60%, PiP 25%, PiZ 15% and Pi null 0%. If normal mean plasma levels for the genotype PiMM are taken as 100%, then levels for PiZZ are 15%, PiSS 60%, PiMZ 57·5% and PiMS 80%. The incidence of these phenotypes in the population is shown in Table 13.

The reason for the decreased liver secretion of these variants is unclear. It is however known that the Z variant, which accumulates in the liver cells, has a subsitution of a glutamic acid residue by glycine. It is probable that this gives rise to a failure of maturation of the protein within the endoplasmic reticulum as there is

TABLE 13
INCIDENCE OF VARIOUS Pi PHENOTYPES IN A UK CAUCASIAN POPULATION

	%
PiM	80·5
PiMS	9·0
PiFM	5·0
PiMZ	3·0
PiFS	0·3
PiS	0·25
PiSZ	0·17
PiFZ	0·10
PiF	0·09
PiZ	0·03
All other phenotypes have a frequency of 0·01% or less	

(Reproduced with permission from A. Milford Ward, 'α_1-Antitrypsin' in *Immunochemistry in Clinical Laboratory Medicine*. A. Milford Ward and J. T. Whicher, Eds. MTP Press, Lancaster. 1978.)

some evidence that the protein in the liver cells has a retained pro-peptide and has incomplete addition of carbohydrate side chains.

Despite the disease associations the gene frequencies for S and Z variants are higher than would be expected for random mutation. It is possible that heterozygotes have a higher fertility due to the decreased inhibition of sperm acrosine and cervical mucolytic enzymes.

Association with Disease. Pulmonary emphysema is associated with phenotypes giving rise to severe deficiencies of α_1-antitrypsin; namely PiZZ and PiSZ though recent evidence suggests an increased incidence of lower lobe emphysema occuring late in life even in PiMS individuals. The normal elastic tissue of the lungs is dependent on elastic fibrils and the ground substance between the alveolar cells. The lung is continually exposed to inhaled particles such as dust and cigarette smoke together with occasional bouts of bacterial infection. All these insults result in phagocytic activity with release of macrophage lysosomal enzymes. In the presence of adequate tissue fluid concentrations of α_1-antitrypsin, the delicate elastic tissue of the lung is protected against such enzymes as elastase. In α_1-antitrypsin deficiency proteolytic damage results in distension of air sacs and loss of elastic recoil—emphysema. Diagnosis is important as the incidence and severity of lung disease can be greatly reduced by avoiding smoking and air-polluted atmosphere. Respiratory tract infection should be treated vigorously.

Some 10–20% of PiZZ infants develop neonatal hepatitis and about one-third of these progress to fatal cirrhosis. A small proportion of PiZZ adults develop cirrhosis. All PiZZ individuals have PAS positive inclusions in the hepatocyte which can be shown by immunofluorescence to be α_1-antitrypsin. It is probable that such liver cells accummulate the protein which they are unable to secrete and are more susceptible to injury succumbing to what could otherwise be mild viral infections. It is also probable that proteolytic damage occurs in the liver as a result of inflammation secondary to infection and, in the absence of α_1-antitrypsin, damage to the collagen and elastic framework of the liver lobule results in the disorganized regeneration typical of cirrhosis.

In the investigation of such cases and their families it is important to realize that simple measurement of α_1-antitrypsin may be inadequate owing to the fact that concomitant infection may give rise to an acute-phase increase in α_1-antitrypsin which masks the deficiencies associated with the phenotypes PiMZ and PiSS. Even in acute illness it is unlikely that PiZZ or PiSZ individuals will attain normal levels. It is possible to use quantitative α_1-antitrypsin measurements as a screening test only if another acute-phase protein such as orosomucoid is measured in parallel.

Simple serum protein electrophoresis is quite adequate for the detection of the severe deficiency phenotypes PiZZ and PiSZ. Agarose electrophoresis gives better resolution of the α_1-region. Phenotyping is now best performed by isoelectric focusing in polyacrylamide or agarose gel.

α_1-antichymotrypsin (MW 68 000)

This protein has a high inhibitory specificity for chymotrypsins *in vitro*. It does not show strong inhibition of any known lysosomal proteases and its role *in vivo* is unclear. It is found in high concentrations in bronchial secretions and it has been suggested that it has a role in protecting mucosal surfaces from proteolytic damage. It is a very rapidly reacting and sensitive acute-phase reactant, only C-reactive protein showing a faster response. It is in this capacity that it is most often measured.

Inter-α-trypsin Inhibitor (MW 160 000)

This is a protease inhibitor of high molecular weight which easily dissociates into fragments which retain inhibitory activity. It is antigenically related to trypsin inhibitors of low molecular weight from bronchial secretions (these exist as monomer MW 22 000 and dimer MW 44 000) and to urinary trypsin inhibitor (MW 77 000). It may thus be the parent molecule for a number of mucosal protease inhibitors. It is a strong inhibitor of sperm acrosine and its role in fertilization is of some interest.

α_2-macroglobulin (MW 725 000)

α_2-macroglobulin is a very important protease inhibitor, synthesised by liver and reticulo-endothelial cells, which binds all known proteases. As with α_1-antitrypsin, limitation of proteolysis by the protease is associated with the inhibitory binding which is achieved by the macroglobulin enfolding the protease molecule and sterically hindering substrate access. It is however of great interest that protease-α_2-macroglobulin complexes still retain proteolytic activity for small substrates and the evidence suggests that these may have a role to play in the inactivation of low molecular weight lymphokines and complement chemotactic factors.[10] α_2-macroglobulin is a large molecule mainly confined to the vascular compartment and experimental evidence suggests that protease may be transferred from α_1-antitrypsin to α_2-macroglobulin. α_2-macroglobulin-protease complexes are rapidly cleared from the circulation with a half-life of about 10 minutes; uncomplexed α_2-macroglobulin has a half-life of 135 hours. Uptake appears to be into reticulo-endothelial cells. Removal of the α_2-macroglobulin-protease complex is far more rapid than that of the α_1-antitrypsin-protease complex which has a half-life of three hours.

α_2-macroglobulin does not behave as an acute-phase reactant. Its level is raised in children prior to adolescence and gives a characteristic appearance to the juvenile electrophoretic strip—see below. Levels increase in pregnancy and in women taking oral contraceptives and may be very raised in nephrotic syndrome (see below). Increased levels have been reported in ataxia telangiectasia, atopic dermatitis and mongolism. Measurement has no established role in clinical practice.

C1-esterase Inhibitor (MW 104 400)

This protein has already been mentioned in the section on complement. Besides inhibition of activated C1 it inhibits plasmin, thrombin and kallikrein. It is an acute phase reactant showing a marked increase in active lymphoid malignancy and it may be useful as a monitor of disease activity. Paradoxically a secondary deficiency may occur in association with the same types of tumour. This can result in the clinical picture of angioneurotic oedema.

Antithrombin III (MW 65 000)

Antithrombin III is an inhibitor of thrombin and in particular of activated factor X. Low levels may be associated with a thrombotic tendency and a genetic deficiency is known. The contraceptive pill causes a reduction in plasma level of about 40%. There has been some interest in the measurement of this protein in users of oral contraceptive to indicate those who may develop thrombosis.[11]

Haptoglobin

The haptoglobins are a family of haemoglobin-binding glycoproteins present in plasma. The haptoglobin molecule binds two molecules of oxyhaemoglobin, the complex being rapidly removed from the circulation by reticulo-endothelial cells. This prevents undue loss of iron by urinary excretion. Haemoglobin is continually released from red cells due to damage as they circulate through small vessels. This is illustrated by the drop in haptoglobin concentration which occurs following severe exercise such as cross-country running. Haemoglobin is also released in areas of inflammation due to capillary damage, stasis and rupture of red cells. It is possible that haptoglobin-haemoglobin complexes possess proteolytic inhibitory activity. It is in this context that haptoglobin may be thought of as a protein involved in inflammation and here may be the reason for its marked acute-phase behaviour.

Structurally, haptoglobin comprises a four-chain molecule with two small alpha-chains and two larger beta-chains which contain the carbohydrate moiety. It shows marked genetic polymorphism and the known mutations fall into three groups. Chromosomal rearrangement with an internal gene duplication has probably resulted in the formation of the α^2-chain from the smaller α^1-chain. Single point mutations at amino-acid 54 on the α^1-chain result in two variant chains α^{1S} and α^{1F}. Alteration in control of the rate of synthesis of either the alpha or beta chains results in reduced haptoglobin synthesis in a manner analogous to the thalassaemias. Such hypohaptoglobinaemia occurs in up to 12% of children among American negros. The common allotypes are the α^1 and α^2 variants of the α-chain which show marked differences in different populations. In Europe the haptoglobin α^1-allotype has a frequency of 0·31–0·45 while it is 0·77 in the Congo and 0·10 in certain areas of India. The result of this is that in Europe the phenotype incidences are Hp 1-1 < Hp 1-2 < Hp 2-2. The presence of the α^2 heavy chain in Hp 1-2 and Hp 2-2 allows polymerization of the molecule to occur with an increase in molecular weight from 85 000 for Hp 1-1 to over 220 000 for Hp 2-2 complexes. The mean plasma level for Hp 1-1 individuals is higher than for Hp 2-2. Hp 2-2 occurs with a higher overall population frequency than Hp 1-1 and as it has been derived from Hp 1-1 it is probable that it confers an evolutionary advantage.

Haemolytic anaemia and liver disease both result in decreased haptoglobin levels and some clinical value may be obtained from detecting these changes. In the absence of inflammation the plasma haptoblobin level is as sensitive as radiochromium-labelled red cell survival in detecting haemolysis. Haptoglobin is an acute-phase reactant and its level is usually elevated in the plasma in trauma, burns, infection, allergic tissue damage, cancer and leukaemia.

Laboratory measurement is beset by the problem that the polymorphic forms Hp1-2 Hp2-2 are underestimated by gel techniques by as much as 20%. Nephelometry is more satisfactory.

Orosomucoid (MW 44 000)

Orosomucoid or α_1-acid glycoprotein is a strong acute-phase reactant present at fairly high concentrations in human plasma. It has a very high carbohydrate content comprising a large number of sialic acid residues which show microheterogeneity due to different sialyl-galactosyl linkages. The protein is synthesized in the liver and the nature of the carbohydrate side chains is affected by circulating oestrogen levels. Plasma orosomucoid shows remarkable heterogeneity due to multiple amino-acid substitutions, a property which is shared only by the immunoglobulins. It is also clear that it shows considerable homology with the variable region of the immunoglobulin light chain, a part of the IgG H-chain and the haptoglobin α-chain, which is itself related to the immunoglobulins. This has led to the suggestion that orosomucoid evolved from the immunoglobulin lineage probably after the duplication of the primitive L-chain but before the formation of the H-chain.

The acute-phase responsiveness of orosomucoid suggests a role in inflammation though as yet no convincing hypothesis for its function has been proposed. It is known to bind propranolol and various Δ^4-3-ketosteroids, particularly progesterone. It is found on platelet membranes and may mediate their adherence to collagen. In clinical practice it is a valuable acute-phase protein to measure in the monitoring of inflammatory disease.

Other Proteins of Inflammation

Blood coagulation is clearly an important part of the inflammatory reaction and it is not surprising that a number of clotting proteins, in particular fibrinogen, are acute phase reactants. Little is known of the changes which occur in plasma kinins such as pre-kallikrein.

TRANSPORT PROTEINS

A number of proteins exist for the purpose of carrying biologically active molecules in the plasma (Table 14). Such transport subserves a number of roles. Molecules may be transported which would otherwise be relatively insoluble in plasma such as bilirubin, lipids and fatty acids. Many substances are biologically inactive when bound to plasma proteins and the bound molecules, being in equilibrium with the free, provide a pool to buffer changes in the free fraction. Toxic substances may also be rendered inactive by protein binding. There is an increasing realization that binding proteins may interact with receptors or enzymes affecting the metabolism of those molecules they carry. For example, apolipoprotein C interacts with lipoprotein lipase, and transferrin may bind to membrane receptors in facilitating the uptake of the iron which it carries.

Albumin (MW 68 000)

Function

Albumin is unique among the major plasma proteins in containing no carbohydrate residues. It is a single polypeptide chain containing a solitary thiol group which, binding a half-cystine, results in the formation of albumin polymers *in vitro*.

Albumin binds a very wide range of substances, in particular lipid-soluble anions. Most of the circulating long-chain fatty acids in plasma are albumin bound with a normal level of 1–2 molecules per molecule of albumin. Other substances which bind are shown in Table 15. Calcium is very weakly bound at 16 sites. A number of drugs, of which a few examples are given, are strongly bound and may displace each other giving rise to dangerous increases of drug activity since only the free drug is active.

Albumin is also important as an intravascular colloid, maintaining the slightly higher oncotic pressure of plasma over tissue fluid but this role may be largely taken on by other proteins in the absence of albumin. Many cells are able to use albumin as a source of amino-acids and take it up by pinocytosis.

Genetic Variation

Twenty-three structural variants of albumin have been described. They are inherited in an autosomal codominant fashion and in the heterozygote give rise to a bisalbuminaemia on electrophoresis (Fig. 17). Both albumin bands are usually of equal staining density as they represent the product of two genes. A number of variants however give rise to dimerization *in vivo* and may give rise to wide albumin bands or double bands of unequal density. None of these 'para-albumins' is associated with disease.

Analbuminaemia has only been described in 11 pedigrees and probably represents a defect in the mechanism controlling albumin synthesis. Small amounts of albumin

TABLE 14
TRANSPORT PROTEINS

Protein	MWt (daltons)	Mean concentration (g/l)	Substance transported
Albumin	68 000	40	(see Table 15)
Transferrin	77 000	2·8	iron
Caeruloplasmin	130 000	0·35	(copper)
Haptoglobin	100 000–400 000	2·0	haemaglobin
Haemopexin	57 000	0·75	haem
Prealbumin	55 000	0·3	vitamin A, thyroxine
Retinol binding protein	21 000	0·045	vitamin A
Thyroxine bining globulin	57 000	0·035	thyroxine
Transcortin	55 700	0·030	corticosteroids
Sex hormone binding globulin	94 000	0·020	androgens and oestrogens
Group component (Gc globulin)	51 000	0·55	vitamin D
Transcobalamin I	60 000–70 000	0·00003	
II	38 000	0·000015	vitamin B_{12}
III	60 000–70 000	0·000025	
Apolipoproteins	see Table 17		

TABLE 15
EXAMPLES OF SUBSTANCES WHICH MAY BE BOUND TO ALBUMIN IN VIVO

Substance	$K_A(M^{-1})$	% bound to albumin in normal plasma
Oleate	$2\cdot6 \times 10^8$	99·9
Palmitate	$6\cdot2 \times 10^7$	99·9
Bilirubin	1×10^8	99·9
L-Thyroxine	$1\cdot6 \times 10^6$	10
Cortisol	5×10^3	30
Aldosterone	$<5 \times 10^3$	60
Urate	3×10^2	15
Cu^{2+}	9×10^6	<5
Ca^{2+}	100	40
Salicylate	4×10^5	40
Warfarin	$2\cdot2 \times 10^5$	97
Phenylbutazone	$2\cdot7 \times 10^5$	99
Clofibrate	$1\cdot3 \times 10^3$	98

Based on Peters, T. Serum albumin in *The Plasma Proteins*. Ed. Putnam, F. W. Academic Press, New York. 1975.

are usually present and affected people suffer no symptoms or at most mild oedema. It is interesting that albumin variants are rare and as heterozygous advantage is probably the most important factor leading to polymorphism it has been suggested that albumin is not essential for life!

Variation in Disease

The factors affecting normal albumin level are its volume of distribution, synthetic rate and catabolic rate. In most disease processes a change in more than one of these factors is operating (Table 16). Changes in albumin level during the menstrual cycle and in pregnancy reflect alterations in the volume of distribution.

Hypoalbuminaemia is a remarkably reliable indication of illness but has little diagnostic specificity. In many diseases decreased albumin synthesis results from malnutrition, and the increased endogenous catabolism associated with injury and pyrexia may decrease the half-life of albumin from the normal 21 days to as little as 7.

Liver disease commonly results in hypoalbuminaemia. This is usually mild but may become severe if recovery is delayed. In cirrhosis the actual rate of hepatic albumin synthesis may be normal or even increased but much of the new albumin enters the hepatic lymph and leaks across the liver capsule directly into the ascitic fluid.

TABLE 16
ALBUMIN CHANGES IN DISEASE

Disease	Synthesis	Loss	Degradation rate	Exchangeable pool
Liver disease	↓	—	↓	↓
Nephrotic syndrome	↑	↑	↑	↑
Protein-losing enteropathy	↑	↑	↑	↓
Malnutrition	↓	—	↓	↓
Acute burn	↑↓	↑	↑	—
Cushing's syndrome	↑↑	—	↑↑	↑
Thyrotoxicosis	↑↑	—	↑↑	↑

Plasma albumin level is thus not a reliable indicator of functional hepatocellular mass.

Increased glomerular permeability results in urinary albumin loss and the nephrotic syndrome. Plasma levels below 20 g/l usually result in oedema. The amount of albuminuria is not a reliable prognostic indicator in nephrotic syndrome, especially in children where 'benign' minimal change nephritis often gives rise to the most severe albuminuria. An idea of the 'selectivity' of the proteinuria such as the IgG clearance:albumin clearance ratio is, however, very useful.

Gastrointestinal protein loss is a common and important cause of hypoalbuminaemia and may result from either inflammatory or neoplastic disease. Protein loss into the gut can be readily estimated by measuring α_1-antitrypsin clearance into the faeces or, less reliably but more simply, the faecal α_1-antitrypsin level. This is possible because α_1-antitrypsin is not degraded by proteases in the gut.[12]

Measurement

Albumin is best measured by immunochemical means as most dye-binding methods overestimate the protein at low levels.[13]

Transferrin (MW 77 000)

Transferrin is the major iron transport protein of plasma. It is synthesized both in the liver and some other tissues such as the reticuloendothelial system. It has two binding sites for iron probably with different affinities; one at the N-terminal and one at the C-terminal end of the molecule.[14] This protein has a central role in iron metabolism, returning iron derived from the catabolism of haemoglobin and other proteins to haemopoietic tissue. Unlike some other transport proteins it is returned to the circulation after unloading its iron at the cell membrane.

The concentration of transferrin in the blood is increased in pregnancy when levels in the last trimester may be higher than those seen in any other condition. Elevated levels are also seen with oral contraceptive use. Iron deficiency results in increased hepatic synthesis with raised serum levels. The more severe and prolonged the iron deficiency, the higher the transferrin level will be.

Transferrin concentration is decreased in malnutrition, liver disease and inflammatory diseases such as rheumatoid arthritis. In this latter group the transferrin level reflects the severity and chronicity of the disease.

In clinical practice transferrin measurement, either immunochemically or as iron binding capacity, has a rôle in assessing the differential diagnosis of anaemia. Iron is usually measured in association with transferrin and the saturation of the protein calculated. Simple iron-deficiency anaemia is usually associated with a raised serum transferrin level with a low saturation whereas anaemia due to failure to incorporate iron into red cells has a normal or low transferrin level with a high saturation.

A high level of saturation with a normal transferrin level is seen in haemochromatosis and other forms of iron overload, and is a useful parameter for diagnosing and monitoring therapy. It has, however, been to some extent superseded by the measurement of serum ferritin.

As with most serum proteins a number of allotypes have been described, all of which are rare. Atransferrinaemia is a rare cause of iron deficiency anaemia which is unresponsive to iron therapy.

Caeruloplasmin (MW 130 000)

Caeruloplasmin is an exquisitely coloured sky-blue protein which carries 8 atoms of copper per molecule. It is synthesized in the liver and binds some 95% of plasma copper, the remainder being attached to albumin.

The function of the protein is very much under debate.[15] Studies with labelled copper show negligible *in vivo* turnover suggesting that caeruloplasmin does not act as a physiological transport protein. Copper is probably transported from the gut to the liver on albumin. It is also clear that copper is attached to caeruloplasmin prior to its secretion from the liver cell and that an increased hepatic copper pool increases caeruloplasmin synthesis. Caeruloplasmin acts a weak oxidant in plasma and it has been suggested that it has a physiological role in the oxidation of ferrous ions at the cell surface to the ferric form for binding to apotransferrin. *In vitro* studies have shown that caeruloplasmin has antioxidant activity inhibiting peroxidation by such important catalysts as iron and ascorbate. It is also possible that the protein may inactivate free radicals produced by phagocytosis thus having a role in inflammation, a suggestion which is supported by its acute-phase behaviour.

Allotypes are rare in Caucasians though at least four have been found in negroes and a fifth in Thailand. The majority of patients with Wilson's disease have low plasma levels of copper and caeruloplasmin. The disease is characterized by copper deposition in tissues, especially the liver, leading to cirrhosis, and the brain, with damage to the basal ganglia. The disease is probably genetically heterogenous but the defect appears to result in a failure of biliary copper excretion with subsequent accumulation in tissues. The evidence suggests that an intracellular copper-binding protein may be at fault, preventing copper from reaching the intracellular pool necessary for biliary excretion and attachment to caeruloplasmin. The low plasma caeruloplasmin level is thus a result of decreased intracellular copper availability.[16]

Low plasma levels of copper or caeruloplasmin are thus an important guide in the diagnosis of Wilson's disease; the demonstration of high levels of copper in the liver by biopsy is also important.

Plasma caeruloplasmin concentration is decreased in severe liver disease, notably in primary biliary cirrhosis and primary-biliary atresia, and also in malabsorbtion. Caeruloplasmin levels are raised 2- to 3-fold in the last trimester of pregnancy and during therapy with oestrogen-containing drugs. It is a slow acute-phase reactant showing increases especially in diseases involving the reticulo-endothelial system, such as Hodgkin's disease where it may be used as an indication of disease activity. Levels are increased in infective and obstructive diseases of the biliary tract probably due to decreased biliary excretion resulting in an increased hepatocellular copper pool with consequent induction of caeruloplasmin synthesis.

Other Transport Proteins

Plasma lipids are transported in macromolecular micellar structures solubized by specific transport proteins.[17] The apolipoproteins may be classified into at least four families, A, B, C, E on the basis of structure and function. They are important not only for transport but, as they interact with both enzymes and cellular receptors, they also play a role in the control of plasma lipid metabolism (Table 17). Their measurement in plasma is not yet of clinical value.

The hormone carrier proteins are listed in Table 14. In general their measurement is of importance in that it allows an assessment of the level of free, and thus biologically active, hormone present in plasma. The role of the retinol binding protein-prealbumin complex in the transport of vitamin A is now attracting considerable interest.[18]

Haemopexin (MW 70 000) binds free haem entering the plasma from the breakdown of haemoglobin released

TABLE 17
THE PROTEINS OF LIPID TRANSPORT

Protein	MWt (daltons)	Proposed function	Comment
Apolipoprotein A-I	28 331 (single chain)	Interacts with lecithin cholesterol acyl transferase (LCAT) binds to phospholopid	major apoprotein of HDL
A-II	17 000 (2 chains)	binds to phospholipid probably micelle former	25% of apoprotein of HDL
Apolipoprotein B	unknown	micelle former interacts with LDL receptor	major apoprotein of LDL insoluble in acqueous media
Apolipoportein C-I	6530	activates LCAT	present especially in VLDL and chylomicrons
C-II	10 000	activates lipoprotein lipase	important in VLDL and chylomicron metabolism
C-III	8764	binds to phospholipid	
Apolipoprotein D	?	?	sometimes called A-III
Apolipoprotein E	33 000	one genetic allotype absent in Type III hyperlipidaemia	present in VLDL

by intravascular haemolysis. The haemopexin-haem complex is taken up by the liver where the iron is bound to intracellular ferritin and the remainder of the haem is converted by haem oxygenase to bilirubin. Decreased serum levels are seen in haemolysis and, as this protein unlike haptoglobin is not an acute-phase reactant, the low levels persist even when those of haptoglobin are normal or raised due to inflammation.

ONCOFETAL AND TISSUE PROTEINS

It seems probable that many tissue proteins are shed into the plasma as a result of cell turnover. This is true for a number of enzymes and some cell membrane proteins such as β_2-microglobulin. It is also clear that some malignant tumours are able to re-express genes normally repressed in mature cells of that tissue. This results in the synthesis, and often the appearance in the plasma, of proteins typical of other adult tissues or fetal cells. Some of the tumour proteins of fetal type, oncofetal proteins, are related to the major histocompatability antigens and thus to β_2-microglobulin. β_2-microglobulin, alpha-fetoprotein and the placental proteins will be briefly described here.

β_2-Microglobulin (MW 11 800)

β_2-microglobulin is a single polypeptide chain of 100 amino-acids devoid of carbohydrate with one intrachain disulphide bridge. It is the light or beta chain of the cell surface HLA antigen (Fig. 14). The heavy chain is

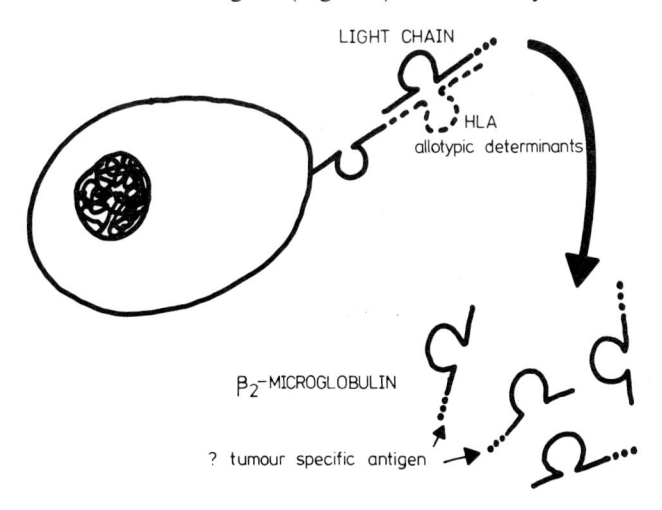

FIG. 14. The histocompatibly antigens consist of two polypeptide chains embedded by the heavy chains in the cell membrane. The light chain is identifiable as β_2-microglobulin in body fluids. β_2-microglobulin and thus HLA from some types of tumour cells may contain specific determinants immunochemically identifiable as tumour specific antigens.

attached to the cell membrane and bears the allotypic determinants which denote HLA specificity.[19] The HLA complex is important in cell recognition and it is probable that β_2-microglobulin controls biosynthesis of the intact molecule in the same way that the immunoglobulin light chain controls biosynthesis of the intact immunoglobulin molecule. Many tumour-specific antigens have been shown to be immunologically cross-reactive with β_2-microglobulin and it has been suggested that they might be regarded as modified histocompatibility antigens.

β_2-microglobulin is synthesized by almost all human cells; tumour cells and lymphocytes have particularly high production rates. The molecule is released from the cell surface and stimulation of lymphocytes has been shown to result in increased release. The molecule is freely filtered at the glomerulus and reabsorbed in the proximal tubule where it is degraded in lysosomes. Excretion thus depends on glomerular filtration rate and high plasma levels are seen in renal failure.

Urinary β_2-microglobulin excretion is a widely used means of assessing renal tubular damage. It has proved valuable in evaluating damage due to nephrotoxic drugs and industrial pollutants such as cadmium. It is probably no more useful than creatinine clearance in predicting transplant rejection.

Plasma levels of β_2-microglobulin closely reflect creatinine clearance but are also affected by increased production rates occurring in some malignant tumours. Measurement may have a role in monitoring tumours particularly those of B-cell origin. Chronic lymphatic leukaemia, B-cell lymphomas and myeloma usually have raised levels. Interestingly, benign monoclonal gammopathies have normal levels suggesting a possible diagnostic role in differentiating these two important conditions. Several inflammatory conditions involving B-cell stimulation have raised plasma β_2-microglobulin levels such as rheumatoid arthritis, systemic lupus erythematosis and Crohn's disease. Plasma β_2-microglobulin levels may be a good marker of disease activity in hepatitis when a fall to normal suggests healing and a persistent high level forebodes the development of chronic hepatitis. It seems clear that the measurement of β_2-microglobulin may have a role in inflammatory disease similar to that of the acute-phase proteins but denoting B-cell activity rather than tissue damage. In all such β_2-microglobulin measurements it is important to relate blood levels to the creatinine concentration to eliminate changes due to variation in glomerular function.

Alpha-fetoprotein (MW 69 000)

Alpha-fetoprotein is present in trace amounts in normal human serum.[20] In the fetus it is produced by the yolk sac, liver and to some extent by the gut. The fetal plasma level is highest at 13 weeks of gestation when it accounts for almost one third of the total plasma proteins and is about one million times higher than the adult level. It falls rapidly towards birth, adult levels being achieved in the first two weeks of life.

The majority of pregnancies affected by open neural tube defects are associated with raised levels of alpha-fetoprotein in amniotic fluid. Determination of amniotic fluid and maternal serum levels in the 15th to 25th week of gestation have proved valuable in screening for such defects (see Chapter 43). Some 30–70% of patients with

TABLE 18
THE PLACENTAL PROTEINS

Name	Abbreviation	M Wt (daltons)	Function
Human chronic gonadotrophin	HCG	40 000	gonadotrophic
Human placental lactogen	HPL	27 000	lactogenic
Schwangerschaft's protein	SP_1	90 000	? immunosuppressive
Pregnancy associated plasma protein A	PAPP-A	750 000	—
Pregnancy associated plasma protein B	PAPP-B	1 000 000	—
Placental protein 5	PP5	36 000	? protease inhibitor

germ cell tumours and 80% of patients with hepatocellular carcinoma have an elevated level of serum alpha-fetoprotein and this has proved useful in the diagnosis and monitoring of such tumours.

Little is known about the function of this protein. It has a number of albumin-like properties which suggest that it may be a transport protein. It also has well established immuno-suppressive effects suggesting a role in maternal-fetal immune regulation.

The Placental and Pregnancy-Associated Proteins

The placenta secretes a number of proteins into the maternal circulation which we suspect convey signals from the fetoplacental unit to the mother (Table 18).[21] Chorionic gonadotrophin and placental lactogen are well known but in the last few years extensive research in this field has revealed several other proteins which appear in the plasma during pregnancy. Only four of these appear to be placental proteins secreted into the maternal circulation.

SP_1 rises rapidly early in pregnancy, the concentration doubling every 2–3 days, and continues to rise more slowly towards term. The change closely resembles that of HPL. PAPP-A has been detected in the 12th week of pregnancy and increases steadily up to term. It has been claimed that a rise in PAPP-A occurs before the onset of pre-eclampsia (see Chapters 42 and 43). PAPP-B is a very large protein with a longer half life than PAPP-A. Little is known of its levels in pregnancy. PP5 unlike the other three proteins is present in placental stroma as well as trophoblast cells and there is a suggestion that it is primarily a tissue protein rather than a secretory product. It has oncofetal characteristics in that it is found in patients with hydatidiform mole and teratoma.

There is as yet no clinical role for the measurement of these proteins though interest is centred around their use in pregnancy testing, in the assessment of fetoplacental wellbeing, and as tumour markers.

THE PLASMA PRECURSORS OF AMYLOID

Amyloid is an amorphous extracellular substance composed of non-branching linear fibrils. It is found in association with a number of conditions (Table 19) and may result in extensive damage to the tissues in which it is deposited. Analysis of the protein content of fibrils has revealed the presence of a number of different proteins which characteristically occur in certain types of amyloid. Precursor proteins for all forms of amyloid have been found circulating in the plasma and the evidence suggests that amyloid deposition occurs as a result of the proteolytic degradation of circulating proteins producing fragments capable of polymerization and deposition in tissue. It is probable that for amyloid to persist there must also be defects in the enzymatic degradation and phagocytic removal of such fibrils. It is thus clear that, biochemically, amyloid represents a heterogenous collection of diseases giving rise to fibril deposition in tissues.

TABLE 19
CLASSIFICATION OF AMYLOID ON THE BASIS OF THE MAJOR PROTEIN COMPONENT

Pathological type	Protein component
Primary amyloid	Light chain fragment (AL protein)
Amyloid associated with B cell tumour	AL protein
Amyloid secondary to infections and immune stimulation	Amyloid A protein (AA protein)
Senile amyloid	Amyloid S protein (AS protein)
Familial forms	
Familial Mediterranean fever	AA protein
Portuguese amyloid	? fragment derived from prealbumin
Localised amyloid	
Thyroid	fragment of thyrocalcitonin
Pancreas	fragment of ? insulin or glucagon?
Skin	

P-component is a protein invariably present as a monor constitutent in the fibrils of all types of amyloid. Electron-microscopically it consists of annular structures made up of five subunits of molecular weight about 27 000. It has been identified as a plasma protein. It is an acute-phase reactant closely related in amino-acid sequence to C-reactive protein. The major component of primary amyloid and of amyloid associated with B-cell tumours is a proteolytically derived fragment of the variable region of the immunoglobulin light chain. In secondary amyloid, on the other hand, AA protein of molecular weight 10 000 is found in the fibrils. This protein is probably a proteolytically produced fragment of circulating plasma AA protein which is an acute-phase reactant also showing marked increases in concentration with increasing age. It has a molecular weight of about 12 000 and easily forms complexes of high molecular weight. Other proteins have been found in specific forms of amyloid such as a fragment derived from prealbumin in Portuguese amyloid. Amyloid may thus be thought of as a disease resulting from the deposition in tissues of a heterogenous group of proteins derived by proteolysis from circulating serum proteins. There is also evidence in the case of AA protein and IgG light chains that certain amino-acid sequences present in the precursor protein may produce fragments particularly likely to polymerize in tissue thus giving some basis for the observed genetic nature of some forms of amyloid. Experimental evidence also suggests that patients with amyloid have a defect in a plasma enzyme capable of degrading the fibrils and an inability to phagocytose the fibrils once deposited in tissue.

The measurement of the serum amyloid components is at the present time of no clinical value except for the detection of monoclonal light chains (Bence-Jones protein) in the diagnosis of AL type amyloid. This is best carried out by electrophoresis and immunofixation of the patient's urine and is of course of considerable importance as it allows diagnosis and thus treatment of the underlying B-cell neoplasm, often resulting in considerable clinical improvement.

THE INTERPRETATION OF ELECTROPHORESIS

Serum and Plasma

Electrophoresis whether performed on cellulose acetate or agarose is of immense value in the preliminary investigation of plasma proteins. Many important abnormalities, such as genetic variants and paraproteins may not be detected by specific protein quantitation alone. Agarose electrophoresis with amido black staining will reveal proteins present in concentrations above 0·1 g/l. Some 10–13 components are usually clearly visible on the electrophoretic separation of serum or plasma (Figs. 15, 16).

Prealbumin. This protein is usually faintly visible in normal individuals, particularly on cellulose acetate separations. Its disappearance is typical of liver cirrhosis, inflammation and malnutrition. High levels are seen in alcoholics.

Albumin. The albumin band shows an increased anodal mobility as a result of the binding of bilirubin, penicillin and acetyl salicylic acid and occasionally due to tryptic activity in acute pancreatitis. Certain variants may give rise to slow or fast bands usually seen in the heterozygote as bisalbuminaemia. Albumin must be decreased by about 30% before a decrease in staining intensity is seen.

Albumin-alpha$_1$-interzone. The staining intensity of this pale area is due to α_1-lipoprotein. A decrease is not easily recognized but occurs in severe inflammation, particularly of the liver while an increase is often seen in

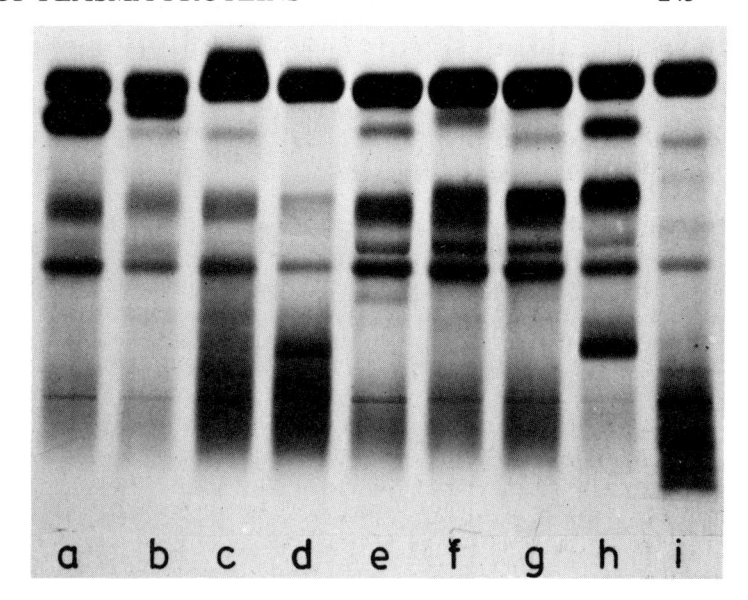

FIG. 15. Examples of serum and plasma electrophoresis. Note the absence of the beta$_2$ band due to C3 activation by freezing and thawing in all samples except (e). Samples (d) and (h) are plasma and contain a fibrinogen band.

(a) Bisalbuminaemia.
(b) Bisalbuminaemia, polymeric variant.
(c) Fast albumin due to bilirubin binding, slightly raised gamma globulin (polyclonal IgG increase) with beta-gamma fusion (IgA increase) in a patient with cirrhosis.
(d) Plasma from a patient with alpha$_1$-antitrypsin deficiency (PiZZ) with early cirrhosis resulting in a polyclonal increase in IgG.
(e) Mild acute phase response with increased alpha$_1$ and alpha$_2$ zones in a patient with immune complex disease. The beta$_2$ band (C3) is visible but fainter than normal due to a low C3 (compare with Fig. 16 (e) and (f)).
(f) Fast alpha$_1$-antitrypsin variant in a patient with a slight acute phase response.
(g) Slow alpha$_1$-antitrypsin variant (PiSS) in a child with an acute phase response.
(h) Raised alpha$_1$, alpha$_2$ and fibrinogen in a patient with inflammation. Hp 1-1 phenotype resulting in a slightly fast alpha$_2$ zone.
(i) Low albumin, slow alpha$_1$-antitrypsin variant, decreased alpha$_2$ and increased gamma globulin with oligoclonal IgG response in a patient with chronic aggressive hepatitis.

Reproduced with permission from J. T. Whicher 'The Interpretation of Electrophoresis', *British Journal of Hospital Medicine*, 1980 in press.

alcoholics, and women during puberty and pregnancy. Very high levels of α-fetoprotein such as occur occasionally in hepatoma may result in a band between the albumin and α_1-bands.

Alpha$_1$-zone. Orosomucoid and α_1-antitrypsin both migrate together but as orosomucoid stains poorly α_1-antitrypsin constitutes the great majority of the α_1-band. Greatly raised levels of orosomucoid do however give rise to an anodal broadening of the band. The acute phase response is characterized by an increase in α_1-antitrypsin and thus of the α_1-band. A selective increase of α_1 is a good indication of liver injury or increased oestrogen levels. An absence of increase with evidence of an acute-phase response in other proteins suggests the presence of a deficient α_1-antitrypsin allotype. Allotypes with altered electrophoretic mobility may be recognizable and homozygous PiZ individuals usually have virtually no visible α_1-band unless there is a marked acute-phase response. Children have lower α_1-antitrypsin levels than adults.

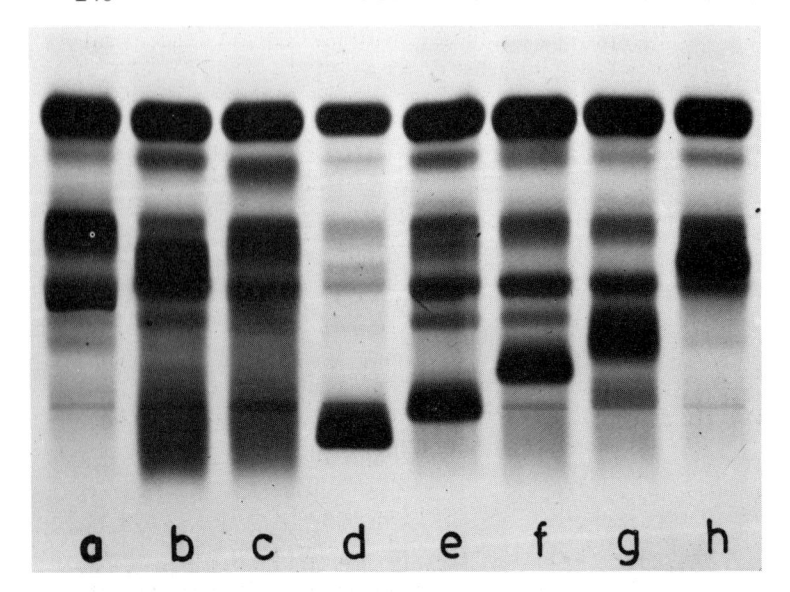

Fig. 16. Examples of Serum Electrophoresis.

(a) Nephrotic syndrome. Prealbumin is visible; albumin is not noticeably decreased. There is a marked increase in the alpha$_2$ zone and in beta lipoprotein. The gamma globulin reflects a hypercatabolic decrease in IgG secondary to the protein-losing state.
(b) Cathodal position and increased density of haptoglobin in a haemolysed sample or as a result of haptoglobin-haemoglobin complexes. Note clearly discernible alpha$_2$-macroglobin at the front of the alpha$_2$ zone where it normally fuses with haptoglobin.
(c) Simple acute phase response with increased alpha$_1$ and alpha$_2$ zones.
(d) Paraprotein in a patient with myeloma. There is immune suppression with a pale gamma area and a decreased albumin. The split beta is due to heterozygosity for a transferrin variant.
(e) IgG paraprotein.
(f) IgG paraprotein.
(g) IgA paraprotein.
(h) IgA paraprotein in the alpha$_2$ region.

Reproduced with permission from J. T. Whicher 'The Interpretation of Electrophoresis', *British Journal of Hospital Medicine*, 1980 in press.

Alpha$_1$-alpha$_2$-interzone. Two very faint bands are usually seen just preceding the main α_2-band on agarose electrophoresis. These represent α_1-antichymotrypsin, inter-α-trypsin inhibitor and Gc globulin. They increase in intensity and coalesce in early acute inflammation as a result of a strong increase in α_1-antichymotrypsin.

Alpha$_2$-zone. This broad band is made up principally of α_2-macroglobulin and haptoglobin. The genetic variants of haptoglobin possess different electrophoretic mobility. Haptoglobin 1-1 which is present in 15% of the population has a faster mobility than α_2-macroglobulin and in fact covers the faint Gc-globulin and α_1-antichymotrypsin bands. Haptoglobin 1-2 and 2-2 migrate to the cathode side of α_2-macroglobulin which is seen as the sharp leading edge of the α_2-zone. Haptoglobin is usually increased as part of the acute-phase response but a normal α_2-zone in the face of a raised α_1-band suggests enhanced haptoglobin catabolism as a result of haemolysis. This is often seen in malignancy and liver disease. Haptoglobin-haemoglobin complexes have a more cathodal mobility than haptoglobin and migrate in the α_2-β_1-interzone where they are frequently seen in samples haemolysed *in vitro*. An increase in the sharp leading edge of the α_2-zone results from a high level of α_2-macroglobulin such as is often seen in normal children, in women either in pregnancy or taking oral contraceptives, and in the nephrotic syndrome.

Alpha$_2$-beta$_1$-interzone. Cold insoluble globulin forms a sharp faint band between the α_2- and β_1-zones. It is precipitated by heparin and is thus not present in heparinised plasma. It is increased in pregnancy. Beta-lipoprotein forms an irregular crenated band anywhere from the α_2- to the β_2-zones. The mobility is affected by concentration and by electrophoretic conditions. It is often visible on agarose but not usually on cellulose acetate when it overlays the α_2-band. An abnormal lipoprotein (lipoprotein x) with low mobility is sometimes seen in the β_1-β_2-region in association with cholestasis.

Beta-zone. Transferrin comprises the β_1-band seen in fresh samples, while native C3 forms the β_2-band. C3 breakdown whether *in vivo* or *in vitro* results in decreased intensity of the β_2-band with an increased intensity of the β_1-band owing to the more anodal mobility of C3c. Splitting of the β_1-band is occasionally seen due to genetic variants of transferrin which are not uncommon. An increased β_1-band is typical of iron-deficiency anaemia while decreased levels are seen in malnutrition and liver disease. β_2 intensity reflects native C3 levels and is therefore often decreased in serum samples and absent in frozen samples where *in vitro* C3 breakdown has occurred. Increased levels occur in the acute-phase response.

Fibrinogen. Plasma samples show a fibrinogen band in the β-γ-region while a faint band in serum samples may be due to the presence of fibrinogen breakdown products formed as a result of allowing the serum to stand on the clot. Increased fibrinogen levels occur as part of the acute phase response.

Gamma-zone. In the normal gamma zone only the immunoglobulins are visible though in pathological conditions other proteins may be seen. IgA has the most anodal mobility of the immunoglobulins and migrates in the β-γ area. IgA deficiency is suggested by pallor in this area and high levels of IgA such as are commonly seen in cirrhosis and rheumatoid arthritis result in increased staining density, the so-called β-γ fusion. Polyclonal IgM increases are difficult to recognize, while increase in IgG may involve all subclasses giving rise to a generalized increase in the γ band, or may involve only fast or slow subclasses. Zones or faint bands in the γ-region are commonly seen in conditions giving rise to oligoclonal immunoglobulin responses; these suggest chronic immune stimulation and are commonly seen in chronic hepatitis and chronic viral infections.

Paraproteins are of immense importance and while usually present in the γ-region may fall anywhere on the separation especially if they complex with other proteins. Substances which may be mistaken for paraproteins are occasionally seen. IgG aggregates produced by freezing and thawing samples and immune complexes produce diffuse bands at the origin. C-reactive protein typically forms a very narrow faint band in the mid-γ region. Lysozyme very rarely reaches high enough concentration in the plasma to be seen as a band cathodal to the slowest γ-band.

Infected Samples. Neuraminidase-producing bacteria contaminating samples may result in desialation of glycoproteins with all fractions except albumin and IgG running more cathodally than usual.

Cerebrospinal Fluid (CSF)

The investigation of CSF and urine proteins will be described in other sections of the book but brief mention will be made here of the electrophoretic appearances (Fig. 17). Ventricular and lumbar CSF differ enormously in protein content as the CSF reaching the lumbar cord has had time to equilibrate with plasma and consequently has a much higher content of plasma proteins. One of the most striking aspects of CSF electrophoresis is the strong pre-albumin band. The α_1-zone is more cathodal than serum as a result of desialation of α_1-antitrypsin by brain neuraminidase. The α_2-zone is faint as high molecular weight α_2-macroglobulin and haptoglobin polymers do not easily enter the CSF. Transferrin is present as two bands, in the native form in the β_1 position and as the desialated 'Tau protein' in the β_2. C3 is not visible in CSF and in fact has a slightly faster mobility than the Tau protein. The γ-region often contains a faint band in the mid-zone but the presence of multiple γ-bands is typical of intrathecal IgG synthesis such as occurs in disseminated sclerosis.

Urine

Glomerular proteinuria results in the passage of increased amounts of proteins with molecular weights above 60 000 into the urine. Thus in nephrotic syndrome albumin and transferrin are usually seen. Tubular damage results in the appearance of proteins not seen on serum electrophoresis such as the α_2-microglobulins and β_2-microglobulin. Bence Jones protein may appear as a band *anywhere* on the electrophoretic separation; it is particularly prone to complex with albumin or α_1-antitrypsin. The acute phase response often results in the appearance of increased amounts of α_1-proteins in the urine as these low molecular weight proteins easily pass the renal glomerulus and may saturate their tubular reabsorbtive mechanisms (Fig. 18).

THE USE OF SPECIFIC PROTEIN MEASUREMENTS IN DISEASE

A combination of electrophoresis and specific protein measurements is of considerable value in the diagnosis and management of many diseases. Limited profiles of groups of proteins such as immunoglobulins and complement components or acute-phase proteins are widely used in clinical laboratories. Some workers recommend the use of more comprehensive profiles particularly in investigating non-specific symptomologies.

Inflammation

Acute-phase protein measurements are useful in the differential diagnosis and management of inflammatory diseases. On electrophoresis the acute phase reaction is seen as an increase in the α_1- and α_2-zones.

In acute generalized inflammation or tissue damage C-reactive protein and anti-chymotrypsin rise rapidly within the first twelve hours and are often associated with a fall in albumin and transferrin of about 10%. This is probably due to loss into inflammatory exudate. Orosomucoid, haptoglobin, fibrinogen and α_1-antitrypsin levels rise 24 to 48 hours after the onset of inflammation. These are followed by a moderate rise in C3, C4, prothrombin, caeruloplasmin and plasminogen

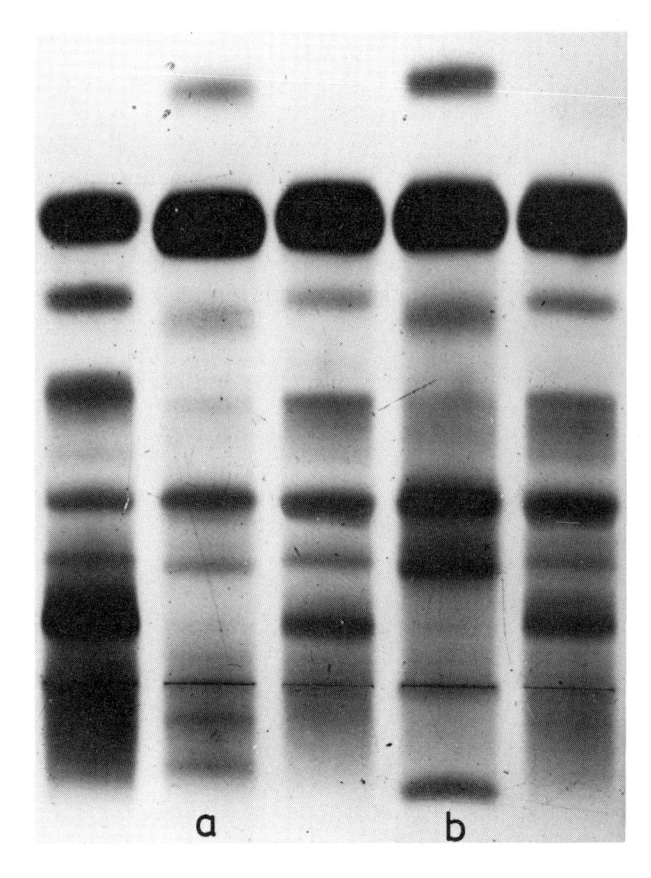

FIG. 17. Electrophoresis of concentrated CSF. Intervening plasma separations have been placed on the plate for comparison. The prealbumin is prominent; the alpha$_1$ is slow due to desialation; the alpha$_2$ is faint; transferrin is in the normal beta$_1$ position; the 'Tau' protein can be seen slightly cathodal to the position of C3 in plasma.

(a) Shows multiple gamma bands due to an oligoclonal IgG response in a patient with disseminated sclerosis.
(b) Shows a single gamma band in a patient with disseminated sclerosis.

Reproduced with permission from J. T. Whicher 'The Interpretation of Electrophoresis', *British Journal of Hospital Medicine*, 1980.

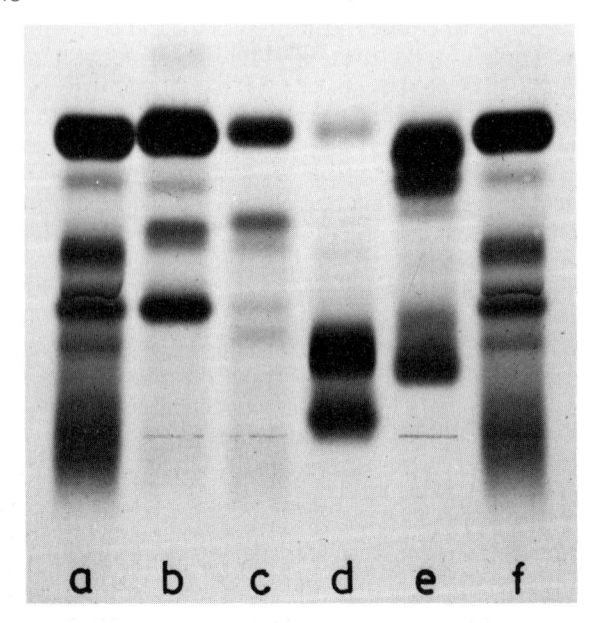

FIG. 18. Electrophoresis of concentrated urine compared with normal serum.

(a) Normal serum.
(b) Glomerular proteinuria with albumin, alpha$_1$, alpha$_2$ and transferrin. Note the fast alpha$_2$ resulting from the passage of non-polymerized haptoglobin molecules into the urine in a patient with the Hp1-1 phenotype. Less alpha$_2$ is seen in nephrotics with the polymerizing phenotypes Hp1-2 and Hp2-2.
(c) Tubular proteinuria. Relatively small albumin band. Split alpha$_2$ due to the presence of alpha$_2$ microglobulins. Trace of transferrin with beta$_2$ microglobulin in the beta$_2$ position, just anodal to the serum C3.
(d) Bence-Jones proteinuria. Two dense bands of Bence-Jones protein in the gamma region.
(e) Factitious proteinuria due to the addition, by the patient, of egg white to the urine. Note that none of the bands have the same mobility as the bands in the serum (f). This expression of the Munchausen syndrome has been seen in a number of cases.
(f) Normal serum.

Reproduced with permission from J. T. Whicher 'The Interpretation of Electrophoresis', *British Journal of Hospital Medicine*, 1980.

four or five days later. If an immune reaction is associated with the inflammation immunoglobulin levels rise 10–14 days later. C-reactive protein has a short half-life and its concentration changes more rapidly than those of other proteins.

The level of some acute-phase proteins may be used as an indication of the severity of inflammation in chronic inflammatory disease with certain provisos. α_1-antitrypsin and antichymotrypsin show little increase in disease associated with vasculitis such as rheumatoid arthritis, possibly due to catabolism of α_1-antitrypsin protease complexes. Chronic inflammation involving the liver is often associated with high levels of α_1-antitrypsin with no orosomucoid increase. A similar pattern is produced by oestrogens or pregnancy. Failure of haptoglobin concentration to increase in chronic inflammation suggests increased intravascular haemolysis such as may occur with hepatic or bone marrow metastases, severe sepsis or autoimmune haemolysis. Children under ten years of age often do not show a haptoglobin response. Normal or decreased fibrinogen levels in inflammation suggest active fibrinolysis such as occurs in disseminated intravascular coagulation. These changes are all reflected in the appearance of the electrophoretic strip.

Immune-complex Disease

The diagnosis of immune-complex disease may be considerably aided by evidence of complement activation or consumption which suggests active intravascular complex deposition. Direct measurements of immune complexes are very valuable. Protein changes such as cryoglobulins, the raised IgA of rheumatoid arthritis or the presence of complement deficiency may give important clues about the aetiology. Disease activity and response to therapy may be monitored by measuring complement breakdown products and acute phase proteins such as orosomucoid or C-reactive protein. There is evidence that C-reactive protein levels reflect the probability of joint destruction in rheumatoid arthritis.

Infectious Disease

Infectious diseases are usually diagnosed and monitored by identification of the offending organisms. However some diseases which may prove difficult to diagnose are associated with useful protein changes. For example, brucellosis, Q fever and Kala Azar are associated with marked increases in IgM whereas most chronic infections produce an IgG response. Immune deficiency and complement deficiencies are important, though rare, causes of chronic or recurrent infection, particularly in childhood. The important immune complex deposition associated with many infections, such as the glomerulonephritis of subacute bacterial endocarditis, may be diagnosed and monitored by complement measurement. C3 breakdown is occasionally a vital clue in the differential diagnosis of bacterial shock in the surgical patient. The rapid diagnosis of infection complicating burns, leukaemia or the post-operative period can be very successfully made by serial measurement of C-reactive protein in such patients.

A raised IgM level in cord blood or in the neonate is a very useful indication of intrauterine or early postnatal infection. It is useful to preserve a sample of cord blood frozen so that in the event of failure to thrive this can then be compared with a sample taken at the time.

Liver Disease

Protein changes may provide useful information in liver disease. Those which commonly occur are shown in Table 20. Electrophoresis is a helpful initial investigation in suspected cirrhosis. The albumin, α_1- and α_2-globulin levels are often decreased while the high IgA and IgG concentrations result in an increased γ-globulin band with β-γ fusion. Primary biliary cirrhosis often shows very high levels of IgM and chronic aggressive hepatitis of IgG; β-γ fusion is often not evident. Alcoholism is associated with high levels of α_1-lipoprotein with increased density of the albumin-α_1 interzone.

TABLE 20
PROTEIN CHANGES IN LIVER DISEASE

Disease	Immunoglobulins	Acute phase proteins	Complement	Other proteins
Hepatitis A	IgM ↑ IgG ↑	↑ α_1-AT	—	—
Hepatitis B	IgG ↑	↑ α_1-AT	activation common	—
Alcoholic cirrhosis	dominant IgA ↑	haptoglobin ↓	—	α-lipoprotein ↑
Chronic aggressive hepatitis	dominant IgG ↑	—	—	—
Primary biliary cirrhosis	dominant IgM ↑	caeruloplasmin ↑ or ↓	—	lipoprotein X
Liver secondaries	—	haptoglobin ↓	—	CEA ↑
Hepatoma	—	—	—	α_1-foetoprotein
Biliary obstruction	—	caeruloplasmin ↑	—	lipoprotein X
Decreased liver cell mass	—	—	—	prealbumin ↓ prothrombin ↓

Gastrointestinal Disease

Protein changes in gastrointestinal disease are of little use diagnostically but may be of considerable value for assessing complications and monitoring therapy. Alpha-chain disease is one of the few conditions in which a specific marker protein is of considerable importance in establishing the diagnosis. The disappearance of free alpha-chains is strong evidence of a successful response to treatment. Carcinoembryonic antigen, while by no means specific for gut tumours, is nevertheless very useful for monitoring the effect of surgery, chemotherapy and radiotherapy.

Excessive protein loss from the gut characterizes many diseases diffusely involving the mucosa. Protein changes in the serum resemble those of the nephrotic syndrome without raised levels of α_2-macroglobulin and β-lipoprotein but with decreased albumin and transferrin. The clearance of α_1-antitrypsin into the faeces may be used to assess this loss.

Many inflammatory gut diseases show changes in acute-phase proteins but these are of limited use in management except in Crohn's disease and ulcerative colitis.

IgA is raised in a number of conditions and a persistent high level in coeliac disease, despite treatment, suggests the possibility of lymphoma or progressive reticular hyperplasia. It is also worth noting that gut disease is common in IgA deficiency and that C1INA deficiency often results in abdominal pain due to mucosal oedema with a history of recurrent laparotomies.

Renal Disease

Glomerular protein loss in renal disease results in characteristic serum changes with increased levels of α_2-macroglobulin and β-lipoprotein with low levels of albumin, transferrin and IgG. Synthesis of α_2-macroglobulin and β-lipoprotein is switched on possibly due to albumin loss and increased levels are attained because these proteins have high molecular weights and are not lost in the urine. The raised β-lipoprotein is the reason for the increased level of plasma cholesterol seen in the nephrotic syndrome. These changes result in the characteristic electrophoretic pattern of decreased albumin and γ-globulin with increased α_2-globulin. Selective protein clearances into the urine are of considerable value in the prognosis and assessment of likely response to steroids in renal disease.[22] This is particularly true of children where the differential diagnosis is simple and renal biopsy is undesirable. An IgG:albumin clearance of less than 0·16 suggests minimal change nephropathy with an excellent steroid response. In adults a selective clearance may be misleading as for example in amyloidosis where steroid treatment is disastrous.

A greater clearance of IgM than would be expected from albumin and IgG measurement is seen in pyelonephritis. The clearance of low molecular weight proteins such as β_2-microglobulin into the urine is of considerable value in the assessment of tubular damage. Serum immunoglobulin changes are not particularly useful owing to a number of complicating factors. Proteinuria results in a catabolic decrease in IgG while uraemia causes a toxic inhibition of immunoglobulin synthesis which affects IgM most. IgA is often raised in pyelonephritis and glomerulonephritis.

Complement measurement is of great importance in distinguishing the serious membrano-proliferative glomerulonephritis from other milder forms.

Neoplasia

A wide range of tumours are now known to produce proteins which can be used with varying degrees of specificity for diagnosis and monitoring.

Secondary host responses to malignant tumours are now of some value in monitoring therapy. Most tumours elicit acute-phase type responses which in a few cases may have remarkable specificity. Measurement of serum caeruloplasmin or copper levels is widely used to monitor therapy and detect relapse in Hodgkin's disease. C1-esterase inhibitor and β_2-microglobulin are useful in other lymphoid neoplasms; β_2-microglobulin may be as useful in myeloma as the paraprotein level. Orosomucoid and haptoglobin are of some value in carcinoma of kidney, bronchus and ovary. It seems clear that serial measurements of a number of proteins will become important tools in the management of malignant disease.

Skin Disease

Many skin diseases are associated with either generalized or localized vasculitis. The detection of immune-complex deposition is clinically important in terms of both aetiology and therapy. Eczema may be associated with IgA deficiency and raised IgE levels are commonly seen in atopic forms. Antigen specific IgE measurement may be of value in assessing the possibility of desentitization. Raised IgA levels are common in many skin diseases but are of little value. The rare condition of lichen myxoedematosis is associated with the presence of IgG lambda paraproteins.

REFERENCES

1. Fudenberg, H. H., Good, R. A., Goodman, H. C., Hitzig, W., Kunkel, H. G., Roitt, I. M., Rosen, F. S., Rowe, D. S., Seligmann, M. and Soothill, J. R. (1971). 'Primary immunodeficiencies,' *Bull. W.H.O.* **45**, 125–142.
2. Bushell, A. C., Whicher, J. T. and Yuille, T. (1979), 'The progressive appearance of multiple Bence Jones proteins and serum paraproteins in a child with immune deficiency,' *Clin. Exp. Immunol.*, **38**, 64–69.
3. Kohn, J. (1979), 'Monoclonal proteins' in *Immunochemistry in Clinical Laboratory Medicine*, pp. 115–126. (Milford Ward, A. and Whicher, J. T. Eds.). Lancaster, U.K.: MTP Press.
4. Kohn, J. (1973), 'The laboratory investigation of paraproteinaemia,' in *Recent Advances in Clinical Pathology*. Series 6, pp. 363–401. (Dyke, S. C. Ed.). Edinburgh and London: Churchill Livingstone.
5. Whicher, J. T., Hawkins, L. and Higginson, J. (1980), 'Clinical applications of immunofixation. A more sensitive technique for the detection of Bence Jones protein,' *J. Clin. Path.*
6. Hobbs, J. R. (1971), 'Immunoglobulins in clinical chemistry,' *Advances in Clinical Chemistry*, **14**, 219–317.
7. Gale, D. S. G., Versey, J. M. B and Hobbs, J. R. (1974), 'Rocket immunoselection for detection of heavy-chain disease,' *Clin. Chem.*, **20**, 1292–1294.
8. Fearon, D. T., Ruddy, S., Schur, P. H. and McCabe, W. R. (1975), 'Activation of the properdin pathway of complement in patients with gram-negative bacteraemia,' *New Eng. J. Med.*, **292**, 937–940.
9. McConkey, B., Crockson, R. A. and Crockson, A. P. (1972), *Q. J. Med.*, New Series XLI, No. 162, 115–125.
10. James, K. (1980), 'Alpha$_2$-macroglobulin and its possible importance in immune systems,' *Trends in Biochemical Sciences*, 43–47.
11. Fagerhol, M. K. and Abildgaard, U. (1970), 'Immunological studies on human antithrombin III. Influence of age, sex and use of oral contraceptives on serum concentrations,' *Scand. J. Haematol.*, **7**, 10–17.
12. Bernier, J. J., Florent, C., Dermazures, C., Aymes, C. and L'Hirondel, C. (1978), 'Diagnosis of protein losing enteropathy by gastrointestinal clearance of α_1-antitrypsin,' *Lancet* II (1978), 763.
13. Slater, L., Carter, P. M. and Hobbs, J. R. (1975), 'Measurement of albumin in sera of patients,' *Ann. Clin. Biochem.*, **12**, 33–40.
14. Leibman, A. and Aisen, P. (1979), 'Distribution of iron between the binding sites of transferrin in serum. Methods and results in normal human subjects.' *Blood*, **53**, 1058–1065.
15. Gutteridge, J. M. C. (1978), 'Caeruloplasmin: a plasma protein, enzyme and antioxidant,' *Ann. Clin. Biochem.*, **15**, 293–296.
16. Delves, H. T. (1976), The clinical value of trace metal measurements,' *Essays in Medical Biochemistry*, **2**, 37–73.
17. Scanu, A. M., Edelstein, C. and Keim, P. (1978). 'Serum lipoproteins,' in *The Plasma Proteins*, (Putnam, F. W. Ed.). New York: Academic Press.
18. Goodman, O. S. (1976), 'Retinol-binding protein, prealbumin and vitamin A transport,' *Prog. Clin. Biol. Res.*, **5**, 313–330.
19. Thompson, O. M., Rauch, J. E., Wetherhead, J. C., Friedlander, P., O'Connor, R., Croner, N., Schuster, J. and Gold, P. (1978), 'Isolation of human tumour-specific antigens associated with β_2-microglobulin,' *Br. J. Cancer*, **37**, 753–775.
20. Norgaard-Pedersen, B. (1976), 'Human alpha-fetoprotein,' *Scand. J. Immunol.*, Suppl. 4.
21. Klopper, A. (1980), 'The new placental proteins,' *Placenta*, **1**, 77–89.
22. Pollak, V. E., First, M. R. and Pesce, A. M. (1974), 'Value of the sieving coefficient in the interpretation of renal protein clearances,' *Nephron*, **13**, 82–94.

SUGGESTIONS FOR FURTHER READING

Protein Metabolism

Ciba Foundation Symposium 9 (new series) (1973), *Protein Turnover*, Amsterdam: North Holland, Elsevier.
Tavill, A. S. and Hoffenberg, R. (1976), 'Turnover of plasma proteins,' in *Structure and Function of Plasma Proteins*, Vol. 2, (Allison, A. C. Ed.). New York and London: Plenum Press.
Schultz, H. E. and Heremans, J. F. (1966), *Molecular Biology of Human Proteins*, Vol. 1. Amsterdam: Elsevier.

Immunoglobulins

Hobbs, J. R. (1971), 'Immunoglobulins in clinical chemistry,' in *Advances in Clinical Chemistry*, Vol. 14, 219–317.
Putnam, F. W. (1977), 'Immunoglobulins I. Structure and Immunoglobulins II. Antibody specificity and genetic control,' in *The Plasma Proteins* (Putnam, F. W. Ed.), Vol. 3. New York: Academic Press.

Complement

Fearon, D. T. and Austen, F. K. (1976), 'The human complement system, biochemistry, biology and pathology,' *Essays in Medical Biochemistry*, **2**, 1–37.
Whicher, J. T. (1978), 'The value of complement assays in clinical chemistry,' *Clin. Chem.*, **24**, 7–22.
Fearon, D. T. and Austen, F. K. (1977), 'Activation of the alternative complement pathway with rabbit erythrocytes by circumvention of the regulatory action of the endogenous control proteins,' *J. Exp. Med.*, **146**, 22–33.

Acute Phase Proteins

Koj, A. (1974), 'Acute phase reactants. Their synthesis, turnover and biological significance,' in *Structure and Function of Plasma Proteins*, Vol. 1, (Allison, A. C. Ed.). New York and London: Plenum Press.
Carrell, R. W. and Owen, M. C. (1979), 'α_1-antitrypsin: structure, variation and disease,' *Essays Med. Biochem.*, **4**, 83–119.
Putnam, F. W. (1975), 'Haptoglobin,' in *The Plasma Proteins*, Vol. 2, (Putnam, F. W. Ed.). New York: Academic Press.

Transport Proteins

Peters, T. Jr. (1975), 'Serum albumin,' in *The Plasma Proteins*, Vol. 1, (Putnam, F. W. Ed.). New York: Academic Press.
Putnam, F. W. (1975), 'Transferrin,' in *The Plasma Proteins*, Vol. 1, (Putnam, F. W. Ed.). New York: Academic Press.

Oncofoetal Proteins

Bagshawe, K. D. and Searle, F. (1977), 'Tumour markers,' *Essays Med. Biochem.*, **3**, 25–73.

Norgaard-Pedersen, B. and Axelsen, N. H. Eds. (1978), 'Carcino-embryonic proteins. Recent progress,' *Scand. J. Immunol. Supplement 8*, Vol. 8.

Amyloid

Rosenthal, C. J. and Franklin, E. C. (1977), 'Amyloidosis and amyloid proteins,' *Recent Advances in Clinical Immunology*, I, 41–77.

Electrophoresis

Jeppsson, J. O., Laurell, C. B. and Franzen, B. (1979), 'Agarose gel electrophoresis,' *Clin. Chem.*, **25**, 629–638.

Thompson, E. J. (1979), 'Immunochemistry of CSF proteins,' in *Immunochemistry in Clinical Laboratory Medicine*, (Milford Ward, A. and Whicher, J. T., Eds.). Lancaster: MTP Press.

Specific Protein Measurement in Disease

Keyser, J. W. (1979), *Human Plasma Proteins. Their Investigation in Pathological Conditions*. Chichester and New York: John Wiley & Sons.

14. THE CLINICAL BIOCHEMISTRY OF BLOOD COAGULATION

DENNIS E. G. AUSTEN

Introduction

General haemostatic mechanism
   General platelet mechanism
   General clotting mechanism
   General fibrinolytic mechanism

Diseases of haemostasis
   Haemophilia and Christmas disease
   von Willebrand's disease
   Fibrinogen deficiency
   Factor XI deficiency
   Factor XII deficiency
   Kinin deficiency
   Factor XIII deficiency
   Antithrombin deficiency
   Combined deficiencies of clotting factors
   Platelet deficiencies
   Defects in fibrinolysis
   Disorders not predominantly associated with defects of clotting factors, platelets or fibrinolysis

Biochemistry of factors and components

Assays of factors and components

INTRODUCTION

There is so much published work which could be included in this chapter that some rigorous discrimination has been inevitable. Selection has been by including that which is important in clinical practice, in treatment of patients and in identification of bleeding disorders. As a result, proportionally less space has been allotted to many studies of molecular structure; these studies tend to be carried out on factors which are present in relatively large amounts in the blood, which lend themselves to complete purification, and which are not the factors usually absent in bleeding disorders. It has not been possible to do justice to all the work reported on platelets and fibrinolysis, as these aspects represent only a modest proportion of bleeding abnormalities; the space allotted to them has been limited accordingly.

GENERAL HAEMOSTATIC MECHANISM

Haemostasis, at the mechanical level, is considered to involve the constriction of damaged blood vessels and the blockage of leaks by a haemostatic plug. The plug is first constructed by aggregating platelets and quickly reinforced by fibrin and this must all take place in a system which provides a control mechanism to restrict blockage only to the wound area and another to dissolve the plug away again once it is no longer necessary. A fascinating inter-play of biochemical reactions is called for, not all of which are understood. More of the general mechanisms are discussed below under the headings of platelets, clotting factors and fibrinolysis.

General Platelet Mechanism

In the complex procedure of building a haemostatic plug, platelets have a series of functions including adhesion to the damaged tissue, aggregation to form the plug and release of essential chemicals. Platelets do not normally adhere to endothelium but when a tissue is damaged and collagen is released they will adhere to the subendothelial tissue which has been exposed. Some adenosine diphosphate (ADP) is released from damaged tissues and even more from platelets which are adhering to subendothelium. When released, it causes platelets to stick to one another, forming aggregates and building up a platelet plug based upon the original injury. Phospholi-

TABLE 1
BASIC PROPERTIES OF THE CLOTTING FACTORS

Factor	Synonyms	Present in human serum	Present in Cohn's fraction	Easily adsorbed by alumina and barium sulphate	Precipitated by ammonium sulphate % saturation
I	Fibrinogen	No	I	No	25
II	Prothrombin	No	III & IV	Yes	50
III	Thromboplastin: Tissue factor				
IV	Calcium				
V	Labile factor: Proaccelerin Accelerator globulin	No	III	No	50
VII	Proconvertin: Stable factor Autoprothrombin I Serum prothrombin conversion accelerator (SPCA) Serum accelerator globulin	Yes	III & IV	Yes	50
VIII	Antihaemophilic globulin (AHG) Antihaemophilic factor A Thromboplastinogen A Platelet cofactor I	No	I	No	33
IX	Christmas factor Plasma thromboplastin component (PTC) Antihaemophilic factor B: Platelet cofactor II Thromboplastinogen B: Autoprothrombin II	Yes	III & IV	Yes	50
X	Stuart Prower factor Autoprothrombin III	Yes	III	Yes	50
XI	Plasma thromboplastin antecedant (PTA) Antihaemophilic factor C	Yes	III & IV	Partly	33
XII	Hageman factor Surface factor	Yes	III & IV	Partly	50
XIII	Fibrin stabilizing factor (FSF) Fibrinase; Laki-Lorand factor	Yes	I	No	33

pid which is also released from platelets (platelet factor III) catalyses reactions of blood clotting factors and the platelet plug becomes reinforced with resultant fibrin giving rise to a firm plug capable of staunching blood flow.

General Clotting Mechanism

The basic pattern of sequential biochemical reactions in blood clotting is well known and extensively described. Synonyms and basic properties of the clotting factors are given in Table 1. In essence the sequential reactions are those shown in Fig. 1 and involve each factor becoming activated and then subsequently activating its successor. At various points, calcium ions and phospholipid are required to obtain the necessary reaction velocities and in addition three of the factors are believed to operate more as co-factors in their relevant steps. These are factors V, VIII and VII but it should be added that this does not preclude their becoming activated as well, while carrying out the clotting functions. Sequential clotting reactions initially involve two paths which meet at the point where they both give rise to activated factor X and, thereafter, a single chain of reactions proceeds from activated factor X until finally a fibrin clot is produced. Of these, that initial path which commences with the activating effect of tissue juices, derived from a wound, is still known by its original name

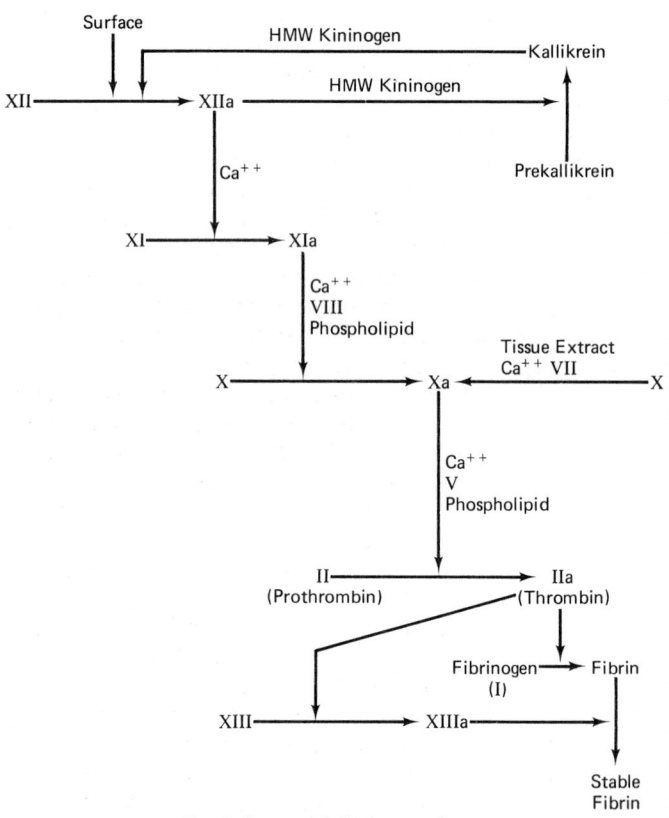

FIG. 1. Sequential clotting reactions.

of the 'extrinsic system'. Likewise, the chain which commences with surface activation of factor XII is known as the 'intrinsic system'. More recently, one section of the clotting scheme which has been further investigated is that concerned with kallikreins. These participate in a somewhat cyclic reaction as shown in Fig. 1 whereby they can become activated by the intrinsic clotting chain and yet, in active form, can themselves activate that same chain. More discussion of this is provided in the relevant sections. Another embellishment is that certain cross-reactions appear to be possible; activated factor XII seems able to participate to some degree in the extrinsic clotting chain and abnormal factor IX can, under certain conditions, affect extrinsic clotting.

The sequence of reactions discussed so far is concerned with activation, clotting and the arrest of haemorrhage but it is also essential that these reactions are contained in the areas where their function is required. It is the function of antithrombin and other naturally occurring inhibitors to consume active components of the clotting process and in this way it is ensured that clotting, once started, does not proceed beyond that area where it is needed to prevent blood loss.

General Fibrinolytic Mechanism

Once bleeding has been staunched by a haemostatic plug, it is necessary as part of the physiological process that fibrin reinforcement in the plug should be dissolved. This process of fibrinolysis is achieved by activation of plasminogen into plasmin, an enzyme whose action is to convert fibrin into soluble components. Plasminogen, the inactive precursor, circulates in blood and has a weak affinity for fibrin. Its affinity becomes stronger when the process of activation has begun and it continues to be absorbed and activated, leading to an accumulation of fibrinolytic plasmin within the fibrin. In the absence of fibrin, any plasmin which might be formed in circulating blood is rapidly inactivated by ever-present inhibitors and in this way the action of plasmin is limited. Most important of the circulating inhibitors are α_2-antiplasmin and to a lesser extent α_2-macroglobulin.

In the fibrinolytic process, activators of plasminogen are divided into two classes dependent upon their source of origin. Intrinsic activators are those derived from precursors (called proactivators) which are themselves circulating in the blood. Extrinsic activators, on the other hand, are released into the circulation from cells which in the main are probably endothelial cells. It appears that intrinsic activators can be formed by two routes, one dependent and one independent of clotting factor XII. In the former route, activated factor XII and the activated complex of high molecular weight kininogen with prekallikrein initiate the process leading to generation of intrinsic activators from their precursors (proactivators). While all of this seems incomplete, there does appear to emerge a fascinating picture of interrelation between clotting and fibrinolytic reactions.

DISEASES OF HAEMOSTASIS

Haemophilia and Christmas Disease

Haemophilia (sometimes called haemophilia A) is the commonest of the severe hereditary bleeding disorders and Christmas disease (sometimes called haemophilia B) is the next most common. These two diseases are clinically indistinguishable; only laboratory assays will differentiate between them. They are both inherited by an X-linked recessive mode and bleeding can be severe and prolonged, even following minor injury, typically into muscles and joints. Bleeding into deep tissues can be apparently spontaneous but when injury is evident the bleeding tends to occur several hours or more after injury. Repeated episodes involving joints, particularly if treatment is delayed, will lead to crippling.

The two diseases can be seen in a range of severities which are fairly well related to the level of factor in the blood. A complete absence of factor VIII or IX results in a very severe diathesis with life-threatening bleeds following minor trauma. A small proportion of factor VIII or IX will, however, decrease the severity enormously and at 25% of the average normal level the patient would probably have few problems, except at surgery (including teeth extractions). The female carrier of haemophilia will, on average, have 50% of normal factor VIII levels but the range between individuals is wide and some can have very little while others have a normal level. These carriers have bleeding problems if their factor VIII or IX is low, roughly as described above for severity of haemophilia and they may suffer from epistaxis and menorrhagia. Treatment of haemophilia and Christmas disease requires replacement therapy with factor concentrates.

Haemophilia can also be acquired by patients who already have some underlying disease such as rheumatoid arthritis, ulcerative colitis or penicillin allergy and it also develops in extremely rare cases after childbirth. In these instances, an antibody develops which destroys factor VIII in the patient's blood. Antibodies also develop in response to replacement therapy in about 6% of haemophiliacs.

von Willebrand's Disease

In this complaint, factor VIII is absent or reduced but it is differentiated from haemophilia by the absence or reduction of factor VIII-related antigen and ristocetin co-factor (these are discussed later). Haemorrhages into muscles and joints are rare in von Willebrand's disease and, instead, bleeding is typically from mucous membranes. Both sexes are affected. Epistaxis is common and can be difficult to control. Gastrointestinal bleeds are also fairly common but haematuria is not a particular feature of the complaint. Easy bruising and excessive bleeding are experienced following trauma and although in these respects the disease might be considered to be milder than haemophilia, some patients have extremely severe and life threatening bleeds particularly through epistaxis. The inheritance pattern is autosomal dominant and treatment is by replacement therapy using fresh frozen plasma, cryoprecipitates or factor VIII concentrates.

Fibrinogen Deficiency

This is rare as a congenital disease and surprisingly the congenital deficiency state is a relatively mild bleeding disorder which usually gives problems only upon surgery or severe trauma. Bleeding into muscles or joints and menorrhagia are rare. It is inherited by an autosomal recessive mechanism and treatment is by replacement therapy using fresh-frozen plasma or fibrinogen concentrates.

Dysfibrinogenaemia is another rare disorder and roughly half the affected people investigated have shown no clinical abnormality. Presumably there are more people with this deficiency state who, being symptomless, have not presented themselves. Out of those families investigated, a proportion had members with a mild bleeding disorder (about 21 families) while, in about 6 families, deficient wound healing was recorded. In other cases a thrombotic tendency was noted (about 6 families, but in 3 of these there were also members with a mild bleeding problem).

Fibrinogen deficiency is also seen as an acquired disorder during intravascular coagulation following the entry into the circulation of some material which triggers the clotting process, probably tissue thromboplastin. In this condition there will usually be an additional reduction in platelet numbers and in the levels of factors II, V and VIII. Occasions when such a condition might arise are in obstetric complications such as premature separation of the placenta, in surgical operations, for example prostatectomy, in chronic illnesses, for example with carcinoma, in snakebite, in some conditions of purpura (e.g. purpura fulminans which is a rare reaction to certain infectious fevers) or as a reaction to a drug. There is profuse bleeding, with epistaxis, melaena, haematemesis and massive bruising. Treatment for intravascular coagulation involves blood replacement, possibly infusion of clotting factors and sometimes heparin therapy. If the prime agent responsible for the intravascular clotting can be removed, rapid improvement is possible and this is the case when the uterus is emptied following premature separation of the placenta.

Factor XI Deficiency

This disease is usually a mild haemorrhagic disorder but a small proportion of patients are moderately affected. Bleeding severity often does not correlate with the level of factor XI in the blood. It is inherited by an autosomal dominant mechanism and most of the patients are of Jewish descent. Replacement therapy using fresh-frozen plasma is usually adequate.

Factor XII Deficiency

Individuals who have a deficiency in factor XII will have prolonged clotting times in laboratory tests of the intrinsic system and, using factor XII deficient plasma, an assay can be carried out. However, these people do not bleed significantly and the sole value of the tests is to avoid confusion with the real haemorrhagic diseases.

Kinin Deficiency

Kinin deficiency states are known under several different names because at the time of their discovery the exact nature of the defect was not understood. In Fitzgerald, Flaujeac or Williams trait it is high molecular weight kininogen which is decreased or missing. Prekallikrein is absent or reduced in Fletcher factor deficiency, an autosomal recessive condition. In general, kinin deficiency does not cause excessive bleeding even though the clotting times recorded in laboratory tests are prolonged.

Factor XIII Deficiency

This is a rare condition with severe bleeding and classically with bleeding from the umbilical stump after birth. Wound healing is slow, giving irregular and retracted scars and intracranial bleeding seems to be a very significant risk. Replacement therapy uses fresh frozen plasma or factor concentrates. Only a few per cent of factor XIII is required for haemostasis and its half-life is relatively long. Because of this, relatively infrequent transfusions are required and prophylaxis is easily possible. Autosomal recessive inheritance has been suggested.

Antithrombin Deficiency

Bleeding caused by excessive antithrombin is so rare that no comments can be made on treatment. Two cases have been reported one of which was apparently due to a molecular variant. On the other hand, there does appear to be a clear correlation between low antithrombin levels and a predisposition to thrombosis. It must be stressed however that it has not yet been proved that a low level of antithrombin *per se* leads to thrombosis, and this would be difficult to demonstrate, since thrombosis itself leads to consumption of antithrombin. Moreover, patients have been described with a predisposition to thrombosis who do not have reduced antithrombin levels.

Reduced levels of antithrombin arise in liver disease, chronic hepatitis, myocardial infarction, post operative states, post prandial hyperlipoproteinemia and in women taking oral contraceptives of high oestrogen level. An hereditary genetic condition of low antithrombin has been seen in a few families; levels as low as 25% of average normal have been reported but about 50% is more usual. Warfarin is generally used in the treatment of these conditions.

Platelet Deficiencies

Platelet abnormalities can be broadly classified into three categories. Platelets can be too few or too many in number, they can fail to adhere to the subendothelium or they can fail to aggregate to form a platelet plug. A list of platelet disorders is given in Table 2. Thrombocytopenia, the condition where platelets are too few in number is the commonest deficiency and most of the cases are those of secondary thrombocytopenia, resulting from the action of drugs, poisons, acute infection, radiation or as a consequence of malignancy or leukaemia. Hereditary thrombocytopenia is very rare.

TABLE 2
PLATELET DISORDERS

Disorder	Platelet abnormality affecting coagulation
Thrombocytopenia	Reduced number of platelets
Thrombasthenia (Glanzmann's disease)	Abnormal ADP aggregation
Storage Pool Deficiency	
(i) Hermansky-Pudlak Syndrome	Defective release reaction
(ii) Wiskott-Aldrich Syndrome	((ii) and (iii) also have thrombocytopenia)
(iii) Chédiak-Higashi Syndrome	
Cyclo-oxygenase deficiency	Defective release reaction
Defective nucleotide metabolism	Defective release reaction
Bernard-Soulier Syndrome	Defective adhesion to subendothelium
Myeloproliferative Disorders	Variable
Renal Disease	Variable
Macroglobulinaemia	Variable
Administration of Aspirin	Defective release reaction
Administration of Penicillin	Variable

Excessive bleeding can be associated with an increased number of platelets as seen in polycythaemia or thrombocythaemia. In addition, haemorrhagic disorders arise where platelets are present in normal numbers but where they have diminished functional ability. Such abnormalities include thrombasthenia, release reaction defects and Bernard-Soulier syndrome and these are discussed later.

Bleeding in platelet disorders is typically from mucous membranes with epistaxis, intestinal or uterine bleeding predominating and petechiae and small diffuse bruises are often seen in the skin. The diseases affect both sexes. Treatment may include infusions of platelet-rich plasma or platelet concentrates and in some cases there will be the need to treat some underlying complaint.

Defects in Fibrinolysis

There is still much discussion on this subject because of the difficulty in distinguishing the results of fibrinolysis from those of diffuse intravascular coagulation (DIC) and this is further complicated because increased fibrinolysis would be a normal sequel to DIC. If there is positive clinical and laboratory evidence for primary fibrinolysis then ε-amino-caproic acid or tranexamic acid is used in treatment but such materials would be contra-indicated for DIC. Successful treatment of increased fibrinolysis has been reported particularly in prostatic carcinoma, prostatectomy and during surgical correction of hypertension in patients with cirrhosis of the liver.

Diseases not Predominantly Associated with Defects of Clotting Factors, Platelets or Fibrinolysis

There is a series of diseases which have not been fully categorized although, in a proportion of them, there does appear to be a defect in the connective tissues in and around the blood vessels which can give rise to a bleeding tendency.

Henoch-Schonlein Purpura is a condition which, in rare cases, can follow an infection, or possibly arise as a reaction to certain foods or sometimes have no obvious cause at all. On the body, the rash is usually below the waist and on the arms and face. Other manifestations are colic, bloody diarrhoea and swollen tender joints, particularly the knees and ankles. The disease usually resolves on its own accord but renal involvement can occasionally be a serious side-effect.

Ehlers-Danlos Syndrome is characterized by hyperextensibility of the joints and hyperelasticity of the skin. Sometimes the eyes reveal blue sclerae. If a bleeding problem is present it is usually mild and consists of easy bruising, bleeding from mucous membranes and bleeding at surgery. Investigations in some patients revealed abnormal collagen which would result in defective platelet adhesion.

Pseudoxanthoma Elasticum involves a loss of skin elasticity with the appearance of skin thickening and nodules. When there is excessive bleeding it is usually from the mucous membranes.

Osteogenesis Imperfecta is marked by brittle bones and blue sclerae. Bleeding is not common but sometimes occurs in the form of easy bruising and bleeding from the mucous membranes. It has been suggested that platelets are sometimes abnormal as well.

Telangiectasia is characterized by vascular lesions which are also sites where bleeding can occur. These lesions can arise in different areas of the body developing during late childhood and can produce serious bleeding or little at all, depending on their position. Haemorrhage is typically from the mucous membranes and epistaxis is the commonest symptom.

Senile Purpura occurs on the arms, backs of the hands and occasionally on the face of old people, usually over seventy years old. A similar condition can arise in middle-aged patients on steroid therapy. No treatment is required in either case.

Scurvy. Bleeding is encountered in scurvy, particularly from the mucous membranes. In patients old enough to have teeth, there is swelling and bleeding of the gums and, in adults, petechiae appear particularly around the hair follicles.

Autoerythrocyte Sensitisation. This is a little understood condition which is found in adult women usually with an emotional disturbance and it has been proposed that they become sensitised against their own red cells, presumably just in the skin area. It recurs at intervals. There is no specific treatment but the disease does tend to become less severe over the years.

BIOCHEMISTRY OF FACTORS AND COMPONENTS

Factor XII

Factor XII is a single chain glycoprotein and estimates of its molecular weight have ranged from 20 000 to 140 000. More recent reports place it between 74 000 and 82 000 and values between 2% and 14% are recorded for the level of carbohydrate in the molecule.

In the sequence of clotting reactions, the role of factor XII is to become activated and then in its turn to activate factor XI and, in addition, in the presence of high molecular weight kininogen to activate prekallikrein. In its active form, factor XII can also initiate the fibrinolytic system; it increases vascular permeability, dilates blood vessels, induces leucocyte migration through blood vessel walls and can activate the first component of complement.

The manner in which factor XII itself is activated in the body is still open to conjecture. Current thoughts favour a mechanism in which the protein becomes absorbed to a negatively charged surface at the site of a wound. However, it appears that it does not spontaneously become activated and probably some unidentified proteolytic enzyme is necessary to bring about conversion.

In vitro, a wide variety of surfaces can absorb factor XII and participate in the reaction, including glass, bentonite, asbestos, collagen, skin, elastin and diatomaceous earths. Ellagic acid, silicic acid and long chain fatty acids are also effective, even when apparently in solution. Surfaces which do not cause significant activation include oils, waxes, silicones, some plastics and vascular endothelium. In a somewhat cyclic manner, as seen previously in Fig. 1, factor XIII can be activated by kallikrein in the presence of high molecular weight kininogen.

Activation of factor XII probably involves proteolytic cleavage at an arginyl-valine bond to produce a two-chain enzyme linked by disulphide bonds. The active site is believed to reside in the light chain.

Factor XI

In the clotting sequence, factor XI is activated by active factor XII in what is believed to be a proteolytic mechanism and, in its turn, it will activate factor IX. Activation of factor XI does not require calcium although the latter is required for the subsequent reaction with factor IX. The molecular weight of factor XI has been reported to be 124 000 and in electrophoresis it migrates in a position between the β- and γ-globulins. It readily adsorbs onto the same materials as factor XII but, when the proportion of adsorbent is limited, only partial adsorption will occur. This is a feature which can help in separating factors XI and XII, although adsorption processes will always lead to some degree of activation.

Factor XI is reported to be a two-chain disulphide-linked molecule which during activation is cleaved, producing a four-chain molecule possessing two light and two heavy chains. Active factor XI possesses esterase activity but it has been shown that this can be separated from its coagulant activity. As a result the manner in which factor XI participates in clotting reactions is still not clear.

Kallikrein and Prekallikrein (Fletcher factor)

Prekallikrein is converted to kallikrein by the action of active factor XII but kallikrein is able to activate factor XII (both reactions requiring the presence of high molecular weight kininogen). The exact significance of this cyclic reaction is not clearly worked out. Kallikrein, once formed, will also activate the fibrinolytic system and in addition it will release bradykinin from high molecular weight kininogen. It has been reported that prekallikrein exists in two forms with molecular weights of 85 000 and 88 000 both of which are activated by active factor XII. Kallikrein has been reported to be a two-chain molecule with its active site in the light chain. Deficiency of prekallikrein has been shown to be synonymous with Fletcher factor deficiency.

High Molecular Weight Kininogen (Fitzgerald, Flaugeac and Williams factors)

This substance has a reported molecular weight of 180 000 and is the co-factor for those kallikrein and factor XII reactions mentioned above. It is also acted upon by kallikrein to yield bradykinin, a nonapeptide with vasoconstrictive properties plus a substance called kinin-free high molecular weight kininogen. Deficiency of high molecular weight kininogen has been shown to be synonymous with Fitzgerald, Flaugeac and Williams traits (separately reported genetic conditions in which abnormal laboratory tests are encountered but in general without excessive bleeding).

Factor IX

Factor IX is a single chain glycoprotein present in normal blood plasma or serum. It is readily adsorbed by inorganic surfaces such as aluminium hydroxide, barium sulphate and calcium phosphate and it can be easily desorbed again. It is difficult to separate from factors II, VII and X and like these three it is produced in the liver by a process which is vitamin K dependent. The electrophoretic mobility has been variously reported and more recent results indicate that its mobility is between that of the α_2-globulins and albumin but that it is increased to nearly that of albumin in the presence of calcium ions. Values of isoelectric point vary between 4·0 or 4·6 for the human factor while a 3·7 is claimed for the bovine material. Molecular weights have been reported between 60 000 and 80 000 for human factor IX while a figure of 55 400 has been given for the bovine material. The same workers recorded carbohydrate contents of 16 to 23% for human and 25·8% for the bovine factor. Factor IX contains γ-carboxy glutamic acid groups as described for factor II (prothrombin).

Factor IX is converted to its active form by the action of factor XI. Calcium ions are necessary in practice although a very slow reaction is possible without them. In the laboratory, strontium, cupric or zinc ions can be substituted for calcium but they are less effective. Barium, magnesium and manganous ions cause only trace activity while cobaltous ions are completely ineffective. Molecular weights of the activated factor have been reported as 50 000 for the human material and 48 000 for bovine. It is believed that activation takes place in two stages. The single chain of factor IX is first cleaved to form an intermediate two-chain disulphide-linked molecule which is then further cleaved in the heavy chain to produce activated factor IX. Heparin and hirudin inhibit its action. Mild reducing agents will reversibly inactivate it and, if thiol blocking agents are additionally introduced, this then becomes irreversible. Disulphide groups are thus indicated. Thiol group reagents on their own do not inactivate the factor, indicating that thiol groups do not play a large part in its action. Unlike active factor X it is not inhibited by soybean trypsin inhibitor.

An interesting feature of factor IX is that it is present in different abnormal molecular forms in some individuals and this can lead to bleeding disorders, classified as variants of Christmas disease. There is also evidence that the molecules in normal subjects are heterogenous to some extent. One of the variants of Christmas disease raises interesting questions on the sequence of clotting reactions. This variant is sometimes classified as haemophilia Bm and here the one-stage prothrombin time is abnormal, providing bovine brain is used as the source of thromboplastin. The simple cascade hypothesis of clotting would not predict such a phenomenon.

Factor VIII

Coagulant factor VIII is found in association with two other substances, factor VIII-related antigen and ristocetin co-factor. Coagulant factor VIII is missing or reduced in haemophilia while factor VIII-related antigen and ristocetin co-factor are not. All three are missing or reduced in von Willebrand's disease. In the laboratory, yet another entity is measured called factor-VIII-clotting antigen (factor VIII: CAg) but it is believed that this is really the molecule of coagulant factor VIII. This clotting antigen is detected even though the molecule has been inactivated in its coagulant action and is found therefore in normal serum as well as in plasma but not in the serum or plasma from haemophiliacs or patients with von Willebrand's disease. These points are summarized in Table 2.

It is now generally agreed that coagulant factor VIII can be separated from ristocetin co-factor and factor VIII-related antigen by several means but there is still considerable disagreement about the extent to which they are attached, if at all, and the strength of that attachment. Separation can be achieved by gel filtration at high or low ionic strengths, by chromatography on poly electrolytes or aminohexyl-Sepharose and by repeated cryoprecipitation. It is less clear whether ristocetin co-factor and factor VIII-related antigen can be separated.

Although a great amount of work has been carried out on factor VIII, not a great deal is known about it. The small amount present in plasma, the difficulties in separation and its labile nature are mainly responsible for this. Moreover, there is a proportion of reported work which is based on wrong assumptions of purity and even more where it is unclear which of the factor VIII-related substances is really being investigated.

Coagulant factor VIII is a glycoprotein which has thiol groups at or near the active clotting site. It is susceptible to both reducing and oxidizing agents. The activity is destroyed by thrombin, plasmin and many other proteotype enzymes and it is therefore absent in normal human serum (factor-VIII-clotting antigen is, however, present). Thrombin, at levels below that required for destruction of activity, will give an apparent activation of factor VIII but quantitative measurement is not possible since the assays involved do not function correctly in these experiments (activation is only demonstrable using the one-stage assay). Trypsin gives the same type of effect as thrombin, while plasmin and chymotrypsin do not. Ancrod and reptilase are reported neither to activate nor destroy factor VIII. No molecular weight change and no new peptides are detected upon this apparent activation.

Electrophoretic mobility of coagulant factor VIII is that of a globulin but different workers have shown it to migrate with the α, β or γ-globulins depending on

TABLE 3
OCCURRENCE OF FACTOR VIII RELATED SUBSTANCES

	Factor VIII clotting activity	Factor VIII clotting antigen	ristocetin co-factor	Factor VIII related antigen
Normal plasma	Yes	Yes	Yes	Yes
Normal serum	No	Yes	Yes	Yes
Severe haemophilic plasma	No	No	Yes	Yes
Severe haemophilic serum	No	No	Yes	Yes
Severe von Willebrand's plasma	No	No	No	No
Severe von Willebrand's serum	No	No	No	No

experimental conditions. Using free-film electrophoresis, it moves with the α-globulins and its mobility is increased in the presence of borates. Coagulant factor VIII is adsorbed slightly by barium sulphate, aluminium hydroxide and bentonite when these adsorbents are in excess. Once adsorbed, it is difficult to remove without damaging the molecule.

The molecular weight of coagulant factor VIII appears to be greater than a million if subjected to gel filtration chromatography but the apparent molecular weight can be reduced by gel filtration at high or low ionic strength, or by chemical reduction followed by gel filtration chromatography under reducing conditions. A typical reagent here would be 1·0 to 1·3 molar sodium chloride or 0·25 molar calcium chloride or, for the reduction reaction, 3×10^{-4} molar dithiothreitol. Estimates of molecular weight after these reactions vary from 40 000 to 340 000. Since factor-VIII-related antigen and ristocetin co-factor are unaffected by these mild conditions, the subsequent gel filtration separates then from coagulant factor VIII. Some workers believe that this phenomenon is the result of factor VIII splitting into sub-units and a variation of this hypothesis is that factor VIII is normally of that sub-unit size but aggregates during blood collection and subsequent handling procedures. Other workers believe that coagulant factor VIII is a small unit which sits as a passenger on a large carrier antigen and from which it can be removed.

If ristocetin co-factor and the antigen are subjected to a higher level of reducing agent (e.g. 3×10^{-3} molar dithiothreitol) and to gel filtration, under the same reducing conditions, then these two substances also show an apparent reduction in molecular weight. Reported estimates vary between 85 000 and 240 000. Size reduction is also seen in polyacrymide gel electrophoresis after chemical reduction and also after attack by proteolytic enzymes.

Factor VII

Factor VII acts in conjunction with tissue thromboplastin to bring about activation of factor X in the so-called extrinsic system. *In vivo*, it is presumed that tissue thromboplastin is provided by damaged tissues at the wound; in the laboratory a saline extract of brain is a crude source of this material. Calcium and phospholipid are also required for the reaction and the latter will usually be present in tissue extracts. In addition, it now appears that factor VII needs to be activated to bring about this reaction. The presence of tissue thromboplastin seems essential for the full activity to appear but some activation can be caused by activated factor XII and by kallikrein.

It is difficult to separate factor VII from factors II, IX and X and its adsorption properties and its production in the liver are the same as described for factors IX and II. On electrophoresis it migrates between the α- and β-globulins. Factor VII has a molecular weight between 53 000 and 63 000 and is believed to be a single chain protein which, when activated, probably by proteolytic action, becomes a two-chain molecule.

Factor X

In the clotting reactions, factor X occupies a unique position since it may be activated by both the so-called intrinsic and extrinsic systems as well as by the venom of the Russell's Viper. Probably the system is more complicated than we presently understand but it is interesting that factor X which has been completely activated by tissue factor plus factor VII cannot be further activated by the venom. When it has become activated, the role of factor X is to activate factor II (prothrombin) to thrombin. This it does in conjunction with phospholipid, calcium and factor V. Factor X is usually found associated with factors II, VII and IX. Its adsorption properties and its production in the liver are the same as described for factors IX and II. It migrates electrophoretically with the α_1-globulins and the isoelectric point has been reported as pH 4·75. its molecular weight has been variously reported between 52 000 and 87 000. Most recent determinations are reported as 59 000. It has a two-chain molecule which is reported to be split by proteolysis at an arginyl-isoleucine bond during activation. A 9000 molecular weight fragment has been detected after activation and molecular weight determinations for the active molecule have ranged between 25 000 and 38 000. Active factor X has esterase activity and both this and its coagulant activity are inhibited by soybean trypsin inhibitor and by DFP. Activity is also destroyed by oxidizing or reducing agents and by carbonyl-group reagents; thiol blocking reagents have no effect.

Factor V

This factor participates in conjunction with factor X in the activation of factor II (prothrombin) to thrombin in a reaction which also requires phospholipid and calcium ions. It is still a matter for discussion, whether factor V is itself activated in the reaction. Estimates of its molecular weight fall mostly between 98 000 or 300 000 and it has been suggested that as much as 25% of the molecule might be a non-protein moiety. It has an isoelectric point of pH 5·2 to 5·3 and its electrophoretic mobility appears to vary with the experimental conditions, especially the nature of the support medium. The factor travels between the α- and β-globulins with paper electrophoresis but in free film experiments it has the mobility of albumin.

Factor V is relatively labile particularly in the case of the human material. Calcium ions seem important to its stabilization and, as a result, oxalate ions and chelating agents such as EDTA aid the destruction of its clotting activity. Human factor V is destroyed by thrombin, hence it is absent in human serum. Bovine factor V being relatively more resistant is present in bovine serum. Thiol-group blocking reagents will also inactivate factor V.

Factor II (Prothrombin) and Thrombin

Prothrombin or factor II is the inactive precursor which is converted into thrombin by activation with factors X and V plus calcium and phospholipid. It is found in conjunction with factors VII, IX and X and it requires chromatographic techniques to separate the four. Even then it can be difficult to achieve complete separation. Adsorption properties of prothrombin are the same as described for factor IX. The association between factors II, VII, IX and X is so close that some workers proposed that they represented one molecule, a prothrombin complex, which cleaved into the four separate entities upon activation. However, the hypothesis is not generally supported today because the molecular weight of factor II is insufficient to house the other three, antibodies to each of the four factors can be separately obtained and deficiency states of each of the four factors have been separately described. Prothrombin is a single chain glycoprotein of molecular weight between 88 000 or 72 000 and about 10–12% of this weight is attributable to a carbohydrate moiety.

Thrombin, the product of activation, is a two-chain molecule linked by disulphide bridging with a molecular weight between 30 000 and 37 000. Its carbohydrate moiety accounts for about 9·7% of its molecular weight; its isoelectric point is pH 5·7. The molecule can apparently sometimes suffer some cleavage during chromatographic separations. Thrombin is a proteolytic enzyme and its role in the clotting reactions is to bring about conversion, of fibrinogen to fibrin. It is reported to do this by splitting arginyl-glycyl bonds. It exhibits esterase activity which can be destroyed, as can the clotting activity, by DFP indicating that serine is an important part of its constitution. If thrombin is acetylated it loses most of its clotting activity but retains its esterolytic action. It has also been claimed that the esterolytic activity can be separated from the coagulant activity. Thrombin converts factor XIII into its active molecule and it can also cause platelet aggregation.

An interesting development in the chemistry of prothrombin is the discovery of γ-carboxyglutamic acid in the molecule. Vitamin K, acting in the liver, is responsible for the introduction of the extra carboxy group into glutamic acid and the new group is a site of calcium and lipid binding and is essential to full clotting activity. The same is true for factors VII, IX and X but the case for prothrombin is more fully elucidated. Bleeding states in which these γ-carboxy groups are missing have been identified and the abnormal prothrombin, without the extra carboxy group, is not adsorbed by precipitated calcium citrate in the same way as is the normal material.

Fibrinogen and Fibrin

Fibrin constitutes the network of a haemostatic clot and fibrinogen is its precursor. The action of thrombin and the chain of events prior to this clotting have already been discussed. Fibrinogen is formed in the liver and is present in normal human blood at a level between 2·5 and 3·5 mg/ml although levels of about 1 mg/ml are usually adequate for haemostasis. It has a molecular weight of about 340 000 and under the electron microscope appears as three spheres connected by two strands. It contains a twin set of 3 polypeptide chains called α (A), β (B) and γ and each of the two sets is reported to have the N-terminal ends of the chains in a so-called 'disulphide knot'. Free thiol groups are not detected and the carbohydrate moiety is relatively small, constituting a few per cent of the molecule (reports vary mostly between 1% and 5%).

When fibrinogen is clotted, the first action of thrombin is release of a fibrin monomer plus fibrinopeptides A and B. The monomer aggregates and then is formed into a network of fibrin with stabilizing cross-links being subsequently introduced by the action of factor XIII.

Factor XIII

This factor is responsible for stabilizing fibrin clots by a transglutaminase reaction. Clots formed in the absence of factor XIII have a much reduced stability which gives rise to a severe bleeding disorder *in vivo*, and which in the laboratory can be easily lysed by such solutions as 30% urea or 2% acetic acid. Factor XIII circulates in blood plasma as an inactive material which, during the clotting process, becomes activated by thrombin. Experimentally, the same activation can be obtained by using trypsin or reptilase. Factor XIII has a molecular weight estimated to be between 320 000 and 350 000 and can be dissociated into subunits with molecular weights reported between 75 000 and 110 000. The factor is shown to be present in platelets and the molecular weight is reported to be 146 000 in this situation. It is not present in white or red cells (some assays may falsely register factor XIII in red cells, presumably by detecting some other transglutaminase activity).

Antithrombin

Several antithrombins have been described and six were originally designated by number. Antithrombin I represents the ability of surfaces such as fibrin or glass to adsorb thrombin. Antithrombin V activity is shown by hypergammaglobulinic blood. Antithrombin VI is the anticoagulant action of fibrinogen-fibrin degradation products. The remaining three antithrombins II, III and IV are now believed to be synonymous and are usually described as antithrombin III. It is a single chain glycoprotein with a molecular weight reported to be between 56 000 and 68 000. Its isoelectric point is pH 5·1. Antithrombin neutralizes thrombin and activated factor X and its action is greatly accelerated by heparin. It has also been reported to inactivate kallikrein, active factor XII and plasmin.

Other protease inhibitors which may be important in haemostasis are α_2-macroglobulin and α_1-antitrypsins. Their molecular weights have been reported as 820 000 and 54 000–respectively. Their importance is uncertain although a severe haemorrhagic disorder (Antithrombin

Pittsburgh) has been attributed to an abnormal variant of α_1-antitrypsin.

Quantities of circulating inhibitors tend to be high. Amounts reported are 0·2 or 0·3 mg/ml antithrombin III and about 3 mg/ml each of α_1-antitrypsin and α_2-macroglobulin.

Antibodies

The commonest acquired inhibitors to blood clotting are specific antibodies to the blood clotting factors. These have been observed in the congenital deficiency states for factors V, VIII, IX and XIII and for fibrinogen and factor VIII-related antigen. Evidence for antibodies in the other congenital factor deficiencies is very sparse and the relative rarity of these diseases would predict this. However, antibodies can also arise as a consequence of some other underlying disease. Specific antibodies to factor II and to factor XI plus factor XII have been separately described in systemic lupus erythematosis. Factor V antibodies have been seen in haemorrhagic pancreatitis and adenocarcinoma of the colon and factor XIII antibodies have developed after therapy with isoniazid. Factor IX antibodies have arisen in collagen disease and there have been many cases of factor VIII antibodies developing in conditions such as rheumatoid arthritis, penicillin allergy, ulcerative colitis; in pregnancy and *post-partum*.

By far the largest number of antibodies are seen to develop in haemophiliacs in response to replacement therapy. Possibly this only reflects the relatively high incidence of haemophilia compared to other bleeding states; about 6% of these patients will have antibodies at some time. They are the best elucidated of all clotting antibodies. Of those which have been investigated nearly all were of the IgG class (IgA and IgM have been reported) and predominantly they appear to be of the IgG_4 subclass. They do not fix complement and are more temperature dependent and slower in their reaction than is normally encountered with antibodies.

Factor IX antibodies also arise in response to replacement therapy in possibly 3% of the factor IX deficient patients. They are less well characterised than the factor VIII equivalent but they do not have the unusual time and temperature response of the latter. Most that have been investigated have been of the IgG class.

Antibodies to fibrinogen have also arisen in afibrinogenaemic patients after replacement therapy but as would be expected they are rare. The diathesis itself is rare and many of those affected only need infrequent infusions of fibrinogen. Anaphylactic reactions have been noted with such antibody patients during therapy but this may only reflect the relatively large amount of circulating fibrinogen compared to other clotting factors.

Other Acquired Inhibitors

Other inhibitors to clotting can occur which are not specifically directed against one clotting factor but which nevertheless interfere with the sequential reactions. The two commonest are that associated with fibrin degradation products (the inhibitor is sometimes given the name antithrombin VI) and that connected with systemic lupus erythematosus. The latter is believed to interfere with activation of prothrombin.

Another material which can interfere with clotting is Protein C. This is a vitamin K dependent protein originally found in bovine plasma. More recently it has been claimed that the human equivalent exists and can interfere with clotting.

Plasminogen and Plasmin

Plasmin is the proteolytic enzyme responsible for fibrinolysis and plasminogen is its inactive precursor. Activation of plasminogen is part of a complicated set of inter-relating reactions which is discussed in the section on fibrinolysis. Plasmin will degrade many proteins apart from fibrinogen, including factor V, factor VIII, ACTH and glucagon. Its action and structure have been likened to those of trypsin and thrombin. Its molecular weight has been estimated as approximately 89 000.

Plasminogen has a marked affinity for fibrinogen and fibrin, so that it is difficult to prepare fibrinogen which does not contain plasminogen as an impurity. Molecular weight estimates have varied between 83 000 and 143 000.

Platelets

Platelets are small cells, devoid of a nucleus and discoid in shape, measuring about 2 to 4 μm in diameter and about 1 μm thick. Most are produced from megakaryocytes in the bone marrow, and have a life span of about 10 days. They are the smallest elements present in blood; they are negatively charged and their normal concentration is between 150 000 and 300 000 per cubic millimetre. A third of their total mass is concentrated in the spleen from where they circulate and exchange freely with the rest in circulation throughout the body. Newly formed platelets are larger than those which have aged in circulation.

The platelet is surrounded by a plasma membrane composed of mainly protein, phospholipid, cholesterol and some carbohydrate. Small channels, described as forming an open canalicular system, connect the platelet interior to the surface and presumably these act as a pathway for platelet secretions to emerge. The plasma membrane is supported by bundles of microtubules and smaller microfilaments are present as well as a dense tubular system.

Within the platelet are different varieties of organelle including several types of granules, mitochondria and dense bodies which are smaller and more electron dense than the granules. Various materials are stored in both the granules and the dense bodies and these include 5-hydroxytryptamine, ADP, ATP, thrombasthenin, lysosomal enzymes, and β-thromboglobulin. Platelets also contain clotting factors as well as platelet **factor 3** and platelet factor 4 and, in addition, clotting factors are non-specifically adsorbed on the membrane.

When exposed to aggregating agents, which *in vivo* are most likely to be thrombin or ADP, there is first a shape change to a more spherical form, followed by aggregation. Reactions follow in which, amongst other substances, ADP, 5-hydroxytryptamine and platelet factor 3 are released. Platelet factor 3 catalyses the clotting reactions and the ADP presumably participates in the second and irreversible phase of aggregation to result finally in a haemostatic plug. The presence of β-thromboglobulin and platelet factor 4 has been extensively monitored on the hypotheses that these indicators of aggregation might be used to predict the thrombotic state.

Prostacyclin (PGI_2) is a substance generated by the vascular wall from arachidonic acid via prostaglandin endoperoxidases and is the most potent endogenous inhibitor of platelet aggregation. The endoperoxides are also converted by platelets into thromboxane A_2, which aggregates platelets, and it is a balance between this, within the platelet, and the prostacyclin in the vascular endothelium which is believed to regulate platelet aggregation.

ASSAYS OF FACTORS AND COMPONENTS

Assays in General—Methods

There are not a large number of choices concerned with the principles of assay. Colorimetric finishes to the tests are an alternative to simple clot observation and improvements are being made all the time. Colorimetric assays for thrombin and, by the same token, antithrombin, can be a good alternative to the old techniques and in the right circumstances can be time saving. Kallikrein assays can also be usefully conducted this way and there is a colorimetric immunoassay for factor VIII-related antigen. For other clotting factors at this time it appears that conventional clotting methods are superior.

For most clotting factors, immuno-assays can be conducted. They can be based on staining precipitates in gels or can be determined by radioactive or colorimetric methods. In any event, these methods complement existing tests rather than replacing them since they record the presence of clotting factor molecules whether they are active in clotting or not. Sometimes immuno-assays are the most convenient or the only ones available, e.g. factor VIII-related antigen, β-thromboglobulin and platelet factor 4.

Having decided that clot observation methods are unavoidable, the next choice is the degree to which automation should be used. Individual decisions will be governed by the experience of the operators involved and the amount of practice they can expect. At present, the overall conclusion seems to be that an automated technique can measure clotting times as precisely, or more so, than can be achieved by the most experienced operator using manual methods. However, automated methods are usually slower to perform and may lead to sample deterioration due to delays between measurements. Hence at the present time automation is more useful for the tests with few manipulations and of short duration, such as one-stage prothrombin times or assays of factors V, VII or X. It is much less useful for assays of the intrinsic system and very difficult to use for two stage assays of factors VIII or IX.

Assays in General—Calculations

Calculation techniques are standard and well known. Graphs of clotting time against dilution are plotted for the test and standard sample on such graph paper as is necessary to provide straight parallel lines. Acceptable limits of straightness and parallelism are somewhat subjective but can be assessed by drawing parallel lines through each point individually and calculating an individual result for each point. The degree of spread can be judged then as acceptable (say $\pm$ 20%) or not. Standard and test samples are normally assayed at three dilutions. The practice of assaying test samples at one dilution only is to be deprecated under normal circumstances.

All these manipulations which are carried out on graph paper can equally well be translated into a technique for computer or calculator. Acceptable limits of straightness and parallelism can be fed into the programme and these limits need only be based upon practical experience. Alternatively a more formal statistical analysis can be used by means of regression analysis followed by analysis of variance; as a result straightness and parallelism are judged as significant or not, taking into account errors between the duplicate clotting times. This approach also allows a statistically derived error limit to be ascribed to the result and is particularly useful in providing an aura of respectibility when there is a commercial or legal interest. However, the use of statistical analysis is not an alternative for good experimental design which should take into account the fact that errors for a single operator performing one set of assays will be less than those obtained between operators, between different days and between assays using different batches of reagent. On average a typical error limit (95% confidence) would be $\pm$ 20% of the level in question if the test involves duplicate clotting times of each of three dilutions for both test and standard sample (i.e. 6 values of clotting time for each).

Preliminary Tests

When confronted with a patient with a suspected deficiency in their haemostatic mechanism, it is obviously desirable to be able to carry out a series of preliminary tests to indicate whether the patient does have a bleeding problem and to give some positive clues regarding the type of deficiency involved. The first action, although not strictly a test, is the recording of bleeding history in the patient and his family, by an experienced clinical

TABLE 4
TESTS TO IDENTIFY A BLEEDING DISORDER

Initial tests
   Bleeding time
   Tourniquet test
   Whole blood clotting time
   Prothrombin consumption index
   Kaolin cephalin clotting time (activated partial thromboplastin time)
   One-stage prothrombin time
   Assay of factor VIII
   Assay of factor IX
   Factor XIII test
   Platelet count
   Blood film

If bleeding history is not explained by initial tests
   Platelet factor III availability
   Platelet aggregation with each of ADP, collagen, adrenaline, ristocetin and animal factor VIII
   Thrombin time
   Fibrinolysis (fresh and cooled samples required)

Specific tests if warranted by initial results
   Specific factor assays
   Inhibitor tests
   Antibody assays
   Fibrin degradation product assay

Tests for when all else has failed
   Antithrombin assay
   Thrombin generation test
   Thromboplastin generation test

TABLE 5
BLEEDING TIME TESTS AND TOURNIQUET TESTS

Disorders which give abnormal results in the bleeding time test:-
   Thrombocytopenia
   Qualitative platelet disorders
   Purpuras
   von Willebrand's disease
   Afibrinogenaemia
   Telangiectasia (if abnormal vessel is punctured)

Disorders which give abnormal results in the tourniquet test:-
   Thrombocytopenia
   Qualitative platelet disorders
   Scurvy

specialist: its importance is overwhelming because there are occasions when positive evidence of bleeding is obtained which no amount of laboratory data can negate, regardless of how negative that data might be. Tests will then be carried out on the patient and on blood samples. Table 4 shows a typical list of tests.

Preliminary Tests on the Patient

Two tests are made directly upon the patient. One is the bleeding time test which records the time taken for a standard skin puncture to stop bleeding and the other is the tourniquet test which examines capillary fragility by recording the number of petechae which arise, following a period in which a sphygmomanometer cuff is applied to the upper arm, at a pressure which restricts venous flow. A list of diseases which might give positive results in these tests is given in Table 5.

In the case of the bleeding time test there are some possible variations; most laboratories seem to use a variation of the Ivy method, where the punctures are made on the forearm. The optimum size of wound is open to discussion. Some procedures involve a pin prick which can be difficult to standardize, while some inflict such a wound that suturing is a common sequel. In the latter technique some distress can be caused, particularly to children, and volunteers to establish the normal range can be conspicuous by their absence. A compromise in wound size might be the best solution.

Preliminary Tests on Whole Blood

Next are the tests which are carried out on whole blood, devoid of anticoagulant; most traditional among these is the whole blood clotting time. Results obtained are of very limited value since the test detects only the grossest of abnormalities. However, it is such a simple test that it seems hardly worthwhile excluding it and the procedure has some value in that it provides a serum sample, without anticoagulant, which can be used in the prothrombin consumption index. This is a test of the intrinsic clotting mechanism plus platelet effectiveness and has been known to detect some obscure clotting deficiencies which are difficult to categorize fully or even to detect by other means.

The whole blood clotting time may be followed up by testing, for example, serial dilutions of blood but although this may have advantage in particular situations it is not generally employed. Another extension of the method is seen in the thromboelastograph, which provides a dynamic plot of clotting progress with time by continuously measuring a function of blood viscosity. Some workers have devoted themselves to this technique and claim significant advantage but, in the routine laboratory, the added complexity and time consumption are not justified by any extra information obtained.

Another investigation which can be carried out on whole blood is that of platelet stickiness. It is not a preliminary test in the sense of giving basic information but it has to be carried out in the early stages while whole blood is still in the syringe. Blood is pumped from the syringe through a short column of tiny glass beads. A platelet count is made before and after this operation to reveal the number which have adhered. Some workers will collect the blood sample directly into heparin in the syringe instead of using the native blood. This is another test which is best performed by someone prepared to devote themselves to the method and rigorously to standardize each component step.

Preliminary Tests on Plasma

Most tests of clotting function are performed on plasma separated from citrated blood and for investigations of clotting factors the blood is centrifuged sufficiently to remove platelets from the plasma. Initial tests most commonly carried out are the one-stage prothrombin time and the activated partial thromboplastin time (kaolin-cephalin clotting time). The latter is a test of the intrinsic clotting system and uses kaolin or an alternative activator to start the reaction chain and a cephalin to mimic the lipid derived from aggregating platelets. The first stage is to incubate the kaolin, cephalin and the test plasma together and it is important to restrict this time of incubation to a short period of say 2 minutes, otherwise the test loses its sensitivity to deficiencies of kinin, in the contact phase of reaction. An abnormally long clotting time in the activated partial thromboplastin time test indicates a factor deficiency or the presence of inhibitory substance in the intrinsic reaction pathway.

In the same way, a deficiency or an inhibitor in the extrinsic pathway is indicated by prolonged clotting times in the one-stage prothrombin time. This test involves a tissue extract, usually of brain, which substitutes for materials released from tissues during injury. The reaction involving tissue extracts is species specific and varying sensitivities to different clotting factors will be obtained by varying the source of brain (usually human, bovine or rabbit). A particular anomaly is sometimes encountered when using bovine brain extracts with plasma from Christmas disease patients and this is discussed in the section on factor IX.

The one-stage prothrombin time is also used to monitor treatment level in anticoagulant therapy. Naturally, a high degree of standardisation is required in this case not only within laboratories but between laboratories and hospitals throughout the country. To this end, a standard brain preparation is available from the National (UK) Reference Laboratory for Anticoagulant Reagents and Control. At one time it was the practice to calibrate one's own extract against the standard and then convert the practical results using a calibration graph. Now that the standard is more freely available, modern practice is to use the standard preparation as the reagent in all tests.

When the activated partial thromboplastin time and the one-stage prothrombin time have been completed on a given plasma sample, one would logically expect to be able to locate any factor deficiency as being either in the intrinsic or extrinsic pathway or else in that area common to both. In most cases this objective is achieved but, there are two severe limitations. First, the two tests are insensitive to factor II, so that a couple of seconds prolongation of the one-stage prothrombin time should not be neglected and a specific assay for prothrombin has to be performed. Second, the activated partial thromboplastin time is not sensitive to factor VIII or IX levels above about 25%. Thus there is a narrow range of factor levels where the patient could have excessive bleeding but the preliminary tests give no indication. The only solution is that all samples should be assayed for factor VIII and IX, regardless of the results of the preliminary tests.

Another of the preliminary tests is the thrombin time, the result of which is a function of fibrinogen content. It can also reflect abnormal fibrinogen or the presence of inhibitors such as fibrin degradation products. It can be performed as a fibrinogen titre by testing serial dilutions, if semi-quantitative results are necessary. There are yet more tests which could be used and some are brought from the archives when it has been impossible to ascribe a specific factor deficiency to a positive bleeding problem. Two which are popular in this respect are the thrombin generation test, which quantitates thrombin generation with time throughout the clotting cycle, and the thromboplastin generation test which can locate a deficiency as being present in plasma, in serum or in both.

Assay of Clotting Factors V, VII, X, XI and XII

Assays of these factors have been grouped together because they are usually performed as simple one-stage assays. Each represents a modification of a simple preliminary test, in that the test sample is diluted and a new reagent is introduced, namely a plasma deficient in that factor to be measured. In this way, the diluted test plasma is the only source of the factor in question and hence clotting time becomes a function of the amount of that factor in the test plasma. The activated partial thromboplastin time test is modified to provide assays of factors VIII, IX, XI, XII and the kinin factors, while the one-stage prothrombin time is modified to provide assays of factors V and VII. Factor X could be assayed similarly but plasma deficient in factor X alone is rare and there is an alternative activating agent, Russell's Viper Venom, which acts directly upon factor X without the need for factor VII. Therefore an assay based upon the venom is used, together with an artificially made plasma which is deficient in factor X (and incidentally deficient in factor VII). Factors VIII and IX can be successfully assayed by the one-stage technique and such assays are popular. However, an alternative method is available and relative advantages and comparisons will be discussed in the next section.

In general, the provision of deficient plasma is a major problem with these assays. Since the test plasma samples are tested in a diluted form, the deficient plasma should ideally contain no measurable quantity of the factor in question, otherwise it could contribute a significant amount of that factor to the assay system.

Assay of Factors VIII and IX

One-stage assays are satisfactory methods of measuring factors VIII and IX and comments in the previous section apply equally well to these cases. However, an alternative technique is available, called the two-stage assay. The basic procedure is to incubate a mixture which is deficient in prothrombin and contains the diluted test sample. In this way a complex of activated factor X develops and rises to a concentration which is a function of the amount of the factor in question, contained by the test sample. The chain of clotting reactions is then allowed to continue by sub-sampling into a mixture containing prothrombin and fibrinogen (usually, normal plasma). Measurement of resultant clotting time becomes a function of the activated factor X which is present, and hence of the factor VIII or IX which was originally present. Measurement of factor VIII differs from that for factor IX in the constitution of the prothrombin-free mixture which is incubated. For factor VIII analysis this contains phospholipid, an activated normal serum sample, factor V and the diluted test sample which has previously been adsorbed, using aluminium hydroxide (adsorbed in the undiluted state). With factor IX analysis, the mixture consists of phospholipid, a source of activated factor XI, serum from a factor IX deficient patient, factors VIII and V and the diluted test sample (not adsorbed). One advantage of the factor VIII assay carried out in this way is immediately apparent: there is no requirement for factor VIII deficient plasma. Less of an advantage exists with factor IX measurement because of the need for deficient serum but requirements for blood with ultra-low factor IX levels are somewhat less stringent when making this serum. As a result, this does amount to an advantage since donors with 1 to 2% of the factor are more plentiful than those with 0%, because the latter are usually treated fairly regularly and for much of the time have the residues of such treatment in their blood.

Since there are two distinct methods of measuring each of these two factors it is not surprising that each method has its supporters, some of whom can become quite uncompromising in their views. It would be true to say that the two-stage factor IX assay has fewer supporters than most because of the added complexities in reagent preparation. In factor VIII analysis both methods are popular. Two-stage assays tend to be more precise, particularly for factor VIII, and have the advantage referred to above concerning factor-deficient reagents. One-stage assays require fewer and simpler reagents. It would be wrong to make further distinctions at this stage because modern investigations suggest that the two assay techniques can give different results, particularly when comparing a plasma sample to a factor concentrate. Clearly this could influence the best choice of assay and it would be premature to draw any more conclusions. Differences are normally less than 20% of the level being considered and so there is no serious effect on patient therapy.

Assay of Factor VIII-Related Antigen and Ristocetin Co-factor

Factor VIII-related antigen and Ristocetin co-factor are closely related to factor VIII though how close this relationship might be is still a matter for discussion. These materials are present in the blood of haemophiliacs and normal subjects while being absent or reduced in the blood of von Willebrand's disease patients (summarized in Table 3). Measuring these levels, together with that of factor VIII activity, will therefore allow the differential diagnosis between haemophilia and von Willebrand's disease. Analysis of factor VIII-related antigen is based, as its name implies, upon an immunological test using a specific antibody. Most usual is the Laurell immunoelectrophoretic assay in which a specific heterologous antibody is included in an agarose gel while the sample, introduced into a small hole, is made to traverse the gel under the influence of an electric field. Peaks of antigen-antibody complex are formed, whose height is a function of antigen content. It is an elegant technique which is accurate for samples containing at least 5% antigen.

Radioactive tests are also available based upon the usual principles of radioimmunoassay and most popular is that in which antibody is adsorbed onto a plastic tube, reacted with test antigen and then a reaction with a radioactive antibody completes the sandwich. As an alternative, the ELISA procedure can be used, when measurement is based on the colorimetric detection of an enzyme coupled to the antibody. Radioactive and colorimetric assays of factor VIII-related antigen are more time consuming and involved than the simple Laurell technique and are normally only justified when measurements are required below the 5% level. They can be used in the detection of some variants of von Willebrand's disease when 'non-parallel lines' are sometimes reported. However, this is only another way of saying that the assay does not function correctly with these samples, so quantitation is hardly advisable. Two-dimensional Laurell techniques will identify von Willebrand's variants.

Assay of ristocetin co-factor is entirely different and based on the premise that normal platelets will be aggregated by ristocetin if the co-factor is present, but will not aggregate in its absence. (Ristocetin is an antibiotic not currently used because of side-effects.) It is necessary to have a correct ristocetin level in order to obtain this differential effect. Two assays are available. In one, the reagents are mixed in an aggregometer which measures light transmission, or scattering, through the sample and aggregation is plotted automatically as a function of time. Plots are compared with those of normal samples to obtain a quantitative result. The other test is to mix the reagents, allow then to react and the aggregated platelets to settle. Unaggregated platelets are then counted in a particle counter, and the percentage aggregation is compared with results using normal samples. It would be premature to state preference here since experiments comparing the two methods are still being carried out.

Assay of Factor VIII Clotting Antigen (Factor VIII: CAg)

This is not to be confused with factor VIII-related antigen, although it is measured in a similar way (see summary in Table 3). The radioactive technique is used in which a radioactive antigen/antibody sandwich is built up upon a plastic tube but the antibody involved here is one specific for factor VIII clotting activity and not factor VIII-related antigen. At the present time the antibodies used are restricted to those of high titre raised, in a haemophiliac, in response to therapy or arising spontaneously in a human subject.

This assay measures a function of the concentration of molecules associated with factor VIII clotting activity and it measures this even if the molecules have been inactivated. This antigen is present in normal serum but not in haemophilic serum. Its measurement has been proposed as a method of detecting haemophilia in a fetus, *in utero*, (since the test can cope with partly clotted samples). This is the subject of much research and it is too early to comment further. It seems likely that fetal blood serum contains rather less of this antigen than adult serum and present interest is to determine whether the method is still accurate, despite this drawback.

Assay of Factor II (Prothrombin)

In principle it is possible to design a one-stage assay for factor II based on the one-stage prothrombin time or the activated partial thromboplastin time test, as described earlier for a series of other factors. In practice, however, such assays are too insensitive, in that measured clotting time does not change sufficiently as factor II concentration is altered. Presumably this is because there are many steps in the clotting sequence and those involving prothrombin cannot be made sufficiently slow in the assay, as to be rate-determining. Logically then, a better assay would be anticipated if fewer steps were present, and this is found to be the case using venom from the taipan or the tiger snake. Taipan venom in the presence of phospholipid activates prothrombin directly, while tiger snake venom is similar but additionally requires factor V to be present. They are good assays whose main problem is that of not giving straight-line plots of clotting time against sample dilution, when drawn on log/log or log/linear paper, except over a restricted range. Problems are only alleviated by plotting reciprocals of sample concentrations and this is a well known method of maximising technician confusion.

An alternative assay for factor II is the so-called two-stage prothrombin time, in which plasma is clotted using tissue extract and the mixture subsampled at intervals into fibrinogen. In this way, a plot of thrombin generation with time can be drawn and the area under this curve, which is a measurement of total thrombin generation, can be compared to that of a normal sample. The advantage of the method is that it demonstrates thrombin generation in the patient's own plasma, in the presence of his own antithrombin. Clotting factor deficiency in the extrinsic chain can also be revealed. It is, however, tedious and time-consuming and cannot be performed satisfactorily on plasma samples which have been frozen, because antithrombin content can vary in storage.

Assay of Fibrinogen (Factor I)

A simple test of qualitative fibrinogen deficiency is the thrombin time test in which clotting time is recorded after adding thrombin. Prolonged clotting times indicate either a low fibrinogen level, an abnormal type of fibrinogen, or that an inhibitory substance is present. An example of a likely inhibitor is that associated with fibrin degradation products, which arise in intravascular coagulation. By using serial dilutions of the test plasma, it is possible to convert this simple test into the so-called fibrinogen titre to give a semi-quantitative result. It is also possible to subsitute reptilase for thrombin in the simple test, when a comparison of results, with those obtained using thrombin, can help distinguish fibrinogen deficiency from an abnormal prothrombin. Additionally, reptilase reagent is unaffected by heparin, should heparinized samples need to be examined.

Quantitative measurement of fibrinogen is also possible and usually involves clotting plasma, washing the clot and finally analysing the amount of protein contained in that clot. Mostly, it is performed by dissolving the clot and following this with one of the conventional analyses for protein in solution.

Not all fibrinogen assays can be performed at leisure. Defibrination, in for example premature separation of the placenta, can be accompanied by catastrophic bleeding calling for speedy blood replacement and a quick confirmation that defibrination is indeed the problem. The laws which seem to govern scientific endeavour will further determine that emergencies will occur at the worst possible moment in a distant area. A simple solution is often possible. Clotting laboratories can provide and maintain small clotting kits in places where problems might occur, such as in obstetric units. Kits would contain tubes for blood collection and the reagents required for determining the thrombin time together with instructions for use by the clinician dealing with the immediate problem.

Assay of Factor XIII

The basis of the factor XIII test is to clot plasma and suspend the clot in a solution such as urea or acetic acid solution. Should the clot dissolve, an abnormality is indicated. Control samples consisting of a normal plasma and a known abnormal plasma are treated identically, for comparison. It would be easy to convert the procedure into a semi-quantitative assay by serially diluting the test plasma in plasma which is deficient in factor XIII and testing the diluted samples. Usually this modification is avoided where possible because of the rarity of the deficient plasma.

Quantitative assays are possible using a radioactive technique based on the measurement of transglutaminase activity but results are not always identical to those of the clot stability assay. Positive results can be obtained in materials, for example lysed red cells, where there is no evidence of activity associated with clot stabilization.

An important precaution when testing for factor XIII is to ensure that the patient has not been given replacement therapy or blood transfusions for some weeks. Because the half life of the factor is relatively long and because 2% of the average normal level is adequate for clot stability, the effects of treatment persist over a long period. Use of a strictly quantitative assay will naturally obviate this restriction to some degree.

Kinin Deficiencies (Fletcher, Flaujeac, Fitzgerald and Williams trait). One stage assays can be devised for these factors in an analagous manner to that described for factors XI and XII but the factor-deficient plasma samples, necessary as reagents, are rare. Subjects with a deficiency in these kinins usually do not experience abnormal bleeding which, in itself, removes some priority from the analyses. The activated partial thromboplastin time is prolonged when some of the kinin factors are reduced, providing a short incubation time of say 2 minutes is employed. By increasing incubation time to say 10 minutes, the prolongation of clotting time is lost and in this way at least a proportion of the kinin deficiencies can be indicated.

Assay of Antithrombin

Antithrombin concentrations are measured, in principle, by incubating the sample with a known amount of thrombin and measuring the thrombin which remains after incubation. Because it is a slow reaction the components must be incubated for about an hour at 37°C. Alternatively heparin may be added which greatly accelerates the reaction. Thrombin present is measured by determining the clotting time when the mixture is added to fibrinogen solution. It is not necessary to calculate absolute thrombin units since a standard plasma sample would be run in parallel and antithrombin expressed as a percentage of average normal level. There is a variation of this method in which the sample is diffused through a gel containing thrombin.

As an alternative, the rate of thrombin inactivation can be measured by subsampling from the reaction mixture (thrombin plus sample) at intervals into fibrinogen and recording the resultant clotting time. Since the reaction rate is increased by heparin, the test can also be converted into a heparin assay. Yet another alternative is to measure the residual thrombin by using a chromogenic substrate and recording the change in colour using a spectrophotometer. Immunological methods are available using specific antisera and these enable molecular concentration to be measured which, as usual, may be distinct from biological activity.

Assay of Inhibitors and Antibodies

Naturally occurring antithrombin has already been discussed but other inhibitors of clotting can present themselves. Some are inhibitors which arise as the result if an underlying disease or clinical condition and these usually interfere with the chain of clotting reactions without reducing the amount of any particular clotting factor. The other category consists of antibodies which are specifically directed against a clotting factor and occur as a direct result of replacement therapy, or else develop spontaneously in a patient with an underlying illness or clinical condition (that patient having previously been normal from the point of view of haemostasis).

Presence of these substances are usually first detected by a prolongation of clotting times in the preliminary activated partial thromboplastin time test and/or one stage prothrombin times. Their inhibitory nature is demonstrated by testing mixtures of patient's plasma with normal plasma, when clotting times will be prolonged by addition of relatively minor proportions of patient's plasma.

Specific antibodies will be revealed when factor assays are performed and the factor in question found to be markedly reduced. Antibody assays can then be carried out and these consist in principle of incubating a patient's plasma with a source of the factor concerned and measuring residual factor, and hence that which has been destroyed. In practice, most antibodies which are encountered are directed against factor VIII, or less usually against factor IX.

Factor IX antibodies appear to have typical antibody characteristics but those directed against factor VIII seem to react by a more time-consuming process, and one which is unusually temperature dependent for an antibody.

Since time of reaction is not an important variable with factor IX antibodies these can be assayed simply on the basis of the principles just outlined. With factor VIII antibodies there are several alternative assays available. Some antibodies can be measured satisfactorily by several methods; some seem to give a different result with each different method that is tried. The picture which is slowly emerging seems to be that a proportion of the antibodies are subject to rather complicated kinetics of reaction. As a result the most consistent results are obtained either at zero time (by extrapolation) when the kinetics have not had time to complicate themselves or else after a long period, of say 4 hours, when the reactions have gone to completion or nearly so. Results at zero time do not have the same practical significance as those at four hours and so the current assay method, based on a four hour incubation of serial dilutions of test plasma with factor VIII, is probably the best available. For a more simple screening assay, the measurement of residual factor VIII after incubation for 1 hour (patient's plasma plus source of factor VIII) is most convenient. Some recent experiments suggest that the source of factor VIII for antibody tests may be important and influence results. Until this is resolved, the best course seems to be to choose a typical sample of the material which will be used to treat the patient as the source of factor VIII to be used in the assay.

Platelet Tests

Measurement of bleeding time is one of the first tests to be carried out when examining a patient for haemostatic defect and a prolonged bleeding time is usually seen in platelet disorders. Counting platelet numbers will identify thrombocytopenia if present. After this, the preliminary tests give strong indications of platelet abnormality when, typically, an abnormal prothrombin-consumption index but a normal activated partial thromboplastin time will be obtained. Both are tests of intrinsic clotting but the prothrombin consumption index is additionally dependent upon platelet function. Tests which follow are a measurement of platelet factor 3 release and tests of aggregation with each of ADP, collagen, adrenalin, ristocetin and an animal factor VIII concentrate. The test of platelet factor 3 is basically an activated partial thromboplastin time in which platelets provide the only source of the cephalin involved. Aggregation tests can be performed simply by mixing platelet rich plasma with the aggregating reagent and timing the appearance of aggregates. Much valuable information can be obtained by visual observation but the usual current procedure is to follow aggregation in the so-called aggregometer, which detects it by light transmission or refraction and provides a continuous plot of aggregation with time. Use of different reagent concentrations allows a family of curves to be built up which can be compared to that of a normal sample. Use of the aggregometer also allows the pattern of aggregation with time to be checked. Typically, when the reagent concentration is correctly chosen, aggregation is seen to be followed by disaggregation and then a 'second phase of aggregation' follows caused by ADP release from the platelets. Absence of this 'second wave' is another indication of abnormality.

By these means, a bleeding disorder can be attributed to platelet abnormality. The results do not predict bleeding severity and laboratory results are viewed in the light of a clinical assessment of bleeding severity. Treatment of bleeding in all platelet disorders is basically the same and so excessive subdivision of the various diseases is not vital. (Treatment of an underlying disease may of course be necessary.) The main divisions (see also Table 2) are as follows:-

Thrombocytopenia. Reduction in the number of platelets which are otherwise normal.

Thrombasthenia. Platelets fail to aggregate with ADP, adrenaline or collagen but aggregate normally with ristocetin or bovine factor VIII. *In vivo*, they would adhere normally to subendothelium but would fail to aggregate beyond a monolayer.

Release Reaction Defects. Platelets aggregate with ADP, adrenaline, ristocetin, or animal factor VIII but fail to aggregate with collagen. In addition, they fail to show a second phase of aggregation with ADP or adrenaline. These defects can be subdivided into storage pool deficiency, cyclo-oxygenase deficiency and defective nucleotide metabolism.

Defective Adhesion. Bernard-Soulier syndrome is the inherited disorder of this type. Platelets are abnormally large and fail to aggregate with ristocetin or animal factor VIII. Aggregation with ADP or collagen is normal.

Platelet Tests of Thrombosis

Assays of β-thromboglobulin and platelet factor IV have been carried out in several centres, based on the hypothesis that their presence in blood, being indicators of platelet aggregation, would possibly indicate an incipient thrombotic state. Some reasonable correlations have been obtained although this is not proof of the prediction possibilities. Mostly these assays have been immunological and there have been considerable difficulties concerning sample collection. Venepuncture techniques are critical and in some centres only certain workers have been able to produce the required results. It would seem that any assay with such limitations is doomed to speedy obsolescence.

Tests of Fibrinolysis

Several precautions in blood sampling are vital for tests of fibrinolytic function. Samples have to be cooled to 4°C immediately they have been taken, and the plasma separated in a refrigerated centrifuge (or ice placed in the buckets of a centrifuge without refrigeration). All initial reagents have to be cooled to 4°C and tests need to be started within 30 minutes of taking the sample. Having obtained a satisfactory sample the following preliminary tests are usually performed:-

Thrombin Time. This has been mentioned before and it is a measure of the clotting time when thrombin is added to plasma. Prolongation indicates a decrease in fibrinogen levels or the presence of fibrin degradation products. A 'serial' modification is also possible in which the plasma is incubated at 37°C and the thrombin times of subsamples are tested at intervals up to 2 hours.

Fibrinogen Titre. This also has been mentioned before and involves clotting serial dilutions of the plasma with thrombin to assess the highest dilution in which clots are visible and, by comparison with a normal sample, to obtain a semi-quantitative estimate of fibrinogen level.

Euglobulin Lysis Time. The principle involved here is to precipitate fibrinolysis activators by dilution in an acidic solution. Without precipitation, a form of equilibrium seems to exist between activators and inhibitors and this limits the sensitivity of any test. The precipitated so-called euglobulin is clotted and the clots observed over a period for lysis.

Fibrin Plate Method. This is a technique in which a buffered solution of fibrinogen is clotted in a petri dish using thrombin. Small spots of a sample are placed on the plate, incubated at 37°C overnight and the areas of lysis on the fibrin plate are a measure of fibrinolytic activity. Samples which are applied are usually euglobulin precipitates; the test is too insensitive using plasma. Results obtained are a measure of plasminogen activator since plasminogen will already be present, bound to the fibrin. The test can be modified to measure plasmin content by heating the plate to 80°C before the test and, by this means, destroying the adherent plasminogen.

Plasminogen Assay. Assay of plasminogen involves separating it from antiplasmin in the plasma sample by acidification. It is then activated to plasmin, by a substance such as streptokinase, and the resultant plasmin is measured by its lytic action upon casein. The plasmin can also be detected by its esterolytic action upon aminoacid esters such as TAMe.

Fibrin Degradation Products. These are usually detected by an immunological test in which red cells or latex particles are coated with an antiserum. The amount of degradation product is indicated by the degree of aggregation of cells or particles obtained when mixed with the sample. There is a similar test in which the sample is mixed with antiserum and the residual antiserum is measured by the aggregation of cells or particles coated with fibrinogen. It will be seen that this test works because the antiserum reacts either with fibrinogen or with fibrin degradation products. By the same token samples for this test must be of well-clotted serum.

FURTHER READING

Austen, D. E. G. and Rhymes, I. L. (1975), *A Laboratory Manual of Blood Coagulation*. Oxford: Blackwell Scientific.

Austen, D. E. G. (1979), 'The structure and function of Factors VIII and IX,' *Clin. Haem.*, **8**, 31–52.

Biggs, R. (Ed.) (1976), *Human Blood Coagulation, Haemostasis and Thrombosis*. Oxford: Blackwell Scientific.

Biggs, R. (Ed.) (1978), *The Treatment of Haemophilia A and B and von Willebrand's Disease*. Oxford: Blackwell Scientific.

Caen, J. P. (1979), 'Blood platelet function', *Nouvelle Revue Francaise d' Hematologie*, **21**, 81–89.

Hardisty, R. M. (1977), 'Disorders of platelet function,' *Brit. med. Bull.*, **33**, 207–212.

Hardisty, R. M. and Ingram, G. I. C., *Bleeding Disorders, Investigation and Management*. Oxford: Blackwell Scientific.

Macfarlane, R. G. (1967), 'Russell's viper venom, 1934–64,' *Brit. J. Haem.*, **13**, 437–451.

Rizza, C. R. (1977), 'Clinical management of haemophilia,' *Brit. med. Bull.*, **33**, 225–230.

Sixma, J. J. and Wester, J. (1977), 'The haemostatic plug,' *Seminars in Haematology*, **14**, 265–299.

SECTION IV

DISORDERS OF THE RENAL TRACT

		PAGE
15.	FUNCTIONS OF THE KIDNEY AND TESTS OF THESE FUNCTIONS: ACUTE AND CHRONIC RENAL FAILURE: URAEMIA	271
16.	EXAMINATION OF URINE	277
17.	UROLITHIASIS	290

15. FUNCTIONS OF THE KIDNEY AND TESTS OF THESE FUNCTIONS. ACUTE AND CHRONIC RENAL FAILURE. URAEMIA

H. A. LEE

Acute renal failure

Renal transplantation

Chronic renal failure
  Proteinuria
  Polyuria
  Glomerular filtration rate
  Tubular function
  Calcium metabolism
  Lipid metabolism

The kidney is one of the principle homeostatic organs in the body. It helps to maintain regularity of body composition by excreting the waste products of metabolism, modifying the acidity of the urine, and regulating electrolyte and water control. Additionally, it has the functions of an endocrine organ with respect to erythropoietin production and bone marrow function, renin production and salt and water and blood pressure control, and the production of 1:24 dihydroxycholecalciferol, the active metabolite of vitamin D, controlling bone formation and structure.

Thus when kidney function is impaired, loss of control of homeostatic mechanisms occurs causing ill health. The impact of such loss of kidney function depends upon the rate at which it occurs, there being fundamental differences between acute and chronic renal failure. Thus, as might be expected, the effects are virtually immediate and life threatening with acute renal failure whereas with chronic renal failure problems occur often over many years and different effects will be seen because of the time duration effect, e.g. on bone metabolism, on fat metabolism.

Clearly one might anticipate that, with such a complex organ as the kidney, many tests must be necessary to assess its function and to decide upon therapeutic measures. Fortunately, in reality, the number of tests are fairly limited and yet give a good guide to treatment.

For the purposes of this chapter I will concentrate upon those tests and their interpretation which help with the diagnosis and treatment of acute and chronic renal failure problems. The emphasis of this chapter is upon biochemical tests, but a brief mention will be made of supportive complementary tests which help establish a diagnosis and point the way to treatment. It is important to remember that biochemical tests alone are no substitution for a comprehensive evaluation of renal function judged by history, clinical examination, examination of urine, radiological investigations, and renal biopsy when necessary. However, they do form a central part to evaluation.

ACUTE RENAL FAILURE

The diagnosis of acute renal failure depends upon an awareness of the condition coupled with early institution of appropriate tests either to ward off incipient acute renal failure or to control established acute renal failure. It is necessary to distinguish between incipient and established acute renal failure for prompt attention to detail can prevent the former proceeding to the latter. Thus any condition leading to hypotension with hypovolaemia, e.g. as a result of plasma, blood or fluid losses, can lead to acute renal failure. Similarly compromise of cardiac function can result in the same end-point. Clinically, one also needs to distinguish between acute and chronic renal failure and look for clues of the latter such as high blood pressure, bone disease, unexplained anaemia, neuropathy and abnormalities of blood lipids. Likewise, awareness of post-renal obstruction leading to anuria must not be forgotten. Finally, the possibility of an acute glomerulonephritis or the ingestion of some toxic drug or a poison must not be overlooked.

Thus if a clinical situation has arisen which can potentially lead to acute renal failure then corrective measures must first and foremost be applied. Thus replacement of blood, plasma, or fluid losses must be undertaken, restoration of cardiac function aimed for and restoration of normal blood pressure achieved. Furthermore, it must not be forgotten there may be a time delay of some six hours between repletion measures having been completed and the establishment of a urine flow of greater than 50 ml/hour.

Incipient renal failure is heralded by a diminution in the formation of bladder urine together with a decreased concentrating ability. The tests referred to below, which imply incipient acute renal failure (IARF) and which may be reversed by appropriate measures, must be done within 48 hours of the onset incident.

Thus the urine volume needs to be measured on an hourly basis and this is one of the rare occasions when a nephrologist would agree it is reasonable to catheterize the bladder. Then urine needs to be collected and examined, firstly under the microscope, to exclude any acute glomerular disease such as acute GN which may be suggested by the presence of casts, particularly granular red cell casts. Likewise, the urine must be tested for protein, bearing in mind that Bence-Jones proteinuria can cause acute renal failure in a hitherto undiagnosed case.

The biochemical tests which are of relevance can be established on random samples of urine and a blood sample taken at the time. Clearly, changes in blood concentrations themselves are not likely to be so precipitate as in themselves to suggest renal impairment. What is important are the ratios between the following (a) urine and plasma urea and (b) plasma and urine osmolality. For, as already implied, in acute renal failure not only is there a decreased urine volume, but the ability of the kidney to produce concentrated urine is also impaired with a consequent retention of waste products in the body resulting fluid retention and acidosis. Normally the urine to plasma urea ratio is 50:1 and the urine to plasma osmolality ratio something in excess of 2·5:1. In IARF the urine:plasma (U:P) urea ratio falls below 15 and the U:P osmolality ratio approximates 1·1:1. Although many other indices have been proposed, in practice none have superseded these two as being the most valuable biochemical tests to undertake. Furthermore, trying to make more elaborate ratios has not proved helpful. Although some advocate that a urinary sodium concentration of above 20 mmol/l also indicates renal tubular damage, this is not such a consistently useful marker. If abnormal ratios are found in an oliguric patient within 48 hours from an acute onset incident, then it is reasonable to treat him with a trial of mannitol to see if a urine flow will be established. From a biochemical standpoint it is important to stress that all the mannitol will achieve is restoration of normal urine volume, but not an immediate change in the ratios referred to. Thus, though the tests are of value in diagnosis of IARF, the only marker of success of a mannitol trial is the improvement in urine volume. Sometimes the delay between the acute onset incident and undertaking tests is too prolonged and thus one is presented with established acute renal failure when the U:P urea ratio is of the order of 10:1 or less and the U:P osmolality ratio nearly 1. Here then the biochemical tests of renal function required are those daily measurements of blood parameters which will indicate (a) the effectiveness of conservative management and (b) when a dialysis procedure may be required.

Thus it is important that daily blood urea and electrolyte concentrations are measured and, preferably, creatinine. It is likewise important to collect daily 24-hour urine volumes and estimate the urea content of these. For with a knowledge of the patient's body weight and the rate of rise of blood urea, then one can establish on the basis of urine volume and its urine urea concentration whether or not the urine volume so produced is sufficient to prevent further rise in blood urea concentration. Graphing such results is very useful in predicting the workload of a renal unit and planning appropriately for so-called prophylactic haemodialysis. Most would agree that the blood urea level should not rise above 35 mmol/l or the bicarbonate level fall below 12 mmol/l or the serum potassium concentration rise above 7 mmol/l. It has been clearly established that the outcome of conservative/haemodialysis treatment for acute renal failure is far better by undertaking prophylactic haemodialyses. Sometimes it will be necessary to measure the plasma potassium concentration twice in a day, for this may rise precipitously in certain patients, e.g. those involved in road traffic accidents with internal injuries and bleeding, and early haemodialysis may be required because of impending, dangerous hyperkalaemia which cannot be controlled by conservative means such as glucose-insulin regimens or by giving calcium resins.

Acute renal failure can last for any period from one week to six weeks during which time haemodialysis and/or peritoneal dialysis is required for control. Furthermore the assessment of the daily rate of rise of the blood urea level helps the clinician to divide patients into so called normo-catabolic (blood urea rise < 6 mmol/l/day), modest catabolic (6–12 mmol/l/day) and hypercatabolic patients (>12 mmol/l/day). This in turn directs whether patients are fed normally, orally or intravenously, or whether they are protein restricted.

In those patients who have not entered a diuretic phase by six weeks then a renal biopsy may be indicated (if not sooner in some cases) to determine whether renal function is likely to return. When a patient enters the diuretic phase, clinical management problems are not over. For then the patient is in danger of severe urinary electrolyte losses because the renal tubules at that stage are still not capable of concentrating urine. Therefore it is still necessary to undertake frequent analyses of blood biochemistry so that deficiencies do not occur. As a rough rule the diuretic phase lasts as long as the oliguric phase.

Careful tests of renal function such as measurement of glomerular filtration rate, and tubular concentrating and diluting ability have shown that it may take up to two years before complete normality is restored to a kidney that has recovered from acute renal failure. However, these refined tests are of little clinical value for the blood biochemistry values in such patients return to normal relatively quickly.

Finally, a few words on the pitfalls of diagnosing acute renal failure. This applies particularly to elderly patients who have undergone surgical procedures. Clearly, as with all organs affected by ageing processes, so the kidney too deteriorates in terms of its concentrating ability. Thus an elderly patient who undergoes an operation becomes relatively catabolic and will produce a solute load, e.g. urea, which exceeds the renal solute excretory capacity even though an adequate urine volume is being passed. This of itself does not imply acute renal failure, but is a reflection of a pathophysiological state. Thus one will find a divergence of the urea:creatinine ratio, but little tendency for the patient to develop hyperkalaemia and only a slight one for a metabolic acidosis. Thus it is important in such patients not to diagnose acute renal failure on the basis of a blood urea concentration. Here it is more important to be aware that attempts are necessary to maintain or increase their usual urine volume, for if neither the urine urea concentration nor the urine volume can be increased in the elderly then, inevitably, retention of waste products must occur. Fortunately, in most situations this is rarely of any clinical significance. Finally it is worth mentioning that certain drugs, e.g. steroids and tetracycline, can cause a considerable increase in the blood urea level without implying renal failure. However, if this is not considered, an osmotic diuresis may occur, by virtue of the rising level of blood urea, leading to quite severe dehydration and in some cases to secondary acute renal failure. Such a situation is deplorable and totally avoidable.

RENAL TRANSPLANTATION

Following renal transplantation, kidneys either diurese immediately or there is a delayed diuresis as the grafted kidney recovers from the effects of ischaemia at donor nephrectomy. Although many tests have been advocated for assessing the function of grafted kidneys and, in particular, whether impending rejection is to occur, at the clinical level very few biochemical tests are required. It must be remembered that many patients being transplanted still have their own kidneys *in situ* which may produce appreciable volumes of dilute urine containing protein. Thus for the renal transplant patient who diureses immediately the most valuable indicators of useful graft function are that the urea and creatinine concentrations decline rapidly in the blood and rise in the urine. Tests of proteinuria are of no value if the patient still has his own kidneys. However, if an anephric patient is transplanted and he subsequently develops proteinuria this may be the harbinger of rejection. For the patient who fails to diurese immediately the problem is that of maintaining him by haemodialysis procedures until the kidney opens up. Here therefore daily measurement of blood urea and creatinine levels alone are sufficient. Further, it is important to stress that the creatinine concentration is the more important since the blood urea will always be artificially raised because of associated immunosuppressant treatment with steroids.

With respect to the diagnosis of rejection in a grafted kidney here again one relies on a profile of clinical examination, changes in blood creatinine with a diminution in urine volume and prompt recourse as necessary to renal biopsy. Although many urinary enzyme tests have been advocated, notably measurement of urinary N-acetyl-glucosaminidase, such tests are subject to alteration by associated antibiotic administration, e.g. gentamicin, and are therefore of limited value. Likewise measurement of urinary β_2-microglobulin has been advocated for diagnosing some forms of renal transplant rejection. However, it must be stressed that the main emphasis in diagnosing graft rejection depends upon changes in blood creatinine concentration with diminishing urine volume. It is also important to have prompt return of the results of requests for these biochemistry tests for clearly hours matter in the treatment of graft rejection and in our laboratory it is customary to have the serum creatinine result within four hours of the sample being sent. Nowadays, miniaturised renography procedures using cadmium telluride crystals and measuring graft blood flow or GFR on a daily basis may offer an even more sensitive method of diagnosing rejection before significant changes in blood creatinine are appreciated. Measurement of urine osmolality and electrolyte concentrations in such patients has not proved to be of significant clinical value.

Finally, from a clinical standpoint, it is also important to be aware of the effects of haemodialysis on the non-diuresing patient and the interpretation of his blood biochemical parameters. Here one has to rely very much on clinical judgement and appreciation of the changing urine volume and the differential clearances, by dialysis, of urea and creatinine.

CHRONIC RENAL FAILURE

The diagnosis of chronic renal failure can be long delayed because kidney function can deteriorate very slowly and the patient adapts extraordinarily well to his changing internal environment. Also it is important to remember that, because the time scale of events in chronic renal failure are so different from acute renal failure, different problems will arise more subtly. For example, changes in bone metabolism and lipid metabolism will occur which are not seen in acute renal failure.

A patient may present for investigation of chronic renal failure as a result of having been found to have proteinuria on routine medical examination, because he is hypertensive, because he has frequency and/or polyuria, because he has normochromic and normocytic anaemia, or because on a random blood sample taken for some other reason his blood urea and creatinine concentrations are found to be elevated. Therefore it might be of value to look at these various indicators separately and to determine which tests are necessary to help in diagnosis and management of renal disease and renal insufficiency leading to chronic renal failure.

Proteinuria

Proteinuria can occur without implying a patient has chronic renal failure or even renal insufficiency as such. Nevertheless, proteinuria is one of the most important indicators of glomerular disease. Firstly orthostatic (benign postural) proteinuria must be excluded, usually by admitting a patient for 24 hours to hospital, to establish whether or not his proteinuria is significant. If proteinuria is confirmed, i.e. it is not postural, then qualitative and quantitative tests should be undertaken.

Although urine protein electrophoresis has had its advocates this technique is of little value with the exception of its ability to detect Bence-Jones protein. Clearly it is important to establish not only the presence of Bence-Jones proteinuria, indicating myeloma, but also the precise nature of the proteinuria which is important for prognosis following treatment. Next it is important to quantitate the proteinuria and this is best done by collecting two consecutive 24-hour urine samples (on an outpatient basis) and measuring the total protein output, the majority of which will be albumin.

However, in paediatric practice in particular, it is valuable to qualitate the proteinuria by measurement of a so-called selectivity index. This measures the ratio of a small protein particle, e.g. transferrin, to a larger protein particle, e.g. IgG. A ratio of greater than 0·2 suggests a non-selective proteinuria which in turn suggests either a proliferative or membranous type of glomerulonephritis. Biopsy alone can determine precisely the nature of the glomerular lesion. Nevertheless, the value of this test is that if a selective proteinuria is found it is likely to indicate a minimal-change glomerulonephritis, the commonest form of glomerulonephritis in childhood which is amenable to steroid treatment. Thus the test is of value in avoiding unnecessary renal biopsies in children. However, the selectivity index is of little value in adults because it does not appear always to stay constant, a factor which invalidates the test.

It is well known that proteinuria will diminish with diminishing renal functioning mass. Therefore an index of leakiness, i.e. an albumin:creatinine ratio, is very valuable. Clearly, diminishing proteinuria could be taken to imply success of treatment whereas in fact it might relate to diminishing functioning renal mass. The albumin:creatinine ratio has been validated both in children and adults and has the merit that random samples of blood and urine can be used. Another pitfall in interpreting proteinuria is that this will diminish if the blood albumin level is dramatically reduced. Thus any interpretation of the amount of proteinuria must be related to plasma albumin concentration. Serum protein electrophoresis may also be helpful in crudely determining the nature of renal disease, e.g. globulin patterns indicating myelomatosis, or showing a raised amount of α_2-globulin with a low albumin and gamma globulin as in the nephrotic syndrome.

Polyuria

Another feature of impaired renal function may be increased frequency of micturition and polyuria. Clearly, here it is important to exclude psychogenic factors and lower urinary tract problems, e.g. small bladder volume. Thus it is often valuable in the first instance to admit such a patient complaining of this symptom to hospital so that the frequency can be verified and individual urine volumes measured, distinguishing between frequency and polyuria.

If true polyuria is confirmed then a number of other disorders need to be excluded before it is concluded that polyuria is due to impairment of renal concentrating ability because of primary renal disease. Generally, polyuria due to renal insufficiency, e.g. chronic glomerulonephritis, chronic pyelonephritis, hypokalaemia, hypercalcaemia is rarely above 5 l/day and associated with elevated blood levels of urea and creatinine. Compulsive water drinkers and patients with diabetes insipidus or nephrogenic diabetes insipidus usually pass more than 5 l/day and have normal blood urea and creatinine concentrations.

Thus compulsive water drinkers need to be excluded; this is best achieved by admission to hospital, noting their personality, their drinking habits and measuring their plasma osmolality which tends to be lower than normal, i.e. around 269 ± 10 instead of 285 ± 14. Furthermore if they are denied access to water then the kidney shows the ability to concentrate and reduce the urine volume passed.

Also it is important to distinguish between diabetes insipidus and nephrogenic diabetes insipidus. This again requires admission to hospital with either a fluid deprivation test or the administration of pitressin tannate. In the overnight water deprivation test the patient must be weighed before commencement of the test, for to lose more than 5% of the total body weight would be dangerous leading to hypotension and collapse. If the patient has true diabetes insipidus then denying him access to water will still result in continuing polyuria and passing dilute urine with increasing plasma osmolality. The administration of pitressin tannate will diminish urine volume with an appropriate increase of urine concentrating ability. If, however, nephrogenic diabetes insipidus is the problem then there is a failure of response to pitressin. Thus the water deprivation test (WDT) investigates the integrity of the supra-optico neurohypophyseal—renal tubule system whereas pitressin tests only the ability of the tubule to respond by concentrating urine. A greater urine osmolality can be achieved with the WDT test than with pitressin (Table 1). Other causes of polyuria such as hypokalaemia secondary to, e.g., laxative abuse, hypercalcaemia, or even to diabetes mellitus, will normally readily be excluded by simple biochemical measurements.

TABLE 1
EFFECTS ON URINE OSMOLALITY mOsm/kg

	Water deprivation test	Pitressin
Compulsive water drinker	750	550
Diabetes insipidus	200	600
Normal	1200	700

If a patient has polyuria then the next important measurement is assessment of the renal functioning mass, i.e. the glomerular filtration rate (GFR). Although experimentally many methods have been used, e.g. inulin clearances, chromium EDTA clearances, vitamin B_{12} etc., the most valuable one for clinical purposes is the creatinine clearance. Although this does not take into account tubular secretion of creatinine nevertheless it is a valuable measurement. Particularly this is so since the blood creatinine concentration can remain within the upper limit of normal whilst at the same time a considerable reduction, up to 50%, of GFR may have taken place. Amongst clinicians and clinical biochemists there is a controversy as to whether one should do a short, two-hour, outpatient creatinine clearance or rely upon 24-hour collections. Both methods are valuable and which is used depends upon the indoctrination of the patients by the clinicians looking after them. My own preference is to measure two consecutive 24-hour urine creatinine clearances, after careful instruction to the patient as to what is required. I prefer this to the outpatient two-hour creatinine clearance for many patients cannot comply with passing urine on demand and the 'washout effect' can be misleading.

Many arguments have been written as to whether significant proteinuria increases tubular secretion of creatinine. Suffice it to say that the evidence would suggest that this is not a significant factor and in any case it makes no difference to the overall interpretation of creatinine clearance for a given patient. All the other methods of creatinine clearance measurements, at least in biochemical terms are not worth the trouble of endeavour. It is likely however that the newer methods of miniaturized renography using radio-isotopic techniques, particularly with diethylene-diamino-penta acetic acid (DTPA) labelled with 99 MTC, will prove a very valuable rapid method of measuring GFR. Using 99 MTC DTPA a good measurement of GFR can be obtained by measuring the clearance of this isotope from another body compartment, e.g. the forearm and obviating the need for plasma and urine sampling. This method has the advantage that it can be performed using a single shot with as little as 1 mc of the isotope.

Measurements of glomerular filtration rate indirectly, by assessing creatinine clearance rate, are valuable in monitoring the progress of patients with chronic renal failure, particularly in patients who may have deteriorating renal function following a renal transplant. Also having an idea of the creatinine clearance rate is valuable with respect to introducing dietary manipulations (*see* Table 2).

TABLE 2
FACTORS INFLUENCING DEGREE OF, AND RESPONSE TO, PROTEIN RESTRICTION IN DIETARY MANAGEMENT OF CRF

(1) GFR 10–15 ml/min—40 g protein diet
(2) GFR 5–10 ml/min—30 g protein diet
(3) GFR   <5 ml/min—strict LPD

Group (1) Blood Urea returns to near normal
Group (2) Blood Urea falls into 6–12 mmol/litre
Group (3) Blood Urea falls into 15–20 mmol/litre

urea/creatinine ratio <27·5

Tubular Function Tests

In chronic renal failure there may be an emphasis on tubular dysfunction as opposed to glomerular dysfunction, yet both lead to a deterioration of renal functioning mass. Although a number of elaborate tests are available, it must be stressed that simple tests will usually distinguish glomerular lesions from tubular lesions. Again, the emphasis on clinical history and symptoms, whether with or without biopsy, cannot be overstressed. Thus with primary glomerular lesions proteinuria is usually greater than 3 g/24 hours as opposed to tubular disorders, e.g. pyelonephritis, analgesic nephropathy where proteinuria tends to be less than 3 g/24 hours. Furthermore, with primary glomerular disorders the tubular concentrating ability is maintained longer than with tubular disorders such as chronic pyelonephritis. Also the tendency towards metabolic acidosis is more prominent with tubular disorders than glomerular disorders, at least in the early stages of renal impairment. As renal function diminishes, however, with a GFR of less than 15 ml/min, the differences become less apparent, for with glomerular diseases the proteinuria will diminish and the tendency to metabolic acidosis will increase. Renal tubular acidosis, of course, is suggested by finding a systemic acidosis with the passage of alkaline urine. The ability of the renal tubules to produce an acidic urine can be challenged by the ammonium chloride test though this is rarely of value in the management of patients with chronic renal failure. This is more of interest in patients with other manifestations of tubular disease, e.g. stone formation.

It is not usually until the GFR has fallen to less than 15 ml/min that dietary procedures are necessary (Table 2) and manipulation of the electrolyte content of the diet may be necessary too. In the early stages of chronic renal failure an increased protein intake is very often required in chronic glomerulonephritis because of the severe proteinuria. However, later the situation may change and such compensation give way to protein restriction. A knowledge of whether the patient has primary renal tubular or glomerular disease can be of value in advanced chronic renal failure, for the tendency to develop bone disease is greater with tubular disorders and the value of bicarbonate therapy likewise more applicable to them.

Chronic Renal Failure and Bone Disorders

In advanced chronic renal failure, renal osteodystrophy not infrequently occurs. This becomes even more of a problem in different parts of the UK when patients become established (if suitable) on regular haemodialysis programme. The advent of renal osteodystrophy is usually suggested by an inappropriately high plasma calcium concentration for the degree of impaired renal function, together with elevated phosphate concentration and raised alkaline phosphatase activity. Such bone complications arise because of the failing ability of the end-stage renal failure kidney to produce the active metabolite of vitamin D, 1:25 DHCC, with associated compensatory increase of parathormone activity in response to hypocalcaemia. Such bone disease may declare itself clinically by bone pain symptoms and be verified by appropriate bone radiology. The importance of making this diagnosis is because nowadays renal osteodystrophy is amenable to treatment by 1α-hydroxycholecalciferol an analogue of vitamin D and can rapidly heal such bone lesions. Clearly, monitoring of blood biochemistry, especially the calcium concentration, is important to avoid overswing effects, i.e. hypercalcaemia from such treatment and serial measurements of alkaline phosphatase activity to determine when it is appropriate to withdraw such treatment.

Chronic Renal Failure and Disorders of Lipid Metabolism

The modern management of many patients with end-stage renal failure includes regular haemodialysis treatment and often subsequent renal transplantation. Whilst in earlier years much emphasis was placed upon modifying protein intake of such patients and adjusting nitrogen metabolism, little attention was given to effects upon lipid metabolism. A majority of patients will develop disorders of lipid metabolism irrespective of the original aetiology of their disease. The commonest lipid disorder is a hypertriglyceridaemia (type IV) and next commonly a type IIb (combined hypercholesterolaemia and hypertriglyceridaemia). Also it has been shown in chronic renal failure that there is a tendency for the high density lipoprotein concentration to decrease, an important factor in promoting lipid-induced atheroma. The causes for such hyperlipidaemia in chronic renal failure are controversial, but the establishment that they exist is important because premature arterial disease in chronic renal failure patients can be prevented by appropriate dietary manipulations. In general terms all patients with chronic renal failure should be advised to take a diet high in polyunsaturated fatty acids, attempting a polyunsaturated:saturated fat ratio of 1·5 or more, with a modest intake of carbohydrate. It is a tragedy that many chronic renal failure patients first maintained by regular haemodialysis and then successfully transplanted should die of premature coronary artery disease because of failure to test for hyperlipidaemia and treat appropriately.

As alluded to initially, limited investigations in chronic renal failure will determine the appropriate treatment and when such treatments should be changed in face of changing renal function. Therefore, depending on the rate of progression in individual patients, the frequency of their outpatient visits will be determined. The management of end-stage renal failure by dietary means depends upon the ability to measure frequently biochemical parameters, to control acidosis with, say, amino-acid supplemented diets, to prevent hypercalcaemia with keto-acid supplemented diets and to treat impending hyperkalaemia with resins if necessary.

None of the renal function tests referred to, however, differentiate between unequally affected kidneys. Here appropriate radiology with modern renography techniques are indicated. I personally feel there is little use in doing individual ureteric catheterization studies and measuring differentially urine sodium and urea concentrations, etc.

One must be aware of the need to interpret differently standard renal function tests when referring to pregnant patients. Here what may be considered for example normal uric acid, urea, and creatinine concentrations can be grossly abnormal in the early stages of pregnancy. Thus to apply the normal figures for the non-pregnant female to the pregnant female can be grossly misleading and lead to errors of interpretation and management.

There can be little doubt that the management of renal failure patients is impossible without the close collaboration of biochemistry departments. Fortunately however, the number of tests required can be minimized and their complexity kept in perspective.

FURTHER READING

Bevan, D. R. (1979), *Renal Function in Anaesthesia and Surgery*. London: Academic Press.
Brochner-Mortensen, J. (1978), 'Routine methods and their reliability for assessment of glomerular filtration in adults,' *Danish Med. Bull.*, **25**, 181–185.
Editorial (1980), 'Acute Renal Failure,' *Brit. med. J.*, **1**, 1333–1335.
Evans, D. B. (1977), 'Disease of the urinary system. Management of chronic renal failure by dialysis and transplantation,' *Brit. med. J.*, **2**, 1585–88.
Goldsmith, H. J. (1972), 'The chemical pathology of renal failure,' *Brit. J. Anaesth.*, **44**, 259–64.
Harrington, J. T. and Cohen, J. J. (1973), 'Clinical disorders of urine concentration and dilution,' *Arch. Intern. Med.*, **131**, 810–16.
Kassirer, J. P. (1971), 'Clinical evaluation of kidney function—glomerular function,' *N. Engl. J. Med.*, **285**, 385–89.
Kassirer, J. P. (1971), 'Clinical evaluation of kidney function—tubular function,' *N. Engl. J. Med.*, **285**, 499–505.
Kerr, D. N. S. (1980), 'The assessment of renal function and detection of urinary abnormalities,' *Medicine* (3rd series), **25**, 1269–1274.
Kerr, D. N. S. and Davison, J. M. (1975), 'The assessment of renal function,' *Brit. J. Hosp. Med.*, **13**, 360–372.
Kleinknecht, D. (1977), 'Acute Renal Failure.' *Medicine* (2nd series), **28**, 1487–1501.
Knapp, M. S., Blamey, R., Cove-Smith, R., Heath, M. (1977), 'Monitoring the function of renal transplants.' *Lancet*, **ii**, 1183–85.
Lavender, S., Hilton, P. J., Jones, N. F. (1969), 'The measurement of glomerular filtration rate in renal disease,' *Lancet*, **ii**, 1216–18.
Lee, H. A. (1977), 'The management of acute renal failure following trauma.' *Brit. J. Anaesth.*, **49**, 697–705.
Lee, H. A. (1978), 'The management of acute renal failure,' pp. 69–97. In *Medical Management of the Critically Ill* (Eds. G. C. Hanson and P. L. Wright). London: Academic Press.
Lindsay, R. M., Linton, A. L., Langland, C. J. (1965), 'Assessment of postoperative renal failure,' *Lancet*, **i**, 978–80.
Linton, A. L. (1980), 'Diagnostic criteria and clinical management of acute renal failure,' pp. 14–36. In *Acute Renal Failure* (Ed. A. Chapman). (Clinics in Critical Care Medicine Vol. 1) Edinburgh: Churchill Livingstone.
Mallick, N. (1980), 'Nephrotic syndrome,' *Medicine* (3rd series), **26**, 1316–1321.
Marks, V. (1973), 'Clinical biochemistry: Assessment of kidney function by biochemical methods,' *Teach-in*, 695–702.
Morgan, D. B., Carver, M. E., Payne, R. B. (1977), 'Plasma creatinine and urea: creatinine ratio in patients with raised plasma urea.' *Brit. Med. J.*, **2**, 929–32.
Roscoe, M. H. (1964), 'Urine in acute and chronic renal failure. *Brit. Med. J.*, **1**, 1084–86.
Sweny, P. (1980), 'Asymptomatic proteinuria.' *Medicine* (3rd series), **26**, 1310–1312.

16. EXAMINATION OF URINE

R. McSWINEY, BRENDA SLAVIN, PATRICIA R. N. KIND, R. V. BROOKS, I. S. MENZIES
AND J. MOUNT

Introduction — R. McSwiney

Routine screening — R. McSwiney
  General principles
  Urine screening in practice
  Further testing

Disorders of the renal tract — R. McSwiney
  Proteinuria
  Water and electrolyte balance
  Hydrogen ion
  Transport defects

Proteinuria — B. Slavin
  Types of proteinuria
  Investigation of proteinuria

Enzyme activity in urine — P. R. N. Kind

Hormones and metabolites in urine — R. V. Brooks
  Urinary free cortisol
  Steroid metabolites
  Female infertility
  Pregnancy monitoring

Sugars in urine (mellituria) — I. S. Menzies
  Types of mellituria
  Detection and identification

Drugs in urine — J. Mount

Metals in urine — R. McSwiney

INTRODUCTION

By the year 1866 the common constituents of urine had been identified and there was a clearly-established distinction between normal and abnormal, but astute clinicians were already using their sense of taste in the testing of urine to differentiate between the diseases diabetes mellitus (sweet) and insipidus (tasteless).

The examination of urine is now mostly performed either to screen for unsuspected diseases (diabetes mellitus and renal disease), or to confirm or eliminate diagnoses, or occasionally to assist in the control of treatment. The advantages of examining urine are numerous. Outstandingly helpful to the analytical chemist are the facts that urine is a concentrate for most substances, perhaps 100 times more concentrated than plasma, and that some blood constituents, which are confusing to the analyst (e.g. red cells an protein), are physiologically reduced in concentration, or removed.

On the physiological front, urine is the dumping ground for many substances. For instance, non-metabolisable sugars of exogenous or endogenous origin are excreted into the urine, and qualitative or quantitative examination can yield diagnostic information about carbohydrate metabolism or about the absorption of such sugars from the intestine and therefore, indirectly, about intestinal function.

Some substances, e.g. morphine, are so rapidly removed from blood that they are virtually undetectable and must be sought through their metabolites in urine. With other substances, the quantity of metabolite is so much greater than that of the original that it is more convenient to estimate the former, especially if concentrated into urine.

In addition, urine in the form of 'timed saves' can be made to act as a convenient carrier and store for the estimation of rates of metabolism or production; if one can reasonably assume that the contents are representative of the original metabolism, then their rate of output is a measure of rates of metabolism. This principle has been used commonly for the estimation of production rates of hormones, notably steroids; indeed the clarification of endocrinological thinking which has occurred during the last 40 years is fundamentally dependent on it. There are fallacies in its application because metabolites may not be representative of production and because of individual variation in metabolic pathways, and the tendency is, therefore, to switch to analysis of the original product in plasma. But the principle still finds useful applications in the diagnosis of phaeochromocytoma (catecholamine metabolites) and carcinoid tumours (5HIAA).

Finally, urine is so familiar to the layman that strange colours, or changes of colour, or unusual odours, or even changes in volume, bring patients for diagnosis. The commonest abnormalities of this kind are blood in urine or exceptional alterations of volume, the other obvious changes being rare. Classical examples are urine, which darkens on exposure to light (alkaptonuria, porphyria, melanoma), or is dark red when passed (myoglobinuria, porphyria again) or has the characteristic odour of 'maple syrup disease'. However, most clinicians see none of these in a lifetime; even in a laboratory, where experience is more concentrated, they are very rare.

Unusual colours are usually due to dyes used to colour sweets or drinks (e.g. eosin), or to injected methylene blue, or to metabolites of beetroot pigment.[12]

It will be seen that, although abnormalities may have been discovered in urine, newer analytical methods of high sensitivity and specificity have abolished some of the original advantages of urine. There is an established trend towards transferring testing and measurement to blood samples because these often give more precise information, and may have other advantages of availability and rapid sampling.

ROUTINE SCREENING

General Principles

Routine screening for unsuspected disease was an established part of hospital medicine before chemistry entered the medical field. Doctors were, and are, trained to make routine checks on the skin, the respiratory system, the heart, the central nervous system, the abdomen and so on, in every case. Obviously, if an abnormality is suspected from previous evidence, for instance the patient's account of symptoms, then the routine examination of a particular system is more detailed. The usefulness of this approach is that the routine examination discloses lesions which would not otherwise be discovered. The approach is that anything which may help the individual patient ought to be done and the clinician will expend time and trouble on the routine aspects of history taking and physical examination just in case useful information should come to light. In the course of admitting a patient to hospital, he will make approximately 100 routine checks, using his own observation, and will then extend the principle to the 'ancillary tests' which require more than personal faculties.

These may be X-rays, haematology, electrocardiograms, plasma chemistry and so on. There must be some limit to the number of such tests which are to be performed in the average patient, but until recent years the question of quantitative efficiency or of cost-effectiveness would not have been raised on the grounds that the individual patient is unique and not average. The value of screening resides in raising the possibility of disease; further tests are required to confirm or deny its presence. A modest proportion of false positives is acceptable as the price of not missing a condition which would affect the patient's welfare; false negatives are undesirable.

Urine Screening in Practice

Routine screening is certain to include chemical investigation of urine. Attention is focused on two conditions (diabetes mellitus and kidney damage) which may be symptomless, but are important because their presence would affect the patient seriously no matter what else might be wrong with him, and sufficiently common to make the effort worthwhile. If glucose is present, the patient is assumed to have diabetes mellitus in the first place and the urine should be tested for ketones; if protein is detected, renal disease is provisionally suspected. Of course, glucose and protein can be present in the urine of people who do not have diabetes mellitus or renal disease.

The basis for routine screening for only these diseases is imposed by the limitations of available chemical tests, not by any limitation in the clinician's wish to know more; urine tests which disclosed, for example, unsuspected cancer or cardiological disturbances would be welcomed.

Routine testing is usually performed outside the clinical chemistry laboratory, but this does not mean that the laboratory should be uninterested; in the hands of clinicians and nurses, the results show approximately 30% of errors. On multiple-test 'sticks' the chemistries for glucose (presumptive evidence for diabetes) and protein (presumptive evidence for kidney damage) are often combined with others: ketones, bile pigment, and pH. These latter subsidiary tests are rarely of value as screening procedures but their importance in selected cases is considerable; they should be performed as tests for confirmation or elimination of diagnoses. It is important to note that the chemistries of 'sticks' may give false positives and that the old 'liquid' tests are usually more reliable.

In addition to the above tests for all hospital patients, there is a special application to patients admitted with acute abdominal pain: porphobilinogen in urine discloses cases of acute intermittent porphyria. The condition is rare, but the diagnosis is crucial to the patient.

Routine urine testing has also been used in babies for the detection of phenylketonuria; it established the value of screening for this condition, but the testing has now been transferred to blood samples because of their greater reliability.

Further Testing

When the clinician has taken the patient's history, performed the physical examination and tested the urine, he is faced with approximately 100 items of data; most are of value, but some are irrelevant to the main problem although appropriate to side issues, and some may be misleading. The situation is almost certainly complicated and needs experience and flair for its solution. In some cases, the diagnosis will be clear already; the clinician may, even then, wish for confirmation or he may need to seek for hidden features of the disease or for complications. In other cases, diagnosis will be provisional and the clinician needs further information to confirm or eliminate various possibilities: he may be able to perform extra tests himself, or he may turn to specialists in other fields.

DISORDERS OF THE RENAL TRACT

Proteinuria

The problems are illustrated by proteinuria. The condition is looked for because (unsuspected) kidney disease is sufficiently common and important for screening to be desirable, and because proteinuria frequently accompanies kidney damage. In principle, absence of protein indicates absence of disease with a sufficiently high probability at this stage of investigation: plasma creatinine or plasma urea concentration and other tests of renal function may be added later.

The clinician is well aware that the simple detection of proteinuria does not necessarily indicate kidney disease. In the first place, the protein may not come from the kidney but from any part of the renal tract or, in women, from the vulva outside the tract. In the second place, even if it came from the kidney it may be normal, as in orthostatic albuminuria, or it may be abnormal protein from elsewhere, as in myelomatosis.

The assumption will be made, however, that proteinuria indicates renal damage unless proved otherwise, and further investigation will be undertaken. The usual first step would be microscopic examination of a deposit from the urine. Casts, which are cylindrical plugs of protein, can only be derived from renal tubules and, therefore, localise the source of protein; they may be normal, so large numbers or the presence of cells embedded in the casts will be looked for.[12]

It may be noted that microscopy used to be a quite usual activity in clinical chemistry laboratories. It may also disclose the presence of excess red blood cells (the most definitive and sensitive indicator of blood in urine), or the white blood cells (indicating inflammation somewhere in the renal tract), or of bacteria or important crystals.[10]

Further guides to diagnosis or therapy may be sought in special cases by laboratory investigation of the type of proteinuria (*see* Proteinuria, p. 281) or of enzymuria (*see* Enzymuria, p. 283-4).

Water, Sodium and Potassium Balance

The normal person achieves metabolic balance of water and sodium by ingesting excesses and excreting surpluses. Physiological mechanisms involve sensors controlling hormonal outputs which in their turn produce reduction or increase of output into urine (*see* Table 1). The mechanisms are the osmolal-antidiuretic

TABLE 1
RENAL TUBULAR FUNCTIONS

Under hormonal control

Water balance—antidiuretic hormone reduces water output.
Sodium balance—aldosterone (and cortisol) reduces output.

Influenced but not controlled by hormones

Potassium balance—aldosterone (and cortisol) increases output.
Phosphate excretion—parathyroid hormone increases output.

Not under hormonal control

Reabsorption and catabolism of some proteins.
Reabsorption of glucose.
Reabsorption of amino acids
Excretion of hydrogen ion.
Synthesis of ammonia.

hormone system for control of water balance and the renin-angiotensin-aldosterone system for control of sodium balance. It may be noted that both mechanisms are highly appropriate to normal living but may be harmful if they operate in diseased people; thus the inappropriate secretion of antidiuretic-hormone is a well-recognised condition, and excess aldosterone secretion is the final trigger for the production of oedema in many cases and is responsible for much ill-health; there is no sensor for the accumulation of sodium and water in physiological concentration and no defence against it. It is also unfortunate that the renal tubules operate by interchanging sodium and potassium because physiological mechanisms and treatment which are designed to alter sodium excretion will inevitably affect potassium also.

The examination of urine is worthwhile in only a few cases of abnormal water, sodium or potassium balance. Considering water first, the problem presents either because an abnormal urinary volume (water output) is noticed by the patient or clinician, or because the clinician has other reasons to suspect excesses or deficiencies. Investigation of renal function will follow and may include urine tests. For instance, if urinary volume is high the problem is whether the patient has an osmotic diuresis due to glycosuria, or lacks antidiuretic-hormone, or is a compulsive water (tea, coffee) drinker, or has renal tubular damage, or has a very rare inherited defect of the renal tubules which renders them unresponsive to ADH. Obviously, tests for glycosuria are necessary and may provide the diagnosis; if they do not, then identification of proteinuria or determinations of the osmolality of urine after an injection of antidiuretic-hormone may help to distinguish between the other conditions; but the diagnosis is not simple.

If the urinary volume is low, the problem is to distinguish between damaged kidneys and normal kidneys responding well to antidiuretic-hormone. The clinician is aided by two facts: (1) normal kidneys cannot produce less than 350 ml of urine per 24 hours, and (2) damaged kidneys cannot concentrate urine. Measurement of urinary volume and osmolality are, therefore, of substantial help in diagnosis; volumes less than 0·25 ml/min (or 350 ml/24 h) indicate damaged kidneys especially if combined with a urine osmolality in the range of 200 to 400 mOsm/kg. Osmolalities over 600 mOsm/kg indicate normality.

We may note that there are formal tests of renal function which depend on the ability of normal kidneys to produce large volumes of dilute urine after a water load or small volumes of concentrated urine during restriction of water intake, as compared with the restricted ability of damaged kidneys to produce anything other than an ultrafiltrate of plasma with the composition and osmolality of plasma. These tests provided useful information in their day but fail when the renal damage is minimal.

Low urinary volume is also an accompaniment of the 'inappropriate ADH syndrome' which follows injury (including surgical operations) in many patients. The body produces more ADH than needed with the result that more water than is necessary is withdrawn from urine; urine is, therefore, more concentrated than necessary and plasma is more dilute. The difficulty is to assess what is 'more than necessary' because so many variables are involved and the diagnosis will probably depend on finding dilute plasma (low plasma sodium concentration or low plasma osmolality) rather than on examination of urine.

Sodium and Potassium Deficiencies

In patients with deficiencies of sodium or potassium or water, examination of urine will be helpful in directing investigation into the right channel. Fundamentally, the clinician's problem is to determine whether the deficiencies are due to insufficient intake or too high an excretion. The general principle is that normal kidneys with normal hormonal control would reduce urinary output to low levels, while renal tubular defects or hormonal abnormalities would have produced the deficiency through high output.

Daily output can easily be measured and the only difficulty is to decide what is an appropriate 'low' level. In sodium deficiency, the metal can be almost entirely removed from urine by the combination of normal kidneys plus normal aldosterone: the output will be less than 5 mmol/24 h. In potassium deficiency, the kidney is less successful in conserving the ion and there is only the incomplete guidance that *if plasma potassium concentration is repeatedly less than 3·0 mmol/l* the urinary potassium should be less than 20 mmol/24 h.

Sodium and Potassium Excess

In patients who have excess sodium or raised plasma potassium concentrations the problem is simpler; the excess must be due to insufficient excretion into urine. Estimation of quantities of these substances in urine is, therefore, unnecessary and attention is turned to a search for evidence of renal disease, errors of hormonal production, cardiac failure, albumin deficiency, hepatic failure, or over-enthusiastic treatment with hypertonic intravenous fluid.

Accumulation of sodium and water, or raised plasma concentration of potassium, may be treated by altering renal tubular function through the use of drugs (diuretics) which increase the output of sodium (which takes water with it) and potassium. In practice, nearly all of the patients on diuretics need to lose sodium and water only. The loss of potassium is an unfortunate side effect, treated by regular doses of potassium salts by mouth; occasionally, the plasma potassium concentration needs to be monitored.

The clinical problem of water and electrolyte balance is illustrated by the condition 'lower nephron necrosis' in which the glomeruli and tubules are destroyed, or nearly so. The remarkable feature of this disaster is that the tubules can recover and regenerate if the patient can be kept alive for a few weeks. Part of the clinical effort is directed to reducing the rate of protein-breakdown but we are concerned with the fact that the clinician who diagnoses the condition has to take over control of water, sodium and potassium balance because the kidneys are incapable.

Diagnosis depends on having a high level of suspicion and on detecting oliguria or low urine water output; for these reasons urinary volume is measured as a screening test in many postoperative patients. Unusually the sign of abnormality is less, not more; normal kidneys cannot excrete less than 350 ml (some clinicians will say 400 to 500 ml) of water per 24 hours, whereas lower nephron necrosis in its first stages may reduce output to nil. The clinician must first reassure himself that the urine collection really represents output and is not confused by errors nor by the accumulation of urine in the patient's bladder. If urine is available, its osmolality will be close to that of plasma (say 300 mOsm/kg) as opposed to the much higher concentration produced by normal kidneys which are responding to the antidiuretic-hormone (say over 600 mOsm/kg). He will probably investigate plasma urea or creatinine concentration and, because of their variability in normal people, may well have the result of yet another screening test—the patient's preoperative value—for comparison. A rapid rise, especially if continued, confirms renal damage.

Control of balance, in the first stages, consists of *not* putting in surpluses, because the normal route of excreting them has no means of increasing the output. Water intake is controlled at a volume which is calculated to allow for insensible losses (say 700 ml/24 h) plus yesterday's urine volume; intakes of sodium and potassium are reduced to nil, because their insensible losses can be neglected. The patient is weighed daily as an indicator of water balance.

This regime continues until the spontaneous regeneration of kidney tissue results in increased glomerular filtration; the tubules are still so damaged that they are virtually unable to alter the filtrate, urinary volume increases and the losses of sodium become significant. It may become worthwhile to determine the urinary loss of sodium each day, and to replace it during the next 24 hours. After a few days or weeks the renal tissue recovers further, glomerular filtration rises, and tubular function begins to recover. The clinician is faced with a patient excreting large volumes of water and sodium, and probably potassium by this time. His new problem is to decide whether the abnormal urine is due to damaged kidneys or is a normal response to his continuing therapy; his recourse is to withdraw a substantial part of the input and look for a fall in the urinary output; he will probably base his judgement on urinary volume rather than on any laboratory test.

Excretion of Hydrogen Ion

Patients may be unable to attain an adequate hydrogen ion gradient across the cells of their renal tubules (renal acidosis) but may be able to compensate by enhanced ammonia output. The ability of renal tubules to excrete hydrogen ion may be tested by examining urinary pH and ammonia output after a standard dose of ammonium chloride.

Transport Defects

There are four clinically notable transport defects in renal tubules. They are failures of reabsorption from the glomerular filtrate, and result in the presence of excess of the affected substance in urine.

Renal glucosuria is harmless and a nuisance because it suggests diabetes mellitus. It is common during pregnancy, and also appears as a rare congenital defect. It is demonstrated by glucosuria with blood glucose concentrations below the normal 'threshold' level of 8 to 9 mmol/l (145–160 mg/100 ml).

Cystinuria is the clinical term for a congenital defect in the reabsorption of the dibasic amino acids, cystine, ornithine, lysine and valine. Cystine is clinically prominent because it is relatively insoluble and crystallises in the pelvis of the kidney, producing very large calculi which may damage the kidney substance by compressing it, or may obstruct urinary flow or may facilitate infections.

Phosphaturia may be due to a rare inherited renal tubular defect; it may result in low plasma concentrations of inorganic phosphate and, therefore, in defective bone mineralisation.

Hypokalaemia is a consequence of a rare congenital defect in reabsorption of potassium.

Combined defects of reabsorption may affect any two or all three of 'cystine', phosphate and potassium. They may or may not be accompanied by a defect of hydrogen ion excretion. The various permutations bear eponymous titles after their discoverers of whom Fanconi may perhaps deserve priority.

PROTEINURIA

Protein in the urine at a greater output than 0·1 g/24 h is considered to be abnormal and it becomes important to establish whether or not there is associated renal disease. Protein losses greater than 0·5 g/24 h must be investigated fully.[5] Prolonged losses may result in a low plasma albumin concentration, which if associated with dependent oedema is known as the 'Nephrotic Syndrome'. The proteins found in normal urine not only include many of those present in plasma but some produced by the urinary tract itself—the most studied being the uromucoid first described by Tam and Horsfall.[21]

Recent work with polyacrylamide gel electrophoresis suggests that even with a quantitatively normal proteinuria, tubular damage and Bence–Jones proteinuria may be present. Measurement of an individual protein such as albumin may better define pathological proteinuria. Barratt, McLaine and Soothill[1] have used this concept as a measure of glomerular dysfunction in children using random urine collections and relating albumin excretion to creatinine concentration.

Types of Proteinuria

Proteinuria can conveniently be classified into:

(i) Loss of plasma proteins not normally present in the glomerular filtrate (Glomerular Proteinuria).
(ii) Saturation of the tubular reabsorption mechanism and/or a failure of this mechanism (Tubular Proteinuria).
(iii) Tissue antigens arising from the renal tract or from other tissues in the body that pass into the renal tract.
(iv) Chyluria.

Common examples are to be found in Table 2.

Glomerular Proteinuria

Glomerular loss of protein may result from an increase in glomerular pressure and flow (e.g. renal vein thrombosis) but is more commonly associated with an alteration in the permeability of the glomerular membrane. Current work in this field suggests that the glomerular wall acts as a negatively charged filter which enhances the filtration of circulating polycations and retards the passage of circulating polyanions.[4] Evidence is now accumulating to suggest that in disease there is a loss of charge and that polyanions, like albumin, readily enter the urinary space; also, that this loss of charge may facilitate the accumulation of larger macromolecules in the mesangium and mesangial matrix production—the ultimate result of which could be glomerular sclerosis, focal or diffuse.

Tubular Proteinuria

Glomerular filtration of serum proteins of molecular weight 40 000 Daltons proceeds with comparative ease and includes fragments of larger serum proteins (e.g. albumin and immunoglobulins), immunoglobulin light chain, β_2-microglobulin, post-gamma globulin, carbonic anhydrase, lysozyme and other enzymes.[17] Normal urine contains only a small amount of these proteins, the greater proportion of them being reabsorbed and partially catabolised by the renal tubules. This process occurs chiefly in the cells of the proximal tubules and it is probable that only certain of these cells may perform this function.

An increase in the production and filtration of low molecular weight proteins by the body (e.g. Bence–Jones protein and lysozyme) will lead to saturation of the tubular reabsorptive mechanism resulting in 'overflow' proteinuria. A high concentration of these proteins may also lead to tubular damage.

Tubular proteinuria associated with a normal filtered load may be an inherited isolated deficiency (e.g. Balkan nephropathy), or it may be associated with other tubular deficiencies (e.g. Fanconi's syndrome) or with tubular damage.

Isolated light chain excretion in the urine has been observed in patients following renal transplant, poisoning with a heavy metal, and gentamycin and rifampin therapy. Children under two years of age have an increase in light chain excretion which may persist in certain families into adult life. The source of these light

TABLE 2
CLASSIFICATION OF PROTEINURIA

Type	Lesion/mechanism	Aetiology	Investigations/comments
Transient			
Exercise	?Increased glomerular permeability		
Postural	?Altered capillary wall and normal reduction in blood flow		Selectivity increment from day to night
General illness	?Inflammatory changes in kidney	Fever, congestive cardiac failure	
Pregnancy	?Renal vascular changes	Toxaemia	
Glomerular			
	Haemodynamic	Renal vein thrombosis	
		Renal artery stenosis	
		Chyluria	
	Increased transfer of proteins across an altered capillary membrane	Glomerulonephritis	
		Collagen vascular disease	
		Infection	
		Diabetes mellitus	
	Low albumin, oedema and proteinuria greater than 3 g/day = Nephrotic Syndrome	Amyloid	80% of patients have a monoclonal protein in serum.
		Myeloma	Monoclonal serum protein, Bence–Jones proteinuria.
		Toxins, e.g. gold, penicillamine	
		Bee strings	
		Neoplasms	
		Congenital	
Tubular			
	Overflow—saturation of the normal tubular mechanism	Bence-Jones	These overflow proteins may also damage the tubular cells
		Myoglobin	
		Lysozyme (leukaemia)	
	Congenital	Fanconi's syndrome	
		Balkan nephropathy	
		Familial light chain excretion	
	Damage to tubular cells:		
	Haemodynamic	Acute tubular necrosis	
	Infection	Pyelonephritis	
	Metabolic	Hypokalaemia	
		Hypomagnesaemia	
		Hypercalcaemia	
		Hyperuricaemia	
	Immunological	Transplant rejection	
	Toxins	Wilson's disease	
		Cystine	
		Cadmium	
Tissue antigens (see text)			
Chyluria (see text)			

chains is probably the plasma but this has not yet been clarified.

β_2-microglobulin measurement appears to be the most sensitive indicator of tubular damage and can be increased even in the presence of normal dip-stix and sulphosalicylic acid screening tests for proteinuria. Although tubular proteinuria does not usually exceed 2 g/day levels as high as 7 g/day may occasionally occur.

Tissue Antigens

Tissue antigens in the urine may arise from the blood, other body organs via the blood stream and the urogenital tract. They are usually glycoproteins and a considerable number are enzymatic in nature.[3]

Tissue antigens excreted by the urogenital tract under normal and pathological conditions are derived from the glomerular basement membrane, renal tubular epithelium, mucous membranes and male accessory glands and testes. The tissue enzymes and Tam–Horsfall protein, which is the major constituent of urinary casts, have aroused most interest.

Chyluria

Chyluria is the presence of lymph in the urine; following a fatty meal the content can be quite spectacular. The lymph gains access via an abnormal communication, between the intestinal and retroperitoneal lymphatic vessels and the urinary tract, which may arise anywhere from the kidney to the urethra. Although the presence of fat in the urine, which is often intermittent, may highlight the condition, it is the protein loss which results in a non-selective nephrotic syndrome which is of most concern. The presence of fibrinogen can also lead to clot formation in the renal tract and bladder. This is a rare condition, except in patients living in filarial-endemic areas although chyluria is a relatively uncommon manifestation of filariasis. Other causes include a primary defect in the lymphatic system, tuberculous peritonitis and neoplasms.[16]

Investigation of Proteinuria

Screening for protein in the urine is usually done by dip-stick procedures. The limitations of these methods are often not appreciated by the clinicians. The problems include:

(i) The assay favours albumin as opposed to globulin. As a consequence other proteins, e.g. Bence–Jones proteins, can be missed and urine containing increased protein other than albumin may be classified as normal.
(ii) The dip-stick can distinguish between protein concentrations less than 0·2 g/l and greater than 3·0 g/l—any greater accuracy requiring quantitative analysis; 0·2 g/l is a concentration which many people would classify as outside the normal range.
(iii) Highly alkaline, buffered urine is well known for giving false positive results.

The sulphosalicylic acid test, which is used as an alternative, can give false positives with X-ray contrast media, tolbutamide, sulphonamides and para-aminosalicylic acid. Penicillin G and ureido-penicillin in doses greater than 0·2 g/day can give false positive results both with the biuret and sulphosalicylic acid methods.

Having established that the patient has proteinuria the next question which arises is whether further investigation is necessary. Transient proteinuria is relatively common following exercise, cold exposure and fever and has been reported following severe stress. The mechanisms are not understood, but are probably due to changes in renal blood flow and in proteinuria associated with fever there may be parenchymal inflammation. The urine should be tested regularly to establish whether the proteinuria is persistent or transient.

Proteinuria in young adults and adolescents is often postural (orthostatic) in nature and may be of no clinical significance. Postural proteinuria is characterised by its absence from early morning specimens, but is present in those collected during waking hours: it may be reduced to within the normal range on standing quietly but can increase to over 1 g/l after exercise. The mechanism is thought to be a combination of an altered capillary wall and associated normal reduction in renal blood flow in the upright position. A small proportion of patients with active glomerulonephritis have a 'postural-like' type of proteinuria and Cameron[5] recommends that all patients with this pattern of proteinuria should have an intravenous urogram. Frey et al.,[11] report that an increment in the selectivity of the proteinuria from day to night is seen in postural proteinuria and not in glomerulonephritis and suggest that this may be a useful test to discriminate between the two conditions.

All patients who have persistent proteinuria must be investigated. A clinical history of sore throat, rashes, joint pains, chronic infection, diabetes, familial nephritis and polycystic kidneys may be an indicator of renal pathology. Nor must it be forgotten that hypercalcaemia (e.g. myeloma, malignancy and sarcoid), hyperuricaemia (e.g. gout and cytotoxic therapy), hypokalaemia and hypomagnesaemia may result in tubular damage with proteinuria. Low serum C3 and C4 levels are found in a number of renal diseases. Hypercholesterolaemia associated with hypoalbuminaemia should raise the possibility of the nephrotic syndrome even in the presence of normal blood urea and creatinine concentrations and clearances, and the urine must be tested for protein. Routine renal function tests when necessary should be supplemented with urine electrophoresis and an assessment of the selectivity of the proteinuria. This latter investigation has limitations but it is still proven to be useful in the investigation of renal disease in children in whom biopsy is to be avoided; it is useful as a parameter in following the course of the disease and occasionally can be used as an indicator of steroid sensitivity in morphological forms of glomerulonephritis which do not usually respond to steroids.[20]

ENZYME ACTIVITY IN URINE

The measurement of the activity of certain enzymes in urine has been used for many years and has proved to be useful as an aid to diagnosis. There are, however, several complicating factors which must be taken into account when evaluating and interpreting the results obtained.

Enzymes present in the urine have different origins. Some plasma enzymes with low molecular weight pass through the glomerulus into the tubular fluid where they may be reabsorbed and metabolised by the tubular cells, and when present in excess will show an increased activity in the urine. The increased activity is mainly due to the increased plasma level and increased load passing through the glomerulus. However, impaired renal tubular function could also be a contributory factor, and likewise impaired filtration could prevent enzyme being excreted, or in the case of glomerular damage enzymes and proteins of high molecuar weight would pass into urine. Other enzymes in the urine but of a high molecular weight originate from the renal tissue itself. There is a continual natural replacement of the renal tubular cells, and as these cells are enzyme rich it is inevitable that they contribute to the normal enzyme excretion pattern. Some forty such enzymes have been assayed.

There have been attempts to distinguish between 'renal tubular' damage and 'parenchymal' damage, by selecting enzymes specific for different areas of the kidney, but the use of such terms has been misleading in some cases.

A further consideration is the presence in the urine of many activators and inhibitors; their varying concentration may have a marked effect on enzyme activity. Dialysis, gel filtration or ultrafiltration have all been used to overcome these problems, and in some cases,

TABLE 3
CLINICAL USES OF URINARY ENZYME ANALYSIS

Enzyme	Origin	Diagnostic use	Analytical procedure	Notes and cautions
Amylase M.W. 45 000	Plasma	Acute pancreatitis, in conjunction with plasma amylase, calculate clearance $$\frac{U\ Amylase}{P\ Amylase} \times \frac{P\ Creatinine}{U\ Creatinine}$$ Normal ratio less than 0·3	Hydrolysis of artificial chromogenic substrate.	Renal disease will give confusing results.
Lysozyme M.W. 17 000	Plasma	Monocytic and monomyelocytic leukaemia for diagnosis, but more for monitoring treatment.	Turbidometric method with suspension of *Micrococcus lysodeikticus*	Renal disease will give confusing results.
N-acetyl-β-glucosaminidase M.W. 140 000	Renal tubular cell lysosomal enzyme	To detect damage to renal tubular cells due to anoxia for any reason. To monitor patients with renal transplants, it has been shown to give two or three days early warning of an episode of rejection in many cases. Raised in other renal conditions, e.g., hypertension, but increase may be quite small and results must be judged with caution.	Colorimetric or fluorometric* substrate hydrolysis. (* very sensitive method.)	Glomerular damage, plasma enzyme will leak through into urine. Drug administration, particularly gentamicin
γ-glutamyl transpeptidase	Renal tubular cell brush border enzyme	Has been claimed to be of use in monitoring renal transplants, but may be useful only during acute tubular necrotic stage which may occur in first few days.	Colorimetric method, reaction-rate using *p*-nitroanilide. β-naphthylide is carcinogenic and should not be used.	Inhibitors in urine, dialysis or high dilution necessary.

where a very sensitive analytical technique is used, a high dilution of the urine has been found to be satisfactory.

There is some evidence that there is a diurnal variation in enzyme excretion which favours the analysis of 24-hour urine saves. However, there may be inactivation of the enzyme over this time and it is not advisable to use preservatives. Some workers advocate overnight (8 hour) or a 6-hour collection from 6 a.m. The enzyme activity may also be expressed in terms of creatinine excretion, thereby making a correction for changes in urine flow. This latter method has the advantage of using random urine samples, and if necessary the result can be available the same day, which may be helpful in monitoring renal function.

It is important to remember that drugs may influence the level of urinary enzymes, giving confusing results. For example, gentamicin causes a marked rise in excretion which may reflect transient renal damage. Some extra-renal diseases such as myocardial infarction may produce changes. The alterations in the excretion pattern need interpreting as well as any increase above the age and sex-related reference ranges.

Table 3 summarises those enzymes that can be of use clinically, and have been shown to have a place in the routine analysis of urine. Admittedly some of these are useful only in specialised cases, but there is a hope that with more careful considerations of the origin and the use of isoenzyme separations, a greater understanding of renal disease might be possible.

Other enzymes have been suggested for investigations of renal disorders, but there has not been very convincing agreement about their role in clinical diagnosis. There have been claims that lactate dehydrogenase and alkaline phosphatase are good indicators of renal tumours, but the claims have not been substantiated by other workers. Alanine amino transpeptidase activity has been used in much the same way as N-acetyl-β-glucosaminidase, but does not have anything like the sensitivity.

HORMONES AND THEIR METABOLITES IN URINE

For many years the very low concentrations of hormones in blood made it impossible to measure them by the available methods and urinary determinations were the only possibility. With the advent of radioimmunoassay the situation has been transformed. The high sensitivity of these techniques has made it possible to measure most hormones in small samples of blood. These techniques have a high productivity and, by frequent sampling it is therefore possible, largely to overcome the uncertainties imposed by episodic and circadian variation of hormone secretion, and even to take advantage of them.

Estimations of concentrations of the protein and polypeptide hormones in the urine have always been of limited value because of ignorance of the processes of their metabolism and excretion. For this reason estimates in serum and plasma have almost completely replaced urine assays for most of these hormones. Although much more is known of the metabolism of the thyroid hormones, urinary excretion of metabolites is of little value in the assessment of thyroid status because of a complicated enterohepatic circulation.

The remaining areas in which urinary assays may offer a significant advantage are:

Urinary free cortisol
Steroid metabolites
Investigations into female infertility
Pregnancy monitoring
Catecholamines
Phenolic substances.

Urinary Free Cortisol

One reason for measuring a hormone concentration is to explain a clinical condition in terms of a raised or lowered biological stimulus to the tissues. In the case of many steroid and thyroid hormones, for which there exist specific binding proteins in the plasma, the concentration measured may depend as much on the concentration of the specific binding protein as on the activity of the endocrine gland which secretes the hormone. Since it is only the free, non-protein-bound, portion of the hormone in the plasma that is biologically active, the total concentration which is measured in the plasma may not be well correlated with the biological activity. The free non-protein-bound hormone in the blood passes into the glomerular filtrate and, although it is mainly passively reabsorbed in the tubule, some passes into the urine. Less than 0·5% of the cortisol secreted is excreted in this way and yet the measurement of this fraction approaches our ideal index for measuring hormonal stimulus to the tissues. It is fairly directly related to the concentration of non-protein-bound hormone in the blood and therefore to the concentration in the ECF. It has the further advantage, when a 24-hour urine collection is used, of giving an integral value for the whole day whereas a blood value applies to a single point in time and will be influenced both by episodic secretion and by circadian variation.[15]

Steroid Metabolites

A slightly different purpose for which a hormone may be measured is to determine the secretory activity of an endocrine gland rather than assess the stimulus to the tissues. For the steroid hormones this has most often been done in the past by measuring the excretion of their major metabolites in the urine (mainly conjugated to glucuronic and sulphuric acids). This may be the major single metabolite of a particular hormone, e.g. pregnanediol glucuronide which accounts for about 20% of the progesterone secreted. Alternatively it may be a group of related metabolites all derived from one hormone as in the case of the 17-oxogenic steriods (17-OGS) which together account for about 50% of secreted cortisol. The 17-OGS have been much used for the assessment of response to dynamic tests, e.g. dexamethasone suppression, synacthen and metyrapone stimulation tests. 17-OGS estimations are still valid for this purpose but measurements of unchanged cortisol in plasma and urine are now increasingly used to assess the first two dynamic tests and even the use of 17-OGS for the metyrapone test is not essential if a radioimmunoassay for ACTH or for 11-deoxycortisol in plasma is available.

A development of the 17-oxogenic steroid method allows the measurement of the 11-oxygenation index.[9] This is useful in the diagnosis of congenital adrenal hyperplasia but in Britain the diagnosis is probably better made with the aid of radioimmunoassay measurements of 17α-hydroxyprogesterone and androstenedione in the plasma.

Investigation of Female Infertility

The changes in the concentrations of oestrogens and progesterone in the blood of their metabolites in the urine during the normal menstrual cycle are large and either can be used for assessment. However, since samples have to be collected every 2 or 3 days for a month or, more, there are considerable practical advantages, especially for the patient, in urine assays, the collections for which can be made at home. The patient collects a 24- (or 8-) hour save in a bottle provided with a 2 (or 1) litre mark, adds tap water to bring the level up to the mark, mixes and transfers a small quantity to a tube which is kept in the refrigerator until the next visit to the hospital. A third alternative to plasma and urine assays is to measure concentrations of hormones in saliva, e.g. progesterone.[22]

Pregnancy Monitoring

The steroid determination required in greatest numbers in current practice is the measurement of oestriol in plasma or urine during pregnancy. Because oestriol is synthesised by the placenta largely from precursors produced by the fetus this concentration is used as an indicator of fetal viability. The greater convenience of the urinary assays must be weighed against the closer temporal correlation of the plasma concentration with the placental production and the elimination of renal factors.[6]

SUGAR IN THE URINE (MELLITURIA)

Excretion of sugar in human urine may be physiological or associated with pathological conditions affecting absorption, metabolism or renal function. Whereas lactose, glucose and certain pentoses can be formed and released by tissue cells, and therefore have an endogenous source, most sugars, including galactose and fructose, appear in the urine only when they have been ingested, or administered by some other route, and are

Recognised Types of Mellituria

Oligosaccharides

Traces of the trisaccharide *raffinose* and the tetrasaccharide *stachyose* may occasionally appear in the urine, derived from vegetables such as beans and artichoke in the diet. An oligosaccharide containing galactose and fructose appears in milligram quantities in the urine of secretors, all of whom are of blood group B, following ingestion of lactose or galactose. Neither of these situations is pathological.

Disaccharides

Small amounts of *lactose* may be present in the urine of healthy individuals following ingestion of milk and milk products which contain this sugar. This dietary mellituria is accentuated when intestinal lactose hydrolysis is impaired (hypolactasia) and especially when intestinal permeability to disaccharide is increased by mucosal damage (coeliac disease, tropical sprue, gastroenteritis). Endogenous lactosuria is seen from the third month of pregnancy, and increases during lactation, being a normal feature of secretory activity of the mammary gland. Abnormal persistence of lactation (galactorrhoea) is also accompanied by lactosuria. *Sucrose* is likewise excreted in small amounts following ingestion of cane sugar, but amounts may become excessive when intestinal disaccharide hydrolysis and permeability are altered in conditions such as coeliac disease and gastroenteritis. Endogenous sucrosuria has been claimed to occur, very rarely, as a complication of pancreatic disease. Two such cases were extensively investigated and reported in the 1930's, but the condition remains an enigma.

Lactulose, either derived from processed milk products or laxative preparations ('Duphalac', 'Gatinar'), is, like lactose and sucrose, excreted in small amounts after ingestion by healthy infants and adults. This sugar has been used as a marker for assessing intestinal permeability by measuring the fraction of an ingested dose excreted in the urine. Unlike other disaccharides, *maltose* is metabolised after reaching the blood stream, and has seldom been found in urine. This is not the case with *isomaltose* which has been found in the urine of casualties with severe trauma, this sugar either being derived from displaced glycogen or infused dextran.

Monosaccharides

Small amounts of *glucose* (less than 10 mg/100 ml, 0·6 mmol/l; 100 mg of 6 mmol/24 h) are constantly present in the urine of healthy individuals. This may increase, as a physiological phenomenon of unknown cause, during the last 6 months of pregnancy. Both situations represent a failure of complete reabsorption by the renal tubules, differing in degree.

Pathological glucosuria may result from impairment of renal tubular reabsorption with normal blood glucose concentrations, or from elevation of blood glucose concentration so that glomerular filtration exceeds the maximum tubular reabsorptive capacity. These are referred to as 'renal' and 'overflow' types of glycosuria, respectively. Estimation of blood glucose concentration at the time of glucosuria is required to distinguish the two types, and usually necessitates a glucose tolerance test with collection of carefully timed consecutive urine samples for analysis. The most important cause of glycosuria is diabetes mellitus, but perhaps the most frequent in hospital practice is intravenous dextrose infusion.

Minor *galactose* excretion is usually present in the urine of healthy persons following ingestion of milk or milk products. This accompanies the absorption of galactose derived from intestinal hydrolysis of lactose, and is therefore minimal or absent in hypolactasic sujects. Renal plasma clearance of galactose is about 30 ml/min, but there is no 'renal threshold'. Pathological galactosuria is not a feature of impaired renal tubular function, but results from conditions in which blood galactose concentrations rise excessively following ingestion of lactose. Thus hepatocellular disease, ethanol ingestion, and especially the inherited defects of galactose metabolism (galactosaemias) are associated with marked galactosuria. If galactosuria, jaundice, fits and cataract suggest uridyl transferase galactosaemia, a galactose-free diet should be instituted without delay, and diagnosis confirmed by erythrocyte enzyme studies. Galactose tolerance tests may be dangerous in this situation.

Traces of *fructose* are regularly excreted in the urine of healthy subjects following ingestion of fruit or sucrose, from which fructose is derived by the action of intestinal sucrase. As with galactose, reabsorption of fructose by the renal tubule is minimal and pathological fructosuria is, therefore, the result of factors that predispose to an excessive rise in blood concentration following ingestion and is not a recognised feature of renal pathology. Marked fructosuria is, therefore, seen in patients with severe hepatocellular disease, and the rare autosomal recessive inherited conditions hereditary fructose intolerance, and essential fructosuria. The latter condition is, like essential pentosuria, completely benign, but may give rise to anxiety through confusion with the glucosuria of diabetes mellitus.

Both *arabinose* and *xylose* are widely distributed in plant tissues, especially as constituents of polysaccharides in gums, resins and fibre. Small amounts derived from the diet are regularly excreted in the urine of healthy subjects, the amount reflecting vegetable intake. D-xylose is employed as a marker for assessing intestinal absorption, interpretation being based on the rise in blood level or timed urinary excretion following a standard oral dose. In contrast to the above two sugars the pentoses L-*xylulose, ribose* and *ribulose* are derived

TABLE 4
EXCRETION OF SUGAR IN THE URINE (MELLITURIA).

	Physiological	Pathological
Stachyose raffinose	Traces, derived from vegetable sources such as soya bean and artichoke may occasionally appear in the urine. (D).	
B-group Oligosaccharide	(galactose ×2 + fucose). Urinary excretion induced in blood-group B secretors by ingestion of galactose or lactose. (DE)	
Lactose	Small amounts usually present, esp. in urine of hypolactasics. (D) Greater amounts during pregnancy and lactation. (E)	1. Lesions of small intestine, esp. when disaccharidase deficiency is associated with increased permeability. (D). 2. When wounds or burns are dressed with lactose containing preparations (e.g. penicillin powder). (Exogenous origin). 3. Galactorrhoea (abnormal persistence of mammary secretion). (E).
Lactulose	Small amounts are excreted in the urine: (a) when lactulose syrup ('Duphalac') is given as a laxative, or for the treatment of hepatic coma. (D) (b) when certain milk preparations, in which minor lactose to lactulose conversion has occurred, are fed to infants. (D)	
Maltose	Small quantities that enter the circulation from the intestine are rapidly metabolised, and rarely reach the urine.	
Sucrose	Small, but variable amounts depending on intake. (D)	1. Endogenous sucrosuria, a very rare and mysterious condition. (E) 2. Lesions of small intestine, esp. when disaccharidase deficiency is associated with increased permeability. (D)
Glucose	Small amount constantly excreted; 30 to 100 mg/24 h (0·18–0·6 mmol/24 h). (E) may become accentuated, or diminished, during pregnancy. (E)	1. Due to hyperglucosaemia. (DE). (diabetes, rapid absorption, intravenous dextrose, etc.). 2. Due to renal tubular defect. (E) (isolated, or part of Fanconi syndrome).
Galactose	Small amounts often present. (D)	1. Galactosaemia (2 types). (D) 2. Severe hepatic disease. (D)
Fructose	Small amounts usually present. (D)	1. Essential fructosuria. (D) 2. Hereditary fructose intolerance. (D) 3. Severe hepatic disease. (D)
Arabinose	Small amounts often present, of vegetable origin. (D)	
Xylose	Small amounts often present (larger excretion when used as absorption test). (D)	
Ribose	Traces usually present. (E)	Increased in some muscle diseases. (E)
L-xylulose	Less than 30 mg/24 h. (E)	1. Essential pentosuria (1–3 g/24 h; 7–20 mmol/24 h) (E) 2. Drug induced (70–300 mg/24 h; 0·5–2·0 mmol/24 h). (E) (amidopyrine, glutethimide, etc.).

E = endogenous. D = dietary, origin; i.e. cannot occur unless sugar in diet

from endogenous sources, trace amounts being excreted in the urine of healthy subjects. Increased excretion of L-xylulose can be induced by administration of drugs that stimulate glucuronic acid metabolism (amidopyrine, glutethimide, etc.), and subjects with the rare inherited autosomal recessive condition, essential pentosuria, which is benign and affects those of Jewish extraction, excrete between 2 g and 4 g of this sugar daily. Ribosuria has been described as a feature of certain muscle diseases, the sugar being derived presumably from catabolised muscle nucleoprotein.

Detection and Identification of Sugar in Urine

Simple screening tests are widely used for the detection of sugar in the urine. Benedict's test for reducing substances, and the more recently introduced convenient modifications such as the Ames Clinitest which includes a thermogenic reaction, will detect significant concentrations of most sugars, but it should be remembered that sugars such as sucrose are non-reducing. Conversely, there are a large number of urinary constituents other than sugars that also have reducing activity, so that these tests are very non-specific.

Tests based on a specific reaction with glucose oxidase have been adapted for use as 'dip-stick' screening procedures specific for glucose (e.g. Ames Clinistix, etc.). They are particularly useful for testing adult urine samples since glucosuria is the most important type of mellituria requiring routine exclusion in this age group.

Many other screening tests have been described, but none is clearly superior for this purpose to the reducing and glucose oxidase tests mentioned. When, especially in the case of an infant, the use of these two complementary tests indicates the presence of a non-glucose reducing substances in the urine, identification is most conveniently obtained by paper or rapid thin-layer chromatography (TLC). The latter procedures have been modified to give reliable qualitative or quantitative analysis of sugars in urine.

For further details refer to Menzies and Seakins.[19]

DRUGS IN URINE

The detection of drugs in biological fluids is thought by some critics to be of little value. In general terms it has been argued that a clinician experienced in treating overdose patients can conclude a successful outcome without ever knowing which drugs were present. It is important to remember that it is the patient who is being treated and not the poison and whilst it may be useful to know the nature of a poison, treatment may not depend on this. Possible exceptions to this are poisoning by tricyclic antidepressants where delayed absorption must be taken into account in order to deal with any cardiac problems which may arise later and poisons for which a specific treatment is required, e.g. paracetamol. After an initial assessment of the patient to classify the severity of the case, time since exposure to the poison, level of consciousness and presence of any complications, treatment is aimed at safeguarding respiration, removing the poison and monitoring fluid balance. To this end the laboratory measurements of blood glucose, plasma urea and electrolytes, and blood gases may be of considerable help to the clinician initially. If procedures such as forced diuresis, haemodialysis or peritoneal dialysis are deemed necessary, then a good case can be made for measuring the blood drug levels. When a clinician does request drug identification and level, plasma or serum is usually the fluid of choice. This topic is dealt with in Chapter 22.

As far as urine is concerned there are several disadvantages. Severely poisoned patients who may or may not be unconscious do not or cannot pass urine, often for long periods of time, in which case a blood sample would yield a faster result. Catheterization, whilst possible, has the potential disadvantage of introducing infection as an added complication. Urine assays are also of little or no value in the acute situation and may even be misleading with possibly the exception of paraquat poisoning. On the positive side, if it is remembered that urine screens are not infallible and that a negative result may indicate insufficient drug for detection rather than its absence, there may be a case for monitoring drugs in urine within 2–3 hours of samples reaching the laboratory. This is an advantage if consideration is given to the fact that there is a high incidence of multiple drug overdose and it is very difficult, even for an experienced clinician, to pinpoint the exact combination of drugs ingested. Plasma is often used for monitoring drug therapy as urine would be inappropriate. However, repetitive urine analysis is often used in monitoring drug-addiction patients, most usefully to indicate whether the methadone therapy has been adhered to and to assess the status of the patient with regard to the ingestion of other addictive drugs.

The detection of drugs in urine should not be undertaken in isolation. Much valuable information can be obtained by direct examination of gastric contents for remains of tablets or capsules or from characteristic smells. Consultation with the clinician may also be helpful. Often the original drug is only found in urine when a large overdose has been taken, the compound having been metabolised, usually to compounds with less pharmacological activity.

Drug metabolism consists of two stages: metabolic transformations whereby drugs may undergo oxidation, reduction or hydrolysis in preparation for the second stage, metabolic conjugation. Conjugation reactions involve the addition of chemical groups to the drugs or their metabolites and involve the formation of glucuronides, sulphates, peptides and thiocyanates, methylation, acetylation and conjugation with glutathione. Although drug metabolites are excreted by various routes, e.g. bile, expired air, sweat, saliva, milk and gastrointestinal secretions, excretion by the kidney occurs by glomerular filtration and tubular secretion, both by passive and active transport. The former is determined by the degree of ionisation and the latter is important for the secretion of acid conjugates such as glucuronides, sulphate esters and glycine conjugates.

Whilst samples of the authentic drug are useful standards when carrying out spot tests or screening schemes, urine from subjects previously known to be positive for a particular drug offers the advantage of providing a positive control which includes drug metabolites. It is equally important to include a known negative control. Quality control is as important in drug detection as in all other areas of clinical chemistry and it is essential that specimens are collected correctly into a clean container with no preservative. The first urine sample after admission is collected, a convenient volume being 100 ml, although even 10 ml may yield useful information. The sample must be properly identified, labelled and accompanied by a fully informative request card. Specimens may be stored at 4°C for a few days but must be deep frozen for longer storage.

A few simple spot tests or preliminary tests carried out on 1 ml aliquots of urine may provide information on the drug or group of drugs ingested.[18] These include tests for salicylates, phenothiazine, imipramine, chloroform and related compounds such as chloral hydrate or dichloralphenazone, paracetamol and paraquat. Senna may also be identified by a simple quantitative test.[13]

To extend the range of drugs that can be identified in urine, more comprehensive screening procedures based on TLC have been developed.[2] This identification is based on an initial separation into acidic drugs (e.g. barburbitures), and basic drugs, (e.g. tricyclic antidepressants), followed by separation and identification of the individual compounds by TLC. Such an identification is made by using selective colour reactions, fluorescence under ultraviolet light and R_f values by comparison with known standards. Screening procedures are also available for benzodiazepines, opiates, methaqualone and phenothiazines which can be detected simultaneously, and for drugs of abuse in urine which include morphine, methadone, codeine, cocaine, amphetamine and quinine.

Various procedures have been used to make the extraction of drugs from urine more specific. These include the use of charcoal and XAD-2 resin. Positive identification of a drug by a TLC screening procedure cannot be made with a single solvent system. It is, therefore, necessary to confirm the identity of the drug by using two or more different solvent systems or by reference to the literature for a suitable confirmatory test.[7]

Quantitation of drugs in urine is rarely required despite the fact that a plethora of methods are in the literature. For the foreseeable future it is likely that most clinical chemistry laboratories will still rely on a combination of spot tests and TLC screening procedures. TLC is the most convenient and rapid chromatographic method of analysis, being cheap and simple to use, although GLC and HPLC are now well established.[8] With the advent of sophisticated reaction-rate and centrifugal analysers and the development in recent years of immunoassay, advanced procedures for the detection and quantitation of drugs in urine using these methods may become more widespread. Radioimmunoassay, established in 1959 by Yalow and Berson, has been applied to urinary drugs of abuse and critically evaluated. More recently radioimmunoassays have become available for benzoylecgonine, a metabolite of cocaine, LSD and other drugs of abuse. Enzyme immunoassays have been developed and studied, and applied to fast centrifugal analysers.[14] A review of developments in toxicology including sections on immunoassay, GC-MS, HPLC and atomic absorption spectroscopy is described by Williams.[23]

METALS IN URINE

Examination of urine of toxic metals such as lead, copper and mercury is a useful procedure almost always combined with blood estimations (see Chapter 21).

In patients with renal calculi, it is worthwhile determining urinary calcium output while the patient is taking a diet with normal calcium content (say 20 mmol or more per 24 hours) and with low normal calcium content (say 10 mmol of calcium per 24 hours). If the reduction of intake reduces calcium output it establishes that the reduction is worth continuing permanently. Otherwise, measurement of urinary calcium output is of little value because the upper end of the normal range is not really definable, because it overlaps almost completely with the range of people with calcium disorders, and because better information is derived more simply from blood (see Chapter 18).

REFERENCES

1. Barratt, T. M., McLaine, P. N. and Soothill, J. F. (1970), 'Albumin excretion as a measure of glomerular dysfunction in children,' *Arch. Dis. Child.*, **45**, 496–501.
2. Berry, D. J. and Grove, J. (1973), 'Emergency toxicological screening for drugs commonly taken in overdose,' *J. Chromat.*, **80**, 205–219.
3. Boss, J. H., Dishon, T., Durst, A. and Rosenmann, E. (1973), 'Tissue antigens excreted in the urine under normal and pathological conditions. *Israel J. Med. Sci.*, **9**, 490–508.
4. Brenner, B. M., Hostetter, T. H. and Humes, H. D. (1978), 'Molecular basis of proteinuria of glomerular origin,' *N. Engl. J. Med.*, **15**, 826–833.
5. Cameron, J. S. (1975), 'Symptomless proteinuria and the nephrotic syndrome,' *Medicine*, **28**:1 (2nd Series), 1527–1537.
6. Chattoraj, S. C. and Wotiz, H. A. (1975), In *Methods in Investigative and Diagnostic Endocrinology*, Vol. 3, Steroid Hormones, p. 23. Dorfman, R. I. (Ed.) Amsterdam: North-Holland.
7. Clarke, E. G. C. (1969), *Isolation and Identification of Drugs*, London: The Pharmaceutical Press.
8. Dixon, P. F., Gray, C. H., Lim, C. K. and Stoll, M. S. (1976), *High Pressure Liquid Chromatography in Clinical Chemistry*. London: Academic Press.
9. Few, J. D. (1968), 'A simple method for the separate estimation of 11-deoxy- and 11-oxygenated 17-hydroxycorticoids in human urine,' *J. Endocrinol.*, **41**, 213–222.
10. Freeman, J. A., Beeler, M. F. (1974), *Laboratory Medicine—Clinical Microscopy*, London: Henry Kimpton.
11. Frey, J. F., Frey, B. M., Koegal, R., Hodler, J. and Wegmueller, E. (1979), 'Selectivity as a clue to diagnosis of postural proteinuria,' *Lancet*, **i**, 343–345.
12. Harrison, G. A. (1957), *Chemical Methods in Clinical Medicine*, London: Churchill Livingstone.
13. Kaspi, T., Royds, R. B. and Turner, P. (1978), 'Qualitative determination of senna in urine,' *Lancet*, **i**, 1162.
14. Lasky, F. D., Ahuja, K. K. and Karmen, A. (1977), 'Enzyme immunoassays with the miniature centrifugal fast analyzer,' *Clin. Chem.*, **23/8**, 1444–1448.
15. Lindholm, J. (1973), 'Studies on some parameters of adrenocortical function,' *Acta Endocrinologica (Kbh)*, **72**, Suppl. 172, 9–153.
16. Lloyd-Davies, R. W., Edwards, J. M. and Kinmonth, J. B. (1967), 'Chyluria—A report of five cases with particular reference to lymphography and direct surgery,' *Brit. J. Urol.*, **39**, 560–570.
17. Manuel, Y., Colle, A., Leclerq, M and Tonnelle, C. (1975), 'Low molecular weight proteinuria,' *Contr. Nephrol.*, **1**, 156–162, Basel: Karger.
18. Meade, B. W., Widdop, B., Blackmore, D. J., Brown, S. S., Curry, A. S., Goulding, R., Higgins, G., Matthew, H. J. S. and Rinsler, M. G. (1972), 'Simple tests to detect poisons. Technical Bulletin No. 24,' *Ann. Clin. Biochem.*, **9**, 35–46.
19. Menzies, I. S. and Seakins, J. W. T. (1976), 'Sugars'. In *Chromatographic and Electrophoretic Techniques*, Vol. 1, p. 183, 4th edn., Smith, I. and Seakins, J. W. T. (Eds.) London: Wm. Heinemann Medical Books.
20. Renner, E., Heinecke, G. and Lange, H. (1975), 'Clinical value of the renal protein clearance determination,' *Contr. Nephrol.*, **1**, 134–142, Basel: Karger.
21. Tam, I. and Horsfall, Jnr. F. L. (1952), 'A mucoprotein derived from human urine which reacts with influenza, mumps and Newcastle disease virus,' *J. Exp. Med.*, **95**, 71–97.
22. Walker, R. F., Read, G. F. and Fahmy, D. R. (1978), 'Salivary progesterone and testosterone concentrations for investigating gonadal function,' *J. Endocrinol.*, **81**, 164–165.
23. Williams, R. L. (1978), 'Developments in the forensic analyses of drugs and poisons,' In *Recent Advances in Clinical Biochemistry*, Alberti, K. G. M. M. (Ed.) London: Churchill Livingstone.

17. UROLITHIASIS

W. C. ALSTON

Introduction
  Urine composition in renal stone formation

Metabolic conditions associated with renal stone disease
  Cystinuria
  Uric acid stone disease

Stone formation due primarily to urinary infection

Calcium stone disease

Idiopathic calcium stone disease

Hyperparathyroidism

Hyperoxaluria
  Primary hyperoxaluria
  Enteric hyperoxaluria

Renal tubular acidosis

INTRODUCTION

It has been estimated that about 2–3% of people in Western countries suffer an attack of renal colic due to renal calculi. Over a follow-up period of a decade it has been observed that the risk of recurrence is between 20% and 50%. The frequency of occurrence of the several types of urinary tract stones varies considerably between countries. However, in Britain, about 80% of calculi are composed of calcium salts, most commonly a mixture of oxalate and phosphate, pure calcium oxalate or calcium phosphate stones are less common. Phosphatic stones, which comprise 28% of the urinary stones in Britain, tend to form in the renal tract of patients who suffer from recurrent urinary tract infections. They consist usually of magnesium ammonium phosphate admixed with calcium phosphate, the latter salt commonly forming the preponderant part of the mixture. Uric acid and cystine stones occur much less commonly, each accounting for about 3% of the total stones in this country. Stones consisting of xanthine or silicates (due to prolonged ingestion of magnesium trisilicate) are very rare indeed. Recently, 2,8-dihydroxyadenine has been identified in a stone which formed in the renal tract of a child with a rare inborn error of metabolism due to the absence of the adenine salvage enzyme, adenine phosphoribosyltransferase.

Urine Composition in Renal Stone Formation

Urinary stone disease is a multifactorial condition. Predisposing factors include metabolic disorders, genetic factors, hormonal imbalance, dietary composition and environmental factors such as the hardness of drinking water. However, stone formation depends ultimately on the formation of abnormal crystalluria, the crystals consisting of one or more of the stone-forming salts and occurring often in the aggregated state. It has been generally accepted that stone formation is favoured by a low urinary volume and that it is inhibited by a high urinary volume. Indeed cystine and pure uric acid stones may be effectively dissolved by having the patient maintain a large fluid intake and hence a high urine volume of high dilution. However, calcium-containing stones cannot be dissolved in this way so investigation is directed towards understanding the pathogenesis of the condition and treatment is aimed at preventing recurrence. In general, the degree of supersaturation of the urine with respect to each salt is the principal factor determining the likelihood of spontaneous crystal formation in urine. In calcium stone formers, the larger and more aggregated the crystals of calcium oxalate and calcium phosphate the greater the risk of recurrent attacks of stone formation and renal colic. In the case of uric acid and cystine stones not only are there excessive amounts of the offending substance present in the urine, but also a consistently acid pH is observed in the urine of recurrent stone-forming patients—a factor which considerably reduces the solubility of these substances. Idiopathic calcium stone formers, on the other hand, in addition to their tendency to excrete higher concentrations of calcium and oxalate than normal, tend to pass a more alkaline urine and the more severe the disease in terms of recurrence the higher is the urinary pH.

The other factor which plays an important role in the risk of urinary stone formation is alteration in the 'inhibitory activity' in the urine. It has been established that certain substances inhibit the formation and growth of crystals. For example magnesium, pyrophosphate and citrate have been identified as inhibitors of calcium phosphate crystal formation while the only significant inhibitors of calcium oxalate crystallisation are the acid mucopolysaccharides. Therefore, the overall risk of urinary stone formation may be viewed as a balance between the degree of supersaturation of the urine with the ionic stone-forming salts and the concentration or activity of the urinary inhibitors.

METABOLIC CONDITIONS ASSOCIATED WITH RENAL STONE DISEASE

Cystinuria

The characteristic appearance of cystine stones has been described as similar to 'maple sugar'. They are moderately radio-opaque, due to their sulphur content, and usually consist almost entirely of cystine crystals. However, in a patient presenting with renal colic, the diagnosis can be readily made by the simple qualitative

Dietary Factors

Clearly, the dietary intake of calcium, oxalate and purines is highly relevant to the interpretation of the 24-hour excretion of these substances and therefore to the investigation of stone-forming subjects. Also certain factors, such as the intake of sodium, carbohydrate and protein, are known to enhance calcium excretion. On the other hand phosphate and oxalate intake tends to reduce the absorption and therefore the urinary excretion of calcium.

Medications

Since ascorbic acid is partly metabolised to oxalate, hyperoxaluria can be caused by the ingestion of large quantities of the vitamin. An excessive intake of vitamins A or D can cause hypercalcaemia and, over a period of time, can be associated with the development of nephrocalcinosis and nephrolithiasis. The milk-alkali syndrome is also known to produce the same results in the long term. In particular, regular ingestion of aluminium hydroxide (Aludrox), formerly used as an antacid, gives rise to total-body phosphate depletion. This in turn causes hypercalciuria which predisposes to stone formation. The diuretic drug acetazolamide, however, causes hypercalciuria secondarily to its main action in producing an 'exogenous' renal tubular acidosis.

Certain conditions are well known to be associated with recurrent calcium stone disease, and well over 90% of the cases of stone disease are associated with the first two conditions listed in Table 1. Of this high proportion, the great majority belongs to the category of idiopathic calcium stone disease.

TABLE 1
SPECIFIC CONDITIONS ASSOCIATED WITH CALCIUM STONE FORMATION

Primary hyperparathyroidism	Milk alkali syndrome
Hyperoxaluria (1) Hereditary (2) Enteric	Drug ingestion-acetazolamide, aludrox
Renal tubular acidosis	Sarcoidosis
Hyperadrenalism	Medullary sponge kidney
Excessive intake of vitamins A, D or C	Immobilisation (Idiopathic)

IDIOPATHIC CALCIUM STONE DISEASE

The diagnostic criteria for this condition have not yet been clearly established. Also it is probable that it is not a homogeneous condition but a generic name covering various subcatagories.

Risk Factors

The main urinary risk factors involved in idiopathic calcium-stone formation may be considered in terms of six factors:

1. Urinary volume.
2. Urinary calcium concentration.
3. Urinary oxalate concentration.
4. Urinary pH.
5. Acid mucopolysaccharide content of the urine.
6. Urinary uric acid level.

In general, calcium oxalate is very insoluble in aqueous solution, as is calcium phosphate, and these facts account for the relatively high incidence of stones containing these salts. Clearly, the smaller the 24-hour urinary volume the greater will be the concentration of these salts. Also, as the urinary pH rises, these salts will become progressively less soluble, thus increasing the risk of the spontaneous formation of calcium oxalate and phosphate crystals and even aggregation of these crystals. However, in all these respects, there is a considerable overlap between normal subjects and stone-forming subjects. Evidently other factors must be operating to change the likelihood of stone formation in patients who suffer from recurrent renal stone formation. These other factors include the inhibitors of crystallisation. The most significant inhibitors of calcium oxalate crystallisation are the acid mucopolysaccharides; the most significant inhibitors of calcium phosphate crystallisation are pyrophosphate and possibly citrate. Roughly half of the male idiopathic calcium-oxalate-stone formers have reduced levels of acid mucopolysaccharide inhibitor and the remainder have a lower effective level owing to an excessive urinary excretion of uric acid. Several theories have been advanced to account for the 'anti-inhibitory' activity of urinary urate, but the most plausible one (already referred to in the section on uric acid stones) proposes that the effective concentration of acid mucopolysaccharides is reduced by being partly adsorbed onto the surface of colloidal particles of urate.

The overall risk of stone formation is then considered to be a balance between the degree of supersaturation of urine with the ionic salts and the inhibitory activity of urine represented by the concentration of acid mucopolysaccharides, pyrophosphate and other inhibitors.

Urinary Volume

The significance of low urinary volume has already been discussed in relation to adaptation to hot environments and climates. Even in temperate climates some stone formers tend to have a low 24-hour urinary volume.

Hypercalciuria

The excretion of calcium varies widely with diet and geographical location. The definition of hypercalciuria adopted in the UK, however, is taken to mean a urinary calcium excretion in excess of 7·5 mmol/24 h for women and in excess of 9·0 mmol/24 h for men. The measurement is usually made with the patient on a 'free diet' and under normal conditions of work and exercise. Four main types of hypercalciuria have been described:

1. Dietary.
2. Absorptive.
3. Renal tubular.
4. Resorptive.

Dietary. Hypercalciuria can result even in normal subjects on a calcium intake above 30 mmol/day (1200 mg/day). Usually this information can be obtained from a dietary history and in such cases the urinary 24-hour calcium excretion falls into the normal range when the patient is equilibrated on a calcium intake of 25 mmol/day or less. With the patient on a calcium intake of 25 mmol/day (1000 mg) a diagnosis of idiopathic hypercalciuria is considered when calcium excretion exceeds 0·1 mmol (4 mg) per kg body weight per day.

Absorptive. Absorptive hypercalciuria is associated with an increased intestinal absorption of calcium with suppression of parathyroid function and excretion of the absorbed calcium into the urine. A proportion of these patients have a low plasma phosphate concentration due to a renal tubular 'leak' of phosphate. As a consequence of this low plasma phosphate level there is an increased synthesis of 1,25-dihydroxycholecalciferol (1,25-diOH D_3) which results in an enhanced absorption of calcium.

Renal Tubular. Another group of patients present with a renal tubular defect in the reabsorption of calcium with secondary hyperparathyroidism. Most of these patients also have a tendency to hypophosphataemia, enhanced synthesis of 1,25-diOHD_3, and enhanced absorption of calcium.

Resorptive. This condition is described as a subtle primary hyperparathyroidism with a tendency to hypophosphataemia, increased synthesis of 1,25-diOH D_3, and increased calcium absorption. These patients have mild hypercalcaemia, i.e. a fasting plasma calcium concentration of 2·55 mmol/l (10·2 mg/100 ml) or more on several occasions. In some cases the hypercalcaemia is intermittent.

If patients have been studied on a 25 mmol/day calcium intake, to exclude those with a purely diet-induced hypercalciuria, they may be assigned to one or other of those groups on the basis of an oral 'calcium tolerance test'. The normal procedure is to put the patient on a low calcium intake (about 10 mmol or 400 mg/day) for 7 to 10 days prior to the test. Following the assessment in the basal fasting state of plasma calcium, calcium excretion and basal parathyroid function the patient is given a standard oral load of calcium (25 mmol or 1000 mg). Over the subsequent 4 hours calcium excretion provides an index of calcium absorption, the calcaemic response an indication of 'calcium tolerance', and the parathyroid response (assessed as changes in nephrogenous cyclic AMP) as a measure of parathyroid suppressibility. Patients with absorptive hypercalciuria show a marked calcaemic response with a greater than normal suppression in nephrogenous cyclic AMP. Those with renal hypercalciuria show a much smaller calcaemic response and a minimal reduction in nephrogenous cyclic AMP, whereas those with subtle primary hyperparathyroidism show a striking intolerance to calcium (a hypercalcaemic response) with only a slight reduction in nephrogenous cyclic AMP.

Hyperoxaluria

The greater importance of mild hyperoxaluria over hypercalciuria in the recurrent idiopathic calcium stone formers is underlined by the demonstration that the severity of the condition in terms of the stone episode rate is strongly related to the urinary excretion of oxalate, but only weakly related to the excretion of calcium. The explanation for the preponderant influence of oxalate excretion lies in the marked effect which it has on the supersaturation of the urine with calcium oxalate. It has recently been demonstrated that the increasing urinary concentration of oxalate bears a strong correlation with calcium oxalate crystalluria in terms of the volume of crystals, the proportion of large crystals, and the aggregation of the crystals.

Both in normal subjects and in idiopathic calcium-stone formers it has recently become clear that a highly significant relationship exists between urinary excretion of oxalate and calcium. This correlation may be ascribed to the fact that most ingested oxalate is bound to calcium in the intestine. The hyperabsorption of calcium, which is a feature of most idiopathic calcium stone formers, leads to a release of bound oxalate which then becomes available for absorption in the lower gastrointestinal tract especially the colon.

Urinary pH

Hyperparathyroid stone formers have, in addition to hypercalciuria, a tendency to an alkaline pH. Also some of the patients with severe recurrent idiopathic calcium stone disease also have a tendency to an alkaline pH. Calcium phosphate solubility decreases as pH of the urine rises, but the solubility of calcium oxalate is affected to a lesser extent. Therefore, pure calcium phosphate stones are suggestive of a persistent alkalinity of the urine.

Inhibitors

As already mentioned, acid mucopolysaccharides tend to inhibit the growth of calcium oxalate crystals and their concentration tends to be reduced in a high percentage of idiopathic calcium-stone formers. Pyrophosphate, on the other hand, inhibits the growth of calcium phosphate crystals as does citrate. These two substances tend to be reduced in concentration in the urine of patients with a urinary tract infection.

Uric Acid Excretion

The uric acid excretion is increased in some patients with idiopathic calcium stone disease and normal in others. The higher the uric acid concentration in urine the greater the risk of spontaneous crystalluria due to its action in lowering the effective acid mucopolysaccharide inhibitor activity.

HYPERPARATHYROIDISM

Recurrent calcium stone disease is a common mode of presentation of patients with primary hyperparathyroidism. On the other hand, calcium-stone disease does not occur in hyperparathyroidism secondary to renal failure or in tertiary hyperparathyroidism, conditions in which calcium excretion in the urine is low. Although the patients with primary hyperparathyroidism and renal stones may have a raised plasma alkaline phosphatase activity and increased urinary hydroxyproline excretion indicating some increased bone osteoclastic activity, it is uncommon for these patients to develop clinically overt bone disease, i.e. osteitis fibrosa cystica.

Risk Factors

The main risk factors predisposing to renal stone development in primary hyperparathyroidism are an increased excretion of calcium and phosphate and an increased urinary pH. The increase in urinary calcium excretion is due to an increased absorption of dietary calcium mediated by an augmented synthesis of 1,25 $diOHD_3$, which in turn is brought about either directly by the action of parathyroid hormone (PTH) or by the low plasma phosphate level activating the enzyme 1-α hydroxylase in the renal tubules. The second factor contributing to the increased calcium excretion is the mobilisation of calcium from bone directly by the plasma PTH. This resorptive action of PTH is evident by the higher than normal calcium:creatinine ratio in urine from fasting subjects and also by the elevated hydroxyproline:creatinine ratio. The alkaline pH commonly observed in these patients is due to the action of PTH in inhibiting bicarbonate reabsorption and the high phosphate excretion is due to the action of PTH in inhibiting the reabsorption of phosphate by the renal tubules.

Oxalate excretion is also found to be increased in hyperparathyroid stone formers and, as in the idiopathic group, it is found to be proportional to calcium excretion; this indicates that the release of oxalate, bound to calcium in the intestine due to calcium hyperabsorption, may also be the mechanism. However, the high excretion of calcium, phosphate and the high urinary pH usually cause calcium phosphate to be the main component of renal stones formed in patients with primary hyperparathyroidism.

HYPEROXALURIA

An increased 24-hour urinary excretion of oxalate can be due either to a primary inborn error of metabolism or to an acquired condition.

Primary Hyperoxaluria

This term is used to include two genetic disorders of glyoxalate metabolism which are characterised by the urinary excretion of excessive amounts of oxalate, usually more than 1·0 mmol/24 h. They are transmitted as autosomal recessive traits, become clinically apparent in early childhood, and include recurrent calcium oxalate nephrolithiasis or nephrocalcinosis or both. These conditions are progressive and lead to death from renal failure in childhood if left untreated.

Primary Hyperoxaluria Type 1 (Glycolic Aciduria). This is associated with a genetic defect in the enzyme α-ketoglutarate: glyoxylate carboligase. As a result of this defect glyoxylate accumulates and is oxidised to oxalate and reduced to glycolate both of which are excreted in the urine in excessive amounts.

Primary-Hyperoxaluria Type 2 (L-Glyceric Aciduria). This is associated with a defect in the enzyme D-glyceric dehydrogenase. Hydroxypyruvate consequently accumulates and is reduced by lactic dehydrogenase to L-glyceric acid. The reduction of hydroxypyruvate to L-glycerate is probably coupled to the oxidation of glyoxylate to oxalate.

Therefore in Type 1, oxalate and glycolate are excreted in excess whereas in Type 2, oxalate and L-glycerate are the main metabolites. Although the urinary excretion of calcium is normal in the early stages of primary hyperoxaluria and low in the later stages (when renal failure supervenes) the excessive excretion of oxalate is the overwhelming risk factor in renal stone formation and nephrocalcinosis.

No specific treatments are available as yet for the management of these very severe inborn errors, but large amounts of pyridoxine are usually given in the Type 1 condition (200–400 mg/24 h) and magnesium oxide has been given in both types with a large fluid intake to maintain a dilute urine.

Enteric Hyperoxaluria

An increased urinary excretion of oxalate associated with oxalate stone disease can also occur in patients with disease of the small intestine, e.g. Crohn's disease, extensive resection of the small bowel and in malabsorptive conditions. Also a high incidence of this complication has been reported in patients who have undergone ileal bypass operations for the treatment of obesity. Chronic pancreatic and biliary disease have also been associated with hyperoxaluria. Although oxalate metabolism and dietary oxalate intake are probably normal, in most of these patients the basic abnormality seems to be a hyperabsorption of dietary oxalate. Also the degree of fat malabsorption in these patients is proportional to the degree of oxalate hyperabsorption and excretion. Therefore it has been postulated that the excess of fatty acids in the intestinal lumen displace calcium from its complex with oxalate thus making more oxalate available for absorption. Thus in these patients urinary excretion of calcium might be normal, but is more commonly reduced due to calcium malabsorption. In addition to the increased excretion of oxalate the other important risk factor is the reduction of urinary 24-hour volume in many of these patients. This may be due to disturbance in small bowel function or resection or the consequence of an ileostomy.

RENAL TUBULAR ACIDOSIS

Renal tubular acidosis (RTA) is a condition in which the patient is unable to acidify urine normally with a consequent renal wastage of bicarbonate. Four types of this disorder have now been described; type 1 (classic or distal), type 2 (proximal), type 3 (a hybrid of distal and proximal types), and type 4 in which the acidosis occurs in association with chronic renal failure. Of these four types only the first is associated with significant renal stone disease, although not a common cause. The primary distal type 1 RTA usually becomes manifest in childhood, and is inherited as an autosomal dominant condition. It can also occur secondarily to other conditions such as drug administration, e.g. gentamicin or phenacetin, to autoimmune disorders such as SLE, and in association with medullary sponge kidney. It may also develop secondarily to prolonged obstructive uropathy.

In children the condition causes severe growth retardation, rickets due to renal phosphate wastage, papillary nephrocalcinosis and calcium phosphate renal stone formation. The tendency towards the last two conditions is due to hypercalciuria and the persistently alkaline urine. The diagnosis is made by the findings of a hyperchloraemic acidosis together with a persistently alkaline urine. In the acquired type, secondary to, for example, obstructive uropathy, the renal tubular acidosis may be incomplete. Then an acid load test may be performed; 0·1 g (1·9 mmol) NH_4Cl per kg body weight is given. Within 6–8 hours after this acid load the normal subject should decrease urine pH below 5·4, and plasma bicarbonate concentrations should not fall.

Although the main aim of treatment is to treat the metabolic acidosis with alkali administration, raising the urinary pH tends to increase the saturation levels of calcium phosphate thus increasing the risk of stone formation. For this reason oral acid treatment, may be tried in patients with less severe metabolic acidosis, especially in those who can decrease the urinary pH below 6·0.

The other conditions described in Table 1 are much less common causes of recurrent calcium stone disease. Dietary and clinical investigation usually provide the clues to making the diagnosis. For example, the anatomical abnormality of medullary sponge kidney will require radiological studies to reveal the characteristic findings. The management of the accompanying hypercalciuria will be as for patients with idiopathic hypercalciuria.

FURTHER READING

1. Broadus, A. E. Thier, S. O. (1979), 'Metabolic basis for renal-stone disease.' *N. Eng. J. Med.*, 300, **15**, 839–845.
2. Pak, C. Y. C., Ohata, M., Lawrence, E. C., Snyder, W. (1974), 'The hypercalciurias: causes, parathyroid functions and diagnostic criteria.' *J. Clin. Invest.* **54**, 2, 387–400.
3. Lemann, Jr., J., Adams, Nancy, D., Gray, R. W. (1979), 'Urinary calcium excretion in human beings.' *N. Eng. J. Med.* 301, **10**, 535–541.
4. Robertson, W. G., Peacock, M. (1980), 'The cause of idiopathic calcium stone disease: hypercalciuria or hyperoxaluria? *Nephron*, **26**, 105–110.
5. Robertson, W. G., Peacock, M., Heyburn, P. J., Marshall, D. H., Clark, P. B. (1978), 'Risk factors in calcium stone disease of the urinary tract.' *Brit. J. Urol.*, 449–454.
6. Wickham, J. E. A. (1979), *Urinary Calculus Disease*. London: Churchill Livingstone.

SECTION V

DISORDERS OF THE SKELETAL SYSTEM

		PAGE
18.	DISORDERS OF CALCIUM AND SKELETAL METABOLISM	299
19.	DISORDERS OF JOINTS	325
20.	DISORDERS OF CONNECTIVE TISSUE	337

18. DISORDERS OF CALCIUM AND SKELETAL METABOLISM

J. A. KANIS, A. D. PATERSON and R. G. G. RUSSELL

Introduction

Physiological aspects
   Distribution and function of calcium
   Calcium balance
   Major regulating hormones
   The influencing hormones
   Calcium transport to and from the extracellular fluid
   Bone
   Integration of organ responses

Disorders of calcium homoestasis
   Disorders of parathyroid secretion
   Disorders of vitamin D metabolism
   Disorders of calcitonin secretion
   Hypercalcaemia
   Hypocalcaemia
   Disorders of bone turnover

INTRODUCTION

The past 20 years has witnessed a rapid growth in knowledge about calcium metabolism, particularly in our understanding of the biochemistry, metabolism, and actions of the calcium-regulating hormones. Not only has this clarified physiological control mechanisms, but it has also resulted in a better understanding of the pathophysiology of several clinical disorders. This in turn has led to a more rational approach to investigating patients with disorders of mineral metabolism.

The purpose of this chapter is to review the pathophysiology of the common disorders of plasma calcium and skeletal homeostasis. The approach used is one of applied physiology and therefore a substantial proportion of the chapter deals with normal physiology. This provides a basis for interpreting the many biochemical, histological and other investigations available to the clinician interested in metabolic bone disease. Of the many tests available, an attempt has been made to distinguish those primarily of physiological and pathophysiological interest from those of diagnostic value. Disorders of collagen metabolism and renal stone disease are dealt with elsewhere in this volume (Chapters 17 and 20).

PHYSIOLOGICAL ASPECTS

Distribution and Function of Calcium

Function of Calcium

Calcium and phosphate are widely distributed throughout living tissues. However, the great majority of calcium and phosphate is found in bone (Table 1) and the ability of the skeleton to turn over this calcium and phosphate is essential for growth, the prevention of and healing of fractures, and the remodelling of the skeleton in response to physiological and pathological stresses. During skeletal growth and remodelling, calcium is transported to bone across the extracellular fluid compartment.

TABLE 1
DISTRIBUTION OF CALCIUM AND PHOSPHATE IN NORMAL HUMAN ADULTS

	Calcium	*Phosphate*
Total body content (for 70 kg adult)	1000–1500 g	700–1000 g (of phosphorus)
Skeleton	98%	85%
Skeletal muscle	0·3%	6%
Skin	0·08%	1%
Liver	0·02%	1%
Central nervous system	0·01%	1%
Other tissues	0·6%	5%
Extracellular fluid	1%	1%

Fifty per cent of the extra-skeletal calcium is found in the extracellular fluid (ECF). The ECF concentration of calcium is critical to maintain normal neuromuscular activity, and a fall in plasma calcium concentration results in tetany and convulsions. Conversely a rise in the concentration of calcium in ECF has many adverse effects (see hypercalcaemia, page 323) including delayed neuromuscular conduction and muscle paralysis.

These two functions of calcium, the maintenance of the skeleton and of the ECF calcium concentration, are closely related and suggest that disorders of the one may induce disorders of the other. This is commonly but not invariably the case. Thus in primary hyperparathyroidism, skeletal disease and disturbed plasma calcium homeostasis commonly coexist. In contrast, in Paget's disease, bone turnover is characteristically increased but plasma calcium is usually normal.

The concentration of calcium within cells is considerably lower than that in the ECF. Cytosol calcium concentrations are generally 100 to 1000 times lower than extracellular. Within cells, however, mitochondria are capable of accumulating large amounts of calcium against electrochemical gradients, to an extent that mitochondrial deposits of insoluble calcium phosphate can form. The activation of many different types of cells by hormones or pharmacological agents is now thought to depend on increases in intracellular calcium concentrations derived from the extracellular fluid or from mitochondria. Many intracellular processes including enzyme activity, cell division and exocytosis are controlled by intracellular calcium, and in particular the calcium-dependent regulatory protein, calmodulin.

Disorders of intracellular calcium homeostasis have not been recognised, perhaps partly because of the difficulties of measuring subcellular calcium concentrations. Hormonal activation is commonly associated with a stimulation of adenylate cyclases specific to the target tissue. This in turn changes the levels of cyclic AMP and intracellular calcium producing further responses within the cell. In some disorders this sequence is lost and the ability of parathyroid hormone (PTH) to stimulate adenylate cyclase and other metabolic events can be used clinically to distinguish various disorders of parathyroid function.

Plasma Calcium

The concentration of plasma calcium in health is maintained within a narrow range (Table 2) despite the large movements of calcium across gut, bone, kidney and cells. Several hormones, including parathyroid hormone and 1,25-dihydroxy vitamin D_3 ($1,25(OH)_2D_3$) appear to regulate the ionised fraction of plasma calcium (approximately 50% of total plasma calcium) by modulating calcium fluxes to and from the extracellular fluid. In turn the secretion rates for these hormones are regulated by the calcium concentration in the extracellular fluid.

TABLE 2
DISTRIBUTION OF PLASMA CALCIUM

Ultrafilterable calcium (53%)	
Ionised calcium	47
Complexed calcium (6%)	
Phosphate	1·5
Citrate	1·5
HCO_3 etc.	3
Protein bound calcium (47%)	
Albumin	37
Globulin	10
Total plasma calcium (2·12–2·6 mmol/l)	100%

Changes in the concentration of plasma ionised calcium are usually accompanied by changes in the total amount of calcium in the extracellular fluid since there is a passive distribution of ionised calcium thoughout the ECF compartment. Within the plasma compartment, however, approximately 40% of the calcium is bound to proteins, mainly albumin, and the binding is pH dependent. Major changes in plasma protein concentration, the presence of abnormal proteins, and large shifts in extracellular hydrogen ion concentration may therefore affect the proportion of total plasma calcium that is bound. A further 5–10% of total plasma calcium is bound to small anions such as citrate, phosphate and bicarbonate and the extent of binding may vary as the concentrations of these anions change. For these reasons the estimation of total plasma calcium may not accurately reflect the ionised calcium concentration.

Changes in the binding of calcium in plasma have some important clinical consequences. Thus, the paraesthesiae found in patients with the hyperventilation syndrome, are due to a decrease in ionised calcium concentration because of alkalosis; but the total plasma calcium level is normal. Also, the infusion of alkali into patients with metabolic acidosis (e.g. in diabetes mellitus and chronic renal failure) may precipitate hypocalcaemic convulsions due to a decrease in ionised calcium without changing total plasma calcium.

In the absence of severe acidosis or alkalosis, the major factor influencing the amount of calcium bound is the quantity of albumin present, since the proportion of calcium which is bound varies little. Failure to account for protein binding may result in the erroneous diagnosis of hypercalcaemia in conditions where there is an increased level or an abnormality of plasma proteins (e.g. dehydration, prolonged venous stasis, myeloma). Also, in hypoproteinaemic states, such as disseminated carcinoma or chronic renal disease, total plasma calcium may be low, though ionised calcium is normal. Similarly, in such disorders the level of total plasma calcium is normal and masks true hypercalcaemia.

The ionized plasma calcium concentration should ideally be measured, since it is the physiologically relevant fraction; methods are available, based on colorimetric reaction with dyes (tetramethyl murexide) or by the use of ion-selective electrodes. They are, however, tedious to perform, and require rapid and anaerobic handling of the sample. Moreover, until recently, the errors of the methods used often exceeded the errors incurred from making appropriate corrections to measured levels of total plasma calcium.

Many formulae have been proposed for predicting the ionised calcium from the total plasma calcium, or 'correcting' the total plasma calcium to a normal protein value. These methods depend on the concurrent measurement of total proteins, albumin or specific gravity of plasma. None of these are entirely satisfactory but a simple correction factor for plasma calcium which is widely used, is to subtract from total plasma calcium 0·02 mmol/l (0·08 mg/dl) for every 1 g/l that the plasma albumin exceeds 40 g/l, provided the sample is drawn without venous stasis. An addition of the same amount is made when the plasma albumin is less than 40 g/l. Many laboratories now report 'corrected' plasma calcium, but it should be realised that these are at best a guide to the ionised calcium concentration.

The small proportion of total plasma calcium (Table 2)

that is complexed with cations such as phosphate, citrate, and bicarbonate should not be ignored. The calcium which is normally filtered by the kidney, includes this complexed calcium as well as ionised calcium, and can be measured by passing plasma through membrane filters which retain the protein-bound calcium. The measurement of ultrafilterable calcium may be of value when assessing the renal handling of calcium particularly in those disorders associated with abnormalities in the complexed fraction of calcium, e.g. disorders of acid-base, phosphate or citrate metabolism. In normal subjects the ultrafilterable fraction of plasma calcium lies between 50 and 60% of the total plasma calcium.

Distribution of Calcium and Phosphate

Most of the body's calcium and phosphate reside in bone (98 and 85% respectively), which is thus a major reservoir for both calcium and phosphate. Unlike calcium, the concentration of which is low in most soft tissues, about 15% of the body's phosphorus lies outside the skeleton as organic phosphate compounds such as nucleic acids, nucleotides, phospholipids and phosphorylated metabolites (Table 1). Extracellular concentrations, and probably intracellular concentrations of phosphate, vary much more than levels of calcium, particularly in response to circadian rhythms and to meals. The measurement of plasma phosphate alone therefore affords less precise information than the measurement of plasma calcium particularly when taken in the non-fasting state, and is best interpreted alongside additional investigations such as estimates of tubular reabsorption for phosphate.

Studies with radioisotopes (calcium and strontium isotopes) have shown that in normal human adults the exchangeable pool of calcium is approximately 1–2% of the total body calcium, in the region of 2 mmols (80 mg) per kg during the first few days after injection. This nevertheless represents approximately 125 mmols, which is a substantial amount considering that the extracellular fluid contains somewhat less than 20 mmols (0.8 g) as ionised calcium. This exchangeable pool of calcium is therefore very important in plasma calcium homeostasis, and movements of calcium between body fluids, cells and surfaces of bone occur continuously.

These large and rapid fluxes of exchangeable calcium. should be distinguished from the movements of calcium which occur in bone as a result of mineralisation and bone resorption. Bone resorption is defined as the complete removal of bone mineral and matrix which occurs during physiological remodelling and is a result of osteoclast activity. This accounts for only a fraction of the total calcium exchange between extracellular fluid and bone, the remainder occurring across the large surface area of osteocytes and their canaliculi without synthesis or destruction of bone matrix. Between 1% and 4% of the adult skeleton is thought to be renewed each year. This process is not uniformly distributed throughout the skeleton, since trabecular bone has a faster turnover than cortical bone. This means that disturbances of bone cell turnover commonly have greater effects in trabecular than cortical bone.

The body is not a closed system with respect to calcium, in the sense that calcium is lost from the body by urinary and intestinal secretion and to a lesser extent in sweat; and enters the body by intestinal absorption and renal tubular reabsorption. The way in which the extracellular fluid concentration for calcium is maintained, and the way in which the body gains and loses calcium from the external environment can be simplified by considering the roles of the major organs responsible for movements of calcium to and from the extracellular fluid; namely gut, kidney and bone.

Calcium Balance

Major Sites of Calcium Flux

The major movements (fluxes) of calcium through gut, bone and kidney in adults are shown in Fig. 1.

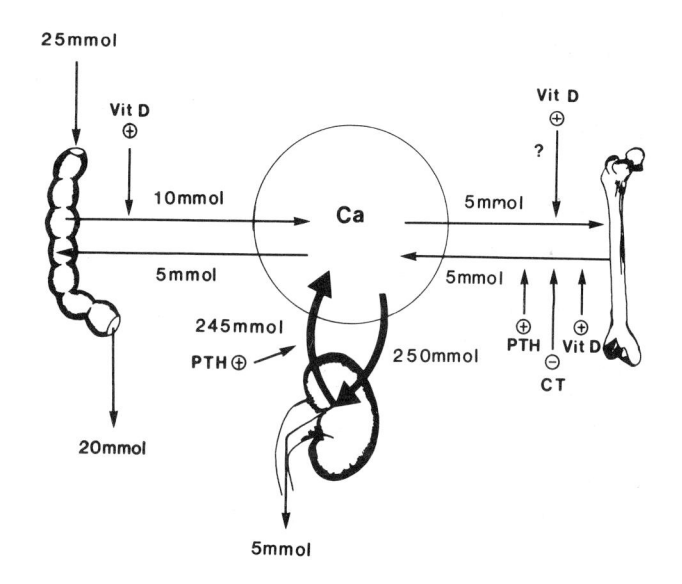

FIG. 1. Major fluxes of calcium (milligrams per day) in a healthy adult. Exchange of calcium in the extracellular fluid occurs with bone, gut and kidney. The net balance for calcium equals the net absorption minus the losses of calcium in faeces and urine, which in a healthy adult is zero. The major fluxes of calcium are regulated by the regulating hormones. PTH increases renal tubular reabsorption of calcium and bone resorption. Calcitonin inhibits bone resorption and vitamin D augments intestinal absorption for calcium. The precise role of vitamin D in augmenting bone resorption and mineralisation *in vivo* is not clear.

Calcium enters the body by intestinal absorption. The true absorption of calcium is greater than the net absorption because some calcium is returned to the gut lumen in biliary, pancreatic and intestinal secretions. Thus, from an average daily dietary intake of 25 mmols (1 g), approximately 10 mmols are absorbed. This is offset by intestinal secretions amounting to approximately 5 mmols (0.2 g) daily, leaving a net transport into the extracellular fluid pool of 5 mmols. True and net fluxes across the gut can be measured by tracer and balance techniques, some of which are now used regularly in clinical practice.

The kidney is a major site for calcium excretion from the body. A large amount of calcium is filtered (Fig. 1) but most of this is reabsorbed so that only 1–3% is excreted into urine. These large fluxes through the kidney to and from the extracellular fluid compartment mean that small changes in renal tubular reabsorption may have profound effects on the extracellular fluid concentration of calcium. Since reabsorption is under hormonal control, principally by parathyroid hormone, this has led to the view that the kidney is a major organ in the regulation of calcium.

Calcium is also lost from the body in sweat. This commonly ignored in metabolic studies since the loss is small and cannot be measured easily. However, losses can be as high as 8 mmols (c. 300 mg) daily under extreme conditions.

In the mature adult who is neither gaining nor losing calcium, bone and soft tissues contribute neither a net gain nor loss of calcium from the extracellular fluid. Thus, the amount of bone resorbed (accounting for approximately 5 mmols (200 mg) of calcium daily) exactly matches the amount formed. This turnover of calcium needs to be distinguished from the rapid and much larger exchange of calcium which occurs within bone and soft tissues.

Calcium Balance

In the healthy adult who is neither gaining nor losing bone, the amount of calcium lost in urine is balanced by an equivalent amount of calcium absorbed by the gut (Fig. 1). A change in the calcium content of the skeleton must clearly result from changes in the net flux of calcium between extracellular fluid and the skeleton. Because plasma calcium is controlled within a close range, and the bulk of calcium is in the skeleton, changes in bone mass are reflected in changes in the external balance for calcium (Fig. 2) rather than in changes in plasma calcium. For example, during growth, where there is a net daily gain of calcium into the skeleton, plasma and intracellular concentrations of calcium are normal. In the long term, therefore, the total body balance for calcium reflects exactly the skeletal balance for calcium (in this case positive balance). Mineral losses begin in middle age (negative balance), particularly in women. Between the age of 25–45 the body is for practical purposes neither gaining nor losing calcium, so that the body's inflow and outflow of calcium are matched.

Calcium balance is measured by the difference between the dietary intake of calcium and the combined faecal and urinary losses. In practice this is time consuming to do and needs the facilities of a metabolic unit. The patient takes a diet constant in calcium and phosphorus which is calculated to be as close as possible to his normal dietary intake of calcium and of as many other constituents as possible. After a 'run-in' period of several days, urine and faeces are collected. Because of the

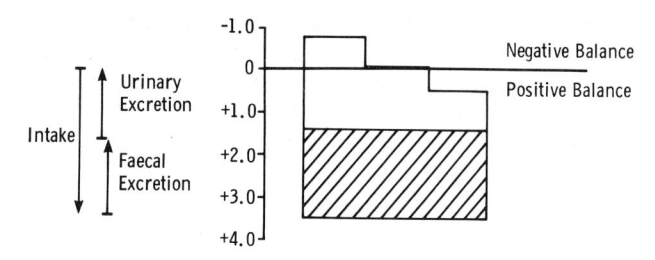

FIG. 2. Metabolic diagram showing conventional method of calculating net balance. The daily intake of the dietary constituent (e.g. calcium) is plotted below the 0 line. From this value the faecal excretion (hatched areas) and the urinary excretion (unshaded areas) are subtracted to give the balance. The figure shows 3 balance periods. In the left hand period the daily faecal output and urinary excretion exceeds the dietary intake and the patient is in negative balance. In contrast, during the balance period on the right of the figure the patient is in positive balance where total excretion is less than dietary intake.

vagaries of bowel habit, it is standard practice to pool collections into periods of between 4–7 days. The accuracy of the measurement is improved by the daily ingestion of a measured amount of an internal marker such as polyethylene glycol, copper thiocyanate or chromium sesquioxide which are all non-absorbable. Calcium losses in the stool can thus be corrected for the measured recoveries of the internal standard. External markers such as carmine red are also commonly administered at the beginning and end of each balance period. The appearance of these dyes can be recognised in the faeces, and this aids timing the faecal output in relation to the corresponding urine collections.

In the best hands the technique has a coefficient of variation of between 5 and 10% of the dietary intake for calcium. This means that it is difficult to be confident of calcium balances which differ by less than 2 to 3 mmols (60–120 mg) daily on repeated estimation. Thus, it is also difficult to be sure that a patient is losing or gaining skeletal mass if the change in bone mass is less than 1–2% per year. Calcium balances are frequently used in clinical research, particularly in the assessment of treatment of metabolic bone disease. In clinical practice they may be useful in determining the activity of disease, for example in osteopetrosis or juvenile osteoporosis, provided the skeletal loss is sufficiently great. A further difficulty arises in the interpretation of balance data. Since a balance is necessarily measured over a relatively short time (weeks) it may be misleading to assume that the balance reflects changes of skeletal significance which occur over years.

Calcium balance is a function of the integrated fluxes across bone, gut and kidney. These fluxes are continually changing and are affected by a variety of factors including several hormones. Thus, the concentration of calcium in the extracellular fluid is set by the relative size of the various fluxes and by the influence of these controlling agents. These hormones can be sub-divided into 'controlling' hormones and 'influencing' hormones. The controlling hormones are the major regulating hormones, parathyroid hormone, calcitonin and the vitamin D metabolites, the secretion of each of which is altered in response to changes in plasma ionised calcium concentrations. The influencing hormones are those other hormones such as thyroid hormones, growth hormone, and adrenal and gonadal steroids, which have effects on calcium metabolism, but whose secretion is determined primarily by factors other than changes in plasma calcium.

Major Regulating Hormones

Parathyroid Hormone (PTH)

In the circulation, PTH consists of several polypeptide fragments which are degraded in the liver and kidney. The major stimulus to the secretion of PTH is a fall in the ionized fraction of plasma calcium. The biological actions of PTH at a variety of target organs serve to increase plasma calcium concentration, which in turn suppresses the secretion of PTH, so that there exists an efficient negative feedback hormonal loop.

Mammalian PTH consists of a single peptide chain, containing 84 amino acids. There are small differences in amino acid sequences between those so far characterised (bovine, porcine and the human hormones; Fig. 3). In common with several other peptide hormones, PTH is synthesised as a pro-hormone which contains an additional 6 amino acids on its N-terminal end. A further precursor form, pre-pro PTH containing a total of 115 amino acids, has been identified in studies *in vitro*. These precursor forms are probably converted to the 84 amino acid peptide before secretion from the gland. The gene structures for both the bovine and human have recently been determined.

Synthesis of the different segments of the polypeptide chain has shown that only the first 32 to 34 amino acids (reading from the N-terminal end) are necessary for biological activity. There is evidence that cleavage occurs naturally, partly in the liver, to produce a short N-terminal biologically active fragment and a larger inactive C-terminal fragment. This cleavage may be necessary for PTH to act on bone. There are also many less well characterised circulating fragments of PTH. The liver and kidney are important sites of degradation and, for example, the C-terminal fragment normally cleared by the kidney may be increased in chronic renal failure, though the circulating biological activity of PTH may be normal. This causes some problems in the interpretation of radioimmunoassay in patients with renal impairment since the C-terminal fragment is the major component measured in many assay systems. Attempts to raise antibodies directed primarily at the amino-terminal portion molecule (the biologically active fragment) have been partially successful in resolving this problem (Fig. 4). Nevertheless, the characteristics of all immunoassays for PTH depend critically upon the immunoreactivity of the antibody, so that each assay system with each antisera must be well characterised at a clinical level before full interpretation of PTH levels are possible.

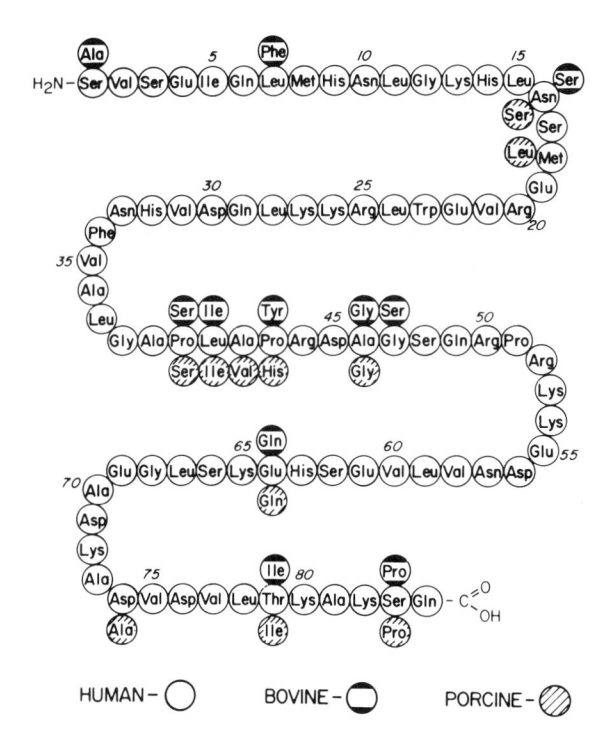

Fig. 3. Amino acid structure of human parathyroid hormone and differences in amino acid sequence in the bovine and porcine hormone.

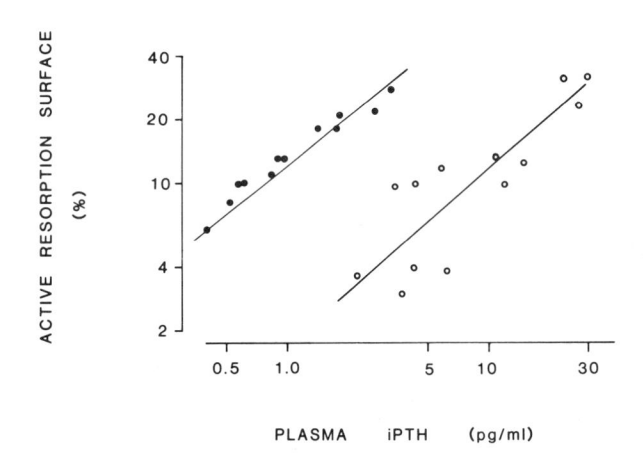

Fig. 4. Plasma immunoreactive levels of parathyroid hormone (iPTH) in patients with end-stage chronic renal failure. Two antisera have been used and values using either assay correlate significantly with known biological effects of PTH (in this case increased bone resorption). However, the values using the 'aminoterminal' assay (closed symbols; left) are 6–10 times lower than those using a 'carboxyterminal' assay (open symbols; right).

Biological assays might also overcome these difficulties but are generally too insensitive or too time consuming. Sensitive biological assays for PTH particularly cytobiochemical assays are being developed though these are not yet established for routine clinical practice. Levels measured by these techniques suggest that levels of biologically active PTH are an order of magnitude less than levels measured by region-specific immunoassays.

The major physiological stimulus to secretion of PTH is a fall in the plasma ionised calcium concentration. Conversely, a rise in plasma ionised calcium above normal suppresses PTH secretion. A host of other factors are known to influence PTH secretion including beta-adrenergic agonists, vitamin-D metabolites, growth hormone and somatostatin, vitamin A, prostaglandins, prolactin, dopamine and other divalent cations such as magnesium and strontium. With the exception of magnesium, the physiological or clinical relevance of these factors is uncertain. In the presence of low levels of magnesium, the secretion of PTH is impaired and this, together with an impaired target organ response to PTH probably accounts for the hypocalcaemia occasionally seen in severe magnesium deficiency in man.

The target organ actions of PTH include effects on bone, kidney and indirectly on gut. PTH acts on the kidney to increase the tubular reabsorption of calcium and to depress the tubular reabsorption of phosphate. This leads to a rise in plasma calcium and a fall in plasma phosphate. PTH also decreases the proximal renal tubular reabsorption for bicarbonate which leads to a hyperchloraemic acidosis (analagous to a proximal renal tubular acidosis). A mild metabolic acidosis is often seen in primary and in secondary hyperparathyroidism. Conversely, a metabolic alkalosis is commonly observed in hypoparathyroidism and in hypercalcaemia due to malignant disease affecting the skeleton. In this latter disorder the alkalosis is partly due to suppression of PTH secretion, but the release of buffer from bone may also contribute. It is possible that the acidosis induced by PTH augments the resorption of bone, thereby raising the level of plasma calcium. PTH has a further important effect on the kidney in stimulating the 1-alpha hydroxylase enzyme responsible for the production of $1,25(OH)_2D_3$, leading to increased intestinal absorption of calcium and possibly to release of calcium from bone. Thus, many of the actions of PTH on the kidney appear either directly or indirectly to increase the extracellular fluid concentration of calcium.

The major effect of PTH on bone is to increase bone resorption, an effect which is readily demonstrated *in vitro*. Both primary and secondary hyperparathyroidism can be associated with obvious radiographic and histological evidence of increased bone resorption. There is experimental evidence, however, that low doses of PTH may increase bone formation and this accords with clinical observation. Thus in kinetic studies, the rate of mineral accretion often matches the rate of bone resorption in primary hyperparathyroidism, suggesting that a main effect of PTH on the skeleton is to cause a redistribution of skeletal calcium and an increase in bone turnover.

There is a great deal of controversy as to whether PTH (and indeed calcitonin) can influence the rapid exchange of calcium that occurs between the extracellular fluid and bone and soft tissue. This rapid transfer may be very important in the minute to minute maintenance of plasma calcium, though it is difficult to envisage this as being important in the long-term regulation of plasma calcium since the net unidirectional flux of calcium across bone and soft tissues must be close to zero.

Calcitonin (CT)

Calcitonin is produced from the C-cells of the thyroid in man. It is a peptide hormone containing 32 amino acid residues with a disulphide bridge between cysteine residues in positions 1 and 7. The entire sequence is essential for biological activity and the gene structure of the human hormone has been elaborated. There are several differences in amino acid composition of the calcitonins from different species and this is associated with different potencies (Fig. 5). Surprisingly, the salmon hormone resembles the human more than other mammalian calcitonins and it is interesting that the salmon hormone is more potent on a molar basis in man than the human hormone itself.

The site of calcitonin secretion in man may not be only from thyroid tissue. The C-cells themselves are derived embryologically from the APUD cell series derived in turn from the neural crest. Non-thyroidal sites for production of calcitonin include the thymus, adrenal, and possibly the pars intermedia of the pituitary gland.

Many agents are known to affect the secretion of calcitonin. These include calcium, gastrointestinal hormones such as cholecystokinin, enteroglucagon and gastrin, beta-adrenergic agents, and alcohol. Since an obvious action of calcitonin is to inhibit bone resorption and thereby lower the level of plasma calcium, it is widely believed that calcitonin is a calcium-regulating hormone with a negative feedback mechanism. Thus the secretion of calcitonin, stimulated by an increase in extracellular fluid concentration of calcium, inhibits bone resorption, and thereby lowers plasma calcium. Conversely, a decrease in plasma calcium concentration reduces the secretion of calcitonin and in turn leads to an increase in calcium release from bone. The actions of calcitonin could, therefore, complement those of parathyroid hormone in maintaining extracellular fluid concentrations of calcium.

The physiological role for calcitonin is not clear. Although the most obvious action of calcitonin in mammals is to inhibit bone resorption and thereby lower plasma calcium, it also has effects at other sites. In the kidney it decreases renal tubular reabsorption of calcium, phosphate, magnesium, potassium, and a variety of other ions, and is a powerful diuretic, though whether these effects are physiological or pharmacological in man is still uncertain. Calcitonin also inhibits the secretion of several gastrointestinal hormones including gastrin, insulin, and glucagons. Once again, the physiological importance of these effects is not clear. One difficulty of ascribing a physiological role to calcitonin is that a deficiency (total thyroidectomy) or excess (medullary carcinoma of the thyroid) of calcitonin is associated with only minor disturbances in skeletal or mineral homeostasis. Athyroidal subjects given thyroid hormones exhibit minor impairments in their ability to handle an oral calcium load in that they excrete more calcium in the

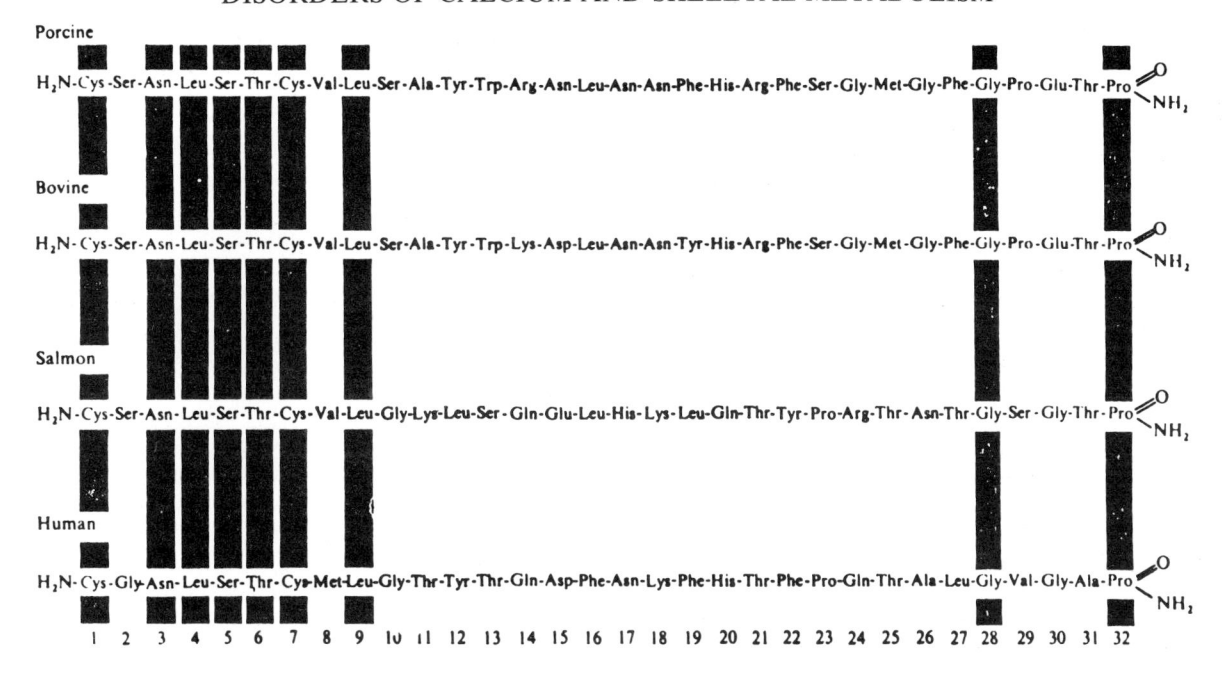

FIG. 5. Amino acid structures of calcitonin from various species. Nine of the 32 amino acids are common to all these species.

urine than do normal controls. This may mean that calcitonin has some role in preventing post-prandial losses of calcium from the body.

A further difficulty in assessing the role of calcitonin metabolism in man arises from the problems of its radioimmunoassay. Like PTH, calcitonin circulates in multiple forms and is degraded in part by the kidney. Many of the immunoreactive fragments are not biologically active and all the assays developed measure some of these fragments. There is much conflict as to how calcitonin secretion responds to various 'physiological' stresses which are partly related to differences in assay technique and differences in the specificity of the antisera used.

Interest has recently been shown in the use of the calcitonin assay as a tumour marker since a variety of non-thyroidal cancers, e.g. breast and bronchus, appear to be associated with raised plasma levels of calcitonin. Once again there are considerable differences in results between centres, and its role in the diagnosis of malignant disease and in following the progress of patients with malignancy is unclear. The assay for calcitonin is, however, useful in the diagnosis of medullary carcinoma of the thyroid and for the detection of family members with the disease in a pre-symptomatic form. A major clinical interest in calcitonin is its use as an inhibitor of bone resorption and turnover in Paget's disease of bone and in hypercalcaemia associated with increased bone resorption.

Vitamin D

Man derives vitamin D_3 (cholecalciferol) from the diet, and from the skin by ultraviolet irradiation of 7-dehydrocholesterol. Vitamin D_2 (ergocalciferol) is a product originally derived from the ultraviolet irradiation of plant sterols and is used to supplement the diet, particularly in margarines. In many respects vitamin D_2 and D_3 are comparable in their metabolism and in their actions.

The photochemical conversion of 7-dehydrocholestrol to vitamin D in skin may be under metabolic control and there is increasing evidence that the conversion occurs more readily in white than dark-skinned races. Both vitamin D and 25 OH-vitamin D are transported in plasma bound to a specific alpha globulin. Vitamin D is fat soluble and absorbed primarily from the duodenum and jejunum into the lymphatic circulation. The vitamin is distributed in fat and muscle. An enterohepatic circulation exists for vitamin D and other metabolites.

Before exerting biological effects, vitamin D undergoes a series of further metabolic conversions (Fig. 6). The first step involves its conversion in the liver to a 25-hydroxylated derivative (25-OHD). This is the major circulating vitamin-D metabolite, and is the metabolite most commonly measured clinically to provide an index of vitamin-D nutritional status. There is a marked seasonal variation in plasma 25-OHD levels with a peak in late summer and a trough in late winter. In winter plasma levels may commonly approach those associated with vitamin D-deficiency states suggesting that in northern Europe both sunlight and dietary intake may be of crucial importance in maintaining vitamin D status.

The second step in the metabolism of vitamin D is its further hydroxylation, mainly in the kidney, to either 1,25-dihydroxy vitamin D_3 ($1,25(OH)_2D_3$) or to $24,25(OH)_2D_3$. Apart from the human placenta and decidua, the kidney is the major, if not the sole site for 1-alpha-hydroxylation, a factor which is of considerable importance in the pathophysiology of renal bone dis-

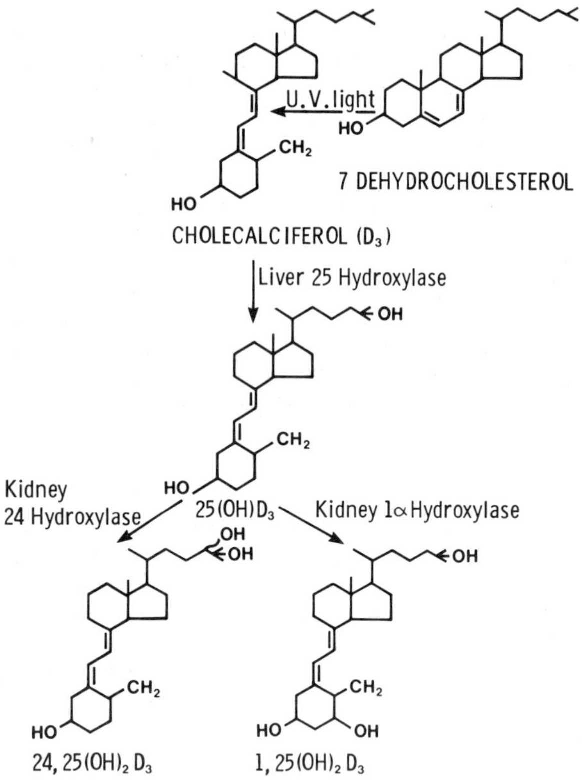

FIG. 6. Steps in the conversion of 7-dehydrocholesterol to vitamin D and its metabolites. The site of synthesis of 25-hydroxy vitamin D_3 is in the liver. The active form of vitamin D ($1,25(OH)_2D_3$) is made in the kidney and placenta. $24,25(OH)_2D_3$ is synthesised in several tissues, but the kidney is probably the major site. The sites of synthesis and function of $25,26(OH)_2D_3$ are unknown.

ease. The renal metabolism of $1,25(OH)_2D_3$ is closely regulated, and its production is favoured under conditions of deficiency of vitamin D, calcium or phosphate. Production of $1,25(OH)_2D_3$ is also augmented by a variety of hormones, including oestradiol, prolactin and growth hormone but it is not clear if this is a direct effect of the hormones on the kidney or if it is mediated by changes in calcium or phosphate concentration.

In many experimental systems studied, $1,25(OH)_2D_3$ appears to have the greatest biological potency. Its principal effects are to increase intestinal absorption of calcium and phosphate and to increase the resorption of calcium from bone. Although lack of vitamin D in man is associated with defective mineralisation of cartilage and bone, the question whether or not vitamin D and its metabolites act directly on bone to increase mineralisation is still unsettled. It is possible that the effects of vitamin D on bone mineralisation are secondary to changes in extracellular fluid concentrations of calcium and phosphate, but this explanation may be too simple to account for all the experimental and clinical observations. Unfortunately there are few experimental systems for studying skeletal mineralisation *in vitro* and the major demonstrable effect of $1,25(OH)_2D_3$ on bone *in vitro* is to increase bone resorption. Apart from vitamin-D toxicity, in which increased bone resorption is well documented, there is surprisingly little evidence that physiological doses of $1,25(OH)_2D_3$ or indeed of any other metabolite of vitamin D, increase bone resorption in man.

From a teleological viewpoint, the action of $1,25(OH)_2D_3$ can be thought of as increasing the availability of calcium and phosphate for mineralisation, or as maintaining plasma levels of calcium and phosphate. $1,25(OH)_2D_3$ can therefore be considered to be the hormonal form of vitamin D in the sense that its secretion from endocrine tissue (the kidney) is controlled by the calcium and phosphate status of the individual, and that the action of this hormone reverses the stimulus to its secretion. The interval between stimulation of $1,25(OH)_2D_3$ synthesis and its effects takes several hours so that the role of vitamin D in calcium homeostasis is to provide regulation over a longer term than the minute-to-minute regulation of plasma calcium that is achieved with PTH or calcitonin.

Receptors for $1,25(OH)_2D_3$ have been found in many other tissues apart from bone and gut. These include skin, breast, salivary, pancreatic and parathyroid tissue. Their physiological significance is not clear but in the case of parathyroid tissue there is some evidence that $1,25(OH)_2D_3$ or possibly one of the other metabolites of vitamin D may also influence parathyroid hormone secretion.

A striking weakness of skeletal muscles, particularly of the pelvic and shoulder girdles are well described features of vitamin D-deficiency. Moreover, myopathy improves rapidly following treatment with vitamin D or one of its metabolites. The mechanisms whereby vitamin D produces an effect on muscle function is unknown but it may involve calcium transfer across the sarcoplasmic reticulum or modifications in the metabolism of troponin C. It is notable that severe phosphate deficiency induced by dietary deprivation or by hyperparathyroidism is associated with muscle weakness, suggesting that the effects of vitamin D could be mediated by hypophosphataemia. However, in the inherited tubular disorder, hypophosphataemic osteomalacia, muscle weakness is characteristically absent even though profound hypophosphataemia is found. There are several other poorly understood trophic effects of vitamin D, particularly on growth, on the maintenance of intestinal mucosa and on the maturation of collagen and cartilage.

The function of $24,25(OH)_2D_3$ in man is unknown. In many experimental systems its production is favoured under conditions which inhibit synthesis of $1,25(OH)_2D_3$ such that a reciprocal relationship is commonly observed between their respective production rates. Thus, under conditions of repletion of vitamin D, calcium or phosphate, $24,25(OH)_2D_3$ it is the major circulating dihydroxy metabolite. A variety of observations in experimental animals and in man suggest that this metabolite may have a physiological role in regulating metabolism in cartilage, bone, or in parathyroid tissue.

Assays for vitamin D metabolites are not generally available with the exception of the competitive protein binding assay for 25-OHD (see appendix for normal ranges). These assays do not generally distinguish between $25(OH)D_2$ and $25(OH)D_3$. Plasma measurements provide a useful index of the adequacy of vitamin D nutrition in the investigation of rickets or osteomalacia. Thus, plasma levels of 25-OHD are low in simple vitamin D-deficient states and in disorders associated with increased destruction or elimination of 25-OHD. In rarer forms of rickets (discussed later), in which the metabolic defect is a failure of production of $1,25(OH)_2D_3$, or is a metabolic error unrelated to vitamin D metabolism, plasma concentrations of 25-OHD are normal. The finding of high levels of 25-OHD may be helpful in an occasional patient presenting with acute hypercalcaemia who withholds a history of excessive dietary intake of vitamin D. Several other vitamin D metabolites have been isolated from plasma and identified, such as 25,26-dihydroxycholecalciferol and a 23,26 lactone derivative. Some of these metabolites appear to have some biological activity but their tissue of origin and metabolic function is not clear.

$1,25(OH)_2D_3$ has a biological half-life measured in hours and a normal plasma concentration of approximately 30 pg/ml in adults. Higher concentrations are observed in growing children. Plasma levels of $25\text{-}OHD_3$ and $24,25(OH)_2D_3$ are 1000-fold and 100-fold greater respectively than those of $1,25(OH)_2D_3$ and both compounds have a long biological half-life in the circulation, perhaps related to their greater affinity for the vitamin D-binding protein. Despite these differences in plasma levels, $1,25(OH)_2D_3$ is considered the active form of vitamin D because it is so much more potent than other derivatives in exerting actions on the target organs.

The discovery that vitamin D must be metabolised to more polar products has led to considerable advances in our understanding of the pathophysiology of a number of disorders, and it might be expected that the assay of the metabolites of 25-OHD, particularly $1,25(OH)_2D_3$, will find a place in clinical practice. Methods of assaying $1,25(OH)_2D_3$ include competitive binding assays using the intestinal cytosol or nuclear receptor protein for $1,25(OH)_2D_3$ and more recently radioimmunoassays using antibodies raised against $1,25(OH)_2D_3$. Both the receptor assays and immunoassays cross react with other vitamin-D compounds; to overcome this problem it is necessary to undertake prior extraction of the plasma followed by chromatographic separation of the metabolites, processes that are time-consuming and laborious. Assay of $1,25(OH)_2D_3$ may eventually have a place in the investigation of disorders of mineral metabolism including sarcoidosis, rare forms of osteomalacia, and disorders of parathyroid secretion.

The Influencing Hormones

From the preceding section, it is clear that the major regulating hormones appear to operate by increasing or decreasing plasma levels of calcium or phosphate in response to changes in these levels. The hormones, PTH, calcitonin, and $1,25(OH)_2D_3$ are all present in the circulation under normal physiological conditions and their secretion might be expected to exert a continuous influence on calcium metabolism and on the rate of bone remodelling. But, since these hormones are regulated by changes in plasma concentrations of calcium or phosphate, their physiological role may be less important in skeletal homeostasis than in maintaining the proper concentrations of these ions in extracellular fluid. Other factors, some of which are at present poorly characterised, influence skeletal metabolism and maintain the integrity of the skeleton in health and under various conditions of stress. On the basis of experimental and clinical observation, several hormones, the secretion rate of which are not primarily governed by extracellular fluid concentrations of calcium or phosphate, are known to be important for normal skeletal metabolism.

Growth Hormone

Growth hormone is best known for its effect on growth of cartilage, an effect which is probably brought about indirectly by growth-hormone-dependent production of somatomedin. There are several known growth factors called somatomedins, only some of which are dependent on growth hormone. Growth hormone also causes an elevation in plasma phosphate by increasing the tubular reabsorption of phosphate in the kidney, and may contribute to the high plasma phosphate found in growing children and in acromegaly. Excess or deficiency of growth hormone is associated with obvious abnormalities in skeletal growth. In acromegaly there is increased periosteal apposition of bone, but, contrary to popular belief, there is no convincing evidence that acromegaly causes osteoporosis.

Thyroid Hormones

Deficiency of thyroid hormones early in life produce the well-known skeletal deformities of cretinism. Before skeletal maturity, thyrotoxicosis may increase longitudinal skeletal growth. In the adult, thyrotoxicosis may be associated with increased bone resorption, hypercalciuria, a raised alkaline phosphatase and occasionally hypercalcaemia. These effects, probably due to direct actions of the thyroid hormones on bone, lead to a tendency for plasma calcium to rise, which in turn leads to depression of secretion of parathyroid hormone and $1,25(OH)_2D_3$.

Adrenal Steroids

The most important effect of glucocorticoids on the skeleton is to regulate growth. Their actions on calcium metabolism are complex and probably involve effects on many target tissues in addition to their effects on metabolism, transport and action of other hormones. In the adult, adrenal insufficiency is not associated with marked skeletal abnormalities but is occasionally accompanied by hypercalcaemia. This is probably due to haemoconcentration (increased albumin concentration) and also to increased renal tubular reabsorption for calcium because of volume depletion.

The effect of glucocorticoid excess in inducing osteoporosis is well established. The pathogenesis for this is far from clear and the relative contributions of diminished intestinal absorption of calcium, diminished bone formation, and increased resorption of bone are not well defined. The ability of corticosteroids to decrease plasma calcium in hypercalcaemia other than that due to primary hyperparathyroidism has been used for many years as a diagnostic aid (see section on hyperparathyroidism), but once again the precise mechanism of action is uncertain. However glucocorticoids are known to inhibit the biosynthis of prostaglandins and of osteoclast activating factor, a product of myeloma cells, both of which may be mediators of tumour-induced bone resorption and hypercalcaemia.

Sex Steroids

Characteristic growth abnormalities are associated with deficiencies of either male or female sex hormones. They appear to play a crucial role in epiphyseal closure and in the so called 'adolescent growth spurt' which precedes this event. They may also influence the amount of calcium present in the skeleton in adult life. In adults the effects of oestrogens are of particular interest because of the loss of bone that occurs in women after the menopause (see osteoporosis). Administration of exogenous oestrogen may slow down this loss and oral contraceptives may reduce bone turnover in premenopausal women. Oestrogen administration lowers plasma calcium slightly, an effect more noticeable in postmenopausal women with hyperparathyroidism or hypercalcaemic women with carcinoma of the breast. It is interesting that hypercalcaemia is also occasionally seen when patients with carcinoma of the breast are treated with tamoxifen, which is an inhibitor of the binding of oestrogen to its receptor. This probably reflects the effects of oestrogen on tumour growth, but it is notable that normal, as well as malignant, breast tissue does have receptors for $1,25(OH)_2D_3$.

Gastrointestinal Hormones

There are many interactions between calcium regulating hormones and gastrointestinal hormones. The relationships between gastrin and calcitonin secretion have been mentioned earlier. Calcium and parathyroid hormone are also involved, and in hyperparathyroidism, gastrin levels are increased, due to the ability of calcium to stimulate gastrin secretion. Of the other gastrointestinal hormones, glucagon is hypocalcaemic due to inhibition of bone resorption, either directly or by stimulating the secretion of calcitonin. Secretin may cause hypercalcaemia, possibly by stimulating release of parathyroid hormone. Insulin is an important hormone for skeletal growth and insulin-dependent diabetics often have diminished skeletal mass. Insulin is one of the few hormones which has been shown to stimulate collagen synthesis *in vitro*.

Osteoclast Activating Factor

Osteoclast activating factor (OAF) is a potent bone-resorbing substance derived from mononuclear leucocytes. There may be more than one type of OAF none of which has been fully identified chemically. Unlike PTH the production of these factors is inhibited by corticosteroids. The physiological importance of OAF is unknown but its production is increased in haematological neoplasias, particularly myeloma, and may be responsible for the hypercalcaemia and bone loss seen in myeloma, which is characteristically sensitive to treatment with corticosteroids.

Prostaglandins

A number of prostaglandins, particularly of the E series resorb bone by directly stimulating osteoclastic activity. There is good experimental evidence that prostaglandins cause the increased bone resorption mediated by certain tumours *in vitro* and in experimental animals. The evidence that prostaglandins are involved in tumour-mediated bone resorption and hypercalcaemia in man is however circumstantial. Prostaglandin synthetase inhibitors have been advocated in the treatment of hypercalcaemia due to metastases in bone, but the general experience is that very few patients respond. Whether or not prostaglandins play a physiological role in skeletal metabolism is also not clear.

From the preceding comments it is clear that the effects of many of these hormones on the skeleton are incompletely understood. The judicious use of assays for these hormones or their metabolites are nevertheless important in investigating certain growth abnormalities, the occasional patient with hypercalcaemia and in excluding the rare but treatable causes of osteoporosis.

Calcium Transport to and from the Extracellular Fluid

With the exception of the pregnant or lactating female, the major fluxes of calcium to and from the extracellular fluid occur across the intestinal mucosa, bone and kidney (Fig. 1). This section describes the extent to which these fluxes are controlled by the major regulating hormones and some of the investigations available to examine these movements of calcium.

Intestine

Unlike the fluxes of calcium between the ECF and bone and kidney, intestinal absorption of calcium is episodic and dependent on an adequate supply of calcium delivered in an available form to the intestinal mucosa. The availability of calcium for absorption depends upon many dietary factors including the presence of phosphate, fatty acids and phytates which bind calcium and render it unavailable for absorption. The influx of calcium depends both on active transport and diffusion processes. Absorption occurs throughout the length of the small intestine and to a lesser extent in the colon. The major sites for active transport are in the duodenum and upper part of the jejunum. However, because the duodenum is relatively short compared with the rest of the gastrointestinal tract, more calcium is probably absorbed at sites distal to the duodenum than within the duodenum itself, at least with normal dietary intakes.

The relationship between the dietary intake and the net calcium absorbed is not linear (Fig. 7). Net absorbtion of calcium increases steeply as the calcium intake

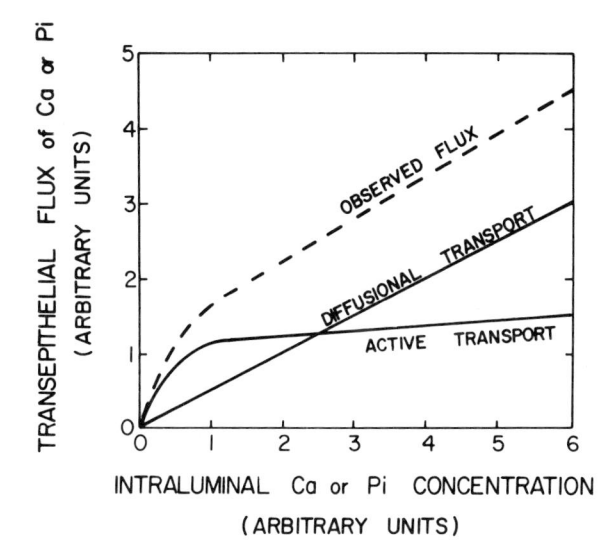

FIG. 7. The active and passive components of intestinal mineral ion transport. The observed relationship between absorption and dietary intake (dashed line) is resolved into active and diffusional transport components. The active transport mechanism is responsible for most of the flux at low intraluminal ion concentrations, whereas diffusion assumes greater quantitative significance at high ion concentrations (from Bringhurst and Potts 1979).

rises from very low amounts, but thereafter flattens off, though does not reach a plateau. This relationship is explicable on the basis of the two known transport systems; the one being an active transport system which is saturable and important at low concentrations of calcium, and the other being a passive diffusion process which is concentration-dependent and is more important at high calcium concentrations in the bowel. This means that the net percentage of dietary calcium absorbed decreases with increasing intake, a fact which is important to take into account when assessing different types of calcium absorption tests (Fig. 8). A clinical consequence of this is that the use of calcium supplements alone is a relatively inefficient way of increasing net intestinal absorption unless the dietary intake of calcium is very low.

At very low intakes of calcium, net absorption may be negative. Calcium delivered into the gut lumen in intestinal secretions may exceed the calcium absorbed. Calcium delivered from intestinal secretions contributes to the so-called 'endogenous faecal calcium' and can be monitored by radioisotope methods in which parenterally administered radiolabelled calcium is recovered in the faeces. In malabsorption syndromes, the endogenous

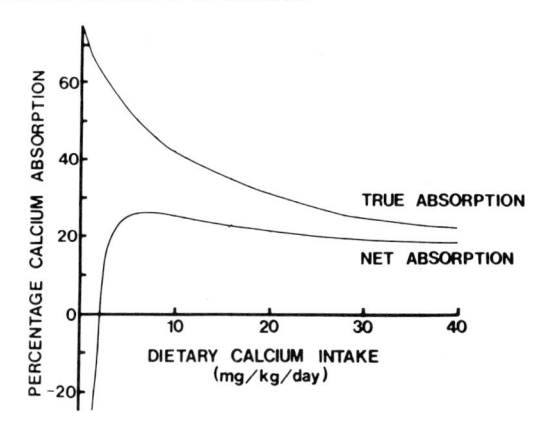

FIG. 8. The relationship between dietary calcium intake and the net and true absorption of calcium expressed as a percentage of the intake (from Nordin, B. E. C. (1976). *Calcium, Phosphate and Magnesium Metabolism*. Edinburgh: Churchill Livingstone).

faecal calcium may appear to rise, but this does not always mean that the amount of calcium secreted in the digestive juices is increased, since the rise is probably due to malabsorption of this digestive juice calcium. Nevertheless, increased calcium secretion has been described in steatorrhoea and adult coeliac disease (gluten-sensitive enteropathy).

Man is able to adapt to variations in dietary calcium so that net absorption remains relatively constant over a fairly wide range of intakes (Fig. 8). This adaptation is slow and occurs over a period of days or weeks rather than hours or minutes. It is now thought that this adaptation is due to changes induced in the synthesis of $1,25(OH)_2D_3$, so that the rate of net intestinal absorption of calcium returns to its original value. Changes in the active transport of calcium are thought to be largely related to changes in the amount of the 1,25-dihydroxy form of vitamin D bound to its intestinal receptor.

The biochemical mechanisms involved in active calcium transport through the intestinal mucosa are not understood in detail, though several features of the system have been identified. $1,25(OH)_2D_3$ appears to stimulate synthesis of a calcium-binding protein in addition to an intestinal alkaline phosphatase, by a mechanism similar to that described for many other steroid hormones. This involves the translocation of vitamin D plus its receptor protein to the intestinal cell nucleus to stimulate the synthesis of messenger RNA and new protein synthesis. The exact function of these proteins is not clear but they could either be involved in the transmucosal transport of calcium or act as buffer to large intracellular shifts in calcium concentration.

There is now good evidence that calcium and phosphate can be absorbed separately from each other, and that $1,25(OH)_2D_3$ has independent effects to enhance the absorption of each. Unlike the absorption of calcium which appears to be closely regulated, the proportion of the dietary phosphate absorbed does not decrease appreciably with increasing intake, but remains fixed at approximately 70% (Fig. 9). This explains in part the

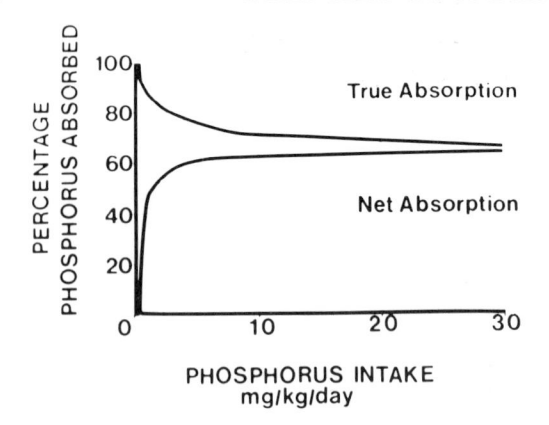

FIG. 9. The relationship between dietary phosphorus intake and the net and true absorption of phosphorus expressed as a percentage of the dietary intake (from Nordin, B. E. C. (1976). *Calcium, Phosphate and Magnesium Metabolism*. Edinburgh: Churchill Livingstone).

changes in plasma phosphate that can occur following a meal and underline the need for studying man under controlled conditions (e.g. fasting) when assessing phosphate homeostasis.

A variety of tests are available to study calcium absorption. Calcium balance provides an index of net calcium transport (Fig. 2). Accurate assessment of unidirectional fluxes of calcium across the gut is dependent upon tracer techniques. A standard method is to compare the plasma appearance and disappearance of an orally administered calcium isotope with the disappearance of an intravenously-given calcium isotope. Absorption is calculated from the difference between the specific activities of the two isotopes in plasma. A simpler technique, but subject to more assumptions, is to follow the appearance and disappearance of a single calcium radioisotope given by mouth. The fractional absorption of calcium will depend on several factors including the availability of the calcium for absorption and the site of absorption. Of great importance is the amount of calcium carrier given with the radioisotope, since a low total calcium dose will present itself for absorption high in the gastrointestinal tract and be more clearly a function of the active transport mechanism. A higher dose of carrier (e.g. 200 to 300 mgs) more closely approximates the calcium provided in a meal and will in addition be available for absorption at sites distal to the duodenum. Clearly the amount of calcium in the diet must also be taken into consideration when assessing the results of calcium absorption tests since, for example, a high fractional absorption of calcium does not necessarily mean that the total amount of calcium absorbed in the diet is high if the diet is severely deficient in calcium, or if calcium is rendered unavailable for absorption by phosphates, phytates, fats or drugs.

A 24-hour urine collection for calcium or phosphate provides an indirect index of intestinal absorption provided it is assumed or known that the net flux of these ions across bone and other tissues is zero. The expression of excretion as a ratio to creatinine helps to adjust for differences in body weight and for incomplete urine collections.

Increased intestinal absorption of calcium is found in several disease states such as hyperparathyroidism, sarcoidosis and idiopathic hypercalciuria, and is thought to be due to increased production of $1,25(OH)_2D_3$. Conversely, malabsorption of calcium is often associated with low levels of $1,25(OH)_2D_3$, for example in hypoparathyroidism, vitamin D deficiency states and in chronic renal failure. Malabsorption of calcium also occurs in disorders in which the target tissue for $1,25(OH)_2D_3$ is defective. A good example of this is in untreated coeliac disease, where the flat intestinal mucosa lacks the villi and enterocytes through which calcium is normally absorbed.

Kidney

The fluxes of calcium across the kidney are much greater than those of the gut (Fig. 1). Renal tubular reabsorption of calcium in the kidney is a complex process, and takes place at several different sites in the nephron. The total amount of calcium reabsorbed can, however, be estimated by subtracting the filtered load of calcium from the renal excretion. The filterable calcium is approximately 60% of total plasma calcium (Table 2), and the filtered load represents the product of the glomerular filtration rate (GFR) and the filterable calcium (approximate 250 mmols (10 g) daily).

In health there is a curvilinear relationship between plasma calcium concentration (an index of filtered load, assuming no change in glomerular filtration rate) and renal excretion of calcium, so that tubular reabsorption often cannot be assessed simply from the calcium clearance.* To take account of variations in glomerular filtration rate, the renal excretion is commonly expressed per litre of glomerular filtration rate (Fig. 10). Any value below the lines depicting the normal relationship would indicate an increase in the net tubular reabsorption of calcium and values above the lines denote decreased net reabsorption. Any value for renal excretion above normal but within the normal range for the relationship between plasma and urine calcium, would indicate an increase in filtered load (gut or bone derived) or a low glomerular filtration rate, but with normal net tubular reabsorption of calcium. Calcium excretion, measured in the fasting state, will reflect more closely calcium derived from bone resorption.

The assessment of renal tubular reabsorption in this way has several limitations, particularly in clinical disorders. Thus, the proportion of total plasma calcium that is filtered is often uncertain in the presence of alkalosis, acidosis or abnormal plasma proteins. Moreover, a change in the relationship normally found between plasma calcium and its renal excretion might be expected in

---

* Calcium clearance as a proportion of GFR can be derived from the urine concentration of calcium (UCa) and creatinine (UCr) and plasma concentrations of ultrafiltrable calcium (PCa) and creatinine (PCr):

$$\frac{UCa \times PCr}{UCr \times PCa}$$

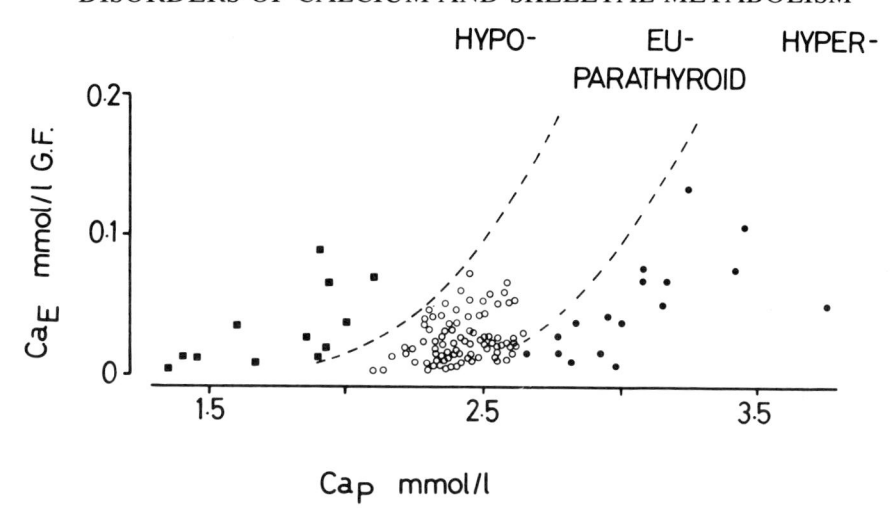

FIG. 10. Relationship between fasting urinary calcium excretion (Ca_E; expressed as mmol/l of glomerular filtrate) and plasma calcium (Ca_p) in health and disorders of parathyroid function. The dashed lines denote the range ($\pm SD$) obtained in normal subjects during calcium infusions. Patients with primary hyperparathyroidism lie on the right of the line describing normal subjects indicating increased renal tubular reabsorption for calcium. In contrast, patients with hypoparathyroidism lie to the left of the normal line indicating decreased renal tubular reabsorption for calcium. Note that the determination of calcium excretion alone (without concurrent measurement of plasma calcium) does not give information concerning renal tubular reabsorption for calcium. Note also that hypoparathyroid patients, and many hyperparathyroid patients have calcium excretion values which are normal (from Kanis J. A. (1980) Etiology and Medical Management of Hypercalcaemia: *Metabolism Bone Disease and Related Research* **2** (3)).

patients with intrinsic renal disease or during saline infusion, dehydration, or treatment with diuretics.

Despite these difficulties, the measurement of calcium excretion together with plasma calcium can help provide information about tubular reabsorption in a steady state, whereas the measurement of calcium excretion alone does not. Whenever a steady state for plasma calcium exists, either in health or disease, calcium must be eliminated from the circulation at the same rate as it enters, whatever the GFR or tubular reabsorption. It follows that hypo- or hypercalciuria do not in themselves indicate disturbances in renal tubular reabsorption, but denote increased or decreased net input into the extracellular fluid due to changes in bone resorption or intestinal absorption. Furthermore, *under steady state conditions*, changes in tubular reabsorption alone will not be reflected by changes in renal excretion rate, but by changes in the plasma calcium concentration.

In health, approximately 97% of the calcium filtered by the kidney is reabsorbed. Many factors influence renal tubular reabsorption, but of the various hormones mentioned PTH is probably the most important under physiological conditions. In mild primary hyperparathyroidism the hypercalcaemia is due mainly to increased renal tubular reabsorption since the fasting calcium excretion is often normal (Fig. 10), whereas an increase would be expected if the net flux of calcium from bone to ECF were increased. Since many patients with primary hyperparathyroidism have increased bone resorption, the normal calcium excretion rate implies an increase in bone formation to match the change in bone resorption. Similarly, in hypoparathyroidism the fasting renal excretion of calcium is commonly normal (Fig. 10), though plasma calcium is low. PTH-mediated changes in plasma calcium are therefore often due to changes in renal tubular reabsorption of calcium.

Plasma phosphate concentration varies more than that of plasma calcium, particularly in response to circadian rhythms and to meals. The level of fasting plasma phosphate is set mainly by the kidney and the measurement of the renal tubular reabsorption of phosphate is often used in clinical diagnosis, such as in hyperparathyroidism. Of the several methods available to calculate phosphate reabsorption, the most appropriate from a physiological point of view is the estimation of *TmP/GFR* (the tubular maximum for phosphate reabsorption per unit of GFR). This examines the relationship between filtered load and renal excretion which, like that for calcium reabsorption, is curvilinear. A nomogram is available for deriving this measurement (Fig. 11).

Phosphate reabsorption is increased by growth hormone, in hypoparathyroid conditions and in phosphate deprivation. It is diminished by calcitonin, in hyperparathyroidism and in several inherited or acquired renal tubular disorders which may also be associated with defects in the reabsorption of glucose, amino acids, or bicarbonate. The calcium and phosphate ions themselves may also influence renal phosphate transport, which makes it difficult to dissociate the direct effects of hormones on tubular resorption from the other metabolic consequences of their secretion.

Both calcium and phosphate excretion are influenced by other factors, notably sodium excretion, extracellular fluid volume, and by the administration of diuretics. There is evidence for both proximal and distal sites of tubular reabsorption and both can be influenced by

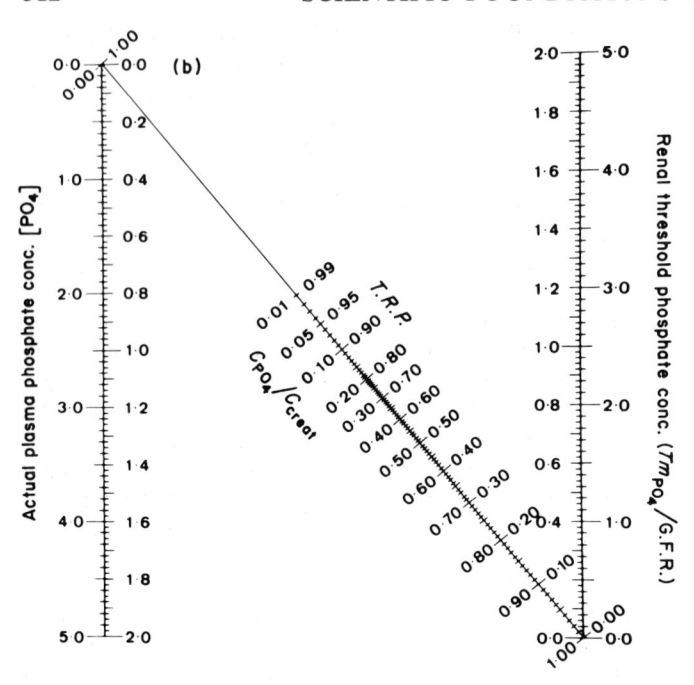

FIG. 11. Nomogram for the derivation of TmP/GFR (estimate of phosphate reabsorption) from simultaneous measurements of tubular reabsorption (TRP) or the ratio of clearance of phosphate to clearance of creatinine (CPO₄/Creat) and plasma phosphate (PO₄). TRP can be calculated from the concentrations of phosphate and creatinine in plasma and creatinine, and urine volume is not required.

$$\frac{\text{Phosphate clearance}}{\text{Creatinine clearance}} = \frac{\text{urine phosphate} \times \text{plasma creatinine}}{\text{plasma phosphate} \times \text{urine creatinine}}$$

A straight line through the appropriate values of plasma phosphate and TRP or phosphate/creatinine clearance passes through the corresponding value of Tmp/GFR. Both TmP/GFR and phosphate are expressed in the same units. The scale in units of the figure are arbitrary (from Walton, R. J. and Bijvoet, O. L. M. *Lancet* ii: 309, 1975).

PTH. Experimental studies show that receptors for PTH exist at multiple locations in the nephron. However, in physiological terms PTH induces changes in phosphate reabsorption which occur predominantly at the proximal convoluted tubule whereas the action of PTH on calcium is probably mainly distal. Infusion of sodium chloride increases the excretion of both calcium and phosphate, an effect which probably contributes to its value in the acute treatment of hypercalcaemia.

The biochemical mechanisms involved in renal transport of calcium and phosphate are not elucidated but the action of PTH on the kidney is known to produce an increase in cortical adenylate cyclase activity, which increases tubular cell and urinary concentrations of 3′, 5′-cyclic adenosine monophosphate (cyclic AMP). It is not known whether this increase in cyclic AMP is the cause of the subsequent changes in phosphate and calcium transport. However, in the clinical disorder, pseudohypoparathyroidism, there is probably a defective receptor mechanism for PTH in kidney (and sometimes bone) so that the administration of PTH does not always produce the normal response of an increased excretion of phosphate and cyclic AMP. This is somewhat analogous to the failure of patients with nephrogenic diabetes insipidus to respond to antidiuretic hormone. The estimation of urinary cyclic AMP is also used in the investigation of hypercalcaemia. Thus, increased excretion of cyclic AMP is seen in primary hyperparathyroidism, but depressed in most conditions where hypercalcaemia is not mediated by PTH. By expressing cyclic AMP excretion so that only the nephrogenous (parathyroid-hormone dependent) component of nucleotide excretion is taken into account, the discriminatory value of the test may be considerably improved. However, a proportion of patients with hypercalcaemia associated with malignant disease, particularly squamous cell carcinoma, but without obvious skeletal metastases may show increased production of nephrogenous cyclic AMP. In some cases this may be due to ectopic production of PTH (so called pseudohyperparathyroidism) but in others there is probably an additional circulating factor which stimulates both renal tubular reabsorption of calcium and the production of cyclic AMP.

It is worth noting that the kidney is the site of degradation for many hormones including parathyroid hormone and calcitonin. Renal function therefore needs to be taken into account when assessing the results of radioimmunoassay, particularly in the case of PTH where the C-terminal fragment normally degraded by the kidney persists in the circulation (see Fig. 4).

Bone

In the adult skeleton, in balance with respect to calcium, the amount of bone resorbed is balanced by an equivalent amount of bone formed (Fig. 1). The processes of mineral accretion and resorption are closely coupled but do not occur in the same anatomical site at the same time. For mineral accretion to occur (mineralisation), new bone matrix (osteoid) must first be formed. Only when bone mineralisation is complete does it undergo resorption releasing calcium, phosphate and collagen degradation products into the ECF. The time taken between the onset of bone formation and the completion of bone resorption in any one site takes many months to be complete. Thus, any factor which influences the rate of mineralisation for example, may have skeletal consequences for many months. In the assessment of treatments for bone disease, particularly osteoporosis, short-term changes in skeletal balance may not necessarily be associated with prolonged changes of clinical significance unless such changes are demonstrated over many months or even years.

The transfer of calcium to and from bone depends on cellular activity. In mature bone three main cell types occur. *Osteoblasts* which form matrix, *osteoclasts* responsible for bone destruction, and *osteocytes*, themselves derived from osteoblasts which become trapped within the bone matrix as maturation proceeds. The osteocytes are interconnected by an extensive canalicular network. This network has a large surface area (one acre in man) and may be responsible for many of the rapid ion fluxes that occur in bone. The origin, life-span, and fate of the various cells in bone is only partially understood, but it is thought that osteoclasts are derived from the monocyte-macrophage lineage of haemopoetic cells, whereas the osteoblasts and osteocytes are derived from mesenchymal cells of the bone marrow stroma. The derivation of bone cells is of clinical interest in the disorder of osteopetrosis, which is characterised by defective bone resorption and the accretion of excessive amounts of calcium in the skeleton. A major abnormality is thought to be the defective production of osteoclasts which is corrected by bone marrow transplantation.

The mineral component of bone is predominantly calcium and phosphate in the form of hydroxyapatite with multiple ionic substitutions. The major constituents of bone matrix are type I collagen which is largely responsible for the tensile strength of bone, and proteoglycans. The mechanisms controlling matrix production are uncertain, though the importance of normal matrix production is apparent from disorders in which it is defective, e.g. in osteogenesis imperfecta or in vitamin-C deficiency in which synthesis of defective collagen results in disturbed bone formation.

Calcification of the collagen matrix is an important step in the transition between matrix production and the formation of mineralised bone. In the skeleton, calcification takes place in two main sites; in epiphyseal cartilage during the growth of long bones, and in bone matrix during intramembranous ossification, and in the remodelling and growth of existing bone. The concentrations of calcium and phosphate in the extracellular fluid are insufficient to initiate the deposition of calcium and phosphate, but can sustain crystal growth once it has started. The first steps in calcification are now thought to take place in or around small membrane-bound vesicles found in the matrix derived from hypertrophic chondrocytes during the maturation of epiphyseal cartilage and possibly from osteoblasts in adult bone.

These vesicles are rich in alkaline phosphatase, an enzyme which has been known for many years to be associated with calcification. Alkaline phosphatase may function in calcification as a component of a membrane pump for calcium and phosphate, or it may be involved in the removal of potential inhibitors of calcification such as inorganic pyrophosphate. The possible importance of alkaline phosphatase in mineralisation is illustrated in hypophosphatasia (a rare and recessively inherited disorder of mineralisation) characterised by low or absent levels of bone-derived alkaline phosphatase. Hypophosphatasia is also characterised by increased concentrations of plasma pyrophosphate. Since pyrophosphate is an inhibitor of the crystal growth of calcium phosphate, it may be responsible for the defective calcification of cartilage and bones in this condition.

Defective mineralisation of adult bone results in increased amounts of osteoid. It is important to remember that the amount of osteoid present in bone is dependent not only upon the rate of mineralisation but also upon the rate of matrix formation. Increased osteoid due to an increased rate of formation occurs in hyperparathyroidism, Paget's disease, fracture repair, and the term osteomalacia is best restricted to those conditions which result in a delay between formation of matrix and its subsequent mineralisation. This occurs in vitamin-D deficiency but is also found in other disorders such as hypophosphataemia, severe acidosis, and aluminium intoxication.

Phosphate plays a critical role in skeletal mineralisation. Low plasma phosphate levels are often associated with osteomalacia and increased levels with the ectopic (extra-skeletal) deposition of calcium phosphate. Phosphate may have specific effects on cells to enhance the uptake of calcium in calcified tissues.

It is widely held that vitamin D or its metabolites are required for normal mineralisation, though there is little direct evidence for such an effect. It is therefore interesting that in many clinical disorders there is a stronger relationship between levels of plasma phosphate and mineralisation of bone, than between levels of the active metabolites of vitamin D and mineralisation. For example, in hypoparathyroidism levels of $1,25(OH)_2D_3$ are classically low though mineralisation of osteoid is usually normal. Conversely, in hypophosphataemic rickets, levels of plasma phosphate are low but levels of $1,25(OH)_2D_3$ are normal. In many of these conditions administration of phosphate alone can improve skeletal calcification. This suggests that $1,25(OH)_2D_3$ may act on mineralisation, predominantly by making calcium and phosphate available for mineralisation. It is possible that the osteomalacia seen in simple vitamin-D deficiency, is partly due to the low plasma phosphate found in this disorder (in part an effect of secondary hyperparathyroidism to reduce TmP/GFR).

The biochemical events occurring during bone resorption are poorly understood. Some resorbing agents such as PTH and prostaglandins can stimulate the production of cyclic AMP in bone but others such as $1,25(OH)_2D_3$ do not. It is interesting that in studies of isolated bone cells *in vitro*, osteoblast-like cells appear to have receptors for $1,25(OH)_2D_3$ and a PTH and prostaglandin-sensitive adenylate cyclase, though these cells themselves are not capable of resorbing bone. It seems likely that there is a complex humoral system of communication between various populations of bone cells which may form the basis of the close coupling between the rates of bone formation and bone resorption. Bone resorption is accompanied by a release of enzymes, such as collagenase and lysosomal enzymes capable of degrading matrix. Osteoclasts, viewed by electron microscopy, possess a ruffled border which is closely applied to the bone surface, and is presumably the site at which removal of bone mineral and matrix occurs. The dissolution of bone matrix is associated with the release of calcium and degradation products of collagen. Proline is a major constituent of the collagen molecule which is hydroxylated to form hydroxyproline in the post-translational stage of collagen biosynthesis. The liberation of hydroxyproline-containing peptides by osteoclasts is therefore an index of collagen degradation and of bone resorption, which can be used clinically in the assessment of metabolic bone disease.

A number of agents are known to inhibit the rate of bone resorption. Administration of calcitonin inhibits bone resorption and decreases the activity and numbers of osteoclasts. Oestrogens also inhibit the rate of bone resorption, though receptors for oestrogens have not yet been convincingly demonstrated in bone cells. Other inhibitors of bone resorption include mithramycin and the diphosphonates. All these agents have been used therapeutically as inhibitors of bone resorption.

Under physiological conditions, resorption of bone is probably under the control of PTH, vitamin D, thyroid hormones, and steroids, but will occur at a basal rate even in the absence of these hormones. Many other agents can stimulate bone resorption under experimental conditions and may be important in disease. These include vitamin A, prostaglandins, particularly of the E series; these latter compounds have been implicated in the hypercalcaemia of malignancy, and in the bone resorption that accompanies rheumatoid arthritis. Myeloma cells and activated lymphocytes can produce materials which cause resorption *in vitro* and which have been designated osteoclast activating factors (OAF).

Coupling of Formation and Resorption. In a variety both of physiological and of pathological states it is clear that there is a remarkably close correlation between the rates of mineral deposition, or bone formation, and mineral resorption. Even though these individual rates may be altered many-fold, the net gains or losses of skeletal mass are minimised by the tight coupling between these rates. This is an important feature in skeletal physiology and therapeutics that is often overlooked. Thus, it is difficult to induce a sustained difference between the rates of mineralisation and resorption, and this is one reason why many potential therapeutic agents have been disappointing in the long-term management of bone diseases, such as senile osteoporosis. Transient dissociations can arise however, for example in the acute loss of bone mineral that occurs in response to immobilisation and in the loss of bone mineral that occurs in malignant disorders, such as myelomatosis. In osteoporosis, where there is a gradual diminution of bone mass, there is clearly an imbalance in the rate of bone formation and rate of bone resorption in favour of net mineral losses. However, there still appears to be a coupling mechanism in the sense that efforts to alter bone resorption, e.g. inhibition by calcitonin, result not only in the inhibition of bone resorption but also the inhibition of bone formation.

Clinical evidence for coupling is readily seen in Paget's disease from the concurrent measurement of alkaline phosphatase and urinary excretion of hydroxyproline. There is a close correlation between these two measurements over a wide range of values (Fig. 12). During treatment of Paget's disease with inhibitors of bone resorption, the earliest response to treatment is a decrease in the urinary excretion of hydroxyproline. This is followed later by a diminution in plasma alkaline phosphatase activity, an index of bone formation. If treatment is sustained, the fall in alkaline phosphatase activity matches the fall in hydroxyproline excretion so that the relationship between the two measurements remains the same as before treatment.

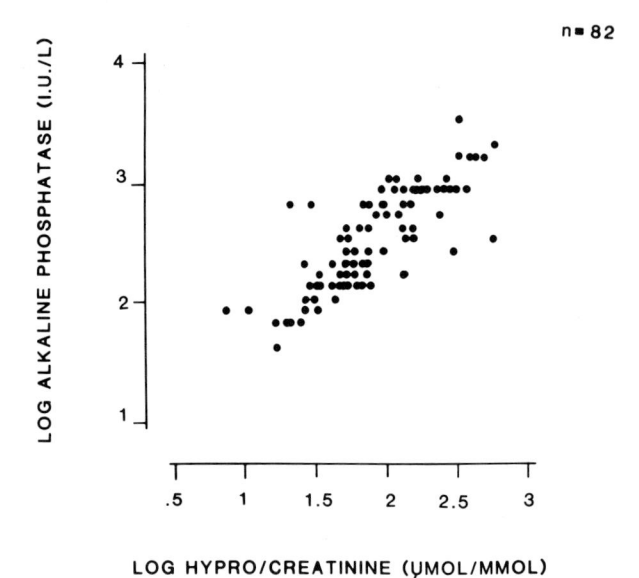

Fig. 12. Relationship between plasma alkaline phosphatase and urinary excretion of hydroxyproline in patients with Paget's disease of bone. Note the logarithmic scales indicating the close relationship between these two measurements over a very wide range of values.

The coupling mechanism illustrates an important difference between the control of skeletal turnover and the regulation of the plasma calcium level. In disorders characterised by increased bone turnover (for example Paget's disease of bone), the rate of bone resorption is high. This, however, is matched by an increased rate of bone formation so that the net contribution of calcium to the extracellular fluid from the skeleton is normal. Thus, under steady state conditions there is no marked disturbance in plasma calcium unless coupling is temporarily disturbed, e.g. during treatment (Fig. 13). Uncoupling may also occur during immobilisation when the rate of bone resorption exceeds formation, so that both hypercalcaemia and hypercalciuria develop.

Bone is not a homogenous structure but consists of cortical (compact) and cancellous (trabecular) bone. On a weight basis there is greater amount of compact than cancellous bone but the surface area (by weight or by volume) in trabecular bone is greater. Since bone turnover occurs at surfaces this has important implications in that disorders of bone turnover are often characterised by more marked changes in trabecular than cortical bone. Examination of trabecular bone may therefore give more sensitive indications of changes in bone remodelling.

An additional feature of bone is its ability to respond to mechanical deformation by changing its structure and shape to counteract stress. One mechanism proposed to account for this involves the generation of small electrical currents within the bone as stresses are applied. Such currents can be demonstrated experimentally and are

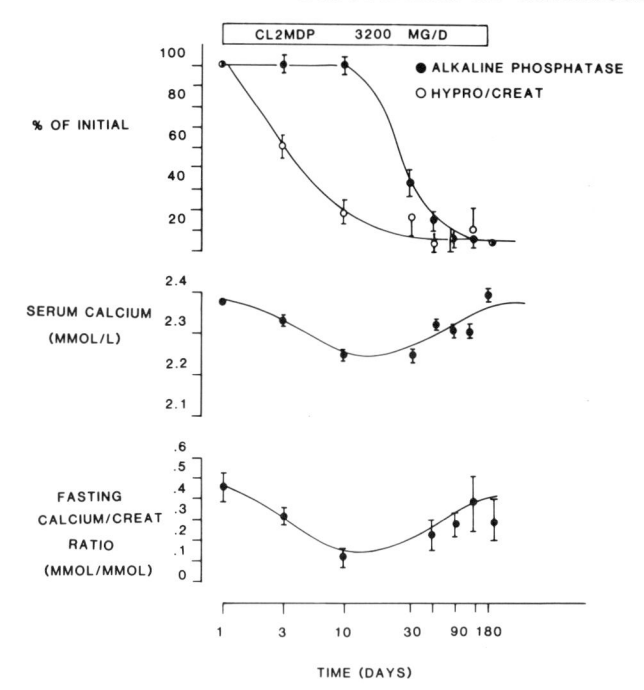

FIG. 13. Biochemical responses to treatment with Clodronate (CL$_2$MDP) an inhibitor of bone resorption in 3 patients with Paget's disease (mean ± SEM). Despite the rapid bone turnover before treatment, plasma calcium and fasting calcium excretion were normal. During the early phase of treatment suppression of bone formation as judged by plasma alkaline phosphatase lagged behind inhibition of bone formation (urinary hydroxyproline/creatinine ratio). During this transient phase when bone formation exceeded resorption, fasting plasma and urinary calcium levels fell until a new steady state was attained when formation matched resorption.

thought to be due to the piezo-electrical properties of the mineral and matrix components.

Finally, it is important to recall that whereas bone cells are intimately related to the accretion and resorption of bone mineral (approximately 5 mmols (200 mg) of calcium per day), this turnover underestimates the true exchange between plasma and bone since exchange of calcium occurs at other sites such as the osteocytic and canalicular surfaces. In the adult, these surfaces have been estimated to constitute an area of more than a thousand square metres, which is 100-fold greater than that of the total trabecular bone surface. This vast exchange area may act as a buffer to calcium stresses imposed on the extracellular fluid. It is unlikely that this rapid exchange contributes to the long-term control of the plasma calcium level or indeed to long-term changes in skeletal mass since in the day-to-day assessment of calcium exchange, the net flux for calcium across these sites must be close to zero.

Assessment of Skeletal Turnover. Skeletal turnover (bone formation, mineralisation, and resorption) can be deduced from kinetic, histological or biochemical tests. Metabolic balance studies give an approximation of the net fluxes of calcium occurring across the skeleton (Fig. 2). They do not however indicate the sizes of the unidirectional fluxes. This can be overcome to some extent by the use of tracer kinetic studies. For example, the rate of bone resorption can be calculated by the release of strontium or calcium isotopes from the skeleton, and together with the balance studies, resorption and formation rates can be calculated separately.

Less direct but more readily measured indices of bone turnover are the estimation in plasma of alkaline phosphatase and the urinary excretion of hydroxyproline (Figs. 12 and 13). Alkaline phosphatase is derived in part from osteoblasts and contributes to the total plasma alkaline phosphatase. It is important to note that plasma alkaline phosphatase is also derived from liver and gut, and that hyperphosphatasia may also be caused by diseases of these systems. Alkaline phosphatase derived from bone has a greater heat stability and somewhat different electrophoretic mobility than that from gut or liver. The concurrent measurement of liver enzymes in plasma (e.g. 5'-nucleotidase) or isoenzyme studies of phosphatase may help to clarify the causes of hyperphosphatasia.

Hydroxyproline, in small peptides derived from collagen degradation can be measured in plasma or urine. As in the case of plasma alkaline phosphatase, there are several limitations which impair interpretation of the results of these estimations. Unless the ingestion of foods rich in hydroxyproline is avoided, high urinary excretion rates may spuriously suggest increased bone resorption. Also a large proportion of the total hydroxyproline in plasma is found in the first component of complement (C1$_q$) and collagen is present in many other structures than bone.

Bone formation, mineralisation, and resorption, can also be assessed from histological studies on bone. Methods exist for estimating the amount of bone undergoing resorption, mineralisation and formation. For practical purposes these techniques are confined to a limited number of centres with an interest in metabolic bone disease. They are nevertheless extremely useful in understanding the pathophysiology of skeletal disorders and their response to therapeutic intervention. They depend heavily on static measurements, for example the proportion of trabecular surfaces occupied by active-looking osteoclasts, but dynamic measurements can also be obtained. The administration of tetracycline prior to bone biopsy results in the deposition of tetracycline at the calcification front. If two courses of tetracycline are administered with a known time interval between them (usually 10–14 days), then the two markers can be seen on sections of the bone biopsy since they fluoresce under ultraviolet light (Fig. 14). In this way the extent and the rates of bone deposition and mineralisation can be measured, and from these the bone formation rate. It is also possible to quantitate the activity of both osteoblasts and osteoclasts, and indeed to calculate the turnover time for bone. Bone biopsy must be necessarily confined to available skeletal sites, usually the ilium and a major limitation of the technique is that inferences made from the biopsy may not reflect events occurring at other skeletal sites, particularly in disorders in which the skeleton is not uniformly disturbed.

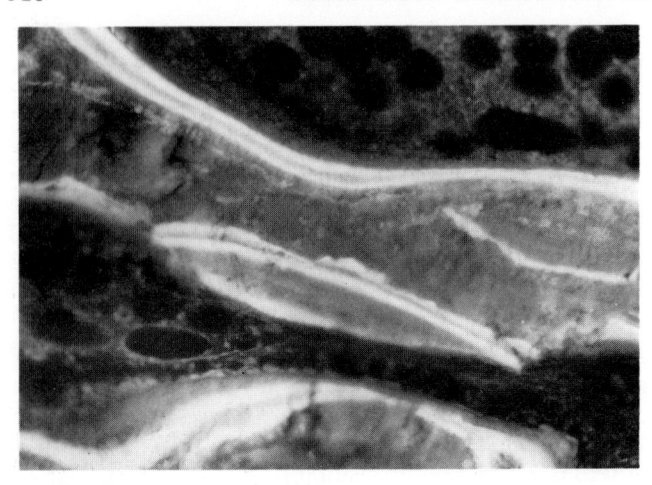

FIG. 14. Section through trabecular bone photographed under ultraviolet light which has been prelabelled with tetracycline. The light bands indicate the two courses of tetracycline given which are separated by a known time interval. This allows the rate and extent of bone formation to be calculated (from P. Meunier).

Despite the many limitations in the assessment of these biochemical indices of bone turnover, there is a surprisingly good correlation between kinetic, histological and biochemical measurements in metabolic bone disease.

Measurement of Skeletal Mass and Loss. A variety of techniques are available for measuring bone loss. With the exception of balance and radioisotope studies, they depend on indirect determinations of skeletal mass at repeated intervals. Conventional balance studies are too insensitive to measure small losses reliably since the magnitude of daily losses of calcium, for example in osteoporosis (say 30 mg per day equivalent to approximately 1% of bone mass per year), is beyond the detection limits of the technique. Moreover, a measurement of skeletal balance over a short period of time may not reflect long-term changes, since bone losses or gains may not occur uniformly with time.

Assessment of skeletal mass can be made from radiographs, but this method is relatively insensitive. Subjective evaluation can be improved by comparing the radiodensity of bone with standards such as aluminium step wedges and with the use of industrial film. In the hip, the loss of femoral trabecular markings on X-rays can give a semi-quantitative index of bone loss (Singh score), and discriminates fairly well between patients with and without femoral neck fracture due to osteoporosis. More precise measurements ($\pm 2\%$) can be made by measuring cortical thickness of long bones. This technique has been widely used in the metacarpal, particularly in the evaluation of osteoporosis. Cortical thickness clearly does not reflect trabecular bone density which may be important, for example, in spinal osteoporosis. These measurements also do not take account of cortical porosity, which is a feature, for example, of primary hyperparathyroidism.

Bone mineral density can be readily measured by photon absorption at peripheral sites. More recently, techniques have been developed to examine central sites such as the spine by photon absorption but are not yet widely available. The measurement of bone mineral density has some advantages over cortical thickness in that for the same degree of precision it provides an index both of cortical and trabecular density. The importance of this is that bone loss is not usually distributed uniformly between cortical and cancellous bone. A recent development has been the use of computerised axial tomography using an X-ray or photon source. This technique has some advantages in that bone density can be computed separately for cortical and trabecular areas, and is likely to offer a significant advance when it becomes more widely available.

Perhaps the most direct method of estimating skeletal mass is from quantitative measurements on bone biopsies (including microradiography which involves taking X-rays of thick (100 micron) slices of bone). For a variety of reasons such methods are very imprecise; but it is important to note that the value of bone biopsy should not be underrated, since it is one of the few reliable ways available of detecting structural abnormalities in bone such as osteomalacia.

In vivo neutron activation analysis is becoming more widely available and gives a precise index of skeletal calcium in the region measured. The technique depends on the neutron bombardment of calcium in the skeleton (and of course in soft tissues). This results in the formation of an unstable calcium isotope (^{49}Ca) which emits gamma irradiation. The amount of calcium in the area under investigation is a function of the radiation emitted. Whole-body neutron-activation analysis provides the most reliable and direct assessment of skeletal calcium, but the method is expensive and not widely available.

A serious limitation of all these techniques, with the exception of whole-body calcium estimation, is that the information derived is necessarily from a restricted part of the body, and may not therefore reflect changes elsewhere in the skeleton. It can be argued that there is often, though not invariably, broad correlation between these various measurements and between sites using the same techniques. But it is hardly surprising that, for example, patients with larger hands may also have more mineral elsewhere. An important limitation of the measurements of skeletal mass is that, although a precision of $\pm 2\%$ can be achieved, this is not sufficiently accurate for many clinical situations. In osteoporosis for example, the rate of bone loss may be less than 2% per year so that on serial measurements it may be difficult to be sure whether or not an individual is losing or gaining bone.

Quality of Bone. The quality of bone can be directly estimated by the use of bone biopsy. Reference has been previously made to quantitative histological techniques that are available, but it is also possible to determine the composition of bone directly, e.g. the calcium: phosphate ratio or the aluminium content, by appropriate analytical methods, including chemical analysis or physical methods such as by energy dispersive X-ray analysis under the electron microscope. Skeletal X-rays have, for many years, provided an important clinical tool for the investigation of skeletal disorders. The radiographic changes which occur in various diseases are beyond the scope of this review but it is important to note that in many disorders, X-rays are a relatively insensitive technique. For example, in osteomalacia occurring in the adult skeleton, radiographic appearances may be quite normal unless also associated with Looser's zones. The architectural detail on radiographs can be improved by the use of fine grain industrial film, which is particularly important in assessing trabecular porosity and the presence of subperiosteal bone resorption. Xeroradiography has limited value in the assessment of skeletal disease but is occasionally useful in revealing bone resorption, occurring, for instance at the distal end of the clavicle, a feature of renal osteodystrophy.

Isotopes such as ^{89}Sr and ^{18}F have been used clinically as scintigraphic agents for detecting regions of increased bone turnover, such as in metastatic tumour deposits. Recently, pyrophosphate, polyphosphates, and diphosphonates, all of which have a high affinity for crystals of hydroxyapatite, have been used as bone scanning agents by linking them to the gamma emitting isotope ^{99m}-technecium (^{99m}Tc) in the presence of stannous ions. Recently, attempts have been made to measure quantitatively the retention of ^{99m}Tc-labelled diphosphonates in the skeleton. These techniques may provide a method of distinguishing patients with osteoporosis alone from those with osteoporosis in combination with osteomalacia, the treatment for which may be quite different. At present these populations can only be adequately separated by examining bone biopsies.

Integration of Organ Responses

It is useful to draw some distinction between the way in which the plasma calcium is set at a particular concentration, and the way in which movements of calcium in and out of extracellular fluid are controlled. The plasma calcium level is set close to a particular value in different individuals in normal and disease states. Deviations from this value are corrected by hormone-induced changes in the relative fluxes of calcium in and out of the extracellular fluid compartment. Alterations in flux rates are therefore monitored and adjusted by the changes in plasma calcium or phosphate concentration they induce. This homeostatic system could operate with the plasma calcium set at any number of different values, with the relative rates of entry and exit of calcium to and from the extracellular fluid being altered as drift occurs from this set point. Thus, in hypo- or hyper-parathyroidism the fluxes of calcium across the gut and in and out of bone may not be greatly different from normal, and external calcium balance can be maintained even though the plasma calcium is set at a markedly abnormal level. The plasma calcium thus provides the point around which adjustments are made.

In considering plasma calcium homeostasis it is useful to consider the steady state and transient changes separately. When the system is disturbed, a steady state no longer exists, and the response which occurs adjusts the system so that a new steady state comes into existence. Deviations of plasma calcium away from its normal value are rapidly corrected by alterations in the secretion of the regulating hormones. In many mammals PTH and calcitonin both respond, but in opposite directions. In man it is uncertain whether or not calcitonin secretion does change in response to changes in plasma calcium concentrations that are within the physiological range. PTH and possibly calcitonin can be considered the fast-acting components of the regulatory system, whereas vitamin D is responsible for adaptation in the longer term. In experimental animals, removal of sources of parathyroid hormone and calcitonin (for example thyroparathyroidectomy in dogs or rats) results in hypocalcaemia (lower set point for plasma calcium) and in a slower than normal return of plasma calcium to starting values in response to acute changes in concentration.

In man, the most important regulator of acute changes in the concentration of calcium is parathyroid hormone, the secretion of which increases within seconds of a fall in plasma calcium, and decreases as the ionised concentration of calcium rises. The rapid control of plasma calcium by PTH is mainly due to its ability to regulate renal tubular reabsorption of calcium and possibly by its effects on calcium exchange between ECF and bone. After parathyroidectomy, the fall in plasma calcium can be largely accounted for by a continued loss of calcium into urine, until a new steady state is achieved in which calcium excretion is the same as its starting value but takes place at a much reduced filtered load of calcium.

A further example of a disruption of the steady state is seen during an infusion of calcium. During a calcium infusion, plasma calcium concentration will begin to rise. If the rate of calcium infusion is constant, levels of plasma calcium will not rise indefinitely but only until the rate of efflux of calcium from the extracellular pool (to bone, kidney, and gut and other tissues) matches the rate of influx. At this point, plasma calcium levels will not rise further, despite continuing the infusion, and a new steady state prevails. In practice, the infusion of calcium will result in the suppression of secretion of parathyroid hormone and a decrease in the renal tubular reabsorption for calcium. This will tend to increase the rate at which a new steady state is achieved. It is notable that the rise in plasma calcium during a calcium infusion is to some extent buffered by the exchange of calcium in bone. Thus, rises or falls in plasma calcium are partially compensated by increased net movements of calcium into or out of bone.

The response to prolonged perturbations brings in contributions from changes in vitamin-D metabolism, and from the intestine and bone. An example is the adaptation that occurs to a change in dietary intake of calcium. If intake is reduced, this will tend to cause a gradual fall in plasma calcium which will increase the secretion of PTH. Apart from its effects on the kidney, a sustained increase in PTH will lead both to enhanced osteoclastic resorption of bone, and to an increase in synthesis of $1,25(OH)_2D_3$. This in turn will enhance intestinal absorption of calcium and resorption of calcium from bone. If the reduction in dietary calcium continues, these changes will act to restore the plasma calcium towards its previous value, and bone formation rate will increase to match the rate of increased bone resorption by the coupling mechanism. The new steady state will consist of a greater efficiency of intestinal absorption of calcium and an increased rate of entry and removal of calcium from bone, so that net balance can be maintained. This response can be seen very clearly in experimental animals and there is good evidence that these adaptive changes also occur in man. When dietary deprivation is so severe that intake can no longer match output, the reserves of calcium in bone are utilised.

DISORDERS OF CALCIUM HOMEOSTASIS

The purpose of this section is not to give a detailed clinical description of disorders of calcium metabolism but to illustrate some of the principles of calcium homeostasis and their application to the investigation of various diseases.

Disorders of Parathyroid Secretion

Hyperparathyroidism

The term hyperparathyroidism is applied to those clinical disorders characterised by an increase in the circulating concentration of parathyroid hormone (Table 3). These can be conveniently classified into primary, secondary, tertiary, and acetopic hyperparathyroidism.

Secondary hyperparathyroidism is due to a rise in PTH because of sustained hypocalcaemia, as, seen for example in vitamin-D deficiency or chronic renal failure. Hypocalcaemia results in the increased secretion of parathyroid hormone and, if sustained, to hyperplasia of the parathyroid glands. The skeletal lesions that result are therefore a reflection of the underlying disorder, the hypocalcaemia, and the high circulating levels of parathyroid hormone. Secondary hyperparathyroidism can be cured by appropriate treatment which restores the plasma calcium concentration to normal (for example vitamin D in simple nutritional deficiency of vitamin D). This, therefore, removes the stimulus to parathyroid overactivity.

Tertiary hyperparathyroidism is a term used to denote those patients with long-standing secondary hyperparathyroidism who develop 'autonomous gland function' and hypercalcaemia. Nowadays this is most commonly seen after renal transplantation but may also be observed in patients with long-standing malabsorption or chronic renal failure. The parathyroid glands are hypertrophied and may show histological evidence of adenomatous transformation. As in the case of many endocrine disorders, the term 'autonomous' secretion is probably a misnomer since parathyroid hormone levels can be suppressed by calcium infusion or further augmented by manoeuvers which lower the level of plasma calcium (e.g. infusion of EDTA). Tertiary hyperparathyroidism therefore implies a change in the set point with respect to PTH secretion for the control of plasma calcium. In contrast to secondary hyperparathyroidism, plasma calcium and PTH are both raised.

Primary hyperparathyroidism is due to the presence of multiple or single adenomata or to hyperplasia of the parathyroid glands. The causes are unknown and it is assumed that the abnormality is intrinsic to the parathyroid glands themselves. Primary hyperparathyroidism is rarely due to carcinoma. As in the case of tertiary hyperparathyroidism, the main biochemical abnormality is an increase in the circulating concentration of both PTH and calcium. There are, however, instances where such patients have a normal total plasma calcium or intermittent hypercalcaemia, sometimes associated with concurrent vitamin-D deficiency. Thus, primary hyperparathyroidism can be defined as a disturbance of the parathyroid glands where circulating levels of PTH are inappropriately high for the prevailing concentrations of plasma calcium. However, both plasma calcium and plasma PTH levels should be interpreted with caution. Thus, total plasma calcium comprises the ionised, complexed and protein-bound fraction and account should be taken of possible abnormalities in protein concentration, pH and in protein binding. Hypercalcaemia itself may impair renal function, and since the kidney is an important site for degradation of PTH, increased levels of PTH in the presence of renal failure may not reflect an increase in the biologically active moiety of PTH (Fig. 4).

In primary hyperparathyroidism, the hypercalcaemia is maintained mainly by a resetting of the renal tubular reabsorption for calcium, so that reabsorption is enhanced for any given filtered load (Fig. 10). Frequently, increased bone resorption and increased intestinal

TABLE 3
CAUSES OF HIGH LEVELS OF PARATHYROID HORMONE IN PLASMA

Primary hyperparathyroidism
   Single or multiple adenoma (common)
   Parathyroid hyperplasia (less common)
   Parathyroid carcinoma (rare)

Secondary hyperparathyroidism
   Vitamin-D-deficiency states
   Drugs causing hypocalcaemia (e.g. phosphate, diphosphonates)
   Renal failure (reduced clearance of PTH fragments, increased secretion or both)
   Medullary carcinoma of thyroid
   Pseudohypoparathyroidism (target organ insensitivity to PTH)

Tertiary hyperparathyroidism
   Prolonged secondary hyperparathyroidism (e.g. chronic bowel or renal disease)
   Transplantation (slow parathyroid involution following restoration of renal function)

Ectopic (or pseudo) hyperparathyroidism
   Carcinoma, e.g. squamous cell carcinoma of bronchus, lymphomas

absorption of calcium (due to stimulation of production of 1,25(OH)$_2$D$_3$) is also found, and this will tend to increase urinary calcium excretion (24-hour excretion). Although resorption of bone is commonly enhanced, so too is bone formation so that the balance for calcium may be normal, but at the expense of an increased bone turnover. Radiographic abnormalities (subperiosteal erosions) are more commonly seen in secondary and tertiary hyperparathyroidism and are due to abnormalities in bone architecture because of increased bone turnover. Calcium balance may be negative particularly in patients with severe bone disease and osteoporosis is occasionally the major radiographic feature. Renal stones are also a common complication of primary hyperparathyroidism. The aetiology of renal stones in hyperparathyroidism is discussed elsewhere (Chapter 17) but is related in part to an increase in the daily amount of calcium excreted (increased intestinal absorption of calcium with or without increased net bone resorption). When urine is examined in the fasting state, the excretion rate for calcium is commonly normal even though urine hydroxyproline excretion is increased (bone resorption = bone formation).

Hypercalcaemia is commonly seen in patients with malignant disease. This may be due to a variety of causes (see page 323) but in some instances may be due to the production of PTH by the tumour itself. This syndrome has been variously described as ectopic hyperparathyroidism or pseudohyperparathyroidism and was once thought to be a common cause for hypercalcaemia in patients with carcinoma. Well-documented examples are exceedingly rare and it is likely that the prevalence of this disorder has been grossly overestimated, perhaps partly due to the inadequacies of the radioimmunoassay for PTH.

The majority of patients with hypercalcaemia have either hyperparathyroidism without overt radiographic evidence of bone disease or non-parathyroid malignant tumours, and the greatest diagnostic difficulties are encountered in distinguishing between these two groups. A variety of tests have been devised to distinguish these disorders, partly because the lack until recently of suitable assays for PTH. The hydrocortisone suppression test has been widely used. Hydrocortisone (usually 40 mg thrice daily for ten days) fails to lower plasma calcium in primary hyperparathyroidism but usually does so in hypercalcaemia due to other causes. With the advent of radioimmunoassay for PTH, this test is becoming redundant but it still has its advocates. It is important to realise that a proportion of patients with primary hyperparathyroidism will suppress their plasma calcium with hydrocortisone, particularly those patients with bone disease. Furthermore, hypercalcaemia in patients with other disorders does not invariably respond to corticosteroids. The actions of corticosteroids on bone metabolism are complex and the basis therefore for the hydrocortisone suppression test is unclear.

Various other tests have been devised to distinguish hypercalcaemia due to hyperparathyroidism from that from other causes. PTH has an effect on the renal handling of phosphorus and this has led to the measurement of indices of tubular reabsorption for phosphate. Many tests are available, though the estimation of TmP/GFR is based on sounder physiological principles than many others. Mention should be made of discriminant function analysis which also has its advocates. One of the major drawbacks of this approach is that the functions need to be calculated in each laboratory. They depend on the simultaneous measurement of several biochemical parameters including phosphate, alkaline phosphatase, chloride, bicarbonate, urea, and erythrocyte sedimentation rate. In the absence of renal impairment, the plasma bicarbonate alone may be helpful. Thus in hyperparathyroidism a metaboloic acidosis is commonly found, reflected as a low plasma bicarbonate, whereas in disorders where PTH secretion is depressed, a metabolic alkalosis is to be expected and frequently observed. Moreover, in hypercalcaemia due to rapid bone destruction, an alkalosis may be accentuated by the release of bicarbonate from bone.

More recently, the urinary excretion of cyclic AMP has been used as a test of parathyroid function. Considerable overlap occurs between patients with hyperparathyroidism and normal subjects, but by expressing the excretion of cAMP as a proportion of creatinine excretion the discrimination is improved. The determination of nephrogenous cyclic AMP improves this differentiation still further (Fig. 15). It is important to recall that a proportion of patients with malignant disease and hypercalcaemia excrete increased amounts of cyclic AMP in the urine despite no evidence for ectopic hyperparathyroidism which decreases the discrimination of the test.

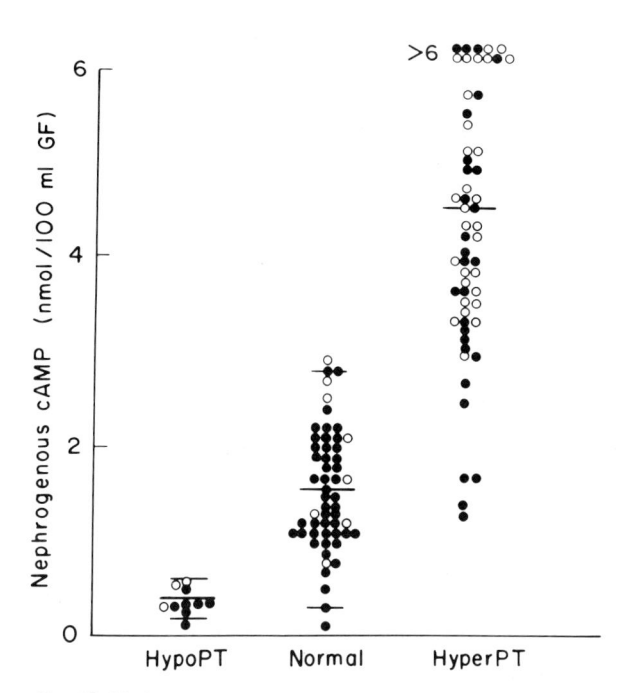

Fig. 15. Nephrogenous cAMP in normal subjects and patients with hypo- and hyperparathyroidism (from A. Broadus et al. (1977) J. Clin. Invest. 60, 771).

Hypoparathyroidism

Decreased secretion of PTH may arise in hypercalcaemic conditions which suppress the secretion of PTH. The term is, however, usually confined to those disorders (Table 4) associated with defective secretion or action of PTH (the latter termed pseudohypoparathyroidism).

TABLE 4
MAJOR CAUSES OF HYPOPARATHYROIDISM

1. Inadequate secretion of PTH
    Surgical—thyroid, parathyroid and radical neck surgery
    Familial
    Sporadic
    Di Giorge syndrome
    Parathyroid disease, e.g. neoplasia, amyloid
2. Suppression of PTH secretion (normal parathyroid glands)
    Neonatal—from maternal hypercalcaemia
    Severe magnesium depletion
    Non-parathyroid induced hypercalcaemia
3. Defective end-organ response to PTH
    Pseudohypoparathyroidism Types I and II

Many of the biochemical and metabolic abnormalities seen in hypoparathyroidism can be understood from an appreciation of the physiology of PTH. Plasma phosphate is raised due to an increase in TmP/GFR. Hypocalcaemia is due in part to a decrease in tubular reabsorption for calcium. PTH stimulates the synthesis of $1,25(OH)_2D_3$ from its precursor, 25-OHD_3, and in hypoparathyroidism circulating concentrations of $1,25(OH)_2D_3$ may be low and account for decreased intestinal absorption of calcium. Hyperphosphataemia may be another reason for low levels of $1,25(OH)_2D_3$ because phosphate suppresses its renal production. It is also thought that hypocalcaemia is due to impaired osteocytic calcium transfer from bone due to deficiency of PTH or $1,25(OH)_2D_3$ or both. The immediate hypocalcaemia, which is seen following parathyroidectomy, may well be explicable in part on this basis but the contribution of osteocytic calcium fluxes under steady state conditions is less clear.

Many of the signs and symptoms of hypoparathyroidism (tetany, calcification of basal ganglia, cataract, etc.) are attributable to hypocalcaemia and hyperphosphataemia.

There exists a group of disorders termed *pseudohypoparathyroidism* in which the secretion of PTH is normal or even increased but some or all of the biochemical features of hypoparathyroidism are present. This is due to a resistance of one or more of the target organ tissues to PTH. There is an association between pseudohypoparathyroidism and characteristic somatic features, including short stature, round face, short neck, shortening of metacarpals and metatarsals. These are not invariably found in pseudohypoparathyroidism and indeed may be present occasionally in idiopathic hypoparathyroidism when present in childhood. Rarely the somatic features of pseudohypoparathyroidism may be found in patients with no impairment of secretion or action of PTH, a condition termed *pseudopseudohypoparathyroidism*.

Simple hypoparathyroidism most commonly results from inadvertent removal of parathyroid glands during thyroid, parathyroid or laryngeal surgery. Idiopathic hypoparathyroidism is much rarer and may be familial, sporadic or be found in combination with other disorders such as pernicious anaemia, Addison's disease or candidiasis. Rarely, hypoparathyroidism may result from infiltration of the parathyroid gland, by malignant disease, haemochromatosis or amyloid. Di Giorge's syndrome refers to a congenital absence of thymic and parathyroid tissue and is associated with immunological deficiencies, cardiovascular anomalies and unusual facies. Hypoparathyroidism has been reported following ^{131}I therapy, but irradiation of the neck appears to be one of the factors predisposing to hyperparathyroidism.

Apart from low or absent levels of PTH, hypocalcaemia and hyperphosphataemia in the presence of normal renal function, a biochemical characteristic of hypoparathyroidism is that patients are responsive to the action of PTH (Fig. 16). The renal excretion of cyclic AMP is low and is augmented by the administration of PTH. Similarly, renal tubular reabsorption for phosphate is acutely decreased following PTH. This forms the basis of the Chase-Aurbach (Ellsworth-Howard) test.

In contrast, pseudohypoparathyroidism (also called PTH-resistant hypoparathyroidism) results from a resistance of target tissues to the action of PTH. Indeed, PTH levels are commonly high, although appropriate for the hypocalcaemia. The resistance to PTH may be partial or complete. Thus during the infusion of PTH, cyclic AMP production is not stimulated (Fig. 16) and in many patients neither is the renal tubular reabsorption for phosphate decreased (pseudohypoparathyroidism type I). Other patients may show normal cAMP responses but the absence of phophaturia (type II), whereas in others responsiveness to PTH may be completely restored by calcium infusion. Not all target tissues need be affected and pseudohypoparathyroidism may be associated with osteitis fibrosa suggesting that the skeleton is sensitive to PTH.

There is a great deal of confusion concerning the definition of pseudopseudohypoparathyroidism, since affected patients may have relatives with pseudohypoparathyroidism and may indeed undergo transitions from hypocalcaemia to normocalcaemia and vice versa.

Mention has been previously made of the hypocalcaemia seen in magnesium deficiency. This may be due to impaired secretion of PTH as well as diminished responsiveness of target tissue to PTH.

Disorders of Vitamin D Metabolism

Excessive Amounts of Vitamin D

Vitamin-D toxicity is a common finding in clinical practice and is usually iatrogenic. Since vitamin D increases calcium absorption and in high doses augments bone resorption, these patients characteristically have hypercalcaemia. Plasma phosphate concentration is

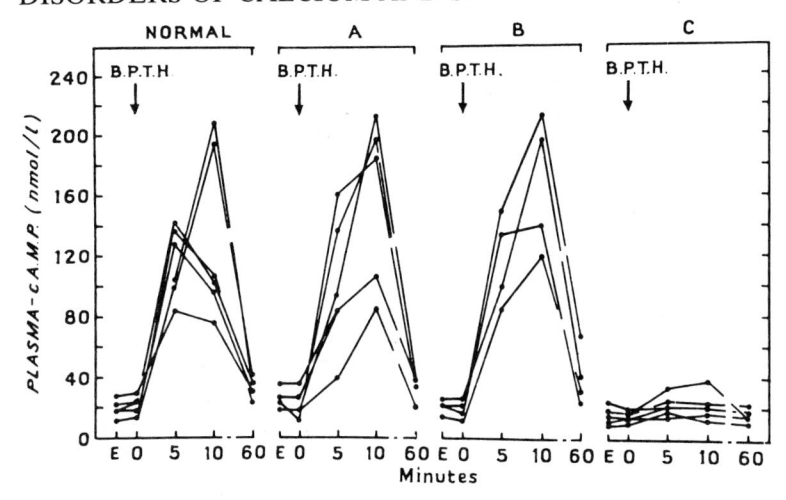

FIG. 16. Plasma cAMP before and up to 60 minutes after an injection of 200 units of bovine PTH (BPTH) in normal subjects and in patients with surgical (A), and idiopathic hypoparathyroidism (B), and pseudohypoparathyroidism (C) (from Tomlinson, S., Hendy, G. N. and O'Riordan, J. L. H. (1976) *Lancet* i, 62).

commonly also elevated due to similar effects on phosphate transport. If over-dosage is prolonged, increased bone resorption causes progressive loss of bone. It is notable that plasma levels of $1,25(OH)_2D_3$ are not raised suggesting that these effects are due to high levels of other metabolites of vitamin D.

Under physiological conditions, the major actions of vitamin D are thought to be due to its conversion to the dihydroxylated metabolites, particularly $1,25(OH)_2D_3$. However, large doses of vitamin D and 25-OHD_3 also induce toxicity, even in patients who are incapable of forming $1,25(OH)_2D_3$ (e.g. anephric patients). Furthermore, in patients with normal renal function and vitamin-D toxicity, plasma levels of $1,25(OH)_2D_3$ are normal.

The production of $1,25(OH)_2D_3$ is under metabolic control. The factors which augment the synthesis of $1,25(OH)_2D_3$ include hypophosphataemia, hypocalcaemia and excessive secretion of PTH. Higher levels of $1,25(OH)_2D_3$ have been reported in patients with primary hyperparathyroidism and in idiopathic renal calculi and may account for the increased calcium absorption found in these disorders. Increased calcium absorption is also seen in acromegaly and may be due to increased secretion of $1,25(OH)_2D_3$. Sarcoidosis is occasionally associated with hypercalcaemia but more commonly with hypercalciuria. This is due to increased intestinal absorption for calcium and in some patients is associated with high levels of $1,25(OH)_2D_3$. It is likely that the $1,25(OH)_2$ vitamin D is produced by the sarcoid tissue itself which may explain the so called vitamin-D sensitivity seen in this disorder.

Defective Production of Vitamin-D Metabolites

The hallmarks of vitamin-D deficiency include defective mineralisation of bone and retardation of growth. Simple vitamin-D deficiency is associated with hypocalcaemia and hypophosphataemia. The hypophosphataemia results from secondary hyperparathyroidism (decreased TmP/GFR) and malabsorption of phosphate. Indeed it is possible that the defective mineralisation of bone is due in part to the phosphate depletion associated with secondary hyperparathyroidism.

It is important to recognise that osteomalacia in adults or rickets in children is not always due to deficiency of vitamin D or its metabolites (Table 5). A common example is X-linked or sporadic hypophosphataemia

TABLE 5
SOME CAUSES OF EXCESS OSTEOID (UNMINERALISED BONE MATRIX)

Vitamin-D deficiency or defective metabolism
   Nutritional deficiency of vitamin D
   Lack of sunlight
   Malabsorption of vitamin D
   Chronic renal failure
   Vitamin-D-dependent rickets

Low plasma phosphate
   Phosphate deficiency
   Renal tubular disorders (vitamin-D-resistant rickets)

Chronic acidosis
   Ureterosigmoidostomy
   Renal tubular acidosis (proximal or distal)

High bone turnover
   Fracture healing
   Paget's disease
   Hyperparathyroidism
   Hyperthyroidism

Drugs
   Anticonvulsants
   Diphosphonates
   Fluoride

Inherited
   Fibrogenesis imperfecta ossium
   Hypophosphatasia

Other unexplained
   Axial osteomalacia

(vitamin-D-resistant rickets) where defective mineralisation of bone is probably due to abnormalities in phosphate transport.

Vitamin-D deficiency may be present despite adequate supplies of vitamin D due to a disturbance in its metabolism. This gives rise to vitamin-D resistance in the sense that the doses of vitamin D necessary to achieve therapeutic effects are greater than the normal physiological requirements. As in the case of osteomalacia, not all vitamin-D resistant disorders are due to defective metabolism of vitamin D. Defective action or production of vitamin-D metabolites may arise in a number of ways (Table 6), by impaired conversion of vitamin D to its natural metabolites or by its impaired target organ responses.

TABLE 6
VITAMIN-D-DEFICIENCY STATES

Reduced availability of vitamin D
    Inadequate sunlight or diet
    Intestinal malabsorption of vitamin D, e.g. malabsorption syndromes, subtotal gastrectomy, ileal bypass surgery

Reduced availability of 25-OHD$_3$
    Liver disease
    Drugs, e.g. phenobarbitone, phenytoin, glutethimide
    Decreased enterohepatic circulation, e.g. malabsorption syndromes and liver disease
    Nephrotic syndrome (urinary losses)

Reduced availability of 1,25(OH)$_2$D$_3$
    Enzyme deficiency (1-alpha-hydroxylase), e.g. Vitamin D dependent Rickets (type 1)
    Enzyme destruction, e.g. chronic renal failure
    Enzyme suppression, e.g. hypoparathyroidism, pseudohypoparathyroidism
    Fanconi syndrome and acidosis
    Tumour-associated osteomalacia (Mesenchymal tumours)
    Itai-Itai disease (cadmium toxicity)

Reduced end-organ response
    Vitamin D-dependent rickets type II
    Anticonvulsants
    ? steroid induced osteoporosis
    Coeliac disease and chronic renal failure (tissue damage to gut)

Uncertain relationship to vitamin D
    X-linked hypophosphataemia
    Diabetes mellitus
    Neonatal hypocalcaemia
    Treatment with corticosteroids
    Osteoporosis

Chronic renal failure, pseudohypoparathyroidism and hypoparathyroidism are disorders in which conversion of 25-hydroxy-vitamin D to 1,25(OH)$_2$D$_3$ is impaired. Plasma levels of 1,25(OH)$_2$D$_3$ are low and physiological quantities of 1,25(OH)$_2$D$_3$ may reverse some of the biochemical abnormalities, whereas pharmacological amounts of vitamin D$_3$ or 25-OHD$_3$ are required to achieve the same response. Perhaps a better example of defective conversion of D$_3$ to 1,25(OH)$_2$D$_3$ is seen in vitamin-D-dependent rickets in which many of the clinical and biochemical features are compatible with a deficient renal 1-alpha hydroxylase enzyme system (pseudo-deficiency rickets). There are patients with similar biochemical characteristics who have normal or high levels of 1,25(OH)$_2$D$_3$—the vitamin-D equivalent of pseudohypoparathyroidism. This rare condition has been described as vitamin-D-dependent rickets type II.

In osteomalacia associated with anticonvulsant drugs, it is claimed though not proven that there is increased metabolic degradation of vitamin D to inactive metabolites because of induction of hepatic microsomal enzymes (including the cytochrome P$_{450}$ system). A more firmly established role for impaired 25-OHD metabolism is seen in the nephrotic syndrome which may be associated with excessive urinary losses of 25-OHD bound to proteins.

Impairment of target organ response to vitamin D probably occurs in coeliac disease (gluten-sensitive enteropathy) where intestinal absorption of calcium is not increased by relatively large doses of 1,25(OH)$_2$D$_3$ even when given intravenously. After effective treatment with a gluten-free diet, a normal sensitivity to 1,25(OH)$_2$D$_3$ is restored. This suggests that the lack of response in the untreated disease is due to damage to the intestinal villi, which are the site of action of vitamin D metabolites on intestinal calcium transport. Impaired target organ responses, both in the intestine and bone may also be responsible for the disorders of mineral metabolism found in anticonvulsant use and corticosteroid treatment.

Low levels of 1,25(OH)$_2$D$_3$ have also been reported in post-menopausal osteoporosis but the importance of this in its pathophysiology is controversial.

The wider availability of the assays for vitamin D and its metabolites will aid in the precise diagnosis of patients with and without vitamin D related osteomalacia. The measurements of 24,25(OH)$_2$D$_3$ and 25,26(OH)$_2$D$_3$ are however not widely available, but the role of these metabolites, both in health and disease, is at present not clear. Low levels of 24,25(OH)$_2$D$_3$ have been reported in anticonvulsant osteomalacia, osteoporosis, and in chronic renal failure.

Disorders of Calcitonin Secretion

Increased secretion of calcitonin occurs in medullary carcinoma of the thyroid, a familial and malignant disorder of the C-cells. This may be associated with other endocrine abnormalities such as phaeochromocytoma, Cushing's syndrome and hyperparathyroidism—the so-called multiple endocrine abnormality (MEA) syndrome. Medullary carcinoma may become evident clinically by the detection of a mass in the thyroid or be suspected from the family history or the presence of associated endocrinopathies. Severe diarrhoea may occur which is probably related to the synthesis of prostaglandins or serotonin. The lack of apparent effect of excessive amounts of calcitonin itself is one of the reasons why the physiological role of calcitonin is uncertain. The diagnosis is established by the presence of high plasma levels of calcitonin and augmented responses to provocative tests of calcitonin secretion (e.g. pentagastrin and calcium infusion).

Few metabolic abnormalities appear to be associated

with calcitonin deficiency. Thus, in athyroidal man given thyroid supplements, calcium metabolism is not markedly disturbed. Defective secretion of calcitonin has been implicated in the pathophysiology of hyperparathyroid bone disease in chronic renal disease and it is possible that osteitis fibrosa in chronic renal failure is in part due to hyperparathyroidism and in part to deficient production of calcitonin.

Hypercalcaemia

Many of the causes of hypercalcaemia have been reviewed previously and are summarised in Table 7. Of

TABLE 7
CAUSES OF HYPERCALCAEMIA

Common
　Artefactual—Hyperproteinaemia due to venous stasis, hyperalbuminaemia (dehydration, i.v. nutrition), hypergammaglobulinaemia (myeloma, sarcoidosis)
　Neoplasia—Carcinoma with skeletal metastases (e.g. breast, lung)
　　Carcinoma without skeletal metastases (? ectopic secretion of humoral agent)
　　Haematological disorders (myeloma, lymphoma)
　Primary hyperparathyroidism
Rare
　'Tertiary' hyperparathyroidism—transplantation, chronic renal failure, malabsorption
　Vitamin-D toxicity
　Vitamin-D 'sensitivity'—sarcoidosis, hypercalcaemia of infancy
　Immobility—Paget's disease
　Milk alkali syndrome
　Thyrotoxicosis
　Thiazide diuretics
　Adrenal failure
　Phaeochromocytoma
　Familial hypocalciuric hypercalcaemia
　Haemodialysis—high dialysate calcium

clinical importance is the ability of hypercalcaemia to alter renal function adversely. Not only may hypercalcaemia cause intrinsic renal damage due to the intrarenal deposition of calcium phosphate, but it also decreases the renal sensitivity to antidiuretic hormone and may therefore result in profound dehydration. Because there is a link between the tubular reabsorption of sodium and that for calcium, dehydration and salt depletion increase renal tubular reabsorption of calcium. All these factors tend to aggravate hypercalcaemia.

The hypercalcaemia associated with malignant disease is complex and arises by several mechanisms. Most commonly this is due to the skeletal destruction of bone associated with widespread metastases. Under such circumstances the secretion of parathyroid hormone is suppressed and the decreased renal tubular reabsorption for calcium thereby induced, has a sparing effect on hypercalcaemia. This may be one reason why patients presenting with hypercalcaemia and skeletal metastases are usually far advanced in the course of the disease.

Certain humoral agents have been associated with osteolytic bone disease due to malignant disorders. These include prostaglandins, osteoclast activating factors (in the case of myeloma) and parathyroid hormone or related substances.

The investigation of patients with hypercalcaemia is usually straightforward, and the diagnosis is commonly reached by clinical assessment and simple biochemical investigations. The majority of patients can be accurately diagnosed by a good history including a full drug history and the simple measurement of plasma calcium, phosphate, creatinine, and an estimate of tubular reabsorption for phosphate such as TmP/GFR. X-rays or bone scans are also most helpful in detecting malignant disease or hyperparathyroid bone disease. The major difficulties arise in those patients without overt skeletal disease and hypercalcaemia. Mention has been previously made of the use of PTH assay, discriminant function analysis, the hydrocortisone suppression test and the measurement of nephrogenous cyclic AMP. These investigations, together with a search for sarcoidosis, myeloma or a drug history usually yield a diagnosis.

Hypocalcaemia

The more common causes of hypocalcaemia are shown in Table 8. The investigation of hypoparathyroidism has been previously discussed, but it is important to

TABLE 8
CAUSES OF HYPOCALCAEMIA

Low plasma albumin or haemodilution
　Malnutrition, liver disease etc.
Vitamin-D-deficiency or resistance
Acute and chronic renal failure
Hypoparathyroidism and pseudohypoparathyroidism
Hypomagnesaemia
Acute pancreatitis
Drugs, e.g. calcitonin, phosphate, diphosphonates, citrate
Carcinoma, particularly of the prostate

emphasise that the distinction between hypoparathyroid and pseudohypoparathyroid states is academic rather than of practical clinical interest since the treatment is the same. The measurement of plasma calcium, phosphate and creatinine, together with a measurement of plasma albumin will distinguish most patients with hypoproteinaemia, vitamin D deficiency, chronic renal failure, and the hypoparathyroid states.

Disorders of Bone Turnover

There are a number of disorders associated with abnormal bone turnover. Several are inherited diseases such as osteogenesis imperfecta, hyperphosphatasia, fibrogenesis imperfecta ossium, various epiphyseal, metaphyseal and diaphyseal dysplasias, and neurofibromatosis. In most of these cases there is no apparent systemic disturbance in calcium metabolism and they are beyond the scope of this chapter.

In *Paget's disease of bone* there is enhanced resorption and formation of bone but the two processes remain coupled so that there is no marked systemic disturbance in calcium metabolism unless the coupling is temporarily disturbed. This may occur during immobilisation when hypercalciuria or hypercalcaemia can develop. There is an association between primary hyperparathyroidism and Paget's disease, and this should be suspected in hypercalcaemic patients. Most of the other complications of Paget's disease arise because the affected bone is structurally abnormal. Bones may be painful, increase in size, become highly vascular, and more liable to deformity and fracture.

The diagnosis is made by the characteristic radiographic findings and the biochemical indices of augmented bone turnover (plasma alkaline phosphatase, urinary hydroxyproline etc.), without disturbances in plasma calcium or phosphate. Occasionally it may be difficult to distinguish localised Paget's disease from fibrous dysplasia or osteoblastic secondary deposits, when bone biopsy may be helpful.

Another important group of disorders of bone turnover include the osteoporoses. *Osteoporosis* is the diminution of bone mass, without detectable changes in the ratio of mineral to non-mineralised matrix. This distinguishes the condition from osteomalacia when the proportion of osteoid to calcified bone is increased. It is important to recognise that osteomalacia and osteoporosis may co-exist since the former is amenable to treatment. A variety of disorders cause osteoporosis, including liver disease, various endocrinopathies and chronic renal failure (Table 9), but is most commonly associated with ageing (senile and post-menopausal osteoporosis).

TABLE 9
SOME CAUSES OF OSTEOPOROSIS (OSTEOPENIA OR THIN BONES)

Primary
    Old age
    Post menopause or post-oöphorectomy
    Idiopathic juvenile osteoporosis
Secondary
    Dietary deficiency of calcium or malabsorption
        Steatorrhoea
        Partial gastrectomy
        Chronic liver disease
    Endocrine
        Hyperparathyroidism
        Hyperthyroidism
        Cushing's syndrome
        Hypogonadism
    Metabolic
        Vitamin-C-deficiency
        Pregnancy
        Osteogenesis imperfecta
    Drugs
        Corticosteroids
        Heparin
        Immobilisation Generalised, e.g. space flight
                          Localised, e.g. after fracture, paraplegia
    Rheumatoid arthritis
    Chronic renal failure or dialysis

In terms of skeletal fluxes, bone loss occurs when the skeletal balance and hence the external balance is negative. This can arise from one of four circumstances.

(a) Normal rate of bone formation but an increased rate of bone resorption.
(b) Reduced rate of formation with a normal rate of resorption.
(c) A decrease in rates of both, but formation more so than resorption.
(d) An increase in both, but resorption more so than formation.

All these mechanisms are probably important in the development of senile and other types of osteoporosis which should therefore not be considered a single (or simple) disorder. Osteoporosis is a difficult condition to study and treat; problems include insufficient knowledge of the population under study, limitations in the methods used, and the small changes in bone metabolism involved. Calcium kinetic studies demonstrate that bone mineralisation and resorption rates fall with age whilst microradiography suggests that bone resorption increases. The calcium kinetic studies are difficult to interpret, since bone formation and resorption are surface phenomena, and the surface of bone available for these events decreases as bone mass is lost. Thus, apparent decreases in formation and resorption may simply be due to a reduction in skeletal mass with the rate of events on individual bone surfaces remaining constant. Similarly, the interpretation of microradiographic data are difficult since they cannot distinguish active resorption (i.e. associated with osteoclasts) from inactive surfaces (i.e. events of the past). Indeed, histological techniques suggest that active bone resorption is increased in only a minority of patients with established osteoporosis, but the extent of the inactive resorbing surface is increased. Active osteoblast surfaces are also usually decreased suggesting that formation, and hence repair of old resorption sites are impaired.

APPENDIX
TYPICAL NORMAL ADULT RANGES FOR SOME SIMPLE BIOCHEMICAL MEASUREMENTS USED IN THE INVESTIGATION OF PATIENTS WITH DISORDERS OF CALCIUM HOMEOSTASIS (M OR F DENOTES SEX)

Measurement	Units	Normal range
Plasma		
Total calcium	mmol/l	2·12–2·60
Ionised calcium	mmol/l	1·10–1·35
Fasting inorganic phosphate	mmol/l	0·6–1·5
Urine		
Calcium	mmol/24 h	M 2·5–10
	mmol/24 h	F 2·5–9·0
Phosphate	mmol/24 h	16–32
Total hydroxyproline	µmol/24 h	M 55–250
	µmol/24 h	F 75–430
Fasting urine		
Calcium/creatinine ratio	mmol/mmol	0·10–0·32
Calcium excretion	mmol/l GF	< 0·04
TmP/GFR	mmol/l	0·8–1·35
Total hydroxyproline/ creatinine	µmol/mmol	< 40
Vitamin D metabolites		
25-OHD	ng/ml	5–50
1,25(OH)$_2$D$_3$	pg/ml	20–40
24,25(OH)$_2$D$_3$	ng/ml	1–5

There are no distinctive biochemical abnormalities found in osteoporosis which are useful in clinical investigation. For example, plasma calcium, phosphate and alkaline phosphatase are invariably normal. Abnormalities in PTH secretion and in vitamin D metabolite production have been described but are of no value in making the diagnosis of osteoporosis in individuals.

It is important to distinguish patients with senile osteoporosis from those with more treatable forms of the syndrome, particularly endocrine disorders such as Cushing's disease and hyperthyroidism (Table 9). Of particular importance is the association of osteomalacia with osteoporosis in the elderly population. The manifestations of osteomalacia are often subtle in the elderly but the clinical consequences are readily treated, unlike the osteoporosis itself.

SUGGESTIONS FOR FURTHER READING

Lawson, D. E. M. (Ed.) (1978). *Vitamin D*. New York: Academic Press.

Kanis, J. A. (Editor), 'Etiology and medical management of hypercalcaemia.' *Metabolic Bone Disease and Related Research*, Vol. 2, No. 3, 1980.

Avioli, L. V. and Krane, S. M. (Ed.) (1977). *Metabolic Bone Disease. Volumes I and II*. New York: Academic Press.

Nordin, B. E. C. (Ed.) (1976). *Calcium, Phosphate and Magnesium Metabolism*. Edinburgh: Churchill Livingstone.

DeLuca, H. F. *Vitamin D Metabolism and Function. Monographs on Endocrinology*. Berlin: Springer-Verlag.

Norman, A. W., v Herrath, D., Grigoleit, H. G., Coburn, J. W., DeLuca, H. F., Mawer, E. B., Suda, T. (Eds) (1979). *Vitamin D. Basic Research and its Clinical Application*. Berlin: De Gruyter.

Vaughan, J. M. (1981). *The Physiology of Bone*. London: Oxford University Press.

Patterson, C. R. (1974). *Metabolic Disorders of Bone*. Oxford: Blackwell Scientific.

Smith, R. (1979). *Biochemical Disorders of the Skeleton*. London: Butterworths.

DeGroot, L. J. (Ed.) (1979). *Endocrinology*. Chapters 40–73. New York: Grune and Stratton.

19. DISORDERS OF JOINTS

ALAN MYLES

Introduction
  Structure of joints

Inflammatory arthritis
  Microbial infections and inflammatory arthritis
  Immunogenetics
  Rheumatoid arthritis
  Sero-negative polyarthritis
  Acute suppurative arthritis

Metabolic joint disease
  Gout
  Calcium pyrophosphate dihydrate deposition disease
  Hypothyroidism
  Hyperparathyroidism
  Ochronosis

Other forms of arthritis
  Arthritis associated with hyperlipoproteinaemia
  Blood diseases
  Sarcoidosis
  Osteoarthrosis

Other methods of investigation
  Synovial fluid analysis
  Crystals
  Synovial biopsy
  Arthroscopy

INTRODUCTION

The term '*arthritis*' is used loosely to describe various diseases of joints. More accurately, it should imply inflammatory joint disease as distinct from degenerative joint disease, commonly known as *osteoarthritis* or more properly as *osteoarthrosis*. This is not a good term because the bony abnormalities are a late and somewhat irrelevant feature of the disease which is primarily a disease of articular cartilage. The words 'osteoarthrosis' and 'osteoarthritis' however, are used throughout the world.

The diagnosis of most forms of joint disease is made mainly on the history and clinical findings. Useful information is gained from radiology and from laboratory investigations. It is however from haematology, histology and immunology rather than biochemistry that most information is gained apart from a relatively small number of conditions which will be discussed in detail.

Structure of Joints

There are two main types of joints: (i) The *synovial joint* (diarthrosis) in which there is a cavity separating two bones which are lined by articular cartilage. The joints are kept in contact by a capsule and by the muscles and tendons which act on the joint. Within the joint capsule there is connective tissue containing a variable amount of fat. The inner lining of the joint is the synovial membrane which is two cells thick. These cells have two functions—one is to produce synovial fluid, the other to phagocytose joint débris or other extraneous material. The fluid has two functions: to lubricate the joint and to provide nutrients to the cartilage. The synovial joints are those which have a considerable range of movement. (ii) At other joints the function is either by cartilage or fibrous tissue. These include the intervertebral discs, the sacroiliac joints and the pubic symphysis. Virtually all other joints are synovial.

INFLAMMATORY ARTHRITIS

Although the synovial joint functions more efficiently than anything man can imitate, it is structurally simple and can therefore undergo only a limited number of pathological changes. Inflammatory arthritis involves hypertrophy of the synovial lining and sub-synovial tissues associated with various patterns of inflammatory cell invasion. The clinical manifestations of this are pain, swelling, stiffness and warmth and differ little with the underlying cause. Thus examination of an individual joint with inflammatory arthritis will seldom lead to a diagnosis, which is based instead on the history, the pattern of joint involvement, the results of laboratory and radiological tests and in some cases the response to therapy.

The Relationship between Microbial Infection and Inflammatory Arthritis

It is postulated that there are four main methods by which microbial infection may lead to inflammatory arthritis.

(i) The presence in the joint of the organism, which is multiplying and causing an inflammatory response and the arthritis is directly the result of this, leading to a *septic arthritis*. Instances of this are infection with *staphylococcus aureus* or more chronically, with the tubercle bacillus.

(ii) The organism may be found within the joint but does not appear to multiply. The arthritis is self-limiting and may result from immune-complex deposition. Examples of this type are the arthritis associated with hepatitis B infection and probably with rubella and secondary syphilis. These are referred to as *post-infective arthritis*.

(iii) Inflammatory arthritis may occur a few weeks after infection, particularly by intestinal bacteria, including salmonella, shigella and yersinia. The organism is not found within the joints in which a polyarthritis occurs, which is usually self-limiting. This is referred to as *'reactive' arthritis*. Patients with this form of arthritis have a high incidence, about 80%, of the histocompatibility antigen HLA B27. It includes the post-dysenteric form of Reiter's disease and the form associated with urogenital infection. A further instance of this group is rheumatic fever which is a reactive polyarthritis associated with the Lancefield group A haemolytic streptococcus.

(iv) Forms of acute, relapsing or chronic polyarthritis in which no organism has been definitely implicated and which may possibly result from a number of different organisms. The most obvious example of this is rheumatoid arthritis, in which a possible explanation is that an infecting organism has affected host cells in such a way that they become antigenic and their progeny are antigenic, thus perpetuating an 'autoimmune disease'.[6]

Immunogenetics

Tissue typing with particular relevance to organ transplantation has lead to the recognition of surface antigens (HLA) on the sixth chromosome. These antigens are present on all nucleated cells and platelets but lymphocytes are most conveniently used, and are found on three loci called A, B and C. D-locus antigens are found mainly on B-lymphocytes. It has long been known that ankylosing spondylitis is a strongly hereditary disease but cannot be characterised in straightforward Mendelian terms. The first diagnostic use of these antigens came from the discovery that over 90% of patients with classical ankylosing spondylitis possessed HLA B27, the normal incidence being about 5%. Subsequently it has been demonstrated that some forms of sero-negative polyarthritis and reactive arthritis have an increased incidence of this antigen, although by no means all. In particular there is a normal incidence in those with rheumatic fever.

HLA B27 is present in about 80% of patients with Reiter's disease, particularly those following dysentery, and also in the reactive arthritis following yersinia, salmonella and shigella infection. There is an increased incidence in juvenile chronic polyarthritis and those that possess HLA B27 tend to develop ankylosing spondylitis.

A normal incidence of A, B and C antigens is found in rheumatoid arthritis, but recently an increased frequency of DW4 has been reported.[1]

Rheumatoid Arthritis

This is the commonest form of polyarthritis, estimated to affect between 3%–5% of the Caucasian population at some stage during their life. The exact incidence is uncertain because it varies from a mild transient arthritis to a severe progressive disease. It has no precise diagnostic features or predictable natural history as occurs with a broken bone or carcinoma. For this reason a classification is accepted using various criteria to divide the condition into possible, probable, definite and classical, which is not only clinically useful but enables the literature on the subject to be comprehensible to other doctors.[16]

Clinical Features

Pain, swelling, stiffness and ultimately deformity, with erosion of cartilage and bone may occur in any synovial joint, but these lesions seldom involve the distal interphalangeal joints or the joints of the spine with the notable exception of the cervical spine. Usually a number of joints are involved and often most of the synovial joints may be affected. It is sometimes described as a symmetrical polyarthritis but true symmetry is very unusual.

Investigations

Radiology. Periarticular osteoporosis is a non-specific finding but in progressive disease erosions occur at the junction of the hyaline cartilage to the bone.

Haematology. Anaemia is a common feature of severe and progressive rheumatoid arthritis. It has many causes and there may be more than one in the individual patient. Sometimes gastrointestinal blood loss is responsible, occuring from many of the anti-inflammatory analgesic drugs that are used, in particular aspirin, phenylbutazone and indomethacin. The corticosteroids, previously thought to be a common cause of blood loss, are very seldom responsible in the small doses that are now used. As with other chronic diseases there is a probability of an increased incidence of peptic ulceration which may itself lead to bleeding. The presence of gastrointestinal bleeding may be established by searching for occult blood in the stools, but can be done with more difficulty by labelling red cells with radioactive chromium. It should be stressed, however, that gastrointestinal bleeding from drugs, particularly from aspirin, may be intermittent. If this is a cause of the anaemia, blood films will show evidence of iron deficiency, the plasma iron concentration being low and the iron binding capacity raised. A more common cause of anaemia, however, is that associated with many chronic diseases in which the marrow is unable to utilise iron to haemoglobinise the red cells. In these, the anaemia is normochromic and normocytic, the plasma iron level is low but in distinction from true iron deficiency the iron binding capacity is also low or normal. Abundant iron is present in the marrow and also is sequestered in the synovial lining cells where it is inaccessible for use. The anaemia is seldom sufficiently severe to have considerable clinical significance, in part because it appears to be related to the severity of the disease. Those severely incapacitated are less in need of a normal haemoglobin level to pursue their usual activities. The anaemia does not respond to iron but tends to improve if the arthritis itself goes into remission. The erythrocyte sedimentation rate (ESR) is usually raised in active arthritis, and, in relatively early disease, is a reasonable guide to disease activity, but clinical examination is often more accurate.[2]

Acute phase proteins, such as C-reactive protein and fibrinogen are more reliable indices of acute inflammation, but are not always in routine use.

Liver Function Tests. About 25% of patients with rheumatoid arthritis have moderately elevated levels of alkaline phosphatase, which can be shown to be of hepatic rather than bone origin by estimations of 5'-nucleotidase or by isoenzyme separation. Gamma-glutamyl transferase activity is moderately elevated in about 75% of patients with active rheumatoid arthritis. Other liver function tests are usually normal in uncomplicated disease, with the exception of the plasma proteins, in which a slight reduction of the albumin and an increase in the globulin content are found. Bilirubin and the amino transferase levels are usually normal. Clinically significant liver disease is extremely uncommon in rheumatoid arthritis and when it occurs is usually the result of amyloidosis or is drug-induced. Minor abnormalities of liver histology have been described. The most abnormal liver function tests occur in those who also have Sjogren's disease.

Immunological Tests

The best recognised immunological abnormality in rheumatoid arthritis is the presence of rheumatoid factor, detected by the latex fixation test, the Rose-Waaler test (differential agglutination titre: DAT) and the sheep cell agglutination test (SCAT). The latex test is usually reported as negative, weakly positive or positive. If a titre is given 1:20 is regarded as the upper range of normal. The upper range of normal for the DAT is 1:16, for the SCAT 1:32.

The latex test gives the highest number of positives but also, therefore, the highest number of false positives. Whichever test is used about 30% of patients with definite rheumatoid arthritis will not be found to have rheumatoid factor present. In the general population about 4% will be positive for one of these tests. This implies that there are more people with positive tests for rheumatoid factor who do not have rheumatoid arthritis than there are patients with rheumatoid arthritis. It is therefore a valuable test in the diagnosis of a patient who has arthritis. If there is no evidence of arthritis, but it is merely used as a screening test, it will have very little value. Positive tests for rheumatoid factor may also occur in other diseases, particularly connective tissue disorders, such as systemic lupus erythematosus, systemic sclerosis and polymylositis. It may also be found in association with chronic active hepatitis, with sarcoidosis and myelomatosis. It may result from chronic infection of which bacterial endocarditis is the most common. Other conditions include chronic tuberculosis, syphilis, leprosy and malaria. Raised titres are commoner in relatives of patients with rheumatoid arthritis and the titre tends to rise with increasing age. Immunoglobulin G rheumatoid factors are now recognised but are less specific and do not yet have any routine clinical use.[13]

About 20% of sero-positive patients with rheumatoid arthritis may have other auto-antibodies, in particular antinuclear factors. They are of no diagnostic significance but tend to be present in those with higher titres of rheumatoid factor, with more severe disease, and with extra-articular manifestations of rheumatoid arthritis particularly vasculitis. DNA-binding is not increased. This and clinical features should avoid any confusion with systemic lupus erethematosus in which antinuclear factors are invariably present, but rheumatoid factor is also commonly present.

Extra-articular Manifestations of Rheumatoid Arthritis

Rheumatoid arthritis is a systemic disease and widespread manifestations occur in a small proportion of patients, more often in men than women for reasons that are not clear, whereas in rheumatoid arthritis with articular manifestations only, women outnumber men by about 3:1. Rheumatoid granuloma formation is found in the eye, particularly the sclera, the pleura, pericardium and in the lungs, in which a form of fibrosing alveolitis occurs. Periarteritis nodosa is also a complication.

Juvenile Chronic Polyarthritis

Unlike adults with rheumatoid arthritis, children, particularly those with an acute systemic onset, may have a marked polymorphonuclear leukocytosis, but rheumatoid factor is nearly always absent. In the small number who possess rheumatoid factor the disease tends to run a more severe course, similar to adult progressive rheumatoid arthritis. Antinuclear antibodies, if present in a high titre, are usually associated with those that have chronic iridocyclitis and should alert the clinical to this possibility. Those having the HLA B27 antigen tend to have mainly large joint involvement and often develop into classical ankylosing spondylitis.

Treatment of Rheumatoid Arthritis

Detailed discussion of the treatment is beyond the scope of this chapter. In early disease rest and the use of anti-inflammatory-analgesic drugs for symptomatic relief are used, although there is little evidence that these measures affect the ultimate prognosis. When progressive disease occurs the use of more specific forms of treatment, including gold, penicillamine, corticosteroids, antimalarials and cytotoxic drugs are used. Surgical treatment, particularly various forms of arthroplasty, are of considerable value in advanced disease affecting a small number of joints. For further details text books of rheumatology should be consulted.[17] Laboratory investigations are of limited value in the assessment and control of treatment in which symptoms and clinical signs are more valuable, although the ESR tends to fall and the haemoglobin level will rise as the disease goes into remission.

Investigations are, however, of considerable importance in monitoring the toxic effects of drugs, particularly the bone-marrow depressant effects of gold, penicillamine and cytotoxic drugs and the nephrotoxic effects of gold and penicillamine.

Sero-negative Polyarthritis

Various forms of inflammatory polyarthritis occur which are not rheumatoid arthritis. The precise nature of some of these remains obscure, but some are associated with other conditions, the more important of which will be considered briefly. In general they do not have any distinct laboratory abnormalities, but the diagnosis is made by the pattern of joint involvement and the association with some other disease. A group remains who are sero-negative, do not appear to be associated with any other condition, in some of whom the arthritis and its progress is very similar to rheumatoid arthritis, although they do not tend to have nodules or to develop the extra-articular complications of rheumatoid arthritis. This group is sometimes referred to as sero-negative rheumatoid arthritis.[19]

Psoriatic Arthritis

An inflammatory arthritis may occur in patients who have psoriasis. It tends to occur in those with the more severe skin lesions, but there is no temporal relationship between the development of arthritis and skin lesions, and the progress of the skin or joint disease does not influence each other. Skin involvement may be confined to the nails in which pitting or dystrophy may be seen. A strong family history of psoriasis in a patient without skin lesions is sometimes accepted as diagnostic in those who have a polyarthritis without features of any other form of arthritis.

There are five main types of arthritis:

1. An arthritis clinically indistinguishable from rheumatoid arthritis but without the presence of rheumatoid nodules or other extra-articular manifestations of rheumatoid arthritis.

2. A deforming type of polyarthritis particularly involving small joints of the hands leading to destructive changes and shortening of the bones. This results in shortening of the fingers and therefore redundancy of the skin, giving rise to a typical clinical appearance known as 'arthritis mutilans'.

3. A form of polyarthritis principally, but not entirely, involving the distal interphalangeal joint of the fingers usually associated with nail involvement by psoriasis.

4. An arthritis predominantly affecting the sacroiliac joints and spine leading to ankylosing spondylitis.

5. An arthritic involvement of a small number of joints, often interphalangeal joints, presenting as an oligoarthritis which may be intermittent.

Laboratory Investigations

1. Rheumatoid factor and other immunological abnormalities are absent but must be estimated to indicate whether there is co-existent rheumatoid arthritis and psoriatic arthritis.

2. Hyperuricaemia, usually of a minor degree, is present in some patients with psoriasis. This is related to the high nucleoprotein turnover in the skin lesions. It has no diagnostic importance but its discovery may

erroneously lead to the diagnosis of gout, particularly in those having the mono- or oligo-articular form.

3. HLA typing has shown that there is an increased incidence of B13 and B17 in uncomplicated psoriasis. There is an increased incidence of B27 in those who have arthritis, almost entirely confined to those with spondylitis.

The prognosis of psoriatic arthritis is somewhat better than that of rheumatoid arthritis. The basis of treatment is similar but clearly drugs which are known to worsen psoriasis should be avoided; this particularly applies to the anti-malarials which may lead to an exfoliative dermatitis.

Ankylosing Spondylitis

This is a form of polyarthritis principally involving the joints of the spine and the sacroiliac joints. It is more common in men than in women and usually begins in early adult life, although it may begin as a form of juvenile polyarthritis. The usual symptoms are of pain and stiffness which is marked first thing in the morning and present in any part of the spine. Rarely, peripheral joints may be involved. The arthritis usually becomes inactive over the course of a few years, but the patient may be left with permanent joint stiffness due to ankylosis of the involved joints and calcification of the ligaments. It does not have the same extra-articular manifestations as rheumatoid arthritis but acute iritis is a common occurence. Cardiac involvement leading to aortic incompetence may occur and a form of pulmonary fibrosis has been described.

Investigations

1. The ESR is usually raised in active disease but has little diagnostic value.
2. Tests for rheumatoid factor and other immunological abnormalities are negative.
3. About 95% of patients have HLA antigens of type B27. This is the strongest known association between an HLA type and disease. The presence of HLA B27 increases the risk of developing ankylosing spondylitis by about 100.

Ankylosing spondylitis may occur in its classical uncomplicated form but also occurs in association with other diseases, particularly with ulcerative colitis, Crohn's disease, psoriasis and Reiter's disease.

Reiter's Disease

The clinical features are an inflammatory polyarthritis mainly affecting large joints associated with acute conjunctivitis or more rarely with iritis. It is usually preceded, one or two weeks previously, by a non-specific urethritis. Low back pain is common. Other features include oral ulceration, balanitis, a form of nail dystrophy and keratodermia blenorrhagica. The symptoms settle down over the course of a few weeks or months. Recurrent attacks occur occasionally and a small number develop chronic manifestations which include ankylosing spondylitis.

Investigations

There are no specific abnormalities. A high ESR is usually present in the early stages. About 70% of those with peripheral arthritis belong to the HLA group B27. Those that develop ankylosing spondilitis have an even higher incidence. In the UK, non-specific urethritis is the commonest cause in which case the condition appears to be almost confined to males. A similar syndrome may occur following infection by salmonella or schigella, particularly in Scandinavia where yersinia infection is also a cause of this condition.

Arthritis in Association with Inflammatory Bowel Disease

Both ulcerative colitis and Crohn's disease may be associated with an inflammatory peripheral polyarthritis mainly affecting large joints, the severity of which is usually related to the severity of the bowel disease. The arthritis usually remits if the bowel disease can be treated satisfactorily. Ankylosing spondylitis may also be associated and may precede the bowel disease. Its progress and prognosis, however, does not depend upon that of the bowel disease, but it progresses as idiopathic ankylosing spondylitis. With peripheral arthritis alone there is no increase over the usual population in the incidence of HLA B27 but the incidence is considerably increased in those who have ankylosing spondylitis.

Acute Suppurative Arthritis

Infection by staphylococcus aureus is considerably more common than involvement by any other organism except in children under the age of two, in whom haemophilus influenzae infection is more common. In most cases the organism reaches the joint by haematogenous spread. More rarely it may result from trauma or infection of neighbouring structures. It is very rarely the result of needle aspiration. Certain factors predispose to infection including diabetes mellitus, rheumatoid arthritis, any debilitating illness and immunodeficiency states including the use of corticosteroids and immunosuppressive drugs.

Usually a single joint is involved but in rheumatoid arthritis and in staphylococcal septicaemia more than one joint may be involved. The clinical findings are of a hot painful swollen joint associated with the usual clinical features of an acute infection. The ESR is usually raised and a polymorphonuclear leukocytosis is usually present but these are not reliable findings in the diagnosis, although they may be helpful in the monitoring of treatment.

Diagnosis

Blood culture and synovial fluid examination are essential. The characteristic findings in the synovial fluid are:

1. The presence of the responsible bacteria.
2. An increased polymorphonuclear leukocyte count usually in excess of 50 000/cm.
3. A low synovial fluid glucose concentration, usually less than half the plasma value. Considerable reliance cannot, however, be placed on this finding in the presence of rheumatoid arthritis in which the synovial fluid glucose level is often less than that of the plasma.

Management

Fortunately, antibiotics freely enter the synovial fluid and the principals of treatment are, therefore, those of any bacterial infection, a suitable antibiotic or combination being given. Purulent synovial fluid is removed by needle aspiration whenever it reaccumulates and the joint is splinted. Surgical drainage is indicated if there is an inadequate response within a few days or the fluid cannot be removed by needle aspiration, which applies particularly to the hip. In a previously undamaged joint, with prompt treatment the prognosis is good.

The ultimate choice of antibiotic will depend upon the organism isolated and its sensitivity and of any known sensitivities to the antibiotics of the individual. The clinical situation may indicate that a particular organism is most likely to be responsible but it will usually be necessary to start with a combination of antibiotics. In view of the prevalence of staphylococcus aureus infection, adequate cover must be given for this organism. Full doses of cloxacillin given parenterally and gentamicin are an appropriate choice until the organism is known. If the patient is known to be pencillin-sensitive, fusidic acid may replace cloxacillin. Close co-operation between clinician and bacteriologist is essential.[4]

METABOLIC JOINT DISEASE

Gout

The inflammatory arthritis which characterises gout results from the deposition of crystals of monosodium urate in the joint cavity. The other important manifestation is the deposition of urate in the kidney and uric acid in the urinary tract. When the plasma becomes supersaturated, urate deposition is likely. Gout therefore occurs in people who have hyperuricaemia, which has been defined as 0·42 mmol/l in males and 0·36 mmol/l in females, when determined by the enzymatic spectrophotometric method. Lower levels are found in pre-pubertal/children, the mean figure being 0·22 mmol/l in both sexes. Normal uric acids however vary with age, sex, ethnic and physical differences.

Hyperuricaemia is defined as primary when the metabolic abnormality has no associated cause.

Pathogenesis of Primary Hyperuricaemia

Uric acid is the end-product of purine metabolism in man, who does not have the enzyme uricase. Endogenous and dietary purine is degraded to hypoxanthine, and subsequently to xanthine and uric acid by the enzyme xanthine oxidase. About one quarter of the daily production of uric acid is excreted into the gut, the remainder is excreted through the kidney. There is no evidence that changes in gastrointestinal loss have any importance in hyperuricaemia. Urate is freely filtered by the glomerulus, is to a large extent reabsorbed, and then actively secreted in the distal renal tubule. Recycling of purines takes place in man, the two most important enzymes involved are phosphoribosyl pyrophosphate (PRPP) synthetase and hypoxanthine-guanine-phosphoribosyl transferase (HGPRT). These enzymes convert hypoxanthine to hypoxanthine-ribose-phosphate and guanine to guanilic acid. Deficiency of HGPRT, allowing increased excretion of uric acid is now well documented and will be considered later. Overactivity of PRPP synthetase has been reported in a few individuals.[9]

On a purine-free diet the average urinary excretion of uric acid is 2·4 mmol/day and is usually less than 3·6 mmol although it can rise to 6 mmol/day on an unrestricted diet. In the great majority of patients with hyperuricaemia however, no underlying metabolic defect can be established. In about 75% of patients with hyperuricaemia, the cause is reduced uric acid excretion, the remainder overproducing uric acid, without any obvious enzymatic cause. There is little doubt than in some patients hyperuricaemia is due to a combination of underexcretion and overproduction.

Secondary Hyperuricaemia

(i) *Renal failure*. Excretion of urate falls in renal failure and there is some relationship between uric acid clearance and inulin clearance. Urate levels do not begin to rise significantly however until the glomerular filtration falls to about 20 ml/minute.

(ii) *High purine diet*. Uric acid excretion is in part related to purine intake, a high intake level leading to high uric acid excretion, which explains some of the racial differences in plasma uric acid levels.

(iii) Overproduction of nucleoprotein accompanies proliferative disorders of the haemopoetic system, including leukaemia, lymphoma, myelomatosis, haemoglobinopathies, sickle cell disease, haemolysis, and polycythaemia of any cause.

(iv) *Drugs*. The antituberculous drug pyrazinamide is a potent inhibitor of uric acid excretion. Many diuretics, including the thiazides, frusemide, ethacrynic acid and chlorthalidone, raise the plasma uric acid level because of a complex action, partly inhibition of tubular secretion, partly increased reabsorption and partly contraction of the extracellular fluid volume.

(v) *Ketosis*. Ketosis arising from starvation or diabetic coma leads to a fall in urate excretion causing hyperuricaemia.

(vi) *Essential hypertension*. Tubular secretion or urate is reduced in some patients with essential hypertension.

(vii) *Lead poisoning.* (Saturine gout). Excessive exposure to lead leads to a renal tubular defect, causing uric acid retention.

(viii) *Hypercalcaemia.* Particularly in hyperparathyroidism, the blood uric acid level tends to rise, not entirely owing to renal failure but also possibly to a specific effect of hypercalcaemia on the renal tubule.

(ix) *Psoriasis.* A moderately increased uric acid level is often found as a result of increased nucleic acid turnover in the skin lesions.

(x) *Lactic acidosis.*

(xi) *Hypothyroidism.*

(xii) *Glucose-6-phosphatase deficiency.*

(xiii) *Lesch-Nyhan Syndrome.* This is a rare X-linked disorder of male subjects due to deficiency of HGPRT. It presents in early childhood with choreoathetosis, spasticity and mental deficiency, associated with a marked tendency to self-mutilation. The plasma uric acid level is raised leading to the development of gout at an early age and an excessive urinary excretion of uric acid. The abnormality of uric acid metabolism can be controlled by allopurinol but this does not have any effect on the neurological changes. It is possible that partial deficiency of HGPRT without a full Lesch-Nyhan Syndrome may be responsible for some cases of gout.

Other Factors

(i) *Genetic.* Gout and hyperuricaemia are clearly familial. A family history will be found in about one-third of patients with gout and about a quarter of asymptomatic relatives will have hyperuricaemia. In view of the many factors involved, however, inheritance is not in a straightforward Mendelian pattern, except in those with a specific abnormality such as HGPRT deficiency.

(ii) *Body weight.* There is a relationship between obesity and hyperuricaemia, which is probably of importance in clinical gout.

(iii) *Social class and intelligence.* Gout and hyperuricaemia are more common in the higher social grades and in those with high intelligence. The precise causes are unknown but could be due to increased purine intake.

(iv) *Alcohol.* The effect of alcohol on uric acid metabolism is complex. Acute intoxication may lead to hyperuricaemia, probably as a result of ketosis and lactic acidosis. Regular alcohol ingestion tends to lead to slightly raised uric acid levels, which could in part be due to the high calorie content of the alcohol leading to obesity. It is also possible that some of those that can afford a high alcohol intake can also afford a high purine diet.

(vi) *Cardiovascular disease.* Uric acid concentration tends to be raised in atherosclerosis, but does not appear to be causally related, the raised level usually being associated with obesity, hypertension and medication.

Clinical Gout

Gouty arthritis (podagra) is about ten times more common in men than in women. The first attack usually occurs in the fourth or fifth decades. It is very uncommon in pre-menopausal women except in cases that are secondary, but with increasing age gout occurs with increasing frequency in women. The first attack is usually an acute inflammatory arthritis affecting the big toe joint. Why this joint is specific to gout is uncertain but it may be due to the decreased solubility of uric acid at low temperatures. Any joint may be involved but it is very unusual for the big toe not to be involved at some stage. The attacks frequently start early in the morning, rapidly become severe and will resolve in the course of about a week. There may be no further attacks, or more commonly attacks will occur every few months, involving one or other toe and sometimes other joints. In a small proportion a more chronic arthritis results, which may be very similar to rheumatoid arthritis. Precipitates of urate occur in the synovial membrane and intra-articular cartilage, but also appear in cartilage elsewhere, such as the ear and in connective tissue and bursae, leading to the characteristic tophi. The olecranon bursa is frequently involved but deposits may also occur in almost any situation, particularly on the Achilles tendon and finger tips. The tophi tend to be pale yellow, but if they discharge, the material which is sodium urate will be white, which should enable it to be distinguished from pus or chronic rheumatoid material.

Uric Acid Calculi. The percentage of patients with gout who develop uric acid calculi differs throughout the world, but it is about 10% in Great Britain. At a pH of 5, uric acid is precipitated in distinction from sodium urate which is the material found at other sites. Some uric acid stones are mixed and may contain calcium, the majority however do not and are therefore radiolucent. They are usually small but may cause renal colic. The tendency to uric acid stone formation can be decreased by a high fluid intake and by taking sodium bicarbonate in order to raise the pH of the urine.

Renal Disease. Renal disease may be secondary to obstruction from uric acid stones or infection related to them. It may also occur as a result of urate in the kidney where it is found particularly in the medulla and pyramids. Vascular nephrosclerosis and chronic pyelonephritis may also occur. Renal involvement is usually associated with hypertension, a mild impairment of renal function and proteinuria.

Diagnosis of Gout. Since gout is due to hyperuricaemia it is clearly necessary to demonstrate a raised plasma uric acid concentration. It has been said that about 1% of patients may have a normal uric acid level at the time of an acute gouty attack. In practice the repeated demonstration of a normal uric acid level will virtually exclude the diagnosis of gout, except in certain circumstances, mainly involving the use of drugs. A patient who is taking uricosuric drugs or a xanthine oxidase inhibitor may have a normal uric acid level. Other drugs, notably phenylbutazone, have a mild uricosuric effect; their use for gouty attacks may lead to a normal plasma uric acid level. Salicylates taken in small doses tend to raise the plasma level of uric acid but, if taken in maximum therapeutic quantities, they have a uricosuric effect and may therefore lower it. The presence of a raised uric acid level however, does not necessarily confirm the diagnosis of gout. Other routine laboratory investigations are usually normal although there may be a slight elevation of the ESR. If renal involvement is present there will be some evidence of renal insufficiency. In chronic gout X-rays may demonstrate erosive changes which, although classically described as being punched-out areas in subchondral bone, have no features which reliably differentiate them from other causes of erosive arthritis.

Twenty-four hour urinary uric acid estimations, particularly when done on a low purine diet, are of value in differentiating those who are oversecretors and those who are undersecreters of uric acid.

The proof of gouty arthritis is by the finding of urate crystals in synovial fluid. Synovial fluid cannot however always be obtained and diagnosis may have to be made on the basis of the history, clinical findings, a persistently raised plasma uric acid and the absence of evidence of any other polyarthritis. It is particularly important to exclude psoriatic arthritis which may be clinically similar and may be associated with a raised plasma level of uric acid. Other conditions which may mimic gout are pseudo-gout, trauma and infection.

Management of Acute Gout

Acute gout is an extremely painful condition, which can be relieved fairly rapidly by prompt treatment. Colchicine is honoured by time but very little else. The usual advice was to start with 1 mg and then give 0·5 mg every two hours until the attack subsides, or more frequently until the onset of severe diarrhoea. It is often effective but no more so than other less unpleasant remedies. It is often supposed to have some diagnostic significance, since most other forms of acute arthritis do not respond to colchicine, but like most other therapeutic tests, it is unreliable. Phenylbutazine is the most effective drug and should be given in a dose of 200 mg, three to four times daily, until the attack subsides, after which it may be rapidly withdrawn over the course of a few days. It should be remembered that this dose is uricosuric and that blood should be taken before starting phenylbutazone if measurement of the uric acid level is to be helpful. Indomethacin is almost as effective taken in a dose of 50 mg, four times daily, but it is more inclined to cause toxic effects on the stomach and central nervous system. Other anti-inflammatory analgesic drugs such as naproxen, 750 mg daily, are useful but probably less effective. If joint fluid is aspirated, more rapid relief may be obtained by giving a local corticosteroid injection at the same time. Systemic corticosteroids or ACTH also provide rapid relief but are seldom needed, except in a small number of patients who do not respond to the anti-inflammatory analgesics or who are known to be intolerant of them.

Long-term Management of Gout

For the patient who has no evidence of renal involvement or stones and only a moderately elevated uric acid concentration, it is necessary to treat only the acute attacks. The patient should be provided with a supply of one of the drugs listed above which has been found to be effective and told to take it as soon as he is aware of the first signs of an acute attack. If the attacks are somewhat more frequent they can sometimes be minimised by taking colchicine regularly in a dose of 0·5 mg, two or three times daily, which seldom causes side-effects. If attacks become frequent some method of lowering the plasma uric acid concentration is desirable. It becomes essential in the following situation:

(i) Gout associated with chronic joint changes or tophi.
(ii) Evidence of renal damage.
(iii) Persistently raised uric acid levels above 0·5 mmol/l.

Drugs Lowering Uric Acid

The aim of either method is to reduce the plasma uric acid to normal levels.[7]

(i) *Uricosuric drugs.* In the United Kingdom probenecid is the drug most commonly used, the dose varying from 1–3 g daily. An alternative drug is sulphinpyrazone, 100 mg three to four times daily. The dose used should be that which maintains the plasma uric acid level within the normal range. Both of these drugs are remarkably free from side-effects. Very occasionally dyspepsia and rashes occur. The nephrotic syndrome has been attributed to the use of probenecid in two cases. Aspirin interferes with their uricosuric action and should not be given concurrently.

(ii) *Xanthine oxidase inhibition.* Allopurinol inhibits the enzyme xanthine oxidase and also suppresses purine biosynthesis at other levels. There is thus an increased excretion of xanthine and hypoxanthine, in the urine. Allopurinol is metabolized to oxipurinol which is itself a weak xanthine oxidase inhibitor. The increased excretion of xanthine and hypoxanthine might be expected to increase the incidence of xanthine and hypoxanthine stones. Fortunately, these substances are more soluble than uric acid. Stone formation has not occurred in the treatment of gout but it has been reported in the Lesch-Nyhan Syndrome and in the treatment of hyperuricaemia associated with lymphosarcoma. Hypoxanthine, xanthine and oxipurinol crystals have been found in skeletal muscle biopsies but do not appear to have any clinical significance. Allopurinol is given in a single daily dose, starting with 100 mg daily and increasing by 100 mg at weekly intervals until the uric acid level is controlled, which it does rapidly. The usual dose is between 300–600 mg daily. Occasional patients can be managed on 200 mg daily; a small proportion require higher doses than 600 mg daily. Allopurinol is well tolerated but may occasionally produce unpleasant skin rashes which unfortunately occur more frequently in those with renal failure in whom allopurinol is particularly useful. The decision on which method of lowering the uric acid level should be used is difficult in the uncomplicated case. Probenecid has been used for considerably longer, appears to be very safe, does not have the metabolic consequences of allopurinol and might therefore be considered to be the drug of first choice, except in particular situations mentioned below. It might be thought that those who are underexcretors of uric acid should go on to uricosuric drugs and those who are overexcretors should have allopurinol, but there is no clear evidence for this assumption in practice. During the first few months of treatment, acute attacks of gout may occur with considerably greater frequency than before. This should be anticipated and suppressed by giving phenylbutazone 300 mg daily, indomethacin 75 mg daily or colchicine 1 mg daily during this time.

Indications for Preferring Allopurinol to Uricosuric drugs

(i) *Extensive tophacious gout.*
(ii) *Gross overproduction of uric acid.*
(iii) *Gout associated with renal failure or renal stones.*
(iv) *Acute uric acid nephropathy.* This occurs when leukemias, lymphomas or large tumour masses are treated by cytotoxic drugs or radiotherapy, leading to excessive breakdown of nuclear protein, which may cause excessively high plasma uric acid levels. A figure of 5·4 mmol/l has been recorded. This should be anticipated by starting allopurinol before treating such patients at high risk.
(v) *Intolerance of uricosuric agents or inability to provide adequate uric acid control.*

In general, allopurinol treatment is more rapidly effective. Gouty deposits and tophi may be expected to disappear and erosive lesions may heal. Unfortunately, there is no evidence that urate nephropathy is improved by the use of drugs which lower the plasma uric acid. In resistant cases allopurinol and uricosuric treatment may be combined.

Other Measures. Gouty patients are often overweight and this itself increases the plasma uric acid concentration and tendency to hypertension. Weight loss should therefore be attempted. Severe dietary restriction of foods with a high purine content and restriction of alcohol intake should achieve a modest reduction in the plasma level of uric acid but is seldom tolerated by the patient or sufficient to be clinically useful. Nevertheless, the patient should be warned that excessive indulgence in either of these may precipitate acute gouty attacks in otherwise well controlled subjects. 'Crash' diets may lead to ketosis and precipitate attacks of gout, and should therefore be avoided.

Symptomless Hyperuricaemia

Some symptomless people have raised plasma uric acid levels without ever developing evidence of acute gouty arthritis or renal involvement. No treatment is required except in those with very high uric acid levels, an arbitrary figure of 0·54 mmol/l being taken, in which case treatment may prevent complications of hyperuricaemia.[10]

Some people like to have regular 'screening tests' and are found to have a moderately elevated uric acid. If they also have some unrelated musculo-skeletal pain such as 'tennis elbow' or a shoulder capsule lesion, they may be submitted to quite unnecessary treatment to lower the uric acid. Presumably those with hyperuricaemia who do not develop manifestations of gouty arthritis or nephropathy possess, by some unknown means, a greater ability to keep uric acid dissolved in the plasma.[18]

Calcium Pyrophosphate Dihydrate Deposition Disease

Calcium pyrophosphate dihydrate (CPPD) may be precipitated in fibro- and hyaline cartilage and may migrate into joints causing a crystal synovitis. The results of this have been referred to as 'pseudo-gout' and chondrocalcinosis, but neither of these terms is adequate. Chondrocalcinosis is a radiological diagnosis, which may not be present in some cases of crystal synovitis and 'pseudo-gout' attacks do not always occur in this condition. The somewhat cumbersome term calcium pyrophosphate deposition disease is, therefore, used. It is associated with the condition listed in Table 1.

TABLE 1
DISEASES ASSOCIATED WITH CPPD DEPOSITION

Commonly	Hyperparathyroidism
	Haemochromatosis
	Gout
Infrequently	Hypothyroidism
	Diabetes mellitus
	Hepatolenticular degeneration
	Ochronosis
	Hypophosphatasia
	Old age

The first described cases of CPPD deposition disease appeared to be familial but this does not now seem to be common. Most cases are not associated with any other discoverable disease. The most definite associations are with haemochromatosis, hyperparathyroidism and gout.

The clinical manifestations are very varied but may be broadly classified into the following:

1. 'Pseudo-gout'. The condition is very similar to gout in that a severe inflammatory arthritis involving a single joint usually occurs. The knee joint, however, is

by far the most commonly involved joint and the large toe is very uncommonly involved. More than one joint may be involved. It may be of sufficient severity to suggest a septic arthritis.

2. A subacute polyarthritis somewhat resembling rheumatoid arthritis clinically but without the distinguishing features of rheumatoid arthritis.

3. Joint symptoms resembling osteoarthrosis but of a more widespread nature involving wrists, metacarpophalangeal joints, shoulders and ankles, in addition to the joints affected by osteoarthrosis. Some of these patients also have attacks of 'pseudo gout'.

4. Changes similar to a neuropathic arthropathy occasionally occur.

Diagnosis

Chondrocalcinosis may be diagnosed radiologically and is most commonly seen in the knee. Calcification in the fibrocartilage of the menisci only is a common finding in old age and is probably not related to CPPD but to deposition of hydroxy-apatite. Hyaline calcification, seen particularly in the lateral radiographs of the knee, is the most common finding. It is also seen in the hip, shoulder, pubic synthesis, the triangular ligament of the wrist and in the intervertebral discs. Erosive changes may occur and are most commonly seen in the metacarpophalangeal joints. If erosive changes are found in these joints in the absence of evidence of obvious inflammatory arthritis CPPD disease should be strongly considered. The diagnosis can be confirmed by identifying cystals of CPPD in synovial fluid, discussed in the section on synovial fluid analysis.

Treatment

The acute attacks of arthritis usually respond to anti-inflammatory analgesics such as phenylbutazone or indomethacin and to joint aspiration and local corticosteroid injection. If an underlying disease is found it may clearly require treatment but this unfortunately does not appear to affect the natural history of CPPD disease.[12]

Hypothyroidism

Cases of hypothyroidism are seen regularly in rheumatological clinics and present in the following ways:

1. Carpal tunnel compression—there is a definitely increased incidence in hypothyroidism. The literature usually relates that the symptoms settle with adequate replacement treatment, but that has not been the author's experience. Initial improvement or apparent cure is usual with local corticosteroid injections but about a third ultimately require surgical treatment.

2. Myopathy—this usually presents with pain and stiffness and some weakness affecting proximal muscles, particularly the thighs. Apart from other tests of thyroid hypofunction, the creatine kinase activity is frequently raised to a moderate degree, and in doubtful cases of polymyositis the diagnosis of hypothyroidism should always be considered. It may be distinguished by muscle biopsy which shows characteristic changes of inflammation in polymyositis, and by electromyography which will usually show fibrillation in addition to an excess of short-duration polyphasic potentials in polymyositis. The myopathy invariably responds to adequate thyroxine replacement.

3. Hypothyroid arthritis—somewhat nonspecific joint symptoms may occur in hypothyroidism which respond to adequate replacement. Sometimes a progressive polyarthritis usually affecting a small number of joints occurs and may be associated with CPPD. The effect of replacement treatment is not well documented but has not, in my experience, affected the arthritis.

A raised plasma uric acid, without evidence of gout, occurs in a small proportion.[5]

Hyperparathyroidism

Hyperparathyroidism may lead to skeletal pain as a result of bone disease, but may also lead to a polyarthritis which is somewhat like rheumatoid arthritis but is sero-negative. More commonly it produces joint disease as a result of CPPD deposition.

Ochronosis

Ochronosis is the description given to the syndrome arising from the absence of the enzyme homogentistic acid oxidase, which is normally present in the liver and kidney. Homogentistic acid is an intermediate metabolite in the breakdown of phenylalanine and tyrosine. The condition is transmitted as an autosomal recessive gene. The absence of homogentistic acid oxidase results in an excessive accumulation of homogentistic acid.

Homogentistic acid is passed in the urine which, if acid, is of a normal colour. If it is alkaline, or alkali is added, the characteristic dark brown appearance occurs. This also occurs slowly after standing due to oxidation.

The main clinical features are of degenerative joint disease, particularly affecting large joints and the spine. Deposits of homogentistic acid are found in both the annulus fibrosis and the nucleus pulposus of the intervertebral discs. Degenerative changes occur, usually in middle age, with narrowing of the joint spaces and posterior osteophyte formation. X-rays show calcification of the intervertebral discs. It has been demonstrated that this is due to the deposition of hydroxy-apatite. Fusion of the vertebral bodies may occur but, unlike ankylosing spondylitis, involvement of the sacroiliac joints does not occur.

The clinical manifestations are the development of a kyphus or kyphosis and limitation of spinal movement. Peripheral joints may be involved, particularly the knee where chondrocalcinosis may be seen. Pigmented deposits also occur in the cartilage of the ears, of the nose and also in the sclera. Skin pigmentation, particularly where there are excessive glands, of a bluish-brown nature may be observed.

The diagnosis is made by the identification of homogentistic acid in the urine, the colour change resulting from this is known as alkaptonuria.

Although it has some striking manifestations and is a well-defined condition, its clinical importance is slight and there are no specific forms of treatment.[14]

OTHER FORMS OF ARTHRITIS

Arthritis Associated with Hyperlipoproteinaemia

Apart from the xanthomata which may appear in tendons, over and near to joints, which are usually painless, a flitting polyarthritis somewhat resembling rheumatic fever may occur with type 2 hyperlipoproteinaemia.[15]

In type IV hyperlipoproteinaemia joint pain and stiffness is not infrequently a symptom and a number of patients develop a persistent synovitis, particularly involving large joints. X-rays may show fairly large periarticular cysts. The synovial fluid or synovial membranes show no diagnostic features. Gradual symptomatic improvement has been reported with appropriate therapy. Type IV hyperlipoproteinaemia is also associated with hyperuricaemia but there has been no evidence of urate crystals found in the synovial fluid.[3,11]

Blood Diseases

Haemophilia

Haemophilia and Christmas Disease commonly cause haemarthrosis which, if repeated, may lead to degenerative joint disease. Acute haemarthrosis may occur as a result of minor trauma or spontaneously, and usually affects large joints, particularly the knees. Those in whom joint involvement is common usually have less than 2% of the normal level of anti-haemophilic globulin. The principles of the treatment are to administer human anti-haemophilic globulin, followed by joint aspiration.

Leukaemia

Acute leukaemia, particularly in children, frequently presents with musculo-skeletal pain. It may present as a polyarthritis and is a frequent source of diagnostic difficulty.

The possibility of gout, secondary to leukaemia, should be considered.

Sarcoidosis

Inflammatory polyarthritis may occur in sarcoidosis, most commonly in the form that is associated with erythema nodosum and pulmonary hilar gland enlargement. It is usually a polyarthritis affecting a small number of large joints settling down without any permanent joint damage over a few weeks or months. It usually requires the use of anti-inflammatory analgesic drugs and, very occasionally, corticosteroids.

Laboratory Investigations

1. The ESR is usually moderately raised.
2. Hypercalcaemia is a very unusual complication of this form of sarcoidosis.
3. Hyperuricaemia has occasionally been recorded.
4. The Kveim test will be positive but is not usually necessary for the diagnosis which is made mainly on the clinical features associated with hilar gland enlargement on chest X-ray.
5. Synovial biopsy. Non-specific inflammatory changes only are found.

Occasionally in chronic sarcoidosis there may be progressive involvement of the joint(s) in which synovial biopsy shows the presence of typical sarcoid granulomata.

Osteoarthrosis

Osteoarthrosis (osteoarthritis) is a degenerative condition of joints primarily affecting cartilage, mainly of weight-bearing joints, including the joints of the spine. If there is an underlying cause which can be recognised it is referred to as secondary osteoarthrosis, otherwise it is known as primary osteoarthrosis.

Secondary osteoarthrosis may arise as a result of a congenital or acquired abnormality of joint structure and in many disorders of bone, cartilage or connective tissue, including all forms of hypermobility. It may result from a severe or repeated trauma. It may follow any acute or chronic inflammatory form of arthritis. It occurs in some metabolic disorders including ochronosis and calcium pyrophosphate crystal disease, considered elsewhere, and also in acromegaly.

There are no laboratory investigations which are helpful in the diagnosis of osteoarthrosis itself.

OTHER METHODS OF INVESTIGATION

Synovial Fluid Analysis

Normal synovial fluid is clear, almost colourless, viscous and of about the consistency of 'golden syrup'. It contains up to 100 cells/cm which are mainly mononuclear; it also contains hyaluronic acid and lubricating glycoproteins. In the presence of inflammation the viscosity decreases and the cell count increases. In virus arthritis the cells remain predominantly mononuclear. In all other forms of inflammatory arthritis the cells are polymorphonuclear leucocytes regardless of whether the cause is a non-infective arthritis or a bacterial arthritis. In bacterial arthritis, however, the cell count is often above 50 000/cm whereas it is usually lower than this in nonbacterial inflammatory arthritis. The two most important uses of synovial fluid analysis are to detect the presence of bacterial arthritis and crystal synovitis. Synovial joints clearly possess effective protection against bacterial infection which explains the very low incidence of septic arthritis induced by needle exploration and the difficulty in cultivating bacteria from septic joints. Meningococci and gonococci are particularly difficult to culture and immediate plating of the synovial fluid is necessary. The normal synovial fluid glucose is approximately the same as the plasma glucose. In the presence of bacterial infection the glucose content of the synovial fluid falls but it also falls in rheumatoid arthritis and is therefore not a very useful test. Recently a raised concentration of lactic acid has been demonstrated in septic arthritis.

The synovial fluid complement level is low in rheumatoid arthritis, suggesting immune-complex deposition, but may be reduced in other forms of inflammatory arthritis. Inclusions in synovial fluid phagocytes are common. The finding of large cells containing a number of phagocytosed neutrophils is suggestive of Reiter's disease.

Rheumatoid factor is frequently present in synovial fluid in rheumatoid arthritis but its presence in seronegative rheumatoid arthritis may help to confirm diagnosis at an early stage. The only other frequently encountered abnormal constituent is blood which may indicate a bleeding disorder, trauma or pigmented villonodular synovitis.

Crystals

With the compensated polarising microscope, calcium pyrophosphate crystals show weakly positive birefringence and tend to be more pleomorphic than urate crystals which are needle shaped. With a first order red compensater they show a faint blue colour against a red background if the axis of the crystal is parallel to the compensater. X-ray diffraction crystallography can also be used, but is not in routine clinical practice. The greater the number of crystals that are found indicates their increasing significance, particularly if they are intracellular. Urate crystals are needle shaped and show a strong negative birefringence and appear blue against a red background when the long axis of the crystal lies parallel to that of the compensater.[8]

Other crystals may be found in synovial fluid but do not have the same precise characteristics. These include cholesterol and cartilage fragments. Calcium oxalate crystals and lithium heparin may be seen if specimens have been put into tubes containing these substances. Slowly absorbed corticosteroids are crystalline and triamcinolone hexacetonide may be similar to CPPD. More recently hydroxy-apatite crystals have been found in osteoarthrosis but are too small to be seen by conventional microscopy.

Synovial Biopsy

Open synovial biopsy has long been used in the diagnosis of arthritis and has been particularly useful in the case of tuberculous arthritis, some other forms of septic arthritis, pigmented villonodular synovitis, and the rare synovial tumours. More recently percutaneous synovial biopsy has been possible, by which small fragments of synovial tissue can be fairly easily obtained using a technique which is only slightly more complicated than a needle aspiration of a joint. It is useful, particularly in the diagnosis of monoarticular arthritis of accessible joints, such as the knee. It may be helpful in distinguishing chronic inflammatory arthritis from other joint disease. The typical features of rheumatoid arthritis are often present early in the disease, although they cannot always be distinguished with certainty from other similar arthritic diseases, such as psoriatic arthritis and Reiter's disease.

Arthroscopy

Arthroscopic examination of the interior of the joint associated with biopsy is an extremely valuable technique and enables abnormal areas to be biopsied with certainty. Unfortunately, at the present time only the knee joint is easily examined by this method.

REFERENCES

1. Albert, D. A. (1977), 'The major histocompatibility complex in man.' *Clin. Rheum. Dis.*, **3**, 175–197.
2. Bennett, R. M. (1977) 'Haematological changes in rheumatoid disease.' *Clin. Rheum. Dis.*, **3**, 433–458.
3. Buckingham, R. H., Bole, G. G. and Basset, D. R. (1975), 'Polyarthritis associated with type IV hyperlipoproteinemia.' *Arch. Int. Med.*, **135**, 286–290.
4. Clarke, J. T. (1978), 'The antibiotic therapy of septic arthritis.' *Clin. Rheum. Dis.*, **4**, 133–153.
5. Dorwart, B. B. and Schumacher, H. R. (1975), 'Joint effusions, chondrocalcinosis and other rheumatic manifestations in hypothyroidism.' *Am. J. Med.*, **59**, 780–790.
6. Dumonde, D. C. and Steward, M. W. (1978), 'Copeman's textbook of the rheumatic diseases.' 5th Ed., pp. 222–252. Edinburgh: Churchill Livingstone.
7. Fox, I. H. (1977), 'The hyperuricaemic agents in the treatment of gout.' *Clin. Rheum. Dis.*, **3**, 145–159.
8. Gatter, R. A. (1977), 'Use of the compensated polarising microscope.' *Clin. Rheum. Dis.*, **3**, 91–103.
9. Kelley, W. N. (1977), 'Inborn errors of purine metabolism.' *Arth. Rheum.*, **20, Suppl.**, 221–228.
10. Klineberg, J. R. (1977), 'The management of asymptomatic hyperuricaemia.' *Clin. Rheum. Dis.*, **3**, 159–169.
11. Lewis, B. (1976), 'The hyperlipidaemias,' p. 220. Oxford: Blackwell Scientific Publications.
12. McCarty, D. J. (1977), 'Calcium pyrophosphate dihydrate crystal deposition disease (Pseudogout syndrome). Clinical aspects.' *Clin. Rheum. Dis.*, **3**, 61–86.
13. Maini, R. N., Glass, D. N. and Scott, J. T. (1977), 'Immunology of the rheumatic disease,' pp. 53–71. London: Edward Arnold.
14. O'Brien, W. M., La Du, D. N. and Bunin, J. J. (1963), 'Biochemical, pathologic and clinical aspects of alcaptonuria, ochronosis and ochronotic arthropathy.' *Amer. J. Med.*, **34**, 813–39.
15. Rooney, P. J., Third, J., Madcur, M. M., Spencer, D. and Dick, W. C. (1978), 'Transient polyarthritis associated with familial hyperbetalipoproteinaemia: a four year follow-up study.' *Quart. J. Med.*, **47**, 249–259.
16. Ropes, M. W., Bennett, G. A., Cobb, S., Jacox, R. and Jessar, R. A. (1959), 'Revision of diagnostic criteria in rheumatoid arthritis.' *Bull. Rheum. Dis.*, **9**, 175.
17. Scott, J. T., (1978), 'Copeman's textbook of the rheumatic diseases.' 5th Ed., pp. 391–508. Edinburgh: Churchill Livingstone.
18. Scott, J. T., (1978), 'Copeman's textbook of the rheumatic diseases.' 5th Ed., pp. 647–683. Edinburgh: Churchill Livingstone.
19. Wright, V. and Moll, J. M. H. (1976), 'Seronegative polyarthritis,' pp. 1–80. Amsterdam, New York, Oxford: North Holland Publishing Company.

20. DISORDERS OF CONNECTIVE TISSUE

ALAN MYLES

Mucopolysaccharidoses

Disorders of collagen and elastin
  Marfan syndrome
  Ehlers-Danlos syndrome
  Pseudoxanthoma elasticum
  Hypermobility syndrome
  Homocystinuria
  Hyperlysinaemia

Other 'connective tissue disorders'
  Systemic lupus erythematosus
  Scleroderma
  Polymyositis and dermatomyositis
  Mixed connective tissue disease

Arteritis
  Polyarteritis nodosa
  Giant cell arteritis

The term *connective tissue disease* usually implies those conditions, predominantly inflammatory, in which involvement of connective tissue is part of the disorder, such as systemic lupus erythematosus. The term might more properly be used to imply those conditions which have an identifiable abnormality of connective tissue only, which can be classified into the following groups:

MUCOPOLYSACCHARIDOSES

These are summarised in Table 1 and are dealt with in more detail in Chapter 33. Some of these have been shown to result from congenital deficiency of lysosomal enzymes which leads to accumulation of the relevant mucopolysaccharide. Attempts to overcome the deficit by infusion of the necessary lysosomal enzyme have been shown to be effective in reducing their accumulation, but are not practical for long-term treatment. Experimental work, however, has shown that the transplantation or injection of normal fibroblasts which are histocompatible may not only lead to biochemical improvement but in some instances induce clinical improvement. This work remains in an experimental state.

DISORDERS OF COLLAGEN AND ELASTIN

There are many of these, some of which have been fully characterised both clinically and biochemically. They are mainly hereditary and there is often a Mendelian dominant pattern.

Marfan Syndrome

This is characterised by an abnormality of body configuration with long extremities and arachnodactyly. The arm span thus exceeds the body height. Skeletal abnormalities include kyphoscoliosis, pectus excavatum, a high arched palate and prognathasism. Dislocation of the lens is a common feature. Dilatation of the ascending aorta occurs and may lead to aortic dissection. It appears to be primarily an abnormality of collagen. The only chemical abnormality that is found is an increased urinary excretion of hydroxyproline.

Ehlers-Danlos Syndrome

This condition is characterised by generalised laxity of the ligaments and hyper-extensibility of the skin which leads to a tendency to bruise into the skin.

Pseudoxanthoma Elasticum

Excessive elastin is present. The skin appears coarse and thickened with a yellowish appearance giving its name pseudoxanthoma. It may be associated with angioid streaks in the retina which may lead to degenerative choroidoretinitis. Early atherosclerosis is common, and as with other diseases of this variety, there is increased laxity of ligaments.

Hypermobility Syndromes

There are many less well-defined syndromes of hypermobility which are not associated with other obvious features. The diagnosis is made on excessive mobility of joints, particularly the thumb, elbow and knees. A minor degree of hypermobility may lead to early degenerative joint disease involving peripheral joints, and also the spine, sometimes known as the 'loose back' syndrome.[3]

Homocystinuria

Homocystinuria is a metabolic defect that has some of the symptoms of Marfan Syndrome. It is due to the hereditary deficiency of the enzyme cystathionine synthetase. It is diagnosed by the detection of homocystine in the urine. It is treated by giving a low methionine diet with which some improvement has been recorded.

Hyperlysinaemia

The main features of this syndrome are hypermobility of joints and mental retardation and it has been shown to be associated with an excess of lysine in the blood, urine and cerebrospinal fluid. It is presumably due to an enzyme defect which has not yet been identified.

OTHER 'CONNECTIVE TISSUE DISORDERS'

In these conditions, the main clinical and pathological effects are on connective tissue structures in which

TABLE 1
MUCOPOLYSACCHARIDOSES. (AFTER MCKUSICK)[5]

	Name of disease	Major clinical features	Genetic pattern	Urinary micro-polysaccharide	Enzyme deficiency
MPS IH	Hurler syndrome	Early clouding of cornea, grave manifestations, death usually before age 10	Homozygous for MPS IH gene	Dermatan sulphate Heparan sulphate	α-L-iduronidase
MPS IS	Scheie syndrome	Stiff joints, cloudy cornea, aortic regurgitation, normal intelligence, ? normal life span	Homozygous for MPS IS gene	Dermatan sulphate Heparan sulphate	α-L-iduronidase
MPS IH/S	Hurler–Scheie compound	Phenotype intermediate between Hurler and Scheie	Genetic compound MPS IH and IS genes	Dermatan sulphate Heparan sulphate	α-L-iduronidase
MPS IIA	Hunter syndrome, severe	No clouding of cornea, milder course than in MPS IH, death usually before age 15	Hemizygous for X-linked gene	Dermatan sulphate Heparan sulphate	Sulpho-iduronide sulphatase
MPS IIB	Hunter syndrome, mild	Survival to 30's to 50's, fair intelligence	Hemizygous for X-linked allele for mild form	Dermatan sulphate Heparan sulphate	Sulpho-iduronide sulphatase
MPS IIIA	Sanfilippo syndrome A	Identical phenotype	Homozygous for Sanfilippo A gene	Heparan sulphate	Heparan sulphate sulphatase
MPS IIIB	Sanfilippo syndrome B	Mild somatic, Severe central nervous system effects	Homozygous for Sanfilippo B (at different locus)	Heparan sulphate	N-acetyl-α-D-glucosaminidase
MPS IV	Morquio syndrome (probably more than one allelic form)	Severe bone changes of distinctive type, cloudy cornea, aortic regurgitation	Homozygous for Morquio gene	Keratan sulphate	? Chondroitin sulphate sulphatase
MPS V	(Vacant (now MPS IS))				
MPS VIA	Maroteaux–Lamy syndrome, severe form	Severe osseous and corneal change, normal intellect	Homozygous for M-L gene	Dermatan sulphate	Arylsulphatase B
MPS VIB	Maroteaux–Lamy, mild form	Mild osseous and corneal change, normal intellect	Homozygous for allele at M-L locus	Dermatan sulphate	Arylsulphatase B
MPS VII	β-glucuronidase deficiency (more than one allelic form?)	Hepatosplenomegaly, dysostosis multiples, white cell inclusions, mental retardation	Homozygous for mutant gene at β-glucuronidase locus	Dermatan sulphate	β-gluconidase

inflammatory and immunological abnormalities are common.

Systemic Lupus Erythematosus

Systemic lupus erythematosus (SLE) may affect almost any tissue. Before precise laboratory tests were available or it had been categorised clinically, it was not surprising that SLE was considered as a possible diagnosis in almost any obscure generalised condition, whereas it is now a disease which can be diagnosed with considerable certainty both on clinical and laboratory evidence. Although not common, it is by no means a rare disease. It affects women about ten times as often as men and usually starts in the child-bearing age group. The precise cause remains uncertain but it is highly probable from family studies that there are strong genetic factors. It is particularly frequently found in negroes of the north American continent. Those that belong to the HLA group B8 have an increased risk of developing SLE and it also appears to be more common in those who have inherited deficiencies of complement. The most striking laboratory abnormalities are immunological and the most specific is the presence of antibodies against double-stranded deoxyribonucleic acid (DNA). In view of the similarities of this antigen with some virus antigens it is conjectured that a virus infection may be the trigger to an autoimmune disease. The high incidence in females, the deleterious effect of pregnancy and of oestrogen-containing contraceptive pills suggests that hormonal influences may be important.

Clinical Manifestations

Cutaneous. There are many cutaneous manifestations of which the butterfly rash on the face is the most well known. Chronic discoid lesions of lupus may occur but do not necessarily imply the presence of systemic disease. Other common and important manifestations include photosensitivity, alopecia and Raynaud's phenomenon. Cutaneous vasculitis may be seen particularly on the hands, and elsewhere in the form of livido reticularis.

Musculo-skeletal. Joint pain or swelling is the most common manifestation of SLE. It usually mimics a mild form of rheumatoid arthritis but it very rarely becomes erosive. A deforming arthritis of the hands with swan neck deformities, ulnar deviation of the metacarpophalangeal joints, and joint subluxation may occur and is the same as the rare post-rheumatic form of arthritis described by Jaccoud. Avascular necrosis of bone, particularly of the head of the femur is not unusual, and is often attributed to corticosteroid therapy, but there have certainly been cases which are unrelated to treatment and may represent vasculitis. A small number have an inflammatory myositis.

Neuro-psychiatric. These complications are increasingly being recognised and are of great importance because they are difficult to diagnose and carry a poor prognosis. The most common presentations are depression, psychosis or fits, although almost any central nervous condition may be simulated. The diagnosis is extremely hard to make, even in a patient who is known to have SLE. The distinction between corticosteroid-induced depression, infection possibly related to corticosteroid and cytotoxic treatment, or actual SLE involvement of the central nervous system may be very difficult.

Renal. Pathologically the kidney may be involved in three main ways. Focal nephritis is the most common lesion and carries a reasonably good prognosis and may be reversible. The membranous type usually presents as the 'nephrotic syndrome'. The diffuse proliferative type carries the worst prognosis and is usually associated with hypertensive disease.

Chest. Recurrent attacks of pleural inflammation are common. Serious impairment of lung function is unusual although most patients with SLE can be demonstrated to have abnormalities of diffusion and ventilation. A striking radiological feature is the progressive elevation of the diaphragms.

Blood. The two most important complications are haemolytic anaemia, which is usually direct Coombs positive, and thrombocytopenic purpura which usually behaves in a similar way to idiopathic thrombocytopenic purpura. Other haematological abnormalities will be detailed under the investigations.

Cardiac. Pericarditis is relatively common but seldom has severe consequences. Myocardial disease may occur but is difficult to diagnose. Valve lesions are uncommon although aortic involvement has been described. The well-known Libman-Sachs endocarditis is rare and has little clinical importance. It is usually diagnosed at autopsy.

Prognosis

The prognosis of SLE is very much better than was originally thought and the fifteen-year survival is about 50%. The main reason for the apparently better prognosis is the increasing recognition of milder forms of SLE as the result of better clinical diagnosis and sophisticated immunological tests which have also lead to improvement in treatment. The mortality occurs mainly in those with renal or neuropsychiatric involvement.

Investigations

A normocytic normochromic anaemia is often present, particularly during exacerbations of the disease. Leukopenia is common and affects both neutrophils and lymphocytes. A mild degree of thrombocytopenia is also frequently present.

The ESR is usually elevated, particularly during active periods of disease, but is not a reliable index.

Minor coagulation defects are frequently reported which appear to be caused by circulating anticoagulants, although their presence seldom leads to a haemorrhagic tendency.

False Positive Tests for Syphilis. These may be present in connective tissue disorders, particularly in SLE. They were frequently present with the older methods which were in use, such as the Wassermann reaction, but false positives have been recorded with all tests in common use.

Immunoglobulins. Polyclonal hyperglobulinaemia is common, the increase mainly being in IgG; cryoglobulins may also be present. Tests for rheumatoid factor are positive in about 40%.

Antinuclear Antibodies (ANA). These are present in virtually all patients with SLE, usually in a titre of >1:20. Weakly positive tests occur in a proportion of normal people and the incidence increases with age. Many drugs may lead to positive tests and are sometimes associated with a syndrome similar to SLE (Table 2). ANA are also found in other connective tissue disorders, particularly systemic sclerosis, Sjogren's syndrome, and a relatively small proportion of patients with rheumatoid arthritis. Organ-specific antibodies are also frequently found in SLE; the pattern of nuclear fluorescence has some value in distinguishing these conditions, the

TABLE 2
DRUGS WHICH CAUSE AN SLE SYNDROME

Hydrallazine	Penicillamine
Procainamide	Thiouracils
Isoniazid	Phenylbutazone
Phenytoin	Quinidine
Chlorpromazine	Reserpine
Oral Contraceptives	Penicillin
Methyldopa	Sulphonamides
L-dopa	

smooth pattern being most commonly seen in SLE. The 'LE' cell test, although frequently positive in SLE is not specific and is no longer used.

The presence of antibodies to double-stranded native DNA is the most specific test, being almost invariably raised in active SLE, and negative in all of the other conditions in which ANA are found.

Extractable nuclear antigen (ENA) which appears to be a mixture of Sm protein and ribonuclear protein is also frequently found.

Serum Complement. C3, C4 and total haemolytic complement are usually measured, low levels nearly always indicate active renal SLE and probably indicate immune complex deposition.

Histopathology

The most characteristic lesion is fibrinoid necrosis, particularly affecting small arteries, arterioles and collagen in various tissues. The haematoxylin body is also found, which represents phagocytosis of nuclei.

Kidney. Focal glomerulonephritis, diffuse proliferative glomerulonephritis, and membranous glomerulonephritis may be found. The well known 'wire loop' appearance occurs in advanced cases only. Electron-microscopy has shown the deposition of material on the endothelial aspect of the glomerular membrane which has been shown to consist of IgG, complement and fibrinogen.

Skin. Involved skin shows classical changes of discoid lupus. In uninvolved skin, immunofluorescent studies have shown the presence of gammaglobulin and complement along the dermal-epidermal junction, particularly in patients with renal disease.

Spleen. The well-known 'onion skin' lesion may be found which consists of perivascular fibrosis around the central arteries, but is seldom of diagnostic use.

Thus in spite of the widespread tissue involvement histopathology plays relatively little part in the diagnosis of SLE.

SLE in Pregnancy

Exacerbations of the disease may occur during or soon after pregnancy but seldom cause any permanent effect provided that renal and cardiac function are normal, and in these cases where there is a reasonably good prognosis, pregnancy is not contraindicated. On the other hand infertility is common in SLE and there is a considerably increased incidence of abortion, which often precedes other manifestations of SLE.

Treatment

In view of the widespread changes and differing features of SLE, treatment has to be considered on an individual basis. In general, excessive exposure to ultraviolet light should be avoided. The use of penicillin and sulphonamides, which may precipitate exacerbations, should be avoided where possible and oral contraceptives should not be used.

Arthritis. Where this is the main feature it is managed along the lines of rheumatoid arthritis using anti-inflammatory analgesic drugs. Corticosteroids are seldom required.

Skin. Skin lesions can often be controlled by the use of antimalarials, and where skin and joint lesions are the main presentation, which is not uncommon, antimalarials are a very helpful form of treatment.

Kidney. Focal nephritis may require no treatment. The nephrotic syndrome is usually treated with corticosteroids, starting with relatively high doses such as 60 mg of prednisolone daily. Diffuse glomerulonephritis presents a major problem. It may be controlled by large doses of prednisolone but side effects become inevitable. The addition of cytotoxic drugs enables the prednisolone dose to be reduced although cytotoxic drugs used alone do not appear to be effective and certainly provide no immediate improvement. Of the drugs commonly used cyclophosphamide appears to be the more effective, but is considerably more toxic than azathioprine. Apart from usual tests of renal function, such as urinalysis, blood urea and creatinine concentrations, and creatinine clearance, the measurement of DNA binding and serum complement are particularly helpful in this situation, the DNA binding being high, and the serum complement concentration low. With improvement they both return towards normal, the serum complement level being the more accurate indicator of disease activity; DNA binding sometimes remains high in spite of little evidence of active disease.

Neuropsychiatric Involvement. Large doses of corticosteroids, such as 100–160 mg prednisolone daily, are frequently used but are not always effective and often cause complications. Cytotoxic drugs do not seem to be particularly helpful.

Blood. Thrombocytopenic purpura and haemolytic anaemia usually require corticosteroid treatment, starting with 40 mg prednisolone daily. A satisfactory response can be expected.

Pleurisy and Pericarditis. If these are causing symptoms small doses of corticosteroids, 10–20 mg prednisolone daily, are usually effective.

Other Methods of Treatment. Plasmaphoresis is being used and appears to have some beneficial effects particularly on renal and neuropsychiatric complications.

In general symptomatic complications of the disease require treatment but there is little evidence that treatment in quiescent periods provides any benefit, although regular follow-up particularly of those with renal disease, who may have few symptoms, is necessary.[2]

SLE has been extensively studied by Dubois.[1]

Scleroderma

Scleroderma is an uncommon condition occurring mainly in women and rarely in children. It is characterised by excessive deposition of collagen, particularly in the skin, but widespread involvement may occur.

Clinical Manifestations

Many classifications are in use which tend to confuse the subject partly because there is considerable overlap.

Localised Scleroderma. Localised areas of thickening of the skin, sometimes of a violaceous colour, is known as morphoea; linear white streaks may also occur. Occasionally widespread skin involvement occurs, but the localised form usually remains a skin disease and there is little risk of development of systemic complications.

Acrosclerosis. This implies involvement of the skin of the extremities which is thickened, pale and stiff and is frequently associated with telangiectasia. Raynaud's phenomenon is very common and may lead to permanent peripheral ischaemia resulting in progressive loss of the pulps of the fingers, resorption of the distal phalanges and in some cases gangrene of the extremities. Calcinosis mainly affecting skin, subcutaneous tissues and occasionally muscle is common.

Progressive Systemic Sclerosis. Progressive systemic sclerosis is usually associated with peripheral changes but is occasionally described without them, in which case it is very hard to diagnose. It is a slowly progressive condition of poor prognosis.

Gastrointestinal Tract. The oesophagus is commonly involved, dysphagia being a frequent complaint. The distal two-thirds of the oesophagus are involved. Abnormalities of motility may be present at an early stage. Fibrotic changes in the oesophagus lead to reflux oesophagitis; hiatal hernia is often present. The lower end of the oesophagus may become stenosed and ultimately the oesophagus becomes an enlarged rigid tube which empties only by gravity.

Involvement of the small bowel may lead to a malabsorption syndrome, in part due to rigidity of the bowel leading to a 'stagnant loop' syndrome.

Kidney. Kidney involvement is a serious complication usually presenting as hypertension which may become malignant.

Heart. Involvement of the myocardium is common. Diffuse myocardial fibrosis may lead to cardiac failure and conduction defects.

Pulmonary Involvement. Diffuse pulmonary fibrosis may occur particularly affecting the lower zones. Pulmonary function tests show restrictive and diffusion defects. Pulmonary hypertension is a late complication.

Joints. A peripheral polyarthritis occurs quite frequently, particularly early in the disease, but is seldom progressive and does not become erosive.

Muscle. Both myopathy and inflammatory myositis occur leading to muscular weakness.

Diagnosis

The history and clinical findings are usually sufficient to enable the diagnosis to be made on these alone.

Investigations

The ESR is moderately raised in less than half and is of little help.

Immunological Tests. Antinuclear antibodies are present in about 50% of cases, the speckled pattern of immunofluorescence being the most characteristic. DNA binding is normal. Rheumatoid factor is present in about 30%. Anti-RNA antibodies have been demonstrated in almost all cases in which they have been sought, but the test is not yet a routine laboratory investigation.

Histopathology. Occasionally skin biopsy is required to confirm the diagnosis. There is increased dermal collagen with low grade perivascular inflammation, and oedema of the dermis. Small blood vessels in the skin and elsewhere may show intimal thickening, increased collagen in the adventitia, and occlusion of the lumen by thrombus. With systemic involvement an increase in collagen in the involved organs is the usual finding although the renal lesion is usually indistinguishable from malignant hypertension.

Treatment

Drug treatment of scleroderma is most unsatisfactory. In view of the effect of penicillamine on collagen cross-linkages it might be expected to be helpful. There is some evidence that it may improve the skin lesions, especially if used early but it has no effect on systemic disease. Other drugs also do not appear to be beneficial for the systemic disease, including corticosteroids and cytotoxic drugs. Corticosteroids may occasionally be needed for the treatment of inflammatory arthritis but this usually responds to anti-inflammatory analgesics. Some drugs may be helpful in the treatment of Raynaud's disease. Reserpine has been used for many years but is of doubtful value. Guanethidine and methyldopa are probably more effective and there have been some favourable reports of the use of griseofulvin. Cervical sympathectomy may be necessary. Oesophageal dilatation in the presence of oseophageal stricture is one of the more satisfactory forms of treatment in this disease.[4]

Polymyositis and Dermatomyositis

Polymyositis is an inflammatory condition of voluntary muscles of unknown cause. Skin involvement occurs in some cases, when it is referred to as dermatomyositis.

Clinical Manifestations

The most common symptoms are weakness and pain, mainly in the proximal muscles, which may also involve the pharyngeal and upper oesophageal musculature leading to dysphagia, and the muscles of respiration. The disease may present acutely with rapidly developing paralysis, which may be life-threatening if the pharangeal and respiratory muscles are involved. Sometimes it presents as a slowly progressive weakness in which case diagnosis is more difficult. It may occur at any age, and is somewhat more common in women. A not infrequent accompaniment is a non-erosive inflammatory polyarthritis, mainly peripheral in distribution, which is not progressive. Raynaud's phenomenon is also a common feature.

Skin Manifestations. The most well known is the erythematous rash on the face and the discolouration of the upper eyelids known as the heliotrope rash. Dusky red coloured papular lesions, sometimes scaly, may appear particularly on bony prominences. Thickening of the skin on the extensor aspect of the metacarpophalangeal joints is commonly seen. The papular lesions may be painful and ulcerate. The skin manifestations may be florid and are often associated with a considerable degree of painful subcutaneous oedema which often leads to diagnostic confusion.

Association with Malignancy. There is a considerably increased incidence of malignancy in association with polymyositis. Fortunately this does not occur in children but the incidence increases with age and is most common in males over the age of 40, particularly in those who have florid skin involvement in which up to 50% may be found to have malignant disease. Carcinoma of the bronchus is by far the most common lesion, but almost any form of malignancy including lymphomas have been described.

Investigations

The ESR is usually raised but a normal value does not exclude the diagnosis.

Activities of various plasma enzymes are raised but the most valuable and specific are creatine kinase and aldolase of which creatine kinase is the most commonly used. A considerable rise in the creatine kinase activity is found in the great majority of cases and parallels the severity of the disease particularly in acute cases. In those with chronic disease and in children, the plasma enzyme levels are less reliable.

The urinary excretion of creatine is also raised. Creatinuria at levels greater than 6% of the creatininuria has been found to be present in active disease, and to be a useful method of monitoring the response to treatment. Hypergammaglobulinaemia is common. Rheumatoid factor and antinuclear antibodies are found in a small number of patients but DNA binding is normal.

Muscle Biopsy. Muscle biopsy confirms the diagnosis, the characteristic features being degeneration of muscle fibres with some evidence of regeneration, infiltration by chronic inflammatory cells, particularly around blood vessels, and phagocytosis of necrotic muscle fibre.

The site and method of biopsy are most important. The deltoid, quadriceps or pectoralis muscles are usually chosen but it is necessary that it should be a weak or painful muscle. Some guidance to the muscle chosen may be given by electromyography but the biopsy should not be done at the precise site at which electromyography needles have been inserted as the needles themselves may cause histological abnormalities.

Electromyography. The characteristic findings are of fibrillation and the presence of an increased number of polyphasic volitional potentials of an abnormally short duration.

Treatment

Since severe polymyositis may be a medical emergency, treatment should be started before the results of specific tests are known. Prednisolone is given in a starting dose of 60 mg daily. Most cases show rapid symptomatic improvement although regeneration of muscle and return of muscular power may be considerably delayed. Much higher doses are needed in a small number of patients if a rapid response is not achieved. Treatment is monitored by the level of the serum creatine kinase and the urine creatine excretion, both of which fall rapidly. High doses of corticosteroids are continued until the creatine kinase activity and urine creatine concentration approach normality, when gradual reduction may be made, based on these levels and on the clinical improvement. It is usually possible to reduce the dose after a few months, to about 10 mg of prednisolone daily. A maintenance dose at this level may be necessary for some years. A small number do not respond to corticosteroids and, in these patients, cytotoxic drugs may be helpful, of which methotrexate appears to be the most valuable. In most cases in which there is an underlying malignancy there is usually some improvement both in the clinical and chemical findings but they are often resistant to treatment. Frequently the malignancy is not obvious at the time of diagnosis but is discovered at some later stage. Failure to respond adequately to treatment should suggest this possibility. The tumour is seldom amenable to treatment.[9]

Mixed Connective Tissue Disease

In some cases, features of systemic lupus erythematosus, polymyositis, and scleroderma co-exist of which arthritis, Raynaud's phenomenon, myositis, and oesophageal involvement are the most common. Unlike systemic lupus erythematosus, renal and neuropsychiatric involvement are absent and skin lesions less common.

Investigations

Anti-nuclear antibodies are frequently present and the speckled pattern of immunofluorescence is the most common finding. The common feature of this condition has been the presence of extractable nuclear antigen (ENA) which appears to be a mixture of ribonucleoprotein and Sm antigen.

Treatment and Prognosis

Considerable controversy remains about whether mixed connective tissue disease is a separate entity. It does, however, appear to respond reasonably well to modest doses of corticosteroids.[8]

ARTERITIS

Polyarteritis Nodosa

Polyarteritis nodosa is a serious but fortunately rare condition which is difficult both to diagnose and to treat.

Clinical Manifestations

These vary considerably but commonly include pyrexia of unknown origin, weight loss, nephritis and hypertension, intestinal infarction, myocardial infarction, asthma, muscular pain and tenderness, and peripheral neuropathy which may be a mononeuritis multiplex. Various skin manifestations occur which are not diagnostic but livido reticularis is probably the most important. Nodules in the skin or on arteries which gave the condition its name are an infrequent feature.

Investigations

The ESR is almost invariably raised. A mild degree of normochromic normocytic anaemia is usually present. A neutrophil leukocytosis is present in over 50%. Eosinophilia occurs in the minority and nearly always in those with pulmonary involvement. Between 20–40% have hepatitis-B surface antigen. In spite of this, clinical evidence of hepatitis is unusual although a minor rise in the plasma levels of hepatic enzymes is not uncommon.

In contrast to many other connective tissue diseases, hyperglobulinaemia or the presence of auto-antibodies or rheumatoid factors is relatively unusual.

A rise in the plasma urea and creatinine concentrations, and a fall in the creatine clearance parallel renal involvement.

Histopathology. The diagnosis can be proved by biopsy of an involved artery, which, in the acute stages shows intense polymorphonuclear leukocyte infiltration which may affect all the layers. At a later stage multiple aneurysm formation occurs leading to the term 'nodosa'. Unfortunately, obviously involved arteries are seldom present. Biopsy of an arbitrarily chosen muscle has been the accepted practice; but unless there is obvious muscular involvement the number of positive biopsies is very small.

Radiology. Renal and coeliac angiography demonstrate the presence of aneurysms in the majority of patients and is now the diagnostic method of choice.[7]

Treatment

Corticosteroids are used, starting with large doses such as 60 mg of prednisolone daily. There is some evidence that the prognosis is improved if early control of the disease can be achieved by high doses. Coexistent hypertension will also require treatment. In those that are resistant to corticosteroids, cytotoxic drugs are usually given but there is at present relatively little evidence of their effectiveness, except in the granulomatous form of arteritis known as Wegener's granulomatosis, in which corticosteroids are relatively unhelpful but cytotoxic drugs, particularly cyclophosphamide, have improved the prognosis considerably.

Giant Cell Arteritis

This condition may present in two main ways: cranial arteritis and polymyalgia rheumatica. Frequently symptoms of both co-exist. It is a condition mainly affecting the elderly and is about four times as common in women as men.

Giant cell arteritis may affect almost any artery but the highest incidence is in those arteries which have a high elastin content, particularly in cranial arteries. It may present as headache and scalp tenderness but any cranial artery may be involved. Less commonly, arteries elsewhere are affected and may lead to the aortic arch syndrome. The most important and feared complication is involvement of the ophthalmic artery which may cause blindness which is almost always permanent.

The main clinical features of polymyalgia rheumatica are summarised in Table 3. Considerable controversy

TABLE 3
COMMON FEATURES OF POLYMYALGIA RHEUMATICA

Elderly patients
Female to male ratio 4:1
Pain and stiffness in central joints and associated muscles
Marked morning stiffness
No muscle weakness or tenderness
Infrequent presence of synovitis
ESR usually >50 mm/hr
Malaise, depression, weight loss
Good response to low-dose corticosteroid treatment

exists about whether giant cell arteritis is the underlying abnormality in all cases.

A form of giant cell arteritis is also found in Takayasu's arteritis, which involves the aorta and its main branches but occurs particularly in young Asiatic females, being one of the causes of 'pulseless disease'. This sometimes occurs in association with cranial arteritis and polymyalgia rheumatica and it is of interest that patients with Takayasu's arteritis often present with a premonitory phase of skeletal pain, which may be indistinguishable from polymyalgia rheumatica.

Investigations

A mild degree of normocytic normochromic anaemia is often present. Occasionally a more severe anaemia occurs and has the characteristics of the anaemia of chronic disease. The plasma iron concentration and iron binding capacity are low, and the marrow shows the presence of abundant iron.

A raised activity of alkaline phosphatase of hepatic origin is found in a small number of cases and liver biopsy has shown the presence of hepatic granulomata in a few cases. Clinical evidence of liver disease, however, is unusual.

A mild degree of hyperglobulinaemia frequently occurs but specific circulating auto-antibodies have not been demonstrated, although in the elderly there is a somewhat higher incidence of low titres of auto-antibodies.

Lymphocyte transformation in response to muscle and arterial antigen was reported and would have proved a useful test, but this work has not been confirmed.

Histopathology. Biopsy of an involved artery, usually the temporal artery, in the acute stage shows a panarteritis which involves particularly the area of the internal elastic lamina which is frequently disrupted. It may be infiltrated with lymphocytes and plasma cells and is the usual situation of the giant cells. Intimal proliferation leads to progressive narrowing of the lumen. At a later stage giant cells and inflammatory cells may not be found but intimal proliferation and disruption of the internal elastic lamina persist.

Temporal Arteriography. Irregularities of the temporal arteries have been demonstrated radiologically, although their value in diagnosis remains uncertain. It can be of value in some doubtful cases and may indicate a suitable site for biopsy, which is important because the arteritis is a patchy process.

Prognosis and Treatment

Corticosteroids are the only effective form of treatment for giant cell arteritis and are necessary to prevent the severe ocular complications. They are also by far the most effective form of treatment of the manifestations of polymyalgia rheumatica. The usual starting dose of prednisolone is 20 mg daily for those with symptoms of cranial arteritis and 10 mg daily for those with polymyalgia rheumatica alone. The subsequent dose is monitored by the symptoms and the ESR. Symptomatic improvement is usually rapid and the ESR falls. The dose used is that which keeps the patient virtually free of symptoms and the ESR at or near to the normal range of less than about 20 mm/lhr. The necessary duration of treatment is impossible to predict and may be for many years.

Associated Conditions

There is a considerably increased incidence of thyroid disease associated with polymyalgia rheumatica, which usually presents as myxoedema. This may cause diagnostic difficulty in view of the muscular pains that are associated with this condition. In these, thyroid auto-antibodies and abnormalities of thyroid function are present.[6]

REFERENCES

1. Dubois, E. L. (1974), *Lupus Erythematosus*. University of Southern California Press.
2. Hughes, G. R. V. (1977), *Connective Tissue Diseases*, pp. 3–59. Oxford: Blackwell Scientific Publications.
3. Kirk, J. A., Answell, B. A. and Bywaters, E. G. L. (1967), 'Hypermobility syndrome'. Ann. Rheum. Dis., 26, 419–425.
4. Leroy, E. C. (1976), *Modern Topics in Rheumatology*, 144–152. London: Heinemann Medical Books Ltd.
5. McKusick, V. A. (1975), *Mendelian Inheritance in Man*, 4th edn. Baltimore: The Johns Hopkins University Press.
6. Mowat, A. G. and Hazleman, B. L. (1974), 'Polymyalgia rheumatica: Clinical study with particular reference to arterial disease'. J. Rheumatol., 1, 190–202.
7. Sack, M., Cassidy, J. T. and Bole, G. G. (1975), 'Prognostic factors in polyarteritis'. J. Rheumatol., 2, 411–421.
8. Sharp, G. C., Irvin, W. S., Tan, E. M., Gould, R. G. and Hoiman, H. R. (1972), 'Mixed connective tissue disease'. Amer. J. Med., 52, 148–157.
9. Walton, J. N. (1974), *Disorders of Voluntary Muscle*, 3rd edn., 614–653. Edinburgh and London: Churchill Livingstone.

SECTION VI

PHARMACOLOGY AND TOXICOLOGY

		PAGE
21.	THE BIOCHEMISTRY AND TOXICOLOGY OF HEAVY METALS	347
22.	LABORATORY INVESTIGATION OF THE POISONED PATIENT	363
23.	REGULATION AND MONITORING OF DRUG THERAPY	379
24.	TRAUMA SHOCK AND SURGERY	400

21. THE BIOCHEMISTRY AND TOXICOLOGY OF HEAVY METALS

ANDREW TAYLOR

Introduction

Toxic metals
  Lead
  Mercury
  Cadmium
  Toxicity of other metals

Essential trace elements
  Zinc
  Copper
  Manganese
  Chromium
  Other essential elements

Other aspects of trace element biology
  Total parenteral nutrition
  Chronic renal failure and maintenance haemodialysis
  Trace elements as pharmacological agents
  Trace elements in the monitoring of disease

Summary

INTRODUCTION

The biological functions and properties of trace elements, or heavy metals (these terms are often used synonymously), are of interest to a wide range of disciplines. In occupational health and toxicology the consequences of gross exposure are studied, while natural requirements, availability and deficiencies of essential elements are of nutritional interest. Physiologists, biochemists and bioinorganic chemists are concerned with mechanisms, e.g. of absorption, and with interactions with other systems. Epidemiological research suggests that there are correlations between morbidity and environmental exposure to certain trace elements, yet a few elements can be used as pharmacological agents. The interests of clinicians and clinical biochemists overlap with all of these specialities, and so occupy a central position, drawing upon each of them and relating one to another.

Prior to the development of sensitive analytical techniques, particularly atomic absorption spectrophotometry, trace elements were largely ignored. The last ten years have seen a radical change and measurements of concentrations in the nanogram per gram range in sample volumes of no more than 50 μl are now relatively straightforward and inexpensive. Together with this increased analytical potential has grown the appreciation of the importance of trace elements in clinical situations other than gross toxicity. In addition, hazards associated with environmental pollution receive considerable public attention so that screening and monitoring programmes are established with the result that, as never before, clinical chemists are required to be conscious of situations where advice and analysis is appropriate.

TOXIC METALS

Most, if not all, metals are toxic, even those known to be essential. In acute toxicity studies lead is less harmful than iron; vanadium is more toxic than mercury. Furthermore, elements that are toxic at trace levels can, at even lower levels, be essential—selenium, for example. The work of Schwarz[1], using rats fed on highly-purified artificial diets deficient in a single trace metal, suggests that arsenic, and possibly cadmium and lead, may be essential at extremely low levels for normal growth and development.

Nevertheless since metals like lead are usually measured when poisoning is suspected, and elements such as zinc are generally investigated in deficiency conditions, there are practical advantages to the conventional classification of trace elements, at the concentrations at which they usually occur within the body, as either essential or toxic.

Lead

Physiology

Lead occurs naturally in the environment of man—in food, drinking water, in the atmosphere and in street dust. The mean food concentration is 135 μg/g (0.65 μmol/g) and drinking water seldom contains more than 10 μg/l (0.05 μmol/l). Thus an adult consuming 1.5 kg of food and 2.5 litres of water ingests approximately 225 μg (1.08 μmol) of lead per day.

Metabolic studies and radioisotope absorption experiments indicate that adults normally absorb about 10% of their daily intake. Children, however, absorb very much more—40–50% of the ingested lead.

Availability of lead for absorption, however, is much more varied than these results suggest. In animals, absorption is inversely related to dietary calcium and phosphorus, zinc and iron. High fat and high protein diets promote absorption while the anion associated with led also affects its uptake.

Approximately 30–50% of inhaled lead is absorbed, either across alveolar membranes or, following transfer of coarse particles from the lung by ciliary action, via the alimentary tract. A typical urban atmosphere contributes 20–30 μg (0·01–0·14 μmol) lead per person per day, i.e. about 10% of total exposure, and is therefore relatively less important than dietary and other sources of lead.

Rabinowitch et al.,[2] described the distribution of lead as approximating to a three-compartmental model. Absorbed lead rapidly associates with erythrocytes (less than 5% of lead in blood is within the plasma). From this first compartment, with a mean life of 35 days, lead passes into the urine (200 nmol/day) or is transferred into the other two compartments. From the second, represented by the soft tissues, a small amount of lead is excreted into nails, hair and alimentary secretions, while the third—primarily the skeleton—serves as the long-term store with a very slow turnover rate and containing 99% of the total body lead.

Lead readily crosses the placenta and is a well-recognised abortifacient.

Toxicity: Industrial

That lead is an extremely versatile and important raw material in very many industries is attested by its continous use for 6000 years.[3] Despite the well-recognised hazards associated with its use, including a number of widespread mass poisoning episodes, no suitable alternative is available and therefore many men and women continue to be occupationally exposed to lead and its compounds.

In England and Wales approximately 20 000 persons employed in industries such as metal smelting, battery and paint manufacture, ship building and ship breaking, construction and demolition work, and petroleum refining, are recognised by the Health and Safety Executive (HSE) as requiring regular examination to safeguard against lead poisoning.

There are, in addition, those who fall outside the regulations within which the HSE act and may therefore present, through their general practitioner, to the hospital without necessarily giving details of their lead exposure.

Signs and Symptoms of Toxicity. The characteristic clinical features of chronic inorganic lead poisoning include weakness and fatigue, anaemia, abdominal pain, deposition of lead as a blue line in the gums, gastrointestinal symptoms, chronic renal failure and a peripheral neuropathy with tremor and weakness.

Lead binds to thiol groups and is therefore an enzyme inhibitor. The activities of erythrocyte pyrimidine nucleotidase and Na^+: K^+-ATPase are reduced in lead poisoning, but the enzymes most studied are those of haem synthesis (Fig. 1) with inhibition of δ-ALA dehydratase, δ-ALA synthetase and ferrochelatase well documented following exposure to lead.[4]

Exposure to organic lead compounds occur in very few occupational processes but, if severe, a syndrome quite distinct from inorganic lead poisoning ensues characterised by an encephalopathy with hallucinations, delusions and convulsions. There is moderate increase in the blood lead concentration although the proportion in the lipid fraction of blood is much higher than in inorganic lead poisoning.

```
glycine + succinyl CoA
        ↓ δ ALA synthase
amino levulinic acid                    ↑ urine
        ↓ δ ALA dehydratase             ↓ blood
porphobilinogen
        ↓
        ↓
coproporphyrinogen III                  ↑ urine
        ↓
        ↓
blood ↑   protoporphyrin IX
        Zn⁺⁺  ╲  ferrochelatase
            Fe⁺⁺
blood ↑ zinc protoporphyrin   haem + globin
                                    ↓
                              haemoglobin    ↓ blood
```

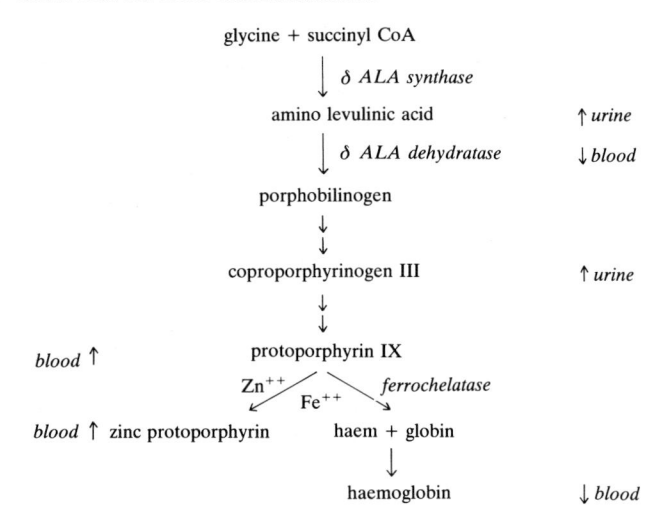

Enzymes with reduced activity due to lead are shown in italics.
Intermediates with altered concentrations are indicated together with body fluid in which assayed.

FIG. 1 *Effect of lead upon haem synthesis*

Laboratory Investigations. Measurement of the blood lead concentration on a properly collected venous blood sample is the best single test of lead absorption, although urinary lead excretion, particularly following stimulation with orally-administered chelating agents, also demonstrates body lead.

Blood lead concentrations in non-occupationally exposed subjects are less than 1·8 μmol/l. Symptomless exposure is indicated by blood lead concentrations of 1·8–3·9 μmol/l. At higher concentrations symptoms may appear and if the blood lead is greater than 6·0 μmol/l they are usually severe. At blood lead concentrations greater than 4·0 μmol/l monitoring of lead workers becomes more intensive and they are removed from occupational exposure to lead.

Tests to assess the effects of lead upon the pathway of haem synthesis have been used as alternatives to the measurement of lead in blood (Fig. 1). Since these tests, unlike the blood lead concentration, are a direct measure of the degree of biological damage wrought by lead they may better demonstrate the state of health of lead workers. Furthermore, contamination of the sample during collection does not affect the result of the tests and some are sufficiently simple to be performed at the collection site.

Increases in urinary coproporphyrin and δ-aminolevulinic acid (ALA) excretion, resulting from the effect of lead on the haem biosynthetic pathway, have been used for many years to monitor lead workers. Since, however, the correlation between either of these tests and the blood lead has repeatedly been shown to be very strong, it is doubtful that they provide any further information than is obtained from the blood lead alone. The lability of coproporphyrins require measurement within a few hours of collection if accurate results are to be obtained. Following the provision of specialist blood lead assay centres in the UK, routine screening using these tests is rapidly being abandoned.

Three-monthly estimations of blood haemoglobin was introduced in 1964 as a statutory test for lead workers but, as with punctate basophylia, alterations are not consistently seen in lead workers until the blood lead levels are greater than 5.5 μmol/l.

Whereas measurement of haemoglobin lacks sensitivity to the presence of lead, measurement of erythrocyte δ-ALA dehydratase is oversensitive. Inhibition of this enzyme occurs at blood lead levels below 2.0 μmol/l and has no role in monitoring occupational exposure.

Following the inhibition by lead of the enzyme ferrochelatase, protoporphyrin IX accumulates in red cells. A simple procedure for fluorometric measurement of free erythrocyte protoporphyrin (FEP) following ethyl acetate extraction was described by Piomelli.[5] Subsequently it was demonstrated that protoporphyrin IX accumulating in the red cell is complexed with zinc and that this zinc protoporphyrin complex (ZPP) can also easily be assayed.

With the development of small, portable dedicated fluorimeters, measurement of ZPP has provoked considerable interest. Piomelli[5] demonstrated excellent logarithmic correlation between protoporphyrin and blood lead and in the USA it is now recommended[6] that, in both occupational exposure and in environmental screening surveys, initial testing should be for erythrocyte protoporphyrin and only when this is increased should be blood lead determination be considered.

Such a trend, however, is not apparent in the UK and while measurement of ZPP will have a role—as a rapid check upon contamination where an unexpectedly high blood lead result is obtained, or as a screen in work with only limited exposure to lead—this test cannot replace blood lead in the monitoring of lead workers (see below).

Lead has been measured in the plasma of occupationally exposed subjects. The concentration is proportional to the total blood lead and despite its availability to other tissues, plasma lead offers no better indication to toxicity than total blood lead.

To determine the test or combination of tests which best predict or reflect morbidity, Irwig et al.,[7] correlated the prevalence of various symptoms with the concentrations of blood and urine lead, urine δ-ALA and packed cell volume in 639 workers receiving considerable occupational lead exposure. The best prediction of morbidity was given by blood lead alone and the effects of lead upon porphyrin metabolism were not related to clinical findings. This work, involving a large number of subjects—25% of whom had blood lead concentrations greater than 5 μmol/l—emphasises the importance of the blood lead in occupational exposure and that other tests should only be considered where it is impracticable to achieve reliable blood lead results. The consequences of recent legislation are that in England and Wales blood lead concentrations will now be measured at intervals related to the level of occupational exposure. A number of workers previously omitted from monitoring programmes will then be included.

Toxicity: Non-industrial

While industrial exposure is responsible for the majority of cases of lead poisoning there are many other sources of lead to which the population in general may be exposed (Table 1). Some of these, e.g. children with

TABLE 1
NON-INDUSTRIAL SOURCES OF LEAD SHOWN TO CAUSE TOXICITY OR TO BE POTENTIAL HEALTH HAZARDS

Domestic water system	Toothpaste tubes
Glazed jugs, bottles, glasses	Sniffing petrol fumes
Kettles and boilers	Herbal and ayurvedic medicines
Embedded bullets	Surma
Indoor firing range	Paints, varnish, polishes
Batteries burned for fuel	Lead-containing opium
Newsprint	Contaminated food

pica (where the behaviour of putting toys, etc., into the mouth is prolonged) who are at risk from paint or dust in old buildings, acidic drinks stored in lead-glazed vessels, lead in petrol exhaust, are well recognised. Some are of incidental interest, being of relevance only when adding to a pre-existing problem of lead exposure. Cases of poisoning attributed to Surma, an Indian eye cosmetic containing up to 90% of lead sulphide, are now being recognised. A related problem concerns unconventional medicines which may be used by some populations. An airline pilot 'treated' in India for diabetes was diagnosed by a doctor in this country as having lead poisoning. His blood lead concentration was 4.2 μmol/l and two ayurvedic medicines which were analysed in our laboratory contained 1.8% and 3.8% of lead. Similar episodes can be expected to occur in other groups.

For reasons which remain obscure the number of reported cases of children with raised blood lead concentrations increases between May and October.

Signs and Symptoms of Toxicity. Symptoms associated with lead poisoning are the same as in industrial exposure, but without the tolerance to lead found in lead workers these patients are affected earlier and more severely. This is especially so in children where encephalophy and death may be the eventual outcome.

Acute poisoning following excessive exposure to lead is extremely uncommon. Weakness, nausea, vomiting and abdominal pain occurred in a group of drug addicts who stole lead-and-opium pills and injected themselves with a suspension of these in water. Those severely affected developed liver failure and reversible renal tubular necrosis while neuropathy with respiratory paralysis was present in one case.

Laboratory Investigations. As with lead workers, the most important test in non-industrial lead poisoning is the measurement of blood lead and wherever a confirmed result of greater than 1·8 μmol/l is discovered, follow-up is essential. This should include an attempt to identify and if possible eliminate the source of exposure, the evaluation of the clinical effects of the lead already absorbed and, where appropriate, initiation of treatment to reduce the body burden.

While the use of chelating agents is essential when the blood lead concentration is dangerously high this is not without risk of producing side-effects and therefore the preferred treatment is removal from the source of exposure. In certain situations, however, removal may be impossible, e.g. where the patient is residing close to a smelting works or where the source of exposure has not been discovered; re-exposure is then inevitable. It is then necessary to assess fully the clinical and biochemical effects of the lead since in these cases too, treatment with chelating agents may be required.

The tests suggested above as alternatives to blood lead are all of value although changes are not always due to lead. Zinc protoporphyrin is elevated in iron-deficiency anaemia to levels comparable to those in lead poisoning. Furthermore, because of the three-month life-span of red cells a change in ZPP will inevitably lag behind absorption or elimination of lead. Inhibition of erythrocyte δ-ALA dehydrase is a very sensitive test of lead exposure and similarly a reduced haemoglobin concentration demonstrates the biochemical effects of lead. A diagnosis of lead poisoning may be suspected by the coincidental finding of low haemoglobin and punctate basophilia during routine haematological investigations of patients not known to be exposed to lead.

Lead poisoning should be considered in the differential diagnosis of unexplained encephalopathy in children. In an urgent situation blood lead can be measured within minutes of receipt of sample and meanwhile radiological examination for lead in the gut and long bones performed.

Lead excreted with the faeces represents the fraction of ingested lead which has not been absorbed, together with the very small amount present in intestinal secretions.[2] Since the diet is the major source for intake of lead, measurement of faecal lead is a valuable determination in children following severe poisoning when blood lead levels may remain high and it is necessary to discover whether this represents re-exposure or equilibration with the lead in bone.

A test not associated with haem synthesis is now available. Nerve conduction velocities in lead exposed subjects (blood lead, 2·0–3·5 μmol/l) are lower than in age-and sex-matched controls, and indicate latent nerve damage without clinical neurological symptoms.

Sub-clinical Lead Exposure

There is now little doubt that residents of areas with increased environmental lead have blood lead concentrations which are greater than those living in other areas. This is particularly evident around lead smelters,[8] but is also observed in subjects living within 100 yards of major roadways or in high traffic density areas. Children in both situations have mean blood lead concentrations which decline with increasing distance between the primary source of lead and the home. Contamination of dust, soil and vegetation, with subsequent ingestion, is probably more important than inhalation as the route of absorption.

Soft acidic water is plumbosolvic and considerable lead can enter the domestic water of homes with lead pipes and/or lead-lined tanks. A positive correlation was found between blood and water lead concentrations of households studied in Glasgow.

Although moderate lead exposure does not produce symptoms of classical lead toxicity it can be recognised by abnormal biochemical findings. It has been suggested that a continuous low-level exposure to lead is responsible for subtle changes in neural function and for other effects, which are summarised as sub-clinical lead toxicity.[9]

With few exceptions, investigations have not shown significantly poorer scores for intelligence or behavioural tests among children exposed to, and having raised concentrations of, blood lead compared with controls matched for age, social and family environment. A similar conclusion was obtained in a study of a group of 47 children who required treatment for lead poisoning and were compared with a sibling of similar age. Landrigan et al.,[8] however, reported that children living for at least twelve of the previous 24 months within 6·6 km of a lead smelter in El Paso, Texas, and who had blood lead concentrations of 1·9–3·3 μmol/l, had poorer results in performance IQ and finger-tapping tests than did normoplumbic children from the same population. Limitations to this study were demonstrated by others[10] and also by themselves. Even their tentative conclusions that lead absorption '*may* result in a subtle but statistically significant impairment in non-verbal cognitive and perceptual motor skills—and sub-clinical impairment in fine motor skills', are not substantiated by other workers who studied children from the same population.

Results from the alternative approach to this question, i.e. measuring the blood lead concentration in children who are mentally handicapped, are more difficult to interpret. A number of reports show raised blood lead concentrations in mentally handicapped children, but this is probably a reflection of the habit of pica which is particularly prevalent in subnormal and hyperactive children.

Beattie et al.,[11] retrospectively measured lead in the drinking water of the houses occupied during pregnancy by the mothers of 77 mentally retarded children and 27 non-retarded matched controls. The probability of mental retardation was significantly increased when water lead exceeded 800 µg/l. Among the deficiencies of this work was the failure to obtain controls matched for maternal age and birth order. Notwithstanding these limitations, excess lead exposure during fetal development remains a possibility in these cases. In a subsequent investigation the same authors measured the concentration of lead in blood samples collected from the same children during the first two weeks of life (collected for phenylketonuria screening). The mean blood lead in the retarded group, 1·25 µmol/l, was significantly higher than in the controls, 1·01 µmol/l. These results emphasise the sensitivity of the fetus to lead, and indicate that attention to household water levels is a priority in high risk areas.

While other associations between high concentrations of lead in water and clinical or laboratory changes have been described, e.g. hypertension and chronic renal insufficiency, it is still the relationship between lead and mental development, which is the most emotive and most discussed aspect of sub-clinical lead toxicity.

Mercury

Mercury rivals lead, with a history of industrial use and recognised toxicity extending over many centuries. Pliny, the Roman historian, was aware of the practice, among those working with compounds of this metal, of using thin bladder skins to prevent inhalation of mercurial dust—an example of an extremely early occupational hygiene procedure.

Mercury, as the metal or its vapour, in inorganic salts or in aryl organic compounds (e.g. phenylmercuric acetate) produces a toxic syndrome that differs considerably, both in its nature and its severity, from that caused by alkyl organic compounds (methyl and ethyl mercury).

Physiology

Exposure to mercury is generally either to methyl mercury or to mercury vapour.

Intestinal absorption of alkyl mercury compounds is virtually total in man. Methyl mercury is present in the environment as a consequence of microbial methylation of inorganic mercury. The normally very low levels are concentrated in fish and other aquatic life; thus total mercury in the range 0·05–0·5 µg/g (0·25–2·5 nmol/g) has been measured in a variety of species including tuna fish, with 80–95% as the methyl compound. Seafood is the main source of exposure to methyl mercury and any containing more than 0·5 µg/g (2·5 nmol/g) is not permitted to be sold. The maximum recommended intake is 200 µg (1·0 µmol) per week, but is usually much less and only populations with a large fish consumption would approach this level. Discharge of industrial effluent into coastal waters around Minamata and Niigata in Japan was responsible for several hundred cases of severe methyl mercury poisoning. Large quantities of fish containing up to 50 µg/g (250 nmol/g) of mercury were eaten and 115 known deaths occurred. Furthermore, at least 22 babies were born with brain damage following fetal exposure to the methyl mercury. Methyl mercury presents a minor occupational hazard. It is applied to seeds, bulbs and tubers as an antifungal agent, and persons handling these have a limited exposure. Treated seeds, however, were responsible in Iraq for thousands of cases of poisoning where imported seed grain was unwittingly used to prepare bread. Similar incidents occurred in Pakistan, Ghana and Guatemala.[12]

Electrical switches, scientific instruments, solders, amalgams, electrodes—in which metallic mercury is employed—are innumerable. As a consequence of its high vapour pressure, volatilisation and exposure is inevitable. Of the inhaled mercury vapour, 70–80% crosses the alveolar membrane and is retained. While fewer in number than a 100 years ago, cases of toxicity continue to be regularly reported.

Other mercury compounds are less important. Elemental mercury is not absorbed in the gut and is innocuous even following intravenous injection or aspiration into the lungs. Aryl compounds are also used as fungicides on seeds, as slimicides in paper mills and in wallpaper pastes, and previously as diuretics. They are relatively non-toxic.

Exposure to inorganic salts is rather more complex. Suspended as dust, inhalation occurs followed (as with lead) by absorption or removal with bronchial secretions. Gastrointestinal absorption depends upon solubility, but does not exceed 20% of the ingested material. In addition to causing considerable irritation absorption undoubtedly occurs through the skin, although the amounts retained have not been determined. Inorganic compounds are used in pigments, for timber preservation, in photography and as antiseptics.

Absorbed mercury is converted to the divalent, mercuric ion. Elemental and aryl compounds are rapidly oxidised in red cells by the catalase–hydrogen peroxide enzyme system. Methyl mercury, however, requires to be demethylated and cleavage of the mercury–carbon bond (possibly accomplished by gut flora) proceeds no faster than about 1% per day. These rates of conversion are relevant to the different patterns of toxicity referred to above.

Methyl mercury is very lipid soluble and rapidly distributes throughout the whole body. Radioactive tracer studies indicate that 10% of the dose goes to the head, with significant amounts also in the liver and kidney. Red cell concentrations are about ten times those in the plasma and account for 5% of the dose. Because of the low plasma concentration little mercury is available to the kidney for filtration and urinary excretion; 90% of elimination is into the faeces via the bile. There is significant transfer across the placenta into the fetus and also into breast milk. A consequence of the mobility of methyl mercury between blood and tissue is that rates of clearance from the blood, the brain and the whole body are very similar. Following a tracer dose of methyl mercury an average half-life of 70 days has been calculated. The concentration of methyl mercury in the blood is the best assessment of both body burden and brain level and at a steady state of exposure the blood level is proportional to intake. Methyl mercury in hair correlates with blood concentration at the time of hair formation and thus a retrospective index of exposure is possible. The total mercury concentration of blood is less than 30 nmol/l except where fish consumption is considerable and may then be 4–5 times this level, although individual values as high as 1000 nmol/l have been reported.[12]

Elemental mercury is initially concentrated in the red cells. Subsequent to oxidation the mercuric ion transfers to the plasma so that 24 hours following exposure the ratio of red cell to plasma mercury is about 2. Thereafter, the concentrations decline with half-lives of around three days.[13] During the first few days following exposure, faecal excretion is greater than urinary loss and a small proportion of the retained mercury is re-excreted from the lungs. There is no correlation between urinary excretion and the plasma concentration, indicating that glomerular filtration is not a predominant pathway for removal of mercury. However, mercury ions gradually redistribute throughout the body and accumulate predominantly in the kidney where tubular secretion into the urine then occurs. An important feature in the metabolism of elemental mercury is that while distribution following oxidation is similar to that of ingested salts, some persists for sufficient time to diffuse from the blood into other tissues. Elemental, atomic mercury is lipophilic and, therefore, crosses the blood-brain barrier and accumulates in neural tissue. Placental passage also occurs, and mercury accumulates in fetal membranes and is found in cord blood at concentrations equivalent to those in maternal blood.[12] The blood mercury concentration is meaningful only following a recent exposure. With the short half-lives referred to above intermittent exposure causes considerable fluctuations in blood levels while with a single exposure, most disappears within a few days.

Determination of urine mercury is the best index of chronic exposure[14] with normal excretion being less than 50 nmol per 24 hours or less than 5·5 μmol per mol of creatinine in randomly collected samples.

Distribution and toxicity of inorganic and organic mercury compounds are modified by inorganic selenium. Retardation of growth caused by oral or parenteral administration of mercury compounds is not observed when selenite is given simultaneously. Protection appears to be afforded by a novel plasma protein to which selenium binds via a sulphydral group and which in turn traps the mercury preventing it from leaving the vascular compartment.

Toxicity

Signs and Symptoms of Toxicity. Soluble mercury salts produce intense inflammation of exposed mucous membranes. There is immediate corrosion and necrosis of the mouth, throat, oesophagus and stomach followed by pain, vomiting and diarrhoea. In severe overwhelming cases death from haemorrhage, shock and circulatory collapse occurs within a few hours. With less severe exposure the primary lesions begin to heal, but further inflammation to the gastrointestinal tract occurs a few days later as secretion of mercuric ions with saliva and colonic fluid commences. Stomatogingivitis, ulcerative colitis and blood loss gradually subside and recovery ensues within a few weeks. Renal tubular necrosis may also be produced which, if not interrupted by death, gradually recovers as the epithelium regenerates, only rarely leaving residual kidney damage. Powders containing calomel (mercurous chloride) were once used to soften the gums of young children cutting teeth and many developed a widespread erythematous rash. Increased urinary mercury excretion has been demonstrated in this condition of Pink Disease (acrodynia), therefore, although not proven, mercury is believed to be responsible.

Toxicity to alkyl mercury compounds has been thoroughly documented following the mass poisonings in Japan and Iraq. Amin-Zaki et al.,[15] have summarised their prolonged observations of children from rural Iraq. Neurological symptoms—ataxia, dysarthria and constriction of the visual fields (the classical signs of organic mercury poisoning)—were most evident. Hyperreflexia was noted in all cases; muscle weakness and parasthaesia, tremor and twitching were also generally present. The severity of poisoning correlated with the blood mercury concentration, which ranged from 2·5–25 μmol/l (normal = less than 30 nmol/l), and was dependent upon the amount of methyl mercury consumed. As in previous reports there was a latent period of up to six weeks with accumulation of mercury before onset of symptoms. Japanese reports of Minamata Bay incidents, where exposure continued for several years with very much higher body burdens being attained, suggested that toxicity was irreversible. In Iraq, however, consumption ceased much earlier and considerable improvement in the condition of these children has slowly taken place.

Much more common than poisoning due to inorganic salts or alkyl mercury compounds is toxicity due to inhalation of metallic mercury vapour and while the hazards associated with the handling and use of this metal are ignored, cases will continue to occur. Very rarely, acute exposure to high concentrations of mercury vapour takes place, producing a diffuse interstitial pneumonitis. Fever, chills, nausea and general malaise develop rapidly, together with tightness in the chest and paroxysmal coughing. Death from lung oedema occurs within three days in very severe cases. Where there is chronic, moderate exposure to mercury vapour symptoms develop insidiously over several weeks or even years. Mercurialism is characterised particularly by muscle tremor and erethism.

Fine muscle tremor of the arms, interfering with writing and other delicate finger work, is an early feature. Facial muscles are often involved and therefore speech may also be affected. In prolonged cases the tremor becomes progressively generalised involving the entire voluntary muscle system. Sudden limb jerking and muscle spasm can develop, with inability to walk or even sit.

Pronounced changes in mood and behaviour with feelings of listlessness and disinterest, gradually leading to withdrawal from social life, form part of the condition of erethism. Subjects also become irritable and over-reactive and tend to sweat or blush readily. These patterns of mood change are noticed by friends and relatives rather than the exposed person himself.

Other symptoms include stomatogingivitis, headache, double-vision, anorexia, diarrhoea and weight loss. Mercurialentis, a brown discolouration of the lens, is a feature of chronic exposure to mercury vapour and is not a sign of toxicity. The intensity of the colour is proportional to length of exposure, but does not interfere with vision.

Current occupational hygiene measures ensure that the full clinical picture of mercurialism is rarely encountered and where a history of exposure is known it is easily recognised. Attention is now centred upon early diagnosis where symptoms are non-specific. Also a problem is the detection of mild mercurialism in patients whose exposure is unrecognised and who present to clinicians with headache, diarrhoea or tremor. Mercurialism is readily reversed by removal from the source of exposure and, if necessary, treatment with chelating agents.

Accounts of nephrotic syndrome following exposure to ammoniated mercury ointments, or to metallic mercury vapour, appear to represent idiosyncratic responses and are not general features of mercury toxicity.

Laboratory Investigations. Conclusive demonstrations of toxicity depend upon measurement of mercury in blood or urine. As mentioned above, the long half-life and good correlation with body and brain burdens make the concentration of mercury in blood the best index of exposure to methyl mercury.

In other situations mercury is rapidly cleared from the blood and determinations in urine are probably more valuable. In cases of toxicity, rates of excretion are very high with milligram amounts being recorded. Lower levels are found during occupational monitoring and there is fluctuation in excretion from day to day. Smith et al.,[14] found excellent correlation between time-weighted exposure and urine mercury excretion in a study of 642 workers manufacturing chlorine, although correlation between excretion and symptoms was less good. Lauwerys and Buchet[16] also showed significant correlation between urine mercury and exposure to vapour. Experience in this laboratory is in agreement with these authors. It has been shown that even the very limited exposure to which dental workers are subjected can be demonstrated in the urine. In early morning urine samples, the ratio of mercury to creatinine is up to three times that found in unexposed subjects. Furthermore, where samples were received from several persons working in a practice, one result was frequently at least 50% higher than the others and was invariably from the person preparing amalgam and working adjacent to the amalgamator.[17]

Collection of urine for mercury analysis must be into containers known to be mercury free. Samples should be acidified and an oxidising agent added to prevent losses of mercury by absorption and volatilisation. Polythene containers cannot be used since mercury vapour diffuses through.

Cadmium

Unlike lead and mercury, industrial exposure to cadmium is a recent phenomenon. Acute exposure can be lethal, but the effects of chronic exposure are misleadingly deceptive with severe, irreversible morbidity suddenly appearing after many years of apparent well-being.

Other than occupationally, food and tobacco represent man's main source of exposure to cadmium. Sufficient is absorbed for the metal to accumulate throughout life with a body burden of 20–50 mg 180–450 μmol), mostly in the kidney, being reached.

Following an environmental tragedy in Japan, where wastes from a cadmium mine contaminated drinking water and growing crops, a unique syndrome with bone pain and osteomalacia was observed in post-menopausal, parous women.

Hypertension can be experimentally induced by prolonged administration of low levels of cadmium to rats and rabbits. There are some data which suggest hypertension and cadmium may be related in man, but this is far from resolved. It has also been proposed that cadmium may be the toxic factor responsible for the increased incidence of coronary heart disease recorded in soft drinking water areas.

Physiology

Shellfish, beef and pork kidney are the main food sources of cadmium contributing to a UK daily intake of 15–30 μg (135–270 nmol) per day. Retention of ingested cadmium varies from 0·7% to 15·6% and is increased at low dietary concentrations of both cadmium and calcium.

Chronic cadmium poisoning was first described as recently as 1950.[18] Zinc and lead ores invariably contain a significant proportion of cadmium, the processes of smelting and refining therefore produce dust and vapour laden with this metal. Electroplating, soldering, brazing and the disposal of industrial waste also generate cadium-rich vapour and dust and it is therefore by inhalation that occupational exposure is effected.

The proportion of cadmium absorbed is variable, but may be as high as 50%.[19] Within industrial locations the concentration of cadmium in the air may be more than 100 times the level in cadmium-free environments and absorption of up to 50 μg (450 nmol) per day takes place. By contrast, less than 1 μg (9 nmol) per day is absorbed by non-exposed subjects, although 20 cigarettes contributes a further 2–4 μg (18–36 nmol) daily.

After absorption, cadmium is transported within the plasma primarily to the liver. There it is complexed with metallothionein, an inducible protein with a molecular weight of 10 000 containing a high proportion (30%) of cysteine residues. This complex is slowly redistributed to the renal cortex via glomerular filtration and tubular reabsorption.

Cadmium levels in whole blood increase following exposure, and then recede as the metal is sequestered into the liver and kidney. A proportion of cadmium entering the liver is excreted as a glutathione complex via biliary secretion and into the faeces. Where synthesis of metallothionein has been induced by recent previous exposure, reduced excretion via this route occurs.

Urinary excretion is no more than 18 nmol per day (unless renal damage has occurred) although there is a small increase with age. Combined biliary and urinary excretion is normally less than 10% of the retained cadmium.

It follows from these very low rates of cadmium excretion, and from its accumulation in the kidney and liver, that overall removal rates are very slow. Few studies have been made in man but from the available data it is apparent that cadmium is virtually fixed within the kidney and only very slowly removed from the liver (half-lives of 17–33, and 7 years respectively).[20]

Toxicity

Antagonistic interactions with certain essential elements are responsible for many of the biochemical effects of cadmium. Metals, especially zinc, can be displaced from enzymes, e.g. carbonic anhydrase. Competition for binding sites is the presumed mechanism here and also in the interference with the metabolism and storage of copper. There is evidence of similar competition between cadmium and both iron and calcium for intestinal transport.[20] Cadmium also inhibits formation of 1,25-dihydroxycholecalciferol, thereby further interfering with calcium metabolism.

Cadmium shares with many metals the property of binding with thiol groups and thereby altering the properties of proteins, particularly enzymes, e.g. alcohol dehydrogenase. Many of the toxic effects of cadmium are limited by the presence of zinc, selenium and thiol compounds.

Signs and Symptoms of Toxicity. 1. Acute. Acute exposure to cadmium oxide vapour causes profound damage to the respiratory system with dyspnoea, bronchitis, pneumonitis and chest pain, irritation to the throat and severe, sometimes fatal pulmonary oedema. Ingestion of cadmium compounds is followed within minutes by gastroenteritis—nausea, vomiting, salivation, diarrhoea, abdominal pain, weakness and headaches being induced. There are few fatalities, possibly due to the emetic effect of high oral cadmium intake.[19]

In animal experiments, disruption of the vascular system of the testes causing necrosis of the germinal epithelium has been produced by acute administration of cadmium. The ovary, placenta and central nervous system have also been shown to be susceptible to huge doses of cadmium. These effects have not been observed in man.

2. Chronic. Inhalation of cadmium-containing fumes over many years eventually precipitates irreversible damage to the kidney, the lungs and possibly bone.

Cadmium as the metallothionein complex is stored within the renal cortex, in this form it is apparently innocuous until what appears to be a critical cortical concentration of 200 μg/g wet weight is achieved. Capacity to store the complex further is overwhelmed and renal tubular damage is manifest with low molecular weight proteinuria, glycosuria, phosphaturia and aminoaciduria. The increase in excretion of low molecular weight proteins is one of the earliest signs of tubular damage and measurement of one of these proteins—β_2 microglobulin—is now used as a diagnostic aid in the early detection of renal dysfunction.[18]

The respiratory effects include pneumonitis with chronic bronchitis and emphysema, dyspnoea and watery nasal discharge. A moderate degree of anaemia and changes in liver function may be produced, but are reversible.

The effects of cadmium upon skeletal tissue are complicated. The osteomalacia of chronic cadmium exposure is generally explained as secondary to the renal damage described above. Inhibition of intestinal absorption of calcium, either by a direct effect upon binding protein or as a consequence of impaired vitamin D metabolism, will accentuate the calcium deficiency.

While skeletal effects are extremely uncommon in cadmium workers, osteomalacia was a consistent feature in the Itai Itai disease in Japan. In this condition the bones became so fragile that considerable pain (Itai Itai literally means ouch ouch) and fractures were caused by very slight pressure. In addition to the osteomalacia, renal manifestations, iron deficiency anaemia, chronic gastritis and enteropathy were observed.

The aetiology of such strikingly severe osteomalacia involves more than just contamination of crops and water by cadmium. The victims were calcium depleted by child-bearing and inadequate diet; a low-fat diet together with limited exposure to sunlight caused vitamin D deficiency. The situation therefore existed, with low dietary calcium and protein, for enhancement of absorption of cadmium into a body ill-equipped to cope with the abuses inflicted.

From various surveys the incidence of carcinoma among men chronically exposed to cadmium at work is reported to be from 0·93% to 8·1%. In 248 workers exposed, the general incidence of carcinoma was not increased, but 4 men with prostatic cancer were seen, which was considerably in excess of the expected 0·58 cases. No significant association has been discovered between respiratory cancer and airborne cadmium. In animals, in common with many unrelated materials, cadmium compounds induced sarcoma development at sites of injection. Interstitial cell carcinoma was found in 68% of rats following cadmium-induced testicular atrophy. However, this could be a consequence of the destruction of the vascular supply and not a primary response to cadmium.

Immunosuppression, induced by administration of cadmium to mice, has not been reported in man.

Laboratory Investigations. Laboratory investigations are required, if undue non-industrial exposure is suspected, to determine the level of exposure in known cases and to assess the body (or renal) burden of cadmium following long-term accumulation.

Interpretation of blood and urinary levels will depend upon the frequency of exposure. Where a steady state of chronic low level exposure obtains, both the urine excretion and blood concentrations of cadmium correlate with the body burden. With exposure, such as to produce urinary levels of 10 nmol/mmol creatinine, there is considerable risk of renal tubular dysfunction. However, when a recent acute exposure to cadmium occurs a surge to much higher levels up to 50 nmol/mmol creatinine may be achieved without associated renal damage. The urinary cadmium is not then an index of body burden. Blood cadmium concentrations reflect recent exposure and tend to fluctuate considerably.[18]

A major problem in occupational monitoring is the early detection, before irreversible changes take place, of renal tubular damage. A sensitive laboratory test is the measurement of urinary β_2 microglobulin. Excretion is positively correlated with the number of years of exposure; normal excretion is less than 350 μg per 24 hours (or 15 μg/mmol creatinine in random samples), whereas men working with cadmium for 40 years may excrete more than 100 times this amount.

Conventional qualitative tests for the presence of protein in urine are positive when β_2 microglobulin is in excess of about 100 μg/mmol creatinine. Routine determination of β_2 microglobulin excretion has not yet been widely applied to the monitoring of cadmium workers and acceptable levels have not been suggested, although these results indicate that excretion should at least be kept to below 100 μg/mmol creatinine.

Determination of renal cadmium by assaying biopsy samples is clearly impractical as a monitoring procedure. A non-invasive technique using neutron-activation analysis for *in vivo* measurement of tissue cadmium concentrations has been developed but has yet to be widely applied. Nevertheless, such a procedure may prove to be more reliable than measurement of cadmium in blood or urine and if small, portable equipment can be developed will be particularly valuable in the future.

Blood levels of metallothionien have not been measured and it remains to be seen whether doing so would be of clinical or occupational value.

Cadmium and Hypertension

Although the mechanism by which the blood pressure is raised is unknown, hypertension can be readily induced in rats and rabbits by long-term, low-level cadmium feeding experiments designed to simulate normal human exposure. Paradoxically, chronic exposure to relatively high amounts of cadmium does not have this effect in animals and hypertension is not a feature of chronic occupational exposure nor of the Itai Itai disease.

The renal cadmium content of hypertensive patients has been reported to be higher than in normotensive controls in some epidemiological studies but not in others. Similarly, measurements of blood or plasma cadmium concentrations in hyper- and normo-tensives are contradictory.

Thus, while the role of cadmium in the aetiology of essential hypertension in man is unsolved, it may yet prove, as suggested by the animal experiments, to be one of many factors responsible for this condition.

Toxicity of Other Metals

Other metals, such as nickel, chromium, beryllium and aluminium, represent occupational hazards with respiratory damage following exposure to fumes or dust being the usual presentation. Absorption through the skin surface or via lesions also occurs. Non-industrial exposure is extremely rare.[21]

ESSENTIAL TRACE ELEMENTS

Modern nutritional research defines an element as essential if it fulfils the following requirements; it occurs in the natural environment and is found in physiological amounts in the diet, it is present in body tissues, and is found in the newborn and/or maternal milk. Recent work, particularly by Schwarz and his colleagues,[1] has extended the number of trace elements known to be essential to animals to 14 with a suggestion that the list is still incomplete.

Most essential elements are functionally important in metalloenzymes, as enzyme co-factors or in porphyrin complexes. Iodine and cobalt are unique in that they appear to be required only for thyroid hormones and vitamin B_{12} respectively. Some trace-element complexes, notably superoxide dismutase but possibly also zinc-containing metalloproteins, afford protection against the destructive potential of highly reactive free-radicals.

Although they are unable to cope with catastrophic amounts, mechanisms exist for the regulation of body burden and blood concentration of essential elements. For most elements, intestinal control is important while for others regulation is at the level of excretion. Further control may be achieved by immobilisation either by storage or by formation of complexes with other elements thus reducing the physiologically active fraction of the total body load.

The following discussion does not cover iron or iodine metabolism which are dealt with elsewhere.

The daily trace element requirement for growing children and for adults is readily achieved by most diets. However, during the refining of sugar and flour, chromium, zinc and possibly other elements are lost and it is apparent that extensive use of refined and convenience foods can produce symptoms of deficiency of these elements. Furthermore, dietary constituents can alter the availability of elements as has been well documented for zinc, with phytate, starch and fibre forming insoluble complexes in the intestinal lumen and preventing absorption. Other factors, such as protein, phosphate, calcium and other trace elements in the diet, not only influence availability but also compete for access to binding sites. Intestinal transport proteins have been demonstrated for zinc and copper (see below), it is therefore likely that similar proteins exist for most or all essential elements.

Within the plasma, iron, copper, zinc and nickel are complexed to specific transport proteins, further fractions may be bound to albumin and to amino acids or small peptides. Chromium occurs in the plasma in 'glucose tolerance factor', a complex containing three amino-acids, nicotinic acid and trivalent chromium. Similar complexes or proteins presumably occur for all essential elements. Pregnancy and oestrogen contraception produce changes in plasma trace element concentrations concomitant with the alterations in circulating plasma proteins.

Only the low molecular weight complexes are filtered at the glomerulus and with the exception of chromium, cobalt, molybdenum and iodine, urinary excretion is not normally the major pathway for elimination. Faecal excretion of essential trace elements represents that secreted into the bile and other intestinal and pacreatic fluids, together with the non-absorbed dietary intake, and varies therefore with food content. Some losses occur with sweating which may be significant under extreme conditions.

TABLE 2
TRACE ELEMENT CONCENTRATIONS IN BLOOD, SERUM AND URINE

Element	Reference range serum plasma	Urine
Aluminium	Less than 1 μmol/l	1·0–2·5 μmol/24 h
Chromium	0·7–6·7 nmol/l	95–235 nmol/24 h
Cobalt	0·3–10·0 nmol/l	0·85–3·25 μmol/l
Copper, men	13·7–24·1 μmol/l	Less than 1·5 μmol/24 h
women	14·0–21·6 μmol/l	
Manganese	6·0–22·0 nmol/l	1·5–43·0 nmol/l
Molybdenum	3·0–12·3 nmol/l	*
Nickel	13·6–88·6 nmol/l	8·5–109 nmol/24 h
Selenium	0·89–2·28 μmol/l	*
Zinc	11·6–19·1 μmol/l	3·0–12·0 μmol/24 h
Blood		
Cadmium	Less than 55 nmol/l	Less than 10 nmol/24 h
Lead	Less than 1·7 μmol/l	Less than 400 nmol/24 h
Mercury	Less than 30 nmol/l	Less than 50 nmol/24 h

* Insufficient data available.

Normal plasma and urinary levels are summarised in Table 2. For a number of elements there is little agreement as to normal values, and those that appear to be the most realistic to date have been included.

Zinc

Changes in Blood Concentration

Most of the circulating zinc, is contained within the red cells (75–85%), a further 3% is present in white cells while the remaining 12–22% is in the plasma. Of this latter fraction, 30–40% is bound to α-macroglobulin, about 1% is complexed with cysteine and histidine and the remainder loosely bound to albumin.[22] Release of zinc from platelets during clotting is suggested to be responsible for serum zinc being an average of 16% higher than in the equivalent plasma sample. No confirmation of the original observation has been published, however, and our laboratory together with others have been unable to confirm this difference. Meals or a glucose solution induce a rapid movement of zinc into tissues with fasting serum levels being restored after two hours. Following prolonged fasting, serum concentrations increase by up to 35%. The serum level is also subject to a diurnal variation which is indirectly related to adrenal steroid metabolism. Peak levels are observed at about 10.00 a.m. and exogenous administration of ACTH results in an increase in serum zinc which is abolished by adrenalectomy.

Serum zinc concentrations fall during the first three days following myocardial infarction and thereafter increase to normal levels by day 10. Lowest levels are correlated with the severity of the infarction and also with the peak serum levels of lactate dehydrogenase. Release of leucocyte endogenous mediator from polymorphonuclear leukocytes in response to infections or tissue damage is rapidly followed by uptake of zinc from serum into the liver. Lowest serum levels are reached after 6–9 hours and then return to basal concentrations. Low serum zinc concentrations occur in certain types of malignancy, e.g. leukaemia and carcinoma of the bronchus, and during pregnancy. In these conditions the serum copper and zinc concentrations vary inversely. Hyperzincaemia (plasma zinc, 45–60 μmol/l) has been observed in one family. The excess zinc was associated with albumin and no apparent clinical abnormality was present.

Decreased Absorption. Zinc deficiency, as a consequence of inadequate diet alone, does not appear to exist. However, in countries such as Iran and Egypt where staple diets are rich in phytate, where there are low water zinc concentrations and where metabolism is complicated by clay geophagia and considerable losses through sweating, effects of gross deficiency are observed (see below).

Artificial feeding regimes frequently contain insufficient zinc to meet metabolic demands. The synthetic diets required for children with phenylketonuria provide only one-third of the amount of zinc present in normal diets. Without supplements these children have significantly lower whole blood zinc concentrations than do normal controls.[10]

Similarly, the profound zinc deficiency which develops in patients receiving prolonged parenteral nutrition is not yet sufficiently well recognised. Trace element status in these patients is complicated by the variable and frequently unknown mineral composition of parenteral foods.

Zinc malabsorption is a feature of damage to or dysfunction of the intestinal mucosa. In coeliac, Crohn's and other gastrointestinal diseases serum zinc concentrations are decreased.

While it has yet to be established in these cases, true deficiency of zinc has been clearly shown in the inherited condition of acrodermatitis enteropathica where the probable lesion is failure to synthesise the mucosal zinc-binding protein (see below).

Increased Excretion. Urinary zinc excretion, usually about 7·5 μmoles per day, is constant for a given adult and is not related to urine volume nor to changes in dietary zinc. Renal zinc clearance is increased by 50% during treatment with diuretics and, although serum concentrations remain unaltered, zinc content of liver is significantly reduced. Chelation therapy for metal toxicity also increases urinary zinc excretion with the development in some cases of metal deficiency. Being associated with albumin, zinc is also lost if there is glomerular damage, e.g. nephrotic syndrome.

Conditions of trauma and stress, such as surgical operations, accidents and burns, are followed by a period of muscle catabolism characterised by muscle atrophy and increased excretion of nitrogen, creatine, potassium, phosphorus, sulphur and magnesium. Muscle forms the major storage site for zinc and during this catabolic phase zinc, bound to cysteine and histidine, is released into the circulation, filtered at the glomerulus and a considerable zincuria ensues. Fell et al.[23] recorded values of over 45 μmoles per 24 hours, following severe burning.

Increased renal clearance of zinc occurs in other catabolic conditions, e.g. alcoholic cirrhosis and thyrotoxicosis, but without evidence of deficiency.

Signs and Symptoms of Zinc Deficiency

The severity of deficiency and its rate of development are reflected in the symptoms produced. Thus the syndrome associated with gross zinc deficiency of rapid onset is quite different from that of chronic dietary insufficiency.

Acute Deficiency. (i) *Zinc repletion* following the considerable urinary losses associated with surgery, etc. is unlikely to be attained during unsupplemented parenteral nutrition. Recovery subsequent to these hypercatabolic phases requires considerable anabolic activity with repair and growth of damaged and wasted tissue. Zinc is essential during the anabolic period and if not replaced the symptoms of acute deficiency ensue. The clinical picture in this situation is characterised by diarrhoea, mental depression, moist, eczematous, pustular dermatitis in the para-nasal, oral and peri-oral regions and marked alopecia. Further crusted skin lesions may be present on the elbow, back, fingers, toes, perineum and scrotum. The symptoms disappear within a few days of commencement of treatment with zinc sulphate.

(ii) *Acrodermatitis enteropathica* is an inborn error of metabolism with autosomal recessive transmission. Symptoms of diarrhoea, alopecia and erosive lesions around the orifices and on the elbows, knees and ankles appear soon after weaning from breast milk and are very similar to those described above.[10] Visual symptoms, e.g. photophobia and an avoidance of central vision, poor growth and malnourished appearance are other signs. Cellular immune deficiency may be apparent although it is uncertain whether this is secondary to prolonged malnutrition. If untreated, acrodermatitis enteropathica is usually fatal within a few years. Until 1973, orally administered diiodohydroxyquinoline was the only successful treatment other than continuation of breast milk.

Moynahan & Barnes[24] first implicated zinc deficiency in this condition. They demonstrated very low (6·0 μmol/l) serum zinc concentrations and a dramatic response to oral zinc therapy.

Impaired intestinal absorption is responsible for the deficiency of zinc. Breast milk, but not cows' milk, contains a zinc binding protein which is similar to a protein present in pancreatic secretions and which is required for zinc absorption. This ligand is found in pancreatic secretions of patients with acrodermatitis enteropathica but its binding of zinc is about 10% of that of the protein from control subjects.

Chronic Deficiency. Experimental zinc deficiency produces growth retardation and testicular atrophy in pigs. A syndrome of iron-deficiency anaemia, hepatosplenomegaly, short stature and hypogonadism was described in Iranian and Egyptian men. The anaemia, but no other symptoms, responded to oral administration of iron and subsequent investigations revealed marked zinc deficiency in these subjects. Treatment of a group of patients with zinc produced increased growth rates, appearance of pubic hair and genital maturation. No such changes were observed in those treated with iron or with an animal protein diet. A similar picture of zinc-deficient hypogonadal dwarfism has been reported in Turkey, Portugal and Morocco. Only male subjects were described in initial reports, but two Iranian women have also been shown to respond to zinc supplementation by sexual maturation and growth.

Investigations of hypothalamo-pituitary-gonadol status have been inconclusive, thus the links between zinc metabolism and endocrine function are not clear.

The aetiology of zinc deficiency in these cases includes binding of zinc by dietary phytate and, where geophagia occurs, by clay and losses in sweat or from chronic bleeding due to intestinal parasites.

Other Signs of Deficiency. In contrast to these florid symptoms, other signs are associated with less severe deficiency. Normal healing of ulcers, wounds and other tissue injuries is retarded in animals made zinc deficient and in patients with low serum zinc concentrations. Pre-surgical zinc supplementation increases wound healing rates in zinc deficient but not in zinc replete subjects. The possibility of deficiency should therefore be considered prior to surgery or where wound healing is delayed, but oral administration of zinc is not indicated as a routine measure.

Cellular immune deficiency associated with malnourished children with thymic atrophy respond to supplementation with zinc. Zinc deficiency is therefore indicated as a cause of immuno-incompetence both here and in acrodermatitis enteropathica.

Mild zinc deficiency reduces and distorts sensitivity to taste and smell. Cirrhotic patients with low serum zinc concentrations are reported to have abnormal dark adaptation which may be due to reduced activity of retinol dehydrogenase—a zinc metalloenzyme.

A potentially more serious feature is the effect of zinc deficiency during pregnancy. Gross congenital abnormalities follow even transitory periods of feeding zinc-deficient diets to rats. There have been no reports of teratogenicity in zinc deficient humans although Burch et al.[22] summarise considerable circumstantial evidence suggesting that such a relationship may exist.

Zinc Toxicity

Ingestion of food contaminated by zinc while stored in galvanised containers causes nausea, vomiting, diarrhoea and fever. Identical symptoms were observed in a woman receiving home haemodialysis against water stored in a galvanised tank. Extreme lethargy, light headedness, staggering and difficulty in writing legibly were experienced by a 16-year-old boy who ingested 12 g of metallic zinc.

Inhalation of zinc oxide fumes causes a pneumonitis that may be fatal.

Assessment of Zinc Status

Serum zinc concentration measured in a sample collected to avoid contamination is the simplest laboratory test of body zinc. However, because of the physiological and pathological changes described above, a low concentration may not necessarily be a definitive indication of zinc deficiency.

Other results, which are indicative of deficiency, are decreased red cell, urine and hair zinc although hyperzincuria will be observed in zinc-losing situations. Analyses of zinc in hair is best performed in short sections of hair if a dynamic picture of changing values within the body is required.

Reduced activity of zinc dependant metalloenzymes, e.g. serum alkaline phosphatase and erythrocyte carbonic anhydrase are indirect measurements of body zinc. Metabolic balance studies with ^{65}Zn provide sensitive indices of zinc status but are rarely necessary in man. Probably the most important demonstration of zinc deficiency is the clinical response to zinc supplementation.

Copper

An adult body contains about 100 mg (1·6 mmol) of copper with highest concentrations in the liver and decreasing amounts in brain, heart and kidney. Like zinc, copper has been recognised as an essential trace element for several decades and about 2·5 mg (40 μmol) is required daily by man. Absorption by intestinal mucosal cells is achieved by at least two mechanisms the most important involving superoxide dismutase and a metallothionein type of protein. Rate of absorption is influenced by other trace elements in the lumen and particularly by molybdenum and sulphate both of which are inhibitory.

Changes in Blood Concentration

Copper circulates in erythrocytes (mainly with superoxide dismutase) and in the plasma. More than 90% of plasma copper is associated with caeruloplasmin. Most of the remainder, representing copper in transit from sites of absorption to tissues, is bound to albumin and a small fraction occurs as amino acid complexes.

Most reports state that normal serum copper concentrations are greater in women 13·7–24·1 μmol/l than in men 14·0–21·6 μmol/l although this difference has been denied by some.

Excretion of copper is almost exclusively via the bile, bound to protein macromolecules and less than 1 μmol appears in the urine each day.

Hypocupraemia can develop following inadequate dietary copper, prolonged parenteral nutrition (particularly in newborn who have an exceptional daily requirement), or malabsorption due to increased intake of molybdenum or sulphate. Deficiency is manifested by anaemia which is responsive to a combination of iron and copper.

Very low serum copper concentrations are features of Wilson's Disease and Menkes Syndrome (see below).

Increased caeruloplasmin synthesis with concomitant rise in serum copper concentration occurs during pregnancy and oestrogen contraception. Interference in biliary excretion of copper may be responsible for the non-specific hypercupraemia of acute and chronic liver disease. Other conditions in which serum concentrations are increased include rheumatoid arthritis, major surgery, myocardial infarction and certain malignant diseases, e.g. acute lymphoblastic leukaemia and Hodgkin's disease (see trace elements and monitoring of disease).

Increased serum copper concentrations have been reported in patients developing delirium tremens or hallucinations during alcohol withdrawal. We have found similar results in patients with episodes of epilepsy.

Copper-containing Enzymes

Caeruloplasmin. Caeruloplasmin, synthesised in the liver, is the transport protein which redistributes copper to other body tissues. It is also an oxidase enzyme and is essential for the mobilisation of iron from ferritin by converting the released ferrous iron to the ferric state which then forms a stable complex with transferrin. Thus copper deficiency can give rise, as has been noted above, to a condition of iron-deficiency anaemia. Because of its role in iron metabolism caeruloplasmin is sometimes referred to as ferroxidase. Other caeruloplasmin oxidase substrates include catecholamines, serotonin and melatonin.

Superoxide Dismutase. Free radicals such as the superoxide ion (O_2^-) are highly reactive and as such are potentially very toxic to adjacent tissue. The superoxide ion is a product of certain enzyme systems, e.g. xanthine oxidase, and to protect vulnerable tissue the copper-containing superoxide dismutase is present in erythrocytes, brain and liver. The O_2^- is converted to oxygen and hydrogen peroxide.

Other copper-dependent enzyme systems include cytochrome oxidase, monoamine oxidase and tyrosinase.

Menkes' Kinky Hair Syndrome

Menkes et al.,[25] first described a syndrome characterised by coarse stubby hair, convulsions and progressive mental retardation, coarse facies, scurvy-like changes in the skeleton and degenerative changes in the internal elastic lamina of blood vessel walls. Hypothermia and blindness may also occur. Kinky hair syndrome is an X-linked recessive condition, the symptoms appearing before 3 months of age. Due to an increased susceptibility to infection, death usually occurs by the age of 3 years.

The structural changes in hair, blood vessels and bone closely resemble abnormalities described in copper-deficient animals and it was demonstrated by Danks[26] that these infants have very low serum concentrations of copper (less than 5.0 μmol/l) and caeruloplasmin. Copper is required for the normal development of each of these tissues. It is in ascorbate oxidase in the bone, in lysyl oxidase (a monoamine oxidase) required for elastin formation and in disulphide bonding in the keratin fibres of hair.

Copper deficiency in these patients is further demonstrated by reduced cytochrome oxidase activity in brain, muscle and liver and by low copper concentrations in liver and brain.

Using ^{64}Cu it has been shown that while the intestinal mucosal cells accumulate copper, only very little transport across the serosal membrane is possible. Following intravenous administration, copper initially enters the red cells but is subsequently taken into the liver where it is retained with a prolonged half-life. It appears that an abnormal metallothionein with altered binding of copper is responsible for immobilisation of the metal in the liver and a similar mechanism would also account for the defective intestinal absorption.

Treatment of patients with copper either parenterally or by intramuscular injection does not reverse pre-existing disease even when therapy commences before the patient is one month old. Normal serum copper and caeruloplasmin concentration can be maintained but either because the copper fails to enter neural tissue or because damage has occurred *in utero*, this treatment is of little benefit.

Wilson's Disease

Wilson's Disease (hepatolenticular degeneration) is a rare inherited disease of copper metabolism. The main features are cirrhosis of the liver, renal tubular dysfunction and progressive degeneration of the central nervous system. A pathognomonic sign is the presence of green–brown Kayser–Fleischer rings in the cornea. Onset of symptoms is usually during the teens, although it has been reported at 6 years of age and is seen earlier in female than in male subjects. Neurological symptoms include poor coordination, a slight tremor, dysarthria, dysphagia and choreoform movements. These signs are however often preceded by hepatic dysfunction which can present itself in a variety of ways, including chronic active hepatitis, acute hepatitis and cryptogenic cirrhosis.

In the serum, low concentrations of copper (< 9 μmol/l) are usually evident (although normal levels exist in a small proportion of cases, particularly those with severe hepatocellular necrosis), while urinary excretion is considerably increased (> 1.5 μmol/24 h). These determinations, together with measurement of caeruloplasmin, are the appropriate biochemical investigations for the diagnosis of Wilson's Disease.

It is suggested that a protein with an increased affinity for copper is present in the liver and synthesis of caeruloplasmin is therefore decreased. Upon saturation of the protein binding sites, excess cellular copper is incorporated into lysosomes. Further hepatic uptake is inhibited with a consequent increase in the amount of copper bound to albumin and small molecular weight compounds. This copper is then distributed to tissues such as the brain, kidney tubules and the cornea.

Irreversible damage to the liver and brain can be prevented if treatment with the chelating agent penicillamine is commenced at an early stage of the disease.

Acute Copper Intoxication

Ingestion of copper salts, either by accident or following suicidal or homicidal intent, produces a complex clinical picture. Nausea, vomiting and diarrhoea may be followed if toxicity is severe, by intravascular haemolysis marked by increased plasma haemoglobin and bilirubin, reticulocytosis and haemoglobinuria. Acute renal failure due to tubular obstruction by haemoglobin, deposition of copper in the kidney and dehydration occurs a few days later and may be fatal despite haemodialysis.

A similar syndrome was recognised some years ago in patients receiving haemodialysis using Cuprophan as the dialysis membrane or dialysis fluid plumbed through copper tubing. Once the cause was determined further cases were prevented.

Manganese

Manganous ion is an activator of many enzymes including RNA polymerase and RNA-dependent DNA polymerase, thus this metal is vital for the synthesis of nucleic acids and protein. Manganese also plays a role in various endocrine systems. The activation of adenylate cyclase by antidiuretic hormone is inhibited by manganese while hepatic storage of the metal is influenced by ACTH and glucocorticoid hormones.

Manganese is present in most body tissues but the highest concentrations are found in mitochondria-rich cells. Very little is excreted in the urine, elimination being via the bile and pancreatic secretions. Under conditions of manganese load intestinal excretion is increased suggesting that homeostatic control of body manganese is achieved at this level rather than at the site of absorption.

Manganese Deficiency

In man, manganese deficiency (with symptoms of dermatitis, weight loss, nausea and vomiting, change of hair colour and hypocholesterolaemia) has been described only in an artificial situation.[22] In animals impaired mucopolysaccharide synthesis causes skeletal deformities, abnormal otolith development is responsible for congenital ataxia while defective sexual development produces sterility. Hypoglycaemia and impaired growth are other features of experimental manganese deficiency.

Manganese Toxicity

Manganese is relatively non-toxic, requiring considerable, long-term exposure, e.g. during mining, to evoke neurological symptoms similar to Wilson's and Parkinson's diseases; psychiatric changes also occur. The tissue manganese concentrations are paradoxically lower in patients with chronic poisoning than in control subjects.

There remains considerable uncertainty as to normal serum levels of manganese,[27] a reflection of the problems in obtaining samples free from contamination and reported changes in concentrations in various pathological states may be illusory.

Chromium

Interest in chromium metabolism began when it was discovered that the activity of glucose tolerance factor (GTF), a principal found in sources such as brewer's yeast and pig kidney, and which prevents the glucose intolerance of rats fed on a 30% Torula yeast diet, was mimicked by complexes of chromium.

GTF is essential for insulin activity possibly forming a complex between insulin and the receptor site, thus disturbances of carbohydrate, protein and lipid metabolism are evident in chromium deficient animals.

Use of refined foods considerably reduces the dietary intake of chromium. Thus it has been suggested that chromium deficiency may be responsible for glucose intolerance in certain groups, e.g. the elderly. Improved glucose tolerance in some individuals following oral chromium supplements supports this hypothesis.

Glinsmann, Feldman and Mertz[28] suggested that the plasma concentration is not an accurate index of chromium status but that the plasma chromium increment 30–120 minutes after an oral glucose load is more meaningful. The magnitude of this rise is dependent upon the level of body chromium and is reduced in patients with impaired glucose tolerance as in some cases of stress or infection, in pregnancy and in maturity onset diabetes. As with many other trace elements the concentrations have been revised in recent years although the post-oral-glucose rise has been confirmed by others. A fasting level of 58 nmol/l (3 ng/ml), with an increment of a further 19 nmol/l (1 ng/ml), was reported by Hambridge.[29] Following the rise in plasma chromium the urinary excretion is increased.

Other Essential Elements

Of other trace elements very little is known concerning their biochemical function, although considerable speculative interest is evident (e.g. silicon and cardiovascular disease). Occupational exposure to fumes of metal compounds, e.g. nickel carbonyl, is a problem during refining and in other industrial processes but there has been little detailed work performed to evaluate the monitoring procedures used. Generally excretion in the urine has been determined although attention is now placed upon blood concentrations in addition to or instead of urine levels.

For detailed reviews of the biochemistry of these elements see:

nickel,[30] cobalt,[31] selenium,[32] molybdenum,[33] silicon, fluorine, tin and vanadium.[34]

Other Aspects of Trace Element Biology

Total Parenteral Nutrition

Subjects with intestinal incompetence following infarction, radiotherapy, surgery, etc., can be treated by prolonged or even permanent total parenteral nutrition. Technical problems associated with intravenous feeding have now been superceded by those attendant upon the gradual development of deficiencies of vitamins, trace elements and other nutrients. These situations arise because trace element requirements, especially during pathological crises, are not known; the trace element content of i.v. foods and supplements are not always available and the symptoms of deficiency (with the exception of zinc) are poorly described.

In this important area of trace element metabolism there is an urgent requirement for the development of methods for multi-element analyses at very low concentrations. Considerable information can also be obtained from veterinary research which for many years has been concerned with recognition of trace element deficiencies in animals.

Chronic Renal Failure and Maintenance Haemodialysis

During a period of routine haemodialysis the body is intimately exposed to far greater volumes of dilute fluid than normally occurs, e.g. through drinking. The consequences of such exposure may be either losses of essential elements or aquisition of toxic elements. It appears, however, that losses probably do not occur. There are no reports of trace element deficiencies in these patients and investigations made during dialysis indicate that only small concentration changes occur which are very soon repaired.

Movement of elements from the dialysis medium into the blood does take place. In addition to the uptake of copper and zinc from dialysis equipment, as referred to earlier, contamination by nickel was discovered at one centre. Recently it has been demonstrated that in areas where aluminium is added to the domestic water supply there is an increased incidence of the syndrome of progressive mental deterioration, osteomalacia and myopathy known as dialysis dementia, among renal patients on home dialysis.[35] Evidence from many dialysis centres where exposure to aluminium has been monitored and then abolished also indicates that aluminium is involved in the aetiology of dialysis dementia.

Kaehny et al.,[36] have shown that plasma is a very efficient sequestering agent, removing aluminium from dialysis fluid and that these patients have increased concentrations of aluminium in the blood and also in cerebral tissues.

Because of the unexpected nature of the contamination in these cases it is not inconceivable that further examples may well arise and metal toxicity must therefore be considered if haemodialysis patients develop bizarre, unexplained symptoms.

Trace Elements as Pharmacological Agents

The history of deliberate administration of trace element compounds probably co-exists with the history of the industrial usage, with mercury (to treat syphyllis), lead (to induce abortion) and arsenic (with homicidal or suicidal intent) as notable examples.

More recently chrysotherapy using soluble gold compounds like sodium aurothiomalate to treat rheumatoid arthritis and cancer chemotherapy with *cis*-complexes of platinum-diamine compounds have had spectacular successes in some patients. Together with the benefit gained from such therapy there is the equally evident toxicity which limits extensive use.

Cobalt salts are sometimes used to stimulate erythropoiesis in the refractory anaemia of chronic renal failure. However, like gold and platinum, the associated toxicity of cobalt makes careful supervision mandatory during such treatment.

Trace Elements in the Monitoring of Disease

In some pathological conditions, apparently unrelated to their normal metabolism, changes in blood concentrations of trace elements have proved valuable in predicting the clinical course or the response to treatment. Thus an increased serum copper concentration is a poor prognostic sign in Hodgkin's disease and in certain other carcinomas, e.g. acute lymphoblastic leukaemia in which concomitant hypozincaemia may also exist.

Copper and caeruloplasmin concentrations in serum from patients with rheumatoid arthritis are similarly increased during the active stages of the disease, returning to normal during remission.

SUMMARY

There are three areas of the biochemistry and toxicology of metals which will almost certainly be prominent among future developments. Investigations of the effects of long-term exposure to metals at levels lower than those associated with 'classical toxicity', which have been initiated for lead and cadmium, will be extended to other elements particularly those of occupational interest. Efforts will be made to determine the metabolism of the essential elements and to recognise clinical syndromes associated with their deficiency. The interactions between trace elements, now recognised as being important in intestinal absorption, will be studied at other sites.

New advances in analytical techniques will undoubtedly play a major role in these developments. The multi-element potential of inductively-coupled plasma emission (ICPE) spectroscopy has yet to be widely applied to biological materials. An exciting concept, still early in its development, is a linking of ICPE with a mass spectroscope detection unit and thus offering, from a limited sample volume, exceptional sensitivity for several dozens of elements.

REFERENCES

1. Schwarz, K. (1977). In *Clinical Chemistry and Chemical Toxicology of Metals*, (Brown, S. S., Ed.). Amsterdam: Elsevier/North Holland.
2. Rabinowitz, M. B., Wetherill, G. W. and Kopple, J. D. (1976), 'Kinetic analysis of lead metabolism in healthy humans', *J. Clin. Invest.*, **58**, 260.
3. Barltrop, D. (1977). In *Clinical Chemistry and Chemical Toxicology of Metals*, (Brown, S. S., Ed.). Amsterdam: Elsevier/North Holland.
4. Clayton, B. (1975), 'Lead: The relation of environment and experimental work', *Br. Med. Bull.*, **31**, 236.
5. Piomelli, S. (1973), 'A micromethod for free erythrocyte porphyrins. The FEP test', *J. Lab. Clin. Med.*, **81**, 932.
6. Center for Disease Control (1978), 'Preventing lead poisoning in young children', *J. Ped.*, **93**, 709.
7. Irwig, L. M., Harrison, W. O., Rocks, P., Webster, I. and Andrew, M. (1978), 'Lead and morbidity: A dose-response relationship', *Lancet*, **ii**, 4.
8. Landrigan, P. J., Baker, E. L., Feldman, R. G., Cox, D. H., Eden, K. V., Orenstein, W. A., Mather, J. A., Yankel, A. J. and Von Lindern, I. H. (1976), 'Increased lead absorption with

anaemia and slowed nerve conduction in children near a lead smelter', *J. Ped.*, **89**, 904.
9. Waldron, M. A. and Stofen, D. (1974), *Sub-clinical Lead Poisoning*. London/New York: Academic Press.
10. Delves, H. T. (1976), 'The clinical value of trace-metal measurements', *Essays Med. Biochem.*, **2**,
11. Beattie, A. D., Moore, M. R., Goldberg, A., Finlayson, M., Graham, J., Mackie, E., Main, J., McLaren, D. A. Murdoch, R. M. and Stewart, G. T. (1975), 'Role of chronic low-level lead exposure on the aetiology of mental retardation', *Lancet*, **i**, 589.
12. Clarkson, T. W. (1977). In *Clinical Chemistry and Chemical Toxicity of Metals*, (Brown, S. S., Ed.). Amsterdam: Elsevier/North Holland.
13. Cherian, M. G., Hursh, J. B., Clarkson, T. W. and Allen, J. (1978), 'Radioactive mercury distribution in biological fluids and excretion in human subjects after inhalation of mercury vapour', *Arch. Environ. Hlth.*, **33**, 109.
14. Smith, R. G., Vorwald, A. J., Patil, L. S. and Mooney, T. F. (1970), 'Effects of exposure to mercury in the manufacture of chlorine', *Am. Ind. Hyg. Ass. J.*, **31**, 687.
15. Amin-Zaki, L., Majeed, M. A., Clarkson, T. W. and Greenwood, M. R. (1978), 'Methyl mercury poisoning in Iraqi children: clinical observations over two years', *Br. Med. J.*, **1**, 613.
16. Lauwerys, R. R. and Buchet, J. P. (1973), 'Occupational exposure to mercury vapors and biological action', *Arch. Environ. Hlth.*, **27**, 65.
17. Marks, V. and Taylor, A. (1979), 'Urinary mercury excretion in dental workers', *Br. Dent. J.*, **146**, 269.
18. Piscator, M. and Pettersson, B. (1977). In *Clinical Chemistry and Chemical Toxicity of Metals*, (Brown, S. S., Ed.). Amsterdam: Elsevier/North Holland.
19. Perry, H. M., Thind, G. S. and Perry, G. F. (1976), 'The biology of cadmium', *Med. Clin. N. Amer.*, **60**, 759.
20. Webb, M. (1975), 'Cadmium,' *Br. Med. Bull.*, **31**, 246.
21. Norval, E. and Butler, L. R. P. (1974), 'Trace metals in man's environment and their determination by atomic absorption spectroscopy', *S. A. Med. J.*, **48**, 2617.
22. Burch, R. E., Hahn, H. K. J. and Sullivan, J. F. (1975), 'Newer aspects of the roles of zinc, manganese and copper in human nutrition', *Clin. Chem.*, **21**, 501.
23. Fell, G. S., Fleck, A., Cuthbertson, D. P., Queen, K., Morrison, C., Bessent, R. C. and Hussain, S. L. (1973), 'Urinary zinc levels as an indication of muscle catabolism', *Lancet*, **i**, 280.
24. Moynahan, E. J. and Barnes, P. M. (1973), 'Zinc deficiency and a synthetic diet for lactose intolerance', *Lancet*, **i**, 676.
25. Menkes, J. H., Alter, M., Steigleder, G. K., Weakley, D. R. and Sung, J. H. (1962), 'A sex-linked recessive disorder with retardation of growth, peculiar hair and focal and cerebellar degeneration', *Pediatrics*, **29**, 764.
26. Danks, D. M., Stevens, B. J., Campbell, P. E., Gillespie, J. M., Walker-Smith, J., Blomfield., J. and Turner, B. (1972), 'Menkes' kinky-hair syndrome', *Lancet*, **i**, 1100.
27. Versieck, J. and Cornelis, R. (1980), 'Normal levels of trace elements in human blood, plasma or serum', *Anal. Chim. Acta*, **116**, 217.
28. Glinsmann, W. H., Feldman, F. J. and Mertz, W. (1966), 'Plasma chromium after glucose administration', *Science*, **152**, 1243.
29. Hambidge, K. M. (1974), 'Chromium nutrition in man', *Am. J. Clin. Nutr.*, **27**, 505.
30. Sunderman, F. W. (1977). In *Clinical Chemistry and Chemical Toxicity of Metals*, (Brown, S. S., Ed.). Amsterdam: Elsevier/North Holland.
31. Taylor, A. and Marks, V. (1978), 'Cobalt: A review', *J. Hum. Nutr.*, **32**, 165.
32. Hoekstra, W. G. (1975), 'Biochemical function of selenium and its relation to vitamin E', *Fed. Proc.*, **34**, 2083.
33. Reinhold, J. G. (1975), 'Trace elements: A selective survey', *Clin. Chem.*, **21**, 476.
34. Nielsen, F. M. and Sandstead, H. H. (1974), 'Are nickel, vanadium, silicon, fluorine and tin essential for man?' *Am. J. Clin. Nutr.*, **27**, 515.
35. Elliott, H. L., Dryburgh, F., Fell, G. S., Sabet, S. and Macdougal, A. I. (1978), 'Aluminium toxicity during regular haemodialysis', *Br. Med. J.*, **i**, 1101.
36. Kaehny, W. D., Alfrey, A. C. Holman, R. E. and Shorr, W. J (1977), 'Aluminium transfer during hemodialysis', *Kidney Intl.*, **12**, 361.

22. LABORATORY INVESTIGATION OF THE POISONED PATIENT

MICHAEL J. STEWART

Introduction
   Drug interactions

The role of the laboratory
   Confirmation of diagnosis
   Prognosis

Treatment
   General
   Gastric lavage and emesis
   Forced diuresis
   Haemodialysis

Interpretation of tests in cases of poisoning
   Blood gas analysis
   Osmolality
   Alterations in electrolyte homeostasis
   Enzyme determinations
   Interference between drugs and chemical tests
   Sample collection
   Summary

Specific applications
   Benzodiazepines
   Carbon monoxide
   Cardioactive drugs
   Ethanol
   Ethylene glycol
   Hypnotic drugs
   Inorganic anions
   Metals
   Methanol
   Narcotics
   Organochlorides
   Organophosphates
   Paracetamol
   Paraquat
   Salicylate
   Tricyclic antidepressants

Conclusions

INTRODUCTION

Poisoned patients account for approximately 10% of all acute admissions to medical wards in British hospitals and a similar pattern is seen in other developed countries. Not all of these patients are there as a result of self-poisoning attempts. Some are poisoned by accident and a minority of cases, including a significant number of children, are due to administration of drugs or poisons by others.

Many poisoning victims die before reaching hospital and many analyses are carried out in departments of Forensic Medicine or Science. Of those who are admitted to hospital approximately 95% are discharged well, the exceptions being those who are moribund on arrival following severe respiratory depression, or who have ingested drugs or poisons which have an irreversible action such as paraquat.

The pattern of poisoning reflects, but lags behind, prescribing patterns and nowhere is it more of a truism that 'common drugs are taken commonly'. Despite this fact there are always single cases of poisoning with uncommon agents which may be difficult to diagnose and manage, and are unfortunately often under-documented.

Although the incidence of acute self-poisoning is either static or falling, there is increasing recognition of the problem of chronic toxicity. This falls into two categories:
(a) Unwitting exposure to toxins
(b) Chronic overdosage with prescribed drugs

The first category will be covered here only in relation to pesticides and lead, but the toxicologist involved in drug analysis must expect to find cases of unwitting chronic overdose as a result of inappropriate therapeutic regimes.

Excess of the majority of drugs will elicit side-effects which are related to the concentration of drug in the plasma. Idiosyncratic side-effects such as agranulocytosis or haemolysis are, however, seldom dose-related and present as acute problems of toxicity.

Common examples of chronic iatrogenic toxicity are the build-up of toxic levels of phenobarbitone and phenytoin in patients on long-term treatment and the development of high concentration of tricyclic antidepressants or digoxin in the elderly as a result of reduced clearance. Chronic lithium toxicity does occur and for this reason patients on this drug for long periods should be occasionally monitored, especially if there are changes in clinical state. Inappropriate dosage during chronic i.v. therapy with agents such as diazepam and chlormethiozole can also give rise to iatrogenic overdose and in these cases quantitative, timed analyses may be required to elicit the full facts.

Drug Interactions

The majority of common drug interactions are known, and data should be available through hospital pharmacies. There may be occasions where an interaction is suspected in the absence of any evidence of dosing with a particular drug; in these cases a simple screen may be of use.

Knowledge of the effects of common drugs when taken in overdose is not always available on the spot and advice may have to be sought outside the receiving hospital. In the UK rapid advice is available through the Regional Poisoning Treatment Centres in London, Cardiff, Birmingham and Edinburgh. Advice on interpretation of laboratory results, prognosis and treatment is available 24 hours a day.

THE ROLE OF THE LABORATORY

It is now accepted that a limited range of drug estimations should be available in all clinical chemistry laboratories and that a wider range of assays, both screening and quantitative determinations should be available in a few selected centres.

The clinical biochemistry laboratory is involved in all serious cases of poisoning by virtue of the derangements which occur in respiratory and renal function and in other major homeostatic mechanisms. In addition, there are a few specific biochemical investigations which may contribute to diagnosis and prognosis and which require to be available on demand. Examples are the estimation of plasma pseudo-cholinesterase in organophosphorus poisoning and measurement of methaemoglobin following poisoning with oxidising agents.

The average laboratory is not equipped to provide a comprehensive screening service in order to confirm or eliminate the presence of all drugs in common use, nor to provide quantitative assays for a large number of these compounds. One study has shown that in only 7% of cases did the results of a comprehensive screening procedure alter the treatment which was contemplated. In addition, screening procedures carried out in laboratories where experience is limited, have led to errors both of omission and commission in the 'identification' of drugs.

Quantitative determinations of drugs are similarly open to difficulties in interpretation, and close liaison between the physician and the analyst is necessary if results are to be correctly interpreted.

Within limits, however, the laboratory should be able to aid the clinician in:
Confirmation of Diagnosis
Prognosis
Monitoring of Treatment

Confirmation of Diagnosis

The physician presented with an unconscious or semiconscious patient has a diagnostic problem which is best solved by a thorough history and clinical examination. Only after this should requests for drug analyses be made. More important in the first instance are glucose analyses since both hypo- and hyperglycaemia can and have been confused with drug overdosage. The presence of ethanol, often in addition to other drugs, should be looked for and an attempt made to obtain details of the amount ingested. With a good examination a guide will be obtained to the drugs ingested in the majority of cases, the exception being children and drug addicts (who are poor historians and may well have purchased adulterated or mixed drugs).

In such cases the priority assays, as in any acute emergency, will be blood gas analyses, glucose/urea and electrolytes. In fitting patients, calcium analyses are mandatory.

Emergency Screening for Drugs

Since treatment of overdose for the vast majority of drugs is conservative, the necessity for screening out of hours is reduced to the resolution of two problems:
1. The differentiation of a drug overdose from some other clinical condition giving the same signs.
2. The positive identification of drugs or poisons for which there is a specific treatment available (Table 1).

TABLE 1
DRUGS AND POISONS FOR WHICH SPECIFIC TREATMENT IS AVAILABLE

Drug/Poison	Treatment	Mechanism
Cyanide	Cobalt tetra-acetate	Chelating Agent
Ethylene Glycol Methanol	Ethanol	Completes for Enzyme Preventing Acid Production
Iron	Desferrioxamine	Specific Chelating Agent
Other Heavy Metals	Dimercaprol and Penicillamine	Chelating Agents
Opiates	Naloxone	Antagonist
Organophosphorus Insecticides	Atropine Pralidoxime	Acetylcholine Antagonist Cholinesterase Reactivator
Paracetamol	N-acetyl cysteine	Inactivates Toxic Intermediate

As an addendum to this last point it should be pointed out that if it is proposed to use a non-specific treatment such as the instillation of charcoal slurry in order to prevent further absorption of the 'drug', then the identification of the drug in question is not necessary since the treatment carries little risk of harm to the patient. If, on the other hand, haemodialysis is contemplated, the identification of the drug should be attempted (if not known from the history) since not all drugs are dialysable and there is a certain risk attached to the procedure. Dialysis may, of course, be instituted in patients with renal failure and in these cases there is a less urgent requirement for drug identification.

The confirmation that a patient's condition is in fact due to drugs is the reason for the majority of requests made for urgent 'drug screening'. Such a request should not be undertaken without discussion with the receiving clinician.

Prognosis

Prognosis in most cases of drug overdose is more related to the general state of the patient on admission and the quality of care than on the levels of drugs detected in the plasma.

There are only a few drugs and poisons for which a

TABLE 2
DRUGS/POISONS FOR WHICH THE PLASMA CONCENTRATION IS PARTICULARLY USEFUL FOR PROGNOSIS OR TREATMENT DECISION

Drug/Poison	Likely effect	Treatment available
Glutethimide	'Cycling' coma	Charcoal slurry Prevents reabsorption
Iron	Gut erosion, Hepatic damage	chelation with Desferrioxamine
Paracetamol	Hepatocellular damage	N-acetyl cysteine
Paraquat	Renal damage, Progressive alveolar degeneration	None
Phenobarbitone	Prolonged coma with Respiratory depression	Forced alkaline diuresis
Quinine/Quinidine	Blindness, Sever arrhythmias	Stellate ganglion block Forced acid diureses
Salicylate	Severe metabolic acidosis	Forced alkaline diuresis

TABLE 3
EXAMPLES OF DRUGS COMMONLY ENCOUNTERED IN OVERDOSE SHOWING HALF-LIVES (AT THERAPEUTIC DOSES) AND PLASMA CONCENTRATIONS ABOVE WHICH TOXICITY WOULD BE EXPECTED

Drug	Half-life (h)	Upper acceptable plasma concentration	Active metabolite
Amitriptyline	20–30 h	800 µg/l	Yes
Barbiturates (Hypnotic)	Varies from 5–40 h	10 mg/l	No
Carbamazepine	5–25 h	12 mg/l	?
Chlorpromazine	20–40 h	700 µg/l	Yes
Chlordiazepoxide	7–28 h	200 µg/l	?
Diazepam	20–90 h	200 µg/l	Yes
Digoxin	24–48 h	3·0 ng/ml	No
Ethanol	Dose dependent	800 mg/l	No
Glutethimide	5–40 h	10 mg/l	?
Imipramine	10–15 h	100 µg/l	Yes
Lithium	16–30 h	2 mmol/l	No
Methaqualone	10–42 h	10 mg/l	No
Nortriptyline	20–40 h	800 µg/l	?
Paracetamol	4 h (in absence of hepatic damage)	Depends upon time after dose	Yes
Phenobarbitone	50–120 h	40 mg/l	No
Phenytoin	20–40 h (adult) 4–10 h (children)	20 mg/l	No
Primidone	Progress best followed by measuring phenobarbitone metabolite		
Quinine/Quinidine	2–12 h	5 mg/l	Yes
Salicylate	Dose dependent	250 mg/l	No
Theophylline	4–9 h (adult)	20 mg/l	Caffeine in Neonates

knowledge of the plasma concentration on admission can indicate a prognosis or a need for specialised treatment. These are summarised in Table 2.

Plasma concentrations of other drugs which are obtained in cases of overdose must be interpreted with caution and a certain amount of basic information is required:

1. What is the accepted range of concentration of this drug which would be found in a patient taking his usual daily dose?

There is a large group of drugs for which this information is not known and only an upper limit, which may be "undetectable" by the method in use, is quoted. A list of drugs commonly encountered in cases of self-poisoning, along with plasma concentrations above which toxic effects would be expected to give in Table 3.

2. Is the patient in the habit of taking the drug in question or is this a single exposure?

The importance of this question is often overlooked, since a naive patient may show a greater response to a concentration of a drug than a patient who is accustomed to that drug.

When answering both of these questions it is pertinent to be aware that the examples given relate to plasma levels, and not to urine or gastric washings, since quantitative information is required.

3. Much useful information may be obtained if more than one analysis is made, with a known time interval between the analyses. An example of the difference in handling of a drug by different patients may be seen in Fig. 1 where twin children each took the same dose of phenytoin during a 'game'. The previously unexposed child showed a much longer half-life than his brother who was receiving phenytoin routinely.

Once equilibration between plasma and tissues has occurred, the normal sequence of events is a fall in plasma drug concentration accompanied by an improvement in the clinical condition of the patient. A fall in plasma level indicates that the detoxification mechanisms of metabolism and elimination are acting to reduce the effects of the drug of interest.

Confirmation of this fact may be necessary in cases where the clinical condition of the patient does not improve at a rate which would be expected. It is for this reason that an on-admission specimen of blood should be obtained from all poisoned patients, since retrospective analysis may give a baseline against which to measure plasma drug concentrations obtained some hours or days later (Fig. 2).

The rate of fall of a drug concentration should approximate to the known biological half-life, some examples of which are given in Table 3. Further figures may be obtained from a variety of sources, but there are many older drugs for which the information is not readily available. Figures for most drugs are now available from the manufacturers. The half-life of a drug may be abnormal in the early stages following overdose due to saturation of the eliminating mechanisms; however such studies are normally requested on the day after ingestion in which case figures obtained in patients receiving therapeutic doses may be used for reference. There are, however, cases in which such a pattern does not occur and it is in such cases that drug analyses can be of great assistance to the clinician.

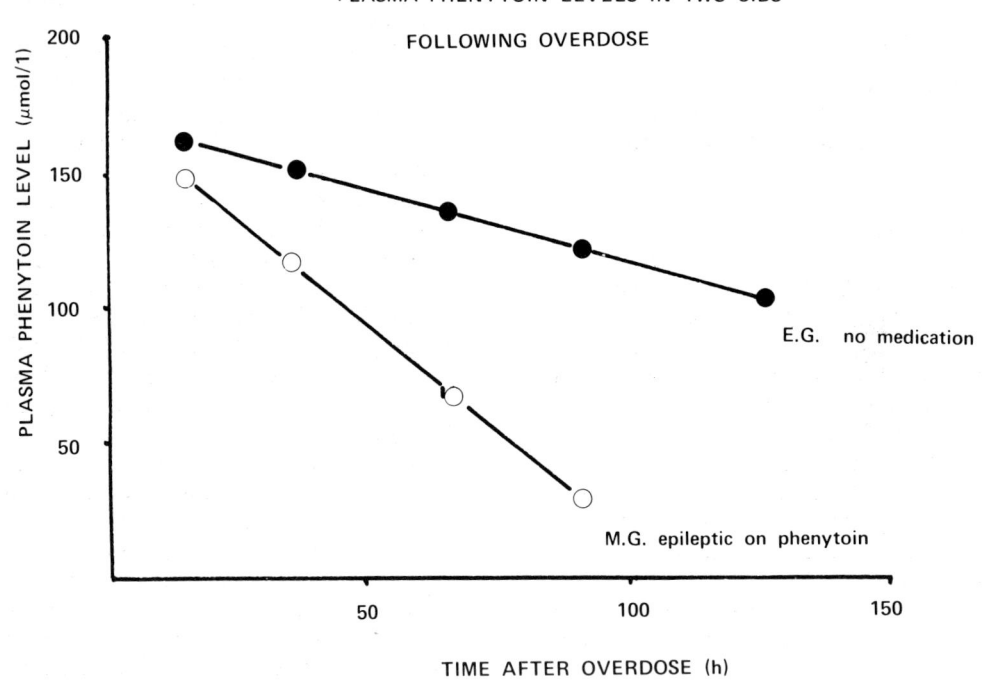

FIG. 1. Plasma phenytoin concentrations in two sibs following overdose with the same number of tablets.

A static or slow-falling drug concentration in plasma may be due to one or more of the following causes:

(a) Reduced renal perfusion, preventing the excretion of water-soluble drugs or metabolites.

(b) Reduced hepatic perfusion, preventing access of drug to metabolising enzymes.

(c) Peripheral hypoxia, reducing the activity of metabolic enzymes.

(d) Physical obstruction to excretion via the renal tubules or bile duct.

A rising plasma drug concentration is less common, but may be observed in the following cases:

(a) In the early stages of alkaline diuresis where plasma pH is high, plasma levels of salicylate or phenobarbitone may rise due to a shift of drug from the tissues into the plasma in advance of its elimination via the kidney.

(b) Some drugs, notably glutethimide, undergo enterohepatic circulation, giving rise to fluctuating plasma levels.

(c) Following periods of haemodialysis or similar active methods of elimination, the plasma drug levels, which may have been reduced to low values, may rise due to re-equilibration with drug in the tissues.

(d) Following inadequate gastric lavage, reabsorption of drugs may take place at a faster rate once blood pressure is restored in hypotensive patients.

(e) Some of the most mystifying cases in which drug levels have remained high or fluctuating for many days have been traced to continued intake by the patient, replenished by relatives or from unwitting absorption of drugs secreted in the small intestine, rectum or vagina. Such causes should be considered when logic fails.

Finally, a seriously poisoned patient may be treated with large volumes of i.v. fluids. It should be remembered that these may give rise to alteration in plasma drug concentrations unless care is taken in the selection of sampling sites and time is allowed for equilibration.

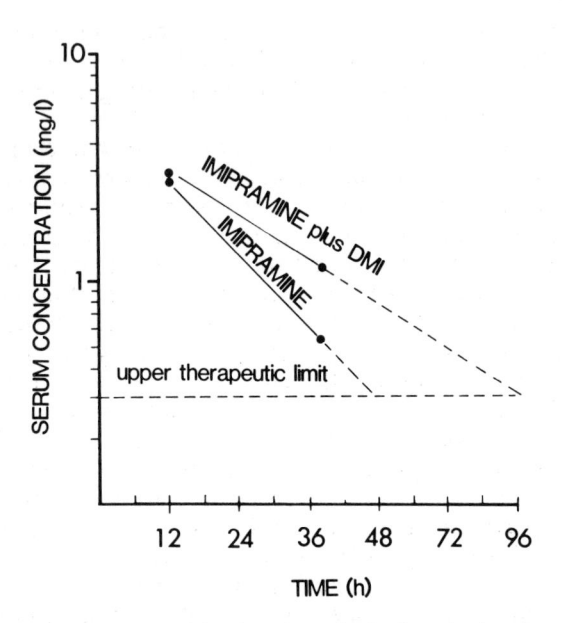

FIG. 2. The use of 2 timed specimens for the determination of the likely recovery time of a patient following an overdose of imipramine. (DMI is an active metabolite).

TREATMENT

General

Regardless of the availability of laboratory support, immediate clinical measures will be instituted in all cases of serious poisoning. Conservative therapy consists of the maintenance of a clear airway and adequate blood pressure.

Correction of cardiac arrhythmias may be necessary where drugs such as trichloral or tricyclic antidepressants have been taken, and assisted ventilation in cases with respiratory depression. The use of the endotracheal tube is mandatory if gastric washout is to be carried out in an unconscious patient and has contributed greatly to the fall in the number of deaths in hospital formerly caused by inhalation of vomit by patients who may have been only mildly poisoned or merely drunk.

Fluid therapy is commonly used in severe cases in order to maintain an adequate hepatic and renal blood flow, without which the elimination of drugs by physiological means cannot proceed.

As with any patient receiving intensive therapy, the regular monitoring of blood gases, urea, electrolytes and potassium is required.

With few exceptions, drugs are eliminated from the body by a combination of hepatic metabolism to more water-soluble metabolites, followed by renal excretion of the metabolites. Some drugs, notably salicylate and paraquat, are excreted by the kidneys with little delay due to metabolism. In addition, alcohol and some volatile solvents may be excreted in part via the lungs. However, the majority of drugs require to be metabolised by the liver to some extent before excretion either in the urine or bile and it is for this reason that maintenance of adequate hepatic blood flow is necessary.

Patients who have been admitted late to hospital may already have a degree of renal impairment following a period of hypotension. Some poisons such as paraquat and heavy metals can lead to the rapid development of acute renal failure, and a similar condition occurs when the exposure of the nephron to large amounts of haemoglobin or myoglobin following haemolysis caused by oxidising agents and other drugs to which individual patients are unusually susceptible.

Since the clinician may not be familiar with the considerable problems that present in cases of poisoning, there is an additional requirement placed on the laboratory for continual monitoring of endogenous constitutents of blood. Since few hospitals have a specialist toxicology service available in parallel with the routine clinical biochemistry service, the question of priorities will often dictate that identification or quantitation of unusual drugs out of normal hours should be attempted only if there is a very good reason for so doing.

Gastric Lavage and Emesis

Except in children, the use of emetics is now discouraged since there is a risk of damage from inhalation of vomit. Where an emetic is required ipecachuana is preferred.

Salt water, used as an emetic, may give rise to severe hypernatraemia, and cases of permanent brain damage as a result of this procedure have been documented. Electrolyte analyses before and after gastric lavage are a useful safeguard.

Gastric lavage is regarded as a necessary procedure in cases of salicylate poisoning and is used commonly in other cases of severe poisoning where there is reasonable suspicion that the stomach contains significant quantities of the drug. Provided that a cuffed endotracheal tube is employed, the procedure is safe in skilled hands and the only common fault is insufficient volume of lavage fluid. Approximately 20 l of water are required for a thorough lavage.

In some cases, such as iron, paraquat and cyanide poisoning (q.v.) specific substances are added to the lavage fluid in order to prevent further absorption.

A considerable body of opinion advocates the use of activated charcoal in lavage fluid, in order to prevent further absorption of drugs. Experience in this country does not confirm the necessity for this procedure in the majority of cases, the exceptions being drugs for which an enterohepatic circulation has been demonstrated, e.g. glutethimide.

Forced Diuresis

This technique is designed to increase the flow of urine and thus enhance the elimination of those drugs which are excreted unchanged by renal excretion. Large volumes of i.v. fluids are given accompanied by diuretics, usually frusemide.

Unfortunately, few drugs fall into this category, most requiring hepatic metabolism which is often the rate-limiting step.

There is a small group of drugs, the elimination of which may be aided by alteration of the urinary pH so as to increase the solubility of the ionised drug and reduce tubular reabsorption.

Sodium bicarbonate will lead to an alkaline urine in which phenobarbitone and salicylate are excreted at an increased rate; potassium supplements are required in order to avoid hypokalaemia and both blood gases and plasma K^+ should be checked during this procedure.

Matthew and Lawson[1] advocated the use of a salicylate cocktail, a mix of NaCl, $NaHCO_3$ and dextrose solutions with potassium supplements. This gives greatly reduced problems with K^+ depletion and leads to a more gradual excretion of salicylate in a safer manner.

The checking of urinary pH during 'forced alkaline diuresis' is a simple but often neglected procedure which can assist the clinician in the balance of fluid therapy. An example of the fluctuations in the excretion of phenobarbitone with urinary pH is given in Figure 3.

Forced acid diuresis, using ammonium chloride solution, has been advocated as a useful adjustment in the treatment of poisoning with basic drugs such as quinine and quinidine, amphetamine and fenfluramine. While

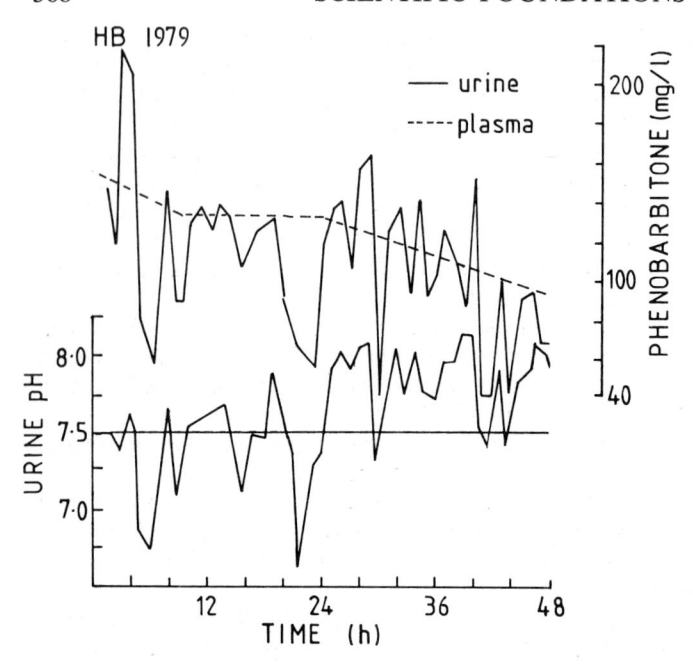

FIG. 3. The effect of urinary pH on phenobarbitone excretion. Urinary phenobarbitone concentrations and urinary pH in 48 hourly samples of urine from a patient with a phenobarbitone overdose treated with NaHCO$_3$, NaCl and 5% Dextrose in rotation.

there is no doubt that the removal of these drugs is enhanced by this procedure, there are risks of the development of a severe metabolic acidosis. Regular blood gas analyses and urinary pH checks are mandatory during this procedure. As there is now widespread experience with haemodialysis, this procedure is becoming safer and is likely to replace forced diuresis with acidification or alkalinisation for all but mild cases.

Haemodialysis

Despite considerable efforts to improve the efficiency of the procedure, haemodialysis is effective in removing only a rather small number of water-soluble drugs from the body, and there is still some morbidity associated with the technique, especially when applied in an emergency.

For this reason it should not be used routinely for the treatment of drug overdose. Drugs which may be removed by haemodialysis are lithium, alcohols, salicylate, phenobarbitone and barbitone, and anions such as bromate, cholorate etc. Although many of these may be eliminated naturally or with the help of forced diuresis, on occasions the priority is to reduce the drug level rapidly and in these cases haemodialysis should be considered.

It is necessary in cases where the normal excretory mechanisms are impaired or forced diuresis is contraindicated, such as in patients who have developed pulmonary oedema or acute renal failure following shock.

The cautious approach to haemodialysis applies also in the main to charcoal haemoperfusion, although this technique is technically simpler. Studies on the amounts of a wide variety of drugs removed by this technique show that relatively small amounts are removed, largely since the majority of the drug is inside the cells and plasma levels available for exchange or adsorption are low.

An example of the effects of haemoperfusion on the plasma concentration of both phenytoin and phenobarbitone in a severe case of poisoning with these drugs is shown in Figure 4.

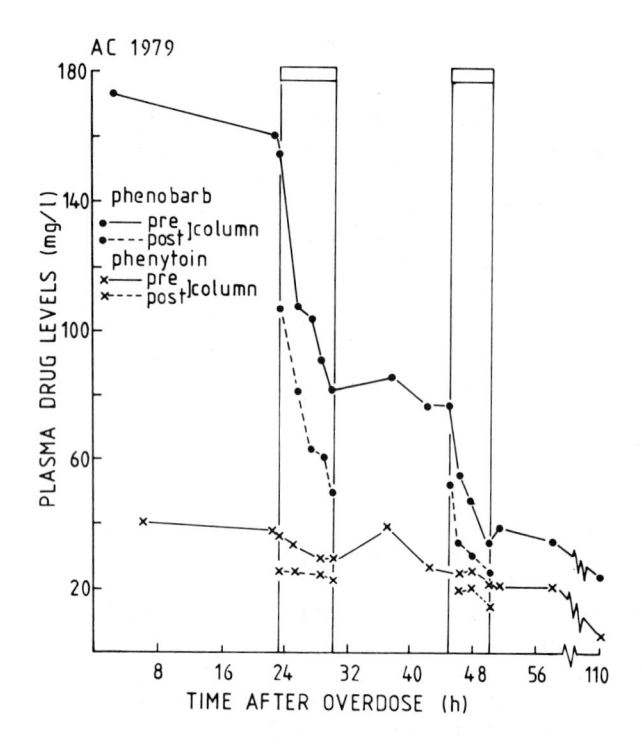

FIG. 4. The different effect of haemodialysis on a polar and non-polar drug. Plasma concentrations of phenobarbitone and phenytoin during two periods of haemodialysis in a patient with an overdose of both drugs.

INTERPRETATION OF SIMPLE CLINICAL CHEMISTRY TESTS IN CASES OF POISONING

Blood Gas Analyses

It is sometimes difficult to place a correct interpretation on the initial results of blood gas analyses. A variety of combinations, from pure respiratory acidosis due to acute respiratory depression to acute metabolic acidosis from methanol or ethylene glycol poisoning, are encountered. The most common finding is a mixed respiratory and metabolic abnormality, the most severe of which may be the mixed respiratory alkalosis and metabolic acidosis of salicylate poisoning. It is necessary for the clinical chemist to become expert at interpreting blood gas results and to be aware of the expected changes in the more common types of overdose. It is also useful to discuss such results with the clinician since in the acute situation little notice may be given to correction of blood gas results for temperature in patients who arrive hypothermic following a period of unconsciousness in an unheated building or outside.

Similarly, correction for the hyperpyrexia observed in patients recovering from barbiturate overdose is seldom applied to their blood gas results. Tables or nomograms for temperature correction of blood gas results should be available in the emergency laboratory.

Once treatment has been instituted, blood gas results need to be carefully correlated with the treatment being given.

Osmolality

With the advent of improved osmometers, increasing use is now being made of osmolality measurements and the question is sometimes posed as to their usefulness in drug overdose. In general, osmolality increases due to the presence of the drugs themselves are relatively rare, the exceptions being salicylate where a change of 5–10 mOs/kg may be observed, or alcohols where the rise may be > 20 mOs/kg. The concentrations of most other drugs in even severe cases are insufficient to have a marked effect on osmolality, although the secondary effect due to production of metabolic acids may be dramatic.

Alterations in Electrolyte Homeostasis

Measurement of sodium concentration may provide a useful aid in the differential diagnosis of the unconscious patient, both hypo- and hyper-natraemia being found on occasions. Potassium levels are most universally raised except in salicylate poisoning, due to underlying acidosis. The need for electrolyte determinations on admission cannot be too highly stressed since without these the alterations observed during and following therapeutic procedures cannot be fully interpreted.

Calcium analyses, while seldom diagnostic, are mandatory in fitting patients. The author has seen several cases of patients given large amounts of diazepam for fits subsequently found to be related to hypocalcaemia.

Serum Enzyme Determinations

The severely poisoned patient may show a variety of non-specific enzyme abnormalities due to the metabolic changes present and also the effect of prolonged hypothermia or immobilisation before admission. It is seldom therefore possible to make use of enzyme analyses in the first 24 hours. The exceptions to this general statement are the use of baseline estimations of AST in paracetamol overdose and of pseudocholinesterase estimations in cases of organophosphorus poisoning.

Interference between Drugs and Clinical Chemistry Tests

Despite volumes written on the interaction of drugs with biochemical tests, there are few of those drugs which are commonly taken in overdose which interact with the common biochemical test procedures. Interference may be by physiological means or by a direct chemical reaction of the drug (or more commonly its metabolites) with the test procedure. Literature references to interactions should be treated with some care since a drug which interferes with one method may interact differently with another. A good example of this is paracetamol, which gives false positive interaction with some methods for blood glucose and false negative answers for another.

For this reason, reports of changes in common clinical chemistry parameters in cases of drug overdose should be interpreted with care and the detailed method used for that analysis should be sought.

In general, simple screening methods and side-room tests are more liable to interference from drugs than more sophisticated methods, e.g. Clinitest® tablets, will show false positive results following ingestion of trichloral compounds or salicylate since these drugs produce metabolites which are reducing agents.

Major interferences are caused by the pharmacological effects or side-effects of the drugs on the metabolites. Thus, narcotic drugs may give rise to increased plasma amylase levels due to spasm of the sphincter of Oddi and overdose with hypoglycaemic drugs may give an expected fall of blood glucose to undetectable levels. Also predictable are the chelation of calcium by oxalate following ingestion of ethylene glycol (q.v.) and the similar dramatic fall in calcium levels following poisoning with fluorides.

Drugs can also interact with the analytical methods for other drugs and drug analysis is bedevilled by non-specific methods. There is a number of compounds which interfere with the Broughton method for barbiturate or the UV screening method for paracetamol but all methods which are required to be definitive require to be checked against not only other common pure drugs but also against their major metabolites.

Sample Collection

There is no ideal body fluid which can be used to answer all clinical questions related to drug ingestion and the choice of specimen is limited by clinical consideration; however some general rules apply.

1. In cases of suspected overdose, the most valuable specimen is the one taken on-admission, and will in an unconscious patient most conveniently be a 10 ml heparinised blood sample. Follow-up specimens are valuable in some cases, but the on-admission specimen gives a baseline against which to consider both the clinical progress of the patient and the results of subsequent assays.

2. Wherever possible, both a urine and a blood sample should be obtained. Urine can often be obtained voluntarily or by gentle compression of the bladder, catheterisation should not be carried out unless there are clinical reasons for so doing. For quantitative analyses, plasma or serum are the fluids of choice.

3. Gastric aspirate has advantages in that the unchanged drug or poison may be found in the aspirate, often in much higher concentrations than in blood or urine. However, care should be taken to ensure that the aspirate is obtained before and not *after* stomach washout.

In an increasing number of cases, the plasma or serum concentration of a drug is being found to be most useful clinically and this trend will continue as methods improve in sensitivity and as information about the pharmacokinetics of more drugs at therapeutic concentrations becomes available.

Summary

From what has been said it will be seen that it is possible to obtain quantitative estimations of the majority of drugs which may be taken, and for a good many poisons.

The interpretation of the results obtained is by no means as simple, but the availability of qualitative results has led to a demand for their production.

A simple rule of thumb for the biochemist involved in clinical toxicology should be 'Do not carry out a quantitative analysis unless either you or the clinician making the request are able to interpret the result'. Using this guideline, over 50% of requests for drug analyses can be averted and much time and effort and money saved. The exception to the rule is where the analyses are contributing to the build-up of information on uncommon or novel drugs, and in such cases research laboratories rely on clinical biochemists to forward samples to central laboratories for analysis.

SPECIFIC APPLICATIONS.

Benzodiazepines

The most commonly encountered drugs in the UK at present are the benzodiazepine group of hypnotics and tranquillisers. These drugs are relatively safe when taken in overdose, and, provided that there is a clear history, there is seldom a requirement for drug analysis as an aid to treatment.

Although coma may last for some days following a severe overdose, there is little respiratory depression and conservative therapy is adequate. The exception is in cases where large amounts of alcohol have also been taken. In such cases respiratory depression may occur and prognosis is poorer.

In a few cases of long-standing coma, a request may be received to confirm that the drug in question is indeed a benzodiazepine in order to allay the fears of the clinician or to avoid other diagnostic procedures. There is a wide range of half-lives among the benzodiazepines and a knowledge of these, coupled with 2 serial analyses can give some idea of the likely time of recovery of consciousness. In such cases, the estimation of benzodiazepines that have one or more active metabolites must be interpreted with care, as the half-lives of the metabolites may differ considerably from that of the parent drug.

Carbon Monoxide

Carbon monoxide poisoning still presents regular problems in the UK, the cause being burning coke in unventilated rooms or deliberate self-exposure to car exhaust fumes.

Carbon monoxide (CO) has an affinity for haemoglobin which is 200–300 times that of oxygen; other haem-containing proteins, such as the cytochromes, are similarly affected.

The clinical diagnosis is seldom in doubt in severe cases and the possibility of cerebral oedema should be investigated by retinal examination. Interpretation of blood carboxyhaemoglobin (HbCO) concentrations are complicated by the fact that many smokers may show concentrations in excess of 10% and that patients are frequently treated with oxygen in the ambulance before admission. Once the patient is removed from the source of CO and breathes air, the half-life for conversion of HbCO to HbO_2 is 320 minutes. The conversion of HbCO to HbO_2 may be augmented by the use of pure oxygen, when the $t_{\frac{1}{2}}$ is reduced to 80 minutes. Hyperbaric oxygen at 3 atmospheres can further reduce this to 23 minutes, but facilities are not commonly available.

Syncope, headaches, nausea and convulsions leading to coma are associated with levels in excess of 30% HbCO. Exchange tranfusion has been advocated but is not recommended unless the patient is moribund. A high proportion of patients who survive severe poisoning show late encephalopathy which is not related to HbCO concentrations.

Carboxyhaemoglobin concentrations are most easily measured using a CO-oximeter. Gas chromatographic methods are also available.

Cardioactive Drugs

Digoxin, quinidine and β-blocking agents are in common use and available in many households, especially among the elderly.

Digoxin

Digoxin poisoning may occur acutely or may build up gradually following an increased therapeutic dose. The clinical presentation is with nausea and vomiting and severe bradycardia occurs. Patients may complain of seeing yellow. Unfortunately, the ECG is not always simple to interpret and a plasma digoxin level may be of assistance. RIA and optical immunoassays are available but it is up to the individual laboratory to decide whether these should be available out of hours. Consultation with the clinician in charge is mandatory.

It is important to realise that, although a digoxin level in excess of 2.5 nmol/l is indicative of toxicity, there is no good relationship between toxicity and plasma level. In particular, since digoxin interferes with the Na^+/K^+ flux at the cell membrane, the potassium level should be measured in all cases. A 'normal' digoxin level may be responsible for toxicity in the presence of hypokalaemia, and since digoxin and diuretics are commonly prescribed together, hypokalaemia is not uncommon and must be excluded, since it is easily treated. Hyperkalaemia, if present, should be corrected with i.m. insulin and glucose infusion.

Quinine and Quinidine

These optical isomers are most commonly encountered following accidental ingestion, especially in children. Both cause severe arrhythmias when present at toxic concentrations (above 20 μmol/l) and in quinine poisoning temporary or permanent blindness may occur. The literature is confusing in relation to levels, since earlier fluorescence methods measured not only the drugs but also their major metabolites.

Confirmation of the presence of one or other of these drugs may be rapidly made by the viewing of urine under a UV lamp at 254 nm where a strong blue fluorescence will be seen. However, since the fluorescence is so strong, a positive reaction will be obtained following ingestion of a bottle of tonic water. Confirmation of toxicity requires qualitative estimation of plasma concentration using a specific method such as HPLC. An enzyme immunoassay method exists for the rapid estimation of quinidine, and at toxic concentration there is sufficient cross-reactivity with quinine for the diagnosis to be confirmed.

Both quinine and quinidine are strongly protein-bound and the elimination rate may be of the order of 2–3 days. Excretion may be enhanced by the use of acid diuresis using ammonium chloride, but this procedure should be carried out only under careful monitoring of acid/base status.

β-blocking Agents

β-blocking agents are now being encountered more frequently in cases of overdose. There are no specific laboratory-based diagnostic tests. Treatment is preferably by large (10 mg) doses of glucagon.

Ethanol

Ethanol is so familiar a poison that it tends to be dismissed in all but the most serious cases of overdose, being present, along with other drugs, in over 60% of all cases of self-poisoning.

The levels of alcohol found in self-poisoned patients may be grossly elevated, up to 5 or 6 times the legal limit for driving (16 mmol/l). In such cases the physician has to decide the relative contribution of alcohol to the patient's state of consciousness and also whether it is necessary to screen for other drugs such as paracetamol.

In many cases the procedure adopted is to wait and observe the clinical condition over a few hours when considerable improvement in the patient's state of consciousness normally occurs. Alcohol has a short half-life and is eliminated with zero order kinetics, thus most patients will show a fall in plasma concentration of between 2 and 5 mmol/l/hr unless there is impairment of hepatic bloodflow.

Recent advances in methodology, in particular the production of reliable fuel-cell based alcohol meters (Alcolyser, Alcometer), make the provision of on-admission alcohol estimates a practical proposition. Alcohol meters may be used by the clinician to determine the blood alcohol level indirectly by analysing air direct from the patient.

By far the best method for the emergency estimation of ethanol in blood is headspace analysis using an alcohol meter. The method correlates well with GLC and the instrumentation is economical, reliable and available for instant use. Analyses are simple and accurate. Enzymatic methods, employing alcohol dehydrogenase, are widely available and may be employed where an alcohol meter or gas chromatograph is not available. Gas chromatography, with direct injection of blood, is a standard method with the advantage that other alcohols may also be rapidly quantitated.

Osmometry is of use where a specific method for alcohol is unavailable. A high osmolality in the absence of either hyperglycaemia or uraemia and with no acidosis is likely to correlate with ethanol concentrations. Where an acidosis is also present then methanol may be implicated, but differentiation from ketoacidosis and lactic acidosis is necessary. Rarely, an individual may show a severe acidosis following ingestion of high concentrations of ethanol alone. The causes are not clear. Hypoglycaemia may occur in children and should be sought. Fructose, which has, in the past, been given as an antidote for ethanol poisoning, can contribute to metabolic acidosis once metabolised and should not be used in treatment. Alcohol is a diuretic and aids its own elimination from the body. Haemodialysis is highly effective and may be required in patients with blood alcohol levels in excess of 50 mmol/l.

The upper legal limit in the UK for ethanol in the blood of vehicle drivers is 16 mmol/l (80 mg/100ml). In severe cases blood ethanol levels may reach more than 60 mmol/l (400 mg/100ml).

Ethylene Glycol

Ethylene glycol used commonly as antifreeze may be taken deliberately by alcoholics as well as by accident by children since most commercial preparations are brightly coloured.

A history is most important since in the absence of suspicion that ethylene glycol has been ingested, treatment may be delayed until too late. The initial presentation may be merely drowsiness, but this rapidly progresses to coma and eventually renal failure. Fitting is a poor prognostic sign.

Ethylene glycol is metabolised via glyoxylate to oxalic acid. The latter complexes calcium and leads to precipitation in the renal tubules with eventual renal failure. There is some delay between ingestion and the appearance of a metabolic acidosis. In severe cases microscopic examination of the urine will reveal calcium oxalate crystals.

Treatment priorities are the maintenance of the plasma calcium concentration, prevention of further metabolism of ethylene glycol and correction of the severe metabolic acidosis. Inhibition of further metabolism may be effected by use of competitive inhibition of the enzyme alcohol dehydrogenase by ethanol. Ethanol should be infused intravenously so as to maintain a blood alcohol level in excess of 22 mmol/l. Once this is achieved there is relatively little risk of oxalate production and therefore hypocalcaemia and renal failure may be avoided. The earlier the alcohol treatment may be started the better, and on no account should the clinician wait until the appearance of a metabolic acidosis. The ethylene glycol may be removed by haemodialysis if necessary, but since some is eliminated unchanged by the kidney, maintenance of a satisfactory urine flow in conjunction with alcohol may prove sufficient.

In patients who present late and in whom metabolic acidosis has developed, the maintenance of pH in the physiological range by use of bicarbonate should be considered; however sodium concentration tends to rise unacceptably before adequate bicarbonate can be given.

Frequent calls for blood gas, sodium, potassium and calcium measurements may well be justified in severe cases, but the prognosis is poor once renal failure has occurred. The presence of ethylene glycol and methanol (present in some brands of antifreeze) may be confirmed by GLC analysis. Quantitative analyses are not required unless dialysis is contemplated.

Poisoning with Hypnotic Drugs

Although this group of drugs once constituted the most common cause of severe self-poisoning, changes in prescribing patterns have drastically reduced the clinical problems. Unfortunately, clinical expertise in dealing with hypnotic overdose is also now less widespread. The drugs prescribed as 'sleeping pills' include *intermediate-acting barbiturates, methaqualone, glutethimide* and *meprobamate, chlormethiazole, trichloral compounds* and a variety of *benzodiazepines*. With the exception of the last two groups of drugs, these are CNS depressants and can lead to severe coma with hypotension, depressed respiration and potentially, hypothermia. Diagnosis of hypnotic poisoning is usually made from the history and on clinical grounds and unless a rapid screening method is available on-call which is at least semiqualitative, the determination of drug levels as an emergency is of little relevance.

Treatment is conservative and only in most severe cases (grade IV coma with no signs of improvement in the first few hours) need effort be made to remove the drugs by external means. Cases of overdose with each of these drugs have been treated by haemodialysis and haemoperfusion and some reduction in the plasma level may be obtained, but the importance of maintaining adequate liver perfusion in order to allow detoxification of the drugs by the normal hepatic metabolic pathways must be stressed. In particular, there is no case for the use of forced alkaline diuresis which is ineffective for this group of compounds: indeed there is a tendency towards the development of pulmonary oedema in cases of *methaqualone* poisoning where forced diuresis has been attempted.

Glutethimide is a particularly complex drug when taken in overdose, since it is metabolised to an active intermediate. Some metabolites are excreted via the bile and reabsorbed from the small bowel, a cycle of events which can lead to fluctuating grades of coma. In such cases the use of activated charcoal is advised in order to prevent reabsorption.

The method of choice for the determination of the *barbiturates, methaqualone* and *glutethimide* is gas chromatography. The glutethimide metabolites show a characteristic pattern on some stationary phases. Hypnotic barbiturates may be rapidly screened by using enzyme immunoassay, but no reagents are readily available for the less common glutethimide and methaqualone.

If analyses are performed, the interpretation of plasma levels must be made with caution. The severity of the case should be assessed by clinical examination of the patient since patients who are chronic users of hypnotics may have induced metabolic pathways and show reduced response to relatively high plasma drug concentration.

Chlormethiazole is currently in use in the UK as a hypnotic with a short half-life (3–5 hr) which is of value in the elderly. The drug is also used for the treatment of delirium tremens, as an intravenous preparation for sedation of aggressive patients, and in pre-eclamptic toxaemia. The oral preparation is dangerous in overdose since it is a central depressant and severe coma and death have been reported. Problems have also been encountered following continuous intravenous infusions. Although in theory chlormethiazole estimations should not be necessary, there is little clinical experience of overdosage with this drug and laboratory support may be helpful. Chlormethiazole is highly volatile and methods which involve evaporation of organic solvents should be avoided. A rapid micromethod is available. The standard forensic UV method also picks up inactive metabolites and should not be used for clinical purposes.

Trichloral compounds, used as hypnotics because of their relatively short half-lives, are uncommonly seen nowadays in overdose cases. Vomiting is common but the clinical problem concerns the bizarre disturbances of cardiac function. Treatment with i.v. propranolol restores the ECG to normal and may need to be repeated until the drug is eliminated.

Chloral hydrate and its related compounds have good and bad features so far as the analyst is concerned. The end product is trichloracetic acid and there are a variety of intermediates. Detection of the end-products is simple and involves heating the plasma or urine with pyridine in the presence of alkali. A red colour in the pyridine layer indicates a drug of this type. The test may be made quantitative in plasma. Care must be taken to avoid false positive reactions from chloroform which can cause contamination if it is being used nearby in the laboratory. Patients receiving trichloral at therapeutic doses (1 g) excrete urine which is strongly positive for reducing substances. Hepatic and renal damage subsequent to overdose have been reported.

The *hypnotic benzodiazepines* are discussed on p. 370.

Poisoning with Inorganic Anions

The ions most commonly encountered in cases of self-poisoning are the oxidising agents, notably *chlorate* from weedkiller, with *bromate, iodate* and *nitrate* being encountered less commonly. *Cyanide* poisoning is also still relatively common as an accidental occurence. Fluoride from rat poison has been taken, with fatal results in some cases, and there is still a small number of reports of overdosage with bromide, used as a sedative.

Chlorate

Chlorate and related oxidising anions exert their toxic effect by oxidation of oxyhaemoglobin to methaemoglobin, with a similar action on the haem moieties of the cytochrome system.

Confirmation in cases of mild poisoning are best made by using the screening test on urine, which consists of the addition of diphenhydramine solution. In severe cases the test will give a blue colour, indicating a positive, in plasma, but this may be masked by the brown colouration of methaemalbumin. The use of spectral methods for the quantitation of methaemoglobin are rather unsatisfactory and provide little more information than that which may be seen with the naked eye. Derivative spectra are more useful than zero order. Modern CO-oximeters will give a quantitative measurement of methaemoglobin. In severe cases the cells will lyse and the haem moiety becomes attached to plasma albumin forming methaemalbumin. In such cases the plasma is likely to be dark brown in colour and may need to be diluted before testing with diphenhydramine for the presence of oxidising agents.

The clinical picture of patients with a muddy skin colour is due to a combination of methaemoglobinaemia and some cyanosis. In cases of nitrate poisoning lysis of the cells is uncommon and treatment is with methylene blue 1–2 mg/kg, repeated as necessary. Oxygen may also be required. Patients treated with methylene blue may show rapid changes of skin colour which should not be a cause for concern.

In more severe cases, where chlorate is involved, lysis of the cells occurs. In such cases methylene blue treatment is less effective and the treatment of choice is exchange transfusion since this both removes the abnormal pigments thus protecting the renal tubules and provides fresh red cells. In addition, the risk of a raised plasma potassium level from haemolysed cells is diminished. Exchange transfusion is preferable to haemodialysis which removes the potassium and anions but does nothing to prevent the deposition of methaemoglobin and methaemalbumin in the renal tubules. In a few cases, haemolysis and methaemoglobin production are the result, not of self-poisoning, but of the idiosyncratic response to a drug at a therapeutic dose.

Bromide

The most common cause of bromism today is the prescription of the hypnotic *Carbromal* to elderly patients. Bromine levels may build up over weeks. Bromide poisoning is serious and the clinical picture may fluctuate due to slow, erratic absorption. Available preparations also contain amylobarbitone and a screen for barbiturates may give a lead. In severe cases a haematological screen is required in order to detect clotting abnormalities.

If high, bromide will falsely elevate plasma chloride concentrations as estimated by standard colorimetric techniques, but the interference is less than stoichiometric. Once suspicion is aroused, estimation of plasma bromide by a colorimetric method may be necessary in order to follow the fall in concentration, which may be slow. Diuresis and peritoneal dialysis are of limited effectiveness, but activated charcoal haemoperfusion has been used with some success.

Cyanide Poisoning

More cases of accidental than suicidal poisoning occur at present, the cause often being the mixing of acid with cyanide waste in drains or waste tips. Cyanide poisoning is often implicated in cases of death by inhalation of smoke from burning plastics.

Patients exposed to cyanide require immediate treatment since there may well be a lag between exposure and development of severe symptoms. Dyspnoea with no cyanosis may be present with vomiting and the gradual loss of consciousness. There is seldom time to confirm diagnosis before treatment.

Two forms of treatment are available. First, conversion of haemoglobin to methaemoglobin using nitrite. The methaemoglobin reacts with further cyanide to form cyanmethaemoglobin which is relatively non-toxic. The cyanide is then converted to thiocyanate by injection of sodium thiosulphate. The initial nitrite administration may be enhanced by the use of amyl nitrite inhalation. Reagents must be fresh. Oxygen therapy is only effective if given with nitrite/thiosulphate.

An alternative form of treatment involves the use of the specific chelating agent 'kelocyanor' but since this compound is toxic in the absence of cyanide and it should not be used in cases where there is doubt.

Identification and estimation of cyanide in blood may be carried out most simply using a Conway diffusion method; however, retrospective analysis of thiocyanate in urine is a simple way of confirming cyanide intake. Blood cyanide levels in smokers can be as high as 150 μg/l but following poisoning may be considerably higher, in the mg/l range. Blood taken for cyanide estimation should be collected into fluoride/oxalate and should not be stored for long periods.

Fluoride Poisoning

While uncommon, this is invariably serious. The fluoride effectively removes calcium from the plasma as may be evidenced by measurement of calcium levels. Calcium estimation is mandatory and active replacement of calcium in massive amounts may be required to control fits. Diuresis should be induced to aid the elimination of the anion, and haemodialysis may be considered but there is often little time and adequate calcium replacement has seldom been given rapidly enough to prevent a fatal outcome.

Poisoning with Metals

Iron

The majority of cases of iron poisoning occur accidentally in young children although on occasion self-poisoning attempts in adults occur. The mortality in severe iron poisoning is high if untreated and urgent efforts must be made to remove the iron from the body. Since the clinical course of severe iron poisoning may show periods of several hours in which there are few symptoms, assessment of severity of iron poisoning requires emergency iron analysis and every laboratory should be equipped to carry out such an estimation. It is probably preferable to use in emergency the method used normally during the day. Where a single analysis is carried out manually, iron-free acid-washed glassware should be kept aside for this eventuality. There is no requirement for emergency estimation of iron-binding capacity, and once treatment has started this measurement can give clinically misleading information. Severe poisoning is indicated if the iron level exceeds 150 μmol/l in an adult for 90 μmol/l in a child

Treatment consists of chelation of the iron both in the gut and the bloodstream using the specific chelating agent desferrioxamine. Some of the agent is left in the stomach. Following lavage, intravenous desferrioxamine is given with i.m. supplement if necessary until plasma iron levels return to acceptable levels (150 μmol/l). The excretion of chelated iron requires an adequate urine flow and in cases where this is not obtained, dialysis may be required. The urine in patients treated with desferrioxamine may appear dark orange due to the chelated iron.

Lithium

The use of lithium as an antidepressant has regained favour now that plasma concentrations may be readily monitored, since the drug has a narrow therapeutic index with plasma concentrations of 0.7–1.3 mmol/l necessary for effective therapy and levels in excess of 1.5 mmol/l associated with toxicity. Concentrations of greater than 3 mmol/l are potentially life-threatening. Most cases of toxicity are due to build up over a short period following a change of dose, development of renal insufficiency or salt depletion. In patients receiving regular therapy, higher concentrations may be tolerated than in those rare cases not previously exposed to the drug.

The symptoms of lithium toxicity are in the main concerned with the CNS, but gastrointestinal problems and in a proportion of patients a nephrogenic diabetes insipidus occurs, which may be refractory to treatment.

Diagnosis and prognosis depend upon estimation of the plasma lithium concentration, the patient with high levels being treated urgently using forced alkaline diuresis or haemodialysis, both of which are effective. During either of these procedures a close watch should be kept on the concentrations of sodium, potassium, and lithium itself. Treatment should be stopped once the plasma lithium concentration falls to within the therapeutic range, since complete removal of the drug may give rise to serious effects in those on long-term therapy. A method for the emergency estimation of lithium should be available in each district general hospital. Flame emission photometry is a satisfactory and simple technique. It is not suitable for the estimation of urinary lithium levels, for which an atomic absorption spectroscopy method should be used.

A proportion of patients receiving lithium therapy develop hypothyroidism and patients receiving the drug who show symptoms of myxoedema should be checked for thyroid status.

Lead

Acute lead poisoning is uncommon but chronic cases occur both from industrial exposure to lead dust and in children who ingest some older paints. Lead analyses should be carried out by a specialist laboratory since the analytical procedure using atomic absorption spectroscopy is difficult to perform accurately without continuous practice. Lead levels above 0.07 mg/100 ml in blood or 15.3 in urine are indicative of excess exposure to lead. A rapid index of lead poisoning may be obtained in any laboratory using a kit method for δ-aminolaevulinic acid which increases if later steps in porphyrin synthesis are blocked by lead. Lead poisoning leads to neurological signs which are the most usual presenting symptoms, but chronic renal damage is a risk. Treatment is by the use of the chelating agents in combination. Again, an adequate urine flow is required for effective treatment. Occupational screening is now mandatory in industries where the risk is recognised.

Mercury

Acute poisoning with mercury is more common than would be expected. The causes are ingestion of mercuric chloride or the incorrect use of mercuric chloride solutions. Chronic poisoning may be caused by elemental mercury which sublimes from spillages, e.g. broken thermometers or dental amalgam. Mercury vapour is toxic when inhaled. A source of this hazard in clinical chemistry laboratories is from thermometers breaking in GLC ovens. The symptoms are neurological and renal damage may occur.

Mercury is estimated by atomic absorption spectrophotometry. Concentrations in excess of 0·5 μmol/l in urine are diagnostic of chronic poisoning.

Treatment of mercury poisoning in cases where the mercuric chloride is ingested is by gastric lavage with sodium formaldehyde sulphoxylate. This reduces the mercuric ion to the less toxic mercurous form. Chelation of the ions using Dimercaprol or N-acetyl D,L-penicillamine is used to remove inorganic mercury.

Aluminium

Aluminium poisoning is now known to be a hazard to patients receiving haemodialysis or chronic ambulatory peritoneal dialysis (CAPD). All water and reagents used in the preparation of such dialysis fluids must be checked for aluminium content before use, or certified to have an Al^{+++} concentration of less than 2 μmol/l. Aluminium encephalopathy is known to occur in patients with plasma Al^{+++} concentrations of greater than 6 μmol/l.

Methanol

Methanol may be taken deliberately by alcoholics, many of whom are remarkably resistant to its effects since they also consume large quantities of ethanol. It may also be ingested accidentally. Methanol, like ethylene glycol, is metabolised by alcohol dehydrogenase to formaldehyde and then to formic acid. It produces a severe and recalcitrant metabolic acidosis which may require large amounts of sodium bicarbonate for correction.

The presenting features are headaches, blurring of vision, dilation of pupils and in severe cases, coma. The diagnostic features in initial laboratory investigation are high plasma osmolality and metabolic acidosis, and since this picture is similar to that obtained in other metabolic disorders, confirmation may be required of the presence of methanol using GLC analysis.

Treatment is by inhibition of metabolism using oral or intravenous ethanol and correction of the metabolic acidosis. Plasma methanol concentrations in excess of 14 mmol/l are an indication for haemodialysis, as is deepening acidosis. Dialysis should be continued until the level falls to 7 mmol/l.

Chronic methanol abuse can lead to blindness and transient and permanent blindness have been described following acute overdose.

Narcotic Overdose

Narcotic overdose in the UK is associated with drug abuse with the exception of *propoxyphene* poisoning which occurs in patients taking a preparation containing paracetamol and propoxyphene. For this reason, paracetamol estimations should be carried out on all patients diagnosed as suffering from narcotic poisoning.

Diagnosis of narcotic poisoning depends upon the clinical picture of coma with respiratory depression and pin-point pupils, both of which respond rapidly to adequate doses of the specific opiate antagonist, *Naloxone*. Naloxone thus contributes to both diagnosis and treatment.

There are no specific laboratory tests required in order to diagnose narcotic poisoning, however amylase levels may be raised due to constriction of the sphincter of Oddi.

Plasma concentrations of narcotics have little significance and urinary screening, if required for identification of the specific drug, should be carried out on a specimen obtained before the start of Naloxone treatment. A variety of other drugs, especially barbiturates, may be present. Specimens obtained from addicts have a high chance of carrying a risk of hepatitis; therefore unnecessary blood samples should not be taken. Patients admitted late after overdose with narcotic agents may well have suffered irreversible cerebral anoxia. In such cases requests for screening for drugs preparatory to the potential donation of organs may be requested. In such cases the analyst should consider the half-lives of the ingested drugs and those drugs (including antibiotics) given after admission to hospital since these may occlude the picture when using methods such as TLC.

Poisonings with Organochlorides, e.g. DDT, pentachlorophenol (PCP)

Cases of poisoning are uncommon in Britain but may be severe when encountered with hepatic and pulmonary damage predominating. There is no specific treatment available unless convulsions are present in which case i.v. diazepam may be used. There is, however, a tendency for low Ca^{++} levels to be encountered. Ca^{++} should be estimated on at least two occasions and the patient treated with calcium gluconate if necessary.

Poisonings with Organophosphorus Compounds

The major problem associated with this group of compounds is inhibition of cholinesterase activity by phosphorylation of the serine residues of the enzyme. Treatment is with atropine and pralidoxine (PAM). The latter competes for the organophosphorus compound, enhances its hydrolysis and regenerates the active enzyme.

Measurement of organophosphorus content of body fluids is unhelpful in diagnosis and treatment but may be required for subsequent enquiries, therefore samples, especially those obtained early, should be kept. Estimation of red cell cholinesterase levels is the most sensitive index; if below 50% of the lower limit of the normal range, treatment should be started even in the absence of symptoms. Most laboratories offer only plasma pseudocholinesterase estimations. These should be available at short notice since treatment needs to be continued, in some cases for days, until levels approach the normal range.

Paracetamol (Acetaminophen)

Poisoning with paracetamol, a mild analgesic available without prescription, is common in the UK and is increasing in the US where it is known as acetaminophen.

There are few clinical signs associated with severe paracetamol overdose in the early stages and confirmation may be required by analysis. The drug is, however, commonly formulated in combination with other drugs, notably the narcotic propoxyphene, and should be sought in cases of self-poisoning with narcotics and in other cases where the exact nature of the poison(s) is not clear.

Paracetamol is toxic via an active metabolite which actively binds to thiol groups. The metabolite is produced to a large extent in the hepatocytes and acute hepatocellular damage is caused if the concentrations of metabolite exceed the available concentrations of glutathione which acts as a scavenger. Individuals with reduced glutathione production due to liver disease or malnourishment and those with induced microsomal oxidation enzymes, who produce relatively higher amounts of metabolite, are more at risk than healthy subjects. Patients with alcohol problems are thus doubly sensitive to paracetamol.

The clinical course following severe paracetamol overdose mirrors that of other causes of acute hepatocellular damage and may progress to hepatic failure and death.

Early death has occurred in those with oesophageal varices or piles due to failure in clotting mechanisms due to the fall-off in production of prothrombin.

Less severe cases may recover and in untreated cases, or those treated late, the estimation of AST and bilirubin concentrations for some days may be of prognostic value.

Provided that diagnosis is made within 10 hours of ingestion of the drug, treatment is now well-established and should be available in district general hospitals in the UK. Intravenous administration of N-acetyl cysteine, which is available commercially, can provide the necessary thiol groups and protect the hepatocytes from damage. Oral therapy has been advocated, but since nausea may be present in a proportion of cases and absorption may be compromised in cases of mixed overdose due to hypotension, the intravenous route is advisable.

The decision on treatment can be made logically only following determination of the plasma paracetamol concentration. If this falls above the 'treatment line' which relates plasma paracetamol concentration to time since ingestion, then treatment is necessary. Patients with concentrations well below the line are not at risk. There are inevitably grey areas associated with uncertainty about drugs and the rule must be to treat if in doubt. Treatment at later than 12 hours has not been shown to be effective, thus reinforcing the requirement for rapid early analysis.

There have been conflicting views on the validity of plasma paracetamol measurement arising from the use of different methods for the measurement. It is important that the method used must detect only paracetamol and not the metabolites, which may be present in high concentrations. The preferred method is HPLC, but the colorimetric method of Glynn and Kendal (1975) is adequate for routine use. An enzymic method is under development which should also be widely applicable and may well become the method of choice for emergency analyses.

Paraquat and Diquat

Poisoning with these two weedkillers (Gramoxone, Weedol, etc) has become relatively common in the UK in recent years. Most of the fatal cases are caused by the agricultural concentrates, which are dark brown liquids resembling the cola soft drinks. The concentrates are extremely toxic, one mouthful being sufficient to cause death, and may also be absorbed through the skin if gloves are not worn, and inhaled with spraying if a mask is not used. The garden herbicides ('Weedol' and 'Pathclear') are in granular form and require some effort to ingest; however, in some suicidal attempts up to three sachets dissolved in water have been taken.

Paraquat is toxic to the lung after a lag period of up to 3 weeks. In severe cases renal failure is an early feature with lung pathology appearing later. Diagnosis of paraquat overdose may be made rapidly using a urine specimen. 10 mg sodium bicarbonate are added to 2 ml urine to render it alkaline, followed by 10 mg sodium nitrite. A strong blue colour indicates the presence of paraquat.

Paraquat is rapidly excreted via the kidney in the early stages following ingestion and even a small dose may give rise to a strong colour reaction in the urine. Prognosis is entirely dependent on the measurement of the level of paraquat in the plasma at a known time after ingestion (Fig. 5). Experience to date indicates that patients with plasma paraquat levels well above the line are unlikely to survive although death may be delayed for days or weeks. Patients with levels well below the line survive with supportive therapy which may include treatment of transient renal failure.

Current treatment consists of gastric lavage using bentonite clay or fullers earth in order to adsorb paraquat. The procedure is cathartic which also aids gut clearance. Forced alkaline diuresis is not effective but prolonged haemodialysis or charcoal haemoperfusion may remove considerable amounts of paraquat if instituted early while the plasma level is still high. Institution of dialysis at later times has not so far proved effective in improving the prognosis.

The methods of choice for the emergency measurement of paraquat are the methods of Knepil, 1977 and Fell, Jarvie and Stewart, 1978, both of which are adequate for prognostic purposes, it being necessary to estimate concentrations down to 10 mg/l. The contribution of diquat is seldom significant, comprising some 14% of the paraquat concentration at all levels (using the colorimetric procedure).

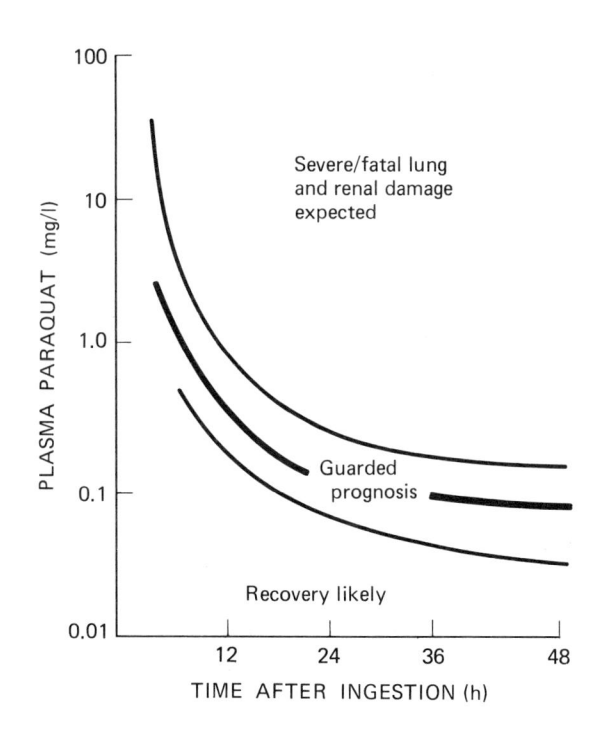

Fig. 5. Prediction time for the likelihood of severe lung damage following Paraquat ingestion.

Salicylate

Salicylate, like paracetamol, is freely available in the UK and represents a still-common cause of severe poisoning in both adults and children with a high mortality (1–7%).

Aspirin, acetylsalicylate, is hydrolysed to salicylic acid which is the major component in plasma. Further metabolism to salicyluric and gentisic acids along with conjugation occurs before excretion in the urine at therapeutic doses. In overdose the binding sites for salicylate arc swamped and salicylate is excreted in the urine, largely unchanged. Excretion does, however, depend on urinary pH.

When used for long-term therapy, e.g. for rheumatoid arthritis the plasma concentration of salicylate may be up to 2·5 mmol/l. Above this concentration, or in previously unexposed patients, salicylate can exert a stimulatory effect on the respiratory centre, leading to hyperventilation which is commonly observed in adult patients who have taken an overdose. However, this takes some time to develop and it is not uncommon for a patient to present with no clinical signs who subsequently turns out to have a high plasma salicylate concentration. In an untreated case the hyperventilation increases in severity leading to a loss of CO_2. Re-equilibration of the bicarbonate buffer system leads to a fall in HCO_2 to extremely low levels, and at the same time there is increasing dehydration due to pure water loss via the lungs. At this stage the blood gas picture in such a patient shows a simple respiratory alkalosis. If treatment is not instituted, the combination of lactic acid production from the diaphragm plus other metabolic acid production caused by the action of salicylate as an uncoupling agent for oxidative phosphorylation causes the development of an increasing metabolic acidosis.

The depleted bicarbonate buffer system is inadequate in the face of the acid load and there develops a severe metabolic acidosis in the presence of an existing respiratory alkalosis. At this stage, loss of consciousness may occur and the prognosis is increasingly poor.

In the absence of a good history the plasma salicylate concentration is necessary for confirmation of diagnosis since the combined metabolic acidosis and respiratory alkalosis present a difficult blood-gas picture and salicylate overdose has been confused with diabetes mellitus, especially since salicylate metabolites are reducing agents which intereferes with Clinitest,® but not with specific tests for glucose.

The plasma salicylate concentration should be related to the putative time of ingestion of the drug. Levels in excess of 5·0 mmol/l should be regarded as serious and levels > 7·5 mmol/l as severe. The colorimetric method of Trinder (1954) is adequate and simple. In these cases gastric lavage is mandatory since there is a tendency for aspirin tablets to form a mass in the pylorus and for continued absorption to occur. In cases where gastric washout is not performed, the plasma concentration may well increase following admission.

The aim of treatment of salicylate poisoning is to hasten the removal of the drug from the body and to replace the bicarbonate buffering capacity and thus restore the pH to normal. Both of these requirements may be met by the administration of alkali.

Forced alkaline diuresis with a cocktail consisting of 1·26% $NaHCO_3$, 0·9% NaCl and 5% dextrose with KCl may be used. This should restore the plasma water and lead to the production of urine of pH > 7.5 without development of hypokalaemia.

In severe cases, monitoring of salicylate, potassium and blood gases may be required until the plasma salicylate concentration falls below 2·5 mmol/l. As with other drugs, if an inadequate flow of urine is obtained despite diuretics, then removal of the drug by haemodialysis or haemoperfusion may be required, but such cases are extremely uncommon.

Tricyclic Antidepressants

Despite the fact that many less toxic antidepressants are available, this group of drugs still give rise to a relatively high proportion of serious poisoning cases. The diagnosis is seldom in doubt since the cholinergic symptoms are well recognised. Cardiac arrhythmias are common and may persist, with a danger of cardiac arrest for some time after recovery of consciousness. For this reason ECG monitoring should be instituted in all serious cases.

Where screening is requested it should be noted that there is a delay of up to 12 hours in the excretion of tricyclic drugs and their metabolites. A urine specimen taken on admission may not show more than a trace.

Analyses of plasma concentrations are now available in some centres using HPLC or immunoassays. In overdose, concentrations in excess of 1 mg/l may be found. This is high enough to give an identifiable spectrum on simple UV analyses of an extract of alkalinised serum. There is a rapid screening test for the presence of *imipramine* and *desmethylimipramine* in plasma and urine, which may be of use diagnostically.

Amitriptyline and *imipramine* are metabolised to their pharmacologically active demethylated derivatives before excretion. It is possible to advise on the necessity for active elimination measures by measuring the concentrations of the unchanged drug and its metabolite over 12 hours. The parent drug concentration should fall accompanied by an initial rise and then a fall by the metabolite.

CONCLUSIONS

The correct laboratory investigation of the poisoned patient requires a considerable knowledge of both the clinical and analytical aspects of toxicology.

There are few drugs and poisons for which specific clinical knowledge is required, but it is in such cases that mistakes may be made because of failure both to appreciate the dangers of a specific toxic mechanism and to treat it with sufficient emergency. Time and effort may be wasted in screening for drugs or performing a qualitative assay when all that is required is an estimate of plasma calcium.

Similarly, insufficient use is made of the simple concept of half-life in order to determine the patient's ability to eliminate a drug and thus determine the likely recovery time or need for intervention with a procedure such as dialysis.

Methods for the rapid and specific determination of drugs and poisons in plasma are improving daily, with HPLC methods in particular providing some exciting possibilities. The clinical chemist faced with the task of assisting clinicians treating poisoned patients has never had more at his disposal and it is the correct application of such methods which is the greatest challenge. The major problem now is to determine whether the work put into the analyses is justified by the clinical problem. Too many analyses are performed, the results of which are never discussed nor acted upon.

Careful attention to the clinical problem and selection of the relevant specimen at the correct time, coupled with the choice of a specific method are the current priorities in this fascinating field of clinical care.

SUGGESTIONS FOR FURTHER READING

Clinical

Matthew, H. and Lawson, A. A. H. (1979). *Treatment of Common Acute Poisonings*, 4th Edn. Edinburgh: Churchill Livingstone.

Goodman, L. S. and Gilman, A. (Eds) (1980). *The Pharmacological Basis of Therapeutics*, 6th Edn. New York: Macmillan.

Doull, J., Klassen, C. D. and Amdur, M. O. (eds) (1981). *Cassaret and Doull's Toxicology*, 2nd Edn. New York: Macmillan.

Proudfoot, A. (1982). *Diagnosis and Management of Acute Poisoning*. Edinburgh: Blackwell.

Analytical

Curry, A. (1975). *Poison Detection in Human Organs*, 3rd Edn. Springfield, Illinois: Charles C Thomas.

Smith, R. V. and Stewart, J. T. (1981). *Textbook of Biopharmaceutic Analysis*. Philadelphia: Lea and Febiger.

Sunshine, I. (1975). *Methodology for Analytical Toxicology*. New York: CRC Press.

Clarke, E. C. G. (Ed) (1969). *Isolation and Identification of Drugs* Vol 1. London: Pharmaceutical Press.

Clarke, E. C. G. (Ed) (1975). *Isolation and Identification of Drugs* Vol. 2. London: Pharmaceutical Press.

23. REGULATION AND MONITORING OF DRUG THERAPY

EMILIO PERUCCA AND ALAN RICHENS

Introduction

Interpretation of serum drug levels
    Pharmacokinetic factors
    Pharmacodynamic factors
    Therapeutic ranges

Uses and abuses of drug monitoring

Salivary drug concentrations

Quality control of drug determinations

Monitoring of particular drugs
    Anti-epileptic drugs
    Lithium
    Tricyclic anti-depressants
    Digoxin
    Anti-dysrhythmic drugs
    β-adrenergic blocking drugs
    Anticoagulants
    Theophylline
    Anti-inflammatory drugs
    Anti-neoplastic drugs
    Aminoglycoside antibiotics

INTRODUCTION

Drugs are given to man to produce pharmacological effects, sometimes on normal tissues—e.g. an opiate analgesic to suppress postoperative pain—sometimes on diseased tissues—e.g. digoxin to stimulate the failing myocardium—and sometimes on invading pathogenic organisms. The physician might reasonably be expected to choose a dose which will achieve an optimal therapeutic response with a minimum risk of adverse effects. If the optimum dose is variable and cannot be predicted beforehand, the physician would be expected to start at a safe dose and increase it gradually until the desired response is obtained. In some diseases, e.g. hypertension, this counsel of perfection is attainable because the physician has an accurate measure, with his sphygmomanometer, of the drug's effect and the dose can be increased until an adequate reduction in blood pressure has been achieved. Sometimes, the subjective response of the patient, e.g. to a hypnotic or analgesic, provides adequate information on the effectiveness of the dose. With some drugs a feedback of information can be generated by measuring a biochemical response such as the reduction in blood sugar produced by a hypoglycaemic drug or the lengthening of the prothrombin time by an oral anticoagulant. As a result of this feedback, the dose can be tailored precisely to the patient's need and this is highlighted by the observation that warfarin dosage varies over a wide range between patients (Fig. 1). With this example, it is, of course, vital to adjust the dose to the prothrombin time if anticoagulation is to be effective and safe.

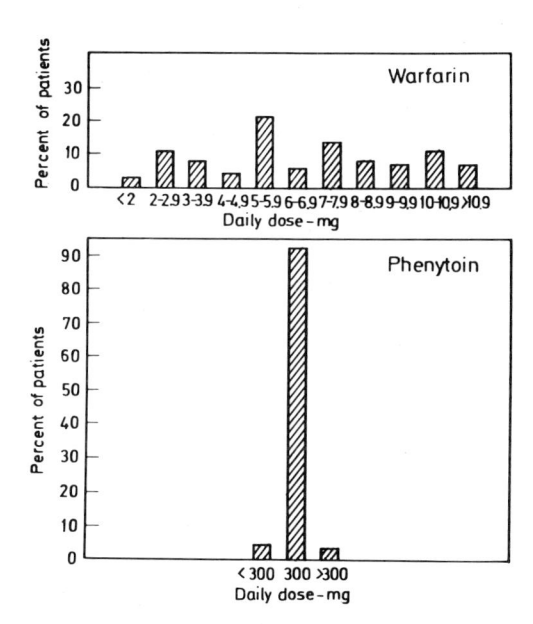

Fig. 1. Distribution of doses of warfarin and phenytoin during chronic therapy at the Massachusetts General Hospital. Each drug has been prescribed for 200 ambulatory patients. (Reproduced from Koch-Weser[10] with permission.)

In contrast, when a drug's effect is more difficult to measure clinically the physician, understandably, tends to play safe by adhering to standard doses, as is illustrated for phenytoin in Fig. 1. Ninety-two per cent of the ambulant epileptic patients studied by Koch-Weser[10] received a dose of 300 mg daily. Plasma level monitoring subsequently showed that only 27% had levels within the accepted therapeutic range of 40–80 μmol/l (10–20 μg/ml) most of the remainder being subtherapeutic (the validity of the concept of a therapeutic range is considered on p. 386). In our own experience, a rather greater proportion of patients were found to have levels within this range on the first occasion on which these had been measured (Fig. 2). Nevertheless, of the 82 patients receiving a dose of 300 mg daily, 27 (33%) had subtherapeutic levels and 17 (21%) toxic levels. Subsequent readjustment of the dose in some of these subjects resulted in improvement of the clinical condition.

Obviously, it is likely that regular monitoring of serum levels will improve drug management by providing information which will allow the physician to tailor the dose until the level is within the optimal range. In this way, satisfactory therapeutic effects will be more likely to be achieved and the patient will be exposed to unrecognised drug intoxication less often. For a number of drugs, evidence in support of this view is now substantial.

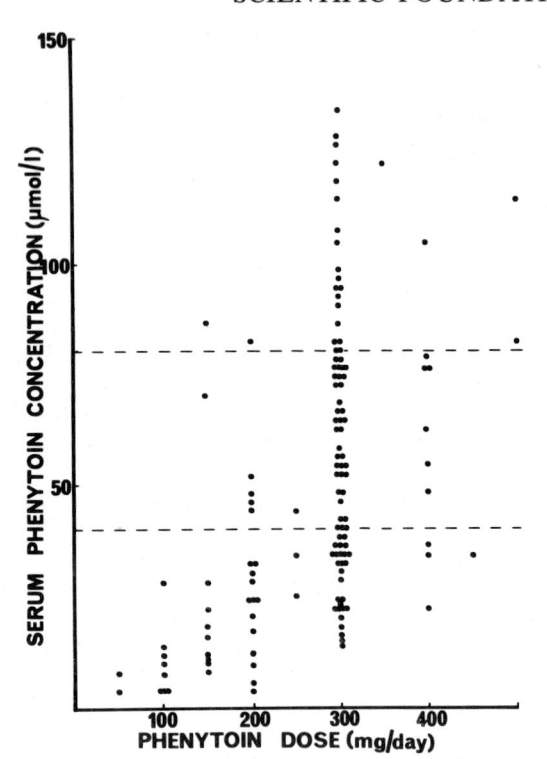

FIG. 2. Serum phenytoin concentrations in 137 chronic epileptic patients on admission to a special centre. With only a few exceptions, phenytoin was being taken in combination with one or more other anti-epileptic drugs. The dashed lines indicate the therapeutic range of serum concentrations. (Note: 4 μmol/l = 1 μg/ml)

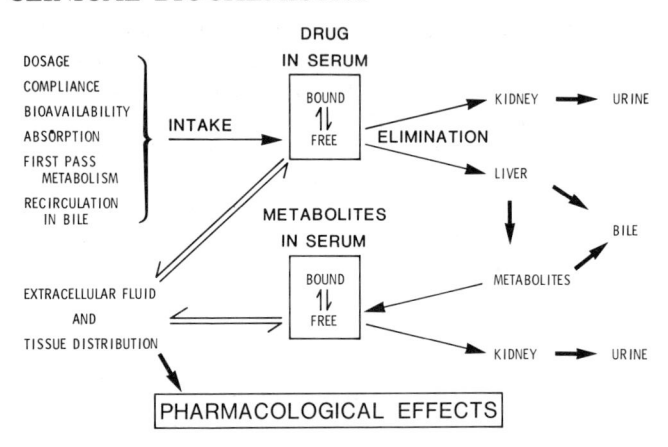

FIG. 3. Factors determining the concentration of a drug and its metabolites in serum, and its pharmacological effects in the tissues.

Serum* is a convenient biological fluid to sample for the measurement of drug concentrations. If we are to monitor many drugs regularly in future it is important that we fully understand the factors responsible for the relationship between dose and serum level, and between serum level and tissue effects. The factors that determine the serum level produced by administration of single or multiple doses of a drug are termed *pharmacokinetic factors* and are summarised in Fig. 3. Intersubject variation in these factors is usually termed *pharmacokinetic variation*, and it is largely this that accounts for differences in drug response between individuals. Particularly important in this respect is the rate of drug elimination, because this is usually the main factor determining the 'steady-state' serum level during chronic administration. (When a state of equilibrium has been achieved between amount of drug administered and amount of drug eliminated 'steady-state' is said to exist, although steadiness of serum levels is only relative. The degree to which the serum level fluctuates throughout a 24-hour period is determined by the rate of absorption, the serum half-life and the frequency of administration, as discussed in more detail below. The object of drug monitoring is usually to estimate the steady-state level produced by chronic therapy rather than the changing levels following single doses.)

The factors that determine the relationship between the serum level and tissue effect of a drug, often called *pharmacodynamic factors*, have been much less studied. Although, generally speaking, the intensity of pharmacological effect tends to be proportional to the concentration of the drug in serum, experience in monitoring drug levels, e.g. of phenytoin, has indicated that considerable variation (which we can term pharmacodynamic variation) occurs in the therapeutic or toxic response to a given serum level. One patient may be severely intoxicated by a level which produces no adverse effect in another. Here is the major limitation of monitoring serum drug levels—precise tailoring of dosage to achieve a level in the 'therapeutic' range eliminates the pharmacokinetic sources of variation in drug response but does not necessarily mean that all patients will show an optimum response and no signs of toxicity. For this reason, monitoring serum levels is not the ultimate answer to regulating therapy. The goal of clinical pharmacology should be accurate measurement of a drug's effect. Nevertheless, this is not always possible in the clinical situation and serum levels are likely to remain of great assistance in many areas of therapeutics. In the sections below, we will consider in more detail the factors which need to be borne in mind in the interpretation of serum level data.

INTERPRETATION OF SERUM DRUG LEVELS

Pharmacokinetic Factors

Dosage

It may seem trite to say that the serum level of a drug is determined by its dose, but it is important to stress that the relationship is not always one of direct proportionality. It would be convenient for the physician if doubling the dose of a drug would double its steady-state serum level, but sometimes this naive concept is not valid. Enzyme systems have a limited ability to handle their

---

* Serum is a more satisfactory fluid for measuring drug levels and therefore, despite the fact that it is strictly correct to refer to *plasma* when describing the physiological *in vivo* state, we will usually refer to *serum levels* rather than *plasma levels* in this chapter. Numerically, these two are identical for all drugs that have been formally studied.

substrates and in therapeutic doses some drugs reach the point of saturation of the enzyme. When this occurs the system will operate in zero order kinetics, i.e. only a fixed quantity (determined by the enzyme's V_{max}) is handled and any amount administered to the patient in excess of this will accumulate. This will give rise to a non-linear relationship between dose and serum level (Fig. 4) which is predictable by using the Michaelis–Menten equation. One of the most important examples of a drug showing saturatable metabolism is phenytoin, and the predicted dose/level relationship has been clearly demonstrated (Fig. 7). The clinical implications of this are fully discussed on page 389. A number of other drugs exhibit saturatable metabolism at therapeutic levels, including dicoumarol, salicylates, phenylbutazone and some barbiturates following overdosage. No doubt further studies will reveal other important examples.

When the therapeutic dose of a drug is far below that required to saturate its metabolism, a linear dose/level relationship would be predicted (assuming that other factors, e.g. absorption and degree of plasma protein binding, are not dose dependent). When this occurs, the drug is said to be metabolised by first order kinetics; under these circumstances an exponential decay in the serum level of the drug occurs following administration of a single dose (once absorption and distribution are complete). A plasma half-life can then be calculated which can be used to predict the time required to reach a steady-state level in the blood on chronic administration of the drug. When a drug's metabolism is saturatable, half-life values should not be calculated because the decay in the serum level is non-exponential following therapeutic doses.

Fig. 4. Dose/serum level relationships for a drug showing saturable metabolism in therapeutic doses, and for another whose dose is well below the point of saturation of the enzyme concerned in its metabolism. (Reproduced from Richens[26] with permission.)

In adults, drugs are usually prescribed in standard doses. However, it has been shown with many drugs that steady-state serum levels are dependent, amongst other things, on body weight. Although it would be better practice to give mg/kg doses routinely, this would complicate prescribing by requiring smaller dosage forms or wider use of linctuses and suspensions, and the improvement in therapy would be small with those drugs (e.g. phenytoin, tricyclic antidepressants, phenothiazines) whose serum levels vary widely from patient to patient. Here, other pharmacokinetic factors are overwhelmingly more important. In neonates and in children, however, adjustment of dose to body weight is better therapeutic practice because the latter varies so much more than in adults. In these age groups, however, further factors must be taken into account, namely (1) a reduced drug-metabolising capacity during the first days of life (particularly in premature newborns) followed by (2) a progressive increase of the metabolic rate to values higher than those observed in adults. As a result, in young children larger doses/kg of body weight are generally required for a satisfactory therapeutic response to be achieved, and calculations based on surface area are preferable as this correlates more closely with metabolic rate.

It is obvious from what has been said that the interpretation of a serum drug level requires knowledge not only of dosage but of age and body weight. The importance of pathological changes in renal and liver function will be dealt with later.

Compliance

Studies in epileptic patients have shown that non-compliance, i.e. failure to take drugs as prescribed, is very common. Up to 50% of patients with tuberculosis fail to take PAS therapy, and in general practice two thirds of psychiatrically-disturbed patients default from drug therapy. The reasons for non-compliance are various and include lack of understanding of the need for treatment; the fear or the occurrence of unpleasant adverse effects (sometimes the patient knows far more about these than does his physician); fear of addiction; an apparent loss of effectiveness of the drug when given long term; and, very often, conflicting medical advice when a patient is under the care of two or more doctors. Factors which have been found to be associated with good compliance include assistance by other members of the family (particularly in the young, old or mentally disabled); fear or personal experience of a recurrence of the disease (when a drug is given prophylactically); and a detailed explanation of the disease, its consequences, and the value of treatment by the physician. Drug level monitoring, frequent outpatient visits, greater severity of the disease, and simplicity of drug regimen are other factors encouraging good compliance.

Because poor compliance is such a widespread problem, precision in drug administration and monitoring is often difficult to achieve and this must be constantly borne in mind when relating drug level to dose. An abrupt rise in level on making an increment in dose may be more from a temporary increase in compliance rather than the revised prescription. Admission to hospital, where drug administration is supervised, is sometimes followed by a rapid increase in drug levels, and/or occasionally the first appearance of drug intoxication.

Bioavailability

Important bioavailability problems have occurred with a number of drugs, particularly digoxin, phenytoin, tetracyclines, hypoglycaemic drugs, coumarin anticoagulants and some anti-inflammatory drugs.

Two aspects of the problem are important for the physician to bear in mind. The first is that if a drug is subject to bioavailability problems, its absorption is likely to be erratic and physiological changes, e.g. in gastrointestinal transit time, may have an important influence on the amount of drug absorbed. The second aspect is that of inequivalence between various formulations of a drug produced by different manufacturers which may result in potentially important changes in drug absorption when patients stabilised on one proprietary preparation are switched over to another.

As far as the interpretation of serum levels is concerned, it is important to bear in mind that with certain drugs a change in formulation can alter the serum level while the prescribed dose remains constant. Drugs whose metabolism is saturatable, e.g. phenytoin, are particularly vulnerable. Any unexplained change in serum level should alert the physician to this possibility.

Absorption and First Pass Metabolism

The amount of an orally administered drug which enters the systemic circulation is determined by its ability to cross the gastrointestinal mucosa, enter the portal blood vessels and pass through the liver without enzymatic destruction. The factors concerned have been less well studied than other aspects of pharmacokinetics, although there is good evidence that reducing gut motility, e.g. by simultaneous administration of a drug with anticholinergic properties, can reduce the rate of absorption. The extent of absorption may concurrently be reduced if, for example, a drug is unstable at the acid pH of the stomach. In the case of poorly soluble compounds such as digoxin, however, a reduction in gut motility may actually increase the extent of absorption by prolonging the time of contact of the drug with the intestinal mucosa. Metoclopramide, which promotes gastric emptying, can have effects opposite to those of anticholinergics. Concurrent intake of food sometimes results in slower rate of absorption and may occasionally cause a drastic reduction in the extent of absorption, as for pivampicillin, isoniazid and tetracyclines. Less well known is the possibility of food enhancing the bioavailability of drugs as in the case of propranolol, metoprolol, hydralazine, camrenone (from spironolactone), nitrofurantoin and hydrochlorothiazide.[15] Bowel disorders such as coeliac disease have also been shown to modify the absorption of some drugs.

Certain drugs are extensively metabolised in the alimentary canal by enzymatic systems localised in gut bacteria (e.g. sulphasalazine) or in the gastrointestinal mucosa (e.g. oestro-progestinic oral contraceptives, chlorpromazine), and their oral bioavailability may be considerably less than unity. The phenomenon by which drug bioavailability is reduced by metabolism in the gut is often referred to as gastrointestinal first-pass effect as opposed to hepatic first-pass effect, which is due to avid metabolic extraction of the drug in passing from hepatic portal to systemic veins following oral administration. In some instances, e.g. nitrates, isoprenaline, lignocaine, the proportion removed by the liver is so great as to make the drug poorly effective by mouth. If large oral doses are given to overcome this effect, considerable differences in serum levels can occur because of interindividual variation in the degree of first pass metabolism. Furthermore, addition of a second drug which modifies metabolism may produce large and even life-threatening changes in bioavailability, e.g. the effect of monoamine oxidase inhibitors on tyramine absorption. The influence of first-pass metabolism upon the pharmacological effect of an orally administered compound may be further complicated when the metabolites produced are pharmacologically active. This is the case for acetylsalicylic acid, chlorpromazine, phenacetin, alprenolol, propranolol, lorcainide and many other drugs. The situation is very relevant to the correct interpretation of serum level data because, under these circumstances, the relationship between the concentration of the parent drug in serum and its pharmacological effect may depend upon the route of administration. In the case of propranolol, for example, it has been shown that the degree of β-adrenergic blockade observed at any given serum concentration of parent drug is greater after oral than after intravenous administration. The difference can be accounted for entirely by the activity of the metabolite 4-hydroxy-propranolol, which accumulates in appreciable amounts in serum only when the oral route of administration is used.

Sometimes, first-pass metabolism may be saturatable so that small oral doses are extensively removed on first-pass, but an increasingly greater proportion reaches the systemic circulation as the dose is raised. This will, of course, produce a non-linear dose/serum concentration relationship as it has been described above (see under *Dosage*). Drugs which show dose-dependent first-pass metabolism include alprenolol, chlormethiazole, lorcainide and, possibly, propranolol.

Recirculation in Bile

Some drugs, e.g. erythromycin, phenolphthalein, are excreted in bile in high concentration in a conjugated form. Release of the drug from its conjugate in the small intestine and subsequent reabsorption is held to account for secondary humps or peaks which are often observed on the serum level profile following single doses. As regards steady-state serum levels, biliary circulation reduces the total body clearance of a drug and produces a higher level than would occur in its absence.

Plasma Protein Binding[21]

Many drugs are highly bound to plasma proteins, especially albumin. The intensity of action of any reversibly acting compound which reaches its site of action by diffusion rather than by active transport is determined by the unbound rather than by the total concentration, because it is only the former that creates the diffusion gradient driving drug molecules across biological membranes. Standard techniques for measuring serum drug levels, however, do not discriminate between drug molecules that are bound to plasma proteins and those that are dissolved in plasma water. If the binding of a particular drug was constant and predictable from one subject to another, then it would be of no importance that both bound and unbound (free) components were measured simultaneously, because the total concentration would accurately reflect the free (pharmacologically active) concentration. In fact, for most of the routinely monitored drugs, the degree of inter-individual variation in binding is relatively small (usually no more than two-fold), provided that the renal and the hepatic functions are normal and no endogenous or exogenous displacing agents are present (see below).

For certain drugs, e.g. clofibrate, naproxen, methadone, prednisolone, quinidine and valproic acid, the binding capacity of plasma proteins is limited and the fraction of drug which is bound decreases as the concentration is raised within the therapeutic range. Under these circumstances, the interpretation of serum level data is complicated by the fact that the increase in concentration of pharmacologically active free drug is greater than expected from the rise in concentration of total drug alone.

One of the major determinants of the free level of a highly bound drug is the plasma concentration of albumin. When the latter is reduced, e.g. in hepatic disease, malnutrition and pregnancy, the proportion of drug bound is smaller, and the measurement of the total drug will underestimate the drug's therapeutic or toxic effects. Renal disease may reduce binding by altering the affinity of albumin for drug molecules. Binding may also be reduced in the neonate, due to the concomitant influence of a number of factors which are only incompletely understood.[17] More rarely, pathological conditions result in an increased degree of plasma protein binding. The best known example of such an occurrence is provided by the enhancement in the binding of chlorpromazine and propranolol in patients with inflammatory disease. The effect is associated with an increased concentration of α_1-glycoprotein to which these cationic drugs are extensively bound.

Two highly bound drugs may compete for the binding sites on the protein, leading to a drug interaction. For drugs characterised by a large volume of distribution (>0·20 l/kg) and restrictive elimination, the interaction is unlikely to produce any significant alteration of pharmacological activity, because the amount of drug displaced from plasma proteins is negligible compared with the total amount of drug already present in the tissues and will undergo elimination in any case. In the case of drugs with a low volume of distribution, however, displacement from protein binding sites may result in increased concentration at the site of action and in consequent enhancement of effect. For example, an important lengthening of the prothrombin time may occur when phenylbutazone is prescribed for a patient stabilised on warfarin. In most cases, the potentiation of pharmacological effect will only be transient because of a compensatory increase in clearance of the displaced drug: this will result in a fall in the concentration of total drug, whereas the concentration of free drug will gradually revert to the pre-interaction value. Under these circumstances, knowledge that a plasma protein binding interaction has occurred is fundamental if serum drug levels are to be interpreted correctly in clinical practice. Both therapeutic and toxic effects will in fact be greater then expected for any given concentration of total drug in serum.

In conditions associated with reduced plasma protein binding, interindividual variation in binding is usually very much greater than in subjects with normal binding capacity. As a result, in these conditions serum levels of total drug will not provide a good indication of free drug concentration and, thereby, pharmacological effect. This principle is illustrated by the observation that in patients with uraemia the normally good correlation between total serum phenytoin levels and effect is lost, even though the correlation between free phenytoin concentration and effect is maintained.[24] Under these conditions it would be desirable to measure the concentration of free drug in serum, but in practice the methods currently available for measuring unbound drug levels are unsatisfactory and are unlikely at present to lead to improved management of drug therapy.

Serum Half-life

The serum half-life of a drug is determined by its clearance (usually hepatic or renal) and by its volume of distribution, i.e. the amount of the drug bound to body tissues. The larger the tissue store, the longer it will take for the clearance mechanism to remove all the drug from the body for a given serum concentration.

A knowledge of the factors determining the serum half-life is essential if a drug is to be used correctly because they will determine the frequency of dosing and the time taken before the serum (and therefore tissue) level reaches steady-state in multiple dosing. It is usually sufficient to administer doses at intervals of one half-life (Table 1) although with some drugs longer intervals are allowable. However, it is important to remember that the serum level just before dosing will be much lower than the peak level after dosing and that 'steady-state' is only relative. Although some drugs have very long half-lives (around 100 hours for phenobarbitone) it is nevertheless usual, for convenience, to administer them

TABLE 1
HALF-LIVES OF SOME DRUGS THAT ARE COMMONLY MEASURED IN SERUM

Drug	Half-life (hours)	Comments
Antipyrine	6–24	Longer with impaired hepatic function.
Carbamazepine	8·5–19 (chronic therapy) 24–46 (single doses)	Autoinduction of metabolism occurs.
Desipramine	12–54	
Diazepam*	15–60	Longer with impaired hepatic function and in the elderly.
Digoxin	30–40	Longer with impaired renal function.
Ethosuximide	60–100	30 hrs in children.
Gentamicin	2–3	Longer with impaired renal function.
Isoniazid	0·5–5	Bimodally distributed.
Lignocaine	1–2	Longer with impaired cardiac and hepatic function.
Lithium	7–20	Longer with impaired renal function.
Nitrazepam	20–40	
Nortriptyline	15–90	
Paracetamol	2–3	Longer after overdose.
Phenobarbitone	53–140	37–73 hrs in children.
Phenytoin	9–100	Dose dependent (see text).
Primidone*	3–13	
Procainamide*	2·5–5	
Propranolol	2–4	
Quinidine	4·5–8·5	
Theophylline	3–9·5	Probably dose-dependent.
Warfarin	15–58	

* Active metabolites with longer half-life produced.

at intervals much shorter than one half-life, and therefore the fluctuation in their serum levels may be small. The rate at which a drug is absorbed will also influence the degree of fluctuation; a slowly absorbed preparation, e.g. carbamazepine, will produce much steadier levels than would be expected by considering only the drug's half-life. Thus, a drug with a serum half-life of 6 hours could be given less frequently if it produces a peak level which is not reached until 4 hours after administration. Drug manufacturers, of course, take this into account when they produce slow-release preparations.

An appreciation of these pharmacokinetic factors is necessary in deciding on the timing of blood samples for routine monitoring purposes. For the example mentioned above, phenobarbitone, a random sample will be entirely satisfactory because the serum level would not be expected to vary by more than 15% with twice daily dosing. With sodium valproate, however, the early morning level may be less than 50% of the peak level and it is therefore necessary to specify the relationship between dosing and sampling. Whether it is better to sample before a dose (the trough level) or at a fixed interval after a dose (the peak level) is disputed. The former is more a measure of the drug's rate of elimination, while the latter is determined more by the rate of absorption. One difficulty in measuring the peak level is that it is likely to occur at different times in different subjects and therefore a standard time interval after dosing will often miss the peak. Measurement of both peak and trough levels is considered to be necessary with some drugs, e.g. antidysrhythmics, so that the dose interval can be adjusted to avoid toxic peak levels and subtherapeutic trough levels.

If serum levels are to be indicative of the intensity of a drug's action, it is necessary that equilibration has taken place between the plasma and tissue compartments. Serum levels measured immediately after an intravenous injection will not reflect the pharmacological response if the drug is still diffusing to its site of action. The assumption is made that the slow fluctuations in serum levels which are seen with multiple dosing of most commonly-measured drugs are mirrored by a similar change in the tissue concentration, i.e. that the plasma:tissue ratio is constant. With highly lipid-soluble drugs this may be valid, but drugs that diffuse more slowly may not satisfy this requirement completely.

Metabolism

For most drugs, particularly lipid soluble ones, elimination from the body is primarily dependent upon metabolism (usually in the liver) to water-soluble metabolites, which are generally pharmacologically inactive and readily excreted in urine. The rate of drug metabolism varies greatly from one subject to another, under the influence of genetic (e.g. the acetylator phenotype), environmental (e.g. the stimulation of oxidative drug metabolism by barbiturates or by cigarette smoking, the inhibition of xanthine oxidase by allopurinol and the inhibition of the intermediary metabolism of alcohol by disulfiram) and pathophysiological factors (e.g. the decrease of drug metabolising capacity in hepatic disease and in old age). Undoubtedly, differences in rate of drug metabolism represent the single most important determinant of interindividual variation in drug response. As a general rule, slow metabolisers require lower maintenance doses of the drug for a satisfactory therapeutic response to be achieved. However, there are a number of exceptions (Table 2). When appreciable amounts of an active metabolite are produced, and the metabolite is eliminated slowly, it may be responsible for part of a drug's action. In some cases (e.g. cyclophosphamide, phenacetin, prednisone) almost all of the pharmacological effect is due to the metabolite. Under these circumstances, an increase in drug metabolising capacity may result in potentiation rather than in reduction of pharmacological effect.

The possibility of metabolites being responsible for a drug's action must be considered when relating serum levels to pharmacological effect. Obviously, measurement of only the parent drug would be inappropriate in this circumstance, and it will be necessary to devise methods for measuring each active substance. Even if this is done, however, it may be difficult to relate the separate levels of these compounds to the overall pharmacological effect unless sound evidence is available on their relative activities, and this is seldom the case.

TABLE 2
SOME IMPORTANT DRUGS WITH ACTIVE METABOLITES

Approved Name	Active metabolite
allopurinol	alloxanthine
amitriptyline	nortriptyline
aspirin	salicylic acid
carbamazepine	carbamazepine-epoxide
chloral hydrate	trichloroethanol
chlordiazepoxide	desmethylchlordiazepoxide
clomipramine	N-desmethylclomipramine
chlorpromazine	many active metabolites
codeine	morphine
cyclophosphamide	many active metabolites
diazepam	nordiazepam
glutethimide	4-hydroxyglutethimide
imipramine	desipramine
methylphenobarbitone	phenobarbitone
nitroprusside	thiocyanate
pethidine	norpethidine
phenacetin	paracetamol
phenylbutazone	oxyphenbutazone
prednisone	prednisolone
primidone	phenobarbitone
	phenylethylmalonamide
procainamide	N-acetylprocainamide
propranolol	4-hydroxypropranolol
spironolactone	canrenone
	canrenoate
sulphasalazine	sulphapyridine

Needless to say, assay methods need to be sufficiently selective to discriminate between a drug and its inactive metabolite if a proper assessment of therapy is to be achieved. If this criterion is not satisfied, an increase in the serum level of a metabolite (e.g. when renal function is impaired) may be interpreted as an accumulation of the parent drug. Even if the metabolites are active, it is preferable to separate them from the parent compound for a full understanding of the relationship between serum levels and pharmacological effects.

Renal Excretion

Numerous drugs and drug metabolites are excreted in urine, either passively (through glomerular filtration) or actively (by tubular secretion). In the presence of impaired renal function, the elimination of these compounds may be drastically reduced. In these patients, careful readjustment of dosage is essential not only for those drugs that are excreted unchanged in urine, but also for those that are fully or partly converted to active metabolites. Individualisation of therapy is usually based on frequent determination of both glomerular filtration rate and serum drug levels. The need for analytical techniques specific for the active substance is obvious under the circumstances.

Pharmacodynamic Factors

If drug level monitoring is to be of value in the clinical situation it is necessary that there should be a good correlation between the serum level and the magnitude of the drug's pharmacological effect. As the latter is determined by the concentration of drug at the tissue receptor sites there needs to be a constant serum:tissue level ratio. In some instances, the whole tissue level of a drug has been measured. For example, phenytoin has been assayed in brain specimens resected from epileptic patients and shows a good correlation with steady-state serum levels, although individual serum:brain concentration ratios show sufficient variation to cast doubt on the simple hypothesis that serum levels will always accurately reflect the anticonvulsant activity of the drug. Whole tissue levels, however, may be misleading because a substantial proportion of a drug in tissues is bound to inactive sites and therefore may not represent the amount available for pharmacological action. A similar problem exists in interpreting serum:tissue level ratios. The concentration gradient which determines the amount of drug diffusing into a tissue is produced not by the total drug in plasma but by the proportion which is not bound to plasma proteins. As discussed above, with some drugs the latter may be small and, furthermore, it may vary from one subject to another. It is likely that the correlation between tissue and serum levels would be improved if based on free rather than total serum levels.

It has been mentioned earlier that some drugs are converted to active metabolites in the liver. In some instances, however, active metabolites may be formed in the target tissue and when this occurs the relationship between serum level and effect may be a poor one. Alpha-methyldopa is a good example of such a drug; its conversion into alpha-methylnoradrenaline and other amine metabolites in central and peripheral noradrenergic neurones is responsible for its hypotensive action. The latter long outlasts measurable levels of the parent compound in serum.

For some drugs a lack of correlation between serum level and therapeutic effect would be predicted from a consideration of their mode of action. Those that have a 'hit-and-run' action, such as cytotoxic drugs, or those that cause prolonged or permanent inactivation of an enzyme system, such as organophosphorus anticholinesterase drugs, would be expected to have disappeared from serum long before their actions had ceased.

Tolerance to a drug's action will also cause a changing relationship between serum level and effect. Marked tolerance is well recognised with opiates, barbiturates and ethyl alcohol, although lesser degrees may occur to a variety of other drugs, e.g. benzodiazepines and indirectly-acting sympathomimetic drugs. It is frequently observed that dose-related adverse effects are often worst immediately after starting a drug, particularly if a loading dose is given in order to achieve a therapeutic serum level as quickly as possible. For instance, a serum phenobarbitone level of 40 μmol/l produced by a single oral dose in a volunteer not previously exposed to the drug may cause much greater sedation that a level of 200 μmol/l resulting from maintenance therapy in an epileptic patient. The difficulty which this causes in defining a therapeutic range of serum levels is considered below.

Therapeutic Ranges of Serum Levels

Before routine monitoring of serum levels will be clinically helpful, the optimum therapeutic range of levels needs to be defined. Ranges derived from retrospective studies are usually unsatisfactory because they merely describe levels which have been produced by standard doses of a drug and these may not be optimum for therapeutic response. Prospective controlled studies in which dosage increments are made during careful monitoring both of serum levels and of clinical response are necessary to define the therapeutic range clearly. However, the selection of patients is critical in such a study. Unless the disease being treated presents only in one form and with a fairly narrow spectrum of severity, the results may not be applicable to all patients with that disease. For example, epilepsy presents in a variety of forms and with varying degrees of severity. A patient with, say, severe temporal lobe epilepsy is likely to respond to a quite different range of serum carbamazepine levels than a patient with infrequent grand mal fits of idiopathic origin. Ideally, a therapeutic range should be defined for all types and severity of disease because one derived from a study which has included an assortment of patients may lead to undertreatment of some patients and excessive medication in others when dosing is based on the monitoring of serum levels. The problems of designing a good prospective study are considerable, and bias may be introduced unless a double-blind technique is used. Ideally, this should involve a fixed number of tablets (a variable mixture of active and placebo tablets), an observer who adjusts drug dosage in the light of serum level monitoring, and a second observer who assesses the patient's clinical response without knowledge of the dosage or serum level.

What has been said so far assumes that the concept of a therapeutic range is valid, i.e. that there is a minimum serum level below which a drug is ineffective and a maximum level above which no further benefit is conferred or toxicity becomes common. The dose-response curve (Fig. 5) is a fundamental tool for the pharmacologist, allowing him to express the relationship between the concentration of a drug in contact with a tissue and the response of the latter to the presence of the drug. The curve is usually sigmoid with the response occurring over a certain concentration range. There is a lower concentration limit, below which little response occurs, and an upper limit at which the response is maximal. If a pharmacological effect in man was produced by an effect on a single population of cells, it might be expected that the dose-response curve (or more strictly the serum concentration-response curve, which excludes pharmacokinetic variation) would be similar to that found in isolated tissues. However, most pharmacological effects in man are produced by actions on several tissues or systems often with different thresholds to a drug's action, and the presence of intact reflexes makes the response to drug administration a complicated one. Propranolol, for instance, when used as an antihypertensive drug probably acts on the heart, peripheral blood vessels and brain to produce its therapeutic effect and it is likely that the dose-response curves for each of these effects are different. When compensatory reflexes come into play the situation is a very complicated one. In treating a diseased patient with drugs we are dealing with a concentration-response relationship which is made up of many individual components, and it may cover a wide range of concentrations and lack the normal sigmoid shape. Therefore, the idealised relationship depicted in Fig. 5 may not be obtainable in practice. Nevertheless, for each patient there is a range of serum levels over which the response passes from threshold to maximum, and a level which is associated with an optimum response. When the number of responders at any given level is plotted, a distribution like that depicted in Fig. 6 will, ideally, be seen, and it should be possible to define the therapeutic range in simple mathematical terms, such as the limits within which 95% of patients respond.

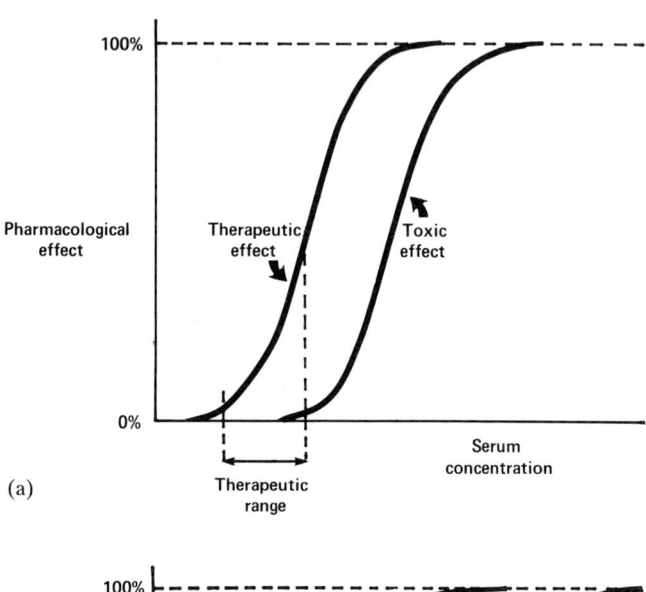

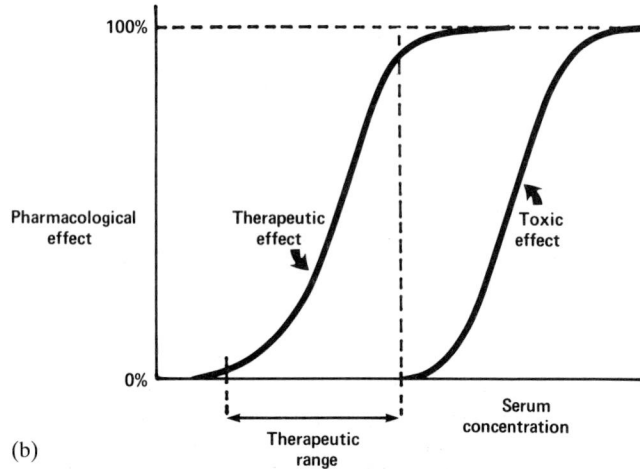

FIG. 5. Hypothetical serum concentration-response curves (a) for a drug with a high therapeutic index (e.g. penicillin) and (b) a drug with a low therapeutic index (e.g. gentamicin). In the first case no appreciable toxicity is seen before the therapeutic effect reaches maximum, and in the second case the therapeutic range of serum concentrations is limited by the development of toxicity. Serum concentration is plotted on a log scale.

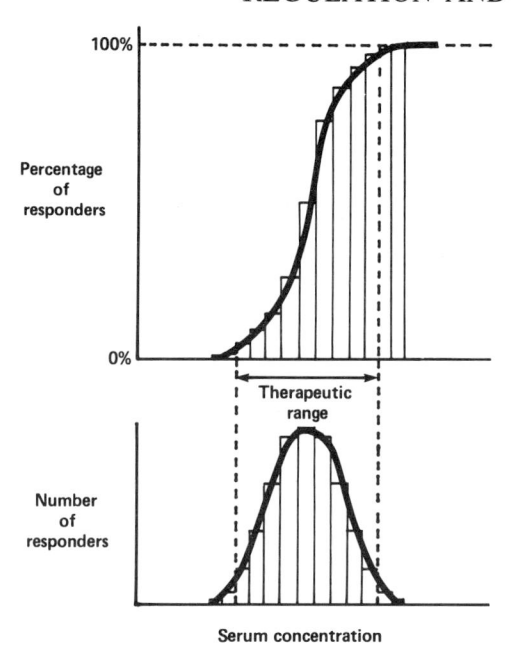

FIG. 6. Cumulative and non-cumulative frequency distribution curves representing the number of patients who show an optimum response at any given serum concentration. Serum concentration is plotted on a log scale. The therapeutic range indicated would apply for a drug with a high therapeutic ratio, and includes 95% of the responders.

Although a definition as precise as this is desirable, it is unlikely to be achieved in practice. However, by carefully defining the type of disease, its severity and perhaps other factors known to determine response to a drug, the optimum range levels at which a given patient is likely to respond might be predictable. A great deal of work in which serum levels and therapeutic effects are related will be necessary before this ideal can be achieved.

USES AND ABUSES OF DRUG MONITORING

Measuring drug levels is time-consuming and expensive and before routine monitoring of drug levels can be justified, the indications for doing so should be carefully considered. As a general rule, drug monitoring can be considered to be of clinical value only when it has been clearly shown that therapeutic effects correlate with the serum level of the drug better than with the prescribed daily dose. This is particularly likely to be so in the following situations:

(i) When there is a wide inter-individual variation in the rate of metabolism of the drug, leading to marked differences in steady-state levels. This can be particularly important in children, in whom differences in body weight and metabolic rate are considerable, and in the elderly.

(ii) When saturation kinetics occur, causing a steep relationship between dose and serum level within the therapeutic range.

(iii) When the therapeutic ratio of a drug is low, i.e. when therapeutic doses are close to toxic doses.

(iv) When a drug is used prophylactically and no other convenient methods are available to assess clinical efficacy.

(v) When signs of toxicity are difficult to recognise clinically, or when signs of overdosage or underdosage are indistinguishable.

(vi) When gastrointestinal, hepatic or renal disease is present causing disturbance of drug absorption, metabolism or excretion.

(vii) When patients are receiving multiple drug therapy with the attendant risk of drug interaction.

(viii) When there is doubt about the patient's reliability in taking his tablets, particularly in the elderly and in the mentally handicapped.

During research studies it is much easier to justify measurement of serum levels. Demonstration of a correlation between the serum level of a drug and its therapeutic effect in a clinical trial is sound evidence that the effect really is due to the drug.

Routine measurement of drug levels is not indicated when the therapeutic range has not been defined by well-designed prospective studies of the relationship between serum level and therapeutic effect. When drugs produce one or more active metabolites a problem arises in the interpretation of the results if only the parent drug is measured. For some drugs a correlation between serum level and effect would not be expected on theoretical grounds, for instance when there is a large inter-individual variation in the degree of serum protein binding (see section below on salivary levels), when a delayed onset of effect or slow recovery of enzyme activity takes place following drug administration (e.g. 'irreversible' anticholinesterase drugs) or when a complex metabolic effect is produced (e.g. cytotoxic drugs). It may be that the plasma profile of many drugs is a misleading representation of their therapeutic effects, and calculating dose intervals from plasma half-lives may result in frequent (and inconvenient) dosing when once-daily dosing is therapeutically satisfactory. There is, for instance, increasing evidence that once daily dosing with propranolol produces satisfactory control of blood pressure despite the short plasma half-life of the drug. Control of angina, however, appears to require more frequent dosing.

Finally, there is always the danger of treating the serum level rather than the patient. With many drugs, measurement of the serum level is invaluable but it is no substitute for careful clinical assessment of the patient's response. If, on clinical grounds, a patient has shown a satisfactory response to therapy without evidence of toxicity, the fact that his serum level may fall outside the approved therapeutic range should be ignored. For anti-epileptic drugs, Booker[5] states the problem in the following terms:

'Unfortunately we have seen occasional adverse effects from the availability of the serum level determination . . . Subjects with low levels have had doses increased, and doses associated with high levels have been decreased. In a few cases, we have seen seizures recur or toxicity develop, so that the treatment of a laboratory value instead of the patient is not always to the benefit of the latter. To say that the fault lies with the physician and not the methodology, while correct, ignores the fact that physicians are human. As such, faced with the increasing complexity and amount of scientific data upon which his practice must be based, the clinician will intuitively turn to simple algorithms that will reduce the complexity of his problems to simple decisions. Unfortunately, these will be increasingly based upon laboratory values. Thus, in the application of population values, i.e. therapeutic and toxic levels, to the treatment of an individual subject, we must always continue to treat the patient and not a laboratory value. Nevertheless, knowledge of the serum levels is a valuable resource for the clinician.'

SALIVARY DRUG CONCENTRATIONS[8]

The unreliability of using total serum drug levels as a measure of pharmacological effect in the presence of wide interindividual variation in plasma protein binding (due to disease or drug interactions) has been discussed above. Under these circumstances it would be desirable to obtain an estimate of free (unbound) pharmacologically active drug.

For certain compounds, e.g. phenytoin, carbamazepine, theophylline, the salivary glands act as a simple dialysis membrane, and the salivary drug concentration is similar to that of the plasma water. For these drugs, it can be argued that measuring salivary levels is superior because differences in binding are allowed for. Furthermore, collecting saliva may be more acceptable to the patient, particularly to children. However, against these arguments must be balanced the greater difficulty of measuring lower concentrations (see section on Quality Control), and the fact that contamination of saliva can occur for up to 3 hours after oral dosing, particularly with syrups, giving artificially high values. Also, for drugs which are largely ionised at physiological values of plasma and salivary pH, considerable fluctuations may occur in the plasma:saliva concentration ratio, leading to seriously inconsistent results. In the opinion of the authors, saliva should not be indiscriminately used for therapeutic drug monitoring.

QUALITY CONTROL OF DRUG DETERMINATIONS

Clinical biochemists have long accepted the need for quality control to check the accuracy of their laboratory results. Drug estimations are generally more difficult than standard clinical biochemistry tests and the need for quality control is therefore much greater. For several years, one of us (AR) has organised the St Bartholomew's Hospital Quality Control Scheme for anti-epileptic drugs and, at the time of writing, specimens are sent to over 200 laboratories throughout Europe.[26] More recently, pilot schemes have been started for theophylline and tricyclic anti-depressants. These schemes have been invaluable in identifying inaccurate methods and in assisting laboratories to set up a reliable assay. For instance, it was obvious at an early stage that spectrophotometric techniques for phenytoin frequently gave results which fell outside the 95% confidence limits, usually above. The reason for this was that these methods were less specific than newer techniques and the presence of other drugs or metabolites frequently caused interference.

More recently, a North American scheme has been set up by Dr Charles Pippenger at Columbia University, New York. Before starting the scheme, Dr Pippenger and his colleagues[22] mailed to each laboratory which offered a service for drug estimations three pooled serum samples containing phenytoin, phenobarbitone, primidone and ethosuximide. By organising distribution centres, they were able to submit specimens to the laboratories as if they had been taken from patients locally, and samples were therefore run without the laboratories knowing that they were under test. The results obtained deserve wide publicity because they highlight the urgent need for quality control of drug estimations. For each specimen the range of results reported was from zero to levels which were so high as to be compatible only with massive overdosage, and the interlaboratory coefficients of variation varied from 38 to 505%: in many cases, these results could have led the clinician to adjust drug dosage in the wrong direction had the specimens come from a real patient.

The message from these studies is that quality control of drug estimations is essential if research studies are to produce meaningful results and if patient care is not to be hazarded. It should be remembered that most of the conventional anti-epileptic drugs are easy to measure and are present in serum in relatively high concentrations. It is interesting to speculate what the outcome of a spot check on, say, propranolol or benzodiazepine levels would be.

MONITORING OF PARTICULAR DRUGS

Anti-epileptic Drugs

For the great majority of epileptic patients long-term drug therapy represents the only practical form of treatment. Once started, therapy will usually continue for at least three years, and in some patients treatment may be lifelong. Although drug therapy is becoming more rational (mainly as a result of serum level monitoring) a substantial proportion of patients require combination therapy and drug interactions are therefore common. Most of the drugs in current use have a low therapeutic ratio and dosage adjustment can therefore be difficult, particularly for phenytoin whose metabolism is saturatable within the therapeutic range. Many of the toxic effects of anti-epileptic drugs are clinically obvious, e.g. ataxia or nystagmus due to phenytoin or sedation caused by phenobarbitone, but some are subtle in their presentation and it may not be immediately obvious that they are drug-induced. For example, bizarre neuropsychiatric symptoms may be caused by phenytoin and barbiturate drugs, and these may be mistaken for progressive deterioration of neurological origin when in fact they are entirely reversible on reducing drug dosage.

All these difficulties in drug management are good reasons for measuring anti-epileptic drug levels and on current evidence the contribution of drug monitoring in this field is greater than in any other, with perhaps the exception of lithium therapy in manic-depressive psychosis and aminoglycoside treatment in patients with renal failure.

Phenytoin[27]

Figure 4 illustrates the predicted relationship between dose and serum level for a drug exhibiting saturatable metabolism. Figure 7 shows the actual relationship found

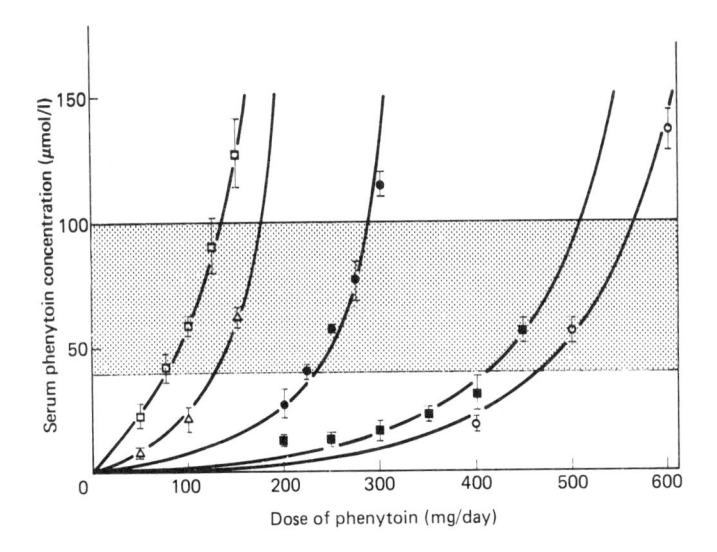

FIG. 7. Dose-serum phenytoin level relationship in five epileptic patients. The steady-state serum concentration was measured at several different doses, each point representing the mean ±S.D. of 3–8 separate estimations. The curves were fitted by computer using the Michaelis-Menten equation. The stippled area represents the therapeutic range used in our laboratory. (Reproduced from Richens and Dunlop[25] with permission.)

in five patients who were studied in detail on several different doses of phenytoin. The patient represented by the left-hand curve was a very slow metaboliser of phenytoin, requiring only 100–125 mg of the drug to achieve a therapeutic level. In contrast, the patient on the right was a fast metaboliser and required 500–550 mg before a therapeutic level was produced. This difference is predominantly genetic in origin and accounts for the wide range of phenytoin levels found in patients who are treated with standard doses without drug monitoring (Fig. 2). Saturation kinetics exaggerate the scatter in serum levels produced by any given dose. The steep dose-level relationship within the therapeutic range gives rise to the situation in which an increment of less than 100 mg will increase the serum level from below the therapeutic range to a level which will produce intoxication. This has a number of important implications for the practical management of the epileptic patient:

(i) If phenytoin therapy is regulated by monitoring serum levels, dose increments should become smaller as the therapeutic range of serum levels is reached. The commonly accepted range is 40–80 μmol/l (10–20 μg/ml) (*see below*) although a proportion of patients with mild epilepsy may be fully controlled at levels below this range. Nevertheless, if control is not achieved at low levels and the physician wishes to increase the dose to obtain a level within this range, the size of the increment should be judged from the steady-state level. A nomogram has been devised for this purpose,[23] although a simple rule-of-thumb can be used: if the serum level is less than 20 μmol/l (5 μg/ml) an increment of 100 mg is generally acceptable, if it is between 20 and 40 μmol/l (5–10 μg/ml) an increment of only 50 mg is permissible, and this should be reduced to a maximum of 25 mg if the level is 40–60 μmol/l (10–15 μg/ml). It should be borne in mind, however, that this simple scheme may overpredict for a slow metaboliser and underpredict for a fast metaboliser.

(ii) Serum concentrations within the therapeutic range are likely to be unstable because a few forgotten tablets or a small change in the amount of drug absorbed will produce a marked change in the level. Similarly, problems of biological availability (which have occurred frequently with phenytoin as it is a poorly-soluble acid) will be exaggerated by the steep relationship between serum level and dose. Only a small increase in biological availability could precipitate intoxication in a patient who previously had a level within the therapeutic range. This was clearly demonstrated when an outbreak of phenytoin intoxication occurred in Australia after the manufacturers of phenytoin capsules changed the excipient from calcium sulphate to lactose.

(iii) Drug interactions with phenytoin are likely to occur frequently and to be of clinical importance. In particular, addition of a drug which inhibits phenytoin metabolism will have an exaggerated effect when the enzyme system is in a critical state of saturation. The higher the initial level of phenytoin the greater will be the effect, and the low therapeutic ratio of phenytoin guarantees a clinically important effect. Sulthiame, isoniazid, chloramphenicol and disulfiram provide typical examples of drugs known to inhibit phenytoin metabolism and to provoke phenytoin intoxication in epileptic patients.

Because of the instability of serum phenytoin levels within the therapeutic range and the ease with which intoxication can be precipitated, the validity of the therapeutic range of serum levels needs critical assessment. A number of prospective studies have indicated that the majority of patients with major epilepsy show an optimum response only when the serum level exceeds 40 μmol/l (10 μg/ml), and as signs of intoxication become common above 80 μmol/l (20 μg/ml), these levels have been set as the limits to the therapeutic range.[13] However, the patients selected for these trials usually had frequent fits and their response to phenytoin does not necessarily reflect the effect of the drug in patients with mild epilepsy. It is generally agreed that many of these latter patients can be adequately controlled by levels below 40 μmol/l (10 μg/ml), especially those who have not had a fit for some years (it seems that good control of fits reduces the likelihood of further fits. No attempt should therefore be made to produce a level within the range of 40–80 μmol/l (10–20 μg/ml) unless it is obvious that an optimum response is not going to be achieved below this range. It is always desirable to keep the dose of phenytoin as low as possible in order to minimise the risks of adverse effects and drug interactions. In our opinion, a good case can be made out for using a therapeutic range of 'up to 80 μmol/l (20 μg/ml)'; if this is done, the physician will be less inclined to make unnecessary increases in dosage.

Despite the high prevalence of adverse effects and the problems which its pharmacokinetics create, phenytoin remains one of the drugs of first choice in major epilepsy. Recent studies have shown that the majority of adult newly-diagnosed patients with epilepsy can be controlled by phenytoin alone provided the dose is tailored to produce optimum serum levels.

Phenobarbitone

The dose-serum level relationship for phenobarbitone appears to be linear within the therapeutic range and therefore the drug lacks the pharmacokinetic problems of phenytoin. Against this, however, must be weighed two major disadvantages. First, phenobarbitone is sedative and there is a growing awareness that this may lead to subtle changes in intellectual function and behaviour. In children, paradoxical effects such as aggressiveness and hyperactivity may also be seen. Second, tolerance occurs to the drug's pharmacological effects on the central nervous system and severe fits or even status epilepticus may ensue when chronic barbiturate treatment is abruptly discontinued. The occurrence of withdrawal fits creates difficulties in the clinical management of the patient because the physician is likely to be unnerved by a sudden increase in fit frequency following a reduction in dosage and is likely to be convinced that the patient needs the higher dose. Had he, in fact, waited for a few weeks, fit frequency would probably have returned to near the original level as the withdrawal effects subsided. Also, the development of tolerance leads to difficulties in interpreting the result of a serum level estimation. It has been said earlier that one requirement for a meaningful interpretation of a drug's effect by monitoring serum levels is that there should be a reasonably constant relationship between the serum level and pharmacological action (i.e. a predictable pharmacodynamic relationship). With phenobarbitone there is not only the expected intersubject variation, but there is, in addition, a progressive change in responsiveness to its central actions within subjects while treatment is continued.

In theory, then, monitoring phenobarbitone levels would not be expected to be very helpful, and this accords with the authors' experience. Some authorities quote a therapeutic range of 45–110 μmol/l (~10–25 μg/ml), despite the fact that a substantial number of patients stabilised on phenobarbitone or primidone therapy without apparent untoward effects have levels exceeding this range. Although sedation may initially be troublesome when large doses are used, it is likely that this effect will diminish progressively with the development of tolerance. Whether tolerance develops not only to the sedative effects of the drug but also to the anticonvulsant action remains to be established. In any case, since serum levels associated with optimum clinical effect seem to vary enormously from one patient to another (and clearly also within the same patient), the value of routine monitoring is bound to be limited.

Primidone

As has been mentioned earlier, primidone is oxidised to two major metabolites, phenobarbitone and phenylethylmalonamide (PEMA). Whereas the serum half-life of the parent drug is on average about 6·5 hours in adults, that of phenobarbitone is about 100 hours (60 hours in children) and of PEMA around 20–30 hours. Although only 25–30% of administered primidone is converted to phenobarbitone, the long half-life of the latter substance causes it to accumulate so that serum phenobarbitone levels are 1·5–4 times higher than primidone levels. Similarly, PEMA accumulates but to a lesser extent than phenobarbitone. There is evidence from animal experiments (although not in man) that both unchanged primidone and PEMA have anticonvulsant activity, so there may be three pharmacologically active substances in serum. Most of the anticonvulsant activity probably rests with phenobarbitone and therefore measuring this substance will give the best estimate of the effect of administered primidone. If, however, the actions of the other two substances are important it will be necessary not only to measure them, but to know their relative anticonvulsant potency compared with phenobarbitone so that the sum total of the therapeutic effect of primidone administration can be assessed. Unfortunately, this information is not available and therefore monitoring levels of unchanged primidone is of no clinical value apart from identifying patients who metabolise the drug relatively slowly and therefore have very high serum concentrations of the parent compound, producing symptoms of toxicity.

Carbamazepine[3]

Carbamazepine is at least as effective as phenytoin and phenobarbitone in grand mal seizures and may be superior to these two in focal epilepsies, in which it is regarded as the agent of choice by some authorities. The absorption of the drug from the gastrointestinal tract is slow and erratic, and this may provide an explanation for the observation that dosage increments sometimes fail to produce a proportional rise in serum levels at steady-state. On chronic administration the half-life of carbamazepine is approximately 8–19 hours but values shorter than these have been described in patients receiving phenytoin or barbiturates in combination, due to induction of carbamazepine metabolism by the latter drugs. Because of the short half-life, serum carbamazepine levels tend to fluctuate considerably during the dosing interval, and the time of sampling in relation to dose is an important variable to be considered when interpreting serum level data.

Carbamazepine is partly metabolised to carbamazepine-10,11-epoxide, that in animal experiments has approximately one third of the anti-epileptic activity of the parent compound. The concentration of the epoxide in serum may approach that of the unmetabolised drug, and it is therefore possible that it may contribute significantly to the overall pharmacological activity. Although for full assessment of the drug's action it would be preferable to monitor the level of both compounds, for practical purposes measuring carbamazepine alone will give a reasonably good guide. Present evidence suggests that serum levels up to 40–50 μmol/l (up to ~10–12 μg/ml) are associated with optimum control of fits in the majority of patients. Due to the existence of large inter-patient differences in the disposition kinetics of the drug, serum level monitoring may prove highly valuable in the optimisation of therapy.

In addition to its anti-epileptic activity, carbamazepine has marked antidiuretic properties. In epileptic patients treated with carbamazepine alone, a good correlation can be found between the drug's serum level and the magnitude of the antidiuretic effect, significant hyponatraemia being relatively common at serum concentration values above 40 μmol/l (Fig. 8). Apart from water intoxication, adverse effects associated with high serum carbamazepine concentrations include drowsiness, incoordination, headache and visual disturbances. These effects are most pronounced at initiation of therapy and tend to subside without any need for a reduction in dosage. This may be explained by the fact that the highest serum concentrations of both carbamazepine and the epoxide are observed usually after 3–4 days and subsequently tend to decrease because of an autoinduction effect.

Ethosuximide

This drug is used primarily for treating petit mal absences in children. The serum half-life in these patients averages 30 hours (60–100 hours in adults) and once-daily dosage may be satisfactory. No active metabolites are produced. In general, intersubject variation in steady-state serum levels is less than for many drugs, but because ethosuximide is used mainly in children, monitoring serum levels helps to select a dose which is appropriate to the age and metabolic needs of the patient. Prospective studies relating serum levels to control of fits indicate that most patients respond optimally within the range of 300–700 μmol/l (~40–100 μg/ml).

Sodium Valproate

More recently introduced, this compound may provide a valuable alternative to ethosuximide in the treatment of petit mal epilepsy. It has a broad spectrum of anti-epileptic activity, although its role in the management of major epilepsy remains to be defined. Valproic acid has a short half-life (5–15 hours) and is extensively (>90%) bound to plasma proteins. Saturation of binding sites occurs at serum concentration values commonly encountered in therapeutic practice and, therefore, the fraction of unbound drug rises with increasing serum concentrations. There is no evidence that monitoring serum valproic acid levels can be valuable in the management of patients with epilepsy. It has been suggested that the serum level of the drug does not reflect the degree of pharmacological activity, possibly due to accumulation of as yet unidentified active metabolites. In one study, the suppression of the photosensitive response following single oral doses of the drug in epileptic patients was achieved only after 3 hours from the attainment of peak serum levels and lasted for up to 5 days after the drug was no longer detectable in blood.[28] These observations correlate with the general impression that twice- or even once-daily dosing produces adequate fit control, contrary to what one might predict from the short half-life of the drug.

In the authors' opinion, routine measurement of serum valproic acid levels should be restricted to those cases in whom assessment of compliance is essential. A number of separate studies have shown that valproic acid displaces phenytoin from plasma proteins and increases by 30–100% the unbound fraction of the latter drug. In patients receiving the two substances in combination, total serum phenytoin levels may grossly underestimate the concentration of free (pharmacologically active) drugs.

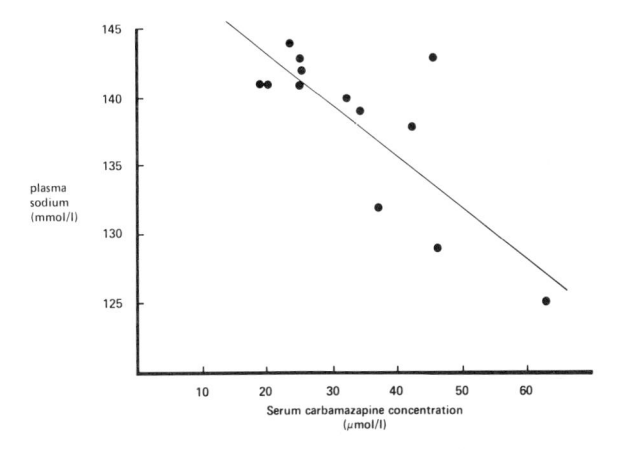

FIG. 8. Relationship between serum carbamazepine concentration and plasma sodium in 13 epileptic patients receiving chronic therapy with carbamazepine alone. (Reproduced from Perucca et al.[20] with permission.)

Other Drugs

A variety of other compounds are used occasionally in the management of epilepsy but apart from benzodiazepine drugs (which are useful in the treatment of epileptic myoclonus and status epilepticus) the drugs discussed above cover virtually all therapeutic needs in this disease. The value of monitoring serum levels of these other compounds has been little studied.

Lithium[1]

Lithium is of proven value as prophylactic therapy in established bipolar manic-depressive illness. Its value in recurrent unipolar depressive illness is less certain, and for the management of acute mania it is preferable to use phenothiazine or butyrophenone drugs in the first instance. Even though lithium is eliminated in the urine, there may be a threefold variation in the plasma lithium levels produced by standard maintenance doses in patients with apparently normal renal function. A high sodium and water intake will enhance the clearance of lithium and will, therefore, influence the maintenance dose required to produce a therapeutic level. The plasma half-life in patients with normal kidney function varies from 7 to 20 hours but it may be considerably longer than this in the presence of renal disease. A twice a day dosing schedule is generally used. Most authorities agree that optimum prophylaxis is achieved with a plasma level of 0·8–1·2 mmol/l, the estimation being performed 12 hours after the last dose. Although the vast majority of potential lithium responders will show a full therapeutic effect at serum concentrations close to the upper limit of the quoted range, clinical experience indicates that such initial high serum levels are associated with side-effects in far too many patients. Typical signs of lithium toxicity include tremor, muscle twitching, giddiness, ataxia, weakness, drowsiness, slurred speech, polyuria and thirst. When initial high dosages are used, some patients will have had such unpleasant experiences that they cannot be persuaded to try lithium again. In order to minimise the risk of adverse effects, a reasonable approach is to aim initially at plasma levels between 0·8 and 1·0 mmol/l, and to increase the dosage subsequently depending upon the clinical response. Patients with initial serum concentrations around 1·20 mmol/l are seriously at risk of developing lithium poisoning, especially during the episodes of dehydration which may result from the diuretic action of the drug. With levels above 3·0 mmol/l confusion, spasticity, convulsions and coma may occur. Obviously serum level monitoring is essential for a safe and effective use of this drug.

Tricyclic Antidepressants[29]

Patients treated with standard doses of tricyclic antidepressants develop markedly different steady-state serum levels, the highest level being approximately thirty times the value of the lowest for both nortriptyline and desipramine. These two compounds are the demethylated derivatives of amitriptyline and imipramine respectively, and have received the greatest attention from the pharmacokinetic point of view because the complication of biotransformation of one active compound into another is avoided. However, both parent compounds appear to show a similar wide variation in serum levels, determined largely by genetic differences in biotransformation rate. In recent years, evidence has accumulated that specific biochemical abnormalities may play an important role in determining the clinical manifestations of depressive disorders. In particular, it has been suggested that patients with endogenous depression can be subdivided into separate groups depending on the relative concentration of the monoamino metabolites 5-hydroxyindol-acetic acid (5-HIAA) and 4-hydroxy-3-methoxyphenyl-glycol (HMPG) in the CSF. Patients with high 5-HIAA (or low HMPG) in the CSF seem to benefit from noradrenaline-uptake inhibitors such as imipramine and nortriptyline, whereas patients with low 5-HIAA (or high HMPG) may do better with a serotonin uptake inhibitor, e.g. clomipramine or amitriptyline. If these observations are confirmed, selection of patients will become a crucial factor to consider when trying to relate serum drug levels to clinical effect.

A critical evaluation of available evidence on the relationship between serum levels of tricyclic antidepressants (particularly nortriptyline) and therapeutic effect indicates that intermediate serum levels are generally associated with maximum improvement, whereas too low or high levels appear to produce less or no clinical benefit. In one study, patients who had failed to respond to high levels were randomised into two groups, one continuing at high levels and the other having the level brought into the therapeutic range by lowering the dose. Whereas the latter group recovered the former did not. More recent studies with amitriptyline[16] and protriptyline suggest that the optimum response to all drugs of this group is also achieved at intermediate serum levels. Failure of therapy at high serum levels is probably due to blockade of postsynaptic monoaminergic receptors, an action that tends to counteract the effect of monoamine uptake blockade (which, in turn, is thought to account for the antidepressive activity of these drugs).

Summarising the available evidence, it would appear that an optimum response to nortriptyline would be expected within the serum level range of 190–570 nmol/l (50–150 ng/ml) whereas for amitriptyline a range of 300–800 nmol/l (approximately 80–220 ng/ml) (combined amitriptyline + nortriptyline) would seem to be appropriate. Although the serum levels of these drugs are low, and may therefore be difficult to measure in a routine laboratory with a consistent degree of precision, regular monitoring may be justified in a selected proportion of patients in whom a firm diagnosis of endogenous depression has been made. The following reasons can be put forward to justify this recommendation: (a) the metabolism of tricyclics and tetracyclines varies widely, making it impossible to use the ingested dose as a means of comparing different patients; (b) they have a narrow therapeutic range; (c) it may be difficult to distinguish therapeutic failure from a dose that is too high or too low; and (d) their therapeutic effects are difficult to assess clinically. However, measuring the serum concentration is no substitute for full clinical assessment of the patient's mood, an elevation of which is the ultimate goal of therapy.

Digoxin[7]

Selection of the correct dose of digitalis and its derivatives has presented problems to almost every physician since William Withering. It has been asserted that if the drug were introduced today as a new preparation, it would be regarded as too dangerous for general use. When a method for measuring serum digoxin concentration became available, it was naturally hoped that the use of the drug would become easier and safer. To what extent have these hopes been fulfilled?

The use of digitalis folia preparations was unsatisfactory because standardisation depended upon a bioassay and because the release of the active glycosides in the gut varied from batch to batch. The purification and standardisation of the major glycoside—digoxin—reduced this variability, although major problems have nevertheless occurred in its bioavailability from oral dosage forms.[6] This was highlighted in England in 1969 when an apparently trivial change in the manufacturing process of Lanoxin reduced its bioavailability. The reintroduction of the original formulation in 1972 was announced by circular and by notices in the medical press in order to minimise the risk of digoxin toxicity. Many subsequent studies have shown marked inequivalence in digoxin preparations from different manufacturers, and this has led to concern about the wisdom of prescribing the drug by its generic name.

Even if the same formulation of digoxin is always used, variation in serum levels produced by a given dose of the drug will still be seen from one patient to another, reflecting other pharmacokinetic differences. The elimination of digoxin is almost entirely by renal excretion, and impairment of this due to advancing age or renal disease will naturally cause a higher serum level on a given maintenance dose. In patients with normal renal function, however, there is a fairly good correlation between dose and serum level (Fig. 9). The serum half-life of the drug is normally 36 hours on average. The therapeutic dose of digoxin is only slightly less than a toxic dose, so using the drug safely requires sound clinical judgement, considerable experience, and not a little luck. A serum level of digoxin that is therapeutic in most patients could be excessive in the presence of hypokalaemia, hypothyroidism, arterial hypoxaemia, myocardial ischaemia or severe coronary heart disease. These disturbances alter the responsiveness of myocardial tissue to the drug.

Until a method for measuring serum digoxin was developed, regulation of dosage depended entirely on the clinical assessment of the therapeutic and toxic effects of the drug. However, the therapeutic effects of digoxin are not always easy to assess clinically. Patients with cardiac failure, for example, almost always receive, in addition to digoxin, other forms of treatment, e.g.

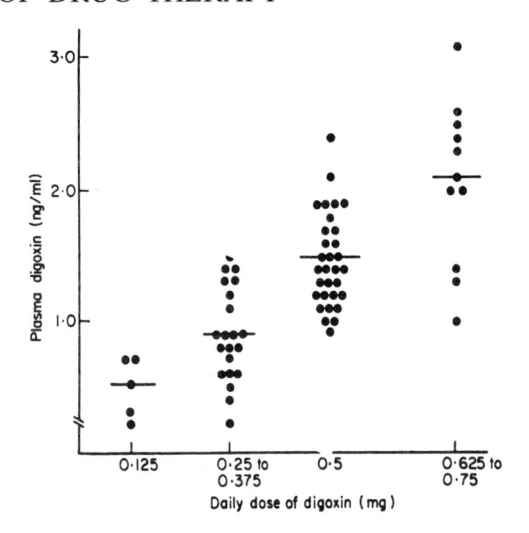

FIG. 9. Plasma digoxin levels plotted against oral dose of digoxin in 68 patients with normal or near-normal renal function. The horizontal bars represent mean values. (Reproduced from Chamberlain et al.[6] with permission.)

diuretics, bronchodilators, oxygen and bed rest. It is therefore difficult to know whether improvement in the patient's condition indicates that a therapeutic dosage of digoxin has been achieved. When the patient is fully recovered, there is no way of predicting whether the dose of the drug could be reduced without ill effect. Clinical assessment of the response to digoxin is somewhat easier in chronic atrial fibrillation. In this condition, the ventricular rate is normally a useful guide to the dose which is required: hence the tradition of withholding the drug when the pulse rate falls below sixty per minute. However, in the presence of thyrotoxicosis, anxiety or inadequately treated cardiac failure, the heart rate may be rapid in spite of toxic serum levels of digoxin. In one study, a poor correlation was found between serum digoxin level and slowing of the heart rate in atrial fibrillation (Fig. 10), unless those patients with a possible atrio-ventricular conduction abnormality, which caused undue sensitivity to digoxin, were excluded.

Since the therapeutic effects of digoxin may not be easily monitored, one is often dependent on the detection of toxic effects of the drug in placing an upper limit on dosage. The idea of establishing a therapeutic dose of digoxin by prescribing a toxic dose and then reducing it is hardly attractive, but nevertheless this is quite often what happens in practice. Toxic effects of digoxin on the heart include brady- and tachy-dysrhythmias of almost all kinds. None of these dysrhythmias is specific to digoxin toxicity and they may equally well be due to the underlying cardiac disease. Symptoms of extra-cardiac digoxin toxicity (e.g. anorexia, nausea, vomiting, disturbance of bowel habit, yellow vision—'xanthopsia'— and mental confusion in the elderly) are also non-specific.

Although it is easy to exaggerate the difficulty of finding the correct dose of digoxin for each patient, the common occurrence of digoxin toxicity is ample evi-

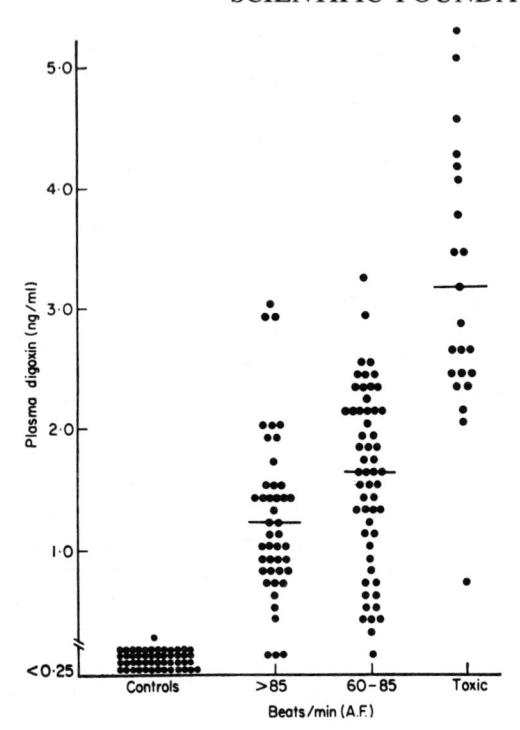

FIG. 10. Plasma digoxin levels in 44 patients with atrial fibrillation and relatively fast heart rates and in 56 patients with atrial fibrillation and slower heart rates. All subjects were normokalaemic. Levels in 22 patients with digoxin toxicity and in 50 control subjects not receiving the drug are shown for comparison. The horizontal bars represent mean values. (Reproduced from Chamberlain et al.[6] with permission.)

dence that clinical methods of monitoring response to the drug are unsatisfactory. The development of a radioimmunoassay technique for measurement of serum digoxin concentration, therefore, has proved to be a considerable advance in drug therapy, although some of the more optimistic hopes have not been fulfilled.[12]

Blood samples for digoxin assay must be taken not less than four hours after the last dose of the drug, in order to avoid the spurious peak levels which occur before this time. Due to the long serum half-life of digoxin (30–40 hours), a sample taken just before dosing is quite acceptable for most purposes. In practice, the greatest difficulty in the interpretation of serum digoxin measurements is the considerable overlap between therapeutic and toxic levels (Fig. 10). The therapeutic range is generally reckoned to be 1·3–3·2 nmol/l (1–2·5 ng/ml) and values of greater than 4·0 nmol/l (~3·0 ng/ml) are regarded as likely to be toxic. However, some patients have undoubted symptoms and signs of digoxin toxicity in spite of having serum levels within the therapeutic range, others may have a satisfactory therapeutic response to digoxin in the presence of 'sub-therapeutic' serum levels, and a few require 'toxic' serum levels in order to produce an adequate therapeutic effect.

Provided that the limitations of digoxin serum level measurements are understood, the technique can be a valuable aid to the physician. First, if serum levels are viewed in the context of clinical history, physical examination and electrocardiographic evidence, they may be helpful in reaching a decision about the patient's state of digitalisation. Second, serum levels may be used to distinguish between failure to take prescribed digoxin and failure to absorb drug which has actually been ingested. If an outpatient whose serum level has remained low in spite of apparently reasonable doses of digoxin is admitted to hospital and given the drug under supervision, the serum level will rise if non-compliance was the cause, but will remain unchanged if pharmacokinetic factors, e.g. failure to absorb the drug, are responsible. Third, if DC cardioversion is required in a digoxin-treated patient, knowledge of the serum level may assist in balancing the risks of the manoeuvre against the hazards of delaying definitive treatment. Fourth, if supraventricular tachycardia develops in a patient on digoxin, a larger dose of the drug may be given if a low serum level of the drug is found, whereas levels within or above the therapeutic range suggest that the drug should be stopped. Finally, when a patient in cardiac failure does not respond to conventional treatment, the digoxin dosage can be adjusted in order to obtain a serum level within the therapeutic range.

Anti-dysrhythmic Drugs

In clinical practice, measurement of serum concentration of drugs has proved more useful in the regulation of therapy with anti-dysrhythmic agents than with any other type of cardio-active drug. This is partly because of the extreme difficulty in assessing the efficacy of anti-dysrhythmic therapy in any other way. These drugs are usually given to prevent supraventricular and ventricular tachycardia or fibrillation, and these dysrhythmias vary in severity from trivial to life-threatening. The apparent abolition of a sporadic dysrhythmia is not a satisfactory criterion of adequate drug therapy: the patient's attacks may be infrequent and unpredictable in occurrence, and a spontaneous remission may be interpreted as a drug response. With potentially lethal dysrhythmias, this approach is clearly out of the question, since the first failure of treatment in a patient may well be the last. Even if failure of treatment results not in death but in a minor episode of illness, each episode may cause discomfort, anxiety, and perhaps admission to hospital.

There are a number of other reasons why measurements of serum levels of anti-dysrhythmic drugs might be expected to be valuable. Pharmacokinetic variation is considerable for these compounds, and therapy may be particularly hazardous in those with impaired renal or hepatic function. The toxic effects of excessive dosage with anti-dysrhythmic drugs may be subtle, and can develop insidiously; heart failure due to myocardial depression is often the only sign, and may easily be wrongly attributed to the underlying heart disease. This means that it is rarely possible to arrive at a therapeutic level of a drug by increasing the dose until toxicity becomes manifest, and then reducing the dose slightly.

In many cases, cardiac dysrhythmias may paradoxically develop as a sign of toxicity. Without the assistance of measurement of the serum level of the drug, the physician may be misled to increase the dose, with potentially disastrous clinical consequences.

There are, however, some important limitations on the value of measuring serum levels of anti-dysrhythmic drugs. First, in order to interpret the results a valid 'therapeutic range' must be established. Ideally this range would correspond to levels found in a large group of successfully treated patients, in whom dysrhythmias have been proven to be abolished on continuous 24-hour ECG recording, and who did not manifest symptoms or signs of drug toxicity. Since it is difficult to obtain this ideal information, therapeutic ranges are generally derived from empirical assessment of drug response in a limited number of patients. In practice this seems to be adequate, but the upper limit of the therapeutic range is likely to be more reliable than the lower as its endpoint is drug toxicity.

It is important to take at least one 'peak' and one 'trough' sample from each patient: this will show whether the size of the dose or the frequency of its administration, or both, should be altered. In addition, as with any other drug, it is unwise to place absolute trust in a single estimation. The assay techniques currently in use do not have the same accuracy and reliability that one expects from routine laboratory procedures. Furthermore, they sometimes fail to measure a major active metabolite and therefore the estimation may not represent the 'total serum anti-dysrhythmic effect' of the drug.

Lignocaine

Lignocaine is used in the acute treatment and prophylaxis of ventricular dysrhythmias, especially those occurring after myocardial infarction. It is almost always administered by the intravenous route, but single intramuscular injections have been used in patients at home who are awaiting transfer to hospital. Oral administration is not used because of a large first-pass effect.

After intravenous administration of a single dose, the serum lignocaine concentration declines rapidly with a half-life of about 8 minutes; this is related to the high lipid solubility of the drug and its large volume of distribution (60–120 litres). Subsequently, the drug is eliminated more slowly mainly by hepatic metabolism, with a serum half-life of 60–120 minutes in normal subjects. The therapeutic range of serum concentrations is 6–25 μmol/l (~1·5–6 μg/ml).

Serum concentrations of lignocaine are rarely measured in clinical practice, and there are several reasons why this is so. The total body clearance of the drug is mainly determined by the rate of liver blood flow, and it is little affected by differences in the degree of drug-metabolising enzyme activity or changes in renal function. The usual mode of administration is a bolus intravenous injection followed by intravenous infusion, so bioavailability problems do not occur and magnitude of dose rather than frequency of administration is the only variable to be considered. Lignocaine therapy is given only to patients under close supervision, and is rarely continued for more than a few days: it is therefore much easier to monitor the pharmacological effect of the drug than with long term oral anti-dysrhythmic therapy. If dysrhythmias fail to respond to the starting dose of lignocaine, the dose is increased progressively until control is achieved. Central nervous system toxicity, including sedation, mental confusion and epileptiform seizures, limits the amount which can be given. Although these effects are unpleasant, they are quickly reversible when the lignocaine infusion is stopped: doses causing central nervous system toxicity do not usually cause depression of myocardial mechanical function. In most clinical situations, decisions about changes in dosage need to be made rapidly. Serum estimations are likely to be of value only if results are fed back to the clinician within minutes, and this is often impractical. If ventricular dysrhythmias fail to respond to maximum doses of lignocaine, a change is immediately made to another drug.

However, there is one clinical situation in which lignocaine cannot be used safely without monitoring the serum levels; this is in severely ill patients with hypotension or cardiogenic shock, in whom lignocaine metabolism may be considerably depressed by the impairment of hepatic perfusion, and in whom excessive drug accumulation may easily occur.

Procainamide[11]

In contrast with lignocaine, procainamide is usually given orally for medium and long-term prophylaxis of dysrhythmias, although the intravenous route is sometimes used in emergency situation. Due to marked inter-patient differences in the pharmacokinetics of the drug, individualisation of dosage is essential if an optimal therapeutic response is to be achieved. The serum half-life of procainamide is only 2·5–4·7 hours, and the therapeutic range of serum levels (17–43 μmol/l, 4–10 μg/ml) is not wide, so unless frequent doses are given many patients will inevitably have alternating toxic and sub-therapeutic levels. Sustained release preparations are available, which help to overcome this problem. Whichever preparation of the drug is used, a knowledge of both 'peak' and 'trough' serum levels is valuable in assessing the adequacy of a dosage regimen, especially in patients whose response to the drug is difficult to evaluate or appears to be abnormal.

About 90% of procainamide administered is excreted unchanged in the urine, so monitoring of serum levels is important in the presence of impaired renal function. Procainamide is metabolised to N-acetyl procainamide, which itself has antidysrhythmic activity. Since the half-life of the metabolite is about twice that of the parent drug, it accumulates in serum during chronic administration and may give an important contribution to the overall antidysrhythmic activity. About one quarter of the UK population are 'slow acetylators' and will therefore have a different proportion of unchanged drug and metabolite in serum compared with that found in 'fast acetylators'. It is therefore important to know whether the technique used in the drug assay is measuring both drug and metabolite, or drug only. Although measurement of both drug and active metabolite will give a better indication of total serum antidysrhythmic activity, it is clear that the pharmacological actions of the two compounds are not simply additive in all rhythm disturbances.

Quinidine

The problems encountered in adjusting quinidine therapy are similar to those described with procainamide. Quinidine is usually given orally for medium and long term prophylaxis of both atrial and ventricular dysrhythmias. Its serum half-life (6 hours) is longer than that of procainamide, but the therapeutic range of serum levels (6–14 μmol/l; 2·3–5 μg/ml) is virtually as narrow. Sustained release preparations are available. Quinidine is oxidised in the liver to metabolites which retain some degree of pharmacological activity. The elimination of both parent drug and metabolites is reduced in cardiac or renal failure. The cardiac toxicity of quinidine develops insidiously, can be serious, and may not be preceded by more benign warning signs of excessive dosage. As with procainamide, quinidine may induce some of the dysrhythmias for which it is usually administered. The therapeutic response to the drug cannot easily be assessed, except possibly with benign recurrent atrial dysrhythmias when the patient's reporting of his own symptoms may be a reliable guide to adequacy of treatment. Measurement of 'peak' and 'trough' levels of quinidine should be routinely performed, especially where the drug is being used in the prophylaxis of malignant ventricular dysrhythmias. The specificity of the analytical technique must be considered if serum quinidine levels are to be interpreted correctly in clinical practice. The therapeutic range quoted above was established by using the Cramer-Isaksson double-extraction fluorimetric assay, and it may be that slightly lower limits apply when more recently developed specific immunological or high-pressure liquid chromatographic methods are used.

Phenytoin

Phenytoin is used both intravenously and orally for the short and long-term suppression of ventricular dysrhythmias. The problems of adjusting dosage are more severe in cardiac patients than in epileptics since failure of treatment may be life-threatening and cardiac dysrhythmias are harder to detect and quantify than epileptic fits. The therapeutic range of serum concentration for suppression of ventricular dysrhythmias is claimed to be 40–72 μmol/l (10–18 μg/ml). Even in concentrations within the therapeutic range, phenytoin has quite marked cardio-depressant effects. Because of the peculiar pharmacokinetics (see page 389) and cardiac toxicity of phenytoin, monitoring of serum levels is essential when the drug is used for long-term prophylaxis in dysrhythmias.

In the acute coronary care situation, however, phenytoin assay is unlikely to be very helpful. The loading dose of phenytoin is 10–15 mg/kg, given in increments of up to 100 mg at five-minute intervals. If this dose is ineffective, another anti-dysrhythmic drug will usually be tried instead. The slow elimination of phenytoin means that maintenance therapy may not be required, since many dysrhythmias occurring after myocardial infarction are transient.

Mexiletine

Mexiletine is used both orally and intravenously for the suppression of ventricular dysrhythmias, particularly those resistant to lignocaine. The terminal half-life of mexiletine is between 11 and 26 hours, although after intravenous bolus administration the serum level declines very rapidly due to redistribution. Elimination of mexiletine is mainly by hepatic metabolism, but some unchanged drug is excreted in urine. The renal elimination of the drug may be enhanced by reduced urinary pH.

The therapeutic range of serum levels is said to be 3·5–15 μmol/l (0·6–2·5 μg/ml), but few laboratories measure the drug. Serum level determination should be useful in differentiating drug resistance of dysrhythmias from inadequate dosage and patient non-compliance.

Disopyramide

Disopyramide is administered orally for the prophylaxis and treatment of both atrial and ventricular dysrhythmias. An intravenous preparation is also available. The serum half-life is 6 hours in normal subjects. A fraction of the administered dose of disopyramide is mono-N-dealkylated in the liver, but the metabolite retains both anti-dysrhythmic and anti-cholinergic activity. Although a therapeutic range of 6–21 μmol/l (2–7 μg/ml) is quoted by some authors, the value of measuring serum levels of the drug requires further evaluation.

In practice, disopyramide has proved remarkably free from serious toxicity, possibly because unpleasant anti-cholinergic effects appear as soon as excessive doses are given. These unwanted effects provide a good indicator of the upper limit of dosage, and cardiac toxicity does not seem an important problem. Measurement of serum levels might be expected to be helpful in patients with impaired renal function; however, in such patients the major metabolite will also suffer impaired elimination, and may not be measured by the assay technique used.

Other Drugs

The value of measuring serum levels of amiodarone, verapamil and some other drugs used in the management of cardiac dysrhythmias has not been adequately assessed to date.

Beta-adrenergic Blocking Drugs

Determination of the plasma level of beta-adrenergic blocking drugs may be useful in the diagnosis of patient non-compliance, but is otherwise of very little value in clinical practice at present. The pharmacological effect of this group of drugs is easily monitored clinically, for example by assessing their effect on exercise heart rate: beta blockade is adequate if the heart rate does not increase beyond 120 beats per minute during maximal or near-maximal effort. Therapeutic effects are also quite easily measured: for example, blood pressure in hypertensive patients, and exercise tolerance or frequency of attacks in patients with angina pectoris. Antidysrhythmic effects are less easily monitored (*see* page 394). The therapeutic index is wide for all beta-blocking drugs, and if the dose is increased progressively, very large amounts can be given safely; this policy is used by some physicians in the treatment of hypertension or schizophrenia. Available evidence suggests that a very high degree of cardiac β-blockade is achieved at levels of propranolol between 290 and 385 nmol/l (75–100 ng/ml) in young volunteers, and that these levels may also be adequate for the prophylaxis of angina. Some patients with ventricular dysrhythmias and low renin hypertension, however, may require serum levels greatly in excess of this. In elderly patients, the tissue sensitivity to the drug is reduced and higher concentrations may therefore be required.

One of the reasons why the therapeutic effects of a beta-blocking drug may not correspond with the measured serum level, is that the presence of active metabolites may not be detected by the assay procedure. For example, orally administered propranolol is extensively metabolised to 4-hydroxypropranolol, which itself has significant beta-blocking activity.

Anticoagulants

The pharmacological effects of anticoagulants are very easily monitored, and enable accurate adjustment of drug dosage. It is therefore rarely necessary to consider measuring the serum concentration of these drugs. Indeed, for a given degree of anticoagulation, serum warfarin concentrations vary four-fold from one patient to another, presumably because of differences in vitamin K intake and capacity to synthesise prothrombin. The dosage of coumarin anticoagulants is adjusted according to the prothrombin time, and the dose of heparin according to the whole blood clotting time or partial thromboplastin time. Warfarin levels may sometimes be useful in distinguishing true 'warfarin resistance' (unusual) from failure to take prescribed medication (common).

Theophylline[18]

Although the use of theophylline and related xanthines in the treatment of asthma decreased with the development of specific β_2-adrenoceptor agonists, growing awareness of their value, particularly when regulated by serum level monitoring, has led to a return in the popularity of these compounds. Their correct use can allow the physician to avoid chronic corticosteroid therapy in most patients. Furthermore, in the management of status asthmaticus, a condition which still carries a high mortality rate, they still have an important place. Individual variation in the rate of metabolism of theophylline is high as evidenced by the wide range of maintenance doses (400–3200 mg daily) required to produce a therapeutic serum level. Furthermore, theophylline has a low therapeutic ratio causing, in addition to nausea and vomiting, serious cardiac and central nervous system toxicity in overdosage, namely agitation, convulsions, cardiac dysrhythmias, hypotension, and sudden death. Usually these effects occur after an intravenous injection which has been given too rapidly, and some become common when the serum level exceeds 100 μmol/l (18 μg/ml). Maximum bronchodilatation occurs in most patients with serum levels in the range of 50–100 μmol/l (9–18 μg/ml), but some subjects show an optimum therapeutic response at levels lower than these.

The serum half-life varies from 3–9.5 hours; it is shorter in cigarette smokers than in non-smokers. Eighty-five per cent of the drug is converted to uric acid derivatives and/or demethylated. There is increasing evidence suggesting that the metabolism of theophylline may be saturable within the therapeutic dose range, and that the dose:serum level relationship therefore may not be linear. Approximately half of the drug in serum is bound to albumin and it has been shown that salivary levels reflect the concentration of unbound drug in serum. This represents an alternative method of monitoring drug levels but it is not recommended because of variations in plasma/saliva concentration ratios which occur with time and between individuals.

When intravenous therapy (as aminophylline) is necessary, an initial dose of 5.6 mg/kg should be given followed, in patients with normal hepatic and cardiac function, by 0.9 mg/kg/hour. The infusion rate can then be adjusted to produce a serum level within the range of 40–100 μmol/l (9–18 μg/ml). Orally, a 500 mg loading dose of theophylline should be given, followed by 300 mg 6-hourly, with subsequent adjustment according to clinical response and serum drug levels. To facilitate interpretation of the results, samples should be taken at the time of the minimal or trough concentration, i.e. prior to the next dose.

Anti-inflammatory Drugs

Aspirin and Salicylic Acid

Salicylates are still considered by many rheumatologists to be the agents of first choice in rheumathoid arthritis. Evidence is available that maintenance of steady-state serum levels of salicylic acid above

1·1 mmol/l (150 μg/ml) is necessary to achieve an optimum therapeutic response, and since toxicity becomes common at levels about 2·2 mmol/l (300 μg/ml) these values have been set as limits for the therapeutic range. Since there is large inter-subject variation in serum levels produced by a fixed dose due to differences in the rate of metabolism, drug interactions and occurrence of autoinduction on repeated dosing, regular monitoring of serum salicylic acid levels may be a valuable supplement to clinical judgement in selected patients. Adjustments of dosage must be made cautiously because salicylic acid, like phenytoin, shows saturation kinetics at therapeutic serum concentrations.

Phenylbutazone

Phenylbutazone is partly metabolised to oxyphenbutazone which retains pharmacological properties similar to those of the parent drug. The therapeutic range of serum phenylbutazone is said to be 195–260 μmol/l (60–80 μg/ml) but more clinical data are required to establish the potential usefulness of monitoring the serum levels of this drug during maintenance therapy.

Gold

Although serum or whole blood levels of gold are sometimes measured in clinical practice, there is no convincing evidence to date that these can be useful in predicting the clinical response to the drug.

Anti-neoplastic Drugs

Three major classes of clinically useful anti-neoplastic agents are currently available: antibiotic DNA intercalators, alkylating agents and anti-metabolites. With most of these drugs, no correlation between serum levels and pharmacological effect has generally been found, and there are several reasons why this may be so. In the first instance, available analytical techniques may not be sufficiently sensitive to measure the serum levels of the drug during the terminal elimination (β) phase (following administration of therapeutic doses in man), and serum levels determined during the distribution phase may not reflect accurately the amount of drug available at the site of action. Second, many of these agents, e.g. 5-fluoro-uracil and cyclophosphamide, are known to act through active metabolites which bind tightly to macromolecular structures in the tumoural tissue. Even when active metabolites are not produced, there is still a tendency for the clinical effect to persist longer than the actual permanence of the drug in the organism, due to the characteristic 'hit and run' action of some of these drugs. Finally, clinically responsive tumours tend to acquire tolerance to the anti-neoplastic effect, resulting in unpredictable changes of the dose-response relationship. Because of these problems monitoring of anti-neoplastic drugs in serum is at present mainly a research exercise, with the possible exception of methotrexate.

Methotrexate

Methotrexate is predominantly used in the treatment of choriocarcinoma, Burkitt's lymphoma, acute lymphatic leukaemia in children, breast cancer and osteogenic sarcoma. The drug acts by inhibiting dihydrofolate reductase, resulting in intracellular depletion of reduced folate and consequent depression of DNA synthesis. It can be given orally, intrathecally and intravenously. As in the case of other anti-neoplastic agents, signs of toxicity can be severe and include myelosuppression, mucositis and renal damage. These effects can be prevented by administering a specific antidote, citrovorum factor (CF), which enters the folate cycle distal to the methotrexate-induced enzymatic block.

Dosage schedules vary considerably according to clinical diagnosis, age, body surface, hepatic and renal function, and specific protocol. The maximum tolerated dose systemically varies from 80 to 900 mg/m^2. In recent years, the observation that CF can promote resumption of DNA synthesis in normal bone marrow, without promoting a similar recovery in tumoral tissue, has led to a new therapeutic approach in which a high dose of methotrexate (up to 3000–30 000 mg/m^2) is administered intravenously, shortly followed by CF rescue by the same route. Evidence is accumulating that, under the circumstances, measurement of serum methotrexate concentration will allow prediction of the risk of toxicity and, thereby, identification of those patients who need extended CF administration, or higher CF dosage. In addition, serum methotrexate levels can be valuable in individualising the time at which CF can be discontinued, resulting in potentially enhanced clinical effectiveness. For further information, the reader is referred to Bleyer.[4]

Aminoglycoside Antibiotics[14,19]

The aminoglycoside antibiotics namely kanamycin, gentamicin, tobramycin, amikacin, streptomycin, sisomicin and netilmicin, are particularly valuable in treating serious Gram-negative infections and tuberculosis. Their major disadvantage, however, is that they have a narrow therapeutic ratio, causing damage to the inner ear or kidney in doses only slightly exceeding those required to combat the infection. Some, e.g. neomycin, are so toxic that they are not normally administered systemically, being used only for surface application or bowel sterilisation (they are very poorly absorbed when given orally).

Aminoglycoside antibiotics are excreted in the urine largely unchanged and therefore the maintenance dose has to be reduced when renal function is impaired. Since it is often necessary to use them in Gram-negative septicaemias accompanying renal failure, a careful assessment of renal function, usually by measuring the creatinine clearance, is imperative. The initial loading dose is determined by the body weight of the patient and will therefore be of the same order whether renal function is normal or reduced. The maintenance dose is reduced in proportion to the creatinine clearance, tables and nomograms being available for this purpose. The predictive value of nomograms can be greatly improved by taking into consideration physiological parameters such as sex, age, lean body mass, and haematocrit. No general agreement exists on which is the most satisfactory mode of administration (intramuscular *vs* intravenous route, bolus injection *vs* continuous infusion).

When renal function is severely impaired, it is essential that the serum level produced by the predicted dose schedule should be monitored. At least two samples should be collected, for the estimation of peak and trough levels respectively. For determination of the peak level, samples should be obtained within 30 min of the intravenous infusion, or one hour after the intramuscular injection, of the aminoglycoside. The sample for the trough level should be collected 30 min before the next dose. The relative contribution of peak and trough concentrations, duration or therapy and total dose administered to the therapeutic and toxic potential of these drugs is not yet completely understood. The reader is referred to Barza and Lauermann[2] for a detailed review of this important topic.

REFERENCES

1. Amdisen, A. (1977), 'Serum level monitoring and clinical pharmacokinetics of lithium.' *Clin. Pharmacokin.*, **2**, 73–92.
2. Barza, M. and Lauermann, M. (1978), 'Why monitor serum levels of gentamicin?' *Clin. Pharmacokin.*, **3**, 202–215.
3. Bertilsson, L. (1978), 'Clinical pharmacokinetics of carbamazepine.' *Clin. Pharmacokin.*, **3**, 128–143.
4. Bleyer, W. A. (1978), 'The clinical pharmacology of methotrexate: new applications of an old drug.' *Cancer*, **41**, 36–51.
5. Booker, H. (1972), 'Phenobarbitol, mephobarbital and metharbital. Relation of plasma levels to clinical control.' In *Antiepileptic Drugs*. (Woodbury, D. M., Penry, J. K. and Schmidt, R. P., Eds), pp. 403–407. New York: Raven Press.
6. Chamberlain, D. A., White, R. J., Howard, M. R. and Smith, T. W. (1970), 'Plasma digoxin concentrations in patients with atrial fibrillation.' *Brit. Med. J.*, **3**, 429–432.
7. Chamberlain, D. A. (1975), 'Digoxin.' In *Advanced Medicine. Topics in Therapeutics 1*, pp. 49–63 (Breckenridge, A. M., Ed.). London: Pitman.
8. Danhof, M. and Breimer, D. D. (1978), 'Therapeutic drug monitoring in saliva.' *Clin. Pharmacokin.*, **3**, 39–57.
9. George, C. F., Fenyvesi, T. and Dollery, C. I. (1973), In *Biological Effects of Drugs in Relation to their Plasma Concentrations*, pp. 123 (Davies, D. S. and Prichard, B. N. C., Eds). Macmillan: London.
10. Koch-Weser, J. (1975), 'The serum level approach to individualization of drug dosage.' *Eur. J. clin. Pharmacol.*, **9**, 1–8.
11. Koch-Weser, J. (1977), 'Serum procainamide levels as therapeutic guides.' *Clin. Pharmacokin.*, **2**, 389–402.
12. Lancet Editorial (1978), 'Digoxin, more problems than solutions.' *Lancet*, **ii**, 1288–1290.
13. Lund, L. (1974), 'Anticonvulsant effects of diphenylhydantoin relative to plasma levels. A prospective 3-year study in ambulant patients with generalized epileptic seizures.' *Arch. Neurol. (Chic.)*, **31**, 289–294.
14. Mawer, G. E. (1976), 'Dosage schedules for aminoglycoside antibiotics.' In *Advanced Medicine. Topics in Therapeutics 2*, pp. 36–50 (Turner, P., Ed.). London: Pitman.
15. Melander, A. (1978), 'Influence of food on the bioavailability of drugs.' *Clin. Pharmacokin.*, **3**, 337–351.
16. Montgomery, S. A., McAuley, R., Rani, S. J., Montgomery, D. B., Braithwaite, R. and Dawlings, S. (1979), 'Amitriptyline plasma concentrations and clinical response.' *Brit. Med. J.*, **1**, 230–231.
17. Morselli, P. L. (1976), 'Clinical pharmacokinetics in neonates.' *Clin. Pharmacokin.*, **1**, 81–98.
18. Ogilvie, R. I. (1978), 'Clinical pharmacokinetics of theophylline.' *Clin. Pharmacokin.*, **3**, 267–293.
19. Pechere, J.-C. and Dugal, R. (1979), 'Clinical pharmacokinetics of aminoglycoside antibiotics.' *Clin. Pharmacokin.*, **4**, 170–199.
20. Perucca, E., Garratt, A., Hebdige, S. and Richens, A. (1978), 'Water intoxication in epileptic patients receiving carbamazepine.' *J. Neurol. Neurosurg. Psych.*, **41**, 713–718.
21. Perucca, E. and Richens, A. (1980), 'Interpretation of drug levels: relevance of plasma protein binding.' In *Drug Concentrations in Neuropsychiatry. (Ciba Foundation Symposium 74)*. Pp. 51–68.
22. Pippenger, C. E., Penry, J. K., White, B. G., Daly, D. D. and Buddington, R. (1976), 'Interlaboratory variability in determination of plasma anti-epileptic drug concentrations.' *Arch. Neurol.*, **33**, 351–355.
23. Rambeck, B., Boenigk, H. E., Dunlop, A., Mullen, P. W., Wadsworth, J. and Richens, A. (1979), 'Predicting phenytoin dose—a revised nomogram.' *Therap. Drug Monitoring* (in press).
24. Reidenberg, M. M. (1977), 'The binding of drugs to plasma proteins and the interpretation of measurements of plasma concentrations of drugs in patients with poor renal function.' *Am. J. Med.*, **62**, 466–470.
25. Richens, A. and Dunlop, A. (1975), 'Serum phenytoin levels in the management of epilepsy.' *Lancet*, **ii**, 247–248.
26. Richens, A. (1978), 'Drug level monitoring—quality and quantity.' *Br. J. clin. Pharmac.*, **5**, 285–288.
27. Richens, A. (1979), 'Clinical pharmacokinetics of phenytoin.' *Clin. Pharmacokin.*, **4**, 153–169.
28. Rowan, A. J., Binnie, C. D., Warfield, C. A., Meinardi, H. and Meijer, J. W. A. (1979), 'The delayed effect of sodium valproate on the photoconvulsive response in man.' *Epilepsia (Amst.)*, **20**, 61–68.
29. Sjöqvist, F. (1975), 'Assessment of antidepressants—pharmacokinetic aspects.' In *Advanced Medicine. Topics in Therapeutics 1*, pp. 198–217 (Breckenridge, A. M., Ed.). London: Pitman.

24. TRAUMA, SHOCK AND SURGERY

A. C. AMES

Introduction
Phasing of the metabolic response to injury
Endocrine changes after trauma
Physiological responses in shock which compensate for hypovolaemia
Local effects of vasoactive substances
Intracellular metabolic changes in shock
Tissue hypoxia and lactic acidosis during shock
Nutritional reserves and energy metabolism
Carbohydrate metabolism
Protein metabolism
Changes in plasma proteins after injury
Fat metabolism
Water metabolism
Sodium and potassium
Magnesium, calcium, phosphate and trace elements
Acid-base disturbance
Pulmonary function in shock
Effects of shock on renal function
Haematological effects of trauma

Introduction

Severe trauma, major surgery, burns or septicaemia can be complicated by 'shock' which is recognised clinically as a combination of pallor, sweating, peripheral cyanosis and hyperventilation with hypotension and tachycardia culminating in oliguria. Hypovolaemic shock follows haemorrhage, the losses of large volumes of fluid caused by major burns, vomiting, diarrhoea and gastrointestinal fistulae or the extravasation of plasma into damaged tissues. Bacteraemic shock may occur in severe gram negative infections but there may be little change in the blood volume. The collapse associated with acute heart failure following myocardial infarction is termed cardiogenic shock.

The physiological and biochemical reactions which are responsible for the shock syndrome occur immediately after injury and are the result of complex but integrated neuroendocrine and cellular responses which are teleologically essential for survival. The main physiological reactions are haemodynamic to compensate for the circulatory failure by redistributing blood from the peripheral tissues to preserve the circulation to vital organs.

If the individual survives the acute injury a much longer period follows in which there is an intense metabolic drive directed towards the mobilisation and utilisation of the endogenuous nutritional reserves along with the conservation of salt and water.

As this may be a time of enforced starvation these processes provide the energy and intermediate substrates necessary for healing and repair.

These responses can be modified or impaired by other factors. For example, in hypercatabolic states which include burns, multiple injuries, peritonitis, septicaemia, pancreatitis and hypercatabolic renal failure the metabolic rate is increased by over 25%. The normal increase in metabolic rate in uncomplicated illness is little more than 10% of the basal rate. Similarly, malnutrition with depleted energy reserves before injury and starvation after injury will impair the chances of survival especially in the debilitated or elderly patient. General anaesthesia, a low environmental temperature and immobilisation also reduce general vitality. In severe injury with prolonged shock the initial physiological and biochemical reactions lose their protective function and become pathological, adversely affecting recovery. This deterioration is more likely to occur if resuscitative measures like blood volume replacement, oxygenation (ventilation) and nutritional support are inadequate.

Appropriate biochemical monitoring together with an understanding of the timing, extent and purpose of these reponses will often enable clinical measures to be taken which will prevent or reverse these undesirable complications.

Phasing of the Metabolic Response to Injury

The response was originally subdivided into an early 'ebb' phase lasting 24–48 hours which corresponded with the period of shock and a longer 'flow' or recovery phase lasting several days or weeks.[13]

The 'flow' phase was subsequently divided into a catabolic phase lasting 3–7 days and a second phase of variable duration which coincided with a period of progressive anabolism.[28]

The 'ebb' or shock phase is one of depressed vitality with a reduction in metabolic rate, oxygen consumption, body temperature and cardiac output. If the injury is severe these effects persist, oxygen consumption declines rapidly and metabolic deterioration culminates in terminal necrobiosis.

In the 'flow' phase heat production and metabolic rate increase together with the activity of the hypothalamic-pituitary-adrenocortical axis. In the initial catabolic period the protein and fat reserves are metabolised to provide energy, a negative nitrogen balance develops and body weight falls. This is followed by a period of anabolism with acceleration of wound healing and restoration of the nutritional reserves (Table 1).

Endocrine Changes after Trauma

Immediately after injury there is an increase in the secretion of catecholamines, adrenaline and noradrenaline. Plasma catecholamines rise transiently and fall rapidly but their urinary excretion remains elevated throughout the 'flow' phase. Catecholamines and stimulation of the sympathetic nervous system (SNS) are

TABLE 1
PHYSIOLOGICAL, ENDOCRINE AND METABOLIC RESPONSES IN RELATION TO THE 'EBB' AND 'FLOW' PHASES AFTER INJURY OR MAJOR SURGERY.

	SHOCK or EBB PHASE	FLOW PHASE early catabolic	FLOW PHASE anabolic
METABOLIC RATE	▼	▲	▲ or N
BODY TEMPERATURE	▼	▲	N
CARDIAC OUTPUT	▼	▲	N
U. CATECHOLAMINES	▲	▲	N
U. 17-OXOGENIC STEROIDS	▲	▲	N
P. INSULIN	▼	▲	▲ or N
P. GLUCAGON	▲	▲	N
P. GROWTH HORMONE	▲	▲	▲ or N
B. GLUCOSE	▲	▲	▲ or N
P. FREE FATTY ACIDS	▲	▲	▼
NITROGEN BALANCE	negative	negative	positive

not only essential for the 'flight or fight' response but play an important central role in the integration of the endocrine response because they stimulate the release of adrenocorticotrophic hormone (ACTH) and glucagon but suppress the secretion of insulin.

Plasma glucagon rises and remains elevated for several days but it participates in a 'positive feed back' system and stimulates catecholamine secretion.

Plasma insulin falls in the acute phase of injury. Insulin suppression is partly due to the dominance of the alpha receptors of the SNS while lacticacidaemia and reduced pancreatic blood flow during shock may both contribute. Insulin levels rise in the 'flow' phase due to stimulation of the SNS adrenergic beta receptors and high levels of glucagon.

Plasma cortisol is transiently raised for 24–48 hours often with large but variable increases in urine-free cortisol depending on the severity and presence of additional stress but the urinary excretion of 17-oxogenic steroids remains elevated for several days after injury. The role of corticosteroids is not clear since the metabolic response of adrenalectomised patients on maintenance dosage of steroids is similar to that of patients with intact adrenals. As they seem neither to initiate nor control the metabolic response corticosteroids may have a permissive function but there is a high mortality after injury in untreated adrenal failure.

Growth hormone production is stimulated during stress and plasma levels remain raised during the 'flow' phase. Although initially responsible for lipolysis it exerts anabolic effects in the presence of insulin during recovery.

After surgery there is an increase in plasma-free thyroxine and tri-iodothyronine levels which persist for several days and coincide with a fall in the thyroxine binding globulin and prealbumin fraction but correlation with increased energy expenditure is equivocal.

The metabolic consequences of the altered endocrine profile will be discussed in subsequent sections. The actions and inter-relationship of these hormones in the mobilisation of nutritional reserves is shown in Fig. 1.

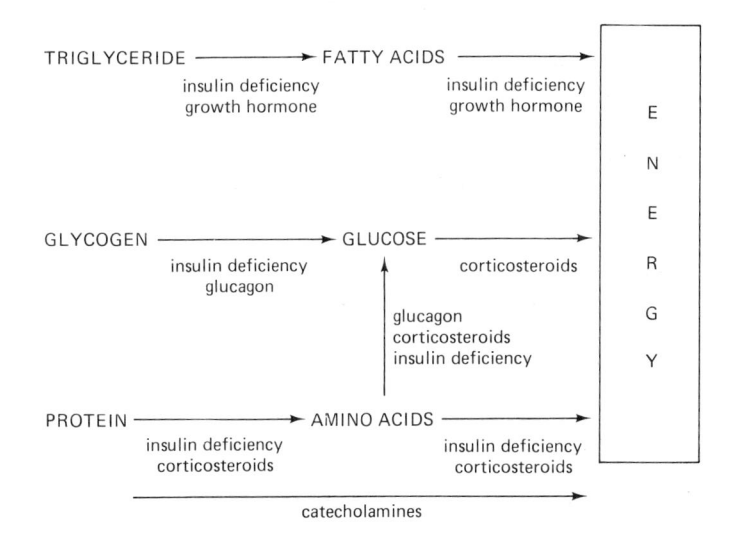

```
TRIGLYCERIDE ─────────► FATTY ACIDS ──────►
                insulin deficiency         insulin deficiency        E
                growth hormone             growth hormone            N
                                                                     E
GLYCOGEN ─────────────► GLUCOSE ──────────►                          R
                insulin deficiency         corticosteroids           G
                glucagon                                             Y
                                           glucagon
                                           corticosteroids
                                           insulin deficiency

PROTEIN ──────────────► AMINO ACIDS ──────►
                insulin deficiency         insulin deficiency
                corticosteroids            corticosteroids
                ─────────────────catecholamines────────────►
```

FIG. 1. The mobilisation of energy reserves by catecholamines indicating the relationship with other hormones which augment catabolism.

Towards the end of the 'flow' phase there is a decline in the activity of the predominantly catabolic catecholamines, corticosteroids and glucagon which allows the anabolic effects of insulin and growth hormone to emerge.

Pain, fear, anaesthesia, hypovolaemia and hyperosmolality associated with injury acting through the hypothalamus stimulate the release of antidiuretic hormone (ADH) from the posterior pituitary for the first 48 hours. Sodium depletion, hypovolaemia and hypotension activate the renin-angiotensin system and aldosterone secretion from the adrenal cortex increases. Both these hormones play an important role in water and sodium conservation (Fig. 2). The endocrine response to trauma has been excellently reviewed.[35,36]

Physiological Responses in Shock which Compensate for Hypovolaemia

Stimulation of the sympathetic nervous system and the release of adrenaline and noradrenaline from the adrenal medulla, causes peripheral vasoconstriction. After haemorrhage the secretion rate of catecholamines is ten to twenty times the basal rate.[34] Cardiac output increases, systemic blood pressure rises and venous return to the heart improves.

Peripheral resistance is also increased by another potent vasoconstrictor, angiotensin II. Hypotension, hypovolaemia and sodium depletion stimulate the release of the proteolytic enzyme renin from the juxtaglomerular apparatus in the kidney.[11] It activates the

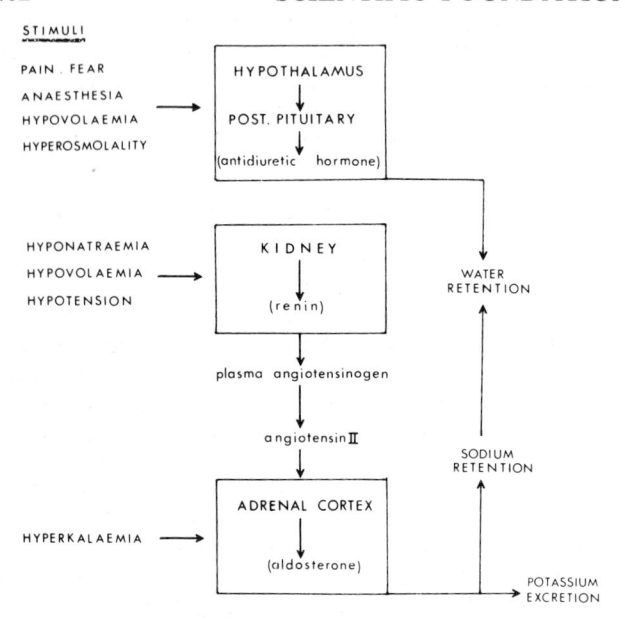

FIG. 2. The stimuli and endocrine responses associated with salt and water conservation after trauma.

alpha globulin precursor angiotensinogen to form angiotensin I which is subsequently converted enzymatically to an octapeptide, angiotensin II.

Catecholamines and angiotensin II reduce the blood flow through the microcirculation and the tissue oxygen consumption falls. Blood is redistributed from the skin, muscles and splanchnic region to vital organs like the heart and brain which are able to maintain their blood flow in spite of a low perfusion pressure by intrinsic vascular autoregulation.

There are several hormone-mediated responses which help to compensate for the fall in blood volume either by inducing a shift of water from the intracellular fluid (ICF) into the extracellular fluid (ECF) or by renal retention of water and sodium.

Selective focal vasoconstriction of the microcirculation by catecholamines reduces the intra-capillary hydrostatic pressure and interstitial fluid enters the vascular compartment.

In haemorrhagic shock Boyd and Mansberger[10] found that plasma osmolality increased. This can persist for several days and the increase is related to the severity of the shock. Hyperglycaemia, a rise in blood urea and sodium retention may contribute to the hyperosmolality which causes a fluid shift from the ICF into the circulation with the osmotic gradient.

There is considerable evidence that this is hormonally mediated[16] because atrial and carotid receptors sensitive to the rate and volume of blood loss stimulate the pituitary via the hypothalamus. ACTH is released and the plasma cortisol level rises after haemorrhage but other hormones like prolactin, growth hormone and vasopressin are probably involved together with glucagon, renin, angiotensin II and aldosterone, following activation of the sympathetic response and liberation of catecholamines.

The tendency to hyperosmolality is counteracted by water retention following the increased activity of ADH on the kidney. In clinical practice the use of intravenous infusions of electrolyte solutions and colloids to restore plasma volume over-rides the compensatory effects of plasma hyperosmolality after injury.

A secondary effect of the fluid shift induced by hyperosmolality of the ECF is a rise in volume and pressure within the interstitial space. It is thought that this assists the return of albumin to the circulation via the lymphatics which raises the capillary oncotic pressure and helps to restore plasma volume.

Local Effects of Vasoactive Substances

The vasopressor action of catecholamines and angiotensin II tends to be antagonised by the vasodilator effects of lactate, histamine, plasma kinins and prostaglandins. As vasoconstriction fades during prolonged shock these substances, by raising the intracapillary pressure and increasing capillary permeability, encourage the passage of fluid into the interstitial space. Mellander and Lewis[27] estimated from animal experiments that the extravasation of fluid could reduce the circulating blood volume in man by 600 ml every hour but would disperse so widely as to be undetectable as oedema.

Histamine is derived from mast cells and histidine following the local activation of histidine decarboxylase and high circulating concentration are often found in bacteraemic shock. Plasma kinins are formed from inactive precursors in the alpha-2 globulin fraction by proteolytic enzymes called kallikreins which are released from damaged tissues and leucocytes. The vasodilator prostaglandins of the E series are released from many tissues after haemorrhage and in bacteraemic shock, but F-prostaglandins cause pulmonary vasoconstriction.

Intracellular Metabolic Changes in Shock

As the oxygen supply to the tissues diminishes the aerobic metabolism of glucose and ultimately pyruvate by the Krebs cycle and electron transport system decreases. The energy requirements are then met by anaerobic glycolysis and lactate accumulates in the cells causing intracellular acidosis. The block in the normal glycolytic pathway causes glucose to diffuse out of the cells.

The change from aerobic to anaerobic glycolysis reduces the generation of adenosinetriphosphate (ATP) within the mitochondria from 30 moles to 2 moles of ATP per mole of glucose which, in turn, reduces the transmembrane potential of muscle cells by 30%. The 'sodium pump' which is ATP-ase dependent and maintains the normal distribution of sodium between the ECF and ICF, fails, causing a net influx of sodium and water into the cells and a corresponding loss of intracellular potassium. It has been calculated that in severe and prolonged shock nearly 2 litres or 10% of the ECF water moves into the cells.[6]

Plasma levels of cyclic AMP (adenosine 3', 5' cyclic phosphate) fall and may be responsible for modifying the effects of insulin, glucagon, catecholamines and possibly corticosteroids on the cells. This may partly explain the phenomenon of 'insulin resistance' after shock and the impaired cellular uptake of glucose.

If cellular metabolism continues to deteriorate there are progressive morphological changes involving cytoplasmic and mitochondrial swelling. As the cell membrane becomes more permeable intracellular lysosomes which contain hydrolytic enzymes diffuse out causing local cell necrosis. These changes can be reversed if the poor tissue perfusion is corrected.

Tissue Hypoxia and Lactic Acidosis During Shock

In hypovolaemic and cardiogenic shock arteriolar vasoconstriction in the peripheral tissues causes oxygen consumption to decrease. As oxygen extraction by the cells increases, the arterio-venous oxygen difference is larger than normal. In the early stages of bacteraemic shock when there is a high cardiac output the A–V oxygen difference is small because pulmonary and splanchnic arterio-venous shunts may open.[8]

Anaerobic glycolysis becomes increasingly important for the generation of ATP but this results in an excessive production of lactate. The normal range of blood lactate is 0·4–1·3 mmol/l (3·6–11·7 mg/100 ml). Lactate homeostasis depends on a balance being maintained between production and removal but this may be severely disturbed during shock because the liver and kidney, which normally remove lactate by gluconeogenesis, actually produce lactate due to impaired hepatic and renal blood flow, resulting in hyperlactataemia.

The origin of the acidosis in hyperlactataemia has been reviewed by Zilva.[37] Anaerobic glycolysis does not produce lactic acid but lactate, ATP and water. Acidosis is caused by the production of hydrogen ions from the hydrolysis of ATP and during the process of gluconeogenesis from lactate.

A useful definition[1] of lactic acidosis is 'Clinical acidosis in an ill patient with a blood lactate persistently above 5 mmol/l (45 mg/100 ml) and an arterial pH of less than 7·25'. Cohen and Woods[12] classified lactic acidosis resulting from hypoxia and shock as Type A to distinguish it from Type B where there is adequate tissue perfusion and oxygenation. Type B occurs in many common disorders like diabetes mellitus, liver disease and after the administration of certain drugs. Intravenous nutrition with fructose, sorbitol or glucose, but without adequate insulin, can cause hyperlactataemia and lactic acidosis,[4] as can ethanol. Their injudicious use in the shocked or postoperative patient may well exacerbate a preexisting hyperlactataemia or acidosis.

Lactic acidosis is a serious complication. There is a progressive rise in mortality as the arterial blood lactate increases[30] with a 90% mortality when it reaches 8·9 mmol/l (80 mg/100 ml). The prognosis is especially poor when lactic acidosis develops in bacteraemic shock. Its presence may be difficult to recognise from the rather non-specific clinical signs and symptoms but a large unexplained plasma 'anion gap' greater than 10 mmol/l may be the first indication. This should be investigated with a specific assay for lactate especially if uraemia and the ketoacidosis of starvation are absent.

Nutritional Reserves and Energy Metabolism

The basal energy requirement of an adult is 7·56 MJ (1,800 Kcals) a day but this is increased by 30% in multiple injuries, 50% in severe infection and 100% or more in severe burns. In the starved and injured patient the hepatic glycogen reserves are rapidly exhausted within twelve hours. The protein reserves, which are mainly muscle but include the small but important reserves in the liver, kidney, pancreas and intestine, etc. are theoretically able to provide 16 000–24 000 Kcals over a twelve day period which is about 20% of the energy expenditure. Protein breakdown provides the amino acids for hepatic gluconeogenesis to meet the requirement for carbohydrate intermediates.

On the other hand, the reserves of triglyceride which are obviously variable have a greater energy potential, containing 109 000–136 000 Kcals, sufficient for a 20–25 day period and providing 75–90% of the energy expenditure. Fatty acids released during the lipolysis of fat cannot be converted into glucose, glycogen or carbohydrate intermediates but can be oxidised to meet the major proportion of the total energy demands. The mobilisation of the energy reserves is under endocrine control and will be discussed in detail in the sections on fat, protein and carbohydrate metabolism. The size of these reserves is an important factor in survival. Patients who lose more than 30% of their normal body weight rapidly have a poor prognosis[23] and a high mortality can be anticipated in those debilitated by age, infection or preoperative malnutrition.

Following injury the energy reserves are quickly depleted unless oral, enteral or parenteral nutrition is given. The route chosen will depend on the presence or absence of gastrointestinal failure. During the 'flow' period the initial phase of catabolism is replaced by one of anabolism which coincides with the restoration of body weight, the replenishment of energy reserves and wound healing.

Carbohydrate Metabolism

Hepatic stores of glycogen are sufficient to supply about 3·36 MJ (800 Kcals) and are rapidly depleted within twelve hours of injury. Catecholamines stimulate hepatic glycogenolysis, activate the Cori cycle whereby muscle glycogen is converted via lactate to glucose in the liver and suppress the secretion of insulin. The resulting rise in blood glucose is maintained by the increased secretion of corticosteroids and glucagon (Fig. 1). Corticosteroids stimulate hepatic gluconeogenesis from alanine and other glucogenic amino acids derived from muscle, and from lactate and pyruvate, but at the same time inhibit the glycolytic enzymes. Glucagon in turn enhances hepatic glycogenolysis and gluconeogenesis while growth hormone, an insulin antagonist, also helps maintain a flow of readily available glucose.

Allison, Prowse and Chamberlain[2] noted the failure of insulin secretion in the acute phase of injury. When intravenous glucose is given the normally expected insulin response does not occur and a diabetic type of glucose intolerance is observed[3] which persists throughout shock. The post-traumatic or post-haemorrhagic hyperglycaemia is usually proportional to the severity of the injury and has been termed 'traumatic diabetes'. This state often persists, however, into the 'flow' phase when the initial inhibition of insulin secretion is replaced by hyperinsulinaemia.

The levels of immunoreactive rise by a factor of two to three times above normal but the abnormal glucose tolerance remains.[17]

The combination of hyperglycaemia and hyperinsulinaemia is consistent with peripheral and visceral cellular 'insulin resistance' and the raised blood glucose is necessary to establish an appropriate glucose gradient to maintain normal cellular uptake of glucose. Large doses of parenteral insulin can overcome this resistance which suggests the presence of either circulating insulin antagonists or a decreased sensitivity of the cell membrane to insulin.

Hepatic gluconeogenesis and muscle amino acid turnover remain active and demonstrate another aspect of 'insulin resistance' because under normal conditions high levels of insulin should inhibit gluconeogenesis and stimulate protein synthesis in muscle. Insulin is normally anabolic and favours protein synthesis.

If hypoglycaemia develops during shock it is probably due to a combination of depleted glycogen stores and inadequate gluconeogenesis coincident with an increased cellular uptake of glucose.

When hypertonic glucose (20% or 50%) forms part of an intravenous nutrition regimen after surgery or trauma, soluble insulin should be given concurrently in a dose which must be adjusted to maintain the blood glucose between 4–8 mmol/l (70–140 mg/100 ml). In these circumstances monitoring of the urine glucose excretion is an unsatisfactory method with which to regulate the dose of insulin.

Protein Metabolism

Protein reserves have been estimated to be able to provide 104 MJ (24 000 Kcals) of energy. Although muscle protein is the principal source after injury or sepsis, visceral protein in the liver, kidney, pancreas and intestine tends to be preserved. This phenomenon is called peripheral to visceral translocation of protein. The exact opposite occurs in starvation because visceral protein is utilised preferentially before muscle[21] and coincides with reduced gluconeogenesis from muscle. After injury, sepsis and starvation 10–14% of the weight loss comes from tissue protein which when converted to lean tissue mass with its high (70%) water content is equivalent to nearly two thirds of the weight lost.[20]

Gluconeogenesis from muscle protein is under endocrine control. Catecholamines convert muscle glycogen to lactic acid which is transported to the liver and converted to glucose (Cori cycle). Insulin deficiency and corticosteroids facilitate the release of amino acids from muscle and in the presence of glucagon stimulate their conversion to glucose in the liver.[32]

Proteolysis mobilises the branched chain amino acids (leucine, isoleucine and valine) in muscle and after oxidation their amino groups combine with pyruvate to form alanine. Alanine, together with three-carbon intermediates like lactate, pyruvate and other amino acids is transported to the liver and raised plasma levels of these substances are associated with increased hepatic gluconeogenic activity. The glucose—three carbon cycle has been called the 'alanine shuttle'[36] and is depicted in Fig. 3.

Increased urea synthesis results from increased oxidative deamination and coincides with a raised urinary non-protein nitrogen (NPN) excretion of which 85% is urea, the remainder being amino acids, ammonia, creatine and creatinine. The normal urine creatine:creatinine ratio is less than 0·1 but is increased after surgery reflecting increased muscle catabolism. In well

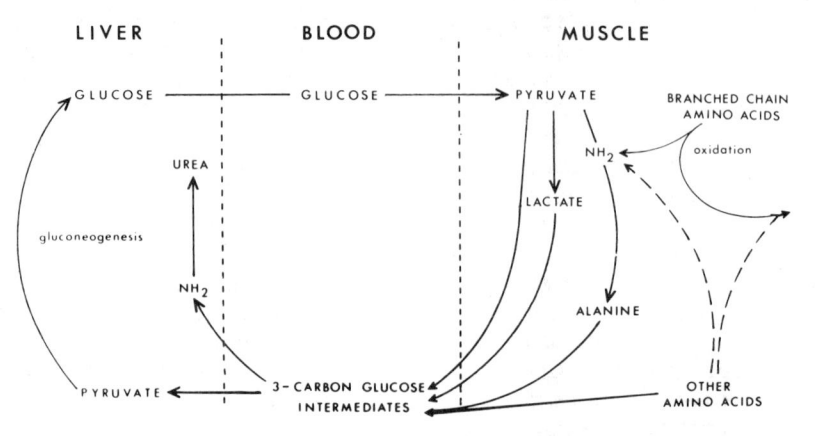

FIG. 3. Hepatic gluconeogenesis from three—carbon glucose intermediates derived from muscle (alanine shuttle).

nourished patients receiving intravenous nutrition, negative nitrogen balance is associated with a high ratio (0·56) while positive nitrogen balance occurred in those with low ratios (0·16) and a poor preoperative nutritional status.[5]

The blood urea also rises and forms about 60% of the total blood NPN. Half of the remainder, is composed of amino acids and alanine is quantitatively the most significant.

If the 'ebb' and early 'flow' phase is compromised by starvation a negative nitrogen balance quickly develops. The severity of the injury bears a direct relationship to the urinary NPN excretion for the greatest losses and the largest negative nitrogen balances occur in hypercatabolic states. The urinary NPN may exceed 25–30 g per day (normal 8–12 g/day). Since 1 g of nitrogen is equivalent to 6.25 g of protein or 25 g of muscle this will involve the substantial loss of about 750 g of muscle a day. The negative nitrogen balance in burns may be even greater due to large plasma exudates from the burned area which may amount to 20% of the total nitrogen loss. The natural course of events in the 'flow' period in uncomplicated trauma with starvation is for the negative nitrogen balance to continue until adequate oral nutrition is possible. Nitrogen balance is then slowly restored from the 7th–10th day onwards.

However, as nutritional support of the critically ill is now regarded as an essential aspect of treatment, positive nitrogen balance can be maintained throughout the flow period either with enteral feeding or intravenous nutrition in those individuals with gastrointestinal failure. The daily nitrogen requirements in the uncomplicated postoperative patient are around 0·21 g N_2/Kg body weight, rising to 0·32 g N_2/Kg in hypercatabolic states. A simple estimate of nitrogen balance is obtained by measuring the 24 hour urine urea excretion in grams, multiplying by 28/60 to convert it into nitrogen and then by 6/5 which corrects for non-urea nitrogen.[24]

24 h urine nitrogen(g) = urine urea(g) × 28/50   **1**

A small correction is necessary for gross proteinuria.

If the blood urea increases in the absence of dehydration it is necessary to correct for nitrogen retention. If urea is assumed to be evenly distributed throughout the body water, which in men is 60% of the body weight, then:

Retained nitrogen(g) = Change in blood urea (g/1/24 h) × 60% body wt (kg) × 28/60   **2**

The total nitrogen produced (g) from protein catabolism is equal to the 24 h urine nitrogen (equation **1**) *plus* the retained nitrogen (equation **2**) and this is equivalent to the amount which is theoretically required to maintain positive nitrogen balance during the next day.

Changes in Plasma Proteins After Injury

Changes in concentration of plasma proteins after injury may be due to shifts of water between the ICF and ECF compartments or protein losses from burns and haemorrhage irrespective of alterations in their rate of synthesis or catabolism.

Albumin

After injury the plasma concentration may fall by 30% around the third to sixth day. This is attributable not only to depression of hepatic synthesis and an increased catabolic rate but to passage into the interstitial fluid which raises the extravascular to intravascular albumin ratio. The latter effect is usually small except after severe burns. Unless the injury is severe there is a progressive but slow restoration to normal levels which may take several weeks.

Globulins

The alpha-1 globulins (alpha-1 antitrypsin, alpha-1 acid glycoproteins) and the alpha-2 globulins (caeruloplasmin, haptoglobin) usually increase by about 50% and 100% respectively on the third day after injury, but there is a fall in the beta globulins. It is postulated that increased hepatic synthesis accounts for the changes in the alpha globulins.

The immunoglobulins IgG, IgA and IgM change little unless in response to severe infection. C-reactive protein increases rapidly by the second day but has little clinical relevance except as a general indication that trauma or sepsis exist.

Fibrinogen

This may rise by 100% soon after injury due to increased hepatic synthesis and remain elevated for many days especially after burns. However, in disseminated intravascular coagulation dangerous hypofibrinogenaemia may occur due to the excessive activity and consumption of blood clotting factors.

Fat Metabolism

Lipolysis of triglyceride in adipose tissue is dependent on catecholamines which activate hormone-dependent lipases and release free fatty acids (FFA) and glycerol into the circulation. The process is enhanced by excess growth hormone and also by insulin deficiency because insulin normally stimulates fat synthesis.

Glycerol is ultimately converted into glucose but FFA either undergo oxidation or are metabolized to ketone bodies. After injury the oxidation rate of FFA is greatly increased but this inhibits further metabolism by the tricarboxylic acid cycle and favours hepatic ketogenesis. Ketonaemia (acetoacetate and 3-hydroxybutyrate) with ketonuria result and this is most marked in starvation.

In severe sepsis fatty acid utilization and ketogenesis are impaired. This is probably due to the high levels of insulin which are found in this situation and inhibit lipolysis.

Most extrahepatic tissues can utilise ketone bodies as

sources of energy and in starvation, by the process of ketoadaptation, the heart and brain obtain a significant proportion of their energy requirements from ketoacids.

In starvation there is evidence that increased ketogenesis can exert a regulatory effect on metabolism.[25] High concentrations of ketone bodies inhibit gluconeogenesis and the peripheral oxidation of glucose. This effectively conserves protein and glucose when metabolic adaptation may be crucial.

Water Metabolism

Accurate fluid balance charts must be kept and scrutinised each day to prevent the serious complications associated with dehydration and water overload after injury or surgery.

It is unusual in these situations to find true water depletion without sodium depletion. Dehydration results from losses of gastrointestinal secretions (vomiting, diarrhoea, gastric aspiration, fistulae, ileostomy), fluid sequestration (paralytic ileus, crush injury), burns and haemorrhage with inadequate replacement. Sweating, diuretics and poor humidification during mechanical ventilation also contribute. When there is a water deficit of about four litres, blood urea and serum protein concentration and haematocrit may increase. Reduction in plasma volume is associated with hypotension and oliguria, intracellular dehydration with loss of tissue turgor and elasticity, and increased thirst, while cerebral dehydration causes mental confusion and coma.

Overhydration is usually iatrogenic and can follow the excessive administration of hypotonic fluids during shock, renal failure, congestive cardiac failure and the early postoperative period when there is increased ADH release. It is occasionally encountered in some head injuries with 'inappropriate' secretion of ADH. The combination of subcutaneous and pulmonary oedema, weakness, confusion, convulsions, hyponatraemia and serum hypo-osmolality suggest the diagnosis.

Sodium and Potassium

After injury the sodium turnover decreases, there is sodium retention and the urinary losses may be reduced to 20–30 mmol by the third day. During recovery there is a sodium and water diuresis. Several factors are responsible for sodium retention, namely aldosterone-mediated renal tubular sodium-potassium exchange, inhibition of the 'third factor' hormone which normally promotes sodium excretion, congestive cardiac failure and positive pressure ventilation. The tendency to hypernatraemia may be masked by simultaneous water retention but it may occur with injudicious infusion of saline, treatment with corticosteroids and some cases of head injury or brain tumour (encephalogenic hypernatraemia). The raised serum osmolality is usually proportional to the increase in serum sodium.

Severe dilutional hyponatraemia (110–120 mmols/l) can be caused by the infusion of large volumes of sodium-free solutions (e.g. 5% glucose), the 'inappropriate' secretion of ADH (e.g. carcinoma of bronchus or pancreas, head injury, severe chest infection), water retention in congestive heart failure and renal failure. This is characterised by a paradoxical natriuresis and serum hypo-osmolality.

Patients with malignancy, liver disease, chronic infection and malnutrition subjected to surgery may develop a moderate dilutional hyponatraemia but sometimes hyponatraemia develops spontaneously in this group and the critically ill. Failure of the cell sodium 'pump' due to hypoxia has been suggested which allows movement of sodium from the ECF into the cells and a corresponding leak of potassium.[15] This has been called the 'sick cell' syndrome. The plasma sodium concentration is usually less than 125 mmol/l, plasma osmolality 260–270 mosm/l with a reduced sodium but increased excretion of potassium (above 80 mmol/day) in the urine.

As the total body sodium content may be normal or even increased, treatment with hypertonic saline is unnecessary and even dangerous. Correction of hypoxia, tissue perfusion and promotion of ion exchange with glucose and insulin which improve cellular metabolism are more beneficial.[22]

Large losses of gastrointestinal fluids after major abdominal surgery can cause severe deficiencies (Table 2). In the early stages the losses from the ECF are compensated by replacement from the exchangeable intracellular reserves, but as these become progressively depleted absolute deficiencies are reflected in low plasma concentrations. Daily measurement of all electrolyte losses in the urine and gastrointestinal secretions facilitates accurate electrolyte replacement and balance to be maintained. Severe hyponatraemia with a plasma sodium concentration of less than 115 mmol/l causes mental disturbances, convulsions and coma.

For several days after trauma there is an increased urinary excretion of potassium of about 120–140 mmols/day which is due to the effect of aldosterone on the kidney and increased losses from injured and catabolising cells. If there are additional losses from gastro-

TABLE 2
NORMAL DAILY VOLUME AND AVERAGE IONIC COMPOSITION IN MILLIMOLES PER LITRE OF GASTROINTESTINAL SECRETIONS

	Na^+	K^+	Cl^-	HCO_3^-	Mg^{++}	Ca^{++}	H^+	Volume (ml)
Gastric	60	10	120	10	0·5	2	70	500–2500
Biliary	145	5	100	40	0·7	2·5	—	700
Pancreatic	140	5	65	110	0·1	1·5	—	800
Ileal	130	11	110	30	0·5	2·0	—	3000
Diarrhoea	75	40	60	50	Variable	Variable	—	—

intestinal secretions, paralytic ileus, or the use of diuretics or corticosteroids, severe potassium depletion and hypokalaemia will develop. This is exacerbated by metabolic alkalosis and with a serum potassium below 2 mmol/l muscle weakness, confusion, ECG changes and cardiac failure may occur.

Hyperkalaemia can be an equally serious complication leading to brachycardia and cardiac arrest. This can be precipitated by renal failure combined with the hypercatabolic effects of sepsis and multiple injuries, massive blood transfusion, adrenal failure and excessive infusion of potassium supplements.

Magnesium, Calcium, Phosphate and Trace Elements

After uncomplicated surgery a transient fall in serum magnesium concentration is followed by a negative magnesium balance which lasts for several days and is related to a reduced food intake.[19] Urinary excretion of magnesium, which is probably derived from hepatic and muscle cell catabolism, rises to a maximum on the third day and slowly declines. The serum magnesium level rapidly returns to normal but magnesium deficiency may result from the continuous losses of magnesium in the drainage of gastrointestinal secretions. As magnesium in these fluids is normally reabsorbed from the ileum and colon massive ileal resection or total colectomy may disturb normal homeostasis and potentiate the development of hypomagnesaemia. If adequate magnesium replacement therapy is given to maintain the serum level between 0·7–1·3 mmol/l (1·8–3·0 mg/100 ml), tetany, muscular weakness, agitation and convulsions will be prevented.

Hypocalcaemia is often found in conjunction with hypomagnesaemia and hypoalbuminaemia. If an ionised calcium measurement is unavailable an absolute reduction in this fraction can be assumed if the serum calcium remains low after a correction for the low albumin is applied. A satisfactory adjustment is to add 0·025 mmol/l (0·1 mg/100 ml) to the measured serum calcium for every gram per litre the plasma albumin is below 40 g/l.[29] Hyperventilation may often precipitate tetany as calcium shifts from the ionised to the protein bound fraction. Conversely, in acidosis the ionised fraction may be normal despite a low total serum calcium. Continuous losses of calcium in the urine and gastrointestinal secretions will contribute towards calcium deficiency.

If postoperative patients are given phosphate-free intravenous solutions there is a transient fall in serum phosphate for two days with a slow return to normal by the fifth day.[31] This may be due to fluid shifts between the ECF and ICF and an increase in urinary phosphate excretion as a result of cellular catabolism.

Persistent hypophosphataemia is observed during intravenous nutrition using hypertonic glucose without phosphate supplements and also during periods of positive nitrogen balance. This probably results from the increased incorporation of phosphate into muscle and liver during anabolism rather than increased urine losses because these are small. Severe hypophosphataemia (0·2–0·3 mmol/l) may cause neuromuscular irritability, confusion and hyperventilation. Equally important but clinically inapparent is the fall in red cell ATP to 15% of normal with depletion of 2,3 diphosphoglycerate. This shifts the oxygen dissociation curve to the left which increases the red cell affinity for oxygen and diminishes the amount of oxygen available to the tissues.[26] An undesirable event particularly in the shocked hypoxic patient.

When there is prolonged gastrointestinal failure after surgery or trauma which necessitates intravenous feeding, deficiencies of the essential biological (trace) elements will arise unless maintenance requirements are given. These include copper, iron, chromium, manganese, cobalt, iodine and zinc among several others.

Zinc deficiency may present clinically with dermatitis, falling hair and impaired wound healing with the plasma zinc level well below the normal range 11·5–17·6 μmol/l (75–115 μg/100 ml). Zinc is essential for the activity of many enzymes and is located within the cells as zinc metalloproteins. Zinc metabolism is altered in response to injury and infection and in the severely catabolic patient there is an increased urinary zinc excretion, maximal after 8–10 days. This comes primarily from muscle catabolism and the zincuria parallels the increased excretion of nitrogen, potassium and creatine.[14] Zinc is excreted in the bile and pancreatic juice and considerable losses can occur in faeces, thus further depleting zinc reserves.

Acid-Base Disturbance

Any acid-base disturbance after injury or shock is most likely to be due to a mixed primary metabolic or respiratory acidosis or alkalosis rather than a single disorder. It may be caused by a combination of any of the precipitating factors listed in Table 3. Compensatory mechanisms may achieve complete or partial correction depending on the magnitude of the initial imbalance but these may be inadequate if renal or pulmonary function is impaired or if postoperative metabolic complications are not recognised and treated.

Interpretation of acid-base laboratory investigations cannot be done accurately in isolation. It is important to take into consideration the clinical condition, general metabolic situation and any treatment the patient may be receiving before assessing or treating alterations in acid-base balance.[7]

Pulmonary Function in Shock

Initial hyperventilation is usually stimulated by metabolic acidosis but it may occur for other unknown reasons. This reduces the arterial carbon dioxide tension ($P\text{co}_2$) and hyperventilation stops when compensation is effected.

Hyperventilation may continue if shock is prolonged and a persistently low arterial $P\text{co}_2$ with a borderline low oxygen tension ($P\text{o}_2$) may be the first sign of incipient pulmonary insufficiency. If this becomes established

TABLE 3
METABOLIC DISTURBANCES AND PREDISPOSING CAUSES OF ACID-BASE IMBALANCE AFTER SHOCK, TRAUMA AND SURGERY

	Metabolic disturbance	*Predisposing cause*
Metabolic Acidosis	Ketonaemia	Starvation Hypercatabolic states Diabetes mellitus
	Hyperlactataemia (Type A)	Shock Congestive cardiac failure Open heart surgery Hypothermia
	Hyperlactataemia (Type B)	Intravenous fructose, sorbitol, ethanol
	H^+ ion retention	Acute and chronic renal failure
	Loss of base	Biliary, pancreatic fistulae Ileostomy (recent)
	Infusion of acid Hyperchloraemia	Amino acid solutions Uretero-sigmoidostomy
Metabolic Alkalosis	Loss of acid	Vomiting, gastric aspiration
	Infusion of base	Sodium bicarbonate Blood and plasma transfusion (citrate)
	Potassium depletion	Gastrointestinal fluid losses Diuretic phase of acute renal failure Diuretics, corticosteroids
Respiratory Acidosis	Retention of CO_2	Bronchial obstruction, pneumothorax Pulmonary oedema, collapse, consolidation Thoracotomy, chest wall injury Medullary depressant drugs, muscle relaxants Inadequate ventilation
Respiratory Alkalosis	Excessive loss of CO_2	Pain, septicaemia Head injury, brain damage Over ventilation

considerable impairment of gas exchange develops with severe hypoxaemia and a high $P{CO_2}$, the initial respiratory alkalosis being replaced by a respiratory acidosis. Increased oxygen extraction by the tissues reduces the oxygen tension in the returning venous blood and if this is shunted through underventilated or oedematous alveoli the arterial $P{O_2}$ will be systematically reduced and the $P{CO_2}$ further increased. When the $P{O_2}$ falls below 60 mm Hg (8 kPa) with room air and the $P{CO_2}$ rises above 55 mm Hg (7·3 kPa) tracheal intubation and assisted ventilation may be necessary.

Respiratory failure in shock has been called the adult respiratory syndrome or 'shock lung'. This is a condition of reduced compliance, pulmonary vascular shunting and an increase in pulmonary vascular resistance. Pulmonary vasoconstriction may be produced by a central reflex mechanism, catecholamines, endotoxins, serotonin, histamine, prostaglandins (F series) and kinins, which may also increase capillary permeability leading to oedema. The lung histologically shows interstitial oedema, diffuse alveolar collapse, bronchopneumonia and hyaline membrane of the alveoli, but these features are not pathognomonic of 'shock lung'.[9]

Additional factors have been frequently associated with the development of this syndrome namely overhydration, alveolar hypoxia with damage of the pulmonary capillaries, oxygen toxicity, intermittent positive pressure ventilation, reduced alveolar surfactant, pulmonary emboli and intravascular coagulation.

The Effects of Shock on Renal Function

Acute renal failure (ARF) after shock or trauma requires prompt diagnosis and careful management. The presenting features are oliguria, which is common but not invariable, and a rising blood urea. Uraemia with polyuria does occur but has a shorter duration and better prognosis than the oliguric form.

Prerenal ARF is usually caused by a sudden reduction in renal blood flow with a fall in glomerular filtration rate due to hypovolaemia and hypotension secondary to haemorrhage, dehydration or burns. The condition is reversible if the circulating blood volume and blood pressure are rapidly restored but if treatment is delayed intrinsic ARF becomes established with the development of vasomotor nephropathy (acute tubular necrosis). Bacteraemic shock, nephrotoxic drugs, acute haemolysis, obstructive jaundice, pancreatitis, disseminated intravascular coagulation, etc., may all be associated with renal failure and potentiate that caused by ischaemia. It is possible that adrenergic stimulation, catecholamines and the release of renin within the renal cortex are responsible for the cortical ischaemia which reduces glomerular filtration.

TABLE 4
INVESTIGATIONS TO DISTINGUISH PRERENAL FROM ESTABLISHED INTRINSIC ACUTE RENAL FAILURE

	Prerenal ARF	Established Intrinsic ARF
Urine osmolality (m. osmol/kg)	> 400	< 400
Osmolality ratio (urine:plasma)	> 1·5:1	< 1·5:1 often 1·1:1
Urea ratio (urine:plasma)	> 10:1	< 10:1 often 4:1
Urine sodium (mmol/l)	< 20	> 30
Effect of intravenous mannitol or diuretic	Diuresis	No diuresis

Table 4 indicates the laboratory investigations that may help in differentiating prerenal and established ARF but they are not unfailingly accurate in every case.

In prerenal ARF there is active conservation of water and salt involving ADH and aldosterone to increase the ECF volume. Consequently, the urine has a high osmolality and urea concentration but a low sodium. In established oliguric ARF the urine:plasma urea and osmolality ratios are low with a high sodium concentration. Intravenous frusemide and mannitol lead to a spontaneous diuresis in prerenal ARF which can usually be reversed by vigorous fluid replacement.

Haematological Effects of Trauma

Tissue injury, endotoxins and noradrenaline cause platelets to release ADP which facilitates their aggregation and initiates haemostatic defence reactions. Occasionally there is inappropriate stimulation of the coagulation cascade system which causes widespread disseminated intravascular coagulation (DIC) and fibrin is deposited in the vessels of the microcirculation. Depending on the intensity and duration of the reaction DIC may have little clinical significance but can become a critical complication if it obstructs the blood flow to vital organs like the kidney, liver and lungs or if it leads to a haemorrhagic tendency. DIC results in abnormal consumption of many clotting factors including fibrinogen and platelets so that a natural protective physiological reaction is replaced by a pathological process which causes diffuse purpura and bleeding into the gastrointestinal and genitourinary tracts and the lungs.

Coincidentally, there is activation of the fibrinolytic system with hyperplasminaemia and the lysis of fibrin produces fibrin degradation products (FDP) which further interfere with coagulation and cause the haemorrhagic diathesis to deteriorate.

The laboratory diagnosis and the pathophysiology has been excellently reviewed by Hamilton, Stalker and Douglas.[18]

REFERENCES

1. Alberti, K. G. M. and Nattrass, M. (1977), 'Lactic acidosis,' *Lancet*, **2**, 25–29.
2. Allison, S. P., Prowse, K. and Chamberlain, M. J. (1967), 'Failure of insulin response to glucose load during operation and after myocardial infarction,' *Lancet*, **1**, 478–481.
3. Allison, S. P., Hinton, P. and Chamberlain, M. J. (1968), 'Intravenous glucose tolerance, insulin and free fatty acid levels in burned patients,' *Lancet*, **2**, 1113–1116.
4. Ames, A. C., Cobbold, S. and Maddock, J. (1975), 'Lactic acidosis complicating treatment of ketosis of labour,' *Brit. Med. J.*, **4**, 611–613.
5. Ames, A. C. and Thomas, A. (1972), 'Nitrogen balance in the postoperative patient receiving parenteral nutrition,' *Ann. Clin. Biochem.*, **9**, 135–140.
6. Baue, A. E. (1976), 'Metabolic abnormalities of shock,' *Surg. Clin. N. America*, **56**, No. 5, 1059–1071. Philadelphia, London, Toronto: W. B. Saunders Company.
7. Beach, F. X. M., Wright, D. M. and Sherwood Jones, E. (1974). In: *Parenteral Nutrition in Acute Metabolic Illness*, pp. 177–195 (Lee, H. A. Ed.), London and New York: Academic Press.
8. Berk, J. L., Hagen, J. F., Maly, G. and Koo, R. (1972), 'The treatment of shock with beta adrenergic blockage,' *Arch. Surg. (Chicago)*, **104**, 46–51.
9. Blaisdell, F. W. (1973), 'Respiratory insufficiency syndrome: clinical and pathological definition,' *J. Trauma*, **13**, 195.
10. Boyd, D. R. and Mansberger, A. R. (1968), 'Serum water and osmolal changes in haemorrhage shock: an experimental and clinical study,' *Am. Surg.*, **34**, 744.
11. Claybaugh, J. R. and Share, L. (1973), 'Vasopressin, renin and cardiovascular responses to slow haemorrhage,' *Amer. J. Physiol.*, **224**, 519.
12. Cohen, R. D. and Woods, H. F. (1976), *Clinical and Biochemical Aspects of Lactic Acidosis*, pp. 77–161. Oxford, London, Edinburgh, Melbourne: Blackwell Scientific.
13. Cuthbertson, D. P. (1942), 'Post shock metabolic response,' *Lancet*, **1**, 433–437.
14. Fell, G. S. and Burns, R. R. (1978), In: *Advances in Parenteral Nutrition*, pp. 241–261; (Johnston, I. D. A. Ed.); Lancaster, England; MTP Press Ltd.
15. Flear, C. T. G. (1970), 'Electrolyte and body water changes after trauma,' *J. Clin. Pth.*, **23**, Suppl. 4, 16–31.
16. Gann, D. S. (1976), 'Endocrine control of plasma protein and volume' *Surg. Clin. N. America*, **56**, No. 5, 1135–1145. Philadelphia, London, Toronto: W. B. Saunders Company.
17. Giddings, A. E. B. (1974), 'The control of plasma glucose in the surgical patient,' *Brit. J. Surg.*, **61**, 787–792.
18. Hamilton, P. J., Stalker, A. L. and Douglas, A. S. (1978), 'Disseminated intravascular coagulation: a review,' *J. Clin. Path.*, **31**, 609–619.
19. Heaton, F. W. (1964), 'Magnesium metabolism in surgical patients,' *Clin. Chim. Acta.*, **9**, 327–333.
20. Kinney, J. M. (1977), In: *Nutritional Aspects of Care of the Critically Ill*. pp. 95–133; (Richards, J. R. and Kinney, J. M. Eds). Edinburgh, London, New York: Churchill Livingstone.
21. Kinney, J. M., Long, C. L. and Duke, J. H. (1970) In: *Energy Metabolism in Trauma*, pp. 103–123, (Porter, R. and Knight, J. Eds). *Ciba Foundation Symposium*, London: J. and A. Churchill.
22. Lancet Editorial, (1974), 'Sick cells and hyponatraemia,' *Lancet*, **1**, 342–343.
23. Lawson, L. J. (1965), 'Parenteral nutrition in surgery,' *Brit. J. Surg.*, **52**, 795–800.
24. Lee, H. A. (1974) In: *Parenteral Nutrition in Acute Metabolic Illness*, pp. 307–331. (Lee, H. A. Ed.). London and New York: Academic Press.
25. Levenson, S. M., Barbul, A. and Seifter, E. (1977). In: *Nutritional Aspects of Care in the Critically Ill*. pp. 3–94, (Richards, J. R. and Kinney, J. M. Eds.). Edinburgh, London, New York: Churchill Livingstone.
26. McConn, R. (1975), 'The oxyhaemoglobin dissociation curve in acute disease,' *Surg. Clin. N. America*, **55**, No. 3, 627–658. Philadelphia, London, Toronto: W. B. Saunders Company.
27. Mellander, S. and Lewis, D. H. (1963), 'Effect of haemorrhagic shock on the reactivity of resistance and capacitance vessels and

on capillary filtration transfer in cat skeletal muscle,' *Circulation Res.*, **13**, 105.
28. Moore, F. D. (1953), 'Bodily changes in surgical convalescence,' *Ann. Surg.*, **137**, 289–315.
29. Payne, R. B., Carver, M. E. and Morgan, D. B. (1979), 'Interpretation of serum total calcium: effects of adjustment of albumin concentration on frequency of abnormal values and on detection of change in the individual,' *J. Clin. Path.*, **32**, 56–60.
30. Peretz, D. I., Scott, H. M., Duff, J., Dossetor, J. B. MacLean, L. D. and McGregor, M. (1965), 'The significance of lacticacidaemia in the shock syndrome,' *Ann. N.Y. Acad. Sci.*, **119**, 1133.
31. Rowlands, B. J. and Giddings, A. E. B. (1976), 'Postoperative hypophosphataemia,' *Lancet*, **2**, 1077.
32. Ryan, N. T. (1976), 'Metabolic adaptations for energy production during trauma and sepsis,' *Surg. Clin. N. America*, **56**, No. 5, 1073–1090. Philadelphia, London, Toronto: W. B. Saunders Company.
33. Sriussadaporn, S. and Cohn, J. N. (1968), 'Regional lactate metabolism in clinical and experimental shock,' *Circulation*, **37**, Suppl. 6, 187.
34. Walker, W. F., Zileil, M. S., Reutter, F. W., Shoemaker, W. C. and Moore, F. D. (1959), 'Adrenomedullary secretion in haemorrhagic shock,' *Amer. J. Physiol.*, **197**, 773.
35. Walker, W. F. and Johnston, I. D. A. (1971), *The Metabolic Basis of Surgical Care*, pp. 83–100. London: Heinemann Medical Books Limited.
36. Wilmore, D. W. (1976), 'Hormonal responses and their effect on metabolism,' *Surg. Clin. N. America*, **56**, No. 5, 999–1018. Philadelphia, London, Toronto: W. B. Saunders Company.
37. Zilva, J. F. (1978), 'The origin of the acidosis in hyperlactataemia,' *Ann. Clin. Biochem*, **15**, 40–43.

SECTION VII

DISORDERS OF THE NEUROMUSCULAR SYSTEM

	Page
25. THE BIOCHEMICAL INVESTIGATION OF COMA AND STUPOR	413
26. PERIPHERAL NEUROPATHY	426
27. THE MYOPATHIES	430
28. CHEMISTRY AND CEREBROSPINAL FLUID IN HEALTH AND DISEASE	445

25. THE BIOCHEMICAL INVESTIGATION OF COMA AND STUPOR

RUDY CAPILDEO, F. R. SMITH and F. CLIFFORD ROSE

Introduction

Terminology

The neurophysiological basis for stupor and coma

The causes of stupor and coma
    Pathophysiological mechanisms

Clinical aspects

Neurological diseases
    Cerebrovascular diseases
    Space-occupying lesions
    Trauma
    Meningitis
    Encephalitis
    Other brain diseases
    Metabolic disorders
    Endocrine disorders
    Toxins
    Other causes

The diagnosis of brain death

Conclusion

INTRODUCTION

Since the monitoring of comatose patients requires the maximal level of nursing and medical care 24 hours a day, all such patients should ideally be admitted to the Intensive Care Unit (ICU) where they can also be offered the benefits of the most sophisticated medical equipment. Apart from those patients requiring artificial ventilation, there is little general agreement as to which type of comatose patient should be accepted to the ICU. Even with the impressive array of life-support equipment now available, the management of the comatose patient remains essentially clinical: (i) to ascertain the cause of coma (ii) to measure the degree of coma (iii) to measure the response to treatment and (iv) to predict the outcome. The ability to maintain homeostasis artificially in the ICU has created the new medical problem of the diagnosis of brain death.

TERMINOLOGY

Before considering definitions of coma and stupor in terms of 'loss of consciousness', the definition of consciousness should first be attempted. This proves to be a difficult task; Ommaya[62] defines normal consciousness as 'that state of awareness in the organism which is characterised by maximum capacity to utilise its sensory input and motor output potential in order to achieve accurate storage and retrieval of events related to contemporary time and space', but Walton[91] states that 'consciousness is a primary element in experience and cannot be defined in terms of anything else'. Both attempts at definition offer little assistance to the clinician but there is an important distinction between content and state of consciousness.

There are many ill-defined terms in current usage to define alterations in the level of consciousness, e.g. drowsy, obtunded, semiconscious, semicomatose, stuporose, comatose. The response to questions, commands and painful stimuli has been traditionally used to delineate these states and, although 'grades of coma' have been devised, there is no sharp division but rather a continuum of these between consciousness and coma. *Coma* can be defined as complete loss of consciousness without response to any stimulus, however vigorous or painful, whilst *stupor* (or semicoma) as complete loss of consciousness with responses only at the reflex level. Concussion has been defined as 'an essentially reversible syndrome without detectable pathology'[24] or 'the loss of consciousness and associated traumatic amnesia which occurs as the consequence of head trauma in the absence of physical damage to the brain'.[92] Although attempts have been made to correlate clinical and pathological findings[63] clinical description is still the best method available for 'altered levels of consciousness' as used in the Glasgow Coma Scale.[83]

THE NEUROPHYSIOLOGICAL BASIS FOR STUPOR AND COMA

The ascending reticular activating system of the brain consists of the central reticular formation situated in the paramedian areas of the lower pons, which radiates upwards to the ventromedial thalamus bilaterally. Depression of this system, for example, by anaesthetics or hypnotics, leads to unconsciousness but both sides of the brain stem must be affected. The lesion itself must be acute in onset or of large extent and lesions of the brain stem below the lower third of the pons do not produce coma.[67] Sleeping and waking can occur in man even after bilateral cerebral hemisphere destruction.

Two other clinical syndromes differentiated from coma and stupor are:

(i) *akinetic mutism*, originally described by Cairns,[12] which is associated with generalised muscular relaxation with apparent lack of response to painful stimuli although the patient's eyes will follow moving objects. The neuroanatomical basis is damage to the central reticular system or its connections, e.g. with the basal ganglia.[67]

(ii) the *'locked-in syndrome'* describes the patient who is conscious and alert, fully aware of his surroundings, but aphonic, anarthric and tetraplegic. Careful examination will reveal that understanding is possible, the patient communicating with various ocular movements or blinking.[36] The neuroanatomical substrate is bilateral destruction of the pons or medulla with sparing of the tegmentum, usually due to infarction.

THE CAUSES OF STUPOR AND COMA

A wide variety of pathological causes can produce stupor or coma. These can be divided into four main groups. *See* (Table 1)

(1) Neurological diseases (3) Endocrine disorders
(2) Metabolic disorders   (4) Toxins

Pathophysiological Mechanisms

Despite the differing aetiologies of the diseases and disorders producing coma, the pathophysiological mechanisms are similar and will be briefly discussed.

Mass Effect

An *acute expanding* intracerebral space-occupying lesion, e.g. an extradural or intracerebral haematoma will produce focal neurological signs according to its site. With continuing rapid expansion stupor and coma are produced by progressive compression of the brain stem so that, at autopsy, multiple small haemorrhages are seen throughout the brain stem. In addition to transtentorial herniation and 'coning' through the foramen magnum, infarction of the corpus callosum can occur due to herniation of the cingulate gyrus under the falx cerebri.

The two important factors producing loss of consciousness are the rate of increase, and the severity, of the intracranial pressure. This *mass effect* must be contrasted with acute lesions, e.g. haemorrhage or infarction, occurring in the pons or mid-brain where even small lesions can cause considerable neurological deficit because of the limited space for expansion.

A *slowly expanding* space-occupying lesion, e.g. a glioma, may not produce any neurological signs initially, particularly if situated in a silent area such as the frontal lobe. Sudden deterioration in the patient's condition is more likely to be due to swelling of brain tissue from cerebral oedema around the tumour than an increase in actual tumour size.

Cerebral oedema can be either localised or generalised and has been classified into three histological types: *vasogenic, cytotoxic* and *interstitial*.[46] *Vasogenic* oedema is associated with the accumulation of fluid in the extracellular space by leakage of plasma and plasma proteins through the blood brain barrier. This type is most commonly seen in cerebral tumour or brain damage associated with acute intracerebral haemorrhage following head injury. Vasogenic oedema often responds to steroid treatment (e.g. dexamethasone). *Cytotoxic* oedema is due to intra- and extra-cellular swelling, usually the result of ischaemia, anoxia or both, and does not respond as well to steroids. The commonest situation is

TABLE 1
DIFFERENTIAL DIAGNOSIS OF COMA

1. *Neurological diseases*

    Cerebrovascular Diseases
    Cerebral infarction
    Cerebral haemorrhage
    Brain stem infarction
    Brain stem haemorrhage
    Subarachnoid haemorrhage
    Hypertensive encephalopathy

    Space-occupying Lesions
    Tumour
    Abscess
    Haematoma—subdural
    —intracranial

    Trauma
    Concussion
    Cerebral contusion
    Haematoma, e.g. extradural
    Fat embolism

    Meningitis
    Bacterial
    Viral
    Tuberculous
    Fungall
    Carcinomatous

    Encephalitis
    e.g. Herpes simplex
        Herpes zoster
        Subacute sclerosing
            panencephalitis

    Diffuse Cerebral Diseases
    e.g. Wernicke's encephalopathy
        Multiple sclerosis
        Central pontine
            myelinosis
        Syphilis

    Epilepsy

    Hysteria

2. *Metabolic disorders*

    Diabetes mellitus
    Renal failure
    Hepatic failure
    Porphyria
    Hypothermia
    Hyperthermia
    Hypoxia
    Hypercapnoea
    Electrolyte disorders, e.g. sodium
                                calcium
                                acid-base balance
    Disseminated intravascular coagulation

3. *Endocrine disorders*
    Hypopituitary coma
    Adreno-cortical failure
    Myxoedema coma
    Parathyroid disorders

4. *Toxins*
    Self-induced poisoning, e.g. alcohol
                                barbiturates
                                methanol
                                carbon monoxide
    Heavy metals   e.g. lead
                                mercury
                                manganese
                                thallium

for both types of oedema to occur, although one type may predominate. *Interstitial* oedema is seen more rarely, e.g. in benign intracranial hypertension.

The sequence of events leading to the development of cerebral oedema is complex but the following events are known to occur (i) failure of the 'blood brain barrier' (ii) failure of cerebral autoregulation (which can be local or involve one or both hemispheres) (iii) alteration of cerebral blood flow. As a result, the brain may not be protected from the systemic blood pressure. Within an ischaemic/anoxic area, there is an accumulation of metabolites, such as lactate, so that the pH will alter and the blood vessels will not respond to changes in arterial P_{CO_2}. Immediately around the ischaemic area, there is a zone of 'luxury perfusion', so that blood may be directed away from the ischaemic area towards normal brain ('steal effect'). Less commonly, blood can be directed towards the ischaemic area, increasing the amount of oedema ('reverse steal effect'). The duration of anoxia and ischaemia, the cerebral blood flow level, the degree of arterial spasm, and the response of ischaemic brain to 're-flow' will also determine the type and amount of cerebral oedema. Other systemic factors are also important, e.g. electrolyte or blood gas disturbances, blood viscosity, polycythaemia, infections or cardio-respiratory problems.

Diffuse Brain Damage

Infections, bacterial or viral, can produce diffuse inflammatory cerebral lesions, associated with cerebral oedema and also cause petechial haemorrhages. Herpes simplex encephalitis, due to virus Type I, can produce clinical signs suggestive of a space-occupying lesion in a temporal lobe (i.e. 'mass effect'). Diffuse brain damage of immediate impact type[38] is associated with injury to white matter throughout the brain and is thought to be responsible for the long-continued alterations in consciousness following complete loss at the moment of impact. At autopsy, the gross pathology of the brain may appear normal and the diffuse nature of the damage found only on histological examination. In stroke, where cerebral ischaemia or infarction is caused by an occlusion of one of the intracerebral arteries, ischaemia is localised and its effects minimised because of 'cross-flow'; studies often show a more generalised disturbance than expected from the anatomical size of the lesion as seen on a CAT scan (computerised axial tomography), the 'non-affected' hemisphere also having areas of reduced blood flow.

Following cardiac arrest, there is total cerebral ischaemia and anoxia. Providing the blood pressure is restored within five minutes, the brain tissue may recover. The areas of brain most in jeopardy are at the boundaries of the anastomoses between the anterior, middle and posterior cerebral arteries so that infarction can occur at these border zones, e.g. occipital infarction, leading to cortical blindness. Metabolic disorders, e.g. hypoglycaemia, can cause similar effects to those seen after severe anoxia. Blood diseases, e.g. leukaemia and idiopathic thrombocytopenic purpura, can also cause diffuse brain damage.

Neurotransmitter Disturbances

Imbalance between dopaminergic and cholinergic mechanisms in the brain has therapeutic implications in various neurological diseases, e.g. the discovery that brain dopamine was depleted in Parkinson's disease lead to the introduction of L-Dopa therapy. The validity of biochemical measurements on human postmortem brain specimens has been considered by Bowen et al.[8] Patients with Huntington's disease and Alzheimer's disease have reduced enzymatic synthesis of gamma-aminobutyric acid (GABA). Increased dopamine-receptor sensitivity possibly due to central dopamine deficiency[15] and GABA lack[66] have been found in brains of schizophrenic patients. In Alzheimer's disease, very low concentrations of choline acetyltransferase and acetylcholinesterase have been found in the brain[23] and, as a result, choline therapy advocated.

The importance of neurotransmitter disturbances in the pathogenesis of coma has yet to be elucidated. In liver disease, various plasma amino acid levels are raised and there is an increase in certain brain amines which could act as false neurotransmitters, mainly by displacing dopamine and noradrenaline: β-hydroxylated phenylethylamines, namely octopamine, and β-phenylethanolamine have been implicated.[27,53] These findings open up the possibility that therapeutic means may be found of restoring low concentrations of neurotransmitters.

L-Dopa therapy has also been tried in hepatic coma (Chapter 7) and in subacute sclerosing panencephalitis (SSPE)[33] and in the latter there may be a metabolic block in neurotransmitter synthesis. A monoamine oxidase inhibitor was also added in low doses which seemed to provide additional benefit to the Levodopa/Carbidopa treatment.

CLINICAL ASPECTS

History

Traditional medical teaching stresses the importance of accurate 'history-taking'. In over 80% of new clinical cases, the diagnosis is made on history alone, the clinical examination and investigations, where appropriate, confirming the initial diagnosis.[34] The 'unexpected' physical sign is only unexpected in that it was not anticipated from the history.

Because of the absence of a full history a different approach is required in the unconscious patient. It may be necessary to institute resuscitation measures, e.g. maintaining the air-way, and a preliminary examination may need to be carried out before taking the history from relatives, neighbours or the ambulance team, e.g. who telephoned for the ambulance?, how was the patient found?, what is known of the patient?, evidence of a recent illness, suicide-attempt and so on. This information is also of medico-legal importance, and a full history must be obtained from the patient once he regains consciousness.

General Assessment of the Unconscious Patient

The first priority is to ensure that the air-way is patent and can be maintained, the second priority to establish the level of coma. Unconsciousness must be distinguished from akinetic mutism, 'locked-in' syndrome (page 414) and hysteria (page 423).

The full general examination is carried out and the patient inspected from 'head to toe' for external signs of bruising and trauma, any evidence that he may have been lying on one side for a long period of time, and possible marks produced by self-inflicted venepuncture. The temperature reading must be accurate and a low-reading thermometer used in cases of hypothermia.

A 'Dextrostix' or similar test performed on a blood sample should exclude hypo- or hyperglycaemia. Following head injury, fluid leading from the nose or ear should be collected and examined by a glucose 'stick' test, or by the laboratory, to detect whether it is cerebrospinal fluid. Central cyanosis indicates cardiorespiratory problems. The cardiac status of the patient, i.e. the nature and quality of the pulse, the presence of hypo- or hypertension, the presence of cardiac murmurs, must be supplemented by an electrocardiogram to determine wheter the patient has had a recent myocardial infarction. If respiratory difficulties are associated with chest injuries resulting from trauma, the possibility of a tension pneumothorax must be considered. An unconscious patient may have aspirated vomitus which will produce upper lobe signs, particularly on the right side. Continuing respiratory problems may necessitate artificial ventilation and this decision will taken in consultation with an experienced anaesthetist. Diabetic ketosis can mimic an acute abdomen but it is important to ensure that the patient has no evidence of intra-abdominal bleeding.

Neurological Grading of Coma

Although the Glasgow Coma Scale was proposed by Teasdale and Jennett[83] for the clinical grading of coma due to head injury, it can be used for all types of coma (Table 2). It consists of three categories of evaluation (i) *eye opening* with four grades ranging from spontaneous to none, (ii) *best verbal response,* with five grades ranging from orientated to no verbal response, and (iii) *best motor response* with six grades from obeying commands to no response. Jennett and Bond[39] have also described outcome measures which supplement the Glasgow Coma Scale.

Simple clinical scales of coma provide a method for monitoring clinical change and for predicting outcome, and the Glasgow Coma Scale has been shown to be of value for both purposes in head injury patients.[40,84]

The importance of the duration and depth of coma as a prognostic sign following an acute stroke is also well recognised.[56,64,69] Assessment of coma 1 hour after a cardiac arrest allows those patients who will die or survive with intellectual damage to be distinguished from those who survive without neurological damage.[94] Outcome following subarachnoid haemorrhage from intracranial aneurysms can be linked with neurological grades[82] using the Botterell grading system.[7]

TABLE 2
GLASGOW COMA SCALE

Eye opening	—spontaneous
	—to speech
	—to pain
	—none
Verbal response	—orientated
	—confused conversation
	—inappropriate words
	—incomprehensive sounds
	—nil
Best motor response	—obeys
	—localises
	—withdraws
	—abnormal flexion
	—extensor response
	—nil

After Teasdale & Jennett.[83]

Neurological Signs

An important distinction must be made between the monitoring of neurological signs, the purpose of which is to detect signs of neurological deterioration, as opposed to coma scales designed to measure and predict outcome.[48] For example, pupillary abnormalities are not a measure of the level of consciousness, although an important neurological finding. Neurological examination of the patient indicates the site and level of the lesion (Table 3). Fundoscopy may indicate the presence

TABLE 3
CORRELATION BETWEEN CLINICAL SIGNS AND LEVELS OF BRAIN FUNCTION

Anatomical Site	Neurological Signs
Cerebral hemispheres	Verbal responses
	Purposive movements
Brain stem	Reflex motor movements:
	Decortication
	Decerebration
Reticular activating system	Eye opening
Midbrain IIIrd CN	Pupil size, reaction to light
Pons Vth and VIIth CN	Corneal reflex
IIIrd VIth CN medial longitudinal fasciculus	Doll's eye manoeuvre and ice-water responses
VIIIth CN	Respiratory rate and pattern
Medulla	Blood pressure

Modified from Coronna and Finklestein[18]

of papilloedema, hypertensive retinopathy, subhyaloid haemorrhages (seen in subarachnoid haemorrhage) or optic atrophy. 'Snail-tracking', so called because of the appearance of the blood vessels, is seen as a preterminal event.

Raised Intracranial Pressure

When coma is associated with the gradual onset of raised intracranial pressure, e.g. due to a cerebral tumour, the patient presents with headache of increasing severity and vomiting, usually in the morning. On examination, papilloedema is typically found. Increasing intracranial pressure in an unconscious patient can be measured by an extradural transducer but this is seldom carried out in clinical practice. Instead, reliance is placed on the physical signs due to brain stem compression. The systemic hypertensive response is association with increased intracranial pressure was first described by Cushing in 1901. The changes preceding this event have been less well described.

The clinical findings in these patients, with fluctuating level of consciousness, rapid changes in respiratory rates, pulse rate, blood pressure and electrocardiographic abnormalities have been associated with high blood levels of catecholamines.[19,60] This is probably due to increased sympathetic activity as a direct result of a hypothalamic disturbance. The hypothalamus could also act through the pituitary-adrenal axis leading to raised plasma cortisol levels, which in turn can potentiate the vascular effects of catecholamines. The pulmonary changes of congestion, haemorrhage and oedema are thought to be mediated by the same mechanism. In an experimental model, Graf and Rossi demonstrated enormous changes in plasma catecholamine levels (35 times normal for dopamine, 145 times for noradrenaline and 1200 times for adrenaline) following raised and sustained intracranial pressure. The effects are likely to be due to distortion of the brain stem with associated functional, structural and vascular changes as much as to the increased intracranial pressure. The 'Cushing response' is probably a preterminal event.

NEUROLOGICAL DISEASES

Cerebrovascular Diseases

A cerebral vascular lesion is a very common cause of coma and the major types of cerebrovascular diseases responsible are (i) *cerebral infarction*, (ii) *cerebral haemorrhage* and (iii) *subarachnoid haemorrhage*.

On clinical grounds alone (i.e. from the history and examination of the patient), cerebral infarction cannot be differentiated from cerebral haemorrhage with any degree of certainty.[22] In contrast, the typical presentation of subarachnoid haemorrhage is the acute onset of severe, occipital headache in a previously fit person, aged between 40 and 60 years. The patient may become drowsy and irritable, lapsing into stupor or coma. The signs of meningism, including the presence of neck stiffness, depend upon the level of consciousness since it is readily detectable when there is only slight alteration of consciousness but absent in the comatose patient. The presence of hemiplegia means that there has also been haemorrhage into a cerebral hemisphere when the clinical picture will be the same as for cerebral haemorrhage.[16]

Following subarachnoid haemorrhage, intense intracerebral arterial spasm may occur which itself can lead to cerebral infarction, even several days after the acute onset of the illness.[82] The high mortality from subarachnoid haemorrhage is due to further bleeding, either in the acute illness, within the first year or several years later.[95]

Cerebral infarction (85% of cases of stroke) is more common than *cerebral haemorrhage* (15% of cases). Alteration of consciousness following infarction usually occurs after 24 hours and is related to the presence of cerebral oedema which increases to reach a peak at day 4, decreasing after this. Occasionally, cerebral oedema can be present for up to 3 weeks. Immediate mortality is related to the extent of the oedema. Following a *cerebral haemorrhage*, there is usually rapid loss of consciousness since the haemorrhage acts as an acute expanding space-occupying lesion. Whereas the overall mortality from cerebral infarction is 40%, the mortality from cerebral haemorrhage is 80%.

Investigations are required to differentiate the major types of cerebrovascular diseases.[13] Cerebrospinal fluid (CSF) examination is still the simplest to perform and will be considered first.

CSF Examination

Although most physicians recognise the importance of CSF examination to confirm the diagnosis of subarachnoid haemorrhage, they are often reluctant to perform this investigation in all stroke patients, partly because of the belief that there is little to be gained by defining the type of stroke, namely infarction or haemorrhage, and partly because of a therapeutic nihilism.[6]

Subarachnoid haemorrhage literally means 'blood in the CSF'. The term is not synonymous with intracerebral aneurysm since these are found in only 60% of cases.[72] The characteristic xanthochromic appearance of centrifuged CSF sample is caused mainly by three substances: oxyhaemoglobin, methaemoglobin and bilirubin.[44] Oxyhaemoglobin is the first to appear within 4–6 hours and the persisting discolouration is then due to bilirubin, which can last 2–3 weeks.[71] Examination of the supernatant fluid will distinguish a true subarachnoid bleed from a traumatic tap. The presence of a large number of erythrocytes may cause the pH of the CSF to fall, thereby stimulating respiration, causing a respiratory alkalosis which can further impair consciousness.[74]

In 80% of *cerebral haemorrhage* cases there is blood in the CSF.[57] False negatives will occur in patients with intracerebral haematoma. Following *cerebral infarction*, the CSF is typically clear, but the CSF protein concentration may be slightly raised during the first 2 weeks (1–2 Gms/l) and an excess of leucocytes may be associated with haemorrhagic infarction.

Direct spectrophotometry of the CSF can distinguish between these pathological types with a far greater degree of accuracy than previously reported and characteristic patterns have been described.[45] In a comparative study of CSF spectrophotometry and computerised axial tomography (CAT) in cerebrovascular disease[78] the former provided a specific diagnosis, i.e. bleeding as opposed to infarction, in almost all cases (95%). CAT scan was less successful (65%). This high success rate depends upon the timing of CSF examination and repeating the lumbar puncture if the initial examination is equivocal. Spectrophotometry will not detect bleeding in normal brain tissue, as opposed to tumour tissue.

The use of serum enzyme estimations, e.g. creatine kinase, aspartate transaminase, lactate dehydrogenase following acute myocardial infarction is well established, and various workers have tried to look for an enzyme marker which would indicate the extent of brain damage following cerebrovascular disease, trauma or infections. Creatine kinase has been estimated for this purpose after brain injury. There are 3 isoenzymes of creatine kinase: muscle (MM-type) heart (MB-type) and brain (BB-type). The heart type (CK–MB) appears to indicate myocardial ischaemia and can be raised even when the total CK level is within normal limits.[54] CK–MB has also been detected in the serum of patients following acute cerebrovascular, traumatic or infectious brain damage.[41] In these patients, ECG abnormalities may be found suggestive of acute myocardial damage. CK–BB has been found in the serum of patients following diffuse brain damage[42,80] Serum total CK measurements have been estimated in patients following *subarachnoid haemorrhage* and the heart-type has also been found.[26] It is probable that the presence of brain-type CK in the serum following acute brain injury carries a poor prognosis[41,42,80] and that the detection of CK–MB also carries a poor prognosis.[42]

Prostaglandins (PG) synthesised in the brain *in vivo*, are actively taken up by the choroid plexus and transported into the venous circulation and can be measured in the CSF. The F-2α type of prostaglandins is a potent vasoconstrictor agent on cerebral arteries whilst the E-type is a vasodilator. In an investigation measuring prostaglandin E2 in CSF from patients following a stroke, a positive correlation was found between the PGE2 level and the severity and clinical outcome of the stroke.[14]

Other Cerebrovascular Diseases

Hypertensive encephalopathy can occur in association with acute and chronic glomerulonephritis, malignant hypertension and eclampsia. The commonest pathological finding is cerebral oedema, probably occurring as a secondary effect. The presenting symptoms are usually convulsions and focal neurological signs such as amaurosis, aphasia and hemiplegia; fundoscopy usually reveals bilateral papilloedema. Attacks can occur following ingestion of tyramine-containing foods in patients taking monoamine oxidase inhibitors and in phaeochromocytoma. Providing the encephalopathy does not complicate renal failure, urinalysis and renal function is usually normal. The CSF is under pressure but CSF examination is otherwise normal, thereby excluding meningitis or subarachnoid haemorrhage.

Space-occupying Lesions

The gradual loss of consciousness due to intracranial tumour, or from a cerebral abscess, has already been mentioned (page 414). Sudden deterioration can be due to haemorrhage into or around a tumour.

At least 5% of patients presenting with acute hemiplegia and clinically diagnosed as 'stroke' will, on further investigation, be found to have an intracerebral tumour.

If an intracerebral space-occupying lesion is suspected, a CAT scan is the investigation of choice. The hazards of lumbar puncture cannot be over-emphasised in this situation and the value of CSF examination has not been established despite several interesting reports. In one study, positive cytology was obtained from preoperative CSF samples in 20% of all cases of single or multiple secondary cerebral tumours. Preoperative CSF cytology was positive in 15.3% of primary cerebral tumours, and in 40% of postoperative CSF samples.[4]

Lysozyme is absent from normal CSF but has been demonstrated in the CSF from patients with primary and secondary central nervous system tumours, the concentrations reflecting the degree of involvement.[61] Lysozyme was also found in infections of the central nervous system.

Dihydrofolate reductase has been found in primary and secondary cerebral tumours which could provide a biochemical rationale for antifolate therapy using methotrexate.[1] Enzymatic evaluation of brain tumours may be a useful diagnostic aid in the surgery of gliomas, since abnormal alanine inhibition of pyruvate kinase (indicating MII-type isoenzyme) has been found in gliomas and meningiomas but not in normal brain tissue.[89] These authors have suggested that the alanine inhibition test could be used to demarcate the tumour at the time of operation.

It is important to monitor the pituitary function of children following treatment of intracranial tumours[73] since growth hormone levels may be drastically reduced, particularly after cranial irradiation.[35]

Trauma

Although there is often a history of 'head injury' in comatose patients, it is always important to consider possible pre-existing illnesses, e.g. did the patient have an acute cerebrovascular or myocardial episode? or was the patient intoxicated? The nature of the underlying pathophysiological disturbance causing coma may be much more difficult to elucidate. Continuing coma may be due to haemorrhage, cerebral oedema or infection. The clinical signs may be complicated by false locating signs due to third or sixth cranial nerve palsies. Injuries to thorax and abdomen complicates management. Dehydration, renal failure, electrolyte and blood-gas abnormalities can lead to further brain damage due to anoxia and ischaemia. Regular monitoring of the plasma urea and electrolyte concentrations is required with blood gas measurements as necessary. Diabetes insipidus is a rare complication following head injury. Accurate observation, monitoring neurological signs, levels of consciousness, pulse, blood pressure, respiratory rate and pattern are vital.

Plain skull X-rays are essential to demonstrate an impacted fracture and this investigation is of medicolegal importance. CAT scanning has revolutionised the management of head injuries since it can rapidly identify localised intracranial lesions. CSF examination is not free of risk. Blood in the CSF may follow a cerebral contusion or cerebral haemorrhage. CSF pressure measurements are best carried out by extradural transducers inserted through a burr-hole. Serum levels of myelin-basic protein (MBP) have been found to be raised immediately after severe head injury, remaining elevated for two weeks. The mean MBP level between two and six days after injury was significantly higher in patients with a poor outcome as compared with those with a good outcome.[85] Myelin basic protein concentration in blood may therefore be valuable in predicting outcome after head injury.

Meningitis

Meningococcal meningitis is the commonest type of acute bacterial meningitis in Britain, affecting all age groups, with the highest mortality occurring at the extremes of life. Even with antibiotic treatment, mortality is 10%. When meningitis leads to coma, the onset is usually subacute and the patient usually complains of intense headache, neck stiffness and there is an associated fever. Kernig's sign is positive (straight-leg raising is limited because of the pain produced by stretching the inflamed meninges). Fulminant meningococcal septicaemia can rapidly lead to coma and death, 3–84 hours after admission.[11]

In suspected cases of meningitis, CSF examination, blood culture and culture of nasopharyngeal secretions must be carried out in order to isolate the organism; this is more difficult if antibiotics have been started before lumbar puncture is performed.[6] If the CSF findings are normal (Gramstain negative, normal cell count) and meningitis is still suspected, the lumbar puncture should be repeated.[76] To increase diagnostic accuracy, countercurrent immunoelectrophoresis (CIE) can be used to detect capsular antigens in the CSF and serum, and has proved successful in the diagnosis of meningococcal, pneumococcal, haemophilus and escherichia coli meningitis. Antigen starts to disappear 48 hours after the initiation of treatment.[47] The typical CSF findings in pyogenic meningitis are a polymorphonuclear leucocytosis, sometimes exceeding 1000 per cu. mm, giving a cloudy appearance to the naked eye, an increase in total protein and a low or absent glucose level. Fibrin degradation products can be increased in the CSF in acute pneumococcal meningitis but are also found in other neurological conditions and therefore are unlikely to be diagnostically significant.[81]

Viral meningitis is common and, apart from mumps, is due mainly to the enteroviruses, especially echovirus and coxsackie virus. The course of the illness is usually benign and is not normally associated with an alteration in the level of consciousness. If this occurs, then it is usually due to an encephalitis. In viral meningitis, the CSF may show an initial increase in polymorphonuclear leucocytes, but mononuclear cells predominate usually by the second week. The glucose level is normal or only slightly depressed. CSF lactate levels have been used

(i) to differentiate bacterial from viral meningitis with levels of over 30 mg/dl being consistent with bacterial meningitis[49]

(ii) to measure response to treatment.[9]

Smith *et al.*[77] have suggested that IgM levels in the CSF may be useful in separating viral from bacterial meningitis, since they are highest in the latter.

γ-aminobutyric acid (GABA) was found in the CSF of patients with bacterial meningitis, glutamic acid and glutamine concentrations were found to be raised in both acute viral and bacterial meningitis.[10]

The main diagnostic problem is in differentiating *tuberculous meningitis* (TBM) from other types of bacterial meningitis when CSF culture is negative. Clinical onset is usually insidious, with a prodrome often lasting 2–3 weeks. Drowsiness and delirium may occur, but lucid intervals are a common feature. In TBM, there is typically a mononuclear response (10–1000 cells per cmmm) in the CSF and the protein level is raised (1–4 gm/l, occasionally higher). The CSF glucose content is low or absent. CSF chloride may be low. CSF lactate levels may take several weeks to return to normal.[49] The bromide partition test (serum and CSF bromide concentrations are measured after 3 days of oral bromide administration or after a single intravenous dose of 2–4 g in 10 mls of sterile distilled water) may be useful. In TBM the ratio serum : CSF approaches unity, whereas in normal subjects, the ratio is 3 : 1. A radioactive bromide test has been developed[52] and a recent report claims a 95% diagnostic accuracy (44 patients) where the serum to CSF ratio below 1.9 : 1 was considered to be highly specific for TBM.[21]

Carcinomatous meningitis also can cause a low glucose level and an abnormal bromide partition ratio.

Encephalitis

The physical signs of viral encephalitis are mainly those of diffuse brain damage, the clinical pattern depending upon the responsible virus. Signs of meningeal irritation may be present (i.e. a meningoencephalitis) and there is often a pleocytosis in the CSF.

The onset of *herpes simplex encephalitis* can be explosive, leading to coma within a few hours. The diagnosis should be suspected from the clinical picture of fits, fever, personality change and localised neurological signs, mimicking a temporal lobe tumour. Changes in serum antibody levels are not always diagnostic but newer techniques with CSF antibodies may prove more useful.[50] Early brain biopsy has been advocated in suspected cases where the virus may be detected by immunofluorescent techniques, electron microscopy, and culture. Periodic sharp wave discharges may be seen on the electroencephalogram but they are not pathognemonic. Untreated, the mortality is 70% and there frequently is severe neurological disability in the survivors.

Herpes zoster encephalitis can arise as an opportunist infection in a patient with disordered immunological responses, e.g. because of a malignant disease, such as leukaemia, or whilst receiving immunosuppressive treatment. The clinical picture is variable but the patient may present with headache, bizarre focal neurological signs or altered level of consciousness and psychotic behaviour.

Subacute sclerosing panencephalitis (SSPE) is due to the long persistence of measles virus after an acute attack. Clinically, 3 stages are recognised, the slow progressive course running from 2 to 18 months. The first stage is associated with mood changes, epileptic attacks and recurrent myoclonic jerking; then, progressive dementia leading to akinetic mutism and finally, decortication. In the CSF, oligoclonal IgG and high titres of measles antibodies can be found. It has been suggested that the detection of virus-specific IgM antibodies in the CSF can be taken as an indication of persistence of virus.[43] The electroencephalogram may be characteristic in SSPE with generalised slow-wave complexes occurring in association with the myoclonic jerks, followed by periods of electrical silence. Brain biopsy confirms the diagnosis in SSPE, and measles virus may be isolated from brain cell cultures.

Other Brain Diseases

Many other diseases, although rare, can cause coma. Deficiency diseases, such as Wernicke's encephalopathy, should be considered in patients with a history of alcoholism, previous neurological disease and poor nutrition. Treatment without vitamin supplements may precipitate coma.[90] In suspected cases, empirical treatment with intravenous thiamine should be given whilst other causes of coma are being investigated. Biochemical investigations may show raised levels of serum pyruvate, reduced erythrocyte transketolase activity and decreased urinary thiamine clearance.

Demyelinating diseases, such as the rare type of *multiple sclerosis* with extensive cerebral demyelination, and diffuse cerebral sclerosis (Schilder's disease) can cause stupor and coma. *Central pontine myelinolysis* is characterised by a rapidly evolving paraparesis or quadriparesis with pseudobulbar signs of dysarthria and dysphagia associated with electrolyte disturbances, mainly severe hyponatraemia.[59] There may be a history of alcohol abuse. *Metachromatic leucodystrophy* is a rare condition usually developing in the second or third year of life and inherited as an autosomal recessive condition. Metachromatic granules may be detected in the urine and aryl-sulphatase estimation in leucocyte suspension may be diagnostic.[65] *Adrenoleucodystrophy* is clinically indistinguishable from Schilder's disease except for the fact that patients show adrenal atrophy so that plasma ACTH and cortisol levels may be helpful. Other rare causes include degenerative or toxic encephalopathies and slow virus disorders such as *multifocal leukoencephalopathy* or *Creutzfeldt–Jakob* disease. Finally, *neurosyphilis* must always be considered. Although new causes, presenting with cerebral parenchymatous disease are rare, routine serological tests must be performed in all cases of unexplained stupor.[51]

Metabolic Disorders

Diabetic Coma

Diabetic coma is normally associated with ketoacidosis, a blood sugar level of over 20 mmol/l and massive ketonuria. Hyperosmolar, non-ketotic diabetic coma occurs in elderly diabetes on hypoglycaemic agents and can be precipitated by thiazide diuretics. A similar syndrome has been described in severe burns. The electrolyte disturbances are a high blood glucose, often exceeding 50 mmol/l, hypernatraemia, severe hypokalaemia and lactic acidosis. Lactic acidosis without ketonuria is also found in cases of severe anoxia or poisoning due to paraldehyde or methyl alcohol. In addition to acute circulatory collapse, the patient in diabetic coma can develop focal epileptic attacks. Absent reflexes may be due to an associated diabetic peripheral neuropathy. There is an increased risk from arterial or venous thrombosis.

Hypoglycaemic Coma

The diagnosis is usually not difficult since the commonest cause is due to an insulin overdose. Glycosuria does not exclude the diagnosis since the urine may have collected before the onset of hypoglycaemia. Spontaneous hypoglycaemia sufficient to produce coma is seen in islet-cell tumours of the pancreas. Hypoglycaemia in infancy may lead to severe brain damage. In adults, other causes of spontaneous hypoglycaemia include liver disease, alcoholism, Addison's disease and hypopituitarism. The difference between diabetic (hyperglycaemic) and hypoglycaemic coma are summarised in Chapter 37. Occasionally, hypoglycaemic coma can present as a 'stroke-like' illness. A full account of the investigation of patients with neuroglycopaenia and spontaneous hypoglycaemia is to be found in the book by Marks and Rose.[55]

Uraemic Coma

Acute and chronic renal failure can lead to coma. The associated electrolyte disturbances include: raised blood urea, creatinine, uric acid, a metabolic acidosis, hyponatraemia, hyperkalaemia, hypocalcaemia and hypermagnasaemia. Fluid retention can reduce serum osmolality below 236 mOsm/1. There is usually a raised blood pyruvate level and cerebral oxygen consumption is reduced. The serum creatinine concentration and the creatinine clearance rate are closely related to renal function.[93] The incidence of uraemic pericarditis is related to the blood uric acid level. Correction of the metabolic acidosis with alkali can further reduce calcium levels, causing tetany. The onset of coma is preceded by headache, vomiting, mental confusion, muscular twitchings and generalised convulsions. Dementia occurring in patients on maintenance haemodialysis has been associated with aluminium toxicity.

Hepatic Coma

The aetiology of hepatic coma is unknown (Chapter 7). One third of patients have cerebral oedema[30] which is not corrected by glycerol therapy.[30] The use of dexamethasone for the treatment of cerebral oedema in these patients carries an increased risk of gastrointestinal haemorrhage. Ammonia levels do not correlate with the onset of stupor since they may be normal or raised in patients with hepatic coma.[75] The encephalopathy may be related to increased concentrations of circulating toxic aromatic amino acids (due to hyperglucagonaemia) which cross the blood brain barrier, and decreased levels of branched-chain amino acids, secondary to hyperinsulinaemia. The failing liver will not be able to metabolise the aromatic amino acids.[79] Octopamine has been postulated in the pathogenesis of encephalopathy.[53] Hypoxia, hypocapnoea and hypotension are often present in decompensated liver disease and probably play a secondary role. The diagnosis is usually straightforward. There is generally a history of liver disease. Coma is usually preceded by asterixis or flapping tremor, and other symptoms and signs of hepatic failure. Coma can be caused by sudden gastrointestinal haemorrhage, high protein foods, intercurrent infection, over-use of diuretics, administration of sedatives or morphine-like drugs, rapid removal of ascitic fluid and hypotension. The glial and neuronal changes in experimental hepatic encephalopathy have been investigated by Diemer.[25] In severe cases, the EEG shows triphasic delta waves and can be used as an additional measure of response to treatment.

Porphyria

Porphyria is a rare cause of coma and is fully considered in Chapter 10.

Hypothermia

Accidental hypothermia can occur after prolonged exposure to cold (fell walkers, mountaineers) or in patients found unconscious at home after stroke or drug overdose. Elderly patients living alone at home, often in unheated rooms, during the winter months, are prone to hypothermia, particularly if their mobility is reduced because of disease such as arthritis or Parkinsonism. Hypothermia can occur in myxoedema and hypopituitarism. Mortality rate in elderly patients is high. Warming should be gradual. Myocardial infarction is a complication. Hypoglycaemia, lactic acidosis, hypoxia and CO_2 retention are common associated findings.

Heat Stroke

Prolonged exertion in hot conditions by, e.g. racing cyclists, marathon runners etc., can lead to hyperpyrexia, cessation of sweating, the rapid onset of coma, convulsions and death. Amphetamine overdose can produce a similar picture. Lesions involving the floor of the third ventricle or the pons (e.g. haemorrhage) can cause hyperpyrexia. Tetanus is a rare cause.

Hypoxia

Plum and Posner[67] divide anoxia into anoxic, anaemic and ischaemic varieties. The neurological picture and outcome is determined by the severity and duration of the hypoxia.

In experimental animals the brain can recover after 30 minutes of hypoxia ($Po_2 < 20$ mm Hg) providing the circulation is maintained.[32]

Anoxic causes include chronic pulmonary disease especially in patients with cor pulmonale, asphyxia (following suffocation), and drowning.

Anaemia reduces the oxygen content of the blood. Carbon monoxide poisoning causes an 'anaemic type' of hypoxia because of the formation of carboxyhaemoglobin. Heart disease, cardiac arrest, and complications of open heart surgery can lead to cerebral *ischaemia* because of poor flow cardiac output. Embolism, disease of main cerebral arteries or smaller cerebral vessels (as in disseminated intravascular coagulation or thrombolic microangiopathy) can also cause cerebral ischaemia and anoxia.

Pulmonary embolism may present with primary neurological features without evidence of paradoxical (right to left) embolisation.[28] Pao_2 is usually below 80 mm Hg and there is often a low $Paco_2$ indicating hyperventilation and physiological shunting. In chronic pulmonary disease, hypoxia is accompanied by an increased $Paco_2$ and the respiratory centre becomes insensitive to changes in arterial CO_2. In this situation the respiratory drive may be determined by the low arterial O_2 tension, and continuous oxygen therapy, by reducing this, can lead to coma. Patients in respiratory failure may exhibit asterixis and myoclonus; papilloedema may be found on fundoscopy. The critical level cannot be established as there is usually concomitant hypercapnoea but may be in the range 40–60 mm Hg. This effect is aggravated by poor cardiac output, anaemia or polycythaemia which can lead to further impairment of consciousness by increasing the degree of cerebral hypoxia.

Electrolyte Disturbances

(i) Sodium. Hyponatraemia below 120 mmol/l leads to confusion and levels below 110 mmol/l to coma. Hyponatraemia seen in hospitalised patients with serious concurrent diseases is usually associated with increased anti-diuretic hormone secretion.[86] Further investigation depends upon the suspected cause (*see* Chapter 1).

Hypernatraemia is seen in hyperosmolar states produced by severe dehydration due either to inadequate fluid intake, e.g. the stroke patient, or to excessive fluid loss, e.g. severe diarrhoea, diabetes insipidus. There is a correlation between the plasma sodium concentration and the levels of consciousness.

(ii) Potassium. Abnormalities in the plasma potassium level do not produce coma but hyperkalaemia is seen in association with a metabolic acidosis which may impair consciousness and the hyperkalaemia itself may cause fatal cardiac dysrhythmias.

Hypokalaemia usually associated with a metabolic acidosis, can cause muscular weakness and paralysis.

(iii) Calcium. Hypercalcaemia as seen in hyperparathyroidism or malignant disease with bone secondaries may produce focal seizures and coma, especially in dehydrated patients.

Hypocalcaemia usually presents with muscle cramps and tetany although the presenting features may be papilloedema or coma.

Acid-base Balance

(i) Metabolic Acidosis severe enough to impair consciousness is seen in diabetes mellitus, renal failure and poisoning due to salicylates, ethyl alcohol, methyl alcohol and rarely ethylene glycol. Lactic acidosis should always be considered in any severely acidotic patient when the commoner causes have been excluded. The diagnosis is usually made when the plasma lactate level is above 5 mmol/l.[2]

(ii) Respiratory Acidosis usually accompanies chronic lung disease but may be found in any condition causing respiratory muscle weakness or brain stem destruction. The correlation between $Paco_2$ and cerebral function is poor. The rate of rise of $Paco_2$ is more important since patients with chronic bronchitis may remain mentally alert even with a $Paco_2$ of 75 mm Hg. Levels above 80 mm Hg are associated with drowsiness. The pH of the cerebro-spinal fluid has been reported to show a better correlation with levels of consciousness.[68]

(iii) Metabolic Alkalosis *per se* does not cause coma.

(iv) Respiratory Alkalosis produces 'light headedness' and limb paraesthesia without loss of consciousness.

Disseminated Intravascular Coagulation (DIC)

Although often associated with severe illness, DIC may present with primary neurological symptoms. Characteristic laboratory abnormalities are prolonged clotting times, thrombocytopenia, hypofibrinogenaemia and raised amounts of fibrinogen-related antigens.

Endocrine Disorders

Diagnostic difficulties arise when a pre-existing endocrine disorder is not suspected.

Hypopituitary Coma

This is usually caused by a number of concurrent factors, namely, reduced adrenal function, hypoglycaemia, hypotension, hyponatraemia, hypothermia, and often precipitated by infection. *Pituitary apoplexy* is the term given to the syndrome of acute pituitary failure due to infarction in a chromophobe adenoma during a period of rapid growth. Rarely, subarachnoid haemorrhage follows. The diagnosis of hypopituitarism depends upon low levels of urinary 17-ketosteroids and plasma cortisol, with a normal response to Synacthen.

Adrenal Cortical Failure

Mild delirium is not uncommon in Addison's disease. Stupor and coma only occur in Addisonian crises. Cerebral oedema and papilloedema have been described. Hypotension, hyponatraemia, hyperkalaemia, hypoglycaemia, lower serum cortisol concentration, diminished excretion of urinary 17-ketosteroids and 17-hydroxycorticoids are the typical findings. The electrocardiogram shows low voltage complexes as does the electroencephalogram. Acute adrenal failure due to meningococcal septicaemia is now rare.

Myxoedema Coma

The characteristic feature is profound hypothermia with temperature readings commonly in the range of 26–31°C. Presentation is usually in the winter months, the myxoedematous patient gradually becoming stuporose and then lapsing into coma. The typical clinical and biochemical findings will be present. The serum triiodothyronine and thyroxine levels will be low and the TSH level will be raised.

Parathyroid Disorders

Parathyroid tumours can cause hypercalcaemia and spontaneous hypoparathyroidism with hypocalcaemia can also cause coma.

Toxins

Self-induced Poisoning

Coma secondary to a drug overdose should always be suspected in younger patients brought into the casualty department on Friday or Saturday nights. 'Overdose' is now a major cause for acute hospital admission. The clinical history is very important. Relatives or friends may indicate that the patient was recently depressed, had made several previous suicide attempts, or was under psychiatric treatment. The nature of the current medication is important and a count of the remaining pills in the bottles and in the patient's room should be carried out. A suicide note may have been left. General examination may reveal flushed face, congested conjuctivae, rapid pulse, low blood pressure and the characteristic odour of *alcohol* on the breath when this is the cause of coma. There may be evidence of *barbiturate blisters* on bony pressure points. If the patient has been lying in one position for many hours, extensive muscle necrosis with high serum creatine kinase levels may be found. Hypothermia and cardiac arrest may occur. Pupil size will vary according to the drug taken, e.g. dilated in alcohol coma and pin-point with opiates. The tendon reflexes may be diminished or absent and the plantar reflexes extensor.

134 deaths were recorded in 4 years in people aged 10–75 years investigated by coroners in Greater London. 75% were under 30 years. *Barbiturate overdose* accounted for over 50% of these deaths.[31] Other drugs commonly taken are aspirin and paracetamol. Less frequently, opiates may be responsible and then there may be signs of self-inflicted venepuncture. Alcohol and sedatives are a common mixture. Diagnosis depends upon the history and identification of the toxin in a blood sample. Investigations must include at least one sample retained for forensic purposes in the event that the patient might succumb. Stomach washings should also be sent for analysis. The blood level of the drug can be estimated on at least two occasions after admission in order to measure the clearance rate when a rough estimate can be made of the total quantity of the drugs taken providing the approximate time of the overdose is known.

The general measures for monitoring and nursing the unconscious patient have been mentioned (page 416). Decisions concerning artificial respiration and increasing elimination of the drug involved by haemodialysis may have to be considered.

In *alcohol intoxication*, serum alcohol levels may be of medico-legal importance. The correlation between serum alcohol levels and altered consciousness is poor but, in general, levels below 2 gm/l usually indicate that alcohol is not the cause for continuing unconsciousness.

Methanol poisoning is occasionally seen in patients who have taken methylated spirits or similar alcohol and there have been outbreaks of poisoning after consumption of home-brewed alcohol. The CAT scan appearance of bilateral symmetrical areas of low attenuation in the putamen is the same as that reported in fatal cases at autopsy.[3]

With the elimination of *carbon monoxide* in the home by the introduction in the UK of natural gas, poisoning attempts using carbon monoxide have decreased. Carboxyhaemoglobin levels between 50–60% have been reported to cause coma and lower levels are associated with headache, vomiting and ataxia.[58] An unusual syndrome (postanoxic encephalopathy) has been described where patients relapse into coma 1–3 weeks after poisoning without any fresh exposure.[29] In fatal cases, the blood is a cherry red colour.

Lead intoxication is still occasionally seen in children and as an occupational disease in plumbers and painters. A sudden rise in the serum level is associated with fits, delirium and coma, with levels in the range of 100–200 μg/ml (normal 10–60 μg/l).

Mercury, manganese and *thallium* poisoning can all cause coma.

Other Causes

Epilepsy

If coma follows epilepsy, there is usually a history of fits before the episode. Examination of the unconscious patient may reveal a bitten tongue and soiled clothing due to incontinence. Focal neurological signs are usually absent apart from extensor plantars. Coma, after a single fit is usually of short duration. After status epilepticus or in the elderly, coma can be protracted and is probably related to cerebral hypoxia and cerebral oedema.

Hysteria

This diagnosis should not be reached by exclusion of other possible causes. A patient feigning coma, possibly after a 'fit' should be carefully examined for physical signs that are inconsistent, particularly in neuroanatomical terms. Deviation of the eyes towards the ground with the patient lying on each side is such a sign and may be found in patients mimicking coma or epilepsy.[37] Serum prolactin concentration rises after a generalised tonic-clonic seizure, being maximal at 15–25 minutes after an attack. Levels are usually above 1000 mU/l.[87] Hysterical blindness and the tests employed to make the diagnosis have been recently reviewed.[88]

THE DIAGNOSIS OF BRAIN DEATH

This is a new problem that has arisen because it is now possible to maintain life even when severe brain damage exists. The decision to switch off a ventilator is a 'straight clinical assessment that the patient is dead though his heart is still beating'.[5] and recommendations to the management of such cases have been published.[17] The conditions for considering the diagnosis of brain death are that the patient must be deeply comatose, with no brain stem reflexes or spontaneous respiration and a positive diagnosis made as to the cause of the brain damage. In particular depressant drugs, hypothermia and metabolic causes must be excluded. Although the diagnosis is clinical, an EEG recording is of medico-legal importance.

CONCLUSION

The management of the stuporose or comatose patient exercises the complete skills of the medical team. Decisions are based on an accurate diagnosis. The approach to the differential diagnosis is essentially clinical using investigations to confirm the diagnosis and monitor change. Coma scales should be used to predict outcome and full clinical and psychological assessments must be carried out on all survivors over an adequate follow-up period.

REFERENCES

1. Abelson, H. T., Fosburg, M., Gorla, L. and Kornbuth, P. (1978), 'Identification of dihydrofolate reductase in human central nervous system tumours,' *Lancet*, **1**, 184–185.
2. Alberti, K. M. M. and Nattrass, M. (1977), 'Lactic Acidosis. An occasional review,' *Lancet*, **2**, 25–29.
3. Aquilonius, S. M., Askmark, H., Enoksson, P., Lundberg, P. O. and Moström, U. (1978), 'Computerised tomography in severe methanol intoxication,' *Brit. Med. J.*, **2**, 929–930.
4. Balhuizen, J. C., Bots, G.Th.A.M., Schaberg, J. L. and Bosman, F. T. (1978), 'Value of cerebrospinal fluid cytology for the diagnosis of malignancies in the central nervous system,' *J. Neurosurg.*, **48**, 747–753.
5. British Medical Journal (1976), 'Leader, Brain death,' **2**, 1157–1158.
6. British Medical Journal (1977), 'Leader, Partially treated pyogenic meningitis,' **1**, 340.
7. Botterell, E. H., Lougheed, W. M., Scott, J. V. and Vandewater, S. L. (1956), 'Hypothermia and interruption of carotid, or carotid and vertebral circulation, in surgical management of intracerebral aneurysms,' *J. Neurosurg.*, **13**, 1–42.
8. Bowen, D. M., Smith, C. B., White, P., Goodhardt, M. J., Spillane, J. A., Flack, R. H. A. and Davison, A. H. (1977), 'Chemical pathology of organic dementias,' *Brain*, **100**, 397–426.
9. Brook, I., Rodriguez, W. J., Controni, G. and Ross, S. (1979), 'CSF lactic acid for differential diagnosis of meningitis,' *Lancet*, **1**, 1035.
10. Buryakova, A. V. and Sytinsky (1975), 'Amino acid composition of cerebrospinal fluid in acute neuroinfections in children,' *Arch. Neurol.*, **32**, 28–31.
11. Cahalane, S. F. and Waters, M. (1975), 'Fulminant meningococcal septicaemia,' *Lancet*, **2**, 120–122.
12. Cairns, H. (1952), 'Disturbances of consciousness with lesions of the brain stem and diencephalon,' *Brain*, **75**, 109–146.
13. Capildeo, R., Haberman, S. and Clifford Rose, F. (1978), 'The definition and classification of stroke,' *Quart. J. Med.*, **47**, 177–196.
14. Carasso, R. L., Vardi, J., Rabay, J. M., Zor, U. and Streifler, M. (1977), 'Measurement of prostaglandin E_2, in cerebrospinal fluid in patients suffering from stroke,' *J. Neurol. Neurosurg. Psych.*, **40**, 967–969.
15. Chouinard, G. and Jones, B. D. (1978), 'Schizophrenia as a dopamine-deficiency disease,' *Lancet*, **2**, 99–100.
16. Collier, J. (1931), 'Cerebral haemorrhage due to causes other than arteriosclerosis,' *Brit. Med. J.*, **2**, 519–521.
17. Conference of the Medical Royal Colleges and their faculties in The United Kingdom (1976), 'Diagnosis of brain death,' *Brit. Med. J.*, **2**, 1187–1188.
18. Coronna, J. J. and Finklestein, S. (1978), 'Neurological symptoms after cardiac arrest,' *Stroke*, **5**, 517–521.
19. Cruickshank, J., Neil-Dwyer, G. and Stott, A. (1974), 'The possible role of catecholamines, corticosteroids and potassium in the production of electrocardiographic abnormalities with subarachnoid haemorrhage,' *Brit. Heart J.*, **36**, 697–706.
20. Cushing, H. (1901), 'Concerning a definite regulatory mechanism of the vasomotor centre which controls blood pressure during cerebral compression,' *Bull. John Hopkins Hosp.*, **12**, 290–292.
21. Da Costa, H., Borker, A. and Lokew, M. (1977), 'Distribution of orally administered Bromine-82 in tubercular meningitis,' *J. Nuc. Med.*, **18**, 123–124.
22. Dalsgaard-Nielsen, T (1955), 'Survey of 1000 cases of apoplexia cerebri,' *Acta Psychiat. Scand.*, **30**, 169–185.
23. Davies, P and Maloney, A. J. F. (1976) 'Selective loss of central cholinergic neurons in Alzheimer's disease,' *Lancet*, **2**, 1403.
24. Denny-Brown, D. and Russell, W. R. (1941), 'Experimental cerebral contusion,' *Brain*, **64**, 93–164.
25. Diemer, N. A. (1978), 'Glial and neuronal changes in experimental hepatic encephalopathy,' *Acta Neurol. Scand. Suppl.*, 71, **58**, Munksgaard, Copenhagen.
26. Fabinyi, G., Hunt, D. and McKinley, L. (1977), 'Myocardial creatine kinase isoenzyme in serum after subarachnoid haemorrhage,' *J. Neurol. Neurosurg. Psych.*, **40**, 818–820.
27. Fisher, J. E. and Baldessarini, R. J. (1921), 'False neurotransmitters and hepatic failure,' *Lancet*, **2**, 75–80.
28. Fred, H. L., Willerson, J. T. and Alexander, J. K. (1967), 'Neurological manifestations of pulmonary thromboembolism,' *Arch. Intern. Med.*, **120**, 33–37.
29. Garland, H. and Pearce, J. (1967), 'Neurological complications of carbon monoxide poisoning,' *Quart. J. Med.*, **36**, 445–455.
30. Gazzard, B., Portmann, B., Murray-Lyon, I. M. and Williams, R. (1975), 'Causes of death in fulminant hepatic failure and relationships to quantitative histological assessment of parenchymal damage,' *Quart. J. Med.*, **44**, 615–627.
31. Ghodse, A. H., Sheehan, M., Stevens, B., Taylor, C. and Edwards, S. (1978), 'Mortality among drug addicts in Greater London,' *Brit. Med. J.*, **2**, 1742–1744.
32. Gray, F. D. Jnr. and Horner, G. J. (1970), 'Survival following extreme hypoxia,' *J.A.M.A.*, **211**, 1815–1817.
33. Halikowski, D. and Piotropawlowska-Weinert, M. (1977), 'Levodopa in subacute sclerosis panencephalitis,' *Lancet*, **2**, 1033.
34. Hampton, J. R., Harrison, M. J. E., Mitchell, J. R. A., Pritchard, J. S. and Seymour, L. (1975), 'Relative contributions of history-taking, physical examination and laboratory investigation to diagnosis and management of medical out-patients,' *Brit. Med. J.*, **2**, 486–489.
35. Harrop, J. S., Davies, T. J., Capra, L. G. and Marks, V. (1975), 'Pituitary function after treatment of intracranial tumours in children,' *Lancet*, **2**, 230–231.
36. Hawkes, C. H. (1974), 'Locked-in Syndrome: Report of seven cases,' *Brit. Med. J.*, **4**, 379–382.
37. Henry, J. A. and Woodruff, G. H. A. (1978), 'A diagnostic sign in states of apparent unconsciousness,' *Lancet*, **2**, 920–921.
38. Hume-Adams, J., Mitchell, D. E., Graham, D. I. and Doyle, D. (1977), 'Diffuse brain damage of immediate impact type,' *Brain*, **100**, 489–502.
39. Jennett, B. and Bond, M. (1975), 'Assessment of outcome after severe brain damage. A practical scale,' *Lancet*, **1**, 480–484.
40. Jennett, B., Teasdale, G. and Galbraith, S. (1977), 'Severe head injuries in three countries,' *J. Neurol. Neurosurg. Psych.*, **40**, 291–298.
41. Kaste, M., Somer, H. and Konttinen, A. (1978), '(i) Heart type creatine kinase isoenzyme (CK MB) in acute cerebral disorders,' *Brit. Heart J.*, **40**, 892–805.
42. Kaste, M., Somer, H. and Konttineh, A. (1978), '(ii) Brain type creatine kinase isoenzyme occurrence in serum in acute cerebral disorders,' *Arch. Neurol.*, **34**, 142–144.
43. Kiessling, W. R., Hall, W. W., Yuhe, L. L. and Meulew, V. ter. (1977) 'Measles-virus-specific immunoglobulin-M response in subacute sclerosing panencephalitis,' *Lancet*, **1**, 324–327.
44. Kjellin, K. G. (1969), 'The finding of xanthochromic compounds in the cerebrospinal fluid,' *J. Neurol. Sci.*, **9**, 597–601.
45. Kjellin, K. G. and Söderström, L. E. (1974), 'Diagnostic significance of CSF spectrophotometry in cerebrovascular diseases,' *J. Neurol. Sci.*, **23**, 359–369.
46. Klatso, I. (1967), 'Neuropathologic aspects of brain oedema,' *J. Neuropath. and Exp. Neurol.*, **26**, 1–15.
47. The Lancet (1976), 'Leader, Diagnosis in meningitis,' **1**, 1277–1278.
48. Langfitt, T. W. (1978), 'Measuring the outcome from head injuries', *J. Neurosurg.*, **48**, 673–678.
49. Lauwers, S. (1978), 'Lactic acid concentration in cerebrospinal fluid and differential diagnosis of meningitis,' *Lancet*, **2**, 163.
50. Levine, D. P., Lauter, L. B. and Learner, A. M. (1978), 'Simultaneous serum and CSF antibodies in herpes simplex virus encephalitis,' *J.A.M.A.*, **240**, 356–360.
51. Luxon, L., Lees, A. J. and Greenwood, R. J. (1974), 'Neurosyphilis today,' *Lancet*, **1**, 90–93.

52. Mandal, B. K., Evans, D. I. K., Ironside, A. G. and Pullan, B. R. (1972), 'Radioactive bromide partition test in differential diagnosis of tuberculous meningitis,' *Brit. Med. J.*, **4**, 413–415.
53. Manghani, K. K., Lunzer, M. R., Billing, B. H. and Sherlock, S. (1975), 'Urinary and serum octopamine in patients with portal systemic encephalopathy,' *Lancet*, **2**, 943–946.
54. Marmor, A., Alpane, G., Keidar, S., Grenadier, E. and Palant, A. 1978), 'The MB isoenzyme of creatine kinase as an indication of severity of myocardial ischaemia,' *Lancet*, **2**, 812–814.
55. Marks, V. and Clifford Rose, F. (1965), *Hypoglycaemia*. Oxford: Blackwell Scientific.
56. Marquardsen, J. (1969), *The Natural History of Acute Cerebrovascular Disease*. Copenhagen: Munksgaard.
57. McKissock, W., Richardson, A. and Walsh, L. (1959), 'Primary intracerebral haemorrhage. Results of surgical treatment in 244 consecutive cases,' *Lancet*, **2**, 683–686.
58. McNally, W. D. (1931), 'Carbon monoxide poisoning,' *Illinois Med. J.*, **59**, 383–388.
59. Messert, B., Orrison, W. W., Hawkins, M. J. and Quaglieri, G. E. (1979), 'Central pontine myelinosis. Considerations on aetiology, diagnosis and treatment,' *Neurology*, **29**, 147–160.
60. Neil-Dwyer, G., Cruickshank, J., Stott, A. and Brice, J. (1974), 'The urinary catecholamines and plasma cortisol levels in patients with subarachnoid haemorrhage,' *J. Neurol. Sci.*, **22**, 375–382.
61. Newman, J., Josephson, A. S., Cacatian A. and Tsang, A. (1974), 'Spinal fluid lysozyme in the diagnosis of central nervous system tumours,' *Lancet*, **2**, 756–756.
62. Ommaya, A. K. (1963), 'Head Injuries: Aspects and Problems,' *Med. Ann. D.C.*, **32**, 18–23.
63. Ommaya, A. K. and Gennarelli, T. A. (1974), 'Cerebral contusion and traumatic unconsciousness,' *Brain*, **97**, 633–654.
64. Oxbiry, J. M., Greenhall, R. B. L. and Grainger, G. M. R. (1975), 'Predicting the outcome of stroke: Acute stage after cerebral infarction,' *Brit. Med. J.*, **3**, 125–127.
65. Percy, A. K. and Brady, R. O. (1968), 'Metachromatic leukodystrophy: Diagnosis with samples of venous blood,' *Science*, **161**, 594–595.
66. Perry, T. L., Kist, S. J., Buchannan, J. and Hansen, S. (1979), 'Aminobutyric acid deficiency in the brains of schizophrenic patients,' *Lancet*, **1**, 237–239.
67. Plum, F. and Posner, J. B. (1972), *Stupor and Coma*. 2nd edition Philadelphia: Davis.
68. Posner, J. B., Swanson, A. G. and Plum, F. (1965), 'Acid-base balance in cerebrospinal fluid,' *Arch. Neurol.*, **12**, 479–496.
69. Rankin, J. (1957), 'Cerebral vascular accident in patients over the age of 60. II Prognosis,' *Scot. Med. J.*, **2**, 200–215.
70. Record, L. O., Chase, R. A., Hughes, R. D., Murray-Lyon, I. M. and Williams, R. (1975), 'Glycerol therapy for cerebral oedema complicating fulminant hepatic failure,' *Lancet*, **1**, 540.
71. Roost, K. T., Pimstone, N. R., Diamond, I. and Schmid, R. (1972), 'The formation of cerebrospinal fluid xanthochromia after subarachnoid haemorrhage. Enzymatic conversion of haemoglobin to bilirubin by the arachnoid and choroid plexus,' *Neurology (Minneap.)*, **22**, 973–977.
72. Sarner, M. and Rose, F. C. (1967), 'Clinical presentation of ruptured intracranial aneurysm,' *J. Neurol. Neurosur. Psych.*, **30**, 67–70.
73. Shalet, S. M., Beardwell, C. G., Morris-Jones, P. H. and Pearson, D. (1975), 'Pituitary function after treatment of intracranial tumours in children,' *Lancet*, **2**, 104–107.
74. Shannon, D. L., Shore, N. and Kazemi, H. (1972), 'Acid-base balance in haemorrhagic cerebrospinal fluid,' *Neurology (Minneap.)*, **22**, 585–589.
75. Sherlock, S. (1975), *Diseases of the Liver and Biliary System*. 5th edition, Oxford: Blackwell Scientific.
76. Smales, D. R. C. and Rutter, N. (1979), 'Difficulties in diagnosing meningococcal meningitis in children,' *Brit. Med. J.*, **1**, 588.
77. Smith, H., Bannister, B. and O'Shea, M. J. (1973), 'Cerebrospinal fluid immunoglobulins in meningitis,' *Lancet*, **2**, 591–593.
78. Söderström, C. E. (1977), 'Diagnostic significance of CSF spectrophotometry and computer tomography in cerebrovascular disease,' *Stroke*, **8**, 606–612.
79. Soeters, P. B. and Fischer, J. E. (1976), 'Insulin, glucagon, aminoacid imbalance and hepatic encephalopathy,' *Lancet*, **2**, 880–882.
80. Somer, H., Kaste, M., Troupp, H. and Konttinen, A. (1975), 'Brain creatine kinase in blood after acute brain injury,' *J. Neurol. Neurosurg. Psych.*, **38**, 572–576.
81. Stuart, J. and Anderson, M. (1978), 'Failure of heparin to alter the outcome of pneumococcal meningitis,' *Brit. Med. J*, **1**, 111.
82. Sundt, T. M. Jnr. and Whisnant, J. D. (1978), 'Subarachnoid haemorrhage from intracranial aneurysms,' *New Eng. J. Med.*, **299**, 116–122.
83. Teasdale, G. and Jennett, B. (1974), 'Assessment of coma and impaired consciousness. A practical scale,' *Lancet*, **2**, 81–84.
84. Teasdale, G., Knill-Jones, R. and Van Der Sande, J. (1978), 'Observer variability in assessing impaired consciousness and coma,' *J. Neurol. Neurosurg. Psych.*, **41**, 603–610.
85. Thomas, D. G. T., Palfreyman, J. W. and Ratcliffe, J. G. (1978), 'Serum myelin basic protein assay in diagnosis and prognosis of patients with head injury,' *Lancet*, **1**, 113–115.
86. Thomas, T. H., Morgan, D. B., Swaminathan, R., Ball, S. F. and Lee, M. R. (1978), 'Severe hyponatraemia,' *Lancet*, **1**, 621–624.
87. Trimble, M. R. (1978), 'Serum prolactin in epilepsy and hysteria,' *Brit. Med. J.*, **2**, 1682.
88. Turner, R. E. (1976), In: *Medical Ophthalmology*. (F. Clifford Rose Ed.), London: Chapman and Hall.
89. Van Veelen, C. W. M., Staal, G. E. J., Verbiest, H. and Vlug, A. M. C. (1977), 'Alenine inhibition of pyruvate kinase in gliomas and meningiomas,' *Lancet*, **2**, 384–385.
90. Wallis, W. E., Willoughby, E. and Baker, P. (1978), 'Coma in the Werniche-Forsakoff syndrome,' *Lancet*, **2**, 400–401.
91. Walton, J. N. (1977), *Brain's Diseases of the Nervous System*, Oxford University Press.
92. Ward, A. A. Jnr. (1966), 'The physiology of concussion,' *Clin. Neurosurg.*, **12**, 95–111.
93. Wardener, H. de (1973), *The Kidney: An Outline of Normal and Abnormal Structure and Function*. 4th edition. Edinburgh: Churchill Livingstone.
94. Willoughby, J. O. and Leach, B. G. (1974), 'Relation of neurological findings after cardiac arrest to outcome,' *Brit. Med. J.*, **3**, 437–439.
95. Winn, H. R., Richardson, A. E., O'Brien, W. and Jane, J. A. (1978), 'The long term prognosis in untreated cerebral aneurysms. 2. Late morbidity and mortality,' *Ann. Neurol.*, **4**, 418–426.

26. PERIPHERAL NEUROPATHY

JOHN P. PATTEN

Introduction

The inherited neuropathies
   The peroneal muscular atrophy group
   Hereditary sensory neuropathies
   Amyloidosis
   Lipoprotein abnormalities
   The lipoidoses

Metabolically triggered neuropathy
   Acute intermittent porphyria

Infective neuropathy
   Leprosy
   Diphtheria
   Guillain-Barré syndrome
   Botulinum toxin poisoning

Toxic neuropathies

Metabolic neuropathies
   Alcoholic neuropathy
   Diabetic neuropathy
   Ischaemic neuropathy
   Vitamin B_{12} deficiency
   Uraemia

Malignant disease
   Myeloma

Laboratory investigation of neuropathy
   Haematological studies
   Biochemical investigations
   Conclusion

INTRODUCTION

Peripheral neuropathy is a common condition with an estimated incidence of 50 cases per million per year. There are nearly 200 recognised causes of peripheral neuropathy, but even following intensive investigation, 40% of cases remain undiagnosed as to the cause. We are here concerned purely with damage to peripheral nerves. The expression 'neuropathy' is sometimes confusingly applied to nerve fibre damage anywhere in the CNS. There are a small number of common and fairly readily identifiable causes (leprosy, diabetes, and alcoholism), but once these conditions have been excluded more difficult and expensive investigations become necessary with a very poor return in establishing the cause and an even greater disappointment in the therapeutic prospects. Even when chemical or pharmacological causes can be identified, and such causes are assuming increasing significance, the prognosis for recovery is often poor.

The pathogenesis of the peripheral nerve damage is not understood. Peripheral nerves consist of motor fibres; a ventral horn cell and its axon, and sensory fibres; the dorsal root ganglion cell and its bipolar fibres from the periphery; and the fibre into the spinal cord. The integrity of the axon is maintained by the cell body with nutrients flowing to the periphery in the axoplasm. Any interference with this process causes a peripheral axonal 'dying back' neuropathy. The motor and sensory cells would seem to have different sensitivity to damage as relatively pure examples of motor or sensory damage occur. The axons are wrapped in a multi-layer of myelin produced by the Schwann cells. Damage to these cells or the myelin itself results in a demyelinating type of neuropathy characterised by extreme slowing of nerve conduction.

The whole subject is extremely difficult to classify in a way that allows any sense to be made of the bewildering complexity of even the simplest clinical features of neuropathy. Classifications can be based on the course (acute, sub-acute, chronic, progressive and relapsing); the distribution (motor, sensory, mixed, autonomic); the causes (infective, chemical, drugs, metabolic, inherited, carcinoma associated); and on the pathological process seen at histology (axonal degeneration, segmental demylination). Unfortunately, no logic emerges even by using multiple classifications. For example, arsenic causes an axonal neuropathy and lead causes segmental demyelination, so that one cannot even categorise similar substances or clinical similarities in the type of neuropathy produced in any sensible way.

Because of this, purely to cover the field comprehensively the writer has chosen to divide the causes into infective, toxic, metabolic etc., although such a classification is only a convenience to ensure that no important causes are missed. Only diffuse peripheral nerve damage will be considered, brachial plexus and single peripheral nerve trunk lesions are excluded and, in view of the nature of this text, emphasis is given to the use of the laboratory in differential diagnosis rather than stressing clinical features, although special peculiarities that may provide important diagnostic clues are mentioned.

TABLE 1
CLASSIFICATION OF PERIPHERAL NEUROPATHY

1. Inherited
2. Metabolically triggered
3. Infective
4. Toxic
5. Metabolic
6. Malignancy

The Inherited Neuropathies

The Peroneal Muscular Atrophy Group

It is generally recognised that the group of inherited neuropathies, including peroneal muscular atrophy

(Charcot-Marie-Tooth Disease) which is a restricted neuropathy, hypertrophic polyneuritis (Dejerine-Sottas disease), a generalised neuropathy, and Friedreich's Ataxia (a generalised neuropathy plus spinal cord and cerebellar damage), are related disorders. Several of these conditions may occur in the same sibship and transitional forms are common. Peronal muscular atrophy maximally affects the leg muscles below mid-thigh and sometimes the distal forearms and hands. Dejerine-Sottas disease affects all nerve roots and peripheral nerves which may become so swollen as to be easily palpable under the skin or even produce spinal cord compression. In Friedreich's ataxia the neuropathy is an early but relatively undramatic feature, the condition being dominated by the ataxic component. There are no specific laboratory tests for these conditions, although nerve conduction studies in this group often reveal extremely slow conduction. The concentration of protein in cerebro-spinal fluid is often elevated and in Dejerine-Sottas disease, in particular, it may reach extremely high levels and interfere with the resorption of CSF. Although a great many metabolic studies have been undertaken in these patients as yet there is no defined metabolic fault or any biochemical test that will distinguish these conditions.

Hereditary Sensory Neuropathies

A heterogenous group of neuropathies, some of which may be present at birth, include hereditary sensory radicular neuropathy (mainly affecting the legs with trophic ulcers and Charcot joints), congenital generalised sensory neuropathy, congenital insensitivity to pain, and Riley-Day Syndrome (neuropathy plus autonomic damage) also known as familial dysautonomia. There are no specific biochemical tests for these conditions and in many instances the patient first comes to medical attention because of repeated unrecognised trauma, trophic ulceration or joint damage as the patients are usually unaware of their unusual disability.

Amyloidosis

There are three varieties of amyloidosis that may affect the peripheral nerves. The first is a fatal neuropathy starting in the third decade and leading to death over a ten year period. To date this has only been recognised in patients of Portuguese extraction. The second milder version, often starts in the arms, mimicking carpal tunnel syndrome, and is associated with vitreous opacities. This type has been recognised in families of Swiss origin. Finally sporadic examples of amyloidosis affecting the peripheral nerves in the absence of underlying disease are recognised with the increasing use of peripheral nerve biopsy, in most instances not associated with systemic amyloidosis. In the familial varieties the amyloid has been shown to be of immunoglobulin origin, but again there are no specific biochemical tests for this disorder.

Lipoprotein Abnormalities

Tangier Disease (familial alpha-lipoprotein deficiency). This is a recently identified form of neuropathy which is mild and tends to show a fluctuating course. It appears to be inherited as an autosomal recessive and the nature of the nerve damage is not understood. The patients so far reported have normal nerve conduction studies and normal peripheral nerve biopsies. The disease is characterised by an extremely low blood cholesterol level with cholesterol esters being stored in the tissues, most strikingly in the tonsils (in which tissue the cholesterol is visible on clinical examination), although the liver, spleen and lymph nodes may be similarly infiltrated. Marrow biopsy may reveal abnormal foam cells, and lipid deposits have been demonstrated in the macrophages on rectal biopsy.

Bassen-Kornzweig Disease (hypo-beta lipoproteinaemia). In this condition the underlying problem appears to be malabsorption of triglycerides from the gut mucosa. Reported cases have shown an association with retinitis pigmentosa, acanthocytosis of red blood cells and degeneration of the dorsal columns, spinocerebellar tracts and peripheral nerves (the distribution of pathology showing similarities to Friedreich's ataxia). The plasma beta-lipoproteins and cholesterol are present at very low concentration and wet preparations of red blood cells show the typical crenated appearance with evidence of haemolysis and elevated reticulocyte count. It has been suggested that the parental use of fat-soluble vitamins in this condition may delay the onset of neurological damage.

The Lipoidoses

These disorders are also known as the leukodystrophies as they are characterised by disturbances of myelin formation in the CNS and peripheral nerves and this is accompanied in many instances by the deposition of lipids in the CNS, hence the other name, the 'lipoidoses'.

Metachromatic Leukodystrophy (sulphatide lipoidoses). This condition, which is the commonest of the leukodystrophies has been shown to be due to a deficiency of the enzyme arylsulphatase A which can usually be demonstrated in the urine. The deposition of the metachromatic granules of abnormal lipid can be shown in the centrifuged deposits of urine and on sural nerve biopsy, enabling the diagnosis to be made in life. The deposition of similar substances in the brain leads to progressive dementia and locomotor dysfunction that dominate the clinical picture of this lethal condition. The other cerebral lipoidoses are dominated by central nervous system damage and peripheral neuropathy is not a diagnostic feature of these conditions.

Refsum's Syndrome (hereditary ataxic neuropathy). In this condition the metabolic defect lies in metabolism of branched chain fatty acids leading to the accumulation of 3, 7, 11, 15-tetramethyl-hexadecanoic acid (phytanic acid) in the tissues. This deposition is mainly in the peripheral nerves and occasionally sufficient to produce palpable swelling. Associated features include retinitis pigmentosa, deafness, and mild ataxia. The course may be an intermittent one and it has been suggested that a diet low in phytol and chlorophyll may prevent exacerbations and deterioration and plasma exchange has been used in some cases. Measurement of the serum phytanic acid level in untreated cases is diagnostic.

Metabolically Triggered Neuropathy

Acute Intermittent Porphyria

Only one of the porphyrias causes neurological problems and that is the acute intermittent type. The metabolic defect in this disorder probably lies in the liver and the onset is often triggered by drugs and chemicals. The attack is characterised by excessive production of delta-amino-laevulinic acid (ALA) and porphobilinogen (the pigment found in the urine). It is thought that ALA mimics gamma amino butyric acid, the central inhibitory transmitter, and that this may be responsible for some of the cerebral features of an attack.

Abdominal pain, vomiting, weight loss and hypertension occur at the onset, quickly followed by mental changes which may mimic almost any known mental disorder, and later epileptic seizures followed by an acute ascending neuropathy, mimicking Guillain-Barré syndrome. It is thought that the neuropathy may be due to inhibition of acetylcholine release as motor features predominate and sensory changes are relatively mild. Cranial nerve involvement is late and a potentially lethal development. Agents that may precipitate an attack of acute intermittent porphyria are listed in Table 2.

TABLE 2
DRUGS PRECIPITATING ATTACKS OF PORPHYRIA

Barbiturates
Sulphonamides
Alphamethyldopa
Levodopa
Grizeofulvin
Dichloralphenazone
Glutethimide
Phenylbutazone
Oestrogens
Chloroquin
Alcohol

The diagnosis can be established by the laboratory with the demonstration of increased levels of porphobilinogen and delta-amino-laevulinic acid in the urine. (See also Chapter 10.)

Infective Neuropathies

Leprosy

On a world wide basis leprosy is probably the commonest cause of neuropathy. The disease is characterised by infiltration of the peripheral nerves producing nodular thickening and local demyelination. This renders nerve trunks extremely vulnerable to pressure; sudden peripheral nerve palsies, a condition known as mononeutritis multiplex, often characterise the disease. In long standing cases diffuse damage to all the peripheral nerves produce widespread evidence of neuropathy. There are no specific biochemical or haematological tests for this condition, although skin biopsy and nerve biopsy studies are diagnostic.

Diphtheria

This is fortunately now a rare cause of peripheral neuropathy, occurring between three weeks and two months after the initial infection. It is thought to be due to diphtheria toxin. The picture is potentially lethal and resembles Guillain-Barré syndrome.

Guillain-Barré Syndrome

This is probably one of the commonest neuropathies occurring as a non-specific neural response to a host of infective triggers including identifiable conditions such as glandular fever. The diagnosis can be quite difficult in a low grade case, but the classical example with the acute onset of rapidly ascending generalised paralysis is readily identifiable. From the laboratory point of view, investigations to identify a specific virus, confirm the diagnosis of glandular fever and, in particular, to exclude hyper- or hypo-kalaemia, which are both capable of mimicking the clinical picture, are extremely important. Diagnostically, an elevated CSF protein concentration in the absence of a cellular response may help to confirm the diagnosis, but unfortunately, the characteristic protein rise may be delayed in some instances and a brisk cellular response may occur in others, so that sometimes the laboratory provides confusing rather than helpful support in the diagnosis of this condition. It is important, however, that these variations from the typical CSF response are well recognised by both the clinicians and the laboratory staff.

Botulinum Toxin Poisoning

This is still a lethal condition, but extremely rare, the recent fatalities in this country being the first since the famous duck paste cases in the early 1920s. Only awareness of this possibility will enable the myasthenia-gravis-like onset (diplopia, dysarthria and dysphagia) to be identified in time to give antitoxin and institute full intensive care and, even more important, quickly to trace others likely to have been exposed to the same food. The laboratory's role lies in identification and typing of the toxin and its course.

Toxic Neuropathies

Although clinical variation in the picture can be used as a diagnostic clue to the type of neuropathy, peripheral nerve damage due to any cause is ultimately going to produce a very similar pattern of disability, but in the field of toxic neuropathies there are some interesting variations. Arsenic typically causes a foot drop (pseudotabes arsenica), lead causes a wrist drop and thallium (used as a cockroach poison) causes very severe pain in the extremities and loss of hair. In all other instances, however, no striking features enable the toxin to be identified. The chemical substances and drugs which have been shown to cause peripheral nerve damage are listed in Table 3 and the toxic levels of various chemical substances are discussed more fully in Chapters 21, 22 and 23.

TABLE 3
CHEMICALS THAT MAY CAUSE PERIPHERAL NEUROPATHY

Acrylamide
Arsenic
Carbon disulphide
Lead
Organophosphates
Thallium
Gold
n-Hexane
Methylbutyl-ketone

DRUGS THAT MAY CAUSE PERIPHERAL NEUROPATHY

Isoniazid	Metronidazole
Nitrofurantoin	Ethambutol
Thalidomide	Hydrallazine
Vincristine	Methaqualone
Clioquinol	Imiprimine
Chloroquine	Amitryptiline
Perhexiline	Indomethacin
Disopyramide	Phenytoin
Clofibrate	Dapsone
Disulfiram	Chlorambucil
Cyanamide	Allopurinol

In most instances a careful history to exclude any possible chemical exposure in the office, factory or home, or possibly even air-borne pollution from nearby factories, and a complete list of any drugs taken by the patient within the past two years, should enable identification of a possible causal agent to be made.

Metabolic Neuropathies

Alcoholic Neuropathy

Alcohol is a major cause of neuropathy and although the peripheral nerve damage responds fairly well to Vitamin B_1 administration, it is far from certain that a simple effect on B_1 metabolism is responsible for the neuropathy. The neuropathy is often characterised by painful peripheral paraesthesiae, often so severe that little motor deficit has a chance to occur. Conversely, some patients may develop an acute Guillain-Barré-like picture with less severe pain. Heavy alcohol intake, combined with a poor diet, should be readily identified and elevation of liver enzyme activities and macrocytosis of red cells are useful screening tests. The pyruvate tolerance test used to be used to identify faults in B_1 metabolism, but it is no longer regarded as a valid test in this situation.

Diabetic Neuropathy

Another very common cause of neuropathy may be found in the maturity-onset type of diabetes where the underlying disease may not be apparent and laboratory assistance in both fasting blood sugar levels and formal glucose tolerance testing is entirely justified in excluding this as a cause of neuropathy in any elderly patient. The CSF protein concentration is usually elevated, but the diabetes itself may be very mild and its actual role in causing neuropathy is uncertain.

Ischaemic Neuropathy

Many patients over 70 develop a low grade neuropathy in the legs. There is considerable suspicion that this is due to poor blood supply and once alcohol, diabetes and B_{12} deficiency have been excluded the diagnosis can be suspected, although underlying malignant disease must remain a possibility until excluded by the passage of time. Polyarteritis nodosa represents a specific form of ischaemic neuropathy with either vascular lesions of whole nerve trunks or diffuse damage to the peripheral nerves developing insidiously. This can be readily identified by nerve biopsy. Eosinophilia and a very high ESR are important clues to the possibility of polyarteritis nodosa.

Vitamin B_{12} Deficiency

This is probably the most frequently suspected cause of neuropathy, although the diagnosis is rarely substantiated. Although the peripheral nerves are affected in B_{12} deficiency the damage to the central pathways dominates the picture. The peripheral paraesthesiae are usually very mild and the physical findings should be diagnostic. In spite of this, vast numbers of unnecessary B_{12} determinations are done each year.

Uraemia

Prior to the advent of dialysis, neuropathy in patients with renal disease was often ascribed (in some cases possibly correctly) to the use of nitrofurantoin. It is now recognised, however, that the uraemic state alone is capable of causing peripheral nerve damage and changes in nerve function can be used as a sensitive index of the efficiency of dialysis in combination with the usual biochemical parameters.

Malignant Disease

For over 30 years the occasional development of a sensori-motor neuropathy as a correlate or even precursor of underlying malignant disease has been recognised. In its truly remote form (i.e. no direct infiltrative neural component) it is very difficult to establish a causal relationship and many studies to define the syndrome and identify metabolic or infective processes secondary to malignancy as a possible cause have failed. The response of the neuropathy to effective treatment of the underlying malignancy is also so variable that no definite support as to the aetiology can be derived. Routine biochemical and haemoatological tests that screen for evidence of underlying dysfunction will obviously play a role in such cases, but are frequently normal. In reported cases there is a predominance of lung cancer and ovarian carcinomas, neither of which will produce typical biochemical or haematological abnormalities.

Myeloma

Myelomatosis may cause a painful neuropathy and carpal tunnel syndrome. Either the disease itself or the secondary amyloidoses derived from the excess immunoglobulins may be responsible in some instances. The sedimentation rate, protein electrophoresis and biochemical evidence of bone, liver and renal disease may all be helpful in establishing the diagnosis. The demonstration of Bence-Jones protein in the urine represents a simple and effective screening test as a preliminary to more extensive biochemical studies.

Laboratory Investigation of Neuropathy

From the point of view of laboratory investigation of peripheral neuropathy the following may be noted.

Haematological Studies

Demonstration of abnormal red cells (acanthocytes, macrocytosis), abnormal white cells (malignancy, lymphoma), high sedimentation rate (myelomatosis, collagen vascular disease) and estimation of the serum and cell folate and B_{12} levels are important in the screening of patients with peripheral neuropathy.

Biochemical Investigations

Many examinations of the urine are of value in the investigation of neuropathy. Evidence of renal failure, porphyria, diabetes, myelomatosis, and a defect in arylsulphatase A in metaochromatic leukodystrophy, may all be found by appropriate urinary examination. Studies of both the chemical and cellular constituents is necessary. Biochemical studies of carbohydrate metabolism and renal and hepatic function are important general screening tests and, in specific instances, estimation of the plasma cholesterol concentration and subsequently lipoprotein analysis are central to the diagnosis of the rare inherited neuropathies associated with disorders of lipid metabolism. Specific tests to detect excess of heavy metals will be requested in appropriate clinical situations. Unfortunately, the chemistry of the cerebral spinal fluid, the body fluid most intimate to the affected tissue, is relatively unhelpful. Although the cerebro-spinal fluid protein concentration is often elevated, it is seemingly a non-specific response and, as yet the use of CSF immunoglobulin studies have not increased the value of CSF examination in the investigation of peripheral nerve disease.

Conclusion

Because of the nebulous and protean manifestations of neuropathy, once a few readily identifiable conditions have been excluded a series of time-consuming, expensive and often unrewarding investigations will be necessary in an attempt to unearth the cause. The clinician and laboratory staff and, most of all, the patient, share the disappointment that so often accompanies even the most diligent search.

27. THE MYOPATHIES

B. P. HUGHES

Introduction
   Classification of the myopathies
   Methods of investigation

Hereditary myopathies
   Non-specific myopathies
   Myopathies with structural muscle abnormalities
   The familial periodic paralyses (FPP)
   Myopathies with specific metabolic defects

Acquired myopathies
   Polymyositis
   Drug myopathies
   Myopathy accompanying general metabolic disturbances
   Nutritional myopathy
   Myasthenia gravis and the myasthenic syndrome

INTRODUCTION

In this chapter we shall be concerned with the role of clinical biochemistry in the diagnosis and management of muscle disorders. It must be admitted from the outset, however, that in the more common hereditary myopathies such as the muscular dystrophies, this role is limited and that biochemical studies will form only one part of the overall assessment of the patient which will include the clinical picture and information derived from other laboratory studies using histological and electrophysiological techniques. Only in certain rare disorders in which a specific biochemical defect has been delineated are biochemical studies of paramount importance.

Classification of the Myopathies

Various general classifications of the myopathies are possible, but for the present purpose we will categorize them as *hereditary* or *acquired*. An alternative would be to divide them into *primary* myopathies, that is disorders in which the cause of the muscle degeneration or dysfunction is within the muscle itself and *secondary*, in which muscle function is disturbed by some external influence, e.g. a toxic substance, endocrine disorder or disease of the motor nerve.

The first alternative was chosen with the intention of avoiding involvement in the controversy as to which myopathies can be classified as primary or secondary. This controversy, although of importance to the basic understanding of muscle disorders, has as yet little impact on the more practical problems with which we are concerned. It will, however, be pertinent to mention briefly the nature of the controversy which particularly concerns the most common of the hereditary myopathies, the muscular dystrophies. These disorders have usually been considered to be disorders of the muscle itself, that is primary myopathies. However, with increasing knowledge of the way in which its specific innervation influences the properties, both biochemical and electrophysiological, of a particular muscle, it was suggested that they might be due to a subtle disturbance of the motor nerve. Although the neurogenic hypothesis has perhaps lost ground recently, other suggestions have invoked vascular or humoral factors to explain muscle degeneration in the muscular dystrophies. Until the aetiology of these disorders has been discovered it is difficult to exclude such factors, and in some myopathies they may indeed play their part in the pathogenesis.

Since the division of the myopathies into the categories of Hereditary and Acquired is too wide to permit consideration of specific types, subdivision is necessary. Although this process will be to some extent arbitrary, the classification shown in Table 1 seems suitable for the present purpose.

Methods of Investigation

Before discussing the various types of myopathy, it is appropriate to discuss what techniques have proved useful for their investigation. Although we are particularly concerned with the role of clinical biochemistry, because this role is limited and must be considered as a part only of the complete assessment of the patient, it is also necessary to mention briefly other laboratory techniques involving histological and electrophysiological procedures.

TABLE 1
A CLASSIFICATION OF THE MYOPATHIES

A —HEREDITARY MYOPATHIES
A1—Non-specific myopathies. These comprise disorders in which there are no specific biochemical, histological or electrophysiological changes in the muscle. The commonest examples are the muscular dystrophies.
A2—Myopathies in which there may be specific and even bizarre structural abnormalities of the muscle fibres or intra-cellular organelles, but in which the relation between these abnormalities and the clinical signs is obscure, and the aetiology of the abnormalities unknown.
A3—Disorders in which there are fairly specific biochemical abnormalities, which however may not be the proximate cause of the muscle signs.
A4—Myopathies with specific biochemical abnormalities probably causally related to the pathogenesis of the disease.
B —ACQUIRED MYOPATHIES
B1—Polymyositis and dermatomyositis.
B2—Drug myopathies
B3—Myopathies accompanying metabolic disturbances such as thyroid dysfunction.
B4—Nutritional myopathy
B5—Myasthenia gravis and the myasthenic syndrome.

Biochemical Methods

Biochemical techniques have been used to study both fluids and muscle biopsies.

Body Fluids—Urine. One of the first biochemical abnormalities reported to be present in muscle disease was an increased excretion of creatine. However, this observation has not proved of any great value, since it appears to be a largely non-specific consequence of diminished creatine uptake by a reduced muscle mass in the presence of a normal rate of creatine synthesis. The resultant elevated creatine excretion is found both in the 'primary' myopathies and in muscle wasting due to denervation. More recently, it was claimed that there was an increased excretion of pentose phosphates, notably ribose phosphate, in the muscular dystrophies. Subsequent studies failed to confirm the original reports. There are also variable degrees of aminoaciduria in the muscular dystrophies but there is no specific excretion pattern associated with any particular disorder. By contrast less hydroxyproline may be excreted, possibly due to increased utilization accompanying connective tissue proliferation.

Body Fluids—Blood. Of much greater practical value has been the assay of constituents of serum or plasma, notably certain enzymes, of which creatine kinase (EC 2.7.3.2.) has proved particularly useful. In a variety of myopathies raised activities of creatine kinase and, to a lesser extent, of certain other enzymes of the muscle cytoplasm are observed in the plasma or serum. Although these changes are not specific, and indeed occur both in Hereditary (Class A) and in Acquired

(Class B) myopathies, they can nevertheless be of considerable assistance in diagnosis if used in conjunction with results of other studies on the patient.

Initially it was considered that raised enzyme activities were not found in muscle degeneration secondary to denervation. However, although later experience indicated that this view was mistaken, it is probably true that elevation of serum enzyme levels are found less frequently and are usually quantitatively less striking in denervation than in the primary myopathies. Their estimation in suspected neurogenic disorders may thus have value when considered as part of the overall assessment of the patient.

Methodology of Enzyme Assay. There are two rather general approaches used for the assay of enzymes of clinical significance. Firstly, activity can be measured by directly following the formation of the product of the reaction when the enzyme is incubated under appropriate conditions in the presence of its substrate. Alternatively, the product of the enzyme action can be coupled to additional enzyme catalysed reactions to produce a more conveniently measured final product. Both approaches can be illustrated by reference to the determination of creatine kinase. The reaction catalysed by this enzyme and written in the most suitable direction for assay is:

$$\text{Creatine phosphate} + \text{ADP} \rightarrow \text{Creatine} + \text{ATP}$$

Creatine can be measured by a suitable colorimetric or fluorimetric method or the system can be coupled to excess of other enzymes, such as hexokinase and glucose-6-phosphate dehydrogenase (G6PDH), which utilize the ATP formed as follows:

$$\text{Glucose} + \text{ATP} \xrightarrow{\textit{Hexokinase}} \text{Glucose-6-phosphate} + \text{ADP}$$

$$\text{Glucose-6-phosphate} + \text{NADP}^+ \xrightarrow{\textit{G6PDH}} \text{6-phosphogluconate} + \text{NADPH} + \text{H}^+$$

The creatine kinase activity is thus measured by the accumulation of NADPH, which can be conveniently followed by recording the optical density of the reaction mixture at 340 nm. A number of similar systems have been devised to assay several enzymes or their substrates by the accumulation of NADPH, or where appropriate NADH, which are both rapid and convenient. A further example is the measurement of blood lactate, a procedure to which we shall refer at a later stage.

Non-Biochemical Methods

Other laboratory investigations are also commonly employed to study muscle disorders; these are morphological and electrophysiological, and consist of light microscopy and electron microscopy techniques and the methods of electromyography.

Light Microscopy. Microscopic examination of muscle biopsies is a standard procedure which, particularly when combined with histochemical methods, can yield valuable information. However, as in the case of biochemical studies many of the abnormalities revealed are non-specific and must be considered together with the sum total of information, from all sources, concerning the patient. In a few rare instances, however, histochemical procedures may yield results of pathognomonic significance.

Electron Microscopy. Electron microscopic studies have shown how the changes seen in the light microscope are related to alterations in cell structures, such as mitochondria, myofilaments and the sarcoplasmic reticulum. To some extent the results of these studies have been disappointing, in as much as the abnormalities seen have proved mainly to be non-specific. However, detection of changes at an early stage when they are not yet advanced enough to alter the light microscopic appearance of the tissue is certainly possible. Early recognition of changes may eventually give valuable insights into the pathogenesis of the muscle disorders even if this potential is still largely unrealized. Moreover, in the case of the A2 myopathies, recognition of specific ultrastructural abnormalities in the muscle fibre may be of considerable diagnostic significance. Again, in Duchenne muscular dystrophy electron microscopy has shown the early appearance of sarcolemmal abnormalities and this has been interpreted as circumstantial evidence in favour of the involvement of a membrane defect in the aetiology of this disorder.

Electromyography. The uses of electrophysiological procedures have largely consisted in helping the differential diagnosis of muscle weakness due to denervation from that due to what have been regarded as primary myopathic conditions. Lest this be thought unduly dismissive, it must be emphasized that electrophysiology has contributed information of fundamental significance to understanding the organization of the muscle in relation to its innervation and to the way in which the nerve initiates and modifies the contractile response of the muscle. Moreover, it is important to realize that even when their clinical applications are relatively minor, the contribution of these various techniques to fundamental knowledge has been very great. Thus detailed knowledge of the structure of the contractile system of muscle rests on the results of electron microscopic studies, while understanding of muscle function at the molecular level derives from basic biochemical investigations. This basic knowledge while perhaps not a sufficient basis for solving clinical problems, is likely to be a necessary condition for doing so.

We shall now consider individual topics in relation to particular myopathic disorders.

A. HEREDITARY MYOPATHIES

In this section we will consider the more common hereditary myopathies without claiming to be comprehensive.

Type A1 Non-specific Myopathies

The muscular dystrophies comprise the commonest group of hereditary muscle disorders. Of these, the most frequently occurring form, and clinically the most severe, is the sex-linked recessive Duchenne muscular dystrophy. We shall use the term 'Duchenne' without qualification, only in connection with the severe muscular dystrophy having a sex-linked recessive mode of inheritance. The term has sometimes been used to describe a severe muscular dystrophy which also affects girls and which is probably inherited as an autosomal recessive trait; however, such use tends to be somewhat confusing.

Duchenne Dystrophy

Apart from being the commonest and most severe form of muscular dystrophy, this disorder is also the one in which biochemical studies of serum enzyme levels have proved to be the most valuable.

Typical features of the disease may be briefly summarized: early onset (3–5 years), rapid progression to severe disablement (unable to walk by age 10) and death in late teens due to respiratory infection or heart failure resulting from the concomitant cardiomyopathy.

Although the activities of a number of enzymes in blood are usually elevated in Duchenne dystrophy, creatine phosphokinase (CPK) has been the most widely studied. Early in the disease serum CPK activity is grossly elevated, but diminishes as time progresses, although even at a late stage seldom returns to normal values. The general relation between serum CPK and the age of the patient is illustrated in Fig. 1.

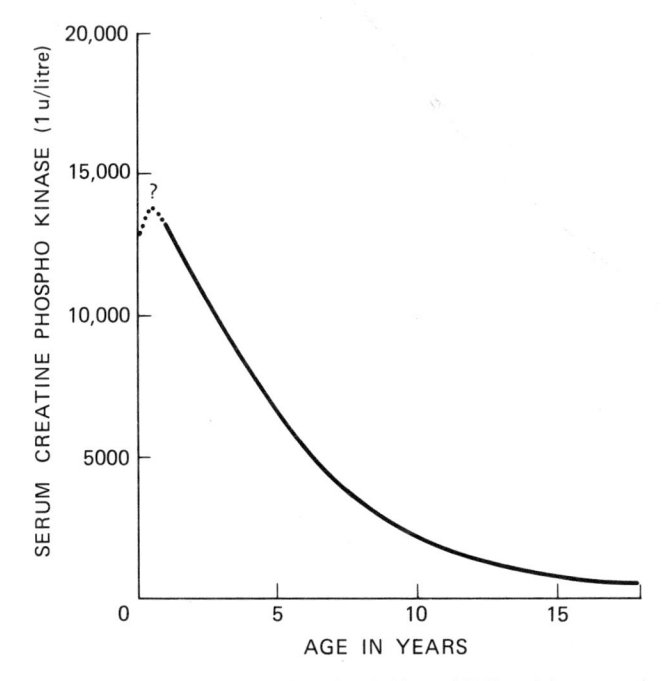

FIG. 1. Relation of serum creatine phosphokinase (CPK) activity to age in patients with Duchenne muscular dystrophy.

From consideration of this relationship, a number of salient points emerge. Thus, although enzyme activity may peak sometime after birth, extremely high values are found in the immediate postnatal period before clinical signs are apparent. This is useful in several ways.

Early Diagnosis. If there is already an affected male in the family and another boy is subsequently born, it will be important to know whether or not he is affected. In the absence of clinical signs, other methods of diagnosis are needed. While it is true that histological changes in the muscle can occur before the onset of clinical weakness, it is clearly an advantage to be able to make a diagnosis from examination of a small blood sample rather than to have to perform a muscle biopsy. The finding of a grossly elevated serum CPK level together with a positive family history will strongly suggest a diagnosis of Duchenne dystrophy, whereas a normal result will provide reasonable grounds for reassurance that the disease is not present.

Antenatal Diagnosis. The fact that the serum CPK level is grossly elevated at birth suggests that it might also be raised *in utero*. Recent technical developments have rendered it possible, albeit not without some risk to the fetus, to sample the fetal blood before birth. By this means antenatal diagnosis might be possible. At present there is some controversy as to the value of this approach, so that further evaluation is required before extensive use of this technically difficult procedure can be justified.

Differential Diagnosis. Although clinical signs are usually absent at birth, there may sometimes be delay in attaining the motor functions, so that if there is a possibility of Duchenne dystrophy, it may be of value to distinguish between this disorder and other conditions which could cause moderate muscle weakness in infancy, such as the more benign denervating disorders, congenital muscular dystrophy and myopathies classified as Type A2 (Table 1). Now although in some or all of these conditions serum CPK elevation is often found, it is seldom as gross as that present in Duchenne dystrophy. Consequently, a grossly elevated serum CPK level will tend to exclude other disorders causing muscle weakness in infancy.

Carrier Detection. If we now consider not the patient, but female members of the family, we find what is arguably the most useful application of serum CPK assay; that of the detection of carriers.

Because it is a sex-linked recessive condition, Duchenne dystrophy is transmitted by the female members of the family, who are usually clinically normal or at most only slightly weak. Nevertheless, it has been found that in a high percentage of cases, probably around 70%, the carriers have somewhat elevated serum CPK activities. Carrier detection may thus be possible with the consequent implications for genetic counselling. Since the degree of serum CPK elevation is often small, care is necessary to avoid both false positives and false negatives, and some factors which relate to this problem must be enumerated.

Effect of Exercise. Vigorous or unusual exercise, especially in a normally sedentary individual can cause a transient elevation in the activities of a number of serum enzymes, including CPK. Consequently, if blood is sampled for enzyme assay following such exercise, there is a risk of a false diagnosis of the disease or of carrier status. In the case of CPK, the maximum of exercise-induced serum activity may occur as much as 18 hours after exercise. It is therefore usual to advise against any undue physical exertion in the period of 48 hours prior to the time at which the blood is sampled. As a further precaution against false positives or negatives, advice on carrier status is usually based on three separate enzyme assays on blood samples whose collection has been spaced at least a week apart.

Distribution of Serum CPK Activities in Normal and Carrier Populations. To attain the best assessment of whether an individual should be assigned to the status of carrier or non-carrier, it is necessary for each laboratory to have sufficient data on normal women and or carriers to permit the distribution of enzyme values in the normal and carrier groups to be determined. In addition, not only will the results of the biochemical or other laboratory investigations be used, but also the probability of an individual being a carrier will be assessed from the family history, and both sources of information will be combined to give the best assessment of whether an individual should be classed as a carrier or non-carrier.

Effect of Pregnancy on Serum CPK. An important factor which may invalidate carrier detection by serum CPK assay is the effect of pregnancy. It has been shown that serum CPK activity is decreased in the early part of pregnancy (up to about 20 weeks) in both carriers and normal women. This observation is relevant to the detection of carriers since comparison of the serum CPK activity of a pregnant carrier with the value for a group of normal non-pregnant women may obviously be misleading. It has been suggested that the difficulty might be avoided by comparing CPK levels in pregnant carriers with those of normal women at the same stage of pregnancy, but this suggestion does not appear to have been followed up and reliable investigation of carrier status by serum CPK activity in pregnant subjects is not possible at present.

Age of Carrier. Since it has been reported that serum CPK activities of carriers tend to fall with increasing age, it has been suggested that the goal of carrier detection is best served by testing young carriers. While this suggestion appears reasonable, it is questionable if there are adequate data to determine the optimum age of testing, and further research on the distribution of serum CPK levels in carriers and normal individuals of different age groups appears desirable.

Other Techniques for Detecting Duchenne Carriers. As alternatives to serum CPK assay a number of other techniques have been proposed, these include quantitative electromyography, examination of muscle biopsies, muscle ribosomal protein synthesis and assay of other serum enzymes such as pyruvate kinase. Although the results of such investigations may indeed reveal abnormalities in carriers, and hence such techniques ought not to be ignored, there does not yet seem to be any convincing evidence that any are superior to serum CPK assay and most are less convenient practically.

Other Sex-linked Muscular Dystrophies

A more benign muscular dystrophy with the same mode of inheritance as Duchenne dystrophy has also been described. The so-called *Becker-type muscular dystrophy* has a much slower progression and consequently the disease may occasionally be transmitted by the male as well as the female members of the family. Serum CPK activity is usually markedly elevated in affected males, although not to the same extent as in Duchenne dystrophy, and is also somewhat raised in a proportion of the female carriers, although because the disorder is less common it is questionable if a reliable estimate can be given of the proportion of carriers who will exhibit serum enzyme abnormalities. Finally, it is not entirely clear whether Becker dystrophy is a homogeneous condition since on clinical grounds further subdivision has been suggested. There is also the question of whether Becker dystrophy arises from a different X-chromosome defect from that in Duchenne dystrophy, or whether a closely linked modifying gene is responsible for the more benign course of the disease. An answer to these and similar questions must probably await discovery of the nature of the primary gene defect in these disorders and its precise location on the X-chromosome.

Limb-Girdle Muscular Dystrophy

Limb-girdle muscular dystrophy, so called because weakness first appears in the muscles of either the upper or the lower limb girdle with subsequent spread to those of the initially unaffected group, is a less well-defined clinical entity than Duchenne dystrophy. However, although it shows a wide variation in severity and age of onset, it is generally a more benign disorder.

Diagnosis is complicated by the fact that certain of the more slowly progressive forms of spinal muscular atrophy may resemble it clinically and may also show histological changes in the muscle usually considered to be of myopathic rather than neurogenic origin. Again the adult form of acid maltase deficiency (see Type A4 myopathies) can present with similar symptoms.

Clinical biochemical investigations are of limited value. Serum CPK activity is usually elevated, but to a lesser extent than in Duchenne dystrophy and moreover does not permit a clear distinction to be made between limb-girdle dystrophy and those forms of spinal muscular atrophy which mimic it clinically. However, if acid maltase deficiency is suspected then such a diagnosis can be confirmed or excluded by measurement of muscle acid maltase activity and muscle glycogen content.

The mode of inheritance, where there is a definitive family history, is usually autosomally recessive; however, a number of cases appear to be sporadic. Also it is not altogether clear whether the severe dystrophy clinically like Duchenne which can affect girls is a particularly severe expression of limb-girdle dystrophy or is a separate entity.

Facio-scapulo-humeral Dystrophy

Facio-scapulo-humeral dystrophy is the most benign of the common muscular dystrophies, and is usually inherited as an autosomal dominant trait. As the name implies, muscles of the face and upper limb girdle are first affected, but weakness may sometimes spread to other muscle groups. There is great variability in severity of symptoms, even within the same family, and some patients may be unaware of any disability.

Serum CPK activity is frequently moderately raised, even when weakness is minimal, so its assay may assist early diagnosis. Although morphological changes in the muscle are not specific they may help to distinguish between facio-scapulo-humeral dystrophy and other disorders giving rise to a similar pattern of weakness. These include a rare form of spinal muscular atrophy and a myopathy or myopathies in which there are structural and biochemical abnormalities in the mitochondria.

Congenital Muscular Dystrophy

This term is applied to a rather ill-defined group of disorders in which muscle weakness is present from early infancy and in which the muscle shows changes considered to indicate a myopathy rather than denervation. The subsequent progression of these disorders may vary from fatal outcome to improvement leading eventually to near normal muscle strength.

Biochemical investigations are seldom useful, but may sometimes exclude a diagnosis of Duchenne dystrophy, although a serum CPK activity as high as in Duchenne dystrophy has been noted in a case of *benign congenital muscular dystrophy*. Investigations of muscle morphology by histochemical or electron microscopic techniques have been more rewarding.

Myotonic Dystrophy

Myotonic dystrophy differs from the myopathies discussed so far in that, besides muscle weakness, myotonia is a prominent symptom and may precede weakness by several years. Furthermore, the disease can best be thought of as a multi-system disorder, which particularly affects muscle, in addition to muscle symptoms, radiological changes in the skull, frontal baldness and testicular atrophy in males, various other endocrine abnormalities, intellectual deterioration and cataract may all be present. Inheritance is usually as an autosomal dominant characteristic and although signs may sometimes be detectable in infancy they are more commonly seen in adult life.

Laboratory Investigations. Serum enzyme estimations are of little value, e.g. serum CPK activity is only elevated in a proportion of cases. However, in view of the multisystem nature of the disease, tests of endocrine function may be useful.

Histological changes in the muscle are mainly nonspecific, but the occurrence of chains of nuclei in longitudinal sections of the muscle fibres may be more prevalent than in the myopathies previously discussed.

Observations of myotonic discharges in the electromyogram, the so-called 'dive-bomber' effect, is a rather characteristic finding. However, it is the overall constellation of changes which must be considered to establish the diagnosis. Investigations may also include non-muscle examination such as slit-lamp testing for detection of early cataract, since in some patients cataract may be the only detectable symptom.

Aetiology. It has been claimed on the basis of some electrophysiological evidence of motor nerve abnormality that myotonic dystrophy is a neurogenic disorder. However, the relation between the reported nerve changes and those in the muscle is obscure and it seems at least as plausible to regard it as a systemic disorder, perhaps of cell membranes, that particularly affects muscle function, but which may also have some effect on nerve.

Myotonia Congenita

In myotonia congenita, myotonia, but not weakness, is present. However, because of the clinical variability of myotonic dystrophy, there has been some dispute as to whether the two disorders are really separate entities. Apart from electrophysiological assessment, laboratory investigations are not particularly helpful. However, clinical and family studies appear to indicate that myotonia congenita is indeed distinct from myotonic dystrophy.

Type A2 Myopathies with Structural Muscle Abnormalities

With the application of electron microscopic and histochemical techniques a number of muscle disorders have been delineated in which there are striking structural abnormalities in the muscle fibre or intra-cellular organelles. Although rare, an increasing number of such disorders have been described frequently amongst disorders formerly classified as congenital muscular dystrophies. We shall consider only a few examples that illustrate the kind of structural abnormalities found and the problems associated with their interpretation.

Central Core Disease

This condition is usually associated with mild and slowly progressive weakness apparent from infancy, the characteristic features being revealed by histochemical and electron microscopic techniques. On examination a high proportion of the patient's muscle fibres contain regions situated more or less centrally, and extending a considerable distance in longitudinal section, that are devoid of glycolytic and oxidative enzyme activity. These regions stain abnormally with various histological reagents, though often not with those most commonly used, e.g. haemotoxylin and eosin. Under the electron microscope the cores contain no mitochondria and myofilament structure is disorganized.

The relation between the structural abnormalities and the muscle weakness is obscure. In some patients who have been followed over a long period there may be a marked increase in the proportion of fibres containing 'cores' without any obvious increase in muscle weakness.

Nemaline Myopathy

Clinically, this condition is similar to Central Core Disease. Aggregates of 'rod' bodies are seen within the fibre particularly at the poles of the nuclei. They are refractile under the light microscope and electron dense under the electron microscope. Because of their morphology, it has been suggested that the 'rods' might be derived from Z-line material, but conclusive evidence to support this suggestion is lacking.

Myotubular Myopathy

In myotubular myopathy the muscle fibres bear a superficial resemblance to myotubes in that the nuclei are centrally located, consequently it has been suggested that there is some developmental failure in this disease. Histochemical methods reveal abnormalities in the central region of the fibres, notably a loss of myosin ATPase perhaps due to a reduced number of myofilaments in the central region. Weakness in early infancy may be severe and progressive.

Mitochondrial Myopathies

There are a number of myopathies in which the muscle mitochondria exhibit various, sometimes bizarre, morphological changes. These may consist of inclusions of an apparently crystalline nature. Such changes can be observed in other myopathies, but less frequently. The nature and origins of these structures is not known, neither is it clear whether they are related to the biochemical changes, such as loosely coupled oxidative phosphorylation which may also be present.

The chief value of these observations lies in defining the various disorders in this group although ultimately they may provide clues to the underlying pathological mechanisms.

Type A3 Myopathies— The Familial Periodic Paralyses (FPP)

This group of disorders is characterized by recurrent paralytic attacks of variable duration and severity which are accompanied by biochemical changes, particularly of tissue and plasma electrolytes.

Three forms can be distinguished according to whether the plasma potassium concentration is raised, lowered or normal during attacks.

Hypokalaemic FPP

In this disease there are attacks of muscle weakness, sometimes profound, which last from a few hours to a few days. Since, however, respiratory muscles and the heart are usually spared, death during an attack is rare. Frequency of attacks tends to be greatest in young adults and diminishes in later life. In a minority of patients there may be a slowly progressive oximal myopathy and if so, the outlook is less favourable.

Biochemical Abnormalities—*Electrolytes.* There is a marked fall of plasma potassium level with no increase in urinary excretion, in fact rather the reverse. This implies a shift of potassium into some other body compartment and indeed a shift into muscle can be demonstrated. However, because of uncertainty about concomitant transfer of water, it is not clear if there is an increase in the muscle concentration of potassium. However, there is undoubtedly an increased concentration difference between muscle and plasma, although electrophysiological measurements reveal depolarization of the muscle fibre membrane rather than the expected hyperpolarization. On correlating the onset of muscle weakness with the fall in plasma potassium level, we find evidence for an abnormal sensitivity. Thus, although plasma potassium may fall as low as 1·5–1·7 mmol/l, weakness may appear at concentrations only slightly less than normal. To account for this observation it has been suggested that movements of other ions such as Na^+ and Cl^- may be involved in the production of weakness. Some support for the involvement of Na^+ is provided by the observed Na^+ retention preceding an attack and the beneficial effects of dietary salt restriction.

Endocrine Changes. Changes in corticosteroid excretion, notably an increase in aldosterone have been reported and it has been suggested that abnormal release of aldosterone with an attendant rise in plasma sodium might explain the symptoms. However, the original observation on aldosterone could not be confirmed so that its significance in the pathogenesis of the disorder is unclear.

Provocative Factors. (1) *Natural.* Rest after exercise, stress and a heavy carbohydrate meal may provoke a paralytic attack. (2) *Artificial.* Administration of glucose plus insulin and of mineralocorticoids such as deoxycorticosterone acetate is also effective.

Morphological Changes. A rather consistent finding is the presence of vacuoles in the muscle, thought by some to be distended elements of the sarcoplasmic reticulum and T-system. Even though the origin of these structural alterations is in question and their significance made rather uncertain by doubt as to whether or not they are present between attacks, they may be involved in the pathogenesis of the slowly progressive myopathy seen in some patients.

Electrophysiological Abnormalities. As muscle weakness develops, changes occur in the shape and duration of the propagated action potential and eventually the muscle becomes unresponsive to either direct or indirect electrical stimulation. By contrast there is no demonstrable abnormality of the motor nerve or neuromuscular junction.

Treatment. Unlike the Type A1 and A2 myopathies (Table 1), various measures have proved useful in treating hypokalaemic FPP.

The most obvious strategem is that of attempting to normalize the plasma K and indeed, episodes of muscle weakness can be terminated or aborted by giving oral potassium.

Various prophylactive measures are also useful:
1. Adherence to a low salt, low carbohydrate diet.
2. Administration of potassium, possibly in a slow release form.
3. Administration of diuretics such as acetazolamide, which promote excretion of both K^+ and Na^+.

Hyperkalaemic FPP and Paramyotonia Congenita

When administration of oral potassium to control hypokalaemic FPP became general practice, it was found that some patients did not improve but actually became worse. Eventually it was realized that they suffered from a different disorder, characterized by a rise in the level of potassium in plasma during onset of muscle weakness, which has been designated hyperkalaemic FPP.

Clinical Features. Attacks are usually of shorter duration than in hypokalaemic FPP and in addition myotonia is present. Both weakness and myotonia are made worse by cold. Although there may again be a slowly progressive myopathy, attacks, starting in infancy tend to become less frequent particularly if strenuous exercise is avoided. Stress and rest after exercise are provocative factors, whereas mild exercise may abort an attack. However, paralysis can be provoked by starvation and aborted, instead of being provoked, by glucose plus insulin.

Biochemical Changes. The most significant change is the rise in plasma potassium and a shift of potassium ions out of the muscle.

Electrophysiological Abnormalities. The muscle fibre membrane is markedly depolarized, but while this change is in the direction expected from the change in plasma potassium, the rise is quantitatively insufficient to explain the degree of depolarization. Moreover, the plasma potassium concentration may remain within normal limits even when weakness is substantial. To explain this apparent abnormal sensitivity to plasma potassium, movements of other ions have been invoked.

Treatment. Termination of an attack with glucose and insulin may be possible and prophylactic administration of diuretics is also useful.

Paramyotonia Congenita. Distinction between hyperkalaemic FPP and paramyotonia congenita has been questioned. The basis for the distinction is the greater sensitivity to cold claimed for paramyotonia congenita, particularly in the production of local weakness, and the fact that mild exercise may increase weakness. However, the argument is unlikely to be finally settled until the underlying metabolic defect is uncovered.

Normokalaemic FPP

A third form of FPP has been reported in which no change in plasma potassium concentration can be demonstrated even in a severe attack. Duration of weakness may be prolonged lasting several weeks, administration of potassium may worsen rather than improve it. Giving NaCl may lead to improvement and successful prophylaxis can be achieved by administration of diuretics to which steroids such as 9-α-fluorohydrocortisone may be added. The basis of these measures seems to be entirely empirical.

Conclusions

Although the observed biochemical abnormalities in the FPPs raise many unanswered questions and may well be secondary to some unknown metabolic defects, they do at least provide a means of distinguishing between the three types and to some extent a guide to appropriate treatment. Consequently, biochemical studies are well worthwhile; likewise electrophysiological measurements have diagnostic value. The value of the morphological observations, namely the presence of vacuoles is perhaps less evident, since rather similar abnormalities occur in all the various forms. In this connection some points, such as the persistence of vacuoles between paralytic attacks, seem to need further study. Such additional information might be of use in predicting which patients are likely to have a progressive proximal myopathy.

Malignant Hyperthermia

Because of its practical importance as a source of preventable anaesthetic emergencies we must briefly mention this recently described disorder.

In malignant hyperthermia there are abnormal and life threatening responses to certain anaesthetics and muscle relaxants, for example halothane and suxamethonium.

These reactions are characterized by a rapidly rising temperature, muscular rigidity and metabolic acidosis.

Treatment is largely symptomatic with vigorous cooling and control of the acidosis being important. In addition, certain muscle relaxants such as dantrolene sodium, may allow an attack to be aborted. It has also been claimed that administration of steroids may be useful, although the rationale of such measures is obscure.

The underlying pathogenesis is thought to involve the presence of raised cytoplasmic Ca^{++} levels, possibly due to impaired sarcoplasmic reticulum Ca^{++} transport or to abnormal release of Ca^{++} from mitochondria. However, the precise nature of the defect remains to be elucidated. Likewise the mode of inheritance whether autosomal recessive or autosomal dominant, with incomplete penetrance, is still in doubt.

Fortunately, in view of the severity of the reaction, it seems likely that individuals at risk may be identified by the abnormal *in vitro* response of their muscle to various agents such as halothane and caffeine.

Type A4 Myopathies with Specific Metabolic Defects

In this group of myopathies there are specific metabolic defects which appear to be the cause of the clinical signs and which usually involve energy generating systems, i.e. ATP formation. Although they are rare it is worth considering a few in detail since they present a paradigm of how biochemical studies can provide a fundamental insight into the pathogenesis of myopathic disorders in a way which, it may be hoped, will eventually become true of the more common hereditary myopathies, such as the muscular dystrophies.

Two chief substrates are used for ATP production: carbohydrate as stored glycogen and lipid as fatty acids.

Glycogen can be used to yield ATP in part at least by anaerobic processes, whereas synthesis of ATP from fatty acids or for the complete utilization of glycogen requires oxygen, i.e. aerobic pathways are involved. Now aerobic metabolism produces ATP from its substrates much more efficiently than the anaerobic pathway, but it cannot function sufficiently rapidly to satisfy peak requirements. It is, however, well suited to satisfy resting requirements or those of sustained sub-maximal activity. Short bursts of maximal activity thus depend additionally on anaerobic glycogen breakdown. The two systems have somewhat different cellular locations, the enzymes of anaerobic glycolysis being found in the cytoplasm, whereas the systems involved in aerobic ATP production are located in the mitochondria. Disorders of energy production could arise as a result of defects in either of these systems, and moreover from what has been stated already the clinical signs would be expected to depend on which system was affected.

It is convenient to consider separately the two main types of defects in energy production:
1. Disorders of anaerobic glycolysis
2. Disorders of aerobic energy production

Disorders of Anaerobic Glycolysis

Myophosphorylase Deficiency. This archetypal disorder of anaerobic glycolysis was first described by McArdle[1] although at that time the precise nature of the enzyme defect was unknown. Subsequently it was shown that the enzyme phosphorylase which together with debrancher enzyme catalyses the first stage of glycogenolysis was absent or present in very small amounts in the muscles of affected individuals. The consequence of this deficit will be appreciated by considering Figure 2, a schematic outline of anaerobic glycolysis. The patient described by

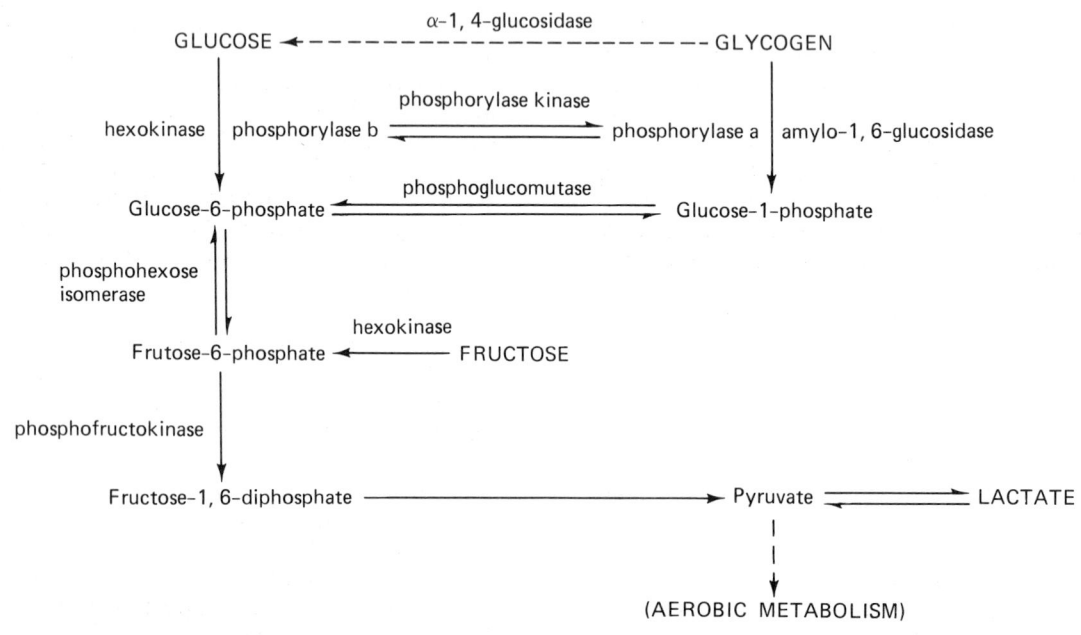

FIG. 2. Simplified representation of anaerobic glycolysis in muscle.

McArdle presented with a disability which can be summarized as follows:

Gentle exercise, such as walking slowly on level ground, could be sustained without discomfort for a considerable period. More strenuous exercise, climbing stairs, running, etc., was quickly halted by severe cramp-like pains in the exercised muscles which remained in a contracted state. In some patients, although not in this case, attempts at vigorous exercise may result in myoglobinuria with the added problem of renal failure.

Figure 2 indicates a number of consequences which result from the position of the enzyme defect:
1. Glucose and fructose which enter the metabolic pathway distal to the block can be utilized.
2. Fatty acid utilization is unaffected.

Catabolism of fatty acids would permit the muscle to maintain itself at rest or at moderate to low activity levels; unimpaired glucose and fructose utilization suggests that an increase in blood fructose or glucose concentrations might increase exercise capability. The first prediction is confirmed, although there are considerable variations in individual exercise capability, even when there is apparently the same degree of enzyme deficiency. These variations do not seem to have received systematic study, although differences in ability to satisfy energy requirements from fatty acids or from blood glucose may provide an explanation. The second prediction seems to be correct in about 50% of patients, and it has been suggested that some of the problems associated with unusual exertion might be mitigated by prophylactic ingestion of oral fructose. However, this suggestion has not received general support.

Methods of Investigation. At this point we can consider the investigations needed to establish the nature of the metabolic defect in patients presenting with a history of exercise–associated muscle pain.

The end product of anaerobic glycolysis is lactate, so that if aerobic metabolism is blocked by occluding the blood supply to the muscle in a normal individual lactate will accumulate. This is the basis of the *ischaemic exercise test*. A patient suspected of having a defect of anaerobic glycolysis can be simply tested by occluding the blood supply to the forearm and getting the subject to exercise the hand, preferably doing a known amount of work. Venous blood is sampled before and after exercise. In a normal subject a marked increase in lactate follows exercise and return to a base line level takes 20–30 minutes. In a patient with myophosphorylase deficiency, no such rise in lactate occurs. Although the results of the ischaemic exercise test indicate the presence or absence of a glycolytic defect, they do not locate the defect on the metabolic pathway. To accomplish this, examination of a muscle biopsy is necessary. The latter can be studied by either:
1. Histochemical staining.
2. A systematic study of the glycolytic activity of a muscle homogenate using different substrates.

Histochemical techniques provide the most rapid method of investigation, but are not well suited to give quantitative information. Moreover, the defect might involve an enzyme for which no satisfactory histochemical technique is available. In the case of myophosphorylase, however, the enzyme can be detected by incubating a biopsy section with excess glucose-1-phosphate and then staining the newly formed glycogen with iodine. In this way phosphorylase in normal muscle can be readily demonstrated and likewise its absence in myophosphorylase deficiency. It should be noted that by using an excess of glucose-1-phosphate the phosphorylase reaction is driven in the reverse of the physiological direction, i.e. towards glycogen synthesis. The fact that glycogen accumulates in the muscle of patients with myophosphorylase deficiency demonstrates that phosphorylase is not involved in glycogen synthesis, but only in its breakdown.

The more general method of locating the position of the glycolytic defect can be illustrated by the results found in another glycolytic defect namely muscle phosphofructokinase deficiency. From Table 2 we see that lactate is formed from fructose-1,6-diphosphate but not from fructose-1-phosphate or glucose-6 phosphate. From this we conclude that there is a deficiency of muscle phosphofructokinase. This conclusion would then be confirmed by direct assay of the deficient enzyme.

TABLE 2
FORMATION OF LACTATE BY HOMOGENATE OF MUSCLE FROM A PATIENT WITH PHOSPHOFRUCTOKINASE DEFICIENCY

Substrate	Lactate formed
Glucose-6-phosphate	–
Fructose-6-phosphate	–
Fructose-1,6-diphosphate	+

Phosphofructokinase Deficiency. Although as anticipated, the clinical signs of muscle phosphofructokinase deficiency are very similar to those of myophosphorylase deficiency, there are some differences. Firstly, utilization of blood glucose or fructose is not possible, so that ingestion of glucose or fructose will not affect exercise capacity, secondly because the muscle form of the enzyme is absent the phosphofructokinase activity of the erythrocytes is reduced. This has the effect of reducing their life span and clinically a mild haemolytic anaemia is observed.

Other Forms of Glycogenosis. *Defects of the Myophosphorylase System.* Phosphorylase can exist in an active and in an inactive form. Interconversion of the two forms is accompanied by transfer of phosphate by a protein kinase whose activity is regulated in turn by a cyclic-AMP (cAMP) regulated kinase. This rather complex mechanism provides a means by which glycolysis can be controlled by release of adrenalin into the circulation. Recently a case of muscle disorder has been reported in which there appeared to be a deficiency of the cAMP activated kinase.

'Debrancher' Enzyme Deficiency. Glycogen consists of a highly branched structure in which glucose units are joined in two different ways: firstly, by 1:4 links to form branches; and secondly, by 1:6 links to form the branch points. Two enzymes are needed to hydrolyse these links; phosphorylase hydrolyses 1:4 links but the enzyme amylo 1:6 glucosidase, or debrancher enzyme, is needed to break the 1:6 links and hence allow further phosphorylase action to occur. In its absence glycogen is only degraded as far as the branch points and material referred to as 'limit dextrin' accumulates. Patients with this particular defect have an impaired ability to break down glycogen. Although the prognosis is generally good, some muscle weakness may develop.

'Brancher' Enzyme Deficiency. The converse of debrancher enzyme deficiency is lack of an enzyme needed to form the 1:6 glucose links amylo-1:4-1:6- transglucosylase. In this disorder there is an accumulation of an abnormal carbohydrate with long chains and few branch points and because the defect is not restricted to skeletal muscle but also involves the liver and heart, prognosis is poor commonly with death in childhood.

Acid Maltase Deficiency. A defect of a rather different kind which affects glycogen metabolism is acid maltase deficiency (see Fig. 2). In this disorder although the normal glycolytic pathway seems to be intact there is a marked accumulation of glycogen, apparently of normal structure, in skeletal muscle and in other tissues, notably the heart and liver. Although it seems reasonable to assume that the disorder is due to the specific enzyme deficiency it is difficult to account for the clinical signs satisfactorily by this means, partly because the normal function of the enzyme is uncertain. Clinically the disorder takes two main forms: firstly, a severe condition affecting infants often with a fatal outcome due to heart and liver involvement; and secondly, a more benign disorder presenting in adult life with symptoms resembling limb-girdle muscular dystrophy.

The diagnosis is established by assay of the enzyme; results of the ischaemic exercise test may be equivocal. Accumulation of glycogen which may be very striking is puzzling since the normal pathway of anaerobic glycolysis is not primarily affected. Consequently the reason for the excessive accumulation is unclear.

Restricted versus Generalized Deficiency. In the last two disorders, besides skeletal muscle, heart and liver are affected, whereas in myophosphorylase deficiency the defect is restricted to muscle. Sometimes as in myophosphorylase deficiency, information on whether the disorder is general or restricted to muscle can be obtained by ascertaining if there is a normal rise in blood glucose in response to adrenalin.

Conclusion. Although the above discussion does not exhaust the list of reported glycolytic disorders, some of the other conditions such as phosphoglucomutase deficiency and a defect at the level of phosphohexose isomerase appear to be less well documented.

Before leaving the glycogenoses, it is worth noting a number of unsolved problems, which can be considered in connection with myophosphorylase deficiency:

Age of Onset. In many cases patients do not experience appreciable disability until their teens or even not until adult life suggesting more or less normal glycolytic activity in infancy or childhood. This is a question which is clearly difficult to investigate since it would require study of at risk individuals over a period of years.

The Nature of the Genetic Defect. In some patients phosphorylase activity may be detectable, but in greatly reduced amounts. At the same time, there may be a considerable amount of protein present with the electrophoretic properties and immunological reactivity of phosphorylase, but lacking enzyme activity. The structural gene for phosphorylase is present but the enzyme has undergone post synthetic modification resulting in absence of enzyme activity. Again even when neither enzyme activity nor phosphorylase-like protein can be detected, this may not necessarily exclude the presence of the structural gene, but may perhaps imply a disorder of its expression.

Some support for this latter suggestion comes from the fact that if muscle from patients with myophosphorylase deficiency is grown *in vitro*, the cultured muscle may show phosphorylase activity. Again, if the muscle is traumatized, the freshly regenerating muscle may show activity *in vivo*. Unfortunately, the significance of these observations is at present obscured by controversy over whether the enzyme observed is of the fetal or adult form.

Muscle ATP levels. Although the most obvious explanation of failure of muscle function in myophosphorylase deficiency or in the other glycogenoses would be a lack of ATP, it has recently been claimed that when muscle function is halted as in myophosphorylase deficiency, tissue ATP levels are not significantly reduced below normal. If this is correct some localized ATP deficit, perhaps at the level of the sarcoplasmic reticulum calcium uptake system, may be implicated without being reflected in overall tissue ATP concentration. Clearly there is considerable scope for further biochemical investigation before a complete understanding of these disorders is attained.

Disorders of Aerobic Metabolism

Introduction. Delineation of specific disorders of the aerobic phase of energy metabolism is a fairly recent development. Again we will refer to a few examples from amongst those which have been most fully characterized so far.

Since the aerobic phase of energy production takes place within the mitochondrion, disorders of aerobic metabolism are disorders of mitochondrial function.

Recently a number of myopathies have been described in which there is evidence for biochemical defects in the mitochondria, which may also show morphological abnormalities such as paracrystalline inclusion bodies similar to those mentioned already in connection with the Type A2 myopathies (Table 1).

The first instance of a disorder, in which a biochemical defect in mitochondria was strongly suggested, was a patient exhibiting nonthyrotoxic hypermetabolism with fairly mild muscle weakness. Mitochondria isolated from the muscle showed loosely coupled oxidative phosphorylation, i.e. although mitochondrial oxygen consumption was high it was not appreciably stimulated by addition of adenosine diphosphate (ADP) to the incubation medium. The site of the defect on the mitochondrial metabolic pathways was not established.

Since then other patients with loosely coupled mitochondrial oxidative phosphorylation have been reported. In these individuals, however, hypermetabolism was not present and again the precise nature of the metabolic defect has not yet been elucidated.

Disorders of Carnitine Metabolism. Amongst the mitochondrial disorders, at least partially characterized biochemically, are disorders of carnitine (β-hydroxy-γ-tetramethyl-amino-butyric acid) metabolism.

In order that long chain fatty acids may be utilized, their coenzyme A-esters must first penetrate the inner mitochondrial membrane. CoA-esters of long chain fatty acids cannot do this in the absence of a suitable carrier system. This is normally provided by carnitine in conjunction with two carnitine acyl transferase enzymes. The processes involved are as follows:
1. Outer surface of inner mitochondrial membrane.

Carnitine acyl transferase I
Acyl-CoA + Carnitine $\rightleftharpoons$ Acyl Carnitine + CoA

2. The acyl carnitine traverses the mitochondrial membrane and on the inner surface acyl-CoA and carnitine are regenerated and the acyl-CoA can then be degraded by β-oxidation eventually to yield ATP

Carnitine acyl transferase II
Acyl Carnitine + CoA $\longrightarrow$ Acyl-CoA + Carnitine

Utilization of long chain fatty acids will be affected if there are defects in this transport system and these can arise either from a lack of carnitine itself or alternatively from the absence of one or both of the transferases.

Carnitine Deficiencies. First consider a deficiency of carnitine itself; this may take two forms; a systemic defect due to some abnormality in hepatic synthesis; or a local muscle deficiency perhaps due to defective uptake from the blood. Patients with both forms of carnitine deficiency have been described and although the clinical and metabolic manifestations are very variable, the systemic form seems to be the more serious and has proved fatal in a number of instances chiefly because of hepatic dysfunction, although cardiac complications may be important.

Clinical and Metabolic Consequences of Systemic Carnitine Deficiency. Lack of carnitine inhibits the oxidative catabolism of long chain fatty acids. Consequently, accumulation of lipid in muscle and other tissues is not unexpected and is a prominent feature of the disorder. Marked muscle weakness, presumably related to a disturbance in the supply of energy derived from fatty acid catabolism, is often present and may be exacerbated by exercise. Pain associated with exercise does not feature prominently. Remission or exacerbation of the muscle symptoms may occur for reasons which are as yet unclear. The general metabolic disturbances which are present seem to be mainly referable to liver rather than muscle dysfunction. They take various forms and at least two can have life-threatening implications.

Metabolic Acidosis. Because fatty acid utilization is blocked, there is an increased reliance on glycolysis as an energy source. Consequently, increased amounts of lactate are produced and this, if not removed, will result in acidosis. Its removal can be achieved by its use as a substrate for hepatic glycogen synthesis or it may enter the tricarboxylic acid cycle via pyruvate. In patients with liver dysfunction hepatic utilization may be reduced. Moreover, since carnitine has been reported to stimulate the pyruvate dehydrogenase system, its absence may cause a diminished transfer of pyruvate to the tricarboxylic acid cycle. Both factors can lead to an accumulation of lactate.

An additional contribution to the production of acidosis is increased cytoplasmic γ-oxidation of fatty acids to give, dicarboxylic acids. Another consequence of increased glycolysis when liver function is impaired is a marked hypoglycaemia. Both hypoglycaemia and acidosis have been reported to occur in an episodic manner in cases of systemic carnitine deficiency and have been associated with rapid onset of coma and death.

Effect of Fasting. Since glycogen stores are rapidly depleted on fasting, reliance on fat as an energy source becomes mandatory, apart from some use of gluconeogenic amino acids from protein catabolism. In a normal individual this leads to an increased production of ketone bodies, which is reflected by increased excretion and raised blood levels; in some patients with carnitine deficiency, by contrast, excretion of ketone bodies is reduced or delayed. Possibly the determining factor is the level of carnitine in the liver if this is sufficiently great and liver function is satisfactory in other respects, then an essentially normal response may be observed. If not, then hepatic fatty acid utilization may be impaired and ketone body production diminished. Cases illustrative of both situations have been reported.

Since fat utilization is impaired, fasting may rapidly deplete glycogen stores and lead to a dangerous degree of hypoglycaemia. This is a point to be noted if it is proposed to use fasting as a provocative stimulus for diagnostic or investigative purposes.

Serum Enzymes. In only some cases is the activity of creatine kinase elevated, even when there is marked muscle weakness. Perhaps of more importance are the high activities of aspartate and alanine amino transferases which may be seen in patients with hepatic involvement.

Other Serum Abnormalities. Abnormalities of serum lipids are an inconstant finding in carnitine deficiency and may take the form of hyperlipidaemia with raised triglycerides or cholesterol. Their significance is obscure.

Therapy. In the case of carnitine deficiency an obvious approach to the problem of treatment is to administer carnitine in the hope of circumventing the metabolic deficit. In some cases striking improvement has resulted with a normalization of blood carnitine levels. However, the relation between changes in blood carnitine levels and clinical effect is obscure, so that there may be a marked clinical improvement in the absence of any striking change in muscle carnitine levels and normalization of blood carnitine may not be accompanied by clinical improvement.

Carnitine Acyl Transferase Deficiency. The second way in which fatty acid utilization can be impaired by a disorder of the carnitine system is a deficiency of either or both of the carnitine acyl transferase enzymes (Equations I + II). The clinical presentation of this disorder is somewhat different from that of a deficiency of carnitine itself. Muscle pain related to prolonged exercise and myoglobinuria are a more common feature of the condition so that it may bear a superficial resemblance to the glycogenoses such as myophosphorylase deficiency. However, the time relation between exercise and symptoms is rather different and there is a rise in lactate after ischaemic exercise. Fasting may cause myoglobinuria and this is an additional reason for caution in using fasting as a provocative stimulus for investigative purposes. As in carnitine deficiency, fasting may not produce a normal ketone body response and serum triglycerides may be elevated. Moreover serum triglycerides may be cleared only slowly after fat loading.

Therapy. There is little evidence that carnitine administration has any beneficial effect; however, administration of steroids has sometimes appeared helpful even if it has no obvious effect on the underlying biochemical deficit. The reason for this benefit is unclear.

Investigative Procedures. The activity of the carnitine acyl transferase enzymes may be measured directly using fatty acids of different chain lengths, or indirectly by measuring the ability of the tissue to oxidize ^{14}C-labelled fatty acids of different chain lengths to $^{14}CO_2$. Both approaches should be employed since the quantitative relation between activity of the carnitine acyl transferases, measured directly, and the ability of the tissue to utilize fatty acids does not seem to have been adequately investigated.

Conclusion. Although it seems reasonable to suppose that the clinical signs observed in disorders of carnitine metabolism result from the biochemical deficits, the mechanisms linking the biochemical abnormalities to the disorders of muscle function are imperfectly understood and require much further study. Despite these limitations there is a considerable role for clinical biochemical investigations directed towards the diagnosis and management of patients with disorders of carnitine metabolism. This may be summarized as follows:
Investigation of the carnitine system.
1. Blood and tissue carnitine levels.
2. Assay of carnitine acyl transferases and investigation of fatty acid utilization.

Study of general metabolic abnormalities.
1. Serum lipids.
2. Serum enzyme profile.
3. Acid-base balance.
4. Blood lactate and pyruvate.
5. Blood glucose.

Monitoring of some of these variables may be vital for the management of crises in which the patient develops severe acidosis or hypoglycaemia, while others, although having less immediate importance, may provide the information needed for a better understanding of these disorders.

Other Mitochondrial Myopathies. Although there are other disorders in which there is increased muscle lipid and loosely coupled mitochondrial oxidative phosphorylation associated with muscle weakness, the site of the biochemical lesion has not in general been located. However, patients have been reported recently whose muscle mitochondria contained abnormally low levels of cytochrome b or cytochrome oxidase. These findings, together with additional studies of mitochondrial respiration, pointed to defects in the mitochondrial electron transport chain. Such defects like the carnitine disorders, would interfere with fatty acid utilization and also with the aerobic phase of carbohydrate metabolism. It is likely that other specific defects in mitochondrial metabolism will eventually come to light and may lead to increased understanding of the relation between biochemical abnormalities and functional disorders of muscle.

B. ACQUIRED MYOPATHIES

Type B1 Polymyositis and Dermatomyositis

Polymyositis is an acquired myopathy in which muscle weakness is usually accompanied by inflammatory changes, although the latter may not be at all prominent. Other histological changes are non-specific and differ little from those seen in the hereditary muscular dystrophies. In *dermatomyositis*, as the name suggests, as well as muscle weakness there may also be skin changes such as a rash or erythaema. Differential diagnosis of polymyositis from the muscular dystrophies can be difficult, but is of great importance because the former, unlike the latter is often a treatable condition.

Therapy and Biochemical Investigation

Treatment of polymyositis with steroids, and more recently with immunosuppressive agents, can frequently prove effective. Such treatment is, however, seldom curative because maintenance on a reduced dosage is usually necessary to prevent relapse.

Biochemical Abnormalities. As in the hereditary myopathies raised serum levels of tissue enzymes, notably CPK, are found and this elevation may be gross, particularly in acute dermatomyositis. Although this abnormality is non-specific it may have predictive value in relation to treatment. Thus, in some patients at least, clinical improvement following treatment is accompanied by a return of serum enzyme activities towards normal values. Moreover this fall in enzyme activities may precede clinical improvement and conversely, a rise may presage a clinical relapse such as can occur if the steroid dose is reduced too rapidly. Consequently, monitoring of serum enzyme activities is an important part of the management of the patient.

Aetiology

The facts that the disease may be favourably influenced by steroids and immunosuppressive drugs, and that there is an association with other collagen vascular disorders, have led to the suggestion that it may be an auto-immune disease. This view has recently received considerable support from observations that circulating anti-skeletal muscle antibodies may be present and that the patient's lymphocytes may exert a cytotoxic effect on cultured skeletal muscle. Finally an association with one of the major histocompatibility antigens HLA-B8 has been claimed.

Various factors have been suggested as 'triggers' of the postulated abnormal immune response including the usually mild infectious illness which may often precede the onset of symptoms. However, there is as yet little detailed knowledge of the pathological mechanisms underlying this condition.

Type B2 Drug Myopathies

A number of therapeutically useful drugs amongst them steroids, some antimalarials and the widely used anti-tumour drug, vincristine, may produce muscle weakness as a side effect. When this happens laboratory investigations do not usually contribute much of interest and the chief practical question is one of recognizing that certain drugs may cause myopathic changes complicating their use. Whether these changes reflect a specific action on muscle or are aspects of a general cytotoxic effect is uncertain.

Type B3 Myopathy Accompanying General Metabolic Disturbances

It is scarcely surprising that general metabolic deficits, for example those arising from abnormalities of pituitary, thyroid or parathyroid function, may sometimes cause muscle weakness. Laboratory investigations have an essential role in delineating the nature and extent of these abnormalities in order to achieve satisfactory management of the patient. Correction of these abnormalities usually reverses the myopathic changes.

Type B4 Nutritional Myopathy

Although in animals muscle weakness can be induced by lack of vitamin E, in man muscle weakness due to vitamin E (α-tocopherol) deficiency is rare. It is only likely to arise in lipid malabsorbtion or severe malnutrition when it may be difficult to distinguish from the other adverse effects of those conditions. At one time, because vitamin E deficiency can cause muscle degeneration, it was thought that a disorder of vitamin E metabolism might underlie the human muscular dystrophies. However, in the absence of evidence to support it, interest in this idea was not maintained.

Type B5 Myasthenia Gravis and the Myasthenic Syndrome

Myasthenia gravis (MG) is a condition characterized by a progressive failure of neuromuscular transmission following repeated voluntary contraction or repetitive artificial stimulation of the motor nerve and has long been recognized as a disorder of the neuromuscular junction.

In neuromuscular transmission, acetyl choline released from the nerve terminal diffuses across the synaptic gap and activates acetyl choline receptor (AChR) molecules on a modified region of the muscle fibre plasma membrane, thus depolarizing the membrane and releasing the propagated action potential which eventually causes the muscle to contract.

In principle, failure of transmission could occur either by failure of acetyl choline release or by impaired response to released transmitter. For a considerable time there was controversy as to whether the defect in MG was in the nerve terminal, i.e. pre-synaptic, or in the muscle fibre membrane, i.e. post-synaptic. A decision between these two possibilities was hindered by the fact that several features of MG are consistent with either mechanism. The problem may be illustrated by considering the action of the therapeutically active anticholinesterases.

Cholinesterase inhibitors

When the released acetyl choline has bound to the AChRs thus depolarizing the muscle fibre membrane and triggering the propagated action potential, it is necessary that the membrane should repolarize so that it may respond to the next stimulus. Removal of the 'surplus' acetyl choline is brought about by the enzyme acetyl cholinesterase. No matter if transmission failure is due to impaired acetyl choline release or to impaired post-synaptic response, we might expect that increasing the life span of the acetyl choline molecules would tend to reverse the transmission failure. Acetyl cholinesterase inhibitors have just this effect and have proved therapeutically very useful in ameliorating the symptoms of MG. However, their mode of action does not distinguish between a pre- or post-synaptic lesion. We may note that these drugs although very efficacious must be used with caution. Too large a dose will cause permanent depolarization of the muscle fibre membrane and again result in transmission failure.

Post-synaptic Origin of MG

The view that the defect in MG is post-synaptic has gradually gained acceptance following advances in basic biochemical knowledge of the AChRs with which the released acetyl choline interacts. Much information has been obtained by studies on the electric organs of certain fish. These contain structures similar in many ways to the neuromuscular junction and, in particular, contain high concentrations of AChR molecules. The AChRs can be extracted by detergents, purified, and characterized by standard biochemical techniques. The isolated AChR protein behaves very much as it does in situ. These studies of the AChR protein have been greatly assisted by the use of certain snake venom neurotoxins, particularly α-bungarotoxin. This polypeptide binds essentially irreversibly to the AChR and because it can be radiolabelled with ^{125}I or tritium without significantly altering its binding properties, it can be used as a marker for locating tissue AChRs by autoradiography, or for following the purification of the isolated protein. The number of AChRs at a neuromuscular junction can be assessed by first locating the position of the junction with cholinesterase histological staining and then measuring the amount of radioactive toxin bound to a known number of junctions.

When normal and MG junctions were compared in this way, it was found that there was a marked reduction in the number of receptor molecules at junctions from myasthenic patients amounting to 70–80% of normal values. Further support for the view that a reduction of functioning receptors could account for the symptoms of MG, comes from the fact that in animals many of the clinical and electro-physiological signs of MG can be reproduced by giving reversible acetyl choline antagonists such as curare. In MG patients there is a greatly increased sensitivity to curare which may be of value for diagnostic purposes.

Aetiology

Because of its association with thymic abnormalities and with other diseases considered to have an immunological basis Simpson[2] suggested that human MG might be an auto-immune disease. Although we will not describe in detail the subsequent development of this idea, it may be useful to summarize the main lines of evidence.

1. Purified AChR from fish electric organs injected into animals produces a response closely similar to that seen in human MG.
2. Electrophysiological signs of myasthenia can be induced in immuno-tolerant mice by prolonged administration of immunoglobulins from MG patients.
3. Circulating antibodies to the AChRs can be demonstrated in the blood of most MG patients. Moreover although the clinical severity does not correlate well with the serum levels of these antibodies, it is claimed that there is a good correlation with the proportion of AChRs to which antibody is bound. Other less specific anti-muscle antibodies may be present but their significance is unknown.

The action of the AChR antibodies is not a simple block since antibody binding does not necessarily prevent subsequent binding of α-bungarotoxin. Two main mechanisms have been suggested for the reduction in the number of AChRs due to auto-antibodies, firstly complement-mediated destruction of the junction and secondly an increased degradation rate consequent upon antibody binding.

We conclude that the features of MG can be explained by an impaired post-synaptic response to acetyl choline due to the action of anti-AChR antibodies. Remaining problems include the following:
1. What triggers the auto-immune response?
2. Are both humoral and cellular immune responses involved, and if so what is their relative importance?

Despite remaining gaps in understanding of its pathogenesis, MG is a good example of the usefulness of combining electrophysiological and biochemical techniques for elucidating the pathogenesis of a neuromuscular disorder.

The Myasthenic Syndrome

Another condition characterized by abnormal muscle fatiguability is the myasthenic syndrome, a disorder often associated with malignancy. The condition differs from MG in a number of respects, namely at certain stimulus frequencies a marked facilitation of the muscle response occurs, the amplitude of the miniature endplate potentials is not reduced, and neither is there a reduction in the number of AChRs at the neuromuscular junction. Anti-AChR antibodies are absent. However, a reduced release of acetyl choline from the nerve terminal has been demonstrated and the condition can be ameliorated by drugs such as quinidine, which facilitate transmitter release. The cause of the reduced acetyl choline release is not understood.

Neonatal Myasthenia

In some infants born to mothers with MG, myasthenic symptoms may be present at birth, usually disappearing quite quickly. This suggests a transplacental transfer of a maternal factor and is quite consistent with the view that circulating anti-AChR antibodies are responsible for the muscle symptoms.

Congenital Myasthenia

Rarely myasthenic symptoms present at birth may persist. The pathogenesis of this condition is little understood but the reported absence of anti-AChR antibodies suggests that it is different from MG.

1. McArdle, B. (1951) 'Myopathy due to defect in muscle glycogen breakdown', *Clin. Sci.* 10, 13–15.
2. Simpson, J. A. (1960) 'Myasthenia gravis; a new hypothesis', *Scot. Med. J.*, 5, 419–436.

28. CHEMISTRY OF THE CEREBROSPINAL FLUID IN HEALTH AND DISEASE

P. T. LASCELLES, E. J. THOMPSON and D. S. WARNER

Introduction and selected clinical aspects
   Bacterial infections
   Viral infections
   Multiple sclerosis
   Intracranial lesions

Chemical examination of the CSF
   Collection
   Inspection
   Routine tests on the CSF
     Glucose
     Chloride
     Total protein
     Tests for syphilis
   Some specialised chemical examinations of the CSF
     Protein fractionation
     Test of integrity of blood-CSF barrier
     Lactic acid
     Enzymes
     Vitamin B_{12} and folic acid
     Amines
     γ-aminobutyric acid (GABA)
     Prostaglandins
     Trace metals
     Drugs
     Identification of CSF rhinorrhoea

Summary

INTRODUCTION AND SELECTED CLINICAL ASPECTS

The cerebrospinal fluid (CSF) is an absolutely clear and colourless fluid found both within the cerebral ventricles and the subarachnoid space. It surrounds and supports the brain, effectively reducing its weight by a factor of six, and serves as a protection against injury. It acts also as a vehicle for the removal of products of metabolism and inflammatory exudates, a rôle which elsewhere in the body is usually ascribed to lymph.

It has a specific gravity of 1·005, a volume of about 130 ml, an average secretion rate of 524 ml/day, and in many respects resembles an ultrafiltrate of blood plasma, though there are reasons for believing that an active process of secretion is involved in its formation. Chemically it differs from plasma in a number of important aspects, especially in its low protein, lipid and calcium content, and in its higher chloride, folate and magnesium content. It is produced in the cerebral ventricles, by a process of filtration and active secretion from the choroid plexuses. CSF, formed in the lateral ventricles, passes into the third ventricle via the Foramen of Monro, backwards through the Aqueduct of Sylvius into the fourth ventricle and thence into the subarachnoid space via the Foramina of Luschka and Magendie to bathe the surfaces of the brain and spinal cord. It lies between the pia mater on the inside and arachnoid mater on the outside, these effectively constituting a vessel for the transport of the CSF. Absorption is via microscopic arachnoid villi into the cerebral venous sinuses.

Mechanical and osmotic barriers exist between blood and the CSF (blood/CSF barrier) and between CSF and brain (CSF/brain barrier) the characteristics of which are frequently altered in disease, leading to changes in CSF composition. For detailed accounts of anatomy and physiology see references[14,35].

In clinical practice CSF is usually sampled from the subarachnoid sac surrounding the cauda equina, below the termination of the spinal cord by means of lumbar puncture (LP). Occasionally it is necessary to carry out cisternal or ventricular puncture.

The indications for LP are:
1. measurement of pressure and collection of CSF for chemical and microscopic examination,
2. introduction of air or contrast media for radiological studies,
3. administration of drugs including local anaesthetics and antibiotics,
4. removal of CSF as a therapeutic measure in patients with benign raised intracranial pressure.

The account which follows is concerned with the first of these headings. Whilst LP is carried out in a wide range of neurological disorders, too numerous to discuss individually, there are a few clinical situations in which the chemical findings are of sufficient interest and importance to merit special comment. For earlier accounts see[9,10,49,53].

Bacterial Infections of the Nervous System

In patients with classical clinical presentations of *acute bacterial meningitis* and typical CSF changes there is rarely a diagnostic problem; the raised CSF pressure, marked polymorphonuclear response, low glucose content and microscopic identification of the organisms being confirmatory.

It is however in the atypical cases, sometimes of longer duration where diagnostic problems present, especially when there is a lymphocytic cell response in the CSF. Depending upon the particular clinical presentation, questions may arise on admission to hospital as to whether or not the diagnosis is one of partially treated bacterial meningitis, cerebral abscess, parameningeal collection of pus, encephalitis, viral meningitis (including benign lymphocytic meningitis), tuberculous meningitis, in which acid-fast bacilli have not been identified by the laboratory, or occasionally fungal infection.

The matter is further complicated by the fact that some cases of viral meningitis exhibit a partial polymorphonuclear response and a low CSF glucose concentration, particularly if the total white cell count is high.

Recently, the estimation of CSF lactate level by gas-liquid chromatography or commercial kits has been advocated in this situation, the finding of lactate levels greater than 3·9 mmol/litre (35 mg/100 ml) point to a bacterial aetiology, and an early fall in response to treatment indicates a good prognosis. Whilst this investigation has been strongly advocated, it has also been pointed out that the 'cut-off' point is not clearly demarcated and significant overlap occurs in the various groups[6,15,18,24].

Attention has also been given to countercurrent immunoelectrophoresis (CIE) of serum and CSF polysaccharides for meningococcal, pneumococcal, haemophilus and Escherichia coli meningitis, with a high-success rate. Meningococcal and pneumococcal meningitis can also be rapidly typed by direct CIE of CSF against specific antisera. Polysaccharide antigens in CSF can also be rapidly detected using antibody-coated latex particles.

Gelation of limulus lysate, the so-called limulus test, is also recommended as giving an almost 100% success rate for the detection of gram-negative bacterial endotoxin.

Quantification of CSF antigen levels has been found to be of some prognostic significance though the persistence of high titres for greater than 48 hours after the commencement of treatment is a stronger index of poor prognosis and may be an indication for a change of therapy[25,55].

Classically *tuberculous meningitis* is associated at an early stage with a lymphocytic response in the CSF (up to 400 cells/μl), a fall in CSF glucose (relative to that of plasma, bearing in mind the increased risk of tuberculosis in diabetic patients and the tendency in any patient with intracranial disease to exhibit decreased glucose tolerance—cerebral hyperglycaemia) and a raised total protein concentration of up to 4·0 g/l. In both of these latter respects tuberculous meningitis differs from benign lymphocytic meningitis. A very low or absent CSF glucose content is rather unusual in tuberculous meningitis and suggests the possibility of pyogenic meningitis (or mixed tuberculous and pyogenic infection) or carcinomatous meningitis. However, it is also reported that a moderate decrease in CSF glucose occurs in some patients with viral meningitis in whom there is a very high CSF lymphocyte count.

Further it must be stressed that a normal CSF glucose, though unusual, does not exclude tuberculous meningitis.

CSF chloride levels are now regarded as noncontributory.

The chemical response of the CSF to antituberculosis treatment is also informative, irrespective of whether drugs are administered intrathecally or not. The glucose level is the first parameter to return to normal but that of protein may remain elevated for many weeks. This is a further differential point from lymphocytic meningitis, where the slightly raised protein level reverts fairly rapidly to normal in the natural course of this disease[43].

Our own experience over several years with the radioactive bromide partition test has been favourable. Although it requires approximately ten days to obtain and administer the isotope and collect the blood and CSF samples, the results so far have been unequivocal in all patients and (in view of the subsequent course of the illness) invariably correct in respect of both positive and negative diagnosis of tuberculous meningitis. An important feature of this test is the fact that commencement of treatment does not interfere with the result.[30,31]

Syphilis of the nervous system is seen clinically over a wide range of age groups. CNS involvement as indicated by CSF changes may occur with all the classical presentations of congenital, primary, secondary or tertiary syphilis. The last group used to present neurologically as meningovascular or spinal syphilis, tabes dorsalis or general paralysis of the insane (GPI), but these classical pictures are now rarely seen in this country, modified neurosyphilis resulting from partial or inadequate previous treatment with penicillin being far commoner[22,26,29]. A full diagnosis must comprise not only the presence of treponemal infection but also indicate whether or not the infection is currently active and, if so, whether it is responding satisfactorily to current therapy; and it is with respect to the latter that examination of the CSF is of particular importance. Generally speaking, serological tests in blood are more sensitive and have higher titres than those in the CSF but are more liable to false positive responses, especially in patients with concurrent auto-immune disease. If all tests in the serum are negative it is unlikely that CSF serology will be positive, particularly late in the disease. Catterall[4] advocates examination of the CSF in every patient in whom the diagnosis of syphilis is suspected as an index of CNS involvement; the prognosis of asymptomatic neurosyphilis being excellent following early adequate treatment. In cases of primary and secondary syphilis the fluid is examined as a test of cure one year after completion of treatment, and, in late syphilis, CSF tests are indicated prior to commencement of treatment as a base-line parameter. Examination of CSF is also advocated in every case of clinical or serological relapse.[4]

Viral Infection

CSF changes in the different forms of specific virus infection is a subject of its own, outside the scope of this article, but brief reference may be made to *herpes simplex encephalitis*.

The majority of cases show changes in the CSF from the outset, there being a predominantly lymphocytic cell response and a moderate rise in total protein content with a tendency for both, but particularly the protein, to rise further with the course of the illness.

There is however no direct correlation between the cell count and the protein level, nor between either of these and the glucose content which is almost invariably normal.

The presence of red cells in the CSF is frequently described in the literature and, in one series, was as high as 40% of cases on first examination with 11% showing xanthochromia.

An important feature however of the CSF examination in herpes simplex encephalitis is that in some 12% of cases the CSF is normal when first examined and that when the diagnosis is eventually established, an initially normal CSF does not necessarily indicate a favourable prognosis.[23,39]

Multiple Sclerosis (MS)

Whilst a slight to moderate increase in the white cell count in CSF is frequently present in the acute phases of MS, changes observed in protein content and type are also most important and informative with respect to activity of the disease process.

A strongly positive Pandy Test in the presence of normal or slightly raised total protein with a paretic type colloidal gold response (indicative of elevated cathodic γ-globulins) in the absence of evidence of treponemal infection is the classical CSF picture in MS.

The presence of two or more protein bands in the γ-globulin region (excluding the γ_1-region) on acrylamide gel electrophoresis constitutes the so-called oligoclonal pattern. Quantitative measurements demonstrate a high γ/β ratio. An oligoclonal pattern is found in 93% of patients with clinically definite MS and in 88% of patients with early probable/latent MS classified according to McDonald.[33] CSF oligoclonal immunoglobulins are an extremely important laboratory diagnostic aid in multiple sclerosis.[50,51]

The relationship between the finding of oligoclonal bands in CSF and abnormalities in visual evoked responses in patients with optic neuritis as an early and isolated presentation of MS is interesting and important. Both tests are positive in greater than 90% of these patients but in our experience in those with mainly brain stem signs and no obvious eye involvement the visual evoked responses are abnormal in approximately 50%, whereas oligoclonal bands are still found in 87% of patients.

The bands alter during the course of the disease with the most marked changes occurring within the first seven days of a relapse.

Oligoclonal patterns are also found in other demyelinating diseases including subacute sclerosing panencephalitis, and occasionally in other disorders of the nervous system. They should therefore not be regarded as being specific for MS.

Intracranial Lesions

Attempts have been made over a number of years to identify markers in the lumbar CSF for *intracranial tumours*.

Lysozyme, which is absent from normal CSF, has been detected in appreciable amounts in the CSF of patients with either primary or secondary tumours of the nervous system.[38] The concentrations were found to reflect the nature of the tumour and the degree of CNS involvement, though other lesions including pyogenic infections also gave rise to increased levels. Non-neoplastic and non-inflammatory lesions did not cause a rise.

In practice the diagnostic importance of this observation appears to be related mainly to metastatic tumours within the CNS and these are known to be associated with an inflammatory reaction. In the case of lymphomatous deposits, which are themselves a potent source of lysozyme, markedly raised levels of lysozyme were detected in CSF.

More recent studies on brain specific proteins, notably astrocyte-specific cerebroprotein, astroprotein (AP) and glial fibrillary acidic protein (GFAP) which have been shown to be immunologically identical, have been encouraging.[21,37,42] In the most recent series of 120 patients with intracranial disease, elevated levels were found in 43% of cases with glial tumours, 16% with non-gliomatous tumours and 21% of non-neoplastic intracranial disorders. In patients with glioblastoma, however, two-thirds showed high levels and this group contained the only few patients with very high levels.[21] Further studies on the correlation between CSF AP levels and the size and location of tumours are awaited.

Biochemical parameters for the semi-quantitative assessment of brain damage resulting from trauma, infection or cerebro-vascular accident have generally been disappointing. Measurements of myelin basic protein in serum following head injury and cerebrovascular accident may prove to be useful in diagnosis and prognosis.[40,41,48] Determination in CSF is less helpful in view of the degrading effect on myelin basic protein of neutral proteinases.[40]

Measurements in CSF of lactate/pyruvate ratio,[8] prostaglandin $F_{2\alpha}$, glutamic oxalacetic transaminase and lactate dehydrogenase enzymes and isoenzymes have each had their advocates, but none has proved to be sufficiently useful to stand the test of time and become part of routine practice.

The estimation of neurotransmitters and their metabolites in CSF, especially γ-aminobutyric acid, 5-hydroxytryptophan, and homovanillic acid, has been extensively discussed in the recent literature in relation to disorders of the extrapyramidal system. The usefulness of these estimations in practice, however, has not been established, due not only to the technical difficulties involved in measuring on a routine basis, the very low levels involved but also to the fact that lumbar fluid levels are a poor index of brain content and turnover. Moreover, very recent work on neurotransmitter-receptor interactions points to the limited interpretations attributable even to estimations of neurotransmitter turnover rates; greater importance is now attached to receptor population density and modulation of transmitter/receptor interactions by endogenous membrane-bound ligands.

ROUTINE CHEMICAL EXAMINATION OF THE CSF

The section which follows discusses in practical terms the individual constituents of CSF.

Collection of CSF

CSF is normally collected by lumbar puncture with the patient in a lateral position. About 6 ml should be collected into three clean, dry, sterile plastic containers and a further 0·5 ml into a small fluoride bottle specially prepared to collect this volume for glucose (with the time of collection clearly marked along with other details) and sent to the laboratories *at once*. Centrifugation should be initiated shortly after receipt, samples from the last plain tube having been taken for cell count and culture. If for any reason, such as problems of transportation, there is delay in processing, the samples should be kept cool, preferably in a refrigerator at +4°C (not frozen in a deep-freeze nor the ice compartment of an ordinary refrigerator). It should be noted that degenerative cell changes become apparent after standing for about 2 hours.

When lumbar puncture is carried out for special investigations more CSF will be required and, in the case of a search for AFB, 10 ml for this alone should be allowed.

Relevant clinical information including details of drug therapy should be supplied on the accompanying request form, together with a note as to whether or not the sample is a first or repeat one. Emergency samples should carry a clear indication as to whom the results should be telephoned. Samples collected in special circumstances, e.g. (1) in the post-operative period, (2) in association with the injection of air for encephalography (collection before the injection only, as the procedure itself raises the cell count and causes an alteration of protein content[16]), (3) following injection of Myodil or (4) following injection of radio-isotopes, should be accompanied by a request form which clearly gives all this information.

When CSF is collected for the radio-active bromide partition test, lumbar puncture should be carried out exactly 48 hours after an oral dose of the radio-active bromide (^{82}Br or ^{77}Br), 2 ml of CSF should be put aside specifically for this investigation and 5 ml of venous blood collected into a dry container at the same time and sent with the CSF to the laboratory.

CSF collected by cisternal or ventricular puncture or by ventricular drain should be labelled accordingly in view of the different interpretation to be placed on the results of analysis.

Whilst it is clear that many factors will influence the timing of non-emergency lumbar punctures, it is in the interest of all parties that samples should *reach the laboratory* as early as possible in the day and, at the latest, to allow one full hour for processing by the 'day' staff.

Inspection

Normal CSF is absolutely clear and colourless. Cloudiness is due to blood, bacteria or white cells (particularly polymorphonuclear leucocytes), and is not caused by a moderate (up to 300 μl) increase in lymphocytes alone.

If blood is present as a result of a traumatic tap during LP the distribution between successively collected specimens will be uneven and centrifugation of freshly-collected samples will reveal a clear supernatant. Following subarachnoid haemorrhage the CSF is usually heavily bloodstained, but it is important to remember that very occasionally aneurysmal bleeding may be localised intracerebrally and that normal CSF may not therefore always exclude this diagnosis. In most cases, however, 'altered' blood in CSF will be present due to subarachnoid haemorrhage and will render the supernatant xanthochromic. Generally this takes about four to six hours to develop after a bleed and clears in about three weeks. Discolouration associated with high circulating levels of bile pigments in the systemic circulation is seen in adults only in very severe cases of obstructive jaundice. A faint yellowish discolouration is also seen in some patients after stroke or in association with high protein content due to glioma, acoustic neuroma, spinal cord tumour resulting in obstructing to circulation of CSF (Froin's syndrome) and also in acute infective polyneuritis (Guillain-Barré syndrome). Fluorescein used in ophthalmology angiography is also occasionally found in CSF, and imparts a bright yellow colour which is intensely fluorescent under ultraviolet light.

The presence of a small fibrin clot developing usually within three hours of collection, especially on standing in the cold, is also indicative of a high protein content but is characteristically seen in association with tuberculous meningitis.

Myodil is occasionally found in CSF specimens and is seen by the naked eye as opalescence or as a whitish opaque deposit and, under the low power of the microscope, as globules which may be mistaken by the inexperienced for red cells.

Routine Tests on CSF

Glucose

Accurate and rapid determination of glucose is one of the most important procedures routinely carried out on the CSF, and great care must be taken by the laboratory to ensure high quality control standards for both the day-time and emergency on-call services.

Estimation is by a glucose oxidase method in specimens collected in sodium fluoride bottles. In the absence of preservative, glycolysis rapidly occurs especially in the presence of bacterial infection or a raised polymorphonuclear cell count. *In vivo* other factors, including a very high lymphocyte count and the presence of malignant cells, contribute towards a low concentration of CSF glucose[34] and, in addition, the plasma glucose concentration has a marked effect on both lowering and raising that of CSF. The importance of measuring plasma glucose is stressed in view of the tendency to non-ketotic hyperglycaemia of 'cerebral' origin in many patients with acute neurological disorders, especially in those on intravenous glucose preparations, and also on account of the increased tendency for tuberculosis to occur in diabetic patients; and it is in the diagnosis of tuberculous meningitis that CSF glucose is particularly important (*vide supra*). Whilst detailed studies of plasma/CSF ratios have been made and reviewed[32] it is useful to remember in day-to-day practice that it is in the presence of CSF glucose levels of less than 2·5 mmol/l (< 45 mg/100 ml) that the fasting plasma : CSF ratio is of particular importance, and that a ratio more than 2·0 can be regarded as definitely abnormal[43].

The results of serial determinations yield important information when considered in conjunction with concomitant measurement of total protein and cell count, and all parameters should be viewed against the clinical response to treatment.

Very low levels of CSF glucose are occasionally associated with purulent meningitis or widespread carcinomatous meningitis, though normal levels of glucose when present should not be regarded as significant evidence against either of these diagnoses.

Chloride

Classically, a decreased chloride concentration in CSF was regarded as helpful in the diagnosis of tuberculous meningitis. Its usefulness in practice is however extremely doubtful, and it is recommended that its determination be discontinued.

Total Protein

Routine measurements of total protein levels in lumbar CSF are important, not only in the diagnosis of neurological disorders but also in following the responses to treatment. Determinations may be made turbidimetrically or colorimetrically though drugs (including salicycates) may interfere with the latter.

In lumbar CSF the upper limit of normal concentration for total protein in adults is 0·4 g/l, though up to 0·7 g/l may be accepted in subjects in the seventh decade and upwards.

Ventricular CSF has a lower protein content than lumbar fluid with an upper limit of normal at 0·1 g/l in middle age.

Contamination of CSF with blood raises the total protein by approximately 0·01 g/l for every 1000 red cells/μl.

The Pandy Test is a useful index of raised total globulin. In the presence of normal or slightly raised total protein, a strongly positive reaction is suggestive of demyelination or neurosyphilis. The test is sensitive and reliable and renders quantitative determination of total globulin unnecessary. If the total protein is greater than 0·7 g/l a positive Pandy is meaningless.

The Nonne-Apelt Test is less satisfactory than the Pandy and should be abandoned.

Visualisation (and sometimes quantitative assay) of individual γ-globulin fractions is of great importance in diagnosis (*vide infra*), but in routine practice very useful information can be obtained from the Lange colloidal gold test provided that meticulous care is taken in preparation of the gold suspension. Some correlation can now be made between these patterns and the presence of individual proteins in the γ-region as determined by electrophoresis. Three types of pattern are classically described, namely, meningitic, luetic and paretic. The meningitic and luetic patterns mainly reflect a frank transudate whilst the paretic reflects a cathodic increase in globulin in excess of transferrin (increased γ/β ratio).

The paretic pattern is the most important abnormal response and is found most commonly in patients with multiple sclerosis (MS); general paralysis of the insane (GPI) is now rare in this country. A paretic Lange test also occurs in some patients with carcinomatous meningitis, subacute sclerosis panencephalitis and torulosis.

Tests for Syphilis

A wide range of laboratory tests of varying specificity and sensitivity is now available for the diagnosis of syphilis. All are indicative of present or past treponemal infection but none will differentiate syphilis from yaws, pinta and bejel.

The literature regarding these tests and their interpretation in CSF is sparse compared with that of serum.

In specialist neurological centres the practice is to test all CSF in parallel with serum and the same test combinations are used; namely:

1. Wassermann complement fixation test (WR),
2. VD reference laboratory flocculation test (VDRL). The tube test is found in our laboratory to be more convenient than the slide test for CSF,
3. Treponema pallidum haemagglutination test (TPHA),
4. Fluorescent treponemal antibody test (FTA(ABS)). This is not carried out if the other tests are negative.

The Reiter's protein complement fixation test (RPCFT) is no longer employed in our laboratory, having been replaced by the TPHA.

The treponema pallidum immobilisation test (TPI) is only occasionally performed now, the main indications being (1) a strong clinical suspicion of syphilis in spite of negative blood and CSF serological tests and (2) when equivocal laboratory results are obtained persistently in a patient in whom the clinical diagnosis is in doubt.

The TPI is a reliable test except in two circumstances: (1) The late stages of syphilis, particularly late congenital syphilis and long-standing taboparesis when there is a scarcity of treponemes and (2) natural infection of the rabbit test serum with treponema cuniculi,[47] though this would normally be taken into account by the laboratory.

Special mention might here be made concerning the VDRL test based on our own experience. Some workers have expressed the opinion that a VDRL test on CSF is of no value. However, of the nine cases of neurosyphilis seen in this hospital during the last year the VDRL was positive in the CSF in five. In two of these patients the first serological indication of response to treatment was a reversion of the CSF VDRL to negative and it is our view that the test is of value.

With respect to false positive tests, in the case of the WR the so-called 'biological false positive responses' in serum are well-documented, but the literature concerning false positives in CSF is again scant. It would appear, however, that the CSF is far less subject to false positive reactions than blood, and in the rare cases where the FTA has been positive, due to a collagen disease, it has been shown that the appearance of the treponemes is one of 'beaded fluorescence' as opposed to a true fluorescence, and the CSF is in reality negative.[49] False positive FTAs have been reported in cases of rheumatoid arthritis and lymphosarcoma.[22]

Up to 25% of patients with collagen disease give a positive serum VDRL and in a few cases the TPHA is also positive,[49] but this has not been described in the CSF. Sight must not be lost of the importance of the colloidal gold test in CSF as an index of treponemal infection, particularly the paretic Lange response in GPI. This finding however is not specific for syphilis, being seen more commonly in demyelinating disorders (especially multiple sclerosis) and also in carcinomatous meningitis and rarely in torulosis.

The most important CSF index of active disease is a raised cell count which is mainly lymphocytic; second in importance to the cell count is a raised total protein concentration and, thirdly, the VD serological tests. By contrast, a positive colloidal gold test is not necessarily an index of current infection, but quantitative measurement of IgM is now considered an important index of response to treatment.[5]

The cell count and raised protein level revert more quickly to normal than the serological tests, with adequate treatment.

In the case of tabes dorsalis the cell count is seldom raised whereas in GPI the number of cells is frequently raised due to an active encephalitis, and the serological and colloidal gold test remain positive—sometimes permanently.

The only quantitative test done at this hospital in routine practice is the VDRL and whereas a titre of one in eight or greater in the blood is invariably diagnostic of active disease the same criterion should not be applied to CSF.

In the presence of positive blood serology, a positive VDRL in CSF should be regarded as indicative of neurosyphilis. If the VDRL is positive in the CSF, it will usually be supported by other positive CSF findings, i.e. WR, TPHA, or FTA.

After successful treatment of early neurosyphilis the tests in CSF return to normal considerably before those in blood, but in late neurosyphilis and especially GPI an abnormal paretic Lange Curve may persist indefinitely. In an untreated patient an entirely normal CSF absolutely excludes GPI, but this does not apply to the late stages of tabes dorsalis.[4]

Some Specialised Chemical Examinations of the CSF

Protein Fractionation

Electrophoresis of normal lumbar CSF on polyacrylamide gel reveals a large number of proteins (many as yet unidentified), ranging in molecular weight from prealbumin (17 000 daltons) at the anode to α_2 macroglobulin (855 000 daltons) in the post-γ-globulin cathodic region.

Some 25 fractions have been identified (see Fig. 1). and, in general, these reflect the pattern in serum, but at a dilution of 1/200. Prealbumin, however, is at a relatively high concentration in lumbar CSF as compared with serum and is even higher in ventricular fluid.

Abnormal CSF frequently contains additional bands and the normal bands are often present in abnormal ratios.[17,27,28,52,54] These are usually expressed in relation to the mid-zone fraction—transferrin. The particular importance of acrylamide gel electrophoresis lies in its ability to give high resolution of abnormal bands in the cathodic γ-globulin region.

A number of distinct patterns can be identified in association with disease processes:

1. Contamination of CSF with Blood This gives rise to increased haemoglobin as well as haemoglobulin-haptoglobin polymers in the slow γ region.

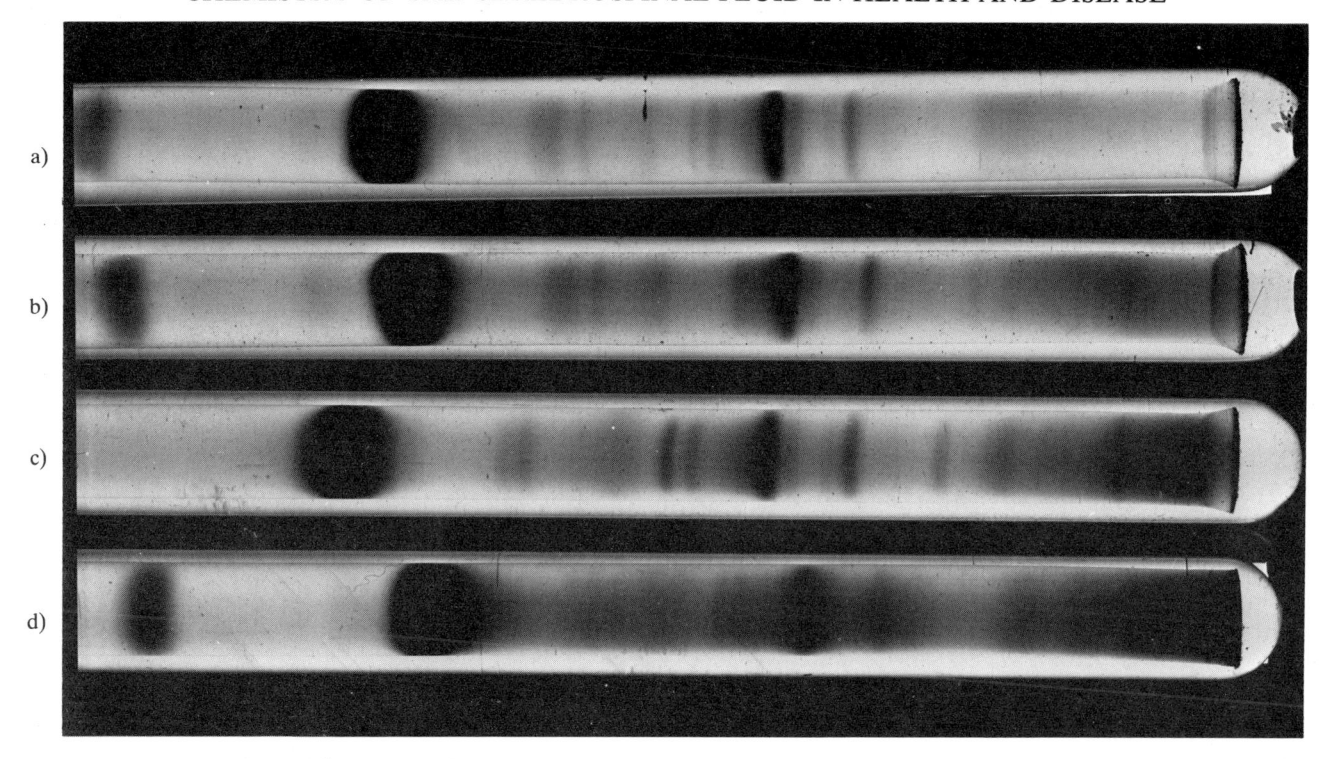

Fig. 1. Polyacrylamide gels from CSF, proteins of four patients. a) Normal b) Abnormal gamma 3, gamma 4, gamma 5 c) Abnormal gamma 4, gamma 5 and post gamma d) Abnormal gamma 4 through post a_2 macroglobulin

2. Transudation of Plasma Proteins High and low molecular weight transudations characterised by the presence of haptoglobin polymers and increased fractions of a_1-antitrypsin and orosomucoid are found in many acute and chronic inflammatory disorders of the nervous system affecting the blood: CSF barrier at the level of the choroid plexus, capillaries of brain parenchyma, meninges or dorsal root ganglia. These patterns are seen therefore in a wide range of disorders including bacterial and viral meningitis, some tumours, strokes, immunoreactive disorders and acute infective polyneuritis (Guillain-Barré syndrome).

Special mention might be made of tuberculous meningitis in which we have found increased fibrinogen degradation products, though for the present this should be regarded as a preliminary observation and not a definite test.

3. Tau Protein Pattern Lysosomal enzymes normally act on CSF transferrin causing partial hydrolysis to the tau protein. Hence any disorder which chronically impairs circulation of the CSF will result in an abnormal tau:β ratio; this pattern is therefore seen in obstructive hydrocephalus and spinal block. Clearly though, conditions associated with increased release of the lysosomal enzymes will also give rise to this pattern which is therefore found in acute and chronic destructive processes of the brain parenchyma. The latter includes Jakob Creutzfeldt disease and some forms of dementia.

4. Oligoclonal Pattern This is by far the most important abnormality from a diagnostic point of view. A typical oligoclonal pattern in lumbar CSF is found in 93% of patients with clinically definite multiple sclerosis and in 88% with early probable disease defined according to the classification of McDonald.[33] It is however not specific for multiple sclerosis (*vide supra*).

Test of Integrity of Blood-CSF Barrier

Many disorders of the nervous system both infective and neoplastic are associated with a breakdown of the normal barriers between blood-CSF and CSF-brain.

In one clinical situation a decrease in the blood-CSF barrier can be exploited to form the basis of an important diagnostic investigation, namely, the bromide partition test, for the differentiation of tuberculous meningitis from benign lymphocytic meningitis.

The definitive diagnosis of tuberculous meningitis depends on identification of acid-fast bacilli in CSF deposit at an early stage. This requires considerable expertise and if organisms are missed, or treatment with antibiotics commences, the diagnosis will be rendered difficult in retrospect. Cytological and chemical changes in CSF, though useful, provide support evidence only.

Thus, although the bromide partition test can be most useful in confirming a tuberculous aetiology, it must be emphasised that it is not specific for tuberculous meningitis and that certain other causes of raised CSF lymphocytic count (e.g. mumps encephalitis) may give a positive result, though clinically it is unlikely that there would be a diagnostic problem in practice in these situations.

Approximately 50 μCi of ^{82}Br or ^{77}Br are given orally to the patient, and exactly 48 hours later a lumbar puncture is performed and a sample of venous blood collected at the same time. Serum and CSF are counted simultaneously for radioactivity till approximately 10 000 counts are recorded from either specimen and the exact ratio of serum : CSF counts calculated. In normal subjects or those with benign lymphocyte meningitis the serum count is higher than the CSF with a ratio greater than 1·6. A ratio of less than 1·6 is indicative of an inappropriately high count in the CSF due to abnormal entry of bromide, and is strongly suggestive of a tuberculous aetiology in the appropriate clinical context. In our own experience the results of this test have always been unequivocal and, in retrospect, correct with this test. It is surprising that it is not employed more extensively, and we would recommend its use in spite of some difficulties sometimes encountered in obtaining the isotope.[30,31]

Lactic Acid

CSF lactate concentration is markedly raised in patients with bacterial and fungal meningitis. Moderately raised levels are also encountered in some patients with viral and tuberculous meningitis, though in these groups some overlap with the normal range occurs.

Several workers have advocated the use of CSF lactate concentration as an aid to diagnosis in the important clinical problem of differentiating partially treated bacterial meningitis from viral meningitis, both conditions being characterised by a lymphocytic response in the CSF. Unfortunately, overlaps occur between the groups, but a cut-off value of 35 mg/100ml (3·9 mmol/l) is useful in distinguishing many cases of viral meningitis which are below this value from bacterial meningitis which are above: some 10% of patients fall on either side, depending upon the length of treatment in the bacterial group.[6,18,24,15]

In an entirely separate context CSF lactate content expressed as a ratio to pyruvate, has been advocated and confirmed as an index of prognosis in patients with head injury.[8] Serial studies emphasised the importance of an estimation within 24 hours of injury, the raised level expressing the effect on CSF of non-respiratory acidosis as a result of lactate efflux from hypoxic cells. In one series, however, a few patients presented with decerebrate rigidity but low lactate levels, due to the presence of a localised lesion in a vital centre. This caused little lacticacidosis but grossly interfered with basic functions, and carried a correspondingly poor prognosis.

Enzymes (including enzymes in cyst fluid)

In spite of considerable work in the late 1950s and early 1960s on enzyme content of CSF, little use is made of these determinations in clinical practice today.

Most attention has been devoted to aspartate aminotransferase (AsT) and lactate dehydrogenase (LDH) (including isoenzyme patterns), although measurements have also been made of phosphohexose isomerase, isocitric dehydrogenase and creatine kinase.[3,11,45]

Sources of CSF enzymes include (1) blood plasma via the choroid plexus and also direct contamination in strokes and subarachnoid haemorrhage, (2) white cells in patients with meningeal irritation, (3) brain parenchyma, (4) synthesis in brain tumours.

Measurement of Michaelis constants have suggested that the source of enzymes is frequently blood rather than brain, though LDH_2 and LDH_3 are recognised by some workers as being of value in determining the presence of destructive lesions in the brain.

Both AsT and LDH activities rise after cerebral infarction, maximal values occurring between three and five days, and lasting for a few days to many months. The finding of an AsT activity greater in CSF than in plasma is more significant than the absolute activity and may be indicative of massive infarction and a bad prognosis.

CSF levels of creatine kinase are raised in Duchenne muscular dystrophy, but add nothing to the significance of serum levels in this condition.

CSF enzymes have proved disappointing in the diagnosis of brain tumours, but examination of enzyme content in cyst fluid of tumours has been more informative.

Further to earlier work of Cumings, a re-appraisal has been made recently of the correlation between malignancy of cystic intracranial tumours and biochemical parameters in the cyst fluid.[44]

Confirmation was established for significantly higher mean levels of cholesterol, alkaline phosphate, LDH, and phosphohexose isomerase (PHI) in tumours in the higher malignancy group (astrocytoma Grades III and IV) than in the lower malignancy group (astrocytoma Grades I and II) but only in the case of LDH and PHI were the mean levels in the high malignancy group significantly above those in the group with benign tumours. Total protein content was helpful from a diagnostic point of view. Biochemical parameters therefore failed to differentiate low malignancy tumours from benign tumours.

Vitamin B_{12} and Folic Acid

Measurements of CSF vitamin B_{12} levels have established that they are exceedingly low compared with those in serum whilst the reverse obtains with folic acid; the CSF concentration being some three times higher than in serum.

The correlation between serum (and in the case of folic acid between red cell content) and CSF concentration is good in health and in most deficiency states. In spite of a great deal of investigative work over many years especially in relation to folic acid,[1] it can now be stated that no useful clinical information can be obtained by measurement of either vitamin B_{12} or folic acid in CSF.[46]

Amines

There is some evidence that the brain content of homovanillic acid (HVA), the terminal metabolite of dopamine, reflects dopaminergic activity, though this is probably an over simplistic view. There is also doubt as to whether or not CSF HVA content is proportional to brain dopamine turnover. Nevertheless there is evidence that, as a group, non-arteriosclerotic Parkinsonian patients with low CSF HVA respond better to levodopa therapy than those with higher levels, though the predictive value of CSF HVA in individual patients is poor.

Abnormal HVA levels in lumbar CSF in other neurological disorders including Huntington's chorea, multiple sclerosis and motor neurone disease have been described but are probably secondary to impaired or disordered movement.[12]

Although evidence exists that ventricular CSF 5-hydroxyindole acetic acid (5-HIAA) content reflects brain 5-hydroxytryptophan (5-HT) turnover, the correlation with lumbar CSF 5-HIAA is not sufficiently reliable to render the latter of practical importance in clinical management of most patients with neurological disorders.[13,19,36]

Some studies, however, are consistent with the view that patients with seizure disorders have a functional deficiency of 5-HT and that certain neurologically useful drugs (including the commonly used anticonvulsants) increase brain amine concentration.[13]

It is also reported that states of anticonvulsant drug intoxication are associated with high lumbar CSF 5-HIAA content.[2]

γ-aminobutyric Acid (GABA)

GABA is the major inhibitory neurotransmitter in the CNS of all mammalian species so far studied, including man. Its membrane binding sites have been demonstrated to possess the properties of postsynaptic receptors.

The density and distribution of these receptors vary between the different regions of the normal brain, and research is currently in progress concerning possible alterations in disease states. A number of methods have been developed for measurement of GABA itself in CSF, with special reference to patients with Huntington's chorea, Parkinson's disease and the pre-senile dementias, but so far these have not proved to be of practical use in day-to-day patient management. A major reason for this is that the very low levels of GABA encountered in lumbar CSF have caused analytical problems, and, in addition, the increase in levels which occur rapidly after sample collection has not always been fully appreciated; thus reported normal and abnormal values in the literature must be interpreted with considerable caution.[20]

Prostaglandins

Prostaglandins (PG) are readily released from brain into CSF in response to disturbance of membranes as a result of trauma, anoxia, stroke, convulsions, or menigoencephalitis. A range of fatty acids is released but most published work has been devoted to $PGF_{2\alpha}$. Prostaglandins are widely distributed in brain so that liberation is of no localising value; however, thromboxane has recently been demonstrated in cerebral cortex, but not cerebellum of experimental animals. The capacity for synthesis is high but resting concentrations are low. In contrast to other tissues, the brain has a very low capacity to degrade prostaglandins so that clearance is into the general circulation via the choroidal and extra colloidal transport mechanisms.[7,56,57,58]

Unfortunately no good data exist in relation to prostaglandin release as an index of progressive tissue damage, e.g. in the active phase of multiple sclerosis though this is an area of continuing research.

Trace Metals

Trace metals analysis in CSF has not to date proved to be of practical importance in diagnosis, even of the rare neurological disorders such as Wilson's disease and Menkes' disease which are known to be associated with abnormal levels of trace metals in brain. This is probably a reflection of the fact that these, and other even rarer disorders, are associated with abnormalities which are primarily of protein binding of the trace metals, and manifestations of these abnormalities are, therefore, more apparent in blood, liver, and other tissues having a high protein content, rather than in CSF.

Drugs

The only drugs measured with any regularity in CSF are gentamicin and the tuberculosis antibiotics, particularly rifampicin, streptomycin and *iso*-nicotinic acid hydrazide. In many patients measurement of blood levels alone will suffice. In our experience, however, there are some patients who, although harbouring sensitive organisms, fail to respond to adequate dosage of antibiotics and, in some of these, assay of CSF antibiotics has demonstrated low levels. Increased dose in these subjects leads to an improved clinical response.

Concentrations of anticonvulsant drugs have also been measured in CSF on occasion, largely on an experimental basis. Levels have been found to reflect the free fraction in plasma. However, on the rare occasions where, in certain patients, an estimate of the free fraction is necessary, this can be obtained more readily by examination of the saliva.

Identification of CSF Rhinorrhoea

A copious discharge of CSF from the nose following head injury will usually be self-evident as a leak of CSF. In cases of doubt, the identification of glucose in a fresh specimen will differentiate CSF from nasal secretion. Real diagnostic problems occur, however, when the leak is minimal and irregular, and in those patients who present only with recurrent meningitis. Unfortunately, it is in these difficult situations that laboratory tests have proved unhelpful. Identification of CSF by use of radioisotopes injected intrathecally has been attempted, but is fraught with practical problems; and difficult cases are often only finally resolved by direct examination and exploration under anaesthesia.

SUMMARY

Chemical and microsopic examination of lumbar CSF has long been established as a procedure essential to the clinical management of patients with disease of the nervous system. Emphasis in the past has been mainly on the confirmation of bacterial and treponemal infections, subarachnoid haemorrhage and stroke. Some specialised tests are of value in differentiating the various cases of CSF lymphocytic pleocytosis.

Modern immunological techniques have greatly enhanced the amount of information obtainable at lumbar puncture and go some way to explaining the changes found in older and empirical tests, such as the colloidal gold response. These specialised techniques are time-consuming, however, and should be seen as complementing and not replacing traditional tests. They are already established as being of particular importance in the very early diagnosis and monitoring of demyelinating diseases, and (more speculatively) in complementing the clinical assessment of intracranial tumours.

Chemical parameters for the assessment of brain damage are at present unsatisfactory.

The estimation of excitatory and inhibitory neurotransmitters and their metabolites in CSF is of theoretical importance; but in spite of much recent research, particularly in the area of brain and CSF amines, their usefulness at present in routine clinical practice has been rather disappointing.

ACKNOWLEDGEMENTS

We should like to thank numerous colleagues for their help and advice and Miss Mavis Shackleton for the preparation of this manuscript.

REFERENCES

1. Botez, M. I. and Reynolds, E. H. (Eds.) (1979), *Folic Acid in Neurology, Psychiatry, and Internal Medicine.* New York: Raven Press.
2. Bowers, M. B. (Jr.) and Reynolds, E. H. (1982), 'Cerebrospinal-fluid folate and acid monoamine metabolites,' *Lancet*, **2**, 1376.
3. Buckell, M. (1968), 'Enzymes in the cerebrospinal fluid,' *Proc. Assoc. clin. Biochem.*, **5**, 33–34.
4. Catterall, R. D. (1977), 'Neurosyphilis,' *Br. J. hosp. Med.*, **17**, 585–604.
5. Catterall, R. D. *Personal communication.*
6. Controni, G., Rodriguez, W. J., Hicks, J. M., Ficke, M., Ross, S., Friedman, G. and Khan, W. (1977). 'Cerebrospinal fluid lactic acid levels in meningitis,' *J. Pediatr.*, **91**, No. 3, 379–384.
7. Cory, H. T., Lascelles, P. T., Millard, B. J., Snedden, W. and Wilson, B. W. (1976), 'Measurement of prostaglandin F2α in human cerebrospinal fluid by single ion monitoring,' *Biomed. Mass Spectrom.*, **3**, 117–121.
8. Crockard, H. A. and Taylor, A. R. (1972), 'Serial CSF lactate/pyruvate values as a guide to prognosis in head injury coma,' *Europ. Neurol.*, **8**, 151–157.
9. Cumings, J. N. (1952), 'The diagnostic value of the examination of the cerebrospinal fluid,' *Postgrad. med. J.*, **28**, 587–591.
10. Cumings, J. N. (1954), 'The cerebrospinal fluid in diagnosis,' *Brit. med. J.*, **1**, 449–456.
11. Cunningham, V. R., Phillips, J. and Field, E. J. (1965), 'Modified dehydrogenase isoenzymes in normal and pathological spinal fluids,' *J. clin. Path.*, **18**, 765–770.
12. Curzon, G. (1975), 'CSF Homovanillic acid: an index of dopaminergic activity,' in *Advances in Neurology.* (Calne, D. B., Chase, T. N. and Barbeau, A. Eds.). Vol. 9, 349–357. New York: Raven Press.
13. Curzon, G. (1978), Serotonin and neurological disease. In *Serotonin in Health and Disease.* (Essman, W. B., Ed.). Vol. 3, *The Central Nervous System,* 403–443. New York: Spectrum.
14. Davson, H. (1967), *Physiology of the Cerebrospinal Fluid.* London: J. and A. Churchill Ltd.
15. D'Souza, E., Mandal, B. K., Hooper, J. and Parker, L. (1978), 'Lactic-acid concentration in cerebrospinal fluid and differential diagnosis of meningitis,' *Lancet*, **2**, 579–580.
16. Dykes, J. R. W. and Stevens, D. L. (1970), 'Alterations in lumbar cerebrospinal fluid protein during air encephalography,' *Brit. med. J.*, **1**, 79–81.
17. Felgenhauer, K. and Hagedorn, D. (1980), 'Two-dimensional separation of human body fluid proteins,' *Clin. Chim. Acta*, **100**, 121–132.
18. Ferguson, I. R. and Tearle, P. V. (1977), 'Gas liquid chromatography in the rapid diagnosis of meningitis,' *J. clin. Path.*, **30**, 1163–1167.
19. Garelis, E., Young, S. N., Lal, S. and Sourkes, T. L. (1974), 'Monamine metabolites in lumbar CSF: the question of their origin in relation to clinical studies.' *Brain Res.*, **79**, 1–8.
20. Hare, T. A., Wood, J. H., Ballenger, J. C. and Post, R. M. (1979), 'γ-aminobutyric acid in human cerebrospinal fluid: normal values,' *Lancet*, **2**, 534–535.
21. Hayakawa, T., Morimoto, K., Ushio, Y., Mori, T., Yoshimine, T., Myoga, A. and Mogami, H. (1980), 'Levels of astroprotein (an astrocyte-specific cerebroprotein) in cerebrospinal fluid of patients with brain tumours,' *J. Neurosurg.*, **52**, 229–233.
22. Hooshmand, H., Escobar, M. R. and Kopf, S. W. (1972), 'Neurosyphilis,' *JAMA*, **219**, (No. 6), 726–729.
23. Illis, L. S. and Gostling, J. V. T. (1972), *Herpes Simplex Encephalitis.* Bristol: Scientechnica (Publishers) Ltd.
24. Lauwers, S. (1978), 'Lactic-acid concentration in cerebrospinal fluid and differential diagnosis of meningitis,' *Lancet*, **2**, 163.
25. Leading article. (1976), 'Diagnosis and prognosis in pyogenic meningitis,' *Lancet*, **1**, 1277–1278.
26. Leading article. (1978), 'Modified neurosyphilis,' *Brit. med. J.*, **2**, 647–648.
27. Link, H. and Tibbling, G. (1977), 'Principles of albumin and IgG analyses in neurological disorders. II. Relation of the concentration of the proteins in serum and cerebrospinal fluid,' *Scand. J. clin. Lab. Invest.*, **37**, 391–396.
28. Link, H. and Tibbling, G. (1977), 'Principles of albumin and IgG analyses in neurological disorders. III. Evaluation of IgG synthesis within the central nervous system in multiple sclerosis,' *Scand. J. clin. Lab. Invest.*, **37**, 397–401.
29. Luxon, L., Lees, A. J. and Greenwood, R. J. (1979), 'Neurosyphilis today,' *Lancet*, **1**, 90–93.
30. Mandal, B. K., Evans, D. I. K., Ironside, A. G. and Pullan, B. R. (1972), 'Radioactive bromide partition test in differential diagnosis of tuberculous meningitis,' *Brit. med. J.*, **4**, 413–415. Correspondence, p. 792; p. 794.
31. Further correspondence appertaining to No. 30, (1975), *Brit. med. J.*, **2**, 560.
32. Marks, V. and Rose, F. C. (1965), *Hypoglycaemia.* Oxford: Blackwell Scientific Publications.
33. McDonald, W. I. and Halliday, A. M. (1977), 'Diagnosis and classification of multiple sclerosis,' *Br. med. Bull.*, **33**, 4–8.
34. Menkes, J. H. (1969), 'The causes for low spinal fluid sugar in bacterial meningitis: another look,' *Pediatrics*, **44**, 1–3.

35. Millen, J. W. and Woollam, D. H. M. (1962), *The Anatomy of the Cerebrospinal Fluid*. London: Oxford University Press.
36. Moir, A. T. B., Ashcroft, G. W., Crawford, T. B. B., Eccleston, D. and Gildberg, H. C. (1970), 'Cerebral metabolites in cerebrospinal fluid as a biochemical approach to the brain,' *Brain*, **93**, Part II, 357–368.
37. Mori, T., Morimoto, K., Hayakawa, T., Ushio, Y., Mogami, H. and Sekiguchi, K. (1978), 'Radioimmunoassay of astroprotein (an astrocyte-specific cerebroprotein) in cerebrospinal fluid and its clinical significance,' *Neurologia medico-chirurg.*, **18**, 25–31.
38. Newman, J., Cacatian, A., Josephson, A. S. and Tsang, A. (1974), 'Spinal-fluid lysozyme in the diagnosis of central-nervous-system tumours,' Preliminary Communication, *Lancet*, **2**, 756–757.
39. Olson, L. C., Buescher, E. L., Artenstein, M. S. and Parkman, P. D. (1967), 'Herpes virus infections of the human central nervous system,' *New Eng. J. Med.*, **227**, 1271–1277.
40. Palfreyman, J. W., Thomas, D. G. T. and Ratcliffe, J. G., (1978), 'Radioimmunoassay of human myelin basic protein in tissue extract, cerebrospinal fluid and serum and its clinical application to patients with head injury,' *Clinica chim. acta*, **82**, 259–270.
41. Palfreyman, J. W., Johnston, R. V. Ratcliffe, J. G., Thomas, D. G. T. and Forbes, C. D. (1979), 'Radioimmunoassay of serum myelin basic protein and its application to patients with cerebrovascular accident,' *Clinica chim. acta*, **92**, 403–409.
42. Palfreyman, J. W., Thomas, D. G. T., Ratcliffe, J. G. and Graham, D. I. (1979). 'Glial fibrillary acidic protein (GFAP),' *J. neurol. Sci.*, **41**, 101–113.
43. Parsons, M. (1979), *Tuberculous Meningitis. A handbook for clinicians*. Oxford: Oxford University Press.
44. Pullicino, P., Thompson, E. J., Moseley, I. F., Zilkha, E. and Shortman, R. C. (1979), 'Cystic intracranial tumours. Cyst fluid, biochemical changes and computerised tomography findings,' *J. Neurol. Sci.*, **44**, 77–85.
45. Rabow, L. (1976), *Lactate dehydrogenase isoenzyme changes after head trauma and in brain tumours*. Sweden: Linköping University Medical Dissertations.
46. Reynolds, E. H. *Personal communication.*
47. Smith, J. L. and Pesetsky, B. R. (1967), 'The current status of *treponema cuniculi*,' *Brit. J. Vener. Dis.*, **43**, 117–127.
48. Thomas, D. G. T., Palfreyman, J. W. and Ratcliffe, J. G. (1978), 'Serum-myelin-basic-protein assay in diagnosis and prognosis of patients with head injury,' *Lancet*, **1**, 113–115.
49. Thompson, E. J., Norman, P. M. and MacDermot, J. (1975), 'The analysis of cerebrospinal fluid,' *Br. J. hosp. Med.*, **14**, 645–652.
50. Thompson, E. J. (1977), 'Laboratory diagnosis of multiple sclerosis: immunological and biochemical aspects,' *Br. med. Bull.*, **33**, No. 1, 28–33.
51. Thompson, E. J., Kaufmann, P., Shortman, R. C., Rudge, P. and McDonald, W. I. (1979), 'Oligoclonal immunoglubulins and plasma cells in spinal fluid of patients with multiple sclerosis,' *Brit. med. J.*, **1**, 16–17.
52. Tibbling, G., Link, H. and Öhman, S. (1977), 'Principles of albumin and IgG analyses in neurological disorders. I. Establishment of reference values,' *Scand. J. clin. Lab. Invest.*, **37**, 385–390.
53. Walton, J. N. (1977), *Brain's diseases of the nervous system*, 8th edn. Oxford: Oxford University Press.
54. Weisner, B. and Bernhardt, W. (1978), 'Protein fractions of lumbar, cisternal and ventricular cerebrospinal fluid,' *J. neurol. Sci.*, **37**, 205–214.
55. Whittle, H. C., Egler, L. J., Tugwell, P. and Greenwood, B. M. (1974), 'Rapid bacteriological diagnosis of pyogenic meningitis by latex agglutination,' *Lancet*, **2**, 619–621.
56. Wolfe, L. S. (1975), 'Possible roles of prostaglandins in the nervous system,' in *Advances in Neurochemistry* (Agranoff, B. W. and Aprison, M. H., Eds.). Vol. 1, 1–49. New York and London: Plenum Press.
57. Wolfe, L. S. (1978), 'Some facts and thoughts on the biosynthesis of prostaglandins and thromboxanes in brain,' in *Advances in Prostaglandin and Thromboxane Research*. (Coceani, F. and Olley, P. M., Eds.). Vol. 4, 215–220. New York: Raven Press.
58. Wolfe, L. S. and Coceani, F. (1979), 'The role of prostaglandins in the central nervous system,' *Ann. Rev. Physiol.*, **41**, 669–684.

SECTION VIII

DISORDERS OF THE CARDIOVASCULAR SYSTEM

		Page
29.	BIOCHEMISTRY OF DEGENERATIVE VASCULAR DISEASE	457
30.	CHEST PAIN AND CARDIAC FAILURE	473
31.	HYPERTENSION	479

29. BIOCHEMISTRY OF DEGENERATIVE VASCULAR DISEASE

G. S. BOYD AND I. F. CRAIG

Introduction
   Normal and atherosclerotic arteries
   Plasma and arterial lipids
   Nature of the atherosclerotic plaque
   Vascular disease in man

Plasma lipoproteins
   Origin and fate
   Chylomicrons
   Chylomicron remnants
   Very-low-density lipoprotein
   Low-density lipoprotein
   High-density lipoprotein
   Lipoprotein-a
   Lipoprotein-x

Lipid exchange factors
   Cholesterol-ester exchange-protein
   Triglyceride exchange-protein
   Phospholipid exchange-protein

Lipoproteinaemias
   Type I
   Type II
   Type III
   Type IV
   Type V
   Other defects of plasma lipoproteins
   Lecithin: cholesterol acyl transferase deficiency
   Diabetes mellitus
   Hypothyroidism

Laboratory investigations into plasma lipoprotein abnormalities
   Visual inspection
   Cholesterol and triglyceride measurements
   Lipoprotein determinations

Conclusions

INTRODUCTION

There are a considerable number of pathological conditions which affect the major blood vessels in man and in experimental animals, but *atherosclerosis* is the most frequent disease of the arterial system. Atherosclerosis is one of the major afflictions of the Western world, being the underlying cause of myocardial infarction and of much of the cerebral vascular disease occurring in Western communities. In this account of the biochemistry of degenerative vascular disease, the emphasis will be on atherosclerosis because of its major significance and also because the aetiology of this condition has been extensively studied.

Normal and Atherosclerotic Arteries

The endothelial surface of normal arteries consists of a smooth layer of cells. These cells form the interface between blood and the other components of the arterial tissue. A typical cross-section of an artery is shown in Fig. 1. The *endothelium* is separated from the *lamina propria* by a thin *basement membrane*. In turn the lamina propria is joined to the *media*, which consists of muscular and elastic tissue. The outermost aspect of the artery is the *adventitia*, and in many arteries the adventitia forms a support for many small blood vessels, the *vasa vasorum*. A portion of the vessel wall derives nutrients directly from the blood bathing the endothelial surface, while the adventitia and the media derive much of their nutrients from the vasa vasorum. The physicochemical characteristics of the arterial intima appear to influence markedly the composition of the intima and the media. Experimental work has shown that large molecules in blood plasma can enter the vessel walls through endothelial cells, presumably by pinocytosis. Similarly, studies have shown that smaller molecules may enter the arterial wall by passing between endothelial cells at cell junctions. Factors that affect pinocytosis may influence the imbibition of selected macromolecules, and factors which influence the adherence of cells to one another may also influence the penetration of blood-born particles into the arterial wall.

When the endothelial surface of large blood vessels such as the aorta is examined, there is frequent evidence of the imbibition of complex lipid-staining material adhering to the internal surface. Such raised areas of microscopic and macroscopic dimensions are frequently referred to as *atherosclerotic plaques*. If a plaque is examined in detail it can be shown to be composed of lipid, blood products such as platelets and other formed elements, smooth muscle cells, fibrin and various other identifiable blood products. Fig. 2 shows a cross-section of a normal artery from a young healthy subject, and, for comparison, a typical atherosclerotic artery from an elderly subject.

Chemical studies of arteries obtained from individuals

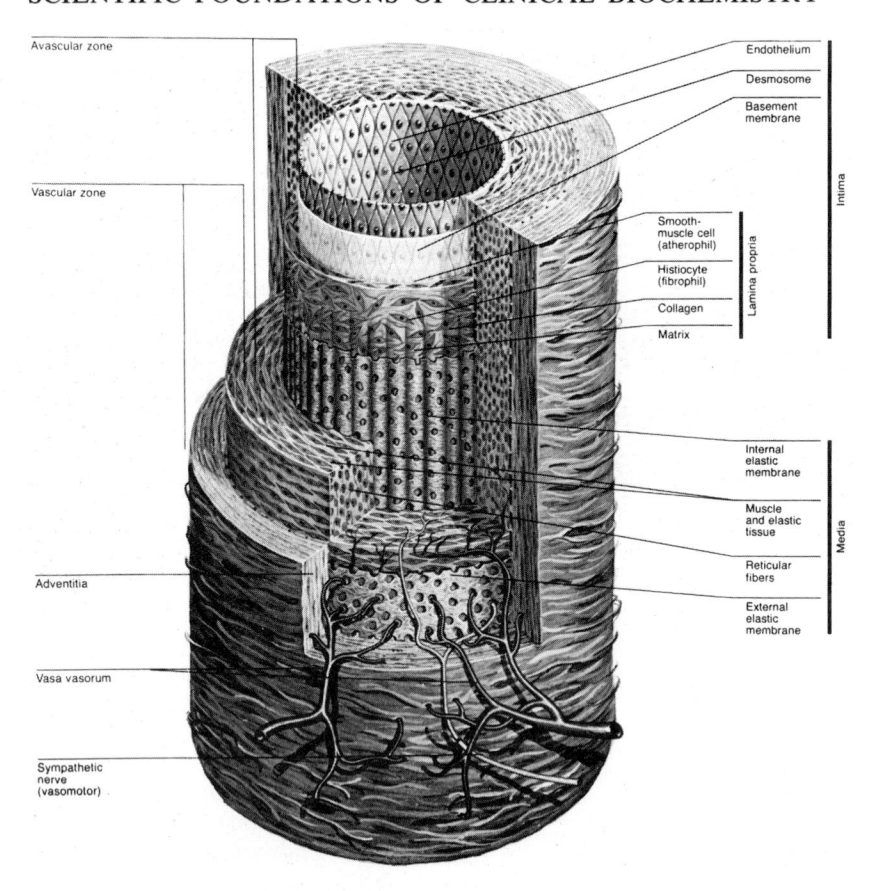

Fig. 1. A diagrammatic cross-section of an artery, showing the relationship between the various components of the arterial tissue.

of different ages show an increase in lipid content which is roughly related to the age of the individual from whom the arteries have been derived. As shown in Table 1, while various common lipids accumulate in arterial tissue the increase in cholesterol ester content is most striking. These changes in the quantity of lipid entrained in the vessel wall may be related to the pathological appraisal of such vessels.

TABLE 1

CHEMICAL COMPOSITION OF ARTERIES WITH VARYING DEGREES OF ATHEROMATOUS INVOLVEMENT.

(note that while there are changes in all the lipid components of the vessel during atheromatosis, the largest percentage change is in the content of esterified cholesterol)

	\multicolumn{4}{c}{Pathological Grading of Severity of Atheromatous Involvement}
	Normal	Slight	Moderate	Severe
Percentage lipid content of sample of arterial tissue	2	4	10	12
Free cholesterol	8	13	16	19
Esterified cholesterol	6	12	30	34
Phospholipids	61	58	42	38
Triglycerides	17	13	10	10

Histological staining procedures applied to the aorta, coronary arteries and the circle of Willis show that the number of identifiable arterial lesions within the intima—the number of atherosclerotic plaques—also increases with advancing years. These facts have tended to direct attention to atherosclerosis or the presence of atherosclerotic plaques in the intimal surface as being closely related to vascular lipid deposition. The atherosclerotic plaque is rich in lipid and the principal lipid found in such lesions is esterified cholesterol. Figure 3 shows a typical atherosclerotic plaque and illustrates the presence of esterified cholesterol, smooth muscle cells, platelets and fibrin within this complex intimal lesion. Since the atherosclerotic plaque is rich in lipid material and the lesion appears on the intimal surface, which is bathed by the circulating blood, it was natural to predict that some of the plasma lipid material might be deposited in the vessel wall by some, perhaps passive, process.

Plasma Lipids and Arterial Lipids

Such a view could be substantiated by experiments on herbivores such as rabbits, which have normally a very low plasma lipid concentration and do not as a rule

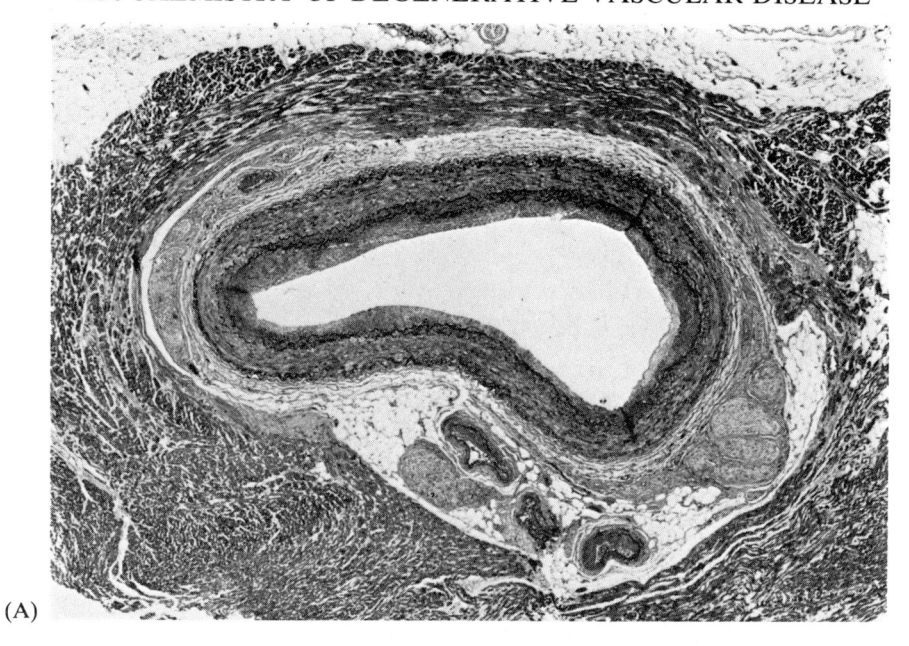

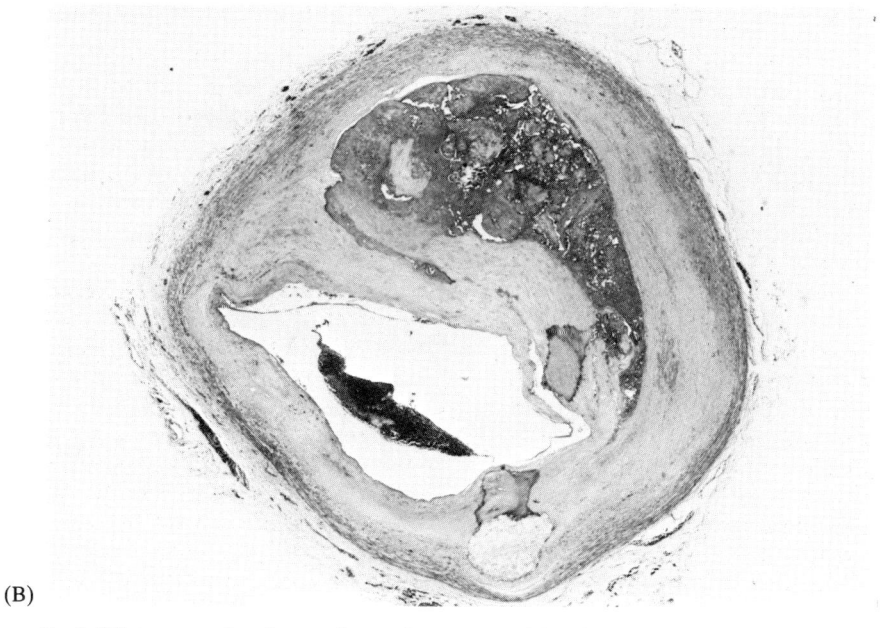

Fig. 2. (A) A cross-section of a normal artery from a young subject, showing the intima relatively free from atheromatous involvement. (B) A cross-section of an artherosclerotic artery, from a middle aged subject showing major atherosclerotic involvement of the vessel.

exhibit spontaneous atherosclerotic plaques. When rabbits are fed an artificial diet containing a substantial cholesterol supplement they show a marked plasma hypercholesterolaemia, and if such treatment is continued for several weeks they then exhibit varying degrees of lipid infiltration of the vascular system such as the appearance of fatty flecks or lipid-laden atherosclerotic plaques within the arterial intima. In general, the degree of lipid infiltration of the vessel wall is roughly related to the magnitude of the hypercholesterolaemia induced in these animals and also to the period of exposure of such animals to this hypercholesterolaemia.

Experimental approaches of this type have been conducted in a great many different animals and birds. In general, confirmation has been obtained that dietetic manipulation leading to elevated plasma cholesterol concentration results in elevated deposition of lipid in

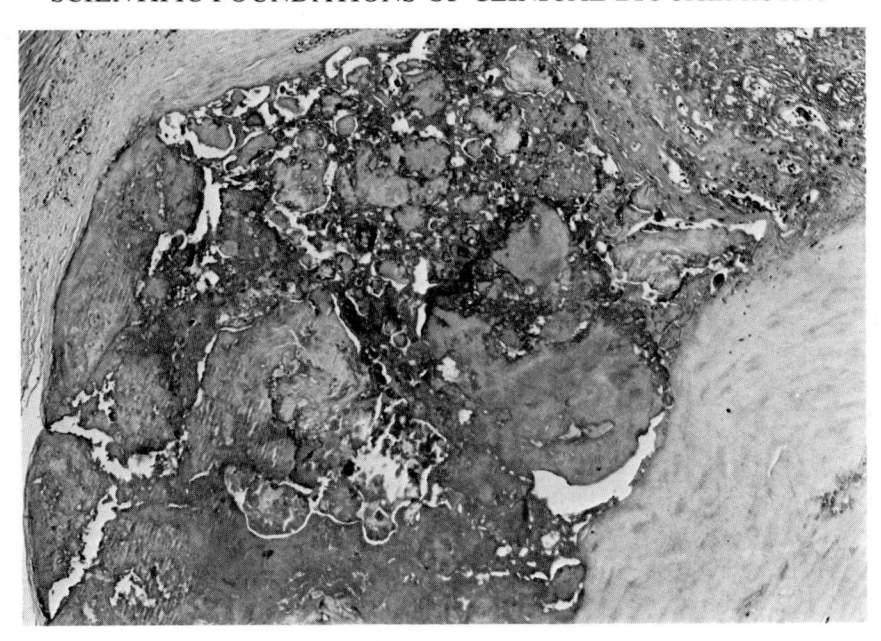

FIG. 3. A high power view of typical atherosclerotic plaque observed in a middle-aged subject.

the arterial system and an increased production of lipid-laden atherosclerotic plaques. But it has not been possible to obtain definitive evidence that artificial lesions produced in experimental animals by dietetic manipulation bear an aetiological similarity to atherosclerotic plaques appearing spontaneously in man.

In view of the ease with which arterial lipid infiltration can be produced in rabbits by induction of plasma hypercholesterolaemia, it is reasonable to deduce that such lipid deposition may be related to atherosclerotic plaque formation. In the cholesterol-fed rabbit there is usually a marked lipid sequestration or retention in the arterial wall.

Various views on the aetiology of atherosclerosis have been actively debated for over half a century; it now seems quite clear to most experimental pathologists, physiologists, biochemists and others that the lipid infiltration theory is perhaps an oversimplification of the complex process which prevails in human atherosclerosis. The atherosclerotic lesion typified by the raised ulcerated atherosclerotic plaque observed in major arteries of middle-aged individuals is indeed a much more complex structure than the lipid deposition observed in rabbits fed a high cholesterol diet for a few weeks. Nevertheless, many of these experimental models which have been produced within recent years by dietetic manipulation have emphasised that the combination of hypercholesterolaemia with certain other physiological or pathological constraints on the organism can result in the appearance of atherosclerotic plaques in experimental animals, and these plaques have in some cases a remarkable similarity to plaques appearing spontaneously in man.

In a great many of these experimental situations, plasma hypercholesterolaemia is a significant and almost invariable factor in the accelerated appearance of the vascular atherosclerotic lesion. Studies over the last 50 years have shown that the lipids in blood plasma occur as a family of *lipoprotein* macromolecules with high molecular weights. As will be discussed later, the evidence has grown in recent years that some of these macromolecules in blood plasma can penetrate into vessel walls through the intimal surface either by pinocytosis or by leakage via cell-cell junctions. These facts reinforce the concept, often referred to as the *ultrafiltration theory*, that lipid infiltration may play a part in predisposing arterial tissue to certain physiological changes, which may result in the ultimate appearance, within and without the vessel, of the structural changes identifiable as atherosclerotic plaques. Plasma lipoproteins may penetrate the basement membrane, and in the media the macromolecules may be subjected to processes akin to ultrafiltration and molecular sieving, as well as to sequestration events similar to ion-exchange chromatography or affinity chromatography. The behaviour and fate of plasma lipoproteins within the vessel wall will be influenced by the size and the charge of such macromolecules. In addition, specific immunological aspects of the plasma lipoproteins may affect their fate in the vessel wall.

The presence of plasma lipoproteins in higher concentration than normal, or the presence of abnormal lipoproteins, may result in these molecules being deposited in whole or in part within the intima or the media of arteries during the process of physiological ultrafiltration through the vessel walls. One of the unsolved problems is the nature of the aberration in the ultrafiltration reaction which predisposes the vessel to lipid deposition as a possible prelude to atherosclerotic plaque formation.

Because of the difficulties associated with attempts to explain atherosclerosis solely on the ultrafiltration theory, a number of other theories have been proposed. In one of these, emphasis is placed upon the appearance of *smooth muscle cells* in the media of arteries. It is known that smooth muscle cells are present in atherosclerotic plaques, and such cells appear to have the ability to synthesise collagen and elastin. It is also known that smooth muscle cells migrate; hence factors which may influence the proliferation of smooth muscle cells or the nutrition of the cells have been extensively studied. Certain plasma lipoproteins are imbibed by such cells, and blood platelets produce a protein which influences the growth and behaviour of smooth muscle cells. It is for this reason that the proliferation of smooth muscle cells within the intima of certain arteries has attracted the attention of experimental pathologists concerned with the aetiology of atherosclerosis. These studies have emphasised that the process of atherosclerosis cannot be attributed simply to the accumulation of lipids in the vessel walls. Much attention is focused at present on the possible role of smooth muscle cells and their proliferation in arteries in the aetiology of the condition which is ultimately identifiable as atheroma.

Nature of the Atherosclerotic Plaque

The pleomorphic appearance of atherosclerotic plaques in man emphasises that a typical vascular lesion is composed of lipid, smooth muscle cells, fibrin, blood platelets and other blood products.

The role of *blood platelets* in the aetiology of vascular disease is now accepted. These small formed elements produced by megakaryocytes in bone marrow, circulate as small oval platelets in blood plasma and have a biological half-life of about 7 days. Blood platelets have a sensitive limiting membrane which confers on them the ability to stick to rough or charged surfaces. During this adhesion process certain components of the blood platelets such as 5-hydroxytryptamine, adenosine diphosphate (ADP), thromboxanes and various other low molecular weight materials may be shed into plasma. ADP and thromboxanes have the ability to influence platelet aggregation. Thus platelets secreting ADP and thromboxane become more sticky and tend to adhere to one another. It is for this reason that some of the events involved in platelets sticking to the rough or charged surfaces within the endothelial surface produce a cascade-like reaction, and these events may well be precursors of thromboembolic episodes.

In addition to ADP and thromboxane, many other physiological factors influence platelet stickiness. Consequently, situations which tend to stretch the arterial intima may bring about a widening of the gap junction between endothelial cells, so that subendothelial materials such as collagen are exposed to blood platelets. The adherence of platelets to collagen may result in the platelets shedding materials such as ADP and thromboxane, and these in turn produce the cascade-like platelet aggregation reaction previously mentioned. It is likely that haemodynamic factors, permeability factors and nutritional factors all contribute to the alterations in the physiology and biochemistry of blood platelets. The stability of these formed elements in blood may control trigger events that can result in thromboembolic cascade reactions.

An artery that is compromised by lipid infiltration or by haemodynamic influences may be predisposed to a platelet-triggered *thromboembolic* episode. The resulting thrombus will be composed of modified plasma proteins such as fibrin, together with formed elements from the blood such as red blood cells, white blood cells and platelets. Such a thrombus is likely to become organised and incorporated into the vascular intima, resulting in a complex plaque containing lipids, platelets, fibrin, etc. It is for this reason that many workers in the field of experimental atherosclerosis subscribe to a view that platelet stickiness, platelet aggregation and platelet disaggregation are significant factors in the aetiology of vascular lesions.

Vascular Disease in Man

It has been known for many years that certain individuals appear to be predisposed to premature cardiovascular occlusive episodes. Metabolic investigations on such individuals confirm the view put forward as a result of animal experimentation that hypercholesterolaemia is a significant factor in premature atherosclerosis and cardiovascular disease in man. As a result of elegant prospective studies performed on large numbers of human volunteers in different centres in the world, it has been possible to identify and quantify some of the physiological and pathophysiological factors which appear to predispose to atheromatous vascular disease in man. Such primary risk factors include *hypercholesterolaemia, hypertension, excessive cigarette smoking* and certain other factors, for example diabetes, obesity, physical inactivity and stressful situations. In this account of the role of the biochemical laboratory in the diagnosis and management of atheromatous vascular disease, emphasis is placed on the hyperlipidaemic state as a primary target for attention.

PLASMA LIPOPROTEINS

In man there are various lipid-containing macromolecules present in blood plasma. As a consequence of the many studies performed over the past 50 years, the physicochemical nature of the plasma lipids is slowly unfolding. In plasma there may be present triglyceride-rich lipid particles of microscopic dimensions called *chylomicrons*. In normal individuals chylomicrons are found only in blood plasma in the postprandial state. There is excellent evidence that chylomicrons originate as a consequence of the absorption of dietary lipids in the gut. As a result of controlled hydrolysis of the dietary triglyceride, absorption occurs of partially degraded triglycerides into the mucosal cells. Much of the triglyceride absorption occurs in the ileum, and this process is dependent upon the presence in the gut of bile salts, which assist in the emulsification of the lipid together with triacylglycerol hydrolase (triglyceride lipase). The resynthesis of triglyceride occurs in the intestinal cells, and the production of spherical triglyceride-rich particles containing specific proteins is a metabolic event occurring in the intestinal mucosa. Chylomicrons are then extruded through the basement membrane of the mucosal cells into lymph, from where the lipid-laden particles are discharged into blood plasma. Within the vascular system the chylomicrons are metabolised by a controlled lipolytic reaction which will be discussed later.

By various physical and chemical means, particularly analytical ultracentrifugation, it has been possible to identify, in blood plasma, lipid particles substantially smaller in diameter than chylomicrons; these particles are triglyceride-rich and have been termed *very-low-density lipoproteins* (VLDL). Present evidence suggests that VLDL particles are synthesised in the gut and in the liver, and that they circulate in the vascular system until degraded like chylomicrons by a controlled lipolytic event. The immediate metabolic fate of VLDL is the hydrolysis of the constituent triglycerides leading to the production of other lipoproteins quite distinct from VLDL, and these are the *low-density liproteins* (LDL). The LDL particles are therefore less rich in triglyceride but contain substantial quantities of free and esterified cholesterol, and phospholipids. LDL particles contain specific apoproteins, mainly apoprotein B, and the stability and the integrity of the lipid-laden LDL macromolecule are dependent upon the presence of such specific apoproteins.

Human blood plasma also contains families of lipoproteins termed *high-density lipoproteins* (HDL). HDL appear to originate in liver, and this class of lipoprotein has a specific function in lipoprotein metabolism. HDL may have a role in transporting or transferring lipids between different lipoproteins and also between different tissues. HDL may also have a role in providing certain specific apoproteins which function as cofactors with certain key enzymes involved in plasma lipoprotein transformations. The approximate chemical composition of chylomicrons, VLDL, LDL and HDL is shown in Table 2.

Origin and Fate of the Plasma Lipoproteins

The plasma lipids, with the exception of free fatty acids which are complexed with albumin, are transported in the blood as lipoproteins. Table 2 shows the major lipoprotein classes and how they differ in the proportions of cholesterol, cholesterol ester, triglyceride, phospholipid and protein. The lipoproteins are further distinguished by the amount and type of specific *apoproteins* that occur in each lipoprotein class. There is a trend towards a higher protein to lipid ratio as the lipoprotein particle size decreases and the hydrated density increases.

The epidemiological correlation of an increased incidence of coronary artery and peripheral vascular diseases with elevated plasma lipid concentrations has stimulated the study of the major plasma lipoprotein classes. There is a relationship between increased plasma concentrations of the lipoproteins containing apo B (VLDL and LDL) and an increased prevalence of vascular disease. However, chylomicrons contain apoprotein B but there is no correlation between plasma chylomicron concentrations and predisposition to vascular disease. Nevertheless, chylomicron degradation products may be harmful to arteries. The HDL class, which contains the lowest ratio of cholesterol to protein, has been extensively studied since epidemiological studies suggest an inverse relationship between plasma levels of HDL and the incidence of atherosclerosis. The working hypothesis has emerged that elevated concentrations in plasma of the cholesterol-rich lipoprotein, LDL, are harmful and predispose to premature atherosclerosis. Conversely, elevated concentrations of the cholesterol-poor lipoprotein, HDL, in plasma may confer protection against premature atherosclerosis.

The plasma lipoproteins are a source of cholesterol for the production of bile acids in the hepatocyte and steroid hormones in the gonads and the adrenal. Plasma lipoproteins may also supply phospholipid and cholesterol for cellular membrane construction, as well as providing reserves of triglyceride for energy production. The plasma lipoproteins transport lipids and carry fat-soluble vitamins, drugs, and hydrophobic materials including some carcinogens.

TABLE 2
LIPID CONTENT OF VARIOUS PLASMA LIPOPROTEIN CLASSES

Lipoprotein Class	Density	Relative Amount (% by weight)				
		Triacyl-glycerol	Cholesterol	Cholesterol ester	Phospholipid	Protein
Chylomicron	0.95	84	2	5	7	2
Chylomicron remnant	1.063	79	3	12	4	2
VLDL	0.95–1.006	70	3	6	11	10
LDL	1.019–1.063	8	8	39	23	22
HDL	1.063–1.21	5	4	15	23	53
HDL_2	1.063–1.125	6	6	16	28	43
HDL_3	1.125–1.21	5	3	12	22	58
HDL_c			10	40	30	20

The study of the origin and metabolic fate of plasma lipoproteins is complicated by the ease of exchange of the surface components between individual lipoproteins and between lipoproteins and cell membranes. In most cell types studied, however, lipoprotein-cell interaction results in uptake of the entire lipoprotein. The chemical composition of the plasma lipoproteins is constantly changing. The site of lipoprotein synthesis is now well established and studies have been made on the mechanism of secretion into plasma. As mentioned previously, the plasma lipoproteins may be modified within the vascular bed before being phagocytosed by specific cell types for degradation by lysosomes. Cholesterol and phospholipid, having some polar groups, are situated in the surface layers of the plasma lipoproteins and can exchange freely between different lipoproteins and between lipoproteins and cell membranes; the non-polar cholesterol ester and triglyceride which lie under the lipoprotein surface layer cannot exchange in this way and require the assistance of specific transport proteins. Phospholipid transport proteins have also been identified in plasma and in bovine liver. The action of these transport proteins complements the metabolic transformation of the lipoprotein mediated by the lipoprotein lipase situated on the vascular endothelium of heart, skeletal muscle, adipose tissue, lung and mammary gland, and by the plasma enzyme lecithin-cholesterol acyltransferase (LCAT), responsible for much of the esterification of cholesterol in plasma.

Several reviews concerned with plasma lipoproteins have been published recently.[1,2,3]

Chylomicrons

After ingestion of a fat-containing meal, the dietary lipids are partly hydrolysed and absorbed into the intestinal mucosa. They are then resynthesised into triacylglycerol (triglyceride), cholesterol ester, and phospholipid. The resulting lipids are combined with specific apoproteins to form the main intestinal lipoproteins, chylomicrons and VLDL. Chylomicrons are spherical triacylglycerol-rich particles found in lymph and ranging in diameter from 0·05–0·5 microns. They consist of a core of triacylglycerol with traces of cholesterol ester surrounded by a monolayer surface film 2·5–3·0 nm wide. The surface film is composed of phospholipid (61%), diacylglycerol (24%), triacylglycerol (1%), cholesterol (3%), monoacylglycerol (0·5%), fatty acids (2%) and proteins (8%). The proteins include apoproteins A, B, C & E, with apo B as the major protein component. The C apoproteins comprise 3 polypeptides, C-I, C-II and C-III, ranging in molecular weight from 7000 to 11 000. Apo C-II activates lipoprotein lipase; chylomicrons isolated from rat intestinal lymph contain apo C-II as the major apoprotein.

As a spherical particle decreases in size the ratio of its surface area to its volume increases. The smaller chylomicrons contain a greater proportion of polar surface components such as phospholipid and protein, and a lesser proportion of non-polar triglyceride.

The apoprotein pattern is independent of chylomicron size, which in turn is a function of the ratio of protein to surface lipid. At high fat absorption rates more triacylglycerol may be packaged in large chylomicrons by a given amount of apoprotein, simply from surface/volume consideration.[4] The fatty acid composition of chylomicron triacylglycerol is dependent on the composition of the diet, but the phospholipids in chylomicrons are less sensitive to alteration by diet.

The metabolic fate of the intestinal lymph chylomicrons is complex. After synthesis and secretion into lymph, the chylomicrons pass into the thoracic duct and thence into the plasma. There are many differences between chylomicrons isolated from lymph and from plasma. The former contain apo C, apo B, apo A-I and apo A-IV as major protein components, while apo A-II and apo E are relatively minor components ($< 5\%$ of the total apoprotein). In the human, plasma chylomicrons contain little apo A-I and have apo B, apo E and apo C as their major apoproteins. Incubation of lymph chylomicrons, with plasma, results in the chylomicrons losing some of their phospholipids and gaining unesterified cholesterol and certain proteins. Cell culture studies suggest that plasma chylomicrons may bind to endothelial cells and be subsequently interiorised and metabolised by a lysosomal pathway. In addition, the chylomicron cholesterol ester may be partially removed by the cells of the vascular surface without interiorisation of the particle, providing the cells with a source of cholesterol. The metabolism of plasma chylomicrons in the rat has been elucidated. The initial process is the removal of most of the triglyceride by lipoprotein lipase. Associated with this lipolytic process is the removal of selected surface components. It is thought that phospholipid is transferred initially to LDL with a subsequent transfer to HDL. Chylomicron apoproteins A-I and A-IV also transfer to the HDL surface, and it is suggested that chylomicron components may be an important source of HDL components. The term *chylomicron remnant* has been used to describe the residual particle left after much of the triacylglycerol contained in newly synthesised chylomicrons has been removed by lipolysis. Thus chylomicron remnants are proportionately enriched in cholesterol.

Chylomicron Remnants

The precise characteristics of the chylomicron remnant particles are dependent on the system used to generate them from chylomicrons. The remnants can be formed by injection of lymph chylomicrons into functionally hepatectomised rats, or by perfusion through isolated organs, or by soluble lipoprotein lipase *in vitro*. There is no evidence that any of the methods produces a discrete monodisperse chylomicron remnant. The median particle diameter is about 90 mμ for the remnants as compared with 200 mμ for unmetabolised chylomicrons, and the sedimentation value (S_f) of most remnant particles is greater than S_f 400.

During chylomicron remnant formation *in vivo* at the

vascular endothelium, the remnants may be repeatedly bound and released so that the existence of a spectrum of remnant particles is to be expected. Particles retaining more than 30% of their initial triacylglycerol content are good substrates for lipoprotein lipase. Intact chylomicrons, VLDL, and their partially lipolysed remnants are believed to compete for lipolytic sites at the endothelial surface. The loss of chylomicron phospholipid during remnant formation involves both the transfer of intact molecules and the formation of lysophosphatidylcholine by phospholipase C activity which is found in association with lipoprotein lipase. An approximate composition of chylomicron remnants is given in Table 2.

The intravenous infusion of chylomicron remnants into normal rats results in a rapid (less than 10 min) removal of most of the injected material. The cholesterol ester and about half of the injected labelled fatty acids are found in the liver. The hepatic uptake of remnants is believed to be a saturatable process, and the internalised remnants reduce hepatic cholesterol synthesis. The chylomicron remnant appears to be a much more effective inhibitor of hepatic cholesterol synthesis than either HDL or LDL, but the reason for this is obscure. It may be due to the chylomicron remnant containing cholesterol from the diet or from *de novo* synthesis in the intestine, whereas much of the cholesterol carried in LDL, HDL and VLDL originates in the liver. It is possible that hepatocytes handle remnants, LDL, and HDL in quite different ways. The fate of HDL and LDL in the hepatocyte may not be immediate lysosomal hydrolysis followed by release of free cholesterol, whereas this seems to be the fate of the remnants.

Chylomicron remnants are almost exclusively endocytosed by the parenchymal cells of the liver. Thus, the factors that cause chylomicron remnants and not chylomicrons to be rapidly taken up by the liver are the subject of much discussion. The loss of certain specific apoproteins by chylomicrons in their conversion to remnants may predispose the latter to hepatic uptake.

The uptake of triglyceride-rich lipoproteins into hepatocytes appears to be regulated by the specific apoproteins of the lipoproteins. Studies with chylomicron remnants prepared *in vitro* and hepatoma cells suggest that C apoproteins of the lipoproteins are involved in the mechanism of uptake by the hepatoma cells. However, in the perfused rat liver the uptake of chylomicron remnants into parenchymal cells appears to be regulated by a receptor in the parenchymal cell membrane that is specific for apoprotein E. This has been conclusively demonstrated by the preparation of artificial lipid dispersions containing purified apo E; in the absence of apoprotein E the dispersion is not phagocytosed, whereas in the presence of apo E there is complete uptake of the protein-lipid complex. In this system the presence of the C apoprotein, specifically apo C-III, is believed to oppose the recognition of the lipoproteins by the receptor. The nascent lymph chylomicrons appear to obtain C apoproteins from HDL in plasma, and this apoprotein may hide the receptor recognition site in the chylomicron from receptors in the vascular bed, promoting the lipolytic metabolism of the chylomicron to the remnant. As the remnants form and the C proteins recycle to HDL, the catalytic activity of lipoprotein lipase is reduced and the recognition site is exposed.

Very Low Density Lipoproteins (VLDL)

In man, VLDL is the major transport vehicle for the removal of endogenously synthesised triacylglycerol from the gut and the liver. The VLDL particles are heterogeneous, their composition varies and they range in size from 30 to 75 nm. The size of the VLDL particle is related directly to the triacylglycerol content and inversely to the phospholipid and protein content. The approximate composition of VLDL is given in Table 2. The ratio of esterified to unesterified cholesterol in VLDL is approximately 2:1. VLDL is synthesised in the intestine and the liver, and while VLDL from these two sites is similar in many respects there are distinct differences, especially in the apoprotein moiety. Newly secreted intestinal VLDL contains apo A-I, apo B and apo A-IV, but no apo E. On the other hand, newly synthesised hepatic VLDL contains apo E, apo B, very small amounts of apo C and little cholesterol ester. Serum VLDL contains much apo C and lesser amounts of apo B. The relative abundance of each of the apoprotein C peptides (C-I, C-II & C-III) varies markedly among individuals and with age. The larger the VLDL particle the greater is its relative content of apoprotein C and the lower its content of apoprotein B. VLDL is considered to be catabolised in the circulation in the following way. The VLDL is converted to LDL by the action of lipoprotein lipase of muscle and adipose tissue. This lipolytic step is possibly also effected by a similar but functionally distinct hepatic lipase. During the conversion of VLDL to LDL, apoprotein C and surface components including cholesterol are transferred to HDL. The initial metabolic clearance rate of VLDL is similar to that of chylomicrons with most of the triacylglycerol being delivered to peripheral tissues. The VLDL remnant particle produced by the action of lipoprotein lipase contains all of the original apoprotein B. In the second step in the catabolism of VLDL, the remnants are taken up by hepatocytes and catabolised further.

Low Density Lipoproteins (LDL)

As stated in the previous section, LDL is derived from VLDL by the action of lipolytic enzymes. Little if any LDL is secreted by the intestine or by the perfused liver. LDL is much more homogeneous with respect to size than chylomicrons of VLDL, having a diameter in the range of 20–25 nm. A typical composition is given in Table 2. The main protein component is apoprotein B, but traces of other proteins are present. Glycerophospholipids and sphingomyelin account for 65% and 25% of the total phospholipid respectively, in which linoleic acid is the major fatty acid. The reported molecular weight of LDL is $2–3.5 \times 10^6$. The metabolic

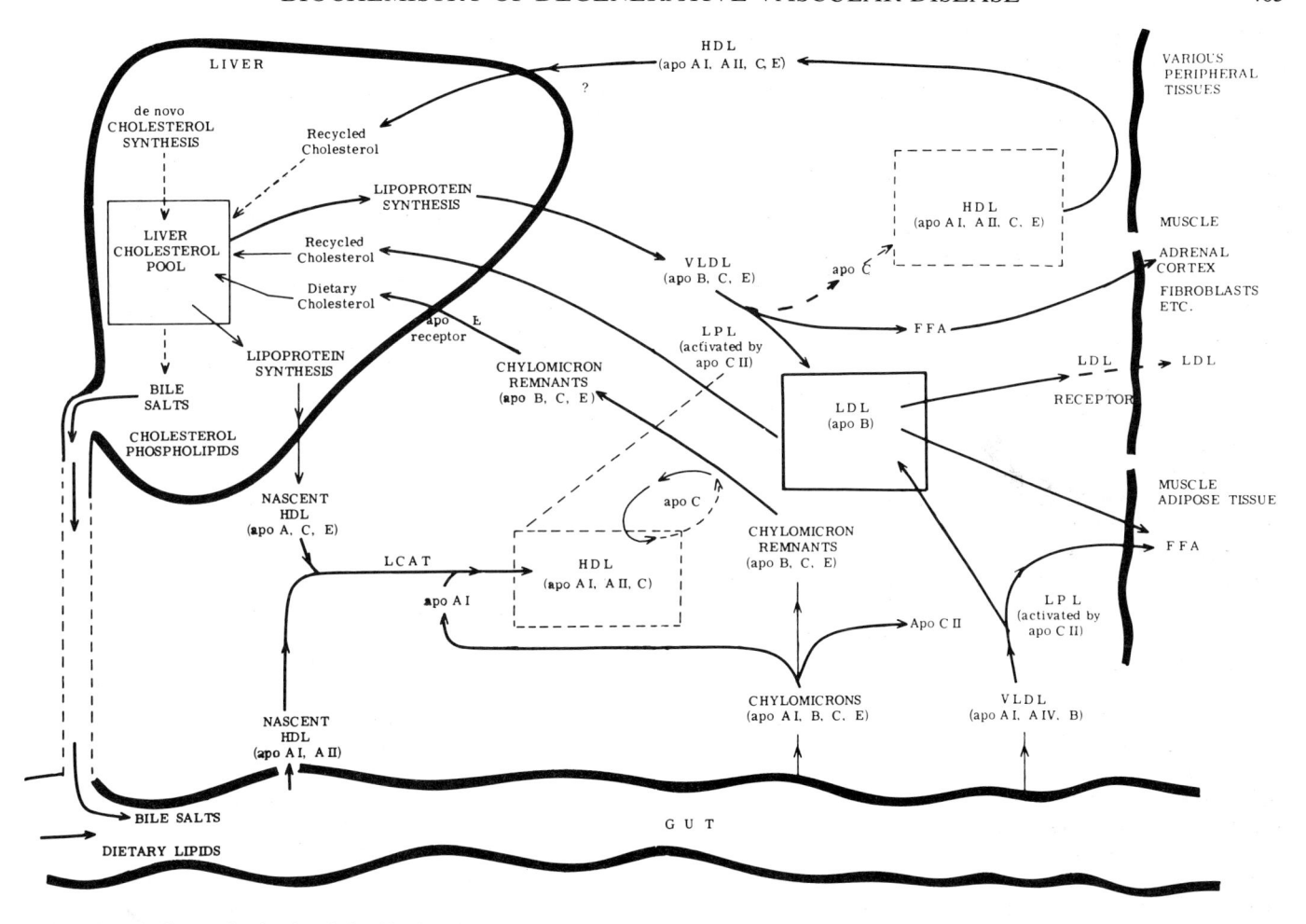

FIG. 4. A schematic diagram showing the relationship of the liver, peripheral tissues and the gut in lipoprotein transformations. In this diagram, the liver and the gut are shown as major sources of nascent HDL and VLDL. Selected peripheral tissues are involved in the uptake of LDL through LDL receptors. The liver is shown as participating in the uptake of chylomicron remnants.

fate of LDL is uncertain. There is disagreement surrounding the relative contribution of the liver and extrahepatic tissues in the catabolism of LDL. As the liver is the major site of cholesterol metabolism and, in man, LDL is the major carrier of plasma cholesterol, it might be reasonable to expect that LDL should be metabolised by the liver. However, comparison of the clearance rates of ^{125}I LDL in the intact rat and in the isolated perfused liver showed that the rate of hepatic LDL catabolism was only 7% of that in the intact rat. Similar experiments conducted in the pig identified the liver as a major site of LDL catabolism. Unfortunately, much of the radioactivity associated with the liver in labelled LDL tissue distribution studies may be due to LDL associated with liver in the extravascular pool and not necessarily to LDL undergoin g degradation. This problem has been partially solved by the use of ^{14}C sucrose covalently linked to LDL. The ^{14}C sucrose is then largely retained in the lysosomes, and cellular accumulation of such ^{14}C sucrose is an index of LDL degradation. However, this technique has been explited only in the study of LDL catabolism in pigs.[5] About 80% of an injected dose of labelled LDL was recovered from the tissues of the pigs 24–48 h after injection. All tissues and organs contained the label. These included the liver, skin, adrenals, spleen, lymph nodes and kidneys. Quantitatively the most important tissues in terms of overall LDL degradation were liver, adipose tissue, muscle and small intestine. The most active tissues in LDL degradation, on a unit-weight basis, were the adrenals, liver and spleen, in that order.. The molecular mechanism involved in the uptake and utilisation of LDL by various extrahepatic tissues has been studied in a number of tissues including ovaries, adrenals and the fibroblast system. Several excellent reviews have appeared recently.[6,7]

Basically, a receptor exists on the fibroblast, which recognises the lysine or arginine moieties of the apoprotein B of LDL. Interaction of LDL with this receptor facilitates internalisation of the LDL and delivers the lipoprotein into secondary lysosomes, where the LDL is degraded and the cholesterol is released as free (non-esterified) cholesterol. The uptake of LDL regulates the number of LDL receptors recycling to the surface and hence limiting the rate of uptake of further LDL molecules. In some types of hypercholesterolaemia, in which the plasma cholesterol concentration may rise due to elevated LDL concentrations, the receptor numbers are drastically reduced, preventing uptake by the normal mechanism. LDL may also be removed from the circulation via the reticuloendothelial system; macrophages appear to contain functional receptors for LDL. These macrophage receptors are slightly different from those of the fibroblasts in that they are not regulated by the uptake of LDL and do not require the LDL to be negatively charged. This could be of importance in the large vessels, as one by-product of thromboxane synthesis is malonaldehyde. It has been suggested that malonaldehyde could react with apoproteins to reduce the negative charge on apoprotein B, thus diminishing the uptake of LDL by fibroblasts but leaving macrophage uptake unaffected. However, there is no indication of how important this process might be quantitatively.

High-Density Lipoproteins (HDL)

HDL contains the highest proportion of protein; it has been shown that these macromolecules can be synthesised in the isolated perfused intestine and in the perfused liver. HDL of hepatic origin contains apoproteins A and C; intestinal HDL contains only apoprotein A. Newly synthesised HDL (nascent HDL) appears as flattened discs stacked like coins, with the protein components orientated around the edge of the discs. Newly synthesised HDL is deficient in non-polar lipids such as cholesterol esters and triacylglycerol, but rich in polar lipids such as cholesterol and phospholipid. Nascent HDL can be found in the plasma of patients with a deficiency of the enzyme lecithin-cholesterol acyltransferase (LCAT). As the LCAT reaction proceeds, the relatively polar unesterified cholesterol is converted to non-polar cholesterol esters. These esters seek a hydrophobic environment and penetrate into the HDL to form a core of neutral lipid. The plasma HDL also acquires triacylglycerol, presumably by an exchange reaction occurring in association with VLDL lipolysis, but the mechanism of this exchange process is still obscure.

Plasma HDL consists of a central non-polar core with an outer polar shell of about 1·4 nm. The core contains cholesterol ester, and the outer shell contains protein and phospholipid. Human HDL obtained by ultracentrifugation contains the apoproteins as shown in Table 3. HDL derived from human plasma may be subdivided into HDL_2 (d 1·067–1·12, diameter 7–10 nm) and HDL_3 (d 1·125–1·21, diameter 4–7 nm). The differences in composition are detailed in Table 3. The previously stated figures for apoprotein composition were for total HDL and are altered as a result of the HDL subfractionation. For example, the mean ratio, by weight, of apoprotein A-I to apoprotein A-II for HDL_2 is 4·75, while the ratio, by weight, of apoprotein A-I to apoprotein A-II for HDL_3 is 3·65. HDL_2 is proportionally richer in apoproteins C and E than HDL_3. Both HDL_2 and HDL_3 are metabolised at similar rates and the HDL apoproteins can exchange within the subfractions. In addition, recent work *in vitro* suggests that the transformation of HDL_3 to HDL_2 occurs in association with the lipolysis of VLDL by lipoprotein lipase.

HDL derived from rat plasma contains a higher concentration of apoprotein E than does human HDL. Apoprotein A-IV is the major apoprotein in rat HDL, and the function of this lipoprotein may be to deliver cholesterol to rat liver in a way similar to the suggested function of LDL in humans. A further HDL species has been isolated from dogs fed cholesterol; it is termed HDL_c and is lipid-rich. The composition of this lipoprotein is detailed in Table 2. It is characterised by the high proportion of apoprotein E and the absence of apoprotein B. It may be isolated over a wide range of densities due to the lipid protein variability in the HDL_c particle. The HDL_c can appear between 1·019 and 1·063 due to this lipid variability and is difficult to separate from LDL. Subfractions of HDL_c are found containing only apo E, and intravenous injection of these particles results in their rapid removal from plasma. This efflux of plasma HDL_c may be achieved by a similar route to that for the uptake of LDL. Cell culture studies have shown that HDL_c and LDL bind to the same cell surface receptors, with the apoprotein-E-rich HDL_c having a higher affinity for the plasma membrane receptor than LDL. This may be associated with a greater number of receptor recognition sites in HDL_c as compared with LDL, or a greater affinity of HDL_c for cell surface receptors.

The fibroblast-binding activity of HDL_c can be accounted for by the presence of apoprotein E in a minor (less than 15%) subfraction of HDL. The conversion of HDL_2 to HDL_c is deduced to occur in healthy human volunteers fed on high cholesterol diets. However, the role of HDL_c in the regulation of cellular cholesterol metabolism *in vivo* remains to be established.

Much has been published recently on the inverse relationship between plasma HDL levels and the incidence of ischaemic heart disease. Premenopausal women, who have a reduced incidence of ischaemic heart disease compared with men of the same age, have

TABLE 3
DISTRIBUTION OF VARIOUS APOPROTEINS IN THE VARIOUS PLASMA LIPOPROTEIN CLASSES

Apoproteins	Percentage of each apoprotein in specific lipoprotein classes				Physiological function
	Lymph chylomicrons	VLDL	LDL	HDL	
A-I	7·4	Trace	—	67	Activates LCAT
A-II	4·2	Trace	—	22	
B	22·5	36·9	98	Trace	
C-I	15	3·3	Trace	1–3	
C-II	15	6·7	Trace	1–3	Activates LPL
C-III	36	39·9	Trace	3–5	
D				+	
E		13	Trace	+	

higher concentrations of HDL_2 in plasma. The plasma HDL_2 concentration can be elevated by physical exercise, by some drugs such as clofibrate, and by increasing plasma lipoprotein lipase by administration of heparin or insulin. These studies tend to confirm the inverse relationship between plasma triacylglycerol concentration and HDL cholesterol concentration.

The precise fate of HDL is unknown. In the rat, the biological half-life of HDL in plasma, as judged by measurements of radiolabelled apoprotein specific activities in HDL, is 10·5 h and the major site of accumulation of radioactivity is in the liver. Both parenchymal and non-parenchymal liver cells are capable of taking up HDL. However, the adrenal cortex is associated with the highest ratio of radioactivity : organ weight. It has been calculated that extrahepatic cells could account for the entire catabolism of HDL, but in man it seems likely that the liver does make a minor contribution.

On the basis of detection of HDL-containing iodine-labelled exchangeable apoprotein, it has been shown that injected HDL remains in human plasma for 6 days and in the body as a whole for 7–10 days. The catabolic rate of HDL in man is enhanced in nephrotic patients, in hypertriglyceridaemic patients, in normal subjects fed on a high carbohydrate diet and in patients with familial HDL deficiency (Tangier disease). The figures quoted earlier depend on the assumption that the clearance of the constituent apoproteins is an accurate index of the metabolism of intact HDL. This may not be strictly correct as the apoprotein constituents (A-I, A-II and C) may turn over at different rates. For example, when the metabolism of HDL was studied by injecting intravenously labelled apoproteins A-I or A-II, the labelled apoprotein associated rapidly with HDL, and apo A-I was cleared from plasma more rapidly than apo A-II. This fact must be considered when equating the metabolism of certain intact lipoproteins with that of their constituent apoproteins. This concept does not apply to apo B in LDL, as radiolabelled apo B appears to be non-exchangeable, so that the entire LDL molecule, lipid and apoprotein B, seem to behave as a single entity within the plasma compartment.

Lipoprotein-a (Lp-a)

Lp-a constitutes a separate class of lipoproteins isolated in the density range that overlaps LDL and HDL_2. This lipoprotein class consists of spherical particles with a lipid composition similar to LDL, but Lp-a is larger than LDL with a molecular weight of 5×10^6 and has a lower lipid to protein ratio than LDL. It is an exception to the usual rule that as lipoproteins decrease in size there is an increase in the ratio of protein to lipid. The protein content of Lp-a is composed of 65% apo B, 15% albumin and 20% of a protein specific to Lp-a. This Lp-a protein is rich in carbohydrate and has a molecular weight of 35–40 000.

The presence of Lp-a in the arterial wall and the ability to precipitate Lp-a from plasma with Ca^{2+} and glycosaminoglycans has suggested that Lp-a is closely associated with vascular disease. However, the plasma concentration of Lp-a in man does not correlate well with age, sex, lipid concentration or the incidence of coronary artery disease.

Family studies of the incidence of Lp-a in the population have shown that this lipoprotein has a high degree of hereditability. Double immuno-diffusion techniques have indicated that 30% of the population of Western Europe have Lp-a in their blood plasma. However, results of similar studies using radioimmunoassay suggest that a much higher proportion of the population have Lp-a in blood plasma. As techniques improve, it may be that Lp-a will be detected in the blood plasma of all subjects. Despite the fact that Lp-a can be detected and its physical properties studied, there is little known of the origin or the metabolic fate of this plasma lipoprotein.

Lipoprotein-x (Lp-x)

Patients with obstructive liver disease possess an abnormal lipoprotein which is often isolated with LDL but does not cross-react with anti-apo B immunoglobulin. It is further differentiated from LDL by its migration on agar gel electrophoresis, since Lp-x migrates at pH 8·6 to the cathode, whereas all other lipoproteins migrate to the anode. Lp-x contains cholesterol and phospholipid in about a 1:1 molar ratio with little non-polar cholesterol ester or triacylglycerol. The composition of Lp-x by weight is phospholipid (65%), cholesterol (25%), cholesterol ester (1%), triacylglycerol (3%) and protein (6%). The protein component of Lp-x is unusual in that it contains a high proportion of albumin (40%), the remainder of the protein consisting of apo C peptides with smaller amounts of apo A-I and apo E. Structurally, Lp-x appears to consist of cholesterol-containing phospholipid bilayers in the form of vesicles 40 nm to 60 nm in diameter and about 10 nm thick. It is assumed that there is an aqueous compartment within the centre of the vesicle which may contain albumin. This has been deduced from the observation that Lp-x will react with anti-albumin immunoglobulin after delipidation but not before.

Although Lp-x is found in the plasma of patients with cholestasis it is also present in other circumstances. For example, Lp-x occurs in plasma in patients with familial LCAT deficiency who have no clinical evidence of liver disease. It is believed that the formation of Lp-x may represent an imbalance between the input of 'surface lipid' into plasma in the form of chylomicrons and the activity of the LCAT enzyme. It has been observed that reduction in the dietary intake of fat reduces the concentration of Lp-x. In patients with cholestasis the source of excess lipid may be derived from increased biosynthesis in liver combined with regurgitation of biliary lipids into plasma. The inverse relationship between LCAT activity and Lp-x concentration is further reinforced by the fact that, *in vitro*, plasma containing Lp-x is a good substrate for assaying the LCAT enzyme activity, and the Lp-x concentration declines as the cholesterol ester concentration increases.

LIPID EXCHANGE FACTORS

The structure, composition and metabolism of plasma lipoproteins are modified by the activity of lipoprotein lipase and LCAT. In addition, there are also a number of recently discovered proteins which effect lipid exchange functions. These include cholesterol ester exchange proteins, triacylglycerol exchange proteins and phospholipid exchange proteins.

Cholesterol-ester Exchange-protein

Free or non-esterified cholesterol exchanges between different lipoprotein classes and also between lipoproteins and cells by tranfer of individual sterol molecules through plasma. On the other hand, the movement of cholesterol ester requires a specific protein, the *cholesterol-ester exchange-protein* (CeEP). Although more than one CeEP has been isolated, they are believed to be glycoproteins and have an isoelectric point at a pH of 5·0–5·2. The precise molecular weight of the purified proteins remains to be resolved, but varies between 8000 and 26 000. The higher molecular weight species is normally isolated from plasma d > 1·25 after removal of lipoproteins by ultracentrifugation. *In vivo* in plasma these proteins are thought to have a greater affinity for HDL than LDL. The lower molecular weight material is isolated from HDL and has similar characteristics to apo D. CeEP occurs in human and rabbit plasma, but is absent from rat plasma. The amount of this material in the plasma of hypercholesterolaemic subjects appears to be five times that found in normal subjects. The absence of CeEP in the rat may explain the observation that in the rat cholesterol esters do not exchange *in vitro* in tissues and body fluids. Furthermore, in man, the fatty acids in cholesterol esters are identical in all plasma lipoproteins, whereas in the rat the composition of the fatty acids in VLDL is similar to that found in the liver but different from that of the other plasma lipoproteins.

The precise role of CeEP is the subject of controversy. *In vitro* incubation of LDL- and HDL-containing labelled cholesterol ester has given opposite results concerning the ability of exchange proteins to catalyse net mass transfer from LDL to HDL. One group observed bidirectional exchange of cholesterol ester between LDL and HDL with no mass transfer, which suggests that *in vivo* the pools of exchangeable cholesterol ester in HDL and LDL are in equilibrium: another group has shown that apo D is the functional exchange protein *in vitro*, and that unidirectional net mass transfer occurs from HDL to LDL and from HDL to VLDL. Such an exchange protein could act to equalise the surface concentration of cholesterol esters on lipoproteins. It is also proposed that an LCAT: apo D: apo A-I complex exists in plasma. A close association of LCAT and a CeEP might prevent product inhibition of the LCAT enzyme and would be consistent with the observed rates of the LCAT system in the presence and absence of this exchange protein.

Similar net mass transfers of cholesterol ester from HDL to VLDL *in vivo* and from LDL to VLDL *in vitro* suggest that the respective pools of lipoprotein cholesterol ester are not in equilibrium. The transfer of cholesterol ester to achieve equilibrium would provide a suitable pathway for the removal of HDL cholesterol via VLDL catabolism *in vivo*. This could explain why there is less esterified cholesterol per particle in LDL than VLDL without having to propose that esterified cholesterol is lost from VLDL during its catabolic conversion to LDL.

Triacylglycerol Exchange-protein (TgEP)

Triacylglycerol is packaged in plasma in chylomicrons of dietary origin or in VLDL of intestinal and hepatic origin. Most of the triacylglycerol is rapidly removed from the circulation by uptake in muscle, liver and adipose tissue, and the rest of the triacylglycerol is transferred to LDL and HDL during the course of lipolysis. This basic system is modified by the action of TgEP. The partially purified glycoprotein has an estimated molecular weight of about 100 000 to 115 000, an isoelectric point of 8–9, and it is normally found in human or rabbit plasma previously cleared of lipoproteins by ultracentrifugation. This protein is absent from rat plasma. In the rabbit *in vivo*, TgEP promotes the bidirectional transfer of triacylglycerol between LDL and HDL and between LDL and VLDL. The rate of transfer exceeds the rate of triacylglycerol removal from plasma. The existence of exchangeable pools of triacylglycerol in LDL and HDL is consistent with the suggestion that triacylglycerol in HDL may not be directly derived from VLDL, as the transfer may involve LDL as an intermediate. *In vitro*, a net mass transfer of VLDL triacylglycerol to HDL has also been described in a process believed to involve an exchange of triacylglycerol for cholesterol ester. This mass transfer has not been identified *in vivo*.

Phospholipid Exchange-protein (PlEP)

An exchange of phospholipid has been shown to occur between individual plasma lipoproteins *in vivo* and *in vitro*. Similarly, phospholipid exchange occurs *in vitro* between plasma lipoproteins, platelets, erythrocyte membranes and a variety of cell membranes and subcellular membranes. The exchange is rapid, with complete equilibrium occurring within 4 h between HDL and LDL phospholipids. In a different system ^{32}P phosphatidylcholine was shown to equilibrate completely between HDL and VLDL within 1 hour. This equilibrium was mediated by a thermolabile factor of molecular weight greater than 100 000 isolated from the d > 1·21 fraction from human plasma.

Recently, a PlEP has been used to modify phospholipid at the surface of LDL. Incubation of saturated dipalmitoyl phosphatidylcholine dispersions with LDL increased the proportion of saturated phospholipid in the LDL surface without affecting the protein or cholesterol content. The use of PlEP to modify the lipoprotein surface should help to elucidate the factors involved in phospholipid transfer. Similarly, its action on plasma lipoproteins containing varying proportions of saturated and unsaturated surface components may give information on the molecular packaging of lipid molecules in plasma lipoproteins.

The existence of lipid exchange proteins in plasma complicates the interpretation of lipoprotein metabolism and suggests possible alternative routes for the catabolism or removal of lipoprotein lipid not involving lipoprotein breakdown.

LIPOPROTEINAEMIAS

The major hyperlipoproteinaemias are shown in Table 4. The definition of the primary hyperlipoproteinaemias is based on the measurement of the plasma lipoproteins. It is primarily a laboratory classification based on statistical values for normal, or the absence of abnormal, lipoproteins and the presence of lipoprotein lipase in plasma after heparin injection. The plasma lipoproteins are defined on a qualitative basis using the mobility in agarose electrophoresis and their flotation rates in ultracentrifugation as means of separating the classes. Certain lipoprotein concentrations are diagnostic of metabolic disorders.

Type II Hyperlipoproteinaemia

Type II plasma hyperlipoproteinaemia is characterised by an increased concentration of plasma LDL and deposition of LDL-derived cholesterol in abnormal sites of the body, especially in tendons and in arteries. Type II hyperlipoproteinaemia is further subdivided into Type II(a) in which the concentration of VLDL is normal and LDL is raised, and Type II(b) in which LDL and VLDL concentrations are both elevated. Type II(a) is inherited and is normally termed familial hypercholesterolaemia. It is transmitted as a dominant trait and recent studies suggest that the cellular defect in Type II(a) familial hypercholesterolaemia is a deficiency (heterozygotes) or absence (homozygotes) of a specific receptor for LDL in peripheral tissues. In the light of recent understanding of the molecular nature of this disorder, it has become apparent that heterozygotes may have one of several mutant alleles at the LDL receptor locus. A detailed description of the various phenotypes is given in a recent review.[8]

Type III Hyperlipoproteinaemia

Fasting chylomicronaemia may be present in Type III hyperlipoproteinaemia but it is not diagnostic of this condition. Type III hyperlipoproteinaemia is defined by the presence of a lipoprotein migrating between LDL and VLDL on electrophoresis of plasma and called

TABLE 4
CLASSIFICATION OF THE MAJOR PLASMA HYPERLIPOPROTEINAEMIAS ENCOUNTERED IN MAN

Hyperlipoproteinaemia classification	PLASMA LIPIDS		PLASMA LIPOPROTEINS				Appearance
	Cholesterol	Triglycerides	Chylomicrons	VLDL	LDL	HDL	
Type I	Normal	Raised	Present	Normal	Normal	Decreased	Milky
Type II(a)	Raised	Normal	Absent	Normal	Raised	Normal	Clear
Type II(b)	Raised	Raised	Absent	Raised	Raised	Normal	Clear
Type III	Raised	Raised	Present	Raised	Raised	Normal	Turbid
Type IV	Raised	Raised	Absent	Raised	Normal	Normal	Turbid
Type V	Raised	Raised	Present	Raised	Normal	Decreased	Milky

Type I Hyperlipoproteinaemia

Chylomicrons must be present in abundance in fasting plasma to establish the so-called Type I or Type V hyperlipoproteinaemias. Type I and Type V are further distinguishable by the complete absence of heparin-releasable lipoprotein lipase in Type I and reduced enzyme activity in Type V. It will be noted that lipoprotein lipase is activated by apoprotein C-II. In some cases Type I hyperlipoproteinaemia can be caused by a deficiency of apo C-II. The plasma triacylglycerol concentrations are returned to normal by infusion of plasma from normal patients or by VLDL from normal subjects as a source of apo C-II to activate lipoprotein lipase.

floating-β lipoprotein. The abnormal component has a density intermediate between VLDL and LDL and is incompletely resolved in the analytical ultracentrifuge. Type III hyperlipoproteinaemia also has a genetic basis; VLDL from normal patients contains three isomorphs of apoprotein E, but in Type III patients apoprotein E-III is lacking. The VLDL occurring in Type III hyperlipoproteinaemia has a ratio of cholesterol to triacylglycerol higher than that found in normal VLDL. In these subjects the rate of VLDL synthesis is higher than that of LDL, implying that a significant proportion of the VLDL in the Type III subjects is removed from the circulation before it can appear as LDL. Further evidence for the uptake of 'surplus' VLDL by extrahepatic tissues comes from results with cell culture. Detailed studies with skin fibroblasts from normal subjects, using VLDL subfractions from the S_f 100–400, 60–100 and 20–60 ultracentrifugation fractions, show that they are all effective in the suppression of cellular cholesterol synthesis, thus supporting the concept that these fractions are taken up by extrahepatic tissues.

Type IV Hyperlipoproteinaemia

This metabolic abnormality is characterised by an elevated concentration of VLDL in plasma. Thus the plasma concentration of triglycerides is raised and as a rule there is a moderate increase in the plasma cholesterol concentration. The plasma obtained from a fasting subject with this lipoproteinaemia is slightly cloudy. This metabolic abnormality is often associated with a defect in carbohydrate metabolism. It is concluded that Type IV hyperlipoproteinaemia is due to overproduction of VLDL.

Type V Hyperlipoproteinaemia

This metabolic abnormality is established by the finding of an elevated plasma chylomicron concentration and an elevated plasma VLDL concentration in a sample of plasma obtained from a fasting subject. In Type V hyperlipoproteinaemia there is a reduced appearance in plasma of the heparin-releasable lipoprotein lipase activity.

Other Defects in Plasma Lipoproteins

In addition to the major plasma lipoproteinaemias there are a number of less common abnormalities of plasma lipoproteins. *Abetalipoproteinaemia* is an identifiable but rare metabolic abnormality in man. It is characterised by an absence of lipoproteins containing apoprotein B with a corresponding decrease in the plasma cholesterol and plasma triglyceride concentrations. There is an associated increase in the plasma concentrations of phosphatidylcholine and sphingomyelin compared to the concentraions of these lipids found in normal subjects. Subjects with this disorder have a defect in the absorption of fat (triglyceride), because they cannot form normal chylomicrons in the gut for export into the lymph and hence into plasma. In addition, these subjects have acanthocytosis, a condition in which the erythrocytes are crenated due to alterations in the red blood cell membranes.

A similar lipoprotein disorder is due an *absence of apoprotein A*. This lipid disorder is termed *Tangier disease* and is characterised by an absence of the apoprotein A-containing lipoproteins, HDL. In this situation there is a decreased plasma cholesterol concentration and an elevated plasma concentration of triglycerides. Usually there is a markedly diminished concentration of LDL in plasma in Tangier disease subjects. There is also histological evidence of cholesterol ester deposition in macrophages, non-vascular smooth muscle cells and in Schwann cells. The deposition of cholesterol esters in the reticuloendothelial system suggests the existence of an alternative route for cholesterol removal from plasma in such subjects.

Lecithin-cholesterol Acyltransferase Deficiency

The enzyme lecithin-cholesterol acyltransferase (LCAT) is synthesised in the liver and secreted into plasma in normal subjects. It has been known for many years that if fresh plasma is allowed to stand at room temperature for several hours, there is a fall in the free cholesterol content and a rise in the esterified cholesterol content. This esterification of the sterol is accomplished by the transference of a fatty acid from the β position of plasma lecithin to the hydroxyl of cholesterol. Much of the cholesterol secreted into plasma as VLDL or as HDL is free or non-esterified, and the subsequent esterification of the sterol occurs in the circulation as a result of the action of the LCAT enzyme. In the absence of this enzyme the plasma free–cholesterol concentration remains high and the plasma esterified cholesterol content remains low. The newly secreted plasma lipoproteins such as HDL remain in an immature or nascent state and there is considerable deposition of the excess non-esterified cholesterol in various tissues.

Diabetes Mellitus

In this condition the lipolytic activity of plasma is often reduced, and the consequent accumulation of VLDL is often associated with the obesity commonly occurring in middle-aged diabetics. Insulin therapy may reduce the plasma concentration of VLDL by increasing the muscle lipolytic activity. Generally, hypertriglyceridaemia in normal weight, non-diabetic patients is considered to be due to an unexplained overproduction of VLDL triacylglycerol, whereas in obese subjects hypertriglyceridaemia seems to be related to a defect in VLDL metabolism.

Hypothyroidism

In hypothyroidism there is a tendency for the plasma VLDL and the plasma LDL concentrations to be raised. The plasma cholesterol concentration is therefore elevated, and on treatment of hypothyroid subjects with thyroxine or tri-iodothyronine at the appropriate dosage the plasma lipoprotein concentrations are normalised. In this case, as in diabetes mellitus, the plasma hyperlipoproteinaemia is of secondary origin.

LABORATORY INVESTIGATIONS OF PLASMA LIPOPROTEIN ABNORMALITIES

The clinical biochemical laboratory may be called upon to determine various lipids and lipoproteins in blood plasma. The ideal situation would be one in which the plasma sample is completely fractionated into all the different plasma lipoprotein classes and the concentration of each of the lipoproteins is determined. There is now excellent evidence suggesting that the major lipoprotein classes are heterogeneous, so that the ideal situation of totally defining all the plasma lipoproteins is not attainable at present. Various workable schemes have been evolved for the separation of plasma lipoprotein classes in clinical biochemical laboratories, and these procedures allow an identification and a quantitation of the plasma chylomicrons, very low density lipoproteins, low density lipoproteins or high density lipoproteins in a small plasma sample. In order that an adequate comparison can be made between results obtained on a patient's sample and results from the 'usual' population it is essential that both types of sample are collected under comparable conditions after a 14-hour fast and 48 hours abstinence from alcohol.

Visual inspection. When a plasma sample is submitted for plasma lipid or plasma lipoprotein determination, it is important to inspect the sample. If the sample is turbid or cloudy it may indicate an elevated plasma chylomicron concentration or an elevated plasma VLDL concentration. Ideally, the sample of plasma should be left for 24 h at 4°C. In gross chylomicronaemia the plasma triglyceride concentration will be raised, causing the original plasma sample to have a milky appearance, and on storage the chylomicrons will float and form a creamy upper layer on the sample. Thus a milky appearance coupled with an elevated plasma triglyceride concentration indicates an elevated chylomicron concentration; this usually means that the sample has been collected from a non-fasting patient; rarely it can indicate Type I hyperlipidaemia.

A plasma sample that is not milky but has a slightly turbid appearance, is frequently associated with an elevated VLDL concentration. VLDL molecules tend to scatter incident light and this accounts for the turbid appearance of such plasma samples. If the sample is left at 4°C for 24 h, the turbidity of the sample remains unchanged. In such a sample the plasma triglyceride concentration will be raised due to the elevated VLDL, which in turn is rich in triglycerides. In such a situation there may also be a slightly elevated plasma cholesterol concentration.

Plasma Cholesterol and Plasma Triglyceride Measurements

By visual inspection of the sample and by leaving the sample at 4°C for about 24 h, and by performing relatively simple chemical manipulations such as the measurement of plasma triglyceride and cholesterol concentrations, a considerable amount of information on the plasma lipoprotein concentrations can be obtained. Plasma cholesterol and triglyceride concentrations can be determined by various automated and enzymatic procedures. The most widely-used automated procedure for the determination of cholesterol is the Technicon M-24 method based upon the use of a manual isopropanol extraction of lipids from plasma. Enzymatic methods are increasingly being used for the determination of cholesterol, e.g. the cholesterol oxidase method which oxidises the 3β-hydroxyl of cholesterol to a ketone and uses oxygen as an electron acceptor; most methods measure by various means the amount of hydrogen peroxide generated during the cholesterol oxidase reaction. Since this enzyme will only attack sterols with a free 3β hydrowyl group, only free or non-esterified cholsesterol in the sample will be measured by this method. It is possible to treat a plasma sample with an enzyme, such as cholesterol-ester hydrolase, and split all the esterified chloesterol to free cholesterol and free fatty acids. Liberated free cholesterol can then be measured by cholesterol oxidase. By simple subtraction it is possible to determine both the non-esterified cholesterol in plasma and the esterified cholesterol content of a plasma sample.

Automated methods are also widely used for the determination of triglycerides in plasma. Some of these techniques use autoanalysis methods following an isopropanol extraction. The basis for most modern methods for the determination of triglyceride is linked to the measurement of glyceride glycerol. The latter is oxidised to formaldehyde and may be measured spectrophotometrically. Glycerol can also be determined using various enzymatic procedures.

Plasma Lipoprotein Determinations

While the appearance of a plasma sample coupled to the measurement of cholesterol and triglyceride concentrations in plasma can give important clues concerning the major lipoproteins in that plasma sample, there is often a requirement to measure the major lipoproteins present in the plasma sample. As discussed previously, the various plasma lipoproteins differ in their hydrated densities and sizes.

Advantage can be taken of these differences, and the plasma lipoproteins may be quantitated by *density-gradient centrifugation* in an analytical ultracentrifuge. However, complete separation of the plasma lipoproteins by density gradient centrifugation requires sequential 18 h centrifugations. Ultracentrifugation procedures are usually restricted to research laboratories because the capital sum required for the purchase of the instruments is high and the procedure is time-consuming and expensive.

A workable scheme for the separation of plasma lipoprotein classes in clinical biochemical laboratories has been evolved. One of the most widely used techniques for lipoprotein examination is *electrophoresis* of the sample using filter paper, cellulose acetate, or agarose gels as the supporting media. As a rule, the best results are obtained using agarose gels. Addition of albumin to the buffer solution often improves the separation of plasma lipoproteins on paper supports. However, complete separation of LDL and VLDL using paper electrophoresis is sometimes difficult.

Having separated the plasma lipoproteins by electrophoresis on cellulose, cellulose acetate or on agarose, the lipoproteins may be detected or quantitated using various stains. The most commonly used stains are Sudan Black, Oil Red 0 and Fat Red 7B. After electrophoretic separation, lipoproteins can be precipitated by the use of polyanions. This has the advantage of providing a rapid measurement and the technique can identify Lp-x which migrates cathodically in agarose, whereas all other lipoproteins move towards the anode at pH 8.6.

Plasma lipoproteins may be separated by *selective precipitation* by polyanions or polycations. A considerable number of different precipitation techniques have been evolved for the separation and measurement of plasma lipoproteins. These involve the use of divalent cations such as manganous, calcium or magnesium ions together with heparin or dextran sulphate. However, EDTA which is necessary as a lipoprotein stabilising agent can interfere with such divalent cation concentration and must be used with care. Furthermore, heparin and manganous ions may interfere with the cholesterol oxidase enzyme used in the determination of free and esterified cholesterol. Polyethylene glycol with a molecular weight of 6000 has been used to precipitate LDL and VLDL selectively, leaving HDL in solution. Alternatively, LDL cholesterol can be directly determined after selective extraction of VLDL and HDL with an Amberlite ion exchange resin and dodecylated polyethylenimine. This selective binding of certain lipoproteins to lipophilic polycations or polyanions is based on hydrophobic interactions between polymers and lipids located on the lipoprotein surface.

Immunochemical methods and *radioimmunoassay techniques* have been applied to the isolation and quantitation of certain specific apoproteins of the plasma lipoproteins. Some apoproteins can be measured by radioimmunoassay or by *rocket immunoelectrophoresis*. However, certain lipoproteins have to be treated with delipidating agents to expose all the antigenic sites of the apoprotein in the lipoprotein before effective immunoassay conditions can be established. Most of these immunoelectrophoresis techniques applied to apoproteins are research-orientated and are not in routine use in clinical biochemical laboratories.

A somewhat cruder, but nevertheless useful, method of estimating lipoprotein concentrations and classifying hyperlipidaemias is that of *nephelometry*. In this technique the property of lipoprotein complexes to scatter light is utilised. By measuring the light-scattering properties of plasma, both before and after filtering out particles of certain size, an estimate can be achieved of the concentrations of the large and the medium-sized lipoprotein complexes. These results, together with the plasma cholesterol concentration, also enable an estimate to be made of the concentrations of small-sized lipoprotein particles. There is a reasonable correspondence of the large sized particles with chylomicrons and of the medium size particles with VLDL. The cholesterol-rich HDL and LDL species are estimated together in the small particle fractions; this inability to distinguish between these two important lipoproteins is the major drawback to nephelometric assay. Nevertheless a knowledge of the plasma cholesterol concentration and a nephelometric pattern is sufficient to classify most hyperlipoproteinaemias and to identify those cases that require a further, more sophisticated, investigation.

CONCLUSIONS

Evidence collected over the last fifty years emphasises that the pathological process of atherosclerosis is associated with the accumulation of cholesterol esters, free cholesterol, phospholipids and certain apoproteins in the arterial intima. Immunological and microanalytical studies strengthen the concept that a sizeable portion of the lipids and lipoproteins accumulating in the arterial intima are derived from the circulating plasma lipoproteins. It is possible that the deposition of such lipids and lipoproteins in the vessel wall may be related to the concentration of specific lipoproteins in plasma. As discussed in this chapter, the various plasma lipoproteins are not independent variables. There is a product-precursor relationship between VLDL and LDL. Similarly, there are important interactions between HDL and chylomicrons as well as various other plasma lipoprotein interactions. In general, elevation of the plasma concentration of VLDL and/or LDL compromises the vascular intima. Conversely, elevation of the plasma concentration of HDL is usually associated with a decreased concentration of VLDL and LDL and a degree of protection against atherosclerosis. These observations are crudely interpreted as indicating that elevated concentrations of plasma VLDL and LDL are atherogenic, whilst elevated concentrations of plasma HDL are antiatherogenic.

Naturally such statements are oversimplifications. Nevertheless, it is on the basis of such concepts that much clinical work affecting biochemical laboratories arises. There are ever-increasing demands on laboratories to measure the concentrations of cholesterol and triglycerides in blood plasma. There are also requests for the determination of VLDL and LDL as sophisticated measurements of the lipid-containing macromolecules found in plasma, whose elevated concentrations are viewed as being of aetiological significance. Lipoprotein studies have naturally led to the introduction of dietary alterations and pharmacological compounds designed to lower the plasma concentrations of VLDL and LDL, or to elevate the plasma concentration of HDL. Such dietetic and drug trials inevitably increase demands on the clinical biochemical laboratory to produce accurate, specific and rapid methods for the determination of the principal plasma lipoprotein classes.

REFERENCES

1. Getz, G. S. and Hay, R. V. (1979). In *Biochemistry of Atherosclerosis*, pp. 151–188, Scanu, A. M., Wissler, R. W., Getz, G. S. (Eds.) New York: Dekker.
2. Shaefer, E. J., Eisenberg, S. and Levy, R. I. (1978) *J. Lipid Res.* **19**, 667–687.

3. Smith, L. C., Pownall, H. J. and Gotto, A. M. (1978) *Ann. Rev. Biochem.* **47**, 751–777.
4. Zilversmit, D. B. (1978). In *Disturbances in Lipid and Lipoprotein Metabolism*, pp. 69–81, Dietschy, J. A., Gotto, A. M. Ontko, J. A. (Eds.) Baltimore: Waverley.
5. Pittman, R. C., Attie, A. D., Carew, T. E. and Steinberg, D. (1979) *Proc. Nat. Acad. Sci. USA* **76**, 5345–5349.
6. Goldstein, J. L., Brown, M. S. and Anderson, R. G. W. (1976–77). In *International Cell Biology*, pp. 634–648, Brinkley, B. R., Porter, K. R. (Eds.) New York: Rockefeller University Press.
7. Stein, Y. and Stein, O. (1979). In *Biochemistry of Atherosclerosis*, pp. 313–344, ed. Scanu, A. M., Wissler, R. W., Getz, G. S. (Eds.) New York: Dekker.
8. Goldstein, J. L. and Brown, M. S. (1979) *Ann. Rev. Genetics* **13**, 259–289.
9. Allen, J. K., Hensley, W. J., Nicholls, A. V. and Whitfield, J. B. (1979) *Clin. Chem.* **25**, 325–327.
10. Heuck, C. C., Middlehiff, G. and Schlierf, G. (1977) *Clin. Chem.* **23**, 1756–1759.

30. CHEST PAIN AND CARDIAC FAILURE

T. FOLEY

Introduction

Acute chest pain
 Myocardial infarction

Acute pleurisy
 Chest wall
 Oesophagitis

Chronic chest pain
 Diagnosis

Heart failure
 Definition
 Treatment

INTRODUCTION

Patients with chest pain present to clinicians in different ways according to whether their pain is acute or longstanding. The aim of this chapter is to show how biochemical tests may help in arriving at a correct diagnosis.

Acute chest pain may arise from any structure in the thorax, including the chest wall. Common causes include damage to the myocardium, pericardium, aorta and great vessels, oesophagus, pleura and rib cage. Pain may also arise from the thoracic spine or from compression of intercostal nerve roots. Inflammation of the trachea can cause pain, but diseases of the lung parenchyma are usually painless unless the pleura is involved. Pain arising from the stomach or gall bladder may also present as chest pain. The differential diagnosis depends, as usual, on a clinical history supplemented by a physical examination and appropriate laboratory and radiological tests. It is impossible to give the absolute, or indeed the relative, frequency of the different organs as sources of chest pain because this varies with time and place and whether the doctor is in hospital or general practice; and whether he is dealing with a general or selected population.

ACUTE CHEST PAIN

Myocardial Infarction

Myocardial infarction is a major source of chest pain and an important cause of morbidity and mortality in most western societies. It presents, as a rule, with pain felt in the centre of the chest, which may extend into the neck and jaw to either, or both, arms or downward into the epigastrium. It is commonly associated with sweating and faintness and the patient may be aware of disturbance in the heart beat, or of shortness of breath. Objectively, the patient may be sweaty if the pain is severe. There is often an alteration in the pulse rate which may be unduly fast, slow or irregular. The jugular venous pressure is commonly raised, though only slightly, and there is often an added heart sound. The blood pressure is generally reduced below the patient's usual level and this can be very helpful, if known.

The diagnosis is usually made on clinical grounds and confirmed by electrocardiography (ECG). In approximately 90% of cases characteristic changes will be found on the ECG. These include an elevation of the ST segments in leads recorded over the infarcted area and a reciprocal depression in leads from other areas. Q waves will be seen if the infarct extends through the full thickness of the myocardium.

It is not sufficiently appreciated that, in a fair proportion of cases, the appearance of ECG changes may be delayed for up to 24 hours and if the diagnosis of myocardial infarction is probable on all other counts it is not refuted by a normal ECG within the first 24 hours. Serial ECG records should always be taken whenever possible. The ECG may be difficult to interpret when the patient has pre-existing heart disease such as a previous myocardial infarct. Under these circumstances biochemical tests play a major role in confirming the diagnosis of myocardial infarction.

The death of heart muscle cells is associated with the release into the circulation of many of their enzymes which can be measured to provide valuable confirmation of the diagnosis of myocardial infarction. The extent of the rise may also give useful information about the severity of the infarct and an indication of the prognosis.[22]

Creatine Kinase (CK)

Amongst the many enzymes that have been used as markers of myocardial infarction, one of the most useful is creatine kinase (CK). This enzyme is present in skeletal tissue and brain as well as in the heart, but occurs in different forms in the various tissues. That found in brain is the BB isoenzyme, and that in the striatal muscle the MM isoenzyme. These are chemically and immunologically distinct from each other and antibodies prepared against one type do not react with the other. CK in heart muscle is the MB variant and shares some of the immunological properties of the other two. Plasma CK activity tends to peak within 24 hours of a myocardial infarct and to return to normal within 72 hours. The sensitivity and, even more, the specificity depends on the method of assay used being increased by confining measurement of activity to the contribution made by the MB isoenzyme.

Total plasma CK activity is markedly increased in patients with muscular dystrophy or inflammatory diseases of the muscle. It is also raised in alcohol intoxication—particularly when associated with delirium tremens—and in diabetes mellitus associated with ketoacidosis, and convulsions. It may also be raised by intramuscular injections and this can lead to diagnostic confusion unless specific isoenyzme measurements are made. A fair proportion of patients with pulmonary embolism show raised CK activity and coronary angiograms may also cause a temporary elevation but, in this case, it is due to MB isoenzyme. Some authors advocate the use of CK-MB activity as a guide to the size and severity of an infarct the larger the lesion the greater the rise in plasma activity.

Aspartate Aminotransferase (AST)

Plasma AST (previously called serum glutamic oxaloacetate transaminase, SGOT) activity typically exceeds the normal range within eight to twelve hours of the onset of myocardial infarction and returns to normal within three to four days. As with all serum enzyme measurements the duration of raised activity depends in part on the highest level attained. Aggress and Kim[1] reviewed 1962 cases of myocardial infarction diagnosed clinically and found true positive elevations of AST in 97%.

AST levels may also be raised in hepatic congestion, primary liver disease, skeletal muscle disorders and shock. Congestive heart failure is associated with a high AST level when the liver is distended. Active myocarditis tends to be associated with a modest rise in AST and pericarditis causes an elevation in perhaps 15% of cases. Twenty-five per cent of patients with pulmonary embolism also show a raised AST activity, but CK levels are normal. Ninety-five per cent of patients who have undergone surgery or suffered major trauma show a rise in AST level as do a high proportion of those given DC counter-shock for cardiac arrhythmias. Patients with diseases of the liver or biliary tract who have received intramuscular injections also have raised AST levels.

Lactate Dehydrogenase (LDH)

Plasma LDH activity typically exceeds the normal range within 24 to 48 hours of the onset of infarction and peaks in three to six days. Thereafter it gradually declines to reach normal values within ten to fourteen days of the onset of the illness. True positive elevations were found in 89% of 282 patients with myocardial infarction diagnosed clinically and are due mainly, if not exclusively, to LDH isoenzymes I and II. These isoenzymes, which have equal activity against both pyruvate and β-hydroxybutyrate, were for some time referred to as β-hydroxybutyrate dehydrogenase (HBD) and thought to be specific for myocardial infarction. High values are, however, found in many disorders apart from myocardial infarction—especially those associated with increased erythropoeitic activity—and HBD measurements have virtually dropped into obsolescence as an aid to the diagnosis of heart disease.

The specificity of plasma LDH assays depends on the population under test. Rises in LDH activity occur in perhaps 30% of patients with congestive cardiac failure; haemolysis, megaloblastic anaemia, leukaemia, acute and chronic liver disease are also associated with raised plasma LDH levels. Pulmonary embolism and neoplastic disease may also cause raised levels resulting in difficulty of interpretation and so may cardiac catheterisation, myocarditis and shock.

The value of plasma enzyme estimation in patients with acute chest pain lies in providing confirmation of a clinical diagnosis of myocardial infarction. They are particularly important when the ECG is difficult to interpret as it often is in cases of previous infarction. Difficulties may arise when there have been intramuscular injections (CK and AST) or when haemolysis has occurred during collection of the sample (LDH). Sobell and his colleagues[23] have suggested that the amount of enzyme released into the blood stream reflects the amount lost from the damaged heart muscle and it has been shown that the height of the maximum rise in enzyme activity correlates with the fall in ejection fraction of the left ventricle after myocardial infarction. The severity and persistence of ventricular rhythm disturbances and morbidity and mortality of patients with myocardial infarction also correlate closely with the height of the rise in enzyme activity.

Myoglobin

Myoglobin is released from damaged heart muscle fibres into the circulation where it can be measured in plasma by radioimmunoassay. Since its concentration in plasma rises more quickly than other markers of myocardial infarction, and since the extent of its rise provides a better measure of infarct size than most other techniques currently available, it is likely to become an increasingly popular investigation in the future.

Blood enzyme estimations have been criticised on the grounds that the rate of disappearance of enzyme from the blood might vary with haemodynamic conditions after infarction.[19] Sobell and coworkers[22] found that, in fact, there was little variation on this score.

Other Tests

Conventional radiology plays little part in the diagnosis of myocardial infarction except in helping to exclude other conditions such as dissection of the aorta which may cause similar symptoms. Radionuclear imaging, particularly with thallium, may provide independent confirmation of the diagnosis.

Acute pericarditis may present with pain similar to that of infarction. It is commonly made worse by deep breathing and usually relieved by sitting up or standing rather than by lying down. The physical signs are commonly a sinus tachycardia, raised venous pressure and an added heart sound. There may also be a friction rub though this is often transitory. If there is a large effusion there may be pulsus paradoxus, i.e. an inspiratory waning of the palpable pulse due to a drop in the pulse pressure. The diagnosis is confirmed by the ECG, which generally shows widespread ST segment elevation in many leads, and generalised T wave inversion. The echocardiogram commonly shows the presence of a small pericardial effusion, and a plain x-ray may show enlargement of the heart shadow. Biochemical tests are generally unhelpful but may reveal a moderate rise in the plasma AST activity.

ACUTE PLEURISY

Acute inflammation of the pleura, from whatever cause, usually causes unilateral discomfort which is made worse by breathing and by movement. There is typically a diminution in the movement of the affected side of the chest. Other findings depend upon the cause.

Biochemical testing is, in general, not helpful. X-ray examination, on the other hand, may be extremely so in disclosing disease of the underlying lung, such as pneumonia, pulmonary infarction, and disease of the pleura itself such as occurs with pleural effusion; lung scanning may reveal evidence of pulmonary embolism.

The clinical history may give valuable clues to the aetiology. A history of recent surgery or trauma to a limb suggests pulmonary infarction; if the predominant features are those of fever, cough and sputum in addition to chest pain, pneumonia is more likely. A positive blood culture may be found in about 30% of cases with pneumonia and lung scanning frequently shows up infarcts that are not apparent radiologically. Nevertheless, the distinction between pulmonary embolism and pneumonia is often far from easy.

Chest Wall

Pain arising from the chest wall tends to be made worse by breathing and by movement. Physical examination may show local tenderness and tenderness on rib compression. X-rays may be helpful in revealing evidence of underlying bone disease, such as fractures, but biochemical tests are generally unhelpful except in cases of multiple myeloma when a paraprotein will be visible on electrophoresis of the plasma proteins.

Pain due to compression of nerve roots may be confused with that of myocardial infarction—particularly if it is bilateral. There may be a history of spinal disease or injury; the pain is often made worse by lifting, straining or coughing. Examination may show evidence of a scoliosis or kyphosis in the thoracic spine; or there may be neurological signs such as a sensory loss. Herpes zoster is a common cause of root pain and may cause difficulty in diagnosis before the vesicles appear.

X-rays, radionuclide imaging or CAT scanning, often help in the diagnosis of chest pain of spinal origin. Biochemical investigations, on the other hand, are seldom helpful except in cases of bone disease due to multiple myeloma, hyperparathyroidism, chronic renal failure, widespread secondary deposits, osteoporosis or osteomalacia; protein electrophoresis, plasma calcium and creatinine levels and the activity of alkaline phosphatase in blood will indicate the presence of one of these conditions.

Oesophagitis

Chest pain due to acute oesophagitis is commonly associated with a history of flatus, acid regurgitation, or heartburn. There may be a history of dysphagia and the pain may be indistinguishable from that of myocardial infarction, except insofar as it is often provoked by bending, leaning forward, or lying down and tends to be worse after meals. The absence of ECG or plasma enzyme changes may be helpful in making the distinction. Barium examination and fibroptic endoscopy may be used to confirm the presence of oesophageal disease.

Peptic ulceration and gall bladder disease may both present with pain in the chest as a dominent feature. Usually, however, there is tenderness in the epigastrium or under the right costal margin and differential diagnosis is not a problem. When disease of the biliary tract is present the patient may be jaundiced and there may be a rise in the plasma alkaline phosphatase and AST activities.

CHRONIC CHEST PAIN

The commonest problem in longstanding chest pain is to distinguish *angina pectoris* from other causes of recurrent pain. Diagnosis of the first named is entirely clinical and relies exclusively on a good clinical history. If the symptoms are genuine the diagnosis stands even when physical examination and investigations are negative. The pain of angina may be felt anywhere between the jaw and the epigastrium and in either arm. When it is felt in the chest it is almost always central. It is generally provoked by exercise and relieved by rest—usually within five minutes. It may also be provoked by emotional excitement or anger.

Diagnosis

Many factors combine to conceal the fact that a patient has angina and there is no totally reliable test to confirm the diagnosis. Rose[18] showed that the simple question 'Do you experience any discomfort in the chest on climbing hills or stairs?' frequently reveals that a patient has angina.

Examination may reveal evidence of anaemia, aortic valve disease, hypertension or hypercholesterolaemia, but it may be entirely normal. Similarly, the resting electrocardiogram during, or after exercise, may provide valuable confirmation, but the latter test is not entirely without risk and is best carried out on a treadmill using a standardized protocol and in the presence of a properly equipped doctor. Biochemical tests provide no direct help in confirming the diagnosis. They may confirm the presence of associated disorders, such as familial hypercholesterolaemia, and coronary sinus lactate estimations carried out during cardiac catheterization may provide confirmation of ischaemia. It might be thought that coronary angiography would provide final confirmation of ischaemia. In general this is so, but the syndrome of angina with normal coronary arteriograms is increasingly being recognised though still not well understood.

HEART FAILURE

Definition

Heart failure is a clinical state (syndrome) characterised by tiredness, shortness of breath on lying down and often by swelling of the legs. It is recognised by tachycardia, a raised jugular venous pressure, and the presence of an added filling sound in diastole. In more advanced failure, the liver is distended and tender and there may be swelling of the legs, the sacral area or the abdomen. Radiologically, the heart is enlarged and the upper lobe veins are unduly prominent; there may be septal lines at the bases of the lungs or there may be frank pulmonary oedema. There are no specific electrocardiographic features.

Physiologically, left ventricular and diastolic pressures are raised and, as a direct result, left atrial and pulmonary capillary pressures are also raised. These can be measured at the bedside by the use of a floating catheter.

The syndrome has many causes, and its symptoms and signs vary with the speed of onset. If the cause is a slowly progressive condition, such as aortic valve disease or systemic hypertension, cardiac failure may be present only during muscular exercise.[21] If, on the other hand, the cause is an acute catastrophe, such as a massive myocardial infarct, the symptoms may be much more severe. Measurements of cardiac output are unhelpful.[3]

Elevation of left ventricular and diastolic pressures bring into play the Frank–Starling mechanism,[24] wherein increased stretch causes an increase in contractility of the myocardium. It also reflexly activates the sympathetic nervous system which causes tachycardia, peripheral vasoconstriction and catecholamine release. There may also be an increase in renin release from the kidney. Local release of catecholamines in the myocardium increases contractility of the heart and is accompanied by an increase in urinary noradrenaline excretion.[4] Myocardial noradrenaline stores are rapidly depleted, both in animals and man,[5] probably due to reduction of tyrosine hydroxylase activity in sympathetic nerve endings in the myocardium.

Stretch receptors in the atria and great veins form the afferent loop of these reflexes;[9,10] their activation normally operates to maintain a constant heart volume by adjusting the heart rate to atrial volume ('venous return'). In cardiac failure they are stretched beyond their normal limits and this seems to be associated with a fall in their discharge rate.[6]

A major biochemical consequence of heart failure is retention of salt and water which tends to expand the plasma volume as well as to increase interstitial fluid volume. Retention is due partly to a fall in the glomerular filtration rate, and partly to an increase in aldosterone secretion due to an increase in activity of the renin–angiotensin–aldosterone system. It seems that inadequate cardiac output causes signal error which tends to increase plasma volume which ought to improve the cardiac output by activating the Frank–Starling mechanism. In fact, the response tends to be excessive and the consequent retention of salt and water leads to congestion and oedema. Pulmonary oedema stiffens the lungs and causes breathlessness and hyperventilation whilst peripheral oedema impairs the functional activity of organs such as the liver whose blood flow is often reduced.[26]

Treatment

Bed Rest and Diuretics

Bed rest, by reducing the metabolic demands of the body, effectively reduces the work of the heart and still has a useful part to play in the treatment of cardiac failure. Nevertheless, since fluid retention is often excessive, and harmful, many patients can benefit by having their symptoms believed by diuretics. Those in common use are:

1. *Thiazides*. These act mainly on the proximal tubule and on the ascending limb of Henle's loop and the cortical diluting segment of the nephron, thereby increasing excretion of salt and water.

2. *Loop diuretics*. This is the collective name given to describe frusemide, ethacrynic acid and bumetanide. These are more powerful in their action than the thiazides and their major effect is on the loop of Henle and early ascending limb of the renal tube.

3. *Aldosterone antagonists.* Spironolactone is a competitive antagonist of aldosterone at the cellular level in a number of tissues. Triamterene and Ameloride, on the other hand, are not true antagonists but do have an opposite action on the distal tubule to that of aldosterone which causes sodium retention, exchanging sodium from the tubular fluid for potassium or hydrogen ions which are excreted instead. The drugs in this third group are not very powerful natriuretics, but are useful when used in conjunction with thiazides or loop diuretics in preventing potassium depletion.

Nicholls and his colleagues[14] studied the effect of diuretic treatment on the renin–angiotensin–aldosterone system in patients with congestive heart failure and identified three phases:

1. In the untreated phase there was a modest increase in activity of the system.
2. In the diuretic phase the plasma aldosterone concentration, paradoxically, became very low in contrast to the brisk rise seen in normal and hypertensive subjects depleted of sodium by diuretics. This paradoxical response is probably due to the very large increase in sodium load reaching the distal tubules which switches off renin release[25] through an action on receptors in the macula densa.
3. In the third stage, as the patient approaches 'dry-weight', the plasma aldosterone concentration rises sharply possibly due to reduction in circulating volume and renal perfusion. It is in this phase that an excessive fall in the circulating volume may occur leading to reduced renal blood flow and consequent increase in the output of renin in an attempt to restore plasma volume and prevent renal failure.

Potassium Depletion

Potassium depletion is common in patients on diuretic treatment and may be associated with a modest fall in plasma potassium concentration. Most studies show a fall in total body potassium which reflects a fall in the lean body mass,[13,15] though there may also be some loss of potassium from the cells.[7,8] The data are difficult to interpret because of varying degrees of wasting and oedema and it has been suggested that height rather than lean body mass should be used as the standard for comparison. Current evidence suggests that hypokalaemia due to secondary hyperaldosteronism produced by diuretic therapy responds better to aldosterone antagonists than to potassium supplements.[23]

Most diuretics in current use cause a fall in plasma potassium concentration.[11] The average fall is less during treatment with frusemide (0·3 mmol/l) than with thiazides (0·6 mmol/l) and is barely influenced by dose or duration of treatment. Plasma potassium concentrations of less than 3 mmol/l are distinctly unusual and the clinical consequences of hypokalaemia in the range 3·0–3·5 mmol/l are probably not very great.

Hyponatraemia

Hyponatraemia is sometimes seen in patients in advanced cardiac failure who have been treated with large doses of diuretics. This could be due to: (1) inappropriate ADH secretion, (2) the sick cell syndrome, i.e. a primary failure of the "sodium pump" mechanism resulting in sodium ions passing from the extracellular to the intracellular fluid, or (3) renal mechanisms preventing the excretion of free water.

In general the evidence suggests that the last mechanism is the most likely since free water excretion depends on an adequate delivery of sodium to the distal tubules. In patients with severe cardiac failure, treated with diuretics, proximal renal tubular reabsorption of sodium may be so great that there is inadequate delivery to the distal tubules to allow free water excretion. This may in turn lead to dilutional hyponatraemia in which water restriction may be required as part of the treatment.

Digitalis is still the main inotropic drug used for the treatment of congestive cardiac failure but, whilst there is no doubt of its value in cases caused by atrial fibrillation, evidence of its value in the longterm treatment of patients in sinus rhythm is scanty. There is increasing awareness of the high incidence of digoxin toxicity, particularly in older people, and that potassium depletion may enhance the incidence of digitalis-induced arrhythmias. The main therapeutic action of digitalis probably depends on blockade of a sodium–potassium–dependent ATPase present in the myocardial cell membrane which is normally responsible for ejecting sodium from the cell in exchange for potassium. After administration of digoxin the concentration of sodium rises locally, within the cell, and alters the membrane potential. This has the secondary effect of increasing the local concentration of calcium, within the active region of the myofibrils, so increasing the force and velocity of contraction.[13]

Vasodilator Treatment

In the last few years the idea of helping the failing heart by reducing the load upon it has become widely accepted. This may be done either by reducing the 'pre-load' or by reducing the 'after-load', the pre-load being the left ventricular filling pressure. Diuretics reduce the pre-load in one way (already discussed) and nitrates achieve it in another by producing veno-dilation; in effect, by enlarging the capacity of the peripheral venous bed, they lower the central venous pressure.

The after-load is the aortic impedance; i.e. aortic pressure divided by the instantaneous flow. The main component of this is the resistance of the peripheral arteriolar bed. Since myocardial oxygen consumption depends on wall tension, which is intraluminal pressure times radius, anything that reduces arterial pressure will reduce the myocardial oxygen requirement. Nitroprusside is one such drug and is used intravenously in acute situations. It dilates both arterioles and veins. It is particularly useful, therefore, in hypertensive crises.

Nitrates act mainly by inducing peripheral veno-dilatation and they, therefore, reduce the pre-load. They are particularly suitable for patients with pulmonary oedema. Hydrallazine, another hypotensive drug, causes arteriolar dilatation as well as having a direct inotropic action upon the heart. It produces a considerable rise in cardiac output with a relatively small fall in the central filling pressure. Like all powerful vasodilators it can cause sodium retention due to renin release. Diazoxide and minoxidil are both powerful vasodilators that have considerable disadvantages. Prozosin appears to be more effective in combating hypertension than in treating cardiac failure.

It is logical to use peripheral vasodilators to treat patients with hypertensive heart failure, but in recent years they have also been found to have a useful place in the treatment of other types of cardiac failure such as those caused by ischaemia, cardiomyopathy and mitral valve disease.[16]

REFERENCES

1. Agress, C. M. and Kim, J. H. (1960) 'Evaluation of enzyme tests in the diagnosis of heart disease,' *American Journal of Cardiology*, **6**, 641–649.
2. Bradley, R. D. (1977) *Studies in Acute Heart Failure*. London: Edward Arnold.
3. Braunwald, E. (1971) 'On the difference between the hearts output and its contractile state,' *Circulation*, **A3**, 171–174.
4. Chidsey, C. A., Braunwauld, E. and Morrow, A. G. (1965) 'Catecholamine excretion and cardiac stores of nor-epinephrine in congestive heart failure,' *American Journal Of Medicine*, **39**, 442–451.
5. De Quattro, V., Nagatsu, T., Mendez, A. and Verska, J. (1973) 'Determinants of cardiac noradrenaline depletion in human congestive failure,' *Cardiovascular Research*, **7**, 344–350.
6. Greenberg, T. T., Richmond, W. M., Stocking, R. A., Gupta, P. D., Meehan, J. P. and Henry, J. P. (1973) 'Impaired atrial receptor responses in dogs with heart failure due to tricuspid insufficiency and pulmonary artery stenosis,' *Circulation Research*, **32**, 424–433.
7. Hamer, J., Knight, R. K., Miall, D. A., Hawkins, L. A. and Dacombe, J. (1976) 'Plasma aldosterone levels in patients with severe, treated congestive heart failure,' *British Heart Journal*, **38**, 534 (abstract).
8. Hamer, J. (1977) 'Heart failure (2).' In *Recent Advances in Cardiology*, Vol. 7, Edinburgh, London and New York (Churchill Livingstone).
9. Linden, R. J. (1972) 'Function of nerves of the heart', *Cardiovascular Research*, **6**, 605–626.
10. Linden, R. J. (1973) 'Function of cardiac receptors,' *Circulation*, **48**, 463–480.
11. Morgan, D. B. and Davidson, C. (1980) 'Hypokalaemia and diuretics, an analysis of publications,' *British Medical Journal*, **1**, 905–908.
12. Nagent de Deuxchaisnes, C. and Mach, R. S. (1974) 'Potassium depletion and potassium supplementation in cardiac failure,' *Lancet*, **i**, 517 (Letter).
13. Nayler, W. C. (1975) 'Ionic basis of contractility relaxation and cardiac failure.' In *Modern Trends in Cardiology*, Vol. 3, p. 154 (Oliver, M. F., Ed.) London: Butterworths.
14. Nicholls, M. G., Espiner, E. A., Donald, R. A. and Hughes, H. (1974) 'Aldosterone and its regulation during diuresis in patients with gross congestive heart failure,' *Clinical Science*, **47**, 301–315.
15. Olesen, B. M. and Valentin, N. (1973) 'Total exchangeable potassium, sodium and chloride in patients with a severe valvular heart disease during preparation for cardiac surgery,' *Scandinavian Journal Of Thoracic and Cardiovascular Surgery*, **7**, 37–44.
16. Opie, L. M. (1980) 'Drugs and the heart. Vasodilating drugs,' *Lancet*, **i**, 966–972.
17. Poole, P. E., Covell, J. W., Levitt, M., Gibb, J. and Brunwald, E. (1967) 'Reduction of cardiac tyrosine hydroxylase activity in experimental congestive heart failure. Its role in the depletion of cardiac nor-epinephrine stores,' *Circulation Research*, **20**, 349–353.
18. Rose, G. A. (1965) 'Ischaemic heart disease: chest pain questionnaire,' *Millbank Memorial Fund Quarterly*, **43**, 32–39.
19. Rowe, and Stamar, (1975)
20. Schartz, A. B. and Swartz, C. D. (1974) 'Dosage of potassium chloride elixir to correct thiazide-induced hypokalaemia,' *Journal of the American Medical Association*, **230**, 702–704.
21. Sharma, B. and Taylor, S. H. (1970) 'Reversible left ventricular failure in angina pectoris,' *Lancet*, **ii**, 902–906.
22. Sobell, B. E., Roberts, R. and Larson, K. B. (1976) 'Considerations in the use of biochemical markers of ischaemic injury,' *Circulation Research*, **38**, (Supplement 1), 99–106.
23. Sobell, B. E., Markham, J. and Roberts, R. (1977) 'Factors influencing enzymatic estimates of infarct size,' *American Journal of Cardiology*, **39**, 130–132.
24. Starling, E. H. (1918) *The Linacre Lecture on the Law of the Heart*. London: Longman.
25. Vander, A. J. (1967) 'Control of renin release,' *Physiological Reviews*, **47**, 359–382.
26. Wade, O. L. and Bishop, J. M. (1962) 'Cardiac output and regional bloodflow,' Oxford: Blackwell Scientific Publications.

31. HYPERTENSION

VINCENT MARKS

Introduction

Symptomatology

Pathophysiology of high blood pressure

   Renin–angiotensin–aldosterone
   Conn's syndrome
   Other steroids

Sodium metabolism

Catecholamines

   Phaeochromocytoma

Vasopressin

Other hormonal causes

INTRODUCTION

Investigation of hypertension is one of the most active and potentially fruitful areas of clinical biochemistry research, for not only is it directly or indirectly one of the most important causes of morbidity and mortality amongst middle-aged individuals in the western world, it is also amenable to therapy. Distinction between primary (or essential) hypertension in which there is no discernible (anatomical) pathology, and secondary hypertension in which there is, provides the basis of a clinically useful classification. However, whilst many causes of hypertension are recognised, in the overwhelming majority of cases, variously put at between 90% and 95%, no primary abnormality, apart from alterations in sodium transport across cell membranes,[9] which may or may not be significant, can be found and treatment is symptomatic. For this reason most investigators and clinicians specialising in the care of hypertensive patients believe that extensive routine investigations are of limited value and are not cost-effective; they should, therefore—unless there are strong indications to the contrary from the history of physical examination—be restricted to a few simple tests (Table 1) which may provide pointers to a remediable cause, should there be one. Such tests may also provide a baseline against which to assess improvement or deterioration in response to time and treatment.

Because of the importance of kidney disease in the pathogenesis of hypertension, examination of the urine for the presence of excess protein, red cells, white cells, casts and microorganisms is essential; glycosuria should also be sought. Measurements of plasma creatinine and possibly urea concentrations provide useful information regarding kidney function. Plasma calcium, potassium and, to a lesser extent, sodium concentrations, may provide the first evidence of hyperparathyroidism or Conn's syndrome respectively. Many centres tend to investigate younger patients, i.e. those under the age of 35 years, more extensively, especially in the absence of a family history but even in this group extensive routine investigation is unwarranted.

SYMPTOMATOLOGY

Hypertension itself is generally symptomless and is discovered either as a result of investigation of one or more of its complications or, more often, as an incidental finding during the course of a routine physical examination. Definition of hypertension is both arbitrary and controversial. It is, however, now generally accepted that a blood pressure persistently above 160 mmHg (21 kPa) systolic, or 105 mmHg (14 kPa) diastolic in a middle-aged man is worth treating; many would set it even lower. Lower values are also accepted as demanding treatment in younger people.

The various clinical syndromes to which the disease gives rise are related to degenerative changes in various organs resulting from arterial degeneration secondary to long continued hypertension. Hypertension predisposes to atherosclerosis, nephrosclerosis, cardiac enlargement and eventually cardiac failure. Coronary thrombosis and cerebral haemorrhage are still common modes of presentation. Hypertension, if untreated, may persist with seemingly little progression for long periods but may at any time—and in any variety—suddenly turn from a relatively benign into a malignant disease which, before effective treatment became available, almost invariably ended fatally within two years of onset.

PATHOPHYSIOLOGY OF HIGH BLOOD PRESSURE

Arterial blood pressure is influenced, in the final analysis, by the balance between the force of expulsion of blood from the heart and the rate of its dissipation, by leakage, through the small arteries, capillaries and veins which between them provide peripheral resistance and a blood storage bed. Although reductions in the force of expulsion of blood by the heart and/or the total volume of the vascular system lead to a reduction of blood pressure the reverse situation does not often occur. Consequently, hypertension is invariably a result of increased peripheral vascular resistance. This is often associated with an expansion of extracellular fluid

TABLE 1
ROUTINE LABORATORY INVESTIGATION OF A PATIENT WITH HYPERTENSION PRIOR TO INITIATION OF TREATMENT

1. Plasma urate; K^+, Ca^{++}, urea, creatinine
2. Urine; protein, glucose
3. ECG
4. Chest X-ray

volume and may occasionally be accompanied by an increase in blood volume.

Peripheral vascular resistance is determined by the bore of the small arteries and arterioles that feed the capillary beds. This is in turn determined by the tone of their musculature which is controlled by nervous and humoral mechanisms.[17] Currently, three major humoral systems, i.e. the renin–angiotensin–aldosterone, vasopressin, and adrenergic systems, are known to be involved in the regulation of peripheral vascular resistance. Each will be dealt with in turn.

Renin–Angiotensin–Aldosterone

Renin is a proteolytic enzyme of approximately 40 000 daltons produced by cells throughout the body but mainly in the kidneys.[1] The secretory cells contain granules and are found in small collections adjacent to the glomeruli, close to both the distal tubule and the afferent and efferent glomeruler arterioles of the corresponding nephron. Known as the macula densa—or more descriptively as the juxtaglomerular apparatus—these cells are uniquely placed to sample the arterial pressure in the kidney and the electrolyte and osmotic composition of distal tubular fluid. Both are probably involved in regulating renin release through an unknown mechanism. Renin is also released in response to sympathetic nerve stimulation independently of any change in blood flow.

Renin liberated into the plasma acts on a glycoprotein of molecular weight approximately 58 000 daltons known as angiotensinogen (renin substrate), which is produced in the liver, to release an inactive decapeptide angiotensin I. This peptide is, in turn, cleaved by a converting enzyme (kininase II) present in the blood and most tissues, especially the lung and kidney itself, into an octapeptide known as angiotensin II. This polypeptide, which was initially recognised for its powerful vasoconstrictor properties, has latterly attracted interest because of (i) its effects on the central nervous system, where it increases sympathetic nervous activity, and (ii) its ability to stimulate aldosterone secretion by the zona glomerulosa cells of the adrenal cortex. Nowadays it is believed that it is through activation of the sympathetic nervous system, rather than by direct action on the arterioles, that angiotensin II exerts its main hypertensive effect.

Increased production of renin is implicated in the production of hypertension in renal artery stenosis (Goldblatt kidney) and possibly other primary diseases of the kidney associated with arterial hypertension. Decreased production of renin occurs when for any reason (other than primary kidney disease) extracellular fluid volume is expanded or plasma osmolality is increased.

Angiotensin

Angiotensin II is formed from angiotensin I under the influence of renin whose activity in the plasma is the rate-limiting step in the production of the active octapeptide, angiotensin II. This polypeptide has an extremely short half-life in the plasma and is, therefore, not readily amenable to meaningful measurement *in vivo*. It can, however, quite readily be measured *in vitro* by radioimmunoassay and its rate of generation in plasma is the basis of a test of plasma renin activity (PRA). Blockade of the conversion of angiotensin I into angiotensin II by the use of orally effective inhibitors of converting enzyme activity, e.g. captopril, provides a useful form of therapy in many patients with hypertension.[19]

Plasma renin

Measurement of plasma renin activity (PRA) is diagnostically useful in a strictly limited number of clinical situations. These are:

(1) Differentiation between primary and secondary hyperaldosteronism when this is in doubt.[13,18]

(2) To indicate whether surgical treatment is likely to benefit a patient with renal artery stenosis and hypertension. This will usually necessitate cannulation of the two renal veins and measuring renin activity in blood collected from each of them individually. High levels of plasma renin activity (PRA) in the effluent from the arterially-constricted kidney and low levels in the other would favour a decision to try and correct the arterial lesion surgically or perform a unilateral nephrectomy.

(3) In helping to decide whether bilateral nephrectomy should be carried out in a patient with hypertension and renal failure necessitating chronic renal dialysis. The presence of increased levels of plasma renin activity favour a decision to carry out nephrectomy but do not necessarily dictate it.

Proper preparation of the patient is essential if a PRA assay is to have any value. The patient must, for example, have been recumbent for at least eight hours before the plasma sample is collected since even a brief period of upright posture may alter the result and makes impossible comparison with either control data or with the 'recumbent' level compared with the level 30 minutes after assuming the upright position. The patient must not have received any diuretic drug or agent affecting the activity of the sympathetic nervous system for at least three weeks prior to the test and should not be taking potassium supplements or drugs such as carbenoxolone that can affect sodium and potassium homeostasis. The diet should have an adequate (100–150 mmol/day) sodium content. Interpretation of the result requires a knowledge of the blood pressure, and the plasma sodium, potassium, urea and aldosterone levels at the time of sampling which, because of diurnal variation, should ordinarily be carried out between 0800 and 0900.

The units used to express renin activity and the methods of measuring it are highly individualistic although attempts to introduce international standards have been made with limited success. Two major types of method are employed—one, which measures the rate of angiotensin I formed by the action of renin on its endogenous substrate during incubation of plasma *in vitro*, is generally referrred to as plasma renin activity (PRA). The other, which measures the enzymic activity of renin in the presence of excess substrate (angiotensinogen) generally obtained from nephrectomised sheep, is often referred to as the plasma renin concentration (PRC). This is, however, a misnomer since it, too, measures 'activity' rather than mass (which could probably only be done by direct immunoassay of renin as a protein; this has not yet been undertaken and would itself give rise to problems since roughly 70–90% of the circulating enzyme is present in the form of an inactive precursor which would almost certainly be measured as immunoreactive renin). Until the difficulties of measurement have been sorted out great care must be exercised by the non-specialist in interpreting plasma renin measurements.

Ordinarily, low plasma renin activity in the presence of raised plasma aldosterone levels is indicative of primary hyperaldosteronism, and high activity of secondary, or kidney-induced, hyperaldosteronism. The occasional occurrence of low levels of plasma renin activity in patients with 'benign' hypertension led to the concept of 'low renin activity hypertension' as a distinct nosological entity, but this is no longer believed to be true.[3,13]

Aldosterone

Aldosterone is the most potent of the endogenous mineralocorticoids and is produced mainly, if not exclusively, by cells in the outer layer, zona glomerulosa, of the adrenal cortex. It differs from other physiologically important adrenal steroids in having an oxygen atom on the C-18 position. This aldehyde group appears to be in equilibrium with an 11–hydroxyfunction to form a cyclic 11,18–hemiacetal form whose unique structure was probably responsible for the delayed discovery of aldosterone (and 18–hydroxycorticosterone) compared with other important adrenosteroids. Although the kidney tubules are the main target organ for aldosterone, the main function of this steroid is to promote exchange of sodium and potassium across cell membranes throughout the body. In the kidneys this leads to reabsorption of sodium and increased excretion of potassium in the urine. In conditions of aldosterone excess, the resulting increase in total body sodium causes hypertension; the loss of potassium leads to all the metabolic consequences of hypokalaemia.

Following its release into the circulation, aldosterone is almost completely metabolised into a variety of compounds which are either excreted in the urine unchanged or as their glucuronides and sulphates. Less than 0.5% of the aldosterone secreted each day is itself excreted unchanged whereas roughly 10% is excreted as aldosterone 18–glucuronide. It is this latter substance which is normally quantitated when urinary 'aldosterone' measurements are carried out.

Circadian variation of plasma aldosterone occurs but is much less obvious than for plasma cortisol; however, marked changes do occur with changes in posture and these are mediated by angiotensin. Assumption of the upright posture leads to a shift in intravascular and extracellular fluid to the lower limbs away from central volume receptors and the kidneys. This in turn causes a rise in plasma renin activity and angiotensin concentration leading to stimulation of aldosterone secretion, hence the importance of controlling posture when clinical measurements of plasma aldosterone are made. ACTH has relatively little effect upon aldosterone secretion by normal or 'hyperplastic' adrenal glands but is thought, paradoxically, to stimulate aldosterone secretion by genuine aldosterone-secreting adrenal adenomas.

Apart from the acute effects of posture, the best known and probably the most physiologically important stimulus to aldosterone secretion is sodium depletion, produced as a result either of reduced intake or of increased output in the urine or other body fluids. Hyperkalaemia is a weak stimulus to aldosterone secretion and hypokalaemia a weak inhibitor.

The effect of aldosterone upon renal tubular function—the exact mechanism of which is still poorly understood—can be largely, if not completely, reversed by a group of substances which behave as competitive aldosterone inhibitors. These drugs, the best known of which is spironolactone (Aldactone), have been extensively used as diuretics in conditions such as cirrhosis and congestive cardiac failure in which secondary hyperaldosteronism contributes to sodium retention and oedema. They have also been used for the treatment of primary hyperaldosteronism especially when this is due to 'idiopathic adrenal hyperplasia'.

Primary Hyperaldosteronism (Conn's Syndrome)

Primary hyperaldosteronism is a rare, but important because curable, cause of hypertension.[18] It is due to a single adrenal tumour in 70–80% of cases and to bilateral adrenal hyperplasia in the rest, though most authors would now deny that the latter is a true clinical entity.[13] Hyperaldosteronism occurs twice as often in women as in men. The patient is usually in middle life and presents as a case of 'hypertension' or with symptoms such as weakness, polyuria, polydypsia and easy fatiguability secondary to potassium depletion. The diagnosis is suggested by the finding of moderate to marked hypokalaemia (plasma potassium below 3.5 mmol/l) for which no other cause can be found (Table 2) and confirmed by finding a high plasma or urinary aldosterone level in the presence of low plasma renin activity. The plasma sodium concentration is usually in the upper half of the reference range but is otherwise unremarkable and the plasma bicarbonate level is generally raised, i.e. hypokalaemic metabolic alkalosis is present. Other biochemical parameters show no consistent pattern.

Although diagnosis is seldom difficult if the investigator has ready access to plasma aldosterone, renin and angiotensin measurements, the rarity of the condition ensures that it is suspected far more often than it is revealed, especially since hypokalaemia, indistinguishable from that associated with primary hyperaldosteron-

TABLE 2
MAIN CAUSES OF HYPOKALAEMIA

Renal wasting
  (i) Hormonal; aldosteronism and other types of mineralocorticoid excess
  (ii) Diuretics (and/or glucocorticoid) therapy
  (iii) Primary kidney (especially tubular) disease

Gastrointestinal loss
  (i) Vomiting
  (ii) Loss through fistulae
  (iii) Chronic diarrhoea; hormonal and purgative-induced
  (iv) Malabsorption
  (v) Secretory tumours of colon and rectum

Maldistribution
  (i) Induced metabolic alkalosis
  (ii) During attacks of familial periodic paralysis (hypokalaemic types only)
  (iii) Rapid reversal of hyperkalaemia with glucose and insulin
  (iv) Intravenous fluid with inadequate K^+ replacement

ism, also occurs in patients with 'benign' hypertension. Since so many factors can influence plasma aldosterone levels, its measurement is not indicated clinically unless the plasma potassium concentration is consistently below 3·7 mmol/l in the absence of obvious cause and is literally less than useless if samples for analysis are collected from unsuitably prepared patients.

Diuretics and all other drugs that affect sodium homeostasis must be discontinued for at least three weeks prior to assessment. Since patients may well be grossly hypokalaemic at presentation, they may require initially substantial potassium supplements, but these should be discontinued for three days before the tests are undertaken. Patients should be receiving 100–300 mmol/day of sodium and 50–100 mmol/day of potassium in their diet at the time the specimens for analysis are collected.

Blood must be taken early in the morning with the patient recumbent. He must not have been allowed to assume the upright posture at any time before the blood samples for analysis are collected.

Traditionally, patients with primary hyperaldosteronism have increased plasma aldosterone and reduced plasma renin activity and angiotensin II levels, but normal levels are not inconsistent with the diagnosis and are indeed observed in upwards of 10% of cases on some occasions.[18]

Though formerly looked upon as variants of a single pathological entity it is now clear that primary hyperaldosteronism due to aldosterone-producing adenomas (APA) and idiopathic adrenal hyperplasia (IAH) are two separate conditions,[13] each with its own appropriate treatment; surgery in the former and spironolactone in the latter. The two conditions can be distinguished from each other only by adrenal vein catheterisation or by use of a combination of tests which depend upon the different behaviour of the two conditions in response to postural changes after extracellular fluid expansion achieved by the use of sodium supplements and fluorocortisone treatment for three days.[13,18]

A rare familial variety of hyperaldosteronism occurs in which treatment with glucocorticoids leads to a reduction in aldosterone secretion, a fall in blood pressure and restoration of normokalaemia. Patients with this abnormality, like those with aldosterone-producing adenomas, but unlike those with 'idiopathic' hyperaldosteronism, have a paradoxical fall in plasma aldosterone after four hours in the upright position. After treatment with dexamethasone the plasma aldosterone response to standing returns to normal (i.e. plasma aldosterone rises instead of falling) presumably because of a rise in plasma renin activity. The cause of the illness, which responds to glucocorticoid therapy, is unknown.

Adrenal vein catheterisation with measurement of the aldosterone:cortisol ratio in the venous effluent from each vein and its comparison with that from the contralateral vein, as well as with that in blood collected from the inferior vena cava, not only permits lateralisation of the tumour, should there be one, but also allows its differentiation from idiopathic adrenal hyperplasia in which the ratio is similar in blood draining both adrenals. Probably the best test of all, however, is adrenal scintigraphy using labelled cholesterol as the marker and/or computerised axial tomography (CAT scanning).

A number of simpler methods for pre-operative differentiation of adrenal adenoma from hyperplasia have been advocated at various times but only quadric analysis and multiple logistic analysis have proved to be consistently reliable and both of these require computer analysis.[3,13,18]

Other Steroids

Hypertension is a characteristic feature of Cushing's syndrome and may provide the first clue to the correct diagnosis which should seldom remain long in doubt once suspected. In a small minority of cases of primary mineralocorticoid excess the steroid mainly responsible for the clinical features is desoxycorticosterone or the more recently discovered 18-hydroxycorticosterone rather than aldosterone which itself may be produced in subnormal amounts. This situation may arise either as the result of an adrenal tumour or as a consequence of an inborn error of steroidogenesis. It occurs characteristically with 11β-hydroxylase deficiency, one of the varieties of congenital adrenal hyperplasia. In addition to virilism (in girls), children afflicted with this disorder have hypertension, normal-to-low plasma potassium and aldosterone concentrations and suppressed plasma renin activity. Plasma and urinary levels of 17α–hydroxyprogesterone and its metabolites are increased and the condition responds to treatment with glucocorticoids. Another even rarer condition in which non-aldosterone mineralocorticoid excess occurs is that caused by 17α–hydroxylase deficiency. This disease is not associated with virilism and generally first comes to light during adolescence. Hypertension with hypokalaemia in a patient with primary amenorrhoea (girls) or pseudohermaphroditism (in boys) is almost pathognomonic and the diagnosis is confirmed by characteristic plasma and urinary steroid findings, i.e. an excess of DOC, 17–hydroxy–DOC, coricosterone and 10–hydroxycorticosterone, together with an almost complete absence of 17α-hydroxysteroids.

The syndromes of endogenous mineralocorticoid excess must be distinguished, clinically, from conditions mimicking them such as those produced by licorice abuse, carbenoxolone, certain nasal sprays and dermatological preparations containing fluorinated steroids with powerful mineralocorticoid activity. All of these conditions are characterised by a combination of hypertension and hypokalaemia but plasma aldosterone levels and renin activity are both low and generally the offending agent can be demonstrated in blood or urine if sought.

PRIMARY ABNORMALITIES OF SODIUM METABOLISM

A relationship between excessive salt consumption and hypertension has been recognised for almost four thousand years[8] and the nature of the association has attracted much attention of late. A defect in Na^+/K^+ ATPase activity[16] with accumulation of greater than normal amounts of intracellular sodium has been demonstrated in patients with hereditary hypertension and/or obesity by a number of different investigators and a similar defect is said also to occur in some of their close blood relatives.[14] De Wardener and his colleagues[7] have recently demonstrated the existence of a non-steroidal natriuretic hormone whose production is defective in persons with hereditary hypertension. This abnormality is apparently related to an inherited inability of the renal tubules to excrete efficiently anything more than a minimal salt load which, in turn, leads to the retention of sodium ions in the body whenever this limited salt excretory capacity is exceeded. This, it is further postulated, can lead to impairment of calcium transport across the plasma membranes of cells which, in the case of the smooth muscle fibres of the arterioles can cause an increase in tone and reduction in intraluminal diameter. Whether such a mechanism is responsible for the production of hypertension by sodium retention secondary to mineralocorticoid excess is unknown.

The possibility that at least some cases of essential hypertension are the result of an inherited abnormality of sodium metabolism is clearly of the utmost clinical importance. Too little factual information is currently available, however, to make profitable further discussion in a work such as this at the present time, but important developments can be confidently predicted over the next few years.

Catecholamines

Adrenaline, possibly the first of the hormones to be studied intensively by pharmacological and biochemical techniques, was characterised initially by its effects upon the cardiovascular system and only much later by its effect upon metabolic processes. Its production and storage by the adrenal medulla and its ability to produce a rapid and dramatic increase in blood pressure was recognised by the end of the nineteenth century but not until 30 years later was it realised that its paroxysmal release by a phaeochromocytoma—the name given to rare tumours of the adrenal medulla—might be a cause of hypertension. Once it had been described, the characteristic clinical syndrome caused by phaeochromocytoma rapidly became known, though only rarely diagnosed during life mainly due to its rarity and the difficulty of diagnosis even when suspected. The situation changed markedly for the better after laboratory procedures for detecting excessive production of adrenaline (and to a lesser extent, noradrenaline) became available during the 1950s.[10]

Production of adrenaline from its immediate precursor, noradrenaline, occurs almost exclusively in the adrenal medulla. It requires the presence of an enzyme, phenylethanolamine N-methyl-transferase (PNMT), which catalyses the conversion of noradrenaline to adrenaline according to the scheme:

$$\begin{array}{c} \text{noradrenaline} \\ + \\ \text{S-adenosyl methionine} \end{array} \xrightarrow{\text{PNMT}} \begin{array}{c} \text{adrenaline} \\ + \\ \text{S-adenosyl homocysteine} \end{array}$$

PNMT is an inducible enzyme which is thought to depend upon the presence of a high concentration of cortisol or other glucocorticoid in the extracellular fluid such as occurs in the adrenal medulla. In man and most other mammals the adrenal medulla receives blood that has already passed through the adrenal cortex and been enriched with adrenal steroids. The inducible nature of the enzyme may account for the facts that chromaffin tissue elsewhere in the body, histologically indistinguishable from that of the adrenal medulla, is only rarely associated with the production and/or storage of adrenaline and that phaeochromocytomas arising in such tissues are characterised by the secretion of noradrenaline and its metabolites rather than of adrenaline and its metabolites. The intriguing suggestion has been made[5] that adrenaline released from the adrenal medulla might, through a complex system of interrelated steps, be involved in the pathogenesis of essential hypertension. This theory is still little more than provocative hypothesis.

Noradrenaline, and its own immediate precursor, dopamine, are produced by enzymic conversion of tyrosine almost exclusively in tissues of nervous origin—including the adrenal medulla. Both they and adrenaline produce their effects upon responsive tissues by interacting with specific receptors located on the surface of the effector cells. Originally looked upon as constituting a single homogeneous class, adrenoceptors are now recognised to be both varied and highly individualistic.[11] The main classes of adrenoceptors are known as α_1, α_2, β_1, and β_2. They are distinct from those responding specifically to dopamine. The pressor effects of adrenaline and noradrenaline are mediated largely through activation of alpha receptors on the surface of smooth muscle fibres of arteries and veins causing them to constrict and additionally in the case of adrenaline, leading to stimulation

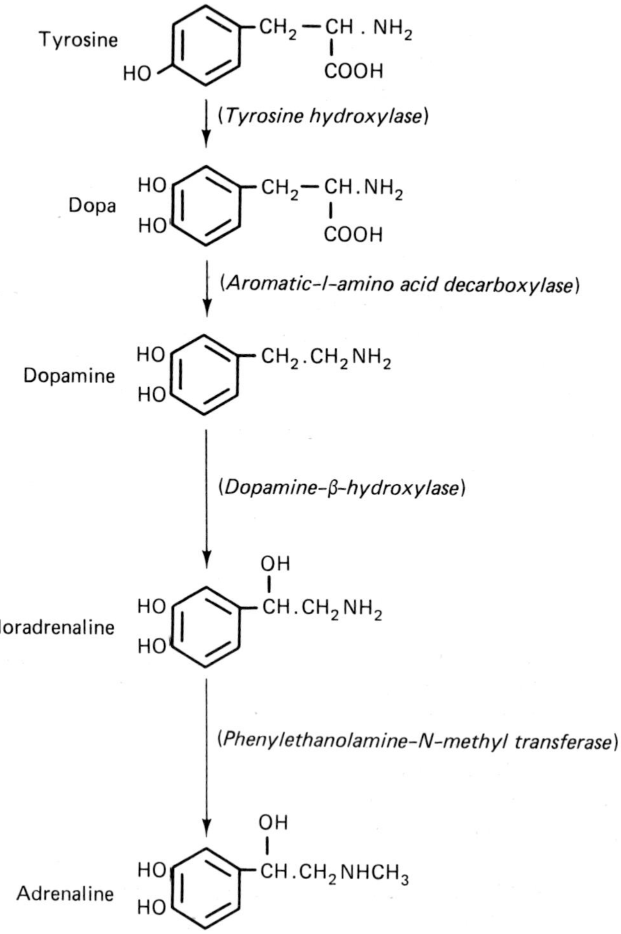

Fig. 1. Biosynthesis of adrenaline and noradrenaline.

of the rate and force of contraction of the heart through activation of β–receptors in the myocardium.[6]

The constricting effect of adrenaline upon the arteries, which is exerted through α–adrenoceptor activation is counteracted, to a greater or lesser extent, by simultaneous activation of β–adrenoceptors on the same or nearby muscle fibres whose action is dilatory. This mutual antagonism does not occur with noradrenaline which has virtually no β–antagonistic properties.

Understanding of the various types of adrenoceptors and development of specific inhibitors (and activators) of their activity has been of immense therapeutic advantage.[11]

Phaeochromocytoma

Phaeochromocytomas are rare tumours with a prevalence of about 1 case per 800 000 of the population. They may occur at any age; 2–5% are familial and/or occur in association with other endocrine abnormalities. Over 80% develop in the adrenal medulla, the remaining 20% in chromaffin tissue elsewhere. The classical symptoms of phaeochromocytoma are intermittent episodes of apprehension, palpitation, sweating, flushing and a sense of epigastric fullness, with or without associated headache, coming on especially commonly after squatting, bending or increased pressure on the abdomen. The episodes, which may last from a few minutes to several hours, are accompanied by a marked rise in blood pressure. More than half the patients subsequently found to be harbouring phaeochromocytoma show few or none of these symptoms, however, and present a clinical picture indistinguishable from that of benign essential hypertension. Pointers to the correct diagnosis, which is made on the basis of laboratory tests, include early onset of non-familial hypertension, a suggestive history and the presence of glycosuria and/or hyperglycaemia.

Considering its rarity, an extremely large number of tests have been used to try and establish a diagnosis of phaeochromocytoma once suspected, but the most reliable[15] is the simplest—namely measurement of adrenaline and noradrenaline metabolites (Fig. 2) in two or more 24-hour urine collections by a sensitive, specific and accurate analytical method.

HPLC using an electrochemical detector is currently the method of choice but GLC and colorimetric methods still enjoy great popularity. The measurement

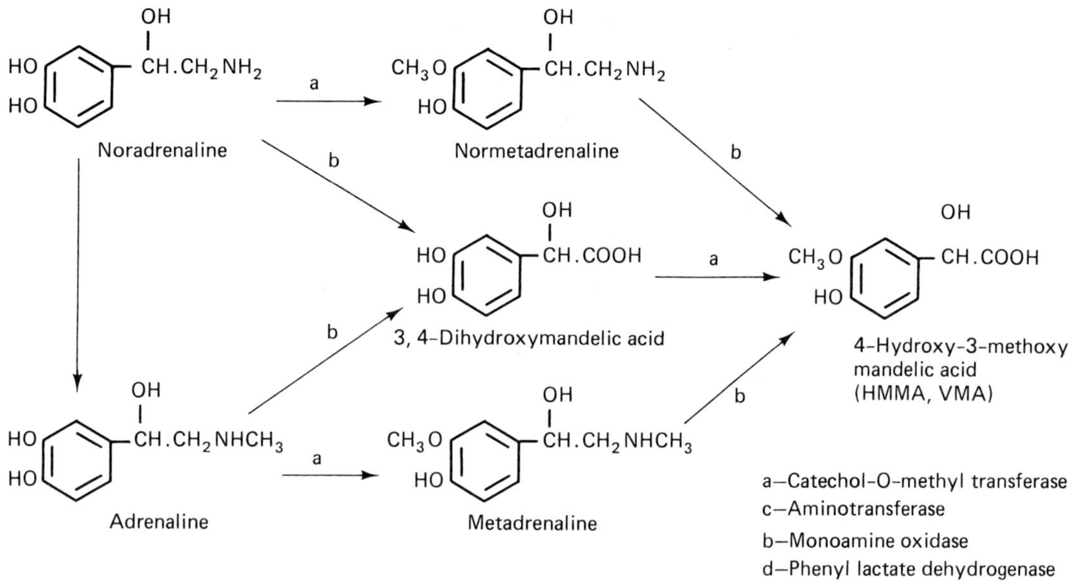

FIG. 2. Metabolic degradation of adrenaline and noradrenaline.

a—Catechol-O-methyl transferase
c—Aminotransferase
b—Monoamine oxidase
d—Phenyl lactate dehydrogenase

of adrenaline and noradrenaline in plasma[4] using an isotope derivatisation technique of HPLC with electrochemical detection has been advocated as being more sensitive and specific but this has been disputed.[15] It is more likely to lead to false negative results in patients with intermittently secreting tumours. Tests based upon the detection and quantifying of catecholamines in urine by fluorescence are obsolete and should not be used, nor should it be necessary to resort to the use of provocative tests such as those employing the intravenous injection of tyramine, histamine or glucagon all of which are unpleasant and potentially extremely dangerous.

Few conditions, apart from phaeochromocytoma, produce an increase in 24-hour urinary excretion of the major catecholamine metabolites, namely metadrenaline and normetadrenaline (often measured as total metadrenaline) and 4-hydroxy, 3-methoxymandelic acid (HMMA). Various antihypertensive drugs can interfere with their assay and/or physiological release—especially if relatively non-specific chromogenic methods are used—and should therefore be discontinued for at least six weeks before samples are collected for analysis if at all possible. It is, however, questionable whether their continued use ever does, in fact, lead to serious diagnostic errors, providing suitably specific analytical techniques are employed.

Values for urinary total metadrenalines in patients with phaeochromocytoma vary from 2–30 times average normal value, the smallest increases generally being observed in patients with intermittent hypertension with or without associated symptoms. Differentiation of adrenal from extra-adrenal phaeochromocytoma on the basis that only the former produce increased amounts of adrenaline, whereas the latter produce mainly noradrenaline, though theoretically possible is unreliable and CAT scanning and scintigraphy are now the recognised methods of choice for preoperative localisation. Either or both of these techniques may be supplemented, in difficult cases, by venous catheterisation to enable the determination of plasma adrenaline concentrations at various sites.[4]

Vasopressin

The antidiuretic hormone of the hypothalamus and posterior pituitary is known officially as vasopressin in recognition of its mode of discovery, which was as a pressor agent, and its standard method of bioassay as a pharmaceutical agent. Chemically it is an octapeptide differing in structure from oxytocin by only two aminoacids. In the pig the arginine residue characteristic of human vasopressin (arginine vasopressin) is substituted by lysine (lysine vasopressin) with remarkably little change in biological activity. The two major stimuli to vasopressin release are a rise in plasma osmolality and a reduction in blood volume and/or blood pressure.[12] The two major stimuli are seemingly independent of each other though both can be looked upon as protecting the cerebral circulation either by conserving water or by reducing the size of the vascular bed in non-essential organs and thereby diverting blood to the brain. Modest dehydration such as that produced by water deprivation for 24 hours is accompanied by a doubling in plasma ADH concentration: this is a modest increase compared with the much greater rises observed in response to stressful stimuli, e.g. hypoglycaemia and haemorrhage.

A possible role for vasopressin in the pathogenesis of hypertension—or even in the maintenance of normal blood pressure—was not seriously entertained until about 1972 when, as a result of various experiments in normal animals and patients with a bizarre syndrome known as primary autonomic insufficiency, the view that it might indeed have such a role gained credibility. This view is once again under serious challenge as the result of (i) the demonstrable absence of hypertension in patients with chronic endogenous vasopressin overproduction occurring as part of the syndrome of inappropriate ADH secretion, (ii) the failure of infusions of exogenous vasopressin to affect blood pressure even when given in quantities sufficient to raise plasma levels to at least those observed in patients with malignant hypertension.[2] Nevertheless, the possibility that vasopressin is important in the maintenance of normal blood pressure under conditions of acute blood volume reduction still deserves serious consideration,[2] since acute haemorrhage is recognised as the most potent stimulus to endogenous vasopressin release.

At the present time, however, the investigation of vasopressin status does not contribute materially to elucidation of the cause of hypertension when, as is usually the case, this is in doubt.

Other Hormonal and Metabolic causes of Hypertension

Hypertension can occur as the presenting feature of several diseases customarily recognized by their more characteristic features. Some of the more important of these are listed in Table 3. The pathological nature of the

TABLE 3
CAUSES OF HYPERTENSION

Essential hypertension
   (no anatomical or pathological cause found)

Kidney disease
   (i) Parenchymal damage
   (ii) Renal artery disease
   (iii) Compression

Endocrine disease
   (i) Phaeochromocytoma
   (ii) Mineralocorticoid excess
   (iii) Glucocorticoid excess
   (vi) Oral contraceptives
   (v) Acromegaly
   (vi) Hyperparathyroidism

Neurogenic hypertension

Toxaemia of pregnancy

Cardiovascular disease

underlying disease will usually be discovered as the result of a careful clinical history and physical examination, and the hypertension can be looked upon as an epiphenomenon, not itself requiring investigation as to cause. Occasionally, however, in hyperparathyroidism, for example, laboratory tests carried out as part of a general diagnostic work-up may provide the first clue to the correct diagnosis.

REFERENCES

1. Atlas, S. A. and Case, D. B. (1981), Renin in essential hypertension. *Clin. Endocrinol. Metab.*, **10**, 537–575.
2. Bartter, F. C. (1981), Vasopressin and blood pressure. *New Engl. J. Med.*, **304**, 1097–1098.
3. Brown, J. J., Lever, A. F., Robertson, J. I. S. *et al.* (1979), Are idiopathic hyperaldosteronism and low renin hypertension variants of essential hypertension? *Ann. Clin. Biochem.*, **16**, 380–388.
4. Brown, M. J., Allison, D. J., Jenner, D. A., Lewis, P. J. and Dollery, C. T. (1981), Increased sensitivity and accuracy of phaeochromocytoma diagnosis achieved by use of plasma-adrenaline estimations and a pentolinium-suppression test. *Lancet*, **i**, 174–177.
5. Brown, M. J. and Macquin, I. (1981), Is adrenaline the cause of essential hypertension? *Lancet*, **ii**, 1079–1081.
6. Cryer, P. E. (1980), Physiology and pathophysiology of the human sympathoadrenal neuroendocrine system. *New Engl. J. Med.*, **303**, 436–444.
7. de Wardener, H. E. (1982), The natriuretic hormone. *Ann. Clin. Biochem.*, **19**, 137–140.
8. Editorial (1981), New evidence linking salt and hypertension. *Br. med. J.*, **282**, 1993–1994.
9. Garay, R. P., Elchozi, J. L., Dagher, G. and Meyer, P. (1980), Laboratory distinction between essential and secondary hypertension by measurement of erythrocyte cation fluxes. *New Engl. J. Med.*, **302**, 769–771.
10. Goldfein, A. (1981), Phaeochromocytoma. *Clin. Endocrinol. Metab.*, **10**, 607–630.
11. Lees, G. M. (1981), A hitch-hiker's guide to the galaxy of adrenoceptors. *Br. med. J.*, **283**, 173–177
12. Martin, J. B., Reichlin, S. and Brown, G. M. (1977), Neural regulation of water and salt metabolism: physiologic function and disease. In: *Clinical Neuroendocrinology*, pp. 63–92. Philadelphia: F. A. Davis.
13. Padfield, P. L., Brown, J. J., Davies, D. L. *et al.* (1981), The myth of idiopathic hyperaldosterone. *Lancet*, **ii**, 83–84.
14. Parfrey, P. S., Vandenberg, M. J., Wright, P. *et al.* (1981), Blood pressure and hormonal changes following alteration in dietary sodium and potassium in mild essential hypertension. *Lancet*, **i**, 59–63.
15. Plouin, P. F., Duclos, J. M., Menard, J., Comoy, E., Bohuon, C. and Alexandre, J. M. (1981), Biochemical tests for diagnosis of phaeochromocytoma: urinary versus plasma determinations. *Br. med. J.*, **282**, 853–854.
16. Sweadner, K. J. and Goldin, S. M. (1980), Active transport of sodium and potassium ions: mechanism, function, and regulation. *New Engl. J. Med.*, **302**, 777–783.
17. Tarazi, R. C. and Gifford, R. W. (1979), Systemic arterial pressure. In: *Pathologic Physiology*, ed. W. A. Sodeman and F. M., pp. 198–299. Sodeman. Philadelphia: W. B. Saunders.
18. Vaughan, N. J. A., Jowett, T. P., Slater, J. D. H. *et al.* (1981), The diagnosis of primary hyperaldosteronism. *Lancet*, **i**, 120–125.
19. Vidt, D. G., Bravo, E. L. and Fouad, F. M. (1982), Drug therapy: captopril. *New Engl. J. Med.*, **306**, 214–219.

FURTHER READING

Biglieri, E. G. and Schambelan, M. (editors) (1981), *Hypertension. Clinics in Endocrinology and Metabolism*, Vol. 10 (No. 3), London: W. B. Saunders.

Mulrow, P. J. (1981). Glucocorticoid-suppressible hyperaldosteronism: a clue to the missing hormone. *New Engl, J. Med.*, **305**, 1012–1014.

Pickering, G. W. (1968), *High Blood Pressure*, 2nd edn. Edinburgh: Churchill Livingstone.

Weinberger, M. H., Grun, C. E., Hollifield, J. W. *et al.* (1979), Primary aldosteronism: diagnosis, localisation and treatment. *Ann. Intern. Med.*, **90**, 386–391.

SECTION IX

NEUROPSYCHIATRIC DISORDERS

	PAGE
32. HEREDITARY AND ACQUIRED MENTAL DEFICIENCY	489
33. THE ROLE OF THE LABORATORY IN SOME PSYCHIATRIC ILLNESSES	523

32. HEREDITARY AND ACQUIRED MENTAL DEFICIENCY

JAN STERN

Introduction
  Size of the problem
  Aetiology

Detection of metabolic disorders with mental handicap
  Technical methods
  Problems, pitfalls and artefacts

Degenerative diseases of the nervous system with mental handicap—lysosomal disorders
  Clinical aspects
  Sphingolipidoses, mucopolysaccharidoses, leucodystrophies
  Approach to diagnosis
  Approaches to treatment

Encephalopathies of extracerebral origin
  Hypoxia and ischaemia
  Hypoglycaemia
  Ventricular haemorrhage
  Prematurity
  Kernicterus and electrolyte disturbances
  Nutritional aspects of mental handicap
  Vitamins, antivitamins and teratogens

Aminoacidurias
  Classification
  Phenylketonuria
  Histidinaemia
  Disorders of tyrosine metabolism
  Non-ketotic hyperglycinaemia
  The hyperammonaemias
  Homocystinuria
  Other aminoacidurias

Other neurometabolic disorders
  Galactosaemia
  Disorders of pyruvate metabolism
  Organic acidurias
  Disorders of purine and pyrimidine metabolism
  Muscular dystrophy
  Hypothyroidism
  Heavy metals and intellectual development

Ethical considerations—genetic counselling
  Problems and dilemmas
  Genetic counselling
  Other medico–ethical issues

INTRODUCTION

A significant proportion of metabolic disease is commonly associated with neuro-psychiatric handicap. While much progress has been made in our understanding of the distortion of metabolic pathways in hereditary metabolic disease, the precise relation of biochemical aberration to mental retardation has often proved elusive. The same is true of environmental factors. Refinements of analytical techniques have made possible the reliable monitoring of nutritional factors, drugs and poisons in body fluids, but again it is frequently difficult to decide if measured levels constitute a threat of injury to the nervous system.

In all cases, it is important to ascertain (1) if the observed biochemical abnormalities do in fact damage the brain (2) by what mechanism and (3) to what extent. Decisions involving massive expenditure on neonatal screening, dietary treatments and public health measures may be based on answers to these questions which can only be given with confidence when reliable biochemical observations are matched by sound clinical and psychological assessments.

Size of the Problem

Mental retardation (synonyms: mental deficiency, mental handicap, subnormality) is a permanent impairment of the intellect sufficiently severe to prejudice normal existence in the community. Mental retardation is thus a social as well as a psychopathological concept; it is a function both of the inadequacy of the brain of an affected individual and of the complexity of the society in which he lives.[44] Contemporary methods of assessing intelligence are based on verbal and non-verbal responses to standardized tasks. The results are expressed as intelligence quotients (IQ) which essentially compare the performance of an individual to that of the population of which he is a member. Commonly used intelligence scales such as the Stanford–Binet or Wechsler are so constructed that the distribution of IQs is roughly normal to conform to the pattern of distribution commonly found for continuously variable biological parameters such as height. Half the population are assigned an IQ in the average range (90–109), a quarter have an IQ of less than 90. Individuals with an IQ 70–50 are classed as mildly retarded, many of them do not need medical care or special educational provisions. They make up approximately 1·8% of the population. Approximately 0·4% of the population are severely subnormal, with an IQ below 50. This group invariably requires special educational provision and often medical care in hospital or the community.

The above figure refers to prevalence, the number of patients living at any time as a fraction of the general population. The incidence at birth of many disorders associated with severe mental retardation is considerably higher than their prevalence because of a differential mortality rate weighted against the severely subnormal.

Aetiology

The relative importance of heredity and environment in shaping intellectual capabilities has been acrimoniously debated since the turn of the century. Here we are only concerned with below average abilities. Traditionally, severe subnormality has been attributed to the effect of 'major genes' or pathological factors in the environment such as intrauterine infection or birth injury. The mildly retarded, on the other hand, were seen as a 'subcultural group' of the general population, their intellectual deficit representing 'biological variation' resulting from the action of 'polygenes', that is a number of genes each producing small but additive effects on intelligence. This approach is not helpful. Genes or environmental factors will impair intelligence if they act at the molecular level to produce structural or metabolic defects which preclude normal function of the brain. Sometimes, genetic disorders, for example tuberous sclerosis, and conditions caused by environmental factors such as rubella embryopathy, may be associated with a whole spectrum of severity of mental defect ranging from virtually normal to the grossest handicap. A distinction between 'subcultural' and 'pathological' mental subnormality cannot be maintained on grounds of aetiology.

This is borne out by neuropathological examination of mentally retarded patients who come to autopsy. The great majority of the severely retarded have gross morphological abnormalities of the brain. In a series in our hospital only 8 out of 191 classified cases were apparently free from structural abnormalities when examined macroscopically and by comparatively straightforward classical histological methods. A significant proportion of the much smaller number of mildly retarded patients who came to post-mortem showed similar but less severe lesions than the lower grade patients.[17]

Genetic and environmental factors may act via common pathological pathways. For example, a high level of unconjugated bilirubin in blood may occur in premature infants, when the enzyme conjugating bilirubin and glucuronic acid is not yet operative, in the Crigler–Najjar syndrome with hereditary absence of the enzyme, or in blood group incompatibility where the conjugating system is overwhelmed by excessive formation of bilirubin. Brain damage with kernicterus (yellow staining of certain formations of the brain) may be seen in all these situations.

Following Scriver, Laberge, Clow and Fraser,[61] we may regard health as a state of equilibrium between the genetic make-up of an individual and environmental forces. Disequilibrium or disease can be placed on a spectrum. At one end are diseases that result from specific, harmful environmental factors acting on the universal genotype, at the other end disease occurs when particular deficient genotypes are exposed to the universal environment. In between are conditions in which disequilibrium results from the interplay of both adverse genetic and environmental factors. This is a highly convenient way of looking at the causes of mental deficiency. Not unnaturally, at least initially conditions at either end of the spectrum attracted most attention: the storage disorders where a genetically-determined enzyme deficiency leads inexorably to the destruction of the cell affected, or classical phenylketonuria where an extracerebral genetic defect inhibits normal intellectual development. At the environmental end of the spectrum are found such conditions as acute lead poisoning or head injury. Yet in a high proportion of cases the neuropsychiatric handicap cannot be attributed to a single cause, be it genetic or environmental[28] (Table 1).

TABLE 1
AETIOLOGICAL FACTORS IN MENTAL RETARDATION[38]

Perinatal disorders	18·5
Chromosome abnormalities	17·5
Metabolic disorders	3·1
Other genetic disorders	2·4
Infections of the CNS	6·6
Malformations	7·5
Injury, poisoning	1·5
Indeterminate causes	42·9
	100%
	n = 2416

It is the result of the interaction of a multiplicity of genetic and environmental factors and it is this interaction which underlies 'biological variation'. To argue the relative importance of genetic and environmental factors in determining biological variation is sterile if not meaningless. What matters is to identify pathogenetic factors, be they environmental or genetic, and to counteract their harmful effects.

The vulnerability of the brain at critical periods of development has been emphasised by Dobbing.[24] The severity and duration of growth-restricting factors and the developmental stage of the brain determine the nature and extent of ultimate deficits rather than the precise nature of the insult. Early in pregnancy, teratogens will produce malformations, in mid-pregnancy neuronal multiplication is at risk. The period from the last trimester to 18 months or 2 years after birth has been called the 'brain growth spurt'. It is the period of glial multiplication, dendritic arborisation, synaptogenesis and of maximum vulnerability to malnutrition and extracerebral inborn errors such as phenylketonuria.[24]

Most neuropathological lesions associated with metabolic disorders are non-specific. Similar patterns of changes may result from different genetically- or environmentally-determined metabolic abnormalities. The repertoire of structural change in the nervous system is rather limited in relation to the great variety of environmental and endogenous harmful factors which may operate via 'common abnormal pathways'.[17] The timing of pathogenetic processes is not always unequivocal. Clinical and biochemical observations are not infallible guides since damage sustained by the brain during early development may only become apparent months or even years later as demands on the nervous system increase and ability for abstract reasoning is tested.[17]

From what has been said it is clear that for any mutation the clinical manifestations may be drastically modified for better or worse by the environment and also by the effects of other genes. Equally important in shaping the disease pattern is the phenomenon of genetic heterogeneity. For mutant genes of large effect, this has nearly always been found when looked for. Genetic heterogeneity may take the form of variable residual enzyme activity. Other things being equal, the higher the residual enzyme activity, the milder the cause of the disorder, and the lower the risk of mental handicap. In other cases, the kinetics or stability of the enzyme may be altered by a mutation. This may result in altered cofactor requirements or increased sensitivity to drugs. Where an enzyme occurs in the form of iso-enzymes these may or may not be under separate genetic control. The clinical manifestations will often differ greatly depending on whether one or other iso-enzyme is involved. For example, deficiency of arylsulphatase A results in a leucodystrophy, that of arylsulphatase B in a mucopolysaccharidosis. Multiple arylsulphatase deficiency combines the clinical features of both a leudodystrophy and a mucopolysaccharidosis and patients are more severely affected than those with a deficiency of a single iso-enzyme.

Sometimes, a mutation will reduce enzyme activity to a few per cent of normal, yet this small amount of enzyme activity may be compatible with normal development. However, those affected may exhibit enhanced vulnerability to hazards such as infections, malnutrition or adverse drug effects (Table 2). Only in the presence of supervening adverse factors is the full clinical picture of the underlying metabolic disorder unmasked. These 'intermittent' inborn errors of metabolism may test severely both clinician and laboratory as often only prompt treatment will prevent death or mental handicap. Between attacks the disorder may or may not be detectable by examination of the body fluids but will always be demonstrable by assay of the appropriate enzyme.

Some of the more important hereditary disorders associated with mental handicap are shown in Table 3. Many more conditions are listed in Crome and Stern,[17] McKusick,[44] Stanbury, Wyngaarden and Frederickson,[66] and Bondy and Rosenberg.[8] All these disorders are rare (Table 4). Estimates of birth frequencies, where available, are inevitably only approximate, sometimes little more than orders of magnitude and always subject to revision in the light of further population epidemiological findings. Even clinicians and biochemists with a special interest in this field are unlikely to see more than a few cases of any of these disorders in the course of their careers. The division into two groups: those 'almost invariably' and those 'occasionally or rarely' associated with mental handicap, is admittedly arbitrary. All one can say is that in general normal intelligence would be unexpected, without treatment, in the great majority of patients with the group disorders listed as 'almost invariably' associated with mental handicap, while the opposite appears to be true of the other group. The extent of mental handicap is not always easy to ascertain. Those inexperienced in the assessment of these patients may be prejudiced by their appearance and history. Both normal and retarded intelligence may be masked by physical incapacity, behaviour disorder or unwillingness to cooperate in psychometric tests.

The situation pertaining to environmental hazards, poisons and teratogens is somewhat different. Here, the concern is what intensity and timing of a metabolic insult to the nervous system is necessary or sufficient to produce lasting mental handicap, and the presence or absence of threshold effects. Some of the more important environmental causes of mental defect are listed in Table 5. They are of interest to the clinical chemist because they are either in themselves metabolic in nature or because they enter into the differential diagnosis of neurometabolic disorder. Many more examples of environmental causes of brain damage will be found in Kirman and Bicknell,[44] Craft,[14] and Menkes.[52]

DETECTION OF METABOLIC DISORDERS CAUSING MENTAL HANDICAP

In the majority of cases, disorders causing mental handicap are diagnosed either by presymptomatic screening of the newborn or by investigation of high risk groups. The logistics and ethics of mass screening have been discussed elsewhere.[5,73] Phenylketonuria is perhaps the best known condition for which mass screening is fully justified both on ethical and economic grounds. Neonatal hypothyroidism, neural tube defects and

TABLE 2
FACTORS WHICH ENHANCE THE VULNERABILITY OF THE NERVOUS SYSTEM IN INBORN ERRORS OF METABOLISM

Malnutrition or inappropriate diet
Anoxia, hypoglycaemia, acidosis, brain oedema
Infection
Drugs
Suboptimal functioning of alternative pathways
Failure of protective mechanisms

Table 3
Hereditary Neurometabolic Disorders and Mental Handicap

	Almost invariably or commonly associated with mental or neurological handicap	Occasionally or rarely associated with neuropsychiatric handicap
Aminoacidurias	Phenylketonuria Homocystinuria Argininosuccinicaciduria Citrullinaemia Hyperammonaemias Hyperglycinaemia Lowe's syndrome Glutathionuria	Histidinaemia Hyperprolinaemias Hydroxyprolinaemia Sarcosinaemia Cystathionaemia Hyperlysinaemias Hartnup disease
Organic acidurias	Maple syrup urine disease Isovaleric acidaemia Methylmalonic acidaemias Glutaric acidaemia 5-oxoprolinuria D-glyceric acidaemia	β-ketothiolase deficiency
Disorders of carbohydrate metabolism	Galactosaemia Glycogen synthetase deficiency Lactic and pyruvic acidaemias	Fructosaemia Von Gierke's disease Fructose diphosphatase deficiency
Storage disorders	Mucopolysaccharidoses Sphingolipidoses Krabbe's disease Metachromatic leucodystrophy Mucolipidoses Wolman's disease Aspartylglucosaminuria Pompe's disease	
Hormonal disorders	Hypothyroidism Pseudohypoparathyroidism	Nephrogenic diabetes insipidus
Miscellaneous disorders	Lesch–Nyhan disease Oroticaciduria Crigler–Najjar syndrome Wilson's disease Menkes' disease Huntington's chorea	Trimethylaminuria
Chromosomal disorders	Autosomal disorders Some sex chromosome disorders	Some sex chromosome disorders

Down's syndrome (trisomy 21) in the offspring of women aged 35 years or over also meet the essential criteria formulated by Wilson and Jungner[73] which should be met before embarking on mass screening. For most other conditions, a convincing case for mass screening has not yet been made out. Often, not enough is known about the range of clinical manifestation, the natural history of the disorder, and in many cases the effectiveness of treatment remains to be demonstrated as required by Wilson and Jungner.[73]

Most cases of neurometabolic disorders are diagnosed in children referred for biochemical assessment because of acute or chronic signs of involvement of the nervous system or because of retarded physical and mental development with or without congenital malformations. Clearly, it is not possible to carry out diagnostic tests for every possible disease on every patient. The need of the laboratory is for a repertory of comparatively simple orienting tests which will indicate the direction further investigations should take and whether and to what specialised centre patient or sample should be referred. It is impossible to devise a system which will not miss the occasional diagnosis particularly in mild or atypical cases or in disorders not previously described. The risk of this will be minimized by cooperation between ward and laboratory, ensuring that samples are collected with care and reach the laboratory promptly and with full details about patient and treatment. Sometimes, clinical signs will suggest a possible diagnosis (Table 6). It must be admitted that some of the signs such as failure to thrive, vomiting, lethargy and acidosis are by no means specific and may assume significance only if, for example, their intensity and failure to respond to routine treatment appear inappropriate in their clinical context. Often, a high index of suspicion on the part of the clinician that he or she is dealing with a neurometabolic disorder is as

TABLE 4
APPROXIMATE ESTIMATES OF THE BIRTH FREQUENCIES OF DISORDERS ASSOCIATED WITH MENTAL HANDICAP[8,52]

Monogenic disorders

Dominant disorders

Huntington's chorea	1:10 000
Neurofibromatosis	1:2 500
Myotonic dystrophy	1:5 000
Tuberous sclerosis	1:30 000

Autosomal recessive disorders

Phenylketonuria	1:12 000
Homocystinuria	1:50 000
Histidinaemia	1:15 000
Maple syrup urine disease	1:250 000
Argininosuccinic aciduria	1:250 000
Galactosaemia	1:75 000
Mucopolysaccharidoses I and III	1:30 000 (combined)
Tay–Sach's disease	1:150 000 (higher in Jews)
Metachromatic leucodystrophy	1:50 000

X-linked disorders

Mucopolysaccharidosis II	1:100 000
Fabry's disease	1:100 000
Menkes' syndrome	1:100 000
Duchenne muscular dystrophy	1:7 000

Endocrine disorders

Neonatal hypothyroidism	1:4 000

Chromosomal disorders

Down's syndrome	1:650
Sex chromosomal disorders	1:550

Congenital malformations

Spina bifida	1:300–1 500

TABLE 5
SELECTED ENVIRONMENTAL CAUSES OF MENTAL HANDICAP[17,52]

Antenatal infections
  Rubella virus
  Cytomegalovirus
  Toxoplasmosis
  Herpes simplex virus

Birth injury

Postnatal infections
  Meningitis
  Subacute sclerosing panencephalitis

Autoimmune and postinfectious disorders
  Pertussis vaccination encephalopathy

Anoxia, asphyxia and hypoglycaemia
Bilirubin toxicity
Electrolyte imbalance
Endocrine imbalance
Malnutrition
Mineral poisons
Drugs and organic poisons
Radiation
Burns

important as any clue. The above considerations apply, of course, with equal force to the investigation of metabolic encephalopathies due to poisons or other adverse environmental factors.

Technical Methods

Chromatography of amino acids, carbohydrates, phenolic acids and purines and pyrimidines on paper or thin layers of cellulose or silica gel[8a,63] is within the scope of most laboratories now that ready-to-use thin layer plates are available commercially. Blood or urine are generally examined, much more information is obtained if both blood and urine are chromatographed at the same time, as it will then be often obvious if excess of a urinary metabolite is an overflow or renal phenomenon. Often, two or three one-dimensional chromatograms visualised by well-selected stains will yield more information than one two-dimensional chromatogram. Chromatography should always be supplemented by urine spot tests.

Many other techniques have proved useful in the detection of neurometabolic disorders. Guthrie bacterial inhibition tests are widely used not only in mass screening of infants for amino acid disorders but also in population genetic surveys and in monitoring the dietary

TABLE 6
CLINICAL PHENOMENA SUGGESTIVE OF INHERITED DISORDERS OF METABOLISM

GENERAL	Positive family history	
	Failure to thrive, slow weight gain, anorexia	
	Unexplained fits	
	Ataxia	
	Dyskinesia-dystonia	
	Persistent jaundice	
	Lethargy, coma	
	Vomiting	
	Unexplained acidosis ('anion gap')	
	Peculiar odour	
	Unusual appearance	
	Unexplained mental retardation	
	Psychotic features	
SPECIFIC	Hepato(spleno)megaly	Storage diseases, galactosaemia
	Vomiting, lethargy, coma	Disorders of Krebs urea cycle (ammonia intoxication)
	Intractable acidosis	Organic acidurias
	Self-mutilation	Lesch–Nyhan syndrome (juvenile hyperuricaemia)
	Cataracts	Some mucopolysaccharidoses, galactosaemia, Lesch–Nyhan syndrome, Lowe's syndrome
	Dislocation of lenses	Homocystinuria, sulphite oxidase deficiency
	Rashes	Phenylketonuria, Hartnup disease, homocystinuria, tryptophanuria
	Hair abnormalities	Argininosuccinic aciduria, Menkes' syndrome, homocystinuria
	Peculiar odour	Organic acidurias

treatment of phenylketonurics. Ion exchange chromatography analysis has been speeded up by the introduction of shorter, narrow bore columns packed with smaller particles and operated at higher flow rates, while sensitivity has been improved by fluorimetric detection of amines and amino acids with fluorescamine and o-phthalaldehyde. Gas–liquid chromatography is the method of choice for the investigation of organic acidurias.[10] For definitive identification of metabolites, it may have to be linked with mass spectrometry. High-pressure liquid chromatography has many applications, particularly in the assay of ultraviolet-absorbing metabolites; its sensitivity and range is being extended by the introduction of new types of detectors. Radio-immunoassay is widely used for the assay of thyroid hormones in the mass screening of neonates.[5]

In neurometabolic disorders the final diagnosis can hardly ever be firmly established without assay of the enzyme affected. Often, it is necessary to ascertain not only the extent of the deficiency of the enzyme but also its organ specificity, cofactor requirements, kinetics and stability. Of particular interest are the lysosomal hydrolases.[32,34] Initially, labelled natural substrates had to be used but fortunately, in many cases it is now possible to substitute synthetic substrates for spectrophotometric or fluorimetric assays. In some lysosomal disorders the enzyme defect is only unmasked after separation by electrophoresis, heat inactivation or pH inactivation of interfering iso-enzymes. Lysosomal enzymes are reasonably stable. Cultured fibroblasts deficient in these and other enzymes survive transport well and may be stored indefinitely in 'cell banks' if frozen with care. Cells may be sent to reference centres irrespective of distance.[39] While most enzyme assays are done on peripheral blood, urine or cultured skin fibroblasts, other tissue fluids have been used, for example hair lysates, saliva and tears.

For the diagnosis of some disorders a liver, muscle, intestinal or peripheral nerve biopsy is necessary. At least in Great Britain, brain biopsies are hardly ever considered justifiable except in some leucodystrophies and cerebral retinal degeneration.[18] Investigation of bone marrow and peripheral blood films is valuable in the storage disorders (vide infra).

Problems, Pitfalls and Artefacts

Mostly, difficulties encountered in this field are of two kinds: problems in interpretation which may be caused by shortcomings in methodology or failure to appreciate the relevance or lack of relevance of biochemical findings; and problems in communication when, for example, appropriate investigations are not initiated because those in charge of a mentally retarded patient do not suspect that they are dealing with a metabolic disease or do not know how to arrange for suitable investigations.

It is widely known that there is a causal relationship between the disturbance of amino acid metabolism and the mental defect in, for example, phenylketonuria, homocystinuria and citrullinaemia. The temptation has, therefore, been to attribute aetiological significance to any abnormal amino acid pattern found in the body fluids of a mentally retarded patient or infant with neurological symptoms. Amino acid abnormalities are in fact produced by a wide variety of both hereditary and environmental factors.[8,63] Many of these abnormalities have no neuropsychiatric significance. Artefacts may be produced by drugs which yield ninhydrin-positive metabolites or interfere with urine screening tests.[41] Urinary excretion patterns of aromatic amino acids and their derivatives depend on the activity of the intestinal flora and are abnormal in protein malnutrition. The excretion of phenolic acids is markedly affected by diet.[63]

Mention must be made of the enhanced urinary excretion of amino acids in the first few weeks of life particularly in premature infants. The amino acids cystine and tyrosine may be detected on routine chromatograms of these infants; these compounds are not normally seen on chromatograms from healthy, older children and adults. Young infants, especially those of low birth weight, also often show elevated blood levels of tyrosine and methionine when on a high protein intake. Diurnal variation and circadian periodicity can in most cases be ignored but may be important when, for example, the ratio of phenylalanine and tyrosine is used for the identification of heterozygotes in phenylketonuria.

Disorders may be missed when the level of abnormal metabolites is too low for available screening tests, or when the pattern characteristic of the disorder only develops with time. Homocystinuria illustrates some of these difficulties. Screening aims to detect elevated blood methionine levels, say, with the Guthrie technique or urinary homcystine excretion by the nitroprusside test. Both methods probably miss the majority of cases. In most affected infants, methionine levels are not sufficiently elevated to be picked up by the Guthrie bacterial test except when the infant has a high protein intake, while the level of homocystine in urine is often below the limit of detection by the nitroprusside test. The situation is made worse by the instability of the sulphur amino acids in alkaline urine and the possible distortion of the sulphur amino acid pattern by bacterial action. Homocystine is unstable in plasma even when frozen. Plasma has to be deproteinized without delay if blood levels are to be estimated. In the mucopolysaccharidoses, bacterial degradation may also invalidate urinary screening tests. Furthermore, mucopolysaccharide excretion is strongly age-dependent and as the mucopolysaccharides which are excreted are partially degraded they may differ in their physicochemical properties from commercially-available standards.[56] Urinary organic acid profiles are also subject to artefacts. For example, benzoic aciduria, a not infrequent finding in mentally retarded children, is commonly caused by bacterial action.

The principal of the Guthrie bacterial inhibition test is simple enough,[5] but great care is needed for reliable results. This applies particularly to the collection of specimens. Use of unsuitable filter paper, accidental contamination with disinfectants or water, incomplete penetration of the blood through the paper and use of ink on the paper are all liable to interfere with the growth of the test organism. Antibiotics may interfere with the test, and autoclaving of the blood spots to deactivate antibiotics and reduce interference by blood pigments, may if carried out carelessly, lead to losses of amino acids. Some antibiotics for example cephaloridine, gentamicin and co-trimoxazole, resist heat deactivation and at sufficiently high blood level no growth may be seen. Such tests must always be repeated by an alternative method or when the infant is no longer on antibiotics. On one occasion an unexpected artefact was produced when an incompletely dried blood sample collected on paper was sent to the laboratory in a non-porous envelope. The sample did not dry out and was infected by a fungus which produced high concentrations of a number of amino acids including phenylalanine, and a strongly positive Guthrie test.

The success of biochemical investigations with mentally handicapped patients depends in no small measure on interdisciplinary cooperation particularly between laboratory staff and those looking after the patients. Increasingly, the mentally retarded live in small units outside hospital in the care of social workers and housemothers rather than doctors and nurses. These workers are often interested in behaviour therapy and rehabilitation rather than metabolic studies, and their training is increasingly based on the behavioural sciences. In these circumstances, knowledge of procedures which are commonplace in a hospital ward often cannot be taken for granted, and detailed instructions and explanations given person to person are essential in all metabolic studies and research. Investigations arranged from a distance by phone or memorandum are unlikely to yield reliable results. A stimulating view of the problems of laboratory investigation of mentally retarded patients has recently been published by Rundle.[58]

DEGENERATIVE DISEASES OF THE NERVOUS SYSTEM WITH MENTAL HANDICAP—LYSOSOMAL DISORDERS

Clinical Aspects

In children, degenerative diseases of the nervous system provide a unique challenge to the biochemist because a significant proportion are genetically determined and therefore, presumably, have a biochemical basis. At the same time, there are few applications of clinical biochemistry in which effective performance is more dependent on close cooperation with other disciplines. First, it is necessary to decide whether or not the condition is progressive. In some cases clinical signs will narrow the diagnostic options. In others, longitudinal observations over weeks or months supplemented by serial psychological assessments and electroencephalographic studies are necessary to demonstrate first slowing, then arrest of acquisition of skills, and finally regression.

Loss of previously acquired milestones is characteristically seen not only in storage disorders but also in many other encephalopathies of diverse origin (Table 7). Conversely, in storage disorders, pathological processes may

TABLE 7
SOME DISORDERS OF THE NERVOUS SYSTEM ASSOCIATED WITH INTELLECTUAL DETERIORATION OR MENTAL HANDICAP[52]

Tumours
Infections (subacute sclerosing panencephalitis)
Autoimmune and postinfectious disorders (Schilder's disease)
Chronic poisoning (lead, organic mercury)
Childhood autism
Heredodegenerative disorders
    Neurocutaneous disorders (tuberous sclerosis, neurofibromatosis)
    Spinal and spinocerebellar degenerations (Friedreich's ataxia)
    Storage disorders (sphingolipidoses, mucopolysaccharidoses, mucolipidoses)
    Leucodystrophies (Krabbe's disease, metachromatic leucodystrophy, adrenoleucodystrophy)
    Huntington's chorea
    Wilson's disease

be active before birth. In these circumstances, perinatal complications are common and their results may mask the progressive deterioration. It must always be remembered that even collectively, the contribution of biochemical factors to neuropsychiatric symptomatology is relatively small compared to, say, antenatal and perinatal injury and some of the hazards listed in Table 5. Epilepsy in most cases has no identifiable biochemical aetiology. Intellectual deterioration may occur after status epilepticus and this involves biochemical mechanisms. Often, however, deterioration is in fact no more than slow learning due to the depressant effects of anticonvulsants, particularly barbiturates. When slow progress is related to age, it appears as a fall in IQ. Very rarely, a degenerative disorder may be responsible for deteriorating performance and behaviour at school, nearly always the causes are social or psychological.

When a degenerative disease is suspected every attempt is made to arrive at a firm diagnosis even if this is unlikely to help the patient so that the parents can be given a reliable estimate of the chance of recurrence in a subsequent child. Some clues may be provided by the fact that the incidence of rare recessive disorders is increased in consanguineous marriages, and that some rare genes have a much increased frequency in certain ethnic groups. Thus in Tay–Sachs disease the gene frequency is 100 times higher in Jews of East European origin than in the rest of the population, in aspartylglucosaminuria the gene frequency is high in Finns. However, if such an ethnic group constitutes a small minority only of the general population then it will contribute a minority of cases of the disorder seen notwithstanding the higher gene frequency. In many European countries the majority of patients with Tay-Sachs disease are not Jewish.

Splingolipidoses, Mucopolysaccharidoses, Leucodystrophies[8,18,21,32,52,66]

A clinical chemist investigating a lysosomal disorder soon comes up against a rather confusing classification and terminology; this has arisen for largely historical reasons. The first of the storage disorders to be identified, the infantile form of cerebromacular degeneration, Tay–Sachs disease, was described in 1881, followed by Gaucher's disease, Niemann–Pick disease and gargoylism or Hurler's disease. Early classification was based on the age of onset and clinicopathological features. Thus, apart from the classicial infantile form of cerebromacular degeneration (Tay–Sachs disease) a late infantile form (Bielschowsky) a juvenile form (Spielmeyer–Vogt) and an adult form (Kufs) have been documented. The late infantile and juvenile forms are sometimes jointly referred to as Batten's or Batten–Mayou disease or amaurotic family idiocy (amaurosis = loss of vision).

From the mid-thirties onwards classification was made more precise by chemical identification of the stored substance. Thus, Tay–Sachs disease proved to be a gangliosidosis, Niemann–Pick disease a sphingomyelinosis and Gaucher's disease a cerebrosidosis. In contrast to these sphingolipidoses, the stored substance in Hurler's disease was found to be an acid glycosaminoglycan (acid mucopolysaccharide, GAG) and this class of disorders was named mucopolysaccharidosis. Some conditions combining the biochemical and clinical features of lipidosis and mucopolysaccharidosis have been termed mucolipidoses. The basic structural unit of the sphingolipids is ceramide the N-acylester of the amino alcohol sphingosine. Ceramide may be covalently linked to phosphoryl choline forming sphingomyelin or to up to four hexoses and three N-acetyl neuraminic acid (sialic acid, NANA) residues giving rise to series of gangliosides (Table 8). The chief lipids of myelin are the phospholipids, sphingomyelin, cerebrosides and cholesterol; nerve cells and their processes contain less phospholipid and more ganglioside. Sphingomyelin is also a constituent of subcellular organelles and the plasma membrane. The acid glycosaminoglycans are macromolecules consisting of repeating units of sulphated hexosamine and hexuronic acid (Table 9). They are found in a variety of tissues particularly the cornea, blood vessels and cartilage where they are an important constituent of ground substance.

A definitive classification has been made possible in many cases with the recognition of the missing enzymes in these disorders. This classification based on enzyme deficiencies does not fit in too well with the earlier terminology. Lack of several distinct enzymes may result in clinically similar conditions, while lack of the same enzyme may produce different clinical manifestations. Invariably, the missing enzymes turned out to be located in the lysosomes. These cytoplasmic organelles normally possess a full range of specific acid hydrolases for the stepwise degradation of the sphingolipids and glycosaminoglycans. Loss of activity at any of the catabolic

TABLE 8
STRUCTURE OF SOME SPHINGOLIPIDS

Glucocerebroside:	Cer-Glc
Galactocerebroside:	Cer-Glc
Ceramide lactoside:	Cer-Glc-Gal
Ceramide trihexoside:	Cer-Glc-Gal-Gal
Haematoside:	Cer-Glc-Gal-NANA
Globoside:	Cer-Glc-Gal-GalNAc
Monosialoganglioside (GM_1):	Cer-Glc-Gal(NANA)-Gal-NAc-Gal
Monosialoganglioside (red cell stroma):	Cer-Glc-Gal-GlcNAc-Gal-NANA
Tay–Sachs ganglioside (GM_2):	Cer-Glc-Gal(NANA)-GalNAc
Asialo–Tay–Sachs ganglioside:	Cer-Glc-Gal-GalNAc
Sulphatide:	Cer-Gal-3-SO_4
Sphingomyelin:	Cer-phosphorylcholine

Cer, ceramide (N-acylsphingosine); Glc, glucose; Gal, galactose; GalNAc, N-acetylgalactosamine; GlcNAc, N-acetylglucosamine; NANA, N-acetylneuraminic acid

TABLE 9
REPEATING UNITS OF THE GLYCOSAMINOGLYCANS

Old name	New name	Abbreviation	Uronic acid	Hexosamine	Sulphate
Hyaluronic acid	Hyaluronic acid	HA	GlcUA	GlcNAc	—
Chondroitin	Chondroitin	Ch	GlcUA	GalNAc	—
Chondroitin sulphate A	Chondroitin-4-sulphate	Ch-4-S	GlcUA	GalNAc	O-SO_4
Chondroitin sulphate C	Chondroitin-6-sulphate	Ch-6-S	GlcUA	GalNAc	O-SO_4
Chondroitin sulphate B	Dermatan sulphate	DS	GlcUA, IdUA	GalNAc	O-SO_4
Heparinate sulphate	Heparin sulphate	HS	GlcUA, IdUA	GlcNAc	O-SO_4, N-SO_4
Heparin	Heparin	Hep	GlcUA	GlcNAc	O-SO_4, N-SO_4
Keratosulphate	Keratan sulphate	KS	—	GlcNAc	O-SO_4

GlcUA, glucuronic acid: IdUA, iduronic acid: GlcNAc, N-acetylglucosamine: GalNAc, N-acetylgalactosamine: KS contains galactose instead of hexuronic acid. Some hexosamine residues in Hep and HS are N-sulphated.

steps is followed by intralysosomal accumulation of the molecule which cannot be broken down, with on occasion, an overspill into the cytoplasm and almost invariably hypertrophy of the lysosomes. The rate of accumulation of stored substances is a function of the quantity of material that has to be degraded in the course of cellular activity. This will vary from tissue to tissue and is a factor in the extent of individual organ involvement. The stored material will be heterogeneous if the bond resistant to hydrolysis occurs in more than one type of molecule. The effect of a hereditary deficiency of a lysosomal enzyme on other lysosomal enzymes is variable, in some cases activity is reduced, in others it may be enhanced several fold. In brain, storage of material leads to mechanical distortion of cells, interferes with their metabolic activity and ends with their destruction. The more important lysosomal disorders are listed in Tables 10 and 11.

In inborn lysosomal disorders as in other errors of metabolism enzyme deficiency may be complete in some cases in others as much as 30% or more of average enzyme activity may be preserved. As pointed out earlier, mutations can result in complete inactivation of an enzyme but also in partial deactivation, reduced stability or altered kinetic properties. In general, low activity is accompanied by early onset and rapid deterioration. Typically, but not invariably, the clinically normal heterozygotes have mean enzyme levels between those for normal and affected homozygotes. Confusingly, the hexosaminidase A level of patients with late infantile onset G_{M2} gangliosidosis may actually be higher than those of the clinically normal heterozygotes of the infantile form. Complications are introduced by the existence of isoenzymes of differing activity and substrate specificity. Thus different β-galactosidases are involved in Landing's and Krabbe's disease. Deficiency of arylsulphatase A results in metachromatic leucodystrophy; deficiency of arylsulphatase B leads to Maroteaux-Lamy disease, a mucopolysaccharidosis. In mucosulphatidosis we find a deficiency of arylsulphatases A, B and C. The patients combine the clinical and pathological features of a leucodystrophy and mucopolysaccharidosis.

Approach to Diagnosis

The overall incidence of lysosomal storage disorders is about 1 in 5000 births in European populations. One case may be seen in the average-sized District Hospital every one or two years. The clinical presentation will in many cases determine the direction biochemical investigations will take. Disturbance of ganglioside metabolism leads to nerve cell destruction expressed by psychomotor retardation, fits, a cherry red spot on the macula and later spasticity and paralysis. Disorders of the metabolism of sulphatide or cerebroside, which are important constituents of myelin, affect the peripheral as well as the central nervous system giving rise to signs of peripheral neuropathy, spasticity and ataxia. Seizures may occur, but only in the later stages of these disorders. In the mucopolysaccharidoses, storage in the skeletal system gradually produces the characteristic appearance and bony changes. Where the stored substance has a high turnover in liver or spleen, visceromegaly is found. Of considerable help are haematological investigations of peripheral blood smears and bone marrow, radiological and electrophysiological studies using recently-developed techniques such as nerve conduction studies, electroretinograms and visual and auditory evoked potential.[18] Thus, reduced nerve conduction velocities are seen in disorders with segmental demyelination, including some leucodystrophies.

TABLE 10
THE MUCOPOLYSACCHARIDOSES[49]

Disorder	Major GAG excreted**	Enzyme defect	Clinical	Mental retardation
MPS IH (Hurler)	DS, HS	α-L-iduronidase	corneal clouding, visceromegaly dysostosis multiplex, heart disease	+++
MPS II (Hunter)*	DS, HS	L-iduronosulphate sulphatase	clear cornea, deafness, otherwise like MPS I but less severe	+++
MPS IIIA (Sanfilippo)	HS, ?DS	Heparan sulphate sulphamidase	mild somatic features, no heart lesions	+++
MPS IIIB (Sanfilippo)	HS, ?DS	α-N-acetylglucosaminidase	clinically indistinguishable from MPS IIIA	+++
MPS IIIC (Sanfilippo)	HS, ?DS	α-glucosaminidase?	clinically indistinguishable from MPS IIIA	+++
MPS IVA (Morquio)	KS	N-acetylgalactosamine-6-sulphatase	corneal clouding, severe bone changes, aortic valve disease	+−
MPS IVB (Morquio)	KS	Keratan sulphate β-galactosidase	corneal clouding, mild dysostosis multiplex	+−
MPS IS (Scheie)***	DS, HS	α-L-iduronidase	corneal clouding, stiff joints dysostosis multiplex, aortic valve disease	+−
MPS VI (Maroteaux-Lamy)*	DS	Arylsulphatase B	corneal clouding, dysostosis multiplex	−
MPS VII (Sly)	DS, HS	β-glucuronidase	corneal clouding? visceromegaly heart disease, dysostosis multiplex	+
MPS VIII (DiFerrante)	KS, HS	Glucosamine-6-sulphate sulphatase	mild dysostosis mutiplex, visceromegaly, coarse hair	+

GAG, glycosaminoglycan: MPS, mucopolysaccharidoses: other symbols as in Table 15.
* Mild and severe forms occur: ** in normal subjects Ch-4-S and Ch-6-S: *** formerly MPS V; the genetic compound MPS IH/S has been described.
MPS II, X-linked recessive, all others autosomal recessive.

TABLE 11
NEUROLIPIDOSES, MUCOLIPIDOSES AND LEUCODYSTROPHIES[18,21,32,52]

NEUROLIPIDOSES

Disorder	Stored substance	Enzyme abnormality	Clinical
Tay–Sachs			
B variant (classical)*	GM_2-ganglioside	hexosaminidase A	Hyperaemia, dementia, epilepsy, paralysis, frequent cherry red spot or macula
O variant (Sandhoff)	GM_2-ganglioside	hexosaminidase A and B	
AB variant	GM_2-ganglioside	hexosaminidases inactive with natural substrate	
Gaucher*	glucocerebroside	glucocerebrosidase	Hepatosplenomegaly, nervous system involvement in infantile but not in late onset chronic variants
Niemann–Pick			
Type A, C, D (Nova Scotia)	sphingomyelin cholesterol	sphingomyelinase in Type A, C and ?D	Hepatosplenomegaly, severity of nervous system involvement A, C, D
Type B (visceral), E	sphingomyelin	sphingomyelinase in Type B and ?E	Visceral involvement only
Fabry	ceramide trihexoside	ceramide trihexosidase	
Wolman*	cholesterol, triglycerides acid lipase		Hepatosplenomegaly, psycho-motor retardation
Pompe*	glycogen	acid α-1,4-glucosidase	Cardiomegaly, hypotonia, cretinoid appearance, mental retardation
Farber	ceramide	acid ceramidase	Hoarse cry, joint deformities, subcutaneous nodules, psycho-motor retardation

MUCOLIPIDOSES†

Disorder	Stored substance	Enzyme abnormality	Clinical
GM_1 gangliosidosis			
Landing	GM_1 ganglioside	β-galactosidase	Features of gargoylism, rapid deterioration, hepatomegaly
Delly	GM_1 ganglioside	β-galactosidase	Mental neurological deterioration, no signs of gargoylism, fits
Fucosidosis			
Type I	oligosaccharides	α-fucosidase	Mental and neurological deterioration, features of gargoylism
Type II	oligosaccharides	α-fucosidase	Milder course than Type I, skin lesions as in Fabry's disease
Mannosidosis‡	oligosaccharides	α-mannosidase	Signs of gargoylism, dementia
Mucosulphatidosis	sulphatide and GAG	sulphatases A, B and C	Psychomotor retardation, dementia, fits
Mucolipidosis I	glycoprotein and glycolipids	α-N-acetylneuraminidase	Features of gargoylism, myoclonus, cherry red spot
Mucolipidosis II	mucopolysaccharides	loss of acid by hydrolases from cells	Psychomotor retardation, features of gargoylism, rapid course
Mucolipidosis III	as for mucolipidosis II but milder	as for mucolipidosis II but milder	As for mucolipidosis II but milder
Mucolipidosis IV	mucopolysaccharides and gangliosides	unknown	Psychomotor retardation athetosis, clouding of cornea

LEUCODYSTROPHIES

Disorder	Stored substance	Enzyme abnormality	Clinical
Krabbe**	galactocerebroside	galactocerebrosidase	Progressive paralysis, dementia, fits, hyperacusis, autosomal recessive
Metachromatic leucodystrophy††	sulphatide	arylsulphatase A	Motor loss, impaired speed spasticity, fits, autosomal recessive
Adrenoleucodystrophy	? C_{26}-C_{30} fatty acids	? fatty acid degradation	Ataxia, fits, adrenal insufficiency, X-linked recessive

* Late onset and milder variants have been described.
† All mucolipidoses are transmitted as autosomal recessive traits.
‡ Mild variants occur.
** May be of early or late infantile onset.
†† May be of late infantile, juvenile or adult onset.

If the patient is dysmorphic (dysmorphic = misshapen) a screening test for excess of urinary glycosaminoglycans will separate the mucopolysaccharidoses in which excretion is increased from the mucolipidoses (oligosaccharidoses). Most screening tests have been based on the polyanionic nature of the glycosaminoglycans which permits metachromatic staining in acid solution with toluidine or alcian blue, or precipitation with cationic detergents such as cetyl pyridinium chloride (CPC). Pennock[56] recommends (1) a turbidity test with CPC (2) quantitative assay of hexuronic acid excretion and (3) identification of abnormal urinary glycosaminoglycans by thin layer chromatography or electrophoresis on cellulose acetate. The oligosaccharides excreted in some of the mucolipidoses are usually looked for by thin layer chromatography.

It has been known for some years that fibroblasts growing in culture incorporate ^{35}S-sulphate into acid glycosaminoglycans. This accumulation is greatly enhanced in fibroblasts from patients with mucopolysaccharidoses. This excessive accumulation, which is due to the inability of the cell to degrade the polymer, may be prevented by supplying secretions from any genotypically different fibroblasts. Genotypes can therefore, be identified by cross-correction experiments even if the precise enzyme defect is not yet known or if suitable enzyme assays are not available. The 'correction factors' also called 'Neufeld factors' after their discoverer, are in fact the lysosomal enzymes deficient in the patient.

In the sphingolipidoses, quantitative analysis of the accumulating substances is possible in urinary sediments. The glycolipids characteristic of Gaucher's, Fabry's and Farber's disease can be measured by HPLC of plasma.[69] However, in all lysosomal disorders, assay of the deficient enzyme is the definitive way of establishing the diagnosis and in favourable circumstances the heterozygous status of a relative. Initially, labelled substrates had to be used but p-nitrophenyl or methylumbelliferyl substrates have been prepared suitable for most requirements. Experimental details will be found, for example, in the publications by Galjaard[32] and Glew and Peters.[34] Enzymes are usually assayed in leucocytes or fibroblasts, occasionally in serum or urine. Most lysosomal disorders can be diagnosed antenatally. Lists of laboratories offering a service are available in the United Kingdom.[47] It is important to communicate early with the chosen centre to find out the form in which material is to be submitted, and what biochemical, family and obstetric information is required. It is considered good practice to send material to two centres who should know of each other.

Electronmicroscopy, supplemented by histochemical studies, is making an increasing contribution to diagnosis, particularly in variants of late infantile and juvenile amaurotic family idiocy in which the enzyme defect has not yet been established, the cerebroretinal degenerations (lipofuscinosis).[18,52,66] Skin biopsies contain amongst other cells axons and Schwann cells. The stored material within these cells exhibits the same characteristic and even pathognomonic patterns on electronmicroscopy as seen in brain biopsies. In the mucopolysaccharidoses, a collection of skeletal abnormalities, known as *dysostosis multiplex*, produces striking radiological changes (Table 10).

Approaches to Treatment

It is a measure of progress in the field of molecular medicine that therapy of what is essentially an intracellular protein deficiency can even be contemplated. It is well known that cells can take up enzymes by endocytosis into pinocytotic veicles which fuse with lysosomes. If the latter contains storage substance the exogenous enzyme can participate in its degradation. The localization of both enzyme and substrate within the lysosomes suggested that lysosomal enzyme deficiency diseases are proper conditions for enzyme replacement therapy.[15] Infusion of leucocytes and plasma have been shown to be effective but the short half life of the enzymes necessitates frequent transfusions with attendant immunological problems. Attempts have been made to prolong enzyme activity and reduce immunogenic reactions by enclosing the enzymes in biodegradable lipid envelopes (liposomes) or erythrocyte ghosts. The practical problems of purification and preparation are formidable.

A more hopeful approach is to provide patients with a continuous source of enzyme by tissue or organ graft. Organ transplants have been biochemically effective in Gaucher's and Fabry's disease but clinically results have been disappointing; the mortality high. More promising are transplants of skin fibroblasts, bone marrow or placental cells.

Additional problems arise where storage occurs primarily in the brain as in Tay–Sachs disease. In animal experiments, infusion of hypertonic mannitol into the external carotid artery has been followed with administration of purified mannosidase. This procedure temporarily altered the blood brain barrier and allowed the exogenous enzyme to enter the brain.[15]

Finally, it should be mentioned that dietary treatment has a place in hereditary neurological disorders. Refsum's disease (heredopathia atactica polyneuritiformis) presents polyneuritis, retinitis pigmentosa, ataxia, progressive paralysis and muscular atrophy. The protein level but not the cell count is increased in the CSF. The disorder is caused by a defect in α-oxidation of fatty acid which results in failure to metabolise phytanic acid which is largely of dietary origin. Patients benefit when placed on a phytol-free diet. A severely restricted diet may lead to weakness and weight loss. A moderately restricted diet in conjunction with plasmapheresis carried out at two- to four-week intervals appears more promising. Plasmapheresis has also been used in Fabry's disease.[54]

Adrenoleucodystrophy[18] is probably caused by a defect in the metabolism of C_{26} fatty acids which are also exclusively of dietary origin. This suggests that the cause of this disorder might also be favourably affected by dietary treatment.[54] While the prospect for the treatment of these degenerative disorders are by no means hopeless, it cannot be gainsaid that in the foreseeable future emphasis will rightly be on early diagnosis, genetic counselling and prevention.[13,32]

ENCEPHALOPATHIES OF EXTRACEREBRAL ORIGIN

Hypoxia and Ischaemia

Hypoxic–ischaemic brain injury is the single most important cause of mental handicap originating in the perinatal period. Other sequelae include fits, and a variety of motor deficits: spasticity, choreoathetosis and ataxia often collectively referred to as cerebral palsy.[71] The brain can be deprived of oxygen either by a diminished oxygen level in the blood—*hypoxaemia*, or by a reduced amount of blood perfusing the tissues—*ischaemia*. Both may occur as the result of *asphyxia*. Hypoxaemia is associated with accelerated uptake of glucose by the brain, increased glycolysis, accelerated production of lactate, reduction in the level of tricarboxylic acid cycle intermediaries and diminished production of the high energy compounds adenosinetriphosphate and phosphocreatine. Glucose and glycogen levels also drop rapidly. Initially, production of lactate lowers the pH, results in vasodilation and increases the supply of substrate. When lactate production 'gets out of hand' toxic levels occur with severe tissue acidosis leading to changes in membrane physiology, breakdown of the blood brain barrier, tissue oedema and widespread destruction of brain tissue. Synthesis of neurotransmitters, particularly acetylcholine, is so sensitive to hypoxia that brain function can be severely affected before any major change in energy metabolism.

The biochemical effects on the nervous system of ischaemia are similar to those of anoxia. In addition, vascular lesions may occur so that when perfusion is restored blood cannot re-enter the ischaemic areas. Commonly, anoxia and ischaemia occur together, in the great majority of cases before and during birth rather than after birth. Clinically, newborn infants are assessed by the Agpar scoring system of heart rate, respiration, muscle tone, reflexes and colour. Fetal hypoxia may be detected by observation of the fetal heart rate and during labour by analysis of fetal blood gases. The risk of mental handicap and neurological sequelae is high if a full term infant has a low Agpar score or neonatal convulsions coupled with severe intrapartum difficulties (e.g. placenta previa, cord around the neck, intrapartum haemorrhage), abnormal fetal monitoring or postnatal difficulties (e.g. respiratory distress, repeated periods of apnea, characteristic EEG abnormalities). Blood pH, pCO_2 and pO_2 when considered alone have not been shown to be of major prognostic value. The important role of hypoxia–ischaemia in the aetiology of neonatal convulsions is illustrated in a recent Swedish study[29] (Table 12). Hypoxia–ischaemia was the probable main aetiological factor in 37 out of 72 infants. Of these, six had died by age one year, six were obviously retarded and six had neurological sequelae without mental handicap. In 12 of the infants, a blood glucose concentration

TABLE 12
PROBABLE AETIOLOGY OF NEONATAL CONVULSIONS IN 77 INFANTS[29]

Probable main cause	Number of infants
Hypoxia	37
Infection	9
Hypoglycaemia	5
Hypocalcaemia	2
Other metabolic causes	2
Unknown causes	22
Total	77

of less than 1·7 mmol/l had been recorded and in three infants a serum calcium below 1·9 mmol/l.

Hypoglycaemia[50]

The brain is the principal organ with an absolute requirement for glucose as a fuel and hypoglycaemia has long been recognized as an important cause of brain damage and mental retardation (Table 13). There is a margin of safety in the supply of glucose to the brain

TABLE 13
SOME AETIOLOGICAL FACTORS IN HYPOGLYCAEMIA IN INFANTS[50]

Environmental
  Alcohol
  Salicylate, paracetamol
  Septicaemia
  Malnutrition
  Reye's syndrome
Endocrine
  Hyperinsulinism
  Hypopituitarism
  Adrenal insufficiency
  Maternal diabetes
Hereditary
  Defects in gluconeogenesis
  Galactosaemia
  Hereditary fructose intolerance
  Most glycogen storage disorders
  Maple syrup urine disease
  Propionic acidaemia
  Methylmalonic acidaemia
Multiple causes
  Transient neonatal hypoglycaemia

when its blood level is normal, half the glucose entering the brain is sufficient for metabolic requirements; if these requirements cannot be met hypoglycaemic coma will ensue. During fasting glucose is mobilized from liver glycogen, and glycerol and free fatty acids from fat. Excess fatty acids are oxidized to ketone bodies which can be used by the brain as a fuel, while glycerol can be transformed to glucose via triosephosphate, the point at which it enters the gluconeogenic pathway. Alanine and other glucogenic amino acids of muscle protein can also be converted to glucose via the gluconeogenic pathway. Deficiency in any enzyme of gluconeogenesis will also predispose to hypoglycaemia. As glucose-6-phosphatase is involved in the mobilization of glucose both from glycogen and by gluconeogenesis, patients suffering from Von Gierkes disease, glucose-6-phosphatase deficiency, are particularly susceptible to hypoglycaemia. Inhibition of gluconeogenesis may also be responsible for the hypoglycaemia frequently seen in infants with organic acidaemias. The ketone bodies acetoacetic and β-hydroxybutyric acids are often thought of in the context of life threatening acidosis in uncontrolled diabetes. However, if availability of carbohydrate is reduced ketone bodies become a major source of energy for the brain, particularly in infants. As with lactic acid, an uncontrolled increase in these strong acids will produce an acute metabolic crisis seen characteristically in the Tildon–Cornblath syndrome, 3-oxoacid CoA-transferase deficiency in which the transformation of acetoacetate to acetyl-CoA is blocked. Massive ketosis results when carbohydrate intake is reduced in affected infants.

In general, risk of brain damage is greater in hypoglycaemia without ketosis. Thus in hyperinsulinism due, for example, to islet hyperplasia, the chances of escaping brain damage by recurrent episodes of hypoglycaemia are not good.[50] On the other hand, episodes of hypoglycaemia with dicarboxylic aciduria and hydroxyaciduria due to mitochondrial medium chain acyl CoA dehydrogenase deficiency do not appear to have resulted in mental retardation in cases with milder forms of this disorder. The infant of a diabetic mother is vulnerable to hypoglycaemia both before and after birth. The complex problems of achieving optimum care for these infants have been reviewed by Farquhar.[30]

So-called ketotic hypoglycaemia is not a disease entity. It is liable to occur when hypoglycaemia is caused by the impaired mobilisation of glucose from glycogen or by failure of gluconeogenic pathways. This can be due to deficiency of a gluconeogenic enzyme, e.g. fructose-1,6-diphosphatase deficiency, but also in infants in whom the gluconeogenic pathway is intact as in severe malnutrition. In some children a fall of the blood glucose to below 1·65 mmol/l and ketosis are accompanied by neurological symptoms while others with comparable biochemical findings are asymptomatic. The symptoms may therefore be due to a relative inability of the brain in symptomatic hypoglycaemia to utilise ketone bodies. Fortunately, most cases recover without mental handicap. The hypoglycaemia in Reye's syndrome is but one fact of a profound derangement of intermediary metabolism (vide infra).

Hypoglycaemia in the neonate, usually defined as a blood glucose level of 1·5 mmol/l or less in full-term infants or 1·1 mmol/l or less in low-birth weight infants, is not uncommon and must be promptly dealt with if the intellectual development of the infant is not to be prejudiced. During the first few hours of independent existence before feeding is established the infant depends on endogenous fuel—glycogen and fat. The low birth weight infant and the infant who is small for gestational age are at a considerable disadvantage. Because of intrauterine malnutrition, glycogen and fat reserves may be low and mobilisation of lipid produce ketone bodies is decreased in small, premature babies because of the immaturity of the enzymes involved. This increases the demand for glucose, and hence the tendency to develop hypoglycaemia. The situation is made worse by the reduced capacity of the transport system carrying glucose into the brain which does not develop its full activity until some time after birth. Infants with congenital malformations or other antenatal insults to the brain often have low birth weight and an excess of perinatal problems which include an enhanced tendency to become hypoglycaemic. A vicious circle is set up whereby pre-existing damage is made worse by episodes of hypoglycaemia.[17]

Ventricular Haemorrhage

In the newborn, particularly in preterm infants (born before 37 completed weeks of gestation) ventricular haemorrhage results not only from circulatory disturbances but is also commonly a consequence of perinatal or postnatal hypoxia, particularly in the respiratory distress syndrome.[71,52] Obviously, the more extensive the haemorrhage the slimmer the chance of survival and of normal intellectual development. Computerised axial tomography is of considerable prognostic value by helping to locate the site of the haemorrhage and assess the extent of the bleeding. Laboratory investigations also provide useful information. Thus, examination of the CSF can provide evidence of recent haemorrhage. The mean red cell count in neonatal CSF is $30/mm^3$ and a count in excess of $200 cells/mm^3$ suggests an intracranial haemorrhage. In neonates, the mean CSF protein level is 0·9 g/l, higher than in older children, in premature neonates it is 1·15 g/l, only values in excess of 1·5 g/l may be considered abnormal. The protein level is further increased by 0·015 g/l for every 1000 fresh $RBCs/mm^3$. Normally, the CSF/blood glucose ratio is 0·48 : 0·6. In neonates this ratio is increased to 0·74; in preterm neonates to 0·81. A ratio in neonates which falls significantly below 0·74 strongly points to the possibility of an intracranial haemorrhage. It is often difficult to decide if blood in the CSF is the result of a haemorrhage or of a traumatic tap. In infants who have had a transfusion of adult blood subsequent to the haemorrhage comparison of the fraction of fetal haemoglobin in the (bloody) CSF and blood will often be diagnostically useful.

By way of contrast, elevation of CSF creatine kinase may reflect cerebral parenchymal damage, while a raised lactate : pyruvate ratio (normal less than 15) which persists for some hours after normal oxygenation has been established, will confirm a clinical diagnosis of hypoxia. Raised blood lactate levels (normally less than 2·8 mmol/l) are also seen in cases of raised intracranial pressure and reduced cerebral blood flow, and consistently in bacterial meningitis, presumably as a consequence of the inflammatory and oedematous changes associated with this condition.

Prematurity

The association between low birth weight (less than 2500 g) and mental handicap has long been known.[17,44] It should be stressed that infants of low birth weight due to intrauterine growth retardation (small for dates infants) should be distinguished from those who are preterm but of normal weight for gestational age. The respiratory distress syndrome and intraventricular haemorrhage are more common in 'genuine' preterm babies; hypoglycaemia in those small for gestational age. The chances of normal intellectual development in small for dates infants are generally less good than for genuine preterm infants of comparable birth weight. The former group includes infants with severe intrauterine malnutrition, chromosomal disorders and congenital malformations. In many of these cases the low birth weight and any perinatal problems are the consequence of an abnormal brain rather than its cause. Prematurity with a birth weight appropriate for gestational age is associated with such factors as cervical incompetence and uterine or placental abnormalities.

For the past thirty years paediatricians have striven to prevent mental handicap in these infants by active management including the correction of metabolic abnormalities. The path of progress has not always been smooth. Occasionally, today's therapeutic advance turns out to be tomorrow's iatrogenic disaster. Thus, deliberate underfeeding of low birth weight infants for fear of inhalation of food in some cases undoubtedly resulted in hypoglycaemia and adversely affected intellectual development. In a number of premature infants with incomplete development of the retinal vasculature, oxygen administered to prevent hypoxia contributed to the development of retrolental fibroplasia, although oxygen excess is not essential in the aetiology of this condition. This is not unknown in infants who have never received oxygen.

Modern techniques have geatly improved the outlook particularly for the 'genuine' premature infant of very low birth weight (less than 1500 g). With continuous recording of arterial oxygen and carbon dioxide tension and careful monitoring of intravenous feeding it is possible to surmount the respiratory and metabolic problems of most of these babies. In leading centres, over 90% of surviving infants are free from major handicap on long-term follow-up. By ultrasound and electroencephalographic techniques it is becoming possible to identify early those infants likely to be left with major handicap. In most cases equipment for blood gas and physiological measurements will be located in the intensive care unit under the control of its director. Apart from monitoring intravenous feeding the clinical chemist will often advise on choice and maintenance of equipment, and supervise quality control.

In reviewing the case histories of infants who faced perinatal crises one is impressed by how resistant the brain can be to severe insult, how well it can often compensate for quite severe metabolic changes and how much nervous function may persist in the presence of quite extensive damage. Usually it is the cumulative effect of several adverse factors each of which acting on its own might well have been overcome by the brain which acting together produce death or lasting handicap.

There is a tendency to attribute any mental handicap to whichever of these factors is best documented. If a nationwide service for the intensive care of very low birth weight infants were provided of a standard comparable to that of the best existing units a significant reduction in the prevalence of mental handicap could be achieved.

Kernicterus and Electrolyte Disturbance

Severe jaundice of the newborn may damage widely scattered areas of the brain, particularly the basal ganglia, brain stem and cerebellum, and the spinal cord. The early lesions are bright yellow, hence the name kernicterus ('Kern' is German for nucleus). Clinically, kernicterus is associated with athetosis, deafness and mental retardation. The pathogenesis of kernicterus is complex: many factors are involved. In general, it occurs when plasma levels of unconjugated bilirubin exceed 340 μmol/l but in premature infants kernicterus may develop at much lower bilirubin levels, while some normal full-term infants may escape handicap with peak levels approaching 500 μmol/l.

Elevated levels of unconjugated bilirubin may arise in a number of ways (Table 14). In the jaundice of the

TABLE 14
SOME FACTORS ENHANCING UNCONJUGATED HYPERBILIRUBINAEMIA IN THE NEWBORN[51]

Bilirubin overproduction
    Infections, septicaemia
    Blood group incompatibility
    Haematoma
    Drugs (vitamin K)
    Erythrocyte enzyme defects
    Haemoglobinopathies

Displacement of bilirubin from albumin
    Drugs (sulphonamides, salicylate)
    Hypoxia, acidosis
    Free fatty acids

Impaired hepatic uptake or conjugation
    Breast milk jaundice
    Drugs (novobiocin, chloramphenicol)
    Hypothyroidism
    Crigler–Najjar syndrome (type I and II)
    Hypoxia, hypoglycaemia, dehydration
    Prematurity

Increased enteric absorption
    Low intestinal obstruction
    Delayed passage of meconium
    Cystic fibrosis of the pancreas

majority of inborn errors such as galactosaemia, fructose intolerance, hepatorenal dysfunction (tyrosinaemia) and α-antitrypsin deficiency bilirubin is predominantly conjugated.

Unconjugated bilirubin occurs in plasma free or bound to albumin. The level of free bilirubin is lower by a factor of over a thousand than the total bilirubin. It is commonly stated that it is the free bilirubin that is responsible for kernicterus. Factors that reduce the binding of bilirubin to albumin thus enhance the risk of kernicterus.

Unfortunately, analytical methods for measuring free bilirubin and its interaction with albumin have not yet been fully validated. Nor is there as yet a generally accepted explanation for the brain damage caused by bilirubin. *In vitro*, bilirubin inhibits many enzymes involved in respiration, oxidative phosphorylation, glycolysis, the tricarboxylic acid cycle, protein synthesis and lipid metabolism.[43] Our understanding of events *in vivo* is incomplete. The blood brain barrier in newborn infants is immature and bilirubin may pass more readily into the brain of neonates. Brain mitochondria are particularly sensitive to respiratory poisons. Bilirubin toxicity may also be enhanced by the absence of ligandin in brain. Ligandin is a cytoplasmic protein present in most other tissues which binds bilirubin and *in vitro* can counteract its effects on respiration and oxidative phosphorylation. As a lipophilic substance, bilirubin binds with phospholipids, particularly sphingomyelin, and has been reported to interact specifically with the gangliosides of neonatal brain. However, the applicability of all these findings to the pathogenesis of kernicterus is conjectural. Antecedent or concomitant damage to the central nervous system and cerebrovascular system may be necessary before the clinical and neuropathological pattern of kernicterus can develop.[35,52]

Severe bilirubinaemia may be counteracted by timely exchange transfusion, intrauterine transfusion, induction of glucuronyl transferase by phenobarbitone and by phototherapy. Rh haemolytic disease of the newborn has been all but eliminated by prophylactic administration of anti-D gammaglobulin immediately after delivery to Rh negative primiparae whose subsequent infants would be at risk from haemolytic disease. In the past two decades the contribution of neonatal hyperbilirubinaemia to the prevalence of mental retardation has dramatically declined.

Improvement in infant feeding and in the management of fluid balance in sick infants has similarly reduced the role of hypernatraemia in the causation of mental handicap. Hyperosmolae dehydration in association with hypernatraemia can lead to intravascular stasis, infarction of brain tissue and haemorrhages. If rehydration is too rapid, oedema may result followed by convulsions some hours after the start of therapy. The importance of brain capillary failure in the pathogenesis of brain oedema and haemorrhage has been stressed by Goldstein.[35]

Neonatal hypocalcaemia occurs either at the age of one week as a result of excessive phosphate intake in milk or, more seriously, during the first two days of life particularly in low birth weight infants often with other perinatal problems, such as hypoxia, maternal diabetes or toxaemia. While the outlook for infants with convulsions due to neonatal hypocalcaemia (serum levels below 1·75 mmol/l) is better than that for infants with hypoglycaemic fits, some are left with handicaps.

Nutritional Aspects of Mental Handicap

The capacity of the adult brain to withstand even extreme and prolonged starvation without irreversible effects on intellectual capacity is axiomatic. Only recently has the vulnerability of the developing brain to malnutrition been demonstrated in animal experiments and this has stimulated numerous epidemiological studies in humans. It is now generally agreed that malnutrition is most dangerous during the brain growth spurt from about midgestation to the second birthday.

A major difficulty has been to isolate the effects of malnutrition as a specific cause of mental handicap from the many other adverse environmental factors with which it is nearly always associated. Also, certain individuals may be specifically vulnerable because of their genetic make-up. Inadequate nutrition can result in reduced brain size, in itself not of great significance, but certain formations, for example, the cerebellum are selectively affected. Biochemical analyses of brains from human infants suffering from intrauterine or postnatal undernutrition have shown alterations in quantities of brain cells, myelin lipids, brain protein and brain mucopolysaccharides. While the reduction in cell number as reflected in DNA analyses is not striking, synaptic ultrastructure seems to be affected as are many enzymes including some involved in neurotransmission such as choline acetyltransferase and glutamate decarboxylase.[3,20] In the developing undernourished brain there is depression of the conversion of glucose carbon into amino acids resulting from a slowing down of the normal increase in amino acids associated with the Krebs cycle, an increasing utilization of ketone bodies and decreased glucose utilization by the brain. Energy reserves in the form of ATP, phosphocreatine glycogen and glucose are not reduced. This implies that the energy-utilising processes have been adjusted to a lower than normal level reflected in the retarded structural, biochemical and functional development of the brain in undernutrition.[3]

The major part of the brain growth spurt takes place after birth and intrauterine malnutrition should thus be correctable to a significant extent in the postnatal period. It is therefore imperative that in small for dates infants the effects of intrauterine malnutrition should not be compounded by postnatal malnutrition. Adequate but not excessive amounts of glucose, lipids and amino acids have to be provided by the appropriate route if the development of these infants is to be safeguarded. The effects of perinatal malnutrition, as those of other forms of brain injury tend to persist if the damage is sufficiently severe or if the biological hazards are reinforced by a poor environment and lack of stimulation.[28]

Malnutrition during pregnancy may in some cases result in low birth weight infants and food supplementation of pregnant women in poor populations has at times been recommended. In a carefully controlled study Rush *et al.*[59] found that supplementing the diet of poor, pregnant women particularly with protein but also with calories unexpectedly resulted in increased fetal loss and few if any good effects. A less concentrated protein supplement with a calorie supplement was less toxic to the fetus. It seems that pregnant women should receive an adequate diet not excessive in any nutrient unless there are clear indications of an increased requirement for a particular nutrient. Indiscriminate supplementation may do more harm than good.

Postnatally, primary nutritional deficiencies as a source of mental retardation are probably extremely rare in Britain except as a form of child abuse. In the Third World they still constitute an important preventable cause of educational underachievement, mental handicap and death in childhood. In children, protein-calorie malnutrition includes a number of disorders caused by a combination of nutritional deficits, infections and other adverse environmental factors. Infantile marasmus is primarily due to calorie insufficiency, too little of an otherwise normal diet. It is characterised by emaciation and growth failure. Kwashiorkor results from a diet containing sufficient calories but inadequate protein. In practice, there is considerable overlap between these two forms of malnutrition and the picture is often complicated by vitamin deficiencies. In severe kwashiorkor there is oedema, dermatitis, hepatomegaly and atrophy of the pancreas and gut mucosa. Disturbance of protein metabolism is reflected in lowered plasma albumin and pre-albumin and reduced levels of the essential amino acids other than phenylalanine and histidine. Levels of the non-essential amino acids are reduced as are those of cholesterol, transferrin and β-lipoprotein, while those of the free fatty acids and triglycerides are increased. Involvement of the pancreas and gut mucosa results in malabsorption. In the marasmic infant biochemical changes are less pronounced.

Neurological abnormalities such as mental changes, hyporeflexia and reduced nerve conduction velocities are commonly seen in protein–calorie malnutrition. While usually insidious in onset, kwashiorkor may present with acute encephalopathy. Tremors and other neurological signs may persist for some time after dietary correction.[52] Many affected children remain mentally retarded. In general, the younger the child during the acute phase, the greater the risk of mental handicap. It must be remembered that in most cases mental handicap is not due to malnutrition alone, but to malnutrition acting synergistically with other adverse factors. Sometimes, malnutrition is the result of brain damage. Severely handicapped children may use unable or unwilling to chew and incapable of tolerating a normal diet. Improvement in their nutritional status results when the consistency of the diet is adjusted to their ability to chew. A detailed discussion of malnutrition and the brain will be found in the publications by Alleyne et al.[1] and Winick.[74]

Vitamins, Antivitamins and Teratogens

Vitamins are involved in enzyme systems as coenzymes and a deficiency, therefore, results in biochemical disturbance which may lead to neurological changes and mental handicap. Beriberi caused by thiamine deficiency may in infants present as an acute encephalopathy, pellagra, nicotinic acid deficiency, may result in dementia, hypocalcaemic tetany with convulsion, may complicate vitamin D deficiency, subdural haematoma may be a result of vitamin C deficiency and vitamin E deficiency can produce neuroaxonal dystrophy. The teratogenic effects of vitamin deficiencies in the fetus have been extensively studied in experimental animals both by selective elimination of vitamins from the diet and the administration of vitamin antagonists. Information on these problems in man is scanty and not always conclusive, but mention must be made of the attempts to use folic acid antagonists as systemic contraceptives some 25 years ago, soon abandoned when some women taking the drug became pregnant and gave birth to infants with congenital malformations and mental retardation. A full discussion of this field will be found in Crome and Stern[17] and Vincken and Bruyn[70].

Postnatally, vitamin imbalance may arise in a number of ways. In Britain, inadequate intake is probably rare as a cause of vitamin deficiency except in alcoholics, psychiatric and particularly psychogeriatric patients. In most individuals with vitamin deficiency states supplementation of the diet up to the required vitamin intake will correct the deficiency state. In some mentally retarded patients, malabsorption is responsible for low blood vitamin levels in spite of adequate dietary intake. A few individuals have a constant specific requirement for a particular vitamin that greatly exceeds the recommended maximum intake. The presence of disease in such individuals requires a constant intake of the vitamin affected in pharmacological doses up to 500 times the normal requirement. In vitamin-dependent individuals this high intake does not produce symptoms of hypervitaminosis. *Vitamin dependency* may arise because of a mutation affecting:

(1) a reaction in the biosynthetic pathway of the coenzyme, provided the block is partial;
(2) the interaction of coenzyme and apoenzyme. Vitamin dependency implies that defects in this interaction can be overcome by high concentrations of coenzyme;
(3) the catalytic site in the apoenzyme producing an enzyme of reduced catalytic activity. Excess of coenzyme may stabilize the apoenzyme, and thereby effectively increase the holoenzyme concentration. Even a small increase in effective concentration may be clinically important.

Vitamin dependency is found in a small minority of

TABLE 15
DRUGS WHICH CAUSE CONGENITAL ABNORMALITIES[64]

Definite
  Thalidomide
  Antimitotic drugs
  Norethisterone
  Alcohol
Probable
  Anticonvulsants
  Warfarin
  Operating theatre environment (? some anaesthetics)
Possible
  Lysergide
  Sex hormones

patients with inborn errors of metabolism which if untreated carry a high risk of death or mental handicap. In these patients a specific and safe treatment is available which is at least as effective as alternative forms of treatment. Some examples of vitamin-dependent syndromes are given in Chap. 6 and on p. 527. In general, however, while there is every reason to encourage diets with a vitamin content close to recommended norms, massive and indiscriminate supplements are useless and potentially harmful: for example, vitamin D excess can lead to hypercalcaemia and renal damage, vitamin A poisoning to neurological symptoms and death. Very exceptionally, a patient with psychotic symptoms may respond to vitamin treatment (see p. 517). However, the treatment of schizophrenic with pharmacological doses of vitamin C proposed some years ago has no sound basis.

Drugs can interfere with prenatal development by disturbing embryogenesis giving rise to *malformations* or by exerting their toxic actions on developing fetal organs resulting in *deformities*. A single teratogen can be responsible for a wide spectrum of defects while the same malformations can be the end result of many pathogenetic processes. In practice, conclusive proof that a drug is teratogenic is elusive even if malformations are present.[64] In retrospective studies it is often difficult to establish when the drug was taken and to pinpoint the all important timing of any teratogenic insult, while prospective studies particularly with low-grade teratogens require very large numbers of pregnant women as the great majority of those taking the drug will escape. Almost any pharmacologically-active drug can be teratogenic given a sufficiently unfavourable enviroment. The timing of the insult may determine the form the malformation takes as much as the nature of the agent in question. A list of drugs which cause congenital malformation is shown in Table 15.

The possibility that maternal alcoholism might harm the fetus was considered by a Select Committee of the House of Commons in 1834 but it is only in the past few years that the role of alcohol as a teratogen has been firmly established. In severe form the fetal alcohol syndrome presents microcephaly, congenital heart disease, pre- and postnatal growth retardation, abnormal brain development and mental retardation.[40,64] The nature of the pathogenetic process is not known. In mildly affected cases there may be no malformations and normal intelligence but behaviour and learning problems. Some risk to the fetus may persist even if pregnancy occurs after the mother has stopped alcohol abuse. There is also an effect of smoking during pregnancy on the intelligence of the offspring. This is associated with a higher risk of the fetus being born small for dates and an increased perinatal death rate. In survivors a four months' lag in reading age was found at age seven years. Smoking is particularly dangerous in mothers whose pregnancies are also at risk for other reasons. Elevated carboxyhaemoglobin levels in the blood of smokers are probably involved in the pathogenetic process.

Antimitotic drugs acting, for example, by folic acid antagonism may be teratogenic and anticonvulsants particularly phenytoin and barbiturate increase the incidence of malformations in pregnant epileptic women. Some effects on the intelligence of the offspring have been observed but failure to control the fits may be more dangerous to the fetus than the anticonvulsants.[67]

Some drugs taken in pregnancy, while failing to cause recognizable malformations, may affect intelligence or behaviour. In particular, drugs which alter the balance of neurotransmitter activity can produce long-lasting functional disturbances in the developing brain. Examples of drugs influencing central neurotransmitter activity are the phenothiazines and adrenergic agonists and antagonists often prescribed in pregnancy. Reliable epidemiological data in man are not available but experimental work suggests that these drugs affect cell proliferation in the developing brain when given to pregnant animals. There is thus reason to believe that some of these drugs might adversely affect the acquisition of cells in the fetus with possible effects on behaviour and intelligence.

Failure of the neural tube to close during the first few weeks of pregnancy can result in a range of defects from spina bifida occulta (no herniation of meninges or neural tissue) to meningocele (herniation of meninges), myelomeningocele (herniation of meninges and neural tissue) and anencephaly. Problems of antenatal detection are discussed in detail elsewhere (p. 672). Here it suffices to say that this is based on assay of serum α-fetoprotein in maternal serum at 16–18 weeks followed in suspected cases by assay of α-fetoprotein in amniotic fluid,[68] usually supplemented by ultrasonic scans. Diagnostic accuracy may be further improved by study of the morphology of amniotic fibroblasts or acetylcholinesterase isoenzymes. Myelomeningocele is almost always associated with some degree of intellectual deficit, more so when it is complicated by hydrocephalus and the Arnold–Chiari malformation (downward displacement of parts of the medulla and cerebellum through the foramen magnum and vertebral canal), and by meningitis and ventriculitis.[17]

The aetiology of neural tube defects is probably multifactorial, both genetic and environmental factors appear involved. In this country the incidence is not uniform varying from 4·5 per 1000 births in Northern Ireland to 1·5 per 1000 births in South-East England. The incidence appears to be increased in communities of Celtic descent, but also in female infants and in social classes IV and V. This suggests that amongst others, nutritional factors might be involved in the aetiology of neural tube defects. To test this hypothesis, Smithells *et al.*[65] gave multivitamin supplements to women who had previously given birth to an infant with a neural tube defect. Such women are known to have a much increased recurrence rate in subsequent pregnancies. Supplements were given from 28 days before conception to at least the date of the second missed period; that is well beyond the time of neural tube closure. Compared to a matched control group, the mothers who had received supplements showed a nearly tenfold reduction in the incidence of neural tube defects. If these results are confirmed, the way will be open to a substantial advance in the prevention of malformation. The supplements given were moderate, little different from the daily allowances recommended by the World Health Organisation. For reasons already discussed massive and indiscriminate supplementation of the diet in pregnancy is unlikely to be beneficial.

AMINOACIDURIAS[8,8a,66]

Classification

Many inherited and acquired diseases with disturbances of amino acid metabolism have been described and in some of these mental retardation has been a constant or frequent manifestation. Aminoacidopathies may be classified as primary, due to an enzymatic defect in intermediary metabolism or amino acid transport or as secondary, due to interference with the metabolism of the liver or kidney by endogenous or exogenous toxic substances. Excessive excretion of amino acids may be an overflow phenomenon when the plasma levels of one or more amino acids are raised and the renal threshold is exceeded, or renal when plasma levels are low or normal but urinary excretion is increased owing to a defect in renal tubular reabsorption. Distinct transport systems exist for the basic amino acids shared in part by cystine, for the acidic amino acids, the neutral amino acids other than glycine, for the β-amino acids and for the imino acids proline and hydroxyproline shared in part by glycine.[8,63] Transport of some amino acids is mediated by two systems: one of low capacity but high specificity; and one of high capacity but low specificity. Transport systems for peptides have also been described. In general, there are no renal mechanisms for the reabsorption of intracellular substances not normally present in plasma. If found in excess, for example, due to a metabolic block, they do not accumulate in plasma but are found in excess in urine, giving rise to the so-called no-threshold aminoacidurias. Defective deamination of amino acids in the liver may go hand in hand with renal tubular defects leading to both overflow and renal aminoacidurias. Again, several substances may accumulate in plasma, some with and some without renal transport mechanisms resulting in a combination of an overflow and no-threshold aminoaciduria. Pure overflow aminoacidurias are usually readily detected by chromatography of plasma, all other aminoacidurias require examination of urine. In practice, it is always best to examine both plasma and urine, preferably collected and analysed at the same time.

Phenylketonuria[4,7,8,33]

The discovery more than 45 years ago of mental retardation associated with well-defined biochemical abnormalities is a landmark in the history of mental deficiency. No other disease in this field has attracted comparable attention. Phenylketonuria is also the first hereditary disease of intermediary metabolism to yield in large measure to specific therapy. The disorder is transmitted as an autosomal recessive trait. Heterozygotes are free from symptoms and are of normal intelligence. The incidence of the disorder in European populations is not uniform; it varies, for example, from 1 in 5000 in Ireland to 1 in 18 000 in South East England. Phenylketonuria is rarer still in non-European populations. No case has yet been found in Finland.

The inborn enzymatic defect in phenylketonuria is the inability to hydroxylate the essential amino acid phenylalanine to tyrosine. This hydroxylation serves as a safety valve preventing the accumulation of phenylalanine when intake exceeds requirements and excretion. Tyrosine is metabolized further via p-hydroxyphenylpyruvic acid and homogentisic acid to carbon dioxide and water. A number of quantitatively less important pathways lead from tyrosine to adrenaline, noradrenaline, thyroxine and melanine. As a result of the metabolic block, phenylalanine accumulates in the body fluids and is diverted into pathways which in normal individuals are of only minor importance. Normally, the blood phenylalanine level is about $100\,\mu mol/l$ or less. In the first few days of life, particularly in preterm infants and in those on a high protein intake the blood phenylalanine may rise and in a few babies exceed $250\,\mu mol/l$. However, these peaks are transient and normal levels are soon regained. In untreated classical phenylketonuria the blood phenylalanine rises over a period of days or weeks and, depending on protein intake usually settles at levels between 1200 and $3000\,\mu mol/l$. In most patients, at blood phenylalanine levels of about $1000\,\mu mol/l$ phenylpyruvic acid formed by transamination is excreted in the urine.

Apart from phenylalanine and phenylpyruvic acid, compounds characteristically excreted in urine include phenyllactic acid, phenylacetic acid and its conjugate phenylacetyglutamine, o-hydroxyphenylacetic acid and γ-glutamylphenylalanine. Classical phenylketonuria is in fact defined biochemically by a blood phenylalanine level persistently in excess of $1\cdot2\,mmol/l$ on at least two occasions, a low or low normal blood tyrosine which does not rise during a phenylalanine load, and the presence in urine of characteristic metabolites.

For activity, phenylalanine hydroxylase requires tetrahydrobiopterin as cofactor, and dihydropteridine reductase to regenerate tetrahydrobiopterin. In the overwhelming majority of cases the mutation affects the phenylalanine hydroxylase moiety of the enzyme system. Very low activity is found in classical phenylketonuria, but variants with residual enzyme activity, moderately raised blood phenylalanine levels (less than $1\,mmol/l$) and normal intellectual development are not uncommon, about one case being found in mass screening for every two cases of classical phenylketonuria. Transient and other atypical variants have also been described.[4,7,8] Patients with deficiency of phenylalanine hydroxylase respond to treatment with a diet low in phenylalanine, provided this is started early, preferably within the first three weeks of life. A strict diet is prescribed when the blood phenylalanine level is persistently above $1\cdot2\,mmol/l$; less restrictive diets are offered for levels between

TABLE 16
VARIANTS OF HYPERPHENYLALANINAEMIA

	Phenylalanine hydroxylase deficiency	Dihydropteridine reductase deficiency	Defects in dihydropteridine synthesis
Tryptophan hydroxylation	mildly affected	severely affected	severely affected
Phenylalanine low diet	normal development	progressive neurological deterioration	progressive neurological deterioration
Dopa, carbidopa 5-hydroxytryptophan	does not respond	most cases respond	most cases respond
Effect of tetrahydrobiopterin	no effect	lowers blood phenylalanine	lowers blood phenylalanine

1·2 mmol/l and 0·6 mmol/l while most paediatricians will not treat an infant with levels below 0·6 mmol/l. Treatment usually aims to achieve blood levels of the order of 0·2 to 0·5 mmol/l, overtreatment may lead to deficiency states.[31]

About 1–3% of cases of classical phenylketonuria are caused by deficiency of dihydropteridine reductase or of tetrahydrobiopterin, due presumably to a deficiency in one of the enzymes catalysing the chain of reactions from guanosine triphosphate to tetrahydrobiopterin.[4,8] In classical phenylketonuria due to phenylalanine hydroxylase, deficiency symptoms other than mental retardation are not too severe. They include eczema, dilution of hair colour, EEG abnormalities and occasionally epilepsy. Patients with mutations involving the pteridine cofactor are neurologically much more severely affected, exhibiting progressive neurological disease with myoclonic seizures, disturbances of tone and posture and difficulties in swallowing. The enzymes hydroxylating tyrosine and tryptophan require the same cofactor as phenylalanine hydroxylase. Failure to form dihydroxyphenylalanine and 5-hydroxytryptophan, the precursors of the neurotransmitters dopamine, noradrenaline adrenaline and serotonin readily explain the neurological signs. In fact, urinary, CSF and brain catecholamines and serotonin are decreased in these variants, which have been termed malignant hyperphenylalaninaemia. Normalisation of the blood phenylalanine level by restriction of phenylalanine intake does not help these patients, but a majority do respond to neurotransmitter replacement with dopa, carbidopa and 5-hydroxytryptophan. Identification of affected infants is not easy. The decrease in urinary catecholamines and 5-hydroxyindoleacetic acid has been used as a screening procedure. HPLC of urinary pteridines is likely to become the method of choice. Administration of tetrahydrobiopterin will decrease blood phenylalanine levels in patients with dihydropteridine reductase deficiency or biopterin deficiencies but not in those with phenylalanine hydroxylase deficiency. This compound is not yet readily available. The variants of hyperphenylalaninaemia are compared in Table 16.

In spite of much research, it is still uncertain how the metabolic derangement in untreated classical phenylketonuria impairs intelligence.[7,33] The search for a unitary pathogenetic mechanism has so far proved abortive. In fact, there is no reason to suppose that pathogenesis is necessarily identical even in two patients with presumably the same mutation (e.g. siblings) or in any one patient at different developmental stages of the brain. An unfavourable environment will further depress the performance of untreated patients.

The benefits of the phenylalanine low diet in classical phenylketonuria are not in doubt. The efficacy of treatment is assessed by monitoring blood phenylalanine levels and correlating biochemical control with intellectual progress. On best evidence, early treatment results in improved intellectual performance amounting to 40–50 points on the appropriate IQ scale but even with best management there is a small deficit of 5–10 IQ points when the performance of patients is compared to that of their parents or unaffected siblings. The reason for this may be delays in diagnosis or deficiencies in treatment—some dietary preparations my be suboptimal in vitamins or trace metals. The strict dietary régime may cause emotional problems in the patient and family with adverse effects on learning. The possibility of slight antenatal damage cannot be excluded. Several isoenzymes of phenylalanine hydroxylase occur in fetal liver and their absence may not be fully compensated by the maternal metabolism. The fetus may also be more vulnerable to minor deficiencies in the maternal diet.

Of the patients on treatment, those whose blood levels are well controlled do better than those whose blood phenylalanine is frequently found to be outside the recommended treatment range. Caution is necessary in interpreting such data. Some patients who do exceptionally well may have residual enzyme activity, making control easier and high blood phenylalanine levels less menacing. Intelligent, caring parents who provide a stimulating home environment are also more likely to succeed with the diet, and the child's superior performance may not just be a consequence of good dietary control. Laboratory control of treatment presents no problems. In most cases, capillary blood collected by health visitors or mothers is assayed for phenylalanine by the Guthrie microbiological inhibition method, fluorimetry or ion exchange chromatography. For fuller information on treatment the reader is referred to Francis.[31]

Until recently, it was believed that treatment could be discontinued once brain development was essentially complete, say, at age eight years. There is now, however, very good though perhaps not conclusive evidence that high blood phenylalanine levels are also harmful in later childhood. Treatment of phenylketonuria should, therefore, be continued beyond age eight years although some relaxation of the diet may probably be tolerated. The dietary treatment is onerous and alternative approaches are being explored. Substance such as β-thienylalanine which reduce renal tubular absorption of phenylalanine are too toxic for clinical use. Supplements of tyrosine and those amino acids which are excluded from the brain by the high level of phenylalanine in the circulation have been suggested, while oral administration of phenylalanine ammonia lyase in capsules has been used in an attempt to metabolise phenylalanine in the gut. None of these approaches has yet been validated in clinical trials.

The phenylketonuric fetus of a heterozygous mother is not exposed to elevated phenylalanine levels during gestation and therefore, the phenylketonuric infant is essentially normal at birth. In the reverse situation of a heterozygous fetus of an untreated phenylketonuric mother the high maternal blood phenylalanine levels can have a devastating effect on the fetus. During the first trimester phenylalanine is teratogenic giving rise amongst others to cardiac and skeletal malformations. Intrauterine growth is retarded. The infants are born with microcephaly, mental retardation and a characteristic facies. Originally it was believed that provided the maternal blood phenylalanine level did not exceed 1–1·2 mmol/l the risk to the fetus was tolerable. In the light of subsequent experiences it is now recommended that the blood levels of a phenylketonuric pregnancy should be controlled at around 0·5 mmol/l starting before conception to avoid the risk of malformations early in pregnancy.[46]

Early treated phenylketonurics are of normal intelligence and will want to marry and have children. There are at present no reliable methods for the diagnosis of affected fetuses and it is therefore, important that heterozygotes should be identifiable amongst prospective marriage partners. The heterozygote frequency in the general population is of the order of 0·02. Heterozygotes may be identified by phenylalanine load tests or more simply, by the phenylalanine tyrosine ratio in a sample collected at noon.[8] Care is necessary as the ratio is altered in pregnant women and in those on oral contraceptives and also by drugs affecting pteridine metabolism such as cotrimoxazole.

Histidinaemia

The biochemical findings in histidinaemia (deficiency of L-histidine ammonia lyase 'histidase') closely parallels those in phenylketonuria. In both disorders minor metabolic pathways assume major significance. In phenylketonuria, phenylalanine accumulates in the body fluids and phenylpyruvate, phenylacetate and phenyllactase are characteristically excreted in the urine. In histidinaemia the transformation of histidine to urocanic acid is blocked, histidine accumulates in the body fluids and imidazolepyruvic acid, imidazolelactic acid and imidazoleacetic acid are found in urine.

To establish harmful effects of histidinaemia on intelligence is of more than theoretical importance since the condition is readily screened for in the neonate with the Guthrie bacterial inhibition assay or some variant of the Scriver test. The diagnosis can be confirmed by assay of histidase in skin homogenates or of urocanic acid in sweat. Biochemical control of the condition is readily achieved with a diet low in histidine. Early cases of histidinaemia had been detected in psychiatric units and a causal link between the disorder and any neuropsychiatric symptoms could not be assumed because of the bias in selection of these patients. To correct for this bias Popkin et al.,[57] reviewed the literature for evidence of handicap in patients diagnosed as histidinaemic amongst siblings of affected propands. They concluded that the likelihood of mental handicap in untreated histidinaemia is at least 40%. On the other hand, Clayton[11] could report that 20 untreated histidinaemic subjects diagnosed by mass screening of the newborn were all normal after follow-up for several years. This should cause no surprise. Perhaps because of the lower histidine content of our diet and more effective renal clearance of histidine, blood levels in untreated histidinaemia tend to be mostly 1 mmol/l or less, often well below 0·7 mmol/l lower than phenylalanine levels in untreated classical phenylketonuria. Few paediatricians would advise a phenylalanine low diet when blood phenylalanine is below 0·7 mmol/l. Also, untreated children with histidinaemia do at least as well as competently treated phenylketonurics; both appear to develop normally. It is therefore difficult to visualise a clinical trial capable of demonstrating the benefits of dietary treatment in histidinaemia. In man, manifestations of maternal histidinaemia, if present at all, are much less severe than those in maternal phenylketonuria. The offspring of a strain of histidinaemic mice have balance defects and behavioural abnormalities but these cannot be solely accounted for by the biochemical defect. In experimental animals, elevated blood histidine levels affect the entry of other amino acids into the brain including that of the neurotransmitter precursors but the effect on the entry of, for example, tryptophan into the brain is much less at blood levels of histidine found typically in histidinaemia than at blood levels of phenylalanine seen in untreated phenylketonuria. Apparently, the distortion of the amino acid pattern in the brain in histidinaemia does not *per se* prejudice normal intellectual function. However, the evidence of Popkin et al.[57] cannot be ignored; an enhanced vulnerability in histidinaemia to neuropsychiatric handicap has to be inferred. Also, we must remember that the adverse affects on the brain may have manifestations other than a deficit in intelligence, for example a susceptibility to epilepsy. Histidinaemia may perhaps be regarded as a form of subclinical malnutrition, which in some animals is associated with a lowered seizure threshold.

Disorders of Tyrosine Metabolism

The relationship between disorders of tyrosine metabolism and intellectual development is complex and not well understood. In hereditary tyrosinaemia (tyrosinosis, deficiency of fumarylacetase and p-hydroxyphenylpyruvic acid oxidase with disturbance of hydroxyphenylpyruvic acid oxidase with disturbance of porphyrin and catecholamine metabolism. Clinically, the disorder may be acute or chronic. It is characterised by severe liver disease, nephropathy with Fanconi syndrome and hyperplasia of the islands of Langerhans. Mental retardation if present is usually mild. Biochemically, there is a marked elevation of serum tyrosine a lesser elevation of serum phenylalanine and sometimes an increase in serum methione, particularly in acute cases, and of α-fetoprotein. In urine, there is a massive excretion of tyrosine derivatives, aminoaciduria, glycosuria, phosphaturia and increased excretion of β-aminolaevulinic acid and catecholamines.[5,8,66]

The first step in the oxidation of tyrosine is transamination by the cytosolic enzyme tyrosine ketoglutarate aminotransferase. Deficiency of this enzyme occurs in the Richner–Hanhart syndrome which presents with keratitis, corneal ulcers, palmoplantar keratosis and mild to severe mental retardation in about half the patients. Liver and kidney function is normal. Excess of tyrosine and its derivatives is found in urine. The disorder responds to treatment with a diet low in tyrosine and phenylalanine. At higher concentrations tyrosine can also be transaminated by a mitochondrial enzyme. The only case of Medes' type tyrosinaemia so far described may have had a defect of the mitochondrial enzyme in the kidney. He was of normal intelligence.

The clinical biochemist is most likely to meet transient neonatal tyrosinaemia due to a temporary deficiency of the enzyme p-hydroxyphenylpyruvate oxidase. This enzyme is low in the fetus and new born, particularly if small for gestational age, is inhibited by excess substrate and requires ascorbic acid or a similar reducing agent for activity. Elevated blood levels of tyrosine (up to 2 mmol/l) are seen in about 1% of newborn infants particularly if of low birth weight with low tissue ascorbic acid levels and on a high protein intake. As these elevated tyrosine levels may persist for days or weeks the question arises if they present a hazard to the developing brain. The infants are symptom free but follow up studies suggest that prolonged elevated blood tyrosine levels may have a deleterious effect on perceptual function later in childhood.[52] Avoidance of high-protein feeding greatly reduces the incidence of neonatal tyrosinaemia. It has been known for some time that excessive intake of tyrosine can be toxic in animals giving rise to skin disorders and ocular changes (cf the Richner–Hanhart syndrome). There is some evidence that the liver microsomal drug-metabolising system is capable of converting tyrosine to a toxic metabolite perhaps a 2,3-epoxide of tyrosine. This drug-metabolising system is quite low in the fetus and newborn animal and thus neonates with transient tyrosinaemia are less at risk than infants with cytosolic transaminase deficiency whose elevated blood tyrosine levels persist indefinitely. Thus, while glycine, glutamate and cystine are more toxic in the neonate than in the older infant, the reverse appears to be true for tyrosine.

The laboratory diagnosis of disorders of tyrosine presents difficulties. Some but not all cases of hepatorenal dysfunction will be identifiable by the presence of excess succinylacetone and δ-aminolaevulinic acid in the urine and elevation of α-fetoprotein in blood. In most cases, however, few conclusions can be drawn from the analysis of tyrosine metabolites in urine. The identification of an enzyme can be exceedingly difficult in a sick infant in whom liver and kidney damage, and possibly hypoglycaemia and malabsorption are superimposed on an inborn error. For example, hereditary fructosaemia and galactosaemia may produce many of the clinical and biochemical features of hepatorenal dysfunction. Many of the biochemical findings in these disorders merely reflect non-specific severe liver damage. Enzyme studies are necessary for a definite diagnosis.[5,8,17]

Non-ketotic Hyperglycinaemia

In man, persistent hyperglycinaemia occurs with ketosis secondary to organic acidurias, such as propionic acidaemia, methylmalonic acidaemia, and β-ketothiolase deficiency. This form of hyperglycinaemia is referred to as ketotic hyperglycinaemia. Hyperglycinaemia without ketosis, nonketotic hyperglycinaemia is associated with deficiency of the glycine cleavage system which transforms glycine and tetrahydrofolate into ammonia, carbon dioxide and 5,10-methylenetetrahydrofolate.

The enzyme system is present in liver and brain; either or both isoenzymes may be affected. If the liver enzyme alone is deficient there are no neurological abnormalities and the EEG is normal, plasma glycine is elevated and CSF glycine is normal. If the brain enzyme is affected a clinical picture is seen of extreme hypotonia, lethargy and myoclonic fits. The EEG is characteristically abnormal exhibiting periodic paroxysmal bursts on an almost flat record. The CSF glycine and CSF/plasma ratio for glycine are elevated. In ketotic hyperglycinaemia, by contrast, both neurological disturbances and EEG abnormalities are non-specific and the CSF/plasma ratio for glycine is normal. The great majority of infants with glycine encephalopathy die during the first few days of life. The survivors are almost invariably severely mentally retarded but if they survive do not necessarily deteriorate further even over periods of years.

The role of glycine as an inhibitory neurotransmitter is now well established. In the rat, both the influx of glycine into the brain and its neurotoxicity are strongly age dependent being much greater in the newborn. The higher influx is required for the highly active protein synthesis in the newborn. The glycine cleavage system may be a safety valve which prevents undue accumulation of glycine at sites where it would interfere with neurotransmission. The rate at which toxic levels build up will then depend on a delicate balance between dietary intake, the active transport system for the amino acid and the cleavage system, a balance which is likely to be particularly vulnerable early in infancy and which may be affected by other genetic, environmental and maturational factors. Strychnine is an antagonist of glycine probably competing for receptor sites on the post-synaptic membrane. Strychnine is being tried in the treatment of this disorder sometimes in association with other measures (Table 17) so far unfortunately with only indifferent success.

TABLE 17
THERAPEUTIC MEASURES PROPOSED FOR NONKETOTIC HYPERGLYCINAEMIA

Protein restriction
Exchange transfusion
Insertion of ventricular shunt
Glycine conjugation with sodium benzoate
Pyridoxine, folate and lipoic acid supplementation
Supply of one-carbon units (N^5-formyltetrahydrofolate, methionine)
Supply of N^5,N^{10}-methylenetetrahydrofolate
Strychnine treatment (competition for glycine receptor sites)
α-Aminoisobutyrate administration (to increase glycine excretion)
α-Methylserine administration (to inhibit endogenous glycine formation from serine)

The Hyperammonaemias

The effects of ammonia on brain metabolism have been studied for nearly 50 years. Ammonia is liberated during convulsions and electroshock, while ammonium salts themselves induce convulsions and coma. By far the most frequent cause of hyperammonaemia in man is acquired liver disease but in the past 25 years many patients have been described who exhibit hyperammonaemia because of a deficiency of one of the enzymes catalysing the conversion of ammonia to urea by the Krebs urea cycle (Table 18). In man, the urea cycle, glutamine formation from glutamate, and formation of glutamate from α-oxoglutarate are the three major mechanisms for the detoxication of ammonia. However, if the urea cycle is inoperative, ammonia detoxication via glutamate is liable to deplete α-oxoglutarate and thereby to reduce the synthesis of high-energy phosphate compounds. Complete blocks in the urea cycle are probably incompatible with life. On the other hand, in some of the inborn errors involving this cycle residual activity may be sufficient for normal development although episodic hyperammonaemic crises may occur in affected subjects, precipitated, for example, by viral infections. The mechanisms by which ammonia produces convulsions and coma are uncertain. Ammonia probably interferes both with post- and pre-synaptic inhibition, and at higher concentrations reduces the rate of synthesis of the neurotransmitters GABA and glutamate, besides depleting the energy stores of the brain.

The differential diagnosis of hyperammonaemia is not easy.[8,45,66] Abnormal amino acid patterns help in the diagnosis of deficiencies of the cytosolic enzymes argininosuccinate synthetase (citrullinaemia), argininosuccinase (argininosuccinicaciduria) and arginase (argininaemia). The diagnosis of deficiency of the mitochondrial enzymes carbamoylphosphate synthetase and ornithine transcarbamylase can only be clinched by enzyme assay on a liver or jejunal biopsy, although in ornithine transcarbamylase deficiency the ratio in serum of this enzyme to alanine transaminase is diagnostically useful. Carbamoylphosphate accumulation occurs in all urea cycle disorders except carbamoylphosphate synthetase deficiency and leads to excessive formation of pyrimidine metabolites, alternative products of carbamoylphosphate metabolism which may be detected in urine. Urine ammonia/urea and ammonia/creatinine ratios are also helpful.[55]

Interpretation of enzyme assays is not straightforward. Urea cycle enzymes show adaptive changes to dietary protein intake and the ornithine transcarbamylase and carbamoylphosphate synthetase will be low in conditions of mitochondrial injury amongst which Reye's encephalopathy must be mentioned.[16]

This syndrome usually follows an acute viral infection but other environmental factors and a genetic predisposition may play a part in the aetiology. Minor non-specific complaints are followed by a change in mental state, vomiting, coma and sometimes convulsions. Liver function tests are abnormal, the blood ammonia is high and the glucose usually low. Liver biopsy reveals microvesicular fatty infiltration. A variety of toxic substances, notably ethanol may produce a similar picture. Mortality may exceed 50% particularly in patients aged less than one year. Some survivors are left with neurological and mental handicap. Treatment is limited to intensive supportive care, the prevention of cerebral oedema and restoration of glucose levels and correction of electrolyte and acid–base balance.

Clinically, it is difficult to distinguish hyperammonaemia due to enzyme deficiency from other conditions such as intracranial haemorrhage or septicaemia which also present with vomiting, lethargy and coma. In fact, hyperammonaemia due to enzyme deficiency is often associated with intracranial haemorrhage. Conversely, sick, anoxic, preterm neonates may develop hyperammonaemia in the absence of any obvious metabolic defect.

Until recently, therapy of disorders of the urea cycle was confined to attempts to reduce the accumulation of

TABLE 18
DIFFERENTIAL DIAGNOSIS OF HYPERAMMONAEMIA[8,45,66]

Primary deficiency of urea cycle enzymes
    carbamoylphosphate synthetase
    ornithine transcarbamylase
    argininosuccinate synthetase (citrullinaemia)
    arginase (argininaemia)

Secondary hyperammonaemia
    ornithinaemia
    lysine intolerance
    nonketotic hyperglycinaemia
    some organic acidurias
    familial protein intolerance with ammonia intoxication
    cerebral atrophic syndrome of Rett
    Reye's syndrome
    diffuse liver failure (poison, infection, infestation, biliary obstruction)
    hypoxia–ischaemia

nitrogenous metabolites by low protein diets with the addition of arginine or of the nitrogen-free analogues of some of the essential amino acids. Some patients benefited, at least temporarily. Other therapeutic approaches are now being explored.[45] Of particular interest are amino acid acylation as a means of increasing waste nitrogen excretion (Table 19). Benzoic acid treatment of

TABLE 19
APPROACHES TO THE TREATMENT OF UREA CYCLE DISORDERS[45]

Dietary measures
    Low protein diet
    Essential amino acid mixtures
    Nitrogen-free analogues of essential amino acids

Promotion of excretion by competitive inhibition of renal tubular reabsorption
    Arginine supplements in argininosuccinicaciduria
    Homoarginine supplements in citrullinaemia and argininaemia

Amino acid acylation to increase nitrogen excretion
    Benzoate loading
    Phenylacetate loading

comatose hyperammonaemic patients has led to clinical improvement and a return of plasma ammonia levels towards normal in a number of cases.

Homocystinuria

Deficiency of cystathionine β-synthetase, the enzyme in the transsulphuration pathway which catalyses the formation of cystathionine from homocysteine and serine is the most common cause of homocystinuria. Clinical signs which point to the diagnosis include long thin limbs, arachnodactyly (long thin digits), chest deformities, a malar flush and dislocation of the lenses with tremor of the iris. These symptoms develop with age; they are not present in the newborn or young infants. Clinically, homocystinuria resembles Marfan's syndrome but the two disorders can be readily distinguished biochemically. In homocystinuria, homocystine is detectable in plasma, the blood methionine level is raised, that of cystine is low. Homocystine which is the disulphide of homocysteine, the mixed disulphide of cysteine and homocysteine, methionine and methionine sulphoxide are excreted in excess in urine. Many other sulphur amino acids have been found in urine.[8,66] They reflect utilization of minor pathways of methionine metabolism or result from transamination or disulphide exchange reactions. Cystathionine occurs in high concentration in human brain. In homocystinuric patients brain cystathionine levels are greatly reduced.

In the majority of patients the enzyme defect appears complete, but a substantial minority have some residual enzyme activity and these patients respond to pharmacological doses of pyridoxine of up to several hundred milligrams a day, both clinically and biochemically. Homocystine disappears from plasma while methionine and cystine levels are normalized. A diet low in methionine and supplemented with cystine is effective in correcting the biochemical abnormalities in all variants of cystathionine synthase deficiency. Treatment results in normal or near normal somatic and intellectual development in the great majority of treated cases but the results are not too easy to interpret. Only about one third of untreated patients are severely retarded, one third are intellectually normal, the rest are mildly retarded. However, even those untreated patients who are intellectually normal or near normal have had abnormal EEGs, epilepsy and psychoses.

Homocysteine can be remethylated to methionine by the enzymes betaine homocysteine methyltransferase and N^5-methyltetrahydrofolate transferase. The latter requires S-adenosylmethionine and cobalamine cofactors. N^5-methyltetrahydrofolate itself is formed from N^5,N^{10}-methylenetetrahydrofolate by N^5,N^{10}-methylenetetrahydrofolate reductase. Deficient activity of N^5-*methyltetrahydrofolate transferase* has been described in several patients with a wide spectrum of clinical manifestations. The deficiency has been due to selective vitamin B_{12} malabsorption or defective intracellular metabolism of cobalamine. In severe forms of this variant CNS lesions resemble those seen in B_{12} deficiency and subacute combined degeneration of the cord. In contrast to the cystathionine synthase defect blood methionine is low. As in B_{12} deficiency, methylmalonic acid is excreted in urine. Cobalamine supplements may benefit these patients.

Decreased N^5,N^{10}-*methylenetetrahydrofolate reductase* is also associated with a wide range of neuropsychiatric symptoms, a low blood methionine level and homocystinuria. Folate administration was followed by dramatic biochemical and clinical improvement in a mildly retarded patient who had a schizophrenia-like psychosis. The three forms of homocystinuria are contrasted in Table 20.

Patients with homocystinuria are liable to die from thromboembolic accidents even at an early age. At autopsy, multiple infarcted areas are found in the brain and these can account for at least part of the intellectual deficits in these disorders. These thromboembolic accidents have been attributed to interference by homocysteine with the stabilization of collagen fibrils. These phenomena occur irrespective of the origin of the homocystine. In fact, the elevated plasma homocysteine levels found in pernicious anaemia may play a part in the pathogenesis of the neurological lesions in this disease.

Other Aminoacidurias

Discussion will be confined to those aminoacidurias, a minority only, with definite adverse effects on brain development. Only one case of *β-alaninaemia* has so far been described.[8] The most likely site of the defect is the enzyme β-alanine α-ketoglutarate aminotransferase. As a result, β-alanine accumulates in the tissues and body fluids. Excess of GABA is found in brain and extracellular fluids including urine. Catabolism of GABA is impaired presumably by competitive inhibition by the excess of β-alanine. GABA has been identified as an inhibitory transmitter similar to glycine, at a number of

TABLE 20
SIMILARITIES AND DIFFERENCES IN THREE FORMS OF HOMOCYSTINURIA[8,66]

	Cystathione synthetase deficiency	N^5-methyltetrahydrofolatehomocysteine methyltransferase deficiency	$N^{5,10}$-methylenetetrahydrofolate reductase deficiency
Mental retardation	common	common	common
Retarded growth	no	common	no
Bone deformities	common	rare	no
Dislocated ocular lenses	common	no	no
Thromboembolic tendency	common	rare	no
Megaloblastic anaemia	no	rare	no
Methylmalonic aciduria	no	yes	no
Plasma and urinary homocystine	increase	increase	increase
Plasma methionine	increase	possible decrease	possible decrease
Plasma and urinary cystathionine	not detected	possible increase	possible increase
SO_4^{2-} after methionine load	decrease	normal	normal
Serum folate	decrease	increase	decrease
Response to vitamin	B_6	B_{12}	folic acid
Severe methionine restriction	helpful	harmful	harmful

TABLE 21
DISORDERS OF THE γ-GLUTAMYL CYCLE[8,66]

Deficient enzyme	Clinical signs	Laboratory findings
γ-Glutamyl transpeptidase	mental retardation, behaviour disorders	excess glutathione in urine and plasma
γ-Glutamylcyclotransferase		
5-Oxoprolinase	? mental retardation in some patients	excess l-5-oxoproline in urine
γ-Glutamylcysteine synthetase*	late onset spinocerebellar degeneration low red cell glutathione	
Glutathione synthetase*	acidosis, mental retardation, neurological deterioration	excess l-5-oxoproline in urine, low red cell glutathione

* In some patients the defect is confined to red cells, when the disorder is characterized by haemolytic anaemia but absence of neurological determination.

synapses. The symptoms in β-alaninaemia which in the case described included somnolence, lethargy, hypotonia, diminished reflexes, convulsions and early death are not unlike those seen in non-ketotic hyperglycinaemia. Clearly, whenever neuro-transmission is upset in a metabolic disorder the prognosis will in general be unfavourable. The diagnosis of β-alaninaemia is unlikely to be difficult. Not only β-alanine but also the β-aminoacids taurine and β-aminoisobutyric acid which share the renal transport system with β-alanine, and GABA will be present in increased amounts in urine. Treatment with pyridoxine might be tried if additional cases are discovered.

Mental retardation has been a feature of all three cases of *γ-glutamyltranspeptidase* deficiency so far identified.[8,76] The enzyme forms part of the γ-glutamyl cycle of Meister[66] formed by the reactions listed in Table 21. It has been postulated that this cycle represents a general though not the sole mechanism for the transport of amino acids across the cell membranes. In fact, patients with γ-glutamyltranspeptidase deficiency have provided no positive evidence for the operation of this cycle in man, since no significant abnormalities of amino acid transport have been observed in kidney or cultured fibroblasts. However, the cycle could have a more restricted but essential function, for example at the blood brain barrier or in certain neurons. The suggestion has been put forward that γ-glutamyltranspeptidase has a role in the transport and metabolism of biologically active peptides in the brain. Peptides are better acceptor substrates for the enzyme than free amino acids. Also, γ-glutamyl amides of histamine, serotonin, dopamine and noradrenaline may be formed by the enzyme from the amines by the transfer of the γ-glutamyl moiety from γ-glutamyl peptides. Deficiency of γ-glutamyltranspeptidase could thus affect the level of serotonin, dopamine and noradrenaline and affect behaviour. Two of the three patients with the disorder have psychiatric problems including severe tension, and aggressive and auto-agressive behaviour. The disorder is readily detectable as glutathione present in excess in urine give a positive nitroprusside test. Neurological symptoms are also found in some patients with deficiency of *5-oxoprolinase*, *γ-glutamyl transferase* and *glutathione synthetase* (pyroglutamic aciduria).[66]

The association of mental retardation and two disorders of imino acid metabolism *hyperprolinaemia type I* (proline oxidase deficiency) and *hydroxyprolinaemia* (hydroxyproline oxidase deficiency) is probably due to chance. A third disorder of imino acid metabolism *hyperprolinaemia type II* is due to deficiency of the enzyme catalysing the second step in the catabolism of the imino acids that is from Δ_1-pyrroline-5-carboxylic acid to glutamric acid and from Δ_1-pyrroline-3-hydroxy-5-carboxylic acid to hydroxyglutamic acid. A substantial proportion but by no means all the patients with hyperprolinaemia type II have had mild mental retardation and epilepsy. A pathogenetic mechanism has not yet been established. The transport system for the imino acids and glycine is distinct from that for other amino acids. Even the very large excess of proline in the body fluids of patients with hyperprolinaemia type II probably does not interfere with the transport of essential amino acid into cells or across the blood brain barrier. This may in part explain why patients are so much less at risk from brain damage than untreated phenylketonurics. Hyperprolinaemia is readily detected on thin-layer chromatograms with an isatin stain. Urine may be screened for Δ_1-pyrroline-5-carboxylic acid with o-aminobenzaldehyde.[8,66]

A number of patients with *hyperlysinaemia* have been described but there is no convincing evidence that hyperlysinaemia *per se* affects intellectual development. The observed biochemical abnormalities have been attributed to deficiency of lysine ketoglutarate dehydrogenase deficiency. They include persistent hyperlysinaemia and hyperlysinuria. In addition, homocitrulline, homoarginine, the N-acetyl lysines, ipiecolic acid and α-aminoadipic acid were identified in urine. This indicates that most metabolic pathways of lysine are operative. The defect in *saccharopinuria* has not yet been established but may involve the enzyme saccharopine dehydrogenase. Two patients so far described were dissimilar but both had neuropsychiatric deficits. It appears then that the nervous system may be adversely affected in several disorders involving amino acids and organic acids of the lysine catabolic pathways. In *α-aminoadipic aciduria* the mitochondrial α-aminoadipate aminotransferase is probably deficient. Only one of the 3 patients with this disorder so far reported was mentally retarded but it may be relevant that α-aminoadipate is toxic to cultured cerebellar cells.

OTHER NEUROMETABOLIC DISORDERS

Galactosaemia

This inborn error was first observed early in this century and many cases have since been described. From mass screening surveys the incidence is probably near 1 in 60 000 births. In classical galactosaemia, symptoms usually begin in the first week of life and include vomiting, failure to thrive, jaundice, signs of progressive liver failure and later cirrhosis. Cataracts develop usually over a period of weeks but some infants show signs of cataract and cirrhosis at birth and these lesions are then not fully reversible. Patients may die rapidly if untreated. Their metabolic disease is often aggravated by infection, typically *Escherichia coli* septicaemia. However, the disorder may also be milder, the patients presenting later with cataracts and mental retardation, while some stay virtually symptomless and mentally normal.

Galactose-1-phosphate uridyltransferase the enzyme affected in galactosaemia exhibits extensive polymorphism (Table 22) and this may explain at least in part, the wide range of clinical manifestations. The metabolic block in galactosaemia is well understood. It results in the accumulation of at least two toxic compounds, galactitol formed from galactose by reduction, and galactose-1-phosphate formed by phosphorylation. Galactitol produces osmotic swelling in the lens and disruption of lens fibres leading to cataract formation, but by itself probably does not produce neuropsychiatric deficits. Galactose-1-phosphate on the other hand is neurotoxic, hepatotoxic and nephrotoxic. Galactosaemia is diagnosed either by mass screening of the newborn using blood collected on filter paper with the Beutler fluorescence test or the Paigen modification of the Guthrie microbiological inhibition test or, in the sick neonate, by following up the findings of a non-glucose reducing substance in urine. The diagnosis has been missed in infants on glucose saline drips when no galactose is excreted in the urine, and in cases where as a result of renal tubular damage glucose as well as galactose are present in urine which then gives a reaction for

TABLE 22
POLYMORPHISM OF GALACTOSE-1-PHOSPHATE URIDYL TRANSFERASE[9]

Variant	Symptoms	Enzyme Activity (%)	Enzyme Electrophoretic mobility	Stability	Gene frequency (estimated)
Classical (Gal)	yes	0–few			0·002–0·005
Negro	sometimes	0*			
Duarte	no	50	fast (3 bands)		0·03–0·08
Rennes	yes	7–10	slow		
Indiana	no	0–40	slow	reduced	
Los Angeles	no	100–140	fast (3 bands)		
Munster	yes	30			
Bern	?	40			0·0009
Chicago fast	no	25	fast	reduced	
Chicago labile	no				

Indiana/Gal, Chicago/Gal and Indiana/Gal compound heterozygotes have shown symptoms.
* 0% activity in red cells; in the negro variant 10% of normal activity is found in liver and intestine.

both glucose and non-glucose reducing substances. The Beutler test is preferred for excluding the disease in sick infants. It shows presence or absence of uridyl transferase and is independent of the nutritional state of the patient. The enzyme is, however, affected by heat and moisture and this can give rise to false positive results. The test will not normally distinguish between complete and partial enzyme deficiencies.

In all cases the diagnosis must be confirmed by quantitative assay of galactose-1-phosphate uridyl transferase. Electrophoretic separation is often necessary to establish the precise genotype. Variants may differ both in electrophoretic mobility and stability (Table 22). Some of the compound heterozygotes have symptoms which are less severe than those in classical galactosaemia and also have markedly reduced transferase levels. It is possible that some patients diagnosed in the past as having mild or atypical galactosaemia were in fact heterozygous for one of the less severe variants of the galactosaemic gene. A full discussion of this topic will be found in Burman et al.[9] and in Bickel et al.[5]

There is no doubt that treatment with a diet as free from galactose as possible is life-saving and results in the reversal of any acute symptoms. In practice, it is extremely difficult to produce a diet which contains no lactose, galactose or galactosides (derived from soya beans) once mixed feeding is introduced. At present it is not certain to what extent treatment can safeguard normal intellectual development. Several points can be made.

(1) In surviving infants delays of less than four months in instituting dietary treatment do not appear to affect intellectual development appreciably.
(2) Even competently treated patients have visiospatial difficulties and intellectual deficits of about 15 IQ points, compared to their parents or unaffected siblings. In some series these differences have been substantially larger. Emotional and behavioural problems are common even in those patients who are of normal or near normal intelligence.
(3) Treatment aims at levels of galactose-1-phosphate in red blood cells of less than 115 μmol/l (30 μg/ml), past infancy patients are often found to have levels in the range 115–200 μmol/l (30–50 μg/l). The correlation between control of red cell galactose-1-phosphate and ultimate intellectual attainment is by no means perfect.
(4) Homozygous galactosaemics have elevated cord blood galactose-1-phosphate levels and the inference can be that the fetus is at risk during pregnancy. Again there appears to be no clear-cut relationship between cord blood galactose-1-phosphate levels and intelligence. By contrast, evidence so far suggests that the heterozygous fetus of a treated galactosaemic mother has an excellent chance of normal development.[9]
(5) There is ample evidence for the biosynthesis of galactose from glucose in man via the pyrophosphorylase pathway. It follows that galactosaemic infants are able to synthetize galactose-1-phosphate from glucose-1-phosphate. This constitutes a possible mechanism of self-intoxication, which at least in some patients might limit the effectiveness of the galactose-free diet.
(6) While in vitro galactose-1-phosphate has been shown to inhibit phosphoglucomutase, glucose-6-phosphate dehydrogenase, and pyrophosphorylase the significance of these findings for in vivo pathogenetic mechanisms is not clear.
(7) Antenatal diagnosis of galactosaemia is possible. In view of the uncertain outlook for affected infants even if treated from birth expert genetic counselling must be available to the parents in conjunction with prenatal diagnostic testing.

Disorders of Pyruvate and Lactate Metabolism

Pyruvate occupies a key position in the metabolism of all animal cells. Even moderate impairment of its metabolism often results in profound biochemical abnormalities in associated pathways, neurological symptoms and mental handicap. Sir Rudolph Peters derived his concept of biochemical lesion from observation of the effects of thiamine deficiency on pyruvate oxidation.

The major pathways of pyruvate utilisation operate by one of four processes: (1) transamination to alanine; (2) reduction to lactate; (3) carboxylation to oxaloacetate and (4) oxidative decarboxylation to acetyl CoA and CO_2. The transamination and reduction are reversible and alanine and lactate may become a source of glucose provided the gluconeogenic pathway is intact. Pyruvate carboxylase is not only the first step in the gluconeogenic pathway but its product oxaloacetate is a precursor of aspartate, a putative neurotransmitter, and may be necessary to prime the Krebs tricarboxylic acid cycle. Oxidative carboxylation is affected by the pyruvate dehydrogenase complex which comprises three catalytic and two regulatory enzymes, thiamine, lipoic acid, co-enzyme A, FAD and NAD. There is relatively little excess of this enzyme system particularly in brain. Even a small reduction in activity may therefore be dangerous. As pyruvate dehydrogenase has a critical role in the central pathway of carbohydrate metabolism mutations resulting in pronounced reduction of activity of this enzyme are probably incompatible with life.

Deficiency of *pyruvate dehydrogenase* may be acquired due, for example, to thiamine deficiency or heavy metal poisoning, or due to the action of any toxin or metabolite which causes mitochondrial damage and a general decrease of mitochondrial enzyme activity.

A number of patients with hereditary deficiency of pyruvate dehydrogenase have been described, mutations affecting one or other of the proteins of the enzyme complex.[6,9] Clinical manifestations correlate with the extent to which overall enzyme activity is reduced. Less than 15% of normal pyruvate dehydrogenase activity is associated with infantile lactic acidosis, severe mental and motor retardation, fits, hypotonia, poor coordination and sometimes optic atrophy. Milder deficiencies with enzyme activities in the range 20–25% of normal have been associated with later onset and milder symptoms, with ataxia as a prominent clinical sign. The largest residual activity ranging from 35–50% has been reported in syndromes of spinocerebellar degeneration, that is a group of hereditary conditions with ataxia and long tract signs. Abnormalities of pyruvate metabolism are detectable in about half the patients with hereditary ataxia such as Friedreich's ataxia. It is difficult to decide if these abnormalities are primary or secondary but at least some of these patients appear to have primary deficiencies of pyruvate dehydrogenase. The most prominent findings in pyruvate carboxylase deficiency are again retarded mental and motor development, fits, hypotonia, acidosis and in some cases hypoglycaemia.[9]

Deficient oxidation of pyruvate leads to impaired production of acetyl CoA, lactic acid and pyruvic acid accumulation and, eventually, lack of ATP. Partial impairment of pyruvate oxidation can compromise brain function long before there is a drop in high-energy compounds by affecting the synthesis of acetyl choline and other neurotransmitters. Deficient carboxylase activity may also interfere with oxidation by impaired priming of the Krebs cycle, and possibly with the synthesis of aspartate and glucose from alanine.[9]

Blood lactate and pyruvate should be measured in patients in whom a disorder of pyruvate metabolism is suspected. Elevation of lactate (more than 1 mmol/l) and of pyruvate (more than 0.1 mmol/l) may only be intermittent even in severely affected patients. In the absence of changes in the NADH/NAD ratio the lactate/pyruvate ratio is generally normal. Pyruvate is always abnormally high one hour after a standard glucose load (more than 0.115 mmol/l). Urinary lactate may be markedly increased particularly during acute episodes. Blood and urine levels of alanine are elevated in some patients. It must be remembered that lactic acidosis is not at all uncommon in sick infants[42] (Table 23). It has been pointed out by Blass[6] that the finding of excess lactate or pyruvate in body fluids indicates no more than that an abnormality exists somewhere in a large area of metabolism which is not specific even for a particular pathway.

Patients with disorders of aminoacid metabolism often show exacerbations when given a protein-containing diet. In contrast, patients with pyruvate dehydrogenase deficiency may deteriorate rapidly on a high carbohydrate diet and those with pyruvate carboxylase deficiency on a high fat diet. At postmortem some patients with disorders of lactate and pyruvate metabolism have shown the neuropathological features of *Leigh's encephalopathy* (subacute necrotising encephalomyelopathy) with a topologically distinct pattern of characteristic lesions in the optic nerves, chiasma and tract, the basal ganglia, cerebellum and spinal cord and, usually most strikingly, in the brain stem.[17] However, this pattern of lesions is not found exclusively in patients with deficiency of pyruvate dehydrogenase and carboxylase but probably represents a common abnormal pathway which can apparently be triggered by several hereditary and possibly also environmental factors.

Assay of pyruvate dehydrogenase and pyruvate carboxylase in small samples of human tissues or cultured fibroblasts is notoriously difficult and there have been discrepancies between the results obtained by research laboratories even when studying the same patient.[9] This has not been a bar to attempts at rational therapy. The pathogenetic process in pyruvate carboxylase deficiency probably operates via the Krebs cycle rather than gluconeogenesis. Attempts have therefore been made to increase the 4-carbon carboxylic acid pool by giving glutamic and aspartic acid with vitamin B_6 supplements to ensure adequate transamination. Beneficial effects have been reported in some but not in other cases. Thiamine in high doses has also been used with or without glutamate or aspartate. Again, only some patients responded. A decrease in blood lactate may occur in some patients by increased activity of pyruvate dehydrogenase or by stimulation of gluconeogenesis. Unfortunately, thiamine, aspartate and glutamate each penetrate the blood brain barrier only slowly. Overall improvement in metabolic state may therefore not extend to the brain.[9]

Pyruvate dehydrogenase deficiency has been treated by a high fat diet to bypass the metabolic block. This diet is adjusted to make the patient ketonaemic but not acidotic and hypoglycaemic, and is supplemented by medium chain triglycerides. Thiamine and lipoic acid have at times been tried, at least one patient responded dramatically to thiamine. This patient appears to have had a true thiamine dependency. Cholinergic agonists such as physostigmine have been used in an attempt to counteract the harmful effects of pyruvate dehydrogenase deficiency or acetylcholine synthesis. Dichloracetate, like thiamine an activator of pyruvate dehydrogenase, is probably too toxic for clinical use. In view of the multiplicity of enzyme defects in these disorders, any one approach to treatment is unlikely to be universally applicable.

Organic Acidurias

Organic acidurias are characterised by the excessive urinary excretion of aromatic or aliphatic carboxylic

TABLE 23
AETIOLOGICAL FACTORS IN LACTIC ACIDOSIS[42]

Environmental
  Shock, cardiopulmonary disease
  Uraemia, liver disease
  Acute infections, septicaemia
  Diabetic ketoacidosis
  Drugs

Hereditary
  Von Gierke's disease
  Fructose 1,6-diphosphatase deficiency
  Methylmalonic acidaemia, propionic acidaemia
  Disorders of pyruvate dehydrogenase complex
  Pyruvate carboxylase deficiency
  Mitochondrial myopathies
  Idiopathic

TABLE 24
SOME ORGANIC ACIDURIAS WITH NEUROPSYCHIATRIC MANIFESTATIONS[10,27]

Disorder	Enzyme defect	Diagnostically useful urinary metabolites
Maple syrup urine disease	Branched chain keto acid decarboxylase	Branched amino and keto acids
Isovaleric acidaemia	Isovaleryl-CoA dehydrogenase	Isovaleric and 3-hydroxyisovaleric acids, isovaleryl glycine
3-Methylcrotonylglycinuria	3-Methylcrotonyl-CoA carboxylase	3-Methylcrotonic and 3-hydroxyisovaleric acids, 3-methylcrotonyl glycine
3-Hydroxy-3-methylglutaric aciduria*	3-Hydroxy-3-methylglutaryl-CoA lyase	3-Hydroxy-3-methylglutaric, 3-methylglutaconic, 3-methylglutaric and 3-hydroxyisovaleric acids
β-Ketothiolase deficiency	β-Ketothiolase	2-Methylacetoacetic and 2-methyl-3-hydroxybutyric acids, glycine
Propionic acidaemia	Propionyl-CoA carboxylase	Propionic, 3-hydroxypropionic and methylcitric acids, glycine
Methyl malonic acidurias (6 variants)	Methylmalonyl-CoA mutase	Methylmalonic, methylcitric, propionic and lactic acids
Holocarboxylase deficiency	Multiple carboxylase deficiency	Propionic, hydroxypropionic, methylcitric and 3-hydroxyisovaleric acids, 3-methylcrotonyl glycine
Pipecolic acidaemia	Pipecolate catabolism	Piperidine-2-carboxylic acid (pipecolic acid)
α-Ketoadipic acidurias	2-Ketoadipic acid decarboxylase	2-Ketoadipic, 2-hydroxyadipic and 2-aminoadipic acids
Glutaric aciduria*	Glutaryl-CoA dehydrogenase	Glutaric, 3-hydroxyglutaric and glutaconic acids
C_6-C_{10}-Dicarboxylic acidurias*	Defect in β-oxidation of fatty acids	Adipic, suberic, sebacic and 5-hydroxycaproic acids
Sidbury syndrome	Green acyl dehydrogenase	Butyric and caproic acids
Glyceric aciduria Type I	D-Glyceric acid dehydrogenase	D-glyceric acid, glycine
Glyceric aciduria Type II	?	D-glyceric acid
Tildon and Cornblath syndrome	Succinyl-CoA: 3-ketoacid CoA transferase	Acetoacetic and 3-hydroxybutyric acids
Pyroglutamic aciduria	Glutathione synthetase	L-2-Pyrrolidine-5-carboxylic acid (pyroglutamic acid)

* Disorder has been described in patients with Reye's syndrome.

acids. Often hereditary, they may or may not be associated with metabolic acidosis and abnormalities of carbohydrate or amino acid metabolism. As first described, many of these disorders were associated with severe neurological symptoms, deterioration of the nervous system and early death. However, some patients with milder forms of the disorders often with somewhat higher residual enzyme activity have survived albeit with mostly severe mental retardation. Information was thus needed on the extent to which organic acidurias contribute on the one hand to perinatal mortality and on the other to the prevalence of mental retardation.

Organic acidurias do not lend themselves readily to screening by paper and thin layer chromatography and progress in this field had to await the advent of gas chromatography mass spectrometry.[10] Using these techniques Watts et al.[72] investigated nearly 2000 severely mentally retarded patients living in institutions. They established, that organic acidurias do not contribute significantly to the prevalence of severe long-term mental handicap. Nearly all the abnormalities in urinary organic acids found in these patients were attributable to drugs, diet or other environmental factors, or to artefacts produced, for example by bacterial contamination.

Table 24 lists organic acidurias which may be responsible for serious disease in early infancy. Despite impressive biochemical heterogeneity the clinical findings are surprisingly similar: failure to feed, vomiting, lethargy, hypotonia, coma, fits. Metabolic acidosis may be compensated or uncompensated and is often accompanied by ketonuria and sometimes also by hypoglycaemia. Many of these findings are seen much more frequently in conditions such as septicaemia, serious perinatal brain injury and poisoning, but also in other rare metabolic disorders, notably those involving enzyme deficiencies of the Krebs urea cycle and some disorders of carbohydrate metabolism.

The analytical problems of this field are formidable. Chalmers and Lawson[10] use preliminary chromatography on DEAE-Sephadex, elution with pyridinium acetate buffer, freeze-drying of the organic acids, conversion to ethyloxime-trimethylsilyl derivatives, gas chromatography and where indicated mass spectrometry. Allowance must be made for the marked age dependence of the urinary organic acid excretion pattern in the neonatal period. Few centres offer these complex and costly investigations and it is important to adhere to careful criteria in referring patients. Valuable diagnostic clues are any of the following: (1) a family history of a previous sibling with similar neonatal findings; (2) acute unexplained illness in the neonatal period or in infancy; (3) metabolic acidosis which fails to respond to treatment with bicarbonate particularly when accompanied by persistent ketonuria and/or hypoglycaemia. Blood gases, urea and electrolytes, liver and renal function tests, calcium, glucose, pyruvate, lactate and ammonia and amino acids (by thin layer chromatography) should be done before going on to more specialized investigations. The incidence of these disorders is not known, and the yield of cases depends on the criteria for referral but in up to 2% of cases examined in specialised centres known inborn errors of metabolism could be identified.[5] A substantial proportion were disorders of organic acids derived from the branched chain amino acids.

A few generalizations may be made about organic acidurias. Impairment of mitochondrial function seems to play an important part in pathogenesis. The toxic effects of short-chain fatty acids are compounded by the consequences of the inhibition of pyruvate and oxoglutarate oxidation. Hyperammonaemia and hyperglycinaemia ('ketotic hyperglycinaemia') are found, for example in methylmalonic acidaemia and propionic acidaemia, presumably as a result of impairment of the mitochondrial enzymes carbamyl phosphate synthetase, ornithine transcarbamylase and glycine-N-acylase, and of the glycine cleavage system. Reye's syndrome may follow severe insult to the mitochondria.

Hypoglycaemia, common in sick neonates, will be exacerbated when gluconeogenesis is inhibited, for example, when pyruvate carboxylase activity is reduced, or where there is failure to utilize ketone bodies, as in the Tildon–Cornblath syndrome, failure to form ketone bodies as in β-hydroxy-β-methylglutaric aciduria, or low rate of formation of ketone bodies as happens in C_6-C_{10}-dicarboxylic aciduria because of decreased production of acetyl CoA resulting from a defect in β-oxidation of fatty acids.

Advances in treatment have not been insignificant. Isovaleric acidaemia responds to a low protein–low leucine diet. Treatment with glycine to conjugate isovaleric acid has been beneficial in some cases. A diet low in branched chain amino acids is life saving in clinical maple syrup urine disease but in most cases does not prevent recurring crises and mental retardation. Patients with partial enzyme deficiency do much better, one patient responded to thiamine.[8] Low protein diets and peritoneal dialysis are commonly only palliative in propionic acidaemia and apoenzyme deficient cases of methylmalonic aciduria. However, four of the six known genotypes have defects of cobalamin metabolism and respond to B_{12} therapy. A minority of patients with carboxylase defects improve when given biotin, most dramatically those with holocarboxylase synthetase deficiency. Treatment with riboflavin may enhance the activity of acyl-CoA dehydrogenase and benefit patients with C_6-C_{10}-carboxylic aciduria.

Always, correction of fluid and electrolyte imbalance, control of the metabolic acidosis, careful attention to the nutritional needs of the patients and vigorous treatment of any infection are mandatory to forestall the rapid, irreversible decline sometimes seen in these disorders. In particular, impairment of mitochondrial function of the endothelial cells of the brain capillaries may result from the combined effects of toxic metabolites, acidosis and infection with consequent failure of the blood brain barrier, brain oedema and irreversible brain damage.[35] A most useful brief guide to the management of newborn babies with suspected metabolic illness has been published by Danks.[19]

Disorders of Purine and Pyrimidine Metabolism

The Lesch–Nyhan syndrome is characterized by gout, choreoathetosis, spasticity, self-mutilation and severe mental retardation. The disease is determined by a gene on the X-chromosome controlling the synthesis of the enzyme hypoxanthine-guanine-phosphoribosyl-transferase. This enzyme plays a vital part in the feedback control of intracellular purine levels. When it is defective there is overproduction of uric acid resulting in a urinary excretion up to eight times normal, while excretion of the purine precursors 4-amino-5-imidazole carboxamide and its riboside is even more abnormal. The excessive uric acid synthesis results in a high incidence of renal lithiasis. This can be controlled by treatment with the xanthine oxidase inhibitor allopurinol.

The unsolved problems of this disorder are the pathogenesis and alleviation of the neurological dysfunction. Of all tissues, brain has the highest HGPRT level, and particularly high activity is found in the basal ganglia. Many of the symptoms of the Lesch–Nyhan syndrome are related to basal ganglia dysfunction. Nyhan et al.[54] suggest that the problem might be an imbalance between serotonergic and dopaminergic or adrenergic influences. Treatment has been tried with 5-hydroxytryptophan in combination with the peripheral decarboxylase inhibitor carbidopa and with imipramine which potentiates the central action of serotonin. Initially most patients show substantial improvement but all develop tolerance to the drugs within one to three months. Treatment with diazepam has been proposed but not yet tried clinically.

The disorder is genetically heterogeneous and the severity of symptoms correlates with the extent of the enzyme defect. Patients with activity a few per cent of normal develop gout at an early age but are otherwise well. Enzyme deficiency is almost complete in the full-blown syndrome, intermediate enzyme levels may be associated with choreoathetosis and spasticity with or without self-mutilation but near normal intelligence.

Assay of the phosphoribosyl transferase on blood collected on filter paper, fibroblasts or amniotic cells is offered by specialized centres which usually also provide an antenatal diagnostic service. A useful screening test is measurement of the molar uric acid/creatinine ratio in urine. This is greater than 2·0 in younger affected children and greater than 1·35 in older affected children. The blood uric acid level is usually but not invariably elevated. The syndrome is discussed in detail in Bondy and Rosenberg.[8]

In recent years the importance of purine and pyrimidine metabolism in the development of the immune system has become apparent. In *purine nucleoside phosphorylase deficiency*[36] severe immunodeficiency disease may be associated in a minority of cases with neurological dysfunction such as developmental delay, tremor, ataxia and tetraparesis. The enzyme catalyses the transformation of inosine to hypoxanthine and guanosine to guanine. In contrast to the Lesch–Nyhan syndrome findings include hypouricosuria, low blood hypoxanthine levels and usually hypouricaemia. Excess purine nucleosides are found in the urine. *Xeroderma pigmentosum*[66] is a disorder in which abnormalities of pigmentation are produced by ultraviolet light. These are liable to become malignant. CNS abnormalities are found in 20–30% of cases. Damage to DNA may be caused by a variety of compounds and by ionizing radiation and this could provide an explanation for the neurological complications. Microcephaly and mental retardation also occurs in up to half the patients with *Fanconi's anaemia*. However, the most important disorder involving defective DNA repair in the field of mental retardation is *ataxia telangiectasia*[36,52] combining cerebellar ataxia, choreoathetosis, mental retardation and teleangiectases (dilatation of capillaries in the conjunctiva, skin and other organs). Complex immunodeficiency and a high incidence of malignancies have been attributed to spontaneous chromosome abnormalities arising from defective DNA repair mechanisms which renders these patients vulnerable to ionizing radiations and other mutagens. The mechanism of the pathological changes in the brain which include atrophy of the cerebellum and posterior columns has not been explained.

A disorder of pyrmidine metabolism *orotic aciduria*[8,66] is of great scientific interest because of its unusual genetic mechanism, and because it is one of the few inborn errors of metabolism responding to treatment based on theoretical considerations. Patients are usually referred because of slow mental and physical development, a severe anaemia refractory to treatment and megaloblastic changes in the bone marrow. Crystals separate in the urine and can be identified as orotic acid. The metabolic block involves deficiency of the two enzymes on the metabolic pathway from orotic acid to uridine-5-phosphate (UMP), orotidine-5-monophosphate pyrophosphorylase and orotidine-5-monophosphate decarboxylase (orotic aciduria type I). Deficiency of decarboxylase with normal pyrophosphorylase activity has been found in one case (orotic aciduria type II). The disorder is expressed in fibroblasts. Those from affected children have only about 4% of normal activity, but this can be stimulated to near normal levels by adding pyrimidine analogues (azauridine, allopurinol). The mutant enzyme has altered electrophoretic mobility and greater thermal lability. It is thought that the deficiency in two enzymes is due to a mutation affecting a protein with two enzymatic activities.

The interruption of pyrimidine nucleotide synthesis creates a nutritional requirement for preformed pyrimidines which has been met by 150 mg uridine per kg body weight per day in some patients. A mixture of cytidylic and uridylic acid may also be effective. Immune responses may be impaired just as in patients with primary defects in purine metabolism. An unusual feature for an inborn error of metabolism is the high incidence of congenital malformations. For a full review of this disorder the reader is referred to chapters in Stanbury *et al.*[66] and Bondy and Rosenberg.[8] Orotic aciduria is also found in disorders of the Krebs urea cycle.

Muscular Dystrophies

Most aspects of muscular disorders in childhood have recently been reviewed by Dubowitz.[25] Mental retardation is frequently found in *myotonic dystrophy (Steinert's disease)*. Some of the retarded patients have congenital malformations of the brain.[17] *Werdnig Hoffman disease* is a neurogenic degeneration of muscle following neuronal decay in the anterior horns of the cord and in the medulla. Most patients die in infancy. In its acute form this disorder may clinically resemble nonketotic hyperglycinaemia.

The association of intellectual impairment and *Duchenne muscular dystrophy* has long been recognised. On the Wechsler scale, the mean IQ of affected boys has been about 85, compared to 105 for normal groups. According to Dubowitz[25] this deficit is not due to factors associated with the patients' physical disability but may be explained by the pleiotropic effects of a single gene involving not only skeletal muscle but also the myocardium and central nervous system. The pathogenic mechanism in Duchenne dystrophy is not known. The intellectual retardation is non-progressive. The clinical chemistry of this disorder is discussed in Chapter 27.

Hypothyroidism

Cretinism occupies a special place in mental deficiency because it was once regarded as the main cause of backwardness. In French and German the terms 'cretin' and 'idiot' are synonymous. The clinical chemistry of the thyroid is described in Chapter 36. Here we are only concerned with the effect of thyroid hypofunction on intellectual development.

Endemic cretinism has largely disappeared in economically advanced countries. It is characterised by retarded growth and development, delayed ossification, mental retardation and other neurological deficits. The condition is thought to result from severe maternal iodine deficiency at an early stage of fetal development but other factors are probably involved.

Mental retardation is associated with several of the rare inborn errors of dyshormonogenesis (Chapter 36) particularly the iodide peroxidase defect, while in the Pendred syndrome, probably an incomplete form of peroxidase deficiency, nerve deafness is the major neuropsychiatric deficit. Mental retardation has also been seen in the dehalogenase defect although there is no reason why this should not be preventable in this variant by very early treatment, and in some cases involving deficits in thyrogloblin synthesis or breakdown.

Goitrous cretinism and mental retardation have been described in the offspring of mothers who had an excessive intake of iodide during pregnancy and in the offspring of thyrotoxic mothers overtreated with antithyroid drugs. However, numerically most important is congenital non-goitrous cretinism associated with thyroid dysgenesis. Congenital hypothyroidism with an estimated incidence of 1 in 4000 births is probably the commonest endocrine disorder in infants.

Work on rats has established that thyroid deficiency in utero or early postnatal life leads to a reduction in brain growth and defective myelination. Cell acquisition is affected, the arborization of dendrites and formation of nerve terminals are retarded. Thyroid hormone is important not only for the formation of the normal number of synapses but also for the development of synaptic organization and of neurotransmitter systems.[3] For example, development of β-adrenergic, GABA and muscarinic cholinergic receptors is affected, not necessarily equally at any stage of development resulting in distortions of the 'chemical wiring' of the brain particularly in the cerebellum.

Unexpectedly, only two cases of hypothyroidism were found amongst 1900 severely retarded institutionalised patients in London.[44] However, some degree of mental handicap is common in congenital hypothyroidism. Early treatment materially improves the outlook for these infants although at least in some cases irreversible injury to the brain has occurred antenatally.[48,75] Often, some functioning thyroid tissue is still present at birth and hypothyroidism develops over a period which may range from days to years.

Treatment with thyroxine should be monitored both by clinical assessment and regular assays of T_4 and TSH. Occasionally children on treatment are found to have normal T_4 but elevated TSH levels. Overtreatment must be avoided as it can result in acceleration of bone age and craniosynostosis.[12]

Heavy Metals and Intellectual Development

Lead

A full discussion of the clinical chemistry of lead will be found on p. 347. Here we confine ourselves to a consideration of how far lead ingestion resulting in blood lead levels not in excess of $2 \cdot 4 \mu\text{mol/l}$ ($50 \mu\text{g}/100\text{ml}$) and not accompanied by recognizable symptoms can impair psychological development and lead to disturbed behaviour. This problem is of more than academic interest since extensive public health measures would have to be instituted if claims about the risk of low-level lead exposure can be substantiated. The methodological problems in this field are formidable and the debate has not always been conducted with scientific detachment.

Until recently most epidemiological investigations have relied on a single estimate of blood lead as the chief or only indicator of the body burden. This approach reflects only recent exposure and provides no guide to its duration or past intoxications. Assays of lead levels in teeth have not always been consistent but do appear to reflect the body burden of lead averaged over a period of time. There still remains the problem of distinguishing shorter periods of high-level exposure from more moderate exposure over more extended periods. Again, a statistically significant association between lead exposure and handicap does not necessarily imply a causal relationship. Children may carry an increased lead burden because of pica (craving for non-food substances such as soil or paint) and this disorder is common in handicapped or disturbed children. Often, in an environment in which there is significant pollution with lead other factors operate which may adversely affect growing children and this may make interpretation of the results more difficult.

After a thorough appraisal of the evidence Rutter[60] concluded that there is impressive but not conclusive evidence for a damaging effect of lead levels hitherto considered harmless.

Mercury

For centuries mercury has been known as an occupational hazard, as a medicine and as a poison. Recently, organic mercury compounds have produced human tragedies on a large scale (p. 351). The oral toxicity of mercury depends on the chemical form of entry. Absorption and toxicity increase in the following order: elemental mercury, mercurous mercury, mercuric mercury and alkylmercury. While the kidney is the organ which is most vulnerable to the toxic effects of mercury, alkylmercurials share with elemental mercury the ability to penetrate biological barriers such as the blood brain and placental barriers. Alkylmercury poisoning gives rise to severe neurological disablement in those affected. Cerebral palsy and mental retardation have been described in infants exposed to these toxic compounds *in utero*.

Copper

Wilson's disease (hepatolenticular degeneration)[52,66] is a rare autosomal recessive neurodegenerative disorder (p. 359). Patients may present with either hepatic or neurological problems.

The diagnosis of Wilson's disease is not difficult once symptoms, either neurological or hepatic, have become established, particularly when Kayer–Fleischer rings are present. Biochemical findings are then usually unequivocal. Problems arise in early hepatic cases when the only abnormality may be a moderately raised ALT, and in differentiating presymptomatic cases from heterozygotes. Measurement of caeruloplasmin and serum copper is often inconclusive. Not only are low levels found in a number of unrelated conditions, but also in up to a quarter of heterozygotes who are about one hundred times more numerous in the general population than homozygotes. Distribution of radiocopper among serum proteins after an oral dose and urinary excretion of radiocopper following an intravenous dose have been recommended as more sensitive tests, but a liver biopsy may be necessary for a definite diagnosis. If started early, treatment with chelating agents is effective.

Menkes' syndrome[52] (kinky hair disease) an X-linked neurodegenerative disorder caused by a defect in copper transport is described on p. 359. Milder cases of the syndrome have been described in which serum copper and caeruloplasmin were near normal. The characteristic hair abnormalities and mental retardation were the signs which pointed to the diagnosis.

Other Metals

Iron deficiency is suspected of causing impaired cognitive and behavioural functioning presumably by decreasing the activity of iron-containing enzymes in brain. In young rats, iron deficiency leads to an elevation of serotonin and 5-hydroxyindoleacetic acid in brain, which is readily reversed by iron injections.

Severe mental retardation as a result of a defect in molybdenum metabolism has been described by Duran *et al.*[26] In man, xanthine oxidase and sulphite oxidase are probably the only two enzymes requiring molybdenum as a cofactor. The clinical and biochemical findings in the patient who was followed over the first two years of her life, were those seen in sulphite oxidase deficiency,[8] and xanthinuria.[8] They included dislocation of the lenses, fits and neurological degeneration, and xanthine stones. Oral ammonium molybdate did not benefit the patient. As the serum molybdenum level was normal the authors surmised that the disorder may involve defective transport of molybdenum into the cell, or disturbance of the incorporation of molybdenum into the enzyme systems.

ETHICAL CONSIDERATION—GENETIC COUNSELLING

Problems and Dilemmas

Not unexpectedly, recent advances in diagnosis, treatment and prevention of both genetically and environmentally determined disorders with mental retardation have given rise to many ethical problems reflected in a voluminous literature. Here we shall confine ourselves to a few points of particular relevance to clinical chemistry.

Some apparent ethical dilemmas are in fact spurious, based on the misunderstanding of the epidemiology of rare recessive disorders. For example, almost everyone involved in treating phenylketonurics is sooner or later reproached for the increased genetic burden created when patients grow up and having been treated are well enough to marry and pass on the harmful gene. In fact, calculations based on actual gene frequencies show clearly that the effect on future generations will be small and that equilibrium will not be reached for many generations.

Reservations must also be raised about investigations of cost-effectiveness of mass-screening and other preventive measures. Some studies which have shown, for example, the economic benefits of screening for phenylketonuria have been valuable.[5] However, the emphasis on cost-savings must not be carried too far. Prevention of mental handicap is worthwhile on ethical grounds, even if not cost-effective. However, resources are finite and there is a very real dilemma in deciding rival claims of measures to prevent mental handicap and other equally worthy causes. The value judgements, subjective feelings, aspirations and prejudices of the public, of the caring professions, and of the legislators determine how this dilemma is faced.

Genetic Counselling

Counselling the parents of a mentally retarded child requires considerable expertise and judgement. Of primary concern to the parents are the limits set by the handicap to their child's intellectual development, the possibilities of treatment and the risk of recurrence in a subsequent child. Some hold that the counsellor should inform rather than advise. In practice, the distinction between information and advice is often blurred, as the decision of the family may be influenced by unconscious bias in the way the information is presented.

There is little doubt that even in cases with a well-defined biochemical aetiology and mode of inheritance genetic counselling should at least initially be the responsibility of the consultant in charge of the case, although he may call in specialist advice in complicated cases. His first task will often be to try to allay the parents' feeling of guilt and self-reproach, which are only too common in their circumstances. He will then go on to explain the nature of the disorder, its prognosis, the scope for treatment and the risk of recurrence.

In dominantly and recessively inherited disorders the recurrence risk is clear cut, 1 in 2 and 1 in 4 respectively, in X-linked recessive conditions the risk of being affected is 1 in 2 for sons, the risk of being a carrier 1 in 2 for daughters. However, in X-linked disorders and even more so in dominantly inherited diseases the mutation rate may be significant. For example, in tuberous sclerosis, a disorder with epilepsy and characteristic skin lesions, which is dominantly transmitted a substantial majority of patients are new mutations, neither parent is affected and there is a negligible recurrence risk. For patients with mental retardation of unknown aetiology ('unclassified mental retardation') it is possible to give the parents an empirical estimate of recurrence which is approximately 3%.[2] The parents may be so confused, upset or angry that they cannot accept or even understand what they are told and several well-spaced counselling sessions may be necessary. Some parents later claim that they are 'kept in the dark' even though the records clearly show that this is not so. Others see specialist after specialist in the hope that an unacceptable diagnosis may be overturned.

The handicapped child and his parents may have frequent contact with the laboratory both for diagnostic tests and for the monitoring of treatment. Collecting specimens from mentally retarded children can be both time consuming and physically and mentally exhausting for staff and parents who are often acutely embarrassed. Some parents may be overawed by senior physicians and find it easier to talk to laboratory staff. A friendly response and conversation on general topics can have a beneficial effect on the parents' relation to the hospital. However, laboratory staff should not discuss diagnosis, prognosis, recurrence risks and management of the patient except when a patient is formally referred for this purpose. Even if they are well informed and what they say is fully in line with the views of the consultant in charge of the case they may be misunderstood and their views quoted out of context or even incorrectly, and therapeutic relationships may be undermined. For more detailed discussion of genetic counselling the reader is referred to Davison and Oakes,[21] Cowie[13] and Galjaard.[32]

Other Medico–ethical Issues

In all investigations of severely retarded patients in research and in clinical trials of new treatment and of drugs it must always be borne in mind that mentally retarded patients are unable to give informed consent, and that parental consent by itself is also not necessarily sufficient if it can be claimed, for example, that agreement was given to gain benefits for the patient. It is therefore, essential to submit full details of any project for approval to the ethical committee of the patients' hospital and to apply for clinical trials certificates to the DHSS where appropriate.

Some research with mentally retarded patients, for example, development of new methods of behaviour therapy or rehabilitation may alleviate the condition of those handicapped here and now but will not, in general, help eliminate the cause or facilitate prevention. Long-term research on aetiology and metabolic disease may benefit future generations but usually does nothing for those patients taking part in the research project. A balance has to be struck: we must neither prejudice the well being of those not yet born by refraining from any research at all on patients who are not likely to benefit personally nor, on the other hand, must we ask handicapped patients to accept unreasonable discomfort or risks for the sake of others when there is no commensurate benefit to themselves.

Laboratory investigations may set a limit to the expectation of life or intellectual potential of a child or fetus and thus critically affect decisions about termination of pregnancy, sterilisation or recommendation for institutional care. The clinical biochemist cannot, therefore, be indifferent to the complex ethical issues involved. Except on religious grounds, the case for aborting a fetus with a fatal storage disorder or autosomal chromosomal abnormality is unanswerable. However, in some hereditary disorders and in some sex chromosome abnormalities those affected may be physically well and only mildly mentally retarded. The decision is then much less clear cut. Can the right of a fetus to live be decided by a figure on an intelligence scale which he may fail to reach? Exposure to a teratogen may carry a 1:10 chance of significant damage to the nervous system. Are we to recommend termination and lose 9 normal fetuses to prevent the birth of one affected one? Clearly, the mother has rights in a situation such as this, but so has, or should have the fetus. In some situations the rights of unaffected siblings must be considered whose prospects may be blighted if parents lavish so much love and attention on the handicapped child that they neglect the rest of the family. These issues have been inconclusively debated at great length by physicians, philosophers and theologians. All one can perhaps say is that flexibility of mind and a sense of humility are desirable qualities in those giving advice, professional arrogance and excessive zeal on the part of the counsellor are often obstacles to sound and humane decisions. Rigid rules, particularly if legally enforced, governing such matters as antenatal detection, termination of pregnancy and sterilisation are liable to cause unnecessary hardship and distress in some of those affected by the decision no matter how well intentioned those formulating the rules. The best safeguard for patients and public is the provision of adequate resources for the handicapped by the community, and access to professional workers with competence, compassion and commonsense.

REFERENCES

1. Alleyne, G. A. O., Hay, R. W., Picou, D. I., Stanfield, J. P. and Whitehead, R. G. (1977). *Protein–Energy Malnutrition*. London: Edward Arnold.
2. Angeli, E. and Kirman, B. H. (1971). Genetic counselling of the family of the mentally retarded child', in Proceedings of the 2nd Congress of the International Association for the Scientific Study of Mental Deficiency (Primrose, D. E. E., ed.), 692–697. Warsaw: Polish Medical Publishers.
3. Balazs, R., Lewis, P. D. and Patel, A. J. (1979). 'Nutritional deficiencies and brain development', in *Human Growth*, Vol. 3 (Falkner, F. and Tanner, J. M., eds), 415–480. London: Baillière Tindall.
4. Bickel, H. (1980). 'Phenylketonuria: past, present, future' (F. P. Hudson Memorial Lecture, Leeds, 1979), *J. Inher. Metabol. Dis.*, **3**, 123–132.
5. Bickel, H. Guthrie, R. and Hammarsen, G. (1980). *Neonatal Screening for Inborn Errors of Metabolism*. Berlin: Springer-Verlag.
6. Blass, J. P. (1979). 'Disorders of pyruvate metabolism', *Neurology*, **29**, 280–286.
7. Blau, K. (1979). 'Phenylalanine hydroxylase deficiency: biochemical physiological and clinical aspects of phenylketonuria and related phenylalaninaemias', in Aromatic Amino Acid Hydroxylases and Mental Diseases (Youdim, M. B. H., ed.), 77–139. Chichester: John Wiley.
8. Bondy, P. K. and Rosenberg, L. E. (1980). *Metabolic Control and Disease*, 8th ed. Philadelphia: Saunders.

8a. Bremer, H. J., Duran, M., Kamerling, J. P., Przyrembel, H. and Wadman, S. K. (1981). *Disturbances of Amino Acid Metabolism: Clinical Chemistry and Diagnosis*. Munich: Urban and Schwarzenberg.
9. Burman, D., Holton, J. B. and Pennock, C. A. (1980). *Inherited Disorders of Carbohydrate Metabolism*. Lancaster: MTP Press.
10. Chalmers, R. A. and Lawson, A. M. (1982). *Organic Acids in Man*, London: Chapman and Hall.
11. Clayton, B. E. (1974). 'Population screening', in *Molecular Variants in Disease* (Raine, D. N., ed.), *Journal of Clinical Pathology*, 145–149.
12. Committee on Drugs (1978). 'Treatment of congenital hypothyroidism', *Pediatrics*, **62**, 413–417.
13. Cowie, V. A. (1977). 'Genetic counselling clinics', in Medico-Social Management of Inherited Metabolic Disease (Raine, D. N., ed.), 103–117. Lancaster: MTP Press.
14. Craft, M. (1979). *Tredgold's Mental Retardation*, 12th ed. London: Baillière Tindall.
15. Crawfurd, M. d'A., Gibbs, D. A. and Watts, R. W. E. (Eds) (1981). *Advances in the Treatment of Inborn Errors of Metabolism*. Chichester: John Wiley.
16. Crocker, J. F. S. (1979). *Reye's Syndrome II*. New York: Grune and Stratton.
17. Crome, L. and Stern, J. (1972). *The Pathology of Mental Retardation*, 2nd ed. Edinburgh: Churchill Livingstone.
18. Crome, L. and Stern, J. (1976). 'Inborn lysosomal enzyme deficiencies', in Greenfield's Neuropathology (Blackwood, W. and Corsellis, J. A. N. eds), 500–580. London: Edward Arnold.
19. Danks, D. M. (1974). 'Management of newborn babies in whom serious metabolic illness is anticipated', *Arch. Dis. Child*, **49**, 576–578.
20. Davison, A. N. (1977). 'The biochemistry of brain development and mental retardation', *Br. J. Psychiatry*, **131**, 565–574.
21. Davison, A. N. and Thompson, R. H. S. (1981). *The Molecular Basis of Neuropathology*. London: Edward Arnold.
22. Davison, B. C. C. and Oakes, M. W. (1976). *Genetic Counselling*, 2nd ed. London: Heinemann.
23. Dickerson, J. W. T. and McGurk, H. (1982). *Brain and Behaviour Development*. Glasgow: Blackie (Surrey University Press).
24. Dobbing, J. (1974). Later development of the brain and its vulnerability, in *Scientific Foundations of Pediatrics*, 565–577 (Davis, J. A. and Dobbing, J., eds). London: Heinemann.
25. Dubowitz, V. (1978). *Muscle Disorders in Childhood*. Philadelphia: Saunders.
26. Duran, M., Beemer, F. A., Heiden, C. v. d., Korteland, J., de Bree, P. K., Brink, M., Wadman, S. K. and Lombeck, I. (1978). 'Combined deficiency of xanthine oxidase and sulphite oxidase: A defect of molybdenum metabolism or transport', *J. Inher. Metabol. Dis.*, **1**, 175–178.
27. Duran, M. and Wadman, S. K. (1981). 'Organic acidurias', in Recent Advances in Clinical Chemistry' (Alberti, K. G. M. M. and Price, C. P., eds), number 2, pp. 103–127. Edinburgh: Churchill Livingstone.
28. Eisenberg, L. (1977). 'Development as a unifying concept in psychiatry', *Br. J. Psychiatry*, **131**, 225–237.
29. Eriksson, M. and Zetterstrom, R. (1979). 'Neonatal convulsions', *Acta. Paediatr. Scand.*, **68**, 807–811.
30. Farquhar, J. W. (1976). 'The infant of the diabetic mother'. *Clin. Endocrinol. Metab.*, **5**, 237–264.
31. Francis, D. E. M. (1982). *Diets for Sick Children*, 4th ed. Oxford: Blackwell.
32. Galjaard, H. (1980). *Genetic Metabolic Diseases*. Amsterdam: Elsevier/North-Holland.
33. Gaull, G. E., Tallan, H. H., Lajtha, A. and Rassin, D. K. (1975). Pathogenesis of brain dysfunction in inborn errors of amino acid metabolism, in Biology of Brain Dysfunction, vol. 3 (Gaull, G. E., ed), 47–143. New York: Plenum Press.
34. Glew, R. H. and Peters, S. P. (1977). *Practical Enzymology of the Sphingolipidoses*. New York: Alan Liss.
35. Goldstein, G. W. (1979). 'Pathogenesis of brain edema and haemorrhage: role of the brain capillary', *Pediatrics*, **64**, 357–360.
36. Güttler, F., Seakins, J. W. T. and Harkness, R. A. (1979). 'Inborn Errors of Immunity and Phagocytosis. Lancaster: MTP Press.
37. Hagberg, B. Kyllerman, M. and Steen, G. (1979). 'Dyskinesia and dystonia in neurometabolic disorders', *Neuropädiatrie*, **10**, 305–320.
38. Hanefeld, F. and König, E. (1974). 'Causes of mental retardation in children. Analysis of 414 cases', *Monatsschr. Kinderheilkd*, **122**, 679–680.
39. Harkness, R. A. Cockburn, F. (1977). *The Cultured Cell and Inherited Metabolic Disease*. Lancaster: MTP Press.
40. Hetzel, B. S. and Smith, R. M. (1981). *Fetal Brain Disorders*. Amsterdam: Elsevier/North-Holland.
41. Hill, A., Casey, R. and Zaleski, W. A. (1976). 'Difficulties and pitfalls in the interpretation of screening tests for the detection of inborn errors of metabolism', *Clin. Chim. Acta.*, **72**, 1–15.
42. Israels, S., Haworth, J. C., Dunn, H. G. and Applegarth, D. A. (1976). 'Lactic acidosis in childhood', *Adv. Paediatr.*, **22**, 267–303.
43. Karp, W. B. (1979). 'Biochemical alterations in neonatal bilirubinaemia and bilirubin encephalopathy: a review', *Pediatrics*, **64**, 361–368.
44. Kirman, B. H. and Bicknell, J. (1975). *Mental Handicap*. Edinburgh: Churchill Livingstone.
45. Koch, R. (ed.) (1981). 'Urea Cycle Symposium', *Pediatrics*, **68**, 271–297 and 446–459.
46. Lenke, R. R. and Levy, H. L. (1980). 'Maternal phenylketonuria and hyperphenylalaninaemia', *N. Engl. J. Med.*, **303**, 1202–1208.
47. McDermott, A. (1980). Prenatal Diagnosis Group Newsletter, PDG Laboratories/Services Information Supplement, 3rd ed., S. W. Regional Cytogenetic Centre, Southmead Hospital, Bristol.
48. MacFaul, R., Dorner, S., Brett, E. M. and Grant, D. B. (1978). 'Neurological abnormalities in patients treated for hypothyroidism from early life', *Arch. Dis. Child.*, **53**, 611–619.
49. McKusick, V. A. (1978). *Mendelian Inheritance in Man*, 5th ed. Baltimore: Johns Hopkins University Press.
50. Marks, V. and Rose, F. C. (1981). Hypoglycaemia, 2nd ed. Oxford: Blackwell.
51. Mathis, R. K., Andres, J. M. and Walker, W. A. (1977). 'Liver disease in infants, part II: hepatic disease states'. *J. Pediatr.*, **90**, 864–880.
52. Menkes, J. H. (1980). *Textbook of Child Neurology*. Philadelphia: Lea and Febiger.
53. Moser, H. W. (1981). 'Recent advances in certain disorders of lipid metabolism', in *Frontiers of Knowledge in Mental Retardation*, Vol. 2 (Mittler, P. and de Jong, J. M., eds), pp. 239–149. Baltimore: University Park Press.
54. Nyhan, W. L., Johnson, H. G., Kaufman, I. A. and Jones, K. L. (1980). 'Serotonergic approaches to the modification of behaviour in the Lesch–Nyhan syndrome', in Applied Research in Mental Retardation, vol. 1, 25–40. New York: Pergamon Press.
55. Palmer, T. and Oberholzer, V. G. (1977). 'Diagnosis of urea cycle disorders', *Ann. Clin. Biochem.*, **14**, 136–138.
56. Pennock, C. A. (1976). 'A review and selection of simple laboratory methods used for the study of glycosaminoglycan excretion and the diagnosis of the mucopolysaccharidoses', *J. Clin. Pathol.*, **29**, 11–123.
57. Popkin, J. S., Clow, C. L., Scriver, C. R. and Grove, J. (1974). 'Is hereditary histidinaemia harmful?' *Lancet*, **i**, 721–722.
58. Rundle, A. T. (1979). 'The use and abuse of the laboratory', in *Tredgold's Mental Retardation*, 12th ed. (Craft, M., ed.). London: Baillière Tindall.
59. Rush, D., Stein, Z. and Susser, M. (1980). 'A randomised controlled trial of prenatal nutritional supplementation in New York City', *Pediatrics*, **65**, 683–697.
60. Rutter, M. (1980). 'Raised lead levels and impaired cognitive/behavioural functioning: a review of the evidence', *Dev. Med. Child Neurol.*, **22** suppl. 1, 1–26.
61. Scriver, C. R., Laberge, C., Clow, C. L. and Fraser, F. C. (1978). 'Genetics and Medicine: an evolving relationship', *Science*, **200**, 946–951.
62. Smith, D. W. (1976). *Recognizable Patterns of Human Malformation*, 2nd ed. Philadelphia: Saunders.

63. Smith, I. and Seakins, J. W. T. (1976). *Chromatographic and Electrophoretic Techniques*, 2 vols, London: Heinemann.
64. Smithells, R. W. (1978). 'Drugs, infections and congenital abnormalities', *Arch. Dis. Child*, **53**, 93–99.
65. Smithells, R. W., Sheppard, S., Schorah, C. J., Seller, M. J., Nevin, N. C., Harris, R., Read, A. P. and Fielding, D. W. (1980). 'Possible prevention of neural-tube defects by periconceptual vitamin supplementation', *Lancet*, **i**, 338–340.
66. Stanbury, J. B., Wyngaarden, J. B. and Fredrickson, D. S. (1978). *Metabolic Basis of Inherited Disease*, 4th ed. New York: McGraw-Hill.
67. Sullivan, F. M. (1979). 'The teratogenic and other toxic effects of drugs in reproduction', in Iatrogenic Disease (D'Arcy, P. F. and Griffin, J. P., eds), 436–458. London: Oxford University Press.
68. UK Collaborative Study (1979). 'Amniotic-fluid alpha-fetoprotein measurement in antenatal diagnosis of anencephaly and open spina bifida in early pregnancy', *Lancet*, **ii**, 651–662.
69. Ullman, M. D. and McCluer, R. H. (1978). 'Quantitative analysis of plasma neutral sphingolipids by high performance liquid chromatography of their perbenzoyl derivatives', *J. Lipid Res.*, **18**, 371–378.
70. Vincken, P. J. and Bruyn, G. W. (1976/7). *Metabolic and Deficiency Diseases of the Nervous System, Handbook of Clinical Neurology*, Vol. 28. Amsterdam: North-Holland.
71. Volpe, J. J. (1981). *Neurology of the Newborn*. Philadelphia: Saunders.
72. Watts, R. W. E., Baraitser, M., Chalmers, R. A. and Purkiss, P. (1980). 'Organic acidurias and aminoacidurias, in the aetiology of long term handicap', *J. Ment. Defic. Res.*, **24**, 257–270.
73. Wilson, J. M. G. and Jungner, G. (1968). *Principals and Practice of Screening for Disease*. Geneva: World Health Organization.
74. Winick, M. (1976). *Malnutrition and Brain Development*. London: Oxford University Press.
75. Wolter, R., Noel, P., DeLock, P. *et al.* (1979). 'Neurophysiological study in treated thyroid dysgenesis', *Acta Paediatr. Scand., Supplement* 277, 41–46.
76. Wright, E. C., Stern, J., Ersser, R. and Patrick, A. D. (1979). 'Glutathionuria: γ-glutamyl transpeptidase deficiency', *J. Inher. Metabol. Dis.*, **2**, 3–7.

33. THE ROLE OF THE LABORATORY IN SOME PSYCHIATRIC ILLNESSES

DAVID M. SHAW

Introduction
  Cerebral dysfunction and some forms of mental illness
  Physical substrate of emotion

Disorders of mood
  Mania
  Depression
  Some hypotheses for affective disorders
  Differential diagnosis
  Maintenance therapy

Schizophrenia
  Differential diagnosis

Organic psychiatric reactions
  Diagnosis of acute organic reaction

Role of the laboratory
  Depression
  Schizophrenia
  Organic reactions

Huntington's chorea

Anxiety states
  Anxiety
  Anxiety and anxiety states
  Assessment

Alcoholism
  Assessment

Laboratory tests in elderly psychiatric patients

Impact of somatic illness on mental state: the endocrine system

Conclusion

INTRODUCTION

In general, in medical conditions, we observe the effects of disease processes on organs and tissues, and the ways in which these various parts of the body influence each other under the conditions of the disease process. Given the large numbers of pathological processes and of different organs and tissues, this makes for a very large number of permutations, so that syndromes of illness can be identified by their own blueprints.

It is true that the brain is the most complex organ in the body, but it is a single organ, and despite the intricacies of its structure, its ways of reacting to organic insult are surprisingly limited—a number of almost stereotyped responses. For instance, a patient suffering from thyrotoxicosis may manifest one of only a few psychiatric symptoms; the patient may become anxious or have 'neurotic' behaviour, develop a depressive illness, or in 'crises' may have an acute brain syndrome. Some of these same few psychological reactions could also be the result of Cushing's syndrome, vitamin deficiencies, etc.

If, therefore, the reactions of the brain to general

metabolic disturbances are relatively non-specific, the history, signs and symptoms, laboratory and other investigations have to be employed together, and are equal in their importance in the diagnosis of somatic illness manifesting itself initially as a psychological disturbance, or of psychiatric illness triggered by organic pathology.

Cerebral Dysfunction and Some Forms of Mental Illness

Psychiatrists are never tired of reiterating the necessity of treating the patient as whole, and as a person with problems rather than 'a case' or 'a problem'. This is their ideal, and includes seeking and treating appropriately those patients in whom the underlying pathology is an organic disorder of the CNS, a somatic illness with secondary affects on the brain, or a biochemical/physiological dysfunction of the brain. At the opposite end of the spectrum are individuals whose main problem, for instance, may be experiential during crucial periods of development, absence of adequate parenting, or other circumstances which have produced a person with maladaptive behaviour and emotional difficulties. In many patients, what is wrong *today* cannot be attributed to single present factors or to neat diagnoses.

Several psychiatric disorders have been selected in this chapter which are primarily either biochemical or organic disorders. In some instances our knowledge of their causes is tentative, in others the underlying processes are understood at least in outline.

Physical Substrate of Emotions

Our knowledge of the physical substrate of emotions is rudimentary. However, until proved otherwise it is assumed that the 'origin' of emotions is likely to be partly in the neocortex, in the perception both of those occurrences in our external and internal environments which trigger these experiences, and of the experience of the emotion evoked. Probably, therefore, the initial efferent and final afferent pathways come from and go to the higher centres.

From thereon our knowledge is less well established. We know however from studies of evolution, that what began as the 'smell brain' in primitive animals developed *pari passu* with the expanding neocortex to form a large complex of brain tissue called 'the limbic system'. This limbic system, which contains the amygdaloid nucleus, hippocampus, limbic cortex, parahippocampal gyrus, fornix, cingulate gyrus, mammillary bodies and which, functionally, should include the hypothalamus, is engaged in the integration of both sensory material and emotional activity. Part of the output from the limbic system, concerned with emotional activity, may be relayed back to the cortex by aminergic neurones, in particular those ascending neuronal pathways with cell bodies in the brain stem and hypothalamus which use noradrenaline, 5 hydroxytryptamine [5HT] or dopamine as transmitter substances; the former two are distributed fairly widely, the latter is more localised to an area in the frontal region.

DISORDERS OF MOOD

Our mood stages have two main facets
1. the level at which it tends to be—its natural set point—so that some people are near the 'average', or tend to be 'above or below the line';
2. mood stability; we all vary in our degrees of stability or instability of mood.

It is beyond the scope of this chapter to discuss the normal variations of mood, except very briefly. Mood is considered to be abnormal when the magnitude of its movement away from 'neutral' is such as to interfere with everyday life and normal functioning, a degree of change which is far commoner than was once generally appreciated.

Mania

The less common disorder of mood is a shift into elation. This may follow various pathological conditions or drugs, but the commonest cause is bipolar affective illness (see below).

The manic individual is happy, elated or even ecstatic. Energy, drive, libido are increased, the patient is 'busy', finds little time to sleep, and may be irritable, demanding, interfering, full of grandiose ideas, and spend money recklessly. Sometimes the person makes unrealistic (and often expensive) plans and schemes, and in their energetic enthusiasm, take others along with them. If more severe, the patient may have religiose delusions or delusions of grandeur. Speech is rapid, thoughts flit from one association to another, the patient is full of puns and makes what are known as 'clang associations'—joining up words on the basis of their similar sounds. Some patients become aggressive.

Depression

As stated above, depression is a normal experience of life which becomes pathological only when its severity and duration are such as to begin to interfere with the patient's life. Its classification is one of considerable and continuing controversy, but as a tentative scheme the following is useful for the purposes of general discussion:

Normal Mood

'Normal' mood has two characteristics, its general set point, characteristic of each individual as determined by constitutional/experiential factors, and its stability. With regard to the latter, some people's mood changes little from hour-to-hour or day-to-day, others with 'mercurial' temperaments alter 'like April showers'. Others change over a longer time, but not to a pathological degree, the so-called cyclothymic individuals. In everybody, elation and depression tend to follow gains and losses respectively (using losses and gains in the widest possible sense). We react with mood change to what is happening to us in our environment and within ourselves.

Depression is common in the puerperium, at the

menopause and in premenstrual tension. Although the symptoms of the latter would be considered pathological on the grounds of severity, it is so common that some authorities do not include it with the abnormal mood states.

Abnormal Mood

Reactive Depression. When reaction to occurrences in the environment is considered disproportionate or over-prolonged, perhaps say following bereavement, or in a reaction to the other 'slings and arrows of outrageous fortune', the person is considered to have a reactive depression. Its origins have not been worked out in detail, but the loss of parents at a young age, up to late adolescence, is one contributing, or perhaps sensitising factor.

Neurotic Depression. Patients in this condition are in a fluctuating, but almost life-long state of reduced mood usually associated with anxiety. It is beyond the scope of this chapter to more than touch on possible causes, but it might be considered as a state of internal psychological conflict, based on experiences in early life, which leads to maladaptive attitudes and behaviour patterns. Traumatic 'life events' may provoke exacerbations of illness of this and other types of depression.

Secondary Depression. By this is meant depression secondary to physical or psychiatric illness:

1. *Somatic illness:*
   (a) Reactive depression to physical illness and disability is extremely common and can be seen in patients on any hospital ward.
   (b) A number of physical disorders cause depression as a manifestation of the illness itself, and clearly, it is this group of illnesses which have to be sought assiduously in the differential diagnosis of depressive illnesses. The means by which these illnesses provoke depressive mood change (or occasionally elation) is not known.
2. *Psychiatric illness.* Depression as a secondary manifestation of other psychiatric illnesses is almost universal. These illnesses will not be considered in detail further here, and are a group of mood changes which have been relatively poorly researched.

Manic-depressive Psychosis (the affective disorders). These are a group of illnesses which have at least two forms—endogenous or recurrrent depression (now usually known as unipolar affective disorder) and manic-depressive illness (now referred to usually as bipolar affective disorder). Patients suffering from the former have attacks of depression only, the latter have phases both of depression and of mania.

It is not known to what extent subgroups exist in unipolar and bipolar disorder, but these conditions are unlikely to be homogeneous.

On the grounds that these illnesses are inherited, that unipolar and bipolar families breed true, the fact that illnesses can be triggered by physical agents (drugs, physical illness, treated by physical methods (drugs and ECT) and attenuated, or prevented by drug treatments (e.g. by lithium salts), they are considered to be disorders of biochemical origin. In other words, the hypothesis is that, whatever is the primary mechanism, it leads to biochemical changes in the brain which, if reversed, lead to disappearance of symptoms.

Affective disorders used to be thought of just as phasic alterations in mood, with periods of normality in between, and other symptoms secondary to the disorder of mood. This is now considered to be an inadequate description of the two groups of disorder.

An alternative description of affective disorders is that they are phasic alterations in vitality, showing themselves psychologically in the cognitive and affective spheres, somatically by changes in drives for food, sex and work, and in both areas by changes in rhythms. In depressive phases cognitive changes include inefficient and often slowed thinking, loss of mental energy, indecision, poor memory and loss of interest. Affective changes are depression and anxiety in the 'down' phase, euphoria and elation in the manic state. Somatic features of phases of depression include loss of interest in food, loss of weight, decreased production of saliva, constipation, loss of libido, inactivity, and changes in body temperature, in endocrine patterns (e.g. cortisol) and in sleep; and there are long term differences in 5HT content of platelets.

This is by no means an exhaustive list.

Some Hypotheses for Affective Disorders

Early observations recorded that mood was elevated by amphetamines which also release noradrenaline and dopamine from noradrenergic and dopaminergic neurones. Depleting these amines with reserpine (at high doses over many months) may lead to the development of severe depression (probably in predisposed subjects). All of the 1st generation antidepressants seemed to enhance the levels of either noradrenaline or of 5-hydroxytryptamine at aminergic synapses. The obvious conclusion was the so-called *amine hypothesis* that the depression or mania were just simple deficiencies or excesses respectively of one or other of these amines. The main contenders were noradrenaline and 5HT.

The view is now thought to be oversimplified. The pharmacological modifications produced by the early antidepressants included increased levels of amine transmitters at the synapses. These changes occurred with great rapidity while, in complete contrast, recovery from depression took from about 4 to 8 weeks to be complete. There was no simple relationship between modifying synaptic amine levels and recovery. This was further emphasised when it was found that acute depletion of all amines, of 5HT alone, or of noradrenaline alone from the brain did not mimic the depressive syndrome in normal controls, nor did potentiation of these amines *in the short term* reverse the depressive syndrome in patients with this disorder. Again, a simple relationship between amine levels and mood states was not a valid hypothesis.

At the same time, it is likely that antidepressant drugs act on aminergic neuronal systems for their therapeutic actions, even if the process *is* slow. Is the primary abnormality of affective illness in these neurones, in a neuronal system in parallel or series with the aminergic cells, or even at some distance away in the nervous system? As yet we do not know.

A number of abnormalities have been identified in depression. For instance, many patients produce excess cortisol, sometimes into the Cushingoid range. There are more and larger than normal secretory bursts, which, instead of being limited to early morning to early afternoon, extend over the whole 24 hours.

In about 40% of patients these increments in production of cortisol cannot be suppressed by dexamethazone, and in the absence of severe medical illness, this is now considered to be diagnostic of 'biological' depression in so far as it is not seen in neurotic or other psychiatric illness. The depressive overproduction of cortisol is blocked by thymoxamine suggesting that it could be mediated at α-adrenoreceptors.

The hypopituitary-pituitary-endocrine gland axis produces most other hormones in normal amounts in depression, but its reactions to some stimuli are blunted in many patients. (For instance, the growth hormone response to hypoglycaemia, the TSH response to TRH.)

Clonidine or desmethylimipramine release growth hormone by an α-adrenoreceptor mediated pathway; this response is muted in many depressives, also suggesting decreased α-receptor activity.

The renin-aldosterone mechanism is abnormal in bipolar illness, with high renin levels, reduced response to posture, and inappropriate production of aldosterone.

Catecholamine metabolism has been examined from a number of aspects. Between 25–60% of the urinary excretion of 3-methoxy-4-hydroxyphenylglycol, [MHPG] a metabolite of noradrenaline, comes from the CNS. Urinary excretion of MHPG tends to be low in unipolar illness, but there is less agreement about urinary production in the bipolar variant. One group claims that low urinary MHPG excretion indicates patients who will respond to imipramine, and high MPHG excretion those who will benefit from amitriptyline.

Many, but not all, reports of the levels of homovillic acid, (HVA—a metabolite of dopamine) in CSF have indicated reduced levels (which may be related to retardation, rather than depression *per se*); accumulation of HVA in CSF tends to be less after probenecid treatment in depressives than in controls (probenecid blocks removal of HVA from the CSF, and serial measures of concentration are taken as a measure of turnover). Platelets may have decreased numbers of α-adrenoreceptor in depression (a finding which is 'echoed' by the reduced growth hormone response to clonidine or desmethylimipramine, see above).

Studies of tryptophan and indolamines have produced findings equally difficult to replicate in some cases.

Urinary tryptamine excretion has been found to be either raised or lowered in different studies of depression, but some kynurenine metabolites tend to be excreted in higher amounts, suggesting increased catabolism of tryptophan. Overall 5HT turnover judging by the urinary excretion of 5-hydroxy indole acetic acid (5HIAA), is normal, but the ratio of urinary 5HIAA: kynurenine is significantly lowered suggesting excess catabolism of tryptophan in relationship to indolamine synthesis. Levels of free and protein-bound tryptophan are normal in most studies but there have been exceptions to this. For instance, one investigation suggested that when psychologically stressed, unipolar depressives tended to have significantly lower plasma tryptophan levels than controls; the many discrepant findings in this area may just be a function of the degree of stress experienced by the patients *before* a test. If it can be shown that unipolar depressives have unusually labile levels of tryptophan which are sensitive to stress, this could prove to be of some interest in the aetiology of this condition.

Tryptophan tolerance tests have suggested that entry of this amino acid to the tissues is low, but when given in excess the amino acid crossed the blood-brain barrier in adequate amounts. However, limited investigations, which need extending, have indicated that there may be concentrations of other amino acids in plasma in these conditions which could compete with tryptophan for transport across the blood-brain barrier.

5HIAA levels in blood, especially after probenecid (which blocks loss of the compound from CSF), tend to be low in some studies, but it is likely that the levels are bimodal. In patients with low 5HIAA in CSF, severity of illness tended to be negatively correlated with concentration.

Finally, post mortem studies, dogged as they are with technical problems, have tended to show reduced levels of brain 5HT and 5HIAA (but with discordant results as well).

Electrolyte and water distribution are altered in both mania and depression, and similarly, the activity of ATPase in red cell membranes is reduced in both conditions.

Other hypotheses have considered changes in energy/transport mechanisms, abnormality of cell permeability, etc. Of the various biochemical changes observed none has proved to be totally specific.

Thus, no test provides an adequate biological marker, and certainly CSF and urinary studies carried out so far have not provided this. There are some indications, however, that perhaps a combination of hypothalamo-pituitary stimulatory tests may be useful in the future, but there is no single distinguishing biochemical investigation which can achieve say 90–95% diagnosis of affective illness.

Differential Diagnosis

Some illnesses can precipitate depression and since they precede the illness, do not come into the diagnostic procedure and the discussion here, except for the purposes of completeness. For instance, operations, virus infections (e.g. influenza), pneumonia, debilitating medical illnesses and also psychological trauma may be followed by depression.

The main somatic precipitants of depression, and of conditions imitating or presenting initially as depression are given in Tables 1 and 2, and are divided rather arbitrarily into the more common and less common factors; a smaller list giving precipitating conditions resembling mania or causing mania are in Table 3.

TABLE 1
THE MORE COMMON SOMATIC CAUSES OF DEPRESSION*

Degenerative	Presenile or senile dementia
	Parkinson's disease
Intracerebral space occupying lesions	Cerebral tumour: Subdural haemorrhage (in fact any intracerebral Space occupying lesion)
Infective	Herpes Zoster: Various bacterial infections e.g. subacute bacterial Endocarditis, glandular fever: Infective hepatitis
Vascular	Cerebrovascular accident
Cardiac	Multi-infarct dementia: Cardiac failure: Coronary thrombosis
Epileptic	There may be a general relationship between epilepsy (esp. temporal lobe) and affective illness: Postictal depression (?)
Metabolic	Deficiency K^+ or Na^+ from any cause: Changes in Ca^{++} levels (see below)
Endocrine	Thyrotoxicosis or hypothyroidism: Menopause: Purperium: Premenstrual tension syndrome. Hypoparathyroidism: Hyperparathyroidism.
Renal	Dialysis Chronic renal failure.
Neoplastic	Carcinoma any site but especially lung and pancreas. Intracerebral primary or secondary growths. Insulinoma.
Vitamin lack	Folate: B_{12}: Thiamine.
Drugs	Reserpine: Antithyroid drugs: Digitalis: Some depot neuroleptics; LSD: anticholinergic drugs. Corticosteroids and ACTH: Possibly oral contraceptives: L-dopa; Physostigmine. Withdrawal of drugs of abuse.

* Excluding conditions in which depression may be obvious sequelae—influenza, typhoid, typhus, major surgery, debilitating medical illness of any kind.

Laboratory Tests Relevant to Affective Illness

The first stage of screening consists of looking for indications for some of the conditions listed in Table 1, and one or two from Table 2. For most depressive patients referred to hospital the initial procedures in-

TABLE 2
THE LESS COMMON SOMATIC CAUSES OF DEPRESSION

CNS	Multiple sclerosis
	Chorea in rheumatic fever
	Low pressure hydrocephaly
Infective	Miliary TB
	Brucellosis
	GPI
Vascular	Temporal arteritis
Metabolic	Porphyria
Endocrine	Cushing's disease (central or peripheral origin)
	Addison's disease
Toxic	Lead
	Wilson's disease (copper)
	Some nerve gases and organo-phosphorous cholinesterase inhibitor insecticides
Anoxic	Sometimes in acclimitisation to high altitude
Nutritional and Vitamin lack	Starvation, any cause general B avitaminosis Pellagra (nicotinamide and other deficiencies) Beri-beri and Wernicke's encephalopathy
Drugs	Excess cortisol ACTH—(may give manic reaction also)

TABLE 3
SOMATIC CAUSES OF MANIA-LIKE CONDITIONS OR MANIA

CNS	Space-occupying lesions
Infective	GPI
Metabolic	Liver failure
	Late stage of pellagra
Endocrine	Rarely in myxoedema
	Hyperthyroidism
	ACTH
	Cortisol
Drugs	Amphetamine abuse
	Tricyclic and other antidepressants
	L-dopa
	Alcohol
Toxic	Tetraethyl lead
	LSD
	Cannabis
	Manganese
	Amphetamine abuse

clude assay of Hb, red cell count and smear, differential white cell count, ESR, plasma electrolytes (including calcium and phosphate), serological tests for syphilis, thyroid profile, fasting or random blood sugar, blood urea and urine tests for sugar, protein and casts. For some patients plasma proteins, serum bilirubin and minimal liver function tests will be done at this stage.

In addition, patients have X-rays of chest and skull, ECG, and in few an EEG.

The procedure beyond this point depends very much on clues from history, physical examination and the results from the initial laboratory tests.

A common situation is a systematic search for the site of a suspected cancer or source of infection, for the causes of any changes in plasma electrolyte levels, or further estimations of hormonal function. A substantial number of patients, particularly the elderly, merit tests for B_{12} or folate deficiency. Since many depressives indulge in bout drinking, it may be worth measuring transketolase activity if alcoholism is suspected.

In general, however, the initial screening procedures suffice for both depression and mania in the majority of patients. The most frequent findings in those presenting via the psychiatric hospital service are thyroid disorders, the presence of neoplasms of the lung, followed perhaps by occult infections (viral or bacterial), and folate/B_{12} deficiency. Only a minority come to psychiatric clinics with physical signs, skull X-rays or EEG findings suggesting undiscovered intracranial pathology, and these are usually referred on to medical or neurological units for further assessment at an early stage.

Maintenance Therapy for Affective Disorders

Many depressive or manic depressive patients are now receiving preventative treatments, usually with the salts of lithium.

Lithium salts have a small therapeutic gap, i.e. the therapeutic and toxic doses are not widely separated, and therefore it is essential to assay the plasma lithium level at regular intervals (it should be noted that the current trend is to use a lower dosage range than hitherto). Ideally, the laboratory procedure should be closely integrated with the lithium clinic so that assays of lithium taken 12 hours after the previous dose can be done 'on the spot'. This allows the patient's dose to be adjusted at the time of attendance to keep the plasma level within the desired range.

Patients on lithium regime require other laboratory checks at regular intervals. About 4% a year become hypothyroid, so that regular tests of thyroid functions—say once or twice yearly—should be done on all patients receiving lithium. Those who develop hypothyroidism may have replacement with the hormone, the level of which will need to be rechecked at intervals.

There is also the problem that lithium may be toxic to the kidney. It certainly seems to cause specific, usually reversible, lesions of the cells of the collecting tubules, but whether or not it causes more serious and irreversible renal damage of other types is still in doubt. Mostly it causes polyuria and failure in ability to concentrate urine, which may return to normal on stopping lithium. However, most lithium clinics examine urine for protein and casts, assay blood urea and do creatinine clearance tests at least once a year on patients maintained on lithium.

SCHIZOPHRENIA

Schizophrenia is an illness of unknown aetiology which has many forms:

1. *Simple:* characterised by lack of drive, initiative and sociability, but with few if any florid symptoms.
2. *Hebephrenia:* where there is disorder of thought processes, auditory hallucinations, and inappropriate affect. When the episodes are recurrent the condition leads to chronic schizophrenia—a defect state with marked deterioration of personality.
3. *Catatonic:* a type of schizophrenia which for unknown reasons has become extremely rare, at least in the Western world.
4. *Paranoid:* a form of the illness accompanied by persecutory or expansive delusions based on disturbing and rather frightening hallucinations. Such individuals are hostile, suspicious, fearful and aggressive. The onset of this form is relatively late in life, and the personality is well preserved compared with other forms of schizophrenic illness.

That the illness is primarily biological in orgin has come from the fact that the illness is inherited. This has been shown by family studies, investigation of monozygotic and dizygotic twins, and the observation that offspring of schizophrenic parents reared away from their biological parents have an increased expectation of illness.

The search for a biological (biochemical) basis for schizophrenia has been a graveyard of forlorn hopes for many years, although there have been encouraging advances recently. Hypotheses have ranged far and wide. In particular it was initially proposed that the disease was caused by endogenous hallucinogens mostly arising from modifications of endogenous indole—or catecholamines. Abnormal proteins in plasma, and abnormal immune systems acting on the brain have been sought.

One of the difficulties in studying this problem is that patients with schizophrenia are very disturbed, very distressed, often depressed and usually anxious, overaroused and perplexed. Many have lived in environments totally different from those of control subjects, and have had powerful drugs, some of which have long-lasting effects. It has been very difficult to distinguish the effects of these secondary processes from those changes going on in perhaps a relatively small population of neurones buried deep in the brain.

There is some evidence that research into schizophrenia is beginning to bear fruit, and that perhaps the problem of schizophrenia is closer to a solution than say that of depression and mania. The starting point was the observation that (1) the antipsychotic activity of neuroleptics such as the phenothiazines, butyrophenones, thioxanthines, and substituted benzamides, is probably associated with postsynaptic blockade of dopaminergic receptors; (2) one isomer of flupenthixol, a thioxanthine, does not block dopaminergic receptors and has minimal antipsychotic activity; the other isomer blocks these receptors and is therapeutically useful for the treatment of acute schizophrenia; (3) the pathway concerned with schizophrenia is not likely to be the nigro-striatal pathway, but the dopaminergic pathway passing from the tegmentum (medial part of the substantia nigra) to the nucleus accumbens and part of the frontal cortex.

The nature of this inherited abnormality is not known, but the success of dopamine receptor-blockers such as the phenothiazines, thioxanthines, butyrophenones and now substituted benzamides in the acute illness and in the prevention of episodes, has focused interest on this area.

Recently, it has been shown in a post mortem study, in an admittedly small series of untreated patients, that their brains has increased binding of radioactively-labelled ligands to dopamine receptors. This, the therapeutic success of neuroleptics, and other studies have put hypersensitivity at dopaminergic synapses in the forefront of contending hypotheses for schizophrenia.

The situation has been complicated by the fact that there is more than one type of dopaminergic receptor; for instance, one type is mediated via the adenylate cyclase–cAMP—system and another does not act directly via this system. Another interesting observation is that the substituted benzamides, with proven antipsychotic activity, attach themselves at relatively low affinity to only a proportion of the binding sites identified by ^{3}H-labelled butyrophenones.

Nevertheless, as stated above, the presence of hypersensitivity in at least a proportion of dopaminergic neurones seems a likely accompaniment of schizophrenia, and may prove to be close to the primary abnormality. The important pathway seems to be the so-called mesolimbic system—the dopaminergic pathways starting in the ventromedial tegmentum (medial part of the substantia nigra) which is distributed to the nucleus accumbens and part of the frontal cortex.

However, to reiterate a point put above, there is still the question as to whether the therapeutic effect of the drugs is a direct one on a pathological process within the dopaminergic pathway as suggested by the limited post mortem data currently available, or is indirect and thus correcting for some other malfunctioning pathway.

Differential Diagnosis

The main illnesses presenting spuriously as schizophrenia are given in Table 4.

A proportion of these come to light from information on the natural history of the illness, family history, previous reports of epilepsy, or illnesses like encephalitis lethargica.

The screening of patients with schizophreniform psychosis starts with assay of Hb, RBC and film, WBC and differential, liver functions tests, thyroid profile, serum tests for syphilis, and chest and skull X-ray. A proportion will need assay for B_{12}, tests for porphyria and rarely plasma cortisol or urinary-free cortisol levels.

If there is suspicion of GPI or multiple sclerosis a lumbar puncture will be helpful, with the proviso that there is no likelihood of a marked increase in intracranial pressure. (This may be associated with papilloedema, but not invariably so.) Lumbar cerebrospinal fluid would indicate the paretic first zone curve in the colloidal gold curve test, increased protein with slight increases in cells, characteristic for GPI and multiple sclerosis. Other tests for syphilis in the CSF may give evidence of active

TABLE 4
CONDITIONS GIVING SCHIZOPHRENIFORM PSYCHOSIS

Degenerative	Schilder's disease
	Early Huntington's Chorea
	Presenile and senile dementia
	Multiple sclerosis
Space occupying lesions	Cerebral tumours in general
	Occasionally in temporal lobe tumours
Infective	GPI
	Post-encephalitica lethargica
Vascular	Cerebral arteriosclerosis
Epilepsy	Temporal lobe epilepsy
Trauma	Post cerebral trauma
Metabolic	Porphyria
	Liver failure
	Wilson's disease
Endocrine	Hyperthyroidism
	Myxoedema
	Cushing's disease
Neoplastic	Insulinoma
Nutritional and Vitamin lack	B_{12} deficiency
Toxic	Alcoholism (alcoholic hallucinosis)
Drugs	Amphetamine
	LSD
	L-dopa
	Excess cortisol
	Cycloserine
Miscellaneous	Narcolepsy

disease. Occasionally neoplastic cells can be demonstrated in carcinomatosis.

Other patients may require cerebral angiography or a CAT scan. Blood or urine should be taken for alcohol or amphetamine levels where indicated.

ORGANIC PSYCHIATRIC REACTIONS

Acute or chronic organic psychiatric reactions are the terms now used to describe disturbances in cerebral function arising from various forms of organic insult to the brain. These include conditions which are reversible or, in terms of our present knowledge, are untreatable. The classification used is only a starting point and framework for further diagnostic exploration. Placing the condition in acute or chronic categories *per se* implies neither treatable nor irreversible pathology.

Acute organic reactions are characterised by impairment of consciousness of varying severity, with fluctuations and often with deterioration at night. Stimulation may lead to temporary improvement in lucidity. Patients with acute organic reactions tend to have severe disturbances of perception, often with visual illusions and hallucinations. There may be florid disturbances in behaviour, and loss of reality leading to vivid phantasies.

Memory disturbances are an outcome of both the fluctuations in level of consciousness and a failure of attention. The individual is disorientated in time and place and, later, person. Restlessness may alternate with slowing of psychomotor activity, with repetitive behaviour and loss of spontaneity. Most individuals are emotionally labile.

Chronic organic reactions are characterised by deterioration in cognitive abilities, usually associated with changes in behaviour and emotion. It occurs mostly in a background of clear consciousness. Chronicity generally implies progressive loss of memory, general disorganisation and deterioration of intellect, and the appearance of odd behaviour.

The patient fails to grasp situations, loses interest, initiative and perseverence, and is easily distracted. Social interaction may decline to almost total withdrawal, and thought processes dwell on the past in a repetitive way.

There are losses of skills and learning capacity, concrete thinking, and rigidity of thought processes. Eventually, with increasing cerebral disorganisation, speech can sink to a level of incomprehensibility.

Diagnosis of Acute Organic Reaction

A list of the main causes of organic reactions are given in Tables 5 and 6. In the diagnosis the initial step is to attempt to get a history of drugs taken recently (prescribed or self-administered). There may be a history of diabetes (the taking of hypoglycaemic drugs, poor compliance with drugs or diet, 'brittle' diabetes, an intercurrent infection, cerebral damage from hypoglycaemic coma). Enquiry should be made of recent operations, whether there were anaesthetic complications and how the patient recovered. The patient may have had a previous cerebrovascular accident. There may have been cerebral damage in the past from a failed suicide attempt, leading to anoxia from intentional or accidental carbon monoxide poisoning, from trauma or from other conditions and circumstances.

Vitamin deficiency may be present due to poor diet, gastrointestinal disease, dietary idiosyncracies, lack of personal care because of incipient dementia or other cause, or self-induced fasting as in anorexia nervosa. There may be indications of rising intracranial pressure, epilepsy, heart failure, venereal disease, endocrine dysfunction, abuse of psychomimetic drugs, or alcoholism.

It is important to enquire whether the patient has been abroad and in contact with tropical diseases. Finally, the knowledge of the patient's home circumstances, hobbies, and previous occupations, may give an indication of possible exposure to heavy metals or other toxic compounds.

Without going into great detail at this point, it is sufficient to say that a careful physical examination will often give the clue as to where to direct futher investigations. Thus there may be evidence of cancer, infection, anaemia, hypo- or hyper-thyroidism, Cushing's disease, Addison's disease, etc.

Dehydration with ketosis will point to uncontrolled diabetes; dehydration without ketosis may indicate uraemia or electrolyte disturbance (e.g. from vomiting, diarrhoea, over use of diuretic drugs, neglect in the elderly, etc.). Muscle twitching may point to uraemia, electrolyte disturbance, hypoglycaemia or hypocalcaemia. The skin, and mouth may give evidence of vitamin deficiency, or poisoning with heavy metals.

The physical examination may indicate the presence of

TABLE 5
CAUSES OF ACUTE ORGANIC REACTIONS. (Taken, with permission, from *Organic Psychiatry*, W. A. Lishman, 1978, p. 1982, Blackwell Scientific Publications)

1.	Degenerative	Presenile or senile dementias complicated by infection, anoxia, etc.
2.	Space occupying lesions	Cerebral tumour, subdural haematoma, cerebral abscess.
3.	Trauma	'Acute post-traumatic psychosis'.
4.	Infection	Encephalitis, meningitis, subacute meningovascular syphilis. Exanthemata, streptococcal infection, septicaemia, pneumonia, influenza, typhoid, typhus, cerebral malaria, trypanosomiasis, rheumatic chorea.
5.	Vascular	Acute cerebral thrombosis or embolism, multi-infarct dementia, transient cerebral ischaemic attack, subarachnoid haemorrhage, hypertensive encephalopathy, systemic lupus erythematosus.
6.	Epileptic	Psychomotor seizures, petit mal status, post-ictal states.
7.	Metabolic	Uraemia, liver disorder, electrolyte disturbances, alkalosis, acidosis, hypercapnia, remote effects of carcinoma, porphyria.
8.	Endocrine	Hyperthyroid crises, myxoedema, Addisonian crises, hypopituitarism, hypo- and hyperparathyroidism, diabetic pre-coma, hypoglycaemia.
9.	Toxic	Alcohol-Wernicke's encephalopathy, delirium tremens. Drugs—barbiturates and other sedatives (including withdrawal), bromides, salicylate intoxication, cannabis, LSD, prescribed medications (antiparkinsonian drugs, scopolamine, tricyclic and MAOI antidepressants, etc.) Others—lead, arsenic, organic mercury compounds, carbon disulphide.
10.	Anoxia	Bronchopneumonia, congestive cardiac failure, cardiac dysrhythmias, silent coronary infarction, silent bleeding, carbon monoxide poisoning, post-anaesthetic.
11.	Vitamin lack	Thiamine (Wernicke's encephalopathy), nicotinic acid (pellagra, acute nicotinic acid deficiency encephalopathy), B_{12} and folic acid deficiency.

raised intracranial pressure, signs of meningial irritation, stigmata of syphilis, signs of an intracerebral space-occupying lesion, cerebrovascular disease or subarachnoid haemorrhage.

The patient may be hypertensive, in cardiac failure, have had a recent coronary, be in a state of cerebral anoxia from acute or chronic respiratory or cardiac diseases, or have the signs of liver failure. The whole body must be examined carefully for neoplastic disease.

At this stage the diagnostic possibilities may be narrowed down, but whatever the clinical findings, some general laboratory tests will be necessary besides more specific ones, because acute organic reactions are conditions *par excellence* where primary and secondary pathology (e.g. diabetes and a urinary infection; carcinoma of the lungs and respiratory infection) may coexist. Indeed the secondary condition may bring the primary lesion to light.

Whatever the probable pathological basis, all patients should have assays of haemoglobin, RBC, WBC and differential count, ESR, serology for syphilis, urine for glucose, protein and casts, and skull and chest X-rays. The further tests are dependent on initial findings, but might include blood urea, serum electrolytes and protein, liver function tests, blood sugar, serum B_{12} and folate, and examination of serum and/or urine for drugs, and heavy metals. The urine may need to be examined for evidence of porphyria.

If the diagnosis is still in doubt, a lumbar puncture may indicate the presence of intracerebral infection (including syphilis) and the patient may also require angiography or CAT scan.

ROLE OF THE LABORATORY

Summary and appraisal of the role of the laboratory in depression, schizophrenia and organic reactions is as follows:

Depression

Depression either as a symptom, an indicator of psychological distress, or as an illness has become more readily recognised and treated by doctors, particularly general practitioners. In part, this is due to the advent of effective antidepressant drugs, and perhaps to the beginnings of an appreciation of what can be achieved in some by cognitive therapy, psychotherapy, etc., as well as by 'social manipulation'.

The result has been that a much more selected group of patients is coming to the attention of psychiatrists, and quite a few of these patients have hidden somatic pathology, some of which will be severe, and some of borderline severity [but still sufficient to trigger affective illness].

With regard to patients presenting to physicians and surgeons with depression as the first indicant of organic disease, it is quite likely that this fact will never be communicated to the laboratory, so that this feature will never become part of the background experience of clinical biochemists.

The notification on laboratory request forms of depressive symptoms will therefore come mostly from the psychiatric services and, as mentioned above, these patients are becoming more and more a highly-selected group. The laboratory plays a most important part in these patients, helping the psychiatrist and physician to sort out the significance of such things as mild hyper- or hypo-thyroidism, the presence and cause of electrolyte disturbances of various kinds (including dehydration secondary to the illness), identifying vitamin deficiencies, detecting infections and malignancies of various kinds, etc.

Schizophrenia

Patients suffering from schizophrenia occupy a very large number of hospital places (perhaps 40% of the beds in the country) and are major users of facilities in day hospitals, rehabilitation centres, hostels, etc. Yet it must be remembered that it is a relatively rare disease, occurring in only about 1% of the population. That the condition accounts for so much in the way of hospital, outpatient and community services is because about half who suffer from an initial attack go on to further episodes and chronic deterioration, and then need continuing provision of one kind or another.

The relative infrequency of the illness means that cases coming for initial diagnosis will be few and far between—one author has suggested that a population of a million may be expected to produce no more than one patient per month to the hospital service in a first attack. This is the 'visible' group—we cannot say how many others come via other services and gravitate into the psychiatric departments at a slightly later stage.

The problem of helping to make the initial diagnosis therefore, is not an everyday problem for a laboratory covering a catchment area of moderate size, but it is a very important one.

It is vital for the patient that trouble is taken to get the correct diagnosis first time whenever this is possible, because, it cannot be emphasised enough, the label schizophrenia should not be applied lightly and without the best evidence available.

Once again laboratory screening and further investigations are essential to distinguish an illness which can be ameliorated but not 'cured' by current treatments, from those conditions which are potentially reversible.

Finally, the laboratory has a vital role in detecting the various forms of drug abuse giving rise to schizophreniform psychoses from the illness itself.

Organic Reactions

Organic reactions pose perhaps the most difficult task for both physician and laboratory because the possibilities are so many, and again the importance lies in distinguishing the treatable from the refractory conditions. With such a wide field from which to choose, the

TABLE 6

CAUSES OF CHRONIC ORGANIC REACTIONS. (Taken, with permission, from *Organic Psychiatry*, W. A. Lishman, 1978, p. 183, Blackwell Scientific Publications)

1.	Degenerative	Senile dementia, arteriosclerotic dementia, Alzheimer's, Pick's, Huntington's, Creuzfeldt–Jakob, normal pressure hydrocephalus, multiple sclerosis, Parkinson's disease, Schilder's, Wilson's, progressive supranuclear palsy, progressive multifocal leucoencephalopathy, progressive myoclonic epilepsy.
2.	Space occupying lesions	Cerebral tumour, subdural haematoma.
3.	Trauma	Post-traumatic dementia.
4.	Infection	General paresis, chronic meningovascular syphilis, subacute and chronic encephalitis.
5.	Vascular	Cerebral arteriosclerosis, 'état lacunaire'.
6.	Epileptic	'Epileptic dementia'.
7.	Metabolic	Uraemia, liver disorder, remote effects of carcinoma.
8.	Endocrine	Myxoedema, Addison's disease, hypopituitarism, hypo- and hyperparathyroidism, hypoglycaemia.
9.	Toxic	'Alcoholic dementia' and Korsakoff psychosis, chronic barbiturate or bromide intoxication, manganese, carbon disulphide.
10.	Anoxia	Anaemia, congestive cardiac failure, chronic pulmonary disease, post-anaesthetic, post carbon-monoxide poisoning, post-cardiac arrest.
11.	Vitamin lack	Lack of thiamine, nicotinic acid, B_{12}, folic acid.

physician has to request some initial 'probes' and then try on clinical grounds and laboratory findings to decide which further investigations are needed to establish a diagnosis.

The second round of investigations (often a combination of laboratory and radiologic tests), may solve many problems, but initial appearances may have been misleading, and first tests may be negative. Then it becomes a combined exercise in detection. Sometimes, the 'crime' goes unsolved.

HUNTINGTON'S CHOREA

Huntington's chorea has been included in this chapter because it is illustrative of one of a number of inherited disorders of the nervous system. It is an inherited illness, based on abnormal biochemistry. The place biochemical investigations have played in this condition range from pure research to the most complex humanitarian and ethical problems posed by attempts to identify carrier individuals, and by attempts to limit spread of the disease.

The condition is present in approximately 5 per 100 000 of the population, and is transmitted by a single autosomal gene of full penetrance. Statistically therefore, half the offspring will be affected. The condition becomes manifest when part of the reproductive life has been passed, usually by 30–50 years of age, and is characterised by choreiform movements, progressive dementia and death within 13–16 years of onset.

Those at risk are only too aware of the chances of becoming ill. However, disturbed behaviour—psychotic, psychopathic, affective illness, alcoholism, and criminality—are common in both carriers and unaffected siblings, and as the carriers begin to deteriorate, they are less likely to take contraceptive precautions.

Attempts to identify carriers have so far been unsuccessful, but have included studies of the EEG, blood groups, finger prints, tremor recording, giving l-dopa (choreiform movements appear but it is not known yet if these will develop into the disease), metals, various lipids and mucopolysaccharides, and urinary amino acids.

A recent study has shown an increase in blood concentration of 5HT, decreased plasma tryptophan and increased platelet monoamine oxidase (MAO) in the condition and also in offspring at risk.

Established illness is associated with some other changes in plasma amino acids, but it is not known if these are primary or secondary.

Post mortem investigations have been fruitful in that they have shown abnormality in gamma amino butyric acid (GABA) in the basal ganglia and reduced glutamic acid decarboxylase activity in the putamen and globus pallidus.

Extensive biochemical studies of affected families have failed so far to identify a peripheral marker separating carrier from non-carrier states. The question is, if one of these tests had been successful in doing so, how would it have been used? With a 50% chance of having the disease, these individuals are subject to enormous psychological strain, and it is difficult to know how one would convey to unaffected siblings their freedom from 'genetic taint', without putting the carrier in the invidious position of guessing that they were destined to end their lives prematurely with an untreatable condition.

That is not to say that research into peripheral markers should cease, only that it makes problems for the clinician. A more profitable approach might be further research into causes and methods of treatment, with further attempts to tackle the thorny problem of containing the spread of the gene until, apart from new mutations, it disappears from the population. However, finding the cause may lead to an ability to detect carriers, so there is no way around the problem.

It is easy to pontificate from the safety of an author's

desk—but the social and other problems of this condition and its management present formidable human problems for the affected individuals, their siblings 'at risk', their families and for society as regards spread of the condition.

ANXIETY STATES

Anxiety

Anxiety is a normal emotional state. Starting from a state of attention and increased alertness and arousal, we can pass from mild unease, apprehension, anxiety, fear, and panic to terror. Anxiety is a normal reaction to one's wellbeing or safety, and only becomes abnormal when it is excessive, prolonged, disproportionate to, or unrelated to any real, or implied threat of harm of any kind. In phobic states anxiety is evoked by a feeling of danger in situations which contain no or minimal danger.

The finding that benzodiazepines were of so much value in the treatment of abnormal anxiety of psychological origin posed some problems for our understanding of the basis of this condition. It had been understood that at least part of the anatomical location of anxiety lay in the limbic system, hypothalamus and reticular formation (in the latter insofar as arousal has a place to play in anxiety).

According to Smythies, the hippocampus has a tonic inhibitory effect on pituitary/adrenal mechanisms, associated with the reaction to 'stress'. The posterior hypothalamus is concerned with fear and aversive reactions. Stimulation of this region in animals gives rise to apparent panic, as can similar treatment of the basolateral amygdala. The septal nuclei tend to inhibit fear reactions.

The amygdala-hippocampal-septal system may exert a push-pull control system, the amygdaloid part promoting, and the hippocampo-septal region reducing, fear. Presumably, the limbic cortex and cortex in general have overall control.

As suggested above, stimulation of the reticular formation enters the picture in that increased activity within it activates the animal, thereby increasing arousal and alerting it, but fear reactions follow if the stimulus is increased.

Compared with these relatively localized pathways and specific connections, the actions of benzodiazepines seem remarkably widespread. They appear to function by unmasking GABA receptors all over the brain, and thereby to give GABA-ergic neurones optimum activity. If it is remembered that these drugs are also muscle relaxants, and anticonvulsant, perhaps their anxiolytic site of action is quite a small part in a localised region of what is a generalised promotion of GABA-ergic activity.

A recent suggestion is that it is the septo-hippocampal pathways and their noradrenergic and serotoninergic efferents which mediate anxiety. The septo-hippocampal system is proposed as a match–mismatch comparator. By this is meant that it monitors current activities, and compares achieved with desired aims. It registers discrepancies (punishment, non-reward, failure) and threats of discrepancies. It brings, current activity to a halt when there is recording of mismatch, and evokes a search—perhaps repetitive checking—for an alternative. Could the septo-hippocampal system be an organ of hesitation and doubt which, if overactive, causes abnormal anxiety? Could benzodiazepines control anxiety by the inhibitory GABA-ergic supply either to noradrenergic neurones which terminate in this region, the lateral septal area, or the hippocampus?

It is a nice theory.

Anxiety and Anxiety States

While anxiety under threat or anticipation of threat to ones integrity/safety as a human being is normal, as stated above, exaggerated or prolonged responses of this kind are not. Nor are periods of anxiety without such external threats.

The commonest problem in patients presenting with anxiety is the anxiety-prone person, the individual whose various endowments and experiences have led to the development of the worrying adult. Anxiety is common in most other psychiatric conditions, especially obsessive-compulsive states, phobias, depression of all kinds, schizophrenia, dementia, subnormality, addiction, or any condition giving rise to confusion or disorientation.

Anxiety gives rise to a whole series of changes in the body mediated via the hypothalamus and the pituitary, the thyroid the adrenal medulla and the autonomic nervous system.

The effects include increases in glucocorticoids and increased production of catecholamines.

Pulse rate is increased, blood pressure is slightly raised and peripheral resistance is diminished. Muscle blood flow is increased and there is an increase of sweating in the palms, axillae and forehead.

Glucose is mobilised and in general the body is prepared for 'flight and fight'.

Assessment of Anxiety

For most individuals the diagnosis of an anxiety state is straightforward. It is where there are atypical features, or borderline physical signs that things become difficult.

The most common request in the doubtful case is for a thyroid profile to confirm or exclude suspected thyrotoxicosis, particularly in the late-middle-aged or elderly, where a minor degree of pathology may not be so obvious clinically. Sporadic anxiety may necessitate a blood sugar or series of blood sugar estimations initially, with more detailed assessment of islet cell function if insulinoma is suspected.

If there is reason to suspect sympathomimetic drugs, they may be sought in the urine, and clinical evidence of the various possible endocrine abnormalities will require the appropriate assays. The main causes of anxiety are given in Table 7, and it can be seen that in most instances the indications for the most relevant laboratory tests are usually fairly evident from clinical observation.

ALCOHOLISM

There is no great degree of agreement about either the causes or nature of 'alcoholism'. The problem is that of those who: (1) suffer physical, psychological or social handicaps through excess input of alcohol, and, (2) possibly a group with common characteristics making them susceptible to addiction to alcohol.

In the *alcohol-dependence syndrome* there is alteration in drinking behaviour which is no longer in accord with social expectation and normal social patterns, there is a need to maintain high blood levels, loss of control of drinking and alcohol craving. Abstinence gives withdrawal symptoms relieved by drinking. More and more alcohol is taken with the development, at first anyway, of increased tolerance.

Alcohol-related disability means that the individual is experiencing physical, mental or social impairment because of alcohol.

Primary alcoholism is a disorder affecting mostly men, who were predominantly normal personalities before alcohol became a problem to them. *Secondary alcoholics* are largely women and the condition is thought of as originating in depression of one kind or another, or are men with antisocial personalities.

Finally, there are a group of ill-defined *binge drinkers* where the input of alcohol is large but intermittent.

It is not certain to what extent there is a genetic factor in the development of alcohol-related problems. There is some evidence of some genetic predisposition in both men and women with perhaps a larger component in the latter.

The deleterious effects of alcohol are psychiatric, physical and social as stated above; the latter will not be discussed here.

Finally, we come to the problem of what alcohol does to make it addictive and there is a paucity of hypotheses. Does it for instance, act mostly as an anxiolytic, or alternatively as a euphoriant, and if so, how? Some of its social response depends on the setting in which it is taken, so it is not invariably euphorthey in its actions, and certainly may not be so when dependence is established.

It has known effects on many biochemical pathways and physiological systems, but there is little in the way of systematic hypotheses to explain its psychotrophic actions.

Assessment of Alcoholism

A brief survey of Table 8 will show how different is the emphasis on alcoholism compared with those discussed in the rest of the chapter.

It is true that much occult alcoholism is missed, and the problem is much more common than is realised by most GP's and clinicians. Measurements of blood alcohol levels are of minimal diagnostic value, but sometimes may have forensic implications. The importance of laboratory investigation is in helping to locate the main site or sites of somatic or neurological damage so that those types of alcohol-induced pathology, that are capable of recovery with suitable treatment or prevention can be identified.

There is no point in going over the various abnormalities one by one because the type of laboratory test is usually indicated by clinical history and examination, and the purpose of the test is to show just how wide an area of psychological/somatic/neurological pathology is encompassed.

Perhaps, however, it is important to stress just two of the complications of alcoholism—*Korasakoff's psychosis* and *Wernicke's encephalopathy*. The former has premonitory signs in the form of presence of some memory deficit before the major process takes over, and the physical signs of Wernicke's encephalopathy should be readily recognisable. These two conditions are something of an emergency because early diagnosis and treatment means all the difference between disabling pathology, and a more or less intact nervous system.

The test required is the red-cell transketolase activity, which if low would indicate the need for rapid reversal of the causative mechanism, lack of thiamine, by injections of B vitamin.

Apart perhaps from suspecting alcoholism in the presence of macrocytosis, no specific investigation is indicated (as mentioned above), only the pursuit of whatever pathology seems appropriate. In most patients nutritional deficiencies, the possibility of liver damage, and the vulnerability of the gastrointestinal tract and possibly pancreas may suggest lines of investigation, but there are no hard and fast rules.

LABORATORY TESTS IN THE ELDERLY WITH PSYCHIATRIC PROBLEMS

Many of the problems associated with the ageing processes have been covered in the sections on acute and chronic organic reactions, depression, and anxiety. Some special mention of old people is merited because of the mistaken view held by many overtly or covertly that 'it is all a waste of time', or that one is dealing with progressive irreversible pathology, and that such facilities that exist ought to be kept for younger age groups.

This view is held on the basis of what is at least a false premise. While it is true that much of old people's disease processes are 'unidirectional', a significant proportion are not. When old people in homes, etc. are given comprehensive screening, it is surprising how much minor reversible pathology is uncovered. The type of conditions commonly found include different forms of anaemia, nutritional problems (such as folate or B_{12} deficiencies), thyroid dysfunction (usually hypothyroidism), diabetes mellitus, various infections, some treatable neoplasms. The list should include also many of the treatable physical causes of mental illness.

Nobody is going to turn the clock back for these

TABLE 7
MAIN CONDITIONS CAUSING ANXIETY

(1) *Somatic*	*Degenerative*	Acute and chronic brain syndromes. Multiple sclerosis.
	CNS	Space occupying lesion After head injury or subarachnoid haemorrhage. Temporal lobe epilepsy.
	Cardiovascular	Heart disease (because of its special psychological significance).
	Neoplastic	Pheochromocytoma Insulinoma
	Metabolic	Reactive hypoglycaemia. Potassium depletion (occasionally). Hypo- and hypercalcaemia.
	Endocrine	Hyperthyroidism. Rarely hypothyroidism. Cushings disease. Premenstrual tension. Menopause. Addisons disease.
	Drugs	Sympathomimetic agents. Hypoglycaemia agents and drugs potentiating them (e.g. MAOI) Excess cortisol. Excess thyroxine.
	Toxic	Anticholinergic drugs. Tetraethyl lead. Caffeinism. LSD: other hallucinogens. Amphetamines and other sympathomimetic agents. Some 'activating' antidepressants agents. Withdrawal: alcohol: opiates: meprobamate, benzodiazepines, barbiturates.
	Vitamin deficiency	Pellagra.
(2) *Psychiatric*	Primary	(a) Anxiety-prone personality. (b) Anxiety states/ anxiety neurosis.
	Secondary	Depressive neurosis (perhaps a variant of condition (b) above). Obsessive compulsive neurosis. Phobic states. Affective disorders. Schizophrenia. Subnormality. Chronic/acute organic reactions. Hypochondriasis of various kinds.
(3) 'Normal' anxiety		Anything which threatens the security of well-being of an individual. Novel, unfamiliar surroundings. Situations where people do not have control over their lives. Situations demanding or seeming to demand more than the individual's capacity.

patients, but the treatment of this kind of physical disorder may mean the difference between independence in reasonable mental health or deterioration in psychiatric illness in the terminal years.

IMPACT OF SOMATIC ILLNESS ON MENTAL STATE: THE ENDOCRINE SYSTEM

The complexity of the structure of the human brain is greater than any known structure, something far beyond our present comprehension, a thing for awe and wonder. Its functions and achievements similarly are so many and intricate as to be beyond our understanding.

Despite this, as discussed above, the responses of the brain to particular metabolic insults are usually surprisingly few and stereotyped, while at the same time being largely unpredictable. For instance, half the patients with Cushing's disease will show minimal psychiatric abnormality, the remainder will show one of several responses, but the nature of the change in mental state and its severity could not be anticipated in advance. Alternatively, in other states, a biochemical disturbance can be followed by almost any psychiatric syndrome.

The responses to endocrine disturbances usually fall into one of the following categories:

TABLE 8
PROBLEMS DUE TO ALCOHOL

Psychiatric and neurological
(a) Intoxication with aggression
(b) 'Black outs' or amnesias—'en bloc', or fragmentary.
(c) Simple withdrawal state
(d) Depression
(e) Suicide
(f) Personality deterioration
(g) Impotence
(h) Delirium tremens
(i) Alcoholic hallucinosis
(j) Intellectual impairment, alcoholic dementia
(k) Withdrawal fits
(l) Alcoholic jealousy
(m) Wernicke's encephalopathy
(n) Korsakoff's psychosis
(o) Marchiafava-Bignami disease (degeneration of the corpus callosum)
(p) Central pontine myelinosis
(q) Cerebellar degeneration
(r) Peripheral neuritis

Somatic	
Degenerative	(g) (j) (m) (n) (o) (p) (q) above
	Cardiomyopathy
	Myopathy
Trauma	Accident/physical injury
Infections	Increased susceptibility to pneumonia and TB
Nutritional	Beri-beri
	Scurvy
	Anaemia secondary to iron, folate, or B_{12} deficiency
Vascular and bone marrow	Secondary anaemia
	Thrombocytopenia
	Macrocytosis
	Bleeding due to lack of Vit. K in cirrhosis of the liver
Alimentary	Gastritis
	Gastric ulcer
	Duodenal ulcer
	Pancreatitis
	Cancer of oesophagus and pharynx
Hepatic	Cirrhosis of liver
Metabolic	Precipitation of gout
Renal	Renal failure secondary to myopathy
Fetal	Excess alcohol intake in mothers can give rise to:
	Mild-moderate mental retardation
	Congenital dislocation of hips
	Congenital heart disease
	Cleft palate
	Small head.
Endocrine	Hypoglycaemia
Methyl alcohol	Retrobulbar neuritis
	Optic atrophy

1. acute or chronic brain reactions,
2. neuroticism—states of anxiety, depression, irritability, unpredictable behaviour, apparent immature or hysterical reactions, lability of mood,
3. various forms of depression, or the mimicking of schizophrenia,
4. loss of energy and drive—apathy, decreased appetite, loss of interest in sex and changes in sleep pattern,
5. disorders of memory,
6. behaviour disorders.

(Paranoid reactions are a frequent feature of these psychiatric disorders.)

Given that individuals may respond in this way, Table 9 gives a list of the more and less common psychiatric manifestations of endocrine disorders. The information overlaps in part with that in the other sections. It is, however, helpful to look at the problem from the other viewpoint, to see just how important are psychiatric manifestations of somatic illness. Indeed many of these conditions present with psychiatric illness before the somatic basis is recognised.

TABLE 9
EFFECTS OF ENDOCRINE DISORDER ON MENTAL STATE

Hormonal system	Abnormality	Psychiatric syndrome or syndromes [relatively common]	[relatively uncommon]
Ant. Pituitary	(1) Hypopituitarism (Simmond's disease)	Depression, emotional lability, irritability, apathy, loss of drive and initiative. Memory impairment In crises, acute brain syndrome.	Paranoid hallucinatory psychosis
	(2) Acromegaly	Apathy, lack of initiative	
Post. Pituitary	Diabetes insipidus	Possibly some depression, irritability and apathy	
Ant. Pituitary/ adrenal cortex	Cushing's disease	Any form of depressive reaction/illness often to point of psychotic depression; emotional lability, irritability. Acute anxiety, Apathy → → stupor	Acute organic reaction Schizophreniform psychosis
Adrenal cortex	Addison's disease	Depression (mild to moderate, various types). Apathy, loss of drive and initiative, mild to moderate chronic brain reaction, memory impairment, Anxiety, Acute brain reaction (in Addisonian crisis)	Psychotic depression
Adrenal medulla	Phaeochromocytoma	Anxiety, fear of impending death, Excitability and confusion.	
Thyroid	Hyperthyroidism	Neurotic behaviour Anxiety state Depression of any degree of severity (often classical affective type) Mania, Acute organic reaction, Schizophreniform psychosis Paranoid reactions	Apathetic state
	Hypothyroidism	Lethargy, retardation, Memory impairment, slowing of all cognitive functions, Apathy. Agitation and aggression, Acute organic reaction, Chronic organic reaction, Schizophreniform illness often with paranoid features Depressive psychosis	
Parathyroid	Hyperparathyroidism	Depression/anergia Tension, irritability, Poor memory and mental slowing Acute brain syndrome (in crisis) Stupor Chronic brain syndrome	
	Hypoparathyroidism	Acute organic reaction, Emotional lability, poor concentration, Chronic brain syndrome, neuroticism, Depression, irritability, social withdrawal.	
Pancreas	Hypoglycaemia from drugs Insulinoma (also hypoglycaemia is a feature of other endocrine disorders)	Anxiety, panic, Depersonalisation, Disinhibition, Aggression, Apathy, social withdrawal Acute brain syndrome	
	Reactive hypoglycaemia	Mild symptoms—anxiety and apprehension	
Sex hormones	Premenstrual tension (basis unknown)	Anxiety, irritability, anger, aggression, depression,	
	Puerperal depression ? rapid withdrawal of sex hormones	Transient depression, Irritability,	
	Menopause	Depression, irritability, loss of drive Emotional lability,	
	Hypogonadism/castration	Tendency to loss of libido and drives, Depression, emotional lability.	

CONCLUSION

By no stretch of the imagination could this chapter be called comprehensive. Choice of topics has included that of four severe psychosyndromes, affective disorders, schizophrenia, acute brain reactions and chronic brain reactions, and has tried to outline the necessary tests needed to pinpoint causative/provocative somatic processes and somatic illnesses mimicking the two former conditions. Acute and chronic brain reactions are important conditions taxing both clinician and laboratory in their diagnosis.

The remaining topics are in part personal choices from what is a vast field of important knowledge.

Additional topics which could have been covered include drugs, environmental toxins, cardiorespiratory disorders, the mental effects of single organ failure (liver, kidneys, etc.), the psychiatric interface with neurological disease, the effects of nutritional deficiencies, etc.

The role of biochemistry in the causation, diagnosis and treatment of psychiatric disease is a large and growing subject.

Note

The author has relied heavily on Dr W. A. Lishman's book *Organic Psychiatry*, Blackwell Scientific Publications, 1978, and, as a start for further reading on the subject, this volume is to be highly recommended.

SECTION X

DISORDERS OF THE ENDOCRINE SYSTEM

	Page
34. INTRODUCTION TO THE ENDOCRINE SYSTEM	541
35. DISORDERS OF THE ANTERIOR PITUITARY GLAND	548
36. THE THYROID GLAND	560
37. ENDOCRINE PANCREAS	583
38. GASTROINTESTINAL HORMONES	606
39. ADRENAL CORTEX	616
40. THE GONADS	633

34. INTRODUCTION TO THE ENDOCRINE SYSTEM

D. L. WILLIAMS

Control of metabolic processess
   Intracellular regulation
   Extracellular integration and control mechanisms
The mechanics of the endocrine system
   The hormones
   Mode or hormone action
   Hormones acting through cyclic nucleotides
   Regulation of hormone action
   Control of hormone production
Investigation of endocrine disease
   Disorders of the endocrine system
   Methods of biochemical investigation
Summary

CONTROL OF METABOLIC PROCESSES

The metabolism of an individual living cell is the sum total of activities of a wide range of separate but inter-dependent biochemical reactions, most of which are catalysed by individual and specific enzymes. Most enzyme-mediated reactions form part of a metabolic pathway so that the substrate used in a particular enzyme reaction is the product of the previous reaction in the pathway, and its product is in turn the substrate for the next reaction in the pathway. By and large, metabolic pathways perform either an anabolic function, changing relatively simple initial substrate compounds into more complex final products, or a catabolic function, breaking down more complex molecules to simple metabolites. Many metabolic pathways are capable of performing either function although frequently the pathway taken during anabolism is slightly different from that taken in reverse during catabolism. The summation of these anabolic and catabolic processes constitute the metabolism of the cell.

Intracellular Regulation

Within the individual cell the complex of catabolic and anabolic processes is regulated by a wide variety of specific control mechanisms and by a number of general factors. These regulatory processes include:

1. *Control of individual enzyme reactions.* The rate of each enzyme reaction is dependent upon the availability of substrate and any co-factors or co-enzymes necessary for the reaction, the negative feed-back effect of high concentrations of product, and the amount of active enzyme available.

2. *Control of metabolic pathways.* In addition to the requirements mentioned above for each individual enzyme, a metabolic pathway may be under some overall control. This is often mediated by a more extensive control of one or more specific enzyme reactions, frequently those at the beginning of the metabolic pathway. Thus, the final product of the pathway may 'allo-sterically' inhibit one of the early enzyme reactions responsible for its eventual synthesis; allo-steric stimulation may also occur, particularly by the mediation of high concentrations of a product of a parallel metabolic pathway. Enzyme activation or inhibition can also be controlled by the concentrations of substances such as AMP, ATP, NAD^+, NADH.

3. *Control of enzyme synthesis.* The activity of some enzymes is dependent, as described above, upon factors which activate or inhibit the enzyme. The activity of other enzymes is controlled by the actual amount of enzyme available and this is controlled in turn by the rate of synthesis of the enzyme. Some substrates have the ability to act as inducers of those enzymes which are required for their subsequent metabolism.

4. *Intracellular organization.* In order for certain metabolic processes to occur it is not only necessary that the required substrate, co-factors, activators, and enzymes are available; they must be available in the appropriate intracellular organelle in which the metabolic process occurs. Some metabolic pathways are initiated in one part of the cell and are completed in another part of the cell. In these cases there are sometimes sophisticated biochemical controls on the passage of metabolites and co-factors across intracellular membranes.

5. *Availability of energy and of substrates.* Most anabolic processes require energy; even catabolic processes producing energy sometimes need to be primed with energy before they can begin to function. All metabolic processes need to obtain their substrates and co-factors and are also dependent upon an adequate supply of amino acids for the synthesis of their enzymes. Thus the overall metabolism of the cell is dependent upon a ready supply of energy, often in mammalian tissues in the form of glucose or fatty acids, and of substrates, amino acids, co-factors and

vitamins, and these substances must be able to be transported, either actively or passively, into the cell across the external cell membrane.

6. *Elimination of waste products.* In addition to individual enzyme reactions being inhibited by their own products, metabolic pathways can give rise to products which, if present in high concentrations, can act as toxic substances which prevent metabolic reactions. If the metabolism of the cell is to continue these by-products must be excreted from the cell and removed from the cell's environment.

Extra-cellular Integration and Control Mechanisms

The types of control mechanisms mentioned above are common to living cells either if they exist as individual animal, plant or microbial cells, or if they are part of a multi-cellular, multi-organ animal or plant. As living organisms evolved from the simple uni-cellular state to more complex forms, certain tissues within the organism took on certain specialised functions. In order that the activities of individual tissues and organs could be co-ordinated for the benefit of the whole living organism it became necessary for more complex control mechanisms to be evolved. Such co-ordinating control mechanisms can be seen in the plant kingdom and in lower animals, but the most sophisticated systems are seen in the mammals where two parallel co-ordinating and control mechanisms have developed, the nervous system and the endocrine system.

By and large the nervous system exerts its effects by means of control of the contraction, relaxation and tone of muscles. Thus it controls both voluntary movement through the skeletal muscles, and also involuntary movement through control by the autonomic nervous system, of for example, the rate and force of cardiac contraction; the tone of arterioles, thereby determining the relative rates of perfusion of various tissues; and the co-ordination of the contraction of certain muscular organs together with the relaxation of associated sphincters in such organs as the gastrointestinal tract, the urinary tract, and the reproductive system. Any control system needs feedback information and in the case of the nervous system this is achieved by means of afferent stimuli carrying information back from the individual organs to the central nervous system.

In contrast the effects of the endocrine system are mediated via the control of metabolic reactions by means of hormones. The term hormone was coined by Bayliss and Starling in 1902[1] and is used to describe chemical substances which are produced in one part of the body, enter the circulation, and are carried to distant organs and tissues to modify their metabolic function.

In addition to the well accepted hormones the term is also now used to describe the active metabolite of vitamin D that is formed in the kidney and then circulates in the blood to have distant effects on the gastrointestinal tract, bone, and kidney tubule; it is also applied to the hypothalamic factors that are synthesised in the nerve cells of the hypothalamus and pass along the nerve axons to be released into a short portal venous system before having their stimulatory or inhibitory effects on cells in the anterior pituitary gland. Feed-back control of the endocrine system may be mediated by the concentration of the hormone itself inhibiting the system responsible for its synthesis and secretion; or the rate of hormone production or activation may be controlled by the level of metabolites produced or utilised as a consequence of the action of the hormone.

It was initially thought that the nervous system and the endocrine system were two independent and parallel control mechanisms which had little or no direct influence on each other. The fact that the adrenal medulla produced the hormone adrenaline as a consequence of autonomic stimulation was looked upon as an interesting physiological oddity. It is now known, however, that there are many other areas where the nervous system and the endocrine system have close and interdependent relationships. In addition to the production of adrenaline and the hypothalamic hormones it seems that the nervous system has an important part to play, for example, in the production of several of the gastrointestinal hormones and it may also have a role in the control of the production of hormones by the pancreatic islets, the gonads, and the macula densa of the kidney.

THE MECHANICS OF THE ENDOCRINE SYSTEM

The Hormones

Types of Hormones. Chemically the hormones fall into three main categories. The simplest are those that are modified amino-acids, thyroxine, adrenaline and possibly some of the other catecholamines such as dopamine. These substances show chemical similarities to nerve transmitters, indeed adrenaline and dopamine probably act both as hormones and as nerve transmitters.

The second group of hormones are polypeptide in nature, and range from relatively simple oligopeptides, such as the tripeptide, thyrotrophin-releasing hormone(TRH), to relatively large proteins such as corticotrophin(ACTH) and growth hormone(hGH). This group includes the APUD(amine precursor uptake and decarboxylation) hormone system–*see* Chapters 35 and 38. The third group is comprised of the steroid hormones, and includes substances such as 1,25-DHCC(1,25-dihydroxy-cholecalciferol) in which one of the rings of the sterol nucleus has been broken.

Site of Synthesis. Classically hormones have been considered to be synthesised in discrete endocrine glands such as the anterior pituitary glands or the thyroid gland. Our increased knowledge of the endocrine system in recent years has however led to the realisation that some hormones are produced by tissues that bear little relationship to these discrete glandular structures. Thus the hypothalamus which produces the pituitary releasing and inhibitory homones is very much part of the nervous system; the kidney produces erythropoieten and 1,25-DHCC from cells in the proximal convoluted tubules and renin from specialised juxtaglomerular cells and closely neighbouring cells at the macula densa; some of the gastrointestinal hormones are produced from cells diffusely scattered in gastrointestinal tissue. Some hormones, particularly the polypeptide and protein hormones, are stored within the cells that synthesise them to be released only in response to a particular triggering mechanism. Many polypeptide hormones are synthesised initially as larger, inert proteins, and are broken down to a more active product either prior to release from storage, as in the case of the formation of insulin from proinsulin, or subsequent to release into the bloodstream, as in the case of the conversion of angiotensinogen to angiotensin I and II within the circulating blood. Other hormones are not stored in large amounts but must be newly synthesised in response to the appropriate stimulation before being released into the blood. Further details of the synthesis of hormones are given in the subsequent chapters in this section.

Transport of Hormones. Many of the smaller hormones, including thyroxine and the steroids, are transported in the blood bound to specific transport proteins such as thyroxine-binding globulin, sex-hormone binding globulin, and cortisol-binding globulin. The protein hormones, having larger molecular weight, do not normally require a transport protein. The physiological and diagnostic importance of these carrier proteins are discussed in the appropriate following chapters.

Mode of Action of the Hormones

Evolution of Hormone Action. An interesting hypothesis of how hormone action evolved is given by Tomkins.[2]

Cellular Effects of Hormones. It has been known for many years that neuro-transmitters at a nerve-nerve synapse or at a neuro-muscular junction perform their function at the recipient neurone or muscle by binding to specific receptor sites on the membrane of the nerve cell or muscle. It was found that the binding of transmitter to the receptor site caused changes in the permeability of the membrane allowing the flux of ions across the membrane, and this secondary action stimulated the consequent polarisation of the muscle cell or nerve cell. It was natural, therefore, to suggest that hormones might work in a similar way by binding to the membranes of the cells which they stimulate. This concept has been shown to be well founded; most of the protein hormones seem to exert the majority of their effects after binding to specific receptor sites on the external cell membrane of tissues which they affect. Steroid hormones, however, appear to exert their actions only after having crossed the external cell membrane. The thyroid hormones appear to work both at the cell membrane and within the cell.

Hormones acting through Cyclic Nucleotides

The Second-messenger Concept. Most hormones that act at the cell membrane appear to do so by stimulating the production within the cell of a chemical substance, the 'second messenger', which is responsible for mediating the effects of the hormone, 'the first messenger'. The work of Sutherland and his colleagues in the 1960s showed that for many hormones the second messenger is a cyclic nucleotide, 3', 5'-cyclic adenosine monophosphate (c-AMP). c-AMP is formed from ATP by the enzyme adenyl cyclase which is situated in association with the receptor site in the cell membrane. Thus, the hormone in the extra-cellular fluid binds to a specific receptor on those tissues that are sensitive to its action, activates adenyl cyclase in the cell membrane, which in turn causes the conversion of ATP to c-AMP within the cell. A list of hormones which depend upon the stimulation of adenyl cyclase and the formation of c-AMP for at least some of their actions is given in Table 1 (taken from

TABLE 1

Hormones that stimulate the cyclic-AMP mechanism
    Catecholamines (β-Receptor)
    Corticotrophin (ACTH)
    FSH
    Glucagon
    Gonadotrophin releasing hormone
    LH
    Lipotropin
    Parathyroid hormone
    Thyrocalcitonin
    Thyrotrophin releasing hormone (TRH)
    Thyrotrophin (TSH)
    Vasopressin (ADH)

Hormones that do not stimulate the cyclic-AMP mechanism
    All steroid hormones
    Thyroxine
    Tri-iodo-thyronine
    Insulin
    Catecholamines (α-Receptor)
    Placental lactogen (HPL)
    Prolactin
    Somatomedin
    Somatostatin
    Growth hormone
    Angiotensin

Baxter and MacLeod[3]). There seems little doubt, however, that cyclic AMP is not the only intracellular second messenger; the binding of some hormones to the receptor site causes a change in flux of certain ions across the membranes; it seems that calcium transport is particularly associated with the binding of some hormones. Guanosine-3', 5'-monophosphate (c-GMP) is also likely to be involved in the mediation of some hormone activities possibly by inhibiting the activity of adenyl cyclase and thereby terminating the hormone action, and possibly also by acting as a second messenger in its own right.

Cyclic AMP produces its effect by stimulating the action of a protein kinase within the cell. The protein kinase then phosphorylates an enzyme protein causing activation or inhibition of that enzyme. Thus c-AMP can be produced within the liver cell by glucagon or adrenaline and this will result in the phosphorylation of at least two proteins via activation of the protein kinase; one protein is the enzyme glycogen synthetase which, when phosphorylated, becomes inactive; the second action is to phosphorylate, and thereby activate, phosphorylase kinase thus starting a short cascade activation of the enzyme liver phosphorylase (*see* Fig. 1). The hormones pharmacology was that catecholamines had at least two activities, called initially α- and β-activities. Subsequent information has led to at least four different types of catecholamine receptors being described. The receptors associated with activation of adenyl cyclase are β-receptors but it is likely also that catecholamines can affect cell metabolism through α-receptor activity. It is suggested that catecholamine binding to the α-receptor site causes increased permeability of the cell membrane to calcium

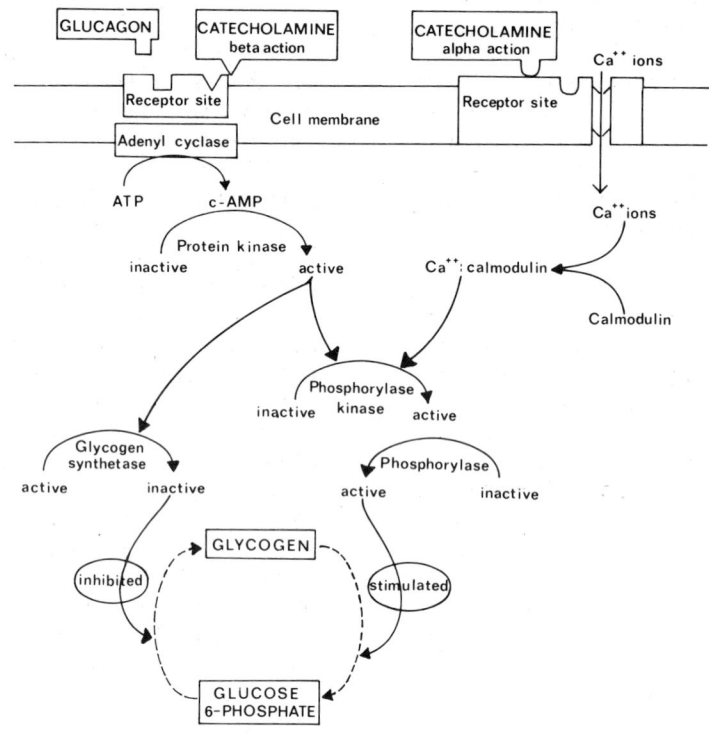

Fig. 1.

thereby inhibit the synthesis of glycogen and enhance its breakdown by this sophisticated control mechanism that requires only one second messenger mediator. Further details of the role of cyclic nucleotides in the control of hormone reactions are given in the reviews by Baxter and MacLeod,[3] Sutherland et al.,[4] Mathason,[5] and Pastan et al.[6]

As can be seen many peptide hormones and adrenaline exert their varied influences on cell metabolism through a very similar mechanism. The essential specificity of hormone action is maintained through the specificity of the hormone-receptor interaction. Thus the hormone binds only to those cells which will respond as required to its presence; only these cells will have receptor sites to bind the particular hormone and to translate that biochemical binding into intra-cellular metabolic activity.

Catecholamines

The action of adrenaline in stimulating c-AMP activity is only one of the ways in which catecholamines can influence cell activity. One of the early discoveries in which thus increases in concentration within the cell.[7] The calcium is in turn thought to exert its effect by combining with a calcium-binding protein, calmodulin.[8] It is further suggested that the calcium-calmodulin entity is then able to activate or inactivate enzyme systems within the cell; thus it is thought that phosphorylase kinase can be activated by this complex. This theory therefore suggests that catecholamines can stimulate the activity of phosphorylase kinase not only via c-AMP β-receptor activity, but also by a calcium-calmodulin α-receptor activity (*see* Fig. 1). It is possible also that the enzyme adenyl cyclase itself is sensitive to changes in calcium ion concentration.

Steroid Hormones

In contrast to the polypeptide hormones, which generally mediate their actions from outside the cell, steroid hormones are known to enter the cell where they bind to certain specific receptors in the cytosol.[9,10] The steroid-receptor complex is then transported to the nuclear chromatin, where it affects the transcription of certain genes, giving rise to the production of specific messenger-RNA's. The messenger-RNA's in their turn cause the synthesis of their associated proteins in the ribosomes. Thus the overall effect of the steroid hormone on the cell is the synthesis of one or more specific proteins. These proteins may be enzymes, transport proteins, binding proteins, or possibly structural proteins, but it is by their secondary effect on cell metabolism that the overall physiological effect of the steroid hormone is mediated. In general each individual steroid hormone affects only a limited number of tissues; the specific tissue which reacts to the hormone is determined by the specificity of the cytoplasmic steroid receptors, and possibly by control at the cellular membrane which may selectively allow only certain steroid hormones to cross. Thus, aldosterone affects specifically those tissues, namely the renal tubule, the gastrointestinal mucosa and the sweat glands, which are responsible for controlling the excretion of sodium and potassium ions from the body. In the kidney tubule it appears that the specific effect of aldosterone is to stimulate the synthesis of a protein or proteins responsible for increasing the transport of sodium ions out of the renal tubular cells into the extra-cellular fluid in exchange for potassium and hydrogen ions; the net effect is to decrease concentration of sodium within the tubular cells and to increase the concentrations of potassium and hydrogen ions; this in turn creates a favourable diffusion gradient between the tubular cell and the tubular contents causing reabsorption of sodium and excretion of potassium and hydrogen ions.

Cortisol appears to affect muscle cells and the liver, stimulating the synthesis of a number of enzymes including those responsible for protein breakdown, the urea cycle, and gluconeogenesis; the major net metabolic effect of cortisol is thus to convert amino acids in proteins to glucose. 1,25-dihydroxy cholecalciferol is known to stimulate the synthesis of a calcium-binding protein in cells of the gastrointestinal mucosa; its net effect is to increase the uptake of calcium from the gastrointestinal tract.

In addition to these actions of steroid hormones in initiating synthesis of certain proteins, it is possible that some effects of steroids are not mediated in this way. Thus, the effect of cortisol in reducing the high concentrations of ACTH in the blood of patients with primary adreno-cortical failure is noticeable within a few minutes of administration and is virtually complete within an hour; this effect is probably too rapid to be accounted for by an effect on pituitary cell DNA, (although it is likely that an effect causing cessation of protein synthesis will probably be more rapid than an effect mediated by stimulation of synthesis). It is also difficult to explain some of the effects of gonadal steroids by means only of changes in the rate of protein synthesis, but this difficulty may merely reflect the complexity of their effects.

Other Hormones

Thyroxine appears to work both by effects on the cellular membrane (although not by a c-AMP stimulated mechanism) and also by stimulating protein synthesis (although not by the same method as steroid hormones); insulin certainly appears to have an effect on cell membrane transport, and may also work through the mediation of a second messenger, possibly calcium, or by inhibiting c-AMP activity. The mechanism of action of these hormones are discussed more fully in Chapters 36 and 37.

Regulation of Hormone Action

It is important for the control of cellular metabolism that the train of events set in motion by the hormone can be brought to a halt when appropriate. The methods of terminating hormone action are not fully understood, but fall into two distinct areas, the removal of the hormone itself and the termination of its effects. Removal of the hormone can be effected in several ways. Binding to receptor sites or receptor proteins is a dynamic equilibrium and the amount of hormone bound to receptors at any one time is a function of the concentration of active hormone in the blood; therefore reduction in the amount of hormone produced can lead, after a finite period of time, to less hormone being bound to receptor. It appears that some polypeptide hormones are ingested by pinocytosis by the cell membrane to which they are attached; it is thought that the polypeptides are then digested by the proteolytic enzymes of intra-cellular lysosomes. Steroid hormones are metabolised within certain cells, particularly in the liver, and then conjugated; these processes reduce their hormonal activity, increase their water solubility, and allow their subsequent excretion. Hormones may also be excreted in the urine, either intact or, as in the cases of catecholamines and steroids, as less active metabolites.

The production of c-AMP within the cell initiated by a polypeptide hormone, is terminated by the enzyme phosphodiesterase which changes c-AMP to AMP. This process can be affected by certain intra-cellular regulators such as calcium ions, nucleotides, possibly including c-GMP, and prostaglandins. The precise effects of these intra-cellular regulators are not fully understood.

Control of Hormone Production

Like any control mechanism, the endocrine system needs feed-back information so that production of its effector substances, the hormones, can be increased or turned off according to the needs of the organism. Feed-back in the nervous system is mediated via the sensory nerves, usually through the appropriate centres in the brain, but occasionally via the shorter, more direct reflex arc through the spinal cord. The afferent pathways in the endocrine system can also employ (i) a long feed-back control system, via the hypothalamus and pituitary, either individually or together, in which the concentration of free circulating hormone produced by the target gland inhibits the production of its own trophic hormone or (ii) a short feed-back loop in which the circulating concentrations of the metabolites, e.g. glucose, calcium and sodium, directly affect the production of those hormones responsible for controlling their concentrations, namely insulin and glucagon, 1,25-DHCC and parathyroid hormone, and renin and aldosterone, respectively.

The short feed-back loop results in relatively constant concentrations of the metabolites, but widely changing concentrations of the hormones. Hormones produced by hypothalamic-pituitary stimulation also show variable blood concentrations, but these are often changes that are rhythmic in nature:

1. Minute to minute fluctuations about a reasonably steady mean value, brought about by the pulsatile nature of the hypothalamic stimulation. These pulses of production of hypothalamic releasing factors may change in frequency with time and result in the more prolonged changes mentioned below. It appears that such pulsatile stimulation is necessary in order to achieve satisfactory pituitary stimulation; thus constant infusion of a dose of gonadotrophin-stimulating hormone can result in inhibition of LH and FSH production, but a pulsatile infusion of the same dose gives satisfactory stimulation.

2. The concentrations of some hormones show a circadian rhythm; thus the levels of prolactin and cortisol are highest in early morning and lowest at midnight.

3. The concentrations of some hormones, especially the female reproductive hormones, show monthly, or even seasonal fluctuation and a few change with age.

4. The concentrations of most hormones exhibit significant change when challenged with the particular stimulation to which they are sensitive.

Further details of these control mechanisms are given in the following chapters.

INVESTIGATION OF ENDOCRINE DISEASE

Disorders of the Endocrine System

Endocrine disorder can arise from abnormality in the gland, the control mechanism, or the target tissue; neoplastic disease can give rise to abnormal 'ectopic' endocrine activity. Thus disorders of the reproductive system can be the result of abnormality in the higher brain centres, the hypothalamus, the pituitary, the gonads, or in the target organs, such as the uterus or the skin. The opportunities for abnormality are somewhat less in those endocrine systems with shorter control systems but even here target organ abnormality should be considered; thus in some forms of diabetes mellitus it seems likely that at least part of the problem is caused by the inability of target tissues to react to adequate circulating concentrations of insulin.

The various parts of the endocrine system are subject to the same wide range of pathological changes that can occur in non-endocrine tissues. These pathological conditions causing endocrine disease include inborn errors of metabolism, congenital defects, infection, inflammation, auto-immune disease, neoplastic change, degenerative disease, and the effects of ischaemia. From a clinical point of view, disease of a particular endocrine system can result in over-activity or under-activity of the gland, or can be associated with no noticeable change in endocrine function. Because of the widespread clinical effects of disorders of the endocrine system, gross endocrine disease is usually diagnosed by the clinical appearance of the patient. However, biochemical and other investigations do play an important part in the management of endocrine disease, and in some cases biochemical abnormalities are the presenting feature; it appears, for instance, that more cases of hyperparathyroidism are now being discovered as a result of routine screening of blood calcium concentrations than by patients presenting with the 'bones, groans, and stones' that are clinical symptoms of the disease.

Methods of Biochemical Investigation

The importance of biochemical and other investigations in the management of endocrine disorders lies in three major areas:

1. In assessing whether a part of the endocrine system is over-active, under-active or has normal activity.

2. In identifying the site of the lesion and its pathology.

3. In monitoring the course and effectiveness of treatment.

The usefulness of the clinical biochemistry laboratory in the investigation of endocrine disease has flowered over the last 20 years since the advent of competitive protein-binding and radio-immunoassays. These assays are both sensitive enough and specific enough to measure individual hormones, trophic hormones and even releasing-hormones in the blood. The widespread availability of such methods has, however, laid a number of traps for the unwary. It is often necessary to ensure that blood samples are taken under correct conditions, e.g. of nutrition, posture or state of rest, and at the appropriate time of day, month, or even season. Many hormones are labile and require that the blood sample is centrifuged under special conditions without delay and that the plasma or serum is stored deep-frozen until the analysis is carried out. Accounts of the collection of specimens for hormone estimation and the effect of biological rhythms on hormone concentrations are given in Chapters 3 and 28 of Volume 1 of this publication. Failure to take these factors into account can lead to misinterpretation of the significance of hormone assay results. The local laboratory should be consulted before requesting any analysis that requires special conditions for sample collection.

It is also useful in the interests of accurate interpretation not only to measure the concentration of the hormone itself, but also to assess the level of its trophic hormone or the concentration of the metabolite that it controls. Thus, poor production of 17β-oestradiol by the ovary can be caused by hypothalamic under-activity, in which case the concentration of FSH will be low, or by ovarian failure in which case the concentration of FSH will be high. Similarly it is difficult to interpret the significance of a concentration of parathyroid hormone in the upper part of the normal range without knowing the concentration of plasma calcium.

It has already been mentioned that several hormones circulate in the bloodstream bound to proteins; the total concentration of hormone may not therefore give a direct indication of the concentration of active hormone as several factors can affect the concentration of binding proteins and thereby distort the result of an assay of total hormone concentration.

The endocrine system lends itself particularly to investigation by means of dynamic function tests[11] rather than by the single estimation of a hormone concentration or of a hormonally controlled metabolite such as calcium or glucose. Thus, the glucose tolerance test, hypoglycaemic stress test, or dexamethazone suppression test can give substantially more information in cases of possible diabetes, growth hormone abnormalities, or Cushing's syndrome than can individual measurements of glucose, growth hormone, or cortisol. Particular use has been made of the stimulation of the pituitary gland by gonadotrophin-releasing hormone (GnRH) or thyrotrophin releasing hormone in the investigation of reproductive or thyroid abnormalities. Thus the response of the pituitary in increasing the concentrations of FSH and LH following the intravenous injection of GnRH has been shown to be useful in the investigation of the delayed onset of puberty, or under-activity of the gonads associated with poor pituitary gonadotrophin drive.

SUMMARY

Our knowledge of the endocrine system has advanced in recent years along with the availability of assays for the measurement of the concentration of circulating hormones and the exploitation of biochemical and physiological techniques to investigate metabolic processes within the cell, at the cell membrane, and between cells. This increase in knowledge and in the sophistication of assay techniques has led to a more systematic and logical approach to the investigation of patients thought to have endocrine disease, and has enabled a more adequate control of treatment. The following chapters deal with some of the more important areas where biochemical investigations can assist in patient management.

Acknowledgement

The author wishes to thank Mrs. Barbara Hunt for typing this chapter and for assistance with the preparation of other chapters.

REFERENCES TO FURTHER READING

1. Baylis, W. M. and Starling, E. H. (1902), 'The mechanism of pancreatic secretion.' *J. Physiol.*, **28**, 325–355.
2. Tomkins, G. M. (1975), 'The Metabolic Code.' *Science*, **189**, 760–3.
3. Baxter, J. D. and MacLeod, K. M. (1980), 'Molecular Basis for Hormone Action.' In *Metabolic Control and Disease* (P. K. Bondy and L. E. Rosenberg, Eds.) W. B. Saunders Co., Philadelphia/London/Toronto pp. 104–160.
4. Sutherland, E. W., Robinson, E. J. and Butcher, R. W. (1968), 'Some aspects of the Biological Role of Adenosine 3',5'-monophosphate (cyclic AMP).' *Circulation*, **33**, 279–306.
5. Mathason, J. A. (1977), 'Cyclic Nucleotides and Nervous System Function.' *Physiological Reviews*, **57**, 157–256.
6. Pastan, I. H., Johnson, G. H. and Anderson, W. B. (1975), 'Role of Cyclic Nucleotides in Growth Control.' *Annual Reviews of Biochemistry*, **44**, 491–522.
7. Rasmussen, H. and Goodman, D. M. (1977), 'Relationships between Calcium and Cyclic Nucleotides in Growth Control in Cell Activation.' *Physiological Reviews*, **57**, 421–509.
8. Wang, J. H. and Wasserman, D. M. (1979), 'Calmodulin and its role in the Second Messenger System.' *Current Topics in Cell Regulation*, **15**, 47–108.
9. Baxter, J. D. and Rousseau, G. G. (1978), *Glucocorticoid Hormone Action*, Springer Verlag, Heidelberg.
10. Gorski, J. and Gannon, F. (1976), 'Current Models of Steroid Hormone Action: a Critical Review.' *Annual Reviews of Physiology*, **38**, 425–450.
11. Marks, V. (1978), 'Laboratory Tests.' In *Scientific Foundations of Clinical Biochemistry Volume 1—Analytical Aspects* (D. L. Williams, R. F. Nunn and V. Marks, Eds.). Heinemann Medical, London pp. 1–12.

35. DISORDERS OF THE ANTERIOR PITUITARY GLAND

JOHN WRIGHT

Introduction

Anatomy
   The hypothalamus
   The pituitary gland

The anterior pituitary hormones
   Measurement of anterior pituitary hormones

Hypothalamic control of anterior pituitary secretion

Physiological control of hypothalamic—pituitary function

Tests of anterior pituitary function

Growth hormone
   Regulation of GH secretion
   Actions of growth hormone
   Tests of GH secretion
   Disorders of growth hormone secretion

INTRODUCTION

It is now half a century since the discovery of the portal capillary connection between the hypothalamus and the anterior pituitary gland, and over twenty-five years since the fundamental studies which established the principle of hypothalamic regulation of anterior pituitary secretion. The past fifteen years have witnessed two important developments: the demonstration of peptide hormone secretion by neural cells within the hypothalamus and the structural identification of a number of these peptides. These advances have not only helped to elucidate the mechanism of hypothalamic-pituitary interaction but have also provided new diagnostic tools for the investigation of pituitary function. The hypothalamic hormones are secreted in response to a multiplicity of neural, chemical and endocrine stimuli. Two hormones (oxytocin and vasopressin) are secreted directly into the systemic circulation from the posterior part of the pituitary gland (which is, in reality, no more than a direct extension of the neural tissue of the hypothalamus) and an understanding of the physiological effects of these peptides has been obtained from study of the responses of remote, accessible structures. The remaining hypothalamic hormones, however, act directly upon the anterior pituitary gland. The inaccessibility and close functional integration of this system have hampered physiological study and have blurred the distinction between hypothalamic and pituitary function which is only now becoming clearly defined.

ANATOMY

The Hypothalamus

The hypothalamus lies at the base of the brain, beneath the thalamus, between the optic chiasma and lamina terminalis anteriorly and the mammillary bodies posteriorly. It forms the floor and part of the lateral walls of the third ventricle up to the level of the hypothalamic sulcus and in the adult human weighs less than 2·5 grams. The stem of the pituitary gland arises from the infundibulum which projects from the under surface of the hypothalamus between the tuber cinereum and the optic chiasma. Within the hypothalamus is a number of aggregations of cell bodies, the hypothalamic nuclei, which, with the exception of the supra-optic nucleus, are poorly defined and indistinct anatomically. The cells within certain of these, particularly the supra-optic, paraventricular, infundibular and ventromedial nuclei, give rise to axons which either terminate in the median eminence or infundibulum, or pass through to the posterior lobe of the pituitary. These are neurosecretory cells and are the source not only of the posterior pituitary hormones (oxytocin and vasopressin), but also the releasing and inhibiting factors which control anterior pituitary function.

The hypothalamus is richly supplied by nerve fibres from all parts of the nervous system. In particular, there are major connections with the limbic system (hippocampus and amygdyla), the globus pallidus, the reticular formation and parts of the midbrain. These centres are among the oldest areas of the brain and are concerned with physiological homeostasis, autonomic function and arousal. The hypothalamus also receives visceral and somatic sensory impulses from spinal nerves via the mamillary bodies, an important olfactory pathway, and inputs from the newest area of the brain, the prefrontal cortex, via the cortico-hypothalamic tracts.

The Pituitary Gland

The pituitary gland (the *hypophysis cerebri*) comprises two distinct parts. The anterior pituitary or *adenohypophysis* consists of the anterior lobe (*pars distalis*) and the *pars tuberalis*, an upward extension which forms a thin cuff of cells around the front and sides of the infundibular stem. The posterior pituitary or *neurohypophysis* consists of the infundibular process (posterior or neural lobe) and the infundibulum. The area which forms the junction between the uppermost part of the infundibulum and the tuber cinereum is frequently referred to as the median eminence.

In man, a distinct intermediate lobe (*pars intermedia*) is found only during fetal life. Functionally and developmentally, the intermediate lobe should be considered as part of the adenohypophysis.

The anterior and posterior parts of the pituitary have quite different embryological origins. The adenohypophysis develops from Rathke's pouch, a midline diverticulum arising from the ectoderm of the primitive buccal cavity which migrates upwards to fuse with a down-

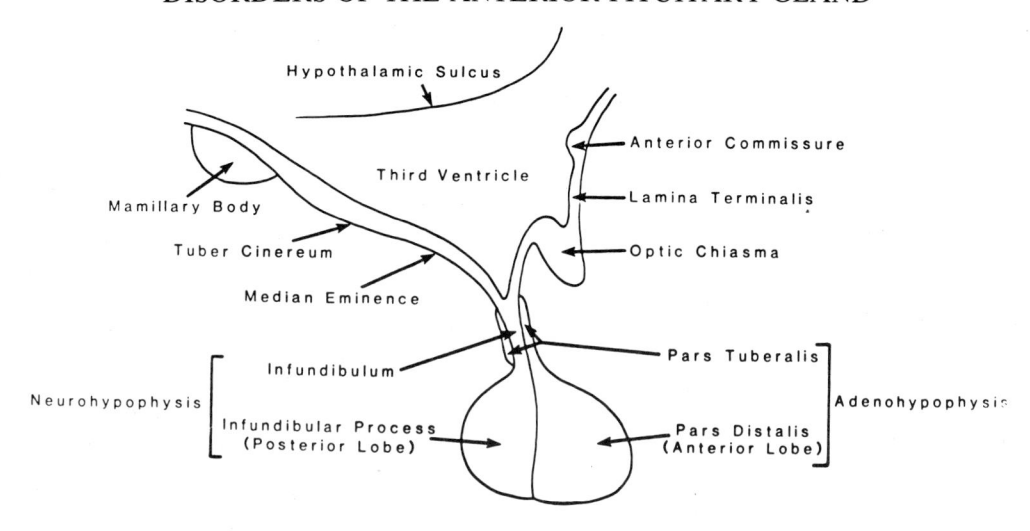

FIG. 1. Sagittal section through the hypothalamus and pituitary (diagrammatic).

growth of neuroectoderm from the hypothalamic part of the midbrain which becomes the posterior lobe of the pituitary. All connection between the anterior lobe and the roof of the pharynx is normally lost but the posterior lobe remains continuous with the ventral hypothalamus via the infundibular stem. The intermediate lobe develops from that portion of the adenohypophysis in contact with the posterior lobe and is separated from the anterior lobe by a cleft which represents the remnant of the lumen of Rathke's pouch.

In the adult, the pituitary measures approximately 12 mm in its transverse diameter and 8 mm anteroposteriorly. The whole gland weighs about 500 mg, increasing considerably during pregnancy largely due to expansion of the anterior lobe.

The pituitary gland occupies the pituitary fossa (*sella turcica*) within the sphenoid bone. The fossa is lined with dura mater which envelops the gland and forms an incomplete covering over the fossa (the diaphragma sellae) through which the pituitary stalk passes. Beneath the pituitary fossa are the sphenoid air spaces and on each side are the cavernous sinuses, each containing the internal carotid artery and the third, fourth, fifth and sixth cranial nerves. The optic chiasma lies on the diaphragma sellae, immediately above the anterior part of the pituitary, where it is vulnerable to compression by supra-sellar expansion of pituitary tumours.

The pituitary gland receives its blood supply from the superior and inferior hypophysial branches of the internal carotid arteries. The superior hypophysial artery supplies branches to the ventral hypothalamus and infundibulum, and a trabecular artery which supplies the lower pituitary stalk. The neural lobe is supplied by the anterior and posterior branches of the inferior hypophysial artery.

The arteries which supply the upper stalk and median eminence are peculiar in emptying into a system of sinusoids which are in intimate contact with the axon terminals from the hypothalamic nuclei. From these sinusoids arise the hypophysial portal vessels which pass down the pituitary stalk to the adenohypophysis which otherwise receives very little direct blood supply. The venous drainage of the pituitary is via short veins which empty into the surrounding dural venous sinuses.

Histology

With the introduction of immunological techniques (immunocytochemistry and immunofluorescence) a more precise classification of anterior pituitary cell types according to their secretory products has been possible. This has largely confirmed the relationships previously established by conventional staining techniques as summarised in Table 1. Staining characteristics are

TABLE 1
CLASSIFICATION OF ANTERIOR PITUITARY CELLS ACCORDING TO HORMONE PRODUCTION

Hormone	Staining reactions			Cytoplasmic Granule size (nm)
	Haematoxylin and Eosin	Mallory's Trichrome	PAS OG PFA AB*	
GH	Eosinophil	Acidophil	Yellow	300–450
PRL	Eosinophil	Acidophil	Yellow	500–750
ACTH and β-LPH	Cyanophil	Basophil	Red	200–400
FSH and LH	Cyanophil	Basophil	Blue	150–300
TSH	Cyanophil	Basophil	Blue	100–200

* PFA AB: Performic acid alcian blue
  PAS OG: Periodic acid—Schiff orange G

determined by the reactions of cytoplasmic secretory granules. Cells which contain few or no granules take up conventional stains poorly and are termed 'chromophobe'; these are commonly seen in pituitary adenomas which secrete prolactin (PRL) or growth hormone (GH) and in which very little hormone is stored prior to discharge.

THE ANTERIOR PITUITARY HORMONES

The hormones of the anterior pituitary fall into three groups: the somatomammotrophic proteins (GH and prolactin), the glycoproteins—thyroid stimulating hormone (TSH), follicle stimulating hormone (FSH), and luteinising hormone (LH)—and adrenocorticotrophic hormone (ACTH) and related polypeptide hormones. The structural characteristics of these hormones are summarised in Table 2.

existence of prolactin as a distinct hormone, separate from GH, in man. In addition, there are considerable structural and physiological similarities between GH, prolactin and placental lactogen, a hormone with somatomammotrophic properties produced by the placenta.

The glycoprotein hormones comprise two separate peptide chains, the α- and β-subunits. The amino acid sequence of the α-chain is identical in all three hormones (and in chorionic gonadotrophin) but there are

TABLE 2
THE HORMONES OF THE ANTERIOR PITUITARY GLAND

Classification	Hormone	Chain length		Molecular weight	Other features
1. Simple proteins	Growth hormone	191		21 700	2 intramolecular S-S bridges
	Prolactin	198		22 500	3 intramolecular S-S bridges
		α	β		
2. Glycoproteins	Thyrotrophin (TSH)	96	113	28 000	Carbohydrate substituents account for 10–20% of molecular weight
	Luteinising hormone (LH)	96	114	29 000	
	Follicle stimulating hormone (FSH)	96	115	32 000	
3. Polypeptides	Adrenocorticotrophin (ACTH)	39		4507	
	β-Lipotrophin (LPH)	91		9500	See Fig. 2
	β-Endorphin	31		3000	

GH and prolactin are closely related peptides; not only is there marked structural similarity between the two hormones but they are secreted by similar cells of the pituitary (*see* Table 1) and there is some overlap of physiological effects. Until recently the immunological similarities between the two molecules and the lack of pure preparations hampered development of a specific radioimmunoassay for prolactin and it was not until 1970 that conclusive evidence was obtained for the

some differences in their carbohydrate content. Functional specificity is determined by the β-chain. LH is identical to interstitial cell stimulating hormone (ICSH).

It is now clear that ACTH is one of a family of related peptides which includes α-MSH, β-MSH, β-LPH, γ-LPH and β-endorphin. These can be divided into two groups on the basis of similarities in primary structure (*see* Fig. 2). Group I includes ACTH and α-MSH both of which share a common heptapeptide core with β-LPH. Group

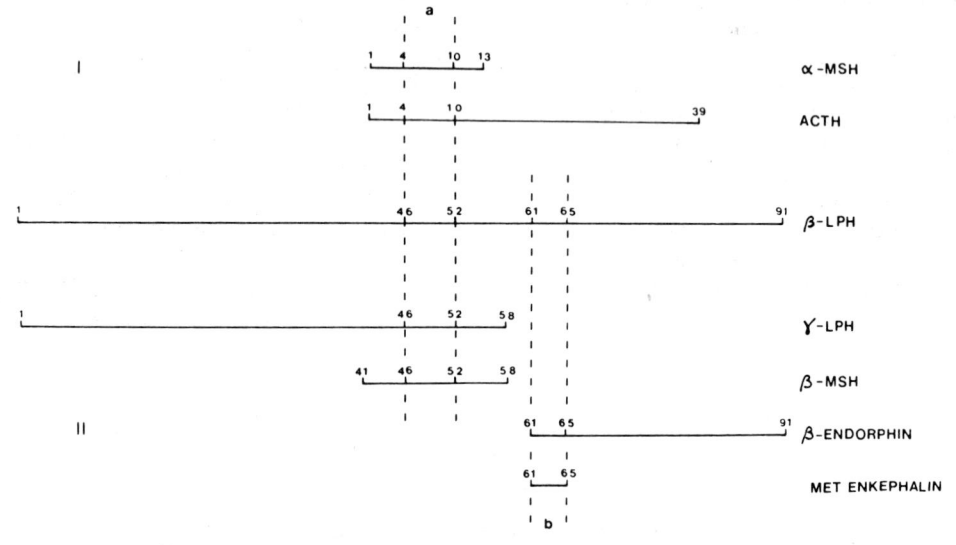

Fig. 2. Structural homologies of the ACTH-related peptides. The numbering of the residues is based on the relationship to ACTH in Group I and to β-LPH in Group II. The vertical dashed lines enclose (a) the common heptapeptide core, Met-Glu-His-Phe-Arg-Trp-Gly and (b) the sequence of Met Enkephalin, Tyr-Gly-Gly-Phe-Met. (Adapted from Prof. L. H. Rees, with permission.)

II includes β-LPH and a number of shorter peptides, which are identical to portions of the β-LPH molecule, and some of which (γ-LPH and β-MSH) also contain the heptapeptide shared by ACTH. This group also includes β-endorphin which is identical to the C-terminal fragment of β-LPH. This structure does not include the common heptapeptide but contains (at residues 61–65 in β-LPH) the structure of Met-enkephalin (see Fig. 3). Of all these peptides, only ACTH, β-LPH and β-endorphin appear to be secreted in significant amounts from the adult pituitary although others may be produced during fetal life. (Earlier demonstrations of β-MSH production were probably due to lack of assay specificity, with cross-reactivity between β-LPH and β-MSH). ACTH, β-LPH and β-endorphin are, under most circumstances, secreted simultaneously and in equimolar amounts and are probably derived from a common precursor molecule or prohormone. It has been suggested that β-endorphin, which possesses powerful central analgaesic properties, may serve to diminish the perception of pain incurred during periods of stress. Although there is no convincing evidence that β-LPH has an important physiological function in its own right, it may have considerable advantage from a laboratory viewpoint. β-LPH is very much more stable than ACTH *in vitro* and, since the two hormones are secreted in parallel, assay of β-LPH may prove a useful and robust alternative to ACTH assay.

Measurement of Anterior Pituitary Hormones

Reliable and specific radioimmunoassay methods are now available for the six major anterior pituitary hormones and these have largely replaced bioassay procedures. Problems of cross-reactivity still exist, particularly with assays for the glycoproteins. These have been partly overcome by antisera directed specifically against β-subunits but in many assays compensation for cross-reactivity is necessary, for example, by addition of an excess of LH to swamp the effect of endogenous LH in the TSH assay. Variations in specificity of antisera make comparison of different assays difficult and for this reason results should always be expressed in terms of established reference preparations.

In physiological situations, there is generally close agreement between radioimmunoassay and bioassay measurements. However, in certain pathological states, in particular when there is ectopic hormone production from a non-endocrine tumour, there may be considerable discrepancy between the two methods due to the production of prohormones, polymers or hormone fragments by tumour cells. In these circumstances, parallel measurement using both immunoassay and bioassay may provide valuable information.

HYPOTHALAMIC CONTROL OF ANTERIOR PITUITARY SECRETION

The synthesis and secretion of anterior pituitary hormones are under the control of individual releasing or inhibitory factors* produced within the hypothalamus. The precise structure of only three of these regulatory factors has yet been determined (see Table 3) but it seems probable that the majority are short-chain peptides. In common with oxytocin and vasopressin, they are synthesised within neurosecretory cells in hypothalamic nuclei and are transported down the axon processes of these cells to the infundibulum. However, whereas the majority of axons transporting vasopressin and oxytocin pass directly to the neurohypophysis, the anterior pituitary regulatory factors are secreted into the sinusoids of the hypophysial portal vessels in the median eminence and pituitary stalk, from where they are carried the short distance to the adenohypophysis in the portal capillaries. Some vasopressin also appears to follow this route but it is not clear whether this plays a significant role in the regulation of anterior pituitary secretion.

There is general agreement that vasopressin and oxytocin are synthesised in the supra-optic and paraventricular nuclei and although the precise anatomical origin of the anterior pituitary regulatory factors is uncertain, experimental evidence suggests that corticotrophin releasing factor is derived from the posterior hypothalamus, TRH from the anterior hypothalamus and GnRH from an intermediate area. High concentrations are demonstrable within the median eminence but this probably represents an accumulation in and around the axon terminals rather than a site of synthesis.

Although there may be both releasing and release-inhibiting factors for several or all of the anterior pituitary hormones, all of these, with the notable exception of prolactin, are under tonic stimulatory control, i.e. it is

---

* By convention, releasing or inhibitory substances whose structure is unknown are called 'factors' and those of known chemical structure 'hormones'. The term 'factor' is used here to describe both collectively.

TABLE 3
PRIMARY STRUCTURE OF HYPOTHALAMIC REGULATORY HORMONES

pGLU-HIS-PRO-NH$_2$
Thyrotrophin releasing hormone

pGLU-HIS-TRP-SER-TRY-GLY-LEU-ARG-PRO-GLY-NH$_2$
Gonadotrophin releasing hormone

H-ALA-GLY-CYS-LYS-ASN-PHE-PHE-TRP-LYS-THR-PHE-THR-SER-CYS
Somatostatin

the releasing factor which is the dominant influence under normal conditions. In contrast, the major influence on prolactin secretion is an inhibitory factor (PIF) and, consequently, hypothalamic damage or pituitary stalk section results in uninhibited hypersecretion of prolactin in association with impaired secretion of all other anterior pituitary hormones. Attempts to isolate a peptide inhibitor of prolactin secretion have been unsuccessful and current evidence indicates that PIF is not a peptide but the neurotransmitter dopamine.

It is now clear that the hypothalamic regulatory factors are not specific in either their anatomical localization or their action on pituitary secretion. Thus, GH release inhibiting hormone (GHRIH, somatostatin) is widely distributed, not only throughout the central nervous system but also the gastro-intestinal tract and the islets of Langerhans in the pancreas, and in addition to its action in inhibiting synthesis and release of GH, it inhibits a variety of endocrine secretions including insulin, glucagon and renin. Similarly, TRH is not entirely specific in its action as it stimulates the release of PRL in addition to TSH although the physiological significance of this is uncertain: the hypersecretion of TSH in hypothyroidism is only occasionally associated with hyperprolactinaemia, and the surge of prolactin secretion in response to suckling is not accompanied by an increase in TSH secretion. The synthesis of both gonadotrophins (LH and FSH) appears to be regulated by a single gonadotrophin releasing hormone (GnRH). The relative responses of LH and FSH to stimulation by GnRH in women of reproductive age probably depend upon the rate of stimulation and up on the circulating levels of ovarian steroids.

As well as providing for translation of a neurological impulse into an endocrine secretion, the hypothalamic-pituitary unit provides amplification of the neuronal signal by means of a cascade system. One inevitable consequence of such a cascade is the pulsatile nature of the secretion both of regulatory factors and pituitary hormones, the rate of hormone secretion being largely determined by the frequency of the pulses of releasing factor passing down the hypophysial portal capillaries.

It is becoming increasingly recognised that a number of diseases previously believed to be due to primary excess or deficiency of pituitary hormones (e.g. Cushing's disease and acromegaly; isolated deficiencies of GH and gonadotrophins) are usually due to over- or under-production of the corresponding hypothalamic releasing factor. As yet, this has few therapeutic implications although a handful of patients with isolated gonadotrophin deficiency have been successfully treated with long-term administration of GnHR.

The Enkephalins

The discovery within the last decade of specific opiate receptors widely distributed throughout the central nervous system has led to a number of exciting developments. The highest concentration of receptors is found not only in those areas of the brain concerned with pain perception such as the thalamus, but also in a number of brain-stem nuclei, in periventricular grey matter including the hypothalamus and, in greatest abundance, in the amygdala. Many of these areas have little to do with pain perception and it is clear that opiate receptors are concerned with a variety of sensory perceptions. Receptors have also been identified within the substantia gelatinosa of the spinal cord and in the myenteric plexus of the gastrointestinal tract. Although the pattern of distribution of opiate receptors has correlated well with the known pharmacological actions of morphine, it seemed improbable that they had evolved to interact with alkaloids from the opium poppy and, indeed, a number of naturally-occurring peptides possessing opiate-like activity and showing high-affinity receptor binding have now been identified. The first to be characterised were the enkephalins (see Fig. 3), two pentapeptides differing only in their C-terminal amino

H-TYR-GLY-GLY-PHE-MET-OH
Methionine–Enkephalin

H-TYR-GLY-GLY-PHE-LEU-OH
Leucine–Enkephalin

FIG. 3. Structure of the Enkephalins.

acid residue, which are closely associated with opiate-receptor neurones throughout the CNS and the gut. It is not yet clear whether the enkephalins themselves act as neurotransmitters or whether they modulate the release of other transmitters at synaptic endings. The relative proportions of Met- and Leu-enkephalin varies in different tissues and the two peptides are probably present in separate neurones and may not have identical actions. Met-enkephalin has also been found in human plasma and preliminary evidence suggests that the circulating pentapeptide may be derived from the adrenal medulla.

The amino acid sequence of Met-enkephalin is also contained within the structure of β-LPH (see Fig. 2), and forms the C-terminal portion of β-endorphin which not only has greater affinity for the opiate receptor but is over 20 times more potent as an analgaesic than the enkephalins themselves. (The generic term 'endorphin' is used to describe those β-LPH fragments containing Met-enkephalin and possessing analgaesic properties.) Both β-LPH and ACTH are also found within the brain and in cerebrospinal fluid. Since their concentration in these tissues is unaffected by hypophysectomy and is normal even in patients with undetectable plasma levels, it seems likely that they are synthesised within the neurones of the brain and then secreted directly into the cerebrospinal fluid.

The physiological function of the enkephalins and endorphins remains uncertain although they are clearly able to modify pain sensation and are probably involved in mediating the analgaesic effects of acupuncture which is associated with a rise in cerebrospinal fluid endorphin levels. The analgaesia produced by both endorphins and acupuncture can be reversed by opiate-antagonists such as naloxone. In addition, a role for endorphins has been suggested in the regulation of mood, temperature and appetite, in alcohol intoxication and in some psychiatric disturbances but much of this remains speculative.

The secretion of a number of pituitary hormones is influenced by enkephalins. GH and prolactin secretion is stimulated while, less consistently, gonadotrophin, ACTH and vasopressin secretions are suppressed. These effects are similar to those produced by morphine and other opiates, and are almost certainly due to modulation of hypothalamic releasing hormones. The physiological significance of these effects is unclear.

PHYSIOLOGICAL CONTROL OF HYPOTHALAMIC—PITUITARY FUNCTION

The major factors influencing hypothalamic-pituitary activity are:

1. Neural influences from other areas of the nervous system.
2. Hypothalamic chemo-receptors.
3. Feedback control by target organ hormones.
4. The pineal gland.

The precise origins of the neurological signals concerned in the regulation of hypothalamic-pituitary secretion are unknown but they probably come from many different areas of the brain and spinal cord, and are perhaps the most important of the regulatory mechanisms. Basal levels and cyclical variations in secretion rates, as well as the responses to a number of stimuli (e.g. stress, sleep) appear to be under neural control but this area of neuro-endocrinology is, as yet, poorly understood.

The hypothalamus contains specific areas sensitive to certain physiological stimuli (e.g. blood glucose concentration, osmolality) which are capable of mediating acute changes in both anterior and posterior pituitary secretions. The anatomical localisation of the majority of these areas is unknown but they are probably closely related to or identical to the nuclei which produce releasing factors in response to the particular stimulus.

With the exception of the positive feedback stimulation of gonadotrophins by high levels of oestrogen, secretion of anterior pituitary trophic hormones (TSH, ACTH, LH and FSH) is inhibited by the corresponding target organ hormone. This negative feedback system is clearly seen to operate both in the presence of primary target organ failure or hypersecretion which result in high or low levels respectively of trophic hormones, and also when target organ hormones are administered exogenously. However, there are clearly a number of situations in which feedback control of hypothalamic-pituitary secretion is over-ridden and it is probably less important physiologically than was previously thought. Nevertheless, feedback control mechanisms may be partly responsible for minute-to-minute control, with the background level of secretion set by supra-hypothalamic influences. In the absence of assays of sufficient sensitivity to measure hypothalamic releasing factors in peripheral blood, it is difficult to determine whether feedback effects are exerted upon the hypothalamus, the pituitary or both, although it seems increasingly likely that, as with other influences, the major effect is upon the hypothalamus. Feedback control may be exerted by target organ hormones (long loop), anterior pituitary hormones (short loop) and possibly by a very short loop system whereby hypothalamic releasing factors act to regulate their own secretion. Neither GH nor prolactin stimulates hormone production by a target organ and neither appears to be controlled by feedback inhibition (although the somatomedins may exert some effect on GH secretion). Interestingly, these are the only two anterior pituitary hormones for which there is good evidence for the existence of a hypothalamic release-inhibiting factor.

The role of the pineal gland in the control of hypothalamic-pituitary function is largely conjectural but it may be important in regulating the changes in gonadotrophin secretion which occur at puberty and in the establishment of secretory rhythms.

TESTS OF ANTERIOR PITUITARY FUNCTION

Anterior pituitary function can be assessed indirectly by measurement of circulating levels of target organ hormones, or directly by measurement of pituitary hormones either in the basal state or following stimulation with synthetic releasing hormones (TRH, GnRH) or an equivalent stimulus (e.g. vasopressin). Additionally, the function of the hypothalamic-anterior pituitary unit can be assessed by measuring the pituitary response to stimuli acting at the level of the hypothalamus. In practice, a combination of all three approaches is generally used. Table 4 summarises those tests in current clinical use.

Measurement of circulating levels of target organ hormones is frequently adequate to exclude absolute-pituitary insufficiency but gives little indication of pituitary reserve and is generally of limited value when pituitary hypersecretion is suspected.

Measurement of basal concentrations of anterior pituitary hormones is of value in confirming primary target organ failure, when elevated levels of the corresponding trophic hormone are found. Stimulation tests generally show an exaggerated, delayed response in these circumstances and add little further information. In contrast, in suspected hypothalamic-pituitary dysfunction, stimulation or suppression tests are usually necessary but the expectation that direct pituitary stimulation tests using synthetic releasing hormones would provide distinction between hypothalamic and anterior pituitary insufficiency has not been entirely fulfilled since the absence of endogenous tonic stimulation of the pituitary due to hypothalamic insufficiency results in a flat response to exogenous stimulation which is frequently indistinguishable from that seen in primary pituitary

TABLE 4

ASSESSMENT OF ANTERIOR PITUITARY FUNCTION. SUMMARY OF INVESTIGATIONAL PROCEDURES. (See text)
(HPT, HPA, HPG: hypothalamic-pituitary-thyroid, -adrenal, -gonadal axes)

Axis	Tests of target organ function	Tests of pituitary function		Tests of hypothalamic-pituitary function	
		Stimulation	Suppression	Stimulation	Suppression
HPT	Thyroid hormones (T3, T4)	TRH	—	—	T3
HPA	Cortisol and other adrenal steroids ACTH/Synacthen stimulation	Vasopressin	—	Hypoglycaemia Metyrapone	Dexamethasone
HPG	Oestradiol, progesterone Testosterone HCG stimulation	GnRH	—	Clomiphene Oestradiol benzoate	—
PRL	—	TRH Metoclopramide Sulpiride	L-Dopa Bromocriptine	Hypoglycaemia	—
GH	—	—	Somatostatin	All stimulation tests (see Table 5)	Hyperglycaemia

disease. More prolonged pituitary stimulation tests should help to resolve this problem.

Individual hypothalamic-pituitary axes may be tested separately but overall pituitary reserve can be conveniently assessed by means of a combined stimulation test using TRH, GnRH and insulin-induced hypoglycaemia as outlined below:

Patient fasted overnight
Zero time: In-dwelling intravenous cannula inserted

    Blood withdrawn for measurement of glucose, GH, cortisol (ACTH), TSH, Prolactin, LH and FSH

    TRH 200 μg
    GnRH 100 μg } given intravenously
    Insulin 0·1 u/kg*

20, 60 minutes: Blood withdrawn for measurement of TSH, LH, FSH, Prolactin
36, 60, 90, 120 minutes: Blood withdrawn for measurement of glucose, GH, cortisol (ACTH)

Prolactin secretion is stimulated by both TRH and hypoglycaemia. In order to distinguish between these responses, insulin should be given at 60 minutes, after completion of the combined TRH and LHRH test.

Interpretation of the GH response is discussed in detail below. Adequate hypoglycaemia (blood glucose falling to less than 2·2 mmol/l) is required to elicit a maximal cortisol response; plasma cortisol should normally rise by at least 160 μmol/l (6 μg/dl) with a peak level of at least 540 μmol/l (20 μg/dl).

The TSH response to TRH shows wide variations in normal subjects. Normal basal TSH values are less than 6 mU/l and may be indistinguishable from zero. The peak response to TRH occurs between 20 and 30 minutes and may be up to 20 mU/l. However, with a sensitive and reliable assay for TSH, a significant rise in TSH above the basal level may be considered normal. In hypothalamic or pituitary insufficiency, the TSH response to TRH is impaired or absent, and a similar response is seen in primary hyperthyroidism. In primary hypothyroidism, the basal TSH level is raised and the response to TRH is both augmented and prolonged.

The gonadotrophin responses to GnRH are generally proportional to the basal values. Pre-pubertally, the rise in FSH exceeds the rise in LH; this ratio is reversed midway through puberty. In adults, there should be a three-fold rise in both gonadotrophins following GnRH. Absent or impaired values are seen with hypothalamic or pituitary disease, including many women with 'functional' or weight-loss associated amenorrhoea; exaggerated and delayed responses are seen in patients with primary gonadal failure.

Further discussion of the hypothalamic-pituitary-adrenal, -thyroid and -gonadal axes are found in Chapters 36, 39 and 40.

GROWTH HORMONE

Regulation of GH Secretion

The major factors which influence GH secretion are summarised in Tables 5a and 5b. In the absence of specific stimuli, circulating levels of GH are low during the day with only minor fluctuations. However, there is a marked surge of GH secretion at night with one or more peaks occurring during early, deep, slow-wave sleep. Serum GH concentrations at these times may increase up to fifty-fold and it is clear that the majority of GH secretory activity takes place during sleep.

GH secretion is stimulated by an increase in the circulating concentration of arginine and of a number of other amino acids following either oral ingestion or intravenous infusion. The GH response to a fall in blood glucose concentration is related to the rate and magnitude of the fall and does not depend on the development

* See 'Tests of GH secretion', below.

of hypoglycaemia. However, the GH response to hypoglycaemia is reliable and quantifiable and remains the reference method for assessment of GH reserve.

The response of GH to neurotransmitters is complex. Acetyl choline, serotonin (5-hydroxy tryptamine), α-adrenergic and dopaminergic agonists have all been shown to stimulate GH secretion. These substances probably act by modifying secretion of GH-releasing factor from the hypothalamus. There is increasing evidence to indicate that serotonin may act as the final common pathway for a number of neurotransmitters, and that a serotoninergic mechanism may be involved in the GH response to hypoglycaemia, exercise, amino acids and slow-wave sleep.

Actions of Growth Hormone

GH possesses both anabolic and catabolic properties; these are summarised below:

1. Anabolic effects
    Stimulation of cellular uptake of amino acids
    Stimulation of RNA and protein synthesis
    Inhibition of protein catabolism and urea synthesis
2. Catabolic effects
    Stimulation of lipolysis
    Inhibition of cellular uptake and phosphorylation of glucose

The effects on protein metabolism and tissue growth are the most important physiologically; GH deficiency in childhood results in severe impairment of skeletal growth but only mild metabolic effects. In cartilage and bone, GH stimulates the synthesis of RNA, collagen, chondroitin sulphate (proteoglycan) and other proteins, resulting in skeletal growth. There is also an increase in protein synthesis in muscle and other tissues but it is not clear whether this contributes to overall somatic growth. *In vitro*, GH itself has been shown to be ineffective in stimulating anabolic processes in cartilage, and these effects are mediated *in vivo* by a group of low molecular weight peptides, the somatomedins and insulin-like growth factors, which are synthesised in the liver and secreted in response to stimulation by GH, insulin and feeding. Initially termed 'sulphation factor', evidence now suggests that there are several related peptides, possessing growth-promoting and insulin-like properties in varying degrees. It is not known whether the somatomedins exert a significant physiological effect on tissues other than cartilage and bone.

The physiological effects of GH are closely interrelated with those of insulin. The anabolic action of GH is synergistic with that of insulin and this action predominates in the postprandial state following stimulation of the release of insulin and GH by dietary carbohydrate and protein respectively. The recent demonstration that gastrin can stimulate GH secretion indicates that the GH response to feeding may be mediated by a humoral gut-hypothalamic-pituitary axis. However, the majority of GH secretion occurs at night and although circulating insulin levels are relatively low at this time, they may be sufficient to facilitate the anabolic action of GH or, alternatively, the insulin-like activity of the somatomedins may fulfil this role. The peak nocturnal secretion of GH coincides with the nadir in the secretion of cortisol which has a catabolic effect on protein metabolism, antagonistic to that of GH.

In contrast, the catabolic actions of GH are seen in the fasting state when insulin levels are low. Lipolysis is stimulated, providing free fatty acids which are the major energy source during fasting; insulin-dependent phosphorylation and utilisation of glucose are impaired, resulting in carbohydrate intolerance in states of GH excess.

The half-life of GH in serum is between twenty and twenty-five minutes, although many of its physiological effects are more sustained, probably due to the action of the somatomedins which have a longer half-life of 3 to 5 hours. GH is largely metabolised by the liver, only a small proportion appearing unchanged in the urine.

Tests of GH Secretion

Basal levels of circulating GH are of limited value in the assessment of patients with suspected pituitary disease. Resting levels are normally low and frequently indistinguishable from those found in hypopituitarism; conversely, the stress of a visit to hospital and venepuncture may be sufficient to raise serum levels into the acromegalic range in nervous subjects. For these reasons, the appropriate stimulation or suppression tests should be used in the investigation of patients with suspected under-production or over-production of GH, respectively (*see* Table 5).

1. Stimulation Tests

a. Insulin-induced hypoglycaemia (Insulin Stress Test)

This remains the standard provocative test for the assessment of pituitary GH reserve, and failure to produce an adequate rise in serum GH in response to satisfactory hypoglycaemia is the major biochemical criterion in the selection of growth-retarded children for treatment with GH. The standard test is performed with the patient resting in bed following an overnight fast. An in-dwelling intravenous cannula is inserted and specimens are obtained for basal GH and glucose assay. The patient is then given a single dose of soluble insulin by intravenous injection. The standard dose is 0·1–0·15 units/kg body weight. If there is a strong possibility of hypopituitarism, a smaller dose (0·05–0·1 u/kg) is used, and in patients with insulin resistance (e.g. obesity, diabetes, acromegaly, Cushing's syndrome) the dose is increased to 0·2 u/kg. Bood specimens are taken at 15, 30, 90, 60 and 120 minutes for measurement of glucose and GH. For the test to be considered satisfactory, the blood glucose level should fall to less than half the basal level or, preferably, symptomatic neuroglycopaenia should be produced.

This procedure is dangerous, especially in children

Table 5A
FACTORS ASSOCIATED WITH STIMULATION OF GH SECRETION IN NORMAL SUBJECTS

1. Sleep*
2. Stress
    Physical (e.g. surgery, exercise*)
    Psychological
3. Changes in circulating levels of metabolic fuels
    Fall in the concentrations of glucose* and free fatty acids
    Rise in amino acid* concentration
4. Biogenic amines
    Dopaminergic agonists (e.g. L-Dopa*, Bromocriptine, Apomorphine)
    Alpha-adrenergic agonists (e.g. Noradrenaline, Clonidine*)
    Beta-adrenergic antagonists (e.g. Propranalol*)
    Serotonin precursors (e.g. 5-Hydroxytryptophan)
5. Peptides
    Glucagon*
6. Oestrogens
7. Opiates
    Morphine and Enkephalin
8. Pentagastrin

Table 5B
FACTORS ASSOCIATED WITH SUPPRESSION OF GH SECRETION IN NORMAL SUBJECTS

1. Somatostatin
2. Changes in circulating levels of metabolic fuels
    Rise in the concentrations of glucose* and free fatty acids
3. Biogenic amines
    Alpha-adrenergic antagonists (e.g. Phentolamine)
    Beta-adrenergic agonists
    Serotonin antagonists (e.g. Cyproheptadine, Methysergide)
4. Corticosteroid excess
5. Hypothyroidism

Those marked with an asterisk have been used in the clinical assessment of GH secretion.

and in patients with pituitary or adrenal insufficiency. The test must be performed in hospital. The patient should not be left alone during the test and a physician should be close at hand. A patent intravenous cannula must be kept in place and the test should be terminated immediately by administration of intravenous glucose if the patient loses consciousness or if any other untoward effects occur. Particular caution should be observed in patients with epilepsy and ischaemic heart disease in whom an alternative stimulation test should be used.

Interpretation of the results is based upon the peak GH concentration reached, assuming that adequate hypoglycaemia was achieved. A peak value of less than 4 mU/l indicates severe GH deficiency; values between 4 and 10 mU/l represent severe partial GH deficiency, and between 10 and 20 mU/l, partial deficiency. A peak value greater than 20 mU/l indicates adequate GH reserve.

Hypoglycaemia also stimulates ACTH and prolactin secretion and, if required, specimens for ACTH, cortisol and prolactin should be collected at the same time as specimens for GH.

b. Prolonged oral glucose tolerance test (OGTT)

In normal subjects, blood GH concentration is suppressed during an OGTT but following the rise in blood glucose concentration, the fall back to normal provokes a rebound rise in GH concentration occurring between 2 and 4 hours after the glucose load. The GH response is less consistent than that provoked by hypoglycaemia but in normal children the serum GH usually exceeds 20 mU/l in at least one sample.

c. Arginine infusion and Bovril tests

Infusion of arginine stimulates GH secretion in normal subjects. After an overnight fast, arginine is given by intravenous infusion in a dose of 0·5 g/kg body weight up to a total dose of 30 g in a volume of 100 ml. Blood samples are collected before and at 30 minute intervals for two hours after the infusion for measurement of GH which usually reaches a peak value between 30 and 60 minutes after the start of the infusion. The test is free of side effects but the GH response is less consistent than the response to hypoglycaemia, particularly in adult men who may require pretreatment with oestrogen (stilboestrol, 0·5–1·0 mg every 12 hours for 48 hours) in order to elicit a maximal response.

In children, the amino acid content of Bovril may elicit a GH response. Blood specimens are taken at 30 minute intervals for two hours following a drink containing Bovril, 14 g/m^2 body surface area. Peak values of GH are generally less than after insulin and some normal children fail to respond adequately. This test has lost favour as a screening procedure.

d. Exercise

Assessment of the GH response to exercise is probably the best screening test in children with suspected GH deficiency. A single blood sample is taken 30 minutes after the start of a 15 minute period of vigorous, preferably standardised, exercise. A serum GH value of 20 mU/l or above at this time indicates adequate GH secretion.

e. L-dopa

The rise in serum GH which follows oral (or intravenous) administration of l-dopa is a useful test of GH reserve and is probably the best alternative in adults in whom an insulin stress test is contra-indicated. Following an overnight fast and collection of basal specimens, l-dopa 500 mg is given orally and further specimens are collected at 30 minute intervals for two hours. In children a smaller dose is used (e.g. 125 mg for body weight less than 15 kg, and 250 mg for body weight of 15 to 30 kg). Concomitant administration of propranalol enhances the GH response to l-dopa. Side effects of l-dopa include transient nausea, occasionally with vomiting. Peak serum GH levels usually occur between 60 and 120 minutes and are similar to those seen with insulin-induced hypoglycaemia.

f. Clonidine

The serum GH response to a single oral dose of Clonidine, 0·15 mg/m^2 surface area, has recently been proposed as an alternative to insulin-hypoglycaemia as a test for GH reserve. Peak values similar to those obtained following insulin occur between 60 and 120 minutes of Clonidine administration. However, this dose of Clonidine may result in severe arterial hypotension and further assessment is required before this is accepted as a standard procedure.

g. Glucagon

Subcutaneous injection of glucagon, 1·0 mg in adults, 0·5 mg in children, provokes a rise in serum GH which is enhanced by simultaneous administration of propranalol. The response to glucagon is delayed, usually starting at around 90 minutes and samples should be taken for a total of 4 hours after injection. For this reason l-dopa is to be preferred as a pharmacological stimulus of GH release.

2. Suppression Tests

a. Glucose tolerance test

In acromegaly and gigantism, basal serum GH levels may be very high, and in the presence of characteristic clinical findings there is frequently little doubt about the diagnosis. However, in many cases, basal levels are only moderately elevated and the biochemical diagnosis depends upon the failure of hyperglycaemia to suppress serum GH. After an overnight fast, a standard 75 g oral glucose tolerance test is performed with specimens for GH collected at 0, 30, 60, 90 and 120 minutes. In normal subjects, serum GH is suppressed to less than 3 mU/l in at least one specimen but in acromegaly and gigantism, serum GH concentration remains above 3 mU/l and there is frequently a paradoxical rise in response to hyperglycaemia. Similar results are obtained with an intravenous glucose tolerance test. In addition to abnormal GH levels, carbohydrate intolerance is found in about 25% of acromegalic patients.

Disorders of Growth Hormone Secretion

Acromegaly and Gigantism

Overproduction of GH in adults results in the syndrome of acromegaly. Rarely, excessive secretion of GH occurs before puberty, prior to closure of the epiphyses of long bones where longitudinal growth occurs, resulting in pituitary gigantism. Although a pituitary adenoma is present in the great majority of cases, the underlying defect may be an abnormality in the hypothalamic control of GH secretion, with pituitary enlargement due to chronic hyperstimulation. In acromegaly there is an increase in the width of long bones and in soft tissue thickness, giving rise to the characteristic spade-like hands and broad feet with thickened heel pad. There is also an increase in the growth of cartilaginous bone which is not limited by epiphyseal closure, resulting in enlargement of the mandible, thickening of the skull and prominence of the supra-orbital ridges which, together with the coarsening of facial features due to soft tissue growth, result in the typical acromegalic appearance. Other features include cardiac enlargement, arterial hypertension, non-toxic goitre, excessive sweating, lethargy, osteoarthritis, osteoporosis, pigmentation and nasal polyps.

The pressure of an expanding pituitary adenoma causes enlargement of the pituitary fossa which is usually evident on a lateral radiograph of the skull. Other local effects of the tumour include headache and visual field defects due to compression of the optic tracts by supra-sellar extension. The characteristic visual field defect is a bi-temporal hemianopia (loss of the outer visual field in both eyes) resulting from compression of the optic chiasma, but any other defect may be encountered.

Other pituitary endocrine deficiencies may occur due to either compression of the hypothalamus by supra-sellar extension of an adenoma, disruption of the hypothalamic-pituitary capillaries or direct damage to pituitary tissue by an expanding tumour. Hypogonadism may result from loss of gonadotrophin secretion but some features (amenorrhoea, oligospermia, loss of libido) may be caused by over-production of prolactin which occurs in about one third of acromegalic patients. In gigantism, hypogonadism delays epiphyseal fusion which prolongs the period of bone growth. Pituitary adenomas occasionally undergo spontaneous infarction (pituitary apoplexy) resulting in varying degrees of pan-hypopituitarism. This is more common with large tumours and may produce a dramatic clinical picture of severe headache, acute visual disturbance, hyperpyrexia and coma.

Rarely, acromegaly occurs as part of a syndrome of multiple endocrine adenomatosis (MEA Type I, Wermer's Syndrome) in association with parathyroid adenomas or hyperplasia, pancreatic islet cell adenomas (β- and non-β-cell) and carcinoid tumours of the gut.

Biochemical Investigations. Failure of serum GH concentration to suppress during a standard glucose tolerance test is characteristic of acromegaly (*see* above). The status of other hypothalamic-pituitary axes should be checked by means of appropriate provocation tests. It is particularly important to identify hyperprolactinaemia as this has therapeutic implications. TRH stimulation frequently produces a rise in blood GH concentration which is not seen in normal subjects.

Other Investigations. The diagnosis of active acromegaly depends upon biochemical criteria but other investigations are essential to determine the size of the pituitary tumour (skull radiography), the extent of suprasellar extension (pneumoencephalography or computerised axial tomography—CAT scan) and the extent of visual field losses. Soft tissue thickness can be assessed by a number of techniques including caliper measurement of skin-fold thickness, hand volume by plethysmography or heel-pad thickness on X-ray. Serial measurements of these parameters are useful in assessing progress of disease or response to treatment.

Treatment. The primary aim of treatment is to reduce the size of the pituitary tumour, particularly in the presence of suprasellar extension, and surgical hypophysectomy, with or without radiotherapy, remains the standard treatment in many centres. Recently, there has been considerable interest in the use of pharmacological agents, particularly bromocriptine which is most effective in those patients with associated hyperprolactinaemia. Although shrinkage of pituitary tumours has been observed on treatment with bromocriptine, primary treatment with surgery or radiotherapy is still recommended in most patients.

Following treatment, re-investigation is essential to assess the reduction in GH secretion and the status of other hypothalamic-pituitary axes. After radiotherapy, GH levels may fall very slowly and may continue to decline for several years, necessitating long-term biochemical follow-up.

GH Deficiency and Pituitary Dwarfism

In childhood, GH deficiency results in short stature due to impaired growth of long bones. This may occur as part of a syndome of panhypopituitarism (*see below*) but in the majority of cases there appears to be congenital failure of the hypothalamic stimulus to GH secretion without evidence of organic disease of either the hypothalamus or pituitary. The condition is commoner in boys, is occasionally familial (with autosomal recessive inheritance) and may be associated with deficiency of other pituitary hormones, particularly gonadotrophins, TSH or ACTH. Intrauterine growth is unaffected and birth size is usually normal, but careful plotting of height characteristically shows a reduced but linear growth rate with height falling below the third percentile by the age of four or five years. There is frequently an increase in subcutaneous fat and consequently most affected children appear short and plump. Bone age as assessed radiologically, is lower than chronological age but usually corresponds to height age. Puberty is often delayed, but in the absence of associated gonadotrophin deficiency, sexual development occurs normally and fertility may be unimpaired. The metabolic effects of GH deficiency are usually minimal and most adults with the condition remain asymptomatic. There is, however, a tendency to fasting hypoglycaemia and an increased sensitivity to insulin can be demonstrated.

GH deficiency accounts for less than 10% of cases of short stature. A summary of other causes is presented in Table 6. Most of those listed are not difficult to distinguish from pituitary dwarfism and therapy with GH plays no part in their management, but in those children in whom there is no apparent cause for short stature, the possibility of GH deficiency should be investigated.

Investigation. A screening test should be performed in all children in whom the diagnosis is suspected and of those available, exercise stimulation is the simplest and least invasive. In hospitalised children, a serum GH concentration measured during the first two hours of sleep is of value. If the GH response to exercise is inadequate and the clinical picture is suggestive of pituitary dwarfism, an insulin stress test should be performed. In a number of conditions, including hypothyroidism, emotional deprivation and malabsorption, the GH response to hypoglycaemia may be impaired but returns to normal on appropriate treatment.

Associated pituitary deficiencies should be excluded by means of the corresponding stimulation tests. However, it is not possible to identify gonadotrophin deficiency before the age of puberty since normal prepubertal values are low and indistinguishable from those found in hypogonadotrophic hypogonadism.

In the very rare but interesting condition of Laron dwarfism, serum GH levels are normal or elevated but somatomedin levels are low resulting in lack of tissue response to GH.

Hypopituitarism

The major causes of hypopituitarism are summarised in Table 7. In many of the conditions listed, the pituitary failure is secondary to hypothalamic damage or disruption of the hypothalamic-pituitary portal system. The clinical picture is very variable and depends on the underlying cause: surgical hypophysectomy is followed by rapid onset of clinical signs, with severe adrenal insufficiency developing within a few days if untreated, but in many conditions the onset is insidious and the full clinical picture may take many years to develop. Surprisingly, this may be the case in hypopituitarism due to apparently dramatic events such as post-partum infarction or severe head injury. In this situation, prolactin, GH and gonadotrophin secretion are usually lost first (failure of lactation and persistent amenorrhoea are the early symptoms of Sheehan's syndrome) with signs of thyroid and adrenal insufficiency often developing slowly. In established panhypopituitarism the features of

TABLE 6
MAJOR CAUSES OF SHORT STATURE

1. Congenital
   Familial short stature
   Constitutional slow growth
2. Chromosomal abnormalities
   Gonadal dysgenesis (Turner's Syndrome)
   Trisomies
3. Intrauterine growth retardation
4. Nutritional
   Malabsorption
   Malnutrition
5. Skeletal
   Osteochondroplasias
6. Chronic systemic disease
   Congenital heart disease
   Chronic renal failure etc.
7. Emotional deprivation
8. Endocrine
   GH deficiency
   Hypothyroidism
   Corticosteroid excess (endogenous or exogenous)
   Congenital adrenal hyperplasia
   Pseudo- and pseudopseudo-hypoparathyroidism
   Precocious puberty
   Somatomedin deficiency (Laron dwarfism)

TABLE 7
CAUSES OF HYPOPITUITARISM

1. Tumours
    Pituitary adenoma
    Craniopharyngioma
    Meningioma
    Glioma
    Pinealoma
    Metastatic carcinoma
2. Trauma
    Head injury
3. Infection
    Basal meningitis
    Tuberculosis
    Syphilis
4. Granulomas
    Sarcoidosis
    Histiocytosis
        Eosinophilic granuloma
        Hand-Schuller-Christian disease
5. Avascular necrosis
    Post-partum (Sheehan's syndrome)
    Infarction of tumour (Pituitary apoplexy)
    Diabetes
6. Idiopathic/Genetic
    Isolated and multiple pituitary hormone deficiencies
7. Iatrogenic
    Surgical hypophysectomy
    External irradiation
    Radioisotope implants
    Suppression by prolonged endocrine therapy (corticosteroids, thyroxine)

thyroid, adrenal and gonadal failure are combined with those of GH and prolactin insufficiency. Symptoms and signs include growth retardation in childhood, lassitude, cold intolerance and constipation; postural hypotension, nausea and abdominal pain; and loss of libido and secondary sex characteristics. The skin is usually pale, dry, fine and wrinkled, and there is thinning of scalp and body hair. Fasting hypoglycaemia may develop and in untreated patients, coma and death may ultimately supervene.

Pituitary damage does not usually result in diabetes insipidus although this may occur with more extensive hypothalamic involvement and occasionally latent diabetes insipidus is precipitated when treatment of hypopituitarism with steroids is instituted. The mechanism of this is unclear but probably concerns the interrelationship between vasopressin and corticotrophin releasing factor.

Biochemical Investigation. The characteristic electrolyte disturbance of primary adrenal insufficiency (hyponatraemia, hyperkalaemia and elevated blood urea) is not usually seen in hypopituitarism in which there is relative preservation of mineralocorticoid function. Suspicion is often raised, biochemically, by the finding of target organ failure without elevation of the corresponding trophic hormone (e.g. hypothyroidism with normal serum TSH concentration), but the diagnosis depends on comprehensive testing of hypothalamic-pituitary function as outlined above. Distinction between a hypothalamic and a pituitary lesion may often be made on the basis of the serum prolactin concentration which is usually raised in hypothalamic disease due to loss of prolactin inhibitory factor, and low in pituitary disease. Additionally, prolonged or repeated stimulation with releasing factors should elicit a response in hypothalamic disease but not in primary pituitary disease.

Other Investigations. Visual field testing and radiological assessment of the pituitary fossa and surrounding area should be performed as in acromegaly. The choice of other investigations will be determined by the suspected nature of the underlying pathological process.

Miscellaneous

Diabetes Mellitus. The GH response to a number of stimuli is increased in both insulin-dependent and non insulin-dependent diabetes, and there is also an increase in the frequency and height of secretory peaks of GH. The importance of this in the pathogenesis and natural history of the disease is unclear. The favourable response of diabetic retinopathy to hypophysectomy was thought to be due, in part at least, to the removal of GH, but the evidence for this belief is inconclusive.

Liver Disease. Basal GH levels are elevated in various forms of liver disease and may show a paradoxical rise after glucose. GH is metabolised by the liver and the elevated levels may be a reflection of decreased breakdown but are possibly due to the increased GH secretion which appears to occur in states of insulin resistance. The high GH levels may be partly due to impaired production of somatomedins in the liver and consequent reduction in negative feedback inhibition of GH secretion.

Alcoholism. In about one quarter of chronic alcoholics, the GH responses to hypoglycaemia is impaired or absent and in a number of these patients, the cortisol response is similarly affected. Abstinence from alcohol results in a return of both GH and cortisol responses. In the majority of alcoholics, this pituitary unresponsiveness to stress is asymptomatic but it is occasionally associated with the development of severe hypoglycaemia.

FURTHER READING

Besser, G. M. (ed.) (1977), 'The hypothalamus and pituitary', *Clin. Endocr. Metab.*, Vol. 6, No. 1.
Locke, W. and Schally, A. V. (Eds.) (1972), *The Hypothalamus and Pituitary in Health and Disease*, Springfield, Illinois: C. C. Thomas.
Martini, L. and Besser, G. M. (Eds.) (1977), *Clinical Neuroendocrinology*. New York: Academic Press.

36. THE THYROID GLAND

D. L. WILLIAMS AND R. GOODBURN

Introduction
   Aspects of thyroid anatomy
   Functions of the gland

Physiology of the thyroid gland
   Hypothalamic-pituitary control
   Synthesis of thyroid hormones
   Release of thyroid hormones
   Circulation of thyroid hormones
   Catabolism and excretion
   Metabolic and physiological effects

Disorders of the thyroid gland
   Pathological changes in thyroid function
   Congenital thyroid disease
   Auto-immune thyroid disease

Clinical aspects of thyroid disease
   Clinical presentation
   Disorders associated with hyperthyroidism
   Disorders associated with hypothyroidism
   Thyroid status in pregnancy
   Thyroid disease in neonates and children
   Thyroid function in the elderly

Investigation of thyroid function
   Introduction
   Thyroid function tests

Summary

INTRODUCTION

Disease of the thyroid gland is common and, although not usually life-threatening, it often causes major changes in the physical and mental well-being of the sufferer. Diagnosis is usually clear-cut. The available methods of treatment are straightforward, well understood and effective. The expenditure of effort and time on the part of clinicians and laboratory staff in ensuring accurate diagnosis and close monitoring of treatment is well rewarded in a noticeable general improvement in the health, both mental and physical, of their patients.

The part played by the clinical biochemistry laboratory in the diagnosis of thyroid disease varies from a relatively minor confirmatory role in the majority of patients, in whom clinical features of the disease are clear, to an important and central role in diagnosis of both adult thyroid disease, when it is mild or at an early stage, and of neo-natal thyroid disease. The role of the clinical biochemistry laboratory is also of central importance in monitoring patients during and after treatment, as changes in thyroid hormone levels will inevitably precede deterioration in the patient's clinical condition. A sound understanding of the physiology and pathology of the thyroid gland will thus enable the clinical biochemist to help his clinical colleagues in the interpretation of thyroid function tests, especially in those cases where the clinical features do not readily lead to the correct diagnosis.

Aspects of Thyroid Anatomy[2,17,18,19]

The thyroid gland arises during embryological development from endodermal tissue in the floor of the pharynx between the first and second pharangeal pouches. At about the sixth week of fetal life, this area, at the rear of the developing tongue, evaginates to form a downward growing tubular duct, the thyroglossal duct, which carries the embryonic thyroid tissue to its normal position where development of the lobes and the isthmus of the gland occurs. The duct degenerates to become a vestigial structure, but rests of thyroid tissue can occasionally be observed at the back of the tongue or in positions along the line of the thyroglossal duct. Pathological changes similar to those observed in the thyroid gland proper can occur in these isolated nodules of glandular tissue. If the gland does not descend, but remains at the back of the tongue, it is termed a 'lingual thyroid' and often exhibits reduced activity.

The adult thyroid gland is a vascular structure of about 25 g in weight, formed from two conical lobes joined by an isthmus and lying anterolaterally to the trachea just below its junction with the larynx.

The gland lies invested in the cervical fascia covered by a thin fibrous capsule. Posterior to the thyroid gland are the parathyroid glands, typically four in number and positioned two behind each lobe; there is, however, great individual variation both in the number and in the positions of the parathyroid glands. The recurrent laryngeal nerves pass on either side posteromedially to the lobes of the thyroid. These nerves innervate the vocal chords and can become damaged either by compression by an enlarged gland or surgically during a thyroidectomy operation; such damage results in hoarseness or complete loss of voice. The gland is not supplied by this nerve but by postganglionic fibres from cervical sympathetic ganglia and also by parasympathetic fibres from the vagus. These nerves branch into plexuses on the blood vessels within the parafollicular spaces, but it is not thought that any branches penetrate between the cells lining the thyroid follicle. It seems unlikely therefore that the thyroid nerves have any secretory activity but purely a vasomotor function.

The thyroid gland consists of follicles (Fig. 1) which are spheroidal groups of cells (follicular cells) forming a single-cell layer surrounding a central core of amorphous colloid composed of thyroglobulin. The follicles range in size from 0·2 to 1 mm in diameter, the larger ones being filled with colloid and surrounded by flattened follicular cells, the smaller ones being more active with less colloid and surrounded by columnar follicular cells. Between the follicles in the 'parafollicular' region, there is a rich capillary blood supply, lymphatic vessels, and specialised parafollicular cells which are now known to produce thyrocalcitonin. It was thought that these thyrocalcitonin-producing cells were limited to the parafollicular spaces, but it is now known that they can also occur interspersed between the thyroxine-producing cells lining the follicle. The gland is one of the most vascular organs in the body receiving its blood supply from the superior and inferior thyroid arteries, which supply 100–125 ml of blood per minute.

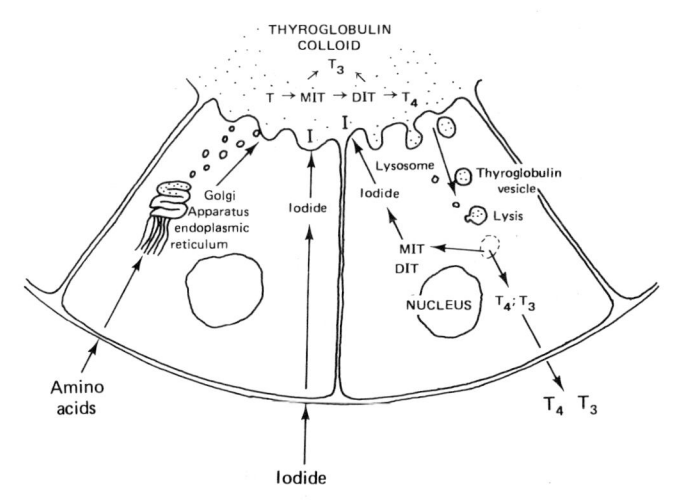

FIG. 1. Diagrammatic representation of thyroid follicular cells. Synthesis of thyroglobulin in the endoplasmic reticulum is indicated on the left, lysis to produce T_3 and T_4 is indicated on the right.

Functions of the Gland

The main functions of the gland are concerned with the trapping of iodine by the follicular cells, the synthesis of thyroid hormones and their storage in the follicular colloid, and their subsequent release into the blood stream. Each of these major functions is dependent upon stimulation by thyroid stimulating hormone (TSH) released by the anterior pituitary gland.

Thyroid Hormones

There are three major hormones produced by the thyroid gland, thyroxine (T_4), tri-iodothyronine (T_3) and reverse tri-iodothyronine (r-T_3). The structures of these hormones are shown in Fig. 2. The thyronine molecule is formed by a condensation reaction between two molecules of tyrosine during which the amino acid side chain of one molecule is lost and an ether (oxygen) bridge is formed between the two aromatic rings. The *in vivo* synthesis of the thyronine molecule is associated with the iodination of tyrosine, initially to form mono- and di-iodotyrosine intermediates, within the thyroid follicle.

Thyroglobulin[3,21]

It is likely that the major source of tyrosine utilised *in vivo* for the synthesis of thyroid hormones is the protein thyroglobulin, the major component of the central colloid of the thyroid follicle. Thyroglobulin is thought to be a 19S protein with a molecular weight of about 660 000. This molecule contains some iodine and further iodination can produce a 27S protein with a molecular weight of well over a million. Biochemical studies on thyroglobulin have been helped by the fact that the thyroid follicle almost entirely consists of thyroglobulin. An important immunological feature which is reflected in some forms of thyroid disease is that thyroglobulin can act as an antigen forming anti-thyroglobulin antibodies.

FIG. 2. Structure of the three major thyroid hormones.

Thyroid-stimulating Hormone (Thyrotrophin)

The synthesis and release of thyroid hormones is under the control of thyroid stimulating hormone (TSH) released from the pituitary gland. This is a glycoprotein secreted by thyrotroph cells in the anterior pituitary gland; it has a molecular weight of 28 000 and is composed of two distinct polypeptide sub-units. The alpha sub-unit is very similar to the alpha sub-units of luteinising hormone, follicle-stimulating hormone, and chorionic gonadotrophin. The beta sub-unit, however, is different from those in each of these other three hormones. Both sub-units are required for biological activity, but it is the beta sub-unit which confers upon TSH its specificity. Higher molecular-weight forms of TSH ('Big' TSH) have been identified in extracts of the anterior pituitary and it is likely that these larger molecules are those initially synthesised, subsequent proteolysis giving rise to the active hormone.

PHYSIOLOGY OF THE THYROID GLAND

Hypothalamic-pituitary Control of Thyroid Activity

The activities of the thyroid gland are stimulated by TSH from the anterior pituitary gland, which in turn is stimulated by thyrotrophin releasing hormone (TRH, a tripeptide-L-pyroglutamyl-L-histidyl-L-prolineamide) from the hypothalamus. Circulating thyroid hormones inhibit this hypothalamic-pituitary activity; by this means the concentrations of thyroid hormones are maintained at relatively constant levels. Details of the hypothalamic-pituitary control mechanism are given in Chapter 35.

Synthesis of Thyroid Hormones[17,18]

The synthesis and secretion of thyroid hormones can be considered in three stages:

1. The trapping of iodide.
2. The oxidation of iodide to iodine, the synthesis of thyroid hormones and their storage in thyroglobulin.
3. The hydrolysis of thyroglobulin and the subsequent secretion of the three thyroid hormones, thyroxine (T_4), tri-iodothyronine (T_3) and (r-T_3).

All three of these stages are stimulated by TSH.

The Trapping of Iodide

Iodide is actively transported from the blood, via the extracellular fluid, across the outer basement membrane of the thyroid follicular cell. The process is an active transport mechanism and is blocked by inhibitors of oxidation and by anoxia; ouabain also blocks the process, indicating that an ATP-ase ionic-pump mechanism, dependent on synthesis of ATP-ase, is involved. Other negatively charged ions of similar size, such as perchlorate, thiocyanate and bromide, compete for this transport mechanism; the radio-isotopic ion pertechnetate ($^{99m}TcO_4^-$) is also transported by this mechanism and has proved a useful label for use in thyroid scanning.

The transport of iodine is relatively slow, equilibrium being reached at about 60 minutes with a concentration gradient of 40:1 in favour of the thyroid cell. In practice, however, a stable equilibrium is not achieved in view of the much faster oxidation of iodide to iodine and its subsequent incorporation into thyroglobulin. Rarely there is a genetic defect in the iodide transport mechanism; in subjects with the defect iodide can still be transported into the cell by passive diffusion, continuation of the process being dependent upon the intracellular utilisation of the iodide; adequate uptake of iodide is achieved only if the plasma concentration of iodide is substantially higher than that normally found. Iodine is of course present in the environment only in limited quantities and there are geographical areas in which there is barely sufficient iodide for people with normal iodide transport mechanisms. These areas would presumably provide insufficient iodide for the rare cases where there is a congenitally defective iodide transport mechanism.

The Oxidation and Organification of Iodide

The next stage in the synthesis of thyroid hormones is a complex of three reactions, the oxidation of iodide to iodine, the iodination of tyrosine residues in the thyroglobulin molecules to mono- and di-iodotyrosines, and the joining of two iodotyrosine residues to form the complete hormones tri-iodothyronine and thyroxine.

Although the first two of these processes can take place independently *in vitro*, it seems likely that *in vivo* they are carried out sequentially on a multi-enzyme complex and that only small amounts of the intermediate can be identified within the thyroid cell.

The Peroxidase Reaction. At least one iodide peroxidase enzyme has been isolated from the thyroid follicular cell. It is present particularly along the apical cell border, although similar peroxidase activity can be found in membranes in other parts of the cell. The peroxidase is a haem-containing protein with an absolute requirement for hydrogen peroxide.

In cell-free systems the peroxidase can be shown to produce molecular iodine:

$$2I^- + H_2O_2 + 2H^+ = I_2 + 2H_2O$$

In the living cell, however, it is probable that the iodination is effected by an intermediate, enzyme-bound ion (such as the iodinium ion, $E^- I^+$) or an enzyme-bound radical ($E - I^.$). The importance of the peroxidase located on the apical membrane of the cell is that the incorporation of iodine into the thyroglobulin molecule occurs in this part of the follicle, utilising either newly synthesised thyroglobulin as it migrates into the colloid core or possibly thyroglobulin which has already reached this storage site. These views have been confirmed by histoautoradiographic studies.

Iodination of Thyroglobulin. Thyroglobulin can be readily iodinated *in vitro* by molecular iodine. It seems likely, however, that the ionic or free radical iodine products from the peroxidase reaction are responsible for *in vivo* iodination. The reaction also requires molecular oxygen.

The products of the reaction are the iodinated tyrosines, MIT and DIT, and the iodinated thyronines, T_4, T_3 and reverse-T_3. The tyrosine molecules that take part in the reaction do so in their covalently bound positions within the thyroglobulin molecule and the final products are released only at a later stage when fragments of the thyroglobulin colloid are hydrolysed within the follicular cell.

It is now known that the formation of the iodothyronines also takes place within the thyroglobulin molecule, by means of a reaction between two iodotyrosine residues on neighbouring polypeptide chains. Thus, tyrosine residues in peptide linkage within the thyroglobulin molecule, having initially been iodinated, participate in an intramolecular rearrangement that is associated with a conformational change of the molecule. An iodinated tyrosine residue in one part of the chain will thus gain an iodinated phenyl ring, linked via an ether bond, while the donor iodotyrosine is converted into a serine residue. The thyroglobulin can thus be thought of as a specialised enzyme; it has a number of active sites; it supplies two of the substrate molecules, the donor and recipient tyrosine residues; it catalyses the reaction and undergoes conformational change; but it fails to release the products of the reaction until the 'enzyme' itself is hydrolysed.

Release of Thyroid Hormones

Hydrolysis of thyroglobulin occurs within the thyroid follicular cell, following endocytosis of small portions of the thyroglobulin colloid by the microvilli that form the apical border of the follicular cell. The endocytosis is stimulated by thyrotrophin. It is likely that newly synthesized thyroglobulin is preferentially used for this purpose, possibly because it is sited nearest to the apical border of the cell. A diagrammatic representation of the site of thyroxine synthesis is given in Fig. 1.

This process whereby small portions of the colloid are pinched off and 'ingested' by the microvilli of the follicular membrane results in small vesicles, the colloid droplets, appearing within the follicular cell. These can be seen within minutes of stimulation by TSH. The droplets migrate towards the basal membrane of the cell and merge with lyzosomes. The lyzosomes contain hydrolytic enzymes including peptidases, which when released into the colloid droplet cause the hydrolysis of thyroglobulin molecules into the constituent amino acids, including the iodinated tyrosines and thyronines. The iodotyrosines undergo de-iodination, by active and specific deiodinases, but the iodothyronines diffuse, probably passively, through the basal membrane into the intracellular space and thence into the blood. Removal of the thyroid hormones by the blood stream, including their binding to thyroxine-binding proteins, means that a constant osmotic gradient exists between the higher concentrations of thyroid hormones intracellularly and the lower ones extracellularly. The process of breakdown of colloid droplets and release of thyroid hormones is relatively rapid, the colloid droplets having a half-life of about 5 minutes.

Circulation of Thyroid Hormones[4,17,18]

On being released into the blood stream both thyroxine and tri-iodothyronine become almost completely bound to carrier proteins. The major proteins involved in binding the thyroid hormones are thyroxine-binding globulin (TBG), which has a high affinity for the thyroid hormones but a relatively low capacity, and albumin and thyroxine-binding pre-albumin (TBPA), which both have a relatively low affinity but a high capacity. Thyroxine-binding globulin is an acidic glycoprotein having a molecular weight of some 63 000 and running between the α-1 and α-2 globulin bands on electrophoresis.

Although pre-albumin and particularly albumin have a much greater binding capacity than thyroxine-binding globulin, the higher affinity of TBG means that, under physiological conditions, about 75% of the circulating thyroxine and tri-iodothyronine is bound to TBG. Of the other two binding proteins, TBPA has a somewhat greater affinity for thyroxine, whereas albumin has a somewhat greater affinity for tri-iodothyronine. These two latter proteins do however play a much more important part in the hereditary disease thyroxine-binding globulin deficiency.

Under normal conditions only about 1 in 3000 molecules of thyroxine is present in the free state, the remainder being bound to thyroxine-binding proteins. Tri-iodothyronine is less strongly bound so that approximately 1 in 300 molecules is free.

The role of thyroxine-binding proteins is two-fold:

1. They do not pass through the renal glomerular membrane, whereas the free thyroid hormones do. Protein binding therefore conserves thyroid hormones, and particularly limits the loss of the scarce element iodine. It is interesting to note in this context that the greater binding affinity of thyroxine means that its turnover rate is only 10% per day whereas that of tri-iodothyronine is some 75% per day.
2. The large circulating pool of bound thyroid hormones is inert but in dynamic equilibrium with the small free fraction and can thus act as an available circulating pool to replenish losses of free hormone as they are taken up into cells, metabolised and/or excreted.

Both bound and free hormones can pass relatively freely between the capillary blood and the extracellular fluid, although this ability is more pronounced in rapidly equilibrating tissues such as liver and kidney and rather less in slowly equilibrating tissues such as muscle and skin. In all tissues, however, it appears that only the free thyroid hormones can penetrate the cellular membrane and enter the cells.

It is thus the free fraction which is responsible for the metabolic effects of the thyroid hormones including the important feed-back effects on the hypothalamus and pituitary gland.

In order to assess the thyroid status of the patient, knowledge of the concentration of free thyroxine and free tri-iodothyronine would be much more helpful than that of the total concentration of each hormone, however accurately these can be measured. This is particularly so because the concentrations of thyroxine-binding proteins, particularly TBG, can be significantly altered by physiological or pathological changes or by the effects of a wide variety of drugs. These changes will result in significant alterations in the total T_4 or T_3 concentrations, but may well not affect the free fractions of these hormones. Factors affecting TBG levels are summarised in Table 1.

TABLE 1
FACTORS AFFECTING THYROXINE-BINDING GLOBULIN (TBG)

Factors Affecting the Concentration of TBG

Rise in TBG concentration	Fall in TBG concentration
Physiological	
Pregnancy	
Neonatal life	
Pathological	
Chronic liver disease (sometimes)	Chronic liver disease (sometimes)
	Active acromegaly
Oestrogen-secreting tumour	Androgen-secreting tumour
Acute intermittent porphyria	Nephrotic syndrome
Acute hepatitis	Protein-calorie malnutrition
Hypothyroidism	Hyperthyroidism
	Cushing's syndrome
	Major illness
Drugs	
Oestrogens	Androgens
Contraceptive pill	Anabolic steroids
Perphenazine	Glucocorticoids (high levels)

Drugs Binding Competitively with TBG
(causing a fall in protein-bound thyroid hormones and thus a fall in total T_4 and T_3 concentrations, but not a fall in free hormone levels)

Salicylates
Phenytoin
Chlorpropamide
Tolbutamide
Diazepam

The strong binding of thyroid hormones in the blood also has a profound effect on the distribution of the hormones between the tissues. The rate of achievement of equilibrium is much slower than with substances of similar size which are not associated with binding proteins to the same extent. Several workers have suggested multi-compartment models to assess the turnover of thyroid hormones. One such model, using four compartments namely blood, gut, fast turnover tissues (liver, kidney) and slow turnover tissues (muscle, brain and skin), has been employed in the sheep. Results suggest that a little more than 20% of thyroxine is in the plasma, a little more than 30% in the fast turnover tissues, about 45% in slow turnover tissues and about 3% in the gut.

The effect of the high degree of binding of thyroid hormones is illustrated by the fact that administration of a single dose of thyroid hormones to hypothyroid patients does not produce a maximal effect on basal metabolic rate until two days afterwards for T_3, or 10 days afterwards for T_4.

Catabolism and Excretion of Thyroid Hormones

The major metabolic reactions undergone by thyroid hormones are de-iodination, conjugation, and deamination/decarboxylation.

De-iodination. There are at least three major de-iodination reactions which iodothyronines and iodotyrosines undergo. De-iodination can occur of either the phenolic ring (3' de-iodination) or the non-phenolic ring (3 de-iodination) of the iodothyronines. The de-iodinases responsible for these reactions have different pH optima (pH 6–7 for the 3' de-iodinase and pH 8–9 for the 3 de-iodinase). The enzymes have not yet been separated however, and it is possible that both actions may be catalysed by the same enzyme working under different conditions. If thyroxine is the substrate the 3' de-iodinase gives tri-iodothyronine as the initial product whereas the 3-iodinase produces rT_3. The specificities of these de-iodinases on T_3 and rT_3 are such that the di-iodothyronine product has one iodine on each ring rather than both on the same ring. It should be noted that when there is reduced activity of the 3' de-iodinase there is a reduced production from thyroxine of T_3; as the same de-iodinase is responsible for the breakdown of rT_3, however, there will be an associated increase in the concentration of rT_3. The contrary situation will occur if there is increased activity of 3' de-iodinase or a reduction of the activity of 3 de-iodinase.

It has been calculated that, under normal circumstances, some 25% of thyroxine secreted by the thyroid gland is converted to T_3, this reaction being responsible for some 80% of the body's production of T_3. Some 40% of thyroxine is converted to rT_3. It can thus be seen that the de-iodination mechanism is responsible not only for initiating the catabolism and excretion of thyroxine, but also for converting it to a more active or a less active product. It is likley, therefore, that the de-iodinase reaction is an important control mechanism for switching-on and switching-off the metabolic effects of the thyroid. It can be compared with the activation of 25-OH vitamin D (see Chapter 18).

De-iodination of mono- and di-iodotyrosines occurs particularly in the thyroid gland, but also in many other tissues. The tyrosine and iodine produced are then part of the general amino-acid and iodine pools respectively.

Conjugation Reactions. Up to about 20% of the thyroid hormones undergo conjugation in the liver. T_4 preferentially forms glucuronide whereas T_3, 3,3' di-iodothyronine and 3' mono-iodothyronine are preferentially conjugated to sulphate (it should be noted that these three compounds all have only a single iodine in the phenolic ring). The conjugates can then be excreted in the bile, although some sulphate conjugate can be found in the systemic blood. There appears to be no enterohepatic circulation of the conjugated hormones.

Deamination/Decarboxylation. As might be expected from the amino-acid nature of the side chain of T_3 and T_4, deamination and decarboxylation reactions can readily occur in the liver. The immediate products are tetra-iodo- or tri-iodo-thyroacetic acids (tetrac and triac). These compounds can be detected in the blood and have some thyroid hormone activity, which is unlikely to be quantitatively important. At least one study has suggested that more than 40% of an injected dose of radio-labelled thyroxine was excreted as thyroacetic acid.

Metabolic and Physiological Effects of Thyroid Hormones

Despite much work over the last quarter of a century the precise roles played by the thyroid hormones are not fully understood. The results of excess or deficiency of the hormones on the body as a whole have been well known for many years and it has been assumed, probably correctly, that many of the pathological effects exhibited in hyper- or hypo-thyroidism reflect an exaggeration or a deficiency of the normal physiological function of the thyroid gland. Thus, it has been considered for many years that the thyroid hormones have a role in the control of growth, this role being reflected in the short stature of the cretin and the gigantism of the pubertal thyrotoxic. The fact that thyroid hormones have a role in controlling metabolic rate is illustrated by the extremes of basal metabolic rate (BMR) which are observed in myoxoedema and thyrotoxicosis. For many years, work has been carried out on the cellular roles of thyroid hormones. Some of this work has tended to confuse the picture because of the use of non-physiological levels of thyroid hormones in *in vitro* studies. The cellular effects are now becoming more fully understood and it should not be long before a complete explanation of the role of thyroid hormones is available, enabling a description of the gross physiological and pathological effects in terms of the cellular biochemistry.

Generalised Effects of Thyroid Hormones

Effects on Growth and Development. It has been known for many years that thyroid hormones are required for the metamorphosis of amphibian animals such as the frog. In the human subject, thyroid hormones are required for growth and development, particularly in the neonate, the child and at puberty. Part of this effect is mediated by thyroid hormone stimulation of the hypothalamus-pituitary to produce growth hormone. Other effects seem to be independent of this ability to stimulate growth hormone production and are probably associated with direct stimulation of protein synthesis at the cellular level. It is possible that this action also involves the production of somatomedins.

Effects on Heat Production. Thyroid hormones cause an increase in the basal metabolic rate, the consumption of oxygen, and the production of heat. A longstanding and plausible theory was that thyroxine uncoupled oxidative phosphorylation, and this was shown in experiments *in vitro*. However, the level of thyroxine (or T_3) needed to elicit this response is thought to be substantially above the levels of thyroxine (or T_3) which occur physiologically. It is currently felt that if uncoupling of oxidative phosphorylation does occur, it is only in the most grossly thyrotoxic patients. It is more likely that the effects of the thyroid on heat production are a summation of a number of intracellular effects of the thyroid hormones, which are described below.

Effects on Neuromuscular Activity. Comparison of myxoedemic and thyrotoxic patients with normal subjects suggests that changes in thyroid hormone concentrations are responsible for changes in the excitability of the central nervous system, the neuromuscular junctions, and the muscles themselves, both splanchnic and somatic. It is probable that these effects are caused by changes in membrane permeability as well as in intracellular metabolism.

Effects of Thyroid Hormones on Metabolic Processes

Effects on Cell Membranes. Both thyroxine and tri-iodothyronine have been shown to increase the uptake of certain amino acids and carbohydrates in isolated cell systems. This effect is a much more rapid effect than some of the others, occurring within minutes, and therefore probably not a consequence of the stimulation of protein synthesis. It is also likely that thyroxine has a definite role *per se* in this respect and does not need first to be converted to tri-iodothyronine. Binding sites for thyroxine have been found on cell membranes, particularly of liver, and it is probable that binding of thyroid hormones to these sites is the trigger to increased permeability.

About half of the increased oxygen consumption produced by thyroid hormones has been shown to be associated with increased activity of $Na^+ K^+$–ATPase. This enzyme is of course responsible for maintaining the extracellular–intracellular differential gradients of sodium and potassium ions. Increased ATPase activity is associated with a great increase in the utilisation of energy and in the production of heat. This activity is likely to be responsible for about half of the increase in oxygen consumption produced by thyroid hormones.

Effects on Protein Synthesis. The administration of thyroid hormones increases the activity of many enzymes, including glycerol 3-phosphate dehydrogenase in mitochondria, NADPH-cytochrome C reductase, malic enzyme, β-hydroxy–β-methylglutaryl coenzyme A reductase, pyruvate carboxylase and phosphoenol pyruvate carboxylase. It is quite likely that it is by stimulating these enzymes, many of which are control enzymes affecting the rate of central metabolic reactions, that some of the effects of thyroid hormones, previously described, might be affected. Similar effects on $Na^+ K^+$–ATPase and on growth hormone synthesis have already been mentioned. In addition it has been known for more than 20 years that thyroid hormones cause increased albumin synthesis in man.

As these stimulatory effects of thyroid hormones are prevented by protein synthesis inhibitors, it is now generally held that many of the actions of thyroid hormones are mediated by stimulation of the synthesis of certain proteins including enzymes and not, for instance, by direct activation of existing enzymes. Studies on the way in which this stimulation might be mediated indicate that thyroid hormones have specific effects both on the nucleus, by regulating transcription, and at the ribosome, by affecting translation. These effects are summarised by Bernal and DeGroot[5] and by Rall.[20] In summary it is obvious that the thyroid hormones have complex effects at the subcellular level on the rate of metabolic processes.

Many of these effects may be mediated by stimulation of the synthesis of specific proteins (and possibly by the inhibition of synthesis of others), but some effects, e.g. the increased uptake of some amino acids by the plasma cell membrane, are likely to be independent of the protein synthesis effects.

DISORDERS OF THE THYROID GLAND

Disorders of the hypothalamic–pituitary–thyroid system can, as in most endocrine systems, result in overactivity or under-activity of the gland, or can cause no change in thyroid function. The gland can be affected by the same spectrum of disease processes that can strike other organs and tissues. Thus changes in function can be brought about by inborn errors of metabolism, congenital defects, infective disease, primary or secondary neoplastic change, toxic damage, auto-immune attack, changes in the control system, and iatrogenic factors, in addition to those ubiquitous pathologies of unknown aetiology.

Pathological Changes in Thyroid Function[1,10,15,17,18,19,23]

The main interests of the clinical biochemistry laboratory are in distinguishing between individuals with normal thyroid function (euthyroidism) and patients with either an under-active (hypothyroidism, myxoedema) or an over-active (hyperthyroidism, thyrotoxicosis) thyroid gland. The causes of changes in thyroid activity are manifold but, as far as the management of an individual patient is concerned, once hypothyroidism, hyperthyroidism or euthyroidism has been established, there is little point in most cases in further investigation of the precise pathological cause of the disease, as the choice of treatment is limited and usually independent of the results of biochemical investigations. Thus the treatment of thyrotoxicosis, by means of surgery, radio-iodine or anti-thyroid drugs, is decided on clinical grounds, such as the age of the patient and the appearance of the gland, and on the results of a thyroid scan, but not on either the degree of elevation of the thyroid hormone concentrations or the detection of thyroid-stimulating antibodies.

Knowledge of the pathological cause of thyroid disease is, however, of general interest and further study of the processes involved can increase our understanding of the aetiology and may, in the long term, result in improved patient management. Also in a small proportion of patients results of further investigations may have an immediate and direct effect on the choice of treatment. For instance, it is thought that a number of clinically hypothyroid patients, who have normal or even elevated levels of thyroxine in blood, associated with sub-normal levels of tri-iodothyronine, have a reduced ability to convert T_4 to T_3 in their tissues; it would obviously be preferable to treat such patients with T_3 rather than T_4.

Specific pathological causes of thyroid disease will not be dealt with in this section, but will be mentioned when the individual diseases are discussed. Two topics, however, have more general application and these will be dealt with here; these are congenital thyroid disease and auto-immune thyroid disease.

Congenital Thyroid Disease[9,14,17,18]

Congenital Hyperthyroidism

Most congenital disease of the thyroid results in diminished activity of the gland, but occasionally hyperthyroidism is observed in a neonate. Usually this is a consequence of a baby being born to a mother who is thyrotoxic and in whose blood there are high levels of thyroid-stimulating antibodies. These antibodies can pass the placental barrier into the fetal circulation and cause prolonged and inappropriate stimulation of the fetal thyroid. The condition is called *neonatal Graves' disease* and is usually self-limiting; in some cases, however, respiratory and cardiac involvement can lead to morbidity and even mortality. Rarely thyrotoxicosis develops in a neonate who has a euthyroid mother and cases have been reported of babies being born with genuine hyperthyroidism, for which infection with rubella has sometimes been blamed.

Occasionally a teratoma of the ovary can contain active thyroid tissue; this congenital tumour, the 'struma ovarii', can become overactive and suppress the activity of the normal gland, thus giving rise to thyrotoxicosis in a patient with an underactive thyroid gland; the toxic manifestations do not usually present until adult life.

Congenital Hypothyroidism

The majority of congenital thyroid disease results in hypothyroidism. The incidence of congenital hypothyroidism varies to some extent from study to study, often being dependent on the population observed; it is likely to be in the order of one case per 3–4000 births. It can result from anatomical or metabolic abnormalities.

Anatomical Defects. There are a numer of uncommon anatomical defects that can cause reduction of thyroid function. There can be congenital absence of the thyroid. Embryological maldevelopment of the gland can result in the whole of the gland, or part of it, being displaced from its normal position; reference has already been made to the lingual thyroid, to thyroglossal cysts, and to thyroid rests. These anatomical anomalies are often associated with reduced thyroid function; they can also be associated with normal thyroid function and it is therefore important to bear them in mind in the investigation of thyroid function in adults, as the abnormally placed tissue is capable of undergoing the same pathological changes as a normally situated gland. Surgical removal of a lingual thyroid usually demands subsequent treatment with thyroid hormone as there may be no other thyroid tissue.

Metabolic Defects. Neonatal hypothyroidism can be caused by a congenital metabolic defect. A number of such defects have been identified and include defects in the iodide concentration mechanism, defects in the iodination mechanism, and a defect in the enzyme which is responsible for deiodinating iodotyrosines.

The term *dyshormonogenesis* covers a number of specific inborn errors of thyroxine metabolism, each of which is of academic interest, but all of which are diagnosed in the same way and treated with exogenous thyroid hormones. These defects include:

1. The inability of the thyroid gland actively to trap iodide. Passive diffusion can, however, take place and, providing that there is a reasonable amount of iodide in the diet, this passive diffusion process can supply adequate iodide to enable the gland to synthesise sufficient thyroid hormone.
2. Deficiency in the oxidation and incorporation of iodide into thyroglobulin. This can be associated with deafness and large, hard, multi-nodular goitre (Pendred's syndrome) or it may be an isolated deficiency. The only well-documented congenital defect associated with poor organification of iodide is that of the enzyme iodide peroxidase, but it does not appear that all patients in this group have a peroxidase deficiency.
3. Defects of thyroglobulin metabolism leading to deficient iodination of tyrosine, deficient conjugation of iodotyrosines to give iodothyronines, and deficiencies of proteolysis have all been postulated, in addition to the possibility of abnormal receptors for the peroxidase molecule. These defects have not yet been conclusively demonstrated although abnormal thyroglobulin, with a high ratio of iodotyrosine:iodothyronine and increased levels of circulating iodoproteins, has been observed.

The importance of these and other causes of neonatal hypothyroidism is that, if undetected, they can cause physical and mental defects, ranging from reduced mental development to the full-blown syndrome of cretinism. Because of the relative frequency of neonatal hypothyroidism it is now the policy in the UK to screen all newborn babies six days after birth to indicate those which may have under-active thyroid function so that, following confirmatory investigations, treatment can be initiated at an early stage and mental retardation thereby prevented. Further details are given in Chapters 32 and 42.

Other Congenital Anomalies

Congenital Deficiency of Thyroxine-binding Globulin. Deficiency of the major thyroid-hormone-binding protein in blood, TBG, is a well-known anomaly which usually has no clinical implications. The level of TBG in individuals carrying a deficient gene ranges from an unmeasurably low concentration in male subjects, to detectable but lower than normal concentrations in affected female subjects. The reason for this difference is that the deficiency is inherited by an X-linked dominant trait. The fact that female carriers have intermediate levels of TBG despite the fact that one of their TBG-carrying genes is both abnormal and dominant, is in accord with the Lyon hypothesis which states that in each female somatic cell one of the two chromosomes is permanently inactivated during early fetal life. This inactivation process, however, is random so that in the single individual some somatic cells will carry a normal active X chromosome whereas others will carry an abnormal inactive X chromosome.

Congenital Thyroxine-binding Globulin Excess. In contrast to thyroxine-binding globulin deficiency some families have been described in which there is an *increased* concentration of thyroxine-binding globulin. There is some doubt as to whether this anomaly is inherited by an autosomal dominant or a sex-linked dominant mode. In either case, however, it is necessary to postulate a defect a little more complicated than in most inborn errors which result in a reduction in the level of protein and not an increase. In inherited TBG excess, it is possible that there is an inherited defect in the synthesis of a repressor gene which is normally responsible for inhibiting TBG synthesis.

In most subjects with either TBG deficiency or TBG excess, there is no associated clinical problem. There is, however, a danger that the associated low blood thyroxine concentration in TBG deficiency or high thyroxine concentration in TBG excess may be picked up in random or screening investigations. A doctor unwise enough to treat patients having either of these conditions without first having assessed their clinical condition, would cause iatrogenic thyrotoxicosis or myxoedema respectively.

Auto-immune Thyroid Disease[11,12,23,24]

In 1956, Roitt *et al.*, showed that antibodies capable of binding thyroglobulin existed in the plasma of patients with certain thyroid diseases. These were true auto-antibodies (*i.e.* they were directed against normal components of the body).

Since these original observations auto-antibodies against a number of thyroid antigens have been demonstrated in association with certain thyroid diseases. These diseases have been termed *auto-immune thyroid disorders* and include the following:

Graves' disease
Hashimoto's disease
Primary myxoedema
Focal lymphocytic thyroiditis

The studies on thyroid auto-antibodies led to the discovery of auto-immunity in association with other diseases (e.g. systemic lupus erythematosis, pernicious anaemia, Addison's disease) and, in a few cases, following damage to some tissues (e.g. antibodies formed to antigens released after vasectomy).

Thyroid auto-immune disease is undoubtedly the most intensively studied of the auto-immune conditions in man. These studies have been useful both as an aid to diagnostic medicine and in helping to elucidate the more widespread phenomenon of auto-immunity.

Characteristics of Auto-immune Diseases

Although many diseases have been said to have an auto-immune aetiology, very few have been shown unequivocally to conform to the strict requirements of this definition which are:

1. The presence of circulating auto-antibodies and/or delayed hypersensitivity reactions directed against normal body components (auto-antigens) must be demonstrated in patients with the disease.
2. The antigens stimulating the auto-immune reaction must be isolated and identified.
3. The auto-immune reaction should be duplicated experimentally in laboratory animals.

Chronic lymphocytic thyroiditis is one of the few so-called auto-immune diseases which have been shown to meet these strict requirements.

Pathological Significance of Thyroid Auto-antibodies

Three major auto-antibody/antigen systems are found in association with the thyroid.

1. Antibodies directed against thyroglobulin.
2. Antibodies directed against microsomal antigen.
3. Thyroid stimulating antibodies.

Additionally there are two systems which appear at present to have little clinical significance:

4. Antibodies directed against the second colloid antigen (CA_2).
5. Antibodies directed against Fragraeus cell-surface antigen.

Incidence of Thyroid Auto-antibodies

Before considering the significance of auto-antibodies in thyroid disease, their occurrence in the blood of normal individuals must be mentioned.

The detection of thyroid auto-antibodies has been facilitated by the introduction of techniques, such as haemagglutination, complement fixation, immunofluorescence, and more recently radio- and enzyme-immunoassays. These techniques are generally extremely sensitive and will ostensibly detect very low levels of auto-antibodies. Some of the methods, however, are prone to non-specific effects, and may show a weakly positive reaction in the absence of specific auto-antibodies. Thus in the most commonly used technique, haemagglutination, a titre of less than 1:20 for thyroglobulin antibodies, and 1:1600 for microsomal antibodies, is considered negative. However, positive reactions with titres well above these levels are seen in some normal individuals. The reported incidence in different studies has varied from 2% to 17·6% of the (normal) population under investigation. This wide variation may be due to regional differences or to difference in methodology. All the studies have shown that the incidence in women is substantially higher than that in men, and that, in women, the incidence increases with age and the disease is particularly prevalent in post-menopausal women.

Thyroid-stimulating antibodies have rarely been reported in normal individuals, but this may be a reflection of the technical complexity and relative insensitivity of methods currently available for their measurement. Receptor assay methods which are now being developed may enable a more accurate assessment of the incidence of thyroid stimulating antibodies in the normal population. In view of the presence in a significant proportion of apparently normal subjects of antibodies against thyroglobulin and thyroid microsomes, their measurement will not give an unequivocal diagnosis of thyroid disease. However, when used in association with other thyroid function tests and other non-biochemical investigations, it will help in the differential diagnosis of simple goitre, Hashimoto's disease, de Quervain's thyroiditis, thyrotoxicosis and thyroid cancer.

Auto-antibodies Directed Against Thyroglobulin

Thyroglobulin is not only found in the intrafollicular colloid, but can also be identified in circulating blood. In some individuals humoral antibodies can be found; the majority of these belong to the IgG class.

Although detection of thyroglobulin auto-antibodies is relatively simple by any of the methods previously mentioned, accurate quantitation is difficult. They are most commonly found in the blood of patients with untreated Hashimoto's disease; increased sensitivity of detection allows earlier diagnosis of this disorder. Thyroglobulin haemagglutination antibody (TGHA) titres of greater than 1:640 indicate the likelihood of Hashimoto's disease (providing there is supporting clinical and biochemical evidence); much higher titres than 1:640 are commonly observed. The absence of thyroglobulin auto-antibodies makes the diagnosis of Hashimoto's disease very unlikely. Positive titres are also observed in the disease described as idiopathic, or primary, myxoedema and suggest a likely aetiological link with Hashimoto's disease. Positive TGHA titres are frequently seen in patients with Graves' disease and it is possible that these patients may in the course of time revert to Hashimoto's disease. The usefulness of TGHA testing in possible thyroid cancer is dubious as the finding of a high titre will not necessarily exclude thyroid neoplasia.

In focal lymphocytic thyroiditis there is also an inconsistent appearance of THGA. High levels are often seen in this disease, but in some cases thyroid lesions are non-progressive, produce no clinical signs, and are associated with low titres of TGHA.

There are usually moderately elevated levels of TGHA in sub-acute (de Quervain's) thyroiditis. This disease is also often associated with the appearance of antibodies to the second colloid antigen (CA_2), which can be shown by immuno-fluorescent microscopy.

Thyroglobulin auto-antibodies are not usually cytotoxic themselves, but are important in the induction of cell-mediated immune processes which lead indirectly to the destruction of thyroid tissue. The cell-mediated immune response is probably induced by sensitisation of the thyroid cells by the IgG auto-antibodies, which then allows their destruction by killer lymphocytes (k-cells). Monitoring of this cell-mediated process is not easy in the laboratory.

Further complications of the auto-antibody–antigen reaction include:

1. The production of immune complexes in the blood of some patients with Hashimoto's disease.
2. The cross-reactivity of the auto-antibody with thyroxine and/or tri-iodothyronine.

This cross-reaction may give rise, when the thyroid hormones are assayed by immunoassay procedures, to results which are not in accord with the clinical condition. This characteristic of antibodies to thyroglobulin in binding thyroxine and tri-iodothyronine has been shown by auto-radiography following addition of radio-active T_3 to blood containing the antibody and also by immuno-fixation and radio-assay techniques.

It is unlikely that a similar cross-reaction also exists with other thyroid auto-antibodies and it is probable that the reason for such cross-reactivity with thyroglobulin is a consequence of the fact that the thyroglobulin molecule contains thyroxine and tri-iodothyronine residues. The prevalence of titres of anti-T_4 and anti-T_3 antibodies sufficient to have a significant effect on the level of circulating thyroid hormones is unlikely to be high, as experience shows that even in the presence of high titres of antithyroglobulin antibodies the thyroxine and tri-iodothyronine concentrations are usually appropriate for the clinical state of the patient.

Auto-antibodies Directed Against Microsomal Antigens

The presence of the second thyroid auto-antigen was first suspected when a positive complement-fixation reaction was observed (using a crude thyroid preparation as the antigen) in the absence of a haemagglutination reaction against thyroglobulin. Immunofluorescent studies showed that the antigen was located in the cytoplasm of the follicular cells and later studies showed it to be a lipoprotein in the membrane of the microvesicles which contain newly-synthesised thyroglobulin.

Anti-microsomal antibodies are also found which are specific to other endocrine organs. The antibody can now be detected by complement fixation, haemagglutination, immunofluorescence, and radio-immunoassay; at best only a semi-quantitative estimate of concentration can be obtained.

The thyroid microsomal antibody has a complement-mediated cytotoxic effect on thyroid tissue and a good correlation has been observed between the presence of microsomal antibody and the degree of hypothyroidism. In one study 83% of patients with overt clinical primary hypothyroidism had anti-microsomal antibodies. The presence of anti-microsomal antibodies is also a characteristic finding in the blood of patients with Graves' disease, in which an elevated anti-microsomal antibody titre (of greater than 1:3200) indicates severe thyrotoxicosis probably associated with concurrent cell damage.

Thyroid-stimulating Antibodies

In 1956 Adams and Purves discovered that serum taken from a patient with Graves' disease was able to mimic the action of thyroid-stimulating hormones when injected into a guinea pig; the duration of action of the serum was much more prolonged than that observed with TSH; the factor responsible for this activity was later called Long Acting Thyroid Stimulator (LATS).

Since that time a number of investigators, using a variety of techniques, have reported several substances which have thyroid-stimulating activities. These substances have been given a rather confusing array of names. All have been shown to be IgG immunoglobulins. A list of some of these stimulating antibodies is given in Table 2. Some have been given names which describe the particular effect by which they are assayed, others have been given more general names. Because none of these compounds have been completely isolated, it is not clear whether they are all distinct entities or whether the analytical methods adopted are measuring a number of different properties of one or more individual substances. Because they each, at least to some extent, stimulate thyroid activity and because they are all immunoglobulin antibodies, the terms thyroid stimulating antibody (TSAb) or thyroid stimulating immunoglobulin (TSI) are useful general names for this class of compound. The former name will be used in the rest of this chapter.

TSAb's are found in almost all (greater than 90%) untreated patients with Graves' disease. They can also be found in subjects who are clinically euthyroid or indeed in patients with Hashimoto's disease, where the stimulatory effect of the TSAb's is presumably counteracted by the damaging effect of other thyroid antibodies. It is also possible that, while TSAb's have stimulatory activity in patients with Graves' disease, their long-term effect is like that of most auto-antibodies, i.e. to damage the tissues against which they are directed.

It is likely that part of the TSAb molecule has a configuration similar to that of the part of the TSH molecule that fits into the receptor site of the thyroid follicular cell; TSAb's may indeed be antibodies against the receptor site itself. This similarity to TSH is sufficient to stimulate the gland, but presumably adjacent parts of the TSAb molecule bind more strongly to the area of cell membrane surrounding the receptor site than does TSH, so that the antibody molecule is less easily displaced or catabolized than TSH.

TABLE 2
THYROID STIMULATING ANTIBODY

Name of Antibody	Abbreviation	Effects	Means of Measurement
Long-acting thyroid stimulator	LATS	Present in the blood of some patients with Graves' disease; Causes prolonged thyroid stimulator activity in guinea pig or mouse bio-assay system.	Bioassay
Long-acting thyroid stimulator protector	LATS-P	Caused anomalies in bio-assays for LATS by preventing binding to the thyroid antigens; later found to have its own thyroid stimulating activity more closely related to the severity of Graves' disease symptoms.	Bioassay
Thyroid stimulating immunoglobulins	TSI	Alternative name for LATS-P	Competitive receptor assay using human thyroid membrane
Human thyroid adenylcyclase stimulator	H-TACS	Stimulates thyroid adenylcyclase	Stimulation of adenylcyclase in human thyroid membranes
Thyrotrophin displacement activity	TDA	Another name given to antibodies which react in the displacement assay mentioned above, possibly not thyroid stimulators.	
Thyroid stimulating antibody	TSAb	In addition to being a generic name for all these substances, this term has been used for substances which increase AMP concentration in thyroid slices.	Increased cyclic AMP in human thyroid slices (2 hr incubation)

Methods of Measuring Thyroid Antibody Levels

A summary of methods available for measuring concentrations of thyroid antibodies is given in Table 3. It is likely that further information about both the biochemistry and clinical importance of thyroid antibodies will follow the wider use of the more recently developed, quantitative and specific methods, such as the micro-ELISA technique.

Incidence of Auto-immune Thyroid Disease

The clinical characteristics of auto-immune thyroid disease are diverse and not always attributable to the auto-immunity. The gland is much more prone to auto-immune disease than other endocrine glands; it also seems to be the only gland in which an auto-immune reaction can commonly cause stimulation, instead of merely destruction. There is an increased incidence of other auto-immune disease in those with auto-immune thyroid disease, particularly diabetes and pernicious anaemia. Individuals with certain histocompatibility antigen (HLA) profiles seem prone to various forms of thyroid disease. Thus there is an association between the susceptibility to sub-acute thyroiditis and the HLA-BW35 antigen, and a higher incidence than expected of Graves' disease in those with HLA-B8 or HLA-Dw3 antigens. Subjects with Turner's syndrome have a higher incidence of thyroid antibodies than the normal population, and it is well known that thyroid antibodies occur more frequently in women than in men.

CLINICAL ASPECTS OF THYROID DISEASES

Clinical Presentation

Patients with thyroid disease can present with the classical symptoms of hyperthyroidism or hypothyroidism, or with visible or palpable abnormality of the gland itself which may or may not be associated with abnormality of endocrine function.

The clinical signs and symptoms of thyroid disease are well and fully presented in most general medical textbooks and in the increasing number of publications devoted to thyroid endocrinology. Description of these clinical features is therefore limited in this chapter to the summaries presented in Tables 6 and 9. Further details can be found in the excellent accounts mentioned in the reference list.[1,7,13,17,18,19,22] Clinical signs and symptoms may be less easy to recognise when the disease is mild or at an early stage, and may also be misleading when thyroid disease presents in the neonate, the child, in pregnancy or in the elderly. Attempts have been made to draw up 'scoresheets', by means of which clinicians give

Table 3
THYROID AUTO-ANTIBODY INVESTIGATION METHODS

Antibody	Detection System	Comments
Thyroglobulin auto-antibody	Immunodiffusion	Lacks sensitivity: useful only in advanced disease. Slow end-point development. Non quantitative.
	Countercurrent immunoelectrophoresis	Similar to immunodiffusion, but rapid end-point development.
	Latex fixation	Superseded by haemagglutination
	Haemagglutination	Now the most commonly used assay. Sensitive semi quantitative and relatively easy to perform, especially in kit form.
	Immunofluorescence	Requires U.V. fluorescence microscopy, and skilled interpretation. Non quantitative, but very sensitive.
	Radioassays	
	(i) Competitive	Most sensitive quantitative assay; requires radio-labelled high titre auto-antibody.
	(ii) Non-competitive	Relatively simple semi-quantitative or quantitative system. Requires radio-labelled thyroglobulin.
	ELISA	Relatively simple semi-quantitative method. Enzyme labels more stable than radio-labels.
Microsomal auto-antibody	Immunofluorescence	As for thyroglobulin autoantibody.
	Complement Fixation	Sensitive semi-quantitative assay, giving an indication of cytotoxicity. Largely superseded by haemagglutination.
	Haemagglutination	As for thyroglobulin auto-antibody, but subject to interference due to contamination of microsomal antigen by thyroglobulin.
	Radioassay	
	(i) Competitive	As for thyroglobulin auto-antibody. Requires radiolabelling of high-titre microsomal auto-antibody.
	(ii) Non-competitive	As for thyroglobulin. Requires radio-labelling of microsomal antigen free of contamination by thyroglobulin.
Second colloid auto-antibody	Immunofluorescence	Seen on immunofluorescence for thyroglobulin auto-antibody. No other assay systems developed.

points if certain signs or symptoms are present and subtract points if these clinical features are absent; the final sum of points indicates the probability of thyroid disease. Experience has indicated that this method is more useful in the assessment of hyperthyroidism than of hypothyroidism. Two of the most well known of the indices, the Wayne index and the Newcastle index, are compared in Table 4.

Disorders Associated with Hyperthyroidism

Hyperthyroidism, the clinical consequence of inappropriately high concentrations of circulating T_4 and T_3, is a relatively common condition; some authorities reserve the term for those diseases giving rise to increased activity of the thyroid gland itself, but not including increased thyroid hormone production from extra-thyroidal tissues; instead they use the term *thyrotoxicosis* to cover the complete list of causes of increased thyroid hormone activity; in this discussion the terms are used synonymously.

It is difficult to establish the precise incidence of hyperthyroidism, but estimates suggest that it is present in about 0·3–0·4% of the population. The incidence in women of reproductive age, the group mainly affected, is substantially higher than this. There is a genetic predisposition to hyperthyroidism, particularly in that associated with Graves' disease, in which female relatives of sufferers have up to an 8% incidence of the disease.

TABLE 4
COMPARISON OF THE WAYNE AND THE NEWCASTLE INDICES FOR THE CLINICAL DIAGNOSIS OF HYPERTHYROIDISM

	WAYNE INDEX		NEWCASTLE INDEX	
	Score if:		Score if:	
	Present	Absent	Present	Absent
Signs				
Pulse rate >90/min	+3		+16	
Pulse rate 80–90/min	0		+8	
Pulse rate <80/min	−3		0	
Atrial fibrillation	+4			
Palpable thyroid	+3	−3	+3	0
Thyroid bruit	+2	−2	+18	0
Exophthalmos	+2	0	+9	0
Lid retraction	+2	0	+2	0
Fine finger tremor	+1	0	+7	0
Hyperkinesis	+4	−2	+4	0
Lid lag	+1	0		
Hands hot	+2	−2		
Hands moist	+1	−1		
Symptoms	Score if present		Score if present	
Increased appetite	+3		+5	
Decreased appetite	−3			
Increased weight	−3			
Decreased weight	+3			
Dyspnoea of effort	+1			
Palpitations	+2			
Tiredness	+2			
Preference for heat	−5			
Preference for cold	+5			
Excess sweating	+3			
Nervousness	+3			
Psychological precipitant			−5	
Frequent checking			−3	
Anticipatory anxiety (severe)			−3	
Age of onset 15–24			0	
25–34			+4	
35–44			+8	
45–54			+12	
55 and over			+16	
Total scores				
Euthyroid	Less than 11		−11 to +23	
Hyperthyroid	Greater than 19		Greater than 40	
Equivocal	11–19		+24 to +39	

Although the list of possible causes of hyperthyroidism is long (Table 5), most cases fall into the categories of Graves' disease, toxic nodular goitre and over-treatment with thyroid hormones. The more common variants of the syndrome are discussed below.

The clinical presentation of hyperthyroidism is a classical one and there is usually no difficulty in making the correct diagnosis in moderate to severe cases; mild or borderline cases are less easy to diagnose and may require a series of investigations. A summary of the major clinical features is given in Table 6.

Graves' Disease

Graves' disease (toxic diffuse goitre, diffuse hyperthyroidism)[22] is the most common cause of hyperthyroidism, being especially so in women in the third and fourth decades. The clinical signs and symptoms (Table 6) are associated with a diffusely enlarged, non-nodular gland. In addition to the typical ophthalmic complications of hyperthyroidism, i.e. the wide-eyed stare, lidlag, and poor convergence, patients with Graves' disease may be affected by characteristic and specific ophthalmic lesions manifested by puffy, itching and inflamed eyelids associated with hyperaemic conjunctivae.

TABLE 5
VARIETIES OF HYPERTHYROIDISM

Graves' disease (toxic diffuse goitre)

Toxic nodular goitre
  Uni-nodular
  Multi-nodular

Thyroiditis
  Sub-acute (de Quervain's) thyroiditis
  Chronic thyroiditis with transient hyperthyroidism

Nodular hyperthyroid goitre due to excess exogenous iodine (Jod–Basedow disease)

Excess exogenous thyroid hormone
  Iatrogenic
  Self-administered

Neoplasia
  Follicular adenoma of thyroid
  Follicular carcinoma of thyroid
  TSH-secreting tumours
    Choriocarcinoma
    Hydatidiform mole
    Ectopic TSH-secreting tumours
    Pituitary TSH-secreting tumour
    (Secondary hyperthyroidism)
  Metastatic thyroid tumours
  Thyroxine-secreting teratoma (Struma ovarii)

TABLE 6
SIGNS AND SYMPTOMS OF HYPERTHYROIDISM
(in approximate order of incidence)

Signs
Tachycardia (occasionally progressing to atrial fibrillation)
Enlarged, palpable thyroid gland
Warm, smooth, moist, skin
Fine finger tremor
Hyper-active movements
Thyroid bruit (rushing sound of blood circulating through the thyroid heard on auscultation)
Eye signs (ranging from wide-eyed stare, mild proptosis and lid-lag to gross ophthalmopathy involving damage to eyelids, cornea and extra-ocular muscles)
Gynaecomastia in male subjects, breast enlargement in women

Symptoms
Nervousness and irritability
Increased sweating, especially palmar and facial
Intolerance to heat, preference for cool weather
Intermittent palpitations and an awareness of increased pulse rate
Weight loss, usually associated with increased appetite
Fatigue and weakness
Menstrual irregularity (usually oligomenorrhoea, occasionally dysfunctional uterine bleeding; reduced fertility)
Diarrhoea and hyper-active bowel

Graves' disease is known to be associated with auto-immune phenomena; in particular, thyroid-stimulating antibodies can be detected in high concentrations in the vast majority of patients with Graves' disease. These features have been mentioned earlier.

Malignant neoplastic change may rarely occur in Graves' disease. It is difficult to know whether there is a genuine increased incidence over the normal population, but one study has suggested evidence of malignancy in 20% of patients with toxic diffuse goitre compared with 0·5% in those with toxic nodular goitre. These figures should, however, be considered in the light of a study in Japan where carcinoma *in situ* was observed in 4% of post mortems carried out on subjects who had not suffered from thyroid disease, and this value rose to 28% if extensive serial sectioning of the glands was carried out. A 'cold' nodule in an otherwise 'hot' diffusely hyperplastic gland, as observed on thyroid scanning, should be taken as a warning sign that there may be an area of malignant change.

Toxic Nodular Goitre

Hyperthyroidism may be associated with one or more localised areas of the thyroid gland having increased endocrine activity; these nodules are usually palpable, exhibit increased activity on thyroid scanning, and are usually autonomous, having 'escaped' from hypothalamic–pituitary control. The increased concentrations of circulating thyroid hormones produced by the nodules are sufficient to cause negative feed-back inhibition of the hypothalamic–pituitary control mechanism. This results in low circulating levels of TSH and therefore lack of stimulation of the normal thyroid tissue. Thyroid scanning will show increased activity in the nodule(s), termed 'hot nodules', with the rest of the gland showing reduced activity. Repeat of the scan following intravenous TSH injection shows that the previously 'cold' tissue is capable of being stimulated to normal activity.

Toxic nodular goitre is observed in a somewhat older population than Graves' disease, although there is a significant overlap in the age of onset. As in Graves' disease there is a much higher incidence of toxic nodular goitre in female than in male subjects. Surveys have indicated that toxic nodular goitre is about half as common as Graves' disease, but the fact that the hyperthyroid symptoms are less severe may result in a less complete identification of subjects with toxic nodular goitre.

The clinical appearance of patients with toxic nodular goitre is variable. The symptoms are usually milder than in Graves' disease, and cardiac arrhythmias, especially atrial fibrillation, may be the only striking clinical sign, particularly in the older age group. The levels of T_4 and T_3 may be in the high–normal or borderline–high ranges, but there is suppression of TSH levels, both unstimulated, and following injection of TRH. More overtly hyperthyroid symptoms can be observed when the dietary iodine intake is increased.

Other Causes of Hyperthyroidism

Sub-acute Thyroiditis. This disease has a number of names, including de Quervain's thyroiditis, and it is likely that it is caused by a viral infection; mumps virus, Echo virus, influenza virus and coxsackie virus have all been implicated.

The clinical features are variable, but there is usually pain and tenderness in the region of the thyroid associated with moderate pyrexia and generalised systemic symptoms such as malaise, weight loss and anorexia. Symptoms of hyperthyroidism often occur in the early stages of the disease with increased nervousness, sweating, intolerance of heat and tachycardia being the most common. These symptoms appear to be associated with release of thyroid hormones from the thyroid gland, although during the course of the initial phase of the disease the thyroid follicles have a reduced uptake of iodine and reduced synthesis of thyroglobulin. It is obvious, if breakdown of thyroid stores is prolonged and/or severe and not accompanied by a re-building of the stores, hypothyroidism will follow. This is often observed, following a brief period in which the patient is euthyroid. The thyroiditis is usually self-limiting, however, and the period of hypothyroidism, either biochemical or clinical, is usually of short duration, following which the gland returns to normal function. The amount of damage done to the gland varies from patient to patient and the average duration of the disease is from 2–5 months. In about 20% of patients less severe recurrences occur which delay complete recovery.

Hashimotos Thyroiditis. (Chronic lymphocytic thyroiditis) is a chronic inflammatory disease of the thyroid gland which is associated with a similar auto-antibody picture to Graves' disease (see below). The aetiology is uncertain but it seems likely that it is an auto-immune disease.

The similarities between the immune phenomena of Graves' disease and Hashimotos disease are given in Table 7, which is taken, with modification, from Volpe.[23]

Hashimoto's lymphocytic thyroiditis is most frequently observed in female subjects and presents as an enlargement of the thyroid gland with relatively mild symptoms. In a minority of patients mild to moderate symptoms of hyperthyroidism can be seen in the initial stages but more frequently the symptoms are those of hypothyroidism, and therefore the disease is described more fully in the section of hypothyroidism.

Iodine-induced Thyrotoxicosis. Increases in the incidence of hyperthyroidism have been noted in areas in which the dietary iodide level, having previously been low, has been deliberately increased, for example by iodinating bread. Subjects particularly prone to develop this complication are those in whom there is already existing thyroid pathology, particularly multi-nodular goitre. There have also been reports of hyperthyroidism occurring in individuals who have been given iodide-

TABLE 7
COMPARISON OF FEATURES IN GRAVES' DISEASE AND HASHIMOTO'S THYROIDITIS

Features	Graves' disease	Hashimoto's thyroiditis
Circulating thyroid antibodies	Almost always present	Almost always present
Circulating thyroid stimulating antibodies	Almost all untreated patients	About 13%
Exophthalmos	About 50%	About 2%
Lymphadenopathy	Common	Rare
Hyper-gammaglobulinaemia	Uncommon	Common
Evidence of cell-mediated immunity	Yes	Yes
Association with other auto-immune diseases	Yes	Yes
Immunoglobulins in thyroid stroma	Yes	Yes
Histocompatability antigen link	HLA-B8, HLA-BW35, HLA-Dw3	?(conflicting evidence)

containing contrast media for radiological studies. The condition is usually mild and remits spontaneously.

Hyperthyroidism Due to Excess Thyrotrophin Production (Secondary hyperthyroidism). It is very rare for hyperthyroidism to be caused by increased production of pituitary TSH. In most of these rare cases a pituitary tumour is observable by radiographic methods. Occasionally, however, there is no distinct tumour, it being thought that the increased TSH secretion comes either from a microadenoma too small to observe or from general pituitary hypersecretion of TSH. Increased production of the alpha TSH sub-unit suggests a pituitary tumour.

Neoplastic Causes of Hyperthyroidism. It is very unusual for hyperthyroidism to be caused by over-activity of a primary thyroid carcinoma; metastases from this primary tumour may, however, produce significant amounts of thyroid hormones, not suppressed by lack of circulating TSH. Removal of such a primary tumour will not cure the hyperthyroid symptoms. Mention has already been made of the excessive production of thyroid hormones by a struma ovarii or by ectopic thyroid tissue.

Inappropriate stimulation of the thyroid resulting in hyperthyroidism can rarely occur by TSH from neoplastic tissue, which is not controlled by the high concentrations of T_3 or T_4 in the blood. Ectopic TSH-like activity can also be produced in patients with trophoblastic tumours (choriocarcinoma, hydatidiform mole) or in some forms of testicular cancer. It is probable that this TSH-like activity is associated with chorionic gonadotrophin, which is produced in large amounts by these tumours, and which has structural similarities to TSH.

Treatment-induced Hyperthyroidism. One of the most common causes of hyperthyroidism is over-enthusiastic treatment of hypothyroidism without adequate monitoring by measurement of TSH and/or thyroid hormones.

Thyrotoxic Crisis

Exceptionally, thyrotoxicosis, usually in Graves' disease or toxic nodular goitre, may become acutely exacerbated, by some precipitating cause such as infection, surgical operation or other trauma, to the severe state of *thyrotoxic crisis*. The symptoms and signs listed in Table 6 are present in an accentuated form, especially tachycardia, fever, sweating and restlessness, and sometimes delirium; later apathy, stupor and coma may occur. The condition is a medical emergency with poor prognosis, and treatment is aimed at correcting the thyrotoxicosis with anti-thyroid drugs, maintaining the cardiac output (as cardiac failure and hypotension can supervene), reducing the pyrexia and correcting body fluid, sodium and glucose levels. Despite these active measures the mortality rate is still of the order of 20%.[17]

Disorders Associated with Hypothyroidism

Hypothyroidism is probably rather less common than hyperthyroidism, occurs most frequently in a somewhat older age group and has a sex incidence ratio of about 6:1 (F:M). There is a long list of possible aetiologies (Table 8) but most patients fall into one or other of the two main categories, chronic lymphocytic thyroiditis or iatrogenic hypothyroidism. The main clinical features are listed in Table 9.

Auto-immune Hypothyroid Disease

Auto-immune Thyroiditis. (Hashimoto's disease, chronic lymphocytic thyroiditis.)[15] This category covers a group of overlapping disorders. Hashimoto's disease presents classically in middle-aged and elderly patients with the symptoms of hypothyroidism and a diffusely enlarged, horseshoe-shaped thyroid gland that has a firm, rubbery consistency. Histologically there is marked fibrosis, loss of thyroid architecture, and an infiltrate of plasma cells. It is associated with high titres of both microsomal and thyroglobulin antibodies.

Also often included in the classification of Hashimoto's disease is a more common but less severe form, presenting at an earlier age and associated with low-normal or mildly sub-normal thyroid hormone concentrations. The goitre is not as firm as in the classical variant and histological examination shows predominantly lymphocytic infiltration with minimal fibrosis.

TABLE 8
CAUSES OF HYPOTHYROIDISM

Congenital anomalies
   Agenesis and maldevelopment
   Ectopic thyroid
   Dyshormonogenesis
   Congenital exposure to iodide or anti-thyroid compound

Auto-immune thyroid disease
   Auto-immune thyroiditis (Hashimoto's disease)
   End-stage Graves' disease

Sub-acute thyroiditis

Iatrogenic hypothyroidism
   Post radiation hypothyroidism
   Post radio-iodine hypothyroidism
   Post-thyroidectomy hypothyroidism
   Excess anti-thyroid drugs (also some food substances)

Infiltrative disease of the thyroid

Endemic hypothyroidism

Disorders of thyroid hormone activity
   Peripheral resistance to thyroid hormones
   Anti-thyroid hormone antibodies
   Impaired conversion T_4 to T_3

Hypothalamic-pituitary hypothyroidism
   Depressed hypothalamic function ('tertiary hypothyroidism')
   Panhypopituitarism
   Selective thyrotrophin deficiency (secondary hypothyroidism)
   Anorexia nervosa

TABLE 9
SIGNS AND SYMPTOMS OF ADULT-ONSET HYPOTHYROIDISM
(in approximate order of incidence)

Dry, coarse, cool skin
Myxoedemic features (coarse skin and lips, puffy eyelids, thick tongue)
Slow, hoarse speech
Thin dry, brittle hair
Undue sensitivity to cold (wearing warm clothes even in warm weather)
Impaired cerebration, poor memory, slow reactions (both mental and physical)
Apathy, listlessness, decreased libido
Slow ankle reflex (both on contraction and especially on relaxation)
Constipation
Weight gain (often associated with decreased appetite)
Peripheral oedema
Slow pulse
Menstrual irregularity (esp. menorrhagia but occasionally amenorrhoea and resulting in impaired fertility)
Deafness
Hyperlipidaemia (giving rise to xanthomata, occlusive vascular disease and its complications, cardiac enlargement and hypertension)
Myxoedema coma (rare, usually in elderly women, associated with hypothermia, poor prognosis)

There are usually high titres of microsomal antibodies, but thyroglobulin antibodies are not a common feature. The incidence of both forms of the disease is much greater in women than in men, the sex difference being more pronounced in the less severe form.

These diseases can co-exist with, or follow, overt Graves' disease and it is tempting to speculate that the many similarities between these diseases (see Table 7) indicate a causal relationship.

End-stage Graves' Disease. That there is a link between Graves' disease and hypothyroidism is undoubted, even if hypothyroidism resulting from radical or over-enthusiastic treatment is excluded. Some patients with Graves' disease require long term anti-thyroid drug therapy, or need repeat partial thyroidectomy because of recurrence of hyperthyroidism; others appear to go into remission, do not require further treatment and exhibit a reduction in anti-thyroid antibodies to normal levels; others, however, pass from clinically overt Graves' disease through a stage of euthyroidism to genuine hypothyroidism; this conversion frequently takes 10–15 years. It is possible that some patients with spontaneous hypothyroidism occurring in middle age have suffered from the sub-clinical Graves' disease. Occasionally patients presenting with hypothyroidism will admit retrospectively to symptoms suggestive of hyperthyroidism occurring some years previously. The ophthalmic complications of Graves' disease are increasingly being reported in patients with hypothyroidism who have not apparently suffered from hyperthyroidism previously.

In contrast some patients who develop Graves' disease do so after experiencing symptoms and showing clinical and biochemical signs of hypothyroidism which may have lasted for some years.

Sub-acute Thyroiditis

As mentioned previously sub-acute thyroiditis, having passed through the acute inflammatory and hyperthyroid phase may revert via euthyroidism to a hypothyroid state. This is usually self-limiting but is sometimes severe enough to warrant temporary treatment with replacement thyroxine.

Endemic Hypothyroidism

Endemic hypothyroidism occurs in populations living in areas where the natural supply of iodide is sub-optimal. It is associated with a much greater incidence of endemic goitre, particularly in mountainous regions of the world such as the Alps, the Himalayas and the Andes. The goitre is of course a physiological response to the low environmental levels of iodide. Within the populations of those regions there is a spread of thyroid function from normal euthyroidism, through borderline hypothyroidism with sub-normal thyroxine levels but clinical euthyroidism, to overt cretinism. Treatment is largely prophylactic, once a region has been identified as being iodine deficient. Additional iodide can be introduced into the diet although care must be taken in initiating such measures that some individuals do not swing suddenly into a thyrotoxic phase.

Disorders of Thyroid Hormone Activity

Resistance to Thyroid Hormone Action. Results of a number of studies have implied that some patients suffer a rare familial metabolic defect in which there is a reduced end-organ response to circulating thyroid hormones, which are usually present in such patients at substantially increased concentrations. These patients do not complain of symptoms or show signs of thyrotoxicosis; indeed they frequently appear, clinically, to be mildly hypothyroid. The reduced response is shown not only in clinical signs and symptoms, but also in a normal or reduced basal metabolic rate, and in the level of TSH, which will exhibit a normal or even elevated response to intravenous injection of TRH. In past years this diagnosis has been arrived at by exclusion of other possible explanations.

Anti-thyroid-hormone Antibodies. The presence of high levels of circulating auto-antibodies which react with thyroxine and tri-iodothyronine could also give rise to similar clinical and biochemical features as discussed below. It would be expected that antibodies to T_3 and T_4 may give rise to spurious changes in total T_3 and T_4 concentrations; the physiologically-active hormone levels could be estimated by direct measurement of the concentrations of the circulating free hormones. Thus it would be expected that this condition could be differentiated from the condition in which there is reduced peripheral response; it is not yet clear that this expectation can be fulfilled. With methods now available, however, the amounts of circulating antibodies can be accurately assayed and there are available some *in vitro* systems for assessing end-organ response.

Impaired Conversion of Thyroxine to Tri-iodothyronine. It is known that tri-iodothyronine has at least five times the biological activity of thyroxine; indeed it may be that thyroxine has no activity *per se* but first requires conversion to tri-iodothyronine before it can exert its biological effects. The ratio of thyroxine: tri-iodothyronine in circulating blood, which is of the order of 50:1, is by no means constant either among normal subjects or among those undergoing thyroid investigation. Whether this is due to individual variation in the ability of tissues to convert thyroxine to tri-iodothyronine, or whether it is due to individual variation in the subsequent metabolism and excretion of the hormones is not known.

It does appear, however, that there is a group of patients in whom the concentration of tri-iodothyronine is inappropriately low when compared with that of thyroxine and some of these patients have the signs and symptoms of hypothyroidism. These findings suggest the possibility of a deficiency of the deiodinase responsible for this inter-conversion. If this is a genuine clinical condition, and if it can be accurately diagnosed, it would obviously be appropriate to treat these patients with tri-iodothyronine rather than with thyroxine.

It is possible that at least in some of these patients there has been a switch of metabolism of thyroxine to reverse tri-iodothyronine. This is known to happen in certain conditions, especially in protein-energy malnutrition, including anorexia nervosa, in postoperative patients and in severe acute illness, including hypothermia.

Drugs and Goitrogens Causing Hypothyroidism

In addition to those drugs which are given therapeutically with the deliberate intention of reducing thyroid activity, a number of other drugs and dietary substances are known to interfere with the production of thyroid hormones. Some of these are listed in Table 10.

TABLE 10
A LIST OF SOME SUBSTANCES KNOWN TO DEPRESS THYROID ACTIVITY

Substance	Mechanism of effect
Drugs	
Lithium	Possible inhibitory effect on the action of cyclic-AMP.
Sulphonamides	Inhibits organification of iodide (noted only in animals).
Sulphonylureas	Blocking iodide uptake (effects extremely rare in commonly used Sulphonylureas, but powerful effect in Carbutamide which is now withdrawn).
Thiocyanate	Blocks uptake and binding of iodide (rarely used).
Food substances	
Soya bean flour	? binds iodide in GI tract and prevents absorption.
Cabbage, turnip and associated vegetables	Contain thioglucosides which are converted to thiocyanate and inhibit iodide uptake.

Disorders of the Hypothalamic–Pituitary Control Mechanism

Reduced Hypothalamic Function is particularly noticeable in subjects who have reduced their dietary intake through personal choice or famine. Thus, in anorexia nervosa borderline-low levels of thyroid hormones, particularly tri-iodothyronine, are observed associated with low concentrations of TSH. That the defect is not in the pituitary gland can be shown by intravenous injection of thyrotrophin-releasing hormone when a satisfactory response, albeit somewhat delayed, can be shown. This may be associated with a similar depression of reproductive function caused by poor pituitary stimulation. The hypothyroidism has been classified as *tertiary hypothyroidism*.

Hypopituitarism. *Secondary hypothyroidism* is one of the associated symptom complexes in panhypopituitarism of whatever cause. Rarely, a condition in which there appears to be an isolated deficiency of TSH has been described. In patients with this deficiency low levels of thyroid hormones would be associated with a low level of TSH unresponsive to TRH stimulation. The most important reason for bearing in mind the possibility of pituitary hypothyroidism is to prevent cretinism. Neonatal screening programmes which depend on the observation of a raised TSH value to diagnose neonatal hypothyroidism will miss the rare patient in whom there is associated hypopituitarism, and it may be several weeks or months before the true diagnosis is made by which time there will have been significant delay in mental development.

Clinical Appearance of Hypothyroidism

Almost irrespective of the cause of the hypothyroidism, the clinical symptoms will fall within the classically described picture of myxoedema. These signs and symptoms are fully described in medical and endocrine text books. They are summarised in Table 9. There is usually no difficulty in making the clinical diagnosis of severe hypothyroidism, but it is now evident that there is a spectrum of disease ranging from mild thyroid hypofunction to gross myxoedema; patients with minimal degrees of disease often do not have clear cut signs and symptoms. It is in these patients that biochemical, immunological and thyroid scanning investigations are particularly helpful to the clinician. One must bear in mind, however, that borderline disease, associated with borderline clinical symptoms, is also frequently accompanied by borderline abnormal results. The possibility of neoplastic change must always be considered but this diagnosis is usually made on the appearance of the thyroid gland and on the results of thyroid scanning.

If thyroid cancer has been excluded, treatment of hypothyroidism is by means of hormone replacement.

Thyroid Status in Pregnancy[6,26]

Because thyroid disease is frequently experienced in women in their reproductive years, the association of thyroid disease in pregnancy is relatively common.

There are a number of findings which might suggest that thyroid activity increases during pregnancy:

1. There is an increased incidence of goitre and of thyroid hyperplasia during pregnancy.
2. The basal metabolic rate is increased.
3. The total thyroxine concentration is increased.
4. The uptake of iodine by the thyroid is increased.

It is likely, however, that these are not true indicators of increased thyroid activity. The increased thyroxine concentration is associated with a substantial increase in the level of thyroxine-binding globulin. Correction of this factor by means of measurement of the free-thyroxine index, T_4:TBG ratio, or free thyroxine concentration, indicate that there are no increases in the amount of free, active thyroid hormones, merely an increase in the protein-bound fraction. The increase in basal metabolic rate is likely to be due to the increased metabolism associated with the feto-placental unit, and the associated haemodynamic circulation of the mother. The fact that the response to injected thyrotrophin-releasing hormone, the TRH test, is exaggerated in pregnancy is also against the theory that the thyroid is over-active in normal pregnancy.

The fetal thyroid develops independently of the state of activity of the maternal thyroid gland in the normal euthyroid pregnancy. The important circulating thyroid hormones, T_3 and T_4, as well as TSH, do not pass across the placenta in significant amounts, in either direction. However, thyroid-stimulating immunoglobulins can pass from a thyrotoxic mother and cause fetal hyperthyroidism. If the maternal hyperthyroidism was not diagnosed prior to the onset of the pregnancy it may be difficult to make the diagnosis as some of the symptoms of thyrotoxicosis can be confused with similar symptoms that occur in a normal pregnancy. The clinician is, therefore, dependent upon accurate assessment of the thyroid state of the pregnant woman, such assessment taking into account the abnormally high levels of TBG which can be expected. Often a patient with mild to moderate hyperthyroidism will cope with the pregnancy satisfactorily, but may relapse into a state of thyrotoxicosis in the postpartum period.

Hypothyroidism is rather less common than thyrotoxicosis in pregnancy, largely because it occurs mainly in a slightly older age group. Hypothyroidism is associated with sub-fertility, but is by no means a contraindication to pregnancy. The rate of miscarriage also seems to be higher in hypothyroid mothers. Biochemical diagnosis of hypothyroidism during pregnancy presents two difficulties: (1) the increased concentration of TBG may push the total thyroxine level into the normal range although the free fraction may be sub-normal, and (2) the exaggerated response to the TRH test in pregnancy may cause the misclassification of the euthyroid pregnant patient into the hypothyroid category.

Whether or not the hypothyroidism affects the development of the fetus is not clear from the literature; however the increased rate of spontaneous abortion is sufficient indication that the hypothyroid mother should be treated during pregnancy and an attempt made to ensure that the total thyroxine level is in the upper part of the normal range.

Postpartum hypothyroidism in the mother is not uncommon. This is sometimes of temporary duration, but occasionally the pregnancy is an event which precipitates incipient hypothyroidism.

Thyroid Disease in Neonates and Children[25]

The incidence of thyroid disease in the young is by no means as common as in the older age-groups. Mention has already been made of the importance of screening for and treating hypothyroidism in neonates, and the topic is more fully discussed in Chapters 32 and 42.

The possibility of hypothyroidism should also be considered in children with abnormally short stature, or with a reduced growth rate. It is interesting to note that the academic school performance of such children, providing the hypothyroidism was not present during the first year of life, may well appear to be better than that of their euthyroid colleagues or may appear to deteriorate somewhat when the euthyroid state is achieved through treatment; this change is particularly noticeable in children who become much more active when treated and is merely a reflection of an increased interest in activities, both mental and physical, outside the somewhat narrower school schedules.

The possibility of hyperthyroidism should always be considered in children who are excessively tall for their age, particularly when this cannot be explained on the basis of parental height, or when there is a sudden, unexpected increase in the rate of linear growth. The effect of hyperthyroidism on a child's height is particularly noticeable at puberty. Clinical examination, backed up with the measurement of thyroid hormone concentrations can prevent excessive, but inappropriate, investigation of possible growth hormone abnormalities.

Thyroid Function in the Elderly[16]

Thyroid disease is more common in the elderly than in the younger population; one report has suggested that thyroid disease is present in over 5% of elderly patients. About three-quarters of thyroid disease in the elderly is hypothyroidism, the remaining quarter being associated with hyperthyroidism. Occasionally atypical presentations occur such as 'apathetic thyrotoxicosis' and rarely *myxoedema coma*, a severe complication with poor prognosis, that may be associated with hypothermia, hypercapnia and hypoglycaemia.

Suggestions have been made that the thyroxine concentration may be a little high in elderly female patients, that the T_3 concentration may be a little low, and that the TSH concentration might be slightly elevated. The explanation of these findings is not clear, but they are likely to be of borderline significance. Nevertheless the high incidence of thyroid disease in the elderly indicates that special attention should be paid by the laboratory to thyroid function tests in this age group.

THE INVESTIGATION OF THYROID FUNCTION

Introduction

The assessment of thyroid function is made, first and foremost, on clinical grounds. A good history and sound examination are sufficient to make a diagnosis in patients with moderate or severe thyroid disease. Laboratory investigations in these cases are useful to give baseline information which can be compared with the results of further investigations during the course of the treatment. Investigations also provide interesting information about the cause of the disease, although this may not be particularly important in deciding on treatment. Investigations are more important in those patients who present with an equivocal picture, although often, if the clinical features are borderline, results of investigations are also borderline. In some studies, but not in all, routine screening for thyroid disease in adults has been shown to be worthwhile (most workers are agreed on the value of neonatal thyroid screening); thyroid disease is common, there are good screening tests, and treatment is relatively simple.

The laboratory investigation of thyroid function has been transformed in the last 10–15 years by the advent of accurate and precise methods for measurement of the thyroid hormones and TSH. Most of the methods at present employed measure the *total* (bound plus free) concentrations of the thyroid hormone and a variety of ingenious procedures have been invoked, with varying degrees of success, to assess, directly or indirectly, the concentrations of the *free* hormones. In recent years practicable methods for the direct measurement of free thyroxine concentration have been developed, and this measurement may soon supplement or even supplant total thyroid hormone measurements, particularly if assay of free hormones are readily automated and inexpensive. The remainder of this section is written on the assumption that, at present, the assay of free T_4 and free T_3 is not routinely available for widespread use in most clinical biochemistry laboratories.

Aims of Investigation

The aims of laboratory investigation of thyroid function are as follows:

1. To use one or two well understood and inexpensive methods for the routine investigation of patients with possible thyroid disease,
2. To use the more sophisticated, difficult, time-consuming investigations only in those patients in whom straightforward tests have not given a satisfactory answer,
3. To minimise the use of invasive techniques or those which cause discomfort or pose threats of danger to patients. In particular, as many patients with thyroid disease are women of reproductive age, the use of *in vivo* radio-isotope methods should be kept to a minimum.

Thyroid Function Tests—Indications and Interpretation

Thyroxine Measurement

It is the authors' opinion that serum total thyroxine concentration is still the best initial measurement on patients with either hypothyroidism or hyperthyroidism. It is also useful in monitoring the progress of treatment. The method is practicable, relatively inexpensive, and can be automated to deal with large numbers of samples. Its major drawbacks are, firstly, that as it is an assay of total hormone concentration a number of patients with abnormal levels of TBG, either from pathological or physiological causes, can be misdiagnosed.

Our experience is that results which cause a patient to be incorrectly categorised are very few, and these can usually be predicted if adequate clinical information is given. The other major drawback of total thyroxine concentration measurements is that they will not pick up a hyperthyroid patient who is suffering from T_3-toxicosis and in whom the thyroxine concentration is in the normal range. It is theoretically possible also that patients with hypothyroidism caused by a low T_3 concentration, but with normal T_4 level, may also be missed.

It is our practice to interpret total thyroxine results, as measured in our laboratory, as follows:

Less than 45 nmol/l (3·5 µg/dl)—Hypothyroid value
   45–55 nmol/l (3·5–4·3 µg/dl) Borderline-hypothyroid value

55–70 nmol/l (4·3–5·5 µg/dl)—Low-normal value
70–135 nmol/l (5·5–10·5 µg/dl)—Normal value
135–145 nmol/l (10·5–11·2 µg/dl)—High-normal value
145–160 nmol/l (11·2–12·4 µg/dl)—Borderline-high value
greater than 160 nmol/l (12·4 µg/dl)—Hyperthyroid value

But these interpretations may be modified in the light of other investigations, e.g. TSH, T_3, done on the same sample, and may also be modified by the clinical appearance of the patient.

TSH Measurements

The normal range for TSH concentrations is not as definite as for thyroxine levels. It is not possible to distinguish between a pathologically low TSH concentration and one which is suppressed by adequate circulating thyroid hormone concentrations. Thus a significant proportion of normal, euthyroid patients will have random TSH concentrations below the limit of detection, usually 1 mU/l, of the methods at present available.

The euthyroid range does however extend above this value. Despite there being comparability of methods of TSH measurement, including the availability of reference TSH anti-serum and standards, there is still inter-laboratory variation in the upper limit of the normal TSH range. We would usually interpret a TSH value of less than 7 mU/l as being normal, one of 7–12 mU/l as being intermediate and worthy of further investigation, and one of greater than 12 mU/l as being abnormally high. Knowledge of the concentration of thyroid hormones does however modify this interpretation: thus an individual with a thyroxine concentration of 120 nmol/l (9 µg/dl) and a TSH of 8 mU/l is likely to be euthyroid, whereas one with a thyroxine concentration of 40 nmol/l (3 µg/dl) and a TSH of 6 mU/l would require further investigation, by means of a TRH test.

The measurement of TSH concentration is useful in the following situations:

1. In patients who have clinical hypothyroidism and whose thyroxine level is significantly below the normal range. This enables distinction to be made between patients with primary hypothyroidism, in whom the TSH value will be significantly elevated, and the very rare cases of secondary hypothyroidism, in whom the TSH concentration will be unmeasurably low. It also acts as a baseline for comparison of future TSH measurements taken once treatment has been started.
2. In patients with borderline sub-normal thyroxine levels, particularly if there are any clinical indications of hypothyroidism. In these patients the TSH level may not be greatly elevated and it may be necessary to carry out a TRH test, or at least to carry out repeat investigations after three months in order to see if there has been any progression of the hypothyroidism. Measurement of thyroid auto-antibodies may also help to confirm incipient hypothyroidism.
3. In monitoring the treatment of hypothyroidism. TSH measurements are particularly useful in monitoring treatment with thyroxine (or tri-iodothyronine) in the early stages, when an appropriate dose is being sought, or subsequently, when it is felt that the treatment dose is not quite adequate. In these cases, when the thyroxine (or tri-iodothyronine) levels are in the lower part of the normal range the TSH value can help to confirm the adequacy or otherwise of the treatment dose. TSH measurement is of little use, however, when the possibility of over-dosage is being considered. In this situation the TSH level is likely to be unmeasurably low, whether the treatment is adequate or excessive.
4. Similarly a single TSH measurement is not of value in the assessment of hyperthyroidism. Multiple TSH measurements are, however, of use in this situation following TRH stimulation.

TRH Test

This is a useful test, of use in the diagnosis both of hyperthyroidism and hypothyroidism. It is classically carried out by injecting 0·2 mg of thyrotrophin-releasing hormone intravenously, having previously taken a basal blood sample. Further blood samples are taken at 20 minutes and 60 minutes after the TRH injection. TSH measurements are carried out on all three samples and the results are compared with those obtained on normal individuals by the laboratory concerned. There is some inter-laboratory variation in basal TSH measurements and this variation is exaggerated when the results of a TRH test are compared between laboratories. The response of TSH to TRH is exaggerated in pregnancy and in patients taking the contraceptive pill; this abnormal response causes difficulty in diagnosing hypothyroidism in such patients.

In general there will be only slight, if any, response in patients with hyperthyroidism but an excessive response, usually starting with a baseline level above the normal range, in patients with primary hypothyroidism. The response in tertiary hypothyroidism, i.e. hypothalamic hypothyroidism, is likely to be somewhat delayed and the peak TSH value often appears in the 60 minute sample rather than the more usual 20 minute peak.

Some workers have suggested that a TRH test should be used routinely in assessing the progress of treatment. We feel that this course of action would entail unnecessary time, expense, and distress to patients; judicious use of TSH and T_4 (or T_3) estimations are sufficient in the vast majority of patients to maintain a euthyroid state.

Measurement of Total Tri-iodothyronine Concentration

The normal range for tri-iodothyronine concentration is also subject to some inter-laboratory variation. Our normal range is 1·3 to 2·8 nmol/l (80–180 ng/dl) with values within 0·3 nmol/l of this range being 'borderline'.

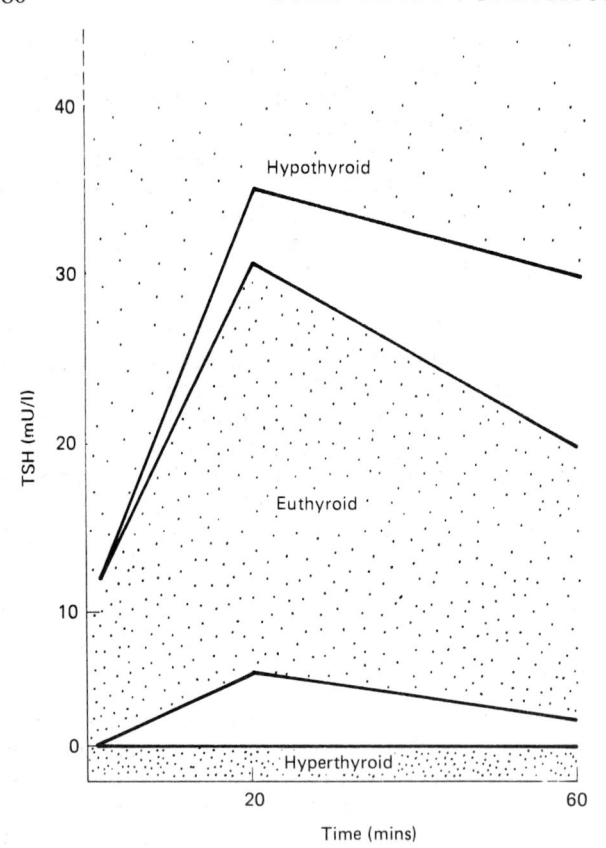

FIG. 3. Typical response ranges of TSH production following injection of 0·2 mg of TRH intravenously.

The major uses for the measurement of T_3 are:

1. To diagnose T_3-toxicosis in patients in whom there are clinical signs or symptoms of hyperthyroidism but in whom the thyroxine concentration is within the normal range.
2. To supplement thyroxine measurements in those patients who show signs of thyrotoxicosis and whose thyroxine levels are in the high-normal or borderline-high regions. Significantly elevated or borderline-elevated T_3 values would support the diagnosis of thyrotoxicosis, but a level well down in the normal range would contra-indicate such a diagnosis.
3. To monitor hypothyroid patients who are being treated with tri-iodothyronine. The T_3 measurement may be supplemented by TSH measurement as mentioned above.
4. The measurement of T_3 may be of value in patients who appear to be hypothyroid but who have normal serum thyroxine concentrations. If in these patients the T_3 concentration is sub-normal it is possible that the patient is unable adequately to convert T_4 to T_3. If, as suggested, T_3 is the active metabolite of the thyroid hormones one can see why such poor conversion may lead to signs of hypothyroidism.

Reverse-T_3

Although various suggestions have been made for using the measurement of reverse-T_3 to assess thyroid function, it does not appear at present that a good case has been made for any particular clinical situation. Reverse-T_3 concentrations are known to rise in certain non-thyroid diseases but this is a non-specific effect and knowledge of them is unlikely to be helpful.

Free-thyroxine Concentration, Free-thyroxine Index, T_3-uptake, Thyroxine Binding Globulin

As mentioned above, a very large percentage of thyroxine and tri-iodothyronine circulates in the plasma bound to thyroxine-binding proteins; it is, however, the free hormone which is likely to have biological effects at the tissue level. Thus, total thyroid hormone measurements in blood do not give a direct estimate of the amount of the active hormones. If the amount of thyroxine-binding proteins and the degree of binding to them did not vary between individuals, the total hormone concentrations would bear a fixed relationship to the free hormone concentrations. Unfortunately this is not so and various factors (see Table 1), alter the amount of binding proteins or interfere with the binding of thyroid hormones to them. Particularly important amongst these are the contraceptive pill and pregnancy, both of which cause a significant increase in the level of thyroxine-binding proteins, thereby increasing the total concentration of thyroid hormones, without necessarily influencing the concentration of the free hormones. An inappropriate diagnosis of thyrotoxicosis or T_3-toxicosis may thus be made in a euthyroid patient.

In order to overcome these difficulties a variety of methods have been investigated to make indirect estimates of the amount of free thyroid hormones. For instance, an estimate of the amount of thyroxine-binding proteins can be achieved with the T_3-uptake test. In recent years this estimate has been superseded by direct measurement of thyroxine-binding globulin. From these results an indirect estimate of the free thyroxine concentration can be made by considering the following equilibrium reaction:

$$T_4 + TBG = T_4 . TBG$$
$$\text{(free)} \qquad \text{(bound)}$$

If the equilibrium constant of this reaction is K, then:

$$K = \frac{(T_4 . TBG)}{(T_4) \times (TBG)}$$

i.e.
$$(T_4) = \frac{1}{K} \times \frac{(T4 \cdot TBG)}{(TBG)}$$

It can be seen that from this statement that the free thyroxine concentration is proportional to the amount of thyroxine bound to TBG (which is almost exactly equal to the total thyroxine concentration) divided by the amount of unbound TBG. This latter amount can be obtained from the T_3-uptake test. The ratio total thyroxine concentration divided by the T_3-uptake per cent is termed *free-thyroxine index*.

Now that the concentration of TBG itself can be measured another ratio, the *total T_4 : TBG ratio*, has also been used as a measure of the amount of free thyroxine. The validity of this value is not explained so easily in terms of the equilibrium, but several workers have suggested that it gives a more clinically appropriate value and causes fewer mistkaes in categorisation of patients especially during pregnancy. It should be noted that the crude reaction kinetics mentioned above are a gross approximation to the true situation. In reality there are at least three binding proteins and the binding coefficients for each are different. Also there is competition for binding sites from T_3 and reverse-T_3, and no account is taken of the effect on the binding constants of the remaining binding sites once some of the sites are already occupied.

These theoretical considerations are likely to prove irrelevant as methods for the direct measurement of the free hormones supersede these indirect estimations.

In our experience it is uncommon for changes in the levels of binding proteins to affect significantly the categorisation of patients. Rarely, however, the clinical appearance of a patient is sufficiently different from the assessment of thyroid status to warrant a full assessment of binding-protein levels.

Interpretation of Unexpected Results

It is the experience of every clinical biochemist and endocrine physician that tests of thyroid function do not always produce the expected results. Sometimes the results do not agree with the patient's presentation, sometimes the results of two tests that should complement each other turn out to be incongruous, at least at first sight. The following discussion is an attempt to explain some of the incongruity.

Erroneous Results. Errors in results can arise in a wide variety of ways. Serum samples are generally preferred, as the fibrinogen in plasma can interfere with the analytical antigen-antibody reaction in some assays, especially that for TSH. The sample must be reasonably fresh when it reaches the laboratory, so that haemolysis can be avoided, and deterioration of the thyroid hormones and the production of potentially interfering artefacts in the sample can be minimised.

The assay methods for thyroid function tests are more complex, involve more manipulation, demand a higher degree of technical skill and are less automated than many other tests carried out in the clinical biochemistry laboratory.

Tests, such as those used for urea and electrolyte determinations, have precise methodology, are fully 'automated' and are frequently linked to computerised data handling systems. This is not usually the case with the radioimmunoassay (or other immunoassay) methods used in thyroid function tests. As a result these tests are liable to greater analytical imprecision, provide more opportunity for operator error, and are subject to more likelihood of mis-identification; they also produce greater inter-laboratory variation.

Some unexpected results are therefore a consequence of errors made in the collection and transport of the sample or in its subsequent analysis. Because abnormal thyroid function frequently needs to be treated for the rest of the patient's life, or demands irreversible surgical or radiation treatment, it is essential that abnormal results are confirmed on a second sample prior to treatment, particularly if the test results are not wholly in accord with the patient's clinical condition.

In addition to genuine errors of this sort, apparently incongruous results can sometimes be obtained in which no error has occurred. These demand interpretation in the light of the patient's condition. Some apparently incongruous results which can be explained are as follows:

High T_4, high T_3. This is the normal result in thyrotoxicosis; in euthyroid subjects however it could indicate raised TBG levels.

High T_4, low or normal T_3. This can also occur in thyrotoxicosis, particularly if there is an associated impaired conversion of T_4 to T_3. It could be observed in elderly thyrotoxic patients or in those with an intercurrent illness. It may also occur if TBPA levels are increased.

Low T_4, normal T_3. These results can occur in mild or early thyroid failure, as the gland attempts to maximise the output of active hormone, in which case TSH and/or a TRH test might give further information. They can also be an indication of iodine deficiency, as the thyroid attempts to make best use of the limited amount of iodine.

Normal T_4, high T_3. This is commonly observed in T_3-toxicosis, or rarely in deliberate or accidental overdosage with T_3. It may also occur in iodine deficiency.

Low T_4, low T_3. In addition to hypothyroidism these results can also be observed in subjects with congenital TBG deficiency and possibly in hypothermia or severe illness in patients who are usually euthyroid.

Low T_4, high TSH. Usual picture in primary hypothyroidism.

Low T_4, low T_3, low TSH. Very low values are associated with secondary (pituitary) hypothyroidism. Moderately reduced thyroid hormone levels with low TSH can also be associated with tertiary (hypothalamic) hypothyroidism. A TRH test would show no rise of TSH in the former case, but a normal, possibly delayed, response in the latter. The picture can also be seen in TBG deficiency in which there would usually be a normal TRH response.

High T_4, high T_3, patient not hyperthyroid. The

possibility of antibodies to T_3 and T_4, or a peripheral tissue resistance to thyroid hormone action should be considered. A TRH test would, in either of these cases, show a normal TSH response, or possibly an elevated response, in contrast to the lack of response in true hyperthyroidism. An excess of TBG, for whatever reason, could give a similar picture, with a normal TRH test response.

Other tests of Thyroid Function

Thyroid Antibody Studies. As mentioned earlier, a significant proportion of thyroid disease is associated with the presence of circulating antibodies to thyroid antigens. Measurement of these, both stimulatory and destructive, can help make the diagnosis when clinical appearance and hormone measurements give equivocal findings. It should be remembered that thyroid auto-antibodies can be found at significant levels in subjects who are clearly euthyroid. Measurement of thyroid-stimulating antibodies is not available routinely at present.

Thyroid Scanning. The use of thyroid scanning is a useful adjunct in the diagnosis of thyroid disease. It is particularly useful in cases of nodular goitre, and also in borderline cases. The subject is, however, beyond the scope of this book.

SUMMARY

The increasing use of the clinical biochemistry laboratory in the investigation of thyroid function, brought about by the development of radioimmunoassays for the hormones involved, has greatly helped in making the diagnosis and monitoring the treatment of patients with thyroid disease. These advances have also led to a greater understanding of the nature of thyroid disease.

In the long run however, diagnosis of thyroid disease is a clinical matter and, in cases where the biochemical results and the clinical features do not completely tally, it is the latter which should determine treatment. A cautionary tale, albeit anecdotal, of one of our patients will suffice to make the point.

The patient was in her mid-fifties when she first presented to her general practitioner with symptoms and signs that he felt to be consistent with mild hypothyroidism. She was referred to a physician who requested thyroid function tests, and was surprised to find that both T_4 and T_3 levels were significantly raised. Repeat tests showed the same results, and both reports stated that the requested TSH estimation was inappropriate, in view of the high levels of thyroid hormones, and was therefore not carried out. Following further discussion the physician persuaded us to carry out a TRH test. This showed, to our surprise, a TSH response a little elevated above that normally observed, and certainly not a hyperthyroid response despite the fact that the thyroid hormone levels continued to be significantly raised. We could not demonstrate the presence of T_4 or T_3 antibodies and therefore concluded that the patient had some peripheral resistance to thyroid hormone action. The physician agreed and advised the general practitioner that it was of no use to treat the patient with thyroxine or tri-iodithyronine as the patient was already producing more than enough already. The practitioner argued that the patient was becoming more hypothyroid and 'needed treatment with thyroxine'; the physician's advice was unchanged. Our opinion was sought and we agreed about the uselessness of giving still more thyroxine. The general practitioner decided, in view of the patient's condition, to take no heed of the advice given and instituted thyroxine therapy. The patient recovered splendidly and is still well three years later!

ACKNOWLEDGEMENT

The authors wish to thank Mrs. Mary Fagence for typing this chapter, and also for her efforts in assisting with the editing of other chapters.

REFERENCES AND SUGGESTIONS FOR FURTHER READING

1. de Visscher, M. (Ed.) (1980), *Comprehensive Endocrinology The Thyroid Gland*. New York: Raven Press.
2. *ibid.* 21–37, Wollman, S. H. 'Structure of the thyroid gland'.
3. *ibid.* 39–79, de Nayer P. and Vassart G. 'Structure and biosynthesis of thyroglobulin'.
4. *ibid.* 81–121, Gergshengorn, M. C., Glinoer, D, and Robbins, J. 'Transport and metabolism of thyroid hormones'.
5. *ibid.* 123–143, Bernal J. and DeGroot L J. 'Mode of action of thyroid hormones'.
6. *ibid.* 215–229, Burrow G. N. 'Thyroid function in relation to age and pregnancy'.
7. *ibid.* 279–362, Mornex, R. and Orgiazzi, J. J. 'Hyperthyroidism'.
8. *ibid.* 377–412, de Visscher, M. and Ingenbleek, Y. 'Hypothyroidism'.
9. *ibid.* 443–487, Salvatore, G. Stanbury, J. B., and Rall, J. E. 'Inherited defects of hormone synthesis'.
10. Evered, D. and Hall, R. (Eds.) (1979), *Clinics in Endocrinology and Metabolism* 8(1), 'Hypothyroidism and Goitre'. New York: W. B. Saunders, Philadelphia, London.
11. *ibid.* 3–20, Lamberg, B.-A. 'The aetiology of hypothyroidism'.
12. *ibid.* 21–28, Tunbridge, W. M. G. 'The epidemiology of hypothyroidism'.
13. *ibid.* 29–38, Hall, R. and Scanlon. M. F. 'Hypothyroidism: Clinical features and complications'.
14. *ibid.* 49–62, Illig, R. 'Congenital hypothyroidism'.
15. *ibid.* 63–80, Doniach, D. Bottazo, G. F. and Russel, R. C. G. 'Goitrous autoimmune thyroiditis 'Hashimoto's disease'.
16. Hodkinson, H. M. (1977). *Biochemical diagnosis in the elderly*. London: Chapman & Hall.
17. Ingbar, S. H. and Woebar, K. A. (1981), 'The thyroid gland' in *Textbook of Endocrinology*, 6th edn, 117–248. (Williams, R. H. Ed.) Philadelphia, London, Toronto: W. B. Saunders Co.
18. Robbins, J., Rall, J. E. and Gordon, P. (1980), 'The thyroid and iodine metabolism'. In *Metabolic Control and Disease* 8th edn, 1325–1389. (Bondy, P. K., Rosenberg, L. E. Eds.) Philadelphia, London, Toronto: W. B. Saunders Co.
19. Werner, S. C. and Ingbar, S. H. (Eds.) (1980), *The Thyroid: a Fundamental and Clinical Text* 4th edition. New York and London: Harper and Row.
20. *ibid.* 138–148, Rall, J. E. 'Mechanism of action of T_4'.

21. van Herle, J., Vassart, G. and Dumont, J. E. (1979) 'Control of thyroglobulin synthesis and secretion'. *New England Journal of Medicine* **301,** 239–246.
22. Volpe, R. (Ed.) (1978), '*Clinics in Endocrinology and Metabolism*', *7(1)*, 'Thyrotoxicosis' Philadelphia, London, Toronto: W. B. Saunders.
23. *ibid.* 3–30, Volpe, R., 'The pathogenesis of Graves' disease: an Overview'.
24. *ibid.* 31–46, McKenzie, J. M., Zakarija, M. and Sato, A. 'Humoral immunity in Graves' disease'.
25. *ibid.* 127–144, Howard, C. P. and Hayles, A. B. 'Hyperthyroidism in children'.
26. *ibid.* 115–126, Barrow, G. N. 'Maternal-fetal considerations in hyperthyroidism'.

37. ENDOCRINE PANCREAS

VINCENT MARKS

Glucose homeostasis
   The glucose pool

Insulin
   Biological properties
   Insulin deficiency and excess
   Insulin in the blood
   C-peptide
   Other blood glucose lowering substances

Glucose inflow
   Absorption from the gut
   Glycogenolysis and gluconeogenesis
   Hormonal control
   Non-hormone factors

Investigation of glucose metabolism in man
Hyperglycaemia
   Diabetes mellitus
   Tests for hyperglycaemia
   Overnight fasting blood glucose
   Oral glucose tolerance tests
   Other tests of glucose tolerance
   Hormone and metabolite assays

Hypoglycaemia
   Classification
   Diagnosis

Inborn errors of metabolism

Insulin is the main, but far from only, hormone of the endocrine pancreas. It has the unique distinction of being the direct or indirect cause of no less than four Nobel Prizes being given: to Banting and Macleod in 1923 for its isolation and purification; to Sanger in 1958 for elucidation of its primary structure; to Dorothy Hodgkin in 1964 for its crystallographic appearance and tertiary structure; and to Rosalyn Yalow in 1979 for the work that she and Solomon Berson put into the development of radioimmunoassay as an analytical system with general application—but which began with the measurement of insulin. The emphasis given to insulin by these scientists and the Nobel Committee reflects its importance in the metabolic economy of the body which will be considered in greater detail below. Suffice, at this stage, to say that insulin is the main anabolic hormone and that it plays a key role in the synthesis of proteins from aminoacids, triglycerides from carbohydrates, and polysaccharides from simpler molecules, as well as exercising profound control over many intermediate metabolic pathways concerned with energy and body-substance production.

GLUCOSE HOMEOSTASIS

The Glucose Pool

There is seldom more than 20 g (0·11 mol) of glucose in the body at any one time and this can be looked upon as constituting a glucose pool to which glucose molecules are constantly being added or subtracted. The glucose pool is confined within the hypothetical 'glucose-space' which, though not anatomically defined, corresponds, more or less exactly, to the extracellular fluid volume of the body plus a small contribution from the intracellular water of the liver and red-blood cells. This is shown diagramatically in Fig. 1, which represents the glucose space as a water cistern and its associated pipes, etc., and the glucose-pool as the water in it.

Since the glucose space remains more or less constant in any individual, the size of the glucose pool corresponds to the height of water in the cistern; i.e. the 'head of pressure', which represents the blood glucose concentration. There is, for all practical purposes, only one route by which new glucose molecules can enter the glucose pool and that is through the liver, having got there either by absorption of a meal containing starch or sugar or by liberation of glucose from preformed glycogen in the liver cells. Once inside the pool, glucose

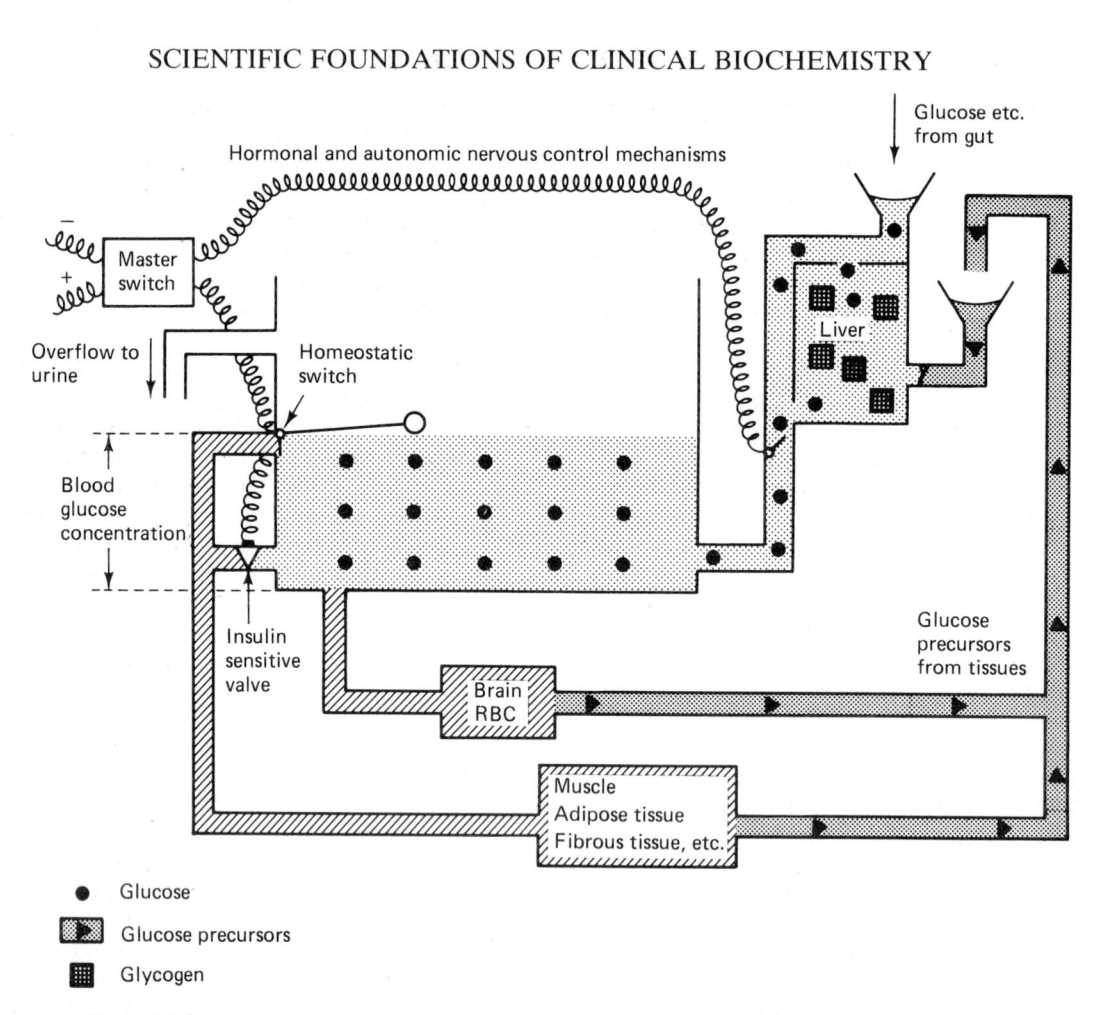

FIG. 1. Schematic representation of glucose pool with autoregulatory controls of inflow and outflow. (Reproduced from Teach-In (June 1972) with permission of the publishers, Update Hospital Publications Ltd.)

molecules can ordinarily leave it only by entering the tissues.

In the fasting subject almost the only drain on the glucose pool is the brain and erythron, i.e. cells which are permeable to glucose even in the absence of insulin. When blood glucose levels rise after a meal they are associated with the release of insulin which in turn permits glucose to leave the glucose pool by gaining entry to the rest of the body tissues. There it is rapidly phosphorylated prior to conversion either into the energy necessary to keep vital processes going or into storage materials such as glycogen and triglycerides. Because entry of glucose into the cells is rate-limiting, intracellular glucose levels are ordinarily very low and do not contribute materially to the size of the glucose pool.

During metabolism many tissues liberate glucose precursors back into the ECF. The most important of these, namely, lactate, pyruvate, glycerol and alanine, are converted by the liver into glucose by a process known as gluconeogenesis. This is almost, but not exactly, the reverse of glycolysis, the process wherein glucose is broken down to pyruvate and lactate. There are, however, four enzymatic steps in the glycolytic pathway that are essentially irreversible but which can be circumvented by participation of enzymes that occur exclusively in the liver and kidneys.

In normal healthy subjects the outflow of glucose from the glucose pool into the tissues is so finely adjusted to the rate of inflow that even the ingestion of a meal containing upwards of 200 g of carbohydrate, i.e. ten times more than the total body glucose pool, seldom causes more than a 50% expansion in size which is reflected by a rise in blood glucose concentration of 2–3 mmol/l at most. Nor, in most healthy adults, does fasting for many days produce more than a modest (10–20%) reduction in glucose pool size, though it may do so in children and pregnant women. The key to this exquisite control of blood glucose concentration is the regulated secretion of insulin which ensures that, as the glucose pool enlarges after ingestion of a meal, additional glucose molecules gain access to the mass of body tissues where they are converted into other substances. Furthermore, under the influence of the increased amount of insulin in the blood, not only do the liver cells cease adding glucose to the glucose pool from their own stores, as they do during fasting, but they actively participate in removing it.

During fasting, the tendency of the blood glucose concentration to fall is counteracted by a decline in insulin secretion. This has the dual effect of switching-off glucose outflow into the tissues and of encouraging glucose inflow from the liver. This latter action compensates for the obligatory loss of glucose from the pool resulting from metabolism in brain and red-blood cells, and prevents the fall in blood glucose concentration that would otherwise inevitably ensue.

It can be seen, from this brief description, that the glucose pool, though seldom varying much in size, is in constant flux, the rate of turnover normally being much greater during, and shortly after, a meal than during fasting when it can become very slow indeed though never ceasing altogether.

Glucose turnover rates are difficult to measure; most methods utilise either radioactive, i.e. ^3H or ^{14}C, or stable, e.g. ^2H or ^{13}C isotope-labelled glucose, and have not, until now, found much application in the investigation and management of individual patients. Investigation of turn-over rates have, however, thrown much light on the homeostatic and pathological mechanisms involved in glucose homeostasis and the possibility cannot be excluded that, with the greater availability of deuterium and ^{13}C-labelled glucose and of cheap mass-spectrometers, their measurement might find a place in the clinical laboratory of the future. Until then, measurement of blood glucose and its response to various procedures provide the main indication of what has gone wrong in patients suffering from disorders of glucose homeostasis.

Changes in blood glucose concentration in an upward direction, i.e. hyperglycaemia, are much commoner, in disease, than those in the downward direction but do not, in themselves, usually cause any acute disturbances in bodily function—at least in the short term. They may, therefore, go unnoticed by the patient unless appropriate measurements are made. Hypoglycaemia, on the other hand, by virtue of its ability to interfere with brain metabolism, leads to dramatic functional changes, which means that it is much less likely to be overlooked even though, because of its comparative infrequency, it may go unsuspected by those unfamiliar with its manifestations. Since both hyper- and hypoglycaemia have many causes neither diagnosis should ever be considered definitive but merely an invitation to further investigation.

INSULIN

Insulin is a polypeptide hormone produced exclusively in the B-cells of the islets of Langerhans. Its structure differs slightly according to the species from which it is isolated. In all mammalian species it contains two peptides known as the A and B chains which contain twenty-one and thirty amino acids respectively. The two peptides are joined to each other by two sulphydryl bridges which are essential for biological activity. Insulin is formed, in the β-granules, by proteolytic cleavage of its precursor, proinsulin. This occurs at two points in the molecule with the production of one molecule of insulin and one of a connecting peptide from which two further amino acids are cleaved, at each end, to produce the (proinsulin) C-peptide. The C-peptide and insulin so formed are stored together in the β-granule until released into the intercellular space in response to an insulinotropic stimulus.

Insulin synthesis and release are independent, though interrelated, processes, but whilst much is known about the latter considerably less is known about the former, which may, in the long run, be the more important of the two. It is generally accepted that blood glucose concentration is extremely important in regulating insulin secretion though, contrary to previous belief, hyperglycaemia itself is a poor stimulus to insulin release. More potent, in this respect, are two polypeptide hormones, GIP and glucagon, whose insulinotropic effects only become manifest, however, in the presence of a slightly raised blood glucose level such as occurs after ingestion of a mixed or carbohydrate-rich meal.[13,16] Certain amino acids can also stimulate insulin secretion under experimental conditions although it is unlikely that this plays a significant role under normal circumstances. β-Hydroxybutyrate and acetoacetate are both capable of stimulating insulin release in some animal species but seemingly have little effect upon insulin secretion in man.

The role of the autonomic nervous system in the overall regulation of insulin secretion is still far from clear. Nevertheless, it is now reasonably certain that inhibition, exerted through the sympathetic branch, is important in reducing insulin secretion during vigorous exercise and that stimulation, via the vagus, plays a part in the production of alimentary insulinaemia in response to food.

Biological Properties

Insulin has many properties in addition to its best known one of lowering the blood glucose concentration when injected into an animal. It is, for example, the main hormone concerned with triglyceride and protein synthesis as well as with glycogen formation. It lowers plasma amino-acid, non-esterified fatty acid (NEFA) and potassium levels, and exerts an inhibitory effect upon gluconeogenesis and glycogenolysis in the liver. The last two actions are the main ways by which insulin, when secreted endogenously or administered parenterally in physiological amounts, lowers blood glucose levels.

Adipose tissue cells are more sensitive to insulin than most others with regard to its ability both to inhibit lipolysis and to encourage glucose uptake. Each of these effects occurs at plasma insulin concentrations close to, or only slightly higher than, those encountered after an overnight fast; considerably higher levels, corresponding to those observed only for brief periods after meals, are necessary to increase glucose uptake by resting muscle.

In order to exert a metabolic effect insulin must first be bound to specific insulin receptors which are found mainly, though probably not exclusively, on the surface of insulin-sensitive cells. Combination of insulin with its receptor activates processes which increase the permeability of the cell plasma-membrane to glucose and effect changes in intracellular metabolism. It is now recognised that both the number and binding characteristics of insulin receptors, which are every bit as important as insulin concentration in determining the extent of insulin activity, can and do change rapidly, sometimes in a matter of only a few hours, in response to changing circumstances. These adaptations, the best studied of which are known as 'up' and 'down' regulation, depending on whether the number of receptors increases or decreases respectively, explain why insulin activity *in vivo*, insofar as it can be judged by blood glucose measurements, bears little relationship to plasma insulin concentration since it represents only one parameter in the equation:

$$\text{Insulin activity} \propto \text{insulin concentration} \times \text{insulin receptors} \times \text{metabolic response}$$

Insulin is normally secreted into the portal circulation resulting in a large, but variable proportion of it being sequestered by the liver. This means that the concentration of insulin in blood perfusing adipose tissue can fall sufficiently low during periods of fasting to permit increased lipolysis, with liberation of NEFA as an alternative fuel to glucose whilst still maintaining a sufficiently high concentration of insulin in the liver to keep gluconeogenesis and glycogenolysis under control.

Proinsulin is normally secreted by healthy B-cells in minute amounts only, except in an exceedingly rare familial condition in which there appears to be a congenital lack of the cleaving enzyme. The prohormone has many of the biological activities of insulin itself, but usually with only one quarter to one tenth the potency of the native hormone. Increased amounts of proinsulin and proinsulin-like components (PLC) circulate in the blood in certain diseases, especially those caused by neoplastic transformation of the B-cells.

Insulin Deficiency and Excess

The catastrophic effect of insulin deficiency upon the glucose economy of the body has been recognised ever since Von Mering and Minkowski first performed a pancreatectomy on a dog and made it diabetic. Subsequently, it became possible, by using selective B-cell poisons, such as alloxan and streptozotocin, to produce permanent insulin deficiency in animals without damaging the exocrine pancreas. Temporary, complete, and virtually specific insulin deficiency can also be produced experimentally, in animals, by injecting them intravenously with an excess of high avidity insulin antiserum. In all cases of acute insulin deficiency the result is the same, namely, the rapid onset of acute diabetes mellitus, characterised by gross hyperglycaemia, glycosuria, ketosis, ketonuria and rapid wasting which is followed, unless reversed by exogenous insulin, by death of the animal.

Interestingly, in some species of birds, lizards and snakes, pancreatectomy lowers, rather than raises, the blood glucose concentration apparently due to loss of the glucagon-secreting cells which in these creatures seem to play a much more important role in the glucose economy of the body than they do in mammals. Even in these glucagon-sensitive species, however, hyperglycaemia due to insulin deficiency can be produced by injecting them with insulin antibodies.

Insulin deficiency occurs in man whenever B-cells are destroyed by disease or removed surgically. However, except in the face of the fiercest challenge, less than 10% of the pancreas is required to maintain normal glucose homeostasis.

A large excess of insulin produces hypoglycaemia regardless of whether the subject has free access to food or not; a moderate excess, on the other hand, may produce nothing more serious than an increase in food intake and obesity. If for any reason food intake is restricted, however, even a very modest increase in insulin production leads to hypoglycaemia mainly through inhibition of glucose release from the liver.

Insulin in the Blood

Though various methods for measuring insulin in blood were tried, none was successful until the introduction of radioimmunoassay by Yalow and Berson in 1960. This revolutionary assay technology not only enabled many measurements to be made on a single individual under a variety of experimental conditions but also extended plasma insulin assays into the clinical domain. Although the quality control of assays was not always very good in the early days, and still often falls far short of perfection, it is generally now possible to obtain clinically meaningful results in any well-equipped clinical biochemistry laboratory should it be necessary.

Most of the circulating immunoreactive insulin (IRI) is identical to the native 51–amino-acid hormone isolated from the pancreas but a small and variable percentage consists of proinsulin-like components (PLC) and insulin fragments which have variable but generally grossly reduced biological activities. The half-life of native insulin in the blood is in the region of three to eight minutes, or shorter. Consequently, plasma insulin levels can, and often do, change extremely rapidly—literally from minute to minute—although in healthy, fasting subjects they remain remarkably constant over long periods. Patients with diseases affecting the endocrine pancreas may, on the other hand, show large and rapid spontaneous fluctuations in plasma IRI levels due to intermittent secretory spurts.

Insulin is inactivated by an insulin-specific protease (insulinase) found in all tissues, especially liver, kidney, muscle and placenta. It is not known, however, whether insulin must be released from binding to its receptors before it can be destroyed or whether inactivation occurs *in situ*. There is no evidence, at the present time, of any disease arising as a result of alterations in insulinase activity, although the possibility cannot be dismissed. Differences in the molecular structure of insulin itself are, however, a rare but recognised cause of diabetes.[10,17]

Circulating autoantibodies to insulin have been observed in people who suffer from a rare type of fasting hypoglycaemia but who have never been exposed to exogenous insulin. In the overwhelming majority of individuals in whom insulin antibodies can be demonstrated, however, there is almost always a history of immunisation with exogenous porcine, or more commonly bovine, insulin, which has been used therapeutically to treat their insulinoprivic diabetes. Autoantibodies to insulin receptors (analogous to those against acetylcholine receptors in myasthenia gravis) are even rarer than insulin autoantibodies and are associated with a bizarre, extremely insulin-resistant type of diabetes mellitus. These antibodies must not be confused with those directed against islet-cells themselves[5] and which are commonly found in insulin-dependent diabetes early, but usually not late, in the course of their illness.

C-Peptide

Though not yet established as a hormone, the (proinsulin) C-peptide already enjoys an important place in diagnostic and investigative clinical biochemistry.[6] Since it is released from the B-cells of the pancreas on an equimolar basis with insulin it can be used as a marker of B-cell activity whenever measurement of insulin itself is impracticable. This occurs, for instance, when insulin antibodies are present in the patient's plasma because of previous immunisation, or when exogenous (porcine) insulin has been used to produce hypoglycaemia as a test of B-cell integrity.

Other Endogenous Substances that Lower Blood Glucose Concentration

Insulin is unique amongst the well-recognised hormones in being the only one that regularly causes a fall in blood glucose concentrations when administered or secreted. It does so both by decreasing glucose inflow from the liver and by increasing uptake by the tissues. The exact contribution made to blood glucose lowering by each component depends upon a variety of factors, the most important of which are the concentration of insulin in blood perfusing the respective organs and their responsiveness.

Strenuous muscular work increases glucose uptake both in normal and diabetic subjects but the way in which it does so is uncertain. Exercise itself undoubtedly increases cellular permeability to glucose in a manner similar to, but distinct from, insulin—thereby permitting its entry into glucose-using cells and increasing the drain on the glucose pool. Exercise also inhibits insulin secretion, probably by activation of the sympathetic nervous system, and this tends to increase rather than decrease the blood glucose concentration. Indeed a small rise in blood glucose concentration is much more common during short bouts of vigorous exercise in healthy subjects than a fall. A fall does occur, however, when, for any reason, insulin secretion or activity does not decline during exercise; this is most likely to happen when the circulating insulin concentration is not under homeostatic control, either because it has been injected—in a diabetic subject, for example—or because it originates from an autonomous tumour.

Although insulin is the main anabolic hormone of the body it has long been recognised that other substances, collectively known as tissue growth factors or somatomedins, are important in the initiation and maintenance of cellular division and growth.[4,20,21,26] A number of these growth factors have been isolated and either partially or completely characterised. Most, if not all, of the somatomedins are produced in the liver. Their formation depends upon the availability of growth hormone which exerts its anabolic effect, largely if not entirely, through them. Most of the somatomedins that have been characterised have only a minor effect upon glucose metabolism though two of them, known as insulin-like growth factors I and II (IGF I and IGF II), have amino-acid compositions and sequences reminiscent of proinsulin. IGF I and IGF II circulate in the plasma at low concentration and, together with a larger molecular weight substance described as non-supressible insulin-like protein (NSILP), constitute what is known as non-suppressible insulin-like activity (NSILA). The physiological function if NSILA, if any, is quite unknown but it has been incriminated in the production of hypoglycaemia in some exceedingly rare patients with neoplastic disease. This must, however, be considered both unproven and unlikely.

Deprivation of food, i.e. a reduction of glucose inflow from the gut, does not ordinarily lead, in healthy adult male subjects, to anything more serious than a modest fall in blood glucose concentration until they are *in extremis* and in danger of imminent death from starvation. This lack of effect of starvation upon blood glucose concentration is due to the almost complete abandonment of glucose as a source of fuel, even by the brain, which, after only a few days' fasting, switches almost entirely to ketones as its source of energy. Under these circumstances the only drain on the glucose pool is the red blood cells which, lacking the capacity to oxidise ketones, are obligatorily glycolytic. Even so the overall amount of glucose they utilise in the course of a day is small since the lactate and pyruvate they produce by glycolysis is resynthesised into glucose in the liver.

Children and young women seem less able than adult men to maintain glucose pool size intact during fasting and, in them, blood glucose levels may fall to well below the conventionally accepted lower limit of normal after 24 to 48 hours without food, though neuroglycopenic symptoms seldom, if ever, develop. Paradoxically, if fasting is continued beyond 48 hours blood glucose levels tend to rise again as ketones and NEFA take over as fuel and drainage of glucose from the pool decreases still further.

GLUCOSE INFLOW

Absorption from the Gut

The average Western diet provides between 150–300 g carbohydrate a day, of which roughly 65% is taken as starch, 25% as sucrose and 10% as other sugars. Starch, sucrose and lactose are hydrolysed into their respective monosaccharides prior to absorption but of the three sugars so produced, i.e. glucose, fructose and galactose, only the first gains more than very limited access to the general circulation. In effect, both fructose and galactose behave as 'first pass' substances, being virtually completely removed from the blood by a single passage through the liver which contains two enzymes, fructokinase and galactose kinase, with high affinities for each of the two sugars respectively.

How large a proportion of the glucose absorbed from the diet is normally taken up by the liver and how much by the peripheral tissues is undecided and controversial. Estimates of hepatic uptake based on results obtained following the ingestion of pure glucose solutions, but which do not necessarily relate to those obtained with mixed meals, have varied from 10–80% depending on the methods of measurement employed. The truth probably lies somewhere in between the two extremes. Equally uncertain is how much of the absorbed glucose is converted into glycogen in the liver and elsewhere, and how much is oxidised immediately to provide energy or converted into other substances such as fats. Estimates made using ^{13}C suggests that about 10% of an orally administered 100 g glucose load is oxidised within three hours of its ingestion, but how this relates to everyday life is unknown since gastric emptying and absorption of carbohydrate is generally much faster after the ingestion of a simple glucose solution than after a mixed meal.

The rate of glucose absorption following ingestion of a meal is determined mainly by the rate of delivery of food into the duodenum. Absorption of carbohydrate (glucose) is, therefore, increased by rapid gastric emptying, such as occurs when isotonic solutions are ingested, or when food is delivered directly into the duodenum or jejunum. In such cases the rate of absorption may be sufficiently high, temporarily, to overwhelm the glucose homeostatic mechanisms leading, in turn, to such a large increase in pool size that glucose 'overflows' into the urine.

The exact duration of the absorptive phase following ingestion of a meal depends upon many factors including its size, composition and physical nature. It is unlikely, however, that the average adult man who eats three meals a day is truly post-absorptive for more than a few hours in any 24 hour period and this mainly between 0200 and 0800 hours. During this time, glucose lost from the pool by utilisation in the tissues is replenished by the liver either by drawing upon its own reserves of glycogen or by releasing new glucose molecules formed from precursors brought to it from the periphery.

Glycogenolysis and Gluconeogenesis

The amount of glycogen in the liver at any one time varies widely and is determined largely by the nature and size of the diet. The average total content, after an overnight fast, is only 44 g (15–80 g) and not much more than this after a meal. Liver glycogen stores may fall, after 36 hours without food, to as low as 4–8 g but do not change much thereafter even if fasting is continued for many days. Parenthetically, it may be noted that the rise in blood glucose concentration evoked by exogenous glucagon reaches its lowest level, virtually zero, after 48–72 hours without food but returns almost to 'normal' if fasting is continued beyond this time.

Glycogen never disappears completely from the liver except *in extremis*; indeed it may be an obligatory intermediate in the production of glucose by the gluconeogenic pathway.

Glycogen synthesis is encouraged by high concentrations of glucose and insulin in the portal blood and glycogen breakdown by the reverse. The two processes are not, however, the opposite of each other. Glycogenolysis depends on the transformation of inactive into active phosphorylase which can occur either as a result of increased cAMP formation from ATP under the influence of adenylate cyclase, or by an alternative method involving stimulation of alpha adrenergic receptors and the entry of calcium ions into the cell. In either case the final product is glucose–1–phosphate which, after undergoing internal rearrangement to glucose–6–phosphate, serves as the substrate for glucose–6–phosphatase. This converts glucose–6–phosphate into inorganic phosphate and glucose which is then free to enter the blood stream and replenish the glucose pool.

Gluconeogenesis, or the formation of new glucose molecules from various precursors, is, for all practical purposes, confined to the liver and, to a lesser extent, the kidneys which assume a major role during prolonged starvation. Although much attention has been paid to the role of specific gluconeogenic enzymes in determining the rate of gluconeogenesis in the liver there seems little doubt that the single most important determinant is the availability of gluconeogenic precursors in the portal blood. Opinions differ as to the relative importance of the various substrates but, in man, lactate is quantitatively the most important especially during exercise. Others in descending order of quantitative importance are alanine, pyruvate, glycerol, glutamate and other amino-acids. The importance of alanine as a gluconeogenic substrate was first recognised following the discovery that its concentration in blood draining the limbs is proportionately much higher than that of other amino-acids and that its clearance by the liver is greater especially during fasting.

Although the extent of its contribution to the formation of new glucose molecules has sometimes been exaggerated, alanine undoubtedly does play an important part in conveying 3–carbon skeletons, derived from glycogen or from other amino-acids in muscle and other tissues, to the liver and kidneys where they can be synthesised into glucose.

Gluconeogenesis is reduced by eating, mainly as a result of increased insulin activity, and enhanced by fasting or anything else that reduces the portal insulin concentration.

Hormonal Control

Adrenaline and Noradrenaline

Adrenaline was the first hormone recognised to have hyperglycaemic properties which are shared, to a certain extent, by noradrenaline. It owes its activity to the dual ability to inhibit insulin secretion and stimulate glycogenolysis by activation of phosphorylase. In liver this leads directly to liberation of glucose into the glucose pool and in muscle to the liberation of lactate and pyruvate which serve as gluconeogenic substrates. The effect upon insulin secretion is ordinarily more important in the regulation of blood glucose than the effect upon glycogenolysis and is mediated mainly, if not exclusively, through the sympathetic nervous innervation of the islets of Langerhans.

For a long time it was thought that adrenaline, like glucagon, exerted its glycogenolytic action in the liver mainly through activation of adenyl cyclase via β–adrenoreceptors. This now seems unlikely in man and some other species in which the glycogenolytic effect involves the entry of calcium ions into the cytosol and which appears to be mediated mainly through activation of α–adrenoreceptors. Regardless of mechanism, the result is the same, namely the conversion of inactive into active phosphorylase and transformation of glycogen into glucose. The discovery of the α–adrenergic, cAMP-independent mode of phosphorylase activation helps reconcile the apparent importance of sympathetic nervous activity in glucose homeostasis with the fact that β–adrenergic blockers do not abolish the hyperglycaemia produced by adrenaline or noradrenaline, nor do they impede recovery from insulin-induced hypoglycaemia in which sympathetic nervous system activation is thought to play a part. Moreover, as Sokal showed over 20 years ago, adrenaline itself almost never circulates in blood under physiological conditions at concentrations sufficiently high to activate hepatic phosphorylase. Noradrenaline, on the other hand, by virtue of its release from nerve terminals which occur in abundance in the liver in intimate connection with hepatocytes, is well suited to a hyperglycaemic role.

Glucagon

Glucagon was discovered as a hyperglycaemic contaminant of the very first insulin preparations but for many years attracted little attention. This changed following the isolation, identification and chemical characterisation of glucagon during the 1950s and, for a short while thereafter, it was thought by some authorities to be almost as important as insulin in the control of glucose metabolism.[3,8] This now seems unlikely in all species except some birds and reptiles in which pancreatectomy leads to hypoglycaemia due to the loss of glucagon secretory capacity, rather than to hyperglycaemia from loss of insulin action.

In man, glucagon is produced exclusively by the A-cells of the islets of Langerhans. They are juxtaposed to the B-cells, which are the source of insulin, and to the D-cells which are the source of pancreatic somatostatin; their secretory product, i.e. glucagon, has a profound effect upon the secretory activity of both of them. The exact way in which the various types of islet cell interact is still largely unknown but there is now very good evidence that glucagon plays an important role in mediating the insulin release stimulated by some amino-acids and intestinal hormones.

Despite a large body of knowledge concerning the pharmacology of glucagon, remarkably little is known about its physiological and pathological roles in man. Amongst the many demonstrable properties of glucagon, those that have attracted most interest are its potent glycogenolytic, gluconeogenic and lipolytic effects (all of which tend to expand the glucose pool) and its apparently paradoxical insulinotropic effect which tends to contract it.

Several substances with immunological and biological properties similar to those of glucagon occur in the gastrointestinal tract but even less is known about their physiology than about true glucagon. They seem not to be directly involved in controlling glucose metabolism but may possibly affect insulin secretion.

Glucagon secretion is stimulated by hypoglycaemia, insulin deficiency, both sympathetic and parasympathetic nervous activity, certain amino-acids and GIP—one of the major insulinotropic gastrointestinal hormones.[13,16] It is inhibited by insulin, hyperglycaemia and somatostatin. Absolute glucagon deficiency is difficult to produce experimentally in mammals but when achieved by injection of anti-glucagon antiserum, is associated with remarkably few, if any, discernible adverse effects apart from an occasional modest lowering of blood glucose concentration. The occurrence of spontaneous glucagon deficiency—without concomitant insulin deficiency—has not been established unequivocally in man, although it has been incriminated as the cause of hypoglycaemia in a child in whom no other lesion was found. Excessive glucagon secretion occurs in all insulinoprivic states, probably due to loss of the inhibitory effect of insulin upon A-cell activity, as well as in the exceedingly rare glucagonoma syndrome in which hyperglycaemia is a common, but not invariable, feature.

Pituitary Hormones

Growth hormone, prolactin and their pregnancy-associated analogue—placental lactogen—exert small but possibly important effects upon glucose homeostasis through poorly understood mechanisms. Their main effect seems to be to produce insulin resistance which, in turn, promotes increased insulin secretion more or less restoring the *status quo* as far as glucose homeostasis is concerned. If, for any reason, the endocrine pancreas cannot respond to the increased need for insulin a condition of impaired glucose tolerance may develop which, if sufficiently severe, may manifest itself as diabetes mellitus. Growth hormone deficiency is associated with impaired insulin secretion in response to an oral glucose challenge but glucose tolerance is normal. This dissociation of effect in which a normal rate of glucose utilisation occurs despite a subnormal rise in plasma insulin is typical of growth hormone deficiency and can be used diagnostically in the investigation of anterior pituitary function.

Adrenocortical Steroids

The naturally occurring glucocorticoid, cortisol, and its more potent synthetic anti-inflammatory analogues such as prednisolone and dexamethasone, are capable, under conditions of pharmacological usage, of increasing glucose pool size, mainly through their ability to promote gluconeogenesis though other mechanisms are also involved. Cortisol promotes mobilisation of NEFA but is not itself ketogenic. In pharmacological doses glucocorticoids produce insulin resistance but their lipolytic effects are completely inhibited by insulin at concentrations similar to those normally present in the blood. Overtreatment with glucocorticoids leads to a characteristic type of obesity and is associated with an increased need for insulin which is normally met by increased pancreatic secretion. Only when this cannot be achieved, because of some defect in B-cell function, does glucose tolerance become impaired.

Non-hormonal Factors

Though usually presented piecemeal for the sake of clarity, the control of glucose pool size is clearly both extremely complex and incompletely understood. How important a role is played by the hypothalamus through its ability to control both insulin and glucagon secretion—not to mention the autonomic nervous system and anterior pituitary hormones—is still largely conjectural; as is the function of the insulinotropic hormones of the gut which are currently attracting so much attention. The importance of fatty acid and ketone body metabolism in the regulation of glucose homeostasis, and *vice versa*, cannot be overemphasised, though detailed discussion would be out of place here. So, too, would consideration of the roles played by K^+, Ca^{++} and, under certain circumstances, Li^+ in regulating glucose metabolism. Nevertheless, it must be noted that moderate and sometimes even gross abnormalities of glucose metabolism are common in what might otherwise be looked upon as asymptomatic reductions in body potassium stores secondary to any one of a variety of diseases.

Of great interest, at the present time, is the nature of the alteration in glucose homeostasis produced by experimental chromium deficiency in animals. There is evidence that organic chromium, especially in a form known as glucose-tolerance-factor which is present in relatively large amounts in brewers' yeast, is involved in sensitizing insulin receptors to insulin, thereby improving glucose tolerance. How it does this is quite unknown.

Attempts to establish a role for chromium deficiency in disorders of glucose metabolism in man have largely been unsuccessful, partly because of the difficulty, hitherto, of measuring chromium in blood and tissues accurately, and partly because the biologically active form of chromium has not yet been identified and quantified.

INVESTIGATION OF GLUCOSE METABOLISM IN MAN

From a clinical point of view it is usual to consider disorders of glucose metabolism as being predominantly associated with either a pathological expansion or contraction of the glucose pool though both may, and often do, occur at different times in the same individual. Moreover, it is quite clear that abnormalities of glucose metabolism exist without any discernible change in total glucose pool size but with the sole exception of hereditary glucose–galactose malabsorption they have, so far, evaded detection—due no doubt mainly to the crudity of investigative methods available for clinical use.[7,23]

HYPERGLYCAEMIA

Abnormal elevations of blood glucose concentration are very common though not necessarily permanent nor pathological. The WHO Expert Committee on Diabetes recently defined diabetes mellitus as a 'state of chronic hyperglycaemia (i.e. the state of having an excess concentration of glucose in the blood) which may result from many environmental and genetic factors often acting jointly' but since blood glucose levels alter throughout the day, even in normal healthy subjects, different criteria must be employed to define hyperglycaemia in the fasting and non-fasting subject. It is extremely useful in clinical practice to distinguish fasting from post-prandial hyperglycaemia since the former almost always indicates the presence of disease requiring treatment, whereas exclusively post-prandial hyperglycaemia alone is usually of less sinister import and often temporary. The main clinical importance of such occasional or temporary post-prandial hyperglycaemia is that it may lead mistakenly to a diagnosis of diabetes mellitus. This should not occur, however, if due attention is paid to the criteria for diagnosis (Table 1) recommended by the WHO Expert Committee and the National Diabetes Data Group.

Diabetes Mellitus

Historical

Soon after it became possible to measure the concentration of sugar in the blood it was established that not only did patients with diabetes mellitus exhibit glycosuria but that they also, invariably, had hyperglycaemia. With further experience it was soon appreciated that certain individuals developed hyperglycaemia and glycosuria only after consuming a glucose-containing drink.

Though some of these individuals were symptomless the majority suffered from some, though seldom all, of the clinical features of diabetes mellitus including glycosuria, polydypsia and polyuria, increased susceptibility to systemic and local infections, gangrene of the limbs, cataracts and kidney disease. Most were fat rather than thin, and showed little tendency to develop ketosis, ketonuria or coma, except when suffering from intercurrent illness, in contrast to patients with classical diabetes mellitus who showed all of these features as well as generally being much younger. By the middle of the present century it had become fashionable, in many quarters, to diagnose all subjects with arbitrarily defined postabsorptive hyperglycaemia as suffering from diabetes mellitus whether or not they had symptoms or other evidence of metabolic disease. In order to prevent perpetuation of the circular argument which had led to this sorry state of affairs, the WHO Expert Committee on Diabetes Mellitus,[25] as well as a mainly American and British group of diabetologists (the National Diabetes Data Group[18]), issued guidelines for the diagnosis of diabetes mellitus based on symptomatology as well as on blood glucose measurements. These acknowledge that a sizeable percentage of people with nothing discernibly wrong with their health and who show little or no propensity to develop any of the complications of diabetes in long-term follow-up studies, nevertheless exhibit blood glucose levels following ingestion of an oral glucose load, that are substantially higher than are observed in the majority of the healthy population but which are lower than those generally observed in patients with clinically manifest diabetes. Such individuals can be described as having impaired glucose tolerance, a condition which is not necessarily indicative of present or future disease (Table 2). Most people with impaired glucose tolerance (now used in a defined technical sense) continue to exhibit impaired glucose tolerance when tested years or even decades later, though some revert towards the norm whilst others develop overt diabetes mellitus.

Diabetic Complications

Acute. Before the insulin era patients with diabetes usually died within five years of the diagnosis, generally from coma. Though much rarer nowadays, coma is still an important cause of morbidity in patients with insulin-dependent (type 1) diabetes and carries a 10–15% mortality rate.

There are one mixed, and three main, types of diabetic coma. The commonest is associated with gross overproduction of ketone bodies and is usually referred to as ketoacidosis. The second commonest is the so-called non-ketotic, or hyperosmolar hypernatraemic variety of diabetic coma and the third is the lactic acidotic type. The latter differs in only one important respect from lactic-acidosis occurring in non-diabetic patients, namely, that the blood glucose concentration is invariably high instead of normal or low. It is also often, but not invariably, a consequence of overtreatment with a biguanide (usually phenformin) which inhibits gluconeogenesis and this permits lactate to accumulate in the blood.

Mixed diabetic syndromes occur but are comparatively rare and usually contain elements of ketotic and lactic acidosis with neither predominating.

Diabetic Ketoacidosis. Profound disturbances of water and electrolyte balance occur in all forms of diabetic coma.[12,19] Though secondary to the disturbances of carbohydrate and fat metabolism, which are the hallmark of diabetes, they are far more dangerous and contribute to a vicious circle (Fig. 2) which, if it is not

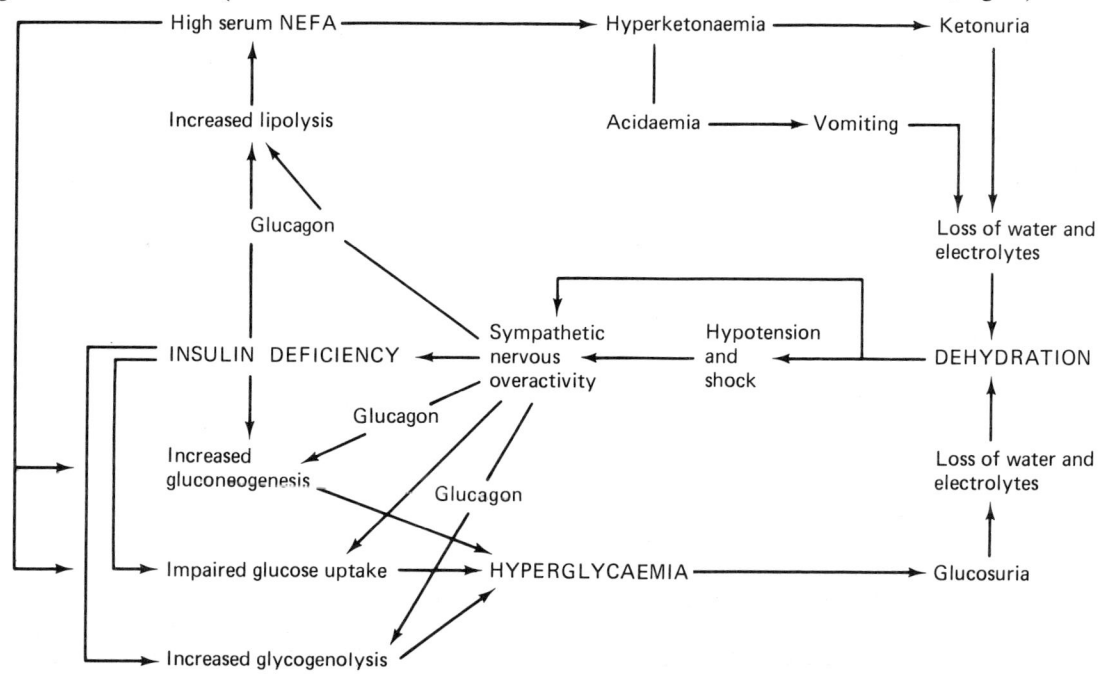

FIG. 2. Schematic representation of the vicious circle predisposing to diabetic ketoacidosis in a patient with embarrassed insulin secretory capacity.

broken rapidly and adequately, may cause death within days or even hours of onset. Metabolic balance studies reveal that the average patient in diabetic ketoacidosis has lost more than 5 litres of fluid or up to 15% of their total body water, by the time the diagnosis is made. Fluid loss through the kidneys is accompanied by large losses of sodium, potassium, phosphorus and magnesium not to mention bicarbonate loss in the form of CO_2 through the lungs. Although the loss of water is relatively greater than that of sodium, plasma sodium levels are generally low by the time the patient is admitted to hospital. Plasma potassium levels, on the other hand, are usually normal or high. This reflects redistribution of the reduced total body potassium between intra- and extracellular fluid compartments as a consequence of changes in hydrogen ion concentration secondary to the accumulation of ketones in the blood and the loss of electrolytes, i.e. sodium and potassium, in the urine. Blood pH is reduced, sometimes to dangerously low levels, and the haematocrit and concentration of plasma proteins are raised due to haemoconcentration. Most of these metabolic derangements arise secondarily to the copious production of urine containing glucose and ketones early in the development of ketoacidosis leading, in turn, to dehydration, activation of the sympathetic nervous system and exacerbation of the metabolic disturbance. By the time the patient is admitted to hospital, reduction in blood volume and renal perfusion may have progressed so far as to cause oliguria and prerenal azotaemia which may, themselves, contribute to the patient's distress.

The aim of treatment in ketoacidosis is to replace lost fluid and electrolytes as rapidly as possible, correct the acidaemia by reducing lipolysis and finally to restore the blood glucose concentration to normal. Only very rarely, i.e. if blood pH falls below 7·0, is it necessary to include alkali in the form of sodium bicarbonate to the therapeutic regime. It is now recognised that too rapid a reduction in blood glucose concentration may produce gross disparities between the osmolalities of the extra- and intracranial compartments and this is suspected of being one of the causes of sudden, unexpected death in patients undergoing seemingly successful correction of ketoacidosis.

The main avoidable complication of treatment of ketoacidosis is the profound fall in plasma potassium concentration that accompanies the reduction in plasma hydrogen ion concentration and restoration of normoglycaemia under the influence of fluid and insulin replacement. Potassium supplements must, therefore, be given, in addition to saline used to replace lost water and sodium, from the very commencement of treatment. They are usually given by continuous intravenous infusion or by regular and frequent intramuscular injections rather than by bolus intravenous injections which can be extremely dangerous because of their deleterious effect upon the heart.

Hyperosmolar Diabetic Coma. In this condition water loss is as great as, and usually greater than, that in ketoacidosis but, because ketosis is minimal, disturbances in potassium and hydrogen ion metabolism are much smaller and hypernatraemia is the rule, rather than the exception as is the case in ketoacidosis. Blood glucose levels often exceed 55 mmol/l (990 mg/100 ml) in contrast to ketoacidosis, in which they generally lie within the range 20–35 mmol/l (360–630 mg/100 ml); plasma ketone levels, on the other hand, seldom exceed 1 mmol/l; haemoconcentration is greater than in ketoacidosis and extracellular fluid volume is proportionately more markedly reduced. Why hyperosmolar rather than ketoacidotic coma develops in some, generally older, subjects and not in others is still poorly understood. One explanation is that in patients with the former condition sufficient insulin secretory capacity remains to restrain lipolysis—one of the precipitating causes of unbridled ketosis—but insufficient to permit free entry of glucose into peripheral tissues where it can be utilised. Another is that the gross hyperglycaemia observed is due, at least in part, to the drinking of glucose-containing fluids, such as lemonade, in a vain attempt to quench the intolerable thirst created by polyuria and glycosuria in a previously unrecognised non-insulin dependent diabetic subject.

Long-term Complications of Diabetes. The advent of insulin and intravenous fluid replacement therapy dramatically reduced the morbidity and mortality from coma in patients with diabetes mellitus which had previously been responsible for death in this disorder, usually within five years of diagnosis. It has, however, done little to reduce the high incidence of chronic complications in diabetes which make it, for example, the commonest cause of blindness in this country and one of the major causes of cataract, nephropathy, kidney failure, peripheral neuropathy and large artery disease. The pathogenic mechanisms are not necessarily the same in all of these conditions and only in the case of diabetic retinitis, and possibly nephropathy, has the causative role of hyperglycaemia itself been established. There is a distinct possibility that 'hyperinsulinaemia' is an aetiological factor in the pathogenesis of large artery disease and that abnormalities of glucose–sorbitol–fructose interconversion play an important role in the pathogenesis of diabetic neuropathy and cataracts.

Diabetic Control

Evidence that hyperglycaemia itself, rather than its underlying cause, is responsible for some or all of the complication of diabetes has been difficult to obtain since, until recently, virtually no independent assessment of 'diabetic control' was available. Reliance upon urinary self-testing as a measure of diabetic control has proved to have been largely misplaced. Currently, however, encouraging results are being obtained with home blood-glucose monitoring schemes based either on the use of battery operated portable colorimeters for measuring the change produced in glucose–oxidase impregnated strips or on the use of dried blood spots on filter paper which are sent to the laboratory for analysis. By using glycosylated haemoglobin (HbA$_1$) as an integrated index of blood glucose concentration over the preceding month, physicians can gain some idea of how effective self-monitoring has been, but this method of control suffers from the disadvantage that it is retrospective rather than prospective.

Glycosylated Haemoglobin. Glycosylation of the haemoglobin molecule occurs through non-enzymatic attachment of glucose to the α–amino groups of lysine at various loci on the A and B chains. The result is a small but detectable change in the physical and immunological properties of the haemoglobin molecule. The glycosylation reaction is essentially irreversible due to internal rearrangement of the glucose molecule after combination with the amino-group. This leads to the formation of a ketoamine linkage which, on cleavage by hydrolysis, yields the ketohexose 2–deoxyfructose whose measurement provides an index of the amount of glycohaemoglobin originally present.

The rate of formation of glycosylated haemoglobin is determined by the law of mass action, i.e. it depends upon the concentration of the two reactants and the length of time they are in contact with each other, but once formed it remains in the blood as long as the red-cell containing it survives. HbA$_1$ is, therefore, a poor index of 'diabetic control' in patients with either anaemia or increased red-cell turnover.

Some of the lysine residues in the haemoglobin molecule are more rapidly and readily glycosylated than others. The most active residue appears to that at the N–amino terminal of the β–chain and its glycosylation produces the product referred to as HbA$_{1C}$. This haemoglobin variant can be separated from other haemoglobins by electrophoresis or gel chromatography and quantified relative to them; alternatively it can be measured by immunoassay, or chemically, after hydrolytic cleavage. None of the methods of measuring glycosylated haemoglobin currently available is wholly satisfactory, though quantitative electrophoresis is probably the simplest and most accurate. There is still no documentary evidence that clinical outcome in diabetic patients is improved by using HbA$_1$ for monitoring them.

Fats

Though diabetes is generally looked upon as a disorder of carbohydrate metabolism, a case could be made for considering it primarily as a disease of fat metabolism in which the disturbances of glucose metabolism observed are secondary to fatty acid and ketone body malhandling.

Non-esterified Fatty Acids. Plasma non-esterified fatty acids (NEFA) are carried in the circulation bound to albumin and originate from adipose tissue as a consequence of intracellular lipolysis. Though normally present in only very small amounts, their rapid turnover and high energy content ensure that they are the main form in which energy is transported around the body. Some of the fatty acids that are liberated by intra-adipocyte lipolysis are re-esterified and returned to stores as triglycerides: the rest, and all of the glycerol, are liberated into the circulation from which they are rapidly cleared by the liver. There the NEFA are either partially or completely oxidised to ketone bodies or carbon dioxide respectively, or are resynthesised into triglycerides and re-exported as very low density lipoproteins (VLDL). What proportion of the NEFA reaching the liver follows each metabolic pathway is determined by many factors, including the availability and supply of other energy yielding substrates, the concentration and activity of various hormones—of which insulin is the most important—but, above all, by the size of NEFA burden which is itself a reflection of the rate of lipolysis in the adipose tissue mass. The quantity of NEFA released by adipose tissue represents the balance struck between intra-adipocyte lipolysis and re-esterification. The former is usually the major determinant, being accelerated by increased adrenergic and other lipolytic hormone activity, and decreased by insulin and some of the prostaglandins.

The plasma half-life of NEFA is extremely short (in the region of one to two minutes) and consequently their concentration often fluctuates wildly, making interpretation of a single determination extremely difficult or impossible. The plasma concentration of NEFA may, for example, more than quadruple in response to the psychogenic stress of a venepuncture or after comparatively mild exercise. More sustained increases in plasma NEFA levels occur in both absolute and relative insulinoprivic states, such as those produced by prolonged fasting, chronic carbohydrate deprivation and, of course, diabetes mellitus. High levels are also associated with increased sympathetic nervous activity such as occurs in anxiety, thyrotoxicosis, phaeochromocytoma and after trauma. In many of these conditions high plasma NEFA are accompanied by a greater or lesser degree of ketonaemia. In the main, plasma NEFA concentrations are more responsive than glucose to small changes in circulating insulin levels due to the greater insulin sensitivity of adipocytes than most other insulin responsive tissues.

Plasma glycerol has a much longer plasma half-life than NEFA and consequently fluctuates less violently in response to short-lived changes in lipolysis rate. It can, therefore, be used clinically as an indicator of the rate of lipolysis in the whole-body adipose tissue mass since, except in liver disease, glycerol clearance is more or less totally dependent upon its plasma concentration. To date, however, neither plasma NEFA nor glycerol measurements have found widespread application in diagnostic clinical biochemistry although they have proved invaluable in helping to elucidate carbohydrate-fat interrelationships in health and disease.

Plasma and Urinary Ketones. Acetoacetate, beta-hydroxybutyrate and acetone, collectively known as ketone bodies or ketones, were discovered in the urine of patients with severe diabetes about a century ago. Their presence in the urine of patients with other diseases attracted little interest until comparatively recently, probably because of the difficulty of measuring them accurately and interpreting the result. In 1962 Williamson and his co-workers introduced the enzymatic techniques, now almost universally employed, for measuring ketones in blood plasma. The availability of this methodology has led to an enormous increase of interest in plasma ketones by physiologists and experimentalists but so far has had relatively little clinical impact.

Ketones are produced from NEFA by a complicated oxidative process, exclusively in the liver. Their concentrations in the plasma can vary a hundredfold or more, even in healthy subjects, and their short plasma half-life and rapid turnover ensures that they are an important source of fuel for many tissues in health as well as in disease. Indeed, they are sometimes preferred to glucose as a source of fuel by the heart, striated muscle, and kidney and, under very special circumstances, even the brain.

The relative proportions of the three ketones in blood and urine vary; in normal subjects beta-hydroxybutyrate and acetoacetate are present in blood in roughly equal amounts and there is virtually no acetone. Ketones are not normally detectable in urine by conventional analytical techniques but often become so in disease. For example, in diabetic ketoacidosis, acetone, which is responsible for the characteristic smell on the breath associated with this disorder, may become the predominant circulating ketone and the ratio of beta-hydroxybutyrate to acetoacetate, normally about 2:1, rises twofold or more.

Changes in this ratio also occur in conditions associated with hypoxia or when intracellular redox potential is shifted towards the reduced or acidic state for any other reason. Similar changes occur during the metabolism of alcohol and explain why ketosis associated with alcohol-induced hypoglycaemia may go undetected by urinalysis since β-hydroxybutyrate—which does not give a positive reaction with Rothera's reagent (Acetest[R]) is excreted in this condition rather than aceto-acetate which does.

The measurement of plasma ketones is still not often used in diagnostic clinical biochemistry. The situation is changing, however, since, like NEFA and glycerol, plasma ketones provide valuable clues as to the level of insulin *activity* in liver and adipose tissue as well as to the balance between the utilisation of carbohydrates and fats as sources of energy under various conditions.

Lactate and Pyruvate

Pyruvic acid is the end product of glycolysis and, under aerobic conditions most of it enters the tricarboxylic (citric) acid cycle. There it is completely metabolised to CO_2 and water and is coupled with the regulated release of energy used to sustain vital processes. Under anaerobic or hypoxic conditions, pyruvic acid is reduced to lactic acid by lactic dehydrogenase using NADH that has been produced from oxidation of glyceraldehyde–3–phosphate earlier in the glycolytic pathway. Regeneration of the NAD^+ by reduction of pyruvate to lactate sustains operation of glycolysis under hypoxic conditions but only at the cost of a shift in the redox potential towards increased acidity; i.e. increased hydrogen ion concentration and an increase in the lactate:pyruvate ratio.

Lactate, unlike pyruvate, is outside the main stream of intracellular metabolism and re-enters it only after re-oxidation to pyruvate. This generally takes place in the liver where lactate is quantitatively the most important substrate for new glucose formation. In plasma the concentration of lactate relative to pyruvate is normally in the region of 10:1 and this ratio reflects the redox potential of the body as a whole but especially that of the liver. It is increased by hypoxia, by impaired hepatic perfusion and by the ingestion of certain substances, the most notable of which are alcohol, the diguanide groups of drugs, e.g. phenformin, and certain polyols, such as sorbitol.

Fructose

Fructose is one of the two main monosaccharides found in fruits; it also constitutes 50% of the sugar present in honey, the rest being glucose. Sucrose, which provides roughly 25% of the carbohydrate content of the average western diet, yields equimolar amounts of glucose and fructose following its hydrolysis in the intestine. The two sugars are absorbed independently of each other through different transport mechanisms; in the exceedingly rare inborn error of metabolism known as glucose–galactose malabsorption, fructose is the only dietary carbohydrate that can be tolerated.

Fructose is cleared from the portal blood after absorption more rapidly and completely than glucose and as a result, systemic blood fructose levels rarely rise above 0·5 mmol/l even after the ingestion of large doses of sucrose, or even fructose itself, except in patients with hepatocellular damage or one of the rare inborn errors of fructose metabolism.

Within the liver, fructose is phosphorylated to fructose–1–phosphate by fructokinase, an enzyme that occurs mainly, though not exclusively, in the liver, kidney and intestinal tract tissue. Fructose–1–phosphate is, itself, the substrate for fructose–1–phosphate aldolase which by cleaving it into dihydroxyacetone phosphate and glyceraldehyde permits the original carbon skeleton of fructose to enter the glycolytic pathway or be converted into glycogen. Fructose is also a substrate for hexokinase, especially in adipose tissue, but only in the almost complete absence of glucose for which the enzyme normally has a much greater affinity. This situation is never likely to occur naturally but the same effect can be produced when insulin activity is lacking since under these circumstances, adipocytes are virtually impermeable to glucose. The ability of fructose to gain entry to, and be metabolised by, adipocytes in the absence of insulin may explain, in part, its greater antiketogenic effect compared with glucose when given to insulinoprivic diabetic animals, although a direct inhibitory effect of fructose upon ketogenesis in the liver is even more important in this context.

Fructose occurs in the body as a normal constituent of semen and is formed from glucose via sorbitol according to the scheme shown in Fig. 3. Transformation takes place in two steps; the first utilises aldose reductase and the second sorbitol dehydrogenase—both of which are present in a number of tissues; especially the lens, liver, seminal vesicles, placenta, brain and peripheral nerves. There is evidence that fructose formation from glucose plays an important part in the aetiology of some of the neural and opthalmic complications of diabetes, due to the high concentration of glucose present in the blood pushing the reactions shown in Fig. 3 far to the right.[9]

$$\text{D-Glucose} + \text{NADPH} + \text{H}^+ \xrightarrow{\text{Aldose reductase}} \text{Sorbitol} + \text{NADP}$$

$$\text{Sorbitol} + \text{NAD} \xrightarrow{\text{Sorbitol dehydrogenase}} \text{D-Fructose} + \text{NADH} + \text{H}^+$$

FIG. 3.

Fructose is the only dietary sugar that does not stimulate insulin secretion when taken by mouth. Nor does it require insulin in order to enter metabolic pathways *in vitro* though, *in vivo*, the administration of fructose to an untreated insulinoprivic subject leads to its almost stoichiometric conversions into glucose. No such conversion occurs, however, in normal or adequately treated insulin-dependent subjects in whom oral fructose produces a smaller, often barely detectable, rise in blood glucose concentration. It can, therefore, be used by diabetic subjects as a non-synthetic sweetener, instead of sucrose, for increasing the palatability of foods, especially since, on a weight basis, fructose is 50% sweeter than sucrose and five times sweeter than sorbitol which was formerly used for this purpose.

Sorbitol

This 6-carbon polyol is normally a trivial constituent of the diet but may be added in substantial amounts to foods prepared for diabetic subjects as a rather poor substitute for sucrose. After absorption it is quickly taken up by the liver where, under the influence of sorbitol dehydrogenase, it is converted into fructose. Thenceforth, it is metabolised along the same lines as described for that of sugar.

Only minute traces of sorbitol and fructose are normally excreted in the urine, but although the amounts are increased in diabetics in direct proportion to the severity of the glycosuria, they have no diagnostic significance. There is, however, some evidence that red-cell sorbitol content can, like HbA_1, be used as an indicator of diabetic control. In addition it may provide information about polyol pathway activity *in vivo* and so throw some light on its role in the pathogenesis of diabetes-associated complications.[9]

Tests for Hyperglycaemia

Pathological hyperglycaemia, unless associated with an abnormality of renal function, is usually accompanied by glycosuria, and this was, until the advent of random blood glucose testing, often the first indication of pathological hyperglycaemia or asymptomatic diabetes.

Urinary Glucose

Roughly 100 grams of glucose are filtered at the glomeruli each day in healthy subjects. Approximately 99·9% of the filtered load is reabsorbed by the renal tubules leaving 50–150 mg to appear in the urine. When the limited, but individually highly variable, capacity of the renal tubules to reabsorb glucose is exceeded, there is a sudden increase in the amount of glucose appearing in the urine with only a very small rise in blood glucose.

Though many methods are available for detecting glycosuria, only two need be considered here. The oldest, and still most popular, is 'Benedict's test' or its modified and grossly simplified version, Clinitest[R]. This method not only detects glucose but also galactose fructose and lactose—should any of these be present and is especially useful, therefore, for screening.[15] The second, simpler, but in some ways less satisfactory, method uses a glucose–oxidase: peroxidase dye-impregnated strip which changes colour when glucose is present. Other reducing sugars do not react in this way. The strips are, therefore, more specific than Benedict's test but this makes them less suitable for screening purposes especially in babies in whom galactosuria and fructosuria are sometimes the first clue to the presence of an inborn error of metabolism. The strips are also much more sensitive to glucose than the older methods and often give a positive response with urinary glucose concentrations as low as 1 mmol/l. This is not really an advantage, though often construed as such, since in a high proportion of cases this degree of glycosuria has no pathological significance. A more serious disadvantage of glucose-oxidase impregnated strips is that they are subject to interference by ascorbate and other reducing substances in the urine and this can produce false negative results in the presence of even moderate glycosuria.

In contrast to Clinistix[R], 'Benedict's test' (Clinitest[R]) is rarely positive unless urinary glucose concentration exceeds 5 mmol/l. A positive test is, therefore, more likely to indicate 'significant glycosuria', i.e. the excretion of more than of 8 mmol (1·45 g) of glucose per day, than mere leakage, i.e. less than 6 mmol (1·08 g) per day, which is normal. Tests for glycosuria should ordinarily be performed on urine collected two to three hours after a meal, and not, as is often done, on the first specimen collected after an overnight fast. A positive Benedict's test on any specimen of urine—but especially one collected from a fasting subject—is an indication for further investigation. In subjects under the age of thirty, it only rarely indicates impairment of glucose tolerance but with increasing age there is, amongst patients with positive urinary 'Benedict's test', a rapidly increasing rise in the incidence of impaired glucose tolerance or frank diabetes. An important use of urinary glucose testing, in the past, was for monitoring the efficacy of dietary and, to a lesser extent, insulin therapy in patients with diabetes. Doubt has, however, recently been cast on the value of this procedure and it is likely to be replaced by home blood-glucose monitoring in the future.

Quantitative measurements of urinary glucose, though possible, are seldom clinically informative and assessment, on the basis of visual inspection, of the 'Benedict's' or Clinitest result provides only a very rough indication of the degree of hyperglycaemia present. It is possible, for example, for a markedly hypoglycaemic subject to pass a grossly glycosuric specimen of urine just before losing consciousness and for this to be misconstrued, by the unwary, as evidence of diabetic, rather than of hypoglycaemic coma. Conversely, it is possible for gross hyperglycaemia to be present with only minimal or no glycosuria, especially in elderly subjects.

Blood Glucose

Measurement of blood glucose is currently the most important investigation in patients suffering, or suspected of suffering, from diseases of carbohydrate metabolism. Many techniques are available and proper interpretation requires a knowledge of which one was used including its limitations, especially as regards specificity, precision and accuracy. Other factors that must be considered are the source of the sample, i.e. whether venous, capillary or arterial, and the time elapsed since the subject last ate. It is also important to know what drugs, if any, the patient was receiving at the time the sample was collected since many can produce aberrant results either by interfering with the analytical procedure or, more commonly, by affecting glucose homeostasis.

Many of the most widely used techniques for measuring blood glucose depend upon the reducing or chromogenic properties of glucose, and are unreliable either when blood glucose concentration is low or when sugars, other than glucose, are present. Tests based on the use of glucose oxidase, glucose dehydrogenase or hexokinase are more dependable though 2–deoxyglucose, a synthetic sugar sometimes used instead of insulin in the modified Hollander test for completeness of selective vagotomy or as a test of adrenomedullary integrity, can act as a substrate for both glucose oxidase and glucose dehydrogenase and give falsely high values. Consequently, under these circumstances, only the hexokinase method can be used with confidence.

The precision and accuracy of glucose measurements can usually be improved by using either plasma or serum, providing that separation from the red-cells has been effected rapidly, but in this event the results are, on average, 14% higher than when the whole blood is employed.

Procedure. Blood for glucose analysis is collected by venepuncture and added to a pot containing a suitable preservative, e.g. sodium fluoride 2 mg/ml, unless separation of the plasma from the red cells is effected immediately. The practice, common in some parts of the world, of allowing blood to clot before separating the serum for glucose analysis can yield falsely low values through glycolysis and should be discouraged. The time and site of collection of the sample should be noted, as should any drugs the patient was receiving; the time of the last meal and the nature of any symptoms that might have been present, especially in cases of suspected hypoglycaemia, should also be recorded.

Interpretation. Random blood glucose measurements are seldom of diagnostic significance except when used to confirm a clinical diagnosis of either diabetes mellitus or of hypoglycaemia which was producing neuroglycopenia at the time the specimen was collected. They do, however, have a role to play as part of the regular monitoring of therapy in patients with diabetes or spontaneous hypoglycaemia.

In healthy subjects, plasma glucose levels lie within the range 3·0–8·0 mmol/l (55–145 mg/100 ml), regardless of relationship to food. Values outside this range are unusual in people under the age of 50 and always warrant further investigation. In people over the age of 60, values up to 10 mmol/l (180 mg/100 ml) are not very uncommon after a meal, and need occasion no concern unless there is additional evidence of impaired glucose tolerance, e.g. glycosuria and/or symptoms of diabetes. Random plasma glucose values greater than 11 mmol/l (200 mg/100 ml) are diagnostic of diabetes, regardless of the patient's age, and further testing, apart from confirmation on at least one further occasion, is not indicated. A random plasma glucose concentration of less than 2·5 mmol/l (45 mg/100 ml) should always be taken seriously, especially if the patient had symptoms at the time the blood was collected. Further analysis, e.g. for insulin, C-peptide, and alcohol, for example, may be indicated, *on that specimen*, in order to elucidate the cause of the hypoglycaemia since it may be difficult or impossible to reproduce it on another occasion. For this reason it is essential when *spontaneous* hypoglycaemia is suspected on clinical grounds, that blood should be collected for confirmation of the diagnosis *before* treatment with glucose is instituted though this is clearly not so important when the hypoglycaemia is iatrogenic.

A plasma glucose concentration of more than 2·8 mmol/l (50 mg/100 ml) in a patient under the age of 60 years and who was symptomatic at the time the specimen was collected makes the diagnosis of neuroglycopenia due to hypoglycaemia extremely unlikely however suggestive the history and prompt the response to treatment might be!

Overnight Fasting Plasma Glucose

The glucose concentration of blood collected after a 12–14 hour fast shows less variation between individuals than blood collected at any other time. There is evidence of a real, though slight, seasonal variation in overnight fasting blood glucose levels; the lowest values occurring in the spring. Average overnight fasting blood glucose levels rise with increasing age, and are especially marked after the age of 50, but there is no clear sex difference.

Glucose levels in venous, capillary and arterial blood are similar to each other in the fasting subject and reflect the size of the body glucose pool; random specimens, on the other hand, may show differences of up to 1·5 mmol/l (25 mg/100 ml) between venous and arterial (capillary) samples due to glucose assimilation in the peripheral tissues under the influence of insulin.

Hyper- and hypoglycaemia are necessarily arbitrary definitions but there is more or less general agreement that overnight fasting plasma glucose levels within the range 3·0–5·5 mmol/l (55–100 mg/100 ml) are clearly 'normal' and that those below 2·5 mmol/l (45 mg/100 ml) or above 8·0 mmol/l (145 mg/100 ml) are clearly abnormal. Overnight values between 2·5–3·0 mmol/l and 5·5–8·0 mmol/l should be viewed with suspicion.

The main diagnostic value of overnight fasting plasma glucose measurements is in the recognition of spontaneous fasting hypoglycaemia and as a guide to the effectiveness of dietary treatment in diabetic subjects. There are, however, those who believe, that diabetes should not be diagnosed—however abnormal the response to an oral glucose load—unless fasting hyperglycaemia can be demonstrated on at least some occasions.

Oral Glucose Tolerance Tests

These have enjoyed a largely undeserved popularity[22] for the diagnosis of diabetes and other metabolic disorders for more than half a century but it is only within the past few years that any consensus as to how the test should be conducted and interpreted has begun to emerge. It is now widely appreciated that formerly far too much significance was attached to minor variations in blood glucose levels during oral glucose tolerance tests, taking into account their notorious unreproducibility and their susceptibility to interference by a variety of different factors, only some of which are capable of regulated control.[24]

The WHO Expert Committee[25] recommends using 75 g of glucose dissolved in 250–300 ml of water, consumed in 5 to 15 minutes, as the standard dose and method of administration, although solutions of partially hydrolysed starch produce the same effect and reduce the incidence of nausea and delayed gastric emptying. Dietary preparation, i.e. the institution of a high carbohydrate diet, is not ordinarily required unless strict carbohydrate restriction, i.e. less than 125 g per day, has been employed for more than three days before the test. Glucose tolerance tests should ordinarily be conducted in the morning after an overnight fast with the patient sitting comfortably in a chair. Smoking should not be permitted but moderate perambulation may be. Urinalysis adds nothing to the diagnostic significance of the test and, though still often advocated, is neither necessary nor indicated except in the identification of patients with an altered renal threshold and even this is of doubtful value.

Measurements of plasma insulin levels, though often performed during oral glucose tolerance tests, have no diagnostic significance nor do measurements of most other hormones and/or metabolites, although they may be useful when oral glucose tolerance tests are carried out for research or for reasons other than the diagnosis of diabetes.

Interpretation (Table 1) of the oral glucose tolerance test rests mainly on the fasting and two-hour post-ingestion blood (or plasma) glucose levels measured by a

TABLE 1
DIAGNOSTIC VALUES FOR ORAL GLUCOSE TOLERANCE TEST UNDER STANDARD CONDITIONS. (LOAD 75 G GLUCOSE IN 250–350 ML OF WATER FOR ADULTS, OR 1·75 G/KG BODY WEIGHT (TO A MAXIMUM OF 75 G) FOR CHILDREN, USING SPECIFIC ENZYMATIC GLUCOSE ASSAY. TWO CLASSES OF RESPONSE ARE IDENTIFIED—DIABETES MELLITUS AND IMPAIRED GLUCOSE TOLERANCE.)

	Glucose concentration		
	Venous whole blood	Capillary whole blood	Venous plasma
Diabetes mellitus Fasting	≥7·0 mmol/l (≥1·2 g/l)	≥7·0 mmol/l (≥1·2 g/l)	≥8·0 mmol/l (≥1·4 g/l)
and/or 2 hours after glucose load	≥10·0 mmol/l (≥1·8 g/l)	≥11·0 mmol/l (≥2·0 g/l)	≥11·0 mmol/l (≥2·0 g/l)
Impaired glucose tolerance Fasting	<7·0 mmol/l (<1·2 g/l)	<7·0 mmol/l (<1·2 g/l)	<8·0 mmol/l (<1·4 g/l)
and 2 hours after glucose load	≥7·0–<10/0 mmol/l (≥1·2–<1·8 g/l)	≥8·0–<11·0 mmol/l (≥1·4–<2·0 g/l)	≥8·0–<11·0 mmol/l (≥1·4–<2·0 g/l)

(Reproduced with permission from WHO Expert Committee on Diabetes Mellitus (2nd Report), WHO, Geneva, 1980)

glucose-specific method.[1,18,25] The Committee further recommends that if the patient has symptoms suggestive of diabetes, it is generally only necessary to measure glucose in a fasting or random blood specimen in order to establish the diagnosis; a random venous plasma glucose concentration of more than 11 mmol/l (200 mg/100 ml) or fasting plasma glucose level of more than 8·0 mmol/l (145 mg/100 ml) if confirmed on one more occasion, is diagnostic of diabetes and oral glucose tolerance testing is contraindicated. Only if the results are equivocal is a glucose tolerance test necessary. A two-hour post-glucose venous plasma glucose concentration of 11 mmol/l, or more, is diagnostic of diabetes; levels below 8 mmol/l are considered normal and those between 8·0 and 11·0 as indicative of *impaired glucose tolerance*.

In the absence of suggestive symptoms at least one additional, abnormal plasma glucose value (e.g. a one-hour post-glucose value of >11 mmol/l) or, preferably, an elevated two-hour or fasting, glucose value on a subsequent occasion, is required to establish a diagnosis of diabetes.

TABLE 2
CLASSIFICATION OF DIABETES MELLITUS AND OTHER CATEGORIES OF GLUCOSE INTOLERANCE

A. Clinical classes
   Diabetes mellitus
      Insulin-dependent type—Type 1
      Non-insulin-dependent type—Type 2
         (a) non-obese
         (b) obese
      Other types including diabetes mellitus associated with certain conditions and syndromes:
         (1) pancreatic disease,
         (2) disease of hormonal etiology,
         (3) drug- or chemical-induced condition,
         (4) insulin receptor abnormalities,
         (5) certain genetic syndromes,
         (6) miscellaneous
   Impaired glucose tolerance
      (a) Non-obese
      (b) Obese
      (c) Impaired glucose tolerance associated with certain conditions and syndromes
   Gestational diabetes

B. Statistical risk classes (subjects with normal glucose tolerance but substantially increased risk of developing diabetes)
   Previous abnormality of glucose tolerance
   Potential abnormality of glucose tolerance

(Reproduced with permission from WHO Expert Committee on Diabetes Mellitus (2nd Report), WHO, Geneva, 1980)

According to the WHO Expert Committee[25] the same criteria can be used to establish or refute a diagnosis of gestational diabetes though others, notably the National Diabetes Data Group,[18] believe that more rigid diagnostic criteria should be employed.

Impaired glucose tolerance usually has no pathological significance though both it and diabetes commonly occur, as an epiphenomenon, in a number of diseases and not infrequently provide the first clue to their existence and subsequent diagnosis. Some of the more common illnesses and conditions in which glucose intolerance occurs are listed in Table 3.

TABLE 3
MAIN CAUSES OF HYPERGLYCAEMIA AND/OR IMPAIRED GLUCOSE TOLERANCE

I Diabetes mellitus
   (i) insulin-dependent types (Type 1)
   (ii) non-insulin dependent types (Type 2)

II Endocrinological disease
   e.g. acromegaly, prolactinoma, Cushing's syndrome, Conn's syndrome, pheochromocytoma, etc.

III Insulin insensitivity and target-organ unresponsiveness
   e.g. obesity, chronic liver disease, renal failure, carcinomatosis and other causes of chronic ill-health, hypokalaemia, hypercalcaemia, lipodystrophy, muscular wasting, etc.

IV Chronic pancreatic disease (may present as diabetes mellitus)

V Insulin receptor autoimmune disease

VI Drugs
   e.g. thiazide diuretics, anti-inflammatory steroids, sympathomimetics, antipsychotics, and others too numerous to mention

Other Tests of Glucose Tolerance

Dissatisfaction with the oral glucose tolerance test as a diagnostic procedure for diabetes and its poor reproducibility and difficulty of interpretation led to many other procedures being introduced over the years but none has found more than minimal acceptance. With the publication of the WHO recommendations for the performance and interpretation of oral glucose tolerance tests all other procedures have become more or less obsolete, except for their use in research.

The Intravenous Glucose Tolerance Test. In this test an amount of glucose, usually 50 ml of a 50% w/v solution, is injected intravenously over the course of two to four minutes, into an overnight fasted individual. Blood is collected through an indwelling venous cannula at ten minute intervals for the next hour. The results of blood glucose analyses obtained between 20 and 60 minutes plotted against time on semilogarithmic graph paper, fall more or less on a straight line, the slope of which, expressed as the glucose assimilation coefficient (K_g), is a measure of glucose uptake by the tissues. K_g is, in large part, determined by the initial insulin secretory response to the rapid rise in blood glucose concentration. Its value is derived from the following formula:

$$K_g = \frac{69 \cdot 3}{t/2}$$

where t/2 is the half-life of glucose in the blood in minutes, and lies between 1·0 and 3·0 in the majority of healthy subjects. Much higher values are occasionally observed in young athletically-fit adults, so that in general there is no upper limit of normal. In patients with diabetes, K_g is always less than 0·9 and usually below 0·5.

The intravenous test offers no diagnostic advantage over the oral glucose tolerance test but, because it can be expressed as a single figure and is less subject to external influences, is useful for following the metabolic response to treatment or passage of time in individuals or, in research, for comparing the glucose tolerance of groups of subjects.

The intravenous glucose test has no place in the diagnosis of hypoglycaemia.

Two-hour Post-prandial Blood-glucose Test. This can be used to reduce the number of oral glucose tolerance tests performed, and consists of measuring the blood glucose concentration 120 ± 10 minutes after ingestion of a meal (or hydrolysed starch mixture) providing not less than 75 g of readily absorbable carbohydrate. A venous plasma glucose concentration of 7 mmol/l (130 mg/100 ml) or less (if confirmed) renders a diagnosis of diabetes mellitus so unlikely that no further action is indicated. A value of greater than 7·0 mmol/l but less than 11·0 mmol/l (200 mg/100 ml), is an indication for performing oral glucose tolerance testing.

Cortisone–Glucose Tolerance Test. Pretreatment of subjects undergoing oral glucose tolerance with cortisone or prednisolone was, at one time, thought to provide useful information with regard to the propensity of subjects with a genetic or other predisposition to glucose intolerance, to go on to develop diabetes mellitus. This expectation has not been borne out by time, and neither the cortisone–glucose tolerance test, nor any of its many variants, has any place in current clinical biochemical practice.

Intravenous Tolbutamide Test. Originally introduced as a rapid diagnostic test for diabetes mellitus, but soon abandoned for this purpose, the intravenous tolbutamide test enjoyed considerable popularity, for a time, as an aid to the diagnosis of spontaneous hypoglycaemia. Even here, however, its use has been largely, if not entirely, superseded by more specific and clinically useful procedures.

Hormone and Metabolite Assays. Despite all expectations, the availability of plasma hormone assays has done comparatively little to advance diagnostic precision, and even less to further effective treatment of diabetes. Plasma insulin levels are often but not always undetectably low in patients with insulin-dependent diabetes and may be 'normal' or even 'high' in those with the non-insulin dependent variety. The diagnostic importance once attached to timing of the 'delayed' insulinaemic response to oral glucose was probably also exaggerated. Much has been made by some authorities of modest differences in plasma glucagon levels and their responses to various stimuli but, whilst possibly having pathological significance, they have proved unhelpful clinically. Measurement of gut hormone levels are of no more than academic interest at the present time and this is still more or less true of the intermediate metabolites that have been investigated, although their measurement in established cases of diabetes is sometimes helpful in monitoring treatment, especially during ketoacidosis or lactic acidosis.

Prognosis in insulin-dependent diabetics is seemingly dependent to a certain extent on the degree, if any, of residual B-cell function. This can now be assessed by measuring the plasma (or possibly urinary) C-peptide response to various insulinotropic stimuli, e.g. glucagon, tolbutamide, arginine or a standard meal, and tests using this principle are beginning to enter clinical practice.

HYPOGLYCAEMIA

Hypoglycaemia is arbitrarily defined as a plasma glucose concentration below 2·5 mmol/l (45 mg/100 ml) which may, or may not, be associated with symptoms.[14] Symptoms, when they occur, are invariably neuroglycopenic in origin (literally shortage of glucose in the neurones).

Four more or less distinct, but not mutually exclusive, neuroglycopenic syndromes can be recognised.

(i) *Acute neuroglycopenia* is the commonest of the four syndromes and results from the rapid onset of hypoglycaemia, such as occurs from overtreatment with soluble insulin, or spontaneously during reactive hypoglycaemia. It begins with a vague sense of ill health, often accompanied by feelings of anxiety, panic and unnaturalness. Palpitations and restlessness are common and objectively there may be tachycardia, facial flushing, sweating, slurring of speech and unsteadiness of gait. These signs and symptoms are transient usually lasting only a few minutes. All, except sweating, which is often exaggerated, are attenuated by pre-treatment with non-selective β-adrenergic blocking agents. Untreated, they usually improve spontaneously as the blood glucose concentration rises under the influence of counter-regulatory homeostatic mechanisms. Rarely, however, they may progress to produce alterations of consciousness, stupor and eventually coma. These symptoms, like those that precede them, can be rapidly and completely reversed by oral or intravenous glucose. Acute neuroglycopenia is associated with an abrupt rise in the plasma concentration of many hormones, e.g. adrenaline, growth hormone, prolactin, cortisol and glucagon, and use has been made of this fact in the recognition and diagnosis of asymptomatic acute neuroglycopenia when it occurs during the night in insulin-treated diabetics. Measurement of the urinary cortisol:creatinine ratio on an early morning urine sample often reveals a marked rise in patients who have experienced 'silent' nocturnal hypoglycaemia compared with normal or non-hypoglycaemic subjects.

(ii) *Subacute neuroglycopenia* is the commonest type of hypoglycaemia encountered in patients suffering from 'spontaneous hypoglycaemia' and results from a gradual fall in blood glucose concentration over a period of an hour or more. The florid signs and symptoms of acute neuroglycopenia, which are largely the result of autonomic nervous activity, are completely lacking. Instead there is a general reduction of spontaneous activity, and conversation and movement become minimal. Sleepiness is an early feature but consciousness is retained until late. Behaviour patterns resembling mild alcoholic intoxication are common and even habitual tasks are performed poorly. Subjectively, discomfort is minimal for the degree of functional impairment. Most episodes abort spontaneously under the influence of counter-regulatory homeostatic mechanisms, but plasma hormone levels rarely rise to anything like the same extent as during acute neuroglycopenia and often are completely normal.

(iii) *Chronic neuroglycopenia* and (iv) *hyperinsulin neuronopathy* are extremely rare clinical variants of neuroglycopenia and are generally only diagnosed after clinical biochemical investigations have revealed the presence of more or less unrelieved hypoglycaemia.

Classification

Many classifications of hypoglycaemia are possible. The most useful clinically is into 'fasting' and 'provoked' (or stimulative) hypoglycaemia (Table 4). A thorough

TABLE 4
CLINICAL CLASSIFICATION OF HYPOGLYCAEMIA
(ACCORDING TO CIRCUMSTANCES OF ITS DEVELOPMENT)

A *Fasting hypoglycaemia*
  I Hyperinsulinism; benign, malignant and multiple insulinomas, microadenomatosis and nesidioblastosis
  II Non-pancreatic tumours
  III Liver and kidney disease
  VI Endocrine diseases: pituitary, adrenal, hypothalamic, etc.
  V Liver glycogen (glycogen storage) disease and other varieties of inborn errors of metabolism
  VI Substrate limitation
  VII Various types of neonatal hypoglycaemia
  VIII Autoimmunity

B *Stimulative hypoglycaemia*
  I Exogenous hypoglycaemic agents; drugs and poisons
  II Essential reactive hypoglycaemia
  III Hereditary fructose intolerance (HFI)
  VI Galactosaemia
  V Alcohol-induced hypoglycaemia

TABLE 5
AETIOLOGICAL CLASSIFICATION OF HYPOGLYCAEMIA

I *Pancreatic causes*
  (a) Insulinoma: benign, malignant, microadenomatosis
  (b) Insular hyperplasia
  (c) Pluriglandular syndrome
  (d) Pancreatitis
  (e) Nesidioblastosis or functional hyperinsulinism

II *Essential reactive hypoglycaemia*
  (a) 'Idiopathic' (rare)
  (b) Alcohol provoked
  (c) Post-gastrectomy

III *Liver and/or kidney disease*
  (a) Hepatocellular disease
  (b) 'End-stage' kidney
  (c) Congestive hepatic disease

VI *Extrapancreatic neoplasia*
  (a) Mesenchymal tumours
  (b) Primary hepatic carcinoma
  (c) Adrenal tumours
  (d) Various carcinomas

V *Endocrine disease*
  (a) Pituitary insufficiency; generalized or specific
  (b) Adrenocortical insufficiency; congenital or acquired
  (c) Hypothyroidism
  (d) Selective hypothalamic insufficiency

VI *Inborn errors of metabolism*
  (a) Glycogenoses (liver glycogen disease)
  (b) Galactosaemia
  (c) Hereditary fructose intolerance (HFI)
  (d) Defective gluconeogenesis

VII *Neonatal hypoglycaemia*
  (a) Infants of diabetic mothers (temporary infantile hyperinsulinism)
    (b) Intrauterine malnutrition
    (c) EMG syndrome

VIII *'Ketotic' hypoglycaemia*
  (a) Secondary to other causes
  (b) 'Idiopathic'

IX *Toxic hypoglycaemia*
  (a) Therapeutic hypoglycaemic agents, e.g. insulin, sulponylureas
  (b) Alcohol
  (c) Drugs
  (d) Poisons

X *Miscellaneous causes*
  (a) Diseases of the nervous system
  (b) Prolonged carbohydrate deprivation (starvation)
  (c) Excessive exercise (especially in combination with certain drugs)
  (d) Lactation (in ungulates)
  (e) Dialysis

and corroborated history, with particular attention to the conditions under which symptoms occur, provides the most important clue to the diagnosis. The occurrence of symptoms before breakfast virtually eliminates essential reactive hypoglycaemia though other types of provoked hypoglycaemia such as that due to alcohol, remain a possibility.

Reactive hypoglycaemia occurs in almost all diseases that are capable of causing fasting hypoglycaemia and should never be diagnosed until they have been excluded by appropriate clinical and laboratory investigations.

Diagnosis

Hypoglycaemia is merely an arbitrary description of the blood glucose concentration and to establish it as the cause of a patient's symptoms is to be committed to identifying its aetiology since this determines the type and nature of the appropriate treatment.

Though hypoglycaemia occurs in many diseases (Table 5), and is often the cause of their presenting symptoms, in comparatively few is it the sole detectable abnormality. Demonstration of a low blood glucose concentration is a *sine qua non* for the initial diagnosis which cannot be made in any other way, however suggestive the symptoms or prompt the relief by food or intravenous glucose might be. An inability to demonstrate hypoglycaemia chemically in a specimen collected whilst the patient was symptomatic virtually excludes the diagnosis.

Usually the patient is asymptomatic when first seen by a physician who can do no more, therefore, than suspect hypoglycaemia as the cause of the symptoms when they do occur. A provisional diagnosis made under these circumstances should always include a note as to whether the hypoglycaemia is likely to be of the 'fasting' or 'stimulative' varieties since this enables investigation to be undertaken in a more rational and orderly manner.

A thorough and detailed history, preferably corroborated by a relative or friend since amnesia is a feature of the disease, is, as always, essential and usually provides the clue to its aetiology.

Apart from measurement of the blood glucose concentration during a symptomatic episode, there is no 'best test' for spontaneous hypoglycaemia. There are, however, many provocative tests, most of which are designed to reveal a propensity to develop hypoglycaemia.

Though not quite the same as establishing that hypoglycaemia is the cause of the patient's (spontaneously occurring) symptoms, they nevertheless often do provide sufficient evidence upon which to make a positive diagnosis.

Prolonged Fast with Exercise Test

Most types of spontaneous hypoglycaemia are provoked by fasting, especially with moderate exercise, and failure to demonstrate hypoglycaemia (which is usually asymptomatic) on any of three occasions after an overnight fast is strong presumptive evidence against a diagnosis of fasting hypoglycaemia. Longer periods of fasting are usually necessary in order to provoke symptoms.

The prolonged fast test is carried out with the patient in hospital under strict supervision in order to prevent him either coming to harm as a result of unobserved hypoglycaemia, or surreptitiously giving himself insulin. Blood is collected at regular intervals and its glucose content measured. Plasma separated from each specimen is stored at $-20\,°C$ and can be analysed for hormones and intermediate metabolites should symptomatic hypoglycaemia develop. The patient is allowed water and non-calorie drinks *ad libitum*, but no food, and should be exercised under supervision and not allowed to lie passively in bed all day.

The test should be terminated after 72 hours, or before if marked symptoms of neuroglycopenia develop. When symptoms do develop, their neuroglycopenic origin can be confirmed, or refuted, by observing the clinical response first to intravenous saline and then to glucose. The former has no effect; the latter produces prompt and complete recovery. An EEG before and after glucose injection provides objective evidence of improvement in cerebral function but is not essential.

Interpretation of the prolonged fasting test is not always simple. Asymptomatic hypoglycaemia with plasma glucose levels as low as 1·7 mmol/l (30 mg/100 ml) or less, is not uncommon even in normal subjects, especially women and children. Normal people do not experience neuroglycopenic symptoms, however, and their plasma insulin and C-peptide levels suppress appropriately. Patients with diseases characterised by fasting hypoglycaemia not only exhibit low blood glucose levels but also experience neuroglycopenia—usually during the first 12–36 hours of the test and always by 72 hours.

A positive response to a prolonged fast test is not diagnostic of any single disease entity. Patients whose hypoglycaemia is due to hyperinsulinism exhibit inappropriately high plasma insulin and C-peptide levels, i.e. plasma levels of more than 10 mU/l and 1 µg/l respectively in the presence of a blood glucose concentration of less than 2·2 mmol/l (40 mg/100 ml)—whilst those with other types of fasting hypoglycaemia do not.

In children, additional information can be gained by measuring plasma β-hydroxybutyrate levels since these are invariably high (>3 mmol/l) in those with 'ketotic hypolycaemia', and low (<1 mmol/l) in those with hyperinsulinism.

Since the fasting test is tedious, for the patient and medical staff alike, and expensive, it has largely been superseded as a screening procedure for insulinoma and for distinguishing between hyperinsulinism and other causes of spontaneous hypoglycaemia by C-peptide suppression tests or, where these are not available, by one or more of the older stimulation tests using glucagon, leucine, tolbutamide or calcium. Substitution of a low carbohydrate, low-energy diet for fasting has gained little popularity as a test for fasting hypoglycaemia except in children, in whom its main use has been to demonstrate 'ketotic' hypoglycaemia.

Insulin Tolerance Tests

Insulin tolerance tests have a dual purpose in the investigation of spontaneous hypoglycaemia. They can be used (a) to reveal a state of insulin sensitivity and/or hypoglycaemic unresponsiveness or, in conjunction with plasma C-peptide assays as a screening test for hyperinsulinism and (b) in conjunction with various plasma and urine hormone measurements as a diagnostic test for hypothalamic-pituitary function and/or integrity of the sympathetic nervous system.

The test is carried out after an overnight fast. An indwelling plastic cannula is inserted into a vein and two resting samples are collected 15 minutes apart and analysed there and then for their glucose content. If either of these reveals hypoglycaemia no further action should be taken. If, as is usually the case, normoglycaemia is present, monocomponent short-acting insulin (0·15 U/kg to a maximum of 15 U) is injected intramuscularly (or 0·1 U/kg intravenously), and blood collected at 30 minute intervals for 2 hours for glucose, C-peptide, GH and cortisol measurements. Patients must *always* be closely supervised during this test and it is a wise precaution to have intravenous hydrocortisone 100 mg available in case glucose alone fails to produce rapid and complete alleviation of symptoms should this be considered necessary. Mild to moderate neuroglycopenic symptoms are common and need occasion no concern. The test should, however, be aborted immediately (but only after a sample of blood for analysis has been collected) by the injection of up to 50 ml of 50% w/v glucose if signs of more severe cerebral involvement, such as changes in consciousness or stupor, begin to appear.

Normal subjects generally experience mild to moderate neuroglycopenic symptoms commencing 20–30 minutes after the insulin injection and lasting 10–30 minutes; these should be recorded. Providing venous plasma glucose levels fall below 2·5 mmol/l (45 mg/100 ml), which is usual, plasma cortisol levels should rise by at least 275 nmol/l, and plasma hGH by 10 µg/l, during the second half of the test if hypothalamic-pituitary function is intact. Blood glucose normally returns to 80% or more of its fasting level within two hours of the injection, the plasma C-peptide concentration falls to less than 1·5 µg/l, and generally below 1·0 µg/l, and urinary catecholamine excretion increases by at least 400%.

Most patients with fasting hypoglycaemia, especially those with hyperinsulinism, non-pancreatic tumours and endocrinopathies, show increased insulin sensitivity and 'hypoglycaemic unresponsiveness'; their plasma glucose levels remaining below 2·5 mmol/l for up to two hours after the insulin injection. A normal result does not exclude 'fasting hypoglycaemia' and many children, especially those with idiopathic ketotic hypoglycaemia, show a normal glucose response to insulin even when, as is often the case, they fail to exhibit the expected rise in plasma or urinary catecholamines.

Patients with liver disease often exhibit insulin resistance and a subnormal, but prolonged, fall in blood glucose concentration whilst those with hypoglycaemia due to hypopituitarism do not show the expected rise in plasma hGH. Patients with adrenocortical dysfunction due to hypothalamic, pituitary or adrenal disease, show no rise in plasma cortisol during induced hypoglycaemia.

Normal subjects and patients with all types of hypoglycaemia, other than those caused by autonomous insulin secretion, show a fall in plasma C-peptide level to below 1·5 µg/l sometime during the course of an insulin tolerance test, providing the plasma glucose concentration falls below 2·5 mmol/l and remains there for at least 20 minutes. Patients with hyperinsulinism generally do not suppress endogenous C-peptide (and insulin) secretion adequately; consequently, failure of plasma C-peptide levels to fall below below 1·5 µg/l during the course of a properly conducted insulin tolerance test provides strong evidence for the diagnosis of hyperinsulinism. The main use of the insulin tolerance test with plasma C-peptide measurements is to reduce, to a bare minimum, the number of patients with reactive and other types of hypoglycaemia who need to be admitted to hospital for a prolonged fast with exercise test.

The commonest cause of a false positive C-peptide suppression test is failure to achieve hypoglycaemia of sufficient degree and duration to permit the plasma C-peptide level to fall below 1·5 µg/l. False negative results may also occur in (i) patients whose hyperinsulinism is predominantly due to proinsulin rather than to insulin, (ii) patients with intermittently secreting tumours, and (iii) those which are partially suppressible.

C-peptide. The C-peptide is formed during the enzymatic conversion of proinsulin into insulin in the beta-granules of the B-cells of the islets of Langerhans. It was first isolated in 1969 and synthesised a few years later. C-peptide remains in the beta-granules alongside the insulin from which it has been cleaved, until both are released into the circulation simultaneously, molecule for molecule, whenever insulin secretion is stimulated.

C-peptides differ markedly from species to species and there is little cross-immunoreactivity between them. Antisera to both natural and synthetic human C-peptides are available and form the basis of a number of different C-peptide immunoassays. There is, however, some confusion about the actual structure of circulating human C-peptide and because of difficulties in its synthesis, the various antisera to human C-peptide and even the C-peptide standards, currently available, differ amongst themselves. Some kits use a 35–amino-acid synthetic 'connecting peptide' as their standard whilst others employ the 31–amino-acid C-peptide. Fortunately, both methods yield similar results.

C-peptide, though released into the circulation stoichiometrically with insulin, is, because of its much longer half-life, invariably present in the plasma in higher molar concentrations than insulin, and the ratio of C-peptide to insulin is inconstant. The high molar concentrations of C-peptide in plasma make it easier to measure accurately, providing correct standards are used, and its longer half-life means that fluctuations in concentration occur more slowly than with insulin. Consequently, a random blood sample analysed for C-peptide content provides a more integrated picture of pancreatic B-cell function than does a single plasma insulin assay which is generally of limited, if any, clinical value except in the presence of documented hypoglycaemia.

Human C-peptide does not cross-react with insulin antibodies. Consequently, it can be measured in the plasma of patients with insulin antibodies produced either as a result of previous exposure to exogenous insulin or because of autoimmunity. More important, from the practical point of view, is the fact that C-peptide can be used as an index of pancreatic B-cell function during insulin-induced hypoglycaemia which is the basis of the 'screening' test for endogenous hyperinsulinism described above.

Although most C-peptide antisera cross-react to a certain extent with proinsulin, this seldom interferes with the analytical result since, except in a few very rare diseases, proinsulin is never present in plasma in more than very low concentrations.

Blood collected for plasma or serum C-peptide assay does not require special preservatives but the plasma or serum should be separated as soon as possible and kept at −20 °C until analysed. This should not be too long delayed, however, as there is evidence that its concentration may decrease by as much as 30% after one year even under ideal storage conditions.

Interpretation of plasma C-peptide assays require the same strict attention to detail as for insulin. Clinically, C-peptide assays are useful *only* when performed on blood collected whilst the patient was hypoglycaemic (except when used to detect residual B-cell function in insulin-dependent diabetes). In the presence of *fasting* hypoglycaemia, a plasma C-peptide concentration of more than 1·5 µg/l is, if confirmed on at least one other occasion, almost pathognomonic of hyperinsulinism. In all other conditions plasma C-peptide levels are low during spontaneous hypoglycaemic episodes. Conversely, a plasma C-peptide concentration of less than 1·0 µg/l under these circumstances makes a diagnosis of endogenous hyperinsulinism unlikely but does not exclude it.

Interpretation of plasma C-peptide measurements in

the non-fasting subject is difficult. Because of its relatively long half-life, compared not only with insulin but also with that of glucose, the plasma C-peptide concentration may not, for example, have had time to fall from a previously high level to below 1·0 µg/l before reactive hypoglycaemia has developed during an oral glucose load test. Too much importance should not, therefore, be attached to an 'inappropriately' raised plasma C-peptide level when hypoglycaemia has resulted from a rapid fall in blood glucose concentration, as can sometimes happen during spontaneous 'reactive' hypoglycaemia.

The demonstration of a low plasma C-peptide level in the presence of hypoglycaemia and a high plasma insulin concentration is virtually diagnostic of insulin-induced factitious hypoglycaemia.

Other Suppression Tests

Before plasma C-peptide assays became available, the only method of demonstrating inappropriate endogenous insulin secretion was either to measure plasma IRI during hypoglycaemia that developed spontaneously during fasting or that was provoked by infusions of alcohol or fish insulin. The last two tests are now obsolete and a somatostatin suppression test for hyperinsulinism, though advocated, was never established as useful and has no place in the clinical laboratory.

A diazoxide suppression test, in which the ability of diazoxide (600 mg in 500 ml saline given intravenously over 1 hour) to suppress insulin secretion and raise the blood glucose concentration, falls into a different category from those described above since it is used in patients in whom the diagnosis of hyperinsulinism has already been made. It is generally used to ascertain whether the patient would respond to therapy with diazoxide in the event that the surgeon was unable to find and remove an insulinoma at operation, since this information will help him decide between performing a subtotal pancreatectomy or a simple biopsy of the pancreas.

Stimulation Tests

A number of substances apart from glucose, stimulate insulin secretion by both normal and abnormal islet tissue. In the past they were used as the basis of various tests for the recognition and differential diagnosis of spontaneous hypoglycaemia, but their continued use for this purpose was rendered obsolete by the introduction of C-peptide suppression tests which are both more sensitive and specific for hyperinsulinism.

Glucagon enjoys a special position amongst the stimulatory tests for hypoglycaemia, however, since it provides information about liver glycogen reserves which is not otherwise readily available. In a hypoglycaemic subject the failure of the (capillary) blood glucose concentration to rise by 1 mmol/l (18 mg/100 ml) or more in response to the intravenous or intramuscular injection of glucagon (30 µg/kg body weight to a maximum of 1 mg) is powerful evidence against a diagnosis of induced or spontaneous hyperinsulinism. In children it is suggestive of an inborn error of metabolism, e.g. glycogenosis type I, or of idiopathic 'ketotic' hypoglycaemia and in adults of alcohol-induced fasting hypoglycaemia or of hypoglycaemia due to liver or endocrine disease.

Extended Oral Glucose Test

This test, often wrongly referred to as the prolonged glucose *tolerance* test, has for many years enjoyed unmerited popularity for the detection and delineation of spontaneous hypoglycaemia. Misinterpretation of the results is responsible for the epidemic of 'non-hypoglycaemia' which is still prevalent in the USA and elsewhere.[14] Moreover, devotion to it, to the exclusion of more appropriate tests, accounts for some of the delays in diagnosis that are experienced by patients with genuine hypoglycaemic disorders.

The only clinical indication for employing an extended oral glucose load test is to provide *limited* support for a diagnosis of essential reactive hypoglycaemia in a patient in whom an insulinoma or other cause of fasting hypoglycaemia is considered unlikely or has been excluded.

The test is normally carried out after an overnight fast though some authors advocate 'afternoon' tests since these often reveal 'reactive hypoglycaemia' when those performed in the morning do not. A soft plastic cannula is inserted into an antecubital, or other convenient vein, and kept patent with 3·8% sodium citrate. The subject is given 100 g hydrolysed starch (liquid glucose, BP, e.g. Hycal^R), dissolved in 300–500 ml of flavoured water to drink over 10–15 minutes. Blood is collected at 30 minute intervals for five hours or can be monitored continuously by automatic analysis. The patient is closely observed throughout the test and the time, nature and duration of any signs or symptoms that develop carefully recorded. The effect upon the symptoms of an intravenous injection of saline, presented to the patient as though it were glucose, may be worth observing, especially if a portable EEG trace is available at the time. Particular attention should be paid to the possibility of hyperventilation. Smoking is not normally permitted during the test and only water may be taken by mouth. Mild physical activity is allowed but should be noted in the record as it may affect the results and influence the interpretation. The blood samples are assayed for glucose and, if the patient becomes symptomatic, for cortisol, but plasma insulin levels are too variable to be diagnostically useful though they may be useful in research. There is insufficient evidence about the clinical significance of changes in plasma GIP and other intestinal hormones in response to oral glucose loads to warrant their measurement clinically, though there is every reason to believe that they will eventually help to throw light upon the aetiology of this rare but interesting condition.

Interpretation of the results of the extended oral glucose test is amongst the most controversial topics in laboratory medicine. Under the test conditions described, most normal subjects experience a modest initial rise followed by a fall in venous blood glucose concentration to below the fasting level. The nadir can occur at any time from 2–5 hours following ingestion of a glucose load, and reach a value as low as 1·7 mmol/l (30 mg/100 ml) in perfectly normal subjects. The median time of the nadir, after a 100 g glucose load, is four hours regardless of whether sampling is intermittent, at half-hourly intervals, or continuous. The drop in blood glucose is less and occurs earlier after smaller doses (e.g. 50 g) and is more pronounced and occurs later after bigger (e.g. 100 g) doses of glucose. Most subjects become bored and restless during the test and many experience vague non-specific symptoms which may be misattributed to neuroglycopenia, because these tend to occur towards the end of the test when blood glucose levels are at their lowest. In most subjects, there is only a very poor correlation between the venous blood glucose concentration and the presence or absence of symptoms. It is my own current practice to consider the combination of rebound hypoglycaemia [arbitrarily defined as a venous plasma glucose concentration of less than 2·5 mmol/l (45 mg/100 ml)], symptoms of neuroglycopenia, a post-nadir rise in plasma cortisol of at least 275 nmol/l accompanied by a rise in plasma GH of 10 μg/l or more, as consistent with, but not proof of, reactive hypoglycaemia. A drop in venous plasma glucose concentration to below 2·5 mmol/l with, or without, the appearance of vague symptoms, but unaccompanied by a rise in plasma cortisol, is considered to be of no particular clinical significance; this restriction has the advantage of limiting the number of patients in whom a tentative diagnosis of essential reactive hypoglycaemia is made to manageable proportions without removing from consideration those in whom reactive hypoglycaemia is a genuine cause of symptoms.

Reactive hypoglycaemia, occurring during an extended oral glucose test, is common in conditions more commonly characterised by fasting hypoglycaemia, e.g. insulinoma, fibrosarcoma and endocrinopathy, and some such 'organic' disorder should always be suspected whenever hypoglycaemia sufficiently severe to cause loss of consciousness occurs.

Contrary to what has been said in the past, neither the shape, magnitude nor duration of the initial hyperglycaemic phase provides help in determining the aetiology of the reactive hypoglycaemia, which can best be determined by careful attention to the clinical history and use of appropriate diagnostic procedures. In particular, a 'flat' oral glucose test is neither evidence of increased glucose tolerance nor, in the absence of strong support from other quarters, of malabsorption, since it is impossible for the blood glucose curve to be flatter (in disease) than it sometimes is in perfectly normal healthy subjects.

There is clearly an urgent need for a more definite test to enable persons who experience genuine symptomatic reactive hypoglycaemia during everyday life to be distinguished from those who do so only under the wholly artificial condition of an extended oral glucose test. The poor reproducibility of the extended oral glucose test in individuals is well recognised but only partially explicable on the basis of changes in diet. It has been known for many years that the consumption of a diet deficient in carbohydrate, such as that often eaten by people with self-diagnosed 'non-hypoglycaemia', predisposes to the development of symptomatic reactive hypoglycaemia during extended oral glucose tests and is the main reason why the test should never be performed in patients whilst they are on a low carbohydrate diet. I completely agree with the Mayo clinic group that 'reliance on the extended oral glucose test for the diagnosis of reactive hypoglycaemia is to be discouraged and should be replaced by documentation of a plasma glucose of less than 2·5 mmol/l (blood glucose 2·2 mmol/l) concurrent with spontaneous symptoms'. Though this is not always easy to achieve, patients can be taught to prick their own fingers during an attack and to collect blood onto filter paper for later analysis in the laboratory. Only if its glucose content is below 2·2 mmol/l should the diagnosis of hypoglycaemia continue to be entertained.

Other Tests

Special tests used for the diagnosis and more particularly the differentiation of spontaneous hypoglycaemia, are usually indicated by the history, physical examination and results of radiological and laboratory investigation. They include measurement of plasma hormone levels in cases of suspected endocrinopathy; plasma lactate levels in cases of lactic acidosis, liver disease and inborn errors of gluconeogenesis; plasma alanine and β–hydroxybutyrate levels in patients with ketotic hypoglycaemia. More specialised procedures are described in appropriate texts.[14]

INBORN ERRORS OF METABOLISM

There are a considerable number of inborn errors of metabolism affecting one or more aspects of carbohydrate metabolism; many produce hypoglycaemia as one of their major clinical manifestations. Amongst the more important of these are hereditary fructose intolerance, galactosaemia and various glycogenoses.

Hereditary fructose intolerance (HFI) is due to an almost complete absence of fructose–1–phosphate aldolase activity and usually manifests itself about the time of weaning when sucrose-containing foods are added to the baby's diet, but if symptoms are mild, as sometimes happens, diagnosis may be delayed until well into adult life. It is suggested by a history of aversion or intolerance to sweet foods, the ingestion of which is typically associated with symptoms of anorexia, nausea, vomiting and profound malaise with, in the more severe cases, alterations in consciousness, coma and even death. Symptoms coincide with, and are partly due to, hypoglycaemia which is produced as a result of inhibition of glycogenolysis and gluconeogenesis. This, in turn, is a consequence of the accumulation of fructose–1–phosphate in the liver whenever fructose is ingested and which is due to an inherited abnormality of fructose–1–phosphate aldolase activity. Fructosuria is usually present if sought but is generally much milder than in benign fructosuria, a rare condition caused by an inherited abnormality of fructokinase which is entirely symptomless.

Diagnosis of HFI is made on the basis of an *intravenous* fructose load test in which fructose, 0·25 g/kg body weight, is given by infusion over a 2–4 minute interval. Blood is collected before and at 10 minute intervals for 1 hour and, if feasible, at half-hour intervals for a further 2 hours and is analysed for glucose, fructose, lactate and inorganic phosphate levels. In normal individuals blood fructose levels fall rapidly after its intravenous injection and plasma glucose levels rise slightly: plasma inorganic phosphate and lactate levels do not change significantly. In HFI fructose levels drop slowly; plasma glucose and inorganic phosphate levels show a moderate to marked decrease and lactate levels rise markedly. In benign fractosuria the results are similar to those observed in healthy subjects, apart from a slow fall in blood fructose and failure of the blood glucose to rise.

Liver biopsy for confirmation of the enzyme defect is rarely necessary or indicated in HFI.

Because fructose is almost completely cleared from the blood during its first passage through the liver, its measurement in peripheral blood after oral administration has been used as a test of liver cell function and more recently as a test of the patency of portal-systemic anastomoses produced surgically for the treatment of portal hypertension.

Galactosaemia, which manifests itself during the first few days of life by failure to thrive, persistent neonatal jaundice, hepatomegaly, galactosuria, aminoaciduria, and a tendency to develop hypoglycaemia after ingestion of lactose-containing foods, must be distinguished from galactokinase deficiency which produces none of the florid features of galactosaemia. If untreated, galactokinase deficiency causes cataracts, sometimes with distressing rapidity, due to the accumulation of galactitol in the lens of eye, during the first few days of life, immediately after initiation of breast feeding.

Galactosaemia can be diagnosed before birth by enzyme analysis on fetal cells harvested by amniocentesis and grown in tissue culture. This investigation is indicated whenever there is a strong family history of galactosaemia since antenatal diagnosis can lead to preventative measures being instituted immediately after birth.

Galactose tolerance tests were used at one time for the diagnosis of galactosaemia but were abandoned as dangerous when safer, more sensitive and specific procedures based on the detection and measurement of red-cell galactose-1-phosphate and/or galactose-1-phosphate uridyl transferase activity became available.

Pentosuria is a benign condition which owes its limited clinical importance to the fact that it was one of the four original disorders to which the name 'inborn error of metabolism' was applied by Garrod. It is identified by the appearance, in the urine, of the reducing sugar l-xylulose, a pentose normally produced in only trace amounts through operation of the metabolic pathway which in most species is responsible for the synthesis of vitamin C from glucose.

REFERENCES AND FURTHER READING

1. Alberti, K. G. M. M. (1980), The World Health Organisation and Diabetes. Editorial, *Diabetologia*, **19**, 169–173.
2. Alberti, K. G. M. M. (1976), *Clinics in Endocrinology and Metabolism: Disorders of Carbohydrate Metabolism Excluding Diabetes*. Eastbourne: W. B. Saunders.
3. Andreani, D., Lefebvre, P. J. and Marks, V. (1980), 'Current views on hypoglycaemia and glucagon.' London: Academic Press.
4. Blundell, T. L. and Humbel, R. E. (1980), 'Hormone families: pancreatic hormones and homologous growth factors.' *Nature*, **287**, 781–787.
5. Doberson, M. J., Scharff, J. E., Ginsberg-Fellner, F. and Notkins, A. L. (1980), 'Cytotoxic autoantibodies to beta cells in the serum of patients with insulin-dependent diabetes mellitus.' *N. Engl. J. Med.*, **303**, 1493–1498.
6. Duckworth, W. C. and Kitabchi, A. E. (1979), 'Measurement of proinsulin and C-peptide in clinical medicine.' In: Special Topics in Endocrinology and Metabolism (ed. M. P. Cohen and P. P. Foa), pp. 55–77. New York: Liss Inc.
7. Felig, P. (1980), 'Disorders of carbohydrate metabolism.' In: Metabolic Control and Disease (ed. P. K. Bondy and L. E. Rosenberg), pp. 276–392. Philadelphia: W. B. Saunders.
8. Foa, P. P., Bajaj, J. S. and Foa, N. L. (1977), *Glucagon: Its Role in Physiology and Medicine*. New York: Springer-Verlag.
9. Gabbay, K. H. (1973), 'The sorbitol pathway and the complications of diabetes.' *N. Engl. J. Med.*, **288**, 831–836.
10. Given, B. D., Mako, M. E., Tager *et al.* (1980), 'Diabetes due to secretion of an abnormal insulin.' *N. Engl. J. Med.*, **302**, 129–139.
11. Goldfine, I. D., Jones, A. L., Hradek, T. T., Wong, K. Y. and Mooney, J. S. (1978), 'Entry of insulin into human cultured lymphocytes: electron microscope autoradiographic analysis.' *Science*, **202**, 760–763.
12. Hockaday, T. D. R. and Alberti, K. G. M. M. (1972), 'Diabetic coma.' *Clinics in Endocrinology and Metabolism*, **1**, 751–788.
13. Marks, V. and Morgan, L. M. (1982), 'Gastrointestinal hormones.' *Molecular Aspects of Biochemistry*, **5**, 225–292.
14. Marks, V. and Rose, F. C. (1981), *Hypoglycaemia*. Oxford: Blackwell Scientific.
15. Marks, V. and Samols, E. (1968), 'Glycosurias other than diabetes mellitus.' In: Carbohydrate Metabolism and its Disorders, Vol. 2 (ed. F. Dickens, P. J. Randle and W. J. Whelan), pp. 338–353. London: Academic Press.
16. Marks, V. and Turner, D. S. (1977), 'The gastrointestinal hormones with particular reference to their role in the regulation of insulin secretion.' *Essays Med. Biochem.*, **3**, 109–152.
17. The New England Journal of Medicine (1980), 'The insulinopathies.' **302**, 165–167.
18. National Diabetes Data Group (1979), 'Classification and diagnosis of diabetes mellitus and other categories of glucose intolerance.' *Diabetes*, **28**, 1039–1057.
19. Oakley, W. G., Pyke, D. A. and Taylor, K. W. (1978), *Diabetes and its Management*, 3rd ed. Oxford: Blackwell Scientific.
20. Phillips, L. S. and Allipopoulou-Sellin, R. (1980), 'Somatomedins.' *N. Engl. J. Med.*, **302**, 371–380 and 438–446.
21. Poffenbarger, P. L. (1979), 'Nonsuppressible insulin-like proteins in health and disease.' In: *Special Topics in Endocrinology and Metabolism, Vol. 1* (ed. M. Cohen and P. P. Foa), pp. 109–140. New York: Liss, Inc.
22. Siperstein, M. D. (1975), 'The glucose tolerance test: a pitfall in the diagnosis of diabetes mellitus.' In: *Advances in Internal Medicine Yearbook*, **20**, 297–323.

23. Sonksen, P. and West, T. E. T. (1978), 'Carbohydrate metabolism and diabetes mellitus.' In: *Recent Advances in Endocrinology and Metabolism* (ed. J. L. H. O'Riordan), pp. 160–187. Edinburgh: Churchill Livingstone.
24. West, K. (1975), 'Substantial differences in the diagnostic criteria used by diabetes experts.' *Diabetes*, **24**, 641–644.
25. World Health Organization (1980), *WHO Expert Committee on Diabetes Mellitus:* Technical Report Series 646, Second Report. WHO, Geneva.
26. Zapf, J., Rinderknecht, E., Humbel, R. E. and Froesch, E. R. (1978), 'Non-suppressible insulin-like activity (NSILA) from human serum: recent accomplishments and their physiologic implications.' *Metabolism*, **27**, 1803–1828.

38. GASTROINTESTINAL HORMONES

VINCENT MARKS

Introduction

Clinical and diagnostic relevance

Gastrin
   Hypergastrinaemia (Zollinger-Ellison syndrome)

Pancreozymin-Cholecystokinin

Glucagon
   Glucagonoma

Secretin

Vasoactive intestinal polypeptide
   HyperVIPaemia (Verner-Morrison syndrome)

Gastric inhibitory polypeptide

Somatostatin
   Somatostatinoma

Mixed and carcinoid tumours

Entero-insular axis

Other hormones
   Motilin
   Pancreatic polypeptide
   Neurotensin

INTRODUCTION

The gastrointestinal tract is far from being an impassive barrier between the exterior and interior of the body with the sole function of facilitating transport of nutrients from the food into the body. It is undoubtedly the largest, and possibly the most complicated of all the endocrine organs of the body. Indeed the term 'hormone' was coined by Bayliss and Starling in 1902 to describe the action of secretin—a product of the duodenum and small intestine—on pancreatic secretion. The study of gastrointestinal hormones was, however, subsequently largely neglected and emphasis was instead given mainly to the neural regulation of intestinal secretion and mobility, its two most important physiological activities apart from the digestion and absorption of food and drink.

It has only been in the past 15 years or so that the gastrointestinal tract itself—as opposed to the pancreas—has been accepted as a major endocrine organ worthy of study by endocrinologists and clinical biochemists. The upsurge of interest was largely due to the development of various techniques in peptide chemistry, in particular the introduction of immunoassay techniques and simplified procedures for isolating, sequencing and synthesizing polypeptides.

With the notable exception of 5–hydroxytryptamine (serotinin)—the hormonal role of which has still not been unequivocally established—all of the gut hormones characterized (including those produced mainly or exclusively by the pancreas) are polypeptides of varying size and heterogeneity. They are produced by cells which possess a common set of cytochemical and ultrastructural characteristics which have led to them being designated by Pearse as APUD cells from their capacity for Amine-Precursor Uptake and/or Decarboxylation.[24]

Many distinct endocrine-type cells have now been identified and are classified on an ultrastructural, histochemical and, increasingly, immunohistological basis. Typically, apart from those that occur clumped together in functional units in the pancreas as islets of Langerhans, they are scattered singly between the gut mucosal cells and are elongated with apical processes reaching and penetrating into the gut lumen; specific secretory granules are generally stored in the basal serosal part of the cell.

The number of biologically-active polypeptides that have either been isolated from the g t or at least partially characterised on the basis of their pharmacological effects, is now very large (Table 1), although only a comparatively small number of them have so far been shown to have clinical relevance. This probably represents the poorly-developed stage of knowledge of gastrointestinal hormones at least as much as to their non-involvement in, or irrelevance to, pathological processes. There is currently widespread belief that many of the so-called gastrointestinal hormones are, in effect,

GASTROINTESTINAL HORMONES

TABLE 1
NOMENCLATURE AND CLASSIFICATION OF MAJOR INTESTINAL HORMONES

Substance	Probable distribution	Main pharmacological activity
i) Gastrin	Antrum and duodenum	Stimulation of gastric acid secretion
(ii) Cholecystokinin (CCK)	Duodenum and jejunum	Stimulation of gall-bladder contraction; stimulation of pancreatic enzyme release
(iii) Secretin	Duodenum and jejunum	Stimulation of pancreatic juice secretion
(iv) Vasoactive intestinal polypeptide (VIP)	Most of gastrointestinal	Vasodilation
(v) Gastric inhibitory polypeptide (GIP)	Duodenum and jejunum	Stimulation of insulin secretion. Inhibition of gastric acid secretion
(vi) Motilin	Duodenum	Stimulation of gastric motor activity
(vii) Glicentin	Duodenum	Stimulation of insulin secretion; glycogenolytic activity
(viii) Chymodenin	Duodenum	Stimulation of chymotrypsin secretion
(ix) Somatostatin-like immunoreactivity	Antrum, duodenum and jejunum	Inhibition of many gastrointestinal and pancreatic hormones
(x) Met/leu enkephalins	Most of gastrointestinal tract	Inhibition of GI motility and secretion
(xi) Urogastrone	Duodenum	Inhibition of gastric acid secretion
(xii) Pancreatic polypeptide (PP)	Pancreas	Inhibition of exocrine pancreatic secretion
(xiii) Neurotensin	Lower part of small intestine	Hypertensive and hyperglycaemic actions
(xiv) PHI	Duodenum and jejunum	Stimulation of insulin, amylase and bicarbonate secretion

TABLE 2
AMINO ACID SEQUENCES OF THE KNOWN GASTROINTESTINAL HORMONES (PORCINE)

Gastrin I
$\quad\quad\quad\quad$ 17 $\quad\quad\quad\quad\quad\quad$ 10 $\quad\quad\quad\quad$ 5 $\quad\quad\quad$ 1
Glu-Gly-Pro-Tyr-Met-Glu-Glu-Glu-Glu-Glu-Ala-Tyr-Gly-Trp-Met-Asp-PheNH$_2$

Cholecystokinin
$\quad$ 33 $\quad\quad$ 30 $\quad\quad\quad\quad\quad\quad\quad\quad$ 20 $\quad\quad\quad\quad\quad\quad\quad\quad\quad\quad$ 10
Lys-Ala-Pro-Ser-Gly-Arg-Val-Ser-Met-Ile-Lys-Ala-Leu-Glu-Ser-Leu-Asp-Pro-Ser-His-Arg-Ile-Ser-Asp-Arg-Asp-
$\quad$ 7 $\quad\quad\quad$ 1
Tyr-Met-Gly-Trp-Met-Asp-Phe-NH$_2$
$\quad$ |
$\quad$ SO$_3$

Secretin
$\quad$ 1 $\quad\quad\quad\quad\quad\quad\quad\quad\quad$ 10 $\quad\quad\quad\quad\quad\quad\quad\quad\quad\quad$ 20 $\quad\quad\quad\quad\quad\quad\quad\quad\quad$ 27
His-Ser-Asp-Gly-Thr-Phe-Thr-Ser-Glu-Leu-Ser-Arg-Leu-Arg-Asp-Ser-Ala-Arg-Leu-Gln-Arg-Leu-Leu-Gln-Gly-Leu-Val-NH$_2$

Glucagon
$\quad$ 1 $\quad\quad\quad\quad\quad\quad\quad\quad\quad$ 10 $\quad\quad\quad\quad\quad\quad\quad\quad\quad\quad$ 20 $\quad\quad\quad\quad\quad\quad\quad\quad\quad$ 29
His-Ser-Gln-Gly-Thr-Phe-Thr-Ser-Asp-Tyr-Ser-Lys-Tyr-Leu-Asp-Ser-Arg-Arg-Ala-Gln-Asp-Phe-Val-Gln-Trp-Leu-Met-Asn-Thr

Vasoactive intestinal polypeptide
$\quad$ 1 $\quad\quad\quad\quad\quad\quad\quad\quad\quad$ 10 $\quad\quad\quad\quad\quad\quad\quad\quad\quad\quad\quad\quad\quad\quad\quad\quad\quad$ 27
His-Ser-Asp-Ala-Val-Phe-Thr-Asp-Asn-Tyr-Thr-Arg-Leu-Arg-Lys-Gln-Met-Ala-Val-Lys-Lys-Tyr-Leu-Asn-Ser-Ile-Asn-NH$_2$

Gastric inhibitory polypeptide
$\quad\quad\quad\quad\quad\quad\quad\quad\quad\quad\quad\quad$ 10 $\quad\quad\quad\quad\quad\quad\quad\quad$ 20
Tyr-Ala-Glu-Gly-Thr-Phe-Ile-Ser-Asp-Tyr-Ser-Ile-Ala-Met-Asp-Lys-Ile-Arg-Gln-Gln-Asp-Phe-Val-Asn-Trp-Leu-Ala-Ala-Gln-
$\quad\quad\quad\quad\quad\quad\quad\quad\quad\quad$ 40 $\quad\quad$ 42
Lys-Gly-Lys-Lys-Ser-Asp-Trp-Lys-His-Asn-Ile-Thr-Gln

Motilin
$\quad$ 1 $\quad\quad\quad\quad\quad\quad\quad\quad$ 10 $\quad\quad\quad\quad\quad\quad\quad$ 20 $\quad$ 22
Pro-Ile-Phe-Thr-Tyr-Gly-Glu-Leu-Gln-Arg-Met-Glu-Glu-Lys-Gln-Arg-Asn-Lys-Gly-Gln

more akin to neurotransmitters than to traditional hormones and that they exert their influence mainly upon adjacent or nearby cells (i.e. they exert a paracrine action) rather than on those at a distance.

All of the GI hormones that have been characterised are (with the exception of insulin, which together with glucagon and pancreatic polypeptide is generally considered in this category even though these are, at least in man, produced exclusively in the pancreas) single chain polypeptides with molecular weights under 5000 daltons; all except gastrin and CCK are basic in character. The primary (Table 2) and even tertiary structures of some of the gastrointestinal hormones resemble each other sufficiently closely to justify grouping them together as 'families' with overlapping pharmacological and possibly even physiological functions. Members of a hormone family are thought to have evolved from a single primitive precursor which presumably had hormonal or neurotransmitter properties.

Most of the gut hormones circulate in a variety of molecular forms of different peptide chain length and, consequently, slightly different biological, physicochemical and above all immunochemical forms. It seems probable that some of the variants represent hormone precursors and others circulating degradation products. From a diagnostic point of view this may be extremely important since, whereas the biological activities (and possibly the clinical significance) of the 'standard' hormones is generally known, the potencies of the other molecular forms are largely unknown; they may be more active, equally or less biologically active than the native hormone.

Clinical and Diagnostic Relevance

Up until now biological and clinical investigators have been concerned mainly with the collection of information, isolation and chemical characterisation of hormones, the investigation of their biological activity and often simply the measurement of circulating hormone levels in various physiological and pathophysiological situations. The clinical and pathological relevance of all but a few GI hormones has proved difficult to identify and, to date, has been concerned almost exclusively with the detection and monitoring of various hormone-secreting tumours of the gastrointestinal tract (Table 3). The special problems inherent in investigating a diffuse endocrine system such as that of the gastrointestinal tract are now realised, however, and are beginning to be overcome. The discovery of the dual nature and localization of many of the gut peptides and their probable role as neurotransmitters has opened up new areas of research. In the meantime I will concentrate in the remainder of this chapter upon those hormones of most importance to the practising clinical biochemist and investigator.

GASTRIN

An antral hormone capable of stimulating gastric acid secretion was first described by Edkins in 1905 but was not really characterised until some 60 years later when its structure was elucidated by Gregory and his co-workers in Liverpool. It was recognized from the very beginning that gastrin could exist in multiple forms.[29] At the present time the unqualified term 'gastrin' is used to describe the 17 amino-acid polypeptide which may (gastrin II) or may not (gastrin I) be sulphated on the tyrosine in the 6–position (counting from the carboxy end); both forms are equally biologically active. Other forms in which gastrin occurs in the gastrointestinal tract and circulation are referred to as 'big gastrin' (G34), 'mini gastrin' (G14) and, according to some, but not all, authors as 'microgastrin' (G4).

G17, G14 and G4 each represent the carboxy terminal of G34 which has the longest half-life in the circulation of all the gastrins and is often the most abundant form found in plasma. The various forms of gastrin share some biological and immunochemical properties but can be distinguished from each other physicochemically. The gastrins are normally produced exclusively by G-cells which are confined to the gastrointestinal tract proper and occur mainly in the gastric antrum and, to a limited extent, the upper duodenum. They may, however, be produced ectopically by any APUD-type tumour, especially those occurring in the pancreas. Their principal

TABLE 3
BIOLOGICAL AND CLINICAL FEATURES OF HORMONE-SECRETING TUMOURS OF THE GI TRACT

Feature	Glucagonoma	Gastrinoma	VIP-oma	Somatostatinoma	Carcinoid
Signs and symptoms	Mild diabetes, skin rash, enlarged liver	Abdominal pain, diarrhoea, GI bleeding epigastric tenderness, heptomegaly	Profuse watery diarrhoea, flushing of skin	Hypoglycaemia or impaired glucose tolerance, hypochlorhydria, steatorrhoea, cholelithiasis	Flushing of skin, diarrhoea, oedema, heart lesions
Male/female ratio	10:90	60:40	25:75	1:2	50:50
Selective radiology	Coeliac axis arteriogram positive in <50% of cases	Peptic ulcer, gastric mucosal hypertrophy	Coeliac axis arteriogram positive in <50% of cases	Coeliac axis arteriogram positive in <50% of cases	
Histopathology	Malignant predominantly A-cell carcinoma	Malignant D-cell tumour (60%), benign (20%), islet hyperplasia (20%)	'Non-β-cell' tumour (80%), islet hyperplasia (20%)	D_1-cell tumour	Malignant carcinoid
Biochemistry	Mild hyperglycemia, hyperinsulinemia, hyperglucagonemia	Resting gastrin hypersecretion and hyperacidity	Hypersecretion of VIP, hyperkalemia, hypochlorhydria, hypercalcemia (40%), impaired glucose tolerance	Hypochlorhydria, somatostatin elevated in portal or peripheral blood	Hypersecretion of 5-hydroxytryptamine (5HT), raised urinary 5-hydroxy-indoleacetic acid (5HIAA)
Diagnosis	Inappropriately high plasma glucagon levels	Excessive gastric acid production and inappropriate hypergastrinemia	High plasma VIP levels	Surgical exploration raised plasma somatotostatin	Raised plasma 5HT, raised urinary 5HIAA

physiological effects are thought to be stimulation of gastric acid and pepsin secretion and trophic effects on the gastric and duodenal mucosa. Almost all the biological activity of gastrin is obtained with the C-terminal tetrapeptide alone and this has led to the synthesis of a 'protected' pentapeptide (Pentagastrin[R]) which has a longer half-life in the plasma than the tetrapeptide and has found widespread use clinically as a means of assessing gastric acid secretion.

Gastrin release is under both neural and hormonal control and is increased by eating a meal. The so-called cephalic phase of gastrin secretion is mediated via the vagus nerves and initiated by the sight, smell and taste of food. Release of gastrin during the gastric phase is caused by distension of the stomach and the presence in it of partially digested protein. All modes of release are inhibited by acidification of the antral mucosa and exaggerated by alkalinisation.

Gastrin is usually present in the plasma in only very low concentrations but can be measured by radioimmunoassay with sufficient reliability to be useful clinically. However, because of the different specificities of the various antisera used against the various circulating forms of gastrin, complete agreement between assayists using different methods is not to be expected. For this reason, and because of the comparative infrequency with which it is required clinically, plasma gastrin assay is probably still best carried out in laboratories specialising in the technique.

Immunoassay results are usually expressed in terms of a G17 standard and overnight fasting values are generally less than 200 ng/l (100 pmol/l). Low plasma gastrin levels are of no diagnostic significance but elevated levels are often useful in the differential diagnosis of chronic abdominal pain, recurrent peptic ulceration and/or chronic diarrhoea, providing gastric hyperacidity can also be demonstrated to be present.

Hypergastrinaemia (Zollinger-Ellison Syndrome)

Zollinger and Ellison[31] described the syndrome of recurrent and intractable peptic ulceration associated with hyperacidity, gastric mucosal hypertrophy and a pancreatic endocrine tumour in 1955—many years before gastrin, the causative agent, was isolated and identified; consequently it is universally known by its eponymous name rather than as hypergastrinism. The syndrome (Z-E) is rare with a probable incidence of less than 1 per million of the population per year. This makes it equally as common as insulinoma and rather more common than most of the other tumours, apart from carcinoids, described in this chapter.

The two most important symptoms of Z-E are abdominal—due to peptic ulcer disease—and chronic diarrhoea which, though less common, is not infrequently the presenting symptom. Men are slightly more often affected than women and most of the patients are middle-aged though no age group is completely exempt.

The illness has an insidious onset in the majority of cases and most of the patients remain undiagnosed for years, although this is changing with greater awareness of its manifestations and the greater availability of plasma gastrin assays. A family history of some form of endocrinopathy is obtained in roughly one third of the patients, many of whom have pluriglandular disease of the so-called multiple endocrine adenomatosis (type I) variety.[30]

A malignant tumour of the pancreas, consisting mainly or exclusively of gastrin-secreting cells, is found in the majority of cases but it is usually slow growing and may, apart from its endocrinological manifestations exerted through secretion of gastrin, have remarkably few adverse effects. In a small percentage of cases the pancreatic tumour appears to be benign, but recent data obtained using plasma gastrin as a tumour marker throw doubt on this suggestion since hypergastrinaemia usually persists or recurs even in those cases in which complete surgical ablation was seemingly effected. In the past most Z-E tumours were large and easily palpable at operation but with improving diagnostic efficiency tumours as small as 2–3 mm in diameter are being revealed as the cause of hypergastrinaemia.[27]

In upwards of 20% of cases the pancreatic lesion causing Z-E is thought to be islet hyperplasia rather than a neoplasm but doubt has recently been cast upon this hypothesis which was based on simple histological rather than on biochemical or immunohistochemical evidence. Islet hyperplasia is a common finding in almost all endocrine lesions of the pancreas and is often associated with an increase in pancreatic polypeptide secretion regardless of the nature of the primary endocrinopathy.

Gastrin is not a normal secretory product of pancreatic islet tissue. Its production by a pancreatic neoplasm is, therefore, an example of ectopic hormone production. Whether hyperplasia or other disease of the antral G-cells, the appropriate source of gastrin, is ever a cause of Z-E and, if so, how commonly, is highly contentious. Current evidence[27] suggests that antral G-cell hyperfunction—with or without hyperplasia—is an even more rare clinical entity than 'classical' Z-E.

Regardless of its aetiology Z-E is characterised clinically by gastric mucosal hypertrophy, which is readily detectable radiologically, and gastric hypersecretion and hyperacidity which persist throughout the night. These were, at one time, used as the main laboratory diagnostic criteria of the disease.

The characteristic biochemical lesion of Z-E is hypergastrinaemia in the presence of a high gastric hydrogen ion concentration, the normal inhibitor of gastrin secretion. Hypergastrinaemia is not itself pathognomonic of Z-E since it occurs in a number of disorders, especially those associated with impaired gastric acid secretion, e.g. pernicious anaemia, atrophic gastritis, vagotomy and/or the overuse of antacids and anticholinergics. Modest hypergastrinaemia occurs also in a small proportion of patients with uncomplicated peptic ulcer disease, especially during treatment with cimetidine or other H_2-receptor antagonists. Hypergastrinaemia may also occur after surgical procedures which by-pass the gastric antrum so that it is no longer bathed in acidic gastric juice (the retained or excluded antrum syndrome) and in pyloric stenosis with severe gastric retention.

The diagnosis of Z-E, once suspected, is seldom long in doubt since hypergastrinaemia is an invariable finding. Attention must, however, be directed towards eliciting the cause of the hypergastrinaemia and establishing whether it is appropriate, i.e. high in the presence of a high antral pH, or inappropriate, i.e. high in the presence of a low antral pH. This can be achieved by measuring the hydrogen ion concentration of aspirated gastric juice and confirming that it comes into physical contact with the pyloric antrum.

Many tests have been advocated for distinguishing between the various causes of hypergastrinaemia, the two most popular depending upon the presumed specific ability of calcium and secretin to stimulate gastrin secretion by gastrin-secreting tumours but not by normal G-cells. Neither of these presumptions has survived critical examination[27,25] and currently the diagnosis of Z-E is best made by repeated measurements of plasma gastrin levels and the demonstration of *inappropriate* hypergastrinaemia. In many of the cases a tumour can be localized in the pancreas by selective arteriography or, where facilities exist, by percutaneous transhepatic pancreatic venous sampling under fluoroscopic control with measurement of plasma gastrin at various sampling points. A high success rate has been claimed for localisation of gastrin-secreting tumours of the pancreas by this technique.[5] This has also been applied, with slightly less success, to the localisation of insulinomas, VIPomas and other endocrine tumours of the pancreas.

The recommended treatment of the Z-E syndrome has undergone several changes since the syndrome was first described. The first, and probably still the most widely practised treatment, was total gastrectomy with or without ablation of the pancreatic tumour. By removing the only known end-organ responsive to gastrin this operation completely alleviated the symptoms caused by hypergastrinaemia, though it left unaltered any subsequent ill-effects caused by inexorable growth of the tumour, not to mention those secondary to total gastrectomy itself. Attempts at curative treatment by removing the pancreatic tumour were uniformly unsuccessful though recent evidence suggests that this might be the treatment of choice in cases in which the tumour is small and still confined to the pancreas. Follow-up studies, with regular monitoring of plasma gastrin levels, are clearly required in all cases of Z-E in which curative pancreatic surgery has been attempted in order to detect relapse as early as possible and prevent reappearance of the full-blown syndrome.

PANCREOZYMIN-CHOLECYSTOKININ

Pancreozymin and cholecystokinin were characterized independently on the basis of their abilities to stimulate enzyme production by the pancreas and contraction of the gall-bladder respectively. It was not until much later, during attempts at their purification and characterisation by Mutt and his co-workers during the 1960s, that they were recognized to be the same substance.[18] It is usually referred to nowadays as cholecystokinin (CCK) on the grounds of precedence.

Several molecular species of CCK are now recognized—the first to have been isolated and characterized being the polypeptide containing 33 amino-acids (CCK-33). The last five C-terminal amino-acids of this and all subsequently CCK variants are identical to those of gastrin with which they have some pharmacological and possibly even physiological properties in common. Most interest now focuses on CCK-8 currently believed to be the most biologically important of the CCK variants.

CCK possesses a sulphated tyrosyl residue (C-7) similar to gastrin II but unlike the latter, sulphation is essential for full expression of the biological, though not the immunological, properties of CCK. Caerulin, an amphibian skin peptide which has CCK-like biological activities shares the common gastrin–CCK pentapeptide and is also sulphated at Tyr–7. Caerulin is used clinically as a stimulant of pancreatic exocrine activity in a test of pancreatic function.

In the gut, CCK is found mainly in endocrine cells of the duodenum which occur mainly in patches rather than evenly distributed. It has, however, been shown by immunohistological techniques that CCK, particularly the C–8 variant, is widely distributed throughout the body, especially in the central nervous system.[26]

The main pharmacological (biological) activities of CCK, in addition to those by which it was discovered, are stimulation of intestinal motility, gastric acid secretion, secretion of Brunner's glands, pancreatic growth and, possibly, insulin release. Its secretion into the circulation is stimulated by the presence of various amino acids and protein hydrolysates, hydrochloric acid and fatty acids in the duodenum.[26] Its concentration in the plasma can be measured by radioimmunoassy and rises after ingestion of a mixed meal. Despite intensive efforts by investigators throughout the world, plasma CCK assays do not, at the present time, yield diagnostically useful information, though in the future they might be expected to do so in disorders of digestion and pancreatic exocrine function.

CCK can be given intravenously, either as the purified porcine C–33 preparation or as caerulin, to test for pancreatic exocrine responsiveness in cases of suspected pancreatic insufficiency. It is usually administered in conjunction with natural or synthetic secretin and the rise in concentration of pancreatic enzymes, e.g. amylase, lipase and trypsin in the plasma, or preferably pancreatic juice obtained by endoscopic intubation, determined as a measure of pancreatic responsiveness.[12,17]

GLUCAGON

Glucagon was named for its ability to raise the blood glucose concentration when injected into fed animals. It was originally obtained as a contaminant of commercial insulin preparations and did not attract much attention from either biologists or clinicians until the mid-1950s when it was first characterized and synthesized.

Its main biological actions are the stimulation of gluconeogenesis and glycogenolysis in the liver, lipolysis in adipocytes, insulin secretion by the B-cells of the pancreas and inhibition of both gastric acid secretion and intestinal motility.[20] Which, if any, of these actions is physiologically important in man is still unknown and there are no recognized ill consequences of glucagon deficiency such as occurs in patients who have undergone total pancreatectomy and receive adequate replacement therapy with insulin and pancreatic enzymes.

Although production of glucagon in man appears to be confined to the A-cells of the pancreas it is produced by similar cells in the stomach in some species. A number of poorly characterized substances which are larger but share many of the immunological determinants of glucagon are, however, produced throughout the intestinal tract of man and other species. They are often referred to as glucagon-like immunoreactants (GLI). One of them, glicentin, has been isolated and characterised but its function, like that of other glucagon cross-reacting materials, is unknown. Their main clinical importance at the present time is their ability to interfere with the results of immunoassays for glucagon unless suitably specific antiserum is used.

Glucagonoma

The syndrome of glucagon excess was first delineated by Mallinson and co-workers in 1974.[21] It usually presents as a dermatological problem in middle-aged or elderly women but also occurs in men. The most characteristic abnormality, apart from high fasting plasma glucagon levels, is a rare skin lesion known to dermatologists as necrolytic migratory erythema. Other common features of the syndrome are stomatitis, normochromic normocytic anaemia, weight loss, hypoalbuminaemia, psychiatric disturbances and a high rate of thromboembolic disease, the last named often being responsible for the patient's death.[3] Diabetes is an inconstant feature and when present is usually mild.

The disease has usually been present for a long time before the diagnosis of hyperglucagonaemia is suspected and subsequently made by demonstrating elevated plasma glucagon levels in the presence of normoglycaemia or modest hyperglycaemia. This may merely reflect unfamiliarity with the syndrome and the difficulty, until recently, of obtaining confirmatory plasma glucagon assays. The causative lesion is almost invariably a malignant pancreatic endocrine tumour which is only very occasionally resectable but, in those rare instances when it is, remission is both rapid and complete unless, and until, it recurs.

SECRETIN

Bayliss and Starling demonstrated that instillation of hydrochloric acid into a denervated loop of small intestine elicited the secretion of pancreatic juice. They suggested that the humoral agent responsible for the effect should be named secretin. Half a century later the hormone was finally isolated in pure form and its amino-acid sequence determined. In spite of this and the enormous amount of work that has subsequently been carried out on it, little more is known about the physiology and pathology of secretin today than when it was first discovered. No diseases directly attributable to diminished, excessive or disorganised secretion of secretin have yet been discovered although they probably do exist if only it was known what should be looked for.

Secretin is produced by the granular S-cells present in the mucosa between the crypts of villi throughout the length of the small intestine with their greatest concentration in the duodenum. It is a strongly basic molecule containing 27 amino-acids. The intact molecule is needed for complete biological activity which appears primarily to stimulate the secretion of water and bicarbonate (but not enzymes) by the pancreas, especially in conjunction with CCK; their actions are highly synergistic, as least under pharmacological conditions. Other biological actions of secretin which have been demonstrated experimentally include stimulation of water and bicarbonate secretion by the Brunner's glands of the intestinal mucosa and insulin release by the B-cells of the pancreas as well as inhibition of gastrin-induced gastric acid secretion and reduction of duodenal motility; whether any of these are physiologically or pathologically significant is unknown.[15]

Plasma secretin concentrations are extremely low but can be measured by radioimmunoassay although contradictory results have been obtained by different investigators under diverse experimental conditions. At present there are no clinical indications for measuring plasma secretin levels. The response to secretin is sometimes used as a test of pancreatic exocrine function,[12] especially in conjunction with CCK, and occasionally in patients suspected of harbouring a gastrinoma. Its use for the latter purpose has, however, declined since it became clear that the discriminatory effect claimed for it was unfounded.[27]

VASOACTIVE INTESTINAL POLYPEPTIDE

Vasoactive intestinal polypeptide (VIP) is a member of the glucagon family of GI hormones and was isolated by Said and Mutt from a side fraction during purification of porcine secretin to which it is even more closely structurally related than to glucagon itself. Initially characterized on the basis of its effects on blood vessels and the circulation, it was soon found to have a wide range of biological actions many of which it shares with other members of the secretin family.[9] Apart from its vasodilator and hypotensive effects VIP inhibits pentagastrin and histamine-stimulated gastric acid secretion and inhibits gastrin secretion. It stimulates water and electrolyte secretion by the exocrine pancreas and colon in a manner similar to that of cholera toxin, i.e. through activation of adenylate cyclase. VIP is a stimulus to glycogenolysis in liver, lipolysis in adipose tissue and insulin secretion by the pancreas. All of these actions may be relevant to its proposed role as a neurotransmitter.

VIP contains 28 amino-acids and shows considerable chemical homology with glucagon, secretin and GIP. Unlike the other members of the 'family', however, it occurs throughout the length of the gastrointestinal tract and is confined to the nervous rather than to the endocrine elements. Indeed, it occurs in relatively large amounts in nerves in various other parts of the body and in the brain. Its role as a hormone has, therefore, been brought into question. Regardless of semantics, however, VIP does occur as a natural constituent in plasma and measurement of its concentration is occasionally helpful in elucidating the cause of severe abdominal pain and/or diarrhoea.

The conditions under which VIP is released into the circulation, and its physiological role, are still poorly understood. It is apparently largely inactivated by a single passage through the liver and this may be the mechanism by which this highly biologically active substance—whose actions are normally confined to well-defined target organs innervated by peptidergic nerves—is removed and denatured when it inadvertently leaks into the general circulation.

The concentration of VIP in plasma can readily be measured by radioimmunoassay, but since measurements for clinical purposes are of limited use and only very rarely indicated, they are best carried out in laboratories specializing in the technique.

HyperVIPaemia (Verner-Morrison Syndrome)

Not long after the first seminal description by Zollinger and Ellison of non-insulin secreting pancreatic tumours producing abdominal disease, it was appreciated by Verner and Morrison[28] that, in a small proportion of patients watery diarrhoea, rather than recurrent peptic ulceration, dominated the clinical picture and that gastric function tests often revealed hypochlorhydria instead of hyperacidity. Differentiation of this syndrome, now often referred to as the Verner-Morrison or Watery Diarrhoea, Hypokalaemia, Achlorhydria (WDHA) syndrome, from Z-E, in which diarrhoea is also often a dominant feature, cannot always be made on clinical grounds alone but may require detailed biochemical and endocrinological examinations.[3]

The likely cause of this disease remained unknown for many years though recognised as being secondary, in the majority of cases, to the secretory product of a non-insulin-secreting pancreatic tumour which was neither gastrin nor glucagon. Though at first wrongly thought to be to gastric inhibitory polypeptide (GIP) it was subsequently shown, mainly by Bloom and his colleagues,[3] that VIP was the main mediator of the syndrome and that the tumours, by analogy with insulinomas, glucagonomas and gastrinomas, were VIPomas. Not all VIPomas, however, cause WDHA nor are all cases of the latter associated with increased plasma VIP levels although many of them are. Moreover, not all of the tumours causing WDHA, whether associated with increased plasma VIP levels or not, arise within the pancreas; some are phaeochromocytomas, ganglioneuromas or bronchial carcinomas.[16] Many of the tumours produce more than one hormone—often three or more—and it becomes difficult to assign a causative role to any single one of them.

The WDHA syndrome is characterised by the production of copious watery stools occurring in explosive bursts with remarkably little cramping, often with long intervals, initially, between attacks. Hypochlorhydria and hypokalaemia (due to loss of potassium in the stool) are both very common and muscular weakness is usual during attacks. Skin flushes and eruptions are also common, as are psychotic episodes. Other biochemical features include hypercalcaemia and impaired glucose tolerance (probably secondary to the hypokalaemia). The condition is three times more common in women than in men and is associated with a metastatic pancreatic tumour in roughly half the cases. The others are associated with apparently benign tumours of the pancreas, simple islet-hyperplasia and ectopic hormone-secreting neoplasms. The endocrinological cause of the syndrome is not known with certainty and is probably not the same in every case. Bloom and Polak have argued cogently for differentiation of WDHA, caused by excessive VIP secretion, from that caused by other agencies. This is justified by analogy with hyperinsulinism which is now recognised to be only one amongst many causes of intractable fasting hypoglycaemia; consequently, VIPoma and hyperVIPaemia can be looked upon as one, but not the only, cause of WDHA. Other causes will undoubtedly emerge in time including increased production of prostaglandins of one type or another or various combinations of gastrointestinal hormones.[14,16]

Diagnosis of VIPoma is made on the basis of a high plasma VIP concentration in a patient with signs and symptoms of WDHA and confirmed radiologically and/or surgically. It must be distinguished, clinically, from the far more common condition of surreptitious purgative-induced diarrhoea and lesions of the bowel itself which are capable of producing intractable diarrhoea, as well as from other endocrinological causes, e.g. carcinoidosis and medullary carcinoma of the thyroid.

Only very rarely is total surgical ablation of a VIPoma possible although, because they are usually so slow growing, removal of tumour bulk often improves symptoms and extends the patient's life.

GASTRIC INHIBITORY POLYPEPTIDE

The existence of a hormone with the ability to inhibit gastric acid secretion and delay gastric emptying and whose secretion was prompted by the ingestion of a fat-rich meal was postulated in 1930 by Kosaka and Lim.[19] They named the hypothetical hormone enterogastrone. In 1969 Brown and co-workers isolated a peptide with enterogastrone activity from a crude intestinal extract and named it Gastric Inhibitory Polypeptide

(GIP), the acronym by which it is now better known.[4] Its amino-acid sequence has been determined and its N-terminal end bears a close resemblance to glucagon and secretin but the C-terminal amino-acid sequence is not common to any other known intestinal polypeptide. Regular GIP consists of 42 amino-acids and has a molecular weight of roughly 5000 daltons. There is, however, evidence of a larger variant of about 8000 daltons whose physiology may or may not be different from that of the smaller variety.

GIP immunoreactive cells are localized mainly in the duodenum and upper jejunum, and to a much lesser extent in the lower bowel. They also, contrary to earlier belief, occur in the pancreatic islets.

Interest in GIP centred exclusively upon its ability to affect gastric acid secretion and the rate of gastric emptying initially but this all changed when it was found to have extremely potent insulin stimulatory properties which only became evident in the presence of mild to moderate hyperglycaemia. This, coupled with the fact that secretion of GIP follows the ingestion and absorption of all the major nutrients, led to its proposal as the main contender for the role of humoral mediator of enteroinsular stimulation[23]—a position it still occupies, though less certainly than once seemed likely—and its renaming as glucose-mediated insulin-releasing polypeptide (GIP).

Whilst it seems likely on circumstantial evidence that abnormalities of GIP secretion are implicated in the pathogenesis of certain metabolic diseases such as non-insulin-dependent diabetes and obesity, no hard evidence has been produced in support of this proposition, despite much searching.[7] Currently, therefore, there are no clinical indications for measuring plasma GIP concentrations, although its interest to researchers in gastrointestinal and metabolic pathology is undiminished.

Somatostatin

Somatostatin is a cyclic tetradecapeptide originally isolated from the hypothalamus by virtue of its ability to inhibit growth hormone secretion by anterior pituitary cells in culture and subsequently found distributed in tissues throughout the body. Indeed, in the rat some 70% of total immunoassayable somatostatin occurs in the gut and a further 5% in the pancreas. Although the chemistry and pharmacological properties of somatostatin have been extensively studied—probably more than any other peptide hormone except insulin—its true physiological role is still enshrouded in mystery.[8] It seems likely that somatostatin acts mainly, if not exclusively, as a paracrine regulator and/or inhibitor of nearby secretory cells whether they be in the pituitary gland, gut or endocrine pancreas. The role of abnormalities of somatostatin secretion in the pathogenesis of disease is the subject of much speculation but there are few objective data. Even the causative role of somatostatin in the somatostatinoma syndrome described below is questionable.

Somatostatin is present in the plasma in at least two major forms, S-14 and S-28, and can be measured by radioimmunoassay. Its concentration varies in response to food and stress and at present is not recognised as having any diagnostic significance except in the recognition and diagnosis of somatostatinoma.

Somatostatinoma

Tumours of the pancreas producing somatostatin as their main—or possibly sole—product were first described a little over five years ago but there is still no clear evidence that they are capable of producing a characteristic syndrome comparable, for example, with insulinoma, VIPoma or gastrinoma. In all of the small number of cases so far described the tumour was malignant and secreted or contained at least one other polypeptide hormone, e.g. calcitonin, corticotrophin or insulin, on immunohistological examination. Plasma somatostatin levels were many tens, or hundreds, of times higher in patients than in control subjects in all cases in which they were examined. Many either had hyperglycaemia or exhibited glucose intolerance though one patient presented with severe and intractable hypoglycaemia. Gall-bladder disease and steatorrhoea were a feature of some of the cases, but not all, and considering the pharmacological effects observed with even tiny amounts of somatostatin, the mildness of the clinical and biochemical abnormalities were more remarkable than their presence.[1]

Mixed and Carcinoid Tumours

All of the peptide-secreting tumours of the gastrointestinal tract, whether they arise in the pancreas or gut itself, present common histological features which have led to them being classed as APUDomas. Differentiation from one another may be difficult if not impossible without recourse to electron microscopy and special staining techniques. A fair proportion of all APUDomas reveal the presence of multiple hormone-containing cell types when subjected to immunohistological examination and many are classified on their histological features alone as carcinoids; indeed, many of them secrete varying amounts of 5-hydroxytryptamine though only a minority manifest this clinically.

Assays carried out on plasma of patients harbouring mixed tumours often reveal the presence of more than one hormone in excess even when this is entirely unsuspected clinically. The reasons for this apparent hormonal silence are unknown.

Carcinoid tumours themselves have been recognised clinically and histologically for longer than any of the other hormone-producing tumours of the gastrointestinal tract. They were first shown to be capable of producing cutaneous flushing, diarrhoea and cardiovascular disease as long ago as 1930 but it was almost a quarter of a century later that the full-blown syndrome of carcinoidosis was recognised and attributed to overproduction of 5-hydroxytryptamine (5HT). This view is now considered somewhat simplistic insofar as 5-hydroxytryptamine, though invariably produced in excess in carcinoidosis, is thought to be responsible for only a minority of the signs and symptoms, the majority of which are due to other hormonal products of the causative tumour.[17] The exact nature of these secretions, and whether they are always the same, is still unknown but the two that have been incriminated most consistently are bradykinin and prostaglandins of one type or another.

Only a tiny proportion of all tumours diagnosed histologically as carcinoids give rise to the clinical syndrome of carcinoidosis: in particular, those arising in the appendix, a favourite site, very rarely do so and consequently were, for a time, considered to be fundamentally different from those developing elsewhere, although this now seems unlikely. What causes the difference is still unknown. Carcinoid tumours developing in the pancreas are particularly likely to contain and secrete one or more polypeptide hormones of gastrointestinal origin and may give rise to some very bizarre syndromes.[11]

Men are more often affected than women and are usually middle-aged or elderly. Intermittent cutaneous flushing is the commonest symptom and is present in over three quarters of the cases. It can often be provoked by alcohol which has been used clinically as a test for the disease.[17] Dilation of veins on the face and the formation of venous telangietases is common, especially over the nose, upper lip and cheek which may give the patient the appearance of being a drunkard. Heart lesions, especially affecting the valves, are common, as is asthma, but it is diarrhoea that is the hallmark of carcinoidosis and usually the most disabling part of the disease. It frequently accompanies or follows episodes of flushing and is characterised by borborygmi, urgency and the frequent, explosive passage of watery stools. It alone amongst the symptoms of carcinoidosis is most likely to be due to excessive 5HT production and may occasionally be relieved by parachlorphenylalanine, an inhibitor of 5HT synthesis. Many patients give a history of having undergone abdominal surgery previously but others have had no such premonitary signs. Enlargement of the liver due to metastases is almost invariably present by the time the patient presents with symptoms but this may be years, or even decades, after the lesion was first discovered.

Diagnosis of the carcinoid syndrome is generally made on the basis of the characteristic clinical picture and excretion, in the urine, of vastly increased amounts of 5HT metabolites, especially 5-hydroxyindoleacetic acid (5HIAA). Various provocative tests utilising alcohol, calcium or adrenaline have been employed but do little to further the diagnosis, although they have helped throw light on the pathogenic mechanisms involved. Plasma 5HT levels are extremely high in patients with carcinoidosis but great care is necessary in order to avoid confusion created by even the slightest degree of thrombolysis, since platelets contain large amounts of 5HT which is liberated into the serum wherever they are damaged.

Treatment of carcinoidosis is essentially symptomatic and aimed at reducing tumour bulk and lessening endocrine effects on the target organs. Because they are usually so slow growing, surgical removal of carcinoid metastases from the liver, when feasible, or interruption of the arterial blood supply, is often followed by remission of the symptoms for up to several years. Alternatively, radiotherapy or cytotoxic drugs, especially 5-fluorouracil, are worthy of trial. The diarrhoea often responds to treatment with methysergide and other 5HT-antagonists or parachlorphenylalanine though the side-effects of the latter drug usually limit its usefulness. Glucocorticoids may help the asthma but so far there is little that can be done to relieve the flushing except avoidance of alcohol should this be implicated in its causation.

The Entero-Insular Axis

The discovery that glucose given by mouth to healthy human subjects led to a much greater release of insulin into the circulation than an equivalent amount given intravenously produced an intensive search for the mechanisms involved. It was early concluded that the main, if not exclusive, mechanism was the release from the gut of one or more humorally active agents which, in conjunction with the mild hyperglycaemia resulting from the absorption of glucose from the gut, stimulated the secretion of glucose by the pancreatic islet B-cells.[22] The role of the autonomic nervous system in augmenting the insulinotropic effect of hyperglycaemia was initially considered to be insignificant, though more recent evidence suggests that this might not be the case. Indeed, current evidence suggests that demedullated nerve fibres innervating the islets via the vagus have an important role in the regulation of insulin secretion.

Though many candidates have been put forward from amongst the gastrointestinal polypeptide hormones as the main factor in mediating intestinal stimulation of insulin release, none fills the bill completely.[22] The leading contender at present is undoubtedly GIP, but GLI, secretin and VIP have all been proposed at one time or another. Dysfunction of the entero-insular axis has been held responsible for certain abnormalities of glucose metabolism such as occur in type II diabetes (non-insulin dependent diabetes) and obesity, but evidence for this is still lacking. The entero-insular axis does, however, provide a fruitful and important field of clinical biochemistry research.

OTHER HORMONES

None of the many GI hormones, apart from those already discussed, has yet been established as unequivocally involved in pathogenic mechanisms or as having diagnostic significance. Nevertheless, at least some of them are of sufficient potential importance to warrant brief discussion.

Motilin

A polypeptide capable of stimulating the contraction of the dog stomach fundus was first isolated from a crude extract of pig gut by Brown and co-workers in 1971 and its structure was determined soon after.[6] It is a single chain polypeptide consisting of 22 amino-acid residues arranged in a sequence unlike any of the other known humorally active GI hormones. It occurs in two main forms in the plasma and tissues: a smaller one corresponding in molecular size to native motilin; a larger one of undetermined composition and behaviour.

The most clearly demonstrable properties of motilin are its ability to stimulate contraction of the gastric musculature and affect the frequency and strength of interdigestive myoelectric complexes. It has been said both to increase and to delay gastric emptying according to the circumstances of its administration and at present its physiological role in man is quite unknown, though it seems reasonable to suppose that it might be involved in regulating gastrointestinal motility. Its concentration in plasma can be measured by radioimmunoassay and shows marked interpersonal differences as well as rhythmical fluctuations, with a periodicity of roughly 90 minutes, throughout the day and night.

Glucose given either by mouth or intravenously produces a rapid fall in plasma motilin concentration, whereas fats and non-absorbable carbohydrates such as xylitol and lactose in lactose-intolerance subjects, produce either no fall or a rise in plasma motilin concentration when they are given by mouth. Diarrhoea, regardless of its casue, is usually associated with high plasma motilin levels but whether these are causal or consequential has not yet been established.

Pancreatic Polypeptide

Polypeptides of roughly similar chemical composition, but with varied biological effects, were isolated from chicken and bovine pancreases by two groups of investigators working independently some 10–15 years ago. These peptides, called, for want of a better term, pancreatic polypeptides (PP), have different immunological properties according to the species from which they are isolated.[10] Immunoassays capable of measuring pancreatic polypeptide in man are available and have led to the accumulation of extensive knowledge with regard to conditions under which pancreatic polypeptide secretion rate is altered, the most important of which is eating. This produces a prompt and large rise in plasma PP which is abolished by vagotomy. So far, however, it has not proved possible to assign a physiological role to pancreatic polypeptide in man or to attribute any illness to abnormalities of its production. The concentration of PP in plasma is raised in many patients with pancreatic APUDomas and its measurement has been suggested as diagnostically useful in this disorder but rejected as insufficiently sensitive and/or specific. Since the rise in plasma concentration of PP which occurs in response to insulin-induced hypoglycaemia is abolished by vagotomy measurement of PP in plasma during an insulin stress test has been proposed as a test of completeness of vagal section following vagotomy for peptic ulceration. The clinical utility of this procedure has not yet been established and currently measurement of plasma PP serves no useful purpose clinically.

Neurotensin

Like somatostatin this 13 amino-acid polypeptide was originally isolated from extracts of hypothalamus and subsequently demonstrated by immunochemical methods to be present in the bowel where it occurs in highest concentration in the ileum.[2] Pharmacologically, neurotensin has powerful hypotensive-vasodilatory properties and stimulates the secretion of insulin, glucagon and gastrin. Its physiological properties are unknown, but its concentration in plasma rises after ingestion of food, especially in patients who experience the dumping syndrome after partial gastrectomy. Whether neurotensin is involved in the pathogenesis of this syndrome is a matter for conjecture. At the present time there are no clear clinical indications for measuring plasma neurotensin levels.

Similar notes to the above could be written about each of the ten or so other biologically active peptides that have been isolated from the gastrointestinal tract, characterised and often sequenced, and to which a hormonal (or paracrine) role has been assigned. No good purpose would be served in doing so, however, and readers are referred to the review articles and books on this subject cited at the end of this chapter.

REFERENCES

1. Axelrod, L., Bush, M. A., Hirsch, H. J. and Loo, S. W. H. (1981), Malignant somatostatinoma: clinical features and metabolic studies. *J. Clin. Endocrinol. Metab.*, **52**, 886–896.
2. Bloom, S. R. and Polak, J. M. (1979), Motilin and Neurotensin: Motilin. *Clin. Endocrinol. Metab.*, **8**, 401–411.
3. Bloom, S. R. and Polak, J. M. (1980), Glucagonomas, VIPomas and somatostatinomas. *Clin. Endocrinol. Metab.*, **9**, 285–297.
4. Brown, J. C., Dryburgh, J. R., Ross, A. and Dupré, J. (1975), Identification and actions of gastric inhibitory polypeptide. *Rec. Prog. Horm. Res.*, **31**, 487–532.
5. Burcharth, F., Stage, J. G., Stadil, F., Jensen, L. I. and Fischermann, K. (1979), Localisation of gastrinomas by transhepatic portal catheterization and gastrin assay. *Gastroenterology*, **77**, 445–450.
6. Chey, W. Y. and Lee, K. Y. (1980), Motilin. *Clin. Gastroenterol.*, **9**, 645–656.
7. Ebert, R. and Creutzfeldt, W. (1978), Aspects of GIP pathology. In: *Gut Hormones* (S. R. Bloom, ed.), pp. 294–300. Edinburgh: Churchill Livingstone.
8. Efendic, S., Hökfelt, T. and Luft, R. (1978), Somatostatin. *Adv. Metab. Disorders*, **9**, 367–424.
9. Fahrenkrug, J. (1980), Vasoactive Intestinal Polypeptide. *Clin. Gastroenterol.*, **9**, 633–643.
10. Floyd, J. C. (1980), Pancreatic polypeptide. *Clin. Gastroenterol.*, **9**, 657–678.
11. Friesen, S. R., Hermreck, A. S. and Mantz, F. A. (1974), Glucagon, gastrin and carcinoid tumors of the duodenum, pancreas and stomach: polypeptide 'Apudomas' of the foregut. *Am. J. Surg.*, **127**, 90–101.
12. Gowenlock, A. H. (1977), Tests of exocrine pancreatic function. *Ann. clin. Biochem.*, **14**, 61–89.
13. Grahame-Smith, D. G. (1974), Natural history and diagnosis of the carcinoid syndrome. *Clin. Gastroenterol.*, **3**, 575–594.
14. Gutniak, M., Rosenqvist, L., Grimelius, L. et al. (1980), Report on a patient with watery diarrhoea syndrome caused by a

pancreatic tumour containing neurotensin, enkephalin and calcitonin. *Acta Med. Scand.*, **208**, 95–100.
15. Häcki, W. H. (1980), Secretin. *Clin. Endocrinol.*, **9**, 609–632.
16. Holst, J. J. (1979), Gut endocrine tumour syndromes. *Clin. Endocrinol. Metab.*, **8**, 413–432.
17. Howat, H. T. and Braganza, J. M. (1979), Assessment of Pancreatic Dysfunction in Man. In: The Exocrine Pancreas (Howat, H. T. and Sarles, H., eds.), pp. 129–175, London: W. B. Saunders.
18. Jorpes, J. E. and Mutt, V. (1973), Secretin and cholecystokinin (CCK). In: *Secretin, Cholecystokinin, Pancreozymin and Gastrin* (Jorpes, J. E. and Mutt, V., eds.), pp. 1–179, Springer Verlag: Berlin.
19. Kosaka, T. and Lim, R. K. S. (1930), Demonstration of the humoral agent in fat inhibition of gastric secretion. *Proc. Soc. Exptl. Biol. Med.*, **27**, 890–896.
20. Lefebvre, P. J. and Unger, R. H. (eds.) (1972), *Glucagon: Molecular Physiology, Clinical and Therapeutic Implications*. Oxford: Pergamon Press.
21. Mallinson, C. N., Bloom, S. R., Warin, A. P., Salmon, P. R. and Cox, B. (1974). A glucagonoma syndrome. *Lancet*, **ii**, 1–5.
22. Marks, V. and Turner, D. S. (1977), The gastrointestinal hormones, with particular reference to their role in the regulation of insulin secretion. *Essays Med. Biochem.*, **3**, 109–152.
23. Morgan, L. M. (1979), Immunoassayable GIP; investigations into its role in carbohydrate metabolism. *Ann. clin. Biochem.*, **16**, 6–14.
24. Pearse, A. G. E. (1966), Common cytochemical properties of cells producing polypeptide hormones, with particular reference to calcitonin and the thyroid C cells. *Vet. Rec.*, **79**, 587–590.
25. Primrose, J. N., Ratcliffe, J. G. and Joffe, S. N. (1980), Assessment of the secretin provocation test in the diagnosis of gastrinoma. *Br. J. Surg.*, **67**, 744–746.
26. Rehfeld, J. F. (1980), Cholecystokinin. *Clin. Gastroenterol.*, **9**, 593–607.
27. Stadil, F. and Stage, J. G. (1979), The Zollinger-Ellison syndrome. *Clin. Endocrinol. Metab.*, **8**, 433–446.
28. Verner, J. V. and Morrison, A. B. (1974), Non-B islet tumours and the syndrome of watery diarrhoea, hypokalaemia and hypochlorhydria. *Clin. Gastroenterol.*, **3**, 595–608.
29. Walsh, J. H. and Lam, S. K. (1980), Physiology and pathology of gastrin. *Clin. Gastroenterol.*, **9**, (3) 567–591.
30. Yamaguchi, K., Kameya, T. and Abe, K. (1980), Multiple endocrine neoplasia: Type I. *Clin. Endocrinol. Metab.*, **9**, 261–284.
31. Zollinger, R. M. and Ellison, E. H. (1955), Primary peptic ulcerations of the jejunum associated with islet-cell tumors of the pancreas. *Am. Surg.*, **142**, 709–723.

FURTHER READING

Bloom, S. R. and Polak, J. M. (Eds.) (1981), *Gut Hormones*, 2nd ed. Edinburgh: Churchill Livingstone.
Bonfils, S. (Ed.) (1974), *Endocrine-Secreting Tumours of the GI Tract*: Clinics in Gastroenterology, Vol. 3.
Buchanan, K. D. (Ed.) (1979), *Gastrointestinal Hormones*: Clinics in Endocrinology and Metabolism, Vol. 8.
Creutzfeldt, W. (Ed.) (1980), *Gastrointestinal Hormones*: Clinics in Gastroenterology, Vol. 9.
Glass, G. B. J. (Ed.) (1980), *Gastrointestinal Hormones*. New York: Raven Press.
Marks, V. and Morgan, L. M. (1982), "Gastrointestinal Hormones" *Molecular Aspects of Medicine*, **5**, 225–292.

39. ADRENAL CORTEX

VINCENT MARKS

Anatomy and embryology
Adrenal steroids
  Glucocorticoids
  Mineralocorticoids
  Adrenal androgens
  Urinary metabolites
Control of adrenocortical activity
Synthetic steroids
Adrenal steroids in blood
  Cortisol
  17α-Hydroxyprogesterone
  Other adrenal steroids
  Cortisol production rate
  Cortisol-binding globulin
Urinary steroids
  Cortisol
Salivary steroids
Plasma ACTH
Dynamic tests of hypothalamic-pituitary-adrenal function
  ACTH stimulation tests
  Insulin hypoglycaemia
  Other stress tests
  Metyrapone test
  Steroid suppression tests
Clinical investigation of adrenocortical dysfunction
  Adrenocortical insufficiency
  Cushing's syndrome
  Nelson's syndrome
  Hypertension
  Adrenogenital virilism
  Congenital adrenal hyperplasia (CAH)
  Feminising adrenal tumours
  Idiopathic hirsutism

ANATOMY AND EMBRYOLOGY

In man the adrenal glands, each weighing approximately 4 g, sit atop the kidneys from which position they derive their earlier name of 'suprarenals'. Each adrenal can be divided anatomically into a larger, outer zone known as the adrenal cortex, and a smaller, inner zone known as the adrenal medulla. The two zones are embryologically, histologically, physiologically and biochemically distinct. The medulla is part of the sympathetic nervous system and exclusively responsible for the secretion of adrenaline in response to appropriate stimulation. It also secretes noradrenaline.

The cortex behaves, in many ways, as though it were three functionally distinct organs all having the capacity, shared only by the gonads, of producing steroid hormones from a cholesterol precursor and secreting them into the blood.[3,16,28,29] The adrenal cortices and gonads are unique amongst endocrine glands in developing from mesoderm rather than from endo- or ectoderm and, with the pancreatic islets, are the only ones that are essential for life.

Both right and left adrenals are copiously supplied with arterial blood from a number of sources including the inferior phrenic artery, the renal arteries and the aorta itself via a number of unnamed vessels. The main venous drainage is different on the two sides; that of the right adrenal is directly into the inferior vena cava whilst that on the left is into the left renal vein. This difference assumes importance when blood from each gland has to be collected individually by catheterisation in order to lateralize a lesion affecting only one of them and which cannot be visualised radiologically.

Blood flow in the adrenals is centripetal which is not without functional significance since there is good evidence that the very high concentration of cortisol in blood perfusing the medulla is responsible for induction, in the chromaffin cells, of the enzyme *phenylethanol-amine-N-methyl transferase* (PNMT) which is capable of effecting synthesis of adrenaline from noradrenaline. This enzyme is, for all practical purposes, confined exclusively to the adrenal medulla and small, circumscribed areas of the brain. Species in which the adrenal medulla and cortex are anatomically separate do not, as a rule, elaborate adrenaline; nor do pheochromocytomas that develop in chromaffin tissue of extra-adrenal origin and which are not, therefore, surrounded by cortisol-secreting tissue.

Histologically the cortex is divided into three zones. The outermost, or zona glomerulosa, is the main producer of aldosterone and the innermost one, or zona reticulosa, of androgens. The middle layer, the zona fasciculata, is normally as large as the other two combined and is the main source of cortisol and corticosterone.

ADRENAL STEROIDS

More than 50 steroids have been isolated from the adrenal cortex. All of them, like their common precursor cholesterol, possess a cyclopentanoperhydrophenanthrene nucleus but differ from each other with respect to their side chains and degree of saturation of their A and B rings. There is no reason to believe that all, or even the majority, of them have hormonal function and many undoubtedly represent intermediate metabolites or artifacts created by separation and purification.

It has long been customary to classify adrenal steroids on the basis of their main biological properties into those that exert their effects mainly upon carbohydrate metabolism, i.e. glucocorticoids, and those that act mainly upon the metabolism of water and electrolytes and which are known as mineralocorticoids. A third group, with mainly androgenic or anabolic properties is also recognized; there is, however, considerable overlap between the biological properties of all three groups, especially in non-mammalian species, and the distinction between them should, therefore, be looked upon more as one of degree than of kind.

Biosynthesis of the adrenal steroids is complicated and, with many alternative routes, it is still not absolutely certain that all subjects utilise the same pathways. The pathway considered, on present evidence, to be the most likely in the majority of subjects, is shown in Fig. 1. All of the biologically active corticosteroids normally *secreted* by the adrenal cortex in man are derivatives of pregnane and are C_{21} steroids carrying two carbon atoms in a side chain at carbon-17. Each has a ketonic group at C_3, a double bond at C_4 and another ketonic group at C_{20}. Many also have a hydroxyl group at C_{17}. The production and secretion of adrenal steroids is largely, though not exclusively, regulated by ACTH which also exercises a trophic effect upon the adrenal cortex as a whole.

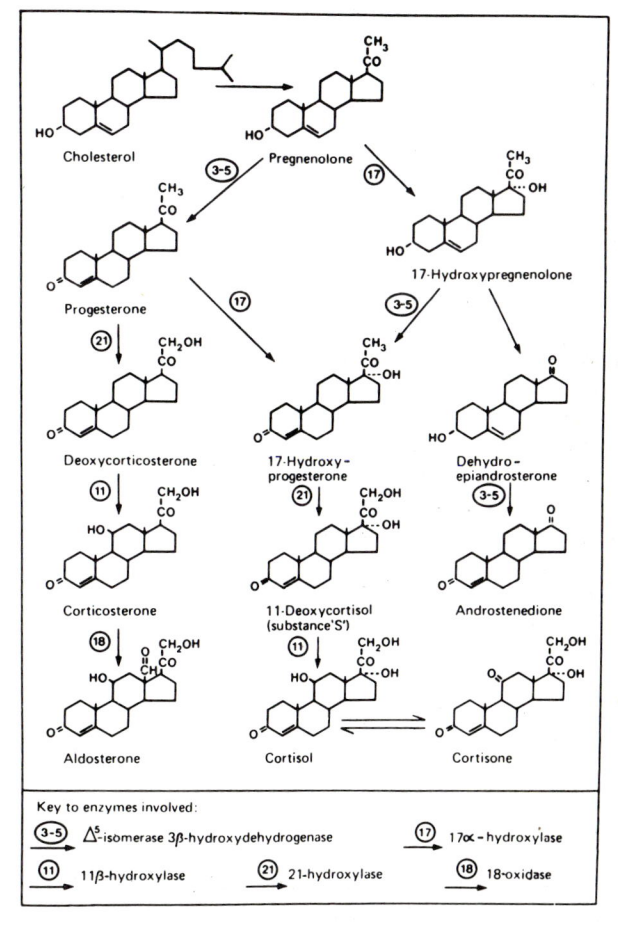

FIG. 1. Biosynthetic routes for adrenocortical steroid hormones. (From James, V. H. T. and Landon, J., *Hypothalamic-Pituitary Adrenal Function Tests*, CIBA Laboratories, Horsham, England, 1976. With permission.)

Glucocorticoids

Cortisol is the main adrenal glucocorticoid produced in man and some, but not all, other mammalian species. It is secreted irregularly, in spurts, at a rate of approximately 20 mg (55 μmol) per day but there are large interindividual variations. It is transported in the plasma mainly bound to a corticosteroid-binding α-globulin, transcortin (often referred to as cortisol-binding globulin: CBG) which has a high binding affinity for cortisol, corticosterone and a few closely related synthetic glucocorticoids, e.g. prednisolone, but few other naturally occurring or synthetic steroids. Conditions prevailing in the blood under ordinary conditions ensure that about 70% of the cortisol present is bound to CBG. Another 20% is bound to albumin which, though present in much higher concentration, has less than 0·01% the binding affinity of CBG for cortisol. The remaining 10% or so of cortisol in the plasma is ultrafilterable and often referred to as 'free' or 'non-protein bound' cortisol. Its concentration, rather than that of the protein-bound fraction, is thought to determine the level of glucocorticoid activity but this is probably an oversimplification.

The cortisol carrying capacity of CBG is limited and, in normal subjects, averages 650 nmol/l and is only rarely more than 780 nmol/l, a value often exceeded during stress and other conditions of enhanced adrenocortical activity. Under these circumstances a much larger proportion of the plasma cortisol exists in the 'free' form, or only loosely bound to albumin, and is more readily accessible to the tissues or filtration at the renal glomeruli.

Cortisone, though a normal constituent of plasma,[12] is usually present at only one tenth the concentration of cortisol ('hydrocortisone') from which it is thought to be formed by metabolism in peripheral tissues. It does not, however, possess biological activity in its own right and must be reconverted into cortisol, in the liver, in order to acquire it. Corticosterone, the main glucocorticoid secreted by the adrenals in rats, rabbits and some other rodents, is normally produced in relatively small amounts by human subjects, its concentration in plasma being approximately one twentieth that of cortisol which it resembles in being tightly bound by CBG.

The average half-life of cortisol in plasma is 70 minutes. It is increased by extremes of age, liver disease and uraemia and decreased by hyperthyroidism. Normally, only a small amount of cortisol, representing less than 20% of that filtered at the glomeruli, is excreted unchanged in the urine. Most of what is filtered is reabsorbed in the distal tubules by an active process which is almost saturated at ordinary plasma cortisol levels and is easily overwhelmed when the filtered load rises only comparatively modestly.

The main route of removal of cortisol from the body is by metabolism in the liver where it is converted into one of many possible metabolites. These are subsequently conjugated into water-soluble glucuronides and sulphates which are then excreted in the urine mainly as 17-oxogenic steroid conjugates. Corticosterone behaves similarly to cortisol but, because of the absence of a 17-hydroxyl group, corticosterone metabolites are not measured as 17-oxogenic steroids.

Although their name implies a special effect upon carbohydrate metabolism, glucocorticoids have profound effects not only upon glucose but also upon fat and protein metabolism behaving, in general, as catabolic rather than as anabolic agents. They encourage proteolysis in peripheral tissue, with liberation of aminoacids, and they promote mobilisation of free fatty acids by adipose tissue, but are not ketogenic. They also affect mineral metabolism, especially corticosterone which has quite marked sodium retaining properties in man.

Mineralocorticoids

Aldosterone is the main mineralocorticoid produced by the adrenal cortex in man but other less potent steroids, e.g. deoxycorticosterone, are also produced in modest amounts except rarely, in disease, when they may assume major importance.

The mineralocorticoids are produced mainly in the zona glomerulosa and, unlike the glucocorticoids and adrenal androgens, are largely independent of hypothalamic-pituitary control. They are, instead, controlled mainly through the renin-angiotensin system which is described elsewhere in this volume.

Adrenal Androgens

Relatively large amounts of steroids with mildly androgenic and anabolic properties are secreted by the adrenal glands and are the major source of androgens in women. The most important are androstenedione (A), dehydroepiandrostenedione (DHA) and its sulphate (DHAS) which are all capable of undergoing conversion, in peripheral tissues, into testosterone and dihydrotestosterone.[22] The adrenal androgens, like cortisol, are largely, if not entirely, under hypothalamic-pituitary control exerted through ACTH. Their importance, under physiological conditions, is difficult to assess since seemingly no untoward effects are produced by their absence. Indeed, they may represent nothing more than obligatory intermediates in adrenal steroid synthesis and degradation, and assume clinical importance only when produced in excess. This view of adrenal androgens is, however, made less likely by the fact that they seem to be produced largely, though not exclusively, by the zona reticulosa, which is histologically a relatively well-defined zone of the adrenal cortex.

Since adrenal androgens and cortisol are both produced in response to adrenal stimulation by ACTH it might be supposed that they would show a constant secretory ratio to each other but this is not so. Indeed, it has been suggested that androgens are produced only in response to dual stimulation by ACTH and some other pituitary trophic hormone, possibly LH, or prolactin but this theory has now largely been discredited.[1] Since adrenal androgens, unlike cortisol, do not exert any negative feedback control on hypothalamic-pituitary-adrenal activity their secretion, under physiological conditions, follows passively that of the glucocorticoids. This does not necessarily happen in diseases affecting the adrenal glands.

Urinary Metabolites

Adrenocortical and gonadal steroids are extensively metabolized in the liver and elsewhere before being excreted in the urine mainly as their glucuronides and, to a lesser extent, sulphates. Both types of conjugate must be hydrolysed by hot strong acid or appropriate enzymes, i.e. β-glucuronidase and sulphatase, before measurements of total urinary steroid excretion can be undertaken. Steroid metabolites, which are generally devoid of important biological actions, are usually classified on a group basis, according to the presence or absence of various functional groups. 17-oxosteroids (17-ketosteroids) possess a ketone group at carbon-17 and react specifically with m-dinitrobenzene in alkaline solution producing the so-called Zimmerman reaction. This produces a blue complex which can be measured spectrophotometrically. Most of the metabolites of the C_{19} adrenal androgens are 17-oxosteroids and were amongst the first class of steroids to be quantitated in urine; unfortunately they provide information relevant to only a limited aspect of adrenocortical function. The major urinary 17-oxosteroids are aetiocholanolone, androsterone, dehydroepiandrosterone (DHA) and their 11-oxygenated derivatives. Measurement of urinary oxosteroids, once extensively used as an index of adrenocortical function, has largely, if not entirely, been superseded by more specific measurements, both in urine and in plasma, of individual steroids.

Many of the C_{21} (glucocorticoid) metabolites can be converted by oxidation with sodium bismuthate (Norymberski), to 17-oxosteroids and, consequently, are often referred to as 17-oxogenic (ketogenic) steroids. Their measurement in urine, once popular as a test of adrenocortical activity, is now obsolete.

Steroids possessing a 17α, 21-dihydroxy, 20-keto group, i.e. most of the cortisol and 11-deoxycortisol metabolites, are capable of reacting with a 2,4-dinitrophenylhydrazone reagent devised by Porter and Silber and are often referred to as 17-hydroxycorticoids. Their measurement, in plasma, was for many years used as the main guide to circulating plasma cortisol levels but has been superseded by more specific methods. The term, 17-hydroxycorticoids, is also used to describe a similar but not identical group of steroids measured in urine by the Norymberski technique but they, too, have been superseded.

In addition to classification on the basis of their reactions with various reagents, urinary steroids can also be classified according to their behaviour in different chromatographic systems. These enable, for example, 11-deoxysteroids to be distinguished from 11-oxysteroids and the metabolites of cortisol from those of its precursors.

Considerable interest has been aroused—especially amongst toxicologists and pharmacologists—by 6β-hydroxycortisol, a naturally occurring metabolite of cortisol whose increased excretion in the urine is a useful marker of enzyme induction in the liver under the influence of a number of enzyme-inducing drugs and toxic compounds. It has, however, attracted comparatively little interest from clinicians and biochemists.

Much of our knowledge of adrenocortical function in disease was obtained using urinary steroid measurements. They suffer from the disadvantage that they provide, at best, indirect information on the nature and quantity of circulating steroids let alone those actually secreted by the adrenals. In order to overcome these problems more attention has been given, in recent years, to the measurement of individual steroids in the blood by colorimetry, fluorimetry and, latterly, by radioimmunoassay, HPLC (high pressure liquid chromatography) and GC-MS (gas chromatography—mass spectrometry). The latter technique, though extremely useful in research, is impracticable for clinical use.

CONTROL OF ADRENOCORTICAL ACTIVITY

The adrenal cortex, at least that part of it concerned with glucocorticoid and androgen production, functions exclusively as part of a complex involving the central nervous system and anterior pituitary gland: the hypothalamic-pituitary-adrenal axis. ACTH is the only known agent by which central control is exercised and, in its absence, adrenocortical activity virtually ceases and the glands atrophy. Mineralocorticoid, i.e. aldosterone, secretion is largely though not wholly independent of ACTH secretion being mainly under the control of the renin-angiotensin system which does not, however, unlike ACTH, appear to exert any trophic effect upon the adrenal glands themselves. The secretion of ACTH is itself under higher nervous control which is channelled through the hypothalamus and is exercised largely by regulated secretion of corticotrophin releasing factor (CRF) which current evidence suggests is a complex, rather than a single substance, one component of which is vasopressin. This octapeptide has long been known as a hypothalamic hormone with antidiuretic and ACTH stimulatory properties but its physiological role as a regulator of ACTH release, though first suggested many years ago, is still disputed.

CRF (whatever it is) is conveyed from the hypothalamus to the anterior pituitary in the hypothalamic-pituitary portal system. Three major mechanisms control its release. The first is a negative feedback control exerted through cortisol (and other glucocorticoid) activity in the plasma; a reduction in plasma cortisol effect leading to a large and rapid increase in ACTH secretion. Advantage is taken of this fact in the metapyrone test of hypothalamic-pituitary-adrenal function in which cortisol production is temporarily blocked by metyrapyrone, a selective inhibitor of 11β-hydroxylase activity. An increase in plasma glucocorticoid activity, on the other hand, produces a decrease in ACTH secretion; this, too, can be exploited clinically in the dexamethasone suppression test of hypothalamic-pituitary-adrenal integrity.

The second control mechanism of hypothalamic-pituitary-adrenal function is mediated through a number of incompletely identified cerebral centres which ensure that ACTH, and consequently cortisol, secretion follows a well-defined circadian rhythm. As a result the highest levels of both these hormones, in plasma, generally occur around 8–10 a.m. and the lowest levels at about midnight. Disturbances of diurnal rhythm, though characteristic of Cushing's disease and endogenous depression, are common in all sorts of diseases but in most of them the disturbances are temporary and probably represent an overriding of general regulation by the third, and possibly most important, mechanism controlling ACTH secretion, namely stimulation by 'stress' of both the physical and psychogenic varieties. The last named control mechanism is thought to be mediated through 'higher-centres', i.e. suprahypothalamic, but its precise anatomical location is unknown. It often remains operational when the other two have been rendered inactive by disease.[13]

SYNTHETIC STEROIDS

Following the discovery that cortisone had a profound, almost miraculous, though infuriatingly temporary, effect upon pain and disability in some cases of rheumatoid arthritis and other inflammatory diseases, an enormous amount of effort was devoted by the pharmaceutical industry to manipulating steroid molecules in order to enhance some of their biological properties and diminish others. In this way, it proved possible to produce compounds with (a) greatly enhanced glucocorticoid but virtually no mineralocorticoid or androgenic properties, e.g. prednisolone, dexamethasone (i.e. typical anti-inflammatory steroids), (b) protein anabolic, but little or no androgenic, mineralo- or glucocorticoid activity, e.g. methandienone and (c) increased mineralocorticoid activity relative to all other steroidal properties, e.g. fludrocortisone.

Members of the various classes of synthetic steroids, especially the anti-inflammatory glucocorticoids, are used extensively in therapy. They assume importance in clinical biochemistry mainly because of their interference with hypothalamic-pituitary-adrenal (HPA) function and other biological processes. The suppressive effect on HPA activity is greatest at about midnight. Advantage is taken of this fact (i) in the treatment of children with congenital adrenogenital virilism who should receive their suppressive doses of cortisone, or other glucocorticoids, as late at night as possible, and (ii) in lessening the adverse effects of steroid therapy for inflammatory diseases by giving them in the morning when their adrenosuppressive effects are minimal.

Some synthetic steroids, notably prednisolone and dexamethasone, are used diagnostically.

ADRENAL STEROIDS IN BLOOD

Cortisol

Plasma glucocorticoids are usually measured nowadays by immunoassay using antisera with greater or lesser specificity for cortisol.[15,34,36] Closely related steroids may contribute to the total measurement but generally to a lesser extent than with previous methods which employed colorimetric or fluorimetric reactions or were based upon competitive protein-binding techniques.[24] As a result, quoted values for total plasma cortisol concentrations tend to be lower now than in the past and this should be taken into account in interpreting results obtained by immunoassay.

Of the many immunoassay methods available those using antisera raised against the 3-O-methoxy-hemisuccinate derivative of cortisol and utilizing either ^{3}H cortisol or a ^{125}I cortisol derivative as label enjoy the greatest popularity. Techniques other than immunoassay are also gaining popularity, especially those employing HPLC, since they confer equal or greater specificity coupled with information about other steroids in the plasma.[40,41] They are, however, relatively slow and tedious to perform and likely, therefore, to be used in situations where they alone can provide the correct analytical answer.

Plasma cortisol levels show a well marked diurnal variation, the highest levels generally occurring around 8–10 a.m. and the lowest around midnight. They do, however, also undergo rapid fluctuations throughout the day even in perfectly healthy subjects under ideal conditions, with peaks and troughs sometimes following each other within minutes in the absence of perceptible reason.

Cortisol is relatively stable in plasma and no special precautions are required during collection of specimens for subsequent assay. It is, however, susceptible to marked and rapid changes in concentration *in vivo*, due mainly to HPA responsiveness to both physical and mental stress, such as many people experience during venesection, but also for reasons stated above. Too much significance should not, therefore, be attached to a single elevated value for plasma cortisol, especially if there is any possibility that the patient was anxious, frightened or subjected to painful venepuncture. This is even more true of plasma ACTH measurements.

The main clinical use of plasma cortisol assays is as an index of the HPA's capacity to respond to a variety of stimuli as part of a test carried out under controlled conditions. They can, however, provide valuable information about the state of HPA function in patients undergoing withdrawal from glucocorticoid therapy.[7] Random plasma cortisol measurements are of some clinical use as evidence—in retrospect—of adrenocortical non-responsiveness during stress and, when collected around midnight, of disturbed circadian rhythm such as occurs in Cushing's syndrome and endogenous depression. Most specimens for cortisol assay are, however, collected between 8 and 10 a.m. when, in most healthy adults, they lie in the range 200–500 nmol/l (7·25–18·2 µg/100 ml).

Because of the key role played by CBG in determining the total amount of cortisol held in the plasma at any particular time, a knowledge of those factors, such as oestrogen status, that themselves determine CBG concentration are important in interpreting plasma cortisol levels. Oestrogens increase, while androgens and/or synthetic anabolic steroids decrease, CBG levels which are also reduced in many hypoproteinaemic conditions as well as in a rare, specific, familial form of CBG deficiency.

Secretion of cortisol is controlled, in health, exclusively by ACTH for which cortisol provides a good bioassay system unless the adrenal cortex is, itself, incapable of responding because of inherited or acquired disease. There is, however, only a poor correlation between plasma ACTH and cortisol levels especially in conditions of moderate to marked stress in which plasma ACTH levels are supramaximal for adrenocortical stimulation. Transient peaks of plasma cortisol, however, can be shown to follow peaks of plasma ACTH under experimental conditions, though timing of the samples is critical.

High plasma cortisol levels, relative to time-related reference values, occur in (i) many acute illnesses; (ii) women on the contraceptive pill and men on oestrogens; in response to (iii) physical or psychological stress; (iv) alcoholism and (v) pregnancy, as well as in Cushing's syndrome and endogenous depression. Low values occur in patients with primary or secondary hypoadrenocorticism and those receiving treatment with anti-inflammatory glucocorticoids providing these do not cross-react in the assay.

17α-Hydroxyprogesterone

This steroid, which is a biosynthetic precursor of cortisol, is not normally present in detectable amounts in plasma but appears in large amounts in patients, generally infants, with either 21- or 11β-hydroxylase deficiencies. 17α-hydroxyprogesterone can be measured in plasma or saliva by immunoassay, using a specific antiserum, or by HPLC, and is currently the method of choice for confirming a suspected diagnosis of adrenogenital virilism, particularly in the first few days of life when urine steroids are especially unreliable. 17α-hydroxyprogesterone can also be used for monitoring the therapeutic response to treatment with glucocorticoids but may be 'too sensitive' and its use for this purpose is likely to be superseded by measurement of plasma androstenedione levels which, like 17α-hydroxyprogesterone, are raised in adrenogenital virilism and depressed by glucocorticoid therapy.

Plasma 17α-hydroxyprogesterone levels, like cortisol, show large diurnal fluctuations. Samples for analysis should ideally be collected between 8 and 10 am, before treatment with glucocorticoids is commenced. Measurements made in neonates before the baby is 2 days old may be misleading, if low, and should, where suspicion persists, be repeated after this time. The plasma concentration of 17α-hydroxyprogesterone reflects the magnitude and severity of the metabolic block and should fall to low or undetectable levels during adequate glucocorticoid replacement therapy. Raised levels are sometimes observed in patients with adrenal tumours causing virilism.

Other Adrenal Steroids

Most, if not all, of clinically significant adrenal steroids that occur in blood (Table 1) are now amenable to measurement by immunoassay, using appropriately specific antisera, either directly on unextracted plasma or more commonly after extraction with, or without, separation by thin layer chromatography or HPLC. Relatively few of the many adrenal steroids that occur in plasma in health and disease have so far found diagnostic clinical applications, although their measurement has helped throw additional light upon pathogenic mechanisms in some cases of idiopathic hirsutism, hypokalaemia and hypertension.

The adrenal glands make the major contribution to androstenedione in the plasma which behaves as a testosterone precursor in certain tissues—notably the skin. Moderately raised plasma androstenedione levels are sometimes observed in women with unexplained hirsutism, which is often associated with the polycystic ovary (Stein-Leventhal) syndrome but are seldom diagnostically useful; higher levels are regularly found in babies and untreated adults with adrenogenital virilism and may be used with, or instead of, 17α-hydroxyprogesterone measurements for diagnosis and monitoring the response to treatment. Like most other steroids of adrenocortical origin, apart from aldosterone and dehydroepiandrosterone sulphate (DHAS), androstenedione secretion shows a well marked diurnal rhythm and its concentration in plasma is depressed by exogenous glucocorticoid therapy.

Dehydroepiendrosterone (DHA) and its sulphate (DHAS), behave as weak androgens and their concentration in plasma is often slightly increased, relative to age-related reference values, in 'idiopathically' hirsute women as well as in those with adrenogenital virilism. They are also said often to be mildly raised in patients with hyperprolactinaemia,[32] regardless of cause, and to fall following restoration of plasma prolactin levels to normal but this has been disputed. Since DHA is derived solely from the adrenal, its suppression by dexamethasone treatment can be used as an index of adrenal autonomy in, for example, the investigation of virilising syndromes.

Though both cortisone and corticosterone are natural constituents of the plasma no diagnostically useful information is currently obtained by measuring them and only in exceedingly rare cases of hypertension due to excess secretion of 11-deoxycorticosterone is measurement of this unusual plasma steroid rewarding.

Cortisol Production Rate

Because of its extensive metabolism *in vivo*, cortisol production rate cannot properly be computed by

TABLE 1
ADRENAL STEROIDS IN BLOOD

Steroid	Mol. wt.	Reference range* Male	Reference range* Female**	Clinical indications for measurement
Androstanediol	288	0·05 nmol (0·02–0·1)		Hirsutism
Androstenedione	286	3·5 nmol (2–11)	4·0 nmol (2–15)	Monitoring glucocorticoid treatment of CAH
Corticosterone	346	11·5 nmol/l (2·5–60)		Cushing's syndrome (rarely)
Cortisol	362	350 nmol/l (200–750)		Cushing's syndrome: Adrenocortical insufficiency (after Synacthen^R); withdrawal from glucocorticoid therapy: test of HPA integrity (during insulin-hypoglycaemia).
DOC	330	2·2 nmol (1·0–3·5)		Hypertension and hypokalaemia
11-Deoxycortisol	346	5·0 nmol/l (2·5–7·5)		During metyrapone tests; 11-hydroxylase deficiency
DHA	289	20 nmol (15–30)		Virilism; adrenal tumours
DHAS	369	5·0 μmol (0·7–11·5)		Hyperprolactinaemia
17α-HP	330	<13 nmol		Congenital adrenal hyperplasia
Testosterone***	228	15 nmol (11–35)	1·6 nmol (0·9–2·7)	Hirsutism; isosexual precocity; feminisation

* Concentration per litre of plasma (in morning). Values vary slightly from one laboratory to another depending upon methodology.
** Early follicular stage of cycle.
*** Sex-difference mainly due to testicular contribution.

DHA = Dehydroandrosterone
DHAS = Dehydroandrosterone sulphate
17α-HP = 17α-hydroxyprogesterone
DHT = Dihydrotestosterone
DCO = Deoxycortisone
CAH = Congenital adrenal hyperplasia

measurement of its excretory products in urine. Instead elaborate isotope dilution techniques must be employed and although these have helped throw considerable light upon the control of adrenocortical function in health and disease they have almost no role in clinical practice. This applies equally to other steroids, whose production rates can be measured by similar methods.

Cortisol Binding Globulin (CBG)

The presence, in the plasma, of transcortin, a high-affinity, low-capacity, glycoprotein capable of binding cortisol and a few closely related steroids, i.e. corticosterone, 17αOH-progesterone, 11-deoxycortisol, progesterone and prednisolone, was first recognised a quarter of a century ago though there is still no clear understanding of its physiological function.[38] It is thought, by some, to provide a plasma pool of cortisol in equilibrium with the 'free' or easily dialysable cortisol fraction; others believe that it may play a role in target orientation whilst yet others look upon it as providing some sort of detoxification mechanism.

In normal subjects transcortin (CBG) has a binding capacity for cortisol of about 650 nmol/l (24 μg/100 ml), a value roughly equivalent to the upper limit of the normal basal plasma cortisol concentration.[35,38] The addition of cortisol to plasma, above this level, leads only to its loose association with albumin; much of it remains free or ultrafilterable. The concentration of transcortin in patients with adrenal insufficiency, Cushing's syndrome, hyperthyroidism, hypothyroidism, hypopituitarism and acromegaly is normal, but it is increased, up to three-fold, in pregnancy and during oestrogen therapy. Its concentration is reduced in many conditions associated with hypoalbuminaemia and, selectively, in a rare familial condition of hereditary transcortin deficiency. Transcortin, in plasma, is usually measured indirectly by determining the cortisol-binding capacity but it can also be measured immunochemically.

URINARY STEROIDS

The 17-oxosteroids (previously known as ketosteroids) represent the main excretory products of adrenal (and to a lesser extent, testicular and ovarian) androgens (Table 2). They were the first group of steroids to become amenable to measurement in the clinical situation and all possess an oxo group at C-17. They are still often used as a screening procedure, especially in conjunction with measurement of urinary 17-oxogenic steroids, in patients suspected of suffering from adrenocortical under- or overactivity.[46] The 17-oxo and 17-oxogenic steroids are produced mainly in the liver by oxidative and other processes involving biotransformations brought about largely by inductible microsomal enzymes. They occur naturally in the urine largely as glucuronides or sulphates, and their excretion, like the secretion of their precursor adrenocortical steroids,

TABLE 2
URINARY ANDRENOSTEROID EXCRETION*

	Method	Male	Female	Clinical indications
Cortisol	RIA	70–250 (nmol)		Suspected Cushing's syndrome
	Fluorimetry**	200–1000 (nmol)		
17-oxosteroids	GLC	15–65 (μmol/l)	10–46	Investigation of suspected adrenocortical dysfunction: metyrapone test.
17-oxogenic steroids	GLC	10–65 (μmol)	5–55	
Androsterone	GLC	5–30 (μmol)	2–20	
Aetiocholanolone	GLC	5–25 (μmol)	5–15	
Cortisol metabolites	GLC	7–55 (μmol)	4–50	
Pregnantriol/THS	GLC	0.5–9.0 (μmol)	0.0–6.0	
11-oxygenation index (11-deoxysteroids/11-oxysteroids)	GLC and/or TLC	<0.5		Congenital adrenocortical hyperplasia (after first week of life)
Cortisol:creatinine ratio	RIA/colorimetry	<20 μmol cortisol/mol creatinine		Adrenocortical overactivity; Cushing's syndrome; nocturnal hypoglycaemia in diabetics.
	Fluorimetry**/colorimetry	<50 μmol cortisol/mol creatinine		

* In 24 hours.
** Non-specific and unreliable.

varies markedly throughout the day. It is customary, therefore, to measure urinary steroids (after hydrolysis) in an aliquot of a 24 hr sample and to express the results in μmol per 24 hr rather than as a concentration.

Many methods are available for measuring urinary steroids and most of them are based upon more or less group-specific colour reactions developed under defined conditions. All of them are susceptible to interference from drugs and other extraneous products which seriously limits their clinical usefulness. More recently introduced methods rely upon GLC or HPLC separation prior to measurement and give lower results than those previously employed. They also yield useful additional information about substitutions in the steroid molecule apart from those occurring on C-17.

Urinary 17-oxosteroid and 17-oxogenic steroid excretions vary markedly with age, sex, height, weight and racial origin as well as with disease status. Interpretation is, therefore, difficult except in cases of gross deviation from normal such as occurs in adrenogenital virilism or advanced Cushing's syndrome; many patients with less advanced disease have results that fall comfortably within the age and sex matched reference range.

Since urinary steroid measurements provide information about the actual quantity of steroids produced by the adrenal that cannot easily be obtained in any other way, except by elaborate isotope-dependent production rate determinations, they still do have a place in the investigation of disease mechanisms—providing they are not overinterpreted. Urinary steroid excretion is, for example, increased, in many cases of simple obesity, largely due to increased metabolic clearance of steroid hormones in the liver, though plasma levels do tend to be higher than average, too. Many drugs that induce hepatic enzymes have a similar effect upon steroid metabolism but do not usually increase urinary steroid excretion though some do interfere chemically with the methods used for measuring them. More important, from a practical point of view, is the ability of many hepatic enzyme-inducing drugs to increase the production and urinary excretion of 6β-hydroxycortisol which enjoys considerable popularity as a marker of hepatic enzyme induction,[31] though it is sometimes excreted in increased amounts by patients with uncomplicated Cushing's syndrome.

Cortisol*

Urinary cortisol reflects the prevailing 'unbound' cortisol level in plasma to which the kidney is exposed and is largely independent of cortisol secretion or destruction rates, liver steroid breakdown or protein binding. It is, however, dependent on glomerular filtration rate (GFR). Providing this is normal, measurement of urinary cortisol excretion is the method of choice for detecting, or confirming, a diagnosis of Cushing's syndrome.[6,8] For this purpose it may not even be necessary to collect urine for 24 hr, but merely to collect an overnight sample and express the result as a cortisol:creatinine ratio. The upper limit of normal found by Chambers and co-workers[8] was 20 μmol cortisol/mol creatinine which is somewhat lower than that observed when cortisol is measured by a non-specific fluorimetric technique. Even more reliable is the integrated concentration of plasma cortisol during a 24-hour period but this is impracticable except in a specially equipped laboratory.[50] Ordinarily very little cortisol is filtered at the glomerulus and even less is excreted in the urine. With modern analytical methods utilising HPLC, and/or radioimmunoassay, it can be shown that most normal subjects excrete less than 0.2 μmol (73 μg) of cortisol per 24 hr. Much higher

* Sometimes referred to as urinary free-cortisol to distinguish it from any cortisol conjugates that might be present. In the absence of protein in the urine none is 'protein bound'.

values used to be recorded with less specific fluorimetric and protein-binding assays which should now be abandoned for this purpose. An interesting application of free urinary cortisol excretion measurements is in the detection of nocturnal hypoglycaemia in insulin-treated diabetics in whom the morning urinary cortisol:creatinine ratio is markedly increased.

SALIVARY STEROIDS

Steroid hormone concentrations in saliva are low compared with those in plasma and their measurement, therefore, requires the use of the highly sensitive methods now available. With these it is possible simply and conveniently to measure cortisol—and other neutral steroids—in small samples of mixed saliva. In addition to the ease with which they can be collected, salivary samples have the advantage that their steroid content accurately reflects the prevailing 'free' (unbound) rather than 'total' plasma hormone level.[37,43,48] The mean resting salivary cortisol level in normal healthy volunteers undergoing a Synacthen[R] test, for example, was only 9 nmol/l (0·3 μg/100 ml), compared with 450 nmol/l (16·4 μg/100 ml), for plasma and showed an eight-fold rather than 3-fold maximum rise 1 h after the injection. This does mean, therefore, that there is not a linear, or any other simple relationship between plasma and salivary cortisol concentrations and that totally new reference ranges will be required before salivary can be substituted for plasma cortisol measurements in clinical diagnostic practice.

Salivary measurements are likely to be particularly valuable in infants and young children and others in whom frequent sampling is required. In addition to cortisol, methods have been developed for measuring 17α-hydroxyprogesterone, progesterone, testosterone, prednisolone and many other steroids in saliva. Serial 17α-hydroxyprogesterone measurements are especially useful in monitoring the response to treatment in children with congenital adrenogenital syndrome.

PLASMA ACTH

ACTH plays a key role in the regulation of adrenocortical function and although it is known to have profound effects upon other tissues such as fat and skin when administered in pharmacological doses, it is doubtful whether this has any physiological significance.

ACTH can be measured in plasma either by radioimmunoassay or by a quantitative cytochemical technique, although the latter is too technically demanding and time-consuming to be used except for research where its greater sensitivity than ordinary immunoassay, makes it especially useful.

Immunoassay of ACTH is complicated and difficult; this is due, in part, to lability of ACTH in plasma and in part to inherent problems of immuno-crossreactivity. Strict attention to detail is essential. ACTH is, for example, rapidly absorbed onto glass and consequently plasma samples for assay must be collected in plastic syringes and containers. Moreover, in order to avoid proteolytic degradation the plasma must be separated from the red cells and frozen to −20°C, or below, within minutes of collection and kept frozen until assay in the laboratory is commenced. β-LPH is more stable in plasma than ACTH, and since it is secreted in equimolar amounts, its measurement may provide a useful and robust alternative to ACTH assay.

Some ACTH antisera used for immunoassays are sensitive to the presence of even small amounts of heparin in the plasma; other anticoagulants, e.g. EDTA, must be used for these assays.

Plasma ACTH measurements have only a very limited clinical application being virtually confined to the *differential* diagnosis of Cushing's syndrome and/or adrenogenital virilism and for monitoring the response to treatment. They are useful in helping to differentiate between primary and secondary adrenocortical insufficiency though this can usually more easily more easily be done by observing the plasma cortisol response to exogenous ACTH.

ACTH secretion, and consequently its concentration in plasma, is even more subject to diurnal variation than is that of cortisol. Peak plasma levels occur early in the morning and nadir levels at about midnight, but as with cortisol—though to an even greater extent—marked fluctuations, both in secretion and plasma levels, occur on a minute to minute basis.

Because of the extreme sensitivity of the hypothalamic-pituitary ACTH secretory mechanism to activation by painful and other physical and psychogenic stresses, blood samples for plasma ACTH assays must always be collected with the minimum of disturbance, preferably through an indwelling plastic venous catheter placed there thirty minutes or so earlier. Failing this, blood may be collected through a single venepuncture providing that it occasioned no pain during its execution.

Aldosterone, the main mineralocorticoid produced by the adrenal glands, together with renin and angiotensin, the chief regulators of its secretion, are considered elsewhere in this volume.

DYNAMIC TESTS OF HYPOTHALAMIC-PITUITARY-ADRENAL FUNCTION

Single measurements of hormones in blood are of limited diagnostic value in the investigation of hypothalamic-pituitary-adrenocortical (HPA) disorders as in most other endocrine diseases. The large diurnal fluctuations that normally occur in plasma ACTH, cortisol and adrenal androgen levels provide useful information about the functional integrity of the HPA axis providing that sampling is properly timed but this is often difficult and inconvenient. In order to overcome some of these difficulties and improve diagnostic efficiency, a number of dynamic function tests have been devised. The more important are described below.

ACTH Stimulation Tests

These were the first and are still amongst the most widely used dynamic tests of adrenocortical function. Used initially for investigating suspected hyper- as well as hypo-adrenocortical function they are used nowadays almost exclusively for the latter.[14,20,21,42]

In the original test described, natural (porcine) ACTH was given by intravenous infusion over an eight-hour period and the adrenocortical response assessed by measuring the increment in urinary 17-oxogenic steroid excretion. More modern versions of the test depend upon measuring the plasma cortisol response to the intramuscular injection of either 0.25 mg of a synthetic ACTH analogue, tetracosactrin (Synacthen[R]), or, more usefully, 1 mg of a long-acting depot preparation of it.

The short test (0.25 mg Synacthen[R]) can easily be performed on outpatients and involves only the collection of blood samples before and thirty minutes after intramuscular (or intravenous) injection and assaying them for cortisol. A normal response, and one that effectively excludes a diagnosis of adrenocortical insufficiency, is an increase in plasma cortisol concentration of more than 200 nmol/l and a peak level, at 30 minutes, of at least 550 nmol/l. Abnormal results occur in a number of conditions apart from adrenocortical insufficiency, although confusion is unlikely to occur providing attention is paid to both the basal and stimulated plasma cortisol levels. An impaired response to the short Synacthen[R] test is extremely common in secondary adrenocortical insufficiency.

The five-hour Synacthen[R] test is performed by measuring plasma cortisol before and at 30, 60 and 300 minutes following the intramuscular injection of 1 mg of long-acting depot Synacthen[R], and provides better discrimination between adrenocortical insufficiency and normal function using the same diagnostic criteria as for the short test. It may still fail to distinguish between primary and secondary adrenocortical failure, however, although this can generally be achieved by repeating the intramuscular injection on four consecutive mornings and comparing the plasma cortisol responses to the first and fourth injections. Patients with secondary insufficiency generally show a markedly greater plasma cortisol response after the fourth compared with the first injection, whereas those with Addison's disease show *no* change or even a deterioration. The distinction between the two varieties of adrenocortical failure can now be made equally accurately and more conveniently, where the facilities exist, by measuring the plasma ACTH concentration at 9 a.m. Patients with primary adrenocortical insufficiency have high plasma ACTH levels and those with secondary failure low or low-normal ACTH levels.

Insulin Hypoglycaemia

All forms of physical and mental stress appear to be capable of stimulating hypothalamic-pituitary-adrenal activity but insulin-induced hypoglycaemia is a particularly effective stimulus and, in most cases, is associated with relatively little discomfort.[42]

The test is usually commenced in the morning after an overnight fast but this is not essential. In most normal-weight subjects it is usually sufficient to give soluble insulin 0.1 U/kg body weight by rapid intravenous injection through an indwelling venous catheter. Blood is collected 10 minutes and 1 minute before the insulin injection and at 20, 30, 60, 90 and 120 minutes afterwards. It is advantageous, if possible, to measure glucose at the bedside during the course of the test so that should neither neuroglycopenia nor hypoglycaemia, i.e. plasma glucose concentration of less than 2.5 mmol/l (4.5 mg/100 ml), appear within 1 h of the initial injection, a further dose of insulin, 0.1–0.2 U/kg depending upon circumstances, can be given. Insulin stress tests should not be performed on patients who are hypoglycaemic before the test commences or on those with overt Addison's disease or hypopituitarism unless special precautions are taken to prevent intractable hypoglycaemia developing. They should be avoided altogether in those with epilepsy or coronary heart disease.

The test, though generally safe, is not totally without danger and it is advisable always to have intravenous glucose (50 ml of 50% V/V glucose in water) and hydrocortisone (100 mg) available for terminating the test should neuroglycopenic symptoms become intolerable or coma supervene; the appearance of mild symptoms, including somnolence, are not a sufficient indication for terminating the test.

Insulin stress tests are often carried out as part of a more general investigation of hypothalamic-pituitary function and can be usefully combined with the measurement of plasma (or urinary) catecholamines and other pituitary hormones, especially if LHRH and TRH are injected simultaneously with insulin.

A normal adrenocortical response to insulin-induced hypoglycaemia is one in which the plasma cortisol concentration rises by at least 200 nmol/l (7.3 μg/100 ml) above baseline to reach a level of 560 nmol/l (20.4 μg/100 ml) or more. Abnormal results occur in a number of disorders including all of those in which there is primary (or secondary) adrenocortical insufficiency. These should ordinarily have been excluded before the test is embarked upon since failure of plasma cortisol to rise under these circumstances provides no useful information about the hypothalamic-pituitary axis. This information can, however, be obtained if plasma ACTH rather than cortisol is measured.

Patients with Cushing's syndrome, whatever its cause, fail to show a rise in plasma cortisol (or ACTH) in response to insulin-induced hypoglycaemia and this can be useful in the differential diagnosis of hypercortisolaemia.

Other Stress Tests

It was at one time not unusual to use either pyrogen or vasopressin to produce physiological stress during which plasma cortisol and/or ACTH could be measured as a test of hypothalamic-pituitary responsiveness.[42] Neither procedure has any obvious advantage over the insulin hypoglycaemia test which is safer and much less unpleasant. Their use has been virtually abandoned except rarely for research purposes.

Metyrapone Test

Metyrapone (Metopyrone[R]) is one of the less toxic agents to have been investigated that is capable of blocking cortisol production by the adrenal cortex. It does so by more or less selectively inhibiting 11-hydroxylation, the final step in the biosynthetic pathway. A consequence of the inhibition is a fall in plasma cortisol concentration which in turn promotes ACTH secretion by activating the negative feedback control mechanism. The rise in plasma ACTH level stimulates steroid biosynthesis with build-up and overflow, into the circulation, of the cortisol precursor 11-deoxycortisol (compound S) which can then be either measured directly in plasma by radioimmunoassay or HPLC, or indirectly, in urine, as 17-oxogenic steroids. 11-deoxycortisol behaves like cortisol in competitive protein binding assays for 'cortisol' that utilise transcortin as the binding agent and in many radioimmunoassays utilising relatively non-specific antisera. It is not measured, however, by assays that depend upon formation of fluorogenic compounds in acid conditions or the better immunoassays. The main use of the metyrapone test is in the investigation of disorders of hypothalamic-pituitary feedback control of adrenocortical activity, which is deranged in some cases of secondary adrenocortical insufficiency, and in the differential diagnosis of Cushing's syndrome.

There are several variants of the metyrapone test[10,18,26,27] depending on whether steroid measurements are made on urine or plasma. Urine tests take longer but are more readily available. Complete 24 h urine collections are made over 4 days—the first two serving as the controls. After completion of the second 24 h urine collection metyrapone, 500–750 mg, is given by mouth *at 4 hourly intervals* for six doses. Urine is collected during and after the administration. All 4 urine samples are measured for their 17-oxogenic steroid content. In normal subjects there is an increase in urinary 17-oxogenic steroid excretion of at least 36 μmol (11 mg) per 24 h on one or both of the test days. Providing primary or secondary adrenocortical failure has been excluded, absence of a rise indicates either an impairment of hypothalamic-pituitary activation by negative feedback control or an inability to increase ACTH secretion due to destruction of the anterior pituitary gland.

A short metyrapone test can be carried out by giving a single oral dose of metyrapone (30 mg/kg body weight) at midnight and collecting blood for cortisol, 11-deoxycortisol and ACTH measurements the following morning at 8–9 a.m. using blood collected the previous morning to provide the basal values used for comparison.[10] Although it has so far been evaluated and proved useful only in the diagnosis and differential diagnosis of primary and secondary adrenocortical insufficiency (due to hypopituitarism and/or long-term glucocorticoid therapy) the short metyrapone test is probably as effective as the standard procedure for differentiating the various kinds of Cushing's syndrome should this still be considered necessary after the more conventional dexamethasone suppression tests have been carried out.

Neither type of metyrapone test has much of a role to play in the clinical investigation of patients with hypothalamic-pituitary-adrenocortical disease, now that plasma hormone assays are readily available. Nevertheless, they are occasionally useful in helping to elucidate the nature of adrenocortical dysfunction in difficult cases especially in patients with iatrogenic adrenocortical insufficiency. Drugs, e.g. phenytoin, that accelerate removal of metyrapone from the plasma by increasing its metabolism in the liver or which have a direct effect upon HPA activity, e.g. exogenous steroids and oestrogens, may invalidate results obtained with metyrapone tests.

Steroid Suppression Tests

The administration of potent glucocorticoids to healthy subjects rapidly leads to suppression of endogenous ACTH secretion which is reflected by a fall in plasma ACTH level and the virtual cessation of cortisol secretion. Dexamethasone is the glucocorticoid most commonly used for this purpose since neither it nor any of its metabolites interferes with the measurement of endogenous adrenocorticoids in urine or plasma.[23,28]

The short dexamethasone suppression test is carried out on outpatients as a screening procedure for Cushing's syndrome. The patient is instructed to take 2 mg of dexamethasone as near to midnight as possible and to attend the laboratory next morning between 8–9 a.m. for blood collection. In normal subjects and those with a hypofunctional HPA axis the plasma cortisol level is depressed by dexamethasone to below 180 nmol/l whereas in patients with Cushing's syndrome and with some types of endogenous depression[9,39] there is little, if any, depression of early morning plasma cortisol levels with this dose of dexamethasone and they generally exceed 300 nmol/l (10·9 μg/100 ml).

The short dexamethasone test is particularly useful in determining the origin of adrenogenital virilism since patients with a congenital defect in adrenocortical biosynthesis normally show virtually complete suppression of plasma 17α-hydroxyprogesterone levels the morning after dexamethasone administration, whereas patients with adrenal tumours show no change in their plasma steroid levels.

If the screening test is positive or clinical suspicion is high, a prolonged dexamethasone test is performed. In the prolonged test, dexamethasone is given six hourly for four days at a dose of 2 mg per day in divided doses for the first two days and 8 mg per day for the second two days. 24-hour urine collections are made on the two days preceding dexamethasone administration and during it. Normal subjects show marked suppression of endogenous adrenocortical steroid production from the very beginning as reflected by a fall in plasma cortisol to below 180 nmol/l (6·5 µg/100 ml), or of urinary 17-oxogenic steroid excretion to below 140 µmol/24 h (5·0 mg/24 h), during the whole period of dexamethasone administration. Patients with Cushing's syndrome due to 'hypothalamic' dysfunction show suppression of adrenocortical function during the second, but not the first half, of the test, while those in whom it is due to an adrenal tumour or secondary to excessive ACTH secretion from a pituitary or ectopic neoplasm usually respond to neither dose of dexamethasone.

False positive results, i.e. failure of cortisol production to suppress, can occur in patients receiving treatment with enzyme-inducing drugs which accelerate dexamethasone metabolism—but this can be circumvented either by using glucocorticoids whose metabolism is not increased by this treatment or by increasing the frequency and dosage of dexamethasone employed.

Dexamethasone suppression tests are less often required now than formerly for diagnosis of adrenocortical disease since plasma ACTH measurements can usually supply the relevant information. They are, however, undergoing extensive evaluation as a means of differentiating between two main varieties of endogenous depression in one of which adrenocortical suppression is normal and in one of which it is not.

CLINICAL INVESTIGATION OF ADRENOCORTICAL DYSFUNCTION

Adrenocortical Insufficiency

Addison's Disease

The commonest cause of primary adrenocortical insufficiency, or Addison's disease, in Britain and most other developed countries, is autoimmune adrenocortical atrophy. Less commonly it is due to other causes of adrenocortical destruction. Other endocrine glands, including the thyroid, ovary and pancreatic islets, may be involved in patients with more generalised autoimmune disease, but in most cases of Addison's disease the adrenal glands alone are affected. In the autoimmune variety of Addison's disease the adrenal medullae are spared and their ability to respond by secreting adrenaline to insulin-induced hypoglycaemia, or to neuroglycopenia induced by 2-deoxyglucose, can be used to distinguish it from the pan-adrenal destruction produced by tuberculosis or adrenal metastases. Autoantibodies to the adrenal cortex can usually be demonstrated by immunochemical methods in the serum of patients with autoimmune Addison's disease but not in the other varieties.

Clinical Presentation and Chemical Pathology

Addison's disease, first recognized almost a century ago, is due to combined deficiency of cortisol and aldosterone and is comparatively rare. Its onset is insidious and in the earliest stages the signs and symptoms are mainly due to cortisol deficiency. They include weakness and easy fatiguability, weight loss, vomiting, diarrhoea and abdominal pain. Pigmentation, which is an early sign of Addison's disease, is due to the melantrophic effects of ACTH whose concentration in plasma is often raised long before overt evidence of cortisol deficiency appears. Amenorrhoea is extremely common and its combination with anorexia, which is nearly always present, may lead to erroneous suspicion of anorexia nervosa. Later, evidence of hypotension and more profound muscular weakness appear. Although the onset of the disease is insidious, the clinical presentation is frequently dramatic. Sometimes patients present in coma due to hypoglycaemia or to collapse from hypovolaemia and/or acute dehydration secondary to sodium and water loss caused by vomiting and/or diarrhoea and the inability to conserve sodium. The loss of adrenal androgen production is associated with few clinical effects except a slight reduction of sexual hair in women which is, however, seldom of sufficient severity to attract attention unless specifically sought by direct questioning. The presence of pigmentation (due to excess ACTH) implies adrenal rather than pituitary failure.

Typically the patient with Addison's disease has a low plasma sodium and high plasma potassium concentrations, the latter being the less common but more characteristic feature. In early cases neither plasma electrolyte level is abnormal.

Other biochemical abnormalities commonly present, especially in the more advanced cases, include hypoglycaemia, azotaemia and hypercreatininaemia due to a decrease in glomerular filtration rate. Urinary sodium excretion is inappropriately high in the presence of hyponatraemia and there is a profound inability to excrete a water load. These two abnormalities were, before the availability of steroid measurements, often used as the basis of a diagnostic test of adrenocortical insufficiency. Nowadays diagnosis can be made earlier, more precisely, and more safely by measuring the adrenocortical response to exogenous ACTH. Total urinary steroid excretion is reduced in patients with Addison's disease, but its measurement is a poor indicator of adrenocortical activity and should be abandoned. Measurement of midnight plasma cortisol levels has no place in the investigation of patients with suspected Addison's disease.

Patients in whom Addison's disease is suspected should be given a short Synacthen[R] test as a screening procedure. A normal plasma cortisol response, i.e. a rise of at least 200 nmol/l (7·3 µg/100 ml) to a level of not less than 550 nmol/l (20 µg/100 ml), virtually eliminates a diagnosis of primary adrenocortical deficiency (i.e. Addison's disease) but is not incompatible with one of secondary adrenocortical insufficiency due to hypopituitarism or other causes of reduced ACTH secretory capacity. Treatment with intravenous saline should not be withheld while this test is carried out and patients who have already begun treatment with steroid replacement therapy can be tested perfectly well, without interrupting therapy. This should, however, be changed—for the duration of the test period—to dexamethasone 0·5 mg and fludrocortisone 0·1 mg every 8 hours.

An abnormal cortisol response to Synacthen strongly supports a tentative diagnosis of adrenocortical deficiency but may not distinguish fully between primary and secondary disease. This is most easily achieved either by measuring the early morning plasma ACTH level or by observing the plasma cortisol response to prolonged (3 days) Synacthen stimulation. In patients with primary adrenocortical insufficiency (Addison's disease) the early morning plasma ACTH level is high and there is seldom a rise of more than 100 nmol/l (3·6 μg/100 ml) in plasma cortisol in response to prolonged Synacthen stimulation. Patients in whom adrenocortical insufficiency is secondary to hypopituitarism or to hypothalamic malfunction—which is itself usually secondary to prolonged anti-inflammatory steroid therapy—have low early-morning plasma ACTH levels and usually exhibit a considerable, i.e. over 200 nmol/l (7·3 μg/100 ml), rise in plasma cortisol after the fourth long-acting synacthen injection.

A clinical diagnosis of adrenocortical deficiency secondary to defective ACTH secretion may be difficult to establish without recourse to insulin tolerance or metyrapone tests. The former is more reliable but must be carried out with caution since patients with hypopituitarism may be unduly sensitive to the hypoglycaemic effects of insulin, especially when there is an associated growth hormone deficiency.

Patients with either selective or non-selective ACTH deficiency fail to show the expected rise in plasma ACTH or cortisol levels. In those in whom the defect is due to hypothalamic rather than anterior pituitary dysfunction, an ACTH response can sometimes be provoked by the infusion of AVP, but this is often unpleasant and rarely necessary since treatment of the two disorders is similar.

ACTH unresponsiveness

This extremely rare disorder is caused by a lack of end organ (i.e. adrenal cortex) response to ACTH.[19] It manifests itself as adrenocortical insufficiency but usually only in response to stress when hypoglycaemia becomes the dominant feature. It is associated with various abnormalities including hyperpigmentation due to high circulating ACTH levels. It must, in the past, have been confused with Addison's disease from which it can, however, be distinguished both on clinical and biochemical grounds since aldosterone secretion is more or less normal in ACTH unresponsiveness but not in primary adrenocortical insufficiency.

Cushing's Syndrome

Cushing's syndrome[2,6,28,45] is invariably a consequence of increase glucocorticoid activity. The commonest cause, in practice, is overtreatment with anti-inflammatory steroids but it also occurs spontaneously due to adrenocortical overactivity secondary to increased hypothalamic-pituitary drive, overproduction of ACTH by a pituitary tumour or an ectopic ACTH-secreting (malignant) neoplasm, or to overproduction of cortisol and other adrenocortical steroids by an autonomous adrenal adenoma or carcinoma. It may also be factitious or a consequence of chronic alcohol abuse (Table 3). Pituitary-dependent Cushing's syndrome—the term used to describe functional overactivity of hypothalamic-pituitary drive and/or oversecretion of ACTH by a pituitary microadenoma—is four times as common in women as in men and has a peak incidence between 30–50 years.

Ectopic ACTH production is commoner in men than in women and is usually due to a malignant tumour composed mainly of APUD cells, most commonly an

TABLE 3
LABORATORY AIDS TO DIAGNOSIS OF CUSHING'S SYNDROME

	Pituitary dependent	Ectopic ACTH	Chronic alcoholism	Adrenal tumour
1. Sex	Usually female	Usually male	Male or female	Male or female
2. Appearance	'Cushingoid'	Often not typically 'Cushingoid'	'Cushingoid'	Mixed Cushingoid /virilism
3. Rapidity of development	Slow	Rapid	Slow	Rapid
4. Hypokalaemia	No	Yes	No	No
5. Cortisoluria	++	++++	±	++
6. Hypercortisolaemia	+	+++	+	+
7. Plasma ACTH	++	+++	±	−
8. Loss of diurnal rhythm	Yes	Yes	Yes/No	Yes
9. Plasma cortisol response to insulin	Absent	Absent	Absent	Absent
10. 11-deoxycortisol response to metyrapone	Exaggerated	Absent		Absent
11. Unusual steroids in urine	Uncommon	Uncommon	Uncommon	Uncommon
12. Dexamethasone suppression				
(i) 2 mg/day	Absent	Absent	Absent	Absent
(ii) 8 mg/day	Present	Absent	Present	Absent
13. Adrenal CAT scan	Bilateral hyperplasia			Unilateral tumour

oat-cell carcinoma of the bronchus. Marked hypokalaemic alkalosis, often without overt somatic manifestations of adrenocortical activity, is a common characteristic of this type of Cushing's syndrome. Non-iatrogenic Cushing's syndrome in children is most often due to an adrenal tumour. Although there are differences in the incidence, nature and severity of signs and symptoms in the various 'types' of Cushing's syndrome these cannot be used to determine the cause in any individual case except in a most general way.

Clinical Presentation and Chemical Pathology

Onset of Cushing's syndrome is generally slow and patients often complain of weakness and tiredness long before the more characteristic features of the disease have developed. Obesity and hirsutism, though common, are seldom gross and hypertension, though almost invariable, is generally modest in severity. Apart from abnormalities of adrenal steroid production there are no simple biochemical or haematological changes that are sufficiently characteristic of Cushing's syndrome to be diagnostically useful. Impairment of glucose tolerance is common but usually mild, manifesting itself as diabetes in less than 25% of cases. The diagnosis of Cushing's syndrome, once suspected on clinical grounds, is provisionally confirmed by demonstrating increased urinary excretion of cortisol (by immunoassay) in a 24 h specimen or relative to creatinine in a random overnight one.[8] Alternatively, a low dose dexamethasone suppression test can be performed (Fig. 2). Normal suppression of endogenous cortisol secretion is strong evidence against the diagnosis, especially if confirmed on one or more occasions. Failure of a single dose of dexamethasone to suppress endogenous cortisol production is an invitation to further investigation. This can usually be done, on an out-patient basis, by measuring the 24 h urinary free-cortisol excretion and plasma cortisol levels during prolonged (2 mg/day) dexamethasone administration and/or insulin-induced hypoglycaemia. Patients who are able to suppress their cortisol production in response to low-level dexamethasone administration or who exhibit a normal plasma cortisol response to hypoglycaemia are extremely unlikely to be suffering from Cushing's syndrome—whereas those who do not suppress adequately probably are, and require further investigation to ascertain its cause. Patients with Cushing's syndrome generally show a disturbed diurnal rhythm of cortisol secretion but so, too, do patients seriously ill from other diseases. An abnormally high midnight cortisol level, i.e. over 400 nmol/l (14·5 μg/100 ml) is not, therefore, diagnostic of Cushing's syndrome though a level below this renders the diagnosis improbable.

Urinary 17-oxogenic steroid excretion is often, but not invariably, raised in patients with Cushing's syndrome but is a poor discriminator as it is increased in many other conditions including simple obesity. 17-oxosteroids

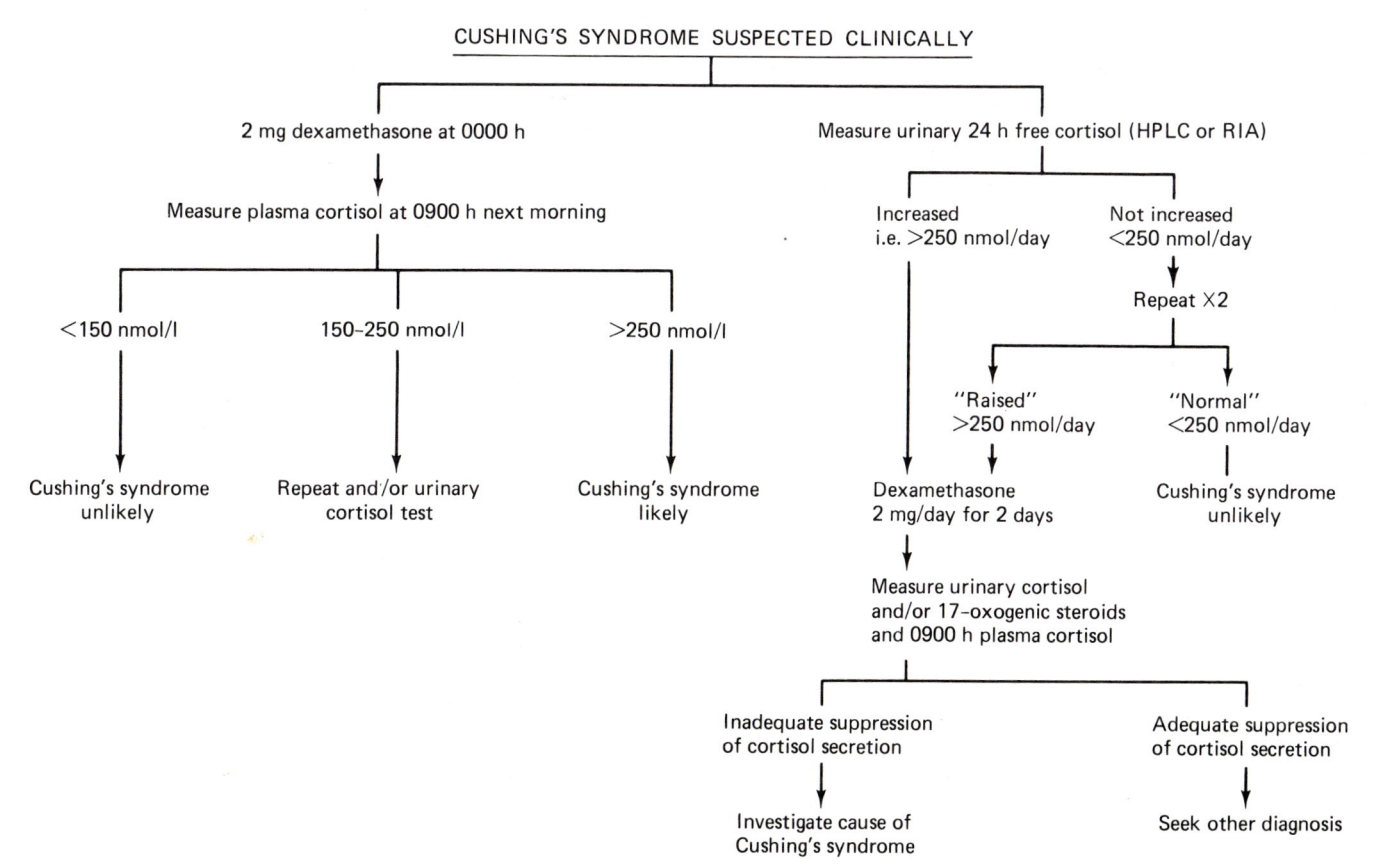

FIG. 2. A flow chart for the laboratory investigation of suspected Cushing's syndrome.

provide no useful information in patients suspected of suffering from Cushing's syndrome.

Plasma ACTH levels are immeasurably low in patients with either factitious or iatrogenic Cushing's syndrome or that due to adrenal tumour; they are 'normal' or, more often, moderately elevated in patients with the pituitary-dependent variety and high or very high in those with ectopic ACTH-secreting tumours. The metyrapone test produces marked increases in plasma ACTH and 11-deoxycortisol levels in patients with pituitary-dependent Cushing's syndrome whilst those suffering from other varieties show no rise in plasma ACTH and only a modest rise in plasma 11-deoxycortisol. This occurs entirely at the expense of cortisol whose concentration in plasma falls, sometimes to dangerously low levels making this test potentially hazardous in patients with tumour-induced Cushing's syndrome.

Further investigations, aimed at localizing the lesion include radiology of the head, chest and other parts of the body, CAT-scan of the adrenal glands and if uncertainty still persists, percutaneous venous catheterisation of the adrenal veins and quantitation of the effluent for cortisol content.

Nelson's Syndrome

Development of intense skin pigmentation and evidence of an expanding intrasellar tumour following adrenalectomy for Cushing's syndrome was first described by Nelson, whose name was applied eponymously to describe the condition by Liddle a few years later.[28] The pigmentation is a consequence of the melanotropic effect of ACTH whose concentration in the plasma is extremely high and nearly always non-suppressible even by high doses of exogenous glucocorticoids. The pituitary lesion is usually—if not invariably—neoplastic and, in a fair proportion of cases is locally invasive, i.e. malignant. It is likely that the syndrome develops, in the majority of cases, as a consequence of interruption of the negative feedback loop which operates, even in patients with Cushing's syndrome, to lessen hypothalamic drive on the ACTH-secreting cells of the anterior pituitary. It is, therefore, a wise precaution to monitor patients treated by adrenalectomy for Cushing's syndrome by measuring their plasma ACTH or β-LPH levels at regular intervals in order to detect, at the earliest possible moment, any increase in pituitary activity which is uncontrolled by steroid replacement therapy.

Hypertension

Adrenocortical dysfunction is an unusual but important, because it is potentially curable, cause of hypertension. In many such cases there is both clinical and biochemical evidence of Cushing's syndrome providing it is sought but in others there are few, if any, pointers to an adrenocortical origin apart from hypokalaemia. Whilst most of the patients falling into the latter category will eventually be shown to have some form of hyperaldosteronism (Conn's syndrome), in a significant minority[4,44] it is deoxycorticosterone (DOC) rather than aldosterone secretion that appears to be primarily at fault. In these patients plasma DOC levels are abnormally high on some occasions, though not on all, and urinary aldosterone excretion is normal. Differentiation from Conn's syndrome may be difficult without recourse to plasma DOC and aldosterone measurements.

Adrenogenital Virilism

Adrenogenital virilism is caused by overproduction of one or more adrenal androgens due to a block in cortisol synthesis (Fig. 1) either as a result of an inherited enzyme defect or an acquired abnormality in an adrenocortical neoplasm.

Congenital Adrenal Hyperplasia (CAH)

There are a number of hereditary disorders of adrenocortical hormone synthesis which are associated with adrenal hyperplasia and produce adrenogenital virilism as their main, if not exclusive, clinical abnormality. Virilization in these syndromes occurs as a result of overproduction of one or more of the weakly androgenic steroids, e.g. DHA, DHAS and androstenedione, which are produced in small amounts even by normal adrenals. All forms of CAH are rare, the commonest being caused by a partial defect in 21-hydroxylation and inherited as an autosomal recessive trait. Other variants, in order of decreasing frequency, are due to greater or less degrees of deficiency in 11β-hydroxylase and 3β-hydroxydehydrogenase activities. The latter disorder produces only minimal virilisation, but is associated with an extremely high mortality rate, as is an even rarer variety of adrenocortical malfunction caused by 20–22 desmolase deficiency. In this condition, not only does virilism *not* occur, but the infant is always of the female phenotype, regardless of its genetic sex due to a failure of androgen synthesis by the gonads as well as by the adrenals. In yet another variety of CAH, that due to 17α-hydroxylase deficiency, boy children are feminised because they cannot make androgens and the girls appear phenotypically normal. Hypertension is common in both sexes.

Salt loss is common in severely affected infants with 21-hydroxylase deficiency and invariably in those with reduced 3β-hydroxydehydrogenase and desmolase activity, and is due to aldosterone deficiency. Patients with 11β-hydroxylase or 17α-hydroxylase deficiency, on the other hand, generally suffer from salt retention due to an increased production of deoxycorticosterone. They are usually hypertensive for the same reason.

The hallmark of all varieties of CAH is a grossly elevated plasma ACTH level in the presence of increased but qualitatively abnormal adrenal steroid production. Diagnosis of adrenogenital virilism due to 21-hydroxylase deficiency is confirmed by demonstrating grossly increased plasma or salivary concentration of 17α-hydroxyprogesterone;[33] the concentration of the adrenal androgens, DHAS, and androstenedione is also increased but is not as diagnostically useful. Both urinary

17-oxo- and 17-oxogenic steroids are increased in amount and pregnanetriol, the main metabolite of 17α-hydroxyprogesterone is excreted in excessive quantities.

Plasma ACTH levels are elevated in all infants with congenital adrenogenital virilism and they, like the adrenal steroid abnormalities, return to normal with cortisol (or prednisolone) replacement therapy.[51] Plasma ACTH, 17α-hydroxyprogesterone or androstendione levels can all be used clinically as a guide to the efficacy of treatment which can often otherwise be difficult to assess until irreversible damage has taken place. Saliva can also be used for monitoring 17α-HP.[49]

Infants with adrenogenital virilism due to defective 11-hydroxylation do *not* invariably show an increase in plasma 17α-hydroxyprogesterone levels, nor do they occur in the three rarer forms of CAH, i.e. those due to desmolase, 3β-hydroxylase and 17α-hydroxylase deficiency. It cannot, therefore, be used as a general screening test for congenital adrenogenital virilism. Since affected children do generally show increased excretion of urinary 17-oxosteroids there is some argument for retaining this non-specific indicator of adrenocortical dysfunction in the laboratory repertoire, if only as a screening test for CAH associated with adrenogenital virilism.

Another test that still enjoys considerable popularity particularly with the advent of GLC or HPLC for measuring urine steroids is calculation of the urinary oxygenation-index which reflects the relative efficiency of the 11β- and 21-hydroxylase systems. It has the avantage[46] that it can be performed on a random sample of urine—providing the baby is more than 8 days old. Normally the ratio of 11-deoxy to 11-oxysteroids is less than 0.5 but in patients with the two commonest forms of adrenogenital virilism, i.e. 21-hydroxylase deficiency and 11β-hydroxylase deficiency the figure is markedly increased.

Virilism Due to Adrenal Tumours

Androgen-secreting adrenal tumours occur in both sexes but are usually more easily recognized in women than in men because of the associated hirsutism, amenorrhoea and virilism. Tumours may be benign or malignant and occur at any age from childhood to senility. Distinction from the commoner ovarian virilising tumours may be difficult but the diagnosis can usually be established by (i) demonstrating greatly increased amounts of the adrenal androgens, DHAS and androstenedione in the plasma, and (ii) their metabolites (measured as 17-oxosteroids) in the urine; (iii) their failure to suppress in response to dexamethasone. In contrast to what is observed in congenital adrenal hyperplasia, plasma ACTH levels are either normal or, more commonly, low.

Adrenocortical tumours often secrete increased amounts of glucocorticoids and their precursors in addition to androgens so that mixed clinical syndromes are common. Differentiation of benign from malignant tumours may be extremely difficult, if not impossible, prior to operation although malignant tumours tend to be more rapid in onset and to secrete larger quantities of unusual steroids than benign ones.

Feminising Adrenal Tumours

A small proportion of both malignant and benign adrenal tumours manifest themselves in men by feminization and in women by interference with cyclic menstrual function. These patients generally have elevated levels of plasma and urinary oestrogens—which can be identified and measured by HPLC and/or immunoassay—as well as of more conventional steroids of adrenocortical origin.

Idiopathic Hirsutism

There is accumulating evidence that at least some cases of idiopathic hirsutism are related to, if not caused by, increased adrenal androgen secretion which is under hypothalamic-pituitary control.[11,30] Plasma DHAS, androstenedione and testosterone levels may all be marginally elevated compared with reference values obtained from non-hirsute women of comparable age. However, the values are generally considerably lower than those seen with virilising tumours and suppress normally in response to dexamethasone. Many of these patients have the clinical features of polycystic ovary syndrome and differentiation may be difficult, if not impossible, by techniques currently available, if indeed there is one since in many cases there is evidence of abnormal steroid production by both the adrenal glands and ovaries. Undoubtedly a tiny percentage of affected subjects have a minimal form of congenital adrenogenital virilism (possibly heterozygotes) and might, therefore, be expected to benefit from glucocorticoid suppression therapy. In the majority of cases, however, no intrinsic biochemical defect can be found to account for the slightly abnormal clinical biochemical findings and glucocorticoid treatment produces dubious benefit.

REFERENCES

1. Anderson, D. C. (1980), 'The adrenal androgen-stimulating hormone does not exist', *Lancet*, **ii**, 453.
2. Aron, D. C., Tyrrell, J. B., Fitzgerald, P. A., Findling, J. W. and Forsham, P. H. (1981), Cushing's syndrome: problems in diagnosis. *Medicine*, **60**, 25-35.
3. Bondy, P. K. (1980), in *Metabolic Control and Disease*, pp. 1427–1499. (Bondy, P. K. and Rosenberg, L. E., Eds.). Philadelphia, Saunders.
4. Brown, J. J., Fraser, R., Love, D. R., Ferriss, J. B., Lever, A. F., Robertson, J. I. S. and Wilson, A. (1972), 'Apparently isolated excess deoxycorticosterone in hypertension', *Lancet*, **ii**, 243–247.
5. Brooks, R. V. (1979), in *The Adrenal Gland*, pp. 67–92. (James, V. H. T., Ed.). New York, Raven Press.
6. Burke, C. W. (1978), in *Recent Advances in Endocrinology and Metabolism*, pp. 61–90. (O'Riordan, J. L. H., Ed.). Edinburgh, Churchill Livingstone.
7. Byyny, R. L. (1976), 'Withdrawal from glucocorticoid therapy', *N. Engl. J. Med.*, **295**, 3–32.

8. Chambers, R. E., Willetts, S. R. and Walters, G. (1980), 'A comparison of urinary free cortisol (RIA) with cortisol metabolites (GLC) for detecting increased cortisol secretion'. *Ann. Clin. Biochem.*, **17**, 233–236.
9. Coppen, A. (1970), in *Progress in Brain Research*, Vol. 32, pp. 336–341. (De Wied, D. and Weijnen, J. A. W. M., Eds.). Amsterdam-London-New York, Elsevier Publishing Company.
10. Dolman, L. I., Nolan, G. and Jubiz, M. (1979), 'Metyrapone test with adrenocorticotrophic levels, separating primary from secondary adrenal insufficiency', *J. Am. Med. Ass.*, **241**, 1251–1253.
11. Ettinger, B., Goldfield, E. B., Burrill, K. C., von Werder, K. and Forsham, P. H. (1973), 'Plasma testosterone stimulation-suppression dynamics in hirsute women: correlation with long-term therapy, *Am. J. Med.*, **54**, 195–200.
12. Few, J. D. and Cashmore, G. C. (1980), 'Plasma cortisone in man: its determination, physiological variation, and significance', *Ann. Clin. Biochem.*, **17**, 227–232.
13. Glass, A. R. and Smith, C. E. (1979), 'Stress-induced cortisol release in hypothalamic hypoadrenalism', *J. Am. Med. Ass.*, **241**, 1612–1613.
14. Greig, W. R., Boyle, J. A., Maxwell, J. D., Lindsay, R. M. and Browning, M. C. K. (1969), 'Criteria for distinguishing normal from subnormal adrenocortical function using the Synacthen test', *Postgrad. med. J.*, **45**, 307–313.
15. Hindawi, R. K., Gaskell, S. J., Read, G. F. and Riad-Fahmy, D. (1980), 'A simple direct solid-phase enzymeimmunoassay for cortisol in plasma', *Ann. Clin. Biochem.*, **17**, 53–59.
16. James, V. H. T. (Ed.), (1979), *The Adrenal Gland*, New York, Raven Press.
17. Jeffcoate, W. J. and Edwards, C. R. W. (1979), in *The Adrenal Gland*, pp. 165–195. (James, V. H. T., Ed.). Raven Press, New York.
18. Jubiz, W., Matsukura, S., Meikle, A. W., Harada, G., West, C. D. and Tyler, F. H. (1970), 'Plasma metyrapone, adrenocorticotropic hormone, cortisol, and deoxycortisol levels: sequential changes during oral and intravenous metyrapone administration', *Arch. Intern. Med.*, **125**, 468–471.
19. Kershnar, A. K., Roe, T. F. and Kogut, M. D. (1972), 'Adrenocorticotropic hormone unresponsiveness: report of a girl with excessive growth and review of 16 reported cases, *J. Ped.*, **80**, 610–619.
20. Keylet, H. and Binder, C. (1973), 'Value of an ACTH test in assessing hypothalamic-pituitary-adrenocortical function in glucocorticoid-treated patients', *Br. med. J.*, **2**, 147–149.
21. Kuhl, W. J. and Lipton, M. A. (1960), 'The diagnosis of adrenocortical disorders by laboratory methods', *N. Engl. J. Med.*, **263**, 128–137.
22. Loriaux, D. L., Malinak, L. R. and Noall, M. W. (1970), 'Contribution of plasma dehydroepiandrosterone sulfate to testosterone in virilized patient with an arrhenoblastoma, *J. Clin. Endocrinol. Metab.*, **31**, 702–704.
23. McHardy-Young, S., Harris, P. W. R., Lessof, M. H. and Lyne, C. (1967), 'Single-dose dexamethasone suppression test for Cushing's syndrome', *Br. Med. J.*, **2**, 740–744.
24. Meikle, A. W., Takiguchi, H., Mizutani, S., Tyler, F. H. and West, C. D. (1969), 'Urinary cortisol excretion determined by competitive protein binding radioassay: A test of adrenal cortical function', *J. Lab. Clin. Med.*, **74**, 803–812.
25. Meikle, A. W. and Tyler, F. H. (1977), 'Potency and duration of action of glucocorticoids: effects of hydrocortisone, prednisone and dexamethasone on human pituitary-adrenal function', *Am. J. Med.*, **63**, 200–207.
26. Melby, J. C. (1971), 'Assessment of adrenocortical function', *N. Engl. J. Med.*, **285**, 735–739.
27. Nattrass, M., Wood, P. J., Smith, J. and Marks, V. (1972), 'Plasma-corticosteroid determinations for assessment of pituitary response to metyrapone', *Lancet*, **ii**, 903–904.
28. Nelson, D. H. (1980), *The Adrenal Cortex: Physiological Function and Disease*, Philadelphia, Saunders.
29. Neville, A. M. and O'Hare, M. J. (1979), in *The Adrenal Gland*, pp. 1–65 (James, V. H. T., Ed.). New York: Raven Press.
30. Oake, R. J., Davies, S. J., McLachlan, M. S. F. and Thomas, J. P. (1974), 'Plasma testosterone in adrenal and ovarian vein blood of hirsute women', *Q. J. Med.*, **43**, 603–613.
31. Pal, S. B. (1978), '6-Hydroxylation of cortisol and urinary 6β-hydroxycortisol', *Metabolism*, **27**, 1003–1011.
32. Parker, L. N., Chang, S. and Odell, W. D. (1978), 'Adrenal androgens in patients with chronic marked elevation of prolactin', *Clin. Endocrinol.*, **8**, 1–5.
33. Pham-Huu-Trung, M. T., Gourmelen, M. and Girard, F. (1973), 'The simultaneous assay of cortisol and 17α-hydroxyprogesterone in the plasma of patients with congenital adrenal hyperplasia', *Acta Endocrinol.*, **74**, 316–330.
34. Pratt, J. J. (1978), 'Steroid immunoassay in clinical chemistry', *Clin. Chem.*, **24**, 1869–1890.
35. Racadot, A., Racadot-Leroy, N., Le Gaillard, F. and Dautrevaux, M. (1975), 'Dosage de la transcortime serique par electroimmunodiffusion', *Clin. Chim. Acta*, **66**, 171–180.
36. Read, G. F., Fahmy, D. R. and Walker, R. F. (1977), 'Determination of cortisol in human plasma by radioimmunoassay: use of the ^{125}I-labelled radioligand', *Ann. Clin. Biochem.*, **14**, 343–349.
37. Riad-Fahmy, D., Read, G. F. and Walker, R. F. (1980), 'Salivary steroid assays for screening endocrine function, *Postgrad. Med. J.*, **56**, 75–78.
38. Rosner, W. (1969), 'Interaction of adrenal and gonadal steroids with proteins in human plasma', *N. Engl. J. Med.*, **281**, 658–665.
39. Schlesser, M. A., Winokur, G. and Sherman, B. M. (1979), 'Genetic subtypes of unipolar primary depressive illness distinguished by hypothalamic-pituitary-adrenal axis activity, *Lancet*, **i**, 739–741.
40. Scott, N. R. and Dixon, P. F. (1979), 'Determination of cortisol in human plasma by reversed-phase higher-performance liquid chromatography', *J. Chromat.*, **164**, 29–34.
41. Scott, N. R., Chakraborty, J. and Marks, V. (1980), 'Determination of prednisolone, prednisone, and cortisol in human plasma by high-performance liquid chromatography, *Anal. Biochem.*, **108**, 266–268.
42. Staub, J. J., Jenkins, J. S., Ratcliffe, J. G. and Landon, J. (1973), 'Comparison of corticotrophin and corticosteroid response to lysone vasopressin, insulin, and pyrogen in man', *Br. Med. J.*, **1**, 267–269.
43. Turkes, A. O., Turkes, A., Joyce, B. G. and Riad-Fahmy, D. (1980), 'A sensitive enzymeimmunoassay with a fluorimetric end-point for the determination of testosterone in female plasma and saliva', *Steroids*, **35**, 89–101.
44. Tan, S. Y., Noth, R. H. and Mulrow, P. J. (1978), 'Deoxycorticosterone and 17-ketosteroids: elevated levels in adult hypertensive patients', *J. Am. Med. Ass.*, **240**, 123–126.
45. Urbanic, R. C. and George, J. M. (1981), 'Cushing's disease—18 years' experience', *Medicine*, **60**, 14–24.
46. Varley, H., Gowenlock, A. H. and Bell, M. (1976), *Practical Clinical Biochemistry*, Vol. 2, Hormones, Vitamins, Drugs and Poisons, 5th edn., London, Heinemann.
47. Walker, M. S. (1977), 'Urinary free 11-hydroxycorticoid ratios in early morning urine samples as an index of adrenal function. *Ann. Clin. Biochem.*, **14**, 203–206.
48. Walker, R. F., Wilson, D. W., Read, G. F and Riad-Fahmy, D. (1980), 'Assessment of testicular function by the radioimmunoassay of testosterone in saliva', *Int. J. Androl.*, **3**, 105–120.
49. Walker, R. F., Read, G. F., Hughes, I. A. and Riad-Fahmy, D. (1979), 'Radioimmunoassay of 17α-hydroxyprogesterone in saliva, parotid fluid, and plasma of congenital adrenal hyperplasma patients', *Clin. Chem.*, **25**, 542–545.
50. Zadik, Z., De Lacerda, L., De Carmargo, L. A. H., Hamilton, B. P., Migeon, C. A. and Kowarski, A. A. (1980), 'Comparative study of urinary 17-hydroxycorticosteroids, urinary free cortisol, and the integrated concentration of plasma cortisol, *J. Clin. Endocrinol. Metab.*, **51**, 1099–1101.
51. Zipf, W. B., Bacon, G. E. and Kelch, R. P. (1980), 'Hormonal and clinical response to prednisone treatment in adolescents with congenital adrenal hyperplasia', *Horm. Res.*, **12**, 206–217.

40. THE GONADS

S. L. JEFFCOATE

Introduction

Normal gonadal function
    Biochemistry of gonadal steroids
    Ovarian function in the adult and its control
    Testicular function in the adult and its control

Principles of measurement of gonadal steroids
    Methods of steroid analysis

Disorders of gonadal function: role of biochemical tests
    The patient as a bioassay
    Disorders of gonadal development
    Disorders of puberty
    The menopause
    Disorders of fertility
    Hirsutism and virilisation
    Where to measure gonadal steroids
    When to measure gonadal steroids
    Which steroids to measure
    Problems of interpretation

INTRODUCTION

In both sexes, the gonads have two functions: the production of gametes and the production of steroid hormones. The purpose of these sex steroids is, broadly speaking, to ensure that the gametes come together at fertilization and that the fertilised ovum finds a suitable environment in which to grow. In both sexes steroidogenesis and gametogenesis are intimately related but in the testis, though not in the ovary, these two functions are anatomically separate.

The anterior pituitary controls both gonadal functions, principally through the secretion of the gonadotrophins, luteinizing hormone (LH) and follicle-stimulating hormone (FSH) though others, prolactin and growth hormone, also affect the gonads. The gonads in turn regulate the hypothalamus and pituitary by means of a complex network of 'feedback loops'. For this reason the gonads are never considered in isolation in clinical practice but as part of a hypothalamo-pituitary-gonadal 'axis', the function of each part of which is dependent on the other parts. Thus this chapter should be read in conjunction with Chapter 35. Furthermore, diseases of other endocrine glands, particularly the adrenal gland which secretes sex steroids in addition to corticosteroids, can affect gonadal function directly, alter the binding or activity of sex steroids, or be associated with clinical appearances similar to those of gonadal disorders. In the sections that follow, I will deal in turn with normal gonadal function and its control, the principles of measurement of gonadal sex steroids and the role of these biochemical tests in the management of disorders of gonadal function.

NORMAL GONADAL FUNCTION

Biochemistry of Gonadal Steroids

Structure and Nomenclature

An appreciation of the chemistry of steroid hormones is undoubtedly an acquired taste, and the presentation of what is disparagingly called 'hydroxylated chickenwire' can be guaranteed to alienate most audiences. This is a pity because a little effort spent in learning the significance of the structure of steroid hormones, into which category all the gonadal sex hormones fall, is repaid by a fuller understanding of the significance of the biochemical tests in health and disease.

The nomenclature of the ring structures is shown in Fig. 1. The basic rules to remember are that: the saturated 'parent' compounds are called '-anes'; compounds with double bonds in the skeleton are called '-enes' and their positions are indicated by the numbers of the lower of the two carbon atoms they join; the presence and position of the carbonyl and hydroxyl groups is also indicated. When multiple substituents are present the rules become more complex.

Reduction at positions 3 and 5 is involved in the metabolism of biologically active sex steroids and produces isomerism, illustrated in Fig. 2. Although the 3-carbonyl group is in the plane of the ring the resulting hydroxyl group can be either on the β-side (indicated by solid bonds ——) or α-side (indicated by broken bonds ----) of the molecule. Similarly reduction of the 4-ene produces either the 5+-ane or β-ane. These isomers have different conformations which are reflected in differences in biological activity. Hydroxyl groups at carbon 20 can also be α or β though in this case it does not refer to the position relative to the ring.

The trivial (i.e. commonly used) and systematic names of the major gonadal steroids and their metabolites are listed in Table 1.

TABLE 1
TRIVIAL NAMES OF THE GONADAL SEX STEROIDS AND SOME OF THEIR METABOLITES

Trivial name	Systematic name
Aetiocholanolone	3α-hydroxy-5β androstan-17-one
Androstenedione	androst-4-ene-3,17-dione
Androsterone	3α-hydroxy-5α androstan-17-one
Dihydrotestosterone (DHT)	17β-hydroxy-5α androstan-3-one
Oestradiol	oestra-1,3,5(10)-triene-3,17β-diol
Oestriol	oestra-1,3,5(10)-triene-3,16α,17β-triol
Oestrone	3-hydroxy-oestra-1,3,5(10)-trien-17-one
Pregnanediol	5β-pregnane-3α,2-α-diol
Progesterone	pregn-4-ene-3,20-dione
Testosterone	17β-hydroxy-androst-4-en-3-one

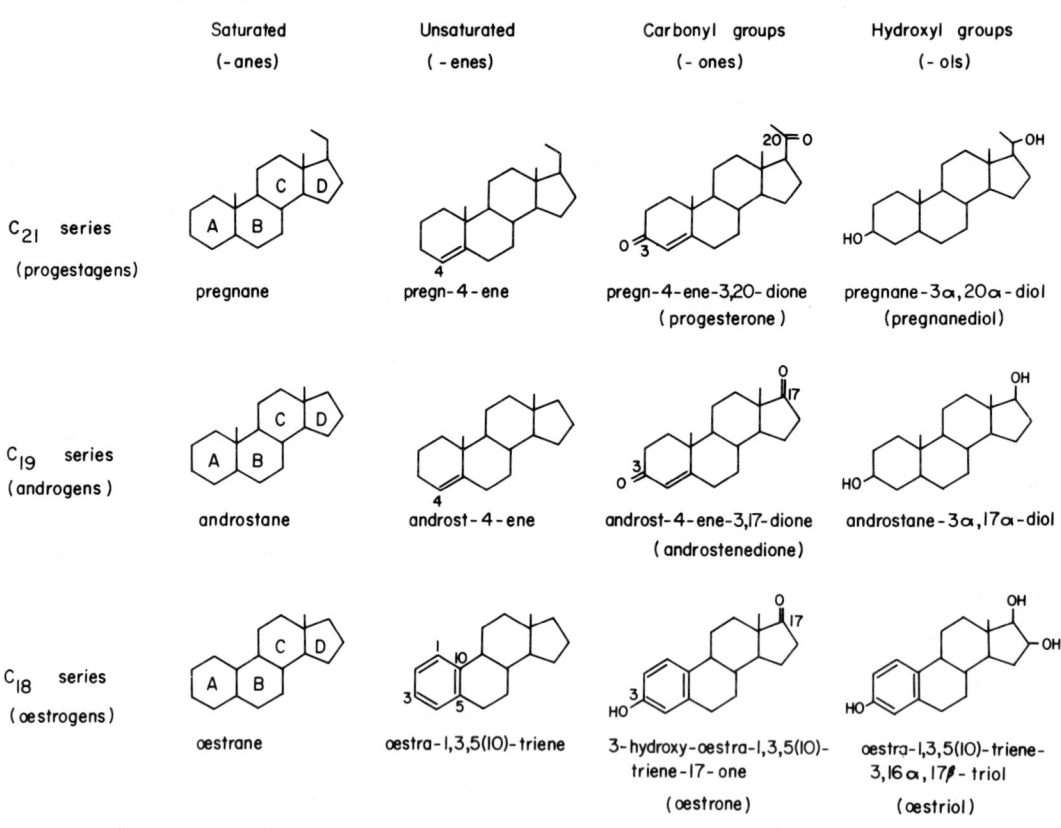

FIG. 1. Structure of nomenclature of some gonadal steroids to illustrate the principles involved.

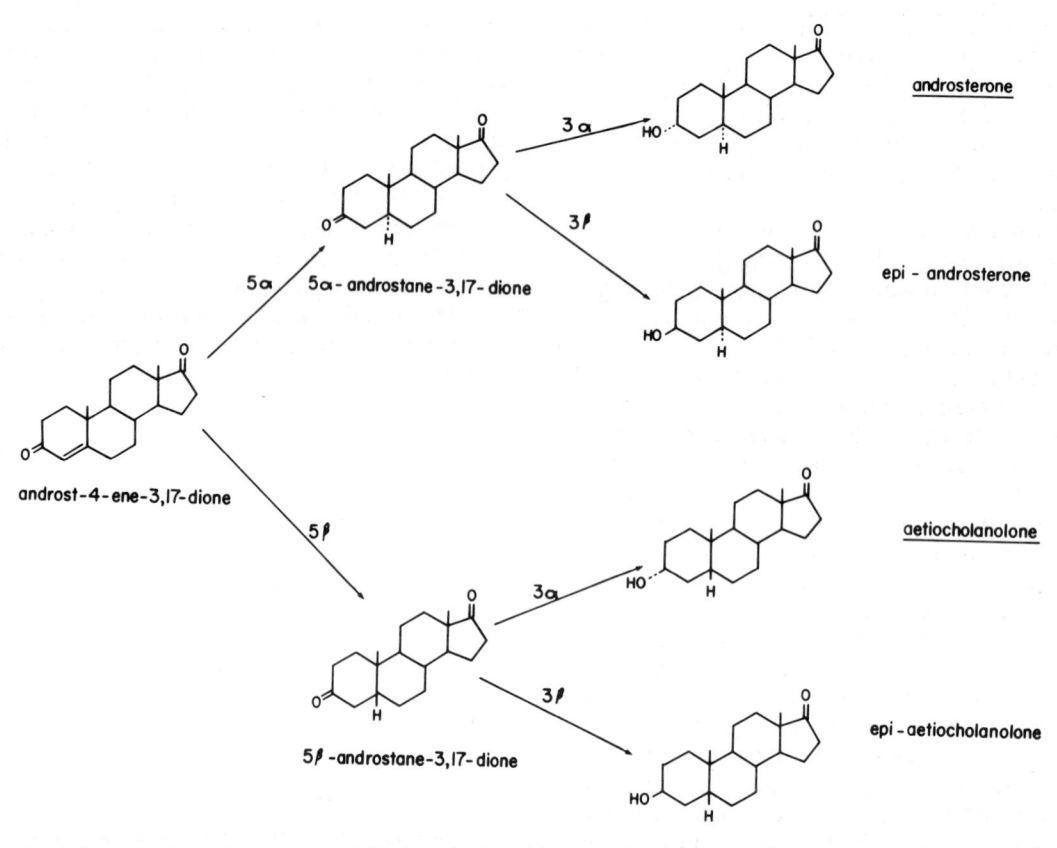

FIG. 2. Isomerism in the C_{19} androstanes following reduction of the A-ring (also seen in the C_{21} pregnanes). Androsterone and aetiocholanolone are the two major 11-deoxy, 17-oxosteroids.

THE GONADS

	Δ_5 Pathway		Δ_4 Pathway	
cholesterol	—(i)→ pregnenolone	—(ii)→	progesterone	→ 20$_\alpha$ and 20$_\beta$ dihydroprogesterone → 16$_\alpha$-hydroxyprogesterone → 6$_\beta$-hydroxyprogesterone
	↓(iii)		↓(iii)	
	17-hydroxypregnenolone	—(ii)→	17-hydroxyprogesterone	
	↓(iv)		↓(iv)	
	dehydroepiandrosterone	—(ii)→	androstenedione	→ oestrone
	↓(v)		↓(v)	⇅
	5-androstenediol	—(ii)→	testosterone	→ oestradiol

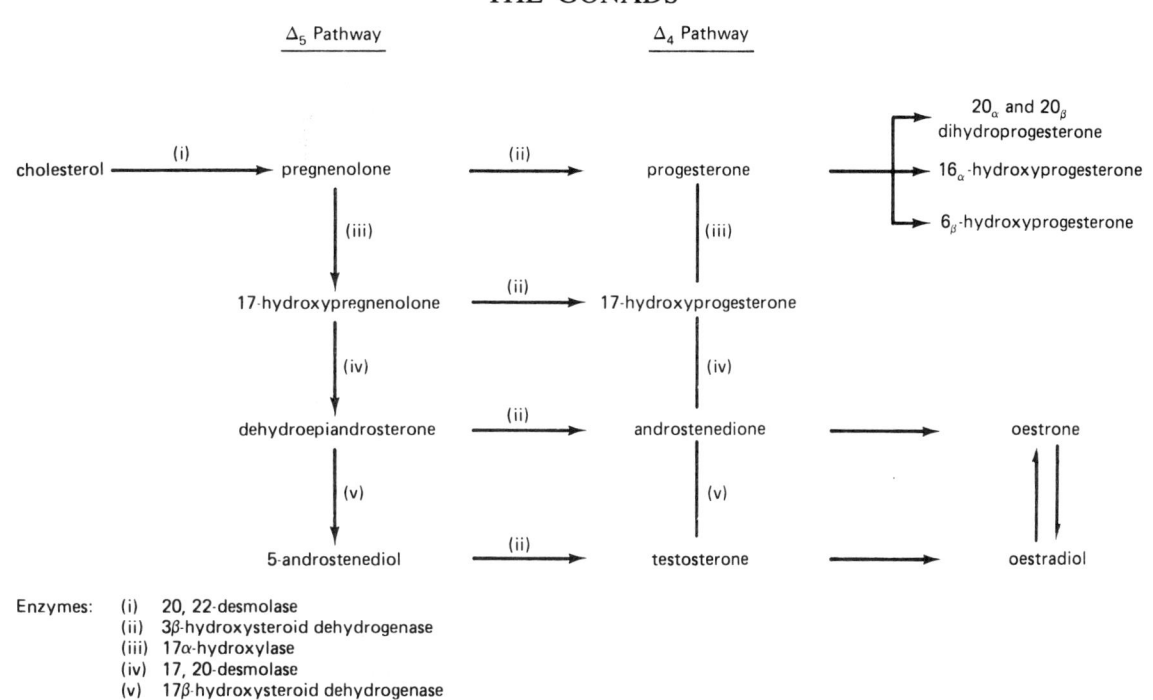

Enzymes: (i)   20, 22-desmolase
          (ii) 3β-hydroxysteroid dehydrogenase
          (iii) 17α-hydroxylase
          (iv) 17, 20-desmolase
          (v)   17β-hydroxysteroid dehydrogenase

FIG. 3. Pathways of biosynthesis of gonadal steroids.

Synthesis and Secretion

The major biosynthetic pathways in the gonads are well established (Fig. 3) and will not be discussed further. All the substances mentioned in Fig. 3 have been shown to be secretory products of either the testis or ovary or both though the 5-androstene series are minor products. In the ovary, several different cell-types contribute to these secretions and their relative importance, which changes during the ovarian cycle, cannot be calculated from the total secretion rate.

The total amount of a steroid entering the circulation in a given time is called the *blood production rate*. It is calculated as the concentration multiplied by the *metabolic clearance rate*. The *secretion rate* is defined as the total amount secreted from the endocrine glands. When the steroid is secreted by only one gland, as is progesterone in the luteal phase of the ovarian cycle, then the blood production rate equals the ovarian secretion rate (Table 2). If it is secreted by two glands, as is androstenedione by the ovary and the adrenal cortex then the secretion rate will be the sum of the contributions of the two glands. Various methods of assessing these contributions in physiological and pathological situations have been devised and their role will be discussed later. Some hormones, such as testosterone, are not only secreted as such but are also produced in peripheral tissues (liver, skin and adipose tissue are the most important sites) from inactive precursors, called prehormones. The total production rate is then the sum of the secretion rates from the endocrine gland *and* this contribution from peripheral non-endocrine tissues. This contribution is important in many physiological and pathological states and will be discussed later.

Binding in Blood and Tissues

All the steroid hormones are bound to some extent to albumin but for some there are also specific binding proteins. The binding to these proteins is of much higher affinity than to albumin but their concentrations in plasma are much less. Thus sex hormone binding globulin (SHBG) binds testosterone with an affinity 50 000 times greater than albumin does but there is about 10 000 times less of it. Because of the relative binding affinities and concentrations of testosterone and proteins, about twice as much testosterone is free (unbound to any protein) and twice as much is albumin-bound in the male compared to the female (Fig. 4). Both these fractions are available for metabolism and there is a similar 2:1 sex ratio in the metabolic clearance rate of testosterone.

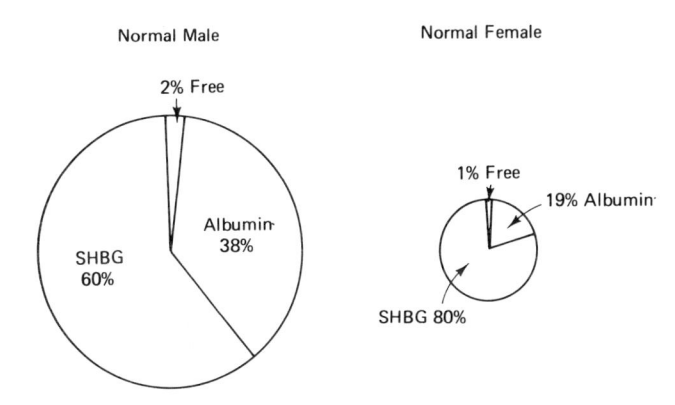

Normal Male: 2% Free, Albumin 38%, SHBG 60%
Normal Female: 1% Free, 19% Albumin, SHBG 80%

FIG. 4. Testosterone and its binding in normal male and normal female plasma. The free and albumin-bound fractions are biologically active and available for metabolism.

Table 2
Production rates and secretion rates of ovarian steroids

	Phase of cycle	Plasma level (nmol/l)	Binding in plasma	Total production rate (μmol/24 h)	% from ovarian secretion (approx)
oestradiol	early follicular	0.2	SHBG and albumin (weak)	0.3	90
	late follicular	1.2		2	100
	mid-luteal	0.6		1	100
oestrone	early follicular	0.15	albumin (weak)	0.4	70
	late follicular	0.75		2	90
	mid-luteal	0.3		0.8	70
progesterone	follicular	1	CBG	3	75
	luteal	30		100	100
20α-dihydroprogesterone	follicular	1	CBG	2	70
	luteal	8		20	80
17α-hydroxyprogesterone	early follicular	0.5	CBG	1.5	50
	late follicular	5		12	85
	mid-luteal	5		12	85
androstenedione	—	5	—	10	40
testosterone	—	1.5	SHBG albumin (weak) CBG (weak)	1	25
dehydro-epi-androsterone	—	15	—	30	15
dihydrotestosterone	—	0.75	SHBG	0.2	20

In certain circumstances, such as hyperthyroidism, pregnancy and during oestrogen therapy, the concentrations of binding globulins and of the corresponding sex steroids are increased. The free (unbound) level remains constant and is thus a better index of steroid 'activity' than is the total concentration.

Sex steroids enter the target cells by simple diffusion. In the cytosol the steroid is bound by a receptor protein of high affinity and specificity: the relative affinity of a series of related steroids for this receptor is proportional to their relative biological potencies. Oestrogens and progesterone act without being further metabolized but testosterone is reduced in the A-ring by 5α-reductase enzymes in the cell to give dihydrotestosterone (DHT) and other active metabolites. These are bound to the cytosol receptor in androgen-sensitive tissues.

Following binding of the steroid to the cytosol receptor a conformational change occurs and the complex passes into the nucleus where messenger RNA synthesis is stimulated. Synthesis of the enzymes appropriate to the particular end-organ cell then follows.

Metabolism and Excretion

After secretion gonadal steroids undergo a complex series of reactions before eventual excretion in the urine. In part, these reactions involve transformation into steroids with biological actions which are enhanced or changed; in part they result in the formation of inactive water-soluble conjugates which are more easily excreted.

Progesterone. Catabolism of progesterone involves three reduction reactions at C-20, C-4,5 and C-3. A number of alternative pathways are possible with a resulting large number of isomers. The major ones are illustrated in Fig. 5. It should be noted that whilst pregnanediol is predominantly derived from secreted progesterone this is not its only source as it also arises from deoxycorticosterone and pregnanolone. By a similar series of reactions 17α-hydroxyprogesterone (predominantly an adrenal metabolite) is converted into 5β-pregnane-3α,17α,20α-triol.

Androgens. C_{19}-steroids are secreted by both the adrenal cortex and the gonads. In the male, the secretion of the other C_{19}-steroids is insignificant compared with that of testosterone from the testis. In the normal female, however, only about 0.5 μmol of testosterone daily is secreted directly—part ovarian, part adrenal—compared with about 70 μmol of the biologically inactive adrenal steroids (dehydroepiandrosterone (DHA) and its sulphate (DHAS) and 11β-hydroxyandrostenedione (11OHΔ—Fig. 6). A further 0.5 μmol daily of testosterone (T) is converted from androstenedione (Δ) which is not only secreted by both adrenal (6 μmol daily) and ovary (3 μmol daily) but is also in part converted from DHA. The weak adrogenicity of these steroids is almost certainly due to their conversion to testosterone in peripheral tissues. The term 'androgen' should thus be used with care; its use to cover all 'C_{19}-steroids' should not be taken to imply procession of intrinsic biological activity.

FIG. 5. Metabolism of progesterone by reduction at C3, C5 and C20. Pregnanediol is the major metabolite.

The metabolites of all these steroids are excreted in urine as 17-oxosteroids (Fig. 6). Most of the androstenedione and testosterone is excreted as the sulphate and glucuronide conjugates of the 11-deoxy-17-oxosteroids, androsterone (A) and aetiocholanolone (E) (*see also* Fig. 2). Some of the DHAS is excreted unchanged and some via androstenedione. 11β-hydroxyandrostenedione (only of adrenal origin) is excreted as a group of 11-oxy-17-oxosteroids (11OA and 11OE) and cortisol also contributes to these. Thus the urinary 17-oxosteroids (17OS) are a complex group, chemically and biologically, being derived from adrenal and gonadal steroids and from androgenic and non-androgenic steroids. In the male, they are a fair reflection of testicular androgen secretion but in the female their measurement is almost useless. Even with a greatly excessive secretion of testosterone, the excretion of 17-oxosteroids may remain within the normal range.

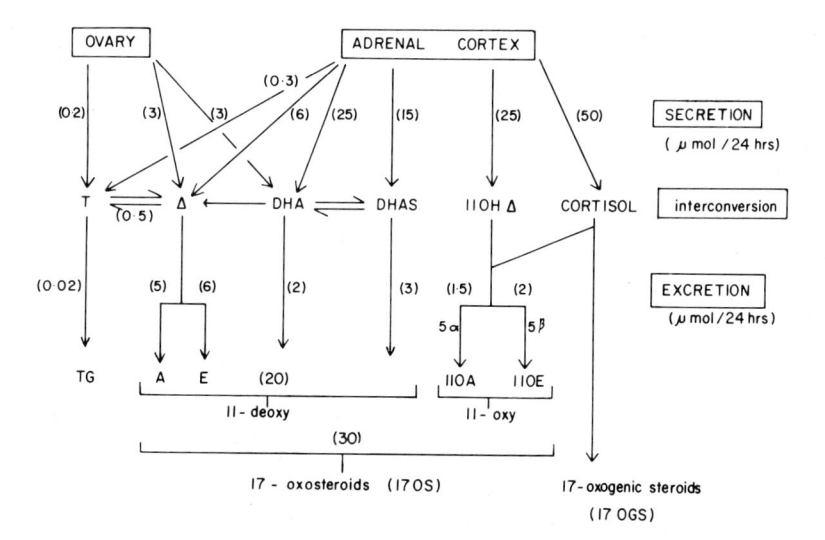

FIG. 6. Interrelationships of the C_{19}-steroids (androgens) in the normal female. The numbers in parenthesis give average daily rates of secretion and excretion (μmol). (T = testosterone, Δ = androstenedione, DHA = dehydroepiandrosterone, TG = testosterone glucanate, A = androsterone, E = aetiocholanolene)

About 1% of testosterone is excreted not as 17-oxosteroid but as the glucuronide of testosterone (TG) itself. This is a better indication of circulating testosterone than 17-OS but in women, a proportion is derived from a separate pool of testosterone glucuronide which is converted from androstenedione in the liver without being released into the circulation as unconjugated active testosterone.

Complex isotopic studies, outside the scope of routine clinical biochemistry, have been used to unravel the complex interrelationships of the C_{19}-steroids in normal subjects and in hirsutism in women.

Morphological Changes

During each menstrual cycle one (rarely more than one) follicle matures, ovulates and gives rise to a corpus luteum (Fig. 8). These complex processes are under the control of pituitary FSH and LH though the exact mechanisms are not clear. The corpus luteum is the major source of sex steroids during the postovulatory (or luteal) phase of the cycle. It lasts for about 14 days, after which it regresses and is replaced by an avascular corpus albicans—unless pregnancy occurs in which case the life of the corpus luteum is maintained by chorionic gonadotrophin.

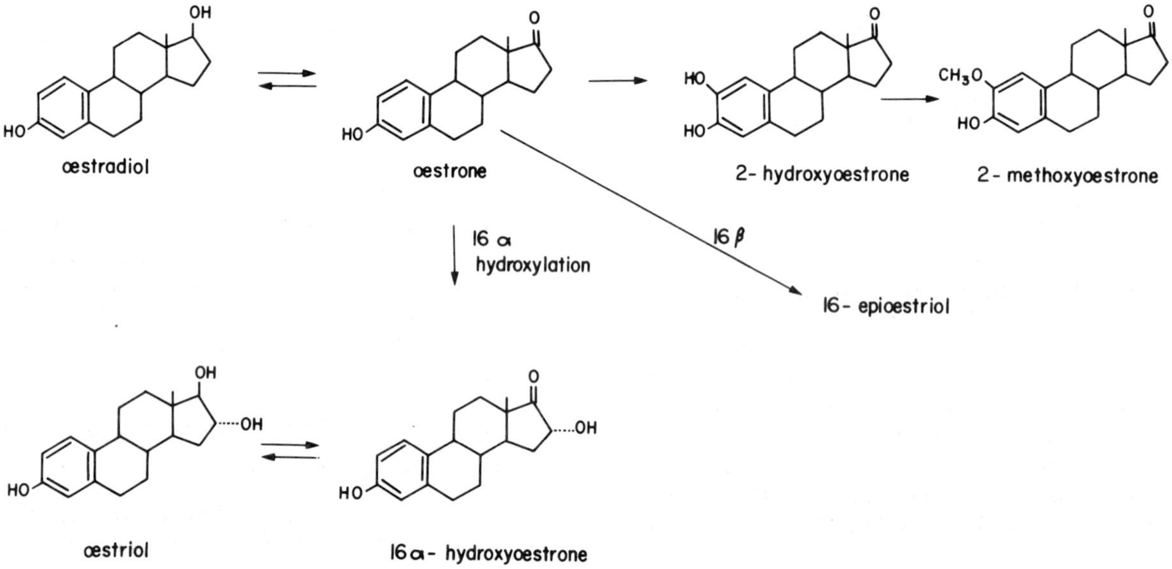

FIG. 7. Metabolism of oestrogens. Only the major pathways are shown. Oestriol is the major metabolite.

Oestrogens. Much of the secreted oestrogen is metabolised and excreted into bile and thence to the faeces. Some is excreted in urine as a complex mixture of glucuronides and sulphates. Metabolic inactivation of the oestrogen can occur at a bewildering number of sites on the molecule (a) oxido-reduction at C_{17} allows conversion of oestrone and oestradiol (b) hydroxylation and subsequent methylation at C-2 produces an important group of methoxy oestrogens (c) hydroxylation or ketone formation occurs also at C-16α (to give oestriol) and at C-6α, C-6β, C-7α, C-14α, C-15α, C-16β, C-18. The most important pathways are illustrated in Fig. 7.

Ovarian Function in the Adult and its Control

Postpubertal human ovaries are complex structures in which different cell types interact to produce sex steroids and an ovum ready for fertilisation in response to cyclical stimulation by pituitary gonadotrophins.

Sex Hormone Changes

These cyclical changes are associated with cyclical changes in sex steroid secretion (Figs. 8 and 9 and Table 2).

Follicular Phase. During the early follicular phase, secretion of ovarian steroids is at a relatively constant and low level. Then as the follicle develops under the influence of FSH, ovarian secretion of oestradiol increases slowly at first then more rapidly to reach a peak on the day before the LH peak. The rise of oestrone is not as great as that of oestradiol and parallels it; only about one third of oestrone is secreted by the ovaries, the remainder is converted peripherally from secreted oestradiol (Fig. 7) and androstenedione. Several days before the LH surge there is also a rise of androstenedione, testosterone, 17α-hydroxyprogesterone and 20α-dihydroprogesterone.

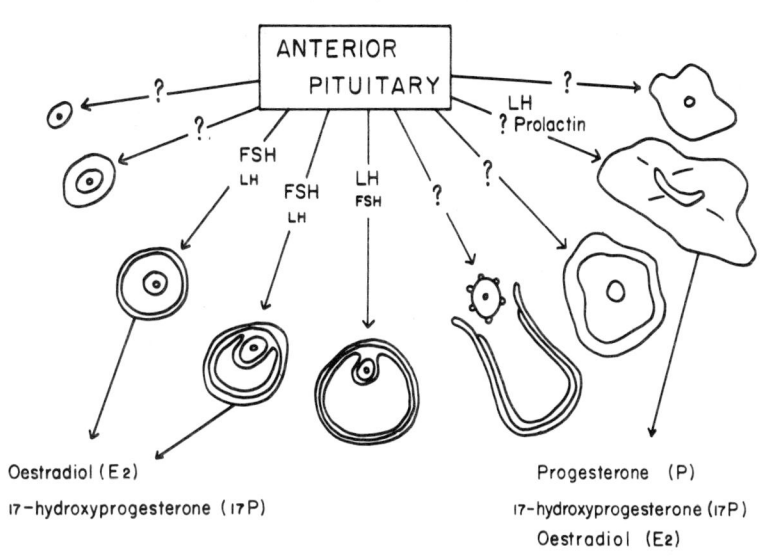

FIG. 8. Cyclical changes in ovarian cell types and steroid secretion (for details see Fig. 9). The role of prolactin in the human is uncertain.

Ovulatory Phase. There is a rapid rise in the level of LH which leads to final maturation of the follicle and ovulation some 24–36 hours later. Before this LH peak there is a sharp fall in the plasma oestradiol level and the plasma progesterone concentration begins to rise.

Luteal Phase. The most important biochemical feature of this phase of the cycle is the marked rise in secretion of progesterone from the corpus luteum which reaches a maximum about 8 days after the LH peak. It is associated with a rise in basal body temperature. There are parallel although smaller increases in 17α-hydroxyprogesterone, oestradiol and oestrone levels. As the steroids increase, LH and FSH gradually decline, the FSH rising again at the end of the luteal phase to initiate the growth of the next follicle.

Testicular Function in the Adult and its Control

Pituitary–Leydig Cell Axis

The interstitial cells of Leydig are located in the connective tissue of the testis between the seminiferous tubules. These cells start to secrete androgens during the seventh week of fetal life and this coincides with the onset of androgen-dependent differentiation of the secondary sexual characteristics.

Testosterone, the major testicular androgen in the adult, is synthesised by the pathways shown in Fig. 3. Nearly all the circulating testosterone in normal men results from testicular secretion. Small amounts of other potent androgens such as dihydrotestosterone are also secreted but their overall contribution is very small. Androstenedione is also secreted but its main importance is as an oestrogen precursor. Progesterone and 17α-hydroxyprogesterone are also secreted in small amounts but their function is unknown. Small amounts of the oestrogens, oestrone and oestradiol are also secreted but the major portion of blood oestrone and oestradiol are derived by peripheral aromatisation from blood androstenedione and testosterone respectively.

The rate of synthesis and secretion of testosterone is dependent upon the rate of LH secretion. This is, in turn, controlled by negative feedback action on the hypothalamus and pituitary so that a fall in testosterone secretion results in increased LH secretion. This stimulates increased testosterone which reduces the LH level. Thus the hypothalamus, pituitary and Leydig cells form

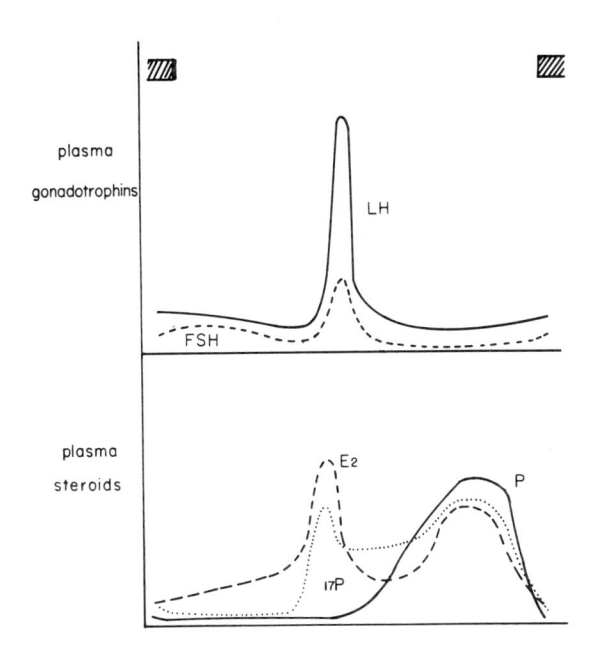

FIG. 9. Cyclical changes in the secretion of pituitary and ovarian hormones. LH = luteinising hormone; FSH = follicle stimulating hormone; E_2 = oestradiol-17β; P = progesterone; 17P = 17-hydroxyprogesterone.

a functional unit which serves to maintain a constant level of testosterone secretion. The exact mechanism of this feedback is not clear but it is certainly complex and may involve aromatisation of testosterone to oestradiol and/or 5α-reduction to dihydrotestosterone and other 5α-androstane metabolites.

Pituitary—Seminiferous Tubule Axis

The major volume of the testis comprises a tightly coiled mass of seminiferous tubules. These are lined by spermatogonia which undergo division into spermatocytes which develop in turn to spermatozoa. A total of about 70 days is required for the complete process. Both LH and FSH are needed; LH indirectly by way of testosterone from the Leydig cells which reaches the tubules by way of intratesticular lymphatics. FSH controls the development of an androgen-binding protein in the Sertoli cells which lie around the outside of the tubules underneath the basement membrane. FSH, like LH, is controlled by a feedback system but unlike LH, steroids appear to play a minor role. A non-steroidal substance of tubular origin (probably from the Sertoli cells) has been postulated. It is called 'inhibin' but little is known about its biochemistry.

PRINCIPLES OF MEASUREMENT OF GONADAL STEROIDS

A detailed consideration of the technical aspects of steroid hormone assay is beyond the scope of this chapter. There are, however, aspects of the methodology and application of the assays which profoundly affect the clinical interpretation of the results and thus it is appropriate to consider them here.

Methods of Steroid Analysis

In the past, the measurement of gonadal steroids in biological fluids has involved extraction and partial purification followed by a more or less specific quantitation system based on a colorimetric or fluorimetric end-point. In some instances the measurements are designed not to be absolutely specific for one steroid but to measure a group of steroids. With the recent rapid extension of protein binding techniques (particularly immunoassay) of high specificity the need for preliminary purification is becoming progressively less.

Separation and Purification Techniques

Since gonadal steroids (except for the conjugated forms) are hydrophobic they can be simply and rapidly removed from aqueous biological media by solvent extraction. This is optimal when the solvent is selected to match the polarity of the steroid of interest thus adding a degree of specificity to the method.

Chromatography of all types has been used over the years for further purification. It has the disadvantage, in the routine laboratory, of being time-consuming and requiring special expertise; the methods are thus prone to error. Added sources of error are associated with variable recoveries from the procedure and the addition of 'blank' material from support media. Because of these, and the progressive refinement of other techniques, chromatographic purification is less and less used. The most common techniques in the field of gonadal steroids are: column partition chromatography on celite; high pressure liquid chromatography; adsorption chromatography on thin layers of silica gel; gel filtration on Sephadex LH 20.

Group Methods

In the past, measurements of groups of related steroids in urine were the only means of chemical assessment of gonadal function. Whilst these are gradually being replaced by simple and specific plasma methods, they are still widely used in laboratories throughout the world and are not without clinical value.

The measurement of 17-oxosteroids (17 OS) by the Zimmermann reaction is widely used as an index of androgen secretion but the urinary 17 OS is a very complex group as described earlier (Fig. 6). This limits the clinical usefulness of the technique.

The number of different oestrogens in urine is also considerable. In non-pregnant subjects the levels, even of the major oestrogens, are low but the measurement of total oestrogens using the Kober reaction has some place where plasma methods are not available for technical or clinical reasons.

Saturation Analysis

The term 'saturation analysis' is generally applied to the assay of substances based on the reversible, non-covalent binding of a ligand by a more-or-less specific binding protein. The sensitivity of these methods is governed by the affinity of this binding, being expressed as an affinity constant or K. These values are of the order of $10^{-9} - 10^{-11}$ litre/mole and the assays are up to a million times more sensitive than the earlier colorimetric methods.

In the early 1960s transport binding proteins were used. Thus SHBG was used to assay testosterone by saturation analysis. Later cytosol receptors, e.g. for oestrogens, were used. All these methods have been largely superseded by immunoassays particularly radioimmunoassays though non-isotopic methods (e.g. with enzyme or fluorescent labels) are becoming increasingly popular.

For all the gonadal steroids of known biological interest specific and sensitive immunoassays in plasma are now widely available.

Where to Measure Gonadal Steroids

Blood Plasma

Most commonly in clinical practice, steroids of gonadal origin are measured in plasma derived from blood obtained from a peripheral vein. Such measurements give an indication of the amount of circulating hormone that is present in the periphery at the time of sampling. There are many difficulties of interpretation (discussed below) but in general these measurements are of clinical relevance (whether they are also of diagnostic value will be discussed later). Steroids are stable in plasma (an exception is the methoxy oestrogen group but these derivatives are not measured routinely) and there are no special problems associated with collection and storage.

In specialised units, increasing use is being made of measurements of steroids in samples obtained by adrenal and/or ovarian venous catheterisation, in an attempt to localise lesions particularly those associated with virilisation (*see* later).

Urine

The major disadvantage of single plasma measurements occurs when there is a significant circadian variation in concentrations. This is not generally a problem with gonadal steroids, but urine measurements do have a place in those laboratories not yet equipped for immunoassay and for clinical situations where a non-invasive approach is advisable, e.g. for repeated determinations in subjects with poor access to clinical staff.

Saliva

Steroids are found in saliva and recent attention has been paid to their measurement there because saliva is an ultrafiltrate of blood plasma and thus only the non-protein bound, or biologically active, fraction appears there. Collection of saliva is also said to be non-invasive though that is a matter of taste!

Other Fluids

Gonadal steroids have been measured in other fluids though only as research projects. Examples are: seminal plasma (as an index of androgen action on the male genital tract); cerebrospinal fluid; ovarian cyst fluid; ovarian and testicular tissue biopsies.

When to Measure Gonadal Steroids

Time of Day

Though there are fluctuations in secretion of gonadal steroids during 24 hours, regular circadian variations are not pronounced and—for clinical purposes—can be ignored.

Time of Cycle

Although there are changes in androgen secretion during the menstrual cycle, these are minor and, for clinical purposes, insignificant. For oestrogens and progestogens, of course, there are marked fluctuations from day to day (Fig. 9) and the samples should be taken and the result interpreted, with this in mind.

Dynamic Tests

As in other areas of endocrinology, single values are often less informative than measurement of the response to an exogenous stimulus. Thus the gonads can be stimulated by exogenous gonadotrophin. In the male, hCG, which acts like LH on the Leydig cell, is given by injection on several days and daily blood samples taken for testosterone measurement. This test helps to distinguish primary from secondary testicular failure. In the female, pituitary (or urine) gonadotrophin preparations may be given in an attempt to induce ovulation. In this situation it is essential that the ovarian response is assessed in order to avoid overstimulation leading possibly to multiple pregnancy.

Which Steroids to Measure

Given the complex mixture of gonadal steroids in the circulation, with their prohormones, metabolites, conjugated forms, the free fraction and that bound to proteins with varying degrees of affinity, and steroids closely related chemically but without or with different biological activities, the decision which steroid to measure in any clinical situation is not always easy.

Oestrogens

Total unconjugated oestradiol in plasma appears to be the measurement of choice for most clinical purposes. In certain situations, e.g. at or after the menopause, oestrone assays are of value. Measurement of oestriol is of little value (except in pregnancy).

Progestogens

In plasma, progesterone can readily be measured by immunoassay and this has largely replaced the measurement of pregnanediol in urine. Plasma 17αOH-progesterone is used as an index of the secretion of adrenal precursors of cortisol rather than of gonadal function. Other natural progestogens are not measured in routine clinical practice. Assays of synthetic progestogens (e.g. norethisterone) are used only in research situations.

Androgens

Here the choice of steroid is more complex. For the reasons mentioned above, urine metabolites are a poor reflection of circulating androgens. Plasma testosterone is commonly measured in both sexes though the diagnostic value of such assays is limited. The 5α-reduced metabolite of testosterone, dihydrotestosterone and/or the androstenediols, can be assayed as an index of end-organ metabolism of testosterone but these have not been widely applied. A group method, based on saturation analysis using SHBG, for all 17β-hydroxy C_{19}-steroids including testosterone, DHT, Δ^5-androstenediol and the androstenediols was used in some laboratories before specific assays for individuals androgenic steroids were used. It is still claimed to have merit in that it measures all biologically active androgens but in my view this advantage is limited by the heterogeneity of the group with respect to their biological potency, tissue of origin and binding and clearance characteristics. Androgen prohormones can also be measured, notably androstenedione.

The choice of analyte is clearly wide and governed by many considerations. As we shall see, from the strictly clinical viewpoint, the management of the individual patient is rarely affected by the ability to measure androgens.

Free and Bound

Although progesterone (to CBG) and oestradiol (to SHBG) are also bound it is the assessment of binding of testosterone to SHBG which is most frequently considered as being of clinical importance. The basis of the argument (analogous to that of thyroid hormone binding to TBG) is the need to monitor the biologically active free fraction rather than the total (free plus bound) which is measured in the routine testosterone assay. There are several approaches to this, though none is widely used in clinical practice.

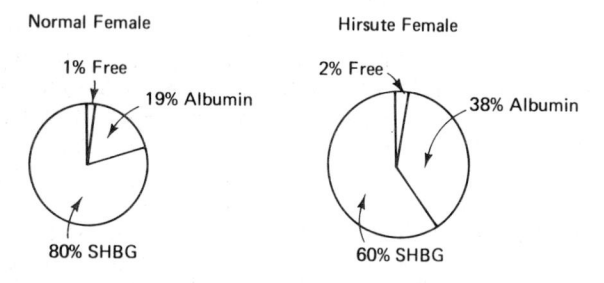

FIG. 10. Testosterone and its binding in normal and hirsute women. (Compare with Fig. 4).

(a) Plasma free testosterone can be measured by techniques such as equilibrium dialysis. The major disadvantages of these methods are: the inconvenience and cost; the difficulty of achieving sensitivity adequate to measure free testosterone concentration in the plasma of female subjects where the assay is most valuable; the fact that it includes albumin-bound testosterone as 'bound' rather than 'free'.

(b) Circulating free testosterone can be assessed indirectly by measurement in 'biological ultrafiltrates'. Thus urine free testosterone and salivary testosterone concentrations reflect the unbound fraction in plasma. These methods suffer from the same disadvantage as the plasma methods except they are simpler. The urine measurement suffers from the severe disadvantage that any hydrolysis of testosterone glucuronide in the urine whilst it is in the bladder (incubating at 37°C) or after voiding will raise the free testosterone concentration.

(c) SHBG binding capacity has been measured by a large number of techniques including equilibrium dialysis, steady-state gel filtration, competitive adsorption (e.g. on florisil), affinity immobilisation, fractional precipitation with ammonium sulphate. This last method, being simple, has been the most widely used but all suffer some degree from interference from albumin binding and problems of dissociation of the SHBG complex.

(d) Direct measurement of the SHBG molecule by radioimmunoassay is being developed in a number of centres but is not yet used in clinical practice.

Other Biochemical Tests

Consideration has only been given here to the measurement of gonadal steroids themselves. As mentioned earlier the assessment of gonadal function involves a wider investigation of the endocrine system as well as a number of ancillary, non-biochemical tests. It is only by considering all the information at his disposal (including clinical details) that a clinician can make appropriate conclusions regarding the diagnosis and treatment of the patient.

Problems of Interpretation

Many of the problems of interpreting the results of gonadal hormone assays are similar to those of clinical biochemistry in general and specific problems have been mentioned elsewhere in this chapter. The following summary is intended to be an aid to intelligent interpretation of sex hormone assay results.

Timing of the Sample

A single value can be interpreted only in terms of other hormonal events, for instance of the ovarian cycle or in the light of treatment with exogenous hormones or other drugs.

Steroid Binding

The degree of binding of 17β-hydroxy androgens and to a lesser extent oestradiol by SHBG is dependent on the amount of SHBG in the circulation and this in turn is dependent on a number of physiological and pathological factors.

(a) Oestrogens increase SHBG levels. Thus the highest levels are seen in pregnancy and during treatment with oestrogens (including oral contraceptives). SHBG levels are lower in postmenopausal women.

(b) Androgens lower SHBG levels. Thus SHBG levels are higher in prepubertal boys than in adult men and rise progressively after the age of 50 years. In hirsutism, SHBG levels are reduced.

(c) Thyroid hormones increase SHBG levels. Thus the neonate and hyperthyroid subjects have raised concentrations of SHBG.

Target Organ Metabolism

Levels of hormones in the circulation do not necessarily reflect the tissue levels in the target organ or their ability to act there and this can sometimes lead to difficulties of interpretation; again androgens provide the best examples. The micro-environment of tissues may contain local high concentrations of steroids. Thus the seminiferous tubules of the testis are bathed by high testosterone concentrations carried there by a local lymphatic circulation.

Once at the cell, binding by the specific cytosol receptor will be one factor governing the activity of the steroid. A classic example is that of testicular feminisation (*see* later) a condition in which androgenisation is absent in spite of adequate circulating testosterone. This is due to an inherited lack of intracellular receptors. This condition represents an extreme case of variations in receptor binding leading to difference in end-organ responsiveness which might exist in lesser degrees in the population.

Differences in enzyme action on the steroid may also be important. Thus 5α-reduction of testosterone to its active metabolites has been shown to vary from tissue to tissue within an individual, from individual to individual, at different stages of reproductive life, and to show familial and racial differences. This is illustrated by the occurrence of much less facial hair in oriental women compared with caucasians despite identical blood testosterone levels.

Reference Ranges

In the assessment of the result on an individual patient, the clinician generally compares it with the values obtained in a normal or reference population. There are several factors that can adversely affect this apparently simple process (note that these apply in other areas of endocrinology and to a greater or lesser extent, throughout clinical biochemistry (ref. to Chapter 29 in Volume I)).

(a) **Methodological Factors** might lead to systematic bias associated with one laboratory. If a patient's result obtained in that laboratory is compared with a normal range established in that laboratory using the same methodology then the results may be clinically valid locally though not necessarily comparable with results obtained elsewhere. Similarly a result on a sample sent to another centre can be compared with a reference range established in that centre. In both these situations, it is important that if any significant change of methodology has been introduced then a new reference range is established (or it is shown that the 'old' values are unaffected by the change).

A more dangerous situation arises when a laboratory is using its own method (which might be a commercial kit, or a method obtained from another centre or one developed *de novo*) but is comparing results with a reference range obtained elsewhere. Use of different reagents and assay procedures means that no two laboratories can expect to obtain the same results.

(b) **Population Factors** are also important and it should be considered whether the reference population is appropriate with respect to such factors as stress, age, sex, race, time of cycle, time of day, diet.

(c) **Statistical Factors** are less well understood by clinicians. Thus the borderline between normality and abnormality is a statistical one and this uncertainty is expressed in most laboratories as the range covering 95% of the normal population (i.e. the mean $\pm$ 2 standard deviations). It follows from this practice that 2.5% of the normal population (i.e. one patient in 40) will have 'raised' levels and a similar number 'lower' levels for statistical reasons alone. This conclusion only follows if the frequency distribution of results from the normal population is gaussian (or 'normal' in the statistical sense). If it is skewed then problems of interpretation, already poorly understood by clinicians, become more pronounced.

How Right is the Result?

A full discussion of the methods of assessment and maintenance of the quality of hormone assay results is beyond the scope of this chapter. Clinicians rarely appreciate (mainly because they haven't been told) the scale of errors that result in imprecision within individual assays and lack of reproducibility of results between assay batches. The amount of information on the assay quality that should be included in the report is arguable but those situations in which clinical decisions are taken on the basis of a result (e.g. of plasma oestradiol in a patient undergoing fertility treatment with gonadotrophins) demand a close relationship between the laboratory and the clinician and full discussion of the likely significance of the result.

DISORDERS OF GONADAL FUNCTION: ROLE OF BIOCHEMICAL TESTS

The role of hormone assay measurements in the assessment of gonadal function depends on whether medicine is considered to be a science or a technology. There is no doubt that the ability to measure these steroids in biological fluids has increased our knowledge of the physiology and endocrine pathology of the human ovary and testis. The impact of this increased scientific knowledge and technical ability on the clinical management of individual patients is, however, arguable. In some instances they are of great help: in others, of none. In all instances the biochemical tests are evaluated within the context of an overall clinical assessment and before considering different gonadal disorders these non-biochemical aspects will be briefly discussed.

The Patient as a Bioassay

Depending on the age of the patient at the time when they occur, abnormalities of gonadal hormone production express themselves in changes in somatic or sexual growth and development or of adult sexual functions that are obvious either to the patient or to the clinician or to both. Thus the patient is responding to his or her own endogenous hormone levels and acting as a 'bioassay subject' giving information about gonadal function from the time of fetal differentiation through puberty to the time of study.

Clinical History

A careful history constructed round the main complaint will help to elucidate the hormonal background to any alteration of sexual development or reproductive function.

Altered sexual development takes the form of: *isosexual precocity* (development which is premature but consistent with the genetic and phenotypic sex of the individual); or *heterosexual precocity* (development of changes associated with the opposite sex); or *delayed puberty*. The clinician in his assessment of these changes must judge them in the light of knowledge about the timing and sequence of changes that take place at normal puberty: there is a wide age range over which these occur. Sexual precocity can be caused by a number of extra-gonadal factors, such as poisoning by hormone-containing pills and creams, and neoplasms of the central nervous system. In delayed puberty there is an absence of secondary sexual characteristics at the appropriate ages. Several syndromes associated with delayed puberty are associated with congenital stigmata and/or a family history.

At any age, signs and symptoms of androgen excess should be sought. The onset of hirsutism or virilization alerts the clinician to a number of possible diagnoses, depending upon the age of onset, family history, rate of progression of the symptoms.

Disturbances of menstruation, including amenorrhoea, are a frequent source of complaint. The history (and physical examination) should attempt to exclude such causes as pregnancy or weight-related amenorrhoea.

Finally, disorders of sexual function are frequently seen in conditions which are not primarily gonadal disorders; e.g. panhypopituitarism, diabetes, obesity.

Physical Examination

The biological effects of gonadal hormones are revealed even more clearly to the observant clinician by the physical examination. This will often make the diagnosis apparent and may obviate the need for extensive biochemical tests. It is not possible to give full details but what follows is a summary of the features the clinician will assess.

(a) Rate of growth is important in establishing the aetiology in several endocrinopathies. Thus sexual precocity initially increases growth rate and then early epiphyseal fusion stunts growth.

(b) Skeletal proportions (e.g. of arm span to overall height) may also be altered.

(c) Nutritional status and body fat distribution.

(d) Physical signs associated with Cushing's syndrome, Addison's disease, thyroid diseases, or congenital gonadal disorders may be seen (e.g. on the skin).

(e) The growth and distribution of hair is carefully assessed. Charts and semi-quantitative scoring systems have been devised for hirsutism. Measuring sebum production has also been used as a means of quantifying androgen production but it has not proved very practicable. Acne is, however, a sensitive clinical indicator of circulating androgens.

(f) Breast examination is important in the assessment of pubertal development in females.

(g) The external genitalia are obvious target organs for the sex hormone action and the state of growth in both sexes indicates the stage of pubertal development. Female genitalia are also sensitive to androgens particularly *in utero*. The most pronounced effects of virilization—labial fusion and clitoral enlargement—are signs of female fetal exposure to androgens.

(h) The internal genitalia also reflect the secretion of sex hormones. In the male the size of the prostate and in the female the size of the ovaries and uterus can be assessed by an experienced clinician. In addition the levels of circulating oestrogen can be as adequately assessed by observing oestrogenization of the vaginal mucosa as by measuring oestradiol levels.

Ancillary Tests

(a) *Vaginal epithelium* is extremely sensitive to oestrogen and microscopic evaluation of vaginal cytology is a valuable adjunct to endocrine investigation. Various indices have been devised to attempt to quantify the oestrogenic stimulation. Serial measurement are most valuable—particularly during menstrual cycles or during a course of fertility treatment.

(b) *Cervical mucus* is another easily examined feature which is controlled by ovarian hormones. In general, secretion is stimulated by oestrogen and inhibited by progesterone. In addition, certain physical and chemical properties show cyclical variations and these can be used to time ovulation and as an index of corpus luteum function.

(c) *Progestogen-induced withdrawal bleed* is a useful test in the amenorrhoeic patient. A positive response indicates an intact pituitary gonadotrophin secretion and adequate ovarian oestrogen secretion and a responsive endometrium. Appropriately used, it obviates the need for plasma oestradiol measurements.

(d) *Bone-age* by X-ray, particularly of the wrist, gives an objective view of the hormonal effects on bone maturation.

(e) *Abnormalities of the sex chromosomes* needs to

be evaluated in cases of apparent disorders of sexual differentiation. A simple technique involves looking at sex chromatin in buccal squames: this gives an indication of the number of X chromosomes. The full karyotype can be obtained from any tissue (normally leucocytes) in culture: special staining techniques can be applied for further analysis of X and Y chromosome structure.

Disorders of Gonadal Development

The existence of the normal male XY karyotype leads to the development of the fetal testis starting at about the 5th week of gestation. This in turn secretes testosterone and other substances (e.g. Müllerian-inhibitory factor) which control the development of the male internal genitalia. The fetal testis is essential for this.

In the female, on the other hand, the XX karyotype leads to the development of the ovary much later, at about the 11th week, and the presence of an ovary is not essential for the development of uterus and tubes. These develop in the absence of any functioning gonad.

Gonadal Dysgenesis (Turner's Syndrome)

In the typical case, there is shortness of stature accompanied by one or more physical abnormalities of which the most frequent are: webbed neck, high-arched palate, short 4th metacarpal, hypoplastic nails, coarctation of the aorta. Thus the typical case is easily diagnosed. The karyotype is 45/XO arising as a result of non-disjunction (incidentally it has been estimated that almost 1% of all human zygotes have this karyotype but less than 3% of these survive till term). The gonads do not develop being present as 'streaks'. As mentioned earlier, ovarian function is not necessary for the development of the female phenotype *in utero* but since the ovarian steroids are not produced at puberty, secondary sexual development and menstruation do not occur. As expected, gonadotrophin levels are very high in the absence of a steroid feedback.

Many variants of the typical case are seen. These arise because of mosaicism (the development of two or more cell lines in one individual). The commonest and mildest form is the XO/XX mosaic but complex karyotypes—including those with Y chromosomes and mixed ovarian/testicular dysgenesis are seen.

Klinefelter's Syndrome

This is the commonest abnormality of testicular development in the male. The sex chromosome abnormality is 47/XXY but many variants are seen. In the classic case there is a failure of pubertal development associated with small testes, azoospermia, gynaecomastia, excessive growth of long bones particularly the legs and some impairment of mental function and social behaviour. Testosterone levels are usually below normal or in the lower part of the normal range; both gonadotrophins are usually elevated. The histological appearance of tubules and Leydig cells is characteristic, so a testicular biopsy can aid in diagnosis.

Pseudohermaphroditism

This term is used to cover a variety of conditions in which the chromosomal and gonadal sex is different from the genital or phenotypic sex. Thus a male pseudohermaphrodite is an individual with testes, a 46/XY karyotype and female appearance.

Male Pseudohermaphroditism results from a failure of virilisation. It can result from deficient androgen synthesis and secretion or a failure of the target organ response.

(a) *Deficient androgen secretion* can result from rare defects involving any of the enzyme steps shown in Fig. 3. A deficiency of the *20–22 desmolase* has been described though it is not usually compatible with life since it involves the adrenal as well as the gonad. A deficiency of *3β-hydroxysteroid dehydrogenase* with a block in the conversion of Δ^5 to Δ^4 steroids also involves the adrenal. Defects of the *17,20-desmolase* and *17β-hydroxysteroid dehydrogenase* have also been described. The former is inherited as an X-linked recessive. Precise diagnosis of all these conditions can only be made in a specialist steroid laboratory with the ability to measure the appropriate metabolites. Stimulation tests with hCG are frequently used as an aid to diagnosis: there is a great increase in the secretion of metabolites preceding the enzyme block (*see* Table 3).

(b) *Defects of androgen action* are more common inherited causes of male pseudohermaphroditism. A number of types have been described which fall into 3 groups.

(i) *Testicular feminisation (complete form)* is the most common. The genotype is XY with an X-linked inheritance and testes are present but the patients have an obvious female body habitus and external genitalia. Plasma testosterone levels are in the normal *male* range (or even above it) and there is evidence for a defect in the cytosol binding protein so that androgens are inactive.

(ii) *Testicular feminisation (incomplete form)*. Clinically these patients resemble the complete type but there is often some virilisation at birth and at puberty. Circulating androgen concentrations are in the normal male range. Many variant forms have been described.

(iii) *Incomplete male pseudohermaphroditism* encompasses two types. Firstly, a spectrum of X-linked recessive disorders, the mildest of which are almost completely masculinised. Again plasma testosterone levels are normal (or even higher than normal) and a variable defect of end-organ responsiveness is postulated. Secondly, an autosomal recessive type which has been shown to involve a deficiency, of 5α-reductase: there are subnormal levels of DHT in plasma and of 5α-metabolites in urine.

TABLE 3
EFFECT OF ENZYME DEFECTS ON SEXUAL DEVELOPMENT

Enzyme	Tissue Affected	Effect on Androgen Secretion	Effect on external genitalia	
			Male	Female
20, 22-D	A, G	↓		normal
3β OH-D	A, G	$\Delta_4 \downarrow; \uparrow \Delta_5$	Pseudohermaphroditism	virilisation
17α-OH	A, G	↓		infantile
18 OH-D	A	no effect	no effect	no effect
21-OH	A	↑	Isosexual precocity	virilisation
11-OH	A	↑		pseudohermaphroditism

Key
A = adrenal; G = gonad; ↓ = decreased; ↑ = increased; Δ_4, Δ_5 (see Fig. 3); 20, 22-D = 20, 22-desmolase; 3β OH-d = 3β-hydroxysteroid dehydrogenase; 17α-OH = 17α-hydroxylase; 18 OH-D = 18 hydroxysteroid dehydrogenase; 21-OH = 21-hydroxylase; 11-OH = 11-hydroxylase.

Female pseudohermaphroditism consists of the presence of the female karyotype 46/XX with varying degrees of masculinisation of the external genitalia. The internal genitalia are normal in all the types and reproductive function is usually possible after appropriate treatment.

Masculinisation of the female fetus is rarely caused by maternal ingestion of androgenic drugs. More usually it is the result of enzyme abnormalities of adrenal steroid biosynthesis (*see* Table 3). These cause masculinisation because the defect of cortisol synthesis leads to hypersecretion of ACTH and a build-up of steroids preceding the block (*see* Chapter 39). The syndrome is commonly called *congenital adrenal hyperplasia* (adrenogenital syndrome in the older literature).

The commonest defect is in the *21-hydroxylase*. Immediately before the block is 17-hydroxyprogesterone and raised levels of this and its urine metabolite, pregnanetriol, are pathognomic of the disease. Secretion of DHA, androstenedione and testosterone is also increased. In milder forms of the disorder androgenisation may be the only abnormality and it may not become apparent until puberty. Stimulation with ACTH and measurement of 17-hydroxyprogesterone is a useful diagnostic test in these cases.

Not so common is the *11-hydroxylase* defect. In addition to excess androgen leading to virilisation there is an excess of 11-deoxycorticosterone (DOC) which causes hypertension. In the urine, 17-oxosteroid excretion is raised although the 11-deoxy fraction is reduced. This forms the basis of a test that is useful in the investigation of the newborn for whom complete urine collections are difficult to obtain. The 11-oxygenation index is a measure of the *ratio* of 11-oxygenated to 11-deoxy steroids.

Management of Disorders of Sexual Development

The earlier in life that errors in sexual differentiation are recognised and treated the better. Whilst the genetic, gonadal, genital and hormonal sex may be of great importance to the clinician, it is the gender identity that is most important to the patient: this is established early in life, certainly before the age of 2 years. Thus all attempts at hormonal or surgical therapy should be carried out early: sex reassignment later in life can be catastrophic. The clinician must also be sensitive to the patient's feelings regarding sexual identity. Thus there is nothing to be gained (and a great deal may be lost) for instance in telling a patient reared as a female that she is 'really a male'.

Disorders of Puberty

Puberty is the state of transition between childhood and full reproductive capacity. It is associated in turn with maturation of the hypothalamus and pituitary, stimulation of the gonads with secretion of sex steroids and development of secondary sexual characteristics, sexual behaviour and fertility. The physical changes have been fully documented (notably by Tanner) and these provide a standard against which patients are assessed clinically.

As puberty develops, the episodic secretion of FSH and LH becomes more apparent with two special features: sleep-entrained LH secretion in both sexes and, in girls, an increase in FSH secretion about 2 years earlier than that of LH. Increased secretion of the gonadal hormones, oestradiol in girls and testosterone in boys follow. There are also increases in the levels of *adrenal* androgens which actually precede those of gonadotrophins. This 'adrenarche' stimulates the growth of pubic and axillary hair in girls.

The factors controlling the timing of puberty are poorly understood. One factor that has received attention recently is that a 'critical body weight' (48 kg) is necessary for the initiation of puberty: a theory that receives support from studies of anorexia nervosa (*see* later).

Delayed Puberty

Because of the wide variation in age of onset, 'delayed puberty' is hard to define: it is important to identify those patients who are destined to undergo normal puberty (constitutional delay) and those who have disorders that may be treatable. The causes fall into groups: those associated with primary gonadal defects and those that are secondary to hypothalamic or pituitary defects.

Primary Gonadal Disorders are characterised biochemically by the finding of raised levels of FSH and LH. The most common forms are those associated with the sex chromosome abnormalities discussed above (Klinefelter's syndrome and gonadal dysgenesis and their many variants). Other causes include treatment with cytotoxic drugs or radiation and congenital absence of the testes (anorchia).

Secondary Gonadal Failure due to gonadotrophin deficiency may be present, but undetected, from birth or may be due to an acquired lesion or trauma. These disorders may be accompanied by defects of other pituitary hormones or involve only LH and FSH (isolated gonadotrophin deficiency). Nearly all tumours (e.g. craniopharyngiomas) are accompanied by other disturbances (e.g. short stature, visual defects). Isolated gonadotrophin deficiency can be familial (the most common form is Kallman's syndrome) or sporadic. Other causes of secondary gonadal failure are: severe intercurrent disease, malnutrition, hypothyroidism.

the youngest end of the normal distribution curve. It is more common in girls and may occur in infancy. Sexual maturation in untreated patients follows the normal pattern.

Tumours of the CNS are the commonest cause in boys. Location is often difficult but there are often localising signs such as headaches or eye defects.

Other Central Causes include the McCune-Albright syndrome, neurofibromatosis, encephalitis and infiltration with granulomas.

hCG-Secreting Tumours cause sexual maturation in boys without activation of the hypothalamic-pituitary pathway. The tumours include pineal teratomas, hepatomas and chorionepitheliomas. The diagnosis depends on specific hCG assays.

Excess Androgen Secretion in congenital adrenal hyperplasia (see Table 3) or from rare adrenal carcino-

TABLE 4
DIFFERENTIAL DIAGNOSIS OF DELAYED PUBERTY

	Stature	Karyotype	Gonadotrophins	LH-RH response	Gonadal steroids
Primary Gonadal Disorders					
1. Gonadal dysgenesis (and variants)	↓	XO and variants	↑	↑	↓
2. Klinefelter's syndrome (and variants)	N or ↑	XXY and variants	↑	↑	N or ↓
Secondary Gonadal Failure					
1. Constitutional delay	↓	N	↓	↓	↓→N
2. CNS tumour	↓	N	↓	↓	↓
3. Isolated gonadotrophin deficiency (including Kallman's syndrome)	N	N	↓	↓	↓
4. Panhypopituitarism	↓↓	N	↓	↓	↓

Key:
N = normal; ↑ = increased; ↓ = decreased (All relative to age)

The Diagnosis of Delayed Puberty involves a careful clinical assessment which will normally lead to a presumptive diagnosis; radiographic and chromosomal studies will be confirmatory in many cases (Table 4). The most difficult differential diagnosis is that between isolated gonadotrophin deficiency and constitutional delayed puberty. Serial measurements of growth, gonadotrophins, oestradiol or testosterone will have prognostic value.

Precocious Puberty

As with delayed puberty, the definition of precocious puberty is made difficult by the wide variation in age of onset of normal puberty. Again, as with delayed puberty, the causes fall into two groups: those associated with excessive gonadotrophin secretion and those in which excess sex steroid secretion is unrelated to gonadotrophins.

Idiopathic Precocious Puberty. These patients are at

mas will cause virilisation in girls and sexual precocity in boys. Biochemically the tumours can be distinguished because they secrete large amounts of DHA and DHA sulphate and androgen secretion is not suppressed by glucocorticoids. In congenital adrenal hypoplasia, raised 17-hydroxyprogesterone and suppressibility with glucocorticoids are the cardinal features.

Excess Oestrogen Secretion causes precocious sexual development in girls. The most common form is the follicular cyst which may become autonomous; granulosa or theca cell tumours are less common. Plasma oestradiol levels may be very high but gonadotrophin levels will be low. Serial oestradiol measurements may be of value in monitoring metastases after surgery.

Diagnosis of Precocious Puberty will rest mainly on clinical evaluation and the results of non-biochemical ancillary tests.

The measurement of LH and FSH (and hCG where appropriate) and of testosterone in boys and oestradiol

TABLE 5
DIFFERENTIAL DIAGNOSIS OF PRECOCIOUS PUBERTY

	Enlargement of gonads	Skull X-ray (etc)	Gonadotrophins	LH-RH response	Steroid levels
1. Idiopathic (male and female)	↑	N	↑	↑	↑ testosterone or oestradiol
2. CNS tumours (male and female)	↑	Abnormal e.g. CAT scan	↑	↑	↑ testosterone or oestradiol
3. hCG-secreting tumours (males)	↑	Abnormal if CNS tumour	↑↑ hCG ↓ LH, FSH	↓	↑ or ↑↑ testosterone
4. Testicular tumours	asymmetrical	N	↓	↓	↑↑ testosterone
5. Ovarian tumours	asymmetrical	N	↓	↓	↑↑ oestradiol
6. Adrenal tumours (males)	↓	N	↓	↓	↑↑ 17 KS and DHA not-suppressible
7. Congenital adrenal hyperplasia (males)	↓	N	↓	↓	↑↑ 17 OHP, DOC or DHA depending on type

Key:
N = normal for age; ↑ = raised to pubertal levels; ↑↑ = raised to adult levels or higher; ↓ = deficient relative to age

in girls are also of primary importance in diagnosis and prognosis (see summary in Table 5).

The Menopause

In both sexes there is a waning of gonadal function in old age but it is in the female that this of clinical importance.

The primary events are a reduction in the responsiveness of the ovary to gonadotrophins, a gradual rise in LH and, particularly, FSH levels and a fall in oestradiol secretion. The postmenopausal ovary continues to secrete some androgen as do the adrenal glands: adrenal androstenedione is an important source (by peripheral aromatisation) of circulating oestrone after the menopause. The levels of oestradiol are significantly inversely correlated with the symptoms of oestrogen lack (e.g. vaginitis) but not with that other symptom of the perimenopausal period—'hot flushes', the cause of which remains unknown.

Oestrogen treatment of these symptoms (and of other features like osteoporosis) has become increasingly popular in recent years. Much needs to be known about the metabolic fate of these oestrogen preparations and the effect of the metabolites on receptors in tissues like the endometrium.

The '*premature menopause*' describes the cessation of menstrual cycles many years before the usual age of the menopause after years of normal function. The reason for this early follicular failure is not known.

There is also a spectrum of patients with small ovaries, relatively resistant to gonadotrophins. This has been called the 'resistant ovary syndrome'.

Disorders of Fertility

About 10% of marriages in the UK are involuntarily childless and as many again have only 1 or 2 children despite the wish to have more. Fertility is a relative state and subfertility can be found in either partner or both. Biochemical tests play almost no part in the management of male infertility but play a major role in establishing the diagnosis and monitoring the treatment of anovulation in the female.

Disorders of Menstruation and Ovulation

Definition. Generally a failure to ovulate is associated with amenorrhoea. In a woman with regular cycles, anovulation is a rare cause of infertility which in this case is likely to be caused by one of a number of physical (e.g. tubal blockage) or chemical (e.g. cervical hostility) factors.

There is no agreement on the length of time that periods are absent before amenorrhoea is diagnosed. The term 'primary amenorrhoea' is applied by clinicians to those who have never menstruated and 'secondary amenorrhoea' to those who previously menstruated. This is unfortunate because it cuts across the standard endocrine classification of diseases as primary in the target organ or secondary to pituitary failure. This distinction frequently confuses clinical biochemists and is illustrated in Table 6.

TABLE 6
SOME CAUSES OF AMENORRHOEA
(illustrating the different uses of the terms 'primary' and 'secondary')

Site of Defect	Clinical Presentation	
	Primary	Secondary
Primary gonadal	Gonadal dysgenesis	Menopause
Secondary to pituitary failure	Inherited gonadotrophin deficiency	Sheehan's syndrome

Clinical Findings. It is important to establish whether secondary sexual characteristics are present or not. The presence and degree of hirsutism is also significant. Some anatomical causes of amenorrhoea will be excluded by physical examination.

Ovarian Failure. All patients in this group have a pathognomic biochemical feature—raised gonadotrophin levels, especially of FSH.

Failure of Hypothalamic-Pituitary Control. Regular ovulation requires a finely integrated pattern of hormone secretion from ovary, hypothalamus and pituitary (*see* above). It is perhaps understandable that this pattern is not infrequently upset.

Thus amenorrhoea is not uncommon in association with the emotional strain of examination, a new job, leaving home etc. Functional disturbance of cyclical hypothalamic release of LH–RH, whilst a theoretical possibility, cannot be assessed biochemically in the human.

The commonest biochemical causes of amenorrhoea and anovulation are associated with abnormal feedback control of the hypothalamus and pituitary. This can occur in many conditions (e.g. obesity, Cushing's syndrome, during administration of oestrogen-containing pills) but the commonest, and most biochemically investigated, cause is the polycystic ovary syndrome.

Polycystic Ovary Syndrome. This is used to describe a heterogeneous group of conditions characterized by ovulatory failure, infertility, hirsutism, obesity (sometimes) and bilateral polycystic ovaries. Clinically there is an overlap with Cushing's syndrome and with virilising ovarian and adrenal tumours. The differential diagnosis of hirsutism is considered in the following section.

In patients with polycystic ovaries LH levels are higher than those of normal women with periods, but the midcycle surge is absent. FSH concentration is generally low or low-normal with little variation. This acyclic LH secretion is thought to be caused by excess circulating oestrone derived from peripheral aromatization of androstenedione of adrenal origin. Many studies have been carried out on steroid synthesis by the ovarian tissue but it is thought that the abnormality starts in the adrenal possibly at the time of adrenarche.

In most cases the diagnosis of the PCO syndrome is not a problem and it should be stressed that biochemical investigation plays little or no part in this process. Raised LH/FSH and E_1/E_2 ratios are typical but not pathognomonic findings.

Monitoring Treatment of Anovulation. The biochemical laboratory plays a very important role in the monitoring of ovulation induction regimes whatever the cause of the anovulation.

Assessment of the follicular response, for instance to exogenous gonadotrophins (FSH is usually administered) is crucial if overstimulation is to be avoided. This is done by serial measurements of either total oestrogens in urine or of plasma oestradiol. The rate of increase is used by the clinician to judge when (or whether) to administer the ovulating dose of hCG. If the rise is too low, the dose of FSH may need to be raised; if it is too fast it may be advisable to avoid ovulation and/or conception and to try a further cycle on a lower dose of FSH.

As mentioned earlier, plasma progesterone measurement is widely used as an index of ovulation having occurred and of corpus luteum function.

Anorexia Nervosa. This bizarre psychiatric state is now recognized as an extreme manifestation of a common cause of amenorrhoea—weight-related amenorrhoea. In these patients there is a regression to a prepubertal state with lower levels of gonadotrophins, particulary LH, and of oestradiol.

Hyperprolactinaemia. Increased secretion of pituitary prolactin is frequently associated with amenorrhoea, though whether this is because of an action on the ovary or at the level of the hypothalamus and pituitary is still unclear.

Hyperprolactinaemia is found in about 15% of patients with amenorrhoea (figures of up to 30 or 40% are quoted but are erroneously high) with or without a pituitary tumour. Galactorrhoea is present in about a third of these patients.

Sheehan's Syndrome. This is the name attached to hypopituitarism resulting from acute necrosis of the anterior pituitary secondary to postpartum haemorrhage or shock. With the modern improvements in obstetric care it is much less common.

'Postpill' Amenorrhoea. Amenorrhoea following discontinuation of oral contraceptive steroids is relatively infrequent and the incidence is higher amongst those with previous menstrual irregularity. Many authorities deny that there is a causal relationship between the pill and the amenorrhoea and in most cases spontaneous resolution occurs.

Other Endocrine Disorders. Anovulation and amenorrhoea are frequently seen in diseases of the adrenal gland such as Cushing's disease and Addison's disease (*see* Chapter 39) and of the thyroid gland (*see* Chapter 36). About 25% of patients with acromegaly have raised prolactin levels associated with menstrual dysfunction.

Male Infertility

In about 40% of infertile couples a male factor is present and semen analysis is an essential feature of the initial investigation. Abnormalities of hormone secretion are rare however and gonadal hormone assays play little or no part in the management of male infertility.

Hirsutism and Virilisation

The difficulties associated with the interpretation of androgen levels have been discussed above. Although DHT and other 5α-reduced metabolites are more active than testosterone their levels reflect peripheral metabolism of testosterone: they play no part in the aetiology of female virilisation. Testosterone is the major circulating androgen and androstenedione is its major precursor. DHA and its sulphate are of negligible importance except in the rare adrenal tumours.

Idiopathic Hirsutism

This most common, usually familial, cause of hirsutism has only mild abnormalities of androgen secretion and metabolism. Although the total plasma testosterone concentration is often normal, SHBG levels are reduced and free testosterone levels raised (Fig. 10). Plasma androstenedione is also often increased. The source of these androgens remains unclear: recent catheterisation studies (*see* below) have suggested the ovary as the major site.

Polycystic Ovary Syndrome

From the androgen view-point there is nothing to distinguish between idiopathic hirsutism and the polycystic ovary (PCO) syndrome. In neither condition does the ability to measure androgens aid in the diagnosis or treatment.

Virilisation

Virilisation (e.g. clitoral hypertrophy) occurs at higher levels of testosterone secretion than hirsutism. It is rare in PCO syndrome and the differential diagnosis usually rests between congenital adrenal hyperplasia (discussed earlier) and a virilising tumour.

The clinical syndromes produced by ovarian and adrenal tumours are indistinguishable though ovarian ones are far more common. Unfortunately only about half of these are palpable clinically. The major product of ovarian virilizing tumours is testosterone. (Note that the great majority of ovarian tumours are non-functional and do not produce hormonal signs and symptoms.)

Adrenal tumours commonly secrete DHA and if large amounts of this 17-oxosteroid are found either in the circulation or urine, an adrenal origin should be sought.

Preoperative localisation of the source of androgen secretion is difficult. Adrenal and ovarian suppression tests, with dexamethasone and oestrogen-containing pills respectively, have been widely applied but are being abandoned as unreliable indicators of androgen source. Selective catheterisation of adrenal and/or ovarian veins with measurement of androgen levels though technically difficult gives a much more accurate localisation and is being increasingly used. Iodocholesterol adrenal scanning and CAT scanning are also possible approaches.

Role of Biochemical Tests in Hirsutism and Virilisation

In summary, androgen measurements are of little or no value in hirsutism and virilisation.

(a) Measurements of 17-oxosteroids are valueless except in the very rare adrenal tumours.

(b) Plasma testosterone concentration may or may not be raised but, in either case, it gives no indication of the underlying pathology.

(c) Localisation with suppression tests has been found to give either ambiguous or erroneous results as often as correct results.

FURTHER READING

Yen, S. S. C. and Jaffe, R. B. (Eds.) (1978), *Reproductive Endocrinology*. W. B. Saunders, Philadelphia/London/Toronto.

Jacobs, H. S. (Ed.) (1979), *Advances in Gynaecological Endocrinology*. Royal College of Obstetricians and Gynaecologists, London.

Ross, G. T. and Lipsett, M. B. (Eds.) (1978), 'Reproductive Endocrinology,' *Clin Endocr. Metab.*, **7**, No. 3.

Makin, H. L. J. (Ed.) (1978), 'Biochemistry of Steroid Hormones,' Oxford: Blackwell.

Sharp, P. J. and Fraser, H. M. (1978), 'Control of Reproduction'. In: *The Endocrine Hypothalamus* (S. L. Jeffcoate and J. S. M. Hutchinson, Eds.). Academic Press, London/New York pp. 271–332.

SECTION XI

CLINICAL BIOCHEMISTRY OF PREGNANCY AND NEONATAL PERIOD

		PAGE
41.	THE CONFIRMATION AND MONITORING OF PREGNANCY	653
42.	PRENATAL DIAGNOSIS OF INHERITED METABOLIC DISEASE, NEURAL TUBE DEFECTS AND LUNG MATURITY	663
43.	PAEDIATRIC CLINICAL CHEMISTRY: Some Special Problems	682

41. THE CONFIRMATION AND MONITORING OF PREGNANCY

G. R. WILSON and A. I. KLOPPER

The confirmation of pregnancy
   Chorionic gonadotrophin
   Detection of hCG in pregnancy
   Pregnancy specific β_1 glycoprotein
   Other biochemical parameters

Biochemical monitoring of feto-placental function
   Plasma or urinary assays
   Steroid assays
   Assay of placental proteins

In order to confirm a pregnancy it is necessary to detect in the body fluids some substance clearly, and if possible exclusively, associated with pregnancy. In this context specificity and sensitivity are of paramount importance and the accuracy of the measurement matters little. On the other hand, reliable quantitative measurement of some product of the pregnancy, is a *sine qua non* for the monitoring of the pregnancy. They are separable activities and should be evaluated separately.

THE CONFIRMATION OF PREGNANCY

In human pregnancy, as soon as the ovum is fertilized it needs to signal this fact to the mother. For the pregnancy to survive, adaptations of maternal physiology such as the diversion of the menstrual cycle from its usual course, have to start at this early stage. It is now 57 years since one of the embryonic signals, chronic gonadotrophin, which enable the mother to recognise that she is pregnant, was first identified.[1] In recent years it has become evident that this process starts while the morula is still being conveyed down the Fallopian tube and, by the time the blastocyst starts to implant, communications with the tube, the endometrium, the myometrium and the ovary have already been established.[25] Some of these signals are not peculiar to pregnancy and most operate at a local level where they cannot be intercepted. Just how soon these fetal signals can be detected is open to question. It has been claimed that hCG can be found at the time of implantation and evidence has been accumulating that it can be detected between then and the due date of the next menstrual period. There is no doubt that with very sensitive methods, embryonic signals can be detected at the time of the first suppressed period or soon after.

Originally, pregnancy tests were based on the belief that products peculiar to pregnancy or at least to trophoblastic tissue could be demonstrated in the blood and urine of pregnant women. That belief has now been eroded by the discovery that the capacity to synthesise certain proteins such as hCG or hPL is not peculiar to trophoblastic tissue but may occur with other primitive cells such as neoplastic tissue or even normal fibroblasts. It is possible that all cells possess the code to make these proteins but that it is suppressed in adult or specialized cells. It is possible that the distinction between trophoblast and other undifferentiated cells in terms of their capacity to make these pregnancy-associated proteins is largely a quantitative one. Embryonic cells simply make more, but there is an overlap with neoplasia. This imposes severe constraints on the biochemical recognition of very early pregnancy. Not only does the test have to be sensitive enough to detect traces of material, it has to be accurate enough to discriminate these from even smaller amounts coming from non-trophoblastic tissue and specific enough to differentiate both from background noise.

Given such a high requirement for specificity, the only technique capable of meeting the demand is some form of immunological recognition of the pregnancy-associated protein. The high sensitivity can be met by radioactivity measurement or the detection of enzyme action. In the present state of the art this means radioimmunoassay with the possible later development of enzyme immunoassays.

A further complication arises from the immunological detection of pregnancy-associated proteins whether the endpoint is measured by radioactivity or by enzyme action. These proteins have structural similarities and therefore antigenic determinants in common with other non-pregnancy maternal proteins. This is most notably so in the case of hCG which has an α chain in common with LH, FSH and TSH. Antibodies directed against the whole molecule of hCG will therefore cross react with the α-chain of these pituitary hormones. At the time of the first missed period an antiserum to hCG cannot discriminate between LH (which is normally present at this stage) and hCG. Fortunately the β-chain of the hCG molecule is unique to this protein and antisera prepared against this fragment are specific for hCG.

If pregnancy tests were required only in normal pregnancy the problem would be straightforward. Unfortunately the most demanding use of pregnancy tests is when some pathology, such as an ectopic pregnancy or a threatened abortion, is present. In these circumstances the production of pregnancy proteins is reduced and here lies the real challenge—to produce a reliable test when it is most needed.

Nonbiochemical techniques for detecting pregnancy have advanced greatly in recent times. Ultrasonic scanning can demonstrate a blastocyst in the uterus soon after the first missed period. Although the fetal heartbeat can only be detected at a stage when the pregnancy is clinically obvious, its demonstration is a secure and reassuring evidence of a live fetus.

Chorionic Gonadotrophin

hCG is a protein produced by the trophoblast and secreted largely into the maternal circulation, the concentration in the fetal blood being much lower. Although it was at first thought to be a product of the cytotrophoblast, more recent immunofluorescent localization studies have shown it to be present in the syncytiotrophoblast.[23] It is a glycoprotein with a molecular size of 45 000–50 000 daltons and a remarkably high carbohydrate content—30% by weight.

hCG like hFSH, hTSH and hLH has two dissimilar amino acid chains, noncovalently linked. These subunits have been designated α and β. The subunits can be dissociated, e.g. in strong urea solutions, and then separated. The amino acid sequence of each subunit has been determined and it transpires that the α-subunit of hCG is very similar to that of the α-subunits of the pituitary hormones, particularly LH. For this reason antisera raised against the whole molecule of hCG or its α-subunit cross react strongly with LH. At first this did not give rise to problems in pregnancy diagnosis as the assays were not sensitive enough to react with the LH concentrations occurring in non-pregnant women. When the tests were done at 6–8 weeks from the last period there was already so much hCG present that false negative results were rare but LH levels were low enough to give rise to false positive results only seldom. When termination of early pregnancy became more common, clinical as well as research pressure for achieving a secure diagnosis of very early pregnancy increased. This led to the development of antisera against the β subunit of hCG and eventually an assay method based on such an antiserum was devised.[32]

Assay Techniques. Bioassays depending on changes induced in the reproductive organs of mice, rabbits, rats or toads are now only part of the history of pregnancy diagnosis. It should not, however be forgotten that the immunological tests which have replaced them are demonstrations only of the presence of hCG as an antigen, not of its biological activity. From the qualitative point of view, i.e. pregnancy detection, this does not matter, but if the object of the test is to measure the presumed activity of hCG in pregnancy maintenance, the quantitative difference between biological and immunological activity may be important.

The immunological approach has been exploited in a variety of modern techniques for pregnancy diagnosis; complement fixation, precipitation, immunodiffusion, immunoelectrophoresis, agglutination inhibition and radioimmunoassay. Complement fixation was introduced by Brody and Carlstrom[3] and proved very useful. For pregnancy detection it has now been largely replaced by simpler methods generally based on agglutination inhibition, which can be put out as cheap, stable pregnancy diagnosis kits. Gel precipitation, which depends on a visible demonstration of the insoluble antigen–antibody complex was brought out at the same time as the complement inhibition technique[24] but has also been largely overtaken by the commercial success of the agglutination inhibition methods. The agglutination inhibition tests stem from the work of Wide and Gemzell[33] who used the agglutination of erythrocytes as an endpoint or of Little[22] who used the agglutination of coated latex particles. The agglutination inhibition tests have a serious disadvantage in that the antiserum used crossreacts with LH. They are therefore deliberately set to work at poor sensitivity so that they can pick up the high levels of hCG obtaining after 6 weeks of pregnancy but not the LH levels of a non-pregnant woman. In the first two weeks after the missed period they are unreliable, which is precisely when accurate diagnosis is most urgently needed. According to Landesman and Saxena[19] at this time the Pregnosticon Test (Organon) showed 53·3% positive results in confirmed pregnancies and 91·4% negative tests in women subsequently found not to be pregnant.

Given the shortcomings of agglutination inhibition tests there remains a substantial use for radioimmunoassay methods. Indeed for research purposes during the first two weeks after the missed period, a radioimmunoassay using antiserum to βhCG is a *sine qua non*. From the clinical point of view it has a serious disadvantage in that the test takes at least 72 hours. There are often situations when the clinician cannot afford to wait that long for an answer. This led to the introduction of a rapid radioreceptor assay.[26] Although sensitive enough to detect hCG at very low levels, the receptor is not specific for hCG and the old problem of crossreaction with LH again arises. Modifications have been designed to cope with the midcycle surge of LH or the high LH levels obtaining in menopausal women, and the originators of the test view it with some enthusiasm[19] but it has not gained wide acceptance in Britain.

Detection of hCG in Pregnancy

There is some controversy, perhaps resulting from methodology, about how soon after fertilisation hCG can be detected. Immunofluorescence experiments suggest that an hCG-like protein could be detected in the two-cell stage of a cleaving mouse embryo removed from the Fallopian tube. Of course it is a long jump from the tube of a mouse to the blood or urine of a woman but it does raise the possibility that the human morula or blastocyst may secrete hCG. The use of the specific hCG-β radioimmunoassay has demonstrated the presence of hCG at the time of initiation of implantation[18] and in peripheral blood it can generally be detected 9–12 days after ovulation.[31] There is little doubt that modern techniques make it possible to diagnose pregnancy before a woman has missed her period. This is a formidable accomplishment when one considers the relative size of the blastocyst (about a pinhead) and the maternal organism throughout which it distributes its signal.

In early pregnancy hCG concentration rises very rapidly. After its first appearance the concentration doubles every 48 hours. This rapid rise is clinically useful both in the detection of pregnancy and in the diagnosis of pregnancy pathology. A test which is doubtful will become positive when repeated if a pregnancy is present and if quantitative measurements do not show a rise the suspicion of a failing gestation is raised.

Clinical Applications

The detection of normal pregnancy at an early stage remains the main clinical application of hCG assay. Of course this is often only a matter of social convenience as the course of events will spell out the answer in a week or two, but early confirmation is a matter of some importance if the question of pregnancy termination arises. This is particularly pertinent to the technique of menstrual regulation by suction curettage. When applied to the diagnosis and management of ectopic pregnancy, threatened abortion and hydatidiform mole, hCG measurements have some limitations. In the case of ectopic pregnancy hCG measurements have largely been replaced by laparoscopic visualisation of the pelvis. As a prognostic measure in threatened abortion hCG measurements have not met with universal acclaim and the diagnosis of incomplete abortion may be obscured by the long half-life of hCG and the persistence of viable placental tissue after fetal death. Although the high levels of hCG associated with hydatidiform mole overlap upper normal values, the use of hCG assays to detect persistent trophoblast tissue after evacuation of a mole or choriocarcinoma is an essential clinical tool.

Pregnancy Specific β_1 Glycoprotein (SP$_1$)

A β_1 globulin present in the blood of pregnant women, but not in non-pregnant women was first described in detail by Tatarinov and Masyukevich.[29] This protein was later isolated and purified from placental extracts and designated as Schwangerschaftsprotein ein (SP$_1$).

The levels of this protein are very low in early pregnancy and can only be detected by sensitive radioimmunoassay methods. Grudzinskas et al.[7] used a method with a sensitivity of 8 μg/l to determine Sp$_1$ levels in early pregnancy. It was originally claimed that SP$_1$ was detected in the plasma of one woman seven days after ovulation, and in two others within 14 days of ovulation. However, it was suggested that the timing of ovulation may have been imprecisely defined, since it was based on one or more of the following criteria: menstrual history, basal body temperature records and plasma progesterone levels. Using a slightly modified assay with a sensitivity of 15 μg/l, it was later found that SP$_1$ appeared in nine subjects 18–23 days after ovulation.[7] In this experiment, ovulation was timed by the mid cycle LH surge. Once detected there was an exponential rise in SP$_1$ levels, the doubling time being 2–3 days. The rate of increase declined slightly about 40 days after the LH surge. βhCG was also measured and found to be detectable 10–16 days after ovulation. The interval between the appearance of βhCG and SP$_1$ was not found to be constant. These clinical studies indicate that the measurement of circulating SP$_1$ may be a useful adjunct in the diagnosis of pregnancy and may be the only biochemical method available when a patient has been treated with hCG to induce ovulation. The similarity in the pattern of increase and the molar concentrations of SP$_1$ and βhCG suggest that the limitations on the use of the SP$_1$ are technical rather than biological.

Other Biochemical Parameters in the Confirmation of Pregnancy

In theory any of the materials specific to the placenta could serve as a means of identifying pregnancy. In practice protein hormones such as placental lactogen or placental enzymes such as heat stable alkaline phosphatase have never caught on. They are expressed too late in pregnancy and are less easy to measure than the simple agglutination techniques which serve for hCG.

The steroid hormones of the fetoplacental unit are of course not specific to it but curiously enough progesterone assays may prove helpful in the diagnosis of pregnancy, particularly when they are done on patients under close survey in a fertility clinic. Finding luteal levels of progesterone does not prove that a patient who has amenorrhoea or is past her period date is pregnant, but proliferative phase values prove that she cannot be pregnant. Equally if plasma progesterone levels remain elevated for more than a fortnight after ovulation has been induced, the likelihood of a pregnancy is high. Indeed persistent luteal levels of progesterone are usually the first evidence of pregnancy in a patient treated with human menopausal gonadotrophin.

BIOCHEMICAL MONITORING OF FETO-PLACENTAL FUNCTION

As with the confirmation of pregnancy, biochemical monitoring of fetal wellbeing involves measuring a fetoplacental product in maternal body fluids. But otherwise it is quite a different kettle of fish. For one thing, fetal monitoring almost always takes place in late pregnancy, when the concentration of fetal products in the maternal fluids is much higher and the sensitivity requirements for monitoring are therefore much less stringent than for the detection of pregnancy. Specificity is also less important for monitoring. When measuring hCG to detect pregnancy it is essential that the technique should pick up only this protein and no other material which may not be peculiar to pregnancy. For monitoring, all that is necessary is that the figures for whatever is being measured should have some bearing on fetal wellbeing; regardless of how many factors may have contributed to those figures. Indeed one of the most successful monitoring methods, the assay of total urinary oestrogen excretion, measures a whole spectrum of oestrogens and the colour endpoint is influenced by a variety of factors other than the oestrogen content of the urine. On the other hand precision, i.e. the variation of replicates in the same assay or of repeated assays on the same sample, is of considerable importance in fetal monitoring. Almost all the substances which might be measured have a large variation from time to time in the same subject and from subject to subject. If there is a coefficient of variation in excess of 10–15% in interassay precision, which has to be allowed for in addition to the physiological variability, it detracts from the value to be ascribed to any assay. Apart from their convenience, the great attraction of automated methods of assay is their precision. They are generally poor in specificity and the results may be far from the scientific truth, but they have the charm of telling the same lie consistently. Accuracy, i.e. the degree to which recoveries of known amounts of material approach 100%, is of more importance in making comparisons with the results of assays from other laboratories or reported in the literature. As long as a laboratory is using its own established norms or doing repeated assays on the same patient, the clinical import of the results is the same, however inaccurate, within limits, the assay might be.

Plasma or Urinary Assays

Although biological fluids such as liquor amnii or saliva in which fetoplacental products may be present are available for assay, in practice the monitoring of fetal wellbeing is done on maternal blood or urine. In the case of blood this involves the decisions whether to use plasma or serum and, for urine, what length of collection time,—spot collections, overnight, 24 hours or 48 hours. All the blood parameters which will be considered in this section can be measured in either plasma or serum and it is a matter of indifference which is used or what anticoagulant is employed to obtain plasma. In the case of urine the argument about collection time or indeed the use of creatinine ratios is unresolved. By and large the trend is toward 24 h urine collections and only these will be considered in the choice between blood and urine.

The pros and cons of assays on blood or urine have been reviewed in detail[13] and will be only briefly summarised here. Plasma assays are one step nearer the fetus and if the object of the assay is to determine the population of biologically active molecules at the target site, there is little doubt that plasma is preferable to urine. But in monitoring pregnancy the purpose of the assay is to assess the rate at which the fetoplacental unit is producing the material, and for this urinary assays may serve as well, or on occasion, better.

In terms of the assay of fetoplacental products, blood is a turbulent pool with many streams flowing in and out, while urine is an endpoint, a reservoir. In general the residence time of any placental product in the plasma compartment is short and may, in the case of some hormones, be only a few minutes. Most products undergo metabolic change between their release from the fetoplacental unit and their appearance in the urine. Therefore, in addition to their physical movement in and out of the plasma compartment, the hormones in blood are in metabolic movement, changing from one molecule to another. When, as is usually the case, one is measuring only one molecular species in a series of related compounds, the amount of this material may be related more to the rate of turnover in the series, than to the secretion rate of the original hormone.

Clinicians are familiar with urinary assays and are apt to regard plasma assays as a more sophisticated version of these. Nothing could be further from the truth. Although it is generally the *concentration* of a fetoplacental product in urine which is measured, the total volume of the urine is known, and what is reported is the total excretion over 24 hours. In the case of plasma assays, all that is known is the *concentration* at the fleeting moment when the blood was drawn. This is a figure which, *inter alia*, is dependent on the volume of the plasma pool which contains the hormone. This varies with the size of the person—a fact which is seldom taken into account in comparing levels in different people. Nor is the plasma volume static from day to day in the same person. This is particularly pertinent to patients who have been vomiting or who are otherwise dehydrated who may have greatly increased concentration by a reduction in plasma volume. Pregnant women may increase their blood volume by 40% during gestation. The increase is variable from one woman to another and may make for differences when the production rate is similar.

Plasma concentrations are the outcome of three variables: rate of inflow, rate of outflow and pool size. In pregnancy monitoring, plasma assays are based on the assumption that the plasma concentration is proportional to the inflow from the fetoplacental unit. This is a gross over-simplification. The inflow is not composed of a single stream from the placenta, nor is there a single channel of outflow via the urine. Most products pass readily from the plasma into adjacent compartments such as the interstitial fluid. In the case of the steroids they are held in specialized pools outwith the plasma such as the body fat and the enterohepatic circulation. Steroid concentration in maternal plasma may be determined, not by the rate of production in the fetoplacental unit but by inflow or outflow into or from these isolated compartments.

The main route of outflow for many fetoplacental products is via urine. This raises the problem of the effect of kidney function on both plasma concentration and urinary excretion. In renal disease the outflow may be impeded, giving rise to a low urinary excretion when fetoplacental production is normal. On the other hand, high blood levels may result from inadequate excretion. The paradoxical situation of low urinary values and high plasma concentration is rare and has not been convincingly documented. Nevertheless it should be kept in mind when reporting results on pregnant women who are particularly liable to impairment of renal function.

In the end a blanket statement governing the choice for all fetoplacental products is not possible. Each material has to be considered in its own right and in the context of the clinical situation in which the assay is being done. This much is clear: plasma assays and urinary outputs are not equivalent; they measure different aspects of pregnancy metabolism. To say one is a better reflection of fetal wellbeing than the other is untrue: each is a mirror held up to a different aspect of the truth, and each reflection is a distortion.

The Normal Range

The traditional way of describing the normal range of any of the biochemical parameters which might be used to assess placental function is as the range covered by twice the standard deviation on either side of the mean value for the stage of gestation. By definition some 95% of all pregnancies would give values within this range. It has already been indicated that there may be a great deal of difference between one normal individual and another. This makes the normal range so large that only the extremes of pathology fall outside it; a situation where the clinician seldom needs biochemical measurements to tell him what is already staring him in the face. Describing a normal range in terms of a standard deviation implies that the values have a Gaussian distribution with roughly as many values above the mean as below it. Inspection of the measurements in the case of almost every material which can be assayed to determine fetal wellbeing shows that this is not the case. The values have a skewed distribution with more below than above the mean. The mean value is pulled up and the range enlarged by a few wild high values. It follows that the normal range is properly described by a logarithmic transformation of values or, for those without a mathematical bent by a centile distribution. Either of these has the effect of raising the lower normal border, thus enlarging the gap between the lowest normal and the highest abnormal values.

Categorisation of Pathology

A great many diseases of pregnancy may put the fetus at hazard. It is unlikely that all diseases will affect the production of a particular fetoplacental product to the same extent, or indeed affect it at all. 'Placental insufficiency' and similar vague diagnostic tags subsume a great many different pathological processes. The value of any particular parameter can be judged only in the context of a particular disease. Thus oestriol production is often markedly depressed by pre-eclamptic toxaemia or retarded fetal growth, but the changes in diabetic pregnancy are variable and there is no notable change at all in fetuses affected by rhesus incompatibility.

The Criteria for Evaluating a Test

It is often possible to demonstrate that the mean value in a group of patients suffering from a particular disease is significantly different from the mean of normal subjects. This is useful in determining which categories of disease should be monitored by which test, but is not of much use to the clinician who wants to know what value to place on a particular result in a single patient. Fetal monitoring may be done in one of three ways. Firstly, as a *diagnostic measure*. Biochemical measurements are of little use for this. There are no changes characteristic of particular diseases and the normal range is so large that only severe disease, easy to diagnose by other means, gives rise to values clearly outwith the normal range. Secondly, as a *control measure* by doing serial assays and judging, not in terms of an arbitrary normal range, but by observing the trend within a single patient. Used in this way biochemical monitoring may yield just the information which enables a clinician to decide when to terminate a pregnancy. In this respect substances like unconjugated oestriol (plasma half-life 6 minutes) serve much better than say SP_1 (plasma half-life 22 hours) as they will respond more quickly to changes in fetoplacental production. Thirdly, measurements may be done as a *screening device*, either to locate an at-risk group in a total population or to determine the degree of fetal hazard in a group known to be at risk, e.g. in terms of an unfavourable obstetric history. Gordon et al.[6] have made some valuable suggestions about assessing the value of any technique for biochemical monitoring when screening a population. They decided that in the group the risk factors for which they wished to screen were perinatal death, a birthweight below the 5th centile for gestation, a major congenital deformity or an Apgar score of 3 or less at 1 minute. If any of these were correctly predicted by, say, a low hPL value, this constituted a true positive (TP). A false positive (FP) resulted when a normal baby was born following a low hPL. If on the other hand both hPL and baby were normal this constituted a true negative (TN) and if hPL were normal but the baby was born with one or other of the risk factors present, this constituted a false negative (FN). The screening value of the biochemical parameter could then be assessed in terms of sensitivity, predictive value and relative risk. Sensitivity was defined as the percentage of true positives among all the affected babies, (TP/TP + FN). The predictive value was the true positive as a percentage of all the positive results (TP/TP + FP) while the relative risk was the ratio of the percentage of positives which were true positives to the percentage of negatives that were false negatives (TP/TP + FP: FN/FN + TN).

This form of analysis of the data makes it possible to make valid comparisons of the usefulness of various parameters in population screening. Automation makes it possible to screen large populations. Such an analysis makes it possible not only to decide which parameter is best in terms of a defined set of risk factors but also how much the results would be improved by a combination of methods. It deserves to be more widely adopted.

Steroid Assays in Biochemical Monitoring

The fetal circulation contains a bewildering variety of steroids. Some of these are probably also secreted by the placenta directly into the maternal circulation; others are transferred from the fetal circulation. Some steroids are metabolized in the placenta during their passage from fetal to maternal circulation, e.g. oestriol sulphate is hydrolysed to the free steroid. Maternal plasma steroids of fetoplacental origin therefore reflect the capacity of the placenta to generate steroids and of its ability to transmit them from the fetal circulation. The picture is further complicated by the fact that, although maternal and fetal steroids are not in equilibrium, tagged steroids injected into the mother soon appear in the fetal circulation and there is clearly a backflow of steroids from mother to fetus.

Interest in steroid assay as a means of monitoring the fetus centres on a trihydroxy-phenolic compound—oestriol. There is a fetal element added to the placental role in the genesis of this oestrogen, and it is the biosynthesis of oestriol which led Egon Diczfalusy to formulate his thesis of the fetus and placenta as discrete parts of a single endocrine unit, the fetoplacental unit.[4] The placenta uses an androgen, dehydroepiandrosterone, as raw material for the synthesis of oestrogens. For the most part the dehydroepiandrosterone comes from the fetal adrenal, although the placenta is able to utilise maternal dehydroepiandrosterone to some extent. The placenta is very poor in the enzyme system which hydroxylates steroids at C16. As oestriol contains a C16-hydroxyl group the placenta has to rely on substrates which have been hydroxylated elsewhere. The C16-hydroxylating capacity of the fetal liver and adrenal is high and 16α-hydroxy-dehydroepiandrosterone of fetal origin is the source from which the placenta synthesizes oestriol. In this way fetal liver and adrenal enzymatic activity are rate limiting steps in the production of oestriol.

The nonpregnant woman also produces oestriol, but does so by metabolism of ovarian oestradiol. In pregnancy, when the fetoplacental unit produces oestriol directly, not via oestradiol, the proportion of oestriol in blood and urine increases greatly. So much so that many methods of oestriol assay make little or no attempt to exclude other oestrogens in the assay.[5,9] Sometimes the fact that a variety of oestrogens contribute (unequally and in combination with other factors) is acknowledged by calling the assay one of 'total oestrogens'. After years of acrimony between proponents of various assay techniques it turns out that the clinical import of the assays is much the same whatever the method and whether the measurement is done on blood or urine. We propose to take them all together under the biochemically questionable title of oestriol assay.

Two physiological characteristics of oestrogen metabolism limit the usefulness of oestriol assays in pregnancy. The first is the natural variability of oestriol levels from time to time. In this respect blood assays have a modest advantage over urinary excretion measurements. From day to day the average coefficient of variation in late pregnancy is 23 per cent[11] whereas in plasma the coefficient of variation is 15 per cent.[17] The second disadvantage of oestriol assay is the large spread of normal values, extending to some 30% on either side of the mean for both urine and blood.[14] This large normal range makes overlap with all but extremes of pathology inevitable and detracts from the usefulness of the assays as a screening device while the variability from time to time adversely affects the reliability of the assays as a control measure in serial assays.

It is essential that any laboratory undertaking oestriol assays should be working to an established normal range, for the stage of gestation. To some extent this is determined by the particular method in use but there is not an overwhelming case for establishing an individual laboratory norm. Accepted normal ranges for each method are in the literature and, provided careful quality control is exercised, the individual laboratory results should not differ substantially. Naturally it is the lower normal limit which is of interest in a clinical context. This will depend on whether the skew nature of the normal distribution has been taken into account. In broad terms the lower limit in late pregnancy for most urinary methods is about 12 mg/24 hours. The total (free plus conjugated) plasma oestriol has a lower normal limit around 400 nmol/l. The lower limit of unconjugated plasma oestriol is around 28 nmol/l.

The usefulness of oestriol assays varies a good deal from one category of obstetric disease to another and can only be assessed in terms of a specific pathology. Oestriol production appears to be related to growth processes in the fetus and the assays are probably most valuable in the management of retarded fetal growth. Unfortunately, although this is a condition which lends itself to definition, e.g. below the 10th centile of birthweight for gestation, it is heterogenous in its composition. Most clinicians would agree that deformity, maternal hypertension, genetic factors and undefined placental malfunction may each lead to retarded fetal growth. Happily oestriol production appears to be more related to the consequence than the cause and there is substantial evidence for reduced oestriol production in retarded growth, regardless of its cause. This is illustrated in Fig. 1 which shows the mean and standard deviation of urinary oestriol excretion in a group of women who each gave birth to a severely growth retarded baby (birthweight below the 5th centile). The values are expressed as percentages of normal for the stage of gestation as measurements were made at different stages of late pregnancy in different women. It can be seen at a glance that oestriol excretion in this group is markedly decreased.

Group comparisons are not, however, very helpful to the clinician who has to make up his mind about individual patients. To some extent the value of the assays will depend on the purpose for which they were done. Figure 2 shows the plasma oestriol concentration at 38 weeks in another group of patients, all of whom produced a baby below the 10th centile of birthweight

PERCENTAGE OESTRIOL EXCRETION IN SEVERE FETAL
GROWTH RETARDATION COMPARED WITH NORMAL MEAN

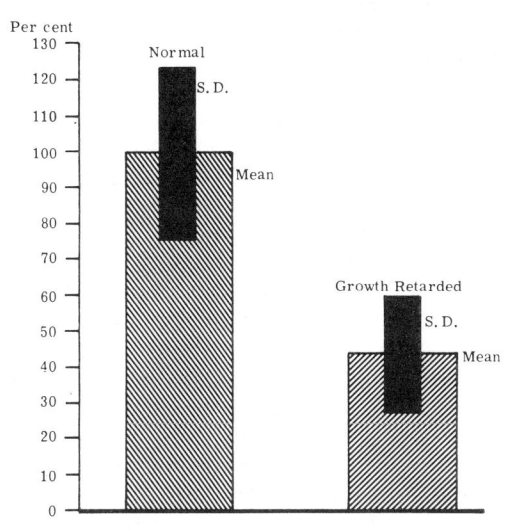

FIG. 1. Urinary oestriol excretion in severe fetal growth retardation expressed as a percentage of the normal mean for the stage of gestation. (From Klopper, A.: The excretion of urinary oestriol in obstetric disease. *Excerpta Medica Int. Congr.* Series 19, p. 546, 1972).

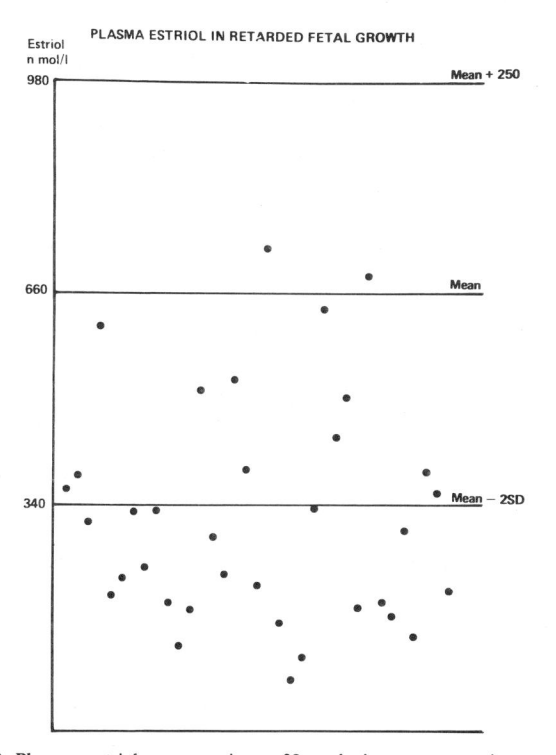

FIG. 2. Plasma oestriol concentration at 38 weeks in women carrying a growth retarded fetus. (From Klopper, A., Jandial, V. and Wilson, G.: Plasma steroid assays in the assessment of placental function. *J. Steroid Biochem.*, **6**, 651, 1975).

would have given a false negative answer in 37 per cent of patients. Of course in practice such tests are not taken in isolation and happily their performance is a good deal better when set sensibly in the context of the whole clinical picture. When considered in terms of a clinically defined risk group, oestriol assays can be helpful not only in defining the likelihood that a patient may have a growth retarded fetus but also by indicating the degree to which the fetus is affected. Figure 3 shows the results of urinary oestriol assay in a group of patients each of whom gave birth to a mildly growth-retarded baby (below the 10th centile). The values tend to cluster round the lower normal border. Comparative results

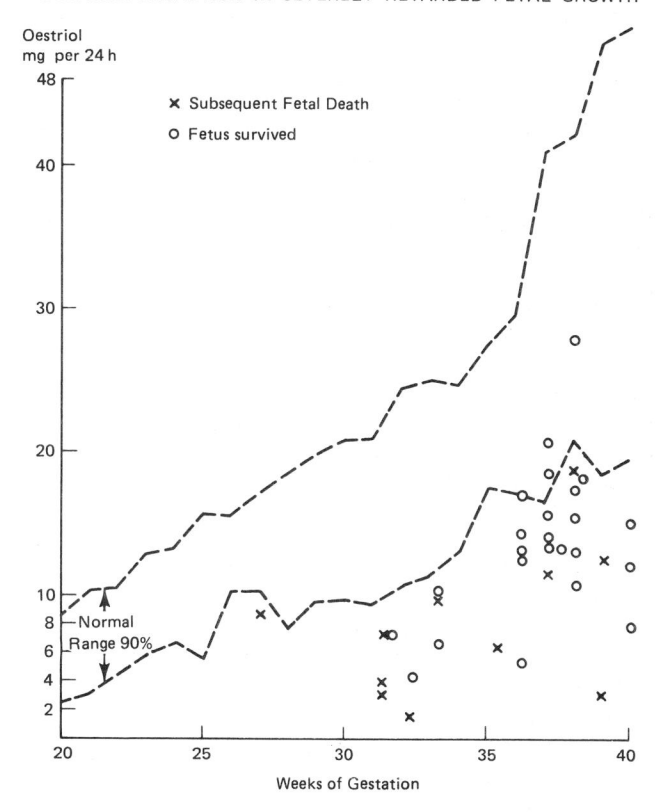

FIG. 3. Urinary oestriol excretion in mildly retarded fetal growth. (From Klopper, A.: Assays of urinary oestriol as a measure of placental function. In Cassano, C. (Ed.) *Research on Steroids*, **2**, 63, Rome, Il Pensiero Scientifico, 1966).

for gestation. Twenty-two of these patients had a plasma oestriol lower than 2 standard deviations below the normal mean, but 13 were within the normal range, 2 being slightly above average. If therefore a depressed plasma oestriol concentration had been used as the sole diagnostic criterion of retarded fetal growth the test

from a group with more severely affected babies (below 5th centile) are shown in Fig. 4. The values are more clearly depressed with a tendency for the lowest values to occur in the most severe group, those which ended in perinatal death.

The value of oestriol estimations in retarded fetal growth lies not so much in being of help in the diagnosis in cases where clinical suspicion is roused, as in the management of the established case. The obstetrician is challenged by the problem of deciding at what point the baby is safer out of, than in, the uterus. In this case serial assays can be very helpful provided two caveats are

OESTROL EXCRETION IN MILDLY RETARDED FETAL GROWTH

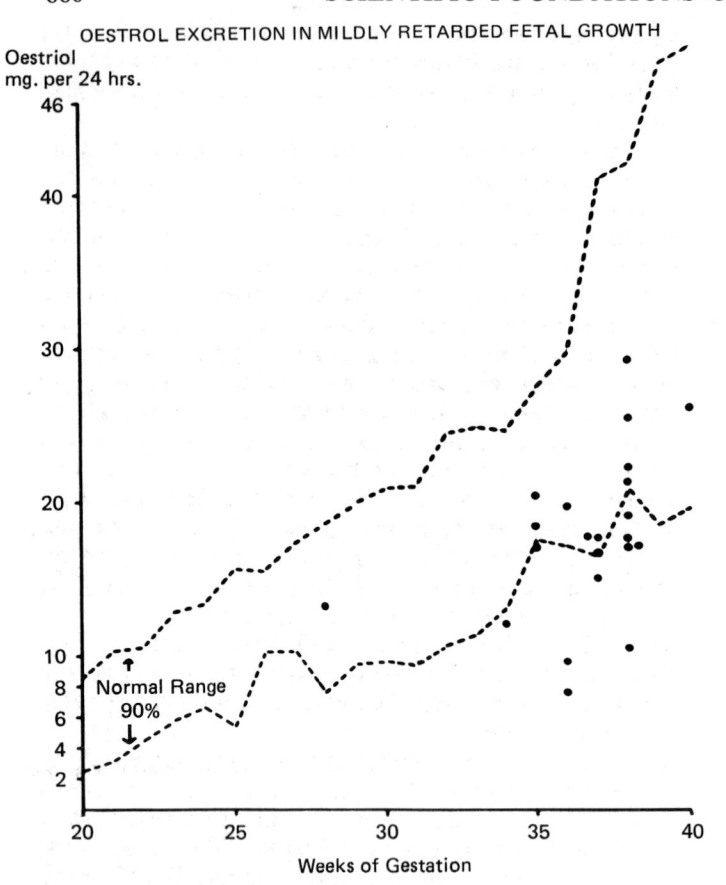

Fig. 4. Urinary oestriol excretion in severely retarded fetal growth. (From Klopper, A.: Assays of urinary oestriol as a measure of placental function. In Cassano, C. (Ed.) *Research on Steroids*, **2**, 63, Rome, Il Pensiero Scientifico, 1966).

borne in mind. The first is the high normal variability (coefficient of variation 15 per cent) and the second is that changes are seldom dramatic. It was noted 17 years ago that oestriol production increases abruptly at about 34 weeks pregnancy.[15] It was speculated that this rise represented a new, purely fetal element, in oestriol biogenesis. It is this surge of oestriol production in late pregnancy which is absent when fetal growth retardation is present, and becomes obvious on serial assay.

The best use of biochemical monitors may not be for a defined risk group but as screening devices for a whole obstetric population. Oestriol has a very short plasma half-life and both urinary and plasma oestriol concentrations are highly variable from time to time. It may therefore not be the best means of screening a population by single assay, although the use of urinary assays for screening has been regarded with considerable enthusiasm by one Australian group.[2] They found that low oestriol excretion occurred in 15 per cent of a large obstetric population and that 21 per cent of the low oestriol group delivered a growth retarded baby while only 3 per cent of the normal oestriol group did so.

Pre-eclamptic toxaemia is another obstetric disease in which oestriol assays are used to monitor the fetal condition. They are of course not needed for diagnosis of the condition and the purpose of the assay is to determine whether complications are likely to occur or to monitor deterioration in the fetal state by serial assays. Beischer and Brown[2] found that normal oestriol excretion, which occurred in 73 out of 103 patients with pre-eclampsia, gave a good prognosis as none of these had any perinatal deaths. On the other hand 7 perinatal deaths occurred among the 30 patients with low oestriol levels. Just how useful serial assays are in patients with pre-eclampsia is difficult to say. The literature abounds with examples but there is an element of selection about these. Illustrative examples of typical oestriol changes in pre-eclampsia are typical of nothing but the prejudices of the person who made the selection.

In obstetric practice the commonest technique for identifying an at-risk group in the clinic population is in terms of obstetric history. Klopper[12] has adduced evidence that relatively early in pregnancy, say 30 weeks, it is possible to sort patients with a poor obstetric history into two groups on the basis of their oestriol output, and that perinatal mortality is concentrated in the group with a poor oestriol output.

Many investigators have addressed themselves to the problem of assessing fetoplacental function in diabetic pregnancy and oestriol assays have featured in this context. Their value is controversial. Although some view them with enthusiasm, diabetic pregnancy is such a protean disease that oestriol production is apt to be variable and it is not advisable to manage pregnancies relying heavily on changes in oestriol level. As far as other conditions for which oestriol assay might be used are concerned, one is faced with an array of decreasing usefulness ranging from prolonged pregnancy on one hand, through antepartum haemorrhage and premature labour to rhesus incompatibility on the other. Oestriol levels bear no relationship to the fetal state when the fetus is affected by Rh isoimmunisation.[16]

When the introduction of radioimmunoassay extended the field of assay from urine to blood, several other steroids, not usually present in urine to any great amount, were considered as means of fetal monitoring. The principal compounds taken into consideration were oestradiol-17β, oestetrol and progesterone. The plasma concentration of oestradiol, like that of oestriol, increases greatly with advancing gestation and at first measurements of this oestrogen seemed promising. In the outcome these hopes were disappointed. The placenta can utilise maternal dehydroepiandrosterone for the synthesis of oestradiol and when, in obstetric disease, the supply of fetal precursor is reduced, the placenta is able to switch to the maternal source. Little justification can therefore be found for the use of plasma oestradiol assays in the follow-up of risk pregnancies.[30] Oestetrol is an oestrogen having hydroxyl groups at both C16 and C15. Hydroxylation at these points in the steroid molecule is a function of fetal liver and it is supposed that oestetrol is entirely a fetal product, being made from oestradiol in the fetal liver. At first high hopes were entertained of oestetrol measurements as a means of assessing the fetal state. Further studies have, however, not been encouraging. It is more difficult to measure than oestriol and seems to have no advantage over it.

Progesterone is in molar terms by much the major steroid produced by the placenta. It is essential for the maintenance of pregnancy and appears to be involved in the onset of labour, both prematurely and at term. It is easy to measure and from time to time investigators have looked at it in the context of a placental function test. The results have been uniformly disappointing. The levels vary a great deal from time to time in normal pregnancy and there may be no change in progesterone concentration even when the fetus is moribund, as long as placental function is maintained. In addition progesterone synthesis by the placenta is not easily disturbed and may be little changed even in the face of extensive placental damage.

If dehydroepiandrosterone sulphate (DHAS) is injected into a non-pregnant subject it is slowly removed and excreted mainly as 17-oxosteroid metabolites in urine. If DHAS is injected into a pregnant subject it is very rapidly removed. This is because the placenta converts it into oestradiol. The plasma oestradiol level rises to 3 or 4 times the pre-injection level within 45 minutes and does not return to baseline for 18 hours. It appears therefore that the injection of DHAS can be used as a dynamic test of the capacity of the placental enzymes to generate oestradiol from its precursor. At first the disappearance rate of DHAS was used as the indicator of placental function[27] but at present most people have followed the lead of the Ulm school in measuring the increase of plasma oestradiol.[20] There have been no reports of untoward effects but recent reports suggest that there may be little change in oestradiol generation even when fetal wellbeing is severely compromised. This may be little more than a question of finding the right kinds of obstetric pathology in which to apply the test. Manipulating the oestrogen concentration of a pregnant woman is an exciting prospect and so is a stress test of the placenta. Provided it can be done without harm to the fetus there may be a place for this test in obstetrics, if not as a placental function test, then as an agent for physiological investigation and possibly therapy.

Assay of Placental Proteins in Fetal Monitoring

The placenta produces an array of proteins which are secreted into the maternal circulation. The trophoblast basement membrane acts as an effective barrier to ingress into the fetal compartment and the concentration of such proteins in the fetal circulation and liquor amnii is much lower than in the maternal blood, in contrast to the findings with steroids. Some of the proteins, notably hCG and hPL, are hormones. Others such as schwangerschaftsprotein 1 (SP_1) or pregnancy associated plasma protein A (PAPP-A) have no known function and cannot be classified. A third group are placental enzymes such as heat stable alkaline phosphatase and cystine aminopeptidase. As placental products all these proteins are candidates as parameters of placental function and therefore indirectly as indices of fetal wellbeing.

Placental lactogen (hPL) has, for a protein, a notably short half-life in the maternal circulation. Fifteen minutes after delivery of the placenta its concentration in the mother's blood has fallen by half and an hour or two after delivery it can no longer be detected. This suggests that plasma levels would respond rapidly to any change in placental production and may make serial assays as a guide to obstetric management attractive. The protein was first isolated by Josimovich and McLaren[10] at a time when radioimmunoassays were already in use and this technique has been employed for hPL assay from the beginning. It is present in late pregnancy in such high concentration that much simpler and less expensive methods such as haemogglutination inhibition, complement fixation, immunodiffusion and immunoelectrophoresis are adequate. hPL assays have, therefore, the additional attraction of being available to centres which do not have available radioactive counting facilities. The normal levels in late pregnancy have varied a good deal from one publication to another. To some extent these differences are due to variation in the standards used for measurement. An agreed international standard is now available and it is accepted that mean levels at term fall between 6 and 10 μg/ml with a standard deviation of about 1·5 μg/ml. The cut-off point (4 μg/ml) suggested by Spellacy, Teoh and Buhi[28] as the lower limit of normal is generally accepted. Gordon et al.[6] screened a whole obstetric population and found that an hPL concentration below the 10th centile (4·3 μg/ml) carried a predictive value of 20 per cent. The relative risk (2·9) indicated that the assay could be used to locate an at risk population. Some of the best studies on the clinical application of hPL assays have used the assays as a means of detecting inevitable abortion in patients presenting with bleeding in early pregnancy.[21] The results are not reliable before 10 weeks gestation but thereafter serial assays will predict with increasing accuracy whether the pregnancy will continue or not.

In pre-eclampsia hPL values generally fall in the normal range. The perinatal death rate and untoward effects such as retarded fetal growth are higher in those eclamptics with a low hPL. The assays are therefore useful in defining a risk group in patients with toxaemia. The assays are less helpful in diabetic pregnancies where the deleterious effect of the diabetes on the placenta may be counteracted by the large mass of trophoblastic tissue. In rhesus incompatibility this effect is more marked. In affected pregnancies hPL tends to be raised above normal even when the fetus is moribund. Plasma hPL is less closely related to fetal weight than oestriol, indeed many investigators have failed to find any correlation between the two. For this reason it is a less reliable indicator of retarded fetal growth, particularly in the absence of hypertension, than oestriol assay.

hPL assays have been used to best effect as a means of screening a whole obstetric population. Of itself a late pregnancy value below 4 µg/ml indicates increased fetal risk and when taken in conjunction with other risk factors such as poor obstetric history, an elderly primigravida, or low oestriol values, hPL assays are very helpful in defining a population in which obstetric pathology is concentrated.

Although hCG is produced throughout pregnancy, the values are low in late pregnancy at the very time when assessment of fetal wellbeing is required. hCG is essential for the maintenance of pregnancy and the production rises steeply in the first 12 weeks of gestation. It might be supposed that to measure a vital product which is increasing rapidly is a sound means of monitoring pregnancy in threatened and recurrent abortion, and many have found it so. A problem arises from the fact that hCG levels only begin to fall when placental function is severely affected. Often the cause of the abortion is primarily fetal in origin and even when it is not, hCG production is not affected until the abortion is well advanced. The assays no longer find favour as a means of prognosis early in the abortion process, but they are useful in determining whether any functioning trophoblast is still present.

Alpha fetoprotein (AFP) is another pregnancy protein the levels of which rise in pregnancy. They reach a plateau at 30 weeks and tend to fall later in the pregnancy, at the very time when rising concentrations would be helpful in assessing the fetal state. Not surprisingly, AFP measurements have not found much application as a parameter of fetoplacental function in late pregnancy. Plasma AFP measurements are now done in all pregnancies at about 16 weeks as a screen for spina bifida, and statements about fetal wellbeing based on this are often made outwith the context of CNS deformity. There is very little evidence to support such statements.

Two new pregnancy-specific proteins of placental origin have been examined as measures of placental function. Schwangerschaftsprotein 1 (SP_1) has a much longer half life than hPL, 22 hours as compared with 15 minutes, and is more likely to be useful in screening than in serial assays for day to day control. Although it has its advocates there is really no substantial evidence to suggest that measuring SP_1 will give information as reliable as hPL or oestriol assays. Its function is unknown and SP_1 levels do not bear a relationship to fetal growth. The concentration in late pregnancy is high, so much so that it can be measured by a simple immunodiffusion technique. If SP_1 assays do turn out to be clinically useful this feature will be very attractive. Assays require no instrumentation and can be done anywhere. The other placental protein, PAPP-A, can be measured equally simply and may be involved in the genesis of pre-eclampsia. Hughes et al.[8] have adduced evidence to show that PAPP-A concentration is raised before the signs of pre-eclampsia become evident. It appears to be part of the immunosuppressive mechanism by which the fetus and placenta escape rejection by the mother. If the initial findings are confirmed, measurements of PAPP-A may come to play an important part in our understanding of the immunology of reproduction.

Heat-stable alkaline phosphatase is a placental isoenzyme which can be distinguished from the liver enzyme by its heat stability. Being a specific placental product, its concentration has occasionally been used as an indicator of placental activity. The concentration ranges widely from one subject to another and from time to time. In spite of some initial enthusiasm the assays have never caught on and are seldom used nowadays. Much the same can be said of other placental enzymes such as cystine aminopeptidase.

Oestriol and hPL assays have an established place in the biochemical monitoring of fetal wellbeing. Other assays are in decline or are unknown quantities. Used with discrimination in appropriate cases oestriol and hPL assays can make a contribution to obstetric practice.

REFERENCES

1. Aschheim, S. and Zondek, B. (1927), 'Hypophysenvorderlappenhormon und Ovarialhormon in Harn von Schwangeren,' *Klinische Wochenschrift*, **6**, 1322–1325.
2. Beischer, N. A. and Brown, J. B. (1972), 'Current status of oestrogen assays in obstetrics and gynecology,' *Obstet. Gynec. Survey*, **27**, 303–343.
3. Brody, S. and Carlstrom, G. (1960), 'Estimation of human chorionic gonadotrophin in biological fluids by complement fixation,' *Lancet*, **ii**, 99–101.
4. Diczfalusy, E. and Mancuso, M. (1969), 'Oestrogen metabolism in pregnancy,' In *Foetus and Placenta*, pp. 191–248 (Klopper, A. and Diczfalusy, E., Eds.). Oxford: Blackwells.
5. Frandsen, V. A. (1963), *The Excretion of Oestriol in Normal Human Pregnancy*. Copenhagen: Munksgaard.
6. Gordon, Y. B., Lewis, J. D., Pendlebury, D. J., Leighton, M. and Gold, J. (1978), 'Is measurement of placental function and maternal weight gain worthwhile?' *Lancet*, **1**, 1001–1003.
7. Grudzinskas, J. G., Lenton, E. A., Gordon, Y. B., Kelso, I. M., Jeffrey, D., Sobowale, O. and Chard, T. (1977), 'Circulating levels of pregnancy specific $β_1$ glucoprotein in early pregnancy,' *Brit. J. Ob. Gyn.*, **84**, 740–746.
8. Hughes, G., Bischof, P., Wilson, G. and Klopper, A. (1980), 'Assay of a placental protein to determine fetal risk,' *Brit. Med. J.*, **1**, 671–673.
9. Ittrich, G. (1958), 'Eine neue Methode zur chemische Bestimmung der oestrogen Hormone im Harn,' *Z. physiol. Chem.*, **312**, 1–7.
10. Josimovich, J. B. and MacLaren, J. A. (1962), 'Presence in the human placenta and term serum of a highly lactogenic substance immunologically related to pituitary growth hormone,' *Endocrinology*, **71**, 209–220.
11. Klopper, A. (1964), 'Variability of urinary sex steroid hormone output in pregnancy,' in *Research on Steroids*, **1**, 119–132. Rome: Il Pensiero Scientifico.
12. Klopper, A. (1968), 'The assessment of placental function in clinical practice,' In *Foetus and Placenta*, pp. 471-555 (Klopper, A. and Diczfalusy, E., Eds.). Oxford: Blackwells.
13. Klopper, A. (1976), 'The choice between assays on blood or on urine,' in *Hormone Assays and their Clinical Application*, pp. 74–86 (Loraine, J. A. and Bell, E. T., Eds.). 4th Edition. Edinburgh: Churchill Livingstone.
14. Klopper, A. (1977), 'Choice of hormone assay in the assessment of fetoplacental function,' in *Endocrinology of Pregnancy*, pp. 350–364. (Fuchs, F. and Klopper, A., Eds.). 2nd Edition. New York: Harper & Row.
15. Klopper, A. and Billewicz, W. (1963), 'Urinary excretion of oestriol and pregnanediol during normal pregnancy,' *J. Obstet. Gynaec. Brit. Cwlth.*, **70**, 1024–1031.
16. Klopper, A. and Stephenson, R. (1966), 'The excretion of oestriol and of pregnanediol in pregnancy complicated by Rh immunisation,' *J. Obstet. Gynaec. Brit. Cwlth.*, **69**, 28.
17. Klopper, A. I., Wilson, G. R. and Masson, G. M. (1977),

'Observations on the variability of plasma estriol,' *Obstet. Gynec.*, **49**, 459–461.
18. Kosada, T. S., Levesque, L. A., Goldstein, D. P. and Taymor, M. L. (1973), 'Early detection of implantation using a radio-immunoassay specific for human chorionic gonadotropin,' *J. Clin. Endocr. Metab.*, **36**, 622–629.
19. Landesman, R. and Saxena, B. (1976), 'Results of the first 1000 radioreceptorassays for the determination of human chorionic gonadotropin,' *Fertil. Steril.*, **27**, 357–368.
20. Lauritzen, Ch., Strecker, J. and Lehmann, W. (1976), 'Dynamic tests of placental function: Some findings on the conversion of DHAS to oestrogens,' in *Plasma Hormone Assays in Evaluation of Fetal Wellbeing*, pp. 113–135 (Klopper, A., Ed.). Edinburgh: Churchill Livingstone.
21. Letchworth, A. T. (1976), 'Human placental lactogen assay as a guide to fetal wellbeing,' pp. 147–173 (Klopper, A., Ed.). Edinburgh: Churchill Livingstone.
22. Little, W. A. (1962), 'Clinical evaluation of the latex flocculation test in pregnancy,' *Am. J. Obstet. Gynecol.*, **84**, 220.
23. Midgley, A. R. and Pierce, G. B. (1962), 'Immunohistochemical localization of human chorionic gonadotrophin,' *J. Expl. Med.*, **115**, 289–294.
24. McKean, C. M. (1960), 'Preparation and use of antisera to human chorionic gonadotropin,' *Am. J. Obstet. Gynecol.*, **80**. 596–600.
25. Psychoyos, A. (1973), 'Endocrine control of egg implantation,' in *Handbook of Physiology*, Vol. II, pp. 187–215 (Greep, R. O. and Astwood, E. B., Eds.). Baltimore: Williams and Wilkins Co.
26. Saxena, B. B., Hassan, S. H., Haour, F. and Gollwitzer, M. S. (1974), *Science*, **184**, 973–976.
27. Siiteri, P. K. and MacDonald, P. C. (1966), 'Placental estrogen biosynthesis during human pregnancy,' *J. Clin. Endocr.*, **26**, 751–761.
28. Spellacy, W. N., Teoh, E. S. and Buhi, W. C. (1970), 'Human chorionic somatomammotropin (HCS) levels prior to fetal death in high risk pregnancies,' *Obstet. Gynec.*, **35**, 685–689.
29. Tatarinov, Y. S. and Masyukevich, V. N. (1970), 'Immunochemical identification of new $beta_1$ globulin in the blood sera of pregnant women,' *Byull. Eksp. Biol. Med.*, **69**, 66–68.
30. Tulchinsky, D. (1976), 'The value of oestrogen assays in obstetric disease,' in *Plasma Hormone Assays in Evaluation of Fetal Wellbeing*, pp. 72–86 (Klopper, A., Ed.). Edinburgh: Churchill Livingstone.
31. Vaitukaitis, J. (1977), 'Human chorionic gonadotropin,' in *Endocrinology of Pregnancy*, pp. 63–75 (Fuchs, R. and Klopper, A., Eds.). 2nd Edition. New York: Harper & Row.
32. Vaitukaitis, J. L., Braunstein, G. D. and Ross, G. T. (1972), 'A radio-immunoassay which specifically measures chorionic gonadotropin in the presence of human luteinizing hormone,' *Am. J. Obstet. Gynecol.*, **113**, 751–760.
33. Wide, L. and Gemzell, C. A. (1960), An immunological pregnancy test. *Acta Endocr. (Kbh)*, **35**, 261–267.

42. PRENATAL DIAGNOSIS OF INHERITED METABOLIC DISEASES, NEURAL TUBE DEFECTS AND LUNG MATURITY

J. B. HOLTON

General introduction
   Possible diagnostic approaches

Prenatal diagnosis of inherited metabolic diseases
   Reasons for prenatal diagnosis
   Principles

Practical aspects of prenatal diagnosis by enzyme assay
   Prelimary studies
   Timing of amniocentesis
   Transport and storage of specimens
   Methods
   Interpretation of results

Neural tube defects
   The need for prenatal diagnosis
   Normal AFP patterns
   Maternal blood AFP screening
   Amniotic fluid AFP
   Conclusion

Assessment of fetal lung maturity
   Principles of assessing lung maturity
   Methods related to surfactant synthesis
   Measurement of cortisol
   Measurement of lipid-positive cells
   Interpretation of results

GENERAL INTRODUCTION

In recent years biochemical tests have become an essential part of ante-natal care and have contributed to the increasing workload in many Clinical Chemistry laboratories. The tests required in pregnancy are of two main types. The first and most frequently requested are the so-called tests of feto-placental function. They examine the viability of the fetus, the placenta and the overall state of the pregnancy. The second kind of test is concerned specifically with the fitness of the fetus for extra-uterine life. This chapter deals with the latter problem.

Post-natal problems that can be diagnosed *in utero* include transient difficulties occurring in the neonatal period like the respiratory distress syndrome (RDS), permanent disorders like the inherited metabolic diseases (IMD) and conditions with a multifactorial etiology such as the neutral tube defects (NTD), spina bifida and anencephaly. The prevention of these conditions has become of far greater significance to society over the past few decades because, with the gradual decline of diseases caused by environmental factors, they now account for a large proportion of the mortality and morbidity in children.

Possible Diagnostic Approaches

There are several ways of obtaining biochemical information about the fetus and these are considered below:

Fetal Blood

Obtaining fetal blood is a highly skilled operation which has to be performed under the visual control of a fetoscope and it is done in only a few places at present. Two techniques have been used.

In the first, blood is aspirated from fetal blood vessels on the surface of the chorionic plate. Unfortunately amniotic fluid or maternal blood is frequently aspirated with the fetal blood. In a rather more elegant method the needle is inserted into a vessel of the umbilical cord, close to the placental surface, where the blood vessels are relatively fixed and large enough to allow collection of up to 1·0 ml of pure fetal blood.

Fetal blood which is contaminated is usually of little value for biochemical tests. If the presence of maternal blood can be demonstrated the sample is obviously not characteristic of the fetus. Large dilutions of the sample with amniotic fluid will be suspected from a low haematocrit and may not matter for tests of a qualitative nature, such as in the diagnosis of the haemoglobinopathies. However, a diluted sample is very unsatisfactory for quantitative tests, in particular the assay of fetal plasma enzymes. Attempts to correct results by estimating the dilutional error from the haematocrit, as in the diagnosis of Duchenne Muscular Dystrophy by plasma creatine phosphokinase assay, are of very doubtful validity.

Fetal loss has occurred in at least 3% of pregnancies following fetal blood sampling but the technique has the advantage that it is the only one in which material is available directly from the fetus. It may be the only method of diagnosing some IMD. However, in view of the high risk it should not be used without a very strong reason, and particularly if a safer approach is available.

Amniotic Fluid

Amniocentesis is the most frequently used method of investigating the fetus and has been employed widely for more than 20 years. The earliest routine application was in predicting the severity of haemolytic disease,[35] an application which demonstrated the limitations and difficulties of amniotic fluid investigations.[26]

Amniotic Fluid Composition and Formation. The composition of amniotic fluid is very similar to that of extracellular fluid. A very small amount of maternal plasma proteins is present in a concentration relative to plasma which is inversely proportional to their molecular weights, with a cut off around 250 000 daltons. Thus liquor probably arises as a transudate of blood plasma with a contribution from both maternal and fetal circulations, because protein markers for both the mother and fetus can be detected in amniotic fluid.[58] The quantitative contribution of each circulation is not known.

Two internal factors seem to modify the composition of the fluid in the latter stages of pregnancy. Fetal urine is believed to be secreted from about the 12th week of pregnancy, but its influence on amniotic fluid composition is noticed mainly in the last trimester, when the osmolality falls rapidly and the creatinine and urea concentrations rise. The other mechanism begins by the fetus swallowing amniotic fluid and then the fluid contents being digested and absorbed in the fetal intestine. This process is particularly important in the turnover of fluid proteins.

The sites of transfer of water and solutes from the mother and fetus into amniotic fluid need to be considered and the ways in which abnormalities in these might occur. On the whole we are concerned only with transport from the fetus. It has been shown that small molecules equilibrate very rapidly and freely between amniotic fluid, maternal and fetal circulations. The transfer of larger molecules is likely to be more difficult and, also, of greater importance. In early pregnancy most external membranes of the fetus, including skin, are permeable to macromolecules, but later effective transfer is probably much more restricted. Many substances are believed to be secreted in fetal urine, including proteins like α-fetoprotein with a MW of 64 000 daltons. In addition, the fetal lung has been shown to secrete large volumes of fluid, and the transfer of this fluid into amniotic fluid is facilitated by respiratory movements in the fetus. Finally, it is possible that the intestinal wall remains relatively permeable to larger molecules.

Some general and some more specific abnormalities which may alter the transfer of substances into amniotic fluid are listed in Table 1. Awareness of these anomalies

TABLE 1
SITES OF TRANSFER OF SUBSTANCES INTO AMNIOTIC FLUID, ABNORMALITIES IN THE SITES, AND HOW THESE WOULD AFFECT FLUID COMPOSITION

Transfer sites	Types of abnormality	Effect on transfer and concentration of constituents
		+ = increase − decrease
Placental membranes	Maldevelopment	+ or −
	Infarction	+
Cord	Exomphalos	+
	Omphalocoele	+
Fetus: skin	Skin aplasia	+
	Increased permeability, Turner's syndrome, Intrauterine death	+
lung and trachea	Atresias	−
	Anoxia, Fallot's tetralogy, Fetal distress	+
GI tract	Atresias	+
	Reduced swallowing	+
kidney	Nephrosis	+
	Agenesis	−
open defects	Spina bifida	+
	Anencephaly	+
	Encephalocoele	+
	Gastroschisis	+

is important since they may be the source of false positive or false negative diagnoses.

One component of amniotic fluid which has not been mentioned is the cellular component. This is utilized in the diagnosis of IMDs and it is important that the cells used are of fetal origin. Most amniotic fluid cells come from the external membranes of the fetus, but some maternal cells may be present and these may pose a problem in the performance of a prenatal diagnosis.

Risk of Amniocentesis. In deciding whether a prenatal investigation should be done one important consideration is the risk attached to obtaining amniotic fluid by transabdominal amniocentesis. This risk is related to gestation. After 20 weeks gestation the risk is usually thought to be negligible. Earlier in the second trimester, when the volume of fluid is lower and the difficulty of amniocentesis correspondingly greater, it seems likely that there could be a finite risk. However, there has been much controversy about the precise level of this risk. Two recent studies have concluded that amniocentesis is not associated with increased fetal loss.[48,57] The most recently published study, carried out in the UK, concluded that there appeared to be a 1 to 1·5% increase in fetal death in women having amniocentesis compared with controls who did not have amniocentesis.[37] In addition there were more neonatal complications and deaths in the amniocentesis group.

It is difficult to see how such different results could be obtained from 3 large studies, but until other data become available it is perhaps best to assume that there is a small increased risk of fetal death due to early amniocentesis of about 1 to 1·5%. There was general agreement that the risk does increase with repeated attempts at amniocentesis, when it is done before 15 weeks gestation and when a blood stained fluid is obtained. It is anticipated that these problems will diminish as experience of amniocentesis increases.

Contamination of Amniotic Fluid. Ideally an amniotic fluid specimen should be clear and free of all contamination, otherwise the validity of the results obtained are in serious doubt. Blood, of fetal and maternal origin, is the most common contaminant at all stages of gestation and can affect analysis directly to give an erroneous result, or may hinder the diagnosis of IMD by preventing the growth of amniotic fluid cells in culture. In some cases it may be relevant to know if the blood is fetal or maternal in deciding whether an analysis would be meaningful. This is usually done by the Kleihauer technique.[4] Towards term the fluid may be contaminated with meconium and occasionally by a variety of antiseptics and obstetric creams.

Maternal Blood and Urine

In some cases maternal blood and urine may be used for prenatal diagnosis, particularly in screening for the NTD. Although there are few difficulties and no risks in obtaining these samples there is a general limitation that they are not in immediate equilibrium with fetal fluids. Therefore, it is to be anticipated that the scope and accuracy of the information they provide would be less.

PRENATAL DIAGNOSIS OF INHERITED METABOLIC DISEASES

Reasons for Prenatal Diagnosis

All but one or two of the conditions which can be diagnosed *in utero* (Table 2) are autosomal recessive in inheritance and parents are discovered to be at risk for the disorder with the birth of an affected child. There is a 1 in 4 chance of subsequent babies being homozygous for the abnormal gene and affected with the disease. Prenatal diagnosis is usually considered in those conditions with a serious outcome and in which there is no successful treatment available. In the event of a positive diagnosis of such a condition the pregnancy may be terminated.

Opinions are widely divided on what is an acceptable degree of handicap in a child and what constitutes a satisfactory form of treatment. Therefore there is no clear distinction between diseases which require to be diagnosed prenatally and those which do not. In counselling parents to consider a prenatal diagnosis the main factors which have to be taken into account are their attitudes towards the abnormalities and reduced life expectancy that their offspring might have, whether they want to terminate the pregnancy if the test is positive and whether they are prepared to accept the small risk of losing a possibly normal fetus because of the amniocentesis. It seems certain that with the changing attitudes of many societies towards abortion and the right of the individual to choose whether to have a handicapped child, the demand for prenatal diagnosis will increase.

The conclusion by some surveys that amniocentesis is not hazardous to the fetus had led to recommendations that prenatal diagnosis might be done simply to provide parents and doctors with prior knowledge of whether the baby is affected. The disagreement about the safety of amniocentesis has been mentioned already but, ignoring this question, another argument against doing prenatal diagnosis for more trivial purposes is the cost. There may be a few diseases however in which therapy can be initiated *in utero* and, in these, a prenatal diagnosis could be justified.

Principles of Prenatal Diagnosis

The biochemical disorder of an IMD is expressed in the body at three different levels, the gene, the primary enzyme defect and the secondary metabolic consequences of the defect; theoretically a disease may be diagnosed by finding an abnormality in any one of these.

The Gene

The fundamental change in the IMD is an altered gene and the ultimate method of diagnosis would be by showing this. Methods of DNA analysis have been applied already to the diagnosis of Sickle Cell Anaemia but not yet to any of the IMD.

The presence of an abnormal gene may also be

TABLE 2
DISORDERS WHICH CAN BE DIAGNOSED PRENATALLY

1. By enzyme assay on cultured cells (relevant enzyme shown in brackets)

   Mucropolysaccharidoses
   IH   Hurler (α-L-iduronidase)
   IS   Scheie (α-L-iduronidase)
   IH/S Hurler/Scheie (α-L-iduronidase)
   II   Hunter (Iduronic acid sulphatase)
   IIIA Sanfillipo A (Heparan-N-sulphatase)
   IIIB Sanfillipo B (N-acetyl-D-glucosaminidase)
   IIIC Sanfillipo C (α-glucosaminidase)
   IVA Morquio C (Galactosamine-6-sulphate sulphatase)
   IVB Morquio B (β-Galactosidase)
   VI   Marateaux-Lamy (Arylsuphatase B)
   VII β-Glucuronidase deficiency (β-Glucuronidase)

   Sphingolipidoses
   GM_1 Gangliosidosis I, generalized (β-Galactosidase A, B and C)
   GM_1 Gangliosidosis II, juvenile (β-Galactosidase B and C)
   GM_2 Gangliosidosis I—Tay Sachs (Hexosaminidase A)
   GM_2 Gangliosidosis II—Sandhoff's (Hexosaminidase A + B)
   GM_2 Gangliosidosis III—Juvenile (Hexosaminidase A + B)
   Metachromatic leukodystrophy (Arylsulphatase A)
   Farber's disease (Ceramidase)
   Gaucher's disease (β-Galactosidase)
   Krabbe's disease (Galactocerebroside β-galactosidase)
   Lactosyl ceramidosis (Lactosyl ceramidase)
   Niemann-Pick disease (Sphingomyelinase)

   Amino acid disorders
   Arginosuccinic aciduria (Arginosuccinase)
   Aspartylglucosaminuria (Aspartylglucosaminidase)
   Citrullinaemia (Arginosuccinate synthetase)
   Cystathioninuria (Cystathionase)
   Histidinaemia (Histidase)
   Homocystinuria (Cystathionine synthetase)
   Hypervalinaemia (Valine transaminase)
   Maple Syrup Urine Disease (Branched chain ketoacid decarboxylase)
   Methylmalonic acidaemia:
      I(B_{12} unresponsive) (Methyl malonyl CoA mutase)
      II(B_{12} responsive) (Various defects in B_{12} metabolism and action)
   Ornithinaemia (Ornithine-α-ketoacid transaminase)
   Ornithine carbamyl transferase deficiency-Hyperammonaemia (Ornithine carbamyl transferase)
   Propionic acidaemia (Ketotic hyperglycinaemia) (Propionyl CoA carboxylase)

   Miscellaneous disorders
   Acatalasia (Catalase)
   Adenine phosphoribosyl transferase deficiency (Adenine phosphoribosyl transferase)
   Adenosine deaminase (Adenosine deaminase)
   Congenital erythropoietic porphyria (Uroporphrinogen III cosynthetase)
   Hypophosphaturia (Alkaline phosphatase)
   Lesch Nyhan syndrome (Hypoxanthine guanine phosphoribosyl transferase)
   Orotic aciduria (Orotidylic pyrophosphorylase and orotidylic decarboxylase)
   Sulphite oxidase deficiency (Sulphite oxidase)
   Refsum's disease (Phytanic acid oxidase)

   Mucolipidoses
   Fucosidosis (α-Fucosidase)
   Mannosidosis (α-Mannosidase)
   Mucolipidosis II—I cell disease (Multiple lysosomal hydrolases)
   Mucolipidosis III—Pseudopolydystrophy (Multiple lysosomal hydrolases)

   Other lysosomal storage diseases
   Acid phosphatase deficiency (Acid phosphatase)
   Cholesterol ester storage disorder (Acid lipase)
   Wolman's disease (Acid lipase)

   Carbohydrate disorders
   Glycogen storage diseases:
     II—Pompe's disease (α-1,4-Glucosidase)
     III—Debrancher deficiency (Amylo-1,6-glucosidase)
     IV—Brancher deficiency (Amylo-1,4 to 1,6-trans glucosidase)
     VI—Phosphorylase kinase deficiency (Phosphorylase kinase)
   Galactokinase deficiency (Galactokinase)
   Galactosaemia (Galactose-1-phosphate uridyl transferase)
   Glucose-6-phosphate dehydrogenase deficiency (Glucose-6-phosphate dehydrogenase)
   Phosphohexose-isomerase deficiency (Phosphohexose isomerase)
   Pyruvate decarboxylase deficiency (Pyruvate decarboxylase)
   Pyruvate dehydrogenase deficiency (Pyruvate dehydrogenase)

2. *Disorders diagnosed prenatally by miscellaneous techniques*
   Cystinosis (Increased accumulation of cystine in cultured cells)
   Adrenogenital syndrome—21 hydroxylase deficiency (Measurement of 17-hydroxyprogesterone in amniotic fluid and by HLA linkage studies)
   Chediak-Higashi disease (Abnormal staining granules in cytoplasm of cultured cells)
   Congenital nephrotic syndrome (Raised α-fetoprotein in amniotic fluid)
   Cystic fibrosis (Proteases in cultured cells and amniotic fluid)
   Duchenne Muscular Dystrophy (Raised creatine phosphokinase in fetal blood)
   Familial hypercholesterolaemia (Failure of feedback inhibition of HMG-CoA reductase by lipoprotein in cultured cells)
   Menkes disease (Cu^{64} incorporation into cultured cells)
   Myotonic dystrophy (Linkage to gene for ABH blood secretor substances)
   Xeroderma pigmentosa (Study of defective DNA repair in cultured cells)

demonstrated indirectly by the method of gene linkage. Studies of a pedigree can establish that in a particular family the abnormal gene for an IMD is always linked in cell division with the gene or genes for some specific marker proteins, most commonly the HLA antigens. By studying the HLA type of the fetus, using cultured amniotic fluid cells, its genotype with respect to the metabolic disease can be inferred. An example of this is the prenatal diagnosis of congenital adrenal hyperplasia.[10]

Since it will be some time before DNA analysis becomes widely used for the diagnosis of IMD, and gene linkage, though it sometimes provides useful corroborative evidence, is limited as far as obtaining a definite answer is concerned, these methods will not be considered in any more detail. However, the attraction of this approach should be appreciated, namely, that the genetic change should be detectable in every type of cell as well as being fundamental to the disease and permanent. Thus, sickle cell anaemia can be diagnosed on cultured amniotic fluid cells and it is not necessary to obtain fetal blood for haemoglobin analysis. DNA analysis could certainly be the answer to the diagnosis of conditions like phenylketonuria, in which the defective enzyme has a very limited expression, mainly to liver cells.

The Enzyme

The enzyme change in an IMD is also fundamental to

the disease and permanent, and is the most reliable, frequently used method of diagnosis. It has the basic limitation that some important enzymes, phenylalanine hydroxylase among them, may not be detectable in clinical material available prenatally, but over sixty diseases have been diagnosed in this way (Table 2). Diagnoses are possible on cellular or plasma enzymes of fetal blood, soluble enzymes of amniotic fluid and the cellular enzymes of the fluid, both by direct analysis or after culturing the cells. All the approaches will be discussed in subsequent sections but enzyme assay on cultured cells will be described most fully because, as will become apparent, it is the method of choice at present.

Secondary Metabolic Consequences

The principle metabolic consequences of a primary enzyme deficiency are: (1) lack of formation of the end product or products subsequent to the enzyme block, (2) accumulation of the immediate precursors of the defective enzyme in the metabolic pathway and (3) occasionally the formation of large amounts of unusual metabolites by conversion of the accumulated precursors. The last two of these effects are the most used in detecting IMD. To make a prenatal diagnosis by this means a qualitative, or quantitative, change in amniotic fluid must be found reflecting the metabolic disorder in the fetus. There are two main limitations to the use of this method in prenatal diagnosis. Firstly, and most important, is the fact that in many IMD the secondary biochemical changes do not develop *in utero* because the accumulation of the precursor compounds is prevented by maternal control of fetal metabolism. Secondly, even if the biochemical disorder develops the accumulated compounds may not be secreted into amniotic fluid, particularly prior to efficient functioning of the fetal kidney.

Amniotic fluid metabolite analysis has been applied in a few diseases but in every case the approach must still be regarded as experimental. In methylmalonicaciduria, large amounts of methylmalonic acid accumulate in amniotic fluid, and the abnormal metabolite is also easily detected in maternal urine.[21] Increased levels of 17 oxosteroids and pregnanetriol in amniotic fluid were reported in congenital adrenal hyperplasia (21-hydroxylase deficiency),[30] but the finding was not consistent in all affected pregnancies. Measurement of 17-hydroxyprogesterone, however, seems a much more promising indication of the enzyme defect.[47]

There is some disagreement regarding the analysis of amniotic fluid glycosaminoglycans for the diagnosis of mucopolysaccharidoses. Some false negative results have been found and Dorfman[13] did not consider the method sound, although the defect certainly develops *in utero*. Greater reliability is claimed for a sensitive method for studying the qualitative pattern of dermatan, chondroitin and heparan sulphates.[25] Finally, the use of amniotic fluid galactitol to diagnose galactosaemia has been described recently.[1]

PRACTICAL ASPECTS OF PRENATAL DIAGNOSIS BY ENZYME ASSAY

Preliminary Studies

Some preliminary investigation of the family is absolutely essential to the proper interpretation of the final result of a prenatal diagnosis. Other tests are less vital, but may be very important in some circumstances.

The essential step in the work-up is confirmation of the diagnosis in the index case of the family. This is essential because usually only a single IMD can be tested in the prenatal diagnosis. If this is the wrong disease the fetus may be incorrectly diagnosed as normal. The original diagnosis must be made by biochemical investigation since a characteristic clinical presentation may have several different causes. Moreover, since similar secondary biochemical changes may arise from more than one enzyme defect, diagnosis based on assay of the primary enzyme is highly desirable. All this emphasises the importance of making a diagnosis prior to the death of any sick child suspected of having a metabolic disorder and, if this is not achieved, to store all specimens and any available post mortem material.

Occasionally prenatal diagnoses are considered when the conditions discussed above are not fulfilled, but in these cases the doubtful validity of the result must be clearly understood. Sometimes it is possible to infer the diagnosis by demonstrating, with reasonable probability, that both parents are heterozygous for the suspected disorder. Unfortunately in many diseases the detection of heterozygotes cannot be achieved with any degree of certainty.

It is very useful in carrying out prenatal tests if the diagnosis in the affected child and its parents has been confirmed by enzyme assays on fibroblast cells grown in culture from skin biopsy material. For many enzymes the activities found on skin fibroblasts are very similar to those of cultured cells from amniotic fluid. Therefore skin fibroblasts from the family may be convenient positive controls in the prenatal diagnosis. A comparison of activities of galactose-1-phosphate uridyl transferase in skin fibroblasts and cultured amniotic fluid cells is shown in Table 3.

A final benefit from preliminary study of the family by assay of skin culture enzymes may be the finding of a variant enzyme which gives misleading results in cultured cells. This may confuse the interpretation of a prenatal diagnosis, particularly if one is unaware of the

TABLE 3
LEVELS OF GALACTOSE-1-PHOSPHATE URIDYL TRANSFERASE ACTIVITY* IN CULTURED CELLS

	Skin fibroblasts			Amniotic fluid cells		
	n	mean	s.d.	n	mean	s.d.
Homozygous normal	16	76.5	12.7	26	78.1	12.2
Normal/galactosaemic heterozygotes	16	32.1	4.2	4	31.0	—

* Enzyme activity is expressed in nmol gal-1-P converted/hr/mg protein.

possibility. This problem is discussed in more detail in the interpretation of the results.

Timing of Amniocentesis

The period in gestation when an amniocentesis is done for diagnosis of an IMD is particularly critical because the test may take up to six weeks to perform, and yet the result should be available before the problem of a late abortion arises. The increased risk of an early amniocentesis may have to be accepted. Amniotic fluid obtained early, particularly before 14 weeks, may also present greater problems in the laboratory because of greater difficulty in establishing a successful cell culture. This is due to fewer cells being present in the amniotic fluid earlier in gestation and also to smaller volumes being available. From the point of view of the actual enzyme assay there appears to be no particular problem in timing, because Nadler[39] showed that the specific activity of many enzymes did not vary in cell cultures established between 10 and 36 weeks gestation. The usual time for amniocentesis is between 14 and 16 weeks gestation.

Transport and Storage of Specimens

Having performed an amniocentesis the specimen may be handled in two ways. It may be sent immediately by the fastest convenient method to the laboratory undertaking the test. This may be a considerable distance, but Niermeijer *et al.*[42] found no particular problems in culturing cells in specimens taking up to 7 days in transportation. Alternatively, if there is a laboratory nearby which does cell culture, the culture may be initiated locally and the cells transported in this form. The main advantage of the latter procedure is that some of the growing cells may be retained in case of loss or damage of the sample during transit. Holton[27] has given a fuller consideration to the problems of transport.

Methods of Prenatal Diagnosis by Enzyme Assay

Prenatal diagnoses have been achieved by assaying the defective enzyme of an IMD using fetal blood, amniotic fluid supernatant and amniotic fluid cells, both cultured and uncultured. The advantages, disadvantages and practical aspects of these techniques will now be considered:

Fetal Blood

Many inherited metabolic diseases have been diagnosed postnatally by enzyme assays on blood, either in plasma or in the cellular constituents. The method has not been applied widely on fetal blood, mainly because of the difficulty and risk of obtaining the sample. The use of fetal blood could usually be justified only if the enzyme involved can be detected in blood and not in amniotic fluid. If blood is employed there is an added advantage that the assay can be done without the delay of cell culture.

The disorders which have been investigated in fetal blood are arginase deficiency,[51] galactosaemia[16] and Tay Sachs disease.[45] There is as yet a shortage of control data on the level of these enzymes in the normal fetus. Hence false positive results could be obtained if the enzyme being assayed is not always present in normal fetal blood of the gestational age studied.

It is pertinent to discuss some attempts at prenatal diagnosis of Duchenne Muscular Dystrophy here, although the principles involved are rather different to those already discussed. Duchenne Muscular Dystrophy is an X-linked recessive disorder whose diagnosis can be confirmed postnatally by the very high plasma levels of creatine phosphokinase which leaks from the dystrophic muscle. A basic enzyme deficiency has not been identified in this disease, but it is assumed that the diagnostic increase in creatine phosphokinase arises as a secondary consequence of an unknown primary defect.

The possibility that an increase in creatine phosphokinase occurs in the plasma of affected male fetuses is being actively explored, but so far with a very discouraging record of more cases being missed than are actually diagnosed. It seems probable that the rise in creatine phosphokinase has not always occurred at the time when prenatal diagnosis is done, but it has also been suggested that the enzyme activity could be inhibited or suppressed by a factor from the maternal circulation. The basis for this proposal is that the mother's creatine phosphokinase level itself is known to be lowered quite significantly throughout the pregnancy. More work is needed on this important and relatively common problem.

Amniotic Fluid Supernatant and Uncultured Amniotic Fluid Cells

These two techniques for prenatal diagnosis are considered together because the source of enzyme in amniotic fluid supernatant is the cells, and the inherent difficulties of both approaches are related. Many enzymes have been located in amniotic fluid cells obtained by centrifuging the fluid at 600 g for 10 min. Unfortunately the concentration of enzyme in the cell sediment will be very variable because the proportion of viable cells may be as low as 20% of the total, and non-viable cells may lose much of their enzyme content. As most of the supernatant enzymes are derived from the same leaky cells it follows that the measure of soluble enzymes must also be subject to wide variations.

An application of the assay of enzymes in non-cultured cells and amniotic fluid supernatant which has proved reasonably reliable is in the diagnosis of Tay Sachs disease. This is essentially because the concentration of the diagnostic enzyme, hexosaminidase A, is measured as a ratio to total hexosaminidase activity (Hexosaminidase A + B). Providing the two enzymes are affected similarly by cell death, determination of the ratio may distinguish the affected fetus, though the absolute concentration of hexosaminidase A may not.

O'Brien et al.[44] found that it was possible to distinguish clearly 6 cases of Tay Sachs disease from normals using the hexosaminidase A/A + B ratio in cultured cells, uncultured cells or supernatant. However, the ratio found in normal subjects was different in the three situations. Cultured cells had the highest ratio and provided the best means of discriminating a case of Tay Sachs disease. Although the use of uncultured cells or supernatant fluid would rarely be accepted as the sole means of diagnosis, it may be a safeguard if problems or artefacts occur in the culture method.

Cultured Amniotic Fluid Cells

The assay of the primary enzyme defect in cultured amniotic fluid cells is almost always the method employed to make a prenatal diagnosis. Alternative procedures may sometimes be used in parallel, usually in an attempt to make a quicker diagnosis, but cell culture is always regarded as the definitive method. In this section the technique of cell culture, the problems of enzyme assay on cultured cells and the interpretation of results will be considered.

Cell Culture Technique. The following is a very brief description of cell culture as applied to amniotic fluid. For a fuller account of the technique the reader is referred to Harnden.[23]

A cell culture may be initiated by resuspending centrifuged amniotic fluid cells in a small volume of culture medium (see Table 4) and placing them in a culture

TABLE 4
CONSTITUENTS OF A TYPICAL CULTURE MEDIUM

Amino acids	Vitamins
Lipids	Nucleic acids
Glucose	Inorganic ions
	Fetal calf serum

Incubated in an atmosphere of CO_2/air (1:20).

flask, or petri dish, maintained at 37° in an atmosphere of 5% CO_2 in air. Alternatively, 3–4 ml of uncentrifuged amniotic fluid is simply diluted with an equal volume of culture medium in the culture vessel. The essential requirement is for the cells to become attached to the bottom of the culture flask, which is designed to present a large surface to the suspended cells. Once the cells are attached, which usually takes 3–4 days, the medium may be renewed at intervals of several days and rapid growth of the cells ensues until a monolayer covers the base of the flask. The layer of cells is described as a 'confluent monolayer' or 'the point of confluency' is said to have been reached.

This initial culture is called the 'primary' culture and can be used for enzyme assay. The cells must be dislodged from the flask bottom with a mixture of trypsin and EDTA solutions (a process described as 'trypsinization') and then transferred to a tube for harvesting by centrifugation. For reasons which become apparent in the subsequent discussion, primary cultures are not highly recommended for the enzyme assay and it is more usual to resuspend and divide the harvested cells, placing them into two fresh culture vessels to begin a new growth cycle.

This division and re-culturing of the cells is known as 'sub-culture' and may be repeated 30 to 40 times with normal amniotic fluid cells before the cell line dies out. The cells in sub-culture usually grow much more rapidly than initially and may reach confluency in 4 to 7 days. It is often found, however, that cells from subjects with a metabolic defect grow more slowly and can be subcultured fewer times before the line dies out.

Enzyme Assays on Cultured Cells. Before performing the enzyme assay on cells harvested from cultures by trypsinization they must be disrupted by osmotic shock, sonication or freezing and thawing. The subsequent enzyme methods are not different in principle from those used for other purposes except that the highest sensitivity is required in order to minimize the number of cells and hence the time taken to grow them. Thus the majority of techniques involve substrates which are radioactively labelled or which yield a highly fluorescent product. Some useful increase in sensitivity can usually be obtained by optimizing the conditions of the assay and, finally, the method is scaled down to the minimum volumes consistent with the required accuracy.

Using the approach just described, which is the one most commonly used, a duplicate assay of enzyme and the protein estimation,[36] to which the enzyme activity is usually related, may be done on about 100 000 cells. A diagnosis may be achieved in under 3 weeks in these circumstances. More sophisticated methods have been developed. Using substrates producing light fluorescent substrates and a microfluorimeter, Galjaard et al.[18] reported diagnostic results using about 100 cells. Autoradiographic techniques have also been combined with microscopy to produce rapid results using few cells.

Use of Primary Cultures for Enzymes Assay. The primary culture is made up of clones of cells which have been derived from a single cell. The number of clones making up the culture will vary depending on the number of viable cells added to the culture vessel, but in the extreme case all the cells of a primary culture and its subcultures may be descended from a single cell. This can give rise to anomalies in enzyme assay and prenatal diagnosis.

Very occasionally a primary culture may be initiated from maternal cells, which is obvious from the misdiagnosis of sex that always occurs in a small proportion of male fetuses. It may also give a false diagnosis of an inherited metabolic disease. Primary cultures derived solely from cells of fetal origin may consist of heterogeneous clones. The predominant cells are usually fibroblast-like, but epithelial cells may also be present. Epithelial cells are morphologically different from fibroblast cells but, more importantly, they have different enzymological characteristics. Even the predominant fibroblast cells are known to be heterogeneous in enzyme composition. For example, only those grown from skin of the genitalia have steroid 5α-reductase activity; an enzyme which is defective in familial incomplete pseudo-hermaphroditism type 2.

One would expect that a primary culture could be most heterogeneous in its cell population and hence the enzyme levels in cells harvested from it would be equally varied (Fig. 1). Gross discrepancies, like the possibility of a maternal cell culture, can be reduced by initiating and assaying several primary cultures, providing enough amniotic fluid cells are available. However, serious errors in diagnosis have been made and reported,[43] and hence the use of primary cultures is not generally favoured.

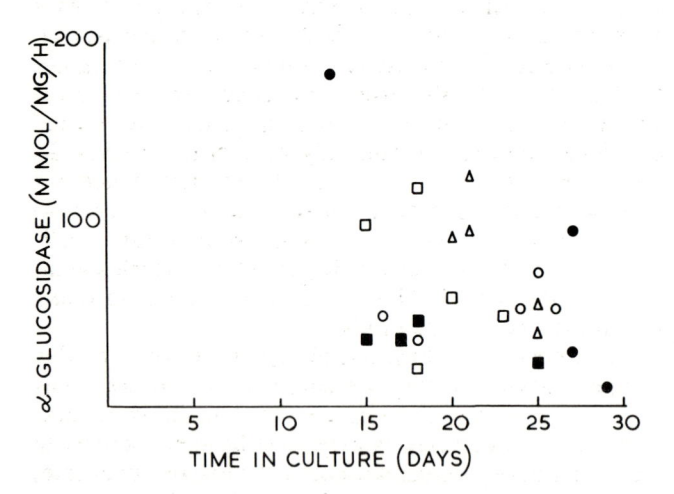

FIG. 1. Variations in the activity of α-glucosidase in primary amniotic fluid cell cultures, expressed in terms of the amount of protein. Results shown with the same symbol indicate primary cultures from the same amniotic fluid (Reproduced from Niermeijer et al.[43])

The Use of Subcultures for Enzyme Assay. By the process of subculture more cells are obtained, but the main purpose is to get a more homogeneous cell population which consists solely of fibroblasts. With this a more consistent pattern of enzyme activity should emerge. This pattern may be modified by conditions in the culture medium like salt composition, energy sources, vitamins, CO_2 concentration and pH and these must be kept constant for a given method.

The specific activity of an enzyme in cultured cells is often a function of the age of the cells. Various patterns of enzyme activity with time after subculture may be found (Fig. 2). This pattern should be reasonably constant for a given enzyme assayed by a particular method, once all the conditions have been established. The optimal time in the culture cycle to harvest the cells can be decided, which is when the enzyme activity is as high as possible but also reasonably constant with time.

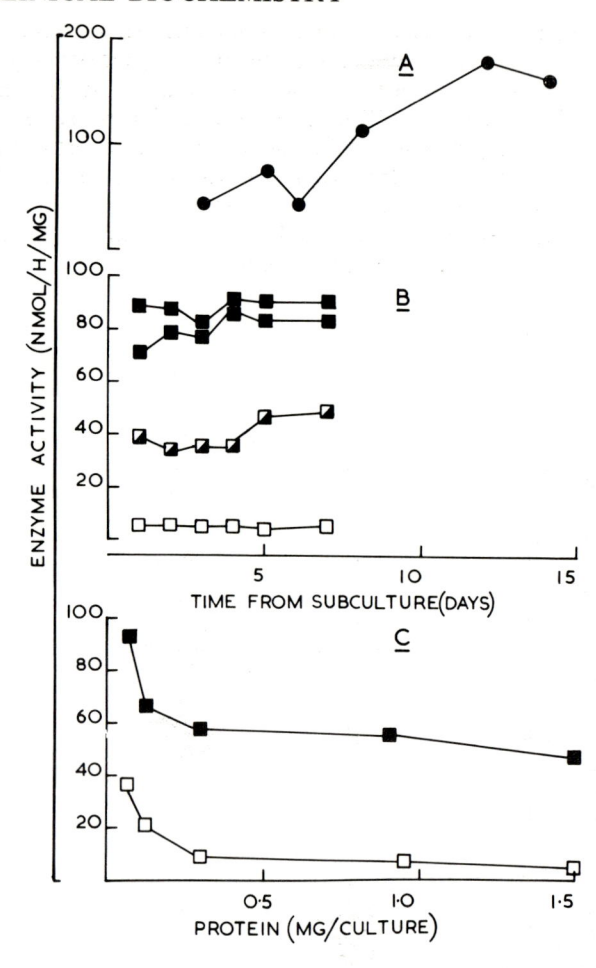

FIG. 2. Different patterns of change of enzyme activity with time in amniotic fluid cell cultures after the first passage. Enzyme activity is expressed per mg of protein in all cases.
A. α-Glucosidase activity assayed from 3 to 15 days after subculture of the primary shown in Fig. 1 (Reproduced from Neimeijer et al.[43]).
B. Galactose-1-phosphate uridyl transferase activity (Gal-1-P converted/h) on days 1 to 7 after subculture of amniotic fluid cells: ■ Normal homozygote cells, ▨ Normal/galactosaemic heterozygote cells, □ Galactosaemic homozygote cells, (Data from the author's laboratory).
C. Galactose-1-phosphate uridyl transferase activity (Gal-1-P converted/h) after subculture, with increasing cell growth measured as mg protein/culture. Symbols as in Fig. 2B. (Reproduced from Fensom and Benson, *Clin. Chim. Acta*, **62**, 1975, 189–194).

Interpretation of the Results of Prenatal Diagnosis

It is clearly the responsibility of the laboratory performing the prenatal diagnostic test to say whether their result indicated that the fetus is or is not affected with the metabolic disease in question. Difficulties in making this decision can be considered under two main headings:

(1) Problems arising from variations of enzyme activity in cell culture.

(2) Problems due to cell culture enzyme activity not representing the true metabolic state in the fetus.

(1) Factors like the heterogeneity of cells, conditions of culture and time of harvesting of cells, all of which

affect enzyme activity in cultured cells, have been discussed in previous sections. Obviously it is necessary to standardize and optimize these conditions as far as possible in each test. With some enzymes it is possible to achieve a very constant and easily interpretable result, so that the actual genotype of the fetus can be predicted with reasonable certainty by comparison with established ranges. Fig. 3 shows how this is done in the prenatal diagnosis of galactosaemia.

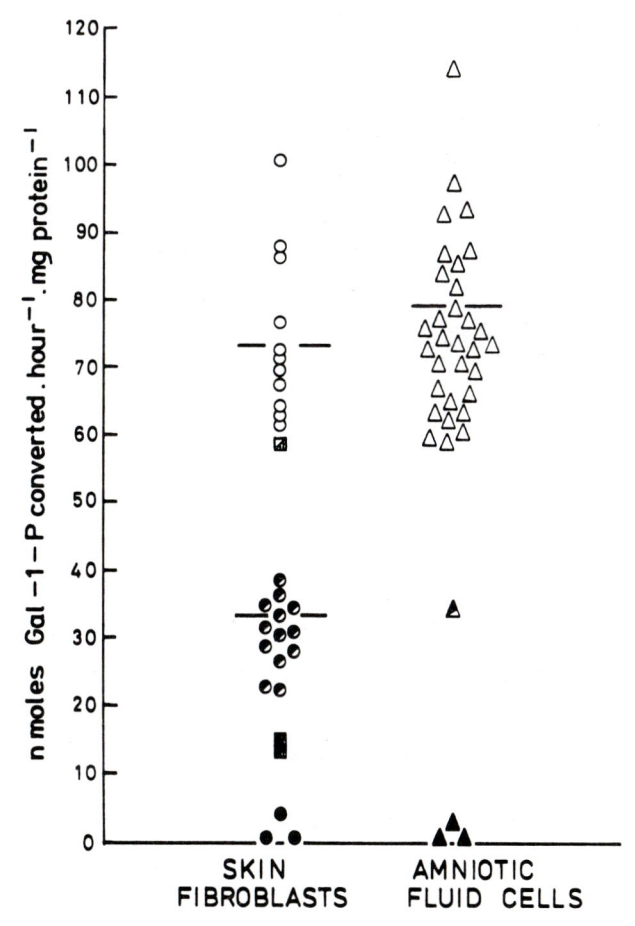

FIG. 3. Galactose-1-phosphate uridyl transferase activity in cultured skin fibroblasts and amniotic fluid cells. ○ △ Normal homozygotes, ◐ ▲ Normal/galactosaemic heterozygotes, ● ▲ Galactosaemic homozygotes, ◨ Normal/Durate heterozygotes, ■ Galactosaemia/Duarte heterozygotes, (Data from the author's laboratory).

There are some enzymes which demonstrate tremendous variation, even between parallel subcultures of the same primary culture. Lysosomal enzymes seem to be exceptionally difficult in this respect, with very great variations between cultures.[61] In these circumstances the interpretation of a result may be extremely difficult. Comparison of ranges of enzyme activity for normal and abnormal cell cultures over a long period may show a complete overlap between the two groups. This problem may be overcome in several ways.

The first is to perform a diagnostic assay within a batch containing a large number of control cell culture samples, normals, affected cases and obligate heterozygotes. Another approach which is useful is to express the enzyme activities relative to the activity of a reference enzyme. Young et al.[61] found that, in spite of wide fluctuations of absolute enzyme activities, the relative levels of lysosomal enzymes were fairly constant. Enzyme activities are usually expressed as a ratio to β-galactosidase activity. Finally, most centres arrange for the cells to be assayed in at least two laboratories and a result would only be accepted if all laboratories agreed on the diagnosis.

The wide variations in the activity of many enzymes in cell culture means that it is particularly difficult to diagnose an affected fetus when a family possesses a gene which codes for a variant enzyme which has intermediate levels of activity. Compound heterozygotes of this gene and the true deficiency gene can have enzyme activities of 10 to 20% of genetically normal individuals and yet be clinically normal themselves. In these circumstances it is necessary to distinguish cultured cells with enzyme activity near zero, in fetuses who would be considered for abortion, from cultured cells with 10 to 20% enzyme activity, in fetuses which should be clinically normal. This is a fine distinction to make with most available methods.

(2) A possiblity which has always to be considered is that the enzyme activity measured in the cultured cells may not represent the enzyme level in body tissues. Unstable enzyme variants may have little net activity in red cells, or body tissues with cells of a low turnover rate, but may have significant activity in cultured cells whose mean age is only a few days. This sort of complication may be anticipated by careful family studies before the prenatal diagnosis is begun.[28]

A false negative result in testing for some metabolic diseases is possible because of contamination of the cell cultures with microorganisms, so that enzyme activity of the foreign organism is measured. The most frequent contaminant is mycoplasma which has been shown to contain hypoxanthine guanine phosphoribosyltransferase, hexosaminidase A and pyruvate dehydrogenase, all of which are enzymes used in the diagnosis of inherited conditions.

Many laboratories will automatically reject cultures which are infected, but this destroys the chance of making what might be a vital diagnosis. An organism may be removed by antibiotic treatment through several subcultures until it is certain that the culture is absolutely clear of infection.

NEURAL TUBE DEFECTS

Introduction—The Need for Prenatal Diagnosis and Screening

Neural tube defects are a major cause of fetal loss, neonatal death and severe handicap in the survivors. The British Isles is generally a high incidence area for the condition, but there are significant regional variations. N. Ireland, W. Scotland and Wales have the highest occurrence and it is least common in the South and East of England. The overall incidence of neural tube defects amongst live and still births lies between 2 and 4 per 1000 but the number of pregnancies affected is much higher, because as many as half of such fetuses may be aborted spontaneously before 28 weeks.

Neural tube defects are due to a failure of embryological development of the central nervous system, the abnormality resulting in a failure of closure of part of the neural tube. The main types of defect are anencephaly, encephalocele and spina bifida. Anencephaly is incompatible with life. Affected babies may be live-born but do not survive beyond the first 24 hours. The other two lesions, involving the head and spine respectively, have varying degrees of anatomical and neurological abnormality.

From the point of view of biochemical detection there is an important distinction between the defects, depending on whether they are 'open' or 'closed' lesions. In closed defects there is usually less neural malformation and the lesion is covered by a normal layer of skin or a very thick membrane. An open defect is usually severe and, as the lesion is not covered by an impermeable barrier, components of fetal extracellular fluid, and neural and other tissues, may 'leak' into the amniotic fluid.

The division into open and closed defects is also related to the clinical severity of the conditions. The prognosis for babies with closed lesions is generally good, but only a very small proportion of those with open defects survive with normal intelligence and minor degrees of physical handicap. Thus the babies which are most affected are the ones which are most likely to show the biochemical changes in amniotic fluid. Cases of closed spina bifida also represent only about 15% of the total of all spinal lesions, or 7% of all neural tube defects, and in future discussion only open spina bifida is considered.

Although the majority of fetuses with neural tube defects abort spontaneously, are still-born, or do not survive the neonatal period, early prenatal detection followed by termination of cases of anencephaly and open spina bifida is considered by many to be a more acceptable way of managing the problem.

A variety of biochemical abnormalities would be expected in amniotic fluid in the presence of an open neural tube defect and some early findings are reviewed by Emery.[15] However, the substance which has proved most valuable in prenatal diagnosis is α-fetoprotein (AFP), a protein which is present in high concentration in the fetus from very early in pregnancy.

Prenatal diagnosis by amniocentesis and estimation of AFP is only justified and possible in a limited number of pregnancies that can be predicted to be a high risk for neural tube malformation. The high risk was originally established by the birth of a previously affected child to the parents, or a family association with neural tube defects (Table 5). Unfortunately this detects only a small number of affected pregnancies because about 90% of infants with neural tube disorders are the first affected children of parents with no family history.

TABLE 5
SOME RISKS OF RECURRENCE OF A NEURAL TUBE DEFECT IN A 'HIGH RISK' AREA

	%
Parents with 1 affected child	5
Parents with 2 affected children	10 to 15
One parent affected	Approx. 3
Parent's niece or nephew affected	1 to 2

In 1973, Brock, Bolton and Monaghan[6] showed that maternal serum AFP levels were increased in pregnancies with fetal neural tube abnormalities and the possibility of a screening test which could be applied in all pregnancies was suggested. This procedure has now been used in many thousands of pregnancies and its problems and limitations as a screening test are well recognized.

In the following sections of this chapter the normal patterns of AFP in fetal serum, amniotic fluid and maternal serum will be considered. The application of AFP analysis to prenatal diagnosis of neural tube disorders will then be considered in the order in which it will now proceed in the majority of cases, i.e. investigation of maternal serum followed by amniocentesis in those cases in which a high risk is indicated.

Normal Patterns of AFP Levels

Fetus

AFP, which has a MW of 64 000 daltons and migrates electrophoretically as an α_1-globulin, originates in the fetal liver and yolk sac, and can be detected in fetal blood from the 4th week of pregnancy. The plasma concentration reaches a maximum of 3 g/l between 12 and 16 weeks and then declines steadily to term (Fig. 4).

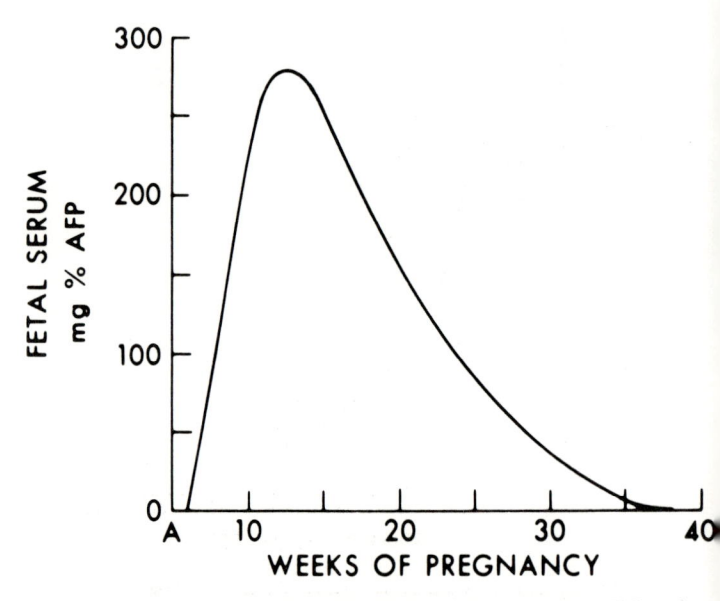

FIG. 4. Mean levels of AFP in normal fetal plasma. (Reproduced from Gitlin and Boesman, *J. Clin. Invest.*, **45**, 1966, 1826–38.)

This gives a false picture of the actual amount of AFP synthesized and it can be calculated that this increases up to the 22nd week, remains almost constant for the next 10 weeks and then decreases from week 32 to term. In fetal plasma the concentration of albumin increases as the level of AFP decreases. There seems to be a reciprocal arrangement between the synthesis of the two proteins by the fetal liver.

Amniotic Fluid

There is a very close parallel between the pattern of change of AFP concentration in amniotic fluid (Fig. 5) with gestation and that of fetal serum, but the actual level in amniotic fluid is only about 1% of that in fetal serum. This constant relationship, which can certainly be demonstrated between about 16 and 32 weeks gestation, could be explained most easily if the protein entered amniotic fluid by a single secretory mechanism which had uniform kinetics over this period. Although permeability of fetal skin and other external membranes may explain the early appearance of AFP in amniotic fluid, it seems almost certain that the most important mechanism in mid-pregnancy is fetal urinary excretion.

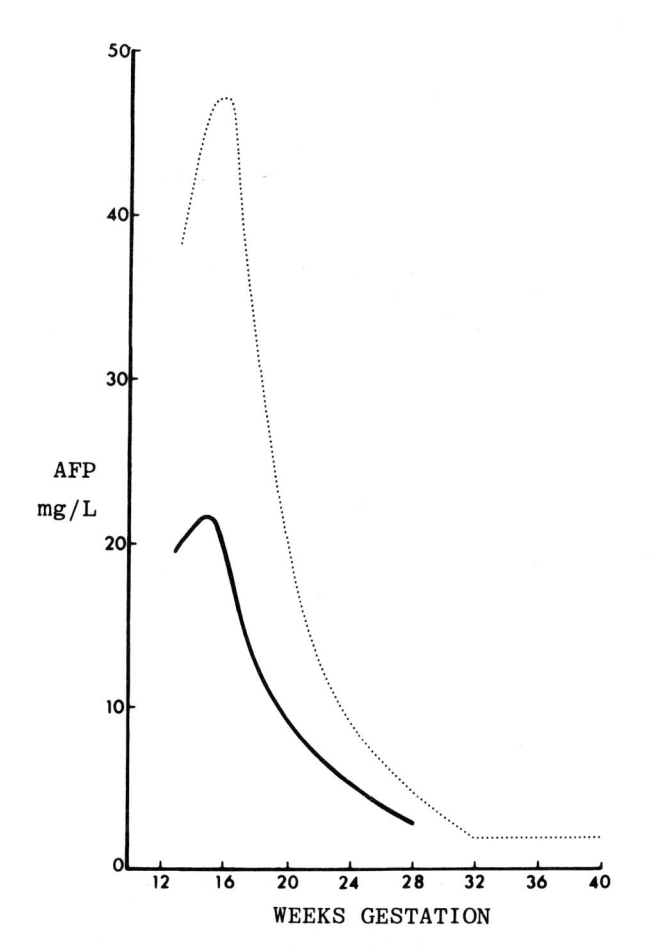

This is consistent with the fact that the AFP concentration in fetal urine follows the same pattern as that in fetal serum and amniotic fluid and that the AFP concentrations in fetal urine and amniotic fluid are very similar until the very last few weeks of pregnancy.

The normal turnover of AFP in amniotic fluid is not known but is of interest if changes in it could lead to altered AFP concentration. Two mechanisms for the turnover have been suggested. First, it is believed that the fetus swallows amniotic fluid and is able to digest protein. Secondly, it is possible that AFP is transferred to the maternal circulation. There is evidence that the former mechanism operates normally but no firm conclusion has been reached regarding the latter. Data on the rate of clearance of AFP when the level has been abnormally raised by a leak of fetal blood indicates that the period over which the AFP concentration falls back to what is considered to be normal is about 4 weeks. It is probable that this involves some passage into the maternal circulation.

Maternal Blood

The pattern of AFP concentration in maternal blood is remarkably different from that in fetal blood, or amniotic fluid, as it goes on increasing to a maximum at around 32 weeks gestation (Fig. 6). Absolute levels in maternal plasma are very much lower than those in fetal plasma by a factor of around 1 to 10 000 (or 1 to 100 compared with levels in amniotic fluid).

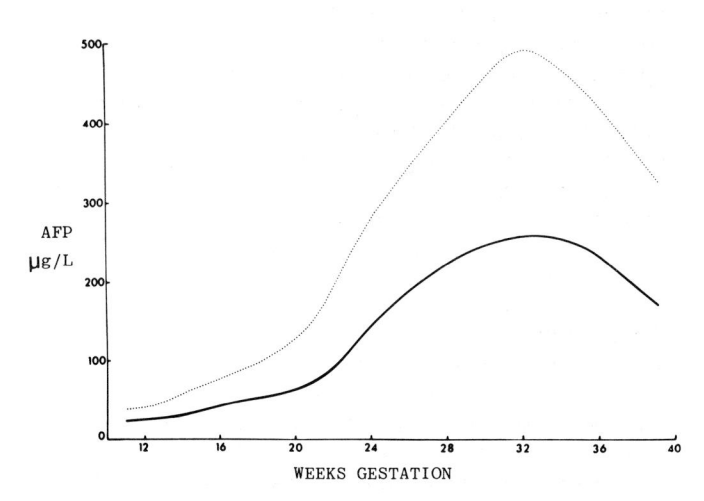

The interesting point is the extent to which the concentration gradient between the compartments is changing, because as fetal plasma and amniotic fluid levels are falling from 16 to 32 weeks gestation, maternal plasma AFP concentrations are steadily rising. The ratio of maternal plasma to fetal plasma or amniotic fluid concentration increases about 100-fold over this period. This suggests that the kinetics of transfer of AFP are altering

FIG. 5. Normal range of amniotic fluid AFP. (Reproduced from Monk and Goldie[38].)
——— mean .... mean + 2 S.D.

FIG. 6. Normal range of maternal serum AFP. (Reproduced from Monk and Goldie[38].)
——— mean .... mean + 2 S.D.

constantly throughout the pregnancy, as the changing concentration gradient could only be partly explained by an increase in the area of transfer which probably occurs as pregnancy progresses.

The source of AFP in maternal serum in normal circumstances is not known, but it would seem more likely to be directly from the fetus because the concentration gradient is much greater. It may also be significant that the level in maternal plasma decreases only when the synthesis of AFP in the fetus declines, at 32 weeks. Recent experiments which show, by a reaction with Concanavalin A, that maternal plasma AFP is qualitatively similar to amniotic fluid AFP over a wide range of normal and abnormal pregnancies with neural tube defects, do not provide any more evidence on this point:[31] they confirm only that abnormally high levels of AFP in amniotic fluid pass into maternal serum.

Maternal Blood AFP Screening

Assay

The sensitivity and precision required to estimate the low levels of AFP in maternal serum can only be achieved using radioimmunoassay. Many methods have been published, one example is that of Monk and Goldie.[38] It will become apparent that very strict attention to the performance of the assay is essential, in respect of both accuracy and precision. Failure to do this can lead, on one hand, to a decrease in the detection rate for neural tube defects and, on the other, to an increase in the number of false positive results in normal pregnancies.

Basis of Screening Procedure

It was clear from the first that maternal serum AFP would not detect every case of open neural tube defect and that the distinction between normal AFP levels and those in affected pregnancies was not absolutely clear cut. Thus the best that can be achieved by screening is to choose a cut off point above which lie a maximum number of abnormal pregnancies and a minimum number of normal pregnancies. By measuring maternal serum AFP at an appropriate time in pregnancy, it is possible to select a group of women with a very high risk of having an open neural tube defect. The details of this approach can be illustrated by the following data from the UK Collaborative Study.[55]

Table 6 shows the detection rate for anencephaly and open spina bifida between 10 and 24 weeks gestation if the 95th percentile is taken as the upper limit of normal for maternal serum AFP. The most consistent results, almost 90% for both types of defect, were obtained between 16 and 18 weeks gestation. This also happens to be a good time to carry out the procedure if the diagnosis is to be made for an early termination of the pregnancy. But using this cut-off point means that 5% of normal pregnancies have a positive screening test. Table 7 shows that, at 16 to 18 weeks, raising the normal limit does not affect the detection of anencephaly substantially, but the detection of open spina bifida may become unacceptably low.

Further consideration of this problem involved using a different statistical method of looking at results. Comparison of results of different laboratories was difficult because of very different normal ranges. The problem was resolved by expressing results as multiples of the normal median value for the same period of gestation. There is an advantage to this device, apart from making it easier to evaluate the composite results of several laboratories. Those who are beginning maternal serum

TABLE 6
SINGLETON PREGNANCIES WITH NEURAL TUBE DEFECTS HAVING MATERNAL SERUM AFP LEVELS EQUAL TO OR GREATER THAN THE NORMAL 95TH PERCENTILE FROM 10 TO 24 WEEKS GESTATION

Week gestation	Anencephaly		Open spina bifida	
	Total	No. detected and percentage	Total	No. detected and percentage
10–12	20	4 (20)	22	4 (18)
13–15	47	34 (72)	35	16 (45)
16–18	51	45 (88)	33	29 (88)
19–21	19	19 (100)	13	7 (54)
22–24	9	7 (78)	5	5 (100)

TABLE 7
SINGLETON PREGNANCIES WITH NEURAL TUBE DEFECTS HAVING MATERNAL SERUM AFP LEVELS EQUAL TO OR GREATER THAN THE NORMAL 95TH TO 99TH PERCENTILE AT 16–18 WEEKS GESTATION

Type of defect	No. of affected pregnancies	Percentage with AFP above normal percentile				
		95th	96th	97th	98th	99th
Anencephaly	51	88	88	86	86	84
Spina bifida	33	88	82	76	76	70

TABLE 8
PERCENTAGE OF PREGNANCIES WITH MATERNAL SERUM AFP LEVELS EQUAL TO OR GREATER THAN SPECIFIED MULTIPLES OF THE NORMAL MEDIAN AT 16 TO 18 WEEKS OF GESTATION

Pregnancy	Cut-off level (multiple of the normal median)				
	2·0	2·5	3·0	3·5	4·0
Singleton anencephalic	90	88	84	82	76
Singleton open spina bifida	91	79	70	64	45
Singleton non-neural tube defect	7·2	3·3	1·4	0·6	0·3
Twin non-neural tube defect	47	26	19	13	11

AFP screening for the first time will find it much quicker to get a reasonable estimate of their normal median than to define their normal range completely. Thus they will be able to interpret results much sooner.

Table 8 shows the percentage of pregnancies with maternal serum AFP levels at 16 to 18 weeks gestation equal to or greater than cut-off levels expressed as multiples of the normal median. This data may then be used to calculate the odds that a maternal serum level AFP greater than various cut-off points is actually associated with a neural tube defect (Table 9). This further calculation depends on the incidence of neural tube defects in the population screened, which would normally lie between 2 and 4 per 1000 live births. It can be seen that a cut-off point at 2·5 times the normal median detects about 90% anencephalics and 80% open spina bifidas, and that a woman with a positive result has between a 1 in 20 and a 1 in 10 chance of having an affected baby.

TABLE 9
ODDS OF A WOMAN WITH A SINGLETON PREGNANCY WITH A SERUM AFP EQUAL OR GREATER THAN THE SPECIFIED CUT-OFF LEVELS AT 16 TO 18 WEEKS GESTATION HAVING A FETUS WITH ANY NEURAL TUBE DEFECT, OR OPEN SPINA BIFIDA

Type of defect and incidence	Odds at cut-off levels (multiples of normal median)				
	2·0	2·5	3·0	3·5	4·0
All defects (incidence 2 per 1000)	1:41	1:21	1:10	1:4	1:3
All defects (incidence 4 per 1000)	1:21	1:10	1:5	1:2	1:1
Open spina bifida (incidence 1 per 1000)	1:79	1:42	1:20	1:9	1:7
Open spina bifida (incidence 2 per 1000)	1:40	1:21	1:10	1:5	1:3

Practical Application of Maternal Serum Screening

A blood specimen is taken, ideally at 16 to 18 weeks gestation. Patients with results above the cut-off point are retested immediately. Probably about 20 to 40% of these repeats will be below the cut-off point and it has been found there is no increased risk of neural tube defect in this 'normal-on-repeat' group. The cause of the high AFP has therefore to be established in the women with persistently raised AFP.

The two most easily established causes of a false positive screening result are twin pregnancy and wrong estimation of dates. These two problems may be investigated by doing an ultrasonic scan. If twin pregnancy is indicated further action is not taken even though a neural tube defect cannot be excluded. Twins are not usually concordant for the defect and it is preferable to allow the pregnancy to continue with one affected fetus than to abort both normal and affected fetuses. If the gestational age has been underestimated it means that the result has been compared with an inappropriately low normal range. When this is corrected these women can be spared further investigation. It has been suggested that there are almost equal numbers of false normal results due to over-estimates of gestational age and this could account for some failures in detection.

The ultrasound scan and a careful consideration of the patient's history may reveal the possibility of one of the many other conditions which can cause a raised maternal serum AFP level (Table 10). In most of these the

TABLE 10
CONDITIONS, OTHER THAN THE NEURAL TUBE DEFECT, ASSOCIATED WITH RAISED AFP IN MATERNAL SERUM AND/OR AMNIOTIC FLUID

Amniocentesis	Oesophageal or duodenal atresia
Fetal distress	Congenital nephrosis
Threatened or missed abortion	Turner's syndrome
Intra-uterine death	Omphalocoele
	Exomphalos
	Rhesus haemolytic disease
	Hydrocephaly
	Fallot's tetralogy
	Skin aplasia

increase in maternal serum AFP occurs in association with, and secondary to, a rise in the level of AFP in amniotic fluid. The causes of some of these abnormalities are discussed by Brock.[5] In other cases, for example in fetal distress, threatened abortion, intra-uterine death and after amniocentesis, it may sometimes be possible for a feto-maternal haemorrhage to occur, causing increased maternal serum AFP, without a significant change in amniotic fluid levels.

After considering all the possible causes of the first raised screening test apart from an actual neural tube defect, and excluding those with normal repeat tests, it is usually possible to eliminate approximately half of the cases. These women will not normally have any further investigations apart from the very rare circumstances where the chance of some other serious congenital defect has been uncovered. If the cut off level selects about 5% positive results this means that amniocentesis is performed on about 2·5% of the total women screened. However, individual centres can improve on this performance. By careful analysis the cut-off point can be raised without lowering the detection rate and by a rigorous search for causes of false positive results also, the amniocentesis rate can be reduced to 0·6% of the total screened.[17] The amniocentesis rate is extremely critical in determining the overall benefits of a screening programme because of the fetal loss associated with it.

Diagnosis of Neural Tube Defects by Amniotic Fluid AFP Assay

Because the levels of AFP are higher, and the neural tube defects are more easily differentiated in amniotic fluid, the method of assay is less critical than in maternal serum. Radioimmunoassay, Laurell 'rocket' electrophoresis, and radial immunodiffusion have all been employed. The first two of these are used most often and are probably more satisfactory for the purpose.

Screening Procedure

The difference between AFP levels in normal pregnancies and those with neural tube defects is much greater in amniotic fluid, and detection rates are higher. The report of the UK Collaborative Study[56] gives a good composite picture of the results obtained in seventeen centres, although the performance of individual laboratories may vary considerably. Once again the useful device of expressing results as multiples of the normal median was used.

Table 11 shows the percentage of pregnancies with and without neural tube defects which have results above the specified limits. The data given has been restricted to 16 to 18 and 19 to 21 weeks gestation and to cut-off levels of 3·0 to 3·5 times the normal median, because these

TABLE 11
SINGLETON PREGNANCIES WITH NEURAL TUBE DEFECTS HAVING RAISED AMNIOTIC FLUID AFP LEVELS, USING CUT OFF LEVELS OF 3·0 AND 3·5 MULTIPLES OF THE NORMAL MEDIAN AT 16 TO 18 AND 19 TO 21 WEEKS GESTATION RESPECTIVELY

Pregnancy	Percentage raised at 16 to 18 weeks	Percentage raised at 19 to 21 weeks
Non-neural tube defect	0·7	1·0
Anencephaly	99	99
Open spina bifida	99	95

periods are when most amniocenteses are likely to be done and the cut-off levels are those recommended in the report. To achieve the same detection limits and specificity in the different periods of gestation the limits have to be changed but, generally speaking, cut-off levels make little difference to detection rates using amniotic fluid, only to the percentage of non-neural tube defect pregnancies with a positive result.

The detection of neural tube defects by use of amniotic fluid AFP levels is almost complete and it is not possible to discern any common cause for the very rare missed cases. About 1% of false positive results occur and it is extremely important to try to recognize these. In fact, published series indicate that it is very unusual for normal fetuses to be terminated.

The most common cause of false positive results is fetal blood contamination and about two thirds of amniotic fluids with a high AFP level are blood-stained. However, only 5% of blood-stained fluids have positive AFP results.

Clearly it is important to distinguish those blood stained fluids containing fetal blood and this may be done using the Kleihauer test.[4] Apart from bloodstaining there are a large number of rarer causes of high AFP and the possibility of one of these must be considered (Table 10). Some are serious congenital defects themselves and termination of the pregnancy may not be inappropriate if one of them is demonstrated.

Alternative Diagnostic Aids

Although the indications are that pregnancy terminations rarely occur in the wrong circumstances, aborting a normal fetus is so unfortunate that other tests for the diagnosis of neural tube defects and to distinguish the false positive result have been sought. Two of these are discussed below.

Smith et al.[50] showed that AFP in the amniotic fluid from abnormal pregnancies was qualitatively different from normal by measuring the percentage of it which had a high affinity for Concanavalin A (Con A). Con A is a plant lectin which bonds glycoproteins terminating with α-D-glucose, α-D-mannose and sterically related structures. In pregnancies affected by a neural tube defect the proportion of AFP which reacted with Con A was much higher. This work was extended by Henderson et al.[24] by showing that normal pregnancies with a raised AFP had qualitatively normal protein as judged by Con A affinity.

The assay of Cholinesterase in amniotic fluid has recently been investigated by a number of groups as an alternative means of diagnosing neural tube defects. It could not improve the detection rates obtained with AFP, but benefit could be gained if the method is not affected by blood contamination. There has been much controversy regarding this latter point and no doubt one important factor is which cholinesterase isoenzyme is being measured by current methods.[11] The most interesting observation has resulted from a study of cholinesterase iso-enzymes in amniotic fluid by polyacrylamide gel electrophoresis.[49] It is claimed that one particular isoenzyme is found only in pregnancies with a fetus affected with a neural tube defect. This would be a very valuable additional technique in the screening laboratory.

Conclusion

An attempt has been made to explain and evaluate the biochemical techniques available for the screening and diagnosis of neural tube defects. The other aspects of this problem, the human and cost benefits are equally important, but are outside the scope of this chapter. For a consideration of these the reader is referred to Chamberlain.[8]

ASSESSMENT OF FETAL LUNG MATURITY AND THE RISK OF RESPIRATORY DISTRESS SYNDROME

Respiratory distress syndrome (RDS) is a major neonatal problem and is strictly a disease of prematurity. Its incidence has been increasing in recent years because of early induction in many pregnancies, in spite of the awareness of the consequences of prematurity. Fortunately mortality from the disease has fallen dramatically in the last decade due to improved methods of treatment. Data reflecting this recent trend are not available generally but in Bristol the mortality of babies with RDS is now only 5% (B. D. Speidel, personal commuication). However, the effort and cost of achieving these results is very considerable and it is better to avoid the complication developing if at all possible.

If an early delivery is being considered the risks of RDS must be taken into account. From studies of incidence in relation to gestation it is possible to get some idea of this risk (Fig. 7). At 32 weeks of gestation, for example, it is very high, but at 37 weeks it can be disregarded. The need for other guidance on the risks of

RDS are two-fold. First, if the gestational age of the fetus lies between the two extremes above, the chances of it developing RDS are about even. A specific indication of the actual risk in a particular pregnancy will be useful in this situation. Secondly, the assessment of the gestational age is usually very unreliable and, therefore, calculation of the risk of RDS from it is also very inaccurate. The use of ultrasound to date pregnancies is more precise and is becoming widely practised. This has reduced partly the dependence on laboratory methods of assessing fetal lung maturity.

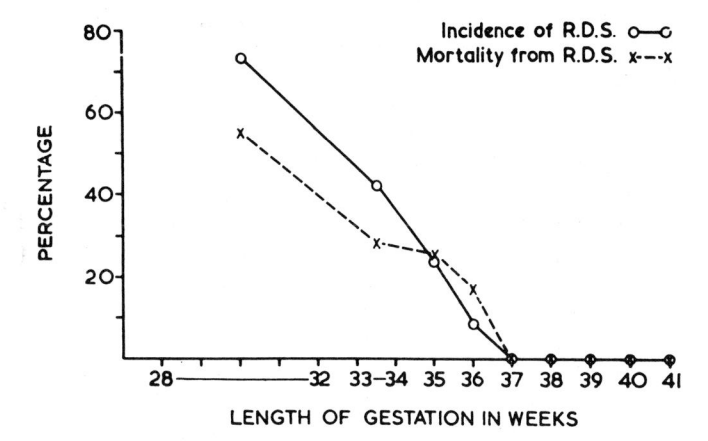

FIG. 7. The probability of the occurrence of, and mortality from, RDS at different gestational ages. (Reproduced from Dunn[14].)

Principles of Assessing Lung Maturity

A wide variety of methods have been used to try to assess the maturity of the fetal lung, or the risk of RDS, in individual pregnancies. Some of these would claim to measure specifically the functional maturity of the fetal lung, whilst others may be no more than tests of general maturity. In fact the theoretical basis of none of the methods has been established with complete certainty but the ideas behind them will be described in this section. Conclusions about whether it is possible to achieve a true assessment of lung maturity will be considered after an examination of the results obtained.

Methods Related to Surfactant Synthesis

The assumption of these methods is that the predominant factor in the determination of lung maturity and occurrence of RDS is the synthesis of surfactant. Surfactant is an extremely effective surface active agent which lines the walls of normal lung alveoli. Its function is to lower the air pressure required to inflate the alveoli and if it is deficient there is a tendency for the lungs to collapse at each expiration. The inspiratory effort needed to reflate the lungs is enormous. These form the principle features of RDS. The significance of surfactant in the etiology of RDS was suggested originally by Avery and Meade[2] and confirmed by others, including Gluck et al.[19] who pioneered the laboratory evaluation of lung maturity.

Surfactant may be isolated from lung homogenates, or from tracheal lavage fluid, and Table 12 shows a typical example of its composition as obtained by King and Clements[32] from the dog. It is composed mainly of lipid, dipalmitoyl lecithin being the major component. King et al.[33] isolated the proteins of surfactant and claimed that one of them, having a MW of 10 000 to 11 000 daltons, is unique to this material. If this is so it is the only component of surfactant which is specific to the lungs.

TABLE 12
COMPOSITION OF SURFACE ACTIVE MATERIAL FROM DOG LUNG[32]

Component	Proportion by weight(%)
Diapalmitoyl lecithin	41
Monoenoic lecithin	25
Protein	9
Cholesterol	8
Phosphatidyl ethanolamine	5
Glycerides	4
Phosphatidyl serine + glycerol	4
Lysolecithin	2
Sphingomyelin	1
Fatty acid	1

It has generally been assumed that the synthesis of the lecithins is the limiting factor in the production of surfactant and two enzyme pathways have been demonstrated which lead to the synthesis of lecithin and other minor phospholipid constituents (Fig. 8). There is no doubt that, of these, the choline incorporation pathway is the most active in the mature lung, and the suggestion by Gluck et al.[20] that the methylation pathway is more important in the immature fetus has not been supported by more recent work. The evidence from work with the rhesus monkey, whose lungs are believed to resemble

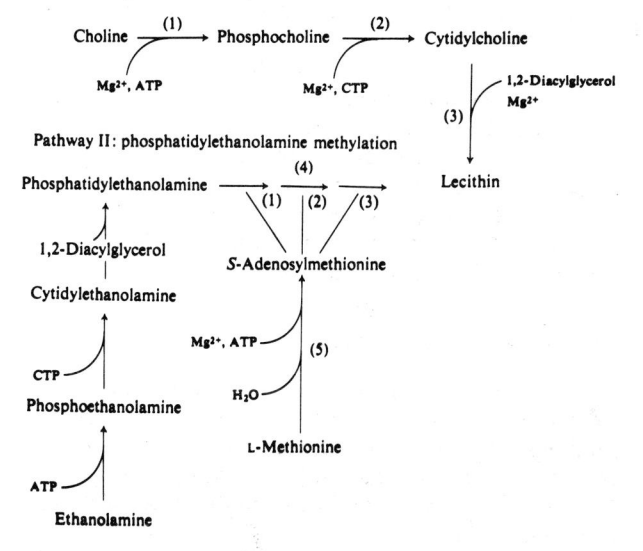

FIG. 8. Pathways of lecithin synthesis.

closely those of the human, is that the choline incorporation pathway is always about 50 times more active than the methylation pathway. Nor has Gluck's claim that palmitoyl myristoyl lecithin is produced in preference to dipalmitoyl lecithin in early fetal life been substantiated. Warren,[59] amongst others, could not find an increased proportion of myristic acid in amniotic fluid from pregnancies early in gestation.

Hallman[22] studied the minor phospholipids of surfactant, phosphatidyl glycerol, phosphatidyl inositol, phosphatidyl ethanolamine and phosphatidyl serine. As the lungs matured changes in the first two of these occurred in amniotic fluid. Early in pregnancy phosphatidyl glycerol was absent and phosphatidyl inositol was very prominent. At 36 weeks the position was reversed and it was claimed that phosphatidyl glycerol became the next most abundant phospholipid to lecithin in the mature lung. A possible application of this work will be mentioned later.

Some surfactant is washed out from the lung by fetal respiratory movements into amniotic fluid and the supposition is that the concentration of surfactant in amniotic fluid reflects its rate of synthesis in the lung. Using lecithin as an indication of surfactant concentration in amniotic fluid, a rapid increase is seen to occur between 34 and 35 weeks of gestation and this is not found in the levels of other phospholipids, like sphingomyelin, which arises principally from sources other than lung (Fig. 9). This increase in lecithin level is presumed to follow closely on a surge of lung surfactant synthesis and the timing corresponds to the period when the lung is maturing, as judged by the rapidly diminishing risk of RDS.

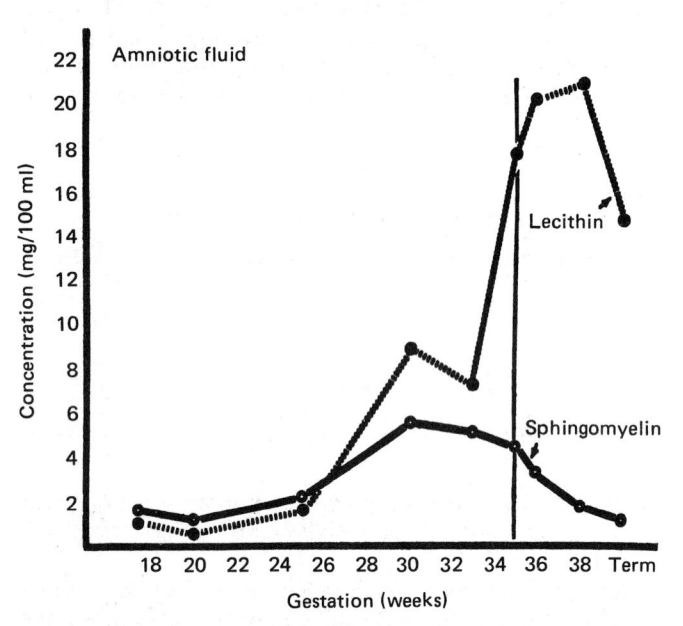

FIG. 9. The mean concentration of sphingomyelin and lecithin in amniotic fluid throughout gestation. (Reproduced from Gluck et al., Amer. J. Obst. Gynaecol., 109, 1971, 440–445.)

Methods of Estimating Surfactant in Amniotic Fluid

Numerous ways of detecting the rise in surfactant concentration in amniotic fluid around 34 to 35 weeks have been published, but few have been validated in clinical practice. There are three basic principles: assessment of a decrease in surface tension, an increase in surface protein or an increase in lecithin. Apart possibly from the assay of surfactant protein, none of the methods can claim to be at all specific to the measurement of surfactant, because the surface active phospholipids, including lecithin, may originate from sources other than the lung.

Surface Tension. The simplest and probably the most widely used method, because it need not involve the laboratory, is based on demonstrating a change in the surface tension lowering property of surfactant in amniotic fluid.[9] The amniotic fluid is diluted progressively with higher concentrations of ethanol to find the greatest dilution at which stable bubble formation occurs. Subjective errors can arise in assessing what are stable bubbles.

Surfactant Protein. King et al.[34] extracted from human endobronchial lavage fluid a protein similar to that which had previously been isolated from dog lung and which was believed to be specific for surfactant. Rabbit antiserum to the proteins has been used for its assay in amniotic fluid by double diffusion precipitation reactions in agarose gel followed by latex fixation.

Lecithin. Many methods have been described which claim to estimate the rise in amniotic fluid lecithin with varying degrees of specificity. Those which are considered most specific involve a preliminary separation of extracted lipids by thin layer chromatography, but it is by no means certain that the lecithin bands contain only this substance.

Following chromatography the lecithin may be extracted and quantitated by a phosphorus method.[41] However, a more rapid and popular technique is to make a semi-quantitative measurement of the lecithin spot on the chromatogram, in comparison with the other major phospholipid spot, sphingomyelin. After charring or staining the phospholopids on the chromatogram, the lecithin/sphingomyelin (L/S) ratio can be estimated by eye, by densitometry or by measuring the area of the spots.

Methods employing a prior isolation of lecithin are rather time consuming and, for this reason, direct analyses on the lipid extract of amniotic fluid have been used. These are probably less specific for lecithin, but are quicker and simpler. Total lipid phosphorus may be estimated, usually after digestion, or glycerophosphatide phosphorus using a sulphuric/periodic acid reagent which should be slightly more specific for lecithin phosphorus. Alternatively, total palmitic acid has been estimated after hydrolysis of the lipid extract, using gas liquid chromatography. Ip et al.[29] attempted to make the palmitic acid estimation more specific to lecithin by extracting the amniotic fluid with a solvent system which preferentially extracts phospholipids rather than triglycerides and free fatty acids.

Methods Measuring Cortisol

There is evidence to suggest that lung maturation is dependent on normal adrenocortical function in the fetus. Babies developing RDS were found to have low cord blood cortisol concentrations[53] and infants dying of RDS smaller adrenals compared to those dying of other causes.[40] The presence of glucocorticoid receptors in fetal lung,[3] and the well established stimulation of lung maturation about 48 hours after the administration of glucocorticoids to the mother,[52] are all consistent with the hypothesis that glucocorticoid activity controls lung maturation.

If the measurement of glucocorticoids in amniotic fluid can serve as a test for lung maturity two conditions must hold. First, there must be a critical level of glucocorticoid activity in the fetus which stimulates, or facilitates, lung maturation. Secondly, amniotic fluid glucocorticoid concentrations must give a good indication of glucocorticoid activity in the fetus.

The test most frequently employed to evaluate this approach is amniotic fluid cortisol, using radioimmunoassay.[46]

Methods Measuring Lipid-positive Cells

The investigation of the proportion of lipid positive cells in amniotic fluid has been used for many years as a test of general fetal maturity. Lipid positive cells are stainable with Nile Blue sulphate and counted under a microscope. Fetal maturity is usually indicated by a proportion of stainable cells over 20%.

There have been only a few attempts to assess amniotic fluid cytology as a test of lung maturity, but some studies suggest that it could have a role in this. Lipid positive cells are thought to arise by exfoliation from the apocrine glands, as the fetal skin matures. An explanation for its application as a lung maturity test depends on the finding that there is a parallel between the maturation of the skin and of other vital organs. In particular maturation of the lipid secreting glands in skin parallels lung maturation, and development of the two processes could be under the same endocrine control. It is also possible that secretions from the mature lung may contribute to the number of lipid positive cells in amniotic fluid.

Interpretation of Results

Although many different methods of assessing lung maturity have been published only a few have been assessed in adequate clinical trials. Most experience has been gained with the bubble stability method and the chromatographic lecithin methods: other techniques have often been assessed only by reference to these. In general, it appears that similar results can be obtained by most methods which measure surfactant and there is little evidence that greater specificity for surfactant, or lecithin, which some techniques are supposed to possess, give any better performance. The principle of the technique being used makes very little difference also to the problems of contamination. Blood and meconium cause the biggest difficulties and they may have the opposite effect on different methods. A result from a contaminated liquor sample should be interpreted with the greatest care.

The ideal lung maturity test would be one having a single critical level which, on one side signified lung maturity, on the other an absolute risk of RDS. Unfortunately this cannot be achieved. In his useful work, Nelson[41] demonstrated a steady decrease in the risk of RDS, from 86·7% to 0·5% as the amniotic fluid lecithin increased (Table 13). As in most series, the majority of

TABLE 13
RISK OF RDS RELATED TO LECITHIN CONCENTRATION IN AMNIOTIC FLUID COLLECTED WITH 72 HOURS OF DELIVERY[41]

Range lecithin concn. (μmol/l)	No. of pregnancies	Risk of RDS (%)
0–7	15	86·7
8–15	16	37·5
16–23	25	20·0
24–31	29	3·4
Over 32	391	0·5

tests were done in pregnancies with mature fetuses and 90% of the tests indicated that there was less than 3% risk of RDS.

Most of the tests which measure surfactant achieve results similar to those described above. A mature fetus is predicted most often with a very high degree of probability. The critical levels indicating lung maturity are approximately 30 μmol/l using quantitative lecithin, 75 μmol/L measuring palmitic acid, or an L/S ratio of 2·0. Only 1–2% of these predictions are likely to be false and these will be pregnancies complicated by certain conditions. Amongst these are maternal diabetes, rhesus haemolytic disease and birth asphyxia. Table 14 shows

TABLE 14
UNEXPECTED RDS SHOWING ASSOCIATED CLINICAL COMPLICATIONS

		Unpredicted cases of RDS			
	Total no. of L/S ratios over 2·0	No. of cases	Associated complications		
Series*			Asphyxia	Diabetes	Rh. disease
Belfast	466	3	1	2	1
Los Angeles	347	13	12	6	—
London (UCH)	108	1	—	1	—
Michigan	not stated	4	—	2	2
Bristol	134	1	—	1	—

* Taken from Whitfield and Sproule[60] with additional data from the author's laboratory.

the numbers of these complications occurring in pregnancies with false positive L/S ratios, in several published series. In maternal diabetes it is even possible to observe a fall in L/S ratio from a mature to an immature result with increasing gestation.

Bustos et al.[7] have measured the levels of other amniotic fluid phospholipids in complicated and uncomplicated pregnancies and claim that phosphatidyl glycerol appears early and phosphatidyl inositol disappears early in complicated pregnancies. This is consistent with a more rapid maturation of the lungs in the presence of complications.

The cause of false mature predictions is not established, but a number of possible mechanisms exist. The lung may have matured normally in these pregnancies but may then be compromised by complications occurring before or during the birth. Another consideration is that the complications which show the phenomenon are usually associated with fetal distress and this is believed to stimulate respiratory movements *in utero*. This could produce an abnormal increase in the transfer of surfactant into amniotic fluid to give a mistaken picture of lung maturity.

Below the critical level for long maturity the test should indicate immaturity, but the risk of RDS varies depending partly on the technique used and the actual result obtained. The bubble stability method lacks value in this situation, an immature result usually indicating only about 20% risk of RDS.[54] Using some lecithin methods different levels of risk can be predicted but in general the problem is that they all produce a large proportion of false predictions of lung immaturity. The detailed interpretation of quantitative lecithin results was shown in Table 13 and the only other technique which has been fully evaluated is the L/S ratio. The actual interpretation of the L/S ratio varies greatly with the centre using the technique (Table 15). In Los

TABLE 15
INCIDENCE OF RDS IN PREGNANCIES WITH L/S RATIOS LESS THAN 2·0

Series	Percentage of babies with RDS when L/S was less than 1·5	Percentage of babies with RDS when L/S was less than 2·0
Belfast[1]	79	33
Bristol[2]	60	33
Cardiff[3]	77	23
Los Angeles[1]	—	63
London (UCH)[1]	—	70

1. From Whitfield and Sproule[60].
2. From author's laboratory.
3. From Turnbull et al.[54].

Angeles and London (UCH) an L/S ratio less than 2·0 was roughly equivalent to one of less than 1·5 in Belfast, Bristol or Cardiff. It is important that a laboratory should assess its performance by a retrospective study of the outcome in tested pregnancies or by comparing results with an established centre.

There are several possible reasons for this difficulty in predicting lung immaturity. First, it is known that the chances of a false result increase as the time interval between the test and delivery lengthens. This is particularly true when the result is close to the critical level indicating lung maturity. The prediction can relate only to the time of the test and there is always a chance that the lung will mature before delivery. Most workers have considered a result to be valid for no longer than 48 hours after amniocentesis. Secondly, the risk of RDS never becomes absolute, no matter how immature the fetus, or how low the synthesis of surfactant. Thus the level of surfactant synthesis in the immature lung is probably not the principle factor determining whether the baby gets RDS. Finally, it is possible that the transfer of surfactant to amniotic fluid is depressed in some immature fetuses and, therefore, the concentration of surfactant may not reflect its synthesis in the lung.

Cortisol levels and lipid-positive cells have not been investigated to anything like the same extent as the surfactant techniques and opinions differ widely as to their value. Some workers have concluded that cortisol is of no use in assessing lung maturity,[46] whilst in one survey cortisol and lipid positive cells were rated better than the L/S ratio or palmitic acid as indicators of the immature lung.[12] However, in this latter evaluation a very high critical L/S value of 3·0 was chosen to indicate lung maturity and the performance of the L/S ratio method was much poorer than in most series.

In conclusion, it appears that most tests which measure amniotic fluid surfactant are able to predict lung maturity with accuracy. An immature result by the bubble stability method is not useful in showing the risk of RDS but other methods which assess lecithin or phospholipid levels may indicate a graded risk, from about 20 to 80%. It would be interesting to compare these results with the probability calculated from an accurate assessment of gestation in order to see whether the test does any more than assess general maturity in these situations.

REFERENCES

1. Allen, J. T., Gillett, M., Holton, J. B., King, G. S. and Pettit, B. R. (1980), 'Evidence of galactosaemia in utero,' *Lancet*, **1**, 603.
2. Avery, M. E. and Meade, J. (1959), 'Surface properties in relation to atelectasis and hyaline membrane disease,' *Amer. J. Dis. Child.*, **97**, 517–523.
3. Ballard, P. L., Baxter, J. D., Higgins, S. J., Bousseau, G. G. (1973), 'Mechanisms of glucocorticoid action: generality of the cytoplasmic receptor system,' *Clin, Res.*, **21**, 289.
4. Betke, K. and Kleihauer, E. (1958), 'Fetaler und bleibender blutfarbstaff in erythrozyhen und erythroblasten von menschlichen feten und neugeboremen,' *Blut.*, **4**, 241–249.
5. Brock, D. J. H. (1976), 'Mechanism by which amniotic fluid alpha-fetoprotein may be assessed in fetal abnormalities,' *Lancet*, **2**, 345–346.
6. Brock, D. J. H., Bolton, A. E. and Monaghan, J. M. (1973), 'Prenatal diagnosis of anencephaly through maternal serum alpha-fetoprotein measurement,' *Lancet*, **2**, 923–924.
7. Bustos, R., Kulovich, M. V., Gluck, L. *et al.* (1979), 'Significance of phosphatidyl glycerol in amniotic fluid in complicated pregnancies,' *Am. J. Obstet. Gynecol.*, **133**, 889–903.
8. Chamberlain, J. (1978), 'Human benefits and costs of a national screening programme for neural tube defects,' *Lancet*, **2**, 1293–1296.

9. Clements, J. A., Platzker, A. C. G., Tierney, D. F., Hobel, C. J., Creasy, R. K., Margolis, A. J., Thibeault, D. W., Tooley, W. H. and Oh, W. (1972), 'Assessment of the risk of the respiratory distress syndrome by a rapid test for surfactant in amniotic fluid,' *N. Eng. J. Med.*, **286**, 1077–1081.
10. Coullin, P., Nicolas, H., Boué, J., Boué, A. (1979), 'HLA typing of amniotic fluid cells applied to prenatal diagnosis of congenital adrenal hyperplasia,' *Lancet*, **1**, 1076.
11. Davies, P., Gosden, C. and Brock, D. J. H. (1979), 'Acetycholinesterase, blood stained amniotic fluid and prenatal diagnosis of neural tube defects,' *Lancet*, **1**, 1303.
12. Doran, T. A., Ford, J. A., Allen, L. C., Wong, P. Y., Benzie, R. J. (1979), 'Amniotic fluid lecithin/sphingomyelin ratio, palmitic acid, palmitic acid/stearic acid ratio, total cortisol, creatinine and percentage of lipid-positive cells in assessment of fetal maturity and fetal pulmonary maturity: A comparison,' *Am. J. Obstet. Gynecol.*, **133**, 302–307.
13. Dorfman, A. (1975), in *Inborn Errors of Skin, Hair and Connective Tissue*, p. 80. (Holton, J. B. and Ireland, J. T., Eds.). Lancaster: Medical and Technical Publishing Co.
14. Dunn, P. M. (1965), 'The respiratory distress syndrome of the newborn: immaturity versus prematurity,' *Arch. Dis. Child.*, **40**, 62–65.
15. Emery, A. E. H. (1973), in *Antenatal Diagnosis of Genetic Disease*, p. 125–130. Edinburgh: Churchill Livingstone.
16. Fensom, A. H., Benson, P. F., Rodeck, C. H., Campbell, S. and Gould, J. D. M. (1979), 'Prenatal diagnosis of a galactosaemia heterozygote by fetal blood enzyme assay,' *Br. Med. J.*, **1**, 21–22.
17. Ferguson-Smith, M. A., May, H. M., Vince, J. D., Robinson, H. P., Rawlinson, H. A., Tait, H. A., Gibson, A. A. M. and Ratcliffe, J. G. (1978), *Lancet*, **1**, 1330–1333.
18. Galjaard, H., Mekes, M., De Josselin De John, J. E. and Niermeijer, M. F. (1973), 'A method for rapid prenatal diagnosis of glycogenosis II (Pompe's disease),' *Clin. Chim. Acta.*, **49**, 361–375.
19. Gluck, L., Kulovich, M. V., Eidelman, A. I., Cordero, L. and Khazin, A. F. (1972), 'Biochemical development of surface activity in mammalian lung IV Pulmonary lecithin synthesis in the human fetus and newborn and etiology of the respiratory distress syndrome,' *Pediat. Res.*, **6**, 81–99.
20. Gluck, L. M., Landowne, R. A. and Kulovich, M. V. (1970), 'Biochemical development of surface activity in mammalian lung III Structural changes in lung lecithin during development of rabbit fetus and newborn,' *Pediatr. Res.*, **4**, 352–364.
21. Gompertz, D., Goodey, P. A., Saudubray, J. M., Charpentier, C. and Chignolle, A. (1974), 'Prenatal diagnosis of methylmalonic aciduria,' *Pediatrics*, **54**, 511–513.
22. Hallman, M., Kulovich, M., Kirkpatrick, E., Sugarman, R. G. and Gluck, L. (1976), 'Phosphatidylinositol and phosphatidylglycerol in amniotic fluid: Indices of lung maturity,' *Amer. J. Obstet. Gynecol.*, **125**, 613–617.
23. Harnden, D. G. (1977), 'Cell biology and cell culture—A review,' in *The Cultured Cell and Inherited Metabolic Disease*, pp. 3–15. (Harkness, R. A. and Cockburn, F. Eds.). Lancaster: MTP Press.
24. Henderson, P., Toftager-Larsen, K. and Nørgaard-Pedersen, B. (1979), 'Concanavallin A reactivity of amniotic fluid alpha-fetoprotein in diagnosis of neural tube defects,' *Lancet*, **2**, 906.
25. Henderson, H. and Whiteman, P. (1976), 'Antenatal diagnosis of Hurler disease,' *Lancet*, **2**, 1024–1025.
26. Holton, J. B. (1977), 'Diagnostic tests on amniotic fluid,' in *Essays in Medical Biochemistry* Vol. 3, pp. 75–107. (Marks, V. and Hales, C. N., Eds.). London: The Biochemical Society and the Association of Clinical Biochemistry.
27. Holton, J. B. (1980), 'Prenatal diagnosis in inherited metabolic disease,' in *Laboratory Investigation of Fetal Disease*. (Barson, A. J., Ed.). Bristol: John Wright.
28. Holton, J. B. and Raymont, C. M. (1979), 'Prenatal diagnosis of classical galactosaemia,' in *Inherited Disorders of Carbohydrate Metabolism*, pp. 141–147. (Burman, D., Holton, J. B. and Pennock, C. A. Eds.). Lancaster: Medical and Technical Publishing Co.
29. Ip, M. P. C., Draisey, T. F., Thibert, R. J., Gagneja, G. L. and Jasey, G. M. (1977), 'Fetal lung maturity as assessed by gas-liquid chromatographic determination of phospholipid palmitic acid in amniotic fluid,' *Clin. Chem.*, **23**, 35–40.
30. Jeffcoate, T. N. A., Fliegner, J. R. H., Russel, S. H., David, J. C. and Wade, A. P. (1965), 'Diagnosis of the adrenogenital syndrome before birth,' *Lancet*, **2**, 553–555.
31. Kelleher, P. C. and Smith, C. J. P. (1979), 'Origin of maternal serum alpha-fetoprotein,' *Lancet*, **2**, 1302–1303.
32. King, R. J. and Clements, J. A. (1972), 'Surface active material from dog lung. II. Composition and physiological correlations,' *Amer. J. Physiol.*, **223**, 715–726.
33. King, R. J., Klass, D. J., Gikas, E. G., Clements, J. A. (1973), 'Isolation of apoproteins from canine surface active material,' *Amer. J. Physiol.*, **224**, 788–795.
34. King, R. J., Ruch, J., Gikas, E. G., Platzker, A. C. G., Creasy, R. K. (1975), 'Appearance of apoproteins of pulmonary surfactant in human amniotic fluid,' *J. Appl. Physiol.*, **39**, 735–741.
35. Liley, A. W. (1961), 'Liquor amnii analysis in the management of pregnancies complicated by rhesus sensitization,' *Amer. J. Obstet. Gynecol.*, **82**, 1359–1365.
36. Lowry, O. H., Roseborough, N. J., Farr, A. L. and Randall, R. J. (1951), *J. Biol. Chem.*, **193**, 265–275.
37. Medical Research Council Working Party on Amniocentesis (1978), 'An assessment of the hazards of amniocentesis,' *Br. J. Obstet. Gynaecol.*, **85, Suppl. 2**, 1–41.
38. Monk, A. M. and Goldie, D. J. (1976), 'The significance of raised maternal serum alpha-fetoprotein levels,' *Br. J. Obstet. Gynaecol.*, **83**, 845–852.
39. Nadler, H. L. (1968), 'Patterns of enzyme development utilizing cultivated human fetal cells derived by amniotic fluid,' *Biochem. Genet.*, **2**, 119–126.
40. Naeye, R. L., Harcke, H. T. and Blanc, W. A. (1971), 'Adrenal gland structure and the development of hyaline membrane disease,' *Pediatrics*, **47**, 650–657.
41. Nelson, G. H. (1975), 'Risk of respiratory distress syndrome as determined by amniotic fluid lecithin concentration,' *Am. J. Obstet. Gynecol.*, **121**, 753-756.
42. Niermeijer, M. J., Halley, D., Sachs, E., Tichelaar-Klepper, C. and Garver, K. L. (1973), 'Transport and storage of amniotic fluid samples for prenatal diagnosis of metabolic diseases,' *Humangenetik*, **20**, 175–178.
43. Niermeijer, M. J., Sachs, E. S., Jahodova, M., Tichelaar-Klepper, C., Kleijer, W. J. and Galjaard, H. (1976), 'Prenatal detection of genetic disorders,' *J. Med. Genet.*, **13**, 182–194.
44. O'Brien, J. S., Okada, S., Fillerup, D. L., Veath, M. L., Adornato, B., Brenner, P. H. and Leroy, J. G. (1971), 'Tay Sachs disease: prenatal disease,' *Science*, **172**, 61–64.
45. Perry, T. B., Hechtman, P. and Chow, J. C. W. (1979), 'Diagnosis of Tay Sachs disease on blood obtained at fetoscopy,' *Lancet*, **1**, 972–973.
46. Perry, L. A., Isherwood, D. M. and Oakey, R. E. (1978), 'Comparison of the concentrations of palmitate and cortisol in amniotic fluid for the prediction of pulmonary maturity,' *Clin. Chim. Acta.*, **83**, 171–176.
47. Pollock, M. S., Levine, L. S., Pang, S., Owens, R. P., Nitowsky, H. M., Maurer, D., New, M. I., Duchon, M., Merkatz, I. R., Sachs, G. and Dupont, B. (1979), 'Prenatal diagnosis of Congenital Adrenal Hyperplasia (21-hydroxylase deficiency) by HLA typing,' *Lancet*, **1**, 1107–1108.
48. Simpson, N. E., Dallaine, L., Miller, J. R., Siminovich, L., Hamerton, J. L., Miller, J., McKeen, C. (1976), 'Prenatal diagnosis of genetic disease in Canada: Report of a collaborative study,' *Canad. Med. Assn. J.*, **115**, 739–748.
49. Smith, A. D. (1979), 'Amniotic fluid acetylcholinestase and neural tube defects: plea for standardisation,' *Lancet*, **2**, 307–308.
50. Smith, C. J., Kelleher, P. C., Bélanger, L. and Dalline, L. (1979), 'Reactivity of amniotic fluid alpha-fetoprotein with concanavallin A in diagnosis of neural tube defects,' *Brit. Med. J.*, **1**, 920–921.
51. Spector, E. B. (1977), 'Properties of human adult and fetal red blood cells arginase: A possible diagnostic test for arginase deficiency,' *Pediat. Res.*, **11**, 464.

52. Spellacy, W. N., Buhi, W. C., Riggall, F. C. and Holsinger, K. L. (1973), *Am. J. Obstet, Gynecol.*, **115**, 261–218.
53. Sybulski, S. and Maughan, G. B. (1976), *Am. J. Obstet. Gynecol.*, **125**, 239–243.
54. Turnbull, A. C., Bhagwanani, S. G. and Fahmy, D. (1974), 'Prenatal detection of the infant at risk', *Proc. Roy. Soc. Med.*, **67**, 243–244.
55. U.K. Collaborative Study on Alpha- Fetoprotein in Relation to Neural Tube Defects (1977), 'Maternal alpha-fetoprotein measurement in antenatal screening for anencephaly and spina bifida in early pregnancy,' *Lancet*, **1**, 1323–1332.
56. U.K. Collaborative Study on Alpha-Fetoprotein in Relation to Neural Tube Defects (1979), 'Amniotic fluid alpha-fetoprotein in antenatal diagnosis of anencephaly and open spina bifida in early pregnancy,' *Lancet*, **2**, 651–662.
57. The United States Institute of Child Health and Human Development (1976), 'Mid-trimester amniocentesis for prenatal diagnosis: Safety and accuracy,' *J. Amer. Med. Assn.*, **236**, 1471–1476.
58. Usategui-Gomez, M. (1974), Immunoglobulins and other protein constituents of amniotic fluid. In *Amniotic Fluid, Physiology, Biochemistry and Clinical Chemistry*, pp. 111–124. (Natelson, S., Scommegna, A. and Epstein, M. B., Eds.). Toronto: John Wiley & Sons.
59. Warren, C. (1973), The estimation of foetal lung maturity, p. 48. M.Sc. Thesis. University of Surrey.
60. Whitfield, C. R. and Sproule, W. B. (1974), Fetal lung maturation. *Br. J. Hosp. Med.*, **12**, 678–690.
61. Young, E., Willcox, P., Whitfield, A. E. and Patrick, A. D. (1975), 'Variability of acid hydrolase activities in cultured skin fibroblasts and amniotic fluid cells,' *J. Med. Gneet.*, **12**, 224–229.

43. PAEDIATRIC CLINICAL CHEMISTRY: Some Special Problems

G. M. ADDISON

Introduction

Special requirements of the paediatric clinical biochemistry laboratory
    Sample size
    Workload differences
    Analytical differences

Reference ranges in paediatrics

Failure to thrive
    Tests of gastro-intestinal function
    Assessment of nutritional status

Neonatal screening
    Neonatal hypothyroid screening
    Screening for cystic fibrosis

Ambiguous genitalia in the neonate

INTRODUCTION

Paediatrics is the medicine of childhood. It is not simply a subspeciality of adult medicine or a part time pursuit for adult physicians. Paediatrics shares much with adult medicine but is clearly distinguished by its concern with both the physical and emotional growth and development of children. The diseases of children show a different pattern from those in adults but affect all the organ systems so that children require amongst other services a comprehensive clinical biochemistry service similar in range of investigation to that provided by laboratories for adult patients. In addition, certain diseases are exclusive to children and laboratory services must be provided to assist paediatricians in the diagnosis and management of these diseases.

The provision of clinical biochemistry services to children is complicated by the fact that children are found in a variety of hospitals or specialized units. These include neonatal wards attached to maternity hospitals, paediatric wards of general hospitals or in specialized childrens hospitals. The clinical chemistry requirements of these different groups is dissimilar. A neonatal unit is concerned with a limited range of problems, respiratory distress syndrome, jaundice, hypoglycaemia, etc. Metabolic diseases often present in neonatal units but, as each disease is rare, only simple screening tests are required. The often complex investigations of inborn errors of metabolism should be done by experienced clinicians and laboratories wherever possible.

The general hospital is the first and often the only place of admission for children who need hospitalization. It needs to provide a wider range of investigations than the neonatal unit. The clinical biochemistry laboratory of a children's hospital will find itself providing services for various subspecialities; these could include nephrology, gastroenterology, endocrinology, oncology, neurology and metabolic medicine as well as surgical specialities such as burns, trauma and cardiac units. No single laboratory is able to provide such a totally comprehensive service for all the specialities without becoming excessively large and inefficient. Paediatric clinical biochemistry services are extended by a network of laboratories with special interests which develop expertise that they make available to others.

It is not possible in a single chapter to attempt to cover the full range of paediatric clinical chemistry. This chapter includes an outline of the needs of a paediatric clinical biochemistry service and gives a few examples of the type of special problems that face a paediatric clinical biochemist. Major areas of paediatric clinical biochemistry must inevitably remain uncovered and emphasis will be placed on areas of uncertainty and controversy in the hope of encouraging more investigation by clinical biochemists to improve the service to the patients.

SPECIAL REQUIREMENTS OF THE PAEDIATRIC CLINICAL BIOCHEMISTRY LABORATORY

Many of the investigations performed on children are similar to those performed on adults and most of the instrumental and analytical techniques are similar. A clinical biochemist working in a paediatric laboratory will need training and background knowledge almost identical to his adult-orientated colleague. Do children then need a special service and need it be separate? In order to answer this question we need to know the special requirements of a paediatric clinical biochemistry service and how these complement or conflict with the needs of adults.

Sample Size

Venepuncture blood samples are difficult to obtain from many children. Taking such samples involves physical restraint of the young patient and emotional upset for all concerned. Furthermore, in the very young child repeated venepunctures will remove a much greater proportion of the child's blood volume than the same amount taken from an adult. For these reasons many analyses are carried out on capillary samples taken by finger or heel prick. This is as skilled a procedure as venepuncture and not without danger, e.g. osteomyelitis following heel prick, but a trained person should be able to collect between 1–2 ml on most occasions in a well hydrated child. Bleeding severely ill children is more difficult and only very small amounts of blood may be obtained. In neonates a high haematocrit will further reduce the amount of serum or plasma obtained. The small specimen volumes require the development of micro methods and the nature of capillary blood affects reference ranges and the interpretation of results. It is important that a record should be kept of the total volume of blood removed (for analytical purposes) from a child particularly in the case of severe or prolonged illness.

The small quantities of plasma or serum make large multi-channel autoanalysers generally unsuitable for much of paediatric work. Various adaptations of this type of equipment have been made but have not had general acceptance in paediatric laboratories. Furthermore these instruments are sometimes too inflexible for paediatric use and require large numbers of samples. Micro techniques are frequently used and have become highly developed in paediatric laboratories and recently these micro methods have become more accepted in adult laboratories. These techniques can be performed manually using a variety of accurate micropipettes or may be semi-automated. Recently developed automated discrete analysers such as the centrifugal analyser can now handle very small plasma samples. They often require modification to reduce the dead volume in the sample cup which in many cases is much larger than the actual volume required for the test. Precautions to prevent sample evaporation may also have to be taken.

In order to avoid blood collection altogether non-invasive techniques are used wherever possible but timed urine collections are as difficult in children (especially girls) as in adults. Random urine samples may provide much information if the analyte is related to the creatinine content.

Workload Differences

Many of the types of analysis carried out in a paediatric laboratory are obviously identical with those in an adult laboratory (electrolytes, urea, blood gases, blood glucose, liver function tests, etc.). Such investigations play an important part in the management of the sick child as well as in establishing a diagnosis. A further battery of tests (sweat test, chromatography of amino acids and sugars, specific tissue enzymes, etc.) are required mainly for diagnosis and almost exclusively by children. These special tests are often time consuming and though small in number constitute a considerable proportion of the workload of the paediatric laboratory.

As well as qualitative differences in the types of test carried out in a paediatric laboratory, there are differences both in time of receipt of the sample in the laboratory and the degree of urgency. Children who are ill show more rapid clinical changes than do adults. For this reason it is often very difficult for the clinician to organize a request the night before so a sample can be taken early in the day. It is therefore difficult to batch tests for one single run a day on large automated equipment. Tests which are often not urgent in adults (e.g. calcium, VMA) may be required urgently in children and such tests may have to be done every day in a paediatric laboratory.

It is the inability of some adult clinical biochemists to recognize these particular requirements of children that has lead to unnecessary friction between paediatricians and some laboratories where there are no separate paediatric laboratory facilities.

Analytical Differences

Differences in concentration of analyte (creatinine) or its nature (alkaline phosphatase isoenzymes) may produce conflict in the choice of analytical technique if a mixed paediatric-adult service is being run. In the case of creatinine many adult laboratories have methods which are insensitive below 50 μmol/l and report difference in steps of 10 μmol. Levels of creatinine in young children are often below 50 μmol/l and changes of 10 μmol represent considerable changes in renal function. For these reasons insensitive methods are unsatisfactory for assessing renal function in children and following progress in an individual patient. Circulating alkaline phosphatase in a child is mainly the bone isoenzyme while in the adult the liver isoenzyme predominates. The choice of assay conditions is therefore particularily important and can have considerable influence on reference ranges, individual result and clinical interpretation.

From these considerations it can be seen that paediatric laboratory services are best separated to some degree from adult services. In the case of the separate children's hospital, an independent laboratory is viable and justifiable. It may have relationships with adult laboratories in nearby hospitals but its relationships are more appropriately with other paediatric laboratories. Because of the wide nature of methods, equipment and tests carried out it is capable of offering a training equivalent in standing to adult laboratories. For the children in general hospitals and neonatal units it is not possible to justify the provision of an independent service within the hospital in particular on the grounds of cost. However, within the general clinical biochemistry services a paediatric sub-department with some autonomy should be set up. It should have at least one paediatrically trained senior member of staff responsible for the service. In view of the acute nature of much paediatric work, the paediatric sub-department could usefully be integrated with the emergency adult work.

REFERENCE RANGES IN PAEDIATRICS

There are major problems in establishing reference ranges for biochemical analyses in children. The first one is the ethical problem of obtaining specimens particularly blood or tissue from 'normal' children. Because of this there are very few surveys of the type so easily carried out with adults. The few surveys that have been carried out have normally dealt with either the neonate or the school child and there is a relative lack of data on the pre-school age group. These surveys have used low numbers which makes statistical analysis difficult and gives rise to occasional conflict in results.

Despite this lack of systematic data there exists a considerable amount of information on paediatric reference ranges which can be found in the literature on disease where appropriate control groups have been studied. Much of this information has been collated by the American Association for Clinical Chemistry and published in a monograph *Paediatric Clinical Chemistry*[19]. This invaluable source of reference besides giving information on paediatric reference ranges gives much useful information and insight into the paediatric laboratory.

The second problem with paediatric reference ranges is the change of the values with age. These changes may reflect the acute adaptation to neonatal life, e.g. plasma calcium, acid-base status and plasma bilirubin; growth of specific tissues or organs, e.g. plasma creatinine and alkaline phosphatase; maturation of function e.g. creatinine clearance and serum immunoglobulins; growth and development of the whole child e.g. gonadotrophins.

Other substances show changes in childhood for reasons which are not well understood. Reference ranges for premature neonates may differ from those in full term infants. The changes of reference values with age make establishing such ranges even more difficult since the numbers required are greater.

Biochemical data should not only be interpreted with respect to reference ranges but also the clinical findings. Several studies on reference ranges in children have shown for some substances statistically significant correlations with age. However a more clinical and less statistical approach suggests that the small differences between age groups may be ignored for the purpose of establishing a reference range for these substances.

On physiological grounds chronological age may not always be the most suitable base for establishing reference ranges. Height, weight, surface area, bone age or some combination of these may be more appropriate. These may serve to narrow the reference range and improve the value of a test. Reference values for renal function tests are usually related both to surface areas and chronological age involving as they do both maturation of function and growth in size. It is preferable in paediatric practice to use m^2 rather than $1.73 \, m^2$ as the base value of surface area. The biochemical changes of puberty are best related to bone age.

Reliable paediatric reference ranges for specimens other than blood are more difficult to obtain. This is obvious with such specimens as tissue biopsies but less so with urine. It is, however, even more difficult to obtain timed collections of urine from children than in adults and to be sure the bladder is completely empty. This has hindered the development of some non-invasive techniques which could replace blood tests, e.g. urine sugar excretion after oral loading tests.

The difficulties in establishing reference ranges for urine due to the problem of collection apply equally to urine samples collected for clinical analysis.

This can be shown in Fig. 1 where creatinine clearance calculated from serum and urine creatinine concentrations and 24 hour urine volume is compared to a calculated creatinine clearance based on the serum creatinine concentration and the patients height. The latter method gives a relatively smooth line showing progression of renal failure while the former shows an undulating curve reflecting poor urine collection even in a well motivated patient.

Other factors than age may influence reference values in children as in adults and it is important to take these into account when using values obtained from the literature. Careful attention must be paid to the nature of the sample tested, especially venous versus capillary blood and to the analytical technique used. In the case of some tests the most notorious of which is the plasma alkaline phosphatase activity there may be no possibility of using literature derived ranges unless an identical method is used. Published paediatric reference ranges for total alkaline phosphatase activity differ by up to a factor of 2 even when supposedly measured using the same method, and greater differences are found when alternative

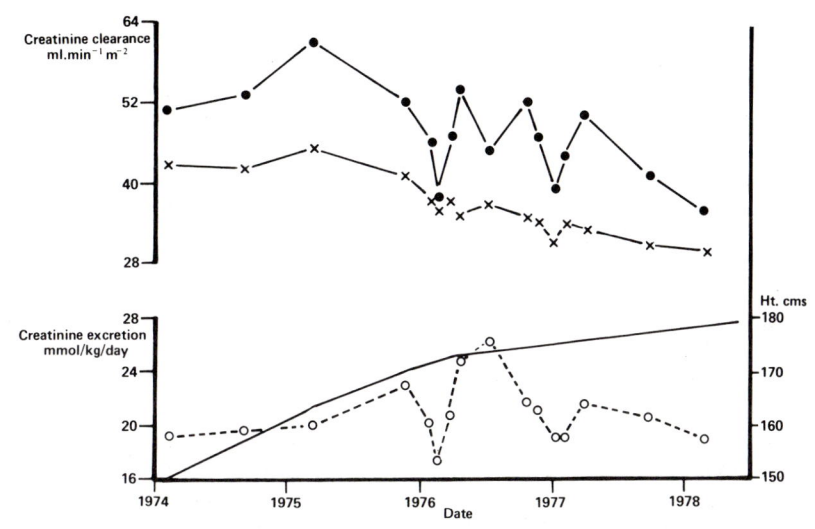

FIG. 1. Creatinine clearance measured by U_{cr} V/S_{cr} (●——●) is compared with the value calculated by the formula Ht × 22/S_{cr} μmol/l (×——×) in a boy with progressive renal failure. The calculated value shows less fluctuation than the measured clearance as variation due to urine collection has been eliminated. The bias between the two values (measured > calculated) is due to systematic over collection of the 24 hour urine specimen by the patient. This can be seen in the high daily creatinine excretion (○-----○).

methods are used. These problems with the measurement of total alkaline phosphatase level in children make its value as a test questionable and alternatives such as the measurement of specific isoenzymes desirable. The interpretation of the isolated raised alkaline phosphatase activity is one of the most common problems in paediatrics. In order to give proper advice each laboratory should establish its own reference range and some improvement could be obtained by relating values to growth velocity where such data is available.

FAILURE TO THRIVE

The term 'failure-to-thrive' is a descriptive term applied to infants who fail to grow appropriately in length or weight for their chronological age. There is no agreed definition and failure-to-thrive has been used in different ways by different authors. Initially its use was confined to patients with nutritional deficiency although it was recognized that malnutrition was not the only cause of poor growth. More specifically in the 1950's the term 'failure-to-thrive' was linked with hypercalcaemia of infancy. Recently with the emphasis on the psychological and emotional development of children 'failure-to-thrive' has been used to refer specifically to those children where growth retardation is due to some disturbance of the child's emotional environment, usually maternal deprivation.

The patients are usually below two years of age and show a failure to gain weight and/or height over a period of time. The failure to gain weight is usually more severe than that of height. The use of this descriptive term implies that there are no signs or symptoms which allow a specific diagnosis to be made. Sometimes children are labelled 'failure-to-thrive' only after exhaustive investigation.

The causes of 'failure-to-thrive' vary geographically. In underdeveloped countries problems of nutrition are most important while in the more 'civilized' developed countries environmental deprivation in some form is the major cause varying between 40–90% in published series. Of the organic causes of 'failure-to-thrive' many require only simple biochemical investigation and only a small number of cases will require a more intensive laboratory work up. The clinical biochemist must know the limitations of his subject and in 'failure-to-thrive' in particular he must not let his biochemical flights of fancy replace an ordered scientific approach to the problem. The evaluation of failure to thrive is best conducted in a planned manner and is not only the province of the clinician and the clinical biochemist but also the nurse, radiologist, bacteriologist, virologist, social worker, psychologist and others will contribute greatly to the assessment of the patient. The child should not be exposed to a battery of random investigations in the hope of a diagnosis turning up.

Children may fail to thrive from birth or develop normally for a period of time and subsequently fail to thrive. In the former group the cause must have existed before or at birth and various placental, fetal and maternal factors are involved. These include genetic and chromosomal factors, congenital abnormalities, intrauterine exposure to infections and drugs (including tobacco and alcohol), endocrine causes and birth trauma. The second group constitutes the major group of children with 'failure-to-thrive'. There is some overlap in the causes of 'failure-to-thrive' between the two groups the main causes are summarized in Table 1.

TABLE 1
AETIOLOGY OF FAILURE-TO-THRIVE

I. *Environmental*
Maternal deprivation: emotional disturbance, socio-economic status; feeding problems.

II. *Organic*

A. *Insufficient net calorie intake*

(1) Inability to suck:- CNS disorders, muscle disorders, respiratory infection, congestive heart disease, etc.
(2) Physical obstruction—cleft lip, hiatus hernia, pyloric stenosis, etc.
(3) Reduced intake:- infection, hypothyroidism.
(4) Vomiting:- air swallowing, metabolic disease, increased intracranial pressure.
(5) Inadequate digestion:- cystic fibrosis, pancreatic deficiency, disaccharidase deficiency.
(6) Inadequate absorption:- gastroenteritis, giardiasis, blind loop, coeliac disease, glucose-galactose malabsorption, biliary atresia; food allergy.
(7) Increased calorie losses:- protein losing enteropathy, diabetes, nephrotic syndrome, aminoaciduria.

B. *Increased calorie requirement*

(1) Muscular activity:- spasticity, congenital heart disease.
(2) Catabolic activity:- chronic infection, hypothyroidism.

C. *Disturbed metabolic function*

(1) Endocrine disorders:- hypothyroidism, adrenal insufficiency, growth hormone deficiency.
(2) Metabolic diseases.
(3) Anaemia.
(4) Hypercalcaemia.
(5) Renal disorders:- pyelonephritis, glomerulonephritis, nephrotic syndrome, renal tubular acidosis.
(6) Lack of specific nutrient.

The investigation of 'failure-to-thrive' should begin with a trial of feeding for up to 10 days in hospital with mother and nurse. If the child puts on adequate weight there is little or no need for any biochemical investigation. At the same time as the trial of feeding, observation of the baby may give clues as to appropriate laboratory investigation. Initial biochemical tests should include the plasma concentrations of urea electrolytes creatinine, calcium, inorganic phosphate and the activity of plasma alkaline phosphatase; the blood sugar level; the urinary excretion of protein, and its amino-acid and sugar content.

Children with a good intake of calories but failure to gain weight in hospital are the group most likely to benefit from the services of the biochemical laboratory. They include children with inadequate digestion or absorption, those with increased calorie requirement or disturbed metabolic function, including inborn errors of metabolism. Children who continue to have a poor intake of food in hospital are more likely to have physical obstruction, neurological or muscular problems, infection etc. Some cases may have endocrine or metabolic disease, e.g. hypothyroidism.

The major problem for the clinical biochemist in 'failure-to-thrive' is the investigation of gastrointestinal function and nutritional status. There is a close interrelationship between the two. The effect of gastrointestinal function on nutritional status is easily understood, however, it should also be appreciated that malnutrition has a direct effect on some gastrointestinal functions. Malnutrition may cause decreases in pancreatic enzymes and intestinal disaccharidases, alterations in bile acid metabolism, steatorrhea and poor absorption of carbohydrates. Therefore in interpreting the results of biochemical investigations it is important to clarify changes which are a primary cause from those which are a secondary effect.

Tests of Gastrointestinal Function

In the investigation of the clinical problem as distinct from research into nutritional problems the emphasis is on measuring gastrointestinal function. (Table 2 lists

TABLE 2
CLINICAL BIOCHEMICAL TEST OF GASTROINTESTINAL FUNCTIONS

I *Tests of enzyme production*

(A) Direct: Pancreatic function tests, (Lundh test meal, CCK-PZ stimulation test); jejunal biopsy with measurement of intestinal disaccharidases; faecal enzymes (chymotrypsin, trypsin).

(B) Indirect: Specific sugar loading tests with measurement in blood, faeces or urine or measurement of breath hydrogen; faecal fat.

II *Tests of absorption*

(A) Direct: Sugar loading tests (xylose, glucose galactose); fat load with measurement of chylomicrons or triglyceride; amino acid loading test; Schilling test.

(B) Indirect: Faecal pH and reducing substances; faecal fat; faecal nitrogen; bile acids.

---

tests of gastrointestinal function useful in cases of failure-to-thrive). Ideally tests should distinguish between failure to digest and failure to absorb. This limits the value of tests such as the faecal fat content; increased excretion of fat being caused by lipase deficiency as well as mucosal damage or inability to form micelles. Such tests are, however, useful for screening patients before more specific tests are carried out. There is some controversy about the value of some of these screening tests and it is probable that none of the tests in use so far has sufficient sensitivity or specificity and several tests may have to be used.

Faecal fat determination and the xylose tolerance test are the two most commonly used screening tests. It has been claimed that faecal fat is the most reliable test of fat absorption. In children with coeliac disease, however, several groups have found that between 30–45% of children do not have steatorrhea. It has been suggested that the cause of the lack of steatorrhea is that the mucosal damage in children is confined to the jejunum and proximal ileum. For this reason various alternative tests of fat absorption have been suggested including fat-loading tests. Measurement of plasma triglycerides either as chylomicrons by nephelometry or directly by standard biochemical analysis have been tried. The increase in plasma triglyceride two hours after a fat load of 2 g neutral fat/kg body weight is used to determine whether fat malabsorption is present. Increases of plasma triglyceride concentration of greater than 0·5 mmol/l or 60 LSI units by nephelometry are accepted as normal. In 32 proven cases of coeliac disease in our laboratory the fat loading test was abnormal in 24 (75%) and the faecal fat in 12 (37%) Fig. 2. Six patients had both tests normal and 2 had steatorrhea but a normal fat load. In another series 92% of patients with coeliac disease have had abnormal fat loading tests. On the other hand patients with other causes of malabsorption and failure-to-thrive may have a normal fat loading test but increased faecal fat. This occurs particularly in patients after gastroenteritis.

	FAT LOADING TEST		
	< 60 LSI units	> 60 LSI units	Total
FAECAL FAT ESTIMATION > 18 mmol/day	12	2	14
FAECAL FAT ESTIMATION < 18 mmol/day	12	6	18
Total	24	8	

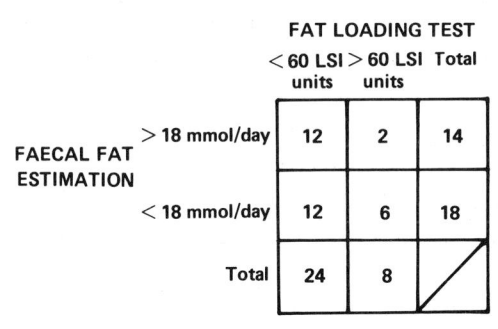

Fig. 2. Comparison of faecal fat estimation and the fat loading test in 32 patients with coeliac disease proven by jejunal biopsy.

Neither the fat loading test nor faecal fat estimations are tests with universal applicability. The choice of test therefore should be made on clinical grounds with knowledge of the probabilities of the eventual diagnosis. Faecal fat estimations give an indication of total fat digestion and absorption whereas the fat loading test using only a 2 hr sample tests the function of the jejunum and proximal ileum. Various other tests of fat absorption have been proposed including the use of radioactively labelled fats (triolein and oleic acid) and the absorption of fat soluble compounds (Vitamin A). None of these tests have gained general acceptance and the use of radioactive isotopes in children should be avoided wherever possible.

The xylose tolerance test has been used as a screening test of sugar malabsorption. In children its use has been limited by the difficulties of obtaining adequately timed urine collections and ensuring complete emptying of the bladder. Errors in collecting the urine are exaggerated by the relatively short time period used. More recently it has been suggested that a blood xylose level of greater than 1·33 mmol/l (20 mg/100 ml) one hour after a 5 g xylose load is highly specific for coeliac disease in children under 30 kg (52 out of 53 cases) and can be used to select patients for intestinal biopsy. Other groups have not been able to confirm this observation and have reported that between 24–35% of patients with proven coeliac disease have a normal blood xylose concentration at one hour. Difficulties in comparing these series are caused by the different doses and means of administration of the xylose load. Xylose absorption is influenced in particular by rate of gastric emptying, intestinal mobility, osmolality of the solution and some drugs. The value of the xylose tolerance test as a screening test remains in doubt until independent series are published using similar techniques which can be compared. On the other hand the xylose tolerance test may prove useful in monitoring changes in the individual patient.

Because of the poor sensitivity and specificity of screening tests of gastrointestinal function it has been suggested that, where a specific test is available and a diagnosis clinically suspected, screening tests should be avoided. This is most strongly advocated in coeliac disease where a jejunal biopsy is used as the specific test. There is little doubt that, where a unit is experienced with obtaining biopsies, this procedure is safe and screening tests can indeed be abandoned. In less experienced units screening tests may still be useful in deciding which children should be referred to a specialist centre but several tests may have to be used.

Some estimation of pancreatic function can be obtained by measuring tryptic activity in faeces. Simple methods such as gelatin digestion have been replaced by more specific methods including spectrophotometric techniques using specific substrates and radioimmunassay. If results are borderline or low a Lundh test meal modified for children or cholecystakinin-pancreozymin (CCK-PZ) test can be carried out with measurement of trypsin, lipase, amylase and bicarbonate secretions by the pancreas. Faecal trypsin estimations have been proposed as a screening test for cystic fibrosis using meconium.

Assessment of Nutritional Status

The assessment of general nutritional status in the individual patient depends more at present on clinical and anthropometric assessment than on biochemical analyses. Although many different alterations in the biochemistry of malnourished children have been reported, these have developed almost entirely from population studies in what might be loosely termed 'community clinical biochemistry'. Difficulties have arisen when attempts have been made to define reliable biochemical markers of general nutritional status to assess the degree of malnutrition and the prognosis for the individual patient. The investigation of the lack of specific nutrients e.g. trace metals and vitamins has proved easier to accomplish especially where some biological action can be measured in vitro.

The problems in assessment of general nutrition, in particular in patients with protein or protein-energy malnutrition, are due to the considerable variability of the measured biochemical parameters in patients with the same degree of clinical malnutrition. Some of the variation is brought about by the complex interrelationship between nutrition and gastrointestinal function, associated diseases and by biochemical or physiological adaptation to dietary changes. Investigations are further handicapped by the small range of type of sample available for analysis (blood, urine and faeces). In addition, in areas of the world where malnutrition is a problem, simple tests which can easily be done in the field are required and these should not rely on expensive high technology.

The search for biochemical tests of nutritional status is based on the not unreasonable hypothesis that biochemical changes will occur before overt tissue damage and clinical malnutrition. The tests listed in Table 3 have

TABLE 3
BIOCHEMICAL ASSESSMENT OF GENERAL NUTRITIONAL STATUS

Plasma proteins	Albumin; Transferrin; Thyroxine binding prealbumin; Retinol binding protein.
Plasma amino acids	Ratio of essential to non-essential amino acids.
Urinary excretion of protein breakdown products	Urea; Hydroxyproline; 3 Methyl Histidine (related to creatinine).
Serum lipids	Cholesterol, Triglycerides Lipoprotein.
Miscellaneous	Creatinine height index.

each in turn been promoted as the answer to the problem of the assessment of protein energy malnutrition only to be shown not to be reliable in all circumstances. It may be that several tests together with multivariate analysis will improve the usefulness of clinical chemistry in this particular field.

For further discussion of nutrition see Chapter 6.

NEONATAL SCREENING

Screening of neonates for certain inborn errors of metabolism and some congenital diseases is based on the expectation that early treatment will in most cases considerably improve the prognosis for the children. These diseases usually occur at very low frequencies (1:16 000 phenylketonuria, 1:60 000 for galactosaemia) or present in a non-specific manner. The detection of these diseases is considerably improved by setting up screening programmes to test all children when suitable biochemical methods have been established. However, the ability simply to detect disease is alone insufficient reason to screen. There must be a quantifiable benefit to the child, its family or the community in order to justify these expensive and often difficult programmes.

Ideally a considerable body of knowledge should exist before screening for a particular disease if certain pitfalls are to be avoided. The pathophysiology of the disease should be well understood and major variants characterized. In some cases variants have only come to light because of the screening programme and some patients have been subjected to unnecessary treatment and emotional trauma. The political pressures to set up preventative medicine programmes make it difficult to set up proper controlled studies of their usefulness before general application. This is particularly true in the neonatal period where screening is often seen as the answer to a clinical problem. One role for the paediatric clinical biochemist is to see that science is not entirely forgotten. Unfortunately some laboratories are over-enthusiastic in their desire to prove their technical ability and adopt screening programmes uncritically soon after they are first suggested in the literature. In times when resources are limited it is even more important that proper evaluation of programmes occurs before general application to large populations. It is usually impossible to do these studies in retrospect or to withdraw a service which may subsequently be shown to be ineffective or at best of marginal effectiveness.

Neonatal Hypothyroid Screening

Congenital hypothyroidism is a disease with multiple aetiologies including athyreosis, maldevelopment or maldescent of the thyroid gland, dyshormonogenesis, iodine deficiency, hypothalamic pituitary disorders and injestion of goitrogens. Untreated or treated later in infancy, the patients are usually mentally retarded. The longer the delay in treatment, the greater the degree of mental retardation appears to be. While this observation is generally true the advocates of neonatal hypothyroid screening failed to explain the exceptions. For instance in one series 10 out of 28 patients treated between 7 and 24 months of age had an IQ of greater than 90, while 5 out of 19 treated before 3 months had an IQ of less than 90. A review of the literature showed that 10 out of 18 cases treated before 6 weeks had an IQ greater than 90 and 2 had an IQ of less than 75; between 7–12 weeks the figures are 17 and 18 out of 47 respectively. Some of these differences reflect the severity of the hypothyroidism. There is evidence that lack of thyroid hormones leads to intrauterine brain damage, some or all of which may be irreversible. This is especially true in cases with total absence of the thyroid gland or dyshormonogenesis.

Another problem with accepting the simple statement that delay in treatment of neonatal hypothyroidism leads to severe mental deficiency is that in all studies reported before screening was adopted the IQ measurements had been taken at a single point in time. Experience in the author's hospital would suggest that at least some patients show improvement in IQ over the years and perhaps only data collected from older children is valid (Table 4). Since screening has been introduced, a considerable body of data is beginning to accummulate on the natural history of neonatal hypothyroidism when treated early and there is little doubt that the patients are doing well. However, we have little data on the natural history of the disease when treated later; such information is almost impossible to obtain retrospectively and with the rapid setting up of screening programmes will never be available.

The argument for screening for neonatal hypothyroidism is reasonably strong but by no means conclusive. The facts are such that economic arguments have to be based on assumptions which may be wrong. (Where are the late treated children with hypothyroidism? Not it appears in large numbers in the mental hospitals). However, there is little doubt that benefit will accrue to many individuals detected in screening programmes.

The complex aetiology of neonatal hypothyroidism makes the choice of test used for screening important. The argument lies essentially between the assay of thyroxine (T4) or thyroid stimulating hormone (TSH).

TABLE 4
NEONATAL HYPOTHYROIDISM IMPROVEMENT OF IQ ON TREATMENT

	Patient 1	Patient 2	Patient 3
Age at diagnosis	1 year	3 years	3½ years
IQ at diagnosis	73*	55	79
IQ 2 years after treatment	85	75	90
IQ 4 years after treatment	93	ND	99
IQ 8 years after treatment	ND	ND	109

ND = No data available.
* IQ at age 2 years 9 months when referred from another hospital.

Reverse triiodothyronine measurement has also been suggested for use in screening but has proved to be of limited value. A screening test should be chosen so that there are no false negatives i.e. high sensitivity. Ideally there should be no false positives either as recall for repeat testing is not only wasteful of effort but disturbing to the family. T4 estimation is a test which will pick up all cases of hypothyroidism, both primary and secondary, provided the right cut-off point is chosen. It will also identify patients with TBG deficiency who will give false positive results. Assay precision is important as values of T4 show a continuous gradation between normal and hypothyroid patients. Recall rates of between 1–5% have been reported for programmes using T4 alone.

TSH measurements will almost inevitably miss secondary hypothyroidism. There is some argument as to whether this is numerically or clinically important. The incidence is about 4% and patients with secondary hypothyroidism often have other associated abnormalities. There is anecdotal evidence that patients detected with TSH deficiency later in life do not have problems of mental retardation. If it is acceptable not to screen for secondary hypothyroidism then TSH becomes almost the ideal test as there is almost complete separation between normals and abnormals. Furthermore, the abnormals have high levels of TSH and this means that there are no problems with assay sensitivity. Recall rates of reported series are between 0·03 and 0·9%; those centres with higher recall rates choosing a lower level of TSH as the cut-off point.

The timing of the test may also have some influence on the choice of the test. Cord blood has been used and has the advantage of being taken at the earliest available time and being available in reasonable quantity. However, cord blood is not often routinely taken now that Rhesus disease can be effectively prevented. There is also much more variability in TSH levels in cord blood. Most screening programmes are based on the use of dried blood spots taken at the time when screening for phenylketonuria and other disorders of aminoacid metabolism is carried out, usually at 3–5 days in North America and Europe and at about 10 days in the United Kingdom. The value of T4 estimation increases as the time of sampling extends as the T4 level falls more rapidly in the patients than in normals, thereby increasing the discrimination. Providing note is taken of the neonatal TSH surge the timing of TSH measurement is not crucial.

The introduction of screening programmes for neonatal hypothyroidism has lead to a considerable increase in our knowledge of the physiology of the thyroid gland during the neonatal period, knowledge which should have been present before screening. It has also led to the identification of at least one additional disorder, that of transient neonatal hypothyroidism. Despite reservations about its over hasty acceptance neonatal hypothyroid screening has on balance proved a worthwhile exercise.

Screening for Cystic Fibrosis

Cystic fibrosis is the most common recessively inherited disorder of children with an incidence of 1:1600. Despite a very large research effort the fundamental defect remains unknown. Diagnosis is based on clinical history together with an abnormal excretion of sodium or chloride in sweat and in 85% of cases evidence of pancreatic disease. Cystic fibrosis has a poor (though improving) prognosis and is very distressing to the child and its family. There has recently been an increasing interest in screening for cystic fibrosis to detect patients before clinical presentation.

Methods proposed for cystic fibrosis screening have included albumin in meconium, trypsin in meconium, trypsin in blood by radioimmunoassay and sweat chloride using an ion-specific electrode. The first method has gained little acceptance due to its poor sensitivity and specificity. The use of the chloride electrode is very labour intensive and probably unsuitable for mass screening especially outside hospital. Screening using trypsin measurements either in meconium or in serum are probably the most useful techniques at present being evaluated. They are unsuitable for patients without pancreatic disease and therefore there will be a minimum of 15% false negatives.

However, the value of screening for cystic fibrosis must remain in doubt until it has been shown that early treatment leads to an improvement in prognosis. Merely advancing the date of diagnosis and thus apparently prolonging survival is useless if the course of the disease and the age of death is unaltered. What evidence there is suggests that there is little improvement in prognosis in patients diagnosed before one year and those diagnosed after one year.

This time is ideal for setting up studies using screening of selected populations to see whether there is any advantage in applying these tests to mass screening.

AMBIGUOUS GENITALIA IN THE NEONATE

The child who presents with ambiguous genitalia in the neonatal period is a paediatric emergency even if the cause is not life threatening. There is an urgent need to establish the child's genetic sex and the aetiology of its ambiguous genitalia in order to allay the parents' anxiety and to establish the most suitable sex (not always the genotypic sex) for rearing the child. The chromosomal sex of the child should be established from leucocyte culture. Buccal smears for sex chromatin are of limited usefulness in neonates as it stains poorly and may be confused with bacterial contamination.

The paediatric clinical biochemistry laboratory has an important role to play in diagnosis with the investigation of the synthesis of the steroid hormones and also the pituitary-gonadal axis. The major causes of ambiguous genitalia are listed in Table 5. Many other rare causes have been reported almost all of which at present do not yield to biochemical investigation. By far the commonest cause is a defect in steroid metabolism or action. The sites of disorders in steroid metabolism are summarized in Fig. 3.

With one exception the causes of ambiguous genitalia in genetic males and females are different. The exception is 3 β-hydroxysteroid dehydrogenase deficiency in which both inadequate virilization of the male and mild virilization of the female can occur. This is thought to be due to partial deficiencies of the enzyme. In the male inadequate testosterone is produced for full virilization while in the female some testosterone or other androgen production is produced by excessive ACTH stimulation of the adrenal cortex and gives rise to mild virilization. On the other hand, the defects in steroid metabolism which give rise to ambiguous genitalia in one sex and are of autosomal recessive inheritance will present in the other sex in some other manner. This is important in the case of salt-losing congenital adrenal hyperplasia. In these cases the biochemical investigation of both sexes will be identical.

TABLE 5
MAJOR CAUSES OF AMBIGUOUS GENITALIA

I *Genetic females (karyotype 46XX)*
(A) Increased endogenous androgen production due to congenital adrenal enzyme deficiencies:- 21 hydroxylase, 11 hydroxylase, 3 β-hydroxysteroid dehydrogenase.
(B) Exogenous androgens and progestogens:- maternal ingestion during pregnancy.
(C) Prominent clitoris of prematurity.

II *Genetic males (karyotype 46XY)*
(A) Decreased endogenous androgen production due to adrenal and/or testicular enzyme deficiencies:- 20–22 desmolase, 3 β-hydroxysteroid dehydrogenase, 17–20 desmolase, 17 α-hydroxysteroid dehydrogenase, 17 reductase.
(B) Defective peripheral androgen metabolism:- 5 α-reductase.
(C) Defective androgen action:- complete and incomplete absence of specific cytosol androgen-binding protein.
(D) Hypospadias with undescended testes.
(E) Maternal injection of progestogens.

III *Genetic mosaicism*
45X/46XY; 46XX/46XY.

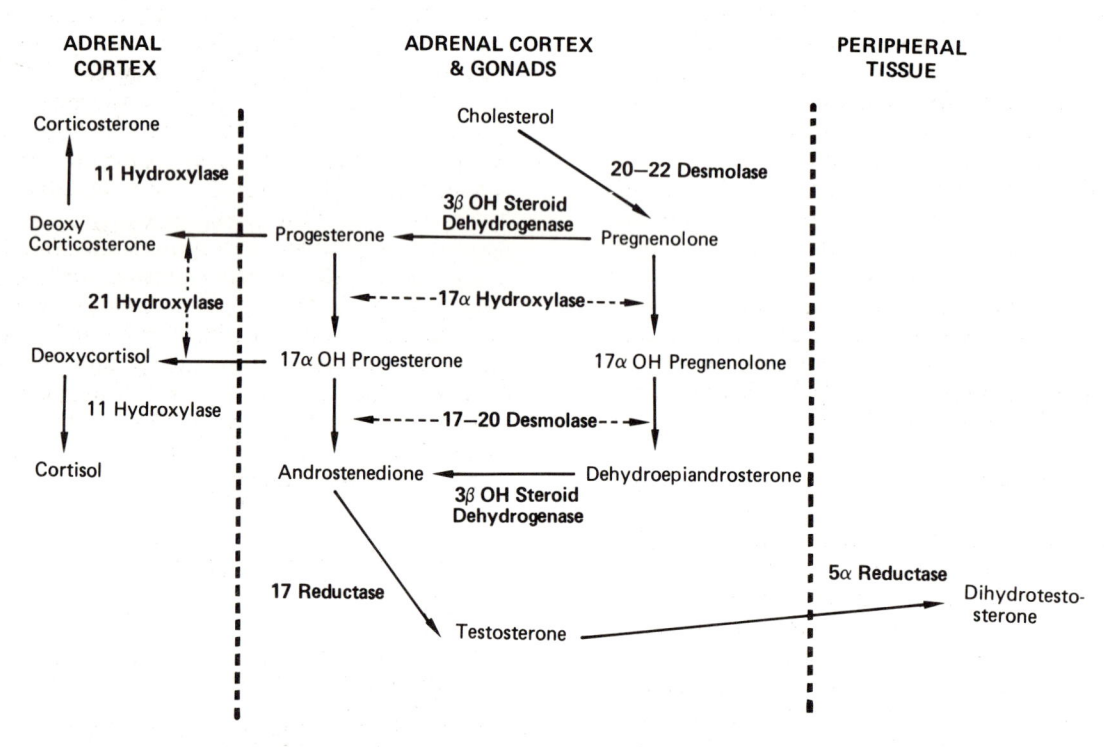

FIG. 3. Disturbances of steroid metabolism can lead to incomplete sexual differentiation and ambiguous genitalia. The figure shows the main steroid pathways which are involved and the enzyme (bold type), deficiencies of which may cause the abnormalities.

TABLE 6
SOME CHANGES IN PLASMA AND URINE STEROIDS IN SYNTHETIC DEFECTS GIVING RISE TO AMBIGUOUS GENITALIA

Defect	21-hydroxylase	11-hydroxylase	17-hydroxylase	3 β-hydroxysteroid dehydrogenase	17-reductase	17–20 desmolase	20–22 desmolase
Urine Ketosteroid	↑↑	↑↑	↓↓	↑	↑	↓↓	↓↓
Pregnanetriol	N→↑↑	N→sl↑		N→↑			N→↓
Oestrogen	↑		↓				
Plasma							
17α-hydroxypregnenolone				↑			
Progesterone	↑		↑				
17α-hydroxyprogesterone	↑↑	N→sl↑					
11-Deoxycorticosterone		↑	↑				
11-Deoxycortisol		↑					
Corticosterone			↑				
Testosterone	↑	↑	N→↓	N→↓	↓	↓	

The biochemical tests involved to establish the diagnosis are the examination of urine and plasma steroids, serum gonadotrophins and in very rare cases enzyme or receptor studies of tissue biopsies (Table 6). Difficulties are often experienced in the neonate and young child in obtaining adequate timed urine collections for quantitative examinations. The presence of abnormal genitalia sometimes exacerbates the difficulty. Urine spot tests e.g. the 11-oxygenation index for the diagnosis of 21-hydroxylase deficiency have been used in an attempt to overcome this problem. Rapid changes in steroid excretion in the neonatal period and the use of non-specific analytical methods can make urine assays in the first few days of life difficult to interpret and false negative results can occur. However the development of relatively specific radioimmunoassays for 17α-hydroxyprogesterone and 11-deoxycortisol (compound S) using small quantities of plasma have essentially replaced the need for urine assays when used for diagnosis of the two commonest defects of steroid synthesis, 21 and 11-hydroxylase deficiencies. Plasma steroids have a further advantage when the child is sick and requires urgent treatment before an adequate urine collection can be made. In these cases a single plasma sample may be taken or the response to exogenous ACTH can be monitored and the enzymic block if any characterized. The ACTH test may even be performed while the child is already on replacement therapy. In rare disorders it is necessary to perform more elaborate studies of plasma and urine steroids, using gas chromatography with or without mass spectroscopy.

Though urine tests have been almost displaced in the diagnosis of congenital adrenal hyperplasia, they retain an important role in the management of the patients. In particular 24-hour excretion of pregnanetriol and 17-oxosteroids reflect the degree of adrenal suppression throughout the day whereas the measurement of plasma steroids, usually 17 α-hydroxyprogesterone and testosterone give information relevant to a short period of time. Recent developments in the assay of steroids in mixed or parotid saliva will allow multiple samples to be taken easily throughout the day. This will facilitate evaluation of the degree of adrenal suppression throughout the day and allow the replacement therapy to be adjusted accordingly. The collection of saliva samples can be done as an out-patient and thus the patient need not be admitted for routine assessments. The control of steroid replacement therapy has to be carefully optimized. If too much cortisol is given then the patient's growth is stunted. On the other hand if too little cortisol is replaced the patient will grow rapidly under the influence of androgens secreted by the unsuppressed adrenal. Rapid advancement of bone age and early fusion of the epiphyses will however result in a loss of potential height. Patients should therefore be evaluated biochemically at frequent intervals.

The diagnosis of ambiguous genitalia in genotypic males due to resistance to androgen action is not within the capability of most laboratories. It involves studies of androgen receptor proteins and the interconversion of testosterone and dihydrotestosterone using biopsy material and cultured fibroblasts. Defects in androgen receptor protein can cause complete or incomplete feminization and as yet it is not possible to separate these two conditions on a biochemical basis. In both cases gonadotrophin secretion is increased and both testosterone and oestrogen production is increased. In deficiency of the enzyme 5 α-reductase which converts testosterone to dihydrotestosterone and gonadotrophin levels are normal as is the plasma testerone concentration, but that of dihydrotestosterone is low. Thus studies of the pituitary gonadal axis may help in establishing the diagnosis but the final diagnosis rests with tissue biopsy studies.

FURTHER READING

General

1. Avery, G. B. (1976), 'Neonatal physiology: basis for the laboratory needs of the nursery' in *The Neonate*, pp. 89–94. New York: John Wiley & Sons.
2. Hicks, J. M. and Young, D. S. (1976), 'Obligations of the paediatric clinical laboratory' in *The Neonate*. pp 325–336. New York: John Wiley & Sons.

3. Hunt, C. E. (1973), 'Capillary blood sampling in the infant: Usefulness and limitations of two methods of sampling,' *Paediatrics*, **51**, 501–506.
4. Mabry, C. C., Roeckel, I. E., Gevedon, R. E., Koepke, J. A. (1968), *Recent Advances in Paediatric Clinical Chemistry*. New York: Grune & Stratton.
5. Meites, S. (1979), 'Skin puncture and blood collecting techniques for infants', *Clin. Chem.*, **25**, 183–189.
6. Meites, S. and Shu-Hui Lin, S. (1980), 'Haemolysis in plasma samples from skin puncture in children,' *Clin. Chem.*, **26**, 987.
7. Michaëlsson, M. and Sjöln. S. (1965), 'Haemolysis in blood samples from newborn infants,' *Acta Paediatr. Scand.*, **54**, 325–330.

Reference Ranges

8. Allansmith, M., McClellan, B. H., Butterworth, M. and Maloney, J. R., (1968), 'The development of immunoglobulin levels in man,' *J. Pediatr.*, **72**, 276–290.
9. de Baare, L., Lewin, J., Sing, H. (1975), 'Ultramicroscale determination of clinical chemical values for blood during the first four days of postnatal life,' *Clin. Chem.*, **21**, 746–750.
10. Belfield, A. and Goldberg, D. M. (1971), 'Normal ranges and diagnostic value of serum 5′ nucleotidase and alkaline phosphatase activities in infancy,' *Arch. Dis. Child.*, **46**, 842–846.
11. Cavallo, L., Margiotta, Kernkamp, C. and Pugliese, G. (1980), 'Serum levels of thyrotropin, thyroxine 3, 3′, 5-triiodothyronine and 3, 3′ 5′-triiodothyronine (reverse T_3) in the first six days of life,' *Acta. Paediatr. Scand.*, **69**, 43–47.
12. Cherian, A. G. and Hill, J. G. (1978), 'Percentile estimates of reference values for fourteen chemical constituents in sera of children and adolescents,' *Amer. J. Clin. Path.*, **69**, 24–31.
13. Clayton, B. E., Jenkins, P. and Round, J. M. (1980), *Paediatric Chemical Pathology: Clinical Tests and Reference Ranges*. Oxford: Blackwell Scientific Publications.
14. Faiman, C., Winter, J. S. D., Reyes, F. I. (1976), 'Patterns of gonadotrophins and gonadal steroids throughout life,' *Clin. Obst. Gynae.*, **3**, 467–483.
15. Gupta, D. (1975), 'Changes in the gonadal and adrenal steroid patterns during puberty,' *Clin. Endocr. Metab.*, **4**, 27–56.
16. Johansson, S. G. O. and Berg, T. (1967), 'Immunoglobulin levels in healthy children,' *Acta Paediatr. Scand.*, **56**, 572–579.
17. Kattwinkel, J., Taussig, L. M., Statland, B. E. and Verter, J. (1973), 'The effects of age on alkaline phosphatase and other serologic liver function tests, in normal subjects and patients with cystic fibrosis,' *J. Pediatr.*, **82**, 234–242.
18. Meites, S., (1975), 'Normal total plasma calcium in the newborn,' *CRC Crit. Rev. Clin. Lab. Sci.*, **6**, 1–18.
19. Meites, S. (Ed.) (1981), *Paediatric Clinical Chemistry*. 2nd edition, Washington D.C.: American Association of Clinical Chemistry.
20. Round, J. M. (1973), 'Plasma calcium, magnesium, phosphorus and alkaline phosphatase levels in normal British school children,' *Brit. Med. J.*, **3**, 137–140.
21. Winter, J. S. D., Hughes, I. A., Reyes, F. I. and Faiman, C. (1975), 'Pituitary-gonadal relations in infancy: patterns of serum gonadal steroid concentration in man from birth to two years of age,' *J. Clin. Endocrinol Metab.*, **42**, 679–686.

Failure-to-Thrive

22. Campbell, A. G. M. (1978), 'Failure to thrive,' in *Textbook of Paediatrics*. Eds. J. O. Forfar and G. A. Arneil. 2nd Edition pp 438–442. Edinburgh: Churchill-Livingstone.
23. Hannaway, P. J. (1970), 'Failure to thrive. A study of 100 infants and children'. *Clin. Paed.*, **9**, 96–99.
24. Komrower, G. M. (1964), 'Failure to thrive,' *Brit. Med. J.*, **2**, 1377–1380.
25. Riley, R. L., Landwirth, J., Kaplan, S. A. and Collipp, P. J. (1968), 'Failure to thrive: An analysis of 83 cases,' *Californian Med.*, **108**, 32–38.
26. Root, A. W. (1976), 'Failure to thrive and problems of growth,' in *The Neonate*. Eds. D. S. Young and J. H. Hicks. pp 157–168. New York: John Wiley & Sons.
27. Smith, C. A. and Berenberg, W. (1970), 'The concept of failure to thrive,' *Paediatrics*, **46**, 661–663.

Gastrointestinal Function Tests

28. Anderson, C. and Burke, V. (1975), *Paediatric Gastroenterology* pp 639–652. Oxford: Blackwell Scientific Publications.
29. Barr, R. G., Perman, J. A., Schoeller, D. A. and Watkins, J. B. (1978), 'Breath tests in paediatric gastrointestinal disorders: New diagnostic opportunities,' *Paediatrics*, **62**, 393–401.
30. Fällström, S. P., Nygren, C. O. and Olegard, R. (1977), 'Plasma triglyceride increase after oral fat load in malabsorption during early childhood,' *Acta. Paediatr. Scand.*, **66**, 111–116.
31. Hamilton, J. R., Lynch, M. J. and Reilly, B. J. (1969), 'Active coeliac disease in childhood: Clinical and laboratory findings of 42 cases,' *Q.J. Med.* **38**, 135–158.
32. Holzel, A. (1967) 'Sugar malabsorption due to deficiencies of dissacharidase activities and of monosaccharide transport,' *Arch. Dis. Child.*, **42**, 341–352.
33. Lamabadusuriya, S. P., Packer, S. and Harries, J. M. (1974), 'Limitations of xylose tolerance test as a screening procedure for coeliac disease,' *Arch. Dis. Child.*, **49**, 244.
34. Larcher, V. F., Shepherd, R., Francis, D. E. M. and Harries, J. T. (1977), 'Protracted diarrhoea in infancy,' *Arch. Dis. Child.*, **52**, 594–605.
35. Newcomer, A. D., McGill, D. B., Thomas, P. J. and Hofmann, A. F. (1975), 'Prospective comparison of indirect methods for detecting lactase deficiency,' *N. Engl. J. Med.*, **293**, 1232–1236.
36. Robards, M. F. (1973), 'Plasma nephelometry after oral fat loading in children with normal and abnormal jejunal morphology,' *Arch. Dis. Child.*, **48**, 656.
37. Rolles, C. J., Kendall, M. J. Nutter, S. and Anderson, C. M. (1973), 'One hour blood xylose screening test for coeliac disease,' *Lancet*, **ii**, 1043–1045.
38. Shmerling, D. H. (1974), 'Screening tests in coeliac disease,' in *Coeliac Disease*. Eds. W. Th, J. M. Hekkens and A. S. Pona. pp 339–345. Leiden: Starfert Kroese.
39. Shmerling, D. M., Forrer, J. C. W. and Prader, A. (1970), 'Faecal fat and nitrogen in healthy children and in children with malabsorption or maldigestion,' *Paediatrics*, **48**, 690–695.
40. Sladen, G. E. and Kumar, P. J. (1973), 'Is the xylose test still a worthwhile investigation?,' *Brit. Med. J.*, **3**, 223–225.

Nutrition

41. Alleyne, G. A. O., Hay, R. W., Picou, D. I., Stanfield, J. P. and Whitehead, R. G. (1977), *Protein-Energy Malnutrition*. pp 54–92 and 150–165. London: Edward Arnold.
42. Ingenbloek, Y., Van Den Schriek, H. G., De Nayer, P., De Visscher, M. (1975), 'Albumin transferrin and thyroxine binding pre-albumin/retinol, binding protein complex in assessment of malnutrition,' *Clinica Chim. Acta.*, **63**, 61–67.
43. Ogunshina, S. O. and Hussain, M. A. (1980), 'Plasma thyroxine binding prealbumin as an index of mild protein energy malnutrition,' *Am. J. Clin. Nutr.*, **33**, 794–800.
44. Reeds, P. J. and Laditan, A. A. O. (1976), 'Serum albumin and transferrin in protein energy malnutrition,' *Brit. J. Nutr.*, **36**, 255–263.
45. Rutishauser, I. H. E. and Whitehead, R. G. (1969), 'Field evaluation of two biochemical tests which may reflect nutritional status in three areas of Uganda,' *Brit. J. Nutr.*, **23**, 1–13.
46. Shetty, P. S., Watrasiewicz, K. E. Jung, R. T. and James, W. P. T. (1979), 'Rapid-turnover transport proteins: An index of subclinical protein-energy malnutrition,' *Lancet*, **ii**, 230–232.
47. Viteri, F. E. and Alvarado, J. (1970), 'The creatinine height index: Its use in the estimation of the degree of protein depletion and repletion in protein calorie malnourished children,' *Paediatrics*, **46**, 696–706.

Screening

48. Bickel, H., Guthrie, R. and Hammersen, G. (Eds.) (1980), *Neonatal Screening for Inborn Errors of Metabolism*. Berlin: Springer-Verlag.

49. Crossley, J. R., Berryman, C. C. and Elliot, R. B. (1977), 'Cystic fibrosis screening in the newborn,' *Lancet*, **ii**, 1093-1095.
50. Fisher, D. A., Dussault, J. H., Foley, T. P., Klein, A. M., Franchi, S., Larson, P. R., Mitchell, M. L., Murphy, W. H., Walfish, P. G. (1979), 'Screening for congenital hypothyroidism. Results of screening one million North American infants,' *J. Pediatr.*, **94**, 700–705.
51. Ginsberg, J., Walfish, P. G., Chopra, I. J. (1978), 'Cord blood reverse T_3 in normal, premature, euthyroid low T_4 and hypothyroid newborns,' *J. Endocrinol Invest.*, **56**, 633–642.
52. Hulse, J. A., Grant, D. B., Clayton, B. E., Lilley, P., Jackson, D., Spracklan, A., Edwards, R. H. W. and Nurse, D. (1980), 'Population screening for neonatal hypothyroidism,' *Brit. Med. J.*, **280**, 675–678.
53. Macfaul, R. and Grant, D. B., (1977), 'Early detection of congenital hypothyroidism,' *Arch. Dis. Child.*, **52**, 81–88.
54. Money, J., Clarke, F. C. and Beck, J. (1978), 'Congenital hypothyroidism and IQ increase, A quarter century follow up'. *J. Pediatr.*, **93**, 432–434.
55. North, A. F. (1974). 'Screening in child health care.' *Pediatrics*, **54**, 608–618.
56. Sackett, D. L. (1975) 'Laboratory screening: A critique,' *Fed. Proc.*, **34**, 2157–2161.
57. Stephan, U., Busch, E. W., Kollberg, H., Hellsing, K. (1975) 'Cystic fibrosis detection by means of a teststrip,' *Paediatrics.* **55**, 35–38.
58. Ten, Kate, L. P., Faenstra-de Gooyer, I., Ploeq-de Groot and Anders, G. J. P. A. (1978), 'Should we screen all newborns for cystic fibrosis?,' *Int. J. Epidemol.*, **7**, 323–334.
59. Walfish, P. G. (1976). 'Evaluation of three thyroid-function screening tests for detecting neonatal hypothyroidism,' *Lancet*, **i**, 1208–1211.

Ambiguous Genitalia

60. Clayton, B. E., Edwards, R. H. W. and Makin, H. L. J. (1971), 'Congenital adrenal hyperplasia and other conditions associated with a raised urinary 11-oxygenation index,' *J. Endocr.*, **50**, 251–265.
61. David, M., Ghali, I., Gillet, P., David, L., Francois, R. and Jeune, M. (1977), 'Management of congenital adrenal hyperplasia by determination of plasma testosterone 17α-hydroxyprogesterone and adrenocorticotropic levels and plasma renin activity: A comparison with the classic method based on urinary steroid determinations', in *Congenital Adrenal Hyperplasia*. Eds. P. A. Lee, L. P. Plotnick, A. A. Kowerski and C. J. Migson. pp 183–194. Baltimore: University Park Press.
62. Gupta, D., Roger, K., Kloman, W. and Attanasid, A. (1977), 'Diagnosis and management of congenital adrenal hyperplasia, in *Congenital Adrenal Hyperplasia*', Eds. P. A. Lee, L. P. Plotnick, A. A. Kowarski and C. J. Migson. pp 283–299. Baltimore: University Park Press.
63. Hughes, I. A. and Winter, J. S. D. (1976), 'The application of a serum 17α-hydroxyprogesterone radioimmunoassay to the diagnosis and management of congenital adrenal hyperplasia', *J. Paediatr.*, **88**, 766–773,
64. Lippe, B. M. (1979), 'Ambiguous Genitalia and 'Pseudohermaphroditism,' *Paed. Clin. N. Amer.*, **26**, 91–106.
65. Lippe, B. M., La Franchi, S. H., Lavin, N., Parlow, A., Coyotupa, J. and Kaplan, S. A. (1974), 'Serum 17-hydroxyprogesterone, progesterone, oestradiol and testosterone in the diagnosis and management of congenital adrenal hyperplasia,') *J. Pediatr.*, **85**, 782–787.
66. Shackleton, C. H. L. and Honour, J. W. (1976), 'Simultaneous estimation of urinary steroids by semi-automated gas chromatography. Investigation of neonatal infants and children with abnormal steroids. *Clin. Chim. Acta.*, **69**, 267–283.
67. Shackleton, C. H., Mitchell, F. L. and Farquhar, J. W. (1972), 'Difficulties in the diagnosis of the adrenogenital syndrome in infancy,' *Paediatrics*, **49**, 198–205.
68. Walker, R. F., Read, G. F., Hughes, I. A. and Riad-Fahmy, D. (1979), 'Radioimmunoassay of 17 α-hydroxyprogesterone in saliva, parotid fluid and plasma of congenital adrenal hyperplasia patients,' *Clin. Chem.*, **25**, 542–545.

INDEX

Abdomen, radiography, 160
Abdominal pain, acute. *See* Acute abdominal pain
Abetalipoproteinaemia, 83, 470
Acetaminophen, 376
Acetazolamide, 293
Acetoacetate, 33–34
Acetyl choline receptor (AChR), 443
Acid-base balance, 165, 422
Acid-base disturbance, 407
Acid maltase deficiency, 440
Acid production, 101
Acid secretion, 67–68
Acid-serum test, 217, 220
Acidosis, 21, 30, 33, 36
Acrodermatitis enteropathica, 357
Acromegaly, 557
ACTH (adrenocorticotrophic hormone), 401, 402, 481, 550, 553, 556, 617, 618, 621, 624, 630, 690, 691
  stimulation tests, 625
  unresponsiveness, 628
Acute abdominal pain, 155–163
  clinical examination, 156–157
  differential diagnosis, 157–158
  history taking, 156
  medical conditions simulating, 162–163
  special investigations, 158
  steps in diagnosis, 156–159
Acute tubular necrosis, 19
Addison's disease, 13, 627
  clinical presentation and chemical pathology, 627
Adenosine diphosphate (ADP), 251, 441, 461
Adenosine monophosphate, cyclic. *See* Cyclic AMP
Adenosine triphosphate (ATP), 402, 403, 438, 440, 543
Adipsia, 61, 65
Adrenal cortex, 616–632
  anatomy, 616
Adrenal cortical failure, 422
Adrenal cortical hormones, 51
Adrenal glands, 616
Adrenal medulla, 616
Adrenal steroids, 307, 617, 620
Adrenal tumours, 631, 650
Adrenaline, 483, 485, 589, 617
Adrenocortical dysfunction, 627–631
Adrenocortical insufficiency, 627
Adrenocortical steroids, 590
Adrenocortical tumours, 631
Adrenocorticotrophic hormone. *See* ACTH
Adrenogenital virilism, 630
Adrenoleucodystrophy, 420, 499
α-adrenoreceptors, 484
β-adrenoreceptors, 484
Aerobic metabolism disorders, 440–442
Affective disorders, 525–526
  differential diagnosis, 526–527
  laboratory tests, 527
  maintenance therapy, 528
Akinetic mutism, 414

ALA (5-Aminolaevulinate), 176–180, 184, 428
ALA-synthetase, 176, 177, 179, 180
Alanine aminotransferase, 142, 151
Albumin, 240, 245, 405, 563
  function, 240
  measurement, 241
  plasma, 144
  structural variants, 240
  variation in disease, 241
Alcohol-dependence syndrome, 534
Alcohol effects, 331
Alcohol intoxication, 423
Alcohol-related disability, 534
Alcoholic liver disease, 151
Alcoholic neuropathy, 429
Alcoholism, 505, 534, 536, 559
Aldactone, 481
Aldosterone
  in hypokalaemia, 19
  tubular unresponsiveness to, 22
Aldosterone antagonists, 477
Aldosterone deficiency, 22
Aldosterone-producing adenomas, 482
Alkaline phosphatases, 142–143
Alkalosis, 19, 40
Alkyl mercury compounds, 351, 352
Allopurinol in gout, 333
Alpha-fetoprotein (AFP), 243, 662
  in amniotic fluid, 673, 675–676
  in maternal blood, 673–675
  in maternal serum, 674
  in normal fetal plasma, 672
  in prenatal diagnosis, 672
  normal patterns of, 672
Alpha-glutamyl transferase, 512
Alpha-methyldopa, 385
Aluminium hydroxide, 293
Aluminium poisoning, 375
Alveolar function, 38
Alzheimer's disease, 415
Ambiguous genitalia, 690
Amenorrhoea, 482, 648
  postpill, 649
Amikacin, 398
Amine hypothesis, 525
Amines in CSF, 453
Amino acid metabolism, 118
Amino acid sequence of oxytocin, 54
Amino-acids, 119, 144, 167–169, 188, 194–196, 198, 208, 494, 506, 517, 607
Aminoacidurias, 506–513
p-Aminobenzoic acid (PABA), 74–75
Aminoglycoside antibiotics, 398–399
5-Aminolaevulinate. *See* ALA
Aminopyrine, 152
Amiodarone, 396
Amitriptyline, 378, 392
Ammonia, urinary, 29
Ammonium in hepatic encephalopathy, 167
Ammonium chloride, 36
Amniocentesis, 664
  risk of, 665
  timing of, 668
  transport and storage of specimens, 668

Amniotic fluid, 664–665
  AFP concentration in, 673
  cholinesterase in, 676
  composition and formation, 664
  contamination of, 665
  cultured cells, 669
  glucocorticoids in, 679
  lecithin in, 678
  lipid-positive cells in, 679
  sphingomyelin, 678
  supernatant, 668
  surfactant in, 678
  uncultured cells, 668
Amyloid plasma precursors, 244–245
Amyloidosis, 427
Anaemia, 122, 211–220, 327
  acquired haemolytic, 216–217
  associated with deficiencies, 212–214
  chronic disorders, 218
  congenital haemolytic, 215–216
  diagnosis of type, 219–220
  haemolytic, 214–216, 220, 339
  infectious causes, 218
  introduction, 212
  iron deficiency, 219
  marrow disorders, 218–219
  mechanical haemolytic, 217
  megaloblastic, 132, 133, 220
  microangiopathic haemolytic, 217
  sideroblastic, 219
Anaerobic glycolysis, 35
  in muscle, disorders of, 438
Androgens, 617, 636, 641, 645, 647
Androstenedione secretion, 621
Anencephaly, 663
Angina pectoris, 397, 475
Angiotensin, 480
Angiotensin II, 50, 61
Anion gap, 15–16, 42
Ankylosing spondylitis, 329
Anorexia nervosa, 97, 126, 649
Anterior pituitary cell types, 549
Anterior pituitary function tests, 553–554
Anterior pituitary gland, 548–559
  anatomy, 548
  histology, 549
Anterior pituitary hormones, 550
  measurement of, 551
Anterior pituitary secretion, hypothalamic control, 551–553
Anthropometry, 120
Antibodies, 259, 266
α_1-Antichymotrypsin, 238
Anticoagulants, 397
Antidiuretic hormone (ADH), 3, 10, 43, 44, 46, 51, 53, 279, 401, 406, 485
  changes in urine concentration and plasma levels, 10
  clinical effect of, 56
  effect on distal tubules and collecting ducts, 53
  effect on renal tubular reabsorption of water, 48
  factors influencing secretions, 3–4

695

INDEX

Antidiuretic hormone (ADH) (contd)
   hyponatraemia without source of ectopic, 12
   in body fluids, 52
   inappropriate secretion of. See Syndrome of inappropriate secretion of ADH (SIADH)
   inhibition of secretion, 12
   mechanisms controlling secretion, 55
   release of, 7–8
Anti-dysrhythmic drugs, 394–396
Anti-epileptic drugs, 388–392
Antiglobulin test, 220
Anti-inflammatory drugs, 397–398
Anti-neoplastic drugs, 398
Antinuclear antibodies (ANA), 339
Antithrombin, 259, 266
Antithrombin III, 239
Antithrombin deficiency, 254
Anti-thyroid-hormone antibodies, 576
α_1-Antitrypsin, 236–238
Anxiety, 525, 532–534
   assessment, 533
   conditions causing, 535
Aplasia, 219
Apolipoproteins, 242
Apoprotein A, 470
Appendicitis, 159
   differential diagnosis, 159–160
   in children, 159
   special investigations, 160
Appetite control, 126
Arginine infusion test, 556
Ariboflavinosis, 132
Arm muscle circumference (AMC), 124
Arnold-Chiari malformation, 505–506
Arterial lipids, 458
Arteries, normal and atherosclerotic, 457
Arteritis, 343–344
Arthritis, 325
   acute suppurative, 329–330
   associated with hyperlipoproteinaemia, 335
   gouty, 331
   inflammatory, 326–330
   other forms of, 335
   post-infective, 326
   reactive, 326
   septic, 326
Arthroscopy, 336
Artificial liver, 164–173
Arylsulphatase A, 491, 497
Arylsulphatase B, 491, 497
Ascorbic acid, 133
Aspartate aminotransferase (AST), 142, 147, 149, 151, 474
Asphyxia, 500
Aspirin, 397
Astroprotein (AP), 447
Ataxia telangiectasia, 518
Atherosclerosis, 457–473
   aetiology of, 460
Atherosclerotic plaque, 457, 459–461
Atropine, 375
Autoerythrocyte sensitisation, 256
Auto-immune disease, 446, 568, 570
   characteristics of, 568
   hypothyroid, 575
Auto-immune thyroiditis, 575

Bacterial infections of the nervous system, 446
Bacterial shock, 235
Barbiturate blisters, 423
Barbiturates, 372
Bassen-Kornzweig disease, 427
B-cell tumours, 229, 230
Becker-type muscular dystrophy, 434
Bence-Jones protein (BJP), 231
Bence-Jones proteinuria, 272
Benedict's test, 287, 595, 596
Benzodiazepines, 270, 372
Beriberi, 131
Bernard-Soulier syndrome, 255, 267
Beta-adrenergic blocking drugs, 397
Beta-blocking agents, 371
Bicarbonate, 25–26
   control by kidney, 27–28
   depleted by buffering, 29
   derived values, 32
   in intestinal tract, 31, 36
   in respiratory acidosis, 38
   reabsorption of, 28
   tubular mechanisms generating, 36
Biguanides, 35
Bile, 75–89
   drug recirculation in, 382
   lipid transport in, 76
Bile acid function, 76
Bile acid metabolism, 75–76, 153–154
   assessment of, 77–78
Bile acids, 145–146, 169
   bacterial deconjugation of, 77
   faecal, 78
   imparied ileal transport, 77
   intraluminal deconjugation, 86
   malabsorption of, 90
   serum estimation, 78
Bile secretion, 75
Biliary cirrhosis, primary, 150
Biliary flow obstruction, 142–144
Biliary secretion, loss of, 100–101
Bilirubin, 151, 502–503
Bilirubin esters, 141–142
Bilirubin estimation, 141
Bilirubin metabolism, 153
Bleeding disorders, 251, 254, 262
Bleeding time, 266
Blood constituents, 174
Blood diseases, 335
Blood examination, 219
Blood gas analyses, 368–369
Blood gas concentrations, 24–43
Blood glucose determinations, 122, 165
Blood platelets, 461
Blood pressure, 479
Blood production rate, 635
Blood tests
   malabsorption, 84
   malnutrition, 84
   myopathies, 431
Blood vessels, 457–473
Blood volume calculation, 5
'Blue bloaters', 40
Body weight, 124
Bohr effect, 191
Bone-age 644
Bone calcium flux, 312–317
Bone disease, 147
Bone disorders, 275–276
Bone marrow, metastatic deposits, 219
Bone quality, 316
Bone resorption, 301
Bone turnover disorders, 323
Bornholm disease, 163
Botulinum toxin poisoning, 428

Bovril tests, 556
Bowel, contamination, 109–110
Brain biopsy, 494
Brain death diagnosis, 423
Brain diseases, 420
Brain growth spurt, 491
Brain injury, 500
Brain reactions, 523, 538
Brain structure, 535
Brancher enzyme deficiency, 440
Breath test
   ^{14}C-glycocholate, 86–87
   ^{14}C-lasctose, 84
   ^{14}C-D xylose, 87
   hydrogen, 84, 87, 109
   liver disease, 152
Bromide partition test, 451
Bromide poisoning, 373
Bromosulphthalein (BSP), 152–153
Buffering, 25–26, 33

Cachexia, cancer, 127
Cadmium toxicity, 353–355
Caeruloplasmin, 150, 242, 358
Caffeine, 438
Calcitonin, 91, 304–306
   secretion disorders, 322
Calcium absorption, 82, 106, 294
Calcium balance, 301–303, 310
Calcium disorders, 299–325
Calcium distribution, 301
Calcium flux, 301
   bone, 312–317
   intestine, 308–309
   kidney, 310–312
Calcium function, 299
Calcium homeostasis disorders, 318–325
Calcium infusion, 69
Calcium levels, 407, 422
Calcium pyrophosphate dihydrate deposition diseases, 333–334
Calcium stone disease, 292–294
Calcium stone formation, 292
Calcium transport, 308
Cancer, protein-energy, malnutrition, 125–126
Cancer cachexia, 127
Carbamazepine, 390
Carbenoxolone, 480
   in hypokalaemia, 19
Carbohydrate absorption, 81
   assessment of, 84
Carbohydrate digestion, 81
Carbohydrate metabolism, 119, 403–404
Carbon dioxide, 25, 26
   arterial blood, 31
   arterial estimations, 42
   arterial levels, 39
   plasma total, 31
Carbon monoxide poisoning, 370, 423
Carbonate dehydratase, 27–31
Carboxyhaemoglobin concentrations, 370
Carbromal, 373
Carcinoembryonic antigen (CEA), 111
Carcinoid syndrome, 111
Carcinoid tumours, 277, 613
Carcinoidosis, 614
Carcinomatosis, 217
Cardiac failure, 476–478
Cardioactive drugs, 370
Cardiovascular disease, 331
Cardiovascular system, 456

Carnitine acyl transferase 441
  deficiency of, 442
Carnitine deficiency, 441, 442
Carnitine metabolism disorders, 441
Carotenaemia, 129
Carpal tunnel syndrome, 430
Casall's collar, 132
Catecholamine metabolism, 277, 526
Catecholamines, 483, 507, 544
Central core disease, 436
Central pontine myelinolysis, 420
Cerebral dysfunction, 524
Cerebral haemorrhage, 417
Cerebral infarction, 417
Cerebral oedema, 414
Cerebrospinal fluid (CSF), 247, 445–455, 526
  blood-brain barrier, 445, 451
  chemical examination, 448–450
  collection of, 448
  contamination, 448–450
  examination, 417–419, 430, 501
  identification of, 454
  in hyperglycinaemia, 509
  inspection, 448
  properties of, 445
Cerebrovascular diseases, 417
Cervical mucus, 644
C1-esterase inhibitor, 239
Cetyl pyridinium chloride (CPC), 499
^{14}C-glycocholate breath test, 78, 86–87
Charcoal haemoperfusion, 170
Charcot-Marie-Tooth disease, 427
Chemoreceptor trigger zone (CTZ), 72
Chest diseases, 163
Chest pain, 473–476
  acute, 473–475
  chronic, 475–476
Chest radiography, 163
Chest wall, pain arising from, 475
Chloral hydrate, 372
Chlorate poisoning, 373
Chloride
  in CSF, 449
  urinary, 30
Chlormethiazole, 372
Cholecystokinin-pancreozymin (CCK), 68, 71, 73, 74, 610, 611
Cholecystokinin-pancreozymin (CCK-PZ) test, 687
Cholestasis, 77, 149
Cholesterol, 82, 145, 146–147
Cholesterol-ester exchange-protein (CeEP), 468
Cholinesterase in amniotic fluid, 676
Cholinesterase inhibitors, 443
Chorionic gonadotrophin (hCG), 653, 654, 661, 662
  assay techniques, 654
  clinical aplications, 655
  detection in pregnancy, 654
Christmas disease, 253, 257, 335
Chromatography, 493, 494
Chromium deficiency, 590
Chromium metabolism, 136, 360
Chronic ambulatory peritoneal dialysis (CAPD), 375
Chronic cold haemagglutinin disease, 216
Chronic lymphocytic thyroiditis, 575
Chylomicron remnants, 463, 465
Chylomicrons, 146, 461, 463
Chyluria, 282
Cigarette smoking, 461
Citrullinaemia, 494

^{14}C-lactose breath test, 84
Climatic effects, 292
Clonidine, 556
Clostridium perfringens, 218
Clotting factors, 144–145, 252–254, 256–267
  assay methods, 261–268
Clotting mechanism, 224, 252–253
Clotting reactions, 252
Coagulation, 144–145, 251–268
Coagulation disorders, 166
Cobalt deficiency, 136
Coeliac disease, 91, 105, 212–213
Collagen disease, 450
Collagen disorders, 337
Collecting ducts, 53, 55–56
Colon, 89–92
  motility, 89
  nutrient absorption, 89
  salt absorption, 89
  water absorption, 89
Colonic fluid transport, 77
Coma, 413–425
  causes of, 414–415
  clinical aspects, 415–417
  definition, 413
  differential diagnosis, 414
  general assessment, 416
  history-taking, 415
  neurological grading, 416
  neurological signs, 416
  neurophysiological basis, 413–414
  trauma, 418–419
  see also Diabetic coma; Endocrine disorders; Hypoglycaemic coma; Hypopituitary coma; Metabolic disorders; Myxoedema Coma; Uraemic coma
Complement, 224, 232–235, 340
  deficiencies, 234
  indicator, 234
  measurements, 233, 235
  physiology, 232
Computerised axial tomography (CAT), 415, 418, 482, 485, 650
Congenital abnormality, drug-induced, 504, 505
Congenital adrenal hyperplasia (CAH), 630–631
Congenital erythropoietic porphyria, 180, 185
Congenital hypothyroidism 688
Congenital muscular dystrophy, 435
Congenital myasthenia, 445
Congestive heart failure, 14
Conjugation reactions, 565
Connective tissue disease, 337
Connective tissue disorders, 337–344
Conn's syndrome, 481
Contaminated small bowel syndrome, 77
Cooperativity, 189
Copper concentration, 358
Copper-containing enzymes, 358
Copper deficiency, 136, 359
Copper metabolism, 136
Copper toxicity, 136, 359, 519
Coproporphyria, 183
  hereditary, 184
Coproporphyrin, 184, 185
Coproporphyrinuria, 182
Corticosteroids, 342–344
  in hypokalaemia, 19
Corticosterone, 617, 621
Cortisol, 545, 617, 620, 621, 623

Cortisol-binding globulin (CRG), 621, 622
Cortisone, 621
Cortisone-glucose tolerance test, 599
Cotrimoxazole, 508
C-peptide, 587, 602–603
Cranial diabetes insipidus, 59, 64
C-reactive protein, 224, 235, 244, 247, 248
Creatine excretion, 431
Creatine kinase, 432, 474
Creatine measurement, 432
Creatine phosphokinase (CPK), 433
Creatinine clearance measurement, 685
Creatinine/height index (CHI), 124
Cretinism, 518, 519
Creutzfeldt-Jakob disease, 420
Crohn's disease, 214, 329, 357
Cryoglobulins, 231
Crystals, 336
Cushing's disease, 325, 535, 620
Cushing's syndrome, 322, 482, 547, 625, 628
  chemical pathology, 629
  clinical presentation, 629
  diagnosis of, 628
Cutaneous manifestations, 338
^{14}C-D-xylose breath test, 87
Cyanide poisoning, 373
Cyanocobalamin, 132
Cyclic AMP, 56, 59, 294, 312, 313, 319, 323, 403, 439, 543, 545
Cyst fluid, 452
Cystic fibrosis, 100, 689
  screening tests, 100
Cystinuria, 281, 290–291
Cytotoxic oedema, 414

DDAVP administration, 64, 65
Deamination/decarboxylation reactions, 565
Debrancher enzyme deficiency, 440
Defective adhesion, 267
Deficiency diseases, 127
Degenerative diseases of nervous system, 495
Degenerative vascular disease, 457–473
Dehydroepiendrosterone (DHA), 621
Dehydroepiendrosterone sulphate (DHAS), 621, 661
De-iodination reactions, 564
Dejerine-Sottas disease, 427
Deletions, 195
Delta-amino-laevulinic acid (ALA), 176–180, 184, 428
Deoxyhaemoglobin, 188, 191
Depression, 524
  laboratory role, 531
  neurotic, 525
  reactive, 525
  secondary, 525
  somatic causes, 527
Dermatomyositis, 342, 442
Desipramine, 392
Desmethylimipramine, 378
17,20-Desmolase, 645
1,25–DHCC, See 1,25-dihydroxy vitamin D_3
Diabetes, 34
  abdominal pain, 163
Diabetes insipidus, 10, 418
  cranial, 59, 64
  nephrogenic, 58, 65
Diabetes mellitus, 10, 19, 48, 57, 570, 559, 590
  classification, 598
  complications, 591
  long-term complications, 592
  monitoring schemes, 592

Diabetic coma, 420, 592
Diabetic ketoacidosis, 591
Diabetic neuropathy, 429
Diagnostic tests, 140
Diarrhoea, 89–92, 111
   investigation of, 90
   occurrence, 89
   osmotic, 91–92
   secretary, 90, 91
Diazoxide in cardiac failure, 478
Dibromosulphthalein (DBSP), 153
Dietary deficiency, 97
Dietary factors, 293, 308
Differential diagnosis
   acute abdominal pain, 157–158
   appendicitis, 159–160
   malabsorption and malnutrition, 95
   osteomalacia and osteoporosis, 130
Diffuse brain damage, 415
Diffuse intravascular coagulation (DIC), 255
Digestive efficiency, 98
Digestive inadequacy, 97–98
Digoxin, 393–394
Digoxin poisoning, 370
Dihydrofolate reductase, 418
2,8-Dihydroxyadenine, 290
Dihydroxyphenylalanine (DOPA), 168
1,25-Dihydroxy vitamin D_3 (1,25$(OH)_2D_3$), 305–307, 309, 313, 319, 320, 321, 322, 542, 546
24,25-Dihydroxy vitamin D_3 (24,25$(OH)_2D_3$), 305–306, 542, 546
2,3-Diphosphoglycerate, 191, 199, 200
Diphtheria, 428
Diphyllobothrium latum, 214
Diquat poisoning, 376
Disaccharides, 286
Disopyramide, 396
Disseminated intravascular coagulation (DIC), 409, 422
Distal tubules, 53, 55–56
Diuresis, 367
Diuretics
   hyponatraemia due to, 13
   in cardiac failure, 476
   in hypokalaemia, 19
DNA, 212, 338, 340, 518
DNA analysis, 666
DNA mapping, 209
Donath-Landsteiner antibody, 216
L-dopa, 556
Dopamine, 483, 524
Down's syndrome, 492
Drug absorption and first pass metabolism, 382
Drug bioavailability, 382
Drug compliance, 381
Drug concentration, 365–366
   salivary, 388
Drug determination, 364
   quality control of, 388
Drug dosage, 380–381
Drug effects
   hypokalaemia, 19
   hyponatraemic, 13
   malabsorption, 108
Drug excretion, 385
Drug interactions, 363
   biochemical tests, 369
Drug metabolism, 152, 384
Drug monitoring, 387–388
Drug myopathies, 443

Drug overdose, 364
Drug recirculation in bile, 382
Drug screening, 364
Drug therapy, 379–399
   see also Serum
Drug toxicity in the elderly, 123
Drugs
   causing congenital abnormalities, 504, 505
   in CSF, 453
   in prenatal development, 505
   in urine, 288–289
   serum levels, 380–387
Duchenne muscular dystrophy, 433–434, 518, 664, 668
Dumping syndromes, 108
Dwarfism, 558
Dysostosis multiplex, 499
Dysfibrinogenaemia, 254
Dyshormonogenesis, 567

Eclampsia, 217
Eczema, 250
EDTA, 472
Ehlers-Danlos syndrome, 255, 337
Ehrlich's reagent, 181
Elastin disorders, 337
Elderly persons, laboratory tests, 534
Electrocardiography (ECG), 473
Electrolyte abnormalities, 436
Electrolyte absorption, small intestine, 80
Electrolyte balance, 1–23, 591
Electrolyte composition of extracellular and intracellular fluid, 1–2
Electrolyte disturbances, 421–422, 502
Electrolyte homeostasis, 369
Electrolyte loss, sources of, 7
Electrolyte secretion, small intestine, 81
Electromyography, 342, 432
Electron microscopy, 432, 499
Electrophoresis, 206, 245–247, 449, 450, 471
   plasma, 245
   serum, 245, 246
   urine, 247
Emesis, 367
Emotions, physical substrate of, 524
Emphysema, 39
Encapsulated charcoals, 170
Encephalitis, 419
Encephalopathies of extracerebral origin, 500
Endemic hypothyroidism, 576
Endocrine changes, 436
   after trauma, 400
Endocrine disorders, 110–111, 422, 536, 537, 540, 546–547, 649
Endocrine pancreas, 583–606
Endocrine system, 535–536, 541–547
Endothelium, 457
Energy metabolism, 120, 403
Enkephalins, 552–553
Enterohepatic circulation, 75–76
Enterohepatic circulation disturbances, 77
Enteroinsular axis, 614
Environmental factors, 292
Enzyme activity in urine, 283–284
Enzyme assay, 432, 667–671
   methods of, 668
   on cultured cells, 669–670
Enzyme change in IMD, 666
Enzyme reactions, control of, 541
Enzyme synthesis, 541
Enzymes
   copper-containing, 358
   faecal, 100, 106

   in CSF, 452
Epilepsy, 423, 495
Erythrocyte sedimentation rate (ESR), 327, 344
Erythrocytes, 27
Erythropoiesis, 211–212
Eschirichia coli septicaemia, 513
Ethanol poisoning, 371
Ethical problems in mental deficiency, 520
Ethosuximide, 391
Ethylene glycol, 36
Ethylene glycol poisoning, 371
Euglena gracilis, 213
Euglobulin lysis time, 267
Exercise test, 601
Extracellular control mechanisms, 542
Extracellular fluid (ECF), 1–2, 299, 308, 317, 402
   changes in composition, 7
   measurement of, 5
Extracellular integration, 542
Extractable nuclear antigen (ENA), 343
Extra-hepatic biliary obstruction, 148

Fabry's disease, 499
Facio-scapulo-humeral dystrophy, 435
Faecal analysis, 99, 105, 109
Faecal enzymes, 100, 106
Faecal fat, 85, 99
   determination, 686
Faecal nitrogen, 99
Faecal protein, 106
Faecal sugars, 106
Failure-to-thrive aetiology, 685–688
Familial alpha-lipoprotein deficiency, 427
Familial periodic paralysis, 19, 436–437
Fanconi's anaemia, 518
Farber's disease, 499
Fasting effects, 441, 587, 601
Fat absorption, 82
   assessment of, 85
   indirect tests, of, 85–86
Fat digestion, 82
Fat loading test, 687
Fat metabolism, 405
Fat stores, 124
Fats, 593
   see also Faecal fats
Fatty acids, 109, 169
   medium-chain, 82
Feminising adrenal tumours, 631
Fertility disorders, 648
Fetal blood, 664, 668
Fetal lung maturity assessment, 676–680
Fetoplacental monitoring, 655–662
Fetoplacental products, 656
Fibrin, 259
   degradation products, 268
   plate method, 267
Fibrinogen, 246, 259, 265, 405
Fibrinogen deficiency, 254
Fibrinogen titre, 267
Fibrinolysis, 253, 267
Fibrinolysis defects, 255
Fibroblasts, 499
Fletcher factor, 256
Fluid balance, 406
   disorders of, 1–23
Fluid input, 45
Fluid output, 45
Fluid retention with oedema, 14
Fluorescent treponemal antibody test (FTA(ABS)), 450

INDEX

Fluoride poisoning, 374
Focal lymphocytic thyroiditis, 568
Folacin, 133
Folate, 83, 133
Folate absorption, assessment of, 86
Folate deficiency, 214
Folic acid, 133, 212, 452, 504
Follicle-stimulating hormone (FSH), 547, 553, 633, 640
Formic acid, 135
Formiminoglutamic acid (FIGLU), 86, 135, 214
Frame-shift error, 196
Frame-shift mutations, 196
Free fatty acids (FFA), 405
Free-thyroxine index, 581
Free water clearance, 48–49
Friedreich's ataxia, 427
Froin's syndrome, 448
Fructose, 35, 286, 594–595
Fructose-1-6-diphosphatase, 36, 439
Fructose-6-phosphate, 439
Fulminant heptic failure (FHF), 164
  aetiology, 164
  charcoal haemoperfusion, 170–171
  clinical features, 164
  initial investigation, 165
  multisystem abnormalities, 165–167
  prognosis, 172
  supportive therapy in, 167
Fusion variants, 195

Galactosaemia, 286, 513–514, 605
  prenatal diagnosis, 671
Galactose, 286
  elimination capacity, 152
Galactose-1-phosphate uridyl transferase, 513, 667
Galactose tolerance tests, 605
Gallbladder function, 75
Gallstones, pathogenesis of, 77
Gamma-aminobutyric acid (GABA), 415, 453, 511–512, 532, 533
Gamma-globulin metabolism, 119
Gamma-glutamyl cycle, 512
Gamma-glutamyl transferase (γGT), 143–144, 147, 151
Gamma-glutamyltranspeptidase deficiency, 512
Gangliosidosis, 497
Gargoylism, 496
Gastrectomy, 108
Gastric emptying, 70–71
  methods for measuring, 71
Gastric inhibitory peptide (GIP), 71
Gastric inhibitory polypeptide (GIP), 612–613
Gastric lavage, 367
Gastric mucosa, 31
Gastric mucosal barrier, 70
Gastric potential difference, 70
Gastric secretion, 101
  non-invasive techniques for measuring, 70
Gastric secretory tests, 68–79
Gastrin, 608
Gastrin levels, 101
Gastrointestinal disease, 249
Gastrointestinal function tests, 686
Gastrointestinal hormones, 308, 606–616
  biological and clinical features, 608
  classification, 607
  clinical and diagnostic relevance, 608
  nomenclature, 607
Gastrointestinal secretions, 18, 406
Gastrointestinal system 66
Gastrointestinal tract, 606
  function of, 67–95
  laboratory assessment, 67–95
Gaucher's disease, 496, 499
Gene change in IMD, 665
Gene mapping, 209
General paralysis of the insane (GPI), 446, 449
Generalized deficiency, 440
Genetic counselling, 520–521
Genetic defects, 440
Genetic deficiencies, 227, 234
Genetic factors, 292
Genetic variation, 237, 240
Genitalia ambiguity, 690
Genito-urinary tract, obstruction of, 57
Gentamicin, 386, 398
Giant cell arteritis, 343–344
Gigantism, 557
Gilbert's disease, 147
Gilbert's syndrome, 148
Glasgow Coma Scale, 413, 416
Glial fibrillary acidic protein (GFAP), 447
Globin chain separation, 208
Globin gene duplication, 192
Globin gene structure, 193–194
Globin monomer, 189
Globin structure, solubility and stability, 189
Globulins, 405
Glomerular filtration rate (GFR), 4, 36, 41, 274–275, 310–311
Glomerular proteinuria, 247, 281
Glomerulonephritis, 275
Glucagon, 557, 589, 610
Glucagon-like immunoreactants (GLI), 611
Glucagonoma, 611
Glucocorticoids, 307–308, 482, 679
Gluconeogenesis, 36, 588
Glucose, 286, 615
  in blood, 596
  in CSF, 449
  overnight fasting plasma, 596–597
  urinary, 595
Glucose absorption, 588
Glucose homeostasis, 583, 590
Glucose inflow, 588
Glucose levels, 587
Glucose metabolism, 590
Glucose pool, 583, 590
Glucose-6-phosphatase deficiency, 501
Glucose-6-phosphate (G-6-P), 35, 439
Glucose-6-phosphate dehydrogenase, 204, 220, 432
Glucose tolerance factor (GTF), 360
Glucose tolerance test, 85, 119, 557, 598
  extended oral, 603
  intravenous, 598
  oral 597
Glucosuria, 287
Glutamate, 29
Glutamic acid, 135
Glutamine, 29
Glutathione synthetase, 512
Glutethimide, 366, 372
L-Glyceric aciduria, 295
Glycogen storage disease, 35
Glycogen synthesis, 588
Glycogenolysis, 588
Glycogenosis, 439
Glycolic aciduria, 295
Glycosaminoglycans, 496
Glycosuria, 595
Glycosylated haemoglobin, 593
Goitre, 573
Gold, 398
Gonadal dysgenesis, 645
Gonadal function, disorders, of, 643–650
Gonadal steriods, 633
  binding in blood and tissues, 635
  in blood plasma, 640
  in saliva, 641
  in urine, 641
  measurement of, 640
  metabolism and excretion, 636
  nomenclature, 633
  problems of interpretation of assays, 642
  reference ranges, 643
  structure, 633
  synthesis and secretion, 635
  target organ metabolism, 643
Gonadotrophin-releasing hormones (GnRH), 547, 553
Gonads, 633–650
  disorders of development, 645
  functions, 633
  physical examination, 644
  primary disorders, 647
  secondary failure, 647
Gout, 330–333
Graves' disease, 568–575
Growth hormone (GH), 307, 550, 554–559
  actions of, 555
  deficiency, 558
  disorders of secretion, 557–559
  regulation of secretion, 554–555
  stimulation tests, 555
  tests of secretion, 555
Guanosine-3', 5'-monophosphate (c-GMP), 543
Guillain-Barré syndrome, 428, 448
Guthrie bacterial inhibitors, 494

Haem, 187–188
Haem biosynthesis, 176–178
  pathway of, 176–177
  regulation of, 177
Haem group, 198
Haem pocket, 188–189, 198
Haem precursors, 178
Haem synthesis, 175–186
Haematological disease, 205
Haematological disturbances, 166
Haematological effects of trauma, 409
Haematological status, 97
Haematology, 204, 327
Haemochromatosis, 150, 211
Haemodialysis, 169, 171–172, 361, 368
Haemoglobin, 26, 27, 194, 450, 593
  diagnosis of unstable, 199
  fetal 203
  glycosylated, 197
  non-genetic variations, 196
  of embryo, fetus, and adult, 192
  unstable 198–199
  variant identification, 208
  variants in man, 196
Haemoglobin geneticds and expression, 192–194
Haemoglobin M, 201
Haemoglobin malfunction, 194
Haemoglobin structure and function, 187–192
Haemoglobin variation and disease, 194–204

INDEX

Haemoglobinaemia, 216, 220
Haemoglobinopathies, 187–210
   laboratory diagnosis, 204–209
Haemoglobinuria, 215, 216, 220
Haemolysate preparation, 205
Haemolysis, 214
Haemolytic uraemic syndrome, 217
Haemopexin, 242
Haemophilia, 253, 335
Haemorrhage, 167
   acute, 214
   chronic, 213, 219
Haemosiderinuria, 217
Haemosiderosis, 211
Haemostasis, 251
Haemostasis diseases, 253–256
Hair morphology, 121
Halothane, 437, 438
Ham's test, 217, 220
Haptoglobins, 239
Hashimoto's disease, 568, 569, 575
Hashimoto's thyroiditis, 574
Hb Lepore, 203
HbA_2 measurement, 205
HbF measurement, 205
Heart failure, *See* Cardiac failure
Heat stroke, 421
Heavy-chain diseases, 231
Heavy metals, 347–362, 519
Heinz bodies, 198, 199, 205, 216
Henderson-Hasselbalch equation, 25, 26
Henle's loop, 2, 3, 17
   water reabsorption, 52–53
Henoch-Schonlein purpura, 255
Hepatic cirrhosis, 14
Hepatic coma, 164, 421
Hepatic encephalopathy, 165, 167
   pathogenesis, 167–170
Hepatic gluconeogenesis, 404
Hepatic malignancy, 151–152
Hepatic urea production, 30
Hepatitis
   acute, 149
   chronic active, 150
Hepatitis B infection, 326
Hepatocellular carcinoma, 151–152
Hepatocellular damage, 142
Hepatosplenomegaly, 128
Heriditary ataxic neuropathy, 427
Hereditary coproporphyria, 180
Hereditary elliptocytosis, 215
Herediatry fructose intolerance (HFI), 604–605
Hereditary Persistence of Fetal Haemoglobin (HPFH), 202
Hereditary spherocytosis, 215
Herpes simplex encephalitis, 419, 447
Herpes zoster, 162
Herpes zoster encephalitis, 420
Heterosexual precocity, 644
Hexosaminidase A, 497, 668
Hill equation, 190
Hill number, 190
Histidinaemia, 508
Histidine, 135
Hirsutism 649, 650
   idiopathic, 650
HLA antigen, 326, 329, 574
Hodgkin's disease, 236, 249, 361
Homeostasis disturbances, 116
Homocystinuria, 337, 494, 511, 512
Homovillic acid (HVA), 526
Hormone assays, 599

Hormones, 526
   acting through cyclic nucleotides, 543
   anterior pituitary gland, 550
   cellular effects, 543
   control of production, 545
   in urine, 284
   influencing, 307–308
   mode of action, 543
   regulating, 303–307
   regulation of action, 545
   site of synthesis, 542
   transport of, 543
   types of 542
Humoral mechanisms, 480
Huntington's chorea, 415, 532
Hurler's disease, 496
Hydrallazine in cardiac failure, 478
Hydrocephalus, 505
Hydrogen breath test, 84, 87, 109
Hydrogen ion balance, 32–41
Hydrogen ion concentrations, 24–43
Hydrogen ion detection, 101
Hydrogen ion disturbances, 41–42
   arterial blood, 32
Hydrogen ion excretion, 280
Hydrogen ion homeostasis, 24
Hydrogen ion overproduction, 34
Hydrogen ion secretion by gastric mucosa, 31
Hydrogen ion status, 31–32
Hydrostatic pressure, 2
5-Hydroxyindole acetic acid (5-HIAA), 277, 392, 453, 526, 614
25-Hydroxylated derivative (25-OHD), 305, 307, 320, 322
4-Hydroxy-3-methoxymandelic acid (HMMA), 485
4-Hydroxy-3-methoxyphenyl-glycol (HMPG), 392
17α-Hydroxyprogesterone, 621
Hydroxyprolinaemia, 512
Hydroxyproline, 315
3β-Hydroxysteroid dehydrogenase, 645
17β-Hydroxysteroid dehydrogenase, 645
5-Hydroxytryptamine (5HT), 111, 169, 453, 461, 524–526, 613–614
5-Hydroxytryptophan 453
Hyperaldosteronism, 14, 481
Hyperammonaemia, 510–511, 517
Hyperbilirubinaemia, 148, 502
   unconjugated, 148
Hyperbolic oxygen dissociation curve, 189
Hypercalcaemia, 58, 152, 294, 308, 323, 422
Hypercalciuria, 293, 294
Hyperchloraemia, 38
Hyperchloraemic acidosis, 36
Hypercholesterolaemia, 276, 461, 459
Hypergastrinaemia, 609
Hyperglobulinaemia, 343
Hyperglycaemia, 402, 404, 590
   causes of, 598
   tests for, 595
Hyperglycinaemia, 517
   non-ketotic, 509–510
Hyperinsulin neuronopathy, 599
Hyperinsulinaemia, 404
Hyperkalaemia, 19, 21–22, 481
Hyperkalaemic periodic paralysis, 22
Hyperlipidaemia, 12, 152
Hyperlipoproteinaemia, 146, 335, 469
   type I, 469
   type II, 469
   type III, 469

   type IV, 470
   type V, 470
Hyperlysinaemia, 337, 513
Hypermobility syndromes, 337
Hypernatraemia, 9–12, 406
   due to excessive sodium intake, 11
   effects of, 11
   therapeutic implications, 11
Hyperosmolal states, 16
Hyperosmolar diabetic coma, 592
Hyperoxaluria, 294, 295
   enteric, 295
   primary, 295
Hyperparathyroidism, 294, 295, 318–319, 322, 486
Hyperphenylalaninaemia, 505
Hyperphosphataemia, 320
Hyperprolactinaemia, 649
Hyperprolinaemia
   type I, 512
   type II, 513
Hyperproteinaemia, 12
HyperVIPaemia, 612
Hypertension, 355, 461, 479–486, 630
   benign, 482
   causes of, 486
   investigation of, 479
   primary, 479
   secondary, 479
   symptomatology, 479
Hypertensive encephalopathy, 418
Hyperthyroidism 325, 572–574
   congenital, 566
   neoplastic causes, 574
   treatment-induced, 574
Hypertonic saline for water intoxication, 14
Hypertriglyceridaemia, 276
Hyperuricaemia, 328, 330–331
   symptomless, 333
Hyperventilation, 407
Hypnotic drug poisoning, 372
Hypoalbuminaemia, 14, 119, 241
Hypoaldosteronism, low-renin, 22
Hypo-beta lipoproteinamia, 427
Hypocalcaemia, 320, 323
Hypochloraemia, 31
Hypodipsia, 61, 65
Hypoglycaemia, 69, 149, 152, 166, 500–501, 517, 585, 586, 599–604, 625
   classification, 600
   diagnosis, 600–601
   fasting, 601, 602
   ketotic, 601
   reactive, 604
Hypoglycaemic coma, 420
Hypokalaemia, 38, 40, 60, 281, 422, 481, 482
   and potassium deficiency, 18–21
   causes of, 18
   effects of, 20
   principles of treatment, 20–21
Hypokalaemic FPP, 436–437
Hyponatraemia, 3, 22, 31, 421, 477
   after injury or surgery, 13
   effects of, 15
   therapeutic implications, 15
   without sources of ectopic ADH, 12–14
Hypo-osmolality, 16
Hypoparathyroidism, 320, 322
Hypophosphataemia, 294
Hypopituitary coma, 422
Hypopituitarism, 558–559, 577
Hypothalamic function, 577

INDEX

Hypothalamic-pituitary-adrenal (HPA) function, 620
Hypothalamic-pituitary-adrenocortical (HPA) disorders, 624
Hypothalamic-pituitary control, 577, 649
Hypothalamic-pituitary function, 553
Hypothalamus, 548
Hypothermia, 421
Hypothyroid disease, auto-immune, 575
Hypothryoid screening, 688
Hypothyroidism, 13, 470, 518–519, 575–577
  causes of, 575
  clinical appearance of, 577
  congenital, 567
  drugs and goitrogens causing, 576
  postpartum, 578
  secondary, 577
  tertiary, 577
Hypovolaemia, 401
Hypoxaemia, 500
Hypoxanthine-guanine phosphoribosyltransferase (HGPRT), 291, 330, 331
Hypoxia, 38, 165, 403, 421, 500
Hysteria, 423

Idiopathic adrenal hyperplasia (IAH), 482
Idiopathic hirsutism, 631
IgA, 150, 220, 225–229, 249, 250
IgD, 226, 227, 229
IgE, 226, 227, 229, 250
IgG, 150, 151, 220, 225–229, 232, 247–249, 569
IgM, 150, 151, 220, 225–229, 231, 232, 248, 249
Imipramine, 366, 378
Immune-complex disease, 234, 248
Immune defence, 107, 224–225
Immunocompetence, 125
Immunogenetics, 326
Immunoglobulins, 224, 339, 569, *see also* under Ig
  classification, 225
  deficiency, 227
  functions of, 225
  normal ranges, 226
  primary deficiencies, 227
  secondary deficiencies, 227
  structure, 224
  synthesis, 225
Immunological tests, 327
Indican excretion, urinary, 86
Indocyanine green (ICG), 153
Indoxyl sulphate, 110
Inductively-coupled plasma emission (ICPE), 361
Infections, 167
Infectious diseases, 248
Infertility, 285, 648
  male, 649
Inflammation, 224, 235, 240, 247
Inflammatory bowel disease, 329
Inherited metabolic diseases (IMD), 663–665
  enzyme change in, 666
  gene alteration, 665
  prenatal diagnosis of, 665–667
Injury
  metabolic response to, 19, 400
  plasma proteins after, 405
Insertions, 195
Insulin, 583, 585–587
  biological properties, 585
  immunoreactive, 586

in blood, 586
Insulin-induced hypoglycaemia, 555, 625
Insulin-like growth factors I and II, 587
Insulin tests, 69, 555, 601
Intelligence quotients (IQ), 489–490, 518, 688
Intermittent idiopathic oedema, 15
Interstitial oedema, 415
Inter-α-trypsin inhibitor, 238
Intestinal lymphagiectasia, 83
Intestinal obstruction, 160–161
  radiography, 161
  special investigations, 161
Intestinal tumours, 111
Intestine, calcium flux, 308–309
Intracellular fluid (ICF), 2, 402
Intracellular regulation, 541
Intracranial lesions, 447
Intracranial pressure, 417
Intracranial tumours, 447
Intravenous hypertonic saline test, 63
Intravenous tolbutamide test, 599
Intrinsic factor, 68, 83, 101–102
Intrinisic factor deficiency (IFD), 102
Iodide oxidation and organification, 562
Iodide trapping, 562
Iodine-induced thyrotoxicosis, 574
Iron absorption, 83, 106
Iron deficiency, 187, 212–213, 219, 520
Iron deficiency anaemia, 219
Iron metabolism, 211–212
Iron poisoning, 374
Ischaemia, 500
Ischaemic exercise test, 439
Ischaemic neuropathy, 429
Iso-enzymes, 491, 494
Isomaltose, 286
Isosexual precocity, 644

Jaundice, 148
Joint disorders, 325
Joint structure, 325–326
Juvenile chronic polyarthritis, 328

Killikrein, 256
Kanamycin, 398
Kayser-Fleischer rings, 150, 359, 519
Kernicterus, 502
Ketoacidosis, 33–34
Ketones, 593, 594
Ketotic hyperglycinaemia, 517
Ketotic hypoglycaemia, 501
Kidney
  bicarbonate control by, 27–28
  calcium flux, 310–312
  functions, 271
Kidney disease, 479
Kinin deficiency, 254, 266
Kinky hair disease, 520
Klinefelter's syndrome, 645
Korasakoff's psychosis, 534
Krabbe's disease, 497
Krebs cycle, 515
Kwashiorkor, 113, 115–117, 120, 132

Lactase activity, 104
Lactase assay, 84
Lactase assessment, 104
Lactase deficiency, 103
Lactase deficiency tests, 84
Lactate, 594
Lactate dehydrogenase (LDH), 474
Lactate formation, 439

Lactate metabolism, 514
Lactate production, 36
Lactic acid, 452
Lactic acidosis, 34, 403, 515
Lactobacillus leichmanii, 213
Lactose, 286, 615
Lactose absorption, 103
Lactose absorption test, 104
Lactose intolerance, 103
Lactose tolerance test, 84
Lactulose, 286
Landing's disease, 497
Lange test, 449
Latex test, 327
L-dopa therapy, 415, 556
Lead exposure risk, 519
Lead poisoning, 163, 181, 187, 347–351, 374, 423
Lean body mass, 124
Lecithin, in amniotic fluid, 678
Lecithin cholesterol acyl transferase (LCAT), 146, 463, 467–470
Lecithin/sphingomyelin (L/S) ratio, 678–680
Lecithin synthesis, 677
Leigh's encephalopathy, 515
Leprosy, 428
Lesch-Nyhan syndrome, 331, 517
Leucine aminopeptidase, 143
Leucocytosis, 161
Leucodystrophy, 491, 496, 498
Leukaemia, 218, 335
Leydig cells, 639
Libman-Sachs endocarditis, 339
Ligandin, 503
Lignocaine, 395
Limb-girdle muscular dystrophy, 434
Limbic system, 524
Lipase, 98
Lipid absorption, 77
Lipid exchange factors, 468
Lipid infiltration, 460
Lipid metabolism disorders, 276
Lipid-positive cells in amniotic fluid, 679
Lipids, 76, 117, 146–147, 242, 458–462
Lipoidoses, 427
Lipoprotein, 117, 146–147, 460
  high density, 462, 464, 466–468, 470–472
  very low density, 82
Lipoprotein-a (Lp-a), 467
Lipoprotein lipases (LPL), 146, 464
Lipoprotein-x (Lp-x), 467
Lipoproteinaemias, 469
β-Lipotrophin (LPH), 550–551
Lithium 392
Lithium carbonate therapy, 58
Lithium poisoning, 363, 374
Lithium salts, 528
Liver, 139–147
  see also Artificial liver
Liver biopsy, 150
Liver cirrhosis, 150
Liver disease, 35, 248, 559
  diagnosis of, 148
  parenchymal, 150–151
  research, 141
  see also Hepatic
Liver failure, charcoal haemoperfusion, 170–172
Liver function tests, 139, 327
Liver status, 96
Liver support devices. *See* Artificial liver
Locked-in syndrome, 414
Long Acting Thyroid Stimulator (LATS), 569

Loop diuretics, 476
Low-density liproteins (LDL), 462, 464–466, 468–472
Lumber puncture (LP), 445
Lundh test, 74
Lundh test meal, 687
Lung, 26
Luteinizing hormone (LH), 550, 553, 633, 639, 640
Lymph nodes, 219
Lymphosarcoma, 450
Lysosomal disorders, 495, 497, 499
Lysozyme, 235, 247, 447

α_2-Macroglobulin, 238
Magnesium absorption, 82
Magnesium levels, 407
Malabsorption
  blood tests, 84
  diagnosis of, 83
  differential diagnosis, 95
  drug-induced, 108
  general information profile, 96
  iatrogenic, 107
  main categories of, 96
  of bile acids, 90
Malaria, 203–204
Malignant disease, 429
Malignant hyperphenylalaninaemia, 507
Malignant hyperthermia, 437–438
Malnutrition
  blood tests, 84
  differential diagnosis, 95
  general information profile, 96
  infantile, 114
  marginal, 114
  pregnancy, 503
  protein energy. *See* Protein-energy malnutrition
Maltose, 286
Manganese deficiency, 136, 360
Manganese intoxication, 136, 360, 423
Manganese role, 360
Mania, 524, 527
Manic-depressive psychosis, 525
Mannitol, 59
Marasmus, 113, 115–116
March haemoglobinuria, 217
Marfan syndrome, 337
Maroteaux-Lamy disease, 497
Marrow disorders, 218–219
Mean cell haemoglobin (MCH), 213, 219
Mean cell volume (MCV), 213, 219
Media, 457
Medico-ethical issues, 521
Medullary cystic disease, 57
Mellituria, 285–288
  recognised types of, 286
Meningitis, 419
  acute bacterial, 446
  See also Tuberculous
Meningococcal meningitis, 419
Menkes' kinky hair syndrome, 359, 520
Menopause, 648
Menstruation disorders, 644, 648
Mental deficiency, 489–523
  aetiology, 490–491
  biochemical investigations, 495
  diagnosis, 497–499
  ethical problems, 520
  genetic and environmental factors, 490–491
  medico-ethical issues, 521
  metabolic, 491–495
  size of problem, 489
  treatment, 499–500
Mental handicap, nutritional aspects, 503–504
Meprobamate, 372
Mercaptans, 169
Mercury toxicity, 351–353, 374–375, 423, 519
Messenger-RNA, 194
Metabolic acidosis, 33, 38, 422, 441, 517
Metabolic alkalosis, 40
Metabolic changes in shock, 402–403
Metabolic clearance rate, 635
Metabolic defects, 438, 567
Metabolic disease, 489
  see also Inherited metabolic diseases
Metabolic disorders, 420, 491–495
  aerobic, 440–442
  carnitine, 441
  inborn, 604
  myopathy in, 443
Metabolic disturbances, 32, 166, 524
Metabolic joint disease, 330–335
Metabolic pathways, 541
Metabolic processes, 541–542
Metabolic response to injury, 19, 400
Metabolic status, 97
Metabolic assays, 599
Metachromatic leucodystrophy, 420, 427, 497
Metadrenaline, 485
Metal poisoning, 347, 374–375
Metals in urine, 289
Methaemoglobinaemia, 200
  acquired, 200
  congenital, 200
  differentiation, 201
Methaemoglobins, 188
  measurement of, 364
Methanol poisoning, 375, 423
Methaqualone, 372
Methionine, 168
Methotrexate, 398
3-Methoxy-4-hydroxyphenyglycol (MHPG), 526
Methylmalonic acid excretion, urinary, 86
Methyl mercury, 351
Metyrapone test, 626
Mexiletine, 396
Microbial infection, 326
β_2-Microglobin, 243
Minerals, in cancer, 127
Minoxidil in cardiac failure, 478
Mitochondrial myopathies, 436, 442
Molecular pathology, 197
Molybdenum metabolism 520
Monitoring tests, 140
Monoamine oxidase (MAO), 532
Monoclonal hypergammaglobulinaemia, 229
Mood disorders, 524–525
Motilin, 614–615
Mucolipidoses, 496, 498
Mucopolysaccharidoses, 337, 491, 496, 497
Mucosal cell dysfunction, 102, 106
Multifocal leukoencephalopathy, 420
Multiple arylsulphatase deficiency, 491
Multiple endocrine abnormality (MEA), 322
Multiple sclerosis, 420, 447, 449
Munchausen syndrome, 163
Muscular dystrophies, 434, 518
Myasthenia gravis (MG), 443–444
Myasthenic syndrome, 444
Mycoplasma pneumoniae infection, 216
Myelofibrosis, 219
Myelomatosis, 218, 231, 430
Myocardinal disease, 339
Myocardial infarction, 473
Myoglobin, 474–475
  structure, 189
Myopathies, 430–451
  acquired, 431, 442–451
  classification, 431
  drugs, 443
  electromyography, 432
  electron microscopy, 432
  hereditary, 431–442
  light microscopy, 432
  metabolic disturbances, 443
  methods of investigation, 431
  primary, 431
Myophosphorylase defects, 438–440
Myotonia congenita, 435
Myotonic dystrophy, 435, 518
Myotubular myopathy, 436
Myxoedema, 344, 566
Myxoedema coma, 422, 578

NADP, NADPH, 216
N^5-Methyltetrahydrofolate transferase, 511
N^5,N^{10}-Methylenetetrahydrofolate reductase, 511
Naloxone, 375
Narcotic overdose, 375
Nelson's syndrome, 630
Nemaline myopathy, 436
Neonatal convulsions, 500
Neonatal Graves' disease, 566
Neonatal hypothyroidism, 491, 567
Neonatal myasthenia, 444
Neonatal screening, 688–689
Nephrogenic diabetes insipidus, 58, 65
Nephrotic syndrome, 14, 241, 247, 339, 353
Nervous system
  bacterial infections of, 446
  degenerative disease of, 495
  vulnerability, 491
Netilmicin, 398
Neufeld factors, 499
Neural tube defects, 491, 506
  alternative diagnostic aids, 676
  amniotic fluid AFP assay, 675–676
  prenatal diagnosis and screening, 671–676
Neuroglycopenia, 599, 625
Neurolipidoses, 498
Neurological diseases, 417–423
Neurometabolic disorders, 492
Neuromuscular system disorders, 412
Neurophysins, 54
Neuropsychiatric disorders, 488
Neurosyphilis, 420
Neurotensin, 615
Neurotransmitter disturbances, 415
Neutral tube defects (NTD), 663
Niacin, 132
Nicotinamide, 132
Nicotine test, 63
Nicotinic acid, 132
Niemann-Pick disease, 496
Nitrates in cardiac failure, 478
Nitrogen balance, 124
Nitrogen excretion, 10
Nitrogen faecal, 99
Nitroprusside in cardiac failure, 477
Non-esterified fatty acids (NEFA), 593, 594
Nonne-Apelt test, 449
Non-protein nitrogen (NPN) excretion, 404–405

INDEX

Nonsense error, 196
Non-suppressible insulin-like activity (NSILA), 587
Non-suppressible insulin-like protein (NSILP), 587
Noradrenaline, 483, 485, 524, 525, 589
Normetadrenaline, 485
Normokalaemic FPP, 437
Nortriptyline, 392
5'-Nucleotidase, 143
Nutrient absorption, colon, 89
Nutritional aspects of mental handicap, 503–504
Nutritional deficiencies in old people, 122–123
Nutritional myopathy, 443
Nutritional reserves, 403
Nutritional status, 96
  assessment in early childhood, 114, 687–688
  normal 137

Oedema
  fluid retention with, 14
  in kwashiorkor, 120
  intermittent idiopathic, 15
Oesophagitis, 475
Oestetrol, 660
Oestradiol-17β, 642, 660
Oestriol, 659, 600, 662
Oestrogens, 638, 641. 647
Oligoclonal hypergammaglobulinaemia, 229
Oligosaccharides, 286
Oncofetal proteins, 243
Oncotic pressure, 2
Oral contraceptive steroids, 649
Oral glucose tolerance test, 556
Organ response integration, 317
Organ transplantation, 326
Organic acidurias, 515–517
Organic psychiatric reactions, 529–531
  cause of, 533
  diagnosis, 530
  laboratory role, 531
Organochloride poisoning, 375
Organophosphorus compounds, 375
Orosmucoid, 239
Orotic aciduria, 518
Osmolal clearance, 48–49
Osmolality, 46–49, 369
Osmometry, 47
Osmoreceptors, 49
Osmotic diuresis, 10, 56, 59
Osmotic gap, 91
Osmotic pressure, 2
Osteoarthritis, 325, 335
Osteoarthrosis, 325, 335
Osteoclast activating factor, 308, 314
Osteogenesis imperfecta, 255
Osteoid excess, 321
Osteomalacia, 129–130, 321
Osteoporosis, 129, 324, 325
Ovarian failure, 648
Ovarian function, 638–639
Ovulation disorders, 648
Ovulation induction regimes, 649
17-Oxogenic steroids, 285
5-Oxoprolinase, 512
Oxygen affinity, 199–200, 206
Oxygen arterial levels, 39
Oxygen transport, 190
Oxyhaemoglobin, 188, 190
Oxytocin, amino acid sequence, 54

Paediatrics, 682–693
  clinical chemistry requirements, 682–683
  failure-to-thrive, aetiology, 685–688
  gastrointestinal function tests, 686
  nutritional status assessment, 687–688
  reference ranges, 684–685
  special requirements of clinical biochemistry laboratory, 683–684
Paget's disease, 147, 314, 323–324
Pancreas, 72–75
Pancreatic carcinoma, 105
Pancreatic cholera syndrome, 91
Pancreatic function
  assessment of, 73–75
  exocrine, 98
  feeding tests, 99
Pancreatic polypeptides, 615
Pancreatic secretion, 72–73, 97
Pancreatitis, 161–162
  acute, 73–74
  chronic, 74–75
Pancreozymin, 610
Pandy test, 449
Paracetamol overdose, 376
Paradoxical breathing, 190
Paraldehyde poisoning, 36
Paramyotonia congenita, 437
Paranoid reactions, 536
Paraproteinaemia, 230
Paraproteins, 229, 231
Paraquat poisoning, 376
Parathyroid disorders, 422
Parathyroid hormone (PTH), 295, 300, 303–304, 307, 311–312, 317–320, 325
Parathyroid secretion disorders, 318
Parenchymal damage, 283
Parenchymal liver disease, 149
Parenteral nutrition, trace elements in, 360
Paroxysmal cold haemoglobinuria, 216
Paroxysmal nocturnal haemoglobinuria, 217
Partial thromboplastin time (PTT), 145
Pellagra, 132
Pendred syndrome, 518
Pentagastrin test, 68
Pentosuria, 605
Pepsinogen secretion, 68
Peptide mapping, 208
Perforated peptic ulcer, 161
Pericarditis, 339, 340
  acute, 475
Peripheral neuropathy, 426–430
  biochemical investigations, 430
  chemicals that may cause, 429
  classification of, 426
  drugs that may cause, 429
  haematological studies, 430
  hereditary sensory, 427
  infective, 428
  inherited, 426–428
  laboratory investigation, 430
  metabolic, 429
  metabolically triggered, 428
  toxic, 428–429
Peroneal muscular atrophy, 426–427
Peroxidase reaction, 562
pH, 25–26, 29, 32, 33, 39, 41, 42, 98, 101, 176, 191, 290–292, 294, 296, 331, 377
Phaeochromocytoma, 277, 484–485
Pharmacodynamic factors, 380, 385
Pharmacokinetic variation, 380
Pharmacology, 346
Phenobarbitone, 363, 390

Phenolic acids, 169
Phanylalanine, 168, 506
Phenylalanine hydroxylase, 507
Phenylbutazone, 398
Phenylethanolamine-N-methyl-transferase, 483, 617
Phenylethylmalonamide, 390
Phenylketonuria, 491, 494, 506–508
Phenylpyruvic acid, 506
Phenytoin, 363, 366, 379, 380, 388–390, 396
Phosphate, 26
  absorption, 82
  distribution, 301
  levels, 407
  urinary, 29, 30, 281
6-Phosphate-dehydrogenase deficiency (G-6-PD), 215–216
Phosphofructokinase deficiency, 439
Phospholipid exchange-protein, 468–469
Phospholipids, 678
Phosphoribosyl pyrophosphate, 330
'Pink puffers', 40
Pituitary dwarfism, 558
Pituitary gland, 548
Pituitary-gonadal axis, 690
Pituitary hormones, 589
Pituitary-Leydig cell axis, 639
Pituitary-seminiferous tubule axis, 640
Placental lactogen (hPL), 653, 661, 662
Placental proteins, 244
Plasma albumin, 121, 144
Plasma amino-acid patterns, 121
Plasma assays, 656
Plasma calcium, 300–301
Plasma cholesterol, 471
Plasma lipids, 458
Plasma lipoproteins, 460–472
Plasma oestriol concentration, 659
Plasma osmolality, 3, 10, 16
Plasma protein binding, 383
Plasma protein concentrations, 223
Plasma protein measurements, 247
Plasma proteins, 26, 121, 125, 144
  abnormalities, 221–251
  after injury, 405
  catabolism, 223
  distribution, 223
  functions of, 224
  inflammation, 224, 235
  investigation of, 245
  metabolism of, 222–224
  neoplasia, 249
  signal, 224
  synthesis, 222–223
  transport, 224, 240–243
  transudation of, 451
Plasma renin activity (PRA) test, 480
Plasma renin concentration, 480, 481
Plasma triglyceride, 471
Plasma volume measurement, 5
Plasmaphoresis, 340
Plasmin, 260
Plasminogen, 260, 267
Plasmodium falciparum, 204, 218
Platelet counts, 166
Platelet deficiencies, 254–255
Platelet disorders, 266
Platelet mechanism, 251
Platelet tests, 266–267
Platelets, 260
Pleurisy, 340
  acute, 475

Point mutation, 194
Poisoning, 363–378
 clinical tests, 268–370
 confirmation of diagnosis, 364
 incidence of, 363
 inorganic anions, 373
 laboratory role, 364
 organochlorides, 375
 prognosis, 364–366
 sample collection, 369–370
 self-induced, 422
 specific applications, 370–378
 statistics, 363
 treatment, 367–368
Polyarteritis nodosa, 343
Polyclonal hypergammaglobulinaemia, 228
Polycystic ovary syndrome, 649–650
Polycythaemia, 200
Polydipsia, 43–65
 clinical groups, 62
 clinical investigation, 63
 iatrogenic, 60
 laboratory investigation of, 62–65
 primary, 60–61
 psychogenic, 60, 64
 secondary, 57–60
Polymyalgia rheumatica, 343–344
Polymyositis, 342, 442
Polyuria, 43–65, 274–275
 clinical groups, 62
 clinical investigation, 63
 laboratory investigation of, 62–65
 primary, 57–60
 secondary, 60–61
Porphobilinogen, 176–178, 180–182, 184, 185
Porphyrias, 163, 175–186, 421
 acute attacks, 179–181
 acute hepatic, 184
 acute intermittent, 184, 428
 classification, 175
 clinical features, 175, 179
 cutanea tarda, 180, 184–185
 diagnosis, 184–185
 drugs precipitating attacks of, 428
 enzyme defects, 178–179
 laboratory differentiation, 181
 laboratory tests, 184–185
 management, 184–185
 skin lesion, 180, 182
 variegate, 184
Porphyrinogens, 177
Porphyrins, 182, 183, 184
Potassium
 balance, 279
 deficiency, 18–21, 280
 depletion, 41, 477
 distribution, 17
 excess, 280
 homeostasis, 480
 in renal tubular cells, 31
 normal dietary intake, 17
 plasma concentration, 21, 406, 422
 renal handling of, 17–18
 space, determination of, 6
 urinary, 20
Pralidoxine, 375
Prealbumin, 245
Prednisolone, 344
 in arteritis, 343
 in SLE, 340
Pre-eclamptic toxaemia, 660
Pregnancy
 biochemical parameters in confirmation

 of, 655
 confirmation of, 653–655
 detection of hCG, 654
 diseases of, 657
 ectopic, 654
 malnutrition during, 503
 monitoring, 285, 655–662
 pathology categorisation, 657
 SLE in, 340
 thyroid disease in, 577
Pregnancy-associated proteins, 244
Pregnancy-specific β_1-glycoprotein (SP_1), 655
Pregnancy tests, 653
 criteria for evaluating, 657
Prekallikrein, 256
Prematurity, 501–502
Prenatal diagnosis, 663–682
 by enzyme assay, 667–671
 interpretation of results, 670
 of inherited metabolic diseases, 665–667
Primary myxoedema, 568
Primidone, 390
Procainamide, 395
Profiling tests, 140, 147
Progesterone, 636, 642, 660, 661
Progestogen-induced withdrawal bleed, 644
Progestogens, 641
Proinsulin, 585
Propoxyphene poisoning, 375
Prostaglandins, 170–171, 308, 313, 453
Protease inhibitors, 236, 238
Protein absorption, 81, 85
Protein assays, 661
Protein-bound compounds, 169
Protein-bound toxins, 171
Protein C, 260
Protein digestion, 81
Protein-energy malnutrition (PEM)
 acute, 118
 aetiology of, 122
 assessment of, 120, 124
 body composition, 117
 clinical features, 115
 development of, 118
 diagnosis, 120
 faecal, 106
 in adults, 122–125
 in children, 113–122
 in CSF, 449
 investigations during treatment, 122
 long-term effects, 118
 malignancy, 125–126
 metabolic disturbances, 116
 mortality, 116
 pathogenesis, 114–115, 122–123
 pathophysiology, 116
 prevalence, 115, 122–123
 prognosis, 116
 psychological aspects, 126
 treatment, 125
Protein fractionation, 450
Protein-losing enteropathy, 107
Protein loss from gut, 85
Protein metabolism, 118, 119, 404
Protein reserves, 404
Protein studies, clinical applications, 222
Protein synthesis, 125, 566
Proteinuria, 273–275, 278, 281–283
 classification, 282
 investigation of, 283
Prothrombin, 259, 265
Prothrombin time (PT), 145, 165, 379

Protoporphyria, 180, 182, 185
Proximal tubule, 52
Prozosin in cardiac failure, 478
Pseudo-gout, 333–334
Pseudo-hermaphroditism, 482, 645
 female, 646
 male, 645
Pseudohyperparathyroidism, 312
Pseudohypoaldosteronism, 22
Pseudohypoparathyroidism 320, 322
Pseudopseudohypoparathyroidism, 320
Pseudoxanthoma elasticum, 255, 337
Psoriatic arthritis, 328
Psychiatric illnesses, 523–538
Puberty
 delayed, 644, 646, 647
 disorders of, 646
 precocious, 647
Pulmonary emphysema, 238
Pulmonary function in shock, 407
Purine metabolism, 517
Purine nucleoside phosphorylase deficiency, 518
Pyloric stenosis, 41
Pyridoxine, 132
Pyrimidine metabolism, 517
Pyruvate, 514, 594
 under-utilization, 35–36
Pyruvate carboxylase, 36
Pyruvate dehydrogenase, 35
Pyruvate dehydrogenase deficiency, 514, 515
Pyruvate kinase deficiency, 215
Pyruvate overproduction, 35

Quinidine, 371, 396
Quinine, 371

Radioallergo-absorption techniques (RAST), 229
Radiography
 abdomen, 160
 chest, 163
 intestinal obstruction, 161
 pancreatitis, 162
Radioimmunoassay techniques, 472
Raffinose, 286
Rapoport-Luebering pathway, 191
Raynaud's phenomenon, 338–339
Reabsorpton of water, 3
Red cells
 abnormalities, 203–204, 215
 autohaemolysis test, 220
 formation, 211
 inclusions, 204
 measurements, 213
 osmotic fragility test, 220
 porphyrin, 183
 protoporphyrin, 183
 regeneration, 215
Refsum's syndrome, 427
Reiter's disease, 326, 329
Reiter's protein complement fixation test (RPCFT), 450
Release reaction defects, 267
Renal disease, 19, 249, 331
Renal disorders, 57
Renal disturbances, 165
Renal failure, 21, 57, 166
 acute, 271–273
 after shock, 408–409
 chronic, 273–276, 322, 361
 incipient acute, 271–272
Renal function, 51

INDEX

Renal function tests, 272
Renal glucosuria, 281
Renal handling of potassium, 17–18
Renal handling of water and salt, 2–5
Renal potassium excretion, 21
Renal status, 96
Renal stone development, 295
Renal stone disease, 290–291
Renal stone formation, 290, 292
Renal tract disorders, 270, 278
Renal transplantation, 57, 273
Renal tubular acidosis, 37, 296
Renal tubular damage, 283
Renal tubular defect, 294
Renal tubular dysfunction, 58
Renal tubular function, 279
Renal tubular function tests, 275
Renal tubular proteinuria, 281
Renal tubular reabsorption, 310
  sodium, 4
Renin, 480
Renin-angiotensin system, 49, 50
Renin-secreting Wilm's tumour, 60
Respiratory acidosis, 38, 422
Respiratory alkalosis, 41, 165, 422
Respiratory centre, 26
Respiratory distress syndrome (RDS), 663, 676–680
Respiratory disturbances, 32
Respiratory dysfunction, 39–40
Respiratory failure in shock, 408
Retinol-binding protein (RBP), 128
Reverse steal effect, 415
Rheumatoid arthritis, 236, 326–328, 450
Rheumatoid factor, 328
Rhinorrhoea, CSF, 453
Riboflavin, 132
Ribose, 286
Ribulose, 286
Richner-Hanhart syndrome, 509
Rickets, 129–130
Ristocetin co-factor, 264
Rocket immunoelectrophoresis, 472
Rose-Waaler test, 327

Saccharopinuria, 513
Salicylate poisoning, 377
Salicylates, 35, 397
  and hydrogen ion balance, 41
Salivary steroids, 624
Salt, renal handling of, 2–5
Salt absorption, colon, 89
Salt consumption, 483
Sarcoidosis, 335
Schilder's disease, 420
Schilling test, 86, 87, 102
Schizophrenia, 528–529
  differential diagnosis, 529
  laboratory role, 531
Schumm's test, 215
Schwangerschaftsprotein 1 (SP_1), 661, 662
Scleroderma, 341–342
Scurvy, 255
Secretin, 71, 73, 74, 611
Secretin test, 70, 74
Selective precipitation, 472
Senile purpura, 255
Sepsis, 59
Sero-negative polyarthritis, 328
Serotonin, 507
Severe combined immunodeficiencies, 227
Sex chromosome abnormalities, 644
Sex hormone binding globulin, 635, 642

Sexual development disorders, 646
Sexual differentiation, 690
Sexual function disorders, 644
Sexual precocity, 644
Sham feeding, 69
Sheehan's syndrome, 558, 649
Sheep cell agglutination test (SCAT), 327
Shock, 400
  metabolic changes in, 402–403
  physiological responses, 401
  pulmonary function in, 407
  renal failure after, 408–409
  respiratory failure in, 408
  tissue hypoxia and lactic acidosis, 403
Sick cell syndrome, 16
Sickle cell anaemia, 163, 187, 666
Sickle cell disease, 197–198, 205
Sickle cell haemoglobin, 205
Sisomicin, 398
Skeletal loss, 316
Skeletal mass, 316
Skeletal metabolism, 299–325
Skeletal system disorders, 298
Skeletal turnover assessment, 315
Skin lesions, porphyrias, 180, 182
Small bowel, bacterial contamination tests, 86
Small bowel resection, 108
Small intestine, 78–79
  bacterial overgrowth tests, 87
  contamination, 77
  electrolyte absorption, 80
  electrolyte secretion, 81
  motility, 87
  transit time, 88
  water absorption, 79
  water secretion, 81
Smooth muscle cells, 461
Sodium
  abnormalities, 5–17
  balance, 279
  bicarbonate, 41
  concentration
    plasma, 11, 15, 406, 421–422
    urine, 17, 30
  deficiency, 280
  depletion, 7–9
    acute hypotensive circulatory failure due to, 9
    circulatory and biochemical changes with, 8
    with water depletion, 8–9
    without water depletion, 7–8
  excess, 280
  homeostasis, 46, 480, 482
  intake, normal dietary, 7
  metabolism, 483
  reabsorption, renal tubular, 4
  retention, 14–15
  space, determination of, 6
  valproate, 391
Somatostatin, 613
Somatostatinoma, 613
Sorbitol, 35, 595
Space-occupying lesions, 418
Sphingolipidoses, 496
Sphingomyelin, amniotic fluid, 678
Sphingomyelinosis, 496
Spina bifida, 663, 675
Spironolactone, 481
Splenomegaly, 219
Stachyose, 286
Stanford-Binet scale, 489
Staphylococcus aureus, 326

Starch gel electrophoresis, 206, 207
Starvation
  in hypokalaemia, 18
  metabolic effects of, 127
Steal effect, 415
Steatorrhoea, 105, 109
Steinert's disease, 518
Steroid analysis, 640, 658
Steroid hormones, 544, 633, 690
Steroid metabolism disorders, 690
Steroid metabolites, 285
Steroid suppression tests, 626
Steroids, urinary, 622
Stimulation tests, 603
Stomach, 67–72
Stool weight measurement, 90
Streptococcus pyogenes, 218
Streptomycin, 398
Stress tests, 625
Stupor, 413–425
  causes of, 414–415
  clinical aspects, 415–417
  definition, 413
  general assessment, 416
  history-taking, 415
  neurophysiological basis, 413–414
Subacute sclerosing panencephalitis, 420
Sub-acute thyroiditis, 573, 576
Subarachnoid haemorrhage, 417, 418, 448
Sucrose, 286
Sugars
  faecal, 106
  in urine, 285–288
Sulphatide lipoidoses, 427
Superoxide dismutase, 359
Suppression tests, 557, 603
Surface tension, 678
Surfactant in amniotic fluid, 678
Surfactant protein, 678
Surfactant synthesis, 677
Surgery, major, 400
Suxamethonium, 437
Sweat test, 100
Syndrome of inappropriate secretion of ADH (SIADH), 9, 12, 55, 61–62, 65
Synovial biopsy, 336
Synovial fluid analysis, 335–336
Synovial joint, 325
Synthetic steroids, 620
Syntocin, in hyponatraemia, 14
Syphilis, 326
  false positive tests, 339
  of the nervous system, 446
  tests for, 449–450
Systemic lupus erythematosus (SLE), 338

Tabes dorsalis, 163
Tangier disease, 427, 470
Taste abnormalities, 126
Tan protein pattern, 451
Tay-Sachs disease, 495, 496, 499, 668, 669
Telangiectasia, 255
Termination mutants, 196
Testicular feminization, 645
Testicular function, 639
Testosterone, 635, 638, 639, 642, 645, 649
Tetany, 40
Tetrahydrobiopterin, 507
Thalassaemia, 187, 192, 195, 196, 201–204
  differentiation traits, 207
  haematology, 201–202
  molecular defects, 202
  pathology, 201–202

INDEX

α-Thalassaemia, 203
β-Thalassaemia, 203, 219
δβ-Thalassaemia, 203
Thallium poisoning, 423
Theophylline monitoring of, 397
Thermoregulation, 120
Thiamine deficiency, 131–132
Thiazides, 476
Thirst, 49
   experimental aspects relating to, 49
   temporary relief by drinking, 49
Thrombasthenia, 267
Thrombin, 259
Thrombin time, 267
Thrombocytopenia, 165, 267
Thrombosis, platelet tests, 267
Thrombotic thrombocytopenic purpura, 217
Thromboxane, 461
Thyroglobulin, 561, 569
   iodination of, 563
Thyroglobulin haemagglutination anti-body (TGHA), 569
Thyroglobulin metabolism, 567
Thyroid antibody levels, 570
Thyroid antibody studies, 582
Thyroid auto-antibodies, 568
Thyroid diseases, 571–578
   clinical presentation, 571
   in the elderly, 578
   in neonates and children, 578
   in pregnancy, 577
Thyroid function assessment, 578–582
Thyroid function tests, 579
   unexpected results, 581
Thyroid gland, 560–583
   anatomy, 560
   auto-immune disease, 568–570
   congenital anomalies, 567
   congenital disease of, 566
   disease of, 560, 566–571
   functions of, 561
   pathological changes, 566
   physiology of, 562
Thyroid hormones, 110, 307, 561–565
   catabolism and excretion, 564
   circulation of, 563
   disorders of, 576
   effects on metabolic processes, 565
   generalised effects of, 565
   metabolic and physiological effects, of, 565
   release of, 563
   resistance to, 576
   synthesis of, 562
Thyroid medullary carcinoma, 991
Thyroid scanning, 582
Thyroid-stimulating antibodies, 568–570
Thyroid stimulating hormone (TSH), 553–579, 689
Thyroid stimulating immunoglobulin (TSI), 570
Thyroiditis, 573–576
Thyrotoxic crisis, 574
Thyrotrophin *see* Thyroid-stimulating hormone
Thyrotrophin-releasing hormone (TRH), 542, 547, 562, 579, 625
Thyroxine, 545, 561, 565, 569, 576, 579, 580, 689
Thyroxine-binding globulin (TBG), 563, 564, 567, 568, 580, 642
Thyroxine-binding pre-albumin (TBPA), 563

Tildon-Cornblath syndrome, 501
Tissue antigens, 282
Tissue growth factors, 587
Tissue proteins, 243
Tissue typing, 326
Tobramycin, 393
TORCH agents, 228
Total body protein, 119
Toxic nodular goitre, 573
Toxins, 422
Trace element deficiency and excess, 135–136
Trace element status assessment, 136
Trace elements, 347, 355–361, 407
   as pharmacological agents, 361
   in monitoring of disease, 361
Trace metals in CSF, 453
Transcellular fluids, composition of, 7
Transcortin, 622
Transferrin, 241–242
Transport defects, 281
Transport proteins. See Plasma proteins
Trauma, 400
   endocrine changes after, 400
   haematological effects, of, 409
   in coma, 418–419
Traumatic diabetes, 404
Treponema pallidum haemagglutination test (TPHA), 450
Treponema pallidum immobilisation test (TPI), 450
Triacylglycerol exchange-protein, 468
Trichloral compounds, 372
Tricyclic antidepressants, 377, 392
Triglyceride, 470
   long-chain, 82
   medium-chain, 82
Tri-iodothyronine, 561, 563, 565, 569, 576
   measurement of, 580
   reverse, 580, 689
Trometamol, 60
Trypsin, 98
Tryptamine, 526
Tryptophan, 169, 526
Tuberculous meningitis (TBM), 419, 446, 451
Turner's syndrome, 645
Tyrosine, 168
   metabolism disorders, 508–509

Ulcerative colitis, 106, 329
Ultrafiltration theory, 460
Uraemia, 429
Uraemic coma, 420
Urea, 60
Urea cycle disorders, 511
Urea metabolism, 119
Ureteric colic, 163
Uric acid, 332, 517, 518
Uric acid calculi, 291, 331
Uric acid excretion, 294
Uricosuric drugs, 332
Uridine-5-phosphate (UMP), 518
Urinary creatinine-height index, 121
Urinary free cortisol, 285
Urinary infection, 292
Urinary oestriol excretion, 659
Urinary volume, 293
Urine
   drugs in, 288–289
   electrophoresis, 247
   enzyme activity in, 283–284
   hormones in, 284
   metals in, 289
   sugar in, 285–288
Urine arginine vasopressin test (AVP), 63
Urine collection, 353
Urine examination, 160, 277–289
Urine osmolality, 16
Urine: plasma urea ratio, 272
Urobilinogen, 142
   excretion, 142
Urolithiasis, 290–296

Vaginal epithelium, 644
Variegate porphyria, 180
Vascular disease, 461
Vasoactive intestinal polypeptide (VIP), 611–612
Vasoactive substances, 402
Vasodilator treatment, 477
Vasogenic oedema, 414
Vasopressin, 485
   in hyponatraemia, 14
VD reference laboratory flocculation test (VDRL), 450
Venepuncture blood samples, 683
Ventricular haemorrhage, 501
Verapamil, 396
Verner-Morrison syndrome, 612
Very-low-density lipoproteins, 462, 464, 469–472
Vital infection, 447
Viral meningitis, 419
Virilisation, 649, 650
Visceral proteins, 125
Vitamin A, 293, 505
Vitamin A deficiency, 128–129
Vitamin B_1, 131
Vitamin B_2, 132
Vitamin B_3, 132
Vitamin B_6, 132
Vitamin B_{12}, 83, 132, 136, 212, 452
   absorption assessment, 86
Vitamin B_{12} deficiency, 101, 132, 212–214, 429
Vitamin C, 133, 505, 605
Vitamin D, 271, 275, 293, 305–307, 317, 321, 505
Vitamin D deficiency, 129–130, 318, 321–322
Vitamin D metabolites, 321–322
Vitamin D_2, 305
Vitamin D_3, 305
Vitamin E deficiency, 130
Vitamn K, 144
Vitamin K deficiency, 130
Vitamin K_1, 131
Vitamin deficiency, 127, 504, 530
Vitamin status assessment, 133–135
Vitamins, 504
   excess, 128
   fat-soluble, 82
   function, 127, 135
   in cancer, 127
   water soluble, 83
Volatile fatty acids, 89
Vomiting, 71–72
Von Gierkes disease, 501
von Willebrand's disease, 253

Warfarin, 379
Warm antibody type haemolytic anaemia, 217
Wassermann complement fixation test (WR), 450
Waste products, elimination, 542

Water
  dynamic aspects, 46
  renal handling of, 2–5
Water absorption, 50–51
  colonic, 51, 77, 89
  small intestinal, 50–51, 79
Water balance, 279, 591
  in infancy, 46
Water composition of the body, 1–2
Water concentration, quantitative aspects, 48
Water content abnormalities, 5–17
Water control, basic principles, 44–46
Water depletion, 9–11
Water deprivation test, 274
Water distribution in body, 44
Water hardness, 292
Water homeostasis, 46
  consequences of imbalance, 56
  physiological control, of, 46–56
Water intake, 44–46
  decreased 56
  excessive, 56
Water intoxication
  hypertonic saline for, 14
  signs of, 15
Water metabolism, 406
Water output, 44–46
Water reabsorption, 52
  in loop of Henle, 52
Water requirement, 44
Water retention, 14–15
  excessive, 12
Water secretion, small intestine, 81
WDHA syndrome, 612
Wechsler scale, 489
Werdnig Hoffman disease, 518
Wernicke-Korsakoff syndrome, 131
Wernicke's encephalopathy, 420, 534
Whipple's disease, 83
Wilm's tumour, renin-secreting, 60
Wilson's disease, 150, 359, 519

Xanthine oxidase inhibition, 332
Xeroderma pigmentosum, 518
Xerophthalmia, 128
Xylitol, 35, 615
Xylose, 286
Xylose absorption test, 103
Xylose tolerance test, 686
D-Xylose excretion test, 85
L-Xylulose, 286, 287

Zinc deficiency, 135, 356–358, 407
Zinc excretion, 357
Zinc intoxication, 136
Zinc malabsorption, 357
Zinc repletion, 357
Zinc status assessment, 358
Zinc toxicity, 358
Zollinger-Ellison syndrome, 69–70, 91, 101, 609–610
Zona fasciculata, 617